T0175770

Bruce L. Bowerman
Miami University

Ruth M. Hummel
JMP

Anne M. Drougas
Dominican University

Kyle B. Moninger
Bowling Green State University

William M. Duckworth
Creighton University

Patrick J. Schur
Miami University

Amy G. Froelich
Iowa State University

Business Statistics and Analytics in Practice

NINTH EDITION

with major contributions by

Steven C. Huchendorf
University of Minnesota

Dawn C. Porter
University of Southern California

Mc
Graw
Hill
Education

BUSINESS STATISTICS AND ANALYTICS IN PRACTICE

Published by McGraw-Hill Education, 2 Penn Plaza, New York, NY 10121. Copyright © 2019 by McGraw-Hill Education. All rights reserved. Printed in the United States of America. No part of this publication may be reproduced or distributed in any form or by any means, or stored in a database or retrieval system, without the prior written consent of McGraw-Hill Education, including, but not limited to, in any network or other electronic storage or transmission, or broadcast for distance learning.

Some ancillaries, including electronic and print components, may not be available to customers outside the United States.

This book is printed on acid-free paper.

13 CPI 24

ISBN 978-1-260-28784-4
MHID 1-260-28784-X

Cover Image: ©*hxdyl/Shutterstock*

All credits appearing on page or at the end of the book are considered to be an extension of the copyright page.

The Internet addresses listed in the text were accurate at the time of publication. The inclusion of a website does not indicate an endorsement by the authors or McGraw-Hill Education, and McGraw-Hill Education does not guarantee the accuracy of the information presented at these sites.

mheducation.com/highered

The McGraw-Hill/Irwin Series in Operations and Decision Sciences

ABOUT THE AUTHORS

Courtesy of Bruce Bowerman

Bruce L. Bowerman Bruce L. Bowerman is emeritus professor of information systems and analytics at Miami University in Oxford, Ohio. He received his Ph.D. degree in statistics from Iowa State University in 1974, and he has over 40 years of experience teaching basic statistics, regression analysis, time series forecasting, survey sampling, and design of experiments to both undergraduate and graduate students. In 1987 Professor Bowerman received an Outstanding Teaching award from the Miami University senior class, and in 1992 he received an Effective Educator award from the Richard T. Farmer School of Business Administration. Together with Richard T. O'Connell, Professor Bowerman has written 25 textbooks. These include *Forecasting, Time Series, and Regression: An Applied Approach* (also coauthored with Anne B. Koehler); *Linear Statistical Models: An Applied Approach*; *Regression Analysis: Unified Concepts, Practical Applications, and Computer Implementation* (also coauthored with Emily S. Murphree); and *Experimental Design: Unified Concepts, Practical Applications, and Computer Implementation* (also coauthored with Emily S. Murphree). The first edition of *Forecasting and Time Series* earned an Outstanding Academic Book award from *Choice* magazine. Professor Bowerman has also published a number of articles in applied stochastic process, time series forecasting, and statistical education. In his spare time, Professor Bowerman enjoys watching movies and sports, playing tennis, and designing houses.

Courtesy of Anne Drougas

Anne Drougas Anne M. Drougas is a Professor of Finance and Quantitative Methods at Dominican University in River Forest, Illinois. Over the course of her academic career, she has received three teaching awards and has developed and taught online and hybrid business statistics and finance courses. Her research is primarily in the areas of corporate finance, simulation, and business analytics with publications in a number of journals including the *Journal of Financial Education* and *Journal of Applied Business and Economics*. She spends her spare time with her family

and serving on the board of directors for Hephzibah House, a social service agency for children in Oak Park, Illinois.

Courtesy of William Duckworth

William Duckworth William M. Duckworth specializes in statistics education and business applications of statistics. His professional affiliations have included the American Statistical Association (ASA), the International Association for Statistical Education (IASE), and the Decision Sciences Institute (DSI). Dr. Duckworth was also a member of the Undergraduate Statistics Education Initiative (USEI), which developed curriculum guidelines for undergraduate programs in statistical science that were officially adopted by the ASA. Dr. Duckworth has published research papers, been an invited speaker at professional meetings, and taught company training workshops, in addition to providing consulting and expert witness services to a variety of companies. During his tenure in the Department of Statistics at Iowa State University, his main responsibility was coordinating, teaching, and improving introductory business statistics courses. Dr. Duckworth currently teaches business analytics to both undergraduate and graduate students in the Heider College of Business at Creighton University.

Courtesy of Amy Froelich

Amy Froelich Amy G. Froelich received her Ph.D. in Statistics from the University of Illinois, Urbana-Champaign, and currently is Associate Professor and Director of Undergraduate Education in the Department of Statistics at Iowa State University. A specialist in undergraduate statistics education, she has taught over 2,700 students at Iowa State in the last 18 years, primarily in introductory statistics, probability and mathematical statistics, and categorical data analysis. Her research in statistics education and psychometrics and educational measurement has appeared in *The American Statistician*, the *Journal of Statistics Education*, *Teaching Statistics*, and the *Journal of Educational Measurement*, and she and her colleagues have received research funding from the National Science Foundation, the U.S. Department of Agriculture, and the U.S. Department of Education. Dr. Froelich has received several teaching and advising awards at Iowa State University and was the 2010 recipient of the Waller Education Award from the American Statistical Association. When not working, she enjoys reading, spending time with her family, and supporting her daughters' extracurricular activities.

Courtesy of Steve Muir/SAS

Ruth Hummel Ruth M. Hummel is an Academic Ambassador with JMP, a division of SAS specializing in desktop software for dynamic data visualization and analysis. As a technical advocate for the use of JMP® in academic settings, she supports professors and instructors who use JMP for teaching and research. She has been teaching and consulting since 2002, when she started her career as a high school math teacher. She has taught high school, undergraduate, and graduate courses in mathematics and statistics, and directed statistical research and analysis in a variety of fields. Ruth holds a Ph.D. in statistics from the Pennsylvania State University.

Courtesy of Kyle Moninger

Kyle Moninger Kyle B. Moninger instructs the Quantitative Business Curriculum at Bowling Green State University in Bowling Green, Ohio. He teaches and plans undergraduate courses in statistics and business calculus, serves on the Quantitative Business Curriculum committee, and supervises the college's math and statistics tutoring center. Kyle has been a visiting instructor three times at Tianjin Polytechnic University in Tianjin, China, and was previously a data scientist at Owens Corning in Toledo, Ohio, where he designed and implemented a corporate training program on business intelligence and analytics.

Courtesy of Pat Schur

Pat Schur Patrick J. Schur is a Senior Clinical Professor in the Department of Information Systems and Analytics in the Farmer School of Business at Miami University in Oxford, Ohio. He received his master's degree in statistics from Purdue University. He has been at Miami University for 11 years, teaching introductory statistics courses and advanced statistics courses including regression modeling, time series modeling, design of experiments, and statistical process control. Before joining Miami University, he worked at Procter & Gamble as a statistical consultant and also worked with multiple startup companies cutting across multiple industries.

AUTHORS' PREVIEW

Business Statistics and Analytics in Practice, Ninth Edition, provides a unique and flexible framework for teaching the introductory course in business statistics. This framework consists of

- A complete presentation of traditional business statistics, with improved discussions of introductory concepts, probability modeling, classical statistical inference (including a much clearer explanation of hypothesis testing), and regression and time series modeling.

- A complete presentation of business analytics, with topic coverage in six optional sections and two optional chapters: a section in Chapter 1 introducing analytics, five sections in Chapters 2 and 3 discussing descriptive analytics, and Chapters 5 and 16 discussing predictive analytics.

- Continuing case studies that facilitate student learning by presenting new concepts in the context of familiar situations.

- Business improvement conclusions—highlighted in yellow and designated by icons ⓑ in the page margins—that explicitly show how statistical analysis leads to practical business decisions.

- Many new exercises.

- Use of Excel (including the Excel add-in MegaStat), Minitab, and JMP to carry out traditional statistical analysis. Use of JMP (and Excel and Minitab where possible) to carry out descriptive and predictive analytics.

We now discuss how these features are implemented in the book's 20 chapters.

Chapters 1, 2, and 3: Introductory concepts. Graphical and numerical descriptive methods. In an improved and simpler Chapter 1 we discuss data, variables, populations, and how to select random and other types of samples. Three case studies—**The Cell Phone Case, The Marketing Research Case,** and **The Car Mileage Case**—are used to illustrate sampling and how samples can be used to make statistical inferences.

In Chapters 2 and 3 we begin to formally discuss the statistical analysis used in making statistical inferences. For example, in Chapter 2 (graphical descriptive methods) we show how to construct a histogram of the car mileages that were sampled in **The Car Mileage**

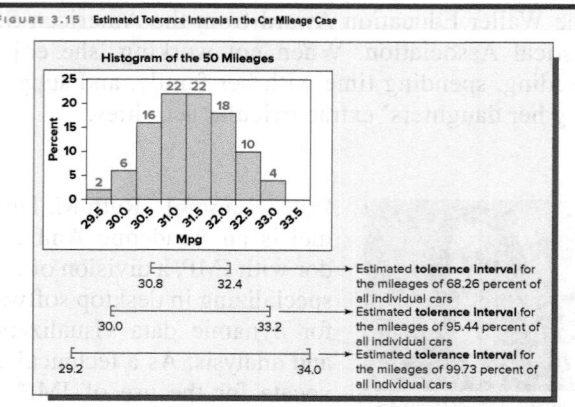

FIGURE 3.15 Estimated Tolerance Intervals in the Car Mileage Case

Case of Chapter 1. In Chapter 3 (numerical descriptive methods) we then use this histogram to help explain the Empirical Rule. As illustrated in Figure 3.15, this rule gives tolerance intervals providing estimates of the "lowest" and "highest" mileages that a new midsize car model should be expected to get in combined city and highway driving.

Chapters 1, 2, and 3: Six optional sections introducing business analytics and data mining and discussing descriptive analytics. In an optional section of Chapter 1 **The Disney Parks Case** introduces how business analytics and data mining are used to analyze big data. This case is then used in an optional section of Chapter 2 to help begin the book's discussion of descriptive analytics. Here, the optional section of Chapter 2 discusses what we call *graphical descriptive analytics,* and four optional sections in Chapter 3 (Part 2 of Chapter 3) discuss what we call *numerical descriptive analytics.* Included in the discussion of graphical descriptive analytics are gauges and dashboards (see Figure 2.35), bullet graphs and treemaps (see the Disney examples in Figures 2.36 and 2.37), and sparklines and data drill-down graphics. Included in the discussion of numerical descriptive analytics are association rules (see Figure 3.25), text mining (see Figure 3.27), hierarchical and k-means cluster analysis (see Figures 3.38 and 3.40), multidimensional scaling (which is part of the cluster analysis section), and factor analysis.

We believe that an early introduction to descriptive analytics will make statistics seem more useful and

FIGURE 2.35   A Dashboard of the Key Performance Indicators for an Airline

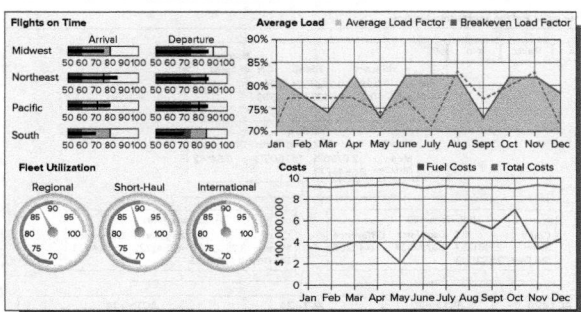

FIGURE 2.36   Excel Output of a Bullet Graph of Disney's Predicted Waiting Times (in minutes) for the Seven Epcot Rides Posted at 3 P.M. on February 21, 2015   DisneyTimes

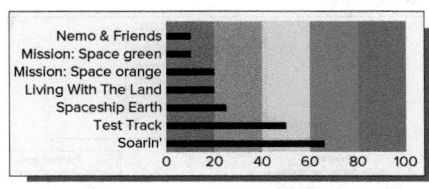

FIGURE 2.37   The Number of Ratings and the Mean Rating for Each of Seven Rides at Epcot (0 = Poor, 1 = Fair, 2 = Good, 3 = Very Good, 4 = Excellent, 5 = Superb) and a JMP Output of a Treemap of the Numbers of Ratings and the Mean Ratings

(b) JMP output of the treemap

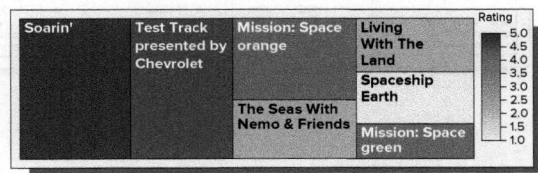

FIGURE 3.25   The JMP Output of an Association Rule Analysis of the DVD Renters Data   DVDRent

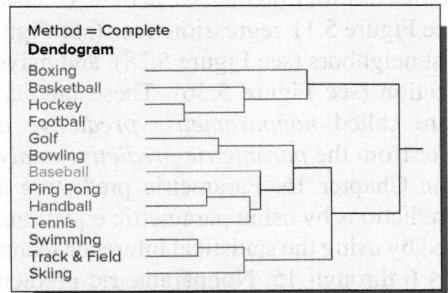

FIGURE 3.27   The JMP Output of Part of a Term and Phrase List and Part of a Word Cloud in the FDA Citations Example   FDACitat

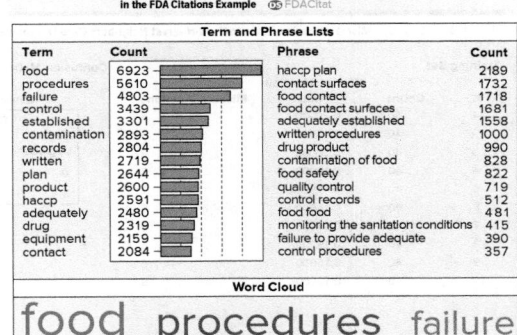

FIGURE 3.38   A Second JMP Output of Hierarchical Clustering of the Sports Perception Data

FIGURE 3.40   The JMP Output of the Biplots in k-Means Clustering of the Sports Perception Data for k = 4, 5, 6, and 7

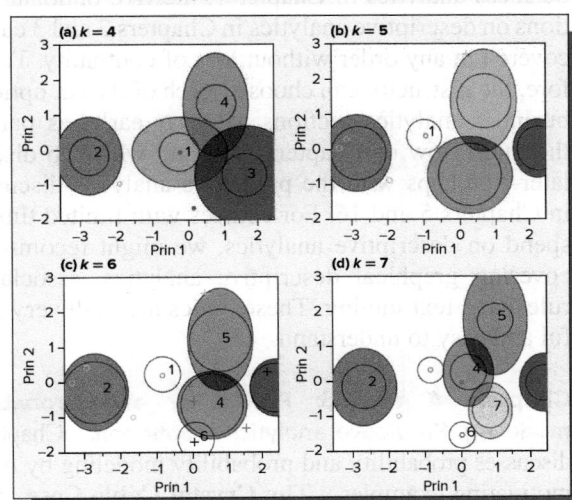

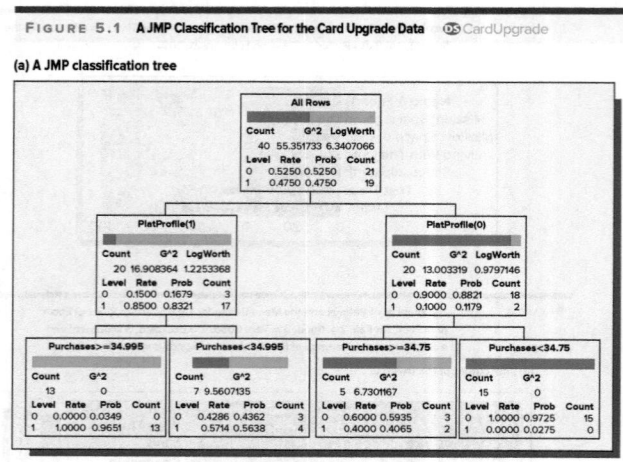

FIGURE 5.1   A JMP Classification Tree for the Card Upgrade Data   DS CardUpgrade

(a) A JMP classification tree

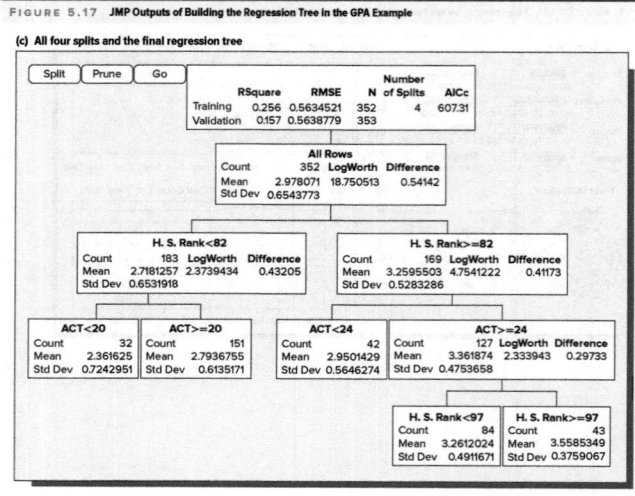

FIGURE 5.17   JMP Outputs of Building the Regression Tree in the GPA Example

(c) All four splits and the final regression tree

FIGURE 5.28   Misclassification Rates in the k-Nearest Neighbors Card Upgrade Example

Training Set

K	Count	Misclassification Rate	Misclassifications
1	40	0.20000	8
2	40	0.12500	5*
3	40	0.20000	8
4	40	0.15000	6
5	40	0.17500	7
6	40	0.15000	6
7	40	0.15000	6
8	40	0.15000	6
9	40	0.15000	6
10	40	0.12500	5

Confusion Matrix for Best K=2

Training Set

Actual Upgrade	Predicted Count 0	1
0	18	3
1	2	17

FIGURE 5.36   Misclassifications in the Naïve Bayes' Card Upgrade Example

Training Set

Count	Misclassification Rate	Misclassifications
40	0.12500	5

Confusion Matrix
Training Set

Actual Upgrade	Predicted Count 0	1
0	17	4
1	1	18

relevant from the beginning and thus motivate students to be more interested in the entire course. However, our presentation gives instructors various choices. This is because, after covering the optional introduction to business analytics in Chapter 1, the five optional sections on descriptive analytics in Chapters 2 and 3 can be covered in any order without loss of continuity. Therefore, the instructor can choose which of the six optional business analytics sections to cover early, as part of the main flow of Chapters 1–3, and which to discuss later—perhaps with the predictive analytics discussed in Chapters 5 and 16. For courses with limited time to spend on descriptive analytics, we might recommend covering graphical descriptive analytics, association rules, and text mining. These topics are both very useful and easy to understand.

Chapters 4 and 5: Probability and probability modeling. Predictive analytics I (optional).   Chapter 4 discusses probability and probability modeling by using motivating examples—**The Crystal Cable Case** and a

real-world example of gender discrimination at a pharmaceutical company—to illustrate the probability rules. Optional Chapter 5 then uses the probability concepts of Chapter 4 and the descriptive statistics of Chapters 2 and 3 to discuss four predictive analytics: classification trees (see Figure 5.1), regression trees (see Figure 5.17). k-nearest neighbors (see Figure 5.28), and naive Bayes' classification (see Figure 5.36). These predictive analytics are called *nonparametric predictive analytics* and differ from the *parametric predictive analytics* discussed in Chapter 16. Parametric predictive analytics make predictions by using parametric equations that are evaluated by using the statistical inference techniques of Chapters 6 through 15. Nonparametric predictive analytics make predictions without using such equations and can be understood (from an applied standpoint) with a background of only descriptive statistics and probability. Chapters 5 and 16 are independent of each other and of the descriptive analytics sections in Chapters 2 and 3. Therefore, the instructor has the option to try to motivate student interest by covering Chapter 5 early, in the

FIGURE 8.1 A Comparison of Individual Car Mileages and Sample Means

(a) A graph of the probability distribution describing the population of six individual car mileages

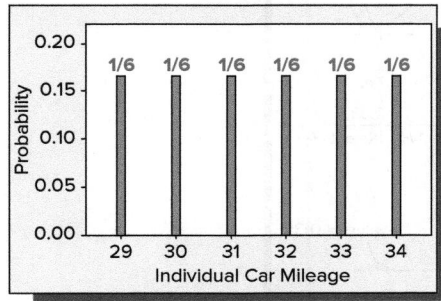

(b) A graph of the probability distribution describing the population of 15 sample means

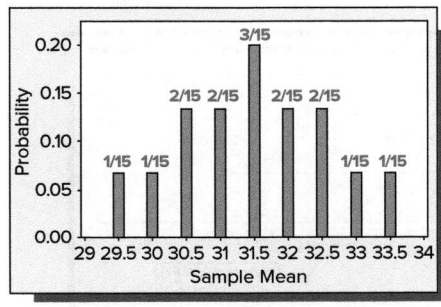

FIGURE 8.5 The Central Limit Theorem Says That the Larger the Sample Size Is, the More Nearly Normally Distributed Is the Population of All Possible Sample Means

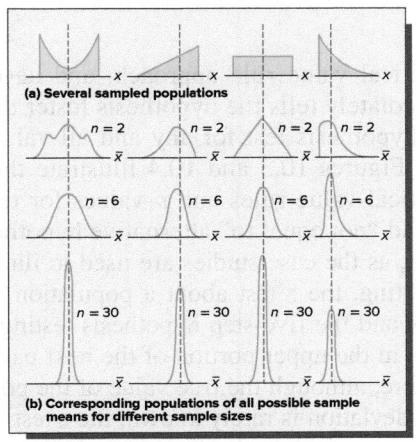

FIGURE 8.3 A Comparison of (1) the Population of All Individual Car Mileages, (2) the Sampling Distribution of the Sample Mean $\bar{x}$ When $n = 5$, and (3) the Sampling Distribution of the Sample Mean $\bar{x}$ When $n = 50$

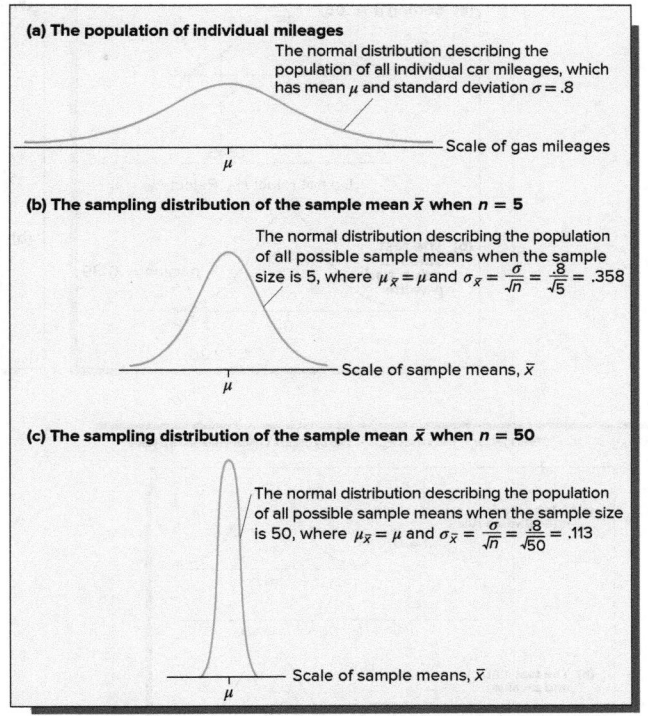

FIGURE 9.2 Three 95 Percent Confidence Intervals for μ

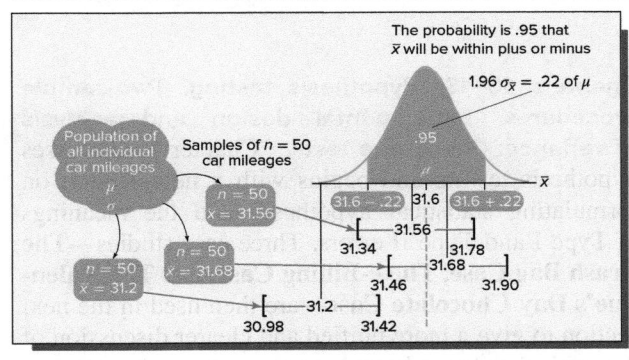

main flow of the course, or wait to cover Chapter 5 until later, perhaps with (before or after) the parametric predictive analytics in Chapter 16. For courses with limited time to spend on nonparametric predictive analytics, we might suggest covering just classification trees and regression trees.

Chapters 6–9: Discrete and continuous probability distributions. Sampling distributions and confidence intervals. Chapters 6 and 7 give discussions of discrete and continuous probability distributions (models)

and feature practical examples illustrating the "rare event approach" to making a statistical inference. In Chapter 8, **The Car Mileage Case** is used to introduce sampling distributions and motivate the Central Limit Theorem (see Figures 8.1, 8.3, and 8.5). In Chapter 9, the automaker in **The Car Mileage Case** uses a confidence interval procedure specified by the Environmental Protection Agency (EPA) to find the EPA estimate of a new midsize model's true mean mileage and determine if the new midsize model deserves a federal tax credit (see Figure 9.2).

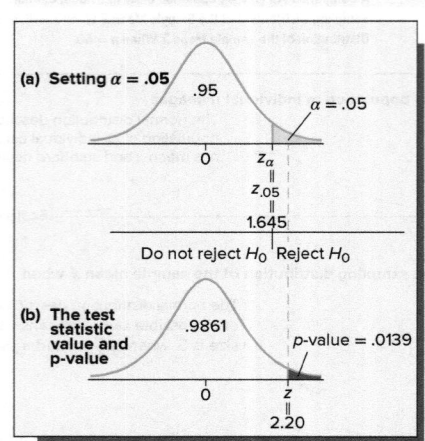

FIGURE 10.1 Testing $H_0: \mu = 50$ versus $H_a: \mu > 50$ by setting α Equal to .05

(a) Setting $\alpha = .05$
.95
$\alpha = .05$
0   z_α
     $\|$
     $z_{.05}$
     $\|$
     1.645
Do not reject H_0 Reject H_0

(b) The test statistic value and p-value
.9861
p-value = .0139
0   z
     $\|$
     2.20

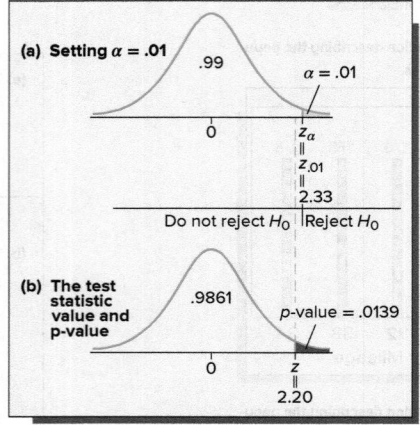

FIGURE 10.2 Testing $H_0: \mu = 50$ versus $H_a: \mu > 50$ by setting α Equal to .01

(a) Setting $\alpha = .01$
.99
$\alpha = .01$
0   z_α
     $\|$
     $z_{.01}$
     $\|$
     2.33
Do not reject H_0 Reject H_0

(b) The test statistic value and p-value
.9861
p-value = .0139
0   z
     $\|$
     2.20

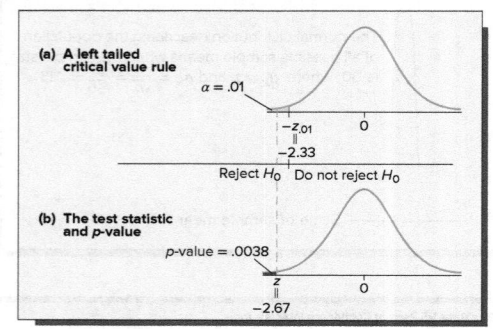

FIGURE 10.3 Testing $H_0: \mu = 19.5$ versus $H_a: \mu < 19.5$ by Using a Critical Value and a p-Value

(a) A left tailed critical value rule
$\alpha = .01$
$-z_{.01}$ 0
$\|$
-2.33
Reject H_0 Do not reject H_0

(b) The test statistic and p-value
p-value = .0038
z 0
$\|$
-2.67

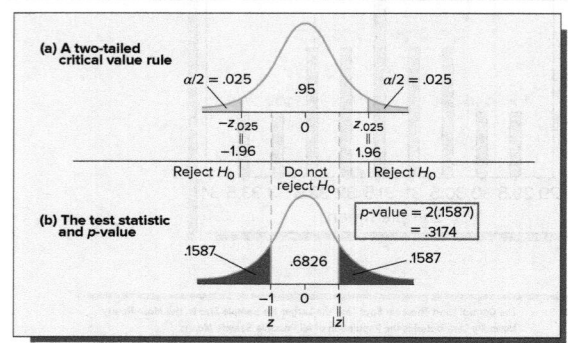

FIGURE 10.4 Testing $H_0: \mu = 330$ versus $H_a: \mu \neq 330$ by Using Critical Values and the p-Value

(a) A two-tailed critical value rule
$\alpha/2 = .025$ .95 $\alpha/2 = .025$
$-z_{.025}$ 0 $z_{.025}$
$\|$ $\|$
-1.96 1.96
Reject H_0 Do not reject H_0 Reject H_0

(b) The test statistic and p-value
p-value = 2(.1587) = .3174
.1587   .6826   .1587
-1 0   1
$\|$ $\|$
z $|z|$

Chapters 10–13: Hypothesis testing. Two-sample procedures. Experimental design and analysis of variance. Chi-square tests. Chapter 10 discusses hypothesis testing and begins with a new section on formulating statistical hypotheses and the meanings of Type I and Type II errors. Three case studies—**The Trash Bag Case, The e-Billing Case,** and **The Valentine's Day Chocolate Case**—are then used in the next section to give a more unified and clearer discussion of the critical value rule and p-value approaches to performing a z test about the population mean. Specifically, for each type of alternative hypothesis, this discussion first illustrates the appropriate critical value rule in a graphical figure and then, in the same graphical figure, shows the appropriate p-value and explains why it is the more informative way to carry out the hypothesis test.

For example, the above Figures 10.1 and 10.2 are presented side-by-side in the text and illustrate testing a "greater than" alternative hypothesis in **The Trash Bag Case.** These figures show the different α's specified by two television networks evaluating a trash bag advertising claim, the different critical values that would have table looked up by an hypothesis tester

using the critical value rule approach, and have the p-value immediately tells the hypothesis tester the results of the hypothesis test for any and all values of α. Similarly, Figures 10.3 and 10.4 illustrate the appropriate critical value rules and p-values for testing "less than" and "not equal to" alternative hypotheses.

In addition, as the case studies are used to illustrate hypothesis testing, the z test about a population mean summary box and the five-step hypothesis testing procedure shown in the upper portion of the next page are developed. Here, although the true value of the population standard deviation is rarely known, the z test about a population mean summary box serves as an easily modifiable model for the book's other more practically useful hypothesis testing summary boxes—for example, for the t test about a population mean summary box and the z test about a population proportion summary box shown in the lower portion of the next page. Moreover, the five-step hypothesis testing procedure emphasizes that to successfully use a hypothesis testing summary box, we simply identify the alternative hypothesis being tested and then looking the summary box for the appropriate critical value rule and/or p-value.

Testing a Hypothesis about a Population Mean When σ Is Known

Null Hypothesis $H_0: \mu = \mu_0$

Test Statistic $z = \dfrac{\bar{x} - \mu_0}{\sigma/\sqrt{n}}$

Assumptions Normal population or Large sample size

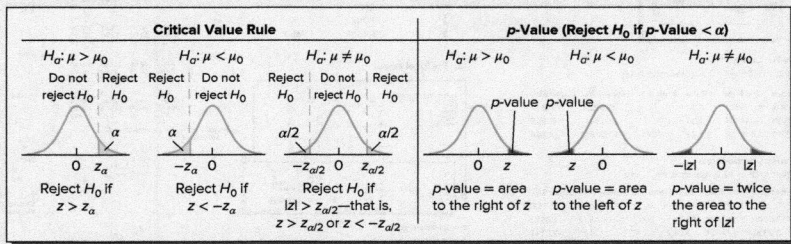

Critical Value Rule			p-Value (Reject H_0 if p-Value < α)		
$H_a: \mu > \mu_0$	$H_a: \mu < \mu_0$	$H_a: \mu \neq \mu_0$	$H_a: \mu > \mu_0$	$H_a: \mu < \mu_0$	$H_a: \mu \neq \mu_0$
Reject H_0 if $z > z_\alpha$	Reject H_0 if $z < -z_\alpha$	Reject H_0 if $|z| > z_{\alpha/2}$—that is, $z > z_{\alpha/2}$ or $z < -z_{\alpha/2}$	p-value = area to the right of z	p-value = area to the left of z	p-value = twice the area to the right of $|z|$

The Five Steps of Hypothesis Testing

1 State the null hypothesis H_0 and the alternative hypothesis H_a.
2 Specify the level of significance α.
3 Plan the sampling procedure and select the test statistic.

Using a critical value rule:

4 Use the summary box to find the critical value rule corresponding to the alternative hypothesis.
5 Collect the sample data, compute the value of the test statistic. and decide whether to reject H_0 by using the critical value rule. Interpret the statistical results.

Using a p-value rule:

4 Collect the sample data and compute the value of the test statistic.
5 Use the summary box to find the p-value corresponding to the alternative hypothesis. Use the computed test statistic value to compute the p-value. Reject H_0 at level of significance α if the p-value is less than α. Interpret the statistical results.

A *t* Test about a Population Mean: σ Unknown

Null Hypothesis $H_0: \mu = \mu_0$

Test Statistic $t = \dfrac{\bar{x} - \mu_0}{s/\sqrt{n}}$ $df = n - 1$

Assumptions Normal population or Large sample size

Critical Value Rule			p-Value (Reject H_0 if p-Value < α)		
$H_a: \mu > \mu_0$	$H_a: \mu < \mu_0$	$H_a: \mu \neq \mu_0$	$H_a: \mu > \mu_0$	$H_a: \mu < \mu_0$	$H_a: \mu \neq \mu_0$
Reject H_0 if $t > t_\alpha$	Reject H_0 if $t < -t_\alpha$	Reject H_0 if $|t| > t_{\alpha/2}$—that is, $t > t_{\alpha/2}$ or $t < -t_{\alpha/2}$	p-value = area to the right of t	p-value = area to the left of t	p-value = twice the area to the right of $|t|$

A Large Sample Test about a Population Proportion

Null Hypothesis $H_0: p = p_0$

Test Statistic $z = \dfrac{\hat{p} - p_0}{\sqrt{\dfrac{p_0(1 - p_0)}{n}}}$

Assumptions[2] $np_0 \geq 5$ and $n(1 - p_0) \geq 5$

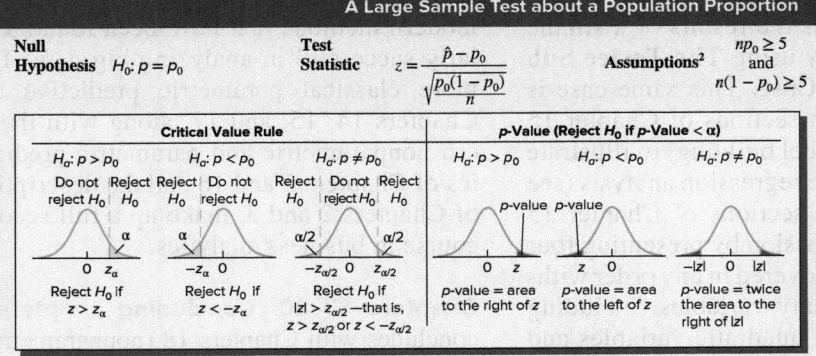

Critical Value Rule			p-Value (Reject H_0 if p-Value < α)		
$H_a: p > p_0$	$H_a: p < p_0$	$H_a: p \neq p_0$	$H_a: p > p_0$	$H_a: p < p_0$	$H_a: p \neq p_0$
Reject H_0 if $z > z_\alpha$	Reject H_0 if $z < -z_\alpha$	Reject H_0 if $|z| > z_{\alpha/2}$—that is, $z > z_{\alpha/2}$ or $z < -z_{\alpha/2}$	p-value = area to the right of z	p-value = area to the left of z	p-value = twice the area to the right of $|z|$

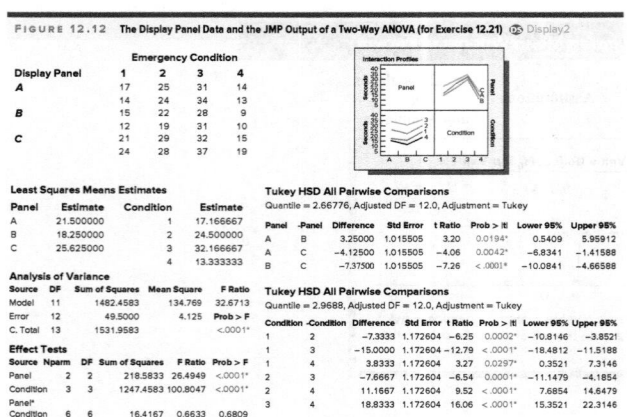

FIGURE 12.12 The Display Panel Data and the JMP Output of a Two-Way ANOVA (for Exercise 12.21) 🖸 Display2

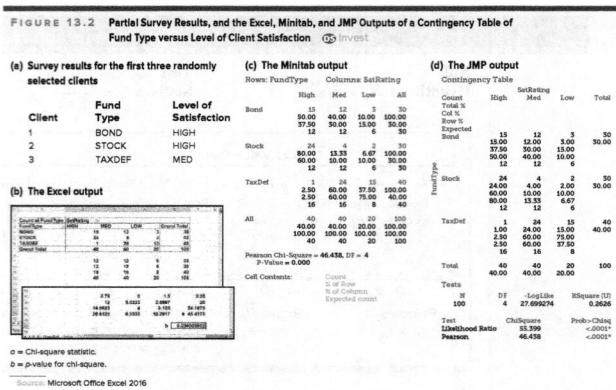

FIGURE 13.2 Partial Survey Results, and the Excel, Minitab, and JMP Outputs of a Contingency Table of Fund Type versus Level of Client Satisfaction 🖸 Invest

Hypothesis testing summary boxes are featured throughout Chapter 10, Chapter 11 (two-sample procedures), Chapter 12 (one-way, randomized block, and two-way analysis of variance), Chapter 13 (chi-square tests of goodness of fit and independence), and the remainder of the book. Furthermore, emphasis is placed throughout on assessing practical importance after testing for statistical significance. For example, as illustrated in Figure 12.12, if an F test finds a significant factor in an analysis of variance, we assess practical importance by finding point estimates of and confidence intervals for the differences in the effects of the different levels of the factor. As another example (see Figure 13.2), if a chi-square test rejects the hypothesis of independence between two variables, we assess practical importance by using the contingency table upon which the chi-square test is based to analyze the nature of the dependence between the variables.

Chapters 14–17: Simple linear regression. Multiple regression and model building. Predictive analytics II (optional). Time series forecasting and index numbers.

Chapter 14 discusses simple linear regression and illustrates the results of a simple linear regression analysis by using **The Tastee Sub Shop** (revenue prediction) **Case.** This same case is then used by the first seven sections of Chapter 15 (multiple regression and model building) to illustrate the results of a basic multiple regression analysis (see Figure 15.4). The last four sections of Chapter 15 continue the regression discussion by presenting four modeling topics that can be covered in any order without loss of continuity: dummy variables (including a discussion of interaction); quadratic variables and quantitative interaction variables; model building and the effects of multicollinearity (including model

building for big data—see Figure 15.31); and residual analysis and diagnosing outlying and influential observations.

With the regression concepts of Chapters 14 and 15 as background, optional Chapter 16 extends these concepts and discusses three parametric predictive analytics: logistic regression (see Figure 16.5), linear discriminate analysis (see Figure 16.12), and neural networks (see Figures 16.17 and 16.19). Moreover, Chapter 17 extends the regression concepts in a different way and discusses time series forecasting methods, including an expanded presentation of exponential smoothing and a new and fuller (but understandable) presentation of the Box–Jenkins methodology.

Note that although we have used the term predictive analytics to refer only to the prediction methods of Chapters 5 and 16, the regression and time series methods of Chapters 14, 15, and 17 all predict (using a parametric equation) values of a response variable and thus are all (parametric) predictive analytics. We have used the term predictive analytics to refer only to the predictive methods of Chapters 5 and 16 because these methods are (for the most part) more modern methods that have been found to be particularly successful in analyzing big data. Together, the more classical parametric predictive analytics of Chapters 14, 15, and 17, along with the more modern nonparametric and parametric predictive analytics of Chapters 5 and 16 and the descriptive analytics of Chapters 2 and 3, make up a full second statistics course in business analytics.

Chapters 18–20: Concluding chapters.

The book concludes with Chapters 18 (nonparametric statistics), Chapter 19 (decision theory), and website Chapter 20 (process improvement using control charts).

FIGURE 15.4 Excel and Minitab Outputs of a Regression Analysis of the Tasty Sub Shop Revenue Data in Table 15.1 Using the Model $y = \beta_0 + \beta_1 x_1 + \beta_2 x_2 + e$

(a) The Excel output

Regression Statistics

Multiple R	0.9905
R Square	0.9810 [8]
Adjusted R Square	0.9756 [9]
Standard Error	36.6856 [7]
Observations	10

ANOVA

	df	SS	MS	F	Significance F
Regression	2	486355.7 [10]	243177.8	180.689 [13]	9.46E-07 [14]
Residual	7	9420.8 [11]	1345.835		
Total	9	495776.5 [12]			

	Coefficients	Standard Error [4]	t Stat [5]	P-value [6]	Lower 95% [19]	Upper 95% [19]
Intercept	125.289 [1]	40.9333	3.06	0.0183	28.4969	222.0807
population	14.1996 [2]	0.9100	15.60	1.07E-06	12.0478	16.3517
bus_rating	22.8107 [3]	5.7692	3.95	0.0055	9.1686	36.4527

(b) The Minitab output

Analysis of Variance

Source	DF	Adj SS	Adj MS	F-Value	P-Value
Regression	2	486356 [10]	243178	180.69 [13]	0.000 [14]
Population	1	327678	327678	243.48	0.000
Bus_Rating	1	21039	21039	15.63	0.006
Error	7	9421 [11]	1346		
Total	9	495777 [12]			

Model Summary

S	R-sq	R-sq(adj)	R-sq(pred)
36.6856 [7]	98.10% [8]	97.56% [9]	96.31%

Coefficients

Term	Coef	SE Coef [4]	T-Value [5]	P-Value [6]	VIF
Constant	125.3 [1]	40.9	3.06	0.018	
Population	14.200 [2]	0.910	15.60	0.000	1.18
Bus_Rating	22.81 [3]	5.77	3.95	0.006	1.18

Regression Equation

Revenue = 125.3 + 14.200 Population + 22.81 Bus_Rating

Variable	Setting	Fit [15]	SE Fit [16]	95% CI [17]	95% PI [18]
Population	47.3	956.606	15.0476	(921.024, 992.188)	(862.844, 1050.37)
Bus_Rating	7				

[1] b_0	[2] b_1	[3] b_2	[4] s_{b_j} = standard error of the estimate b_j	[5] t statistics	[6] p-values for t statistics	[7] s = standard error
[8] R^2	[9] Adjusted R^2	[10] Explained variation	[11] SSE = Unexplained variation	[12] Total variation	[13] F(model) statistic	
[14] p-value for F(model)	[15] $\hat{y}$ = point prediction when x_1 = 47.3 and x_2 = 7	[16] $s_{\hat{y}}$ = standard error of the estimate $\hat{y}$				
[17] 95% confidence interval when x_1 = 47.3 and x_2 = 7	[18] 95% prediction interval when x_1 = 47.3 and x_2 = 7	[19] 95% confidence interval for β_j				

FIGURE 15.31 The JMP Output of the Potential Quantitative Independent Variables, Including the Dummy Variables, and Forward Selection with Simultaneous Validation in the Used Toyota Corolla Sales Price Example

(a) The Variables

Age_08_04
Mfg_Month
Mfg_Year
KM

Fuel_Type[CNG&Petrol-Diesel]
Fuel_Type[CNG-Petrol]
HP
Met_Color[0-1]
Color[White&Beige&Violet&Green&Red-Blue&Black&Silver&Grey&Yellow]
Color[White&Beige-Violet&Green&Red]
Color[White-Beige]
Color[Violet&Green-Red]
Color[Blue&Black&Silver-Grey&Yellow]

Color[Blue-Black&Silver]
Color[Black-Silver]
Color[Grey-Yellow]
Automatic[0-1]
CC
Doors
Cylinders
Gears
Quarterly_Tax
Weight
Mfg_Guarantee
BOVAG_Guarantee
Guarantee_Period
ABS
Airbag_1
Airbag_2
Airco
Automatic_airco
Boardcomputer
CD_Player
Central_Lock
Powered_Windows
Power_Steering
Radio
Mistlamps
Sport_Model
Backseat_Divider
Metallic_Rim
Radio_cassette
Tow_Bar

(b) Forward selection with simultaneous validation

Step History

Step	Parameter	Action	"Sig Prob"	RSquare	RSquare Validation	P
1	Mfg_Year	Entered	0.0000	0.7834	0.7831	2
2	Automatic_airco	Entered	0.0000	0.8326	0.8339	3
3	HP	Entered	0.0000	0.8569	0.8392	4
4	KM	Entered	0.0000	0.8675	0.8472	5
5	Weight	Entered	0.0000	0.8865	0.8895	6
6	Powered_Windows	Entered	0.0000	0.8896	0.8933	7
7	Quarterly_Tax	Entered	0.0000	0.8930	0.8942	8
8	Guarantee_Period	Entered	0.0000	0.8961	0.8940	9
9	BOVAG_Guarantee	Entered	0.0000	0.8993	0.8956	10
10	Color[White&Beige&Violet&Green&Red-Blue&Black&Silver&Grey&Yellow]	Entered	0.0001	0.9011	0.8965	11
11	Sport_Model	Entered	0.0003	0.9026	0.8975	12
12	Fuel_Type[CNG-Petrol]	Entered	0.0015	0.9041	0.9008	14
13	Boardcomputer	Entered	0.0014	0.9053	0.9003	15
14	ABS	Entered	0.0110	0.9060	0.9006	16
15	Age_08_04	Entered	0.0209	0.9066	0.9010	17
16	Automatic[0-1]	Entered	0.0185	0.9072	0.9005	18
17	Metallic_Rim	Entered	0.0171	0.9078	0.9007	19
18	Airco	Entered	0.0585	0.9082	0.9010	20
19	Mfg_Guarantee	Entered	0.0712	0.9086	0.9023	21
20	Backseat_Divider	Entered	0.0614	0.9089	0.9029	22
21	Color[Blue&Black&Silver-Grey&Yellow]	Entered	0.0912	0.9092	0.9027	23
22	Central_Lock	Entered	0.0833	0.9096	0.9017	24
23	Doors	Entered	0.1099	0.9098	0.9015	25
24	Color[White&Beige-Violet&Green&Red]	Entered	0.1189	0.9101	0.9020	26
25	Tow_Bar	Entered	0.1151	0.9104	0.9029	27
26	Airbag_1	Entered	0.3079	0.9105	0.9025	28
27	Color[Grey-Yellow]	Entered	0.3464	0.9106	0.9024	29
28	CD_Player	Entered	0.4018	0.9107	0.9029	30
29	Airbag_2	Entered	0.3797	0.9107	0.9030	31
30	Color[Violet&Green-Red]	Entered	0.4227	0.9108	0.9032	32
31	CC	Entered	0.5240	0.9108	0.9034	33
32	Met_Color[0-1]	Entered	0.5621	0.9109	0.9036	34
33	Gears	Entered	0.6239	0.9109	0.9035	35
34	Color[Violet-Green]	Entered	0.6316	0.9109	0.9036	36
35	Mistlamps	Entered	0.6821	0.9110	0.9034	37
36	Color[Black-Silver]	Entered	0.7004	0.9110	0.9032	39
37	Power_Steering	Entered	0.8402	0.9110	0.9032	40
38	Radio	Entered	0.9038	0.9110	0.9032	41
39	Mfg_Month	Entered	0.9997	0.9110	0.9032	42
40	Best	Specific	.	0.9109	0.9036	35

FIGURE 16.5 JMP Output of a Logistic Regression of the Credit Card Upgrade Data

Whole Model Test

Model	-LogLikelihood	DF	ChiSquare	Prob>ChiSq
Difference	18.492960	2	36.98592	<.0001*
Full	9.182906			
Reduced	27.675866			

RSquare (U)　0.6682

Lack Of Fit

Source	DF	-LogLikelihood	ChiSquare
Lack Of Fit	37	9.1829064	18.36581
Saturated	39	0.0000000	Prob>ChiSq
Fitted	2	9.1829064	0.9956

Parameter Estimates

Term	Estimate	Std Error	ChiSquare	Prob>ChiSq	Lower 95%	Upper 95%
Intercept	-9.9524146	4.0818938	5.94	0.0148*	-20.989597	-4.1146535
Purchases	0.20983066	0.0907079	5.35	0.0207*	0.07153188	0.44477918
PlatProfile	4.14786846	1.6101814	6.64	0.0100*	1.64251771	8.57651011

Effect Likelihood Ratio Tests

Source	Nparm	DF	L-R ChiSquare	Prob>ChiSq
Purchases	1	1	11.5458696	0.0007*
PlatProfile	1	1	12.8442451	0.0003*

Odds Ratios

For UpGrade odds of 1 versus 0
Tests and confidence intervals on odds ratios are likelihood ratio based.

Unit Odds Ratios

Per unit change in regressor

Term	Odds Ratio	Lower 95%	Upper 95%	Reciprocal
Purchases	1.233469	1.074152	1.560146	0.8107215
PlatProfile	63.29893	5.168165	5305.557	0.0157981

FIGURE 16.12 The Group Means and p-Values for Test 1 and Test 2

Group Means

Count	Group	Test 1	Test 2	Column	F Ratio	Prob>F
20	0	84.750000	79.100000	Test 1	27.384	0.0000056
23	1	92.434783	84.782609	Test 2	2.369	0.1316083
43	All	88.860465	82.139535			

FIGURE 16.17 The Single Layer Perceptron

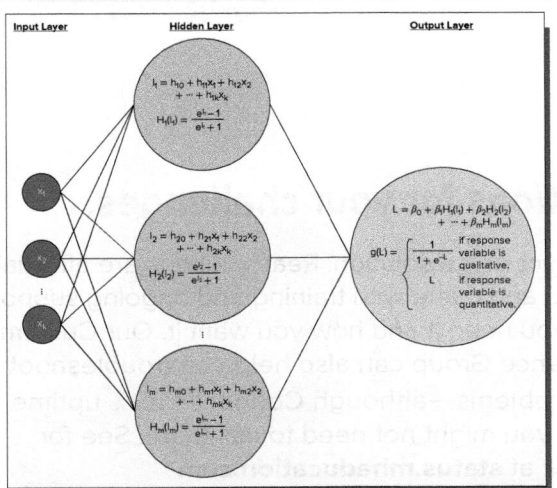

FIGURE 16.19 JMP Output of Neural Network Estimation for the Credit Card Upgrade Data ⊙ CardUpgrade

Neural　Validation Column: Validation　Model NTanH(3)

Estimates

Parameter	Estimate
H1_1:Purchases	-0.00498
H1_1:PlatProfile:0	0.393252
H1_1:Intercept	0.32598
H1_2:Purchases	-0.05685
H1_2:PlatProfile:0	0.219115
H1_2:Intercept	1.712453
H1_3:Purchases	-0.01804
H1_3:PlatProfile:0	0.556931
H1_3:Intercept	0.651032
Upgrade(0):H1_1	2.299942
Upgrade(0):H1_2	4.493357
Upgrade(0):H1_3	4.160657
Upgrade(0):Intercept	0.25162

Upgrade	Purchases	PlatProfile	Validation	H1_1	H1_2	H1_3	Probability (Upgrade==0)	Probability (Upgrade==1)	Most Likely Upgrade
41	42.571	1		-0.138671214	-0.432804209	-0.324738223	0.0334660985	0.9665339015	1
42	51.835	0		0.2266252687	-0.468027018	0.1355958367	0.3173449846	0.6826550154	1

Students—study more efficiently, retain more and achieve better outcomes. Instructors—focus on what you love—teaching.

SUCCESSFUL SEMESTERS INCLUDE CONNECT

FOR INSTRUCTORS

You're in the driver's seat.

Want to build your own course? No problem. Prefer to use our turnkey, prebuilt course? Easy. Want to make changes throughout the semester? Sure. And you'll save time with Connect's auto-grading too.

65%
Less Time Grading

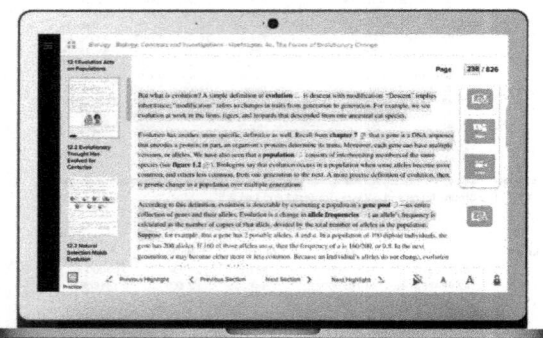

They'll thank you for it.

Adaptive study resources like SmartBook® help your students be better prepared in less time. You can transform your class time from dull definitions to dynamic debates. Hear from your peers about the benefits of Connect at **www.mheducation.com/highered/connect**

Make it simple, make it affordable.

Connect makes it easy with seamless integration using any of the major Learning Management Systems—Blackboard®, Canvas, and D2L, among others—to let you organize your course in one convenient location. Give your students access to digital materials at a discount with our inclusive access program. Ask your McGraw-Hill representative for more information.

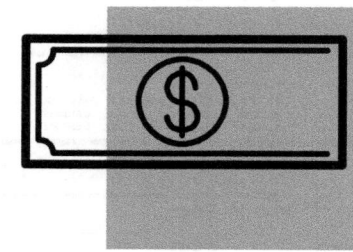

©Hill Street Studios/Tobin Rogers/Blend Images LLC

Solutions for your challenges.

A product isn't a solution. Real solutions are affordable, reliable, and come with training and ongoing support when you need it and how you want it. Our Customer Experience Group can also help you troubleshoot tech problems—although Connect's 99% uptime means you might not need to call them. See for yourself at **status.mheducation.com**

FOR STUDENTS

Effective, efficient studying.

Connect helps you be more productive with your study time and get better grades using tools like SmartBook, which highlights key concepts and creates a personalized study plan. Connect sets you up for success, so you walk into class with confidence and walk out with better grades.

©Shutterstock/wavebreakmedia

" I really liked this app—it made it easy to study when you don't have your text-book in front of you. **"**

- Jordan Cunningham,
Eastern Washington University

Study anytime, anywhere.

Download the free ReadAnywhere app and access your online eBook when it's convenient, even if you're offline. And since the app automatically syncs with your eBook in Connect, all of your notes are available every time you open it. Find out more at **www.mheducation.com/readanywhere**

No surprises.

The Connect Calendar and Reports tools keep you on track with the work you need to get done and your assignment scores. Life gets busy; Connect tools help you keep learning through it all.

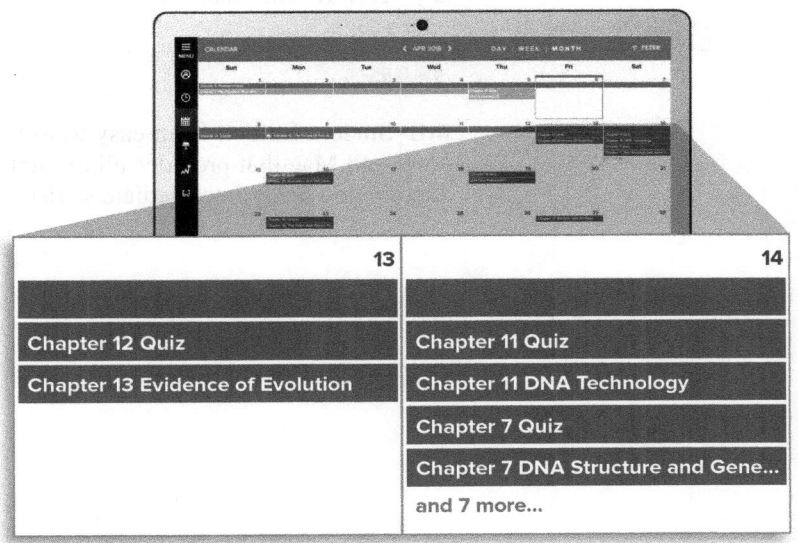

13	14
Chapter 12 Quiz	Chapter 11 Quiz
Chapter 13 Evidence of Evolution	Chapter 11 DNA Technology
	Chapter 7 Quiz
	Chapter 7 DNA Structure and Gene...
	and 7 more...

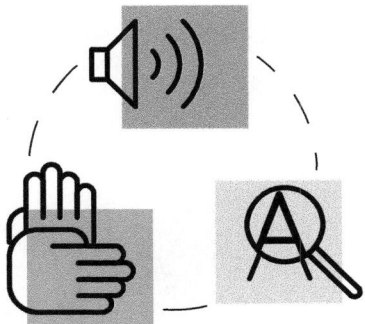

Learning for everyone.

McGraw-Hill works directly with Accessibility Services Departments and faculty to meet the learning needs of all students. Please contact your Accessibility Services office and ask them to email accessibility@mheducation.com, or visit **www.mheducation.com/about/accessibility.html** for more information.

ADDITIONAL RESOURCES

MEGASTAT® FOR MICROSOFT EXCEL® (AND EXCEL: MAC)

MegaStat is a full-featured Excel add-in by J. B. Orris of Butler University that is available with this text. It performs statistical analyses within an Excel workbook. It does basic functions such as descriptive statistics, frequency distributions, and probability calculations, as well as hypothesis testing, ANOVA, and regression.

MegaStat output is carefully formatted. Ease-of-use features include AutoExpand for quick data selection and Auto Label detect. Since MegaStat is easy to use, students can focus on learning statistics without being distracted by the software. MegaStat is always available from Excel's main menu. Selecting a menu item pops up a dialog box. MegaStat works with all recent versions of Excel. For more information, go to **mhhe.com/megastat**.

MINITAB®

Minitab® Student Version 18 is available to help students solve the business statistics exercises in the text. This software is available in the student version and can be packaged with any McGraw-Hill business statistics text.

JMP®

JMP Student Edition is an easy-to-use streamlined version of JMP software for both Windows and Mac that provides all the statistical analysis and graphical tools covered in introductory and many intermediate statistics courses.

CHAPTER-BY-CHAPTER REVISIONS FOR 9TH EDITION:

Chapter 1

- Improved and simpler introduction to business statistics
- Completely rewritten and much improved introduction to business analytics
- Appendix on using JMP added

Chapter 2

- JMP examples and exercises added
- Appendix on using JMP added

Chapter 3

- Improved discussion of association rules
- New section on text mining and latent semantic analysis added
- Much improved discussions of hierarchical clustering, k-means clustering, and multidimensional scaling; Improved discussion of factor analysis
- JMP examples and exercises added
- Appendix on using JMP added

Chapter 4

- Improved discussion of probability modeling

Chapter 5

- A new chapter on "nonparametric" prediction analytics—classification trees, regression trees, k-nearest neighbors, and naive Bayes' classification
- JMP examples and exercises added
- Appendix on using JMP added

Chapter 6 (formerly 5)

- JMP examples and exercises added
- Appendix on using JMP added

Chapter 7 (formerly 6)

- JMP examples and exercises added
- Appendix on using JMP added

Chapter 8 (formerly 7)

- No significant changes

Chapter 9 (formerly 8)

- JMP examples and exercises added
- Appendix on using JMP added

Chapter 10 (formerly 9)

- Improved section on formulating statistical hypotheses and the meanings of Type I and Type II errors
- Much improved (more unified, simpler, clearer, and shorter) explanation of using critical value rules and p-values to test hypotheses
- JMP examples and exercises added
- Appendix on using JMP added

Chapter 11 (formerly 10)

- JMP examples and exercises added
- Appendix on using JMP added

Chapter 12 (formerly 11)

- JMP examples and exercises added
- Appendix on using JMP added

Chapter 13 (formerly 12)

- JMP examples and exercises added
- Appendix on using JMP added

Chapter 14 (formerly 13)

- JMP examples and exercises added
- Appendix on using JMP added

Chapter 15 (formerly 14)

- New section on model building for big data added
- Improved discussion of diagnosing outlying and influential observations
- JMP examples and exercises added
- Appendix on using JMP added

Chapter 16

- A new chapter on "parametric" predictive analytics–logistic regression, linear discriminate analysis, and neural networks
- JMP examples and exercises added
- Appendix on using JMP added

Chapter 17 (formerly 15)

- Expanded discussion of exponential smoothing
- New and fuller (but understandable) discussion of the Box—Jenkins methodology
- JMP examples and exercises added
- Appendix on using JMP added

Chapter 18 (formerly 17)

- JMP examples and exercises added
- Appendix on using JMP added

Chapter 19 (formerly 18)

- No significant changes

Chapter 20 (on website–formerly 16)

- Appendix on using JMP added

ACKNOWLEDGMENTS

We wish to thank many people who have helped to make this book a reality. As indicated on the title page, we thank Professor Steven C. Huchendorf, University of Minnesota, and Dawn C. Porter, University of Southern California, for major contributions to this book. We also thank former co-author Emily Murphree for all of her excellent work on previous editions.

We wish to thank the people at McGraw-Hill for their dedication to this book. These people include senior brand manager Noelle Bathurst, who is extremely helpful to the authors; senior development editor Tobi Philips who has shown great dedication to the improvement of this book; content project manager Fran Simon, who has very capably and diligently guided this book through its production and who has been a tremendous help to the authors; and our former executive editor Steve Scheutz, who always greatly supported our books. We also thank lead product developer Michelle Janicek

for her tremendous help in developing this new edition; our former executive editor Scott Isenberg for the tremendous help he has given us in developing all of our McGraw-Hill business statistics books; and our former executive editor Dick Hercher, who persuaded us to publish with McGraw-Hill.

We wish to thank Larry White, Eastern Illinois University, for revising the Test Bank and Vickie Fry, Indiana University Bloomington, for revising the PowerPoints and reviewing the Test Bank. Additional thanks to our co-authors for their work above and beyond the revision of the text: Patrick Schur, Miami University, for his work developing learning resources; Kyle Moninger, Bowling Green State University, for updating and revising the LearnSmart content; and Amy Froelich, Iowa State University, for developing the Solutions Manual. Most importantly, we wish to thank our families for their acceptance, unconditional love, and support.

DEDICATION

Bruce L. Bowerman

— Richard T. O'Connell, my best friend and wonderful co-author and person, whom I miss so much.

— Herbert T. David, my brilliant and kind Ph.D. advisor, who generously helped so many Ph.D. students.

— All of my loved ones.

Anne Drougas

I would like to dedicate this book to my parents, Arthur and Mary, and my sister, Cathy, who have been the greatest blessings in my life, my Ph.D. dissertation advisor, John, and Steve.

William Duckworth

To my wife, Shelia, and children—Billy, Kim, and Andrew: You have been supportive, helpful, and patient during this project and I thank you.

Amy Froelich

To my husband, Jim, and daughters, Sarah and Jamie.

To my Mom and Dad and brother, Scott.

Ruth Hummel:

To Bruce, for his friendship and the many interesting statistical and non-statistical conversations we've had.

Kyle Moninger

Dedicated to my family. They are everything to me.

Pat Schur

To my wife, children, and grandchildren

    — Lorie

    — Andy, Manda, and Angie

    — Cooper, Chloe, Emma, Gracie, and Nolan

BRIEF CONTENTS

CONTENTS

©Tetra Images/Alamy

CHAPTER 1

An Introduction to Business Statistics and Analytics

Learning Objectives

After mastering the material in this chapter, you will be able to:

LO1-1 Define a variable.

LO1-2 Describe the difference between a quantitative variable and a qualitative variable.

LO1-3 Describe the difference between cross-sectional data and time series data.

LO1-4 Construct and interpret a time series (runs) plot.

LO1-5 Identify the different types of data sources: existing data sources, experimental studies, and observational studies.

LO1-6 Explain the basic ideas of data warehousing and big data.

LO1-7 Describe the difference between a population and a sample.

LO1-8 Distinguish between descriptive statistics and statistical inference.

LO1-9 Explain the concept of random sampling and select a random sample.

LO1-10 Explain some of the uses of business analytics and data mining (Optional).

LO1-11 Identify the ratio, interval, ordinal, and nominative scales of measurement (Optional).

LO1-12 Describe the basic ideas of stratified random, cluster, and systematic sampling (Optional).

LO1-13 Describe basic types of survey questions, survey procedures, and sources of error (Optional).

Chapter Outline

1.1 Data

1.2 Data Sources, Data Warehousing, and Big Data

1.3 Populations, Samples, and Traditional Statistics

1.4 Random Sampling and Three Case Studies That Illustrate Statistical Inference

1.5 Business Analytics and Data Mining (Optional)

1.6 Ratio, Interval, Ordinal, and Nominative Scales of Measurement (Optional)

1.7 Stratified Random, Cluster, and Systematic Sampling (Optional)

1.8 More about Surveys and Errors in Survey Sampling (Optional)

he subject of statistics involves the study of how to collect, analyze, and interpret data. **Data are facts and figures from which conclusions can be drawn**. Such conclusions are important to the decision making of many professions and organizations. For example, **economists** use conclusions drawn from the latest data on unemployment and inflation to help the government make policy decisions. **Financial planners** use recent trends in stock market prices and economic conditions to make investment decisions. **Accountants** use **sample data** concerning a company's *actual sales revenues* to assess whether the company's *claimed sales revenues* are valid. **Marketing professionals** and **data miners** help businesses decide which products to develop and market and which consumers to target in marketing campaigns by using data that reveal consumer preferences. **Production supervisors** use manufacturing data to evaluate, control, and improve product quality. **Politicians** rely on data from public opinion polls to formulate legislation and to devise campaign strategies. **Physicians and hospitals** use data on the effectiveness of drugs and surgical procedures to provide patients with the best possible treatment.

In this chapter we begin to see how we collect and analyze data. As we proceed through the chapter, we introduce several case studies. These case studies (and others to be introduced later) are revisited throughout later chapters as we learn the statistical methods needed to analyze them. Briefly, we will begin to study four cases:

The Cell Phone Case: A bank estimates its cellular phone costs and decides whether to outsource management of its wireless resources by studying the calling patterns of its employees.

The Marketing Research Case: A beverage company investigates consumer reaction to a new bottle design for one of its popular soft drinks.

The Car Mileage Case: To determine if it qualifies for a federal tax credit based on fuel economy, an automaker studies the gas mileage of its new midsize model.

The Disney Parks Case: Walt Disney World Parks and Resorts in Orlando, Florida, manages Disney parks worldwide and uses data gathered from its guests to give these guests a more "magical" experience and increase Disney revenues and profits.

1.1 Data

LO1-1
Define a variable.

Data sets, elements, and variables

We have said that data are facts and figures from which conclusions can be drawn. Together, the data that are collected for a particular study are referred to as a **data set.** For example, Table 1.1 is a data set that gives information about the new homes sold in a Florida luxury home development over a recent three-month period. Potential home buyers could choose either the "Diamond" or the "Ruby" home model design and could have the home built on either a lake lot or a treed lot (with no water access).

In order to understand the data in Table 1.1, note that any data set provides information about some group of individual **elements,** which may be people, objects, events, or other entities. The information that a data set provides about its elements usually describes one or more characteristics of these elements.

Any characteristic of an element is called a **variable.**

TABLE 1.1 **A Data Set Describing Five Home Sales** DS HomeSales

Home	Model Design	Lot Type	List Price	Selling Price
1	Diamond	Lake	$494,000	$494,000
2	Ruby	Treed	$447,000	$398,000
3	Diamond	Treed	$494,000	$440,000
4	Diamond	Treed	$494,000	$469,000
5	Ruby	Lake	$447,000	$447,000

LO1-2

Describe the difference between a quantitative variable and a qualitative variable.

TABLE 1.2
2016 MLB Payrolls
ⒹⓈ MLB

Team	2016 Payroll
Los Angeles Dodgers	223
New York Yankees	213
Boston Red Sox	182
Detroit Tigers	172
San Francisco Giants	166
Washington Nationals	166
Los Angeles Angels	146
Texas Rangers	144
Philadelphia Phillies	133
Toronto Blue Jays	126
Seattle Mariners	123
St. Louis Cardinals	120
Cincinnati Reds	117
Chicago Cubs	117
Baltimore Orioles	116
Kansas City Royals	113
San Diego Padres	113
Minnesota Twins	108
New York Mets	100
Chicago White Sox	99
Milwaukee Brewers	99
Colorado Rockies	98
Atlanta Braves	88
Cleveland Indians	86
Pittsburgh Pirates	86
Miami Marlins	85
Oakland Athletics	80
Tampa Bay Rays	74
Arizona Diamondbacks	71
Houston Astros	69

Source: www.stevetheump.com, January 15, 2017.

For the data set in Table 1.1, each sold home is an element, and four variables are used to describe the homes. These variables are (1) the home model design, (2) the type of lot on which the home was built, (3) the list (asking) price, and (4) the (actual) selling price. Moreover, each home model design came with "everything included"—specifically, a complete, luxury interior package and a choice (at no price difference) of one of three different architectural exteriors. The builder made the list price of each home solely dependent on the model design. However, the builder gave various price reductions for homes built on treed lots.

The data in Table 1.1 are real (with some minor changes to protect privacy) and were provided by a business executive—a friend of the authors—who recently received a promotion and needed to move to central Florida. While searching for a new home, the executive and his family visited the luxury home community and decided they wanted to purchase a Diamond model on a treed lot. The list price of this home was $494,000, but the developer offered to sell it for an "incentive" price of $469,000. Intuitively, the incentive price's $25,000 savings off list price seemed like a good deal. However, the executive resisted making an immediate decision. Instead, he decided to collect data on the selling prices of new homes recently sold in the community and use the data to assess whether the developer might accept a lower offer. In order to collect "relevant data," the executive talked to local real estate professionals and learned that new homes sold in the community during the previous three months were a good indicator of current home value. Using real estate sales records, the executive also learned that five of the community's new homes had sold in the previous three months. The data given in Table 1.1 are the data that the executive collected about these five homes.

When the business executive examined Table 1.1, he noted that homes on lake lots had sold at their list price, but homes on treed lots had not. Because the executive and his family wished to purchase a Diamond model on a treed lot, the executive also noted that two Diamond models on treed lots had sold in the previous three months. One of these Diamond models had sold for the incentive price of $469,000, but the other had sold for a lower price of $440,000. Hoping to pay the lower price for his family's new home, the executive offered $440,000 for the Diamond model on the treed lot. Initially, the home builder turned down this offer, but two days later the builder called back and accepted the offer. The executive had used data to buy the new home for $54,000 less than the list price and $29,000 less than the incentive price!

Quantitative and qualitative variables

For any variable describing an element in a data set, we carry out a **measurement** to assign a value of the variable to the element. For example, in the real estate example, real estate sales records gave the actual selling price of each home to the nearest dollar. As another example, a credit card company might measure the time it takes for a cardholder's bill to be paid to the nearest day. Or, as a third example, an automaker might measure the gasoline mileage obtained by a car in city driving to the nearest one-tenth of a mile per gallon by conducting a mileage test on a driving course prescribed by the Environmental Protection Agency (EPA). If the possible values of a variable are numbers that represent quantities (that is, "how much" or "how many"), then the variable is said to be **quantitative.** For example, (1) the actual selling price of a home, (2) the payment time of a bill, (3) the gasoline mileage of a car, and (4) the 2016 payroll of a Major League Baseball team are all quantitative variables. Considering the last example, Table 1.2 gives the 2016 payroll (in millions of dollars) for each of the 30 Major League Baseball (MLB) teams. Moreover, Figure 1.1 portrays the team payrolls as a **dot plot.** In this plot, each team payroll is shown as a dot located on the real number

FIGURE 1.1   A Dot Plot of 2016 MLB Payrolls (Payroll Is a Quantitative Variable)

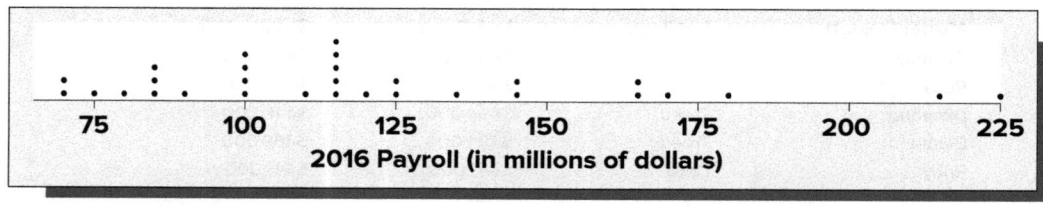

FIGURE 1.2   The Ten Most Popular Car Colors in the World for 2012 (Car Color Is a Qualitative Variable)

FIGURE 1.3   Time Series Plot of the Average Basic Cable Rates in the U.S. from 1999 to 2009   DS BasicCable

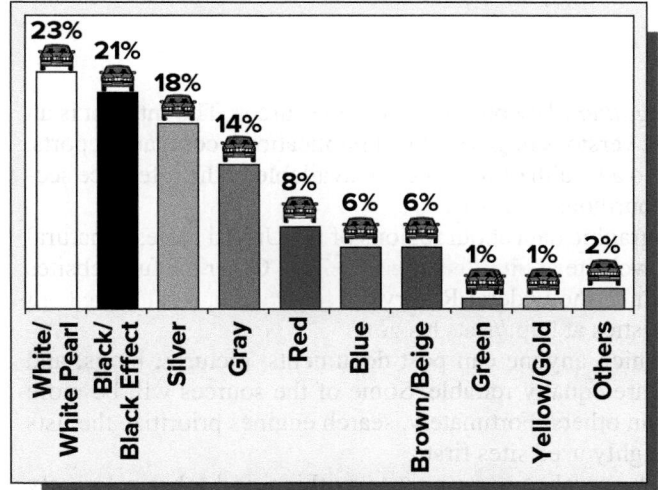
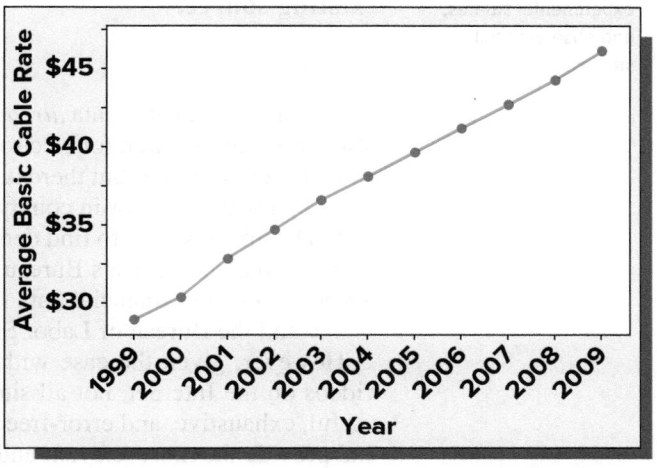

Source: Author created using data from http://autoweek.com/article/car-life/dupont-color-survey-puts-white-and-black-top-again (accessed September 12, 2013).

TABLE 1.3   The Average Basic Cable Rates in the U.S. from 1999 to 2009   DS BasicCable

Year	1999	2000	2001	2002	2003	2004	2005	2006	2007	2008	2009
Cable Rate	$ 28.92	30.37	32.87	34.71	36.59	38.14	39.63	41.17	42.72	44.28	46.13

Source: U.S. Energy Information Administration, http://www.eia.gov/.

line—for example, the rightmost dot represents the payroll for the Los Angeles Dodgers. In general, the values of a quantitative variable are numbers on the real line. In contrast, if we simply record into which of several categories an element falls, then the variable is said to be **qualitative** or **categorical.** Examples of categorical variables include (1) a person's gender, (2) whether a person who purchases a product is satisfied with the product, (3) the type of lot on which a home is built, and (4) the color of a car.[1] Figure 1.2 illustrates the categories we might use for the qualitative variable "car color." This figure is a **bar chart** showing the 10 most popular (worldwide) car colors for 2012 and the percentages of cars having these colors.

Cross-sectional and time series data

Some statistical techniques are used to analyze *cross-sectional data,* while others are used to analyze *time series data.* **Cross-sectional data** are data collected at the same or approximately the same point in time. For example, suppose that a bank wishes to analyze last month's cell phone bills for its employees. Then, because the cell phone costs given by these bills are for different employees in the same month, the cell phone costs are cross-sectional data. **Time series data** are data collected over different time periods. For example, Table 1.3 presents the average basic cable television rate in the United States for each of the years 1999 to 2009. Figure 1.3 is a **time series plot**—also called a **runs plot**—of these data. Here we plot each cable rate on the vertical scale versus its corresponding time index (year) on the horizontal scale. For instance, the first cable rate ($28.92) is plotted versus 1999, the second cable rate ($30.37) is plotted versus 2000, and so forth. Examining the time series plot, we see that the cable rates increased substantially from 1999 to 2009. Finally, because the five homes in Table 1.1 were sold over a three-month period that represented a relatively stable real estate market, we can consider the data in Table 1.1 to essentially be cross-sectional data.

LO1-3
Describe the difference between cross-sectional data and time series data.

LO1-4
Construct and interpret a time series (runs) plot.

[1]Optional Section 1.6 discusses two types of quantitative variables (ratio and interval) and two types of qualitative variables (ordinal and nominative).

LO1-5
Identify the different
types of data sources:
existing data sources,
experimental studies,
and observational
studies.

1.2 Data Sources, Data Warehousing, and Big Data

Primary data are data collected by an individual or business directly through planned **experimentation** or **observation. Secondary data** are data taken from an **existing source.**

Existing sources

Sometimes we can use data *already gathered* by public or private sources. The Internet is an obvious place to search for electronic versions of government publications, company reports, and business journals, but there is also a wealth of information available in the reference section of a good library or in county courthouse records.

If a business wishes to find demographic data about regions of the United States, a natural source is the U.S. Census Bureau's website at http://www.census.gov. Other useful websites for economic and financial data include the Federal Reserve at http://research.stlouisfed.org /fred2/ and the Bureau of Labor Statistics at http://stats.bls.gov/.

However, given the ease with which anyone can post documents, pictures, blogs, and videos on the Internet, not all sites are equally reliable. Some of the sources will be more useful, exhaustive, and error-free than others. Fortunately, search engines prioritize the lists and provide the most relevant and highly used sites first.

Obviously, performing such web searches costs next to nothing and takes relatively little time, but the tradeoff is that we are also limited in terms of the type of information we are able to find. Another option may be to use a private data source. Most companies keep and use employee records and information about their customers, products, processes (inventory, payroll, manufacturing, and accounting), and advertising results. If we have no affiliation with these companies, however, these data may be difficult to obtain.

Another alternative would be to contact a data collection agency, which typically incurs some kind of cost. You can either buy subscriptions or purchase individual company financial reports from agencies like Bloomberg and Dow Jones & Company. If you need to collect specific information, some companies, such as ACNielsen and Information Resources, Inc., can be hired to collect the information for a fee. Moreover, no matter what existing source you take data from, it is important to assess how reliable the data are by determing how, when, and where the data were collected.

Experimental and observational studies

There are many instances when the data we need are not readily available from a public or private source. In cases like these, we need to collect the data ourselves. Suppose we work for a beverage company and want to assess consumer reactions to a new bottled water. Because the water has not been marketed yet, we may choose to conduct taste tests, focus groups, or some other market research. As another example, when projecting political election results, telephone surveys and exit polls are commonly used to obtain the information needed to predict voting trends. New drugs for fighting disease are tested by collecting data under carefully controlled and monitored experimental conditions. In many marketing, political, and medical situations of these sorts, companies sometimes hire outside consultants or statisticians to help them obtain appropriate data. Regardless of whether newly minted data are gathered in-house or by paid outsiders, this type of data collection requires much more time, effort, and expense than are needed when data can be found from public or private sources.

When initiating a study, we first define our variable of interest, or **response variable.** Other variables, typically called **factors,** that may be related to the response variable of interest will also be measured. When we are able to set or manipulate the values of these factors, we have an **experimental study.** For example, a pharmaceutical company might wish to determine the most appropriate daily dose of a cholesterol-lowering drug for patients having cholesterol levels that are too high. The company can perform an experiment in which one

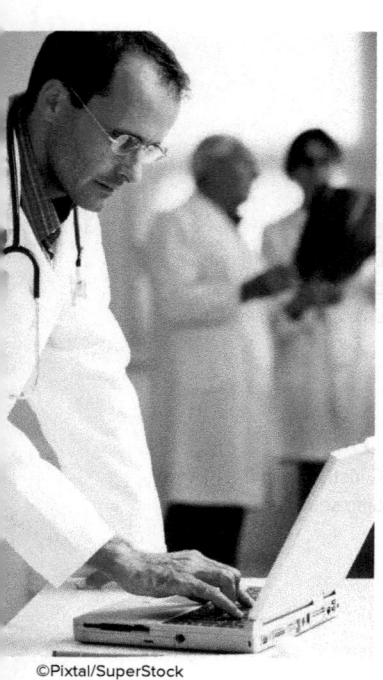

©Pixtal/SuperStock

sample of patients receives a placebo; a second sample receives some low dose; a third a higher dose; and so forth. This is an experiment because the company controls the amount of drug each group receives. The optimal daily dose can be determined by analyzing the patients' responses to the different dosage levels given.

When analysts are unable to control the factors of interest, the study is **observational.** In studies of diet and cholesterol, patients' diets are not under the analyst's control. Patients are often unwilling or unable to follow prescribed diets; doctors might simply ask patients what they eat and then look for associations between the factor *diet* and the response variable *cholesterol level.*

Asking people what they eat is an example of performing a **survey.** In general, people in a survey are asked questions about their behaviors, opinions, beliefs, and other characteristics. For instance, shoppers at a mall might be asked to fill out a short questionnaire which seeks their opinions about a new bottled water. In other observational studies, we might simply observe the behavior of people. For example, we might observe the behavior of shoppers as they look at a store display, or we might observe the interactions between students and teachers.

Transactional data, data warehousing, and big data

With the increased use of online purchasing and with increased competition, businesses have become more aggressive about collecting information concerning customer transactions. Every time a customer makes an online purchase, more information is obtained than just the details of the purchase itself. For example, the web pages searched before making the purchase and the times that the customer spent looking at the different web pages are recorded. Similarly, when a customer makes an in-store purchase, store clerks often ask for the customer's address, zip code, e-mail address, and telephone number. By studying past customer behavior and pertinent demographic information, businesses hope to accurately predict customer response to different marketing approaches and leverage these predictions into increased revenues and profits. Dramatic advances in data capture, data transmission, and data storage capabilities are enabling organizations to integrate various databases into **data warehouses.** *Data warehousing* is defined as a process of centralized data management and retrieval and has as its ideal objective the creation and maintenance of a central repository for all of an organization's data. The huge capacity of data warehouses has given rise to the term **big data,** which refers to massive amounts of data, often collected at very fast rates in real time and in different forms and sometimes needing quick preliminary analysis for effective business decision making.

LO1-6
Explain the basic ideas of data warehousing and big data.

 EXAMPLE 1.1 The Disney Parks Case: Improving Visitor Experiences

Annually, approximately 100 million visitors spend time in Walt Disney parks around the world. These visitors could generate a lot of data, and in 2013, Walt Disney World Parks and Resorts introduced the wireless-tracking wristband *MagicBand* in Walt Disney World in Orlando, Florida.

The MagicBands are linked to a credit card and serve as a park entry pass and hotel room key. They are part of the *McMagic+* system and wearing a band is completely voluntary. In addition to expediting sales transactions and hotel room access in the Disney theme parks, MagicBands provide visitors with easier access to FastPass lines for Disney rides and attractions. Each visitor to a Disney theme park may choose a FastPass for three rides or attractions per day. A FastPass allows a visitor to enter a line where there is virtually no waiting time. The McMagic+ system automatically programs a visitor's FastPass selections into his or her MagicBand. As shown by the photo, a visitor simply places the MagicBand on his or her wrist next to a FastPass entry reader and is immediately admitted to the ride or attraction.

In return, the McMagic+ system allows Disney to collect massive amounts of valuable data like real-time location, purchase history, riding patterns, and audience analysis and

©Bob Croslin

segmentation data. For example, the data tell Disney the types and ages of people who like specific attractions. To store, process, analyze and visualize all the data, Disney has constructed a gigantic data warehouse and a big data analysis platform. The data analysis allows Disney to improve daily park operations (by having the right numbers of staff on hand for the number of visitors currently in the park); to improve visitor experiences when choosing their "next" ride (by having large displays showing the waiting times for the park's rides); to improve its attraction offerings; and to tailor its marketing messages to different types of visitors.

Finally, although it collects massive amounts of data, Disney is very ethical in protecting the privacy of its visitors. First, as previously stated, visitors can choose not to wear a MagicBand. Moreover, visitors who do decide to wear one have control over the quantities of data collected, stored, and shared. Visitors can use a menu to specify whether Disney can send them personalized offers during or after their park visit. Parents also have to opt in before the characters in the park can address their children by name or use other personal information stored in the MagicBands.

Exercises for Sections 1.1 and 1.2

CONCEPTS   connect

1.1 Define what we mean by a *variable,* and explain the difference between a quantitative variable and a qualitative (categorical) variable.

1.2 Below we list several variables. Which of these variables are quantitative and which are qualitative? Explain.
  a The dollar amount on an accounts receivable invoice.
  b The net profit for a company in 2017.
  c The stock exchange on which a company's stock is traded.
  d The national debt of the United States in 2017.
  e The advertising medium (radio, television, or print) used to promote a product.

1.3 (1) Discuss the difference between cross-sectional data and time series data. (2) If we record the total number of cars sold in 2017 by each of 10 car salespeople, are the data cross-sectional or time series data? (3) If we record the total number of cars sold by a particular car salesperson in each of the years 2013, 2014, 2015, 2016, and 2017, are the data cross-sectional or time series data?

1.4 Consider a medical study that is being performed to test the effect of smoking on lung cancer. Two groups of subjects are identified; one group has lung cancer and the other one doesn't. Both are asked to fill out a questionnaire containing questions about their age, sex, occupation, and number of cigarettes smoked per day. (1) What is the response variable? (2) Which are the factors? (3) What type of study is this (experimental or observational)?

1.5 What is a data warehouse? What does the term *big data* mean?

METHODS AND APPLICATIONS

1.6 Consider the five homes in Table 1.1. What do you think you would have to pay for a Ruby model on a treed lot?

1.7 Consider the five homes in Table 1.1. What do you think you would have to pay for a Diamond model on a lake lot? For a Ruby model on a lake lot?

1.8 The number of Bismark X-12 electronic calculators sold at Smith's Department Stores over the past 24 months have been: 197, 211, 203, 247, 239, 269, 308, 262, 258, 256, 261, 288, 296, 276, 305, 308, 356, 393, 363, 386, 443, 308, 358, and 384. Make a time series plot of these data. That is, plot 197 versus month 1, 211 versus month 2, and so forth. What does the time series plot tell you? DS CalcSale

1.3 Populations, Samples, and Traditional Statistics

LO1-7

Describe the difference between a population and a sample.

We often collect data in order to study a population.

> A **population** is the set of all elements about which we wish to draw conclusions.

Examples of populations include (1) all of last year's graduates of Dartmouth College's Master of Business Administration program, (2) all current MasterCard cardholders, and (3) all Buick LaCrosses that have been or will be produced this year.

We usually focus on studying one or more variables describing the population elements. If we carry out a measurement to assign a value of a variable to each and every population element, we have a *population of measurements* (sometimes called *observations*). If the population is small, it is reasonable to do this. For instance, if 150 students graduated last year from the Dartmouth College MBA program, it might be feasible to survey the graduates and to record all of their starting salaries. In general:

> If we examine all of the population measurements, we say that we are conducting a **census** of the population.

Often the population that we wish to study is very large, and it is too time-consuming or costly to conduct a census. In such a situation, we select and analyze a subset (or portion) of the population elements.

> A **sample** is a subset of the elements of a population.

For example, suppose that 8,742 students graduated last year from a large state university. It would probably be too time-consuming to take a census of the population of all of their starting salaries. Therefore, we would select a sample of graduates, and we would obtain and record their starting salaries. When we measure a characteristic of the elements in a sample, we have a **sample of measurements.**

We often wish to describe a population or sample.

LO1-8
Distinguish between descriptive statistics and statistical inference.

> **Descriptive statistics** is the science of describing the important aspects of a set of measurements.

As an example, if we are studying a set of starting salaries, we might wish to describe (1) what a typical salary might be and (2) how much the salaries vary from each other.

When the population of interest is small and we can conduct a census of the population, we will be able to directly describe the important aspects of the population measurements. However, if the population is large and we need to select a sample from it, then we use what we call **statistical inference.**

> **Statistical inference** is the science of using a sample of measurements to make generalizations about the important aspects of a population of measurements.

For instance, we might use the starting salaries recorded for a sample of the 8,742 students who graduated last year from a large state university to *estimate* the typical starting salary and the variation of the starting salaries for the entire population of the 8,742 graduates. Or General Motors might use a sample of Buick LaCrosses produced this year to estimate the typical EPA combined city and highway driving mileage and the variation of these mileages for all LaCrosses that have been or will be produced this year.

What we might call **traditional statistics** consists of a set of concepts and techniques that are used to describe populations and samples and to make statistical inferences about populations by using samples. Much of this book is devoted to traditional statistics, and in the next section we will discuss **random sampling** (or approximately random sampling). However, traditional statistics is sometimes not sufficient to analyze big data, which (we recall) refers to massive amounts of data often collected at very fast rates in real time and sometimes needing quick preliminary analysis for effective business decision making. For this reason, two related extensions of traditional statistics—**business analytics** and **data mining**—have been developed to help analyze big data. In optional Section 1.5 we will begin to discuss business analytics and data mining. As one example of using business analytics, we will see how Disney uses the large amount of data it collects every day concerning the riding patterns of its visitors. These data are used to keep its visitors informed of the current waiting times for different rides, which helps patrons select the next ride to go on or attraction to attend.

1.4 Random Sampling and Three Case Studies That Illustrate Statistical Inference

Random sampling

If the information contained in a sample is to accurately reflect the population under study, the sample should be **randomly selected** from the population. To intuitively illustrate random sampling, suppose that a small company employs 15 people and wishes to randomly select two of them to attend a convention. To make the random selections, we number the employees from 1 to 15, and we place in a hat 15 identical slips of paper numbered from 1 to 15. We thoroughly mix the slips of paper in the hat and, blindfolded, choose one. The number on the chosen slip of paper identifies the first randomly selected employee. Then, still blindfolded, we choose another slip of paper from the hat. The number on the second slip identifies the second randomly selected employee.

Of course, when the population is large, it is not practical to randomly select slips of paper from a hat. For instance, experience has shown that thoroughly mixing slips of paper (or the like) can be difficult. Further, dealing with many identical slips of paper would be cumbersome and time-consuming. For these reasons, statisticians have developed more efficient and accurate methods for selecting a random sample. To discuss these methods we let n denote the number of elements in a sample. We call n the **sample size.** We now define a random sample of n elements and explain how to select such a sample.[2]

> 1   If we select n elements from a population in such a way that every set of n elements in the population has the same chance of being selected, then the n elements we select are said to be a **random sample.**
>
> 2   In order to select a random sample of n elements from a population, we make n *random selections*—one at a time—from the population. On each **random selection,** we give every element remaining in the population for that selection the same chance of being chosen.

In making random selections from a population, we can sample *with or without replacement.* If we **sample with replacement,** we place the element chosen on any particular selection back into the population. Thus, we give this element a chance to be chosen on any succeeding selection. If we **sample without replacement,** we do not place the element chosen on a particular selection back into the population. Thus, we do not give this element a chance to be chosen on any succeeding selection. **It is best to sample without replacement.** Intuitively, this is because choosing the sample without replacement guarantees that all of the elements in the sample will be different, and thus we will have the fullest possible look at the population.

We now introduce three case studies that illustrate (1) the need for a random (or approximately random) sample, (2) how to select the needed sample, and (3) the use of the sample in making statistical inferences.

Ⓒ **EXAMPLE 1.2** The Cell Phone Case: Reducing Cellular Phone Costs

Part 1: The Cost of Company Cell Phone Use   Rising cell phone costs have forced companies having large numbers of cellular users to hire services to manage their cellular and other wireless resources. These cellular management services use sophisticated software and mathematical models to choose cost-efficient cell phone plans for their clients. One such firm, mindWireless of Austin, Texas, specializes in automated wireless cost management.

---

[2]Actually, there are several different kinds of random samples. The type we will define is sometimes called a *simple random sample*. For brevity's sake, however, we will use the term *random sample*.

According to Kevin Whitehurst, co-founder of mindWireless, cell phone carriers count on *overage*—using more minutes than one's plan allows—and *underage*—using fewer minutes than those already paid for—to deliver almost half of their revenues.[3] As a result, a company's typical cost of cell phone use can be excessive—18 cents per minute or more. However, Mr. Whitehurst explains that by using mindWireless automated cost management to select calling plans, this cost can be reduced to 12 cents per minute or less.

In this case we consider a bank that wishes to decide whether to hire a cellular management service to choose its employees' calling plans. While the bank has over 10,000 employees on many different types of calling plans, a cellular management service suggests that by studying the calling patterns of cellular users on 500-minute-per-month plans, the bank can accurately assess whether its cell phone costs can be substantially reduced. The bank has 2,136 employees on a variety of 500-minute-per-month plans with different basic monthly rates, different overage charges, and different additional charges for long distance and roaming. It would be extremely time-consuming to analyze in detail the cell phone bills of all 2,136 employees. Therefore, the bank will estimate its cellular costs for the 500-minute plans by analyzing last month's cell phone bills for a *random sample* of 100 employees on these plans.[4]

Part 2: Selecting a Random Sample   The first step in selecting a random sample is to obtain a numbered list of the population elements. This list is called a **frame.** Then we can use a *random number table* or *computer-generated random numbers* to make random selections from the numbered list. Therefore, in order to select a random sample of 100 employees from the population of 2,136 employees on 500-minute-per-month cell phone plans, the bank will make a numbered list of the 2,136 employees on 500-minute plans. The bank can then use a **random number table,** such as Table 1.4(a), to select the random sample. To see how this is done, note that any single-digit number in the table has been chosen in such a way that any of the single-digit numbers between 0 and 9 had the same chance of being chosen. For this reason, we say that any single-digit number in the table is a **random number** between 0 and 9. Similarly, any two-digit number in the table is a random number between 00 and 99, any three-digit number in the table is a random number between 000 and 999, and so forth. Note that the table entries are segmented into groups of five to make the table easier to read. Because the total number of employees on 500-minute cell phone plans (2,136) is a four-digit number, we arbitrarily select any set of four digits in the table (we have circled these digits). This number, which is 0511, identifies the first randomly selected employee. Then, moving in any direction from the 0511 (up, down, right, or left—it does not matter which), we select additional sets of four digits. These succeeding sets of digits identify additional randomly selected employees. Here we arbitrarily move down from 0511 in the table. The first seven sets of four digits we obtain are

0511   7156   0285   4461   3990   4919   1915

(See Table 1.4(a)—these numbers are enclosed in a rectangle.) Because there are no employees numbered 7156, 4461, 3990, or 4919 (remember only 2,136 employees are on 500-minute plans), we ignore these numbers. This implies that the first three randomly selected employees are those numbered 0511, 0285, and 1915. Continuing this procedure, we can obtain the entire random sample of 100 employees. Notice that, because we are sampling without replacement, we should ignore any set of four digits previously selected from the random number table.

While using a random number table is one way to select a random sample, this approach has a disadvantage that is illustrated by the current situation. Specifically, because most four-digit random numbers are not between 0001 and 2136, obtaining 100 different, four-digit random numbers between 0001 and 2136 will require ignoring a large number of random numbers in the random number table, and we will in fact need to use a random number table that is larger than Table 1.4(a). Although larger random number tables are readily available in books of mathematical and statistical tables, a good alternative is to use a computer

---

[3]The authors would like to thank Kevin Whitehurst for help in developing this case.

[4]In Chapter 9 we will discuss how to plan the *sample size*—the number of elements (for example, 100) that should be included in a sample. Throughout this book we will take large enough samples to allow us to make reasonably accurate statistical inferences.

TABLE 1.4   **Random Numbers**

(a) A portion of a random number table

33276	85590	79936	56865	05859	90106	78188
03427	90511	69445	18663	72695	52180	90322
92737	27156	33488	36320	17617	30015	74952
85689	20285	52267	67689	93394	01511	89868
08178	74461	13916	47564	81056	97735	90707
51259	63990	16308	60756	92144	49442	40719
60268	44919	19885	55322	44819	01188	55157
94904	01915	04146	18594	29852	71585	64951
58586	17752	14513	83149	98736	23495	35749
09998	19509	06691	76988	13602	51851	58104
14346	61666	30168	90229	04734	59193	32812
74103	15227	25306	76468	26384	58151	44592
24200	64161	38005	94342	28728	35806	22851
87308	07684	00256	45834	15398	46557	18510
07351	86679	92420	60952	61280	50001	94953

(b) Minitab output of 100 different, four-digit random numbers between 1 and 2136

705	1131	169	1703	1709	609
1990	766	1286	1977	222	43
1007	1902	1209	2091	1742	1152
111	69	2049	1448	659	338
1732	1650	7	388	613	1477
838	272	1227	154	18	320
1053	1466	2087	265	2107	1992
582	1787	2098	1581	397	1099
757	1699	567	1255	1959	407
354	1567	1533	1097	1299	277
663	40	585	1486	1021	532
1629	182	372	1144	1569	1981
1332	1500	743	1262	1759	955
1832	378	728	1102	667	1885
514	1128	1046	116	1160	1333
831	2036	918	1535	660	
928	1257	1468	503	468	

software package, which can generate random numbers that are between whatever values we specify. For example, Table 1.4(b) gives the Minitab output of 100 different, four-digit random numbers that are between 0001 and 2136 (note that the "leading 0's" are not included in these four-digit numbers). If used, the random numbers in Table 1.4(b) would identify the 100 employees that form the random sample. For example, the first three randomly selected employees would be employees 705, 1990, and 1007.

Finally, note that computer software packages sometimes generate the same random number twice and thus are sampling with replacement. Because we wished to randomly select 100 employees without replacement, we had Minitab generate more than 100 (actually, 110) random numbers. We then ignored the repeated random numbers to obtain the 100 different random numbers in Table 1.4(b).

Part 3: A Random Sample and Inference   When the random sample of 100 employees is chosen, the number of cellular minutes used by each sampled employee during last month (the employee's *cellular usage*) is found and recorded. The 100 cellular-usage figures are given in Table 1.5. Looking at this table, we can see that there is substantial overage and underage—many employees used far more than 500 minutes, while many others failed to use all of the 500 minutes allowed by their plan. In Chapter 3 we will use these 100 usage figures to estimate the bank's cellular costs and decide whether the bank should hire a cellular management service.

TABLE 1.5   **A Sample of Cellular Usages (in Minutes) for 100 Randomly Selected Employees**
    DS CellUse

75	485	37	547	753	93	897	694	797	477
654	578	504	670	490	225	509	247	597	173
496	553	0	198	507	157	672	296	774	479
0	822	705	814	20	513	546	801	721	273
879	433	420	521	648	41	528	359	367	948
511	704	535	585	341	530	216	512	491	0
542	562	49	505	461	496	241	624	885	259
571	338	503	529	737	444	372	555	290	830
719	120	468	730	853	18	479	144	24	513
482	683	212	418	399	376	323	173	669	611

ⒸEXAMPLE 1.3 The Marketing Research Case: Rating a Bottle Design

Part 1: Rating a Bottle Design   The design of a package or bottle can have an important effect on a company's bottom line. In this case a brand group wishes to research consumer reaction to a new bottle design for a popular soft drink. Because it is impossible to show the new bottle design to "all consumers," the brand group will use the *mall intercept method* to select a sample of 60 consumers. On a particular Saturday, the brand group will choose a shopping mall and a sampling time so that shoppers at the mall during the sampling time are a representative cross-section of all consumers. Then, shoppers will be intercepted as they walk past a designated location, will be shown the new bottle, and will be asked to rate the bottle image. For each consumer interviewed, a bottle image **composite score** will be found by adding the consumer's numerical responses to the five questions shown in Figure 1.4. It follows that the minimum possible bottle image composite score is 5 (resulting from a response of 1 on all five questions) and the maximum possible bottle image composite score is 35 (resulting from a response of 7 on all five questions). Furthermore, experience has shown that the smallest acceptable bottle image composite score for a successful bottle design is 25.

Part 2: Selecting an Approximately Random Sample   Because it is not possible to list and number all of the shoppers who will be at the mall on this Saturday, we cannot select a random sample of these shoppers. However, we can select an *approximately* random sample of these shoppers. To see one way to do this, note that there are 6 ten-minute intervals during each hour, and thus there are 60 ten-minute intervals during the 10-hour period from 10 A.M. to 8 P.M.—the time when the shopping mall is open. Therefore, one way to select an approximately random sample is to choose a particular location at the mall that most shoppers will walk by and then randomly select—at the beginning of each ten-minute period—one of the first shoppers who walks by the location. Here, although we could randomly select one person from any reasonable number of shoppers who walk by, we will (arbitrarily) randomly select one of the first five shoppers who walk by. For example, starting in the upper left-hand corner of Table 1.4(a) and proceeding down the first column, note that the first three random numbers between 1 and 5 are 3, 5, and 1. This implies that (1) at 10 A.M. we would select the 3rd customer who walks by; (2) at 10:10 A.M. we would select the 5th shopper who walks by; (3) at 10:20 A.M. we would select the 1st customer who walks by, and so forth. Furthermore, assume that the composite score ratings of the new bottle design that would be given by all shoppers at the mall on the Saturday are representative of the composite score ratings that would be given by all possible consumers. It then follows that the composite score ratings given by the 60 sampled shoppers can be regarded as an approximately random sample that can be used to make statistical inferences about the population of all possible consumer composite score ratings.

Part 3: The Approximately Random Sample and Inference   When the brand group uses the mall intercept method to interview a sample of 60 shoppers at a mall on a particular Saturday, the 60 bottle image composite scores in Table 1.6 are obtained. Because these scores

---

FIGURE 1.4   **The Bottle Design Survey Instrument**

Please circle the response that most accurately describes whether you agree or disagree with each statement about the bottle you have examined.

Statement	Strongly Disagree						Strongly Agree
The size of this bottle is convenient.	1	2	3	4	5	6	7
The contoured shape of this bottle is easy to handle.	1	2	3	4	5	6	7
The label on this bottle is easy to read.	1	2	3	4	5	6	7
This bottle is easy to open.	1	2	3	4	5	6	7
Based on its overall appeal, I like this bottle design.	1	2	3	4	5	6	7

TABLE 1.6   A Sample of Bottle Design Ratings (Composite Scores for a Sample of 60 Shoppers)
DS Design

34	33	33	29	26	33	28	25	32	33
32	25	27	33	22	27	32	33	32	29
24	30	20	34	31	32	30	35	33	31
32	28	30	31	31	33	29	27	34	31
31	28	33	31	32	28	26	29	32	34
32	30	34	32	30	30	32	31	29	33

vary from a minimum of 20 to a maximum of 35, we might infer that *most* consumers would rate the new bottle design between 20 and 35. Furthermore, 57 of the 60 composite scores are at least 25. Therefore, we might estimate that a proportion of $57/60 = .95$ (that is, 95 percent) of all consumers would give the bottle design a composite score of at least 25. In future chapters we will further analyze the composite scores.

Processes

Sometimes we are interested in studying the population of all of the elements that will be or could potentially be produced by a *process*.

> A **process** is a sequence of operations that takes inputs (labor, materials, methods, machines, and so on) and turns them into outputs (products, services, and the like).

Processes produce output *over time*. For example, this year's Buick LaCrosse manufacturing process produces LaCrosses over time. Early in the model year, General Motors might wish to study the population of the city driving mileages of all Buick LaCrosses that will be produced during the model year. Or, even more hypothetically, General Motors might wish to study the population of the city driving mileages of all LaCrosses that could *potentially* be produced by this model year's manufacturing process. The first population is called a **finite population** because only a finite number of cars will be produced during the year. The second population is called an **infinite population** because the manufacturing process that produces this year's model could in theory always be used to build "one more car." That is, theoretically there is no limit to the number of cars that could be produced by this year's process. There are a multitude of other examples of finite or infinite hypothetical populations. For instance, we might study the population of all waiting times that will or could potentially be experienced by patients of a hospital emergency room. Or we might study the population of all the amounts of grape jelly that will be or could potentially be dispensed into 16-ounce jars by an automated filling machine. To study a population of potential process observations, we sample the process—often at equally spaced time points—over time.

Ⓒ **EXAMPLE 1.4** The Car Mileage Case: Estimating Mileage

Part 1: Auto Fuel Economy   Personal budgets, national energy security, and the global environment are all affected by our gasoline consumption. Hybrid and electric cars are a vital part of a long-term strategy to reduce our nation's gasoline consumption. However, until use of these cars is more widespread and affordable, the most effective way to conserve gasoline is to design gasoline-powered cars that are more fuel efficient.[5] In the short term, "that will give you the biggest bang for your buck," says David Friedman, research director of the Union of Concerned Scientists' Clean Vehicle Program.[6]

In this case study we consider a tax credit offered by the federal government to automakers for improving the fuel economy of gasoline-powered midsize cars. According to *The Fuel Economy Guide—2017 Model Year,* virtually every gasoline-powered midsize car equipped with an automatic transmission and a six-cylinder engine has an EPA combined city and

---

[5,6]Bryan Walsh, "Plugged In," *Time,* September 29, 2008 (see page 56).

TABLE 1.7		A Sample of 50 Mileages			
30.8	30.8	32.1	32.3	32.7	Note: Time
31.7	30.4	31.4	32.7	31.4	order is given
30.1	32.5	30.8	31.2	31.8	by reading
31.6	30.3	32.8	30.7	31.9	down the
32.1	31.3	31.9	31.7	33.0	columns from
33.3	32.1	31.4	31.4	31.5	left to right.
31.3	32.5	32.4	32.2	31.6	
31.0	31.8	31.0	31.5	30.6	
32.0	30.5	29.8	31.7	32.3	
32.4	30.5	31.1	30.7	31.4	

FIGURE 1.5   A Time Series Plot of the 50 Mileages

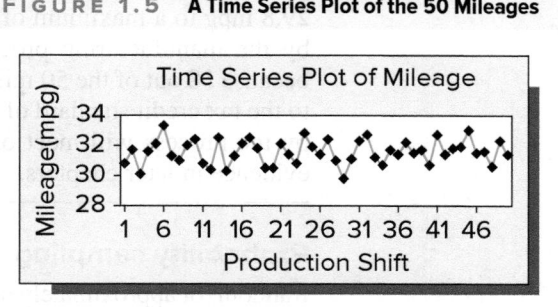

highway mileage estimate of 26 miles per gallon (mpg) or less.[7] As a matter of fact, when this book was written in 2017, the mileage leader in this category was the Nissan Altima, which registered a combined city and highway mileage of 26 mpg. While fuel economy has seen improvement in almost all car categories, the EPA has concluded that an additional 5 mpg increase in fuel economy is significant and feasible.[8] Therefore, suppose that the government has decided to offer the tax credit to any automaker selling a midsize model with an automatic transmission and a six-cylinder engine that achieves an EPA combined city and highway mileage estimate of at least 31 mpg.

Part 2: Sampling a Process   Consider an automaker that has recently introduced a new midsize model with an automatic transmission and a six-cylinder engine and wishes to demonstrate that this new model qualifies for the tax credit. In order to study the population of all cars of this type that will or could potentially be produced, the automaker will choose a sample of 50 of these cars. The manufacturer's production operation runs 8-hour shifts, with 100 midsize cars produced on each shift. When the production process has been fine-tuned and all start-up problems have been identified and corrected, the automaker will select one car at random from each of 50 consecutive production shifts. Once selected, each car is to be subjected to an EPA test that determines the EPA combined city and highway mileage of the car.

To randomly select a car from a particular production shift, we number the 100 cars produced on the shift from 00 to 99 and use a random number table or a computer software package to obtain a random number between 00 and 99. For example, starting in the upper left-hand corner of Table 1.4(a) and proceeding down the two leftmost columns, we see that the first three random numbers between 00 and 99 are 33, 3, and 92. This implies that we would select car 33 from the first production shift, car 3 from the second production shift, car 92 from the third production shift, and so forth. Moreover, because a new group of 100 cars is produced on each production shift, repeated random numbers would not be discarded. For example, if the 15th and 29th random numbers are both 7, we would select the 7th car from the 15th production shift and the 7th car from the 29th production shift.

Part 3: The Sample and Inference   Suppose that when the 50 cars are selected and tested, the sample of 50 EPA combined mileages shown in Table 1.7 is obtained. A time series plot of the mileages is given in Figure 1.5. Examining this plot, we see that, although the mileages vary over time, they do not seem to vary in any unusual way. For example, the mileages do not tend to either decrease or increase (as did the basic cable rates in Figure 1.3) over time. This intuitively verifies that the midsize car manufacturing process is producing consistent car mileages over time, and thus we can regard the 50 mileages as an approximately random sample that can be used to make statistical inferences about the population of all

---

[7]The "26 miles per gallon (mpg) or less" figure relates to midsize cars with an automatic transmission *and* at least a six-cylinder, 3.5-liter engine. Therefore, when we refer to a midsize car with an automatic transmission in future discussions, we are assuming that the midsize car also has at least a six-cylinder, 3.5-liter engine.

[8]The authors wish to thank Jeff Alson of the EPA for this information.

possible midsize car mileages.[9] Therefore, because the 50 mileages vary from a minimum of 29.8 mpg to a maximum of 33.3 mpg, we might conclude that most midsize cars produced by the manufacturing process will obtain between 29.8 mpg and 33.3 mpg. Moreover, because 38 out of the 50 mileages—or 76 percent of the mileages—are greater than or equal to the tax credit standard of 31 mpg, we have some evidence that the "typical car" produced by the process will meet or exceed the tax credit standard. We will further evaluate this evidence in later chapters.

Probability sampling

Random (or approximately random) sampling—as well as the more advanced kinds of sampling discussed in optional Section 1.7—are types of *probability sampling.* In general, **probability sampling** is sampling where we know the chance (or probability) that each element in the population will be included in the sample. If we employ probability sampling, the sample obtained can be used to make valid statistical inferences about the sampled population. However, if we do not employ probability sampling, we cannot make valid statistical inferences.

One type of sampling that is not probability sampling is **convenience sampling,** where we select elements because they are easy or convenient to sample. For example, if we select people to interview because they look "nice" or "pleasant," we are using convenience sampling. Another example of convenience sampling is the use of **voluntary response samples,** which are frequently employed by television and radio stations and newspaper columnists. In such samples, participants self-select—that is, whoever wishes to participate does so (usually expressing some opinion). These samples overrepresent people with strong (usually negative) opinions. For example, the advice columnist Ann Landers once asked her readers, "If you had it to do over again, would you have children?" Of the nearly 10,000 parents who *voluntarily* responded, 70 percent said that they would not. A probability sample taken a few months later found that 91 percent of parents would have children again.

Another type of sampling that is not probability sampling is **judgment sampling,** where a person who is extremely knowledgeable about the population under consideration selects population elements that he or she feels are most representative of the population. Because the quality of the sample depends upon the judgment of the person selecting the sample, it is dangerous to use the sample to make statistical inferences about the population.

To conclude this section, we consider a classic example where two types of sampling errors doomed a sample's ability to make valid statistical inferences. This example occurred prior to the presidential election of 1936, when the *Literary Digest* predicted that Alf Landon would defeat Franklin D. Roosevelt by a margin of 57 percent to 43 percent. Instead, Roosevelt won the election in a landslide. *Literary Digest*'s first error was to send out sample ballots (actually, 10 million ballots) to people who were mainly selected from the *Digest*'s subscription list and from telephone directories. In 1936 the country had not yet recovered from the Great Depression, and many unemployed and low-income people did not have phones or subscribe to the *Digest*. The *Digest*'s sampling procedure excluded these people, who overwhelmingly voted for Roosevelt. Second, only 2.3 million ballots were returned, resulting in the sample being a voluntary response survey. At the same time, George Gallup, founder of the Gallup Poll, was beginning to establish his survey business. He used a probability sample to correctly predict Roosevelt's victory. In optional Section 1.8 we discuss various issues related to designing surveys and more about the errors that can occur in survey samples.

Ethical guidelines for statistical practice

The American Statistical Association, the leading U.S. professional statistical association, has developed the report "Ethical Guidelines for Statistical Practice."[10] This report provides information that helps statistical practitioners to consistently use ethical statistical practices

---

[9]In Chapter 20 (on the website) we will discuss more precisely how to assess whether a process is operating consistently over time.

[10]American Statistical Association, "Ethical Guidelines for Statistical Practice," 1999.

and that helps users of statistical information avoid being misled by unethical statistical practices. Unethical statistical practices can take a variety of forms, including:

- **Improper sampling**   Purposely selecting a biased sample—for example, using a non-random sampling procedure that overrepresents population elements supporting a desired conclusion or that underrepresents population elements not supporting the desired conclusion—is unethical. In addition, discarding already sampled population elements that do not support the desired conclusion is unethical. More will be said about proper and improper sampling later in this chapter.

- **Misleading charts, graphs, and descriptive measures**   In Section 2.7, we will present an example of how misleading charts and graphs can distort the perception of changes in salaries over time. Using misleading charts or graphs to make the salary changes seem much larger or much smaller than they really are is unethical. In Section 3.1, we will present an example illustrating that many populations of individual or household incomes contain a small percentage of very high incomes. These very high incomes make the *population mean income* substantially larger than the *population median income*. In this situation we will see that the population median income is a better measure of the typical income in the population. Using the population mean income to give an inflated perception of the typical income in the population is unethical.

- **Inappropriate statistical analysis or inappropriate interpretation of statistical results**   The American Statistical Association report emphasizes that selecting many different samples and running many different tests can eventually (by random chance alone) produce a result that makes a desired conclusion seem to be true, when the conclusion really isn't true. Therefore, continuing to sample and run tests until a desired conclusion is obtained and not reporting previously obtained results that do not support the desired conclusion is unethical. Furthermore, we should always report our sampling procedure and sample size and give an estimate of the reliability of our statistical results. Estimating this reliability will be discussed in Chapter 8 and beyond.

The above examples are just an introduction to the important topic of unethical statistical practices. The American Statistical Association report contains 67 guidelines organized into eight areas involving general professionalism and ethical responsibilities. These include responsibilities to clients, to research team colleagues, to research subjects, and to other statisticians, as well as responsibilities in publications and testimony and responsibilities of those who employ statistical practitioners.

Exercises for Sections 1.3 and 1.4

CONCEPTS

1.9 (1) Define a *population*. (2) Give an example of a population that you might study when you start your career after graduating from college. (3) Explain the difference between a census and a sample.

1.10 Explain each of the following terms:
   a Descriptive statistics.   **c** Random sample.
   b Statistical inference.   **d** Process.

1.11 Explain why sampling without replacement is preferred to sampling with replacement.

METHODS AND APPLICATIONS

1.12 In the page margin, we list 15 companies that have historically performed well in the food, drink, and tobacco industries. Consider the random numbers given in the random number table of Table 1.4(a). Starting in the upper left corner of Table 1.4(a) and moving down the two leftmost columns, we see that the first three two-digit numbers obtained are: 33, 03, and 92. Starting with these three random numbers, and moving down the two leftmost columns of Table 1.4(a) to find more two-digit random numbers, use Table 1.4(a) to randomly select five of these companies to be interviewed in detail about their business strategies. Hint: Note that we have numbered the companies from 1 to 15.

Companies:
1 Altria Group
2 PepsiCo
3 Coca-Cola
4 Archer Daniels
5 Anheuser-Bush
6 General Mills
7 Sara Lee
8 Coca-Cola Enterprises
9 Reynolds American
10 Kellogg
11 ConAgra Foods
12 HJ Heinz
13 Campbell Soup
14 Pepsi Bottling Group
15 Tyson Foods

FIGURE 1.6   The Video Game Satisfaction Survey Instrument

Statement	Strongly Disagree						Strongly Agree
The game console of the XYZ-Box is well designed.	1	2	3	4	5	6	7
The game controller of the XYZ-Box is easy to handle.	1	2	3	4	5	6	7
The XYZ-Box has high-quality graphics capabilities.	1	2	3	4	5	6	7
The XYZ-Box has high-quality audio capabilities.	1	2	3	4	5	6	7
The XYZ-Box serves as a complete entertainment center.	1	2	3	4	5	6	7
There is a large selection of XYZ-Box games to choose from.	1	2	3	4	5	6	7
I am totally satisfied with my XYZ-Box game system.	1	2	3	4	5	6	7

1.13 THE VIDEO GAME SATISFACTION RATING CASE DS VideoGame

A company that produces and markets video game systems wishes to assess its customers' level of satisfaction with a relatively new model, the XYZ-Box. In the six months since the introduction of the model, the company has received 73,219 warranty registrations from purchasers. The company will randomly select 65 of these registrations and will conduct telephone interviews with the purchasers. Specifically, each purchaser will be asked to state his or her level of agreement with each of the seven statements listed on the survey instrument given in Figure 1.6. Here, the level of agreement for each statement is measured on a 7-point Likert scale. Purchaser satisfaction will be measured by adding the purchaser's responses to the seven statements. It follows that for each consumer the minimum composite score possible is 7 and the maximum is 49. Furthermore, experience has shown that a purchaser of a video game system is "very satisfied" if his or her composite score is at least 42.

a Assume that the warranty registrations are numbered from 1 to 73,219 in a computer. Starting in the upper left corner of Table 1.4(a) and moving down the five leftmost columns, we see that the first three five-digit numbers obtained are 33276, 03427, and 92737. Starting with these three random numbers and moving down the five leftmost columns of Table 1.4(a) to find more five-digit random numbers, use Table 1.4(a) to randomly select the numbers of the first 10 warranty registrations to be included in the sample of 65 registrations.

b Suppose that when the 65 customers are interviewed, their composite scores are as given in Table 1.8. Using the largest and smallest observations in the data, estimate limits between which most of the 73,219 composite scores would fall. Also, estimate the proportion of the 73,219 composite scores that would be at least 42.

1.14 THE BANK CUSTOMER WAITING TIME CASE
DS WaitTime

A bank manager has developed a new system to reduce the time customers spend waiting to be served by tellers during peak business hours. Typical waiting times during peak business hours

TABLE 1.8   Composite Scores for the Video Game Satisfaction Rating Case DS VideoGame

39	44	46	44	44
45	42	45	44	42
38	46	45	45	47
42	40	46	44	43
42	47	43	46	45
41	44	47	48	
38	43	43	44	
42	45	41	41	
46	45	40	45	
44	40	43	44	
40	46	44	44	
39	41	41	44	
40	43	38	46	
42	39	43	39	
45	43	36	41	

under the current system are roughly 9 to 10 minutes. The bank manager hopes that the new system will lower typical waiting times to less than 6 minutes and wishes to evaluate the new system. When the new system is operating consistently over time, the bank manager decides to select a sample of 100 customers that need teller service during peak business hours. Specifically, for each of 100 peak business hours, the first customer that starts waiting for teller service at or after a randomly selected time during the hour will be chosen.

a Consider the peak business hours from 2:00 P.M. to 2:59 P.M., from 3:00 P.M. to 3:59 P.M., from 4:00 P.M. to 4:59 P.M., and from 5:00 P.M. to 5:59 P.M. on a particular day. Also, assume that a computer software system generates the following four random numbers between 00 and 59: 32, 00, 18, and 47. This implies that the randomly selected times during the first three peak business hours are 2:32 P.M., 3:00 P.M., and 4:18 P.M. What is the randomly selected time during the fourth peak business hour?

b When each customer is chosen, the number of minutes the customer spends waiting for teller service is recorded. The 100 waiting times that are observed are given in Table 1.9. Using the largest and smallest observations in the data,

TABLE 1.9		Waiting Times (in Minutes) for the Bank Customer Waiting Time Case DS WaitTime				
1.6	6.2	3.2	5.6	7.9	6.1	7.2
6.6	5.4	6.5	4.4	1.1	3.8	7.3
5.6	4.9	2.3	4.5	7.2	10.7	4.1
5.1	5.4	8.7	6.7	2.9	7.5	6.7
3.9	.8	4.7	8.1	9.1	7.0	3.5
4.6	2.5	3.6	4.3	7.7	5.3	6.3
6.5	8.3	2.7	2.2	4.0	4.5	4.3
6.4	6.1	3.7	5.8	1.4	4.5	3.8
8.6	6.3	.4	8.6	7.8	1.8	5.1
4.2	6.8	10.2	2.0	5.2	3.7	5.5
5.8	9.8	2.8	8.0	8.4	4.0	
3.4	2.9	11.6	9.5	6.3	5.7	
9.3	10.9	4.3	1.3	4.4	2.4	
7.4	4.7	3.1	4.8	5.2	9.2	
1.8	3.9	5.8	9.9	7.4	5.0	

TABLE 1.10	Trash Bag Breaking Strengths DS TrashBag	
61.1	63.8	52.1
51.8	50.5	55.4
52.2	47.3	49.1
42.1	54.5	52.1
51.9	50.0	55.0
56.5	44.6	59.9
47.1	58.9	57.2
49.4	51.9	52.7
46.7	57.7	55.8
46.5	41.7	49.5

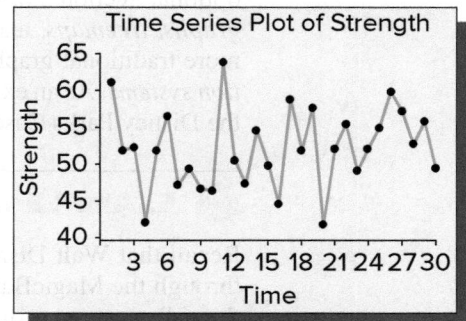

estimate limits between which the waiting times of most of the customers arriving during peak business hours would be. Also, estimate the proportion of waiting times of customers arriving during peak business hours that are less than 6 minutes.

1.15 In an article entitled "Turned Off" in the June 2–4, 1995, issue of *USA Weekend*, Don Olmsted and Gigi Anders reported results of a survey where readers were invited to write in and express their opinions about sex and violence on television. The results showed that 96 percent of respondents were very or somewhat concerned about sex on TV, and 97 percent of respondents were very or somewhat concerned about violence on TV. Do you think that these results could be generalized to all television viewers in 1995? Why or why not?

1.16   **THE TRASH BAG CASE**[11]   DS TrashBag

A company that produces and markets trash bags has developed an improved 30-gallon bag. The new bag is produced using a specially formulated plastic that is both stronger and more biodegradable than previously used plastics, and the company wishes to evaluate the strength of this bag. The *breaking strength* of a trash bag is considered to be the amount (in pounds) of a representative trash mix that when loaded into a bag suspended in the air will cause the bag to sustain significant damage (such as ripping or tearing). The company has decided to select a sample of 30 of the new trash bags. For each of 30 consecutive hours, the first trash bag produced at or after a randomly selected time during the hour is chosen. The bag is then subjected to a *breaking strength test*. The 30 breaking strengths obtained are given in Table 1.10. Using the largest and smallest observations in the data, estimate limits between which the breaking strengths of most trash bags would fall. Assume that the trash bag manufacturing process is operating consistently over time.

1.5   Business Analytics and Data Mining (Optional)

LO1-10

Explain some of the uses of business analytics and data mining (Optional).

Big data, which sometimes needs quick (sometimes almost real-time) analysis for effective business decision making and which may be too massive to be analyzed by traditional statistical methods, has resulted in an extension of traditional statistics called *business analytics*. In general, **business analytics** might be defined as the use of traditional and newly developed statistical methods, advances in information systems, and techniques from *management science* to continuously and iteratively explore and investigate past business performance, with the purpose of gaining insight and improving business planning and operations. There are three broad categories of business analytics: *descriptive analytics, predictive analytics*, and *prescriptive analytics*.

Descriptive analytics

Descriptive analytics are graphical and numerical methods used to find and visualize patterns, associations, anomalies, and other relationships in data sets, with the purpose of

---

[11]This case is based on conversations by the authors with several employees working for a leading producer of trash bags. For purposes of confidentiality, we have withheld the company's name.

business improvement. In the next two subsections we will introduce what we call **graphical descriptive analytics** and **numerical descriptive analytics**.

Graphical descriptive analytics

In previous examples we have illustrated using dot plots, bar charts, and time series plots to graphically display data. These and other traditional methods for displaying data are fully discussed in Sections 2.1 through 2.7 of Chapter 2. The methods discussed in these sections, and more recently developed statistical display techniques designed to take advantage of the dramatic advances in data capture, transmission, and storage, make up the toolset of *graphical descriptive analytics*. **Graphical descriptive analytics** uses the traditional and/or newer graphics to present to executives (and sometimes customers) easy-to-understand visual summaries of up-to-the minute information concerning the operational status of a business. In optional Section 2.8, we will discuss some of the new graphics, which include *gauges, bullet graphs, treemaps,* and *sparklines*. We will also see how they are used with each other and more traditional graphics to form analytic *dashboards*, which are part of *executive informa-tion systems*. As an example of one of the new graphics—the *bullet graph*—we again consider the Disney Parks Case.

C **EXAMPLE 1.5** The Disney Parks Case: Predicting Ride Waiting Times

Recall that Walt Disney World Orlando collects massive amounts of data from its visitors through the MagicBands they wear. Because these data include real-time location data and the riding patterns of Disney visitors, they allow Disney to continuously predict the wait-ing times for each ride or attraction in its parks. This prediction is done by using the data to estimate the visitor arrival rate to the line at each ride or attraction. Then it uses statistics and a management science technique called **queuing theory** to predict the current waiting time.

One Walt Disney World Orlando park—Epcot Center—consists of the World Showcase, which features the cultures of different countries, other non-ride attractions, and (at the time that one of this book's authors visited the park) seven main rides. Near each ride, Disney posts its current predicted waiting time and also summarizes its predictions for all seven rides on large display screens located throughout the park. On February 21, 2015, one of this book's authors spent the day at Epcot and recorded the predicted waiting times (in minutes) for the seven rides. We summarize the predicted times posted at approximately 3 P.M. in the **bullet graph** in Figure 1.7. Note that the colors in the bullet graph range from dark green to red to signify short (0 to 20 minute) to very long (80 to 100 minute) predicted wait times. For each ride, a black horizontal bar representing Disney's predicted wait extends into the colors. Rather than using a bullet graph on its display screens, Disney flashes its predictions for the different rides—one at a time—on the display screens. This display method requires visitors to remember previously flashed times and thus complicates choosing the next ride. We think that Disney would be better off using a display, such as a bullet graph, that simulta-neously posts all of the predicted waiting times. In whatever manner the times are displayed,

---

F I G U R E 1 . 7   **A Bullet Graph of Disney's Predicted Waiting Times (in minutes) for the Seven Epcot Rides Posted at 3 P.M. on February 21, 2015**   Ⓓ DisneyTimes

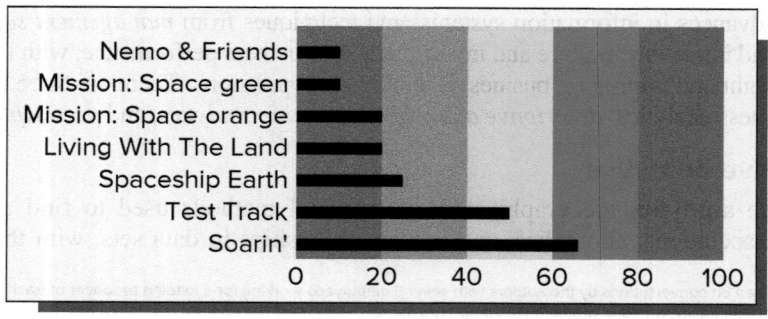

however, they provide important information to park visitors. This is because Epcot visitors choose the three rides or attractions on which they'll gain FastPass access from certain categories, and on any given day a visitor can choose only one of Epcot's two most popular rides—Soarin' and Test Track. By viewing the posted predicted waiting times for the rides, visitors can assess whether the current predicted waiting time for a popular rides is short enough to get in line.

Finally, note that continuously monitoring predicted waiting times for the seven rides helps not only visitors but Disney management. For example, if predicted waits for rides are frequently very long, Disney might see the need to add more popular rides or attractions to its parks. As a matter of fact, Channel 13 News in Orlando reported on March 6, 2015, that Disney had announced plans to add a third "theatre" for Soarin' (a virtual ride) in order to shorten long visitor waiting times. On June 17, 2016, the third theatre for Soarin' was completed and available to park visitors.

BI

Numerical descriptive analytics

In Sections 3.1 through 3.6 of Chapter 3 we discuss methods of traditional numerical descriptive statistics—for example, using means, standard deviations, and correlation coefficients. These methods are part of the toolset of **numerical descriptive analytics,** which extends these methods and consists of the topic areas of *association learning, text mining, cluster analysis,* and *factor analysis.* We now briefly describe these topic areas, which are discussed in optional Sections 3.7 through 3.10 of Chapter 3.

Association learning   This involves identifying items that tend to co-occur and finding the rules that describe their co-occurrence. For example, a supermarket chain once found that men who buy baby diapers on Thursdays also tend to buy beer on Thursdays (possibly in anticipation of watching sports on television over the weekend). This led the chain to display beer near the baby aisle in its stores. As another example, Netflix might find that customers who rent fictional dramas also tend to rent historical documentaries or that some customers will rent almost any type of movie that stars a particular actor or actress. Disney might find that visitors who spend more time at the Magic Kingdom also tend to buy Disney cartoon character clothing. Disney might also find that visitors who stay in more luxurious Disney hotels also tend to play golf on Disney courses and take cruises on the Disney Cruise Line. These types of findings are used for targeting coupons, deals, or advertising to the right potential customers.

Text mining   Text mining is the science of discovering knowledge, insights, and patterns from a collection of textual documents or databases. Using *latent semantic analysis*, we can analyze the relationships between a collection of documents and the words they contain to produce a set of key concepts or factors related to the documents and words. For example, a Food and Drug Administration (FDA) administrator might use a large number of FDA drug and safety citations issued to businesses and other organizations to determine the key issues underlying the citations. Businesses and organizations can then hopefully be successfully encouraged to improve their performances in the areas related to the issues, or a company that makes pet products might analyze a set of texts from pet owners about their pets to devise advertising about their products that would appeal to the pet owners.

Cluster analysis   This involves finding natural groupings, or clusters, within data without having to prespecify a set of categories. For example, Michaels Arts and Crafts might use cluster detection and customer purchase records to define different groupings of customers that reveal different buying behaviors and tastes. A financial analyst might use cluster detection to define different groupings of stocks based on the past history of stock price fluctuations. A sports marketing analyst might use cluster detection and sports fan perceptions concerning the attributes of different sports to define groupings of sports that would lead to determining why some sports are more popular than others. A waterfront developer might use cluster detection and restaurant goers' perceptions of 10 types of cultural foods to define groupings of the cultural foods that would lead to the best choice of a limited number of different cultural restaurants for the waterfront development.

Factor analysis   This involves starting with a large number of correlated variables and finding fewer underlying, uncorrelated factors that describe the "essential aspects" of the large number of correlated variables. For example, a sales manager might rank job applicants on 15 attributes and then use factor analysis to find that the underlying factors that describe the essential aspects of these 15 variables are "extroverted personality," "experience," "agreeable personality," "academic ability," and "appearance." The manager of a large discount store might have shoppers rate the store on 29 attributes and use factor analysis to find that the underlying factors that describe the essential aspects of the 29 attributes are "good, friendly service," "price level," "attractiveness," spaciousness," and "size." Reducing the large number of variables to fewer underlying factors helps a business focus its activities and strategies.

It is important to note that the optional sections or descriptive analytics in Chapters 2 and 3 (Sections 2.8, 3.7, 3.8, 3.9, and 3.10) cannot only be skipped entirely but also can be read in any order without loss of continuity. Therefore, the reader has the option to choose which of these sections to study in the main flow of the course and which to study later, perhaps with Chapters 5 and 16 on predictive analytics. For readers who wish a short, easy-to-understand, and motivating introduction to descriptive analytics, we might suggest studying Sections 2.8, 3.7, and 3.8 on graphical descriptive analytics, association rules, and text mining.

Predictive analytics

Predictive analytics are methods used to predict values of a **response variable** (for example, sales of a product) on the basis of one or more **predictor variables** (for example, price and advertising expenditure). Predictive analytics determine how to make predictions by finding a relationship between previously observed values of the response variable and corresponding previously observed values of the predictor variable(s). Here, the previously observed values of the response variable are said to guide, or supervise, the learning of how to make predictions of future values of the response variable, and therefore predictive analytics are called **supervised learning** techniques. In contrast, although the descriptive analytics discussed in the previous subsection detect patterns and relationships in data, there is not a particular response variable we are trying to predict, and thus these descriptive analytics are called **unsupervised learning** techniques. The response variable we wish to predict when using predictive analytics can be quantitative (for example, sales of a product) or qualitative. If the response variable is qualitative, we use predictive analytics to perform **classification,** which assigns items to specified categories or classes. For example, a bank might study the financial records and mortgage payment histories of previous borrowers to predict whether a new mortgage applicant should be classified as a future successful or unsuccessful mortgage payer. As another example, a spam filter might use classification methodology and the differences in word usage between previously analyzed legitimate and spam e-mails to predict whether or not an incoming e-mail is spam.

Predictive analytics fall into two classes—**nonparametric predictive analytics** and **parametric predictive analytics.** Essentially, **parametric predictive analytics** find a mathematical equation that relates the response variable to the predictor variable(s) and involves unknown parameters that must be estimated and evaluated by using sample data. Parametric predictive analytics include *classical linear regression, logistic regression, discriminate analysis, neural networks*, and *time series forecasting*. Because evaluating the parameters in the equations obtained when using these predictive analytics requires knowledge of formal statistical inference, which is discussed in Chapters 6 through 10 (along with probability distributions) and extended in Chapters 11 through 13, we do not study parametric predictive analytics until Chapters 14 through 17. On the other hand, nonparametric predictive analytics make predictions by using a relationship between the response variable and the predictor variables that is not expressed in terms of a mathematical equation involving parameters. These analytics include *decision trees (classification and regression trees),* *k-nearest neighbors,* and *naive Bayes' classification.* Furthermore, the practical applications of these analytics in business can be understood with only the background of descriptive statistics and probability that is provided in Chapters 2, 3, and 4. Thus, we give an early and hopefully easy-to-understand and motivating discussion of nonparametric predictive analytics in Chapter 5. The reader thus has the option to study these analytics early in the course or nearer the time of studying parametric predictive analytics in Chapters 14 through 17.

Data mining and prescriptive analytics

Data mining is the process of discovering useful knowledge in extremely large data sets (big data). For each data mining project, data mining first uses computer science algorithms and information system techniques to extract the data needed for the project from a data source (for example, a data warehouse); data mining then uses descriptive analytics and/or predictive analytics to analyze these data. It is estimated that for any data mining project, approximately 65 percent to 90 percent of the time is spent in *data preparation*—checking, correcting, reconciling inconsistencies in, and otherwise "cleaning" the data. Also, whereas descriptive and predictive analytics might be most useful to decision makers when used with data mining, these methods can also be important, as we will see, when analyzing smaller data sets. Whatever the size of the data set being analyzed, however, it is important to turn the knowledge obtained into an optimal course of action that will lead to business improvement.

    Prescriptive analytics are techniques that combine external and internal constraints (for example, the state of the economy and a company's debt relative to its equity) with results from descriptive or predictive analytics (for example, the probability that an investment being considered by the company would be successful) to recommend an optimal course of action. Prescriptive analytics include *decision theory methods* (see Chapter 19), *linear optimization, nonlinear optimization,* and *simulation.* While we will not discuss the last three of these analytics in this book (see any book on *management science* or *operations research*), we will intuitively use results from descriptive and predictive analytics to suggest business improvement courses of action.

Exercises for Section 1.5

CONCEPTS

1.17 Why are predictive analytics supervised learning techniques?

1.18 Why are descriptive analytics unsupervised learning techniques?

1.19 What is data mining?

1.20 What are prescriptive analytics?

1.6 Ratio, Interval, Ordinal, and Nominative Scales of Measurement (Optional)

LO1-11
Identify the ratio, interval, ordinal, and nominative scales of measurement (Optional).

In Section 1.1 we said that a variable is **quantitative** if its possible values are *numbers that represent quantities* (that is, "how much" or "how many"). In general, a quantitative variable is measured on a scale having a *fixed unit of measurement* between its possible values. For example, if we measure employees' salaries to the nearest dollar, then one dollar is the fixed unit of measurement between different employees' salaries. There are two types of quantitative variables: **ratio** and **interval.** A **ratio variable** is a quantitative variable measured on a scale such that ratios of its values are meaningful and there is an inherently defined zero value. Variables such as salary, height, weight, time, and distance are ratio variables. For example, a distance of zero miles is "no distance at all," and a town that is 30 miles away is "twice as far" as a town that is 15 miles away.

    An **interval variable** is a quantitative variable where ratios of its values are not meaningful and there is not an inherently defined zero value. Temperature (on the Fahrenheit scale) is an interval variable. For example, zero degrees Fahrenheit does not represent "no heat at all," just that it is very cold. Thus, there is no inherently defined zero value. Furthermore, ratios of temperatures are not meaningful. For example, it makes no sense to say that 60° is twice as warm as 30°. In practice, there are very few interval variables other than temperature. Almost all quantitative variables are ratio variables.

In Section 1.1 we also said that if we simply record into which of several categories a population (or sample) unit falls, then the variable is **qualitative** (or **categorical**). There are two types of qualitative variables: **ordinal** and **nominative.** An **ordinal variable** is a qualitative variable for which there is a meaningful *ordering,* or *ranking,* of the categories. The measurements of an ordinal variable may be nonnumerical or numerical. For example, a student may be asked to rate the teaching effectiveness of a college professor as excellent, good, average, poor, or unsatisfactory. Here, one category is higher than the next one; that is, "excellent" is a higher rating than "good," "good" is a higher rating than "average," and so on. Therefore, teaching effectiveness is an ordinal variable having nonnumerical measurements. On the other hand, if (as is often done) we substitute the numbers 4, 3, 2, 1, and 0 for the ratings excellent through unsatisfactory, then teaching effectiveness is an ordinal variable having numerical measurements.

In practice, both numbers and associated words are often presented to respondents asked to rate a person or item. When numbers are used, statisticians debate whether the ordinal variable is "somewhat quantitative." For example, statisticians who claim that teaching effectiveness rated as 4, 3, 2, 1, or 0 is *not* somewhat quantitative argue that the difference between 4 (excellent) and 3 (good) may not be the same as the difference between 3 (good) and 2 (average). Other statisticians argue that as soon as respondents (students) see equally spaced numbers (even though the numbers are described by words), their responses are affected enough to make the variable (teaching effectiveness) somewhat quantitative. Generally speaking, the specific words associated with the numbers probably substantially affect whether an ordinal variable may be considered somewhat quantitative. It is important to note, however, that in practice numerical ordinal ratings are often analyzed as though they are quantitative. Specifically, various arithmetic operations (as discussed in Chapters 2 through 19) are often performed on numerical ordinal ratings. For example, a professor's teaching effectiveness average and a student's grade point average are calculated.

To conclude this section, we consider the second type of qualitative variable. A **nominative variable** is a qualitative variable for which there is no meaningful ordering, or ranking, of the categories. A person's gender, the color of a car, and an employee's state of residence are nominative variables.

Exercises for Section 1.6

CONCEPTS ![Mc Graw Hill Education] connect

1.21 Discuss the difference between a ratio variable and an interval variable.

1.22 Discuss the difference between an ordinal variable and a nominative variable.

METHODS AND APPLICATIONS

1.23 Classify each of the following qualitative variables as ordinal or nominative. Explain your answers.

Qualitative Variable	Categories				
Statistics course letter grade	A	B	C	D	F
Door choice on *Let's Make A Deal*	Door #1	Door #2	Door #3		
Television show classifications	TV-G	TV-PG	TV-14	TV-MA	
Personal computer ownership	Yes	No			
Restaurant rating	*****	****	***	**	*

Qualitative Variable	Categories
Income tax filing status	Married filing jointly; Married filing separately; Single; Head of household; Qualifying widow(er)

1.24 Classify each of the following qualitative variables as ordinal or nominative. Explain your answers.

Qualitative Variable	Categories									
Personal computer operating system	Windows XP; Windows Vista; Windows 7; Windows 8; Windows 10									
Motion picture classifications	G	PG	PG-13	R	NC-17	X				
Level of education	Elementary; Middle school; High school; College; Graduate school									
Rankings of the top 10 college football teams	1	2	3	4	5	6	7	8	9	10
Exchange on which a stock is traded	AMEX	NYSE	NASDAQ	Other						
Zip code	45056	90015	etc.							

1.7 Stratified Random, Cluster, and Systematic Sampling (Optional)

LO1-12
Describe the basic ideas of stratified random, cluster, and systematic sampling (Optional).

Random sampling is not the only kind of sampling. Methods for obtaining a sample are called **sampling designs,** and the sample we take is sometimes called a **sample survey.** In this section we explain three sampling designs that are alternatives to random sampling— **stratified random sampling, cluster sampling,** and **systematic sampling.**

One common sampling design involves separately sampling important groups within a population. Then, the samples are combined to form the entire sample. This approach is the idea behind **stratified random sampling.**

> In order to select a **stratified random sample,** we divide the population into nonoverlapping groups of similar elements (people, objects, etc.). These groups are called **strata.** Then a random sample is selected from each stratum, and these samples are combined to form the full sample.

It is wise to stratify when the population consists of two or more groups that differ with respect to the variable of interest. For instance, consumers could be divided into strata based on gender, age, ethnic group, or income.

As an example, suppose that a department store chain proposes to open a new store in a location that would serve customers who live in a geographical region that consists of (1) an industrial city, (2) a suburban community, and (3) a rural area. In order to assess the potential profitability of the proposed store, the chain wishes to study the incomes of all households in the region. In addition, the chain wishes to estimate the proportion and the total number of households whose members would be likely to shop at the store. The department store chain feels that the industrial city, the suburban community, and the rural area differ with respect to income and the store's potential desirability. Therefore, it uses these subpopulations as strata and takes a stratified random sample.

Taking a stratified sample can be advantageous because such a sample takes advantage of the fact that elements in the same stratum are similar to each other. It follows that a stratified sample can provide more accurate information than a random sample of the same size. As a simple example, if all of the elements in each stratum were exactly the same, then examining only one element in each stratum would allow us to describe the entire population. Furthermore, stratification can make a sample easier (or possible) to select. Recall that, in order to take a random sample, we must have a list, or **frame** of all of the population elements. Although a frame might not exist for the overall population, a frame might exist for each stratum. For example, suppose nearly all the households in the department store's geographical region have telephones. Although there might not be a telephone directory for the overall geographical region, there might be separate telephone directories for the industrial city, the suburb, and the rural area. For more discussion of stratified random sampling, see Mendenhall, Schaeffer, and Ott (1986).

Sometimes it is advantageous to select a sample in stages. This is a common practice when selecting a sample from a very large geographical region. In such a case, a frame often does not exist. For instance, there is no single list of all registered voters in the United States. There is also no single list of all households in the United States. In this kind of situation, we can use **multistage cluster sampling.** To illustrate this procedure, suppose we wish to take a sample of registered voters from all registered voters in the United States. We might proceed as follows:

Stage 1: Randomly select a sample of counties from all of the counties in the United States.

Stage 2: Randomly select a sample of townships from each county selected in Stage 1.

Stage 3: Randomly select a sample of voting precincts from each township selected in Stage 2.

Stage 4: Randomly select a sample of registered voters from each voting precinct selected in Stage 3.

We use the term *cluster sampling* to describe this type of sampling because at each stage we "cluster" the voters into subpopulations. For instance, in Stage 1 we cluster the voters into counties, and in Stage 2 we cluster the voters in each selected county into townships. Also, notice that the random sampling at each stage can be carried out because there are lists of (1) all counties in the United States, (2) all townships in each county, (3) all voting precincts in each township, and (4) all registered voters in each voting precinct.

As another example, consider sampling the households in the United States. We might use Stages 1 and 2 above to select counties and townships within the selected counties. Then, if there is a telephone directory of the households in each township, we can randomly sample households from each selected township by using its telephone directory. Because *most* households today have telephones, and telephone directories are readily available, most national polls are now conducted by telephone. Further, polling organizations have recognized that many households are giving up landline phones, and have developed ways to sample households that only have cell phones.

It is sometimes a good idea to combine stratification with multistage cluster sampling. For example, suppose a national polling organization wants to estimate the proportion of all registered voters who favor a particular presidential candidate. Because the presidential preferences of voters might tend to vary by geographical region, the polling organization might divide the United States into regions (say, Eastern, Midwestern, Southern, and Western regions). The polling organization might then use these regions as strata, and might take a multistage cluster sample from each stratum (region).

The analysis of data produced by multistage cluster sampling can be quite complicated. For a more detailed discussion of cluster sampling, see Mendenhall, Schaeffer, and Ott (1986).

In order to select a random sample, we must number the elements in a frame of all the population elements. Then we use a random number table (or a random number generator on a computer) to make the selections. However, numbering all the population elements can be quite time-consuming. Moreover, random sampling is used in the various stages of many complex sampling designs (requiring the numbering of numerous populations). Therefore, it is useful to have an alternative to random sampling. One such alternative is called **systematic sampling.** In order to systematically select a sample of n elements without replacement from a frame of N elements, we divide N by n and round the result down to the nearest whole number. Calling the rounded result ℓ, we then randomly select one element from the first ℓ elements in the frame—this is the first element in the systematic sample. The remaining elements in the sample are obtained by selecting every ℓth element following the first (randomly selected) element. For example, suppose we wish to sample a population of $N = 14{,}327$ allergists to investigate how often they have prescribed a particular drug during the last year. A medical society has a directory listing the 14,327 allergists, and we wish to draw a systematic sample of 500 allergists from this frame. Here we compute 14,327/500 = 28.654, which is 28 when rounded down. Therefore, we number the first 28 allergists in the directory from 1 to 28, and we use a random number table to randomly select one of the first 28 allergists. Suppose we select allergist number 19. We interview allergist 19 and every 28th allergist in the frame thereafter, so we choose allergists 19, 47, 75, and so forth until we obtain our sample of 500 allergists. In this scheme, we must number the first 28 allergists, but we do not have to number the rest because we can "count off" every 28th allergist in the directory. Alternatively, we can measure the approximate amount of space in the directory that it takes to list 28 allergists. This measurement can then be used to select every 28th allergist.

Exercises for Section 1.7

CONCEPTS   **connect**

1.25 When is it appropriate to use stratified random sampling? What are strata, and how should strata be selected?

1.26 When is cluster sampling used? Why do we describe this type of sampling by using the term *cluster*?

1.27 Explain how to take a systematic sample of 100 companies from the 1,853 companies that are members of an industry trade association.

1.28 Explain how a stratified random sample is selected. Discuss how you might define the strata to survey student opinion on a proposal to charge all students a $100 fee for a new university-run bus system that will provide transportation between off-campus apartments and campus locations.

1.29 Marketing researchers often use city blocks as clusters in cluster sampling. Using this fact, explain how

a market researcher might use multistage cluster sampling to select a sample of consumers from all cities having a population of more than 10,000 in a large state having many such cities.

1.8 More about Surveys and Errors in Survey Sampling (Optional)

LO1-13
Describe basic types of survey questions, survey procedures, and sources of error (Optional).

We have seen in Section 1.2 that people in surveys are asked questions about their behaviors, opinions, beliefs, and other characteristics. In this section we discuss various issues related to designing surveys and the errors that can occur in survey sampling.

Types of survey questions

Survey instruments can use **dichotomous** ("yes or no"), **multiple-choice,** or **open-ended** questions. Each type of question has its benefits and drawbacks. Dichotomous questions are usually clearly stated, can be answered quickly, and yield data that are easily analyzed. However, the information gathered may be limited by this two-option format. If we limit voters to expressing support or disapproval for stem-cell research, we may not learn the nuanced reasoning that voters use in weighing the merits and moral issues involved. Similarly, in today's heterogeneous world, it would be unusual to use a dichotomous question to categorize a person's religious preferences. Asking whether respondents are Christian or non-Christian (or to use any other two categories like Jewish or non-Jewish; Muslim or non-Muslim) is certain to make some people feel their religion is being slighted. In addition, this is a crude and unenlightening way to learn about religious preferences.

Multiple-choice questions can assume several different forms. Sometimes respondents are asked to choose a response from a list (for example, possible answers to the religion question could be Jewish, Christian, Muslim, Hindu, Agnostic, or Other). Other times, respondents are asked to choose an answer from a numerical range. We could ask the question:

"In your opinion, how important are SAT scores to a college student's success?"

Not important at all   1   2   3   4   5   Extremely important

These numerical responses are usually summarized and reported in terms of the average response, whose size tells us something about the perceived importance. The Zagat restaurant survey (www.zagat.com) asks diners to rate restaurants' food, décor, and service, each on a scale of 1 to 30 points, with a 30 representing an incredible level of satisfaction. Although the Zagat scale has an unusually wide range of possible ratings, the concept is the same as in the more common 5-point scale.

Open-ended questions typically provide the most honest and complete information because there are no suggested answers to divert or bias a person's response. This kind of question is often found on instructor evaluation forms distributed at the end of a college course. College students at Georgetown University are asked the open-ended question, "What comments would you give to the instructor?" The responses provide the instructor feedback that may be missing from the initial part of the teaching evaluation survey, which consists of numerical multiple-choice ratings of various aspects of the course. While these numerical ratings can be used to compare instructors and courses, there are no easy comparisons of the diverse responses instructors receive to the open-ended question. In fact, these responses are often seen only by the instructor and are useful, constructive tools for the teacher despite the fact they cannot be readily summarized.

Survey questionnaires must be carefully constructed so they do not inadvertently bias the results. Because survey design is such a difficult and sensitive process, it is not uncommon for a pilot survey to be taken before a lot of time, effort, and financing go into collecting a large amount of data. Pilot surveys are similar to the beta version of a new electronic product; they are tested out with a smaller group of people to work out the "kinks" before being used on a larger scale. Determination of the sample size for the final survey is an important process for many reasons. If the sample size is too large, resources may be wasted during the data collection. On the other hand, not collecting enough data for a meaningful analysis will obviously be detrimental to the study. Fortunately, there are several formulas that will help decide how large a sample should be, depending on the goal of the study and various other factors.

Types of surveys

There are several different survey types, and we will explore just a few of them. The **phone survey** is particularly well-known (and often despised). A phone survey is inexpensive and usually conducted by callers who have very little training. Because of this and the impersonal nature of the medium, the respondent may misunderstand some of the questions. A further drawback is that some people cannot be reached and that others may refuse to answer some or all of the questions. Phone surveys are thus particularly prone to have a low **response rate.**

The **response rate** is the proportion of all people whom we attempt to contact that actually respond to a survey. A low response rate can destroy the validity of a survey's results.

It can be difficult to collect good data from unsolicited phone calls because many of us resent the interruption. The calls often come at inopportune times, intruding on a meal or arriving just when we have climbed a ladder with a full can of paint. No wonder we may fantasize about turning the tables on the callers and calling *them* when it is least convenient.

Numerous complaints have been filed with the Federal Trade Commission (FTC) about the glut of marketing and survey telephone calls to private residences. The National Do Not Call Registry was created as the culmination of a comprehensive, three-year review of the Telemarketing Sales Rule (TSR) (www.ftc.gov/donotcall/). This legislation allows people to enroll their phone numbers on a website so as to prevent most marketers from calling them.

Self-administered surveys, or **mail surveys,** are also very inexpensive to conduct. However, these also have their drawbacks. Often, recipients will choose not to reply unless they receive some kind of financial incentive or other reward. Generally, after an initial mailing, the response rate will fall between 20 and 30 percent. Response rates can be raised with successive follow-up reminders, and after three contacts, they might reach between 65 and 75 percent. Unfortunately, the entire process can take significantly longer than a phone survey would.

Web-based surveys have become increasingly popular, but they suffer from the same problems as mail surveys. In addition, as with phone surveys, respondents may record their true reactions incorrectly because they have misunderstood some of the questions posed.

A personal interview provides more control over the survey process. People selected for interviews are more likely to respond because the questions are being asked by someone face-to-face. Questions are less likely to be misunderstood because the people conducting the interviews are typically trained employees who can clear up any confusion arising during the process. On the other hand, interviewers can potentially "lead" a respondent by body language, which signals approval or disapproval of certain sorts of answers. They can also prompt certain replies by providing too much information. **Mall surveys** are examples of personal interviews. Interviewers approach shoppers as they pass by and ask them to answer the survey questions. Response rates around 50 percent are typical. Personal interviews are more costly than mail or phone surveys. Obviously, the objective of the study will be important in deciding upon the survey type employed.

Errors occurring in surveys

In general, the goal of a survey is to obtain accurate information from a group, or sample, that is representative of the entire population of interest. We are trying to estimate some aspect (numerical descriptor) of the entire population from a subset of the population. This is not an easy task, and there are many pitfalls. First and foremost, the *target population* must be well defined and a *sample frame* must be chosen.

The **target population** is the entire population of interest to us in a particular study.

Are we intending to estimate the average starting salary of students graduating from any college? Or from four-year colleges? Or from business schools? Or from a particular business school?

> The **sample frame** is a list of sampling elements (people or things) from which the sample will be selected. It should closely agree with the target population.

Consider a study to estimate the average starting salary of students who have graduated from the business school at Miami University of Ohio over the last five years; the target population is obviously that particular group of graduates. A sample frame could be the Miami University Alumni Association's roster of business school graduates for the past five years. Although it will not be a perfect replication of the target population, it is a reasonable frame.

We now discuss two general classes of survey errors: **errors of nonobservation** and **errors of observation.** From the sample frame, units are randomly chosen to be part of the sample. Simply by virtue of the fact that we are taking a sample instead of a census, we are susceptible to *sampling error.*

> **Sampling error** is the difference between a numerical descriptor of the population and the corresponding descriptor of the sample.

Sampling error occurs because our information is incomplete. We observe only the portion of the population included in the sample while the remainder is obscured. Suppose, for example, we wanted to know about the heights of 13-year-old boys. There is extreme variation in boys' heights at this age. Even if we could overcome the logistical problems of choosing a random sample of 20 boys, there is nothing to guarantee the sample will accurately reflect heights at this age. By sheer luck of the draw, our sample could include a higher proportion of tall boys than appears in the population. We would then overestimate average height at this age (to the chagrin of the shorter boys). Although samples tend to look more similar to their parent populations as the sample sizes increase, we should always keep in mind that sample characteristics and population characteristics are not the same.

If a sample frame is not identical to the target population, we will suffer from an *error of coverage.*

> **Undercoverage** occurs when some population elements are excluded from the process of selecting the sample.

Undercoverage was part of the problem dooming the *Literary Digest* Poll of 1936. Although millions of Americans were included in the poll, the large sample size could not rescue the poll results. The sample represented those who could afford phone service and magazine subscriptions in the lean Depression years, but in excluding everyone else, it failed to yield an honest picture of the entire American populace. Undercoverage often occurs when we do not have a complete, accurate list of all the population elements. If we select our sample from an incomplete list, like a telephone directory or a list of all Internet subscribers in a region, we automatically eliminate those who cannot afford phone or Internet service. Even today, 7 to 8 percent of the people in the United States do not own telephones. Low-income people are often underrepresented in surveys. If underrepresented groups differ from the rest of the population with respect to the characteristic under study, the survey results will be biased.

Often, pollsters cannot find all the people they intend to survey, and sometimes people who are found will refuse to answer the questions posed. Both of these are examples of the **nonresponse** problem. Unfortunately, there may be an association between how difficult it is to find and elicit responses from people and the type of answers they give.

> **Nonresponse** occurs whenever some of the individuals who were supposed to be included in the sample are not.

For example, universities often conduct surveys to learn how graduates have fared in the workplace. The alumnus who has risen through the corporate ranks is more likely to have a current address on file with his alumni office and to be willing to share career information than a classmate who has foundered professionally. We should be politely skeptical about reports touting the average salaries of graduates of various university programs. In some surveys, 35 percent or more of the selected individuals cannot be contacted—even when several callbacks are made. In such cases, other participants are often substituted for those who cannot be contacted. If the substitutes and the originally selected participants differ with respect to the characteristic under study, the survey will be biased. Furthermore, people who will answer highly sensitive, personal, or embarrassing questions might be very different from those who will not.

As discussed in Section 1.4, the opinions of those who bother to complete a voluntary response survey may be dramatically different from those who do not. (Recall the Ann Landers question about having children.) The viewer voting on the television show *American Idol* is another illustration of **selection bias,** because only those who are interested in the outcome of the show will bother to phone in or text message their votes. The results of the voting are not representative of the performance ratings the country would give as a whole.

Errors of observation occur when data values are recorded incorrectly. Such errors can be caused by the data collector (the interviewer), the survey instrument, the respondent, or the data collection process. For instance, the manner in which a question is asked can influence the response. Or, the order in which questions appear on a questionnaire can influence the survey results. Or, the data collection method (telephone interview, questionnaire, personal interview, or direct observation) can influence the results. A **recording error** occurs when either the respondent or interviewer incorrectly marks an answer. Once data are collected from a survey, the results are often entered into a computer for statistical analysis. When transferring data from a survey form to a spreadsheet program like Excel, Minitab, or JMP, there is potential for entering them incorrectly. Before the survey is administered, the questions need to be very carefully worded so that there is little chance of misinterpretation. A poorly framed question might yield results that lead to unwarranted decisions. Scaled questions are particularly susceptible to this type of error. Consider the question "How would you rate this course?" Without a proper explanation, the respondent may not know whether "1" or "5" is the best.

If the survey instrument contains highly sensitive questions and respondents feel compelled to answer, they may not tell the truth. This is especially true in personal interviews. We then have what is called **response bias.** A surprising number of people are reluctant to be candid about what they like to read or watch on television. People tend to overreport "good" activities like reading respected newspapers and underreport their "bad" activities like delighting in the *National Enquirer*'s stories of alien abductions and celebrity meltdowns. Imagine, then, the difficulty in getting honest answers about people's gambling habits, drug use, or sexual histories. Response bias can also occur when respondents are asked slanted questions whose wording influences the answer received. For example, consider the following question:

Which of the following best describes your views on gun control?

1　The government should take away our guns, leaving us defenseless against heavily armed criminals.

2　We have the right to keep and bear arms.

This question is biased toward eliciting a response against gun control.

Exercises for Section 1.8

CONCEPTS

1.30　Explain:
　　a　Three types of surveys and discuss their advantages and disadvantages.
　　b　Three types of survey questions and discuss their advantages and disadvantages.

1.31　Explain each of the following terms:
　　a　Undercoverage.　b　Nonresponse.　c　Response bias.

1.32　A market research firm sends out a web-based survey to assess the impact of advertisements placed on a search engine's results page. About 65 percent of the surveys were answered and sent back. What types of errors are possible in this scenario?

Chapter Summary

We began this chapter by discussing **data.** We learned that the data that are collected for a particular study are referred to as a **data set,** and we learned that **elements** are the entities described by a data set. In order to determine what information we need about a group of elements, we define important **variables,** or characteristics, describing the elements. **Quantitative variables** are variables that use numbers to measure quantities (that is, "how much" or "how many") and **qualitative, or categorical, variables** simply record into which of several categories an element falls.

We next discussed the difference between cross-sectional data and time series data. **Cross-sectional data** are data collected at the same or approximately the same point in time. **Time series data** are data collected over different time periods, and we saw that time series data are often depicted by using a **time series plot.**

Next we learned about data sources. **Primary data** are collected by an individual through personally planned experimentation or observation, while **secondary data** are taken from existing sources. We discussed some readily available existing data sources, and we learned the difference between *experimental* and *observational* studies. We found that a study is **experimental** when we are able to set or manipulate the factors that may be related to the response variable and that a study is **observational** when we are unable to control the factors of interest. We learned that with the increased use of online purchasing and with increased competition, businesses have become more aggressive about collecting information about customer transactions. Dramatic advances in data capture, data transmission, and data storage capabilities are enabling organizations to integrate various databases into **data warehouses.** The term **big data** refers to massive amounts of data, often collected at very fast rates in real time and in different forms and sometimes needing quick preliminary analysis for effective business decision making.

We often collect data to study a **population,** which is the set of all elements about which we wish to draw conclusions. We saw that, because many populations are too large to examine in their entirety, we frequently study a population by selecting a **sample,** which is a subset of the population elements. We saw that we often wish to describe a population or sample and

that **descriptive statistics** is the science of describing the important aspects of a population or sample. We also learned that if a population is large and we need to select a sample from it, we use what is called **statistical inference,** which is the science of using a sample to make generalizations about the important aspects of a population.

Next we learned that if the information contained in a sample is to accurately represent the population, then the sample should be **randomly selected** from the population. In Section 1.4 we formally defined a **random sample,** and we studied three cases that introduced how we can take a random (or approximately random) sample. The methods we illustrated included using a **random number table** and using **computer-generated random numbers.** We also learned that we often wish to sample a **process** over time, and we illustrated how such sampling might be done. Finally, Section 1.4 presented some ethical guidelines for statistical practice.

We concluded this chapter with four optional sections. In optional Section 1.5 we learned that big data has resulted in an extension of traditional statistics called **business analytics,** and we introduced some basic ideas of **descriptive analytics, predictive analytics, data mining,** and **prescriptive analytics.** In optional Section 1.6, we considered different types of quantitative and qualitative variables. We learned that there are two types of **quantitative variables—ratio variables,** which are measured on a scale such that ratios of its values are meaningful and there is an inherently defined zero value, and **interval variables,** for which ratios are not meaningful and there is no inherently defined zero value. We also saw that there are two types of **qualitative variables—ordinal variables,** for which there is a meaningful ordering of the categories, and **nominative variables,** for which there is no meaningful ordering of the categories. Optional Section 1.7 introduced several advanced sampling designs: **stratified random sampling, cluster sampling,** and **systematic sampling.** Finally, optional Section 1.8 discussed more about sample surveys. Topics included **types of surveys** (such as phone surveys, mail surveys, and mall surveys), types of **survey questions** (such as dichotomous, multiple-choice, and open-ended questions), and **survey errors** (such as sampling error, error due to undercoverage, and error due to nonresponse).

Glossary of Terms

big data: Massive amounts of data, often collected at very fast rates in real time and in different forms and sometimes needing quick preliminary analysis for effective business decision making.

business analytics: The use of traditional and newly developed statistical methods, advances in information systems, and techniques from *management science* to continuously and iteratively explore and investigate past business performance, with

the purpose of gaining insight and improving business planning and operations.

categorical (qualitative) variable: A variable having values that indicate into which of several categories a population element belongs.

census: An examination of all the elements in a population.

cluster sampling (multistage cluster sampling): A sampling design in which we sequentially cluster population elements into subpopulations.

convenience sampling: Sampling where we select elements because they are easy or convenient to sample.

cross-sectional data: Data collected at the same or approximately the same point in time.

data: Facts and figures from which conclusions can be drawn.

data mining: The process of discovering useful knowledge in extremely large data sets.

data set: Facts and figures, taken together, that are collected for a statistical study.

data warehousing: A process of centralized data management and retrieval that has as its ideal objective the creation and maintenance of a central repository for all of an organization's data.

descriptive analytics: The use of traditional and more recently developed statistical graphics to present to executives (and sometimes customers) easy-to-understand visual summaries of up-to-the-minute information concerning the operational status of a business.

descriptive statistics: The science of describing the important aspects of a set of measurements.

element: A person, object, or other entity about which we wish to draw a conclusion.

errors of nonobservation: Sampling error related to population elements that are not observed.

errors of observation: Sampling error that occurs when the data collected in a survey differs from the truth.

experimental study: A statistical study in which the analyst is able to set or manipulate the values of the factors.

factor: A variable that may be related to the response variable.

finite population: A population that contains a finite number of elements.

frame: A list of all of the population elements.

infinite population: A population that is defined so that there is no limit to the number of elements that could potentially belong to the population.

interval variable: A quantitative variable such that ratios of its values are not meaningful and for which there is not an inherently defined zero value.

judgment sampling: Sampling where an expert selects population elements that he/she feels are representative of the population.

measurement: The process of assigning a value of a variable to an element in a population or sample.

nominative variable: A qualitative variable for which there is no meaningful ordering, or ranking, of the categories.

nonresponse: A situation in which population elements selected to participate in a survey do not respond to the survey instrument.

observational study: A statistical study in which the analyst is not able to control the values of the factors.

ordinal variable: A qualitative variable for which there is a meaningful ordering or ranking of the categories.

predictive analytics: Methods used to find anomalies, patterns, and associations in data sets, with the purpose of predicting future outcomes. The applications of predictive analytics include *anomaly (outlier) detection, association learning, classification, cluster detection, prediction,* and *factor analysis.*

prescriptive analytics: The use of internal and external variables, along with the predictions obtained from predictive analytics, to recommend one or more courses of action.

population: The set of all elements about which we wish to draw conclusions.

probability sampling: Sampling where we know the chance (probability) that each population element will be included in the sample.

process: A sequence of operations that takes inputs and turns them into outputs.

qualitative (categorical) variable: A variable having values that indicate into which of several categories a population element belongs.

quantitative variable: A variable having values that are numbers representing quantities.

random number table: A table containing random digits that is often used to select a random sample.

random sample: A sample selected in such a way that every set of n elements in the population has the same chance of being selected.

ratio variable: A quantitative variable such that ratios of its values are meaningful and for which there is an inherently defined zero value.

response bias: Bias in the results obtained when carrying out a statistical study that is related to how survey participants answer the survey questions.

response rate: The proportion of all people whom we attempt to contact that actually respond to a survey.

response variable: A variable of interest that we wish to study.

sample: A subset of the elements in a population.

sample frame: A list of sampling elements from which a sample will be selected. It should closely agree with the target population.

sampling error: The difference between the value of a sample statistic and the population parameter; it occurs because not all of the elements in the population have been measured.

sampling with replacement: A sampling procedure in which we place any element that has been chosen back into the population to give the element a chance to be chosen on succeeding selections.

sampling without replacement: A sampling procedure in which we do not place previously selected elements back into the population and, therefore, do not give these elements a chance to be chosen on succeeding selections.

selection bias: Bias in the results obtained when carrying out a statistical study that is related to how survey participants are selected.

statistical inference: The science of using a sample of measurements to make generalizations about the important aspects of a population.

strata: The subpopulations in a stratified sampling design.

stratified random sampling: A sampling design in which we divide a population into nonoverlapping subpopulations and then select a random sample from each subpopulation (stratum).

survey: An instrument employed to collect data.

systematic sample: A sample taken by moving systematically through the population. For instance, we might randomly select one of the first 200 population elements and then systematically sample every 200th population element thereafter.

target population: The entire population of interest in a statistical study.
time series data: Data collected over different time periods.
time series plot (runs plot): A plot of time series data versus time.

undercoverage: A situation in sampling in which some groups of population elements are underrepresented.
variable: A characteristic of a population or sample element.
voluntary response sample: Sampling in which the sample participants self-select.

Supplementary Exercises

1.33   **THE COFFEE TEMPERATURE CASE**   DS Coffee

According to the website of the American Association for Justice,[12] Stella Liebeck of Albuquerque, New Mexico, was severely burned

by McDonald's coffee in February 1992. Liebeck, who received third-degree burns over 6 percent of her body, was awarded $160,000 in compensatory damages and $480,000 in punitive damages. A postverdict investigation revealed that the coffee temperature at the local Albuquerque McDonald's had dropped from about 185°F before the trial to about 158° after the trial.

This case concerns coffee temperatures at a fast-food restaurant. Because of the possibility of future litigation and to possibly improve

©iStockphoto/Getty Images

the coffee's taste, the restaurant wishes to study the temperature of the coffee it serves. To do this, the restaurant personnel measure the temperature of the coffee being dispensed (in degrees Fahrenheit) at a randomly selected time during each of the 24 half-hour periods from 8 A.M. to 7:30 P.M. on a given day. This is then repeated on a second day, giving the 48 coffee temperatures in Table 1.11. Make a time series plot of the coffee temperatures, and assuming process consistency, estimate limits between which most of the coffee temperatures at the restaurant would fall.

1.34   In the article "Accelerating Improvement" published in *Quality Progress,* Gaudard, Coates, and Freeman describe a restaurant that caters to business travelers and has a self-service breakfast buffet. Interested in customer satisfaction, the manager conducts a survey over a three-week period and finds that the main customer complaint is having to wait too long to be seated. On each day from September 11 to October 1, a problem-solving team records the percentage of patrons who must wait more than one minute to be seated. A time series plot of the daily percentages is shown in Figure 1.8.[13] What does the time series plot tell us about how to improve the waiting time situation?

TABLE 1.11   **The Coffee Temperatures for Exercise 1.33**   DS Coffee

154°F	156	158	166
165	151	160	158
148	161	153	173
157	157	161	162
160	154	160	155
157	159	158	150
152	155	169	165
149	153	163	154
171	173	146	160
168	164	167	162
165	161	162	159
164	151	159	166

Note: Time order is given by reading down the columns from left to right.

FIGURE 1.8   **Time Series Plot of Daily Percentages of Customers Waiting More Than One Minute to Be Seated (for Exercise 1.34)**

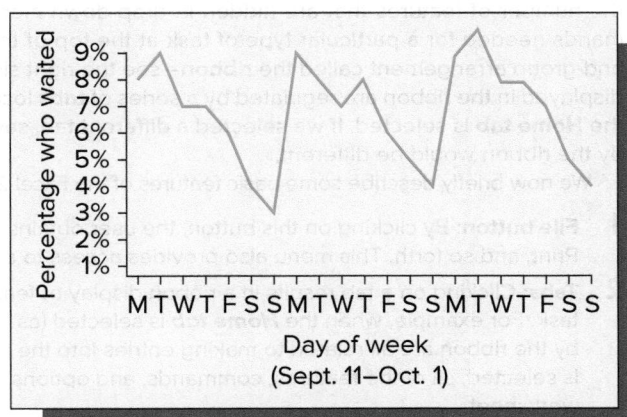

[12]American Association for Justice, June 16, 2006.

[13]Source 1.8 is M. Gaudard, R. Coates, and L. Freeman, "Accelerating Improvement," *Quality Progress,* October 1991, pp. 81–88. © 1991 American Society for Quality Control. Used with permission.

1.35 ■ **Internet Exercise**

The website maintained by the U.S. Census Bureau provides a multitude of social, economic, and government data (http://www.census.gov/). Go to the U.S. Census Bureau website and find the table of Consumer Price Indexes by Major Groups 1990 to 2010 (www.census.gov /compendia/statab/2012/tables/12s0725.pdf). Construct time series plots of (1) the price index for all items over time (years), (2) the price index for food over time, (3) the price index for energy over time, and (4) the price index for housing over time. For each time series plot, describe apparent trends in the price index.

Excel, Minitab, and JMP for Statistics

In this book we use three types of software to carry out statistical analysis—Microsoft Excel 2016 (including the Excel add in MegaStat), Minitab 18, and JMP Pro 14. Excel is, of course, a general purpose electronic spreadsheet program and analytical tool. The analysis ToolPak in Excel includes many procedures for performing various kinds of basic statistical analyses. The Excel add-in package Megastat is specifically designed for performing basic and intermediate-level statistical analysis in the Excel spreadsheet environment. Minitab and JMP are computer packages designed expressly for conducting basic, intermediate, and advanced statistical analysis. They are widely used at many colleges and universities and in a large number of business organizations. The principal advantage of Excel is that, because of its broad acceptance among students and professionals as a multipurpose analytical tool, it is both well known and widely available. The advantages of special-purpose statistical software packages like Minitab and JMP are that they provide a far wider range of statistical procedures than either Excel or MegaStat and give the experienced analyst a range of options to better control the analysis. The advantages of MegaStat include (1) its ability to perform a number of intermediate-level statistical calculations that are not automatically done by the procedures in the Excel ToolPak and (2) features that make it easier to use than Excel for a wide variety of statistical analyses. In addition, the output obtained by using MegaStat is automatically placed in a standard Excel spreadsheet and can be edited by using any of the features in Excel. MegaStat can be copied from the book's website.

Commonly used features of Excel 2016, Minitab 18, and JMP Pro 14 are presented in this chapter along with an initial application—the construction of a time series plot of the gas mileages in Table 1.7. Descriptions of these packages then continue throughout this book. The book's website shows how to use MegaStat.

Appendix 1.1 ■ Getting Started with Excel

Because Excel 2016 may be new to some readers, we will begin by describing some characteristics of the Excel 2016 window. Versions of Excel prior to 2007 employed many drop-down menus. This meant that many features were "hidden" from the user, which resulted in a steep learning curve for beginners. Beginning with Excel 2010, Microsoft tried to reduce the number of features that are hidden in drop-down menus. Therefore, Excel 2016 displays all of the applicable commands needed for a particular type of task at the top of the Excel window. These commands are represented by a tab-and-group arrangement called the **ribbon**—see the right side of the illustration of an Excel 2016 window. The commands displayed in the ribbon are regulated by a series of **tabs** located near the top of the ribbon. For example, in the illustration, the **Home tab** is selected. If we selected a different tab, say, for example, the **Page Layout tab**, the commands displayed by the ribbon would be different.

We now briefly describe some basic features of the Excel 2016 window:

1 **File button:** By clicking on this button, the user obtains a menu of often used commands—for example, Open, Save, Print, and so forth. This menu also provides access to a large number of Excel options settings.

2 **Tabs:** Clicking on a tab results in a ribbon display of features, commands, and options related to a particular type of task. For example, when the *Home tab* is selected (as in the figure), the features, commands, and options displayed by the ribbon are all related to making entries into the Excel worksheet. As another example, if the *Formulas tab* is selected, all of the features, commands, and options displayed in the ribbon relate to using formulas in the Excel worksheet.

3 **Quick access toolbar:** This toolbar displays buttons that provide shortcuts to often-used commands. Initially, this toolbar displays Save, Undo, and Redo buttons. The user can customize this toolbar by adding shortcut buttons for other commands (such as New, Open, Quick Print, and so forth). This can be done by clicking on the arrow button directly to the right of the Quick Access toolbar and by making selections from the "Customize" drop-down menu that appears.

Tabs (Home Tab Selected)

File Button   Quick Access Toolbar   Title Bar   Maximize, Minimize, and Close Buttons   Ribbon

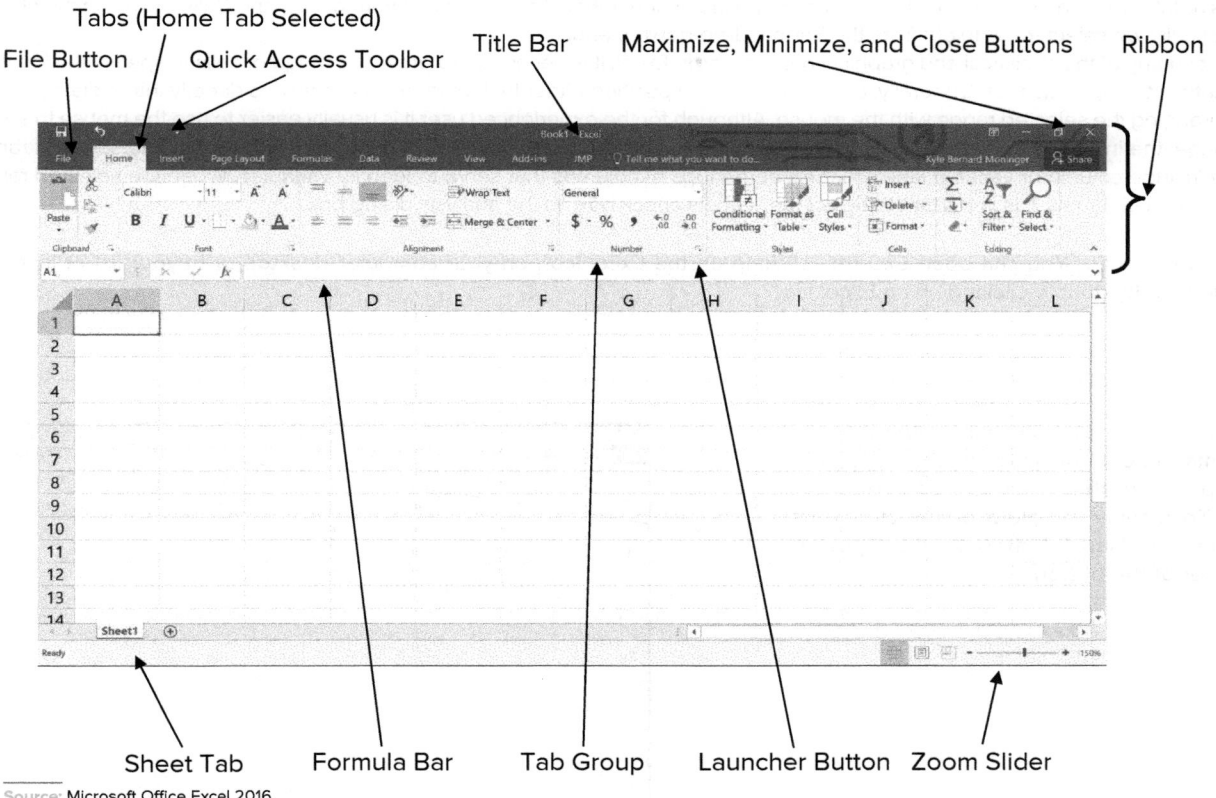

Sheet Tab   Formula Bar   Tab Group   Launcher Button   Zoom Slider

Source: Microsoft Office Excel 2016

4 Title bar: This bar shows the name of the currently active workbook and contains the Quick Access Toolbar as well as the Maximize, Minimize, and Close buttons.

5 Ribbon: A grouping of toolbars, tabs, commands, and features related to performing a particular kind of task—for example, making entries into the Excel spreadsheet. The particular features displayed in the ribbon are controlled by selecting a *Tab*. If the user is working in the spreadsheet workspace and wishes to reduce the number of features displayed by the ribbon, this can be done by right-clicking on the ribbon and by selecting "Minimize the Ribbon." We will often minimize the ribbon in the Excel appendices of this book in order to focus attention on operations being performed and results being displayed in the Excel spreadsheet.

6 Sheet tabs: These tabs show the name of each sheet in the Excel workbook. When the user clicks a sheet tab, the selected sheet becomes active and is displayed in the Excel spreadsheet. The name of a sheet can be changed by double-clicking on the appropriate sheet tab and by entering the new name.

7 Formula bar: When a worksheet cell is selected, the formula bar displays the current content of the cell. If the cell content is defined by a formula, the defining formula is displayed in the formula bar.

8 Tab group: This is a labeled grouping of commands and features related to performing a particular type of task.

9 Launcher button: Some of the tab groups have a launcher button—for example, the Clipboard, Font, Alignment, and Number tab groups each have such a button. Clicking on the launcher button opens a dialog box or task pane related to performing operations in the tab group.

10 **Zoom slider:** By moving this slider right or left, the cells in the Excel spreadsheet can be enlarged or reduced in size.

We now take a look at some features of Excel that are common to many analyses. When the instructions call for a sequence of selections, the sequence will be presented in the following form:

Select Home : Format : Row Height

This notation indicates that we first select the Home tab on the ribbon, then we select Format from the Cells Group on the ribbon, and finally we select Row Height from the Format drop-down menu.

For many of the statistical and graphical procedures in Excel, it is necessary to provide a range of cells to specify the location of data in the spreadsheet. Generally, the range may be specified either by typing the cell locations directly into a dialog box or by dragging the selected range with the mouse. Although for the experienced user it is usually easier to use the mouse to select a range, the instructions that follow will, for precision and clarity, specify ranges by typing in cell locations. The selected range may include column or variable labels—labels at the tops of columns that serve to identify variables. When the selected range includes such labels, it is important to select the "Labels check box" in the analysis dialog box.

Opening Excel   You can open Excel by clicking on the Excel icon on your computer desktop or in your list of installed options or by double-clicking on an Excel file.

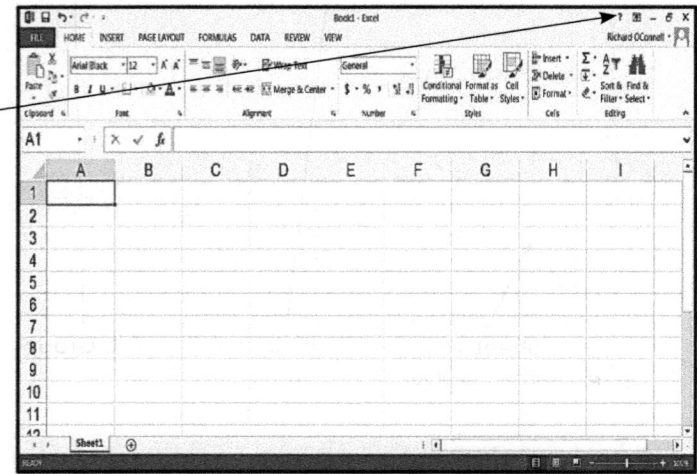

Help resources Like most Windows programs, Excel includes online help via a Help Menu that includes search capability. To display the Help Menu, click on the "Question Mark" button in the upperright corner of the ribbon.

Source: Microsoft Office Excel 2016

Entering data (entering the gas mileages in Table 1.7) from the keyboard (data file: GasMiles. xlsx):

- In a new Excel workbook, click on cell A1 in Sheet 1 and type a label—that is, a variable name—say, Mileage, for the gasoline mileages.

- Beginning in cell A2 (directly under the column label Mileage), type the mileages from Table 1.7 down the column, pressing the Enter key following each entry.

Source: Microsoft Office Excel 2016

Saving data (saving the gasoline mileage data):

- To begin, click on the **File** button and select **Save As.**

- In the "Save As" dialog box, select the destination drive and folder. Here we have selected a file folder called New Data Files in Rick's System folder on the local C drive.

- Enter the desired file name in the "File name" window. In this case we have chosen the name GasMiles.

- Select Excel Workbook in the "Save as type" window.

- Click the **Save** button in the "Save As" dialog box.

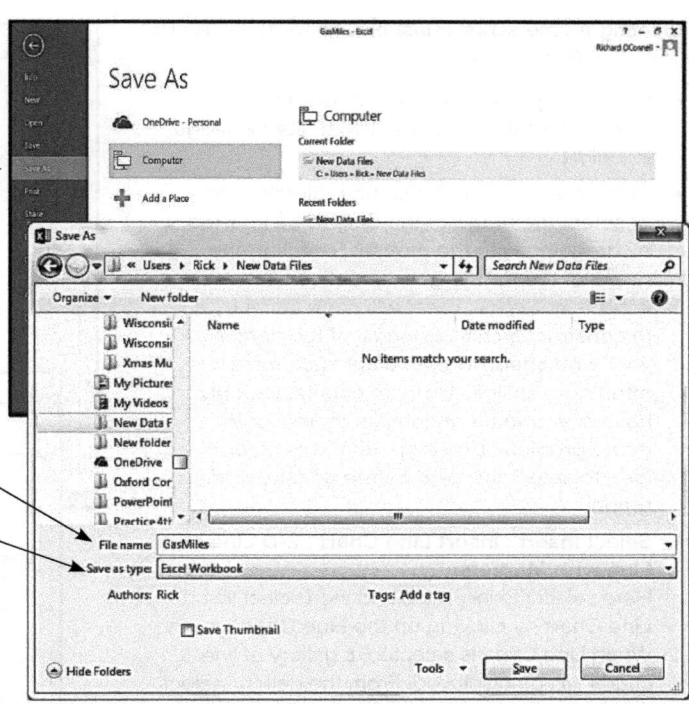

Source: Microsoft Office Excel 2016

Retrieving an Excel spreadsheet containing the gasoline mileages in Table 1.7 (data file: GasMiles.xlsx):

- Select **File : Open : Computer : Browse**
  That is, click on the File button and then select Open, Computer, and Browse.

- In the Open dialog box, select the desired source drive, folder, and file. Here we have selected the GasMiles file in a folder named New Data Files in Rick's System folder on the local C drive. When you select the desired file by clicking on it, the file name will be shown in the "File name" window.

- Click the Open button in the Open dialog box.

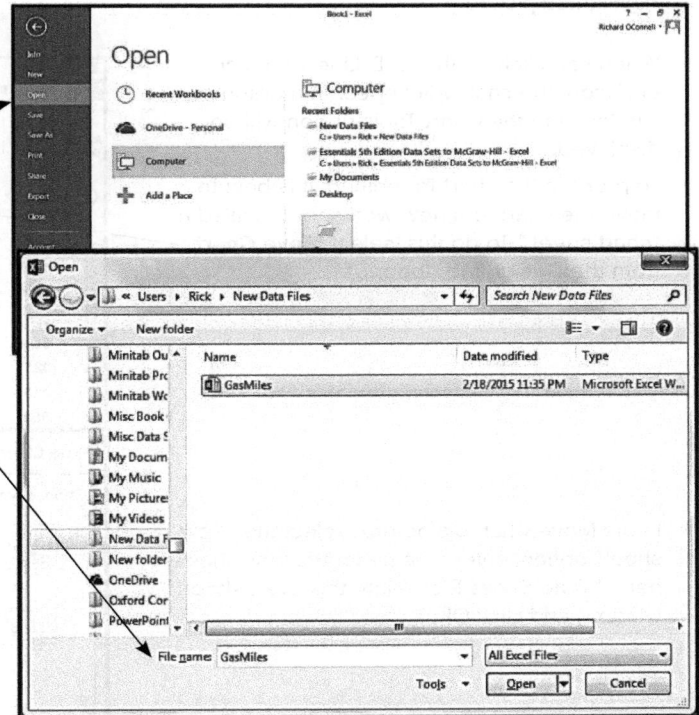

Source: Microsoft Office Excel 2016

Creating a time series (runs) plot similar to Figure 1.5 (data file: GasMiles.xlsx):

- Enter the gasoline mileage data into column A of the worksheet with label Mileage in cell A1.

- Click on any cell in the column of mileages, or select the range of the data to be charted by dragging with the mouse. Selecting the range of the data is good practice because, if this is not done, Excel will sometimes try to construct a chart using all of the data in your worksheet. The result of such a chart is often nonsensical. Here, of course, we only have one column of data, so there would be no problem. But, in general, it is a good idea to select the data before constructing a graph.

- Select **Insert : Insert Line Chart : 2-D Line : Line with Markers**
  Here select the Insert tab and then select Insert Line Chart by clicking on the Line Chart icon. When Line Chart is selected, a gallery of line charts will be displayed. From the gallery, select the desired chart—in this case a 2-D Line chart with markers. The proper chart can be selected by looking at the sample pictures. As an alternative, if the cursor is hovered over a picture, a descriptive "tool tip" of the chart type will be displayed. In this case, the "Line with Markers" tool tip was obtained by hovering the cursor over the highlighted picture.

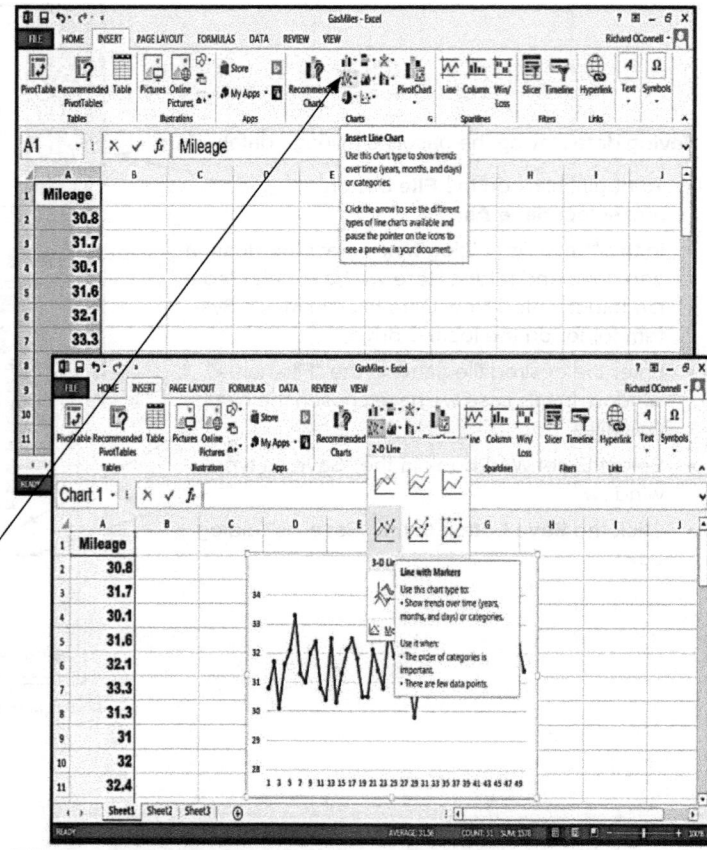

Source: Microsoft Office Excel 2016

- When you click on the "2-D Line with Markers" icon, the chart will appear in a graphics window and the Chart Tools ribbon will be displayed.

- To prepare the chart for editing, it is best to move the chart to a new worksheet—called a "chart sheet." To do this, select **Move Chart** from the Design tab ribbon.

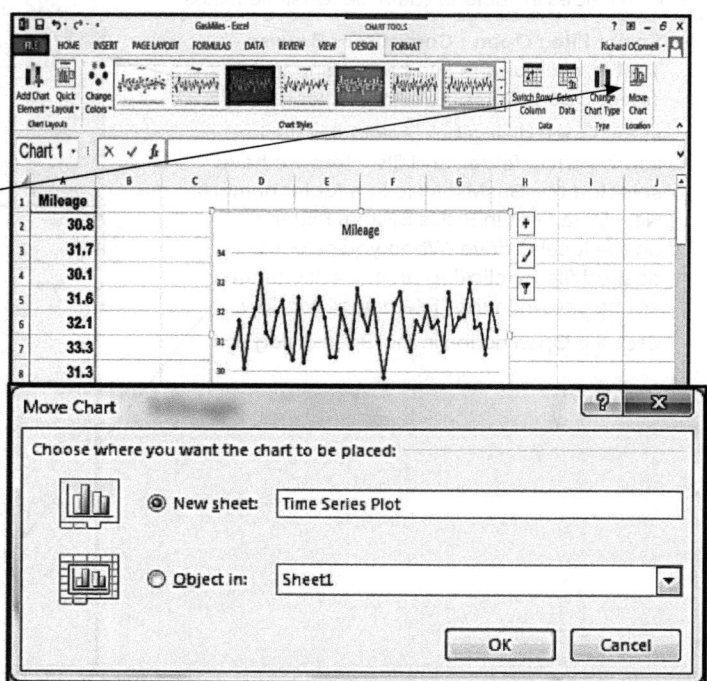

Source: Microsoft Office Excel 2016

- In the Move Chart dialog box, select the "New sheet" option, enter a name for the new sheet—here, "Time Series Plot"—into the "New sheet" window, and click OK.

- Here we show an edited time series plot. This revised chart was constructed from the original time series plot created by Excel using various options like those illustrated above. This chart can be copied directly from the worksheet (simply right-click on the graph and select Copy from the pop-up menu) and can then be pasted into a word-processing document.

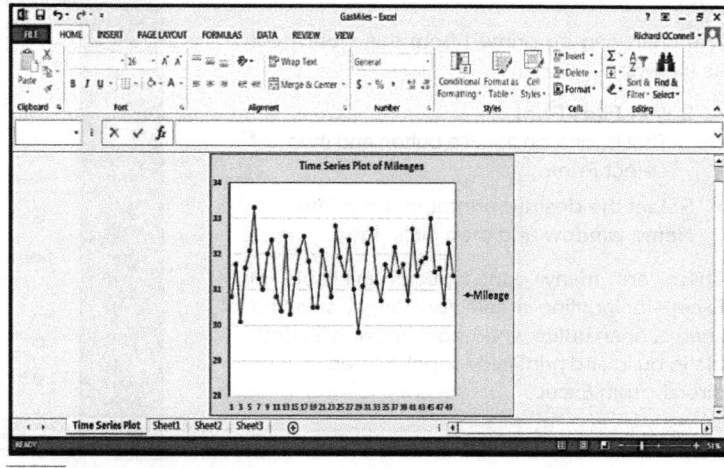

Source: Microsoft Office Excel 2016

- The Chart Tools ribbon will be displayed and the chart will be placed in a chart sheet format that is more convenient for editing.

- The chart can be edited in several ways. You can select a chart style from the gallery of styles shown on the ribbon. Or, by clicking on the icons that appear in the upper right corner of the chart, you can edit many chart elements such as axes, axis labels, data labels (numbers that label the data points), and so forth. Chart styles and the data points that are displayed can be edited by clicking on the second and third icons in the upper right corner of the chart. Colors and other chart attributes can be edited by making various selections from the ribbon.

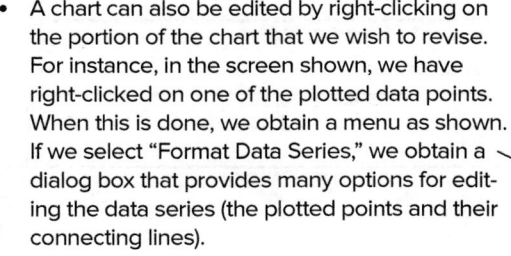

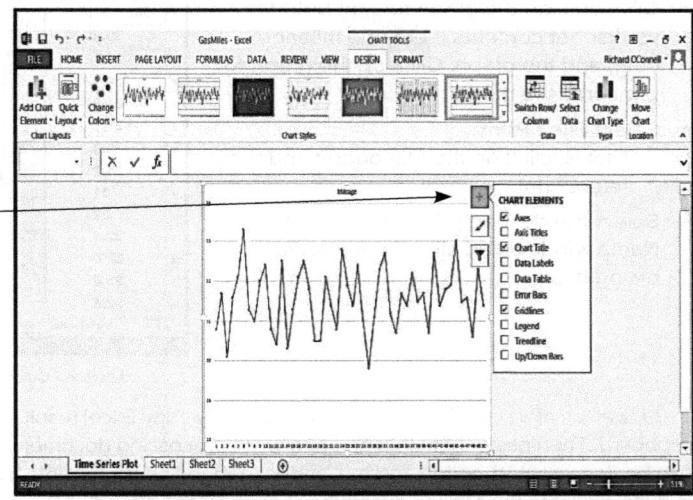

Source: Microsoft Office Excel 2016

- A chart can also be edited by right-clicking on the portion of the chart that we wish to revise. For instance, in the screen shown, we have right-clicked on one of the plotted data points. When this is done, we obtain a menu as shown. If we select "Format Data Series," we obtain a dialog box that provides many options for editing the data series (the plotted points and their connecting lines).

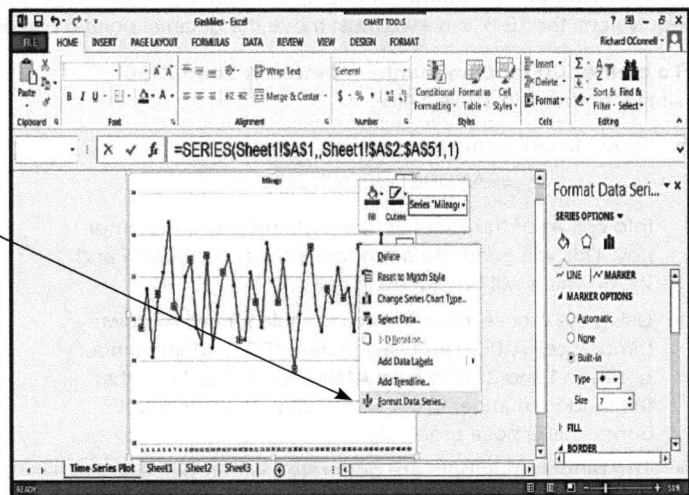

Source: Microsoft Office Excel 2016

The chart can be printed from this worksheet as follows:

- Select **File : Print**
    That is, click on the File button and then select Print.
- Select the desired printer in the Printer Name window and then click Print.

There are many print options available in Excel—for printing a selected range, selected sheets, or an entire workbook—making it possible to build and print fairly sophisticated reports directly from Excel.

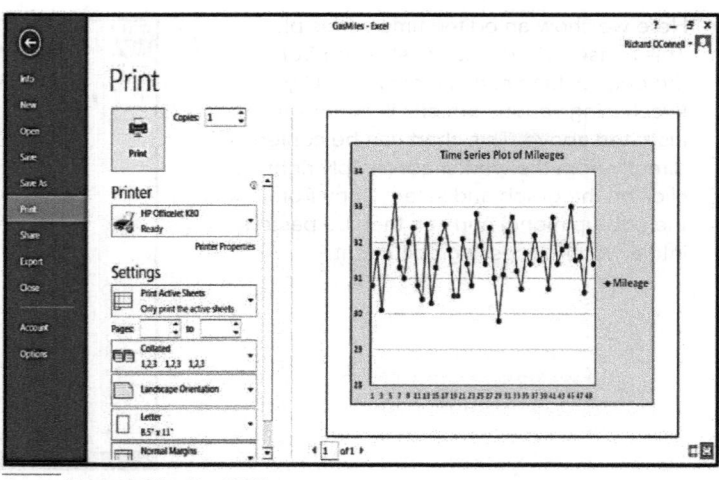

Source: Microsoft Office Excel 2016

Printing a spreadsheet with an embedded graph:

- Click outside the graph to print both the worksheet contents (here, the mileage data) and the graph. Click on the graph to print only the graph.
- Select **File : Print**
    That is, click on the File button and then select Print.
- Select the desired printer in the Printer Name window and click OK in the Print dialog box.

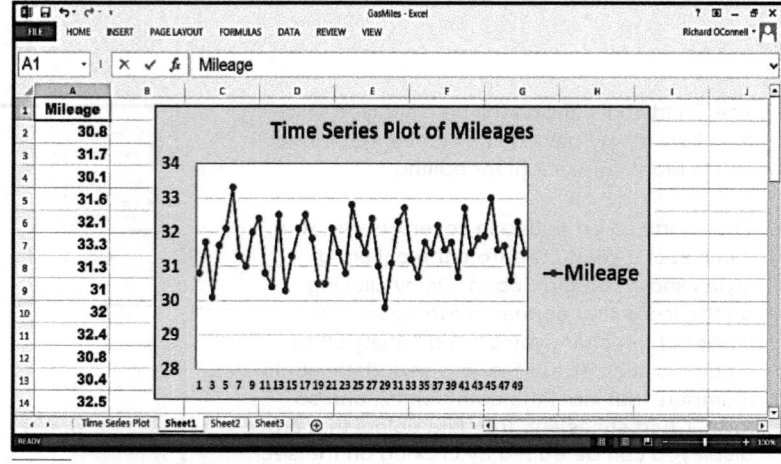

Source: Microsoft Office Excel 2016

Including Excel output in reports   You can easily copy Excel results—selected spreadsheet ranges and graphs—to the Windows clipboard. Then paste them into an open word-processing document. Once copied to a word-processing document, Excel results can be documented, edited, resized, and rearranged as desired into a cohesive record of your analysis. The cut-and-paste process is quite similar to the Minitab examples at the end of Appendix 1.2.

Calculated results   As we proceed through this book, you will see that Excel often expresses calculated results that are fractions in **scientific notation.** For example, Excel might express the results of a calculation as 7.77 E-6. To get the decimal point equivalent, the "E-6" says we must move the decimal point 6 places to the left. This would give us the fraction .00000777.

To create 100 random numbers between 1 and 2136 similar to those in Table 1.4(b).

- Type the cell formula

    =RANDBETWEEN(1, 2136)

    into cell A1 of the Excel worksheet and press the enter key. This will generate a random integer between 1 and 2136, which will be placed in cell A1.
- Using the mouse, copy the cell formula for cell A1 down through cell A100. This will generate 100 random numbers between 1 and 2136 in cells A1 through A100. (Note that the random number in cell A1 will change when this is done. This is not a problem.)
- The random numbers are generated with replacement. Repeated numbers would be skipped if the random numbers were being used to sample without replacement.

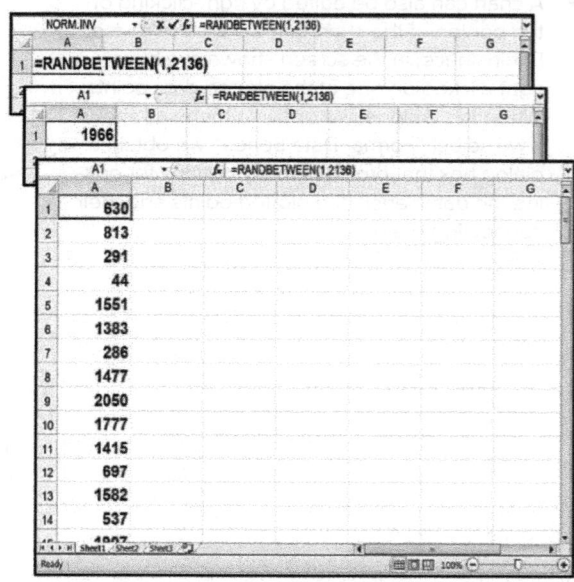

Source: Microsoft Office Excel 2016

Appendix 1.2 ■ Getting Started with Minitab

We begin by looking at some features of Minitab that are common to most analyses. When the instructions call for a sequence of selections from a series of menus, the sequence will be presented in the following form:

Stat : Basic Statistics : Descriptive Statistics

This notation indicates that you first select Stat from the Minitab menu bar, next select Basic Statistics from the Stat pull-down menu, and finally select Descriptive Statistics from the Basic Statistics pull-down menu.

Opening Minitab   You can open Minitab by clicking on the Minitab icon on your computer desktop or in your list of installed options or by double-clicking on a Minitab file.

After you open Minitab, the display is partitioned into two working windows.

- The "Session window" is the area where Minitab commands and basic output are displayed. ——
- The "Data window" is an Excel-like worksheet where data can be entered and edited. ⟍

Help resources Like most Windows programs, Minitab offers online help via a Help Menu. Help includes standard Contents and Search entries as well as Tutorials introducing Minitab concepts and walk-throughs of typical Minitab sessions. A StatGuide explains how to interpret statistical tables and graphs in a practical, straightforward way.

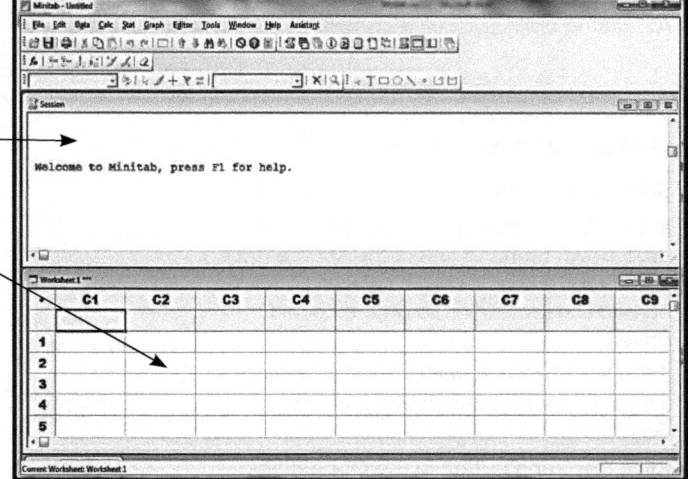

Source: Minitab 18

Entering data (entering the gasoline mileage data in Table 1.7) from the keyboard:

- In the Data window, click on the cell directly below C1 and type a name for the variable— say, Mileage—and press the Enter key.
- Starting in row 1 under column C1, type the values for the variable (gasoline mileages from Table 1.7) down the column, pressing the Enter key after each entry.

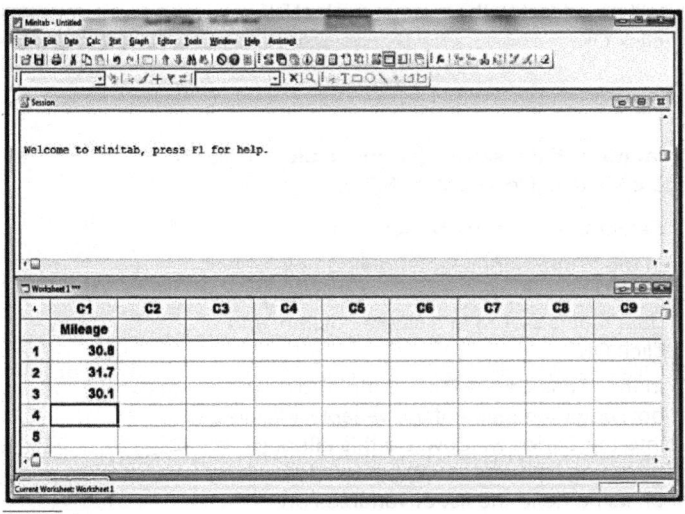

Source: Minitab 18

Saving data (saving the gasoline mileage data):

- Select **File : Save Project As**
- In the "Save Worksheet As" dialog box (the project here is a worksheet), use the "Save in" drop-down menu to select the destination drive and folder. (Here we have selected a folder named Data Files on the Local C drive.)
- Enter the desired file name in the File name box. Here we have chosen the name GasMiles. Minitab will automatically add the extension .mpj.
- Click the Save button in the "Save Project As" dialog box.

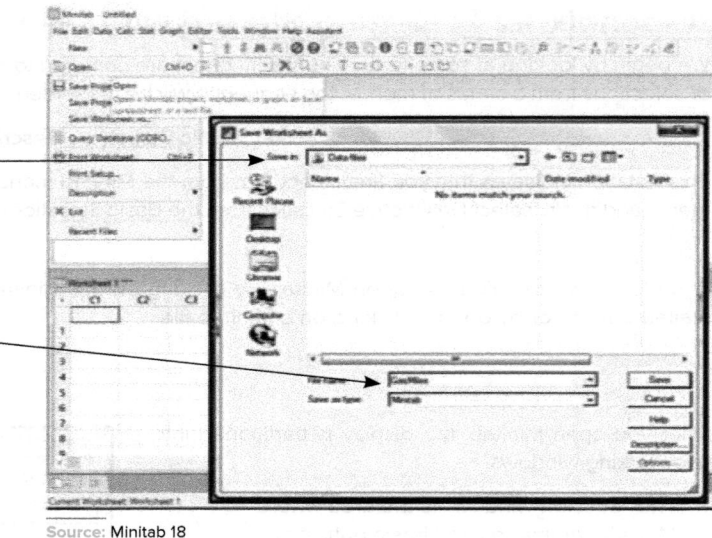

Source: Minitab 18

Retrieving a Minitab worksheet

- Select **File : Open**
- In the "Open" dialog box, use the "Look in" drop-down menu to select the source drive and folder. (Here we have selected a folder named Datasets Minitab on the Local C drive.)
- Enter the desired name in the File name box. (Here we have chosen the worksheet GasMiles.)
- Click the Open button in the Open dialog box.
- Minitab may display a dialog box with the message "A copy of the contents of this file will be added to the current project." If so, click OK.

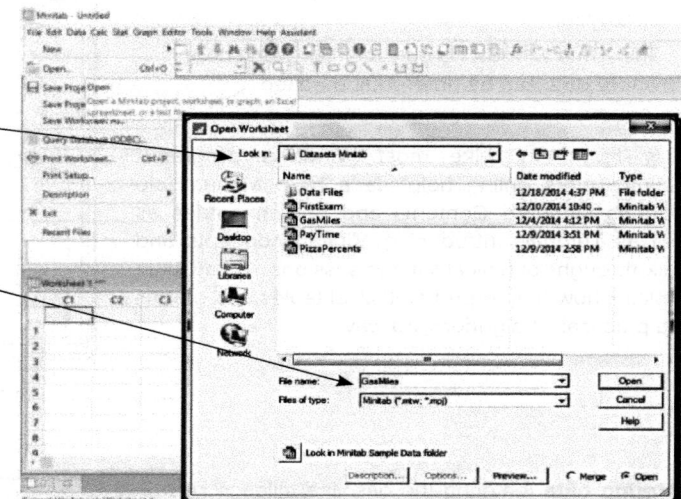

Source: Minitab 18

Creating a time series (or runs) plot similar to Figure 1.5 (data file: GasMiles.MTW):

- Select **Graph : Time Series Plot**
- In the "Time Series Plots" dialog box, select Simple, which produces a time series plot of data that is stored in a single column, and click OK.
- In the "Time Series Plot—Simple" dialog box, enter the name of the variable, Mileage, into the Series window. Do this either (1) by typing its name, or (2) by double-clicking on its name in the list of variables on the left side of the dialog box. Here, this list consists of the single variable Mileage in column C1.
- Click OK in the "Time Series Plot—Simple" dialog box.

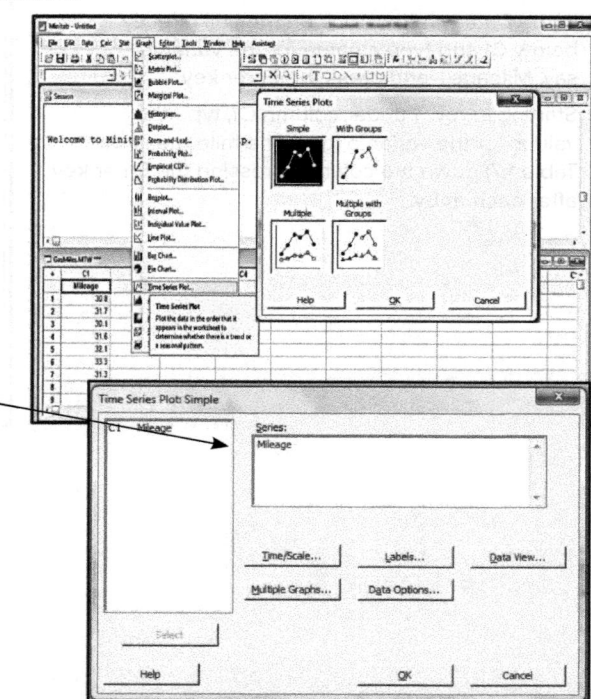

Source: Minitab 18

- The time series plot will appear in a graphics window.

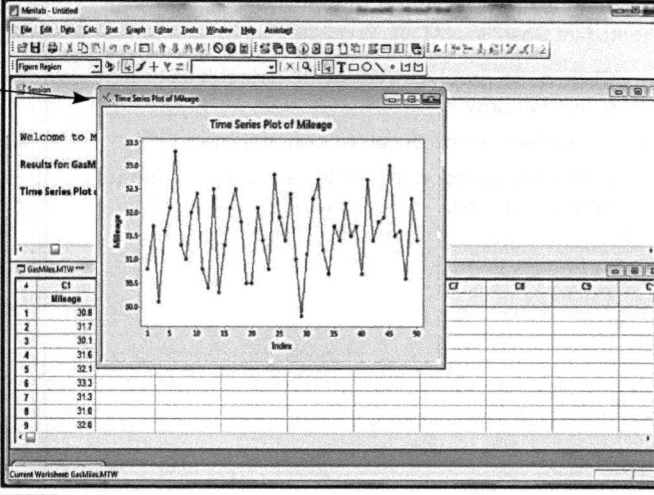

Source: Minitab 18

- The graph can be edited by right-clicking on the portion you wish to edit. For instance, here we have right-clicked on the data region.

- Selecting "Edit Data Region" from the pop-up window yields a dialog box allowing you to edit this region. The *x* and *y* scales, *x* and *y* axis labels, figure region, and so forth can all be edited by right-clicking on the appropriate portion of the graph.

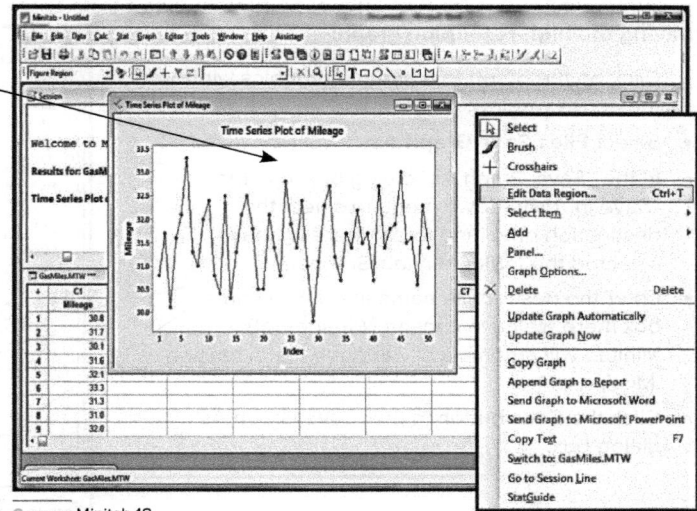

Source: Minitab 18

- For instance, after right-clicking on the data region and selecting "Edit Data Region" from the pop-up menu, we can use the Attributes tab on the Edit Data Region menu to customize the data region. Here we have chosen to change the background color to yellow.

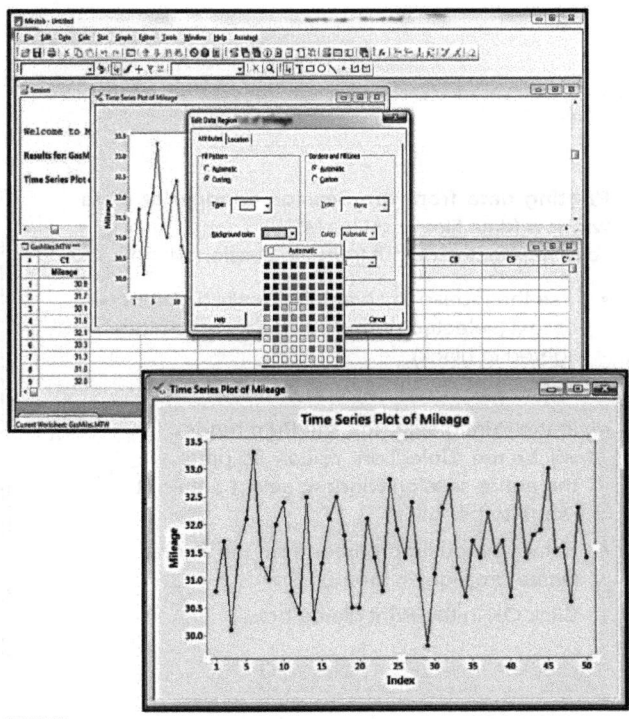

Source: Minitab 18

Printing a high-resolution graph similar to Figure 1.5 (data file: GasMiles.MTW):

- Click in the graphics window to select it as the active window.
- Select **File : Print Graph** to print the graph.
- Select the appropriate printer and click OK in the Print dialog box.

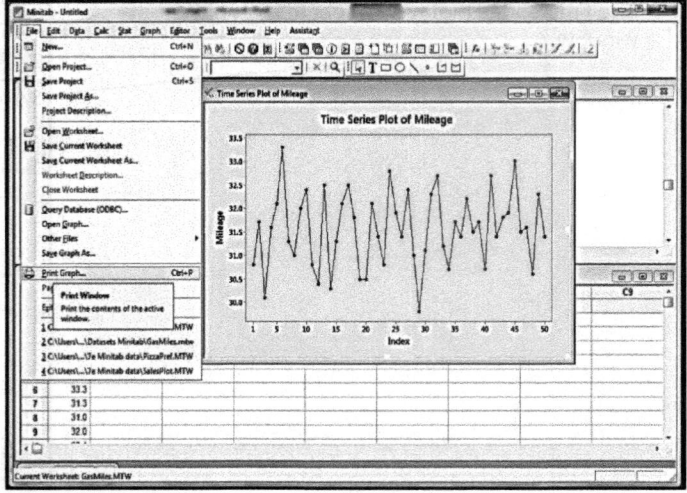

Source: Minitab 18

Saving the high-resolution graph:

- Click on the graph to make the graphics window the active window.
- Select **File : Save Graph As**
- In the "Save Graph As" dialog box, use the "Save in" drop-down menu to select the destination drive and folder (here we have selected the folder Minitab Graphs).
- Enter the desired file name in the File name box (here we have chosen MileagePlot). Minitab will automatically add the file extension .MGF.
- Click the Save button in the "Save Graph As" dialog box.

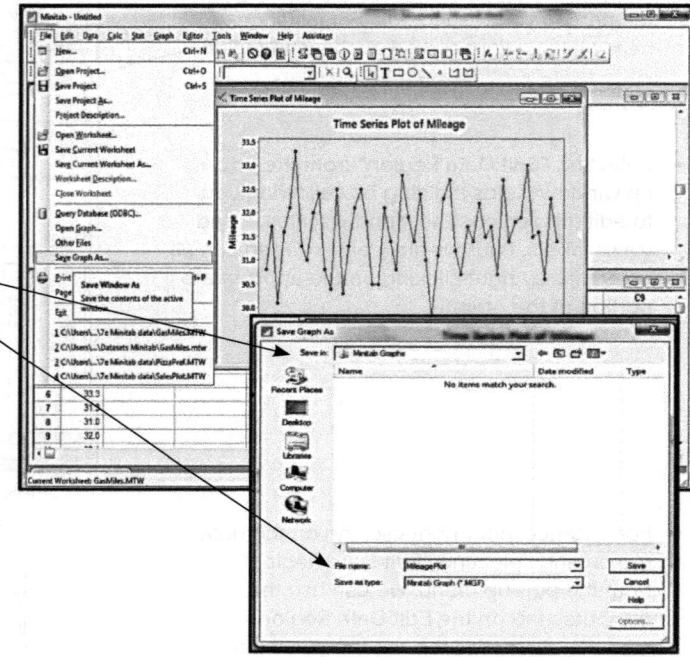

Source: Minitab 18

Printing data from the Session window or Data window (data file: GasMiles.MTW):
To print selected output from the Session window:

- Use the mouse to select the desired output or text (selected output will be reverse-high-lighted in black).
- Select **File : Print Session Window**
- In the Print dialog box, the Print range will be the "Selection" option. To print the entire session window, select the Print range to be "All."
- Select the desired printer from the Printer Name drop-down menu.
- Click OK in the Print dialog box.

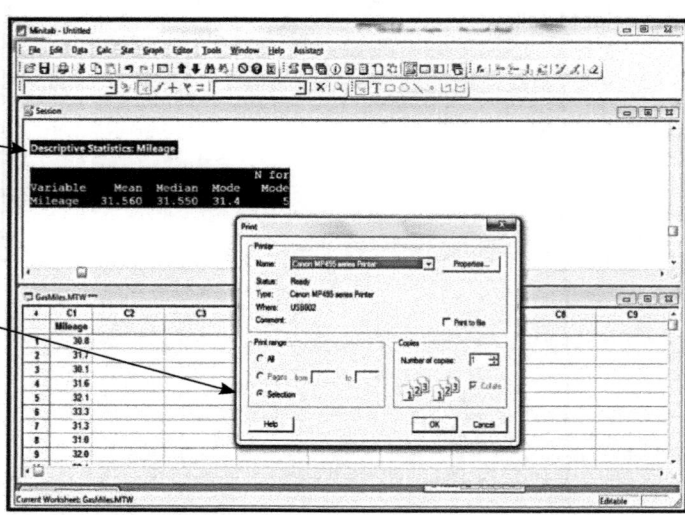

Source: Minitab 18

To print the contents of the Data window (that is, to print the Minitab worksheet):

- Click in the Data window to select it as active
- Select **File : Print Worksheet**
- Make selections as desired in the Data Window Print Options dialog box; add a title in the Title window if desired; and click OK.
- Select the desired printer from the Printer Name drop-down menu and click OK.

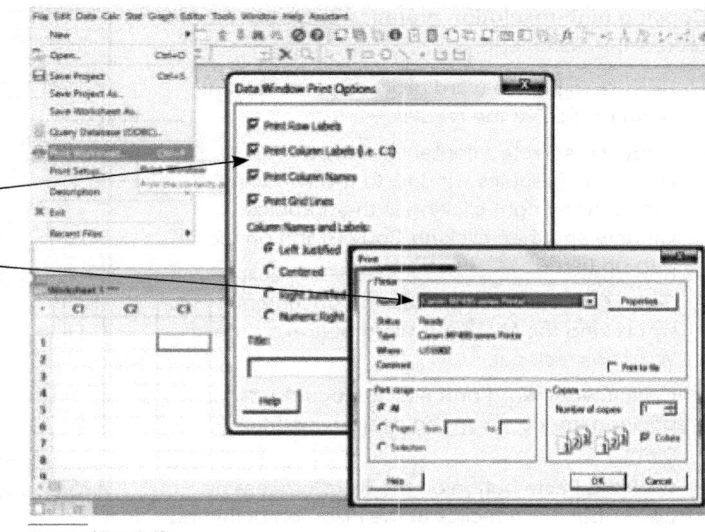

Source: Minitab 18

Including Minitab output in reports The preceding examples show how to print various types of output directly from Minitab. Printing is a useful way to capture a hard copy record of a result. However, you may prefer to combine your results with narrative documentation to create a report that can be saved and printed as a seamless unit. You can do this by copying selected Minitab results to the Windows clipboard and then pasting them into your favorite word processor. Once copied to a word processor document, Minitab results can be annotated, edited, resized, and rearranged into a cohesive record of your analysis. The following sequence of screens illustrates this process.

Copying session window output to a word processing document:

- Be sure to have a word processing document open to receive the results.
- Use the scroll bar on the right side of the Session window to locate the results you wish to copy and drag the mouse to select the desired output.
- Copy the selected output to the Windows clipboard by clicking on the Edit icon on the Minitab toolbar and then choosing Copy or by right-clicking on the selected text and then selecting Copy from the pop-up menu.
- Switch to your word processing document by clicking the Microsoft Word button on the Windows task bar.

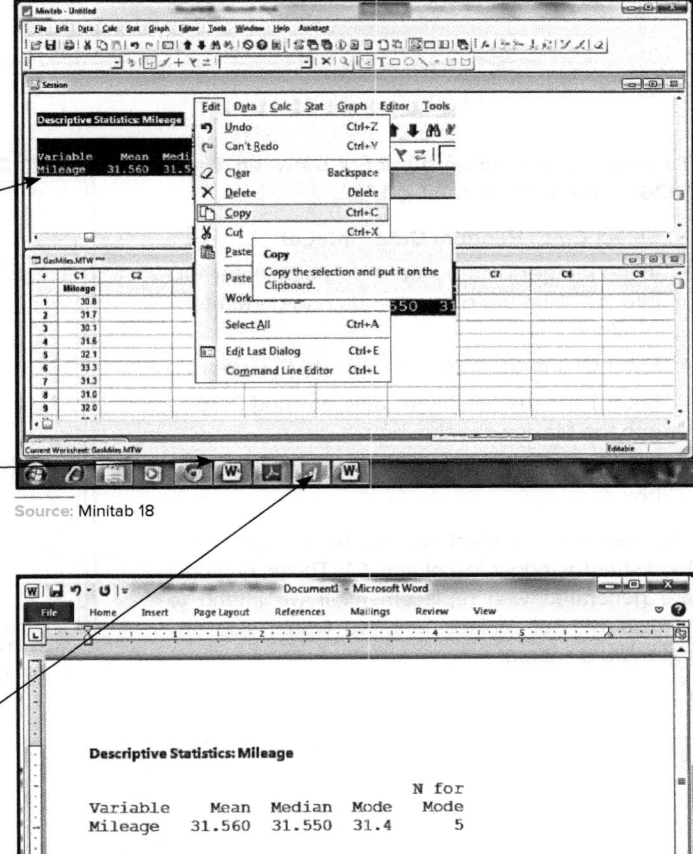

Source: Minitab 18

- Click in your word processing document to position the cursor at the desired insertion point.
- Click the Paste button on the word processing power bar or right-click at the insertion point and select Paste from the pop-up menu.
- Return to your Minitab session by clicking the Minitab button on the Windows task bar.

Source: Minitab 18

Copying high-resolution graphics output to a word processing document:

- Be sure to have a word processing document open to receive the results.
- Copy the selected contents of the high-resolution graphics window to the Windows clipboard by right-clicking in the graphics window and then clicking Copy Graph on the pop-up menu.
- Switch to your word processing document by clicking the Microsoft Word button on the Windows task bar.
- Click in your word processing document to position the cursor at the desired insertion point.
- Click the Paste button on the word processing power bar or right-click at the insertion point and select Paste from the pop-up menu.
- Return to your Minitab session by clicking the Minitab button on the Windows task bar.

Results Here is how the copied results might appear in Microsoft Word. These results can be edited, resized, and combined with your own additional documentation to create a cohesive record of your analysis.

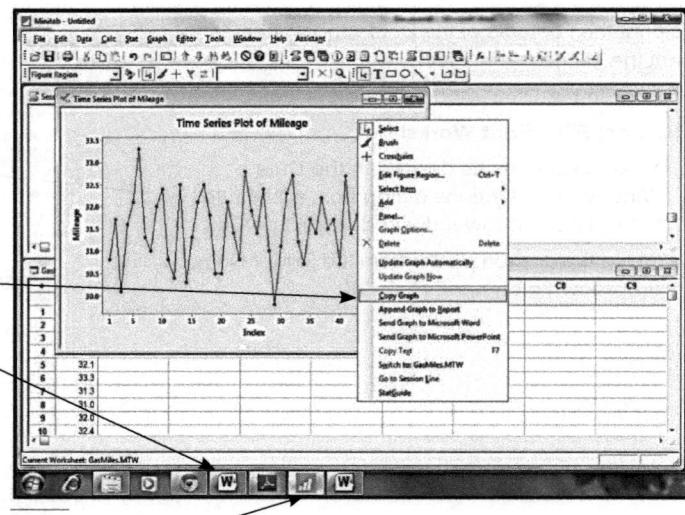

Source: Minitab 18

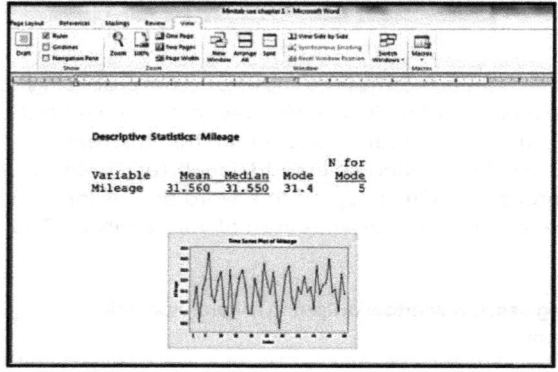

Source: Minitab 18

To create 100 random numbers between 1 and 2136 similar to those in Table 1.4(b):

- Select **Calc : Random Data : Integer**
- In the Integer Distribution dialog box, enter 100 into the "Number of rows of data to generate" window.
- Enter C1 into the "Store in column(s)" window.
- Enter 1 into the Minimum value box and 2136 into the Maximum value box.
- Click OK in the Integer Distribution dialog box.

The 100 random numbers will be placed in the Worksheet window in column C1. These numbers are generated with replacement. If we intend to sample without replacement, we would skip any repeated numbers.

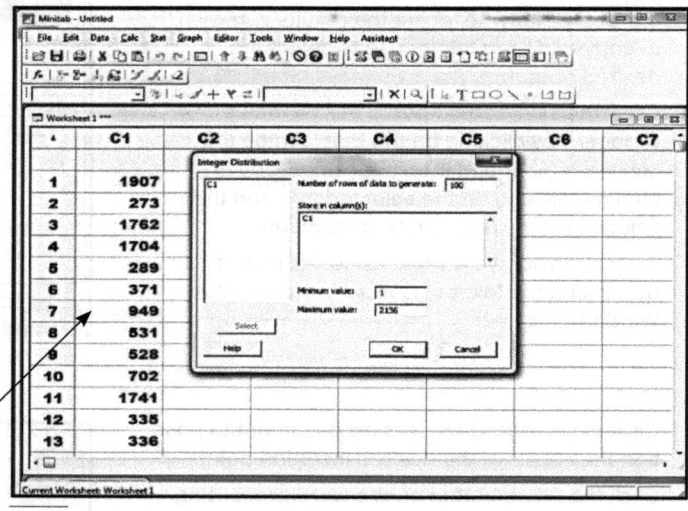

Source: Minitab 18

Appendix 1.3 ■ Getting Started with JMP®

JMP is a commercially available desktop statistical software package for Mac and Windows environments. If you are a student or professor, you may have already have access to JMP through your school.

There are three versions of JMP software available: the professional JMP Pro, the standard JMP, and the JMP Student Edition. Some of the advanced business analytics tools that are covered in this book will require the JMP Pro version, so we will show JMP Pro throughout the book. To learn more about the features available only in JMP Pro, go to https://www.jmp.com/en_us /software/predictive-analytics-software/key-features-of-jmp-pro.html. To compare the features in the JMP Student Edition to the standard and Pro versions of JMP, see the "JMP Student Edition Comparison Chart" available at https://www.jmp.com/en_be /academic/jmp-student-edition.html.

Opening JMP You can open JMP by clicking on the JMP icon on your computer desktop or in your list of installed applications or by double-clicking on a JMP file.

Upon opening JMP, you will see the **Tip of the Day** pop-up window and, beneath that, the **JMP Home Window**. If you are using JMP on a Mac, you will also see the **JMP Starter** window.

- **Tip of the Day** gives helpful hints on using JMP.
- The **JMP Home Window** displays recently used files and open data tables and windows.
- The **JMP Starter** window (click on **View : JMP Starter** on Windows to open) provides short-cuts for using JMP, including opening files and accessing JMP analyses.

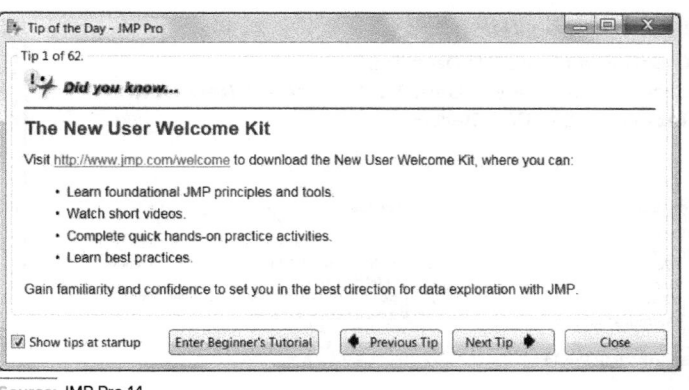

Source: JMP Pro 14

The JMP menus, across the top, can be used to perform JMP functions.

The JMP menu bar:
Although you can launch the JMP platforms from the **JMP Starter** window, we will show all analyses launched from the JMP menu bar.

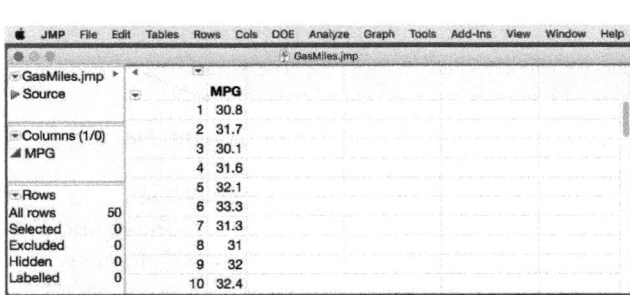

JMP Menu bar on Mac

Source: JMP Pro 14

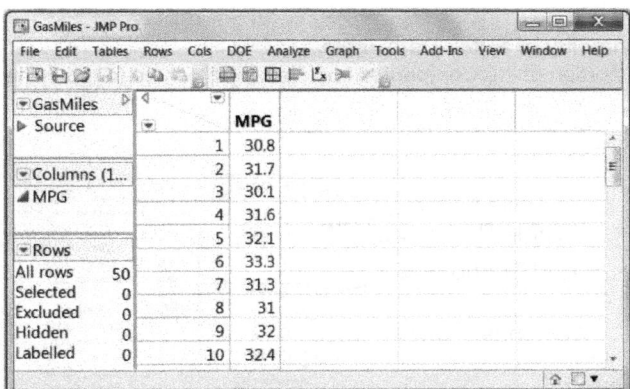

JMP Menu bar on Windows

Source: JMP Pro 14

Along the top of your screen you will see the menu bar for JMP. If you are using JMP on a Mac, this menu bar will be unattached to the open JMP windows and will instead appear at the top of your computer screen. If you are using JMP on Windows, the menu bar will be attached to the top of any open JMP window. Note that, by default, JMP on Windows will hide the menu bar when the JMP window is small. To make the menu bar reappear, click **Alt,** or hover your mouse on the report window where the menu would normally appear. You can also change this default preference to keep the menu bar permanently visible regardless of the window size. To make this change, go to **File : Preferences : Windows Specific** and change **Auto-hide menus and toolbars** from **Based on window size** to **Never**.

JMP Tools

The JMP tools, available in the Tools menu and as a toolbar located beneath the menu bar, provides many shortcuts and helpful tools.

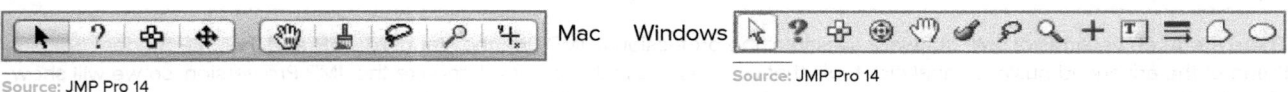

Source: JMP Pro 14   Mac   Windows   Source: JMP Pro 14

The default tool is the Arrow, or Cursor. To change tools, click on the icon from the toolbar or select from the Tools menu. Your curser will change from the arrow to the shape of the tool selected, and the active tool will be highlighted in the toolbar.

Data tables in JMP

Creating a New JMP Data Table

Click on **File : New : Data Table** or select **New Data Table** from the **JMP Starter**.

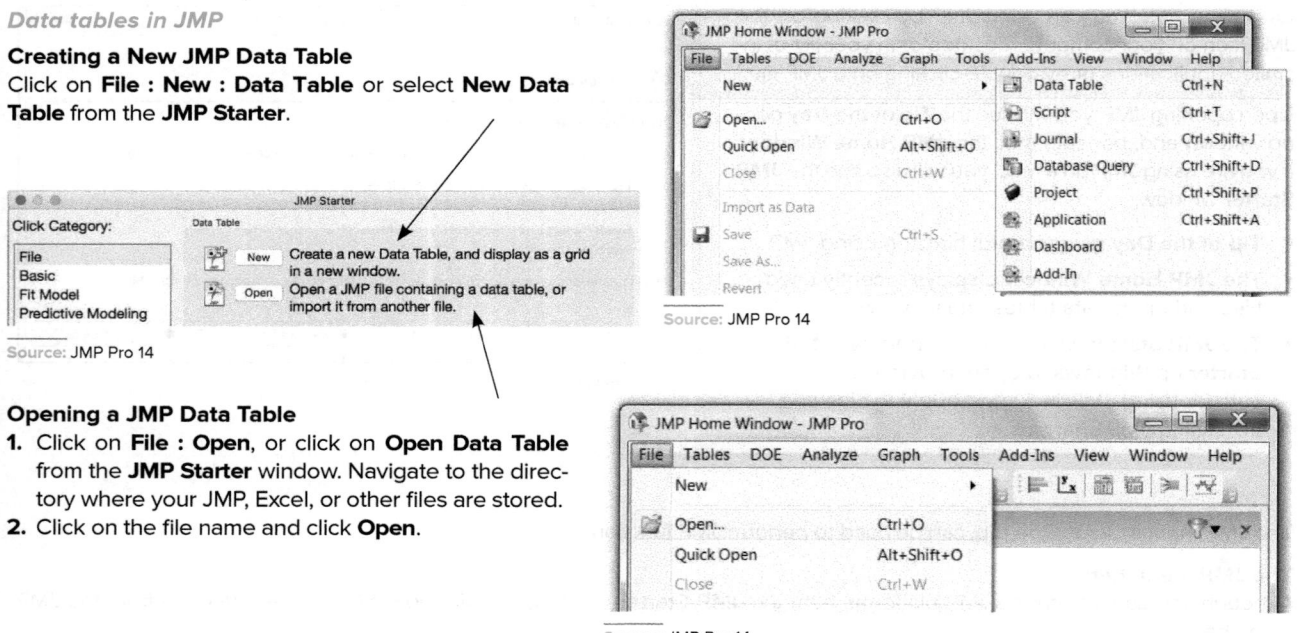

Source: JMP Pro 14

Opening a JMP Data Table

1. Click on **File : Open**, or click on **Open Data Table** from the **JMP Starter** window. Navigate to the directory where your JMP, Excel, or other files are stored.
2. Click on the file name and click **Open**.

Source: JMP Pro 14

Anatomy of a JMP Data Table

A portion of the **Companies.jmp** file, which can be found in the JMP **Help : Sample Data Library**, is shown.

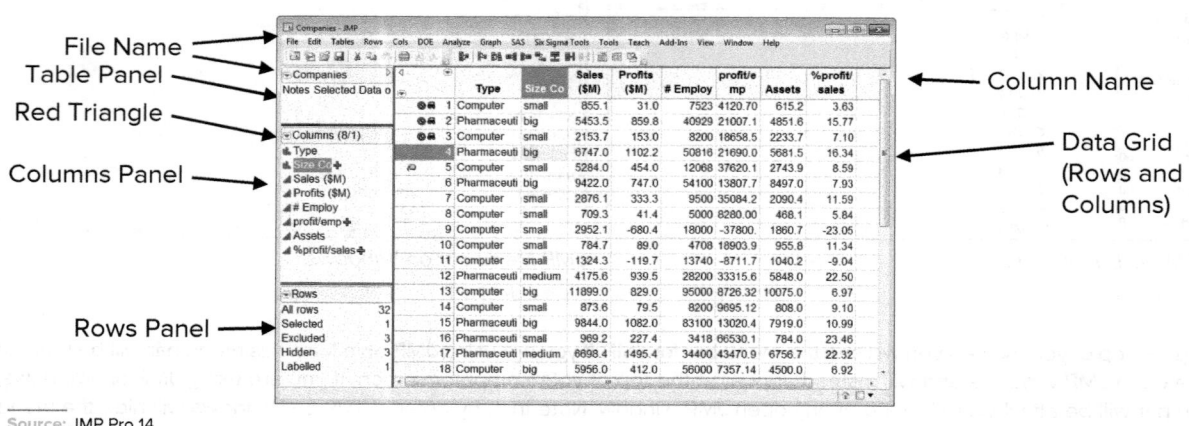

Source: JMP Pro 14

Tables Panel

- The data table name
- A list of table properties and scripts

> — Double-click on properties to display or open.

Columns Panel

- The number of columns
- The number of selected columns
- Column names
- The modeling type for each column
- Column properties

> — This data table has eight columns, and one column has been selected.
> — Two columns are nominal (red bars).
> — Six columns are continuous (blue triangles).
> — There are no ordinal columns (green bars).
> — Three columns have stored formulas (the plus sign after the column name—click on the plus sign to display the formula).

Rows Panel

- Rows
- Selected rows
- Hidden rows (with a mask)
- Excluded rows (with a "don't" sign)
- Labeled rows (with a tag)

> — This data table has 32 rows, or observations.
> — One row has been selected (row 4).
> — Three rows have been both excluded and hidden (rows 1—3).
> — Hidden rows will not display on graphs.
> — Excluded rows will not be included in most future analyses.
> — One row has been labeled (row 5).

Notes: **Red triangles** are used throughout JMP to access other commands, and **gray triangles** are used to minimize display areas. **Right-click** in different regions of the data table (or graphs) for additional options. For additional information, see the book *Discovering JMP* (under **Help : Books**).

Creating a time series (or runs) plot

- To plot one continuous variable against the observation order, select **Graph : Overlay Plot**. (Note: Graph Builder can also be used, but this method will require a column variable for the time or observation order.)
- Drag the **Miles** variable name from the "Select Column" list on the left into the "Cast Selected Columns into Roles" box, or just click on the variable name to select it and click the Y button.
- Click **OK** to create the plot.

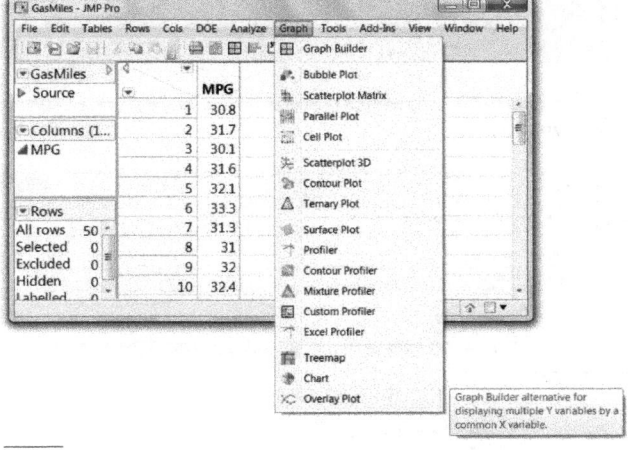

Source: JMP Pro 14

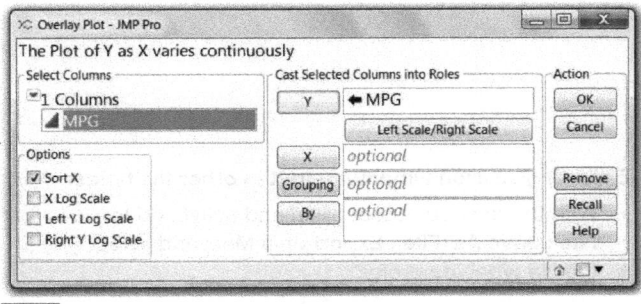

Source: JMP Pro 14

- Click on the **red triangle** to explore options to change the plot appearance. For example, you can select **Connect Thru Missing** to draw a line to connect the points, and you can deselect **Y Options : Show Points** to take away the large points.

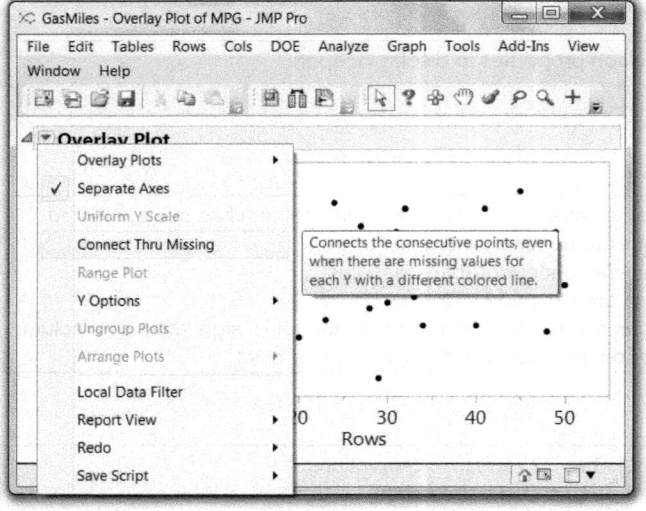

Source: JMP Pro 14

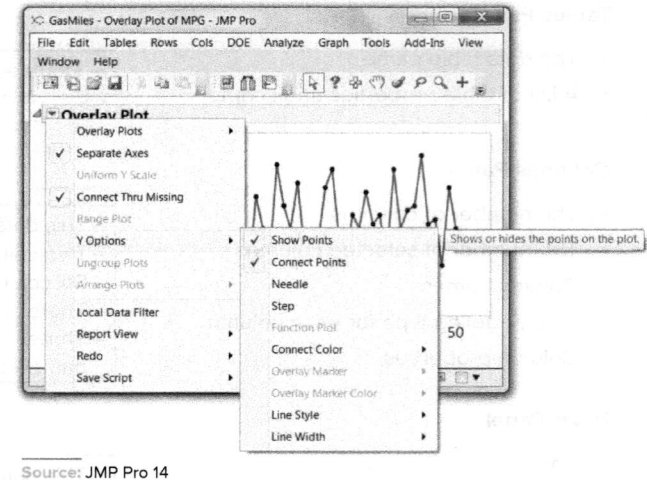

Source: JMP Pro 14

Including JMP output in reports

Copying session window output to a word processing document:

- Open the word processing document to which you will be adding JMP results.

- Use the selection tool (⊕) to highlight the JMP output you would like to copy. Go to **Edit : Copy** or right-click and select **Copy** or use **ctrl-c** to copy the selection. Switch your cursor into the word processing document to make the document active; paste the output by using the **ctrl-v** keyboard shortcut, by right-clicking and selecting **Paste**, or by pasting using the menus of your document processing application.

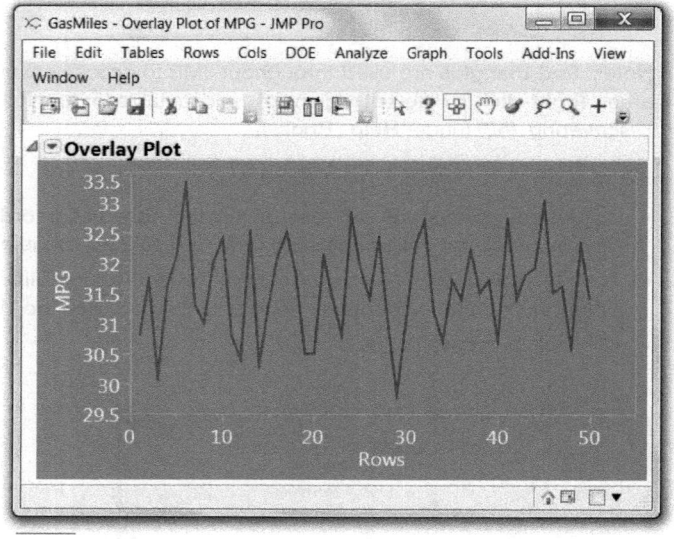

Source: JMP Pro 14

Exporting session window output in other file types:

- With the desired output open and active, go to **File : Save As** (**File : Export** on a Mac) and select the file type you prefer.

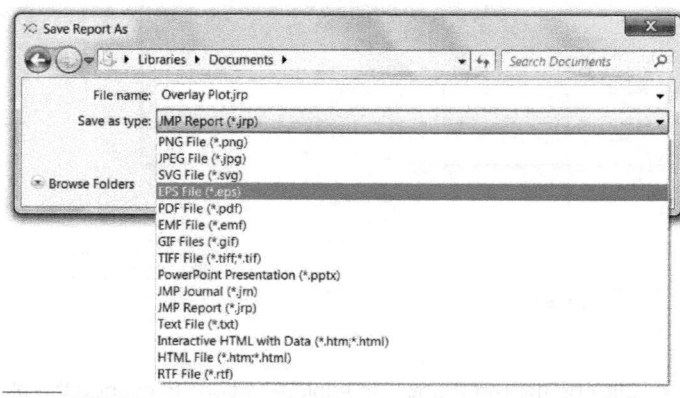

Source: JMP Pro 14

Creating random numbers

To create 100 random numbers between 1 and 2136 similar to those in Table 1.4(b)

- From an open JMP data table, select **Cols : New Column**.

- Change the field in **Initialize Data** to **Random**. Keep the Number of rows as 100. Change the Maximum value to **2136** and click **OK**.

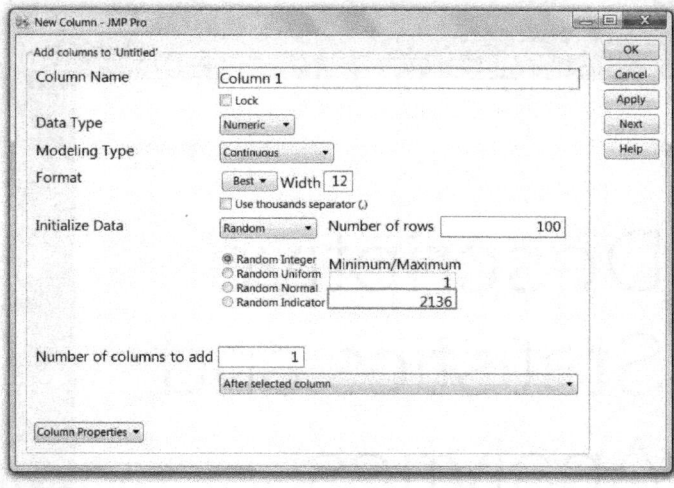

Source: JMP Pro 14

Resources for help

JMP Learning Library

Basic instructions on how to get the most out of JMP, including one-page overviews, step-by-step tutorials, and short videos, are available at www.jmp.com/learn. Topics include importing and opening data, saving results and graphics, and many statistical method topics like *t*-tests, regression, and ANOVA.

Hover Help

When selecting analytic options within JMP, hover help is available to provide a short description of the option. Simply hover your mouse above the selectable option, and JMP will display a description.

Question Mark Tool

Help – The help tool or question mark, available in the JMP toolbar, accesses the JMP help system. Select the help tool, then click on an area of a data table or report on which you need assistance. Context-sensitive help tells about the items located near the location of your click.

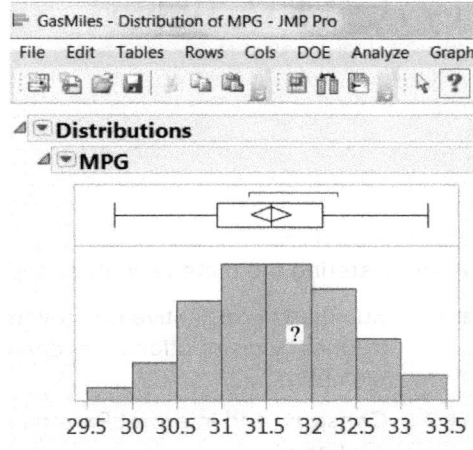

Source: JMP Pro 14

Help menu

The **Help** menu provides many resources to help you get started:

- Searchable documentation (Help Contents, Search the Help, Help Index and Books),

- A summary of new features,

- Tutorials,

- Sample Data,

- Indexes of statistical terms and JMP scripting functions, and more.

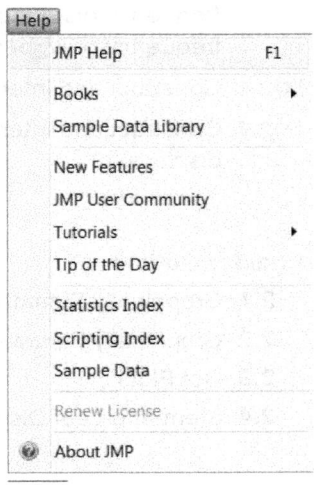

Design Elements: (CD): ©Comstock Images/Alamy; (All Others): ©McGraw-Hill Education

Source: JMP Pro 14

## CHAPTER 2

Descriptive Statistics and Analytics: Tabular and Graphical Methods

©Mlenny Photography/Getty Images

Learning Objectives

After mastering the material in this chapter, you will be able to:

LO2-1 Summarize qualitative data by using frequency distributions, bar charts, and pie charts.

LO2-2 Construct and interpret Pareto charts (Optional).

LO2-3 Summarize quantitative data by using frequency distributions, histograms, frequency polygons, and ogives.

LO2-4 Construct and interpret dot plots.

LO2-5 Construct and interpret stem-and-leaf displays.

LO2-6 Examine the relationships between variables by using contingency tables (Optional).

LO2-7 Examine the relationships between variables by using scatter plots (Optional).

LO2-8 Recognize misleading graphs and charts (Optional).

LO2-9 Construct and interpret gauges, bullet graphs, treemaps, and sparklines (Optional).

Chapter Outline

2.1 Graphically Summarizing Qualitative Data

2.2 Graphically Summarizing Quantitative Data

2.3 Dot Plots

2.4 Stem-and-Leaf Displays

2.5 Contingency Tables (Optional)

2.6 Scatter Plots (Optional)

2.7 Misleading Graphs and Charts (Optional)

2.8 Graphical Descriptive Analytics (Optional)

n Chapter 1 we saw that although we can sometimes take a census of an entire population, we often must randomly select a sample from a population. When we have taken a census or a sample, we typically wish to describe the observed data set. In particular, we describe a sample in order to make inferences about the sampled population.

In this chapter we begin to study **descriptive statistics** and **analytics,** which is the science of describing and visually depicting the important characteristics of a data set. The techniques of descriptive statistics and analytics include **tabular and graphical methods,** which are discussed in this chapter, and **numerical methods,**

which are discussed in Chapter 3. We will see that, in practice, the methods of this chapter and the methods of Chapter 3 are used together to describe data. We will also see that the methods used to describe quantitative data differ somewhat from the methods used to describe qualitative data. Finally, we will see that there are methods—both graphical and numerical—for studying the relationships between variables.

We will illustrate the methods of this chapter by describing the cell phone usages, bottle design ratings, and car mileages introduced in the cases of Chapter 1, and we will further consider the Disney Parks Case. In addition, we introduce two new cases:

The e-Billing Case: A management consulting firm assesses how effectively a new electronic billing system reduces bill payment times.

The Brokerage Firm Case: A financial broker examines whether customer satisfaction depends upon the type of investment product purchased.

2.1 Graphically Summarizing Qualitative Data

LO2-1
Summarize qualitative data by using frequency distributions, bar charts, and pie charts.

Frequency distributions

When data are qualitative, we use names to identify the different categories (or classes). Often we summarize qualitative data by using a frequency distribution.

> A **frequency distribution** is a table that summarizes the number (or **frequency**) of items in each of several nonoverlapping classes.

EXAMPLE 2.1 Describing Pizza Preferences

A business entrepreneur plans to open a pizza restaurant in a college town. There are currently six pizza restaurants in town: four chain restaurants—Domino's Pizza, Little Caesars Pizza, Papa John's Pizza, and Pizza Hut—and two local establishments—Bruno's Pizza and Will's Uptown Pizza. Before developing a basic pizza recipe (crust ingredients, sauce ingredients, and so forth), the entrepreneur wishes to study the pizza preferences of the college students in town. In order to do this, the entrepreneur selects a random sample of 50 students enrolled in the local college and asks each sampled student to name his or her favorite among the six pizza places in town. The survey results are given in Table 2.1.

©Karen Mower/Getty Images

Part 1: Studying Pizza Preferences by Using a Frequency Distribution Unfortunately, the raw data in Table 2.1 do not reveal much useful information about the pattern of pizza preferences. In order to summarize the data in a more useful way, we can construct a frequency distribution. To do this we simply count the number of times each of the six pizza restaurants appears in Table 2.1. We find that Bruno's appears 8 times, Domino's appears 2 times, Little Caesars appears 9 times, Papa John's appears 19 times, Pizza Hut appears 4 times, and Will's Uptown Pizza appears 8 times. The frequency distribution for the pizza preferences is given in Table 2.2—a list of each of the six restaurants along with their corresponding counts (or **frequencies**). The frequency distribution shows us how the preferences are distributed among the six restaurants. The purpose of the frequency

TABLE 2.1   Pizza Preferences of 50 College Students   ⒹⓈ PizzaPref

Little Caesars	Papa John's	Bruno's	Papa John's	Domino's
Papa John's	Will's Uptown	Papa John's	Pizza Hut	Little Caesars
Pizza Hut	Little Caesars	Will's Uptown	Little Caesars	Bruno's
Papa John's	Bruno's	Papa John's	Will's Uptown	Papa John's
Bruno's	Papa John's	Little Caesars	Papa John's	Little Caesars
Papa John's	Little Caesars	Bruno's	Will's Uptown	Papa John's
Will's Uptown	Papa John's	Will's Uptown	Bruno's	Papa John's
Papa John's	Domino's	Papa John's	Pizza Hut	Will's Uptown
Will's Uptown	Bruno's	Pizza Hut	Papa John's	Papa John's
Little Caesars	Papa John's	Little Caesars	Papa John's	Bruno's

TABLE 2.2   A Frequency Distribution of Pizza Preferences   ⒹⓈ PizzaFreq

Restaurant	Frequency
Bruno's	8
Domino's	2
Little Caesars	9
Papa John's	19
Pizza Hut	4
Will's Uptown	8
	50

TABLE 2.3   Relative Frequency and Percent Frequency Distributions for the Pizza Preference Data   ⒹⓈ PizzaPercents

Restaurant	Relative Frequency	Percent Frequency
Bruno's	8/50 = .16	16%
Domino's	.04	4%
Little Caesars	.18	18%
Papa John's	.38	38%
Pizza Hut	.08	8%
Will's Uptown	.16	16%
	1.0	100%

distribution is to make the data easier to understand. Certainly, looking at the frequency distribution in Table 2.2 is more informative than looking at the raw data in Table 2.1. We see that Papa John's is the most popular restaurant, and that Papa John's is roughly twice as popular as each of the next three runners-up—Bruno's, Little Caesars, and Will's. Finally, Pizza Hut and Domino's are the least preferred restaurants.

When we wish to summarize the proportion (or fraction) of items in each class, we employ the **relative frequency** for each class. If the data set consists of n observations, we define the relative frequency of a class as follows:

$$\textbf{Relative frequency of a class} = \frac{\text{frequency of the class}}{n}$$

This quantity is simply the fraction of items in the class. Further, we can obtain the **percent frequency** of a class by multiplying the relative frequency by 100.

Table 2.3 gives a relative frequency distribution and a percent frequency distribution of the pizza preference data. A **relative frequency distribution** is a table that lists the relative frequency for each class, and a **percent frequency distribution** lists the percent frequency for each class. Looking at Table 2.3, we see that the relative frequency for Bruno's pizza is $8/50 = .16$ and that (from the percent frequency distribution) 16% of the sampled students preferred Bruno's pizza. Similarly, the relative frequency for Papa John's pizza is $19/50 = .38$ and 38% of the sampled students preferred Papa John's pizza. Finally, the sum of the relative frequencies in the relative frequency distribution equals 1.0, and the sum of the percent frequencies in the percent frequency distribution equals 100%. These facts are true for all relative frequency and percent frequency distributions.

Part 2: Studying Pizza Preferences by Using Bar Charts and Pie Charts   A **bar chart** is a graphic that depicts a frequency, relative frequency, or percent frequency distribution. For example, Figure 2.1 gives an Excel bar chart of the pizza preference data. On the horizontal axis we have placed a label for each class (restaurant), while the vertical axis measures frequencies. To construct the bar chart, Excel draws a bar (of fixed width) corresponding to each class label.

FIGURE 2.1   **Excel Bar Chart of the Pizza Preference Data**

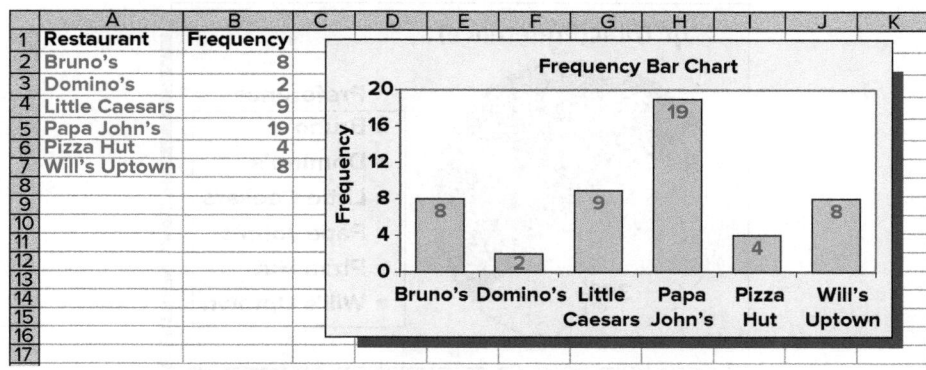

FIGURE 2.2   **Minitab Percent Bar Chart of the Pizza Preference Data**

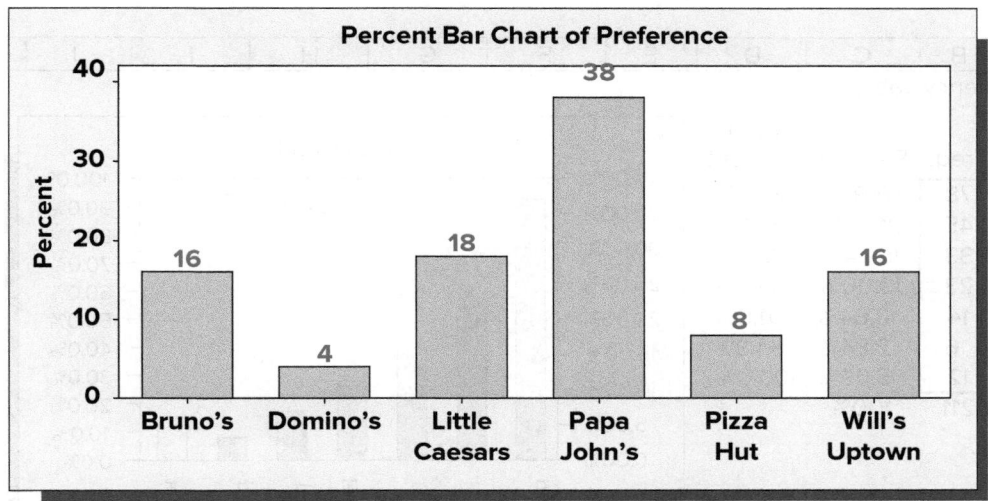

Each bar is drawn so that its height equals the frequency corresponding to its label. Because the height of each bar is a frequency, we refer to Figure 2.1 as a **frequency bar chart.** Notice that there are gaps between the bars. When data are qualitative, the bars should always be separated by gaps in order to indicate that each class is separate from the others. The bar chart in Figure 2.1 clearly illustrates that, for example, Papa John's pizza is preferred by more sampled students than any other restaurant and Domino's pizza is least preferred by the sampled students.

If desired, the bar heights can represent relative frequencies or percent frequencies. For instance, Figure 2.2 is a Minitab **percent bar chart** for the pizza preference data. Here the heights of the bars are the percentages given in the percent frequency distribution of Table 2.3. Lastly, the bars in Figures 2.1 and 2.2 have been positioned vertically. Because of this, these bar charts are called **vertical bar charts.** However, sometimes bar charts are constructed with horizontal bars and are called **horizontal bar charts.**

A **pie chart** is another graphic that can be used to depict a frequency distribution. When constructing a pie chart, we first draw a circle to represent the entire data set. We then divide the circle into sectors or "pie slices" based on the relative frequencies of the classes. For example, remembering that a circle consists of 360 degrees, Bruno's Pizza (which has relative frequency .16) is assigned a pie slice that consists of .16(360) = 57.6 degrees. Similarly, Papa John's Pizza (with relative frequency .38) is assigned a pie slice having .38(360) = 136.8 degrees. The resulting pie chart (constructed using JMP) is shown in Figure 2.3 on the next page. Here we have labeled the pie slices using the percent frequencies. The pie slices can also be labeled using frequencies or relative frequencies.

FIGURE 2.3   JMP Pie Chart of the Pizza Preference Data

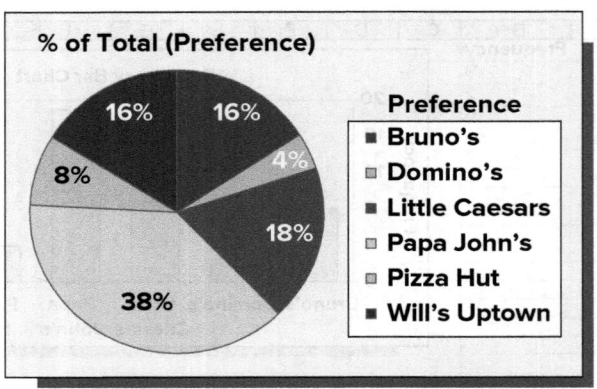

FIGURE 2.4   Excel Frequency Table and Pareto Chart of Labeling Defects   DS Labels

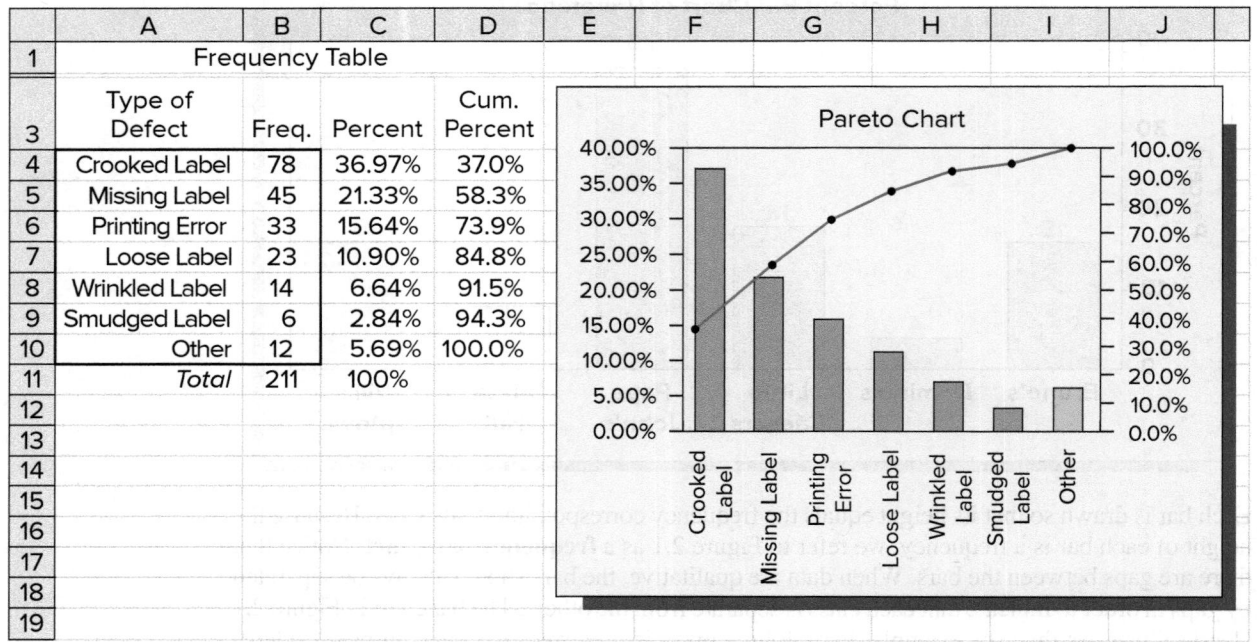

	A	B	C	D
1	Frequency Table			
3	Type of Defect	Freq.	Percent	Cum. Percent
4	Crooked Label	78	36.97%	37.0%
5	Missing Label	45	21.33%	58.3%
6	Printing Error	33	15.64%	73.9%
7	Loose Label	23	10.90%	84.8%
8	Wrinkled Label	14	6.64%	91.5%
9	Smudged Label	6	2.84%	94.3%
10	Other	12	5.69%	100.0%
11	Total	211	100%	

The Pareto chart (Optional)

LO2-2
Construct and interpret Pareto charts (Optional).

Pareto charts are used to help identify important quality problems and opportunities for process improvement. By using these charts we can prioritize problem-solving activities. The Pareto chart is named for Vilfredo Pareto (1848–1923), an Italian economist. Pareto suggested that, in many economies, most of the wealth is held by a small minority of the population. It has been found that the **"Pareto principle"** often applies to defects. That is, only a few defect types account for most of a product's quality problems.

To illustrate the use of Pareto charts, suppose that a jelly producer wishes to evaluate the labels being placed on 16-ounce jars of grape jelly. Every day for two weeks, all defective labels found on inspection are classified by type of defect. If a label has more than one defect, the type of defect that is most noticeable is recorded. The Excel output in Figure 2.4 presents the frequencies and percentages of the types of defects observed over the two-week period.

In general, the first step in setting up a **Pareto chart** summarizing data concerning types of defects (or categories) is to construct a frequency table like the one in Figure 2.4. Defects or categories should be listed at the left of the table in *decreasing order by frequencies*—the defect with the highest frequency will be at the top of the table, the defect with the second-highest frequency below the first, and so forth. If an "other" category is employed, it should be placed at the bottom of the table. The "other" category should not make up 50 percent or more of the total of the frequencies, and the frequency for the "other" category should not exceed the frequency for the defect at the top of the table. If the frequency for the "other" category is too high, data should be collected so that the "other" category can be broken down into new categories. Once the frequency and the percentage for each category are determined, a cumulative percentage for each category is computed. As illustrated in Figure 2.4, the cumulative percentage for a particular category is the sum of the percentages corresponding to the particular category and the categories that are above that category in the table.

A Pareto chart is simply a bar chart having the different kinds of defects or problems listed on the horizontal scale. The heights of the bars on the vertical scale typically represent the frequency of occurrence (or the percentage of occurrence) for each defect or problem. The bars are arranged in decreasing height from left to right. Thus, the most frequent defect will be at the far left, the next most frequent defect to its right, and so forth. If an "other" category is employed, its bar is placed at the far right. The Pareto chart for the labeling defects data is given in Figure 2.4. Here the heights of the bars represent the percentages of occurrences for the different labeling defects, and the vertical scale on the far left corresponds to these percentages. The chart graphically illustrates that crooked labels, missing labels, and printing errors are the most frequent labeling defects.

As is also illustrated in Figure 2.4, a Pareto chart is sometimes augmented by plotting a **cumulative percentage point** for each bar in the Pareto chart. The vertical coordinate of this cumulative percentage point equals the cumulative percentage in the frequency table corresponding to the bar. The cumulative percentage points corresponding to the different bars are connected by line segments, and a vertical scale corresponding to the cumulative percentages is placed on the far right. Examining the cumulative percentage points in Figure 2.4, we see that crooked and missing labels make up 58.3 percent of the labeling defects and that crooked labels, missing labels, and printing errors make up 73.9 percent of the labeling defects.

Technical note

The Pareto chart in Figure 2.4 illustrates using an "other" category which combines defect types having low frequencies into a single class. In general, when we employ a frequency distribution, a bar chart, or a pie chart and we encounter classes having small class frequencies, it is common practice to combine the classes into a single "other" category. Classes having frequencies of 5 percent or less are usually handled this way.

Exercises for Section 2.1

CONCEPTS connect

2.1 Explain the purpose behind constructing a frequency or relative frequency distribution.

2.2 Explain how to compute the relative frequency and percent frequency for each class if you are given a frequency distribution.

2.3 Find an example of a pie chart or bar chart in a newspaper or magazine. Copy it, and hand it in with a written analysis of the information conveyed by the chart.

METHODS AND APPLICATIONS

2.4 A multiple choice question on an exam has four possible responses—(a), (b), (c), and (d). When 250 students take the exam, 100 give response (a), 25 give response (b), 75 give response (c), and 50 give response (d).

a Write out the frequency distribution, relative frequency distribution, and percent frequency distribution for these responses.

b Construct a bar chart for these data using frequencies.

2.5 Consider constructing a pie chart for the exam question responses in Exercise 2.4.

a How many degrees (out of 360) would be assigned to the "pie slice" for response (a)?

b How many degrees would be assigned to the "pie slice" for response (b)?

c Construct the pie chart for the exam question responses.

2.6 Consider the partial relative frequency distribution of consumer preferences for four products— W, X, Y, and Z—that is shown on the next page.

a Find the relative frequency for product X.

b If 500 consumers were surveyed, give the frequency distribution for these data.

c Construct a percent frequency bar chart for these data.

d If we wish to depict these data using a pie chart, find how many degrees (out of 360) should be assigned to each of products W, X, Y, and Z. Then construct the pie chart.

Product	Relative Frequency
W	.15
X	—
Y	.36
Z	.28

2.7 Below we give the overall dining experience ratings (Outstanding, Very Good, Good, Average, or Poor) of 30 randomly selected patrons at a restaurant on a Saturday evening.   DS RestRating

Outstanding	Good	Very Good
Outstanding	Outstanding	Outstanding
Very Good	Outstanding	Outstanding
Outstanding	Good	Very Good
Good	Very Good	Outstanding
Very Good	Outstanding	Good
Very Good	Very Good	Average
Outstanding	Outstanding	Very Good
Outstanding	Very Good	Outstanding
Very Good	Good	Outstanding

a Find the frequency distribution and relative frequency distribution for these data.

b Construct a percentage bar chart for these data.

c Construct a percentage pie chart for these data.

2.8 Fifty randomly selected adults who follow professional sports were asked to name their favorite professional sports league. The results are as follows where MLB = Major League Baseball, MLS = Major League Soccer, NBA = National Basketball Association, NFL = National Football League, and NHL = National Hockey League.   DS ProfSports

NFL	NBA	NFL	MLB	MLB
MLB	NFL	MLB	NBA	NBA
NBA	NFL	NHL	NFL	MLS
NHL	MLB	NHL	NFL	NFL
MLS	NFL	MLB	NBA	NFL
NHL	NFL	NFL	MLS	MLB
NFL	NFL	NFL	NHL	NBA
NFL	MLB	NFL	MLB	NFL
NFL	MLB	NFL	NBA	NFL
NFL	MLB	NBA	NFL	NFL

a Find the frequency distribution, relative frequency distribution, and percent frequency distribution for these data.

b Construct a frequency bar chart for these data.

c Construct a pie chart for these data.

d Which professional sports league is most popular with these 50 adults? Which is least popular?

2.9 The National Automobile Dealers Association (NADA) publishes *AutoExec* magazine, which annually reports on new vehicle sales and market shares by manufacturer. As given on the *AutoExec* magazine website in May 2006, new vehicle market shares in the United States for 2005 were as follows:[1] Chrysler/Dodge/Jeep 13.6%, Ford 18.3%, GM 26.3%, Japanese (Toyota/Honda/Nissan) 28.3%, other imports 13.5%. Construct a percent frequency bar chart and a percentage pie chart for the 2005 auto market shares.   DS AutoShares05

2.10 Figure 2.5 gives a percentage pie chart of new vehicle market shares in the United States for 2014 as given by GoodCarBadCar.net. Use this pie chart and your results from Exercise 2.9 to write an analysis explaining how new vehicle market shares in the United States have changed from 2005 to 2014.   DS AutoShares14

2.11 On January 11, 2005, the Gallup Organization released the results of a poll investigating how many Americans have private health insurance. The results showed that among Americans making less than $30,000 per year, 33% had private insurance, 50% were covered by Medicare/Medicaid, and 17% had no health insurance, while among Americans making $75,000 or more per year, 87% had private insurance, 9% were covered by Medicare/Medicaid, and 4% had no health insurance.[2] Use bar and pie charts to compare health coverage of the two income groups.

2.12 In an article in *Quality Progress,* Barbara A. Cleary reports on improvements made in a software supplier's responses to customer calls. In this article, the author states:

> In an effort to improve its response time for these important customer-support calls, an inbound telephone inquiry team was formed at PQ Systems, Inc., a software and training organization in Dayton, Ohio. The team found that 88 percent of the customers' calls were already being answered immediately by the technical support group, but those who had to be called back had to wait an average of 56.6 minutes. No customer complaints had been registered, but the team believed that this response rate could be improved.

As part of its improvement process, the company studied the disposition of complete and incomplete calls to its technical support analysts. A call is considered complete if the customer's problem has been resolved; otherwise the call is incomplete. Figure 2.6 shows a Pareto chart analysis for the incomplete customer calls.

a What percentage of incomplete calls required "more investigation" by the analyst or "administrative help"?

b What percentage of incomplete calls actually presented a "new problem"?

c In light of your answers to *a* and *b*, can you make a suggestion?

[1]Source: www.autoexecmag.com, May 15, 2006.
[2]Source: http://news.gallup.com/poll/123149/cost-is-foremost-healthcare-issue-for-americans.aspx.

FIGURE 2.5   A Pie Chart of U.S. Automobile Sales in 2014 as Given by GoodCarBadCar.net (for Exercise 2.10)
DS AutoShares14

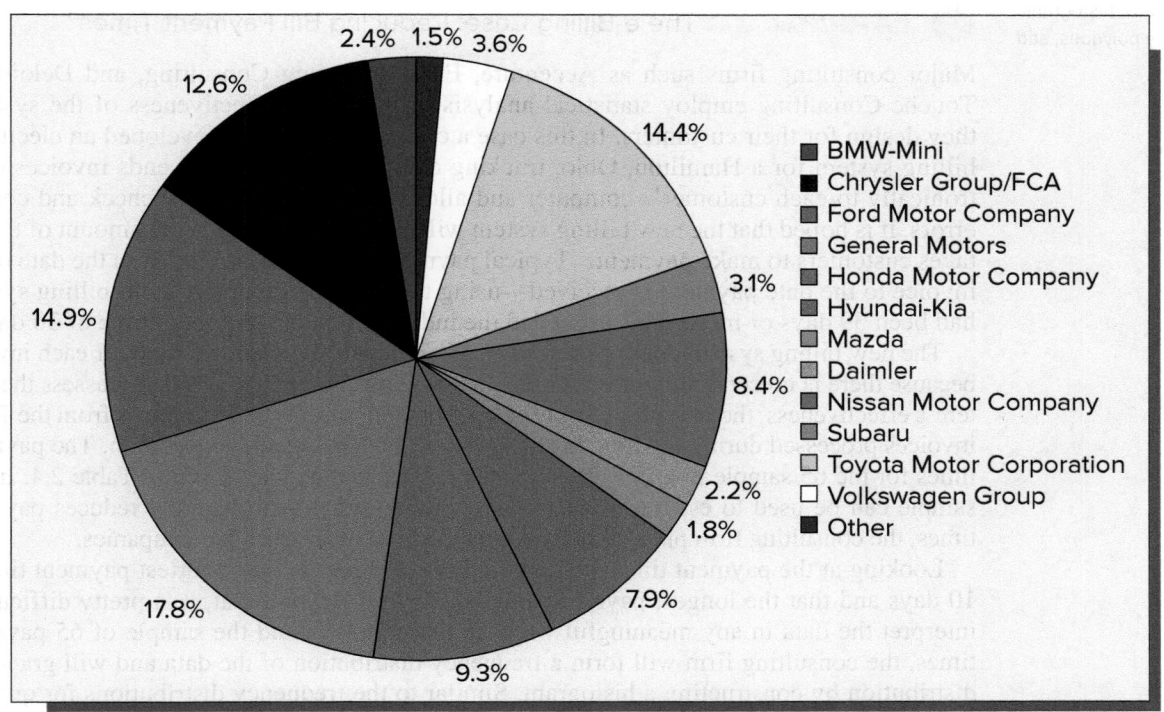

■ BMW-Mini
■ Chrysler Group/FCA
■ Ford Motor Company
■ General Motors
■ Honda Motor Company
■ Hyundai-Kia
■ Mazda
■ Daimler
■ Nissan Motor Company
■ Subaru
□ Toyota Motor Corporation
□ Volkswagen Group
■ Other

Source: Author created using data from http://www.goodcarbadcar.net/search/label/US%20Auto%20Sales%20By%20Brand?maxresults=5 (accessed January 14, 2015).

FIGURE 2.6   A Pareto Chart for Incomplete Customer Calls (for Exercise 2.12)

Required customer to get more data   29.17%
Required more investigation by us   28.12%
Callbacks   21.87%
Required development assistance   12.50%
Required administrative help   4.17%
Actually is a new problem   4.17%

Source: Author created using data from B. A. Cleary, "Company Cares about Customers' Calls," *Quality Progress* (November 1993), pp. 60–73. American Society for Quality Control, 1993.

2.2 Graphically Summarizing Quantitative Data

Frequency distributions and histograms

We often need to summarize and describe the shape of the distribution of a population or sample of **measurements.** Such data are often summarized by grouping the measurements

LO2-3
Summarize quantitative
data by using frequency
distributions, histograms,
frequency polygons, and
ogives.

into the classes of a frequency distribution and by displaying the data in the form of a **histogram.** We explain how to construct a histogram in the following example.

 EXAMPLE 2.2 The e-Billing Case: Reducing Bill Payment Times[3]

Major consulting firms such as Accenture, Ernst & Young Consulting, and Deloitte & Touche Consulting employ statistical analysis to assess the effectiveness of the systems they design for their customers. In this case a consulting firm has developed an electronic billing system for a Hamilton, Ohio, trucking company. The system sends invoices electronically to each customer's computer and allows customers to easily check and correct errors. It is hoped that the new billing system will substantially reduce the amount of time it takes customers to make payments. Typical payment times—measured from the date on an invoice to the date payment is received—using the trucking company's old billing system had been 39 days or more. This exceeded the industry standard payment time of 30 days.

The new billing system does not automatically compute the payment time for each invoice because there is no continuing need for this information. Therefore, in order to assess the system's effectiveness, the consulting firm selects a random sample of 65 invoices from the 7,823 invoices processed during the first three months of the new system's operation. The payment times for the 65 sample invoices are manually determined and are given in Table 2.4. If this sample can be used to establish that the new billing system substantially reduces payment times, the consulting firm plans to market the system to other trucking companies.

Looking at the payment times in Table 2.4, we can see that the shortest payment time is 10 days and that the longest payment time is 29 days. Beyond that, it is pretty difficult to interpret the data in any meaningful way. To better understand the sample of 65 payment times, the consulting firm will form a frequency distribution of the data and will graph the distribution by constructing a histogram. Similar to the frequency distributions for qualitative data we studied in Section 2.1, the frequency distribution will divide the payment times into classes and will tell us how many of the payment times are in each class.

Step 1: Find the Number of Classes   One rule for finding an appropriate number of classes says that the number of classes should be the smallest whole number K that makes the quantity 2^K greater than the number of measurements in the data set. For the payment time data we have 65 measurements. Because $2^6 = 64$ is less than 65 and $2^7 = 128$ is greater than 65, we should use $K = 7$ classes. Table 2.5 gives the appropriate number of classes (determined by the 2^K rule) to use for data sets of various sizes.

Step 2: Find the Class Length   We find the length of each class by computing

$$\text{approximate class length} = \frac{\text{largest measurement} - \text{smallest measurement}}{\text{number of classes}}$$

Because the largest and smallest payment times in Table 2.4 are 29 days and 10 days, the approximate class length is $(29 - 10)/7 = 2.7143$. To obtain a simpler final class length, we round this value. Commonly, the approximate class length is rounded up to the precision of the data measurements (that is, increased to the next number that has the same number

TABLE 2.4   A Sample of Payment Times (in Days) for 65 Randomly Selected Invoices
DS PayTime

22	29	16	15	18	17	12	13	17	16	15
19	17	10	21	15	14	17	18	12	20	14
16	15	16	20	22	14	25	19	23	15	19
18	23	22	16	16	19	13	18	24	24	26
13	18	17	15	24	15	17	14	18	17	21
16	21	25	19	20	27	16	17	16	21	

[3]This case is based on a real problem encountered by a company that employs one of the authors' former students. For purposes of confidentiality, we have withheld the company's name.

TABLE 2.5	Recommended Number of Classes for Data Sets of *n* Measurements*
Number of Classes	**Size, *n*, of the Data Set**
2	$1 \le n < 4$
3	$4 \le n < 8$
4	$8 \le n < 16$
5	$16 \le n < 32$
6	$32 \le n < 64$
7	$64 \le n < 128$
8	$128 \le n < 256$
9	$256 \le n < 528$
10	$528 \le n < 1056$

*For completeness' sake we have included all values of $n \ge 1$ in this table. However, we do not recommend constructing a histogram with fewer than 16 measurements.

TABLE 2.6	Seven Nonoverlapping Classes for a Frequency Distribution of the 65 Payment Times
Class 1	10 days and less than 13 days
Class 2	13 days and less than 16 days
Class 3	16 days and less than 19 days
Class 4	19 days and less than 22 days
Class 5	22 days and less than 25 days
Class 6	25 days and less than 28 days
Class 7	28 days and less than 31 days

of decimal places as the data measurements). For instance, because the payment times are measured to the nearest day, we round 2.7143 days up to 3 days.

Step 3: Form Nonoverlapping Classes of Equal Width We can form the classes of the frequency distribution by defining the **boundaries** of the classes. To find the first class boundary, we find the smallest payment time in Table 2.4, which is 10 days. This value is the lower boundary of the first class. Adding the class length of 3 to this lower boundary, we obtain $10 + 3 = 13$, which is the upper boundary of the first class and the lower boundary of the second class. Similarly, the upper boundary of the second class and the lower boundary of the third class equals $13 + 3 = 16$. Continuing in this fashion, the lower boundaries of the remaining classes are 19, 22, 25, and 28. Adding the class length 3 to the lower boundary of the last class gives us the upper boundary of the last class, 31. These boundaries define seven nonoverlapping classes for the frequency distribution. We summarize these classes in Table 2.6. For instance, the first class—10 days and less than 13 days—includes the payment times 10, 11, and 12 days; the second class—13 days and less than 16 days—includes the payment times 13, 14, and 15 days; and so forth. Notice that the largest *observed* payment time—29 days—is contained in the last class. In cases where the largest measurement is not contained in the last class, we simply add another class. Generally speaking, the guidelines we have given for forming classes are not inflexible rules. Rather, they are intended to help us find reasonable classes. Finally, the method we have used for forming classes results in classes of equal length. Generally, forming classes of equal length will make it easier to appropriately interpret the frequency distribution.

Step 4: Tally and Count the Number of Measurements in Each Class Having formed the classes, we now count the number of measurements that fall into each class. To do this, it is convenient to tally the measurements. We simply list the classes, examine the payment times in Table 2.4 one at a time, and record a tally mark corresponding to a particular class each time we encounter a measurement that falls in that class. For example, because the first four payment times in Table 2.4 are 22, 19, 16, and 18, the first four tally marks are shown below. Here, for brevity, we express the class "10 days and less than 13 days" as "10 < 13" and use similar notation for the other classes.

Class	First 4 Tally Marks	All 65 Tally Marks	Frequency
10 < 13		III	3
13 < 16		IƝƝ IƝƝ IIII	14
16 < 19	II	IƝƝ IƝƝ IƝƝ IƝƝ III	23
19 < 22	I	IƝƝ IƝƝ II	12
22 < 25	I	IƝƝ III	8
25 < 28		IIII	4
28 < 31		I	1

TABLE 2.7   Frequency Distributions of the 65 Payment Times

Class	Frequency	Relative Frequency	Percent Frequency
10 < 13	3	3/65 = .0462	4.62%
13 < 16	14	14/65 = .2154	21.54
16 < 19	23	.3538	35.38
19 < 22	12	.1846	18.46
22 < 25	8	.1231	12.31
25 < 28	4	.0615	6.15
28 < 31	1	.0154	1.54

FIGURE 2.7   A Frequency Histogram of the 65 Payment Times

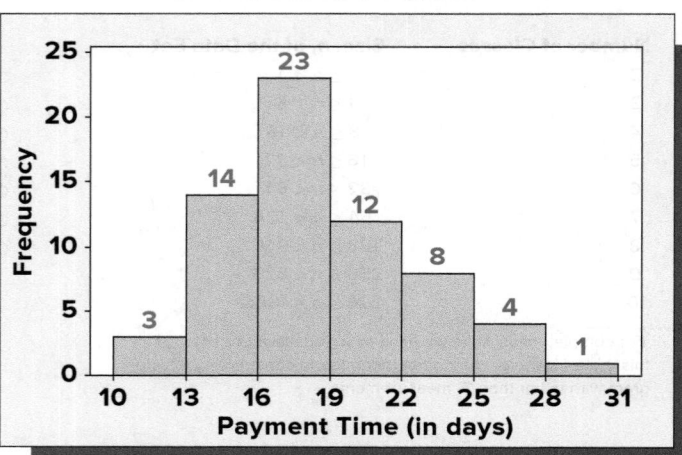

After examining all 65 payment times, we have recorded 65 tally marks. We find the **frequency** for each class by counting the number of tally marks recorded for the class. For instance, counting the number of tally marks for the class "13 < 16," we obtain the frequency 14 for this class. The frequencies for all seven classes are summarized in Table 2.7. This summary is the **frequency distribution** for the 65 payment times. Table 2.7 also gives the *relative frequency* and the *percent frequency* for each of the seven classes. The **relative frequency** of a class is the proportion (fraction) of the total number of measurements that are in the class. For example, there are 14 payment times in the second class, so its relative frequency is 14/65 = .2154. This says that the proportion of the 65 payment times that are in the second class is .2154, or, equivalently, that 100(.2154)% = 21.54% of the payment times are in the second class. A list of all of the classes—along with each class relative frequency—is called a **relative frequency distribution.** A list of all of the classes—along with each class percent frequency—is called a **percent frequency distribution.**

Step 5: Graph the Histogram   We can graphically portray the distribution of payment times by drawing a **histogram.** The histogram can be constructed using the frequency, relative frequency, or percent frequency distribution. To set up the histogram, we draw rectangles that correspond to the classes. The base of the rectangle corresponding to a class represents the payment times in the class. The height of the rectangle can represent the class frequency, relative frequency, or percent frequency.

We have drawn a **frequency histogram** of the 65 payment times in Figure 2.7. The first (leftmost) rectangle, or "bar," of the histogram represents the payment times 10, 11, and 12. Looking at Figure 2.7, we see that the base of this rectangle is drawn from the lower boundary (10) of the first class in the frequency distribution of payment times to the lower boundary (13) of the second class. The height of this rectangle tells us that the frequency of the first class is 3. The second histogram rectangle represents payment times 13, 14, and 15. Its base is drawn from the lower boundary (13) of the second class to the lower boundary (16) of the third class, and its height tells us that the frequency of the second class is 14. The other histogram bars are constructed similarly. Notice that there are no gaps between the adjacent rectangles in the histogram. Here, although the payment times have been recorded to the nearest whole day, the fact that the histogram bars touch each other emphasizes that a payment time could (in theory) be any number on the horizontal axis. In general, histograms are drawn so that adjacent bars touch each other.

Looking at the frequency distribution in Table 2.7 and the frequency histogram in Figure 2.7, we can describe the payment times:

1   None of the payment times exceeds the industry standard of 30 days. (Actually, all of the payment times are less than 30—remember the largest payment time is 29 days.)

2   The payment times are concentrated between 13 and 24 days (57 of the 65, or (57/65) × 100 = 87.69%, of the payment times are in this range).

FIGURE 2.8   A Percent Frequency Histogram of the 65 Payment Times

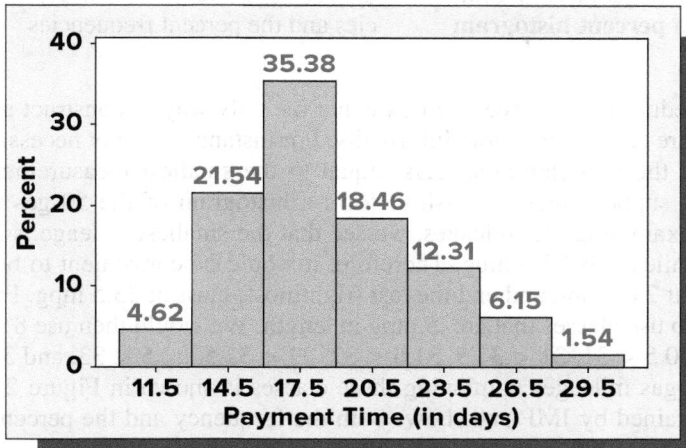

3   More payment times are in the class "16 < 19" than are in any other class (23 payment times are in this class).

Notice that the frequency distribution and histogram allow us to make some helpful conclusions about the payment times, whereas looking at the raw data (the payment times in Table 2.4) did not.

A **relative frequency histogram** and a **percent frequency histogram** of the payment times would both be drawn like Figure 2.7 except that the heights of the rectangles would represent, respectively, the relative frequencies and the percent frequencies in Table 2.7. For example, Figure 2.8 gives a percent frequency histogram of the payment times. This histogram also illustrates that we sometimes label the classes on the horizontal axis using the **class midpoints.** Each class midpoint is exactly halfway between the boundaries of its class. For instance, the midpoint of the first class, 11.5, is halfway between the class boundaries 10 and 13. The midpoint of the second class, 14.5, is halfway between the class boundaries 13 and 16. The other class midpoints are found similarly. The percent frequency distribution of Figure 2.8 tells us that 21.54% of the payment times are in the second class (which has midpoint 14.5 and represents the payment times 13, 14, and 15).

In the following box we summarize the steps needed to set up a frequency distribution and histogram:

Constructing Frequency Distributions and Histograms

1   Find the number of classes. Generally, the number of classes K should equal the smallest whole number that makes the quantity 2^K greater than the total number of measurements n (see Table 2.5).

2   Compute the approximate class length:

$$\frac{\text{largest measurement} - \text{smallest measurement}}{K}$$

Often the final class length is obtained by rounding this value up to the same level of precision as the data.

3   Form nonoverlapping classes of equal length. Form the classes by finding the **class boundaries.** The lower boundary of the first class is the smallest measurement in the data set. Add the class length to this boundary to obtain the next boundary. Successive boundaries are found by repeatedly adding the class length until the upper boundary of the last (Kth) class is found.

4   Tally and count the number of measurements in each class. The **frequency** for each class is the count of the number of measurements in the class. The **relative frequency** for each class is the fraction of measurements in the class. The **percent frequency** for each class is its relative frequency multiplied by 100%.

5   Graph the histogram. To draw a **frequency histogram,** plot each frequency as the height of a rectangle

positioned over its corresponding class. Use the class boundaries to separate adjacent rectangles. A **relative frequency histogram** and a **percent histogram** are graphed in the same way except that the heights of the rectangles are, respectively, the relative frequencies and the percent frequencies.

The procedure in the preceding box is not the only way to construct a histogram. Often, histograms are constructed more informally. For instance, it is not necessary to set the lower boundary of the first (leftmost) class equal to the smallest measurement in the data. As an example, suppose that we wish to form a histogram of the 50 gas mileages given in Table 1.7. Examining the mileages, we see that the smallest mileage is 29.8 mpg and that the largest mileage is 33.3 mpg. Therefore, it would be convenient to begin the first (leftmost) class at 29.5 mpg and end the last (rightmost) class at 33.5 mpg. Further, it would be reasonable to use classes that are .5 mpg in length. We would then use 8 classes: 29.5 < 30, 30 < 30.5, 30.5 < 31, 31 < 31.5, 31.5 < 32, 32 < 32.5, 32.5 < 33, and 33 < 33.5. A histogram of the gas mileages employing these classes is shown in Figure 2.9. This histogram has been obtained by JMP and shows both the frequency and the percent frequency of the gas mileages in each class.

Sometimes it is desirable to let the nature of the problem determine the histogram classes. For example, to construct a histogram describing the ages of the residents in a city, it might be reasonable to use classes having 10-year lengths (that is, under 10 years, 10–19 years, 20–29 years, 30–39 years, and so on).

Notice that in our examples we have used classes having equal class lengths. In general, it is best to use equal class lengths whenever the raw data (that is, all the actual measurements) are available. However, sometimes histograms are formed with unequal class lengths—particularly when we are using published data as a source. Economic data and data in the social sciences are often published in the form of frequency distributions having unequal class lengths. Dealing with this kind of data is discussed in Exercises 2.26 and 2.27. Also discussed in these exercises is how to deal with **open-ended** classes. For example, if we are constructing a histogram describing the yearly incomes of U.S. households, an open-ended class could be households earning over $500,000 per year.

As indicated by the JMP histogram in Figure 2.9, we can use software packages such as Excel, JMP, and Minitab to construct histograms. Each of these packages will automatically define histogram classes for the user. However, these automatically defined classes will not necessarily be the same as those that would be obtained using the manual method we have

FIGURE 2.9   **A JMP Histogram of the Gas Mileages: The Gas Mileage Distribution Is Symmetrical and Mound Shaped**

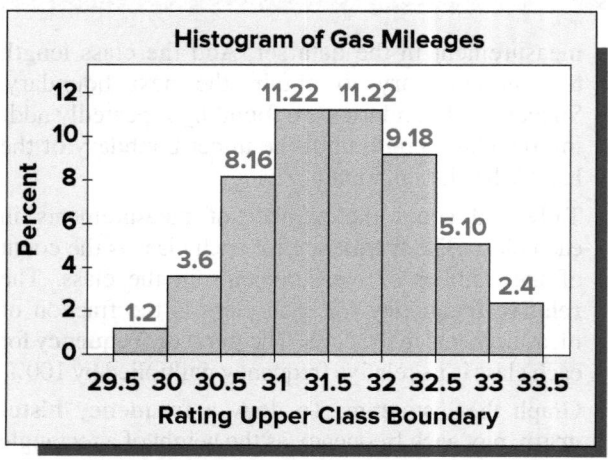

FIGURE 2.10   **A Minitab Frequency Histogram of the Payment Times with Automatic Classes: The Payment Time Distribution Is Skewed to the Right**

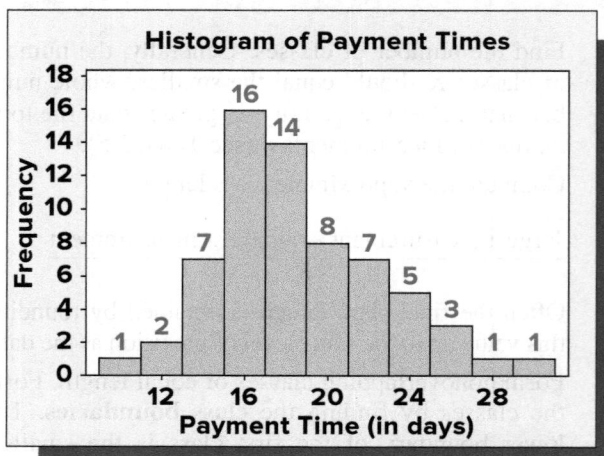

FIGURE 2.11   **An Excel Frequency Histogram of the Bottle Design Ratings: The Distribution of Ratings Is Skewed to the Left**

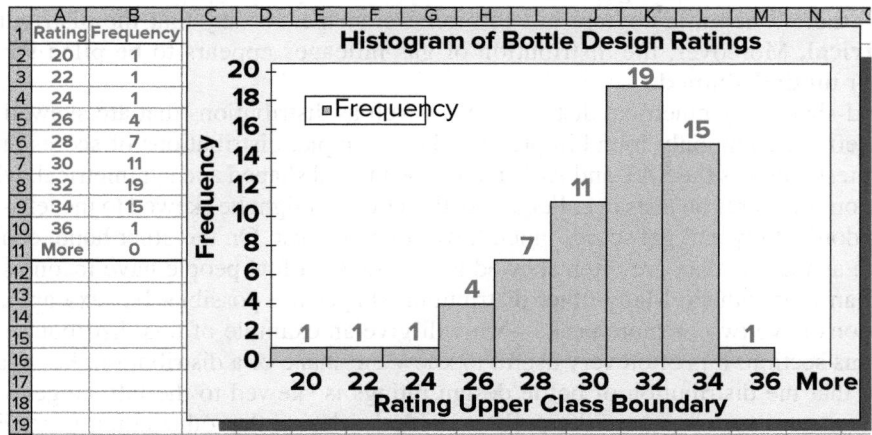

previously described. Furthermore, the packages define classes by using different methods. (Descriptions of how the classes are defined can often be found in help menus.) For example, in addition to Figure 2.9, consider Figure 2.10, which gives a Minitab frequency histogram of the payment times in Table 2.4. Here, Minitab has defined 11 classes and has labeled 5 of the classes on the horizontal axis using midpoints (12, 16, 20, 24, 28). It is easy to see that the midpoints of the unlabeled classes are 10, 14, 18, 22, 26, and 30. Moreover, the boundaries of the first class are 9 and 11, the boundaries of the second class are 11 and 13, and so forth. Minitab counts frequencies as we have previously described. For instance, one payment time is at least 9 and less than 11, two payment times are at least 11 and less than 13, seven payment times are at least 13 and less than 15, and so forth.

Figure 2.11 gives an Excel frequency distribution and histogram of the bottle design ratings in Table 1.6. Excel labels histogram classes using their upper class boundaries. For example, the first class has an upper class boundary equal to the smallest rating of 20 and contains only this smallest rating. The boundaries of the second class are 20 and 22, the boundaries of the third class are 22 and 24, and so forth. The last class corresponds to ratings more than 36. Excel's method for counting frequencies differs from that of Minitab (and, therefore, also differs from the way we counted frequencies by hand in Example 2.2). Excel assigns a frequency to a particular class by counting the number of measurements that are greater than the lower boundary of the class and less than or equal to the upper boundary of the class. For example, one bottle design rating is greater than 20 and less than or equal to (that is, at most) 22. Similarly, 15 bottle design ratings are greater than 32 and at most 34.

In Figure 2.10 we have used Minitab to automatically form histogram classes. It is also possible to force software packages to form histogram classes that are defined by the user. We explain how to do this in the appendices at the end of this chapter. Because Excel does not always automatically define acceptable classes, the classes in Figure 2.11 are a modification of Excel's automatic classes. We also explain this modification in the appendices at the end of this chapter.

Some common distribution shapes

We often graph a frequency distribution in the form of a histogram in order to visualize the *shape* of the distribution. If we look at the histogram of payment times in Figure 2.10, we see that the right tail of the histogram is longer than the left tail. When a histogram has this general shape, we say that the distribution is **skewed to the right.** Here the long right tail tells us that a few of the payment times are somewhat longer than the rest. If we look at the histogram of bottle design ratings in Figure 2.11, we see that the left tail of the histogram is much longer than the right tail. When a histogram has this general shape, we say that the distribution is **skewed to the left.** Here the long tail to the left tells us that, while most of the bottle design ratings are concentrated above 25 or so, a few of the ratings are lower than the rest. Finally, looking at the histogram of gas mileages in Figure 2.9, we see that the right and left tails of the histogram appear to be mirror images of each other. When a histogram has this general shape, we say that the distribution is **symmetrical.** Moreover, the distribution of gas mileages appears to be piled up in the middle or **mound shaped.**

Mound-shaped, symmetrical distributions as well as distributions that are skewed to the right or left are commonly found in practice. For example, distributions of scores on standardized tests such as the SAT and ACT tend to be mound shaped and symmetrical, whereas distributions of scores on tests in college statistics courses might be skewed to the left—a few students don't study and get scores much lower than the rest. On the other hand, economic data such as income data are often skewed to the right—a few people have incomes much higher than most others. Many other distribution shapes are possible. For example, some distributions have two or more peaks—we will give an example of this distribution shape later in this section. It is often very useful to know the shape of a distribution. For example, knowing that the distribution of bottle design ratings is skewed to the left suggests that a few consumers may have noticed a problem with the design that others didn't see. Further investigation into why these consumers gave the design low ratings might allow the company to improve the design.

Frequency polygons

Another graphical display that can be used to depict a frequency distribution is a **frequency polygon.** To construct this graphic, we plot a point above each class midpoint at a height equal to the frequency of the class—the height can also be the class relative frequency or class percent frequency if so desired. Then we connect the points with line segments. As we will demonstrate in the following example, this kind of graphic can be particularly useful when we wish to compare two or more distributions.

EXAMPLE 2.3 Comparing Two Grade Distributions

Table 2.8 lists (in increasing order) the scores earned on the first exam by the 40 students in a business statistics course taught by one of the authors several semesters ago. Figure 2.12 gives a percent frequency polygon for these exam scores. Because exam scores are often reported by using 10-point grade ranges (for instance, 80 to 90 percent), we have defined the following classes: 30 < 40, 40 < 50, 50 < 60, 60 < 70, 70 < 80, 80 < 90, and

TABLE 2.8		Exam Scores for the First Exam Given in a Statistics Class DS FirstExam		
32	63	69	85	91
45	64	69	86	92
50	64	72	87	92
56	65	76	87	93
58	66	78	88	93
60	67	81	89	94
61	67	83	90	96
61	68	83	90	98

FIGURE 2.12   A Percent Frequency Polygon of the Exam Scores

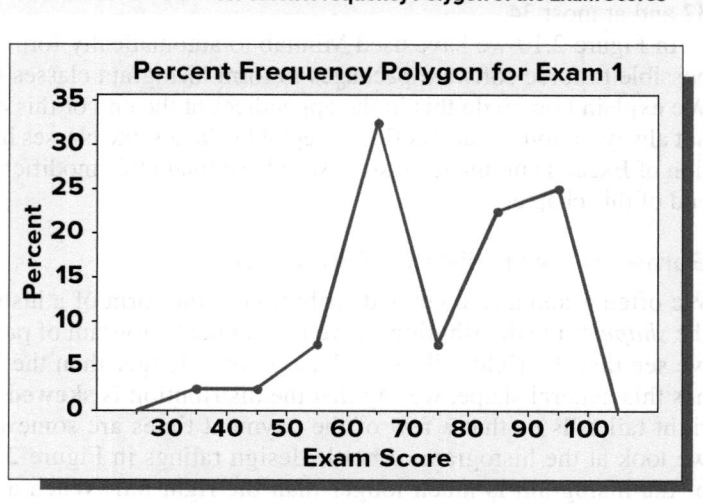

TABLE 2.9	Exam Scores for the Second Statistics Exam—after a New Attendance Policy ⒹⓈ SecondExam			
55	74	80	87	93
62	74	82	88	94
63	74	83	89	94
66	75	84	90	95
67	76	85	91	97
67	77	86	91	99
71	77	86	92	
73	78	87	93	

FIGURE 2.13   Percent Frequency Polygons of the Scores on the First Two Exams in a Statistics Course

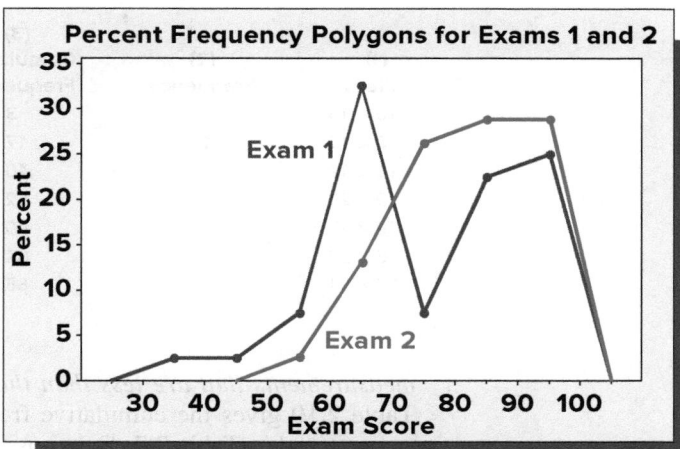

90 < 100. This is an example of letting the situation determine the classes of a frequency distribution, which is common practice when the situation naturally defines classes. The points that form the polygon have been plotted corresponding to the midpoints of the classes (35, 45, 55, 65, 75, 85, 95). Each point is plotted at a height that equals the percentage of exam scores in its class. For instance, because 10 of the 40 scores are at least 90 and less than 100, the plot point corresponding to the class midpoint 95 is plotted at a height of 25 percent.

Looking at Figure 2.12, we see that there is a concentration of scores in the 85 to 95 range and another concentration of scores around 65. In addition, the distribution of scores is somewhat skewed to the left—a few students had scores (in the 30s and 40s) that were quite a bit lower than the rest.

This is an example of a distribution having two peaks. When a distribution has multiple peaks, finding the reason for the different peaks often provides useful information. The reason for the two-peaked distribution of exam scores was that some students were not attending class regularly. Students who received scores in the 60s and below admitted that they were cutting class, whereas students who received higher scores were attending class on a regular basis.

After identifying the reason for the concentration of lower scores, the instructor established an attendance policy that forced students to attend every class—any student who missed a class was to be dropped from the course. Table 2.9 presents the scores on the second exam—after the new attendance policy. Figure 2.13 presents (and allows us to compare) the percent frequency polygons for both exams. We see that the polygon for the second exam is single peaked—the attendance policy[4] eliminated the concentration of scores in the 60s, although the scores are still somewhat skewed to the left.

Cumulative distributions and ogives

Another way to summarize a distribution is to construct a **cumulative distribution.** To do this, we use the same number of classes, the same class lengths, and the same class boundaries that we have used for the frequency distribution of a data set. However, in order to construct a **cumulative frequency distribution,** we record for each class *the number of*

---

[4]Other explanations are possible. For instance, all of the students who did poorly on the first exam might have studied harder for the second exam. However, the instructor's 30 years of teaching experience suggest that attendance was the critical factor.

TABLE 2.10 A Frequency Distribution, Cumulative Frequency Distribution, Cumulative Relative Frequency Distribution, and Cumulative Percent Frequency Distribution for the Payment Time Data

(1) Class	(2) Frequency	(3) Cumulative Frequency	(4) Cumulative Relative Frequency	(5) Cumulative Percent Frequency
10 < 13	3	3	3/65 = .0462	4.62%
13 < 16	14	17	17/65 = .2615	26.15
16 < 19	23	40	.6154	61.54
19 < 22	12	52	.8000	80.00
22 < 25	8	60	.9231	92.31
25 < 28	4	64	.9846	98.46
28 < 31	1	65	1.0000	100.00

measurements that are less than the upper boundary of the class. To illustrate this idea, Table 2.10 gives the cumulative frequency distribution of the payment time distribution summarized in Table 2.7. Columns (1) and (2) in this table give the frequency distribution of the payment times. Column (3) gives the **cumulative frequency** for each class. To see how these values are obtained, the cumulative frequency for the class 10 < 13 is the number of payment times less than 13. This is obviously the frequency for the class 10 < 13, which is 3. The cumulative frequency for the class 13 < 16 is the number of payment times less than 16, which is obtained by adding the frequencies for the first two classes—that is, 3 + 14 = 17. The cumulative frequency for the class 16 < 19 is the number of payment times less than 19—that is, 3 + 14 + 23 = 40. We see that, in general, a cumulative frequency is obtained by summing the frequencies of all classes representing values less than the upper boundary of the class.

Column (4) gives the **cumulative relative frequency distribution** for each class, which is obtained by summing the relative frequencies of all classes representing values less than the upper boundary of the class. Or, more simply, this value can be found by dividing the cumulative frequency for the class by the total number of measurements in the data set. For instance, the cumulative relative frequency for the class 19 < 22 is 52/65 = .8. Column (5) gives the **cumulative percent frequency distribution** for each class, which is obtained by summing the percent frequencies of all classes representing values less than the upper boundary of the class. More simply, this value can be found by multiplying the cumulative relative frequency of a class by 100. For instance, the cumulative percent frequency for the class 19 < 22 is .8 × (100) = 80 percent.

As an example of interpreting Table 2.10, 60 of the 65 payment times are 24 days or less, or, equivalently, 92.31 percent of the payment times (or a fraction of .9231 of the payment times) are 24 days or less. Also, notice that the last entry in the cumulative frequency distribution is the total number of measurements (here, 65 payment times). In addition, the last entry in the cumulative relative frequency distribution is 1.0 and the last entry in the cumulative percent frequency distribution is 100%. In general, for any data set, these last entries will be, respectively, the total number of measurements, 1.0, and 100%.

An **ogive** (pronounced "oh-jive") is a graph of a cumulative distribution. To construct a frequency ogive, we plot a point above each upper class boundary at a height equal to the cumulative frequency of the class. We then connect the plotted points with line segments. A similar graph can be drawn using the cumulative relative frequencies or the cumulative percent frequencies. As an example, Figure 2.14 gives a percent frequency ogive of the payment times. Looking at this figure, we see that, for instance, a little more than 25 percent (actually, 26.15 percent according to Table 2.10) of the payment times are less than 16 days, while 80 percent of the payment times are less than 22 days. Also notice that we have completed the ogive by plotting an additional point at the lower boundary of the first (leftmost) class at a height equal to zero. This depicts the fact that none of the payment times is less than 10 days. Finally, the ogive graphically shows that all (100 percent) of the payment times are less than 31 days.

FIGURE 2.14   **A Percent Frequency Ogive of the Payment Times**

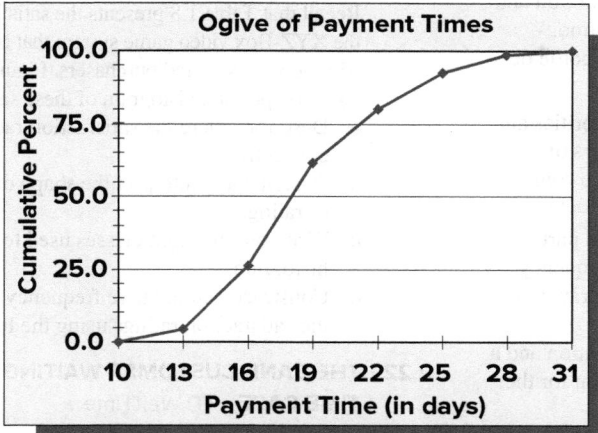

Exercises for Section 2.2

CONCEPTS **connect**

2.13 Explain:
  a Why we construct a frequency distribution and a histogram for a data set.
  b The difference between a frequency histogram and a frequency polygon.
  c The difference between a frequency polygon and a frequency ogive.

2.14 When constructing a frequency distribution and histogram, explain how to find:
  a The frequency for a class.
  b The relative frequency for a class.
  c The percent frequency for a class.

2.15 Explain what each of the following distribution shapes looks like. Then draw a picture that illustrates each shape.
  a Symmetrical and mound shaped
  b Double peaked
  c Skewed to the right
  d Skewed to the left

METHODS AND APPLICATIONS

2.16 Consider the data in the table to the right. **DS** HistoData

36	46	40	38
39	40	38	38
36	42	33	34
35	34	37	37
36	41	22	17
20	36	33	25
19	42	28	38

  a Find the number of classes needed to construct a histogram.
  b Find the class length.
  c Define nonoverlapping classes for a frequency distribution.
  d Tally the number of values in each class and develop a frequency distribution.
  e Draw a frequency histogram for these data.
  f Develop a percent frequency distribution.

2.17 Consider the frequency distribution of exam scores in the page margin above. **DS** GradeDist

  a Develop a relative frequency distribution and a percent frequency distribution.
  b Develop a cumulative frequency distribution and a cumulative percent frequency distribution.
  c Draw a percent frequency polygon.
  d Draw a percent frequency ogive.

Class	Frequency
50 < 60	2
60 < 70	5
70 < 80	14
80 < 90	17
90 < 100	12

DS GradeDist

THE MARKETING RESEARCH CASE

Recall that 60 randomly selected shoppers have rated a new bottle design for a popular soft drink. The data are given below. **DS** Design

34	33	33	29	26
32	25	27	33	22
24	30	20	34	31
32	28	30	31	31
31	28	33	31	32
32	30	34	32	30
33	28	25	32	33
27	32	33	32	29
32	30	35	33	31
33	29	27	34	31
28	26	29	32	34
30	32	31	29	33

Use these data to work Exercises 2.18 and 2.19.

2.18 **a** Find the number of classes that should be used to construct a frequency distribution and histogram for the bottle design ratings.
  b If we round up to the nearest whole rating point, show that we should employ a class length equal to 3.
  c Define the nonoverlapping classes for a frequency distribution.
  d Tally the number of ratings in each class and develop a frequency distribution.
  e Draw the frequency histogram for the ratings data, and describe the distribution shape. **DS** Design

2.19 **a** Construct a relative frequency distribution and a percent frequency distribution for the bottle design ratings.

b Construct a cumulative frequency distribution and a cumulative percent frequency distribution.

c Draw a percent frequency ogive for the bottle design ratings.   DS *Design*

2.20 Table 2.11 gives the 25 most powerful celebrities and their annual earnings as ranked by the editors of *Forbes* magazine and as listed on the Forbes.com website on January 14, 2015.   DS *PowerCeleb*

a Remove Dr. Dre from the data set in this part as well as in parts *b* and *c*. Develop a frequency distribution for the remaining data and draw a frequency histogram.

b Develop a cumulative frequency distribution and a cumulative percent frequency distribution for the altered data set.

c Draw a percent frequency ogive for the altered data set.

d Reinsert Dr. Dre into the data set and develop a frequency distribution for the original earnings data.

2.21 THE VIDEO GAME SATISFACTION RATING CASE   DS *VideoGame*

Recall that Table 1.8 presents the satisfaction ratings for the XYZ-Box video game system that have been given by 65 randomly selected purchasers. Figure 2.15 gives the Excel output of a histogram of these satisfaction ratings.

a Describe where the satisfaction ratings seem to be concentrated.

b Describe and interpret the shape of the distribution of ratings.

c Write out the eight classes used to construct this histogram.

d Construct a cumulative frequency distribution of the satisfaction ratings using the histogram classes.

2.22 THE BANK CUSTOMER WAITING TIME CASE   DS *WaitTime*

Recall that Table 1.9 presents the waiting times for teller service during peak business hours of 100 randomly selected bank customers.

TABLE 2.11   **The 25 Most Powerful Celebrities as Rated by *Forbes* Magazine (for Exercise 2.20)**   DS PowerCeleb

Power Ranking	Celebrity Name	Earnings ($ mil)	Power Ranking	Celebrity Name	Earnings ($ mil)
1	Beyoncé Knowles	115	14	Bruno Mars	60
2	LeBron James	72	15	Kobe Bryant	62
3	Dr. Dre	620	16	Roger Federer	55
4	Oprah Winfrey	82	17	Miley Cyrus	36
5	Ellen DeGeneres	70	18	Taylor Swift	64
6	Jay-Z	60	19	Lady Gaga	33
7	Floyd Mayweather	105	20	Kanye West	30
8	Rihanna	48	21	Calvin Harris	66
9	Katy Perry	40	22	Tiger Woods	61
10	Robert Downey Jr.	75	23	Dwayne Johnson	52
11	Steven Spielberg	100	24	Rafael Nadal	45
12	Jennifer Lawrence	34	25	Bruce Springsteen	81
13	Bon Jovi	82			

Source: http://www.forbes.com/celebrities/list/#tab:overall (accessed January 14, 2015).

FIGURE 2.15   **Excel Frequency Histogram of the 65 Satisfaction Ratings (for Exercise 2.21)**

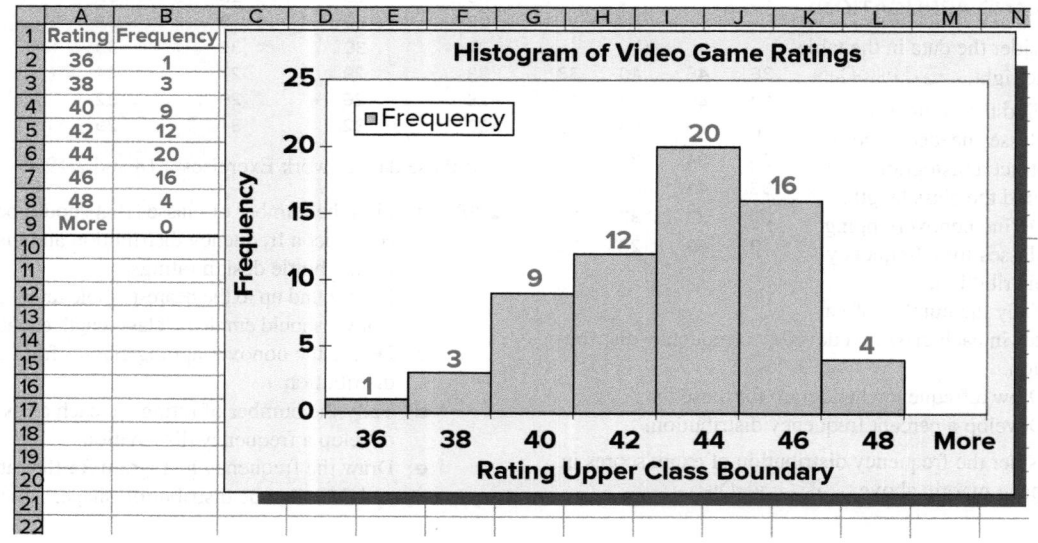

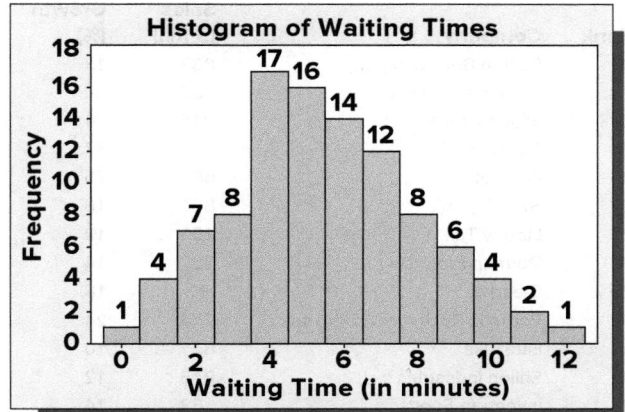

FIGURE 2.16   Minitab Frequency Histogram of the 100 Waiting Times (for Exercise 2.22)

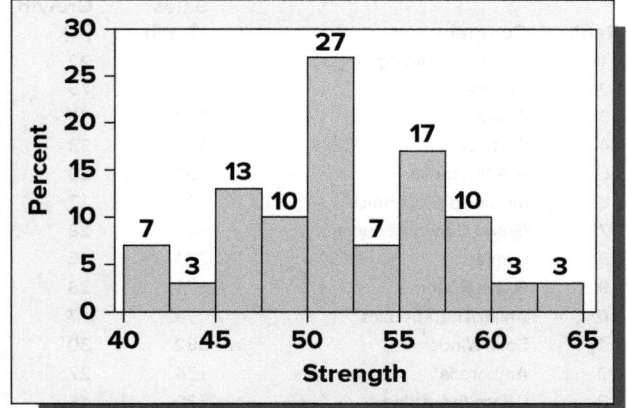

FIGURE 2.17   JMP Percent Frequency Histogram of the 30 Breaking Strengths (for Exercise 2.23)

TABLE 2.12   Major League Baseball Team Valuations and Revenues as Given on the Forbes.com Website on January 14, 2015 (for Exercise 2.24) DS MLBTeams

Rank	Team	Value ($ mil)	Revenue ($ mil)	Rank	Team	Value ($ mil)	Revenue ($ mil)
1	New York Yankees	2500	461	16	Baltimore Orioles	620	198
2	Los Angeles Dodgers	2000	293	17	San Diego Padres	615	207
3	Boston Red Sox	1500	357	18	Toronto Blue Jays	610	218
4	Chicago Cubs	1200	266	19	Minnesota Twins	605	221
5	San Francisco Giants	1000	316	20	Cincinnati Reds	600	209
6	Philadelphia Phillies	975	265	21	Arizona Diamondbacks	585	192
7	Texas Rangers	825	257	22	Colorado Rockies	575	197
8	St. Louis Cardinals	820	283	23	Pittsburgh Pirates	572	204
9	New York Mets	800	238	24	Cleveland Indians	570	196
10	Los Angeles Angels	775	253	25	Milwaukee Brewers	565	197
11	Atlanta Braves	730	253	26	Houston Astros	530	186
12	Seattle Mariners	710	210	27	Miami Marlins	500	159
13	Washington Nationals	700	244	28	Oakland Athletics	495	187
14	Chicago White Sox	695	210	29	Kansas City Royals	490	178
15	Detroit Tigers	680	262	30	Tampa Bay Rays	455	181

Source: http://www.forbes.com/mlb-valuations/ (accessed January 14, 2015).

Figure 2.16 above gives the Minitab output of a histogram of these waiting times that has been constructed using automatic classes.

a Describe where the waiting times seem to be concentrated.

b Describe and interpret the shape of the distribution of waiting times.

c What is the class length that has been automatically defined by Minitab?

d Write out the automatically defined classes and construct a cumulative percent frequency distribution of the waiting times using these classes.

2.23   THE TRASH BAG CASE DS TrashBag

Recall that Table 1.10 presents the breaking strengths of 30 trash bags selected during a 30-hour pilot production run. Figure 2.17 gives a percent frequency histogram of these breaking strengths.

a Describe where the breaking strengths seem to be concentrated.

b Describe and interpret the shape of the distribution of breaking strengths.

2.24 Table 2.12 gives the 2014 franchise value and revenues for each of the 30 teams in Major League Baseball as reported by *Forbes* magazine and as listed on the Forbes.com website on January 14, 2015. DS MLBTeams

a Develop a frequency distribution and a frequency histogram for the 30 team values. Then describe the distribution of team values.

b Develop a percent frequency distribution and a percent frequency histogram for the 30 team

TABLE 2.13 **America's Top 40 Best Small Companies of 2014 as Rated by *Forbes* Magazine (for Exercise 2.25)**
DS SmallComp

Rank	Company	Sales ($ mil)	Sales Growth (%)	Rank	Company	Sales ($ mil)	Sales Growth (%)
1	US Silica Holdings	680	22	21	Boston Beer Company	837	13
2	Gentherm	754	75	22	Nathan's Famous	87	11
3	EPAM Systems	633	32	23	SL Industries	215	4
4	Proto Labs	185	33	24	Cirrus Logic	712	37
5	IPG Photonics	701	29	25	Fortinet	685	25
6	Methode Electronics	824	12	26	Syntel	881	16
7	Grand Canyon Education	641	28	27	Liberty Tax	159	19
8	Lannett	274	15	28	Dorman Products	727	14
9	Sturm Ruger	676	28	29	Clearfield	63	18
10	Anika Therapeutics	99	17	30	Portfolio Recovery Associates	774	24
11	SolarWinds	382	30	31	Littelfuse	827	10
12	Ambarella	174	27	32	Shiloh Industries	815	12
13	Cavco Industries	539	46	33	Inventure Foods	253	14
14	Air Methods	958	14	34	Winnebago Industries	913	14
15	Calavo Growers	772	15	35	Chuy's Holdings	223	33
16	Myriad Genetics	778	19	36	NIC	259	19
17	Manhattan Associates	451	7	37	Core Molding Technologies	162	10
18	Medifast	331	28	38	Stamps.com	131	10
19	Winmark	57	10	39	Multi-Color	743	24
20	Astronics	511	14	40	Alliance Fiber OPtic Products	94	14

Source: http://www.forbes.com/best-small-companies/list/ (accessed January 14, 2015).

revenues. Then describe the distribution of team revenues.

c Draw a percent frequency polygon for the 30 team values.

2.25 Table 2.13 gives America's top 40 best small companies of 2014 as rated on the Forbes.com website on January 14, 2015. DS SmallComp

a Develop a frequency distribution and a frequency histogram for the sales values. Describe the distribution of these sales values.

b Develop a percent frequency histogram for the sales growth values and then describe this distribution.

Exercises 2.26 and 2.27 relate to the following situation. ISO 9000 is a series of international standards for quality assurance management systems. CEEM Information Services presents the results of a Quality Systems Update/Deloitte & Touche survey of ISO 9000–registered companies conducted in July 1993.[5] Included in the results is a summary of the total annual savings associated with ISO 9000 implementation for surveyed companies. The findings (in the form of a frequency distribution of ISO 9000 savings) are given to the right of Exercise 2.26. Notice that the classes in this distribution have unequal lengths and that there is an open-ended class (>$500K).

2.26 To construct a histogram for these data, we select one of the classes as a base. It is often convenient to choose the shortest class as the base (although it is not necessary to do so). Using this choice, the 0 to $10K class is the base. This means that we will draw a rectangle over the 0 to $10K class having a height equal to 162 (the frequency given for this class in the published data). Because the other classes are longer than the base, the heights of the rectangles above these classes will be adjusted. To do this we employ a rule that says that the area of a rectangle positioned over a particular class should represent the relative proportion of measurements in the class. Therefore, we proceed as follows. The length of the $10K to 25K class differs from the base class by a factor of $(25 - 10)/(10 - 0) = 3/2$, and, therefore, we make the height of the rectangle over the $10K to 25K

DS ISO9000

Annual Savings	Number of Companies
0 to $10K	162
$10K to $25K	62
$25K to $50K	53
$50K to $100K	60
$100K to $150K	24
$150K to $200K	19
$200K to $250K	22
$250K to $500K	21
(>$500K)	37

Note: (K = 1000)

[5]Source: CEEM Information Services, Fairfax, Virginia. *Is ISO 9000 for you?*

class equal to $(2/3)(62) = 41.333$. Similarly, the length of the \$25K to 50K class differs from the length of the base class by a factor of $(50 - 25)/(10 - 0) = 5/2$, and, therefore, we make the height of the rectangle over the \$25K to 50K class equal to $(2/5)(53) = 21.2$.

  ⒹⓈ ISO9000

a Use the procedure just outlined to find the heights of the rectangles drawn over all the other classes (with the exception of the open-ended class, >\$500K).

b Draw the appropriate rectangles over the classes (except for >\$500K).

2.27 To complete the histogram from Exercise 2.26, we place an asterisk (∗) to the right of \$500K on the scale of measurements and note "37" next to the ∗ to indicate 37 companies saved more than \$500K. Complete the histogram by doing this.   **ⒹⓈ** ISO9000

2.3 Dot Plots

A very simple graph that can be used to summarize a data set is called a **dot plot.** To make a dot plot we draw a horizontal axis that spans the range of the measurements in the data set. We then place dots above the horizontal axis to represent the measurements. As an example, Figure 2.18(a) shows a dot plot of the exam scores in Table 2.8. Remember, these are the scores for the first exam given before implementing a strict attendance policy. The horizontal axis spans exam scores from 30 to 100. Each dot above the axis represents an exam score. For instance, the two dots above the score of 90 tell us that two students received a 90 on the exam. The dot plot shows us that there are two concentrations of scores—those in the 80s and 90s and those in the 60s. Figure 2.18(b) gives a dot plot of the scores on the second exam (which was given after imposing the attendance policy). As did the percent frequency polygon for Exam 2 in Figure 2.13, this second dot plot shows that the attendance policy eliminated the concentration of scores in the 60s.

    Dot plots are useful for detecting **outliers,** which are unusually large or small observations that are well separated from the remaining observations. For example, the dot plot for exam 1 indicates that the score 32 seems unusually low. How we handle an outlier depends on its cause. If the outlier results from a measurement error or an error in recording or processing the data, it should be corrected. If such an outlier cannot be corrected, it should be discarded. If an outlier is not the result of an error in measuring or recording the data, its cause may reveal important information. For example, the outlying exam score of 32 convinced the author that the student needed a tutor. After working with a tutor, the student showed considerable improvement on Exam 2. A more precise way to detect outliers is presented in Section 3.3.

LO2-4

Construct and interpret dot plots.

FIGURE 2.18   **Comparing Exam Scores Using Dot Plots**

(a) Dot Plot of the Scores on Exam 1: Before Attendance Policy

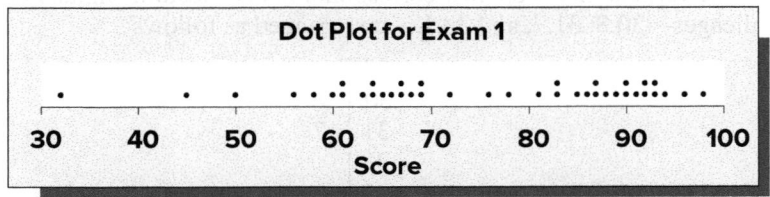

(b) Dot Plot of the Scores on Exam 2: After Attendance Policy

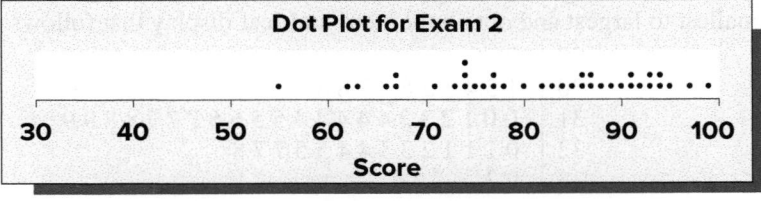

Exercises for Section 2.3

CONCEPTS 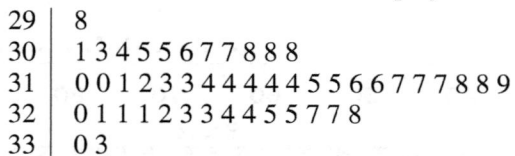 connect

2.28 When we construct a dot plot, what does the horizontal axis represent? What does each dot represent?

2.29 If a data set consists of 1,000 measurements, would you summarize the data set using a histogram or a dot plot? Explain.

METHODS AND APPLICATIONS

2.30 The following data consist of the number of students who were absent in a professor's statistics class each day during the last month. ⒟Ⓢ AbsenceData

2	0	3	1	2
1	10	6	2	2
5	8	0	1	4
0	3	6	0	1

Construct a dot plot of these data, and then describe the distribution of absences.

2.31 The following are the revenue growth rates for 30 fast-growing companies. ⒟Ⓢ RevGrowth30

93%	43%	91%	49%	70%
33%	40%	60%	35%	51%
87%	46%	38%	30%	33%
44%	71%	70%	52%	59%
48%	39%	61%	25%	87%
43%	29%	38%	60%	32%

Develop a dot plot for these data and describe the distribution of revenue growth rates.

2.32 The yearly home run totals for Babe Ruth during his career as a New York Yankee are as follows (the totals are arranged in increasing order): 22, 25, 34, 35, 41, 41, 46, 46, 46, 47, 49, 54, 54, 59, 60. Construct a dot plot for these data and then describe the distribution of home run totals. ⒟Ⓢ RuthsHomers

2.4 Stem-and-Leaf Displays

LO2-5
Construct and interpret stem-and-leaf displays.

Another simple graph that can be used to quickly summarize a data set is called a **stem-and-leaf display.** This kind of graph places the measurements in order from smallest to largest, and allows the analyst to simultaneously see all of the measurements in the data set and see the shape of the data set's distribution.

Ⓒ **EXAMPLE 2.4** The Car Mileage Case: Estimating Mileage

Table 2.14 presents the sample of 50 gas mileages for the new midsize model previously introduced in Chapter 1. To develop a stem-and-leaf display, we note that the sample mileages range from 29.8 to 33.3 and we place the leading digits of these mileages—the whole numbers 29, 30, 31, 32, and 33—in a column on the left side of a vertical line as follows.

```
29 |
30 |
31 |
32 |
33 |
```

This vertical arrangement of leading digits forms the **stem** of the display. Next, we pass through the mileages in Table 2.14 one at a time and place each last digit (the tenths place) to the right of the vertical line in the row corresponding to its leading digits. For instance, the first three mileages—30.8, 31.7, and 30.1—are arranged as follows:

```
29 |
30 | 8 1
31 | 7
32 |
33 |
```

We form the **leaves** of the display by continuing this procedure as we pass through all 50 mileages. After recording the last digit for each of the mileages, we sort the digits in each row from smallest to largest and obtain the stem-and-leaf display that follows:

```
29 | 8
30 | 1 3 4 5 5 6 7 7 8 8 8
31 | 0 0 1 2 3 3 4 4 4 4 4 5 5 6 6 7 7 7 8 8 9 9
32 | 0 1 1 1 2 3 3 4 4 5 5 7 7 8
33 | 0 3
```

©vadimguzhva/Getty Images

TABLE 2.14   A Sample of 50 Mileages for a New Midsize Model   Ⓓ🖢 GasMiles

30.8	30.8	32.1	32.3	32.7
31.7	30.4	31.4	32.7	31.4
30.1	32.5	30.8	31.2	31.8
31.6	30.3	32.8	30.7	31.9
32.1	31.3	31.9	31.7	33.0
33.3	32.1	31.4	31.4	31.5
31.3	32.5	32.4	32.2	31.6
31.0	31.8	31.0	31.5	30.6
32.0	30.5	29.8	31.7	32.3
32.4	30.5	31.1	30.7	31.4

As we have said, the numbers to the left of the vertical line form the stem of the display. Each number to the right of the vertical line is a leaf. Each combination of a stem value and a leaf value represents a measurement in the data set. For instance, the first row in the display

$$29 \mid 8$$

tells us that the first two digits are 29 and that the last (tenth place) digit is 8—that is, this combination represents the mileage 29.8 mpg. Similarly, the last row

$$33 \mid 0\ 3$$

represents the mileages 33.0 mpg and 33.3 mpg.

The entire stem-and-leaf display portrays the overall distribution of the sample mileages. It groups the mileages into classes, and it graphically illustrates how many mileages are in each class, as well as how the mileages are distributed within each class. The first class corresponds to the stem 29 and consists of the mileages from 29.0 to 29.9. There is one mileage—29.8—in this class. The second class corresponds to the stem 30 and consists of the mileages from 30.0 to 30.9. There are 11 mileages in this class. Similarly, the third, fourth, and fifth classes correspond to the stems 31, 32, and 33 and contain, respectively, 22 mileages, 14 mileages, and 2 mileages. Moreover, the stem-and-leaf display shows that the distribution of mileages is quite symmetrical. To see this, imagine turning the stem-and-leaf display on its side so that the vertical line becomes a horizontal number line. We see that the display now resembles a symmetrically shaped histogram. However, *the stem-and-leaf display is advantageous because it allows us to actually see the measurements in the data set* in addition to the distribution's shape.

When constructing a stem-and-leaf display, there are no rules that dictate the number of stem values (rows) that should be used. If we feel that the display has collapsed the mileages too closely together, we can stretch the display by assigning each set of leading digits to two or more rows. This is called *splitting the stems*. For example, in the following stem-and-leaf display of the mileages the first (uppermost) stem value of 30 is used to represent mileages between 30.0 and 30.4. The second stem value of 30 is used to represent mileages between 30.5 and 30.9.

```
29 │ 8
30 │ 1 3 4
30 │ 5 5 6 7 7 8 8 8
31 │ 0 0 1 2 3 3 4 4 4 4 4
31 │ 5 5 6 6 7 7 7 8 8 9 9
32 │ 0 1 1 1 2 3 3 4 4
32 │ 5 5 7 7 8
33 │ 0 3
```

FIGURE 2.19   **Minitab Stem-and-Leaf Display of the 50 Mileages**

```
Stem-and-Leaf Display: Mpg
Stem-and-leaf of Mpg N = 50
Leaf unit = 0.10
   1    29   8
   4    30   134
  12    30   55677888
  23    31   00123344444
 (11)   31   55667778899
  16    32   011123344
   7    32   55778
   2    33   03
```

Notice that, in this particular case, splitting the stems produces a display that seems to more clearly reveal the symmetrical shape of the distribution of mileages.

Most statistical software packages can be used to construct stem-and-leaf displays. Figure 2.19 gives a Minitab stem-and-leaf display of the 50 sample mileages. This output has been obtained by splitting the stems—Minitab produced this display automatically. Minitab also provides an additional column of numbers (on the left) that provides information about how many mileages are in the various rows. For example, if we look at the Minitab output, the 11 (in parentheses) tells us that there are 11 mileages between 31.5 mpg and 31.9 mpg. The 12 (no parentheses) tells us that a total of 12 mileages are at or below 30.9 mpg, while the 7 tells us that a total of 7 mileages are at or above 32.5 mpg.

It is possible to construct a stem-and-leaf display from measurements containing any number of digits. To see how this can be done, consider the following data which consists of the number of DVD players sold by an electronics manufacturer for each of the last 12 months.

13,502	15,932	14,739	15,249	14,312	17,111	DVDPlayers
19,010	16,121	16,708	17,886	15,665	16,475	

To construct a stem-and-leaf display, we will use only the first three digits of each sales value and we will define leaf values consisting of one digit. The stem will consist of the values 13, 14, 15, 16, 17, 18, and 19 (which represent thousands of units sold). Each leaf will represent the remaining three digits rounded to the nearest 100 units sold. For example, 13,502 will be represented by placing the leaf value 5 in the row corresponding to 13. To express the fact that the leaf 5 represents 500, we say that the **leaf unit** is 100. Using this procedure, we obtain the following stem-and-leaf display:

```
          Leaf unit = 100
          13 | 5
          14 | 3 7
          15 | 2 7 9
          16 | 1 5 7
          17 | 1 9
          18 |
          19 | 0
```

The standard practice of always using a single digit for each leaf allows us to construct a stem-and-leaf display for measurements having any number of digits as long as we appropriately define a leaf unit. However, it is not possible to recover the original measurements from such a display. If we do not have the original measurements, the best we can do is to approximate them by multiplying the digits in the display by the leaf unit. For instance, the measurements in the row corresponding to the stem value 17 can be approximated to be $171 \times (100) = 17{,}100$ and $179 \times (100) = 17{,}900$. In general, leaf units can be any power of

10 such as 0.1, 1, 10, 100, 1000, and so on. If no leaf unit is given for a stem-and-leaf display, we assume its value is 1.0.

We summarize how to set up a stem-and-leaf display in the following box:

Constructing a Stem-and-Leaf Display

1　Decide what units will be used for the stems and the leaves. Each leaf must be a single digit and the stem values will consist of appropriate leading digits. As a general rule, there should be between 5 and 20 stem values.

2　Place the stem values in a column to the left of a vertical line with the smallest value at the top of the column and the largest value at the bottom.

3　To the right of the vertical line, enter the leaf for each measurement into the row corresponding to the proper stem value. Each leaf should be a single digit—these can be rounded values that were originally more than one digit if we are using an appropriately defined leaf unit.

4　Rearrange the leaves so that they are in increasing order from left to right.

If we wish to compare two distributions, it is convenient to construct a **back-to-back stem-and-leaf display.** Figure 2.20 presents a back-to-back stem-and-leaf display for the previously discussed exam scores. The left side of the display summarizes the scores for the first exam. Remember, this exam was given before implementing a strict attendance policy. The right side of the display summarizes the scores for the second exam (which was given after imposing the attendance policy). Looking at the left side of the display, we see that for the first exam there are two concentrations of scores—those in the 80s and 90s and those in the 60s. The right side of the display shows that the attendance policy eliminated the concentration of scores in the 60s and illustrates that the scores on Exam 2 are almost single peaked and somewhat skewed to the left.

Stem-and-leaf displays are useful for detecting **outliers,** which are unusually large or small observations that are well separated from the remaining observations. For example, the stem-and-leaf display for Exam 1 indicates that the score 32 seems unusually low. How we handle an outlier depends on its cause. If the outlier results from a measurement error or an error in recording or processing the data, it should be corrected. If such an outlier cannot be corrected, it should be discarded. If an outlier is not the result of an error in measuring or recording the data, its cause may reveal important information. For example, the outlying exam score of 32 convinced the author that the student needed a tutor. After working with a tutor, the student showed considerable improvement on Exam 2. A more precise way to detect outliers is presented in Section 3.3.

FIGURE 2.20　A Back-to-Back Stem-and-Leaf Display of the Exam Scores

Exam 1		Exam 2
2	3	
	3	
	4	
5	4	
0	5	
8 6	5	5
4 4 3 1 1 0	6	2 3
9 9 8 7 7 6 5	6	6 7 7
2	7	1 3 4 4 4
8 6	7	5 6 7 7 8
3 3 1	8	0 2 3 4
9 8 7 7 6 5	8	5 6 6 7 7 8 9
4 3 3 2 2 1 0 0	9	0 1 1 2 3 3 4 4
8 6	9	5 7 9

Exercises for Section 2.4

CONCEPTS connect

2.33 Explain the difference between a histogram and a stem-and-leaf display.

2.34 What are the advantages of using a stem-and-leaf display?

2.35 If a data set consists of 1,000 measurements, would you summarize the data set by using a stem-and-leaf display or a histogram? Explain.

METHODS AND APPLICATIONS

2.36 The following data consist of the revenue growth rates (in percent) for a group of 20 firms. Construct a stem-and-leaf display for these data.
DS RevGrowth20

36	59	42	65	91
30	55	33	63	70
32	56	28	49	51
44	42	83	53	43

2.37 The following data consist of the profit margins (in percent) for a group of 20 firms. Construct a stem-and-leaf display for these data. DS ProfitMar20

25.2	16.1	22.2	15.2	14.1
16.4	13.9	10.4	13.8	14.9
15.2	14.4	15.9	10.4	14.0
16.1	15.8	13.2	16.8	12.6

2.38 The following data consist of the sales figures (in millions of dollars) for a group of 20 firms. Construct a stem-and-leaf display for these data. Use a leaf unit equal to 100. DS Sales20

6835	1973	2820	5358	1233
3517	1449	2384	1376	1725
3291	2707	3291	2675	3707
6047	7903	4616	1541	4189

2.39 **THE e-BILLING CASE AND THE MARKETING RESEARCH CASE**

Figure 2.21 gives JMP stem-and-leaf displays of the payment times in Table 2.4 and of the bottle design ratings in Table 1.6. Describe the shapes of the two displays. DS Paytime DS Design

2.40 **THE TRASH BAG CASE** DS TrashBag

Figure 2.22 gives a Minitab stem-and-leaf display of the sample of 30 breaking strengths in the trash bag case. Use the stem-and-leaf display to describe the distribution of breaking strengths.

2.41 Babe Ruth's record of 60 home runs in a single year was broken by Roger Maris, who hit 61 home runs in 1961. The yearly home run totals for Ruth in his career as a New York Yankee are (arranged in increasing order) 22, 25, 34, 35, 41, 41, 46, 46, 46, 47, 49, 54, 54, 59, and 60. The yearly home run totals for Maris over his career in the American League are

---

FIGURE 2.21 **JMP Stem-and-Leaf Displays of the Payment Times and Bottle Design Ratings (for Exercise 2.39)**

Payment Times

Stem	Leaf	Count
2	9	1
2	64	2
2	44455	5
2	22233	5
2	0001111	7
1	88888899999	11
1	66666666677777777	17
1	44445555555	11
1	22333	5
1	0	1

Note: Whereas Minitab puts the smallest data value at the top of a stem-and-leaf display, JMP puts the largest data value at the top of a stem-and-leaf display.

Ratings

Stem	Leaf	Count
35	0	1
34	00000	5
33	0000000000	10
32	0000000000	11
31	00000000	8
30	000000	6
29	00000	5
28	0000	4
27	000	3
26	00	2
25	00	2
24	0	1
23		
22	0	1
21		
20	0	1

FIGURE 2.22 **A Minitab Stem-and-Leaf Display of the 30 Breaking Strengths (for Exercise 2.40)**

Stem-and-leaf plot for strength

Stem-and-leaf of strength   N = 30
Leaf Unit = 1.0

1	4	1
2	4	2
3	4	4
7	4	6677
10	4	999
15	5	00111
15	5	2222
11	5	4555
7	5	677
4	5	89
2	6	1
1	6	3

TABLE 2.15 **Waiting Times (in Minutes) for the Bank Customer Waiting Time Case (for Exercise 2.42)** DS WaitTime

1.6	6.2	3.2	5.6	7.9	6.1	7.2
6.6	5.4	6.5	4.4	1.1	3.8	7.3
5.6	4.9	2.3	4.5	7.2	10.7	4.1
5.1	5.4	8.7	6.7	2.9	7.5	6.7
3.9	.8	4.7	8.1	9.1	7.0	3.5
4.6	2.5	3.6	4.3	7.7	5.3	6.3
6.5	8.3	2.7	2.2	4.0	4.5	4.3
6.4	6.1	3.7	5.8	1.4	4.5	3.8
8.6	6.3	.4	8.6	7.8	1.8	5.1
4.2	6.8	10.2	2.0	5.2	3.7	5.5
5.8	9.8	2.8	8.0	8.4	4.0	
3.4	2.9	11.6	9.5	6.3	5.7	
9.3	10.9	4.3	1.3	4.4	2.4	
7.4	4.7	3.1	4.8	5.2	9.2	
1.8	3.9	5.8	9.9	7.4	5.0	

(arranged in increasing order) 8, 13, 14, 16, 23, 26, 28, 33, 39, and 61. Compare Ruth's and Maris's home run totals by constructing a back-to-back stem-and-leaf display. What would you conclude about Maris's record-breaking year?

DS HomeRuns

2.42 THE BANK CUSTOMER WAITING TIME CASE DS WaitTime

Table 2.15 reproduces the 100 waiting times for teller service that were originally given in Table 1.9.

a Construct a stem-and-leaf display of the waiting times.

b Describe the distribution of the waiting times.

2.43 THE VIDEO GAME SATISFACTION RATING CASE DS VideoGame

Recall that 65 purchasers have participated in a survey and have rated the XYZ-Box video game system. The composite ratings that have been obtained are as follows:

39	45	38	42	42	41
38	42	46	44	40	39
40	42	45	44	42	46
40	47	44	43	45	45
40	46	41	43	39	43
46	45	45	46	43	47
43	41	40	43	44	41
38	43	36	44	44	45
44	46	48	44	41	45
44	44	44	46	39	41
44	42	47	43	45	

a Construct a stem-and-leaf display for the 65 composite ratings. Hint: Each whole number rating can be written with an "implied tenth place" of zero. For instance, 39 can be written as 39.0. Use the implied zeros as the leaf values and the whole numbers 36, 37, 38, 39, etc., as the stem values.

b Describe the distribution of composite ratings.

c If we consider a purchaser to be "very satisfied" if his or her composite score is at least 42, can we say that almost all purchasers of the XYZ-Box video game system are "very satisfied"?

2.5 Contingency Tables (Optional)

Previous sections in this chapter have presented methods for summarizing data for a single variable. Often, however, we wish to use statistics to study possible relationships between several variables. In this section we present a simple way to study the relationship between two variables. Crosstabulation is a process that classifies data on two dimensions. This process results in a table that is called a **contingency table.** Such a table consists of rows and columns—the rows classify the data according to one dimension and the columns classify the data according to a second dimension. Together, the rows and columns represent all possibilities (or *contingencies*).

LO2-6
Examine the relationships between variables by using contingency tables (Optional).

 EXAMPLE 2.5 The Brokerage Firm Case: Studying Client Satisfaction

An investment broker sells several kinds of investment products—a stock fund, a bond fund, and a tax-deferred annuity. The broker wishes to study whether client satisfaction with its products and services depends on the type of investment product purchased. To do this, 100 of the broker's clients are randomly selected from the population of clients who have purchased shares in exactly one of the funds. The broker records the fund type purchased by each client and has one of its investment counselors personally contact the client. When contacted, the client is asked to rate his or her level of satisfaction with the purchased fund as high, medium, or low. The resulting data are given in Table 2.16.

Looking at the raw data in Table 2.16, it is difficult to see whether the level of client satisfaction varies depending on the fund type. We can look at the data in an organized way by constructing a contingency table. A crosstabulation of fund type versus level of client satisfaction is shown in Table 2.17. The classification categories for the two variables are defined along the left and top margins of the table. The three row labels—bond fund, stock fund, and tax-deferred annuity—define the three fund categories and are given in the left table margin. The three column labels—high, medium, and low—define the three levels of client satisfaction and are given along the top table margin. Each row and column combination, that is, each fund type and level of satisfaction combination, defines what we call a "cell" in the table. Because each of the randomly selected clients has invested in exactly one fund type and has reported exactly one level of satisfaction, each client can be placed in a particular cell in the contingency table. For example, because client number 1 in Table 2.16 has invested in the bond fund and reports a high level of client satisfaction, client number 1 can be placed in the upper left cell of the table (the cell defined by the Bond Fund row and High Satisfaction column).

©Hill Street Studios/Blend Images LLC

TABLE 2.16   **Results of a Customer Satisfaction Survey Given to 100 Randomly Selected Clients Who Invest in One of Three Fund Types—a Bond Fund, a Stock Fund, or a Tax-Deferred Annuity**   Invest

Client	Fund Type	Level of Satisfaction	Client	Fund Type	Level of Satisfaction	Client	Fund Type	Level of Satisfaction
1	BOND	HIGH	35	STOCK	HIGH	69	BOND	MED
2	STOCK	HIGH	36	BOND	MED	70	TAXDEF	MED
3	TAXDEF	MED	37	TAXDEF	MED	71	TAXDEF	MED
4	TAXDEF	MED	38	TAXDEF	LOW	72	BOND	HIGH
5	STOCK	LOW	39	STOCK	HIGH	73	TAXDEF	MED
6	STOCK	HIGH	40	TAXDEF	MED	74	TAXDEF	LOW
7	STOCK	HIGH	41	BOND	HIGH	75	STOCK	HIGH
8	BOND	MED	42	BOND	HIGH	76	BOND	HIGH
9	TAXDEF	LOW	43	BOND	LOW	77	TAXDEF	LOW
10	TAXDEF	LOW	44	TAXDEF	LOW	78	BOND	MED
11	STOCK	MED	45	STOCK	HIGH	79	STOCK	HIGH
12	BOND	LOW	46	BOND	HIGH	80	STOCK	HIGH
13	STOCK	HIGH	47	BOND	MED	81	BOND	MED
14	TAXDEF	MED	48	STOCK	HIGH	82	TAXDEF	MED
15	TAXDEF	MED	49	TAXDEF	MED	83	BOND	HIGH
16	TAXDEF	LOW	50	TAXDEF	MED	84	STOCK	MED
17	STOCK	HIGH	51	STOCK	HIGH	85	STOCK	HIGH
18	BOND	HIGH	52	TAXDEF	MED	86	BOND	MED
19	BOND	MED	53	STOCK	HIGH	87	TAXDEF	MED
20	TAXDEF	MED	54	TAXDEF	MED	88	TAXDEF	LOW
21	TAXDEF	MED	55	STOCK	LOW	89	STOCK	HIGH
22	BOND	HIGH	56	BOND	HIGH	90	TAXDEF	MED
23	TAXDEF	MED	57	STOCK	HIGH	91	BOND	HIGH
24	TAXDEF	LOW	58	BOND	MED	92	TAXDEF	HIGH
25	STOCK	HIGH	59	TAXDEF	LOW	93	TAXDEF	LOW
26	BOND	HIGH	60	TAXDEF	LOW	94	TAXDEF	LOW
27	TAXDEF	LOW	61	STOCK	MED	95	STOCK	HIGH
28	BOND	MED	62	BOND	LOW	96	BOND	HIGH
29	STOCK	HIGH	63	STOCK	HIGH	97	BOND	MED
30	STOCK	HIGH	64	TAXDEF	MED	98	STOCK	HIGH
31	BOND	MED	65	TAXDEF	MED	99	TAXDEF	MED
32	TAXDEF	MED	66	TAXDEF	LOW	100	TAXDEF	MED
33	BOND	HIGH	67	STOCK	HIGH			
34	STOCK	MED	68	BOND	HIGH			

TABLE 2.17   A Contingency Table of Fund Type versus Level of Client Satisfaction

Fund Type	Level of Satisfaction			
	High	Medium	Low	Total
Bond Fund	15	12	3	30
Stock Fund	24	4	2	30
Tax-Deferred Annuity	1	24	15	40
Total	40	40	20	100

We fill in the cells in the table by moving through the 100 randomly selected clients and by tabulating the number of clients who can be placed in each cell. For instance, moving through the 100 clients results in placing 15 clients in the "bond fund—high" cell, 12 clients in the "bond fund—medium" cell, and so forth. The counts in the cells are called the **cell frequencies.** In Table 2.17 these frequencies tell us that 15 clients invested in the bond fund and reported a high level of satisfaction, 4 clients invested in the stock fund and reported a medium level of satisfaction, and so forth.

The far right column in the table (labeled Total) is obtained by summing the cell frequencies across the rows. For instance, these totals tell us that $15 + 12 + 3 = 30$ clients invested in the bond fund, $24 + 4 + 2 = 30$ clients invested in the stock fund, and $1 + 24 + 15 = 40$ clients invested in the tax-deferred annuity. These **row totals** provide a frequency distribution for the different fund types. By dividing the row totals by the total of 100 clients surveyed, we can obtain relative frequencies; and by multiplying each relative frequency by 100, we can obtain percent frequencies. That is, we can obtain the frequency, relative frequency, and percent frequency distributions for fund type as follows:

Fund Type	Frequency	Relative Frequency	Percent Frequency
Bond fund	30	30/100 = .30	.30 (100) = 30%
Stock fund	30	30/100 = .30	.30 (100) = 30%
Tax-deferred annuity	40	40/100 = .40	.40 (100) = 40%
	100		

We see that 30 percent of the clients invested in the bond fund, 30 percent invested in the stock fund, and 40 percent invested in the tax-deferred annuity.

The bottom row in the table (labeled Total) is obtained by summing the cell frequencies down the columns. For instance, these totals tell us that $15 + 24 + 1 = 40$ clients reported a high level of satisfaction, $12 + 4 + 24 = 40$ clients reported a medium level of satisfaction, and $3 + 2 + 15 = 20$ clients reported a low level of satisfaction. These **column totals** provide a frequency distribution for the different satisfaction levels (see below). By dividing the column totals by the total of 100 clients surveyed, we can obtain relative frequencies, and by multiplying each relative frequency by 100, we can obtain percent frequencies. That is, we can obtain the frequency, relative frequency, and percent frequency distributions for level of satisfaction as follows:

Level of Satisfaction	Frequency	Relative Frequency	Percent Frequency
High	40	40/100 = .40	.40 (100) = 40%
Medium	40	40/100 = .40	.40 (100) = 40%
Low	20	20/100 = .20	.20 (100) = 20%
	100		

We see that 40 percent of all clients reported high satisfaction, 40 percent reported medium satisfaction, and 20 percent reported low satisfaction.

We have seen that the totals in the margins of the contingency table give us frequency distributions that provide information about each of the variables *fund type* and *level of client satisfaction*. However, the main purpose of constructing the table is to investigate possible relationships *between* these variables. Looking at Table 2.17, we see that clients who have invested in the stock fund seem to be highly satisfied and that those who have invested

T A B L E 2.18   Row Percentages for Each Fund Type

Fund Type	Level of Satisfaction			
	High	Medium	Low	Total
Bond Fund	50%	40%	10%	100%
Stock Fund	80%	13.33%	6.67%	100%
Tax-Deferred Annuity	2.5%	60%	37.5%	100%

in the bond fund seem to have a high to medium level of satisfaction. However, clients who have invested in the tax-deferred annuity seem to be less satisfied.

One good way to investigate relationships such as these is to compute **row percentages** and **column percentages.** We compute row percentages by dividing each cell's frequency by its corresponding row total and by expressing the resulting fraction as a percentage. For instance, the row percentage for the upper left-hand cell (bond fund and high level of satisfaction) in Table 2.17 is $(15/30) \times 100\% = 50\%$. Similarly, column percentages are computed by dividing each cell's frequency by its corresponding column total and by expressing the resulting fraction as a percentage. For example, the column percentage for the upper left-hand cell in Table 2.17 is $(15/40) \times 100\% = 37.5\%$. Table 2.18 summarizes all of the row percentages for the different fund types in Table 2.17. We see that each row in Table 2.18 gives a percent frequency distribution of level of client satisfaction given a particular fund type.

For example, the first row in Table 2.18 gives a percent frequency distribution of client satisfaction for investors who have purchased shares in the bond fund. We see that 50 percent of bond fund investors report high satisfaction, while 40 percent of these investors report medium satisfaction, and only 10 percent report low satisfaction. The other rows in Table 2.18 provide percent frequency distributions of client satisfaction for stock fund and annuity purchasers.

All three percent frequency distributions of client satisfaction—for the bond fund, the stock fund, and the tax-deferred annuity—are illustrated using bar charts in Figure 2.23. In this figure, the bar heights for each chart are the respective row percentages in Table 2.18. For example, these distributions tell us that 80 percent of stock fund investors report high satisfaction, while 97.5 percent of tax-deferred annuity purchasers report medium or low satisfaction. Looking at the entire table of row percentages (or the bar charts in Figure 2.23), we might conclude that stock fund investors are highly satisfied, that bond fund investors are quite satisfied (but, somewhat less so than stock fund investors), and that tax-deferred annuity purchasers are less satisfied than either stock fund or bond fund investors. In general, row percentages and column percentages help us to quantify relationships such as these.

In the investment example, we have crosstabulated two qualitative variables. We can also crosstabulate a quantitative variable versus a qualitative variable or two quantitative variables against each other. If we are crosstabulating a quantitative variable, we often define categories by using appropriate ranges. For example, if we wished to crosstabulate level of education (grade school, high school, college, graduate school) versus income, we might define income classes $0–$50,000, $50,001–$100,000, $100,001–$150,000, and above $150,000.

F I G U R E 2.23   Bar Charts Illustrating Percent Frequency Distributions of Client Satisfaction as Given by the Row Percentages for the Three Fund Types in Table 2.18

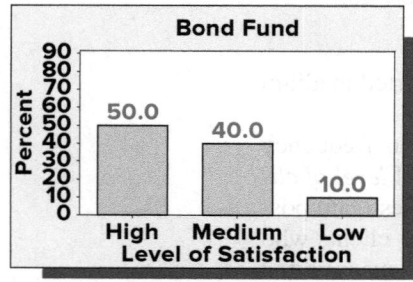
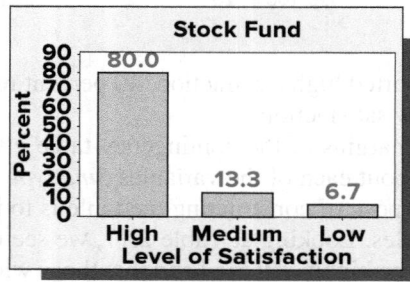
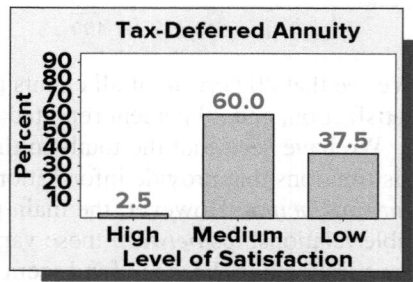

Exercises for Section 2.5

CONCEPTS ▣ connect

2.44 Explain the purpose behind constructing a contingency table.

2.45 A contingency table consists of several "cells." Explain how we fill the cells in the table.

2.46 Explain how to compute (1) the row percentages for a contingency table, and (2) the column percentages. What information is provided by the row percentages in a particular row of the table? What information is provided by the column percentages in a particular column of the table?

METHODS AND APPLICATIONS

Exercises 2.47 through 2.49 are based on the following situation:

The marketing department at the Rola-Cola Bottling Company is investigating the attitudes and preferences of consumers toward Rola-Cola and a competing soft drink, Koka-Cola. Forty randomly selected shoppers are given a "blind taste test" and are asked to give their cola preferences. The results are given in Table 2.19—each shopper's preference, Rola-Cola or Koka-Cola, is revealed to the shopper only after he or she has tasted both brands without knowing which cola is which. In addition, each survey participant is asked to answer three more questions: (1) Have you previously purchased Rola-Cola: Yes or No? (2) What is your sweetness preference for cola drinks: very sweet, sweet, or not so sweet? (3) How many 12-packs of cola drinks does your family consume in a typical month? These responses are also given in Table 2.19.

2.47 Construct a contingency table using cola preference (Rola or Koka) as the row variable and Rola-Cola purchase history (Yes or No) as the

column variable. Based on the table, answer the following: ⒟ ColaSurvey

a How many shoppers who preferred Rola-Cola in the blind taste test had previously purchased Rola-Cola?

b How many shoppers who preferred Koka-Cola in the blind taste test had not previously purchased Rola-Cola?

c What kind of relationship, if any, seems to exist between cola preference and Rola-Cola purchase history?

2.48 Construct a contingency table using cola preference (Rola or Koka) as the row variable and sweetness preference (very sweet, sweet, or not so sweet) as the column variable. Based on the table, answer the following: ⒟ ColaSurvey

a How many shoppers who preferred Rola-Cola in the blind taste test said that they preferred a cola drink to be either very sweet or sweet?

b How many shoppers who preferred Koka-Cola in the blind taste test said that they preferred a cola drink to be not so sweet?

c What kind of relationship, if any, seems to exist between cola preference and sweetness preference?

2.49 Construct a contingency table using cola preference (Rola or Koka) as the row variable and the number of 12-packs consumed in a typical month (categories 0 through 5, 6 through 10, and more than 10) as the column variable. Based on the table, answer the following: ⒟ ColaSurvey

a How many shoppers who preferred Rola-Cola in the blind taste test purchase 10 or fewer 12-packs of cola drinks in a typical month?

TABLE 2.19   **Rola-Cola Bottling Company Survey Results** ⒟ ColaSurvey

Shopper	Cola Preference	Previously Purchased?	Sweetness Preference	Monthly Cola Consumption	Shopper	Cola Preference	Previously Purchased?	Sweetness Preference	Monthly Cola Consumption
1	Koka	No	Very Sweet	4	21	Koka	No	Very Sweet	4
2	Rola	Yes	Sweet	8	22	Rola	Yes	Not So Sweet	9
3	Koka	No	Not So Sweet	2	23	Rola	Yes	Not So Sweet	3
4	Rola	Yes	Sweet	10	24	Koka	No	Not So Sweet	2
5	Rola	No	Very Sweet	7	25	Koka	No	Sweet	5
6	Rola	Yes	Not So Sweet	6	26	Rola	Yes	Very Sweet	7
7	Koka	No	Very Sweet	4	27	Koka	No	Very Sweet	7
8	Rola	No	Very Sweet	3	28	Rola	Yes	Sweet	8
9	Koka	No	Sweet	3	29	Rola	Yes	Not So Sweet	6
10	Rola	No	Very Sweet	5	30	Koka	No	Not So Sweet	3
11	Rola	Yes	Sweet	7	31	Koka	Yes	Sweet	10
12	Rola	Yes	Not So Sweet	13	32	Rola	Yes	Very Sweet	8
13	Rola	Yes	Very Sweet	6	33	Koka	Yes	Sweet	4
14	Koka	No	Very Sweet	2	34	Rola	No	Sweet	5
15	Koka	No	Not So Sweet	7	35	Rola	Yes	Not So Sweet	3
16	Rola	Yes	Sweet	9	36	Koka	No	Very Sweet	11
17	Koka	No	Not So Sweet	1	37	Rola	Yes	Not So Sweet	9
18	Rola	Yes	Very Sweet	5	38	Rola	No	Very Sweet	6
19	Rola	No	Sweet	4	39	Koka	No	Not So Sweet	2
20	Rola	No	Sweet	12	40	Rola	Yes	Sweet	5

b How many shoppers who preferred Koka-Cola in the blind taste test purchase 6 or more 12-packs of cola drinks in a typical month?

c What kind of relationship, if any, seems to exist between cola preference and cola consumption in a typical month?

2.50 A marketing research firm wishes to study the relationship between wine consumption and whether a person likes to watch professional tennis on television. One hundred randomly selected people are asked whether they drink wine and whether they watch tennis. The following results are obtained:
DS WineCons

	Watch Tennis	Do Not Watch Tennis	Total
Drink Wine	16	24	40
Do Not Drink Wine	4	56	60
Total	20	80	100

DS WineCons

a What percentage of those surveyed both watch tennis and drink wine? What percentage of those surveyed do neither?

b Using the survey data, construct a table of row percentages.

c Using the survey data, construct a table of column percentages.

d What kind of relationship, if any, seems to exist between whether or not a person watches tennis and whether or not a person drinks wine?

e Illustrate your conclusion of part *d* by plotting bar charts of appropriate column percentages for people who watch tennis and for people who do not watch tennis.

2.51 In a survey of 1,000 randomly selected U.S. citizens aged 21 years or older, 721 believed that the amount of violent television programming had increased over the past 10 years, 454 believed that the overall quality of television programming had gotten worse over the past 10 years, and 362 believed both.

a Use this information to fill in the contingency table below.

	TV Violence Increased	TV Violence Not Increased	Total
TV Quality Worse			
TV Quality Not Worse			
Total			

b Using the completed contingency table, construct a table of row percentages.

c Using the completed contingency table, construct a table of column percentages.

d What kind of relationship, if any, seems to exist between whether a person believed that TV violence had increased over the past 10 years and whether a person believed that the overall quality of TV programming had gotten worse over the past 10 years?

e Illustrate your answer to part *d* by constructing bar charts of appropriate row percentages.

2.52 In a Gallup Lifestyle Poll concerning American tipping attitudes, the Gallup News service (on January 8, 2007) reported results that allow construction of two contingency tables given below. The first table uses row percentages to investigate a possible relationship between recommended tip percentage and income level, and the second table uses column percentages to investigate a possible relationship between whether or not a customer has ever left a restaurant without tipping because of bad service and the customer's recommended tip percentage. **DS** TipPercent **DS** LeftTip

a Using the first table, construct a percentage bar chart of recommended tip percentage for each of the three income ranges. Interpret the results.
DS TipPercent

b Using the second table, construct a percentage bar chart of the categories "Yes, have left without tipping" and "No, have not left without tipping" for each of the three appropriate tip percentage categories. Interpret the results. **DS** LeftTip

A table of row percentages for Exercise 2.52a

	Appropriate Tip Percent[*]				
Income	Less than 15%	15%	16–19%	20% or more	Total
Less than $30,000	28.41%	42.04%	1.14%	28.41%	100%
$30,000 through $74,999	15.31%	42.86%	6.12%	35.71%	100%
$75,000 or more	8.16%	32.66%	9.18%	50.00%	100%

[*]Among those surveyed having an opinion.

DS TipPercent

A table of column percentages for Exercise 2.52b

	Tip less than 15%	Tip 15% through 19%	Tip 20% or more
Yes, have left without tipping	64%	50%	35%
No, have not left without tipping	36%	50%	65%
Total	100%	100%	100%

 LeftTip

2.6 Scatter Plots (Optional)

We often study relationships between variables by using graphical methods. A simple graph that can be used to study the relationship between two variables is called a **scatter plot.** As an example, suppose that a marketing manager wishes to investigate the relationship between the sales volume (in thousands of units) of a product and the amount spent (in units of $10,000) on advertising the product. To do this, the marketing manager randomly selects 10 sales regions having equal sales potential. The manager assigns a different level of advertising expenditure for January 2016 to each sales region as shown in Table 2.20. At the end of the month, the sales volume for each region is recorded as also shown in Table 2.20.

LO2-7
Examine the relationships between variables by using scatter plots (Optional).

A scatter plot of these data is given in Figure 2.24. To construct this plot, we place the variable advertising expenditure (denoted x) on the horizontal axis and we place the variable sales volume (denoted y) on the vertical axis. For the first sales region, advertising expenditure equals 5 and sales volume equals 89. We plot the point with coordinates $x = 5$ and $y = 89$ on the scatter plot to represent this sales region. Points for the other sales regions are plotted similarly. The scatter plot shows that there is a positive relationship between advertising expenditure and sales volume—that is, higher values of sales volume are associated with higher levels of advertising expenditure.

We have drawn a straight line through the plotted points of the scatter plot to represent the relationship between advertising expenditure and sales volume. We often do this when the relationship between two variables appears to be a **straight line,** or **linear, relationship.** Of course, the relationship between x and y in Figure 2.24 is not perfectly linear—not all of the points in the scatter plot are exactly on the line. Nevertheless, because the relationship between x and y appears to be approximately linear, it seems reasonable to represent the general relationship between these variables using a straight line. In future chapters we will explain ways to quantify such a relationship—that is, describe such a relationship numerically. Moreover, not all linear relationships between two variables x and y are positive linear relationships (that is, have a positive slope). For example, Table 2.21 gives the average hourly outdoor temperature (x) in a city during a week and the city's natural gas consumption (y) during the week for each of the previous eight weeks. The temperature readings are expressed in degrees Fahrenheit and the natural gas consumptions are expressed in millions of cubic feet of natural gas. The scatter plot in Figure 2.25 shows that there is a

TABLE 2.20   Advertising Expenditure and Sales Volume for Ten Sales Regions   ⓓⓢ SalesPlot

Sales Region	1	2	3	4	5	6	7	8	9	10
Advertising Expenditure, x	5	6	7	8	9	10	11	12	13	14
Sales Volume, y	89	87	98	110	103	114	116	110	126	130

FIGURE 2.24   A Scatter Plot of Sales Volume versus Advertising Expenditure

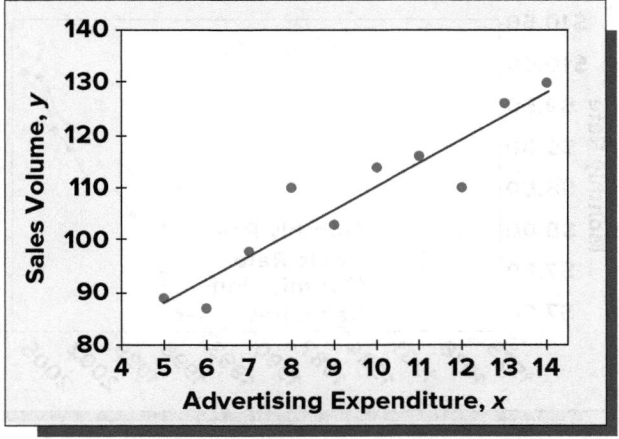

FIGURE 2.25   A Scatter Plot of Natural Gas Consumption versus Temperature

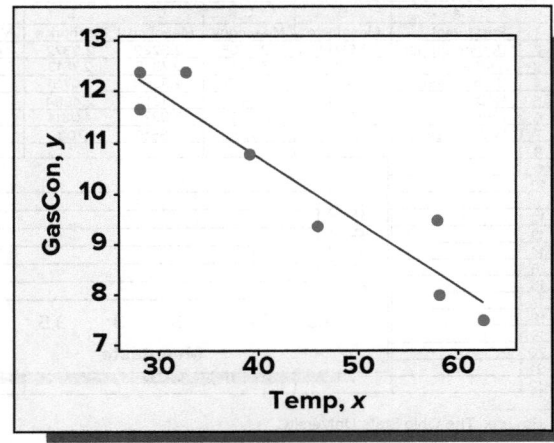

FIGURE 2.26

Little or No Relationship between x and y

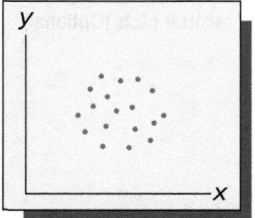

TABLE 2.21   The Natural Gas Consumption Data   DS GasCon1

Week	1	2	3	4	5	6	7	8
Temperature, x	28.0	28.0	32.5	39.0	45.9	57.8	58.1	62.5
Natural Gas Consumption, y	12.4	11.7	12.4	10.8	9.4	9.5	8.0	7.5

negative linear relationship between x and y—that is, as average hourly temperature in the city increases, the city's natural gas consumption decreases in a linear fashion. Finally, not all relationships are linear. In Chapter 15 we will consider how to represent and quantify curved relationships, and, as illustrated in Figure 2.26, there are situations in which two variables x and y do not appear to have any relationship.

To conclude this section, recall from Chapter 1 that a **time series plot** (also called a **runs plot**) is a plot of individual process measurements versus time. This implies that a time series plot is a scatter plot, where values of a process variable are plotted on the vertical axis versus corresponding values of time on the horizontal axis.

Exercises for Section 2.6

CONCEPTS   **connect**

Copiers, x	Minutes, y
4	109
2	58
5	138
7	189
1	37
3	82
4	103
5	134
2	68
4	112
6	154

DS SrvcTime

2.53 Explain the purpose for constructing a scatter plot of y versus x.

2.54 Discuss the relationship between a scatter plot and a time series plot.

METHODS AND APPLICATIONS

2.55   THE SERVICE TIME CASE
DS SrvcTime

Accu-Copiers, Inc., sells and services the Accu-500 copying machine. To obtain information about the time it takes to perform routine service, Accu-Copiers has collected the data in the page margin for 11 service calls. Here, x denotes the

number of copiers serviced and y denotes the number of minutes required to perform service on the service call. Construct a scatter plot of y versus x and interpret what the plot says.

2.56   THE FAST-FOOD RESTAURANT RATING CASE
DS FastFood

Figure 2.27 presents the ratings given by 406 randomly selected individuals of six fast-food restaurants on the basis of taste, convenience, familiarity, and price. The data were collected by researchers at The Ohio State University in the early 1990s. Here, 1 is the best rating and 6 the worst. In addition, each individual ranked the restaurants from 1 through 6 on the basis of overall preference. Interpret the Excel scatter plot, and construct and interpret other relevant scatter plots.

2.57 Figure 2.28 gives a time series plot of the average U.S. monthly pay cable TV rate (for premium services) for each year from 1975 to 2005. Figure 2.28 also gives a

FIGURE 2.27   Fast-Food Restaurant Data and a Scatter Plot   DS FastFood

	A	B	C	D	E	F
1	Restaurant	Meantaste	Meanconv	Meanfam	Meanprice	Meanpref
2	Borden Burger	3.5659	2.7005	2.5282	2.9372	4.2552
3	Hardee's	3.329	3.3483	2.7345	2.7513	4.0911
4	Burger King	2.4231	2.7377	2.3368	3.0761	3.0052
5	McDonald's	2.0895	1.938	1.4619	2.4884	2.2429
6	Wendy's	1.9661	2.892	2.3376	4.0814	2.5351
7	White Castle	3.8061	3.7242	2.6515	1.708	4.7812

Scatter plot: Meanpref (y-axis, 0 to 6) versus Meantaste (x-axis, 1.5 to 4).

Source: The Ohio State University.

FIGURE 2.28   Time Series Plots for Exercise 2.57   DS PayTVRates

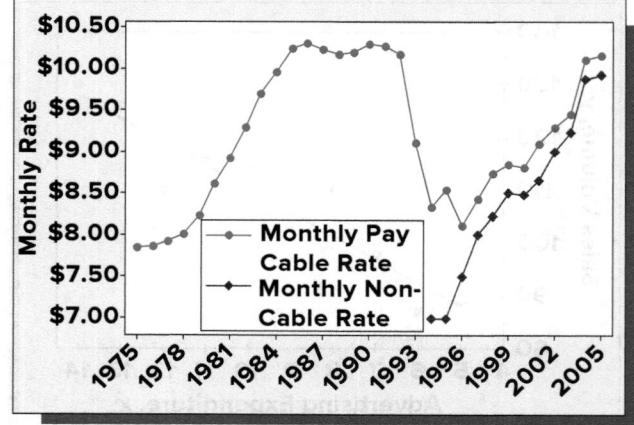

time series plot of the average monthly non-cable (mostly satellite) TV rate (for premium services) for each year from 1994 to 2005.[6] Satellite TV became a serious competitor to cable TV in the early 1990s. Does it appear

that the emergence of satellite TV had an influence on cable TV rates? What happened after satellite TV became more established in the marketplace?
DS PayTVRates

2.7 Misleading Graphs and Charts (Optional)

LO2-8
Recognize misleading graphs and charts (Optional).

The statistical analyst's goal should be to present the most accurate and truthful portrayal of a data set that is possible. Such a presentation allows managers using the analysis to make informed decisions. However, it is possible to construct statistical summaries that are misleading. Although we do not advocate using misleading statistics, you should be aware of some of the ways statistical graphs and charts can be manipulated in order to distort the truth. By knowing what to look for, you can avoid being misled by a (we hope) small number of unscrupulous practitioners.

As an example, suppose that the nurses at a large hospital will soon vote on a proposal to join a union. Both the union organizers and the hospital administration plan to distribute recent salary statistics to the entire nursing staff. Suppose that the mean nurses' salary at the hospital and the mean nurses' salary increase at the hospital (expressed as a percentage) for each of the last four years are as follows:

Year	Mean Salary	Mean Salary Increase (Percent)
1	$60,000	3.0%
2	61,600	4.0
3	63,500	4.5
4	66,100	6.0

©GLG3/Digital Vision/Getty Images

The hospital administration does not want the nurses to unionize and, therefore, hopes to convince the nurses that substantial progress has been made to increase salaries without a union. On the other hand, the union organizers wish to portray the salary increases as minimal so that the nurses will feel the need to unionize.

Figure 2.29 gives two bar charts of the mean nurses' salaries at the hospital for each of the last four years. Notice that in Figure 2.29(a) the administration has started the vertical scale of the bar chart at a salary of $58,000 by using a *scale break* (⌇). Alternatively, the chart

FIGURE 2.29   Two Bar Charts of the Mean Nurses' Salaries at a Large Hospital for the Last Four Years

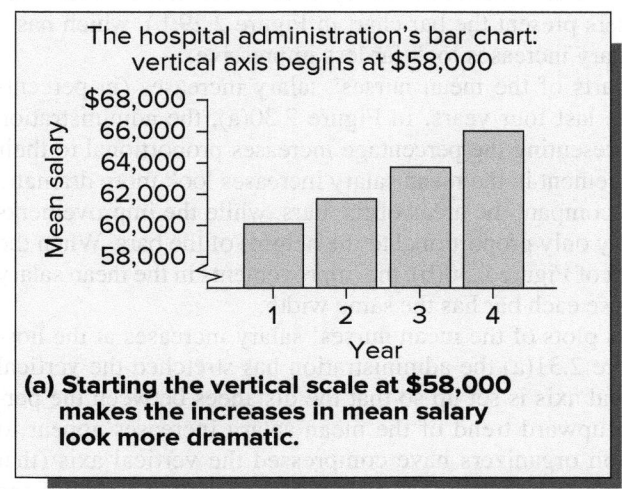

(a) Starting the vertical scale at $58,000 makes the increases in mean salary look more dramatic.

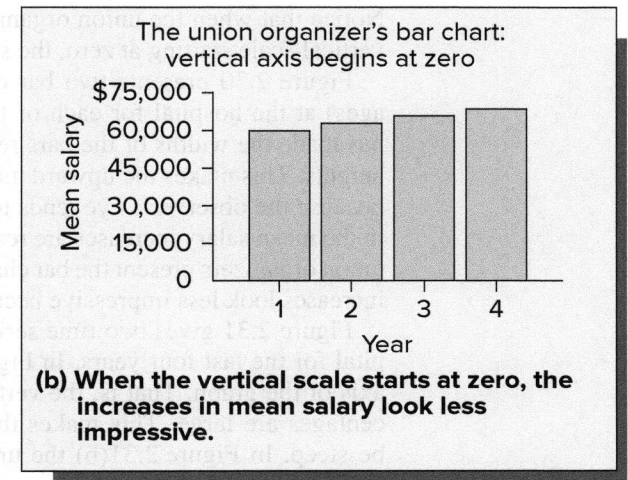

(b) When the vertical scale starts at zero, the increases in mean salary look less impressive.

[6]The time series data for this exercise are on the website for this book.

FIGURE 2.30   **Two Bar Charts of the Mean Nurses' Salary Increases at a Large Hospital for the Last Four Years**

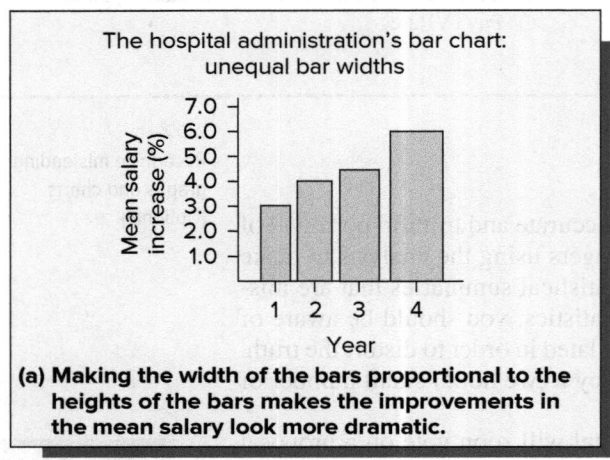

(a) Making the width of the bars proportional to the heights of the bars makes the improvements in the mean salary look more dramatic.

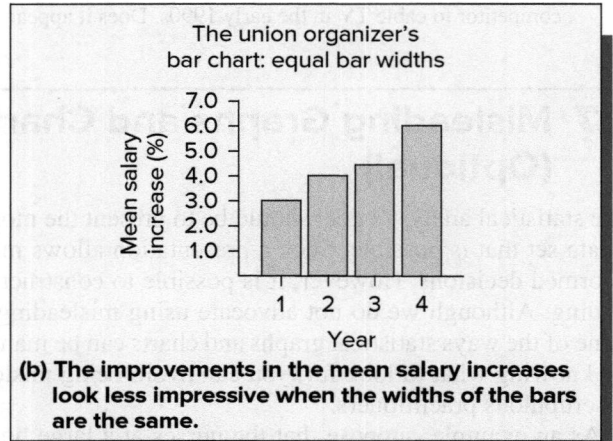

(b) The improvements in the mean salary increases look less impressive when the widths of the bars are the same.

FIGURE 2.31   **Two Time Series Plots of the Mean Nurses' Salary Increases at a Large Hospital for the Last Four Years**

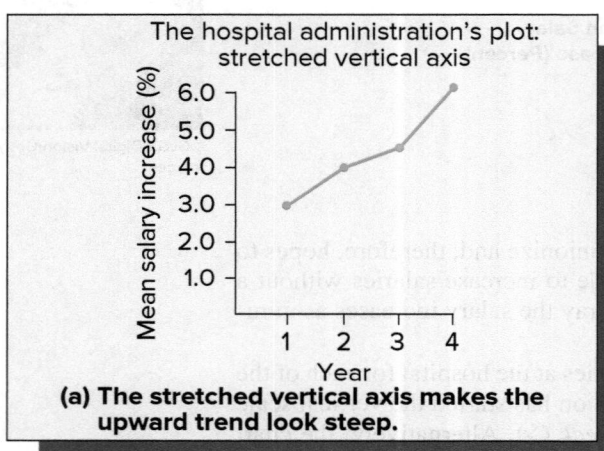

(a) The stretched vertical axis makes the upward trend look steep.

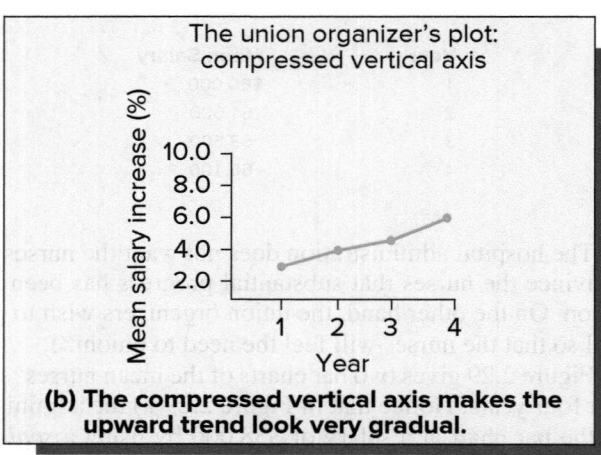

(b) The compressed vertical axis makes the upward trend look very gradual.

could be set up without the scale break by simply starting the vertical scale at $58,000. Starting the vertical scale at a value far above zero makes the salary increases look more dramatic. Notice that when the union organizers present the bar chart in Figure 2.29(b), which has a vertical scale starting at zero, the salary increases look far less impressive.

Figure 2.30 presents two bar charts of the mean nurses' salary increases (in percentages) at the hospital for each of the last four years. In Figure 2.30(a), the administration has made the widths of the bars representing the percentage increases proportional to their heights. This makes the upward movement in the mean salary increases look more dramatic because the observer's eye tends to compare the areas of the bars, while the improvements in the mean salary increases are really only proportional to the heights of the bars. When the union organizers present the bar chart of Figure 2.30(b), the improvements in the mean salary increases look less impressive because each bar has the same width.

Figure 2.31 gives two time series plots of the mean nurses' salary increases at the hospital for the last four years. In Figure 2.31(a) the administration has stretched the vertical axis of the graph. That is, the vertical axis is set up so that the distances between the percentages are large. This makes the upward trend of the mean salary increases appear to be steep. In Figure 2.31(b) the union organizers have compressed the vertical axis (that is, the distances between the percentages are small). This makes the upward trend of the mean salary increases appear to be gradual. As we will see in the exercises, stretching and compressing the *horizontal* axis in a time series plot can also greatly affect the impression given by the plot.

It is also possible to create totally different interpretations of the same statistical summary by simply using different labeling or captions. For example, consider the bar chart of mean nurses' salary increases in Figure 2.30(b). To create a favorable interpretation, the hospital administration might use the caption "Salary Increase Is Higher for the Fourth Year in a Row." On the other hand, the union organizers might create a negative impression by using the caption "Salary Increase Fails to Reach 10% for Fourth Straight Year."

In summary, it is important to carefully study any statistical summary so that you will not be misled. Look for manipulations such as stretched or compressed axes on graphs, axes that do not begin at zero, bar charts with bars of varying widths, and biased captions. Doing these things will help you to see the truth and to make well-informed decisions.

Exercises for Section 2.7

CONCEPTS ![McGraw Hill Education] **connect**

2.58 When we construct a bar chart or graph, what is the effect of starting the vertical axis at a value that is far above zero? Explain.

2.59 Find an example of a misleading use of statistics in a newspaper, magazine, corporate annual report, or other source. Then explain why your example is misleading.

METHODS AND APPLICATIONS

2.60 Figure 2.32 gives two more time series plots of the previously discussed mean nurses' salary increases. In Figure 2.32(a) the hospital administration has compressed the horizontal axis. In Figure 2.32(b) the union organizers have stretched the horizontal axis. Discuss the different impressions given by the two time series plots.

2.61 In the article "How to Display Data Badly" in the May 1984 issue of *The American Statistician*, Howard Wainer presents a *stacked bar chart* of the number of public and private elementary schools (1929–1970). This bar chart is given in Figure 2.33. Wainer also gives a line graph of the number of private elementary schools (1930–1970). This graph is shown in Figure 2.34.

   a Looking at the bar chart of Figure 2.33, does there appear to be an increasing trend in the number of private elementary schools from 1930 to 1970?

   b Looking at the line graph of Figure 2.34, does there appear to be an increasing trend in the number of private elementary schools from 1930 to 1970?

   c Which portrayal of the data do you think is more appropriate? Explain why.

   d Is either portrayal of the data entirely appropriate? Explain.

FIGURE 2.32   **Two Time Series Plots of the Mean Nurses' Salary Increases at a Large Hospital for the Last Four Years (for Exercise 2.60)**

(a) The administration's plot: compressed horizontal axis

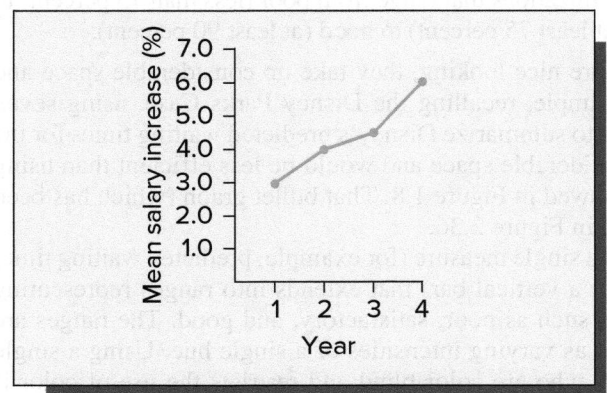

(b) The union organizer's plot: stretched horizontal axis

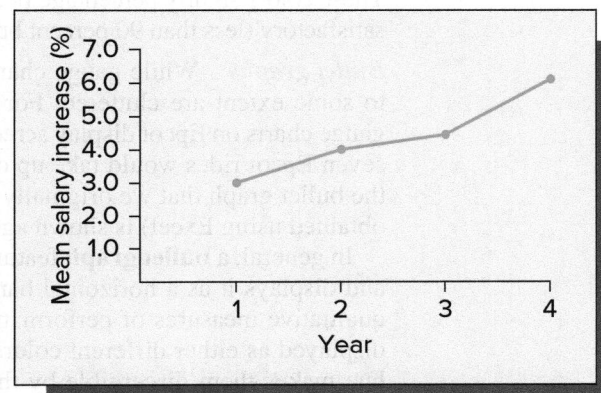

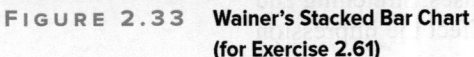

FIGURE 2.33   Wainer's Stacked Bar Chart (for Exercise 2.61)

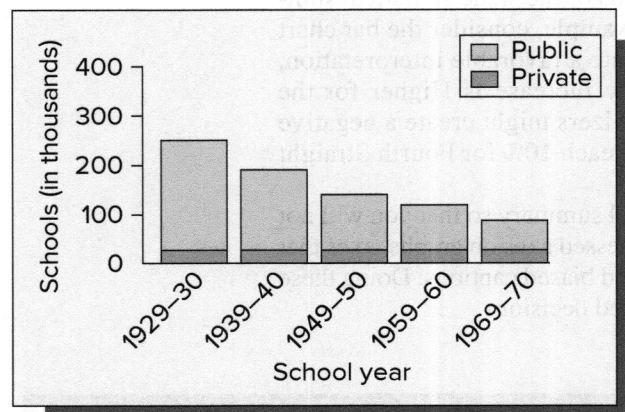

FIGURE 2.34   Wainer's Line Graph (for Exercise 2.61)

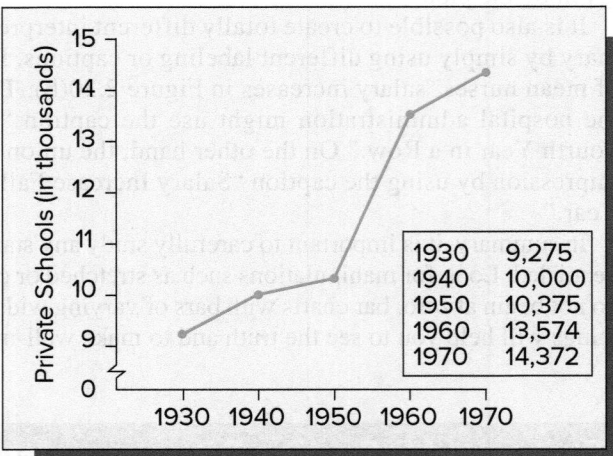

1930	9,275
1940	10,000
1950	10,375
1960	13,574
1970	14,372

Source: Data from H. Wainer. "How to Display Data Badly," *The American Statistician*. May 1984, pp. 137–147. American Statistical Association, 1984.

2.8 Graphical Descriptive Analytics (Optional)

LO2-9

Construct and interpret gauges, bullet graphs, treemaps, and sparklines (Optional).

In Section 1.5 we said that **graphical descriptive analytics** uses traditional and more recently developed graphics to present to executives (and sometimes customers) easy-to-understand visual summaries of up-to-the-minute information concerning the operational status of a business. In this section we will discuss some of the more recently developed graphics used by graphical descriptive analytics, which include *gauges, bullet graphs, treemaps,* and *sparklines*. In addition, we will see how they are used with each other and more traditional graphics to form analytic *dashboards*, which are part of *executive information systems*. We will also briefly discuss *data discovery*, which involves, in it simplest form, *data drill down*.

Dashboards and gauges   An analytic **dashboard** provides a graphical presentation of the current status and historical trends of a business's key performance indicators. The term *dashboard* originates from the automotive dashboard, which helps a driver monitor a car's key functions. Figure 2.35 shows a dashboard that graphically portrays some key performance indicators for a (fictitious) airline for last year. In the lower left-hand portion of the dashboard we see three **gauges** depicting the percentage utilizations of the regional, short-haul, and international fleets of the airline. In general, a gauge (chart) allows us to visualize data in a way that is similar to a real-life speedometer needle on an automobile. The outer scale of the gauge is often color coded to provide additional performance information. For example, note that the colors on the outer scale of the gauges in Figure 2.35 range from red to dark green to light green. These colors signify percentage fleet utilizations that range from poor (less than 75 percent) to satisfactory (less than 90 percent but at least 75 percent) to good (at least 90 percent).

Bullet graphs   While gauge charts are nice looking, they take up considerable space and to some extent are cluttered. For example, recalling the Disney Parks Case, using seven gauge charts on Epcot display screens to summarize Disney's predicted waiting times for the seven Epcot rides would take up considerable space and would be less efficient than using the bullet graph that we originally showed in Figure 1.8. That bullet graph (which has been obtained using Excel) is shown again in Figure 2.36.

In general, a **bullet graph** features a single measure (for example, predicted waiting time) and displays it as a horizontal bar (or a vertical bar) that extends into ranges representing qualitative measures of performance, such as poor, satisfactory, and good. The ranges are displayed as either different colors or as varying intensities of a single hue. Using a single hue makes them discernible by those who are color blind and restricts the use of color if the bullet graph is part of a dashboard that we don't want to look too "busy." Many bullet

FIGURE 2.35   **A Dashboard of the Key Performance Indicators for an Airline**

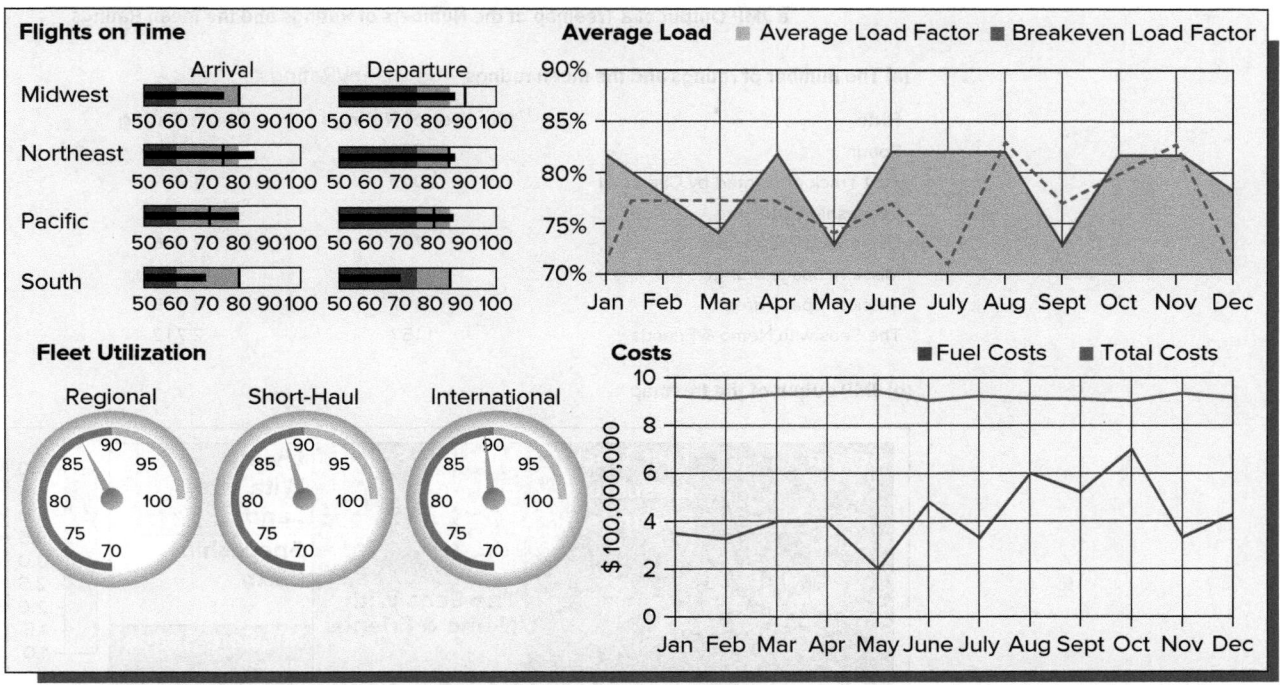

FIGURE 2.36   **Excel Output of a Bullet Graph of Disney's Predicted Waiting Times (in minutes) for the Seven Epcot Rides Posted at 3 P.M. on February 21, 2015**   Ⓓ DisneyTimes

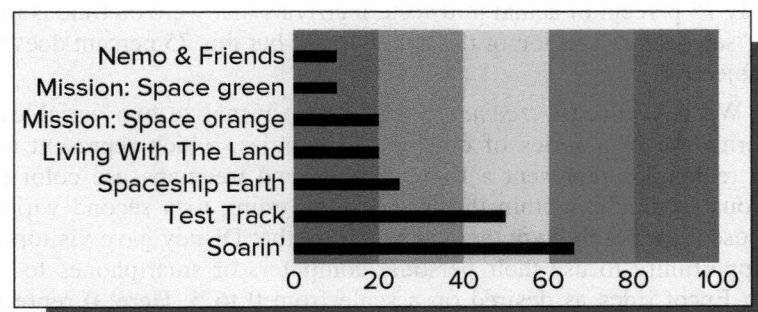

graphs compare the single primary measure to a target, or objective, which is represented by a symbol on the bullet graph. The bullet graph of Disney's predicted waiting times uses five colors ranging from dark green to red and signifying short (0 to 20 minutes) to very long (80 to 100 minutes) predicted waiting times. This bullet graph does not compare the predicted waiting times to an objective. However, the bullet graphs located in the upper left of the dashboard in Figure 2.35 (representing the percentages of on-time arrivals and departures for the airline) do display objectives represented by short vertical black lines. For example, consider the bullet graphs representing the percentages of on-time arrivals and departures in the Midwest, which are shown below.

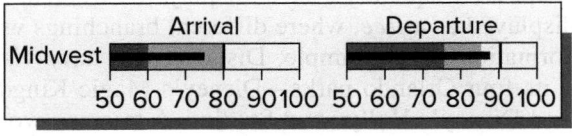

FIGURE 2.37   The Number of Ratings and the Mean Rating for Each of Seven Rides at Epcot
(0 = Poor, 1 = Fair, 2 = Good, 3 = Very Good, 4 = Excellent, 5 = Superb) and
a JMP Output of a Treemap of the Numbers of Ratings and the Mean Ratings

(a) The number of ratings and the mean ratings   DisneyRatings

Ride	Number of Ratings	Mean Rating
Soarin'	2572	4.815
Test Track presented by Chevrolet	2045	4.247
Spaceship Earth	697	1.319
Living With The Land	725	2.186
Mission: Space orange	1589	3.408
Mission: Space green	467	3.116
The Seas with Nemo & Friends	1157	2.712

(b) JMP output of the treemap

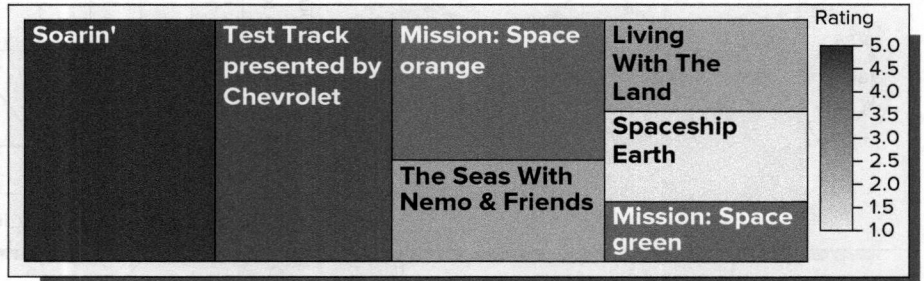

The airline's objective was to have 80 percent of midwestern arrivals be on time. The approximately 75 percent of actual midwestern arrivals that were on time is in the airline's light brown "satisfactory" region of the bullet graph, but this 75 percent does not reach the 80 percent objective.

Treemaps   We next discuss *treemaps*, which help visualize two variables. **Treemaps** display information in a series of clustered rectangles, which represent a whole. The sizes of the rectangles represent a first variable, and treemaps use color to characterize the various rectangles within the treemap according to a second variable. For example, suppose (as a purely hypothetical example) that Disney gave visitors at Epcot the voluntary opportunity to use their personal computers or smartphones to rate as many of the seven Epcot rides as desired on a scale from 0 to 5. Here, 0 represents "poor," 1 represents "fair," 2 represents "good," 3 represents "very good," 4 represents "excellent," and 5 represents "superb." Figure 2.37(a) gives the number of ratings and the mean rating for each ride on a particular day. (These data are completely fictitious.) Figure 2.37(b) shows the JMP output of a treemap, where the size and color of the rectangle for a particular ride represent, respectively, the total number of ratings and the mean rating for the ride. The colors range from dark blue (signifying a mean rating near the "superb," or 5, level) to white (signifying a mean rating near the "fair," or 1, level), as shown by the color scale on the treemap. Note that six of the seven rides are rated to be at least "good," four of the seven rides are rated to be at least "very good," and one ride is rated as "fair." Many treemaps use a larger range of colors (ranging, say, from dark green to red), but we had JMP use the range of colors shown in Figure 2.37(b) to construct the Epcot ride treemap. Also, note that treemaps are frequently used to display *hierarchical* information (information that could be displayed as a *tree*, where different branchings would be used to show the hierarchical information). For example, Disney could have visitors voluntarily rate the rides in each of its four Orlando parks—Disney's Magic Kingdom, Epcot, Disney's Animal Kingdom, and Disney's Hollywood Studios. A treemap would be constructed by

	A	B	C	D	E	F	G	H
1	Monthly Closing Price							
2			Jan.	Feb.	Mar.	Apr.	May	June
3	Stock 1		103.91	98.74	87.42	85.97	75.62	71.10
4	Stock 2		213.14	218.84	201.76	197.41	191.10	181.49
5	Stock 3		59.34	65.28	61.14	68.97	73.42	75.81
6	Stock 4		90.72	95.51	98.41	99.95	98.78	106.20
7	Stock 5		325.26	311.10	314.75	286.15	276.24	259.88

breaking a large rectangle into four smaller rectangles representing the four parks and by then breaking each smaller rectangle representing a park into even smaller rectangles representing the rides in the park. Similarly, the U.S. budget could be displayed as a treemap, where a large rectangle is broken into smaller rectangles representing the major segments of the budget—national defense, Social Security, Medicare, and so forth. Then each smaller rectangle representing a major segment of the budget is broken into even smaller rectangles representing different parts of the major budget segment.

Sparklines   Figure 2.38 shows *sparklines* depicting the monthly closing prices of five stocks over six months. In general, a **sparkline,** another of the new descriptive analytics graphics, is a line chart that presents the general shape of the variation (usually over time) in some measurement, such as temperature or a stock price. A sparkline is typically drawn without axes or coordinates and made small enough to be embedded in text. Sparklines are often grouped together so that comparative variations can be seen. Whereas most charts are designed to show considerable information and are set off from the main text, sparklines are intended to be succinct and located where they are discussed. Therefore, if the sparklines in Figure 2.38 were located in the text of a report comparing stocks, they would probably be made smaller than shown in Figure 2.38, and the actual monthly closing prices of the stocks used to obtain the sparklines would probably not be shown next to the sparklines.

Data discovery   To conclude this section, we briefly discuss **data discovery** methods, which allow decision makers to interactively view data and make preliminary analyses. One simple version of data discovery is **data drill down,** which reveals more detailed data that underlie a higher-level summary. For example, an online presentation of the airline dashboard in Figure 2.35 might allow airline executives to drill down by clicking the gauge which shows that approximately 89 percent of the airline's international fleet was utilized over the year to see the bar chart in Figure 2.39. This bar chart depicts the *quarterly* percentage utilizations of the airline's international fleet over the year. The bar chart suggests that the airline might offer international travel specials in quarter 1 (Winter) and quarter 4 (Fall) to increase international travel and thus increase the percentage of the international fleet utilized in those quarters. Drill down can be done at multiple levels. For example, Disney executives might wish to have weekly reports of total merchandise sales at its four Orlando parks, which can be drilled down to reveal total merchandise sales at each park, which can be further drilled down to reveal total merchandise sales at each "land" in each park, which can be yet further drilled down to reveal total merchandise sales in each store in each land in each park.

FIGURE 2.39　Drill Down: The Quarterly Percentage Utilizations of the International Fleet of the Airline

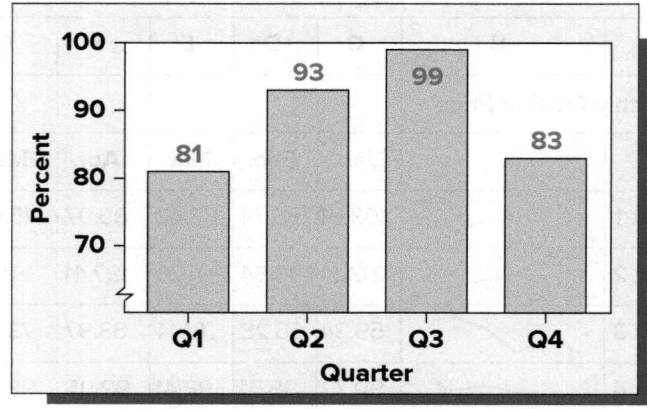

Exercises for Section 2.8

CONCEPTS

2.62 Discuss how each of the following is used: gauge, bullet graph, treemap, sparkline.

2.63 What is data drill down?

METHODS AND APPLICATIONS

2.64 Interpret the bullet graphs in Figure 2.40.

2.65 In the treemap in Figure 2.41, dark blue represents the lowest ozone level and bright red represents the highest ozone level. Where is the ozone level higher, Chicago or New York City?

FIGURE 2.40　Bullet Graphs of Year-to-Date Key Performance Indicators for a Company

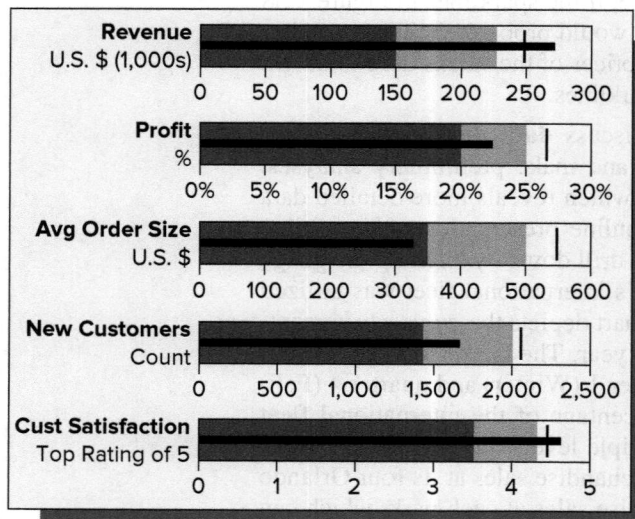

Source: Adapted from Few, Stephen (2006). *Information Dashboard Design: The Effective Visual Communication of Data.* https://en.wikipedia.org/wiki/Bullet_graph (accessed February 18, 2015).

2.66 At 5 P.M. on February 21, 2015, the waiting times posted at Epcot for Nemo & Friends, Mission: Space green, Mission: Space orange, Living With The Land, Spaceship Earth, Test Track, and Soarin' were, respectively, 10, 15, 20, 15, 20, 60, and 85 minutes. Use Excel (see the Excel appendix for this chapter) to construct a gauge for each waiting time and to construct a bullet graph similar to Figure 2.36 for the waiting times. ⒹⓈ DisneyTimes2

2.67 Suppose on the day that the numbers of ratings and mean ratings for the seven Epcot rides in Figure 2.37(a) were obtained, data drill down shows that the numbers of ratings and the mean ratings given by children ages 10 to 17 are as follows. (Note that these data are completely fictitious.) ⒹⓈ DisneyRatings2.

Ride	Number of Ratings	Mean Rating
Living With The Land	135	1.563
Mission: Space green	197	3.475
Mission: Space orange	871	4.394
Soarin'	1194	4.724
Spaceship Earth	246	1.217
Test Track presented by Chevrolet	911	4.412
The Seas with Nemo & Friends	541	3.122

Use Excel (see the Excel appendix) to construct a treemap of this information.

2.68 Suppose that the closing prices at the end of July for the five stocks in Figure 2.38 were, respectively, 65.20, 174.88, 78.25, 110.44, and 248.67. Use Excel (see the Excel appendix) to construct sparklines of the closing prices of the five stocks from January to July. ⒹⓈ StockSpark2

FIGURE 2.41   **Treemap of the Population Sizes and Ozone Levels of 52 U.S. Cities (for Exercise 2.65)**

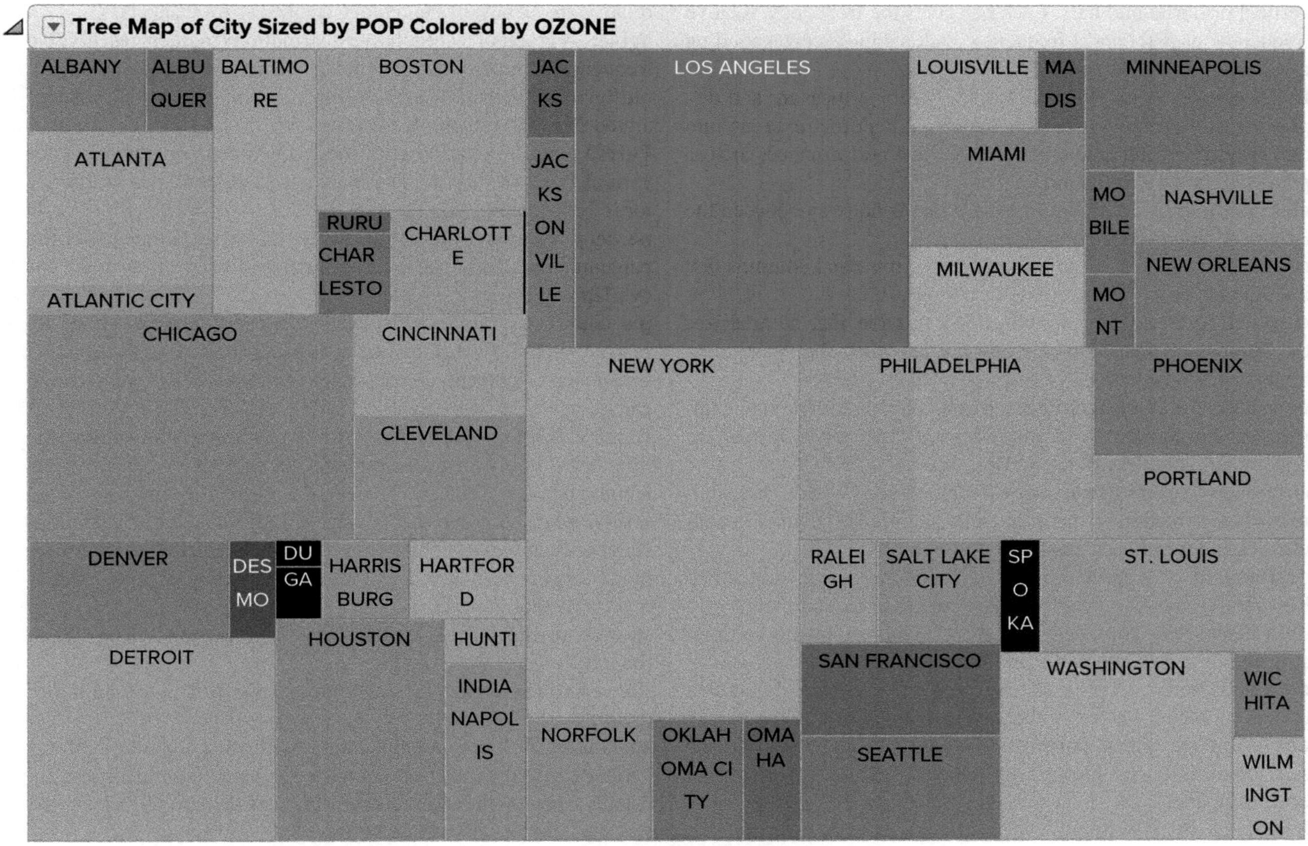

Source: http://www.jmp.com/support/help/Example_of_Treemaps.shtml **(accessed February 17, 2015).**

Chapter Summary

We began this chapter by explaining how to summarize qualitative data. We learned that we often summarize this type of data in a table that is called a **frequency distribution.** Such a table gives the **frequency, relative frequency,** or **percent frequency** of items that are contained in each of several nonoverlapping classes or categories. We also learned that we can summarize qualitative data in graphical form by using **bar charts** and **pie charts** and that qualitative quality data are often summarized using a special bar chart called a **Pareto chart.** We continued in Section 2.2 by discussing how to graphically portray quantitative data. In particular, we explained how to summarize such data by using frequency distributions and histograms. We saw that a **histogram** can be constructed using frequencies, relative frequencies, or percentages, and that we often construct histograms using statistical software such as Minitab or the analysis toolpak in Excel. We used histograms to describe the shape of a distribution and we saw that distributions are sometimes **mound shaped and symmetrical,** but that a distribution can also be **skewed** (**to the right** or **to the left**). We also learned that a frequency distribution can be graphed by using a **frequency polygon** and that a graph of a **cumulative frequency distribution** is called an **ogive.** In Sections 2.3 and 2.4 we showed how to summarize relatively small data sets by using **dot plots**

and **stem-and-leaf displays.** These graphics allow us to see all of the measurements in a data set and to (simultaneously) see the shape of the data set's distribution. Next, we learned about how to describe the relationship between two variables. First, in optional Section 2.5 we explained how to construct and interpret a **contingency table,** which classifies data on two dimensions using a table that consists of rows and columns. Then, in optional Section 2.6 we showed how to construct a **scatter plot.** Here, we plot numerical values of one variable on a horizontal axis versus numerical values of another variable on a vertical axis. We saw that we often use such a plot to look at possible straight-line relationships between the variables. In optional Section 2.7 we learned about misleading graphs and charts. In particular, we pointed out several graphical tricks to watch for. By careful analysis of a graph or chart, one can avoid being misled. Finally, in optional Section 2.8 we discussed some of the more recently developed graphics used by graphical descriptive analytics. These include **gauges, bullet graphs, treemaps**, and **sparklines.** Further, we saw how they are used with each other and more traditional graphics to form analytic **dashboards**, which are part of *executive information systems.* We also briefly discussed **data discovery,** which involves in its simplest form **data drill down.**

Glossary of Terms

bar chart: A graphical display of data in categories made up of vertical or horizontal bars. Each bar gives the frequency, relative frequency, or percentage frequency of items in its corresponding category.

bullet graph: A graphic that features a single measure and displays it as a horizontal bar (or a vertical bar) that extends into ranges representing qualitative measures of performance, such as poor, satisfactory, and good.

class midpoint: The point in a class that is halfway between the lower and upper class boundaries.

contingency table: A table consisting of rows and columns that is used to classify data on two dimensions.

cumulative frequency distribution: A table that summarizes the number of measurements that are less than the upper class boundary of each class.

cumulative percent frequency distribution: A table that summarizes the percentage of measurements that are less than the upper class boundary of each class.

cumulative relative frequency distribution: A table that summarizes the fraction of measurements that are less than the upper class boundary of each class.

dashboard: A graphical presentation of the current status and historical trends of a business's key performance indicators.

data discovery: A process that allows decision makers to interactively view data and make preliminary analyses.

data drill down: A process that reveals more detailed data that underlie a higher-level summary.

dot plot: A graphical portrayal of a data set that shows the data set's distribution by plotting individual measurements above a horizontal axis.

frequency distribution: A table that summarizes the number of items (or measurements) in each of several nonoverlapping classes.

frequency polygon: A graphical display in which we plot points representing each class frequency (or relative frequency or percent frequency) above their corresponding class midpoints and connect the points with line segments.

gauge: A graphic that allows us to visualize data in a way that is similar to a real-life speedometer needle on an automobile.

histogram: A graphical display of a frequency distribution, relative frequency distribution, or percentage frequency distribution.

It divides measurements into classes and graphs the frequency, relative frequency, or percentage frequency for each class.

ogive: A graph of a cumulative distribution (frequencies, relative frequencies, or percent frequencies may be used).

outlier: An unusually large or small observation that is well separated from the remaining observations.

Pareto chart: A bar chart of the frequencies or percentages for various types of defects. These are used to identify opportunities for improvement.

percent frequency distribution: A table that summarizes the percentage of items (or measurements) in each of several non-overlapping classes.

pie chart: A graphical display of data in categories made up of "pie slices." Each pie slice represents the frequency, relative frequency, or percentage frequency of items in its corresponding category.

relative frequency distribution: A table that summarizes the fraction of items (or measurements) in each of several nonoverlapping classes.

scatter plot: A graph that is used to study the possible relationship between two variables y and x. The observed values of y are plotted on the vertical axis versus corresponding observed values of x on the horizontal axis.

skewed to the left: A distribution shape having a long tail to the left.

skewed to the right: A distribution shape having a long tail to the right.

sparkline: A line chart that presents the general shape of the variation (usually over time) in some measurement. Sparklines are usually drawn without axes or coordinates and are intended to be succinct and located where they are discussed (in text).

stem-and-leaf display: A graphical portrayal of a data set that shows the data set's distribution by using stems consisting of leading digits and leaves consisting of trailing digits.

symmetrical distribution: A distribution shape having right and left sides that are "mirror images" of each other.

treemap: A graphic that displays information in a series of clustered rectangles, which represent a whole. The sizes of the rectangles represent a first variable, and treemaps use color to characterize the various rectangles within the treemap according to a second variable.

Important Formulas and Graphics

Frequency distribution: pages 53, 62, 63

Relative frequency: pages 54, 62

Percent frequency: pages 54, 62

Bar chart: pages 54, 55

Pie chart: page 55

Pareto chart: page 56

Histogram: pages 60, 62

Frequency polygon: page 66

Cumulative distribution: page 67

Ogive: page 68

Dot plot: page 73

Stem-and-leaf display: page 74

Contingency table: page 79

Scatter plot: page 85

Time series plot: page 86

Gauge: page 90

Bullet graph: page 90

Treemap: page 92

Sparkline: page 93

Supplementary Exercises

connect | **2.69** At the end of 2011 Chrysler Motors was trying to decide whether or not to discontinue production of a Jeep model introduced in 2002—the Jeep Liberty. Although Liberty sales had generally declined since 2007 with the cancellation of one Jeep model and the introduction of three new Jeep models, Liberty sales had continued to be good (and the Liberty name had retained a positive image) at some Jeep dealerships. Suppose that the owner of one such dealership in Cincinnati, Ohio, wished to present Chrysler with evidence that Liberty sales were still an important part of total sales at his dealership. To do this, the owner decided to compare his dealership's Liberty sales in 2011 with those in 2006. Recently summarized sales data show that in 2011 the owner's dealership sold 30 Jeep Compasses, 50 Jeep Grand Cherokees, 45 Jeep Libertys, 40 Jeep Patriots, 28 Jeep Wranglers, and 35 Jeep Wrangler Unlimiteds. Such summarized sales data were not available for 2006, but raw sales data were found. Denoting the four Jeep models sold in 2006 (Commander, Grand Cherokee, Liberty, and Wrangler) as C, G, L, and W, these raw sales data are shown in Table 2.22. Construct percent bar charts of (1) Jeep sales in 2011 and (2) Jeep sales in 2006. Compare the charts and write a short report supporting the owner's position that in 2011 Liberty sales were still an important part of sales at his dealership. **DS** JeepSales

> Exercises 2.70 through 2.77 are based on the data in Table 2.23. This table gives the results of the J.D. Power initial quality study of 2014 automobiles. Each model is rated on overall mechanical quality and overall design quality on a scale from "among the best" to "the rest" (see the Scoring Legend). **DS** JDPower

2.70 Develop a frequency distribution of the overall mechanical quality ratings. Describe the distribution. **DS** JDPower

2.71 Develop a relative frequency distribution of the overall design quality ratings. Describe the distribution. **DS** JDPower

2.72 Construct a percentage bar chart of the overall mechanical quality ratings for each of the following: automobiles of United States origin; automobiles of Pacific Rim origin (Japan/Korea); and automobiles of European origin (Germany/Italy/Great Britain/Sweden). Compare the three distributions in a written report. **DS** JDPower

2.73 Construct a percentage pie chart of the overall design quality ratings for each of the following: automobiles of United States origin; automobiles of Pacific Rim origin (Japan/Korea); and automobiles of European origin (Germany/Italy/Great Britain/Sweden). Compare the three distributions in a written report. **DS** JDPower

2.74 Construct a contingency table of automobile origin versus overall mechanical quality rating. Set up rows corresponding to the United States, the Pacific Rim (Japan/Korea), and Europe (Germany/Italy/Great Britain/Sweden), and set up columns corresponding to the ratings "among the best" through "the rest." Describe any apparent relationship between origin and overall mechanical quality rating. **DS** JDPower

2.75 Develop a table of row percentages for the contingency table you set up in Exercise 2.74. Using these row percentages, construct a percentage frequency distribution of overall mechanical quality rating for each of the United States, the Pacific Rim, and Europe. Illustrate these three frequency distributions using percent bar charts and compare the distributions in a written report. **DS** JDPower

2.76 Construct a contingency table of automobile origin versus overall design quality rating. Set up rows corresponding to the United States, the Pacific Rim (Japan/Korea), and Europe (Germany/Italy/Great Britain/Sweden), and set up columns corresponding to the ratings "among the best" through "the rest." Describe any apparent relationship between origin and overall design quality. **DS** JDPower

2.77 Develop a table of row percentages for the contingency table you set up in Exercise 2.76. Using these row percentages, construct a percentage frequency distribution of overall design quality rating for each of the United States, the Pacific Rim, and Europe. Illustrate these three frequency distributions using percentage pie charts and compare the distributions in a written report. **DS** JDPower

TABLE 2.22 **2006 Sales at a Greater Cincinnati Jeep Dealership** **DS** JeepSales

W	L	L	W	G	C	C	L	C	L	G	W	C	L	L	C	C	G	L	L	W	C	G	G	C	L
L	L	G	L	C	C	G	C	C	G	C	L	W	W	G	G	W	G	C	W	W	G	L	L	G
G	L	G	C	C	C	C	C	G	G	L	G	G	L	L	L	C	C	G	C	L	G	G	G	L
L	G	L	L	G	L	C	W	G	L	G	L	G	G	G	C	G	W	G	L	L	L	C	C	L
G	L	C	L	C	L	L	L	C	G	L	C	L	W	L	W	G	W	C	W	C	W	C	L	C
C	G	C	C	C	C	C	C	C	G	C	C	W	G	C	G	L	L	L	C	L	L	G	G	G	L
L	L	C	G	L	C	C	L	L	G	G	L	L	L	C	G	L	C	L	W	L	L	C	G	C
G	G	G	L	C	L	L	G	L	C	C	L	G	W	W	W	C	C	C	G	G	L	G	C	G
C	L	L	G	G	L	W	W	L	C	C	C	G	W	C	C	W	L	G	W	L	L	L	G	G
G	W	L	L	C	G	C	C	W	C	L	L	L	G	G	W	L	L	C	L	G	G	W	G	G

TABLE 2.23   Results of the J.D. Power Initial Quality Study of 2014 Automobiles (for Exercises 2.70–2.77)   JDPower

Company	Country of Origin	Overall Quality Mechanical	Overall Quality Design	Company	Country of Origin	Overall Quality Mechanical	Overall Quality Design
Acura	Japan	3	3	Lexus	Japan	5	3
Audi	Germany	4	3	Lincoln	United States	4	3
BMW	Germany	4	3	Mazda	Japan	3	2
Buick	United States	4	3	Mercedes-Benz	Germany	3	3
Cadillac	United States	4	3	MINI	Great Britain	4	2
Chevrolet	United States	4	3	Mitsubishi	Japan	2	3
Chrysler	United States	4	4	Nissan	Japan	3	3
Dodge	United States	2	3	Porsche	Germany	2	5
Fiat	Italy	2	2	Ram	United States	2	4
Ford	United States	3	3	Scion	Japan	3	3
GMC	United States	3	3	Subaru	Japan	3	3
Honda	Japan	3	3	Toyota	Japan	3	3
Hyundai	Korea	4	4	Volkswagen	Germany	3	3
Infiniti	Japan	3	3	Volvo	Sweden	3	3
Jaguar	Great Britain	3	5				
Jeep	United States	3	3				
Kia	Korea	3	3				
Land Rover	Great Britain	2	3				

Scoring Legend

5 — Among the best	3 — About average	
4 — Better than most	2 — The rest	

Source: http://ratings.jdpower.com/automotive/ratings/909201795/2014-Initial+Quality+Study/index.htm (accessed January 16, 2015).

THE CIGARETTE ADVERTISEMENT CASE   ModelAge

In an article in the *Journal of Marketing,* Mazis, Ringold, Perry, and Denman discuss the perceived ages of models in cigarette advertisements.[7] To quote the authors:

> Most relevant to our study is the Cigarette Advertiser's Code, initiated by the tobacco industry in 1964. The code contains nine advertising principles related to young people, including the following provision (*Advertising Age* 1964): "Natural persons depicted as smokers in cigarette advertising shall be at least 25 years of age and shall not be dressed or otherwise made to appear to be less than 25 years of age."

Tobacco industry representatives have steadfastly maintained that code provisions are still being observed. A 1988 Tobacco Institute publication, "Three Decades of Initiatives by a Responsible Cigarette Industry," refers to the industry code as prohibiting advertising and promotion "directed at young people" and as "requiring that models in advertising must be, and must appear to be, at least 25 years old." John R. Nelson, Vice President of Corporate Affairs for Philip Morris, wrote, "We employ only adult models in our advertising who not only are but *look* over 25." However, industry critics have charged that current cigarette advertising campaigns use unusually young-looking models, thereby violating the voluntary industry code.

Suppose that a sample of 50 people is randomly selected at a shopping mall. Each person in the sample is shown a typical cigarette advertisement and is asked to estimate the age of the model in the ad. The 50 perceived age estimates so obtained are given in the right column. Use these data to do Exercises 2.78 through 2.81.

26	30	23	27	27
31	28	24	26	29
30	28	25	31	22
28	26	24	30	27
29	32	27	17	30
32	28	19	25	29
27	28	17	28	21
29	18	27	29	23
25	26	28	20	24
27	21	29	26	28

2.78   Consider constructing a frequency distribution and histogram for the perceived age estimates.   ModelAge

a Develop a frequency distribution, a relative frequency distribution, and a percent frequency distribution for the perceived age estimates using eight classes each of length 2. Note that, while the procedure presented in Section 2.2 would tell us to use six classes, in this case we get a more informative frequency distribution by using eight classes.

b Draw a percent frequency histogram for the perceived age estimates.

c Describe the shape of the distribution of perceived age estimates.

2.79   Construct a percent frequency polygon of the perceived age estimates. Use the classes of Exercise 2.78. ModelAge

2.80   Construct a dot plot of the perceived age estimates and describe the shape of the distribution. What percentage of

[7]Source: M. B. Mazis, D. J. Ringold, E. S. Perry, and D. W. Denman, "Perceived Age and Attractiveness of Models in Cigarette Advertisements," *Journal of Marketing* 56 (January 1992), pp. 22–37.

the perceived ages are below the industry's code provision of 25 years old? Do you think that this percentage is too high? **DS** ModelAge

2.81 Using the frequency distribution you developed in Exercise 2.78, develop: **DS** ModelAge

  a A cumulative frequency distribution.
  b A cumulative relative frequency distribution.
  c A cumulative percent frequency distribution.
  d A percent frequency ogive of the perceived age estimates.
  e How many perceived age estimates are 28 or less?
  f What percentage of perceived age estimates are 22 or less?

2.82 Table 2.24 presents data concerning the largest U.S. charities in 2014 as rated on the Forbes.com website on January 14, 2015. **DS** Charities

  a Construct a percent frequency histogram of each of (1) the charities' private support figures, (2) the charities' total revenues, and (3) the charities' fundraising efficiencies.
  b Describe the shape of each histogram.

2.83 The price/earnings ratio of a firm is a multiplier applied to a firm's earnings per share (EPS) to determine the value of the firm's common stock. For instance, if a firm's earnings per share is $5, and if its price/earnings ratio (or P/E ratio) is 10, then the market value of each share of common stock is ($5)(10) = $50. To quote Stanley B. Block and Geoffrey A. Hirt in their book *Foundations of Financial Management*:[8]

> The P/E ratio indicates expectations about the future of a company. Firms expected to provide returns greater than those for the market in general with equal or less risk often have P/E ratios higher than the market P/E ratio.

TABLE 2.24 Data Concerning the Largest U.S. Charities in 2014 as Rated by *Forbes* Magazine (for Exercise 2.82)
DS Charities

Name of Charity	Private Support ($ mil)	Total Revenue ($ mil)	Fundraising Efficiency (%)	Name of Charity	Private Support ($ mil)	Total Revenue ($ mil)	Fundraising Efficiency (%)
United Way	3,870	4,266	91	Lutheran Services in America	373	20,980	81
Salvation Army	2,080	4,316	90	Boy Scouts of America	362	1,240	87
Feeding America	1,855	1,894	99	MAP International	346	349	99
Task Force for Global Health	1,575	1,609	100	Step Up for Students	332	333	100
American National Red Cross	1,079	3,504	82	CARE USA	320	492	93
Food for the Poor	1,023	1,030	97	American Jewish Joint Distribution Committee	312	393	96
Goodwill Industries International	975	5,178	97	Good 360	310	310	100
YMCA of the USA	939	6,612	87	Mayo Clinic	310	3,899	90
American Cancer Society	871	935	77	Leukemia & Lymphoma Society	279	290	83
St. Jude Children's Research Hospital	869	1,287	83	Project HOPE	273	287	97
World Vision	795	981	86	Dana-Farber Cancer Institute	261	1,057	92
Boys & Girls Clubs of America	766	1,686	88	Planned Parenthood Federation of America	315	1,210	81
Catholic Charities USA	715	4,337	89	Metropolitan Museum of Art	259	632	96
Compassion International	657	660	90	Cross International	257	257	98
AmeriCares Foundation	620	622	99	Operation Blessing International Relief & Development.	255	267	99
Habitat for Humanity International	606	1,665	81	Make-A-Wish Foundation of America	253	265	86
United States Fund for UNICEF	587	593	94	Population Services International	251	609	99
Catholic Medical Mission Board	513	527	99	Alzheimer's Association	243	285	83
Campus Crusade for Christ	504	544	92	Catholic Relief Services	244	641	89
American Heart Association	502	663	85	National Multiple Sclerosis Society	243	256	85
Nature Conservancy	500	881	82	Brother's Brother Foundation	243	244	100
Save the Children Federation	455	658	94	Chronic Disease Fund	238	238	100
Direct Relief	450	450	100	San Francisco Museum of Modern Art	236	243	98
Feed the Children	442	454	93	Marine Toys for Tots Foundation	235	238	97
Samaritan's Purse	426	461	92				
Memorial Sloan-Kettering Cancer Center	389	3,386	87				

Source: Data from http://www.forbes.com/top-charities/list/ (accessed January 14, 2015).

[8]Source: Excerpt from S. B. Block and G. A. Hirt, *Foundations of Financial Management*, p. 28. McGraw-Hill Companies, Inc., 1994.

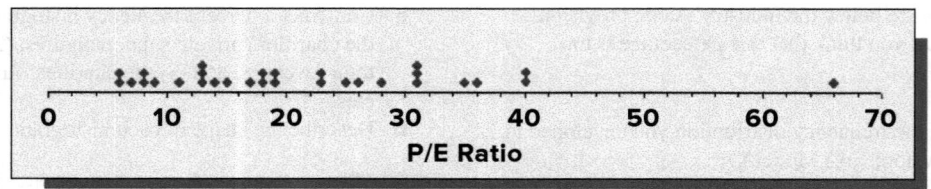

In the figure above we give a dot plot of the P/E ratios for 30 fast-growing companies. Describe the distribution of the P/E ratios.

2.84 A basketball player practices free throws by taking 25 shots each day, and he records the number of shots missed each day in order to track his progress. The numbers of shots missed on days 1 through 30 are, respectively, 17, 15, 16, 18, 14, 15, 13, 12, 10, 11, 11, 10, 9, 10, 9, 9, 9, 10, 8, 10, 6, 8, 9, 8, 7, 9, 8, 7, 5, 8. **DS** FreeThrw

a Construct a stem-and-leaf display and a time series plot of the numbers of missed shots.

b Do you think that the stem-and-leaf display is representative of the numbers of shots that the player will miss on future days? Why or why not?

2.85 In the Fall 1993 issue of *VALIC Investment Digest,* the Variable Annuity Life Insurance Company used pie charts to help give the following description of an investment strategy called **rebalancing:**

Once you've established your ideal asset allocation mix, many experts recommend that you review your portfolio at least once a year to make sure your portfolio remains consistent with your preselected asset allocation mix. This practice is referred to as *rebalancing.*

For example, let's assume a moderate asset allocation mix of 50 percent equities funds, 40 percent bond funds, and 10 percent cash-equivalent funds. The chart [see Figure 2.42] based on data provided by Ibbotson, a major investment and consulting firm, illustrates how rebalancing works. Using the Standard & Poor's 500 Index,

the Salomon Brothers Long-Term High-Grade Corporate Bond Index, and the U.S. 30-day Treasury bill average as a cash-equivalent rate, our hypothetical portfolio balance on 12/31/90 is $10,000. One year later the account had grown to $12,380. By the end of 1991, the allocation had changed to 52.7%/38.7%/8.5%. The third pie chart illustrates how the account was once again rebalanced to return to a 50%/40%/10% asset allocation mix.

Rebalancing has the potential for more than merely helping diversify your portfolio. By continually returning to your original asset allocation, it is possible to avoid exposure to more risk than you previously decided you were willing to assume.

a Suppose you control a $100,000 portfolio and have decided to maintain an asset allocation mix of 60 percent stock funds, 30 percent bond funds, and 10 percent government securities. Draw a pie chart illustrating your portfolio (like the ones in Figure 2.42).

b Over the next year your stock funds earn a return of 30 percent, your bond funds earn a return of 15 percent, and your government securities earn a return of 6 percent. Calculate the end-of-year values of your stock funds, bond funds, and government securities. After calculating the end-of-year value of your entire portfolio, determine the asset allocation mix (percent stock funds, percent bond funds, and percent government securities) of your portfolio before rebalancing. Finally, draw an end-of-year pie chart of your portfolio before rebalancing.

c Rebalance your portfolio. That is, determine how much of the portfolio's end-of-year value must be invested in stock

FIGURE 2.42   **Using Pie Charts to Illustrate Portfolio Rebalancing (for Exercise 2.85)**

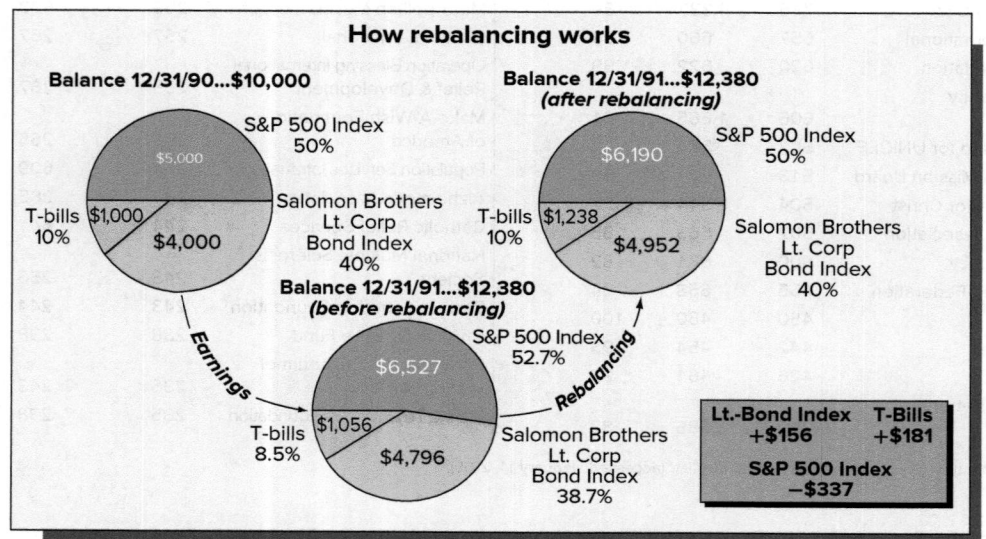

Source: The Variable Annuity Life Insurance Company, *VALIC* 6, no. 4 (Fall 1993).

funds, bond funds, and government securities in order to restore your original asset allocation mix of 60 percent stock funds, 30 percent bond funds, and 10 percent government securities. Draw a pie chart of your portfolio after rebalancing.

2.86   Figure 2.43 was used in various Chevrolet magazine advertisements in 1997 to compare the overall resale values of Chevrolet, Dodge, and Ford trucks in the years from 1990 to 1997. What is somewhat misleading about this graph?

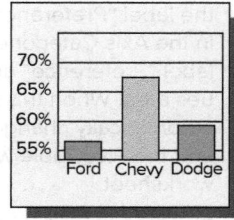

FIGURE 2.43
A Graph Comparing the Resale Values of Chevy, Dodge, and Ford Trucks

Source: General Motors Corporation.

2.87   Internet Exercise

The Gallup Organization provides market research and consulting services around the world. Gallup publishes the Gallup Poll, a widely recognized barometer of American and international opinion. The Gallup website provides access to many recent Gallup studies. Although a subscription is needed to access the entire site, many articles about recent Gallup Poll results can be accessed free of charge. To find poll results, go to the Gallup home page (http://www.gallup.com/). The poll results are presented using a variety of statistical summaries and graphics that we have learned about in this chapter.

a   Go to the Gallup Organization website and access several of the articles presenting recent poll results. Find and print examples of some of the statistical summaries and graphics that we studied in this chapter. Then write a summary describing which statistical methods and graphics seem to be used most frequently by Gallup when presenting poll results.

b   Read the results of a Gallup poll that you find to be of particular interest and summarize (in your own words) its most important conclusions. Cite the statistical evidence in the article that you believe most clearly backs up each conclusion.

c   By searching the web, or by searching other sources (such as newspapers and magazines), find an example of a misleading statistical summary or graphic. Print or copy the misleading example and write a paragraph describing why you think the summary or graphic is misleading.

Appendix 2.1 ■ Tabular and Graphical Methods Using Excel

The instructions in this section begin by describing the entry of data into an Excel spreadsheet. Alternatively, the data may be downloaded from this book's website. The appropriate data file name is given at the top of each instruction block. Please refer to Appendix 1.1 for further information about entering data, saving data, and printing results in Excel.

Construct a frequency distribution and frequency bar chart of pizza preferences as in Table 2.2 and Figure 2.1 (data file: PizzaPref.xlsx):

• Enter the pizza preference data in Table 2.1 into column A with label Preference in cell A1.

We obtain the frequency distribution and bar chart by forming what is called a **PivotChart**. This is done as follows:

• Select **Insert : PivotChart : PivotChart & PivotTable**

• In the Create PivotTable dialog box, click "Select a table or range."

• Enter the range of the data to be analyzed into the Table/Range window. Here we have entered the range of the pizza preference data A1:A51, that is, the entries in rows 1 through 51 in column A. The easiest way to do this is to click in the Table/Range window and to then use the mouse to drag the cursor from cell A1 through cell A51.

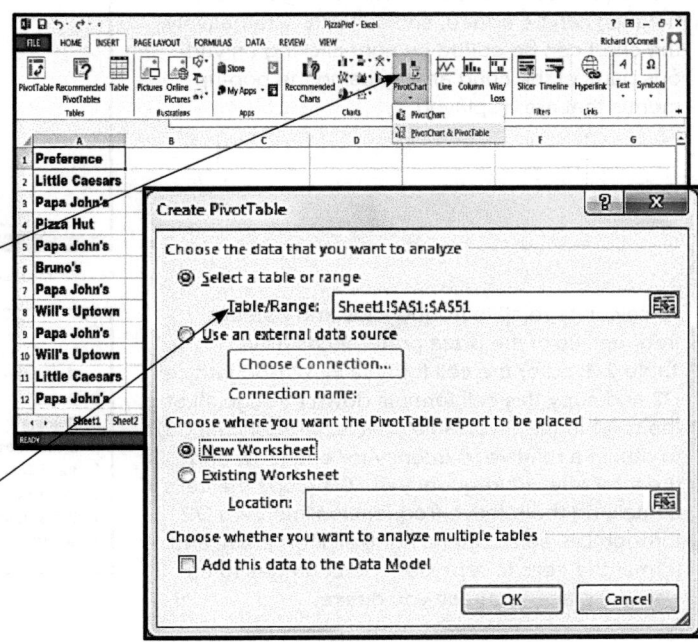

Source: Microsoft Office Excel 2016

- Select "New Worksheet" to have the PivotChart and PivotTable output displayed in a new worksheet.
- Click OK in the Create PivotTable dialog box.
- In the **PivotChart Fields task pane,** place a checkmark in the checkbox to the left of the column label "Preference"—when you do this, the label "Preference" will also be placed in the Axis Categories area. Also drag the label "Preference" and drop it into the Σ Values area. When this is done, the label will automatically change to "Count of Preference" and the PivotTable will be displayed in the new worksheet.

  The PivotChart (bar chart) will also be displayed in a graphics window in the new worksheet.

  Note: You may need to close the PivotChart Fields task pane in order to see the bar chart.

- As demonstrated in Appendix 1.1, move the bar chart to a new worksheet before editing.

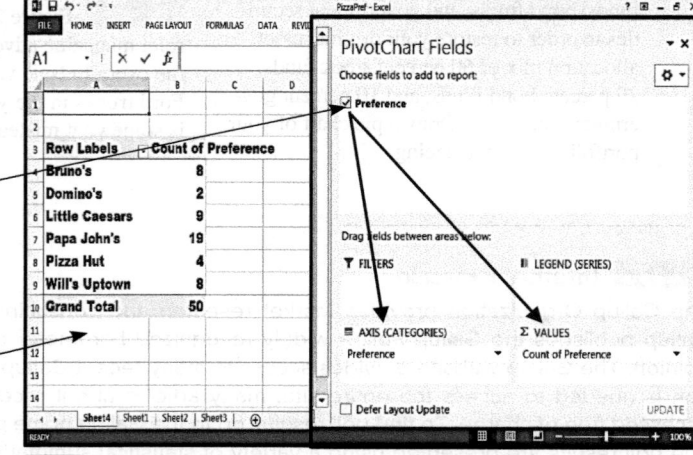

Source: Microsoft Office Excel 2016

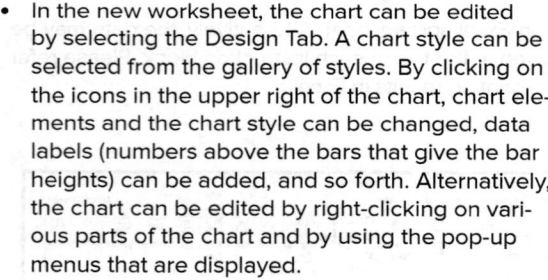

Source: Microsoft Office Excel 2016

- In the new worksheet, the chart can be edited by selecting the Design Tab. A chart style can be selected from the gallery of styles. By clicking on the icons in the upper right of the chart, chart elements and the chart style can be changed, data labels (numbers above the bars that give the bar heights) can be added, and so forth. Alternatively, the chart can be edited by right-clicking on various parts of the chart and by using the pop-up menus that are displayed.

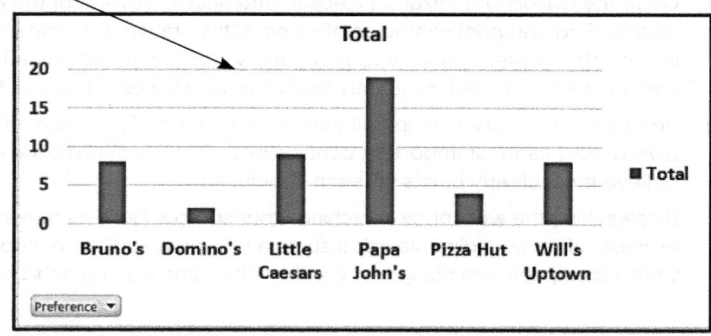

Source: Microsoft Office Excel 2016

- To calculate relative frequencies and percent frequencies of the pizza preferences as in Table 2.3, enter the cell formula = B2/B$8 into cell C2 and copy this cell formula down through all of the rows in the PivotTable (that is, through cell C8) to obtain a relative frequency for each row and the total relative frequency of 1.00. Copy the cells containing the relative frequencies into cells D2 through D8, select them, right-click on them, and format the cells to represent percentages to the decimal place accuracy you desire.

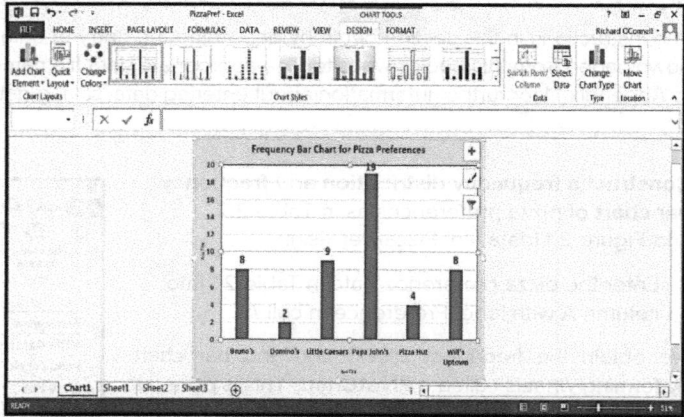

Source: Microsoft Office Excel 2016

Construct a frequency bar chart of pizza preferences from a frequency distribution (data file: PizzaFreq.xlsx):

- Enter the frequency distribution of pizza preferences in Table 2.2 as shown in the screen with the various restaurant identifiers in column A (with label Restaurant) and with the corresponding frequencies in column B (with label Frequency).

- Select the entire data set using the mouse.

- Select **Insert : PivotChart : PivotChart**

- In the Create PivotChart dialog box, click "Select a table or range."

- Enter the entire range of the data into the Table/Range window. Here we have entered the range of the frequency distribution, that is, A1:B7. Again, the easiest way to do this is to click in the Table/Range window and then to select the range of the data with the mouse.

- Select "New Worksheet" to have the PivotChart (and a Pivot Table) displayed in a new worksheet and click OK in the Create PivotChart dialog box.

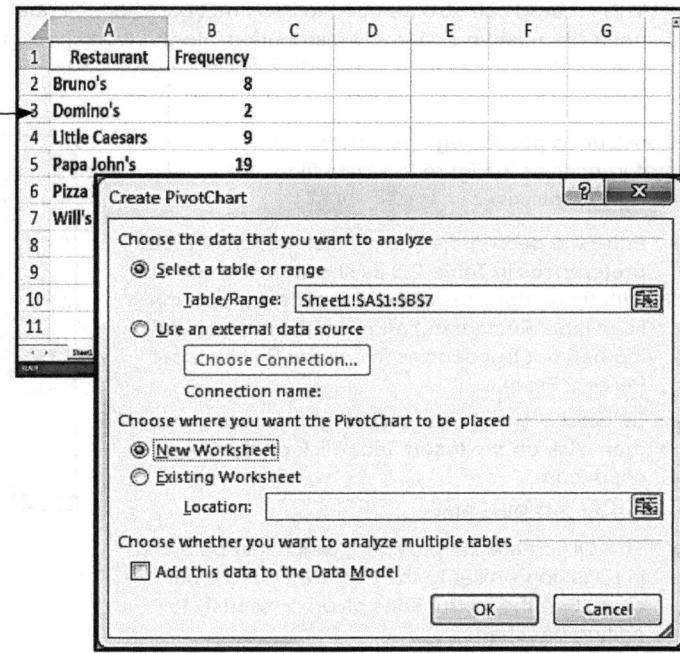

Source: Microsoft Office Excel 2016

- Place checkmarks in the checkboxes to the left of the field names Restaurant and Frequency. When this is done, the field name Restaurant will be placed in the Axis area, "Sum of Frequency" will be placed in the Σ Values area, a Pivot Table will be placed in the new worksheet, and the Pivot Chart (bar chart) will be constructed. You may need to close the PivotChart Fields pane to see the chart. The chart can be moved to a new sheet and edited as previously described.

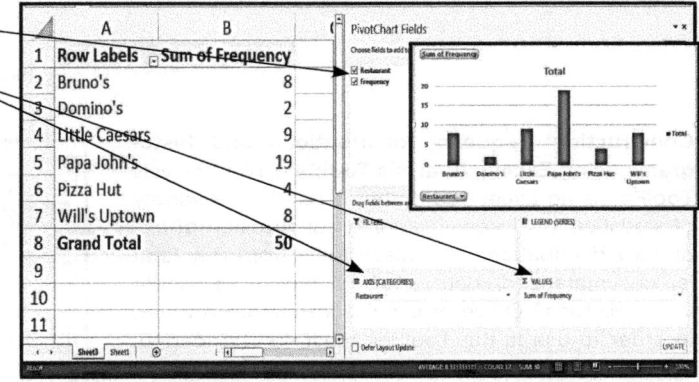

Source: Microsoft Office Excel 2016

- **A bar chart can also be constructed without using a Pivot Table.** To do this, select the entire frequency distribution with the mouse and then click on the **Insert** tab. Click on the (vertical) bar chart icon and select

  2-D Column : Clustered Column

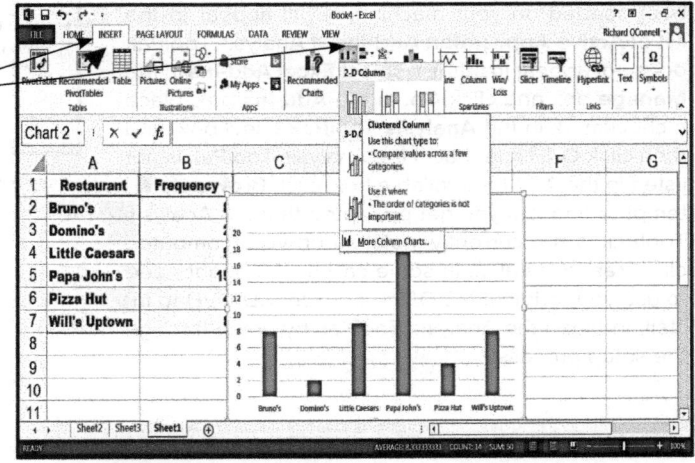

Source: Microsoft Office Excel 2016

- This method can also be used to construct bar charts of relative frequency and percent frequency distributions. Simply enter the relative or percent distribution into the worksheet, select the entire distribution with the mouse, and make the same selections as in the preceding bullet.

Construct a percentage pie chart of pizza preferences as in Figure 2.3 (data file: PizzaPercents.xlsx):

- Enter the percent frequency distribution of pizza preferences in Table 2.3 as shown in the screen with the various restaurant identifiers in column A (with label Restaurant) and with the corresponding percent frequencies in column B (with label Percent Freq).

- Select the entire data set using the mouse and then click on the **Insert** Tab. Click on the pie chart icon.

- Select **2-D Pie : Pie**

- This will create the pie chart, which can be edited in a fashion similar to the way we edited a bar chart. See the instructions given previously for editing bar charts.

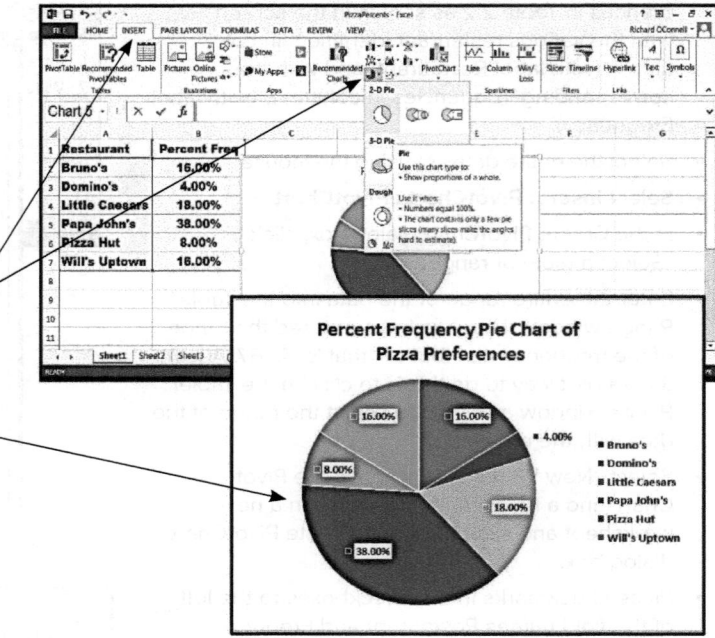

Source: Microsoft Office Excel 2016

Constructing frequency distributions and histograms using Excel's Analysis ToolPak: The Analysis ToolPak is an Excel add-in that is used for a variety of statistical analyses—including construction of frequency distributions and histograms from raw (that is, unsummarized) data. The ToolPak is available when Microsoft Office or Excel is installed. However, in order to use it, the ToolPak must first be loaded. To see if the Analysis ToolPak has been loaded on your computer, click the **Microsoft File Button,** click **Options,** and finally click **Add-Ins.** If the ToolPak has been loaded on your machine, it will appear in the list of **Active Application Add-Ins.** If Analysis ToolPak does not appear in this list, select **Excel Add-Ins** in the **Manage** box and click **Go.** In the **Add-Ins** box, place a checkmark in the **Analysis ToolPak** checkbox, and then click **OK.** Note that, if the Analysis ToolPak is not listed in the Add-Ins available box, click **Browse** to attempt to find it. If you get prompted that the Analysis ToolPak is not currently installed on your computer, click **Yes** to install it. In some cases, you might need to use your original MS Office or Excel CD/DVD to install and load the Analysis ToolPak by going through the setup process.

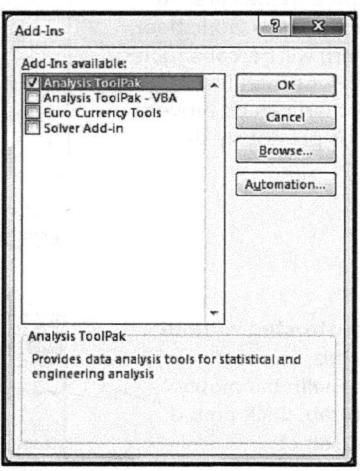

Source: Microsoft Office Excel 2016

Constructing a frequency histogram of the bottle design ratings as in Figure 2.11 (data file: Design .xlsx):

- Enter the 60 bottle design ratings in Table 1.6 into Column A with label Rating in cell A1.

- Select **Data : Data Analysis**

- In the Data Analysis dialog box, select Histogram in the Analysis Tools window and click OK.

- In the Histogram dialog box, click in the Input Range window and select the range of the data A1:A61 into the Input Range window by dragging the mouse from cell A1 through cell A61.

- Place a checkmark in the Labels checkbox.

- Under "Output options," select "New Worksheet Ply."

- Enter a name for the new worksheet in the New Worksheet Ply window—here, Histogram 1.

- Place a checkmark in the Chart Output checkbox.

- Click OK in the Histogram dialog box.

- Notice that we are leaving the Bin Range window blank. This will cause Excel to define automatic classes for the frequency distribution and histogram. However, because Excel's automatic classes are often not appropriate, we will revise these automatic classes as explained below.

- The frequency distribution will be displayed in the new worksheet and the histogram will be displayed in a graphics window.

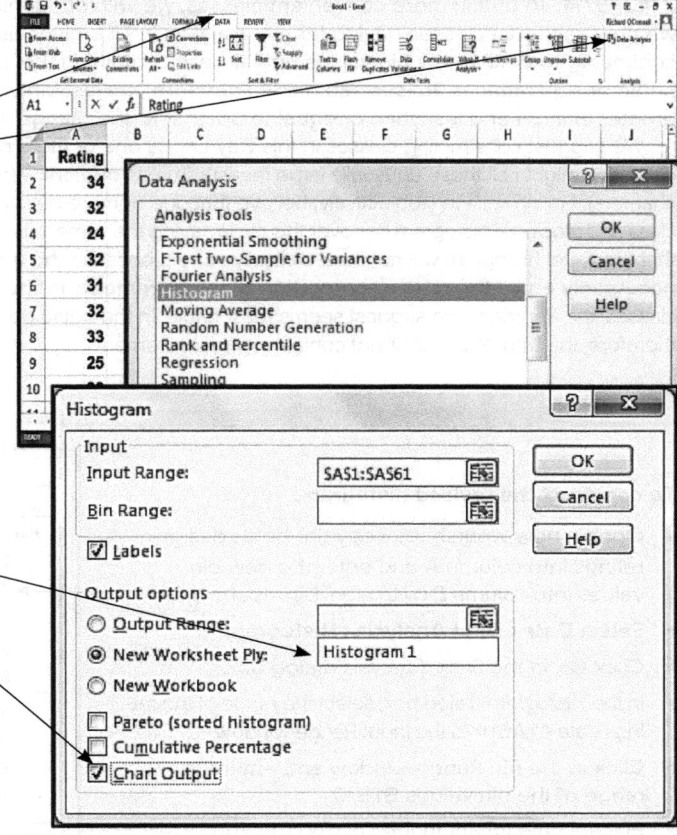

Source: Microsoft Office Excel 2016

Notice that Excel defines what it calls bins when constructing the histogram. The bins define the automatic classes for the histogram. The bins that are automatically defined by Excel are often cumbersome—the bins in this example are certainly inconvenient for display purposes! Although one might be tempted to simply round the bin values, we have found that the rounded bin values can produce an unacceptable histogram with unequal class lengths. (Whether this happens depends on the particular bin values in a given situation.)

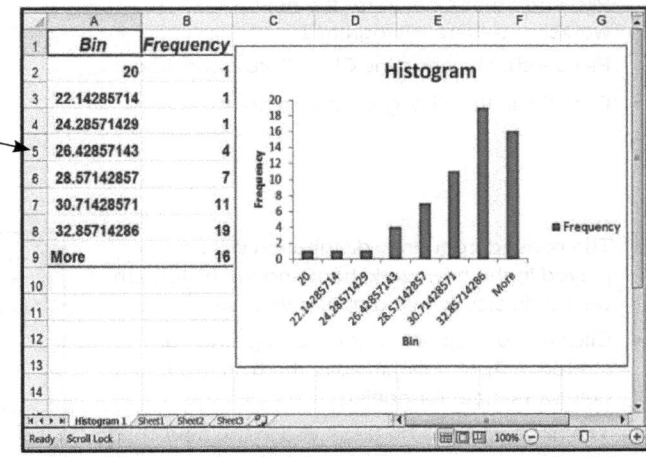

Source: Microsoft Office Excel 2016

To obtain more acceptable results, **we recommend defining new bin values that are roughly based on the automatic bin values.** We can do this as follows. First, we note that the smallest bin value is 20 and that this bin value is expressed using the same decimal place accuracy as the original data (recall that the bottle design ratings are all whole numbers). Remembering that Excel obtains a cell frequency by counting the number of measurements that are less than or equal to the upper class boundary and greater than the lower class boundary, the first class contains bottle design ratings less than or equal to 20. Based on the authors' experience, the first automatic bin value given by Excel is expressed to the same decimal place accuracy as the data being analyzed. However, if the smallest bin value were to be expressed using more decimal places than the original data, then **we suggest rounding it down to the decimal place accuracy of the original data being analyzed.** Frankly, the authors are not sure that this would ever need to be done—it was not necessary in any of the examples we have tried. Next, find the class length of the Excel automatic classes and round it to a convenient value. For the bottle design ratings, using the first and second bin values in the screen, the class length is 22.14285714 − 20, which equals

2.14285714. To obtain more convenient classes, we will round this value to 2. Starting at the first automatic bin value of 20, we now construct classes having length equal to 2. This gives us new bin values of 20, 22, 24, 26, and so on. We suggest continuing to define new bin values until a class containing the largest measurement in the data is found. Here, the largest bottle design rating is 35 (see Table 1.6). Therefore, the last bin value is 36, which says that the last class will contain ratings greater than 34 and less than or equal to 36, that is, the ratings 35 and 36.

We suggest constructing classes in this way unless one or more measurements are unusually large compared to the rest of the data—we might call these unusually large measurements **outliers.** We will discuss outliers more thoroughly in Chapter 3 (and in later chapters). For now, if we (subjectively) believe that one or more outliers exist, we suggest placing these measurements in the "more" class and placing a histogram bar over this class having the same class length as the other bars. In such a situation, we must recognize that the Excel histogram will not be technically correct because **the area of the bar (or rectangle) above the "more" class will not necessarily equal the relative proportion of measurements in the class.** Nevertheless, given the way Excel constructs histogram classes, the approach we suggest seems reasonable. In the bottle design situation, the largest rating of 35 is not unusually large and, therefore, the "more" class will not contain any measurements.

To construct the revised histogram:

- Open a new worksheet, copy the bottle design ratings into column A and enter the new bin values into column B (with label Bin) as shown.

- Select **Data : Data Analysis : Histogram**

- Click OK in the Data Analysis dialog box.

- In the Histogram dialog box, select the range of the ratings data A1:A61 into the Input Range window.

- Click in the Bin Range window and enter the range of the bin values B1:B10.

- Place a checkmark in the Labels checkbox.

- Under "Output options," select "New Worksheet Ply" and enter a name for the new worksheet—here Histogram 2.

- Place a checkmark in the Chart Output checkbox.

- Click OK in the Histogram dialog box.

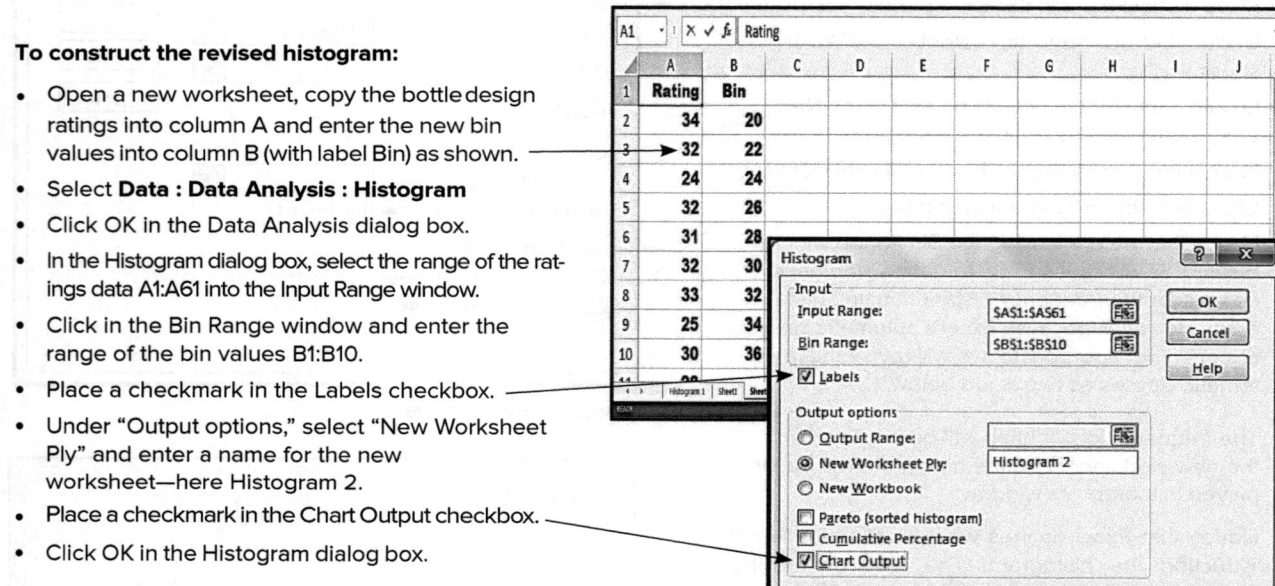

Source: Microsoft Office Excel 2016

- The revised frequency distribution will be displayed in the new worksheet and the histogram will be displayed in a graphics window.

- Click in the graphics window and (as demonstrated in Appendix 1.1) move the histogram to a new worksheet for editing.

- The histogram will be displayed in the new chart sheet in a much larger format that makes it easier to carry out editing.

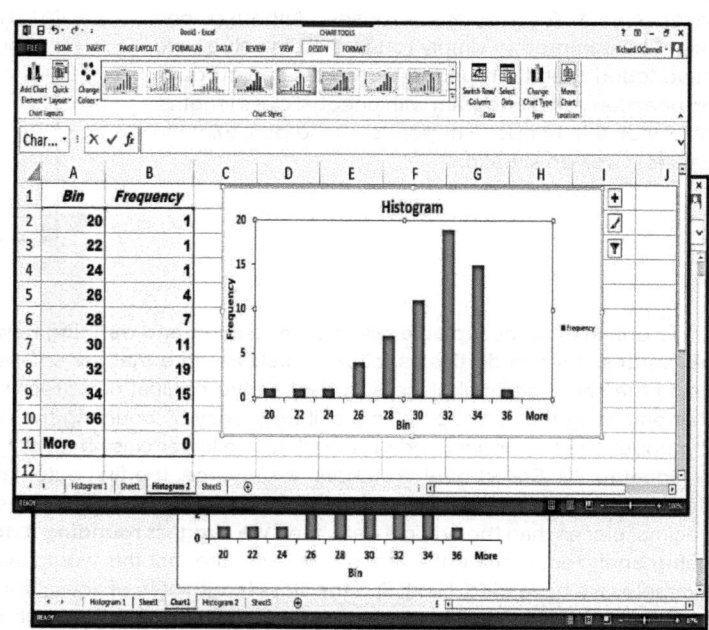

Source: Microsoft Office Excel 2016

To remove the gaps between the histogram bars:

- Right-click on one of the histogram bars and select Format Data Series from the pop-up window.

- Set the gap width to zero by moving the gap width slider to 0%.

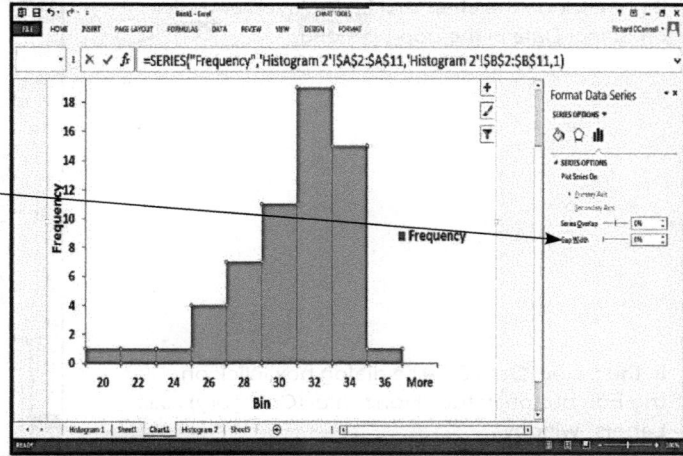

Source: Microsoft Office Excel 2016

- By selecting the Chart Tools tab, the histogram can be edited in many ways. This can also be done by right-clicking on various portions of the histogram and by making desired pop-up menu selections.

- To obtain **data labels** (the numbers on the tops of the bars that indicate the bar heights), right click on one of the histogram bars and select "Add data labels" from the pop-up menu.

After final editing, the histogram might look like the one illustrated in Figure 2.11.

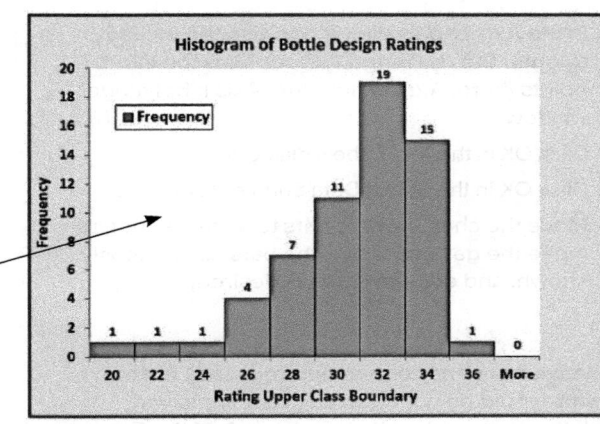

Source: Microsoft Office Excel 2016

Constructing a frequency histogram of bottle design ratings from summarized data:

- Enter the **midpoints** of the frequency distribution classes into column A with label Midpoint and enter the class frequencies into column B with label Frequency.

- Use the mouse to select the cell range that contains the frequencies (here, cells B2 through B10).

- Select **Insert : Column : 2-D Column (Clustered Column)**

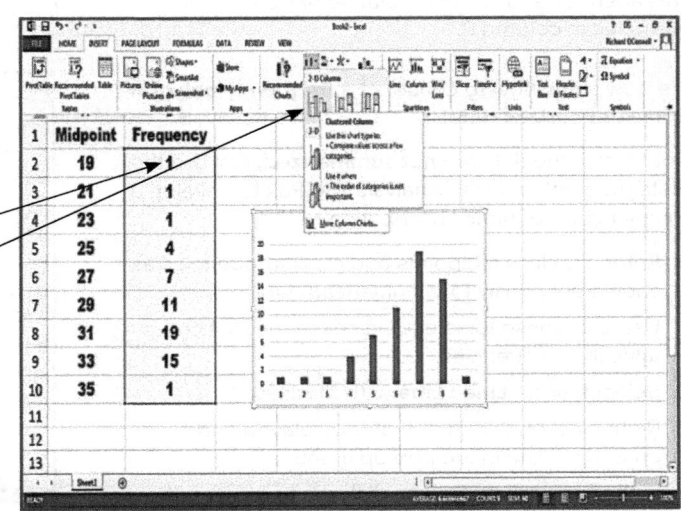

Source: Microsoft Office Excel 2016

- Right-click on the chart that is displayed and click on Select Data in the pop-up menu.

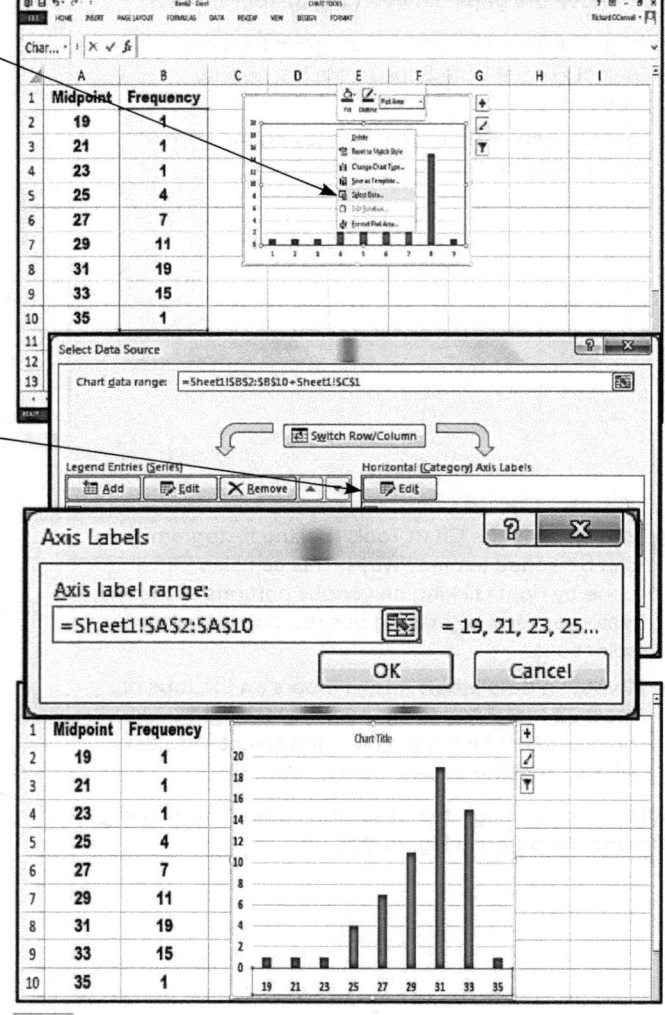

- In the Select Data Source dialog box, click on the Edit button in the "Horizontal (Category) Axis Labels" window.

- In the Axis Labels dialog box, use the mouse to enter the cell range that contains the midpoints (here, A2:A10) into the "Axis label range" window.
- Click OK in the Axis Labels dialog box.
- Click OK in the Select Data Source dialog box.
- Move the chart that appears to a chart sheet, remove the gaps between the bars as previously shown, and edit the chart as desired.

Relative frequency or percent frequency histograms would be constructed in the same way with the class midpoints in column A of the Excel spreadsheet and with the relative or percent frequencies in column B.

We now show how to construct a **frequency polygon from summarized data.**

Note that, if the data are **not summarized,** first use the Histogram option in the Analysis ToolPak to develop a summarized frequency distribution.

- Enter the class midpoints and class frequencies as shown previously for summarized data.
- Use the mouse to select the cell range that contains the frequencies.
- Select **Insert : Line : Line with Markers**
- Right-click on the chart that is displayed and click on Select Data in the pop-up menu.
- In the Select Data Source dialog box, click on the Edit button in the "Horizontal (Category) Axis Labels" window.
- In the Axis Labels dialog box, use the mouse to enter the cell range that contains the midpoints into the "Axis label range" window.
- Click OK in the Axis Labels dialog box.
- Click OK in the Select Data Source dialog box.
- Move the chart that appears to a chart sheet, and edit the chart as desired.

Source: Microsoft Office Excel 2016

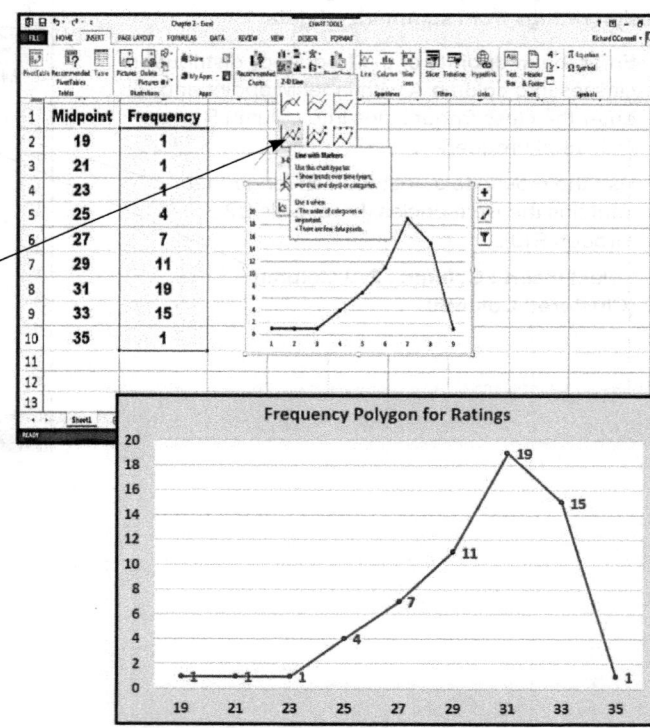

Source: Microsoft Office Excel 2016

To construct a percent frequency ogive for the bottle design rating distribution (data file: Design.xlsx):

Follow the instructions for constructing a histogram by using the Analysis ToolPak with changes as follows:

- In the Histogram dialog box, place a checkmark in the Cumulative Percentage checkbox.

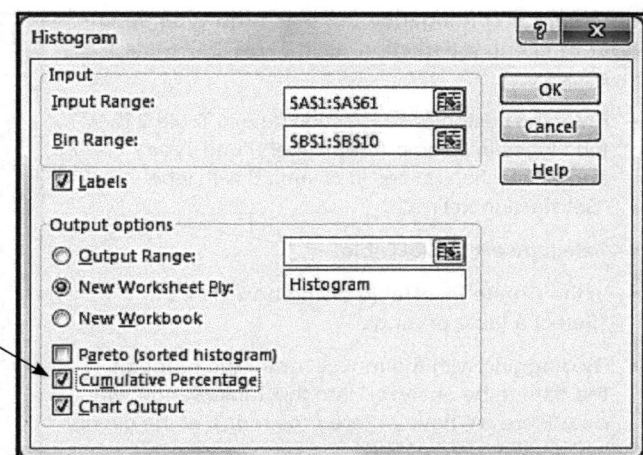

Source: Microsoft Office Excel 2016

- After moving the histogram to a chart sheet, right-click on any histogram bar.

- Select "Format Data Series" from the pop-up menu.

- In the "Format Data Series" task pane, (1) select Fill from the list of "Series Options" and select "No fill" from the list of Fill options; (2) select Border from the list of "Series Options" and select "No line" from the list of Border options; (3) close the Format Data Series task pane.

- Click on the chart to remove the histogram bars.

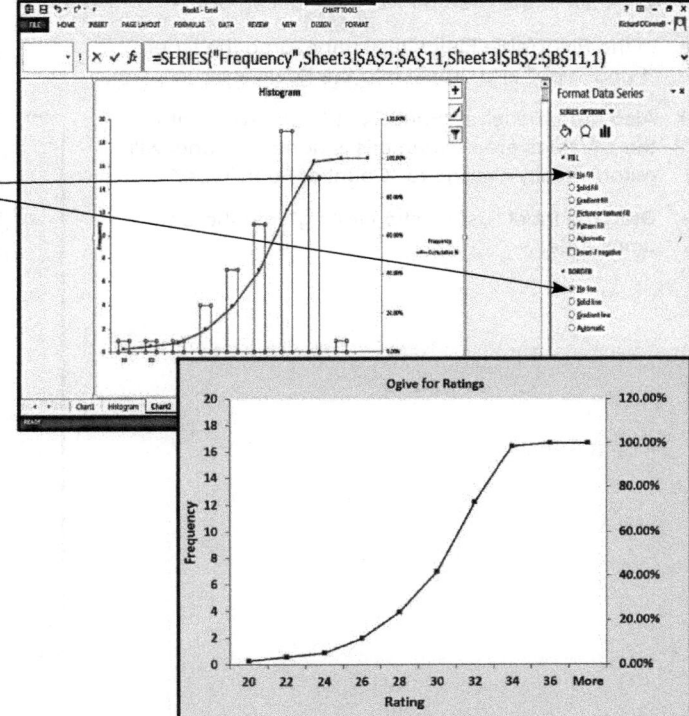

Source: Microsoft Office Excel 2016

Construct a contingency table of fund type versus level of client satisfaction as in Table 2.17 (data file: Invest.xlsx):

- Enter the customer satisfaction data in Table 2.16—fund types in column A with label "Fund Type" and satisfaction ratings in column B with label "Satisfaction Rating."

- Select **Insert : PivotTable**

- In the Create PivotTable dialog box, click "Select a table or range."

- By dragging with the mouse, enter the range of the data to be analyzed into the Table/Range window. Here we have entered the range of the client satisfaction data A1:B101.

- Select the New Worksheet option to place the PivotTable in a new worksheet.

- Click OK in the Create PivotTable dialog box.

- In the PivotTable Fields task pane, drag the label "Fund Type" and drop it into the Rows area.

- Also drag the label "Fund Type" and drop it into the Σ Values area. When this is done, the label will automatically change to "Count of Fund Type."

- Drag the label "Satisfaction Rating" into the Columns area.

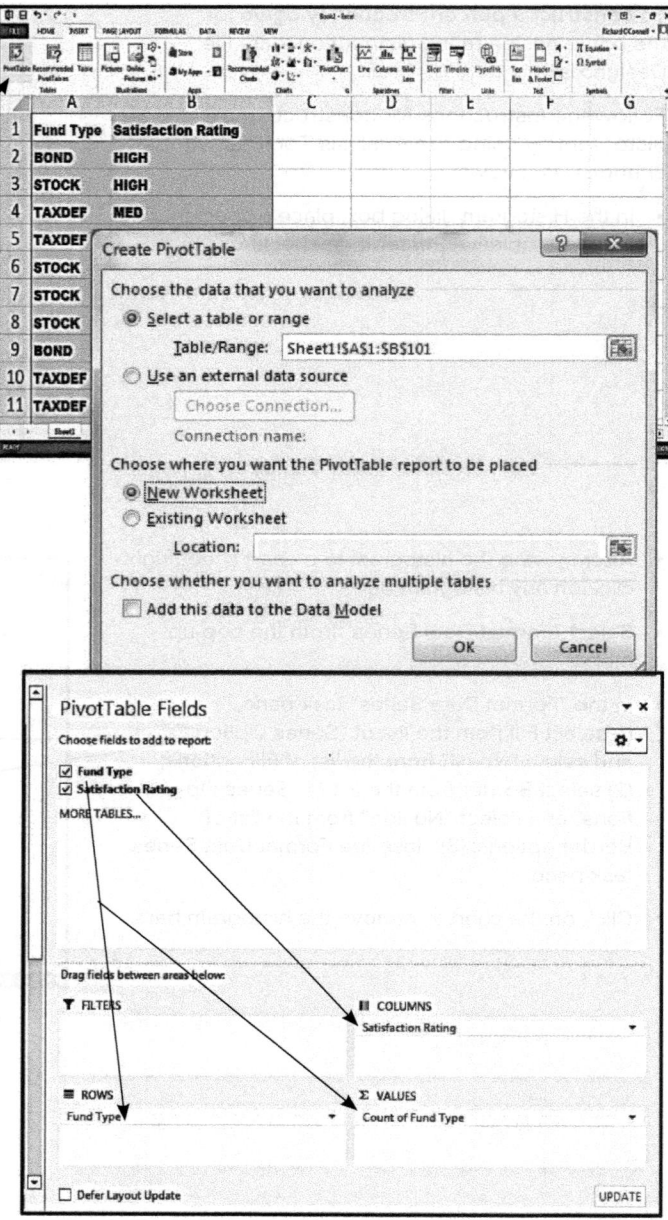

Source: Microsoft Office Excel 2016

- The PivotTable will be created and placed in a new worksheet.

- Now right-click inside the PivotTable and select PivotTable Options from the pop-up menu.

- In the PivotTable Options dialog box, select the Totals & Filters tab and make sure that a checkmark has been placed in each of the "Show grand totals for rows" and the "Show grand totals for columns" checkboxes.

- Select the Layout & Format tab, place a checkmark in the "For empty cells show" checkbox, and enter 0 (the number zero) into its corresponding window. (For the customer satisfaction data, none of the cell frequencies equal zero, but, in general, this setting should be made to prevent empty cells from being left blank in the contingency table.)

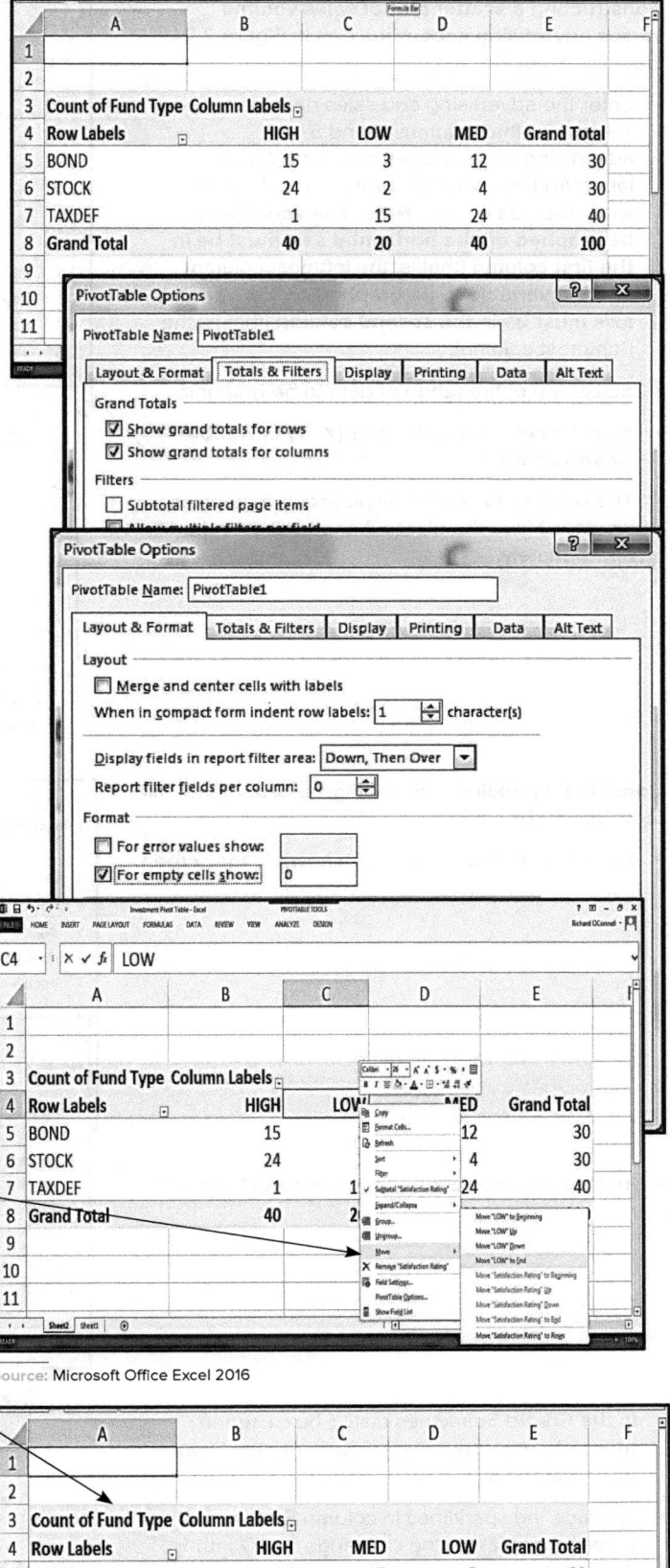

- To change the order of the column labels from the default alphabetical ordering (High, Low, Medium) to the more logical ordering of High, Medium, Low, right-click on LOW, select Move from the pop-up menu, and select "Move LOW to End."

- The contingency table is now complete.

Source: Microsoft Office Excel 2016

	A	B	C	D	E	F
1						
2						
3	Count of Fund Type	Column Labels				
4	Row Labels	HIGH	MED	LOW	Grand Total	
5	BOND	15	12	3	30	
6	STOCK	24	4	2	30	
7	TAXDEF	1	24	15	40	
8	Grand Total	40	40	20	100	

Source: Microsoft Office Excel 2016

Constructing a scatter plot of sales volume versus advertising expenditure as in Figure 2.24 (data file: SalesPlot.xlsx):

- Enter the advertising and sales data in Table 2.20 into columns A and B—advertising expenditures in column A with label "Ad Exp" and sales values in column B with label "Sales Vol." **Note: The variable to be graphed on the horizontal axis must be in the first column** (that is, the leftmost column) and **the variable to be graphed on the vertical axis must be in the second column** (that is, the rightmost column).

- Select the entire range of data to be graphed.

- Select **Insert : Insert Scatter (X, Y) or Bubble Chart : Scatter**

- The scatter plot will be displayed in a graphics window. Move the plot to a chart sheet and edit appropriately.

Source: Microsoft Office Excel 2016

Construct sparklines as in Figure 2.38 (data file: StockSpark.xlxs):

- Enter the stock price data as shown in the screen.

- In the Sparklines Group, select **Insert : Line**

- In the Create Sparklines dialog box, use the mouse to enter the range C3:H7 into the Data Range window.

- To place the sparklines in column B of the spreadsheet, enter the cell range B3:B7 into the Location Range window.

- Click OK in the Create Sparklines dialog box to have the sparklines created in cells B3 through B7.

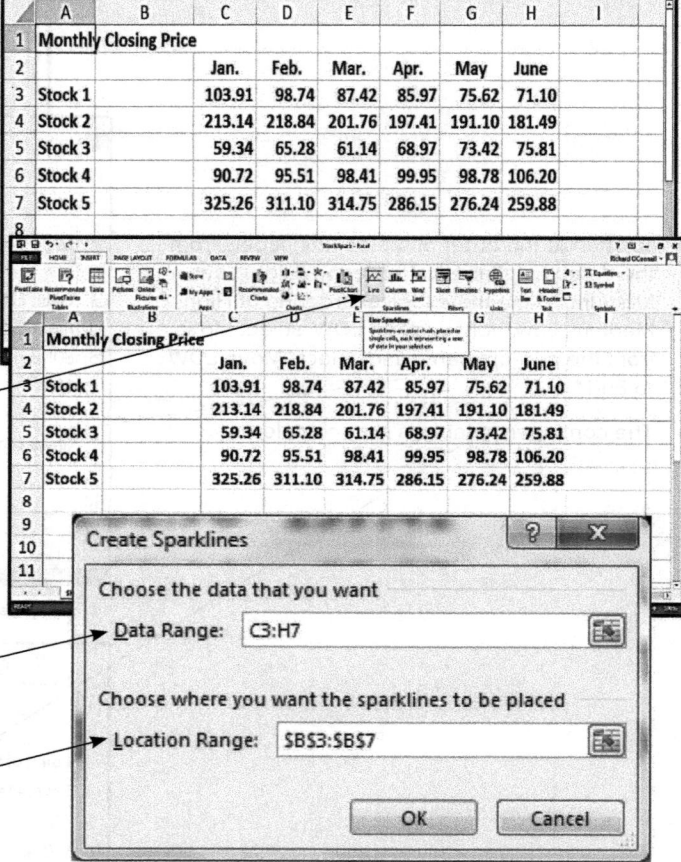

Source: Microsoft Office Excel 2016

- To edit the sparklines, select cell B3 (or any other cell that contains a sparkline). Then select the Design tab in Sparkline Tools.

- For example, in the Show Group, place check-marks in the High Point, Low Point, and Last Point checkboxes to have these points marked on the sparklines. Select Marker Color to choose colors for the plotted points. Many other sparkline attributes can be edited in a similar fashion.

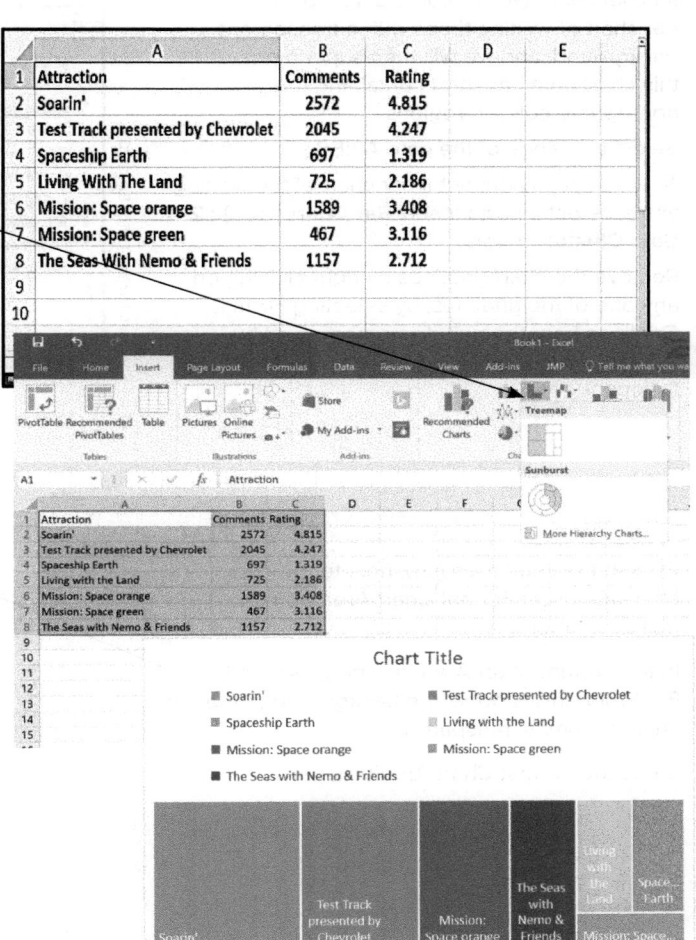

Source: Microsoft Office Excel 2016

Construct a treemap as in Figure 2.37(b):

- Enter the Disney rating data in Figure 2.37(a) (data file: DisneyRatings.xlsx), as shown in the screen.

- Select the data, click on the "Select Hierarchy Chart" button and select treemap from the drop-down list.

- The treemap will appear on the worksheet.

- The color scheme of the treemap can be edited by clicking on the Design tab in the toolbar and by selecting a color scheme from the gallery that is displayed. Unfortunately, Excel does not (at the time of the writing of this book) allow the use of different shades of the same color or a scheme of different colors that would meaningfully charac-terize the rectangles in the treemap according to changes in the value of a second variable.

Source: Microsoft Office Excel 2016

Construct a Bullet Graph of waiting times as in Figure 2.36 (data file: DisneyTimes .xlsx).

- Enter the waiting time data as shown in the screen with the attraction names in column A with label Attraction and the waiting times in column B with label Wait Time.

	A	B	C	D	E	F	G
1	Attraction	Wait Time					
2	Soarin'	65		20			
3	Test Track	50		20			
4	Spaceship Earth	25		20			
5	Living With The Land	20		20			
6	Mission: Space orange	20		20			
7	Mission: Space green	10					
8	Nemo & Friends	10					
9							
10							
11							

Source: Microsoft Office Excel 2016

- The Excel chart galleries do not automatically construct a bullet chart. Therefore, to construct a bullet chart, we will construct a horizontal bar chart of waiting times with a transparent background, and we will then superimpose this chart on a bar chart containing the appropriate colored regions.

- Select the range of the data A2:B8.

- To construct a horizontal bar chart of the waiting times, select **Insert : Insert Bar Chart (icon) : 2-D Bar : Clustered Bar.**

- Remove the chart gridlines by right-clicking on any one of the gridlines, by selecting Format Gridlines from the pull-down menu, and by selecting No Line in the Format Major Gridlines task pane.

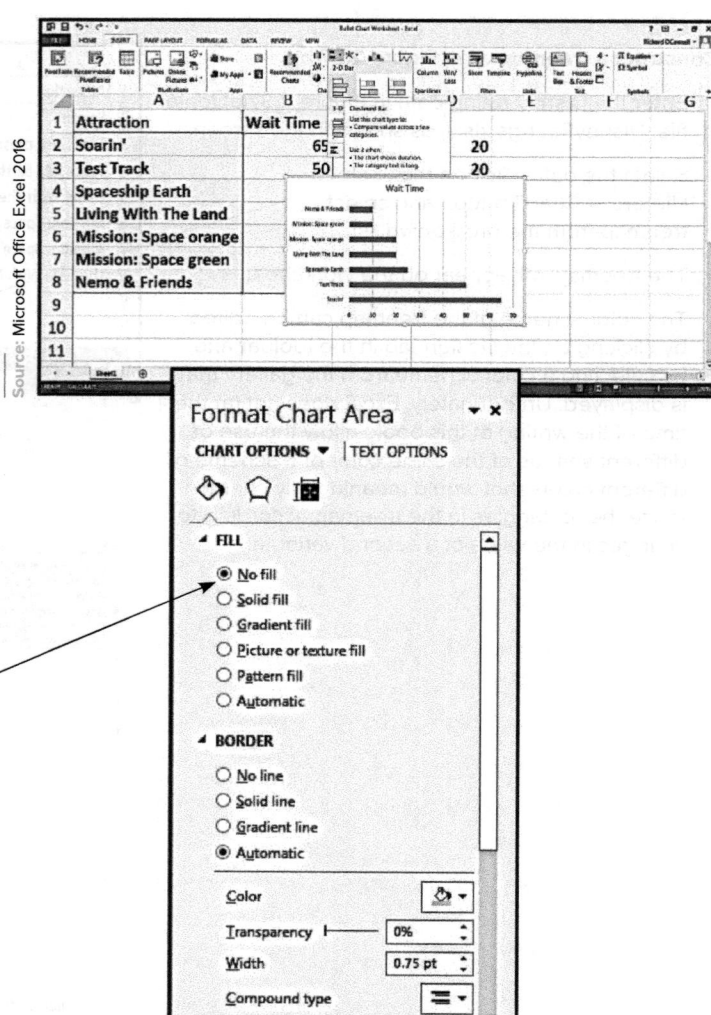

Source: Microsoft Office Excel 2016

- Click in the chart area (say, next to the chart title) and select Format Chart Area from the pull-down menu.

- In the Format Chart Area task pane, select No Fill. When this is done, the background of the bar chart becomes transparent.

- Close the Format Chart Area task pane.

Source: Microsoft Office Excel 2016

- Right-click on any one of the chart bars and select Format Data Series from the pull-down menu.

- In the Format Data Series task pane, click on the Fill & Line (paint bucket) icon and select Solid Fill.

- Then click on the color button and select Black to color the chart bars black (or use dark blue if you wish).

- Close the Format Data Series task pane.

- Select the chart title and delete it.

- We now have a horizontal bar chart of the waiting times having dark bars and a transparent background.

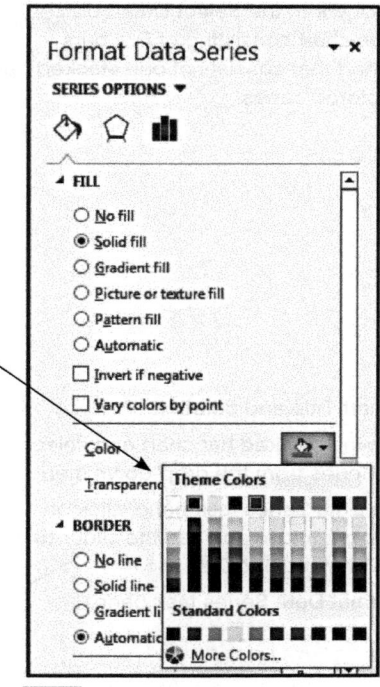

Source: Microsoft Office Excel 2016

- We will now construct a bar chart consisting of colored zones that will serve as the background for the bullet graph.

- Because the categories (good, OK, etc.) that describe the waiting times are 20 units long, enter the value 20 into each cell from D2 to D6. These values have been entered in the first screen of this section on bullet charts.

- Select the cell range D2:D6 using the mouse.

- Select **Insert : Insert Bar Chart (icon) : 2-D Bar : Stacked Bar**

- In the new bar chart that is constructed, remove the gridlines as previously shown.

- Right-click in the Chart Area (say, next to the Chart Title) and choose Select Data from the drop-down menu.

- In the Select Data Source dialog box, click on "Switch Row/Column."

- Click OK in the Select Data Source dialog box.

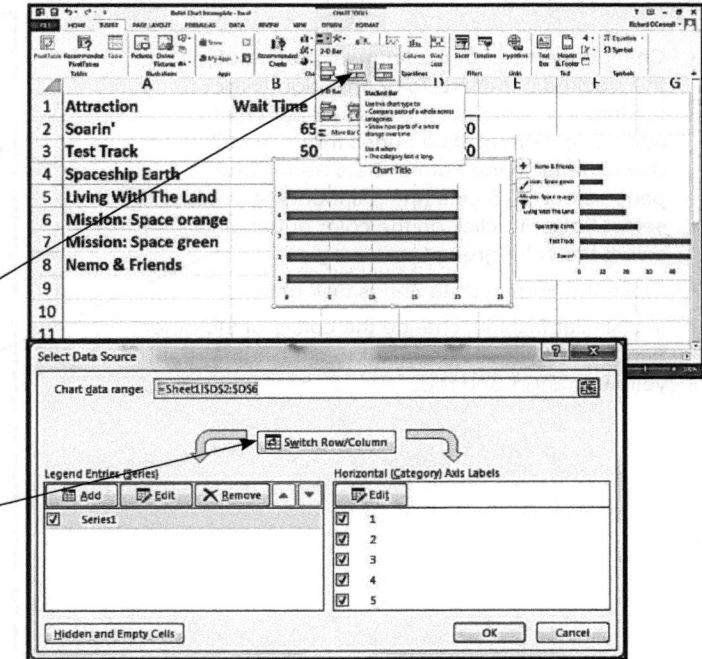

Source: Microsoft Office Excel 2016

- When you click OK in the Select Data Source dialog box, the chart consisting of five bars becomes a chart that consists of one stacked bar having five colored zones.

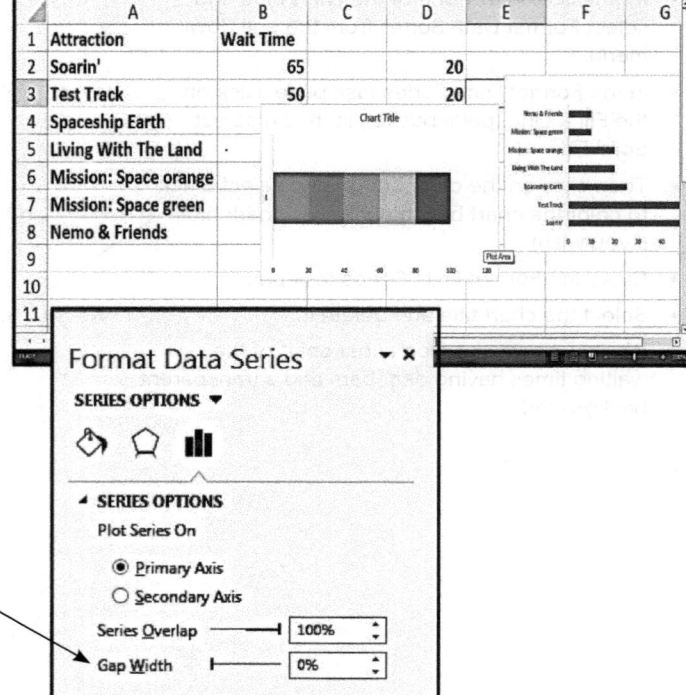

- Select the Chart Title and delete it.
- Right-click on the stacked bar chart and select Format Data Series from the drop-down menu.
- In the Format Data Series task pane, click on the small histogram icon and use the slider to set the gap width equal to 0%.
- Close the Format Data Series task pane.

Source: Microsoft Office Excel 2016

- Change the color of the leftmost portion of the stacked bar chart. To do this, right-click on the leftmost region of the stacked bar chart and select Format Data Series from the pull-down menu. In the Format Data Series task pane select Fill & Line (the paint bucket icon), select Solid Fill, click on the color button and select the color green for this region.
- Close the Format Data Series task pane.
- In a similar fashion, change the colors of the other bar chart regions to (from left to right) light green, yellow, orange, and red.

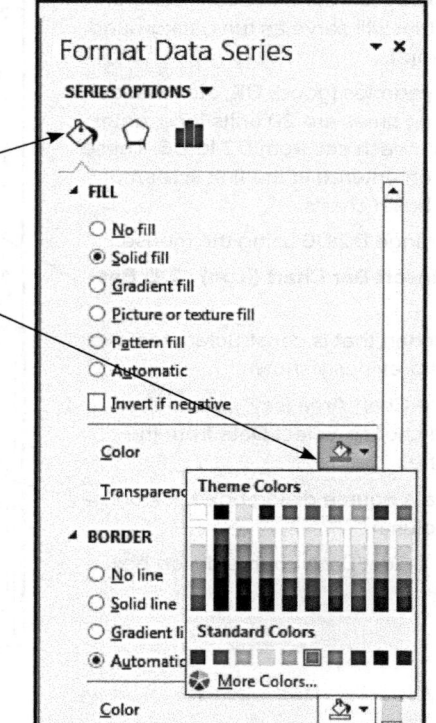

Source: Microsoft Office Excel 2016

- Right-click on the horizontal axis of the stacked bar chart and select Format Axis from the drop-down menu.

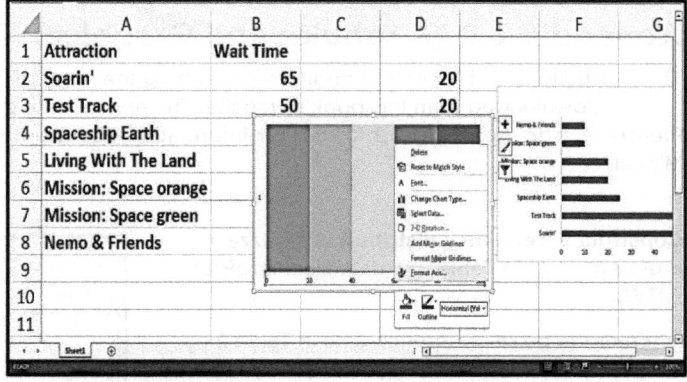

Source: Microsoft Office Excel 2016

- In the Format Axis task pane, select Axis Options and enter 100.0 into the (Bounds) Maximum window.
- Close the Format Axis task pane.
- In a similar fashion, edit the axis of the unstacked bar chart so that the maximum bound equals 100.

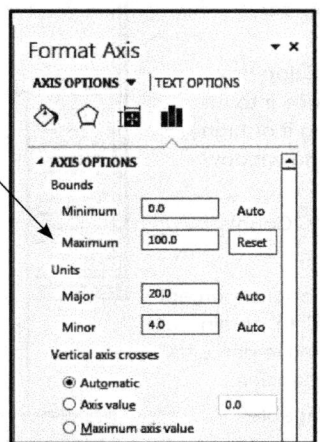

Source: Microsoft Office Excel 2016

- Click on the border of the stacked chart with the colored regions and select Send To Back.
- Drag the unstacked bar chart with the transparent background over the stacked chart with the colored regions and adjust the charts so that the zero horizontal axis values coincide. Then adjust the size of the charts to get the rest of the axis labels to coincide.

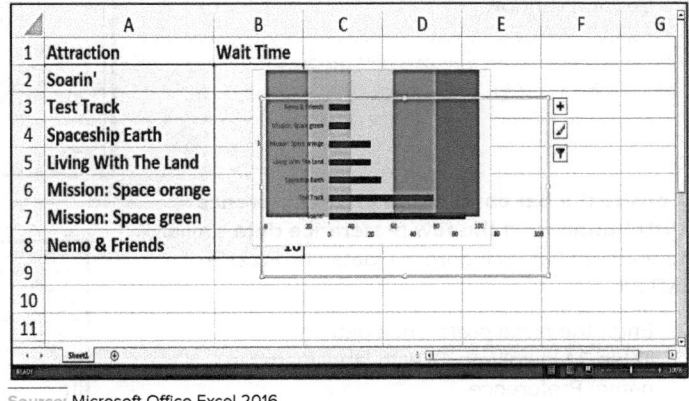

Source: Microsoft Office Excel 2016

- When the axes of the two charts coincide, the bullet graph is complete.

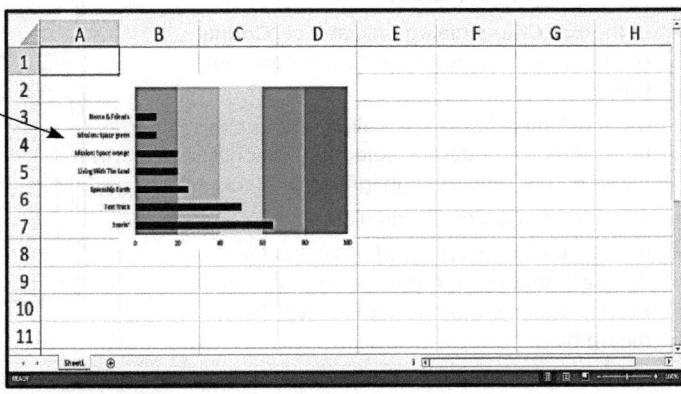

Source: Microsoft Office Excel 2016

Appendix 2.2 ■ Tabular and Graphical Methods Using Minitab

The instructions in this section begin by describing the entry of data into the Minitab data window. Alternatively, the data may be downloaded from this book's website. The appropriate data file name is given at the top of each instruction block. Please refer to Appendix 1.2 for further information about entering data, saving data, and printing results when using Minitab.

Construct a frequency distribution of pizza preferences as in Table 2.2 (data file: PizzaPref .MTW):

- Enter the pizza preference data in Table 2.1 into column C1 with label (variable name) Preference.

- Select **Stat : Tables : Tally Individual Variables**

- In the Tally Individual Variables dialog box, enter the variable name Preference into the Variables window either by typing it or highlighting Preference in the left-hand window and then clicking Select.

- Place a checkmark in the Display "Counts" checkbox to obtain frequencies.

  We would check "Percents" to obtain percent frequencies, "Cumulative counts" to obtain cumulative frequencies, and "Cumulative percents" to obtain cumulative percent frequencies.

- Click OK in the Tally Individual Variables dialog box.

- The frequency distribution is displayed in the Session window.

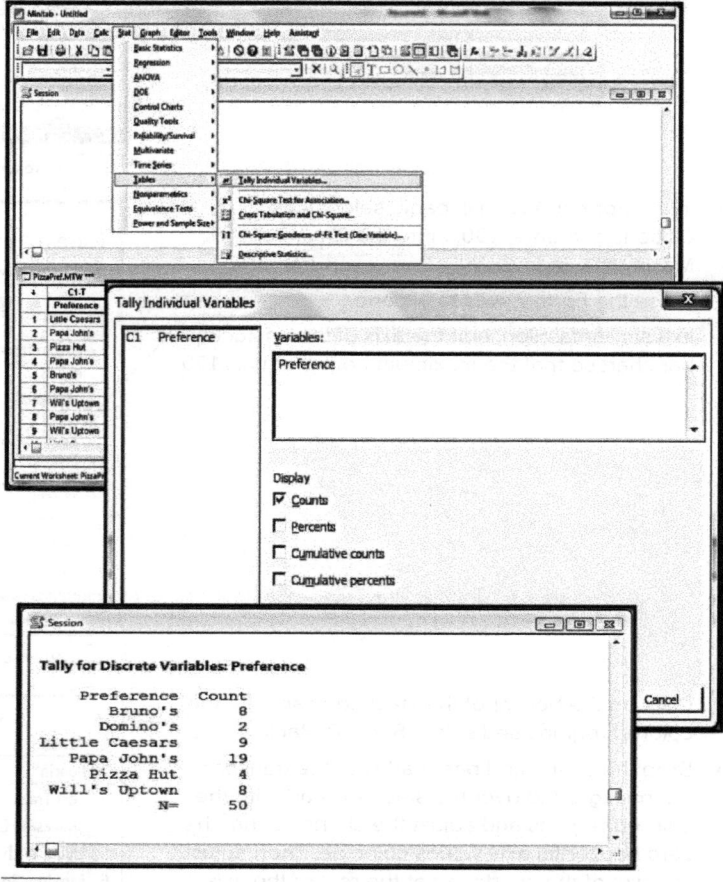

Source: Minitab 18

Construct a bar chart of the pizza preference distribution from the raw preference data similar to the bar chart in Figure 2.1 (data file: PizzaPref .MTW):

- Enter the pizza preference data in Table 2.1 in column C1 with label (variable name) Preference.

- Select **Graph : Bar Chart**

- In the Bar Charts dialog box, select "Counts of unique values" from the "Bars represent" pull-down menu.

- Select "Simple" from the gallery of bar chart types (this is the default selection, which is indicated by the reverse highlighting in black).

- Click OK in the Bar Charts dialog box.

- In the "Bar Chart: Count of unique values, Simple" dialog box, enter the variable name Preference into the "Categorical variables" dialog box.

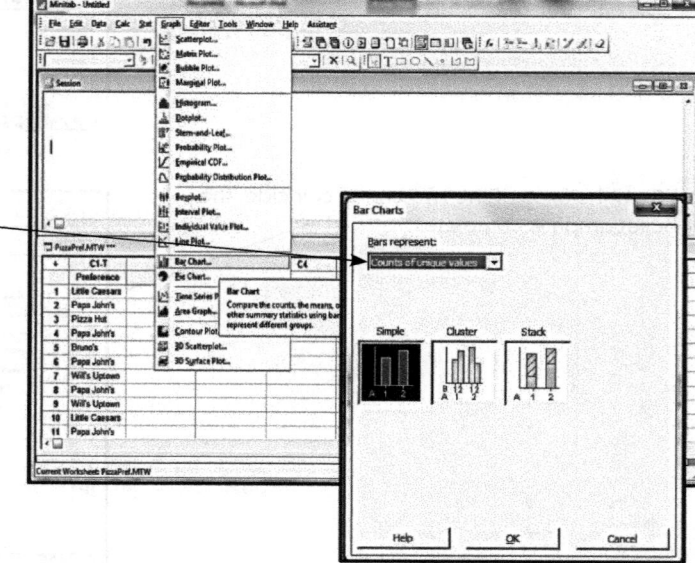

Source: Minitab 18

- To obtain Data labels (numbers above the bars indicating the heights and frequencies of the bars), click on the Labels... button.
- In the "Bar Chart Labels" dialog box, click on the Data Labels tab and select "Use y-value labels." This will produce data labels that equal the category frequencies.
- Click OK in the "Bar Chart Labels" dialog box.
- Click OK in the "Bar Chart Counts of unique values, Simple" dialog box.

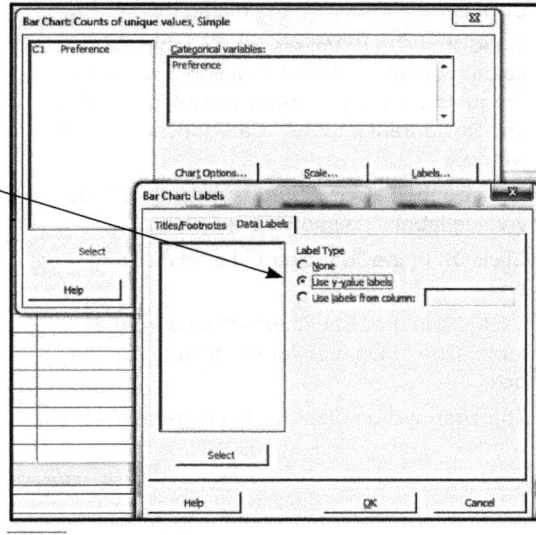

Source: Minitab 18

- The bar chart will be displayed in a graphics window. The chart may be edited by right-clicking on various portions of the chart and by using the pop-up menus that appear. See Appendix 1.2 for more details.
- If you prefer to create a percent frequency bar chart (as in Figure 2.2) rather than a frequency bar chart, click on the Chart Options... button and select "Show Y as Percent" in the Bar Chart— Options dialog box.

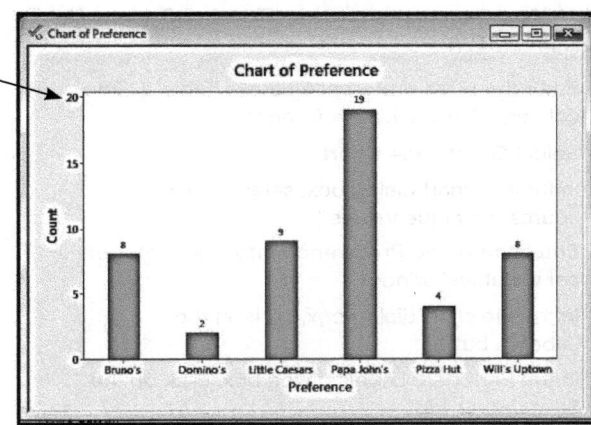

Source: Minitab 18

Construct a bar chart from the tabular frequency distribution of pizza preferences in Table 2.2 (data file: PizzaFreq.MTW):

- Enter the pizza preference data in table form with unique restaurants in column C1 and their associated frequencies in column C2.
- Select **Graph : Bar Chart**
- In the Bar Charts dialog box, select "Values from a table" in the "Bars represent" pull-down menu.
- Select "One column of values, Simple" from the gallery of bar types.

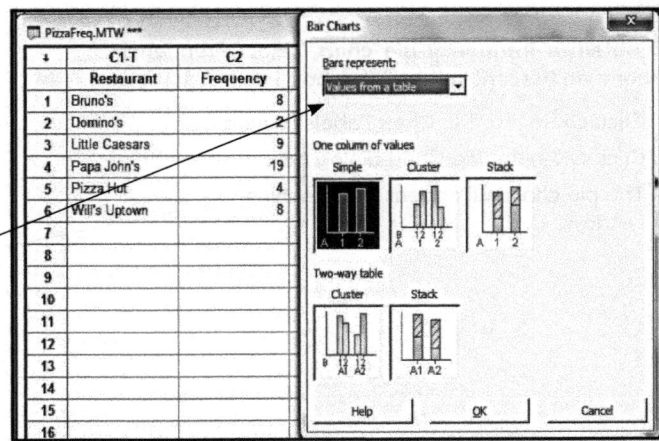

Source: Minitab 18

- Click OK in the Bar Charts dialog box.
- In the "Bar Chart: Values from a table, One column of values, Simple" dialog box, enter Frequency into the "Graph variables" window and Restaurant into the "Categorical variable" window.
- Click on the Labels... button and select "Use y-value labels" as shown previously.
- Click OK in the "Bar Chart—Labels" dialog box.
- Click OK in the "Bar Chart—Values from a table, One column of values, Simple" dialog box.
- The chart will be displayed in a graphics window.

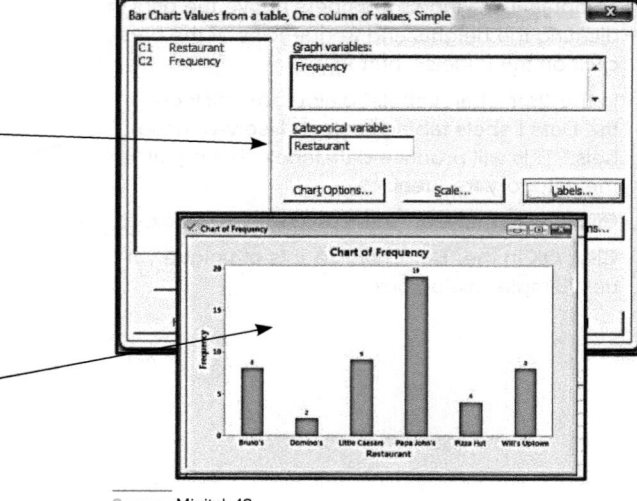

Source: Minitab 18

Construct a pie chart of pizza preference percentages similar to that shown in Figure 2.3 (data file: PizzaPref.MTW)

- Enter the pizza preference data in Table 2.1 into column C1 with label Preference.
- Select **Graph : Pie Chart**
- In the Pie chart dialog box, select "Chart counts of unique values."
- Enter the name Preference into the "Categorical variables" window.
- In the Pie chart dialog box, click on the Labels... button.
- In the Pie Chart: Labels dialog box, click on the Slice Labels tab.
- Place checkmarks in the Category name, Percent, and "Draw a line from label to slice" checkboxes.

To obtain a frequency pie chart, select Frequency rather than Percent in this dialog box.

- Click OK in the "Pie Chart Labels" dialog box.
- Click OK in the Pie Chart dialog box.
- The pie chart will appear in a graphics window.

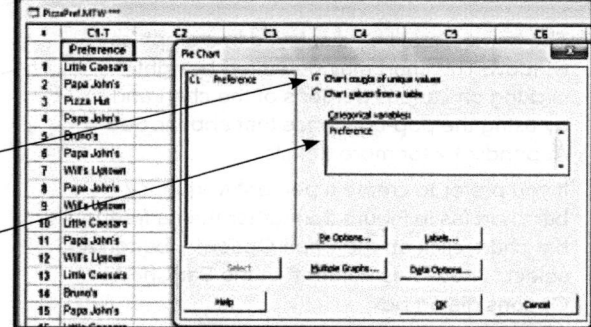

Source: Minitab 18

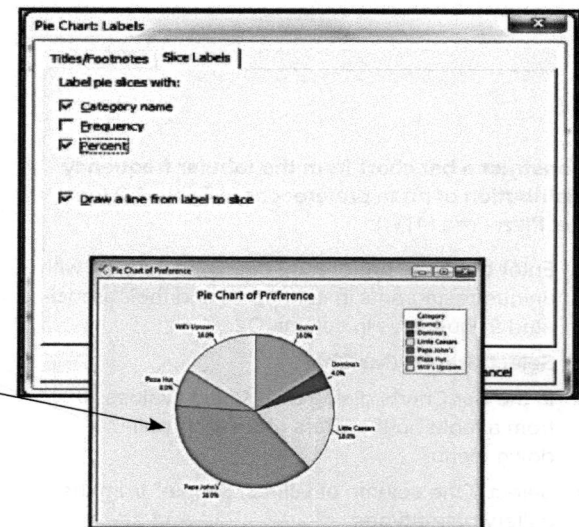

Source: Minitab 18

Construct a pie chart from the tabular distribution of pizza preferences in Table 2.3 (data file: PizzaPercents .MTW):

- Enter the pizza preference percentage distribution from Table 2.3 or from the data file.
- Select **Graph : Pie Chart**
- In the Pie Chart dialog box, select "Chart values from a table."
- Enter Restaurant into the "Categorical variable" window.
- Enter 'Percent Freq' into the "Summary variables" window. The single quotes are required because this variable has a two-word name.
- Continue by following the previous directions for adding data labels and for generating the pie chart.

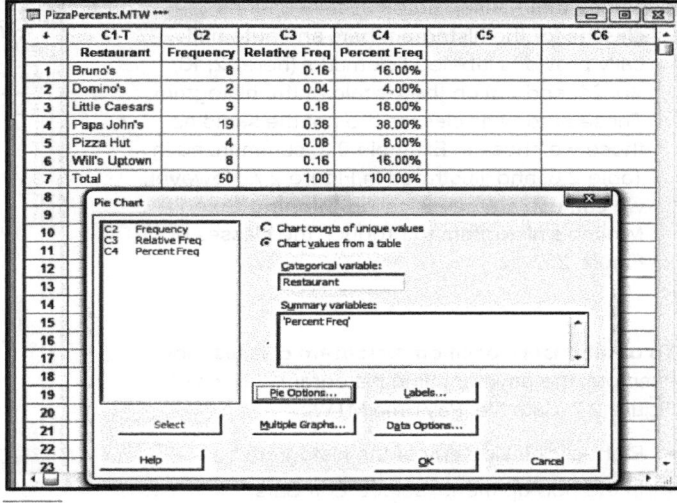

Source: Minitab 18

Construct a frequency histogram of the payment times as in Figure 2.7 (data file: PayTime.MTW):

- Enter the payment time data from Table 2.4 into column C1.
- Select **Graph : Histogram**
- In the Histogram dialog box, select Simple from the gallery of histogram types and click OK.
- In the "Histogram: Simple" dialog box, enter the variable name PayTime into the Graph variables window and click on the Scale button.
- In the "Histogram: Scale" dialog box, click on the "Y-Scale Type" tab and select **Frequency** to obtain a frequency histogram. We would select **Percent** to request a percent frequency histogram. Then click OK in the "Histogram: Scale" dialog box.

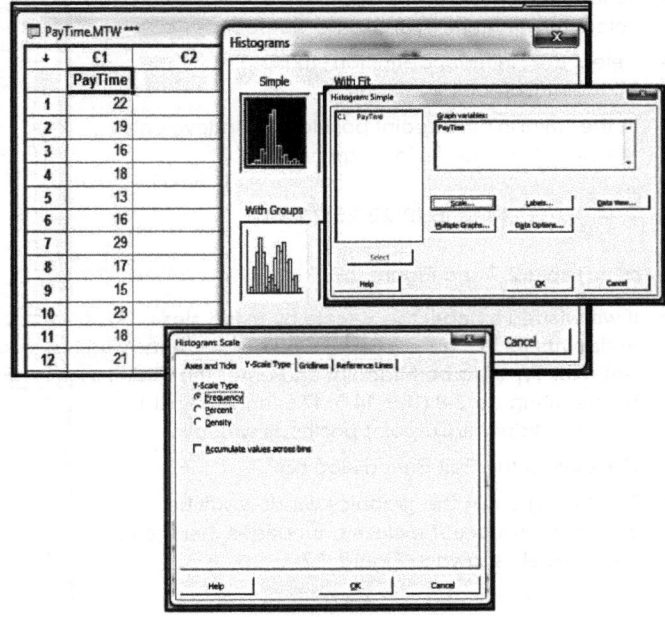

Source: Minitab 18

- Data labels are requested in the same way as we have shown for bar charts. Click on the Labels... button in the "Histogram: Simple" dialog box. In the "Histogram: Labels" dialog box, click on the Data Labels tab and select "Use y-value labels." Then click OK in the "Histogram: Labels" dialog box.
- To create the histogram, click OK in the "Histogram: Simple" dialog box.
- The histogram will appear in a graphics window and can be edited, printed, or copied and pasted into a word processing document as described in Appendix 1.2.

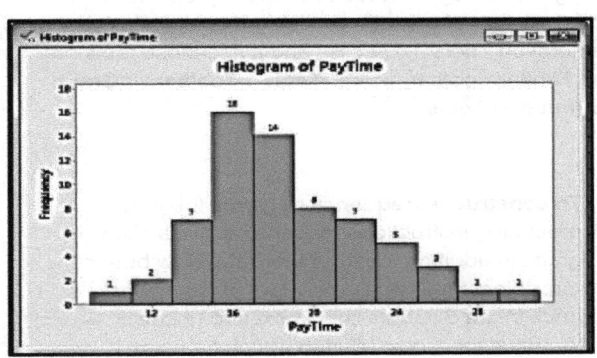

Source: Minitab 18

- Notice that Minitab automatically defines classes for the histogram bars and automatically provides labeled tick marks (here 12, 16, 20, 24, and 28) on the x-scale of the histogram. These automatic classes are not the same as those we chose in Example 2.2, summarized in Table 2.6, and illustrated in Figure 2.7. However, we can use a process called "**binning**" to edit Minitab's histogram to produce the classes of Figure 2.7.

To obtain user-specified histogram classes—for example, the payment time histogram classes of Figure 2.7 (data file: PayTime.MTW):

- Right-click inside any of the histogram bars.
- In the pop-up menu, select "Edit Bars."
- In the "Edit Bars" dialog box, select the Binning tab.
- To label the x-scale by using class boundaries, select the "Interval Type" to be Cutpoint.
- Select the "Interval Definition" to be Midpoint/Cutpoint positions.
- In the "Midpoint/Cutpoint positions" window, enter the class boundaries (or cutpoints)

        10 13 16 19 22 25 28 31

as in Table 2.7 and Figure 2.7.

- If we wished to label the x-scale by using class midpoints as in Figure 2.8, we would select the "Interval Type" to be Midpoint and enter the midpoints of Figure 2.8 (11.5, 14.5, 17.5, and so forth) into the Midpoint/Cutpoint positions window.
- Click OK in the Edit Bars dialog box.
- The histogram in the graphics window will be edited to produce the class boundaries, bars, and x-axis labels shown in Figure 2.7.

Frequency Polygons and Ogives: Minitab does not have automatic procedures for constructing frequency polygons and ogives. However, these graphics can be constructed quite easily by using the Minitab Graph Annotation Tools. To access and place these tools on the Minitab toolbar, select **Tools : Toolbars : Graph Annotation Tools.**

- **To construct a frequency polygon**, follow the preceding instructions for constructing a histogram. In addition, click on the Data View button; select the Data Display tab; place a checkmark in the Symbols checkbox; and uncheck the Bars checkbox. Minitab will plot a point above each histogram class rather than a bar. Now select the polygon tool

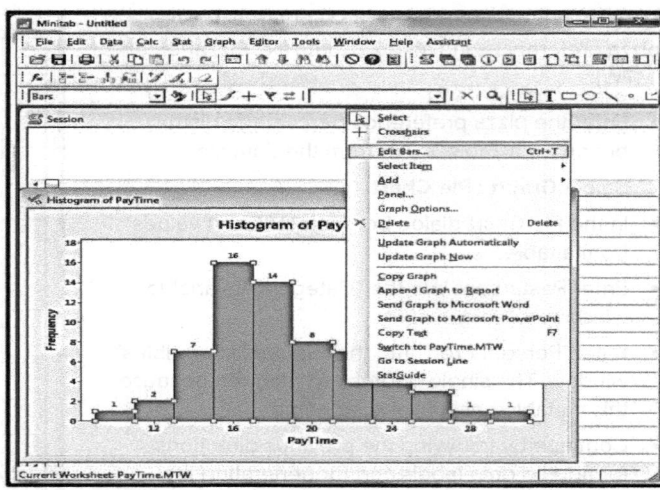
Source: Minitab 18

Source: Minitab 18

Source: Minitab 18

Source: Minitab 18

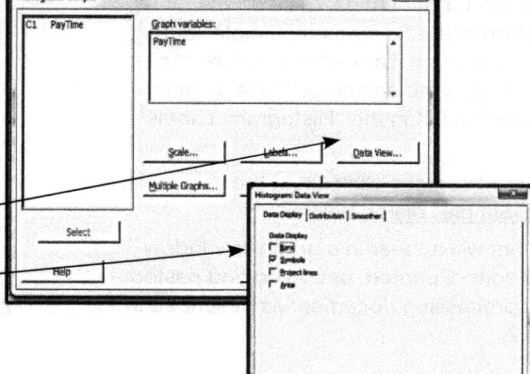

from the Graph Annotation Tools toolbar and draw connecting lines to form the polygon. Instructions for using the polygon tool can be found in the Minitab help resources.

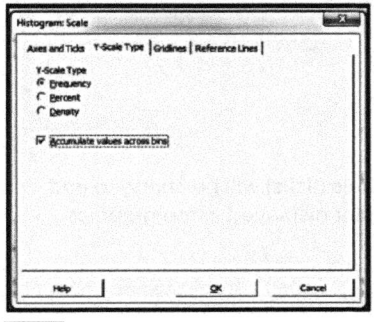

Source: Minitab 18

- **To construct an ogive**, follow the instructions for constructing a frequency polygon. In addition, click on the Scale button; select the "Y-Scale Type" tab; and check the "Accumulate values across bins" checkbox. This will result in a plot of cumulative frequencies above each histogram class. Now select the polyline tool

from the Graph Annotation Tools toolbar and draw connecting lines to form the ogive. Instructions for using the polyline tool can be found in the Minitab help resources.

Source: Minitab 18

Construct a dot plot of the exam scores as in Figure 2.18(a) (data file: FirstExam.MTW):

- Enter the scores for exam 1 in Table 2.8 into column C1 with variable name 'Exam 1.'
- Select **Graph : Dot Plot**
- In the Dotplots dialog box, select "One Y, Simple" from the gallery of dot plots.
- Click OK in the Dotplots dialog box.
- In the "Dotplot: One Y, Simple" dialog box, enter 'Exam 1' into the "Graph variables" window. Be sure to include the single quotes.
- Click OK in the "Dotplot: One Y, Simple" dialog box.
- The dotplot will be displayed in a graphics window.

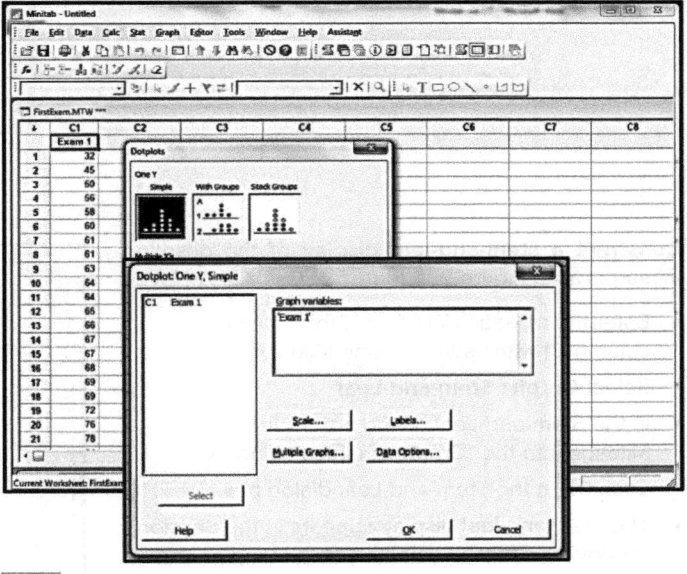

Source: Minitab 18

- To change the *x*-axis labels (or ticks), right click on any of the existing labels (for instance, the 45) and select "Edit X Scale..." from the pop-up menu.

- In the Edit Scale dialog box, select the Scale tab and select "Positon of Ticks" as the "Major Tick Positions" setting.

- Enter the desired ticks (30 40 50 60 70 80 90 100) into the "Positon of Ticks" window and click OK in the Edit Scale dialog box.

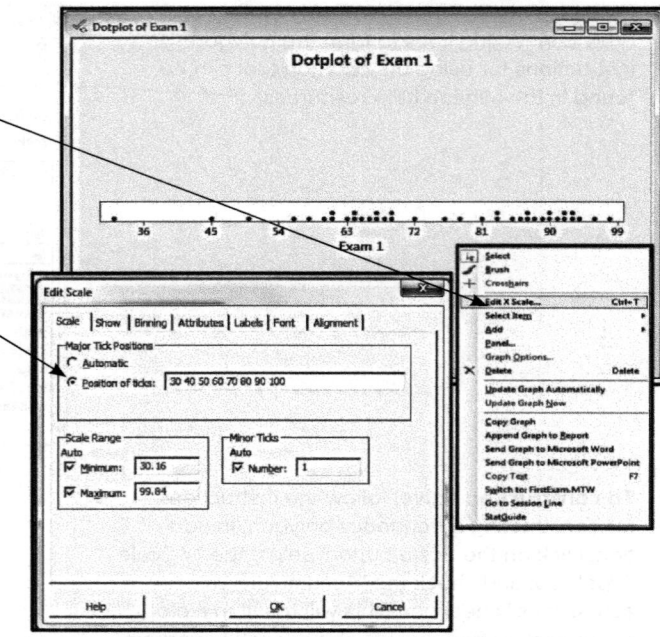

Source: Minitab 18

- The *x*-axis labels (ticks) will be changed and the new dot plot displayed in the graphics window.

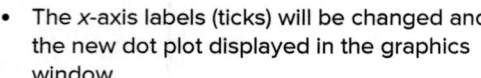

Source: Minitab 18

Construct a stem-and-leaf display of the gasoline mileages as in Figure 2.19 (data file: GasMiles.MTW):

- Enter the mileage data from Table 2.14 into column C1 with variable name Mileage.

- Select **Graph : Stem-and-Leaf**

- In the Stem-and-Leaf dialog box, enter Mileage into the "Graph variables" window.

- Click OK in the Stem-and-Leaf dialog box.

- The stem-and-leaf display appears in the Session window and can be selected for printing or copied and pasted into a word processing document. (See Appendix 1.2.)

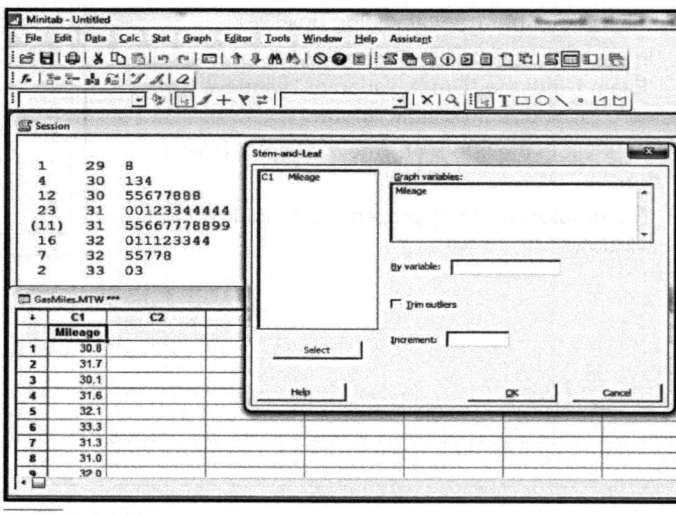

Source: Minitab 18

Construct a contingency table of fund type versus level of client satisfaction as in Table 2.17 (data file: Invest.MTW):

- Enter the client satisfaction data from Table 2.16 with client number in column C1 with variable name Client and with fund type and satisfaction rating in columns C2 and C3 with variable names Fund and Satisfaction.

The default ordering for different levels of each categorical variable in the contingency table will be alphabetical (BOND, STOCK, TAXDEF for Fund and HIGH, LOW, MED for Satisfaction). To change the ordering to HIGH, MED, LOW for Satisfaction:

- Click on any cell in column C3 (Satisfaction).
- Select **Editor : Column : Value order**
- In the "Value Order for C3 (Satisfaction)" dialog box, select the "User-specified order" option.
- In the "Define and order (one value per line)" window, specify the order HIGH, MED, LOW.
- Click OK in the "Value Order for C3 (Satisfaction)" dialog box.

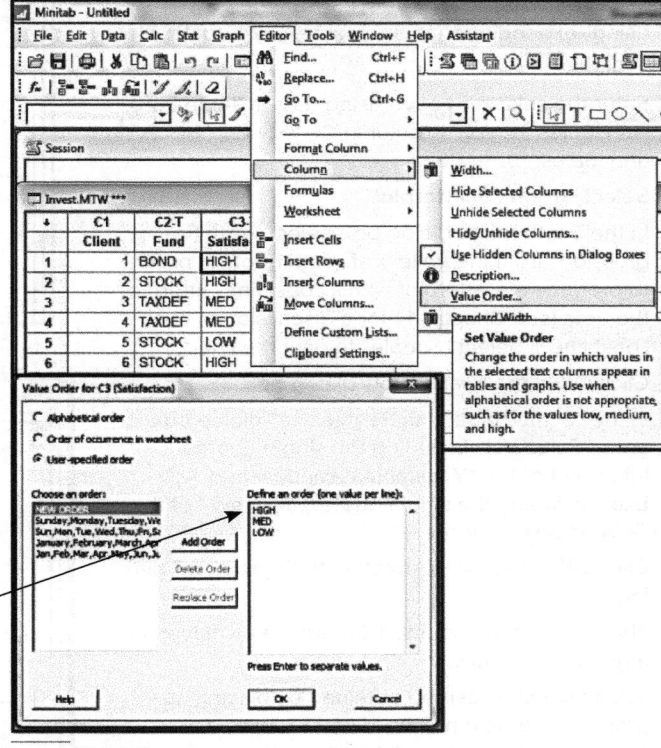

Source: Minitab 18

To construct the contingency table:

- Select **Stat : Tables : Cross Tabulation and Chi-Square**
- In the "Cross Tabulation and Chi-Square" dialog box, use the "Raw data" option and enter Fund into the "Rows" window and Satisfaction into the "Columns" window.
- Check "Counts" in the Display choices. Check "Row percents" or "Column percents" if you wish to produce a table with row or column percentages.
- Click OK in the "Cross Tabulation and Chi-Square" dialog box to obtain results in the session window.

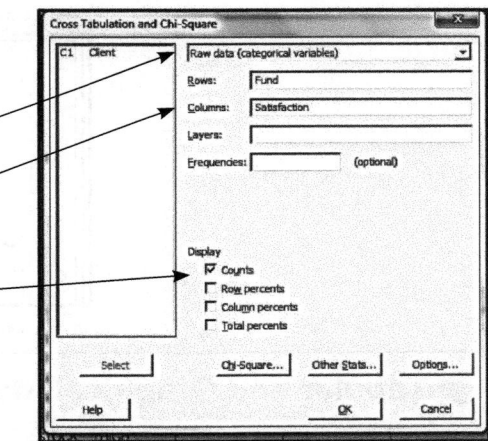

Source: Minitab 18

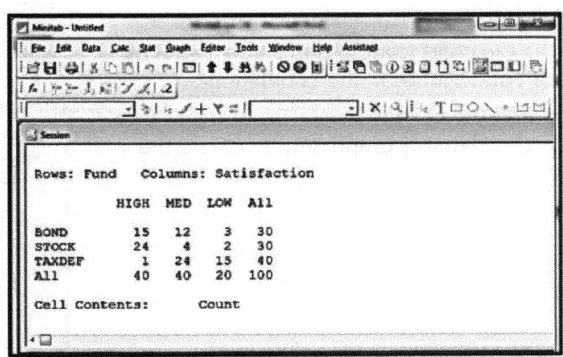

Rows: Fund   Columns: Satisfaction

	HIGH	MED	LOW	All
BOND	15	12	3	30
STOCK	24	4	2	30
TAXDEF	1	24	15	40
All	40	40	20	100

Cell Contents: Count

Source: Minitab 18

Construct a scatter plot of sales volume versus advertising expenditure as in Figure 2.24 (data file: SalesPlot.MTW):

- Enter the sales and advertising data in Table 2.20 with 'Sales Region' in column C1, 'Adv Exp' in C2, and 'Sales Vol' in C3.

- Select **Graph : Scatterplot**

- In the Scatterplots dialog box, select "With Regression" from the gallery of scatterplots in order to produce a scatterplot with a "best line" fitted to the data (see Chapter 14 for a discussion of this "best line"). Select "Simple" to omit the line.

- Click OK in the Scatterplots dialog box.

- In the "Scatterplot: With Regression" dialog box, enter 'Sales Vol' (including the single quotes) into row 1 of the "Y variables" window and 'Adv Exp' (including the single quotes) into row 1 of the "X variables" window.

- Click OK in the "Scatterplot: With Regression" dialog box.

- The scatter plot and fitted line will be displayed in a graphics window.

- Additional plots can be obtained by placing appropriate variable names in other rows in the "Y variables" and "X variables" windows.

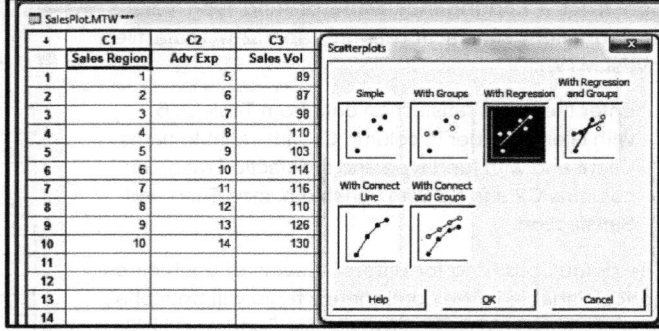

Source: Minitab 18

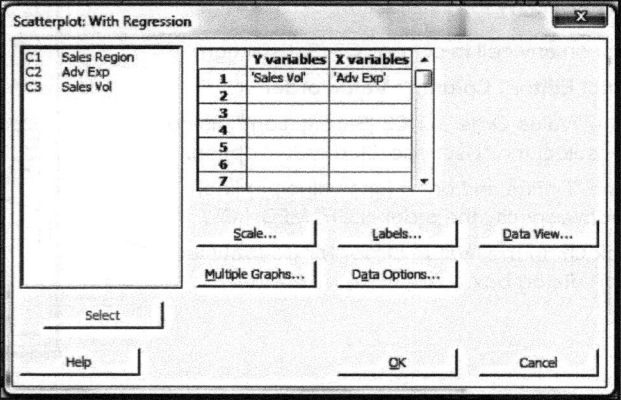

Source: Minitab 18

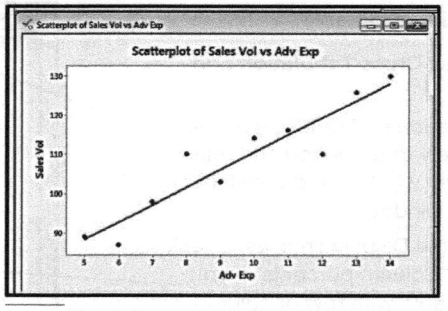

Source: Minitab 18

Appendix 2.3 ■ Tabular and Graphical Methods Using JMP

Pie chart in Figure 2.3 (data file: PizzaPref.jmp):

- After opening the data file, select **Graph : Graph Builder.**

- Click on the pie chart icon 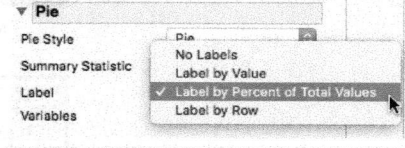 and select **Label by Percent of Total Values** in the **Label** menu.

- Drag and drop the column name "Preference" into the center of the empty graph area to create the pie chart.

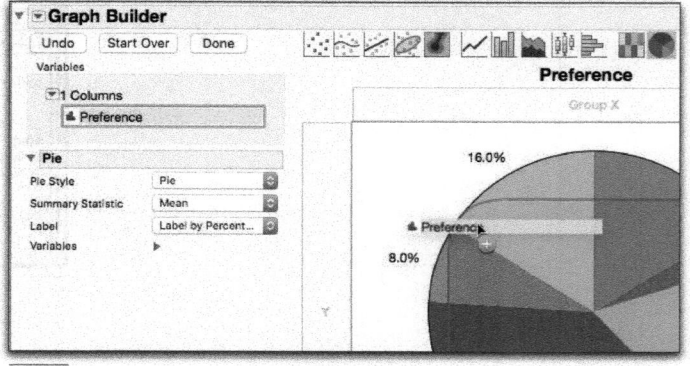

Source: JMP Pro 14

Frequency distributions and **bar charts** (data file: PizzaPref.jmp):

- After opening the data file, select **Analyze : Distribution**.

- In the Distribution dialog box, select the "Preference" column and click the "Y, Columns" button and click "OK."

- From the red triangle menu beside "Preference" select **Histogram Options : Show Percents, Histogram Options : Show Counts,** and **Display Options : Horizontal Layout.**

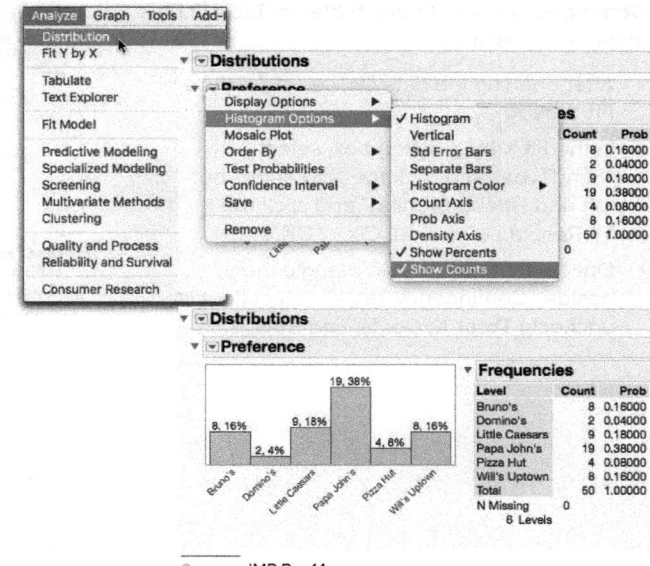

Source: JMP Pro 14

Histogram in Figure 2.9 (data file: GasMiles.jmp):

- Follow the exact same steps as above (for frequency distributions and bar charts with the PizzaPref data) using the MPG column to obtain a histogram like the one in Figure 2.9.

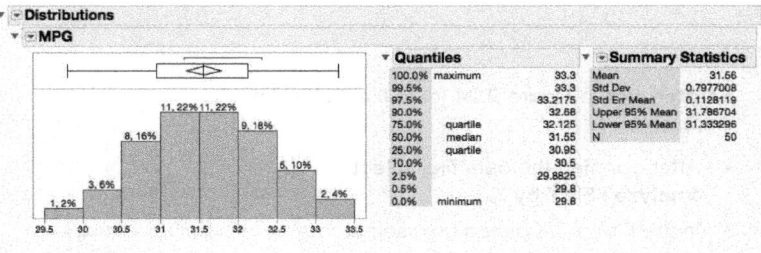

Source: JMP Pro 14

Stem-and-leaf display in Figure 2.21 (data file: PayTime.jmp):

- After opening the data file, select **Analyze : Distribution**.

- In the Distribution dialog box, select the "PayTime" column and click the "Y, Columns" button and click "OK."

- From the red triangle menu beside "PayTime" select **Stem and Leaf.**

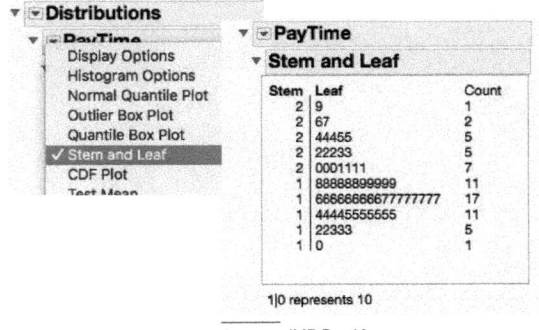

Source: JMP Pro 14

Contingency table (pivot table) in Table 2.17 (data file: Invest.jmp):

- After opening the data file, select **Analyze : Fit Y by X.**

- In the Fit Y by X dialog box, select "FundType" and click the "X, Factor" button and select "SRating" and click the "Y, Response" button. Click "OK."

- Optionally, from the red triangle menu beside "Contingency Table" select (that is, uncheck) **Total %, Col %,** and **Row %.**

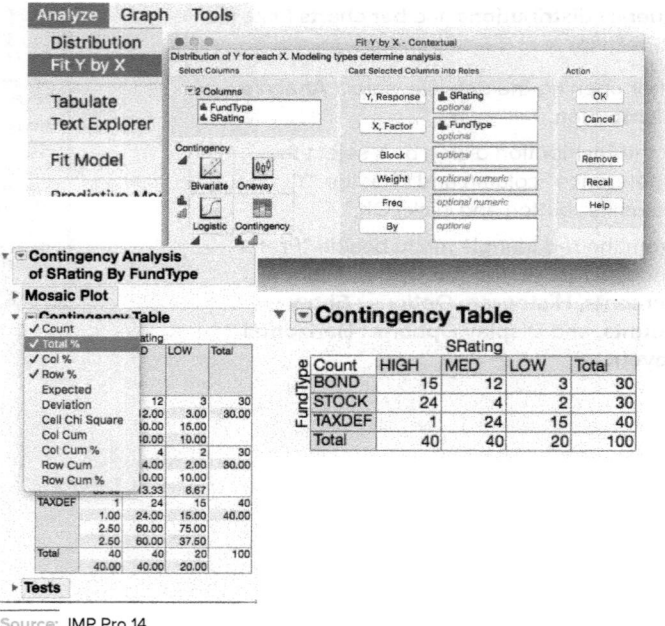

Contingency Table

		SRating		
Count	HIGH	MED	LOW	Total
BOND	15	12	3	30
STOCK	24	4	2	30
TAXDEF	1	24	15	40
Total	40	40	20	100

Source: JMP Pro 14

Scatter plot in Figure 2.24 (data file: SalesPlot .jmp):

- After opening the data file, select **Analyze : Fit Y by X.**

- In the Fit Y by X dialog box, select "Advertising Expenditure" and click the "X, Factor" button and select "Sales Volume" and click the "Y, Response" button. Click "OK."

- Optionally, to obtain the line in Figure 2.24, from the red triangle menu beside "Bivariate Fit of Sales Volume by Advertising Expenditure" select **Fit Line.**

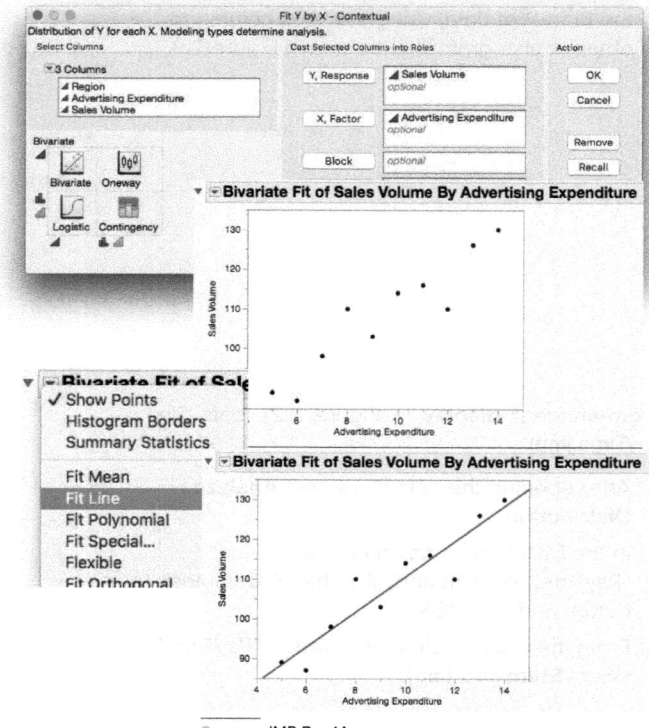

Source: JMP Pro 14

Treemap in Figure 2.37(b) (data file: Disney Ratings.jmp):

- After opening the data file, select **Graph : Graph Builder.**

- Click on the Treemap icon 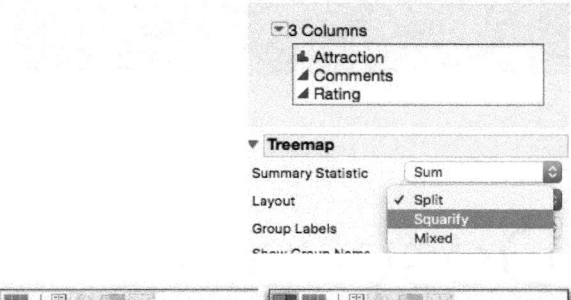 (to the right of the pie chart icon) and select **Squarify** in the **Layout** menu.

- Drag and drop the column names as follows: "Attraction" into the center of the empty graph area, "Comments" into the "Size" drop zone, and "Rating" into the "Color" drop zone.

- From the red triangle menu beside "Graph Builder" select **Continuous Color Theme…** and under "Sequential" click on the second color gradient (which is a "White to Blue" theme). Click "OK."

- Optionally, resize the window (make it horizontally wider) to obtain the exact treemap in Figure 2.37(b).

- Optionally, from the red triangle menu beside "Graph Builder" select (that is, uncheck) **Show Control Panel.**

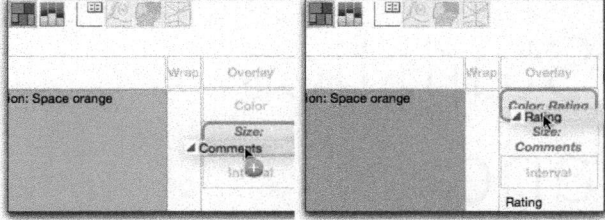

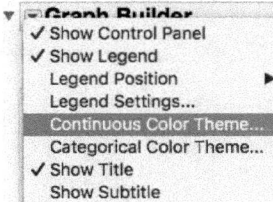

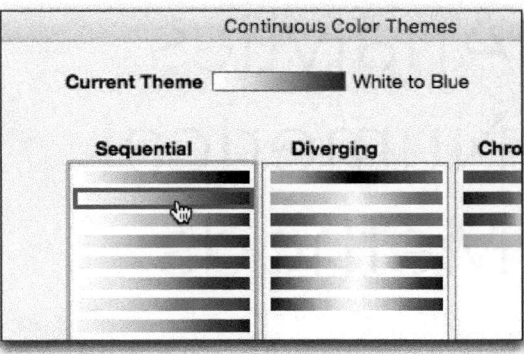

Source: JMP Pro 14

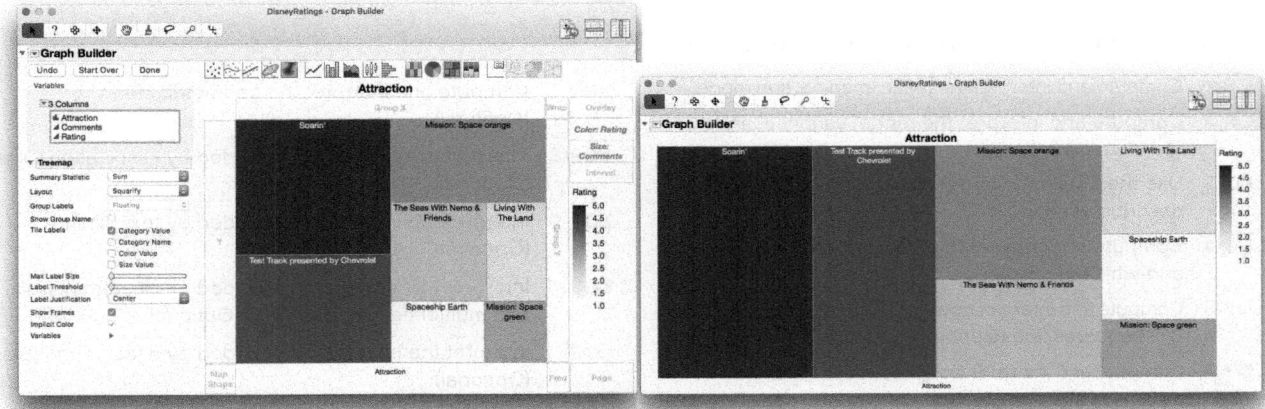

Source: JMP Pro 14

Design Elements: (CD): ©Comstock Images/Alamy; (All Others): ©McGraw-Hill Education

©Hero Images/Getty Images

CHAPTER 3

Descriptive Statistics and Analytics: Numerical Methods

Learning Objectives

After mastering the material in this chapter, you will be able to:

LO3-1 Compute and interpret the mean, median, and mode.

LO3-2 Compute and interpret the range, variance, and standard deviation.

LO3-3 Use the Empirical Rule and Chebyshev's Theorem to describe variation.

LO3-4 Compute and interpret percentiles, quartiles, and box-and-whiskers displays.

LO3-5 Compute and interpret covariance, correlation, and the least squares line (Optional).

LO3-6 Compute and interpret weighted means and the mean and standard deviation of grouped data (Optional).

LO3-7 Compute and interpret the geometric mean (Optional).

LO3-8 Interpret the information provided by association rules (Optional).

LO3-9 Interpret the information provided by text mining (Optional).

LO3-10 Interpret the information provided by cluster analysis and multidimensional scaling (Optional).

LO3-11 Interpret the information provided by a factor analysis (Optional).

Chapter Outline

Part 1 Numerical Methods of Descriptive Statistics
- **3.1** Describing Central Tendency
- **3.2** Measures of Variation
- **3.3** Percentiles, Quartiles, and Box-and-Whiskers Displays
- **3.4** Covariance, Correlation, and the Least Squares Line (Optional)
- **3.5** Weighted Means and Grouped Data (Optional)
- **3.6** The Geometric Mean (Optional)

Part 2 Numerical Descriptive Analytics (Optional)
- **3.7** Association Rules (Optional)
- **3.8** Text Mining (Optional)
- **3.9** Cluster Analysis and Multidimensional Scaling (Optional)
- **3.10** Factor Analysis (Optional and Requires Section 3.4)

Note: The reader may study Sections 3.7, 3.8, 3.9, and 3.10 in any order without loss of continuity.

130

n this chapter we study numerical methods for describing the important aspects of a set of measurements. If the measurements are values of a quantitative variable, we often describe (1) what a typical measurement might be and (2) how the measurements vary, or differ, from each other. For example, in the car mileage case we might estimate (1) a typical EPA gas mileage for the new midsize model and (2) how the EPA mileages vary from car to car. Or, in the marketing research case, we might estimate (1) a typical bottle design rating and (2) how the bottle design ratings vary from consumer to consumer.

Taken together, the graphical displays of Chapter 2 and the numerical methods of this chapter give us a basic understanding of the important aspects of a set of measurements. We will illustrate this by continuing to analyze the car mileages, payment times, bottle design ratings, and cell phone usages introduced in Chapters 1 and 2. We will also use the numerical methods of this chapter to give an optional introduction to four important techniques of numerical descriptive analytics: *association rules*, *text mining*, *cluster analysis*, and *factor analysis*.

3.1 Describing Central Tendency

The mean, median, and mode

In addition to describing the shape of the distribution of a sample or population of measurements, we also describe the data set's **central tendency.** A measure of central tendency represents the *center* or *middle* of the data. Sometimes we think of a measure of central tendency as a *typical value.* However, as we will see, not all measures of central tendency are necessarily typical values.

One important measure of central tendency for a population of measurements is the **population mean.** We define it as follows:

> The **population mean,** which is denoted μ and pronounced *mew,* is the average of the population measurements.

LO3-1
Compute and interpret the mean, median, and mode.

More precisely, the population mean is calculated by adding all the population measurements and then dividing the resulting sum by the number of population measurements. For instance, suppose that Chris is a college junior majoring in business. This semester Chris is taking five classes and the numbers of students enrolled in the classes (that is, the class sizes) are as follows:

Class	Class Size 🔵 ClassSizes
Business Law	60
Finance	41
International Studies	15
Management	30
Marketing	34

The mean μ of this population of class sizes is

$$\mu = \frac{60 + 41 + 15 + 30 + 34}{5} = \frac{180}{5} = 36$$

Because this population of five class sizes is small, it is possible to compute the population mean. Often, however, a population is very large and we cannot obtain a measurement for each population element. Therefore, we cannot compute the population mean. In such a case, we must estimate the population mean by using a sample of measurements.

In order to understand how to estimate a population mean, we must realize that the population mean is a **population parameter.**

> A **population parameter** is a number calculated using the population measurements that describes some aspect of the population. That is, a population parameter is a descriptive measure of the population.

There are many population parameters, and we discuss several of them in this chapter. The simplest way to estimate a population parameter is to make a **point estimate,** which is a one-number estimate of the value of the population parameter. Although a point estimate is a guess of a population parameter's value, it should not be a *blind guess*. Rather, it should be an educated guess based on sample data. One sensible way to find a point estimate of a population parameter is to use a **sample statistic.**

> A **sample statistic** is a number calculated using the sample measurements that describes some aspect of the sample. That is, a sample statistic is a descriptive measure of the sample.

The sample statistic that we use to estimate the population mean is the **sample mean,** which is denoted as $\bar{x}$ (pronounced *x bar*) and is the average of the sample measurements.

In order to write a formula for the sample mean, we employ the letter n to represent the number of sample measurements, and we refer to n as the **sample size.** Furthermore, we denote the sample measurements as $x_1, x_2, \ldots, x_n$. Here x_1 is the first sample measurement, x_2 is the second sample measurement, and so forth. We denote the last sample measurement as x_n. Moreover, when we write formulas we often use *summation notation* for convenience. For instance, we write the sum of the sample measurements

$$x_1 + x_2 + \cdots + x_n$$

as $\sum_{i=1}^{n} x_i$. Here the symbol Σ simply tells us to add the terms that follow the symbol. The term x_i is a generic (or representative) observation in our data set, and the $i = 1$ and the n indicate where to start and stop summing. Thus

$$\sum_{i=1}^{n} x_i = x_1 + x_2 + \cdots + x_n$$

We define the sample mean as follows:

> The **sample mean** $\bar{x}$ is defined to be
>
> $$\bar{x} = \frac{\sum_{i=1}^{n} x_i}{n} = \frac{x_1 + x_2 + \cdots + x_n}{n}$$
>
> and is the **point estimate of the population mean** μ.

©monkeybusinessimages/Getty Images RF

 EXAMPLE 3.1 The Car Mileage Case: Estimating Mileage

In order to offer its tax credit, the federal government has decided to define the "typical" EPA combined city and highway mileage for a car model as the mean μ of the population of EPA combined mileages that would be obtained by all cars of this type. Here, using the mean to represent a typical value is probably reasonable. We know that some individual cars will get mileages that are lower than the mean and some will get mileages that are above it. However, because there will be many thousands of these cars on the road, the mean mileage obtained by these cars is probably a reasonable way to represent the model's overall fuel economy. Therefore, the government will offer its tax credit to any automaker selling a midsize model equipped with an automatic transmission that achieves a mean EPA combined mileage of at least 31 mpg.

To demonstrate that its new midsize model qualifies for the tax credit, the automaker in this case study wishes to use the sample of 50 mileages in Table 3.1 to estimate μ, the model's mean mileage. Before calculating the mean of the entire sample of 50 mileages, we will illustrate the formulas involved by calculating the mean of the first five of these mileages.

TABLE 3.1   The Sample of 50 Mileages in the Car Mileage Case

(a) A sample of 50 mileages   DS GasMiles

30.8	30.8	32.1	32.3	32.7
31.7	30.4	31.4	32.7	31.4
30.1	32.5	30.8	31.2	31.8
31.6	30.3	32.8	30.7	31.9
32.1	31.3	31.9	31.7	33.0
33.3	32.1	31.4	31.4	31.5
31.3	32.5	32.4	32.2	31.6
31.0	31.8	31.0	31.5	30.6
32.0	30.5	29.8	31.7	32.3
32.4	30.5	31.1	30.7	31.4

(b) Fuel economy ratings for the 2017 Nissan Altima

■ 2017 Nissan Altima 6 cyl, 3.5 L, Automatic (AV-S7)

	Regular Gasoline		
©Darren Brode/Shutterstock	**26** combined city/highway	**MPG** 22 city	32 highway
	3.8 gals/100 miles		

Source: http://www.fueleconomy.gov (accessed June 10, 2017).

Table 3.1 tells us that $x_1 = 30.8$, $x_2 = 31.7$, $x_3 = 30.1$, $x_4 = 31.6$, and $x_5 = 32.1$, so the sum of the first five mileages is

$$\sum_{i=1}^{5} x_i = x_1 + x_2 + x_3 + x_4 + x_5$$
$$= 30.8 + 31.7 + 30.1 + 31.6 + 32.1 = 156.3$$

Therefore, the mean of the first five mileages is

$$\bar{x} = \frac{\sum_{i=1}^{5} x_i}{5} = \frac{156.3}{5} = 31.26$$

Of course, intuitively, we are likely to obtain a more accurate point estimate of the population mean by using all of the available sample information. The sum of all 50 mileages can be verified to be

$$\sum_{i=1}^{50} x_i = x_1 + x_2 + \cdots + x_{50} = 30.8 + 31.7 + \cdots + 31.4 = 1578$$

Therefore, the mean of the sample of 50 mileages is

$$\bar{x} = \frac{\sum_{i=1}^{50} x_i}{50} = \frac{1578}{50} = 31.56$$

This point estimate says we estimate that the mean mileage that would be obtained by all of the new midsize cars that will or could potentially be produced this year is 31.56 mpg. Unless we are extremely lucky, however, there will be **sampling error.** That is, the point estimate $\bar{x} = 31.56$ mpg, which is the average of the sample of fifty randomly selected mileages, will probably not exactly equal the population mean μ, which is the average mileage that would be obtained by all cars. Therefore, although $\bar{x} = 31.56$ provides some evidence that μ is at least 31 and thus that the automaker should get the tax credit, it does not provide definitive evidence. In later chapters, we discuss how to assess the *reliability* of the sample mean and how to use a measure of reliability to decide whether sample information provides definitive evidence.

BI

The point estimate obtained from the sample of 50 mileages that we have just calculated is similar to the combined city and highway mileage estimate that would be reported by the EPA in *The Fuel Economy Guide.* Table 3.1(b) presents the fuel economy ratings for the 2017 Nissan Altima as given on the fueleconomy.gov website. Here we see that the estimated gas mileage for the Nissan Altima in combined city and highway driving is 26 mpg. Recall from Chapter 1 that the Nissan Altima is the mileage leader among midsize cars having an automatic transmission and a 6-cylinder, 3.5-liter engine.

Another descriptive measure of the central tendency of a population or a sample of measurements is the **median.** Intuitively, the median divides a population or sample into two roughly equal parts. We calculate the median, which is denoted M_d, as follows:

Consider a population or a sample of measurements, and arrange the measurements in increasing order. The **median, M_d,** is found as follows:

1 If the number of measurements is odd, the median is the middlemost measurement in the ordering.

2 If the number of measurements is even, the median is the average of the two middlemost measurements in the ordering.

For example, recall that Chris's five classes have sizes 60, 41, 15, 30, and 34. To find the median of this population of class sizes, we arrange the class sizes in increasing order as follows:

$$15 \quad 30 \quad \textcircled{34} \quad 41 \quad 60$$

Because the number of class sizes is odd, the median of the population of class sizes is the middlemost class size in the ordering. Therefore, the median is 34 students (it is circled).

As another example, suppose that in the middle of the semester Chris decides to take an additional class—a sprint class in individual exercise. If the individual exercise class has 30 students, then the sizes of Chris's six classes are (arranged in increasing order):

$$15 \quad 30 \quad \textcircled{30} \quad \textcircled{34} \quad 41 \quad 60$$

Because the number of classes is even, the median of the population of class sizes is the average of the two middlemost class sizes, which are circled. Therefore, the median is $(30 + 34)/2 = 32$ students. Note that, although two of Chris's classes have the same size, 30 students, each observation is listed separately (that is, 30 is listed twice) when we arrange the observations in increasing order.

As a third example, if we arrange the sample of 50 mileages in Table 3.1 in increasing order, we find that the two middlemost mileages—the 25th and 26th mileages—are 31.5 and 31.6. It follows that the median of the sample is 31.55. Therefore, we estimate that the median mileage that would be obtained by all of the new midsize cars that will or could potentially be produced this year is 31.55 mpg. The Excel output in Figure 3.1 shows this median mileage, as well as the previously calculated mean mileage of 31.56 mpg. Other quantities given on the output will be discussed later in this chapter.

A third measure of the central tendency of a population or sample is the **mode,** which is denoted M_o.

The **mode,** M_o, of a population or sample of measurements is the measurement that occurs most frequently.

For example, the mode of Chris's six class sizes is 30. This is because more classes (two) have a size of 30 than any other size. Sometimes the highest frequency occurs at more than one measurement. When this happens, two or more modes exist. When exactly two modes exist, we say the data are *bimodal.* When more than two modes exist, we say the data are *multimodal.* If data are presented in classes (such as in a frequency or percent histogram), the class having the highest frequency or percent is called the *modal class.* For example, Figure 3.2 shows a histogram of the car mileages that has two modal classes—the class from 31.0 mpg to 31.5 mpg and the class from 31.5 mpg to 32.0 mpg. Because the mileage 31.5 is

FIGURE 3.1   **Excel Output of Statistics Describing the 50 Mileages**

Mileage

Mean	31.56
Standard Error	0.1128
Median	31.55
Mode	31.4
Standard Deviation	0.7977
Sample Variance	0.6363
Kurtosis	−0.5112
Skewness	−0.0342
Range	3.5
Minimum	29.8
Maximum	33.3
Sum	1578
Count	50

FIGURE 3.2   **A Percent Histogram Describing the 50 Mileages**

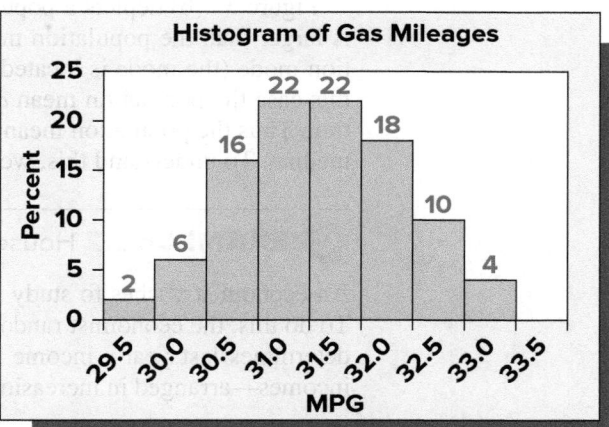

in the middle of the modal classes, we might estimate that the population mode for the new midsize model is 31.5 mpg. Or, alternatively, because the Excel output in Figure 3.1 tells us that the mode of the sample of 50 mileages is 31.4 mpg (it can be verified that this mileage occurs five times in Table 3.1), we might estimate that the population mode is 31.4 mpg. Obviously, these two estimates are somewhat contradictory. In general, it can be difficult to define a reliable method for estimating the population mode. Therefore, although it can be informative to report the modal class or classes in a frequency or percent histogram, the mean or median is used more often than the mode when we wish to describe a data set's central tendency by using a single number. Finally, the mode is a useful descriptor of qualitative data. For example, we have seen in Chapter 2 that the most preferred pizza restaurant in the college town was Papa John's, which was preferred by 38 percent of the college students.

Comparing the mean, median, and mode

Often we construct a histogram for a sample to make inferences about the shape of the sampled population. When we do this, it can be useful to "smooth out" the histogram and use the resulting *relative frequency curve* to describe the shape of the population. Relative frequency curves can have many shapes. Three common shapes are illustrated in Figure 3.3. Part (a) of this figure depicts a population described by a symmetrical relative frequency curve. For such a population, the mean (μ), median (M_d), and mode (M_o) are all equal. Note that in this case all three of these quantities are located under the highest point of the curve. It follows that when the frequency distribution of a sample of measurements is approximately symmetrical, then the sample mean, median, and mode will be nearly the same. For instance,

FIGURE 3.3   **Typical Relationships among the Mean μ, the Median M_d, and the Mode M_o**

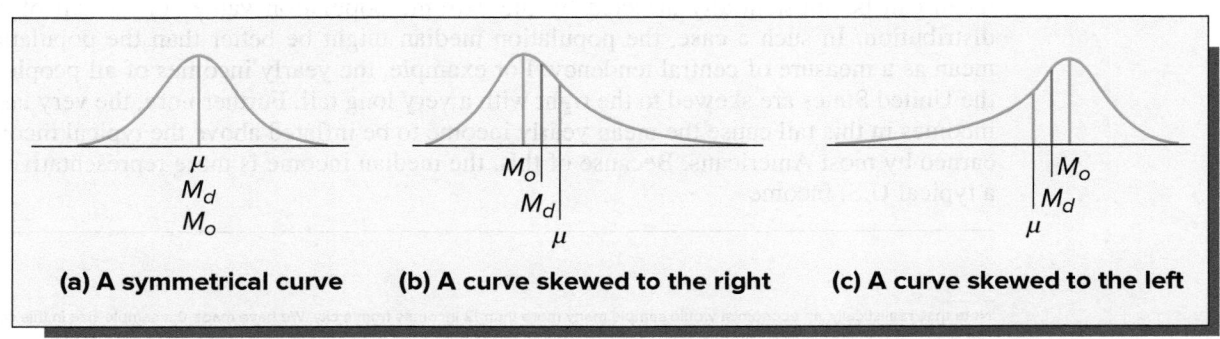

consider the sample of 50 mileages in Table 3.1. Because the histogram of these mileages in Figure 3.2 is approximately symmetrical, the mean—31.56—and the median—31.55—of the mileages are approximately equal to each other.

Figure 3.3(b) depicts a population that is skewed to the right. Here the population mean is larger than the population median, and the population median is larger than the population mode (the mode is located under the highest point of the relative frequency curve). In this case the population mean *averages in* the large values in the upper tail of the distribution. Thus the population mean is more affected by these large values than is the population median. To understand this, we consider the following example.

Ⓒ **EXAMPLE 3.2** Household Incomes

An economist wishes to study the distribution of household incomes in a Midwestern city. To do this, the economist randomly selects a sample of $n = 12$ households from the city and determines last year's income for each household.[1] The resulting sample of 12 household incomes—arranged in increasing order—is as follows (the incomes are expressed in dollars):

 Incomes

7,524	11,070	18,211	26,817	36,551	41,286
49,312	57,283	72,814	90,416	135,540	190,250

Because the number of incomes is even, the median of the incomes is the average of the two middlemost incomes, which are enclosed in ovals. Therefore, the median is (41,286 + 49,312)/2 = $45,299. The mean of the incomes is the sum of the incomes, 737,074, divided by 12, or $61,423 (rounded to the nearest dollar). Here, the mean has been affected by averaging in the large incomes $135,540 and $190,250 and thus is larger than the median. The median is said to be resistant to these large incomes because the value of the median is affected only by the position of these large incomes in the ordered list of incomes, not by the exact sizes of the incomes. For example, if the largest income were smaller—say $150,000—the median would remain the same but the mean would decrease. If the largest income were larger—say $300,000—the median would also remain the same but the mean would increase. Therefore, the median is resistant to large values but the mean is not. Similarly, the median is resistant to values that are much smaller than most of the measurements. In general, we say that **the median is resistant to extreme values.**

Figure 3.3(c) depicts a population that is skewed to the left. Here the population mean is smaller than the population median, and the population median is smaller than the population mode. In this case the population mean *averages in* the small values in the lower tail of the distribution, and the mean is more affected by these small values than is the median. For instance, in a survey several years ago of 20 Decision Sciences graduates at Miami University, 18 of the graduates had obtained employment in business consulting that paid a mean salary of about $43,000. One of the graduates had become a Christian missionary and listed his salary as $8,500, and another graduate was working for his hometown bank and listed his salary as $10,500. The two lower salaries decreased the overall mean salary to about $39,650, which was below the median salary of about $43,000.

When a population is skewed to the right or left with a very long tail, the population mean can be substantially affected by the extreme population values in the tail of the distribution. In such a case, the population median might be better than the population mean as a measure of central tendency. For example, the yearly incomes of all people in the United States are skewed to the right with a very long tail. Furthermore, the very large incomes in this tail cause the mean yearly income to be inflated above the typical income earned by most Americans. Because of this, the median income is more representative of a typical U.S. income.

---

[1]Note that, realistically, an economist would sample many more than 12 incomes from a city. We have made the sample size in this case small so that we can simply illustrate various ideas throughout this chapter.

When a population is symmetrical or not highly skewed, then the population mean and the population median are either equal or roughly equal, and both provide a good measure of the population central tendency. In this situation, we usually make inferences about the population mean because much of statistical theory is based on the mean rather than the median.

Ⓒ EXAMPLE 3.3 The Marketing Research Case: Rating a Bottle Design

The JMP output in Figure 3.4 tells us that the mean and the median of the sample of 60 bottle design ratings are 30.35 and 31, respectively. Because the histogram of the bottle design ratings in Figure 3.5 is not highly skewed to the left, the sample mean is not much less than the sample median. Therefore, using the mean as our measure of central tendency, we estimate that the mean rating of the new bottle design that would be given by all consumers is 30.35. This is considerably higher than the minimum standard of 25 for a successful bottle design.

Ⓒ EXAMPLE 3.4 The e-Billing Case: Reducing Bill Payment Times

The Minitab output in Figure 3.6 gives a histogram of the 65 payment times, and the Minitab output in Figure 3.7 tells us that the mean and the median of the payment times are 18.108 days and 17 days, respectively. Because the histogram is not highly skewed to the right, the sample mean is not much greater than the sample median. Therefore, using the mean as our measure of central tendency, we estimate that the mean payment time of all bills using the new billing system is 18.108 days. This is substantially less than the typical payment time of 39 days that had been experienced using the old billing system.

FIGURE 3.4   **JMP Output of Statistics Describing the 60 Bottle Design Ratings**

FIGURE 3.5   **JMP Histogram of the 60 Bottle Design Ratings**

Summary Statistics	
Mean	30.35
Std Dev	3.1072632
Std Err Mean	0.401146
N	60
Sum	1821
Variance	9.6550847
Minimum	20
Maximum	35
Median	31
Mode	32
Range	15
Interquarile Range	4

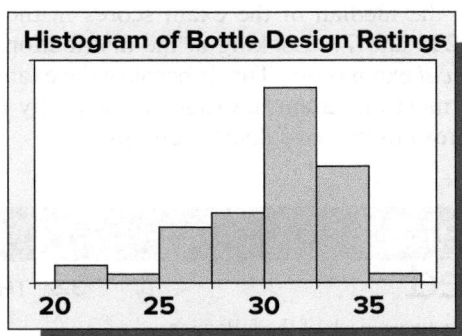

FIGURE 3.6   **Minitab Frequency Histogram of the 65 Payment Times**

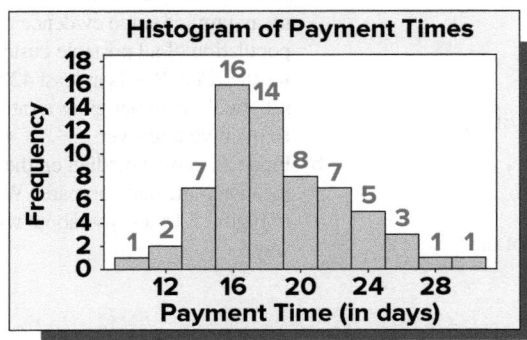

Minutes Used

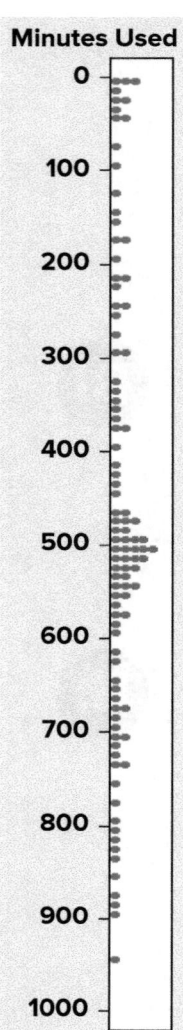

F I G U R E 3 . 7   **Minitab Output of Statistics Describing the 65 Payment Times**

Variable	Count	Mean	StDev	Variance
PayTime	65	18.108	3.961	15.691

Variable	Minimum	Q1	Median	Q3	Maximum	Range
PayTime	10.000	15.000	17.000	21.000	29.000	19.000

Ⓒ EXAMPLE 3.5 The Cell Phone Case: Reducing Cellular Phone Costs

Suppose that a cellular management service tells the bank that if its cellular cost per minute for the random sample of 100 bank employees is over 18 cents per minute, the bank will benefit from automated cellular management of its calling plans. Last month's cellular usages for the 100 randomly selected employees are given in Table 1.5, and a dot plot of these usages is given in the page margin. If we add the usages together, we find that the 100 employees used a total of 46,625 minutes. Furthermore, the total cellular cost incurred by the 100 employees is found to be \$9,317 (this total includes base costs, overage costs, long distance, and roaming). This works out to an average of \$9,317/46,625 = \$.1998, or 19.98 cents per minute. Because this average cellular cost per minute exceeds 18 cents per minute, the bank will hire the cellular management service to manage its calling plans.

BI

To conclude this section, note that the mean and the median convey useful information about a population having a relative frequency curve with a sufficiently regular shape. For instance, the mean and median would be useful in describing the mound-shaped, or single-peaked, distributions in Figure 3.3. However, these measures of central tendency do not adequately describe a double-peaked distribution. For example, the mean and the median of the exam scores in the double-peaked distribution of Figure 2.12 are 75.225 and 77. Looking at the distribution, neither the mean nor the median represents a *typical* exam score. This is because the exam scores really have *no central value*. In this case the most important message conveyed by the double-peaked distribution is that the exam scores fall into two distinct groups.

Exercises for Section 3.1

CONCEPTS ▦ **connect**

3.1 Explain the difference between each of the following:
   a A population parameter and its point estimate.
   b A population mean and a corresponding sample mean.

3.2 Explain how the population mean, median, and mode compare when the population's relative frequency curve is
   a Symmetrical.
   b Skewed with a tail to the left.
   c Skewed with a tail to the right.

METHODS AND APPLICATIONS

3.3 Calculate the mean, median, and mode of each of the following populations of numbers:
   a 9, 8, 10, 10, 12, 6, 11, 10, 12, 8
   b 110, 120, 70, 90, 90, 100, 80, 130, 140

3.4 Calculate the mean, median, and mode for each of the following populations of numbers:
   a 17, 23, 19, 20, 25, 18, 22, 15, 21, 20
   b 505, 497, 501, 500, 507, 510, 501

3.5 THE VIDEO GAME SATISFACTION RATING CASE
   Ⓓ VideoGame

Recall that Table 1.8 presents the satisfaction ratings for the XYZ-Box game system that have been given by 65 randomly selected purchasers. Figures 3.8 and 3.11 give the Minitab and Excel outputs of statistics describing the 65 satisfaction ratings.

   a Find the sample mean on the outputs. Does the sample mean provide some evidence that the mean of the population of all possible customer satisfaction ratings for the XYZ-Box is at least 42? (Recall that a "very satisfied" customer gives a rating that is at least 42.) Explain your answer.
   b Find the sample median on the outputs. How do the mean and median compare? What does the histogram in Figure 2.15 tell you about why they compare this way?

FIGURE 3.8   Minitab Output of Statistics Describing the 65 Satisfaction Ratings (for Exercise 3.5)

Variable	Count	Mean	StDev	Variance		
Ratings	65	42.954	2.642	6.982		

Variable	Minimum	Q1	Median	Q3	Maximum	Range
Ratings	36.000	41.000	43.000	45.000	48.000	12.000

FIGURE 3.9   Minitab Output of Statistics Describing the 100 Waiting Times (for Exercise 3.6)

Variable	Count	Mean	StDev	Variance		
WaitTime	100	5.460	2.475	6.128		

Variable	Minimum	Q1	Median	Q3	Maximum	Range
WaitTime	0.400	3.800	5.250	7.200	11.600	11.200

3.6   THE BANK CUSTOMER WAITING TIME CASE

DS WaitTime

Recall that Table 1.9 presents the waiting times for teller service during peak business hours of 100 randomly selected bank customers. Figures 3.9 and 3.12 give the Minitab and Excel outputs of statistics describing the 100 waiting times.

a Find the sample mean on the outputs. Does the sample mean provide some evidence that the mean of the population of all possible customer waiting times during peak business hours is less than six minutes (as is desired by the bank manager)? Explain your answer.

b Find the sample median on the outputs. How do the mean and median compare? What does the histogram in Figure 2.16 tell you about why they compare this way?

3.7   THE TRASH BAG CASE   **DS** TrashBag

Consider the trash bag problem. Suppose that an independent laboratory has tested 30-gallon trash bags and has found that none of the 30-gallon bags currently on the market has a mean breaking strength of 50 pounds or more. On the basis of these results, the producer of the new, improved trash bag feels sure that its 30-gallon bag will be the strongest such bag on the market if the new trash bag's mean breaking strength can be shown to be at least 50 pounds. Recall that Table 1.10 presents the breaking strengths of 30 trash bags of the new type that were selected during a 30-hour pilot production

run. Figure 3.10 gives the Minitab output of statistics describing the 30 breaking strengths.

a Find the sample mean on the output. Does the sample mean provide some evidence that the mean of the population of all possible trash bag breaking strengths is at least 50 pounds? Explain your answer.

b Find the sample median on the output. How do the mean and median compare? What does the histogram in Figure 2.17 tell you about why they compare this way?

3.8 Lauren is a college sophomore majoring in business. This semester Lauren is taking courses in accounting, economics, management information systems, public speaking, and statistics. The sizes of these classes are, respectively, 350, 45, 35, 25, and 40. Find the mean and the median of the class sizes. What is a better measure of Lauren's "typical class size"—the mean or the median? Explain.

In the National Basketball Association (NBA) lockout of 2011, the owners of NBA teams wished to change the existing collective bargaining agreement with the NBA Players Association. The owners wanted a "hard salary cap" restricting the size of team payrolls. This would allow "smaller market teams" having less revenue to be (1) financially profitable and (2) competitive (in terms of wins and losses) with the "larger market teams." The NBA owners also wanted the players to agree to take less than the 57 percent share of team revenues that they had been receiving. The players opposed these changes. Table 3.2 gives, for each NBA team, the team's 2009–2010 revenue, player expenses (including benefits and bonuses), and operating income as given on the Forbes.com website on October 26, 2011.

FIGURE 3.10   Minitab Output of Statistics Describing the 30 Breaking Strengths (for Exercise 3.7)

Variable	N	Mean	SE Mean	StDev		
Strength	30	52.167	0.985	5.394		

Variable	Minimum	Q1	Median	Q3	Maximum	Range
Strength	41.700	48.650	52.000	55.975	63.800	22.100

FIGURE 3.11   Excel Output of Satisfaction Rating
                Statistics (for Exercise 3.5)

Ratings	
Mean	42.954
Standard Error	0.3277
Median	43
Mode	44
Standard Deviation	2.6424
Sample Variance	6.9822
Kurtosis	−0.3922
Skewness	−0.4466
Range	12
Minimum	36
Maximum	48
Sum	2792
Count	65

FIGURE 3.12   Excel Output of Waiting Time Statistics
                (for Exercise 3.6)

WaitTime	
Mean	5.46
Standard Error	0.2475
Median	5.25
Mode	5.8
Standard Deviation	2.4755
Sample Variance	6.1279
Kurtosis	−0.4050
Skewness	0.2504
Range	11.2
Minimum	0.4
Maximum	11.6
Sum	546
Count	100

Here, the operating income of a team is basically the team's profit (that is, the team's revenue minus the team's expenses—including player expenses, arena expenses, etc.—but not including some interest, depreciation, and tax expenses). Use the data in Table 3.2 to do Exercises 3.9 through 3.15.

3.9 Construct a histogram and a stem-and-leaf display of the teams' revenues.   DS NBAIncome

3.10 Compute the mean team revenue and the median team revenue, and explain the difference between the values of these statistics.   DS NBAIncome

3.11 Construct a histogram and a stem-and-leaf display of the teams' player expenses.   DS NBAIncome

3.12 Compute the mean team player expense and the median team player expense and explain the difference between the values of these statistics.   DS NBAIncome

3.13 Construct a histogram and a stem-and-leaf display of the teams' operating incomes.   DS NBAIncome

3.14 Compute the mean team operating income and the median team operating income, and explain the difference between the values of these statistics.   DS NBAIncome

3.15 The mean team operating income is the operating income each NBA team would receive if the NBA

TABLE 3.2   National Basketball Association Team Revenues, Player Expenses, and Operating Incomes as Given on the Forbes.com Website on October 26, 2011   DS NBAIncome

Rank	Team	Revenue ($mil)	Player Expenses ($mil)	Operating Income ($mil)	Rank	Team	Revenue ($mil)	Player Expenses ($mil)	Operating Income ($mil)
1	New York Knicks	226	86	64.0	16	Utah Jazz	121	76	−3.9
2	Los Angeles Lakers	214	91	33.4	17	Philadelphia 76ers	110	69	−1.2
3	Chicago Bulls	169	74	51.3	18	Oklahoma City Thunder	118	62	22.6
4	Boston Celtics	151	88	4.2	19	Washington Wizards	107	73	−5.2
5	Houston Rockets	153	67	35.9	20	Denver Nuggets	113	79	−11.7
6	Dallas Mavericks	146	81	−7.8	21	New Jersey Nets	89	64	−10.2
7	Miami Heat	124	78	−5.9	22	Los Angeles Clippers	102	62	11.0
8	Phoenix Suns	147	69	20.4	23	Atlanta Hawks	105	70	−7.3
9	San Antonio Spurs	135	84	−4.7	24	Sacramento Kings	103	72	−9.8
10	Toronto Raptors	138	72	25.3	25	Charlotte Bobcats	98	73	−20.0
11	Orlando Magic	108	86	−23.1	26	New Orleans Hornets	100	74	−5.9
12	Golden State Warriors	119	70	14.3	27	Indiana Pacers	95	71	−16.9
13	Detroit Pistons	147	64	31.8	28	Memphis Grizzlies	92	59	−2.6
14	Portland Trail Blazers	127	64	10.7	29	Minnesota Timberwolves	95	67	−6.7
15	Cleveland Cavaliers	161	90	2.6	30	Milwaukee Bucks	92	69	−2.0

Source: http://www.forbes.com/lists/2011/32/basketball-valuations-11.

owners divided the total of their operating incomes equally among the 30 NBA teams. (Of course, some of the owners might object to dividing their operating incomes equally among the teams.) 🖙 NBAIncome

a   How would the players use the mean team operating income to justify their position opposing a hard salary cap?

b   Use Table 3.2 to find the number of NBA teams that made money (that is, had a positive operating income) and the number of teams that lost money (had a negative operating income).

c   How would the owners use the results of part *b* and the median team operating income to justify their desire for a hard salary cap?

3.2 Measures of Variation

LO3-2
Compute and interpret the range, variance, and standard deviation.

Range, variance, and standard deviation

In addition to estimating a population's central tendency, it is important to estimate the **variation** of the population's individual values. For example, Figure 3.13 shows two histograms. Each portrays the distribution of 20 repair times (in days) for personal computers at a major service center. Because the mean (and median and mode) of each distribution equals four days, the measures of central tendency do not indicate any difference between the American and National Service Centers. However, the repair times for the American Service Center are clustered quite closely together, whereas the repair times for the National Service Center are spread farther apart (the repair time might be as little as one day, but could also be as long as seven days). Therefore, we need measures of variation to express how the two distributions differ.

One way to measure the variation of a set of measurements is to calculate the *range*.

> Consider a population or a sample of measurements. The **range** of the measurements is the largest measurement minus the smallest measurement.

In Figure 3.13, the smallest and largest repair times for the American Service Center are three days and five days; therefore, the range is $5 - 3 = 2$ days. On the other hand, the range for the National Service Center is $7 - 1 = 6$ days. The National Service Center's larger range indicates that this service center's repair times exhibit more variation.

In general, the range is not the best measure of a data set's variation. One reason is that it is based on only the smallest and largest measurements in the data set and therefore may reflect an extreme measurement that is not entirely representative of the data set's variation. For example, in the marketing research case, the smallest and largest ratings in the sample of 60 bottle design ratings are 20 and 35. However, to simply estimate that most bottle design ratings are between 20 and 35 misses the fact that 57, or 95 percent, of the 60 ratings are at least as large as the minimum rating of 25 for a successful bottle design. In general, to

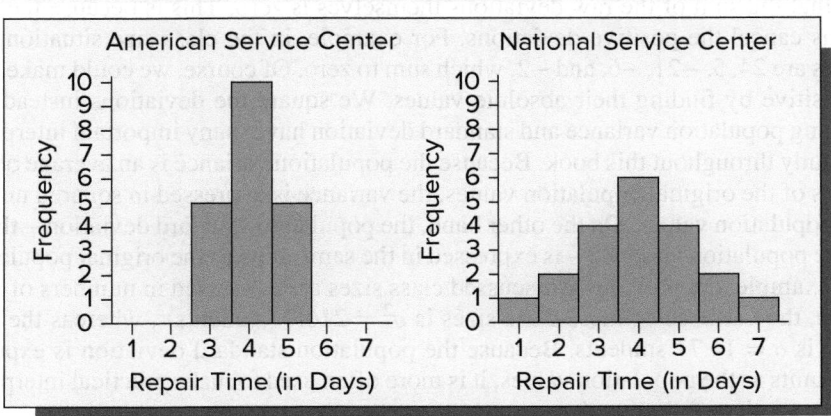

FIGURE 3.13   **Repair Times for Personal Computers at Two Service Centers**

fully describe a population's variation, it is useful to estimate intervals that contain *different percentages* (for example, 70 percent, 95 percent, or almost 100 percent) of the individual population values. To estimate such intervals, we use the **population variance** and the **population standard deviation.**

The Population Variance and Standard Deviation

The **population variance** σ^2 (pronounced *sigma squared*) is the average of the squared deviations of the individual population measurements from the population mean μ.

   The **population standard deviation** σ (pronounced *sigma*) is the positive square root of the population variance.

For example, consider again the population of Chris's class sizes this semester. These class sizes are 60, 41, 15, 30, and 34. To calculate the variance and standard deviation of these class sizes, we first calculate the population mean to be

$$\mu = \frac{60 + 41 + 15 + 30 + 34}{5} = \frac{180}{5} = 36$$

Next, we calculate the deviations of the individual population measurements from the population mean $\mu = 36$ as follows:

$$(60 - 36) = 24 \quad (41 - 36) = 5 \quad (15 - 36) = -21 \quad (30 - 36) = -6 \quad (34 - 36) = -2$$

Then we compute the sum of the squares of these deviations:

$$(24)^2 + (5)^2 + (-21)^2 + (-6)^2 + (-2)^2 = 576 + 25 + 441 + 36 + 4 = 1,082$$

Finally, we calculate the population variance σ^2, the average of the squared deviations, by dividing the sum of the squared deviations, 1,082, by the number of squared deviations, 5. That is, σ^2 equals $1,082/5 = 216.4$. Furthermore, this implies that the population standard deviation σ (the positive square root of σ^2) is $\sqrt{216.4} = 14.71$.

   To see that the variance and standard deviation measure the variation, or spread, of the individual population measurements, suppose that the measurements are spread far apart. Then, many measurements will be far from the mean μ, many of the squared deviations from the mean will be large, and the sum of squared deviations will be large. It follows that the average of the squared deviations—the population variance—will be relatively large. On the other hand, if the population measurements are clustered closely together, many measurements will be close to μ, many of the squared deviations from the mean will be small, and the average of the squared deviations—the population variance—will be small. Therefore, the more spread out the population measurements, the larger is the population variance, and the larger is the population standard deviation.

   To further understand the population variance and standard deviation, note that one reason we square the deviations of the individual population measurements from the population mean is that the sum of the raw deviations themselves is zero. This is because the negative deviations cancel the positive deviations. For example, in the class size situation, the raw deviations are 24, 5, −21, −6, and −2, which sum to zero. Of course, we could make the deviations positive by finding their absolute values. We square the deviations instead because the resulting population variance and standard deviation have many important interpretations that we study throughout this book. Because the population variance is an average of squared deviations of the original population values, the variance is expressed in squared units of the original population values. On the other hand, the population standard deviation—the square root of the population variance—is expressed in the same units as the original population values. For example, the previously discussed class sizes are expressed in numbers of students. Therefore, the variance of these class sizes is $\sigma^2 = 216.4$ (students)^2, whereas the standard deviation is $\sigma = 14.71$ students. Because the population standard deviation is expressed in the same units as the population values, it is more often used to make practical interpretations about the variation of these values.

When a population is too large to measure all the population units, we estimate the population variance and the population standard deviation by the **sample variance** and the **sample standard deviation.** We calculate the sample variance by dividing the sum of the squared deviations of the sample measurements from the sample mean by $n - 1$, the sample size minus one. Although we might intuitively think that we should divide by n rather than $n - 1$, it can be shown that dividing by n tends to produce an estimate of the population variance that is too small. On the other hand, dividing by $n - 1$ tends to produce a larger estimate that we will show in Chapter 8 is more appropriate. Therefore, we obtain:

The Sample Variance and the Sample Standard Deviation

The **sample variance** s^2 (pronounced s *squared*) is defined to be

$$s^2 = \frac{\sum_{i=1}^{n}(x_i - \bar{x})^2}{n - 1} = \frac{(x_1 - \bar{x})^2 + (x_2 - \bar{x})^2 + \cdots + (x_n - \bar{x})^2}{n - 1}$$

and is the **point estimate of the population variance** σ^2.

The **sample standard deviation** $s = \sqrt{s^2}$ is the positive square root of the sample variance and is the **point estimate of the population standard deviation** σ.

Ⓒ **EXAMPLE 3.6** The Car Mileage Case: Estimating Mileage

To illustrate the calculation of the sample variance and standard deviation, we begin by considering the first five mileages in Table 3.1: $x_1 = 30.8$, $x_2 = 31.7$, $x_3 = 30.1$, $x_4 = 31.6$, and $x_5 = 32.1$. Because the mean of these five mileages is $\bar{x} = 31.26$ it follows that

$$\sum_{i=1}^{5}(x_i - \bar{x})^2 = (x_1 - \bar{x})^2 + (x_2 - \bar{x})^2 + (x_3 - \bar{x})^2 + (x_4 - \bar{x})^2 + (x_5 - \bar{x})^2$$
$$= (30.8 - 31.26)^2 + (31.7 - 31.26)^2 + (30.1 - 31.26)^2$$
$$+ (31.6 - 31.26)^2 + (32.1 - 31.26)^2$$
$$= (-.46)^2 + (.44)^2 + (-1.16)^2 + (.34)^2 + (.84)^2$$
$$= 2.572$$

Therefore, the variance and the standard deviation of the sample of the first five mileages are

$$s^2 = \frac{2.572}{5 - 1} = .643 \qquad \text{and} \qquad s = \sqrt{.643} = .8019$$

Of course, intuitively, we are likely to obtain more accurate point estimates of the population variance and standard deviation by using all the available sample information. Recall that the mean of all 50 mileages is $\bar{x} = 31.56$. Using this sample mean, it can be verified that

$$\sum_{i=1}^{50}(x_i - \bar{x})^2 = (x_1 - \bar{x})^2 + (x_2 - \bar{x})^2 + \cdots + (x_{50} - \bar{x})^2$$
$$= (30.8 - 31.56)^2 + (31.7 - 31.56)^2 + \cdots + (31.4 - 31.56)^2$$
$$= (-.76)^2 + (.14)^2 + \cdots + (-.16)^2$$
$$= 31.18$$

Therefore, the variance and the standard deviation of the sample of 50 mileages are

$$s^2 = \frac{31.18}{50 - 1} = .6363 \qquad \text{and} \qquad s = \sqrt{.6363} = .7977.$$

Notice that the Excel output in Figure 3.1 gives these quantities. Here $s^2 = .6363$ and $s = .7977$ are the point estimates of the variance, σ^2, and the standard deviation, σ, of the population of the mileages of all the cars that will be or could potentially be produced. Furthermore, the sample standard deviation is expressed in the same units (that is, miles per gallon) as the sample values. Therefore $s = .7977$ mpg.

Before explaining how we can use s^2 and s in a practical way, we present a formula that makes it easier to compute s^2. This formula is useful when we are using a handheld calculator that is not equipped with a statistics mode to compute s^2.

The **sample variance** can be calculated using the *computational formula*

$$s^2 = \frac{1}{n-1}\left[\sum_{i=1}^{n} x_i^2 - \frac{\left(\sum_{i=1}^{n} x_i\right)^2}{n}\right]$$

ⓒ EXAMPLE 3.7 The e-Billing Case: Reducing Bill Payment Times

Consider the sample of 65 payment times in Table 2.4. Using these data, it can be verified that

$$\sum_{i=1}^{65} x_i = x_1 + x_2 + \cdots + x_{65} = 22 + 19 + \cdots + 21 = 1{,}177 \quad \text{and}$$

$$\sum_{i=1}^{65} x_i^2 = x_1^2 + x_2^2 + \cdots + x_{65}^2 = (22)^2 + (19)^2 + \cdots + (21)^2 = 22{,}317$$

Therefore,

$$s^2 = \frac{1}{(65-1)}\left[22{,}317 - \frac{(1{,}177)^2}{65}\right] = \frac{1{,}004.2464}{64} = 15.69135$$

and $s = \sqrt{s^2} = \sqrt{15.69135} = 3.9612$ days. Note that the Minitab output in Figure 3.7 gives these results in slightly rounded form.

LO3-3

Use the Empirical Rule and Chebyshev's Theorem to describe variation.

A practical interpretation of the standard deviation: The Empirical Rule

One type of relative frequency curve describing a population is the **normal curve,** which is discussed in Chapter 7. The normal curve is a symmetrical, bell-shaped curve and is illustrated in Figure 3.14. If a population is described by a normal curve, we say that the population is **normally distributed,** and the following result can be shown to hold.

The Empirical Rule for a Normally Distributed Population

If a population has **mean μ** and **standard deviation σ** and is **described by a normal curve,** then, as illustrated in Figure 3.14,

1 68.26 percent of the population measurements are within (plus or minus) one standard deviation of the mean and thus lie in the interval $[\mu - \sigma, \mu + \sigma] = [\mu \pm \sigma]$

2 95.44 percent of the population measurements are within (plus or minus) two standard deviations of the mean and thus lie in the interval $[\mu - 2\sigma, \mu + 2\sigma] = [\mu \pm 2\sigma]$

3 99.73 percent of the population measurements are within (plus or minus) three standard deviations of the mean and thus lie in the interval $[\mu - 3\sigma, \mu + 3\sigma] = [\mu \pm 3\sigma]$

In general, an interval that contains a specified percentage of the individual measurements in a population is called a **tolerance interval.** It follows that the one, two, and three standard deviation intervals around μ given in (1), (2), and (3) are tolerance intervals containing,

FIGURE 3.14   The Empirical Rule and Tolerance Intervals for a Normally Distributed Population

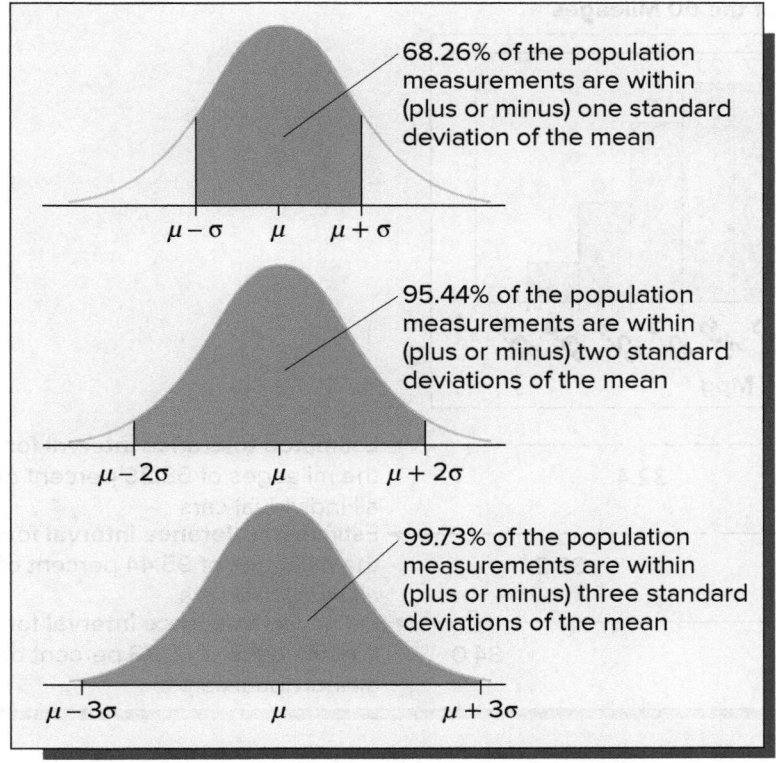

respectively, 68.26 percent, 95.44 percent, and 99.73 percent of the measurements in a normally distributed population. Often we interpret the *three-sigma interval* $[\mu \pm 3\sigma]$ to be a tolerance interval that contains *almost all* of the measurements in a normally distributed population. Of course, we usually do not know the true values of μ and σ. Therefore, we must estimate the tolerance intervals by replacing μ and σ in these intervals by the mean $\bar{x}$ and standard deviation s of a sample that has been randomly selected from the normally distributed population.

C EXAMPLE 3.8 The Car Mileage Case: Estimating Mileage

Again consider the sample of 50 mileages. We have seen that $\bar{x} = 31.56$ and $s = .7977$ for this sample are the point estimates of the mean μ and the standard deviation σ of the population of all mileages. Furthermore, the histogram of the 50 mileages in Figure 3.15 suggests that the population of all mileages is normally distributed. To illustrate the Empirical Rule more simply, we will round $\bar{x}$ to 31.6 and s to .8. It follows that, using the interval

1   $[\bar{x} \pm s] = [31.6 \pm .8] = [31.6 - .8, 31.6 + .8] = [30.8, 32.4]$, we estimate that 68.26 percent of all individual cars will obtain mileages between 30.8 mpg and 32.4 mpg.

2   $[\bar{x} \pm 2s] = [31.6 \pm 2(.8)] = [31.6 \pm 1.6] = [30.0, 33.2]$, we estimate that 95.44 percent of all individual cars will obtain mileages between 30.0 mpg and 33.2 mpg.

3   $[\bar{x} \pm 3s] = [31.6 \pm 3(.8)] = [31.6 \pm 2.4] = [29.2, 34.0]$, we estimate that 99.73 percent of all individual cars will obtain mileages between 29.2 mpg and 34.0 mpg.

FIGURE 3.15   **Estimated Tolerance Intervals in the Car Mileage Case**

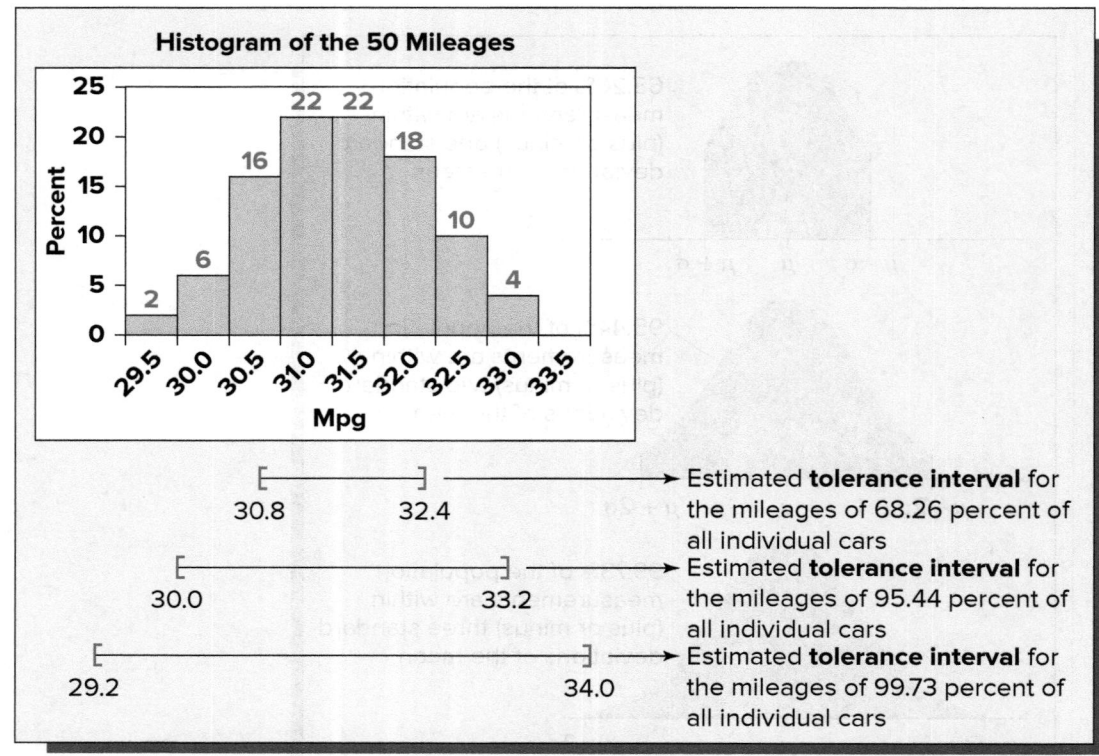

Figure 3.15 depicts these estimated tolerance intervals, which are shown below the histogram. Because the difference between the upper and lower limits of each estimated tolerance interval is fairly small, we might conclude that the variability of the individual car mileages around the estimated mean mileage of 31.6 mpg is fairly small. Furthermore, the interval $[\bar{x} \pm 3s] = [29.2, 34.0]$ implies that almost any individual car that a customer might purchase this year will obtain a mileage between 29.2 mpg and 34.0 mpg.

Before continuing, recall that we have rounded $\bar{x}$ and s to one decimal point accuracy in order to simplify our initial example of the Empirical Rule. If, instead, we calculate the Empirical Rule intervals by using $\bar{x} = 31.56$ and $s = .7977$ and then round the interval endpoints to one decimal place accuracy at the end of the calculations, we obtain the same intervals as obtained above. In general, however, rounding intermediate calculated results can lead to inaccurate final results. Because of this, throughout this book we will avoid greatly rounding intermediate results.

We next note that if we actually count the number of the 50 mileages in Table 3.1 that are contained in each of the intervals $[\bar{x} \pm s] = [30.8, 32.4]$, $[\bar{x} \pm 2s] = [30.0, 33.2]$, and $[\bar{x} \pm 3s] = [29.2, 34.0]$, we find that these intervals contain, respectively, 34, 48, and 50 of the 50 mileages. The corresponding sample percentages—68 percent, 96 percent, and 100 percent—are close to the theoretical percentages—68.26 percent, 95.44 percent, and 99.73 percent—that apply to a normally distributed population. This is further evidence that the population of all mileages is (approximately) normally distributed and thus that the Empirical Rule holds for this population.

To conclude this example, we note that the automaker has studied the combined city and highway mileages of the new model because the federal tax credit is based on these combined mileages. When reporting fuel economy estimates for a particular car model to the public, however, the EPA realizes that the proportions of city and highway driving vary from purchaser to purchaser. Therefore, the EPA reports both a combined mileage estimate and separate city and highway mileage estimates to the public (see Table 3.1(b)).

©Blend Images/Getty Images RF

Skewness and the Empirical Rule

The Empirical Rule holds for normally distributed populations. In addition:

> The Empirical Rule also approximately holds for populations having **mound-shaped (single-peaked) distributions that are not very skewed to the right or left.**

In some situations, the skewness of a mound-shaped distribution can make it tricky to know whether to use the Empirical Rule. This will be investigated in the end-of-section exercises. When a distribution seems to be too skewed for the Empirical Rule to hold, it is probably best to describe the distribution's variation by using **percentiles,** which are discussed in the next section.

Chebyshev's Theorem

If we fear that the Empirical Rule does not hold for a particular population, we can consider using **Chebyshev's Theorem** to find an interval that contains a specified percentage of the individual measurements in the population. Although Chebyshev's Theorem technically applies to any population, we will see that it is not as practically useful as we might hope.

Chebyshev's Theorem

Consider any population that has mean μ and standard deviation σ. Then, for any value of k greater than 1, at least $100(1 - 1/k^2)\%$ of the population measurements lie in the interval $[\mu \pm k\sigma]$.

For example, if we choose k equal to 2, then at least $100(1 - 1/2^2)\% = 100(3/4)\% = 75\%$ of the population measurements lie in the interval $[\mu \pm 2\sigma]$. As another example, if we choose k equal to 3, then at least $100(1 - 1/3^2)\% = 100(8/9)\% = 88.89\%$ of the population measurements lie in the interval $[\mu \pm 3\sigma]$. As yet a third example, suppose that we wish to find an interval containing at least 99.73 percent of all population measurements. Here we would set $100(1 - 1/k^2)\%$ equal to 99.73%, which implies that $(1 - 1/k^2) = .9973$. If we solve for k, we find that $k = 19.25$. This says that at least 99.73 percent of all population measurements lie in the interval $[\mu \pm 19.25\sigma]$. Unless σ is extremely small, this interval will be so long that it will tell us very little about where the population measurements lie. We conclude that Chebyshev's Theorem can help us find an interval that contains a reasonably high percentage (such as 75 percent or 88.89 percent) of all population measurements. However, unless σ is extremely small, Chebyshev's Theorem will not provide a useful interval that contains almost all (say, 99.73 percent) of the population measurements.

Although Chebyshev's Theorem technically applies to any population, it is only of practical use when analyzing a **non-mound-shaped** (for example, a double-peaked) **population that is not *very* skewed to the right or left.** Why is this? First, **we would not use Chebyshev's Theorem to describe a mound-shaped population that is not very skewed because we can use the Empirical Rule** to do this. In fact, the Empirical Rule is better for such a population because it gives us a shorter interval that will contain a given percentage of measurements. For example, if the Empirical Rule can be used to describe a population, the interval $[\mu \pm 3\sigma]$ will contain 99.73 percent of all measurements. On the other hand, if we use Chebyshev's Theorem, the interval $[\mu \pm 19.25\sigma]$ is needed. As another example, the Empirical Rule tells us that 95.44 percent of all measurements lie in the interval $[\mu \pm 2\sigma]$, whereas Chebyshev's Theorem tells us only that at least 75 percent of all measurements lie in this interval.

It is also not appropriate to use Chebyshev's Theorem—or any other result making use of the population standard deviation σ—to describe a population that is very skewed. This is because, if a population is very skewed, the measurements in the long tail to the left or right will inflate σ. This implies that tolerance intervals calculated using σ will be too long to be useful. In this case, it is best to measure variation by using **percentiles,** which are discussed in the next section.

z-scores

We can determine the relative location of any value in a population or sample by using the mean and standard deviation to compute the value's z-score. For any value x in a population or sample, the z-**score** corresponding to x is defined as follows:

z-score:

$$z = \frac{x - \text{mean}}{\text{standard deviation}}$$

The z-score, which is also called the *standardized value,* is the number of standard deviations that x is from the mean. A positive z-score says that x is above (greater than) the mean, while a negative z-score says that x is below (less than) the mean. For instance, a z-score equal to 2.3 says that x is 2.3 standard deviations above the mean. Similarly, a z-score equal to −1.68 says that x is 1.68 standard deviations below the mean. A z-score equal to zero says that x equals the mean.

A z-score indicates the relative location of a value within a population or sample. For example, below we calculate the z-scores for each of the profit margins for five competing companies in a particular industry. For these five companies, the mean profit margin is 10 percent and the standard deviation is 3.406 percent.

Company	Profit margin, x	x − mean	z-score
1	8%	8 − 10 = −2	−2/3.406 = −.59
2	10	10 − 10 = 0	0/3.406 = 0
3	15	15 − 10 = 5	5/3.406 = 1.47
4	12	12 − 10 = 2	2/3.406 = .59
5	5	5 − 10 = −5	−5/3.406 = −1.47

These z-scores tell us that the profit margin for Company 3 is the farthest above the mean. More specifically, this profit margin is 1.47 standard deviations above the mean. The profit margin for Company 5 is the farthest below the mean—it is 1.47 standard deviations below the mean. Because the z-score for Company 2 equals zero, its profit margin equals the mean.

Values in two different populations or samples having the same z-score are the same number of standard deviations from their respective means and, therefore, have the same relative locations. For example, suppose that the mean score on the midterm exam for students in Section A of a statistics course is 65 and the standard deviation of the scores is 10. Meanwhile, the mean score on the same exam for students in Section B is 80 and the standard deviation is 5. A student in Section A who scores an 85 and a student in Section B who scores a 90 have the same relative locations within their respective sections because their z-scores, (85 − 65)/10 = 2 and (90 − 80)/5 = 2, are equal.

The coefficient of variation

Sometimes we need to measure the size of the standard deviation of a population or sample relative to the size of the population or sample mean. The **coefficient of variation,** which makes this comparison, is defined for a population or sample as follows:

$$\text{coefficient of variation} = \frac{\text{standard deviation}}{\text{mean}} \times 100$$

The coefficient of variation compares populations or samples having different means and different standard deviations. For example, suppose that the mean yearly return for a particular stock fund, which we call Stock Fund 1, is 10.39 percent with a standard deviation of 16.18 percent, while the mean yearly return for another stock fund, which we call Stock Fund 2, is 7.7 percent with a standard deviation of 13.82 percent. It follows that the coefficient of variation for Stock Fund 1 is (16.18/10.39) × 100 = 155.73, and that the coefficient of variation for Stock Fund 2 is $(13.82/7.7) \times 100 = 179.48$. This tells us that, for Stock Fund 1, the standard deviation is 155.73 percent of the value of its mean yearly return. For Stock Fund 2, the standard deviation is 179.48 percent of the value of its mean yearly return.

In the context of situations like the stock fund comparison, the coefficient of variation is often used as a measure of *risk* because it measures the variation of the returns (the standard deviation) relative to the size of the mean return. For instance, although Stock Fund 2 has a smaller standard deviation than does Stock Fund 1 (13.82 percent compared to 16.18 percent), Stock Fund 2 has a higher coefficient of variation than does Stock Fund 1 (179.48 versus 155.73). This says that, *relative to the mean return,* the variation in returns for Stock Fund 2 is higher. That is, we would conclude that investing in Stock Fund 2 is riskier than investing in Stock Fund 1.

Exercises for Section 3.2

CONCEPTS ▣ **connect**

3.16 Define the range, variance, and standard deviation for a population.

3.17 Discuss how the variance and the standard deviation measure variation.

3.18 The Empirical Rule for a normally distributed population and Chebyshev's Theorem have the same basic purpose. In your own words, explain what this purpose is.

METHODS AND APPLICATIONS

3.19 Consider the following population of five numbers: 5, 8, 10, 12, 15. Calculate the range, variance, and standard deviation of this population.

3.20 Table 3.3 gives spending data for the football programs at the 10 universities that spent the most money on football in 2012. Calculate the population range, variance, and standard deviation of the 10 total football spending figures. Do the same for the 10 spending per scholarship player figures. **DS** Football

3.21 Consider Exercise 3.20. **DS** Football

    a Compute and interpret the *z*-score for each total spending figure.

    b Compute and interpret the *z*-score for each spending per scholarship player figure.

3.22 In order to control costs, a company wishes to study the amount of money its sales force spends entertaining clients. The following is a random sample of six entertainment expenses (dinner costs for four people) from expense reports submitted by members of the sales force. **DS** DinnerCost

    $357 $332 $309 $345 $325 $339

    a Calculate $\bar{x}$, s^2, and s for the expense data. In addition, show that the two different formulas for calculating s^2 give the same result.

    b Assuming that the distribution of entertainment expenses is approximately normally distributed, calculate estimates of tolerance intervals containing 68.26 percent, 95.44 percent, and 99.73 percent of all entertainment expenses by the sales force.

    c If a member of the sales force submits an entertainment expense (dinner cost for four) of $390, should this expense be considered unusually high (and possibly worthy of investigation by the company)? Explain your answer.

    d Compute and interpret the *z*-score for each of the six entertainment expenses.

3.23 **THE TRASH BAG CASE** **DS** TrashBag

The mean and the standard deviation of the sample of 30 trash bag breaking strengths are $\bar{x} = 52.167$ and $s = 5.394$.

    a What does the histogram in Figure 2.17 say about whether the Empirical Rule should be used to describe the trash bag breaking strengths?

Rank	School	2012 Total Spending ($ mil)	Spending per Scholarship Player ($)
1	The Ohio State University	34.36	400,000
2	University of Alabama	37.77	360,000
3	Auburn University	33.33	303,000
4	University of Wisconsin	24.23	285,000
5	University of Arkansas	24.33	283,000
6	Oklahoma State University	26.24	279,000
7	Virginia Tech	24.72	275,000
8	University of Arizona	24.12	274,000
9	University of Florida	23.25	273,000
10	University of Michigan	23.64	272,000

TABLE 3.3 Football Spending Data for 2012's Top 10 College Football Spenders (for Exercises 3.20 and 3.21) **DS** Football

Source: *Journal-News* (Hamilton, Ohio), January 11, 2015.

b   Use the Empirical Rule to calculate estimates of tolerance intervals containing 68.26 percent, 95.44 percent, and 99.73 percent of all possible trash bag breaking strengths.

c   Does the estimate of a tolerance interval containing 99.73 percent of all breaking strengths provide evidence that almost any bag a customer might purchase will have a breaking strength that exceeds 45 pounds? Explain your answer.

d   How do the percentages of the 30 breaking strengths in Table 1.10 that actually fall into the intervals $[\bar{x} \pm s]$, $[\bar{x} \pm 2s]$, and $[\bar{x} \pm 3s]$ compare to those given by the Empirical Rule? Do these comparisons indicate that the statistical inferences you made in parts b and c are reasonably valid?

3.24   THE BANK CUSTOMER WAITING TIME CASE
DS WaitTime

The mean and the standard deviation of the sample of 100 bank customer waiting times are $\bar{x} = 5.46$ and $s = 2.475$.

a   What does the histogram in Figure 2.16 say about whether the Empirical Rule should be used to describe the bank customer waiting times?

b   Use the Empirical Rule to calculate estimates of tolerance intervals containing 68.26 percent, 95.44 percent, and 99.73 percent of all possible bank customer waiting times.

c   Does the estimate of a tolerance interval containing 68.26 percent of all waiting times provide evidence that at least two-thirds of all customers will have to wait less than eight minutes for service? Explain your answer.

d   How do the percentages of the 100 waiting times in Table 1.9 that actually fall into the intervals $[\bar{x} \pm s]$,

$[\bar{x} \pm 2s]$, and $[\bar{x} \pm 3s]$ compare to those given by the Empirical Rule? Do these comparisons indicate that the statistical inferences you made in parts b and c are reasonably valid?

3.25   THE VIDEO GAME SATISFACTION RATING CASE   DS VideoGame

The mean and the standard deviation of the sample of 65 customer satisfaction ratings are $\bar{x} = 42.95$ and $s = 2.6424$.

a   What does the histogram in Figure 2.15 say about whether the Empirical Rule should be used to describe the satisfaction ratings?

b   Use the Empirical Rule to calculate estimates of tolerance intervals containing 68.26 percent, 95.44 percent, and 99.73 percent of all possible satisfaction ratings.

c   Does the estimate of a tolerance interval containing 99.73 percent of all satisfaction ratings provide evidence that 99.73 percent of all customers will give a satisfaction rating for the XYZ-Box game system that is at least 35 (the minimal rating of a "satisfied" customer)? Explain your answer.

d   How do the percentages of the 65 customer satisfaction ratings in Table 1.8 that actually fall into the intervals $[\bar{x} \pm s]$, $[\bar{x} \pm 2s]$, and $[\bar{x} \pm 3s]$ compare to those given by the Empirical Rule? Do these comparisons indicate that the statistical inferences you made in parts b and c are reasonably valid?

3.26   Consider the 63 automatic teller machine (ATM) transaction times given in Table 3.4 below.

a   Construct a histogram (or a stem-and-leaf display) for the 63 ATM transaction times. Describe the shape of the distribution of transaction times.   DS ATMTime

b   When we compute the sample mean and sample standard deviation for the transaction times, we

TABLE 3.4   ATM Transaction Times (in Seconds) for 63 Withdrawals   DS ATMTime

Transaction	Time	Transaction	Time	Transaction	Time
1	32	22	34	43	37
2	32	23	32	44	32
3	41	24	34	45	33
4	51	25	35	46	33
5	42	26	33	47	40
6	39	27	42	48	35
7	33	28	46	49	33
8	43	29	52	50	39
9	35	30	36	51	34
10	33	31	37	52	34
11	33	32	32	53	33
12	32	33	39	54	38
13	42	34	36	55	41
14	34	35	41	56	34
15	37	36	32	57	35
16	37	37	33	58	35
17	33	38	34	59	37
18	35	39	38	60	39
19	40	40	32	61	44
20	36	41	35	62	40
21	32	42	33	63	39

find that $\bar{x} = 36.56$ and $s = 4.475$. Compute each of the intervals $[\bar{x} \pm s]$, $[\bar{x} \pm 2s]$, and $[\bar{x} \pm 3s]$. Then count the number of transaction times that actually fall into each interval and find the percentage of transaction times that actually fall into each interval.

    c   How do the percentages of transaction times that fall into the intervals $[\bar{x} \pm s]$, $[\bar{x} \pm 2s]$, and $[\bar{x} \pm 3s]$ compare to those given by the Empirical Rule? How do the percentages of transaction times that fall into the intervals $[\bar{x} \pm 2s]$ and $[\bar{x} \pm 3s]$ compare to those given by Chebyshev's Theorem?

    d   Explain why the Empirical Rule does not describe the transaction times extremely well.

3.27   Consider three stock funds, which we will call Stock Funds 1, 2, and 3. Suppose that Stock Fund 1 has a mean yearly return of 10.93 percent with a standard deviation of 41.96 percent, Stock Fund 2 has a mean yearly return of 13 percent with a standard deviation of 9.36 percent, and Stock Fund 3 has a mean yearly return of 34.45 percent with a standard deviation of 41.16 percent.

    a   For each fund, find an interval in which you would expect 95.44 percent of all yearly returns to fall. Assume returns are normally distributed.

    b   Using the intervals you computed in part a, compare the three funds with respect to average yearly returns and with respect to variability of returns.

    c   Calculate the coefficient of variation for each fund, and use your results to compare the funds with respect to risk. Which fund is riskier?

3.3 Percentiles, Quartiles, and Box-and-Whiskers Displays

LO3-4

Compute and interpret percentiles, quartiles, and box-and-whiskers displays.

Percentiles, quartiles, and five-number displays

In this section we consider **percentiles** and their applications. We begin by defining the *p*th percentile.

> For a set of measurements arranged in increasing order, the *p*th **percentile** is a value such that p percent of the measurements fall at or below the value, and $(100 - p)$ percent of the measurements fall at or above the value.

There are various procedures for calculating percentiles. **One procedure for calculating the *p*th percentile for a set of *n* measurements uses the following three steps:**

Step 1: Arrange the measurements in increasing order.
Step 2: Calculate the index

$$i = \left(\frac{p}{100}\right)n$$

Step 3: (a) If i is not an integer, round up to obtain the next integer greater than i. This integer denotes the position of the *p*th percentile in the ordered arrangement.

        (b) If i is an integer, the *p*th percentile is the average of the measurements in positions i and $i + 1$ in the ordered arrangement.

To illustrate the calculation and interpretation of percentiles, recall in the household income situation that an economist has randomly selected a sample of $n = 12$ households from a Midwestern city and has determined last year's income for each household. In order to assess the variation of the population of household incomes in the city, we will calculate various percentiles for the sample of incomes. Specifically, we will calculate the 10th, 25th, 50th, 75th, and 90th percentiles of these incomes. The first step is to arrange the incomes in increasing order as follows:

7,524	11,070	18,211	26,817	36,551	41,286
49,312	57,283	72,814	90,416	135,540	190,250

To find the 10th percentile, we calculate (in step 2) the index

$$i = \left(\frac{p}{100}\right)n = \left(\frac{10}{100}\right)12 = 1.2$$

Because $i = 1.2$ is not an integer, step 3(a) says to round $i = 1.2$ up to 2. It follows that the 10th percentile is the income in position 2 in the ordered arrangement—that is, 11,070. To find the 25th percentile, we calculate the index

$$i = \left(\frac{p}{100}\right)n = \left(\frac{25}{100}\right)12 = 3$$

Because $i = 3$ is an integer, step 3(b) says that the 25th percentile is the average of the incomes in positions 3 and 4 in the ordered arrangement—that is, $(18{,}211 + 26{,}817)/2 = 22{,}514$. To find the 50th percentile, we calculate the index

$$i = \left(\frac{p}{100}\right) n = \left(\frac{50}{100}\right) 12 = 6$$

Because $i = 6$ is an integer, step 3(b) says that the 50th percentile is the average of the incomes in positions 6 and 7 in the ordered arrangement—that is, $(41{,}286 + 49{,}312)/2 = 45{,}299$. To find the 75th percentile, we calculate the index

$$i = \left(\frac{p}{100}\right) n = \left(\frac{75}{100}\right) 12 = 9$$

Because $i = 9$ is an integer, step 3(b) says that the 75th percentile is the average of the incomes in positions 9 and 10 in the ordered arrangement—that is, $(72{,}814 + 90{,}416)/2 = 81{,}615$. To find the 90th percentile, we calculate the index

$$i = \left(\frac{p}{100}\right) n = \left(\frac{90}{100}\right) 12 = 10.8$$

Because $i = 10.8$ is not an integer, step 3(a) says to round $i = 10.8$ up to 11. It follows that the 90th percentile is the income in position 11 in the ordered arrangement—that is, 135,540.

One appealing way to describe the variation of a set of measurements is to divide the data into four parts, each containing approximately 25 percent of the measurements. This can be done by defining the *first, second,* and *third quartiles* as follows:

The **first quartile,** denoted Q_1, is the **25th percentile.**
The **second quartile** (or **median**), denoted M_d, is the **50th percentile.**
The **third quartile,** denoted Q_3, is the **75th percentile.**

Note that the second quartile is simply another name for the median. Furthermore, the procedure we have described here that is used to find the 50th percentile (second quartile) will always give the same result as the previously described procedure (see Section 3.1) for finding the median. To illustrate how the quartiles divide a set of measurements into four parts, consider the following display of the sampled incomes, which shows the first quartile (the 25th percentile), $Q_1 = 22{,}514$, the median (the 50th percentile), $M_d = 45{,}299$, and the third quartile (the 75th percentile), $Q_3 = 81{,}615$:

7,524	11,070	18,211		26,817	36,551	41,286	

$Q_1 = 22{,}514$ $\qquad\qquad\qquad\qquad\qquad\qquad$ $M_d = 45{,}299$

49,312	57,283	72,814		90,416	135,540	190,250

$Q_3 = 81{,}615$

Using the quartiles, we estimate that for the household incomes in the Midwestern city: (1) 25 percent of the incomes are less than or equal to $22,514, (2) 25 percent of the incomes are between $22,514 and $45,299, (3) 25 percent of the incomes are between $45,299 and $81,615, and (4) 25 percent of the incomes are greater than or equal to $81,615. In addition, to assess some of the lowest and highest incomes, the 10th percentile estimates that 10 percent of the incomes are less than or equal to $11,070, and the 90th percentile estimates that 10 percent of the incomes are greater than or equal to $135,540.

We sometimes describe a set of measurements by using a **five-number summary.** The summary consists of (1) the smallest measurement; (2) the first quartile, Q_1; (3) the median, M_d; (4) the third quartile, Q_3; and (5) the largest measurement. It is easy to graphically depict a five-number summary. For example, we have seen that for the 12 household incomes, the smallest income is $7,524, $Q_1 = \$22{,}514$, $M_d = \$45{,}299$, $Q_3 = \$81{,}615$, and the largest income is $190,250. A graphical depiction of this five-number summary is shown in the page margin. Notice that we have drawn a vertical line extending from the smallest income to the largest income. In addition, a rectangle is drawn that extends from Q_1 to Q_3, and a horizontal line is drawn to indicate the location of the median. The summary divides the incomes into four parts, with the middle 50 percent of the incomes depicted by the rectangle. The summary indicates that the largest 25 percent of the incomes is much more spread out than the smallest 25 percent of the incomes and that the second-largest 25 percent of the incomes is more spread out than the second-smallest 25 percent of the incomes. Overall, the summary indicates that the incomes are fairly highly skewed toward the larger incomes.

In general, unless percentiles correspond to very high or very low percentages, they are resistant (like the median) to extreme values. For example, the 75th percentile of the household incomes would remain $81,615 even if the largest income ($190,250) were, instead, $7,000,000. On the other hand, the standard deviation in this situation would increase. In general, if a population is highly skewed to the right or left, the standard deviation is so large that using it to describe variation does not provide much useful information. For example, the standard deviation of the 12 household incomes is inflated by the large incomes $135,540 and $190,250 and can be calculated to be $54,567. Because the mean of the 12 incomes is $61,423, Chebyshev's Theorem says that we estimate that at least 75 percent of all household incomes in the city are in the interval $[\bar{x} \pm 2s] = [61,423 \pm 2(54,567)] = [-47,711, 170,557]$; that is, are $170,557 or less. This is much less informative than using the 75th percentile, which estimates that 75 percent of all household incomes are less than or equal to $81,615. In general, if a population is highly skewed to the right or left, it can be best to describe the variation of the population by using various percentiles. This is what we did when we estimated the variation of the household incomes in the city by using the 10th, 25th, 50th, 75th, and 90th percentiles of the 12 sampled incomes and when we depicted this variation by using the five-number summary. Using other percentiles can also be informative. For example, the Bureau of the Census sometimes assesses the variation of all household incomes in the United States by using the 20th, 40th, 60th, and 80th percentiles of these incomes.

We next define the **interquartile range,** denoted IQR, to be the difference between the third quartile Q_3 and the first quartile Q_1. That is, $IQR = Q_3 - Q_1$. This quantity can be interpreted as the length of the interval that contains the *middle 50 percent* of the measurements. For instance, the interquartile range of the 12 household incomes is $Q_3 - Q_1 = 81,615 - 22,514 = 59,101$. This says that we estimate that the middle 50 percent of all household incomes fall within a range that is $59,101 long.

The procedure we have presented for calculating the first and third quartiles is not the only procedure for computing these quantities. In fact, several procedures exist, and, for example, different statistical computer packages use several somewhat different methods for computing the quartiles. These different procedures sometimes obtain different results, but the overall objective is always to divide the data into four equal parts.

Box-and-whiskers displays (box plots)

A more sophisticated modification of the graphical five-number summary is called a **box-and-whiskers display** (sometimes called a **box plot**). Such a display is constructed by using Q_1, M_d, Q_3, and the interquartile range. As an example, suppose that 20 randomly selected customers give the following satisfaction ratings (on a scale of 1 to 10) for a DVD recorder:

1   3   5   5   7   8   8   8   8   8   9   9   9   9   9   10   10   10   10   DS DVDSat

The Minitab output in Figure 3.16 says that for these ratings $Q_1 = 7.25$, $M_d = 8$, $Q_3 = 9$, and $IQR = Q_3 - Q_1 = 9 - 7.25 = 1.75$. To construct a box-and-whiskers display, we first draw a box that extends from Q_1 to Q_3. As shown in Figure 3.17, for the satisfaction ratings data this box extends from $Q_1 = 7.25$ to $Q_3 = 9$. The box contains the middle 50 percent of the data set. Next a vertical line is drawn through the box at the value of the median M_d. This line divides the data set into two roughly equal parts. We next define what we call the **lower** and **upper limits.** The **lower limit** is located $1.5 \times IQR$ below Q_1 and the **upper limit** is located $1.5 \times IQR$ above Q_3. For the satisfaction ratings data, these limits are

$$Q_1 - 1.5(IQR) = 7.25 - 1.5(1.75) = 4.625 \quad \text{and} \quad Q_3 + 1.5(IQR) = 9 + 1.5(1.75) = 11.625$$

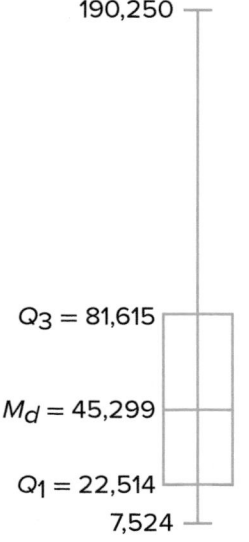

Household Income Five-Number Summary

190,250

$Q_3 = 81,615$

$M_d = 45,299$

$Q_1 = 22,514$

7,524

FIGURE 3.16   **Minitab Output of Statistics Describing the 20 Satisfaction Ratings**

Variable	Count	Mean	StDev	Range
Rating	20	7.700	2.430	9.000

Variable	Minimum	Q1	Median	Q3	Maximum
Rating	1.000	7.250	8.000	9.000	10.000

FIGURE 3.17   **Constructing a Box Plot of the Satisfaction Ratings**

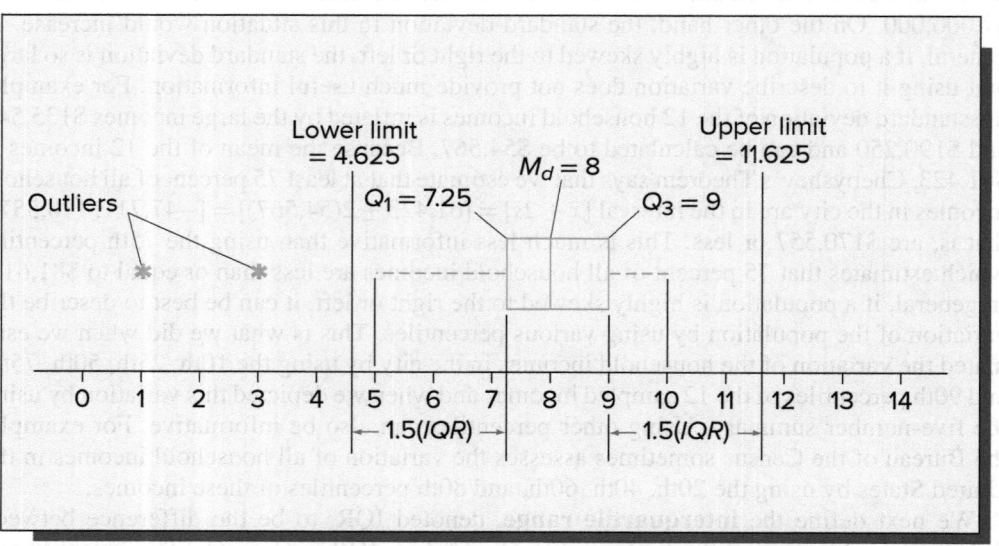

The lower and upper limits help us to draw the plot's **whiskers:** dashed lines extending below Q_1 and above Q_3 (as in Figure 3.17). One whisker is drawn from Q_1 to the smallest measurement between the lower and upper limits. For the satisfaction ratings data, this whisker extends from $Q_1 = 7.25$ down to 5, because 5 is the smallest rating between the lower and upper limits 4.625 and 11.625. The other whisker is drawn from Q_3 to the largest measurement between the lower and upper limits. For the satisfaction ratings data, this whisker extends from $Q_3 = 9$ up to 10, because 10 is the largest rating between the lower and upper limits 4.625 and 11.625. The lower and upper limits are also used to identify *outliers*. An **outlier** is a measurement that is separated from (that is, different from) most of the other measurements in the data set. A measurement that is less than the lower limit or greater than the upper limit is considered to be an outlier. We indicate the location of an outlier by plotting this measurement with the symbol *. For the satisfaction rating data, the ratings 1 and 3 are outliers because they are less than the lower limit 4.625. Figure 3.18 gives the Minitab output of a box-and-whiskers plot of the satisfacting ratings.

We now summarize how to construct a box-and-whiskers plot.

FIGURE 3.18   **Minitab Output of a Box Plot of the Satisfaction Ratings**

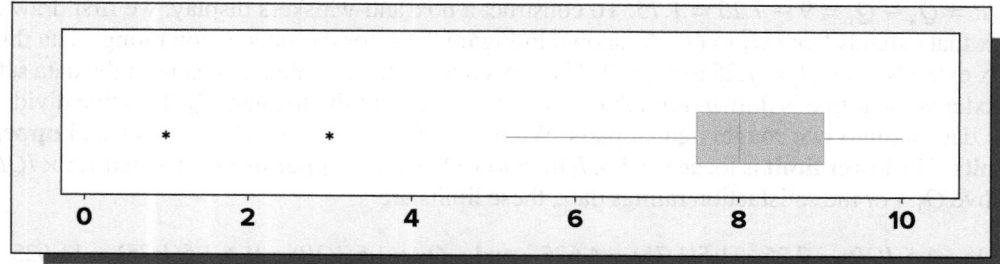

Constructing a Box-and-Whiskers Display (Box Plot)

1 Draw a **box** that extends from the first quartile Q_1 to the third quartile Q_3. Also draw a vertical line through the box located at the median M_d.

2 Determine the values of the **lower** and **upper limits.** The **lower limit** is located $1.5 \times IQR$ below Q_1 and the **upper limit** is located $1.5 \times IQR$ above Q_3. That is, the lower and upper limits are

$$Q_1 - 1.5(IQR) \quad \text{and} \quad Q_3 + 1.5(IQR)$$

3 Draw **whiskers** as dashed lines that extend below Q_1 and above Q_3. Draw one whisker from Q_1 to the *smallest* measurement that is between the lower and upper limits. Draw the other whisker from Q_3 to the *largest* measurement that is between the lower and upper limits.

4 A measurement that is less than the lower limit or greater than the upper limit is an **outlier.** Plot each outlier using the symbol *.

When interpreting a box-and-whiskers display, keep several points in mind. First, the box (between Q_1 and Q_3) contains the middle 50 percent of the data. Second, the median (which is inside the box) divides the data into two roughly equal parts. Third, if one of the whiskers is longer than the other, the data set is probably skewed in the direction of the longer whisker. Last, observations designated as outliers should be investigated. Understanding the root causes behind the outlying observations will often provide useful information. For instance, understanding why two of the satisfaction ratings in the box plot of Figure 3.18 are substantially lower than the great majority of the ratings may suggest actions that can improve the DVD recorder manufacturer's product and/or service. Outliers can also be caused by inaccurate measuring, reporting, or plotting of the data. Such possibilities should be investigated, and incorrect data should be adjusted or eliminated.

Graphical five-number summaries and box-and-whiskers displays are perhaps best used to compare different sets of measurements. We demonstrate this use of such displays in the following example.

Ⓒ EXAMPLE 3.9 The Standard and Poor's 500 Case

Figure 3.19 shows box plots of the percentage returns of stocks on the Standard and Poor's 500 (S&P 500) for different time horizons of investment. Figure 3.19(a) compares a 1-year time horizon with a 3-year time horizon. We see that there is a 25 percent chance of a negative return (loss) for the 3-year horizon and a 25 percent chance of earning more than 50 percent on the principal during the three years. Figures 3.19(b) and (c) compare a 5-year time horizon with a

FIGURE 3.19   **Box Plots of the Percentage Returns of Stocks on the S&P 500 for Different Time Horizons of Investment**

(a) 1-year versus 3-year

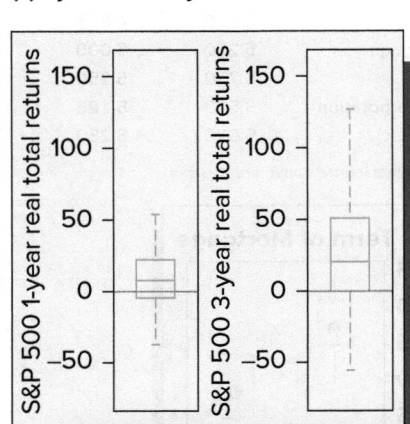

(b) 5-year versus 10-year

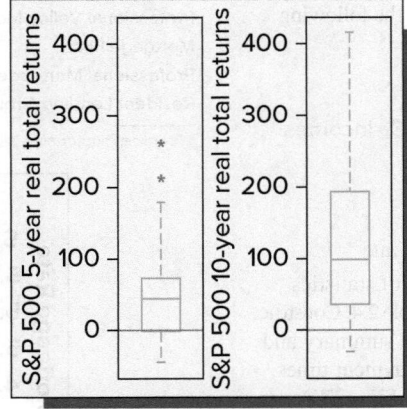

(c) 10-year versus 20-year

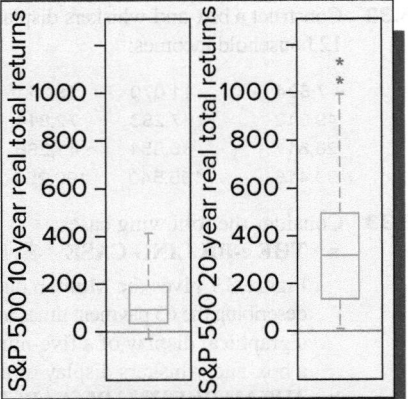

Data from Global Financial Data

Source: http//junkcharts.typepad.com/junk_charts/boxplot/.

10-year time horizon and a 10-year time horizon with a 20-year time horizon. We see that there is still a positive chance of a loss for the 10-year horizon, but the median return for the 10-year horizon almost doubles the principal (a 100 percent return, which is about 8 percent per year compounded). With a 20-year horizon, there is virtually no chance of a loss, and there were two positive outlying returns of over 1000 percent (about 13 percent per year compounded).

Exercises for Section 3.3

CONCEPTS ■ connect

3.28 Explain each of the following in your own words: a percentile; the first quartile, Q_1; the third quartile, Q_3; and the interquartile range, IQR.

3.29 Discuss how a box-and-whiskers display is used to identify outliers.

METHODS AND APPLICATIONS

3.30 Recall that 20 randomly selected customers give the following satisfaction ratings (on a scale of 1 to 10) for a DVD recorder: 1, 3, 5, 5, 7, 8, 8, 8, 8, 8, 8, 9, 9, 9, 9, 9, 10, 10, 10, 10.   ⓄS DVDSat

   a Using the technique discussed earlier, find the first quartile, median, and third quartile for these data.

   b Do you obtain the same values for the first quartile and the third quartile that are shown on the Minitab output in Figure 3.16?

   c Using your results, construct a graphical display of a five-number summary and a box-and-whiskers display.

3.31 Thirteen internists in the Midwest are randomly selected, and each internist is asked to report last year's income. The incomes obtained (in thousands of dollars) are 152, 144, 162, 154, 146, 241, 127, 141, 171, 177, 138, 132, 192. Find   ⓄS DrSalary

   a The 90th percentile.
   b The median.
   c The first quartile.
   d The third quartile.
   e The 10th percentile.
   f The interquartile range.
   g Develop a graphical display of a five-number summary and a box-and-whiskers display.

3.32 Construct a box-and-whiskers display of the following 12 household incomes:

7,524	11,070	18,211
49,312	57,283	72,814
26,817	36,551	41,286
90,416	135,540	190,250

ⓄS Incomes

3.33 Consider the following cases:

   a THE e-BILLING CASE   ⓄS PayTime

   Figure 3.7 gives the Minitab output of statistics describing the 65 payment times in Table 2.4. Construct a graphical display of a five-number summary and a box-and-whiskers display of the payment times.

   b THE MARKETING RESEARCH CASE   ⓄS Design

   Consider the 60 bottle design ratings in Table 1.6. The smallest rating is 20, $Q_1 = 29$, $M_d = 31$, $Q_3 = 33$, and the largest rating is 35. Construct a graphical

display of this five-number summary and verify that a box-and-whiskers display of the bottle design ratings is as shown in the JMP output of Figure 3.20.

   c Discuss the difference between the skewness of the 65 payment times and the skewness of the 60 bottle design ratings.

3.34 Figure 3.21 gives graphical displays of five-number summaries of a large company's employee salaries for different salary grades, with a comparison of salaries for males and females and a comparison of actual salaries to the prescribed salary ranges for the salary grades.

   a What inequities between males and females are apparent? Explain.

   b How closely are the prescribed salary ranges being observed? Justify your answer.

3.35 On its website, the *Statesman Journal* newspaper (Salem, Oregon, 2005) reports mortgage loan interest rates for 30-year and 15-year fixed-rate mortgage loans for a number of Willamette Valley lending institutions. Of interest is whether there is any systematic difference between 30-year rates and 15-year rates (expressed as annual percentage rate or APR). The table below displays the 30-year rate and the 15-year rate for each of nine lending institutions. The figure below the data gives side-by-side Minitab box-and-whiskers plots of the 30-year rates and the 15-year rates. Interpret the plots by comparing the central tendencies and variabilities of the 15-and 30-year rates.   ⓄS Mortgage

Lending Institution	30-Year	15-Year
Blue Ribbon Home Mortgage	5.375	4.750
Coast To Coast Mortgage Lending	5.250	4.750
Community Mortgage Services Inc.	5.000	4.500
Liberty Mortgage	5.375	4.875
Jim Morrison's MBI	5.250	4.875
Professional Valley Mortgage	5.250	5.000
Mortgage First	5.750	5.250
Professional Mortgage Corporation	5.500	5.125
Resident Lending Group Inc.	5.625	5.250

Source: http://online.statesmanjournal.com/mortrates.cfm.

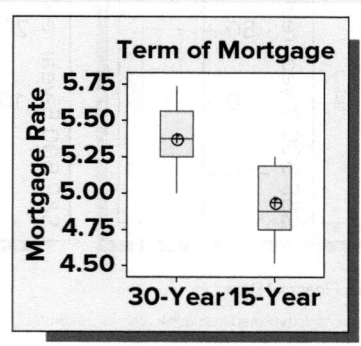

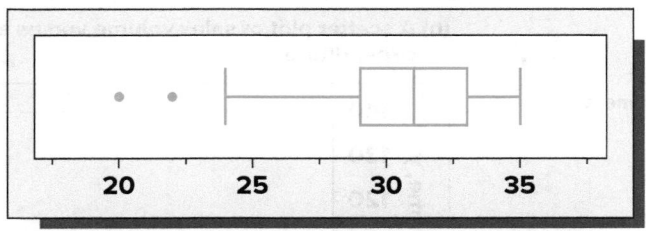

FIGURE 3.21   **Employee Salaries by Salary Grade and Gender (for Exercise 3.34)**

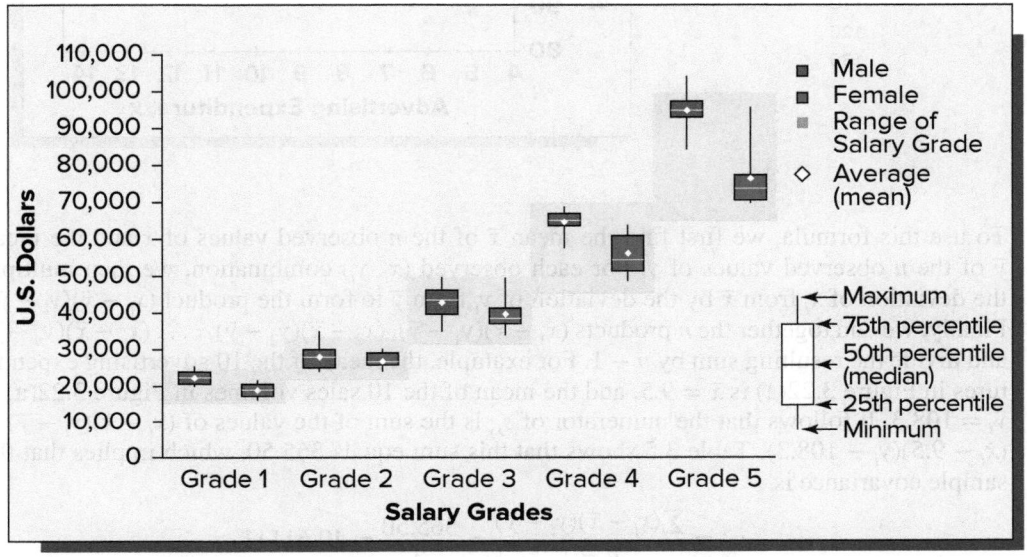

Source: www.perceptualedqe.com/articles/dmreview/boxes_of_insight.pdf.

3.4 Covariance, Correlation, and the Least Squares Line (Optional)

LO3-5
Compute and interpret covariance, correlation, and the least squares line (Optional).

In Section 2.6 we discussed how to use a scatter plot to explore the relationship between two variables x and y. To construct a scatter plot, a sample of n pairs of values of x and y—(x_1, y_1), $(x_2, y_2), \ldots, (x_n, y_n)$—is collected. Then, each value of y is plotted against the corresponding value of x. If the plot points seem to fluctuate around a straight line, we say that there is a **linear relationship** between x and y. For example, suppose that 10 sales regions of equal sales potential for a company were randomly selected. The advertising expenditures (in units of \$10,000) in these 10 sales regions were purposely set in July of last year at the values given in the second column of Figure 3.22(a). The sales volumes (in units of \$10,000) were then recorded for the 10 sales regions and are given in the third column of Figure 3.22(a). A scatter plot of sales volume, y, versus advertising expenditure, x, is given in Figure 3.22(b) and shows a linear relationship between x and y.

A measure of the **strength of the linear relationship** between x and y is the **covariance.** The **sample covariance** is calculated by using the sample of n pairs of observed values of x and y.

The **sample covariance** is denoted as s_{xy} and is defined as follows:

$$s_{xy} = \frac{\sum_{i=1}^{n}(x_i - \bar{x})(y_i - \bar{y})}{n-1}$$

FIGURE 3.22   The Sales Volume Data, and a Scatter Plot

(a) The sales volume data ⒹⓈ SalesPlot

Sales Region	Advertising Expenditure, x	Sales Volume, y
1	5	89
2	6	87
3	7	98
4	8	110
5	9	103
6	10	114
7	11	116
8	12	110
9	13	126
10	14	130

(b) A scatter plot of sales volume versus advertising expenditure

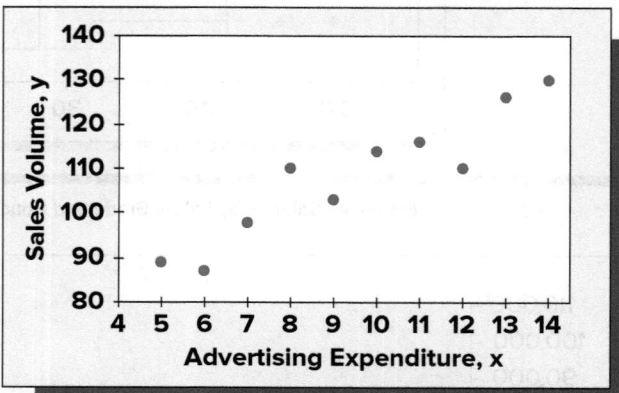

To use this formula, we first find the mean $\bar{x}$ of the n observed values of x and the mean $\bar{y}$ of the n observed values of y. For each observed (x_i, y_i) combination, we then multiply the deviation of x_i from $\bar{x}$ by the deviation of y_i from $\bar{y}$ to form the product $(x_i - \bar{x})(y_i - \bar{y})$. Finally, we add together the n products $(x_1 - \bar{x})(y_1 - \bar{y}), (x_2 - \bar{x})(y_2 - \bar{y}), \ldots, (x_n - \bar{x})(y_n - \bar{y})$ and divide the resulting sum by $n - 1$. For example, the mean of the 10 advertising expenditures in Figure 3.22(a) is $\bar{x} = 9.5$, and the mean of the 10 sales volumes in Figure 3.22(a) is $\bar{y}_i = 108.3$. It follows that the numerator of s_{xy} is the sum of the values of $(x_i - \bar{x})(y_i - \bar{y}) = (x_i - 9.5)(y_i - 108.3)$. Table 3.5 shows that this sum equals 365.50, which implies that the sample covariance is

$$s_{xy} = \frac{\Sigma(x_i - \bar{x})(y_i - \bar{y})}{n - 1} = \frac{365.50}{9} = 40.61111$$

To interpret the covariance, consider Figure 3.23(a). This figure shows the scatter plot of Figure 3.22(b) with a vertical blue line drawn at $\bar{x} = 9.5$ and a horizontal red line drawn at $\bar{y} = 108.3$. The lines divide the scatter plot into four quadrants. Points in quadrant I correspond to x_i greater than $\bar{x}$ and y_i greater than $\bar{y}$ and thus give a value of $(x_i - \bar{x})(y_i - \bar{y})$ greater than 0. Points in quadrant III correspond to x_i less than $\bar{x}$ and y_i less than $\bar{y}$ and thus also give a value of $(x_i - \bar{x})(y_i - \bar{y})$ greater than 0. It follows that if s_{xy} is positive, the points having the greatest influence on $\Sigma(x_i - \bar{x})(y_i - \bar{y})$ and thus on s_{xy} must be in quadrants I and III. Therefore, a positive value of s_{xy} (as in the sales volume example) indicates a positive linear relationship between x and y. That is, as x increases, y increases.

TABLE 3.5   The Calculation of the Numerator of s_{xy}

x_i	y_i	$(x_i - 9.5)$	$(y_i - 108.3)$	$(x_i - 9.5)(y_i - 108.3)$
5	89	−4.5	−19.3	86.85
6	87	−3.5	−21.3	74.55
7	98	−2.5	−10.3	25.75
8	110	−1.5	1.7	−2.55
9	103	−0.5	−5.3	2.65
10	114	0.5	5.7	2.85
11	116	1.5	7.7	11.55
12	110	2.5	1.7	4.25
13	126	3.5	17.7	61.95
14	130	4.5	21.7	97.65
Totals 95	1083	0	0	365.50

If we further consider Figure 3.23(a), we see that points in quadrant II correspond to x_i less than $\bar{x}$ and y_i greater than $\bar{y}$ and thus give a value of $(x_i - \bar{x})(y_i - \bar{y})$ less than 0. Points in quadrant IV correspond to x_i greater than $\bar{x}$ and y_i less than $\bar{y}$ and thus also give a value of $(x_i - \bar{x})(y_i - \bar{y})$ less than 0. It follows that if s_{xy} is negative, the points having the greatest influence on $\Sigma(x_i - \bar{x})(y_i - \bar{y})$ and thus on s_{xy} must be in quadrants II and IV. Therefore, a negative value of s_{xy} indicates a negative linear relationship between x and y. That is, as x increases, y decreases, as shown in Figure 3.23(b). For example, a negative linear relationship might exist between average hourly outdoor temperature (x) in a city during a week and the city's natural gas consumption (y) during the week. That is, as the average hourly outdoor temperature increases, the city's natural gas consumption would decrease. Finally, note that if s_{xy} is near zero, the (x_i, y_i) points would be fairly evenly distributed across all four quadrants. This would indicate little or no linear relationship between x and y, as shown in Figure 3.23(c).

From the previous discussion, it might seem that a large positive value for the covariance indicates that x and y have a strong positive linear relationship and a very negative value for the covariance indicates that x and y have a strong negative linear relationship. However, one problem with using the covariance as a measure of the strength of the linear relationship between x and y is that the value of the covariance depends on the units in which x and y are measured. A measure of the strength of the linear relationship between x and y that does not depend on the units in which x and y are measured is the **correlation coefficient.**

The **sample correlation coefficient** is denoted as r and is defined as follows:

$$r = \frac{s_{xy}}{s_x s_y}$$

Here, s_{xy} is the previously defined sample covariance, s_x is the sample standard deviation of the sample of x values, and s_y is the sample standard deviation of the sample of y values.

For the sales volume data:

$$s_x = \sqrt{\frac{\sum_{i=1}^{10}(x_i - \bar{x})^2}{9}} = 3.02765 \qquad \text{and} \qquad s_y = \sqrt{\frac{\sum_{i=1}^{10}(y_i - \bar{y})^2}{9}} = 14.30656$$

FIGURE 3.23 **Interpretation of the Sample Covariance**

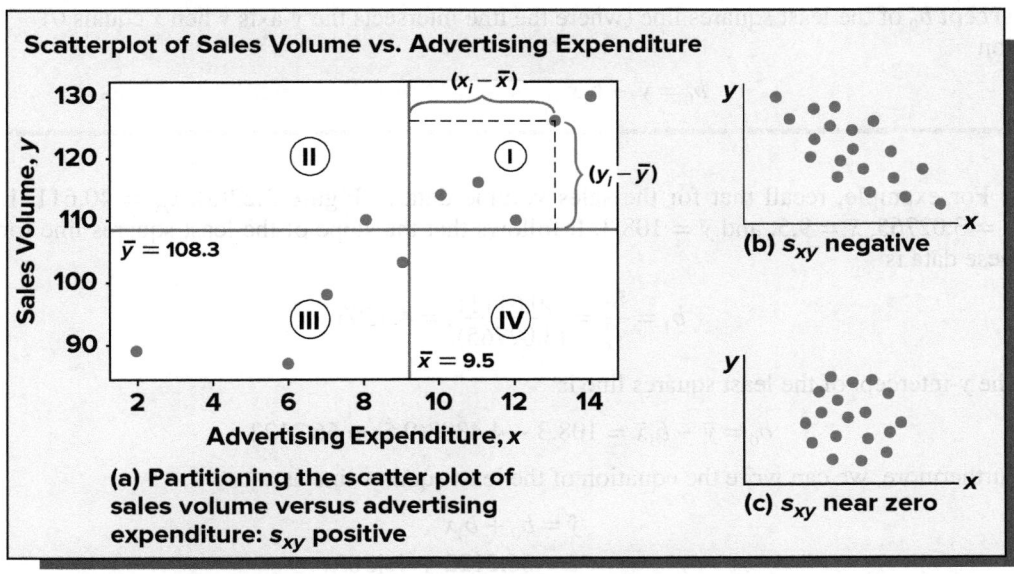

Scatterplot of Sales Volume vs. Advertising Expenditure

(a) Partitioning the scatter plot of sales volume versus advertising expenditure: s_{xy} positive

(b) s_{xy} negative

(c) s_{xy} near zero

Therefore, the sample correlation coefficient is

$$r = \frac{s_{xy}}{s_x s_y} = \frac{40.61111}{(3.02765)(14.30656)} = .93757$$

It can be shown that the sample correlation coefficient r is always between -1 and 1. A value of r near 0 implies little linear relationship between x and y. A value of r close to 1 says that x and y have a strong tendency to move together in a straight-line fashion with a positive slope and, therefore, that x and y are highly related and **positively correlated.** A value of r close to -1 says that x and y have a strong tendency to move together in a straight-line fashion with a negative slope and, therefore, that x and y are highly related and **negatively correlated.** Note that if $r = 1$, the (x, y) points fall exactly on a positively sloped straight line, and, if $r = -1$, the (x, y) points fall exactly on a negatively sloped straight line. For example, because $r = .93757$ in the sales volume example, we conclude that advertising expenditure (x) and sales volume (y) have a strong tendency to move together in a straight-line fashion with a positive slope. That is, x and y have a strong positive linear relationship.

We next note that the sample covariance s_{xy} is the point estimate of the **population covariance,** which we denote as σ_{xy}, and the sample correlation coefficient r is the point estimate of the **population correlation coefficient,** which we denote as ρ (pronounced *row*). To define σ_{xy} and ρ, let μ_x and σ_x denote the mean and the standard deviation of the population of all possible x values, and let μ_y and σ_y denote the mean and the standard deviation of the population of all possible y values. Then, σ_{xy} is the average of all possible values of $(x - \mu_x)(y - \mu_y)$, and ρ equals $\sigma_{xy}/(\sigma_x \sigma_y)$. Similar to r, ρ is always between -1 and 1.

After establishing that a strong positive or a strong negative linear relationship exists between two variables x and y, we might wish to predict y on the basis of x. This can be done by drawing a straight line through a scatter plot of the observed data. Unfortunately, however, if different people *visually* drew lines through the scatter plot, their lines would probably differ from each other. What we need is the "best line" that can be drawn through the scatter plot. Although there are various definitions of what this best line is, one of the most useful best lines is the *least squares line*. The least squares line will be discussed in detail in Chapter 14. For now, we will say that, intuitively, the **least squares line** is the line that minimizes the sum of the squared vertical distances between the points on the scatter plot and the line.

It can be shown that the **slope b_1** (defined as rise/run) of the least squares line is given by the equation

$$b_1 = \frac{s_{xy}}{s_x^2}$$

In addition, the **y-intercept b_0** of the least squares line (where the line intersects the y-axis when x equals 0) is given by the equation

$$b_0 = \bar{y} - b_1 \bar{x}$$

For example, recall that for the sales volume data in Figure 3.22(a), $s_{xy} = 40.61111$, $s_x = 3.02765$, $\bar{x} = 9.5$, and $\bar{y} = 108.3$. It follows that the slope of the least squares line for these data is

$$b_1 = \frac{s_{xy}}{s_x^2} = \frac{40.61111}{(3.02765)^2} = 4.4303$$

The y-intercept of the least squares line is

$$b_0 = \bar{y} - b_1 \bar{x} = 108.3 - 4.4303(9.5) = 66.2122$$

Furthermore, we can write the equation of the least squares line as

$$\hat{y} = b_0 + b_1 x$$
$$= 66.2122 + 4.4303x$$

FIGURE 3.24   The Least Squares Line for the Sales Volume Data

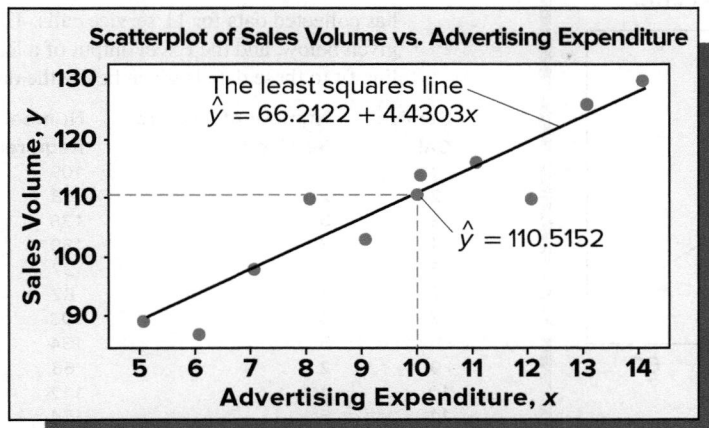

Here, because we will use the line to predict y on the basis of x, we call $\hat{y}$ (pronounced y *hat*) **the predicted value of y** when the advertising expenditure is x. For example, suppose that we will spend $100,000 on advertising in a sales region in July of a future year. Because an advertising expenditure of $100,000 corresponds to an x of 10, a prediction of sales volume in July of the future year is (see Figure 3.24):

$$\hat{y} = 66.2122 + 4.4303(10)$$
$$= 110.5152 \text{ (that is, \$1,105,152)}$$

Is this prediction likely to be accurate? If the least squares line developed from last July's data applies to the future July, then, because the sample correlation coefficient $r = .93757$ is fairly close to 1, we might hope that the prediction will be reasonably accurate. However, we will see in Chapter 14 that a sample correlation coefficient near 1 does not necessarily mean that the least squares line will predict accurately. We will also study (in Chapter 14) better ways to assess the potential accuracy of a prediction.

Exercises for Section 3.4

CONCEPTS ▓ connect

3.36 Discuss what the covariance and the correlation coefficient say about the linear relationship between two variables x and y.

3.37 Discuss how the least squares line is used to predict y on the basis of x.

METHODS AND APPLICATIONS

3.38 THE NATURAL GAS CONSUMPTION CASE
⑤ GasCon1

In the table to the right we give the average hourly outdoor temperature (x) in a city during a week and the city's natural gas consumption (y) during the week for each of eight weeks (the temperature readings are expressed in degrees Fahrenheit and the natural gas consumptions are expressed in millions

of cubic feet of natural gas—denoted MMcf). The output on the next page is obtained when Minitab is used to fit a least squares line to the natural gas consumption data.

Week	Average Hourly Temperature, x (°F)	Weekly Natural Gas Consumption, y (MMcf)
1	28.0	12.4
2	28.0	11.7
3	32.5	12.4
4	39.0	10.8
5	45.9	9.4
6	57.8	9.5
7	58.1	8.0
8	62.5	7.5

⑤ GasCon1

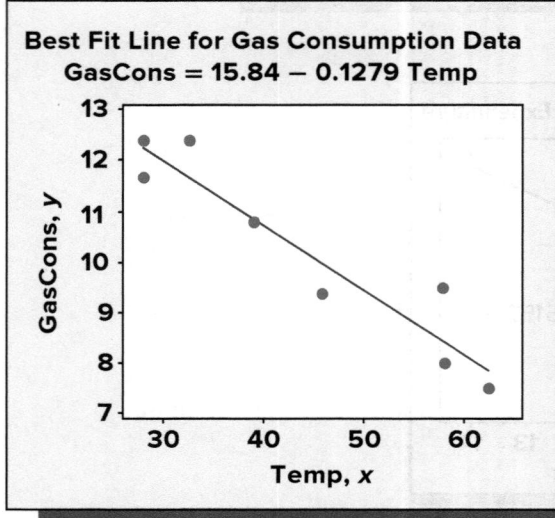

It can be shown that for the gas consumption data:

$\bar{x} = 43.98 \qquad \bar{y} = 10.2125$

$$\sum_{i=1}^{8}(x_i - \bar{x})^2 = 1404.355 \qquad \sum_{i=1}^{8}(y_i - \bar{y})^2 = 25.549$$

$$\sum_{i=1}^{8}(x_i - \bar{x})(y_i - \bar{y}) = -179.6475$$

a Calculate s_{xy}, s_x, s_y, and r.
b Using the formulas for the slope and y-intercept, calculate (within rounding) the values $b_1 = -.1279$ and $b_0 = 15.84$ on the Minitab output.
c Find a prediction of the natural gas consumption during a week when the average hourly temperature is 40° Fahrenheit.

3.39   THE SERVICE TIME CASE   🖵 SrvcTime

Accu-Copiers, Inc., sells and services the Accu-500 copying machine. As part of its standard service

contract, the company agrees to perform routine service on this copier. To obtain information about the time it takes to perform routine service, Accu-Copiers has collected data for 11 service calls. The data are given below, and the Excel output of a least squares line fit to these data is given below the data.

Service Call	Number of Copiers Serviced, x	Number of Minutes Required, y
1	4	109
2	2	58
3	5	138
4	7	189
5	1	37
6	3	82
7	4	103
8	5	134
9	2	68
10	4	112
11	6	154

🖵 SrvcTime

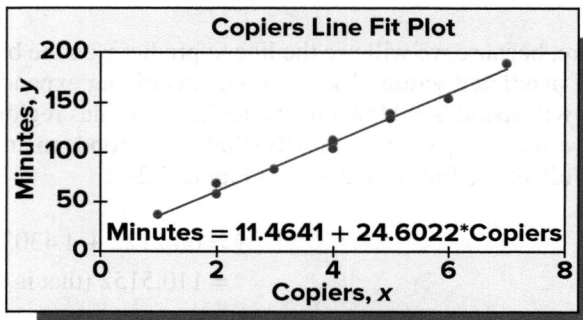

a The sample correlation coefficient r can be calculated to equal .9952 for the service time data. What does this value of r say about the relationship between x and y?
b Predict the service time for a future service call on which five copiers will be serviced.

LO3-6
Compute and interpret weighted means and the mean and standard deviation of grouped data (Optional).

3.5 Weighted Means and Grouped Data (Optional)

Weighted means

In Section 3.1 we studied the mean, which is an important measure of central tendency. In order to calculate a mean, we sum the population (or sample) measurements, and then divide this sum by the number of measurements in the population (or sample). When we do this, each measurement counts equally. That is, each measurement is given the same importance or weight.

Sometimes it makes sense to give different measurements unequal weights. In such a case, a measurement's weight reflects its importance, and the mean calculated using the unequal weights is called a **weighted mean.**

We calculate a weighted mean by multiplying each measurement by its weight, summing the resulting products, and dividing the resulting sum by the sum of the weights:

Weighted Mean

The weighted mean equals

$$\frac{\sum w_i x_i}{\sum w_i}$$

where
x_i = the value of the ith measurement
w_i = the weight applied to the ith measurement

Such a quantity can be computed for a population of measurements or for a sample of measurements.

In order to illustrate the need for a weighted mean and the required calculations, suppose that an investor obtained the following percentage returns on different amounts invested in four stock funds:

Stock Fund	Amount Invested	Percentage Return
1	$50,000	9.2%
2	$10,000	12.8%
3	$10,000	−3.3%
4	$30,000	6.1%

DS StockRtn

If we wish to compute a mean percentage return for the total of $100,000 invested, we should use a weighted mean. This is because each of the four percentage returns applies to a different amount invested. For example, the return 9.2 percent applies to $50,000 invested and thus should count more heavily than the return 6.1 percent, which applies to $30,000 invested.

The percentage return measurements are $x_1 = 9.2$ percent, $x_2 = 12.8$ percent, $x_3 = -3.3$ percent, and $x_4 = 6.1$ percent, and the weights applied to these measurements are $w_1 = \$50,000$, $w_2 = \$10,000$, $w_3 = \$10,000$, and $w_4 = \$30,000$. That is, we are weighting the percentage returns by the amounts invested. The weighted mean is computed as follows:

$$\mu = \frac{50,000(9.2) + 10,000(12.8) + 10,000(-3.3) + 30,000(6.1)}{50,000 + 10,000 + 10,000 + 30,000}$$

$$= \frac{738,000}{100,000} = 7.38\%$$

In this case the unweighted mean of the four percentage returns is 6.2 percent. Therefore, the unweighted mean understates the percentage return for the total of $100,000 invested.

The weights chosen for calculating a weighted mean will vary depending on the situation. For example, in order to compute the grade point average of a student, we would weight the grades A(4), B(3), C(2), D(1), and F(0) by the number of hours of A, the number of hours of B, the number of hours of C, and so forth. Or, in order to compute a mean profit margin for a company consisting of several divisions, the profit margins for the different divisions might be weighted by the sales volumes of the divisions. Again, the idea is to choose weights that represent the relative importance of the measurements in the population or sample.

Descriptive statistics for grouped data

We usually calculate measures of central tendency and variability using the individual measurements in a population or sample. However, sometimes the only data available are in the form of a frequency distribution or a histogram. For example, newspapers and magazines often summarize data using frequency distributions and histograms without giving the individual measurements in a data set. Data summarized in frequency distribution or histogram form are often called **grouped data.** In this section we show how to compute descriptive statistics for such data.

Suppose we are given a frequency distribution summarizing a sample of 65 customer satisfaction ratings for a consumer product.

Satisfaction Rating	Frequency
36–38	4
39–41	15
42–44	25
45–47	19
48–50	2

DS SatRatings

Because we do not know each of the 65 individual satisfaction ratings, we cannot compute an exact value for the mean satisfaction rating. However, we can calculate an approximation of this mean. In order to do this, we let the midpoint of each class represent the measurements in the class. When we do this, we are really assuming that the average of the measurements in each class equals the class midpoint. Letting M_i denote the midpoint of class i, and letting f_i denote the frequency of class i, we compute the mean by calculating a weighted mean of the class midpoints using the class frequencies as the weights. The logic here is that if f_i measurements are included in class i, then the midpoint of class i should count f_i times in the weighted mean. In this case, the sum of the weights equals the sum of the class frequencies, which equals the sample size. Therefore, we obtain the following equation for the sample mean of grouped data:

Sample Mean for Grouped Data

$$\bar{x} = \frac{\Sigma f_i M_i}{\Sigma f_i} = \frac{\Sigma f_i M_i}{n}$$

where
f_i = the frequency for class i
M_i = the midpoint for class i
$n = \Sigma f_i$ = the sample size

Table 3.6 summarizes the calculation of the mean satisfaction rating for the previously given frequency distribution of satisfaction ratings. Note that in this table each midpoint is halfway between its corresponding class limits. For example, for the first class $M_1 = (36 + 38)/2 = 37$. We find that the sample mean satisfaction rating is approximately 43.

We can also compute an approximation of the sample variance for grouped data. Recall that when we compute the sample variance using individual measurements, we compute the squared deviation from the sample mean $(x_i - \bar{x})^2$ for each individual measurement x_i and then sum the squared deviations. For grouped data, we do not know each of the x_i values. Because of this, we again let the class midpoint M_i represent each measurement in class i. It follows that we compute the squared deviation $(M_i - \bar{x})^2$ for each class and then sum these squares, weighting each squared deviation by its corresponding class frequency f_i. That is, we approximate $\Sigma(x_i - \bar{x})^2$ by using $\Sigma f_i(M_i - \bar{x})^2$. Finally, we obtain the sample variance for the grouped data by dividing this quantity by the sample size minus 1. We summarize this calculation in the following box:

Sample Variance for Grouped Data

$$s^2 = \frac{\Sigma f_i(M_i - \bar{x})^2}{n-1}$$

where $\bar{x}$ is the sample mean for the grouped data.

Table 3.7 illustrates calculating the sample variance of the previously given frequency distribution of satisfaction ratings. We find that the sample variance s^2 is approximately 8.15625 and, therefore, that the sample standard deviation s is approximately $\sqrt{8.15625} = 2.8559$.

Finally, although we have illustrated calculating the mean and variance for grouped data in the context of a sample, similar calculations can be done for a population of measurements.

TABLE 3.6   Calculating the Sample Mean Satisfaction Rating

Satisfaction Rating	Frequency (f_i)	Class Midpoint (M_i)	$f_i M_i$
36–38	4	37	4(37) = 148
39–41	15	40	15(40) = 600
42–44	25	43	25(43) = 1,075
45–47	19	46	19(46) = 874
48–50	2	49	2(49) = 98
	$n = 65$		2,795

$$\bar{x} = \frac{\Sigma f_i M_i}{n} = \frac{2,795}{65} = 43$$

TABLE 3.7 Calculating the Sample Variance of the Satisfaction Ratings

Satisfaction Rating	Frequency f_i	Class Midpoint M_i	Deviation $(M_i - \bar{x})$	Squared Deviation $(M_i - \bar{x})^2$	$f_i(M_i - \bar{x})^2$
36–38	4	37	$37 - 43 = -6$	36	$4(36) = 144$
39–41	15	40	$40 - 43 = -3$	9	$15(9) = 135$
42–44	25	43	$43 - 43 = 0$	0	$25(0) = 0$
45–47	19	46	$46 - 43 = 3$	9	$19(9) = 171$
48–50	2	49	$49 - 43 = 6$	36	$2(36) = 72$
	65				$\sum f_i(M_i - \bar{x})^2 = 522$

$$s^2 = \text{sample variance} = \frac{\sum f_i(M_i - \bar{x})^2}{n - 1} = \frac{522}{65 - 1} = 8.15625$$

If we let N be the size of the population, the grouped data formulas for the population mean and variance are given in the following box:

Population Mean for Grouped Data

$$\mu = \frac{\sum f_i M_i}{N}$$

Population Variance for Grouped Data

$$\sigma^2 = \frac{\sum f_i (M_i - \mu)^2}{N}$$

Exercises for Section 3.5

CONCEPTS connect

3.40 Consider calculating a student's grade point average using a scale where 4.0 represents an A and 0.0 represents an F. Explain why the grade point average is a weighted mean. What are the x_i values? What are the weights?

3.41 When we perform grouped data calculations, we represent the measurements in a class by using the midpoint of the class. Explain the assumption that is being made when we do this.

3.42 When we compute the mean, variance, and standard deviation using grouped data, the results obtained are approximations of the population (or sample) mean, variance, and standard deviation. Explain why this is true.

METHODS AND APPLICATIONS

3.43 Sound City sells the TrueSound-XL, a top-of-the-line satellite car radio. Over the last 100 weeks, Sound City has sold no radios in three of the weeks, one radio in 20 of the weeks, two radios in 50 of the weeks, three radios in 20 of the weeks, four radios in 5 of the weeks, and five radios in 2 of the weeks. The following table summarizes this information. **DS** TrueSound

Number of Radios Sold	Number of Weeks Having the Sales Amount
0	3
1	20
2	50
3	20
4	5
5	2

Compute a weighted mean that measures the average number of radios sold per week over the 100 weeks.

3.44 The following table gives a summary of the grades received by a student for the first 64 semester hours of university coursework. The table gives the number of semester hours of A, B, C, D, and F earned by the student among the 64 hours. **DS** Grades

Grade	Number of Hours
A (that is, 4.00)	18
B (that is, 3.00)	36
C (that is, 2.00)	7
D (that is, 1.00)	3
F (that is, 0.00)	0

a By assigning the numerical values, 4.00, 3.00, 2.00, 1.00, and 0.00 to the grades A, B, C, D, and F (as shown), compute the student's grade point average for the first 64 semester hours of coursework.

b Why is this a weighted average?

3.45 The following frequency distribution summarizes the weights of 195 fish caught by anglers participating in a professional bass fishing tournament. **DS** BassWeights

Weight (Pounds)	Frequency
1–3	53
4–6	118
7–9	21
10–12	3

a Calculate the (approximate) sample mean for these data.

b Calculate the (approximate) sample variance for these data.

3.46 The following is a frequency distribution summarizing earnings per share (EPS) growth data for the 30 fastest-growing firms as given on *Fortune* magazine's website on January 14, 2015.   Ⓓ EPSGrowth

EPS Growth (Percent)	Frequency
37<87	15
87<137	7
137<187	6
187<237	1
237<287	1

Source: http://fortune.com/100-fastest-growing-companies/ (accessed January 14, 2015).

Calculate the (approximate) population mean, variance, and standard deviation for these data.

3.47 The Data and Story Library website (a website devoted to applications of statistics) gives a histogram of the ages of a sample of 60 CEOs. We present the data in the form of a frequency distribution below.
Ⓓ CEOAges

Age (Years)	Frequency
28–32	1
33–37	3
38–42	3
43–47	13
48–52	14
53–57	12
58–62	9
63–67	1
68–72	3
73–77	1

Source: http://lib.stat.cmu.edu/DASL/Stories/ceo.html (accessed January 14, 2015).

Calculate the (approximate) sample mean, variance, and standard deviation of these data.

LO3-7

Compute and interpret the geometric mean (Optional).

3.6 The Geometric Mean (Optional)

In Section 3.1 we defined the mean to be the average of a set of population or sample measurements. This mean is sometimes referred to as the arithmetic mean. While very useful, the arithmetic mean is not a good measure of the rate of change exhibited by a variable over time. To see this, consider the rate at which the value of an investment changes—its rate of return. Suppose that an initial investment of \$10,000 increases in value to \$20,000 at the end of one year and then decreases in value to its original \$10,000 value after two years. The rate of return for the first year, R_1, is

$$R_1 = \left(\frac{20{,}000 - 10{,}000}{10{,}000}\right) \times 100\% = 100\%$$

and the rate of return for the second year, R_2, is

$$R_2 = \left(\frac{10{,}000 - 20{,}000}{20{,}000}\right) \times 100\% = -50\%$$

Although the value of the investment at the beginning and end of the two-year period is the same, the arithmetic mean of the yearly rates of return is $(R_1 + R_2)/2 = (100\% + (-50\%))/2 = 25\%$. This arithmetic mean does not communicate the fact that the value of the investment is unchanged at the end of the two years.

To remedy this situation, we define the **geometric mean** of the returns to be **the constant return R_g that yields the same wealth at the end of the investment period as do the actual returns.** In our example, this says that if we express R_g, R_1, and R_2 as decimal fractions (here $R_1 = 1$ and $R_2 = -.5$),

$$(1 + R_g)^2 \times 10{,}000 = (1 + R_1)(1 + R_2) \times 10{,}000$$

or

$$R_g = \sqrt{(1 + R_1)(1 + R_2)} - 1$$

$$= \sqrt{(1 + 1)(1 + (-.5))} - 1$$

$$= \sqrt{1} - 1 = 0$$

Therefore, the geometric mean R_g expresses the fact that the value of the investment is unchanged after two years.

In general, if $R_1, R_2, \ldots, R_n$ are returns (expressed in decimal form) over n time periods:

The **geometric mean** of the returns $R_1, R_2, \ldots, R_n$ is

$$R_g = \sqrt[n]{(1 + R_1)(1 + R_2) \ldots (1 + R_n)} - 1$$

and the ending value of an initial investment ℓ experiencing returns $R_1, R_2, \ldots, R_n$ is $\ell(1 + R_g)^n$.

As another example, suppose that in year 3 our investment's value increases to $25,000, which says that the rate of return for year 3 (expressed as a percentage) is

$$R_3 = \left(\frac{25,000 - 10,000}{10,000}\right) \times 100\%$$
$$= 150\%$$

Because (expressed as decimals) $R_1 = 1$, $R_2 = -.5$, and $R_3 = 1.5$, the geometric mean return at the end of year 3 is

$$R_g = \sqrt[3]{(1 + 1)(1 + (-.5))(1 + 1.5)} - 1$$
$$= 1.3572 - 1$$
$$= .3572$$

and the value of the investment after 3 years is

$$10,000\,(1 + .3572)^3 = \$25,000$$

Exercises for Section 3.6

CONCEPTS ⬛ connect

3.48 In words, explain the interpretation of the geometric mean return for an investment.

3.49 If we know the initial value of an investment and its geometric mean return over a period of years, can we compute the ending value of the investment? If so, how?

METHODS AND APPLICATIONS

3.50 Suppose that a company's sales were $5,000,000 three years ago. Since that time sales have grown at annual rates of 10 percent, −10 percent, and 25 percent.
  a Find the geometric mean growth rate of sales over this three-year period.
  b Find the ending value of sales after this three-year period.

3.51 Suppose that a company's sales were $1,000,000 four years ago and are $4,000,000 at the end of the four years. Find the geometric mean growth rate of sales.

3.52 The following table gives the value of the Dow Jones Industrial Average (DJIA), NASDAQ, and the S&P 500 on the first day of trading for the years 2010 through 2013. ⓄⓈ StockIndex

Year	DJIA	NASDAQ	S&P 500
2010	10,583.96	2308.42	1132.99
2011	11,670.75	2691.52	1271.87
2012	12,397.38	2648.72	1277.06
2013	13,412.55	3112.26	1462.42

Source: http://www.davemanuel.com/where-did-the -djia-nasdaq-sp500-trade-on.php (accessed January 14, 2015).

  a For each stock index, compute the rate of return from 2010 to 2011, from 2011 to 2012, and from 2012 to 2013.
  b Calculate the geometric mean rate of return for each stock index for the period from 2010 to 2013.
  c Suppose that an investment of $100,000 is made in 2010 and that the portfolio performs with returns equal to those of the DJIA. What is the investment worth in 2013?
  d Repeat part c for the NASDAQ and the S&P 500.

3.53 Refer to Exercise 3.52. The values of the DJIA on the first day of trading in 2005, 2006, 2007, 2008, and 2009 were 10,729.43, 10,847.41, 12,474.52, 13,043.96, and 9,034.69.
  a Calculate the geometric mean rate of return for the DJIA from 2005 to 2013.
  b If an investment of $100,000 is made in 2005 and the portfolio performs with returns equal to those of the DJIA, what is the investment worth in 2013?

**PART 2
Numerical
Descriptive Analytics
(Optional)**

C H A P T E R   3

LO3-8
Interpret the information
provided by association
rules (Optional).

3.7 Association Rules (Optional)

Association learning (also called **association rule mining**) identifies items that tend to co-occur and finds rules, called **association rules,** that describe their co-occurrence. Sophisticated versions of association rules are used by companies such as Amazon and Netflix to make product and movie recommendations to their customers. There are various formal definitions that can be used to describe how association rules are obtained. We will take a more informal approach and illustrate how association rules are obtained by using an example. Suppose that Flickers.com rents movie DVDs online. An analyst at Flickers has decided to use association rules to tailor further DVD recommendations to customers' histories. She categorizes DVDs into genres like comedy, drama, classics, science fiction, and so forth. Consider comedy. Suppose that in the last month, Flickers has added six new comedies to its catalog. We label them A–F. Also consider 10 customers who rent fairly heavily in the comedy genre. (We are keeping the example small for clarity.) Table 3.8 shows which of the six new comedies each of the 10 customers has rented and "liked." Note that Flickers has customers rate the movies they rent because it wishes to make recommendations to future customers on the basis of movies that previous customers have both rented and liked. That way customers will trust its recommendations. Therefore, when in the following discussions we refer to a DVD that has been rented, we mean that the DVD has been rented and liked.

In order to study the associations between the DVDs that the 10 comedy enthusiasts have rented, the analyst starts with pairs of DVDs rented and chooses a 50 percent "threshold of support" for a pair of DVDs to qualify as a basis for future recommendations. To understand what we mean by a 50 percent threshold of support, note that the analyst first tallies how many of the customers have rented each potential pair of movies. There are 15 such pairs, but some are unpopular. For example, no one rented the pair BF and only one person rented the pair DE. The only pairs appearing five or more times among the 10 customers are AB, AC, BC, and CE (occurring 5, 6, 7, and 5 times each). These four meet the 50 percent threshold when we compare the number of times the pair was rented to 10, the number of customers being considered.

Next, the analyst considers the association between renting one of these pairs and renting a third DVD. For example, the most popular pair is BC. If a customer has rented BC, what other DVD is a good recommendation for that person's tastes? If you look at the rightmost column, you will see that four of the seven customers who rented BC also rented E. (These four are highlighted in red.) While this is a majority, the ratio is not high enough to convince the analyst to use the resources required to recommend E to future renters of B and C. To make a recommendation, the analyst has decided to require a "confidence" percentage of 80 percent. Because $(4/7)100\% = 57\%$, she will not recommend E to future renters of B and C. However, when she considers the pair AB, she discovers that every customer who rented the pair AB also rented C. (These five are highlighted in green.) This is $(5/5)100\% = 100\%$.

TABLE 3.8 **The DVDs Rented by 10 Customers and Development of Association Rules Based on Pairs of DVDs Rented** DVDRent

	DVDs	Pairs	Triples
Customer 1	A, C, D, F	AC, AD, AF, CD, CF, DF	ACD, ACF, ADF, CDF
Customer 2	A, B, C, D	AB, AC, AD, BC, BD, CD	ABC, ABD, ACF, BCD
Customer 3	C, E, F	CE, CF, EF	CEF
Customer 4	A, B, C	AB, AC, BC	ABC
Customer 5	B, C, E	BC, BE, CE	BCE
Customer 6	A, B, C, E	AB, AC, AE, BC, BE, CE	ABC, ABE, ACE, BCE
Customer 7	A, E	AE	none
Customer 8	A, B, C, E	AB, AC, AE, BC, BE, CE	ABC, ABE, ACE, BCE
Customer 9	B, C, D, E	BC, BD, BE, CD, CE, DE	BCD, BCE, BDE, CDE
Customer 10	A, B, C	AB, AC, BC	ABC

TABLE 3.9 **The DVDs Rented by 10 Customers and Development of Association Rules Based on Single DVDs Rented**

	DVDs	Pairs	Triples
Customer 1	A, C, D, F	AC, AD, AF, CD, CF, DF	ACD, ACF, ADF, CDF
Customer 2	A, B, C, D	AB, AC, AD, BC, BD, CD	ABC, ABD, ACF, BCD
Customer 3	C, E, F	CE, CF, EF	CEF
Customer 4	A, B, C	AB, AC, BC	ABC
Customer 5	B, C, E	BC, BE, CE	BCE
Customer 6	A, B, C, E	AB, AC, AE, BC, BE, CE	ABC, ABE, ACE, BCE
Customer 7	A, E	AE	none
Customer 8	A, B, C, E	AB, AC, AE, BC, BE, CE	ABC, ABE, ACE, BCE
Customer 9	B, C, D, E	BC, BD, BE, CD, CE, DE	BCD, BCE, BDE, CDE
Customer 10	A, B, C	AB, AC, BC	ABC

Similarly, of the six customers who rented the pair AC, five also rented B. This is $(5/6)100\% = 83\%$. Finally, of the five customers who rented the pair CE, four also rented B. This just meets the 80 percent standard. Therefore, given the analyst's choices of (1) using a 50 percent support threshold for pairs to qualify for the association analysis and (2) requiring a confidence percentage of 80 percent, she will make the following recommendations to future Flickers customers in the comedy genre:

1   Recommend C to renters of A and B.
2   Recommend B to renters of A and C.
3   Recommend B to renters of C and E.

To continue this example, suppose the analyst next considers basing recommendations on single DVDs rented. She keeps the 50 percent threshold for support and the requirement of an 80 percent confidence percentage. Therefore, she begins by tallying the number of times each DVD has been rented. The only two not making the 50 percent cut are D, with three renters, and F, with just two. Next, she considers the association between renting one of these DVDs and renting a second. If we consider Table 3.9, we see that of the seven customers who rented A, five (in red) also rented B. Because $(5/7)100\% = 71\%$, this is insufficient association to warrant recommending B to renters of A. However, of the seven customers who rented A, six (in green) also rented C. Because $(6/7)100\% = 86\%$, this is high enough to warrant recommending C to renters of A. Continuing in this fashion, we see that all seven renters of B also rented, C; and five of six renters of E also rented C. No other associations meet the 80 percent requirement. Thus the analyst will make the following recommendations to future customers in the comedy genre based on single DVDs rented: a Recommend C to renters of A; b Recommend C to renters of B; c Recommend C to renters of E.

Recall from the analysis of pairs of DVDs that the analyst decided to recommend C to renters of A and B; B to renters of A and C; and B to renters of C and E. The last of these recommendations would not be made using the software package JMP if we used the analyst's specified support percentage of 50 percent and specified confidence percentage of 80 percent. To understand this, note that JMP calls the recommending movie(s) the **Condition** and the recommended movies(s) the **Consequent.** Moreover, JMP requires that the specified support percentage hold for both the condition and the combination of the condition and consequent. In the recommendation of B to renters of C and E, the condition CE has a support percentage of 50 percent (5 out of 10 customers rented C and E). However, the combination of the condition and consequent, BCE, has a support percentage of only 40 percent (4 out of 10 customers rented B and C and E) and thus does not meet the 50 percent threshold. With this in mind, the JMP output of an association rule analysis of the DVD renters data using a specified support percentage of 50 percent and a specified confidence percentage of 80 percent is given in Figure 3.25. On this output, the **Confidence** for a recommendation of y based on x is computed by dividing the **Support** for x & y by the **Support** for x and expressing the result as a percentage. For example, consider

FIGURE 3.25   **The JMP Output of an Association Rule Analysis of the DVD Renters Data** (DS) DVDRent

Association Analysis

Frequent Item Sets

Item Set	Support	N Items
{C}	90%	1
{A}	70%	1
{B}	70%	1
{B, C}	70%	2
{E}	60%	1
{A, C}	60%	2
{A, B}	50%	2
{C, E}	50%	2
{A, B, C}	50%	3

Rules

Rule

Condition	Consequent	Confidence	Lift
A	C	86%	0.952
B	C	100%	1.111
E	C	83%	0.926
A, B	C	100%	1.111
A, C	B	83%	1.19

the recommendation of B to renters of A & C. The **Support** for A & C is 60% (6 customers rented A and C), and the **Support** for A & C & B is 50% (5 customers rented A and C and B). Therefore, the **Confidence** is (50%/60%)100% = 83% (actually, 83.33%).

The **Lift** (ratio) for a recommendation on the output is the confidence percentage for the recommendation divided by the support percentage for the consequent of the recommendation. For example, for the recommendation of B to renters of A and C, the confidence percentage is 83% and the support percentage for the consequent B is 70%. Therefore, the lift ratio is 83%/70% = 1.19. This lift ratio is greater than 1 and estimates that a renter of A and C is 19 percent more likely to rent B than is a randomly selected customer in the comedy genre. On the other hand, the lift ratio for recommending C to a renter of E is 83%/90% = .926. This estimates that a renter of E is 7.4% less likely to rent C than is a randomly selected customer in the comedy genre. Only a movie recommendation rule having a lift ratio greater than 1 is better at identifying customers likely to respond positively to the recommendation than no rule at all.

In the real world, companies such as Amazon and Netflix sell or rent thousands or even millions of items and find association rules based on millions of customers. In order to make obtaining meaningful association rules manageable, these companies break products for which they are obtaining association rules into various categories (for example, comedies or thrillers) and hierarchies (for example, a hierarchy related to how new the product is).

Exercises for Section 3.7

CONCEPTS

3.54   What is the purpose of association rules?

3.55   Discuss the meanings of the terms *support percentage, confidence percentage*, and *lift ratio*.

METHODS AND APPLICATIONS

3.56   In the JMP output of Figure 3.25, show how the lift ratio of 1.111(rounded) for the recommendation of C to renters of B has been calculated. Interpret this lift ratio.

3.57   The JMP output of an association rule analysis of the DVD renters data using a support percentage of 40 percent, a confidence percentage of 70 percent, and a lift ratio of .8 is shown in Figure 3.26.
(DS) DVDRent

a   Summarize the recommendations based on a lift ratio greater than 1.

b   Consider the recommendation of DVD B based on having rented C & E. (**1**) Identify and interpret the support for C & E. Do the same for the support for C & E & B. (**2**) Show how the Confidence of 80% has been calculated. (**3**) Show how the Lift Ratio of 1.143 (rounded) has been calculated.

3.58   Suppose that the analyst at Flickers wishes to do an association rule analysis for seven new science fiction movies that Flickers has recently added to its catalog. We label these movies A–G, and Table 3.10 shows which of the movies 10 customers have rented and liked. Perform the association rule analysis using a support percentage of 30%, a confidence percentage of 60%, and a lift ratio of .8.

FIGURE 3.26   **The JMP Output of an Association Rule Analysis of the DVD Renters Data Using a Support Percentage of 40%, a Confidence Percentage of 70%, and a Lift Ratio of .8**
DVDRent

Association Analysis

Frequent Item Sets

Item Set	Support	N Items
{C}	90%	1
{A}	70%	1
{B}	70%	1
{B, C}	70%	2
{E}	60%	1
{A, C}	60%	2
{A, B}	50%	2
{C, E}	50%	2
{A, B, C}	50%	3
{B, E}	40%	2
{B, C, E}	40%	3

Rules

Rule Condition	Consequent	Confidence	Lift
A	B	71%	1.02
B	A	71%	1.02
A	C	86%	0.952
B	C	100%	1.111
C	B	78%	1.111
E	C	83%	0.926
A	B, C	71%	1.02
B	A, C	71%	1.19
A, B	C	100%	1.111
A, C	B	83%	1.19
B, C	A	71%	1.02
B, E	C	100%	1.111
C, E	B	80%	1.143

TABLE 3.10   **Science Fiction Movie DVD's Rented by 10 Customers**   DVDRent

DVDs		DVDs	
Customer 1	A, B, C, D, F, G	Customer 6	B, F, G
Customer 2	B, E, F, G	Customer 7	A, D, F, G
Customer 3	A, C, E, F	Customer 8	D, E, F
Customer 4	B, C, F, G	Customer 9	A, D, E
Customer 5	A, C, E, F, G	Customer 10	A, B, C, F

3.8 Text Mining (Optional)

LO3-9
Interpret the information provided by text mining (Optional).

Text mining is the science of discovering knowledge, insights, and patterns from a collection of textual documents or databases. To illustrate using JMP to carry out text mining, we consider 23,848 food and drug safety citations issued to various businesses and organizations by the Food and Drug Administration (FDA). Basic text mining analysis of these citations by JMP produces the *term* (or *word*) and *phrase lists* in Figure 3.27. These lists give the number of the 23,848 citations that contain the most frequently occurring terms and phrases. Figure 3.27 also shows a **word cloud**, where the relative sizes of the words indicate the relative frequencies of the citations containing these words.

Suppose that an FDA administrator wishes to understand the key issues or factors in organizations and businesses that cause citations to be issued. If we arrange—using an educated guess—the largest six words in the term list and word cloud into a phrase (adding—the word "to"), we might obtain the phrase "failure to control established food contamination procedures." To get a better idea of the key issues or factors that cause the citations to be issued, we can use **latent semantic analysis.** This technique analyzes the relationships between a set of documents and the words they contain to produce a set of key factors (or issues or concepts) related to the documents and words. This is done by producing a count of the most frequently occurring words in each document and using advanced statistical methods beyond the scope of this book to define the key factors. Latent semantic analysis will group (or cluster) the documents around the key factors best described by those documents. By reading the documents clustered around the key factors, we can understand what the factors are. Figure 3.28 shows the JMP output of a geometric display

FIGURE 3.27   The JMP Output of Part of a Term and Phrase List and Part of a Word Cloud
in the FDA Citations Example   DS FDACit

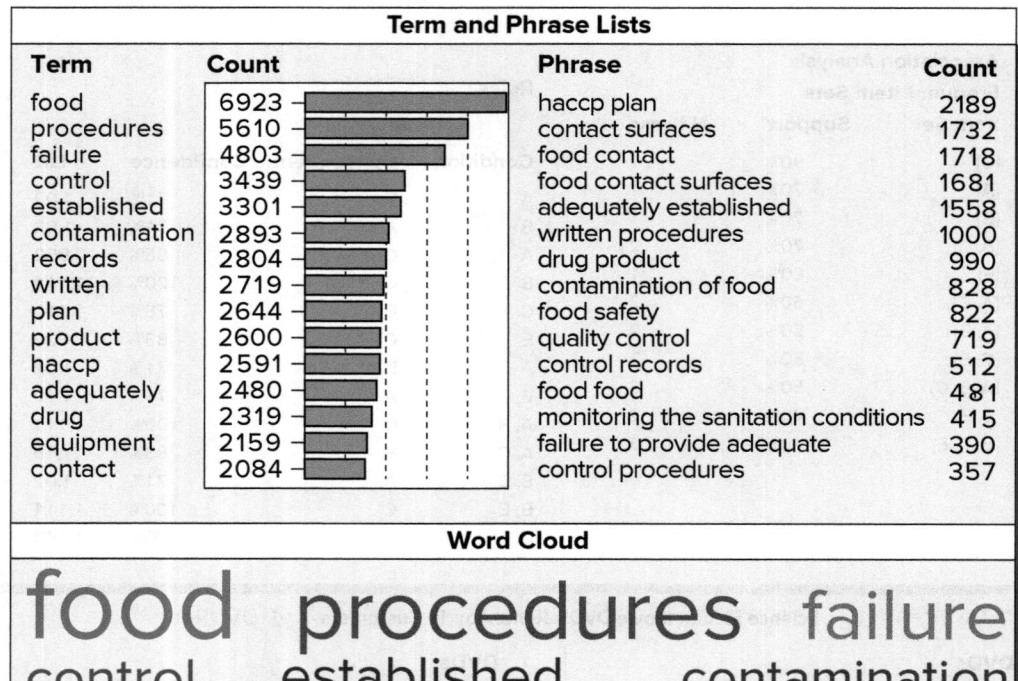

Term and Phrase Lists

Term	Count		Phrase	Count
food	6923		haccp plan	2189
procedures	5610		contact surfaces	1732
failure	4803		food contact	1718
control	3439		food contact surfaces	1681
established	3301		adequately established	1558
contamination	2893		written procedures	1000
records	2804		drug product	990
written	2719		contamination of food	828
plan	2644		food safety	822
product	2600		quality control	719
haccp	2591		control records	512
adequately	2480		food food	481
drug	2319		monitoring the sanitation conditions	415
equipment	2159		failure to provide adequate	390
contact	2084		control procedures	357

Word Cloud

food procedures failure control established contamination records written plan product haccp adequately drug equipment contact ensure surfaces quality appropriate monitoring process used manner production

FIGURE 3.28   The JMP Output of a Geometric Display of Two Key Factors Related to the
FDA Citations   DS FDACit

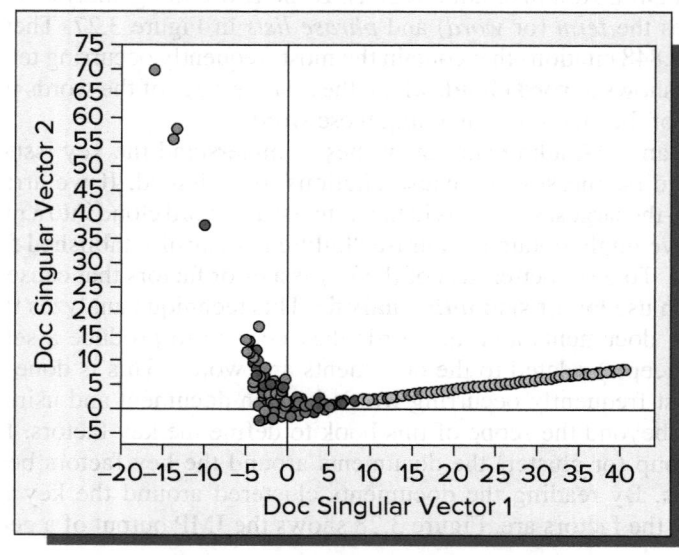

of two key factors (see *Doc Singular Vector 1* and *Doc Singular Vector 2*) related to the 23,848 FDA citations and the clustering of the citations describing the factors around the factors. Figure 3.29 shows a JMP output of some citations clustered around the first factor. Reading these citations, we might describe the first factor as "failure to monitor, maintain, and keep records concerning sanitary conditions." Figure 3.30 shows a JMP output of some citations clustered around the second factor. Reading these citations, we

---

FIGURE 3.29   **The JMP Output of Some Citations Clustered around the First Factor in the FDA Citations Examples**

You are not monitoring the sanitation conditions and practices with sufficient frequency to assure conformance with Current Good Manufacturing Practices including safety of water that comes into contact with food or food contact surfaces, including water used to manufacture ice, condition and cleanliness of food contact surfaces, prevention of cross-contamination from insanitary objects, protection of food, food packaging material, and food contact surfaces from adulteration, proper labeling, storage and use of toxic chemicals, and control of employee health conditions. [12]

You do not have or have not implemented a sanitation standard operating procedure that addresses sanitation conditions and practices before, during and after processing. [103]

You do not always maintain sanitation standard operating procedure records that document the monitoring of conditions and practices during processing. [104]

GMP training is not conducted on a continuing basis and with sufficient frequency to assure that employees remain familiar with CGMP requirements applicable to them. [173]

You are not maintaining sanitation control records that document monitoring and corrections of sanitation deficiencies for safety of water that comes into contact with food or food contact surfaces, including water used to manufacture ice, condition and cleanliness of food contact surfaces, prevention of cross-contamination from insanitary objects, maintenance of hand washing, hand sanitizing, and toilet facilities, protection of food, food packaging material, and food contact surfaces from adulteration, proper labeling, storage and use of toxic chemicals, control of employee health conditions, and exclusion of pests. [206]

You are not maintaining sanitation control records that document corrections of sanitation deficiencies for maintenance of hand washing, hand sanitizing, and toilet facilities. [211]

Failure to properly identify and store toxic cleaning compounds, sanitizing agents, and pesticide chemicals in a manner that protects against contamination of food, food-contact surfaces, and food-packaging materials. [266]

You are not maintaining sanitation control records that document monitoring for safety of water that comes into contact with food or food contact surfaces, including water used to manufacture ice, protection of food, food packaging material, and food contact surfaces from adulteration, proper labeling, storage and use of toxic chemicals, control of employee health conditions, and exclusion of pests. [347]

Proper precautions to protect food, food-contact surfaces, and food-packaging materials from contamination with filth and extraneous material cannot be taken because of deficiencies in plant construction. [370]

You are not monitoring the sanitation conditions and practices with sufficient frequency to assure conformance with Current Good Manufacturing Practices including protection of food, food packaging material, and food contact surfaces from adulteration and exclusion of pests. [418]

FIGURE 3.30   **The JMP Output of Some Citations Clustered around the Second Factor in the FDA Citations Example**

An MDR report was not submitted within 30 days of receiving or otherwise becoming aware of information that reasonably suggests that a marketed device may have caused or contributed to a death or serious injury. [27]

An MDR report was not submitted within 30 days of receiving or otherwise becoming aware of information that reasonably suggests that a marketed device may have caused or contributed to a death or serious injury. [79]

Follow-up reports were not submitted within 15 calendar days of receipt of new information concerning post marketing 15-day reports. [86]

An NDA-Field Alert Report was not submitted within three working days of receipt of information concerning bacteriological contamination and significant chemical, physical, or other change or deterioration in a distributed drug product. [89]

An initial establishment registration was not submitted within 30 days after starting an operation requiring registration. [219]

might describe the second factor as "failure to file reports to the FDA on time." By preemptively emphasizing the importance of these issues to businesses and organizations, the FDA can hopefully be successful in motivating businesses and organizations to improve their performance in the areas related to these issues.

Exercises for Section 3.8

CONCEPTS

3.59   What is text mining?

3.60   How is latent semantic analysis used?

METHODS AND APPLICATIONS

3.61   A company that makes and sells pet products has randomly selected 194 Internet posts from cat and dog owners concerning their pets. Figure 3.31 and 3.32 show JMP outputs of term and phrase lists and of a word cloud for these 194 posts. Figure 3.33 shows the

JMP output of a geometrical display of two key factors related to the 194 posts and the clustering of posts describing these key factors around the key factors. Figure 3.34 shows a JMP output of some posts clustered around the first factors, and Figure 3.35 shows a JMP output of some posts clustered around the second factor. By reading the posts, describe conceptually the two factors and discuss how the pet product company might use this information to make television commercials advertising its products that would appeal to cat and dog owners.

FIGURE 3.31   **The JMP Output of Part of a Term and Phrase List in the Pet Posts Exercise** Ⓓⓢ PetPosts

Term and Phrase Lists

Term	Count	Phrase	Count
cat	51	video of the cat	5
dogs	48	sit in my lap	4
dog	42	dog barks	3
cats	17	duck hunting	3
lap	14	funny video	3
barking	12	great job	3
video	11	last week	3
sit	10	stop barking	3
walk	10	cat in my lap	2
mice	9	dogs do a great	2
just	8	hunting with the dogs	2
like	8	sled in the winter	2
take	8	video of a cat	2
bark	7	around the block	2
huskies	7	lap and purr	2
time	7	take the huskies	2

FIGURE 3.32   **The JMP Output of a Word Cloud in the Pet Posts Exercise** Ⓓⓢ PetPosts

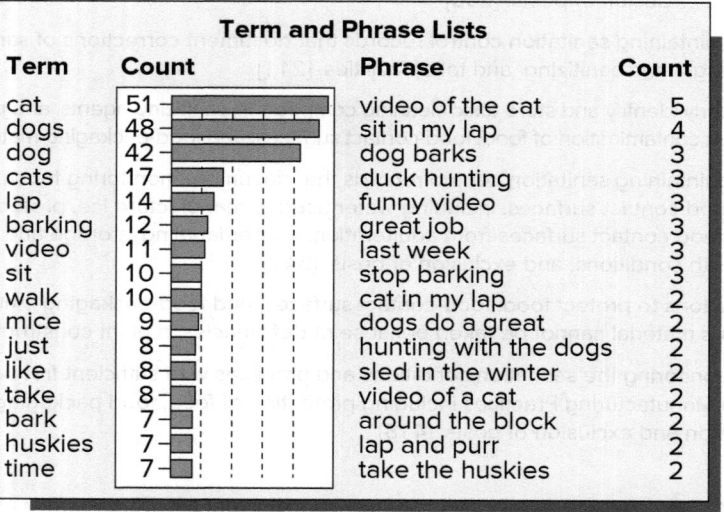

Word Cloud

cat dogs dog cats lap barking video sit walk mice just like take bark huskies time always cattle day every funny house hunting one sheep sled winter around barks go keep last two videos allergic away barn catch couch door getting great job jumped love made now squirrels trying dog food cat food

FIGURE 3.33   **The JMP Output of a Geometric Display of Two Key Factors in the Pet Posts Exercise**

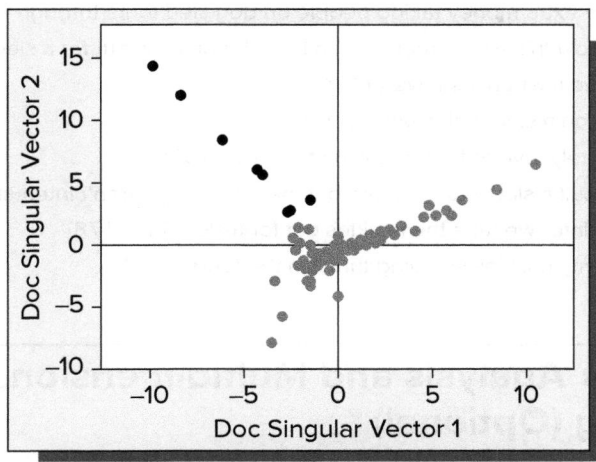

FIGURE 3.34   **The JMP Output of Some Posts Clustered around the First Factor in the Pet Posts Exercise**

The cat purrs and rubs against my leg when she wants to sit in my lap. [11]

My cat loves to sit in my lap and purr. [13]

That video of the cat knocking over the TV was hilarious. [21]

The cat just wants to sit on my lap and be pet all the time. [26]

The cats jump into my lap as soon as I sit down. [33]

Fluffy cats are my favorite. [35]

My cat tries to sit in my lap and purr while I am working on the computer. [46]

There was this funny video of a cat trying to jump into someones lap, but fell into the pool instead. [56]

The cats always purr loudly when they want to be petted. [76]

My kids are always trying to make videos of the cats doing funny things. [78]

The cat jumps out of my lap when he hears someone shake that box of cat food. [79]

We don't have fluffy cats anymore because I am allergic to them. [88]

The cats are always trying to sit in my lap while I am sitting down to dinner. [93]

I made a funny video of the cat chasing the dog down the stairs. [95]

That video of the cat jumping from the car into the window was amazing. [100]

My cat is so weird, he swims in the pool sometimes. [119]

It's like every time I sit down the cat jumps into my lap. [127]

I wish the cats would catch those squirrels living in the attic. [129]

The cat jumped out of the box just when I started recording the video. [137]

The funny cat video where the cats jumped through the window right into the bathtub was hilarious. [142]

Last year I bought two cats so that I can make cat videos. [148]

We made this funny video of the cat trying to climb the wall to chase a laser pointer. [153]

FIGURE 3.35   **The JMP Output of Some Posts Clustered around the Second Factor in the Pet Posts Exercise**

In the winter I make extra money taking people on dog sled tours through the park. [2]

There is enough snow this winter that we can take the huskies out for a sled ride. [103]

The huskies just love it when it snows. [117]

I take the dogs out on a sled in the winter. [136]

The huskies absolutely love to be out pulling the sled. [160]

My huskies like to pull a sled in the winter, but the other dogs aren't interested. [172]

When it snows in winter we take the huskies out for a sled ride. [178]

In the winter I take my huskies sledding through the forest. [191]

LO3-10

Interpret the information provided by cluster analysis and multidimensional scaling (Optional).

3.9 Cluster Analysis and Multidimensional Scaling (Optional)

Cluster analysis seeks to find natural groupings, or clusters, within data without having to prespecify a set of categories, as is done in classification. Two of the most common methods of clustering are **hierarchical clustering** and *k*-means clustering. We will begin by discussing hierarchical clustering, and then we will consider *k*-means clustering, which is often used in data mining.

Hierarchical clustering and multidimensional scaling

A recent analysis of baseball's popularity by Gary Gillette, editor of the *2006 ESPN Baseball Encyclopedia*, has led him to say in various online articles and interviews that "baseball is not nearly as popular now as it once was." Not all analysts agree with this conclusion, but what is undeniable is that another summer sport—tennis—is on life support in terms of American sports fan interest. Television ratings for the U.S. Open Tennis Tournament are a fraction of what they were in the era of Chris Evert, Jimmy Conners, and John McEnroe. Suppose that representatives of both baseball and tennis have approached a sports marketing analyst and have asked for suggestions about how to increase American sports fan interest in their sports. The analyst begins by randomly selecting 45 American sports fans and has them give their current perceptions of various sports. Specifically, the analyst has each of the 45 American sports fans give each of boxing (BX), basketball (BK), golf (G), swimming (SW), skiing (SK), baseball (BB), ping pong (PP), hockey (HK), handball (H), track and field (TF), bowling (BW), tennis (T), and football (F) an integer rating of 1 to 7 on six scales: fast moving (1) versus slow moving (7); complicated rules (1) versus simple rules (7); team sport (1) versus individual sport (7); easy to play (1) versus hard to play (7); noncontact (1) versus contact (7); competition against opponent (1) versus competition against standard (7). The first two rows of Table 3.11 present a particular sports fan's ratings of boxing and basketball on each of the six scales, and Table 3.12 presents the average rating by all 45 sports fans of each sport on each of the six scales.

TABLE 3.11   **A Particular Sports Fan's Ratings of Boxing and Basketball**

Sport	(1) Fast Mvg. (7) Slow Mvg.	(1) Compl. (7) Simple	(1) Team (7) Indv.	(1) Easy to Play (7) Hard to Play	(1) Ncon. (7) Con.	(1) Comp Opp. (7) Comp Std.	
Boxing	3	5	7	4	6	1	Distance
Basketball	2	3	2	4	4	2	$= \sqrt{(1)^2 + (2)^2 + (5)^2 + (0)^2 + (2)^2 + (-1)^2}$
Paired Difference	1	2	5	0	2	−1	$= \sqrt{35} = 5.9161$

Source: of Tables 3.11, 3.12, 3.13, Figure 3.36: Data from D. M. Levine, "Nonmetric Multidimensional Scaling and Hierarchical Clustering: Procedures for the Investigation of the Perception of Sports," *Research Quarterly*, vol. 48 (1977), pp. 341–348.

TABLE 3.12   **Average Rating of Each Sport on Each of the Six Scales** DS SportsRatings

Sport	(1) Fast Mvg. (7) Slow Mvg.	(1) Compl. (7) Simple	(1) Team (7) Indv.	(1) Easy to Play (7) Hard to Play	(1) Ncon. (7) Con.	(1) Comp Opp. (7) Comp Std.
Boxing	3.07	4.62	6.62	4.78	6.02	1.73
Basketball	1.84*	3.78	1.56*	3.82	4.89*	2.27*
Golf	6.13	4.49	6.58	3.84	1.82	4.11
Swimming	2.87	5.02	5.29	3.64	2.22	4.36
Skiing	2.13	4.60	5.96	5.22	2.51	4.71
Baseball	4.78✓	4.18	2.16✓	3.33	3.60✓	2.67✓
Ping pong	3.18	5.13	5.38	2.91	2.04	2.20
Hockey	1.71*	3.22	1.82*	5.04	5.96*	2.49*
Handball	2.53	4.67	4.78	3.71	2.78	2.31
Track & field	2.82	4.38	4.47	3.84	2.89	3.82
Bowling	5.07	5.16	5.40	3.11	1.60	3.73
Tennis	2.89✗	3.78	5.47✗	4.09	2.16✗	2.42✗
Football	2.42*	2.76	1.44*	5.00	6.47*	2.33*

To better understand the perceptions of the 13 sports, we will *cluster* them into groups. The first step in doing this is to consider the **distance** between each pair of sports for each sports fan. For example, to calculate the distance between boxing and basketball for the sports fan whose ratings are given in Table 3.11, we calculate the paired difference between the ratings on each of the six scales, square each paired difference, sum the six squared paired differences, and find the square root of this sum. The resulting distance is 5.9161. A distance for each sports fan for each pair of sports can be found, and then an *average distance* over the 45 sports fans for each pair of sports can be calculated. Statistical software packages do this, but these packages sometimes standardize the individual ratings before calculating the distances. We will not discuss the various ways in which such standardization can be done. Rather, we note that Table 3.13 presents a matrix containing the average distance over the 45 sports fans for each pair of sports, and we note that this matrix has been obtained by using a software package that uses a standardization procedure. There are many different approaches to using the average distances to cluster the sports. We will discuss one approach—the **hierarchical, complete linkage approach. Hierarchical clustering** implies that once two sports are clustered together at a particular stage, they are considered to be permanently joined and cannot be separated into different clusters at a later stage. **Complete linkage** bases the merger of two clusters of sports (either cluster of which can be an individual sport) on the **maximum distance** between sports in the clusters. For example, because Table 3.13 shows that the smallest average distance is the average distance between football and hockey, which is 2.20, football and hockey are clustered together in the first stage of

TABLE 3.13   **A Matrix Containing the Average Distances** DS SportDis

Sport	BX	BK	G	SK	SW	BB	PP	HK	H	TF	BW	T
BK	3.85											
G	4.33	4.88										
SK	3.80	4.05	3.73									
SW	3.81	3.81	3.56	2.84								
BB	4.12	3.15	3.83	4.16	3.60							
PP	3.74	3.56	3.61	3.67	2.72	3.41						
HK	3.85	2.58	5.11	4.02	4.17	3.49	4.27					
H	3.41	3.24	3.92	3.25	2.80	3.34	2.58	3.52				
TF	3.81	3.36	3.88	3.20	2.84	3.37	3.06	3.72	2.75			
BW	4.07	4.23	2.72	3.75	2.89	3.32	2.87	4.58	3.13	3.26		
T	3.49	3.32	3.59	3.19	2.82	3.25	2.54	3.58	2.33	2.72	2.85	
F	3.86	2.51	5.15	4.38	4.41	3.43	4.35	2.20	3.68	3.84	4.67	3.69

FIGURE 3.36   A Tree Diagram Showing Clustering of the 13 Sports

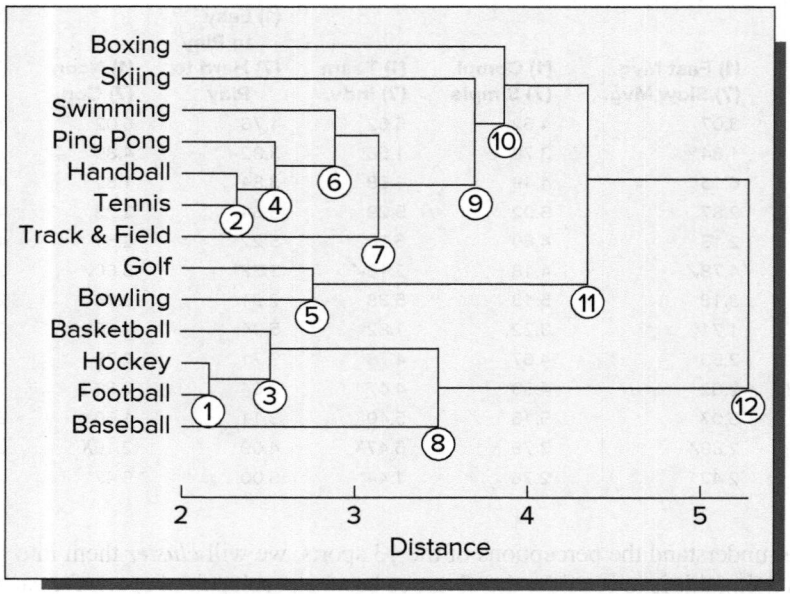

clustering. (See the tree diagram in Figure 3.36.) Because the second smallest average distance is the average distance between tennis and handball, which is 2.33, tennis and handball are clustered together in the second stage of clustering. The third smallest average distance is the average distance between football and basketball, which is 2.51, but football has already been clustered with hockey. The average distance between hockey and basketball is 2.58, and so the average distance between basketball and the cluster containing football and hockey is 2.58—the maximum of 2.51 and 2.58. This average distance is equal to the average distance between ping pong and the cluster containing tennis and handball, which (as shown in Table 3.13) is the maximum of 2.54 and 2.58—that is, 2.58. There is no other average distance as small as 2.58. Furthermore, note that the distance between basketball and football is 2.51, whereas the distance between ping pong and tennis is 2.54. Therefore, we will break the "tie" between the two average distances of 2.58 by adding basketball to the cluster containing football and hockey in the third stage of clustering. Then, we add ping pong to the cluster containing tennis and handball in the fourth stage of clustering. Figure 3.36 shows the results of all 12 stages of clustering, and Figure 3.37 shows Minitab and JMP outputs of the 12 stages of clustering. Note that the "tree diagrams" shown in the outputs are sometimes called **dendograms.**

At the end of seven stages of clustering, six clusters have been formed. They are:

Cluster 1: Boxing   **Cluster 4:** Golf, Bowling

Cluster 2: Skiing   **Cluster 5:** Basketball, Hockey, Football

Cluster 3: Swimming, Ping Pong, Handball,   **Cluster 6:** Baseball
Tennis, Track and Field

Cluster 5 consists of basketball, hockey, and football. Examining the means describing the perceptions of these sports in Table 3.12, we see that basketball, hockey, and football are alike in that they are rated (see the asterisks) (1) as being of "high action," which combines being "fast moving" with being "contact oriented," and (2) as being "team oriented," which combines being a "team sport" with featuring "competition against an opponent." Noting that football and basketball are very popular sports and hockey is gaining in popularity, we might conclude that "fast-moving, team-oriented" sports are currently popular. Further examining the means in Table 3.12, we see that baseball is rated (see the checks) as being fairly "slow moving" and not as "team oriented" as basketball, hockey, and football. It might be that high baseball player salaries, free agency, frequent player moves, and the inability of small market teams to compete make baseball seem less team oriented to fans. Perhaps more revenue

FIGURE 3.37   **Minitab and JMP Outputs of Hierachical Clustering of the Sports Perception Data**

(a) The Minitab output

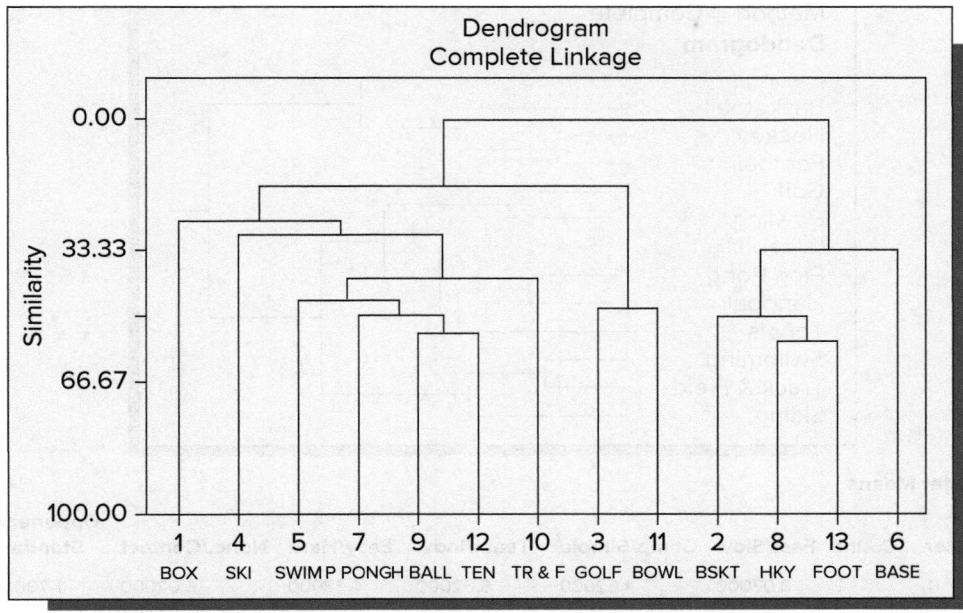

(b) The JMP output

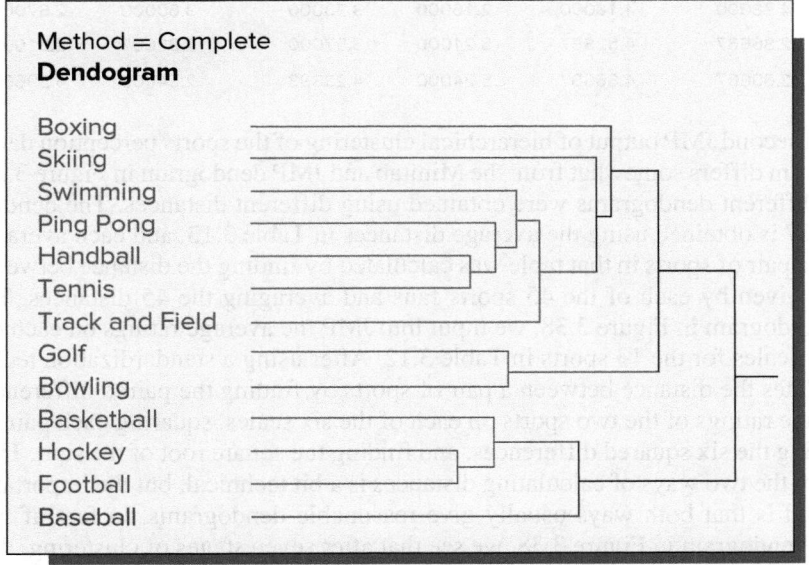

sharing between small and large market teams and including more teams in the end-of-season playoffs would help fans feel more positive about their teams and cause baseball to gain in general popularity. Also, perhaps rule changes to speed up the pace of the game would make baseball seem faster moving and make it more popular. As for tennis, the means in Table 3.12 show this sport is rated (see the **X**'s) as being not team oriented (which is fairly obvious) and as being slower moving than basketball, hockey, and football. It might be that power tennis (partially due to new tennis racquet technologies), which has resulted in somewhat shorter rallies and made it impossible for any major player to use a "charge the net" game, has made tennis seem less action-oriented to American sports fans. Perhaps limiting the power of tennis racquets would allow smaller, exciting players (like Jimmy Connors and John McEnroe of the 1980s) and players that use "charge the net" games (like John McEnroe) to be more competitive. This might make tennis seem more action-oriented.

FIGURE 3.38   A Second JMP Output of Hierarchical Clustering of the Sports Perception Data

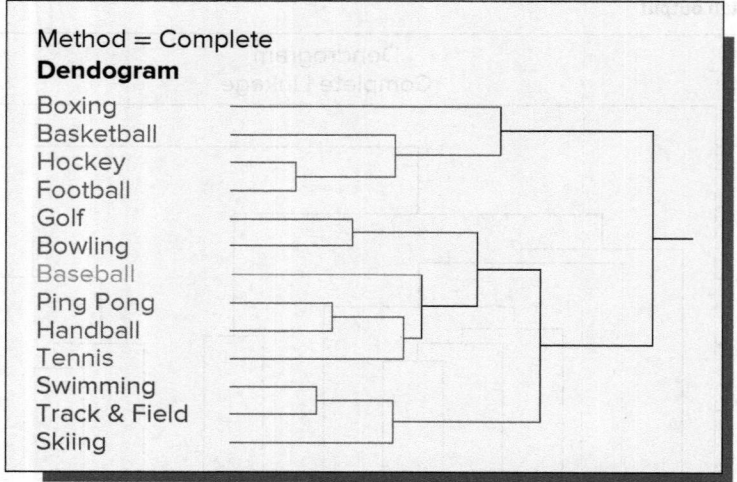

Cluster Means

Cluster	Count	Fast/Slow	Comp/Simple	Team/Indv.	Easy/Hard	Nonc./Contact	Opponent/ Standard
1	1	3.07000	4.62000	6.62000	4.78000	6.02000	1.73000
2	3	1.99000	3.25333	1.60667	4.62000	5.77333	2.36333
3	2	5.60000	4.82500	5.99000	3.47500	1.71000	3.92000
4	1	4.78000	4.18000	2.16000	3.33000	3.60000	2.67000
5	3	2.86667	4.52667	5.21000	3.57000	2.32667	2.31000
6	3	2.60667	4.66667	5.24000	4.23333	2.54000	4.29667

Figure 3.38 is a second JMP output of hierarchical clustering of the sports perception data. This JMP dendogram differs somewhat from the Minitab and JMP dendogram in Figure 3.37 because the two different dendograms were obtained using different distances. The dendogram in Figure 3.37 is obtained using the average distances in Table 3.13, and each average distance between a pair of sports in that table was calculated by finding the distance between the pair of sports given by each of the 45 sports fans and averaging the 45 distances. To obtain the JMP dendogram in Figure 3.38, we input into JMP the average ratings on each of the six perception scales for the 13 sports in Table 3.12. After using a standardization technique, JMP calculates the distance between a pair of sports by finding the paired difference between the average ratings of the two sports on each of the six scales, squaring each paired difference, summing the six squared differences, and finding the square root of the sum. The difference between the two ways of calculating distances is a bit technical, but the important thing to understand is that both ways usually give reasonable dendograms. In fact, if we examine the JMP dendogram in Figure 3.38, we see that after seven stages of clustering, the six clusters formed are as follows: Cluster 1: boxing; Cluster 2: swimming, track and field, skiing; Cluster 3: ping pong, handball, tennis; Cluster 4: golf, bowling; Cluster 5: basketball, hockey, football; Cluster 6: baseball. Here, Clusters 1, 4, 5, and 6 are identical to Clusters 1, 4, 5, and 6 obtained from Figure 3.37. Moreover, Clusters 2 and 3 do not differ in a meaningful way from Clusters 2 and 3 obtained from Figure 3.37. Overall, we would reach the same practical conclusions using either dendogram.

We next note that in Figure 3.39 we present a two-dimensional graph that is the result of a procedure called **multidimensional scaling.** To understand this procedure, note that, because each sport is represented by six ratings, each sport exists geometrically as a point in six-dimensional space. Multidimensional scaling uses the *relative average distances* between the sports in the six-dimensional space and attempts to find points in a lesser-dimensional space that approximately have the same relative average distances between them. In this example we illustrate mapping the six-dimensional space into a two-dimensional space, because a two-dimensional

FIGURE 3.39   The JMP Output of Multidimensional Scaling of the Sports Perception Data

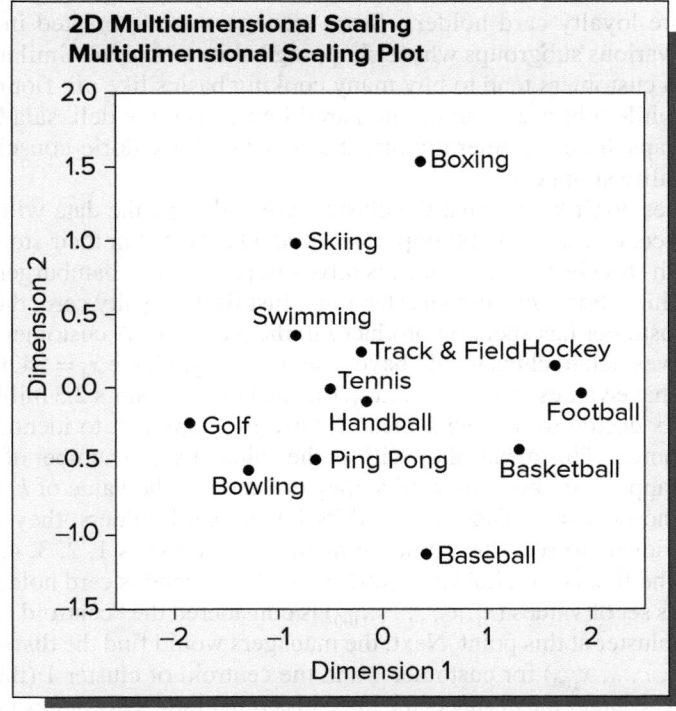

space allows us to most easily interpret the results of multidimensional scaling—that is, to study the location of the sports relative to each other and thereby determine the overall factors or dimensions that appear to separate the sports. Figure 3.39 gives the output of multidimensional scaling that is obtained by using JMP. (We will not discuss the numerical procedure used to actually carry out the multidimensional scaling.) By comparing the sports on the right of Dimension 1 with the sports on the left, and by using the average ratings in Table 3.12, we see that Dimension 1 probably represents the factor "team-oriented versus individual." By comparing the sports near the top of Dimension 2 with the sports near the bottom, and by using the average ratings in Table 3.12, we see that Dimension 2 probably represents the factor "degree of action," which combines "fast-moving/slow-moving" aspects with "contact/noncontact" aspects. Also, note that the two clusters that have been formed at the end of 11 stages of clustering in Figure 3.38 support the existence of the "team-oriented versus individual" factor. In summary, the two-dimensional graph in Figure 3.39 shows that two of the main factors underlying the perceptions of sports are "degree of action" and "team-oriented versus individual." We therefore have graphical confirmation of similar conclusions we reached from analyzing the previously discussed six clusters and thus possibly for the recommendations we made for improving the popularities of baseball and tennis.

k-means clustering

In the previously discussed hierarchical clustering, we start with each item (for example, sport) being its own cluster. Then, the two items closest to each other are merged into a single cluster, and the process continues until all items are merged into a single cluster. In **k-means clustering**, we set the number of clusters (k) at the start of the process. Items are then assigned to clusters in an iterative process that seeks to make the means (or **centroids**) of the k clusters as different as possible. In hierarchical clustering, once an item is assigned to a cluster, it never leaves the cluster. In k-means clustering, an item is often reassigned to a different cluster later in the process. It is generally recommended that hierarchical clustering not be used with more than 250 items. In a large data mining project, where we might be clustering millions of customers, k-means clustering is often used.

We will illustrate k-means clustering by using a real data mining project. For confidentiality purposes, we will consider a fictional grocery chain. However, the conclusions reached are real. Consider, then, the Just Right grocery chain, which has 2.3 million store loyalty card holders. Store managers are interested in clustering their customers into various subgroups whose shopping habits tend to be similar. They expect to find that certain customers tend to buy many cooking basics like oil, flour, eggs, rice, and raw chickens, while others are buying prepared items from the deli, salad bar, and frozen food aisle. Perhaps there are other important categories like calorie-conscious, vegetarian, or premium-quality shoppers.

The executives don't know what the clusters are and hope the data will enlighten them. They choose to concentrate on 100 important products offered in their stores. Suppose that product 1 is fresh strawberries, product 2 is olive oil, product 3 is hamburger buns, and product 4 is potato chips. For each customer having a Just Right loyalty card, they will know the amount x_i the customer has spent on product i in the past year. A customer who never buys olive oil but craves potato chips would have $x_2 = 0$ but might have $x_4 = \$420$. The data set in question would have values of $x_1, x_2, \ldots, x_{100}$ for each of the chain's 2.3 million card holders.

The managers decide to use the k-means clustering approach to identify homogeneous clusters of customers. This means they will set the value of k, the number of desired clusters. (If they are unhappy with the final results, they can modify the value of k, as we shall see.) Suppose they choose $k = 5$. Then, from all its loyalty card holders, they would randomly select five individuals to serve as founding members of clusters 1, 2, 3, 4, and 5. Suppose that the first of the five is card holder 12,867, while the second is card holder 498,015. Each cluster founder's set of values $(x_1, x_2, \ldots, x_{100})$ is considered the "centroid" or average set of values for their cluster at this point. Next, the managers would find the **distance** between the values of $(x_1, x_2, \ldots, x_{100})$ for customer 1 and the centroid of cluster 1 (the values for customer 12,867). The larger the distance, the more dissimilar these two shoppers' decisions seem to be. They would continue by computing the distance between the values of $(x_1, x_2, \ldots, x_{100})$ for customer 1 and the centroid of cluster 2 (the values for customer 498,015). Suppose that after computing similar distances between customer 1's values and those of the centroids of clusters 3, 4, and 5, they find that the smallest of the five distances is that from the centroid of cluster 4. Then customer 1 would be placed in cluster 4 in this initial stage of the analysis. In a similar fashion, every other customer who is not a founder would be assigned to a cluster on the basis of their distances from the five centroids. Thus, at the end of stage one of the cluster analysis, each of the 2.3 million customers has been assigned to one of five clusters.

In stage two, new centroids are computed for each cluster by finding the mean values of $(x_1, x_2, \ldots, x_{100})$ for all members of the clusters. Therefore, the set of numbers tells us the mean amount spent by cluster members on fresh strawberries, olive oil, hamburger buns, and so forth. Then the classification process is repeated all over again. The distance between each card holder's (including the initial five founders') spending values and each cluster centroid is computed. The customer is assigned to the cluster for which the distance is smallest. At stage three, the cluster centroids are recomputed for the new mix of members, and the process of computing distances and making new cluster assignments is repeated.

The process keeps repeating until it "converges," which means the cluster assignments become stable from one stage to the next. When this happens, the variability of the values of $(x_1, x_2, \ldots, x_{100})$ within the same cluster should be small in comparison to the variability between the cluster centroids. This is because we should have similarities among members of the same cluster and differences between the clusters themselves. Occasionally, no stable solution is found and the entire process is repeated with a different choice for k, the number of clusters. If several different values of k lead to stable cluster assignments, one way to decide on a "good" k is to consider the ratio of the within-cluster variability to the between-cluster variability. Smaller values of the ratio are better, at least up to a point. As k increases, the ratio will necessarily decrease. This is because more clusters allow more differentiation between groups and more homogeneity within clusters. Of course, you don't want k to become too large or the clusters may be so fragmented that they may be difficult to interpret. Therefore, considering the ratio as a characteristic of k, the last value of k that shows real improvement over its predecessor is often chosen as the k to be used.

In the Just Right grocery case, the clustering process did converge and the managers were happy with their choice of $k = 5$. They studied the characteristics of the cluster centroids to tease out practical interpretations of the five groups. Two of the five had particularly compelling interpretations. One was the "fresh food lovers." They bought lots of organic foods, fresh fruits and vegetables, and both salad ingredients and prepared salads. The other was the "convenience junkies." They purchased many easy-to-prepare or already prepared products. Because the grocery chain could distinguish the customers belonging to these groups, they could target them appropriately. The fresh food lovers received advertising and special offers centered on the stores' year-round stock of high-quality fresh produce and organic foods. The convenience junkies received ads and offers touting the stores' in-house line of frozen foods, deli offerings, and speed of checkout. The chain saw substantial increases in per customer spending as a result of these targeted marketing campaigns.

JMP carries out k-means clustering and helps the user choose the appropriate number of clusters k by iteratively trying different values of k specified by the user. As a very useful option, JMP presents *biplots* picturing the clusters (in two dimensions) that help the user evaluate the within-cluster variability compared to the between-cluster variability for each value of k. Figure 3.40 shows the biplots for $k = 4$, 5, 6, and 7 clusters when we have JMP try values of k from 1 to 7 in performing a k-means cluster analysis of the sports perception data. Examining the biplots, we conclude that the within-cluster variability is too large for $k = 4$ and that when $k = 7$ the clusters seem too fragmented. The choice of the best k seems to be 5 or 6. For these values of k, Figure 3.41 gives the cluster memberships. The

FIGURE 3.40   **The JMP Output of the Biplots in *k*-Means Clustering of the Sports Perception Data for *k* = 4, 5, 6, and 7**

FIGURE 3.41   The JMP Output of Cluster Memberships in *k*-Means Clustering of the Sports Perception Data
             for *k* = 5 and 6

	Sport	Fast/ Slow	Comp/ Simple	Team/ Indv.	Easy/ Hard	Nonc./ Contact	Opponent/ Standard	Cluster	Distance	Cluster 2	Distance 2
1	Boxing	3.07	4.62	6.62	4.78	6.02	1.73	1	0	1	0
2	Basketball	1.84	3.78	1.56	3.82	4.89	2.27	2	1.9067983206	2	1.9067983206
3	Golf	6.13	4.49	6.58	3.84	1.82	4.11	3	0.7352206817	3	0.7352206817
4	Swimming	2.87	5.02	5.29	3.64	2.22	4.36	5	0.922769955	5	0.922769955
5	Skiing	2.13	4.6	5.96	5.22	2.51	4.71	5	2.1447343982	5	2.1447343982
6	Baseball	4.78	4.18	2.16	3.33	3.6	2.67	4	3.1246810518	6	0
7	Ping Pong	3.18	5.13	5.38	2.91	2.04	2.2	4	1.9258641139	4	1.5408552895
8	Hockey	1.71	3.22	1.82	5.04	5.96	2.49	2	0.3930960263	2	0.3930960263
9	Handball	2.53	4.67	4.78	3.71	2.78	2.31	4	0.5882292355	4	0.2529497922
10	Track & Field	2.82	4.38	4.47	3.84	2.89	3.82	5	0.8820181373	5	0.8820181373
11	Bowling	5.07	5.16	5.4	3.11	1.6	3.73	3	0.7352206817	3	0.7352206817
12	Tennis	2.89	3.78	5.47	4.09	2.16	2.42	4	1.87473886	4	1.5534667771
13	Football	2.42	2.76	1.44	5	6.47	2.33	2	0.9770784332	2	0.9770784332

clusters are the same, except that when $k = 5$, cluster 4 contains baseball, ping pong, handball, and tennis, and when $k = 6$, cluster 4 contains ping pong, handball, and tennis, and the new cluster—cluster 6—contains baseball. Looking at the biplots when $k = 5$ and 6, and comparing the standard deviations of the perceptions of the sports in cluster 4 when $k = 5$ and 6 (see Figure 3.42), we conclude that the perceptions in cluster 4 are substantially less variable when baseball is broken out into its own cluster. Therefore, we choose $k = 6$. Examining the means of the perceptions of the sports in the different clusters (these means are given in both Figures 3.41 and 3.42), we reach conclusions similar to the conclusions reached using hierarchical clustering: (1) basketball, hockey, and football (the members of cluster 2) are rated as being of "high action" and "team oriented"; (2) baseball (the one member of cluster 6) is rated as being "slow moving" and not as "team oriented" as basketball, hockey, and football; and (3) ping pong, handball, and tennis (the members of cluster 4) are rated as being not "team oriented" (which is obvious) and not as "fast moving" as basketball, hockey, and football.

FIGURE 3.42   The Means and Standard Deviations of the Clusters in *k*-Means Clustering of the
             Sports Perception Data for *k* = 5 and 6

(a) *k* = 5
Cluster Means

Cluster	Fast/Slow	Comp/Simple	Team/Indv.	Easy/Hard	Nonc./Contact	Opponent/ Standard
1	3.07	4.62	6.62	4.78	6.02	1.73
2	1.99	3.25333333	1.60666667	4.62	5.77333333	2.36333333
3	5.6	4.825	5.99	3.475	1.71	3.92
4	3.345	4.44	4.4475	3.51	2.645	2.4
5	2.60666667	4.66666667	5.24	4.23333333	2.54	4.29666667

FIGURE 3.42   *(Continued)*

Cluster Standard Deviations

Cluster	Fast/Slow	Comp/Simple	Team/Indv.	Easy/Hard	Nonc./Contact	Opponent/Standard
1	0	0	0	0	0	0
2	0.30865299	0.4170798	0.15860503	0.56592108	0.65839873	0.09285592
3	0.53	0.335	0.59	0.365	0.11	0.19
4	0.85989825	0.50798622	1.3470593	0.4384062	0.61876894	0.17421251
5	0.33767177	0.26549744	0.60931656	0.70244019	0.27434771	0.36609046

(b) $k = 6$

Cluster Means

Cluster	Fast/Slow	Comp/Simple	Team/Indv.	Easy/Hard	Nonc./Contact	Opponent/Standard
1	3.07	4.62	6.62	4.78	6.02	1.73
2	1.99	3.25333333	1.60666667	4.62	5.77333333	2.36333333
3	5.6	4.825	5.99	3.475	1.71	3.92
4	2.86666667	4.52666667	5.21	3.57	2.32666667	2.31
5	2.60666667	4.66666667	5.24	4.23333333	2.54	4.29666667
6	4.78	4.18	2.16	3.33	3.6	2.67

Cluster Standard Deviations

Cluster	Fast/Slow	Comp/Simple	Team/Indv.	Easy/Hard	Nonc./Contact	Opponent/Standard
1	0	0	0	0	0	0
2	0.30865299	0.4170798	0.15860503	0.56592108	0.65839873	0.09285592
3	0.53	0.335	0.59	0.365	0.11	0.19
4	0.26587382	0.56037686	0.30626786	0.49179942	0.32427697	0.08981462
5	0.33767177	0.26549744	0.60931656	0.70244019	0.27434771	0.36609046
6	0	0	0	0	0	0

Exercises for Section 3.9

CONCEPTS

3.62 Explain how hierarchical clustering and *k*-means clustering differ with respect to reassigning an item to a different cluster.

METHODS AND APPLICATIONS

3.63 In this exercise we consider a marketing research study concerning the similarities and differences between the 10 types of food shown in Table 3.14. Each type of food was given an integer rating of 1 to 7 by 50 randomly selected restaurant-goers on three scales: bland (1) versus spicy (7), light (1) versus heavy (7), and low calories (1) versus high calories (7). Table 3.14 gives the average value for each of the food types on the three scales.

Figures 3.43 and 3.44 present the results of a hierarchical cluster analysis and multidimensional scaling of the 10 food types. Figures 3.45, 3.46, and 3.47 present the results of a *k*-means cluster analysis of the 10 food types. Here, the results for $k = 4$, 5, and 6 clusters are shown.   **DS** FoodTypes

a   Discuss why the two axes in Figure 3.44 may be interpreted as "Asian versus western" and "spicy versus bland."

b   Using all of the results shown, discuss the similarities and differences between the food types.

c   Suppose that you are in charge of choosing restaurants to be included in a new riverfront development that initially will include a limited number of restaurants. Discuss how you might use the information in this exercise to do this.

TABLE 3.14 Average Ratings of the Food Types on Three Scales ⓓⓢ FoodTypes

Food	Spicy/Bland	Heavy/Light	High/Low Calories
Japanese (JPN)	2.8	3.2	3.4
Cantonese (CNT)	2.6	5.3	5.4
Szechuan (SCH)	6.6	3.6	3.0
French (FR)	3.5	4.5	5.1
Mexican (MEX)	6.4	4.3	4.3
Mandarin (MAN)	3.4	4.1	4.2
American (AMR)	2.3	5.8	5.7
Spanish (SPN)	4.7	5.4	4.9
Italian (ITL)	4.6	6.0	6.2
Greek (GRK)	5.3	4.7	6.0

Source: Data from Mark L. Berenson, David M. Levine, and Mathew Goldstein, *Intermediate Statistical Methods and Applications, A Computer Package Approach* (Prentice Hall, 1983).

FIGURE 3.43 JMP Output of Hierachical Clustering of the 10 Food Types

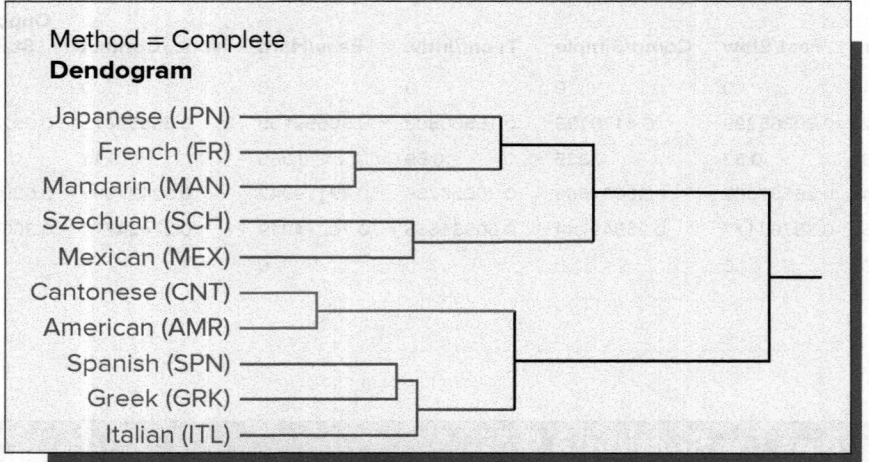

Cluster Means

Cluster	Count	Spicy/Bland	Heavy/Light	High/Low Calories
1	1	2.80000	3.20000	3.40000
2	2	3.45000	4.30000	4.65000
3	2	6.50000	3.95000	3.65000
4	2	2.45000	5.55000	5.55000
5	3	4.86667	5.36667	5.70000

FIGURE 3.44 The JMP Output of Multidimensional Scaling of the 10 Food Types

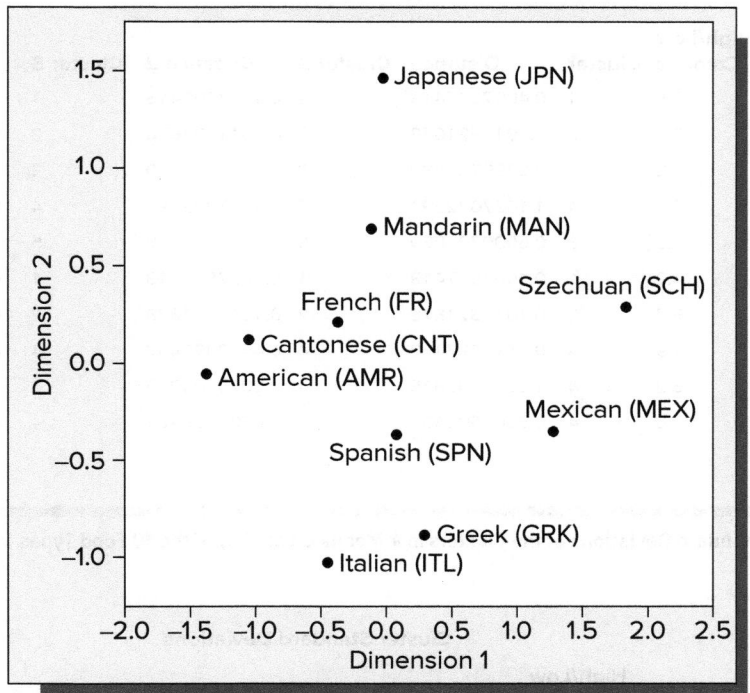

FIGURE 3.45 The JMP Output of the Biplots in *k*-Means Clustering of the 10 Food Types for *k* = 4, 5, and 6

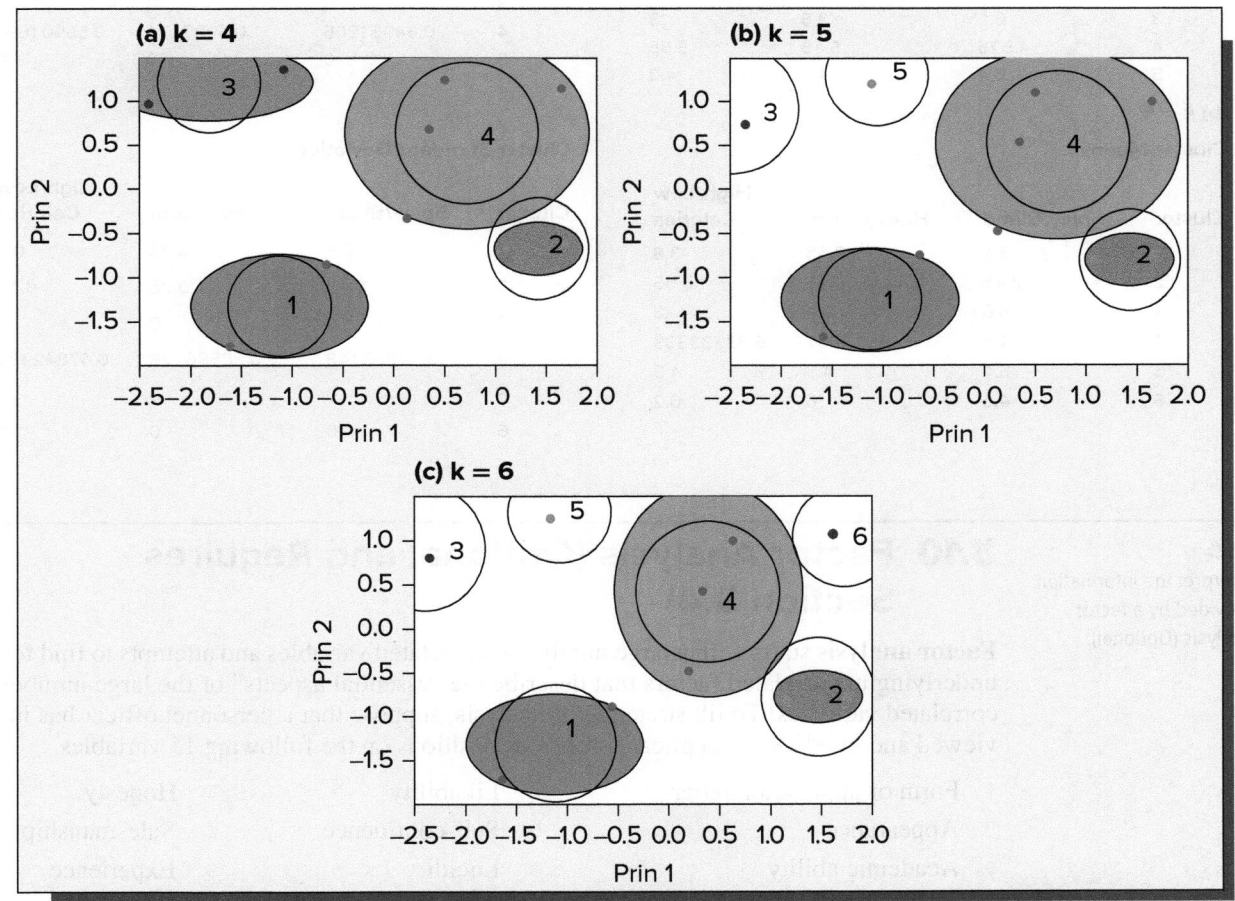

FIGURE 3.46   The JMP Output of Cluster Memberships in *k*-Means Clustering of the 10 Food Types

Food	Spicy/ Bland	Heavy/ Light	High/Low Calories	Cluster	Distance	Cluster 2	Distance 2	Cluster 3	Distance 3
1 Japanese (JPN)	2.8	3.2	3.4	1	0.4097597413	1	0.4097597413	1	0.4097597413
2 Cantonese (CNT)	2.6	5.3	5.4	2	0.1011621678	2	0.1011621678	2	0.1011621678
3 Szechuan (SCH)	6.6	3.6	3	3	0.5090733889	3	0	3	0
4 French (FR)	3.5	4.5	5.1	4	1.1037022991	4	1.1037022991	4	0.6215932311
5 Mexican (MEX)	6.4	4.3	4.3	3	0.5090733889	5	0	5	0
6 Mandarin (MAN)	3.4	4.1	4.2	1	0.4097597413	1	0.4097597413	1	0.4097597413
7 American (AMR)	2.3	5.8	5.7	2	0.1011621678	2	0.1011621678	2	0.1011621678
8 Spanish (SPN)	4.7	5.4	4.9	4	0.4482097432	4	0.4482097432	4	0.5075987511
9 Italian (ITL)	4.6	6	6.2	4	1.2023100975	4	1.2023100975	6	0
10 Greek (GRK)	5.3	4.7	6	4	0.6602606401	4	0.6602606401	4	0.6822106676

FIGURE 3.47   The Means and Standard Deviations of the Clusters in *k*-Means Clustering of the 10 Food Types for *k* = 5 and 6

(a) *k* = 5

Cluster Means

Cluster	Spicy/Bland	Heavy/Light	High/Low Calories
1	3.1	3.65	3.8
2	2.45	5.55	5.55
3	6.6	3.6	3
4	4.525	5.15	5.55
5	6.4	4.3	4.3

Cluster Standard Deviations

Cluster	Spicy/Bland	Heavy/Light	High/Low Calories
1	0.3	0.45	0.4
2	0.15	0.25	0.15
3	0	0	0
4	0.64951905	0.5937171	0.55901699
5	0	0	0

(b) *k* = 6

Cluster Means

Cluster	Spicy/Bland	Heavy/Light	High/Low Calories
1	3.1	3.65	3.8
2	2.45	5.55	5.55
3	6.6	3.6	3
4	4.5	4.86666667	5.33333333
5	6.4	4.3	4.3
6	4.6	6	6.2

Cluster Standard Deviations

Cluster	Spicy/Bland	Heavy/Light	High/Low Calories
1	0.3	0.45	0.4
2	0.15	0.25	0.15
3	0	0	0
4	0.74833148	0.38586123	0.47842334
5	0	0	0
6	0	0	0

LO3-11
Interpret the information provided by a factor analysis (Optional).

3.10 Factor Analysis (Optional and Requires Section 3.4)

Factor analysis starts with a large number of correlated variables and attempts to find fewer underlying uncorrelated factors that describe the "essential aspects" of the large number of correlated variables. To illustrate factor analysis, suppose that a personnel officer has interviewed and rated 48 job applicants for sales positions on the following 15 variables.

1   Form of application letter
2   Appearance
3   Academic ability
4   Likability
5   Self-confidence
6   Lucidity
7   Honesty
8   Salesmanship
9   Experience

10	Drive	12	Grasp	14	Keenness to join
11	Ambition	13	Potential	15	Suitability

In order to better understand the relationships between the 15 variables, the personnel officer will use factor analysis. The first step in factor analysis is to **standardize** each variable. A variable is standardized by calculating the mean and the standard deviation of the 48 values of the variable and then subtracting from each value the mean and dividing the resulting difference by the standard deviation. The variance of the values of each standardized variable can be shown to be equal to 1, and the pairwise correlations (that is, correlation coefficients) between the standardized variables can be shown to be equal to the pairwise correlations between the original variables. Although we will not give the 48 values of each of the 15 variables,[2] we present in Table 3.15 a matrix containing the pairwise correlations of these variables. Considering the matrix, we note that there are so many fairly large pairwise correlations that it is difficult to understand the relationship among the 15 variables. When we use factor analysis, we determine whether there are **uncorrelated factors,** fewer in number than 15, that (1) explain a large percentage of the total variation in the 15 variables and (2) help us to understand the relationships among and the "essential aspects" of the 15 variables.

To find the desired factors, we first find what are called **principal components.** The *first principal component* is the *composite* of the 15 standardized variables that explains the highest percentage of the total of the variances of these variables. The Minitab output in Figure 3.48 tells us that the first principal component is

$$y_{(1)} = .447x_1 + .583x_2 + .109x_3 + \cdots + .646x_{15}$$

where $x_1, x_2, \ldots, x_{15}$ denote the 15 standardized variables. Here, the coefficient multiplied by each x_i is called the *factor loading* of $y_{(1)}$ on x_i and can be shown to equal the pairwise correlation between $y_{(1)}$ and x_i. For example, the factor loading .583 says that the pairwise correlation between $y_{(1)}$ and x_2 is .583. The Minitab output also tells us that the variance of the 48 values of $y_{(1)}$ is 7.5040. Furthermore, because the sum of the variances of the 15 standardized variables is 15, the Minitab output tells us that the variance of $y_{(1)}$ explains $(7.5040/15)100\% = 50\%$ of the total variation in the standardized variables. Similarly, the Minitab output shows the *second principal component,* which has a variance of 2.0615 and explains $(2.0615/15)100\% = 13.7\%$ of the total variation in the standardized variables. In all, there are 15 principal components that are uncorrelated with each other and explain a cumulative percentage of 100 percent of the total variation in the 15 variables. Also, note that

TABLE 3.15 A Matrix of Pairwise Correlations for the Applicant Data

Variable	1	2	3	4	5	6	7	8	9	10	11	12	13	14	15
1	1.00	.24	.04	.31	.09	.23	−.11	.27	.55	.35	.28	.34	.37	.47	.59
2		1.00	.12	.38	.43	.37	.35	.48	.14	.34	.55	.51	.51	.28	.38
3			1.00	.00	.00	.08	−.03	.05	.27	.09	.04	.20	.29	−.32	.14
4				1.00	.30	.48	.65	.35	.14	.39	.35	.50	.61	.69	.33
5					1.00	.81	.41	.82	.02	.70	.84	.72	.67	.48	.25
6						1.00	.36	.83	.15	.70	.76	.88	.78	.53	.42
7							1.00	.23	−.16	.28	.21	.39	.42	.45	.00
8								1.00	.23	.81	.86	.77	.73	.55	.55
9									1.00	.34	.20	.30	.35	.21	.69
10										1.00	.78	.71	.79	.61	.62
11											1.00	.78	.77	.55	.43
12												1.00	.88	.55	.53
13													1.00	.54	.57
14														1.00	.40
15															1.00

Source: Data from Maurice Kendall, *Multivariate Analysis,* 2nd ed. Charles Griffin & Company Ltd., of London and High Wycombe, 1980.

___
[2]See Sir Maurice Kendall, *Multivariate Analysis,* 2nd ed. (London: Griffin, 1980).

FIGURE 3.48   Minitab Output of a Factor Analysis of the Applicant Data (7 Factors Used)

Principal Component Factor Analysis of the Correlation Matrix

Unrotated Factor Loadings and Communalities

Variable	Factor1	Factor2	Factor3	Factor4	Factor5	Factor6	Factor7	Communality
Var 1	0.447	−0.619	−0.376	−0.121	−0.102	0.425	−0.085	0.937
Var 2	0.583	0.050	0.020	0.282	−0.752	−0.033	−0.003	0.988
Var 3	0.109	−0.339	0.494	0.714	0.181	0.161	−0.182	0.973
Var 4	0.617	0.181	−0.580	0.357	0.099	0.078	0.057	0.896
Var 5	0.798	0.356	0.299	−0.179	−0.000	0.004	−0.066	0.890
Var 6	0.867	0.185	0.184	−0.069	0.178	0.117	0.301	0.961
Var 7	0.433	0.582	−0.360	0.446	0.061	−0.216	−0.065	0.909
Var 8	0.882	0.056	0.248	−0.228	−0.030	−0.063	−0.010	0.900
Var 9	0.365	−0.794	−0.093	0.074	0.090	−0.260	0.068	0.859
Var 10	0.863	−0.069	0.100	−0.166	0.176	−0.175	−0.297	0.936
Var 11	0.872	0.098	0.256	−0.209	−0.137	0.076	−0.125	0.919
Var 12	0.908	0.030	0.135	0.097	0.064	0.102	0.247	0.928
Var 13	0.913	−0.032	0.073	0.218	0.105	0.047	0.004	0.901
Var 14	0.710	0.115	−0.558	−0.235	0.101	0.059	−0.144	0.919
Var 15	0.646	−0.604	−0.107	−0.029	−0.064	−0.293	0.105	0.895
Variance	7.5040	2.0615	1.4677	1.2091	0.7414	0.4840	0.3441	13.8118
% Var	0.500	0.137	0.098	0.081	0.049	0.032	0.023	0.921

Rotated Factor Loadings and Communalities

Varimax Rotation

Variable	Factor1	Factor2	Factor3	Factor4	Factor5	Factor6	Factor7	Communality
Var 1	0.124	−0.042	−0.428	−0.005	0.853✓	−0.094	0.015	0.937
Var 2	0.326	−0.212	−0.117	0.056	0.077	−0.902✓	0.011	0.988
Var 3	0.054	0.028	−0.134	0.974✓	−0.012	−0.039	0.000	0.973
Var 4	0.221	−0.858✓	−0.131	−0.012	0.265	−0.100	0.115	0.896
Var 5	0.911✓	−0.154	0.083	−0.042	−0.051	−0.139	−0.069	0.890
Var 6	0.879✓	−0.257	−0.101	0.017	0.059	0.003	0.328	0.961
Var 7	0.202	−0.876✓	0.134	0.001	−0.230	−0.160	−0.070	0.909
Var 8	0.901✓	−0.078	−0.220	−0.060	0.056	−0.161	−0.045	0.900
Var 9	0.065	0.030	−0.887✓	0.163	0.201	0.012	−0.002	0.859
Var 10	0.797✓	−0.209	−0.359	0.023	0.106	0.026	−0.340	0.936
Var 11	0.894✓	−0.060	−0.087	−0.018	0.166	−0.260	−0.113	0.919
Var 12	0.797✓	−0.306	−0.236	0.141	0.129	−0.149	0.291	0.928
Var 13	0.730✓	−0.404	−0.290	0.265	0.162	−0.141	0.061	0.901
Var 14	0.459	−0.567	−0.170	−0.386	0.425	0.038	−0.162	0.919
Var 15	0.340	−0.076	−0.843	−0.010	0.184	−0.171	0.007	0.895
Variance	5.5634	2.2804	2.1176	1.2251	1.2029	1.0467	0.3758	13.8118
% Var	0.371	0.152	0.141	0.082	0.080	0.070	0.025	0.921

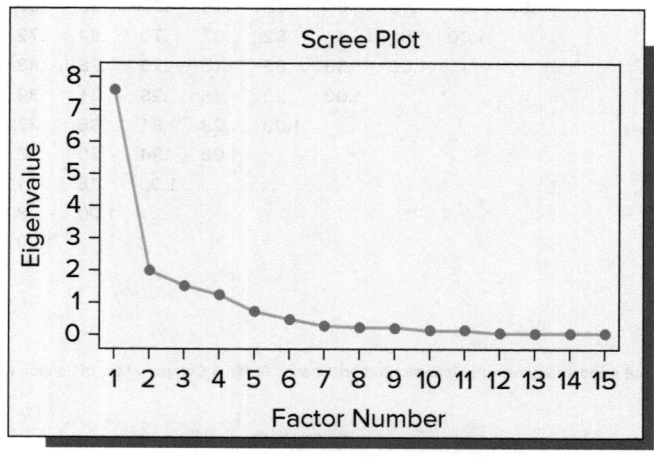

the variance of a particular principal component can be shown to equal the sum of the squared pairwise correlations between the principal component and the 15 standardized variables. For example, examining the first column of pairwise correlations in the upper portion of Figure 3.48, it follows that the variance of the first principal component is

$$(.447)^2 + (.583)^2 + \cdots + (.646)^2 = 7.5040$$

Although the Minitab output shows the percentage of the total variation explained by each of the 15 principal components, it shows only 7 of these principal components. The reason is that, because we wish to obtain final factors that are fewer in number than the number of original variables, we have instructed Minitab to retain 7 principal components for "further study." The choice of 7 principal components, while somewhat arbitrary, is based on the belief that 7 principal components will explain a high percentage of the total variation in the 15 variables. The Minitab output tells us that this choice is reasonable—the first 7 principal components explain 92.1 percent of the total variation in the 15 variables. The reason that we need to further study the 7 principal components is that, in general, principal components tend to be correlated with many of the standardized variables (see the factor loadings on the Minitab output) and thus tend to be difficult to interpret in a meaningful way. For this reason, we **rotate** the 7 principal components by using **Varimax rotation.** This technique attempts to find final uncorrelated factors each of which loads highly on (that is, is strongly correlated with) a *limited number* of the 15 original standardized variables and loads as low as possible on the rest of the standardized variables. The Minitab output shows the results of the Varimax rotation. Examining the check marks that we have placed on the output, we see that Factor 1 loads heavily on variables 5 (self-confidence), 6 (lucidity), 8 (salesmanship), 10 (drive), 11 (ambition), 12 (grasp), and 13 (potential). Therefore, Factor 1 might be interpreted as an "extroverted personality" dimension. Factor 2 loads heavily on variables 4 (likability) and 7 (honesty). Therefore, Factor 2 might be interpreted as an "agreeable personality" dimension. (Note that the negative correlations −.858 and −.876 imply that the higher the likability and honesty are for a job applicant, the more negative is the value of Factor 2 for the job applicant. Therefore, the more negative the value of Factor 2 is for a job applicant, the more "agreeable" is his or her personality.) Similarly, Factors 3 through 7 might be interpreted as the following dimensions: Factor 3, "experience"; Factor 4, "academic ability"; Factor 5, "form of application letter"; Factor 6, "appearance"; Factor 7, no discernible dimension. Note that, although variable 14 (keenness to join) does not load heavily on any factor, its correlation of −.567 with Factor 2 might mean that it should be interpreted to be part of the agreeable personality dimension.

We next note that the **communality** to the right of each variable in Figure 3.48 is the percentage of the variance of the variable that is explained by the 7 factors. The communality for each variable can be shown to equal the sum of the squared pairwise correlations between the variable and the 7 factors. For example, examining the first row of pairwise correlations in the lower portion of Figure 3.48 (which describes the rotated factors), it follows that the communality for factor 1 is

$$(.124)^2 + (-.042)^2 + \cdots + (.015)^2 = .937$$

All of the communalities in Figure 3.48 seem high. However, some statisticians might say that we have retained too many factors. To understand this, note that the upper portion of Figure 3.48 tells us that the sum of the variances of the first seven unrotated factors is

$$7.5040 + 2.0615 + 1.4677 + 1.2091 + .7414 + .4840 + .3441 = 13.8118$$

This variance is $(13.8118/15)100\% = 92.1\%$ of the sum of the variances of the 15 standardized variables. Some statisticians would suggest that we retain a factor only if its variance exceeds 1, the variance of each standardized variable. If we do this, we would retain 4 factors, because the variance of the fourth factor is 1.2091 and the variance of the fifth factor is .7414. The bottom of Figure 3.48 shows the Minitab output of a **scree plot** of each unrotated factor variance (or **eigenvalue**) versus the factor number. This plot graphically illustrates that the variances (eigenvalues) of only the first four factors are greater than 1, suggesting that we retain only the first four factors.

FIGURE 3.49   Minitab Output of a Factor Analysis of the Applicant Data (4 Factors Used)

Principal Component Factor Analysis of the Correlation Matrix

Unrotated Factor Loadings and Communalities

Variable	Factor1	Factor2	Factor3	Factor4	Communality
Var 1	0.447	−0.619	−0.376	−0.121	0.739
Var 2	0.583	0.050	0.020	0.282	0.422
Var 3	0.109	−0.339	0.494	0.714	0.881
Var 4	0.617	0.181	−0.580	0.357	0.877
Var 5	0.798	0.356	0.299	−0.179	0.885
Var 6	0.867	0.185	0.184	−0.069	0.825
Var 7	0.433	0.582	−0.360	0.446	0.855
Var 8	0.882	0.056	0.248	−0.228	0.895
Var 9	0.365	−0.794	−0.093	0.074	0.779
Var 10	0.863	−0.069	0.100	−0.166	0.787
Var 11	0.872	0.098	0.256	−0.209	0.879
Var 12	0.908	0.030	0.135	0.097	0.852
Var 13	0.913	−0.032	0.073	0.218	0.888
Var 14	0.710	0.115	−0.558	−0.235	0.884
Var 15	0.646	−0.604	−0.107	−0.029	0.794
Variance	7.5040	2.0615	1.4677	1.2091	12.2423
% Var	0.500	0.137	0.098	0.081	0.816

Rotated Factor Loadings and Communalities

Varimax Rotation

Variable	Factor1	Factor2	Factor3	Factor4	Communality
Var 1	0.114	−0.833✓	−0.111	−0.138	0.739
Var 2	0.440	−0.150	−0.394	0.226	0.422
Var 3	0.061	−0.127	−0.006	0.928✓	0.881
Var 4	0.216	−0.247	−0.874✓	−0.081	0.877
Var 5	0.919✓	0.104	−0.162	−0.062	0.885
Var 6	0.864✓	−0.102	−0.259	0.006	0.825
Var 7	0.217	0.246	−0.864✓	0.003	0.855
Var 8	0.918✓	−0.206	−0.088	−0.049	0.895
Var 9	0.085	−0.849✓	0.055	0.219	0.779
Var 10	0.796✓	−0.354	−0.160	−0.050	0.787
Var 11	0.916✓	−0.163	−0.105	−0.042	0.879
Var 12	0.804✓	−0.259	−0.340	0.152	0.852
Var 13	0.739✓	−0.329	−0.425	0.230	0.888
Var 14	0.436	−0.364	−0.541	−0.519	0.884
Var 15	0.379	−0.798✓	−0.078	0.082	0.794
Variance	5.7455	2.7351	2.4140	1.3478	12.2423
% Var	0.383	0.182	0.161	0.090	0.816

Figure 3.49 gives the Minitab output obtained by using 4 factors. Examining the check marks that we have placed on the output, we see that Factors 1 through 4 might be interpreted as follows: Factor 1, "extroverted personality"; Factor 2, "experience"; Factor 3, "agreeable personality"; Factor 4, "academic ability." Variable 2 (appearance) does not load heavily on any factor and thus is its own factor, as Factor 6 on the Minitab output in Figure 3.48 indicated is true. Variable 1 (form of application letter) loads heavily on Factor 2 ("experience"). In summary, there is not much difference between the 7-factor and 4-factor solutions. We might therefore conclude that the 15 variables can be reduced to the following five uncorrelated factors: "extroverted personality," "experience," "agreeable personality," "academic ability," and "appearance." This conclusion helps the personnel officer focus on the "essential characteristics" of a job applicant. Moreover, if a company analyst wishes at a later date to use predictive analytics to predict sales performance on the basis of the characteristics of salespeople, the analyst can simplify the prediction modeling procedure by using the five uncorrelated factors instead of the original 15 correlated variables as potential predictor variables.

To conclude this section, we note that whereas Minitab will use a correlation matrix or the raw data as input for performing a factor analysis, JMP (at the time of the writing of this text) requires the raw data. We will illustrate using JMP to carry out a factor analysis in Exercises 3.68 and 3.69.

Exercises for Section 3.10

CONCEPTS

3.64 What is the purpose of factor analysis?

3.65 Why does factor analysis do factor rotation?

METHODS AND APPLICATIONS

3.66 In *Applied Multivariate Techniques* (John Wiley and Sons, 1996), Subhash Sharma considers a study in which 143 respondents rated three brands of laundry detergent on 12 product attributes using a 5-point Likert scale. The 12 product attributes are:

V1:   Gentle to natural fabrics   V7:   Makes colors bright
V2:   Won't harm colors   V8:   Removes grease stains
V3:   Won't harm synthetics   V9:   Good for greasy oil
V4:   Safe for lingerie   V10: Pleasant fragrance
V5:   Strong, powerful   V11: Removes collar soil
V6:   Gets dirt out   V12: Removes stubborn stains

Table 3.16 is a matrix containing the pairwise correlations between the variables, and Figure 3.50 is a software package output of a factor analysis of the detergent data. **(1)** Examining the "EIGENVALUE" column in Figure 3.50, and recalling that the eigenvalue of a factor is the variance of the factor, explain why the analyst chose to retain two factors. **(2)** Discuss why Factor 1 can be interpreted to be the ability of the detergent to clean clothes. **(3)** Discuss why Factor 2 can be interpreted to be the mildness of the detergent.

3.67 Table 3.17 shows the output of a factor analysis of the ratings of 82 respondents who were asked to evaluate a particular discount store on 29 attributes using a 7-point Likert scale. Interpret and give names to the five factors.

TABLE 3.16   **Correlation Matrix for Detergent Data**

	V1	V2	V3	V4	V5	V6	V7	V8	V9	V10	V11	V12
V1	1.00000	0.41901	0.51840	0.56641	0.18122	0.17454	0.23034	0.30647	0.24051	0.21192	0.27443	0.20694
V2	0.41901	1.00000	0.57599	0.49886	0.18666	0.24648	0.22907	0.22526	0.21967	0.25879	0.32132	0.25853
V3	0.51840	0.57599	1.00000	0.64325	0.29080	0.34428	0.41083	0.34028	0.32854	0.38828	0.39433	0.36712
V4	0.56641	0.49886	0.64325	1.00000	0.38360	0.39637	0.37699	0.40391	0.42337	0.36564	0.33691	0.36734
V5	0.18122	0.18666	0.29080	0.38360	1.00000	0.57915	0.59400	0.67623	0.69269	0.43873	0.55485	0.65261
V6	0.17454	0.24648	0.34428	0.39637	0.57915	1.00000	0.57756	0.70103	0.62280	0.62174	0.59855	0.57845
V7	0.23034	0.22907	0.41083	0.37699	0.59400	0.57756	1.00000	0.67682	0.68445	0.54175	0.78361	0.63889
V8	0.30647	0.22526	0.34028	0.40391	0.67623	0.70103	0.67682	1.00000	0.69813	0.68589	0.71115	0.71891
V9	0.24051	0.21967	0.32854	0.42337	0.69269	0.62280	0.68445	0.69813	1.00000	0.58579	0.64637	0.69111
V10	0.21192	0.25879	0.38828	0.36564	0.43873	0.62174	0.54175	0.68589	0.58579	1.00000	0.62250	0.63494
V11	0.27443	0.32132	0.39433	0.33691	0.55485	0.59855	0.78361	0.71115	0.64637	0.62250	1.00000	0.63973
V12	0.20694	0.25853	0.36712	0.36734	0.65261	0.57845	0.63889	0.71891	0.69111	0.63494	0.63973	1.00000

Source: Data from Subhash Sharma, *Applied Multivariate Techniques* (New York: Wiley, 1996).

FIGURE 3.50   **A Software Package Output of a Factor Analysis of the Detergent Data**

	INITIAL STATISTICS:						ROTATED FACTOR MATRIX:		
VARIABLE	COMMUNALITY	*	FACTOR	EIGENVALUE	PCT OF VAR	CUM PCT		FACTOR 1	FACTOR 2
V1	.42052	*	1	6.30111	52.5	52.5	VI	.12289	.65101
V2	.39947	*	2	1.81757	15.1	67.7	V2	.13900	.64781
V3	.56533	*	3	.66416	5.5	73.2	V3	.24971	.78587
V4	.56605	*	4	.57155	4.8	78.0	V4	.29387	.74118
V5	.60467	*	5	.55995	4.7	82.6	V5	.73261	.15469
V6	.57927	*	6	.44517	3.7	86.3	V6	.73241	.20401
V7	.69711	*	7	.41667	3.5	89.8	V7	.77455	.22464
V8	.74574	*	8	.32554	2.7	92.5	V8	.85701	.20629
V9	.66607	*	9	.27189	2.3	94.8	V9	.80879	.19538
V10	.59287	*	10	.25690	2.1	96.9	V10	.69326	.23923
V11	.71281	*	11	.19159	1.6	98.5	V11	.77604	.25024
V12	.64409	*	12	.17789	1.5	100.0	V12	.79240	.19822

TABLE 3.17   Factor Analysis of the Discount Store Data

Scale	I	II	III	IV	V	Communality
1. Good service	.79	−.15	.06	.12	.07	.67
2. Helpful salespersons	.75	−.03	.04	.13	.31	.68
3. Friendly personnel	.74	−.07	.17	.09	−.14	.61
4. Clean	.59	−.31	.34	.15	−.25	.65
5. Pleasant store to shop in	.58	−.15	.48	.26	.10	.67
6. Easy to return purchases	.56	−.23	.13	−.03	−.03	.39
7. Too many clerks	.53	−.00	.02	.23	.37	.47
8. Attracts upper-class customers	.46	−.06	.25	−.00	.17	.31
9. Convenient location	.36	−.30	−.02	−.19	.03	.26
10. High-quality products	.34	−.27	.31	.12	.25	.36
11. Good buys on products	.02	−.88	.09	.10	.03	.79
12. Low prices	−.03	−.74	.14	.00	.13	.59
13. Good specials	.35	−.67	−.05	.10	.14	.60
14. Good sales on products	.30	−.67	.01	−.08	.16	.57
15. Reasonable value for price	.17	−.52	.11	−.02	−.03	.36
16. Good store	.41	−.47	.47	.12	.11	.63
17. Low-pressure salespersons	−.20	−.30	−.28	−.03	−.05	.18
18. Bright store	−.02	−.10	.75	.26	−.05	.61
19. Attractive store	.19	.03	.67	.34	.24	.66
20. Good displays	.33	−.15	.61	.15	−.20	.57
21. Unlimited selections of products	.09	.00	.29	−.03	.00	.09
22. Spacious shopping	.00	.20	.00	.70	.10	.54
23. Easy to find items you want	.36	−.16	.10	.57	.01	.49
24. Well-organized layout	−.02	−.05	.25	.54	−.17	.39
25. Well-spaced merchandise	.20	.15	.27	.52	.16	.43
26. Neat	.38	−.12	.45	.49	−.34	.72
27. Big store	−.20	.15	.06	.07	−.65	.49
28. Ads frequently seen by you	.03	−.20	.07	.09	.42	.23
29. Fast checkout	.30	−.16	.00	.25	−.33	.28
Percentage of variance explained	16	12	9	8	5	
Cumulative variance explained	16	28	37	45	50	

Source: Data from David A. Aaker, V. Kumar, and George S. Dax, *Marketing Research*, 6th ed. (New York: Wiley, 1998).

3.68 Figure 3.51 shows the JMP output of 10 colleges. From a data set containing 1302 colleges and 22 variables describing these colleges. We will ignore the first two variables (*State* and *Public/Private*), and when we use JMP, we obtain Figures 3.52 and 3.53, which show the JMP output of the first 10 principal components and the variances (eigenvalues) of these components. The first five principal components have a variance greater than 1. However, after trying a number of principal components from 3 to 10 and rotating these factors, we conclude that 6 factors best describe the essential aspects of the variables describing the colleges. Figure 3.54 shows the unrotated factor loadings when 6 factors are retained, and Figure 3.55 shows the rotated factor loadings, final communality estimates, and variance explained by each factor when 6 factors are retained. Five of these rotated factors—factors 1, 2, 3, 5, and 6—have clear interpretations. Interpret these rotated factors.

3.69 Interpret the meanings of the communality estimates in Figure 3.55.

FIGURE 3.51 **The JMP Output of Part of the College Data** ⓄⓈColleges

	College Name	State	Public (1)/ Private (2)	Math SAT	Verbal SAT	ACT	# appli. rec'd	# appl. accepted	# new stud. enrolled	% new stud. from top 10%	% new stud. from top 25%
1	Alaska Pacific University	AK	2	490	482	20	193	146	55	16	44
2	University of Alaska at Fairbanks	AK	1	499	462	22	1852	1427	928	.	.
3	University of Alaska Southeast	AK	1	.	.	.	146	117	89	4	24
4	University of Alaska at Anchorage	AK	1	459	422	20	2065	1598	1162	.	.
5	Alabama Agri. & Mech. Univ.	AL	1	.	.	17	2817	1920	984	.	.
6	Faulkner University	AL	2	.	.	20	345	320	179	.	27
7	University of Montevallo	AL	1	.	.	21	1351	892	570	18	78
8	Alabama State University	AL	1	.	.	.	4639	3272	1278	.	.
9	Auburn University-Main Campus	AL	1	575	501	24	7548	6791	3070	25	57
10	Birmingham-Southern College	AL	2	575	525	26	805	588	287	67	88

	# FT undergrad	# PT undergrad	in-state tuition	out-of-state tuition	room	board	add. fees	estim. book costs	estim. personal $	% fac. w/PHD
1	249	869	7560	7560	1620	2500	130	800	1500	76
2	3885	4519	1742	5226	1800	1790	155	650	2304	67
3	492	1849	1742	5226	2514	2250	34	500	1162	39
4	6209	10537	1742	5226	2600	2520	114	580	1260	48
5	3958	305	1700	3400	1108	1442	155	500	850	53
6	1367	578	5600	5600	1550	1700	300	350	.	52
7	2385	331	2220	4440	.	.	124	300	600	72
8	4051	405	1500	3000	1960	.	84	500	.	48
9	16262	1716	2100	6300	.	.	.	600	1908	85
10	1376	207	11660	11660	2050	2430	120	400	900	74

	stud./fac. ratio	Graduation rate
1	11.9	15
2	10	.
3	9.5	39
4	13.7	.
5	14.3	40
6	32.8	55
7	18.9	51
8	18.7	15
9	16.7	69
10	14	72

FIGURE 3.52   The JMP Output of the First 10 Principal Components

Eigenvectors

	Prin1	Prin2	Prin3	Prin4	Prin5	Prin6	Prin7	Prin8	Prin9	Prin10
Math SAT	**0.34423**	0.03401	**−0.24616**	0.07439	−0.05898	**0.09788**	0.01241	0.01551	**−0.13753**	0.07245
Verbal SAT	**0.33842**	−0.03971	**−0.23563**	0.06675	−0.06513	**0.11122**	−0.00937	0.03193	**−0.13172**	0.08301
ACT	**0.33584**	−0.01294	**−0.20487**	0.04069	−0.05765	**0.15521**	0.03641	0.05262	**−0.09975**	0.07353
# appli. rec'd	0.16719	**0.37003**	0.11308	−0.11292	0.07286	**−0.19701**	**−0.16248**	0.07203	**0.00017**	−0.01818
# appl. accepted	0.12390	**0.40030**	0.13874	−0.13484	0.02783	**−0.18291**	**−0.16255**	0.07578	0.06523	−0.08685
# new stud. enrolled	0.07901	**0.42575**	0.06267	−0.06935	−0.08671	**−0.14076**	**−0.15035**	0.02786	−0.02378	−0.06405
% new stud. from top 10%	**0.33657**	0.03083	−0.16408	0.05912	0.12823	0.00183	0.07276	−0.00990	**−0.14474**	0.01836
% new stud. from top 25%	**0.32860**	0.04191	−0.20006	0.06533	0.05453	0.04074	0.05988	0.02385	**−0.23982**	−0.03208
# FT undergrad	0.06717	**0.43546**	0.06036	−0.03646	−0.10403	−0.08966	−0.08527	−0.00502	−0.02457	−0.03704
# PT undergrad	−0.04625	**0.29618**	**0.15152**	**0.17586**	**−0.39430**	**0.24242**	0.16161	**−0.36101**	−0.07642	**0.56108**
in-state tuition	**0.25443**	**−0.27729**	**0.17467**	−0.01834	−0.08854	−0.11229	−0.08546	0.07691	**0.27088**	0.08963
out-of-state tuition	**0.29743**	**−0.18106**	**0.19634**	−0.04100	−0.12014	−0.09651	−0.04610	0.02658	**0.33810**	0.07321
room	0.12602	−0.07304	**0.59231**	0.04597	0.08592	0.15407	0.19716	0.16783	**−0.57381**	**−0.31610**
board	0.20605	−0.09473	**0.51514**	0.00877	−0.11945	0.16043	0.05557	0.02905	0.12831	0.23328
add. fees	0.05730	0.16378	0.06651	**−0.20936**	**0.78311**	0.15942	**0.31922**	−0.14481	0.16549	0.29737
estim. book costs	0.06321	0.05028	0.11445	**0.68973**	0.30394	0.24343	**−0.50908**	−0.18452	0.17107	−0.14856
estim. personal $	−0.07416	0.16006	−0.02668	**0.58422**	0.00917	−0.31428	**0.49971**	**0.49042**	0.16788	0.08309
% fac. w/PHD	**0.23784**	0.13214	−0.02236	−0.06547	−0.17289	0.23021	**0.40529**	−0.24665	**0.45486**	**−0.56579**
stud./fac. ratio	**−0.15203**	0.16677	−0.07823	−0.15228	−0.06714	**0.67271**	−0.13298	**0.61791**	0.16299	0.02766
Graduation rate	**0.26969**	−0.12001	0.04790	−0.14156	0.07475	−0.18174	−0.19137	**0.28184**	0.11110	0.21658

FIGURE 3.53   The Variances of the First 10 Principal Components

Summary Plots

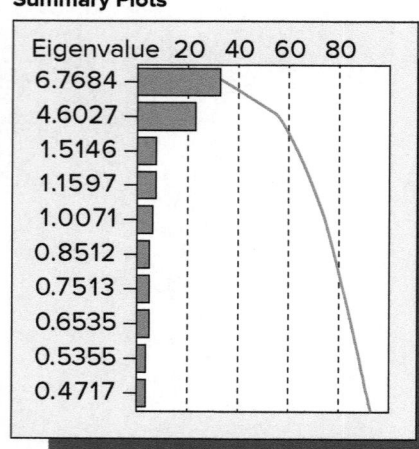

Eigenvalue	20 40 60 80
6.7684	
4.6027	
1.5146	
1.1597	
1.0071	
0.8512	
0.7513	
0.6535	
0.5355	
0.4717	

FIGURE 3.54   The Unrotated Factor Loadings When 6 Factors Are Retained

Unrotated Factor Loading

	Factor1	Factor2	Factor3	Factor4	Factor5	Factor6
Math SAT	**0.900487**	0.055929	−0.317891	0.107706	−0.046837	0.039209
Verbal SAT	**0.879502**	−0.101146	−0.292279	0.099219	−0.076600	0.053447
ACT	**0.868334**	−0.044193	−0.246630	0.092846	−0.053367	0.127148
# appli. rec'd	**0.443350**	**0.789294**	0.179156	−0.212424	0.002863	−0.047950
# appl. accepted	0.332719	**0.861968**	0.220265	−0.181628	−0.077771	0.000656
# new stud. enrolled	0.216142	**0.914336**	0.098844	−0.003307	−0.136127	−0.066316
% new stud. from top 10%	**0.874611**	0.049404	−0.200708	−0.089709	0.214439	−0.128753
% new stud. from top 25%	**0.848561**	0.072932	−0.246027	−0.025976	0.157723	−0.087760
# FT undergrad	0.185637	**0.935723**	0.082826	0.079424	−0.099199	−0.044926
# PT undergrad	−0.106926	**0.575443**	0.097012	**0.404556**	−0.003083	0.024893
in-state tuition	**0.659727**	**−0.608972**	**0.318316**	−0.029810	**−0.153346**	−0.154690
out-of-state tuition	**0.773081**	**−0.402349**	**0.347079**	0.024648	**−0.195873**	−0.064306
room	**0.295148**	−0.140460	**0.457322**	0.100159	**0.240845**	0.094707
board	**0.500459**	−0.193030	**0.492871**	0.173679	**0.142168**	0.116991
add. fees	0.137476	0.302107	0.032394	−0.261089	**0.291662**	0.167096
estim. book costs	0.145777	0.083782	0.030162	0.153297	**0.224355**	−0.176736
estim. personal $	−0.168807	0.288766	−0.057396	0.169469	**0.130664**	−0.239653
% fac. w/PHD	**0.582592**	0.245021	0.013825	0.138301	−0.014727	0.188762
stud./fac. ratio	**−0.354374**	0.312332	−0.083840	0.040518	−0.038583	0.192523
Graduation rate	**0.660019**	−0.243713	0.094762	−0.182515	−0.058888	0.040287

FIGURE 3.55   **The Rotated Factor Loadings, Final Communality Estimates, and Variance Explained by Each Factor When 6 Factors Are Retained**

Rotated Factor Loading

	Factor 1	Factor 2	Factor 3	Factor 4	Factor 5	Factor 6
Math SAT	**0.946663**	0.145748	0.104633	0.035647	−0.019067	0.019917
Verbal SAT	**0.923419**	0.016137	0.154021	0.073825	−0.061917	−0.042191
ACT	**0.894433**	0.066747	0.186678	0.016591	−0.007189	−0.069123
# appli. rec'd	**0.241130**	**0.872205**	0.061392	0.086314	**0.252321**	0.072811
# appl. accepted	0.132090	**0.939004**	0.030766	0.006623	**0.201719**	0.019680
# new stud. enrolled	0.102394	**0.927880**	−0.113575	−0.114125	0.049273	0.125061
% new stud. from top 10%	**0.826755**	0.135032	0.172063	0.262055	0.209324	0.185539
% new stud. from top 25%	**0.836556**	0.142195	0.127756	0.181380	0.153984	0.161802
# FT undergrad	0.092639	**0.916445**	−0.116987	−0.200030	0.042140	0.176461
# PT undergrad	−0.114647	**0.475175**	−0.029396	**−0.435594**	−0.118870	**0.269447**
in-state tuition	**0.433389**	**−0.258460**	**0.624946**	**0.471008**	−0.257902	−0.148991
out-of-state tuition	**0.524360**	−0.044093	**0.656671**	0.354875	−0.245391	−0.169837
room	0.056962	−0.010230	**0.609330**	0.007148	0.113173	0.074169
board	0.232765	0.012088	**0.734663**	0.014671	0.004804	0.020266
add. fees	0.069171	0.247369	0.028262	−0.012331	**0.474741**	0.008173
estim. book costs	0.105944	0.054885	0.105161	0.006657	0.033134	**0.328349**
estim. personal $	−0.134432	0.173000	−0.170650	−0.081811	−0.035127	**0.365175**
% fac. w/PHD	**0.533446**	**0.311864**	0.212391	−0.157763	0.048414	−0.021153
stud./fac. ratio	−0.259032	0.170447	−0.260257	**−0.313382**	0.069483	−0.058627
Graduation rate	**0.521900**	−0.020473	**0.363105**	**0.309547**	0.045369	−0.199706

Final Communality Estimates

Math SAT	0.93039
Verbal SAT	0.88775
ACT	0.84442
# appli. rec'd	0.89907
# appl. accepted	0.94125
# new stud. enrolled	0.91544
% new stud. from top 10%	0.87828
% new stud. from top 25%	0.81916
# FT undergrad	0.93507
# PT undergrad	0.51627
in-state tuition	0.95574
out-of-state tuition	0.92311
room	0.39299
board	0.59470
add. fees	029237
estim. book costs	0.13425
estim. personal $	0.21840
% fac. w/PHD	0.45461
stud./fac. ratio	0.27036
Graduation rate	0.54240

Variance Explained by Each Factor

Factor	Variance	Percent	Cum Percent
Factor1	5.2184	26.092	26.092
Factor2	3.9234	19.617	45.709
Factor3	2.1661	10.830	56.539
Factor4	0.9327	4.663	61.202
Factor5	0.5710	2.855	64.057
Factor6	0.5346	2.673	66.730

Chapter Summary

We began this chapter by presenting and comparing several measures of **central tendency.** We defined the **population mean** and we saw how to estimate the population mean by using a **sample mean.** We also defined the **median** and **mode,** and we compared the mean, median, and mode for symmetrical distributions and for distributions that are skewed to the right or left. We then studied measures of **variation** (or *spread*). We defined the **range, variance,** and **standard deviation,** and we saw how to estimate a population variance and standard deviation by using a sample. We learned that a good way to interpret the standard deviation when a population is (approximately) normally distributed is to use the **Empirical Rule,** and we studied **Chebyshev's Theorem,** which gives us intervals containing reasonably large fractions of the population units no matter what the population's shape might be. We also saw that, when a data set is highly skewed, it is best

to use **percentiles** and **quartiles** to measure variation, and we learned how to construct a **box-and-whiskers plot** by using the quartiles.

After learning how to measure and depict central tendency and variability, we presented various optional topics. First, we discussed several numerical measures of the relationship between two variables. These included the **covariance,** the **correlation coefficient,** and the **least squares line.** We then introduced the concept of a **weighted mean** and also explained how to compute descriptive statistics for **grouped data.** In addition, we showed how to calculate the **geometric mean** and demonstrated its interpretation. Finally, we used the numerical methods of this chapter to give an introduction to four important techniques of **descriptive analytics: association rules, text mining, cluster analysis,** and **factor analysis.**

Glossary of Terms

association rules: Rules to identify items that tend to co-occur and to describe their co-occurence.

box-and-whiskers display (box plot): A graphical portrayal of a data set that depicts both the central tendency and variability of the data. It is constructed using Q_1, M_d, and Q_3.

central tendency: A term referring to the middle of a population or sample of measurements.

Chebyshev's Theorem: A theorem that (for any population) allows us to find an interval that contains a specified percentage of the individual measurements in the population.

cluster analysis: The detection of natural groupings, or clusters, within data without having to prespecify a set of categories. Two of the most common methods of clustering are hierarchical clustering and k-means clustering.

coefficient of variation: A quantity that measures the variation of a population or sample relative to its mean.

correlation coefficient: A numerical measure of the linear relationship between two variables that is between −1 and 1.

covariance: A numerical measure of the linear relationship between two variables that depends upon the units in which the variables are measured.

Empirical Rule: For a normally distributed population, this rule tells us that 68.26 percent, 95.44 percent, and 99.73 percent, respectively, of the population measurements are within one, two, and three standard deviations of the population mean.

factor analysis: A technique that starts with a large number of correlated variables and attempts to find fewer underlying, uncorrelated factors that describe the "essential aspects" of the large number of correlated variables.

first quartile (denoted Q_1): A value below which approximately 25 percent of the measurements lie; the 25th percentile.

geometric mean: The constant return (or rate of change) that yields the same wealth at the end of several time periods as do actual returns.

grouped data: Data presented in the form of a frequency distribution or a histogram.

interquartile range (denoted IQR): The difference between the third quartile and the first quartile (that is, $Q_3 - Q_1$).

least squares line: The line that minimizes the sum of the squared vertical differences between points on a scatter plot and the line.

lower and upper limits (in a box-and-whiskers display): Points located $1.5 \times IQR$ below Q_1 and $1.5 \times IQR$ above Q_3.

measure of variation: A descriptive measure of the spread of the values in a population or sample.

median (denoted M_d): A measure of central tendency that divides a population or sample into two roughly equal parts.

mode (denoted M_o): The measurement in a sample or a population that occurs most frequently.

mound-shaped: Description of a relative frequency curve that is "piled up in the middle."

normal curve: A bell-shaped, symmetrical relative frequency curve. We will present the exact equation that gives this curve in Chapter 7.

outlier (in a box-and-whiskers display): A measurement less than the lower limit or greater than the upper limit.

percentile: The value such that a specified percentage of the measurements in a population or sample fall at or below it.

point estimate: A one-number estimate for the value of a population parameter.

population mean (denoted μ): The average of a population of measurements.

population parameter: A descriptive measure of a population. It is calculated using the population measurements.

population standard deviation (denoted σ): The positive square root of the population variance. It is a measure of the variation of the population measurements.

population variance (denoted σ^2): The average of the squared deviations of the individual population measurements from the population mean. It is a measure of the variation of the population measurements.

range: The difference between the largest and smallest measurements in a population or sample. It is a simple measure of variation.

sample mean (denoted $\bar{x}$): The average of the measurements in a sample. It is the point estimate of the population mean.

sample size (denoted n): The number of measurements in a sample.

sample standard deviation (denoted s): The positive square root of the sample variance. It is the point estimate of the population standard deviation.

sample statistic: A descriptive measure of a sample. It is calculated from the measurements in the sample.

sample variance (denoted s^2): A measure of the variation of the sample measurements. It is the point estimate of the population variance.

text mining: The science of discovering knowledge, insights, and patterns from a collection of textual documents or databases.

third quartile (denoted Q_3): A value below which approximately 75 percent of the measurements lie; the 75th percentile.

tolerance interval: An interval of numbers that contains a specified percentage of the individual measurements in a population.

weighted mean: A mean where different measurements are given different weights based on their importance.

z-score (of a measurement): The number of standard deviations that a measurement is from the mean. This quantity indicates the relative location of a measurement within its distribution.

Important Formulas and Graphics

The population mean, μ: page 131

The sample mean, $\bar{x}$: page 132

The median: page 134

The mode: page 134

Supplementary Exercises

connect **3.70** In the book *Modern Statistical Quality Control and Improvement*, Nicholas R. Farnum presents data concerning the elapsed times from the completion of medical lab tests until the results are recorded on patients' charts. Table 3.18 gives the times it took (in hours) to deliver and chart the results of 84 lab tests over one week. Use the techniques of this and the previous chapter to determine if there are some deliveries with excessively long waiting times. Which deliveries might be investigated in order to discover reasons behind unusually long delays? **DS** LabTest

3.71 Figure 3.56 gives five-number summaries comparing the base yearly salaries of employees in marketing and employees in research for a large company. Interpret these summaries.

3.72 **THE INVESTMENT CASE** **DS** InvestRet

The Fall 1995 issue of *Investment Digest*, a publication of the Variable Annuity Life Insurance Company of Houston, Texas,

TABLE 3.18	Elapsed Time (in Hours) for Completing and Delivering Medical Lab Tests DS LabTest		
6.1	8.7	1.1	4.0
2.1	3.9	2.2	5.0
2.1	7.1	4.3	8.8
3.5	1.2	3.2	1.3
1.3	9.3	4.2	7.3
5.7	6.5	4.4	16.2
1.3	1.3	3.0	2.7
15.7	4.9	2.0	5.2
3.9	13.9	1.8	2.2
8.4	5.2	11.9	3.0
24.0	24.5	24.8	24.0
1.7	4.4	2.5	16.2
17.8	2.9	4.0	6.7
5.3	8.3	2.8	5.2
17.5	1.1	3.0	8.3
1.2	1.1	4.5	4.4
5.0	2.6	12.7	5.7
4.7	5.1	2.6	1.6
3.4	8.1	2.4	16.7
4.8	1.7	1.9	12.1
9.1	5.6	13.0	6.4

Source: Data from N. R. Farnum, *Modern Statistical Quality Control and Improvement*, p. 55. Brooks/Cole.

FIGURE 3.56 **Five-Number Summaries Comparing Base Salaries in Marketing and Base Salaries in Research**

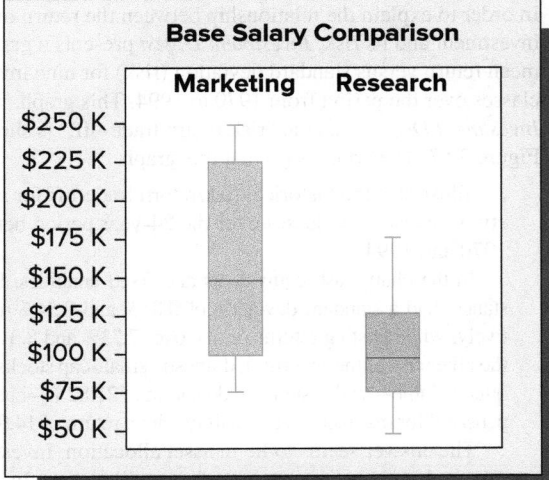

Source: http://nelsontouchconsulting.wordpress.com /2011/01/07/behold-the-box-plot/.

FIGURE 3.57   The Risk/Return Trade-Off (for Exercise 3.72)

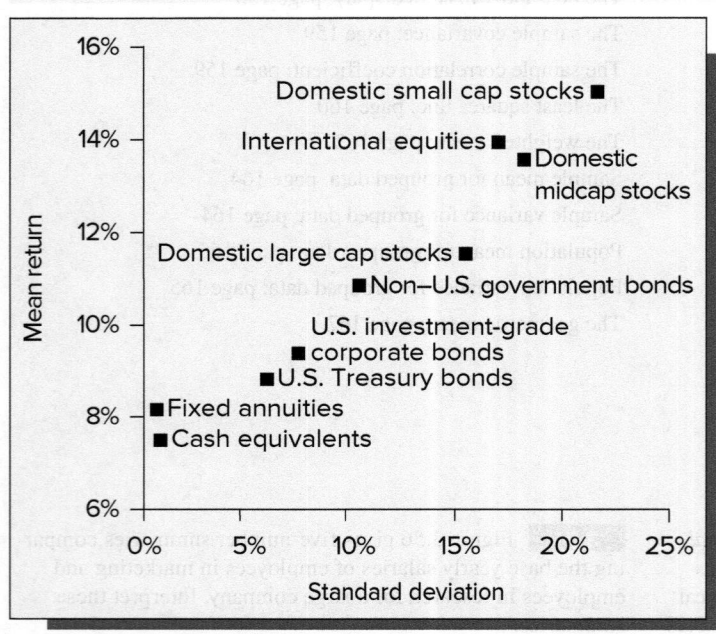

Source: Based on The Variable Annuity Life Insurance Company, VALIC 9 (1995), no. 3.

TABLE 3.19   **Mean Return and Standard Deviation for Nine Investment Classes (for Exercise 3.72)**
ⒹⓈ InvestRet

Investment Class	Mean Return	Standard Deviation
Fixed annuities	8.31%	.54%
Cash equivalents	7.73	.81
U.S. Treasury bonds	8.80	5.98
U.S. investment-grade corporate bonds	9.33	7.92
Non–U.S. government bonds	10.95	10.47
Domestic large cap stocks	11.71	15.30
International equities	14.02	17.16
Domestic midcap stocks	13.64	18.19
Domestic small cap stocks	14.93	21.82

discusses the importance of portfolio diversification for long-term investors. The article states:

> While it is true that investment experts generally advise long-term investors to invest in variable investments, they also agree that the key to any sound investment portfolio is diversification. That is, investing in a variety of investments with differing levels of historical return and risk.
>
> Investment risk is often measured in terms of the volatility of an investment over time. When volatility, sometimes referred to as *standard deviation,* increases, so too does the level of return. Conversely, as risk (standard deviation) declines, so too do returns.

In order to explain the relationship between the return on an investment and its risk, *Investment Digest* presents a graph of mean return versus standard deviation (risk) for nine investment classes over the period from 1970 to 1994. This graph, which *Investment Digest* calls the "risk/return trade-off," is shown in Figure 3.57. The article says that this graph

> . . . illustrates the historical risk/return trade-off for a variety of investment classes over the 24-year period between 1970 and 1994.
>
> In the chart, cash equivalents and fixed annuities, for instance, had a standard deviation of 0.81% and 0.54% respectively, while posting returns of just over 7.73% and 8.31%. At the other end of the spectrum, domestic small-cap stocks were quite volatile—with a standard deviation of 21.82%—but compensated for that increased volatility with a return of 14.93%.
>
> The answer seems to lie in asset allocation. Investment experts know the importance of asset allocation. In a nutshell, asset allocation is a method of creating a diversified portfolio of investments that minimize historical risk and maximize potential returns to help you meet your retirement goals and needs.

Suppose that, by reading off the graph of Figure 3.57, we obtain the mean return and standard deviation combinations for the various investment classes as shown in Table 3.19.

Further suppose that future returns in each investment class will behave as they have from 1970 to 1994. That is, for each investment class, regard the mean return and standard deviation in Table 3.19 as the population mean and the population standard deviation of all possible future returns. Then do the following:

a   Assuming that future returns for the various investment classes are mound-shaped, for each investment class compute intervals that will contain approximately 68.26 percent and 99.73 percent of all future returns.

b   Making no assumptions about the population shapes of future returns, for each investment class compute intervals that will contain at least 75 percent and at least 88.89 percent of all future returns.

c   Assuming that future returns are mound-shaped, find

   (1) An estimate of the maximum return that might be realized for each investment class.

   (2) An estimate of the minimum return (or maximum loss) that might be realized for each investment class.

d   Assuming that future returns are mound-shaped, which two investment classes have the highest estimated maximum returns? What are the estimated minimum returns (maximum losses) for these investment classes?

e   Assuming that future returns are mound-shaped, which two investment classes have the smallest estimated maximum returns? What are the estimated minimum returns for these investment classes?

3.73   Table 3.20 gives data concerning America's 30 largest private companies of 2014 as rated by *Forbes* magazine.

a   Construct a box-and-whiskers display of the large company revenues.

b   Construct a box-and-whiskers display of the large company numbers of employees.

c   Interpret the displays of parts *a* and *b*.   ⒹⓈ LargeComp

TABLE 3.20   America's 30 Largest Private Companies of 2014 as Rated by *Forbes* Magazine DS LargeComp

Rank	Company	Revenue ($bil)	Employees	Rank	Company	Revenue ($bil)	Employees
1	Cargill	134.90	143,000	16	Enterprise Holdings	17.80	83,369
2	Koch Industries	115.00	100,000	17	Cumberland Gulf Group	17.00^e	7,200
3	Dell	57.20^e	111,300	18	Cox Enterprises	15.90	50,000
4	Bechtel	39.40	53,000	19	Meijer	15.00^e	74,000
5	PricewaterhouseCoopers	34.00	195,000	20	Performance Food Group	13.78^2	12,000
6	Mars	33.00	75,000	21	Fidelity Investments	13.63	41,000
7	Pilot Flying J	32.09	21,000	22	Toys "R" Us	12.54	67,000
8	Publix Super Markets	28.92	166,000	23	JM Family Enterprises	12.50	4,000
9	Ernst & Young	27.40	190,000	24	Platinum Equity	12.00^e	50,000
10	C&S Wholesale Grocers	25.00	14,000	25	Kiewit Corporation	11.83	33,200
11	Albertsons	23.00^e	127,000	26	Amway	11.80	21,000
12	Reyes Holdings	23.00	16,100	27	QuikTrip	11.45	15,126
13	Love's Travel Stops & Country Stores	22.65	10,500	28	Southern Wine & Spirits	11.40	14,000
				29	Trammo	11.32	455
14	US Foods	22.20^e	25,000	30	HJ Heinz	11.20^e	30,000
15	HE Butt Grocery	20.00^e	76,000				

e = *Forbes* estimate
2 = Company-provided estimate

Source: http://www.forbes.com/largest-private-companies/list/ (accessed January 14, 2015).

3.74   THE FLORIDA POOL HOME CASE DS PoolHome

In Florida, real estate agents refer to homes having a swimming pool as *pool homes*. In this case, Sunshine Pools Inc. markets and installs pools throughout the state of Florida. The company wishes to estimate the percentage of a pool's cost that can be recouped by a buyer when he or she sells the home. For instance, if a homeowner buys a pool for which the current purchase price is $30,000 and then sells the home in the current real estate market for $20,000 more than the homeowner would get if the home did not have a pool, the homeowner has recouped (20,000/30,000) × 100% = 66.67% of the pool's cost. To make this estimate, the company randomly selects 80 homes from all of the homes sold in a Florida city (over the last six months) having a size between 2,000 and 3,500 square feet. For each sampled home, the following data are collected: selling

price (in thousands of dollars); square footage; the number of bathrooms; a niceness rating (expressed as an integer from 1 to 7 and assigned by a real estate agent); and whether or not the home has a pool (1 = yes, 0 = no). The data are given on the next page in Table 3.21. Figure 3.58 on the next page gives descriptive statistics for the 43 homes having a pool and for the 37 homes that do not have a pool.

a   Using Figure 3.58, compare the mean selling prices of the homes having a pool and the homes that do not have a pool.

b   Using these data, and assuming that the average current purchase price of the pools in the sample is $32,500, estimate the percentage of a pool's cost that can be recouped when the home is sold.

c   The comparison you made in part *a* could be misleading. Noting that different homes have different square footages, numbers of bathrooms, and niceness ratings, explain why.

3.75   Internet Exercise

The Data and Story Library (DASL) houses a rich collection of data sets useful for teaching and learning statistics, from a variety of sources, contributed primarily by university faculty members. DASL can be reached through the BSC by clicking on the Data Bases button in the BSC home screen and by then clicking on the Data and Story Library link. The DASL can also be reached directly using the url http://lib.stat.cmu.edu/DASL/. The objective of this exercise is to retrieve a data set of chief executive officer salaries and to construct selected graphical and numerical statistical summaries of the data.

a   From the McGraw-Hill/Irwin Business Statistics Center Data Bases page, go to the DASL website and select "List all topics." From the Stories by Topic page, select Economics, then CEO Salaries to reach the CEO Salaries story. From the CEO Salaries story page, select the Datafile Name: CEO Salaries to reach the data set page. The data set includes the ages and salaries (save for a single missing observation) for a sample of 60 CEOs. Capture these observations and copy them into an Excel or Minitab worksheet. This data capture can be accomplished in a number of ways. One simple approach is to use simple copy and paste procedures from the DASL data set to Excel or Minitab. DS CEOSal

b   Use your choice of statistical software to create graphical and numerical summaries of the CEO Salaries data and use these summaries to describe the data. In Excel, create a histogram of salaries and generate descriptive statistics. In Minitab, create a histogram, stem-and-leaf display, box plot, and descriptive statistics. Offer your observations about typical salary level, the variation in salaries, and the shape of the distribution of CEO salaries.

TABLE 3.21   The Florida Pool Home Data (for Exercise 3.74)   DS PoolHome

Home	Price ($1000s)	Size (Sq Feet)	Number of Bathrooms	Niceness Rating	Pool? yes=1; no=0	Home	Price ($1000s)	Size (Sq Feet)	Number of Bathrooms	Niceness Rating	Pool? yes=1; no=0
1	260.9	2666	2.5	7	0	41	285.6	2761	3.0	6	1
2	337.3	3418	3.5	6	1	42	216.1	2880	2.5	2	0
3	268.4	2945	2.0	5	1	43	261.3	3426	3.0	1	1
4	242.2	2942	2.5	3	1	44	236.4	2895	2.5	2	1
5	255.2	2798	3.0	3	1	45	267.5	2726	3.0	7	0
6	205.7	2210	2.5	2	0	46	220.2	2930	2.5	2	0
7	249.5	2209	2.0	7	0	47	300.1	3013	2.5	6	1
8	193.6	2465	2.5	1	0	48	260.0	2675	2.0	6	0
9	242.7	2955	2.0	4	1	49	277.5	2874	3.5	6	1
10	244.5	2722	2.5	5	0	50	274.9	2765	2.5	4	1
11	184.2	2590	2.5	1	0	51	259.8	3020	3.5	2	1
12	325.7	3138	3.5	7	1	52	235.0	2887	2.5	1	1
13	266.1	2713	2.0	7	0	53	191.4	2032	2.0	3	0
14	166.0	2284	2.5	2	0	54	228.5	2698	2.5	4	0
15	330.7	3140	3.5	6	1	55	266.6	2847	3.0	2	1
16	289.1	3205	2.5	3	1	56	233.0	2639	3.0	3	0
17	268.8	2721	2.5	6	1	57	343.4	3431	4.0	5	1
18	276.7	3245	2.5	2	1	58	334.0	3485	3.5	5	1
19	222.4	2464	3.0	3	1	59	289.7	2991	2.5	6	1
20	241.5	2993	2.5	1	0	60	228.4	2482	2.5	2	0
21	307.9	2647	3.5	6	1	61	233.4	2712	2.5	1	1
22	223.5	2670	2.5	4	0	62	275.7	3103	2.5	2	1
23	231.1	2895	2.5	3	0	63	290.8	3124	2.5	3	1
24	216.5	2643	2.5	3	0	64	230.8	2906	2.5	2	0
25	205.5	2915	2.0	1	0	65	310.1	3398	4.0	4	1
26	258.3	2800	3.5	2	1	66	247.9	3028	3.0	4	0
27	227.6	2557	2.5	3	1	67	249.9	2761	2.0	5	0
28	255.4	2805	2.0	3	1	68	220.5	2842	3.0	3	0
29	235.7	2878	2.5	4	0	69	226.2	2666	2.5	6	0
30	285.1	2795	3.0	7	1	70	313.7	2744	2.5	7	1
31	284.8	2748	2.5	7	1	71	210.1	2508	2.5	4	0
32	193.7	2256	2.5	2	0	72	244.9	2480	2.5	5	0
33	247.5	2659	2.5	2	1	73	235.8	2986	2.5	4	0
34	274.8	3241	3.5	4	1	74	263.2	2753	2.5	7	1
35	264.4	3166	3.0	3	1	75	280.2	2522	2.5	6	1
36	204.1	2466	2.0	4	0	76	290.8	2808	2.5	7	1
37	273.9	2945	2.5	5	1	77	235.4	2616	2.5	3	0
38	238.5	2727	3.0	1	1	78	190.3	2603	2.5	2	0
39	274.4	3141	4.0	4	1	79	234.4	2804	2.5	4	0
40	259.6	2552	2.0	7	1	80	238.7	2851	2.5	5	0

FIGURE 3.58   Descriptive Statistics for Homes with and without Pools (for Exercise 3.74)

Descriptive Statistics (Homes with Pools)	Price	Descriptive Statistics (Homes without Pools)	Price
count	43	count	37
mean	276.056	mean	226.900
sample variance	937.821	sample variance	609.902
sample standard deviation	30.624	sample standard deviation	24.696
minimum	222.4	minimum	166
maximum	343.4	maximum	267.5
range	121	range	101.5

Appendix 3.1 ■ Numerical Descriptive Statistics Using Excel

The instructions in this section begin by describing the entry of data into an Excel worksheet. Alternatively, the data may be downloaded from this book's website. The appropriate data file name is given at the top of each instruction block. Please refer to Appendix 1.1 for further information about entering data, saving data, and printing results when using Excel.

Numerical descriptive statistics for the bottle design ratings as in Figure 3.4 (data file: Design.xlsx):

- Enter the bottle design ratings data into column A with the label Rating in cell A1 and with the 60 design ratings from Table 1.6 in cells A2 to A61.
- Select **Data : Data Analysis : Descriptive Statistics.**
- Click OK in the Data Analysis dialog box.
- In the Descriptive Statistics dialog box, enter the range for the data, A1:A61, into the "Input Range" window.
- Check the "Labels in first row" checkbox.
- Click in the "Output Range" window and enter the desired cell location for the upper left corner of the output, say cell C1.
- Check the "Summary statistics" checkbox.
- Click OK in the Descriptive Statistics dialog box.
- The descriptive statistics summary will appear in cells C1:D15. Drag the column C border to reveal complete labels for all statistics.

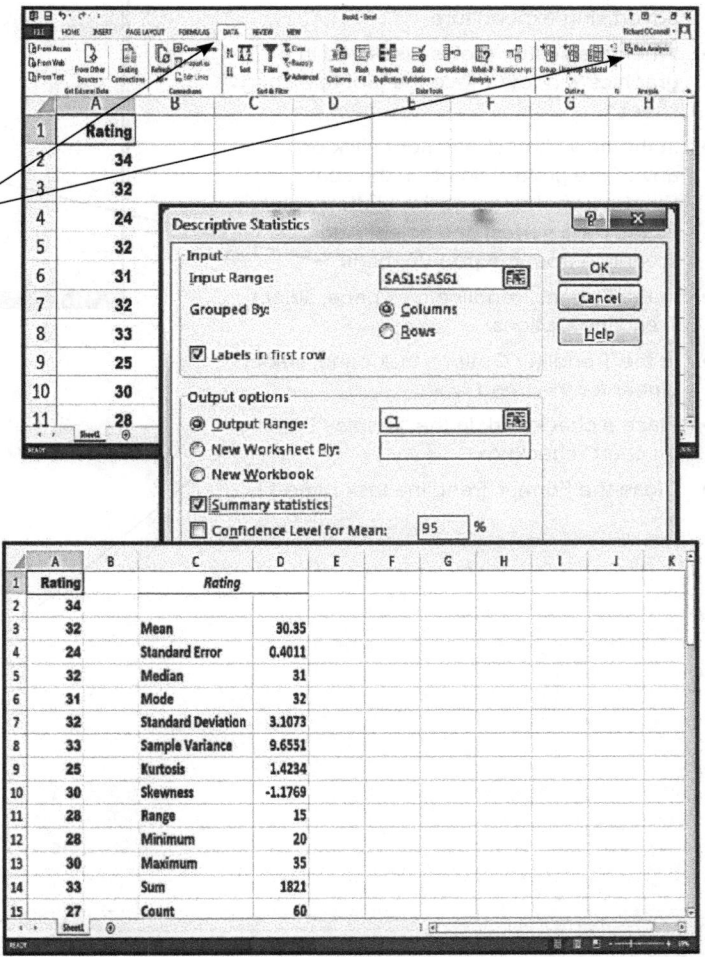

Source: Microsoft Office Excel 2016

Least squares line, correlation, and covariance for the sales volume data in Figure 3.22(a) (datafile: SalesPlot.xlsx):

To compute the equation of the least squares line:

- Follow the directions in Appendix 2.1 for constructing a scatter plot of sales volume versus advertising expenditure.

- When the scatter plot is displayed in a graphics window, move the plot to a chart sheet.

- In the new chart sheet, right-click on any of the plotted points in the scatter plot (Excel refers to the plotted points as the **data series**) and select Add Trendline from the pop-up menu.

- In the Format Trendline task pane, select Trendline Options.

- In the Trendline Options task pane, select Linear for the trend type.

- Place a checkmark in the "Display Equation on chart" checkbox.

- Close the Format Trendline task pane.

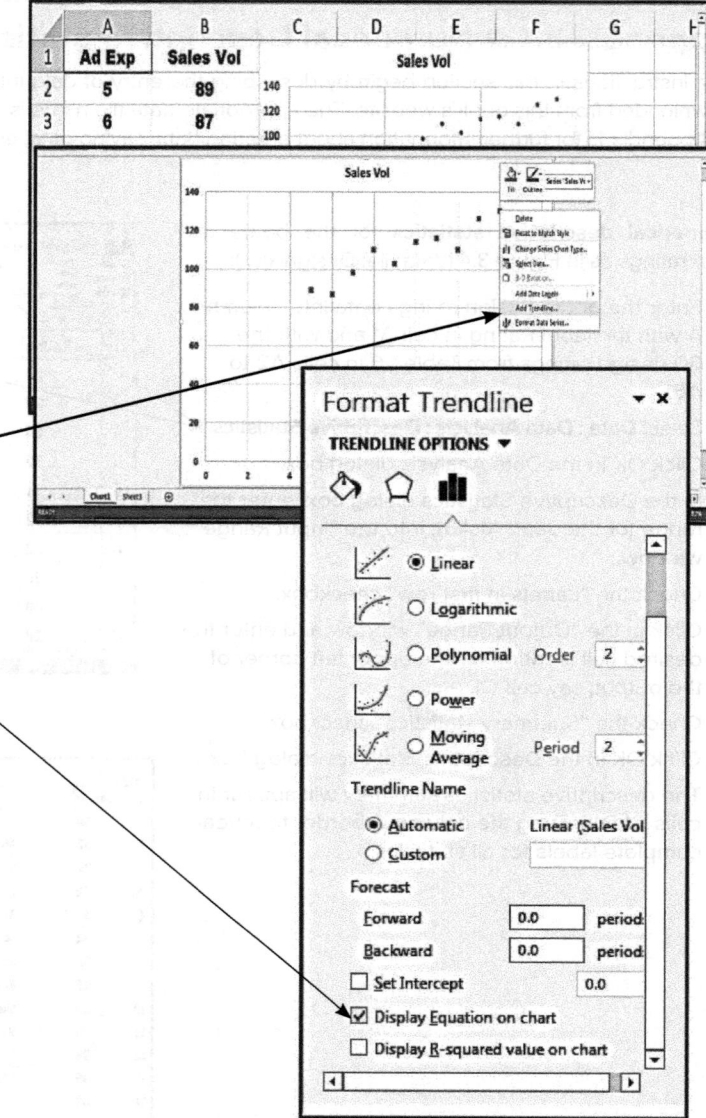

Source: Microsoft Office Excel 2016

- The Trendline equation will be displayed in the scatter plot and the chart can then be edited appropriately.

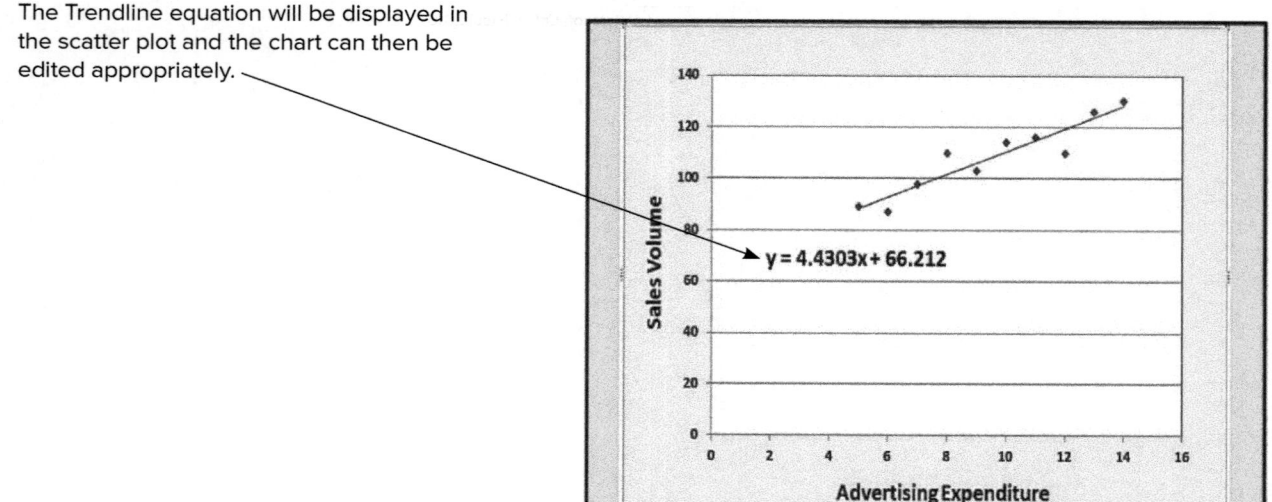

$y = 4.4303x + 66.212$

Source: Microsoft Office Excel 2016

To compute the sample covariance between sales volume and advertising expenditure:

- Enter the advertising and sales data in Figure 3.22(a) into columns A and B—advertising expenditures in column A with label "Ad Exp" and sales values in column B with label "Sales Vol."

- Select **Data : Data Analysis : Covariance.**

- Click OK in the Data Analysis dialog box.

- In the Covariance dialog box, enter the range of the data, A1:B11 into the Input Range window.

- Select "Grouped By: Columns" if this is not already the selection.

- Place a checkmark in the "Labels in first row" checkbox.

- Under "Output options," select Output Range and enter the cell location for the upper left corner of the output, say A13, in the Output Range window.

- Click OK in the Covariance dialog box.

The Excel ToolPak Covariance routine calculates the population covariance. This quantity is the value in cell B15 (=36.55). To compute the sample covariance from this value, we will multiply by $n/(n-1)$ where n is the sample size. In this situation, the sample size equals 10. Therefore, we can compute the sample covariance as follows:

- Type the label "Sample Covariance" in cell E14.

- In cell E15 write the cell formula = (10/9)*B15 and type enter.

- The sample covariance (=40.61111) is the result in cell E15.

To compute the sample correlation coefficient between sales volume and advertising expenditure:

- Select **Data : Data Analysis : Correlation.**

- In the correlation dialog box, enter the range of the data, A1:B11 into the Input Range window.

- Select "Grouped By: Columns" if this is not already the selection.

- Place a checkmark in the "Labels in first row" checkbox.

- Under Output options, select "New Worksheet Ply" to have the output placed in a new work- sheet and enter the name Output for the new worksheet.

- Click OK in the Correlation dialog box.

- The sample correlation coefficient (=0.93757) is displayed in the Output worksheet.

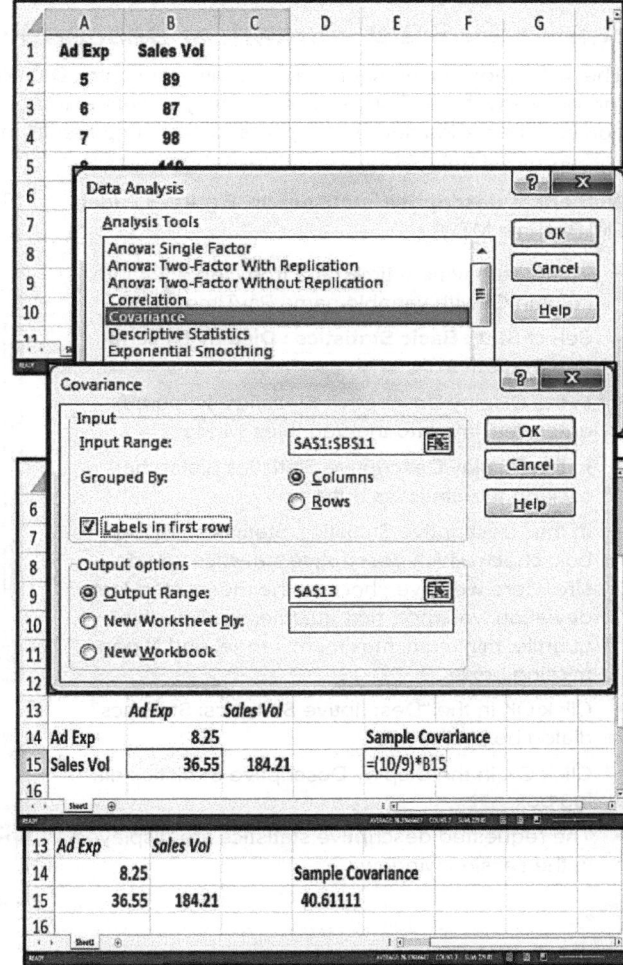

Source: Microsoft Office Excel 2016

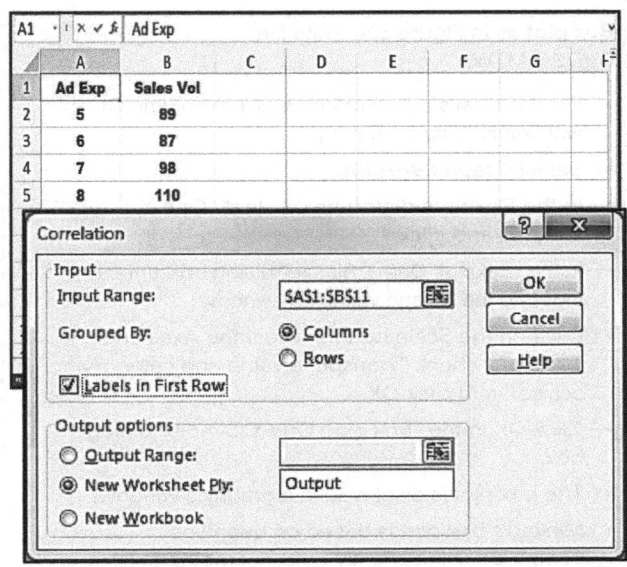

Source: Microsoft Office Excel 2016

Source: Microsoft Office Excel 2016

Appendix 3.2 ■ Numerical Descriptive Statistics Using Minitab

The instructions in this section begin with data entry. Data can be typed into the Minitab data window or downloaded from this book's website. Data file names are given at the top of each instruction block. Refer to Appendix 1.2 for further information about entering data, saving data, and printing results in Minitab.

Numerical descriptive statistics in Figure 3.7 (data file: PayTime.MTW)

• Enter the payment time data from Table 2.4 into column C1 with variable name PayTime.

• Select **Stat : Basic Statistics : Display Descriptive Statistics...**

• In the Display Descriptive Statistics dialog box, enter PayTime into the Variables window.

• In the Display Descriptive Statistics dialog box, click on the Statistics button.

• In the "Descriptive Statistics: Statistics" dialog box, check which descriptive statistics you desire. Here we have checked the mean, standard deviation, variance, first quartile, median, third quartile, minimum, maximum, range, and N non-missing boxes.

• Click OK in the "Descriptive Statistics: Statistics" dialog box.

• Click OK in the Display Descriptive Statistics dialog box.

• The requested descriptive statistics are displayed in the session window.

Source: Minitab 18

Box plot as in Figure 3.18 (data file: DVDSat.MTW):

• Enter the satisfaction rating data into column C1 with variable name Ratings.

• Select **Graph : Boxplot.**

• In the Boxplots dialog box, select "One Y, Simple" and click OK.

• In the "Boxplot: One Y, Simple" dialog box, enter Ratings into the "Graph variables" window.

• Click on the Scale button, select the Axes and Ticks tab, check "Transpose value and category scales," and click OK.

• Click OK in the "Box plot: One Y, Simple" dialog box.

• The box plot is displayed in a graphics window.

• Minitab's box plot is based on quantiles computed using slightly different methods from those presented in Section 3.3 of this book. Consult the Minitab help menu for a precise description of the methods it uses.

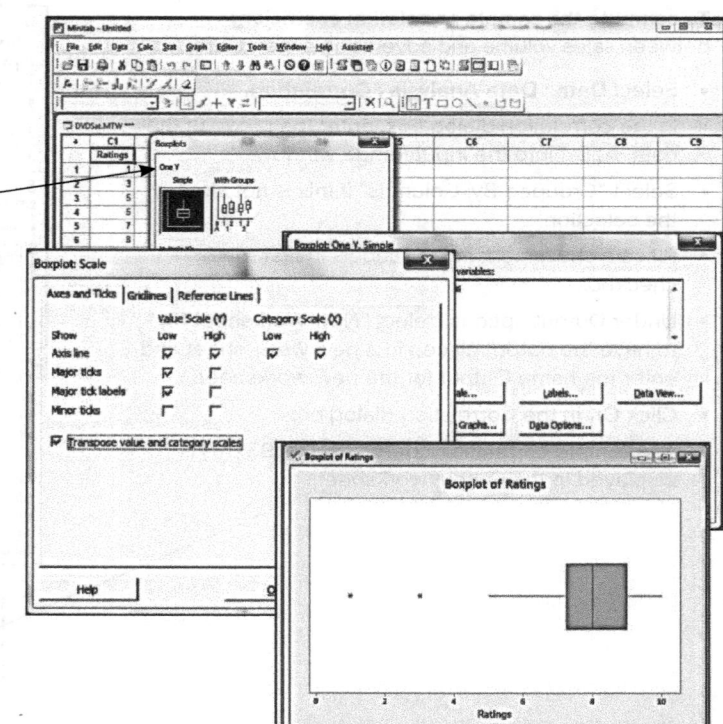

Source: Minitab 18

Least squares line, correlation, and covariance for the sales volume data as in Section 3.4 (data file: SalesPlot.MTW):

To compute the equation of the least squares line:

- Enter the sales and advertising data in Figure 3.22(a) with Sales Region in column C1, Adv Exp in C2, and Sales Vol in C3.
- Select **Stat : Regression : Fitted Line Plot.**
- In the Fitted Line Plot dialog box, enter 'Sales Vol' (including the single quotes) into the "Response (Y)" window.
- Enter 'Adv Exp' (including the single quotes) into the "Predictor (X)" window.
- Select Linear for the "Type of Regression Model."
- Click OK in the Fitted Line Plot dialog box.

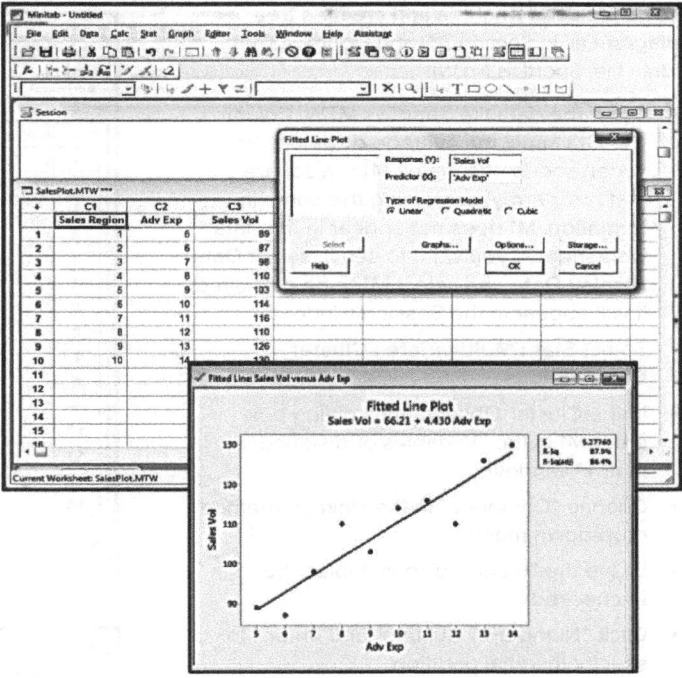

Source: Minitab 18

To compute the sample correlation coefficient:

- **Select Stat : Basic Statistics : Correlation.**
- In the Correlation dialog box, enter 'Adv Exp' and 'Sales Vol' (including the single quotes) into the Variables window.
- Choose "Pearson correlation" in the Method drop-down menu.
- Remove the check from the "Display p-values" box (or keep it checked if desired; we will learn about p-values in later chapters).
- Click OK in the Correlation dialog box.
- The correlation coefficient will be displayed in the session window.

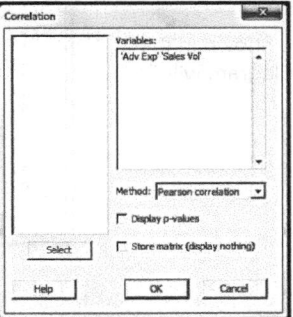
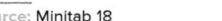

Source: Minitab 18

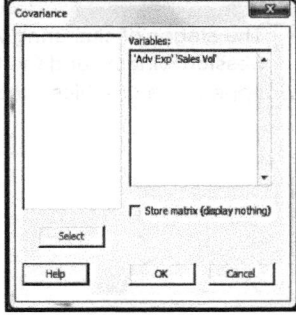

Source: Minitab 18

To compute the sample covariance:

- Select **Stat : Basic Statistics : Covariance.**
- In the Covariance dialog box, enter 'Adv Exp' and 'Sales Vol' (including the single quotes) into the Variables window.
- Click OK in the Covariance dialog box.
- The covariance will be displayed in the session window. [The covariance between advertising expenditure and sales volume is 40.61111; the other two numbers displayed are the sample variances of advertising expenditure (9.16667) and of sales volume (204.67778).]

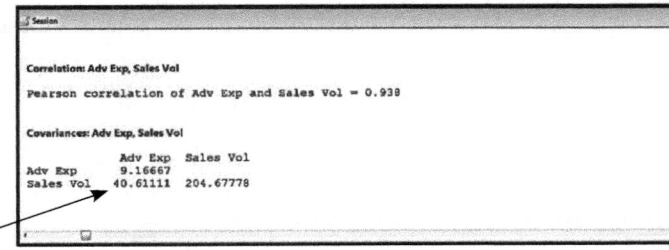

Source: Minitab 18

To do a cluster analysis and create a tree diagram as in Section 3.9 and Figure 3.37(a) (data file: SportDis.MTW):

- Open the data file SportDis.MTW. Columns C1—C13 show the average distances between sports; the matrix M1 is a square matrix (or array) containing this same information. M1 does not appear in the data worksheet. If you want to see it, select **Data: Display Data** and select M1 to be displayed. It will appear in the Session window.

- Select **Stat : Multivariate : Cluster Observations.**

- In the Cluster Observations dialog box, enter M1 in the "Variables or distance matrix" window.

- Choose "Complete" in the Linkage method drop-down menu.

- Leave the "Standardize variables" box unchecked.

- Click "Number of clusters" and enter 1 to specify the final partition.

- Check the box to request the dendrogram (tree) be given.

- Click OK in the Cluster Observations dialog box.

- The stages of clustering will be shown in the Session window and the tree diagram will appear in a graphics window.

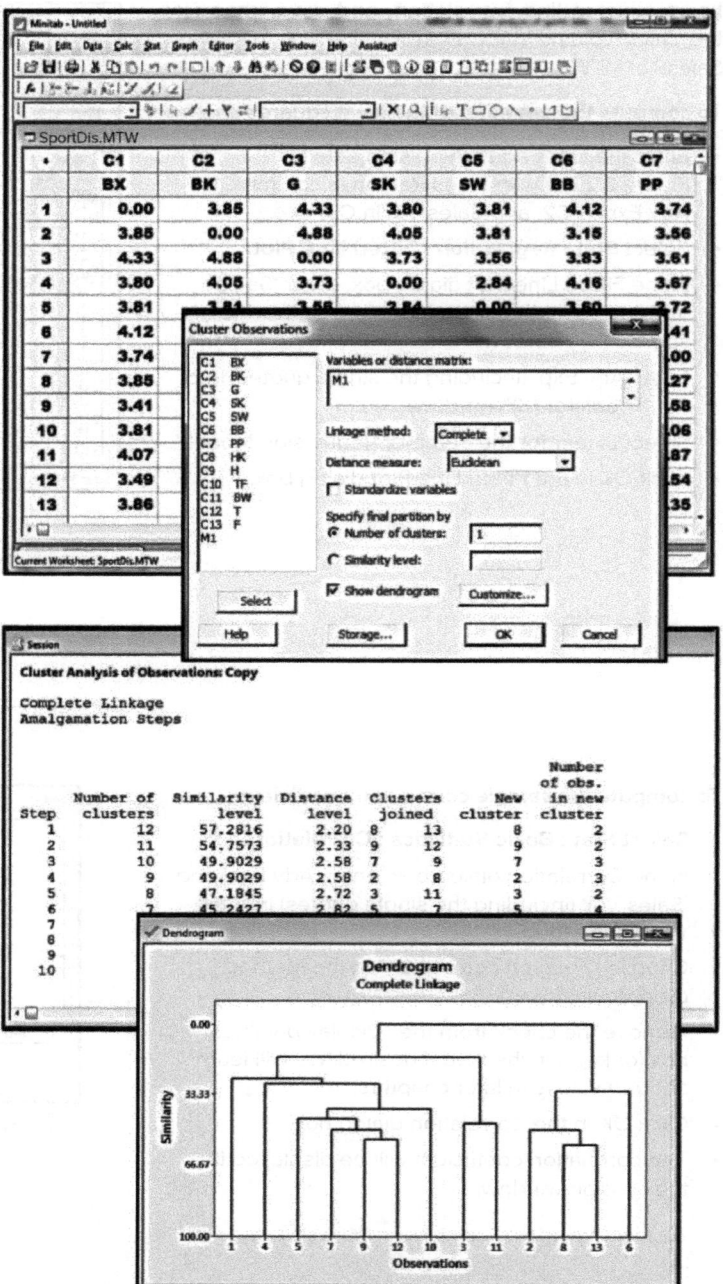

Source: Minitab 18

Perform a factor analysis of the applicant data as in Figure 3.48 (data file: Applicant.MTW):

- Open the file Applicant.MTW stored on this book's website.
- The data window contains columns of pairwise correlations, as in Table 3.15. The worksheet also contains M1, a matrix version of these 15 columns, but it is not visible. You can request to see it by selecting **Data: Display Data** and then entering M1 in the "Columns, constants, and matrices to display" window.
- Select **Stat : Multivariate : Factor Analysis.**
- In the Factor Analysis dialog box, leave the "Variables" window blank, enter 7 as the number of factors to extract, select Principal components as the "Method of Extraction," and select Varimax as the "Type of Rotation."

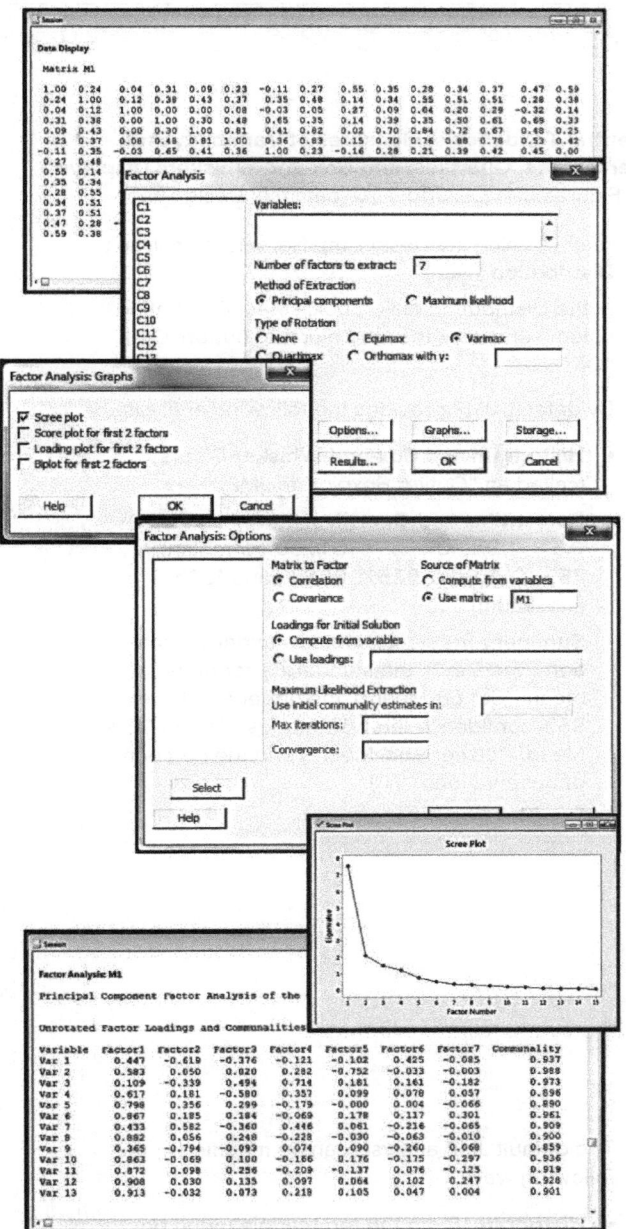

- Click on the Graphs . . . button in the Factor Analysis dialog box and check "Scree plot" in the "Factor Analysis: Graphs" dialog box. Click OK.
- Click on the Options . . . button in the Factor Analysis dialog box.
- In the "Factor Analysis: Options" dialog box, choose "Correlation" as the Matrix, to Factor; enter M1 in the "Use matrix" window as the Source of Matrix, and choose "Compute from variables" as the Loadings for Initial Solution.
- Click OK in the "Factor Analysis: Options" dialog box and also in the Factor Analysis dialog box.
- The Scree plot will appear in a graphics window and the factor analysis will be in the Session window.

Source: Minitab 18

Appendix 3.3 ■ Descriptive Statistics and Analytics: Numerical Methods Using JMP

Central Tendencies, Measures of Variability and Percentiles, Quartiles, and Box-and-Whiskers Displays in Figures 3.4, 3.5, **and** 3.20 (data file: Design.jmp)

- After opening the Design.jmp file, select **Analyze : Distribution**.

- In the Distribution dialog box, select the "Rating" column and click the "Y, Columns" button. Click "OK."

- By default, JMP provides the following analysis.

  - **Histogram** and **Box-and-Whisker Display** (called an "Outlier Boxplot" in JMP).

  - **Percentiles** and **Quartiles**: 0% (minimum), 0.5%, 2.5%, 10%, 25% (Q1), 50% (median), 75% (Q3), 90%, 97.5%, 99.5%, and 100% (maximum).

  - **Summary Statistics**: Mean, standard deviation ("Std Dev"), the standard error of the mean ("Std Err Mean"), the upper and lower 95% confidence interval bounds ("Upper 95% Mean," "Lower 95% Mean") and the number of observations ("N").

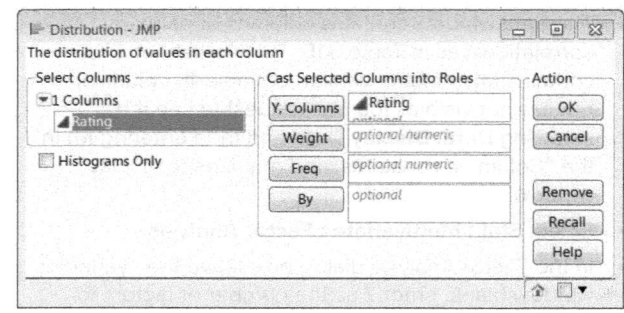

Source: JMP Pro 14

Quantiles

100.0%	maximum	35
99.5%		35
97.5%		34.475
90.0%		33.9
75.0%	quartile	33
50.0%	median	31
25.0%	quartile	29
10.0%		26
2.5%		21.05
0.5%		20
0.0%	minimum	20

Summary Statistics

Mean	30.35
Std Dev	3.1072632
Std Err Mean	0.401146
Upper 95% Mean	31.152691
Lower 95% Mean	29.547309
N	60

Source: JMP Pro 14

- The default JMP analysis can be modified in the following ways:

  - **Histogram**: From the red triangle menu beside "Rating," select "Histogram Options". To change from a vertical to horizontal presentation, unselect "Vertical." Other common options are to add a "Count Axis" to the histogram, "Show Counts" or "Show Percents" for each bar in the histogram, or to change the "Histogram Color" from the default JMP color.

  - **Box-and-Whisker Display**: The default outlier boxplot in JMP includes extra information (the diamond and the red interval). To remove, right-click anywhere in the boxplot and select "Customize" from the menu. In the Customize Graph dialog box, click "Box Plot" on the left side of the window. Unselect "Confidence Diamond" and "Shortest Half." Click "OK."

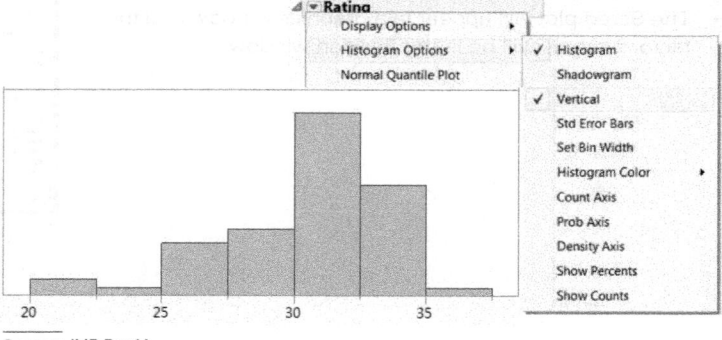

Source: JMP Pro 14

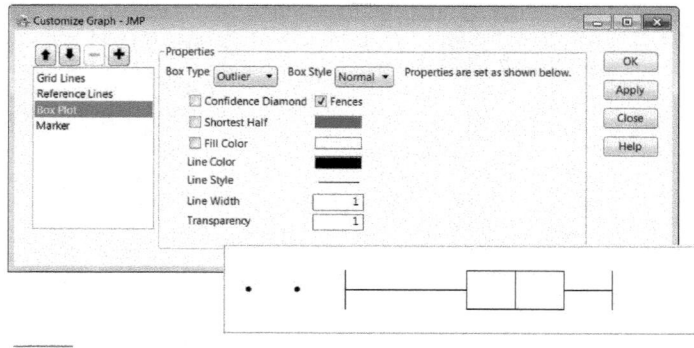

Source: JMP Pro 14

- **Summary Statistics**: From the red triangle menu beside "Summary Statistics," select "Customize Summary Statistics." In the Customize Summary Statistics dialog box, select the set of desired statistics: "Mean, Std Dev, N, Sum, Variance, Minimum, Maximum, Median, Mode, Range, Interquartile Range." Click "OK."

Summary Statistics

Customize Summary Statistics	
Show All Modes	
Upper 95% Mean	31.152691
Lower 95% Mean	29.547309
N	60

Summary Statistics

Mean	30.35
Std Dev	3.1072632
Std Err Mean	0.401146
N	60
Sum	1821
Variance	9.6550847
Minimum	20
Maximum	35
Median	31
Mode	32
Range	15
Interquartile Range	4

Source: JMP Pro 14

Covariance, Correlation, and the Least Squares Line in Figure 3.24 (data file: SalesPlot.jmp):

- After opening the SalesPlot.jmp file, select **Analyze : Fit Y by X**.

- In the Fit Y by X dialog box, select the "Sales Volume" column and click "Y, Response" and select the "Advertising Expenditure" and click "X, Factor." Click "OK."

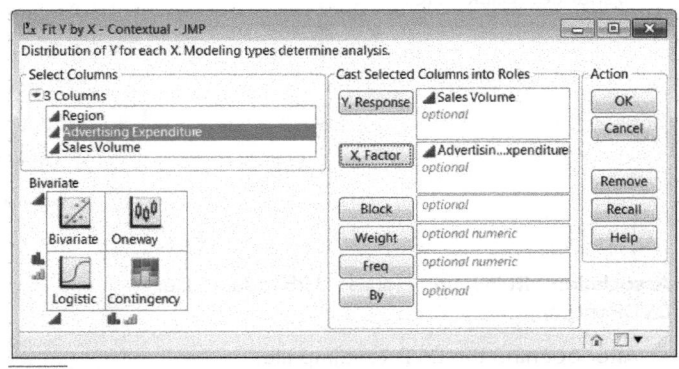

Source: JMP Pro 14

- By default, JMP will generate a **Scatterplot**. Additional analyses can be added.

  - **Sample Covariance and Sample Correlation Coefficient**: From the red triangle menu beside "Bivariate Fit of Sales Volume by Advertising Expenditure," select "Summary Statistics."

  - **Least Squares Line**: From the red triangle menu beside "Bivariate Fit of Sales Volume by Advertising Expenditure," select "Fit Line." The least squares line is added to the scatterplot. The equation of the least squares line—the predicted value of Sales Volume from Advertising Expenditure— appears in the output. The equation includes the y-intercept b_0 (66.2121) and the slope b_1 (4.4303). These values can vary slightly depending on the use of rounding in the calculations.

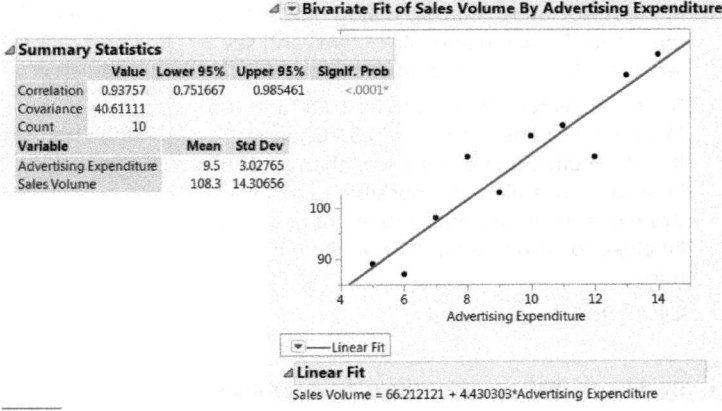

Source: JMP Pro 14

Weighted Means and Grouped Data in Tables 3.6 and 3.7 (data file: SatRatings.jmp):

- After opening the SatRatings.jmp file, select **Analyze : Distribution**.

- In the Distribution dialog box, select the "Class Midpoint" column and click "Y, Columns" and select the "Frequency" column and click "Freq." Click "OK."

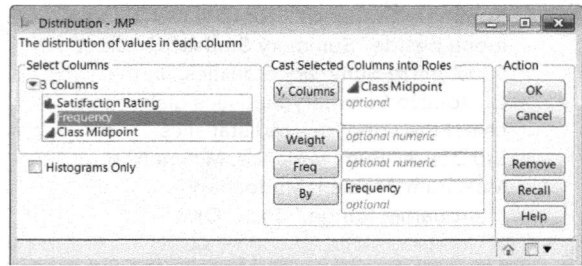

- **Sample Mean and Variance for Grouped Data**: From the red triangle menu beside "Summary Statistics," select "Customize Summary Statistics." In the Customize Summary Statistics dialog box, select the desired statistics "Mean," "Std Dev," "N," "Sum," and "Variance." Click "OK."

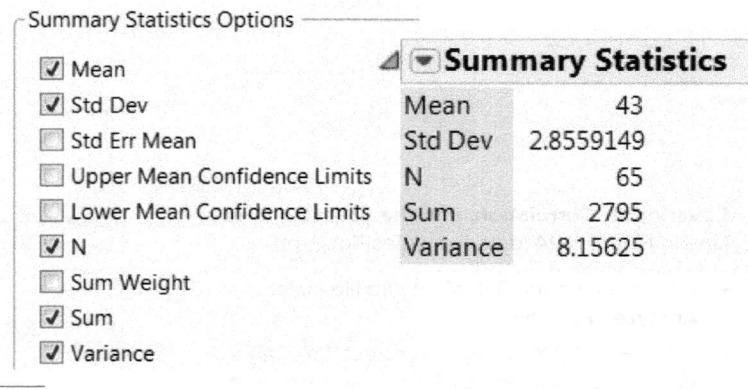

Source: JMP Pro 14

Association Rules in Figure 3.25 (data file: DVDRent.jmp):

- After opening the DVDRent.jmp file, select **Analyze : Screening : Association Analysis**.

- In the Association Analysis dialog box, select the "Movie" column and click "Item" and select the "Customer" column and click "ID."

- Set the Minimum Support to be "0.5" and the Minimum Confidence to be "0.8." Change the Minimum Lift to a value less than "0.926" in order to see all of the possible rules in this example. In general, you might use a lift close to zero in order to see all possible rules.

- Click "OK."

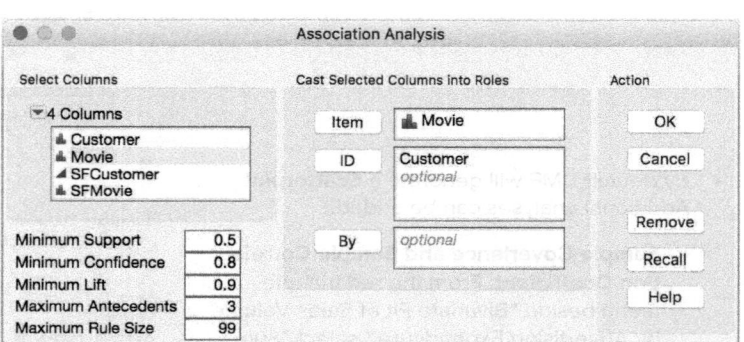

- By default, JMP provides the following tables of results:

  - **Frequent Item Sets**: the list of item sets whose support (proportion of occurrences) exceeds the minimum value specified in the launch window.

  - **Rules**: Association rules that meet the minimum support, minimum confidence, minimum lift, maximum antecedents, and maximum rule size requirements specified in the launch dialog.

- Optional: To see the list of items selected for each customer, from the red triangle menu beside "Association Analysis," select "Transaction Listing."

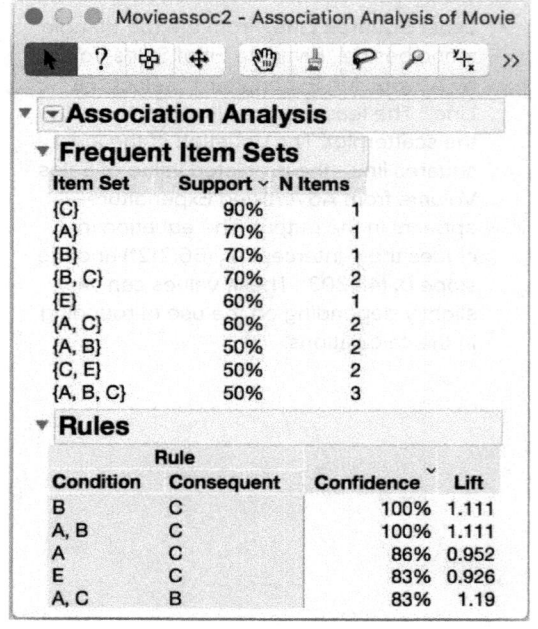

Source: JMP Pro 14

Text Mining in Figures 3.27 and 3.28 (data file: FDACit.jmp):

- After opening the FDACit.jmp file, select **Analyze : Text Explorer**.
- In the Text Explorer dialog box, select the "Citation Description" column and click "Text Columns." Change the "Maximum Words per Phrase," "Maximum Number of Phrases," and other options if desired.
- Click "OK."

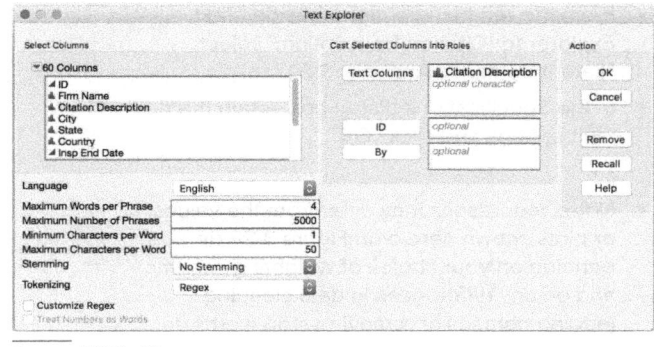

Source: JMP Pro 14

- The complete term and phrase lists for the data is shown to the right (Figure 3.27 gives a partial list). Many text preparation options, including stemming and parsing, are available from the red triangle menu beside "Text Explorer for Citation Description." Additional options are available from the "Term and Phrase Lists."

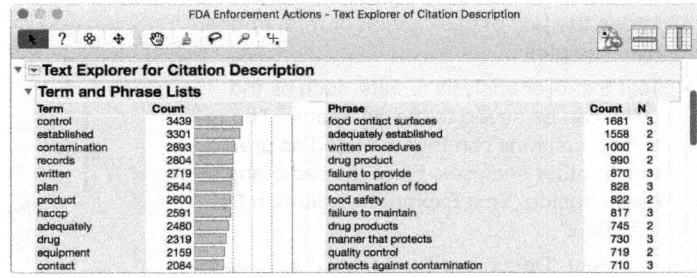

Source: JMP Pro 14

- To remove terms or phrases, select the terms (or phrases), right-click, and select "Add Stop Word."
- To add phrases to the term list, select the phrases, right-click, and select "Add Phrase."
- To combine terms or change the values, select the terms, right-click, and select "Recode."

- From the red triangle menu beside "Text Explorer for Citation Description," select **Display Options : Show Word Cloud**.
- From the red triangle menu next to "Word Cloud," select an alternate layout, such as "Centered," or choose a Coloring, such as "Arbitrary Colors" (as shown to the right).
- Right-click on a term in the Term or Phrase Lists or on a word in the word cloud and select Show Text to see the citations that include this term.

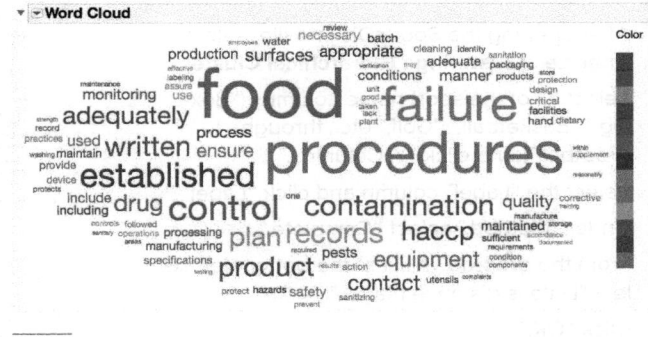

Source: JMP Pro 14

- Text analysis involves transforming prepared text data into a **document term matrix (DTM)**. Each row in the DTM corresponds to a document (a cell in a JMP data table), and each column in the DTM corresponds to a term. The DTM is then used as an input in all analyses.
- Two main analysis options are available from the red triangle menu beside Text Explorer for Citation Description:
  - **Latent Class Analysis** groups documents into clusters of similar documents.
  - **Latent Semantic Analysis** is a dimension reduction technique similar to principal components analysis. This method will reduce the dimension of the DTM by applying singular value decomposition (SVD).

- From the red triangle menu beside "Text Explorer for Citation Description," select "Latent Semantic Analysis, SVD."
- In the Specifications dialog box, retain the default options and click "OK."

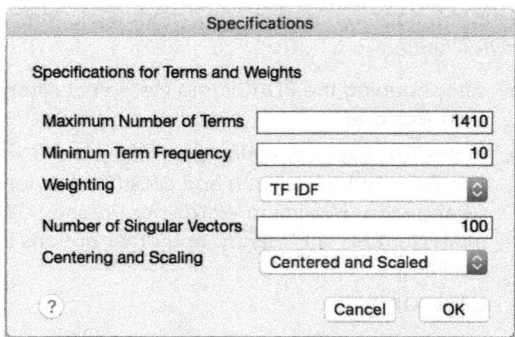

- Note: Your results may differ from the values or plots shown here or in Figure 3.28 depending on your choice of weighting scheme and on any differences in data cleaning (adding phrases or removing stop words, for example).
- In the plots, hover your mouse over a point to see the text of this citation or term appear over the plot.
- Text Explorer analysis results, such as the DTM, can be saved to the data table. The resulting columns can then be used as predictors in other analyses. From the red triangle menu beside "Text Explorer for Citation Description,"
  - Select "Document Term Matrix" to save information that corresponds to documents.
  - Select "Save Term Table" to save information that corresponds to terms, without respect to the specific documents.

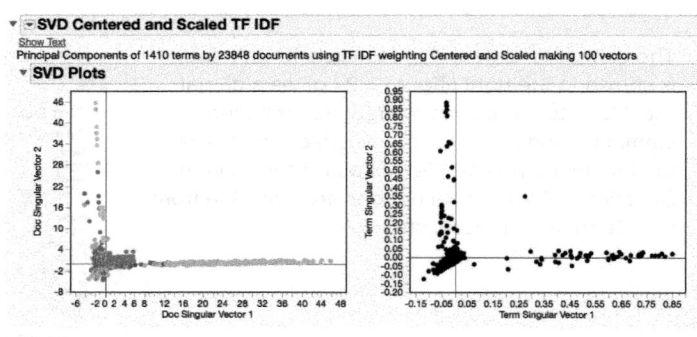

Source: JMP Pro 14

Cluster Analysis in Figure 3.37(b) (data file: SportsDis.jmp):

- After opening the SportsDis.jmp file, select **Analyze : Clustering : Hierarchical Cluster**.
- Select all of the sport name columns ("Boxing," "Basketball," "Golf," etc., through "Football") and click "Y, Columns."
- Select the "Label" column and click "Label."
- Under "Method," select "Complete."
- From the drop-down list at the bottom left select "Data is distance matrix."
- Click "OK."
- A description of more cluster analysis options is given in the next example.

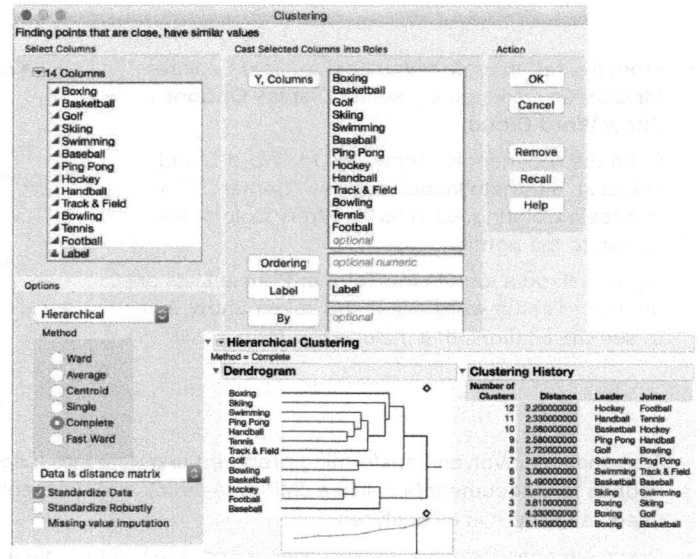

Source: JMP Pro 14

Cluster Analysis in Figure 3.38 (data file: SportsRatings.jmp):

- After opening the SportsRatings.jmp file, select **Analyze : Clustering : Hierarchical Cluster**.
- Select all columns except Sport ("Fast/Slow," "Comp/Simple," etc.) and click "Y, Columns."
- Select the "Sport" column and click "Label."
- Under "Method," select "Complete."
- Click "OK."

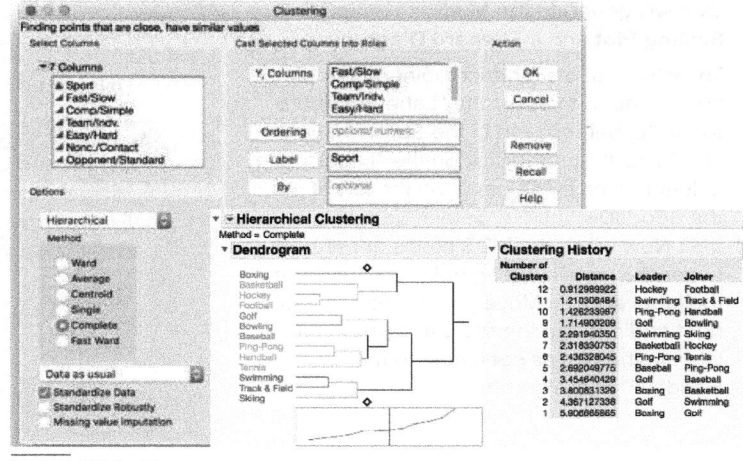

Source: JMP Pro 14

JMP will generate:

- A **dendrogram**, showing the clusters formed at each step.
- A **scree plot**, showing the distance bridged each step.
- The **clustering history**, giving cluster statistics for each step. To reveal, click on the small gray disclosure triangle to the left of "Clustering History."
- To dynamically change the number of clusters, click and drag one of the **black diamonds** left or right. Slide to the left until there are six clusters.
- From the red triangle menu beside "Hierarchical Clustering," select "Color Clusters" to color code the clusters in the dendrogram.
- From the red triangle menu beside "Hierarchical Clustering," select "Cluster Summary." The Cluster Summary will show the Cluster Means in a table as in Figure 3.38 and as a parallel plot.

Multidimensional Scaling in Figure 3.39 (data file: SportsDis.jmp):

- After opening the SportsDis.jmp file, select **Analyze : Multivariate Methods : Multidimensional Scaling**.
- Select all of the sport variable columns and click "Y Columns."
- Keep the default "Data Format" as "Distance Matrix."
- Keep the default "Set Dimensions" as "2."
- Click "OK."

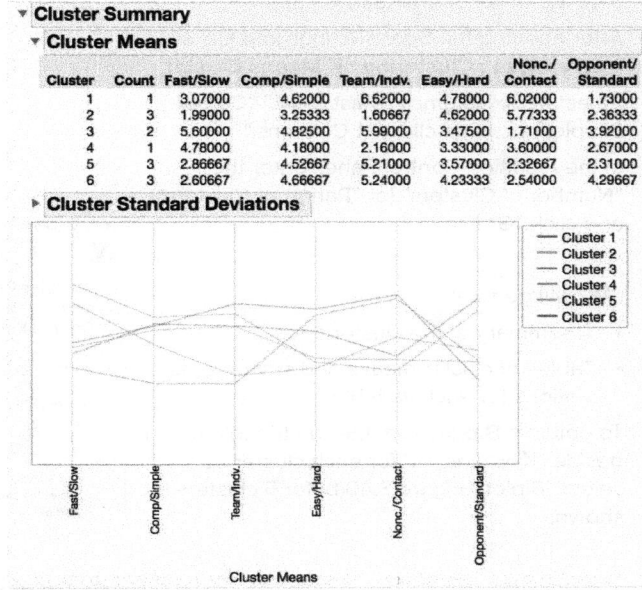

Cluster Summary

Cluster Means

Cluster	Count	Fast/Slow	Comp/Simple	Team/Indv.	Easy/Hard	Nonc./Contact	Opponent/Standard
1	1	3.07000	4.62000	6.62000	4.78000	6.02000	1.73000
2	3	1.99000	3.25333	1.60667	4.62000	5.77333	2.36333
3	2	5.60000	4.82500	5.99000	3.47500	1.71000	3.92000
4	1	4.78000	4.18000	2.16000	3.33000	3.60000	2.67000
5	3	2.86667	4.52667	5.21000	3.57000	2.32667	2.31000
6	3	2.60667	4.66667	5.24000	4.23333	2.54000	4.29667

Cluster Standard Deviations

Source: JMP Pro 14

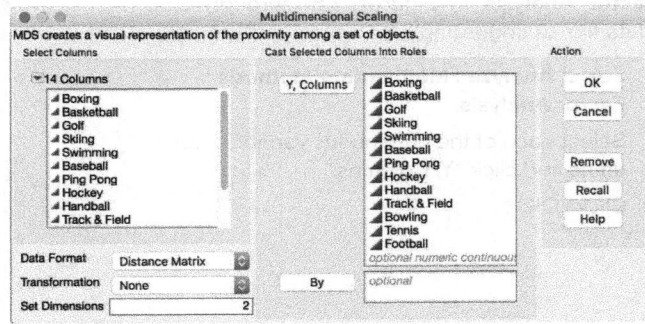

Source: JMP Pro 14

- JMP will generate the **Multidimensional Scaling Plot** and a **Shepard Diagram**.
- To include labels for each point on the plot, you will need to apply the "Label" attribute to the "Label" column in the SportsDis.jmp file. To do this, right-click on the "Label" column name in the panel on the left of the SportsDis.jmp file. Select "Label/Unlabel." Now when the MDS plot is generated, hover the mouse over a point until you see the label appear. Move your mouse to the right and click on the red pin to keep that label in the plot as seen next to the "Label: Handball" in the graph.

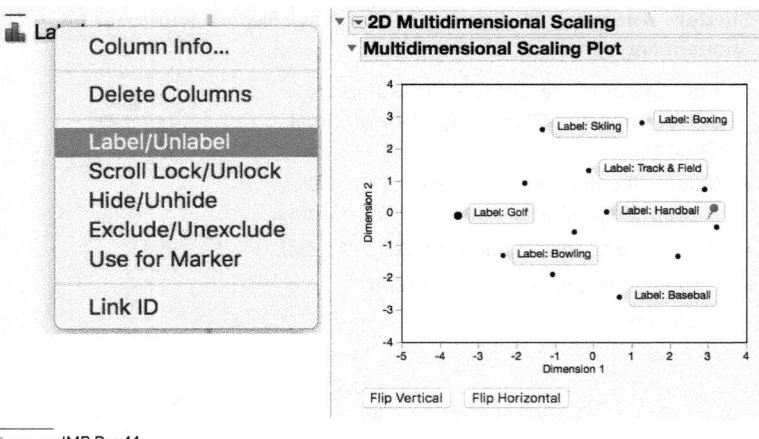

Source: JMP Pro 14

- **k-Means Clustering** in Figures 3.40, 3.41, and 3.42 (data file: SportsRatings.jmp):
- After opening the SportsRatings.jmp file, select **Analyze : Clustering : K Means Cluster**.
- Select all six columns ("Fast/Slow," "Comp/ Simple," etc.) and click "Y, Columns."
- In the resulting Control Panel, enter the "Number of Clusters" (or "Range of Clusters") and click "Go."

- JMP will generate:
  - A summary of the cluster sizes.
  - Tables of cluster means and standard deviations for each variable.
- To obtain a **Biplot**, from the red triangle menu beside "K Means . . ." for each cluster size, select "Biplot." Figure 3.40(b) for 5 clusters is shown.

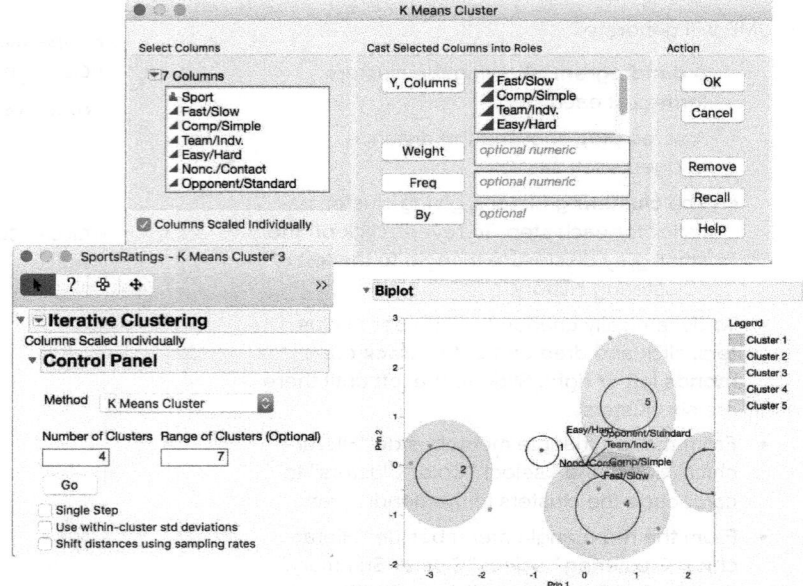

Source: JMP Pro 14

Factor Analysis in Figures 3.51 through 3.55 (data file: Colleges.jmp):

- Select **Analyze : Multivariate Methods : Factor Analysis**.
- Select each of the continuous variables columns and click "Y, Columns."
- Click "OK."

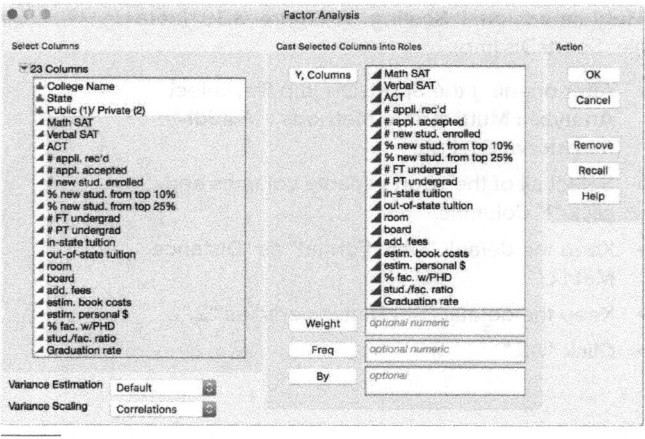

Source: JMP Pro 14

- JMP displays an **Eigenvalue** report and a **Scree Plot**.

- In the "Model Launch" area, under "Factoring Method," select "Principal Axis" and under "Prior Commonality," select "Common Factor Analysis." Enter "6" for the "Number of Factors" and select "Unrotated" for the "Rotation method."

- Click "OK."

- The following results are provided (additional options are available under the red triangle menu beside "Factor Analysis on Correlations . . ."):

  - Final communality estimates
  - The variance explained by each factor
  - Significance tests for the factor analysis
  - Unrotated Factor Loadings (as shown)
  - Factor loading plot

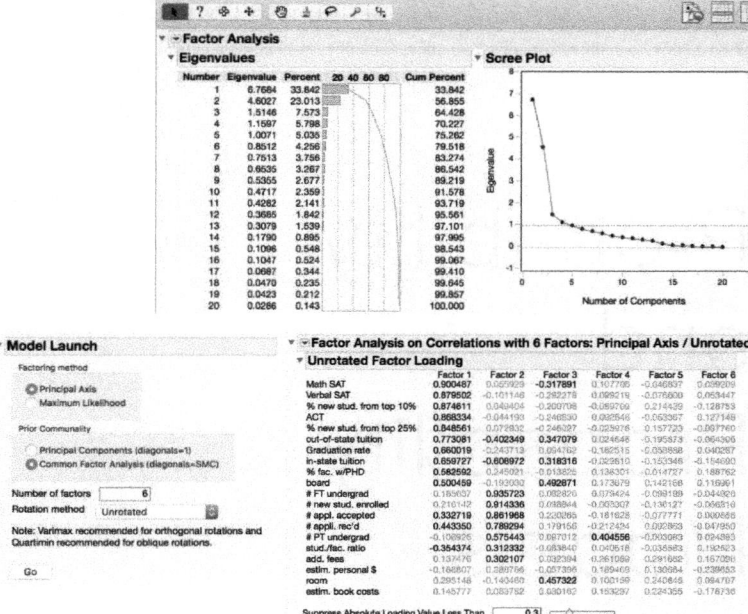

Source: JMP Pro 14

- To rotate the factors using the Varimax rotation, click on the small gray disclosure triangle beside "Model Launch" to open the area and change "Rotation method" to "Varimax." Click "OK."

- To save the factor scores as new columns to the data table, click on the red triangle menu beside "Factor Analysis on Correlations with 6 Factors: Principal Axis/Varimax" and select "Save Rotated Components."

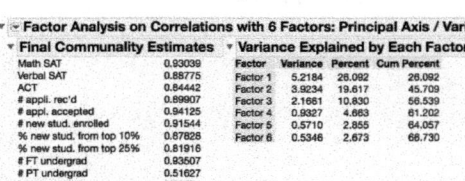

Note: Factor Analysis can also be accessed from the **Principal Components** platform.

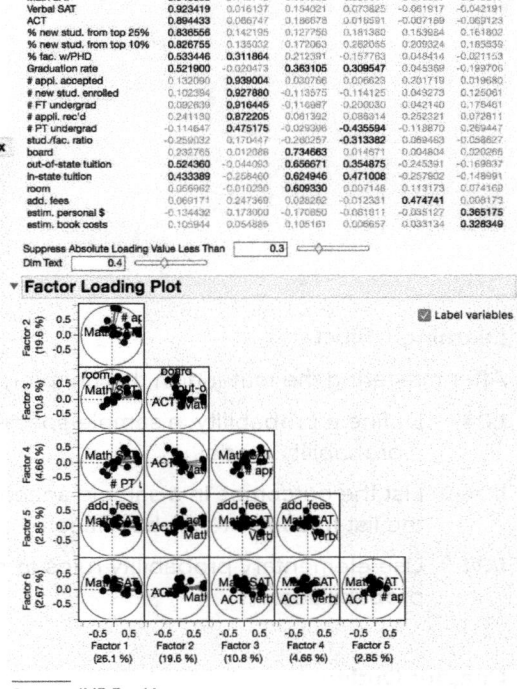

Source: JMP Pro 14

Design Elements: (CD): ©Comstock Images/Alamy; (All Others): ©McGraw-Hill Education

## CHAPTER 4

Probability and Probability Models

©Corbis/Glow Images

Learning Objectives

After mastering the material in this chapter, you will be able to:

LO4-1 Define a probability, a sample space, and a probability model.

LO4-2 List the outcomes in a sample space and use the list to compute probabilities.

LO4-3 Use elementary probability rules to compute probabilities.

LO4-4 Compute conditional probabilities and assess independence.

LO4-5 Use Bayes' Theorem to update prior probabilities to posterior probabilities (Optional).

LO4-6 Use some elementary counting rules to compute probabilities (Optional).

Chapter Outline

I n Chapter 3 we explained how to use sample statistics as point estimates of population parameters. Starting in Chapter 8, we will focus on using sample statistics to make more sophisticated **statistical inferences** about population parameters. We will see that these statistical inferences are generalizations—based on calculating **probabilities**—about population parameters. In this chapter and in Chapters 6 and 7 we present the fundamental concepts about probability that are needed to understand how we make such statistical inferences. We begin our discussions in this chapter by considering rules for calculating probabilities.

In order to illustrate some of the concepts in this chapter, we will introduce a new case.

The Crystal Cable Case: A cable company uses probability to assess the market penetration of its television and Internet services.

4.1 Probability, Sample Spaces, and Probability Models

LO4-1
Define a probability, a sample space, and a probability model.

An introduction to probability and sample spaces

We use the concept of **probability** to deal with uncertainty. Intuitively, the probability of an event is a number that measures the chance, or likelihood, that the event will occur. For instance, the probability that your favorite football team will win its next game measures the likelihood of a victory. The probability of an event is always a number between 0 and 1. The closer an event's probability is to 1, the higher is the likelihood that the event will occur; the closer the event's probability is to 0, the smaller is the likelihood that the event will occur. For example, if you believe that the probability that your favorite football team will win its next game is .95, then you are almost sure that your team will win. However, if you believe that the probability of victory is only .10, then you have very little confidence that your team will win.

When performing statistical studies, we sometimes collect data by **performing a controlled experiment.** For instance, we might purposely vary the operating conditions of a manufacturing process in order to study the effects of these changes on the process output. Alternatively, we sometimes obtain data by **observing uncontrolled events.** For example, we might observe the closing price of a share of General Motors' stock every day for 30 trading days. In order to simplify our terminology, we will use the word *experiment* to refer to either method of data collection. We now formally define an experiment and the *sample space* of an experiment.

> An **experiment** is any process of observation that has an uncertain outcome. The **sample space** of an experiment is the set of all possible outcomes for the experiment. The possible outcomes are sometimes called **experimental outcomes** or **sample space outcomes.**

When specifying the sample space of an experiment, we must define the sample space outcomes so that on any single repetition of the experiment, one and only one sample space outcome will occur. For example, if we consider the experiment of tossing a coin and observing whether the upward face of the coin shows as a "head" or a "tail," then the sample space consists of the outcomes "head" and "tail." If we consider the experiment of rolling a die and observing the number of dots showing on the upward face of the die, then the sample space consists of the outcomes 1, 2, 3, 4, 5, and 6. If we consider the experiment of subjecting an automobile to a "pass-fail" tailpipe emissions test, then the sample space consists of the outcomes "pass" and "fail."

Assigning probabilities to sample space outcomes

We often wish to assign probabilities to sample space outcomes. This is usually done by using one of three methods: the *classical method*, the *relative frequency method*, or the *subjective method.* Regardless of the method used, **probabilities must be assigned to the sample space outcomes so that two conditions are met:**[1]

> 1   The probability assigned to each sample space outcome must be between 0 and 1. That is, if E represents a sample space outcome and if $P(E)$ represents the probability of this outcome, then $0 \leq P(E) \leq 1$.
>
> 2   The probabilities of all of the sample space outcomes must sum to 1.

The **classical method** of assigning probabilities can be used when the sample space outcomes are equally likely. For example, consider the experiment of tossing a fair coin. Here, there are *two* equally likely sample space outcomes—head (H) and tail (T). Therefore, logic suggests that the probability of observing a head, denoted $P(H)$, is $1/2 = .5$, and that the probability of observing a tail, denoted $P(T)$, is also $1/2 = .5$. Notice that each probability is between 0 and 1. Furthermore, because H and T are all of the sample space outcomes, $P(H) + P(T) = 1$. In general, if there are N equally likely sample space outcomes, the probability assigned to each sample space outcome is $1/N$. To illustrate this, consider the experiment of rolling a fair die. It would seem reasonable to think that the six sample space outcomes 1, 2, 3, 4, 5, and 6 are equally likely, and thus each outcome is assigned a probability of $1/6$. If $P(1)$ denotes the probability that one dot appears on the upward face of the die, then $P(1) = 1/6$. Similarly, $P(2) = 1/6$, $P(3) = 1/6$, $P(4) = 1/6$, $P(5) = 1/6$, and $P(6) = 1/6$.

Before discussing the *relative frequency method* for assigning probabilities, we note that probability is often interpreted to be a **long-run relative frequency.** To illustrate this, consider tossing a fair coin—a coin such that the probability of its upward face showing as a head is .5. If we get 6 heads in the first 10 tosses, then the relative frequency, or fraction, of heads is $6/10 = .6$. If we get 47 heads in the first 100 tosses, the relative frequency of heads is $47/100 = .47$. If we get 5,067 heads in the first 10,000 tosses, the relative frequency of heads is $5,067/10,000 = .5067$. Note that the relative frequency of heads is approaching (that is, getting closer to) .5. The long run relative frequency interpretation of probability says that, if we tossed the coin an indefinitely large number of times (that is, a number of times *approaching infinity*), the relative frequency of heads obtained would approach .5. Of course, in actuality it is impossible to toss a coin (or perform any experiment) an indefinitely large number of times. Therefore, a relative frequency interpretation of probability is a mathematical idealization. To summarize, suppose that E is a sample space outcome that might occur when a particular experiment is performed. Then the probability that E will occur, $P(E)$, can be interpreted to be the number that would be approached by the relative frequency of E if we performed the experiment an indefinitely large number of times. It follows that we often think of a probability in terms of the percentage of the time the sample space outcome would occur in many repetitions of the experiment. For instance, when we say that the probability of obtaining a head when we toss a coin is .5, we are saying that, when we repeatedly toss the coin an indefinitely large number of times, we will obtain a head on 50 percent of the repetitions.

Sometimes it is either difficult or impossible to use the classical method to assign probabilities. Because we can often make a relative frequency interpretation of probability, we can estimate a probability by performing the experiment in which an outcome might occur many times. Then, we estimate the probability of the outcome to be the proportion of the time that the outcome occurs during the many repetitions of the experiment. For example, to estimate the probability that a randomly selected consumer prefers Coca-Cola to all other soft drinks, we perform an experiment in which we ask a randomly selected consumer for his or her preference. There are two possible sample space outcomes: "prefers Coca-Cola" and "does not prefer Coca-Cola." However, we have no reason to believe that these sample space outcomes are equally likely, so we cannot use the classical method. We might perform the experiment, say, 1,000 times by surveying 1,000 randomly selected consumers. Then, if 140 of those surveyed said that they prefer Coca-Cola, we would estimate the probability that a randomly selected consumer prefers Coca-Cola to all other soft drinks to be $140/1,000 = .14$. This is an example of the **relative frequency method** of assigning probability.

---

[1]These conditions assume that the sample space consists of either a finite number of sample space outcomes or an infinite number of sample space outcomes that can be counted or listed (such as the sequence of all nonnegative integers 0, 1, 2, 3, . . .).

If we cannot perform the experiment many times, we might estimate the probability by using our previous experience with similar situations, intuition, or special expertise that we may possess. For example, a company president might estimate the probability of success for a one-time business venture to be .7. Here, on the basis of knowledge of the success of previous similar ventures, the opinions of company personnel, and other pertinent information, the president believes that there is a 70 percent chance the venture will be successful.

When we use experience, intuitive judgment, or expertise to assess a probability, we call this the **subjective method** of assigning probability. Such a probability (called a **subjective probability**) may or may not have a relative frequency interpretation. For instance, when the company president estimates that the probability of a successful business venture is .7, this may mean that, if business conditions similar to those that are about to be encountered could be repeated many times, then the business venture would be successful in 70 percent of the repetitions. Or the president may not be thinking in relative frequency terms but rather may consider the venture a "one-shot" proposition. We will discuss some other subjective probabilities later. However, the interpretations of statistical inferences we will explain in later chapters are based on the relative frequency interpretation of probability. For this reason, we will concentrate on this interpretation.

Probability models

Throughout this chapter and the rest of this book, we will use what are called *probability models* to calculate the probabilities that various events of interest will occur. We define a probability model as follows:

A **probability model** is a mathematical representation of a random phenomenon.

One type of random phenomenon is an experiment, which we have said is any process of observation that has an uncertain outcome. The probability model describing an experiment consists of (1) the sample space of the experiment and (2) a procedure for calculating probabilities concerning the sample space outcomes. In this chapter we will use the probability model describing an experiment to find rules for calculating the probabilities of various simple and complex events related to the experiment.

In Chapters 6 and 7, we will consider another type of random phenomenon called a *random variable*. A **random variable** is a variable whose value is numeric and is determined by the outcome of an experiment. For example, in Chapter 6 we will consider the experiment of selling the TrueSound-XL satellite car radio at a store during a future week, and we will define a random variable that equals the number of TrueSound-XL radios that will be sold at the store during the future week. As another example, in Chapter 7 we will consider randomly selecting a new midsize car, and we will define a random variable that equals the EPA combined city and highway driving mileage that will be obtained by this car when it is tested. The probability model describing a random variable is called a **probability distribution** and consists of (1) a specification of the possible values of the random variable and (2) a table, graph, or formula that can be used to calculate probabilities concerning the values that the random variable might equal. For example, we might calculate the probability that at least two TrueSound-XL radios will be sold at the store during the future week.

There are two types of probability distributions: **discrete probability distributions** (discussed in Chapter 6) and **continuous probability distributions** (discussed in Chapter 7). Two important discrete distributions are the *binomial distribution* and the *Poisson distribution*, which are sometimes called (respectively) the *binomial model* and the *Poisson model*. Two important continuous distributions are the *normal distribution* and the *exponential distribution*, which are sometimes called (respectively) the *normal model* and the *exponential model*. Every probability model—and thus every probability distribution—is based on one or more assumptions concerning the random phenomenon that the model describes. For example, the probability models discussed in this chapter are sometimes based on sample spaces that are assumed to have equally likely sample space outcomes. As another example, probability models are sometimes based on sample spaces that have sample space outcomes associated with the possible occurrences of *independent events*. As a third example, the binomial distribution of Chapter 6 is based on the assumption of a *binomial experiment*. Similar to physical models that describe the motions of planetary bodies, the assumptions made by a probability model will not capture all of the nuances of the random phenomenon that the probability model is describing. However, if these assumptions capture the important aspects of the random phenomenon, the probability model should provide reasonably accurate probabilities concerning the random phenomenon.

4.2 Probability and Events

In the first section of this chapter, we have informally talked about events. We now give the formal definition of an event.

> An **event** is a set of one or more sample space outcomes.

For example, if we consider the experiment of tossing a fair die, the event "at least five spots will show on the upward face of the die" consists of the sample space outcomes 5 and 6. That is, the event "at least five spots will show on the upward face of the die" will occur if and only if one of the sample space outcomes 5 or 6 occurs.

To find the probability that an event will occur, we can use the following result.[2]

> The **probability of an event** is the **sum of the probabilities of the sample space outcomes** that correspond to the event.

As an example, we have seen that if we consider the experiment of tossing a fair die, then the sample space outcomes 5 and 6 correspond to the occurrence of the event "at least five spots will show on the upward face of the die." Therefore, the probability of this event is

$$P(5) + P(6) = \frac{1}{6} + \frac{1}{6} = \frac{2}{6} = \frac{1}{3}$$

EXAMPLE 4.1 Boys and Girls

A newly married couple plans to have two children. Naturally, they are curious about whether their children will be boys or girls. Therefore, we consider the experiment of having two children. In order to find the sample space of this experiment, we let B denote that a child is a boy and G denote that a child is a girl. Then, it is useful to construct the **tree diagram** shown in Figure 4.1. This diagram pictures the experiment as a two-step process—having the first child, which could be either a boy or a girl (B or G), and then having the second child, which could also be either a boy or a girl (B or G). Each branch of the tree leads to a sample space outcome. These outcomes are listed at the right ends of the branches. We see that there are four sample space outcomes. Therefore, the sample space (that is, the set of all the sample space outcomes) is

$$BB \qquad BG \qquad GB \qquad GG$$

In order to consider the probabilities of these outcomes, suppose that boys and girls are equally likely each time a child is born. Intuitively, this says that each of the sample space outcomes is equally likely. That is, this implies that

$$P(BB) = P(BG) = P(GB) = P(GG) = \frac{1}{4}$$

Therefore:

1   The probability that the couple will have two boys is

$$P(BB) = \frac{1}{4}$$

because two boys will be born if and only if the sample space outcome BB occurs.

---

[2]This result assumes that the sample space consists of either a finite number of sample space outcomes or an infinite number of sample space outcomes that can be counted or listed.

FIGURE 4.1   A Tree Diagram of the Genders of Two Children

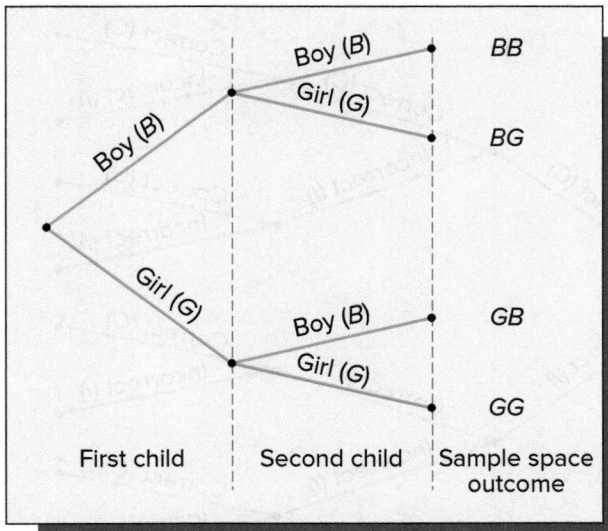

2   The probability that the couple will have one boy and one girl is

$$P(BG) + P(GB) = \frac{1}{4} + \frac{1}{4} = \frac{1}{2}$$

because one boy and one girl will be born if and only if one of the sample space outcomes *BG* or *GB* occurs.

3   The probability that the couple will have two girls is

$$P(GG) = \frac{1}{4}$$

because two girls will be born if and only if the sample space outcome *GG* occurs.

4   The probability that the couple will have at least one girl is

$$P(BG) + P(GB) + P(GG) = \frac{1}{4} + \frac{1}{4} + \frac{1}{4} = \frac{3}{4}$$

because at least one girl will be born if and only if one of the sample space outcomes *BG*, *GB*, or *GG* occurs.

EXAMPLE 4.2 Pop Quizzes

A student takes a pop quiz that consists of three true–false questions. If we consider our experiment to be answering the three questions, each question can be answered correctly or incorrectly. We will let *C* denote answering a question correctly and *I* denote answering a question incorrectly. Figure 4.2 depicts a tree diagram of the sample space outcomes for the experiment. The diagram portrays the experiment as a three-step process—answering the first question (correctly or incorrectly, that is, *C* or *I*), answering the second question, and answering the third question. The tree diagram has eight different branches, and the eight sample space outcomes are listed at the ends of the branches. We see that the sample space is

CCC   CCI   CIC   CII

ICC   ICI   IIC   III

Next, suppose that the student was totally unprepared for the quiz and had to blindly guess the answer to each question. That is, the student had a 50–50 chance (or .5 probability) of

FIGURE 4.2   A Tree Diagram of Answering Three True–False Questions

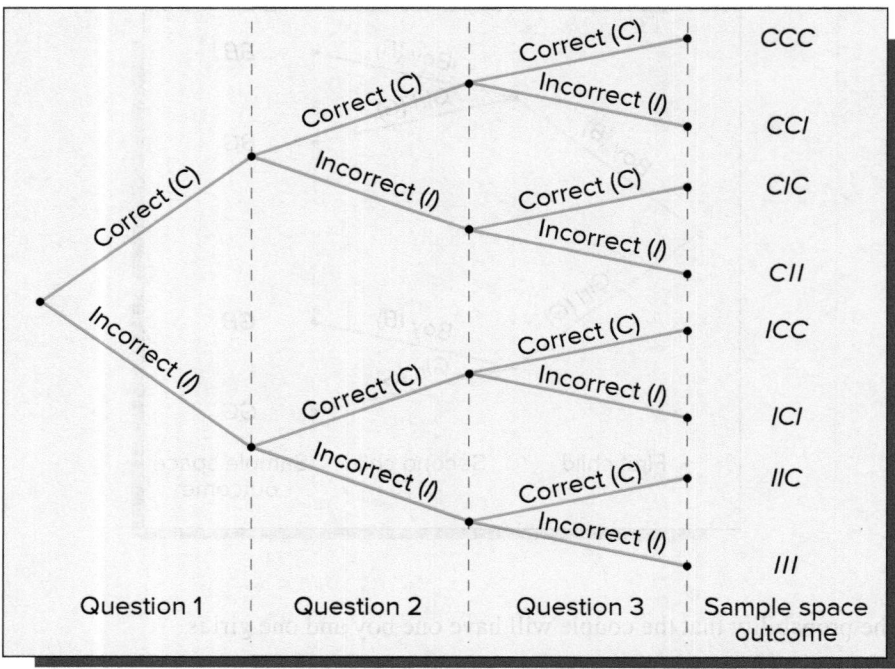

correctly answering each question. Intuitively, this would say that each of the eight sample space outcomes is equally likely to occur. That is,

$$P(CCC) = P(CCI) = \cdots = P(III) = \frac{1}{8}$$

Therefore:

1   The probability that the student will get all three questions correct is

$$P(CCC) = \frac{1}{8}$$

2   The probability that the student will get exactly two questions correct is

$$P(CCI) + P(CIC) + P(ICC) = \frac{1}{8} + \frac{1}{8} + \frac{1}{8} = \frac{3}{8}$$

because two questions will be answered correctly if and only if one of the sample space outcomes *CCI*, *CIC*, or *ICC* occurs.

3   The probability that the student will get exactly one question correct is

$$P(CII) + P(ICI) + P(IIC) = \frac{1}{8} + \frac{1}{8} + \frac{1}{8} = \frac{3}{8}$$

because one question will be answered correctly if and only if one of the sample space outcomes *CII*, *ICI*, or *IIC* occurs.

4   The probability that the student will get all three questions incorrect is

$$P(III) = \frac{1}{8}$$

5   The probability that the student will get at least two questions correct is

$$P(CCC) + P(CCI) + P(CIC) + P(ICC) = \frac{1}{8} + \frac{1}{8} + \frac{1}{8} + \frac{1}{8} = \frac{1}{2}$$

because the student will get at least two questions correct if and only if one of the sample space outcomes *CCC*, *CCI*, *CIC*, or *ICC* occurs.

Notice that in the true–false question situation, we find that, for instance, the probability that the student will get exactly one question correct equals the ratio

$$\frac{\text{the number of sample space outcomes resulting in one correct answer}}{\text{the total number of sample space outcomes}} = \frac{3}{8}$$

In general, when a sample space is finite we can use the following method for computing the probability of an event.

If all of the sample space outcomes are equally likely, then the probability that an event will occur is equal to the ratio

$$\frac{\text{the number of sample space outcomes that correspond to the event}}{\text{the total number of sample space outcomes}}$$

It is important to emphasize, however, that we can use this rule only when all of the sample space outcomes are equally likely (as they are in the true–false question situation). If the sample space outcomes are not equally likely, the rule may give an incorrect probability.

EXAMPLE 4.3 Choosing a CEO

A company is choosing a new chief executive officer (CEO). It has narrowed the list of candidates to four finalists (identified by last name only)—Adams, Chung, Hill, and Rankin. If we consider our experiment to be making a final choice of the company's CEO, then the experiment's sample space consists of the four possible outcomes:

$A \equiv$ Adams will be chosen as CEO.

$C \equiv$ Chung will be chosen as CEO.

$H \equiv$ Hill will be chosen as CEO.

$R \equiv$ Rankin will be chosen as CEO.

Next, suppose that industry analysts feel (subjectively) that the probabilities that Adams, Chung, Hill, and Rankin will be chosen as CEO are .1, .2, .5, and .2, respectively. That is, in probability notation

$$P(A) = .1 \qquad P(C) = .2 \qquad P(H) = .5 \qquad \text{and} \qquad P(R) = .2$$

Also, suppose only Adams and Hill are internal candidates (they already work for the company). Letting *INT* denote the event that "an internal candidate will be selected for the CEO position," then *INT* consists of the sample space outcomes A and H (that is, *INT* will occur if and only if either of the sample space outcomes A or H occurs). It follows that $P(INT) = P(A) + P(H) = .1 + .5 = .6$. This says that the probability that an internal candidate will be chosen to be CEO is .6.

Finally, it is important to understand that if we had ignored the fact that sample space outcomes are not equally likely, we might have tried to calculate $P(INT)$ as follows:

$$P(INT) = \frac{\text{the number of internal candidates}}{\text{the total number of candidates}} = \frac{2}{4} = .5$$

This result would be incorrect. Because the sample space outcomes are not equally likely, we have seen that the correct value of $P(INT)$ is .6, not .5.

EXAMPLE 4.4 The Crystal Cable Case: Market Penetration

Like all companies, cable companies send shareholders reports on their profits, dividends, and return on equity. They often supplement this information with some metrics unique to the cable business. To construct one such metric, a cable company can compare the number of households it actually serves to the number of households its current transmission lines could reach (without extending the lines). The number of households that the cable company's lines could reach is called its number of **cable passings,** while the ratio of the number of households the cable company actually serves to its number of cable passings is called the company's **cable penetration.** There are various types of cable penetrations—one for cable television, one for cable Internet, one for cable phone, and others. Moreover, a cable penetration is a probability, and interpreting it as such will help us to better understand various techniques to be discussed in the next section. For example, in a recent quarterly report, Crystal Cable reported that it had 12.4 million cable television customers and 27.4 million cable passings.[3] Consider randomly selecting one of Crystal's cable passings. That is, consider selecting one cable passing by giving each and every cable passing the same chance of being selected. Let A be the event that the randomly selected cable passing has Crystal's cable television service. Then, because the sample space of this experiment consists of 27.4 million equally likely sample space outcomes (cable passings), it follows that

$$P(A) = \frac{\text{the number of cable passings that have Crystal's cable television service}}{\text{the total number of cable passings}}$$

$$= \frac{12.4 \text{ million}}{27.4 \text{ million}}$$

$$= .45$$

This probability is Crystal's cable television penetration and says that the probability, that a randomly selected cable passing has Crystal's cable television service is .45. That is, 45 percent of Crystal's cable passings have Crystal's cable television service.

To conclude this section, we note that in optional Section 4.6 we discuss several *counting rules* that can be used to count the number of sample space outcomes in an experiment. These rules are particularly useful when there are many sample space outcomes and thus these outcomes are difficult to list.

Exercises for Sections 4.1 and 4.2

CONCEPTS connect

4.1 Define the following terms: *experiment, event, probability, sample space.*

4.2 Explain the properties that must be satisfied by a probability.

METHODS AND APPLICATIONS

4.3 Two randomly selected grocery store patrons are each asked to take a blind taste test and to then state which of three diet colas (marked as A, B, or C) he or she prefers.
  a Draw a tree diagram depicting the sample space outcomes for the test results.
  b List the sample space outcomes that correspond to each of the following events:
  (1) Both patrons prefer diet cola A.
  (2) The two patrons prefer the same diet cola.
  (3) The two patrons prefer different diet colas.
  (4) Diet cola A is preferred by at least one of the two patrons.
  (5) Neither of the patrons prefers diet cola C.

  c Assuming that all sample space outcomes are equally likely, find the probability of each of the events given in part *b*.

4.4 Suppose that a couple will have three children. Letting B denote a boy and G denote a girl:
  a Draw a tree diagram depicting the sample space outcomes for this experiment.
  b List the sample space outcomes that correspond to each of the following events:
  (1) All three children will have the same gender.
  (2) Exactly two of the three children will be girls.
  (3) Exactly one of the three children will be a girl.
  (4) None of the three children will be a girl.
  c Assuming that all sample space outcomes are equally likely, find the probability of each of the events given in part *b*.

4.5 Four people will enter an automobile showroom, and each will either purchase a car (P) or not purchase a car (N).
  a Draw a tree diagram depicting the sample space of all possible purchase decisions that could potentially be made by the four people.

---
[3]Although these numbers are hypothetical, they are similar to results actually found in Time Warner Cable's quarterly reports. See www.TimeWarnerCable.com. Click on Investor Relations.

b List the sample space outcomes that correspond to each of the following events:
  (1) Exactly three people will purchase a car.
  (2) Two or fewer people will purchase a car.
  (3) One or more people will purchase a car.
  (4) All four people will make the same purchase decision.

c Assuming that all sample space outcomes are equally likely, find the probability of each of the events given in part *b*.

4.6 The U.S. Census Bureau compiles data on family income and summarizes its findings in *Current Population Reports*. The table to the right is a frequency distribution of the annual incomes for a random sample of U.S. families. Find an estimate of the probability that a randomly selected U.S. family has an income between $60,000 and $199,999.   🅳🅢 FamIncomes

🅳🅢 FamIncomes

Income	Frequency (in thousands)
Under $20,000	11,470
$20,000–$39,999	17,572
$40,000–$59,999	14,534
$60,000–$79,999	11,410
$80,000–$99,999	7,535
$100,000–$199,999	11,197
$200,000 and above	2,280
	75,998

4.7 Let A, B, C, D, and E be sample space outcomes forming a sample space. Suppose that $P(A) = .2$, $P(B) = .15$, $P(C) = .3$, and $P(D) = .2$. What is $P(E)$? Explain how you got your answer.

4.3 Some Elementary Probability Rules

We can often calculate probabilities by using formulas called **probability rules.** We will begin by presenting the simplest probability rule: the *rule of complements*. To start, we define the complement of an event:

LO4-3
Use elementary probability rules to compute probabilities.

Given an event A, the **complement of A** is the event consisting of all sample space outcomes that do not correspond to the occurrence of A. The complement of A is denoted $\overline{A}$. Furthermore, $P(\overline{A})$ denotes **the probability that A will not occur.**

Figure 4.3 is a **Venn diagram** depicting the complement $\overline{A}$ of an event A. In any probability situation, either an event A or its complement $\overline{A}$ must occur. Therefore, we have

$$P(A) + P(\overline{A}) = 1$$

This implies the following result:

FIGURE 4.3
The Complement of an Event (the Shaded Region Is $\overline{A}$, the Complement of A)

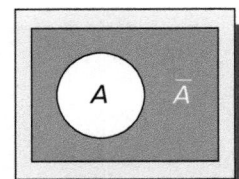

The Rule of Complements
Consider an event A. Then, **the probability that A *will not occur*** is

$$P(\overline{A}) = 1 - P(A)$$

EXAMPLE 4.5 The Crystal Cable Case: Market Penetration

Recall from Example 4.4 that the probability that a randomly selected cable passing has Crystal's cable television service is .45. It follows that the probability of the complement of this event (that is, the probability that a randomly selected cable passing does not have Crystal's cable television service) is $1 - .45 = .55$.

We next define the *intersection* of two events. Consider performing an experiment a single time. Then:

Given two events A and B, the **intersection of A and B** is the event that occurs if both A and B simultaneously occur. The intersection is denoted by $A \cap B$. Furthermore, $P(A \cap B)$ denotes **the probability that *both A and B will simultaneously occur.***

EXAMPLE 4.6 The Crystal Cable Case: Market Penetration

Recall from Example 4.4 that Crystal Cable has 27.4 million cable passings. Consider randomly selecting one of these cable passings, and define the following events:

$A \equiv$ the randomly selected cable passing has Crystal's cable television service.

$\overline{A} \equiv$ the randomly selected cable passing does not have Crystal's cable television service.

$B \equiv$ the randomly selected cable passing has Crystal's cable Internet service.

$\overline{B} \equiv$ the randomly selected cable passing does not have Crystal's cable Internet service.

$A \cap B \equiv$ the randomly selected cable passing has both Crystal's cable television service and Crystal's cable Internet service.

$A \cap \overline{B} \equiv$ the randomly selected cable passing has Crystal's cable television service and does not have Crystal's cable Internet service.

$\overline{A} \cap B \equiv$ the randomly selected cable passing does not have Crystal's cable television service and does have Crystal's cable Internet service.

$\overline{A} \cap \overline{B} \equiv$ the randomly selected cable passing does not have Crystal's cable television service and does not have Crystal's cable Internet service.

Table 4.1 is a *contingency table* that summarizes Crystal's cable passings. Using this table, we can calculate the following probabilities, each of which describes some aspect of Crystal's cable penetrations:

1 Because 12.4 million out of 27.4 million cable passings have Crystal's cable television service, A, then

$$P(A) = \frac{12.4}{27.4} = .45$$

This says that 45 percent of Crystal's cable passings have Crystal's cable television service (as previously seen in Example 4.4).

2 Because 9.8 million out of 27.4 million cable passings have Crystal's cable Internet service, B, then

$$P(B) = \frac{9.8}{27.4} = .36$$

This says that 36 percent of Crystal's cable passings have Crystal's cable Internet service.

3 Because 6.5 million out of 27.4 million cable passings have Crystal's cable television service and Crystal's cable Internet service, $A \cap B$, then

$$P(A \cap B) = \frac{6.5}{27.4} = .24$$

This says that 24 percent of Crystal's cable passings have both Crystal's cable television service and Crystal's cable Internet service.

TABLE 4.1 **A Contingency Table Summarizing Crystal's Cable Television and Internet Penetration (Figures in Millions of Cable Passings)**

Events	Has Cable Internet Service, B	Does Not Have Cable Internet Service, $\overline{B}$	Total
Has Cable Television Service, A	6.5	5.9	12.4
Does Not Have Cable Television Service, $\overline{A}$	3.3	11.7	15.0
Total	9.8	17.6	27.4

4   Because 5.9 million out of 27.4 million cable passings have Crystal's cable television service, but do not have Crystal's cable Internet service, $A \cap \overline{B}$, then

$$P(A \cap \overline{B}) = \frac{5.9}{27.4} = .22$$

This says that 22 percent of Crystal's cable passings have only Crystal's cable television service.

5   Because 3.3 million out of 27.4 million cable passings do not have Crystal's cable television service, but do have Crystal's cable Internet service, $\overline{A} \cap B$, then

$$P(\overline{A} \cap B) = \frac{3.3}{27.4} = .12$$

This says that 12 percent of Crystal's cable passings have only Crystal's cable Internet service.

6   Because 11.7 million out of 27.4 million cable passings do not have Crystal's cable television service and do not have Crystal's cable Internet service, $\overline{A} \cap \overline{B}$, then

$$P(\overline{A} \cap \overline{B}) = \frac{11.7}{27.4} = .43$$

This says that 43 percent of Crystal's cable passings have neither Crystal's cable television service nor Crystal's cable Internet service.

We next consider the *union* of two events. Again consider performing an experiment a single time. Then:

Given two events A and B, the **union of A and B** is the event that occurs if A or B (or both) occur. The union is denoted $A \cup B$. Furthermore, $P(A \cup B)$ denotes **the probability that A or B (or both) will occur.**

EXAMPLE 4.7 The Crystal Cable Case: Market Penetration

Consider randomly selecting one of Crystal's 27.4 million cable passings, and define the event

$A \cup B \equiv$ the randomly selected cable passing has Crystal's cable television service or Crystal's cable Internet service (or both)—that is, has at least one of the two services.

Looking at Table 4.1, we see that the cable passings that have Crystal's cable television service or Crystal's cable Internet service are (1) the 5.9 million cable passings that have only Crystal's cable television service, $A \cap \overline{B}$, (2) the 3.3 million cable passings that have only Crystal's cable Internet service, $\overline{A} \cap B$, and (3) the 6.5 million cable passings that have both Crystal's cable television service and Crystal's cable Internet service, $A \cap B$. Therefore, because a total of 15.7 million cable passings have Crystal's cable television service or Crystal's cable Internet service (or both), it follows that

$$P(A \cup B) = \frac{15.7}{27.4} = .57$$

This says that the probability that the randomly selected cable passing has Crystal's cable television service or Crystal's cable Internet service (or both) is .57. That is, 57 percent of Crystal's cable passings have Crystal's cable television service or Crystal's cable Internet service (or both). Notice that $P(A \cup B) = .57$ does not equal

$$P(A) + P(B) = .45 + .36 = .81$$

Logically, the reason for this is that both $P(A) = .45$ and $P(B) = .36$ count the 24 percent of the cable passings that have both Crystal's cable television service and Crystal's cable Internet service. Therefore, the sum of $P(A)$ and $P(B)$ counts this 24 percent of the cable passings once too often. It follows that if we subtract $P(A \cap B) = .24$ from the sum of $P(A)$ and $P(B)$, then we will obtain $P(A \cup B)$. That is,

$$P(A \cup B) = P(A) + P(B) - P(A \cap B)$$
$$= .45 + .36 - .24 = .57$$

Noting that Figure 4.4 shows **Venn diagrams** depicting the events A, B, $A \cap B$, and $A \cup B$, we have the following general result:

The Addition Rule

Let A and B be events. Then, **the probability that A or B (or both) will occur** is

$$P(A \cup B) = P(A) + P(B) - P(A \cap B)$$

The reasoning behind this result has been illustrated at the end of Example 4.7. Similarly, the Venn diagrams in Figure 4.4 show that when we compute $P(A) + P(B)$, we are counting each of the sample space outcomes in $A \cap B$ twice. We correct for this by subtracting $P(A \cap B)$.

We next define the idea of *mutually exclusive events:*

Mutually Exclusive Events

Two events A and B are **mutually exclusive** if they have no sample space outcomes in common. In this case, the events A and B cannot occur simultaneously, and thus

$$P(A \cap B) = 0$$

Noting that Figure 4.5 is a Venn diagram depicting two mutually exclusive events, we consider the following example.

EXAMPLE 4.8 Selecting Playing Cards

Consider randomly selecting a card from a standard deck of 52 playing cards. We define the following events:

$J \equiv$ the randomly selected card is a jack.

$Q \equiv$ the randomly selected card is a queen.

$R \equiv$ the randomly selected card is a red card (that is, a diamond or a heart).

Because there is no card that is both a jack and a queen, the events J and Q are mutually exclusive. On the other hand, there are two cards that are both jacks and red cards—the jack of diamonds and the jack of hearts—so the events J and R are not mutually exclusive.

We have seen that for any two events A and B, the probability that A or B (or both) will occur is

$$P(A \cup B) = P(A) + P(B) - P(A \cap B)$$

FIGURE 4.4   Venn Diagrams Depicting the Events *A*, *B*, *A ∩ B*, and *A ∪ B*

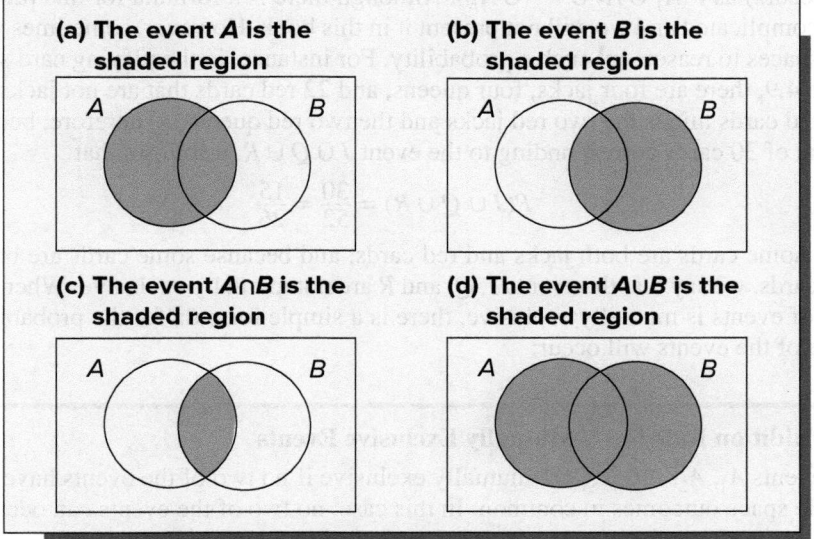

(a) The event *A* is the shaded region

(b) The event *B* is the shaded region

(c) The event *A∩B* is the shaded region

(d) The event *A∪B* is the shaded region

FIGURE 4.5
Two Mutually Exclusive Events

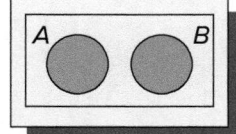

Therefore, when calculating $P(A \cup B)$, we should always subtract $P(A \cap B)$ from the sum of $P(A)$ and $P(B)$. However, when A and B are mutually exclusive, $P(A \cap B)$ equals 0. Therefore, in this case—and only in this case—we have the following:

The Addition Rule for Two Mutually Exclusive Events

Let A and B be **mutually exclusive** events. Then, **the probability that *A or B* will occur** is

$$P(A \cup B) = P(A) + P(B)$$

EXAMPLE 4.9 Selecting Playing Cards

Again consider randomly selecting a card from a standard deck of 52 playing cards, and define the events

$J \equiv$ the randomly selected card is a jack.

$Q \equiv$ the randomly selected card is a queen.

$R \equiv$ the randomly selected card is a red card (a diamond or a heart).

Because there are four jacks, four queens, and 26 red cards, we have $P(J) = \frac{4}{52}$, $P(Q) = \frac{4}{52}$, and $P(R) = \frac{26}{52}$. Furthermore, because there is no card that is both a jack and a queen, the events J and Q are mutually exclusive and thus $P(J \cap Q) = 0$. It follows that the probability that the randomly selected card is a jack or a queen is

$$P(J \cup Q) = P(J) + P(Q)$$
$$= \frac{4}{52} + \frac{4}{52} = \frac{8}{52} = \frac{2}{13}$$

Because there are two cards that are both jacks and red cards—the jack of diamonds and the jack of hearts—the events J and R are not mutually exclusive. Therefore, the probability that the randomly selected card is a jack or a red card is

$$P(J \cup R) = P(J) + P(R) - P(J \cap R)$$
$$= \frac{4}{52} + \frac{26}{52} - \frac{2}{52} = \frac{28}{52} = \frac{7}{13}$$

We now consider an arbitrary group of events—$A_1, A_2, \ldots, A_N$. We will denote the probability that A_1 or A_2 or $\cdots$ or A_N occurs (that is, the probability that at least one of the events occurs) as $P(A_1 \cup A_2 \cup \cdots \cup A_N)$. Although there is a formula for this probability, it is quite complicated and we will not present it in this book. However, sometimes we can use sample spaces to reason out such a probability. For instance, in the playing card situation of Example 4.9, there are four jacks, four queens, and 22 red cards that are not jacks or queens (the 26 red cards minus the two red jacks and the two red queens). Therefore, because there are a total of 30 cards corresponding to the event $J \cup Q \cup R$, it follows that

$$P(J \cup Q \cup R) = \frac{30}{52} = \frac{15}{26}$$

Because some cards are both jacks and red cards, and because some cards are both queens and red cards, we say that the events J, Q, and R are not mutually exclusive. When, however, a group of events is mutually exclusive, there is a simple formula for the probability that at least one of the events will occur:

The Addition Rule for N Mutually Exclusive Events

The events $A_1, A_2, \ldots, A_N$ are mutually exclusive if no two of the events have any sample space outcomes in common. In this case, no two of the events can occur simultaneously, and

$$P(A_1 \cup A_2 \cup \cdots \cup A_N) = P(A_1) + P(A_2) + \cdots + P(A_N)$$

As an example of using this formula, again consider the playing card situation and the events J and Q. If we define the event

$$K \equiv \text{the randomly selected card is a king}$$

then the events J, Q, and K are mutually exclusive. Therefore,

$$P(J \cup Q \cup K) = P(J) + P(Q) + P(K)$$
$$= \frac{4}{52} + \frac{4}{52} + \frac{4}{52} = \frac{12}{52} = \frac{3}{13}$$

Exercises for Section 4.3

CONCEPTS   connect

4.8   Explain what it means for two events to be mutually exclusive; for N events.

4.9   If A and B are events, define (in words) $\bar{A}$, $A \cup B$, $A \cap B$, and $\bar{A} \cap \bar{B}$.

METHODS AND APPLICATIONS

4.10   Consider a standard deck of 52 playing cards, a randomly selected card from the deck, and the following events:

$$R = \text{red} \quad B = \text{black} \quad A = \text{ace}$$
$$N = \text{nine} \quad D = \text{diamond} \quad C = \text{club}$$

   a   Describe the sample space outcomes that correspond to each of these events.

   b   For each of the following pairs of events, indicate whether the events are mutually exclusive. In each case, if you think the events are mutually exclusive, explain why the events have no common sample space outcomes. If you think the events are

not mutually exclusive, list the sample space outcomes that are common to both events.

    (1) R and A (3) A and N (5) D and C

    (2) R and C (4) N and C

4.11   The following contingency table summarizes the number of students at a college who have a MasterCard or a Visa credit card.

	Have Visa	Do Not Have Visa	Total
Have MasterCard	1,000	1,500	2,500
Do not have MasterCard	3,000	4,500	7,500
Total	4,000	6,000	10,000

   a   Find the probability that a randomly selected student

    (1)   Has a MasterCard.

    (2)   Has a Visa.

    (3)   Has both credit cards.

TABLE 4.2   **Results of a Concept Study for a New Wine Cooler (for Exercises 4.14 and 4.15)**   ⒟Ⓢ WineCooler

Rating	Total	Gender		Age Group		
		Male	**Female**	**21–24**	**25–34**	**35–49**
Extremely appealing (5)	151	68	83	48	66	37
(4)	91	51	40	36	36	19
(3)	36	21	15	9	12	15
(2)	13	7	6	4	6	3
Not at all appealing (1)	9	3	6	4	3	2

Source: W. R. Dillon, T. J. Madden, and N. H. Firtle, *Essentials of Marketing Research* (Burr Ridge, IL: Richard D. Irwin, Inc., 1993), p. 390.

b   Find the probability that a randomly selected student
   (1)   Has a MasterCard or a Visa.
   (2)   Has neither credit card.
   (3)   Has exactly one of the two credit cards.

4.12   The card game of Euchre employs a deck that consists of all four of each of the aces, kings, queens, jacks, tens, and nines (one of each suit—clubs, diamonds, spades, and hearts). Find the probability that a randomly selected card from a Euchre deck is (**1**) a jack (*J*), (**2**) a spade (*S*), (**3**) a jack or an ace (*A*), (**4**) a jack or a spade. (**5**) Are the events *J* and *A* mutually exclusive? Why or why not? (**6**) Are *J* and *S* mutually exclusive? Why or why not?

4.13   Each month a brokerage house studies various companies and rates each company's stock as being either "low risk" or "moderate to high risk." In a recent report, the brokerage house summarized its findings about 15 aerospace companies and 25 food retailers in the following table:

Company Type	Low Risk	Moderate to High Risk
Aerospace company	6	9
Food retailer	15	10

If we randomly select one of the total of 40 companies, find
a   The probability that the company is a food retailer.
b   The probability that the company's stock is "low risk."
c   The probability that the company's stock is "moderate to high risk."

d   The probability that the company is a food retailer and has a stock that is "low risk."
e   The probability that the company is a food retailer or has a stock that is "low risk."

4.14   In the book *Essentials of Marketing Research,* William R. Dillon, Thomas J. Madden, and Neil H. Firtle present the results of a concept study for a new wine cooler. Three hundred consumers between 21 and 49 years old were randomly selected. After sampling the new beverage, each was asked to rate the appeal of the phrase

    Not sweet like wine coolers, not filling like beer, and more refreshing than wine or mixed drinks

as it relates to the new wine cooler. The rating was made on a scale from 1 to 5, with 5 representing "extremely appealing" and with 1 representing "not at all appealing." The results obtained are given in Table 4.2. Estimate the probability that a randomly selected 21 to 49-year-old consumer
a   Would give the phrase a rating of 5.
b   Would give the phrase a rating of 3 or higher.
c   Is in the 21–24 age group; the 25–34 age group; the 35–49 age group.
d   Is a male who gives the phrase a rating of 4.
e   Is a 35- to 49-year-old who gives the phrase a rating of 1.   ⒟Ⓢ WineCooler

4.15   In Exercise 4.14 estimate the probability that a randomly selected 21- to 49-year-old consumer is a 25- to 49-year-old who gives the phrase a rating of 5.   ⒟Ⓢ WineCooler

4.4 Conditional Probability and Independence

LO4-4
Compute conditional probabilities and assess independence.

Conditional probability

In Table 4.3 we repeat Table 4.1 summarizing data concerning Crystal Cable's 27.4 million cable passings. Suppose that we randomly select a cable passing and that the chosen cable passing reports that it has Crystal's cable Internet service. Given this new information, we wish to find the probability that the cable passing has Crystal's cable television service. This new probability is called a **conditional probability.**

> The **probability of the event *A*, given the condition that the event *B* has occurred,** is written as $P(A \mid B)$—pronounced "the probability of *A* given *B*." We often refer to such a probability as the **conditional probability of *A* given *B*.**

TABLE 4.3   A Contingency Table Summarizing Crystal's Cable Television and Internet Penetration (Figures in Millions of Cable Passings)

Events	Has Cable Internet Service, B	Does Not Have Cable Internet Service, $\overline{B}$	Total
Has Cable Television Service, A	6.5	5.9	12.4
Does Not Have Cable Television Service, $\overline{A}$	3.3	11.7	15.0
Total	9.8	17.6	27.4

In order to find the conditional probability that a randomly selected cable passing has Crystal's cable television service, given that it has Crystal's cable Internet service, notice that if we know that the randomly selected cable passing has Crystal's cable Internet service, we know that we are considering one of Crystal's 9.8 million cable Internet customers (see Table 4.3). That is, we are now considering what we might call a **reduced sample space** of Crystal's 9.8 million cable Internet customers. Because 6.5 million of these 9.8 million cable Internet customers also have Crystal's cable television service, we have

$$P(A\,|\,B) = \frac{6.5}{9.8} = .66$$

This says that the probability that the randomly selected cable passing has Crystal's cable television service, given that it has Crystal's cable Internet service, is .66. That is, 66 percent of Crystal's cable Internet customers also have Crystal's cable television service.

Next, suppose that we randomly select another cable passing from Crystal's 27.4 million cable passings, and suppose that this newly chosen cable passing reports that it has Crystal's cable television service. We now wish to find the probability that this cable passing has Crystal's cable Internet service. We write this new probability as $P(B\,|\,A)$. If we know that the randomly selected cable passing has Crystal's cable television service, we know that we are considering a reduced sample space of Crystal's 12.4 million cable television customers (see Table 4.3). Because 6.5 million of these 12.4 million cable television customers also have Crystal's cable Internet service, we have

$$P(B\,|\,A) = \frac{6.5}{12.4} = .52$$

This says that the probability that the randomly selected cable passing has Crystal's cable Internet service, given that it has Crystal's cable television service, is .52. That is, 52 percent of Crystal's cable television customers also have Crystal's cable Internet service.

If we divide both the numerator and denominator of each of the conditional probabilities $P(A\,|\,B)$ and $P(B\,|\,A)$ by 27.4, we obtain

$$P(A\,|\,B) = \frac{6.5}{9.8} = \frac{6.5/27.4}{9.8/27.4} = \frac{P(A \cap B)}{P(B)}$$

$$P(B\,|\,A) = \frac{6.5}{12.4} = \frac{6.5/27.4}{12.4/27.4} = \frac{P(A \cap B)}{P(A)}$$

We express these conditional probabilities in terms of $P(A)$, $P(B)$, and $P(A \cap B)$ in order to obtain a more general formula for a conditional probability. We need a more general formula because, although we can use the reduced sample space approach we have demonstrated to find conditional probabilities when all of the sample space outcomes are equally likely, this approach may not give correct results when the sample space outcomes are *not* equally likely. We now give expressions for conditional probability that are valid for any sample space.

Conditional Probability

1 The **conditional probability of the event A given that the event B has occurred** is written $P(A \mid B)$ and is defined to be

$$P(A \mid B) = \frac{P(A \cap B)}{P(B)}$$

Here we assume that $P(B)$ is greater than 0.

2 The **conditional probability of the event B given that the event A has occurred** is written $P(B \mid A)$ and is defined to be

$$P(B \mid A) = \frac{P(A \cap B)}{P(A)}$$

Here we assume that $P(A)$ is greater than 0.

If we multiply both sides of the equation

$$P(A \mid B) = \frac{P(A \cap B)}{P(B)}$$

by $P(B)$, we obtain the equation

$$P(A \cap B) = P(B)P(A \mid B)$$

Similarly, if we multiply both sides of the equation

$$P(B \mid A) = \frac{P(A \cap B)}{P(A)}$$

by $P(A)$, we obtain the equation

$$P(A \cap B) = P(A)P(B \mid A)$$

In summary, we now have two equations that can be used to calculate $P(A \cap B)$. These equations are often referred to as the **general multiplication rule** for probabilities.

The General Multiplication Rule—Two Ways to Calculate $P(A \cap B)$

Given any two events A and B,

$$P(A \cap B) = P(A)P(B \mid A)$$
$$= P(B)P(A \mid B)$$

C **EXAMPLE 4.10** Gender Issues at a Pharmaceutical Company

At a large pharmaceutical company, 52 percent of the sales representatives are women, and 44 percent of the sales representatives having a management position are women. (There are various types of management positions in the sales division of the pharmaceutical company.) Given that 25 percent of the sales representatives have a management position, we wish to find

- The percentage of the sales representatives that have a management position and are women.
- The percentage of the female sales representatives that have a management position.
- The percentage of the sales representatives that have a management position and are men.
- The percentage of the male sales representatives that have a management position.

In order to find these percentages, consider randomly selecting one of the sales representatives. Then, let W denote the event that the randomly selected sales representative is a woman, and let M denote the event that the randomly selected sales representative is a man. Also, let MGT denote the event that the randomly selected sales representative has a management position. The information given at the beginning of this example says that 52 percent of the sales representatives are women and 44 percent of the sales representatives having a management position are women. This implies that $P(W) = .52$ and that $P(W \mid MGT) = .44$.

The information given at the beginning of this example also says that 25 percent of the sales representatives have a management position. This implies that $P(MGT) = .25$. To find the percentage of the sales representatives that have a management position and are women, we find $P(MGT \cap W)$. The general multiplication rule tells us that

$$P(MGT \cap W) = P(MGT)P(W|MGT) = P(W)P(MGT|W)$$

Although we know that $P(W) = .52$, we do not know $P(MGT|W)$. Therefore, we cannot calculate $P(MGT \cap W)$ as $P(W)P(MGT|W)$. However, because we know that $P(MGT) = .25$ and $P(W|MGT) = .44$, we can calculate

$$P(MGT \cap W) = P(MGT)P(W|MGT) = (.25)(.44) = .11$$

This says that 11 percent of the sales representatives have a management position and are women. Moreover,

$$P(MGT|W) = \frac{P(MGT \cap W)}{P(W)} = \frac{.11}{.52} = .2115$$

This says that 21.15 percent of the female sales representatives have a management position.

To find the percentage of the sales representatives that have a management position and are men, we find $P(MGT \cap M)$. Because we know that 52 percent of the sales representatives are women, the rule of complements tells us that 48 percent of the sales representatives are men. That is, $P(M) = .48$. We also know that 44 percent of the sales representatives having a management position are women. It follows (by an extension of the rule of complements) that 56 percent of the sales representatives having a management position are men. That is, $P(M|MGT) = .56$. Using the fact that $P(MGT) = .25$, the general multiplication rule implies that

$$P(MGT \cap M) = P(MGT)P(M|MGT) = (.25)(.56) = .14$$

This says that 14 percent of the sales representatives have a management position and are men. Moreover,

$$P(MGT|M) = \frac{P(MGT \cap M)}{P(M)} = \frac{.14}{.48} = .2917$$

This says that 29.17 percent of the male sales representatives have a management position.

We have seen that $P(MGT) = .25$, while $P(MGT|W) = .2115$. Because $P(MGT|W)$ is less than $P(MGT)$, the probability that a randomly selected sales representative will have a management position is smaller if we know that the sales representative is a woman than it is if we have no knowledge of the sales representative's gender. Another way to see this is to recall that $P(MGT|M) = .2917$. Because $P(MGT|W) = .2115$ is less than $P(MGT|M) = .2917$, the probability that a randomly selected sales representative will have a management position is smaller if the sales representative is a woman than it is if the sales representative is a man.

Independence

In Example 4.10 the probability of the event MGT is influenced by whether the event W occurs. In such a case, we say that the events MGT and W are **dependent.** If $P(MGT|W)$ were equal to $P(MGT)$, then the probability of the event MGT would not be influenced by whether W occurs. In this case we would say that the events MGT and W are **independent.** This leads to the following definition:

Independent Events

Two events A and B are **independent** if and only if

1   $P(A|B) = P(A)$ or, equivalently,

2   $P(B|A) = P(B)$

Here we assume that $P(A)$ and $P(B)$ are greater than 0.

(C) EXAMPLE 4.11 Gender Issues at a Pharmaceutical Company

Recall that 52 percent of the pharmaceutical company's sales representatives are women. If 52 percent of the sales representatives having a management position were also women, then $P(W|MGT)$ would equal $P(W) = .52$. Moreover, recalling that $P(MGT) = .25$, it would follow that

$$P(MGT|W) = \frac{P(MGT \cap W)}{P(W)} = \frac{P(MGT)P(W|MGT)}{P(W)} = \frac{P(MGT)(.52)}{.52} = P(MGT)$$

That is, 25 percent of the female sales representatives—as well as 25 percent of all of the sales representatives—would have a management position. Of course, this independence is only hypothetical. The actual pharmaceutical company data led us to conclude that MGT and W are dependent. Specifically, because $P(MGT|W) = .2115$ is less than $P(MGT) = .25$ and $P(MGT|M) = .2917$, we conclude that women are less likely to have a management position at the pharmaceutical company. Looking at this another way, note that the ratio of $P(MGT|M) = .2917$ to $P(MGT|W) = .2115$ is $.2917/.2115 = 1.3792$. This says that **the probability that a randomly selected sales representative will have a management position is 37.92 percent higher if the sales representative is a man than it is if the sales representative is a woman.** Moreover, this conclusion describes the actual employment conditions that existed at Novartis Pharmaceutical Company from 2002 to 2007.[4] In the largest gender discrimination case ever to go to trial, Sanford, Wittels, and Heisler LLP used data implying the conclusion above—along with evidence of salary inequities and women being subjected to a hostile and sexist work environment—to successfully represent a class of 5,600 female sales representatives against Novartis. On May 19, 2010, a federal jury awarded $250 million to the class. The award was the largest ever in an employment discrimination case, and in November 2010 a final settlement agreement between Novartis and its female sales representatives was reached.

©Chris Schmidt/Getty Images RF

If the occurrences of the events A and B have nothing to do with each other, then we know that A and B are independent events. This implies that $P(A|B)$ equals $P(A)$ and that $P(B|A)$ equals $P(B)$. Recall that the general multiplication rule tells us that, for any two events A and B, we can say that $P(A \cap B) = P(A)P(B|A)$. Therefore, if $P(B|A)$ equals $P(B)$, it follows that $P(A \cap B) = P(A)P(B)$.

This equation is called the **multiplication rule for independent events.** To summarize:

The Multiplication Rule for Two Independent Events

If A and B are **independent events,** then

$$P(A \cap B) = P(A)P(B)$$

As a simple example, let CW denote the event that your favorite college football team wins its first game next season, and let PW denote the event that your favorite professional football team wins its first game next season. Suppose you believe that $P(CW) = .6$ and $P(PW) = .6$. Then, because the outcomes of a college football game and a professional football game would probably have nothing to do with each other, it is reasonable to assume that CW and PW are independent events. It follows that

$$P(CW \cap PW) = P(CW)P(PW) = (.6)(.6) = .36$$

This probability might seem surprisingly low. That is, because you believe that each of your teams has a 60 percent chance of winning, you might feel reasonably confident that both your college and professional teams will win their first game. Yet the chance of this happening is really only .36!

[4]**Source:** http://www.bononilawgroup.com/blog/2010/07/women-win-a-bias-suit-against-novartis.shtml.

Next, consider a group of events $A_1, A_2, \ldots, A_N$. Intuitively, the events $A_1, A_2, \ldots, A_N$ are independent if the occurrences of these events have nothing to do with each other. Denoting the probability that all of these events will simultaneously occur as $P(A_1 \cap A_2 \cap \cdots \cap A_N)$, we have the following:

The Multiplication Rule for N Independent Events

If $A_1, A_2, \ldots, A_N$ are independent events, then

$$P(A_1 \cap A_2 \cap \cdots \cap A_N) = P(A_1)P(A_2) \cdots P(A_N)$$

C   **EXAMPLE 4.12** An Application of the Independence Rule: Customer Service

This example is based on a real situation encountered by a major producer and marketer of consumer products. The company assessed the service it provides by surveying the attitudes of its customers regarding 10 different aspects of customer service—order filled correctly, billing amount on invoice correct, delivery made on time, and so forth. When the survey results were analyzed, the company was dismayed to learn that only 59 percent of the survey participants indicated that they were satisfied with all 10 aspects of the company's service. Upon investigation, each of the 10 departments responsible for the aspects of service considered in the study insisted that it satisfied its customers 95 percent of the time. That is, each department claimed that its error rate was only 5 percent. Company executives were confused and felt that there was a substantial discrepancy between the survey results and the claims of the departments providing the services. However, a company statistician pointed out that there was no discrepancy. To understand this, consider randomly selecting a customer from among the survey participants, and define 10 events (corresponding to the 10 aspects of service studied):

$A_1 \equiv$ the customer is satisfied that the order is filled correctly (aspect 1).

$A_2 \equiv$ the customer is satisfied that the billing amount on the invoice is correct (aspect 2).

$\vdots$

$A_{10} \equiv$ the customer is satisfied that the delivery is made on time (aspect 10).

Also, define the event

$S \equiv$ the customer is satisfied with all 10 aspects of customer service.

Because 10 different departments are responsible for the 10 aspects of service being studied, it is reasonable to assume that all 10 aspects of service are independent of each other. For instance, billing amounts would be independent of delivery times. Therefore, $A_1, A_2, \ldots, A_{10}$ are independent events, and

$$P(S) = P(A_1 \cap A_2 \cap \cdots \cap A_{10})$$
$$= P(A_1) P(A_2) \cdots P(A_{10})$$

If, as the departments claim, each department satisfies its customers 95 percent of the time, then the probability that the customer is satisfied with all 10 aspects is

$$P(S) = (.95)(.95) \cdots (.95) = (.95)^{10} = .5987$$

This result is almost identical to the 59 percent satisfaction rate reported by the survey participants.

If the company wants to increase the percentage of its customers who are satisfied with all 10 aspects of service, it must improve the quality of service provided by the 10 departments. For example, to satisfy 95 percent of its customers with all 10 aspects of service, the company must require each department to raise the fraction of the time it satisfies its customers to x, where x is such that $(x)^{10} = .95$. It follows that

$$x = (.95)^{\frac{1}{10}} = .9949$$

and that each department must satisfy its customers 99.49 percent of the time (rather than the current 95 percent of the time).

Exercises for Section 4.4

CONCEPTS ■ connect

4.16 Give an example of a conditional probability that would be of interest to you.

4.17 Explain what it means for two events to be independent.

METHODS AND APPLICATIONS

4.18 The following contingency table summarizes the number of students at a college who have a MasterCard and/or a Visa credit card.

	Have Visa	Do Not Have Visa	Total
Have MasterCard	1,000	1,500	2,500
Do Not Have MasterCard	3,000	4,500	7,500
Total	4,000	6,000	10,000

a Find the proportion of MasterCard holders who have Visa cards. Interpret and write this proportion as a conditional probability.

b Find the proportion of Visa cardholders who have MasterCards. Interpret and write this proportion as a conditional probability.

c Are the events *having a MasterCard* and *having a Visa* independent? Justify your answer.

4.19 Each month a brokerage house studies various companies and rates each company's stock as being either "low risk" or "moderate to high risk." In a recent report, the brokerage house summarized its findings about 15 aerospace companies and 25 food retailers in the following table:

Company Type	Low Risk	Moderate to High Risk
Aerospace company	6	9
Food retailer	15	10

If we randomly select one of the total of 40 companies, find

a The probability that the company's stock is moderate to high risk given that the firm is an aerospace company.

b The probability that the company's stock is moderate to high risk given that the firm is a food retailer.

c Determine if the events *the firm is a food retailer* and *the firm's stock is low risk* are independent. Explain.

4.20 John and Jane are married. The probability that John watches a certain television show is .4. The probability that Jane watches the show is .5. The probability that John watches the show, given that Jane does, is .7.

a Find the probability that both John and Jane watch the show.

b Find the probability that Jane watches the show, given that John does.

c Do John and Jane watch the show independently of each other? Justify your answer.

4.21 In Exercise 4.20, find the probability that either John or Jane watches the show.

4.22 In the July 29, 2001, issue of the *Journal News* (Hamilton, Ohio), Lynn Elber of the Associated Press reported that "while 40 percent of American families own a television set with a V-chip installed to block designated programs with sex and violence, only 17 percent of those parents use the device."[5]

a Use the report's results to find an estimate of the probability that a randomly selected American family has used a V-chip to block programs containing sex and violence.

b According to the report, more than 50 percent of parents have used the TV rating system (TV-14, etc.) to control their children's TV viewing. How does this compare to the percentage using the V-chip?

4.23 According to the Associated Press report (in Exercise 4.22), 47 percent of parents who have purchased TV sets after V-chips became standard equipment in January 2000 are aware that their sets have V-chips, and of those who are aware of the option, 36 percent have programmed their V-chips. Using these results, find an estimate of the probability that a randomly selected parent who has bought a TV set since January 2000 has programmed the V-chip.

4.24 Fifteen percent of the employees in a company have managerial positions, and 25 percent of the employees in the company have MBA degrees. Also, 60 percent of the managers have MBA degrees. Using the probability formulas,

a Find the proportion of employees who are managers and have MBA degrees.

b Find the proportion of MBAs who are managers.

[5]Source: *Journal News* (Hamilton, Ohio), July 29, 2001, p. C5.

c Are the events *being a manager* and *having an MBA* independent? Justify your answer.

4.25 In Exercise 4.24, find the proportion of employees who either have MBAs or are managers.

4.26 Consider Exercise 4.14. Using the results in Table 4.2, estimate the probability that a randomly selected 21 to 49-year-old consumer would:
ⒹⓈ WineCooler

a Give the phrase a rating of 4 or 5 given that the consumer is male; give the phrase a rating of 4 or 5 given that the consumer is female. Based on these results, is the appeal of the phrase among males much different from the appeal of the phrase among females? Explain.

b Give the phrase a rating of 4 or 5, **(1)** given that the consumer is in the 21–24 age group; **(2)** given that the consumer is in the 25–34 age group; **(3)** given that the consumer is in the 35–49 age group. **(4)** Based on these results, which age group finds the phrase most appealing? Least appealing?

4.27 In a survey of 100 insurance claims, 40 are fire claims (*FIRE*), 16 of which are fraudulent (*FRAUD*). Also, there are a total of 40 fraudulent claims.

a Construct a contingency table summarizing the claims data. Use the pairs of events *FIRE* and $\overline{FIRE}$, *FRAUD* and $\overline{FRAUD}$.

b What proportion of the fire claims are fraudulent?

c Are the events *a claim is fraudulent* and *a claim is a fire claim* independent? Use your probability of part *b* to prove your answer.

4.28 Recall from Exercise 4.3 that two randomly selected customers are each asked to take a blind taste test and then to state which of three diet colas (marked as *A*, *B*, or *C*) he or she prefers. Suppose that cola *A*'s distributor claims that 80 percent of all people prefer cola *A* and that only 10 percent prefer each of colas *B* and *C*.

a Assuming that the distributor's claim is true and that the two taste test participants make independent cola preference decisions, find the probability of each sample space outcome.

b Find the probability that neither taste test participant will prefer cola *A*.

c If, when the taste test is carried out, neither participant prefers cola *A*, use the probability you computed in part *b* to decide whether the distributor's claim seems valid. Explain.

4.29 A sprinkler system inside an office building has two types of activation devices, *D*1 and *D*2, which operate independently. When there is a fire, if either device operates correctly, the sprinkler system is turned on. In case of fire, the probability that *D*1 operates correctly is .95, and the probability that *D*2 operates correctly is .92. Find the probability that

a Both *D*1 and *D*2 will operate correctly.

b The sprinkler system will turn on.

c The sprinkler system will fail.

4.30 A product is assembled using 10 different components, each of which must meet specifications for five different quality characteristics. Suppose that there is a .9973 probability that each individual specification will be met. Assuming that all 50 specifications are met independently, find the probability that the product meets all 50 specifications.

4.31 In Exercise 4.30, suppose that we wish to have a 99.73 percent chance that all 50 specifications will be met. If each specification will have the same chance of being met, how large must we make the probability of meeting each individual specification?

4.32 **GENDER ISSUES AT A DISCOUNT CHAIN**

Suppose that 65 percent of a discount chain's employees are women and 33 percent of the discount chain's employees having a management position are women. If 25 percent of the discount chain's employees have a management position, what percentage of the discount chain's female employees have a management position?

4.33 In a murder trial in Los Angeles, the prosecution claims that the defendant was cut on the left middle finger at the murder scene, but the defendant claims the cut occurred in Chicago, the day after the murders had been committed. Because the defendant is a sports celebrity, many people noticed him before he reached Chicago. Twenty-two people saw him casually, one person on the plane to Chicago carefully studied his hands looking for a championship ring, and another person stood with him as he signed autographs and drove him from the airport to the hotel. None of these 24 people saw a cut on the defendant's finger. If in fact he was not cut at all, it would be extremely unlikely that he left blood at the murder scene.

a Because a person casually meeting the defendant would not be looking for a cut, assume that the probability is .9 that such a person would not have seen the cut, even if it was there. Furthermore, assume that the person who carefully looked at the defendant's hands had a .5 probability of not seeing the cut even if it was there and that the person who drove the defendant from the airport to the hotel had a .6 probability of not seeing the cut even if it was there. Given these assumptions, and also assuming that all 24 people looked at the defendant independently of each other, what is the probability that none of the 24 people would have seen the cut, even if it was there?

b What is the probability that at least one of the 24 people would have seen the cut if it was there?

c Given the result of part *b* and given the fact that none of the 24 people saw a cut, do you think the defendant had a cut on his hand before he reached Chicago?

d How might we estimate what the assumed probabilities in part *a* would actually be? (Note: This would not be easy.)

4.5 Bayes' Theorem (Optional)

Sometimes we have an initial or **prior probability** that an event will occur. Then, based on new information, we revise the prior probability to what is called a **posterior probability.** This revision can be done by using a theorem called **Bayes' Theorem.**

LO4-5
Use Bayes' Theorem to update prior probabilities to posterior probabilities (Optional).

EXAMPLE 4.13 Should HIV Testing Be Mandatory?

HIV (Human Immunodeficiency Virus) is the virus that causes AIDS. Although many have proposed mandatory testing for HIV, statisticians have frequently spoken against such proposals. In this example, we use Bayes' Theorem to see why.

Let HIV represent the event that a randomly selected American has the HIV virus, and let $\overline{HIV}$ represent the event that a randomly selected American does not have this virus. Because it is estimated that .6 percent of the American population have the HIV virus, $P(HIV) = .006$ and $P(\overline{HIV}) = .994$. A diagnostic test is used to attempt to detect whether a person has HIV. According to historical data, 99.9 percent of people with HIV receive a positive (POS) result when this test is administered, while 1 percent of people who do not have HIV receive a positive result. That is, $P(POS\,|\,HIV) = .999$ and $P(POS\,|\,\overline{HIV}) = .01$. If we administer the test to a randomly selected American (who may or may not have HIV) and the person receives a positive test result, what is the probability that the person actually has HIV? This probability is

$$P(HIV\,|\,POS) = \frac{P(HIV \cap POS)}{P(POS)}$$

The idea behind Bayes' Theorem is that we can find $P(HIV\,|\,POS)$ by thinking as follows. A person will receive a positive result (POS) if the person receives a positive result and actually has HIV—that is, ($HIV \cap POS$)—or if the person receives a positive result and actually does not have HIV—that is, ($\overline{HIV} \cap POS$). Therefore,

$$P(POS) = P(HIV \cap POS) + P(\overline{HIV} \cap POS)$$

This implies that

$$P(HIV\,|\,POS) = \frac{P(HIV \cap POS)}{P(POS)}$$

$$= \frac{P(HIV \cap POS)}{P(HIV \cap POS) + P(\overline{HIV} \cap POS)}$$

$$= \frac{P(HIV)P(POS\,|\,HIV)}{P(HIV)P(POS\,|\,HIV) + P(\overline{HIV})P(POS\,|\,\overline{HIV})}$$

$$= \frac{.006(.999)}{.006(.999) + (.994)(.01)} = .38$$

This probability says that, if all Americans were given a test for HIV, only 38 percent of the people who get a positive result would actually have HIV. That is, 62 percent of Americans identified as having HIV would actually be free of the virus! The reason for this rather surprising result is that, because so few people actually have HIV, the majority of people who test positive are people who are free of HIV and, therefore, erroneously test positive. This is why statisticians have spoken against proposals for mandatory HIV testing.

In the preceding example, there were two *states of nature*—HIV and $\overline{HIV}$—and two outcomes of the diagnostic test—POS and $\overline{POS}$. In general, there might be any number of states of nature and any number of experimental outcomes. This leads to a general statement of Bayes' Theorem.

Bayes' Theorem

Let $S_1, S_2, \ldots, S_k$ be k mutually exclusive states of nature, one of which must be true, and suppose that $P(S_1), P(S_2), \ldots, P(S_k)$ are the prior probabilities of these states of nature. Also, let E be a particular outcome of an experiment designed to help determine which state of nature is really true. Then, the **posterior probability** of a particular state of nature, say S_i, given the experimental outcome E, is

$$P(S_i|E) = \frac{P(S_i \cap E)}{P(E)} = \frac{P(S_i)P(E|S_i)}{P(E)}$$

where

$$P(E) = P(S_1 \cap E) + P(S_2 \cap E) + \cdots + P(S_k \cap E)$$

$$= P(S_1)P(E|S_1) + P(S_2)P(E|S_2) + \cdots + P(S_k)P(E|S_k)$$

Specifically, if there are two mutually exclusive states of nature, S_1 and S_2, one of which must be true, then

$$P(S_i|E) = \frac{P(S_i)P(E|S_i)}{P(S_1)P(E|S_1) + P(S_2)P(E|S_2)}$$

We have illustrated Bayes' Theorem when there are two states of nature in Example 4.13. In the next example, we consider three states of nature.

Ⓒ EXAMPLE 4.14 The Oil Drilling Case: Site Selection

An oil company is attempting to decide whether to drill for oil on a particular site. There are three possible states of nature:

1.   No oil (state of nature S_1, which we will denote as *none*)
2.   Some oil (state of nature S_2, which we will denote as *some*)
3.   Much oil (state of nature S_3, which we will denote as *much*)

Based on experience and knowledge concerning the site's geological characteristics, the oil company feels that the prior probabilities of these states of nature are as follows:

$$P(S_1 \equiv none) = .7 \qquad P(S_2 \equiv some) = .2 \qquad P(S_3 \equiv much) = .1$$

In order to obtain more information about the potential drilling site, the oil company can perform a seismic experiment, which has three readings—low, medium, and high. Moreover, information exists concerning the accuracy of the seismic experiment. The company's historical records tell us that

1.   Of 100 past sites that were drilled and produced no oil, 4 sites gave a high reading. Therefore,

$$P(high|none) = \frac{4}{100} = .04$$

2.   Of 400 past sites that were drilled and produced some oil, 8 sites gave a high reading. Therefore,

$$P(high|some) = \frac{8}{400} = .02$$

©Digital Vision/Getty Images RF

3   Of 300 past sites that were drilled and produced much oil, 288 sites gave a high reading. Therefore,

$$P(\text{high} \mid \text{much}) = \frac{288}{300} = .96$$

Intuitively, these conditional probabilities tell us that sites that produce no oil or some oil seldom give a high reading, while sites that produce much oil often give a high reading.

Now, suppose that when the company performs the seismic experiment on the site in question, it obtains a high reading. The previously given conditional probabilities suggest that, given this new information, the company might feel that the likelihood of much oil is higher than its prior probability $P(\text{much}) = .1$, and that the likelihoods of some oil and no oil are lower than the prior probabilities $P(\text{some}) = .2$ and $P(\text{none}) = .7$. To be more specific, we wish to *revise the prior probabilities* of no, some, and much oil to what we call *posterior probabilities.* We can do this by using Bayes' Theorem as follows.

If we wish to compute $P(\text{none} \mid \text{high})$, we first calculate

$$\begin{aligned} P(\text{high}) &= P(\text{none} \cap \text{high}) + P(\text{some} \cap \text{high}) + P(\text{much} \cap \text{high}) \\ &= P(\text{none})P(\text{high} \mid \text{none}) + P(\text{some})P(\text{high} \mid \text{some}) + P(\text{much})P(\text{high} \mid \text{much}) \\ &= (.7)(.04) + (.2)(.02) + (.1)(.96) = .128 \end{aligned}$$

Then Bayes' Theorem says that

$$P(\text{none} \mid \text{high}) = \frac{P(\text{none} \cap \text{high})}{P(\text{high})} = \frac{P(\text{none})P(\text{high} \mid \text{none})}{P(\text{high})} = \frac{.7(.04)}{.128} = .21875$$

Similarly, we can compute $P(\text{some} \mid \text{high})$ and $P(\text{much} \mid \text{high})$ as follows.

$$P(\text{some} \mid \text{high}) = \frac{P(\text{some} \cap \text{high})}{P(\text{high})} = \frac{P(\text{some})P(\text{high} \mid \text{some})}{P(\text{high})} = \frac{.2(.02)}{.128} = .03125$$

$$P(\text{much} \mid \text{high}) = \frac{P(\text{much} \cap \text{high})}{P(\text{high})} = \frac{P(\text{much})P(\text{high} \mid \text{much})}{P(\text{high})} = \frac{.1(.96)}{.128} = .75$$

These revised probabilities tell us that, given that the seismic experiment gives a high reading, the revised probabilities of no, some, and much oil are .21875, .03125, and .75, respectively.

Because the posterior probability of much oil is .75, we might conclude that we should drill on the oil site. However, this decision should also be based on economic considerations. The science of **decision theory** provides various criteria for making such a decision. An introduction to decision theory can be found in Chapter 19.

In this section we have only introduced Bayes' Theorem. There is an entire subject called **Bayesian statistics,** which uses Bayes' Theorem to update prior belief about a probability or population parameter to posterior belief. The use of Bayesian statistics is controversial in the case where the prior belief is largely based on subjective considerations, because many statisticians do not believe that we should base decisions on subjective considerations. Realistically, however, we all do this in our daily lives. For example, how each person viewed the evidence in the O. J. Simpson murder trial had a great deal to do with the person's prior beliefs about both O. J. Simpson and the police.

Exercises for Section 4.5

CONCEPTS ■ connect

4.34 What is a prior probability? What is a posterior probability?

4.35 Explain the purpose behind using Bayes' Theorem.

METHODS AND APPLICATIONS

4.36 Suppose that A_1, A_2, and B are events where A_1 and A_2 are mutually exclusive and

$$P(A_1) = .8 \quad P(B|A_1) = .1$$
$$P(A_2) = .2 \quad P(B|A_2) = .3$$

Use this information to find $P(A_1|B)$ and $P(A_2|B)$.

4.37 Suppose that A_1, A_2, A_3, and B are events where A_1, A_2, and A_3 are mutually exclusive and

$$P(A_1) = .2 \quad P(A_2) = .5 \quad P(A_3) = .3$$
$$P(B|A_1) = .02 \quad P(B|A_2) = .05 \quad P(B|A_3) = .04$$

Use this information to find $P(A_1|B)$, $P(A_2|B)$ and $P(A_3|B)$.

4.38 Again consider the diagnostic test for HIV discussed in Example 4.13 and recall that $P(POS|HIV) = .999$ and $P(POS|\overline{HIV}) = .01$, where POS denotes a positive test result. Assuming that the percentage of people who have HIV is 1 percent, recalculate the probability that a randomly selected person has HIV, given that his or her test result is positive.

4.39 A department store is considering a new credit policy to try to reduce the number of customers defaulting on payments. A suggestion is made to discontinue credit to any customer who has been one week or more late with his/her payment at least twice. Past records show 95 percent of defaults were late at least twice. Also, 3 percent of all customers default, and 30 percent of those who have not defaulted have had at least two late payments.
   a Find the probability that a customer with at least two late payments will default.
   b Based on part *a*, should the policy be adopted? Explain.

4.40 A company administers an "aptitude test for managers" to aid in selecting new management trainees. Prior experience suggests that 60 percent of all applicants for management trainee positions would be successful if they were hired. Furthermore, past experience with the aptitude test indicates that 85 percent of applicants who turn out to be successful managers pass the test and 90 percent of applicants who do not turn out to be successful managers fail the test.
   a If an applicant passes the "aptitude test for managers," what is the probability that the applicant will succeed in a management position?
   b Based on your answer to part *a*, do you think that the "aptitude test for managers" is a valuable way to screen applicants for management trainee positions? Explain.

4.41 THE OIL DRILLING CASE

Recall that the prior probabilities of no oil (*none*), some oil (*some*), and much oil (*much*) are:

$$P(\text{none}) = .7 \quad P(\text{some}) = .2 \quad P(\text{much}) = .1$$

Of 100 past sites that were drilled and produced no oil, 5 gave a medium reading. Of the 400 past sites that were drilled and produced some oil, 376 gave a medium reading. Of the 300 past sites that were drilled and produced much oil, 9 gave a medium reading. This implies that the conditional probabilities of a medium reading (medium) given no oil, some oil, and much oil are:

$$P(\text{medium}|\text{none}) = \frac{5}{100} = .05$$

$$P(\text{medium}|\text{some}) = \frac{376}{400} = .94$$

$$P(\text{medium}|\text{much}) = \frac{9}{300} = .03$$

Calculate the posterior probabilities of no, some, and much oil, given a medium reading.

4.42 THE OIL DRILLING CASE

Of 100 past sites that were drilled and produced no oil, 91 gave a low reading. Of the 400 past sites that were drilled and produced some oil, 16 gave a low reading. Of the 300 past sites that were drilled and produced much oil, 3 gave a low reading. Calculate the posterior probabilities of no, some, and much oil, given a low reading.

4.43 Three data entry specialists enter requisitions into a computer. Specialist 1 processes 30 percent of the requisitions, specialist 2 processes 45 percent, and specialist 3 processes 25 percent. The proportions of incorrectly entered requisitions by data entry specialists 1, 2, and 3 are .03, .05, and .02, respectively. Suppose that a random requisition is found to have been incorrectly entered. What is the probability that it was processed by data entry specialist 1? By data entry specialist 2? By data entry specialist 3?

4.44 A truth serum given to a suspect is known to be 90 percent reliable when the person is guilty and 99 percent reliable when the person is innocent. In other words, 10 percent of the guilty are judged innocent by the serum and 1 percent of the innocent are judged guilty. If the suspect was selected from a group of suspects of which only 5 percent are guilty of having committed a crime, and the serum indicates that the suspect is guilty of having committed a crime, what is the probability that the suspect is innocent?

4.6 Counting Rules (Optional)

LO4-6
Use some elementary
counting rules to
compute probabilities
(Optional).

Consider the situation in Example 4.2 in which a student takes a pop quiz that consists of three true–false questions. If we consider our experiment to be answering the three questions, each question can be answered correctly or incorrectly. We will let C denote answering a question correctly and I denote answering a question incorrectly. Figure 4.6 depicts a tree diagram of the sample space outcomes for the experiment. The diagram portrays the experiment as a three-step process—answering the first question (correctly or incorrectly, that is, C or I), answering the second question (correctly or incorrectly, that is, C or I), and answering the third question (correctly or incorrectly, that is, C or I). The tree diagram has eight different branches, and the eight distinct sample space outcomes are listed at the ends of the branches.

In general, a rule that is helpful in determining the number of experimental outcomes in a multiple-step experiment is as follows:

A Counting Rule for Multiple-Step Experiments

If an experiment can be described as a sequence of k steps in which there are n_1 possible outcomes on the first step, n_2 possible outcomes on the second step, and so on, then the total number of experimental outcomes is given by $(n_1)(n_2) \cdots (n_k)$.

FIGURE 4.6   A Tree Diagram of Answering Three True–False Questions

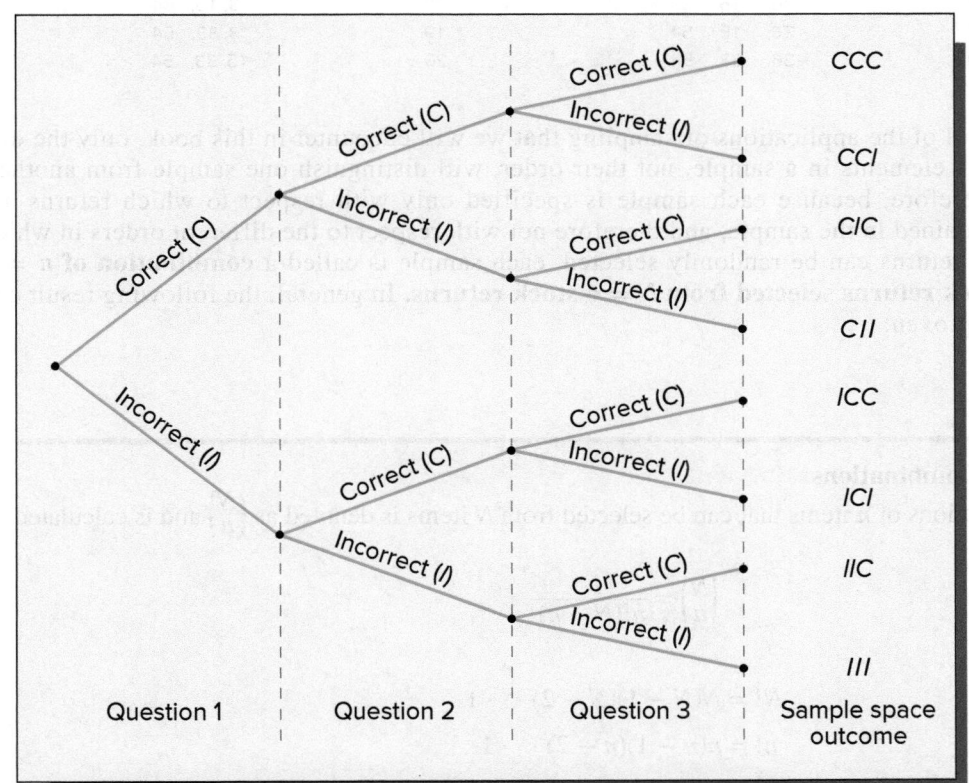

For example, the pop quiz example consists of three steps in which there are $n_1 = 2$ possible outcomes on the first step, $n_2 = 2$ possible outcomes on the second step, and $n_3 = 2$ possible outcomes on the third step. Therefore, the total number of experimental outcomes is $(n_1)(n_2)(n_3) = (2)(2)(2) = 8$, as is shown in Figure 4.6. Now suppose the student takes a pop quiz consisting of five true–false questions. Then, there are $(n_1)(n_2)(n_3)(n_4)(n_5) = (2)(2)(2)(2)(2) = 32$ experimental outcomes. If the student is totally unprepared for the quiz and has to blindly guess the answer to each question, the 32 experimental outcomes might be considered to be equally likely. Therefore, because only one of these outcomes corresponds to all five questions being answered correctly, the probability that the student will answer all five questions correctly is $1/32$.

As another example, suppose a bank has three branches; each branch has two departments, and each department has four employees. One employee is to be randomly selected to go to a convention. Because there are $(n_1)(n_2)(n_3) = (3)(2)(4) = 24$ employees, the probability that a particular one will be randomly selected is $1/24$.

Next, consider the population of last year's percentage returns for six high-risk stocks. This population consists of the percentage returns -36, -15, 3, 15, 33, and 54 (which we have arranged in increasing order). Now consider randomly selecting without replacement a sample of $n = 3$ stock returns from the population of six stock returns. Below we list the 20 distinct samples of $n = 3$ returns that can be obtained:

Sample	n = 3 Returns in Sample	Sample	n = 3 Returns in Sample
1	−36, −15, 3	11	−15, 3, 15
2	−36, −15, 15	12	−15, 3, 33
3	−36, −15, 33	13	−15, 3, 54
4	−36, −15, 54	14	−15, 15, 33
5	−36, 3, 15	15	−15, 15, 54
6	−36, 3, 33	16	−15, 33, 54
7	−36, 3, 54	17	3, 15, 33
8	−36, 15, 33	18	3, 15, 54
9	−36, 15, 54	19	3, 33, 54
10	−36, 33, 54	20	15, 33, 54

In all of the applications of sampling that we will encounter in this book, only the distinct elements in a sample, not their order, will distinguish one sample from another. Therefore, because each sample is specified only with respect to which returns are contained in the sample, and therefore not with respect to the different orders in which the returns can be randomly selected, each sample is called a **combination of $n = 3$ stock returns selected from $N = 6$ stock returns.** In general, the following result can be proven:

A Counting Rule for Combinations

The number of combinations of n items that can be selected from N items is denoted as $\binom{N}{n}$ and is calculated using the formula

$$\binom{N}{n} = \frac{N!}{n!(N-n)!}$$

where

$$N! = N(N-1)(N-2) \cdots 1$$

$$n! = n(n-1)(n-2) \cdots 1$$

Here, for example, $N!$ is pronounced "N factorial." Moreover, zero factorial (that is, $0!$) is defined to equal 1.

For example, the number of combinations of $n = 3$ stock returns that can be selected from the six previously discussed stock returns is

$$\binom{6}{3} = \frac{6!}{3!\,(6-3)!} = \frac{6!}{3!3!} = \frac{6 \cdot 5 \cdot 4 \,(3 \cdot 2 \cdot 1)}{(3 \cdot 2 \cdot 1)\,(3 \cdot 2 \cdot 1)} = 20$$

The 20 combinations are listed on the previous page. As another example, the Ohio lottery system uses the random selection of 6 numbers from a group of 47 numbers to determine each week's lottery winner. There are

$$\binom{47}{6} = \frac{47!}{6!\,(47-6)!} = \frac{47 \cdot 46 \cdot 45 \cdot 44 \cdot 43 \cdot 42 \,(41!)}{6 \cdot 5 \cdot 4 \cdot 3 \cdot 2 \cdot 1 \,(41!)} = 10{,}737{,}573$$

combinations of 6 numbers that can be selected from 47 numbers. Therefore, if you buy a lottery ticket and pick six numbers, the probability that this ticket will win the lottery is 1/10,737,573.

Exercises for Section 4.6

CONCEPTS connect

4.45 Explain why counting rules are useful.

4.46 Explain when it is appropriate to use the counting rule for multiple-step experiments.

4.47 Explain when it is appropriate to use the counting rule for combinations.

METHODS AND APPLICATIONS

4.48 A credit union has two branches; each branch has two departments, and each department has four employees. How many total people does the credit union employ? If you work for the credit union, and one employee is randomly selected to go to a convention, what is the probability that you will be chosen?

4.49 Construct a tree diagram (like Figure 4.6) for the situation described in Exercise 4.48.

4.50 How many combinations of two high-risk stocks could you randomly select from eight high-risk stocks? If

you did this, what is the probability that you would obtain the two highest-returning stocks?

4.51 A pop quiz consists of three true–false questions and three multiple-choice questions. Each multiple-choice question has five possible answers. If a student blindly guesses the answer to every question, what is the probability that the student will correctly answer all six questions?

4.52 A company employs eight people and plans to select a group of three of these employees to receive advanced training. How many ways can the group of three employees be selected?

4.53 The company of Exercise 4.52 employs Mr. Withrow, Mr. Church, Ms. David, Ms. Henry, Mr. Fielding, Mr. Smithson, Ms. Penny, and Mr. Butler. If the three employees who will receive advanced training are selected at random, what is the probability that Mr. Church, Ms. Henry, and Mr. Butler will be selected for advanced training?

Chapter Summary

In this chapter we studied **probability** and the idea of a **probability model.** We began by defining an **event** to be an experimental outcome that may or may not occur and by defining the **probability of an event** to be a number that measures the likelihood that the event will occur. We learned that a probability is often interpreted as a **long-run relative frequency,** and we saw that probabilities can be found by examining **sample spaces** and by using **probability rules.** We learned several important probability rules—**addition rules, multiplication rules,** and **the rule of**

complements. We also studied a special kind of probability called a **conditional probability,** which is the probability that one event will occur given that another event has occurred, and we used probabilities to define **independent events.** We concluded this chapter by studying two optional topics. The first of these was **Bayes' Theorem,** which can be used to update a **prior** probability to a **posterior** probability based on receiving new information. Second, we studied **counting rules** that are helpful when we wish to count sample space outcomes.

Glossary of Terms

Bayes' Theorem: A theorem (formula) that is used to compute posterior probabilities by revising prior probabilities.

Bayesian statistics: An area of statistics that uses Bayes' Theorem to update prior belief about a probability or population parameter to posterior belief.

classical method (for assigning probability): A method of assigning probabilities that can be used when all of the sample space outcomes are equally likely.

complement (of an event): If A is an event, the complement of A is the event that A will not occur.

conditional probability: The probability that one event will occur given that we know that another event has occurred.

decision theory: An approach that helps decision makers to make intelligent choices.

dependent events: When the probability of one event is influenced by whether another event occurs, the events are said to be dependent.

event: A set of one or more sample space outcomes.

experiment: A process of observation that has an uncertain outcome.

independent events: When the probability of one event is not influenced by whether another event occurs, the events are said to be independent.

intersection of events A and B: The event that occurs if both event A and event B simultaneously occur. This is denoted $A \cap B$.

mutually exclusive events: Events that have no sample space outcomes in common, and, therefore, cannot occur simultaneously.

posterior probability: A revised probability obtained by updating a prior probability after receiving new information.

prior probability: The initial probability that an event will occur.

probability (of an event): A number that measures the chance, or likelihood, that an event will occur when an experiment is carried out.

probability model: A mathematical representation of a random phenomenon.

relative frequency method (for assigning probability): A method of estimating a probability by performing an experiment (in which an outcome of interest might occur) many times.

sample space: The set of all possible experimental outcomes (sample space outcomes).

sample space outcome: A distinct outcome of an experiment (that is, an element in the sample space).

subjective method (for assigning probability): Using experience, intuition, or expertise to assess the probability of an event.

subjective probability: A probability assessment that is based on experience, intuitive judgment, or expertise.

tree diagram: A diagram that depicts an experiment as a sequence of steps and is useful when listing all the sample space outcomes for the experiment.

union of events A and B: The event that occurs when event A or event B (or both) occur. This is denoted $A \cup B$.

Important Formulas

Probabilities when all sample space outcomes are equally likely: page 225

The rule of complements: page 227

The addition rule for two events: page 230

The addition rule for two mutually exclusive events: page 231

The addition rule for N mutually exclusive events: page 232

Conditional probability: pages 233, 235

The general multiplication rule: page 235

Independence: page 236

The multiplication rule for two independent events: page 237

The multiplication rule for N independent events: page 238

Bayes' Theorem: page 242

Counting rule for multiple-step experiments: page 245

Counting rule for combinations: page 246

Supplementary Exercises   connect

Exercises 4.54 through 4.57 are based on the following situation: An investor holds two stocks, each of which can rise (R), remain unchanged (U), or decline (D) on any particular day.

4.54 Construct a tree diagram showing all possible combined movements for both stocks on a particular day (for instance, RR, RD, and so on, where the first letter denotes the movement of the first stock, and the second letter denotes the movement of the second stock).

4.55 If all outcomes are equally likely, find the probability that both stocks rise; that both stocks decline; that exactly one stock declines.

4.56 Find the probabilities you found in Exercise 4.55 by assuming that for each stock $P(R) = .6$, $P(U) = .1$, and

$P(D) = .3$, and assuming that the two stocks move independently.

4.57   Assume that for the first stock (on a particular day)

$$P(R) = .4, \; P(U) = .2, \; P(D) = .4$$

and that for the second stock (on a particular day)

$$P(R) = .8, \; P(U) = .1, \; P(D) = .1$$

Assuming that these stocks move independently, find the probability that both stocks decline; the probability that exactly one stock rises; the probability that exactly one stock is unchanged; the probability that both stocks rise.

The Bureau of Labor Statistics reports on a variety of employment statistics. *College Enrollment and Work Activity of 2004 High School Graduates* provides information on high school graduates by gender, by race, and by labor force participation as of October 2004.[6] (All numbers are in thousands.) The following two tables provide sample information on the "Labor force status of persons 16 to 24 years old by educational attainment and gender, October 2004." Using the information contained in the tables, do Exercises 4.58 through 4.62.   DS LabForce

Women, Age 16 to 24	Civilian Labor Force		Not in Labor Force	Row Total
	Employed	Unemployed		
< High School	662	205	759	1,626
HS degree	2,050	334	881	3,265
Some college	1,352	126	321	1,799
Bachelor's degree or more	921	55	105	1,081
Column Total	4,985	720	2,066	7,771

Men, Age 16 to 24	Civilian Labor Force		Not in Labor Force	Row Total
	Employed	Unemployed		
< High School	1,334	334	472	2,140
HS degree	3,110	429	438	3,977
Some college	1,425	106	126	1,657
Bachelor's degree or more	708	37	38	783
Column Total	6,577	906	1,074	8,557

4.58   Find an estimate of the probability that a randomly selected female aged 16 to 24 is in the civilian labor force, if she has a high school degree.   DS LabForce

4.59   Find an estimate of the probability that a randomly selected female aged 16 to 24 is in the civilian labor force, if she has a bachelor's degree or more.   DS LabForce

4.60   Find an estimate of the probability that a randomly selected female aged 16 to 24 is employed, if she is in the civilian labor force and has a high school degree.   DS LabForce

4.61   Find an estimate of the probability that a randomly selected female aged 16 to 24 is employed, if she is in the civilian labor force and has a bachelor's degree or more.   DS LabForce

4.62   Repeat Exercises 4.58 through 4.61 for a randomly selected male aged 16 to 24. In general, do the tables imply that labor force status and employment status depend upon educational attainment? Explain your answer.   DS LabForce

Suppose that in a survey of 1,000 U.S. residents, 721 residents believed that the amount of violent television programming had increased over the past 10 years, 454 residents believed that the overall quality of television programming had decreased over the past 10 years, and 362 residents believed both. Use this information to do Exercises 4.63 through 4.69.

4.63   What proportion of the 1,000 U.S. residents believed that the amount of violent programming had increased over the past 10 years?

4.64   What proportion of the 1,000 U.S. residents believed that the overall quality of programming had decreased over the past 10 years?

4.65   What proportion of the 1,000 U.S. residents believed that both the amount of violent programming had increased and the overall quality of programming had decreased over the past 10 years?

4.66   What proportion of the 1,000 U.S. residents believed that either the amount of violent programming had increased or the overall quality of programming had decreased over the past 10 years?

4.67   What proportion of the U.S. residents who believed that the amount of violent programming had increased, believed that the overall quality of programming had decreased?

4.68   What proportion of the U.S. residents who believed that the overall quality of programming had decreased, believed that the amount of violent programming had increased?

4.69   What sort of dependence seems to exist between whether U.S. residents believed that the amount of violent programming had increased and whether U.S. residents believed that the overall quality of programming had decreased? Explain your answer.

4.70   Enterprise Industries has been running a television advertisement for Fresh liquid laundry detergent. When a survey was conducted, 21 percent of the individuals surveyed had purchased Fresh, 41 percent of the individuals surveyed had recalled seeing the advertisement, and 13 percent of the individuals surveyed had purchased Fresh and recalled seeing the advertisement.

[6]Source: *College Enrollment and Work Activity of 2004 High School Graduates*, Table 2, "Labor force status of persons 16 to 24 years old by school enrollment, educational attainment, sex, race, and Hispanic or Latino ethnicity, October 2004," www.bls.gov.

a What proportion of the individuals surveyed who recalled seeing the advertisement had purchased Fresh?

b Based on your answer to part *a*, does the advertisement seem to have been effective? Explain.

4.71 A company employs 400 salespeople. Of these, 83 received a bonus last year, 100 attended a special sales training program at the beginning of last year, and 42 both attended the special sales training program and received a bonus. (Note: The bonus was based totally on sales performance.)

a What proportion of the 400 salespeople received a bonus last year?

b What proportion of the 400 salespeople attended the special sales training program at the beginning of last year?

c What proportion of the 400 salespeople both attended the special sales training program and received a bonus?

d What proportion of the salespeople who attended the special sales training program received a bonus?

e Based on your answers to parts *a* and *d*, does the special sales training program seem to have been effective? Explain your answer.

4.72 On any given day, the probability that the Ohio River at Cincinnati is polluted by a carbon tetrachloride spill is .10. Each day, a test is conducted to determine whether the river is polluted by carbon tetrachloride. This test has proved correct 80 percent of the time. Suppose that on a particular day the test indicates carbon tetrachloride pollution. What is the probability that such pollution actually exists?

4.73 In the book *Making Hard Decisions: An Introduction to Decision Analysis,* Robert T. Clemen presents an example in which he discusses the 1982 John Hinckley trial. In describing the case, Clemen says:

In 1982 John Hinckley was on trial, accused of having attempted to kill President Reagan. During Hinckley's trial, Dr. Daniel R. Weinberger told the court that when individuals diagnosed as schizophrenics were given computerized axial tomography (CAT) scans, the scans showed brain atrophy in 30% of the cases compared with only 2% of the scans done on normal people. Hinckley's defense attorney wanted to introduce as evidence Hinckley's CAT scan, which showed brain atrophy. The defense argued that the presence of atrophy strengthened the case that Hinckley suffered from mental illness.

a Approximately 1.5 percent of the people in the United States suffer from schizophrenia. If we consider the prior probability of schizophrenia to be .015, use the information given to find the probability that a person has schizophrenia given that a person's CAT scan shows brain atrophy.

b John Hinckley's CAT scan showed brain atrophy. Discuss whether your answer to part *a* helps or hurts the case that Hinckley suffered from mental illness.

c It can be argued that .015 is not a reasonable prior probability of schizophrenia. This is because .015 is the probability that a randomly selected U.S. citizen has schizophrenia. However, John Hinckley was not a randomly selected U.S. citizen. Rather, he was accused of attempting to assassinate the president. Therefore, it

might be reasonable to assess a higher prior probability of schizophrenia. Suppose you are a juror who believes there is only a 10 percent chance that Hinckley suffers from schizophrenia. Using .10 as the prior probability of schizophrenia, find the probability that a person has schizophrenia given that a person's CAT scan shows brain atrophy.

d If you are a juror with a prior probability of .10 that John Hinckley suffers from schizophrenia and given your answer to part *c*, does the fact that Hinckley's CAT scan showed brain atrophy help the case that Hinckley suffered from mental illness?

e If you are a juror with a prior probability of .25 that Hinckley suffers from schizophrenia, find the probability of schizophrenia given that Hinckley's CAT scan showed brain atrophy. In this situation, how strong is the case that Hinckley suffered from mental illness?

4.74 Below we give two contingency tables of data from reports submitted by airlines to the U.S. Department of Transportation. The data concern the numbers of on-time and delayed flights for Alaska Airlines and America West Airlines at five major airports. ⑤ AirDelays

⑤ AirDelays	Alaska Airlines		
	On Time	**Delayed**	**Total**
Los Angeles	497	62	559
Phoenix	221	12	233
San Diego	212	20	232
San Francisco	503	102	605
Seattle	1,841	305	2,146
Total	3,274	501	3,775
	America West		
	On Time	**Delayed**	**Total**
Los Angeles	694	117	811
Phoenix	4,840	415	5,255
San Diego	383	65	448
San Francisco	320	129	449
Seattle	201	61	262
Total	6,438	787	7,225

Source: A. Barnett, "How Numbers Can Trick You," *Technology Review,* October 1994, pp. 38–45. MIT Technology Review, 1994.

a What percentage of all Alaska Airlines flights were delayed? That is, use the data to estimate the probability that an Alaska Airlines flight will be delayed. Do the same for America West Airlines. Which airline does best overall?

b For Alaska Airlines, find the percentage of delayed flights at each airport. That is, use the data to estimate each of the probabilities P(delayed|Los Angeles), P(delayed|Phoenix), and so on. Then do the same for America West Airlines. Which airline does best at each individual airport?

c We find that America West Airlines does worse at every airport, yet America West does best overall. This seems impossible, but it is true! By looking carefully at the data, explain how this can happen. Hint: Consider the weather in Phoenix and Seattle. (This exercise is an example of what is called *Simpson's paradox.*)

4.75 **Internet Exercise** DS CDCData

What is the age, gender, and ethnic composition of U.S. college students? As background for its 1995 study of college students and their risk behaviors, the Centers for Disease Control and Prevention collected selected demographic data—age, gender, and ethnicity—about college students. A report on the 1995 National Health Risk Behavior Survey can be found at the CDC website by going directly to http://www.cdc.gov/mmwr/preview/mmwrhtml/00049859.htm. This report includes a large number of tables, the first of which summarizes the demographic information for the sample of $n = 4609$ college students. An excerpt from Table 1 is given on the right.

Using conditional probabilities, discuss (a) the dependence between age and gender and (b) the dependence between age and ethnicity for U.S. college students.
DS CDCData

TABLE 1. Demographic Characteristics of Undergraduate College Students Aged >=18 Years, by Age Group — United States, National College Health Risk Behavior Survey, 1995

Category	Total (%)	Age Group (%) 18–24 Years	>=25 Years
Total	--	63.6	36.4
Sex			
Female	55.5	52.0	61.8
Male	44.5	48.0	38.2
Race/ethnicity			
White	72.8	70.9	76.1
Black	10.3	10.5	9.6
Hispanic	7.1	6.9	7.4
Other	9.9	11.7	6.9

Design Elements: (CD): ©Comstock Images/Alamy; (All Others): ©McGraw-Hill Education

Predictive Analytics I: Trees, k-Nearest Neighbors, Naive Bayes', and Ensemble Estimates

©Purestock/SuperStock

Learning Objectives

After mastering the material in this chapter, you will be able to:

LO5-1 Interpret the information provided by classification trees.

LO5-2 Interpret the information provided by regression trees.

LO5-3 Interpret the information provided by k-nearest neighbors.

LO5-4 Interpret the information provided by naive Bayes' classification.

LO5-5 Interpret the information provided by ensemble models.

Chapter Outline

As discussed in Chapter 1, predictive analytics are methods used to predict a response variable (for example, sales of a product) on the basis of one or more predictor variables (for example, price and advertising expenditure). Predictive analytics determine how to make predictions by finding a relationship between previously observed values of the response variable and corresponding previously observed values of the predictor variables.

Predictive analytics fall into two classes—parametric predictive analytics and nonparametric predictive analytics. Essentially, parametric predictive analytics find a mathematical equation that relates the response variable to the predictor variables and involves unknown parameters that must be estimated and evaluated by using sample data. Parametric predictive analytics include simple and multiple linear regression, time series forecasting, logistic regression, neural networks,

and discriminate analysis. Because the evaluation of the parameters in the equations obtained by these analytics requires knowledge of formal statistical inference (which is discussed in Chapters 7 through 10 and extended in Chapters 11 through 13), we do not study parametric predictive analytics until Chapters 14 through 17.

Nonparametric predictive analytics make predictions by using a relationship between the response variable and the predictor variables that is not expressed in terms of a mathematical equation involving parameters. These analytics include decision trees (classification and regression trees), k-nearest neighbors, and naive Bayes' classification, and in this chapter we study these analytics. It is important to note that this chapter is optional. Therefore, the reader has the option to study the material in this chapter at this point in the course or to study it when studying parametric predictive analytics.

5.1 Decision Trees I: Classification Trees

LO5-1
Interpret the information provided by classification trees.

In this section we begin to study making predictions by using what are called **decision trees.** If we are predicting a quantitative response variable (for example, product demand, weekly natural gas consumption, or grade point average[1]), we call the decision tree a **regression tree.** If we are predicting a qualitative, or categorical, response variable (for example, whether or not a Silver credit card holder will upgrade to a Platinum card), we call the decision tree a **classification tree.** One reason for this name is that we can use a classification tree to do **classification,** which we have said (in Chapter 1) involves assigning items (for example, Silver credit card holders) to prespecified categories, or classes. Specifically, if we can use a classification tree to estimate the probability that a particular item will fall into a class, then we might assign the item to the class if the estimated probability that the item will fall into the class is "high enough."

Because classification trees are somewhat easier to understand than regression trees, we will begin by explaining classification trees. Figure 5.1(a) shows an example of a classification tree, which has been obtained by using JMP to analyze the credit card upgrade data in Figure 5.1(b). To understand these data and the tree, suppose that a bank wishes to predict whether or not an existing holder of its Silver credit card will upgrade, for an annual fee, to its Platinum credit card. To do this, the bank carries out a pilot study that randomly selects 40 of its existing Silver card holders and offers each Silver card holder an upgrade to its Platinum card. The data in Figure 5.1(b) are the results of the study. Here, the response variable Upgrade equals 1 if the Silver card holder decided to upgrade and 0 otherwise. Moreover, the predictor variable Purchases is last year's purchases (in thousands of dollars) by the Silver card holder, and the predictor variable PlatProfile equals 1 if the Silver card holder conforms to the bank's *Platinum profile* and 0 otherwise. Here, the term *Platinum profile* is the name we are using for a real concept used by a major credit card company with which we have spoken.[2] To explain this concept, we assume that when an individual applies for one of the bank's credit cards, the bank has the person fill out a questionnaire related to how he or she will use the credit card. This helps the bank assess which type of credit card might best meet the needs of the individual. For example, a Platinum card might be best for someone who travels frequently and for whom the card's extra reward points and luxury travel perks justify paying the card's annual fee. Based on the questionnaire filled out by the individual, the bank assesses whether the Silver or

---

[1]Readers who have studied optional Section 1.6 will recall that grade point average may be considered to be "somewhat quantitative."

[2]Although the idea of a Platinum profile comes solely from our discussions with a real credit card company, the original motivation for a credit card upgrade example comes from an example in Berenson, Levine, and Szabat (2015).

FIGURE 5.1   A JMP Classification Tree for the Card Upgrade Data   ⊙ CardUpgrade

(a) A JMP classification tree

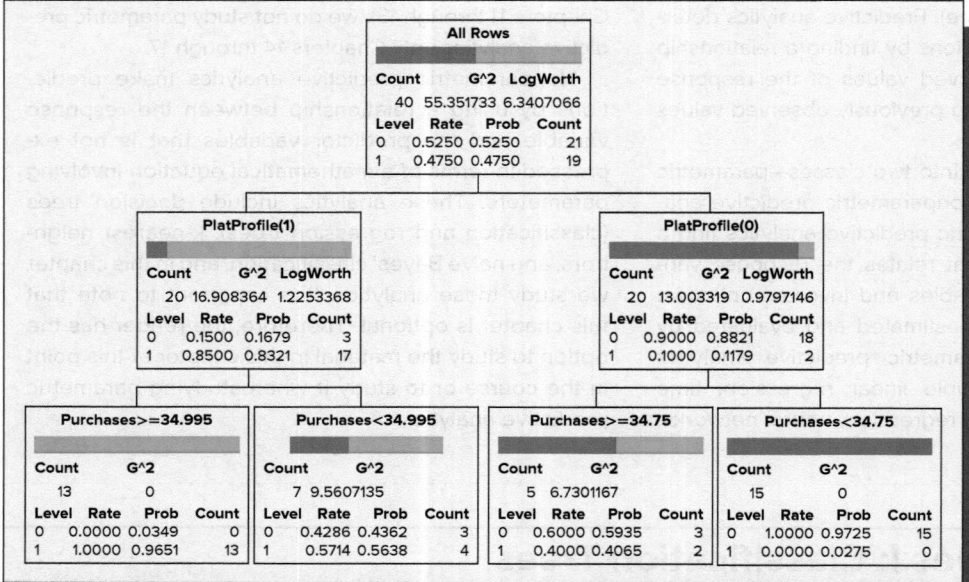

(b) The card upgrade data

Upgrade	Purchases	Plat-Profile
0	7.471	0
0	21.142	0
1	39.925	1
1	32.450	1
1	48.950	1
0	28.520	1
0	7.822	0
0	26.548	0
1	48.831	1
0	17.584	0
1	49.820	1
1	50.450	0
0	28.175	0
0	16.200	0
1	52.978	1
1	58.945	1
1	40.075	1
1	42.380	0
1	38.110	1
1	26.185	1
0	52.810	0
1	34.521	1
0	34.750	0
1	46.254	1
0	24.811	0
0	4.792	0
1	55.920	1
0	38.620	0
0	12.742	0
0	31.950	0
1	51.211	1
1	30.920	1
0	23.527	0
0	30.225	0
0	28.387	1
0	27.480	0
1	41.950	1
1	34.995	1
0	34.964	1
0	7.998	0

Platinum card might best meet his or her needs and explains (with a recommendation) the advantages and disadvantages of both cards. Because of the Platinum card's annual fee, many customers whose needs might be best met by the Platinum card choose the Silver card instead. However, the bank keeps track of each Silver card holder's annual purchases and the consistency with which the Silver card is used. Periodically, the bank offers some Silver card holders an upgrade (for the annual fee) to the Platinum card. A Silver card holder who has used the card consistently over the past two years and whose original questionnaire indicates that his or her needs might be best met by a Platinum card is said by the bank to *conform* to its *Platinum profile*. In this case, the predictor variable Plat-Profile for the holder is set equal to 1. If a Silver card holder does not conform to the bank's Platinum profile, PlatProfile is set equal to 0. Of course, a Silver card holder's needs might change over time, and thus, in addition to using the predictor variable PlatProfile, the bank will also use the previously defined predictor variable Purchases (last year's purchases) to predict whether a Silver card holder will upgrade. Note that the bank will only consider Silver card holders who have an excellent payment history. Also, note that in sampling the 40 existing Silver card holders, the bank purposely randomly selected 20 Silver card holders who conformed to the Platinum profile and randomly selected 20 Silver card holders who did not conform to the Platinum profile. In addition, note that Upgrade and PlatProfile, each of which equals 1 or 0, are examples of what is called a (1, 0) **dummy variable.** This is a quantitative variable used to represent a qualitative variable. Dummy variables are discussed in more detail in Chapter 15.

The classification tree modeling procedure tries to find predictor variables that distinguish existing Silver card holders who upgrade from those who do not. Neither Excel nor Minitab has a classification tree modeling procedure, but—as shown in Figure 5.1(a)—JMP does. To understand the JMP output in Figure 5.1(a), note that the top of the output shows the "trunk" of the decision tree and that 19 of the 40 sampled Silver card holders upgraded. That is, considering the 40 sampled Silver card holders, a **sample proportion** $\hat{p} = 19/40 = .4750$ (47.50 percent) of these Silver card holders upgraded. The sample proportion $\hat{p} = .4750$ is shown on the output as a *Rate* of .4750. The values of *Prob* on the output will be explained later.

The JMP classification tree procedure first examines each of the potential predictor variables Purchases and PlatProfile and every possible way of splitting the values of each variable into two groups. Specifically, it looks at Purchases, and for every value of Purchases, it calculates (1) the proportion of Silver card holders with purchases greater than or equal to that value who upgraded and (2) the proportion of Silver card holders with purchases less than that value who upgraded. It also looks at PlatProfile and calculates (3) the proportion of Silver card holders conforming to the Platinum profile (PlatProfile = 1) who upgraded and (4) the proportion of Silver card holders not conforming to the Platinum profile (PlatProfile = 0) who upgraded. The classification tree procedure then chooses the combination of a predictor variable and a split point that, intuitively, produces the greatest difference between the proportion of Silver card holders who upgraded and the proportion of Silver card holders who did not upgrade. Technically JMP determines this "greatest difference" by finding the combination of a predictor variable and a split point that gives the largest reduction in what is called the **LogWorth criterion.** The definition of this criterion is very theoretical and will not be discussed in this book. However, as shown in the following portion of the classification tree output in Figure 5.1(a), the combination of a predictor variable and a split point that JMP chooses for the first split is the predictor variable PlatProfile and the split point 1:

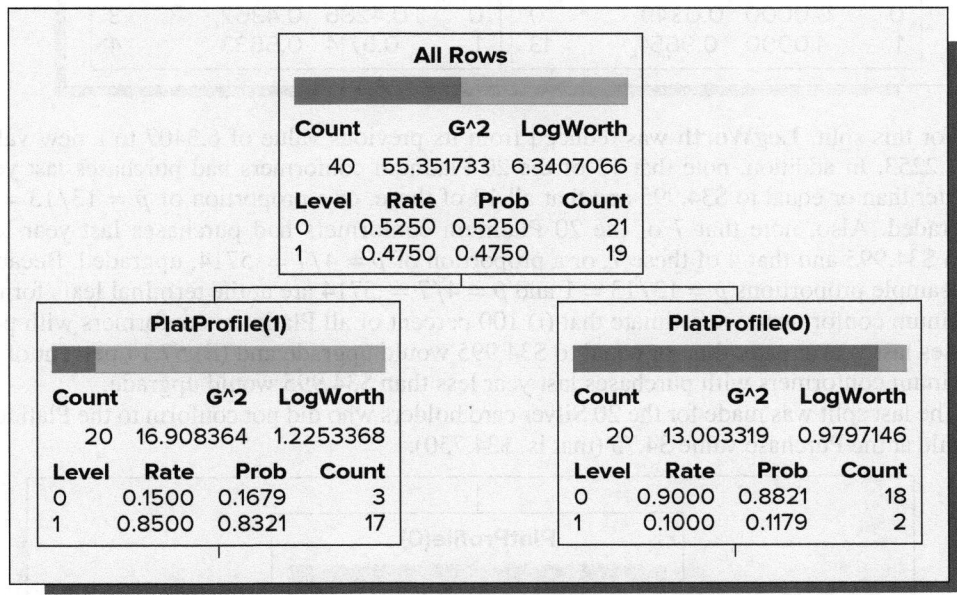

LogWorth for this split is 6.3407. In addition, note that out of the 20 Silver card holders who conformed to the Platinum profile (PlatProfile = 1), 17 of these 20, or a proportion of $\hat{p} = 17/20 = .8500$ (85 percent), upgraded. Also, out of the 20 Silver card holders who did not conform to the Platinum profile (PlatProfile = 0), 2 of these 20, or a proportion of $\hat{p} = 2/20 = .10$ (10 percent), upgraded.

After JMP finds the first split, it continues searching again on the two resulting groups, finding the next best combination of a predictor variable and a split point—the combination that gives the largest reduction in the LogWorth of 6.3407. JMP continues in this manner but stops splitting at a *leaf* when either (1) the next split at the leaf would produce a new leaf that had a *count* (see the output) that is less than a specified minimum split size value (JMP uses a default of 5 for this value, but the user can choose a different specified value), or (2) the proportion $\hat{p}$ of the number of Silver card holders at the leaf who have upgraded is either 1 or 0 (in which case no more splitting is possible and we call the leaf *pure*). When JMP has stopped splitting at a leaf, we call the leaf a *terminal leaf*, and JMP gives an estimate of the upgrade proportion (or percentage) for all Silver card holders with values of Purchases and PlatProfile corresponding to the terminal leaf.

To be specific, the next split was made for the 20 Silver card holders who conformed to the Platinum profile at the Purchase value 34.995 (that is, $34.995):

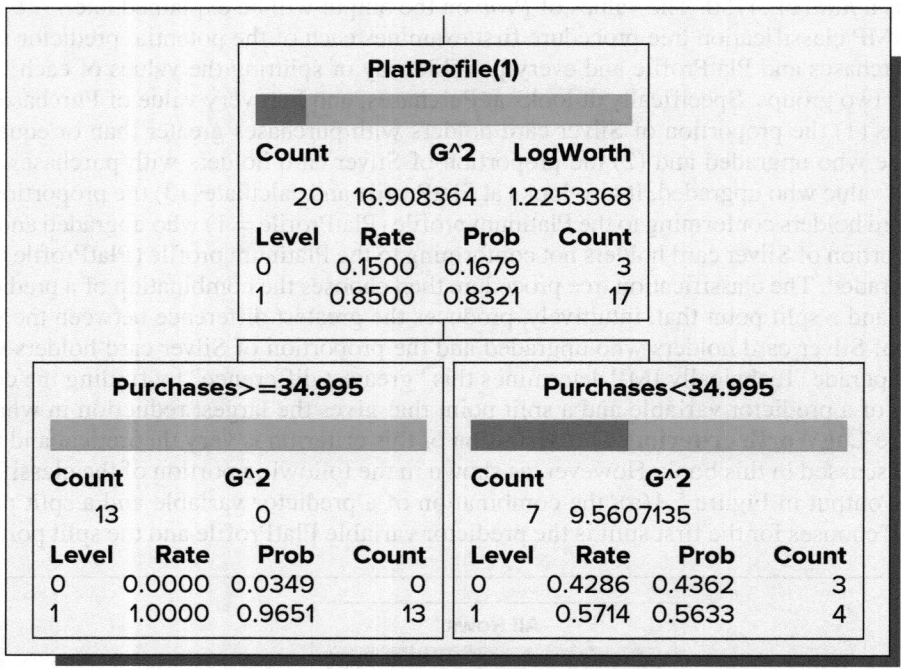

For this split, **LogWorth** was reduced from its previous value of 6.3407 to a new value of 1.2253. In addition, note that 13 of the 20 Platinum conformers had purchases last year greater than or equal to $34,995 and that all 13 of these, or a proportion of $\hat{p} = 13/13 = 1$, upgraded. Also, note that 7 of the 20 Platinum conformers had purchases last year less than $34,995 and that 4 of these 7, or a proportion of $\hat{p} = 4/7 = .5714$, upgraded. Because the sample proportions $\hat{p} = 13/13 = 1$ and $\hat{p} = 4/7 = .5714$ are at the terminal leafs for the Platinum conformers, we estimate that (i) 100 percent of all Platinum conformers with purchases last year greater than or equal to $34,995 would upgrade and (ii) 57.14 percent of all Platinum conformers with purchases last year less than $34,995 would upgrade.

The last split was made for the 20 Silver card holders who did not conform to the Platinum profile at the Purchase value 34.75 (that is, $34,750):

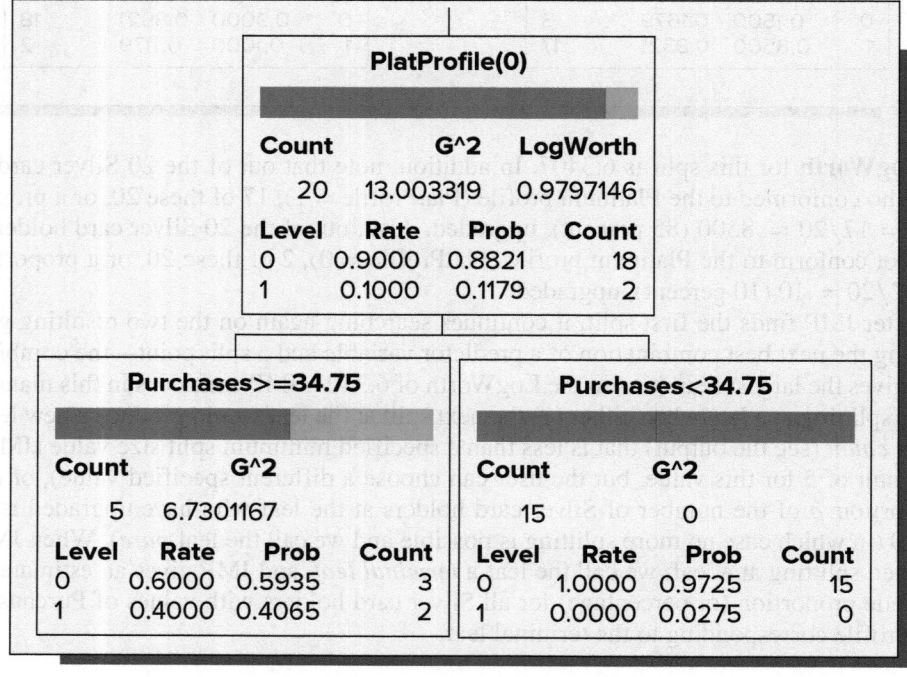

For this split, **LogWorth** was reduced from its previous value of 1.2253 to a new value of .9797. In addition, note that 5 of the 20 Platinum nonconformers had purchases last year greater than or equal to \$34,750 and that 2 of these 5, or a proportion of $\hat{p} = 2/5 = .40$, upgraded. Also, note that 15 of the 20 Platinum nonconformers had purchases last year less than \$34,750 and that 0 of these 15, or a proportion of $\hat{p} = 0/15 = 0$, upgraded. Because the sample proportions $\hat{p} = 2/5 = .40$ and $\hat{p} = 0/15 = 0$ are at the terminal leafs for the Platinum nonconformers, we estimate that (iii) 40 percent of all Platinum nonconformers with purchases last year greater than or equal to \$34,750 would upgrade and (iv) 0 percent (that is, none) of all Platinum nonconformers with purchases last year less than \$34,750 would upgrade.

In the previous two paragraphs, we have used the sample proportions $\hat{p} = 1$, $\hat{p} = .5714$ $\hat{p} = .4$, and $\hat{p} = 0$ to obtain the estimated upgrade percentages 100 percent, 57.14 percent, 40 percent, and 0 percent corresponding to the four terminal leafs. Each of these sample proportions (that is, each *Rate* on the JMP output) is also an estimate of the *probability* of a Silver card holder upgrading to a Platinum card. Moreover, JMP modifies each sample proportion to obtain a more sophisticated estimate (called a *Prob* on the JMP output) of the probability of a Silver card holder upgrading to a Platinum card. To make this modification, JMP uses Bayes' Theorem. Although we have discussed Bayes' Theorem in Chapter 4, exactly how JMP uses it to calculate each *Prob* on the JMP output is beyond the scope of this book. However, note that the *Probs* corresponding to *Level 1* at the four terminal leafs on the JMP output are .9651, .5638, .4065, and .0275. Therefore, JMP estimates that (i) the upgrade probability for a Platinum conformer with purchases last year greater than or equal to \$34,995 is .9651; (ii) the upgrade probability for a Platinum conformer with purchases last year less than \$34,995 is .5638; (iii) the upgrade probability for a Platinum nonconformer with purchases last year greater than or equal to \$34,750 is .4065; and (iv) the upgrade probability for a Platinum nonconformer with purchases last year less than \$34,750 is .0275. We conclude that Platinum conformers, who have upgrade probability estimates of .9651 and .5638 depending on their purchases last year, are the most likely to upgrade and thus probably should be sent an upgrade offer. In addition, Platinum nonconformers with purchases last year greater than or equal to \$34,750 have a .4065 probability of upgrading and thus might—depending on the cost of sending the upgrade offer—also be sent the offer.

Figure 5.2 shows the JMP *Leaf Report* that summarizes the upgrade probability estimates .9651, .5638, .4065 and .0275 that correspond to the four combinations of values of PlatProfile and Purchases at the terminal leafs (see the *Leaf Labels*). Figure 5.3 shows the JMP output of the upgrade probability estimate [see the column headed *Prob (Upgrade==1)*] and *Leaf Number Formula* (that is, the number of the *Leaf Label* giving the upgrade probability estimate) for each of the 40 Silver card holders in the observed sample and for 2 Silver card holders (holders 41 and 42) to whom the bank might send an upgrade offer. JMP classifies each Silver card holder as a 1 (that is, as an upgrader) if the upgrade probability estimate for the holder is at least .5 and as a 0 otherwise. The classifications are given in Figure 5.3 under the column labeled *Most Likely Upgrade*. If we compare

BI

FIGURE 5.2   **The JMP Leaf Report for the Card Upgrade Tree**

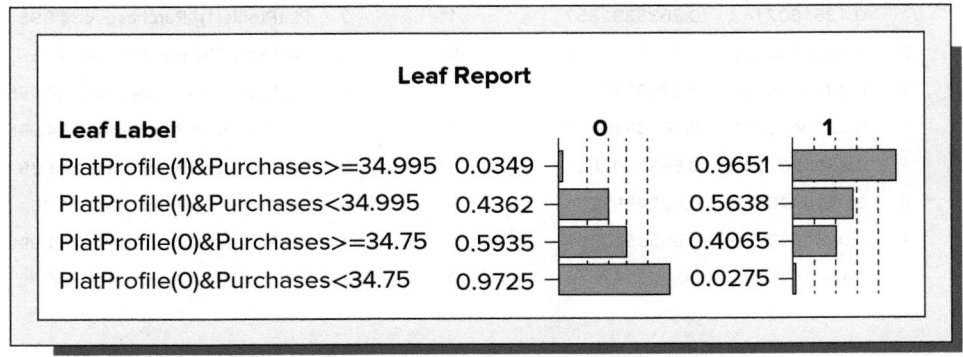

Leaf Report

Leaf Label		0	1
PlatProfile(1)&Purchases>=34.995	0.0349		0.9651
PlatProfile(1)&Purchases<34.995	0.4362		0.5638
PlatProfile(0)&Purchases>=34.75	0.5935		0.4065
PlatProfile(0)&Purchases<34.75	0.9725		0.0275

FIGURE 5.3   **The JMP Upgrade Probability Estimates and Classifications Using the Card Upgrade Tree**

	UpGrade	Purchases	PlatProfile	Prob (UpGrade==0)	Prob (UpGrade==1)	Most Likely UpGrade	Leaf Number Formula	Leaf Label Formula
1	0	7.471	0	0.9725446429	0.0274553571	0	4	PlatProfile(0)&Purchases<34.75
2	0	21.142	0	0.9725446429	0.0274553571	0	4	PlatProfile(0)&Purchases<34.75
3	1	39.925	1	0.0349489796	0.9650510204	1	1	PlatProfile(1)&Purchases>=34.995
4	1	32.45	1	0.4361607143	0.5638392857	1	2	PlatProfile(1)&Purchases<34.995
5	1	48.95	1	0.0349489796	0.9650510204	1	1	PlatProfile(1)&Purchases>=34.995
6	0	28.52	1	0.4361607143	0.5638392857	1*	2	PlatProfile(1)&Purchases<34.995
7	0	7.822	0	0.9725446429	0.0274553571	0	4	PlatProfile(0)&Purchases<34.75
8	0	26.548	0	0.9725446429	0.0274553571	0	4	PlatProfile(0)&Purchases<34.75
9	1	48.831	1	0.0349489796	0.9650510204	1	1	PlatProfile(1)&Purchases>=34.995
10	0	17.584	0	0.9725446429	0.0274553571	0	4	PlatProfile(0)&Purchases<34.75
11	1	49.82	1	0.0349489796	0.9650510204	1	1	PlatProfile(1)&Purchases>=34.995
12	1	50.45	0	0.593452381	0.406547619	0*	3	PlatProfile(0)&Purchases>=34.75
13	0	28.175	0	0.9725446429	0.0274553571	0	4	PlatProfile(0)&Purchases<34.75
14	0	16.2	0	0.9725446429	0.0274553571	0	4	PlatProfile(0)&Purchases<34.75
15	1	52.978	1	0.0349489796	0.9650510204	1	1	PlatProfile(1)&Purchases>=34.995
16	1	58.945	1	0.0349489796	0.9650510204	1	1	PlatProfile(1)&Purchases>=34.995
17	1	40.075	1	0.0349489796	0.9650510204	1	1	PlatProfile(1)&Purchases>=34.995
18	1	42.38	0	0.593452381	0.406547619	0*	3	PlatProfile(0)&Purchases>=34.75
19	1	38.11	0	0.0349489796	0.9650510204	1	1	PlatProfile(1)&Purchases>=34.995
20	1	26.185	1	0.4361607143	0.5638392857	1	2	PlatProfile(1)&Purchases<34.995
21	0	52.81	0	0.593452381	0.406547619	0	3	PlatProfile(0)&Purchases>=34.75
22	1	34.521	1	0.4361607143	0.5638392857	1	2	PlatProfile(1)&Purchases<34.995
23	0	34.75	0	0.593452381	0.406547619	0	3	PlatProfile(0)&Purchases>=34.75
24	1	46.254	1	0.0349489796	0.9650510204	1	1	PlatProfile(1)&Purchases>=34.995
25	0	24.811	0	0.9725446429	0.0274553571	0	4	PlatProfile(0)&Purchases<34.75
26	0	4.792	0	0.9725446429	0.0274553571	0	4	PlatProfile(0)&Purchases<34.75
27	1	55.92	0	0.0349489796	0.9650510204	1	1	PlatProfile(1)&Purchases>=34.995
28	0	38.62	0	0.593452381	0.406547619	0	3	PlatProfile(0)&Purchases>=34.75
29	0	12.742	0	0.9725446429	0.0274553571	0	4	PlatProfile(0)&Purchases<34.75
30	0	31.95	0	0.9725446429	0.0274553571	0	4	PlatProfile(0)&Purchases<34.75
31	1	51.211	1	0.0349489796	0.9650510204	1	1	PlatProfile(1)&Purchases>=34.995
32	1	30.92	1	0.4361607143	0.5638392857	1	2	PlatProfile(1)&Purchases<34.995
33	0	23.527	0	0.9725446429	0.0274553571	0	4	PlatProfile(0)&Purchases<34.75
34	0	30.225	0	0.9725446429	0.0274553571	0	4	PlatProfile(0)&Purchases<34.75
35	0	28.387	1	0.4361607143	0.5638392857	1*	2	PlatProfile(1)&Purchases<34.995
36	0	27.48	0	0.9725446429	0.0274553571	0	4	PlatProfile(0)&Purchases<34.75
37	1	41.95	1	0.0349489796	0.9650510204	1	1	PlatProfile(1)&Purchases>=34.995
38	1	34.995	1	0.0349489796	0.9650510204	1	1	PlatProfile(1)&Purchases>=34.995
39	0	34.964	1	0.4361607143	0.5638392857	1*	2	PlatProfile(1)&Purchases<34.995
40	0	7.998	0	0.9725446429	0.0274553571	0	4	PlatProfile(0)&Purchases<34.75
41		42.571	1	0.0349489796	0.9650510204	1	1	PlatProfile(1)&Purchases>=34.995
42		51.835	0	0.593452381	0.406547619	0	3	PlatProfile(0)&Purchases>=34.75

FIGURE 5.4   The JMP Fit Details for the Card Upgrade Tree

Fit Details

Measure	Training	Definition
Entropy RSquare	0.6738	1-Loglike(model)/Loglike(0)
Generalized RSquare	0.8092	$(1-(L(0)/L(model))^{\wedge}(2/n))/(1-L(0)^{\wedge}(2/n))$
Mean -Log p	0.2257	$\Sigma -Log(\rho[j])/n$
RMSE	0.2712	$\sqrt{} \Sigma (y[j]-\rho[j])^2/n$
Mean Abs Dev	0.1677	$\Sigma |y[j]-\rho[j]|/n$
Misclassification Rate	0.1250	$\Sigma (\rho[j] \neq \rho Max)/n$
N	40	n

Confusion Matrix
Training

Actual	Predicted Count	
UpGrade	0	1
0	18	3
1	2	17

the *Upgrade* column with the *Most Likely Upgrade* column, we see that 5 classifications (see the asterisks) out of 40 classifications are inaccurate. This gives a *Misclassification Rate* of 5/40 = .1250, which is shown in the JMP output of Figure 5.4, as part of the *Fit Details*. Also shown as part of the *Fit Details* is a *Confusion Matrix*, which says that (1) out of 21 Silver card holders who are 0's (non-upgraders), 18 are accurately classified as 0's and 3 are inaccurately classified as 1's, for a non-upgrader misclassification rate of 3/21 = .1429; and (2) out of 19 Silver card holders who are 1's (upgraders), 17 are accurately classified as 1's and 2 are inaccurately classified as 0's, for an upgrader misclassification rate of 2/19 = .1053.

Although the confusion matrix is a good indicator of the classification tree's accuracy, another useful indicator is the *Entropy RSquare* value of .6738 in the fit details. We did not show this value in the original classification tree presented in Figure 5.1(a), but JMP gives *Entropy RSquare* at the top of each decision tree it presents and calls it simply *RSquare*. For example, JMP gives the *Entropy RSquare* value of .6738 as .674 in the following form at the top of the classification tree in Figure 5.1(a):

RSquare	N	Number of Splits
0.674	40	3

The exact definition of *Entropy RSquare* is beyond the scope of this book. However, it will be used in later discussions to help choose more complex decision trees, and it may intuitively be thought of as the square of the *simple correlation coefficient* between the 40 observed 0 and 1 upgrade values in the *Upgrade* column of Figure 5.4 and the 40 corresponding upgrade probability estimates in the *Prob (Upgrade==1)* column in Figure 5.4. As discussed in optional Section 3.4 of Chapter 3, the simple correlation coefficient between two variables is a measure of the linear relationship between the variables and is always between -1 and 1. This implies that the square of the simple correlation coefficient is between 0 and 1. The nearer that the square of the simple correlation coefficient is to 1, the stronger is the linear relationship between the two variables. If we use a computer software package, or the formula in Section 3.4, to compute the simple correlation coefficient between the upgrade values and the upgrade probability estimates, we find that this simple correlation coefficient is .8413. The square of this coefficient is .7078, which is close to the *Entropy RSquare* value of .6738. Given the fact that there are only two upgrade values (0 and 1), the *Entropy RSquare* value of .6738 is fairly large. This, along with the fact that the *Confusion Matrix* indicates that the upgrade probability estimates are classifying the *Upgrade* values fairly accurately, indicates that we can have reasonable confidence in the ability of the classification tree to find accurate upgrade probability estimates and classifications for Silver card holders to whom the bank is considering sending an upgrade offer. Specifically, examining Silver card holders 41 and 42, we see that (1) the upgrade probability estimate for a Platinum conformer with purchases of $42,571 is .9651 (see the first *Leaf Label* in Figure 5.2), and thus this Silver card holder is classified as a 1 (an upgrader); and (2) the upgrade probability estimates for a Platinum nonconformer with purchases of $51,835 is .4065 (see the third *Leaf Label* in Figure 5.2), and thus this Silver card holder is classified as a 0 (a non-upgrader).

When the number of observations in a decision tree study is small to medium sized (as in the credit card upgrade example), it is reasonable to use the entire data set to continue to make splits until the previously discussed termination criterion based on a specified minimum split size or leaf purity is reached. However, if the number of observations in the study is very large, doing this might well "overfit" the data, resulting in an overly complex tree that fits the data too closely and thus fails to capture the real data patterns that would help to accurately predict and/or classify future observations. One approach to avoid over-fitting is to divide the entire data set into a **training data set,** which is used to make splits based on the specified minimum split size/leaf purity termination criterion, and a **valida-tion data set,** which is used as an additional criterion to decide when to stop making splits. Specifically, as the training data set is used to make splits and thus to obtain a sequence of more complex trees, the trees are used to predict the response variable values in both the training data set and the validation data set. It can be shown that the *RSquare* describing a tree's predictions of the training data set's response variable values will continue to increase as more splits are made. However, if (as often occurs) the tree being built reaches a point where the *RSquare* describing the tree's predictions of the validation set's response variable values is not exceeded in ten additional steps of making splits, the tree is "pruned back" to the tree that existed before validation *RSquare* stopped increasing. In general, the percentages of the entire data set that are used as the training data set and the validation data set differ in different applications, but usually at least 50 percent of the entire data set is used as the training data set.

To illustrate using a validation data set, we consider a cell phone company that has randomly selected a sample of 3,332 observations concerning customer values of the response variable Churn and 17 predictor variables. Here, Churn equals True if a customer churned—left the cell phone company for another cell phone company—and equals False otherwise. The 17 predictor variables describe aspects of the customer's account and recent cell phone bill and are summarized across the top of Figure 5.5, which is a JMP output of 12 observations in the data and part of a JMP classification tree analysis. Because the sample of 3,332 observations is very large, we begin by having JMP form a *Validation* column (shown in Figure 5.5) by randomly selecting 67 percent (or 2,221) of the observations as the training data set and 33 percent (or 1,111) of the observations as the validation data set. We also (somewhat arbitrarily) specify a minimum split size of 20 (in the exercises the reader will use a default value of 5), and Figure 5.6 shows the final classification tree obtained by JMP, when JMP uses the procedure discussed in the previous paragraph to find a tree. Figure 5.7 shows the JMP leaf report that summarizes the churn probability estimates corresponding to the 13 combinations of the predictor variable values at the terminal leafs of the classification tree. Note that Figure 5.5 gives the JMP churn probability estimate [see the column headed *Probability (Churn==True)*] and the number of the leaf label giving this estimate for each of the customers shown in this figure. JMP classifies each customer as a *True* (the customer will churn) if the customer's churn probability estimate is at least .5 and as a *False* otherwise. The classifications are given in Figure 5.5 under the column headed *Most Likely Upgrade*. Of the 7 training set customers shown in Figure 5.5, 2 are misclassified, and of the 5 validation set customers, 1 is misclassified. The confusion matrix in Figure 5.8 shows that (1) 134 out of 2,221 training data set customers are misclassified, for a training set misclassification rate of 134/2221 = .0603; and (2) 85 out of 1,111 validation set customers are misclassified, for a validation set misclassification rate of 85/1111 = .0765. While these misclassification rates are fairly small, the confusion matrix shows that (1) 89 of 322 churners in the training set are misclassified, for a churner training set misclassification rate of 89/322 = .2764; and (2) 54 out of 161 churners in the validation set are misclassified, for a churner validation set misclassification rate of 54/161 = .3354. That is, the classification tree does not classify the churners as accurately as the non-churners. However, given that—logically—the churners would be more difficult

FIGURE 5.5   The JMP Churn Probability Estimates and Classifications for 12 Customers Using the Churn Tree ⑤Churn

	Churn	AcctLength	IntlPlan	VMPlan	NVMailMsgs	DayMinutes	DayCalls	DayCharge	EveMinutes	EveCalls	EveCharges
1	False	107	no	yes	26	161.6	123	27.47	195.5	103	16.62
2	False	137	no	no	0	243.4	114	41.38	121.2	110	10.3
3	False	84	yes	no	0	299.4	71	50.9	61.9	88	5.26
4	False	75	yes	no	0	166.7	113	28.34	148.3	122	12.61
10	True	65	no	no	0	129.1	137	21.95	228.5	83	19.42
11	False	74	no	no	0	187.7	127	31.91	163.4	148	13.89
12	False	168	no	no	0	128.8	96	21.9	104.9	71	8.92
13	False	95	no	no	0	156.6	88	26.62	247.6	75	21.05
14	False	62	no	no	0	120.7	70	20.52	307.2	76	26.11
15	True	161	no	no	0	332.9	67	56.59	317.8	97	27.01
16	False	85	no	yes	27	196.4	139	33.39	280.9	90	23.88
99	True	77	no	no	0	251.8	72	42.81	205.7	126	17.48

	NightMin	NightCalls	NightCharge	IntlMin	IntlCalls	IntlCharge	NCust ServiceCalls	Validation	Prob (Churn==False)	Prob (Churn==True)	Most Likely Churn	Leaf Number Formula
1	254.4	103	11.45	13.7	3	3.7	1	Training	0.9722670632	0.0277329368	False	13
2	162.6	104	7.32	12.2	5	3.29	0	Training	0.9346342468	0.0653657532	False	12
3	196.9	89	8.86	6.6	7	1.78	2	Training	0.4924075637	0.5075924363	True*	2
4	186.9	121	8.41	10.1	3	2.73	3	Validation	0.9644147429	0.0355852571	False	9
10	208.8	111	9.4	12.7	6	3.43	4	Training	0.2867232952	0.7132767048	True	5
11	196	94	8.82	9.1	5	2.46	0	Training	0.9722670632	0.0277329368	False	13
12	141.1	128	6.35	11.2	2	3.02	1	Validation	0.9722670632	0.0277329368	False	13
13	192.3	115	8.65	12.3	5	3.32	3	Validation	0.9722670632	0.0277329368	False	13
14	203	99	9.14	13.1	6	3.54	4	Training	0.2867232952	0.7132767048	True*	5
15	160.6	128	7.23	5.4	9	1.46	4	Validation	0.9613822935	0.0386177065	True	1
16	89.3	75	4.02	13.8	4	3.73	1	Training	0.9722670632	0.0277329368	False	13
99	275.2	109	12.38	9.8	7	2.65	2	Validation	0.9346342468	0.0653657532	False*	12

FIGURE 5.6   The JMP Churn Classification Tree

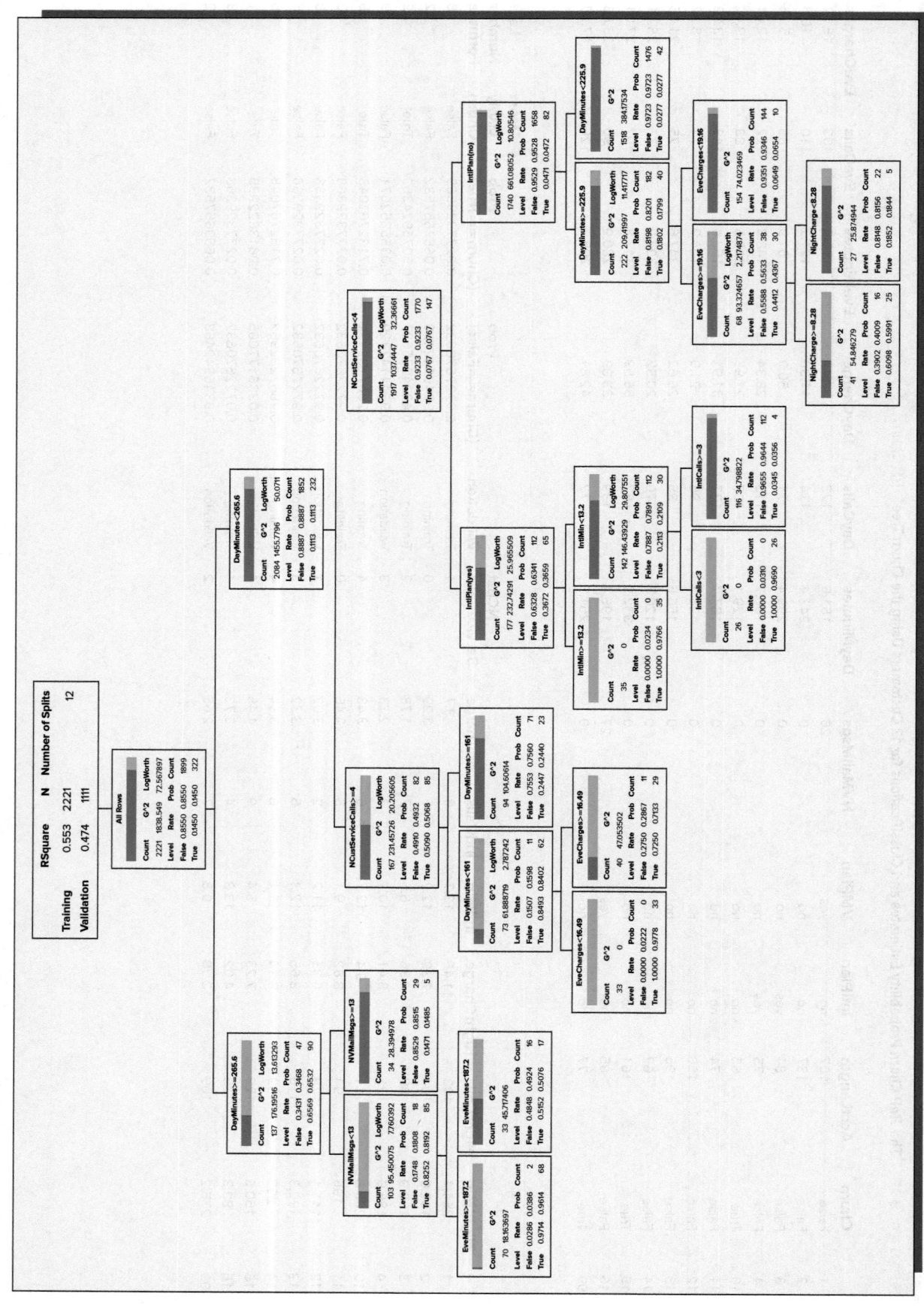

FIGURE 5.7   **The JMP Leaf Report for the Churn Tree**

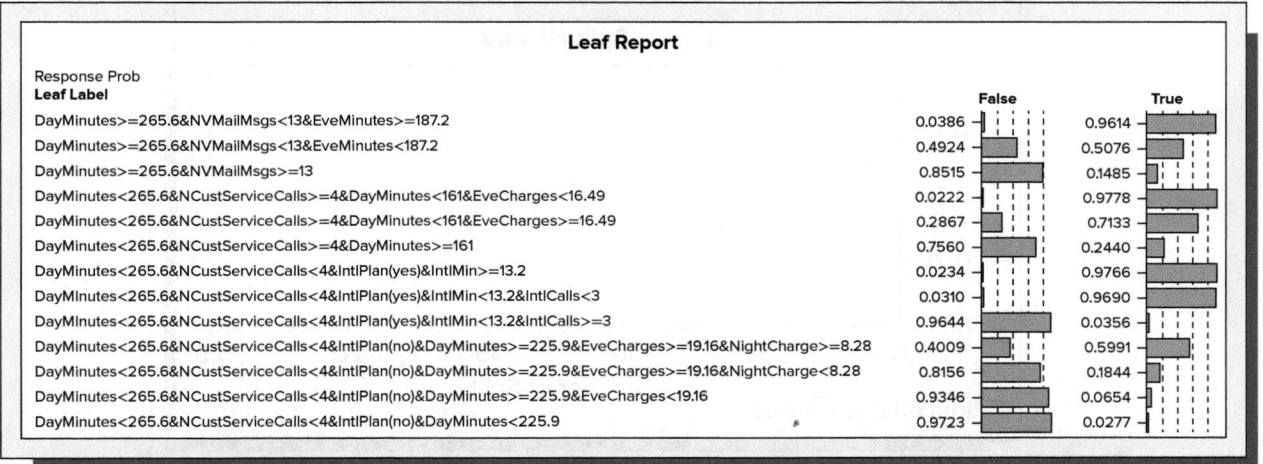

FIGURE 5.8   **The JMP Fit Details for the Churn Tree**

Fit Details

Measure	Training	Validation	Definition
Entropy RSquare	0.5525	0.4742	1-Loglike(model)/Loglike(0)
Generalized RSquare	0.6520	0.5766	$(1-(L(0)/L(model))^{\wedge}(2/n))/(1-L(0)^{\wedge}(2/n))$
Mean-Log p	0.1852	0.2176	Σ -Log(ρ[j])/n
RMSE	0.2203	0.2419	$\sqrt{\Sigma (y[j]-\rho[j])^2/n}$
Mean Abs Dev	0.0986	0.1116	Σ |y[j]-ρ[j]|/n
Misclassification Rate	0.0603	0.0765	Σ (ρ[j] $\neq$ ρMax)/n
N	2221	1111	n

Confusion Matrix

Training				Validation			
		Predicted				Predicted	
Actual		Count		Actual		Count	
Churn	False	True		Churn	False	True	
False	1854	45		False	919	31	
True	89	233		True	54	107	

to detect, we might conclude that the tree is reasonably accurate in classifying churners and non-churners.

To see how JMP obtained the final classification tree, consider Figure 5.9. This figure shows that for the first 12 splits, *RSquare* increases for both the training data set and validation data set. At the 13th split, *RSquare* increases for the training data set (it can be verified that the increase is from .5525 to .5561) but decreases for the validation data set (it can be verified that the decrease is from .4742 to .4704). Moreover, Figure 5.9 shows that after 22 splits *RSquare* for the validation data set never increases beyond the value of .4742 that it had after 12 splits. Therefore, the tree is pruned back to what it was after 12 splits—the tree shown in Figure 5.6.

The purpose of building the tree is to classify customers who have just been sent a cell phone bill, so that the cell phone company can take early action to convince customers who are classified as churners not to churn. Customer 3333 in Figure 5.10 is a customer who has

FIGURE 5.9   The JMP Split History for the Churn Tree

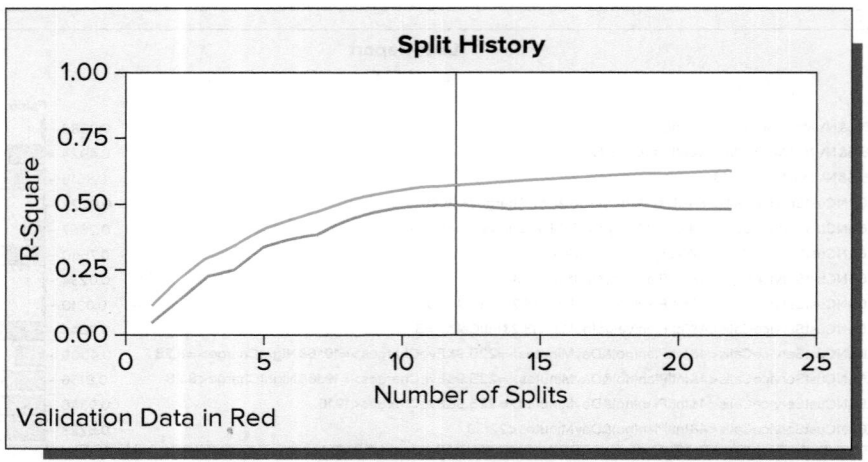

FIGURE 5.10   The JMP Churn Probability Estimate and Classification for Customer 3333 Using the Churn Tree

	Churn	AcctLength	IntlPlan	VMPlan	NVMailMsgs	DayMinutes	DayCalls	DayCharge	EveMinutes	EveCalls	EveCharges
3333		87	no	yes	10	153	110	28.35	190.7	104	18.75

	NightMin	NightCalls	NightCharge	IntlMin	IntlCalls	IntlCharge	NCustServiceCalls	Prob(Churn==False)
3333	205	81	8.81	11.7	3	3.87	5	0.2867232952

	Prob(Churn==True)	Most Likely Churn 4	Leaf Number Formula	Leaf Label Formula
3333	0.7132767048	True	5	DayMinutes<265.6&NCustServiceCalls>=4&DayMinutes<161&EveCharges>=16.49

just been sent a cell phone bill. It can be verified, as shown in Figure 5.10, that the values of the 17 predictor variables for customer 3333 correspond to the fifth *Leaf Label* in Figure 5.2, and therefore the classification tree estimates that the probability that customer 3333 will churn is .7133. It follows that customer 3333 is classified as a churner, and thus the cell phone company will take action to convince this customer not to churn.

Exercises for Section 5.1

CONCEPTS

5.1   What prediction does a classification tree give and how is this prediction used to classify?

METHODS AND APPLICATIONS

5.2   Consider the classification trees in Figures 5.11(a) and (b).
   a   Use the mortgage default tree in Figure 5.11(a) to estimate the percentage of all mortgage applicants

with a household income of $65,000 and debt of $20,000 who would default.
   b   Salford Systems (see Salford-systems.com) describes a study it helped carry out at UC San Diego to predict the probability that a heart attack survivor would have a second heart attack within 30 days. The study measured the values of over 100 variables concerning demographics, medical history,

FIGURE 5.11   **Classification Trees, Regression Trees, and Data for Exercises 5.2 and 5.3**

(a) Mortgage default tree

(b) Second heart attack tree

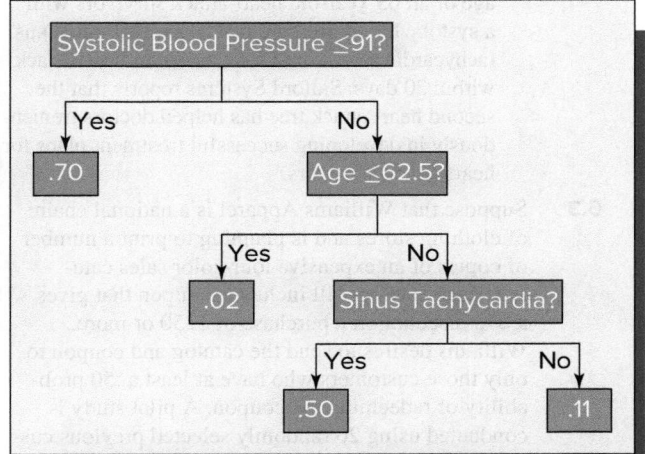

(c) The coupon redemption

    data **DS** CoupRedemp

Coupon	Purchases	Card
0	10.52	0
0	14.89	0
0	15.75	1
1	29.27	1
1	48.15	1
1	51.40	0
1	29.89	1
0	21.23	0
0	17.55	1
0	46.61	0
0	28.01	0
0	15.97	0
1	50.10	1
1	51.62	0
1	27.48	1
0	4.95	0
1	43.25	1
0	35.04	1
0	28.94	1
1	53.67	1
0	13.24	0
1	51.54	1
1	42.05	1
0	29.71	0
0	39.24	0
1	58.38	1
0	23.75	0
0	23.89	0
0	50.94	0
1	36.37	1
1	34.40	1
1	38.01	1
1	55.87	1
0	21.46	0
1	40.75	1
1	46.52	1
0	34.93	0
0	31.97	0
0	42.28	0
0	26.28	0

(d) JMP output of a classification tree for the coupon redemption data (for Exercise 5.3)

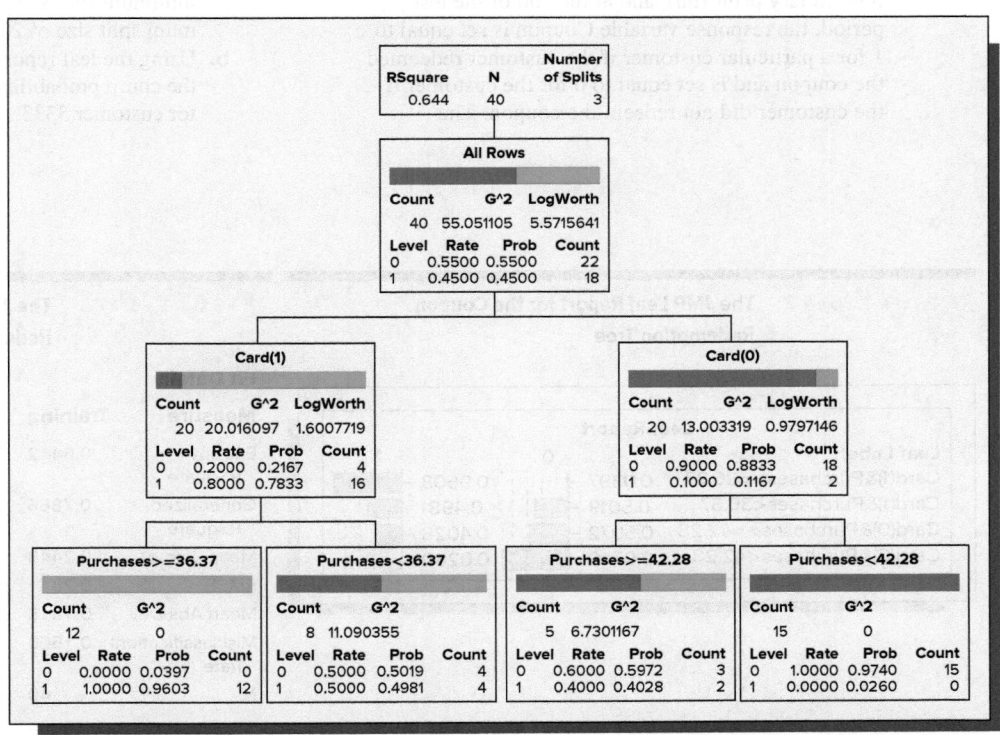

and lab results for heart attack survivors. From Salford Systems' description of the results, we have constructed the second heart attack tree in Figure 5.11(b). Use the tree to estimate the percentage of all 65-year-old heart attack survivors with a systolic blood pressure of 120 and with no sinus tachycardia who would have a second heart attack within 30 days. Salford Systems reports that the second heart attack tree has helped doctors tremendously in developing successful treatment plans for heart attack survivors.

5.3 Suppose that Williams Apparel is a national chain of clothing stores and is planning to print a number of copies of an expensive four-color sales catalog. Each catalog will include a coupon that gives a $70 discount on a purchase of $250 or more. Williams desires to send the catalog and coupon to only those customers who have at least a .50 probability of redeeming the coupon. A pilot study is conducted using 20 randomly selected previous customers who have a Williams credit card (Card=1) and 20 randomly selected previous customers who do not have a Williams credit card (Card=0). Each of the 40 randomly selected previous customers is sent a copy of the catalog and coupon (from a small preliminary print run), and at the end of the test period, the response variable Coupon is set equal to 1 for a particular customer if the customer redeemed the coupon and is set equal to 0 for the customer if the customer did not redeem the coupon. The

values of the response variable Coupon are given for the 40 customers in Figure 5.11(c), along with the corresponding values of the predictor variable Card and of the predictor variable Purchases (last year's purchases, in hundreds of dollars).[3] Figure 5.11(d) gives the JMP classification tree analyzing the data in Figure 5.11(c) using the default minimum split size of 5. Figures 5.12 and 5.13 give the leaf report and fit details for the classification tree, and Figure 5.14 gives the coupon redemption probability estimates and classifications for the 40 sampled customers the 40 samples customers. **DS** CoupRedemp

a. Interpret the confusion matrix in Figure 5.13.

b. Using the classification tree or leaf report, find (to four decimal places) the coupon redemption probability estimates and classifications for customers 41 and 42. Which customers will be sent a catalog?

5.4 Figure 5.15 gives a partial JMP classification tree analysis of the churn data using the default minimum split size of 5.

a. Using the fit details and confusion matrix, compare the fit and misclassification rates (including the churner misclassification rates) when using a minimum split size of 5 and when using a minimum split size of 20.

b. Using the leaf report, find (to four decimal places) the churn probability estimate and classification for customer 3333.

FIGURE 5.12   **The JMP Leaf Report for the Coupon Redemption Tree**

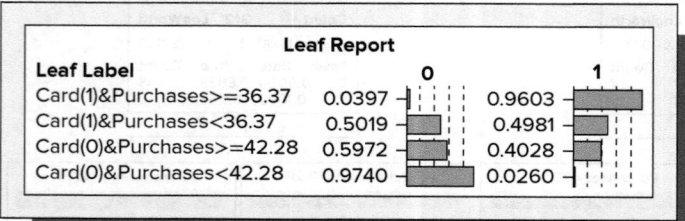

Leaf Report			
Leaf Label		**0**	**1**
Card(1)&Purchases>=36.37		0.0397	0.9603
Card(1)&Purchases<36.37		0.5019	0.4981
Card(0)&Purchases>=42.28		0.5972	0.4028
Card(0)&Purchases<42.28		0.9740	0.0260

FIGURE 5.13   **The JMP Fit Details for the Coupon Redemption Tree**

Fit Details

Measure	Training	Definition
Entropy RSquare	0.6442	1-Loglike(model)/Loglike(0)
Generalized RSquare	0.7866	$(1-(L(0)/L(model))^{(2/n)})/(1-L(0)^{(2/n)})$
Mean -Log p	0.2448	Σ -Log(ρ[j])/n
RMSE	0.2841	$\sqrt{\Sigma\ (y[j]-\rho[j])^2/n}$
Mean Abs Dev	0.1818	$\Sigma\ |y[j]-\rho[j]|/n$
Misclassification Rate	0.1500	$\Sigma\ (\rho[j] \neq \rho Max)/n$
N	40	n

Confusion Matrix

Training

Actual	Predicted Count	
Coupon	**0**	**1**
0	22	0
1	6	12

---

[3]The idea for this exercise was obtained from an example in Anderson, Sweeney, Williams, Camm, and Cochran (2014). The data in Figure 5.11 (c) are our data.

FIGURE 5.14   **The JMP Coupon Redemption Probability Estimates and Classifications Using the Coupon Redemption Tree**

	Coupon	Purchases	Card	Prob (Coupon==0)	Prob (Coupon==1)	Most Likely Coupon	Leaf Number	Formula	Leaf Label Formula
1	0	10.52	0	0.9739583333	0.0260416667	0	4	1	Card(0)&Purchases<42.28
2	0	14.89	0	0.9739583333	0.0260416667	0	4	2	Card(0)&Purchases<42.28
3	0	15.75	1	0.5018518519	0.4981481481	0	2	3	Card(1)&Purchases<36.37
4	1	29.27	1	0.5018518519	0.4981481481	0	2	4	Card(1)&Purchases<36.37
5	1	48.15	1	0.0397435897	0.9602564103	1	1	5	Card(1)&Purchases>=36.37
6	1	51.40	0	0.5972222222	0.4027777778	0	3	6	Card(0)&Purchases>=42.28
7	1	29.89	1	0.5018518519	0.4981481481	0	2	7	Card(1)&Purchases<36.37
8	0	21.23	0	0.9739583333	0.0260416667	0	4	8	Card(0)&Purchases<42.28
9	0	17.55	1	0.5018518519	0.4981481481	0	2	9	Card(1)&Purchases<36.37
10	0	46.60	0	0.5972222222	0.4027777778	0	3	10	Card(0)&Purchases>=42.28
11	0	28.01	0	0.9739583333	0.0260416667	0	4	11	Card(0)&Purchases<42.28
12	0	15.97	0	0.9739583333	0.0260416667	0	4	12	Card(0)&Purchases<42.28
13	1	50.10	1	0.0397435897	0.9602564103	1	1	13	Card(1)&Purchases>=36.37
14	1	51.62	0	0.5972222222	0.4027777778	0	3	14	Card(0)&Purchases>=42.28
15	1	27.48	1	0.5018518519	0.4981481481	0	2	15	Card(1)&Purchases<36.37
16	0	4.95	0	0.9739583333	0.0260416667	0	4	16	Card(0)&Purchases<42.28
17	1	43.25	1	0.0397435897	0.9602564103	1	1	17	Card(1)&Purchases>=36.37
18	0	35.04	1	0.5018518519	0.4981481481	0	2	18	Card(1)&Purchases<36.37
19	0	28.94	1	0.5018518519	0.4981481481	0	2	19	Card(1)&Purchases<36.37
20	1	53.67	1	0.0397435897	0.9602564103	1	1	20	Card(1)&Purchases >=36.37
21	0	13.24	0	0.9739583333	0.0260416667	0	4	21	Card(0)&Purchases<42.28
22	1	51.54	1	0.0397435897	0.9602564103	1	1	22	Card(1)&Purchases>=36.37
23	1	42.05	1	0.0397435897	0.9602564103	1	1	23	Card(1)&Purchases>=36.37
24	0	29.71	0	0.9739583333	0.0260416667	0	4	24	Card(0)&Purchases<42.28
25	0	39.24	0	0.9739583333	0.0260416667	0	4	25	Card(0)&Purchases<42.28
26	1	58.38	1	0.0397435897	0.9602564103	1	1	26	Card(1)&Purchases>=36.37
27	0	23.75	0	0.9739583333	0.0260416667	0	4	27	Card(0)&Purchases<42.28
28	0	23.89	0	0.9739583333	0.0260416667	0	4	28	Card(0)&Purchases<42.28
29	0	50.94	0	0.5972222222	0.4027777778	0	3	29	Card(0)&Purchases>=42.28
30	1	36.37	1	0.0397435897	0.9602564103	1	1	30	Card(1)&Purchases>=36.37
31	1	34.40	1	0.5018518519	0.4981481481	0	2	31	Card(1)&Purchases<36.37
32	1	38.01	1	0.0397435897	0.9602564103	1	1	32	Card(1)&Purchases>=36.37
33	1	55.87	1	0.0397435897	0.9602564103	1	1	33	Card(1)&Purchases>=36.37
34	0	21.46	0	0.9739583333	0.0260416667	0	4	34	Card(0)&Purchases<42.28
35	1	40.75	1	0.0397435897	0.9602564103	1	1	35	Card(1)&Purchases>=36.37
36	1	46.52	1	0.0397435897	0.9602564103	1	1	36	Card(1)&Purchases>=36.37
37	0	34.93	0	0.9739583333	0.0260416667	0	4	37	Card(0)&Purchases<42.28
38	0	31.97	0	0.9739583333	0.0260416667	0	4	38	Card(0)&Purchases<42.28
39	0	42.28	0	0.5972222222	0.4027777778	0	3	39	Card(0)&Purchases>=42.28
40	0	26.28	0	0.9739583333	0.0260416667	0	4	40	Card(0)&Purchases<42.28
41		43.97	1				1	41	Card(1)&Purchases>=36.37
42		52.48	0				3	42	Card(0)&Purchases>=42.28

FIGURE 5.15 **Partial JMP Classification Tree Analysis of the Churn Data Using a Specified Minimum Split Size of 5**

Leaf Report

Response Prob

Leaf Label	False	True
DayMinutes>=265.6&NVMailMsgs<13&EveMinutes>=187.2	0.0386	0.9614
DayMinutes>=265.6&NVMailMsgs<13&EveMinutes<187.2&NightCharge>=9.64	0.1698	0.8302
DayMinutes>=265.6&NVMailMsgs<13&EveMinutes<187.2&NightCharge<9.64	0.7746	0.2254
DayMinutes>=265.6&NVMailMsgs>=13&IntlPlan(yes)	0.4013	0.5987
DayMinutes>=265.6&NVMailMsgs>=13&IntlPlan(no)	0.9589	0.0411
DayMinutes<265.6&NCustServiceCalls>=4&DayMinutes<161&EveCharges<19.84&VMPlan(no)	0.0146	0.9854
DayMinutes<265.6&NCustServiceCalls>=4&DayMinutes<161&EveCharges<19.84&VMPlan(yes)	0.2836	0.7164
DayMinutes<265.6&NCustServiceCalls>=4&DayMinutes<161&EveCharges>=19.84	0.5472	0.4528
DayMinutes<265.6&NCustServiceCalls>=4&DayMinutes>=161&EveMinutes<135.2	0.3179	0.6821
DayMinutes<265.6&NCustServiceCalls>=4&DayMinutes>=161&EveMinutes>=135.2&DayMinutes<175.8&EveMinutes<212.5	0.1995	0.8005
DayMinutes<265.6&NCustServiceCalls>=4&DayMinutes>=161&EveMinutes>=135.2&DayMinutes<175.8&EveMinutes>=212.5	0.9852	0.0148
DayMinutes<265.6&NCustServiceCalls>=4&DayMinutes>=161&EveMinutes>=135.2&DayMinutes>=175.8	0.9279	0.0721
DayMinutes<265.6&NCustServiceCalls<4&IntlPlan(yes)&tIntlCharge>=3.56	0.0234	0.9766
DayMinutes<265.6&NCustServiceCalls<4&IntlPlan(yes)&IntlCharge<3.56&IntlCalls<3	0.0310	0.9690
DayMinutes<265.6&NCustServiceCalls<4&IntlPlan(yes)&IntlCharge<3.56&IntlCalls>=3	0.9644	0.0356
DayMinutes<265.6&NCustServiceCalls<4&IntlPlan(no)&DayMinutes>=225.9&EveCharges>=19.16&VMPlan(no)&NightMin>=218	0.0945	0.9055
DayMinutes<265.6&NCustServiceCalls<4&IntlPlan(no)&DayMinutes>=225.9&EveCharges>=19.16&VMPlan(no)&NightMin<218	0.6186	0.3814
DayMinutes<265.6&NCustServiceCalls<4&IntlPlan(no)&DayMinutes>=225.9&EveCharges>=19.16&VMPlan(yes)	0.9419	0.0581
DayMinutes<265.6&NCustServiceCalls<4&IntlPlan(no)&DayMinutes>=225.9&EveCharges<19.16	0.9346	0.0654
DayMinutes<265.6&NCustServiceCalls<4&]ntlPlan(no)&DayMinutes<225.9	0.9723	0.0277

Fit Details

Measure	Training	Validation	Definition
Entropy RSquare	0.6145	0.5173	1-Loglike(model)/Loglike(0)
Generalized RSquare	0.7082	0.6187	$(1-(L(0)/L(model))^{(2/n)})/(1-L(0)^{(2/n)})$
Mean -Log p	0.1596	0.1997	Σ -Log($\rho[j]$)/n
RMSE	0.1970	0.2274	$\sqrt{\Sigma (y[j]- \rho[j])^2/n}$
Mean Abs Dev	0.0800	0.0955	Σ |y[j]- $\rho[j]$|/n
Misclassification Rate	0.0450	0.0648	$\Sigma (\rho[j] \neq \rho Max)/n$
N	2221	1111	n

Confusion Matrix

Training

Actual	Predicted Count	
Churn	False	True
False	1884	15
True	85	237

Validation

Actual	Predicted Count	
Churn	False	True
False	940	10
True	62	99

	Prob(Churn==False) 2	Prob(Churn==True) 2	Most Likely Churn 3	Leaf Number Formula
3333				7

	Leaf Label Formula
3333	DayMinutes<265.6&NCustServiceCalls>=4&DayMinutes<161&EveCharges<19.84&VMPlan(yes)

Source: William Mendenhall and Terry Sincich, *A Second Course in Business Statistics: Regression Analysis,* Fourth edition, ©1993. Reprinted with permission of Prentice Hall.

TABLE 5.1 The Hiring Status Data DS Gender

Hiring Status y	Education x_1, years	Experience x_2, years	Gender x_3	Hiring Status y	Education x_1, years	Experience x_2, years	Gender x_3
0	6	2	0	1	4	5	1
0	4	0	1	0	6	4	0
1	6	6	1	0	8	0	1
1	6	3	1	1	6	1	1
0	4	1	0	0	4	7	0
1	8	3	0	0	4	1	1
0	4	2	1	0	4	5	0
0	4	4	0	0	6	0	1
0	6	1	0	1	8	5	1
1	8	10	0	0	4	9	0
0	4	2	1	0	8	1	0
0	8	5	0	0	6	1	1
0	4	2	0	1	4	10	1
0	6	7	0	1	6	12	0

5.5 Mendenhall and Sincich (1993) present data that can be used to investigate allegations of gender discrimination in the hiring practices of a particular firm. These data are given in Table 5.1. In this table, y is a dummy variable that equals 1 if a potential employee was hired and 0 otherwise; x_1 is the number of years of education of the potential employee; x_2 is the number of years of experience of the potential employee; and x_3 is a dummy variable that equals 1 if the potential employee was a male and 0 if the potential employee was a female.

a. Calling the variables y, x_1, x_2, and x_3, respectively, Hire, Education, Experience, and Gender, use JMP to find a classification tree describing the data in Table 5.1 when the minimum split size is the default value of 5.

b. Find a point estimate of the probability that a potential employee having 4 years of education and 5 years of experience will be hired if (1) the potential employee is a male and (2) if the potential employee is a female.

c. Use the results in *a* and *b* to discuss whether there has been gender discrimination in the hiring practices of the firm.

5.2 Decision Trees II: Regression Trees

LO5-2
Interpret the information provided by regression trees.

To discuss regression trees, we consider an example presented by Kutner, Nachtsheim, Neter, and Li (2005). In this example, a college admissions officer wishes to predict a college applicant's grade point average at the end of the freshman year (GPA, also denoted as y) on the basis of the applicant's ACT entrance test score (ACT) and the applicant's high school rank (H. S. Rank). The high school rank is the percentile at which the student stands in his or her graduating class at the time of application. The admissions officer will use a random sample of 705 previously admitted applicants to build a regression tree that can be used to predict the GPAs of future applicants. As a preliminary step in the analysis, we randomly select 50 percent of the 705 applicants (actually, 352 applicants) as the training data set and 50 percent of the applicants (353 applicants) as the validation data set. Figure 5.16 shows the GPA, ACT, and H. S. Rank for some of the applicants in each of these data sets. (We assume that the admissions officer has arranged the applicants in each data set in order of increasing GPA.)

JMP's first step in finding a regression tree to predict future GPAs is to compute the mean of the 352 GPAs in the training data set. This mean is the initial prediction of any GPA in the training data set or validation data set and, as shown in Figure 5.17(a), is 2.978071. To assess the accuracy of 2.978071 or other predictions based on a regression

FIGURE 5.16   The JMP Output of Some Training Data and Some Validation Data in the GPA
Example   ⊙Ⓢ GPA

(a) Some training data

	GPA	ACT	H. S. Rank
1	1.25	19	74
2	1.32	23	95
38	3.15	25	73
39	3.15	25	95
350	3.933	27	97
351	4	29	97
352	4	32	99

(b) Some validation data

	GPA	ACT	H. S. Rank
353	0.98	20	61
354	1.13	20	84
384	2.71	25	79
385	2.73	21	63
703	3.956	29	97
704	4	26	98
705	4	29	97

tree developed by using the training data set, we use the prediction(s) to calculate three quantities—MSE, RMSE, and RSquare—for the GPAs in each of the training data set and the validation data set. Here, (1) MSE is the mean of the squared deviations between the GPAs in the data set being considered (training or validation) and the predictions based on the training data set of these GPAs, (2) RMSE is the square root of MSE, and (3) RSquare is (intuitively) the square of the simple correlation coefficient between the GPAs in the data set being considered and the predictions based on the training data set of these GPAs. As illustrated in Figure 5.17(a), if we use the training data set mean of 2.978071 as the prediction of any GPA in the training or validation data set, then RMSE is .6534 for the training data set and .6142 for the validation data set. Moreover, RSquare is 0 for each data set.

In the second step of finding a regression tree, JMP examines each of the predictor variables ACT and H. S. Rank and every possible way of splitting the values of each predictor variable into two groups. The procedure then chooses the combination of a predictor variable and a split point that gives the biggest reduction in LogWorth (which can be shown to be approximately equivalent to finding the combination of a predictor variable and a split point that gives the biggest reduction in RMSE). Examining Figure 5.17(b), we see that the procedure chooses the predictor variable H. S. Rank and the split point 82. Also, the mean of the 183 training data set GPAs corresponding to a H. S. Rank less than 82 is 2.7181, and this mean is the prediction of any GPA in the training or validation data set that corresponds to a H. S. Rank less than 82. Moreover, the mean of the 169 training data set GPAs corresponding to a H. S. Rank greater than or equal to 82 is 3.2596, and this mean is the prediction of any GPA in the training or validation data set that corresponds to a H. S. Rank greater than or equal to 82. Using these predictions, RMSE is .5948 for the training data set and .5811 for the validation data set. In addition RSquare is .171 for the training data set and .105 for the validation data set, and LogWorth is 18.7505.

As shown in Figure 5.17(c), the JMP regression tree procedure continues to find split points, where the next split point gives the biggest reduction in LogWorth. In step 3, the split point is an ACT of 24 (which reduces LogWorth to 4.7541), in step 4 the split point is an ACT of 20 (which reduces LogWorth to 2.3739), and in step 5 the split point is a H. S. Rank of 97 (which reduces LogWorth to 2.3339). Figure 5.18 graphically illustrates that RSquare increases for both the training and validation data sets up through step 5 (that is, up through the fourth split, which is at a H. S. Rank of 97). However, at the fifth split (which, although not shown, is at a H. S. Rank of 99), RSquare decreases (from .157 to .154) for the validation data set, and RSquare does not increase above .157 again for the validation data set up through the fourteenth split. Therefore, the final regression tree is the tree obtained after the fourth split and shown in Figure 5.17. The leaf report in Figure 5.19 shows the predictions for the various combinations of ACT and H. S. Rank. For example, Figure 5.20 shows that the prediction of GPA for a new applicant with an ACT of 26 and a H. S. Rank of 91 corresponds to the fourth leaf label and equals 3.2612.

FIGURE 5.17   JMP Outputs of Building the Regression Tree in the GPA Example

(a) The mean of the 352 GPAs in the training data set

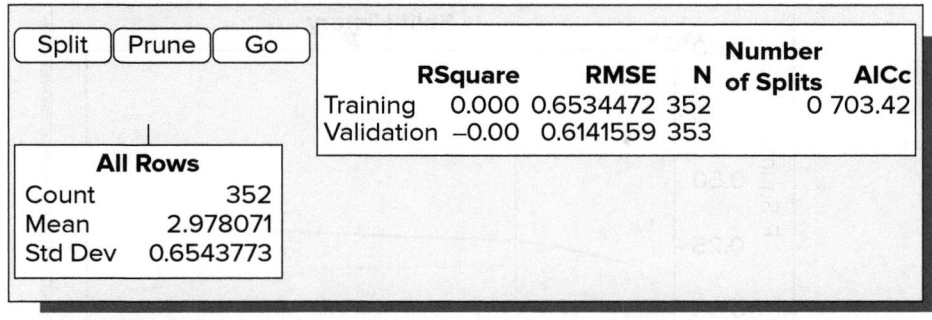

(b) The first split

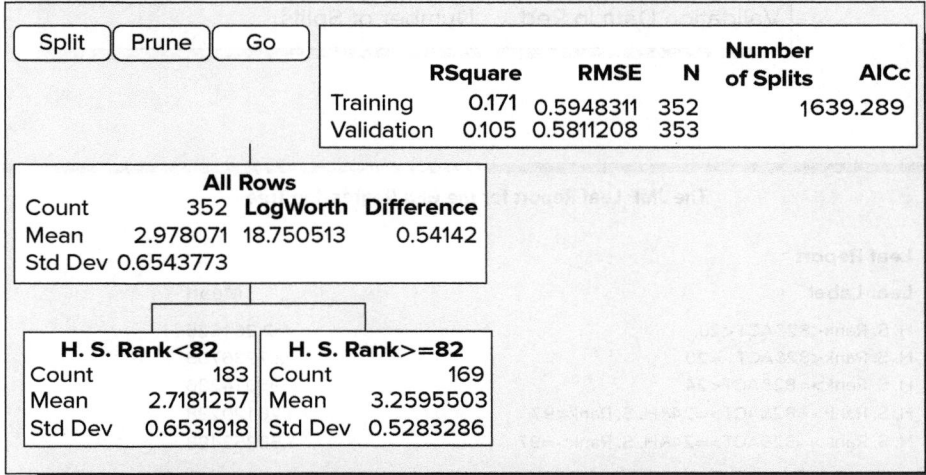

(c) All four splits and the final regression tree

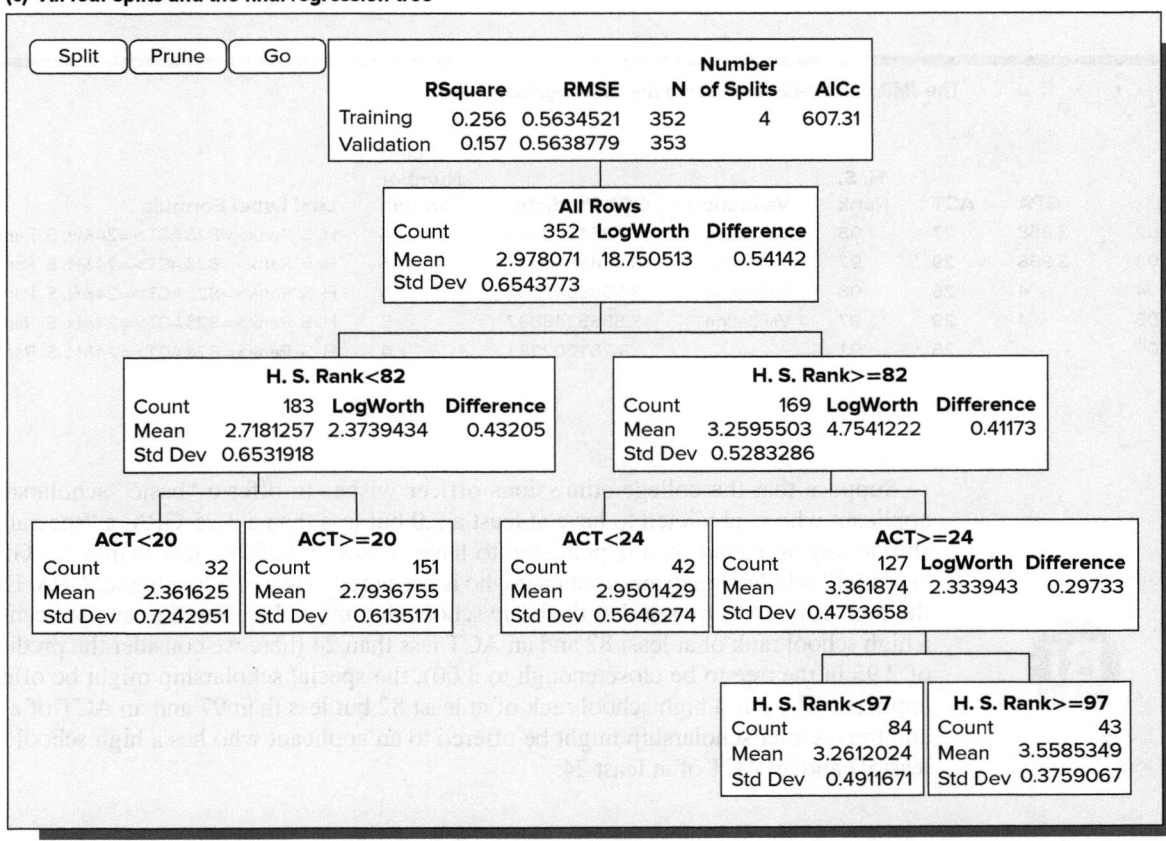

FIGURE 5.18   The JMP Split History for the GPA Regression Tree

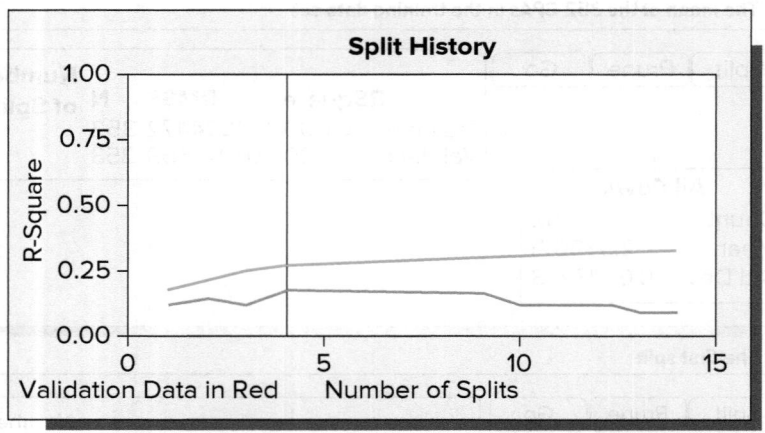

FIGURE 5.19   The JMP Leaf Report for the GPA Regression Tree

Leaf Report

Leaf Label	Mean	Count
H. S. Rank<82&ACT<20	2.361625	32
H. S. Rank<82&ACT>=20	2.7936755	151
H. S. Rank>=82&ACT<24	2.95014286	42
H. S. Rank>=82&ACT>=24&H. S. Rank<97	3.26120238	84
H. S. Rank>=82&ACT>=24&H. S. Rank>=97	3.55853488	43

FIGURE 5.20   The JMP GPA Predictions Using the GPA Regression Tree

	GPA	ACT	H. S. Rank	Validation	GPA Predictor	Leaf Number Formula	Leaf Label Formula
702	3.858	27	96	Validation	3.261202381	4	H. S. Rank>=82&ACT>=24&H. S. Rank<97
703	3.956	29	97	Validation	3.5585348837	5	H. S. Rank>=82&ACT>=24&H. S. Rank>=97
704	4	26	98	Validation	3.5585348837	5	H. S. Rank>=82&ACT>=24&H. S. Rank>=97
705	4	29	97	Validation	3.5585348837	5	H. S. Rank>=82&ACT>=24&H. S. Rank>=97
706	.	26	91	.	3.261202381	4	H. S. Rank>=82&ACT>=24&H. S. Rank<97

Suppose that the college admissions officer wishes to offer a "basic" scholarship to any applicant who is predicted to have at least a 3.0 but less than a 3.25 GPA, a "special" scholarship to any applicant who is predicted to have at least a 3.25 but less than a 3.5 GPA, and a "premier" scholarship to any applicant who is predicted to have a 3.5 or higher GPA. Examining the regression tree, we see that the basic scholarship might be offered to an applicant who has a high school rank of at least 82 and an ACT less than 24 (here we consider the predicted GPA of 2.95 in the tree to be close enough to 3.00); the special scholarship might be offered to an applicant who has a high school rank of at least 82 but less than 97 and an ACT of at least 24; and the premier scholarship might be offered to an applicant who has a high school rank of at least 97 and an ACT of at least 24.

Exercises for Section 5.2

CONCEPTS

5.6 How does a regression tree use the response variable values corresponding to a terminal leaf to make a prediction?

METHODS AND APPLICATIONS

5.7 **THE FRESH DETERGENT CASE**
DS Fresh2

Enterprise Industries produces Fresh, a brand of liquid laundry detergent. In order to manage its inventory more effectively and make revenue projections, the company would like to better predict demand for Fresh. To develop a prediction model, the company has gathered data concerning demand for Fresh over the last 30 sales periods (each sales period is defined to be a four-week period). The demand data are presented in Table 5.2. For each sales period, let

y = the demand for the large size bottle of Fresh (in hundreds of thousands of bottles) in the sales period (Demand)

x_1 = the price (in dollars) of Fresh as offered by Enterprise Industries in the sales period

x_2 = the average industry price (in dollars) of competitors' similar detergents in the sales period

x_3 = Enterprise Industries' advertising expenditure (in hundreds of thousands of dollars) to promote Fresh in the sales period (AdvExp)

x_4 = the difference between the average industry price (in dollars) of competitors' similar detergents and the price (in dollars) of Fresh as offered by Enterprise Industries in the sales period (PriceDif). (Note that $x_4 = x_2 - x_1$).

Figures 5.21, 5.22, and 5.23 show the JMP outputs of a regression tree analysis of the Fresh demand data, where the response variable is Demand and the predictor variables are AdvExp and PriceDif. The default minimum split size of 5 was used. Find the JMP regression tree prediction of demand for Fresh in Future sales periods 31 and 32.

5.8 A Toyota dealer wished to predict the sales prices of used Toyota Corollas and gathered 1,436 observations concerning sales prices of used Corollas and the values of 36 predictor variables. Sixty percent of the observations (actually 862 observations) were randomly selected as the training data set, and the remaining 40 percent of the observations

TABLE 5.2 **Historical Data, Including Price Differences, Concerning Demand for Fresh Detergent** DS Fresh2

Sales Period	Price for Fresh, x_1 (Dollars)	Average Industry Price, x_2 (Dollars)	Price Difference, $x_4 = x_2 - x_1$ (Dollars)	Advertising Expenditure for Fresh, x_3 (Hundreds of Thousands of Dollars)	Demand for Fresh, y (Hundreds of Thousands of Bottles)
1	3.85	3.80	−.05	5.50	7.38
2	3.75	4.00	.25	6.75	8.51
3	3.70	4.30	.60	7.25	9.52
4	3.70	3.70	0	5.50	7.50
5	3.60	3.85	.25	7.00	9.33
6	3.60	3.80	.20	6.50	8.28
7	3.60	3.75	.15	6.75	8.75
8	3.80	3.85	.05	5.25	7.87
9	3.80	3.65	−.15	5.25	7.10
10	3.85	4.00	.15	6.00	8.00
11	3.90	4.10	.20	6.50	7.89
12	3.90	4.00	.10	6.25	8.15
13	3.70	4.10	.40	7.00	9.10
14	3.75	4.20	.45	6.90	8.86
15	3.75	4.10	.35	6.80	8.90
16	3.80	4.10	.30	6.80	8.87
17	3.70	4.20	.50	7.10	9.26
18	3.80	4.30	.50	7.00	9.00
19	3.70	4.10	.40	6.80	8.75
20	3.80	3.75	−.05	6.50	7.95
21	3.80	3.75	−.05	6.25	7.65
22	3.75	3.65	−.10	6.00	7.27
23	3.70	3.90	.20	6.50	8.00
24	3.55	3.65	.10	7.00	8.50
25	3.60	4.10	.50	6.80	8.75
26	3.65	4.25	.60	6.80	9.21
27	3.70	3.65	−.05	6.50	8.27
28	3.75	3.75	0	5.75	7.67
29	3.80	3.85	.05	5.80	7.93
30	3.70	4.25	.55	6.80	9.26

(574 observations) were used as the validation data set. A portion of the data, as well as part of a JMP regression tree analysis of the data, are shown in Figures 5.24 and 5.25. The default minimum split size of 5 was used. DS ToyotaCorolla

a. Using the information given in Figures 5.24 and 5.25, find the prediction of the sales price of used Toyota Corolla 1437, which has just been obtained by the Toyota dealer and has not yet been sold.

b. Redo the regression trees analysis and part a using a specified minimum split size of 20. Compare the results you obtain with the results obtained using a minimum split size of 5.

FIGURE 5.21   The JMP Output of a Regression Tree for the Fresh Detergent Demand Data   DS Fresh2

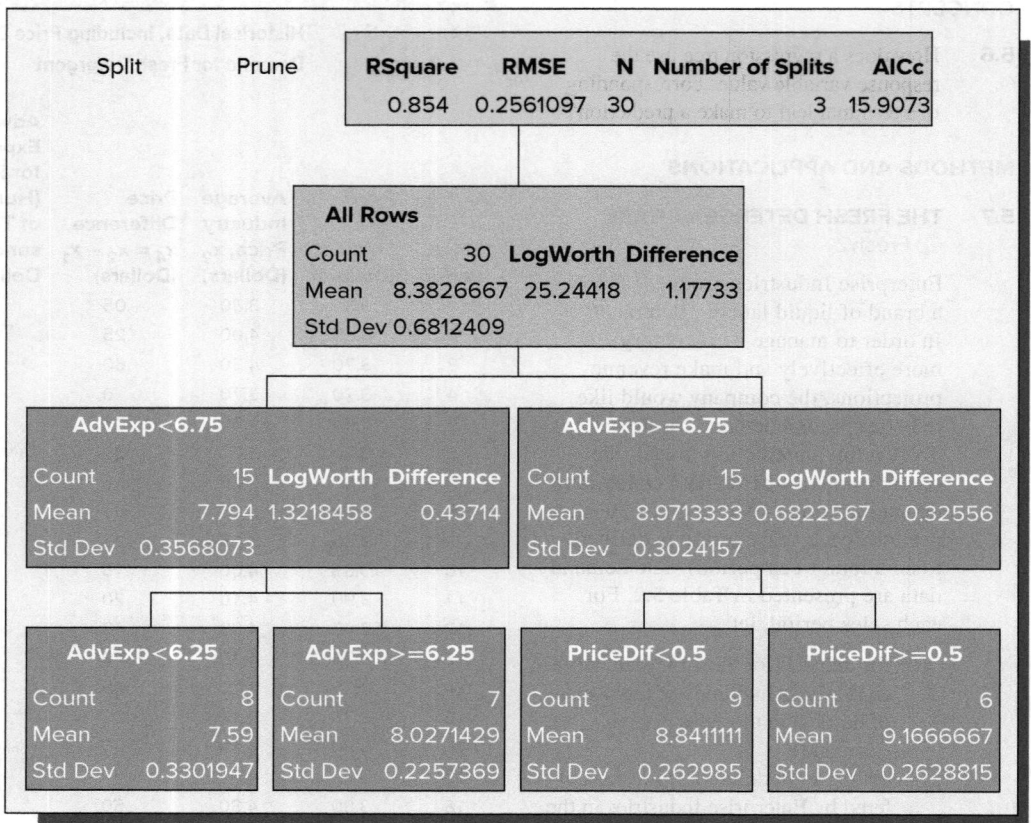

	RSquare	RMSE	N	Number of Splits	AICc
	0.854	0.2561097	30	3	15.9073

All Rows

Count	30	LogWorth	Difference
Mean	8.3826667	25.24418	1.17733
Std Dev	0.6812409		

AdvExp<6.75

Count	15	LogWorth	Difference
Mean	7.794	1.3218458	0.43714
Std Dev	0.3568073		

AdvExp>=6.75

Count	15	LogWorth	Difference
Mean	8.9713333	0.6822567	0.32556
Std Dev	0.3024157		

AdvExp<6.25

Count	8
Mean	7.59
Std Dev	0.3301947

AdvExp>=6.25

Count	7
Mean	8.0271429
Std Dev	0.2257369

PriceDif<0.5

Count	9
Mean	8.8411111
Std Dev	0.262985

PriceDif>=0.5

Count	6
Mean	9.1666667
Std Dev	0.2628815

FIGURE 5.22   The JMP Leaf Report for the Fresh Detergent Regression Tree

Leaf Report

Leaf Label	Mean	Count
AdvExp<6.75&AdvExp<6.25	7.59	8
AdvExp<6.75&AdvExp>=6.25	8.02714286	7
AdvExp>=6.75&PriceDif<0.5	8.84111111	9
AdvExp>=6.75&PriceDif>=0.5	9.16666667	6

FIGURE 5.23   The JMP Predictions of Demand in Periods 31 and 32 Using the Fresh Detergent Regression Tree

	PriceDif	AdvExp	Demand	Demand Predictor	Leaf Number Formula	Leaf Label Formula
31	0.3	6.9	.		3	AdvExp>=6.75&PriceDif<0.5
32	0.1	6.4	.		2	AdvExp<6.75&AdvExp>=6.25

F I G U R E   5 . 2 4   **The JMP Output of a Portion of the Used Toyota Corolla Sales Price Data and of Part of a Regression Tree Analysis of These Data**

	Model	Price	Age_08_04	Mfg_Month	Mfg_Year	KM	Fuel_Type	HP	Met_Color
1	TOYOTA Corolla 2.0 D4D HATCHB TERRA 2/3-Doors	13500	23	10	2002	46986	Diesel	90	1
2	TOYOTA Corolla 2.0 D4D HATCHB TERRA 2/3-Doors	13750	23	10	2002	72937	Diesel	90	1
3	TOYOTA Corolla 2.0 D4D HATCHB TERRA 2/3-Doors	13950	24	9	2002	41711	Diesel	90	1
8	TOYOTA Corolla 2.0 D4D 90 3DR TERRA 2/3-Doors	18600	30	3	2002	75889	Diesel	90	1
9	TOYOTA Corolla 1800 T SPORT VVT I 2/3-Doors	21500	27	6	2002	19700	Petrol	192	0
10	TOYOTA Corolla 1.9 D HATCHB TERRA 2/3-Doors	12950	23	10	2002	71138	Diesel	69	0
1437	TOYOTA Corolla 2.0 D4D 90 3DR TERRA 2/3-Doors	16900	27	6	2002	94612	Diesel	90	1

	Color	Automatic	CC	Doors	Cylinders	Gears	Quarterly_Tax	Weight	Mfg_Guarantee	BOVAG_Guarantee	Guarantee_Period
1	Blue	0	2000	3	4	5	210	1165	0	1	3
2	Silver	0	2000	3	4	5	210	1165	0	1	3
3	Blue	0	2000	3	4	5	210	1165	1	1	3
8	Grey	0	2000	3	4	5	210	1245	1	1	3
9	Red	0	1800	3	4	5	100	1185	0	1	3
10	Blue	0	1900	3	4	5	185	1105	0	1	3
1437	Grey	0	2000	3	4	5	210	1245	0	1	3

	ABS	Airbag_1	Airbag_2	Airco	Automatic_airco	Boardcomputer	CD_Player	Central_Lock	Powered_Windows
1	1	1	1	0	0	1	0	1	1
2	1	1	1	1	0	1	1	1	0
3	1	1	1	0	0	1	0	0	0
8	1	1	1	1	0	1	1	1	1
9	1	1	0	1	0	0	0	1	1
10	1	1	1	1	0	1	0	0	0
1437	1	1	1	1	0	1	0	1	1

	Power_Steering	Radio	Mistlamps	Sport_Model	Backseat_Divider	Metallic_Rim	Radio_cassette	Tow_Bar	Validation
1	1	0	0	0	1	0	0	0	Validation
2	1	0	0	0	1	0	0	0	Validation
3	1	0	0	0	1	0	0	0	Training
8	1	0	0	0	1	0	0	0	Training
9	1	1	0	0	0	1	1	0	Training
10	1	0	0	0	1	0	0	0	Validation
1437	1	0	0	1	1	0	0	0	Training

	RSquare	RMSE	N	Number of Splits	AICc
Training	0.933	938.52912	862	47	14349.9
Validation	0.853	1387.5365	574		

	Validation	Price Residual	Price Predicted	Leaf Number	Leaf Label
1437				43	Age_08_04<33&Weight>=1125&Metallic_Rim<1&Mfg_Year<2003&Doors<5

Source of the used Toyota Corolla sales price data: Gayle Shmaeli, Peter Broce, Mia Stephens, and Nitin Patel, *Data Mining for Business Analytics* (Hoboken, NJ: John Wiley & Sons, 2017).

Leaf Report

Leaf Label	Mean
Age_08_04>=33&Age_08_04>=57&Age_08_04>=69&KM>=145996&ABS<1	5811.25
Age_08_04>=33&Age_08_04>=57&Age_08_04>=69&KM>=145996&ABS>=1&Fuel_Type(Petrol)	6200
Age_08_04>=33&Age_08_04>=57&Age_08_04>=69&KM>=145996&ABS>=1&Fuel_Type(Diesel)	7050
Age_08_04>=33&Age_08_04>=57&Age_08_04>=69&KM<145996&Quarterly_Tax<85&Mfg_Guarantee<1	7687.84884
Age_08_04>=33&Age_08_04>=57&Age_08_04>=69&KM<145996&Quarterly_Tax<85&Mfg_Guarantee>=1&Doors<5	7993.71429
Age_08_04>=33&Age_08_04>=57&Age_08_04>=69&KM<145996&Quarterly_Tax<85&Mfg_Guarantee>=1&Doors>=5	8840
Age_08_04>=33&Age_08_04>=57&Age_08_04>=69&KM<145996&Quarterly_Tax>=85&Tow_Bar>=1	8112.47368
Age_08_04>=33&Age_08_04>=57&Age_08_04>=69&KM<145996&Quarterly_Tax>=85&Tow_Bar<&BOVAG_Guarantee<1	7607
Age_08_04>=33&Age_08_04>=57&Age_08_04>=69&KM<145996&Quarterly_Tax>=85&Tow_Bar<1&BOVAG_Guarantee>=1	8720.51282
Age_08_04>=33&Age_08_04>=57&Age_08_04<69&KM>=135337&HP<110	7166.77778
⋮	⋮
Age_08_04<33&Weight<1125&Age_08_04>=19&Central_Lock<1	14390.8333
Age_08_04<33&Weight<1125&Age_08_04>=19&Central_Lock>=1&Weight<1115	15989.5833
Age_08_04<33&Weight<1125&Age_08_04>=19&Central_Lock>=1&Weight>=1115	17504.375
Age_08_04<33&Weight<1125&Age_08_04<19	18286.4
Age_08_04<33&Weight>=1125&Metallic_Rim<1&Mfg_Year<2003&Doors<5	16270
Age_08_04<33&Weight>=1125&Metallic_Rim<1&Mfg_Year<2003&Doors>=5	17934.5385
Age_08_04<33&Weight>=1125&Metallic_Rim<1&Mfg_Year>=2003&Quarterly_Tax<234	19468.75
Age_08_04<33&Weight>=1125&Metallic_Rim<1&Mfg_Year>=2003&Quarterly_Tax>=234	21810
Age_08_04<33&Weight>=1125&Metallic_Rim>=1&Mfg_Month>=4	21228.125
Age_08_04<33&Weight>=1125&Metallic_Rim>=1&Mfg_Month<4	22982.2222

LO5-3
Interpret the information provided by k-nearest neighbors.

5.3 k-Nearest Neighbors

Classifying a qualitative response variable

To explain the k-nearest neighbors approach to classification, note that the JMP output in Figure 5.26 shows again the 40 credit card upgrade observations, and consider Silver card holder 22, who had purchases last year of \$34,521 (Purchases = 34.521) and is a Platinum conformer (PlatProfile = 1). If we did not know that Silver card holder 22 had upgraded (Upgrade = 1) and wished to predict whether Silver card holder 22 would upgrade, a very simple way to do this would be to search through the data and find the Silver card holders that had values of Purchases and PlatProfile that are the closest, or nearest, to the Purchases and PlatProfile for Silver card holder 22. We would then predict that Silver card holder 22 would upgrade if a majority of the "nearest neighbors" to Silver card holder 22 upgraded. Figure 5.26 shows the four nearest neighbors to each Silver card holder. As shown there, the four nearest neighbors to Silver card holder 22 are (1) Silver card holder 39 (see *RowNear 1*), who is the first (or overall) nearest neighbor and has Purchases = 34.964 and PlatProfile = 1; (2) Silver card holder 38 (see *RowNear 2*), who is the second nearest neighbor and has Purchases = 34.995 and PlatProfile = 1; (3) Silver card holder 4 (see *RowNear 3*), who is the third nearest neighbor and has Purchases = 32.450 and PlatProfile = 1; and (4) Silver card holder 19 (see *RowNear 4*), who is the fourth nearest neighbor and has Purchases = 38.110 and PlatProfile = 1. In general, JMP determines the nearest neighbors to an observation by measuring the *distance* between the set of predictor variable values for that observation and the set of predictor variable values for every other observation. We discuss measuring such a distance in our discussion of cluster analysis in Section 3.9 of Chapter 3.

If we use the first nearest neighbor, Silver card holder 39, to classify Silver card holder 22, then, since *Upgrade* = 0 for holder 39 (holder 39 did not upgrade), we predict that holder 22 would also not upgrade. That is, as shown in the column headed *Predicted Upgrade 1* in Figure 5.27, JMP

FIGURE 5.26 The First, Second, Third, and Fourth Nearest Neighbors in the Card Upgrade Example

	UpGrade	Purchases	PlatProfile	RowNear 1	RowNear 2	RowNear 3	RowNear 4
1	0	7.471	0	7	40	26	29
2	0	21.142	0	33	10	25	14
3	1	39.925	1	17	19	37	38
4	1	32.450	1	32	22	39	38
5	1	48.950	1	9	11	31	24
6	0	28.520	1	35	20	32	4
7	0	7.822	0	40	1	26	29
8	0	26.548	0	36	13	25	33
9	1	48.831	1	5	11	31	24
10	0	17.584	0	14	2	29	33
11	1	49.820	1	5	9	31	15
12	1	50.450	0	21	18	28	23
13	0	28.175	0	36	8	34	25
14	0	16.200	0	10	29	2	33
15	1	52.978	1	31	27	11	5
16	1	58.945	1	27	15	31	11
17	1	40.075	1	3	37	19	38
18	1	42.380	0	28	23	12	21
19	1	38.110	1	3	17	38	39
20	1	26.185	1	35	6	32	4
21	0	52.810	0	12	18	28	23
22	1	34.521	1	39	38	4	19
23	0	34.750	0	30	28	34	13
24	1	46.254	1	9	5	11	37
25	0	24.811	0	33	8	36	13
26	0	4.792	0	1	7	40	29
27	1	55.920	1	15	16	31	11
28	0	38.620	0	18	23	30	34
29	0	12.742	0	14	40	10	7
30	0	31.950	0	34	23	13	36
31	1	51.211	1	11	15	5	9
32	1	30.920	1	4	6	35	22
33	0	23.527	0	25	2	8	36
34	0	30.225	0	30	13	36	8
35	0	28.387	1	6	20	32	4
36	0	27.480	0	13	8	25	34
37	1	41.950	1	17	3	19	24
38	1	34.995	1	39	22	4	19
39	0	34.964	1	38	22	4	19
40	0	7.998	0	7	1	26	29
41		42.571	1	.	.	.	.
42		51.835	0	.	.	.	.

FIGURE 5.27   **Classification Using the Nearest, Two Nearest, Three Nearest, and Four Nearest Neighbors in the Card Upgrade Example**

	UpGrade	Predicted UpGrade 1	Predicted UpGrade 2	Predicted UpGrade 3	Predicted UpGrade 4	Predicted Formula UpGrade 2
1	0	0	0	0	0	0
2	0	0	0	0	0	0
3	1	1	1	1	1	1
4	1	1	1	1	1	1
5	1	1	1	1	1	1
6	0	0	0	1 *	1 *	0
7	0	0	0	0	0	0
8	0	0	0	0	0	0
9	1	1	1	1	1	1
10	0	0	0	0	0	0
11	1	1	1	1	1	1
12	1	0 *	1	0 *	0 *	0
13	0	0	0	0	0	0
14	0	0	0	0	0	0
15	1	1	1	1	1	1
16	1	1	1	1	1	1
17	1	1	1	1	1	1
18	1	0 *	0 *	0 *	0 *	0
19	1	1	1	1	1	1
20	1	0 *	0 *	0 *	1	0
21	0	1 *	1 *	1 *	0	0
22	1	0 *	1	1	1	0
23	0	0	0	0	0	0
24	1	1	1	1	1	1
25	0	0	0	0	0	0
26	0	0	0	0	0	0
27	1	1	1	1	1	1
28	0	1 *	0	0	0	0
29	0	0	0	0	0	0
30	0	0	0	0	0	0
31	1	1	1	1	1	1
32	1	1	1	0 *	0 *	1
33	0	0	0	0	0	0
34	0	0	0	0	0	0
35	0	0	1 *	1 *	1 *	0
36	0	0	0	0	0	0
37	1	1	1	1	1	1
38	1	0 *	1	1	1	0
39	0	1 *	1 *	1 *	1 *	0
40	0	0	0	0	0	0
41						1
42						0

FIGURE 5.28   Misclassification Rates in the k-Nearest Neighbors Card Upgrade Example

Training Set

K	Count	Misclassification Rate	Misclassifications
1	40	0.20000	8
2	40	0.12500	5*
3	40	0.20000	8
4	40	0.15000	6
5	40	0.17500	7
6	40	0.15000	6
7	40	0.15000	6
8	40	0.15000	6
9	40	0.15000	6
10	40	0.12500	5

Confusion Matrix for Best K=2

Training Set

Actual Upgrade	Predicted Count 0	1
0	18	3
1	2	17

classifies holder 22 as a 0 (a non-upgrader), which is a misclassification, since holder 22 did upgrade (*Upgrade* = 1). If we use the first two nearest neighbors, Silver card holders 39 and 38, to classify Silver card holder 22, then, since *Upgrade* = 0 for holder 39 and *Upgrade* = 1 for holder 38, there is a "tie," in that the two nearest neighbors consist of an equal number of upgrades (1) and non-upgrades (1). In such a case, JMP randomly selects a classification to break the tie, and, as shown in the column headed *Predicted Upgrade 2*, correctly classifies Silver card holder 22 a 1 (an upgrader). If we use the first three nearest neighbors, Silver card holders 39, 38, and 4, who have *Upgrade* values of 0, 1, and 1, then the majority of the three nearest neighbors are upgraders. Thus, as shown in the column headed *Predicted Upgrade 3*, JMP correctly classifies Silver card holder 22 as a 1 (an upgrader). If we use the first four nearest neighbors, Silver card holders 39, 38, 4, and 19, who have *Upgrade* values of 0, 1, 1, and 1, then the majority of the first four nearest neighbors are upgraders. Therefore, as shown in the column headed *Predicted Upgrade 4*, JMP correctly classifies Silver card holder 22 as a 1 (an upgrader).

Examining the asterisks in the column headed *Predicted Upgrades 1, 2, 3*, and *4* in Figure 5.27, we see that if we use the 1, 2, 3, and 4 nearest neighbors to classify all 40 observed Silver card holders, then there are, respectively, 8, 5, 8, and 6 misclassifications, for misclassification rates of .2, .125, .20, and .15. Figure 5.28 shows a summary of misclassification rates for using k-nearest neighbors, where we have arbitrarily tried k values from 1 to 10. Because k = 2 gives the smallest misclassification rate, we will classify Silver card holders to whom the bank is considering sending an upgrade offer by using the two nearest neighbors. For example, as shown in the column headed *Predicted Formula Upgrade 2*, JMP classifies Silver card holders 41 and 42 as, respectively, a 1 (an upgrader) and a 0 (a non-upgrader). (Because the column headed *Predicted Formula Upgrade 2* uses *all* 40 previously observed Silver card holders to determine the nearest neighbors to any Silver card holder, the nearest neighbor to any previously observed Silver card holder will be that Silver card holder. It makes no sense to classify a previously observed Silver card holder in that way, so the classifications of previously observed Silver card holders in *Predicted Formula Upgrade 2* should be ignored.)

To understand the validation approach to k-nearest neighbors, consider the churn example and data in Figure 5.5. To implement the validation approach, JMP finds the k-nearest neighbors from the 2,221 cell phone customers in the training data set to each cell phone customer in the training data set and to each cell phone customer in the validation data set. (Of course, remembering our discussion at the end of the previous paragraph, JMP does not allow a cell phone customer in the training data set to be the nearest neighbor to itself.) JMP then classifies each cell phone customer in the training data set and each cell phone customer in the validation data set as a churner (True) or non-churner (False) by using the k-nearest

FIGURE 5.29   Misclassification Rates and Classification in the k-Nearest Neighbors Churn Example

Training Set

Misclassification

K	Count	Rate	Misclassifications
1	2221	0.13643	303
2	2221	0.14273	317
3	2221	0.11076	246
4	2221	0.11571	257
5	2221	0.10986	244 *
6	2221	0.11436	254
7	2221	0.11076	246
8	2221	0.11301	251
9	2221	0.11436	254
10	2221	0.11661	259

Validation Set

Misclassification

K	Count	Rate	Misclassifications
1	1111	0.14941	166
2	1111	0.14581	162
3	1111	0.11521	128
4	1111	0.11431	127
5	1111	0.11611	129
6	1111	0.11251	125 *
7	1111	0.11701	130
8	1111	0.11431	127
9	1111	0.11611	129
10	1111	0.11341	126

Confusion Matrix for Best K=6

Training Set

Actual Churn	Predicted Count	
	False	True
False	1885	14
True	240	82

Validation Set

Actual Churn	Predicted Count	
	False	True
False	941	9
True	116	45

	Predicted Formula Churn 6
3333	False

neighbors from the training data set to this cell phone customer. JMP chooses the optimal value of k to be the value of k that minimizes the misclassification rate in the validation data set. Figure 5.29 shows that the optimal value of k is 6, and that, using the 6 nearest neighbors to cell phone customer 3333 (who has just been sent a cell phone bill), JMP classifies this customer as a non-churner (a False). Note, however, that although the misclassification rate in the validation data set is fairly low (.11251), the confusion matrix shows that the 6 nearest neighbors classify churners very inaccurately—much more inaccurately than the classification tree of Figure 5.6 classified churners (see the classification tree's confusion matrix in Figure 5.8). This demonstrates that some classification techniques are much better than others when analyzing a particular data set.

Predicting a quantitative response variable

The procedure for using the k-nearest neighbors to predict a quantitative response variable is the same as the procedure for classifying a qualitative response variable, except that we predict the quantitative response variable by averaging the response variable values for the k-nearest neighbors. For example, recall that a college admissions officer wishes to predict the GPA of a future applicant on the basis of the applicant's ACT and H. S. Rank and that the officer will use 352 previously admitted applicants as the training data set and 353 previously admitted applicants as the validation data set.

To implement the validation approach to k-nearest neighbors, JMP finds the k-nearest neighbors (in terms of ACT and H. S. Rank) from the 352 applicants in the training data set to each applicant in the training data set and to each applicant in the validation data set. JMP then predicts the GPA of each applicant in the training data set and of each applicant in the validation data set by finding the mean of the GPAs of the k-nearest neighbors from the training data set to that applicant. JMP chooses the optimal value of k to be the value of k that minimizes RMSE (the square root of the mean of the squared deviations of the predicted GPAs from the observed GPAs) for the applicants in the validation data set. Figure 5.30

FIGURE 5.30   RMSE Values and Prediction Using k-Nearest Neighbors and Validation in the GPA Example

Training Set

K	Count	RMSE	SSE	K	Count	RMSE	SSE
1	352	0.85758	258.876	1	353	0.81706	235.657
2	352	0.73294	189.094	2	353	0.68765	166.918
3	352	0.67404	159.926	3	353	0.62740	138.952
4	352	0.65309	150.135	4	353	0.60973	131.233
5	352	0.62839	138.996	5	353	0.60512	129.256
6	352	0.62355	136.862	6	353	0.58980	122.795
7	352	0.61641	133.746	7	353	0.57891	118.303
8	352	0.61585	133.503	8	353	0.57222	115.585
9	352	0.61444	132.893	9	353	0.56997	114.678
10	352	0.61077	131.309	10	353	0.56723	113.577
11	352	0.60729	129.819	11	353	0.56697	113.474
12	352	0.60397	128.402	12	353	0.56479	112.602
13	352	0.60054	126.949	13	353	0.56042	110.867
14	352	0.59700	125.454	14	353	0.56090	111.056
15	352	0.59494	124.59	15	353	0.56079	111.013
16	352	0.59279	123.694	16	353	0.55317	108.017
17	352	0.59278	123.689	17	353	0.55439	108.493
18	352	0.59034	122.671	18	353	0.55172	107.45
19	352	0.59177	123.267	19	353	0.54726	105.721
20	352	0.59239	123.528	20	353	0.54914	106.45
21	352	0.58755	121.516	21	353	0.54766	105.874
22	352	0.58586	120.817	22	353	0.54586	105.181
23	352	0.58498	120.453	23	353	0.54318	104.15
24	352	0.58436	120.201	24	353	0.54054	103.14
25	352	0.58428	120.169	25	353	0.54039	103.085
26	352	0.58539	120.625	26	353	0.53955	102.762
27	352	0.58547	120.658	27	353	0.53799	102.171
28	352	0.58508	120.495	28	353	0.53542	101.197
29	352	0.58356	119.871	29	353	0.53451	100.854 *
30	352	0.58359	119.883	30	353	0.53605	101.436
31	352	0.58459	120.296	31	353	0.53741	101.95
32	352	0.58385	119.992	32	353	0.53735	101.927
33	352	0.58354	119.862*	33	353	0.53837	102.313
34	352	0.58279	119.556	34	353	0.53862	102.409
35	352	0.58239	119.392*	35	353	0.53869	102.437
36	352	0.58314	119.699	36	353	0.53807	102.2
37	352	0.58477	120.368	37	353	0.53671	101.686
38	352	0.58429	120.169	38	353	0.53657	101.632
39	352	0.58406	120.076	39	353	0.53644	101.58
40	352	0.58493	120.435	40	353	0.53722	101.879

	GPA	ACT	H. S. Rank	Validation	Predicted Formula GPA 29
702	3.858	27	96	Validation	3.4790344828
703	3.956	29	97	Validation	3.4474482759
704	4	26	98	Validation	3.4684482759
705	4	29	97	Validation	3.4474482759
706	.	26	91	.	3.3264137931

shows that the optimal value of k is 29 and that the prediction of the GPA for a new applicant (applicant 706) having an ACT of 26 and a H. S. Rank of 91 is 3.3264. This prediction has been calculated by averaging the GPAs of the 29 nearest previously admitted applicants in the training data set to this new applicant.

Exercises for Section 5.3

CONCEPTS

5.9   Discuss how k-nearest neighbors classifies a qualitative response variable.

5.10   Discuss how k-nearest neighbors predicts a quantitative response variable.

METHODS AND APPLICATIONS

5.11   Figure 5.31 gives the misclassification rates when k-nearest neighbors is used to classify the coupon redemption data shown in Figure 5.32.
  a   What is the optimal value of k?
  b   Discuss how JMP obtained the classification of 1 for customer 4. This classification is given in the column headed *Predicted Coupon 3*.
  c   Find and report the classifications for customers 41 and 42.

5.12   **THE FRESH DETERGENT CASE**    Fresh2
Figure 5.33 gives the JMP output obtained when k-nearest neighbors is applied to the Fresh detergent demand data.
  a   Using the 6 nearest neighbors to sales period 1 (observation 1), show how the predicted demand 7.55667 (that is 755,667 bottles) for sales period 1 has been calculated.
  b   Find and report the predicted demands for future sales periods 31 and 32.

5.13   Figure 5.34 gives the JMP output obtained when k-nearest neighbors is applied to the used Toyota Corolla sales price data in Figure 5.24. Discuss how the predicted sales price of $14,978.57 for used Toyota Corolla 1437 has been calculated.

FIGURE 5.31   Misclassification Rates in the k-Nearest Neighbors Coupon Redemption Exercise

Training Set

Misclassification

K	Count	Rate	Misclassifications
1	40	0.17500	7
2	40	0.17500	7
3	40	0.12500	5*
4	40	0.15000	6
5	40	0.15000	6
6	40	0.15000	6
7	40	0.15000	6
8	40	0.15000	6
9	40	0.12500	5
10	40	0.12500	5

Confusion Matrix for Best K=3

Training Set

Actual Coupon	Predicted Count 0	1
0	19	3
1	2	16

FIGURE 5.32   Classification Using the Three Nearest Neighbors in the Coupon Redemption Exercise

	Coupon	Purchases	Card	RowNear 1	RowNear 2	RowNear 3	Predicted Coupon 3	Predicted Formula Coupon 3
1	0	10.52	0	21	2	12	0	0
2	0	14.89	0	12	21	1	0	0
3	0	15.75	1	9	15	19	0	0
4	1	29.27	1	19	7	15	1	1
5	1	48.15	1	36	13	22	1	1
6	1	51.40	0	14	29	10	0	1
7	1	29.89	1	4	19	15	1	1
8	0	21.23	0	34	27	28	0	0
9	0	17.55	1	3	15	19	0	0
10	0	46.60	0	39	29	6	0	0
11	0	28.01	0	24	40	38	0	0
12	0	15.97	0	2	21	8	0	0
13	1	50.10	1	22	5	20	1	1
14	1	51.62	0	6	29	10	0	1
15	1	27.48	1	19	4	7	1	1
16	0	4.95	0	1	21	2	0	0
17	1	43.25	1	23	35	36	1	1
18	0	35.04	1	31	30	32	1	1
19	0	28.94	1	4	7	15	1	1
20	1	53.67	1	22	33	13	1	1
21	0	13.24	0	2	1	12	0	0
22	1	51.54	1	13	20	5	1	1
23	1	42.05	1	17	35	32	1	1
24	0	29.71	0	11	38	40	0	0
25	0	39.24	0	39	37	38	0	0
26	1	58.38	1	33	20	22	1	1
27	0	23.75	0	28	34	8	0	0
28	0	23.89	0	27	40	34	0	0
29	0	50.94	0	6	14	10	1	1
30	1	36.37	1	18	32	31	1	1
31	1	34.40	1	18	30	32	1	1
32	1	38.01	1	30	35	18	1	1
33	1	55.87	1	20	26	22	1	1
34	0	21.46	0	8	27	28	0	0
35	1	40.75	1	23	17	32	1	1
36	1	46.52	1	5	17	13	1	1
37	0	34.93	0	38	25	24	0	0
38	0	31.97	0	24	37	11	0	0
39	0	42.28	0	25	10	37	0	0
40	0	26.28	0	11	28	27	0	0
41		43.97	1	.	.	.		1
42		52.48	0	.	.	.		1

FIGURE 5.33　The 6 Nearest Neighbors to Observation 1, RMSE Values, and Predictions in the k-Nearest Neighbors Fresh Detergent Demand Exercise

	RowNear 1	RowNear 2	RowNear 3
1	4	28	9

	RowNear 4	RowNear 5	RowNear 6
	8	29	22

K	Count	RMSE	SSE
1	30	0.29283	2.5724
2	30	0.31029	2.88835
3	30	0.29055	2.53259
4	30	0.31371	2.95246
5	30	0.30562	2.80211
6	30	0.27685	2.29938*
7	30	0.28662	2.46448
8	30	0.30393	2.77111

	PriceDif	AdvExp	Demand	Predicted Demand 6	Predicted Formula Demand 6
1	−0.05	5.50	7.38	7.5566666667	7.575
2	0.25	6.75	8.51	8.67	8.7733333333
3	0.60	7.25	9.52	9.0566666667	9.185
4	0.00	5.50	7.50	7.5366666667	7.575
5	0.25	7.00	9.33	8.7716666667	8.91
6	0.20	6.50	8.28	8.3616666667	8.2633333333
7	0.15	6.75	8.75	8.4183333333	8.3216666667
8	0.05	5.25	7.87	7.5966666667	7.575
9	−0.15	5.25	7.10	7.6033333333	7.575
10	0.15	6.00	8.00	7.9866666667	7.9866666667
11	0.20	6.50	7.89	8.4266666667	8.2633333333
12	0.10	6.25	8.15	7.9616666667	7.995
13	0.40	7.00	9.10	8.94	8.9783333333
14	0.45	6.90	8.86	8.96	8.9533333333
15	0.35	6.80	8.90	8.9033333333	8.8316666667
16	0.30	6.80	8.87	8.89	8.91
17	0.50	7.10	9.26	9.0816666667	9.0816666667
18	0.50	7.00	9.00	8.9966666667	9.0383333333
19	0.40	6.80	8.75	8.9133333333	8.8716666667
20	−0.05	6.50	7.95	8.0983333333	8.0066666667
21	−0.05	6.25	7.65	7.8733333333	7.87
22	−0.10	6.00	7.27	7.8083333333	7.6416666667
23	0.20	6.50	8.00	8.4083333333	8.2633333333
24	0.10	7.00	8.50	8.605	8.7066666667
25	0.50	6.80	8.75	9.0566666667	8.9716666667
26	0.60	6.80	9.21	9.1083333333	9.0566666667
27	−0.05	6.50	8.27	8.045	8.0066666667
28	0.00	5.75	7.67	7.6216666667	7.625
29	0.05	5.80	7.93	7.6616666667	7.625
30	0.55	6.80	9.26	8.9716666667	9.0566666667
31	0.30	6.90	.	.	8.91
32	0.10	6.40	.	.	8.17

FIGURE 5.34   **RMSE Values and Prediction in the Used Toyota Corollas Sales Price k-Nearest Neighbors Exercise**

Training Set

K	Count	RMSE	SSE
1	862	1689.4	2.46e+9
2	862	1523.6	2e+9
3	862	1448.1	1.81e+9
4	862	1429.3	1.76e+9
5	862	1414.2	1.72e+9 *
6	862	1438.5	1.78e+9
7	862	1468.1	1.86e+9
8	862	1474.4	1.87e+9
9	862	1471.7	1.87e+9
10	862	1486.8	1.91e+9
11	862	1506.9	1.96e+9
12	862	1513.2	1.97e+9
13	862	1519.2	1.99e+9
14	862	1530.0	2.02e+9
15	862	1540.2	2.04e+9

Validation Set

K	Count	RMSE	SSE
1	574	1761.9	1.78e+9
2	574	1621.4	1.51e+9
3	574	1528.9	1.34e+9
4	574	1532.5	1.35e+9
5	574	1476.4	1.25e+9
6	574	1433.5	1.18e+9
7	574	1429.5	1.17e+9 *
8	574	1447.4	1.2e+9
9	574	1456.9	1.22e+9
10	574	1471.0	1.24e+9
11	574	1484.8	1.27e+9
12	574	1486.2	1.27e+9
13	574	1503.8	1.3e+9
14	574	1514.7	1.32e+9
15	574	1526.2	1.34e+9

Predicted Formula Price 7	
1437	14978.571429

5.4 Naive Bayes' Classification

LO5-4
Interpret the information provided by naive Bayes' classification.

The naive Bayes' classification procedure uses a "naive" version of Bayes' Theorem to classify observations. To introduce this procedure, suppose that we randomly select a Silver card holder from the 40 Silver card holders in Figure 5.1(b). Let the events U, U^c, HP, and PP denote the events that the randomly selected Silver card holder, respectively, upgraded, did not upgrade, had a high purchase amount last year [a purchase amount greater than the median amount of 34.873 ($34,873)], and conformed to the bank's Platinum profile. The data in Figure 5.1(b) show that:

(1) 19 out of 40 Silver card holders upgraded, or $P(U) = 19/40$.

(2) 21 out of 40 Silver card holders did not upgrade, or $P(U^c) = 21/40$.

(3) 15 out of 19 upgraders had a high purchase amount last year, or $P(HP|U) = 15/19$.

(4) 3 out of 21 non-upgraders had a high purchase amount last year, or $P(HP|U^c) = 3/21$.

(5) 17 out of 19 upgraders conformed to the Platinum profile, or $P(PP|U) = 17/19$.

(6) 3 out of 21 non-upgraders conformed to the Platinum profile, or $P(PP|U^c) = 3/21$.

(7) 13 out of 19 upgraders had high purchases last year and conformed to the Platinum profile, or $P(HP \cap PP|U) = 13/19$.

(8) 1 out of 21 non-upgraders had high purchases last year and conformed to the Platinum profile, or $P(HP \cap PP|U^c) = 1/21$.

(9) 13 out of 14 Silver card holders who had high purchases last year and conformed to the Platinum profile upgraded, or $P(U|HP \cap PP) = 13/14$.

The last probability, $P(U|HP \cap PP) = 13/14 = .92857$, is an estimate of the upgrade probability for a Silver card holder who had high purchases last year and conformed to the Platinum profile. Moreover, we can obtain this probability exactly by using the full version of Bayes' Theorem and approximately by using a naive version of Bayes' Theorem. Since we know that $P(U|HP \cap PP) = .92857$, the reader might ask why we would wish to use Bayes' Theorem to find it. The answer to this question will be explained after we demonstrate the use of Bayes' Theorem. First, then, using the previously calculated probabilities in (1), (2), (7), and (8), the full version of Bayes' Theorem says that

$$P(U|HP \cap PP) = \frac{P(U)\,P(HP \cap PP\,|\,U)}{P(U)\,P(HP \cap PP|U) + P(U^c)\,P(HP \cap PP\,|\,U^c)}$$

$$= \frac{(19/40)(13/19)}{(19/40)(13/19) + (21/40)(1/21)}$$

$$= \frac{(13/40)}{(13/40) + (1/40)}$$

$$= \frac{13}{14}$$

To introduce the naive version of Bayes' Theorem, note that

- The previously found probabilities (3) and (5) imply that $P(HP|U)\,P(PP|U) = (15/19)$ $(17/19) = .70637$, which is not exactly equal to, but close to, the previously found probability (7) of $P(HP \cap PP|U) = 13/19 = .68421$.

- The previously found probabilities (4) and (6) imply that $P(HP|U^c)\,P(PP|U^c) = (3/21)$ $(3/21) = .02041$, which is not exactly equal to, but close to, the previously found probability (8) of $P(HP \cap PP|U^c) = 1/21 = .04762$.

If HP and PP were statistically independent for upgraders and statistically independent for non-upgraders, then $P(HP \cap PP|U)$ would exactly equal $P(HP|U)\,P(PP|U)$ and $P(HP \cap PP|U^c)$ would exactly equal $P(HP|U^c)\,P(PP|U^c)$, and that is what naive Bayes' Theorem assumes. The fact that these probabilities approximately hold implies that we can approximately calculate $P(U|HP \cap PP) = 13/14 = .92857$ as follows:

$$P(U|HP \cap PP) = \frac{P(U)\,P(HP \cap PP|U)}{P(U)\,P(HP \cap PP|U) + P(U^c)\,P(HP \cap PP|U^c)}$$

$$\approx \frac{P(U)\,P(HP|U)\,P(PP|U)}{P(U)\,P(HP|U)\,P(PP|U) + P(U^c)\,P(HP|U^c)\,P(PP|U^c)}$$

$$= \frac{(19/40)(17/19)(15/19)}{(19/40)(17/19)(15/19) + (21/40)(3/21)(3/21)}$$

$$= .96906$$

In general, suppose that we wish to find the probability that a randomly selected observation described by the values $x_1, x_2, \ldots x_k$ of k predictor variables will fall into a particular category (for example, will upgrade or churn). Letting C denote the event that the observation will fall into particular category and C^c denote the event that the observation will not fall into the particular category, Bayes' Theorem says that

$$P(C|x_1 \cap x_2 \cap \ldots \cap x_k) = \frac{P(C)\,P(x_1 \cap x_2 \cap \ldots \cap x_k|C)}{P(C)\,P(x_1 \cap x_2 \cap \ldots \cap x_k|C) + P(C^c)\,P(x_1 \cap x_2 \cap \ldots \cap x_k|C^k)}$$

Naive Bayes' Theorem assumes that the events that the predictor variables take the values x_1, $x_2, \ldots, x_k$ are statistically independent for observations that fall into the particular category and statistically independent for observations that do not fall into the particular category. Therefore, naive Bayes' Theorem says that

$$P(C|x_1 \cap x_2 \cap \ldots \cap x_k) = \frac{P(C)P(x_1|C)P(x_2|C)\ldots P(x_k|C)}{P(C)P(x_1|C)P(x_2|C)\ldots P(x_k|C) + P(C^c)P(x_1|C^c)P(x_2|C^c)\ldots P(x_k|C^c)}$$

To explain why we need naive Bayes' Theorem to estimate $P(C|x_1, x_2, \ldots, x_k)$, note in the credit card upgrade example that because there were 14 Silver card holders that had high purchases last year and conformed to the Platinum profile, we were able to use the 13 of these that upgraded to directly estimate that $P(U|HP \cap PP) = 13/14$. However, if there are a large number of predictor variables (there are 17 in the cell phone churn example), there might be very few (and possibly no) observations—even in a large data set—that have a *combination* of predictor variable values that *match* the *particular combination* of predictor variable values $x_1, x_2, \ldots, x_k$ for which we wish to estimate $P(C| x_1 \cap x_2 \cap \ldots \cap x_k)$. In this case we would not be able to directly estimate $P(C|x_1 \cap x_2 \cap \ldots \cap x_k)$. In addition, we would also not be able to use the full version of Bayes' Theorem to estimate $P(C|x_1 \cap x_2 \cap \ldots \cap x_k)$. This is because doing this would require estimating $P(x_1 \cap x_2 \cap \ldots \cap x_k|C)$, which cannot be directly done if there are very few (and possibly no) observations in category C that have a *combination* of predictor variable values that *match* the *combination* $x_1, x_2, \ldots, x_k$. However, naive Bayes' Theorem provides a solution to this problem. To understand this, note in the credit card upgrade example that, while there are 13 out of 19 upgraders who had high purchases last year *and* conformed to the Platinum profile, there are *more* upgraders—15 out of 19 upgraders—who had high purchases last year—and there are *more* upgraders—17 out of 19 upgraders—who conformed to the Platinum profile. In general, there might be very few (and possibly no) observations in category C having a combination of predictor variable values matching the combination $x_1, x_2, \ldots, x_k$ and thus allowing us to estimate $P(x_1 \cap x_2 \cap \ldots \cap x_k|C)$. However, there well might be, for $j = 1, 2, \ldots, k$, enough observations in category C that have a value of the single j th predictor variable that matches x_j and and thus allow us to estimate $P(x_j|C)$. Therefore, we would be able to estimate $P(x_1|C) \ P(x_2|C) \ldots P(x_k|C)$ and thus use naive Bayes' Theorem.

Before showing the use of JMP in carrying out naive Bayes classification, note that there might be virtually no matches in the observed data to the value of a *quantitative* predictor variable (for example, $42,571) that is part of the combination of predictor variable values describing the observation that we wish to classify. Therefore, JMP "bins" the values of any quantitative predictor variable into categories. This is, in fact, what we did in the first example of this section when we binned last year's purchase amounts into two categories—purchase amounts greater than the median amount of 34,873 ($34,873) and purchase amounts less than or equal to this median amount. With this in mind, JMP will carry out naive Bayes' classification with or without validation. When JMP does not use validation, it uses the entire data set to calculate the naive Bayes' probabilities. For example, Figure 5.35 shows non-validation naive Bayes' classification analysis using the credit card upgrade data. A Silver card holder is classified as a 1 (an upgrader) if the naive Bayes' upgrade probability estimate for this card holder (see the column headed *Naive Prob 1*) is at least .5. The naive Bayes' classifications are given in the column headed *Naive Predicted Formula Upgrade*. As shown by the asterisks in Figure 5.35, there are 5 naive Bayes' misclassifications and thus a misclassification error rate of $5/40 = .125$ (see Figure 5.36). The bottom of Figure 5.35 shows that the naive Bayes' upgrade probability estimates for Silver card holders 41 and 42 are, respectively, .9581 and .6084, and thus both card holders are classified as upgraders.

When JMP uses a validation approach, it uses the data in the training data set as the basis for calculating naive Bayes' probabilities that describe and classify the observations in both the training and validation data sets. For example, Figure 5.37 shows that when we apply naive Bayes' classification analysis to the cell phone churn data, the training data set error rate is .11706 and the validation data set error rate is .11881. These are higher than the training and validation data set error rates of .0603 and .1116 given in Figure 5.8 for the classification tree describing the churn data and about the same as

FIGURE 5.35   Naive Bayes' Classification in the Card Upgrade Example

	UpGrade	Purchases	PlatProfile	Naive Prob 0	Naive Prob 1	Naive Predicted Formula UpGrade
1	0	7.471	0	0.9998288767	0.0001711233	0
2	0	21.142	0	0.9917850447	0.0082149553	0
3	1	39.925	1	0.0604854522	0.9395145478	1
4	1	32.45	0	0.1889667659	0.8110332341	1
5	1	48.95	0	0.0195352914	0.9804647086	1
6	0	28.52	0	0.3393707907	0.6606292093	1 *
7	0	7.822	0	0.999808689	0.000191311	0
8	0	26.548	0	0.9713275502	0.0286724498	0
9	1	48.831	1	0.0197811876	0.9802188124	1
10	0	17.584	0	0.996702733	0.003297267	0
11	1	49.82	1	0.0178648833	0.9821351167	1
12	1	50.45	0	0.4235912448	0.5764087552	1
13	0	28.175	0	0.9596688542	0.0403311458	0
14	0	16.2	0	0.9977302834	0.0022697166	0
15	1	52.978	1	0.0133202148	0.9866797852	1
16	1	58.945	1	0.0087537895	0.9912462105	1
17	1	40.075	1	0.0591976304	0.9408023696	1
18	1	42.38	0	0.6590858818	0.3409141182	0 *
19	1	38.11	1	0.0789372955	0.9210627045	1
20	1	26.185	1	0.4603323545	0.5396676455	1
21	0	52.81	0	0.3709623056	0.6290376944	1 *
22	1	34.521	1	0.1368550223	0.8631449777	1
23	0	34.75	0	0.8675551976	0.1324448024	0
24	1	46.254	1	0.026365046	0.973634954	1
25	0	24.811	0	0.9804452394	0.0195547606	0
26	0	4.792	0	0.999928445	0.000071555	0
27	1	55.92	1	0.0105917343	0.9894082657	1
28	0	38.62	0	0.7726667964	0.2273332036	0
29	0	12.742	0	0.9991455495	0.0008544505	0
30	0	31.95	0	0.9169695507	0.0830304493	0
31	1	51.211	1	0.0156040083	0.9843959917	1
32	1	30.92	1	0.2389754918	0.7610245082	1
33	0	23.527	0	0.9854355836	0.0145644164	0
34	0	30.225	0	0.939550729	0.060449271	0
35	0	28.387	1	0.3457071078	0.6542928922	1 *
36	0	27.48	0	0.9650652269	0.0349347731	0
37	1	41.95	1	0.0455510798	0.9544489202	1
38	1	34.995	1	0.1271184401	0.8728815599	1
39	0	34.964	1	0.1277327917	0.8722672083	1 *
40	0	7.998	0	0.9997977348	0.0002022652	0
41		42.571	1	0.0418904939	0.9581095061	1
42		51.835		0.3915908597	0.6084091403	1

FIGURE 5.36   **Misclassifications in the Naive Bayes' Card Upgrade Example**

Training Set

Misclassification

Count	Rate	Misclassifications
40	0.12500	5

Confusion Matrix
Training Set

Actual Upgrade	Predicted Count	
	0	1
0	17	4
1	1	18

FIGURE 5.37   **Naive Bayes' Misclassifications and Prediction in the Churn Exercise with Validation**

Training Set

Misclassification

Count	Rate	Misclassifications
2221	0.11706	260

Validation Set

Misclassification

Count	Rate	Misclassifications
1111	0.11881	132

Confusion Matrix

Training Set

Actual Churn	Predicted Count	
	False	True
False	1818	81
True	179	143

Validation Set

Actual Churn	Predicted Count	
	False	True
False	903	47
True	85	76

	Naive Prob False	Naive Prob True	Naive Predicted Churn
3333	0.4727135145	0.5272864855	True

the training and validation set error rates of .11436 and .11251 given in Figure 5.29 for the k-nearest neighbor results describing the churn data. Furthermore, the confusion matrix in Figure 5.37 shows that the naïve Bayes' approach classifies the churners less accurately than the classification tree approach (see Figure 5.8) but more accurately than the k-nearest neighbors approach (see Figure 5.29). The naïve Bayes' churn probability estimate for cell phone customer 3333 (who has just been sent a cell phone bill) is .5273. Therefore, the naïve Bayes' approach classifies customer 3333 as a churner (as did the classification tree approach).

Exercises for Section 5.4

CONCEPTS

5.14 How does naive Bayes' Theorem differ from the usual Bayes' Theorem?

METHODS AND APPLICATIONS

5.15 Figures 5.38 and 5.39 are JMP outputs of a naive Bayes' classification analysis of the coupon redemption data shown in Figure 5.38.

a Find and report the naive Bayes' coupon redemption probability estimates for customers 41 and 42.

b How are customers 41 and 42 classified?

FIGURE 5.38   Naive Bayes' Classification in the Coupon Redemption Exercise

	Coupon	Purchases	Card	Naive Prob 0	Naive Prob 1	Naive Predicted Formula Coupon
1	0	10.52	0	0.9991316454	0.0008683546	0
2	0	14.89	0	0.9974782494	0.0025217506	0
3	0	15.75	1	0.9124979935	0.0875020065	0
4	1	29.27	1	0.3915205079	0.6084794921	1
5	1	48.15	1	0.0416680074	0.9583319926	1
6	1	51.40	0	0.4939099007	0.5060900993	1
7	1	29.89	1	0.3655092209	0.6344907791	1
8	0	21.23	0	0.9896854692	0.0103145308	0
9	0	17.55	1	0.8735216192	0.1264783808	0
10	0	46.60	0	0.6148342156	0.3851657844	0
11	0	28.01	0	0.9616866992	0.0383133008	0
12	0	15.97	0	0.9967566198	0.0032433802	0
13	1	50.10	1	0.0344097523	0.9655902477	1
14	1	51.62	0	0.4888129668	0.5111870332	1
15	1	27.48	1	0.4717385347	0.5282614653	1
16	0	4.95	0	0.9998000396	0.0001999604	0
17	1	43.25	1	0.0710955405	0.9289044595	1
18	0	35.04	1	0.195885098	0.804114902	1
19	0	28.94	1	0.4057791818	0.5942208182	1
20	1	53.67	1	0.0250843437	0.9749156563	1
21	0	13.24	0	0.9982987148	0.0017012852	0
22	1	51.54	1	0.0301262134	0.9698737866	1
23	1	42.05	1	0.0818709133	0.9181290867	1
24	0	29.71	0	0.9485748914	0.0514251086	0
25	0	39.24	0	0.8017085656	0.1982914344	0
26	1	58.38	1	0.0177471555	0.9822528445	1
27	0	23.75	0	0.9827823924	0.0172176076	0
28	0	23.89	0	0.9822999359	0.0177000641	0
29	0	50.94	0	0.5047227516	0.4952772484	0
30	1	36.37	1	0.1655294829	0.8344705171	1
31	1	34.40	1	0.2123371265	0.7876628735	1
32	1	38.01	1	0.1345343059	0.8654656941	1
33	1	55.87	1	0.0211234699	0.9788765301	1
34	0	21.46	0	0.9891791523	0.0108208477	0
35	1	40.75	1	0.0957534988	0.9042465012	1
36	1	46.52	1	0.0493707255	0.9506292745	1
37	0	34.93	0	0.8848818518	0.1151181482	0
38	0	31.97	0	0.9257238037	0.0742761963	0
39	0	42.28	0	0.7285927506	0.2714072494	0
40	0	26.28	0	0.9720225284	0.0279774716	0
41		43.97	1	0.0654393672	0.9345606328	1
42		52.48	0	0.4693726801	0.5306273199	1

FIGURE 5.39   **Misclassifications in the Coupon Redemption Exercise**

Training Set

Misclassification

Count	Rate	Misclassifications
40	0.05000	2

Confusion Matrix

Training Set

Actual Coupon	Predicted Count 0	1
0	20	2
1	0	18

5.5 An Introduction to Ensemble Estimates

LO5-5
Interpret the information provided by ensemble models.

Table 5.3 summarizes the upgrade probability estimates and classifications for Silver card holders 41 and 42 obtained by using the previously discussed classification tree, k-nearest neighbors, and naive Bayes' techniques and also by using three additional techniques to be discussed in Chapter 16—logistic regression, discriminate analysis, and neural networks. Because different techniques will give different upgrade probability estimates and might give different classifications for the same Silver card holder, we might determine a final classification by using the predominant classification given by the techniques and a final upgrade probability estimate by averaging the five estimates given. Looking at Table 5.3, it is clear that Silver card holder 41 should be classified as an upgrader (since all six techniques classify Silver card holder 41 as an upgrader). In addition, the average of the five upgrade probability estimates for Silver card holder 41 is .9667. On the other hand, three techniques classify Silver card holder 42 as a non-upgrader and three techniques classify Silver card holder 42 as an upgrader. The average of the five upgrade probability estimates for Silver card holder 42 is .5606, which is slightly greater than .5, so we might classify Silver card holder 42 as an upgrader.

In calculating the ensemble upgrade probability estimates of the previous paragraph, we calculated a simple average of the individual upgrade probability estimates for each Silver card holder. If we have different trusts in different techniques (as perhaps indicated by the historical *RSquare* values, misclassification rates, and confusion matrices obtained using the different techniques), we might use a weighted average of the different results given by the different techniques. When predicting a quantitative response variable (see Exercise 5.17), this trust might be assessed by considering the historical *RSquare* and *RMSE* values obtained using the different techniques. In the real world, ensemble models have proven to be very effective.

TABLE 5.3   **Upgrade Probability Estimates and Classifications for Silver Card Holders 41 and 42 Using Six Techniques**

	Classification Tree	k-Nearest Neighbors	Naive Bayes'	Logistic Regression	Discriminate Analysis	Neural Networks
Upgrade Probability Estimate for Holder 41	.96505	–	.95811	.95804	.98593	.96653
Classification for Holder 41	1	1	1	1	1	1
Upgrade Probability Estimate for Holder 42	.40655	–	.60841	.71589	.38970	.68266
Classification for Holder 42	0	0	1	1	0	1

Exercises for Section 5.5

CONCEPTS

5.16   Discuss how ensemble estimates are made.

METHODS AND APPLICATIONS

5.17   Consider the GPA example. We have seen that regression trees and k-nearest neighbors give predicted GPAs of 3.2612 and 3.3264 for new applicant 706, and in Chapter 16 we will see that neural networks give a predicted GPA of 3.1866 for this applicant. By using simple averaging, find an ensemble estimate of the GPA for new applicant 706.

Chapter Summary

Nonparametric predictive analytics make predictions by using a relationship between the response variable and the predictor variables that is not expressed in terms of a mathematical equation involving parameters. In this chapter we have discussed four nonparametric predictive analytics—**classification trees, regression trees, k-nearest neighbors**, and **naive Bayes' classification**. A **classification or regression tree** splits values of the predictor variables off into branches that end in terminal leafs. For each observation having predictor variable values corresponding to the predictor variable values in a terminal leaf, (1) a classification tree's terminal leaf estimates the probabilities that the observation will fall into the various classes being considered, and (2) a regression tree's terminal leaf predicts the observation's response variable value. The third analytic we have studied—**k-nearest neighbors**—finds the k observations having predictor variable values that are the k smallest *distances* from the predictor variable values of an observation that we wish to classify or predict. The procedure then uses these k-nearest neighbors to do the classification or prediction. The fourth analytic—**naive Bayes' classification**—uses a naive version of Bayes' Theorem to estimate the probabilities that an observation will fall into the various classes being considered. Of course, if we use different classification or prediction analytics to analyze the same data set, the different analytics will usually give different probability estimates or response variable predictions. Therefore, it can be useful to form an **ensemble estimate**, which combines (for example, averages) the estimates or predictions obtained from different analytics to arrive at an overall result.

Glossary of Terms

confusion matrix: A matrix that summarizes a classification analytic's success in classifying observations in the training data set and/or validation data set.

training data: The portion of the data used to fit the analytic.

validation data: The portion of the data used to assess how well the analytic fitted to the training data fits data different from the training data.

Important Graphics

classification tree: pages 254, 255 and 256

regression tree: page 272

Supplementary Exercises

5.18 The personnel director of a firm has developed two tests to determine whether potential employees would perform successfully in a particular position. To help estimate the usefulness of the tests, the director gives both tests to 43 employees who currently hold the position. Figure 5.40 gives the scores of each employee on both tests and indicates whether the employee is currently performing successfully or unsuccessfully in the position. If the employee is performing successfully, we set the variable *Group* equal to 1; if the employee is performing unsuccessfully, we set *Group* equal to 0.

a Using the nonvalidation approach, perform a classification tree analysis of these data. Interpret the confusion matrix and estimate the probability that a potential employee who scores a 93 on Test 1 and an 84 on Test 2 will perform successfully in the position.

b Repeat part a using the validation approach and the validation column in Figure 5.40.

c Compare your results from a. and b.

5.19 Do Exercise 5.18 using k-nearest neighbors classification.

5.20 Do Exercise 5.18 using naive Bayes' classification.

5.21 Compare the nonvalidation approach results in Exercises 5.18, 5.19, and 5.20 and then compare the validation approach results in these exercises.

5.22 Find an ensemble estimate of the probability of success for the potential employee described in Exercise 5.18, first using the nonvalidation approach and then using the validation approach.

FIGURE 5.40   **The JMP Output of the Performance Data** DS PerfTest

	Group	Test 1	Test 2	Validation
1	1	96	85	Validation
2	1	96	88	Training
3	1	91	81	Validation
4	1	95	78	Validation
5	1	92	85	Validation
6	1	93	87	Training
7	1	98	84	Training
8	1	92	82	Training
9	1	97	89	Validation
10	1	95	96	Validation
11	1	99	93	Validation
12	1	89	90	Training
13	1	94	90	Training
14	1	92	94	Training
15	1	94	84	Training
16	1	90	92	Training
17	1	91	70	Training
18	1	90	81	Training
19	1	86	81	Training
20	1	90	76	Training
21	1	91	79	Training
22	1	88	83	Training
23	1	87	82	Training
24	0	93	74	Training
25	0	90	84	Training
26	0	91	81	Validation
27	0	91	78	Training
28	0	88	78	Validation
29	0	86	86	Training
30	0	79	81	Validation
31	0	83	84	Training
32	0	79	77	Validation
33	0	88	75	Training
34	0	81	85	Validation
35	0	85	83	Validation
36	0	82	72	Validation
37	0	82	81	Validation
38	0	81	77	Validation
39	0	86	76	Training
40	0	81	84	Validation
41	0	85	78	Training
42	0	83	77	Training
43	0	81	71	Training

Appendix 5.1 ■ Predictive Analytics Using JMP

Classification trees in Figures 5.1–5.4 (data file: CardUpgrade.jmp):

- After opening the CardUpgrade.jmp file, select **Analyze : Predictive Modeling : Partition.**
- In the Partition dialog box, select the "UpGrade" column and click "Y, Response" and select the "Purchases" and "PlatProfile" columns and click "X, Factor."
- Under "Method," select "Decision Tree."
- Click "OK."

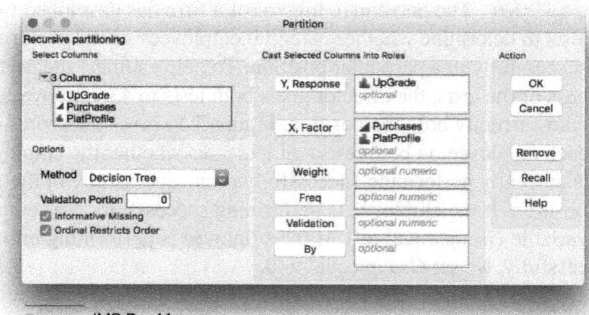

Source: JMP Pro 14

- By default, JMP first provides the overall summary of the partitioning for All Rows.
- Optional: To show the probability, in each condition at each step, for a customer to upgrade, select "Display Options : Show Split Prob" and "Display Options : Show Split Count" from the red triangle menu beside "Partition for UpGrade."
- Click "Split" several times to perform some partitioning steps.

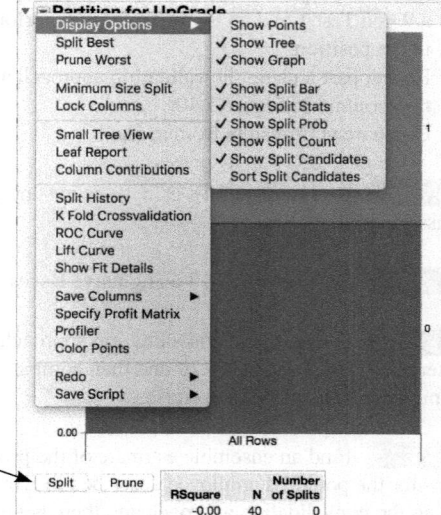

Source: JMP Pro 14

- To generate the leaf report from Figure 5.2, click on the red triangle menu beside "Partition for UpGrade" and select "Leaf Report."
- To save predictions to the data table, from the red triangle menu beside "Partition for UpGrade," select "Save Columns : Save Prediction Formula." The three columns "Prob(UpGrade==0)," "Prob(UpGrade==1)," and "Most Likely UpGrade" are saved to the file "CardUpgrade.jmp," with associated formulas.
- To save the leaf number and label to the data table, from the red triangle menu beside "Partition for UpGrade," select "Save Columns : Save Leaf Number Formula" and "Save Columns : Save Leaf Label Formula." The two columns "Leaf Number Formula" and "Leaf Label Formula" are saved to the file "CardUpgrade.jmp," with associated formulas.
- When prediction columns with associated formulas are saved to a JMP data table, these formulas can be applied to new values of the predictors. In the "CardUpgrade.jmp" file, enter "42.571" for "Purchases" and "1" for "PlatProfile" into Row 41. Similarly, enter "51.835" for "Puchases" and "0" for "PlatProfile" into Row 42. JMP will automatically predict the five formula values for these new rows.

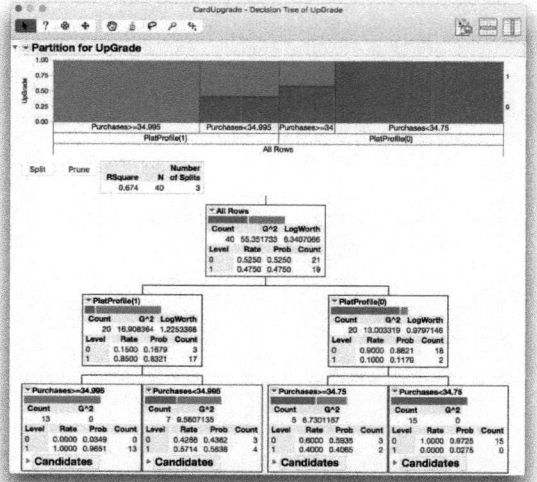

Source: JMP Pro 14

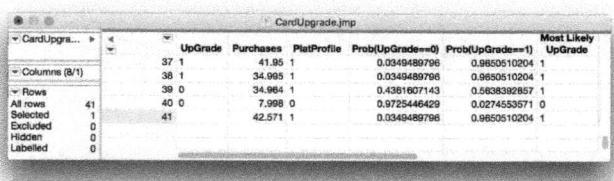

Source: JMP Pro 14

- From the red triangle menu beside "Partition for UpGrade," select "Show Fit Details" to generate the information shown in Figure 5.4.

Classification trees with validation in Figures 5.5–5.10 (data file: Churn.jmp):

- After opening the Churn.jmp file, create the validation column by selecting **Analyze : Predictive Modeling : Make Validation Column**.
- In the Make Validation Column dialog box
  - Change the proportion of data to allocate to the "Training Set" to "0.67."
  - Change the proportion of data to allocate to the "Validation Set" to "0.33."
  - Choose the "Formula Random" method to apply these proportions randomly to all the data.

Note: JMP will randomly determine which category each data point will be assigned to, according to the probabilities specified. Due to the random variation, the validation column may have very slightly different proportions or counts after assigning the categories.

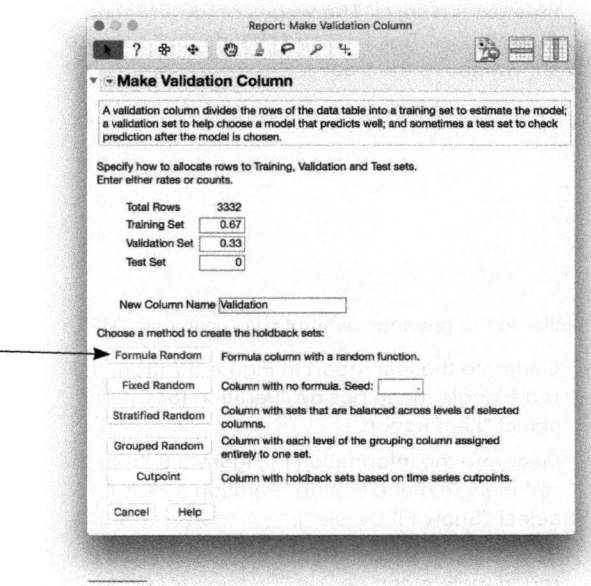

Source: JMP Pro 14

- To create the classification tree model, select **Analyze : Predictive Modeling : Partition**.
- In the Partition dialog box
  - Select the "Churn" column and click "Y, Response."
  - Select the 17 predictor variable columns, from "AcctLength" to "NCustServiceCalls," and click "X, Factor."
  - Select the "Validation" column and click "Validation."
  - Click "OK."

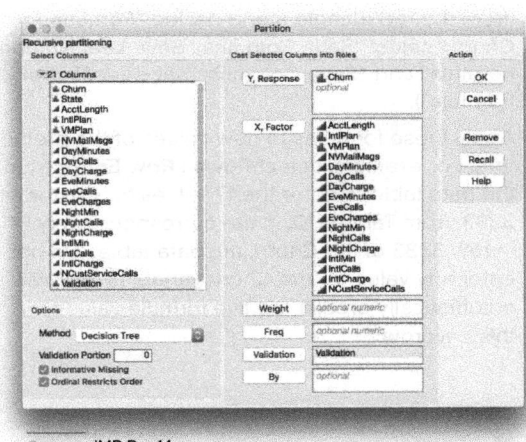

Source: JMP Pro 14

- From the red triangle menu beside "Partition for Churn," select "Minimum Split Size." Enter "20" in into the prompt box to change from JMP default of "5." Click "OK."
- Optional: Similar to the previous example, to show the probabilities and counts in each condition, select "Display Options : Show Split Prob" and "Display Options : Show Split Count" from the red triangle menu beside "Partition for Churn."
- To perform some partitioning steps, click "Split" several times, or click "Go" to have JMP automatically split until the validation R-Square is better than the validation R-Square that would be obtained by the next 10 splits.

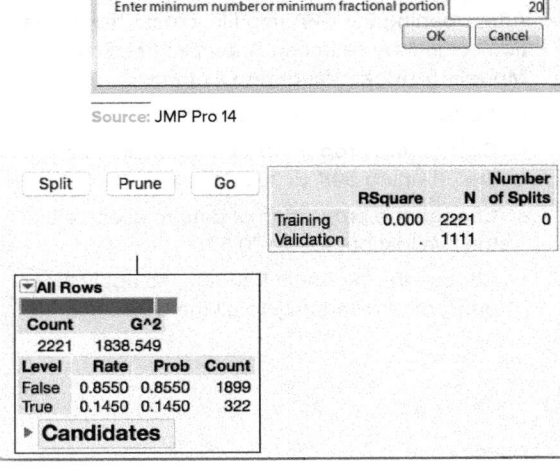

Source: JMP Pro 14

Source: JMP Pro 14

- Split History is automatically generated and confirms that this validation R-Square was maximized at the selected number of splits in comparison to the next ten splits. The values of the "RSquare" for both the "Training" and "Validation" sets are shown above the tree.

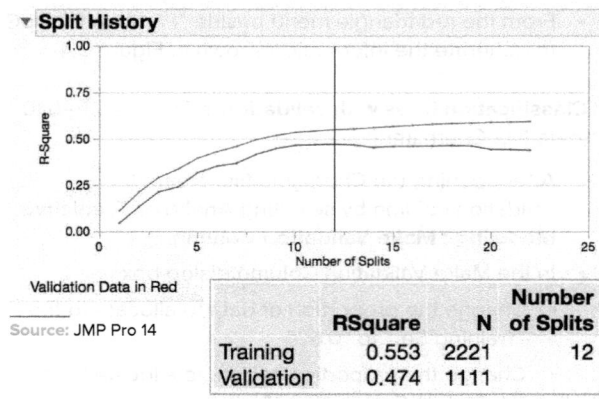

Split History

Validation Data in Red

	RSquare	N	Number of Splits
Training	0.553	2221	12
Validation	0.474	1111	

Source: JMP Pro 14

Similar to the previous example, you can use JMP to

- Generate the leaf report in Figure 5.7 (from the red triangle menu beside "Partition for Churn," select "Leaf Report").
- Generate the information in Figure 5.8 (from the red triangle menu beside "Partition for Churn," select "Show Fit Details").
- Save the predictions to the data table (from the red triangle menu beside "Partition for Churn," select "Save Columns : Save Prediction Formula").
- Save the leaf number and label to the data table (from the red triangle menu beside "Partition for Churn," select "Save Columns : Save Leaf Number Formula" and "Save Columns : Save Leaf Label Formula").
- Apply these formulas to new values of the predictors. For example, select **Rows : Row Editor** from the data table to enter the 17 values for Customer 3333, from Table 5.10, in the corresponding cells in row 3333 of the Churn.jmp data table. (Do not enter any value for the Churn variable.) JMP will automatically predict the five formula values for this new row.

Row Editor for Churn.jmp — Row: 3333

Churn	
AcctLength	87
IntlPlan	no
VMPlan	yes
NVMailMsgs	10
DayMinutes	153
DayCalls	110
DayCharge	28.35
EveMinutes	190.7
EveCalls	104
EveCharges	18.75
NightMin	205
NightCalls	81
NightCharge	8.81
IntlMin	11.7
IntlCalls	3
IntlCharge	3.87
NCustServiceCalls	5
Validation	.
Prob(Churn==False)	0.2867232952
Prob(Churn==True)	0.7132767048
Most Likely Churn	True
Leaf Number Formula	5
Leaf Label Formula	DayMinutes<265.6&NCustServiceCalls>=4&DayMinutes<161&EveCharges>=16.49

JMP Pro 14

Regression trees with validation in Figures 5.16–5.20 (data file: GPA.jmp):

- After opening the GPA.jmp file, create the validation column by selecting **Analyze : Predictive Modeling : Make Validation Column**.
- In the Make Validation Column dialog box
  - Change the proportion of data to allocate to the "Training Set" to "0.5."
  - Change the proportion of data to allocate to the "Validation Set" to "0.5."
  - Choose the "Formula Random" to apply these proportions randomly to all the data.

Report: Make Validation Column - JMP

Make Validation Column

A validation column divides the rows of the data table into a training set to estimate the model; a validation set to help choose a model that predicts well; and sometimes a test set to check prediction after the model is chosen.

Specify how to allocate rows to Training, Validation and Test sets. Enter either rates or counts.

Total Rows	706
Training Set	0.5
Validation Set	0.5
Test Set	0

New Column Name Validation

Choose a method to create the holdback sets:

Formula Random	Formula column with a random function.
Fixed Random	Column with no formula. Seed:
Stratified Random	Column with sets that are balanced across levels of selected columns.
Grouped Random	Column with each level of the grouping column assigned entirely to one set.
Cutpoint	Column with holdback sets based on time series cutpoints.

Cancel   Help

- To sort the data set by the "Validation" group and then by "GPA," select **Tables : Sort**.
- In the Sort dialog box, select the "Validation" and "GPA" columns and click "By."
- Select either the "Replace table" option or enter a name for the sorted data table in the Output table name input box.
- Click "OK."

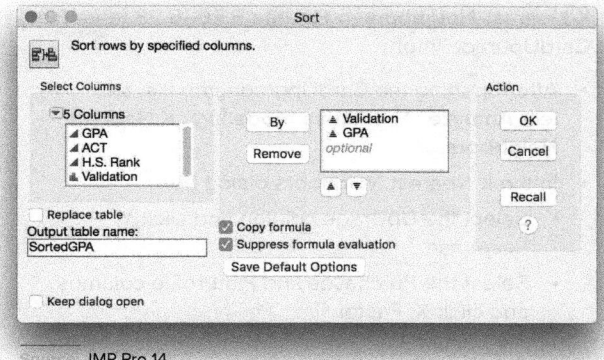

Source: JMP Pro 14

- To create the regression tree model, select **Analyze : Predictive Modeling : Partition**.
- In the Partition dialog box
  - Select the "GPA" column and click "Y, Response."
  - Select the "ACT" and "H.S. Rank" columns and click "X, Factor."
  - Select the "Validation" column and click "Validation."
  - Click "OK."
- Click "Split" once, then click "Prune" in order to see the RMSE appear in the table for "Number of Splits = 0," as in Figure 5.17(a).

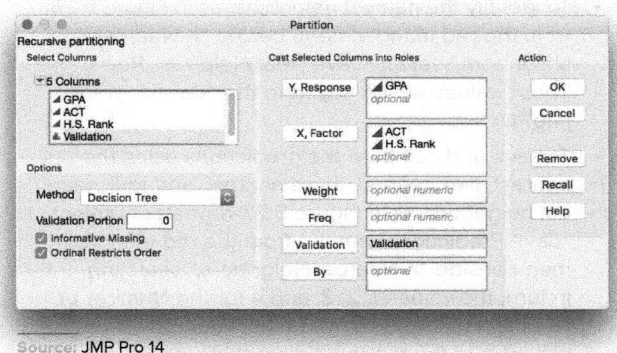

Source: JMP Pro 14

Similar to the previous examples, you can use JMP to

- Perform some partitioning steps (click "Split" several times).
- Automatically split (click "Go").
- Generate the split history from Figure 5.18 (from the red triangle menu beside "Partition for GPA," select "Split History").
- Generate the leaf report from Figure 5.19 (from the red triangle menu beside "Partition for GPA," select "Leaf Report").
- Save the predicted values, the leaf number, and the leaf label to the data table, as in Figure 5.20 (from the red triangle menu beside "Partition for GPA," select the formula options: "Save Prediction Formula," "Save Leaf Number Formula," and "Save Leaf Label Formula").
- Apply these formulas to new values of the predictors. For example, in row 706 of the GPA.jmp data table, enter "26" for "ACT" and "91" for H.S. Rank. (Do not enter any value for the GPA variable.) JMP will automatically predict the values for any formula column for this new row.

K-Nearest Neighbors in Figures 5.26–5.28 (data file: CardUpgrade.jmp):

- After opening the CardUpgrade.jmp file, select **Analyze : Predictive Modeling : K Nearest Neighbors**.
- In the K Nearest Neighbors dialog box
  - Select the UpGrade column and click Y, Response.
  - Select the Purchases and PlatProfile columns and click X, Factor.
  - Click "OK."

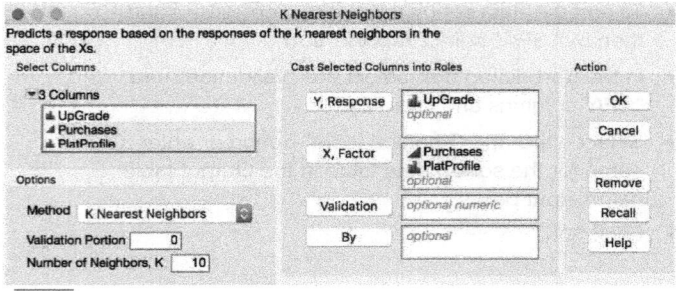

Source: JMP Pro 14

By default, JMP first provides the output as in Figure 5.28.

- To identify the nearest neighbors, as in Figure 5.26, from the red triangle menu beside "K Nearest Neighbors," select "Save Near Neighbor Rows." These values will be saved to the "CardUpgrade.jmp" file.
- To save and compare the predictions using the nearest, two nearest, three nearest, and four nearest neighbors, as in Figure 5.27, repeatedly select "Save Prediction Formula" from the red triangle menu beside "K Nearest Neighbors," selecting, in turn, the values 1, 2, 3, and 4 for the Number of Neighbors, K. Keep in mind that your results may vary due to the random tie breaking, as described in the chapter.
- After calculating the nearest neighbors and the predicted values, add rows 41 and 42 to the CardUpgrade.jmp file, with "Purchases" = 42.571 and "PlatProfile" = 1 and "Purchases" = 51.835 and "PlatProfile" = 0, respectively. The formula columns will update with the new predictions.
- Note that you should not include these extra prediction rows until *after* you save the nearest neighbors and create the prediction formulas. If those extra rows are included before these steps, JMP will include these rows as candidates for the nearest neighboring rows and the results will be affected.

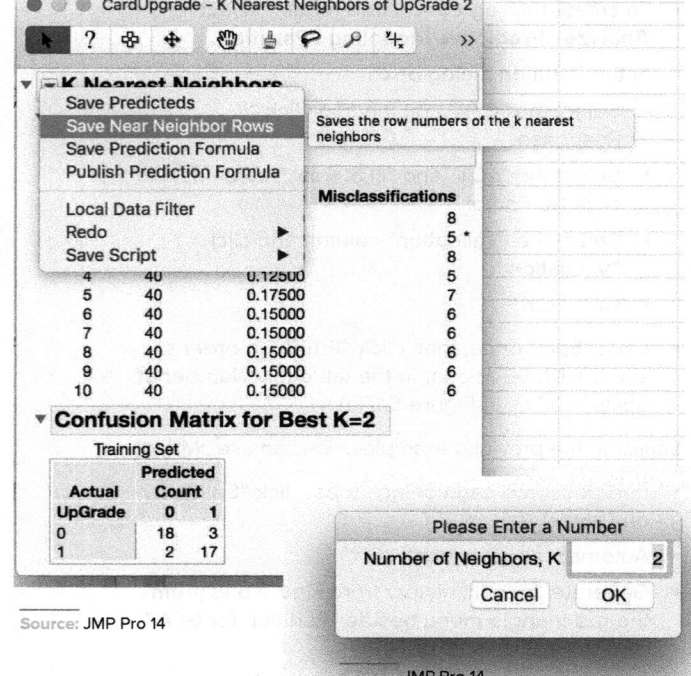

Source: JMP Pro 14

K-Nearest Neighbors with validation in Figure 5.29 (data file: Churn.jmp):

- After opening the Churn data file in JMP, select **Analyze : Predictive Modeling : K Nearest Neighbors**.
- In the K Nearest Neighbors dialog box
  - Select the column "Churn" and click "Y, Response."
  - Select the 17 predictor variable columns, from AcctLength to NCustServiceCalls, and click "X, Factor."
  - Select the "Validation" column and click "Validation."
  - Click "OK."

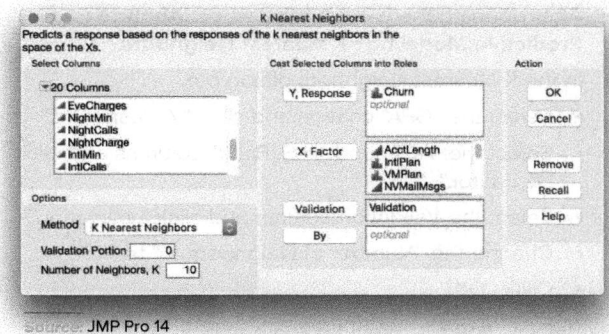

Source: JMP Pro 14

By default, JMP first provides the output as in Figure 5.29. Keep in mind that your results may vary.

- To identify the nearest neighbors, from the red tri-angle menu beside "K Nearest Neighbors," select "Save Near Neighbor Rows."

- To save the predictions with the formula to the data table, select "Save Prediction Formula" from the red triangle menu beside "K Nearest Neighbors." Use the recommended default value or enter a different number for the "Number of Neighbors, K," in the pop-up dialog box.

- To find the prediction for hypothetical row 3333, enter the 17 values for Customer 3333, from Table 5.10, in the corresponding cells in row 3333 of the Churn.jmp data table. (Do not enter any value for the Churn variable.) JMP will automatically fill in the prediction column for this new row. As before, you can use the Row Editor to enter these values (select **Rows : Row Editor** from the Churn.jmp file).

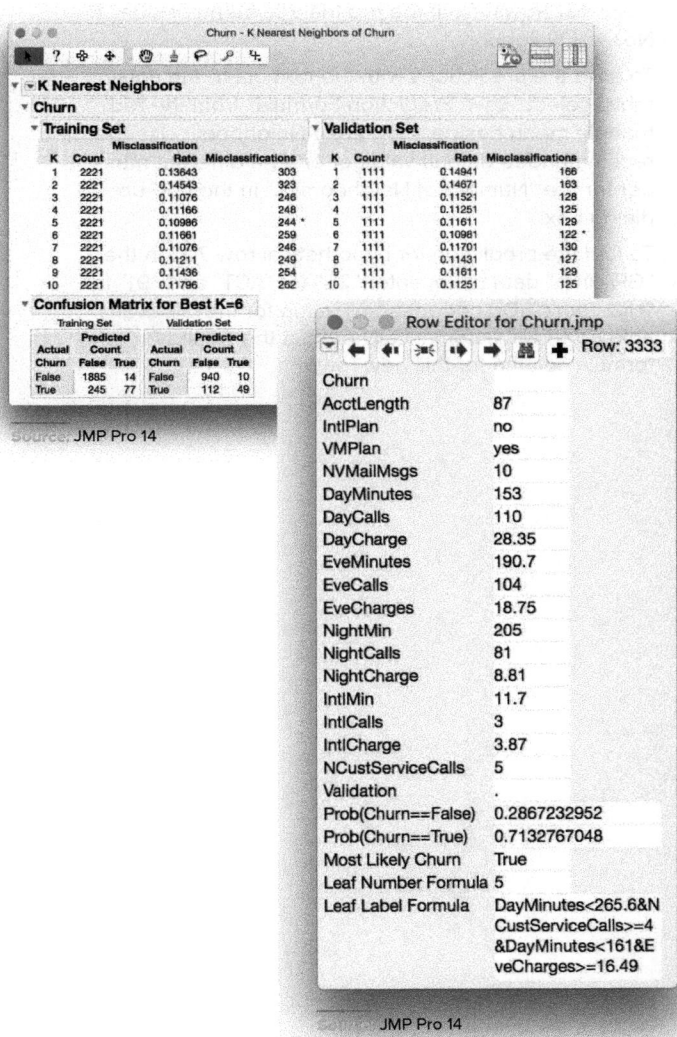

Source: JMP Pro 14

K-Nearest Neighbors with validation in Figure 5.30 (data file: GPA.jmp):

- After opening the GPA.jmp file, select **Analyze : Predictive Modeling : K Nearest Neighbors**.

- In the K Nearest Neighbors dialog box
  - Select the "GPA" column and click "Y, Response."
  - Select the "ACT" and "H.S. Rank" columns and click "X, Factor."
  - Select the Validation column and click "Validation."
  - Change the "Number of Neighbors, K" to "40."
  - Click "OK."

By default, JMP first provides the output as in Figure 5.30. Keep in mind that your results may vary.

- To identify the nearest neighbors, select "Save Near Neighbor Rows" from the red triangle menu beside "K Nearest Neighbors."

- To save the predictions with the formula to the data table, select "Save Prediction Formula" from the red triangle menu beside "K Nearest Neighbors." Use the recommended default value or enter a different number for the "Number of Neighbors, K" in the pop-up dialog box.

- To find the prediction for hypothetical row 706 in the "GPA.jmp" data table, enter "26" for "ACT" and "91" for "H.S. Rank." (Do not enter any value for the GPA variable.) JMP will automatically predict the values for any formula column for this new row.

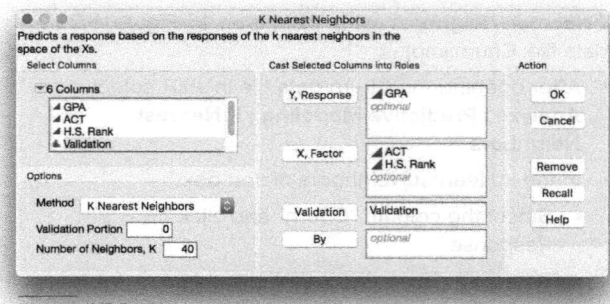

Source: JMP Pro 14

GPA - K Nearest Neighbors of GPA

▼ K Nearest Neighbors
▼ GPA

▼ Training Set					▼ Validation Set			
K	**Count**	**RMSE**	**SSE**		**K**	**Count**	**RMSE**	**SSE**
1	352	0.85758	258.876		1	353	0.81706	235.657
2	352	0.73294	189.094		2	353	0.68765	166.918
3	352	0.67404	159.926		3	353	0.62740	138.952
4	352	0.65309	150.135		4	353	0.60973	131.233
5	352	0.62839	138.996		5	353	0.60512	129.256
6	352	0.62355	136.862		6	353	0.58980	122.795
7	352	0.61641	133.746		7	353	0.57891	118.303
8	352	0.61585	133.503		8	353	0.57222	115.585
9	352	0.61444	132.893		9	353	0.56997	114.678
10	352	0.61077	131.309		10	353	0.56723	113.577
11	352	0.60729	129.819		11	353	0.56697	113.474
12	352	0.60397	128.402		12	353	0.56479	112.602
13	352	0.60054	126.949		13	353	0.56042	110.867
14	352	0.59700	125.454		14	353	0.56090	111.056
15	352	0.59494	124.59		15	353	0.56079	111.013
16	352	0.59279	123.694		16	353	0.55317	108.017
17	352	0.59278	123.689		17	353	0.55439	108.493
18	352	0.59034	122.671		18	353	0.55172	107.45
19	352	0.59177	123.267		19	353	0.54726	105.721
20	352	0.59239	123.528		20	353	0.54914	106.45
21	352	0.58755	121.516		21	353	0.54766	105.874
22	352	0.58586	120.817		22	353	0.54586	105.181
23	352	0.58498	120.453		23	353	0.54318	104.15
24	352	0.58436	120.201		24	353	0.54054	103.14
25	352	0.58428	120.169		25	353	0.54039	103.085
26	352	0.58539	120.625		26	353	0.53955	102.762
27	352	0.58547	120.658		27	353	0.53799	102.171
28	352	0.58508	120.495		28	353	0.53542	101.197
29	352	0.58356	119.871		29	353	0.53451	100.854 *
30	352	0.58359	119.883		30	353	0.53605	101.436
31	352	0.58459	120.296		31	353	0.53741	101.95
32	352	0.58385	119.992		32	353	0.53735	101.927
33	352	0.58354	119.862		33	353	0.53837	102.313
34	352	0.58279	119.556		34	353	0.53862	102.409
35	352	0.58239	119.392 *		35	353	0.53869	102.437
36	352	0.58314	119.699		36	353	0.53807	102.2
37	352	0.58477	120.368		37	353	0.53671	101.686
38	352	0.58429	120.169		38	353	0.53657	101.632
39	352	0.58406	120.076		39	353	0.53644	101.58
40	352	0.58493	120.435		40	353	0.53722	101.879

JMP Pro 14

Naive Bayes' Classification in Figures 5.35–5.36 (data file: CardUpgrade.jmp):

- After opening the CardUpgrade.jmp file, select **Analyze : Predictive Modeling : Naive Bayes**.

- In the Naive Bayes dialog box
  - Select the "UpGrade" column and click "Y, Response."
  - Select the Purchases and PlatProfile column and click "X, Factor."
  - Click "OK."

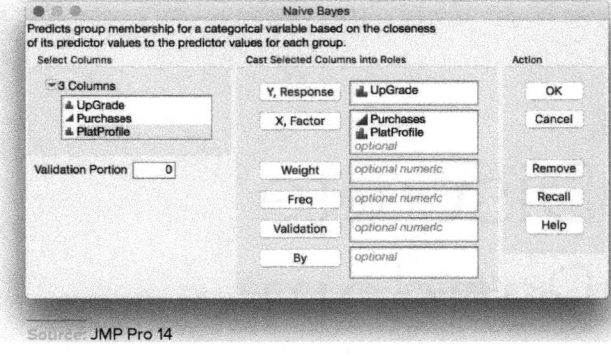

Source: JMP Pro 14

- The misclassification rate and count and the confusion matrix are displayed by default, along with the ROC curve.

- To save the probabilities and predictions with formulas to the data table, select "Save Probability Formula" from the red triangle menu beside "Naive Bayes."

- Add the hypothetical rows 41 and 42 to the "CardUpgrade.jmp" file, with "Purchases" = 42.571 and "PlatProfile" = 1 and "Purchases" = 51.835 and "PlatProfile" = 0, respectively. The formula columns will update with the new predictions.

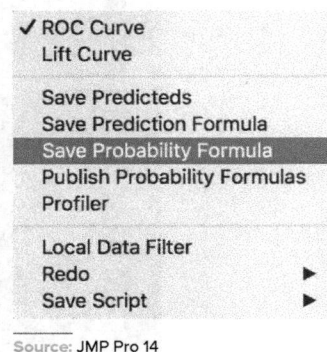

Source: JMP Pro 14

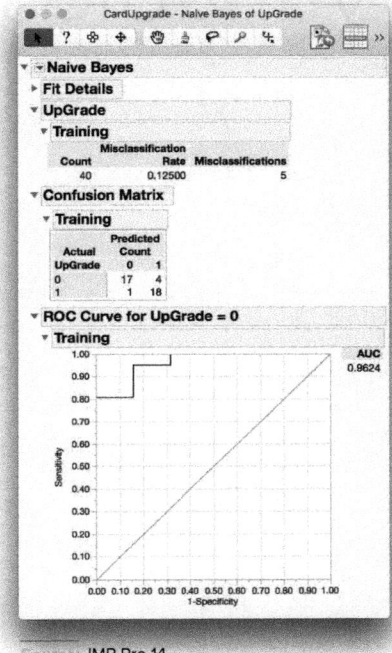

Source: JMP Pro 14

Naive Bayes' Classification with validation in Figure 5.37 (data file: Churn.jmp):

- After opening the Churn data file in JMP, select **Analyze : Predictive Modeling : Naive Bayes.**

- In the Naive Bayes dialog box
  - Select the Churn column and click "Y, Response."
  - Select the 17 predictor variable columns from "AcctLength" to "NCustServiceCalls" and click "X, Factor."
  - Select the "Validation" column and click "Validation."
  - Click "OK."

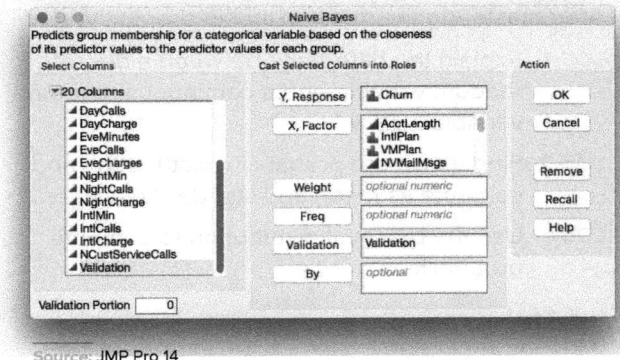

Source: JMP Pro 14

The misclassification rate and count and the confusion matrix are displayed by default, along with the ROC curve.

Design Elements: (CD): ©Comstock Images/Alamy; (All Others): ©McGraw-Hill Education

## CHAPTER 6

Discrete Random Variables

©Sorbis/Shutterstock

Learning Objectives

After mastering the material in this chapter, you will be able to:

LO6-1 Explain the difference between a discrete random variable and a continuous random variable.

LO6-2 Find a discrete probability distribution and compute its mean and standard deviation.

LO6-3 Use the binomial distribution to compute probabilities.

LO6-4 Use the Poisson distribution to compute probabilities (Optional).

LO6-5 Use the hypergeometric distribution to compute probabilities (Optional).

LO6-6 Compute and understand the covariance between two random variables (Optional).

Chapter Outline

e often use what we call **random variables** to describe the important aspects of the outcomes of experiments. In this chapter we introduce two important types of random variables—**discrete random variables** and **continuous random variables**—and learn how to find probabilities concerning discrete random variables. As one application, we will begin to see how to use probabilities concerning discrete random variables to make statistical inferences about populations.

6.1 Two Types of Random Variables

LO6-1
Explain the difference between a discrete random variable and a continuous random variable.

We begin with the definition of a random variable:

> A **random variable** is a variable whose value is numerical and is determined by the outcome of an experiment. A random variable assigns one and only one numerical value to each experimental outcome.

Before an experiment is carried out, its outcome is uncertain. It follows that, because a random variable assigns a number to each experimental outcome, a random variable can be thought of as *representing an uncertain numerical outcome.*

To illustrate the idea of a random variable, suppose that Sound City sells and installs car stereo systems. One of Sound City's most popular stereo systems is the TrueSound-XL, a top-of-the-line satellite car radio. Consider (the experiment of) selling the TrueSound-XL radio at the Sound City store during a particular week. If we let x denote the number of radios sold during the week, then x is a random variable. That is, looked at before the week, the number of radios x that will be sold is uncertain, and, therefore, x is a random variable.

Notice that x, the number of TrueSound-XL radios sold in a week, might be 0 or 1 or 2 or 3, and so forth. In general, when the possible values of a random variable can be counted or listed, we say that the random variable is a **discrete random variable.** That is, a discrete random variable may assume a finite number of possible values or the possible values may take the form of a *countable* sequence or list such as 0, 1, 2, 3, 4, . . . (a *countably infinite* list).

Some other examples of discrete random variables are

1   The number, x, of the next three customers entering a store who will make a purchase. Here x could be 0, 1, 2, or 3.

2   The number, x, of four patients taking a new antibiotic who experience gastrointestinal distress as a side effect. Here x could be 0, 1, 2, 3, or 4.

3   The number, x, of television sets in a sample of 8 five-year-old television sets that have not needed a single repair. Here x could be any of the values 0, 1, 2, 3, 4, 5, 6, 7, or 8.

4   The number, x, of major fires in a large city in the next two months. Here x could be 0, 1, 2, 3, and so forth (there is no definite maximum number of fires).

5   The number, x, of dirt specks in a one-square-yard sheet of plastic wrap. Here x could be 0, 1, 2, 3, and so forth (there is no definite maximum number of dirt specks).

The values of the random variables described in examples 1, 2, and 3 are countable and finite. In contrast, the values of the random variables described in 4 and 5 are countable and infinite (or countably infinite lists). For example, in theory there is no limit to the number of major fires that could occur in a city in two months.

Not all random variables have values that are countable. When a random variable may assume any numerical value in one or more intervals on the real number line, then we say that the random variable is a **continuous random variable.**

Ⓒ EXAMPLE 6.1 The Car Mileage Case: A Continuous Random Variable

Consider the car mileage situation that we have discussed in Chapters 1–3. The EPA combined city and highway mileage, x, of a randomly selected midsize car is a continuous random variable. This is because, although we have measured mileages to the nearest one-tenth of a mile per gallon, technically speaking, the potential mileages that might be obtained correspond (starting at, perhaps, 26 mpg) to an interval of numbers on the real line. We cannot count or list the numbers in such an interval because they are infinitesimally close together. That is, given any two numbers in an interval on the real line, there is always another number between them. To understand this, try listing the mileages starting with 26 mpg. Would the next mileage be 26.1 mpg? No, because we could obtain a mileage of 26.05 mpg. Would 26.05 mpg be the next mileage? No, because we could obtain a mileage of 26.025 mpg. We could continue this line of reasoning indefinitely. That is, whatever value we would try to list as the *next mileage*, there would always be another mileage between this *next mileage* and 26 mpg.

Some other examples of continuous random variables are

1   The temperature (in degrees Fahrenheit) of a cup of coffee served at a McDonald's restaurant.

2   The weight (in ounces) of strawberry preserves dispensed by an automatic filling machine into a 16-ounce jar.

3   The time (in seconds) that a customer in a store must wait to receive a credit card authorization.

4   The interest rate (in percent) charged for mortgage loans at a bank.

Exercises for Section 6.1

CONCEPTS ▇ **connect**

6.1   Explain the concept of a random variable.

6.2   Explain how the values of a discrete random variable differ from the values of a continuous random variable.

6.3   Classify each of the following random variables as discrete or continuous:

    a x = the number of girls born to a couple who will have three children.

b x = the number of defects found on an automobile at final inspection.

c x = the weight (in ounces) of the sandwich meat placed on a submarine sandwich.

d x = the number of incorrect lab procedures conducted at a hospital during a particular week.

e x = the number of customers served during a given day at a drive-through window.

f x = the time needed by a clerk to complete a task.

g x = the temperature of a pizza oven at a particular time.

LO6-2
Find a discrete probability distribution and compute its mean and standard deviation.

6.2 Discrete Probability Distributions

The value assumed by a random variable depends on the outcome of an experiment. Because the outcome of the experiment will be uncertain, the value assumed by the random variable will also be uncertain. However, we can model the random variable by finding, or estimating, its **probability distribution,** which describes how probabilities are distributed over values of the random variable. If a random variable is discrete, we find, or estimate, the probabilities that are associated with the different specific values that the random variable can take on.

> The **probability distribution** of a discrete random variable is called a **discrete probability distribution** and is a table, graph, or formula that gives the probability associated with each possible value that the random variable can assume.

We denote the probability distribution of the discrete random variable x as $p(x)$. As we will demonstrate in Section 6.3 (which discusses the *binomial distribution*), we can sometimes use the sample space of an experiment and probability rules to find the probability distribution of a discrete random variable. In other situations, we collect data that will allow us to estimate the probabilities in a discrete probability distribution.

EXAMPLE 6.2 The Sound City Case: Selling TrueSound-XL Radios

Recall that Sound City sells the TrueSound-XL car radio, and define the random variable x to be the number of such radios sold in a particular week at Sound City. Also, suppose that Sound City has kept historical records of TrueSound-XL sales during the last 100 weeks.

TABLE 6.1	An Estimate (Based on 100 Weeks of Historical Data) of the Probability Distribution of x, the Number of TrueSound-XL Radios Sold at Sound City in a Week
x, Number of Radios Sold	*p(x)*, the Probability of *x*
0	$p(0) = P(x = 0) = 3/100 = .03$
1	$p(1) = P(x = 1) = 20/100 = .20$
2	$p(2) = P(x = 2) = 50/100 = .50$
3	$p(3) = P(x = 3) = 20/100 = .20$
4	$p(4) = P(x = 4) = 5/100 = .05$
5	$p(5) = P(x = 5) = 2/100 = .02$

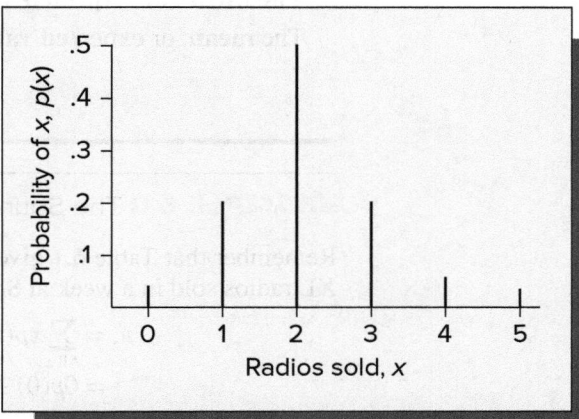

FIGURE 6.1 A Graph of the Probability Distribution of *x*, the Number of TrueSound-XL Radios Sold at Sound City in a Week

These records show 3 weeks with no radios sold, 20 weeks with one radio sold, 50 weeks with two radios sold, 20 weeks with three radios sold, 5 weeks with four radios sold, and 2 weeks with five radios sold. Because the records show 3 of 100 weeks with no radios sold, we estimate that $p(0) = P(x = 0)$ is $3/100 = .03$. That is, we estimate that the probability of no radios being sold during a week is .03. Similarly, because the records show 20 of 100 weeks with one radio sold, we estimate that $p(1) = P(x = 1)$ is $20/100 = .20$. That is, we estimate that the probability of exactly one radio being sold during a week is .20. Continuing in this way for the other values of the random variable *x*, we can estimate $p(2) = P(x = 2), p(3) = P(x = 3), p(4) = P(x = 4)$, and $p(5) = P(x = 5)$. Table 6.1 gives the entire estimated probability distribution of the number of TrueSound-XL radios sold at Sound City during a week, and Figure 6.1 shows a graph of this distribution. Moreover, such a probability distribution helps us to more easily calculate probabilities about events related to a random variable. For example, the probability of at least two radios being sold at Sound City during a week [that is, $P(x \geq 2)$] is $p(2) + p(3) + p(4) + p(5) = .50 + .20 + .05 + .02 = .77$. This says that we estimate that in 77 percent of all weeks, at least two TrueSound-XL radios will be sold at Sound City.

Finally, note that using historical sales data to obtain the estimated probabilities in Table 6.1 is reasonable if the TrueSound-XL radio sales process is stable over time. This means that the number of radios sold weekly does not exhibit any long-term upward or downward trends and is not seasonal (that is, radio sales are not higher at one time of the year than at others).

In general, a discrete probability distribution $p(x)$ must satisfy two conditions:

Properties of a Discrete Probability Distribution $p(x)$

A **discrete probability distribution** $p(x)$ must be such that

1 $p(x) \geq 0$ for each value of *x*

2 $\sum_{\text{All } x} p(x) = 1$

The first of these conditions says that each probability in a probability distribution must be zero or positive. The second condition says that the probabilities in a probability distribution must sum to 1. Looking at the probability distribution illustrated in Table 6.1, we can see that these properties are satisfied.

Suppose that the experiment described by a random variable *x* is repeated an indefinitely large number of times. If the values of the random variable *x* observed on the repetitions are recorded, we would obtain the population of all possible observed values of the random variable *x*. This population has a mean, which we denote as μ_x and which we sometimes call the

expected value of x. In order to calculate μ_x, we multiply each value of x by its probability $p(x)$ and then sum the resulting products over all possible values of x.

The Mean, or Expected Value, of a Discrete Random Variable

The **mean**, or **expected value**, of a discrete random variable x is

$$\mu_x = \sum_{\text{All } x} xp(x)$$

EXAMPLE 6.3 The Sound City Case: Selling TrueSound-XL Radios

Remember that Table 6.1 gives the probability distribution of x, the number of TrueSound-XL radios sold in a week at Sound City. Using this distribution, it follows that

$$\mu_x = \sum_{\text{All } x} xp(x)$$
$$= 0p(0) + 1p(1) + 2p(2) + 3p(3) + 4p(4) + 5p(5)$$
$$= 0(.03) + 1(.20) + 2(.50) + 3(.20) + 4(.05) + 5(.02)$$
$$= 2.1$$

To see that such a calculation gives the mean of all possible observed values of x, recall from Example 6.2 that the probability distribution in Table 6.1 was estimated from historical records of TrueSound-XL sales during the last 100 weeks. Also recall that these historical records tell us that during the last 100 weeks Sound City sold

1   Zero radios in 3 of the 100 weeks, for a total of $0(3) = 0$ radios
2   One radio in 20 of the 100 weeks, for a total of $1(20) = 20$ radios
3   Two radios in 50 of the 100 weeks, for a total of $2(50) = 100$ radios
4   Three radios in 20 of the 100 weeks, for a total of $3(20) = 60$ radios
5   Four radios in 5 of the 100 weeks, for a total of $4(5) = 20$ radios
6   Five radios in 2 of the 100 weeks, for a total of $5(2) = 10$ radios

In other words, Sound City sold a total of

$$0 + 20 + 100 + 60 + 20 + 10 = 210 \text{ radios}$$

in 100 weeks, or an average of $210/100 = 2.1$ radios per week. Now, the average

$$\frac{210}{100} = \frac{0 + 20 + 100 + 60 + 20 + 10}{100}$$

can be written as

$$\frac{0(3) + 1(20) + 2(50) + 3(20) + 4(5) + 5(2)}{100}$$

which can be rewritten as

$$0\left(\frac{3}{100}\right) + 1\left(\frac{20}{100}\right) + 2\left(\frac{50}{100}\right) + 3\left(\frac{20}{100}\right) + 4\left(\frac{5}{100}\right) + 5\left(\frac{2}{100}\right)$$
$$= 0(.03) + 1(.20) + 2(.50) + 3(.20) + 4(.05) + 5(.02)$$

which is $\mu_x = 2.1$. That is, if observed sales values occur with relative frequencies equal to those specified by the probability distribution in Table 6.1, then the average number of radios sold per week is equal to the expected value of x and is 2.1 radios.

Of course, if we observe radio sales for another 100 weeks, the relative frequencies of the observed sales values would not (unless we are very lucky) be exactly as specified by the estimated probabilities in Table 6.1. Rather, the observed relative frequencies would differ slightly from the estimated probabilities in Table 6.1, and the average number of radios sold per week would not exactly equal $\mu_x = 2.1$ (although the average would likely be close). However, the point is this: If the probability distribution in Table 6.1 were the true probability distribution of weekly radio sales, and if we were to observe radio sales for an indefinitely large number of weeks, then we would observe sales values with relative frequencies that are exactly equal to those specified by the probabilities in Table 6.1. In this case, when we calculate the expected value of x to be $\mu_x = 2.1$, we are saying that *in the long run* (that is, over an indefinitely large number of weeks) Sound City would average selling 2.1 TrueSound-XL radios per week.

EXAMPLE 6.4 The Life Insurance Case: Setting a Policy Premium

An insurance company sells a $20,000 whole life insurance policy for an annual premium of $300. Actuarial tables show that a person who would be sold such a policy with this premium has a .001 probability of death during a year. Let x be a random variable representing the insurance company's profit made on one of these policies during a year. The probability distribution of x is

©BJI/Blue Jean Images/Getty Images RF

x, Profit	p(x), Probability of x
$300 (if the policyholder lives)	.999
$300 − $20,000 = − $19,700	.001
(a $19,700 loss if the policyholder dies)	

The expected value of x (expected profit per year) is

$$\mu_x = \$300(.999) + (-\$19{,}700)(.001)$$
$$= \$280$$

This says that if the insurance company sells a very large number of these policies, it will average a profit of $280 per policy per year. Because insurance companies actually do sell large numbers of policies, it is reasonable for these companies to make profitability decisions based on expected values.

Next, suppose that we wish to find the premium that the insurance company must charge for a $20,000 policy if the company wishes the average profit per policy per year to be greater than $0. If we let *prem* denote the premium the company will charge, then the probability distribution of the company's yearly profit x is

x, Profit	p(x), Probability of x
prem (if policyholder lives)	.999
prem − $20,000 (if policyholder dies)	.001

The expected value of x (expected profit per year) is

$$\mu_x = prem(.999) + (prem - 20{,}000)(.001)$$
$$= prem - 20$$

In order for this expected profit to be greater than zero, the premium must be greater than $20. If, as previously stated, the company charges $300 for such a policy, the $280 charged in excess of the needed $20 compensates the company for commissions paid to salespeople, administrative costs, dividends paid to investors, and other expenses.

In general, it is reasonable to base decisions on an expected value if we perform the experiment related to the decision (for example, if we sell the life insurance policy) many times. If we do not (for instance, if we perform the experiment only once), then it may not be a good idea to base decisions on the expected value. For example, it might not be wise for you—as an individual—to sell one person a $20,000 life insurance policy for a premium of $300. To see this, again consider the probability distribution of yearly profit:

x, Profit	p(x), Probability of x
$300 (if policyholder lives)	.999
$300 − $20,000 = −$19,700 (if policyholder dies)	.001

and recall that the expected profit per year is \$280. However, because you are selling only one policy, you will not receive the \$280. You will either gain \$300 (with probability .999) or you will lose \$19,700 (with probability .001). Although the decision is personal, and although the chance of losing \$19,700 is very small, many people would not risk such a loss when the potential gain is only \$300.

Just as the population of all possible observed values of a discrete random variable x has a mean μ_x, this population also has a variance σ_x^2 and a standard deviation σ_x. Recall that the variance of a population is the average of the squared deviations of the different population values from the population mean. To find σ_x^2, we calculate $(x - \mu_x)^2$ for each value of x, multiply $(x - \mu_x)^2$ by the probability $p(x)$, and sum the resulting products over all possible values of x.

The Variance and Standard Deviation of a Discrete Random Variable

The **variance** of a discrete random variable x is

$$\sigma_x^2 = \sum_{\text{All } x} (x - \mu_x)^2 p(x)$$

The **standard deviation** of x is the positive square root of the variance of x. That is,

$$\sigma_x = \sqrt{\sigma_x^2}$$

EXAMPLE 6.5 The Sound City Case: Selling TrueSound-XL Radios

Table 6.1 gives the probability distribution of x, the number of TrueSound-XL radios sold in a week at Sound City. Remembering that we have calculated μ_x (in Example 6.3) to be 2.1 radios, it follows that

$$
\begin{aligned}
\sigma_x^2 &= \sum_{\text{All } x} (x - \mu_x)^2 p(x) \\
&= (0 - 2.1)^2 p(0) + (1 - 2.1)^2 p(1) + (2 - 2.1)^2 p(2) + (3 - 2.1)^2 p(3) \\
&\quad + (4 - 2.1)^2 p(4) + (5 - 2.1)^2 p(5) \\
&= (4.41)(.03) + (1.21)(.20) + (.01)(.50) + (.81)(.20) + (3.61)(.05) + (8.41)(.02) \\
&= .89
\end{aligned}
$$

and that the standard deviation of x is $\sigma_x = \sqrt{.89} = .9434$ radios. To make one interpretation of a standard deviation of .9434 radios, suppose that Sound City sells another top-of-the-line satellite car radio called the ClearTone-400. If the ClearTone-400 also has mean weekly sales of 2.1 radios, and if the standard deviation of the ClearTone-400's weekly sales is 1.2254 radios, we would conclude that there is more variability in the weekly sales of the ClearTone-400 than in the weekly sales of the TrueSound-XL.

In Chapter 3 we considered the percentage of measurements in a population that are within (plus or minus) one, two, or three standard deviations of the mean of the population. Similarly, we can consider the probability that a random variable x will be within (plus or minus) one, two, or three standard deviations of the mean of the random variable. For example, consider the probability distribution in Table 6.1 of x, the number of TrueSound-XL radios sold in a week at Sound City. Also, recall that $\mu_x = 2.1$ and $\sigma_x = .9434$. If (for instance) we wish to find the probability that x will be within (plus or minus) two standard deviations of μ_x, then we need to find the probability that x will lie in the interval

$$
\begin{aligned}
[\mu_x \pm 2\sigma_x] &= [2.1 \pm 2(.9434)] \\
&= [.2132, 3.9868]
\end{aligned}
$$

As illustrated in Figure 6.2, there are three values of x ($x = 1$, $x = 2$, and $x = 3$) that lie in the interval [.2132, 3.9868]. Therefore, the probability that x will lie in the interval [.2132, 3.9868] is the probability that x will equal 1 or 2 or 3, which is $p(1) + p(2) + p(3) = .20 + .50 + .20 = .90$. This says that in 90 percent of all weeks, the number of TrueSound-XL radios sold at Sound City will be within (plus or minus) two standard deviations of the mean weekly sales of the TrueSound-XL radio at Sound City.

FIGURE 6.2   The Interval $[\mu_x \pm 2\sigma_x]$ for the Probability Distribution Describing TrueSound-XL Radio Sales (see Table 6.1)

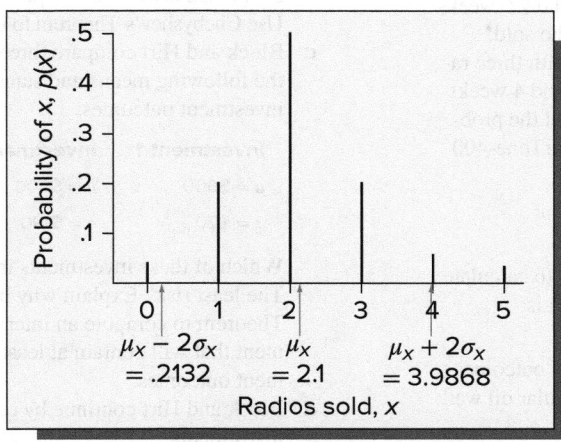

In general, consider any random variable with mean μ_x and standard deviation σ_x. Then, Chebyshev's Theorem (see Chapter 3) tells us that, for any value of k that is greater than 1, the probability is at least $1 - 1/k^2$ that x will be within (plus or minus) k standard deviations of μ_x and thus will lie in the interval $[\mu_x \pm k\sigma_x]$. For example, setting k equal to 2, the probability is at least $1 - 1/2^2 = 1 - 1/4 = 3/4$ that x will lie in the interval $[\mu_x \pm 2\sigma_x]$. Setting k equal to 3, the probability is at least $1 - 1/3^2 = 1 - 1/9 = 8/9$ that x will lie in the interval $[\mu_x \pm 3\sigma_x]$. If (as in the Sound City situation) we have the probability distribution of x, we can calculate exact probabilities, and thus we do not need the approximate probabilities given by Chebyshev's Theorem. However, in some situations we know the values of μ_x and σ_x, but we do not have the probability distribution of x. In such situations the approximate Chebyshev's probabilities can be quite useful. For example, let x be a random variable representing the return on a particular investment, and suppose that an investment prospectus tells us that, based on historical data and current market trends, the investment return has a mean (or expected value) of $\mu_x = \$1,000$ and a standard deviation of $\sigma_x = \$100$. It then follows from Chebyshev's Theorem that the probability is at least $8/9$ that the investment return will lie in the interval $[\mu_x \pm 3\sigma_x] = [1000 \pm 3(100)] = [700, 1300]$. That is, the probability is fairly high that the investment will have a minimum return of $700 and a maximum return of $1300.

In the next several sections, we will see that a probability distribution $p(x)$ is sometimes specified by using a formula. As a simple example of this, suppose that a random variable x is equally likely to assume any one of n possible values. In this case we say that x is described by the discrete **uniform distribution** and we specify $p(x)$ by using the formula $p(x) = 1/n$. For example, if we roll a fair die and x denotes the number of spots that show on the upward face of the die, then x is uniformly distributed and $p(x) = 1/6$ for $x = 1, 2, 3, 4, 5,$ and 6. As another example, if historical sales records show that a Chevrolet dealership is equally likely to sell 0, 1, 2, or 3 Chevy Malibus in a given week, and if x denotes the number of Chevy Malibus that the dealership sells in a week, then x is uniformly distributed and $p(x) = 1/4$ for $x = 0, 1, 2,$ and 3.

Exercises for Section 6.2

CONCEPTS   connect

6.4 What is a discrete probability distribution? Explain in your own words.

6.5 What conditions must be satisfied by the probabilities in a discrete probability distribution? Explain what these conditions mean.

6.6 Describe how to compute the mean (or expected value) of a discrete random variable, and interpret what this quantity tells us about the observed values of the random variable.

6.7 Describe how to compute the standard deviation of a discrete random variable, and interpret what this quantity tells us about the observed values of the random variable.

METHODS AND APPLICATIONS

6.8 Recall from Example 6.5 that Sound City also sells the ClearTone-400 satellite car radio. For this radio, historical sales records over the last 100 weeks show 6 weeks with no radios sold, 30 weeks with one radio sold, 30 weeks with two radios sold, 20 weeks with three radios sold, 10 weeks with four radios sold, and 4 weeks with five radios sold. Estimate and write out the probability distribution of x, the number of ClearTone-400 radios sold at Sound City during a week.

6.9 Use the estimated probability distribution in Exercise 6.8 to calculate μ_x, σ_x^2, and σ_x.

6.10 Use your answers to Exercises 6.8 and 6.9 to calculate the probabilities that x will lie in the intervals $[\mu_x \pm \sigma_x]$, $[\mu_x \pm 2\sigma_x]$, and $[\mu_x \pm 3\sigma_x]$.

6.11 The following table summarizes investment outcomes and corresponding probabilities for a particular oil well:

x = the outcome in $	$p(x)$
−$40,000 (no oil)	.25
10,000 (some oil)	.7
70,000 (much oil)	.05

a Graph $p(x)$; that is, graph the probability distribution of x.

b Find the expected monetary outcome. Mark this value on your graph of part a. Then interpret this value.

c Calculate the standard deviation of x.

6.12 In the book *Foundations of Financial Management* (7th ed.), Stanley B. Block and Geoffrey A. Hirt discuss risk measurement for investments. Block and Hirt present an investment with the possible outcomes and associated probabilities given in Table 6.2. The authors go on to say that the probabilities

> may be based on past experience, industry ratios and trends, interviews with company executives, and sophisticated simulation techniques. The probability values may be easy to determine for the introduction of a mechanical stamping process in which the manufacturer has 10 years of past data, but difficult to assess for a new product in a foreign market. **DS** OutcomeDist

a Use the probability distribution in Table 6.2 to calculate the expected value (mean) and the standard deviation of the investment outcomes. Interpret the expected value.

b Block and Hirt interpret the standard deviation of the investment outcomes as follows: "Generally, the larger the standard deviation (or spread of outcomes), the greater is the risk." Explain why this makes sense. Use Chebyshev's Theorem to illustrate your point.

c Block and Hirt compare three investments having the following means and standard deviations of the investment outcomes:

Investment 1	Investment 2	Investment 3
$\mu = \$600$	$\mu = \$600$	$\mu = \$600$
$\sigma = \$20$	$\sigma = \$190$	$\sigma = \$300$

Which of these investments involves the most risk? The least risk? Explain why by using Chebyshev's Theorem to compute an interval for each investment that will contain at least 8/9 of the investment outcomes.

d Block and Hirt continue by comparing two more investments:

Investment A	Investment B
$\mu = \$6,000$	$\mu = \$600$
$\sigma = \$600$	$\sigma = \$190$

The authors explain that Investment A

> appears to have a high standard deviation, but not when related to the expected value of the distribution. A standard deviation of $600 on an investment with an expected value of $6,000 may indicate less risk than a standard deviation of $190 on an investment with an expected value of only $600.
>
> We can eliminate the size difficulty by developing a third measure, the **coefficient of variation** (V). This term calls for nothing more difficult than dividing the standard deviation of an investment by the expected value. Generally, the larger the coefficient of variation, the greater is the risk.

$$\text{Coefficient of variation } (V) = \frac{\sigma}{\mu}$$

Calculate the coefficient of variation for investments A and B. Which investment carries the greater risk?

e Calculate the coefficient of variation for investments 1, 2, and 3 in part c. Based on the coefficient of variation, which investment involves the most risk? The least risk? Do we obtain the same results as we did by comparing standard deviations (in part c)? Why?

6.13 An insurance company will insure a $50,000 diamond for its full value against theft at a premium of $400 per year. Suppose that the probability that the diamond will be stolen is .005, and let x denote the insurance company's profit.

a Set up the probability distribution of the random variable x.

b Calculate the insurance company's expected profit.

c Find the premium that the insurance company should charge if it wants its expected profit to be $1,000.

TABLE 6.2 **Probability Distribution of Outcomes for an Investment** **DS** OutcomeDist

Outcome	Probability of Outcome	Assumptions
$300	.2	Pessimistic
600	.6	Moderately successful
900	.2	Optimistic

Source: S. B. Block and G. A. Hirt, *Foundations of Financial Management*, 7th ed., p. 378. McGraw-Hill Companies, Inc., 1994.

FIGURE 6.3   A Tree Diagram of Two Project Choices

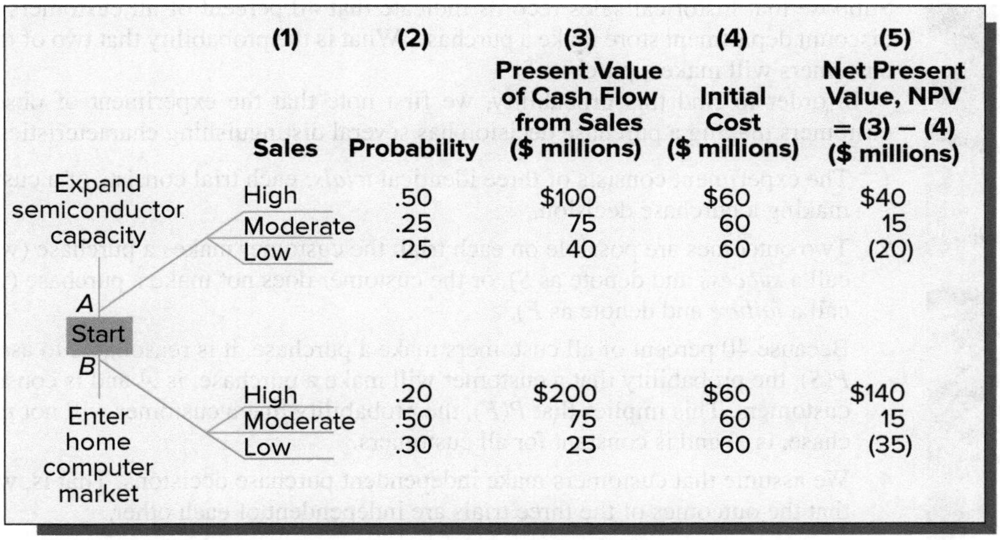

	(1) Sales	(2) Probability	(3) Present Value of Cash Flow from Sales ($ millions)	(4) Initial Cost ($ millions)	(5) Net Present Value, NPV = (3) − (4) ($ millions)
Expand semiconductor capacity	High	.50	$100	$60	$40
	Moderate	.25	75	60	15
	Low	.25	40	60	(20)
A Start B					
Enter home computer market	High	.20	$200	$60	$140
	Moderate	.50	75	60	15
	Low	.30	25	60	(35)

Source: S. B. Block and G. A. Hirt, *Foundations of Financial Management*, 7th ed., p. 387. McGraw-Hill Companies, Inc., 1994.

6.14 In the book *Foundations of Financial Management* (7th ed.), Stanley B. Block and Geoffrey A. Hirt discuss a semiconductor firm that is considering two choices: (1) expanding the production of semiconductors for sale to end users or (2) entering the highly competitive home computer market. The cost of both projects is $60 million, but the net present value of the cash flows from sales and the risks are different.

Figure 6.3 gives a tree diagram of the project choices. The tree diagram gives a probability distribution of expected sales for each project. It also gives the present value of cash flows from sales and the net present value (NPV = present value of cash flow from sales minus initial cost) corresponding to each sales alternative. Note that figures in parentheses denote losses.

a For each project choice, calculate the expected net present value.
b For each project choice, calculate the variance and standard deviation of the net present value.
c Calculate the coefficient of variation for each project choice. See Exercise 6.12d for a discussion of the coefficient of variation.
d Which project has the higher expected net present value?
e Which project carries the least risk? Explain.
f In your opinion, which project should be undertaken? Justify your answer.

6.15 Five thousand raffle tickets are to be sold at $10 each to benefit a local community group. The prizes, the

number of each prize to be given away, and the dollar value of winnings for each prize are as follows:
DS Raffle

Prize	Number to Be Given Away	Dollar Value
Automobile	1	$20,000
Entertainment center	2	3,000 each
DVD recorder	5	400 each
Gift certificate	50	20 each

DS Raffle

If you buy one ticket, calculate your expected winnings. (Form the probability distribution of x = your dollar winnings, and remember to subtract the cost of your ticket.)

6.16 A survey conducted by a song rating service finds that the percentages of listeners *familiar* with "Poker Face" by Lady Gaga who would give the song ratings of 5, 4, 3, 2, and 1 are, respectively, 43 percent, 21 percent, 22 percent, 7 percent, and 7 percent. Assign the numerical values 1, 2, 3, 4, and 5 to the (qualitative) ratings 1, 2, 3, 4, and 5 and find an estimate of the probability distribution of x = this song's rating by a randomly selected listener who is familiar with the song.

6.17 In Exercise 6.16,
a Find the expected value of the estimated probability distribution.
b Interpret the meaning of this expected value in terms of all possible "Poker Face" listeners.

6.3 The Binomial Distribution

LO6-3
Use the binomial distribution to compute probabilities.

In this section we discuss what is perhaps the most important discrete probability distribution—the **binomial distribution** (or **binomial model**). We begin with an example.

EXAMPLE 6.6 Purchases at a Discount Store

©UpperCut Images/Getty Images

Suppose that historical sales records indicate that 40 percent of all customers who enter a discount department store make a purchase. What is the probability that two of the next three customers will make a purchase?

In order to find this probability, we first note that the experiment of observing three customers making a purchase decision has several distinguishing characteristics:

1   The experiment consists of three identical *trials;* each trial consists of a customer making a purchase decision.

2   Two outcomes are possible on each trial: the customer makes a purchase (which we call a *success* and denote as *S*), or the customer does not make a purchase (which we call a *failure* and denote as *F*).

3   Because 40 percent of all customers make a purchase, it is reasonable to assume that $P(S)$, the probability that a customer will make a purchase, is .4 and is constant for all customers. This implies that $P(F)$, the probability that a customer will not make a purchase, is .6 and is constant for all customers.

4   We assume that customers make independent purchase decisions. That is, we assume that the outcomes of the three trials are independent of each other.

Using the tree diagram in Figure 6.4, we can find the sample space of the experiment of three customers making a purchase decision. As shown in the tree diagram, the sample space of the experiment consists of the following eight sample space outcomes:

SSS	FSS
SSF	FSF
SFS	FFS
SFF	FFF

FIGURE 6.4   A Tree Diagram of Three Customers Making a Purchase Decision

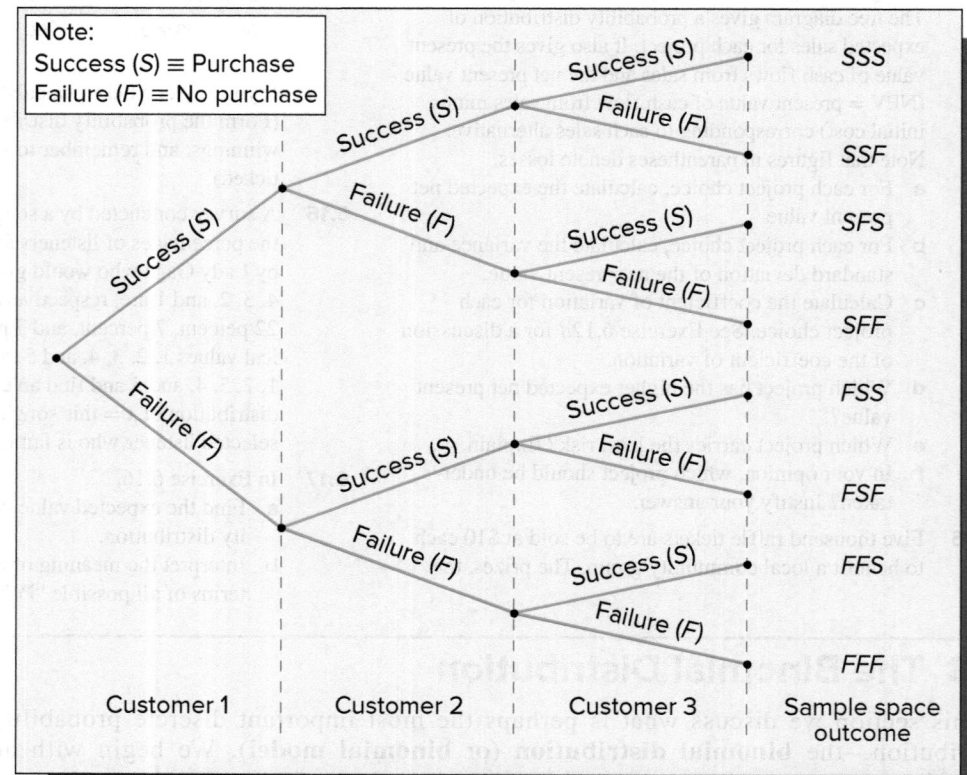

Here the sample space outcome *SSS* represents the first customer making a purchase, the second customer making a purchase, and the third customer making a purchase. On the other hand, the sample space outcome *SFS* represents the first customer making a purchase, the second customer not making a purchase, and the third customer making a purchase. In addition, each sample space outcome corresponds to a specific number of customers (out of three customers) making a purchase. For example, the sample space outcome *SSS* corresponds to all three customers making a purchase. Each of the sample space outcomes *SSF*, *SFS*, and *FSS* corresponds to two out of three customers making a purchase. Each of the sample space outcomes *SFF*, *FSF,* and *FFS* corresponds to one out of three customers making a purchase. Finally, the sample space outcome *FFF* corresponds to none of the three customers making a purchase.

To find the probability that two out of the next three customers will make a purchase (the probability asked for at the beginning of this example), we consider the sample space outcomes *SSF*, *SFS*, and *FSS*. Because the trials (individual customer purchase decisions) are independent, we can multiply the probabilities associated with the different trial outcomes (each of which is *S* or *F*) to find the probability of a sequence of customer outcomes:

$$P(SSF) = P(S)P(S)P(F) = (.4)(.4)(.6) = (.4)^2(.6)$$
$$P(SFS) = P(S)P(F)P(S) = (.4)(.6)(.4) = (.4)^2(.6)$$
$$P(FSS) = P(F)P(S)P(S) = (.6)(.4)(.4) = (.4)^2(.6)$$

It follows that the probability that two out of the next three customers will make a purchase is

$$P(SSF) + P(SFS) + P(FSS)$$
$$= (.4)^2(.6) + (.4)^2(.6) + (.4)^2(.6)$$
$$= 3(.4)^2(.6) = .288$$

We can now generalize the previous result and find the probability that *x* of the next *n* customers will make a purchase. Here we will assume that *p* is the probability that a customer will make a purchase, $q = 1 - p$ is the probability that a customer will not make a purchase, and that purchase decisions (trials) are independent. To generalize the probability that two out of the next three customers will make a purchase, which equals $3(.4)^2(.6)$, we note that

1　The 3 in this expression is the number of sample space outcomes (*SSF*, *SFS*, and *FSS*) that correspond to the event "two out of the next three customers will make a purchase." Note that this number equals the number of ways we can arrange two successes among the three trials.

2　The .4 is *p*, the probability that a customer will make a purchase.

3　The .6 is $q = 1 - p$, the probability that a customer will not make a purchase.

Therefore, the probability that two of the next three customers will make a purchase is

$$\left(\begin{array}{c} \text{The number of ways} \\ \text{to arrange 2 successes} \\ \text{among 3 trials} \end{array} \right) p^2 q^1$$

Now, notice that, although each of the sample space outcomes *SSF*, *SFS*, and *FSS* represents a different arrangement of the two successes among the three trials, each of these sample space outcomes consists of two successes and one failure. For this reason, the probability of each of these sample space outcomes equals $(.4)^2(.6)^1 = p^2 q^1$. It follows that *p* is raised to a power that equals the number of successes (2) in the three trials, and *q* is raised to a power that equals the number of failures (1) in the three trials.

In general, each sample space outcome describing the occurrence of *x* successes (purchases) in *n* trials represents a different arrangement of *x* successes in *n* trials. However, each outcome consists of *x* successes and $n - x$ failures. Therefore, the probability of each sample space outcome is $p^x q^{n-x}$. It follows by analogy that the probability that x of the next n trials will be successes (purchases) is

$$\left(\begin{array}{c} \text{The number of ways} \\ \text{to arrange } x \text{ successes} \\ \text{among } n \text{ trials} \end{array} \right) p^x q^{n-x}$$

We can use the expression we have just arrived at to compute the probability of x successes in the next n trials if we can find a way to calculate the number of ways to arrange x successes among n trials. It can be shown that:

The number of ways to arrange x successes among n trials equals

$$\frac{n!}{x!(n-x)!}$$

where $n!$ is pronounced "n factorial" and is calculated as $n! = n(n-1)(n-2) \cdots (1)$ and where (by definition) $0! = 1$.

For instance, using this formula, we can see that the number of ways to arrange $x = 2$ successes among $n = 3$ trials equals

$$\frac{n!}{x!\,(n-x)!} = \frac{3!}{2!\,(3-2)!} = \frac{3!}{2!\,1!} = \frac{3 \cdot 2 \cdot 1}{2 \cdot 1 \cdot 1} = 3$$

Of course, we have previously seen that the three ways to arrange $x = 2$ successes among $n = 3$ trials are *SSF*, *SFS*, and *FSS*.

Using the preceding formula, we obtain the following general result:

The Binomial Distribution (The Binomial Model)

A **binomial experiment** has the following characteristics:

1 The experiment consists of n *identical trials.*
2 Each trial results in a **success** or a **failure**.
3 The probability of a success on any trial is p and remains constant from trial to trial. This implies that the probability of failure, q, on any trial is $1 - p$ and remains constant from trial to trial.
4 The trials are **independent** (that is, the results of the trials have nothing to do with each other).

Furthermore, if we define the random variable

   $x = $ the total number of successes in n trials of a binomial experiment

then we call x a **binomial random variable**, and the probability of obtaining x successes in n trials is

$$p(x) = \frac{n!}{x!\,(n-x)!} p^x q^{n-x}$$

Noting that we sometimes refer to the formula for $p(x)$ as the **binomial formula,** we illustrate the use of this formula in the following example.

EXAMPLE 6.7 Purchases at a Discount Store

Consider the discount department store situation discussed in Example 6.6. In order to find the probability that three of the next five customers will make purchases, we calculate

$$p(3) = \frac{5!}{3!\,(5-3)!}\,(.4)^3(.6)^{5-3} = \frac{5!}{3!\,2!}\,(.4)^3(.6)^2$$

$$= \frac{5 \cdot 4 \cdot 3 \cdot 2 \cdot 1}{(3 \cdot 2 \cdot 1)(2 \cdot 1)}\,(.4)^3(.6)^2$$

$$= 10(.064)(.36)$$

$$= .2304$$

Here we see that

1 $\frac{5!}{3!(5-3)!} = 10$ is the number of ways to arrange three successes among five trials. For instance, two ways to do this are described by the sample space outcomes *SSSFF* and *SFSSF*. There are eight other ways.

2 $(.4)^3(.6)^2$ is the probability of any sample space outcome consisting of three successes and two failures.

Thus far we have shown how to calculate binomial probabilities. We next give several examples that illustrate some practical applications of the binomial distribution. As we demonstrate in the first example, the term *success* does not necessarily refer to a *desirable* experimental outcome. Rather, it refers to an outcome that we wish to investigate.

EXAMPLE 6.8 The Phe-Mycin Case: Drug Side Effects

Antibiotics occasionally cause nausea as a side effect. A major drug company has developed a new antibiotic called Phe-Mycin. The company claims that, at most, 10 percent of all patients treated with Phe-Mycin would experience nausea as a side effect of taking the drug. Suppose that we randomly select $n = 4$ patients and treat them with Phe-Mycin. Each patient will either experience nausea (which we arbitrarily call a success) or will not experience nausea (a failure). We will assume that p, the true probability that a patient will experience nausea as a side effect, is .10, the maximum value of p claimed by the drug company. Furthermore, it is reasonable to assume that patients' reactions to the drug would be independent of each other. Let x denote the number of patients among the four who will experience nausea as a side effect. It follows that x is a binomial random variable, which can take on any of the potential values 0, 1, 2, 3, or 4. That is, anywhere between none of the patients and all four of the patients could potentially experience nausea as a side effect. Furthermore, we can calculate the probability associated with each possible value of x as shown in Table 6.3. For instance, the probability that none of the four randomly selected patients will experience nausea is

$$p(0) = P(x = 0) = \frac{4!}{0!\,(4-0)!}\,(.1)^0(.9)^{4-0}$$

$$= \frac{4!}{0!\,4!}\,(.1)^0(.9)^4$$

$$= \frac{4!}{(1)(4!)}\,(1)(.9)^4$$

$$= (.9)^4 = .6561$$

Because Table 6.3 lists each possible value of x and also gives the probability of each value, we say that this table gives the **binomial probability distribution of x.**

The binomial probabilities given in Table 6.3 need not be hand calculated. Excel, JMP, and Minitab can be used to calculate binomial probabilities. For instance, Figure 6.5(a) gives the Excel output of the binomial probability distribution listed in Table 6.3.[1] Figure 6.5(b) shows a graph of this distribution.

---

[1]As we will see in this chapter's appendixes, we can use Minitab and JMP to obtain output of the binomial distribution that is essentially identical to the output given by Excel.

TABLE 6.3	The Binomial Probability Distribution of x, the Number of Four Randomly Selected Patients Who Will Experience Nausea as a Side Effect of Being Treated with Phe-Mycin
x (Number Who Experience Nausea)	$p(x) = \dfrac{n!}{x!\,(n-x)!}\, p^x(1-p)^{n-x}$
0	$p(0) = P(x = 0) = \dfrac{4!}{0!\,(4-0)!}\,(.1)^0(.9)^{4-0} = .6561$
1	$p(1) = P(x = 1) = \dfrac{4!}{1!\,(4-1)!}\,(.1)^1(.9)^{4-1} = .2916$
2	$p(2) = P(x = 2) = \dfrac{4!}{2!\,(4-2)!}\,(.1)^2(.9)^{4-2} = .0486$
3	$p(3) = P(x = 3) = \dfrac{4!}{3!\,(4-3)!}\,(.1)^3(.9)^{4-3} = .0036$
4	$p(4) = P(x = 4) = \dfrac{4!}{4!\,(4-4)!}\,(.1)^4(.9)^{4-4} = .0001$

FIGURE 6.5   The Binomial Probability Distribution with $p = .10$ and $n = 4$

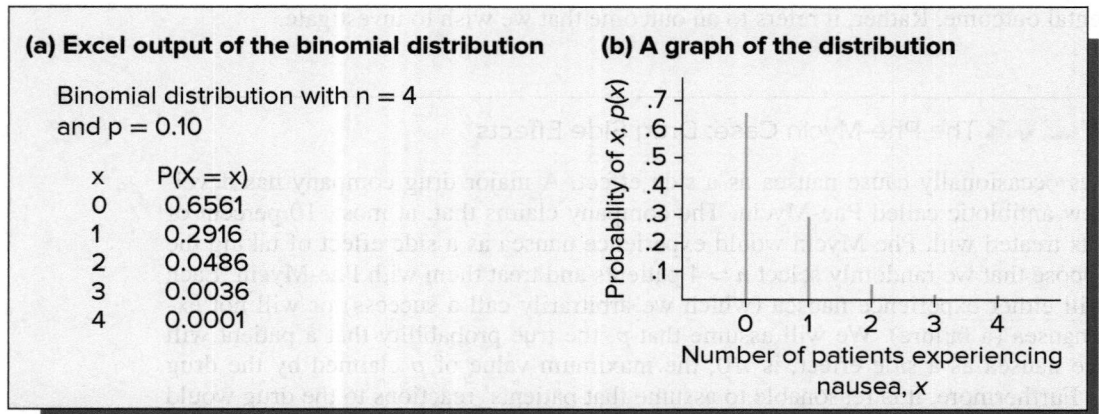

(a) Excel output of the binomial distribution

Binomial distribution with n = 4
and p = 0.10

x	P(X = x)
0	0.6561
1	0.2916
2	0.0486
3	0.0036
4	0.0001

(b) A graph of the distribution

In order to interpret these binomial probabilities, consider administering the antibiotic Phe-Mycin to all possible samples of four randomly selected patients. Then, for example,

$$P(x = 0) = 0.6561$$

says that none of the four sampled patients would experience nausea in 65.61 percent of all possible samples. Furthermore, as another example,

$$P(x = 3) = 0.0036$$

says that three out of the four sampled patients would experience nausea in only .36 percent of all possible samples.

Another way to avoid hand calculating binomial probabilities is to use **binomial tables**, which have been constructed to give the probability of x successes in n trials. A table of binomial probabilities is given in Table A.1. A portion of this table is reproduced in Table 6.4(a) and (b). Part (a) of this table gives binomial probabilities corresponding to $n = 4$ trials. Values of p, the probability of success, are listed across the top of the table (ranging from $p = .05$ to $p = .50$ in steps of .05), and more values of p (ranging from $p = .50$ to $p = .95$ in steps of .05) are listed across the bottom of the table. When the value of p being considered is one of those across the top of the table, values of x (the number of successes in four trials) are listed down the left side of the table. For instance, to find the probabilities

TABLE 6.4   A Portion of a Binomial Probability Table

(a) A Table for $n = 4$ Trials

Values of p (.05 to .50)

	↓ .05	.10	.15	.20	.25	.30	.35	.40	.45	.50	
	0 .8145	.6561	.5220	.4096	.3164	.2401	.1785	.1296	.0915	.0625	**4**
	1 .1715	.2916	.3685	.4096	.4219	.4116	.3845	.3456	.2995	.2500	**3**
Number of	**2** .0135	.0486	.0975	.1536	.2109	.2646	.3105	.3456	.3675	.3750	**2 Number of**
Successes	**3** .0005	.0036	.0115	.0256	.0469	.0756	.1115	.1536	.2005	.2500	**1 Successes**
	4 .0000	.0001	.0005	.0016	.0039	.0081	.0150	.0256	.0410	.0625	**0**
	.95	.90	.85	.80	.75	.70	.65	.60	.55	.50	↑

Values of p (.50 to .95)

(b) A Table for $n = 8$ trials

Values of p (.05 to .50)

	↓ .05	.10	.15	.20	.25	.30	.35	.40	.45	.50	
	0 .6634	.4305	.2725	.1678	.1001	.0576	.0319	.0168	.0084	.0039	**8**
	1 .2793	.3826	.3847	.3355	.2670	.1977	.1373	.0896	.0548	.0313	**7**
	2 .0515	.1488	.2376	.2936	.3115	.2965	.2587	.2090	.1569	.1094	**6**
Number of	**3** .0054	.0331	.0839	.1468	.2076	.2541	.2786	.2787	.2568	.2188	**5 Number of**
Successes	**4** .0004	.0046	.0185	.0459	.0865	.1361	.1875	.2322	.2627	.2734	**4 Successes**
	5 .0000	.0004	.0026	.0092	.0231	.0467	.0808	.1239	.1719	.2188	**3**
	6 .0000	.0000	.0002	.0011	.0038	.0100	.0217	.0413	.0703	.1094	**2**
	7 .0000	.0000	.0000	.0001	.0004	.0012	.0033	.0079	.0164	.0313	**1**
	8 .0000	.0000	.0000	.0000	.0000	.0001	.0002	.0007	.0017	.0039	**0**
	.95	.90	.85	.80	.75	.70	.65	.60	.55	.50	↑

Values of p (.50 to .95)

that we have computed in Table 6.3, we look in part (a) of Table 6.4 ($n = 4$) and read down the column labeled .10. Remembering that the values of x are on the left side of the table because $p = .10$ is on top of the table, we find the probabilities in Table 6.3 (they are shaded). For example, the probability that none of four patients will experience nausea is $p(0) = .6561$, the probability that one of the four patients will experience nausea is $p(1) = .2916$ and so forth. If the value of p is across the bottom of the table, then we read the values of x from the right side of the table. As an example, if p equals .70, then the probability of three successes in four trials is $p(3) = .4116$ (we have shaded this probability).

EXAMPLE 6.9 The Phe-Mycin Case: Drug Side Effects

Suppose that we wish to investigate whether p, the probability that a patient will experience nausea as a side effect of taking Phe-Mycin, is greater than .10, the maximum value of p claimed by the drug company. This assessment will be made by assuming, for the sake of argument, that p equals .10, and by using sample information to weigh the evidence against this assumption and in favor of the conclusion that p is greater than .10. Suppose that when a sample of $n = 4$ randomly selected patients is treated with Phe-Mycin, three of the four patients experience nausea. Because the fraction of patients in the sample that experience nausea is $3/4 = .75$, which is far greater than .10, we have some evidence contradicting the assumption that p equals .10. To evaluate the strength of this evidence, we calculate the probability that at least 3 out of 4 randomly selected patients would experience nausea as a side effect if, in fact, p equals .10. Using the binomial probabilities in Table 6.4(a), and realizing that the events $x = 3$ and $x = 4$ are mutually exclusive, we have

$$P(x \geq 3) = P(x = 3 \text{ or } x = 4)$$
$$= P(x = 3) + P(x = 4)$$
$$= .0036 + .0001$$
$$= .0037$$

This probability says that, if p equals .10, then in only .37 percent of all possible samples of four randomly selected patients would at least three of the four patients experience nausea as a side effect. This implies that, if we are to believe that p equals .10, then we must believe that we have observed a sample result that is so rare that it can be described as a 37 in 10,000 chance. Because observing such a result is very unlikely, we have very strong evidence that p does not equal .10 and is, in fact, greater than .10.

Next, suppose that we consider what our conclusion would have been if only one of the four randomly selected patients had experienced nausea. Because the sample fraction of patients who experienced nausea is $1/4 = .25$, which is greater than .10, we would have some evidence to contradict the assumption that p equals .10. To evaluate the strength of this evidence, we calculate the probability that at least one out of four randomly selected patients would experience nausea as a side effect of being treated with Phe-Mycin if, in fact, p equals .10. Using the binomial probabilities in Table 6.4(a), we have

$$P(x \geq 1) = P(x = 1 \text{ or } x = 2 \text{ or } x = 3 \text{ or } x = 4)$$
$$= P(x = 1) + P(x = 2) + P(x = 3) + P(x = 4)$$
$$= .2916 + .0486 + .0036 + .0001$$
$$= .3439$$

This probability says that, if p equals .10, then in 34.39 percent of all possible samples of four randomly selected patients, at least one of the four patients would experience nausea. Because it is not particularly difficult to believe that a 34.39 percent chance has occurred, we would not have much evidence against the claim that p equals .10.

Example 6.9 illustrates what is sometimes called the **rare event approach to making a statistical inference.** The idea of this approach is that if the probability of an observed sample result under a given assumption is *small,* then we have *strong evidence* that the assumption is false. Although there are no strict rules, many statisticians judge the probability of an observed sample result to be small if it is less than .05. The logic behind this will be explained more fully in Chapter 10.

EXAMPLE 6.10 The ColorSmart-5000 Case: TV Repairs

The manufacturer of the ColorSmart-5000 television set claims that 95 percent of its sets last at least five years without requiring a single repair. Suppose that we will contact $n = 8$ randomly selected ColorSmart-5000 purchasers five years after they purchased their sets. Each purchaser's set will have needed no repairs (a success) or will have been repaired at least once (a failure). We will assume that p, the true probability that a purchaser's television set will require no repairs within five years, is .95, as claimed by the manufacturer. Furthermore, it is reasonable to believe that the repair records of the purchasers' sets are independent of each other. Let x denote the number of the $n = 8$ randomly selected sets that will have lasted at least five years without a single repair. Then x is a binomial random variable that can take on any of the potential values 0, 1, 2, 3, 4, 5, 6, 7, or 8. The binomial distribution of x is shown in the page margin. Here we have obtained these probabilities from Table 6.4(b). To use Table 6.4(b), we look at the column corresponding to $p = .95$. Because $p = .95$ is listed at the bottom of the table, we read the values of x and their corresponding probabilities from bottom to top (we have shaded the probabilities). Notice that the values of x are listed on the right side of the table. Alternatively, we could use the binomial probability formula to find the binomial probabilities. For example, assuming that p equals .95, the probability that 5 out of 8 randomly selected television sets would last at least five years without a single repair is

$$p(5) = \frac{8!}{5!(8-5)!}(.95)^5(.05)^{8-5} = .0054$$

x	$p(x)$
0	$p(0) = .0000$
1	$p(1) = .0000$
2	$p(2) = .0000$
3	$p(3) = .0000$
4	$p(4) = .0004$
5	$p(5) = .0054$
6	$p(6) = .0515$
7	$p(7) = .2793$
8	$p(8) = .6634$

The page margin shows first the Minitab output of the binomial distribution with $p = .95$ and $n = 8$ and then the JMP output of this distribution.

Now, suppose that when we actually contact eight randomly selected purchasers, we find that five out of the eight television sets owned by these purchasers have lasted at least five years without a single repair. Because the sample fraction, $5/8 = .625$, of television sets needing no repairs is less than .95, we have some evidence contradicting the manufacturer's claim that p equals .95. To evaluate the strength of this evidence, we will calculate the probability that five or fewer of the eight randomly selected televisions would last five years without a single repair if, in fact, p equals .95. Using the appropriate binomial probabilities, we have

$$P(x \le 5) = P(x = 5 \text{ or } x = 4 \text{ or } x = 3 \text{ or } x = 2 \text{ or } x = 1 \text{ or } x = 0)$$
$$= P(x = 5) + P(x = 4) + P(x = 3) + P(x = 2) + P(x = 1) + P(x = 0)$$
$$= .0054 + .0004 + .0000 + .0000 + .0000 + .0000$$
$$= .0058$$

This probability says that, if p equals .95, then in only .58 percent of all possible samples of eight randomly selected ColorSmart-5000 televisions would five or fewer of the eight televisions last five years without a single repair. Therefore, if we are to believe that p equals .95, we must believe that a 58 in 10,000 chance has occurred. Because it is difficult to believe that such a small chance has occurred, we have strong evidence that p does not equal .95, and is, in fact, less than .95.

*Binomial with
n = 8 and
p = 0.95

x	P (X = x)
3	0.0000
4	0.0004
5	0.0054
6	0.0515
7	0.2793
8	0.6634

*Probabilities for x = 0, 1, and 2 are not listed because each has a probability that is approximately zero.

x	P (X = x)
0	3.90625e-11
1	5.9375e-9
2	3.9484375e-7
3	0.0000150041
4	0.0003563465
5	0.0054164666
6	0.0514564323
7	0.2793349184
8	0.6634204313

In Examples 6.8 and 6.10 we have illustrated binomial distributions with different values of n and p. The values of n and p are often called the **parameters** of the binomial distribution. Figure 6.6 shows several different binomial distributions. We see that, depending on the parameters, a binomial distribution can be skewed to the right, skewed to the left, or symmetrical.

We next consider calculating the mean, variance, and standard deviation of a binomial random variable. If we place the binomial probability formula into the expressions (given in Section 6.2) for the mean and variance of a discrete random variable, we can derive formulas

FIGURE 6.6   **Several Binomial Distributions**

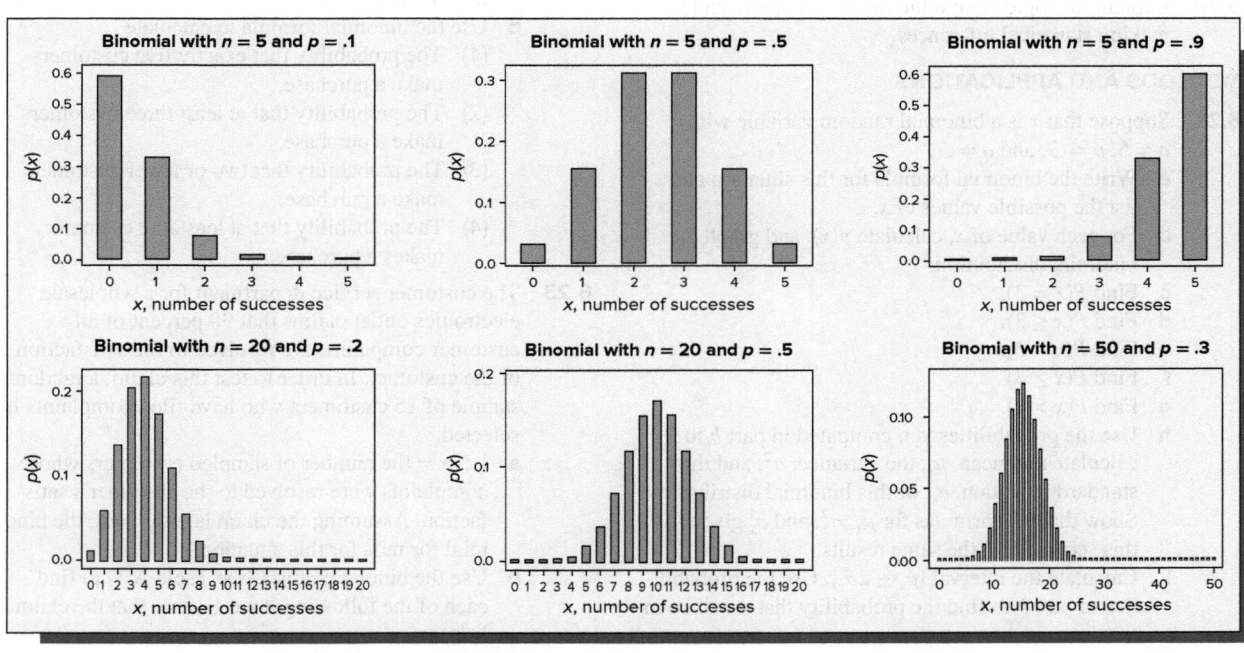

that allow us to easily compute μ_x, σ_x^2, and σ_x for a binomial random variable. Omitting the details of the derivation, we have the following results:

The Mean, Variance, and Standard Deviation of a Binomial Random Variable

If x is a binomial random variable, then

$$\mu_x = np \qquad \sigma_x^2 = npq \qquad \sigma_x = \sqrt{npq}$$

where n is the number of trials, p is the probability of success on each trial, and $q = 1 - p$ is the probability of failure on each trial.

As a simple example, again consider the television manufacturer, and recall that x is the number of eight randomly selected ColorSmart-5000 televisions that last five years without a single repair. If the manufacturer's claim that p equals .95 is true (which implies that q equals $1 - p = 1 - .95 = .05$), it follows that

$$\mu_x = np = 8(.95) = 7.6$$
$$\sigma_x^2 = npq = 8(.95)(.05) = .38$$
$$\sigma_x = \sqrt{npq} = \sqrt{.38} = .6164$$

In order to interpret $\mu_x = 7.6$, suppose that we were to randomly select all possible samples of eight ColorSmart-5000 televisions and record the number of sets in each sample that last five years without a repair. If we averaged all of our results, we would find that the average number of sets per sample that last five years without a repair is equal to 7.6.

Exercises for Section 6.3

CONCEPTS ≣ **connect**

6.18 List the four characteristics of a binomial experiment.

6.19 Suppose that x is a binomial random variable. Explain what the values of x represent. That is, how are the values of x defined?

6.20 Explain the logic behind the rare event approach to making statistical inferences.

METHODS AND APPLICATIONS

6.21 Suppose that x is a binomial random variable with $n = 5$, $p = .3$, and $q = .7$.
  a Write the binomial formula for this situation and list the possible values of x.
  b For each value of x, calculate $p(x)$, and graph the binomial distribution.
  c Find $P(x = 3)$.
  d Find $P(x \leq 3)$.
  e Find $P(x < 3)$.
  f Find $P(x \geq 4)$.
  g Find $P(x > 2)$.
  h Use the probabilities you computed in part b to calculate the mean, μ_x, the variance, σ_x^2, and the standard deviation, σ_x, of this binomial distribution. Show that the formulas for μ_x, σ_x^2, and σ_x given in this section give the same results.
  i Calculate the interval $[\mu_x \pm 2\sigma_x]$. Use the probabilities of part b to find the probability that x will be in this interval.

6.22 Thirty percent of all customers who enter a store will make a purchase. Suppose that six customers enter the store and that these customers make independent purchase decisions.
  a Let $x =$ the number of the six customers who will make a purchase. Write the binomial formula for this situation.
  b Use the binomial formula to calculate
    (1) The probability that exactly five customers make a purchase.
    (2) The probability that at least three customers make a purchase.
    (3) The probability that two or fewer customers make a purchase.
    (4) The probability that at least one customer makes a purchase.

6.23 The customer service department for a wholesale electronics outlet claims that 90 percent of all customer complaints are resolved to the satisfaction of the customer. In order to test this claim, a random sample of 15 customers who have filed complaints is selected.
  a Let $x =$ the number of sampled customers whose complaints were resolved to the customer's satisfaction. Assuming the claim is true, write the binomial formula for this situation.
  b Use the binomial tables (see Table A.1) to find each of the following if we assume that the claim is true:

(1) $P(x \le 13)$.
(2) $P(x > 10)$.
(3) $P(x \ge 14)$.
(4) $P(9 \le x \le 12)$.
(5) $P(x \le 9)$.

c   Suppose that of the 15 customers selected, 9 have had their complaints resolved satisfactorily. Using part b, do you believe the claim of 90 percent satisfaction? Explain.

6.24 The United States Golf Association requires that the weight of a golf ball must not exceed 1.62 oz. The association periodically checks golf balls sold in the United States by sampling specific brands stocked by pro shops. Suppose that a manufacturer claims that no more than 1 percent of its brand of golf balls exceed 1.62 oz. in weight. Suppose that 24 of this manufacturer's golf balls are randomly selected, and let x denote the number of the 24 randomly selected golf balls that exceed 1.62 oz. Figure 6.7 gives part of an Excel output of the binomial distribution with $n = 24$, $p = .01$, and $q = .99$. (Note that, because $P(X = x) = .0000$ for values of x from 6 to 24, we omit these probabilities.) Use this output to

a   Find $P(x = 0)$, that is, find the probability that none of the randomly selected golf balls exceeds 1.62 oz. in weight.

b   Find the probability that at least one of the randomly selected golf balls exceeds 1.62 oz. in weight.

c   Find $P(x \le 3)$.

d   Find $P(x \ge 2)$.

e   Suppose that 2 of the 24 randomly selected golf balls are found to exceed 1.62 oz. Using your result from part d, do you believe the claim that no more than 1 percent of this brand of golf balls exceed 1.62 oz. in weight?

6.25 An industry representative claims that 50 percent of all satellite dish owners subscribe to at least one premium movie channel. In an attempt to justify this claim, the representative will poll a randomly selected sample of dish owners.

a   Suppose that the representative's claim is true, and suppose that a sample of four dish owners is randomly selected. Assuming independence, use an appropriate formula to compute

(1) The probability that none of the dish owners in the sample subscribes to at least one premium movie channel.

(2) The probability that more than two dish owners in the sample subscribe to at least one premium movie channel.

b   Suppose that the representative's claim is true, and suppose that a sample of 20 dish owners is randomly selected. Assuming independence, what is the probability that

(1) Nine or fewer dish owners in the sample subscribe to at least one premium movie channel?

(2) More than 11 dish owners in the sample subscribe to at least one premium movie channel?

(3) Fewer than five dish owners in the sample subscribe to at least one premium movie channel?

c   Suppose that, when we survey 20 randomly selected dish owners, we find that 4 of the dish owners actually subscribe to at least one premium movie channel. Using a probability you found in this exercise as the basis for your answer, do you believe the industry representative's claim? Explain.

6.26 For each of the following, calculate μ_x, σ_x^2, and σ_x by using the formulas given in this section. Then (1) interpret the meaning of μ_x, and (2) find the probability that x falls in the interval $[\mu_x \pm 2\sigma_x]$.

a   The situation of Exercise 6.22, where $x =$ the number of the six customers who will make a purchase.

b   The situation of Exercise 6.23, where $x =$ the number of 15 sampled customers whose complaints were resolved to the customer's satisfaction.

c   The situation of Exercise 6.24, where $x =$ the number of the 24 randomly selected golf balls that exceed 1.62 oz. in weight.

6.27 The January 1986 mission of the Space Shuttle *Challenger* was the 25th such shuttle mission. It was unsuccessful due to an explosion caused by an O-ring seal failure.

a   According to NASA, the probability of such a failure in a single mission was 1/60,000. Using this value of p and assuming all missions are independent, calculate the probability of no mission failures in 25 attempts. Then calculate the probability of at least one mission failure in 25 attempts.

b   According to a study conducted for the Air Force, the probability of such a failure in a single mission was 1/35. Recalculate the probability of no mission failures in 25 attempts and the probability of at least one mission failure in 25 attempts.

c   Based on your answers to parts a and b, which value of p seems more likely to be true? Explain.

d   How small must p be made in order to ensure that the probability of no mission failures in 25 attempts is .999?

FIGURE 6.7   **Excel Output of the Binomial Distribution with $n = 24$, $p = .01$, and $q = .99$**

Binomial distribution with $n = 24$ and $p = 0.01$

x	$P(X = x)$
0	0.7857
1	0.1905
2	0.0221
3	0.0016
4	0.0001
5	0.0000

LO6-4
Use the Poisson
distribution to compute
probabilities (Optional).

6.4 The Poisson Distribution (Optional)

We now discuss a discrete random variable that describes the number of occurrences of an event over a specified interval of time or space. For instance, we might wish to describe (1) the number of customers who arrive at the checkout counters of a grocery store in one hour, or (2) the number of major fires in a city during the next two months, or (3) the number of dirt specks found in one square yard of plastic wrap.

Such a random variable can often be described by a **Poisson distribution** (or **Poisson model**). We describe this distribution and give two assumptions needed for its use in the following box:

The Poisson Distribution (The Poisson Model)

Consider the number of times an event occurs over an interval of time or space, and assume that

1 The probability of the event's occurrence is the same for any two intervals of equal length, and

2 Whether the event occurs in any interval is independent of whether the event occurs in any other nonoverlapping interval.

Then, the probability that the event will occur x times in a *specified interval* is

$$p(x) = \frac{e^{-\mu}\mu^x}{x!}$$

Here μ is the mean (or expected) number of occurrences of the event in the *specified interval*, and $e = 2.71828\ldots$ is the base of Napierian logarithms.

In theory, there is no limit to how large x might be. That is, theoretically speaking, the event under consideration could occur an indefinitely large number of times during any specified interval. This says that a **Poisson random variable** might take on any of the values 0, 1, 2, 3, . . . and so forth. We will now look at an example.

EXAMPLE 6.11 The Air Safety Case: Traffic Control Errors

In an article in the August 15, 1998, edition of the *Journal News* (Hamilton, Ohio),[2] the Associated Press reported that the Cleveland Air Route Traffic Control Center, the busiest in the nation for guiding planes on cross-country routes, had experienced an unusually high number of errors since the end of July. An error occurs when controllers direct flights either within five miles of each other horizontally, or within 2,000 feet vertically at a height of 18,000 feet or more (the standard is 1,000 feet vertically at heights less than 18,000 feet). The controllers' union blamed the errors on a staff shortage, whereas the Federal Aviation Administration (FAA) claimed that the cause was improved error reporting and an unusual number of thunderstorms.

Suppose that an air traffic control center has been averaging 20.8 errors per year and that the center experiences 3 errors in a week. The FAA must decide whether this occurrence is unusual enough to warrant an investigation as to the causes of the (possible) increase in errors. To investigate this possibility, we will find the probability distribution of x, the number of errors in a week, when we assume that the center is still averaging 20.8 errors per year.

Arbitrarily choosing a time unit of one week, the average (or expected) number of errors per week is 20.8/52 = .4. Therefore, we can use the Poisson formula (note that the Poisson assumptions are probably satisfied) to calculate the probability of no errors in a week to be

$$p(0) = P(x = 0) = \frac{e^{-\mu}\mu^0}{0!} = \frac{e^{-.4}(.4)^0}{1} = .6703$$

---
[2]F. J. Frommer, "Errors on the Rise at Traffic Control Center in Ohio," *Journal News*, August 15, 1998.

TABLE 6.5   **A Portion of a Poisson Probability Table**

x, Number of Occurrences	μ, Mean Number of Occurrences									
	.1	.2	.3	.4	.5	.6	.7	.8	.9	1.0
0	.9048	.8187	.7408	.6703	.6065	.5488	.4966	.4493	.4066	.3679
1	.0905	.1637	.2222	.2681	.3033	.3293	.3476	.3595	.3659	.3679
2	.0045	.0164	.0333	.0536	.0758	.0988	.1217	.1438	.1647	.1839
3	.0002	.0011	.0033	.0072	.0126	.0198	.0284	.0383	.0494	.0613
4	.0000	.0001	.0003	.0007	.0016	.0030	.0050	.0077	.0111	.0153
5	.0000	.0000	.0000	.0001	.0002	.0004	.0007	.0012	.0020	.0031
6	.0000	.0000	.0000	.0000	.0000	.0000	.0001	.0002	.0003	.0005

Source: Data from Brooks/Cole, 1991.

FIGURE 6.8   **The Poisson Probability Distribution with $\mu = .4$**

(a) The Minitab output		(b) The JMP output		(c) A graph of the distribution
x	P (X = x)	x	P (X = x)	
0	0.6703	0	0.670320046	
1	0.2681	1	0.2681280184	
2	0.0536	2	0.0536256037	
3	0.0072	3	0.0071500805	
4	0.0007	4	0.000715008	
5	0.0001	5	0.0000572006	
6	0.0000	6	3.8133763e-6	

Similarly, the probability of three errors in a week is

$$p(3) = P(x = 3) = \frac{e^{-.4}(.4)^3}{3!} = \frac{e^{-.4}(.4)^3}{3 \cdot 2 \cdot 1} = .0072$$

As with the binomial distribution, tables have been constructed that give Poisson probabilities. A table of these probabilities is given in Table A.2. A portion of this table is reproduced in Table 6.5. In this table, values of the mean number of occurrences, μ, are listed across the top of the table, and values of x (the number of occurrences) are listed down the left side of the table. In order to use the table in the traffic control situation, we look at the column in Table 6.5 corresponding to .4, and we find the probabilities of 0, 1, 2, 3, 4, 5, and 6 errors (we have shaded these probabilities). For instance, the probability of one error in a week is .2681. Also, note that the probability of any number of errors greater than 6 is so small that it is not listed in the table. Figures 6.8(a) and (b) give Minitab and JMP outputs of the Poisson distribution of x, the number of errors in a week. (Excel gives similar outputs.) The Poisson distribution is graphed in Figure 6.8(c).

Next, recall that there have been three errors at the air traffic control center in the last week. This is considerably more errors than .4, the expected number of errors assuming the center is still averaging 20.8 errors per year. Therefore, we have some evidence to contradict this assumption. To evaluate the strength of this evidence, we calculate the probability that at least three errors will occur in a week if, in fact, μ equals .4. Using the Poisson probabilities in Figure 6.8(a), we obtain

$$P(x \geq 3) = p(3) + p(4) + p(5) + p(6) = .0072 + .0007 + .0001 + .0000 = .008$$

This probability says that, if the center is averaging 20.8 errors per year, then there would be three or more errors in a week in only .8 percent of all weeks. That is, if we are to believe that the control center is averaging 20.8 errors per year, then we must believe that an 8 in 1,000 chance has occurred. Because it is very difficult to believe that such a rare event has occurred, we have strong evidence that the average number of errors per week has increased. Therefore, an investigation by the FAA into the reasons for such an increase is probably justified.

EXAMPLE 6.12 Errors in Computer Code

In the book *Modern Statistical Quality Control and Improvement,* Nicholas R. Farnum (1994) presents an example dealing with the quality of computer software. In the example, Farnum measures software quality by monitoring the number of errors per 1,000 lines of computer code.

Suppose that the number of errors per 1,000 lines of computer code is described by a Poisson distribution with a mean of four errors per 1,000 lines of code. If we wish to find the probability of obtaining eight errors in 2,500 lines of computer code, we must adjust the mean of the Poisson distribution. To do this, we arbitrarily choose a *space unit* of one line of code, and we note that a mean of four errors per 1,000 lines of code is equivalent to $4/1,000$ of an error per line of code. Therefore, the mean number of errors per 2,500 lines of code is $(4/1,000)(2,500) = 10$. It follows that

$$p(8) = \frac{e^{-\mu}\mu^8}{8!} = \frac{e^{-10}10^8}{8!} = .1126$$

The mean, μ, is often called the *parameter* of the Poisson distribution. Figure 6.9 shows several Poisson distributions. We see that, depending on its parameter (mean), a Poisson distribution can be very skewed to the right or can be quite symmetrical.

Finally, if we place the Poisson probability formula into the general expressions (of Section 6.2) for μ_x, σ_x^2, and σ_x, we can derive formulas for calculating the mean, variance, and standard deviation of a Poisson distribution:

The Mean, Variance, and Standard Deviation of a Poisson Random Variable

Suppose that x is a **Poisson random variable**. If μ is the average number of occurrences of an event over the specified interval of time or space of interest, then

$$\mu_x = \mu \qquad \sigma_x^2 = \mu \qquad \sigma_x = \sqrt{\mu}$$

FIGURE 6.9   Several Poisson Distributions

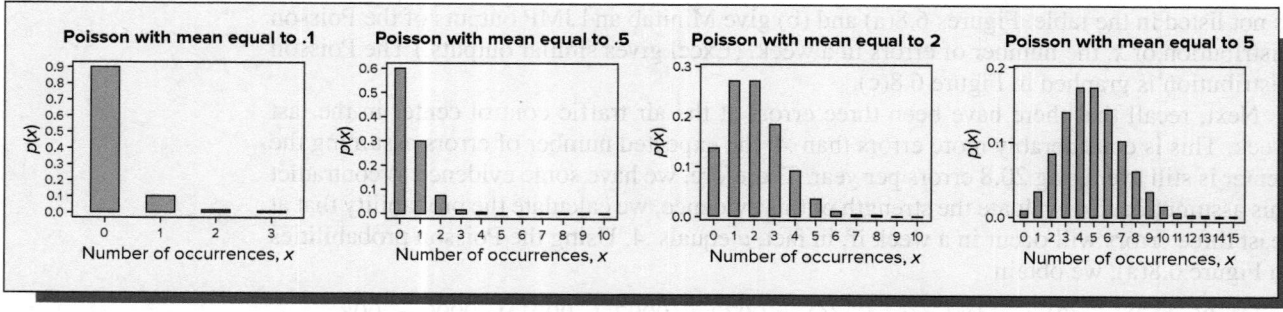

Here we see that both the mean and the variance of a Poisson random variable equal the average number of occurrences μ of the event of interest over the specified interval of time or space. For example, in the air traffic control situation, the Poisson distribution of x, the number of errors at the air traffic control center in a week, has a mean of $\mu_x = .4$ and a standard deviation of $\sigma_x = \sqrt{.4} = .6325$.

Exercises for Section 6.4

CONCEPTS ▦ **connect**

6.28 What do the possible values of a Poisson random variable x represent?

6.29 Explain the assumptions that must be satisfied when a Poisson distribution adequately describes a random variable x.

METHODS AND APPLICATIONS

6.30 Suppose that x has a Poisson distribution with $\mu = 2$.
  a Write the Poisson formula and describe the possible values of x.
  b Starting with the smallest possible value of x, calculate $p(x)$ for each value of x until $p(x)$ becomes smaller than .001.
  c Graph the Poisson distribution using your results of b.
  d Find $P(x = 2)$.   e Find $P(x \leq 4)$.
  f Find $P(x < 4)$.   g Find $P(x \geq 1)$ and $P(x > 2)$.
  h Find $P(1 \leq x \leq 4)$.   i Find $P(2 < x < 5)$.
  j Find $P(2 \leq x < 6)$.

6.31 Suppose that x has a Poisson distribution with $\mu = 2$.
  a Use the formulas given in this section to compute the mean, μ_x, variance, σ_x^2, and standard deviation, σ_x.
  b Calculate the intervals $[\mu_x \pm 2\sigma_x]$ and $[\mu_x \pm 3\sigma_x]$. Then use the probabilities you calculated in Exercise 6.30 to find the probability that x will be inside each of these intervals.

6.32 A bank manager wishes to provide prompt service for customers at the bank's drive-up window. The bank currently can serve up to 10 customers per 15-minute period without significant delay. The average arrival rate is 7 customers per 15-minute period. Let x denote the number of customers arriving per 15-minute period. Assuming x has a Poisson distribution:
  a Find the probability that 10 customers will arrive in a particular 15-minute period.
  b Find the probability that 10 or fewer customers will arrive in a particular 15-minute period.
  c Find the probability that there will be a significant delay at the drive-up window. That is, find the probability that more than 10 customers will arrive during a particular 15-minute period.

6.33 A telephone company's goal is to have no more than five monthly line failures on any 100 miles of line. The company currently experiences an average of two monthly line failures per 50 miles of line. Let x denote the number of monthly line failures per 100 miles of line. Assuming x has a Poisson distribution:
  a Find the probability that the company will meet its goal on a particular 100 miles of line.
  b Find the probability that the company will not meet its goal on a particular 100 miles of line.
  c Find the probability that the company will have no more than five monthly failures on a particular 200 miles of line.

6.34 A local law enforcement agency claims that the number of times that a patrol car passes through a particular neighborhood follows a Poisson process with a mean of three times per nightly shift. Let x denote the number of times that a patrol car passes through the neighborhood during a nightly shift.
  a Calculate the probability that no patrol cars pass through the neighborhood during a nightly shift.
  b Suppose that during a randomly selected night shift no patrol cars pass through the neighborhood. Based on your answer in part a, do you believe the agency's claim? Explain.
  c Assuming that nightly shifts are independent and assuming that the agency's claim is correct, find the probability that exactly one patrol car will pass through the neighborhood on each of four consecutive nights.

6.35 When the number of trials, n, is large, binomial probability tables may not be available. Furthermore, if a computer is not available, hand calculations will be tedious. As an alternative, the Poisson distribution can be used to approximate the binomial distribution when n is large and p is small. Here the mean of the Poisson distribution is taken to be $\mu = np$. That is, when n is large and p is small, we can use the Poisson formula with $\mu = np$ to calculate binomial probabilities, and we will obtain results close to those we would obtain by using the binomial formula. A common rule is to use this approximation when $n/p \geq 500$.

  To illustrate this approximation, in the movie *Coma*, a young female intern at a Boston hospital was very upset when her friend, a young nurse, went into a coma during routine anesthesia at the hospital. Upon investigation, she found that 10 of the last 30,000 healthy patients at the hospital had gone into comas during routine anesthesias. When she confronted the hospital administrator with this fact and the fact that the national average was 6 out of 100,000 healthy patients going into comas during routine anesthesias, the administrator replied that

10 out of 30,000 was still quite small and thus not that unusual.

    a Use the Poisson distribution to approximate the probability that 10 or more of 30,000 healthy patients would slip into comas during routine anesthesias, if in fact the true average at the hospital was 6 in 100,000. Hint: $\mu = np = 30,000(6/100,000) = 1.8$.

    b Given the hospital's record and part a, what conclusion would you draw about the hospital's medical practices regarding anesthesia?

(Note: It turned out that the hospital administrator was part of a conspiracy to sell body parts and was purposely putting healthy adults into comas during routine anesthesias. If the intern had taken a statistics course, she could have avoided a great deal of danger.)

6.36 Suppose that an automobile parts wholesaler claims that .5 percent of the car batteries in a shipment are defective. A random sample of 200 batteries is taken, and four are found to be defective. **(1)** Use the Poisson approximation discussed in Exercise 6.35 to find the probability that four or more car batteries in a random sample of 200 such batteries would be found to be defective, if in fact the wholesaler's claim is true. **(2)** Do you believe the claim? Explain.

LO6-5

Use the hypergeometric distribution to compute probabilities (Optional).

6.5 The Hypergeometric Distribution (Optional)

The Hypergeometric Distribution (The Hypergeometric Model)

Suppose that a population consists of N items and that r of these items are *successes* and $(N - r)$ of these items are *failures*. If we randomly select n of the N items **without replacement**, it can be shown that the probability that x of the n randomly selected items will be successes is given by the **hypergeometric probability formula**

$$p(x) = \frac{\binom{r}{x}\binom{N-r}{n-x}}{\binom{N}{n}}$$

Here $\binom{r}{x}$ is the number of ways x successes can be selected from the total of r successes in the population, $\binom{N-r}{n-x}$ is the number of ways $n - x$ failures can be selected from the total of $N - r$ failures in the population, and $\binom{N}{n}$ is the number of ways a sample of size n can be selected from a population of size N.

    To demonstrate the calculations, suppose that a population of $N = 6$ stocks consists of $r = 4$ stocks that are destined to give positive returns (that is, there are $r = 4$ *successes*) and $N - r = 6 - 4 = 2$ stocks that are destined to give negative returns (that is, there are $N - r = 2$ *failures*). Also suppose that we randomly select $n = 3$ of the six stocks in the population without replacement and that we define x to be the number of the three randomly selected stocks that will give a positive return. Then, for example, the probability that $x = 2$ is

$$P(x = 2) = \frac{\binom{r}{x}\binom{N-r}{n-x}}{\binom{N}{n}} = \frac{\binom{4}{2}\binom{2}{1}}{\binom{6}{3}} = \frac{\left(\frac{4!}{2!\,2!}\right)\left(\frac{2!}{1!\,1!}\right)}{\left(\frac{6!}{3!\,3!}\right)} = \frac{(6)(2)}{20} = .6$$

Similarly, the probability that $x = 3$ is

$$P(x = 3) = \frac{\binom{4}{3}\binom{2}{0}}{\binom{6}{3}} = \frac{\left(\frac{4!}{3!\,1!}\right)\left(\frac{2!}{0!\,2!}\right)}{\left(\frac{6!}{3!\,3!}\right)} = \frac{(4)(1)}{20} = .2$$

It follows that the probability that at least two of the three randomly selected stocks will give a positive return is $P(x = 2) + P(x = 3) = .6 + .2 = .8$.

    If we place the hypergeometric probability formula into the general expressions (of Section 6.2) for μ_x and σ_x^2, we can derive formulas for the mean and variance of the hypergeometric distribution.

The Mean and Variance of a Hypergeometric Random Variable

Suppose that x is a hypergeometric random variable. Then

$$\mu_x = n\left(\frac{r}{N}\right) \qquad \text{and} \qquad \sigma_x^2 = n\left(\frac{r}{N}\right)\left(1 - \frac{r}{N}\right)\left(\frac{N-n}{N-1}\right)$$

In the previous example, we have $N = 6$, $r = 4$, and $n = 3$. It follows that

$$\mu_x = n\left(\frac{r}{N}\right) = 3\left(\frac{4}{6}\right) = 2, \qquad \text{and}$$

$$\sigma_x^2 = n\left(\frac{r}{N}\right)\left(1 - \frac{r}{N}\right)\left(\frac{N-n}{N-1}\right) = 3\left(\frac{4}{6}\right)\left(1 - \frac{4}{6}\right)\left(\frac{6-3}{6-1}\right) = .4$$

and that the standard deviation $\sigma_x = \sqrt{.4} = .6325$.

To conclude this section, note that, on the first random selection from the population of N items, the probability of a success is r/N. Because we are making selections *without replacement,* the probability of a success changes as we continue to make selections. However, **if the population size N is "much larger" than the sample size n (say, at least 20 times as large),** then making the selections will not substantially change the probability of a success. In this case, we can assume that the probability of a success stays essentially constant from selection to selection, and the different selections are essentially independent of each other. Therefore, **we can approximate the hypergeometric distribution by the binomial distribution.** That is, we can compute probabilities about the hypergeometric random variable x by using the easier binomial probability formula

$$p(x) = \frac{n!}{x!(n-x)!}\, p^x(1-p)^{n-x} = \frac{n!}{x!\,(n-x)!}\left(\frac{r}{N}\right)^x\left(1 - \frac{r}{N}\right)^{n-x}$$

where the binomial probability of success equals r/N. Exercise 6.43 illustrates this.

Exercises for Section 6.5

CONCEPTS   **connect**

6.37 In the context of the hypergeometric distribution, explain the meanings of N, r, and n.

6.38 When can a hypergeometric distribution be approximated by a binomial distribution? Explain carefully what this means.

METHODS AND APPLICATIONS

6.39 Suppose that x has a hypergeometric distribution with $N = 8$, $r = 5$, and $n = 4$. Find

  a $P(x = 0)$.   **e** $P(x = 4)$.
  b $P(x = 1)$.   **f** $P(x \geq 2)$.
  c $P(x = 2)$.   **g** $P(x < 3)$.
  d $P(x = 3)$.   **h** $P(x > 1)$.

6.40 Suppose that x has a hypergeometric distribution with $N = 10$, $r = 4$, and $n = 3$.

  a Write out the probability distribution of x.
  b Find the mean μ_x, variance σ_x^2, and standard deviation σ_x of this distribution.

6.41 Among 12 metal parts produced in a machine shop, 3 are defective. If a random sample of three of these metal parts is selected, find

  a The probability that this sample will contain at least two defectives.
  b The probability that this sample will contain at most one defective.

6.42 Suppose that you purchase (randomly select) 3 TV sets from a production run of 10 TV sets. Of the 10 TV sets, 9 are destined to last at least five years without needing a single repair. What is the probability that all three of your TV sets will last at least five years without needing a single repair?

6.43 Suppose that you own a car dealership and purchase (randomly select) 7 cars of a certain make from a production run of 200 cars. Of the 200 cars, 160 are destined to last at least five years without needing a major repair. Set up an expression using the hypergeometric distribution for the probability that at least 6 of your 7 cars will last at least five years without needing a major repair. Then, using the binomial tables (see Table A.1), approximate this probability by using the binomial distribution. What justifies the approximation? Hint: $p = r/N = 160/200 = .8$.

LO6-6

Compute and understand the covariance between two random variables (Optional).

6.6 Joint Distributions and the Covariance (Optional)

Below we present (1) the probability distribution of x, the yearly proportional return for stock A; (2) the probability distribution of y, the yearly proportional return for stock B; and (3) the **joint probability distribution of (x, y),** the joint yearly proportional returns for stocks A and B [note that we have obtained the data below from Pfaffenberger and Patterson (1987)].

x	$p(x)$	y	$p(y)$
−0.10	0.400	−0.15	0.300
0.05	0.125	−0.05	0.200
0.15	0.100	0.12	0.150
0.38	0.375	0.46	0.350

$\mu_x = .124$ $\mu_y = .124$

$\sigma_x^2 = .0454$ $\sigma_y^2 = .0681$

$\sigma_x = .2131$ $\sigma_y = .2610$

Joint Distribution of (x, y)

Stock B Return, y	Stock A Return, x −0.10	0.05	0.15	0.38
−0.15	0.025	0.025	0.025	0.225
−0.05	0.075	0.025	0.025	0.075
0.12	0.050	0.025	0.025	0.050
0.46	0.250	0.050	0.025	0.025

To explain the joint probability distribution, note that the probability of .250 enclosed in the rectangle is the probability that in a given year the return for stock A will be −.10 and the return for stock B will be .46. The probability of .225 enclosed in the oval is the probability that in a given year the return for stock A will be .38 and the return for stock B will be −.15. Intuitively, these two rather large probabilities say that (1) a negative return x for stock A tends to be associated with a highly positive return y for stock B, and (2) a highly positive return x for stock A tends to be associated with a negative return y for stock B. To further measure the association between x and y, we can calculate the *covariance* between x and y. To do this, we calculate $(x - \mu_x)(y - \mu_y) = (x - .124)(y - .124)$ for each combination of values of x and y. Then, we multiply each $(x - \mu_x)(y - \mu_y)$ value by the probability $p(x, y)$ of the (x, y) combination of values and add up the quantities that we obtain. The resulting number is the **covariance,** denoted σ_{xy}^2. For example, for the combination of values $x = -.10$ and $y = .46$, we calculate

$$(x - \mu_x)(y - \mu_y)\, p(x, y) = (-.10 - .124)(.46 - .124)(.250) = -.0188$$

Doing this for all combinations of (x, y) values and adding up the resulting quantities, we find that the covariance is −.0318. In general, a negative covariance says that as x increases, y tends to decrease in a linear fashion. A positive covariance says that as x increases, y tends to increase in a linear fashion.

In this situation, the covariance helps us to understand the importance of investment diversification. If we invest all of our money in stock A, we have seen that $\mu_x = .124$ and $\sigma_x = .2131$. If we invest all of our money in stock B, we have seen that $\mu_y = .124$ and $\sigma_y = .2610$. If we invest half of our money in stock A and half of our money in stock B, the return for the portfolio is $P = .5x + .5y$. To find the expected value of the portfolio return, we need to use a *property of expected values*. This property says if a and b are constants, and if x and y are random variables, then

$$\mu_{(ax+by)} = a\mu_x + b\mu_y$$

Therefore,

$$\mu_P = \mu_{(.5x+.5y)} = .5\mu_x + .5\mu_y = .5(.124) + .5(.124) = .124$$

To find the variance of the portfolio return, we must use a *property of variances*. In general, if x and y have a covariance σ_{xy}^2, and a and b are constants, then

$$\sigma_{(ax+by)}^2 = a^2\sigma_x^2 + b^2\sigma_y^2 + 2ab\sigma_{xy}^2$$

Therefore,

$$\sigma_P^2 = \sigma_{(.5x+.5y)}^2 = (.5)^2\sigma_x^2 + (.5)^2\sigma_y^2 + 2(.5)(.5)\sigma_{xy}^2$$

$$= (.5)^2(.0454) + (.5)^2(.0681) + 2(.5)(.5)(-.0318) = .012475$$

and $\sigma_P = \sqrt{.012475} = .1117$. Note that, because $\mu_P = .124$ equals $\mu_x = .124$ and $\mu_y = .124$, the portfolio has the same expected return as either stock A or B. However, because $\sigma_P = .1117$ is less than $\sigma_x = .2131$ and $\sigma_y = .2610$, the portfolio is a less risky investment. In other words, diversification *can* reduce risk. Note, however, that the reason that σ_P is *less* than σ_x and σ_y is that $\sigma_{xy}^2 = -.0318$ *is negative*. Intuitively, this says that the two stocks tend to balance each other's returns. However, if the covariance between the returns of two stocks is positive, σ_P can be larger than σ_x and/or σ_y. The student will demonstrate this in Exercise 6.46.

Next, note that a measure of linear association between x and y that is unitless and always between -1 and 1 is the **correlation coefficient,** denoted ρ. We define ρ as follows:

The **correlation coefficient** between x and y is $\rho = \sigma_{xy}^2/\sigma_x\,\sigma_y$.

For the stock return example, ρ equals $(-.0318)/((.2131)(.2610)) = -.5717$.

To conclude this section, we summarize four properties of expected values and variances that we will use in optional Section 8.3 to derive some important facts about the sample mean:

Property 1: If a is a constant and x is a random variable, $\mu_{ax} = a\mu_x$.

Property 2: If $x_1, x_2, \ldots, x_n$ are random variables, $\mu_{(x_1+x_2+\cdots+x_n)} = \mu_{x_1} + \mu_{x_2} + \cdots + \mu_{x_n}$.

Property 3: If a is a constant and x is a random variable, $\sigma_{ax}^2 = a^2\sigma_x^2$.

Property 4: If $x_1, x_2, \ldots, x_n$ are statistically independent random variables (that is, if the value taken by any one of these independent variables is in no way associated with the value taken by any other of these random variables), then the covariance between any two of these random variables is zero and $\sigma_{(x_1+x_2+\cdots+x_n)}^2 = \sigma_{x_1}^2 + \sigma_{x_2}^2 + \cdots + \sigma_{x_n}^2$.

Exercises for Section 6.6

CONCEPTS

6.44 Explain the meaning of a negative covariance.

6.45 Explain the meaning of a positive covariance.

METHODS AND APPLICATIONS

6.46 Let x be the yearly proportional return for stock C, and let y be the yearly proportional return for stock D. If $\mu_x = .11$, $\mu_y = .09$, $\sigma_x = .17$, $\sigma_y = .17$, and $\sigma_{xy}^2 = .0412$, find the mean and standard deviation of the portfolio return $P = .5x + .5y$. Discuss the risk of the portfolio.

6.47 In the table in the right column we give a joint probability table for two utility bonds where the random variable x represents the percentage return for bond 1 and the random variable y represents the percentage return for bond 2. In this table, probabilities associated with values of x are given in the row labeled $p(x)$ and probabilities associated with values of y are given in the column labeled $p(y)$. For example, $P(x = 10) = .38$ and $P(y = 10) = .40$. The entries inside the body of the table are joint probabilities—for instance, the

			x			
y	8	9	10	11	12	p(y)
8	.03	.04	.03	.00	.00	.10
9	.04	.06	.06	.04	.00	.20
10	.02	.08	.20	.08	.02	.40
11	.00	.04	.06	.06	.04	.20
12	.00	.00	.03	.04	.03	.10
p(x)	.09	.22	.38	.22	.09	

Source: David K. Hildebrand and Lyman Ott, *Statistical Thinking for Managers,* 2nd edition (Boston, MA: Duxbury Press, 1987), p. 101.

probability that x equals 9 and y equals 10 is .08. Use the table to do the following:

a Calculate μ_x, σ_x, μ_y, σ_y, and σ_{xy}^2.

b Calculate the variance and standard deviation of a portfolio in which 50 percent of the money is used to buy bond 1 and 50 percent is used to buy bond 2. That is, find σ_P^2 and σ_P, where $P = .5x + .5y$. Discuss the risk of the portfolio.

Chapter Summary

In this chapter we began our study of **random variables.** We learned that **a random variable represents an uncertain numerical outcome.** We also learned that a random variable whose values can be listed is called a **discrete random variable,** while the values of a **continuous random variable** correspond to one or more intervals on the real number line. We saw that a **probability distribution** of a discrete random variable is a table, graph, or formula that gives the probability associated with each of the random variable's possible values. We also discussed several descriptive measures of a discrete random variable—its **mean** (or **expected value**), its **variance,** and its **standard deviation.** We continued this chapter by studying two important, commonly used discrete probability distributions—the **binomial distribution** and the **Poisson distribution**—and we demonstrated how these distributions can be used to make statistical inferences. Finally, we studied a third important discrete probability distribution, the **hypergeometric distribution,** and we discussed **joint distributions** and the **covariance.**

Glossary of Terms

binomial distribution: The probability distribution that describes a binomial random variable.

binomial experiment: An experiment that consists of n independent, identical trials, each of which results in either a success or a failure and is such that the probability of success on any trial is the same.

binomial random variable: A random variable that is defined to be the total number of successes in n trials of a binomial experiment.

binomial tables: Tables in which we can look up binomial probabilities.

continuous random variable: A random variable whose values correspond to one or more intervals of numbers on the real number line.

correlation coefficient: A unitless measure of the linear relationship between two random variables.

covariance: A non-unitless measure of the linear relationship between two random variables.

discrete random variable: A random variable whose values can be counted or listed.

expected value (or **mean**) **of a random variable:** The mean of the population of all possible observed values of a random variable. That is, the long-run average value obtained if values of a random variable are observed a (theoretically) infinite number of times.

hypergeometric distribution: The probability distribution that describes a hypergeometric random variable.

hypergeometric random variable: A random variable that is defined to be the number of successes obtained in a random sample selected without replacement from a finite population of N elements that contains r successes and $N - r$ failures.

joint probability distribution of (x, y): A probability distribution that assigns probabilities to all combinations of values of x and y.

mean of a random variable: The mean of the population of all possible observed values of a random variable. That is, the long-run average value of the random variable.

Poisson distribution: The probability distribution that describes a Poisson random variable.

Poisson random variable: A discrete random variable that can often be used to describe the number of occurrences of an event over a specified interval of time or space.

probability distribution of a discrete random variable: A table, graph, or formula that gives the probability associated with each of the discrete random variable's values (also called a discrete probability distribution).

random variable: A variable that assumes numerical values that are determined by the outcome of an experiment. That is, a variable that represents an uncertain numerical outcome.

standard deviation (of a random variable): The standard deviation of the population of all possible observed values of a random variable. It measures the spread of the population of all possible observed values of the random variable.

variance (of a random variable): The variance of the population of all possible observed values of a random variable. It measures the spread of the population of all possible observed values of the random variable.

Important Formulas

Properties of a discrete probability distribution: page 305

The mean (expected value) of a discrete random variable: page 306

Variance and standard deviation of a discrete random variable: page 308

Binomial probability formula: page 314

Mean, variance, and standard deviation of a binomial random variable: page 320

Poisson probability formula: page 322

Mean, variance, and standard deviation of a Poisson random variable: page 324

Hypergeometric probability formula: page 326

Mean and variance of a hypergeometric random variable: page 327

Covariance of x and y: page 328

Correlation coefficient between x and y: page 329

Properties of expected values and variances: page 329

Supplementary Exercises

6.48 A rock concert promoter has scheduled an outdoor concert on July 4th. If it does not rain, the promoter will make $30,000. If it does rain, the promoter will lose $15,000 in guarantees made to the band and other expenses. The probability of rain on the 4th is .4.

a What is the promoter's expected profit? Is the expected profit a reasonable decision criterion? Explain.

b In order to break even, how much should an insurance company charge to insure the promoter's full losses? Explain your answer.

6.49 The demand (in number of copies per day) for a city newspaper, x, has historically been 50,000, 70,000, 90,000, 110,000, or 130,000 with the respective probabilities .1, .25, .4, .2, and .05.

a Graph the probability distribution of x.

b Find the expected demand. Interpret this value, and label it on the graph of part *a*.

c Using Chebyshev's Theorem, find the minimum percentage of all possible daily demand values that will fall in the interval $[\mu_x \pm 2\sigma_x]$.

d Calculate the interval $[\mu_x \pm 2\sigma_x]$. Illustrate this interval on the graph of part *a*. According to the probability distribution of demand x previously given, what percentage of all possible daily demand values fall in the interval $[\mu_x \pm 2\sigma_x]$?

6.50 United Medicine, Inc., claims that a drug, Viro, significantly relieves the symptoms of a certain viral infection for 80 percent of all patients. Suppose that this drug is given to eight randomly selected patients who have been diagnosed with the viral infection.

a Let x equal the number of the eight randomly selected patients whose symptoms are significantly relieved. What distribution describes the random variable x? Explain.

b Assuming that the company's claim is correct, find $P(x \le 3)$.

c Suppose that of the eight randomly selected patients, three have had their symptoms significantly relieved by Viro. Based on the probability in part *b*, would you believe the claim of United Medicine, Inc.? Explain.

6.51 A consumer advocate claims that 80 percent of cable television subscribers are not satisfied with their cable service. In an attempt to justify this claim, a randomly selected sample of cable subscribers will be polled on this issue.

a Suppose that the advocate's claim is true, and suppose that a random sample of five cable subscribers is selected. Assuming independence, use an appropriate formula to compute the probability that four or more subscribers in the sample are not satisfied with their service.

b Suppose that the advocate's claim is true, and suppose that a random sample of 25 cable subscribers is selected. Assuming independence, use a computer to find

   (1) The probability that 15 or fewer subscribers in the sample are not satisfied with their service.

   (2) The probability that more than 20 subscribers in the sample are not satisfied with their service.

   (3) The probability that between 20 and 24 (inclusive) subscribers in the sample are not satisfied with their service.

   (4) The probability that exactly 24 subscribers in the sample are not satisfied with their service.

c Suppose that when we survey 25 randomly selected cable television subscribers, we find that 15 are actually not satisfied with their service. Using a probability you found in this exercise as the basis for your answer, do you believe the consumer advocate's claim? Explain.

6.52 A retail store has implemented procedures aimed at reducing the number of bad checks cashed by its cashiers. The store's goal is to cash no more than eight bad checks per week. The average number of bad checks cashed is three per week. Let x denote the number of bad checks cashed per week. Assuming that x has a Poisson distribution:

a Find the probability that the store's cashiers will not cash any bad checks in a particular week.

b Find the probability that the store will meet its goal during a particular week.

c Find the probability that the store will not meet its goal during a particular week.

d Find the probability that the store's cashiers will cash no more than 10 bad checks per two-week period.

e Find the probability that the store's cashiers will cash no more than five bad checks per three-week period.

6.53 Suppose that the number of accidents occurring in an industrial plant is described by a Poisson process with an average of 1.5 accidents every three months. During the last three months, four accidents occurred.

a Find the probability that no accidents will occur during the current three-month period.

b Find the probability that fewer accidents will occur during the current three-month period than occurred during the last three-month period.

c Find the probability that no more than 12 accidents will occur during a particular year.

d Find the probability that no accidents will occur during a particular year.

6.54 A high-security government installation has installed four security systems to detect attempted break-ins. The four security systems operate independently of each other, and each has a .85 probability of detecting an attempted break-in. Assume an attempted break-in occurs. Use the binomial distribution to find the probability that at least one of the four security systems will detect it.

6.55 A new stain removal product claims to completely remove the stains on 90 percent of all stained garments. Assume that the product will be tested on 20 randomly selected stained garments, and let x denote the number of these garments from which the stains will be completely removed. Use the binomial distribution to find $P(x \le 13)$ if the stain removal product's claim is correct. If x actually turns out to be 13, what do you think of the claim?

6.56 Consider Exercise 6.55, and find $P(x \leq 17)$ if the stain removal product's claim is correct. If x actually turns out to be 17, what do you think of the claim?

6.57 A state has averaged one small business failure per week over the past several years. Let x denote the number of small business failures in the next eight weeks. Use the Poisson distribution to find $P(x \geq 17)$ if the mean number of small business failures remains what it has been. If x actually turns out to be 17, what does this imply?

6.58 A candy company claims that its new chocolate almond bar averages 10 almonds per bar. Let x denote the number of almonds in the next bar that you buy. Use the Poisson distribution to find $P(x \leq 4)$ if the candy company's claim is correct. If x actually turns out to be 4, what do you think of the claim?

6.59 Consider Exercise 6.58, and find $P(x \leq 8)$ if the candy company's claim is true. If x actually turns out to be 8, what do you think of the claim?

Appendix 6.1 ■ Binomial, Poisson, and Hypergeometric Probabilities Using Excel

Binomial probabilities in Figure 6.5(a):

- Enter the title, "Binomial with n = 4 and p = 0.10," in the cell location where you wish to place the binomial results. We have placed the title beginning in cell A2 (any other choice will do).
- In cell A3, enter the heading, x.
- Enter the values 0 through 4 in cells A4 through A8.
- In cell B3, enter the heading P(X = x).
- Click in cell B4 (this is where the first binomial probability will be placed). Click on the Insert Function button f_x.
- In the Insert Function dialog box, select Statistical from the "Or select a category:" menu, select BINOM.DIST from the "Select a function:" menu, and click OK.

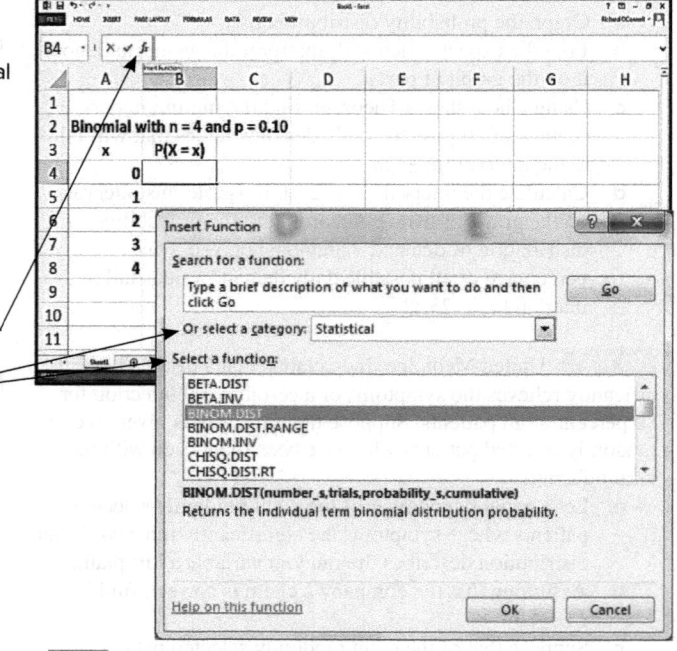

Source: Microsoft Office Excel 2016

- In the BINOM.DIST Function Arguments dialog box, enter the cell location A4 (this cell contains the value for which the first binomial probability will be calculated) in the "Number_s" window.
- Enter the value 4 in the Trials window.
- Enter the value 0.10 in the "Probability_s" window.
- Enter the value 0 in the Cumulative window.
- Click OK in the BINOM.DIST Function Arguments dialog box.

Source: Microsoft Office Excel 2016

- When you click OK, the calculated result (0.6561) will appear in cell B4. Double-click the drag handle (in the lower right corner) of cell B4 to automatically extend the cell formula to cells B5 through B8.
- The remaining probabilities will be placed in cells B5 through B8.

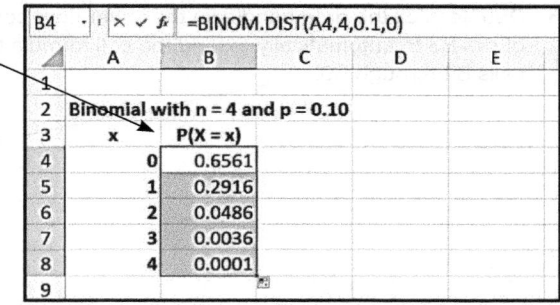

B4		× ✓ fx	=BINOM.DIST(A4,4,0.1,0)		
	A	B	C	D	E
1					
2	Binomial with n = 4 and p = 0.10				
3	x	P(X = x)			
4	0	0.6561			
5	1	0.2916			
6	2	0.0486			
7	3	0.0036			
8	4	0.0001			
9					

Source: Microsoft Office Excel 2016

To obtain **hypergeometric probabilities**:

Enter data as above, click the Insert Function button, and then select HYPGEOM.DIST from the "Select a function" menu. In the Function Arguments dialog box:

- Enter the location of the initial number of successes in the Sample_s window.
- Enter the size of the sample in the Number_sample window.
- Enter the number of successes in the population in the Population_s window.
- Enter the size of the population in the Number_pop window.

Then click OK and proceed as above to compute the probabilities.

Poisson probabilities similar to Figure 6.8(a):

- Enter the title "Poisson with mean = 0.40" in the cell location where you wish to place the Poisson results. Here we have placed the title beginning in cell A1 (any other choice will do).
- In cell A2, enter the heading, x.
- Enter the values 0 through 6 in cells A3 through A9.
- In cell B2, enter the heading, P(X = x).
- Click in cell B3 (this is where the first Poisson probability will be placed). Click on the Insert Function button f_x on the Excel toolbar.
- In the Insert Function dialog box, select Statistical from the "Or select a category" menu, select POISSON.DIST from the "Select a function:" menu, and click OK.

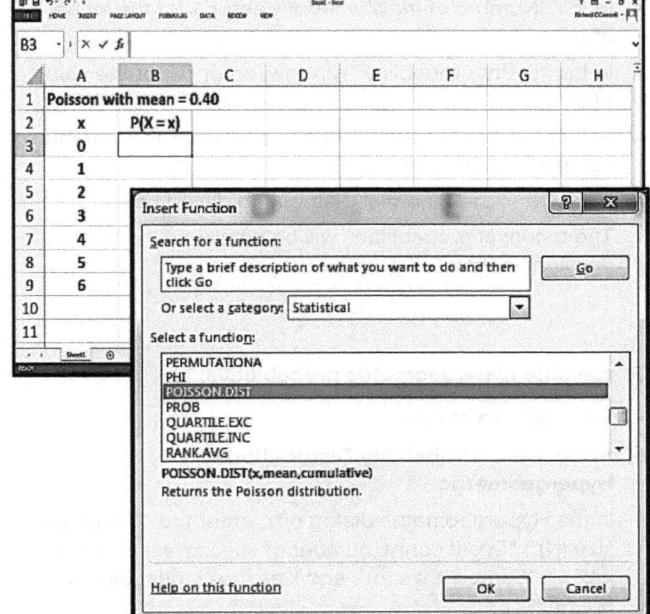

Source: Microsoft Office Excel 2016

- In the POISSON.DIST Function Arguments dialog box, enter the cell location A3 (this cell contains the value for which the first Poisson probability will be calculated) in the "X" window.
- Enter the value 0.40 in the Mean window.
- Enter the value 0 in the Cumulative window.
- Click OK in the POISSON.DIST Function Arguments dialog box.
- The calculated result for the probability of 0 events will appear in cell B3.

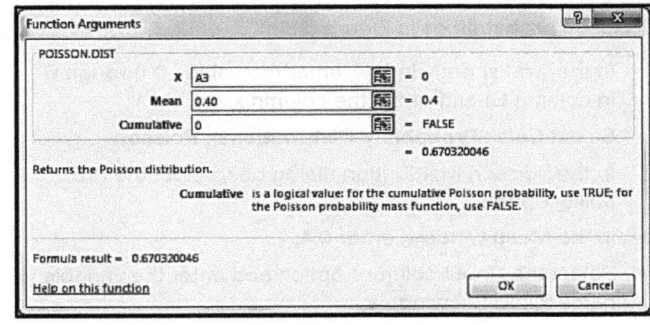

Source: Microsoft Office Excel 2016

- Double-click the drag handle (in the lower right corner) of cell B3 to automatically extend the cell formula to cells B4 through B9.

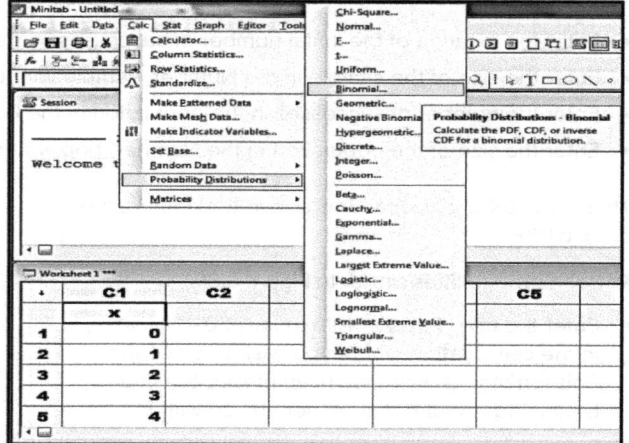

Source: Microsoft Office Excel 2016

Appendix 6.2 ■ Binomial, Poisson, and Hypergeometric Probabilities Using Minitab

Binomial probabilities similar to Figure 6.5(a):

- In the worksheet window, enter the values 0 through 4 in column C1 and name the column x.
- Select **Calc : Probability Distributions : Binomial**
- In the Binomial Distribution dialog box, select the Probability option.
- In the "Number of trials" window, enter 4 for the value of *n*.
- In the "Event probability" window, enter 0.1 for the value of *p*.
- Select the "Input column" option and enter the variable name x into the window.
- Click OK in the Binomial Distribution dialog box.
- The binomial probabilities will be displayed in the Session window.

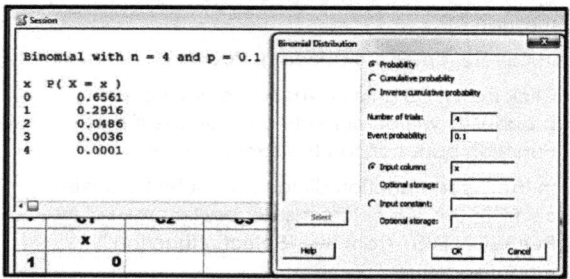

Source: Minitab 18

To compute **hypergeometric probabilities:**

- Enter data as above.
- Select **Calc : Probability Distributions : Hypergeometric**
- In the Hypergeometric dialog box, enter the "Population size (N)"; "Event count (number of successes) in population"; "Sample Size (n)"; and x as the "Input column" option.
- Click OK to obtain the probabilities in the Session window.

Poisson probabilities in Figure 6.8(a):

- In the worksheet window, enter the values 0 through 6 in column C1 and name the column x.
- Select **Calc : Probability Distributions : Poisson**
- In the Poisson Distribution dialog box, select the Probability option.
- In the Mean window, enter 0.4.
- Select the "Input column" option and enter the variable name x into the window.
- Click OK in the Poisson Distribution dialog box.
- The Poisson probabilities will be displayed in the Session window.

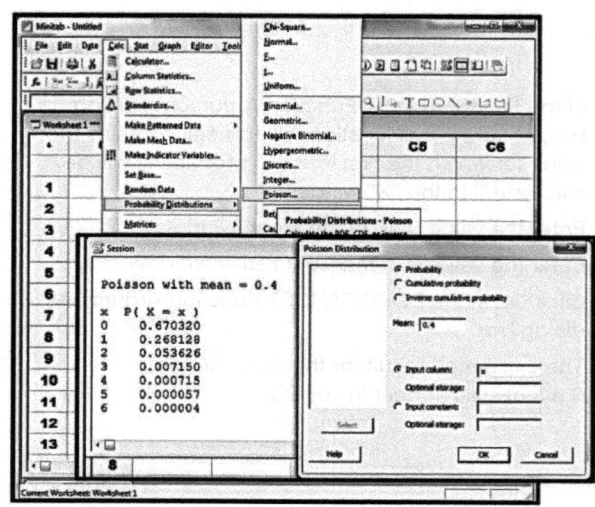

Source: Minitab 18

Appendix 6.3 ■ Binomial, Poisson, and Hypergeometric Probabilities Using JMP

Using the Data Table

- From the JMP main menu, select **File : New : Data Table**.

- Right click on the **Column 1** heading. Select "Column Info."

- In the Column 1 dialog box, change "Column Name" to "x." Click "OK."

- Enter the values of x by clicking on the first cell in the "x" column and entering the values sequentially.

- From the JMP main menu, select **Cols: New Column**.

- In the New Column dialog box, change "Column Name" to "P(X = x)."

- At the bottom left of the New Column dialog box, click on the "Column Properties" drop down menu. From this menu, select the first option, "Formula."

- Follow the directions below to calculate probabilities for the three distributions.

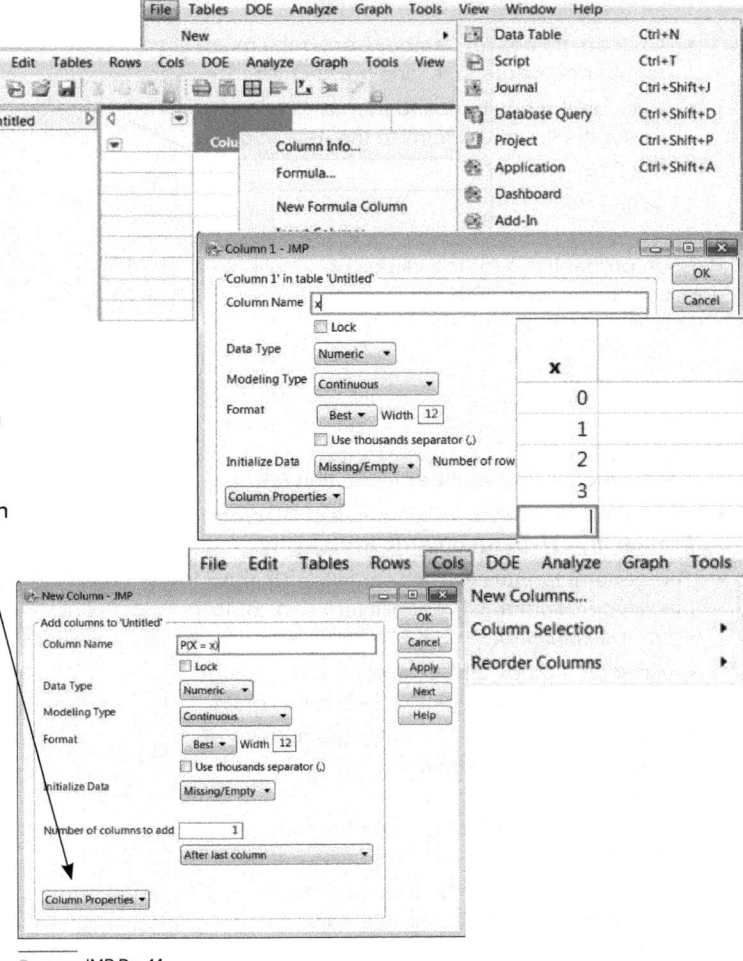

Source: JMP Pro 14

- **Binomial Probabilities** in Example 6.10:

  - On the left side of the window, select **Discrete Probability : Binomial Probability**.

  - Enter the values for "p," "n," and "k":

    - Click on the box for "p," enter the probability of success on each trial "0.95," and press Enter.

    - Click on the box for "n," enter the number of trials "8," and press Enter.

    - Click on the box for k and then click on "x" under the list of Columns in the middle of the page.

    - Click "OK."

  - In the New Column dialog box, click "OK." The binomial probabilities for the values of "x" will appear in the column "P(X = x)."

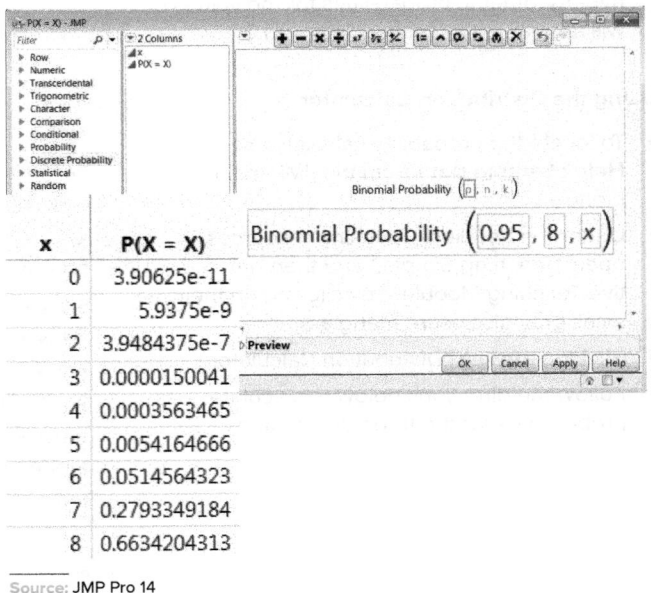

x	P(X = X)
0	3.90625e-11
1	5.9375e-9
2	3.9484375e-7
3	0.0000150041
4	0.0003563465
5	0.0054164666
6	0.0514564323
7	0.2793349184
8	0.6634204313

Source: JMP Pro 14

- **Poisson Probabilities** in Figure 6.8(b):
  - On the left side of the window, select **Discrete Probability : Poisson Probability**.
  - Enter the values for "lambda" and "k":
    - Click on the box for "lambda," enter the mean number of events "0.4," and press Enter.
    - Click on the box for k and then click on "x" under the list of Columns in the middle of the page.
    - Click "OK."
- In the New Column dialog box, click "OK." The Poisson probabilities for the values of "x" will appear in the column "P(X = x)."

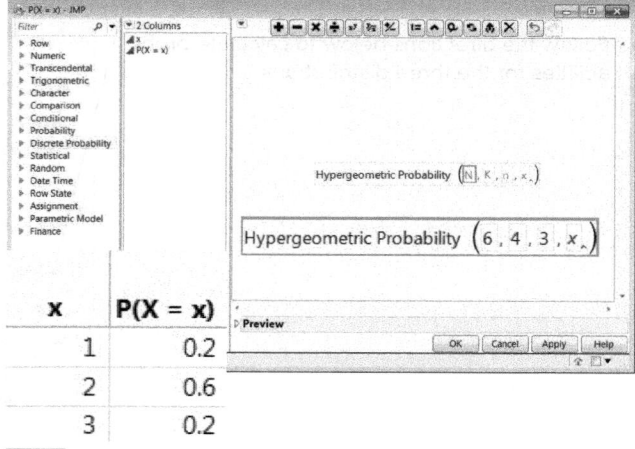

x	P(X = X)
0	0.670320046
1	0.2681280184
2	0.0536256037
3	0.0071500805
4	0.000715008
5	0.0000572006
6	3.8133763e-6

Source: JMP Pro 14

- **Hypergeometric Probabilities** in Section 6.5:
  - On the left side of the window, select **Discrete Probability : Hypergeometric Probability**.
  - The general formula will appear in the formula box: Hypergeometric Probability (N, K, n, x). To enter the values for "N," "K," "n," and "x":
    - Click on the box for "N," enter the number of items in the population "6," and press Enter.
    - Click on the box for "K," enter the number of successes in the population "4," and press Enter.
    - Click on the box for "n," enter the number items in the sample "3," and press Enter.
    - Click on the box for "x" and then click on "x" under the list of Columns in the middle of the page.
    - Click "OK."
- In the New Column dialog box, click "OK." The hypergeometric probabilities for the values of "x" will appear in the column "P(X = x)."

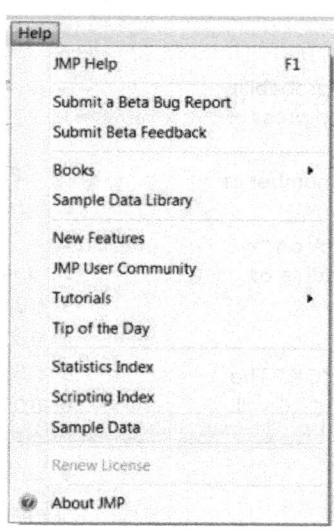

x	P(X = x)
1	0.2
2	0.6
3	0.2

Source: JMP Pro 14

Using the Distribution Calculator

- To locate the probability calculator, select **Help : Sample Data** from the JMP main menu.
- Under the heading "Teaching resources," open "Teaching Scripts," and then "Interactive Teaching Modules" by clicking on their small gray disclosure triangles.
- Click on the link "Distribution Calculator."
- Follow the directions below to calculate probabilities for the three distributions.

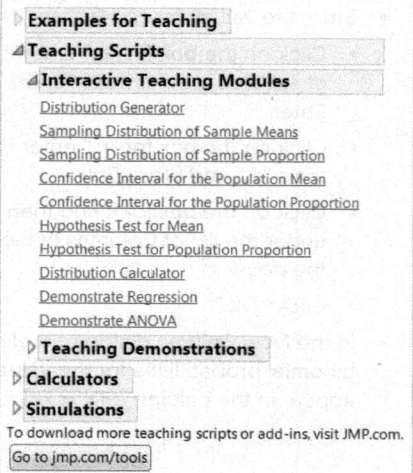

Source: JMP Pro 14

- **Binomial Probabilities** in Example 6.9:
  - Under "Distribution Characteristics":
    - Select "Binomial" from "Distribution" menu.
    - Enter "0.1" for "P(Success)."
    - Enter "4" for "N."
  - To calculate P(X ≥ 3), under "Calculations":
    - Select "P(X > q)."
    - Enter "2" in the "Value:" box.
  - The probability = 0.0037 is given.

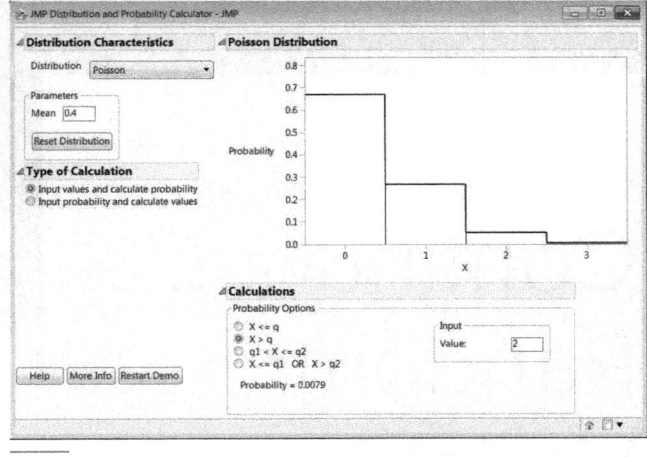

Source: JMP Pro 14

- **Poisson Probabilities** in Example 6.11:
  - Under "Distribution Characteristics":
    - Select "Poisson" from "Distribution" menu.
    - Enter "0.4" for "Mean."
  - To calculate P(X ≥ 3), under "Calculations":
    - Select "P(X > q)."
    - Enter "2" in the "Value:" box.
  - The probability = 0.0079 is given. The difference between this value and 0.008 is due to differences in rounding precision.

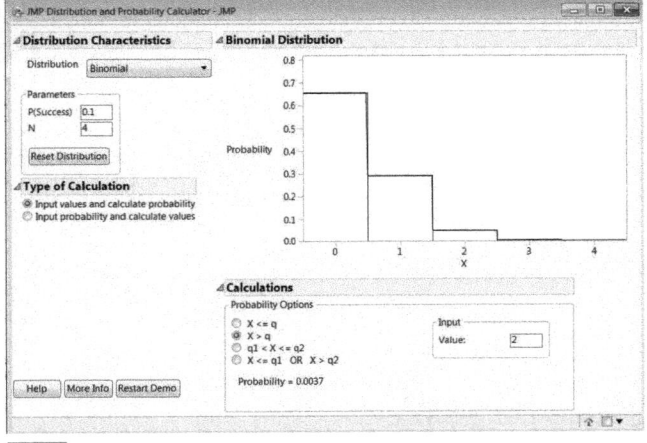

Source: JMP Pro 14

- **Hypergeometric Probabilities** in Section 6.5:
  - Under "Distribution Characteristics":
    - Select "Hypergeometric" from "Distribution" menu.
    - Enter "3" for "Sample Size."
    - Enter "4" for "Number of Items."
    - Enter "6" for "Population Size."
  - To calculate P(X ≥ 2), under "Calculations":
    - Select "P(X > q)."
    - Enter "1" in the "Value:" box.
  - The probability = 0.8 is given.

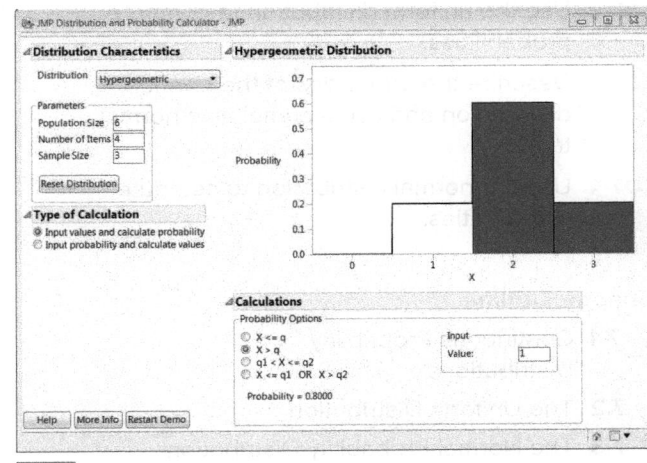

Source: JMP Pro 14

Design Elements: (CD): ©Comstock Images/Alamy; (All Others): ©McGraw-Hill Education

CHAPTER 7

Continuous Random Variables

©MarioGuti/Getty Images

Learning Objectives

After mastering the material in this chapter, you will be able to:

LO7-1 Define a continuous probability distribution and explain how it is used.

LO7-2 Use the uniform distribution to compute probabilities.

LO7-3 Describe the properties of the normal distribution and use a cumulative normal table.

LO7-4 Use the normal distribution to compute probabilities.

LO7-5 Find population values that correspond to specified normal distribution probabilities.

LO7-6 Use the normal distribution to approximate binomial probabilities (Optional).

LO7-7 Use the exponential distribution to compute probabilities (Optional).

LO7-8 Use a normal probability plot to help decide whether data come from a normal distribution (Optional).

Chapter Outline

7.1 Continuous Probability Distributions

7.2 The Uniform Distribution

7.3 The Normal Probability Distribution

7.4 Approximating the Binomial Distribution by Using the Normal Distribution (Optional)

7.5 The Exponential Distribution (Optional)

7.6 The Normal Probability Plot (Optional)

n Chapter 6 we defined discrete and continuous random variables. We also discussed discrete probability distributions, which are used to compute the probabilities of values of discrete random variables. In this chapter we discuss **continuous probability distributions.** These are used to find probabilities concerning continuous random variables. We begin by explaining the general idea behind a

continuous probability distribution. Then we present three important continuous distributions—the **uniform, normal,** and **exponential distributions.** We also study when and how the normal distribution can be used to approximate the binomial distribution (which was discussed in Chapter 6).

We will illustrate the concepts in this chapter by using two cases:

The Car Mileage Case: A competitor claims that its midsize car gets better mileage than an automaker's new midsize model. The automaker uses sample information and a probability based on the normal distribution to provide strong evidence that the competitor's claim is false.

The Coffee Temperature Case: A fast-food restaurant uses the normal distribution to estimate the proportion of coffee it serves that has a temperature (in degrees Fahrenheit) inside the range 153° to 167°, the customer requirement for best-tasting coffee.

7.1 Continuous Probability Distributions

LO7-1
Define a continuous probability distribution and explain how it is used.

We have said in Section 6.1 that when a random variable may assume any numerical value in one or more intervals on the real number line, then the random variable is called a **continuous random variable.** For example, as discussed in Section 6.1, the EPA combined city and highway mileage of a randomly selected midsize car is a continuous random variable. Furthermore, the temperature (in degrees Fahrenheit) of a randomly selected cup of coffee at a fast-food restaurant is also a continuous random variable. We often wish to compute probabilities about the range of values that a continuous random variable x might attain. For example, suppose that marketing research done by a fast-food restaurant indicates that coffee tastes best if its temperature is between 153° F and 167° F. The restaurant might then wish to find the probability that x, the temperature of a randomly selected cup of coffee at the restaurant, will be between 153° and 167°. This probability would represent the proportion of coffee served by the restaurant that has a temperature between 153° and 167°. Moreover, one minus this probability would represent the proportion of coffee served by the restaurant that has a temperature outside the range 153° to 167°.

In general, to compute probabilities concerning a continuous random variable x, we model x by using a **continuous probability distribution,** which assigns probabilities to **intervals of values.** To understand this idea, suppose that $f(x)$ is a continuous function of the numbers on the real line, and consider the continuous curve that results when $f(x)$ is graphed. Such a curve is illustrated in the figure in the page margin. Then:

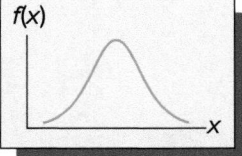

Continuous Probability Distributions

The curve $f(x)$ is the **continuous probability distribution** of the random variable x if the probability that x will be in a specified interval of numbers is the area under the curve $f(x)$ corresponding to the interval. Sometimes we refer to a continuous probability distribution as a **probability curve** or as a **probability density function.**

In this chapter we will study three continuous probability distributions—the *uniform, normal,* and *exponential* distributions. As an example of using a continuous probability distribution to describe a random variable, suppose that the fast-food restaurant will study the temperature of the coffee being dispensed at one of its locations. A temperature measurement is taken at a randomly selected time during each of the 24 half-hour periods from

FIGURE 7.1   The Coffee Temperature Data, Histogram, and Normal Curve ⦿ Coffee

(a) The Coffee Temperature Data (Time Order Is Given by Reading Across and Then Down.)

154° F	165	148	157	160	157	152	149	171	168	165	164	156	151	161	157
154	159	155	153	173	164	161	151	158	160	153	161	160	158	169	163
146	167	162	159	166	158	173	162	155	150	165	154	160	162	159	166

(b) The Histogram

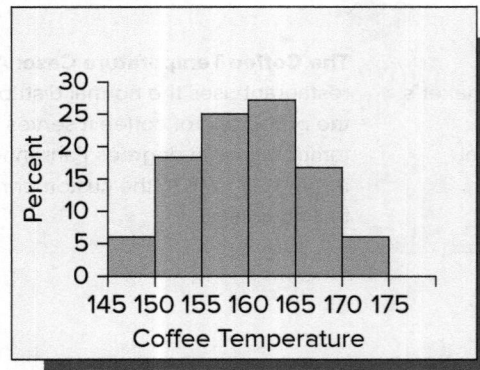

(c) The Normal Curve

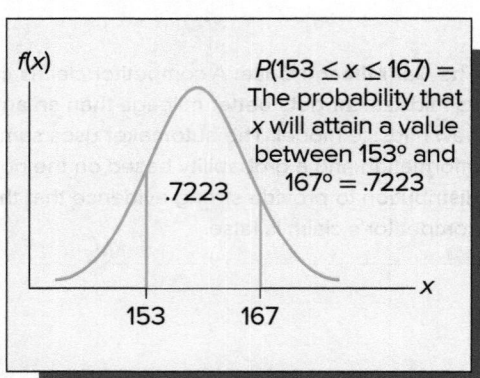

8 A.M. to 7:30 P.M. on a given day. This is then repeated on a second day, giving the 48 coffee temperatures in Figure 7.1(a). Figure 7.1(b) shows a percent frequency histogram of the coffee temperatures. If we were to smooth out the histogram with a continuous curve, we would get a curve similar to the symmetrical and bell-shaped curve in Figure 7.1(c). One continuous probability distribution that graphs as a symmetrical and bell-shaped curve is the *normal probability distribution* (or *normal curve*). Because the coffee temperature histogram looks like a normal curve, it is reasonable to conclude that x, the temperature of a randomly selected cup of coffee at the fast-food restaurant, is described by a normal probability distribution. It follows that the probability that x will be between 153° and 167° is the area under the coffee temperature normal curve between 153 and 167. In Section 7.3, where we discuss the normal curve in detail, we will find that this area is .7223. That is, in probability notation: $P(153 \leq x \leq 167) = .7223$ (see the blue area in Figure 7.1(c)). In conclusion, we estimate that 72.23 percent of the coffee served at the restaurant is within the range of temperatures that is best and 27.77 percent of the coffee served is not in this range. If management wishes a very high percentage of the coffee served to taste best, it must improve the coffee-making process by better controlling temperatures.

We now present some general properties of a continuous probability distribution. We know that any probability is 0 or positive, and we also know that the probability assigned to all possible values of x must be 1. It follows that, similar to the conditions required for a discrete probability distribution, a probability curve must satisfy the following:

Properties of a Continuous Probability Distribution

The **continuous probability distribution** (or **probability curve**) $f(x)$ of a random variable x must satisfy the following two conditions:

1   $f(x) \geq 0$ for any value of x.

2   The total area under the curve $f(x)$ is equal to 1.

We have seen that to calculate a probability concerning a continuous random variable, we must compute an appropriate area under the curve $f(x)$. Because there is no area under a continuous curve at a single point, or number, on the real line, the probability that a continuous random variable x will equal a single numerical value is always equal to 0. It follows that if $[a, b]$ denotes an arbitrary interval of numbers on the real line, then $P(x = a) = 0$ and $P(x = b) = 0$. Therefore, $P(a \leq x \leq b)$ equals $P(a < x < b)$ because each of the interval endpoints a and b has a probability that is equal to 0.

7.2 The Uniform Distribution

LO7-2
Use the uniform distribution to compute probabilities.

Suppose that over a period of several days the manager of a large hotel has recorded the waiting times of 1,000 people waiting for an elevator in the lobby at dinnertime (5:00 P.M. to 7:00 P.M.). The observed waiting times range from zero to four minutes. Furthermore, when the waiting times are arranged into a histogram, the bars making up the histogram have approximately equal heights, giving the histogram a rectangular appearance. This implies that the relative frequencies of all waiting times from zero to four minutes are about the same. Therefore, it is reasonable to use the continuous *uniform distribution* to describe the random variable x, the amount of time a randomly selected hotel patron spends waiting for an elevator. The equation describing the uniform distribution in this situation is

$$f(x) = \begin{cases} \dfrac{1}{4} & \text{for } 0 \leq x \leq 4 \\ 0 & \text{otherwise} \end{cases}$$

Noting that this equation is graphed in Figure 7.2(a), suppose that the hotel manager feels that an elevator waiting time of 2.5 minutes or more is unacceptably long. Therefore, to find the probability that a hotel patron will wait too long, the manager wishes to find the probability that a randomly selected patron will spend at least 2.5 minutes waiting for an elevator. This probability is the area under the curve $f(x)$ that corresponds to the interval $[2.5, 4]$. As shown in Figure 7.2(a), this probability is the area of a rectangle having a base equal to $4 - 2.5 = 1.5$ and a height equal to $1/4$. That is,

$$P(x \geq 2.5) = P(2.5 \leq x \leq 4) = \text{base} \times \text{height} = 1.5 \times \frac{1}{4} = .375$$

This says that 37.5 percent of all hotel patrons will spend at least 2.5 minutes waiting for an elevator at dinnertime. Based on this result, the hotel manager would probably decide

FIGURE 7.2   The Uniform Distribution

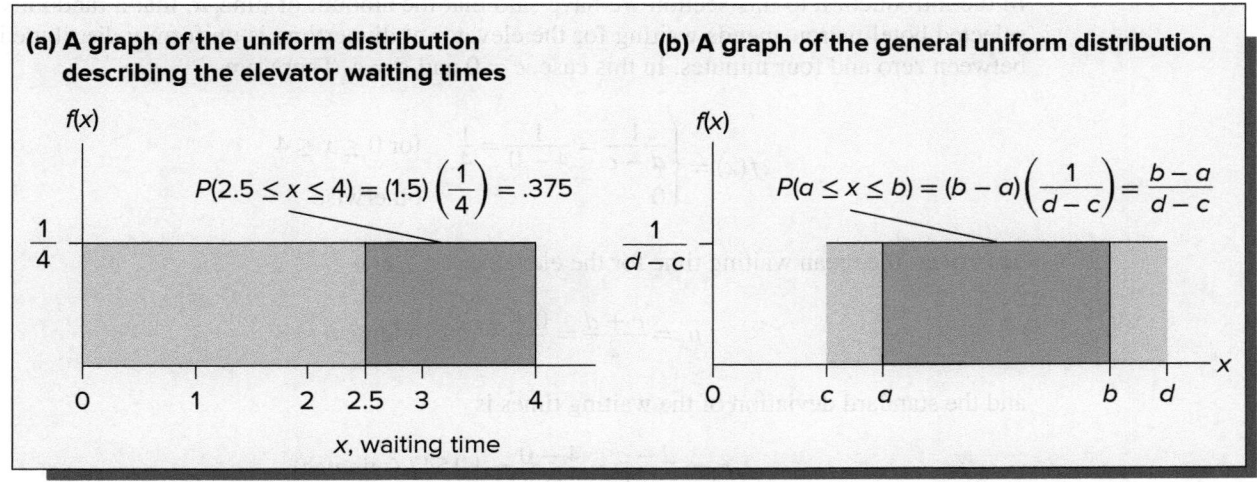

that too many patrons are waiting too long for an elevator and that action should be taken to reduce elevator waiting times.

In general, the equation that describes the **uniform distribution** (or **uniform model**) is given in the following box and is graphed in Figure 7.2(b).

The Uniform Distribution (The Uniform Model)

If c and d are numbers on the real line, the equation describing the **uniform distribution** is

$$f(x) = \begin{cases} \dfrac{1}{d-c} & \text{for } c \leq x \leq d \\ 0 & \text{otherwise} \end{cases}$$

Furthermore, the mean and the standard deviation of the population of all possible observed values of a random variable x that has a uniform distribution are

$$\mu_x = \frac{c+d}{2} \quad \text{and} \quad \sigma_x = \frac{d-c}{\sqrt{12}}$$

Notice that the total area under the uniform distribution is the area of a rectangle having a base equal to $(d-c)$ and a height equal to $1/(d-c)$. Therefore, the probability curve's total area is

$$\text{base} \times \text{height} = (d-c)\left(\frac{1}{d-c}\right) = 1$$

(remember that the total area under any continuous probability curve must equal 1). Furthermore, if a and b are numbers that are as illustrated in Figure 7.2(b), then the probability that x will be between a and b is the area of a rectangle with base $(b-a)$ and height $1/(d-c)$. That is,

$$P(a \leq x \leq b) = \text{base} \times \text{height} = (b-a)\left(\frac{1}{d-c}\right) = \frac{b-a}{d-c}$$

Ⓒ EXAMPLE 7.1 Elevator Waiting Times

In the introduction to this section we have said that the amount of time, x, that a randomly selected hotel patron spends waiting for the elevator at dinnertime is uniformly distributed between zero and four minutes. In this case, $c = 0$ and $d = 4$. Therefore,

$$f(x) = \begin{cases} \dfrac{1}{d-c} = \dfrac{1}{4-0} = \dfrac{1}{4} & \text{for } 0 \leq x \leq 4 \\ 0 & \text{otherwise} \end{cases}$$

Moreover, the mean waiting time for the elevator is

$$\mu_x = \frac{c+d}{2} = \frac{0+4}{2} = 2 \text{ (minutes)}$$

and the standard deviation of the waiting times is

$$\sigma_x = \frac{d-c}{\sqrt{12}} = \frac{4-0}{\sqrt{12}} = 1.1547 \text{ (minutes)}$$

Therefore, noting that $\mu_x - \sigma_x = 2 - 1.1547 = .8453$ and that $\mu_x + \sigma_x = 2 + 1.1547 = 3.1547$, the probability that the waiting time of a randomly selected patron will be within (plus or minus) one standard deviation of the mean waiting time is

$$P(.8453 \leq x \leq 3.1547) = (3.1547 - .8453) \times \frac{1}{4}$$

$$= .57735$$

Exercises for Sections 7.1 and 7.2

CONCEPTS connect

7.1 A discrete probability distribution assigns probabilities to individual values. To what are probabilities assigned by a continuous probability distribution?

7.2 How do we use the continuous probability distribution (or probability curve) of a random variable x to find probabilities? Explain.

7.3 What two properties must be satisfied by a continuous probability distribution (or probability curve)?

7.4 Is the height of a probability curve over a given point a probability? Explain.

7.5 When is it appropriate to use the uniform distribution to describe a random variable x?

METHODS AND APPLICATIONS

7.6 Suppose that the random variable x has a uniform distribution with $c = 2$ and $d = 8$.
  a Write the formula for the probability curve of x, and write an interval that gives the possible values of x.
  b Graph the probability curve of x.
  c Find $P(3 \leq x \leq 5)$.
  d Find $P(1.5 \leq x \leq 6.5)$.
  e Calculate the mean μ_x, variance σ_x^2, and standard deviation σ_x.
  f Calculate the interval $[\mu_x \pm 2\sigma_x]$. What is the probability that x will be in this interval?

7.7 Consider the figure given below. Find the value h that makes the function $f(x)$ a valid continuous probability distribution.

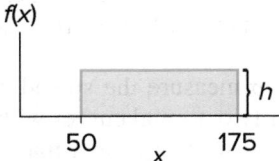

7.8 Assume that the waiting time x for an elevator is uniformly distributed between zero and six minutes.
  a Write the formula for the probability curve of x.
  b Graph the probability curve of x.
  c Find $P(2 \leq x \leq 4)$.
  d Find $P(3 \leq x \leq 6)$.
  e Find $P(\{0 \leq x \leq 2\}$ or $\{5 \leq x \leq 6\})$.

7.9 Refer to Exercise 7.8.
  a Calculate the mean, μ_x, the variance, σ_x^2, and the standard deviation, σ_x.
  b Find the probability that the waiting time of a randomly selected patron will be within one standard deviation of the mean.

7.10 Consider the figure given below. Find the value k that makes the function $f(x)$ a valid continuous probability distribution.

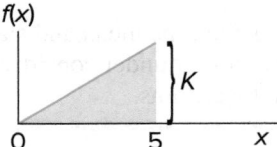

7.11 Suppose that an airline quotes a flight time of 2 hours, 10 minutes between two cities. Furthermore, suppose that historical flight records indicate that the actual flight time between the two cities, x, is uniformly distributed between 2 hours and 2 hours, 20 minutes. Letting the time unit be one minute,
  a Write the formula for the probability curve of x.
  b Graph the probability curve of x.
  c Find $P(125 \leq x \leq 135)$.
  d Find the probability that a randomly selected flight between the two cities will be at least five minutes late.

7.12 Refer to Exercise 7.11.
  a Calculate the mean flight time and the standard deviation of the flight time.
  b Find the probability that the flight time will be within one standard deviation of the mean.

7.13 Consider the figure given below. Find the value c that makes the function $f(x)$ a valid continuous probability distribution.

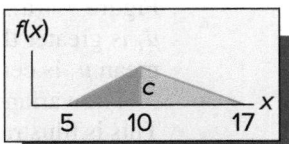

7.14 A weather forecaster predicts that the May rainfall in a local area will be between three and six inches but has no idea where within the interval the amount will be. Let x be the amount of May rainfall in the local area, and assume that x is uniformly distributed over the interval three to six inches.
  a Write the formula for the probability curve of x and graph this probability curve.
  b What is the probability that May rainfall will be at least four inches? At least five inches?

7.15 Refer to Exercise 7.14 and find the probability that the observed May rainfall will fall within two standard deviations of the mean May rainfall.

LO7-3
Describe the properties
of the normal distribution
and use a cumulative
normal table.

7.3 The Normal Probability Distribution

The normal curve

The bell-shaped appearance of the **normal probability distribution** (or **normal model**) is illustrated in Figure 7.3. The equation that defines this normal curve is given in the following box:

The Normal Probability Distribution (The Normal Model)

The **normal probability distribution** is defined by the equation

$$f(x) = \frac{1}{\sigma\sqrt{2\pi}}\, e^{-\frac{1}{2}\left(\frac{x-\mu}{\sigma}\right)^2} \quad \text{for all values of } x \text{ on the real line}$$

Here μ and σ are the mean and standard deviation of the population of all possible observed values of the random variable x under consideration. Furthermore, $\pi = 3.14159\ldots$, and $e = 2.71828\ldots$ is the base of Napierian logarithms.

FIGURE 7.3

The Normal

Probability Curve

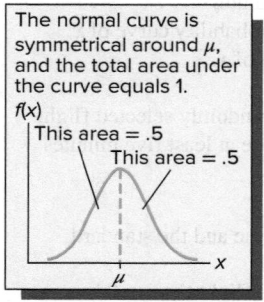

The normal curve is symmetrical around μ, and the total area under the curve equals 1.

$f(x)$

This area = .5

This area = .5

Although this equation looks very intimidating, we will not use it to find areas (and thus probabilities) under the normal curve. Instead, we will use a *normal curve table*. What is important to know for now is that the normal probability distribution has several important properties:

1   There is an entire family of normal probability distributions; the specific shape of each normal distribution is determined by its mean μ and its standard deviation σ.

2   The highest point on the normal curve is located at the mean, which is also the median and the mode of the distribution.

3   The normal distribution is symmetrical: The curve's shape to the left of the mean is the mirror image of its shape to the right of the mean.

4   The tails of the normal curve extend to infinity in both directions and never touch the horizontal axis. However, the tails get close enough to the horizontal axis quickly enough to ensure that the total area under the normal curve equals 1.

5   Because the normal curve is symmetrical, the area under the normal curve to the right of the mean (μ) equals the area under the normal curve to the left of the mean, and each of these areas equals .5 (see Figure 7.3).

Intuitively, the mean μ positions the normal curve on the real line. This is illustrated in Figure 7.4(a). This figure shows two normal curves with different means μ_1 and μ_2 (where μ_1 is greater than μ_2) and with equal standard deviations. We see that the normal curve with mean μ_1 is centered farther to the right.

The variance σ^2 and the standard deviation σ measure the spread of the normal curve. This is illustrated in Figure 7.4(b), which shows two normal curves with the same mean and two different standard deviations σ_1 and σ_2. Because σ_1 is greater than σ_2, the normal curve with standard deviation σ_1 is more spread out (flatter) than the normal curve with standard deviation σ_2. In general, larger standard deviations result in normal curves that are flatter and more spread out, while smaller standard deviations result in normal curves that have higher peaks and are less spread out.

Suppose that a random variable x is described by a normal probability distribution (or is, as we say, **normally distributed**) with mean μ and standard deviation σ. If a and b are numbers on the real line, we consider the probability that x will be between a and b. That is, we consider

$$P(a \leq x \leq b)$$

which equals the area under the normal curve with mean μ and standard deviation σ corresponding to the interval $[a, b]$. Such an area is depicted in Figure 7.5. We soon explain how to find such areas using a statistical table called a **normal table.** For now, we emphasize three important areas under a normal curve. These areas form the basis for the **Empirical Rule**

FIGURE 7.4   How the Mean μ and Standard Deviation σ Affect the Position and Shape of a Normal Probability Curve

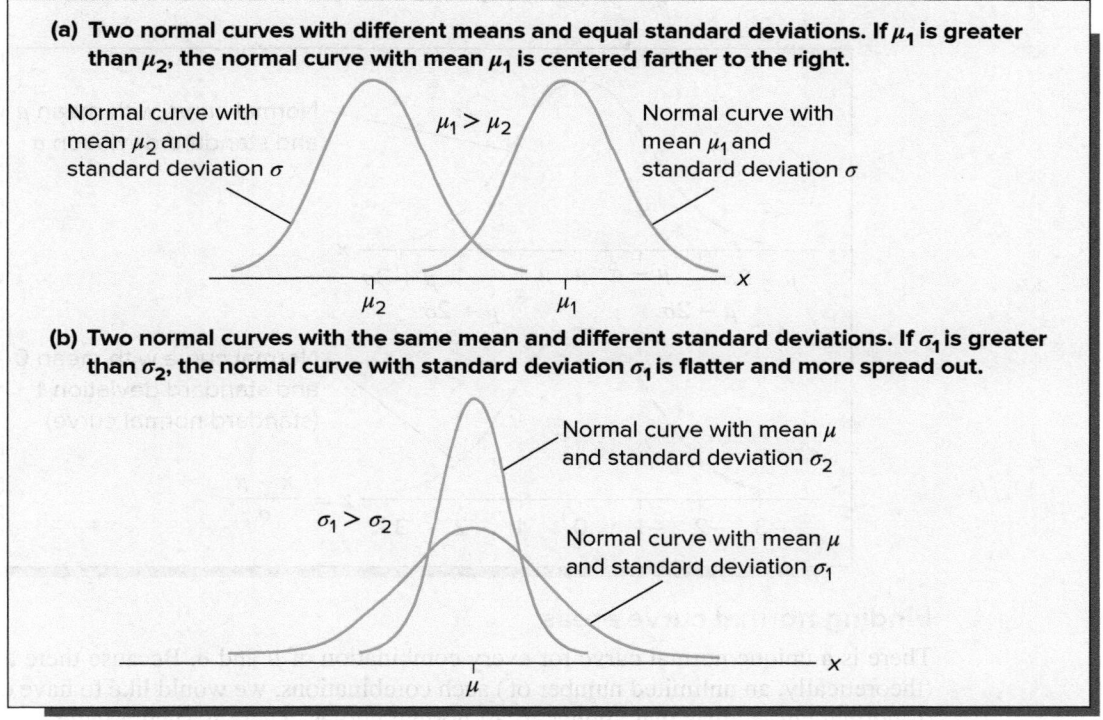

(a) Two normal curves with different means and equal standard deviations. If μ_1 is greater than μ_2, the normal curve with mean μ_1 is centered farther to the right.

Normal curve with mean μ_2 and standard deviation σ

$\mu_1 > \mu_2$

Normal curve with mean μ_1 and standard deviation σ

(b) Two normal curves with the same mean and different standard deviations. If σ_1 is greater than σ_2, the normal curve with standard deviation σ_1 is flatter and more spread out.

Normal curve with mean μ and standard deviation σ_2

$\sigma_1 > \sigma_2$

Normal curve with mean μ and standard deviation σ_1

FIGURE 7.5   An Area under a Normal Curve Corresponding to the Interval $[a, b]$

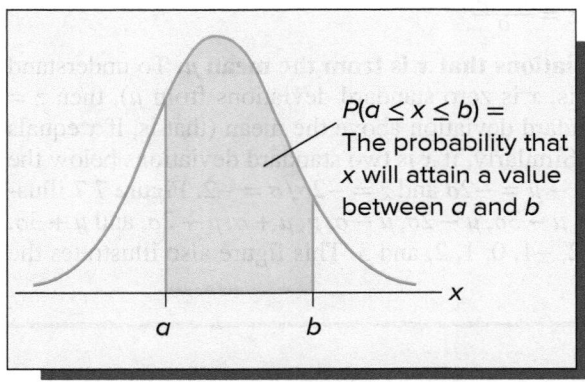

$P(a \leq x \leq b) =$ The probability that x will attain a value between a and b

FIGURE 7.6   Three Important Percentages Concerning a Normally Distributed Random Variable x with Mean μ and Standard Deviation σ

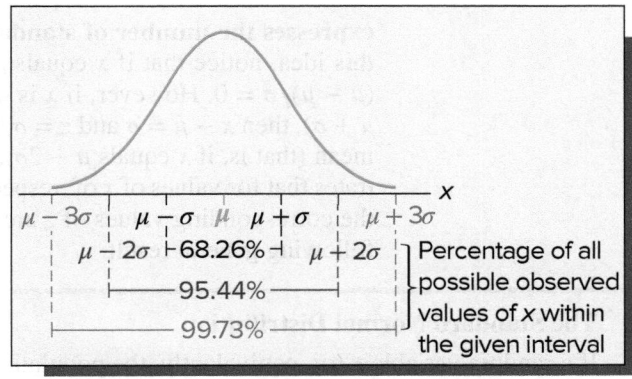

$\mu - 3\sigma$ $\mu - \sigma$ μ $\mu + \sigma$ $\mu + 3\sigma$

$\mu - 2\sigma$ ⊢68.26%⊣ $\mu + 2\sigma$ Percentage of all possible observed values of x within the given interval

95.44%

99.73%

for a normally distributed population. Specifically, if x is normally distributed with mean μ and standard deviation σ, it can be shown (using a normal table) that, as illustrated in Figure 7.6:

Three Important Areas under the Normal Curve

1 $P(\mu - \sigma \leq x \leq \mu + \sigma) = .6826$

This means that 68.26 percent of all possible observed values of x are within (plus or minus) one standard deviation of μ.

2 $P(\mu - 2\sigma \leq x \leq \mu + 2\sigma) = .9544$

This means that 95.44 percent of all possible observed

values of x are within (plus or minus) two standard deviations of μ.

3 $P(\mu - 3\sigma \leq x \leq \mu + 3\sigma) = .9973$

This means that 99.73 percent of all possible observed values of x are within (plus or minus) three standard deviations of μ.

FIGURE 7.7   If x Is Normally Distributed with Mean μ and Standard Deviation σ, Then
$z = \dfrac{x - \mu}{\sigma}$ Is Normally Distributed with Mean 0 and Standard Deviation 1

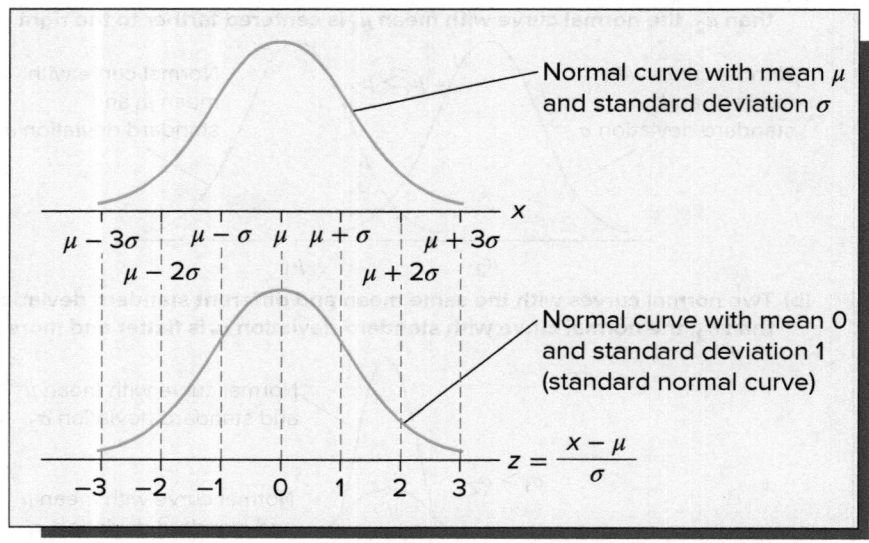

Finding normal curve areas

There is a unique normal curve for every combination of μ and σ. Because there are many (theoretically, an unlimited number of) such combinations, we would like to have one table of normal curve areas that applies to all normal curves. There is such a table, and we can use it by thinking in terms of how many standard deviations a value of interest is from the mean. Specifically, consider a random variable x that is normally distributed with mean μ and standard deviation σ. Then the random variable

$$z = \frac{x - \mu}{\sigma}$$

expresses the number of standard deviations that x is from the mean μ. To understand this idea, notice that if x equals μ (that is, x is zero standard deviations from μ), then $z = (\mu - \mu)/\sigma = 0$. However, if x is one standard deviation above the mean (that is, if x equals $\mu + \sigma$), then $x - \mu = \sigma$ and $z = \sigma/\sigma = 1$. Similarly, if x is two standard deviations below the mean (that is, if x equals $\mu - 2\sigma$), then $x - \mu = -2\sigma$ and $z = -2\sigma/\sigma = -2$. Figure 7.7 illustrates that for values of x of, respectively, $\mu - 3\sigma, \mu - 2\sigma, \mu - \sigma, \mu, \mu + \sigma, \mu + 2\sigma$, and $\mu + 3\sigma$, the corresponding values of z are $-3, -2, -1, 0, 1, 2$, and 3. This figure also illustrates the following general result:

The Standard Normal Distribution

If a random variable x (or, equivalently, the population of all possible observed values of x) is normally distributed with mean μ and standard deviation σ, then the random variable

$$z = \frac{x - \mu}{\sigma}$$

(or, equivalently, the population of all possible observed values of z) is normally distributed with mean 0 and standard deviation 1. A normal distribution (or curve) with mean 0 and standard deviation 1 is called a **standard normal distribution** (or **curve**).

Table A.3 is a table of *cumulative* areas under the standard normal curve. This table is called a *cumulative normal table*, and it is reproduced as Table 7.1. Specifically,

The **cumulative normal table** gives, for many different values of z, the area under the standard normal curve to the left of z.

TABLE 7.1   **Cumulative Areas under the Standard Normal Curve**

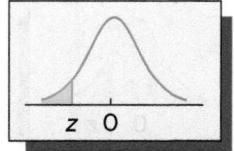

z	0.00	0.01	0.02	0.03	0.04	0.05	0.06	0.07	0.08	0.09
−3.9	0.00005	0.00005	0.00004	0.00004	0.00004	0.00004	0.00004	0.00004	0.00003	0.00003
−3.8	0.00007	0.00007	0.00007	0.00006	0.00006	0.00006	0.00006	0.00005	0.00005	0.00005
−3.7	0.00011	0.00010	0.00010	0.00010	0.00009	0.00009	0.00008	0.00008	0.00008	0.00008
−3.6	0.00016	0.00015	0.00015	0.00014	0.00014	0.00013	0.00013	0.00012	0.00012	0.00011
−3.5	0.00023	0.00022	0.00022	0.00021	0.00020	0.00019	0.00019	0.00018	0.00017	0.00017
−3.4	0.00034	0.00032	0.00031	0.00030	0.00029	0.00028	0.00027	0.00026	0.00025	0.00024
−3.3	0.00048	0.00047	0.00045	0.00043	0.00042	0.00040	0.00039	0.00038	0.00036	0.00035
−3.2	0.00069	0.00066	0.00064	0.00062	0.00060	0.00058	0.00056	0.00054	0.00052	0.00050
−3.1	0.00097	0.00094	0.00090	0.00087	0.00084	0.00082	0.00079	0.00076	0.00074	0.00071
−3.0	0.00135	0.00131	0.00126	0.00122	0.00118	0.00114	0.00111	0.00107	0.00103	0.00100
−2.9	0.0019	0.0018	0.0018	0.0017	0.0016	0.0016	0.0015	0.0015	0.0014	0.0014
−2.8	0.0026	0.0025	0.0024	0.0023	0.0023	0.0022	0.0021	0.0021	0.0020	0.0019
−2.7	0.0035	0.0034	0.0033	0.0032	0.0031	0.0030	0.0029	0.0028	0.0027	0.0026
−2.6	0.0047	0.0045	0.0044	0.0043	0.0041	0.0040	0.0039	0.0038	0.0037	0.0036
−2.5	0.0062	0.0060	0.0059	0.0057	0.0055	0.0054	0.0052	0.0051	0.0049	0.0048
−2.4	0.0082	0.0080	0.0078	0.0075	0.0073	0.0071	0.0069	0.0068	0.0066	0.0064
−2.3	0.0107	0.0104	0.0102	0.0099	0.0096	0.0094	0.0091	0.0089	0.0087	0.0084
−2.2	0.0139	0.0136	0.0132	0.0129	0.0125	0.0122	0.0119	0.0116	0.0113	0.0110
−2.1	0.0179	0.0174	0.0170	0.0166	0.0162	0.0158	0.0154	0.0150	0.0146	0.0143
−2.0	0.0228	0.0222	0.0217	0.0212	0.0207	0.0202	0.0197	0.0192	0.0188	0.0183
−1.9	0.0287	0.0281	0.0274	0.0268	0.0262	0.0256	0.0250	0.0244	0.0239	0.0233
−1.8	0.0359	0.0351	0.0344	0.0336	0.0329	0.0322	0.0314	0.0307	0.0301	0.0294
−1.7	0.0446	0.0436	0.0427	0.0418	0.0409	0.0401	0.0392	0.0384	0.0375	0.0367
−1.6	0.0548	0.0537	0.0526	0.0516	0.0505	0.0495	0.0485	0.0475	0.0465	0.0455
−1.5	0.0668	0.0655	0.0643	0.0630	0.0618	0.0606	0.0594	0.0582	0.0571	0.0559
−1.4	0.0808	0.0793	0.0778	0.0764	0.0749	0.0735	0.0721	0.0708	0.0694	0.0681
−1.3	0.0968	0.0951	0.0934	0.0918	0.0901	0.0885	0.0869	0.0853	0.0838	0.0823
−1.2	0.1151	0.1131	0.1112	0.1093	0.1075	0.1056	0.1038	0.1020	0.1003	0.0985
−1.1	0.1357	0.1335	0.1314	0.1292	0.1271	0.1251	0.1230	0.1210	0.1190	0.1170
−1.0	0.1587	0.1562	0.1539	0.1515	0.1492	0.1469	0.1446	0.1423	0.1401	0.1379
−0.9	0.1841	0.1814	0.1788	0.1762	0.1736	0.1711	0.1685	0.1660	0.1635	0.1611
−0.8	0.2119	0.2090	0.2061	0.2033	0.2005	0.1977	0.1949	0.1922	0.1894	0.1867
−0.7	0.2420	0.2389	0.2358	0.2327	0.2296	0.2266	0.2236	0.2206	0.2177	0.2148
−0.6	0.2743	0.2709	0.2676	0.2643	0.2611	0.2578	0.2546	0.2514	0.2482	0.2451
−0.5	0.3085	0.3050	0.3015	0.2981	0.2946	0.2912	0.2877	0.2843	0.2810	0.2776
−0.4	0.3446	0.3409	0.3372	0.3336	0.3300	0.3264	0.3228	0.3192	0.3156	0.3121
−0.3	0.3821	0.3783	0.3745	0.3707	0.3669	0.3632	0.3594	0.3557	0.3520	0.3483
−0.2	0.4207	0.4168	0.4129	0.4090	0.4052	0.4013	0.3974	0.3936	0.3897	0.3859
−0.1	0.4602	0.4562	0.4522	0.4483	0.4443	0.4404	0.4364	0.4325	0.4286	0.4247
−0.0	0.5000	0.4960	0.4920	0.4880	0.4840	0.4801	0.4761	0.4721	0.4681	0.4641

(Table Continues)

Two such areas are shown next to Table 7.1—one with a negative z value and one with a positive z value. The values of z in the cumulative normal table range from −3.99 to 3.99 in increments of .01. As can be seen from Table 7.1, values of z accurate to the nearest tenth are given in the far left column (headed z) of the table. Further graduations to the nearest hundredth (.00, .01, .02, ... , .09) are given across the top of the table. The areas under the normal curve are given in the body of the table, accurate to four (or sometimes five) decimal places.

As an example, suppose that we wish to find the area under the standard normal curve to the left of a z value of 2.00. This area is illustrated in Figure 7.8. To find this area, we skip past the portion of the normal table on this page (which contains negative z values) and go to the portion of the normal table (which contains positive z values). We start at the top of the leftmost column and scan down the column until we find the z value 2.0—see the red arrow. We now scan across the row in the table corresponding to the z value 2.0 until we find the column corresponding to the heading .00. The desired area (which we have shaded blue) is in the row corresponding to the z value 2.0 and in the column headed .00. This area, which equals .9772, is the probability that the

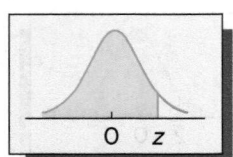

TABLE 7.1   **Cumulative Areas under the Standard Normal Curve (*Continued*)**

z	0.00	0.01	0.02	0.03	0.04	0.05	0.06	0.07	0.08	0.09
0.0	0.5000	0.5040	0.5080	0.5120	0.5160	0.5199	0.5239	0.5279	0.5319	0.5359
0.1	0.5398	0.5438	0.5478	0.5517	0.5557	0.5596	0.5636	0.5675	0.5714	0.5753
0.2	0.5793	0.5832	0.5871	0.5910	0.5948	0.5987	0.6026	0.6064	0.6103	0.6141
0.3	0.6179	0.6217	0.6255	0.6293	0.6331	0.6368	0.6406	0.6443	0.6480	0.6517
0.4	0.6554	0.6591	0.6628	0.6664	0.6700	0.6736	0.6772	0.6808	0.6844	0.6879
0.5	0.6915	0.6950	0.6985	0.7019	0.7054	0.7088	0.7123	0.7157	0.7190	0.7224
0.6	0.7257	0.7291	0.7324	0.7357	0.7389	0.7422	0.7454	0.7486	0.7518	0.7549
0.7	0.7580	0.7611	0.7642	0.7673	0.7704	0.7734	0.7764	0.7794	0.7823	0.7852
0.8	0.7881	0.7910	0.7939	0.7967	0.7995	0.8023	0.8051	0.8078	0.8106	0.8133
0.9	0.8159	0.8186	0.8212	0.8238	0.8264	0.8289	0.8315	0.8340	0.8365	0.8389
1.0	0.8413	0.8438	0.8461	0.8485	0.8508	0.8531	0.8554	0.8577	0.8599	0.8621
1.1	0.8643	0.8665	0.8686	0.8708	0.8729	0.8749	0.8770	0.8790	0.8810	0.8830
1.2	0.8849	0.8869	0.8888	0.8907	0.8925	0.8944	0.8962	0.8980	0.8997	0.9015
1.3	0.9032	0.9049	0.9066	0.9082	0.9099	0.9115	0.9131	0.9147	0.9162	0.9177
1.4	0.9192	0.9207	0.9222	0.9236	0.9251	0.9265	0.9279	0.9292	0.9306	0.9319
1.5	0.9332	0.9345	0.9357	0.9370	0.9382	0.9394	0.9406	0.9418	0.9429	0.9441
1.6	0.9452	0.9463	0.9474	0.9484	0.9495	0.9505	0.9515	0.9525	0.9535	0.9545
1.7	0.9554	0.9564	0.9573	0.9582	0.9591	0.9599	0.9608	0.9616	0.9625	0.9633
1.8	0.9641	0.9649	0.9656	0.9664	0.9671	0.9678	0.9686	0.9693	0.9699	0.9706
1.9	0.9713	0.9719	0.9726	0.9732	0.9738	0.9744	0.9750	0.9756	0.9761	0.9767
2.0	0.9772	0.9778	0.9783	0.9788	0.9793	0.9798	0.9803	0.9808	0.9812	0.9817
2.1	0.9821	0.9826	0.9830	0.9834	0.9838	0.9842	0.9846	0.9850	0.9854	0.9857
2.2	0.9861	0.9864	0.9868	0.9871	0.9875	0.9878	0.9881	0.9884	0.9887	0.9890
2.3	0.9893	0.9896	0.9898	0.9901	0.9904	0.9906	0.9909	0.9911	0.9913	0.9916
2.4	0.9918	0.9920	0.9922	0.9925	0.9927	0.9929	0.9931	0.9932	0.9934	0.9936
2.5	0.9938	0.9940	0.9941	0.9943	0.9945	0.9946	0.9948	0.9949	0.9951	0.9952
2.6	0.9953	0.9955	0.9956	0.9957	0.9959	0.9960	0.9961	0.9962	0.9963	0.9964
2.7	0.9965	0.9966	0.9967	0.9968	0.9969	0.9970	0.9971	0.9972	0.9973	0.9974
2.8	0.9974	0.9975	0.9976	0.9977	0.9977	0.9978	0.9979	0.9979	0.9980	0.9981
2.9	0.9981	0.9982	0.9982	0.9983	0.9984	0.9984	0.9985	0.9985	0.9986	0.9986
3.0	0.99865	0.99869	0.99874	0.99878	0.99882	0.99886	0.99889	0.99893	0.99897	0.99900
3.1	0.99903	0.99906	0.99910	0.99913	0.99916	0.99918	0.99921	0.99924	0.99926	0.99929
3.2	0.99931	0.99934	0.99936	0.99938	0.99940	0.99942	0.99944	0.99946	0.99948	0.99950
3.3	0.99952	0.99953	0.99955	0.99957	0.99958	0.99960	0.99961	0.99962	0.99964	0.99965
3.4	0.99966	0.99968	0.99969	0.99970	0.99971	0.99972	0.99973	0.99974	0.99975	0.99976
3.5	0.99977	0.99978	0.99978	0.99979	0.99980	0.99981	0.99981	0.99982	0.99983	0.99983
3.6	0.99984	0.99985	0.99985	0.99986	0.99986	0.99987	0.99987	0.99988	0.99988	0.99989
3.7	0.99989	0.99990	0.99990	0.99990	0.99991	0.99991	0.99992	0.99992	0.99992	0.99992
3.8	0.99993	0.99993	0.99993	0.99994	0.99994	0.99994	0.99994	0.99995	0.99995	0.99995
3.9	0.99995	0.99995	0.99996	0.99996	0.99996	0.99996	0.99996	0.99996	0.99997	0.99997

random variable z will be less than or equal to 2.00. That is, we have found that $P(z \leq 2) =$ .9772. Note that, because there is no area under the normal curve at a single value of z, there is no difference between $P(z \leq 2)$ and $P(z < 2)$. As another example, the area under the standard normal curve to the left of the z value 1.25 is found in the row corresponding to 1.2 and in the column corresponding to .05. We find that this area (also shaded blue) is .8944. That is, $P(z \leq 1.25) = .8944$ (see Figure 7.9).

We now show how to use the cumulative normal table to find several other kinds of normal curve areas. First, suppose that we wish to find the area under the standard normal curve to the right of a z value of 2—that is, we wish to find $P(z \geq 2)$. This area is illustrated in Figure 7.10 and is called a **right-hand tail area.** Because the total area under the normal curve equals 1, the area under the curve to the right of 2 equals 1 minus the area under the curve to the left of 2. Because Table 7.1 tells us that the area under the standard normal curve to the left of 2 is .9772, the area under the standard normal curve to the right of 2 is $1 - .9772 = .0228$. Said in an equivalent fashion, because $P(z \leq 2) = .9772$, it follows that $P(z \geq 2) = 1 - P(z \leq 2) = 1 - .9772 = .0228$.

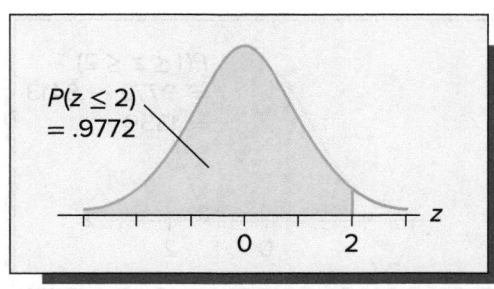

FIGURE 7.8   Finding P(z ≤ 2)

$P(z \leq 2)$
$= .9772$

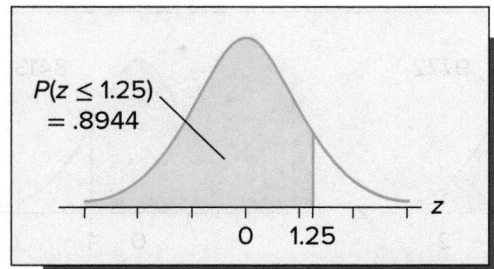

FIGURE 7.9   Finding P(z ≤ 1.25)

$P(z \leq 1.25)$
$= .8944$

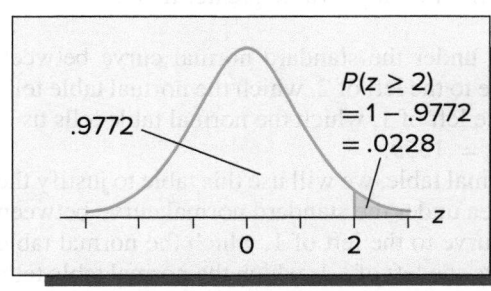

FIGURE 7.10   Finding P(z ≥ 2)

$P(z \geq 2)$
$= 1 - .9772$
$= .0228$

.9772

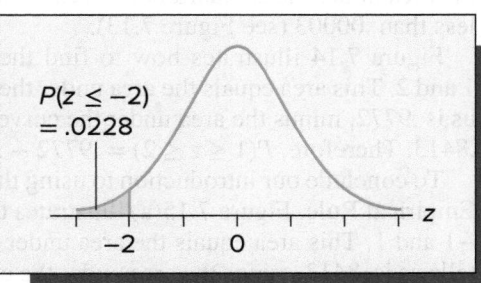

FIGURE 7.11   Finding P(z ≤ −2)

$P(z \leq -2)$
$= .0228$

Next, suppose that we wish to find the area under the standard normal curve to the left of a z value of -2. That is, we wish to find $P(z \leq -2)$. This area is illustrated in Figure 7.11 and is called a **left-hand tail area.** The needed area is found in the row of the cumulative normal table corresponding to -2.0 and in the column headed by .00. We find that $P(z \leq -2) = .0228$. Notice that the area under the standard normal curve to the left of -2 is equal to the area under this curve to the right of 2. This is true because of the symmetry of the normal curve.

Figure 7.12 illustrates how to find the area under the standard normal curve to the right of -2. Because the total area under the normal curve equals 1, the area under the curve to the right of -2 equals 1 minus the area under the curve to the left of -2. Because Table 7.1 tells us that the area under the standard normal curve to the left of -2 is .0228, the area under the standard normal curve to the right of -2 is $1 - .0228 = .9772$. That is, because $P(z \leq -2) = .0228$, it follows that $P(z \geq -2) = 1 - P(z \leq -2) = 1 - .0228 = .9772$.

The smallest z value in Table 7.1 is -3.99, and the table tells us that the area under the standard normal curve to the left of -3.99 is .00003 (see Figure 7.13). Therefore, if we wish to find the area under the standard normal curve to the left of any z value less than -3.99, the

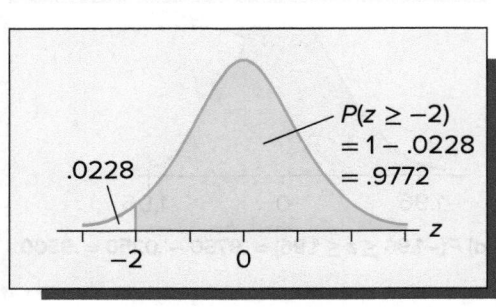

FIGURE 7.12   Finding P(z ≥ −2)

$P(z \geq -2)$
$= 1 - .0228$
$= .9772$

.0228

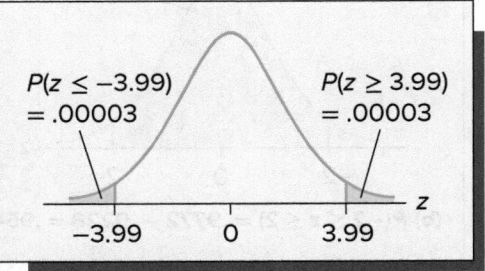

FIGURE 7.13   Finding P(z ≤ −3.99)

$P(z \leq -3.99)$
$= .00003$

$P(z \geq 3.99)$
$= .00003$

FIGURE 7.14   Calculating $P(1 \leq z \leq 2)$

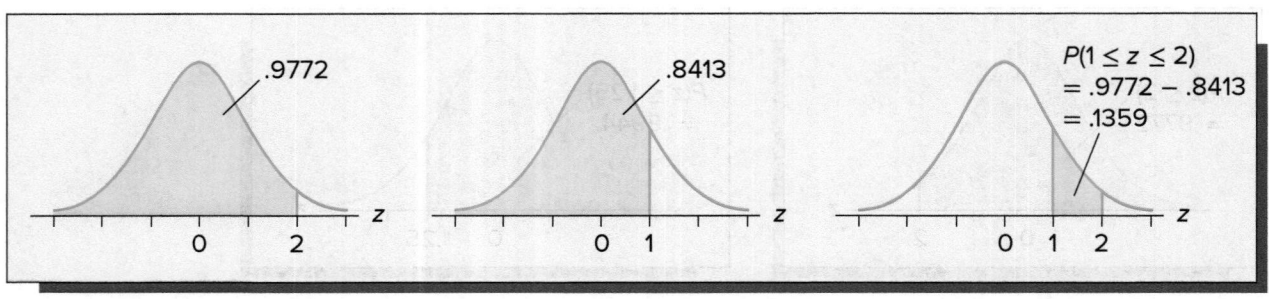

most we can say (without using a computer) is that this area is less than .00003. Similarly, the area under the standard normal curve to the right of any z value greater than 3.99 is also less than .00003 (see Figure 7.13).

Figure 7.14 illustrates how to find the area under the standard normal curve between 1 and 2. This area equals the area under the curve to the left of 2, which the normal table tells us is .9772, minus the area under the curve to the left of 1, which the normal table tells us is .8413. Therefore, $P(1 \leq z \leq 2) = .9772 - .8413 = .1359$.

To conclude our introduction to using the normal table, we will use this table to justify the Empirical Rule. Figure 7.15(a) illustrates the area under the standard normal curve between -1 and 1. This area equals the area under the curve to the left of 1, which the normal table tells us is .8413, minus the area under the curve to the left of -1, which the normal table tells us is .1587. Therefore, $P(-1 \leq z \leq 1) = .8413 - .1587 = .6826$. Now, suppose that a random variable x is normally distributed with mean μ and standard deviation σ, and remember that z is the number of standard deviations σ that x is from μ. It follows that when we say that $P(-1 \leq z \leq 1)$ equals .6826, we are saying that 68.26 percent of all possible observed values of x are between a point that is one standard deviation below μ (where z equals -1) and a point

FIGURE 7.15   Some Areas under the Standard Normal Curve

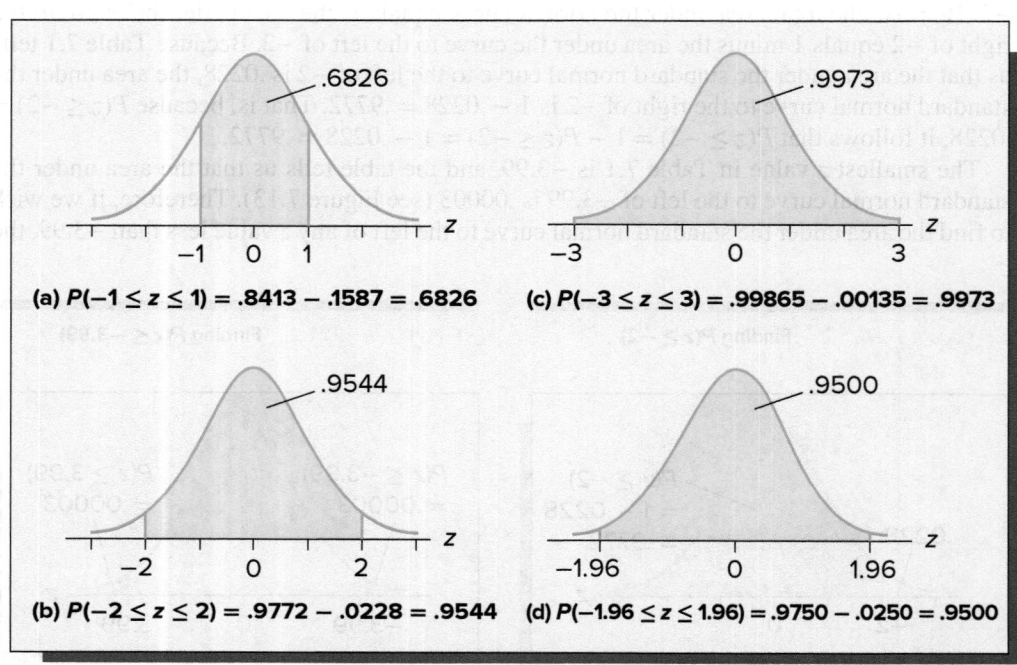

(a) $P(-1 \leq z \leq 1) = .8413 - .1587 = .6826$

(c) $P(-3 \leq z \leq 3) = .99865 - .00135 = .9973$

(b) $P(-2 \leq z \leq 2) = .9772 - .0228 = .9544$

(d) $P(-1.96 \leq z \leq 1.96) = .9750 - .0250 = .9500$

that is one standard deviation above μ (where z equals 1). That is, 68.26 percent of all possible observed values of x are within (plus or minus) one standard deviation of the mean μ.

Figure 7.15(b) illustrates the area under the standard normal curve between −2 and 2. This area equals the area under the curve to the left of 2, which the normal table tells us is .9772, minus the area under the curve to the left of −2, which the normal table tells us is .0228. Therefore, $P(-2 \leq z \leq 2) = .9772 - .0228 = .9544$. That is, 95.44 percent of all possible observed values of x are within (plus or minus) two standard deviations of the mean μ.

Figure 7.15(c) illustrates the area under the standard normal curve between −3 and 3. This area equals the area under the curve to the left of 3, which the normal table tells us is .99865, minus the area under the curve to the left of −3, which the normal table tells us is .00135. Therefore, $P(-3 \leq z \leq 3) = .99865 - .00135 = .9973$. That is, 99.73 percent of all possible observed values of x are within (plus or minus) three standard deviations of the mean μ.

Although the Empirical Rule gives the percentages of all possible values of a normally distributed random variable x that are within one, two, and three standard deviations of the mean μ, we can use the normal table to find the percentage of all possible values of x that are within any particular number of standard deviations of μ. For example, in later chapters we will need to know the percentage of all possible values of x that are within plus or minus 1.96 standard deviations of μ. Figure 7.15(d) illustrates the area under the standard normal curve between −1.96 and 1.96. This area equals the area under the curve to the left of 1.96, which the normal table tells us is .9750, minus the area under the curve to the left of −1.96, which the table tells us is .0250. Therefore, $P(-1.96 \leq z \leq 1.96) = .9750 - .0250 = .9500$. That is, 95 percent of all possible values of x are within plus or minus 1.96 standard deviations of the mean μ.

Some practical applications

We have seen how to use z values and the normal table to find areas under the standard normal curve. However, most practical problems are not stated in such terms. We now consider an example in which we must restate the problem in terms of the standard normal random variable z before using the normal table.

LO7-4
Use the normal distribution to compute probabilities.

Ⓒ **EXAMPLE 7.2** The Car Mileage Case: Estimating Mileage

Recall from previous chapters that an automaker has recently introduced a new midsize model and that we have used the sample of 50 mileages to estimate that the population of mileages of all cars of this type is normally distributed with a mean mileage equal to 31.56 mpg and a standard deviation equal to .798 mpg. Suppose that a competing automaker produces a midsize model that is somewhat smaller and less powerful than the new midsize model. The competitor claims, however, that its midsize model gets better mileages. Specifically, the competitor claims that the mileages of all its midsize cars are normally distributed with a mean mileage μ equal to 33 mpg and a standard deviation σ equal to .7 mpg. In the next example we consider one way to investigate the validity of this claim. In this example we assume that the claim is true, and we calculate the probability that the mileage, x, of a randomly selected competing midsize car will be between 32 mpg and 35 mpg. That is, we wish to find $P(32 \leq x \leq 35)$. As illustrated in Figure 7.16, this probability is the area between 32 and 35 under the normal curve having mean $\mu = 33$ and standard deviation $\sigma = .7$. In order to use the normal table, we must restate the problem in terms of the standard normal random variable z by computing z **values** corresponding to 32 mpg and 35 mpg. The z value corresponding to 32 is

©Adam Gault/Getty Images RF

$$z = \frac{x - \mu}{\sigma} = \frac{32 - 33}{.7} = \frac{-1}{.7} = -1.43$$

which says that the mileage 32 is 1.43 standard deviations below the mean $\mu = 33$. The z value corresponding to 35 is

$$z = \frac{x - \mu}{\sigma} = \frac{35 - 33}{.7} = \frac{2}{.7} = 2.86$$

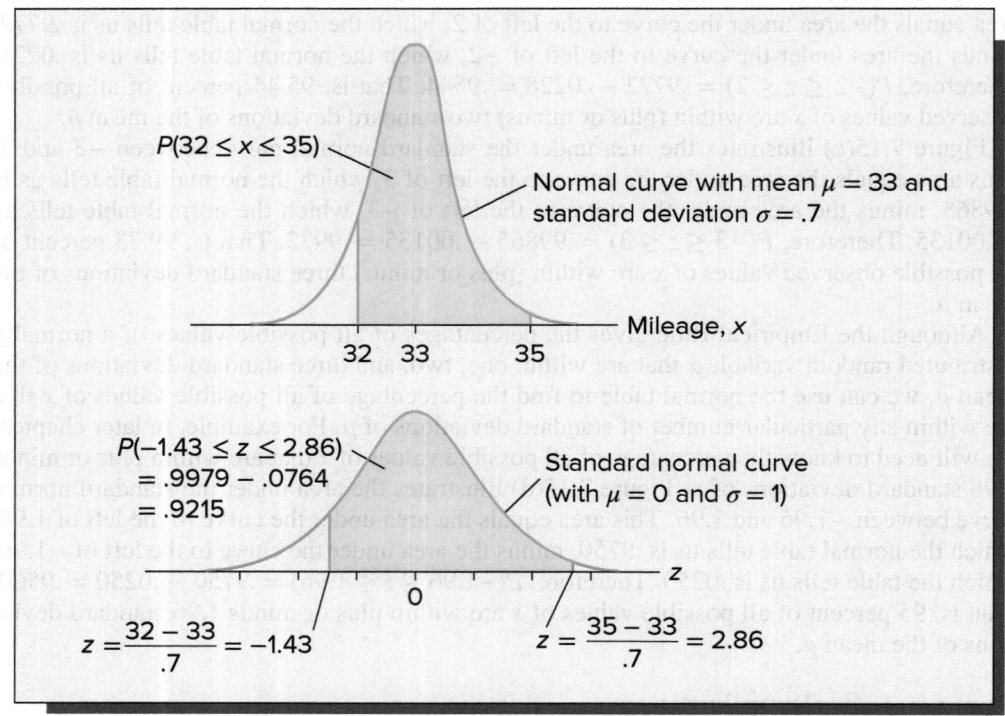

FIGURE 7.16   Finding $P(32 \leq x \leq 35)$ When $\mu = 33$ and $\sigma = .7$ by Using a Normal Table

which says that the mileage 35 is 2.86 standard deviations above the mean $\mu = 33$. Looking at Figure 7.16, we see that the area between 32 and 35 under the normal curve having mean $\mu = 33$ and standard deviation $\sigma = .7$ equals the area between -1.43 and 2.86 under the standard normal curve. This equals the area under the standard normal curve to the left of 2.86, which the normal table tells us is .9979, minus the area under the standard normal curve to the left of -1.43, which the normal table tells us is .0764. We summarize this result as follows:

$$P(32 \leq x \leq 35) = P\left(\frac{32 - 33}{.7} \leq \frac{x - \mu}{\sigma} \leq \frac{35 - 33}{.7}\right)$$

$$= P(-1.43 \leq z \leq 2.86) = .9979 - .0764 = .9215$$

This probability says that, if the competing automaker's claim is valid, then 92.15 percent of all of its midsize cars will get mileages between 32 mpg and 35 mpg.

Example 7.2 illustrates the general procedure for finding a probability about a normally distributed random variable x. We summarize this procedure in the following box:

Finding Normal Probabilities

1 Formulate the problem in terms of the random variable x.

2 Calculate relevant z values and restate the problem in terms of the standard normal random variable

$$z = \frac{x - \mu}{\sigma}$$

3 Find the required area under the standard normal curve by using the normal table.

4 Note that it is always useful to draw a picture illustrating the needed area before using the normal table.

EXAMPLE 7.3 The Car Mileage Case: Estimating Mileage

Recall from Example 7.2 that the competing automaker claims that the population of mileages of all its midsize cars is normally distributed with mean $\mu = 33$ and standard deviation $\sigma = .7$. Suppose that an independent testing agency randomly selects one of these cars and finds that it gets a mileage of 31.2 mpg when tested as prescribed by the EPA. Because the sample mileage of 31.2 mpg is *less than* the claimed mean $\mu = 33$, we have some evidence that contradicts the competing automaker's claim. To evaluate the strength of this evidence, we will calculate the probability that the mileage, x, of a randomly selected midsize car would be *less than or equal to* 31.2 if, in fact, the competing automaker's claim is true. To calculate $P(x \leq 31.2)$ under the assumption that the claim is true, we find the area to the left of 31.2 under the normal curve with mean $\mu = 33$ and standard deviation $\sigma = .7$ (see Figure 7.17). In order to use the normal table, we must find the z value corresponding to 31.2. This z value is

$$z = \frac{x - \mu}{\sigma} = \frac{31.2 - 33}{.7} = -2.57$$

which says that the mileage 31.2 is 2.57 standard deviations below the mean mileage $\mu = 33$. Looking at Figure 7.17, we see that the area to the left of 31.2 under the normal curve having mean $\mu = 33$ and standard deviation $\sigma = .7$ equals the area to the left of -2.57 under the standard normal curve. The normal table tells us that the area under the standard normal curve to the left of -2.57 is .0051, as shown in Figure 7.17. It follows that we can summarize our calculations as follows:

$$P(x \leq 31.2) = P\left(\frac{x - \mu}{\sigma} \leq \frac{31.2 - 33}{.7}\right)$$

$$= P(z \leq -2.57) = .0051$$

This probability says that, if the competing automaker's claim is valid, then only 51 in 10,000 cars would obtain a mileage of less than or equal to 31.2 mpg. Because it is very difficult to believe that a 51 in 10,000 chance has occurred, we have very strong evidence against the competing automaker's claim. It is probably true that μ is less than 33 and/or σ is greater than .7 and/or the population of all mileages is not normally distributed.

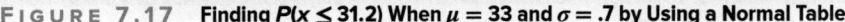

FIGURE 7.17   **Finding $P(x \leq 31.2)$ When $\mu = 33$ and $\sigma = .7$ by Using a Normal Table**

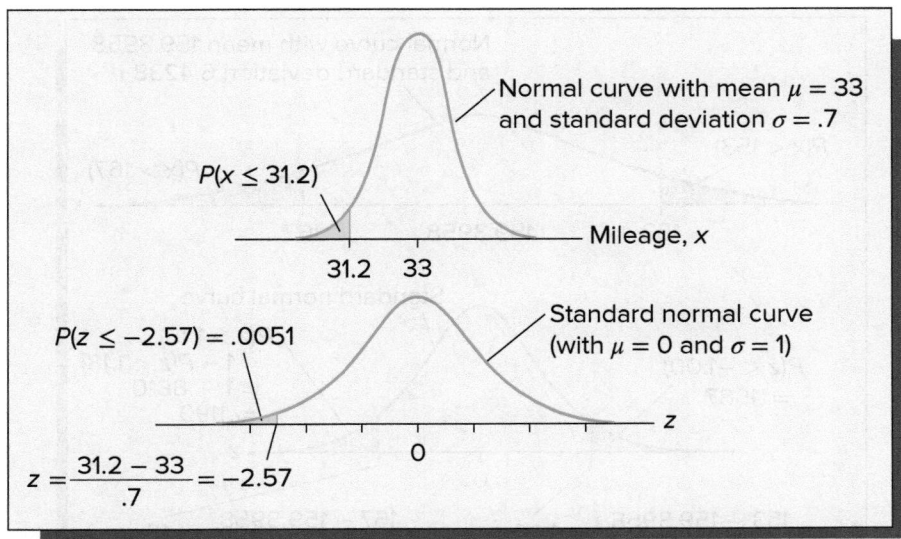

©Purestock/SuperStock RF

 EXAMPLE 7.4 The Coffee Temperature Case: Meeting Customer Requirements

Recall that marketing research done by a fast-food restaurant indicates that coffee tastes best if its temperature is between 153°F and 167°F. The restaurant has sampled the coffee it serves and observed the 48 temperature readings in Figure 7.1(a). The temperature readings have a mean $\bar{x} = 159.3958$ and a standard deviation $s = 6.4238$ and are described by a bell-shaped histogram. Using $\bar{x}$ and s as point estimates of the mean μ and the standard deviation σ of the population of all possible coffee temperatures, we wish to calculate the probability that x, the temperature of a randomly selected cup of coffee, is outside the customer require-ments for best-tasting coffee (that is, less than 153° or greater than 167°). In order to com-pute the probability $P(x < 153$ or $x > 167)$, we compute the z values

$$z = \frac{153 - 159.3958}{6.4238} = -1.00 \quad \text{and} \quad z = \frac{167 - 159.3958}{6.4238} = 1.18$$

Because the events $\{x < 153\}$ and $\{x > 167\}$ are mutually exclusive, we have

$$P(x < 153 \text{ or } x > 167) = P(x < 153) + P(x > 167)$$
$$= P(z < -1.00) + P(z > 1.18)$$
$$= .1587 + .1190 = .2777$$

This calculation is illustrated in Figure 7.18. The probability of .2777 implies that 27.77 percent of the coffee temperatures do not meet customer requirements and 72.23 percent of the coffee temperatures do meet these requirements. If management wishes a very high percentage of its coffee temperatures to meet customer requirements, the coffee-making process must be improved.

LO7-5

Find population values that correspond to specified normal distribution probabilities.

Finding a point on the horizontal axis under a normal curve

In order to use many of the formulas given in later chapters, we must be able to find the z value so that the tail area to the right of z under the standard normal curve is a particular value. For instance, we might need to find the z value so that the tail area to the right of z under the standard normal curve is .025. This z value is denoted $z_{.025}$, and we illustrate $z_{.025}$ in Figure 7.19(a). We refer to $z_{.025}$ as **the point on the horizontal axis under the standard**

FIGURE 7.18   Finding $P(x < 153$ or $x > 167)$ in the Coffee Temperature Case

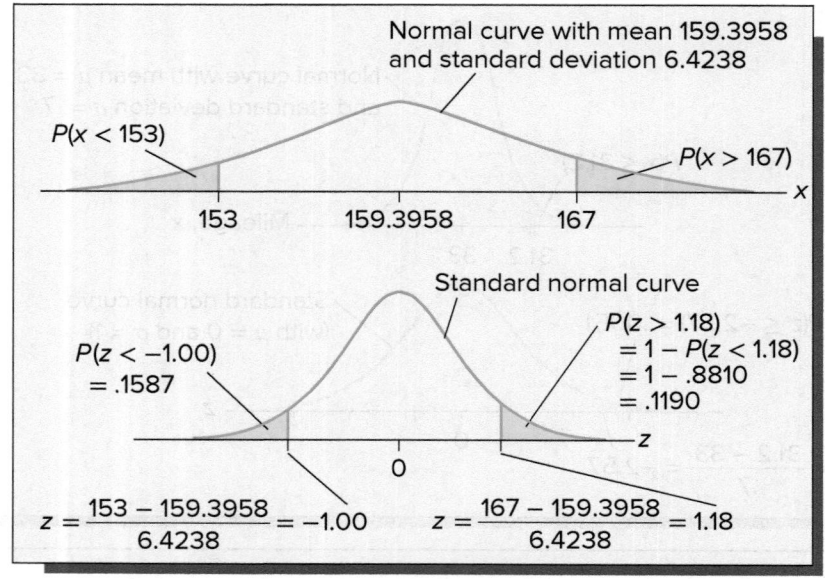

FIGURE 7.19   The Point $z_{.025} = 1.96$

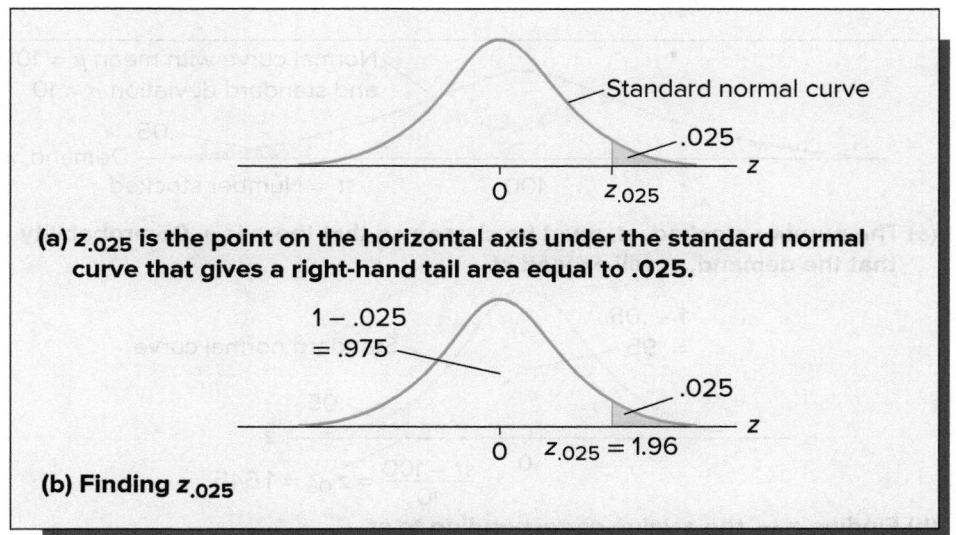

(a) $z_{.025}$ is the point on the horizontal axis under the standard normal curve that gives a right-hand tail area equal to .025.

(b) Finding $z_{.025}$

normal curve that gives a right-hand tail area equal to .025. It is easy to use the cumulative normal table to find such a point. For instance, in order to find $z_{.025}$, we note from Figure 7.19(b) that the area under the standard normal curve to the left of $z_{.025}$ equals .975. Remembering that areas under the standard normal curve to the left of z are the four-digit (or five-digit) numbers given in the body of Table 7.1, we scan the body of the table and find the area .9750. We have shaded this area in Table 7.1, and we note that the area .9750 is in the row corresponding to a z of 1.9 and in the column headed by .06. It follows that the z value corresponding to .9750 is 1.96. Because the z value 1.96 gives an area under the standard normal curve to its left that equals .975, it also gives a right-hand tail area equal to .025. Therefore, $z_{.025} = 1.96$.

In general, **we let z_α denote the point on the horizontal axis under the standard normal curve that gives a right-hand tail area equal to α.** With this definition in mind, we consider the following example.

Ⓒ EXAMPLE 7.5 The DVD Case: Managing Inventory

A large discount store sells 50 packs of HX-150 blank DVDs and receives a shipment every Monday. Historical sales records indicate that the weekly demand, x, for these 50 packs is normally distributed with a mean of $\mu = 100$ and a standard deviation of $\sigma = 10$. How many 50 packs should be stocked at the beginning of a week so that there is only a 5 percent chance that the store will run short during the week?

If we let st equal the number of 50 packs that will be stocked, then st must be chosen to allow only a .05 probability that weekly demand, x, will exceed st. That is, st must be chosen so that

$$P(x > st) = .05$$

Figure 7.20(a) shows that the number stocked, st, is located under the right-hand tail of the normal curve having mean $\mu = 100$ and standard deviation $\sigma = 10$. In order to find st, we need to determine how many standard deviations st must be above the mean in order to give a right-hand tail area that is equal to .05.

The z value corresponding to st is

$$z = \frac{st - \mu}{\sigma} = \frac{st - 100}{10}$$

and this z value is the number of standard deviations that st is from μ. This z value is illustrated in Figure 7.20(b), and it is the point on the horizontal axis under the standard normal curve that gives a right-hand tail area equal to .05. That is, the z value corresponding to st is $z_{.05}$. Because the area under the standard normal curve to the left of $z_{.05}$ is

FIGURE 7.20   **Finding the Number of 50 Packs of DVDs Stocked, *st*, so That *P(x > st)* = .05 When $\mu = 100$ and $\sigma = 10$**

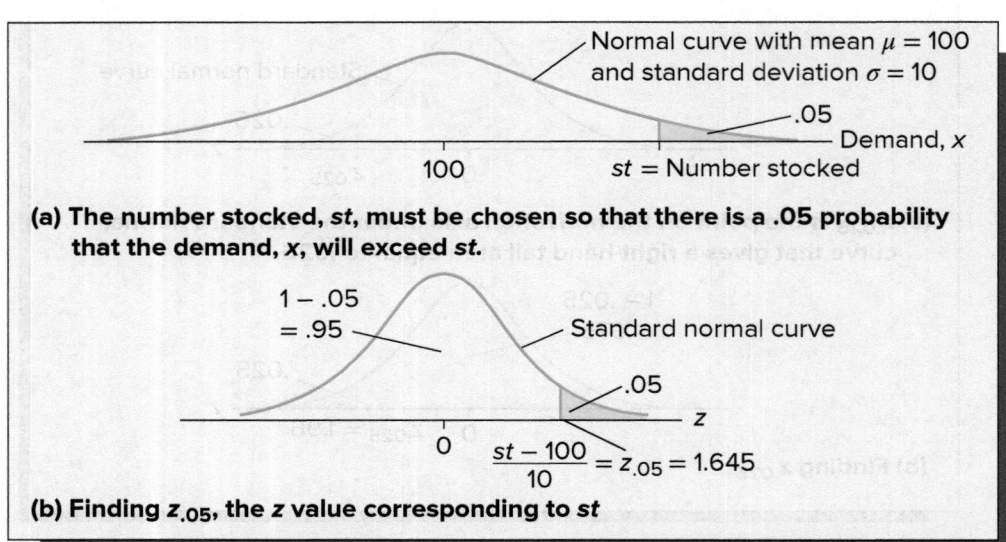

(a) The number stocked, *st*, must be chosen so that there is a .05 probability that the demand, *x*, will exceed *st*.

(b) Finding $z_{.05}$, the *z* value corresponding to *st*

$1 - .05 = .95$—see Figure 7.20(b)—we look for .95 in the body of the normal table. In Table 7.1, we see that the areas closest to .95 are .9495, which has a corresponding z value of 1.64, and .9505, which has a corresponding z value of 1.65. Although it would probably be sufficient to use either of these z values, we will (because it is easy to do so) interpolate halfway between them and assume that $z_{.05}$ equals 1.645. To find *st*, we solve the equation

$$\frac{st - 100}{10} = 1.645$$

for *st*. Doing this yields

$$st - 100 = 1.645(10)$$

or

$$st = 100 + 1.645(10) = 116.45$$

BI

This last equation says that *st* is 1.645 standard deviations ($\sigma = 10$) above the mean ($\mu = 100$). Rounding $st = 116.45$ up so that the store's chances of running short will be *no more* than 5 percent, the store should stock 117 of the 50 packs at the beginning of each week.

Sometimes we need to find the point on the horizontal axis under the standard normal curve that gives a particular **left-hand tail area** (say, for instance, an area of .025). Looking at Figure 7.21,

FIGURE 7.21   **The *z* Value $-z_{.025} = -1.96$ Gives a Left-Hand Tail Area of .025 under the Standard Normal Curve**

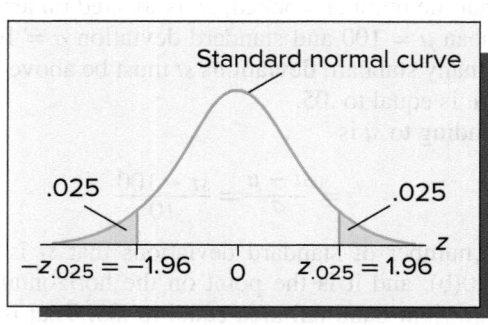

it is easy to see that, if, for instance, we want a left-hand tail area of .025, the needed z value is $-z_{.025}$, where $z_{.025}$ gives a right-hand tail area equal to .025. To find $-z_{.025}$, we look for .025 in the body of the normal table and find that the z value corresponding to .025 is -1.96. Therefore, $-z_{.025} = -1.96$. In general, $-z_\alpha$ **is the point on the horizontal axis under the standard normal curve that gives a left-hand tail area equal to α.**

Ⓒ **EXAMPLE 7.6** Setting a Guarantee Period

Extensive testing indicates that the lifetime of the Everlast automobile battery is normally distributed with a mean of $\mu = 60$ months and a standard deviation of $\sigma = 6$ months. The Everlast's manufacturer has decided to offer a free replacement battery to any purchaser whose Everlast battery does not last at least as long as the minimum lifetime specified in its guarantee. How can the manufacturer establish the guarantee period so that only 1 percent of the batteries will need to be replaced free of charge?

If the battery will be guaranteed to last l months, l must be chosen to allow only a .01 probability that the lifetime, x, of an Everlast battery will be less than l. That is, we must choose l so that

$$P(x < l) = .01$$

Figure 7.22(a) shows that the guarantee period, l, is located under the left-hand tail of the normal curve having mean $\mu = 60$ and standard deviation $\sigma = 6$. In order to find l, we need to determine how many standard deviations l must be below the mean in order to give a left-hand tail area that equals .01. The z value corresponding to l is

$$z = \frac{l - \mu}{\sigma} = \frac{l - 60}{6}$$

and this z value is the number of standard deviations that l is from μ. This z value is illustrated in Figure 7.22(b), and it is the point on the horizontal axis under the standard normal curve that gives a left-hand tail area equal to .01. That is, the z value corresponding to l is $-z_{.01}$. To find $-z_{.01}$, we look for .01 in the body of the normal table. Doing this, we see that the area closest to .01 is .0099, which has a corresponding z value of -2.33. Therefore, $-z_{.01}$ is (roughly) -2.33. To find l, we solve the equation

$$\frac{l - 60}{6} = -2.33$$

for l. Doing this yields

$$l - 60 = -2.33(6)$$

or

$$l = 60 - 2.33(6) = 46.02$$

FIGURE 7.22　**Finding the Guarantee Period, l, so That $P(x < l) = .01$ When $\mu = 60$ and $\sigma = 6$**

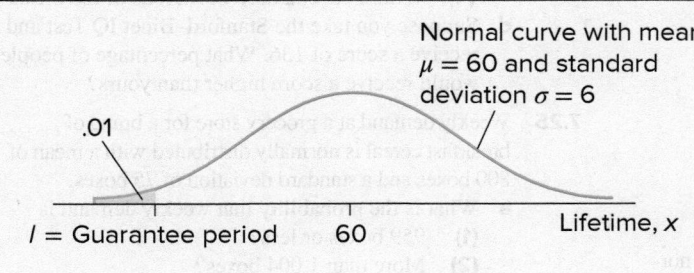

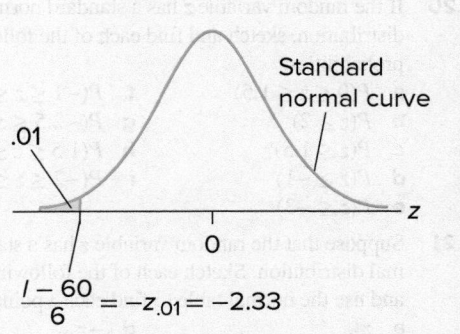

(a) The guarantee period, l, must be chosen so that there is a .01 probability that the lifetime, x, will be less than l.

(b) Finding $-z_{.01}$, the z value corresponding to l

Note that this last equation says that l is 2.33 standard deviations ($\sigma = 6$) below the mean ($\mu = 60$). Rounding $l = 46.02$ down so that *no more* than 1 percent of the batteries will need to be replaced free of charge, it seems reasonable to guarantee the Everlast battery to last 46 months.

Whenever we use a normal table to find a z point corresponding to a particular normal curve area, we will use the *halfway interpolation* procedure illustrated in Example 7.5 if the area we are looking for is exactly halfway between two areas in the table. Otherwise, as illustrated in Example 7.6, we will use the z value corresponding to the area in the table that is closest to the desired area.

Exercises for Section 7.3

CONCEPTS ■ **connect**

7.16 Explain what the mean, μ, tells us about a normal curve, and explain what the standard deviation, σ, tells us about a normal curve.

7.17 Explain how to compute the z value corresponding to a value of a normally distributed random variable. What does the z value tell us about the value of the random variable?

METHODS AND APPLICATIONS

7.18 In each case, sketch the two specified normal curves on the same set of axes:
  a A normal curve with $\mu = 20$ and $\sigma = 3$, and a normal curve with $\mu = 20$ and $\sigma = 6$.
  b A normal curve with $\mu = 20$ and $\sigma = 3$, and a normal curve with $\mu = 30$ and $\sigma = 3$.
  c A normal curve with $\mu = 100$ and $\sigma = 10$, and a normal curve with $\mu = 200$ and $\sigma = 20$.

7.19 Let x be a normally distributed random variable having mean $\mu = 30$ and standard deviation $\sigma = 5$. Find the z value for each of the following observed values of x:
  a $x = 25$ **d** $x = 40$
  b $x = 15$ **e** $x = 50$
  c $x = 30$

In each case, explain what the z value tells us about how the observed value of x compares to the mean, μ.

7.20 If the random variable z has a standard normal distribution, sketch and find each of the following probabilities:
  a $P(0 \leq z \leq 1.5)$ **f** $P(-1 \leq z \leq 1)$
  b $P(z \geq 2)$ **g** $P(-2.5 \leq z \leq .5)$
  c $P(z \leq 1.5)$ **h** $P(1.5 \leq z \leq 2)$
  d $P(z \geq -1)$ **i** $P(-2 \leq z \leq -.5)$
  e $P(z \leq -3)$

7.21 Suppose that the random variable z has a standard normal distribution. Sketch each of the following z points, and use the normal table to find each z point.
  a $z_{.01}$ **d** $-z_{.01}$
  b $z_{.05}$ **e** $-z_{.05}$
  c $z_{.02}$ **f** $-z_{.10}$

7.22 Suppose that the random variable x is normally distributed with mean $\mu = 1{,}000$ and standard deviation $\sigma = 100$.

Sketch and find each of the following probabilities:
  a $P(1{,}000 \leq x \leq 1{,}200)$ **e** $P(x \leq 700)$
  b $P(x > 1{,}257)$ **f** $P(812 \leq x \leq 913)$
  c $P(x < 1{,}035)$ **g** $P(x > 891)$
  d $P(857 \leq x \leq 1{,}183)$ **h** $P(1{,}050 \leq x \leq 1{,}250)$

7.23 Suppose that the random variable x is normally distributed with mean $\mu = 500$ and standard deviation $\sigma = 100$. For each of the following, use the normal table to find the needed value k. In each case, draw a sketch.
  a $P(x \geq k) = .025$ **f** $P(x > k) = .95$
  b $P(x \geq k) = .05$ **g** $P(x \leq k) = .975$
  c $P(x < k) = .025$ **h** $P(x \geq k) = .0228$
  d $P(x \leq k) = .015$ **i** $P(x > k) = .9772$
  e $P(x < k) = .985$

7.24 Stanford–Binet IQ Test scores are normally distributed with a mean score of 100 and a standard deviation of 16.
  a Sketch the distribution of Stanford–Binet IQ Test scores.
  b Write the equation that gives the z score corresponding to a Stanford–Binet IQ Test score. Sketch the distribution of such z scores.
  c Find the probability that a randomly selected person has an IQ test score
    (1) Over 140.
    (2) Under 88.
    (3) Between 72 and 128.
    (4) Within 1.5 standard deviations of the mean.
  d Suppose you take the Stanford–Binet IQ Test and receive a score of 136. What percentage of people would receive a score higher than yours?

7.25 Weekly demand at a grocery store for a brand of breakfast cereal is normally distributed with a mean of 800 boxes and a standard deviation of 75 boxes.
  a What is the probability that weekly demand is '
    (1) 959 boxes or less?
    (2) More than 1,004 boxes?
    (3) Less than 650 boxes or greater than 950 boxes?
  b The store orders cereal from a distributor weekly. How many boxes should the store order for a week to have only a 2.5 percent chance of running short of this brand of cereal during the week?

7.26 The lifetimes of a particular brand of DVD player are normally distributed with a mean of eight years and a standard deviation of six months. Find each of the following probabilities where x denotes the lifetime in years. In each case, sketch the probability.

a $P(7 \leq x \leq 9)$ e $P(x \leq 7)$
b $P(8.5 \leq x \leq 9.5)$ f $P(x \geq 7)$
c $P(6.5 \leq x \leq 7.5)$ g $P(x \leq 10)$
d $P(x \geq 8)$ h $P(x > 10)$

7.27 United Motors claims that one of its cars, the Starbird 300, gets city driving mileages that are normally distributed with a mean of 30 mpg and a standard deviation of 1 mpg. Let x denote the city driving mileage of a randomly selected Starbird 300.

a Assuming that United Motors' claim is correct, find $P(x \leq 27)$.

b If you purchase (randomly select) a Starbird 300 and your car gets 27 mpg in city driving, what do you think of United Motors' claim? Explain your answer.

7.28 An investment broker reports that the yearly returns on common stocks are approximately normally distributed with a mean return of 12.4 percent and a standard deviation of 20.6 percent. On the other hand, the firm reports that the yearly returns on tax-free municipal bonds are approximately normally distributed with a mean return of 5.2 percent and a standard deviation of 8.6 percent. Find the probability that a randomly selected

a Common stock will give a positive yearly return.

b Tax-free municipal bond will give a positive yearly return.

c Common stock will give more than a 10 percent return.

d Tax-free municipal bond will give more than a 10 percent return.

e Common stock will give a loss of at least 10 percent.

f Tax-free municipal bond will give a loss of at least 10 percent.

7.29 A filling process is supposed to fill jars with 16 ounces of grape jelly. Specifications state that each jar must contain between 15.95 ounces and 16.05 ounces. A jar is selected from the process every half hour until a sample of 100 jars is obtained. When the fills of the jars are measured, it is found that $\bar{x} = 16.0024$ and $s = .02454$. Using $\bar{x}$ and s as point estimates of μ and σ, estimate the probability that a randomly selected jar will have a fill, x, that is out of specification. Assume that the process is stable and that the population of all jar fills is normally distributed.

7.30 A tire company has developed a new type of steel-belted radial tire. Extensive testing indicates the population of mileages obtained by all tires of this new type is normally distributed with a mean of 40,000 miles and a standard deviation of 4,000 miles. The company wishes to offer a guarantee providing a discount on a new set of tires if the original tires purchased do not exceed the mileage stated in the guarantee. What should the guaranteed mileage be if the tire company desires that no more than 2 percent of the tires will fail to meet the guaranteed mileage?

7.31 Recall from Exercise 7.28 that yearly returns on common stocks are normally distributed with a

mean of 12.4 percent and a standard deviation of 20.6 percent.

a (1) What percentage of yearly returns are at or below the 10th percentile of the distribution of yearly returns? (2) What percentage are at or above the 10th percentile? (3) Find the 10th percentile of the distribution of yearly returns.

b Find the first quartile, Q_1, and the third quartile, Q_3, of the distribution of yearly returns.

7.32 Two students take a college entrance exam known to have a normal distribution of scores. The students receive raw scores of 63 and 93, which correspond to z scores (often called the standardized scores) of -1 and 1.5, respectively. Find the mean and standard deviation of the distribution of raw exam scores.

7.33 In the book *Advanced Managerial Accounting*, Robert P. Magee discusses monitoring cost variances. A *cost variance* is the difference between a budgeted cost and an actual cost. Magee considers weekly monitoring of the cost variances of two manufacturing processes, Process A and Process B. One individual monitors both processes and each week receives a weekly cost variance report for each process. The individual has decided to investigate the weekly cost variance for a particular process (to determine whether or not the process is out of control) when its weekly cost variance is too high. To this end, a weekly cost variance will be investigated if it exceeds $2,500.

a When Process A is in control, its potential weekly cost variances are normally distributed with a mean of $0 and a standard deviation of $5,000. When Process B is in control, its potential weekly cost variances are normally distributed with a mean of $0 and a standard deviation of $10,000. For each process, find the probability that a weekly cost variance will be investigated (that is, will exceed $2,500) even though the process is in control. Which in-control process will be investigated more often?

b When Process A is out of control, its potential weekly cost variances are normally distributed with a mean of $7,500 and a standard deviation of $5,000. When Process B is out of control, its potential weekly cost variances are normally distributed with a mean of $7,500 and a standard deviation of $10,000. For each process, find the probability that a weekly cost variance will be investigated (that is, will exceed $2,500) when the process is out of control. Which out-of-control process will be investigated more often?

c If both Processes A and B are almost always in control, which process will be investigated more often?

d Suppose that we wish to reduce the probability that Process B will be investigated (when it is in control) to .3085. (1) What cost variance investigation policy should be used? That is, how large a cost variance should trigger an investigation? (2) Using this new policy, what is the probability that an out-of-control cost variance for Process B will be investigated?

7.34 Suppose that yearly health care expenses for a family of four are normally distributed with a mean expense equal to $3,000 and a standard deviation of $500. An insurance company has decided to offer a health insurance premium reduction if a policyholder's health care expenses do not exceed a specified dollar amount. What dollar amount should be established if the insurance company wants families having the lowest 33 percent of yearly health care expenses to be eligible for the premium reduction?

7.35 Suppose that the 33rd percentile of a normal distribution is equal to 656 and that the 97.5th percentile of this normal distribution is 896. Find the mean μ and the standard deviation σ of the normal distribution. Hint: Sketch these percentiles.

LO7-6

Use the normal distribution to approximate binomial probabilities (Optional).

7.4 Approximating the Binomial Distribution by Using the Normal Distribution (Optional)

Recall that Figure 6.6 illustrates several binomial distributions. In general, we can see that as n gets larger and as p gets closer to .5, the graph of a binomial distribution tends to have the symmetrical, bell-shaped appearance of a normal curve. It follows that, under conditions given in the following box, we can approximate the binomial distribution by using a normal distribution.

The Normal Approximation of the Binomial Distribution

Consider a binomial random variable x, where n is the number of trials performed and p is the probability of success on each trial. If n and p have values so that $np \geq 5$ and $n(1 - p) \geq 5$, then x is approximately normally distributed with mean $\mu = np$ and standard deviation $\sigma = \sqrt{npq}$, where $q = 1 - p$.

This approximation is often useful because binomial tables for large values of n are often unavailable. The conditions $np \geq 5$ and $n(1 - p) \geq 5$ must be met in order for the approximation to be appropriate. Note that if p is near 0 or near 1, then n must be larger for a good approximation, while if p is near .5, then n need not be as large.[1] We illustrate exactly how to do the approximation in the examples to follow.

EXAMPLE 7.7 The Continuity Correction

Consider the binomial random variable x with $n = 50$ trials and probability of success $p = .5$. Suppose we want to use the normal approximation to this binomial distribution to compute the probability of 23 successes in the 50 trials. That is, we wish to compute $P(x = 23)$. Because $np = (50)(.5) = 25$ is at least 5, and $n(1 - p) = 50(1 - .5) = 25$ is also at least 5, we can appropriately use the approximation. Moreover, we can approximate the binomial distribution of x by using a normal distribution with mean $\mu = np = 50(.5) = 25$ and standard deviation $\sigma = \sqrt{npq} = \sqrt{50(.5)(1 - .5)} = 3.5355$.

In order to compute the needed probability, we must make a **continuity correction.** This is because a discrete distribution (the binomial) is being approximated by a continuous distribution (the normal). Because there is no area under a normal curve at the single point $x = 23$, we must assign an area under the normal curve to the binomial outcome $x = 23$. It is logical to assign the area corresponding to the interval from 22.5 to 23.5 to the integer outcome $x = 23$. That is, the area under the normal curve corresponding to all values within .5 of a unit of the integer outcome $x = 23$ is assigned to the value $x = 23$. So we approximate

---

[1] As an alternative to the rule that both np and $n(1 - p)$ must be at least 5, some statisticians suggest using the more conservative rule that both np and $n(1 - p)$ must be at least 10.

FIGURE 7.23   **Approximating the Binomial Probability $P(x = 23)$ by Using the Normal Curve When $\mu = np = 25$ and $\sigma = \sqrt{npq} = 3.5355$**

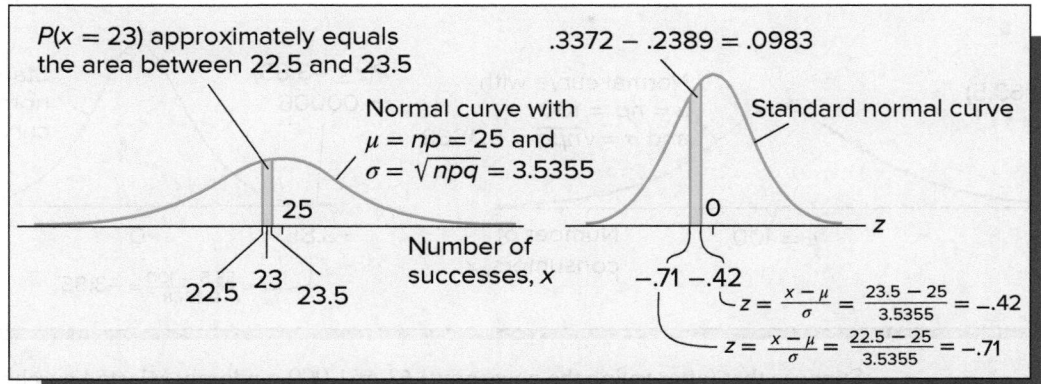

the binomial probability $P(x = 23)$ by calculating the normal curve area $P(22.5 \leq x \leq 23.5)$. This area is illustrated in Figure 7.23. Calculating the z values

$$z = \frac{22.5 - 25}{3.5355} = -.71 \quad \text{and} \quad z = \frac{23.5 - 25}{3.5355} = -.42$$

we find that $P(22.5 \leq x \leq 23.5) = P(-.71 \leq z \leq -.42) = .3372 - .2389 = .0983$. Therefore, we estimate that the binomial probability $P(x = 23)$ is .0983.

Making the proper continuity correction can sometimes be tricky. A good way to approach this is to list the numbers of successes that are included in the event for which the binomial probability is being calculated. Then assign the appropriate area under the normal curve to each number of successes in the list. Putting these areas together gives the normal curve area that must be calculated. For example, again consider the binomial random variable x with $n = 50$ and $p = .5$. If we wish to find $P(27 \leq x \leq 29)$, then the event $27 \leq x \leq 29$ includes 27, 28, and 29 successes. Because we assign the areas under the normal curve corresponding to the intervals [26.5, 27.5], [27.5, 28.5], and [28.5, 29.5] to the values 27, 28, and 29, respectively, then the area to be found under the normal curve is $P(26.5 \leq x \leq 29.5)$. Table 7.2 gives several other examples.

Ⓒ **EXAMPLE 7.8** The Cheese Spread Case: Improving Profitability

A food processing company markets a soft cheese spread that is sold in a plastic container with an "easy pour" spout. Although this spout works extremely well and is popular with consumers, it is expensive to produce. Because of the spout's high cost, the company has developed a new, less expensive spout. While the new, cheaper spout may alienate some purchasers, a company study shows that its introduction will increase profits if fewer than 10 percent of the cheese spread's current purchasers are lost. That is, if we let p be the true proportion of all current purchasers who would stop buying the cheese spread if the new spout were used, profits will increase as long as p is less than .10.

TABLE 7.2   **Several Examples of the Continuity Correction ($n = 50$)**

Binomial Probability	Numbers of Successes Included in Event	Normal Curve Area (with Continuity Correction)
$P(25 < x \leq 30)$	26, 27, 28, 29, 30	$P(25.5 \leq x \leq 30.5)$
$P(x \leq 27)$	0, 1, 2, . . . , 26, 27	$P(x \leq 27.5)$
$P(x > 30)$	31, 32, 33, . . . , 50	$P(x \geq 30.5)$
$P(27 < x < 31)$	28, 29, 30	$P(27.5 \leq x \leq 30.5)$

FIGURE 7.24   **Approximating the Binomial Probability $P(x \leq 63)$ by Using the Normal Curve When $\mu = np = 100$ and $\sigma = \sqrt{npq} = 9.4868$**

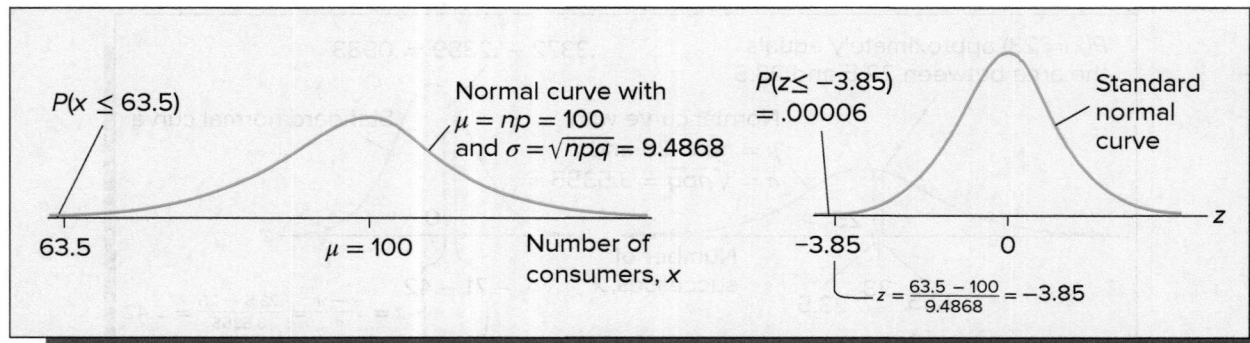

Suppose that (after trying the new spout) 63 of 1,000 randomly selected purchasers say that they would stop buying the cheese spread if the new spout were used. To assess whether p is less than .10, we will assume for the sake of argument that p equals .10, and we will use the sample information to weigh the evidence against this assumption and in favor of the conclusion that p is less than .10. Let the random variable x represent the number of the 1,000 purchasers who say they would stop buying the cheese spread. Assuming that p equals .10, then x is a binomial random variable with $n = 1,000$ and $p = .10$. Because the sample result of 63 is less than $\mu = np = 1,000(.1) = 100$, the expected value of x when p equals .10, we have some evidence to contradict the assumption that p equals .10. To evaluate the strength of this evidence, we calculate the probability that *63 or fewer* of the 1,000 randomly selected purchasers would say that they would stop buying the cheese spread if the new spout were used if, in fact, p equals .10.

Because both $np = 1,000(.10) = 100$ and $n(1 - p) = 1,000(1 - .10) = 900$ are at least 5, we can use the normal approximation to the binomial distribution to compute the needed probability. The appropriate normal curve has mean $\mu = np = 1,000(.10) = 100$ and standard deviation $\sigma = \sqrt{npq} = \sqrt{1,000(.10)(1 - .10)} = 9.4868$. In order to make the continuity correction, we note that the discrete value $x = 63$ is assigned the area under the normal curve corresponding to the interval from 62.5 to 63.5. It follows that the binomial probability $P(x \leq 63)$ is approximated by the normal probability $P(x \leq 63.5)$. This is illustrated in Figure 7.24. Calculating the z value for 63.5 to be

$$z = \frac{63.5 - 100}{9.4868} = -3.85$$

we find that

$$P(x \leq 63.5) = P(z \leq -3.85)$$

Using the normal table, we find that the area under the standard normal curve to the left of -3.85 is .00006. This says that, if p equals .10, then in only 6 in 100,000 of all possible random samples of 1,000 purchasers would 63 or fewer say they would stop buying the cheese spread if the new spout were used. Because it is very difficult to believe that such a small chance (a .00006 chance) has occurred, we have very strong evidence that p does not equal .10 and is, in fact, less than .10. Therefore, it seems that using the new spout will be profitable.

Exercises for Section 7.4

CONCEPTS ▪ **connect**

7.36   Explain why it might be convenient to approximate binomial probabilities by using areas under an appropriate normal curve.

7.37   Under what condition may we use the normal approximation to the binomial distribution?

7.38   Explain how we make a continuity correction. Why is a continuity correction needed when we approximate a binomial distribution by a normal distribution?

METHODS AND APPLICATIONS

7.39   Suppose that x has a binomial distribution with $n = 200$ and $p = .4$.

a Show that the normal approximation to the binomial can appropriately be used to calculate probabilities about x.

b Make continuity corrections for each of the following, and then use the normal approximation to the binomial to find each probability:

    (1) $P(x = 80)$
    (2) $P(x \le 95)$
    (3) $P(x < 65)$
    (4) $P(x \ge 100)$
    (5) $P(x > 100)$

7.40 Repeat Exercise 7.39 with $n = 200$ and $p = .5$.

7.41 An advertising agency conducted an ad campaign aimed at making consumers in an eastern state aware of a new product. Upon completion of the campaign, the agency claimed that 20 percent of consumers in the state had become aware of the product. The product's distributor surveyed 1,000 consumers in the state and found that 150 were aware of the product.

a Assuming that the ad agency's claim is true:

    (1) Verify that we may use the normal approximation to the binomial.
    (2) Calculate the mean, μ, and the standard deviation, σ, we should use in the normal approximation.
    (3) Find the probability that 150 or fewer consumers in a random sample of 1,000 consumers would be aware of the product.

b Should the distributor believe the ad agency's claim? Explain.

7.42 In order to gain additional information about respondents, some marketing researchers have used ultraviolet ink to precode questionnaires that promise confidentiality to respondents. Of 205 randomly selected marketing researchers who participated in an actual survey, 117 said that they disapprove of this practice. Suppose that, before the survey was taken, a marketing manager claimed that at least 65 percent of all marketing researchers would disapprove of the practice.

a Assuming that the manager's claim is correct, calculate the probability that 117 or fewer of 205 randomly selected marketing researchers would disapprove of the practice. Use the normal approximation to the binomial.

b Based on your result of part a, do you believe the marketing manager's claim? Explain.

7.43 When a store uses electronic article surveillance (EAS) to combat shoplifting, it places a small sensor on each item of merchandise. When an item is legitimately purchased, the sales clerk is supposed to remove the sensor to prevent an alarm from sounding as the customer exits the store. In an actual survey of 250 consumers, 40 said that if they were to set off an EAS alarm because store personnel (mistakenly) failed to deactivate merchandise, they would never shop at that store again. A company marketing the alarm system claimed that no more than 5 percent of all consumers would say that they would never shop at that store again if they were subjected to a false alarm.

a Assuming that the company's claim is valid, use the normal approximation to the binomial to calculate the probability that at least 40 of the 250 randomly selected consumers would say that they would never shop at that store again if they were subjected to a false alarm.

b Do you believe the company's claim based on your answer to part a? Explain.

7.44 A department store will place a sale item in a special display for a one-day sale. Previous experience suggests that 20 percent of all customers who pass such a special display will purchase the item. If 2,000 customers will pass the display on the day of the sale, and if a one-item-per-customer limit is placed on the sale item, how many units of the sale item should the store stock in order to have at most a 1 percent chance of running short of the item on the day of the sale? Assume here that customers make independent purchase decisions.

7.5 The Exponential Distribution (Optional)

LO7-7
Use the exponential distribution to compute probabilities (Optional).

In Example 6.11, we considered an air traffic control center where controllers occasionally misdirect pilots onto flight paths dangerously close to those of other aircraft. We found that the number of these controller errors in a given time period has a Poisson distribution and that the control center is averaging 20.8 errors per year. However, rather than focusing on the number of errors occurring in a given time period, we could study the time elapsing between successive errors. If we let x denote the number of weeks elapsing between successive errors, then x is a continuous random variable that is described by what is called the *exponential distribution*. Moreover, because the control center is averaging 20.8 errors per year, the center is averaging a mean, denoted λ, of $20.8/52 = .4$ error per week (λ is pronounced *lambda*). It follows that the control center is averaging a mean of $52/20.8 = 2.5$ (that is, $1/\lambda = 1/.4 = 2.5$) weeks between successive errors.

In general, if the number of events occurring per unit of time or space (for example, the number of controller errors per week or the number of imperfections per square yard of cloth) has a Poisson distribution with mean λ, then the number of units, x, of time or space between successive events has an *exponential distribution* with mean $1/\lambda$. The equation of

FIGURE 7.25   A Graph of the Exponential Distribution $f(x) = \lambda e^{-\lambda x}$

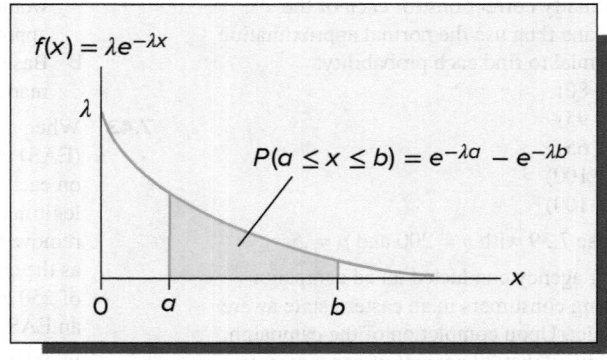

the probability curve describing the **exponential distribution** (or **exponential model**) is given in the following formula box.

The Exponential Distribution (The Exponential Model)

If x is described by an exponential distribution with mean $1/\lambda$, then the equation of the probability curve describing x is

$$f(x) = \begin{cases} \lambda e^{-\lambda x} & \text{for } x \geq 0 \\ 0 & \text{otherwise} \end{cases}$$

Using this probability curve, it can be shown that

$$P(a \leq x \leq b) = e^{-\lambda a} - e^{-\lambda b}$$

In particular, because $e^0 = 1$ and $e^{-\infty} = 0$, this implies that

$$P(x \leq b) = 1 - e^{-\lambda b} \quad \text{and} \quad P(x \geq a) = e^{-\lambda a}$$

Furthermore, both the mean and the standard deviation of the population of all possible observed values of a random variable x that has an exponential distribution are equal to $1/\lambda$. That is, $\mu_x = \sigma_x = 1/\lambda$.

The graph of the equation describing the exponential distribution and the probability $P(a \leq x \leq b)$ where x is described by this exponential distribution are illustrated in Figure 7.25.

We illustrate the use of the exponential distribution in the following examples.

EXAMPLE 7.9 The Air Safety Case: Traffic Control Errors

We have seen in the air traffic control example that the control center is averaging $\lambda = .4$ error per week and $1/\lambda = 1/.4 = 2.5$ weeks between successive errors. Letting x denote the number of weeks elapsing between successive errors, the equation of the exponential distribution describing x is $f(x) = \lambda e^{-\lambda x} = .4e^{-.4x}$. For example, the probability that the time between successive errors will be between 1 and 2 weeks is

$$P(1 \leq x \leq 2) = e^{-\lambda a} - e^{-\lambda b} = e^{-\lambda(1)} - e^{-\lambda(2)}$$
$$= e^{-.4(1)} - e^{-.4(2)} = e^{-.4} - e^{-.8}$$
$$= .6703 - .4493 = .221$$

EXAMPLE 7.10 Emergency Room Arrivals

Suppose that the number of people who arrive at a hospital emergency room during a given time period has a Poisson distribution. It follows that the time, x, between successive arrivals of people to the emergency room has an exponential distribution. Furthermore, historical records indicate that the mean time between successive arrivals of people to the emergency room is seven minutes. Therefore, $\mu_x = 1/\lambda = 7$, which implies that $\lambda = 1/7 = .14286$. Noting that $\sigma_x = 1/\lambda = 7$, it follows that

$$\mu_x - \sigma_x = 7 - 7 = 0 \quad \text{and} \quad \mu_x + \sigma_x = 7 + 7 = 14$$

Therefore, the probability that the time between successive arrivals of people to the emergency room will be within (plus or minus) one standard deviation of the mean interarrival time is

$$P(0 \leq x \leq 14) = e^{-\lambda a} - e^{-\lambda b}$$
$$= e^{-(.14286)(0)} - e^{-(.14286)(14)}$$
$$= 1 - .1353$$
$$= .8647$$

To conclude this section we note that the exponential and related Poisson distributions are useful in analyzing waiting lines, or **queues.** In general, **queuing theory** attempts to determine the number of servers (for example, doctors in an emergency room) that strikes an optimal balance between the time customers wait for service and the cost of providing service. The reader is referred to any textbook on management science or operations research for a discussion of queueing theory.

Exercises for Section 7.5

CONCEPTS   connect

7.45 Give two examples of situations in which the exponential distribution might be used appropriately. In each case, define the random variable having an exponential distribution.

7.46 State the formula for the exponential probability curve. Define each symbol in the formula.

7.47 Explain the relationship between the Poisson and exponential distributions.

METHODS AND APPLICATIONS

7.48 Suppose that the random variable x has an exponential distribution with $\lambda = 2$.
  a Write the formula for the exponential probability curve of x. What are the possible values of x?
  b Sketch the probability curve.
  c Find $P(x \leq 1)$.
  d Find $P(.25 \leq x \leq 1)$.
  e Find $P(x \geq 2)$.
  f Calculate the mean, μ_x, the variance, σ_x^2, and the standard deviation, σ_x, of the exponential distribution of x.
  g Find the probability that x will be in the interval $[\mu_x \pm 2\sigma_x]$.

7.49 Repeat Exercise 7.48 with $\lambda = 3$.

7.50 Recall in Exercise 6.32 that the number of customer arrivals at a bank's drive-up window in a 15-minute period is Poisson distributed with a mean of seven customer arrivals per 15-minute period. Define the random variable x to be the time (in minutes) between successive customer arrivals at the bank's drive-up window.
  a Write the formula for the exponential probability curve of x.
  b Sketch the probability curve of x.
  c Find the probability that the time between arrivals is
    (1) Between one and two minutes.
    (2) Less than one minute.
    (3) More than three minutes.
    (4) Between ½ and 3½ minutes.
  d Calculate μ_x, σ_x^2, and σ_x.
  e Find the probability that the time between arrivals falls within one standard deviation of the mean; within two standard deviations of the mean.

7.51 The length of a particular telemarketing phone call, x, has an exponential distribution with mean equal to 1.5 minutes.
  a Write the formula for the exponential probability curve of x.
  b Sketch the probability curve of x.
  c Find the probability that the length of a randomly selected call will be
    (1) No more than three minutes.
    (2) Between one and two minutes.
    (3) More than four minutes.
    (4) Less than 30 seconds.

7.52 The maintenance department in a factory claims that the number of breakdowns of a particular machine follows a Poisson distribution with a mean of two breakdowns every 500 hours. Let x denote the time (in hours) between successive breakdowns.
  a Find λ and μ_x.
  b Write the formula for the exponential probability curve of x.
  c Sketch the probability curve.
  d Assuming that the maintenance department's claim is true, find the probability that the time between successive breakdowns is at most five hours.
  e Assuming that the maintenance department's claim is true, find the probability that the time between successive breakdowns is between 100 and 300 hours.
  f Suppose that the machine breaks down five hours after its most recent breakdown. Based on your answer to part d, do you believe the maintenance department's claim? Explain.

7.53 Suppose that the number of accidents occurring in an industrial plant is described by a Poisson distribution with an average of one accident per month. Let x denote the time (in months) between successive accidents.

  a Find the probability that the time between successive accidents is

(1) More than two months.
(2) Between one and two months.
(3) Less than one week (¼ of a month).

  b Suppose that an accident occurs less than one week after the plant's most recent accident. Would you consider this event unusual enough to warrant special investigation? Explain.

LO7-8

Use a normal probability plot to help decide whether data come from a normal distribution (Optional).

7.6 The Normal Probability Plot (Optional)

The **normal probability plot** is a graphic that is used to visually check whether sample data come from a normal distribution. In order to illustrate the construction and interpretation of a normal probability plot, consider the e-billing case and suppose that the trucking company operates in three regions of the country—the north, central, and south regions. In each region, 24 invoices are randomly selected and the payment time for each sampled invoice is found. The payment times obtained in the north region are given in the page margin. To construct a normal probability plot of these payment times, we first arrange the payment times in order from smallest to largest. The ordered payment times are shown in column (1) of Table 7.3. Next, for each ordered payment time, we compute the quantity $i/(n + 1)$, where i denotes the observation's position in the ordered list of data and n denotes the sample size. For instance, for the first and second ordered payment times, we compute $1/(24 + 1) = 1/25 = .04$ and $2/(24 + 1) = 2/25 = .08$. Similarly, for the last (24th) ordered payment time, we compute $24/(24 + 1) = 24/25 = .96$. The positions ($i$ values) of all 24 payment times are given in column (2) of Table 7.3, and the corresponding values of $i/(n + 1)$ are given in column (3) of this table. We continue by computing what is called the **standardized normal quantile value** for each ordered payment time. This value (denoted O_i) is the z value that gives an area of $i/(n + 1)$ to its left under the standard normal curve. Figure 7.26 illustrates finding O_1. Specifically, O_1—the standardized normal quantile value corresponding to the first ordered residual—is the z value that gives an area of $1/(24 + 1) = .04$ to its left under the standard normal curve. Looking up a cumulative normal curve area of .04 in Table 7.1, the z value (to two decimal places) that gives a cumulative area closest to .04 is $O_1 = -1.75$ (see Figure 7.26). Similarly, O_2 is the z value that gives an area of $2/(24 + 1) = .08$ to its left under the standard normal curve. Looking up a cumulative normal curve area of .08 in Table 7.1, the z value (to two decimal places) that gives a cumulative area closest to .08 is $O_2 = -1.41$. The standardized normal quantile values corresponding to all 24 ordered payment times are given in column (4) of Table 7.3. Finally, we obtain the **normal probability plot** by plotting the 24 ordered payment times on the vertical axis versus the corresponding standardized normal quantile values (O_i values) on the horizontal axis. Figure 7.27 gives an Excel add-in (MegaStat) output of this normal probability plot.

In order to interpret the normal plot, notice that, although the areas in column (3) of Table 7.3 (that is, the $i/(n + 1)$ values: .04, .08, .12, etc.) are equally spaced, the z values corresponding to these areas are not equally spaced. Because of the mound-shaped nature of the standard normal curve, the negative z values get closer together as they get closer to the mean ($z = 0$) and the positive z values get farther apart as they get farther from the mean (more positive). If the distances between the payment times behave the same way as the distances between the z values—that is, if the distances between the payment times are proportional to the distances between the z values—then the normal probability plot will be a straight line. This would suggest that the payment times are normally distributed. Examining Figure 7.27, the normal probability plot for the payment times from the north region is approximately a straight line and, therefore, it is reasonable to assume that these payment times are approximately normally distributed.

In the page margin we give the ordered payment times for the central region, and Figure 7.28 plots these values versus the standardized normal quantile values in column (4) of Table 7.3. The resulting normal probability plot for the central region has a nonlinear appearance. The plot points rise more steeply at first and then continue to increase at

North Region Payment Times

26	28
27	26
21	21
22	32
22	23
23	24
27	25
20	15
22	17
29	19
18	34
24	30

DS NorthPay

Ordered Central Region Payment Times	Ordered South Region Payment Times
7	19
12	19
15	19
18	20
19	20
20	21
21	21
22	22
22	23
23	23
24	24
24	24
25	25
25	26
25	27
26	27
26	28
27	29
28	31
28	33
29	36
29	39
29	44
29	50

DS CentralPay
DS SouthPay

TABLE 7.3   **Calculations for Normal Probability Plots in the e-billing Example**

Ordered North Region Payment Times Column (1)	Observation Number (i) Column (2)	Area $i/(n+1)$ Column (3)	z value O_i Column (4)
15	1	0.04	−1.75
17	2	0.08	−1.41
18	3	0.12	−1.18
19	4	0.16	−0.99
20	5	0.2	−0.84
21	6	0.24	−0.71
21	7	0.28	−0.58
22	8	0.32	−0.47
22	9	0.36	−0.36
22	10	0.4	−0.25
23	11	0.44	−0.15
23	12	0.48	−0.05
24	13	0.52	0.05
24	14	0.56	0.15
25	15	0.6	0.25
26	16	0.64	0.36
26	17	0.68	0.47
27	18	0.72	0.58
27	19	0.76	0.71
28	20	0.8	0.84
29	21	0.84	0.99
30	22	0.88	1.18
32	23	0.92	1.41
34	24	0.96	1.75

FIGURE 7.26   **The Standardized Normal Quantile Value O_1**

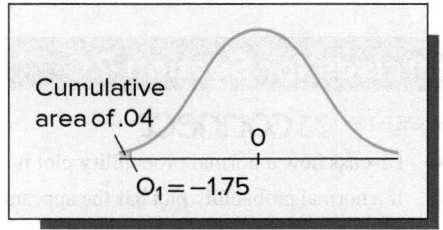

Cumulative area of .04

$O_1 = −1.75$

FIGURE 7.27   **Excel Add-in (MegaStat) Normal Probability Plot for the North Region: Approximate Normality**

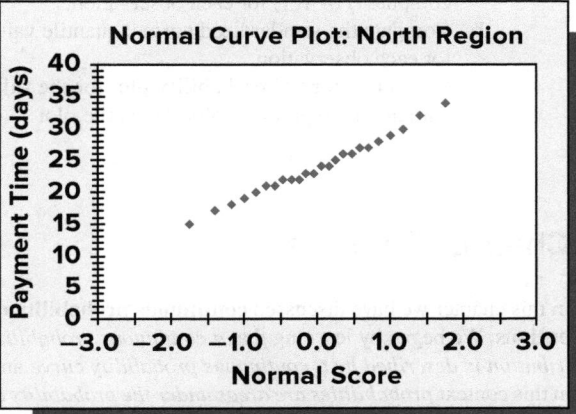

a decreasing rate. This pattern indicates that the payment times for the central region are skewed to the left. Here the rapidly rising points at the beginning of the plot are due to the payment times being farther apart in the left tail of the distribution. In the page margin we also give the ordered payment times for the south region, and Figure 7.29 gives the normal probability plot for this region. This plot also has a nonlinear appearance. The points rise slowly at first and then increase at an increasing rate. This pattern indicates

FIGURE 7.28   **Excel Add-in (MegaStat) Normal Probability Plot for the Central Region: Data Skewed to the Left**

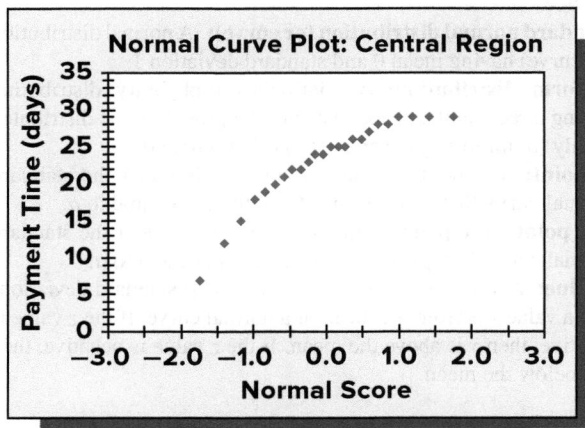

FIGURE 7.29   **Excel Add-in (MegaStat) Normal Probability Plot for the South Region: Data Skewed to the Right**

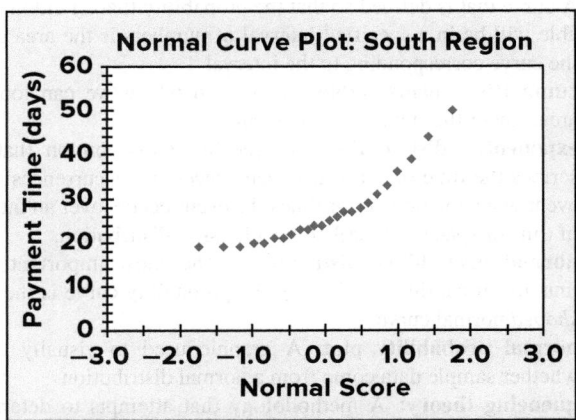

that the payment times for the south region are skewed to the right. Here the rapidly rising points on the right side of the plot are due to the payment times being farther apart in the right tail of the distribution.

Exercises for Section 7.6

CONCEPTS ▦ **connect**

7.54 Discuss how a normal probability plot is constructed.

7.55 If a normal probability plot has the appearance of a straight line, what should we conclude?

METHODS AND APPLICATIONS

7.56 Consider the sample of 12 incomes given in Example 3.2.

    a Sort the income data from smallest to largest, and compute $i/(n + 1)$ for each observation.

    b Compute the standardized normal quantile value O_i for each observation.

    c Graph the normal probability plot for the salary data and interpret this plot. Does the plot

indicate that the data are skewed? Explain. ⓄⓈ Incomes

7.57 Consider the 20 DVD satisfaction ratings given in the file DVDSat. Construct a normal probability plot for these data and interpret the plot. ⓄⓈ DVDSat

7.58 A normal probability plot can be constructed using Minitab. Use the selections Stat: Basic Statistics: Normality test, and select the data to be analyzed. Although the Minitab plot is slightly different from the plot outlined in this section, its interpretation is the same. Use Minitab to construct a normal probability plot of the gas mileage data in Table 3.1. Interpret the plot. ⓄⓈ GasMiles

Chapter Summary

In this chapter we have discussed **continuous probability distributions.** We began by learning that *a continuous probability distribution is described by a continuous probability curve* and that in this context *probabilities are areas under the probability curve.* We next studied two important continuous probability distributions—**the uniform distribution and the normal distribution.** In particular, we concentrated on the normal distribution, which is the most important continuous probability distribution. We learned about the properties of the normal curve, and we saw how to use **a normal table** to find various areas under a normal curve.

We then demonstrated how we can use a normal curve probability to make a statistical inference. We continued with an optional section that explained how we can use a normal curve probability to approximate a binomial probability. Then we presented an optional section that discussed another important continuous probability distribution—**the exponential distribution,** and we saw how this distribution is related to the Poisson distribution. Finally, we concluded this chapter with an optional section that explained how to use a **normal probability plot** to decide whether data come from a normal distribution.

Glossary of Terms

continuous probability distribution (or **probability curve**): A curve that is defined so that the probability that a random variable will be in a specified interval of numbers is the area under the curve corresponding to the interval.

cumulative normal table: A table in which we can look up areas under the standard normal curve.

exponential distribution: A probability distribution that describes the time or space between successive occurrences of an event when the number of times the event occurs over an interval of time or space is described by a Poisson distribution.

normal probability distribution: The most important continuous probability distribution. Its probability curve is the *bell-shaped* normal curve.

normal probability plot: A graphic used to visually check whether sample data come from a normal distribution.

queueing theory: A methodology that attempts to determine the number of servers that strikes an optimal balance between

the time customers wait for service and the cost of providing service.

standard normal distribution (or curve): A normal distribution (or curve) having mean 0 and standard deviation 1.

uniform distribution: A continuous probability distribution having a rectangular shape that says the probability is distributed evenly (or uniformly) over an interval of numbers.

z_α **point:** The point on the horizontal axis under the standard normal curve that gives a right-hand tail area equal to α.

$-z_\alpha$ **point:** The point on the horizontal axis under the standard normal curve that gives a left-hand tail area equal to α.

z **value:** A value that tells us the number of standard deviations that a value x is from the mean of a normal curve. If the z value is positive, then x is above the mean. If the z value is negative, then x is below the mean.

Important Formulas and Graphics

The uniform probability curve: page 342

Mean and standard deviation of a uniform distribution: page 342

The normal probability curve: page 344

z values: pages 346, 351

Finding normal probabilities: page 352

Normal approximation to the binomial distribution: page 360

The exponential probability curve: page 364

Mean and standard deviation of an exponential distribution: page 364

Constructing a normal probability plot: pages 366–368

Supplementary Exercises

7.59 In a bottle-filling process, the amount of drink injected into 16 oz bottles is normally distributed with a mean of 16 oz and a standard deviation of .02 oz. Bottles containing less than 15.95 oz do not meet the bottler's quality standard. What percentage of filled bottles do not meet the standard?

7.60 In a murder trial in Los Angeles, a shoe expert stated that the range of heights of men with a size 12 shoe is 71 inches to 76 inches. Suppose the heights of all men wearing size 12 shoes are normally distributed with a mean of 73.5 inches and a standard deviation of 1 inch. What is the probability that a randomly selected man who wears a size 12 shoe
a Has a height outside the range 71 inches to 76 inches?
b Is 74 inches or taller?
c Is shorter than 70.5 inches?

7.61 In the movie *Forrest Gump*, the public school required an IQ of at least 80 for admittance.
a If IQ test scores are normally distributed with mean 100 and standard deviation 16, what percentage of people would qualify for admittance to the school?
b If the public school wishes 95 percent of all children to qualify for admittance, what minimum IQ test score should be required for admittance?

7.62 The amount of sales tax paid on a purchase is rounded to the nearest cent. Assume that the round-off error is uniformly distributed in the interval −.5 to .5 cent.
a Write the formula for the probability curve describing the round-off error.
b Graph the probability curve describing the round-off error.
c What is the probability that the round-off error exceeds .3 cent or is less than −.3 cent?
d What is the probability that the round-off error exceeds .1 cent or is less than −.1 cent?
e Find the mean and the standard deviation of the round-off error.
f Find the probability that the round-off error will be within one standard deviation of the mean.

7.63 A *consensus forecast* is the average of a large number of individual analysts' forecasts. Suppose the individual forecasts for a particular interest rate are normally distributed with a mean of 5.0 percent and a standard deviation of 1.2 percent. A single analyst is randomly selected. Find the probability that his/her forecast is
a At least 3.5 percent.
b At most 6 percent.
c Between 3.5 percent and 6 percent.

7.64 Recall from Exercise 7.63 that individual forecasts of a particular interest rate are normally distributed with a mean of 5 percent and a standard deviation of 1.2 percent.
a What percentage of individual forecasts are at or below the 10th percentile of the distribution of forecasts? What percentage are at or above the 10th percentile? Find the 10th percentile of the distribution of individual forecasts.
b Find the first quartile, Q_1, and the third quartile, Q_3, of the distribution of individual forecasts.

7.65 The scores on the entrance exam at a well-known, exclusive law school are normally distributed with a mean score of 200 and a standard deviation equal to 50. At what value should the lowest passing score be set if the school wishes only 2.5 percent of those taking the test to pass?

7.66 A machine is used to cut a metal automobile part to its desired length. The machine can be set so that the mean length of the part will be any value that is desired. The standard deviation of the lengths always runs at .02 inch. Where should the mean be set if we want only .4 percent of the parts cut by the machine to be shorter than 15 inches long?

7.67 A motel accepts 325 reservations for 300 rooms on July 1, expecting 10 percent no-shows on average from past records. Use the normal approximation to the binomial to find the probability that all guests who arrive on July 1 will receive a room.

7.68 Suppose a software company finds that the number of errors in its software per 1,000 lines of code is described by a Poisson distribution. Furthermore, it is found that there is an average of four errors per 1,000 lines of code. Letting x denote the number of lines of code between successive errors:
a Find the probability that there will be at least 400 lines of code between successive errors in the company's software.
b Find the probability that there will be no more than 100 lines of code between successive errors in the company's software.

7.69 THE INVESTMENT CASE InvestRet
For each investment class in Table 3.18, assume that future returns are normally distributed with the population mean and standard deviation given in Table 3.18. Based on this assumption:
a For each investment class, find the probability of a return that is less than zero (that is, find the probability of a loss). Is your answer reasonable for all investment classes? Explain.

b For each investment class, find the probability of a return that is
   (1) Greater than 5 percent.
   (2) Greater than 10 percent.
   (3) Greater than 20 percent.
   (4) Greater than 50 percent.
c For which investment classes is the probability of a return greater than 50 percent essentially zero? For which investment classes is the probability of such a return greater than 1 percent? Greater than 5 percent?
d For which investment classes is the probability of a loss essentially zero? For which investment classes is the probability of a loss greater than 1 percent? Greater than 10 percent? Greater than 20 percent?

7.70   The daily water consumption for an Ohio community is normally distributed with a mean consumption of 800,000 gallons and a standard deviation of 80,000 gallons. The community water system will experience a noticeable drop in water pressure when the daily water consumption exceeds 984,000 gallons. What is the probability of experiencing such a drop in water pressure?

7.71   Suppose the times required for a cable company to fix cable problems in its customers' homes are uniformly distributed between 10 minutes and 25 minutes. What is the probability that a randomly selected cable repair visit will take at least 15 minutes?

7.72   Suppose the waiting time to get food after placing an order at a fast-food restaurant is exponentially distributed with a mean of 60 seconds. If a randomly selected customer orders food at the restaurant, what is the probability that the customer will wait at least
a 90 seconds?
b Two minutes?

7.73   Net interest margin—often referred to as *spread*—is the difference between the rate banks pay on deposits and the rate they charge for loans. Suppose that the net interest margins for all U.S. banks are normally distributed with a mean of 4.15 percent and a standard deviation of .5 percent.

a Find the probability that a randomly selected U.S. bank will have a net interest margin that exceeds 5.40 percent.
b Find the probability that a randomly selected U.S. bank will have a net interest margin less than 4.40 percent.
c A bank wants its net interest margin to be less than the net interest margins of 95 percent of all U.S. banks. Where should the bank's net interest margin be set?

7.74   In an article in *Advertising Age*, Nancy Giges studies global spending patterns. Giges presents data concerning the percentage of adults in various countries who have purchased various consumer items (such as soft drinks, athletic footwear, blue jeans, beer, and so on) in the past three months.
a Suppose we wish to justify the claim that fewer than 50 percent of adults in Germany have purchased blue jeans in the past three months. The survey reported by Giges found that 45 percent of the respondents in Germany had purchased blue jeans in the past three months.

   Assume that a random sample of 400 German adults was employed, and let p be the proportion of all German adults who have purchased blue jeans in the past three months. If, for the sake of argument, we assume that $p = .5$, use the normal approximation to the binomial distribution to calculate the probability that 45 percent or fewer of 400 randomly selected German adults would have purchased blue jeans in the past three months. Note: Because 45 percent of 400 is 180, you should calculate the probability that 180 or fewer of 400 randomly selected German adults would have purchased blue jeans in the past three months.
b Based on the probability you computed in part *a*, would you conclude that p is really less than .5? That is, would you conclude that fewer than 50 percent of adults in Germany have purchased blue jeans in the past three months? Explain.

7.75   Assume that the ages for first marriages are normally distributed with a mean of 26 years and a standard deviation of 4 years. What is the probability that a person getting married for the first time is in his or her twenties?

Appendix 7.1 ■ Normal Distribution Using Excel

Normal probability $P(X \le 31.2)$ in Example 7.3:

- Click in the cell where you wish to place the answer. Here we have clicked in cell A11. Then select the Insert Function button f_x from the Excel toolbar.

- In the Insert Function dialog box, select Statistical from the "Or select a category:" menu, select NORM.DIST from the "Select a function:" menu, and click OK.

- In the NORM.DIST: Function Arguments dialog box, enter the value 31.2 in the X window.

- Enter the value 33 in the Mean window.

- Enter the value 0.7 in the Standard_dev window.

- Enter the value 1 in the Cumulative window.

- Click OK in the NORM.DIST: Function Arguments dialog box.

- When you click OK in this dialog box, the answer will be placed in cell A11.

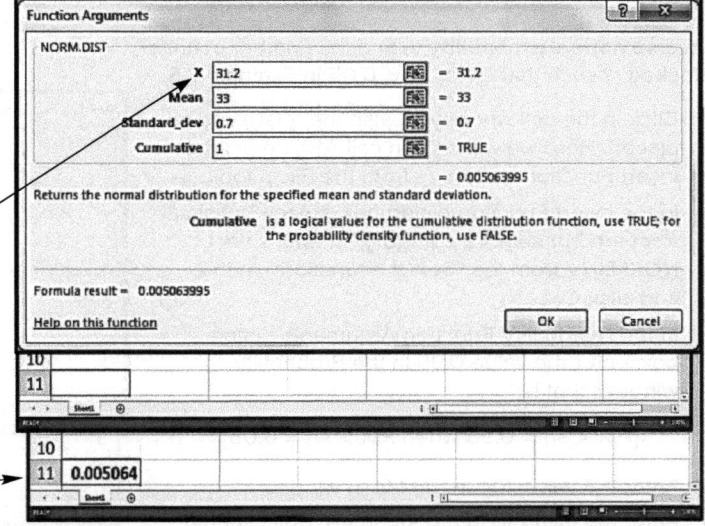

Source: Microsoft Office Excel 2016

Normal probability $P(X < 153$ or $X > 167)$ in Example 7.4):

- Enter the headings—x, P(X < x), P(X > x)—in the spreadsheet where you wish the results to be placed. Here we will enter these headings in cells A12, B12, and C12. The calculated results will be placed below the headings.

- In cells A13 and A14, enter the values 153 and 167.

- Click in cell B13 and select the Insert Function button f_x from the Excel toolbar.

- In the Insert Function dialog box, select Statistical from the "Or select a category:" menu, select NORM.DIST from the "Select a function:" menu, and click OK.

- In the NORM.DIST Function Arguments dialog box, enter the cell location A13 in the X window.

- Enter the value 159.3958 in the Mean window.

- Enter the value 6.4238 in the Standard_dev window.

- Enter the value 1 in the Cumulative window.

- Click OK in the NORM.DIST Function Arguments dialog box.

- When you click OK, the result for $P(X < 153)$ will be placed in cell B13. Double-click the drag-handle (in the lower right corner) of cell B13 to automatically extend the cell formula of B13 through cell B14.

- In cells C13 and C14, enter the formulas =1−B13 and =1−B14. The results for $P(X > 153)$ and $P(X > 167)$ will be placed in cells C13 and C14.

- In cell E14, enter the formula =B13+C14.

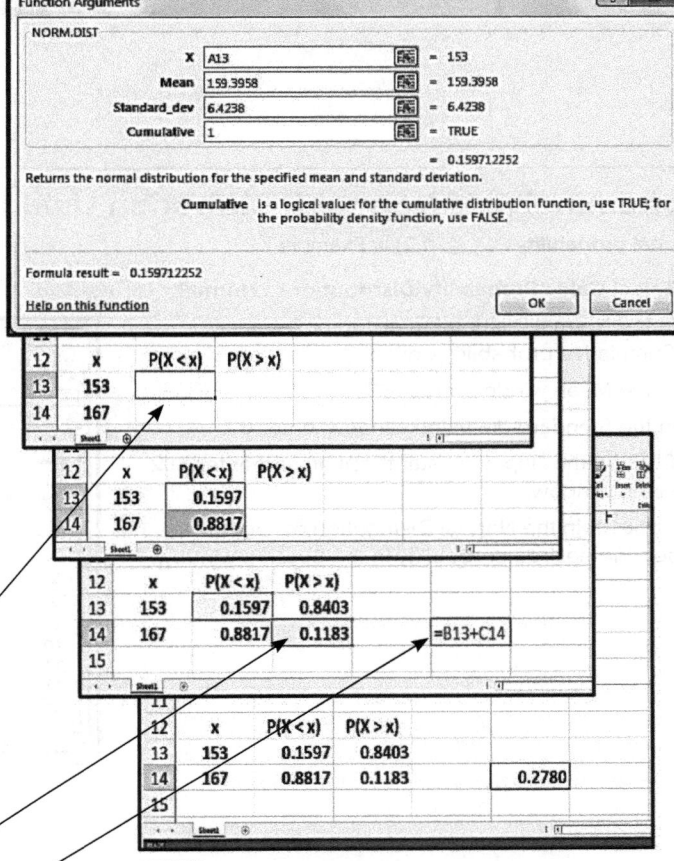

Source: Microsoft Office Excel 2016

The desired probability is in cell E14, the sum of the lower-tail probability for 153 and the upper-tail probability for 167. This value differs slightly from the value in Example 7.4 because Excel carries out probability calculations to higher precision than can be achieved using normal probability tables.

Inverse normal probability to find the number of units stocked *st* such that $P(X > st) = 0.05$ in Example 7.5:

- Click in the cell where you wish the answer to be placed. Here we will click in cell A10. Select the Insert Function button $\boxed{f_x}$ from the Excel toolbar.

- In the Insert Function dialog box, select Statistical from the "Or select a category:" menu, select NORM.INV from the "Select a function:" menu, and click OK.

- In the NORM.INV Function Arguments dialog box, enter the value 0.95 in the Probability window; that is,

    $$[P(X < st) = 0.95 \text{ when } P(X > st) = 0.05]$$

- Enter the value 100 in the Mean window.
- Enter the value 10 in the Standard_dev window.
- Click OK in the NORM.INV Function Arguments dialog box.
- When you click OK, the answer is placed in cell A10.

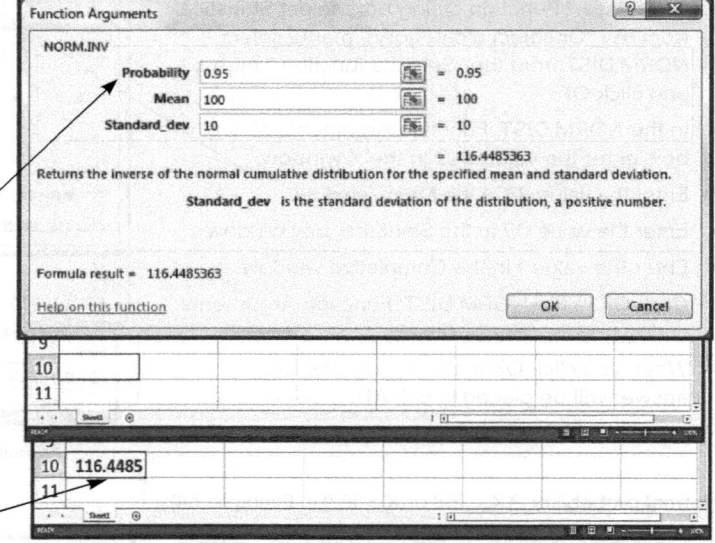

Source: Microsoft Office Excel 2016

Appendix 7.2 ■ Normal Distribution Using Minitab

Normal probability $P(X \leq 31.2)$ in Example 7.3:

- Select **Calc : Probability Distributions : Normal**
- In the Normal Distribution dialog box, select the Cumulative probability option.
- In the Mean window, enter 33.
- In the Standard deviation window, enter 0.7.
- Click on the "Input constant" option and enter 31.2 in the window.
- Click OK in the Normal Distribution dialog box to see the desired probability in the Session window.

Source: Minitab 18

Normal probability $P(X < 153$ or $X > 167)$ in Example 7.4:

- In columns C1–C3, enter the variable names x, $P(X < x)$, and $P(X > x)$.
- In column C1, enter the values 153 and 167.
- Select **Calc : Probability Distributions : Normal**
- In the Normal Distribution dialog box, select the Cumulative probability option.
- In the Mean window, enter 159.3958.
- In the Standard deviation window, enter 6.4238.
- Click on the "Input column" option, enter x in the window, and enter 'P(X < x)' in the "Optional storage" window.
- Click OK in the Normal Distribution dialog box.
- Select **Calc : Calculator**.
- In the Calculator dialog box, enter 'P(X > x)' in the "Store result in variable" window.
- Enter 1 − 'P(X > x)' in the Expression window.
- Click OK in the Calculator dialog box.

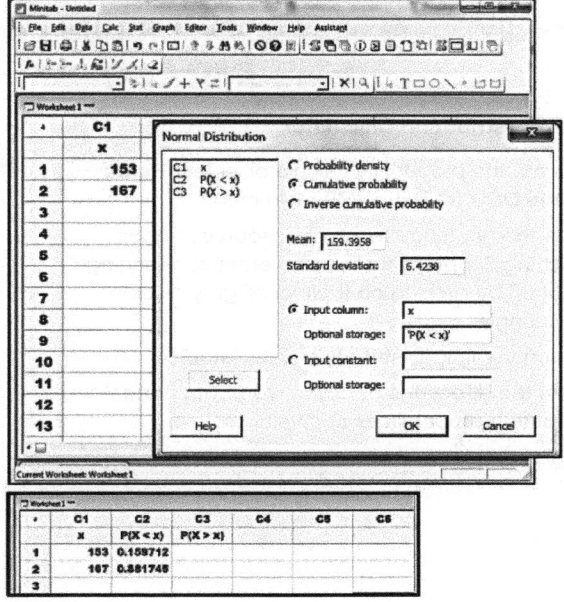

Source: Minitab 18

The desired probability is the sum of the lower-tail probability for 153 and the upper-tail probability for 167 or $0.159712 + 0.118255 = 0.277967$. This value differs slightly from the value in Example 7.4 because Minitab carries out the calculations to higher precision than what can be achieved using normal probability tables.

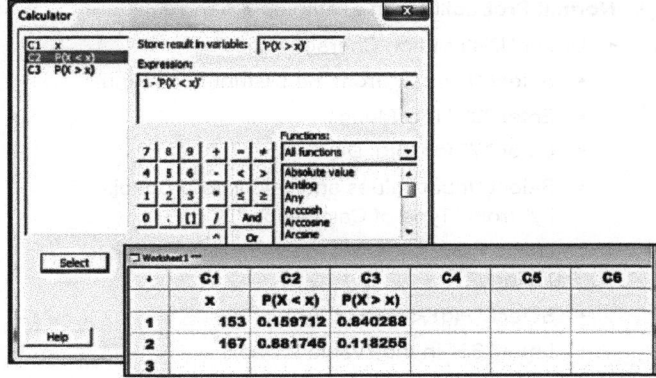

Source: Minitab 18

Inverse normal probability to find *st*, the number of units stocked, such that $P(X > st) = .05$ in Example 7.5:

- Select **Calc : Probability Distributions : Normal**
- In the Normal Distribution dialog box, select the Inverse cumulative probability option.
- In the Mean window, enter 100.
- In the Standard deviation window, enter 10.
- Click the "Input constant" option and enter 0.95 in the window. That is, $P(X \le st) = 0.95$ when $P(X > st) = 0.05$.
- Click OK in the Normal Distribution dialog box to see the value of *st* in the Session window.

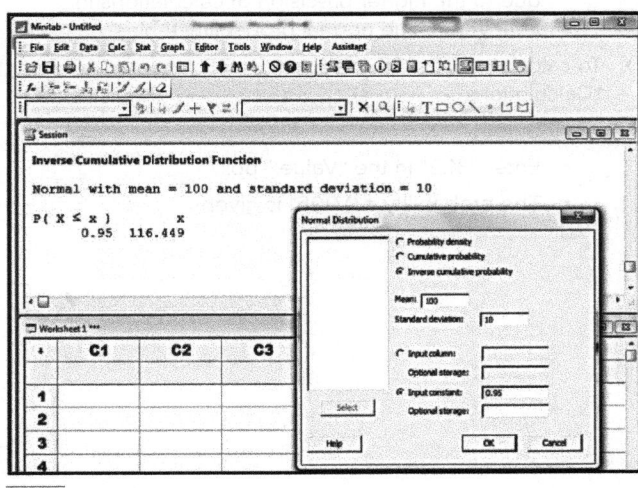

Source: Minitab 18

Appendix 7.3 ■ Normal Distribution and Normal Probability Plot Using JMP

Using the Distribution Calculator

- To locate the probability calculator, select **Help : Sample Data** from the JMP main menu.
- Under the heading "Teaching resources," open "Teaching Scripts," and then "Interactive Teaching Modules" by clicking on their small gray disclosure triangles.
- Click on the link "Distribution Calculator."
- Follow the directions below to work with normal distribution probabilities and values.

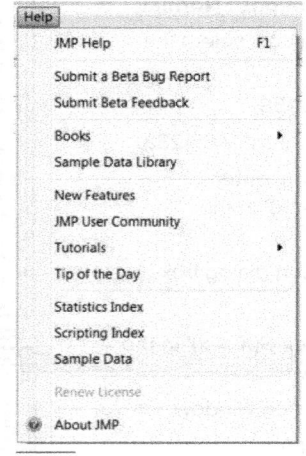
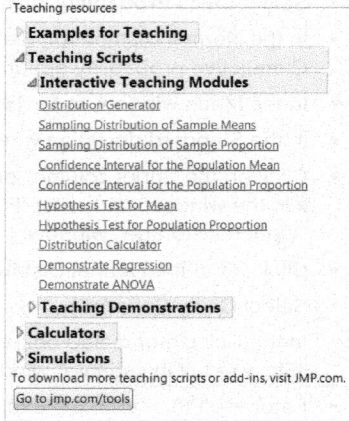

Source: JMP Pro 14

- **Normal Probabilities** in Examples 7.2 and 7.3:
  - Under "Distribution Characteristics":
    - Select "Normal" from the Distribution menu.
    - Enter "33" for "Mean."
    - Enter ".7" for "Std. Dev."
    - Select "Input values and calculate probability" from "Type of Calculation."
- To calculate P(32 ≤ X ≤ 35) in Example 7.2, under "Calculations":
    - Select "P(q1 < X <= q2)."
    - Enter "35" in the "Value 2:" box.
    - Enter "32" in the "Value 1:" box.
    - The probability = 0.9213 is given. This value differs slightly from the value in Example 7.2 due to the higher precision of calculations in JMP versus the normal probability tables.
- To calculate P(X ≤ 31.2) in Example 7.3, under "Calculations":
    - Select "P(X <= q)."
    - Enter "31.2" in the "Value:" box.
    - The probability = 0.0051 is given.

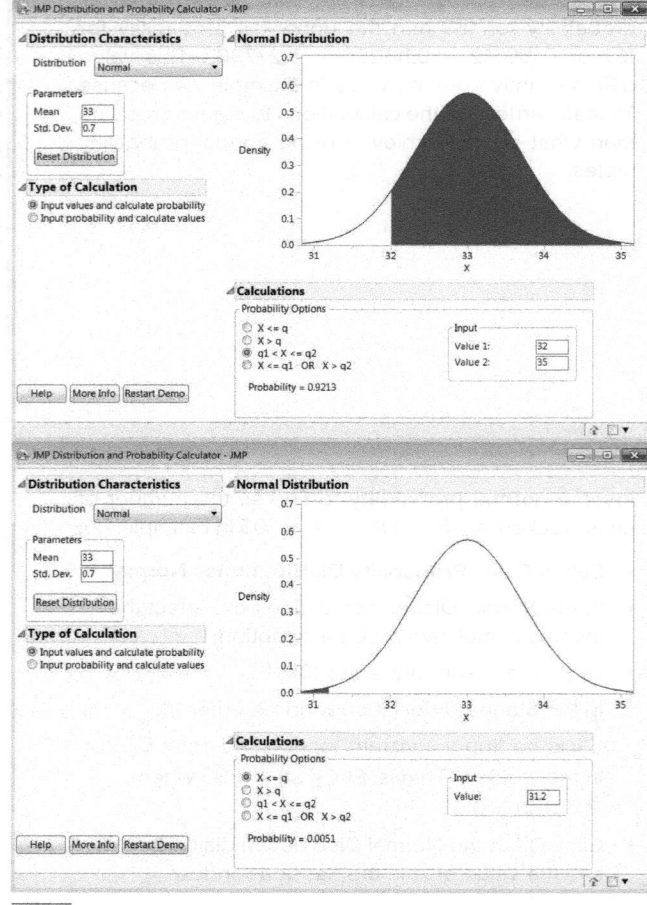

Source: JMP Pro 14

- **Inverse Normal Probability** in Example 7.5:
  - Under "Distribution Characteristics":
    - Select "Normal" from the Distribution menu.
    - Enter "100" for "Mean."
    - Enter "10" for "Std. Dev."
    - Select "Input probability and calculate values" from "Type of Calculation."
  - Under "Calculations":
    - Select "Right tail probability."
    - Enter "0.05" in the "Probability:" box.
- The value = 116.4485 is given. This value differs slightly from the value in Example 7.5 due to the higher precision of calculations in JMP versus the normal probability tables.

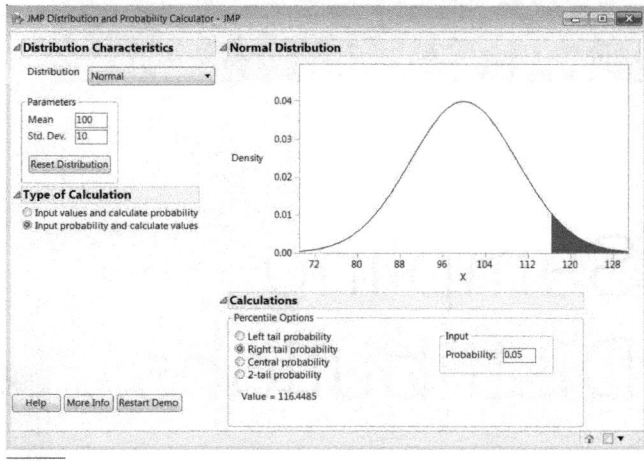

Source: JMP Pro 14

- **Normal Probability Plots** in Figure 7.27 (data file: NorthPay):
  - After opening the NorthPay.jmp file, select **Analyze : Distribution**.
  - In the Distribution dialog box, select the "Payment Time" column and click the "Y, Columns" button. Click "OK."
  - From the red triangle menu beside "Payment Time," select "Display Options : Horizontal Layout."
  - From the red triangle menu beside "Payment Time," select "Normal Quantile Plot."
  - The variable "Payment Time" is on the horizontal axis and the corresponding "Normal Score" is on the vertical axis. When the points follow the red "Diagonal Reference Line," the distribution shows approximately normality. The red curves indicate 95% confidence bounds for the points in the normal probability plot. If the data come from a normal distribution, 95% of all points will be within these bounds. The bounds will be wider for smaller data sets and smaller for larger data sets.

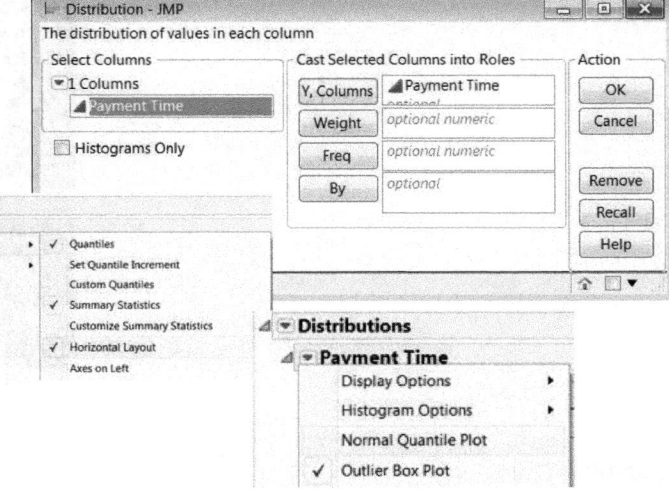

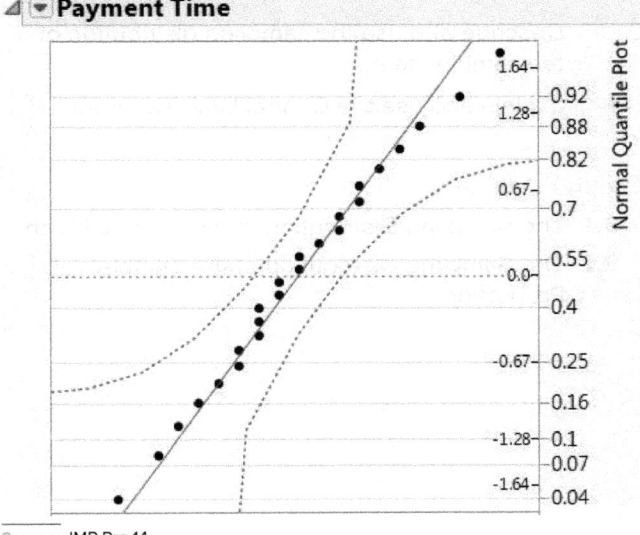

Source: JMP Pro 14

Design Elements: (CD): ©Comstock Images/Alamy; (All Others): ©McGraw-Hill Education

## CHAPTER 8

Sampling Distributions

©Hybrid Images/Getty Images RF

Learning Objectives

After mastering the material in this chapter, you will be able to:

LO8-1 Describe and use the sampling distribution of the sample mean.

LO8-2 Explain and use the Central Limit Theorem.

LO8-3 Describe and use the sampling distribution of the sample proportion.

Chapter Outline

n this chapter we discuss two probability distributions that are related to random sampling. To understand these distributions, note that if we select a random sample, then we use the sample mean as the point estimate of the population mean and the sample proportion as the point estimate of the population proportion. Two probability distributions that help us assess how accurate the sample mean and sample proportion are likely to be as point estimates are the sampling distribution of the sample

mean and the sampling distribution of the sample proportion. In Section 8.1 we discuss the first of these sampling distributions, and in Section 8.2 we discuss the second of these sampling distributions. Finally, in optional Section 8.3 we derive some important properties of the sampling distribution of the sample mean.

We illustrate the ideas of this chapter by employing three previously introduced cases:

In the **Car Mileage Case,** we use the sampling distribution of the sample mean to provide evidence that the midsize model's mean mileage exceeds 31 mpg and that the midsize model should be awarded a tax credit.

In the **e-Billing Case,** we use the sampling distribution of the sample mean to provide evidence that the new

billing system has reduced the mean bill payment time by more than 50 percent.

In the **Cheese Spread Case**, we use the sampling distribution of the sample proportion to provide evidence that a change in the packaging of a soft cheese spread will improve profitability.

8.1 The Sampling Distribution of the Sample Mean

LO8-1
Describe
and use the sampling
distribution of the
sample mean.

Introductory ideas and basic properties

Suppose that we are about to randomly select a sample of n elements (for example, cars) from a population of elements. Also, suppose that for each sampled element we will measure the value of a characteristic of interest. (For example, we might measure the mileage of each sampled car.) Before we actually select the sample, there are many different samples of n elements and corresponding measurements that we might potentially obtain. Because different samples of measurements generally have different sample means, there are many different sample means that we might potentially obtain. It follows that, *before we draw the sample, the sample mean $\bar{x}$ is a random variable,* and we will see that we can use the random variable $\bar{x}$ to make statistical inferences about the population mean μ. To make these inferences, we first model $\bar{x}$ by finding its *sampling distribution.*

> The **sampling distribution of the sample mean $\bar{x}$** is the probability distribution of the population of all possible sample means that could be obtained from all possible samples of the same size.

In order to illustrate the sampling distribution of the sample mean, we begin with an example that is based on the authors' conversations with University Chrysler/Jeep of Oxford, Ohio. To keep the example simple, we have used simplified car mileages to help explain the concepts.

 EXAMPLE 8.1 The Car Mileage Case: Estimating Mean Mileage

This is the first year that the automaker has offered its new midsize model for sale to the public. However, last year the automaker made six preproduction cars of this new model. Two of these six cars were randomly selected for testing, and the other four were sent to auto shows at which the new model was introduced to the news media and the public. As is standard industry practice, the automaker did not test the four auto show cars before or during the five months these auto shows were held, because testing can potentially harm the appearance of the cars.

In order to obtain a preliminary estimate—to be reported at the auto shows—of the midsize model's combined city and highway driving mileage, the automaker subjected the two cars selected for testing to the EPA mileage test. When this was done, the cars obtained mileages of 30 mpg and 32 mpg. The mean of this sample of mileages is

$$\bar{x} = \frac{30 + 32}{2} = 31 \text{ mpg}$$

This sample mean is the point estimate of the mean mileage μ for the population of six preproduction cars and is the preliminary mileage estimate for the new midsize model that was reported at the auto shows.

When the auto shows were over, the automaker decided to further study the new midsize model by subjecting the four auto show cars to various tests. When the EPA mileage test was performed, the four cars obtained mileages of 29 mpg, 31 mpg, 33 mpg, and 34 mpg. Thus, the mileages obtained by the six preproduction cars were 29 mpg, 30 mpg, 31 mpg, 32 mpg, 33 mpg, and 34 mpg. The probability distribution of this population of six individual car mileages is given in Table 8.1 and graphed in Figure 8.1(a). The mean of the population of

TABLE 8.1   **A Probability Distribution Describing the Population of Six Individual Car Mileages**

Individual Car Mileage	29	30	31	32	33	34
Probability	1/6	1/6	1/6	1/6	1/6	1/6

FIGURE 8.1   **A Comparison of Individual Car Mileages and Sample Means**

(a) A graph of the probability distribution describing the population of six individual car mileages

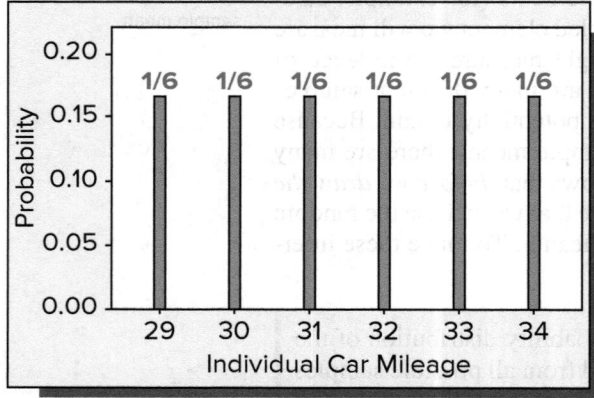

(b) A graph of the probability distribution describing the population of 15 sample means

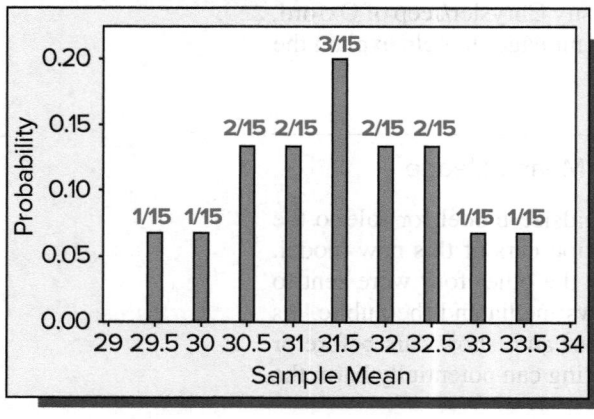

TABLE 8.2   **The Population of Sample Means**

(a) The population of the 15 samples of $n = 2$ car mileages and corresponding sample means

Sample	Car Mileages	Sample Mean
1	29, 30	29.5
2	29, 31	30
3	29, 32	30.5
4	29, 33	31
5	29, 34	31.5
6	30, 31	30.5
7	30, 32	31
8	30, 33	31.5
9	30, 34	32
10	31, 32	31.5
11	31, 33	32
12	31, 34	32.5
13	32, 33	32.5
14	32, 34	33
15	33, 34	33.5

(b) A probability distribution describing the population of 15 sample means: the sampling distribution of the sample mean

Sample Mean	Frequency	Probability
29.5	1	1/15
30	1	1/15
30.5	2	2/15
31	2	2/15
31.5	3	3/15
32	2	2/15
32.5	2	2/15
33	1	1/15
33.5	1	1/15

car mileages is

$$\mu = \frac{29 + 30 + 31 + 32 + 33 + 34}{6} = 31.5 \text{ mpg}$$

Note that the point estimate $\bar{x} = 31$ mpg that was reported at the auto shows is .5 mpg less than the true population mean μ of 31.5 mpg. Of course, different samples of two cars and corresponding mileages would have given different sample means. There are, in total, 15 samples of two mileages that could have been obtained by randomly selecting two cars from the population of six cars and subjecting the cars to the EPA mileage test. These samples correspond to the 15 combinations of two mileages that can be selected from the six mileages: 29, 30, 31, 32, 33, and 34. The samples are given, along with their means, in Table 8.2(a).

In order to find the probability distribution of the population of sample means, note that different sample means correspond to different numbers of samples. For example, because the sample mean of 31 mpg corresponds to 2 out of 15 samples—the sample (29, 33) and the sample (30, 32)—the probability of obtaining a sample mean of 31 mpg is 2/15. If we analyze all of the sample means in a similar fashion, we find that the probability distribution of the population of sample means is as given in Table 8.2(b). This distribution is the *sampling distribution of the sample mean*. A graph of this distribution is shown in Figure 8.1(b) and illustrates the accuracies of the different possible sample means as point estimates of the population mean. For example, whereas 3 out of 15 sample means exactly equal the population mean of 31.5 mpg, other sample means differ from the population mean by amounts varying from .5 mpg to 2 mpg.

©Darren Brode/Shutterstock

As illustrated in Example 8.1, one of the purposes of the sampling distribution of the sample mean is to tell us how accurate the sample mean is likely to be as a point estimate of the population mean. Because the population of six individual car mileages in Example 8.1 is small, we were able (after the auto shows were over) to test all six cars, determine the values of the six car mileages, and calculate the population mean mileage. Often, however, the population of individual measurements under consideration is very large—either a large finite population or an infinite population. In this case, it would be impractical or impossible to determine the values of all of the population measurements and calculate the population mean. Instead, we randomly select a sample of individual measurements from the population and use the mean of this sample as the point estimate of the population mean. Moreover, although it would be impractical or impossible to list all of the many (perhaps trillions of) different possible sample means that could be obtained if the sampled population is very large, statisticians know various theoretical properties about the sampling distribution of these sample means. Some of these theoretical properties are intuitively illustrated by the sampling distribution of the 15 sample means in Example 8.1. Specifically, suppose that we will randomly select a sample of n individual measurements from a population of individual measurements having mean μ and standard deviation σ. Then, it can be shown that:

- **In many situations, the distribution of the population of all possible sample means looks, at least roughly, like a normal curve.** For example, consider Figure 8.1. This figure shows that, while the distribution of the population of six individual car mileages is a uniform distribution, the distribution of the population of 15 sample means has a somewhat bell-shaped appearance. Noting, however, that this rough bell-shaped appearance is not extremely close to the appearance of a normal curve, we wish to know when the distribution of all possible sample means is exactly or approximately normally distributed. Answers to this question will begin on the next page.

- **The mean, $\mu_{\bar{x}}$, of the population of all possible sample means is equal to μ, the mean of the population from which we will select the sample.** For example, the mean, $\mu_{\bar{x}}$, of the population of 15 sample means in Table 8.2(a) can be calculated by adding up the 15 sample means, which gives 472.5, and dividing by 15. That is, $\mu_{\bar{x}} = 472.5/15 = 31.5$, which is the same as μ, the mean of the population of six individual car mileages in Table 8.1. Furthermore, because $\mu_{\bar{x}}$ equals μ, we call the sample mean an **unbiased point estimate** of the population mean. This unbiasedness property says that, although most of the possible sample means that we might obtain are either above or below the population mean, there is no systematic tendency for the sample mean to overestimate or underestimate the population mean. That is, although we will randomly select only one sample, the unbiased sample mean is "correct on the average" in all possible samples.

- **The standard deviation, $\sigma_{\bar{x}}$, of the population of all posssible sample means is less than σ, the standard deviation of the population from which we will select the sample.** This is illustrated in Figure 8.1, which shows that the distribution of all possible sample means is less spread out than the distribution of all individual car mileages. Intuitively, we see that $\sigma_{\bar{x}}$ is smaller than σ because each possible sample mean is an average of n measurements (n equals 2 in Table 8.2). Thus, **each sample mean *averages out* high and low sample measurements and can be expected to be closer to the population mean μ than many of the individual population measurements would be.** It follows that the different possible sample means are more closely clustered around μ than are the individual population measurements.

- **If the population from which we will select the sample is normally distributed, then for any sample size n the population of all possible sample means is normally distributed.** For example, consider the population of the mileages of all of the new midsize cars that could potentially be produced by this year's manufacturing process. As discussed in Chapter 1, we consider this population to be an infinite population because the automaker could always make "one more car." Moreover, assume that (as will be verified in a later example) this infinite population of all individual car mileages is normally distributed (see the top curve in Figure 8.2), and assume that the automaker will randomly select a sample of $n = 5$ cars, test them as prescribed by the EPA, and

FIGURE 8.2   **The Normally Distributed Population of All Individual Car Mileages and the Normally Distributed Population of All Possible Sample Means**

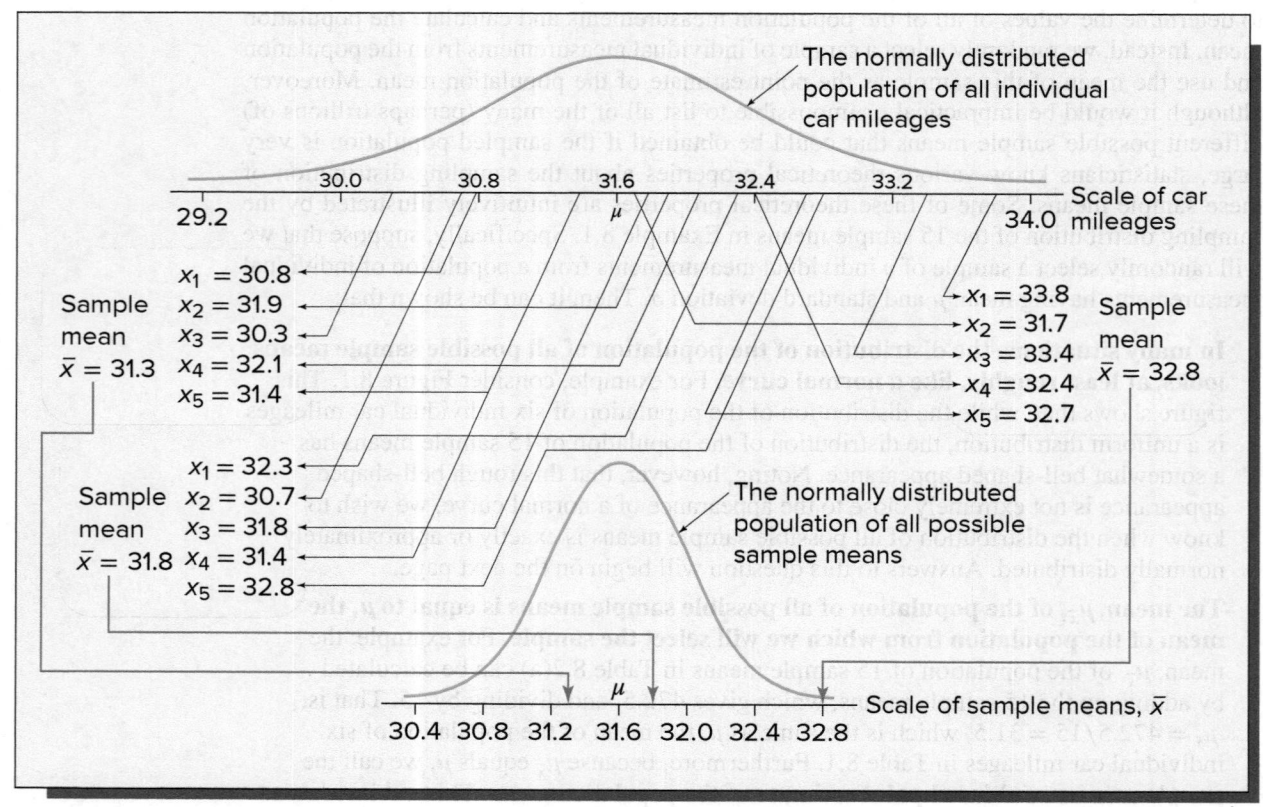

calculate the mean of the resulting sample mileages. It then follows that the population of all possible sample means that the automaker might obtain is normally distributed. This is illustrated in Figure 8.2 (see the bottom curve), which also depicts the unbiasedness of the sample mean $\bar{x}$ as a point estimate of the population mean μ. Specifically, note that the normally distributed population of all possible sample means is centered over μ, the mean of the normally distributed population of all individual car mileages. This says that, although most of the possible sample means that the automaker might obtain are either above or below the true population mean μ, the mean of all of the possible sample means that the automaker might obtain, $\mu_{\bar{x}}$, is equal to μ. To make Figure 8.2 easier to understand, we hypothetically assume that the true value of the population mean mileage μ is 31.6 mpg (this is slightly different from the 31.5 mpg population mean mileage for last year's six preproduction cars). Of course, the true value of μ is really unknown. Our objective is to estimate μ, and to do this effectively, it is important to know more about $\sigma_{\bar{x}}$, the standard deviation of the population of all possible sample means. We will see that having a formula for $\sigma_{\bar{x}}$ will help us to choose a sample size n that is likely to make the sample mean $\bar{x}$ an accurate point estimate of the population mean μ. That is, although Figure 8.2 is based on selecting a sample of $n = 5$ car mileages, perhaps we should select a larger sample of, say, 50 or more car mileages. The following summary box gives a formula for $\sigma_{\bar{x}}$ and also summarizes other previously discussed facts about the probability distribution of the population of all possible sample means.

The Sampling Distribution of $\bar{x}$

Assume that the population from which we will randomly select a sample of n measurements has mean μ and standard deviation σ. Then, the population of all possible sample means

1　Has a normal distribution, if the sampled population has a normal distribution.

2　Has mean $\mu_{\bar{x}} = \mu$.

3　Has standard deviation $\sigma_{\bar{x}} = \dfrac{\sigma}{\sqrt{n}}$.

The formula for $\sigma_{\bar{x}}$ in (3) holds exactly if the sampled population is infinite. If the sampled population is finite, this formula holds approximately under conditions to be discussed at the end of this section.

Stated equivalently, **the sampling distribution of $\bar{x}$ has mean $\mu_{\bar{x}} = \mu$, has standard deviation $\sigma_{\bar{x}} = \sigma/\sqrt{n}$ (if the sampled population is infinite), and is a normal distribution (if the sampled population has a normal distribution).**[1]

The third result in the summary box says that, if the sampled population is infinite, then $\sigma_{\bar{x}} = \sigma/\sqrt{n}$. In words, $\sigma_{\bar{x}}$, the standard deviation of the population of all possible sample means, equals σ, the standard deviation of the sampled population, divided by the square root of the sample size n. It follows that, if the sample size n is greater than 1, then $\sigma_{\bar{x}} = \sigma/\sqrt{n}$ is smaller than σ. This is illustrated in Figure 8.2, where the sample size n is 5. Specifically, note that the normally distributed population of all possible sample means is less spread out than the normally distributed population of all individual car mileages. Furthermore, the formula $\sigma_{\bar{x}} = \sigma/\sqrt{n}$ says that $\sigma_{\bar{x}}$ decreases as n increases. That is, intuitively, when the sample size is larger, each possible sample averages more observations. Therefore, the resulting different possible sample means will differ from each other by less and thus will become more closely clustered around the population mean. It follows that, if we take a larger sample, we are more likely to obtain a sample mean that is near the population mean.

In order to better see how $\sigma_{\bar{x}} = \sigma/\sqrt{n}$ decreases as the sample size n increases, we will compute some values of $\sigma_{\bar{x}}$ in the context of the car mileage case. To do this, we will assume that, although we do not know the true value of the population mean μ, we do know the true value of the population standard deviation σ. Here, knowledge of σ might be based on theory or history related to the population under consideration. For example, because the automaker

[1]In optional Section 8.3 we derive the formulas $\mu_{\bar{x}} = \mu$ and $\sigma_{\bar{x}} = \sigma/\sqrt{n}$.

has been working to improve gas mileages, we cannot assume that we know the true value of the population mean mileage μ for the new midsize model. However, engineering data might indicate that the spread of individual car mileages for the automaker's midsize cars is the same from model to model and year to year. Therefore, if the mileages for previous models had a standard deviation equal to .8 mpg, it might be reasonable to assume that the standard deviation of the mileages for the new model will also equal .8 mpg. Such an assumption would, of course, be questionable, and in most real-world situations there would probably not be an actual basis for knowing σ. However, assuming that σ is known will help us to illustrate sampling distributions, and in later chapters we will see what to do when σ is unknown.

Ⓒ EXAMPLE 8.2 The Car Mileage Case: Estimating Mean Mileage

Part 1: Basic Concepts   Consider the infinite population of the mileages of all of the new midsize cars that could potentially be produced by this year's manufacturing process. If we assume that this population is normally distributed with mean μ and standard deviation

FIGURE 8.3   **A Comparison of (1) the Population of All Individual Car Mileages, (2) the Sampling Distribution of the Sample Mean $\bar{x}$ When $n = 5$, and (3) the Sampling Distribution of the Sample Mean $\bar{x}$ When $n = 50$**

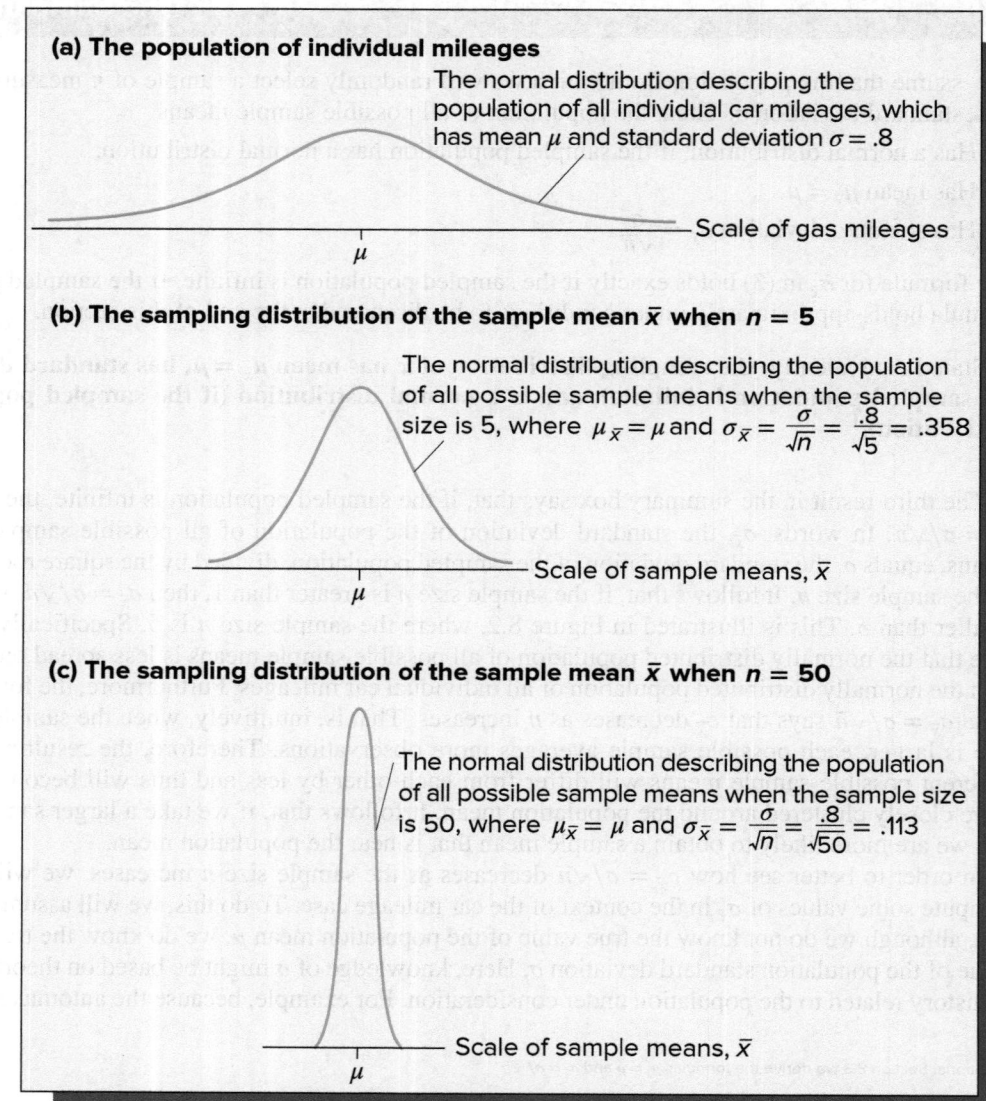

$\sigma = .8$ (see Figure 8.3(a)), and if the automaker will randomly select a sample of n cars and test them as prescribed by the EPA, it follows that the population of all possible sample means is normally distributed with mean $\mu_{\bar{x}} = \mu$ and standard deviation $\sigma_{\bar{x}} = \sigma/\sqrt{n} = .8/\sqrt{n}$. In order to show that a larger sample is more likely to give a more accurate point estimate $\bar{x}$ of μ, compare taking a sample of size $n = 5$ with taking a sample of size $n = 50$. If $n = 5$, then

$$\sigma_{\bar{x}} = \frac{\sigma}{\sqrt{n}} = \frac{.8}{\sqrt{5}} = .358$$

and it follows (by the Empirical Rule) that 95.44 percent of all possible sample means are within plus or minus $2\sigma_{\bar{x}} = 2(.358) = .716$ mpg of the population mean μ. If $n = 50$, then

$$\sigma_{\bar{x}} = \frac{\sigma}{\sqrt{n}} = \frac{.8}{\sqrt{50}} = .113$$

and it follows that 95.44 percent of all possible sample means are within plus or minus $2\sigma_{\bar{x}} = 2(.113) = .226$ mpg of the population mean μ. Therefore, if $n = 50$, the different possible sample means that the automaker might obtain will be more closely clustered around μ than they will be if $n = 5$ (see Figures 8.3(b) and (c)). This implies that the larger sample of size $n = 50$ is more likely to give a sample mean $\bar{x}$ that is near μ.

Part 2: Statistical Inference　　Recall from Chapter 3 that the automaker has randomly selected a sample of $n = 50$ mileages, which has mean $\bar{x} = 31.56$. We now ask the following question: If the population mean mileage μ exactly equals 31 mpg (the minimum standard for the tax credit), what is the probability of observing a sample mean mileage that is greater than or equal to 31.56 mpg? To find this probability, recall from Chapter 2 that a histogram of the 50 mileages indicates that the population of all individual mileages is normally distributed. Assuming that the population standard deviation σ is known to equal .8 mpg, it follows that the sampling distribution of the sample mean $\bar{x}$ is a normal distribution, with mean $\mu_{\bar{x}} = \mu$ and standard deviation $\sigma_{\bar{x}} = \sigma/\sqrt{n} = .8/\sqrt{50} = .113$. Therefore,

$$P(\bar{x} \geq 31.56 \quad \text{if} \quad \mu = 31) = P\left(z \geq \frac{31.56 - \mu_{\bar{x}}}{\sigma_{\bar{x}}}\right) = P\left(z \geq \frac{31.56 - 31}{.113}\right)$$

$$= P(z \geq 4.96)$$

To find $P(z \geq 4.96)$, notice that the largest z value given in Table A.3 is 3.99, which gives a right-hand tail area of .00003. Therefore, because $P(z \geq 3.99) = .00003$, it follows

FIGURE 8.4　　**The Probability That $\bar{x} \geq 31.56$ When $\mu = 31$ in the Car Mileage Case**

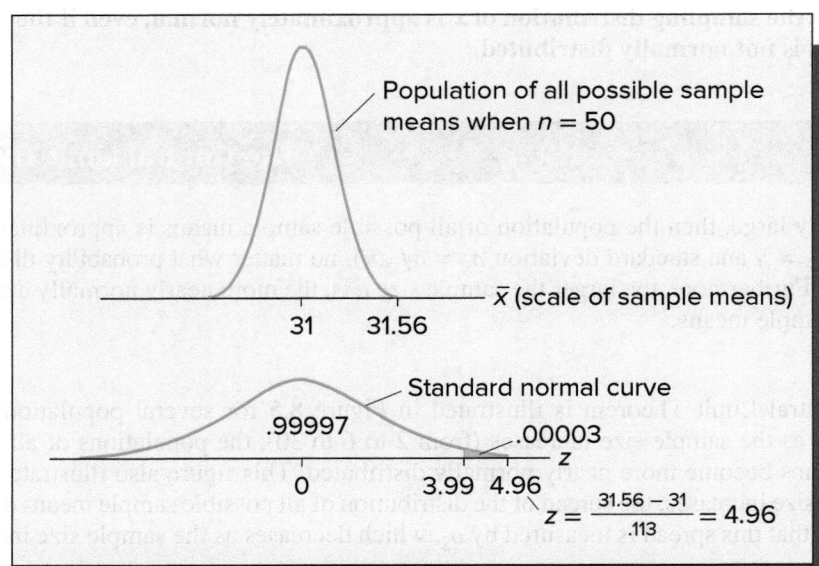

BI

that $P(z \geq 4.96)$ is less than .00003 (see Figure 8.4). The fact that this probability is less than .00003 says that, if μ equals 31, then fewer than 3 in 100,000 of all possible sample means are at least as large as the sample mean $\bar{x} = 31.56$ that we have actually observed. Therefore, if we are to believe that μ equals 31, then we must believe that we have observed a sample mean that can be described as a smaller than 3 in 100,000 chance. Because it is extremely difficult to believe that such a small chance would occur, we have extremely strong evidence that μ does not equal 31 and that μ is, in fact, larger than 31. This evidence would probably convince the federal government that the midsize model's mean mileage μ exceeds 31 mpg and thus that the midsize model deserves the tax credit.

To conclude this subsection, it is important to make two comments. First, the formula $\sigma_{\bar{x}} = \sigma/\sqrt{n}$ follows, in theory, from the formula for $\sigma_{\bar{x}}^2$, the variance of the population of all possible sample means. The formula for $\sigma_{\bar{x}}^2$ is $\sigma_{\bar{x}}^2 = \sigma^2/n$. Second, in addition to holding exactly if the sampled population is infinite, **the formula $\sigma_{\bar{x}} = \sigma/\sqrt{n}$ holds approximately if the sampled population is finite and much larger than (say, at least 20 times) the size of the sample.** For example, if we define the population of the mileages of all new midsize cars to be the population of the mileages of all cars that will actually be produced this year, then the population is finite. However, the population would be very large—certainly at least as large as 20 times any reasonable sample size. For example, if the automaker produces 100,000 new midsize cars this year, and if we randomly select a sample of $n = 50$ of these cars, then the population size of 100,000 is more than 20 times the sample size of 50. It follows that, even though the population is finite and thus the formula $\sigma_{\bar{x}} = \sigma/\sqrt{n}$ would not hold exactly, this formula would hold approximately. The exact formula for $\sigma_{\bar{x}}$ when the sampled population is finite is given in a technical note at the end of this section. It is important to use this exact formula if the sampled population is finite and less than 20 times the size of the sample. **However, with the exception of the populations considered in the technical note and in Section 9.5, we will assume that all of the remaining populations to be discussed in this book are either infinite or finite and at least 20 times the size of the sample. Therefore, it will be appropriate to use the formula $\sigma_{\bar{x}} = \sigma/\sqrt{n}$.**

Sampling a nonnormally distributed population: The Central Limit Theorem

LO8-2
Explain and use the Central Limit Theorem.

We now consider what can be said about the sampling distribution of $\bar{x}$ when the sampled population is not normally distributed. First, as previously stated, the fact that $\mu_{\bar{x}} = \mu$ is still true. Second, as also previously stated, the formula $\sigma_{\bar{x}} = \sigma/\sqrt{n}$ is exactly correct if the sampled population is infinite and is approximately correct if the sampled population is finite and much larger than (say, at least 20 times as large as) the sample size. Third, an extremely important result called the **Central Limit Theorem** tells us that, **if the sample size n is large, then the sampling distribution of $\bar{x}$ is approximately normal, even if the sampled population is not normally distributed.**

The Central Limit Theorem

If the sample size n is sufficiently large, then the population of all possible sample means is approximately normally distributed (with mean $\mu_{\bar{x}} = \mu$ and standard deviation $\sigma_{\bar{x}} = \sigma/\sqrt{n}$), no matter what probability distribution describes the sampled population. Furthermore, the larger the sample size n is, the more nearly normally distributed is the population of all possible sample means.

The Central Limit Theorem is illustrated in Figure 8.5 for several population shapes. Notice that as the sample size increases (from 2 to 6 to 30), the populations of all possible sample means become more nearly normally distributed. This figure also illustrates that, as the sample size increases, the spread of the distribution of all possible sample means decreases (remember that this spread is measured by $\sigma_{\bar{x}}$, which decreases as the sample size increases).

FIGURE 8.5   **The Central Limit Theorem Says That the Larger the Sample Size Is, the More Nearly Normally Distributed Is the Population of All Possible Sample Means**

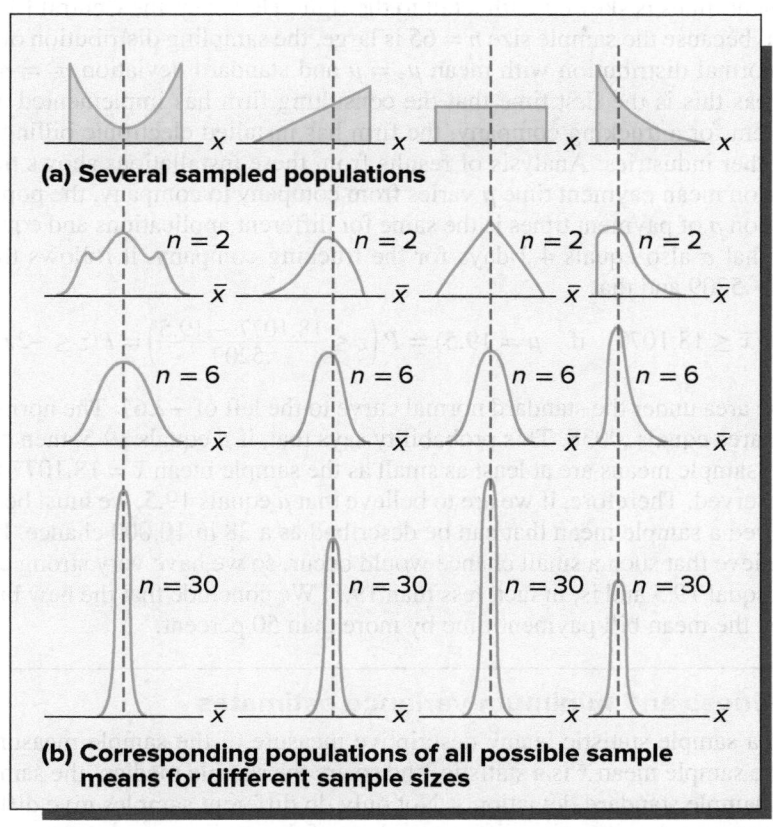

(a) Several sampled populations

(b) Corresponding populations of all possible sample means for different sample sizes

How large must the sample size be for the sampling distribution of $\bar{x}$ to be approximately normal? In general, the more skewed the probability distribution of the sampled population, the larger the sample size must be for the population of all possible sample means to be approximately normally distributed. For some sampled populations, particularly those described by symmetric distributions, the population of all possible sample means is approximately normally distributed for a fairly small sample size. In addition, studies indicate that, **if the sample size is at least 30, then for most sampled populations the population of all possible sample means is approximately normally distributed.** In this book, whenever the sample size n is at least 30, we will assume that the sampling distribution of $\bar{x}$ is approximately a normal distribution. Of course, if the sampled population is exactly normally distributed, the sampling distribution of $\bar{x}$ is exactly normal for any sample size.

Ⓒ **EXAMPLE 8.3** The e-Billing Case: Reducing Mean Bill Payment Time

Recall that a management consulting firm has installed a new computer-based electronic billing system in a Hamilton, Ohio, trucking company. Because of the previously discussed advantages of the new billing system, and because the trucking company's clients are receptive to using this system, the management consulting firm believes that the new system will reduce the mean bill payment time by more than 50 percent. The mean payment time using the old billing system was approximately equal to, but no less than, 39 days. Therefore, if μ denotes the new mean payment time, the consulting firm believes that μ will be less than 19.5 days. To assess whether μ is less than 19.5 days, the consulting firm has randomly selected a sample of $n = 65$ invoices processed using the new billing system and has determined the payment times for these invoices. The mean of the 65 payment times is $\bar{x} = 18.1077$ days, which is less than 19.5 days. Therefore, we ask the following question: If

©shotbydave/Getty Images RF

the population mean payment time is 19.5 days, what is the probability of observing a sample mean payment time that is less than or equal to 18.1077 days? To find this probability, recall from Chapter 2 that a histogram of the 65 payment times indicates that the population of all payment times is skewed with a tail to the right. However, the Central Limit Theorem tells us that, because the sample size $n = 65$ is large, the sampling distribution of $\bar{x}$ is approximately a normal distribution with mean $\mu_{\bar{x}} = \mu$ and standard deviation $\sigma_{\bar{x}} = \sigma/\sqrt{n}$. Moreover, whereas this is the first time that the consulting firm has implemented an electronic billing system for a trucking company, the firm has installed electronic billing systems for clients in other industries. Analysis of results from these installations shows that, although the population mean payment time μ varies from company to company, the population standard deviation σ of payment times is the same for different applications and equals 4.2 days. Assuming that σ also equals 4.2 days for the trucking company, it follows that $\sigma_{\bar{x}}$ equals $4.2/\sqrt{65} = .5209$ and that

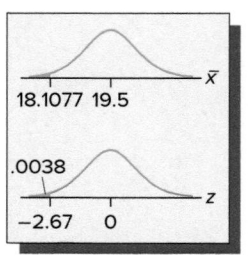

$$P(\bar{x} \leq 18.1077 \quad \text{if} \quad \mu = 19.5) = P\left(z \leq \frac{18.1077 - 19.5}{.5209}\right) = P(z \leq -2.67)$$

which is the area under the standard normal curve to the left of -2.67. The normal table tells us that this area equals .0038. This probability says that, if μ equals 19.5, then only .0038 of all possible sample means are at least as small as the sample mean $\bar{x} = 18.1077$ that we have actually observed. Therefore, if we are to believe that μ equals 19.5, we must believe that we have observed a sample mean that can be described as a 38 in 10,000 chance. It is very difficult to believe that such a small chance would occur, so we have very strong evidence that μ does not equal 19.5 and is, in fact, less than 19.5. We conclude that the new billing system has reduced the mean bill payment time by more than 50 percent.

Unbiasedness and minimum-variance estimates

Recall that a sample statistic is any descriptive measure of the sample measurements. For instance, the sample mean $\bar{x}$ is a statistic, and so are the sample median, the sample variance s^2, and the sample standard deviation s. Not only do different samples give different values of $\bar{x}$, different samples also give different values of the median, s^2, s, or any other statistic. It follows that, *before we draw the sample, any sample statistic is a random variable,* and

> The **sampling distribution of a sample statistic** is the probability distribution of the population of all possible values of the sample statistic.

In general, we wish to estimate a population parameter by using a sample statistic that is what we call an *unbiased point estimate* of the parameter.

> A sample statistic is an **unbiased point estimate** of a population parameter if the mean of the population of all possible values of the sample statistic equals the population parameter.

For example, we use the sample mean $\bar{x}$ as the point estimate of the population mean μ because **$\bar{x}$ is an unbiased point estimate of μ.** That is, $\mu_{\bar{x}} = \mu$. In words, the average of all the different possible sample means (that we could obtain from all the different possible samples) equals μ.

Although we want a sample statistic to be an unbiased point estimate of the population parameter of interest, we also want the statistic to have a small standard deviation (and variance). That is, we wish the different possible values of the sample statistic to be closely clustered around the population parameter. If this is the case, when we actually randomly select one sample and compute the sample statistic, its value is likely to be close to the value of the population parameter. Furthermore, some general results apply to estimating the mean μ of a *normally distributed population*. In this situation, it can be shown that both the sample mean and the sample median are unbiased point estimates of μ. In fact, there are many unbiased point estimates of μ. However, it can be shown that the variance of the population of all possible sample means is smaller than the variance of the population of all possible values of any other unbiased point estimate of μ. For this reason, **we call the sample mean**

a minimum-variance unbiased point estimate of μ. When we use the sample mean as the point estimate of μ, we are more likely to obtain a point estimate close to μ than if we used any other unbiased sample statistic as the point estimate of μ. This is one reason why we use the sample mean as the point estimate of the population mean.

We next consider estimating the population variance σ^2. It can be shown that if the sampled population is infinite, then s^2 **is an unbiased point estimate of** σ^2. That is, the average of all the different possible sample variances that we could obtain (from all the different possible samples) is equal to σ^2. This is why we use a divisor equal to $n-1$ rather than n when we estimate σ^2. It can be shown that, if we used n as the divisor when estimating σ^2, we would not obtain an unbiased point estimate of σ^2. When the population is finite, s^2 may be regarded as an approximately unbiased estimate of σ^2 as long as the population is fairly large (which is usually the case).

It would seem logical to think that, because s^2 is an unbiased point estimate of σ^2, s should be an unbiased point estimate of σ. This seems plausible, but it is not the case. There is no easy way to calculate an unbiased point estimate of σ. Because of this, the usual practice is to use s as the point estimate of σ (even though it is not an unbiased estimate).

This ends our discussion of the theory of point estimation. It suffices to say that in this book we estimate population parameters by using sample statistics that statisticians generally agree are best. Whenever possible, these sample statistics are unbiased point estimates and have small variances.

Technical Note: If we randomly select a sample of size n without replacement from a finite population of size N, then it can be shown that $\sigma_{\bar{x}} = (\sigma/\sqrt{n})\sqrt{(N-n)/(N-1)}$, where the quantity $\sqrt{(N-n)/(N-1)}$ is called the **finite population multiplier.** If the size of the sampled population is at least 20 times the size of the sample (that is, **if $N \geq 20n$), then the finite population multiplier is approximately equal to one, and $\sigma_{\bar{x}}$ approximately equals $\sigma/\sqrt{n}$.** However, if the population size N is smaller than 20 times the size of the sample, then the finite population multiplier is substantially less than one, and we must include this multiplier in the calculation of $\sigma_{\bar{x}}$. For instance, in Example 8.1, where the standard deviation σ of the population of $N=6$ car mileages can be calculated to be 1.7078, and where $N=6$ is only three times the sample size $n=2$, it follows that

$$\sigma_{\bar{x}} = \frac{\sigma}{\sqrt{n}}\sqrt{\frac{N-n}{N-1}} = \left(\frac{1.7078}{\sqrt{2}}\right)\sqrt{\frac{6-2}{6-1}} = 1.2076(.8944) = 1.08$$

We will see how this formula can be used to make statistical inferences in Section 9.5.

Exercises for Section 8.1

CONCEPTS

8.1 The sampling distribution of the sample mean $\bar{x}$ is the probability distribution of a population. Describe this population.

8.2 What does the Central Limit Theorem tell us about the sampling distribution of the sample mean?

METHODS AND APPLICATIONS

8.3 Suppose that we will take a random sample of size n from a population having mean μ and standard deviation σ. For each of the following situations, find the mean, variance, and standard deviation of the sampling distribution of the sample mean $\bar{x}$:
   a $\mu = 10, \quad \sigma = 2, \quad n = 25$
   b $\mu = 500, \quad \sigma = .5, \quad n = 100$
   c $\mu = 3, \quad \sigma = .1, \quad n = 4$
   d $\mu = 100, \quad \sigma = 1, \quad n = 1,600$

8.4 For each situation in Exercise 8.3, find an interval that contains (approximately or exactly) 99.73 percent of

all the possible sample means. In which cases must we assume that the population is normally distributed? Why?

8.5 Suppose that we will randomly select a sample of 64 measurements from a population having a mean equal to 20 and a standard deviation equal to 4.
   a Describe the shape of the sampling distribution of the sample mean $\bar{x}$. Do we need to make any assumptions about the shape of the population? Why or why not?
   b Find the mean and the standard deviation of the sampling distribution of the sample mean $\bar{x}$.
   c Calculate the probability that we will obtain a sample mean greater than 21; that is, calculate $P(\bar{x} > 21)$. Hint: Find the z value corresponding to 21 by using $\mu_{\bar{x}}$ and $\sigma_{\bar{x}}$ because we wish to calculate a probability about $\bar{x}$. Then sketch the sampling distribution and the probability.
   d Calculate the probability that we will obtain a sample mean less than 19.385; that is, calculate $P(\bar{x} < 19.385)$.

THE GAME SHOW CASE

Exercises 8.6 through 8.10 are based on the following situation.

Congratulations! You have just won the question-and-answer portion of a popular game show and will now be given an opportunity to select a grand prize. The game show host shows you a large revolving drum containing four identical white envelopes that have been thoroughly mixed in the drum. Each of the envelopes contains one of four checks made out for grand prizes of 20, 40, 60, and 80 thousand dollars. Usually, a contestant reaches into the drum, selects an envelope, and receives the grand prize in the envelope. Tonight, however, is a special night. You will be given the choice of either selecting one envelope or selecting two envelopes and receiving the average of the grand prizes in the two envelopes. If you select one envelope, the probability is 1/4 that you will receive any one of the individual grand prizes 20, 40, 60, and 80 thousand dollars. To see what could happen if you select two envelopes, do Exercises 8.6 through 8.10.

8.6 There are six combinations, or samples, of two grand prizes that can be randomly selected from the four grand prizes 20, 40, 60, and 80 thousand dollars. Four of these samples are (20, 40), (20, 60), (20, 80), and (40, 60). Find the other two samples.

8.7 Find the mean of each sample in Exercise 8.6.

8.8 Find the probability distribution of the population of six sample mean grand prizes.

8.9 If you select two envelopes, what is the probability that you will receive a sample mean grand prize of at least 50 thousand dollars?

8.10 Compare the probability distribution of the four individual grand prizes with the probability distribution of the six sample mean grand prizes. Would you select one or two envelopes? Why? Note: There is no single correct answer. It is a matter of opinion.

8.11 THE BANK CUSTOMER WAITING TIME CASE
    DS WaitTime

Recall that the bank manager wants to show that the new system reduces typical customer waiting times to less than six minutes. One way to do this is to demonstrate that the mean of the population of all customer waiting times is less than 6. Letting this mean be μ, in this exercise we wish to investigate whether the sample of 100 waiting times provides evidence to support the claim that μ is less than 6.

For the sake of argument, we will begin by assuming that μ equals 6, and we will then attempt to use the sample to contradict this assumption in favor of the conclusion that μ is less than 6. Recall that the mean of the sample of 100 waiting times is $\bar{x} = 5.46$ and assume that σ, the standard deviation of the population of all customer waiting times, is known to be 2.47.

a Consider the population of all possible sample means obtained from random samples of 100 waiting times. What is the shape of this population of sample means? That is, what is the shape of the sampling distribution of $\bar{x}$? Why is this true?

b Find the mean and standard deviation of the population of all possible sample means when we assume that μ equals 6.

c The sample mean that we have actually observed is $\bar{x} = 5.46$. Assuming that μ equals 6, find the probability of observing a sample mean that is less than or equal to $\bar{x} = 5.46$.

d If μ equals 6, what percentage of all possible sample means are less than or equal to 5.46? Because we have actually observed a sample mean of $\bar{x} = 5.46$, is it more reasonable to believe that (1) μ equals 6 and we have observed one of the sample means that is less than or equal to 5.46 when μ equals 6, or (2) that we have observed a sample mean less than or equal to 5.46 because μ is less than 6? Explain. What do you conclude about whether the new system has reduced the typical customer waiting time to less than six minutes?

8.12 THE VIDEO GAME SATISFACTION RATING CASE
    DS VideoGame

Recall that a customer is considered to be very satisfied with his or her XYZ Box video game system if the customer's composite score on the survey instrument is at least 42. One way to show that customers are typically very satisfied is to show that the mean of the population of all satisfaction ratings is at least 42. Letting this mean be μ, in this exercise we wish to investigate whether the sample of 65 satisfaction ratings provides evidence to support the claim that μ exceeds 42 (and, therefore, is at least 42).

For the sake of argument, we begin by assuming that μ equals 42, and we then attempt to use the sample to contradict this assumption in favor of the conclusion that μ exceeds 42. Recall that the mean of the sample of 65 satisfaction ratings is $\bar{x} = 42.95$, and assume that σ, the standard deviation of the population of all satisfaction ratings, is known to be 2.64.

a Consider the sampling distribution of $\bar{x}$ for random samples of 65 customer satisfaction ratings. Use the properties of this sampling distribution to find the probability of observing a sample mean greater than or equal to 42.95 when we assume that μ equals 42.

b If μ equals 42, what percentage of all possible sample means are greater than or equal to 42.95? Because we have actually observed a sample mean of $\bar{x} = 42.95$, is it more reasonable to believe that (1) μ equals 42 and we have observed a sample mean that is greater than or equal to 42.95 when μ equals 42, or (2) that we have observed a sample mean that is greater than or equal to 42.95 because μ is greater than 42? Explain. What do you conclude about whether customers are typically very satisfied with the XYZ Box video game system?

8.13 In an article in the *Journal of Management,* Joseph Martocchio studied and estimated the costs of employee absences. Based on a sample of 176 blue-collar workers, Martocchio estimated that the mean amount of *paid* time lost during a three-month period was 1.4 days per employee with a standard deviation of 1.3 days. Martocchio also estimated that the mean amount of *unpaid* time lost during a three-month period was 1.0 day per employee with a standard deviation of 1.8 days.

Suppose we randomly select a sample of 100 blue-collar workers. Based on Martocchio's estimates:

a What is the probability that the average amount of *paid* time lost during a three-month period for the 100 blue-collar workers will exceed 1.5 days? Assume σ equals 1.3 days.

b What is the probability that the average amount of *unpaid* time lost during a three-month period for the 100 blue-collar workers will exceed 1.5 days? Assume σ equals 1.8 days.

c Suppose we randomly select a sample of 100 blue-collar workers, and suppose the sample mean amount of unpaid time lost during a three-month period actually exceeds 1.5 days. Would it be reasonable to conclude that the mean amount of unpaid time lost has increased above the previously estimated 1.0 day? Explain. Assume σ still equals 1.8 days.

8.2 The Sampling Distribution of the Sample Proportion

LO8-3

Describe and use the sampling distribution of the sample proportion.

A food processing company markets a soft cheese spread that is sold in a plastic container with an "easy pour" spout. Although this spout works extremely well and is popular with consumers, it is expensive to produce. Because of the spout's high cost, the company has developed a new, less expensive spout. While the new, cheaper spout may alienate some purchasers, a company study shows that its introduction will increase profits if fewer than 10 percent of the cheese spread's current purchasers are lost. That is, if we let p be the true proportion of all current purchasers who would stop buying the cheese spread if the new spout were used, profits will increase as long as p is less than .10.

Suppose that (after trying the new spout) 63 of 1,000 randomly selected purchasers say that they would stop buying the cheese spread if the new spout were used. The point estimate of the population proportion p is the sample proportion $\hat{p} = 63/1,000 = .063$. This sample proportion says that we estimate that 6.3 percent of all current purchasers would stop buying the cheese spread if the new spout were used. Because $\hat{p}$ equals .063, we have some evidence that the population proportion p is less than .10. In order to determine the strength of this evidence, we need to consider the sampling distribution of $\hat{p}$. In general, assume that we will randomly select a sample of n elements from a population, and assume that a proportion p of all the elements in the population fall into a particular category (for instance, the category of consumers who would stop buying the cheese spread). Before we actually select the sample, there are many different samples of n elements that we might potentially obtain. The number of elements that fall into the category in question will vary from sample to sample, so the sample proportion of elements falling into the category will also vary from sample to sample. For example, if three possible random samples of 1,000 soft cheese spread purchasers had, respectively, 63, 58, and 65 purchasers say that they would stop buying the cheese spread if the new spout were used, then the sample proportions given by the three samples would be $\hat{p} = 63/1000 = .063$, $\hat{p} = 58/1000 = .058$, and $\hat{p} = 65/1000 = .065$. In general, before we randomly select the sample, there are many different possible sample proportions that we might obtain, and thus the sample proportion $\hat{p}$ is a random variable. In the following box we give the properties of the probability distribution of this random variable, which is called **the sampling distribution of the sample proportion $\hat{p}$.**

The Sampling Distribution of the Sample Proportion $\hat{p}$

The population of all possible sample proportions

1 Approximately has a normal distribution, if the sample size n is large.

2 Has mean $\mu_{\hat{p}} = p$.

3 Has standard deviation $\sigma_{\hat{p}} = \sqrt{\dfrac{p(1-p)}{n}}$.

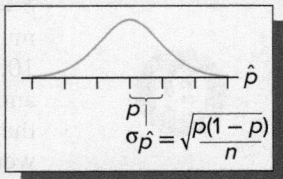

Stated equivalently, the sampling distribution of $\hat{p}$ has mean $\mu_{\hat{p}} = p$, has standard deviation $\sigma_{\hat{p}} = \sqrt{p(1-p)/n}$, and is approximately a normal distribution (if the sample size n is large).

Property 1 in the box says that, if n is large, then the population of all possible sample proportions approximately has a normal distribution. Here, it can be shown that **n should be considered large if both np and $n(1 - p)$ are at least 5.**[2] Property 2, which says that $\mu_{\hat{p}} = p$, is valid for any sample size and tells us that $\hat{p}$ is an unbiased estimate of p. That is, although the sample proportion $\hat{p}$ that we calculate probably does not equal p, the average of all the different sample proportions that we could have calculated (from all the different possible samples) is equal to p. Property 3, which says that

$$\sigma_{\hat{p}} = \sqrt{\frac{p(1-p)}{n}}$$

is exactly correct if the sampled population is infinite and is approximately correct if the sampled population is finite and much larger than (say, at least 20 times as large as) the sample size. Property 3 tells us that the standard deviation of the population of all possible sample proportions decreases as the sample size increases. That is, the larger n is, the more closely clustered are all the different sample proportions around the true population proportion. Finally, note that the formula for $\sigma_{\hat{p}}$ follows, in theory, from the formula for $\sigma_{\hat{p}}^2$, the variance of the population of all possible sample proportions. The formula for $\sigma_{\hat{p}}^2$ is $\sigma_{\hat{p}}^2 = p(1 - p)/n$.

ⓒ **EXAMPLE 8.4** The Cheese Spread Case: Improving Profitability

In the cheese spread situation, the food processing company must decide whether p, the proportion of all current purchasers who would stop buying the cheese spread if the new spout were used, is less than .10. In order to do this, remember that when 1,000 purchasers of the cheese spread are randomly selected, 63 of these purchasers say they would stop buying the cheese spread if the new spout were used. Noting that the sample proportion $\hat{p} = .063$ is less than .10, we ask the following question. If the true population proportion is .10, what is the probability of observing a sample proportion that is less than or equal to .063?

If p equals .10, we can assume that the sampling distribution of $\hat{p}$ is approximately a normal distribution because both $np = 1,000(.10) = 100$ and $n(1 - p) = 1,000(1 - .10) = 900$ are at least 5. Furthermore, the mean and standard deviation of the sampling distribution of $\hat{p}$ are $\mu_{\hat{p}} = p = .10$ and

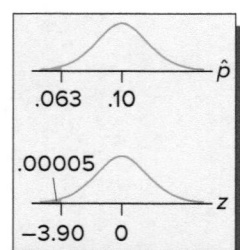

$$\sigma_{\hat{p}} = \sqrt{\frac{p(1-p)}{n}} = \sqrt{\frac{(.10)(.90)}{1,000}} = .0094868$$

Therefore,

$$P(\hat{p} \leq .063 \quad \text{if} \quad p = .10) = P\left(z \leq \frac{.063 - \mu_{\hat{p}}}{\sigma_{\hat{p}}}\right) = P\left(z \leq \frac{.063 - .10}{.0094868}\right)$$

$$= P(z \leq -3.90)$$

which is the area under the standard normal curve to the left of -3.90. The normal table tells us that this area equals .00005. This probability says that, if p equals .10, then only 5 in 100,000 of all possible sample proportions are at least as small as the sample proportion $\hat{p} = .063$ that we have actually observed. That is, if we are to believe that p equals .10, we must believe that we have observed a sample proportion that can be described as a 5 in 100,000 chance. It follows that we have extremely strong evidence that p does not equal .10 and is, in fact, less than .10. In other words, we have extremely strong evidence that fewer than 10 percent of current purchasers would stop buying the cheese spread if the new spout were used. It seems that introducing the new spout will be profitable.

---

[2]Some statisticians suggest using the more conservative rule that both np and $n(1 - p)$ must be at least 10.

Exercises for Section 8.2

CONCEPTS   ■ connect

8.14 The sampling distribution of $\hat{p}$ is the probability distribution of a population. Describe this population.

8.15 If the sample size n is large, the sampling distribution of $\hat{p}$ is approximately a normal distribution. What condition must be satisfied to guarantee that n is large enough to say that $\hat{p}$ is approximately normally distributed?

8.16 Write formulas that express the central tendency and variability of the population of all possible sample proportions. Explain what each of these formulas means in your own words.

8.17 Describe the effect of increasing the sample size on the population of all possible sample proportions.

METHODS AND APPLICATIONS

8.18 In each of the following cases, determine whether the sample size n is large enough to say that the sampling distribution of $\hat{p}$ is a normal distribution.
    a $p = .4,$ $n = 100$ **d** $p = .8,$ $n = 400$
    b $p = .1,$ $n = 10$ **e** $p = .98,$ $n = 1{,}000$
    c $p = .1,$ $n = 50$ **f** $p = .99,$ $n = 400$

8.19 In each of the following cases, find the mean, variance, and standard deviation of the sampling distribution of the sample proportion $\hat{p}$.
    a $p = .5,$ $n = 250$ **c** $p = .8,$ $n = 400$
    b $p = .1,$ $n = 100$ **d** $p = .98,$ $n = 1{,}000$

8.20 For each situation in Exercise 8.19, find an interval that contains approximately 95.44 percent of all the possible sample proportions.

8.21 Suppose that we will randomly select a sample of $n = 100$ elements from a population and that we will compute the sample proportion $\hat{p}$ of these elements that fall into a category of interest. If the true population proportion p equals .9:
    a Describe the shape of the sampling distribution of $\hat{p}$. Why can we validly describe the shape?
    b Find the mean and the standard deviation of the sampling distribution of $\hat{p}$.
    c Find $P(\hat{p} \geq .96)$.
    d Find $P(.855 \leq \hat{p} \leq .945)$.
    e Find $P(\hat{p} \leq .915)$.

8.22 On February 8, 2002, the Gallup Organization released the results of a poll concerning American attitudes toward the 19th Winter Olympic Games in Salt Lake City, Utah. The poll results were based on telephone interviews with a randomly selected national sample of 1,011 adults, 18 years and older, conducted February 4–6, 2002.
    a Suppose we wish to use the poll's results to justify the claim that more than 30 percent of Americans (18 years or older) say that figure skating is their favorite Winter Olympic event. The poll actually found that 32 percent of respondents reported that

figure skating was their favorite event.[3] If, for the sake of argument, we assume that 30 percent of Americans (18 years or older) say figure skating is their favorite event (that is, $p = .3$), calculate the probability of observing a sample proportion of .32 or more; that is, calculate $P(\hat{p} \geq .32)$.
    b Based on the probability you computed in part a, would you conclude that more than 30 percent of Americans (18 years or older) say that figure skating is their favorite Winter Olympic event?

8.23 *Quality Progress* reports on improvements in customer satisfaction and loyalty made by Bank of America. A key measure of customer satisfaction is the response (on a scale from 1 to 10) to the question: "Considering all the business you do with Bank of America, what is your overall satisfaction with Bank of America?" Here, a response of 9 or 10 represents "customer delight."
    a Historically, the percentage of Bank of America customers expressing customer delight has been 48 percent. Suppose that we wish to use the results of a survey of 350 Bank of America customers to justify the claim that more than 48 percent of all current Bank of America customers would express customer delight. The survey finds that 189 of 350 randomly selected Bank of America customers express customer delight. If, for the sake of argument, we assume that the proportion of customer delight is $p = .48$, calculate the probability of observing a sample proportion greater than or equal to $189/350 = .54$. That is, calculate $P(\hat{p} \geq .54)$.
    b Based on the probability you computed in part a, would you conclude that more than 48 percent of current Bank of America customers express customer delight? Explain.

8.24 Again consider the survey of 350 Bank of America customers discussed in Exercise 8.23, and assume that 48 percent of Bank of America customers would currently express customer delight. That is, assume $p = .48$. Find:
    a The probability that the sample proportion obtained from the sample of 350 Bank of America customers would be within three percentage points of the population proportion. That is, find $P(.45 \leq \hat{p} \leq .51)$.
    b The probability that the sample proportion obtained from the sample of 350 Bank of America customers would be within six percentage points of the population proportion. That is, find $P(.42 \leq \hat{p} \leq .54)$.

8.25 Based on your results in Exercise 8.24, would it be reasonable to state that the survey's "margin of error" is $\pm$ 3 percentage points? $\pm$ 6 percentage points? Explain.

[3]Source: The Gallup Organization, http://news.gallup.com/poll/5305/Figure-Skating-Tops-List-Americans-Favorite-Winter-Olympic-Events.aspx. February 13, 2002.

8.26 An article in *Fortune* magazine discussed "outsourcing." According to the article, outsourcing is "the assignment of critical, but noncore, business functions to outside specialists." This allows a company to immediately bring operations up to best-in-world standards while avoiding huge capital investments. The article included the results of a poll of business executives addressing the benefits of outsourcing.

   a Suppose we wish to use the poll's results to justify the claim that fewer than 20 percent of business executives feel that the benefits of outsourcing are either "less or much less than expected." The poll actually found that 15 percent of the respondents felt that the benefits of outsourcing were either "less or much less than expected." If 1,000 randomly selected business executives were polled, and if for the sake of argument, we assume that 20 percent of all business executives feel that the benefits of outsourcing are either less or much less than expected (that is, $p = .20$), calculate the probability of observing a sample proportion of .15 or less. That is, calculate $P(\hat{p} \leq .15)$.

   b Based on the probability you computed in part *a*, would you conclude that fewer than 20 percent of business

executives feel that the benefits of outsourcing are either "less or much less than expected"? Explain.

8.27 *Fortune* magazine reported the results of a survey on executive training that was conducted by the Association of Executive Search Consultants. The survey showed that 75 percent of 300 polled CEOs believe that companies should have "fast-track training programs" for developing managerial talent.

   a Suppose we wish to use the results of this survey to justify the claim that more than 70 percent of CEOs believe that companies should have fast-track training programs. Assuming that the 300 surveyed CEOs were randomly selected, and assuming, for the sake of argument, that 70 percent of CEOs believe that companies should have fast-track training programs (that is, $p = .70$), calculate the probability of observing a sample proportion of .75 or more. That is, calculate $P(\hat{p} \geq .75)$.

   b Based on the probability you computed in part *a*, would you conclude that more than 70 percent of CEOs believe that companies should have fast-track training programs? Explain.

8.3 Derivation of the Mean and the Variance of the Sample Mean (Optional)

Before we randomly select the sample values $x_1, x_2, \ldots, x_n$ from a population having mean μ and variance σ^2, we note that, for $i = 1, 2, \ldots, n$, the ith sample value x_i is a random variable that can potentially be any of the values in the population. Moreover, it can be proven (and is intuitive) that

1. The mean (or expected value) of x_i, denoted μ_{x_i}, is μ, the mean of the population from which x_i will be randomly selected. That is, $\mu_{x_1} = \mu_{x_2} = \cdots = \mu_{x_n} = \mu$.

2. The variance of x_i, denoted $\sigma_{x_i}^2$, is σ^2, the variance of the population from which x_i will be randomly selected. That is, $\sigma_{x_1}^2 = \sigma_{x_2}^2 = \cdots = \sigma_{x_n}^2 = \sigma^2$.

If we consider the sample mean $\bar{x} = \sum_{i=1}^{n} x_i/n$, then we can prove that $\mu_{\bar{x}} = \mu$ by using the following two properties of the mean discussed in Section 6.6:

Property 1: If a is a fixed number, $\mu_{ax} = a\mu_x$

Property 2: $\mu_{(x_1 + x_2 + \cdots + x_n)} = \mu_{x_1} + \mu_{x_2} + \cdots + \mu_{x_n}$

The proof that $\mu_{\bar{x}} = \mu$ is as follows:

$$\mu_{\bar{x}} = \mu\left(\sum_{i=1}^{n} x_i/n\right)$$

$$= \frac{1}{n}\mu\left(\sum_{i=1}^{n} x_i\right) \qquad \text{(see Property 1)}$$

$$= \frac{1}{n}\mu_{(x_1+x_2+\cdots+x_n)}$$

$$= \frac{1}{n}(\mu_{x_1} + \mu_{x_2} + \cdots + \mu_{x_n}) \qquad \text{(see Property 2)}$$

$$= \frac{1}{n}(\mu + \mu + \cdots + \mu) = \frac{n\mu}{n} = \mu$$

We can prove that $\sigma_{\bar{x}}^2 = \sigma^2/n$ by using the following two properties of the variance discussed in Section 6.6:

Property 3: If a is a fixed number, $\sigma_{ax}^2 = a^2\sigma_x^2$

Property 4: If $x_1, x_2, \ldots, x_n$ are statistically independent, $\sigma^2(x_1+x_2+\cdots+x_n) = \sigma_{x_1}^2 + \sigma_{x_2}^2 + \cdots + \sigma_{x_n}^2$

The proof that $\sigma_{\bar{x}}^2 = \sigma^2/n$ is as follows:

$$\sigma_{\bar{x}}^2 = \sigma^2\left(\sum_{i=1}^{n} x_i/n\right) = \left(\frac{1}{n}\right)^2 \sigma^2\left(\sum_{i=1}^{n} x_i\right) \qquad \text{(see Property 3)}$$

$$= \frac{1}{n^2}\,\sigma^2(x_1+x_2+\cdots+x_n)$$

$$= \frac{1}{n^2}\,(\sigma_{x_1}^2 + \sigma_{x_2}^2 + \cdots + \sigma_{x_n}^2) \qquad \text{(see Property 4)}$$

$$= \frac{1}{n^2}\,(\sigma^2 + \sigma^2 + \cdots + \sigma^2) = \frac{n\sigma^2}{n^2} = \frac{\sigma^2}{n}$$

Note that we can use Property 4 if $x_1, x_2, \ldots, x_n$ are independent random variables. In general, $x_1, x_2, \ldots, x_n$ are independent if we are drawing these sample values from an infinite population. When we select a sample from an infinite population, a population value obtained on one selection can also be obtained on any other selection. This is because, when the population is infinite, there are an infinite number of repetitions of each population value. Therefore, because a value obtained on one selection is not precluded from being obtained on any other selection, the selections and thus $x_1, x_2, \ldots, x_n$ are statistically independent. Furthermore, this statistical independence approximately holds if the population size is much larger than (say, at least 20 times as large as) the sample size. Therefore, in this case $\sigma_{\bar{x}}^2 = \sigma^2/n$ is approximately correct.

Chapter Summary

We began this chapter by discussing sampling distributions. A **sampling distribution** is the probability distribution that describes the population of all possible values of a sample statistic. In this chapter we studied the properties of two important sampling distributions—the sampling distribution of the sample mean, $\bar{x}$, and the sampling distribution of the sample proportion, $\hat{p}$.

Because different samples that can be randomly selected from a population give different sample means, there is a population of sample means corresponding to a particular sample size. The probability distribution describing the population of all possible sample means is called the **sampling distribution of the sample mean, $\bar{x}$.** We studied the properties of this sampling distribution when the sampled population is and is not normally distributed. We found that, when the sampled population has a normal distribution, then the sampling distribution of the sample mean is a normal distribution. Furthermore, the **Central Limit Theorem** tells us that, if the sampled population is not normally distributed, then the sampling distribution of the sample mean is approximately a normal distribution when the sample size is large (at least 30). We also saw that the mean of the sampling distribution of $\bar{x}$ always equals the mean of the sampled population, and we presented formulas for the variance and the standard deviation of this sampling distribution. Finally, we explained that the sample mean is a **minimum-variance unbiased point estimate** of the mean of a normally distributed population.

We also studied the properties of the **sampling distribution of the sample proportion $\hat{p}$.** We found that, if the sample size is large, then this sampling distribution is approximately a normal distribution, and we gave a rule for determining whether the sample size is large. We found that the mean of the sampling distribution of $\hat{p}$ is the population proportion p, and we gave formulas for the variance and the standard deviation of this sampling distribution.

Throughout our discussions of sampling distributions, we demonstrated that knowing the properties of sampling distributions can help us make statistical inferences about population parameters. In fact, we will see that the properties of various sampling distributions provide the foundation for most of the techniques to be discussed in future chapters. In the last (optional) section of this chapter, we derived the mean and variance of the sampling distribution of the sample mean $\bar{x}$.

Glossary of Terms

Central Limit Theorem: A theorem telling us that when the sample size n is sufficiently large, then the population of all possible sample means is approximately normally distributed no matter what probability distribution describes the sampled population.
minimum-variance unbiased point estimate: An unbiased point estimate of a population parameter having a variance that is smaller than the variance of any other unbiased point estimate of the parameter.
sampling distribution of a sample statistic: The probability distribution of the population of all possible values of the sample statistic.

sampling distribution of the sample mean $\bar{x}$: The probability distribution of the population of all possible sample means obtained from samples of a particular size n.
sampling distribution of the sample proportion $\hat{p}$: The probability distribution of the population of all possible sample proportions obtained from samples of a particular size n.
unbiased point estimate: A sample statistic is an unbiased point estimate of a population parameter if the mean of the population of all possible values of the sample statistic equals the population parameter.

Important Results and Formulas

The sampling distribution of the sample mean: pages 381 and 384
   when a population is normally distributed: page 381

Central Limit Theorem: page 384

The sampling distribution of the sample proportion: page 389

Supplementary Exercises

8.28 **THE TRASH BAG CASE** 🖵 TrashBag
Recall that the trash bag manufacturer has concluded that its new 30-gallon bag will be the strongest such bag on the market if its mean breaking strength is at least 50 pounds. In order to provide statistical evidence that the mean breaking strength of the new bag is at least 50 pounds, the manufacturer randomly selects a sample of n bags and calculates the mean $\bar{x}$ of the breaking strengths of these bags. If the sample mean so obtained is at least 50 pounds, this provides some evidence that the mean breaking strength of all new bags is at least 50 pounds.

Suppose that (unknown to the manufacturer) the breaking strengths of the new 30-gallon bag are normally distributed with a mean of $\mu = 50.6$ pounds and a standard deviation of $\sigma = 1.62$ pounds.
a Find an interval containing 95.44 percent of all possible sample means if the sample size employed is $n = 5$.
b Find an interval containing 95.44 percent of all possible sample means if the sample size employed is $n = 40$.
c If the trash bag manufacturer hopes to obtain a sample mean that is at least 50 pounds (so that it can provide evidence that the population mean breaking strength of the new bags is at least 50), which sample size ($n = 5$ or $n = 40$) would be best? Explain why.

8.29 **THE STOCK RETURN CASE**
The year 1987 featured extreme volatility on the stock market, including a loss of over 20 percent of the market's value on a single day. Figure 8.6(a) shows the percent frequency histogram of the percentage returns for the entire year 1987 for the population of all 1,815 stocks listed on the New York Stock Exchange. The mean and the standard deviation of the population of percentage returns are −3.5 percent and 26 percent, respectively. Consider drawing a random sample of $n = 5$ stocks from the population of 1,815 stocks and calculating the mean return, $\bar{x}$, of the sampled stocks. If we use a computer, we can generate all the different samples of five stocks that can

be obtained (there are trillions of such samples) and calculate the corresponding sample mean returns. A percent frequency histogram describing the population of all possible sample mean returns is given in Figure 8.6(b). Comparing Figures 8.6(a) and (b), we see that, although the histogram of individual stock returns and the histogram of sample mean returns are both bell-shaped and centered over the same mean of −3.5 percent, the histogram of sample mean returns looks *less spread out* than the histogram of individual returns. A sample of 5 stocks is a portfolio of stocks, where the average return of the 5 stocks is the portfolio's return if we invest equal amounts of money in each of the 5 stocks. Because the sample mean returns are less spread out than the individual stock returns, we have illustrated that diversification reduces risk. **(1)** Find the standard deviation of the population of all sample mean returns, and **(2)** assuming that this population is normally distributed, find an interval that contains 95.44 percent of all sample mean returns.

8.30 Suppose that we wish to assess whether more than 60 percent of all U.S. households in a particular income class bought life insurance last year. That is, we wish to assess whether p, the proportion of all U.S. households in the income class that bought life insurance last year, exceeds .60. Assume that an insurance survey is based on 1,000 randomly selected U.S. households in the income class and that 640 of these households bought life insurance last year.
a Assuming that p equals .60 and the sample size is 1,000, what is the probability of observing a sample proportion that is at least .64?
b Based on your answer in part a, do you think more than 60 percent of all U.S. households in the income class bought life insurance last year? Explain.

8.31 A computer supply house receives a large shipment of flash drives each week. Past experience has shown that the number of flaws (bad sectors) per flash drive is either 0, 1, 2, or 3 with probabilities .65, .2, .1, and .05, respectively.

FIGURE 8.6 The New York Stock Exchange in 1987: A Comparison of Individual Stock Returns and Sample Mean Returns

(a) The percent frequency histogram describing the population of individual stock returns

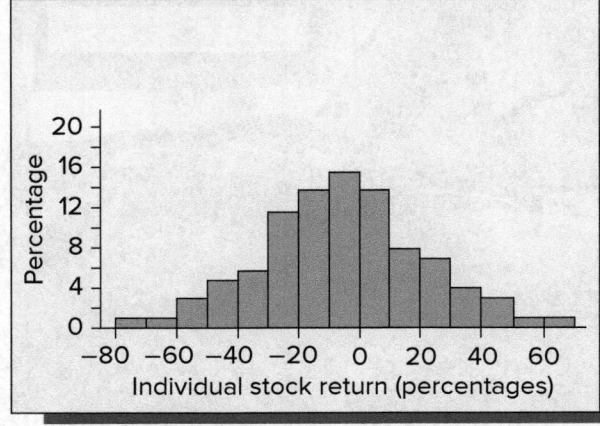

(b) The percent frequency histogram describing the population of all possible sample mean returns when $n = 5$

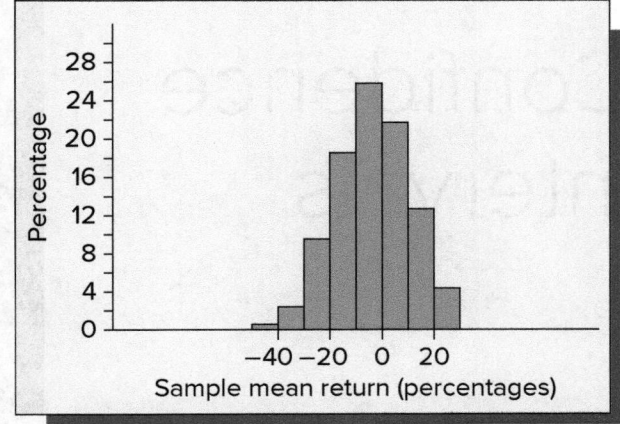

Source: Based on John K. Ford, "A Method for Grading 1987 Stock Recommendations," *The American Association of Individual Investors Journal*, March 1988, pp. 16–17.

a   Calculate the mean and standard deviation of the number of flaws per flash drive.

b   Suppose that we randomly select a sample of 100 flash drives. Describe the shape of the sampling distribution of the sample mean $\bar{x}$. Then compute the mean and the standard deviation of the sampling distribution of $\bar{x}$.

c   Sketch the sampling distribution of the sample mean $\bar{x}$ and compare it to the distribution describing the number of flaws on a single flash drive.

d   The supply house's managers are worried that the flash drives being received have an excessive number of flaws. Because of this, a random sample of 100 flash drives is drawn from each shipment and the shipment is rejected (sent back to the supplier) if the average number of flaws per flash drive for the 100 sample drives is greater than .75. Suppose that the mean number of flaws per flash drive for this week's entire shipment is actually .55. What is the probability that this shipment will be rejected and sent back to the supplier?

8.32   Each day a manufacturing plant receives a large shipment of drums of Chemical ZX-900. These drums are supposed to have a mean fill of 50 gallons, while the fills have a standard deviation known to be .6 gallon.

a   Suppose that the mean fill for the shipment is actually 50 gallons. If we draw a random sample of 100 drums from the shipment, what is the probability that the average fill for the 100 drums is between 49.88 gallons and 50.12 gallons?

b   The plant manager is worried that the drums of Chemical ZX-900 are underfilled. Because of this, she decides to draw a sample of 100 drums from each daily shipment and will reject the shipment (send it back to the supplier) if the average fill for the 100 drums is less than 49.85 gallons. Suppose that a shipment that actually has a mean fill of 50 gallons is received. What is the probability that this shipment will be rejected and sent back to the supplier?

Design Elements: (CD): ©Comstock Images/Alamy; (All Others): ©McGraw-Hill Education

## CHAPTER 9

Confidence Intervals

©Aaron Kiley Photography

Learning Objectives

After mastering the material in this chapter, you will be able to:

LO9-1 Calculate and interpret a z-based confidence interval for a population mean when σ is known.

LO9-2 Describe the properties of the t distribution and use a t table.

LO9-3 Calculate and interpret a t-based confidence interval for a population mean when σ is unknown.

LO9-4 Determine the appropriate sample size when estimating a population mean.

LO9-5 Calculate and interpret a large sample confidence interval for a population proportion.

LO9-6 Determine the appropriate sample size when estimating a population proportion.

LO9-7 Find and interpret confidence intervals for parameters of finite populations (Optional).

Chapter Outline

e have seen that the sample mean is the point estimate of the population mean and the sample proportion is the point estimate of the population proportion. In general, although a point estimate is a reasonable one-number estimate of a population parameter (mean, proportion, or the like), the point estimate will not—unless we are extremely lucky—equal the true value of the population parameter.

In this chapter we study how to use a **confidence interval** to estimate a population parameter. A confidence interval for a population parameter is an interval, or range of numbers, constructed around the point estimate so that we are very sure, or confident, that the true value of the population parameter is inside the interval.

By computing such an interval, we estimate—with confidence—the possible values that a population parameter might equal. This, in turn, can help us to assess—with confidence—whether a particular business improvement has been made or is needed.

In order to illustrate confidence intervals, we revisit several cases introduced in earlier chapters and also introduce some new cases. For example:

In the **Car Mileage Case,** we use a confidence interval to provide strong evidence that the mean EPA combined city and highway mileage for the automaker's new midsize model meets the tax credit standard of 31 mpg.

In the **e-Billing Case,** we use a confidence interval to more completely assess the reduction in mean payment time that was achieved by the new billing system.

In the **Cheese Spread Case,** we use a confidence interval to provide strong evidence that fewer than 10 percent of all current purchasers will stop buying the cheese spread if the new spout is used, and, therefore, that it is reasonable to use the new spout.

9.1 *z*-Based Confidence Intervals for a Population Mean: σ Known

LO9-1
Calculate and interpret a *z*-based confidence interval for a population mean when σ is known.

An introduction to confidence intervals for a population mean

In the *car mileage case*, we have seen that an automaker has introduced a new midsize model and wishes to estimate the mean EPA combined city and highway mileage, μ, that would be obtained by all cars of this type. In order to estimate μ, the automaker has conducted EPA mileage tests on a random sample of 50 of its new midsize cars and has obtained the sample of mileages in Table 1.7. The mean of this sample of mileages, which is $\bar{x} = 31.56$ mpg, is the point estimate of μ. However, a sample mean will not—unless we are extremely lucky—equal the true value of a population mean. Therefore, the sample mean of 31.56 mpg does not, by itself, provide us with any confidence about the true value of the population mean μ. One way to estimate μ with confidence is to calculate a *confidence interval* for this mean.

A **confidence interval** for a population mean is an interval constructed around the sample mean so that we are reasonably sure, or confident, that this interval contains the population mean. Any confidence interval for a population mean is based on what is called a **confidence level.** This confidence level is a percentage (for example, 95 percent or 99 percent) that expresses how confident we are that the confidence interval contains the population mean. In order to explain the exact meaning of a confidence level, we will begin in the car mileage case by finding and interpreting a confidence interval for a population mean that is based on the most commonly used confidence level—the 95 percent level. Then we will generalize our discussion and show how to find and interpret a confidence interval that is based on any confidence level.

Before the automaker selected the sample of $n = 50$ new midsize cars and tested them as prescribed by the EPA, there were many samples of 50 cars and corresponding mileages that the automaker might have obtained. Because different samples generally have different sample means, we consider the probability distribution of the population of all possible

FIGURE 9.1   The Sampling Distribution of the Sample Mean

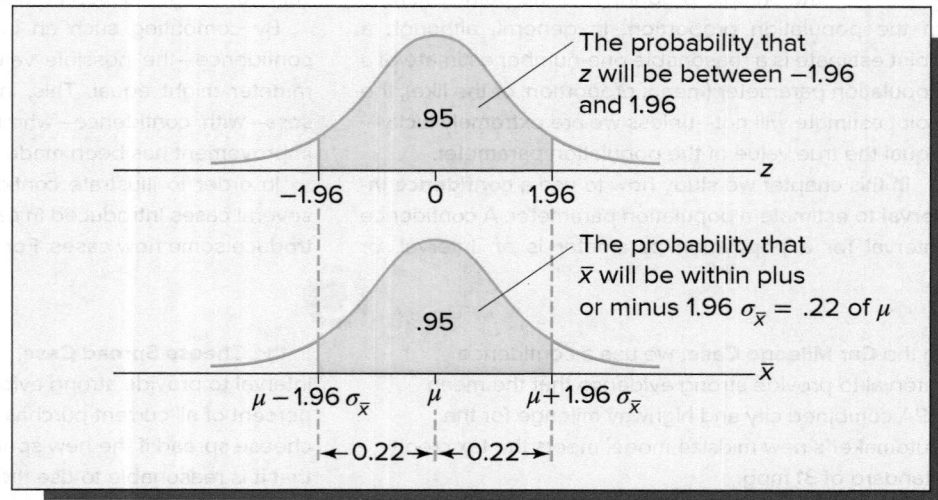

sample means that would be obtained from all possible samples of $n = 50$ car mileages. In Chapter 8 we have seen that such a probability distribution is called the sampling distribution of the sample mean, and we have studied various properties of sampling distributions. Several of these properties tell us that, if the population of all individual midsize car mileages is normally distributed with mean μ and standard deviation σ, then for any sample size n the sampling distribution of the sample mean is a normal distribution with mean $\mu_{\bar{x}} = \mu$ and standard deviation $\sigma_{\bar{x}} = \sigma/\sqrt{n}$. Moreover, because the sample size n is 50, and because we can assume that the true value of the population standard deviation σ is 0.8 mpg (as discussed in Chapter 8), it follows that

$$\sigma_{\bar{x}} = \frac{\sigma}{\sqrt{n}} = \frac{.8}{\sqrt{50}} = .113$$

This allows us to reason as follows:

1   In Chapter 7 (see Figure 7.15) we have seen that the area under the standard normal curve between -1.96 and 1.96 is .95 (see Figure 9.1). This .95 area says that 95 percent of the values in any normally distributed population are within plus or minus 1.96 standard deviations of the mean of the population. Because the population of all possible sample means is normally distributed with mean μ and standard deviation $\sigma_{\bar{x}} = .113$, it follows that 95 percent of all possible sample means are within plus or minus $1.96\sigma_{\bar{x}} = 1.96(.113) = .22$ of the population mean μ. That is, **considered before we select the sample,** the probability is .95 that the sample mean $\bar{x}$ will be within plus or minus $1.96\sigma_{\bar{x}} = .22$ of the population mean μ (again, see Figure 9.1).

2   Saying

$\bar{x}$ will be within plus or minus .22 of μ

is the same as saying

$\bar{x}$ will be such that the interval $[\bar{x} \pm .22]$ contains μ

To understand this, consider Figure 9.2. This figure illustrates three possible samples of 50 mileages and the means of these samples. Also, this figure assumes that (unknown to any human being) the true value of the population mean μ is 31.6. Then, as illustrated in Figure 9.2, because the sample mean $\bar{x} = 31.56$ is within .22 of $\mu = 31.6$, the interval $[31.56 \pm .22] = [31.34, 31.78]$ contains μ. Similarly, because the sample mean $\bar{x} = 31.68$ is within .22 of $\mu = 31.6$, the interval $[31.68 \pm .22] = [31.46, 31.90]$ contains μ. However, because the sample mean $\bar{x} = 31.2$ is not within .22 of $\mu = 31.6$, the interval $[31.2 \pm .22] = [30.98, 31.42]$ does not contain μ.

FIGURE 9.2   Three 95 Percent Confidence Intervals for μ

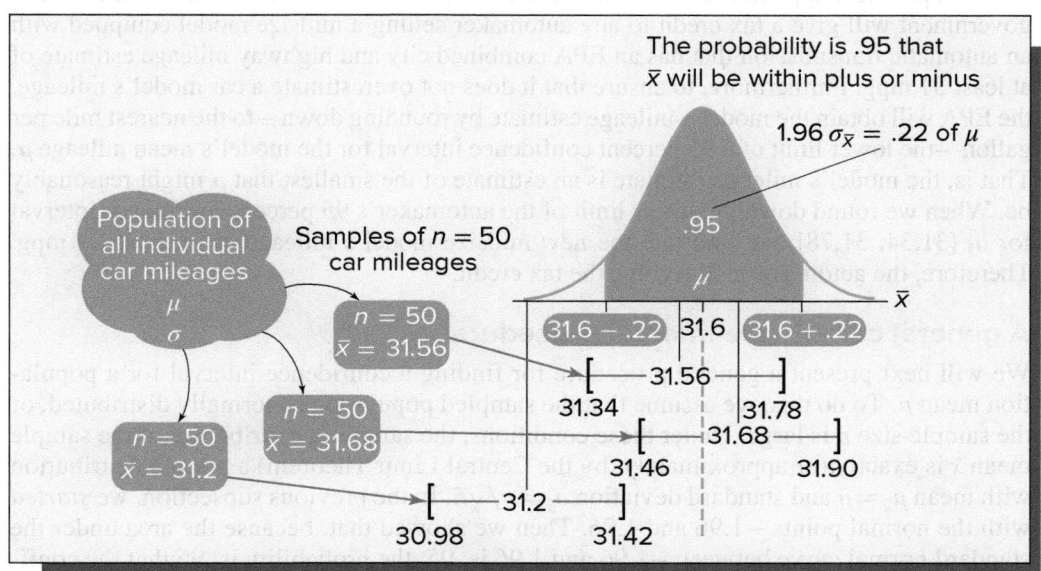

3   In statement 1 we showed that the probability is .95 that the sample mean $\bar{x}$ will be within plus or minus $1.96\sigma_{\bar{x}} = .22$ of the population mean μ. In statement 2 we showed that $\bar{x}$ being within plus or minus .22 of μ is the same as the interval $[\bar{x} \pm .22]$ containing μ. Combining these results, we see that the probability is .95 that the sample mean $\bar{x}$ will be such that the interval

$$[\bar{x} \pm 1.96\sigma_{\bar{x}}] = [\bar{x} \pm .22]$$

contains the population mean μ.

A 95 percent confidence interval for μ

Statement 3 says that, **before we randomly select the sample,** there is a .95 probability that we will obtain an interval $[\bar{x} \pm .22]$ that contains the population mean μ. In other words, 95 percent of all intervals that we might obtain contain μ, and 5 percent of these intervals do not contain μ. For this reason, we call the interval $[\bar{x} \pm .22]$ a **95 percent confidence interval for μ.** To better understand this interval, we must realize that, **when we actually select the sample,** we will observe one particular sample from the extremely large number of possible samples. Therefore, we will obtain one particular confidence interval from the extremely large number of possible confidence intervals. For example, recall that when the automaker randomly selected the sample of $n = 50$ cars and tested them as prescribed by the EPA, the automaker obtained the sample of 50 mileages given in Table 1.7. The mean of this sample is $\bar{x} =$ 31.56 mpg, and a histogram constructed using this sample (see Figure 2.9) indicates that the population of all individual car mileages is normally distributed. It follows that a 95 percent confidence interval for the population mean mileage μ of the new midsize model is

$$[\bar{x} \pm .22] = [31.56 \pm .22]$$

$$= [31.34, 31.78]$$

Because we do not know the true value of μ, we do not know for sure whether this interval contains μ. However, we are 95 percent confident that this interval contains μ. That is, we are 95 percent confident that μ is between 31.34 mpg and 31.78 mpg. What we mean by "95 percent confident" is that we hope that the confidence interval [31.34, 31.78] is one of the 95 percent of all confidence intervals that contain μ and not one of the 5 percent of all confidence intervals that do not contain μ. Here, we say that 95 percent is the **confidence level** associated with the confidence interval.

A practical application

To see a practical application of the automaker's confidence interval, recall that the federal government will give a tax credit to any automaker selling a midsize model equipped with an automatic transmission that has an EPA combined city and highway mileage estimate of at least 31 mpg. Furthermore, to ensure that it does not overestimate a car model's mileage, the EPA will obtain the model's mileage estimate by rounding down—to the nearest mile per gallon—the lower limit of a 95 percent confidence interval for the model's mean mileage μ. That is, the model's mileage estimate is an estimate of the smallest that μ might reasonably be. When we round down the lower limit of the automaker's 95 percent confidence interval for μ, [31.34, 31.78], we find that the new midsize model's mileage estimate is 31 mpg. Therefore, the automaker will receive the tax credit.[1]

A general confidence interval procedure

We will next present a general procedure for finding a confidence interval for a population mean μ. To do this, we assume that the sampled population is normally distributed, or the sample size n is large. Under these conditions, the sampling distribution of the sample mean $\bar{x}$ is exactly (or approximately, by the Central Limit Theorem) a normal distribution with mean $\mu_{\bar{x}} = \mu$ and standard deviation $\sigma_{\bar{x}} = \sigma/\sqrt{n}$. In the previous subsection, we *started* with the normal points -1.96 and 1.96. Then we showed that, because the area under the standard normal curve between -1.96 and 1.96 is .95, the probability is .95 that the confidence interval $[\bar{x} \pm 1.96\sigma_{\bar{x}}]$ will contain the population mean. Usually, we do not start with two normal points, but rather we start by choosing the probability (for example, .95 or .99) that the confidence interval will contain the population mean. This probability is called the **confidence coefficient.** Next, we find the normal points that have a symmetrical area between them under the standard normal curve that is equal to the confidence coefficient. Then, using $\bar{x}$, $\sigma_{\bar{x}}$, and the normal points, we find the confidence interval that is based on the confidence coefficient. To illustrate this, we will start with a confidence coefficient of .95 and use the following three-step procedure to find the appropriate normal points and the corresponding 95 percent confidence interval for the population mean:

Step 1: As illustrated in Figure 9.3, place a symmetrical area of .95 under the standard normal curve and find the area in the normal curve tails beyond the .95 area. Because the entire area under the standard normal curve is 1, the area in both normal curve tails is $1 - .95 = .05$, and the area in each tail is .025.

Step 2: Find the normal point $z_{.025}$ that gives a right-hand tail area under the standard normal curve equal to .025, and find the normal point $-z_{.025}$ that gives a left-hand tail area under the curve equal to .025. As shown in Figure 9.3, the area under the standard normal curve between $-z_{.025}$ and $z_{.025}$ is .95, and the area under this curve to the left of $z_{.025}$ is .975. Looking up a cumulative area of .975 in Table A.3 or in Table 9.1 (which shows a portion of Table A.3), we find that $z_{.025} = 1.96$.

Step 3: Form the following 95 percent confidence interval for the population mean:

$$[\bar{x} \pm z_{.025}\, \sigma_{\bar{x}}] = \left[\bar{x} \pm 1.96\frac{\sigma}{\sqrt{n}}\right]$$

If all possible samples were used to calculate this interval, then 95 percent of the resulting intervals would contain the population mean. For example, recall in the car mileage case that $n = 50$, $\bar{x} = 31.56$, and $\sigma = .8$. Therefore, we can directly calculate the previously obtained 95 percent confidence interval for the midsize model's mean mileage μ as follows:

$$\left[\bar{x} \pm 1.96\frac{\sigma}{\sqrt{n}}\right] = \left[31.56 \pm 1.96\frac{.8}{\sqrt{50}}\right]$$
$$= [31.56 \pm .22]$$
$$= [31.34, 31.78]$$

---

[1]This example is based on the authors' conversations with the EPA. However, there are approaches for showing that μ is at least 31 mpg that differ from the approach that uses the confidence interval [31.34, 31.78]. Some of these other approaches are discussed in Chapter 10, but the EPA tells the authors that it would use the confidence interval approach described here.

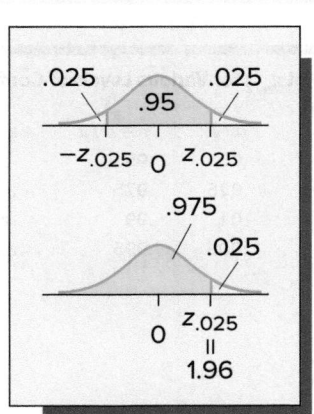

FIGURE 9.3   The Point $z_{.025}$

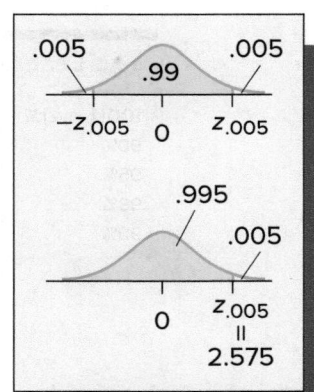

FIGURE 9.4   The Point $z_{.005}$

TABLE 9.1   Cumulative Areas under the Standard Normal Curve

z	0.00	0.01	0.02	0.03	0.04	0.05	0.06	0.07	0.08	0.09
1.5	0.9332	0.9345	0.9357	0.9370	0.9382	0.9394	0.9406	0.9418	0.9429	0.9441
1.6	0.9452	0.9463	0.9474	0.9484	0.9495	0.9505	0.9515	0.9525	0.9535	0.9545
1.7	0.9554	0.9564	0.9573	0.9582	0.9591	0.9599	0.9608	0.9616	0.9625	0.9633
1.8	0.9641	0.9649	0.9656	0.9664	0.9671	0.9678	0.9686	0.9693	0.9699	0.9706
1.9	0.9713	0.9719	0.9726	0.9732	0.9738	0.9744	0.9750	0.9756	0.9761	0.9767
2.0	0.9772	0.9778	0.9783	0.9788	0.9793	0.9798	0.9803	0.9808	0.9812	0.9817
2.1	0.9821	0.9826	0.9830	0.9834	0.9838	0.9842	0.9846	0.9850	0.9854	0.9857
2.2	0.9861	0.9864	0.9868	0.9871	0.9875	0.9878	0.9881	0.9884	0.9887	0.9890
2.3	0.9893	0.9896	0.9898	0.9901	0.9904	0.9906	0.9909	0.9911	0.9913	0.9916
2.4	0.9918	0.9920	0.9922	0.9925	0.9927	0.9929	0.9931	0.9932	0.9934	0.9936
2.5	0.9938	0.9940	0.9941	0.9943	0.9945	0.9946	0.9948	0.9949	0.9951	0.9952

We next start with a confidence coefficient of .99 and find the corresponding 99 percent confidence interval for the population mean:

Step 1: As illustrated in Figure 9.4, place a symmetrical area of .99 under the standard normal curve, and find the area in the normal curve tails beyond the .99 area. Because the entire area under the standard normal curve is 1, the area in both normal curve tails is $1 - .99 = .01$, and the area in each tail is .005.

Step 2: Find the normal point $z_{.005}$ that gives a right-hand tail area under the standard normal curve equal to .005, and find the normal point $-z_{.005}$ that gives a left-hand tail area under the curve equal to .005. As shown in Figure 9.4, the area under the standard normal curve between $-z_{.005}$ and $z_{.005}$ is .99, and the area under this curve to the left of $z_{.005}$ is .995. Looking up a cumulative area of .995 in Table A.3 or in Table 9.1, we find that $z_{.005} = 2.575$.

Step 3: Form the following 99 percent confidence interval for the population mean:

$$\left[\bar{x} \pm z_{.005}\, \sigma_{\bar{x}} \right] = \left[\bar{x} \pm 2.575 \frac{\sigma}{\sqrt{n}} \right]$$

If all possible samples were used to calculate this interval, then 99 percent of the resulting intervals would contain the population mean. For example, in the car mileage case, a 99 percent confidence interval for the midsize model's mean mileage μ is

$$\left[\bar{x} \pm 2.575 \frac{\sigma}{\sqrt{n}} \right] = \left[31.56 \pm 2.575 \frac{.8}{\sqrt{50}} \right]$$

$$= [31.56 \pm .29]$$

$$= [31.27, 31.85]$$

FIGURE 9.5 **The Point $z_{\alpha/2}$**

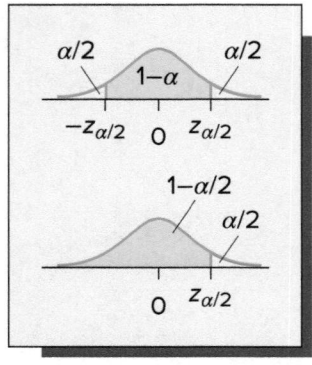

TABLE 9.2 **The Normal Point $z_{\alpha/2}$ for Various Levels of Confidence**

$100(1-\alpha)\%$	$1-\alpha$	α	$\alpha/2$	$1-\alpha/2$	$z_{\alpha/2}$
90%	.90	.10	.05	.95	$z_{.05} = 1.645$
95%	.95	.05	.025	.975	$z_{.025} = 1.96$
98%	.98	.02	.01	.99	$z_{.01} = 2.33$
99%	.99	.01	.005	.995	$z_{.005} = 2.575$

To compare the 95 percent confidence interval $[\bar{x} \pm 1.96(\sigma/\sqrt{n})]$ with the 99 percent confidence interval $[\bar{x} \pm 2.575(\sigma/\sqrt{n})]$, note that each of these confidence intervals can be expressed in the form $[\bar{x} \pm$ margin of error]. Here, for a given level of confidence, the **margin of error** expresses the farthest that the sample mean $\bar{x}$ might be from the population mean μ. Moreover, the margin of error $2.575(\sigma/\sqrt{n})$ used to compute the 99 percent interval is larger than the margin of error $1.96(\sigma/\sqrt{n})$ used to compute the 95 percent interval. Therefore, the 99 percent interval is the longer of these intervals. **In general, increasing the confidence level (1) has the advantage of making us more confident that μ is contained in the confidence interval, but (2) has the disadvantage of increasing the margin of error and thus providing a less precise estimate of the true value of μ.** Frequently, 95 percent confidence intervals are used to make conclusions. If conclusions based on stronger evidence are desired, 99 percent intervals are sometimes used. In the car mileage case, the fairly large sample size of $n = 50$ produces a 99 percent margin of error, $2.575(.8/\sqrt{50}) = .29$, that is not much larger than the 95 percent margin of error, $1.96(.8/\sqrt{50}) = .22$. Therefore, the 99 percent confidence interval for μ, $[31.56 \pm .29] = [31.27, 31.85]$, is not much longer than the 95 percent confidence interval for μ, $[31.56 \pm .22] = [31.34, 31.78]$.

In general, we let α denote the probability that a confidence interval for a population mean will *not* contain the population mean. This implies that $1 - \alpha$ is the probability that the confidence interval will contain the population mean. In order to find a confidence interval for a population mean that is based on a confidence coefficient of $1 - \alpha$ (that is, a **$100(1 - \alpha)$ percent confidence interval** for the population mean), we do the following:

Step 1: As illustrated in Figure 9.5, place a symmetrical area of $1 - \alpha$ under the standard normal curve, and find the area in the normal curve tails beyond the $1 - \alpha$ area. Because the entire area under the standard normal curve is 1, the combined areas in the normal curve tails are α, and the area in each tail is $\alpha/2$.

Step 2: Find the normal point $z_{\alpha/2}$ that gives a right-hand tail area under the standard normal curve equal to $\alpha/2$, and find the normal point $-z_{\alpha/2}$ that gives a left-hand tail area under this curve equal to $\alpha/2$. As shown in Figure 9.5, the area under the standard normal curve between $-z_{\alpha/2}$ and $z_{\alpha/2}$ is $(1 - \alpha)$, and the area under this curve to the left of $z_{\alpha/2}$ is $1 - \alpha/2$. **This implies that we can find $z_{\alpha/2}$ by looking up a cumulative area of $1 - \alpha/2$ in Table A.3.**

Step 3: Form the following $100(1 - \alpha)$ percent confidence interval for the population mean:

$$[\bar{x} \pm z_{\alpha/2}\,\sigma_{\bar{x}}] = \left[\bar{x} \pm z_{\alpha/2}\,\frac{\sigma}{\sqrt{n}}\right]$$

If all possible samples were used to calculate this interval, then $100(1 - \alpha)$ percent of the resulting intervals would contain the population mean. Moreover, we call **$100(1 - \alpha)$ percent** the **confidence level** associated with the confidence interval.

The general formula that we just obtained for a $100(1 - \alpha)$ percent confidence interval for a population mean implies that we now have a formal way to find the normal point corresponding to a particular level of confidence. Specifically, we can set $100(1 - \alpha)$ percent equal to the particular level of confidence, solve for α, and use a cumulative normal table to find the normal point $z_{\alpha/2}$ corresponding to the cumulative area $1 - \alpha/2$. For example, suppose that we wish to find a 99 percent confidence interval for the population mean. Then, because $100(1 - \alpha)$ percent equals 99 percent, it follows that $1 - \alpha = .99$, $\alpha = .01$, $\alpha/2 = .005$, and $1 - \alpha/2 = .995$. Looking up .995 in a cumulative normal table, we find that $z_{\alpha/2} = z_{.005} = 2.575$. This normal point is the same normal point that we previously found using the three-step procedure and the normal curves illustrated in Figure 9.4. Table 9.2 summarizes finding the values of $z_{\alpha/2}$ for different values of the confidence level $100(1 - \alpha)$ percent.

The following box summarizes the formula used in calculating a $100(1 - \alpha)$ percent confidence interval for a population mean μ. This interval is based on the normal distribution and assumes that the true value of σ is known. If (as is usually the case) σ is not known, we can use a confidence interval for μ discussed in the next section.

A Confidence Interval for a Population Mean μ: σ Known

Suppose that the sampled population is normally distributed with mean μ and standard deviation σ. Then a $100(1 - \alpha)$ **percent confidence interval for μ** is

$$\left[\bar{x} \pm z_{\alpha/2} \frac{\sigma}{\sqrt{n}} \right] = \left[\bar{x} - z_{\alpha/2} \frac{\sigma}{\sqrt{n}}, \quad \bar{x} + z_{\alpha/2} \frac{\sigma}{\sqrt{n}} \right]$$

Here, $z_{\alpha/2}$ is the normal point that gives a right-hand tail area under the standard normal curve of $\alpha/2$. The normal point $z_{\alpha/2}$ can be found by looking up a cumulative area of $1 - \alpha/2$ in Table A.3. This confidence interval is also approximately valid for nonnormal populations if the sample size is large (at least 30).

ⓒ EXAMPLE 9.1 The e-Billing Case: Reducing Mean Bill Payment Time

Recall that a management consulting firm has installed a new computer-based electronic billing system in a Hamilton, Ohio, trucking company. The population mean payment time using the trucking company's old billing system was approximately equal to, but no less than, 39 days. In order to assess whether the population mean payment time, μ, using the new billing system is substantially less than 39 days, the consulting firm will use the sample of $n = 65$ payment times in Table 2.4 to find a 99 percent confidence interval for μ. The mean of the 65 payment times is $\bar{x} = 18.1077$ days, and we will assume that the true value of the population standard deviation σ for the new billing system is 4.2 days (as discussed in Chapter 8). Then, because we previously showed that the normal point corresponding to 99 percent confidence is $z_{\alpha/2} = z_{.005} = 2.575$, a 99 percent confidence interval for μ is

$$\left[\bar{x} \pm z_{.005} \frac{\sigma}{\sqrt{n}} \right] = \left[18.1077 \pm 2.575 \frac{4.2}{\sqrt{65}} \right]$$

$$= [18.1077 \pm 1.3414]$$

$$= [16.8, 19.4]$$

Recalling that the mean payment time using the old billing system is 39 days, this interval says that we are 99 percent confident that the population mean payment time using the new billing system is between 16.8 days and 19.4 days. Therefore, we are 99 percent confident that the new billing system reduces the mean payment time by at most $39 - 16.8 = 22.2$ days and by at least $39 - 19.4 = 19.6$ days.

Exercises for Section 9.1

CONCEPTS ■ connect

9.1 Explain why it is important to calculate a confidence interval.

9.2 Explain the meaning of the term "95 percent confidence."

9.3 Under what conditions is the confidence interval $[\bar{x} \pm z_{\alpha/2}\,(\sigma/\sqrt{n})]$ for μ valid?

9.4 For a fixed sample size, what happens to a confidence interval for μ when we increase the level of confidence?

9.5 For a fixed level of confidence, what happens to a confidence interval for μ when we increase the sample size?

METHODS AND APPLICATIONS

9.6 Suppose that, for a sample of size $n = 100$ measurements, we find that $\bar{x} = 50$. Assuming that σ equals 2, calculate confidence intervals for the population mean μ with the following confidence levels:

    a 95%   **b** 99%   **c** 97%
    d 80%   **e** 99.73%   **f** 92%

9.7 **THE TRASH BAG CASE** ⓓⓢ TrashBag

Consider the trash bag problem. Suppose that an independent laboratory has tested trash bags and has found that no 30-gallon bags that are currently on the market have a mean breaking strength of 50 pounds or more. On the basis of these results, the producer of the new, improved trash bag feels sure that its 30-gallon bag will be the strongest such bag on the market if the new trash bag's mean breaking strength can be shown to be at least 50 pounds. The mean of the sample of 40 trash bag breaking strengths in Table 1.10 is 50.575. If we let μ denote the mean of the breaking strengths of all possible trash bags of the new type and assume that the population standard deviation equals 1.65:

    a Calculate 95 percent and 99 percent confidence intervals for μ.

    b Using the 95 percent confidence interval, can we be 95 percent confident that μ is at least 50 pounds? Explain.

    c Using the 99 percent confidence interval, can we be 99 percent confident that μ is at least 50 pounds? Explain.

    d Based on your answers to parts b and c, how convinced are you that the new 30-gallon trash bag is the strongest such bag on the market?

9.8 **THE BANK CUSTOMER WAITING TIME CASE** ⓓⓢ WaitTime

Recall that a bank manager has developed a new system to reduce the time customers spend waiting to be served by tellers during peak business hours. The mean waiting time during peak business hours under the current system is roughly 9 to 10 minutes. The bank manager hopes that the new system will have a mean waiting time that is less than six minutes. The mean of the sample of 100 bank customer waiting times in Table 1.9 is 5.46. If we let μ denote the mean of all possible bank customer waiting times using the new system and assume that the population standard deviation equals 2.47:

    a Calculate 95 percent and 99 percent confidence intervals for μ.

    b Using the 95 percent confidence interval, can the bank manager be 95 percent confident that μ is less than six minutes? Explain.

    c Using the 99 percent confidence interval, can the bank manager be 99 percent confident that μ is less than six minutes? Explain.

    d Based on your answers to parts b and c, how convinced are you that the new mean waiting time is less than six minutes?

9.9 **THE VIDEO GAME SATISFACTION RATING CASE** ⓓⓢ VideoGame

The mean of the sample of 65 customer satisfaction ratings in Table 1.8 is 42.95. If we let μ denote the mean of all possible customer satisfaction ratings for the XYZ Box video game system, and assume that the population standard deviation equals 2.64:

    a Calculate 95 percent and 99 percent confidence intervals for μ.

    b Using the 95 percent confidence interval, can we be 95 percent confident that μ is at least 42 (recall that a very satisfied customer gives a rating of at least 42)? Explain.

    c Using the 99 percent confidence interval, can we be 99 percent confident that μ is at least 42? Explain.

    d Based on your answers to parts b and c, how convinced are you that the mean satisfaction rating is at least 42?

9.10 In an article in *Marketing Science*, Silk and Berndt investigate the output of advertising agencies. They describe ad agency output by finding the shares of dollar billing volume coming from various media categories such as network television, spot television, newspapers, radio, and so forth.

    a Suppose that a random sample of 400 U.S. advertising agencies gives an average percentage share of billing volume from network television equal to 7.46 percent, and assume that the population standard deviation equals 1.42 percent. Calculate a 95 percent confidence interval for the mean percentage share of billing volume from network television for the population of all U.S. advertising agencies.

    b Suppose that a random sample of 400 U.S. advertising agencies gives an average percentage share of billing volume from spot television commercials equal to 12.44 percent, and assume that the population standard deviation equals 1.55 percent. Calculate a 95 percent confidence interval for the mean percentage share of billing volume from spot television commercials for the population of all U.S. advertising agencies.

c Compare the confidence intervals in parts *a* and *b*. Does it appear that the mean percentage share of billing volume from spot television commercials for all U.S. advertising agencies is greater than the mean percentage share of billing volume from network television for all U.S. advertising agencies? Explain.

9.11 In an article in *Accounting and Business Research*, Carslaw and Kaplan investigate factors that influence "audit delay" for firms in New Zealand. Audit delay, which is defined to be the length of time (in days) from a company's financial year-end to the date of the auditor's report, has been found to affect the market reaction to the report. This is because late reports often seem to be associated with lower returns and early reports often seem to be associated with higher returns.

Carslaw and Kaplan investigated audit delay for two kinds of public companies—owner-controlled and manager-controlled companies. Here a company is considered to be owner controlled if 30 percent or more of the common stock is controlled by a single outside investor (an investor not part of the management group or board of directors). Otherwise, a company is considered manager controlled. It was felt that the type of control influences audit delay. To quote Carslaw and Kaplan:

> Large external investors, having an acute need for timely information, may be expected to pressure the company and auditor to start and to complete the audit as rapidly as practicable.

a Suppose that a random sample of 100 public owner-controlled companies in New Zealand is found to give a mean audit delay of 82.6 days, and assume that the population standard deviation equals 33 days. Calculate a 95 percent confidence interval for the population mean audit delay for all public owner-controlled companies in New Zealand.

b Suppose that a random sample of 100 public manager-controlled companies in New Zealand is found to give a mean audit delay of 93 days, and assume that the population standard deviation equals 37 days. Calculate a 95 percent confidence interval for the population mean audit delay for all public manager-controlled companies in New Zealand.

c Use the confidence intervals you computed in parts *a* and *b* to compare the mean audit delay for all public owner-controlled companies versus that of all public manager-controlled companies. How do the means compare? Explain.

9.12 In an article in the *Journal of Marketing*, Bayus studied the differences between "early replacement buyers" and "late replacement buyers" in making consumer durable good replacement purchases. Early replacement buyers are consumers who replace a product during the early part of its lifetime, while late replacement buyers make replacement purchases late in the product's lifetime. In particular, Bayus studied automobile replacement purchases. Consumers who traded in cars with ages of zero to three years and mileages of no more than 35,000 miles were classified as early replacement buyers. Consumers who traded in cars with ages of seven or more years and mileages of more than 73,000 miles were classified as late replacement buyers. Bayus compared the two groups of buyers with respect to demographic variables such as income, education, age, and so forth. He also compared the two groups with respect to the amount of search activity in the replacement purchase process. Variables compared included the number of dealers visited, the time spent gathering information, and the time spent visiting dealers.

a Suppose that a random sample of 800 early replacement buyers yields a mean number of dealers visited equal to 3.3, and assume that the population standard deviation equals .71. Calculate a 99 percent confidence interval for the population mean number of dealers visited by all early replacement buyers.

b Suppose that a random sample of 500 late replacement buyers yields a mean number of dealers visited equal to 4.3, and assume that the population standard deviation equals .66. Calculate a 99 percent confidence interval for the population mean number of dealers visited by all late replacement buyers.

c Use the confidence intervals you computed in parts *a* and *b* to compare the mean number of dealers visited by all early replacement buyers with the mean number of dealers visited by all late replacement buyers. How do the means compare? Explain.

9.2 *t*-Based Confidence Intervals for a Population Mean: σ Unknown

LO9-2

Describe the properties of the *t* distribution and use a *t* table.

If we do not know σ (which is usually the case), we can use the sample standard deviation *s* to help construct a confidence interval for μ. The interval is based on the sampling distribution of

$$t = \frac{\bar{x} - \mu}{s/\sqrt{n}}$$

If the sampled population is normally distributed, then for any sample size *n* this sampling distribution is what is called a ***t* distribution.**

FIGURE 9.6   **As the Number of Degrees of Freedom Increases, the Spread of the *t* Distribution Decreases and the *t* Curve Approaches the Standard Normal Curve**

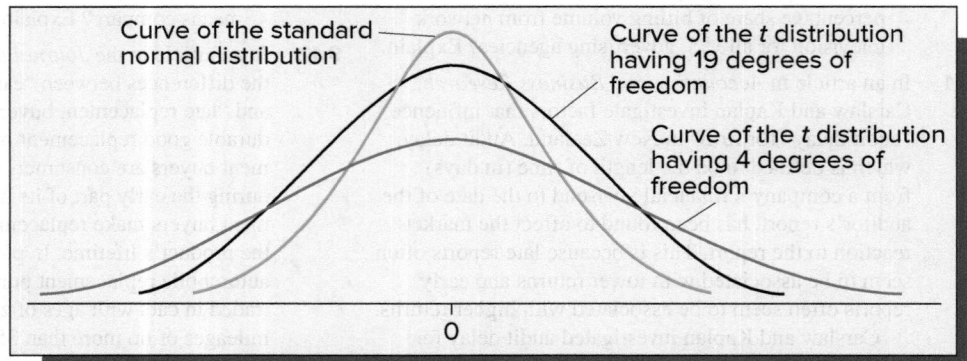

The curve of the *t* distribution has a shape similar to that of the standard normal curve. Two *t* curves and a standard normal curve are illustrated in Figure 9.6. A *t* curve is symmetrical about zero, which is the mean of any *t* distribution. However, the *t* distribution is more spread out, or variable, than the standard normal distribution. Because the *t* statistic above is a function of two random variables, $\bar{x}$ and s, it is logical that the sampling distribution of this statistic is more variable than the sampling distribution of the z statistic, which is a function of only one random variable, $\bar{x}$. The exact spread, or standard deviation, of the *t* distribution depends on a parameter that is called the **number of degrees of freedom (denoted *df*)**. The number of degrees of freedom *df* varies depending on the problem. In the present situation the sampling distribution of *t* has a number of degrees of freedom that equals the sample size minus 1. We say that this sampling distribution is a ***t* distribution with $n - 1$ degrees of freedom.** As the sample size n (and thus the number of degrees of freedom) increases, the spread of the *t* distribution decreases (see Figure 9.6). Furthermore, as the number of degrees of freedom approaches infinity, the curve of the *t* distribution approaches (that is, becomes shaped more and more like) the curve of the standard normal distribution.

In order to use the *t* distribution, we employ a ***t* point that is denoted t_α.** As illustrated in Figure 9.7, t_α **is the point on the horizontal axis under the curve of the *t* distribution that gives a right-hand tail area equal to α.** The value of t_α in a particular situation depends upon the right-hand tail area α and the number of degrees of freedom of the *t* distribution. Values of t_α are tabulated in a ***t* table.** Such a table is given in Table A.4 of Appendix A and a portion of Table A.4 is reproduced in this chapter as Table 9.3. In this *t* table, the rows correspond to the different numbers of degrees of freedom (which are denoted as *df*). The values of *df* are listed down the left side of the table, while the columns designate the right-hand tail area α. For example, suppose we wish to find the *t* point that gives a right-hand tail area of .025 under a *t* curve having $df = 14$ degrees of freedom. To do this, we look in Table 9.3 at the row labeled 14 and the column labeled $t_{.025}$. We find that this $t_{.025}$ point is 2.145 (also see Figure 9.8). Similarly, when there are $df = 14$ degrees of freedom, we find that $t_{.005} = 2.977$ (see Table 9.3 and Figure 9.9).

Table 9.3 gives *t* points for degrees of freedom *df* from 1 to 30. The table also gives *t* points for 40, 60, 120, and an infinite number of degrees of freedom. Looking at this table, it is useful to realize that **the normal points giving the various right-hand tail areas are listed in the row of the *t* table corresponding to an infinite (∞) number of degrees of freedom.** Looking at the row corresponding to ∞, we see that, for example, $z_{.025} = 1.96$. Therefore, we can use this row in the *t* table as an alternative to using the normal table when we need to find normal points (such as $z_{\alpha/2}$ in Section 9.1).

FIGURE 9.7   **An Example of a *t* Point Giving a Specified Right-Hand Tail Area (This *t* Point Gives a Right-Hand Tail Area Equal to α)**

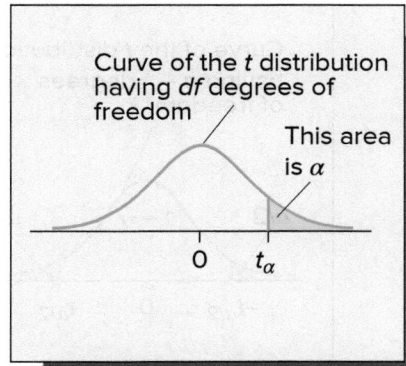

FIGURE 9.8   **The *t* Point Giving a Right-Hand Tail Area of .025 under the *t* Curve Having 14 Degrees of Freedom: $t_{.025} = 2.145$**

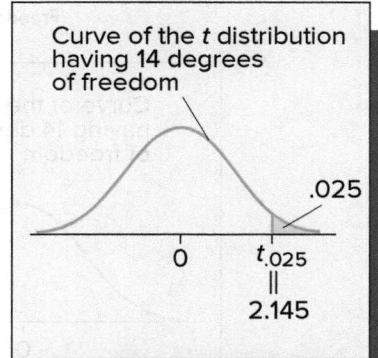

TABLE 9.3   **A *t* Table**

df	$t_{.10}$	$t_{.05}$	$t_{.025}$	$t_{.01}$	$t_{.005}$	$t_{.001}$	$t_{.0005}$
1	3.078	6.314	12.706	31.821	63.657	318.31	636.62
2	1.886	2.920	4.303	6.965	9.925	22.326	31.598
3	1.638	2.353	3.182	4.541	5.841	10.213	12.924
4	1.533	2.132	2.776	3.747	4.604	7.173	8.610
5	1.476	2.015	2.571	3.365	4.032	5.893	6.869
6	1.440	1.943	2.447	3.143	3.707	5.208	5.959
7	1.415	1.895	2.365	2.998	3.499	4.785	5.408
8	1.397	1.860	2.306	2.896	3.355	4.501	5.041
9	1.383	1.833	2.262	2.821	3.250	4.297	4.781
10	1.372	1.812	2.228	2.764	3.169	4.144	4.587
11	1.363	1.796	2.201	2.718	3.106	4.025	4.437
12	1.356	1.782	2.179	2.681	3.055	3.930	4.318
13	1.350	1.771	2.160	2.650	3.012	3.852	4.221
14	1.345	1.761	2.145	2.624	2.977	3.787	4.140
15	1.341	1.753	2.131	2.602	2.947	3.733	4.073
16	1.337	1.746	2.120	2.583	2.921	3.686	4.015
17	1.333	1.740	2.110	2.567	2.898	3.646	3.965
18	1.330	1.734	2.101	2.552	2.878	3.610	3.922
19	1.328	1.729	2.093	2.539	2.861	3.579	3.883
20	1.325	1.725	2.086	2.528	2.845	3.552	3.850
21	1.323	1.721	2.080	2.518	2.831	3.527	3.819
22	1.321	1.717	2.074	2.508	2.819	3.505	3.792
23	1.319	1.714	2.069	2.500	2.807	3.485	3.767
24	1.318	1.711	2.064	2.492	2.797	3.467	3.745
25	1.316	1.708	2.060	2.485	2.787	3.450	3.725
26	1.315	1.706	2.056	2.479	2.779	3.435	3.707
27	1.314	1.703	2.052	2.473	2.771	3.421	3.690
28	1.313	1.701	2.048	2.467	2.763	3.408	3.674
29	1.311	1.699	2.045	2.462	2.756	3.396	3.659
30	1.310	1.697	2.042	2.457	2.750	3.385	3.646
40	1.303	1.684	2.021	2.423	2.704	3.307	3.551
60	1.296	1.671	2.000	2.390	2.660	3.232	3.460
120	1.289	1.658	1.980	2.358	2.617	3.160	3.373
∞	1.282	1.645	1.960	2.326	2.576	3.090	3.291

Source: Data from E. S. Pearson and H. O. Hartley eds., *The Biometrika Tables for Statisticians* 1, 3d ed. (Biometrika, 1966).

FIGURE 9.9 The *t* Point Giving a Right-Hand Tail Area of .005 under the *t* Curve Having 14 Degrees of Freedom: $t_{.005} = 2.977$

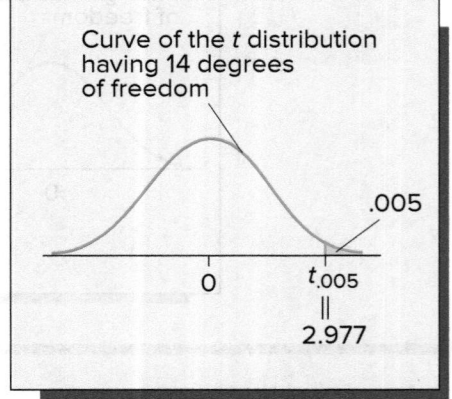

FIGURE 9.10 The Point $t_{\alpha/2}$ with $n - 1$ Degrees of Freedom

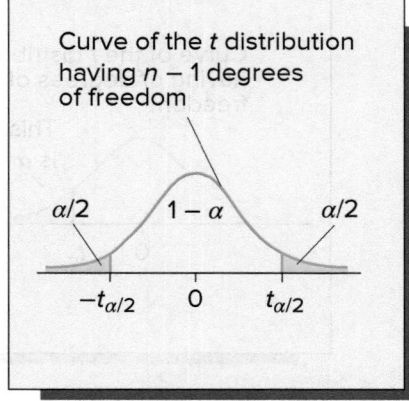

LO9-3

Calculate and interpret a *t*-based confidence interval for a population mean when σ is unknown.

Table A.4 of Appendix A gives *t* points for values of *df* from 1 to 100. We can use a computer to find *t* points based on values of *df* greater than 100. Alternatively, because a *t* curve based on more than 100 degrees of freedom is approximately the shape of the standard normal curve, *t* points based on values of *df* greater than 100 can be approximated by their corresponding *z* points. That is, **when performing hand calculations, it is reasonable to approximate values of t_α by z_α when *df* is greater than 100.**

We now present the formula for a $100(1 - \alpha)$ percent confidence interval for a population mean μ based on the *t* distribution.

A *t*-Based Confidence Interval for a Population Mean μ: σ Unknown

If the sampled population is normally distributed with mean μ, then a **$100(1 - \alpha)$ percent confidence interval for μ** is

$$\left[\bar{x} \pm t_{\alpha/2} \frac{s}{\sqrt{n}} \right]$$

Here *s* is the sample standard deviation, $t_{\alpha/2}$ is the *t* point giving a right-hand tail area of $\alpha/2$ under the *t* curve having $n - 1$ degrees of freedom, and *n* is the sample size. This confidence interval is also approximately valid for nonnormal populations if the sample size is large (at least 30).

Before presenting an example, we need to make a few comments. First, it has been shown that, even if the sample size is not large, this confidence interval is approximately valid for many populations that are not exactly normally distributed. In particular, this interval is approximately valid for a mound-shaped, or single-peaked, population, even if the population is somewhat skewed to the right or left. Second, this interval employs the point $t_{\alpha/2}$, which as shown in Figure 9.10, gives a right-hand tail area equal to $\alpha/2$ under the *t* curve having $n - 1$ degrees of freedom. Here $\alpha/2$ is determined from the desired confidence level $100(1 - \alpha)$ percent.

 EXAMPLE 9.2 The Commercial Loan Case: Mean Debt-to-Equity Ratio

One measure of a company's financial health is its *debt-to-equity ratio*. This quantity is defined to be the ratio of the company's corporate debt to the company's equity. If this ratio is too high, it is one indication of financial instability. For obvious reasons, banks often monitor the financial health of companies to which they have extended commercial loans. Suppose that, in order to reduce risk, a large bank has decided to initiate a policy limiting the mean debt-to-equity ratio for its portfolio of commercial loans to being less than 1.5. In order to estimate the mean debt-to-equity ratio of its (current) commercial loan portfolio, the bank randomly selects a sample of 15 of its commercial loan accounts. Audits of these companies result in the following debt-to-equity ratios:

1.31	1.05	1.45	1.21	1.19
1.78	1.37	1.41	1.22	1.11
1.46	1.33	1.29	1.32	1.65

DS DebtEq

A stem-and-leaf display of these ratios is given in the page margin and looks reasonably bell-shaped and symmetrical. Furthermore, the sample mean and standard deviation of the ratios can be calculated to be $\bar{x} = 1.3433$ and $s = .1921$.

```
1.0 | 5
1.1 | 1 9
1.2 | 1 2 9
1.3 | 1 2 3 7
1.4 | 1 5 6
1.5 |
1.6 | 5
1.7 | 8
```

Suppose the bank wishes to calculate a 95 percent confidence interval for the loan portfolio's mean debt-to-equity ratio, μ. The reasonably bell-shaped and symmetrical stem-and-leaf display in the page margin implies that the population of all debt-to-equity ratios is (approximately) normally distributed. Thus, we can base the confidence interval on the *t* distribution. Because the bank has taken a sample of size $n = 15$, we have $n - 1 = 15 - 1 = 14$ degrees of freedom, and the level of confidence $100(1 - \alpha)$ percent = 95 percent implies that $1 - \alpha = .95$ and $\alpha = .05$. Therefore, we use the *t* point $t_{\alpha/2} = t_{.05/2} = t_{.025}$, which is the *t* point giving a right-hand tail area of .025 under the *t* curve having 14 degrees of freedom. This *t* point is illustrated in the figure below:

```
        df = 14
  .025          .025
          .95
 -t.025   0   t.025
              ‖
            2.145
```

Using Table 9.3, we find that $t_{.025}$ with 14 degrees of freedom is 2.145. It follows that the 95 percent confidence interval for μ is

$$\left[\bar{x} \pm t_{.025} \frac{s}{\sqrt{n}} \right] = \left[1.3433 \pm 2.145 \frac{.1921}{\sqrt{15}} \right]$$

$$= [1.3433 \pm 0.1064]$$

$$= [1.2369, 1.4497]$$

This interval says the bank is 95 percent confident that the mean debt-to-equity ratio for its portfolio of commercial loan accounts is between 1.2369 and 1.4497. Based on this interval, the bank has strong evidence that the portfolio's mean ratio is less than 1.5 (or that the bank is in compliance with its new policy).

Recall that in the two cases discussed in Section 9.1, we calculated *z*-based confidence intervals for μ by assuming that the population standard deviation σ is known. If σ is actually not known (which would probably be true), we should compute *t*-based confidence intervals. Furthermore, recall that in each of these cases the sample size is large (at least 30). As stated in the summary box, **if the sample size is large, the *t*-based confidence interval for μ is approximately valid even if the sampled population is not normally distributed.** Therefore, consider the car mileage case and the sample of 50 mileages in Table 1.7, which has mean $\bar{x} = 31.56$ and standard deviation $s = .7977$.

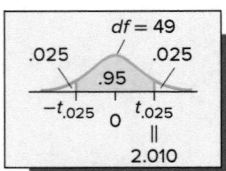

The 95 percent t-based confidence interval for the population mean mileage μ of the new midsize model is

$$\left[\bar{x} \pm t_{.025} \frac{s}{\sqrt{n}}\right] = \left[31.56 \pm 2.010 \frac{.7977}{\sqrt{50}}\right] = [31.33, 31.79]$$

where $t_{.025} = 2.010$ is based on $n - 1 = 50 - 1 = 49$ degrees of freedom—see Table A.4. This interval says we are 95 percent confident that the model's population mean mileage μ is between 31.33 mpg and 31.79 mpg. Based on this interval, the model's EPA mileage estimate is 31 mpg, and the automaker will receive the tax credit.

As another example, the sample of 65 payment times in Table 2.4 has mean $\bar{x} = 18.1077$ and standard deviation $s = 3.9612$. The 99 percent t-based confidence interval for the population mean payment time using the new electronic billing system is

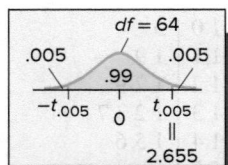

$$\left[\bar{x} \pm t_{.005} \frac{s}{\sqrt{n}}\right] = \left[18.1077 \pm 2.655 \frac{3.9612}{\sqrt{65}}\right] = [16.8, 19.4]$$

where $t_{.005} = 2.655$ is based on $n - 1 = 65 - 1 = 64$ degrees of freedom—see Table A.4. Recalling that the mean payment time using the old billing system is 39 days, the interval says that we are 99 percent confident that the population mean payment time using the new billing system is between 16.8 days and 19.4 days. Therefore, we are 99 percent confident that the new billing system reduces the mean payment time by at most 22.2 days and by at least 19.6 days.

C **EXAMPLE 9.3** The Marketing Research Case: Rating a Bottle Design

Recall that a brand group is considering a new bottle design for a popular soft drink and that Table 1.6 gives a random sample of $n = 60$ consumer ratings of this new bottle design. Let μ denote the mean rating of the new bottle design that would be given by all consumers. In order to assess whether μ exceeds the minimum standard composite score of 25 for a successful bottle design, the brand group will calculate a 95 percent confidence interval for μ. The mean and the standard deviation of the 60 bottle design ratings are $\bar{x} = 30.35$ and $s = 3.1073$. It follows that a 95 percent confidence interval for μ is

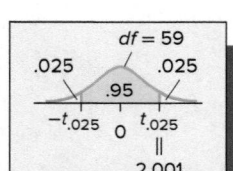

$$\left[\bar{x} \pm t_{.025} \frac{s}{\sqrt{n}}\right] = \left[30.35 \pm 2.001 \frac{3.1073}{\sqrt{60}}\right] = [29.5, 31.2]$$

where $t_{.025} = 2.001$ is based on $n - 1 = 60 - 1 = 59$ degrees of freedom—see Table A.4. Because the interval says we are 95 percent confident that the population mean rating of the new bottle design is between 29.5 and 31.2, we are 95 percent confident that this mean rating exceeds the minimum standard of 25 by at least 4.5 points and by at most 6.2 points.

Confidence intervals for μ can be computed using Excel, JMP, and Minitab. For example, the Minitab and JMP outputs in Figure 9.11 tell us that the t-based 95 percent confidence interval for the mean debt-to-equity ratio is [1.2370, 1.4497]. This result is, within rounding, the same interval calculated in Example 9.2. The Minitab output also gives the sample mean

FIGURE 9.11 Minitab and JMP Outputs of a t-Based 95 Percent Confidence Interval for the Mean Debt-to-Equity Ratio

(a) The Minitab Output

Variable	N	Mean	StDev
Ratio	15	1.3433	0.1921

SE Mean	95% CI
0.0496	(1.2370, 1.4497)

(b) The JMP Output

Parameter	Estimate	Lower CI	Upper CI
Mean	1.343333	1.236962	1.449704

FIGURE 9.12   The Excel Output for the Debt-to-Equity Ratio Example

	A	B	C
1	*Descriptive Statistics*		
2			
3	Mean	1.3433	
4	Standard Error	0.0496	
5	Median	1.32	
6	Mode	#N/A	
7	Standard Deviation	0.1921	
8	Sample Variance	0.0369	
9	Kurtosis	0.8334	
10	Skewness	0.8050	
11	Range	0.73	
12	Minimum	1.05	
13	Maximum	1.78	
14	Sum	20.15	
15	Count	15	
16	Confidence Level(95.0%)	0.1064	

$\bar{x} = 1.3433$, as well as the sample standard deviation $s = .1921$ and the quantity $s/\sqrt{n} = .0496$, which is called the **standard error of the estimate** $\bar{x}$ and denoted "SE Mean" on the Minitab output. Figure 9.12 gives the Excel output of the information needed to calculate the *t*-based 95 percent confidence interval for the mean debt-to-equity ratio. If we consider the Excel output, we see that $\bar{x} = 1.3433$ (see "Mean"), $s = .1921$ (see "Standard Deviation"), $s/\sqrt{n} = .0496$ (see "Standard Error"), and $t_{.025}(s/\sqrt{n}) = .1064$ [see "Confidence Level (95.0%)"]. The interval, which must be hand calculated (or calculated by using Excel cell formulas), is $[1.3433 \pm .1064] = [1.2369, 1.4497]$.

Finally, if the sample size *n* is small and the sampled population is not mound-shaped or is highly skewed, then the *t*-based confidence interval for the population mean might not be valid. In this case we can use a **nonparametric method**—a method that makes no assumption about the shape of the sampled population and is valid for any sample size. Nonparametric methods are discussed in Chapter 18.

Making the correct interpretations: The difference between a confidence interval and a tolerance interval

Recall in the car mileage case that the mean and the standard deviation of the sample of 50 mileages are $\bar{x} = 31.56$ and $s = .7977$. Also, we have seen on the previous page that a 95 percent *t*-based confidence interval for the mean, μ, of the mileages of all individual cars is $[\bar{x} \pm 2.010(s/\sqrt{50})] = [31.33, 31.79]$. A correct interpretation of this confidence interval says that we are 95 percent confident that the population mean mileage μ is between 31.33 mpg and 31.79 mpg. An incorrect interpretation—and one that is sometimes made by beginning statistics students—is that we estimate that 95 percent of all individual cars would (if tested) get between 31.33 mpg and 31.79 mpg. In general, an interval that contains a specified percentage of the individual measurements in a population is a *tolerance interval* (as previously discussed in Chapter 3). A tolerance interval is of the form $[\mu \pm z_{\alpha/2}\sigma]$ if the population of all individual measurements is normally distributed. For example, the histogram in Figure 9.13 suggests that the population of the mileages of all individual cars is normally distributed. Therefore, consider estimating a tolerance interval that contains the mileages of 95 percent of all individual cars. To do this, we first find the normal point $z_{\alpha/2}$ such that the area under the standard normal curve between $-z_{\alpha/2}$ and $z_{\alpha/2}$ is .95. As shown in the page margin, the appropriate normal point $z_{\alpha/2}$ is $z_{.025} = 1.96$. Estimating μ and σ in the tolerance interval $[\mu \pm 1.96\sigma]$ by $\bar{x} = 31.56$ and $s = .7977$, we then obtain an estimated tolerance interval of $[31.56 \pm 1.96(.7977)] = [30.0, 33.1]$. This estimated tolerance interval implies that approximately 95 percent of all individual cars would (if tested) get between 30.0 mpg and 33.1 mpg. Furthermore, Figure 9.13 shows that the estimated tolerance interval, which is meant to contain the *many mileages* that would be obtained by 95 percent of all individual cars, is much longer than the 95 percent confidence interval, which is meant to contain the *single population mean* μ.

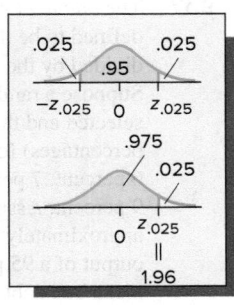

FIGURE 9.13   A Comparison of a Confidence Interval and a Tolerance Interval

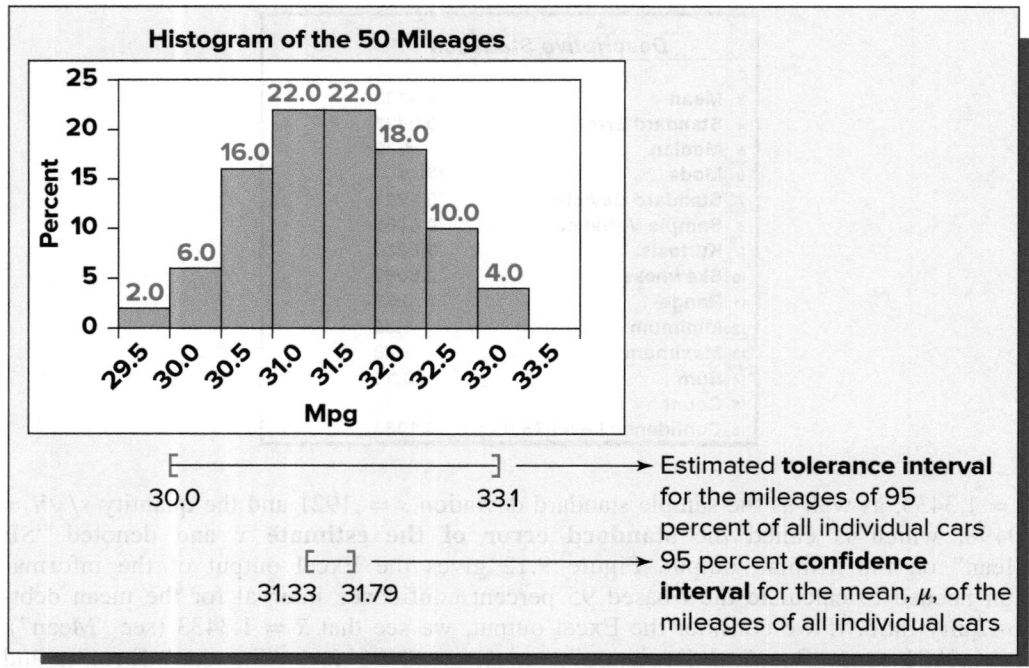

Estimated **tolerance interval** for the mileages of 95 percent of all individual cars

95 percent **confidence interval** for the mean, μ, of the mileages of all individual cars

Exercises for Section 9.2

CONCEPTS ▧ connect

9.13 Explain how each of the following changes as *the number of degrees of freedom* describing a *t* curve *increases*:
   a The standard deviation of the *t* curve.
   b The points t_α and $t_{\alpha/2}$.

9.14 Discuss when it is appropriate to use the *t*-based confidence interval for μ.

METHODS AND APPLICATIONS

9.15 Using Table A.4, find $t_{.100}$, $t_{.025}$, and $t_{.001}$ based on 11 degrees of freedom. Also, find these *t* points based on 6 degrees of freedom.

9.16 Suppose that for a sample of $n = 11$ measurements, we find that $\bar{x} = 72$ and $s = 5$. Assuming normality, compute confidence intervals for the population mean μ with the following levels of confidence:
   a 95%   **b** 99%   **c** 80%   **d** 90%
   e 98%   **f** 99.8%

9.17 The *bad debt ratio* for a financial institution is defined to be the dollar value of loans defaulted divided by the total dollar value of all loans made. Suppose a random sample of seven Ohio banks is selected and that the bad debt ratios (written as percentages) for these banks are 7 percent, 4 percent, 6 percent, 7 percent, 5 percent, 4 percent, and 9 percent. Assuming the bad debt ratios are approximately normally distributed, the Minitab output of a 95 percent confidence interval for the mean bad debt ratio of all Ohio banks is given

below. Using the sample mean and sample standard deviation on the Minitab output, demonstrate the calculation of the 95 percent confidence interval, and calculate a 99 percent confidence interval for the population mean debt-to-equity ratio.

ⒹⓈ BadDebt

Variable	N	Mean	StDev	SE Mean	95% CI
D-Ratio	7	6.00000	1.82574	0.69007	(4.31147, 7.68853)

9.18 In Exercise 9.17, suppose bank officials claim that the mean bad debt ratio for all banks in the Midwest region is 3.5 percent and that the mean bad debt ratio for all Ohio banks is higher. Using the 95 percent confidence interval (given by Minitab), can we be 95 percent confident that this claim is true? Using the 99 percent confidence interval you calculated, can we be 99 percent confident that this claim is true? Explain.

9.19 Air traffic controllers have the crucial task of ensuring that aircraft don't collide. To do this, they must quickly discern when two planes are about to enter the same air space at the same time. They are aided by video display panels that track the aircraft in their sector and alert the controller when two flight paths are about to converge. The display panel currently in use has a mean "alert time" of 15 seconds. (The alert time is the time elapsing between the instant when two aircraft enter into a collision course and when a controller initiates a call to reroute the planes.) According to Ralph Rudd, a supervisor of air traffic controllers at the Greater Cincinnati International Airport, a new display panel has been developed that uses artificial intelligence to project a plane's current flight path into the future. This new panel provides air traffic controllers with an earlier warning that a collision is likely. It is hoped that the mean "alert time," μ, for the new panel is less than 8 seconds. In order to test the new panel, 15 randomly selected air traffic controllers are trained to use the panel and their alert times for a simulated collision course are recorded. The sample alert times (in seconds) are: 7.2, 7.5, 8.0, 6.8, 7.2, 8.4, 5.3, 7.3, 7.6, 7.1, 9.4, 6.4, 7.9, 6.2, 8.7. Using the facts that the sample mean and sample standard deviation are 7.4 and 1.026, respectively, **(1)** Find a 95 percent confidence interval for the population mean alert time, μ, for the new panel. Assume normality. **(2)** Can we be 95 percent confident that μ is less than 8 seconds? Explain.   AlertTimes

©Monty Rakusen/Getty Images RF

9.20   Whole Foods is an all-natural grocery chain that has 50,000 square foot stores, up from the industry average of 34,000 square feet. Sales per square foot of super-markets average just under $400 per square foot, as re-ported by *USA Today* in an article on "*A Whole New Ballgame in Grocery Shopping.*" Suppose that sales per square foot in the most recent fiscal year are recorded for a random sample of 10 Whole Foods supermarkets. The data (sales dollars per square foot) are as follows: 854, 858, 801, 892, 849, 807, 894, 863, 829, 815. Using the facts that the sample mean and sample standard de-viation are 846.2 and 32.866, respectively, **(1)** Find a 95 percent confidence interval for the population mean sales dollars per square foot for all Whole Foods super-markets during the most recent fiscal year. Assume normality. **(2)** Are we 95 percent confident that this population mean is greater than $800, the historical av-erage for Whole Foods? Explain.   WholeFoods

9.21   A production supervisor at a major chemical company wishes to determine whether a new catalyst, catalyst XA-100, increases the mean hourly yield of a chemical pro-cess beyond the current mean hourly yield, which is known to be roughly equal to, but no more than, 750 pounds per hour. To test the new catalyst, five trial runs using catalyst XA-100 are made. The resulting yields for the trial runs (in pounds per hour) are 801, 814, 784, 836, and 820. Assuming that all factors affecting yields of the process have been held as constant as possible during the test runs, it is reasonable to regard the five yields ob-tained using the new catalyst as a random sample from the population of all possible yields that would be obtained by using the new catalyst. Furthermore, we will assume that this population is approximately normally distributed.   ChemYield

FIGURE 9.14   **Excel Output for Exercise 9.21**

	A	B	C
1	*Descriptive Statistics for Yield*		
2			
3	Mean	811	
4	Standard Error	8.7864	
5	Median	814	
6	Mode	#N/A	
7	Standard Deviation	19.6469	
8	Sample Variance	386	
9	Kurtosis	-0.1247	
10	Skewness	-0.2364	
11	Range	52	
12	Minimum	784	
13	Maximum	836	
14	Sum	4055	
15	Count	5	
16	Confidence Level(95.0%)	24.3948	

a   Using the Excel output in Figure 9.14, find a 95 percent confidence interval for the mean of all possible yields obtained using catalyst XA-100.

b   Based on the confidence interval, can we be 95 percent confident that the population mean yield using catalyst XA-100 exceeds 750 pounds per hour? Explain.

9.22   **THE VIDEO GAME SATISFACTION RATING**
     CASE   VideoGame

The mean and the standard deviation of the sample of 65 customer satisfaction ratings in Table 1.8 are 42.95 and 2.6424, respectively. **(1)** Calculate a *t*-based 95 percent confidence interval for μ, the mean of all possible customer satisfaction ratings for the XYZ-Box video game system. **(2)** Are we 95 percent confident that μ is at least 42, the minimal rating given by a very satisfied customer? Explain.

9.23   **THE BANK CUSTOMER WAITING TIME**
     CASE   WaitTime

The mean and the standard deviation of the sample of 100 bank customer waiting times in Table 1.9 are 5.46 and 2.475, respectively. **(1)** Calculate a *t*-based 95 percent confidence interval for μ, the mean of all possible bank customer waiting times using the new system. **(2)** Are we 95 percent confident that μ is less than six minutes? Explain.

9.24   **THE TRASH BAG CASE**   TrashBag

The mean and the standard deviation of the sample of 30 trash bag breaking strengths in Table 1.10 are 52.167 and 5.394, respectively.

a   Calculate a *t*-based 95 percent confidence interval for μ, the mean of the breaking strengths of all pos-sible new trash bags. Are we 95 percent confident that μ is at least 50 pounds? Explain.

b   Assuming that the population of all individual trash bag breaking strengths is normally distributed, estimate tolerance intervals of the form $[\mu \pm z_{\alpha/2}\sigma]$ that contain **(1)** 95 percent of all individual trash bag breaking strengths and **(2)** 99.8 percent of all indi-vidual trash bag breaking strengths.

9.3 Sample Size Determination

LO9-4
Determine the appropriate sample size when estimating a population mean.

In Section 9.1 we used a sample of 50 mileages to construct a 95 percent confidence interval for the midsize model's mean mileage μ. The size of this sample was not arbitrary—it was planned. To understand this, suppose that before the automaker selected the random sample of 50 mileages, it randomly selected the following sample of five mileages: 30.7, 31.9, 30.3, 32.0, and 31.6. This sample has mean $\bar{x} = 31.3$. Assuming that the population of all mileages is normally distributed and that the population standard deviation σ is known to equal .8, it follows that a 95 percent confidence interval for μ is

$$\left[\bar{x} \pm z_{.025}\frac{\sigma}{\sqrt{n}}\right] = \left[31.3 \pm 1.96\frac{.8}{\sqrt{5}}\right]$$

$$= [31.3 \pm .701]$$

$$= [30.6, 32.0]$$

Although the sample mean $\bar{x} = 31.3$ is at least 31, the lower limit of the 95 percent confidence interval for μ is less than 31. Therefore, the midsize model's EPA mileage estimate would be 30 mpg, and the automaker would not receive its tax credit. One reason that the lower limit of this 95 percent interval is less than 31 is that the sample size of 5 is not large enough to make the interval's margin of error

$$z_{.025}\frac{\sigma}{\sqrt{n}} = 1.96\frac{.8}{\sqrt{5}} = .701$$

small enough. We can attempt to make the margin of error in the interval smaller by increasing the sample size. If we feel that the mean $\bar{x}$ of the larger sample will be at least 31.3 mpg (the mean of the small sample we have already taken), then the lower limit of a $100(1 - \alpha)$ percent confidence interval for μ will be at least 31 if the margin of error is .3 or less.

We will now explain how to find the size of the sample that will be needed to make the margin of error in a confidence interval for μ as small as we wish. In order to develop a formula for the needed sample size, we will initially assume that we know σ. Then, if the population is normally distributed or the sample size is large, the z-based $100(1 - \alpha)$ percent confidence interval for μ is $[\bar{x} \pm z_{\alpha/2}(\sigma/\sqrt{n})]$. To find the needed sample size, we set $z_{\alpha/2}(\sigma/\sqrt{n})$ equal to the desired margin of error and solve for n. Letting E denote the desired margin of error, we obtain

$$z_{\alpha/2}\frac{\sigma}{\sqrt{n}} = E$$

Multiplying both sides of this equation by $\sqrt{n}$ and dividing both sides by E, we obtain

$$\sqrt{n} = \frac{z_{\alpha/2}\sigma}{E}$$

Squaring both sides of this result gives us the formula for n.

Determining the Sample Size for a Confidence Interval for μ : σ Known

A sample of size

$$n = \left(\frac{z_{\alpha/2}\sigma}{E}\right)^2$$

makes the margin of error in a $100(1 - \alpha)$ percent confidence interval for μ equal to E. That is, this sample size makes us $100(1 - \alpha)$ percent confident that $\bar{x}$ is within E units of μ. If the calculated value of n is not a whole number, round this value up to the next whole number (so that the margin of error is at least as small as desired).

If we consider the formula for the sample size n, it intuitively follows that the value E is the farthest that the user is willing to allow $\bar{x}$ to be from μ at a given level of confidence, and the normal point $z_{\alpha/2}$ follows directly from the given level of confidence. Furthermore, because the population standard deviation σ is in the numerator of the formula for n, it follows that the more variable that the individual population measurements are, the larger is the sample size needed to estimate μ with a specified accuracy.

In order to use this formula for n, we must either know σ (which is unlikely) or we must compute an estimate of σ. We first consider the case where we know σ. For example, suppose in the car mileage situation we wish to find the sample size that is needed to make the margin of error in a 95 percent confidence interval for μ equal to .3. Assuming that σ is known to equal .8, and using $z_{.025} = 1.96$, the appropriate sample size is

$$n = \left(\frac{z_{.025}\sigma}{E}\right)^2 = \left(\frac{1.96(.8)}{.3}\right)^2 = 27.32$$

Rounding up, we would employ a sample of size 28.

In most real situations, of course, we do not know the true value of σ. If σ is not known, we often estimate σ by using a preliminary sample. In this case we modify the above formula for n by replacing σ by the standard deviation s of the preliminary sample and by replacing $z_{\alpha/2}$ by $t_{\alpha/2}$. This approach usually gives a sample at least as large as we need. Thus, we obtain:

$$n = \left(\frac{t_{\alpha/2}s}{E}\right)^2$$

where the number of degrees of freedom for the $t_{\alpha/2}$ point is the size of the preliminary sample minus 1.

Intuitively, using $t_{\alpha/2}$ compensates for the fact that the preliminary sample's value of s might underestimate σ, and, therefore, give a sample size that is too small.

C EXAMPLE 9.4 The Car Mileage Case: Estimating Mean Mileage

Suppose that in the car mileage situation we wish to find the sample size that is needed to make the margin of error in a 95 percent confidence interval for μ equal to .3. Assuming we do not know σ, we regard the previously discussed sample of five mileages as a preliminary sample. Therefore, we replace σ by the standard deviation of the preliminary sample, which can be calculated to be $s = .7583$, and we replace $z_{\alpha/2} = z_{.025} = 1.96$ by $t_{.025} = 2.776$, which is based on $5 - 1 = 4$ degrees of freedom. We find that the appropriate sample size is

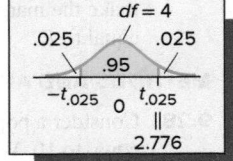

$$n = \left(\frac{t_{.025}s}{E}\right)^2 = \left(\frac{2.776(.7583)}{.3}\right)^2 = 49.24$$

Rounding up, we employ a sample of size 50.

When we make the margin of error in our 95 percent confidence interval for μ equal to .3, we can say we are 95 percent confident that the sample mean $\bar{x}$ is within .3 of μ. To understand this, suppose the true value of μ is 31.6. Recalling that the mean of the sample of 50 mileages is $\bar{x} = 31.56$, we see that this sample mean is within .3 of μ. Other samples of 50 mileages would give different sample means that would be different distances from μ. When we say that our sample of 50 mileages makes us 95 percent confident that $\bar{x}$ is within .3 of μ, we mean that **95 percent of all possible sample means based on 50 mileages are within .3 of μ** and 5 percent of such sample means are not.

C **EXAMPLE 9.5** The Car Mileage Case: Estimating Mean Mileage

To see that the sample of 50 mileages has actually produced a 95 percent confidence interval with a margin of error that is as small as we requested, recall that the 50 mileages have mean $\bar{x} = 31.56$ and standard deviation $s = .7977$. Therefore, the t-based 95 percent confidence interval for μ is

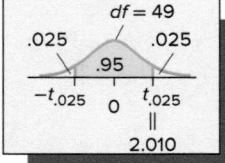

$$\left[\bar{x} \pm t_{.025}\frac{s}{\sqrt{n}}\right] = \left[31.56 \pm 2.010\frac{.7977}{\sqrt{50}}\right]$$

$$= [31.56 \pm .227]$$

$$= [31.33, 31.79]$$

where $t_{.025} = 2.010$ is based on $n - 1 = 50 - 1 = 49$ degrees of freedom—see Table A.4. We see that the margin of error in this interval is .227, which is smaller than the .3 we asked for. Furthermore, as the automaker had hoped, the sample mean $\bar{x} = 31.56$ of the sample of 50 mileages turned out to be at least 31.3. Therefore, because the margin of error is less than .3, the lower limit of the 95 percent confidence interval is above 31 mpg, and the midsize model's EPA mileage estimate is 31 mpg. Because of this, the automaker will receive its tax credit.

Finally, sometimes we do not know σ and we do not have a preliminary sample that can be used to estimate σ. In this case it can be shown that, if we can make a reasonable guess of the range of the population being studied, then a conservatively large estimate of σ is this estimated range divided by 4. For example, if the automaker's design engineers feel that almost all of its midsize cars should get mileages within a range of 5 mpg, then a conservatively large estimate of σ is $5/4 = 1.25$ mpg. When employing such an estimate of σ, it is sufficient to use the z-based sample size formula $n = (z_{\alpha/2}\sigma/E)^2$, because a conservatively large estimate of σ will give us a conservatively large sample size.

Exercises for Section 9.3

CONCEPTS ■ **connect**

9.25 Explain what is meant by the margin of error for a confidence interval. What error are we talking about in the context of an interval for μ?

9.26 Explain exactly what we mean when we say that a sample of size n makes us 99 percent confident that $\bar{x}$ is within E units of μ.

9.27 Why do we often need to take a preliminary sample when determining the size of the sample needed to make the margin of error of a confidence interval equal to E?

METHODS AND APPLICATIONS

9.28 Consider a population having a standard deviation equal to 10. We wish to estimate the mean of this population.
  a How large a random sample is needed to construct a 95 percent confidence interval for the mean of this population with a margin of error equal to 1?
  b Suppose that we now take a random sample of the size we have determined in part *a*. If we obtain a sample mean equal to 295, calculate the 95 percent confidence interval for the population mean. What is the interval's margin of error?

9.29 Referring to Exercise 9.11*a*, assume that the population standard deviation equals 33. How large a random

sample of public owner-controlled companies is needed to make us
  a 95 percent confident that $\bar{x}$, the sample mean audit delay, is within a margin of error of four days of μ, the population mean audit delay?
  b 99 percent confident that $\bar{x}$ is within a margin of error of four days of μ?

9.30 Referring to Exercise 9.12*b*, assume that the population standard deviation equals .66. How large a sample of late replacement buyers is needed to make us
  a 99 percent confident that $\bar{x}$, the sample mean number of dealers visited, is within a margin of error of .04 of μ, the population mean number of dealers visited?
  b 99.73 percent confident that $\bar{x}$ is within a margin of error of .05 of μ?

9.31 Referring to Exercise 9.21, regard the sample of five trial runs (which has standard deviation 19.65) as a preliminary sample. Determine the number of trial runs of the chemical process needed to make us
  a 95 percent confident that $\bar{x}$, the sample mean hourly yield, is within a margin of error of eight pounds of the population mean hourly yield μ when catalyst XA-100 is used.
  b 99 percent confident that $\bar{x}$ is within a margin of error of five pounds of μ. **DS** ChemYield

9.32 Referring to Exercise 9.20, regard the sample of 10 sales figures (which has standard deviation 32.866) as a preliminary sample. How large a sample of sales figures is needed to make us 95 percent confident that $\bar{x}$, the sample mean sales dollars per square foot, is within a margin of error of $10 of μ, the population mean sales dollars per square foot for all Whole Foods supermarkets? Ⓓ WholeFoods

9.33 THE AIR SAFETY CASE Ⓓ AlertTimes

Referring to Exercise 9.19, regard the sample of 15 alert times (which has standard deviation 1.026) as a preliminary sample. Determine the sample size needed to make us 95 percent confident that $\bar{x}$, the sample mean alert time, is within a margin of error of .3 second of μ, the population mean alert time using the new display panel.

9.4 Confidence Intervals for a Population Proportion

LO9-5

Calculate and interpret a large sample confidence interval for a population proportion.

In Chapter 8, the soft cheese spread producer decided to replace its current spout with the new spout if p, the true proportion of all current purchasers who would stop buying the cheese spread if the new spout were used, is less than .10. Suppose that when 1,000 current purchasers are randomly selected and are asked to try the new spout, 63 say they would stop buying the spread if the new spout were used. The point estimate of the population proportion p is the sample proportion $\hat{p} = 63/1,000 = .063$. This sample proportion says we estimate that 6.3 percent of all current purchasers would stop buying the cheese spread if the new spout were used. Because $\hat{p}$ equals .063, we have some evidence that p is less than .10.

In order to see if there is strong evidence that p is less than .10, we can calculate a confidence interval for p. As explained in Chapter 8, if the sample size n is large, then the sampling distribution of the sample proportion $\hat{p}$ is approximately a normal distribution with mean $\mu_{\hat{p}} = p$ and standard deviation $\sigma_{\hat{p}} = \sqrt{p(1-p)/n}$. By using the same logic we used in developing confidence intervals for μ, it follows that a $100(1-\alpha)$ percent confidence interval for p is

$$\left[\hat{p} \pm z_{\alpha/2}\sqrt{\frac{p(1-p)}{n}}\right]$$

Estimating $p(1-p)$ by $\hat{p}(1-\hat{p})$, it follows that a $100(1-\alpha)$ percent confidence interval for p can be calculated as summarized below.

A Large Sample Confidence Interval for a Population Proportion p

If the sample size n is large, **a $100(1-\alpha)$ percent confidence interval for the population proportion p** is

$$\left[\hat{p} \pm z_{\alpha/2}\sqrt{\frac{\hat{p}(1-\hat{p})}{n}}\right]$$

Here n should be considered large if both $n\hat{p}$ and $n(1-\hat{p})$ are at least 5.[2]

Ⓒ EXAMPLE 9.6 The Cheese Spread Case: Improving Profitability

Suppose that the cheese spread producer wishes to calculate a 99 percent confidence interval for p, the population proportion of purchasers who would stop buying the cheese spread if the new spout were used. To determine whether the sample size $n = 1,000$ is large enough to enable us to use the confidence interval formula just given, recall that the point estimate of p is $\hat{p} = 63/1,000 = .063$. Therefore, because $n\hat{p} = 1,000(.063) = 63$ and

---

[2]Some statisticians suggest using the more conservative rule that both $n\hat{p}$ and $n(1-\hat{p})$ must be at least 10. Furthermore, because $\hat{p}(1-\hat{p})/(n-1)$ is an unbiased point estimate of $p(1-p)/n$, a more correct $100(1-\alpha)$ percent confidence interval for p is $\left[\hat{p} \pm z_{\alpha/2}\sqrt{\hat{p}(1-\hat{p})/(n-1)}\right]$. Computer studies and careful theory suggest that an even more accurate $100(1-\alpha)$ percent confidence interval for p is $\left[\tilde{p} \pm z_{\alpha/2}\sqrt{\tilde{p}(1-\tilde{p})/(n+4)}\right]$. Here $\tilde{p} = (x+2)/(n+4)$, where x is the number of the n sample elements that fall into the category being studied (for example, the number of the 1,000 sampled customers who say that they would stop buying the cheese spread if the new spout were used). The estimate $\tilde{p}$ was proposed by Edwin Wilson in 1927 but was rarely used until recently.

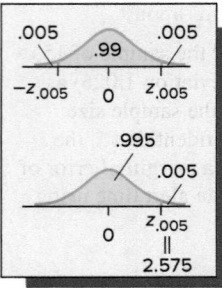

BI

$n(1 - \hat{p}) = 1,000(.937) = 937$ are both greater than 5, we can use the confidence interval formula. It follows that the 99 percent confidence interval for p is

$$\left[\hat{p} \pm z_{.025}\sqrt{\frac{\hat{p}(1 - \hat{p})}{n}}\right] = \left[.063 \pm 2.575\sqrt{\frac{(.063)(.937)}{1000}}\right]$$

$$= [.063 \pm .0198]$$

$$= [.0432, .0828]$$

This interval says that we are 99 percent confident that between 4.32 percent and 8.28 percent of all current purchasers would stop buying the cheese spread if the new spout were used. Moreover, because the upper limit of the 99 percent confidence interval is less than .10, we have very strong evidence that the true proportion p of all current purchasers who would stop buying the cheese spread is less than .10. Based on this result, it seems reasonable to use the new spout. Below we show the JMP output of the 99 percent confidence interval for p.

Number of Successes	63	Confidence Interval	
Sample Size	1000	Confidence Level	0.99
		Result	Value
		Estimated Proportion	0.063
		Lower Limit	0.04321
		Upper Limit	0.08279

In the cheese spread example, a sample of 1,000 purchasers gives us a 99 percent confidence interval for p that has a margin of error of .0198 and a 95 percent confidence interval for p that has a margin of error of .0151. Both of these error margins are reasonably small. Generally, however, quite a large sample is needed in order to make the margin of error in a confidence interval for p reasonably small. The next two examples demonstrate that a sample size of 200, which most people would consider quite large, does not necessarily give a 95 percent confidence interval for p with a small margin of error.

EXAMPLE 9.7 The Phe-Mycin Case: Drug Side Effects

Antibiotics occasionally cause nausea as a side effect. Scientists working for a major drug company have developed a new antibiotic called Phe-Mycin. The company wishes to estimate p, the proportion of all patients who would experience nausea as a side effect when being treated with Phe-Mycin. Suppose that a sample of 200 patients is randomly selected. When these patients are treated with Phe-Mycin, 35 patients experience nausea. The point estimate of the population proportion p is the sample proportion $\hat{p} = 35/200 = .175$. This sample proportion says that we estimate that 17.5 percent of all patients would experience nausea as a side effect of taking Phe-Mycin. Furthermore, because $n\hat{p} = 200(.175) = 35$ and $n(1 - \hat{p}) = 200(.825) = 165$ are both at least 5, we can use the previously given formula to calculate a confidence interval for p. Doing this, we find that a 95 percent confidence interval for p is

$$\left[\hat{p} \pm z_{.025}\sqrt{\frac{\hat{p}(1 - \hat{p})}{n}}\right] = \left[.175 \pm 1.96\sqrt{\frac{(.175)(.825)}{200}}\right]$$

$$= [.175 \pm .053]$$

$$= [.122, .228]$$

This interval says we are 95 percent confident that between 12.2 percent and 22.8 percent of all patients would experience nausea as a side effect of taking Phe-Mycin. Notice that the margin of error (.053) in this interval is rather large. Therefore, this interval is fairly long, and it does not provide a very precise estimate of p.

(C) EXAMPLE 9.8 The Marketing Ethics Case: Confidentiality

In the book *Essentials of Marketing Research*, William R. Dillon, Thomas J. Madden, and Neil H. Firtle discuss a survey of marketing professionals, the results of which were originally published by Ishmael P. Akoah and Edward A. Riordan in the *Journal of Marketing Research*. In the study, randomly selected marketing researchers were presented with various scenarios involving ethical issues such as confidentiality, conflict of interest, and social acceptability. The marketing researchers were asked to indicate whether they approved or disapproved of the actions described in each scenario. For instance, one scenario that involved the issue of confidentiality was described as follows:

> **Use of ultraviolet ink**   A project director went to the marketing research director's office and requested permission to use an ultraviolet ink to precode a questionnaire for a mail survey. The project director pointed out that although the cover letter promised confidentiality, respondent identification was needed to permit adequate crosstabulations of the data. The marketing research director gave approval.

Of the 205 marketing researchers who participated in the survey, 117 said they disapproved of the actions taken in the scenario. It follows that a point estimate of p, the proportion of all marketing researchers who disapprove of the actions taken in the scenario, is $\hat{p} = 117/205 = .5707$. Furthermore, because $n\hat{p} = 205(.5707) = 117$ and $n(1 - \hat{p}) = 205(.4293) = 88$ are both at least 5, a 95 percent confidence interval for p is

$$\left[\hat{p} \pm z_{.025} \sqrt{\frac{\hat{p}(1 - \hat{p})}{n}} \right] = \left[.5707 \pm 1.96 \sqrt{\frac{(.5707)(.4293)}{205}} \right]$$

$$= [.5707 \pm .0678]$$

$$= [.5029, .6385]$$

This interval says we are 95 percent confident that between 50.29 percent and 63.85 percent of all marketing researchers disapprove of the actions taken in the ultraviolet ink scenario. Notice that because the margin of error (.0678) in this interval is rather large, this interval does not provide a very precise estimate of p. Below we show the Minitab output of this interval.

CI for One Proportion

X	N	Sample p	95% CI
117	205	0.570732	(0.502975, 0.638488)

In order to find the size of the sample needed to estimate a population proportion, we consider the theoretically correct interval

$$\left[\hat{p} \pm z_{\alpha/2} \sqrt{\frac{p(1 - p)}{n}} \right]$$

To obtain the sample size needed to make the margin of error in this interval equal to E, we set

$$z_{\alpha/2} \sqrt{\frac{p(1 - p)}{n}} = E$$

and solve for n. When we do this, we get the following result:

LO9-6
Determine the appropriate sample size when estimating a population proportion.

Determining the Sample Size for a Confidence Interval for p

A sample of size

$$n = p(1 - p) \left(\frac{z_{\alpha/2}}{E} \right)^2$$

makes the margin of error in a $100(1 - \alpha)$ percent confidence interval for p equal to E. That is, this sample size makes us $100(1 - \alpha)$ percent confident that $\hat{p}$ is within E units of p. If the calculated value of n is not a whole number, round this value up to the next whole number.

FIGURE 9.15   **The Graph of $p(1-p)$ versus p**

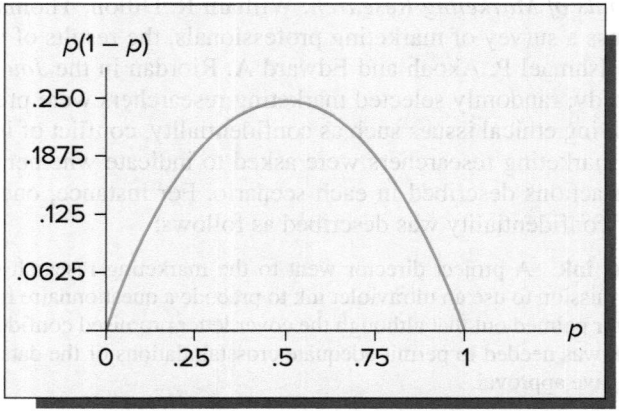

Looking at this formula, we see that the larger $p(1-p)$ is, the larger n will be. To make sure n is large enough, consider Figure 9.15, which is a graph of $p(1-p)$ versus p. This figure shows that $p(1-p)$ equals .25 when p equals .5. Furthermore, $p(1-p)$ is never larger than .25. Therefore, if the true value of p could be near .5, we should set $p(1-p)$ equal to .25. This will ensure that n is as large as needed to make the margin of error as small as desired. For example, suppose we wish to estimate the proportion p of all registered voters who currently favor a particular candidate for President of the United States. If this candidate is the nominee of a major political party, or if the candidate enjoys broad popularity for some other reason, then p could be near .5. Furthermore, suppose we wish to make the margin of error in a 95 percent confidence interval for p equal to .02. If the sample to be taken is random, it should consist of

$$n = p(1-p)\left(\frac{z_{\alpha/2}}{E}\right)^2 = .25\left(\frac{1.96}{.02}\right)^2 = 2{,}401$$

registered voters. In reality, a list of all registered voters in the United States is not available to polling organizations. Therefore, it is not feasible to take a (technically correct) random sample of registered voters. For this reason, polling organizations actually employ other (more complicated) kinds of samples. We have explained some of the basic ideas behind these more complex samples in optional Section 1.7. For now, we consider the samples taken by polling organizations to be approximately random. Suppose, then, that when the sample of voters is actually taken, the proportion $\hat{p}$ of sampled voters who favor the candidate turns out to be greater than .52. It follows, because the sample is large enough to make the margin of error in a 95 percent confidence interval for p equal to .02, that the lower limit of such an interval is greater than .50. This says we have strong evidence that a majority of all registered voters favor the candidate. For instance, if the sample proportion $\hat{p}$ equals .53, we are 95 percent confident that the proportion of all registered voters who favor the candidate is between .51 and .55.

Major polling organizations conduct public opinion polls concerning many kinds of issues. While making the margin of error in a 95 percent confidence interval for p equal to .02 requires a sample size of 2,401, making the margin of error in such an interval equal to .03 requires a sample size of only

$$n = p(1-p)\left(\frac{z_{\alpha/2}}{E}\right)^2 = .25\left(\frac{1.96}{.03}\right)^2 = 1{,}067.1$$

or 1,068 (rounding up). Of course, these calculations assume that the proportion p being estimated could be near .5. However, for any value of p, increasing the margin of error from .02 to .03 substantially decreases the needed sample size and thus saves considerable time and money. For this reason, although the most accurate public opinion polls use a margin of error of .02, the vast majority of public opinion polls use a margin of error of .03 or larger.

FIGURE 9.16   **As p Gets Closer to .5, $p(1 - p)$ Increases**

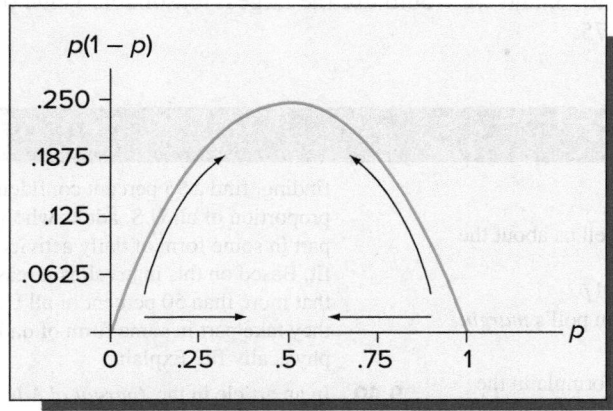

When the news media report the results of a public opinion poll, they express the margin of error in a 95 percent confidence interval for p *in percentage points*. For instance, if the margin of error is .03, the media would say the poll's margin of error is 3 percentage points. The media seldom report the level of confidence, but almost all polling results are based on 95 percent confidence. Sometimes the media make a vague reference to the level of confidence. For instance, if the margin of error is 3 percentage points, the media might say that "the sample result will be within 3 percentage points of the population value in 19 out of 20 samples." Here the "19 out of 20 samples" is a reference to the level of confidence, which is $100(19/20) = 100(.95) = 95$ percent.

As an example, suppose a news report says a recent poll finds that 34 percent of the public favors military intervention in an international crisis, and suppose the poll's margin of error is reported to be 3 percentage points. This means the sample taken is large enough to make us 95 percent confident that the sample proportion $\hat{p} = .34$ is within .03 (that is, 3 percentage points) of the true proportion p of the entire public that favors military intervention. That is, we are 95 percent confident that p is between .31 and .37.

If the population proportion we are estimating is substantially different from .5, setting p equal to .5 will give a sample size that is much larger than is needed. In this case, we should use our intuition or previous sample information (along with Figure 9.16) to determine the largest reasonable value for $p(1 - p)$. Figure 9.16 implies that as p gets closer to .5, $p(1 - p)$ increases. It follows that $p(1 - p)$ is maximized by the reasonable value of p that is closest to .5. Therefore, **when we are estimating a proportion that is substantially different from .5, we use the reasonable value of p that is closest to .5 to calculate the sample size needed to obtain a specified margin of error.**

EXAMPLE 9.9 The Phe-Mycin Case: Drug Side Effects

Again consider estimating the proportion of all patients who would experience nausea as a side effect of taking the new antibiotic Phe-Mycin. Suppose the drug company wishes to find the size of the random sample that is needed in order to obtain a 2 percent margin of error with 95 percent confidence. In Example 9.7 we employed a sample of 200 patients to compute a 95 percent confidence interval for p. This interval, which is [.122, .228], makes us very confident that p is between .122 and .228. Because .228 is the reasonable value of p that is closest to .5, the largest reasonable value of $p(1 - p)$ is $.228(1 - .228) = .1760$, and thus the drug company should take a sample of

$$n = p(1 - p)\left(\frac{z_{\alpha/2}}{E}\right)^2 = .1760\left(\frac{1.96}{.02}\right)^2 = 1{,}691 \text{ (rounded up)}$$

patients.

Finally, as a last example of choosing p for sample size calculations, suppose that experience indicates that a population proportion p is at least .75. Then, .75 is the reasonable value of p that is closest to .5, and we would use the largest reasonable value of $p(1 - p)$, which is $.75(1 - .75) = .1875$.

Exercises for Section 9.4

CONCEPTS ▌**connect**

9.34 **a** What does a population proportion tell us about the population?
   b Explain the difference between p and $\hat{p}$.
   c What is meant when a public opinion poll's *margin of error* is 3 percent?

9.35 Suppose we are using the sample size formula in the box on page 419 to find the sample size needed to make the margin of error in a confidence interval for p equal to E. In each of the following situations, explain what value of p would be used in the formula for finding n:
   a We have no idea what value p is—it could be any value between 0 and 1.
   b Past experience tells us that p is no more than .3.
   c Past experience tells us that p is at least .8.

METHODS AND APPLICATIONS

9.36 In each of the following cases, determine whether the sample size n is large enough to use the large sample formula presented in the box on page 417 to compute a confidence interval for p.
   a $\hat{p} = .1$, $n = 30$ **d** $\hat{p} = .8$, $n = 400$
   b $\hat{p} = .1$, $n = 100$ **e** $\hat{p} = .9$, $n = 30$
   c $\hat{p} = .5$, $n = 50$ **f** $\hat{p} = .99$, $n = 200$

9.37 In each of the following cases, compute 95 percent, 98 percent, and 99 percent confidence intervals for the population proportion p.
   a $\hat{p} = .4$ and $n = 100$ **c** $\hat{p} = .9$ and $n = 100$
   b $\hat{p} = .1$ and $n = 300$ **d** $\hat{p} = .6$ and $n = 50$

9.38 In a news story distributed by the *Washington Post*, Lew Sichelman reports that a substantial fraction of mortgage loans that go into default within the first year of the mortgage were approved on the basis of falsified applications. For instance, loan applicants often exaggerate their income or fail to declare debts. Suppose that a random sample of 1,000 mortgage loans that were defaulted within the first year reveals that 410 of these loans were approved on the basis of falsified applications.
   a Find a point estimate of and a 95 percent confidence interval for p, the proportion of all first-year defaults that are approved on the basis of falsified applications.
   b Based on your interval, what is a reasonable estimate of the minimum percentage of all first-year defaults that are approved on the basis of falsified applications?

9.39 Suppose that 60 percent of 1,000 randomly selected U.S. adults say that they take part in some form of daily activity to keep physically fit. Based on this

finding, find a 95 percent confidence interval for the proportion of all U.S. adults who would say they take part in some form of daily activity to keep physically fit. Based on this interval, is it reasonable to conclude that more than 50 percent of all U.S. adults would say they take part in some form of daily activity to keep physically fit? Explain.

9.40 In an article in the *Journal of Advertising,* Weinberger and Spotts compare the use of humor in television ads in the United States and the United Kingdom. They found that a substantially greater percentage of U.K. ads use humor.
   a Suppose that a random sample of 400 television ads in the United Kingdom reveals that 142 of these ads use humor. Find a point estimate of and a 95 percent confidence interval for the proportion of all U.K. television ads that use humor.
   b Suppose a random sample of 500 television ads in the United States reveals that 122 of these ads use humor. Find a point estimate of and a 95 percent confidence interval for the proportion of all U.S. television ads that use humor.
   c Do the confidence intervals you computed in parts *a* and *b* suggest that a greater percentage of U.K. ads use humor? Explain. How might an ad agency use this information?

9.41 **THE MARKETING ETHICS CASE: CONFLICT OF INTEREST**

Consider the marketing ethics case described in Example 9.8. One of the scenarios presented to the 205 marketing researchers is as follows:

> A marketing testing firm to which X company gives most of its business recently went public. The marketing research director of X company had been looking for a good investment and proceeded to buy some $20,000 of their stock. The firm continues as X company's leading supplier for testing.

Of the 205 marketing researchers who participated in the ethics survey, 111 said that they disapproved of the actions taken in the scenario. Use this sample result to show that the 95 percent confidence interval for the proportion of all marketing researchers who disapprove of the actions taken in the conflict of interest scenario is as given in the Minitab output below. Interpret this interval.

CI for One Proportion

X	N	Sample p	95% CI
111	205	0.541463	(0.473254, 0.609673)

9.42 On the basis of the confidence interval given in Exercise 9.41, is there convincing evidence that a majority of all marketing researchers disapprove of the actions taken in the conflict of interest scenario? Explain.

9.43 In an article in *CA Magazine,* Neil Fitzgerald surveyed Scottish business customers concerning their satisfaction with aspects of their banking relationships. Fitzgerald reports that, in 418 telephone interviews conducted by George Street Research, 67 percent of the respondents gave their banks a high rating for overall satisfaction.

　　a Assuming that the sample is randomly selected, calculate a 99 percent confidence interval for the proportion of all Scottish business customers who give their banks a high rating for overall satisfaction.

　　b Based on this interval, can we be 99 percent confident that more than 60 percent of all Scottish business customers give their banks a high rating for overall satisfaction? Explain.

9.44 The manufacturer of the ColorSmart-5000 television set claims 95 percent of its sets last at least five years without needing a single repair. In order to test this claim, a consumer group randomly selects 400 consumers who have owned a ColorSmart-5000 television set for five years. Of these 400 consumers, 316 say their ColorSmart-5000 television sets did not need a repair, whereas 84 say their ColorSmart-5000 television sets did need at least one repair.

　　a Find a 99 percent confidence interval for the proportion of all ColorSmart-5000 television sets that have lasted at least five years without needing a single repair.

　　b Does this confidence interval provide strong evidence that the percentage of all ColorSmart-5000 television sets that last at least five years without a single repair is less than the 95 percent claimed by the manufacturer? Explain.

9.45 *Consumer Reports* (January 2005) indicates that profit margins on extended warranties are much greater than on the purchase of most products.[3] In this exercise we consider a major electronics retailer that wishes to increase the proportion of customers who buy extended warranties on digital cameras. Historically, 20 percent of digital camera customers have purchased the retailer's extended warranty. To increase this percentage, the retailer has decided to offer a new warranty that is less expensive and more comprehensive. Suppose that three months after starting to offer the new warranty, a random sample of 500 customer sales invoices shows that 152 out of 500 digital camera customers purchased the new warranty. Find a 95 percent confidence interval for the proportion of all digital camera customers who have purchased the new warranty. Are we 95 percent confident that this proportion exceeds .20? Explain.

9.46 Consider Exercise 9.39 and suppose we wish to find the sample size n needed in order to be 95 percent confident that $\hat{p}$, the sample proportion of respondents who said they took part in some sort of daily activity to keep physically fit, is within a margin of error of .02 of p, the proportion of all U.S. adults who say that they take part in such activity. In order to find an appropriate value for $p(1 - p)$, note that the 95 percent confidence interval for p that you calculated in Exercise 9.39 was [.57, .63]. This indicates that the reasonable value for p that is closest to .5 is .57, and thus the largest reasonable value for $p(1 - p)$ is .57(1 − .57) = .2451. Calculate the required sample size n.

9.47 Referring to Exercise 9.44, determine the sample size needed in order to be 99 percent confident that $\hat{p}$, the sample proportion of ColorSmart-5000 television sets that last at least five years without a single repair, is within a margin of error of .03 of p, the proportion of all sets that last at least five years without a single repair.

9.48 Suppose we conduct a poll to estimate the proportion of voters who favor a major presidential candidate. Assuming that 50 percent of the electorate could be in favor of the candidate, determine the sample size needed so that we are 95 percent confident that $\hat{p}$, the sample proportion of voters who favor the candidate, is within a margin of error of .01 of p, the proportion of all voters who are in favor of the candidate.

9.5 Confidence Intervals for Parameters of Finite Populations (Optional)

LO9-7

Find and interpret confidence intervals for parameters of finite populations (Optional).

Random sampling

Companies in financial trouble have sometimes falsified their accounts receivable invoices in order to mislead stockholders. For this reason, independent auditors are often asked to estimate a company's true total sales for a given period. To illustrate this, consider a company that sells home theaters. The company accumulated 2,418 sales invoices last year. The total of the sales amounts listed on these invoices (that is, the total sales claimed by the company)

---

is $5,127,492.17. In order to estimate the true total sales for last year, an independent auditor randomly selects 242 of the invoices without replacement and determines the actual sales amounts by contacting the purchasers. The mean and the standard deviation of the actual sales amounts for these invoices are $\bar{x} = \$1{,}843.93$ and $s = \$516.42$. To use this sample information to estimate the company's true total sales, note that the mean μ of a finite population is the **population total** τ (pronounced *tau*), which is the sum of the values of all of the population measurements, divided by the number, N, of population measurements. That is, we have $\mu = \tau/N$, which implies that $\tau = N\mu$. Because a point estimate of the population mean μ is the sample mean $\bar{x}$, it follows that *a point estimate of the population total τ is $N\bar{x}$.* In the context of estimating the true total sales τ for the home theater company's population of $N = 2{,}418$ sales invoices, the sample mean invoice amount is $\bar{x} = \$1{,}843.93$, and thus the point estimate of τ is $N\bar{x} = 2{,}418(\$1{,}843.93) = \$4{,}458{,}622.74$. This point estimate is considerably lower than the claimed total sales of $5,127,492.17. However, we cannot expect the point estimate of τ to exactly equal the true total sales, so we need to calculate a confidence interval for τ before drawing any unwarranted conclusions.

In general, consider randomly selecting without replacement a large sample of n measurements from a finite population consisting of N measurements and having mean μ and standard deviation σ. It can then be shown that the sampling distribution of the sample mean $\bar{x}$ is approximately a normal distribution with mean $\mu_{\bar{x}} = \mu$ and standard deviation $\sigma_{\bar{x}} = (\sigma/\sqrt{n})\sqrt{(N-n)/(N-1)}$. Estimating σ by the sample standard deviation s, it can also be shown that the appropriate point estimate of $\sigma_{\bar{x}}$ is $(s/\sqrt{n})\sqrt{(N-n)/N}$, which is used in the following result.

Confidence Intervals for the Population Mean and Population Total for a Finite Population

Suppose we randomly select a sample of n measurements **without replacement from a finite population of N measurements.** Then, if n is large (say, at least 30)

1 **A $100(1-\alpha)$ percent confidence interval for the population mean μ is**

$$\left[\bar{x} \pm z_{\alpha/2}\,\frac{s}{\sqrt{n}}\,\sqrt{\frac{N-n}{N}}\,\right]$$

2 **A $100(1-\alpha)$ percent confidence interval for the population total τ** is found by multiplying the lower and upper limits of the $100(1-\alpha)$ percent confidence interval for μ by N.

The quantity $\sqrt{(N-n)/N}$ in the confidence intervals for μ and τ is called the **finite population correction** and is always less than 1. If the population size N is much larger than the sample size n, then the finite population correction is only slightly less than 1. For example, if we randomly select (without replacement) a sample of 1,000 from a population of 1 million, then the finite population correction is $\sqrt{(1{,}000{,}000 - 1{,}000)/1{,}000{,}000} = .9995$. In such a case, the finite population correction is not far enough below 1 to meaningfully shorten the confidence intervals for μ and τ and thus can be set equal to 1. However, **if the population size N is not much larger than the sample size n (say, if n is more than 5 percent of N), then the finite population correction is substantially less than 1 and should be included** in the confidence interval calculations.

EXAMPLE 9.10 The Home Theater Case: Auditing Total Sales

Recall that when the independent auditor randomly selects a sample of $n = 242$ invoices, the mean and standard deviation of the actual sales amounts for these invoices are $\bar{x} = 1{,}843.93$ and $s = 516.42$. Here the sample size $n = 242$ is $(242/2{,}418) \times 100 = 10.008$ percent of the population size $N = 2{,}418$. Because n is more than 5 percent of N, we should include the finite population correction in our confidence interval calculations. It follows that a 95 percent confidence interval for the true mean sales amount μ per invoice is

$$\left[\bar{x} \pm z_{.025} \frac{s}{\sqrt{n}} \sqrt{\frac{N-n}{N}} \right] = \left[1{,}843.93 \pm 1.96 \frac{516.42}{\sqrt{242}} \sqrt{\frac{2{,}418 - 242}{2{,}418}} \right]$$

$$= [1{,}843.93 \pm 61.723812]$$

or [\$1,782.21, \$1,905.65]. Moreover, multiplying the lower and upper limits of this interval by $N = 2{,}418$, we find that a 95 percent confidence interval for the true total sales τ is [1,782.21(2,418), 1,905.65(2,418)], or [\$4,309,383.80, \$4,607,861.70]. Because the upper limit of this interval is more than \$500,000 below the claimed total sales amount of \$5,127,492.17, we have strong evidence that the claimed total sales amount is overstated.

We sometimes estimate the total number, τ, of population units that fall into a particular category. For instance, the auditor of Example 9.10 might wish to estimate the total number of the 2,418 invoices having incorrect sales amounts. Here the proportion, p, of the population units that fall into a particular category is the total number, τ, of population units that fall into the category divided by the number, N, of population units. That is, $p = \tau/N$, which implies that $\tau = Np$. Therefore, because a point estimate of the population proportion p is the sample proportion $\hat{p}$, a point estimate of the population total τ is $N\hat{p}$. For example, suppose that 34 of the 242 sampled invoices have incorrect sales amounts. Because the sample proportion is $\hat{p} = 34/242 = .1405$, a point estimate of the total number of the 2,418 invoices that have incorrect sales amounts is $N\hat{p} = 2{,}418(.1405) = 339.729$.

Confidence Intervals for the Proportion of and Total Number of Units in a Category When Sampling a Finite Population

Suppose that we randomly select a sample of n units **without replacement from a finite population of N units.** Then, if n is large

1 A $100(1-\alpha)$ percent confidence interval for the population proportion p is

$$\left[\hat{p} \pm z_{\alpha/2} \sqrt{\frac{\hat{p}(1-\hat{p})}{n} \left(\frac{N-n}{N} \right)} \right]$$

2 A $100(1-\alpha)$ percent confidence interval for the population total τ is found by multiplying the lower and upper limits of the $100(1-\alpha)$ percent confidence interval for p by N.

EXAMPLE 9.11 The Home Theater Case: Auditing Sales Invoices

Recall that we found that 34 of the 242 sampled invoices have incorrect sales amounts. Because $\hat{p} = 34/242 = .1405$, a 95 percent confidence interval for the true proportion of the 2,418 invoices that have incorrect sales amounts is

$$\left[\hat{p} \pm z_{.025} \sqrt{\frac{\hat{p}(1-\hat{p})}{n} \left(\frac{N-n}{N} \right)} \right] = \left[.1405 \pm 1.96 \sqrt{\frac{(.1405)(.8595)}{242} \left(\frac{2{,}418 - 242}{2{,}418} \right)} \right]$$

$$= [.1405 \pm .0416]$$

or [.0989, .1821]. Moreover, multiplying the lower and upper limits of this interval by $N = 2{,}418$, we find that a 95 percent confidence interval for the true total number, τ, of the 2,418 sales invoices that have incorrect sales amounts is [.0989(2,418), .1821(2,418)], or [239.14, 440.32]. Therefore, we are 95 percent confident that between (roughly) 239 and 440 of the 2,418 invoices have incorrect sales amounts.

Exercises for Section 9.5

CONCEPTS

9.49 Define a population total. Give an example of a population total that a business might estimate.

9.50 Explain why the finite population correction $\sqrt{(N-n)/N}$ is unnecessary when the sample size is less than 5 percent of the population size. Give an example using numbers.

METHODS AND APPLICATIONS

9.51 A retailer that sells audio and video equipment accumulated 10,451 sales invoices during the previous year. The total of the sales amounts listed on these invoices (that is, the total sales claimed by the company) is $6,384,675. In order to estimate the true total sales for last year, an independent auditor randomly selects 350 of the invoices and determines the actual sales amounts by contacting the purchasers. The mean and the standard deviation of the 350 sampled sales amounts are $532 and $168, respectively.

   a Find point estimates of and 95 percent confidence intervals for **(1)** μ, the true mean sales amount per invoice for the 10,451 invoices, and **(2)** τ, the true total sales amount for last year.

   b What does the interval for τ say about the company's claim that the true total sales were $6,384,675? Explain.

9.52 A company's manager is considering simplification of a travel voucher form. In order to assess the costs associated with erroneous travel vouchers, the manager must estimate the total number of such vouchers that were filled out incorrectly in the last month. In a random sample of 100 vouchers drawn without replacement from the 1,323 travel vouchers submitted in the last month, 31 vouchers were filled out incorrectly.

   a Find point estimates of and 95 percent confidence intervals for **(1)** p, the true proportion of travel vouchers that were filled out incorrectly in the last month and **(2)** τ, the total number of vouchers filled out incorrectly in the last month.

   b If it costs the company $10 to correct an erroneous travel voucher, find a reasonable estimate of the minimum cost of correcting all of last month's erroneous travel vouchers. Would it be worthwhile to spend $5,000 to design a simplified travel voucher that could be used for at least a year? Explain.

Chapter Summary

In this chapter we discussed **confidence intervals** for population **means** and **proportions**. First, we studied how to compute a confidence interval for a **population mean.** We saw that when the population standard deviation σ is known, we can use the **normal distribution** to compute a confidence interval for a population mean. When σ is not known, if the population is normally distributed (or at least mound-shaped) or if the sample size n is large, we use the t **distribution** to compute this interval. We also studied how to find the size of the sample needed if we wish to compute a confidence interval for a mean with a prespecified *confidence level* and with a prespecified *margin of*

error. Figure 9.17 is a flowchart summarizing our discussions concerning how to compute an appropriate confidence interval for a population mean.

   Next we saw that we are often interested in estimating the proportion of population units falling into a category of interest. We showed how to compute a large sample confidence interval for a **population proportion,** and we saw how to find the sample size needed to estimate a population proportion. We concluded this chapter with an optional section that discusses how to estimate means, proportions, and totals for finite populations.

Glossary of Terms

confidence coefficient: The (before sampling) probability that a confidence interval for a population parameter will contain the population parameter.

confidence interval: An interval of numbers computed so that we can be very confident (say, 95 percent confident) that a population parameter is contained in the interval.

confidence level: The percentage of time that a confidence interval would contain a population parameter if all possible samples were used to calculate the interval.

degrees of freedom (for a t curve): A parameter that describes the exact spread of the curve of a t distribution.

margin of error: The quantity that is added to and subtracted from a point estimate of a population parameter to obtain a confidence interval for the parameter.

population total: The sum of the values of all the population measurements.

standard error of the estimate $\bar{x}$: The point estimate of $\sigma_{\bar{x}}$.

t distribution: A commonly used continuous probability distribution that is described by a distribution curve similar to a normal curve. The t curve is symmetrical about zero and is more spread out than a standard normal curve.

t point, t_α: The point on the horizontal axis under a t curve that gives a right-hand tail area equal to α.

t table: A table of t point values listed according to the area in the tail of the t curve and according to values of the degrees of freedom.

FIGURE 9.17   Computing an Appropriate Confidence Interval for a Population Mean

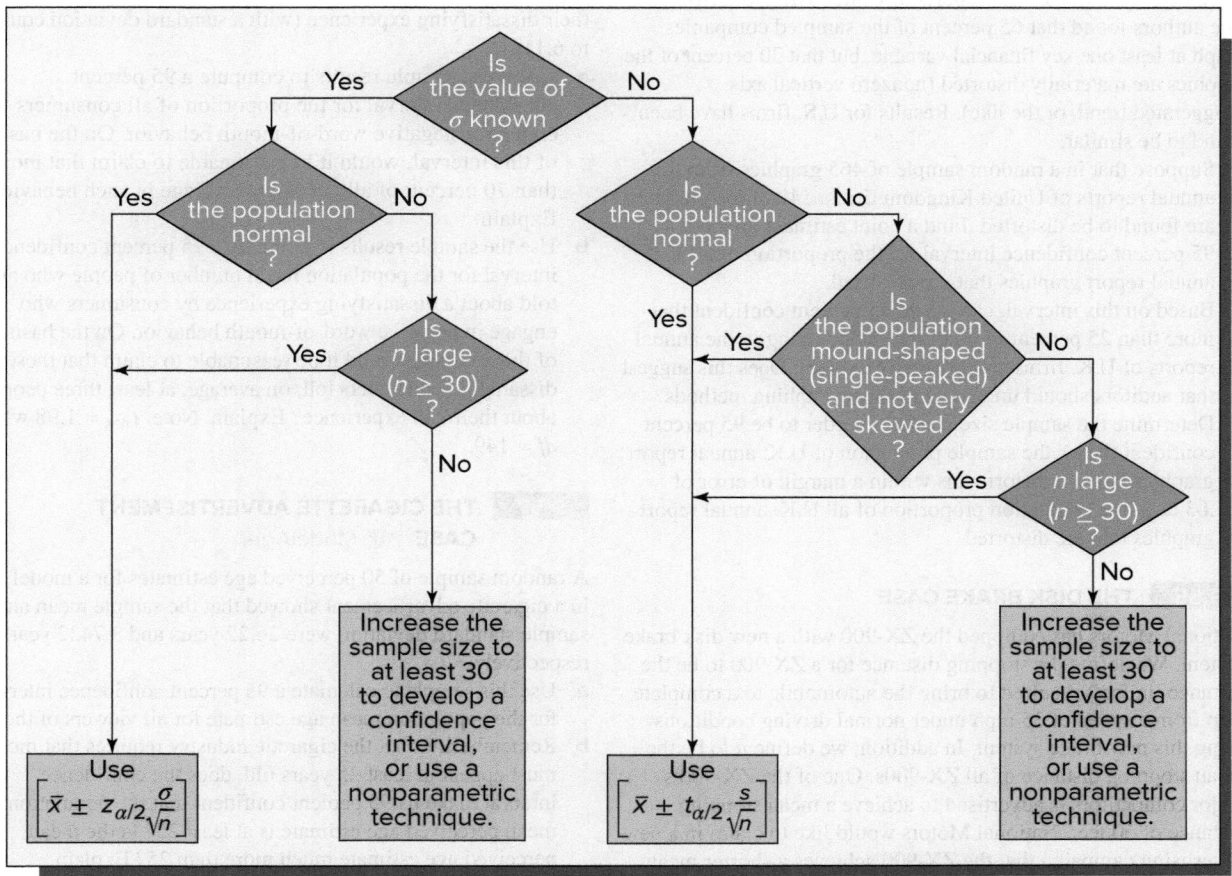

Important Formulas

A z-based confidence interval for a population mean μ with σ known: page 403

A t-based confidence interval for a population mean μ with σ unknown: page 408

Sample size when estimating μ: pages 414 and 415

A large sample confidence interval for a population proportion p: page 417

Sample size when estimating p: page 419

Confidence intervals for a population mean and population total (finite populations): page 424

Confidence intervals for the proportion of and total number of units in a category (finite populations): page 425

Supplementary Exercises

connect **9.53** In an article in the *Journal of Accounting Research*, Ashton, Willingham, and Elliott studied audit delay (the length of time from a company's fiscal year-end to the date of the auditor's report) for industrial and financial companies. In the study, a random sample of 250 industrial companies yielded a mean audit delay of 68.04 days with a standard deviation of 35.72 days, while a random sample of 238 financial companies yielded a mean audit delay of 56.74 days with a standard deviation of 34.87 days. Use these sample results to do the following:

a Calculate a 95 percent confidence interval for the mean audit delay for all industrial companies. Note: $t_{.025} = 1.97$ when $df = 249$.

b Calculate a 95 percent confidence interval for the mean audit delay for all financial companies. Note: $t_{.025} = 1.97$ when $df = 237$.

c By comparing the 95 percent confidence intervals you calculated in parts *a* and *b*, is there strong evidence that the mean audit delay for all financial companies is shorter than the mean audit delay for all industrial companies? Explain.

9.54   In an article in *Accounting and Business Research*, Beattie and Jones investigate the use and abuse of graphic presentations in the annual reports of United Kingdom firms. The authors found that 65 percent of the sampled companies graph at least one key financial variable, but that 30 percent of the graphics are materially distorted (nonzero vertical axis, exaggerated trend, or the like). Results for U.S. firms have been found to be similar.

a Suppose that in a random sample of 465 graphics from the annual reports of United Kingdom firms, 142 of the graphics are found to be distorted. Find a point estimate of and a 95 percent confidence interval for the proportion of all U.K. annual report graphics that are distorted.

b Based on this interval, can we be 95 percent confident that more than 25 percent of all graphics appearing in the annual reports of U.K. firms are distorted? Explain. Does this suggest that auditors should understand proper graphing methods?

c Determine the sample size needed in order to be 95 percent confident that $\hat{p}$, the sample proportion of U.K. annual report graphics that are distorted, is within a margin of error of .03 of p, the population proportion of all U.K. annual report graphics that are distorted.

9.55   THE DISK BRAKE CASE

National Motors has equipped the ZX-900 with a new disk brake system. We define the stopping distance for a ZX-900 to be the distance (in feet) required to bring the automobile to a complete stop from a speed of 35 mph under normal driving conditions using this new brake system. In addition, we define μ to be the mean stopping distance of all ZX-900s. One of the ZX-900's major competitors is advertised to achieve a mean stopping distance of 60 feet. National Motors would like to claim in a new advertising campaign that the ZX-900 achieves a shorter mean stopping distance.

Suppose that National Motors randomly selects a sample of 81 ZX-900s. The company records the stopping distance of each automobile and calculates the mean and standard deviation of the sample of 81 stopping distances to be 57.8 ft and 6.02 ft., respectively.

a Calculate a 95 percent confidence interval for μ. Can National Motors be 95 percent confident that μ is less than 60 ft? Explain.

b Using the sample of 81 stopping distances as a preliminary sample, find the sample size necessary to make National Motors 95 percent confident that $\bar{x}$ is within a margin of error of one foot of μ.

9.56   In an article in the *Journal of Retailing*, J. G. Blodgett, D. H. Granbois, and R. G. Walters investigated negative word-of-mouth consumer behavior. In a random sample of 201 consumers, 150 reported that they engaged in negative

word-of-mouth behavior (for instance, they vowed never to patronize a retailer again). In addition, the 150 respondents who engaged in such behavior, on average, told 4.88 people about their dissatisfying experience (with a standard deviation equal to 6.11).

a Use these sample results to compute a 95 percent confidence interval for the proportion of all consumers who engage in negative word-of-mouth behavior. On the basis of this interval, would it be reasonable to claim that more than 70 percent of all consumers engage in such behavior? Explain.

b Use the sample results to compute a 95 percent confidence interval for the population mean number of people who are told about a dissatisfying experience by consumers who engage in negative word-of-mouth behavior. On the basis of this interval, would it be reasonable to claim that these dissatisfied consumers tell, on average, at least three people about their bad experience? Explain. Note: $t_{.025} = 1.98$ when $df = 149$.

9.57   THE CIGARETTE ADVERTISEMENT CASE   DS ModelAge

A random sample of 50 perceived age estimates for a model in a cigarette advertisement showed that the sample mean and sample standard deviation were 26.22 years and 3.7432 years, respectively.

a Use this sample to calculate a 95 percent confidence interval for the population mean age estimate for all viewers of the ad.

b Remembering that the cigarette industry requires that models must appear at least 25 years old, does the confidence interval make us 95 percent confident that the population mean perceived age estimate is at least 25? Is the mean perceived age estimate much more than 25? Explain.

9.58   How safe are child car seats? *Consumer Reports* (May 2005) tested the safety of child car seats in 30 mph crashes. They found "slim safety margins" for some child car seats. Suppose that *Consumer Reports* simulates the safety of the market-leading child car seat. Their test consists of placing the maximum claimed weight in the car seat and simulating crashes at higher and higher miles per hour until a problem occurs. The following data identify the speed at which a problem with the car seat (such as the strap breaking, seat shell cracked, strap adjuster broke, detached from base, etc.) first appeared: 31.0, 29.4, 30.4, 28.9, 29.7, 30.1, 32.3, 31.7, 35.4, 29.1, 31.2, 30.2. Using the facts that the sample mean and sample standard deviation are 30.7833 and 1.7862, respectively, find a 95 percent confidence interval for the population mean speed at which a problem with the car seat first appears. Assume normality. Are we 95 percent confident that this population mean is at least 30 mph? Explain.   DS CarSeat

Appendix 9.1 ■ Confidence Intervals Using Excel

Confidence interval for a population mean in Figure 9.12 (data file: DebtEq.xlsx):

- Enter the debt-to-equity ratio data from Example 9.2 into cells A2 to A16 with the label Ratio in cell A1.

- Select **Data : Data Analysis : Descriptive Statistics.**

- Click OK in the Data Analysis dialog box.

- In the Descriptive Statistics dialog box, enter A1:A16 into the Input Range window.

- Place a checkmark in the "Labels in first row" checkbox.

- Under output options, select "New Worksheet Ply" to have the output placed in a new worksheet and enter the name Output for the new worksheet.

- Place checkmarks in the Summary Statistics and "Confidence Level for Mean" checkboxes. This produces a *t*-based margin of error for a confidence interval.

- Type 95 in the "Confidence Level for Mean" box.

- Click OK in the Descriptive Statistics dialog box.

- A descriptive statistics summary will be displayed in cells A3 through B16 in the Output worksheet. Drag the column borders to reveal complete labels for all of the descriptive statistics.

- Type the heading "95% Confidence Interval" into cells D13 through E13.

- Compute the lower bound of the interval by typing the cell formula =B3 − B16 into cell D14. This subtracts the margin of error of the interval (labeled "Confidence Level (95.0%)") from the sample mean.

- Compute the upper bound of the interval by typing the formula =B3 + B16 into cell E14.

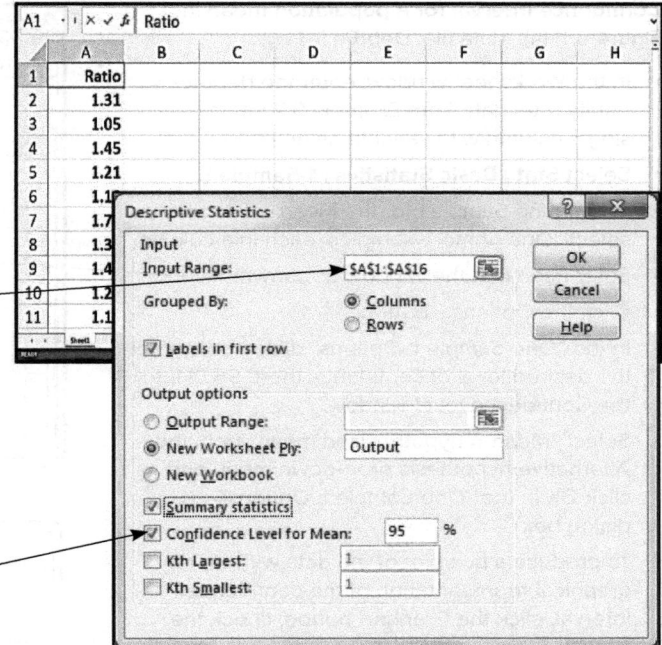

Source: Microsoft Office Excel 2016

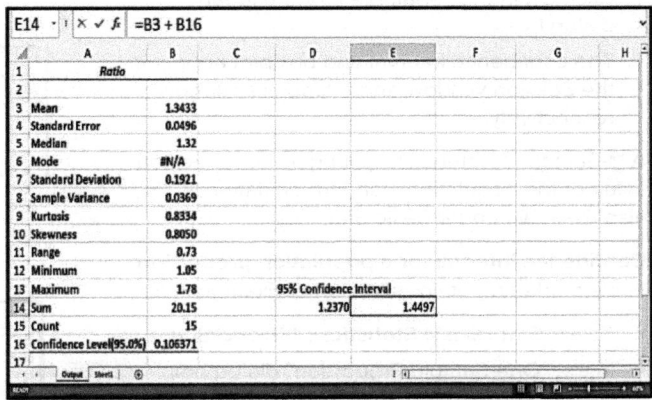

Source: Microsoft Office Excel 2016

Appendix 9.2 ■ Confidence Intervals Using Minitab

Confidence interval for a population mean in Figure 9.11 (a): (data file: DebtEq.MTW):

- In the Worksheet window, enter the debt-to-equity ratio data from Example 9.2 into a single column with variable name Ratio.

- Select **Stat : Basic Statistics : 1-Sample t.**

- In the "One-Sample t for the Mean" dialog box, select "One or more samples, each in a column."

- Enter Ratio into the window as shown.

- Click the Options... button.

- In the "One-Sample t: Options" dialog box, enter the desired level of confidence (here 95.0) into the Confidence level window.

- Select "Mean ≠ hypothesized mean" from the Alternative hypothesis drop-down menu and click OK in the "One-Sample t: Options" dialog box.

- To produce a box plot of the data with a graphical representation of the confidence interval, click the Graphs... button; check the "Boxplot" box; and click OK in the "One-Sample t: Graphs" dialog box.

- Click OK in the "1-Sample t for the Mean" dialog box.

- The confidence interval and box plot appear in the Session window and a graphics window respectively.

A "1-Sample Z" interval requiring a user-specified value of the population standard deviation is also available under Basic Statistics.

Confidence interval for a population proportion in the marketing ethics situation of Example 9.8.

- Select **Stat : Basic Statistics : 1 Proportion**

- In the "One-Sample Proportion" dialog box, select "Summarized data" from the drop-down menu.

- Enter the number of events (here 117) and the number of trials (here 205) in the appropriate windows.

- Click on the Options... button.

- In the "One-Sample Proportion: Options" dialog box, enter the desired level of confidence (here 95.0) and choose "Proportion ≠ hypoth-esized proportion" as the Alternative hypothesis.

- Choose "Normal approximation" as the Method.

- Click OK in the "One-Sample Proportion: Options" dialog box.

- Click OK in the "One-Sample Proportion" dialog box.

- The confidence interval will be displayed in the Session window.

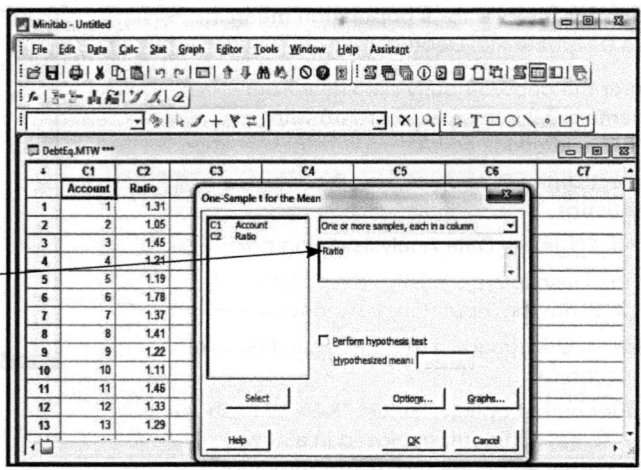

Source: Minitab 18

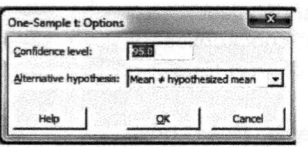

Source: Minitab 18

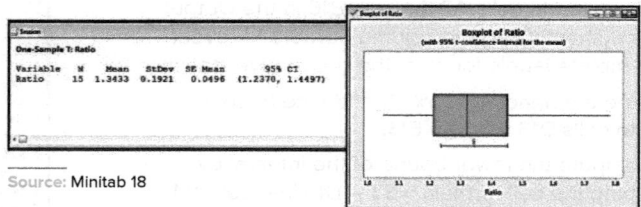

Source: Minitab 18

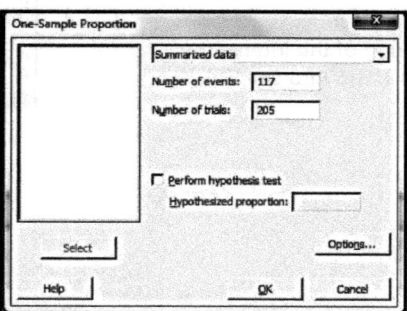

Source: Minitab 18

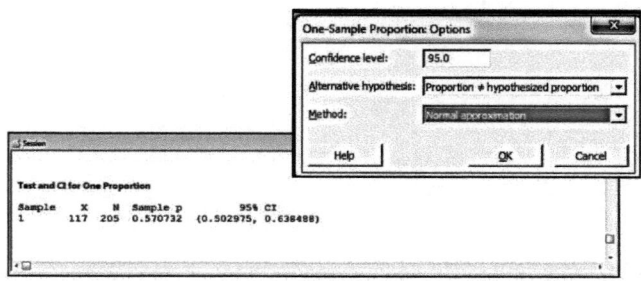

Source: Minitab 18

Appendix 9.3 ■ Confidence Intervals Using JMP

Confidence Interval for a Mean using raw data in Figure 9.11 (data file: DebtEq.jmp):

- After opening the DebtEq.jmp file, select **Analyze: Distribution**.

- Select the "Debt To Equity" column and click "Y, Columns." Click "OK."

- From the red triangle menu beside "Debt To Equity," select "Confidence Interval" and "0.95." The confidence interval for the mean is in the first row.

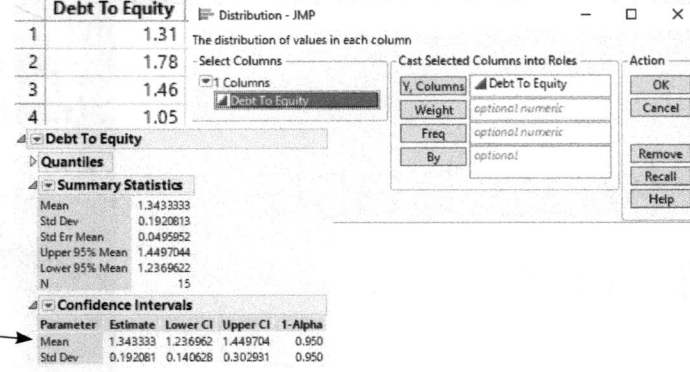

Source: JMP Pro 14

Confidence Interval for a Mean using summary statistics in Exercise 9.16:

- Select **Help: Sample Data**.

- Expand the **Calculators** area under "Teaching resources" by clicking on the grey triangle next to **Calculators**. Select "Confidence Interval for One Mean." Select "Summary Statistics" and click "OK."

- Select "t" under "Choose Interval Type." Enter the summary statistics under the "Summary Information" heading as shown. Enter the "Confidence Level" of "0.95." The Lower and Upper Limits will be automatically calculated under "Result."

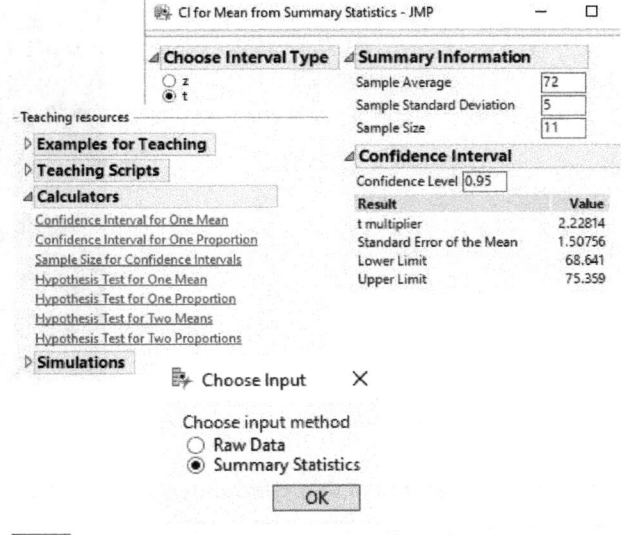

Source: JMP Pro 14

Confidence Interval for a Proportion in Example 9.6:

- Select **Help: Sample Data**

- Expand the **Calculators** area under "Teaching resources" by clicking on the grey triangle next to **Calculators**. Select "Confidence Interval for One Proportion." Select "Summary Statistics" and click "OK."

- Select "Normal Approximation" under "Sample Statistic." Enter the summary statistics under the "Summary Information" heading as shown. Enter the "Confidence Level" of "0.99." The Lower and Upper Limits will be automatically calculated under "Result."

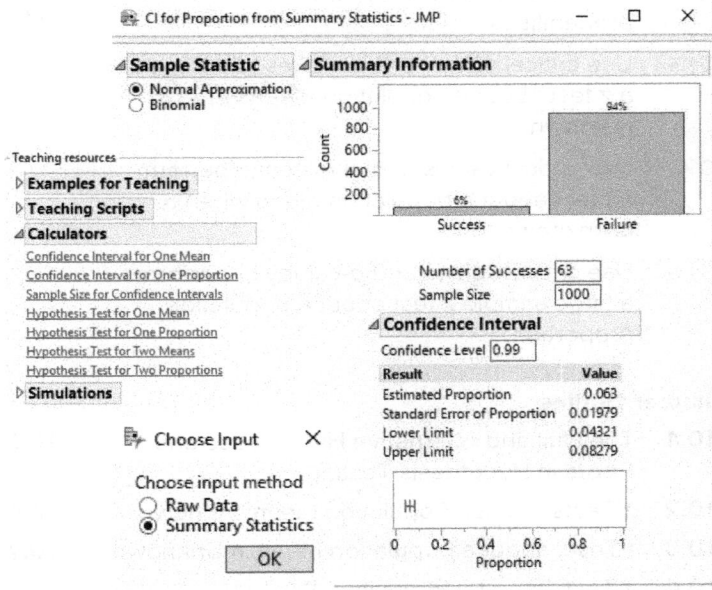

Source: JMP Pro 14

Design Elements: (CD): ©Comstock Images/Alamy; (All Others): ©McGraw-Hill Education

## CHAPTER 10

Hypothesis Testing

©George Doyle/Getty Images

Learning Objectives

After mastering the material in this chapter, you will be able to:

LO10-1 Set up appropriate null and alternative hypotheses.

LO10-2 Describe Type I and Type II errors and their probabilities.

LO10-3 Use critical values and p-values to perform a z test about a population mean when σ is known.

LO10-4 Use critical values and p-values to perform a t test about a population mean when σ is unknown.

LO10-5 Use critical values and p-values to perform a large sample z test about a population proportion.

LO10-6 Calculate Type II error probabilities and the power of a test, and determine sample size (Optional).

LO10-7 Describe the properties of the chi-square distribution and use a chi-square table.

LO10-8 Use the chi-square distribution to make statistical inferences about a population variance (Optional).

Chapter Outline

10.1 The Null and Alternative Hypotheses and Errors in Hypothesis Testing

10.2 z Tests about a Population Mean: σ Known

10.3 t Tests about a Population Mean: σ Unknown

10.4 z Tests about a Population Proportion

10.5 Type II Error Probabilities and Sample Size Determination (Optional)

10.6 The Chi-Square Distribution

10.7 Statistical Inference for a Population Variance (Optional)

Hypothesis testing is a statistical procedure used to provide evidence in favor of some statement (called a *hypothesis*). For instance, hypothesis testing might be used to assess whether a population parameter, such as a population mean, differs from a specified standard or previous value. In this chapter

we discuss testing hypotheses about population means, proportions, and variances.

In order to illustrate how hypothesis testing works, we revisit several cases introduced in previous chapters and also introduce some new cases:

The e-Billing Case: The consulting firm uses hypothesis testing to provide strong evidence that the new electronic billing system has reduced the mean payment time by more than 50 percent.

The Cheese Spread Case: The cheese spread producer uses hypothesis testing to supply extremely strong evidence that fewer than 10 percent of all current purchasers would stop buying the cheese spread if the new spout were used.

The Trash Bag Case: A marketer of trash bags uses hypothesis testing to support its claim that the mean breaking strength of its new trash bag is greater than 50 pounds. As a result, a television network approves use of this claim in a commercial.

The Valentine's Day Chocolate Case: A candy company uses hypothesis testing to assess whether it is reasonable to plan for a 10 percent increase in sales of its special Valentine box of assorted chocolates.

10.1 The Null and Alternative Hypotheses and Errors in Hypothesis Testing

LO10-1
Set up appropriate null and alternative hypotheses.

One of the authors' former students is employed by a major television network in the standards and practices division. One of the division's responsibilities is to reduce the chances that advertisers will make false claims in commercials run on the network. Our former student reports that the network uses a statistical methodology called **hypothesis testing** to do this.

In hypothesis testing we choose between two competing statements. For example, when in doubt about the truth of an advertising claim, the television network must decide whether the advertising claim is true or false. We call one of the competing statements the *null hypothesis* and the other the *alternative hypothesis*. Hypothesis testing does not treat these hypotheses evenhandedly. Therefore, it is crucial to decide which of the competing statements is the null hypothesis and which is the alternative hypothesis. In the following summary box, we define the differing roles of the null and alternative hypotheses.

The Null Hypothesis and the Alternative Hypothesis

In hypothesis testing:

1   The **null hypothesis,** denoted H_0, is the statement being tested. The null hypothesis is given the **benefit of the doubt** and is not rejected unless there is convincing sample evidence that it is false.

2   The **alternative hypothesis,** denoted H_a, is a statement that is assigned the **burden of proof.** The alternative hypothesis is accepted only if there is convincing sample evidence that it is true.

The meaning of the term *convincing sample evidence* will be discussed as we proceed through this chapter. For now, it is intuitively helpful to know that the differing roles of the null and alternative hypotheses reflect the philosophy adopted in an American criminal court. In court, the null hypothesis is that the defendant is innocent while the alternative hypothesis is that the defendant is guilty. The judge or jury must assume that the defendant is innocent (or that H_0 is true) and will find the defendant guilty (or reject H_0) only if the prosecutor builds a strong case for the defendant's guilt that is "beyond a reasonable doubt." Understanding this philosophy helps us to place many of the hypothesis testing situations that we will encounter into one of two categories.

1. The alternative hypothesis as a research hypothesis

In some situations an entrepreneur, pioneering scientist, or business has developed a new product, service, process, or other innovation that is intended to replace an existing standard. Here the innovation is generally considered to be a potential improvement over the existing standard, and the developer wants to claim that the innovation is better than the standard. The developer's claim is called a **research hypothesis.** Because the developer is trying to convince consumers (or businesses, or other scientists) that the innovation is superior to the existing standard, the burden of proof should be on the developer of the innovation. Thus, the research hypothesis should be the alternative hypothesis. The null hypothesis is then the opposite of H_a, and says that the innovation is not better than the existing standard. Only if sample data provide convincing evidence that the innovation is better will H_0 be rejected and will the developer have statistical support to claim that the innovation is superior to the existing standard. Notice that in this type of hypothesis testing scenario *it is easiest to establish the alternative hypothesis H_a first and to then form the null hypothesis as the opposite of H_a.* We now consider two examples. The first illustrates that many of the advertising claims evaluated by television networks are research hypotheses concerning potential product or service improvements.

(C) EXAMPLE 10.1 The Trash Bag Case: Testing An Advertising Claim

©Aluma Images/Getty Images RF

An important attribute of a trash bag is its *breaking strength*. This is the amount of a representative trash mix (in pounds) that, when loaded into the bag suspended in the air, will cause the bag to rip, drip, or tear. In this example we consider a trash bag manufacturer that is known for selling inexpensive but quality trash bags and has been making a 30-gallon bag (a commonly purchased size) that has a mean breaking strength of 40 pounds. A 40-pound mean breaking strength is typical for many brands of 30-gallon bags, but the manufacturer has just designed a new 30-gallon bag that it believes has a mean breaking strength that is greater than 50 pounds, the advertised mean breaking strength of the strongest 30-gallon bag currently on the market.

The new trash bag is made with a specially formulated plastic developed by the manufacturer that is stronger and also more biodegradable than other plastics. The stronger plastic allows the new bag's thickness to be reduced, and the resulting cost swings will enable the manufacturer to lower its already low price and make the new bag the least expensive 30-gallon bag on the market. The trash bag manufacturer wishes to run a commercial on television networks advertising that the new bag is the least expensive, most environmentally friendly, and strongest 30-gallon trash bag on the market. Networks are convinced of the price and environmental advantages of the new bag by price comparisons and the chemical composition of the new bag's plastic. However, the basis of the advertisement that the new bag is the strongest on the market—the mean breaking strength μ of the new bag—is in question, and thus the claim that μ is greater than 50 pounds is a research hypothesis that is assigned the burden of proof. It follows that the statement that *μ is greater than 50 pounds will be the alternative hypothesis, H_a,* and the opposite statement that *μ is less than or equal to 50 pounds will be the null hypothesis, H_0.* (Note that the *null hypothesis* says that the μ is *not* greater than 50 pounds.) We summarize the null and alternative hypotheses by saying that we are testing H_0: $\mu \leq 50$ versus H_a: $\mu > 50$. A network will run the manufacturer's commercial if a random sample of n new bags provides sufficient evidence to reject H_0: $\mu \leq 50$ in favor of H_a: $\mu > 50$.

EXAMPLE 10.2 The e-Billing Case: Reducing Mean Bill Payment Time

Recall that a management consulting firm has installed a new computer-based electronic billing system for a Hamilton, Ohio, trucking company. Because of the system's advantages, and because the trucking company's clients are receptive to using this system, the management consulting firm believes that the new system will reduce the mean bill payment time by more than 50 percent. The mean payment time using the old billing system was approximately equal to, but no less than, 39 days. Therefore, if μ denotes the mean payment time using the new system, the consulting firm believes and wishes to show that μ *is less than 19.5 days.* The statement that μ is less than 19.5 days is a research hypothesis that claims a large reduction in mean bill payment time and is assigned the burden of proof. It follows the statement that μ *is less than 19.5 days will be the alternative hypothesis, H_a,* and the opposite statement that μ *is greater than or equal to 19.5 days will be the null hypothesis, H_0.* The consulting firm will randomly select a sample of invoices and will determine if their payment times provide sufficient evidence to reject $H_0: \mu \geq 19.5$ in favor of $H_a: \mu < 19.5$. If such evidence exists, the consulting firm will conclude that the new electronic billing system has reduced the Hamilton trucking company's mean bill payment time by more than 50 percent. This conclusion will be used to help demonstrate the benefits of the new billing system both to the Hamilton company and to other trucking companies that are considering using such a system.

Although we have initially said that research hypotheses often represent potential product, service, or process improvements, there are other types of research hypotheses. For example, the hypothesis that an existing standard (such as a popular and frequently prescribed medicine) is less effective than originally thought is a research hypothesis that would be assigned the burden of proof. As another example, consider the cheese spread case, and let p be the proportion of current purchasers who would stop buying the cheese spread if the new spout were used. Recalling that profits will increase only if p is less than .1, the statement p *is less than .1* is a research hypothesis that would be assigned the burden of proof. Therefore, the alternative hypothesis H_a would say that p *is less than .1,* and the null hypothesis H_0 would say that p *is greater than or equal to .1.*

2. The null hypothesis as a hypothesis about a successful, ongoing process

The second kind of hypothesis testing situation involves an ongoing process which has functioned acceptably for a reasonable period of time. Here the null hypothesis is a statement that assumes the process is continuing to function acceptably, while the alternative hypothesis says the process has gone awry. For example, suppose that an auto parts supplier has been producing an auto part with an intended mean diameter of 2 inches. The supplier wishes to periodically assess whether the auto part production process is in fact producing auto parts with the intended mean diameter. When the process is found to be producing an unacceptable mean diameter, it will be readjusted to correct the situation. Suppose that past experience indicates that the process has been consistently successful in producing parts with the intended mean diameter. Then, if failure of one of these parts would not be catastrophic (for instance, would not lead to injury or death), it is reasonable to give the production process the benefit of the doubt. Therefore, letting μ be the mean diameter of the parts made by the process, the null hypothesis H_0 will state that μ equals 2 inches. The alternative hypothesis H_a will say that μ does not equal 2 inches (that is, the process is producing parts with a mean diameter that is too large or too small). The auto parts supplier will *readjust the process* if sample data provide strong evidence against $H_0: \mu = 2$ and in favor of $H_a: \mu \neq 2$. Note that in this type of hypothesis testing scenario, *it is easiest to establish the null hypothesis first and to then form the alternative hypothesis as the opposite of H_0.*

As another example, suppose that a coffee producer is selling cans that are labeled as containing three pounds of coffee, and also suppose that the Federal Trade Commission (FTC) wishes to assess whether the mean amount of coffee μ in all such cans is at least three pounds. If the FTC has received no complaints or other information indicating that μ is less than 3, it is reasonable to give the coffee producer the benefit of the doubt. Therefore, the FTC will take action against the producer only if sample data provide convincing evidence against H_0: $\mu \geq 3$ and in favor of H_a: $\mu < 3$. A third example follows.

©Jill Fromer/Getty Images RF

Ⓒ EXAMPLE 10.3 The Valentine's Day Chocolate Case:[1] Production Planning

A candy company markets a special 18-ounce box of assorted chocolates to large retail stores each year for Valentine's Day. This year's assortment features especially tempting chocolates nestled in an attractive new box. The company has a long history in the candy business, and its experience has enabled it to accurately forecast product demand for years. Because this year's box is particularly special, the company projects that sales of the Valentine box will be 10 percent higher than last year.

Before the beginning of the Valentine's Day sales season, the candy company sends information to its client stores about its upcoming Valentine assortment. This information includes a product description and a preview of the advertising displays that will be provided to encourage sales. Each client retail store then responds with a single (nonreturnable) order to meet its anticipated customer demand for the Valentine box. Last year, retail stores ordered an average of 300 boxes per store. If the projected 10 percent increase in sales is correct, this year's mean order quantity, μ, for client stores will equal 330 boxes per store.

Because the company has accurately projected product demand for many years, its ongoing forecasting process has been functioning acceptably. However, because accurate forecasting is important (the company does not want to produce more Valentine's Day boxes than it can sell, nor does it want to fail to satisfy demand), the company needs to assess whether it is reasonable to plan for a 10 percent increase in sales. To do this by using a hypothesis test, the null hypothesis says that the forecasting process is continuing to make accurate predictions. That is, the null hypothesis is H_0: $\mu = 330$. The alternative hypothesis says that the forecasting process has gone awry, that is, H_a: $\mu \neq 330$.

To perform the hypothesis test, the candy company will randomly select a sample of its client retail stores. These stores will receive an early mailing that promotes the upcoming Valentine box. Each store will be asked to report the number of boxes it expects to order. If the sample data do not allow the company to reject H_0: $\mu = 330$, it will plan its production based on a 10 percent sales increase. However, if there is sufficient evidence to reject H_0, the company will change its production plans.

We next summarize the sets of null and alternative hypotheses that we have thus far considered in the trash bag, e-billing, and Valentine's Day chocolate cases:

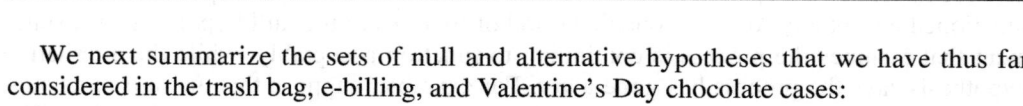

$$H_0: \mu \leq 50 \qquad H_0: \mu \geq 19.5 \qquad H_0: \mu = 330$$
$$\text{versus} \qquad \text{versus} \qquad \text{versus}$$

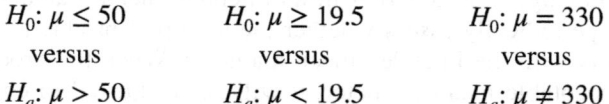

$$H_a: \mu > 50 \qquad H_a: \mu < 19.5 \qquad H_a: \mu \neq 330$$

The alternative hypothesis H_a: $\mu > 50$ is called a **one-sided, greater than alternative** hypothesis, whereas H_a: $\mu < 19.5$ is called a **one-sided, less than alternative** hypothesis, and H_a: $\mu \neq 330$ is called a **two-sided, not equal to alternative** hypothesis. Many of the alternative hypotheses we consider in this book are one of these three types. Also, note that each null hypothesis we have considered involves an *equality*. For example, the null hypothesis H_0: $\mu \leq 50$ says that μ is either less than or *equal to* 50. We will see that, in general, the approach we use to test a null hypothesis versus an alternative hypothesis requires that the null hypothesis involve an equality. For this reason, **we always formulate the null and alternative hypotheses so that the null hypothesis involves an equality.**

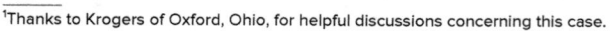

[1]Thanks to Krogers of Oxford, Ohio, for helpful discussions concerning this case.

Sample evidence and the legal system analogy

Recall from the beginning of this section that hypothesis testing rejects a null hypothesis H_0 in favor of an alternative hypothesis H_0 only if there is convincing sample evidence that H_0 is false and H_a is true. Also, recall that this is similar to our legal system, which rejects the innocence of the accused only if the evidence of guilt is beyond a reasonable doubt. For example, consider the trash bag case. Because the mean $\bar{x}$ of the breaking strengths of a random sample of new trash bags is the point estimate of the population mean breaking strength μ, in Section 10.2 we will use $\bar{x}$ to test $H_0: \mu \le 50$ versus $H_a: \mu > 50$. Moreover, we will reject $H_0: \mu \le 50$ in favor of $H_a: \mu > 50$ only if $\bar{x}$ is enough greater than 50 to show beyond a reasonable doubt that $H_0: \mu \le 50$ is false and $H_a: \mu > 50$ is true. An $\bar{x}$ only slightly greater than 50 might not be convincing enough to reject $H_0: \mu \le 50$, but such an $\bar{x}$ would provide a small amount of evidence against $H_0: \mu \le 50$ and certainly would not provide convincing evidence to support accepting $H_0: \mu \le 50$. In general, if the sample evidence is not strong enough to convince us to reject a null hypothesis H_0, **we do not say that we accept H_0. Rather we say that we do not reject H_0.** Again, this is similar to our legal system, where the lack of evidence of guilt beyond a reasonable doubt results in a verdict of **not guilty,** but does not prove that the accused is innocent.

Type I and Type II errors and their probabilities

To determine exactly how much statistical evidence is required to reject H_0, we consider the errors and the correct decisions that can be made in hypothesis testing. These errors and correct decisions, as well as their implications in the trash bag advertising example, are summarized in Tables 10.1 and 10.2. Across the top of each table are listed the two possible "states of nature." Either $H_0: \mu \le 50$ is true, which says the manufacturer's claim that μ is greater than 50 is false, or H_0 is false, which says the claim is true. Down the left side of each table are listed the two possible decisions we can make in the hypothesis test. Using the sample data, we will either reject $H_0: \mu \le 50$, which implies that the claim will be advertised, or we will not reject H_0, which implies that the claim will not be advertised.

In general, the two types of errors that can be made in hypothesis testing are defined as follows:

LO10-2
Describe Type I and Type II errors and their probabilities.

Type I and Type II Errors

If we reject H_0 when it is true, this is a **Type I error.**
If we do not reject H_0 when it is false, this is a **Type II error.**

As can be seen by comparing Tables 10.1 and 10.2, if we commit a Type I error, we will advertise a false claim. If we commit a Type II error, we will fail to advertise a true claim.

TABLE 10.1   Type I and Type II Errors

	State of Nature	
Decision	$H_0: \mu \le 50$ **True**	$H_0: \mu \le 50$ **False**
Reject $H_0: \mu \le 50$	Type I error	Correct decision
Do not reject $H_0: \mu \le 50$	Correct decision	Type II error

TABLE 10.2   The Implications of Type I and Type II Errors in the Trash Bag Example

	State of Nature	
Decision	**Claim False**	**Claim True**
Advertise the claim	Advertise a false claim	Advertise a true claim
Do not advertise the claim	Do not advertise a false claim	Do not advertise a true claim

We now let the symbol α (pronounced **alpha**) **denote the probability of a Type I error,** and we let β (pronounced **beta**) **denote the probability of a Type II error.** Obviously, we would like both α and β to be small. A common (but not the only) procedure is to base a hypothesis test on taking a sample of a fixed size (for example, $n = 30$ trash bags.) and on setting α equal to a small prespecified value. Setting α low means there is only a small chance of rejecting H_0 when it is true. This implies that we are requiring convincing evidence against H_0 before we reject it.

We sometimes choose α as high as .10, but we usually choose α to be between .05 and .01. A frequent choice for α is .05. However, the choice of a value for α in a particular hypothesis test will depend on various factors, such as the relative plausibilities of the null and alternative hypotheses and the relative seriousnesses of making Type I and Type II errors. For example, in the trash bag case it is plausible that the mean breaking strength of the new trash bag is greater than 50 pounds and making a Type I error (advertising the claim that the new trash bag is the strongest on the market when it is not) might be regarded as of average seriousness when it comes to evaluating advertising claims. Therefore, many television networks might choose the value of .05 for α. On the other hand, other networks might be more risk averse when facing the possibility of advertising a false trash bag strength claim and choose a smaller value for α–say, one of the values .04 or .03 or .02 or .01. Unfortunately, for a fixed sample size, the smaller a network sets α, the larger is β, the probability of a Type II error (not advertising the claim that the new trash bag is the strongest on the market when it is). It follows that a network should not set α smaller than it feels is necessary to guard against making a Type I error.

As another example, suppose a pharmaceutical company wishes to advertise that it has developed an effective treatment for a disease that has been very resistant to treatment. Such a claim is (perhaps) difficult to believe. Moreover, if the claim is false, patients suffering from the disease would be subjected to false hope, needless expense, and possibly serious side effects. In such a case, it might be reasonable for most networks to set α at .01—or even less—because this would lower the chance of advertising the claim if it is false. Unless committing a Type I error is extremely serious (it might be in the pharmaceutical example), we usually do not set α lower than .01 because doing so often leads to an unacceptably large value of β.

Exercises for Section 10.1

CONCEPTS

10.1   Define each of the following: Type I error, α, Type II error, β.

10.2   When testing a hypothesis, why don't we set the probability of a Type I error to be extremely small? Explain.

METHODS AND APPLICATIONS

10.3   **THE VIDEO GAME SATISFACTION RATING CASE**   DS VideoGame

Recall that "very satisfied" customers give the XYZ-Box video game system a rating that is at least 42. Suppose that the manufacturer of the XYZ-Box wishes to use the 65 satisfaction ratings to provide evidence supporting the claim that the mean composite satisfaction rating for the XYZ-Box exceeds 42.

a   Letting μ represent the mean composite satisfaction rating for the XYZ-Box, set up the null and alternative hypotheses needed if we wish to attempt to provide evidence supporting the claim that μ exceeds 42. Regard the claim as a research hypothesis.

b   In the context of this situation, interpret making a Type I error; interpret making a Type II error.

10.4   **THE BANK CUSTOMER WAITING TIME CASE**   DS WaitTime

Recall that a bank manager has developed a new system to reduce the time customers spend waiting for teller service during peak hours. The manager hopes the new system will reduce waiting times from the current 9 to 10 minutes to less than 6 minutes.

Suppose the manager wishes to use the 100 waiting times to support the claim that the mean waiting time under the new system is shorter than six minutes.

a   Letting μ represent the mean waiting time under the new system, set up the null and alternative hypotheses needed if we wish to attempt to provide evidence supporting the claim that μ is shorter than six minutes.

b   In the context of this situation, interpret making a Type I error; interpret making a Type II error.

10.5   The Crown Bottling Company has just installed a new bottling process that will fill 16-ounce bottles of the popular Crown Classic Cola soft drink. Both overfilling and underfilling bottles are undesirable: Underfilling leads to customer complaints and overfilling costs the company considerable money. In order to verify that the filler is set up correctly, the company wishes to see whether the mean bottle fill, μ, is close to the target fill of 16 ounces. To this end, a random sample of 36 filled bottles is selected from the output of a test filler run. If the sample results cast a substantial amount of doubt on the hypothesis that the mean bottle fill is the desired 16 ounces, then the filler's initial setup will be readjusted.

a The bottling company wants to set up a hypothesis test so that the filler will be readjusted if the null hypothesis is rejected. Set up the null and alternative hypotheses for this hypothesis test.

b In the context of this situation, interpret making a Type I error; interpret making a Type II error.

10.2 z Tests about a Population Mean: σ Known

LO10-3
Use critical values and p-values to perform a z test about a population mean when σ is known.

A critical value rule for testing a "greater than" alternative hypothesis

In this subsection we use the trash bag case to illustrate what is called the **critical value rule approach** to hypothesis testing. We will show that this approach can be carried out in five steps, the first two of which have been introduced in Section 10.1.

Step 1: State the null hypothesis H_0 and the alternative hypothesis H_a.   In Section 10.1 we have seen that a television network will run the commercial stating that the new trash bag is the strongest on the market if a random sample of new trash bags provides sufficient evidence to reject $H_0: \mu \leq 50$ in favor of $H_a: \mu > 50$. Here, μ is the mean breaking strength of the population of all new trash bags. In order to test $H_0: \mu \leq 50$ versus $H_a: \mu > 50$, we will test the modified null hypothesis $H_0: \mu = 50$ versus $H_a: \mu > 50$. The idea here is that if there is sufficient evidence to reject the hypothesis that μ equals 50 in favor of $\mu > 50$, then there is certainly also sufficient evidence to reject the hypothesis that μ is less than or equal to 50.

Step 2: Specify a value for α, the probability of a Type I error for the hypothesis test.   Suppose that a particular television network that we will call network 1 has specified how much evidence is sufficient to reject $H_0: \mu = 50$ in favor of $H_a: \mu > 50$ by specifying a value of .05 for α. A Type I error—rejecting $H_0: \mu = 50$ when H_0 is true—would result in network 1 advertising the claim that the new trash bag is the strongest on the market when it is not. Therefore, network 1 is allowing a 5 percent chance of advertising that the new bag is the strongest when it is not.

Step 3: Plan the sampling procedure and select the test statistic.   In order to test $H_0: \mu = 50$ versus $H_a: \mu > 50$, we will randomly select a sample of $n = 30$ new trash bags and calculate the mean $\bar{x}$ of the breaking strengths of these bags. Because the sample size $n = 30$ is large, the Central Limit Theorem tells us that the sampling distribution of all possible sample means that could obtained from all possible samples is (at least approximately) a normal distribution having mean $\mu_{\bar{x}} = \mu$ and standard deviation $\sigma_{\bar{x}} = \sigma/\sqrt{n}$. Here, σ is the standard deviation of the breaking strengths of the population of all new trash bags. It follows that, if $H_0: \mu = 50$ is true, the distribution of all possible values of

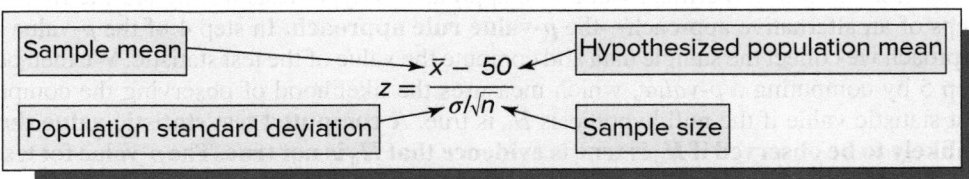

is (at least approximately) a standard normal distribution. We use z as the **test statistic** for testing $H_0: \mu = 50$ versus $H_a: \mu > 50$. A positive value of z results when $\bar{x}$ is greater than 50. This implies that μ might be greater than 50 and thus provides evidence to support rejecting $H_0: \mu = 50$ in favor of $H_a: \mu > 50$.

Step 4: Determine the critical value rule for deciding whether to reject H_0.   In order to decide whether the value of the test statistic z computed from the sample that will be observed is large enough to reject H_0: $\mu = 50$ in favor of H_a: $\mu > 50$, we can use a **critical value rule,** which says to do the following:

- Place the probability of a Type I error, α, in the right-hand tail of the standard normal curve and use the normal table (see Table A.3) to find the normal point z_α. Here z_α, which we call a **critical value,** is the point on the horizontal axis under the standard normal curve that gives a right-hand tail area equal to α [see Figure 10.1(a)].
- **Reject H_0: $\mu = 50$ in favor of H_a: $\mu > 50$ if and only if the computed value of the test statistic z is greater than the critical value z_α.**

The critical value for network 1's α of .05 is $z_\alpha = z_{.05}$. Looking up a cumulative area of .95 in Table A.3, we find that $z_{.05} = 1.645$ [see Figure 10.1(a)]. It follows that the critical value rule tells network 1 to reject H_0: $\mu = 50$ if and only if the computed value of the test statistic z is greater than the critical value $z_{.05} = 1.645$. Moreover, the α of .05 means that, if H_0: $\mu = 50$ is true, then only 5 percent of all possible values of z are greater than $z_{.05} = 1.645$ and thus would cause network 1 to incorrectly reject H_0 (commit a Type I error).

Step 5 : Collect the sample data, compute the value of the test statistic, and decide whether to reject H_0 by using the critical value rule. Interpret the statistical results.   Suppose that when the sample of $n = 30$ new trash bags is randomly selected, the mean of the breaking strengths is calculated to be $\bar{x} = 52.167$ pounds. Also suppose that the trash bag manufacturer has improved its trash bags multiple times in the past and has found that the population standard deviation σ of individual trash bag breaking strengths remains constant for each update and equals 5.4 pounds. Assuming that σ is 5.4 pounds for the new trash bag, the value of the test statistic is

$$z = \frac{\bar{x} - 50}{\sigma/\sqrt{n}} = \frac{52.167 - 50}{5.4/\sqrt{30}} = \frac{2.167}{.9859} = 2.20$$

Because $z = 2.20$ is greater than the critical value $z_{.05} = 1.645$, the critical value rule tells network 1 to reject H_0: $\mu = 50$ in favor of H_a: $\mu > 50$ and thus run the trash bag commercial. Furthermore, consider a second network—network 2—that has decided to set α equal to .01. Looking up a cumulative area of .99 in Table A.3, we find that $z_\alpha = z_{.01} = 2.33$ [see Figure 10.2(a)]. Because $z = 2.20$ is less than $z_{.01} = 2.33$, the critical value rule tells network 2 to not reject H_0: $\mu = 50$ in favor of H_0: $\mu > 50$ and thus not run the trash bag commercial. Here, although setting α equal to .01 reduced the chances of making a Type I error (advertising a false trash bag strength claim), network 2 has ended up not rejecting H_0 and thus might have committed a Type II error (not advertising a true trash bag strength claim).

A *p*-value rule for testing a "greater than" alternative hypothesis

Steps 1, 2 and 3 of the critical value rule approach to hypothesis testing are the first three steps of an alternative approach—the ***p*-value rule approach.** In step 4 of the *p*-value rule approach we collect the sample data and compute the value of the test statistic. We then begin step 5 by computing a *p-value,* which measures the likelihood of observing the computed test statistic value if the null hypothesis H_0 is true. **A computed test statistic value that is unlikely to be observed if H_0 is true is evidence that H_0 is not true.** The *p*-value for testing a null hypothesis H_0 versus an alternative hypothesis H_a is defined as follows:

The ***p*-value** is the probability, computed assuming that the null hypothesis H_0 is true, of observing a value of the test statistic that is at least as contradictory to H_0 and supportive of H_a as the value computed from the sample data.

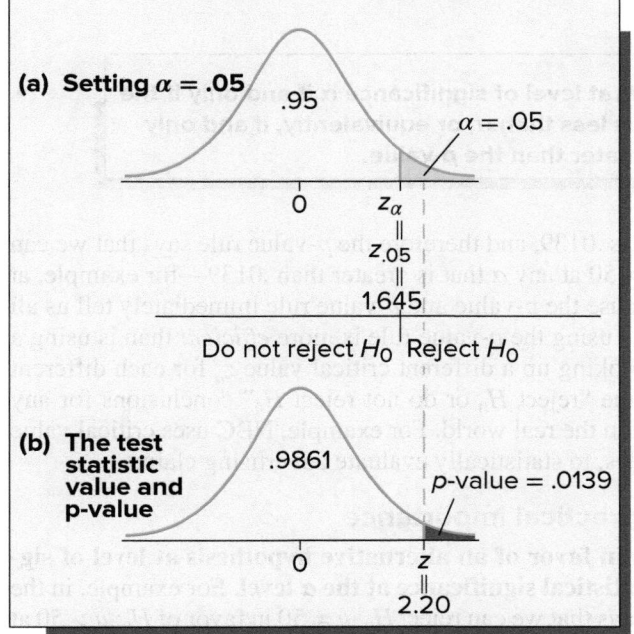

FIGURE 10.1 Testing H_0: $\mu = 50$ versus H_a: $\mu > 50$ by setting α Equal to .05

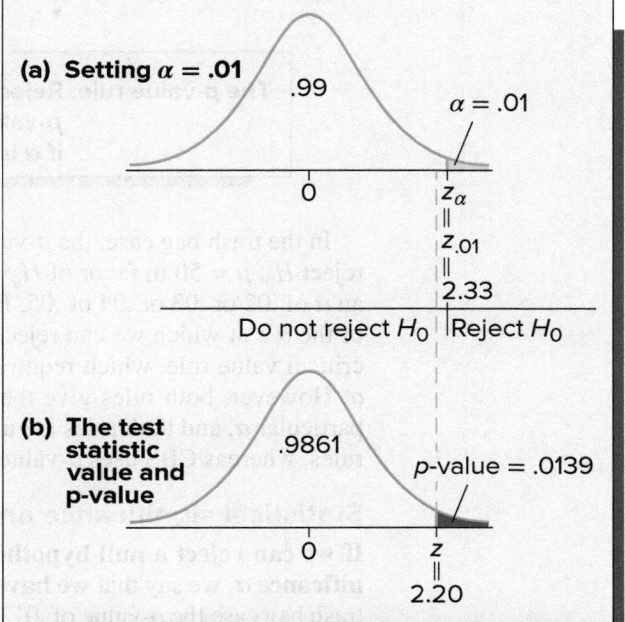

FIGURE 10.2 Testing H_0: $\mu = 50$ versus H_a: $\mu > 50$ by setting α Equal to .01

In the trash bag case, the computed test statistic value $z = 2.20$ is based on a sample mean of $\bar{x} = 52.167$, which is greater than 50. Therefore, to some extent $z = 2.20$ contradicts H_0: $\mu = 50$ and supports H_a: $\mu > 50$. A test statistic value that is at least as contradictory to H_0: $\mu = 50$ as $z = 2.20$ would be based on a sample mean that is at least as large as 52.167. But this is a test statistic value that would be greater than or equal to $z = 2.20$. It follows that the *p*-value is the probability, computed assuming that H_0: $\mu = 50$ is true, of observing a test statistic value that is **greater than or equal to $z = 2.20$.** Because the distribution of all possible values of the test statistic is a standard normal distribution if H_0: $\mu = 50$ is true, this *p*-value can be calculated as follows:

> **The *p*-value for testing H_0: $\mu = 50$ versus H_a: $\mu > 50$ is the area under the standard normal curve to the right of $z = 2.20$.**

As illustrated in Figure 10.1(b), the area under the standard normal curve to the right of $z = 2.20$ (and thus the *p*-value) is $1 - .9861 = .0139$ (see Table A.3). The *p*-value of .0139 says that, if H_0: $\mu = 50$ is true, then only 1.39 percent of all possible test statistic values are at least as large and contradictory to H_0 as the value $z = 2.20$. That is, if we are to believe that H_0 is true, we must believe that we have observed a test statistic value that can be described as having a 1.39 percent chance of occurring. It is difficult to believe that we have observed such a small chance. Moreover, the *p*-value of .0139 tells us for any α—without having to find the numerical value of the critical value z_α—whether the computed test statistic value $z = 2.20$ is greater than z_α and thus whether the critical value rule says to reject H_0: $\mu = 50$. To see this, compare Figures 10.1 (a) and (b). Doing this, we see that because the *p*-value of .0139 (the red area) is less than the α of .05 (the blue area), the test statistic value $z = 2.20$ is greater than the critical value $z_{.05}$, and thus the critical value rule says to reject H_0: $\mu = 50$. Also, compare Figures 10.2 (a) and (b). Doing this, we see that because the *p*-value of .0139 (the red area) is greater than the α of .01 (the blue area), the test statistic value $z = 2.20$ is less than the critical value $z_{.01}$, and thus the critical value rule says to not reject H_0: $\mu = 50$. Furthermore, generalizing what is shown in Figures 10.1 and 10.2, we conclude that as long as the *p*-value of .0139 is less than α, the computed test statistic value $z = 2.20$ will be greater than the critical value z_α, and thus the critical value rule will say to reject H_0: $\mu =$ 50. Noting that the **probability of a Type I error** for a hypothesis test is called the **level of significance** of the test, it follows that we have the following **p-value rule** (which we will see holds for all hypothesis tests):

> **The p-value rule: Reject H_0 at level of significance α if and only if the p-value is less than α, or equivalently, if and only if α is greater than the p-value.**

In the trash bag case, the p-value is .0139, and therefore the p-value rule says that we can reject H_0: $\mu = 50$ in favor of H_a: $\mu > 50$ at any α that is greater than .0139—for example, at an α of .02 or .03 or .04 or .05. Because the p-value and p-value rule immediately tell us all of the α's at which we can reject H_0, using the p-value rule is more *efficient* than is using a critical value rule, which requires looking up a different critical value z_α for each different α. However, both rules give the same "reject H_0 or do not reject H_0" conclusions for any particular α, and both rules are used in the real world. For example, NBC uses critical value rules, whereas CBS uses p-value rules, to statistically evaluate advertising claims.

Statistical significance and practical importance

If we can reject a null hypothesis in favor of an alternative hypothesis at level of significance α, we say that **we have statistical significance at the α level**. For example, in the trash bag case the p-value of .0139 says that we can reject H_0: $\mu = 50$ in favor of H_0: $\mu > 50$ at any α greater than .0139, and therefore we have statistical significance at any α level greater than .0139. In addition, this statistical significance is **practically important** to the trash bag manufacturer because it means that any television network specifying an α greater than .0139 will conclude that the new trash bag is the strongest on the market and thus run the trash bag commercial. However, to help consumers decide whether they think that the new trash bag being the strongest on the market is practically important, networks will require that the point estimate of the new bag's mean breaking strength, $\bar{x} = 52.167$, be included (in rounded form.) in the commercial. The $\bar{x}$ of 52.167 estimates that the new trash bag is the strongest on the market by only about 2 pounds. Of course, the new bag's over 50 pound mean breaking is still the largest on the market and will be practically important to many consumers who will feel that, because the new bag is also the least expensive and most environmental friendly bag on the market, it is definitely worth purchasing. On the other hand, to consumers who are looking only for a bag that holds much more than 50 pounds of trash, the statistical conclusions will not be practically important. Notice that, in general, assessing the practical importance of a statistically significant result is somewhat subjective and aided by estimating the parameter (for example, mean) being tested.

Measuring the weight of evidence against the null hypothesis

In addition to deciding whether H_0: $\mu = 50$ can be rejected in favor of H_a: $\mu > 50$ at each television network's chosen α, the trash bag manufacturer would almost certainly wish to know exactly how much evidence there is that its new trash bag is the strongest on the market. To measure this evidence, recall that the p-value for testing a null hypothesis H_0 is the probability of observing the computed test statistic value, or a test statistic value more contradictory to H_0, if H_0 is true. The smaller the p-value is, the less likely we were to have observed the computed test statistic value if H_0 is true, and thus the stronger is the evidence that H_0 is not true. Although there are really no sharp borders between different **weights of evidence,** some statisticians consider the potential values .10, .05, .01, and .001 of the p-value to represent, respectively, mildly strong evidence, strong evidence, very strong evidence, and extremely strong evidence against H_0. For example, the p-value of .0139 in the trash bag case is somewhat less than .05 and slightly more than .01. Therefore, we might conclude that there is somewhat more than strong evidence, but not quite very strong evidence, that H_0: $\mu = 50$ is false and thus that the new trash bag is the strongest on the market.

A general *z* test procedure

Consider the e-billing case. We have seen that in order to study whether the new electronic billing system reduces the mean payment time by more than 50 percent, the management consulting firm wishes to test $H_0: \mu \geq 19.5$ versus $H_a: \mu < 19.5$. Similar to the trash bag case, where we tested $H_0: \mu \leq 50$ versus $H_a: \mu > 50$ by testing the modified null hypothesis $H_0: \mu = 50$ versus $H_a: \mu > 50$, in the e-billing case we will test $H_0: \mu \geq 19.5$ versus $H_a: \mu < 19.5$ by testing the modified null hypothesis $H_0: \mu = 19.5$ versus $H_a: \mu < 19.5$. In general, let μ_0 be a specific number such as 50 or 19.5. Then, in order to test the alternative hypothesis $H_a: \mu > \mu_0$ or the alternative hypothesis $H_a: \mu < \mu_0$, we modify the null hypothesis to be $H_0: \mu = \mu_0$. The summary box below gives the critical value rule and the *p*-value rule for testing $H_0: \mu = \mu_0$ versus any one of the alternative hypotheses $H_a: \mu > \mu_0$, $H_a: \mu < \mu_0$, or $H_a: \mu \neq \mu_0$. (As an example of $H_a: \mu \neq \mu_0$, recall in the Valentine's Day chocolate case that we wish to test $H_0: \mu = 330$ versus $H_a: \mu \neq 330$.)

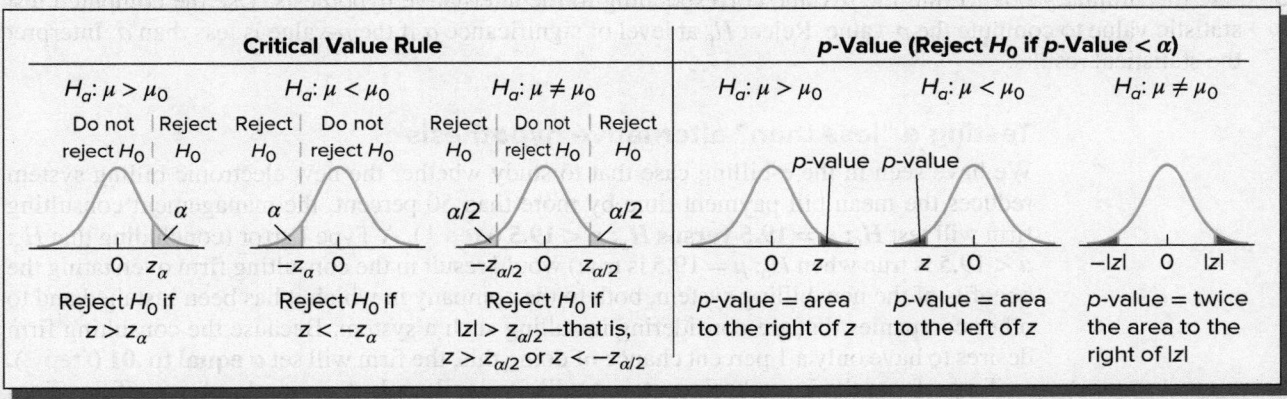

Testing a Hypothesis about a Population Mean When σ Is Known

Null Hypothesis $H_0: \mu = \mu_0$ **Test Statistic** $z = \dfrac{\bar{x} - \mu_0}{\sigma/\sqrt{n}}$ **Assumptions**   Normal population or Large sample size

The procedures given in the summary box are based on the assumption that either the sampled population is normally distributed or the sample size is large. Under this assumption, the sampling distribution of $z = (\bar{x} - \mu_0)/(\sigma/\sqrt{n})$ is exactly or approximately a standard normal distribution if $H_0: \mu = \mu_0$ is true, and thus we use z as the test statistic for performing the desired hypothesis test. Such a test is called a **z test** and requires that the *true value of the population standard deviation σ is known*. To use the summary box to test a particular alternative hypothesis, we identify the type of alternative hypothesis, and then we look in the summary box under the appropriate heading—either $H_a: \mu > \mu_0$, $H_a: \mu < \mu_0$, or $H_a: \mu \neq \mu_0$—to find the appropriate critical value rule and *p*-value. Examining the summary box, we see that the appropriate critical value rule and *p*-value will be either **right tailed, left tailed,** or **two tailed.** Furthermore, note that for any of the alternative hypotheses, the *p*-value rule says to reject $H_0: \mu = \mu_0$ if and only if the *p*-value is less than α. However, the critical value rule, as well as the *p*-value used in the *p*-value rule, depend upon the alternative hypothesis. For example, consider testing $H_0: \mu = 50$ versus $H_a: \mu > 50$ in the trash bag case. Because $H_a: \mu > 50$ is of the form $H_a: \mu > \mu_0$, we look in the summary box under the critical value rule heading $H_a: \mu > \mu_0$ and find that the appropriate critical value rule is a **right-tailed critical value rule.** This rule, which is the critical value rule that we have previously used, says to reject $H_0: \mu = 50$ in favor of $H_a: \mu > 50$ if and only if the computed test statistic value z is greater than the critical value z_α. Similarly, looking in the summary box under the *p*-value heading $H_a: \mu > \mu_0$, the appropriate *p*-value is a **right-tailed *p*-value.** This *p*-value, which we have previously computed, is the area under the standard normal curve to the right of the computed test statistic value z. In the next two subsections we will discuss using the critical value rules and p-values in the summary box to test a "less than" alternative hypothesis (H_a: $\mu < \mu_0$) and a "not equal to" alternative hypothesis (H_a: $\mu \neq \mu_0$). Moreover, throughout this book we will (formally or informally) use the five steps below to implement the critical value and p-value approaches to hypothesis testing.

The Five Steps of Hypothesis Testing

1   State the null hypothesis H_0 and the alternative hypothesis H_a.
2   Specify the level of significance α.
3   Plan the sampling procedure and select the test statistic.

Using a critical value rule:

4   Use the summary box to find the critical value rule corresponding to the alternative hypothesis.
5   Collect the sample data, compute the value of the test statistic. and decide whether to reject H_0 by using the critical value rule. Interpret the statistical results.

Using a p-value rule:

4   Collect the sample data and compute the value of the test statistic.
5   Use the summary box to find the p-value corresponding to the alternative hypothesis. Use the computed test statistic value to compute the p-value. Reject H_0 at level of significance α if the p-value is less than α. Interpret the statistical results.

Testing a "less than" alternative hypothesis

We have seen in the e-billing case that to study whether the new electronic billing system reduces the mean bill payment time by more than 50 percent, the management consulting firm will test H_0: $\mu = 19.5$ versus H_a: $\mu < 19.5$ (step 1). A Type I error (concluding that H_a: $\mu < 19.5$ is true when H_0: $\mu = 19.5$ is true) would result in the consulting firm overstating the benefits of the new billing system, both to the company in which it has been installed and to other companies that are considering installing such a system. Because the consulting firm desires to have only a 1 percent chance of doing this, the firm will set α **equal to .01** (step 2).

To perform the hypothesis test, we will randomly select a sample of $n = 65$ invoices paid using the new billing system and calculate the mean $\bar{x}$ of the payment times of these invoices. Then, because the sample size is large, we will utilize the **test statistic in the summary box** (step 3):

$$z = \frac{\bar{x} - 19.5}{\sigma/\sqrt{n}}$$

A value of the test statistic z that is less than zero results when $\bar{x}$ is less than 19.5. This implies that μ might be less than 19.5 and thus provides evidence to support rejecting H_0: $\mu = 19.5$ in favor of H_a: $\mu < 19.5$. To decide how much less than zero the value of the test statistic must be to reject H_0 in favor of H_a at level of significance α, we note that H_a: $\mu < 19.5$ is of the form H_a: $\mu < \mu_0$, and we look in the summary box under the critical value rule heading H_a: $\mu < \mu_0$. The critical value rule that we find is a **left-tailed critical value rule** and says to do the following:

- Place the probability of a Type I error, α, in the left-hand tail of the standard normal curve and use the normal table to find the critical value $-z_\alpha$. Here $-z_\alpha$ is the negative of the normal point z_α. That is, $-z_\alpha$ is the point on the horizontal axis under the standard normal curve that gives a left-hand tail area equal to α.

- **Reject H_0: $\mu = 19.5$ in favor of H_a: $\mu < 19.5$ if and only if the computed value of the test statistic z is less than the critical value $-z_\alpha$ (step 4).** Because α equals .01, the critical value $-z_\alpha$ is $-z_{.01} = -2.33$ [see Table A.3 and Figure 10.3(a)].

FIGURE 10.3 Testing $H_0: \mu = 19.5$ versus $H_a: \mu < 19.5$ by Using a Critical Value and a p-Value

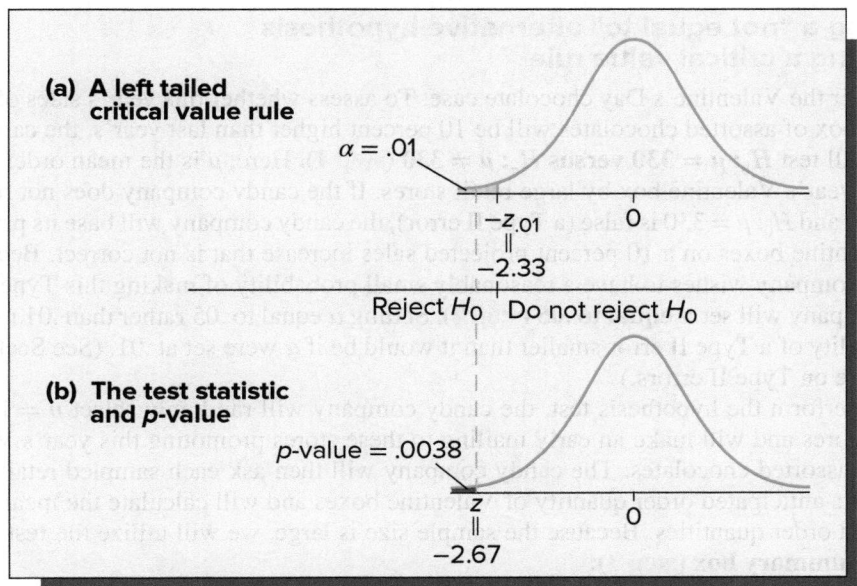

When the sample of $n = 65$ invoices is randomly selected, the mean of the payment times of these invoices is calculated to be $\bar{x} = 18.1077$ days. Assuming that the population standard deviation σ of payment times for the new electronic billing system is 4.2 days (as discussed in Example 8.3), the **value of the test statistic** is

$$z = \frac{\bar{x} - 19.5}{\sigma/\sqrt{n}} = \frac{18.1077 - 19.5}{4.2/\sqrt{65}} = -2.67$$

Because $z = -2.67$ is less than the critical value $-z_{.01} = -2.33$, we can reject H_0: $\mu = 19.5$ in favor of $H_a: \mu < 19.5$ by setting α equal to .01 (step 5). Therefore, we conclude (at an α of .01) that the population mean payment time for the new electronic billing system is less than 19.5 days. This, along with the fact that the sample mean $\bar{x} = 18.1077$ is slightly less than 19.5, implies that it is reasonable for the management consulting firm to conclude (and claim) that the new electronic billing system has reduced the population mean payment time by slightly more than 50 percent.

To find the p-value for the hypothesis test, note that the computed test statistic value $z = -2.67$ is based on a sample mean of $\bar{x} = 18.1077$, which is less than 19.5. Therefore, to some extent $z = -2.67$ contradicts $H_0: \mu = 19.5$ and supports $H_a: \mu < 19.5$. A test statistic value that is at least as contradictory to $H_0: \mu = 19.5$ as $z = -2.67$ would be based on a sample mean that is as small as or smaller than 18.1077. But this is a test statistic value that would be less than or equal to $z = -2.67$. It follows that the p-value is the probability, computed assuming that $H_0: \mu = 19.5$ is true, of observing a test statistic value that is *less than or equal to* $z = -2.67$. As shown in the summary box under the p-value heading H_a: $\mu < \mu_0$, this p-value can be calculated as follows:

> **The p-value for testing $H_0: \mu = 19.5$ versus $H_a: \mu < 19.5$ is the area under the standard normal curve to the left of $z = -2.67$.**

Using Table A.3, we find that this area—and thus the p-value—is .0038, as shown in Figure 10.3(b). The p-value of .0038 says that if $H_0: \mu = 19.5$ is true, then only 38 in 10,000 of all possible test statistic values are at least as negative and contradictory to H_0 as the value $z = -2.67$. Moreover, recall that the management consulting firm has set α equal to .01. **Because the p-value of .0038 is less than the α of .01, we can reject $H_0: \mu = 19.5$ in favor of $H_a: \mu < 19.5$ by setting α equal to .01** (step 5). In addition, because the p-value of

.0038 is less than .01 but not less than .001, we have very strong evidence, but not extremely strong evidence, that H_0 is false and H_a is true.

Testing a "not equal to" alternative hypothesis by using a critical value rule

Consider the Valentine's Day chocolate case. To assess whether this year's sales of its Valentine box of assorted chocolates will be 10 percent higher than last year's, the candy company will test $H_0: \mu = 330$ versus $H_a: \mu \neq 330$ (step 1). Here, μ is the mean order quantity of this year's Valentine box by large retail stores. If the candy company does not reject $H_0: \mu = 330$ and $H_0: \mu = 330$ is false (a Type II error), the candy company will base its production of Valentine boxes on a 10 percent projected sales increase that is not correct. Because the candy company wishes to have a reasonably small probability of making this Type II error, the company will set α equal to .05 (step 2). Setting α equal to .05 rather than .01 makes the probability of a Type II error smaller than it would be if α were set at .01. (See Section 10.5 for more on Type II errors.)

To perform the hypothesis test, the candy company will randomly select $n = 100$ large retail stores and will make an early mailing to these stores promoting this year's Valentine box of assorted chocolates. The candy company will then ask each sampled retail store to report its anticipated order quantity of Valentine boxes and will calculate the mean $\bar{x}$ of the reported order quantities. Because the sample size is large, we will utilize the **test statistic in the summary box** (step 3):

$$z = \frac{\bar{x} - 330}{\sigma / \sqrt{n}}$$

A value of the test statistic that is greater than 0 results when $\bar{x}$ is greater than 330. This implies that μ might be greater than 330 and thus provides evidence to support rejecting $H_0: \mu = 330$ in favor of $H_a: \mu \neq 330$. Similarly, a value of the test statistic that is less than 0 results when $\bar{x}$ is less than 330. This implies that μ might be less than 330 and thus also provides evidence to support rejecting $H_0: \mu = 330$ in favor of $H_a: \mu \neq 330$. To decide how different from zero (positive or negative) the value of the test statistic must be in order to reject H_0 in favor of H_a at level of significance α, we note that $H_a: \mu \neq 330$ is of the form $H_a: \mu \neq \mu_0$, and we look in the summary box under the critical value rule heading $H_0: \mu \neq \mu_0$. The critical value rule that we find is a **two-tailed critical value rule** and says to do the following:

- Divide the probability of a Type I error, α, into two equal parts, and place the area $\alpha/2$ in the right-hand tail of the standard normal curve and the area $\alpha/2$ in the left-hand tail of the standard normal curve. Then use the normal table to find the critical values $z_{\alpha/2}$ and $-z_{\alpha/2}$. Here $z_{\alpha/2}$ is the point on the horizontal axis under the standard normal curve that gives a right-hand tail area equal to $\alpha/2$, and $-z_{\alpha/2}$ is the point giving a left-hand tail area equal to $\alpha/2$.
- Reject $H_0: \mu = 330$ in favor of $H_a: \mu \neq 330$ if and only if the computed value of the test statistic z is greater than the critical value $z_{\alpha/2}$ or less than the critical value $-z_{\alpha/2}$ (step 4). Note that this is equivalent to saying that we should **reject H_0 if and only if the absolute value of the computed value of the test statistic z is greater than the critical value $z_{\alpha/2}$.** Because α equals .05, the critical values are [see Table A.3 and Figure 10.4(a)]

$$z_{\alpha/2} = z_{.05/2} = z_{.025} = 1.96 \quad \text{and} \quad -z_{\alpha/2} = -z_{.025} = -1.96.$$

When the sample of $n = 100$ large retail stores is randomly selected, the mean of their reported order quantities is calculated to be $\bar{x} = 326$ boxes. Assuming that the population standard deviation σ of large retail store order quantities for this year's Valentine box will be 40 boxes (the same as it was for previous years' Valentine boxes), the value of the test statistic is

$$z = \frac{\bar{x} - 330}{\sigma / \sqrt{n}} = \frac{326 - 330}{40 / \sqrt{100}} = -1$$

FIGURE 10.4   Testing $H_0: \mu = 330$ versus $H_a: \mu \neq 330$ by Using Critical Values and the p-Value

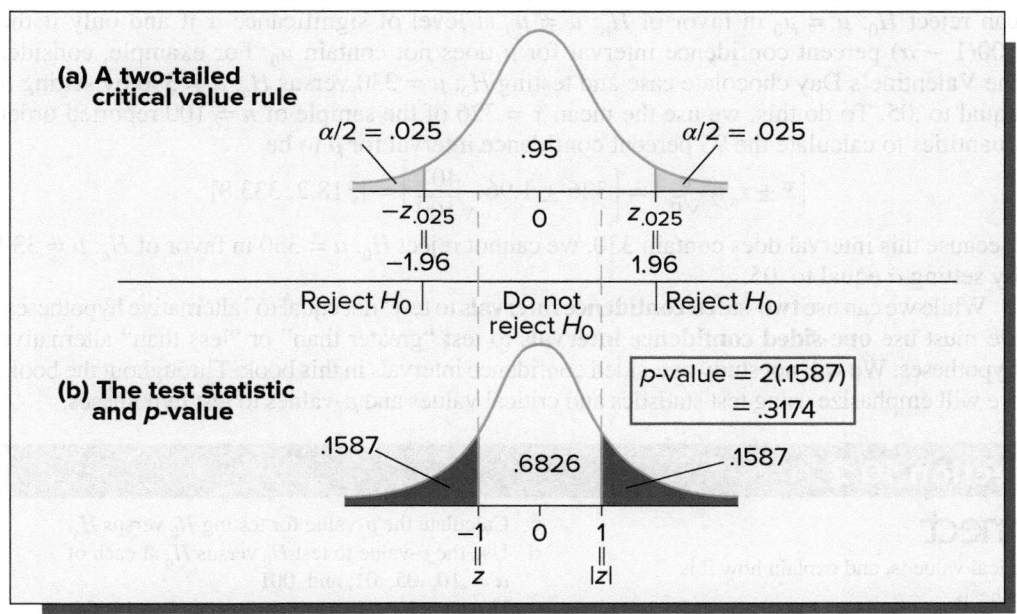

Because $z = -1$ is between the critical values $-z_{.025} = -1.96$ and $z_{.025} = 1.96$ (or, equivalently, because $|z| = 1$ is less than $z_{.025} = 1.96$), we cannot reject $H_0: \mu = 330$ in favor of $H_a: \mu \neq 330$ by setting α equal to .05 (step 5). Therefore, we cannot conclude (at an α of .05) that the population mean order quantity of this year's Valentine box by large retail stores will differ from 330 boxes. It follows that the candy company will base its production of Valentine boxes on the 10 percent projected sales increase.

To find the p-value for the hypothesis test, note that the value of the test statistic computed from the sample data is $z = -1$. Because the alternative hypothesis $H_a: \mu \neq 330$ says that μ might be greater or less than 330, both positive and negative test statistic values contradict $H_0: \mu = 330$ and support $H_a: \mu \neq 330$. It follows that a value of the test statistic that is at least as contradictory to H_0 and supportive of H_a as $z = -1$ is a value of the test statistic that is greater than or equal to 1 or less than or equal to -1. Therefore, the p-value is the probability, computed assuming that $H_0: \mu = 330$ is true, of observing a value of the test statistic that is greater than or equal to 1 or less than or equal to -1. As illustrated in Figure 10.4(b), this p-value equals the area under the standard normal curve to the right of 1, plus the area under this curve to the left of -1. But, by the symmetry of the normal curve, the sum of these two areas, and thus the p-value, is as follows:

> The p-value for testing $H_0: \mu = 330$ versus $H_a: \mu \neq 330$ is twice the area under the standard normal curve to the right of $|z| = 1$, the absolute value of the test statistic.

Note that this p-value is shown in the summary box under the p-value heading $H_a: \mu \neq \mu_0$. Because the area under the standard normal curve to the right of $|z| = 1$ is $1 - .8413 = .1587$ (see Table A.3), the p-value is $2(.1587) = .3174$. The p-value of .3174 says that, if $H_0: \mu = 330$ is true, then 31.74 percent of all possible test statistic values are at least as contradictory to H_0 as $z = -1$. Moreover, recall that the candy company has set α equal to .05. **Because the p-value of .3174 is greater than the α of .05, we cannot reject $H_0: \mu = 330$ in favor of $H_a: \mu \neq 330$ by setting α equal to .05 (step 5).** In fact, the p-value of .3174 provides very little evidence against H_0 and in favor of H_a.

Using confidence intervals to test hypotheses

Confidence intervals can be used to test hypotheses. Specifically, it can be proven that we can reject H_0: $\mu = \mu_0$ in favor of H_a: $\mu \neq \mu_0$ at level of significance α if and only if the $100(1 - \alpha)$ percent confidence interval for μ does not contain μ_0. For example, consider the Valentine's Day chocolate case and testing H_0: $\mu = 330$ versus H_a: $\mu \neq 330$ by setting α equal to .05. To do this, we use the mean $\bar{x} = 326$ of the sample of $n = 100$ reported order quantities to calculate the 95 percent confidence interval for μ to be

$$\left[\bar{x} \pm z_{\alpha/2} \frac{\sigma}{\sqrt{n}}\right] = \left[326 \pm 1.96 \frac{40}{\sqrt{100}}\right] = [318.2, 333.8]$$

Because this interval does contain 330, we cannot reject H_0: $\mu = 330$ in favor of H_a: $\mu \neq 330$ by setting α equal to .05.

While we can use **two-sided confidence intervals** to test "not equal to" alternative hypotheses, we must use **one-sided confidence intervals** to test "greater than" or "less than" alternative hypotheses. We will not study one-sided confidence intervals in this book. Throughout the book, we will emphasize using test statistics and critical values and p-values to test hypotheses.

Exercises for Section 10.2

CONCEPTS   connect

10.6 Explain what a critical value is, and explain how it is used to test a hypothesis.

10.7 Explain what a p-value is, and explain how it is used to test a hypothesis.

METHODS AND APPLICATIONS

10.8 Suppose that we wish to test H_0: $\mu = 80$ versus H_a: $\mu > 80$, where the population standard deviation is known to equal 20. Also, suppose that a sample of 100 measurements randomly selected from the population has a mean equal to 85.
   a Calculate the value of the test statistic z.
   b By comparing z with a critical value, test H_0 versus H_a at $\alpha = .05$.
   c Calculate the p-value for testing H_0 versus H_a.
   d Use the p-value to test H_0 versus H_a at each of $\alpha = .10, .05, .01,$ and .001.

10.9 In Exercise 10.8, how much evidence is there that H_0: $\mu = 80$ is false and H_a: $\mu > 80$ is true?

10.10 Suppose that we wish to test H_0: $\mu = 20$ versus H_a: $\mu < 20$, where the population standard deviation is known to equal 7. Also, suppose that a sample of 49 measurements randomly selected from the population has a mean equal to 18.
   a Calculate the value of the test statistic z.
   b By comparing z with a critical value, test H_0 versus H_a at $\alpha = .01$.
   c Calculate the p-value for testing H_0 versus H_a.
   d Use the p-value to test H_0 versus H_a at each of $\alpha = .10, .05, .01,$ and .001.

10.11 In Exercise 10.10, how much evidence is there that H_0: $\mu = 20$ is false and H_a: $\mu < 20$ is true?

10.12 Suppose that we wish to test H_0: $\mu = 40$ versus H_a: $\mu \neq 40$, where the population standard deviation is known to equal 18. Also, suppose that a sample of 81 measurements randomly selected from the population has a mean equal to 35.
   a Calculate the value of the test statistic z.
   b By comparing z with critical values, test H_0 versus H_a at $\alpha = .05$.

   c Calculate the p-value for testing H_0 versus H_a.
   d Use the p-value to test H_0 versus H_a at each of $\alpha = .10, .05, .01,$ and .001.
   e How much evidence is there that H_0: $\mu = 40$ is false and H_a: $\mu \neq 40$ is true?

10.13 THE VIDEO GAME SATISFACTION RATING CASE VideoGame

Recall that "very satisfied" customers give the XYZ-Box video game system a rating that is at least 42. Suppose that the manufacturer of the XYZ-Box wishes to use the random sample of 65 satisfaction ratings to provide evidence supporting the claim that the mean composite satisfaction rating for the XYZ-Box exceeds 42.
   a Letting μ represent the mean composite satisfaction rating for the XYZ-Box, set up the null hypothesis H_0 and the alternative hypothesis H_a needed if we wish to attempt to provide evidence supporting the claim that μ exceeds 42.
   b The random sample of 65 satisfaction ratings yields a sample mean of 42.954. Assuming that the population standard deviation equals 2.64, use critical values to test H_0 versus H_a at each of $\alpha = .10, .05, .01,$ and .001.
   c Using the information in part b, calculate the p-value and use it to test H_0 versus H_a at each of $\alpha = .10, .05, .01,$ and .001.
   d How much evidence is there that the mean composite satisfaction rating exceeds 42?

10.14 THE BANK CUSTOMER WAITING TIME CASE WaitTime

Recall that a bank manager has developed a new system to reduce the time customers spend waiting for teller service during peak hours. The manager hopes the new system will reduce waiting times from the current 9 to 10 minutes to less than 6 minutes.

Suppose the manager wishes to use the random sample of 100 waiting times to support the claim that the mean waiting time under the new system is shorter than six minutes.

a Letting μ represent the mean waiting time under the new system, set up the null and alternative hypotheses needed if we wish to attempt to provide evidence supporting the claim that μ is shorter than six minutes.

b The random sample of 100 waiting times yields a sample mean of 5.46 minutes. Assuming that the population standard deviation equals 2.47 minutes, use critical values to test H_0 versus H_a at each of $\alpha = .10, .05, .01,$ and .001.

c Using the information in part *b*, calculate the *p*-value and use it to test H_0 versus H_a at each of $\alpha = .10, .05, .01,$ and .001.

d How much evidence is there that the new system has reduced the mean waiting time to below six minutes?

10.15 Consolidated Power, a large electric power utility, has just built a modern nuclear power plant. This plant discharges wastewater that is allowed to flow into the Atlantic Ocean. The Environmental Protection Agency (EPA) has ordered that the wastewater may not be excessively warm so that thermal pollution of the marine environment near the plant can be avoided. Because of this order, the wastewater is allowed to cool in specially constructed ponds and is then released into the ocean. This cooling system works properly if the mean temperature of wastewater discharged is 60°F or cooler. Consolidated Power is required to monitor the temperature of the wastewater. A sample of 100 temperature readings will be obtained each day, and if the sample results cast a substantial amount of doubt on the hypothesis that the cooling system is working properly (the mean temperature of wastewater discharged is 60°F or cooler), then the plant must be shut down and appropriate actions must be taken to correct the problem.

a Consolidated Power wishes to set up a hypothesis test so that the power plant will be shut down when the null hypothesis is rejected. Set up the null hypothesis H_0 and the alternative hypothesis H_a that should be used.

b Suppose that Consolidated Power decides to use a level of significance of $\alpha = .05$, and suppose a random sample of 100 temperature readings is obtained. If the sample mean of the 100 temperature readings is 60.482, test H_0 versus H_a and determine whether the power plant should be shut down and the cooling system repaired. Perform the hypothesis test by using a critical value and a *p*-value. Assume that the population standard deviation equals 2.

10.16 Do part *b* of Exercise 10.15 if the sample mean equals 60.262.

10.17 Do part *b* of Exercise 10.15 if the sample mean equals 60.618.

10.18 An automobile parts supplier owns a machine that produces a cylindrical engine part. This part is supposed to have an outside diameter of three inches. Parts with diameters that are too small or too large do not meet customer requirements and must be rejected. Lately, the company has experienced problems meeting customer requirements. The technical staff feels that the mean diameter produced by the machine is off target. In order to verify this, a special study will randomly sample

40 parts produced by the machine. The 40 sampled parts will be measured, and if the results obtained cast a substantial amount of doubt on the hypothesis that the mean diameter equals the target value of three inches, the company will assign a problem-solving team to intensively search for the causes of the problem.

a The parts supplier wishes to set up a hypothesis test so that the problem-solving team will be assigned when the null hypothesis is rejected. Set up the null and alternative hypotheses for this situation.

b A sample of 40 parts yields a sample mean diameter of 3.006 inches. Assuming that the population standard deviation equals .016: **(1)** Use a critical value to test H_0 versus H_a by setting α equal to .05. **(2)** Should the problem-solving team be assigned? **(3)** Use a *p*-value to test H_0 versus H_a with $\alpha = .05$.

10.19 The Crown Bottling Company has just installed a new bottling process that will fill 16-ounce bottles of the popular Crown Classic Cola soft drink. Both overfilling and underfilling bottles are undesirable: Underfilling leads to customer complaints and overfilling costs the company considerable money. In order to verify that the filler is set up correctly, the company wishes to see whether the mean bottle fill, μ, is close to the target fill of 16 ounces. To this end, a random sample of 36 filled bottles is selected from the output of a test filler run. If the sample results cast a substantial amount of doubt on the hypothesis that the mean bottle fill is the desired 16 ounces, then the filler's initial setup will be readjusted.

a The bottling company wants to set up a hypothesis test so that the filler will be readjusted if the null hypothesis is rejected. Set up the null and alternative hypotheses for this hypothesis test.

b Suppose that Crown Bottling Company decides to use a level of significance of $\alpha = .01$, and suppose a random sample of 36 bottle fills is obtained from a test run of the filler. For each of the following four sample means—$\bar{x} = 16.05, \bar{x} = 15.96, \bar{x} = 16.02,$ and $\bar{x} = 15.94$—determine whether the filler's initial setup should be readjusted. In each case, use **(1)** a critical value, **(2)** a *p*-value, and **(3)** a confidence interval. Assume that the population standard deviation equals .1 oz.

10.20 THE DISK BRAKE CASE

National Motors has equipped the ZX-900 with a new disk brake system. We define μ to be the mean stopping distance (from a speed of 35 mph) of all ZX-900s. National Motors would like to claim that the ZX-900 achieves a shorter mean stopping distance than the 60 ft claimed by a competitor.

a Set up the null and alternative hypotheses needed to support National Motors' claim.

b A television network will allow National Motors to advertise its claim if the appropriate null hypothesis can be rejected at $\alpha = .05$. If a random sample of 81 ZX-900s has a mean stopping distance of 57.8 ft, will National Motors be allowed to advertise the claim? Assume that the population standard deviation equals 6.02 ft and justify your answer using both a critical value and a *p*-value.

LO10-4

Use critical values and *p*-values to perform a *t* test about a population mean when σ is unknown.

10.3 *t* Tests about a Population Mean: σ Unknown

If we do not know σ (which is usually the case), we can base a hypothesis test about μ on the sampling distribution of

$$\frac{\bar{x} - \mu}{s/\sqrt{n}}$$

If the sampled population is normally distributed (or if the sample size is large—at least 30), then this sampling distribution is exactly (or approximately) a ***t* distribution having $n - 1$ degrees of freedom.** This leads to the following results:

A *t* Test about a Population Mean: σ Unknown

| Null Hypothesis $H_0: \mu = \mu_0$ | Test Statistic $t = \dfrac{\bar{x} - \mu_0}{s/\sqrt{n}}$ | $df = n - 1$ | Assumptions | Normal population or Large sample size |

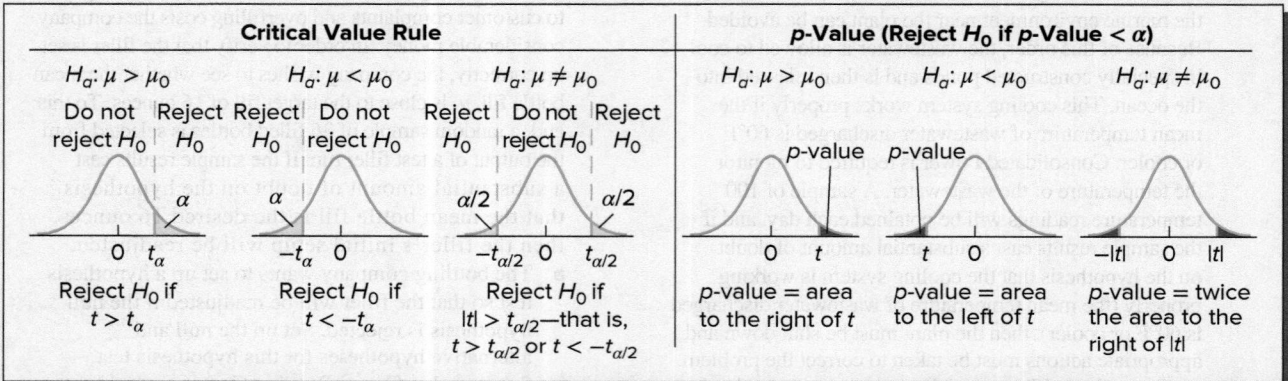

Critical Value Rule			*p*-Value (Reject H_0 if *p*-Value < α)		
$H_a: \mu > \mu_0$	$H_a: \mu < \mu_0$	$H_a: \mu \neq \mu_0$	$H_a: \mu > \mu_0$	$H_a: \mu < \mu_0$	$H_a: \mu \neq \mu_0$
Do not reject H_0 / Reject H_0	Reject H_0 / Do not reject H_0	Reject H_0 / Do not reject H_0 / Reject H_0	*p*-value	*p*-value	
Reject H_0 if $t > t_\alpha$	Reject H_0 if $t < -t_\alpha$	Reject H_0 if $\lvert t \rvert > t_{\alpha/2}$—that is, $t > t_{\alpha/2}$ or $t < -t_{\alpha/2}$	*p*-value = area to the right of *t*	*p*-value = area to the left of *t*	*p*-value = twice the area to the right of $\lvert t \rvert$

Stem-and-leaf display (page margin):

```
1.0 | 5
1.1 | 1 9
1.2 | 1 2 9
1.3 | 1 2 3 7
1.4 | 1 5 6
1.5 |
1.6 | 5
1.7 | 8
```

DS **DebtEq**

C **EXAMPLE 10.4** The Commercial Loan Case: Mean Debt-to-Equity Ratio

One measure of a company's financial health is its *debt-to-equity ratio*. This quantity is defined to be the ratio of the company's corporate debt to the company's equity. If this ratio is too high, it is one indication of financial instability. For obvious reasons, banks often monitor the financial health of companies to which they have extended commercial loans. Suppose that, in order to reduce risk, a large bank has decided to initiate a policy limiting the mean debt-to-equity ratio for its portfolio of commercial loans to being less than 1.5. In order to assess whether the mean debt-to-equity ratio μ of its (current) commercial loan portfolio is less than 1.5, the bank will test the **null hypothesis $H_0: \mu = 1.5$ versus the alternative hypothesis $H_a: \mu < 1.5$.** In this situation, a Type I error (rejecting $H_0: \mu = 1.5$ when $H_0: \mu = 1.5$ is true) would result in the bank concluding that the mean debt-to-equity ratio of its commercial loan portfolio is less than 1.5 when it is not. Because the bank wishes to be very sure that it does not commit this Type I error, it will test H_0 versus H_a by using a **.01 level of significance.** To perform the hypothesis test, the bank randomly selects a sample of 15 of its commercial loan accounts. Audits of these companies result in the following debt-to-equity ratios (arranged in increasing order): 1.05, 1.11, 1.19, 1.21, 1.22, 1.29, 1.31, 1.32, 1.33, 1.37, 1.41, 1.45, 1.46, 1.65, and 1.78. The mound-shaped stem-and-leaf display of these ratios is given in the page margin and indicates that the population of all debt-to-equity ratios is (approximately) normally distributed. It follows that it is appropriate to calculate the value of the **test statistic *t* in the summary box.** Furthermore, because the alternative hypothesis $H_a: \mu < 1.5$ says to use

FIGURE 10.5   Testing H_0: $\mu = 1.5$ versus H_a: $\mu < 1.5$ by Using a Critical Value and the *p*-Value

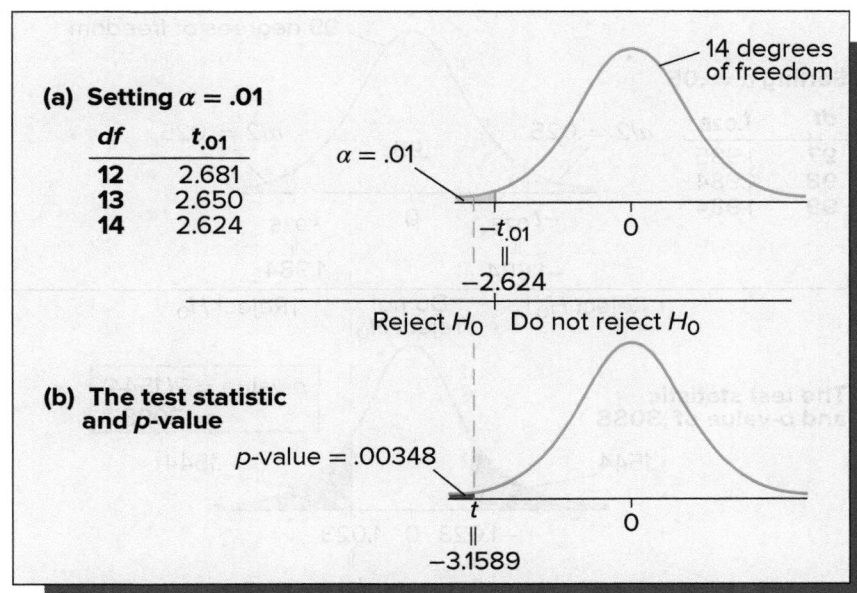

Test of mu = 1.5 vs < 1.5

Variable	N	Mean	StDev	SE Mean	95% Upper Bound	T	P
Ratio	15	1.3433	0.1921	0.0496	1.4307	−3.16	0.003

the left-tailed critical value rule in the summary box, we should **reject H_0: $\mu = 1.5$ if the value of *t* is less than the critical value $-t_\alpha = -t_{.01} = -2.624$.** Here, $-t_{.01} = -2.624$ is based on $n - 1 = 15 - 1 = 14$ degrees of freedom (see Table A.4), and this critical value is illustrated in Figure 10.5(a). The mean and the standard deviation of the random sample of $n = 15$ debt-to-equity ratios are $\bar{x} = 1.3433$ and $s = .1921$. This implies that the **value of the test statistic** is

$$t = \frac{\bar{x} - 1.5}{s/\sqrt{n}} = \frac{1.3433 - 1.5}{.1921/\sqrt{15}} = -3.1589$$

Because $t = -3.1589$ is less than $-t_{.01} = -2.624$, we reject H_0: $\mu = 1.5$ in favor of H_a: $\mu < 1.5$. That is, we conclude (at an α of .01) that the population mean debt-to-equity ratio of the bank's commercial loan portfolio is less than 1.5. This, along with the fact that the sample mean $\bar{x} = 1.3433$ is slightly less than 1.5, implies that it is reasonable for the bank to conclude that the population mean debt-to-equity ratio of its commercial loan portfolio is slightly less than 1.5.

The *p*-value for testing H_0: $\mu = 1.5$ versus H_a: $\mu < 1.5$ is the left-tailed *p*-value in the summary box: the area under the curve of the *t* distribution having 14 degrees of freedom to the left of $t = -3.1589$. Tables of *t* points (such as Table A.4) are not complete enough to give such areas for most *t* statistic values, so we use computer software packages to calculate *p*-values that are based on the *t* distribution. For example, Excel tells us that the *p*-value for testing H_0: $\mu = 1.5$ versus H_a: $\mu < 1.5$ is .00348, which is shown in Figure 10.5(b) and is rounded to .003 on the Minitab output at the bottom of Figure 10.5. The JMP output on the page margin rounds this *p*-value to .0035. The *p*-value of .00348 says that if we are to believe that H_0 is true, we must believe that we have observed a test statistic value that can be described as having a 348 in 100,000 chance of occurring. Moreover, because the *p*-value of .00348 is between .01 and .001, we have very strong evidence, but not extremely strong evidence, that H_0: $\mu = 1.5$ is false and H_a: $\mu < 1.5$ is true. That is, we have very strong evidence that the mean debt-to-equity ratio of the bank's commercial loan portfolio is less than 1.5.

Test Mean

Hypothesized Value	1.5
Actual Estimate	1.34333
DF	14
Std Dev	0.19208

t Test	
Test Statistic	-3.1589
Prob > |t|	0.0070*
Prob > t	0.9965
Prob < t	0.0035*

FIGURE 10.6   Testing $H_0: \mu = 330$ versus $H_a: \mu \neq 330$ by Using Critical Values and the p-Value

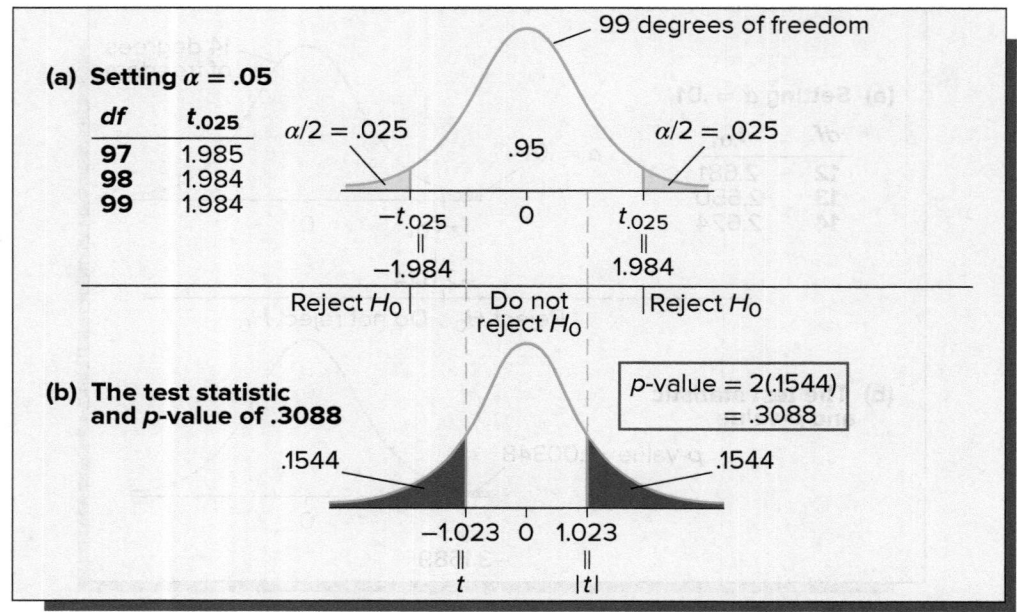

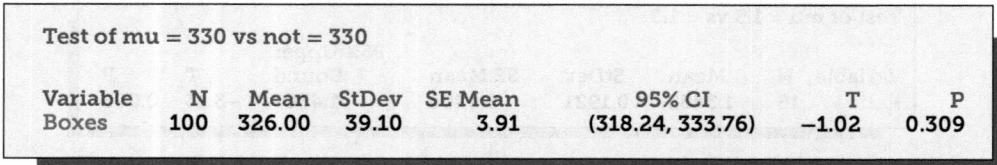

Variable	N	Mean	StDev	SE Mean	95% CI	T	P
Boxes	100	326.00	39.10	3.91	(318.24, 333.76)	−1.02	0.309

Test of mu = 330 vs not = 330

   Recall that in three cases discussed in Section 10.2 we tested hypotheses by assuming that the population standard deviation σ is known and by using z tests. If σ is actually not known in these cases (which would probably be true), we should test the hypotheses under consideration by using t tests. Furthermore, recall that in each case the sample size is large (at least 30). **In general, it can be shown that if the sample size is large, the t test is approximately valid even if the sampled population is not normally distributed (or mound shaped).** Therefore, consider the Valentine's Day chocolate case and testing $H_0: \mu = 330$ versus H_a: $\mu \neq 330$ at the .05 level of significance. To perform the hypothesis test, assume that we will randomly select $n = 100$ large retail stores and use their anticipated order quantities to calculate the value of the **test statistic t in the summary box.** Then, because the alternative hypothesis $H_a: \mu \neq 330$ says to use the two-tailed critical value rule in the summary box, we will **reject $H_0: \mu = 330$ if the absolute value of t is greater than $t_{\alpha/2} = t_{.025} = 1.984$ (based on $n - 1 = 99$ degrees of freedom)**—see Figure 10.6(a). Suppose that when the sample is randomly selected, the mean and the standard deviation of the $n = 100$ reported order quantities are calculated to be $\bar{x} = 326$ and $s = 39.1$. The **value of the test statistic** is

$$t = \frac{\bar{x} - 330}{s/\sqrt{n}} = \frac{326 - 330}{39.1/\sqrt{100}} = -1.023$$

Because $|t| = 1.023$ is less than $t_{.025} = 1.984$, we cannot reject $H_0: \mu = 330$ by setting α equal to .05. It follows that we cannot conclude (at an α of .05) that this year's population mean order quantity of the Valentine box by large retail stores will differ from 330 boxes. Therefore, the candy company will base its production of Valentine boxes on the 10 percent projected sales increase. The p-value for the hypothesis test is the two-tailed p-value in the summary box: twice the area under the t distribution curve having 99 degrees of freedom to the right of $|t| = 1.023$. Using a computer, we find that this p-value is .3088 (see the Minitab output in Figure 10.6(b) and the JMP output on the page margin). This p-value provides little evidence against $H_0: \mu = 330$ and in favor of $H_a: \mu \neq 330$. In Exercises 10.24 and 10.25, the reader will assume that σ is unknown in the e-billing and trash bag cases and will perform appropriate t tests in these cases.

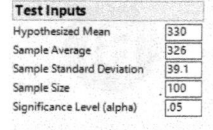

Test Inputs	
Hypothesized Mean	330
Sample Average	326
Sample Standard Deviation	39.1
Sample Size	100
Significance Level (alpha)	.05

Test Results	
Result	Value
t-score	-1.023
t Critical Values	+/- 1.9842
Observed Significance (p-value)	0.3088
Fail to Reject Null Hypothesis	

To conclude this section, note that if the sampled population is not approximately normally distributed and the sample size is not large, it might be appropriate to use a **nonparametric test about the population median** (see Chapter 18).

Exercises for Section 10.3

CONCEPTS connect

10.21 What assumptions must be met in order to carry out a *t* test about a population mean?

10.22 How do we decide whether to use a *z* test or a *t* test when testing a hypothesis about a population mean?

METHODS AND APPLICATIONS

10.23 Suppose that a random sample of nine measurements from a normally distributed population gives a sample mean of 2.57 and a sample standard deviation of .3. Use critical values to test H_0: $\mu = 3$ versus H_a: $\mu \neq 3$ using levels of significance $\alpha = .10$, $\alpha = .05$, $\alpha = .01$, and $\alpha = .001$.

10.24 Consider the e-billing case. The mean and the standard deviation of the sample of 65 payment times are 18.1077 and 3.9612, respectively. **(1)** Test H_0: $\mu = 19.5$ versus H_a: $\mu < 19.5$ by setting α equal to .01 and using a critical value rule. **(2)** Interpret the (computer calculated) *p*-value of .0031 for the test. That is, explain how much evidence there is that H_0 is false.
 PayTime

10.25 Consider the trash bag case. The mean and the standard deviation of the sample of 30 trash bag breaking strengths are 52.167 and 5.394, respectively. **(1)** Test H_0: $\mu = 50$ versus H_a: $\mu > 50$ by setting α equal to .05 and using a critical value rule. **(2)** Interpret the (computer calculated) *p*-value of .018 for the test. That is, explain how much evidence there is that H_0 is false.   TrashBag

10.26 THE AIR SAFETY CASE   AlertTimes

Recall that it is hoped that the mean alert time, μ, using the new display panel is less than eight seconds. **(1)** Formulate the null hypothesis H_0 and the alternative hypothesis H_a that would be used to attempt to provide evidence that μ is less than eight seconds. **(2)** Discuss the meanings of a Type I error and a Type II error in this situation. **(3)** The mean and the standard deviation of the sample of 15 alert times are 7.4 and 1.0261, respectively. Test H_0 versus H_a by setting α equal to .05 and using a critical value. Assume that the population of all alert times using the new display panel is approximately normally distributed. **(4)** Interpret the *p*-value of .02 for the test.

10.27 The *bad debt ratio* for a financial institution is defined to be the dollar value of loans defaulted divided by the total dollar value of all loans made. Suppose that a random sample of seven Ohio banks is selected and that the bad debt ratios (written as percentages) for these banks are 7%, 4%, 6%, 7%, 5%, 4%, and 9%.
 BadDebt

a Banking officials claim that the mean bad debt ratio for all Midwestern banks is 3.5 percent and that the mean bad debt ratio for Ohio banks is higher. **(1)** Set up the null and alternative hypotheses needed to attempt to provide evidence supporting the claim that the mean bad debt ratio for Ohio banks exceeds 3.5 percent. **(2)** Discuss the meanings of a Type I error and a Type II error in this situation.

b Assuming that bad debt ratios for Ohio banks are approximately normally distributed: **(1)** Use a critical value and the given sample information to test the hypotheses you set up in part *a* by setting α equal to .01. **(2)** Interpret the *p*-value of .006 for the test.

10.28 How might practical importance be defined for the situation in Exercise 10.27?

10.29 THE VIDEO GAME SATISFACTION RATING CASE   VideoGame

Recall that "very satisfied" customers give the XYZ-Box video game system a composite satisfaction rating that is at least 42.

a Letting μ represent the mean composite satisfaction rating for the XYZ-Box, set up the null and alternative hypotheses needed if we wish to attempt to provide evidence supporting the claim that μ exceeds 42.

b The mean and the standard deviation of a sample of 65 customer satisfaction ratings are 42.95 and 2.6424, respectively. **(1)** Use a critical value to test the hypotheses you set up in part *a* by setting α equal to .01. **(2)** Interpret the *p*-value of .0025 for the test.

10.30 THE BANK CUSTOMER WAITING TIME CASE
 WaitTime

Recall that a bank manager has developed a new system to reduce the time customers spend waiting for teller service during peak hours. The manager hopes the new system will reduce waiting times from the current 9 to 10 minutes to less than 6 minutes.

a Letting μ represent the mean waiting time under the new system, set up the null and alternative hypotheses needed if we wish to attempt to provide evidence supporting the claim that μ is shorter than 6 minutes.

b The mean and the standard deviation of a sample of 100 bank customer waiting times are 5.46 and 2.475, respectively. **(1)** Use a critical value to test the hypotheses you set up in part *a* by setting α equal to .05. **(2)** Interpret the *p*-value of .0158 for the test.

10.31 Consider a chemical company that wishes to determine whether a new catalyst, catalyst XA-100, changes the mean hourly yield of its chemical process from the historical process mean of 750 pounds per hour. When five trial runs are made using the new catalyst, the following yields (in pounds per hour) are recorded: 801, 814, 784, 836, and 820. 🟢 ChemYield

    a Letting μ be the mean of all possible yields using the new catalyst, set up the null and alternative hypotheses needed if we wish to attempt to provide evidence that μ differs from 750 pounds.

    b The mean and the standard deviation of the sample of 5 catalyst yields are 811 and 19.647, respectively. **(1)** Using a critical value and assuming approximate normality, test the hypotheses you set up in part *a* by setting α equal to .01. **(2)** The *p*-value for the hypothesis test is given in the Excel output to the right. Interpret this *p*-value.

t-statistic
6.942585
p-value
0.002261

10.32 Recall from Exercise 9.12 that Bayus (1991) studied the mean numbers of auto dealers visited by early and late replacement buyers. **(1)** Letting μ be the mean number of dealers visited by all late replacement buyers, set up the null and alternative hypotheses needed if we wish to attempt to provide evidence that μ differs from 4 dealers. **(2)** A random sample of 100 late replacement buyers gives a mean and a standard deviation of the number of dealers visited of 4.32 and .67, respectively. Use critical values to test the hypotheses you set up by setting α equal to .10, .05, .01, and .001. **(3)** Do we estimate that μ is less than 4 or greater than 4?

10.33 In 1991 the average interest rate charged by U.S. credit card issuers was 18.8 percent. Since that time, there has been a proliferation of new credit cards affiliated with retail stores, oil companies, alumni associations, professional sports teams, and so on. A financial officer wishes to study whether the increased competition in the credit card business has reduced interest rates. To do this, the officer will test a hypothesis about the current mean interest rate, μ, charged by all U.S. credit card issuers. To perform the hypothesis test, the officer randomly selects $n = 15$ credit cards and obtains the following interest rates (arranged in increasing order): 14.0, 14.6, 15.3, 15.6, 15.8, 16.4, 16.6, 17.0, 17.3, 17.6, 17.8, 18.1, 18.4, 18.7, and 19.2. A stem-and-leaf display of the interest rates is given at the top right, and the Minitab and Excel outputs for testing $H_0: \mu = 18.8$ versus $H_a: \mu < 18.8$ is given below.

14	06
15	368
16	46
17	0368
18	147
19	2

🟢 CreditCd

    a Set up the null and alternative hypotheses needed to provide evidence that mean interest rates have decreased since 1991.

    b Use the Minitab and Excel outputs and critical values to test the hypotheses you set up in part *a* at the .05, .01, and .001 levels of significance.

    c Use the Minitab and Excel outputs and a *p*-value to test the hypotheses you set up in part *a* at the .05, .01, and .001 levels of significance.

    d Based on your results in parts *b* and *c*, how much evidence is there that mean interest rates have decreased since 1991?

Test of mu = 18.8 vs < 18.8

Variable	N	Mean	StDev	SE Mean	T	P
Rate	15	16.8267	1.5378	0.3971	−4.97	0.000

t-statistic
−4.97
p-value
0.000103

LO10-5

Use critical values and p-values to perform a large sample z test about a population proportion.

10.4 *z* Tests about a Population Proportion

In this section we study a large sample hypothesis test about a population proportion (that is, about the fraction of population elements that possess some characteristic). We begin with an example.

🅲 **EXAMPLE 10.5** The Cheese Spread Case: Improving Profitability

Recall that the cheese spread producer has decided that replacing the current spout with the new spout is profitable only if p, the true proportion of all current purchasers who would stop buying the cheese spread if the new spout were used, is less than .10. The producer feels that it is unwise to change the spout unless it has very strong evidence that p is less than .10. Therefore, the spout will be changed if and only if the null hypothesis $H_0: p = .10$ can be rejected in favor of the alternative hypothesis $H_a: p < .10$ at the .01 level of significance.

In order to see how to test this kind of hypothesis, remember that when n is large, the sampling distribution of

$$\frac{\hat{p} - p}{\sqrt{\frac{p(1-p)}{n}}}$$

is approximately a standard normal distribution. Let p_0 denote a specified value between 0 and 1 (its exact value will depend on the problem), and consider testing the null hypothesis $H_0: p = p_0$. We then have the following result:

A Large Sample Test about a Population Proportion

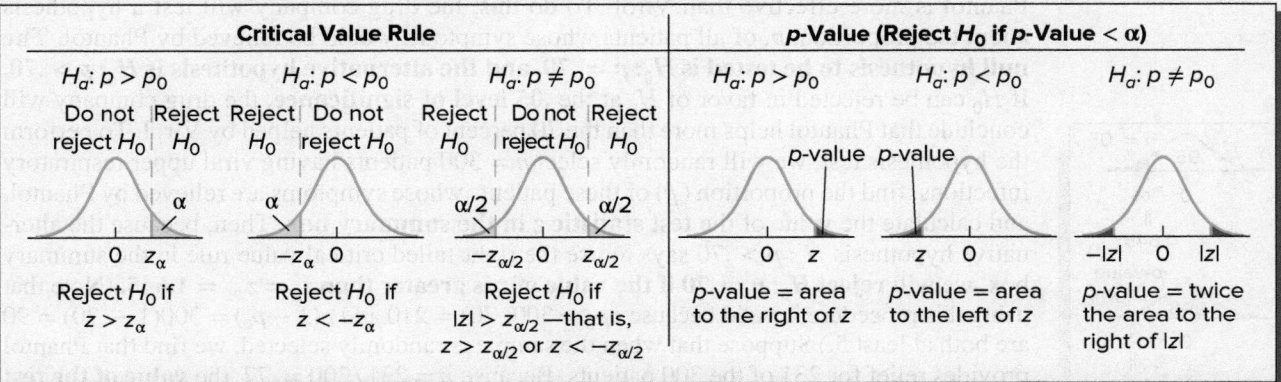

Null Hypothesis $H_0: p = p_0$

Test Statistic $z = \dfrac{\hat{p} - p_0}{\sqrt{\dfrac{p_0(1-p_0)}{n}}}$

Assumptions[2] $np_0 \geq 5$ and $n(1-p_0) \geq 5$

Critical Value Rule	p-Value (Reject H_0 if p-Value $< \alpha$)
$H_a: p > p_0$ — Do not reject H_0 / Reject H_0 — Reject H_0 if $z > z_\alpha$	$H_a: p > p_0$ — p-value = area to the right of z
$H_a: p < p_0$ — Reject H_0 / Do not reject H_0 — Reject H_0 if $z < -z_\alpha$	$H_a: p < p_0$ — p-value = area to the left of z
$H_a: p \neq p_0$ — Reject H_0 / Do not reject H_0 / Reject H_0 — Reject H_0 if $|z| > z_{\alpha/2}$—that is, $z > z_{\alpha/2}$ or $z < -z_{\alpha/2}$	$H_a: p \neq p_0$ — p-value = twice the area to the right of $|z|$

C EXAMPLE 10.6 The Cheese Spread Case: Improving Profitability

We have seen that the cheese spread producer wishes to test $H_0: p = .10$ versus $H_a: p < .10$, where p is the proportion of all current purchasers who would stop buying the cheese spread if the new spout were used. The producer will use the new spout if H_0 can be rejected in favor of H_a at the **.01 level of significance.** To perform the hypothesis test, we will randomly select $n = 1{,}000$ current purchasers of the cheese spread, find the proportion ($\hat{p}$) of these purchasers who would stop buying the cheese spread if the new spout were used, and calculate the value of the **test statistic z in the summary box.** Then, because the alternative hypothesis $H_a: p < .10$ says to use the left-tailed critical value rule in the summary box, we will **reject $H_0: p = .10$ if the value of z is less than** $-z_\alpha = -z_{.01} = -2.33$. (Note that using this procedure is valid because $np_0 = 1{,}000(.10) = 100$ and $n(1-p_0) = 1{,}000(1-.10) = 900$ are both at least 5.) Suppose that when the sample is randomly selected, we find that 63 of the 1,000 current purchasers say they would stop buying the cheese spread if the new spout were used. Because $\hat{p} = 63/1{,}000 = .063$, the **value of the test statistic** is

$$z = \frac{\hat{p} - p_0}{\sqrt{\frac{p_0(1-p_0)}{n}}} = \frac{.063 - .10}{\sqrt{\frac{.10(1-.10)}{1{,}000}}} = -3.90$$

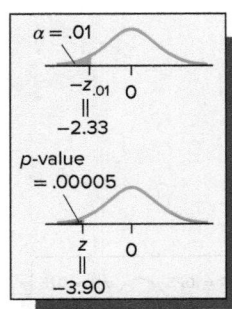

Because $z = -3.90$ is less than $-z_{.01} = -2.33$, we reject $H_0: p = .10$ in favor of $H_a: p < .10$. That is, we conclude (at an α of .01) that the proportion of all current purchasers who would stop buying the cheese spread if the new spout were used is less than .10. It follows that the company will use the new spout. Furthermore, the point estimate $\hat{p} = .063$ says we estimate that 6.3 percent of all current customers would stop buying the cheese spread if the new spout were used.

BI

---

[2]Some statisticians suggest using the more conservative rule that both np_0 and $n(1-p_0)$ must be at least 10.

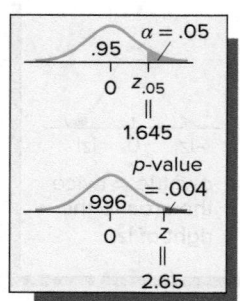

Test Inputs

Hypothesized Proportion	.01
Number of Successes	63
Sample Size	1000
Significance Level (alpha)	.01

Test Results

Result	Value
Estimated Proportion	0.063
z Critical Values	-2.3263
Test Statistic	-3.9001
Observed Significance (p-Value)	0
Reject Null Hypothesis	

Although the cheese spread producer has made its decision by setting α equal to a single, pre-chosen value (.01), it would probably also wish to know the weight of evidence against H_0 and in favor of H_a. The p-value for testing H_0: $p = .10$ versus H_a: $p < .10$ is the left-tailed p-value in the summary box: the area under the standard normal curve to the left of $z = -3.90$. Table A.3 tells us that this area is .00005, which is rounded to 0 in the JMP output on the page margin. Because this p-value is less than .001, we have extremely strong evidence that H_a: $p < .10$ is true. That is, we have extremely strong evidence that fewer than 10 percent of all current purchasers would stop buying the cheese spread if the new spout were used.

EXAMPLE 10.7 The Phantol Case: Testing the Effectiveness of a Drug

Recent medical research has sought to develop drugs that lessen the severity and duration of viral infections. Virol, a relatively new drug, has been shown to provide relief for 70 percent of all patients suffering from viral upper respiratory infections. A major drug company is developing a competing drug called Phantol. The drug company wishes to investigate whether Phantol is more effective than Virol. To do this, the drug company will test a hypothesis about the proportion, p, of all patients whose symptoms would be relieved by Phantol. **The null hypothesis to be tested is H_0: $p = .70$, and the alternative hypothesis is H_a: $p > .70$.** If H_0 can be rejected in favor of H_a at the **.05 level of significance,** the drug company will conclude that Phantol helps more than the 70 percent of patients helped by Virol. To perform the hypothesis test, we will randomly select $n = 300$ patients having viral upper respiratory infections, find the proportion ($\hat{p}$) of these patients whose symptoms are relieved by Phantol, and calculate the value of the **test statistic z in the summary box.** Then, because the alternative hypothesis H_a: $p > .70$ says to use the right-tailed critical value rule in the summary box, we will **reject H_0: $p = .70$ if the value of z is greater than $z_\alpha = z_{.05} = 1.645$.** (Note that using this procedure is valid because $np_0 = 300(.70) = 210$ and $n(1 - p_0) = 300(1 - .70) = 90$ are both at least 5.) Suppose that when the sample is randomly selected, we find that Phantol provides relief for 231 of the 300 patients. Because $\hat{p} = 231/300 = .77$, the **value of the test statistic** is

$$z = \frac{\hat{p} - p_0}{\sqrt{\frac{p_0(1 - p_0)}{n}}} = \frac{.77 - .70}{\sqrt{\frac{(.70)(1 - .70)}{300}}} = 2.65$$

Because $z = 2.65$ is greater than $z_{.05} = 1.645$, we reject H_0: $p = .70$ in favor of H_a: $p > .70$. That is, we conclude (at an α of .05) that Phantol will provide relief for more than 70 percent of all patients suffering from viral upper respiratory infections. More specifically, the point estimate $\hat{p} = .77$ of p says that we estimate that Phantol will provide relief for 77 percent of all such patients. Comparing this estimate to the 70 percent of all patients whose symptoms are relieved by Virol, we conclude that Phantol is somewhat more effective.

The p-value for testing H_0: $p = .70$ versus H_a: $p > .70$ is the right-tailed p-value in the summary box: the area under the standard normal curve to the right of $z = 2.65$. This p-value is $(1.0 - .9960) = .004$ (see Table A.3), and it provides very strong evidence against H_0: $p = .70$ and in favor of H_a: $p > .70$. That is, we have very strong evidence that Phantol will provide relief for more than 70 percent of all patients suffering from viral upper respiratory infections.

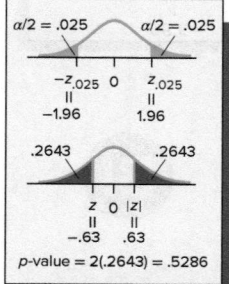

Ⓒ EXAMPLE 10.8 The Electronic Article Surveillance Case: False Alarms

A sports equipment discount store is considering installing an electronic article surveillance device and is concerned about the proportion, p, of all consumers who would never shop in the store again if the store subjected them to a false alarm. Suppose that industry data for general discount stores says that 15 percent of all consumers say that they would never shop in a store again if the store subjected them to a false alarm. To determine whether this percentage is different for the sports equipment discount store, the store will test the **null hypothesis H_0: $p = .15$ versus the alternative hypothesis H_a: $p \neq .15$** at the **.05 level of significance.** To perform the hypothesis test, the store will randomly select

$n = 500$ consumers, find the proportion $\hat{p}$ of these consumers who say that they would never shop in the store again if the store subjected them to a false alarm, and calculate the value of the **test statistic z in the summary box.** Then, because the alternative hypothesis $H_a: p \neq .15$ says to use the two-tailed critical value rule in the summary box, we will **reject $H_0: p = .15$ if $|z|$, the absolute value of the test statistic z, is greater than $z_{\alpha/2} = z_{.025} = 1.96$.** (Note that using this procedure is valid because $np_0 = (500)(.15) = 75$ and $n(1 - p_0) = (500)(1 - .15) = 425$ are both at least 5.) Suppose that when the sample is randomly selected, we find that 70 out of 500 consumers say that they would never shop in the store again if the store subjected them to a false alarm. Because $\hat{p} = 70/500 = .14$, the **value of the test statistic** is

$$z = \frac{\hat{p} - p_0}{\sqrt{\dfrac{p_0(1 - p_0)}{n}}} = \frac{.14 - .15}{\sqrt{\dfrac{.15(1 - .15)}{500}}} = -.63$$

Because $|z| = .63$ is less than $z_{.025} = 1.96$, we cannot reject $H_0: p = .15$ in favor of H_a: $p \neq .15$. That is, we cannot conclude (at an α of .05) that the percentage of all people who would never shop in the sports discount store again if the store subjected them to a false alarm differs from the general discount store percentage of 15 percent.

The p-value for testing $H_0: p = .15$ versus $H_a: p \neq .15$ is the two-tailed p-value in the summary box: twice the area under the standard normal curve to the right of $|z| = .63$. Because the area under the standard normal curve to the right of $|z| = .63$ is $(1 - .7357) = .2643$ (see Table A.3), the p-value is $2(.2643) = .5286$. As can be seen on the output below, Minitab calculates this p-value to be 0.531 (this value is slightly more accurate than our (hand- and table-) calculated result).

BI

Test of p = 0.15 vs p not = 0.15					
X	N	Sample p	95% CI	Z-Value	P-Value
70	500	0.14	(0.1096, 0.1704)	−0.63	0.531

This p-value is large and provides little evidence against $H_0: p = .15$ and in favor of H_a: $p \neq .15$. That is, we have little evidence that the percentage of people who would never shop in the sports discount store again if the store subjected them to a false alarm differs from the general discount store percentage of 15 percent.

Exercises for Section 10.4

CONCEPTS   connect

10.34 If we wish to test a hypothesis to provide evidence supporting the claim that fewer than 5 percent of the units produced by a process are defective, formulate the null and alternative hypotheses.

10.35 What condition must be satisfied in order for the methods of this section to be appropriate?

METHODS AND APPLICATIONS

10.36 Suppose we test $H_0: p = .3$ versus $H_a: p \neq .3$ and that a random sample of $n = 100$ gives a sample proportion $\hat{p} = .20$.
   a Test H_0 versus H_a at the .01 level of significance by using critical values. What do you conclude?
   b Find the p-value for this test.
   c Use the p-value to test H_0 versus H_a by setting α equal to .10, .05, .01, and .001. What do you conclude at each value of α?

10.37 THE MARKETING ETHICS CASE: CONFLICT OF INTEREST

Recall that a conflict of interest scenario was presented to a sample of 205 marketing researchers and that 111 of these researchers disapproved of the actions taken.
   a Let p be the proportion of all marketing researchers who disapprove of the actions taken in the conflict of interest scenario. Set up the null and alternative hypotheses needed to attempt to provide evidence supporting the claim that a majority (more than 50 percent) of all marketing researchers disapprove of the actions taken.
   b Assuming that the sample of 205 marketing researchers has been randomly selected, use critical values and the previously given sample information to test the hypotheses you set up in part a at the .10, .05, .01, and .001 levels of significance. How much evidence is there that a majority of all marketing researchers disapprove of the actions taken?

c   Suppose a random sample of 1,000 marketing researchers reveals that 540 of the researchers disapprove of the actions taken in the conflict of interest scenario. Use critical values to determine how much evidence there is that a majority of all marketing researchers disapprove of the actions taken.

d   Note that in parts *b* and *c* the sample proportion $\hat{p}$ is (essentially) the same. Explain why the results of the hypothesis tests in parts *b* and *c* differ.

10.38 Last year, television station WXYZ's share of the 11 P.M. news audience was approximately equal to, but no greater than, 25 percent. The station's management believes that the current audience share is higher than last year's 25 percent share. In an attempt to substanti-ate this belief, the station surveyed a random sample of 400 11 P.M. news viewers and found that 146 watched WXYZ.

a   Let *p* be the current proportion of all 11 P.M. news viewers who watch WXYZ. Set up the null and alternative hypotheses needed to attempt to provide evidence supporting the claim that the current audience share for WXYZ is higher than last year's 25 percent share.

b   Use critical values and the Minitab output (at the bottom of the page) to test the hypotheses you set up in part *a* at the .10, .05, .01, and .001 levels of significance. How much evidence is there that the current audience share is higher than last year's 25 percent share?

c   Find the *p*-value for the hypothesis test in part *b*. Use the *p*-value to carry out the test by setting α equal to .10, .05, .01, and .001. Interpret your results.

d   Do you think that the result of the station's survey has practical importance? Why or why not?

10.39 In the book *Essentials of Marketing Research*, William R. Dillon, Thomas J. Madden, and Neil H. Firtle discuss a marketing research proposal to study day-after recall for a brand of mouthwash. To quote the authors:

> The ad agency has developed a TV ad for the introduction of the mouthwash. The objective of the ad is to create awareness of the brand. The objective of this research is to evaluate the awareness generated by the ad measured by aided- and unaided-recall scores.
>
> A minimum of 200 respondents who claim to have watched the TV show in which the ad was aired the night before will be contacted by telephone in 20 cities.
>
> The study will provide information on the inci-dence of unaided and aided recall.

Suppose a random sample of 200 respondents shows that 46 of the people interviewed were able to recall the commercial without any prompting (unaided recall).

a   In order for the ad to be considered successful, the percentage of unaided recall must be above the category norm for a TV commercial for the product class. If this norm is 18 percent, set up the null and alternative hypotheses needed to attempt to provide evidence that the ad is successful.

b   Use the previously given sample information to: (**1**) Compute the *p*-value for the hypothesis test you set up in part *a*. (**2**) Use the *p*-value to carry out the test by setting α equal to .10, .05, .01, and .001. (**3**) How much evidence is there that the TV commercial is successful?

c   Do you think the result of the ad agency's survey has practical importance? Explain your opinion.

10.40 An airline's data indicate that 50 percent of people who begin the online process of booking a flight never complete the process and pay for the flight. To reduce this percentage, the airline is considering changing its website so that the entire booking process, including flight and seat selection and payment, can be done on two simple pages rather than the current four pages. A random sample of 300 customers who begin the book-ing process are exposed to the new system, and 117 of them do not complete the process. (**1**) Formulate the null and alternative hypotheses needed to attempt to provide evidence that the new system has reduced the noncompletion percentage. (**2**) Use critical values and a *p*-value to perform the hypothesis test by setting α equal to .10, .05, .01, and .001.

10.41 Suppose that a national survey finds that 73 percent of restaurant employees say that work stress has a negative impact on their personal lives. A random sample of 200 employees of a large restaurant chain finds that 141 employees say that work stress has a negative impact on their personal lives. (**1**) Formulate the null and alternative hypotheses needed to attempt to provide evidence that the percentage of work-stressed employees for the restaurant chain differs from the national percentage. (**2**) Use critical values and a *p*-value to perform the hypothesis test by setting α equal to .10, .05, .01, and .001.

10.42 The manufacturer of the ColorSmart-5000 television set claims that 95 percent of its sets last at least five years without needing a single repair. In order to test this claim, a consumer group randomly selects 400 consumers who have owned a ColorSmart-5000 television set for five years. Of these 400 consumers, 316 say that their ColorSmart-5000 television sets did not need repair, while 84 say that their ColorSmart-5000 television sets did need at least one repair.

a   Letting *p* be the proportion of ColorSmart-5000 television sets that last five years without a single repair, set up the null and alternative hypotheses

Test of p = 0.25 vs p > 0.25

Sample	X	N	Sample p	Z-Value	P-Value
1	146	400	0.365000	5.31	0.000

that the consumer group should use to attempt to show that the manufacturer's claim is false.

b   Use critical values and the previously given sample information to test the hypotheses you set up in part a by setting α equal to .10, .05, .01, and .001. How

much evidence is there that the manufacturer's claim is false?

c   Do you think the results of the consumer group's survey have practical importance? Explain your opinion.

10.5 Type II Error Probabilities and Sample Size Determination (Optional)

LO10-6

Calculate Type II error probabilities and the power of a test, and determine sample size (Optional).

As we have seen, we often take action (for example, advertise a claim) on the basis of having rejected the null hypothesis. In this case, we know the chances that the action has been taken erroneously because we have prespecified α, the probability of rejecting a true null hypothesis. However, sometimes we must act (for example, decide how many Valentine's Day boxes of chocolates to produce) on the basis of *not* rejecting the null hypothesis. If we must do this, it is best to know the probability of not rejecting a false null hypothesis (a Type II error). If this probability is not small enough, we may change the hypothesis testing procedure. In order to discuss this further, we must first see how to compute the probability of a Type II error.

As an example, the Federal Trade Commission (FTC) often tests claims that companies make about their products. Suppose coffee is being sold in cans that are labeled as containing three pounds, and also suppose that the FTC wishes to determine if the mean amount of coffee μ in all such cans is at least three pounds. To do this, the FTC tests H_0: $\mu \geq 3$ (or $\mu = 3$) versus H_a: $\mu < 3$ by setting $\alpha = .05$. Suppose that a sample of 35 coffee cans yields $\bar{x} = 2.9973$. Assuming that σ is known to equal .0147, we see that because

$$z = \frac{2.9973 - 3}{.0147/\sqrt{35}} = -1.08$$

is not less than $-z_{.05} = -1.645$, we cannot reject H_0: $\mu \geq 3$ by setting $\alpha = .05$. Because we cannot reject H_0, we cannot have committed a Type I error, which is the error of rejecting a true H_0. However, we might have committed a Type II error, which is the error of not rejecting a false H_0. Therefore, before we make a final conclusion about μ, we should calculate the probability of a Type II error.

A Type II error is not rejecting H_0: $\mu \geq 3$ when H_0 is false. Because any value of μ that is less than 3 makes H_0 false, there is a different Type II error (and, therefore, a different Type II error probability) associated with each value of μ that is less than 3. In order to demonstrate how to calculate these probabilities, we will calculate the probability of not rejecting H_0: $\mu \geq 3$ when in fact μ equals 2.995. This is the probability of failing to detect an average underfill of .005 pound. For a fixed sample size (for example, $n = 35$ coffee can fills), the value of β, the probability of a Type II error, depends upon how we set α, the probability of a Type I error. Because we have set $\alpha = .05$, we reject H_0 if

$$\frac{\bar{x} - 3}{\sigma/\sqrt{n}} < -z_{.05}$$

or, equivalently, if

$$\bar{x} < 3 - z_{.05}\frac{\sigma}{\sqrt{n}} = 3 - 1.645\frac{.0147}{\sqrt{35}} = 2.9959126$$

Therefore, we do not reject H_0 if $\bar{x} \geq 2.9959126$. It follows that β, the probability of not rejecting H_0: $\mu \geq 3$ when μ equals 2.995, is

$$\beta = P(\bar{x} \geq 2.9959126 \text{ when } \mu = 2.995)$$

$$= P\left(z \geq \frac{2.9959126 - 2.995}{.0147/\sqrt{35}}\right)$$

$$= P(z \geq .37) = 1 - .6443 = .3557$$

FIGURE 10.7   Calculating β When μ Equals 2.995

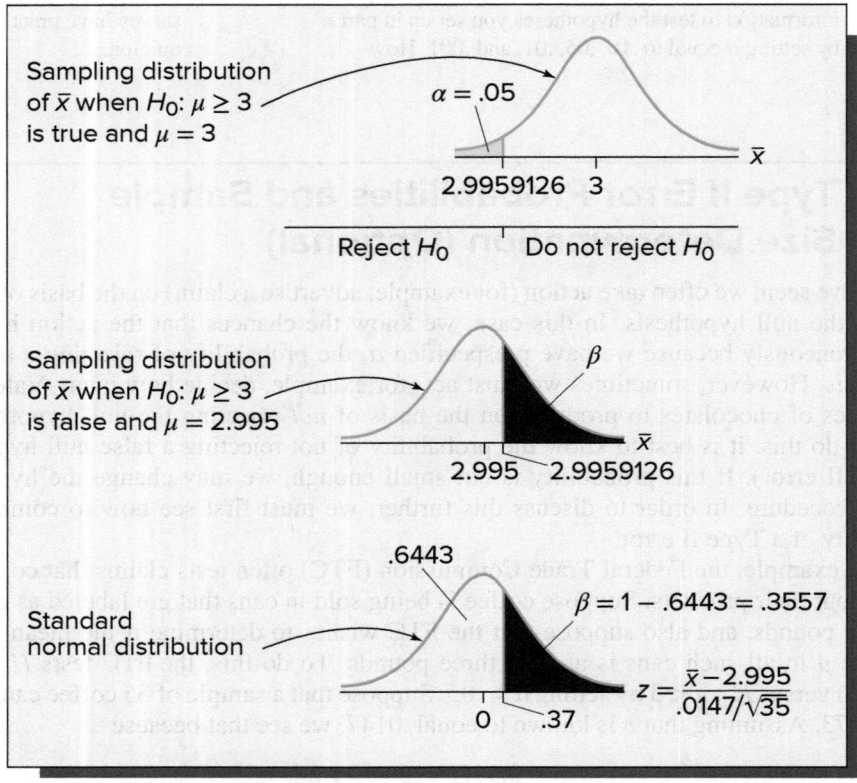

This calculation is illustrated in Figure 10.7. Similarly, it follows that β, the probability of not rejecting H_0: $\mu \geq 3$ when μ equals 2.99, is

$$\beta = P(\bar{x} \geq 2.9959126 \text{ when } \mu = 2.99)$$

$$= P\left(z \geq \frac{2.9959126 - 2.99}{.0147/\sqrt{35}}\right)$$

$$= P(z \geq 2.38) = 1 - .9913 = .0087$$

It also follows that β, the probability of not rejecting H_0: $\mu \geq 3$ when μ equals 2.985, is

$$\beta = P(\bar{x} \geq 2.9959126 \text{ when } \mu = 2.985)$$

$$= P\left(z \geq \frac{2.9959126 - 2.985}{.0147/\sqrt{35}}\right)$$

$$= P(z \geq 4.39)$$

This probability is less than .00003 (because z is greater than 3.99).

In Figure 10.8 we illustrate the values of β that we have calculated. Notice that the closer an alternative value of μ is to 3 (the value specified by H_0: $\mu = 3$), the larger is the associated value of β. Although alternative values of μ that are closer to 3 have larger associated probabilities of Type II errors, these values of μ have associated Type II errors with less serious consequences. For example, we are more likely not to reject H_0: $\mu = 3$ when $\mu = 2.995$ ($\beta = .3557$) than we are not to reject H_0: $\mu = 3$ when $\mu = 2.99$ ($\beta = .0087$). However, not rejecting H_0: $\mu = 3$ when $\mu = 2.995$, which means that we are failing to detect an average underfill of .005 pound, is less serious than not rejecting H_0: $\mu = 3$ when $\mu = 2.99$, which means that we are failing to detect a larger average underfill of .01 pound. In order to decide whether a particular hypothesis test adequately controls the probability of a Type II error, we must determine which Type II errors are serious, and then we must decide whether the probabilities of these errors are small enough. For example, suppose that the FTC and the

FIGURE 10.8   How β Changes as the Alternative Value of μ Changes

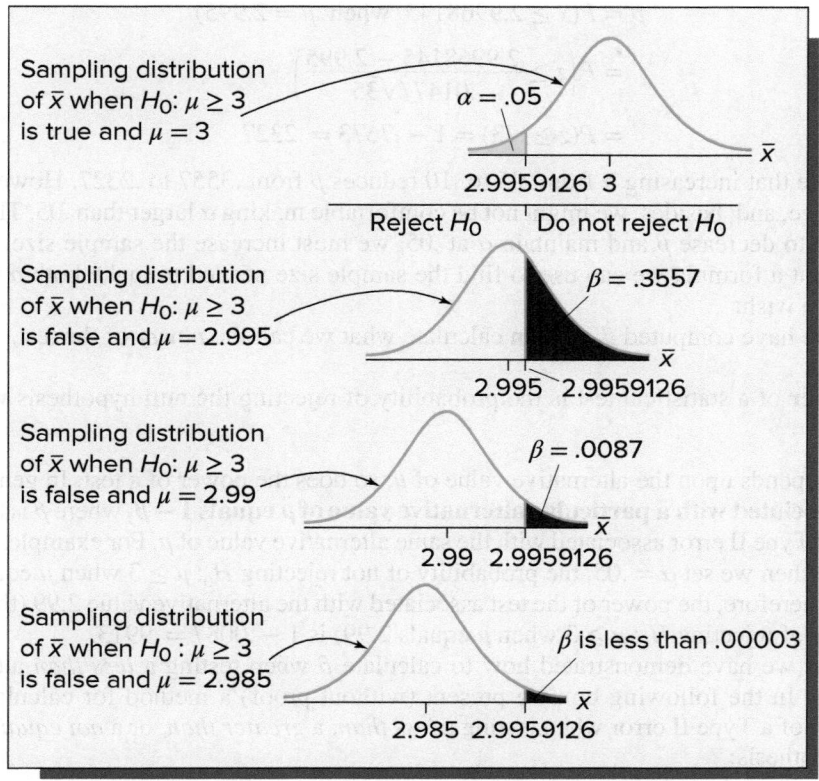

coffee producer agree that failing to reject $H_0: \mu = 3$ when μ equals 2.99 is a serious error, but that failing to reject $H_0: \mu = 3$ when μ equals 2.995 is not a particularly serious error. Then, because the probability of not rejecting $H_0: \mu = 3$ when μ equals 2.99 is .0087, which is quite small, we might decide that the hypothesis test adequately controls the probability of a Type II error. To understand the implication of this, recall that the sample of 35 coffee cans, which has $\bar{x} = 2.9973$, does not provide enough evidence to reject $H_0: \mu \geq 3$ by setting $\alpha = .05$. We have just shown that the probability that we have failed to detect a serious underfill is quite small (.0087), so the FTC might decide that no action should be taken against the coffee producer. Of course, this decision should also be based on the variability of the fills of the individual cans. Because $\bar{x} = 2.9973$ and $\sigma = .0147$, we estimate that 99.73 percent of all individual coffee can fills are contained in the interval $[\bar{x} \pm 3\sigma] = [2.9973 \pm 3(.0147)] = [2.9532, 3.0414]$. If the FTC believes it is reasonable to accept fills as low as (but no lower than) 2.9532 pounds, this evidence also suggests that no action against the coffee producer is needed.

Suppose, instead, that the FTC and the coffee producer had agreed that failing to reject $H_0: \mu \geq 3$ when μ equals 2.995 is a serious mistake. The probability of this Type II error is .3557, which is large. Therefore, we might conclude that the hypothesis test is not adequately controlling the probability of a serious Type II error. In this case, we have two possible courses of action. First, we have previously said that, for a fixed sample size, the lower we set α, the higher is β, and the higher we set α, the lower is β. Therefore, if we keep the sample size fixed at $n = 35$ coffee cans, we can reduce β by increasing α. To demonstrate this, suppose we increase α to .10. In this case we reject H_0 if

$$\frac{\bar{x} - 3}{\sigma/\sqrt{n}} < -z_{.10}$$

or, equivalently, if

$$\bar{x} < 3 - z_{.10}\frac{\sigma}{\sqrt{n}} = 3 - 1.282\frac{.0147}{\sqrt{35}} = 2.9968145$$

Therefore, we do not reject H_0 if $\bar{x} \geq 2.9968145$. It follows that β, the probability of not rejecting H_0: $\mu \geq 3$ when μ equals 2.995, is

$$\beta = P(\bar{x} \geq 2.9968145 \text{ when } \mu = 2.995)$$

$$= P\left(z \geq \frac{2.9968145 - 2.995}{.0147/\sqrt{35}}\right)$$

$$= P(z \geq .73) = 1 - .7673 = .2327$$

We thus see that increasing α from .05 to .10 reduces β from .3557 to .2327. However, β is still too large, and, besides, we might not be comfortable making α larger than .05. Therefore, if we wish to decrease β and maintain α at .05, we must increase the sample size. We will soon present a formula we can use to find the sample size needed to make both α and β as small as we wish.

Once we have computed β, we can calculate what we call the *power* of the test.

> The **power** of a statistical test is the probability of rejecting the null hypothesis when it is false.

Just as β depends upon the alternative value of μ, so does the power of a test. In general, **the power associated with a particular alternative value of μ equals $1 - \beta$,** where β is the probability of a Type II error associated with the same alternative value of μ. For example, we have seen that, when we set $\alpha = .05$, the probability of not rejecting H_0: $\mu \geq 3$ when μ equals 2.99 is .0087. Therefore, the power of the test associated with the alternative value 2.99 (that is, the probability of rejecting H_0: $\mu \geq 3$ when μ equals 2.99) is $1 - .0087 = .9913$.

Thus far we have demonstrated how to calculate β when testing a *less than* alternative hypothesis. In the following box we present (without proof) a method for calculating the probability of a Type II error when testing a *less than*, a *greater than*, or a *not equal to* alternative hypothesis:

Calculating the Probability of a Type II Error

Assume that the sampled population is normally distributed, or that a large sample will be taken. Consider testing H_0: $\mu = \mu_0$ versus one of H_a: $\mu > \mu_0$, H_a: $\mu < \mu_0$, or H_a: $\mu \neq \mu_0$. Then, if we set the probability of a Type I error equal to α and randomly select a sample of size n, the probability, β, of a Type II error corresponding to the alternative value μ_a of μ is (exactly or approximately) equal to the area under the standard normal curve to the left of

$$z^* - \frac{|\mu_0 - \mu_a|}{\sigma/\sqrt{n}}$$

Here z^* equals z_α if the alternative hypothesis is one-sided ($\mu > \mu_0$ or $\mu < \mu_0$), in which case the method for calculating β is exact. Furthermore, z^* equals $z_{\alpha/2}$ if the alternative hypothesis is two-sided ($\mu \neq \mu_0$), in which case the method for calculating β is approximate.

© EXAMPLE 10.9 The Valentine's Day Chocolate Case: Production Planning

In the Valentine's Day chocolate case we are testing H_0: $\mu = 330$ versus H_a: $\mu \neq 330$ by setting $\alpha = .05$. We have seen that the mean of the reported order quantities of a random sample of $n = 100$ large retail stores is $\bar{x} = 326$. Assuming that σ equals 40, it follows that because

$$z = \frac{326 - 330}{40/\sqrt{100}} = -1$$

is between $-z_{.025} = -1.96$ and $z_{.025} = 1.96$, we cannot reject H_0: $\mu = 330$ by setting $\alpha = .05$. Because we cannot reject H_0, we might have committed a Type II error. Suppose that the candy company decides that failing to reject H_0: $\mu = 330$ when μ differs from 330 by as many as 15 Valentine boxes (that is, when μ is 315 or 345) is a serious Type II error. Because we have set α equal to .05, β for the alternative value $\mu_a = 315$ (that is, the probability of not rejecting $H_0: \mu = 330$ when μ equals 315) is the area under the standard normal curve to the left of

$$z^* - \frac{|\mu_0 - \mu_a|}{\sigma/\sqrt{n}} = z_{.025} - \frac{|\mu_0 - \mu_a|}{\sigma/\sqrt{n}}$$

$$= 1.96 - \frac{|330 - 315|}{40/\sqrt{100}}$$

$$= -1.79$$

Here $z^* = z_{\alpha/2} = z_{.05/2} = z_{.025}$ because the alternative hypothesis ($\mu \neq 330$) is two-sided. The area under the standard normal curve to the left of -1.79 is .0367. Therefore, β for the alternative value $\mu_a = 315$ is .0367. Similarly, it can be verified that β for the alternative value $\mu_a = 345$ is .0367. It follows, because we cannot reject $H_0: \mu = 330$ by setting $\alpha = .05$, and because we have just shown that there is a reasonably small (.0367) probability that we have failed to detect a serious (that is, a 15 Valentine box) deviation of μ from 330, that it is reasonable for the candy company to base this year's production of Valentine boxes on the projected mean order quantity of 330 boxes per large retail store.

To calculate β for the alternative value $\mu_a = 315$ directly (that is, without using a formula), note that $\sigma_{\bar{x}} = \sigma/\sqrt{n} = 40/\sqrt{100} = 4$. We will fail to reject $H_0: \mu = 330$ when $\alpha = .05$ if $-1.96 \leq (\bar{x} - 330)/4 \leq 1.96$. Algebra shows that this event is equivalent to the event $322.16 \leq \bar{x} \leq 337.84$. Therefore, the probability that we will fail to reject $H_0: \mu = 330$ when $\mu = 315$ is the probability that $322.16 \leq \bar{x} \leq 337.84$ when $\mu = 315$ (and $\sigma_{\bar{x}} = 4$). This probability is the area under the standard normal curve between $(322.16 - 315)/4 = -5.71$ and $(337.84 - 315)/4 = -1.79$. Because there is virtually no area under the standard normal curve to the left of -5.71, the desired probability is approximately the area under the standard normal curve to the left of -1.79. This area is .0367, the same result obtained using the formula.

In the following box we present (without proof) a formula that tells us the sample size needed to make both the probability of a Type I error and the probability of a Type II error as small as we wish:

Calculating the Sample Size Needed to Achieve Specified Values of α and β

Assume that the sampled population is normally distributed, or that a large sample will be taken. Consider testing $H_0: \mu = \mu_0$ versus one of $H_a: \mu > \mu_0$, $H_a: \mu < \mu_0$, or $H_a: \mu \neq \mu_0$. Then, in order to make the probability of a Type I error equal to α and the probability of a Type II error corresponding to the alternative value μ_a of μ equal to β, we should take a sample of size

$$n = \frac{(z^* + z_\beta)^2 \sigma^2}{(\mu_0 - \mu_a)^2}$$

Here z^* equals z_α if the alternative hypothesis is one-sided ($\mu > \mu_0$ or $\mu < \mu_0$), and z^* equals $z_{\alpha/2}$ if the alternative hypothesis is two-sided ($\mu \neq \mu_0$). Also, z_β is the point on the scale of the standard normal curve that gives a right-hand tail area equal to β.

Ⓒ **EXAMPLE 10.10** Finding A Sample Size

Although we sometimes set both α and β equal to the same value, we do not always do this. For example, again consider the Valentine's Day chocolate case, in which we are testing $H_0: \mu = 330$ versus $H_a: \mu \neq 330$. Suppose that the candy company decides that failing to reject $H_0: \mu = 330$ when μ differs from 330 by as many as 15 Valentine boxes (that is, when μ is 315 or 345) is a serious Type II error. Furthermore, suppose that it is also decided that this Type II error is more serious than a Type I error. Therefore, α will be set equal to .05 and β for the alternative value $\mu_a = 315$ (or $\mu_a = 345$) of μ will be set equal to .01. It follows that the candy company should take a sample of size

$$n = \frac{(z^* + z_\beta)^2 \sigma^2}{(\mu_0 - \mu_a)^2} = \frac{(z_{\alpha/2} + z_\beta)^2 \sigma^2}{(\mu_0 - \mu_a)^2}$$

$$= \frac{(z_{.025} + z_{.01})^2 \sigma^2}{(\mu_0 - \mu_a)^2}$$

$$= \frac{(1.96 + 2.326)^2 (40)^2}{(330 - 315)^2}$$

$$= 130.62 = 131 \text{ (rounding up)}$$

Here, $z^* = z_{\alpha/2} = z_{.05/2} = z_{.025} = 1.96$ because the alternative hypothesis ($\mu \neq 330$) is two-sided, and $z_\beta = z_{.01} = 2.326$ (see the bottom row of the t table, Table 8.3).

As another example, consider the coffee fill example and suppose we wish to test $H_0: \mu \geq 3$ (or $\mu = 3$) versus $H_a: \mu < 3$. If we wish α to be .05 and β for the alternative value $\mu_a = 2.995$ of μ to be .05, we should take a sample of size

$$n = \frac{(z^* + z_\beta)^2 \sigma^2}{(\mu_0 - \mu_a)^2} = \frac{(z_\alpha + z_\beta)^2 \sigma^2}{(\mu_0 - \mu_a)^2}$$

$$= \frac{(z_{.05} + z_{.05})^2 \sigma^2}{(\mu_0 - \mu_a)^2} = \frac{(1.645 + 1.645)^2 (.0147)^2}{(3 - 2.995)^2}$$

$$= 93.5592 = 94 \text{ (rounding up)}$$

Here, $z^* = z_\alpha = z_{.05} = 1.645$ because the alternative hypothesis ($\mu < 3$) is one-sided, and $z_\beta = z_{.05} = 1.645$.

To conclude this section, we point out that the methods we have presented for calculating the probability of a Type II error and determining sample size can be extended to other hypothesis tests that utilize the normal distribution. We will not, however, present the extensions in this book.

Exercises for Section 10.5

CONCEPTS 📘 **connect**

10.43 Explain what is meant by
   a A serious Type II error.
   b The power of a statistical test.

10.44 In general, do we want the power corresponding to a serious Type II error to be near 0 or near 1? Explain.

METHODS AND APPLICATIONS

10.45 Again consider the Consolidated Power wastewater situation. Remember that the power plant will be shut down and corrective action will be taken on the cooling system if the null hypothesis $H_0: \mu \leq 60$ is rejected in favor of $H_a: \mu > 60$. In this exercise we calculate probabilities of various Type II errors in the context of this situation.
   a Recall that Consolidated Power's hypothesis test is based on a sample of $n = 100$ temperature readings and assume that σ equals 2. If the power company sets $\alpha = .025$, calculate the probability of a Type II error for each of the following alternative values of μ: 60.1, 60.2, 60.3, 60.4, 60.5, 60.6, 60.7, 60.8, 60.9, 61.

   b If we want the probability of making a Type II error when μ equals 60.5 to be very small, is Consolidated Power's hypothesis test adequate? Explain why or why not. If not, and if we wish to maintain the value of α at .025, what must be done?

   c The **power curve** for a statistical test is a plot of the power = $1 - \beta$ on the vertical axis versus values of μ that make the null hypothesis false on the horizontal axis. Plot the power curve for Consolidated Power's test of $H_0: \mu \leq 60$ versus $H_a: \mu > 60$ by plotting power = $1 - \beta$ for each of the alternative values of μ in part a. What happens to the power of the test as the alternative value of μ moves away from 60?

10.46 Again consider the automobile parts supplier situation. Remember that a problem-solving team will be assigned to rectify the process producing the cylindrical engine parts if the null hypothesis $H_0: \mu = 3$ is rejected in favor of $H_a: \mu \neq 3$. In this exercise we calculate probabilities of various Type II errors in the context of this situation.

a Suppose that the parts supplier's hypothesis test is based on a sample of $n = 100$ diameters and that σ equals .023. If the parts supplier sets $\alpha = .05$, calculate the probability of a Type II error for each of the following alternative values of μ: 2.990, 2.995, 3.005, 3.010.

b If we want both the probabilities of making a Type II error when μ equals 2.995 and when μ equals 3.005 to be very small, is the parts supplier's hypothesis test adequate? Explain why or why not. If not, and if we wish to maintain the value of α at .05, what must be done?

c Plot the power of the test versus the alternative values of μ in part a. What happens to the power of the test as the alternative value of μ moves away from 3?

10.47 In the Consolidated Power hypothesis test of $H_0: \mu \leq 60$ versus $H_a: \mu > 60$ (as discussed in Exercise 10.45) find the sample size needed to make the probability of a Type I error equal to .025 and the probability of a Type II error corresponding to the alternative value $\mu_a = 60.5$ equal to .025. Here, assume σ equals 2.

10.48 In the automobile parts supplier's hypothesis test of $H_0: \mu = 3$ versus $H_a: \mu \neq 3$ (as discussed in Exercise 10.46) find the sample size needed to make the probability of a Type I error equal to .05 and the probability of a Type II error corresponding to the alternative value $\mu_a = 3.005$ equal to .05. Here, assume σ equals .023.

10.6 The Chi-Square Distribution

Sometimes we can make statistical inferences by using the **chi-square distribution.** The probability curve of the χ^2 (pronounced *chi-square*) distribution is skewed to the right. Moreover, the exact shape of this probability curve depends on a parameter that is called the **number of degrees of freedom** (denoted *df*). Figure 10.9 illustrates chi-square distributions having 2, 5, and 10 degrees of freedom.

In order to use the chi-square distribution, we employ a **chi-square point,** which is denoted χ_α^2. As illustrated in the upper portion of Figure 10.10, χ_α^2 is the point on the horizontal axis under the curve of the chi-square distribution that gives a right-hand tail area equal to α. The value of χ_α^2 in a particular situation depends on the right-hand tail area α and the number of degrees of freedom (*df*) of the chi-square distribution. Values of χ_α^2 are tabulated in a **chi-square table.** Such a table is given in Table A.5 of Appendix A; a portion of this table is reproduced as Table 10.3 on the next page. Looking at the chi-square table, the rows correspond to the appropriate number of degrees of freedom (values of which are listed down the left side of the table), while the columns designate the right-hand tail area α. For example, suppose we wish to find the chi-square point that gives a right-hand tail area of .05 under a chi-square curve having 5 degrees of freedom. To do this, we look in Table 10.3 at the row labeled 5 and the column labeled $\chi_{.05}^2$. We find that this $\chi_{.05}^2$ point is 11.0705 (see the lower portion of Figure 10.10).

LO10-7

Describe the properties of the chi-square distribution and use a chi-square table.

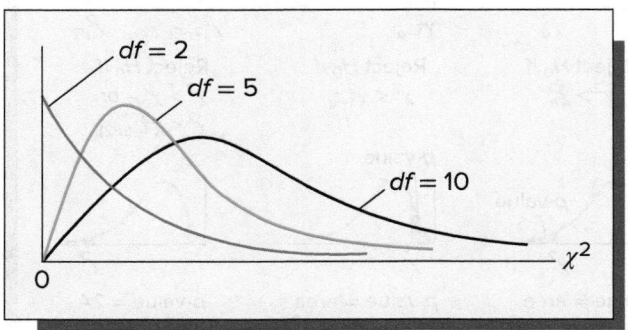

FIGURE 10.9 **Chi-Square Distributions with 2, 5, and 10 Degrees of Freedom**

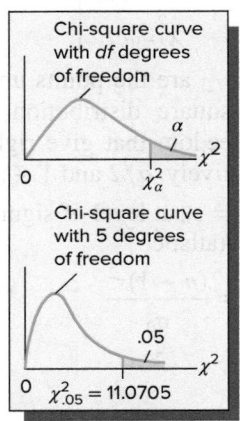

FIGURE 10.10 **Chi-Square Points**

TABLE 10.3   A Portion of the Chi-Square Table

Degrees of Freedom (df)	$\chi^2_{.10}$	$\chi^2_{.05}$	$\chi^2_{.025}$	$\chi^2_{.01}$	$\chi^2_{.005}$
1	2.70554	3.84146	5.02389	6.63490	7.87944
2	4.60517	5.99147	7.37776	9.21034	10.5966
3	6.25139	7.81473	9.34840	11.3449	12.8381
4	7.77944	9.48773	11.1433	13.2767	14.8602
5	9.23635	11.0705	12.8325	15.0863	16.7496
6	10.6446	12.5916	14.4494	16.8119	18.5476

LO10-8
Use the chi-square distribution to make statistical inferences about a population variance (Optional).

10.7 Statistical Inference for a Population Variance (Optional)

A jelly and jam producer has a filling process that is supposed to fill jars with 16 ounces of grape jelly. Through long experience with the filling process, the producer knows that the population of all fills produced by the process is normally distributed with a mean of 16 ounces, a variance of .000625, and a standard deviation of .025 ounce. Using the Empirical Rule, it follows that 99.73 percent of all jar fills produced by the process are in the tolerance interval $[16 \pm 3(.025)] = [15.925, 16.075]$. In order to be competitive with the tightest specifications in the jelly and jam industry, the producer has decided that at least 99.73 percent of all jar fills must be between 15.95 ounces and 16.05 ounces. Because the tolerance limits [15.925, 16.075] of the current filling process are not inside the specification limits [15.95, 16.05], the jelly and jam producer designs a new filling process that will hopefully reduce the variance of the jar fills. A random sample of $n = 30$ jars filled by the new process is selected, and the mean and the variance of the corresponding jar fills are found to be $\bar{x} = 16$ ounces and $s^2 = .000121$. In order to attempt to show that the variance, σ^2, of the population of all jar fills that would be produced by the new process is less than .000625, we can use the following result:

Statistical Inference for a Population Variance

Suppose that s^2 is the variance of a sample of n measurements randomly selected from a normally distributed population having variance σ^2. The sampling distribution of the statistic $(n - 1)s^2/\sigma^2$ is a chi-square distribution having $n - 1$ degrees of freedom. This implies that

1 A $100(1 - \alpha)$ **percent confidence interval for** σ^2 is

$$\left[\frac{(n - 1)s^2}{\chi^2_{\alpha/2}}, \frac{(n - 1)s^2}{\chi^2_{1-(\alpha/2)}} \right]$$

Here $\chi^2_{\alpha/2}$ and $\chi^2_{1-(\alpha/2)}$ are the points under the curve of the chi-square distribution having $n - 1$ degrees of freedom that give right-hand tail areas of, respectively, $\alpha/2$ and $1 - (\alpha/2)$.

2 We can test $H_0: \sigma^2 = \sigma_0^2$ at level of significance α by using the test statistic

$$\chi^2 = \frac{(n - 1)s^2}{\sigma_0^2}$$

and by using the critical value rule or p-value that is positioned under the appropriate alternative hypothesis. **We reject H_0 if the p-value is less than α.**

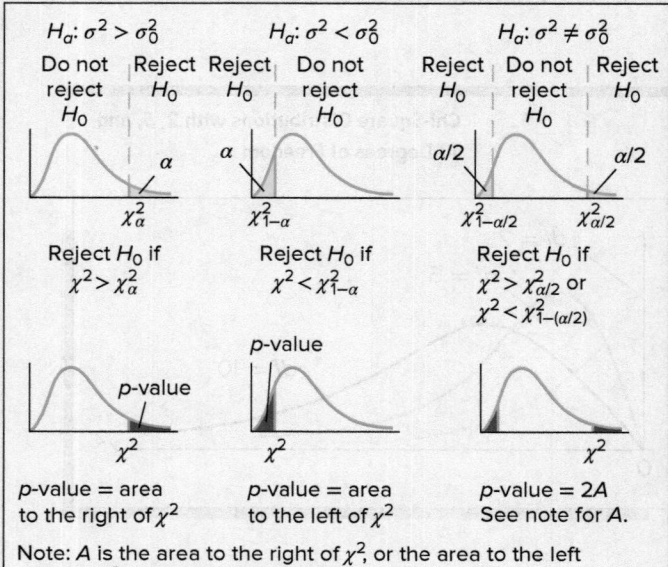

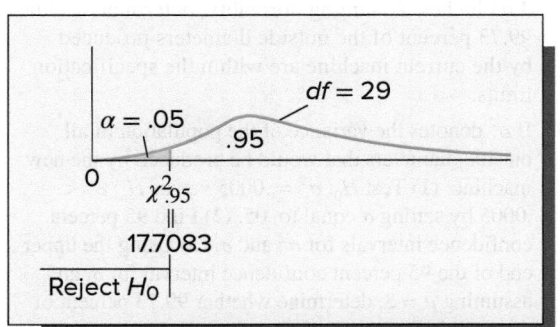

FIGURE 10.11 Testing H_0: $\sigma^2 = .000625$ versus H_a: $\sigma^2 < .000625$ by Setting $\alpha = .05$

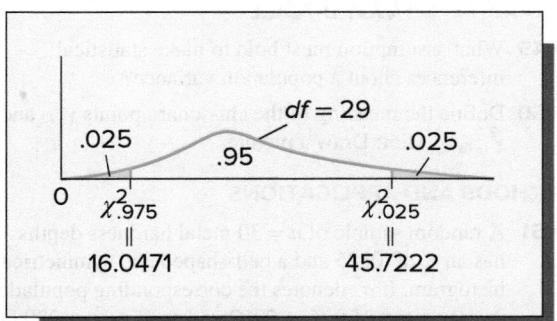

FIGURE 10.12 The Chi-Square Points $\chi^2_{.025}$ and $\chi^2_{.975}$

The assumption that the sampled population is normally distributed must hold fairly closely for the statistical inferences just given about σ^2 to be valid. When we check this assumption in the jar fill situation, we find that a histogram (not given here) of the sample of $n = 30$ jar fills is bell-shaped and symmetrical. In order to assess whether the jar fill variance σ^2 for the new filling process is less than .000625, we test **the null hypothesis H_0: $\sigma^2 = .000625$ versus the alternative hypothesis H_a: $\sigma^2 < .000625$.** If H_0 can be rejected in favor of H_a at the **.05 level of significance,** we will conclude that the new process has reduced the variance of the jar fills. Because the histogram of the sample of $n = 30$ jar fills is bell-shaped and symmetrical, **the appropriate test statistic is given in the summary box.** Furthermore, because the alternative hypothesis H_a: $\sigma^2 < .000625$ says to use the left-tailed critical value rule in the summary box, we will **reject H_0: $\sigma^2 = .000625$ if the value of χ^2 is less than the critical value $\chi^2_{1-\alpha} = \chi^2_{.95} = 17.7083$** (see Table A.5). Here $\chi^2_{.95} = 17.7083$ is based on $n - 1 = 30 - 1 = 29$ degrees of freedom, and this critical value is illustrated in Figure 10.11. Because the sample variance is $s^2 = .000121$, the **value of the test statistic** is

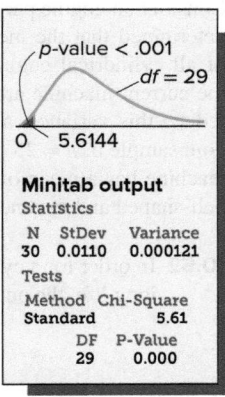

p-value < .001
$df = 29$
0   5.6144

Minitab output
Statistics

N	StDev	Variance
30	0.0110	0.000121

Tests

Method	Chi-Square
Standard	5.61

DF	P-Value
29	0.000

$$\chi^2 = \frac{(n-1)s^2}{\sigma_0^2} = \frac{(29)(.000121)}{.000625} = 5.6144$$

Because $\chi^2 = 5.6144$ is less than $\chi^2_{.95} = 17.7083$, we reject H_0: $\sigma^2 = .000625$ in favor of H_a: $\sigma^2 < .000625$. That is, we conclude (at an α of .05) that the new process has reduced the population variance of the jar fills. Moreover, the p-value for testing H_0: $\sigma^2 = .000625$ versus H_a: $\sigma^2 < .000625$ is the left-tailed p-value in the summary box: the area under the chi-square distribution curve having 29 degrees of freedom to the left of $\chi^2 = 5.6144$. Examining the figure in the page margin, we see that this p-value is less than .001. Therefore, we have extremely strong evidence that the population jar fill variance has been reduced.

In order to compute a 95 percent confidence interval for σ^2, we note that $\alpha = .05$. Therefore, $\chi^2_{\alpha/2}$ is $\chi^2_{.025}$ and $\chi^2_{1-(\alpha/2)}$ is $\chi^2_{.975}$. Table A.5 tells us that these points (based on $n - 1 = 29$ degrees of freedom) are $\chi^2_{.025} = 45.7222$ and $\chi^2_{.975} = 16.0471$ (see Figure 10.12). It follows that a 95 percent confidence interval for σ^2 is

$$\left[\frac{(n-1)s^2}{\chi^2_{\alpha/2}}, \frac{(n-1)s^2}{\chi^2_{1-(\alpha/2)}} \right] = \left[\frac{(29)(.000121)}{45.7222}, \frac{(29)(.000121)}{16.0471} \right]$$

$$= [.000076746, .00021867]$$

Moreover, taking square roots of both ends of this interval, we find that a 95 percent confidence interval for σ is [.008760, .01479]. The upper end of this interval, .01479, is an estimate of the largest that σ for the new process might reasonably be. Recalling that the estimate of the population mean jar fill μ for the new process is $\bar{x} = 16$, and using .01479 as the estimate of σ, we estimate that (at the worst) 99.73 percent of all jar fills that would be produced with the new process are in the tolerance interval $[16 \pm 3(.01479)] = [15.956, 16.044]$. This tolerance interval is narrower than the tolerance interval of [15.925, 16.075] for the old jar filling process and is within the specification limits of [15.95, 16.05].

Exercises for Sections 10.6 and 10.7

CONCEPTS ▧ connect

10.49 What assumption must hold to make statistical inferences about a population variance?

10.50 Define the meaning of the chi-square points $\chi^2_{\alpha/2}$ and $\chi^2_{1-(\alpha/2)}$. Hint: Draw a picture.

METHODS AND APPLICATIONS

10.51 A random sample of $n = 30$ metal hardness depths has an s^2 of .0885 and a bell-shaped and symmetrical histogram. If σ^2 denotes the corresponding population variance, test $H_0: \sigma^2 = .2209$ versus $H_a: \sigma^2 < .2209$ by setting α equal to .05.

Exercises 10.52 and 10.53 relate to the following situation: Consider an engine parts supplier and suppose the supplier has determined that the mean and the variance of the population of all cylindrical engine part outside diameters produced by the current machine are, respectively, 3 inches and .0005. To reduce this variance, a new machine is designed, and a random sample of $n = 25$ outside diameters produced by this new machine has a mean of 3 inches, a variance of .00014, and a bell-shaped and symmetrical histogram.

10.52 In order for a cylindrical engine part to give an engine long life, the outside diameter of the part must be between the specification limits of 2.95 inches and 3.05 inches. Assuming normality, determine whether 99.73 percent of the outside diameters produced by the current machine are within the specification limits.

10.53 If σ^2 denotes the variance of the population of all outside diameters that would be produced by the new machine: **(1)** Test $H_0: \sigma^2 = .0005$ versus $H_a: \sigma^2 < .0005$ by setting α equal to .05. **(2)** Find 95 percent confidence intervals for σ^2 and σ. **(3)** Using the upper end of the 95 percent confidence interval for σ, and assuming $\mu = 3$, determine whether 99.73 percent of the outside diameters produced by the new machine are within the specification limits.

10.54 A manufacturer of coffee vending machines has designed a new, less expensive machine. The current machine is known to dispense (into cups) an average of 6 fl. oz., with a standard deviation of .2 fl. oz. When the new machine is tested using 15 cups, the mean and the standard deviation of the fills are found to be 6 fl. oz. and .214 fl. oz. Test $H_0: \sigma = .2$ versus $H_a: \sigma \neq .2$ at levels of significance .05 and .01. Assume normality.

10.55 In Exercise 10.54, test $H_0: \sigma = .2$ versus $H_a: \sigma > .2$ at levels of significance .05 and .01.

Chapter Summary

We began this chapter by learning about the two hypotheses that make up the structure of a hypothesis test. The **null hypothesis** is the statement being tested. The null hypothesis is given the benefit of the doubt and is not rejected unless there is convincing sample evidence that it is false. The **alternative hypothesis** is a statement that is assigned the burden of proof. It is accepted only if there is convincing sample evidence that it is true. In some situations, we begin by formulating the alternative hypothesis as a **research hypothesis.** We also learned that two types of errors can be made in a hypothesis test. A **Type I error** occurs when we reject a true null hypothesis, and a **Type II error** occurs when we do not reject a false null hypothesis.

We studied two commonly used ways to conduct a hypothesis test. The first involves comparing the value of a test statistic with what is called a **critical value,** and the second employs what is called a **p-value.** The p-value measures the weight of evidence against the null hypothesis. The smaller the p-value, the more we doubt the null hypothesis.

The specific hypothesis tests we covered in this chapter all dealt with a hypothesis about one population parameter. First, we studied a test about a **population mean** that is based on the assumption that the population standard deviation σ **is known.** This test employs the **normal distribution.** Second, we studied a test about a population mean that assumes that σ **is unknown.** We learned that this test is based on the **t distribution.** Figure 10.13 presents a flowchart summarizing how to select an appropriate test statistic to test a hypothesis about a population mean. Then we presented a test about a **population proportion** that is based on the **normal distribution.** Next (in optional Section 10.5) we studied Type II error probabilities, and we showed how we can find the sample size needed to make both the probability of a Type I error and the probability of a serious Type II error as small as we wish. We concluded this chapter by discussing (in Section 10.6 and optional Section 10.7) the **chi-square distribution** and its use in making statistical inferences about a **population variance.**

Glossary of Terms

alternative hypothesis: A statement that is assigned the burden of proof. It is accepted only if there is convincing sample evidence that it is true.

chi-square distribution: A useful continuous probability distribution. Its probability curve is skewed to the right, and the exact shape of the probability curve depends on the number of degrees of freedom associated with the curve.

FIGURE 10.13 Selecting an Appropriate Test Statistic to Test a Hypothesis about a Population Mean

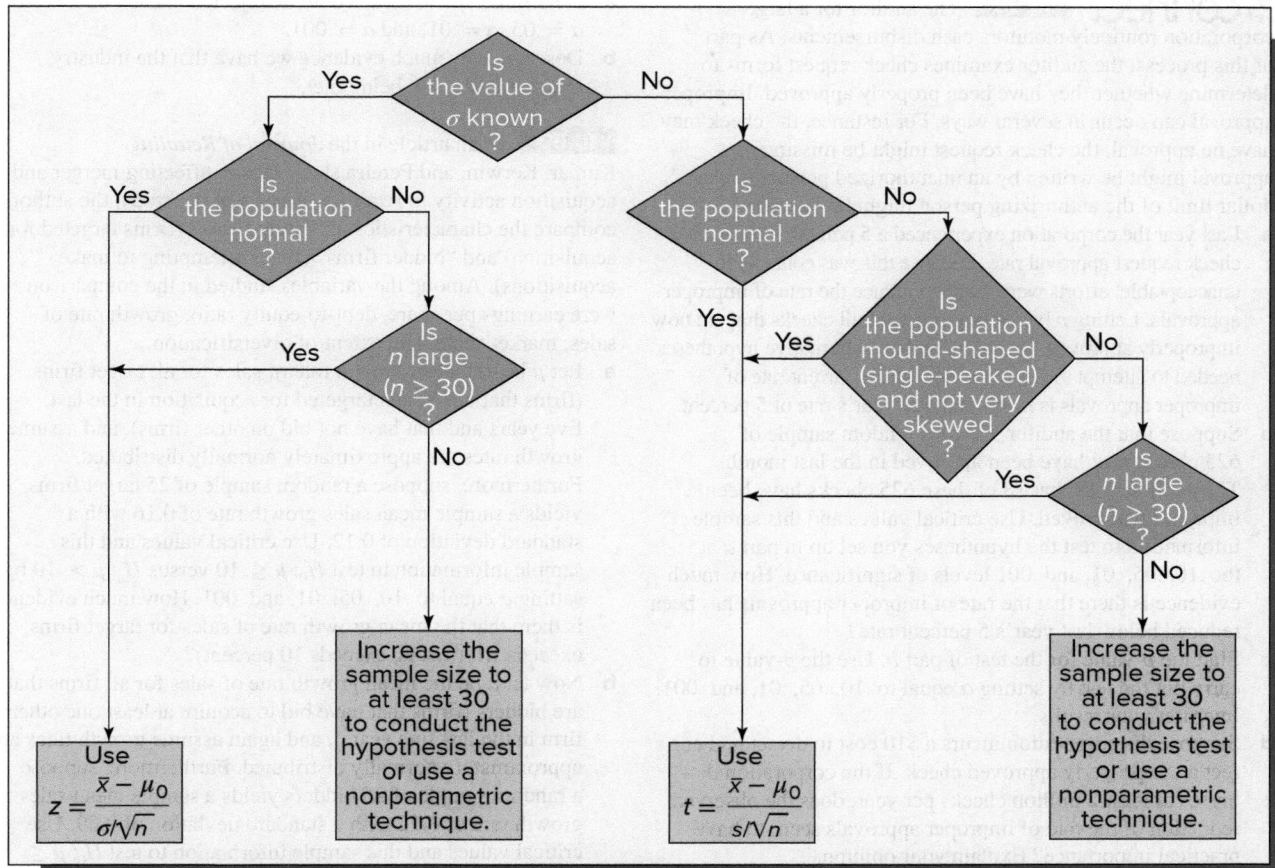

critical value: The value of the test statistic is compared with a critical value in order to decide whether the null hypothesis can be rejected.

greater than alternative: An alternative hypothesis that is stated as a *greater than* (>) inequality.

less than alternative: An alternative hypothesis that is stated as a *less than* (<) inequality.

level of significance: The probability of making a Type I error when performing a hypothesis test.

not equal to alternative: An alternative hypothesis that is stated as a *not equal to* (≠) inequality.

null hypothesis: The statement being tested in a hypothesis test. It is given the benefit of the doubt.

power (of a statistical test): The probability of rejecting the null hypothesis when it is false.

p-value (probability value): The probability, computed assuming that the null hypothesis H_0 is true, of observing a value of the test statistic that is at least as contradictory to H_0 and supportive of H_a as the value actually computed from the sample data. The p-value measures how much doubt is cast on the null hypothesis by the sample data. The smaller the p-value, the more we doubt the null hypothesis.

test statistic: A statistic computed from sample data in a hypothesis test. It is either compared with a critical value or used to compute a p-value.

two-sided alternative hypothesis: An alternative hypothesis that is stated as a *not equal to* (≠) inequality.

Type I error: Rejecting a true null hypothesis.

Type II error: Failing to reject a false null hypothesis.

Important Formulas and Tests

Hypothesis testing steps: page 444

A hypothesis test about a population mean (σ known): page 443

A t test about a population mean (σ unknown): page 450

A large sample hypothesis test about a population proportion: page 455

Calculating the probability of a Type II error: page 462

Sample size determination to achieve specified values of α and β: page 463

Statistical inference about a population variance: page 466

Supplementary Exercises

■connect **10.56** The auditor for a large corporation routinely monitors cash disbursements. As part of this process, the auditor examines check request forms to determine whether they have been properly approved. Improper approval can occur in several ways. For instance, the check may have no approval, the check request might be missing, the approval might be written by an unauthorized person, or the dollar limit of the authorizing person might be exceeded.

a Last year the corporation experienced a 5 percent improper check request approval rate. Because this was considered unacceptable, efforts were made to reduce the rate of improper approvals. Letting p be the proportion of all checks that are now improperly approved, set up the null and alternative hypotheses needed to attempt to demonstrate that the current rate of improper approvals is lower than last year's rate of 5 percent.

b Suppose that the auditor selects a random sample of 625 checks that have been approved in the last month. The auditor finds that 18 of these 625 checks have been improperly approved. Use critical values and this sample information to test the hypotheses you set up in part a at the .10, .05, .01, and .001 levels of significance. How much evidence is there that the rate of improper approvals has been reduced below last year's 5 percent rate?

c Find the p-value for the test of part b. Use the p-value to carry out the test by setting α equal to .10, .05, .01, and .001. Interpret your results.

d Suppose the corporation incurs a $10 cost to detect and correct an improperly approved check. If the corporation disburses at least 2 million checks per year, does the observed reduction of the rate of improper approvals seem to have practical importance? Explain your opinion.

10.57 **THE CIGARETTE ADVERTISEMENT CASE** ⓓ ModelAge

Recall that the cigarette industry requires that models in cigarette ads must appear to be at least 25 years old. Also recall that a sample of 50 people is randomly selected at a shopping mall. Each person in the sample is shown a "typical cigarette ad" and is asked to estimate the age of the model in the ad.

a Let μ be the mean perceived age estimate for all viewers of the ad, and suppose we consider the industry requirement to be met if μ is at least 25. Set up the null and alternative hypotheses needed to attempt to show that the industry requirement is not being met.

b Suppose that a random sample of 50 perceived age estimates gives a mean of 23.663 years and a standard deviation of 3.596 years. Use these sample data and critical values to test the hypotheses of part a at the .10, .05, .01, and .001 levels of significance.

c How much evidence do we have that the industry requirement is not being met?

d Do you think that this result has practical importance? Explain your opinion.

10.58 **THE CIGARETTE ADVERTISEMENT CASE** ⓓ ModelAge

Consider the cigarette ad situation discussed in Exercise 10.57. Using the sample information given in that exercise, the p-value for testing H_0 versus H_a can be calculated to be .0057.

a Determine whether H_0 would be rejected at each of $\alpha = .10$, $\alpha = .05$, $\alpha = .01$, and $\alpha = .001$.

b Describe how much evidence we have that the industry requirement is not being met.

10.59 In an article in the *Journal of Retailing*, Kumar, Kerwin, and Pereira study factors affecting merger and acquisition activity in retailing. As part of the study, the authors compare the characteristics of "target firms" (firms targeted for acquisition) and "bidder firms" (firms attempting to make acquisitions). Among the variables studied in the comparison were earnings per share, debt-to-equity ratio, growth rate of sales, market share, and extent of diversification.

a Let μ be the mean growth rate of sales for all target firms (firms that have been targeted for acquisition in the last five years and that have not bid on other firms), and assume growth rates are approximately normally distributed. Furthermore, suppose a random sample of 25 target firms yields a sample mean sales growth rate of 0.16 with a standard deviation of 0.12. Use critical values and this sample information to test $H_0: \mu \le .10$ versus $H_a: \mu > .10$ by setting α equal to .10, .05, .01, and .001. How much evidence is there that the mean growth rate of sales for target firms exceeds .10 (that is, exceeds 10 percent)?

b Now let μ be the mean growth rate of sales for all firms that are bidders (firms that have bid to acquire at least one other firm in the last five years), and again assume growth rates are approximately normally distributed. Furthermore, suppose a random sample of 25 bidders yields a sample mean sales growth rate of 0.12 with a standard deviation of 0.09. Use critical values and this sample information to test $H_0: \mu \le .10$ versus $H_a: \mu > .10$ by setting α equal to .10, .05, .01, and .001. How much evidence is there that the mean growth rate of sales for bidders exceeds .10 (that is, exceeds 10 percent)?

10.60 A consumer electronics firm has developed a new type of remote control button that is designed to operate longer before becoming intermittent. A random sample of 35 of the new buttons is selected and each is tested in continuous operation until becoming intermittent. The resulting lifetimes are found to have a sample mean of 1,241.2 hours and a sample standard deviation of 110.8.

a Independent tests reveal that the mean lifetime (in continuous operation) of the best remote control button on the market is 1,200 hours. Letting μ be the mean lifetime of the population of all new remote control buttons that will or could potentially be produced, set up the null and alternative hypotheses needed to attempt to provide evidence that the new button's mean lifetime exceeds the mean lifetime of the best remote button currently on the market.

b Using the previously given sample results, use critical values to test the hypotheses you set up in part a by setting α equal to .10, .05, .01, and .001. What do you conclude for each value of α?

c Suppose that a sample mean of 1,241.2 and a sample standard deviation of 110.8 had been obtained by testing a sample of 100 buttons. Use critical values to test the hypotheses you set up in part a by setting α equal to .10, .05, .01, and .001. Which sample (the sample of 35 or the sample of 100) gives a more statistically significant result? That is, which sample provides stronger evidence that H_a is true?

d If we define practical importance to mean that μ exceeds 1,200 by an amount that would be clearly noticeable to most consumers, do you think that the results of parts *b* and *c* have practical importance? Explain why the samples of 35 and 100 both indicate the same degree of practical importance.

e Suppose that further research and development effort improves the new remote control button and that a random sample of 35 buttons gives a sample mean of 1,524.6 hours and a sample standard deviation of 102.8 hours. Test your hypotheses of part *a* by setting α equal to .10, .05, .01, and .001.

(1) Do we have a highly statistically significant result? Explain.

(2) Do you think we have a practically important result? Explain.

10.61 Again consider the remote control button lifetime situation discussed in Exercise 10.60. Using the sample information given in the introduction to Exercise 10.60, the *p*-value for testing H_0 versus H_a can be calculated to be .0174.

a Determine whether H_0 would be rejected at each of $\alpha = .10$, $\alpha = .05$, $\alpha = .01$, and $\alpha = .001$.

b Describe how much evidence we have that the new button's mean lifetime exceeds the mean lifetime of the best remote button currently on the market.

10.62 Several industries located along the Ohio River discharge a toxic substance called carbon tetrachloride into the river. The state Environmental Protection Agency monitors the amount of carbon tetrachloride pollution in the river. Specifically, the agency requires that the carbon tetrachloride contamination must average no more than 10 parts per million. In order to monitor the carbon tetrachloride contamination in the river, the agency takes a daily sample of 100 pollution readings at a specified location. If the mean carbon tetrachloride reading for this sample casts substantial doubt on the hypothesis that the average amount of carbon tetrachloride contamination in the river is at most 10 parts per million, the agency must issue a shutdown order. In the event of such a shutdown order, industrial plants along the river must be closed until the carbon tetrachloride contamination is reduced to a more acceptable level. Assume that the state Environmental Protection Agency decides to issue a shutdown order if a sample of 100 pollution readings implies that $H_0: \mu \leq 10$ can be rejected in favor of $H_a: \mu > 10$ by setting $\alpha = .01$. If σ equals 2, calculate the probability of a Type II error for each of the following alternative values of μ: 10.1, 10.2, 10.3, 10.4, 10.5, 10.6, 10.7, 10.8, 10.9, and 11.0.

10.63 *Consumer Reports* (January 2005) indicates that profit margins on extended warranties are much greater than on the purchase of most products.[3] In this exercise we consider a major electronics retailer that wishes to increase the proportion of customers who buy extended warranties on digital cameras. Historically, 20 percent of digital camera customers have purchased the retailer's extended warranty. To increase this percentage, the retailer has decided to offer a new warranty that is less expensive and more comprehensive. Suppose that three months after starting to offer the new warranty, a random sample of 500 customer sales invoices shows that 152 out of 500 digital camera customers purchased the new warranty. **(1)** Letting *p* denote the proportion of all digital camera customers who have purchased the new warranty, calculate the *p*-value for testing $H_0: p = .20$ versus $H_a: p > .20$. **(2)** How much evidence is there that *p* exceeds .20? **(3)** Does the difference between $\hat{p}$ and .2 seem to be practically important? Explain your opinion.

[3]*Consumer Reports*, January 2005, page 51.

Appendix 10.1 ■ One-Sample Hypothesis Testing Using Excel

Hypothesis test for a population mean in Exercise 10.33 (data file: CreditCd.xlsx):

The Data Analysis ToolPak in Excel does not explicitly provide for one-sample tests of hypotheses. A one-sample test can be conducted using the Descriptive Statistics component of the Analysis ToolPak and a few additional computations using Excel.

Descriptive statistics:

- Enter the interest rate data from Exercise 10.33 (into cells A2:A16 with the label Rate in cell A1.
- Select **Data : Data Analysis : Descriptive Statistics.**
- Click OK in the Data Analysis dialog box.
- In the Descriptive Statistics dialog box, enter A1:A16 into the Input Range box.
- Place checkmarks in the "Labels in first row" and Summary Statistics checkboxes.
- Under output options, select "New Worksheet Ply" and enter Output for the worksheet's name.

- Click OK in the Descriptive Statistics dialog box.

The resulting block of descriptive statistics is displayed in the Output worksheet and entries needed to carry out the *t* test have been entered into the cell range D3:E6.

Computation of the test statistic and *p*-value:

- In cell E7, type the formula
  =(E3-E4)/(E5/SQRT(E6))
  to compute the test statistic t (= −4.970).
- Click on cell E8 and then select the Insert Function button f_x on the Excel toolbar.
- In the Insert Function dialog box, select Statistical from the "Or select a category:" menu, select T.DIST from the "Select a function:" menu, and click OK.
- In the T.DIST Function Arguments dialog box, enter E7 into the X window.
- Enter 14 into the Deg_freedom window and Enter 1 into the Cumulative window.
- Click OK in the T.DIST Function Arguments dialog box.
- The *p*-value for the test (=0.000103) will be placed in cell E8.

The T.DIST function returns the left tail area under the appropriate *t* curve. Because we are testing a "less than" alternative hypothesis in this example, the desired *p*-value is a left-tail area. If we are testing a **"greater than" alternative,** the *p*-value (which is a right-tail area) is found by using a cell formula to subtract the left-tail area provided by Excel from one. In the case of a **"not equal to" alternative,** we use the T.DIST function to find the area under the *t* curve to the left of the absolute value (abs) of the *t* statistic. We then use cell formula(s) to subtract this area from one and to multiply the resulting right-tail area by two.

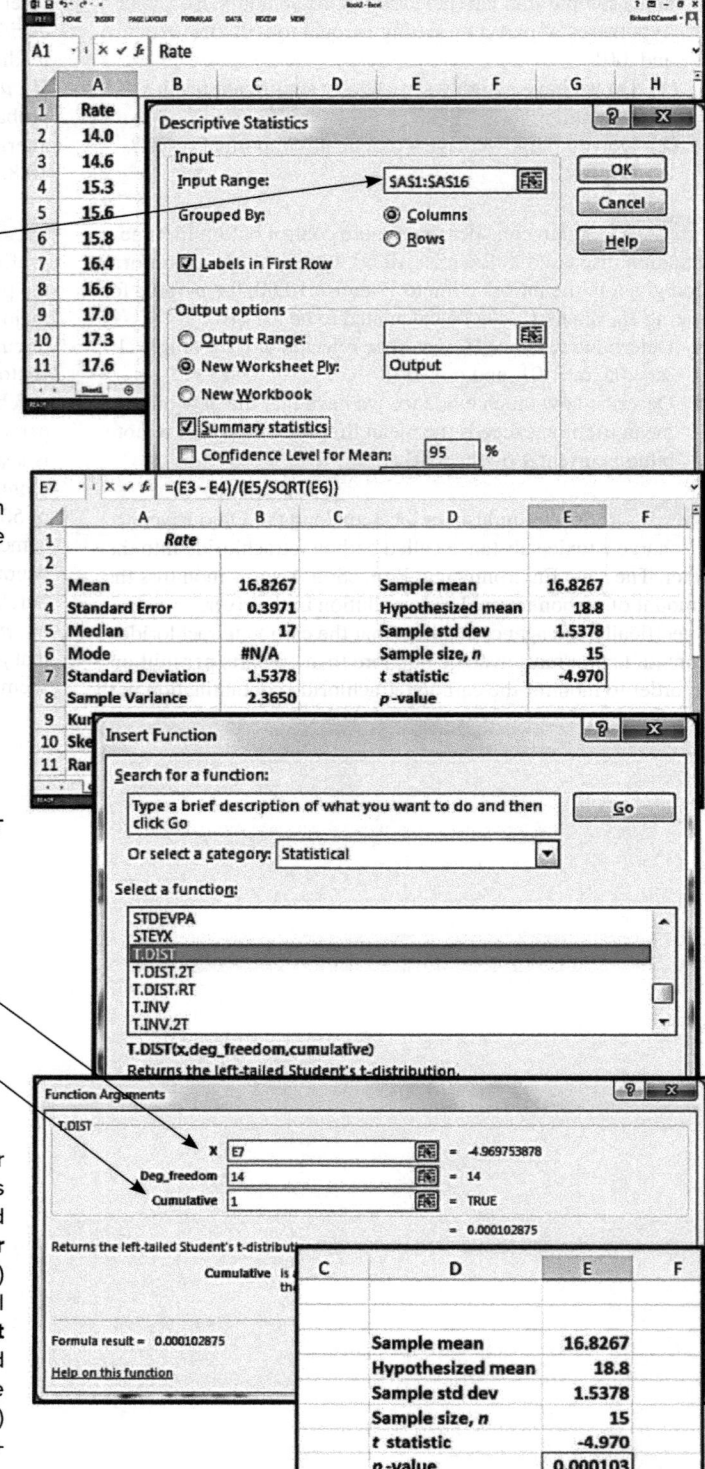

Appendix 10.2 ■ One-Sample Hypothesis Testing Using Minitab

Hypothesis test for a population mean in Exercise 10.33 (data file: CreditCd.MTW):

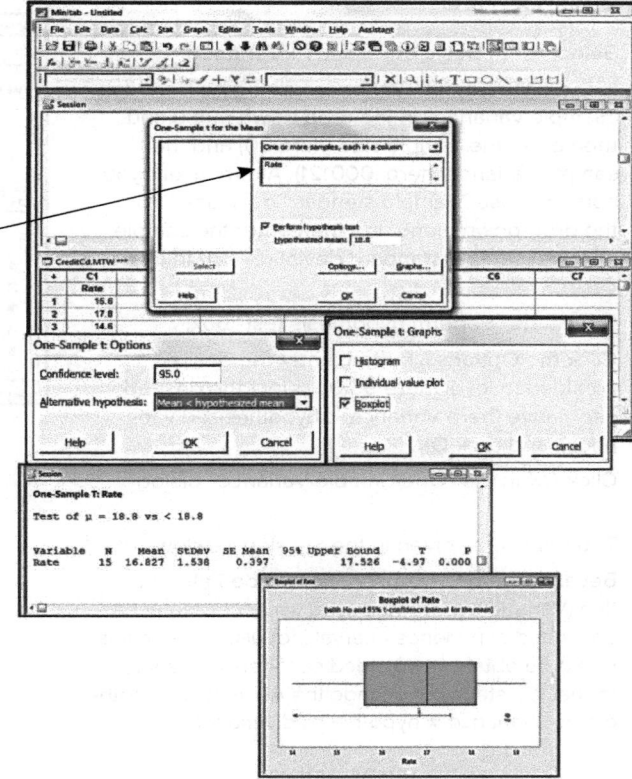

- In the Worksheet window, enter the interest rate data from Exercise 10.33 into a single column with variable name Rate.

- **Select Stat : Basic Statistics : 1-Sample t.**

- In the "One-Sample t for the Mean" dialog box, select "One or more samples, each in a column."

- Enter Rate into the next window as shown.

- Check the "Perform hypothesis test" box and enter 18.8 as the Hypothesized mean.

- Click on the Options... button.

- In the "One-Sample t: Options" dialog box, enter the desired level of confidence (here 95.0) into the Confidence level window.

- Select the desired alternative (here "Mean < hypothesized mean") from the drop-down menu and click OK in the "One-Sample t: Options" dialog box.

- To produce a box plot of the data with a graphical representation of the hypothesis test, click the Graphs... button; check the "Boxplot" box; and click OK in the "One-Sample t: Graphs" dialog box.

- Click OK in the "One-Sample t for the Mean" dialog box.

- The *t* test results are given in the Session window and the box plot is displayed in a graphics window.

Source: Minitab 18

Hypothesis test for a population proportion in Exercise 10.38.

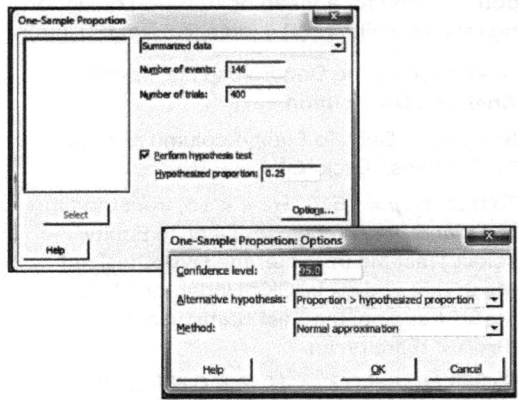

- **Select Stat : Basic Statistics : 1 Proportion**

- In the "One-Sample Proportion" dialog box, select "Summarized data."

- Enter the number of events (here 146) and the number of trials (here 400).

- Check the "Perform hypothesis test" box and enter the Hypothesized proportion (here 0.25).

- Click on the Options... button.

- In the "One-Sample Proportion: Options" dialog box, select the desired alternative (here "Proportion > hypothesized proportion") from the drop-down menu.

- Choose "Normal approximation" from the "Method" drop-down menu.

- Click OK in the "One-Sample Proportion: Options" and in the "One-Sample Proportion" dialog boxes.

- The test results appear in the Session window.

Source: Minitab 18

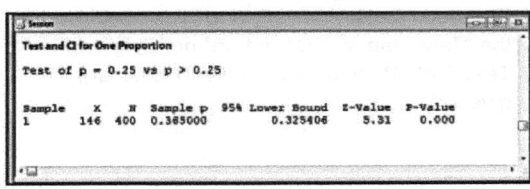

Source: Minitab 18

Hypothesis test and confidence intervals for a population variance in the example in Section 10.7, where the hypothesized variance and standard deviation are .000625 and .025.

- **Select Stat : Basic Statistics : 1 Variance**
- In the "One-Sample Variance" dialog box, choose "Sample variance" in the drop-down menu and then enter the sample size (here 30) and the sample variance (here .000121). Alternatively, you could choose "Sample standard deviation" from the drop-down menu and then enter the sample size and sample standard deviation (.011 in this case).
- Check the "Perform hypothesis test" box.
- Click the Options... button; enter the desired confidence level (here 95.0); select the desired alternative (here Variance < hypothesized variance); and click OK.
- Click OK in the "One-Sample Variance" dialog box.
- The results are given in the Session window.
- Because we have chosen a one-sided alternative (Variance < hypothesized variance), we get a one-sided confidence interval (not discussed in this book). To obtain a two-sided confidence interval, repeat the steps but change the Alternative hypothesis to "Variance ≠ hypothesized variance."

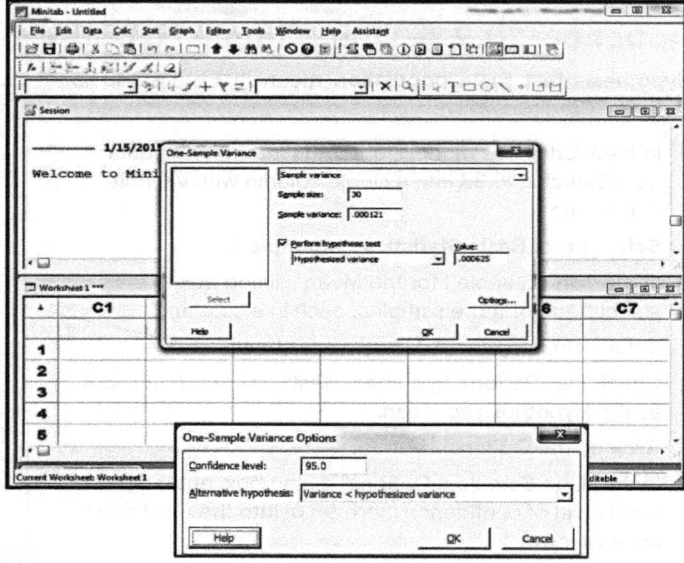

Source: Minitab 18

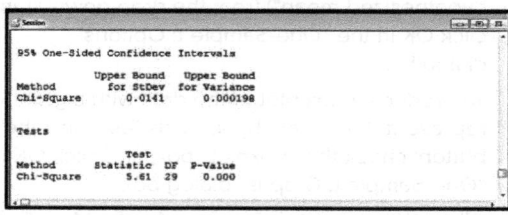

Source: Minitab 18

Appendix 10.3 ■ Hypothesis Testing Using JMP

Hypothesis test for a mean or standard deviation using raw data: Figure 10.4 (data file: DebtEq.jmp):

- After opening the DebtEq.jmp file, select **Analyze : Distribution**.
- Select the "Debt To Equity" column and click "Y, Columns." Click "OK."
- To test H_0: $\mu = 1.5$ vs. H_a: $\mu < 1.5$, from the red triangle menu beside "Debt To Equity," select "Test Mean." Enter the **Hypothesized Mean** (1.5) and click "OK." Under the "Test Mean" heading, the **Test Statistic** and p-value **(Prob < t)** are given.
- To test H_0: $\sigma = 0.3$ vs. H_a: $\sigma < 0.3$, from the red triangle menu beside "Debt to Equity," select "Test Std Dev." Enter the **Hypothesized Standard Deviation** (0.3) and click "OK." Under the "Test Standard Deviation" heading, the **Test Statistic** and p-value **(Prob < Chisq)** are given.

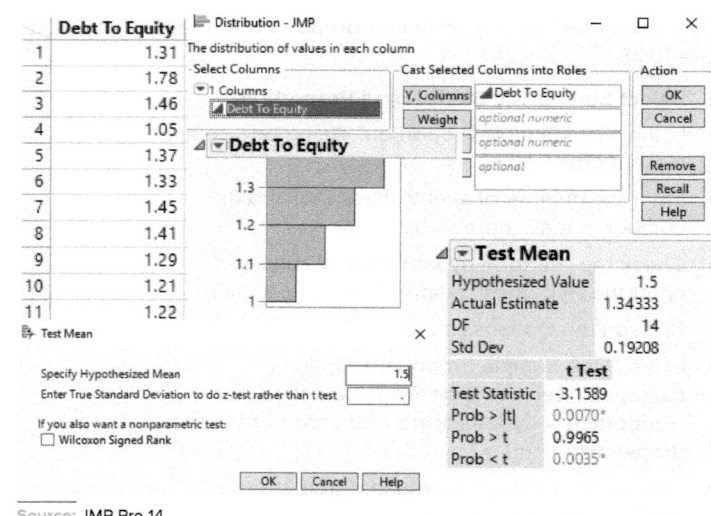

Source: JMP Pro 14

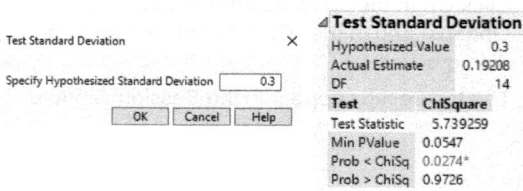

Source: JMP Pro 14

Hypothesis test for a mean using summary statistics: Testing H_0: $\mu = 330$ vs H_a: $\mu \neq 330$ in Section 10.3:

- Select **Help : Sample Data**.

- Expand the **Calculators** area under "Teaching resources" by clicking on the grey triangle next to **Calculators**. Select "Hypothesis Test for One Mean." Select "Summary Statistics" and click "OK."

- Select "*t*-test" under "Choose Type of Test." Choose the type of the alternative hypothesis ("two-tailed"). Enter the summary statistics under the "Test Inputs" heading. The test statistic ("*t*-score") and *p*-value will be automatically calculated in "Test Results."

- You can also check the "Reveal Decision" box to show whether or not the null hypothesis should be rejected.

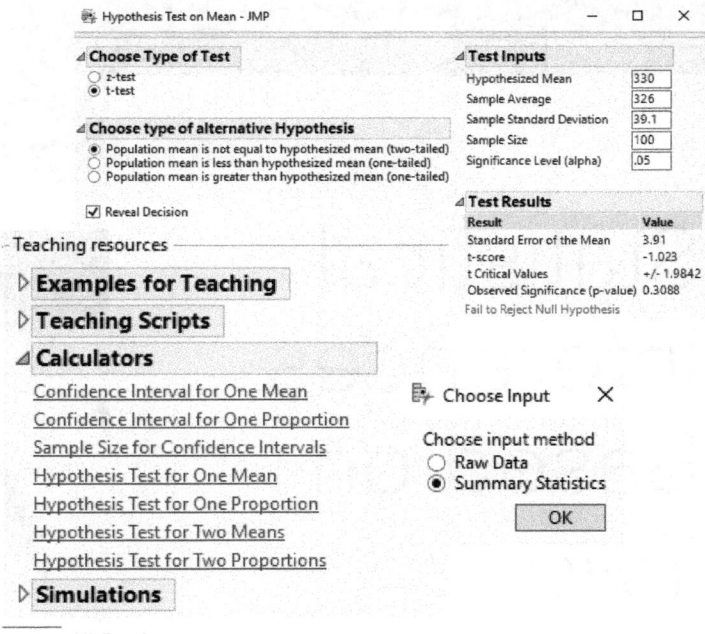

Source: JMP Pro 14

Hypothesis test for a proportion: Testing H_0: $p = .1$ vs H_a: $p < .1$ in Section 10.6:

- Select **Help : Sample Data**.

- Expand the **Calculators** area by clicking on the gray triangle next to **Calculators**. Select "Hypothesis Test for One Proportion." Select "Summary Statistics" and click "OK."

- Choose the type of the alternative hypothesis ("less than"—"one tailed." Enter the summary statistics under the "Test Inputs" heading. The "Test Statistic" and "*p*-value" will be automatically calculated in "Test Results."

- You can also check the Reveal Decision box to show whether or not the null hypothesis should be rejected.

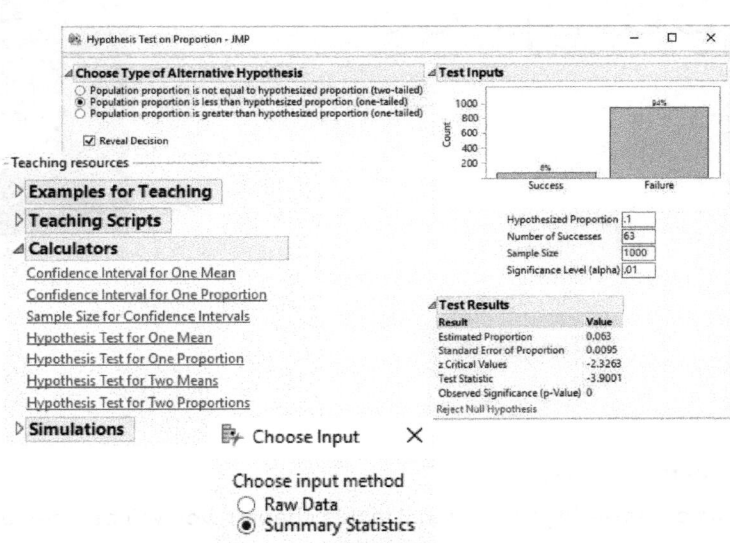

Source: JMP Pro 14

Design Elements: (CD): ©Comstock Images/Alamy; (All Others): ©McGraw-Hill Education

## CHAPTER 11

Statistical Inferences Based on Two Samples

©Chris Ryan/Media Bakery RF

Learning Objectives

After mastering the material in this chapter, you will be able to:

LO11-1 Compare two population means when the samples are independent.

LO11-2 Recognize when data come from independent samples and when they are paired.

LO11-3 Compare two population means when the data are paired.

LO11-4 Compare two population proportions using large independent samples.

LO11-5 Describe the properties of the F distribution and use an F table.

LO11-6 Compare two population variances when the samples are independent.

Chapter Outline

usiness improvement often requires making comparisons. For example, to increase consumer awareness of a product or service, it might be necessary to compare different types of advertising campaigns. Or to offer more profitable investments to its customers, an investment firm might compare the profitability of different investment portfolios. As a third example, a manufacturer might compare different production methods in order to minimize or eliminate out-of-specification product.

In this chapter we discuss using confidence intervals and hypothesis tests to **compare two populations.** Specifically, we compare two population means, two population variances, and two population proportions. For instance, to compare two population means, say μ_1 and μ_2, we consider the difference between these means, $\mu_1 - \mu_2$. If, for example, we use a confidence interval or hypothesis test to conclude that $\mu_1 - \mu_2$ is a positive number, then we conclude that μ_1 is greater than μ_2. On the other hand, if a confidence interval or hypothesis test shows that $\mu_1 - \mu_2$ is a negative number, then we conclude that μ_1 is less than μ_2. As another example, if we wish to compare two population variances, say σ_1^2 and σ_2^2, we might test the null hypothesis that the two population variances are equal versus the alternative hypothesis that the two population variances are not equal.

We explain many of this chapter's methods in the context of three new cases:

The Catalyst Comparison Case: The production supervisor at a chemical plant uses confidence intervals and hypothesis tests for the difference between two population means to determine which of two catalysts maximizes the hourly yield of a chemical process. By maximizing yield, the plant increases its productivity and improves its profitability.

The Auto Insurance Case: In order to reduce the costs of automobile accident claims, an insurance company uses confidence intervals and hypothesis tests for the difference between two population means to compare repair cost estimates for damaged cars at two different garages.

The Test Market Case: An advertising agency is test marketing a new product by using one advertising campaign in Des Moines, Iowa, and a different campaign in Toledo, Ohio. The agency uses confidence intervals and hypothesis tests for the difference between two population proportions to compare the effectiveness of the two advertising campaigns.

11.1 Comparing Two Population Means by Using Independent Samples

LO11-1

Compare two population means when the samples are independent.

A bank manager has developed a new system to reduce the time customers spend waiting to be served by tellers during peak business hours. We let μ_1 denote the population mean customer waiting time during peak business hours under the current system. To estimate μ_1, the manager randomly selects $n_1 = 100$ customers and records the length of time each customer spends waiting for service. The manager finds that the mean and the variance of the waiting times for these 100 customers are $\bar{x}_1 = 8.79$ minutes and $s_1^2 = 4.8237$. We let μ_2 denote the population mean customer waiting time during peak business hours for the new system. During a trial run, the manager finds that the mean and the variance of the waiting times for a random sample of $n_2 = 100$ customers are $\bar{x}_2 = 5.14$ minutes and $s_2^2 = 1.7927$.

In order to compare μ_1 and μ_2, the manager estimates $\mu_1 - \mu_2$, the difference between μ_1 and μ_2. Intuitively, a logical point estimate of $\mu_1 - \mu_2$ is the difference between the sample means

$$\bar{x}_1 - \bar{x}_2 = 8.79 - 5.14 = 3.65 \text{ minutes}$$

This says we estimate that the current population mean waiting time is 3.65 minutes longer than the population mean waiting time under the new system. That is, we estimate that the new system reduces the mean waiting time by 3.65 minutes.

To compute a confidence interval for $\mu_1 - \mu_2$ (or to test a hypothesis about $\mu_1 - \mu_2$), we need to know the properties of the sampling distribution of $\bar{x}_1 - \bar{x}_2$. To understand this

©Ryan McVay/Getty Images RF

477

FIGURE 11.1   **The Sampling Distribution of $\bar{x}_1 - \bar{x}_2$ Has Mean $\mu_1 - \mu_2$ and Standard Deviation $\sigma_{\bar{x}_1-\bar{x}_2}$**

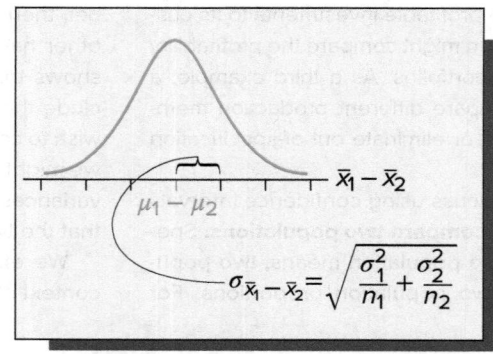

sampling distribution, consider randomly selecting a sample[1] of n_1 measurements from a population having mean μ_1 and variance σ_1^2. Let $\bar{x}_1$ be the mean of this sample. Also consider randomly selecting a sample of n_2 measurements from another population having mean μ_2 and variance σ_2^2. Let $\bar{x}_2$ be the mean of this sample. Different samples from the first population would give different values of $\bar{x}_1$, and different samples from the second population would give different values of $\bar{x}_2$—so different pairs of samples from the two populations would give different values of $\bar{x}_1 - \bar{x}_2$. In the following box we describe the **sampling distribution of $\bar{x}_1 - \bar{x}_2$**, which is the probability distribution of all possible values of $\bar{x}_1 - \bar{x}_2$. Here we assume that the randomly selected samples from the two populations are independent of each other. This means that there is no relationship between the measurements in one sample and the measurements in the other sample. In such a case, we say that we are performing an **independent samples experiment.**

The Sampling Distribution of $\bar{x}_1 - \bar{x}_2$

If the randomly selected samples are **independent** of each other, then the population of all possible values of $\bar{x}_1 - \bar{x}_2$

1   Has a normal distribution if each sampled population has a normal distribution, or has approximately a normal distribution if the sampled populations are not normally distributed and each of the sample sizes n_1 and n_2 is large.

2   Has mean $\mu_{\bar{x}_1-\bar{x}_2} = \mu_1 - \mu_2$

3   Has standard deviation $\sigma_{\bar{x}_1-\bar{x}_2} = \sqrt{\dfrac{\sigma_1^2}{n_1} + \dfrac{\sigma_2^2}{n_2}}$

Figure 11.1 illustrates the sampling distribution of $\bar{x}_1 - \bar{x}_2$. Using this sampling distribution, we can find a confidence interval for and test a hypothesis about $\mu_1 - \mu_2$ by using the normal distribution. However, the interval and test assume that the true values of the population variances σ_1^2 and σ_2^2 are known, which is very unlikely. Therefore, we will estimate σ_1^2 and σ_2^2 by using s_1^2 and s_2^2, the variances of the samples randomly selected from the populations being compared, and base a confidence interval and a hypothesis test on the t distribution. There are two approaches to doing this. The first approach gives theoretically correct confidence intervals and hypothesis tests but assumes that the population variances σ_1^2 and σ_2^2 are equal. The second approach does not require that σ_1^2 and σ_2^2 are equal but gives only approximately correct confidence intervals and hypothesis tests. In the bank customer waiting time situation, the sample variances are $s_1^2 = 4.8237$ and $s_2^2 = 1.7927$.

---

[1]Each sample in this chapter is a *random* sample. As has been our practice throughout this book, for brevity we sometimes refer to "random samples" as "samples."

The difference in these sample variances makes it questionable to assume that the population variances are equal. More will be said later about deciding whether we can assume that two population variances are equal and about choosing between the two t distribution approaches in a particular situation. For now, we will first consider the case where the population variances σ_1^2 and σ_2^2 can be assumed to be equal. Denoting the common value of these variances as σ^2, it follows that

$$\sigma_{\bar{x}_1 - \bar{x}_2} = \sqrt{\frac{\sigma_1^2}{n_1} + \frac{\sigma_2^2}{n_2}} = \sqrt{\frac{\sigma^2}{n_1} + \frac{\sigma^2}{n_2}} = \sqrt{\sigma^2 \left(\frac{1}{n_1} + \frac{1}{n_2} \right)}$$

Because we are assuming that $\sigma_1^2 = \sigma_2^2 = \sigma^2$, we do not need separate estimates of σ_1^2 and σ_2^2. Instead, we combine the results of the two independent random samples to compute a single estimate of σ^2. This estimate is called the *pooled estimate* of σ^2, and it is a weighted average of the two sample variances s_1^2 and s_2^2. Denoting the pooled estimate as s_p^2, it is computed using the formula

$$s_p^2 = \frac{(n_1 - 1)s_1^2 + (n_2 - 1)s_2^2}{n_1 + n_2 - 2}$$

Using s_p^2, the estimate of $\sigma_{\bar{x}_1 - \bar{x}_2}$ is

$$\sqrt{s_p^2 \left(\frac{1}{n_1} + \frac{1}{n_2} \right)}$$

and we form the statistic

$$\frac{(\bar{x}_1 - \bar{x}_2) - (\mu_1 - \mu_2)}{\sqrt{s_p^2 \left(\frac{1}{n_1} + \frac{1}{n_2} \right)}}$$

It can be shown that, if we have randomly selected independent samples from two normally distributed populations having equal variances, then the sampling distribution of this statistic is a t distribution having $(n_1 + n_2 - 2)$ degrees of freedom. Therefore, we can obtain the following confidence interval for $\mu_1 - \mu_2$:

A t-Based Confidence Interval for the Difference between Two Population Means: Equal Variances

Suppose we have randomly selected independent samples from two normally distributed populations having equal variances. Then, a $100(1 - \alpha)$ percent confidence interval for $\mu_1 - \mu_2$ is

$$\left[(\bar{x}_1 - \bar{x}_2) \pm t_{\alpha/2} \sqrt{s_p^2 \left(\frac{1}{n_1} + \frac{1}{n_2} \right)} \right] \quad \text{where} \quad s_p^2 = \frac{(n_1 - 1)s_1^2 + (n_2 - 1)s_2^2}{n_1 + n_2 - 2}$$

and $t_{\alpha/2}$ is based on $(n_1 + n_2 - 2)$ degrees of freedom.

Ⓒ **EXAMPLE 11.1** The Catalyst Comparison Case: Process Improvement

A production supervisor at a major chemical company must determine which of two catalysts, catalyst XA-100 or catalyst ZB-200, maximizes the hourly yield of a chemical process. In order to compare the mean hourly yields obtained by using the two catalysts, the supervisor runs the process using each catalyst for five one-hour periods. The resulting yields (in pounds per hour) for each catalyst, along with the means, variances, and box plots[2] of the

---

[2]All of the box plots presented in this chapter and in Chapter 12 have been obtained using Minitab.

TABLE 11.1 **Yields of a Chemical Process Obtained Using Two Catalysts** ⟲ Catalyst

Catalyst XA-100	Catalyst ZB-200
801	752
814	718
784	776
836	742
820	763
$\bar{x}_1 = 811$	$\bar{x}_2 = 750.2$
$s_1^2 = 386$	$s_2^2 = 484.2$

Boxplot of XA-100, ZB-200

yields, are given in Table 11.1. Assuming that all other factors affecting yields of the process have been held as constant as possible during the test runs, it seems reasonable to regard the five observed yields for each catalyst as a random sample from the population of all possible hourly yields for the catalyst. Furthermore, because the sample variances $s_1^2 = 386$ and $s_2^2 = 484.2$ do not differ substantially (notice that $s_1 = 19.65$ and $s_2 = 22.00$ differ by even less), it might be reasonable to conclude that the population variances are approximately equal.[3] It follows that the pooled estimate

$$s_p^2 = \frac{(n_1 - 1)s_1^2 + (n_2 - 1)s_2^2}{n_1 + n_2 - 2}$$

$$= \frac{(5 - 1)(386) + (5 - 1)(484.2)}{5 + 5 - 2} = 435.1$$

is a point estimate of the common variance σ^2.

We define μ_1 as the mean hourly yield obtained by using catalyst XA-100, and we define μ_2 as the mean hourly yield obtained by using catalyst ZB-200. If the populations of all possible hourly yields for the catalysts are normally distributed, then a 95 percent confidence interval for $\mu_1 - \mu_2$ is

$$\left[(\bar{x}_1 - \bar{x}_2) \pm t_{.025} \sqrt{s_p^2 \left(\frac{1}{n_1} + \frac{1}{n_2} \right)} \right]$$

$$= \left[(811 - 750.2) \pm 2.306 \sqrt{435.1 \left(\frac{1}{5} + \frac{1}{5} \right)} \right]$$

$$= [60.8 \pm 30.4217]$$

$$= [30.38, 91.22]$$

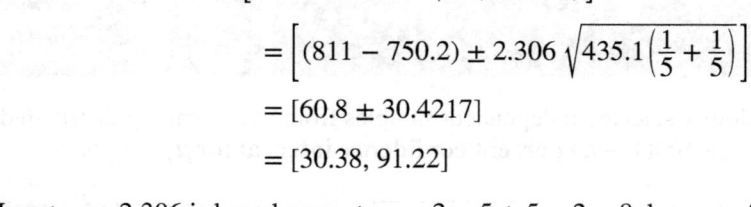

Here $t_{.025} = 2.306$ is based on $n_1 + n_2 - 2 = 5 + 5 - 2 = 8$ degrees of freedom. This interval tells us that we are 95 percent confident that the mean hourly yield obtained by using catalyst XA-100 is between 30.38 and 91.22 pounds higher than the mean hourly yield obtained by using catalyst ZB-200.

Suppose we wish to test a hypothesis about $\mu_1 - \mu_2$. In the following box we describe how this can be done. Here we test the null hypothesis $H_0: \mu_1 - \mu_2 = D_0$, where D_0 is a number whose value varies depending on the situation. Often D_0 will be the number 0. In such a case, the null hypothesis $H_0: \mu_1 - \mu_2 = 0$ says there is **no difference** between the population means μ_1 and μ_2. In this case, each alternative hypothesis in the box implies that the population means μ_1 and μ_2 differ in a particular way.

---

[3]We describe how to test the equality of two variances in Section 11.5 (although, as we will explain, this test has drawbacks).

A t Test about the Difference between Two Population Means: Equal Variances

Null Hypothesis	Test Statistic	Assumptions
$H_0: \mu_1 - \mu_2 = D_0$	$t = \dfrac{(\bar{x}_1 - \bar{x}_2) - D_0}{\sqrt{s_p^2 \left(\dfrac{1}{n_1} + \dfrac{1}{n_2}\right)}}$	Independent samples and Equal variances and either Normal populations or Large sample sizes

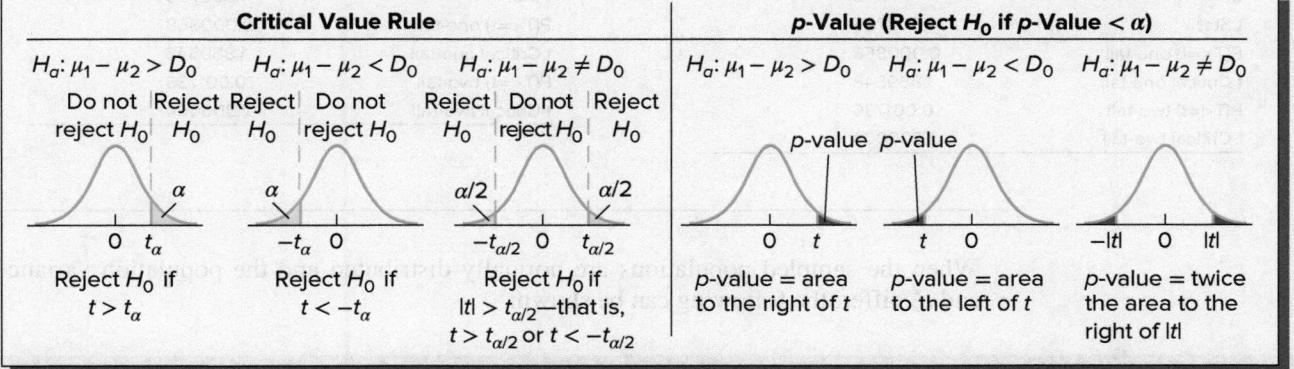

Here t_α, $t_{\alpha/2}$, and the p-values are based on $n_1 + n_2 - 2$ degrees of freedom.

C **EXAMPLE 11.2** The Catalyst Comparison Case: Process Improvement

In order to compare the mean hourly yields obtained by using catalysts XA-100 and ZB-200, we will test $H_0: \mu_1 - \mu_2 = 0$ versus $H_a: \mu_1 - \mu_2 \neq 0$ at the **.05 level of significance.** To perform the hypothesis test, we will use the sample information in Table 11.1 to calculate the value of the **test statistic t in the summary box.** Then, because $H_a: \mu_1 - \mu_2 \neq 0$ says to use the two-tailed critical value rule in the summary box, we will **reject $H_0: \mu_1 - \mu_2 = 0$ if the absolute value of t is greater than $t_{\alpha/2} = t_{.025} = 2.306$.** Here the $t_{\alpha/2}$ point is based on $n_1 + n_2 - 2 = 5 + 5 - 2 = 8$ degrees of freedom. Using the data in Table 11.1, the **value of the test statistic** is

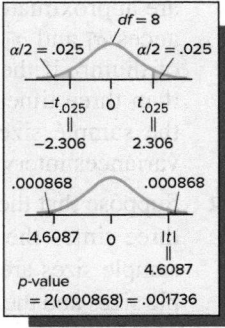

$$t = \frac{(\bar{x}_1 - \bar{x}_2) - D_0}{\sqrt{s_p^2 \left(\dfrac{1}{n_1} + \dfrac{1}{n_2}\right)}} = \frac{(811 - 750.2) - 0}{\sqrt{435.1\left(\dfrac{1}{5} + \dfrac{1}{5}\right)}} = 4.6087$$

Because $|t| = 4.6087$ is greater than $t_{.025} = 2.306$, we can reject $H_0: \mu_1 - \mu_2 = 0$ in favor of $H_a: \mu_1 - \mu_2 \neq 0$. We conclude (at an α of .05) that the mean hourly yields obtained by using the two catalysts differ. Furthermore, the point estimate $\bar{x}_1 - \bar{x}_2 = 811 - 750.2 = 60.8$ says we estimate that the mean hourly yield obtained by using catalyst XA-100 is 60.8 pounds higher than the mean hourly yield obtained by using catalyst ZB-200.

Figure 11.2(a) gives the Excel output for using the equal variance t statistic to test $H_0: \mu_1 - \mu_2 = 0$ versus $H_a: \mu_1 - \mu_2 \neq 0$. The output tells us that $t = 4.6087$ and that the associated p-value is .001736. This p-value is the two-tailed p-value in the summary box: twice the area under the t distribution curve having 8 degrees of freedom to the right of $|t| = 4.6087$. The very small p-value of .001736 tells us that we have very strong evidence against $H_0: \mu_1 - \mu_2 = 0$ and in favor of $H_a: \mu_1 - \mu_2 \neq 0$. In other words, we have very strong evidence that the mean hourly yields obtained by using the two catalysts differ. (Note that in Figure 11.2(b) we give the Excel output for using an *unequal variances t statistic,* which is discussed on the following pages, to perform the hypothesis test.)

FIGURE 11.2   Excel Outputs for Testing the Equality of Means in the Catalyst Comparison Case

(a) The Excel Output Assuming Equal Variances

t-Test: Two-Sample Assuming Equal Variances

	XA-100	ZB-200
Mean	811	750.2
Variance	386	484.2
Observations	5	5
Pooled Variance	435.1	
Hypothesized Mean Diff	0	
df	8	
t Stat	4.608706	
P(T<=t) one-tail	0.000868	
t Critical one-tail	1.859548	
P(T<=t) two-tail	0.001736	
t Critical two-tail	2.306004	

(b) The Excel Output Assuming Unequal Variances

t-Test: Two-Sample Assuming Unequal Variances

	XA-100	ZB-200
Mean	811	750.2
Variance	386	484.2
Observations	5	5
Hypothesized Mean Diff	0	
df	8	
t Stat	4.608706	
P(T<=t) one-tail	0.000868	
t Critical one-tail	1.859548	
P(T<=t) two-tail	0.001736	
t Critical two-tail	2.306004	

When the sampled populations are normally distributed and the population variances σ_1^2 and σ_2^2 differ, the following can be shown.

t-Based Confidence Intervals for $\mu_1 - \mu_2$, and t Tests about $\mu_1 - \mu_2$: Unequal Variances

1 When the sample sizes n_1 and n_2 are equal, the "equal variances" t-based confidence interval and hypothesis test given in the preceding two boxes are approximately valid even if the population variances σ_1^2 and σ_2^2 differ substantially. As a rough rule of thumb, if the larger sample variance is not more than three times the smaller sample variance when the sample sizes are equal, we can use the equal variances interval and test.

2 Suppose that the larger sample variance is more than three times the smaller sample variance when the sample sizes are equal or suppose that both the sample sizes and the sample variances differ substantially. Then, we can use an approximate procedure that is sometimes called an "unequal variances" procedure. This procedure says that an **approximate $100(1 - \alpha)$ percent confidence interval for $\mu_1 - \mu_2$ is**

$$\left[(\bar{x}_1 - \bar{x}_2) \pm t_{\alpha/2} \sqrt{\frac{s_1^2}{n_1} + \frac{s_2^2}{n_2}} \right]$$

Furthermore, we can test $H_0: \mu_1 - \mu_2 = D_0$ by using the test statistic

$$t = \frac{(\bar{x}_1 - \bar{x}_2) - D_0}{\sqrt{\frac{s_1^2}{n_1} + \frac{s_2^2}{n_2}}}$$

and by using the previously given critical value rules and p-values.

For both the interval and the test, the degrees of freedom are equal to

$$df = \frac{\left(s_1^2/n_1 + s_2^2/n_2\right)^2}{\dfrac{\left(s_1^2/n_1\right)^2}{n_1 - 1} + \dfrac{\left(s_2^2/n_2\right)^2}{n_2 - 1}}$$

Here, if df is not a whole number, we can round df down to the next smallest whole number.

In general, both the "equal variances" and the "unequal variances" procedures have been shown to be approximately valid when the sampled populations are only approximately normally distributed (say, if they are mound-shaped). Furthermore, although the above summary box might seem to imply that we should use the unequal variances procedure only if we cannot use the equal variances procedure, this is not necessarily true. In fact, because the unequal variances procedure can be shown to be a very accurate approximation whether or not the population variances are equal and for most sample sizes (here, both n_1 and n_2 should be at least 5), **many statisticians believe that it is best to use the unequal variances procedure in almost every situation.** If each of n_1 and n_2 is large (at least 30), both the equal variances procedure and the unequal variances procedure are approximately valid, no matter what probability distributions describe the sampled populations.

To illustrate the unequal variances procedure, consider the bank customer waiting time situation, and recall that $\mu_1 - \mu_2$ is the difference between the mean customer waiting time under the current system and the mean customer waiting time under the new system. Because of cost considerations, the bank manager wants to implement the new system only if it reduces the mean waiting time by more than three minutes. Therefore, the manager will test the **null hypothesis H_0: $\mu_1 - \mu_2 = 3$ versus the alternative hypothesis H_a: $\mu_1 - \mu_2 > 3$.** If H_0 can be rejected in favor of H_a at the **.05 level of significance,** the manager will implement the new system. Recall that a random sample of $n_1 = 100$ waiting times observed under the current system gives a sample mean $\bar{x}_1 = 8.79$ and a sample variance $s_1^2 = 4.8237$. Also, recall that a random sample of $n_2 = 100$ waiting times observed during the trial run of the new system yields a sample mean $\bar{x}_2 = 5.14$ and a sample variance $s_2^2 = 1.7927$. Because each sample is large, we can use the **unequal variances test statistic t in the summary box.** The degrees of freedom for this statistic are

$$df = \frac{\left(s_1^2/n_1 + s_2^2/n_2\right)^2}{\dfrac{\left(s_1^2/n_1\right)^2}{n_1 - 1} + \dfrac{\left(s_2^2/n_2\right)^2}{n_2 - 1}}$$

$$= \frac{[(4.8237/100) + (1.7927/100)]^2}{\dfrac{(4.8237/100)^2}{99} + \dfrac{(1.7927/100)^2}{99}}$$

$$= 163.657$$

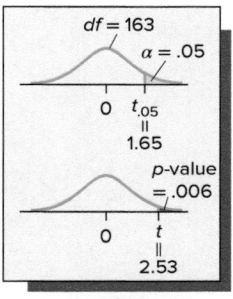

which we will round down to 163. Therefore, because H_a: $\mu_1 - \mu_2 > 3$ says to use the right-tailed critical value rule in the summary box, we will **reject H_0: $\mu_1 - \mu_2 = 3$ if the value of the test statistic t is greater than $t_\alpha = t_{.05} = 1.65$** (which is based on 163 degrees of freedom and has been found using a computer). Using the sample data, the **value of the test statistic** is

$$t = \frac{(\bar{x}_1 - \bar{x}_2) - 3}{\sqrt{\dfrac{s_1^2}{n_1} + \dfrac{s_2^2}{n_2}}} = \frac{(8.79 - 5.14) - 3}{\sqrt{\dfrac{4.8237}{100} + \dfrac{1.7927}{100}}} = \frac{.65}{.25722} = 2.53$$

Because $t = 2.53$ is greater than $t_{.05} = 1.65$, we reject H_0: $\mu_1 - \mu_2 = 3$ in favor of H_a: $\mu_1 - \mu_2 > 3$. We conclude (at an α of .05) that $\mu_1 - \mu_2$ is greater than 3 and, therefore, that the new system reduces the population mean customer waiting time by more than 3 minutes. Therefore, the bank manager will implement the new system. Furthermore, the point estimate $\bar{x}_1 - \bar{x}_2 = 3.65$ says that we estimate that the new system reduces mean waiting time by 3.65 minutes.

The JMP output on the page margin and the Minitab output in Figure 11.3 show the results of using the unequal variances procedure to test H_0: $\mu_1 - \mu_2 = 3$ versus H_a: $\mu_1 - \mu_2 > 3$. The outputs tell us that $t = 2.53$ and that the associated p-value is .006. This p-value is the right-tailed p-value in the summary box: the area under the t distribution curve having 163 degrees of freedom to the right of $t = 2.53$. The very small p-value of .006 tells us that we have very strong evidence against H_0: $\mu_1 - \mu_2 = 3$ and in favor of H_a: $\mu_1 - \mu_2 > 3$. That is, we have very strong evidence that $\mu_1 - \mu_2$ is greater than 3 and, therefore, that the new system reduces the mean customer waiting time by more than 3 minutes. To find a 95 percent confidence interval for $\mu_1 - \mu_2$, note that we can use a computer to find that $t_{.025}$ based on 163 degrees of freedom is 1.97. It follows that the 95 percent confidence interval for $\mu_1 - \mu_2$ is

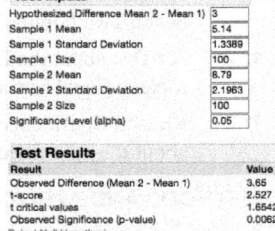

Test Inputs

Hypothesized Difference Mean 2 - Mean 1)	3
Sample 1 Mean	5.14
Sample 1 Standard Deviation	1.3389
Sample 1 Size	100
Sample 2 Mean	8.79
Sample 2 Standard Deviation	2.1963
Sample 2 Size	100
Significance Level (alpha)	0.05

Test Results

Result	Value
Observed Difference (Mean 2 - Mean 1)	3.65
t-score	2.527
t critical values	1.6542
Observed Significance (p-value)	0.0062
Reject Null Hypothesis	

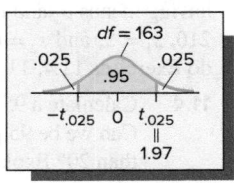

$$\left[(\bar{x}_1 - \bar{x}_2) \pm t_{.025} \sqrt{\frac{s_1^2}{n_1} + \frac{s_2^2}{n_2}} \right] = \left[(8.79 - 5.14) \pm 1.97 \sqrt{\frac{4.8237}{100} + \frac{1.7927}{100}} \right]$$

$$= [3.65 \pm .50792] = [3.14, 4.16]$$

FIGURE 11.3   Minitab Output of the Unequal Variances
              Procedure for the Bank Customer Waiting
              Time Case

Two-Sample T-Test and CI

Sample	N	Mean	StDev	SE Mean
Current	100	8.79	2.20	0.22
New	100	5.14	1.34	0.13

Difference = mu(1) - mu(2)
Estimate for difference: 3.650
95% lower bound for difference: 3.224
T Test of difference = 3 (vs >):
   T-Value = 2.53   P-Value = 0.006   DF = 163

FIGURE 11.4   Minitab Output of the Unequal Variances
              Procedure for the Catalyst Comparison
              Case

Two-Sample T-Test and CI: XA-100, ZB-200

	N	Mean	StDev	SE Mean
XA-100	5	811.0	19.6	8.8
ZB-200	5	750.2	22.0	9.8

Difference = mu (XA-100) - mu (ZB-200)
Estimate for difference: 60.8000
95% CI for difference: (29.6049, 91.9951)
T Test of difference = 0 (vs not =):
   T-Value = 4.61   P-Value = 0.002   DF = 7

This interval says that we are 95 percent confident that the new system reduces the mean of all customer waiting times by between 3.14 minutes and 4.16 minutes.

In general, the degrees of freedom for the unequal variances procedure will always be less than or equal to $n_1 + n_2 - 2$, the degrees of freedom for the equal variances procedure. For example, if we use the unequal variances procedure to analyze the catalyst comparison data in Table 11.1, we can calculate df to be 7.9. This is slightly less than $n_1 + n_2 - 2 = 5 + 5 - 2 = 8$, the degrees of freedom for the equal variances procedure. Figure 11.2(b) gives the Excel output, and Figure 11.4 gives the Minitab output, of the unequal variances analysis of the catalyst comparison data. Note that the Excel unequal variances procedure rounds $df = 7.9$ up to 8 and obtains the same results as did the equal variances procedure (see Figure 11.2(a)). On the other hand, Minitab rounds $df = 7.9$ down to 7 and finds that a 95 percent confidence interval for $\mu_1 - \mu_2$ is [29.6049, 91.9951]. Minitab also finds that the test statistic for testing $H_0: \mu_1 - \mu_2 = 0$ versus $H_a: \mu_1 - \mu_2 \neq 0$ is $t = 4.61$ and that the associated p-value is .002. These results do not differ by much from the results given by the equal variances procedure.

To conclude this section, it is important to point out that if the sample sizes n_1 and n_2 are not large (at least 30), and if we fear that the sampled populations might be far from normally distributed, we can use a **nonparametric method.** One nonparametric method for comparing populations when using independent samples is the **Wilcoxon rank sum test.** This test is discussed in Chapter 18.

Exercises for Section 11.1

CONCEPTS   connect

For each of the formulas in the boxes named below, list all of the assumptions that must be satisfied in order to validly use the formula.

11.1   A t-Based Confidence Interval for the Difference between Two Population Means: Equal Variances

11.2   A t Test about the Difference between Two Population Means: Equal Variances

11.3 t-Based Confidence Intervals for $\mu_1 - \mu_2$, and t Tests about $\mu_1 - \mu_2$: Unequal Variances

METHODS AND APPLICATIONS

Suppose we have taken independent, random samples of sizes $n_1 = 7$ and $n_2 = 7$ from two normally distributed populations having means μ_1 and μ_2, and suppose we obtain $\bar{x}_1 = 240$, $\bar{x}_2 = 210$, $s_1 = 5$, and $s_2 = 6$. Using the equal variances procedure, do Exercises 11.4, 11.5, and 11.6.

11.4   Calculate a 95 percent confidence interval for $\mu_1 - \mu_2$. Can we be 95 percent confident that $\mu_1 - \mu_2$ is greater than 20? Explain why we can use the equal variances procedure here.

11.5   Use critical values to test the null hypothesis $H_0: \mu_1 - \mu_2 \leq 20$ versus the alternative hypothesis $H_a: \mu_1 - \mu_2 > 20$ by setting α equal to .10, .05, .01, and .001. How much evidence is there that the difference between μ_1 and μ_2 exceeds 20?

11.6   Use critical values to test the null hypothesis $H_0: \mu_1 - \mu_2 = 20$ versus the alternative hypothesis $H_a: \mu_1 - \mu_2 \neq 20$ by setting α equal to .10, .05, .01, and .001. How much evidence is there that the difference between μ_1 and μ_2 is not equal to 20?

11.7   Repeat Exercises 11.4 through 11.6 using the unequal variances procedure. Compare your results to those obtained using the equal variances procedure.

11.8   An article in *Fortune* magazine reported on the rapid rise of fees and expenses charged by mutual funds. Assuming that stock fund expenses and municipal bond fund expenses are each approximately normally distributed, suppose a random sample of 12 stock funds gives a mean annual expense of 1.63 percent with a standard deviation of .31 percent, and an independent random sample of 12 municipal bond funds gives a mean annual expense of 0.89 percent with a standard deviation of .23 percent. Let μ_1 be the mean annual expense for stock funds, and let μ_2 be the mean annual expense for municipal bond funds. Do parts *a*, *b*, and *c* by using the equal variances procedure. Then repeat *a*, *b*, and *c* using the unequal variances procedure. Compare your results.

a Set up the null and alternative hypotheses needed to attempt to establish that the mean annual expense for stock funds is larger than the mean annual expense for municipal bond funds. Test these hypotheses at the .05 level of significance. What do you conclude?

b Set up the null and alternative hypotheses needed to attempt to establish that the mean annual expense for stock funds exceeds the mean annual expense for municipal bond funds by more than .5 percent. Test these hypotheses at the .05 level of significance. What do you conclude?

c Calculate a 95 percent confidence interval for the difference between the mean annual expenses for stock funds and municipal bond funds. Can we be 95 percent confident that the mean annual expense for stock funds exceeds that for municipal bond funds by more than .5 percent? Explain.

11.9 In the book *Business Research Methods,* Donald R. Cooper and C. William Emory (1995) discuss a manager who wishes to compare the effectiveness of two methods for training new salespeople. The authors describe the situation as follows:

The company selects 22 sales trainees who are randomly divided into two experimental groups—one receives type *A* and the other type *B* training. The salespeople are then assigned and managed without regard to the training they have received. At the year's end, the manager reviews the performances of salespeople in these groups and finds the following results:

	A Group	B Group
Average Weekly Sales	$1,500	$1,300
Standard Deviation	225	251

a Set up the null and alternative hypotheses needed to attempt to establish that type *A* training results in higher mean weekly sales than does type *B* training.

b Because different sales trainees are assigned to the two experimental groups, it is reasonable to believe that the two samples are independent. Assuming that the normality assumption holds, and using the equal variances procedure, test the hypotheses you set up in

part *a* at levels of significance .10, .05, .01, and .001. How much evidence is there that type *A* training produces results that are superior to those of type *B*?

c Use the equal variances procedure to calculate a 95 percent confidence interval for the difference between the mean weekly sales obtained when type *A* training is used and the mean weekly sales obtained when type *B* training is used. Interpret this interval.

11.10 A marketing research firm wishes to compare the prices charged by two supermarket chains—Miller's and Albert's. The research firm, using a standardized one-week shopping plan (grocery list), makes identical purchases at 10 of each chain's stores. The stores for each chain are randomly selected, and all purchases are made during a single week.

The shopping expenses obtained at the two chains, along with box plots of the expenses, are as follows:
Ⓓ ShopExp

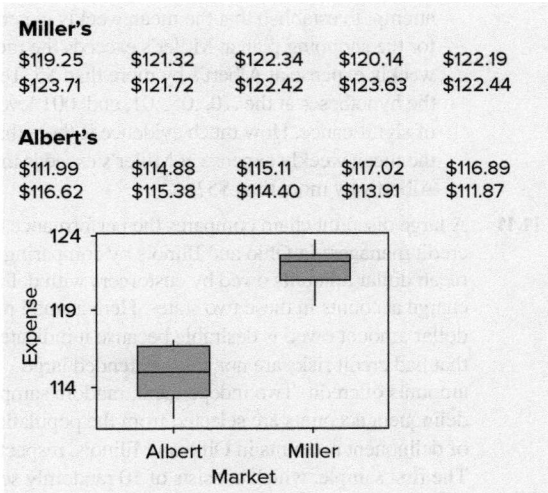

Miller's				
$119.25	$121.32	$122.34	$120.14	$122.19
$123.71	$121.72	$122.42	$123.63	$122.44

Albert's				
$111.99	$114.88	$115.11	$117.02	$116.89
$116.62	$115.38	$114.40	$113.91	$111.87

Because the stores in each sample are different stores in different chains, it is reasonable to assume that the samples are independent, and we assume that weekly expenses at each chain are normally distributed.

a Letting μ_M be the mean weekly expense for the shopping plan at Miller's, and letting μ_A be the mean weekly expense for the shopping plan at Albert's, Figure 11.5 gives the Minitab output of the test of $H_0: \mu_M - \mu_A = 0$ (that is, there is no difference between μ_M and μ_A) versus $H_a: \mu_M - \mu_A \neq 0$ (that is,

FIGURE 11.5 **Minitab Output of Testing the Equality of Mean Weekly Expenses at Miller's and Albert's Supermarket Chains (for Exercise 11.10)**

```
Two-sample T for Millers vs Alberts

           N    Mean    StDev    SE Mean
Millers   10   121.92    1.40      0.44
Alberts   10   114.81    1.84      0.58

Difference = mu(Millers) - mu(Alberts)   Estimate for difference: 7.10900
95% CI for difference: (5.57350, 8.64450)
T-Test of diff = 0 (vs not =): T-Value = 9.73   P-Value = 0.000   DF = 18
Both use Pooled StDev = 1.6343
```

μ_M and μ_A differ). Note that Minitab has employed the equal variances procedure. Use the sample data to show that $\bar{x}_M = 121.92$, $s_M = 1.40$, $\bar{x}_A = 114.81$, $s_A = 1.84$, and $t = 9.73$.

b Using the t statistic given on the output and critical values, test H_0 versus H_a by setting α equal to .10, .05, .01, and .001. How much evidence is there that the mean weekly expenses at Miller's and Albert's differ?

c Figure 11.5 gives the p-value for testing $H_0: \mu_M - \mu_A = 0$ versus $H_a: \mu_M - \mu_A \neq 0$. Use the p-value to test H_0 versus H_a by setting α equal to .10, .05, .01, and .001. How much evidence is there that the mean weekly expenses at Miller's and Albert's differ?

d Figure 11.5 gives a 95 percent confidence interval for $\mu_M - \mu_A$. Use this confidence interval to describe the size of the difference between the mean weekly expenses at Miller's and Albert's. Do you think that these means differ in a practically important way?

e Set up the null and alternative hypotheses needed to attempt to establish that the mean weekly expense for the shopping plan at Miller's exceeds the mean weekly expense at Albert's by more than $5. Test the hypotheses at the .10, .05, .01, and .001 levels of significance. How much evidence is there that the mean weekly expense at Miller's exceeds that at Albert's by more than $5?

11.11 A large discount chain compares the performance of its credit managers in Ohio and Illinois by comparing the mean dollar amounts owed by customers with delinquent charge accounts in these two states. Here a small mean dollar amount owed is desirable because it indicates that bad credit risks are not being extended large amounts of credit. Two independent, random samples of delinquent accounts are selected from the populations of delinquent accounts in Ohio and Illinois, respectively. The first sample, which consists of 10 randomly selected delinquent accounts in Ohio, gives a mean dollar amount of $524 with a standard deviation of $68. The second sample, which consists of 20 randomly selected delinquent accounts in Illinois, gives a mean dollar amount of $473 with a standard deviation of $22.

a Set up the null and alternative hypotheses needed to test whether there is a difference between the population mean dollar amounts owed by customers with delinquent charge accounts in Ohio and Illinois.

b Figure 11.6 gives the Minitab output of using the unequal variances procedure to test the equality of mean dollar amounts owed by customers with delinquent charge accounts in Ohio and Illinois. Assuming that the normality assumption holds, test the hypotheses you set up in part a by setting α equal to .10, .05, .01, and .001. How much evidence is there that the mean dollar amounts owed in Ohio and Illinois differ?

c Assuming that the normality assumption holds, calculate a 95 percent confidence interval for the difference between the mean dollar amounts owed in Ohio and Illinois. Based on this interval, do you think that these mean dollar amounts differ in a practically important way?

11.12 A loan officer compares the interest rates for 48-month fixed-rate auto loans and 48-month variable-rate auto loans. Two independent, random samples of auto loan rates are selected. A sample of five 48-month variable-rate auto loans had the following loan rates: ⊙S AutoLoan

2.6%   3.07%   2.872%   3.24%   3.15%

while a sample of five 48-month fixed-rate auto loans had loan rates as follows:

4.032%   3.85%   4.385%   3.75%   4.16%

a Set up the null and alternative hypotheses needed to determine whether the mean rates for 48-month variable-rate and fixed-rate auto loans differ.

b Figure 11.7 gives the JMP output of using the equal variances procedure to test the hypotheses you set up in part a. Assuming that the normality and equal variances assumptions hold, use the JMP output and critical values to test these hypotheses by setting α equal to .10, .05, .01, and .001. How much evidence is there that the mean rates for 48-month fixed- and variable-rate auto loans differ?

c Figure 11.7 gives the p-value for testing the hypotheses you set up in part a. Use the p-value to test these hypotheses by setting α equal to .10, .05, .01, and .001. How much evidence is there that the mean rates for 48-month fixed- and variable-rate auto loans differ?

FIGURE 11.6   Minitab Output of Testing the Equality of Mean Dollar Amounts Owed for Ohio and Illinois (for Exercise 11.11)

Two-Sample T-Test and CI

Sample	N	Mean	StDev	SE Mean
Ohio	10	524.0	68.0	22
Illinois	20	473.0	22.0	4.9

Difference = mu(1) - mu(2)
Estimate for difference: 51.0
95% CI for difference: (1.1, 100.9)
T-Test of difference = 0 (vs not =):
   T-Value = 2.31   P-Value = 0.046   DF = 9

FIGURE 11.7   JMP Output of Testing the Equality of Mean Loan Rates for Variable and Fixed 48-Month Auto Loans (for Exercise 11.12)

Means and Std Deviations

Level	Number	Mean	Std Dev
Fixed	5	4.03540	0.251785
Variable	5	2.98640	0.255176

t Test
Variable-Fixed
Assuming equal variances

Difference	−1.0490	t Ratio	−6.54321
Std Err Dif	0.1603	DF	8
Upper CL Dif	−0.6793	Prob > ltl	0.0002*
Lower CL Dif	−1.4187	Prob > t	0.9999
Confidence	0.95	Prob < t	<.0001*

d Calculate a 95 percent confidence interval for the difference between the mean rates for fixed- and variable-rate 48-month auto loans. Can we be 95 percent confident that the difference between these means exceeds .4 percent? Explain.

e Use a hypothesis test to establish that the difference between the mean rates for fixed- and variable-rate 48-month auto loans exceeds .4 percent. Use α equal to .05.

11.2 Paired Difference Experiments

LO11-2
Recognize when data come from independent samples and when they are paired.

Ⓒ EXAMPLE 11.3 The Auto Insurance Case: Comparing Mean Repair Costs

Home State Casualty, specializing in automobile insurance, wishes to compare the repair costs of moderately damaged cars (repair costs between $700 and $1,400) at two garages. One way to study these costs would be to take two independent samples (here we arbitrarily assume that each sample is of size $n = 7$). First we would randomly select seven moderately damaged cars that have recently been in accidents. Each of these cars would be taken to the first garage (garage 1), and repair cost estimates would be obtained. Then we would randomly select seven *different* moderately damaged cars, and repair cost estimates for these cars would be obtained at the second garage (garage 2). This sampling procedure would give us independent samples because the cars taken to garage 1 differ from those taken to garage 2. However, because the repair costs for moderately damaged cars can range from $700 to $1,400, there can be substantial differences in damages to moderately damaged cars. These differences might tend to conceal any real differences between repair costs at the two garages. For example, suppose the repair cost estimates for the cars taken to garage 1 are higher than those for the cars taken to garage 2. This difference might exist because garage 1 charges customers more for repair work than does garage 2. However, the difference could also arise because the cars taken to garage 1 are more severely damaged than the cars taken to garage 2.

To overcome this difficulty, we can perform a **paired difference experiment.** Here we could randomly select one sample of $n = 7$ moderately damaged cars. The cars in this sample would be taken to both garages, and a repair cost estimate for each car would be obtained at each garage. The advantage of the paired difference experiment is that the repair cost estimates at the two garages are obtained for the same cars. Thus, any true differences in the repair cost estimates would not be concealed by possible differences in the severity of damages to the cars.

©Westend61/Getty Images RF

Suppose that when we perform the paired difference experiment, we obtain the repair cost estimates in Table 11.2 (these estimates are given in units of $100). To analyze these data,

TABLE 11.2   **A Sample of $n = 7$ Paired Differences of the Repair Cost Estimates at Garages 1 and 2 (Cost Estimates in Hundreds of Dollars)**   Ⓓ Ⓢ Repair

Sample of $n = 7$ Damaged Cars	Repair Cost Estimates at Garage 1	Repair Cost Estimates at Garage 2	Sample of $n = 7$ Paired Differences
Car 1	$ 7.1	$ 7.9	$d_1 = -.8$
Car 2	9.0	10.1	$d_2 = -1.1$
Car 3	11.0	12.2	$d_3 = -1.2$
Car 4	8.9	8.8	$d_4 = .1$
Car 5	9.9	10.4	$d_5 = -.5$
Car 6	9.1	9.8	$d_6 = -.7$
Car 7	10.3	11.7	$d_7 = -1.4$
	$\bar{x}_1 = 9.329$	$\bar{x}_2 = 10.129$	$\bar{d} = -.8 = \bar{x}_1 - \bar{x}_2$
			$s_d^2 = .2533$
			$s_d = .5033$

we calculate the difference between the repair cost estimates at the two garages for each car. The resulting **paired differences** are given in the last column of Table 11.2. The mean of the sample of $n = 7$ paired differences is

$$\bar{d} = \frac{-.8 + (-1.1) + (-1.2) + \cdots + (-1.4)}{7} = -.8$$

which equals the difference between the sample means of the repair cost estimates at the two garages

$$\bar{x}_1 - \bar{x}_2 = 9.329 - 10.129 = -.8$$

Furthermore, $\bar{d} = -.8$ (that is, $-\$80$) is the point estimate of

$$\mu_d = \mu_1 - \mu_2$$

the mean of the population of all possible paired differences of the repair cost estimates (for all possible moderately damaged cars) at garages 1 and 2 (which is equivalent to μ_1, the mean of all possible repair cost estimates at garage 1, minus μ_2, the mean of all possible repair cost estimates at garage 2). This says we estimate that the mean of all possible repair cost estimates at garage 1 is $\$80$ less than the mean of all possible repair cost estimates at garage 2.

In addition, the variance and standard deviation of the sample of $n = 7$ paired differences

$$s_d^2 = \frac{\sum_{i=1}^{7}(d_i - \bar{d})^2}{7 - 1} = .2533$$

and

$$s_d = \sqrt{.2533} = .5033$$

are the point estimates of σ_d^2 and σ_d, the variance and standard deviation of the population of all possible paired differences.

LO11-3

Compare two population means when the data are paired.

In general, suppose we wish to compare two population means, μ_1 and μ_2. Also suppose that we have obtained two different measurements (for example, repair cost estimates) on the same n units (for example, cars), and suppose we have calculated the n paired differences between these measurements. Let $\bar{d}$ and s_d be the mean and the standard deviation of these n paired differences. If it is reasonable to assume that the paired differences have been randomly selected from a normally distributed (or at least mound-shaped) population of paired differences with mean μ_d and standard deviation σ_d, then the sampling distribution of

$$\frac{\bar{d} - \mu_d}{s_d / \sqrt{n}}$$

is a t distribution having $n - 1$ degrees of freedom. This implies that we have the following confidence interval for μ_d:

A Confidence Interval for the Mean, μ_d, of a Population of Paired Differences

Let μ_d be the mean of a **normally distributed population of paired differences,** and let $\bar{d}$ and s_d be the mean and standard deviation of a sample of n paired differences that have been randomly selected from the population. Then, a **$100(1 - \alpha)$ percent confidence** interval for $\mu_d = \mu_1 - \mu_2$ is

$$\left[\bar{d} \pm t_{\alpha/2} \frac{s_d}{\sqrt{n}}\right]$$

Here $t_{\alpha/2}$ is based on $(n - 1)$ degrees of freedom.

Ⓒ EXAMPLE 11.4 The Auto Insurance Case: Comparing Mean Repair Costs

Using the data in Table 11.2, and assuming that the population of paired repair cost differences is normally distributed, a 95 percent confidence interval for $\mu_d = \mu_1 - \mu_2$ is

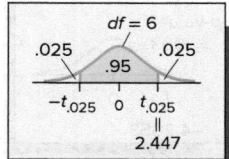

$$\left[\bar{d} \pm t_{.025} \frac{s_d}{\sqrt{n}} \right] = \left[-.8 \pm 2.447 \frac{.5033}{\sqrt{7}} \right]$$

$$= [-.8 \pm .4654]$$

$$= [-1.2654, -.3346]$$

Here $t_{.025} = 2.447$ is based on $n - 1 = 7 - 1 = 6$ degrees of freedom. This interval says that Home State Casualty can be 95 percent confident that μ_d, the mean of all possible paired differences of the repair cost estimates at garages 1 and 2, is between $-\$126.54$ and $-\$33.46$. That is, we are 95 percent confident that μ_1, the mean of all possible repair cost estimates at garage 1, is between \$126.54 and \$33.46 less than μ_2, the mean of all possible repair cost estimates at garage 2.

We can also test a hypothesis about μ_d, the mean of a population of paired differences. We show how to test the null hypothesis

$$H_0: \mu_d = D_0$$

in the following box. Here the value of the constant D_0 depends on the particular problem. Often D_0 equals 0, and the null hypothesis $H_0: \mu_d = 0$ says that μ_1 and μ_2 do not differ.

Testing a Hypothesis about the Mean, μ_d, of a Population of Paired Differences

Null Hypothesis $H_0: \mu_d = D_0$ **Test Statistic** $t = \dfrac{\bar{d} - D_0}{s_d/\sqrt{n}}$ $df = n - 1$ **Assumptions** Normal population of paired differences or Large sample size

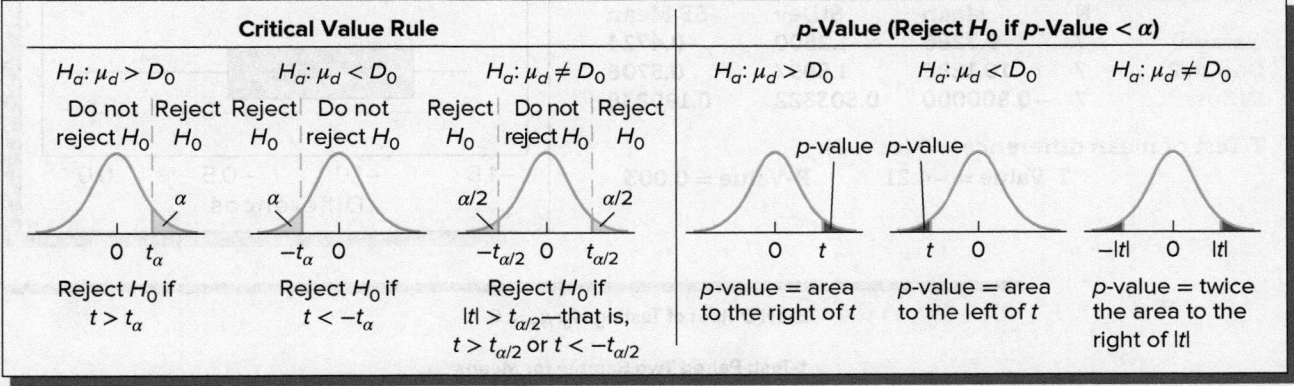

Ⓒ EXAMPLE 11.5 The Auto Insurance Case: Comparing Mean Repair Costs

Home State Casualty currently contracts to have moderately damaged cars repaired at garage 2. However, a local insurance agent suggests that garage 1 provides less expensive repair service that is of equal quality. Because it has done business with garage 2 for years, Home State has decided to give some of its repair business to garage 1 only if it has very strong evidence that μ_1, the mean repair cost estimate at garage 1, is smaller than μ_2, the

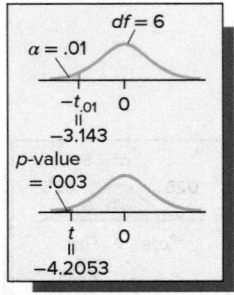

mean repair cost estimate at garage 2—that is, if $\mu_d = \mu_1 - \mu_2$ is less than zero. Therefore, we will test $H_0: \mu_d = 0$ or, equivalently, $H_0: \mu_1 - \mu_2 = 0$, versus $H_a: \mu_d < 0$ or, equivalently, $H_a: \mu_1 - \mu_2 < 0$, at the .01 level of significance. To perform the hypothesis test, we will use the sample data in Table 11.2 to calculate the value of the test statistic t in the summary box. Because $H_a: \mu_d < 0$ says to use the left-tailed critical value rule in the summary box, we will reject $H_0: \mu_d = 0$ if the value of t is less than $-t_\alpha = -t_{.01} = -3.143$. Here the t_α point is based on $n - 1 = 7 - 1 = 6$ degrees of freedom. Using the data in Table 11.2, the value of the test statistic is

$$t = \frac{\bar{d} - D_0}{s_d/\sqrt{n}} = \frac{-.8 - 0}{.5033/\sqrt{7}} = -4.2053$$

Because $t = -4.2053$ is less than $-t_{.01} = -3.143$, we can reject $H_0: \mu_d = 0$ in favor of $H_a: \mu_d < 0$. We conclude (at an α of .01) that μ_1, the mean repair cost estimate at garage 1, is less than μ_2, the mean repair cost estimate at garage 2. As a result, Home State will give some of its repair business to garage 1. Furthermore, Figure 10.8 gives the Minitab output of this hypothesis test and shows us that the p-value for the test is .003. This p-value for testing $H_0: \mu_d = 0$ versus $H_a: \mu_d < 0$ is the left-tailed p-value in the summary box: the area under the t distribution curve having 6 degrees of freedom to the left of $t = -4.2053$. The very small p-value of .003 says that we have very strong evidence that H_0 should be rejected and that μ_1 is less than μ_2.

Figure 11.9 shows the Excel output for testing $H_0: \mu_d = 0$ versus $H_a: \mu_d < 0$ (the "one-tail" test) and for testing $H_0: \mu_d = 0$ versus $H_a: \mu_d \neq 0$ (the "two-tail" test). The Excel p-value for testing $H_0: \mu_d = 0$ versus $H_a: \mu_d < 0$ is .002826, which in the rounded form .003 is the same as the Minitab p-value. This very small p-value tells us that Home State has very strong evidence that the mean repair cost at garage 1 is less than the mean repair cost at garage 2. The Excel p-value for testing $H_0: \mu_d = 0$ versus $H_a: \mu_d \neq 0$ is the two-tailed p-value in the summary box and equals .005653.

FIGURE 11.8   Minitab Output of Testing $H_0: \mu_d = 0$ versus $H_a: \mu_d < 0$

Paired T for Garage1 – Garage2

	N	Mean	StDev	SE Mean
Garage1	7	9.3286	1.2500	0.4724
Garage2	7	10.1286	1.5097	0.5706
Difference	7	−0.800000	0.503322	0.190238

T Test of mean difference = 0 (vs < 0):

T-Value = −4.21   P-Value = 0.003

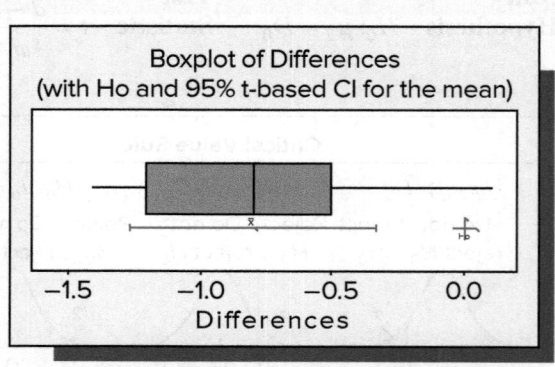

Boxplot of Differences
(with Ho and 95% t-based CI for the mean)

FIGURE 11.9   Excel Output of Testing $H_0: \mu_d = 0$

t-Test: Paired Two Sample for Means

	Garage1	Garage2
Mean	9.328571	10.12857
Variance	1.562381	2.279048
Observations	7	7
Pearson Correlation	0.950744	
Hypothesized Mean	0	
df	6	
t Stat	−4.20526	
P(T<=t) one-tail	0.002826	
t Critical one-tail	1.943181	
P(T<=t) two-tail	0.005653	
t Critical two-tail	2.446914	

In general, an experiment in which we have obtained two different measurements on the same n units is called a **paired difference experiment.** The idea of this type of experiment is to remove the variability due to the variable (for example, the amount of damage to a car) on which the observations are paired. In many situations, a paired difference experiment will provide more information than an independent samples experiment. As another example, suppose that we wish to assess which of two different machines produces a higher hourly output. If we randomly select 10 machine operators and randomly assign 5 of these operators to test machine 1 and the others to test machine 2, we would be performing an independent samples experiment. This is because different machine operators test machines 1 and 2. However, any difference in machine outputs could be obscured by differences in the abilities of the machine operators. For instance, if the observed hourly outputs are higher for machine 1 than for machine 2, we might not be able to tell whether this is due to (1) the superiority of machine 1 or (2) the possible higher skill level of the operators who tested machine 1. Because of this, it might be better to randomly select five machine operators, thoroughly train each operator to use both machines, and have each operator test both machines. We would then be *pairing on the machine operator,* and this would remove the variability due to the differing abilities of the operators.

The formulas we have given for analyzing a paired difference experiment are based on the t distribution. These formulas assume that the population of all possible paired differences is normally distributed (or at least mound-shaped). If the sample size is large (say, at least 30), the t-based interval and tests of this section are approximately valid no matter what the shape of the population of all possible paired differences. If the sample size is small, and if we fear that the population of all paired differences might be far from normally distributed, we can use a nonparametric method. One nonparametric method for comparing two populations when using a paired difference experiment is the **Wilcoxon signed ranks test.** This nonparametric test is discussed in Chapter 18.

Exercises for Section 11.2

CONCEPTS

11.13 Explain how a paired difference experiment differs from an independent samples experiment in terms of how the data for these experiments are collected.

11.14 Why is a paired difference experiment sometimes more informative than an independent samples experiment? Give an example of a situation in which a paired difference experiment might be advantageous.

11.15 Suppose a company wishes to compare the hourly output of its employees before and after vacations. Explain how you would collect data for a paired difference experiment to make this comparison.

METHODS AND APPLICATIONS

11.16 Suppose a sample of 49 paired differences that have been randomly selected from a normally distributed population of paired differences yields a sample mean of $\bar{d} = 5$ and a sample standard deviation of $s_d = 7$.
 a Calculate a 95 percent confidence interval for $\mu_d = \mu_1 - \mu_2$. Can we be 95 percent confident that the difference between μ_1 and μ_2 is not equal to 0?

 b Test the null hypothesis $H_0: \mu_d = 0$ versus the alternative hypothesis $H_a: \mu_d \neq 0$ by setting α equal to .10, .05, .01, and .001. How much evidence is there that μ_d differs from 0? What does this say about how μ_1 and μ_2 compare?
 c The p-value for testing $H_0: \mu_d \leq 3$ versus $H_a: \mu_d > 3$ equals .0256. Use the p-value to test these hypotheses with α equal to .10, .05, .01, and .001. How much evidence is there that μ_d exceeds 3? What does this say about the size of the difference between μ_1 and μ_2?

11.17 Suppose a sample of 11 paired differences that has been randomly selected from a normally distributed population of paired differences yields a sample mean of 103.5 and a sample standard deviation of 5.
 a Calculate 95 percent and 99 percent confidence intervals for $\mu_d = \mu_1 - \mu_2$.
 b Test the null hypothesis $H_0: \mu_d \leq 100$ versus $H_a: \mu_d > 100$ by setting α equal to .05 and .01. How much evidence is there that $\mu_d = \mu_1 - \mu_2$ exceeds 100?
 c Test the null hypothesis $H_0: \mu_d \geq 110$ versus $H_a: \mu_d < 110$ by setting α equal to .05 and .01. How much evidence is there that $\mu_d = \mu_1 - \mu_2$ is less than 110?

| TABLE 11.3 | Preexposure and Postexposure Attitude Scores (for Exercise 11.18) DS AdStudy |

Subject	Preexposure Attitudes (A_1)	Postexposure Attitudes (A_2)	Attitude Change (d_i)
1	50	53	3
2	25	27	2
3	30	38	8
4	50	55	5
5	60	61	1
6	80	85	5
7	45	45	0
8	30	31	1
9	65	72	7
10	70	78	8

Source: W. R. Dillon, T. J. Madden, and N. H. Firtle, *Essentials of Marketing Research* (Burr Ridge, IL: Richard D. Irwin, 1993), p. 435. McGraw-Hill Companies, Inc., 1993.

| TABLE 11.4 | Average Account Ages in 2015 and 2016 for 10 Randomly Selected Accounts (for Exercise 11.20) DS AcctAge |

Account	Average Age of Account in 2017 (Days)	Average Age of Account in 2016 (Days)
1	27	35
2	19	24
3	40	47
4	30	28
5	33	41
6	25	33
7	31	35
8	29	51
9	15	18
10	21	28

11.18 In the book *Essentials of Marketing Research,* William R. Dillon, Thomas J. Madden, and Neil H. Firtle (1993) present preexposure and postexposure attitude scores from an advertising study involving 10 respondents. The data for the experiment are given in Table 11.3. Assuming that the differences between pairs of postexposure and preexposure scores are normally distributed: DS AdStudy

a   Set up the null and alternative hypotheses needed to attempt to establish that the advertisement increases the mean attitude score (that is, that the mean postexposure attitude score is higher than the mean preexposure attitude score).

b   Test the hypotheses you set up in part *a* at the .10, .05, .01, and .001 levels of significance. How much evidence is there that the advertisement increases the mean attitude score?

c   Estimate the minimum difference between the mean postexposure attitude score and the mean preexposure attitude score. Justify your answer.

11.19 National Paper Company must purchase a new machine for producing cardboard boxes. The company must choose between two machines. The machines produce boxes of equal quality, so the company will choose the machine that produces (on average) the most boxes. It is known that there are substantial differences in the abilities of the company's machine operators. Therefore, National Paper has decided to compare the machines using a paired difference experiment. Suppose that eight randomly selected machine operators produce boxes for one hour using machine 1 and for one hour using machine 2, with the following results: DS BoxYield

	Machine Operator			
	1	2	3	4
Machine 1	53	60	58	48
Machine 2	50	55	56	44

	Machine Operator			
	5	6	7	8
Machine 1	46	54	62	49
Machine 2	45	50	57	47

a   Assuming normality, perform a hypothesis test to determine whether there is a difference between the mean hourly outputs of the two machines. Use $\alpha = .05$.

b   Estimate the minimum and maximum differences between the mean outputs of the two machines. Justify your answer.

11.20 During 2017 a company implemented a number of policies aimed at reducing the ages of its customers' accounts. In order to assess the effectiveness of these measures, the company randomly selects 10 customer accounts. The average age of each account is determined for the years 2016 and 2017. These data are given in Table 11.4. Assuming that the population of paired differences between the average ages in 2017 and 2016 is normally distributed, DS AcctAge

a   Set up the null and alternative hypotheses needed to establish that the mean average account age has been reduced by the company's new policies.

b   Figure 11.10 gives the JMP output needed to test the hypotheses of part *a*. Use critical values to test these hypotheses by setting α equal to .10, .05, .01, and .001. How much evidence is there that the mean average account age has been reduced?

c   Figure 11.10 gives the *p*-value for testing the hypotheses of part *a*. Use the *p*-value to test these hypotheses by setting α equal to .10, .05, .01, and .001. How much evidence is there that the mean average account age has been reduced?

d   Calculate a 95 percent confidence interval for the mean difference in the average account ages between 2016 and 2017. Estimate the minimum reduction in the mean average account ages from 2016 to 2017.

11.21 Do students reduce study time in classes where they achieve a higher midterm score? In a *Journal of Economic Education* article (Winter 2005), Gregory Krohn and Catherine O'Connor studied student effort and performance in a class over a semester. In an intermediate macroeconomics course, they found that "students respond to higher midterm scores by

FIGURE 11.10   **JMP Output of a Paired Difference Analysis of the Account Age Data (for Exercise 11.20)**

Matched Pairs

Difference: Average age of account in
2016 (days)-Average age of account in 2017

Average age of account in 2016 (days)	34	t-Ratio	3.61211
Average age of account in 2017	27	DF	9
Mean Difference	7	Prob > ItI	0.0056*
Std Error	1.93793	Prob > t	0.0028*
Upper 95%	11.3839	Prob < t	0.9972
Lower 95%	2.61611		
N		10	
Correlation		0.80459	

TABLE 11.5   **Weekly Study Time Data for Students Who Perform Well on the MidTerm**   StudyTime

Students	1	2	3	4	5	6	7	8
Before	15	14	17	17	19	14	13	16
After	9	9	11	10	19	10	14	10

reducing the number of hours they subsequently allocate to studying for the course."[4] Suppose that a random sample of $n = 8$ students who performed well on the midterm exam was taken and weekly study times before and after the exam were compared. The resulting data are given in Table 11.5. Assume that the population of all possible paired differences is normally distributed.

a   Set up the null and alternative hypotheses to test whether there is a difference in the population mean study time before and after the midterm exam.

b   In the right column we present the Minitab output for the paired differences test. Use the output and critical values to test the hypotheses at the .10, .05, and .01

levels of significance. Has the population mean study time changed?

c   Use the *p*-value to test the hypotheses at the .10, .05, and .01 levels of significance. How much evidence is there against the null hypothesis?

Paired T-Test and CI: StudyBefore, StudyAfter

Paired T for StudyBefore - StudyAfter

	N	Mean	StDev	SE Mean
StudyBefore	8	15.6250	1.9955	0.7055
StudyAfter	8	11.5000	3.4226	1.2101
Difference	8	4.12500	2.99702	1.05961

95% CI for mean difference: (1.61943, 6.63057)
T-Test of mean difference = 0 (vs not = 0):
T-Value = 3.89   P-Value = 0.006

11.3 Comparing Two Population Proportions by Using Large, Independent Samples

Ⓒ EXAMPLE 11.6 The Test Market Case: Comparing Advertising Media

LO11-4
Compare two population proportions using large independent samples.

Suppose a new product was test marketed in the Des Moines, Iowa, and Toledo, Ohio, metropolitan areas. Equal amounts of money were spent on advertising in the two areas. However, different advertising media were employed in the two areas. Advertising in the

[4]Source: "Student Effort and Performance over the Semester," *Journal of Economic Education*, Winter 2005, pages 3–28.

Des Moines area was done entirely on television, while advertising in the Toledo area consisted of a mixture of television, radio, newspaper, and magazine ads. Two months after the advertising campaigns commenced, surveys are taken to estimate consumer awareness of the product. In the Des Moines area, 631 out of 1,000 randomly selected consumers are aware of the product, while in the Toledo area 798 out of 1,000 randomly selected consumers are aware of the product. We define p_1 to be the proportion of all consumers in the Des Moines area who are aware of the product and p_2 to be the proportion of all consumers in the Toledo area who are aware of the product. It follows that, because the sample proportions of consumers who are aware of the product in the Des Moines and Toledo areas are

$$\hat{p}_1 = \frac{631}{1,000} = .631$$

and

$$\hat{p}_2 = \frac{798}{1,000} = .798$$

then a point estimate of $p_1 - p_2$ is

$$\hat{p}_1 - \hat{p}_2 = .631 - .798 = -.167$$

This says we estimate that p_1 is .167 less than p_2. That is, we estimate that the percentage of all consumers who are aware of the product in the Toledo area is 16.7 percentage points higher than the percentage in the Des Moines area.

In order to find a confidence interval for and to carry out a hypothesis test about $p_1 - p_2$, we need to know the properties of the **sampling distribution of $\hat{p}_1 - \hat{p}_2$.** In general, therefore, consider randomly selecting n_1 elements from a population, and assume that a proportion p_1 of all the elements in the population fall into a particular category. Let $\hat{p}_1$ denote the proportion of elements in the sample that fall into the category. Also, consider randomly selecting a sample of n_2 elements from a second population, and assume that a proportion p_2 of all the elements in this population fall into the particular category. Let $\hat{p}_2$ denote the proportion of elements in the second sample that fall into the category.

The Sampling Distribution of $\hat{p}_1 - \hat{p}_2$

If the randomly selected samples are independent of each other, then the population of all possible values of $\hat{p}_1 - \hat{p}_2$

1   Approximately has a normal distribution if each of the sample sizes n_1 and n_2 is large. Here n_1 and n_2 are large enough if $n_1 p_1$, $n_1(1 - p_1)$, $n_2 p_2$, and $n_2(1 - p_2)$ are all at least 5.

2   Has mean $\mu_{\hat{p}_1 - \hat{p}_2} = p_1 - p_2$

3   Has standard deviation $\sigma_{\hat{p}_1 - \hat{p}_2} = \sqrt{\dfrac{p_1(1 - p_1)}{n_1} + \dfrac{p_2(1 - p_2)}{n_2}}$

If we estimate p_1 by $\hat{p}_1$ and p_2 by $\hat{p}_2$ in the expression for $\sigma_{\hat{p}_1 - \hat{p}_2}$, then the sampling distribution of $\hat{p}_1 - \hat{p}_2$ implies the following $100(1 - \alpha)$ percent confidence interval for $p_1 - p_2$.

A Large Sample Confidence Interval for the Difference between Two Population Proportions[5]

Suppose we randomly select a sample of size n_1 from a population, and let $\hat{p}_1$ denote the proportion of elements in this sample that fall into a category of interest. Also suppose we randomly select a sample of size n_2 from another population, and let $\hat{p}_2$ denote the proportion of elements in this second sample that fall into the category of interest. Then, if each of the sample sizes n_1 and n_2 is large (n_1 and n_2 are considered to be large if $n_1\hat{p}_1$, $n_1(1-\hat{p}_1)$, $n_2\hat{p}_2$, and $n_2(1-\hat{p}_2)$ are all at least 5), and if the random samples are independent of each other, a **100(1 − α) percent confidence interval for $p_1 - p_2$** is

$$\left[(\hat{p}_1 - \hat{p}_2) \pm z_{\alpha/2} \sqrt{\frac{\hat{p}_1(1-\hat{p}_1)}{n_1} + \frac{\hat{p}_2(1-\hat{p}_2)}{n_2}} \right]$$

© **EXAMPLE 11.7** The Test Market Case: Comparing Advertising Media

Recall that in the advertising media situation described at the beginning of this section, 631 of 1,000 randomly selected consumers in Des Moines are aware of the new product, while 798 of 1,000 randomly selected consumers in Toledo are aware of the new product. Also recall that

$$\hat{p}_1 = \frac{631}{1,000} = .631$$

and

$$\hat{p}_2 = \frac{798}{1,000} = .798$$

Because $n_1\hat{p}_1 = 1,000(.631) = 631$, $n_1(1-\hat{p}_1) = 1,000(1-.631) = 369$, $n_2\hat{p}_2 = 1,000(.798) = 798$, and $n_2(1-\hat{p}_2) = 1,000(1-.798) = 202$ are all at least 5, both n_1 and n_2 can be considered large. It follows that a 95 percent confidence interval for $p_1 - p_2$ is

$$\left[(\hat{p}_1 - \hat{p}_2) \pm z_{.025} \sqrt{\frac{\hat{p}_1(1-\hat{p}_1)}{n_1} + \frac{\hat{p}_2(1-\hat{p}_2)}{n_2}} \right]$$

$$= \left[(.631 - .798) \pm 1.96 \sqrt{\frac{(.631)(.369)}{1,000} + \frac{(.798)(.202)}{1,000}} \right]$$

$$= [-.167 \pm .0389]$$

$$= [-.2059, -.1281]$$

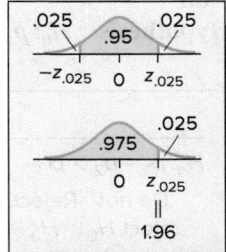

This interval says we are 95 percent confident that p_1, the proportion of all consumers in the Des Moines area who are aware of the product, is between .2059 and .1281 less than p_2, the proportion of all consumers in the Toledo area who are aware of the product. Thus, we have substantial evidence that advertising the new product by using a mixture of television, radio, newspaper, and magazine ads (as in Toledo) is more effective than spending an equal amount of money on television commercials only.

BI

---

[5]More correctly, because $\hat{p}_1(1-\hat{p}_1)/(n_1-1)$ and $\hat{p}_2(1-\hat{p}_2)/(n_2-1)$ are unbiased point estimates of $p_1(1-p_1)/n_1$ and $p_2(1-p_2)/n_2$, a point estimate of $\sigma_{\hat{p}_1 - \hat{p}_2}$ is

$$s_{\hat{p}_1 - \hat{p}_2} = \sqrt{\frac{\hat{p}_1(1-\hat{p}_1)}{n_1-1} + \frac{\hat{p}_2(1-\hat{p}_2)}{n_2-1}}$$

and a 100(1 − α) percent confidence interval for $p_1 - p_2$ is $[(\hat{p}_1 - \hat{p}_2) \pm z_{\alpha/2}s_{\hat{p}_1 - \hat{p}_2}]$. Because both n_1 and n_2 are large, there is little difference between the interval obtained by using this formula and those obtained by using the formula in the box above.

To test the null hypothesis H_0: $p_1 - p_2 = D_0$, we use the test statistic

$$z = \frac{(\hat{p}_1 - \hat{p}_2) - D_0}{\sigma_{\hat{p}_1 - \hat{p}_2}}$$

A commonly employed special case of this hypothesis test is obtained by setting D_0 equal to 0. In this case, the null hypothesis H_0: $p_1 - p_2 = 0$ says there is **no difference** between the population proportions p_1 and p_2. When $D_0 = 0$, the best estimate of the common population proportion $p = p_1 = p_2$ is obtained by computing

$$\hat{p} = \frac{\text{the total number of elements in the two samples that fall into the category of interest}}{\text{the total number of elements in the two samples}}$$

Therefore, the point estimate of $\sigma_{\hat{p}_1 - \hat{p}_2}$ is

$$s_{\hat{p}_1 - \hat{p}_2} = \sqrt{\frac{\hat{p}(1 - \hat{p})}{n_1} + \frac{\hat{p}(1 - \hat{p})}{n_2}}$$

$$= \sqrt{\hat{p}(1 - \hat{p})\left(\frac{1}{n_1} + \frac{1}{n_2}\right)}$$

For the case where $D_0 \neq 0$, the point estimate of $\sigma_{\hat{p}_1 - \hat{p}_2}$ is obtained by estimating p_1 by $\hat{p}_1$ and p_2 by $\hat{p}_2$. With these facts in mind, we present the following procedure for testing H_0: $p_1 - p_2 = D_0$:

A Hypothesis Test about the Difference between Two Population Proportions

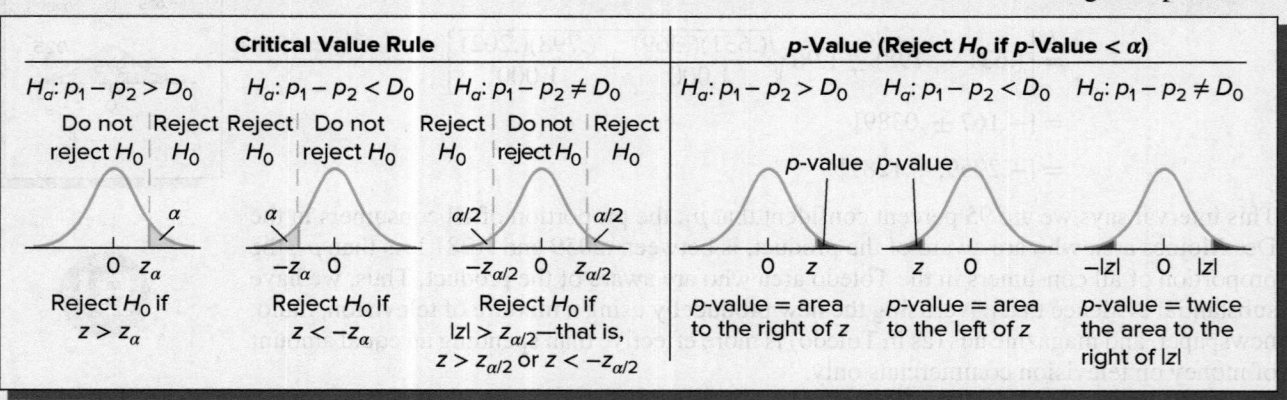

Null Hypothesis H_0: $p_1 - p_2 = D_0$ **Test Statistic** $z = \frac{(\hat{p}_1 - \hat{p}_2) - D_0}{\sigma_{\hat{p}_1 - \hat{p}_2}}$ **Assumptions**   Independent samples and Large sample sizes

Note:

1 If $D_0 = 0$, we estimate $\sigma_{\hat{p}_1 - \hat{p}_2}$ by

$$s_{\hat{p}_1 - \hat{p}_2} = \sqrt{\hat{p}(1 - \hat{p})\left(\frac{1}{n_1} + \frac{1}{n_2}\right)}$$

2 If $D_0 \neq 0$, we estimate $\sigma_{\hat{p}_1 - \hat{p}_2}$ by

$$s_{\hat{p}_1 - \hat{p}_2} = \sqrt{\frac{\hat{p}_1(1 - \hat{p}_1)}{n_1} + \frac{\hat{p}_2(1 - \hat{p}_2)}{n_2}}$$

Ⓒ **EXAMPLE 11.8** The Test Market Case: Comparing Advertising Media

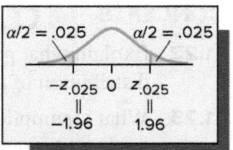

Recall that p_1 is the proportion of all consumers in the Des Moines area who are aware of the new product and that p_2 is the proportion of all consumers in the Toledo area who are aware of the new product. To test for the equality of these proportions, we will test H_0: $p_1 - p_2 = 0$ versus H_a: $p_1 - p_2 \neq 0$ at the **.05 level of significance.** Because both of the Des Moines and Toledo samples are large (see Example 11.7), we will calculate the value of the **test statistic z in the summary box** (where $D_0 = 0$). Because H_a: $p_1 - p_2 \neq 0$ says to use the two-tailed critical value rule in the summary box, we will **reject H_0: $p_1 - p_2 = 0$ if the absolute value of z is greater than $z_\alpha/2 = z_{.05/2} = z_{.025} = 1.96$.** Because 631 out of 1,000 randomly selected Des Moines residents were aware of the product and 798 out of 1,000 randomly selected Toledo residents were aware of the product, the estimate of $p = p_1 = p_2$ is

$$\hat{p} = \frac{631 + 798}{1,000 + 1,000} = \frac{1,429}{2,000} = .7145$$

and the **value of the test statistic is**

$$z = \frac{(\hat{p}_1 - \hat{p}_2) - D_0}{\sqrt{\hat{p}(1 - \hat{p})\left(\frac{1}{n_1} + \frac{1}{n_2}\right)}} = \frac{(.631 - .798) - 0}{\sqrt{(.7145)(.2855)\left(\frac{1}{1,000} + \frac{1}{1,000}\right)}} = \frac{-.167}{.0202} = -8.2673$$

Because $|z| = 8.2673$ is greater than 1.96, we can reject H_0: $p_1 - p_2 = 0$ in favor of H_a: $p_1 - p_2 \neq 0$. We conclude (at an α of .05) that the proportions of all consumers who are aware of the product in Des Moines and Toledo differ. Furthermore, the point estimate $\hat{p}_1 - \hat{p}_2 = .631 - .798 = -.167$ says we estimate that the percentage of all consumers who are aware of the product in Toledo is 16.7 percentage points higher than the percentage of all consumers who are aware of the product in Des Moines. The p-value for testing H_0: $p_1 - p_2 = 0$ versus H_a: $p_1 - p_2 \neq 0$ is the two-tailed p-value in the summary box: twice the area under the standard normal curve to the right of $|z| = 8.2673$. Because the area under the standard normal curve to the right of 3.99 is .00003, the p-value for testing H_0 is less than $2(.00003) = .00006$. It follows that we have extremely strong evidence that H_0: $p_1 - p_2 = 0$ should be rejected in favor of H_a: $p_1 - p_2 \neq 0$. That is, this small p-value provides extremely strong evidence that p_1 and p_2 differ. Figure 11.11 presents the Minitab and JMP outputs of the hypothesis test of H_0: $p_1 - p_2 = 0$ versus H_a: $p_1 - p_2 \neq 0$ and of a 95 percent confidence interval for $p_1 - p_2$. Note that the Minitab output gives a value of the test statistic z (that is, the value -8.41) that is slightly different from the value -8.2673 calculated above. The reason is that, even though we are testing H_0: $p_1 - p_2 = 0$, Minitab uses the second formula in the summary box (rather than the first formula) to calculate $s_{\hat{p}_1 - \hat{p}_2}$.

FIGURE 11.11   **Minitab and JMP Outputs of Statistical Inference in the Test Market Case**

(a) The Minitab output

Test and CI for Two Proportions

Sample	X	N	Sample p
1	631	1000	0.631000
2	798	1000	0.798000

Difference = p(1) − p(2)
Estimate for difference: −0.167
95% CI for difference: (−0.205906, −0.128094)
Test of difference = 0 (vs not = 0):
         Z = −8.41, P-value = 0.000

(b) The JMP output

Test Inputs

Hypothesized Difference (p2-p1)	0
Sample 1 Count (x1)	798
Sample 1 Size (n1)	1000
Sample 2 Count (x2)	631
Sample 1 Size (n2)	1000
Significance Level (alpha)	0.05

Test Results

Result	Value
Sample 1 Proportion	0.798
Sample 2 Proportion	0.631
Difference in Proportions (p2-p1)	−0.167
z-score	−8.2679
z Critical Value(s)	+/−1.96
Observed Significance (p-value)	<.0001
Reject Null Hypothesis	

Exercises for Section 11.3

CONCEPTS ![connect logo]

11.22 Explain what population is described by the sampling distribution of $\hat{p}_1 - \hat{p}_2$.

11.23 What assumptions must be satisfied in order to use the methods presented in this section?

METHODS AND APPLICATIONS

In Exercises 11.24 through 11.26 we assume that we have selected two independent random samples from populations having proportions p_1 and p_2 and that $\hat{p}_1 = 800/1{,}000 = .8$ and $\hat{p}_2 = 950/1{,}000 = .95$.

11.24 Calculate a 95 percent confidence interval for $p_1 - p_2$. Interpret this interval. Can we be 95 percent confident that $p_1 - p_2$ is less than 0? That is, can we be 95 percent confident that p_1 is less than p_2? Explain.

11.25 Test $H_0: p_1 - p_2 = 0$ versus $H_a: p_1 - p_2 \neq 0$ by using critical values and by setting α equal to .10, .05, .01, and .001. How much evidence is there that p_1 and p_2 differ? Explain.

11.26 Test $H_0: p_1 - p_2 \geq -.12$ versus $H_a: p_1 - p_2 < -.12$ by using a p-value and by setting α equal to .10, .05, .01, and .001. How much evidence is there that p_2 exceeds p_1 by more than .12? Explain.

11.27 In an article in the *Journal of Advertising*, Weinberger and Spotts compare the use of humor in television ads in the United States and in the United Kingdom. Suppose that independent random samples of television ads are taken in the two countries. A random sample of 400 television ads in the United Kingdom reveals that 142 use humor, while a random sample of 500 television ads in the United States reveals that 122 use humor.

   a Set up the null and alternative hypotheses needed to determine whether the proportion of ads using humor in the United Kingdom differs from the proportion of ads using humor in the United States.

   b Test the hypotheses you set up in part *a* by using critical values and by setting α equal to .10, .05, .01, and .001. How much evidence is there that the proportions of U.K. and U.S. ads using humor are different?

   c Set up the hypotheses needed to attempt to establish that the difference between the proportions of U.K. and U.S. ads using humor is more than .05 (five percentage points). Test these hypotheses by using a p-value and by setting α equal to .10, .05, .01, and .001. How much evidence is there that the difference between the proportions exceeds .05?

   d Calculate a 95 percent confidence interval for the difference between the proportion of U.K. ads using humor and the proportion of U.S. ads using humor. Interpret this interval. Can we be 95 percent confident that the proportion of U.K. ads using humor is greater than the proportion of U.S. ads using humor?

11.28 In the book *Essentials of Marketing Research,* William R. Dillon, Thomas J. Madden, and Neil H. Firtle discuss a research proposal in which a telephone company wants to determine whether the appeal of a new security system varies between homeowners and renters. Independent samples of 140 homeowners and 60 renters are randomly selected. Each respondent views a TV pilot in which a test ad for the new security system is embedded twice. Afterward, each respondent is interviewed to find out whether he or she would purchase the security system.

   Results show that 25 out of the 140 homeowners definitely would buy the security system, while 9 out of the 60 renters definitely would buy the system.

   a Letting p_1 be the proportion of homeowners who would buy the security system, and letting p_2 be the proportion of renters who would buy the security system, set up the null and alternative hypotheses needed to determine whether the proportion of homeowners who would buy the security system differs from the proportion of renters who would buy the security system.

   b Find the test statistic z and the p-value for testing the hypotheses of part *a*. Use the p-value to test the hypotheses with α equal to .10, .05, .01, and .001. How much evidence is there that the proportions of homeowners and renters differ?

   c Calculate a 95 percent confidence interval for the difference between the proportions of homeowners and renters who would buy the security system. On the basis of this interval, can we be 95 percent confident that these proportions differ? Explain.

11.29 In the book *Cases in Finance,* Nunnally and Plath present a case in which the estimated percentage of uncollectible accounts varies with the age of the account. Here the age of an unpaid account is the number of days elapsed since the invoice date.

   An accountant believes that the percentage of accounts that will be uncollectible increases as the ages of the accounts increase. To test this theory, the accountant randomly selects independent samples of 500 accounts with ages between 31 and 60 days and 500 accounts with ages between 61 and 90 days from the accounts receivable ledger dated one year ago. When the sampled accounts are examined, it is found that 10 of the 500 accounts with ages between 31 and 60 days were eventually classified as uncollectible, while 27 of the 500 accounts with ages between 61 and 90 days were eventually classified as uncollectible. Let p_1 be the proportion of accounts with ages between 31 and 60 days that will be uncollectible, and let p_2 be the proportion of accounts with ages between 61 and 90 days that will be uncollectible.

Test and CI for Two Proportions

Sample	X	N	Sample p
1 (31 to 60 days)	10	500	0.020000
2 (61 to 90 days	27	500	0.054000

Difference = p(1) − p(2)
Estimate for difference: −0.034

95% CI for difference: (−0.0573036, −0.0106964)
Test for difference = 0 (vs not = 0): Z = −2.85   P-Value = 0.004

a Use the Minitab output above to determine how much evidence there is that we should reject H_0: $p_1 - p_2 = 0$ in favor of H_a: $p_1 - p_2 \neq 0$.

b Identify a 95 percent confidence interval for $p_1 - p_2$, and estimate the smallest that the difference between p_1 and p_2 might be.

11.30 On January 7, 2000, the Gallup Organization released the results of a poll comparing the lifestyles of today with yesteryear. The survey results were based on telephone interviews with a randomly selected national sample of 1,031 adults, 18 years and older, conducted December 20–21, 1999. The poll asked several questions and compared the 1999 responses with the responses given in polls taken in previous years. Below we summarize some of the poll's results.[6]

Percentage of respondents who

1	Had taken a vacation lasting six days or more within the last 12 months:	December 1999 42%	December 1968 62%
2	Took part in some sort of daily activity to keep physically fit:	December 1999 60%	September 1977 48%
3	Watched TV more than four hours on an average weekday:	December 1999 28%	April 1981 25%

Assuming that each poll was based on a randomly selected national sample of 1,031 adults and that the samples in different years are independent,

a Let p_1 be the December 1999 population proportion of U.S. adults who had taken a vacation lasting six days or more within the last 12 months, and let p_2 be the December 1968 population proportion who had taken such a vacation. Calculate a 99 percent confidence interval for the difference between p_1 and p_2. Interpret what this interval says about how these population proportions differ.

b Let p_1 be the December 1999 population proportion of U.S. adults who took part in some sort of daily activity to keep physically fit, and let p_2 be the September 1977 population proportion who did the same. Carry out a hypothesis test to attempt to justify that the proportion who took part in such daily activity increased from September 1977 to December 1999. Use $\alpha = .05$ and explain your result.

c Let p_1 be the December 1999 population proportion of U.S. adults who watched TV more than four hours on an average weekday, and let p_2 be the April 1981 population proportion who did the same. Carry out a hypothesis test to determine whether these population proportions differ. Use $\alpha = .05$ and interpret the result of your test.

11.4 The *F* Distribution

LO11-5

Describe the properties of the *F* distribution and use an *F* table.

In this and upcoming chapters we will make statistical inferences by using what is called an **F distribution.** In general, as illustrated in Figure 11.12, the curve of the *F* distribution is skewed to the right. Moreover, the exact shape of this curve depends on two parameters that are called the **numerator degrees of freedom (denoted df_1)** and the **denominator degrees of freedom (denoted df_2).** In order to use the *F* distribution, we employ an **F point,** which is denoted F_α. As illustrated in Figure 11.12(a), F_α **is the point on the horizontal axis under the curve of the *F* distribution that gives a right-hand tail area equal to α.** The value of F_α in a particular situation depends on the size of the right-hand tail area (the size of α) and on the numerator degrees of freedom (df_1) and the denominator degrees of freedom (df_2). Values of F_α are given in an **F table.** Tables A.6, A.7, A.8, and A.9 give values of $F_{.10}$, $F_{.05}$, $F_{.025}$, and $F_{.01}$, respectively. Each table tabulates values of F_α according to the appropriate numerator degrees of freedom (values listed across the top of the table) and the appropriate denominator degrees of freedom (values listed down the left side of the table). A portion

---

[6]Source: www.gallup.com/, The Gallup Poll, December 30, 1999.

FIGURE 11.12   **F Distribution Curves and F Points**

(a) The point F_α corresponding to df_1 and df_2 degrees of freedom

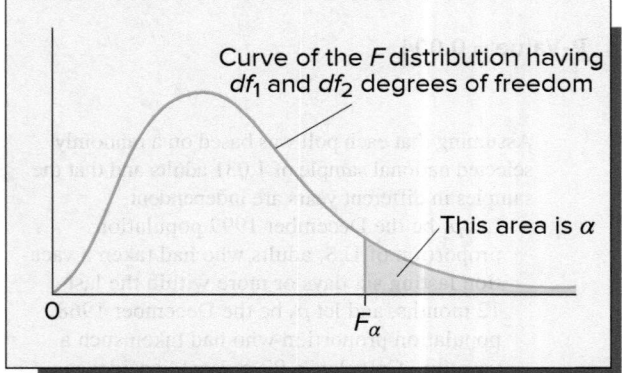

(b) The point $F_{.05}$ corresponding to 4 and 7 degrees of freedom

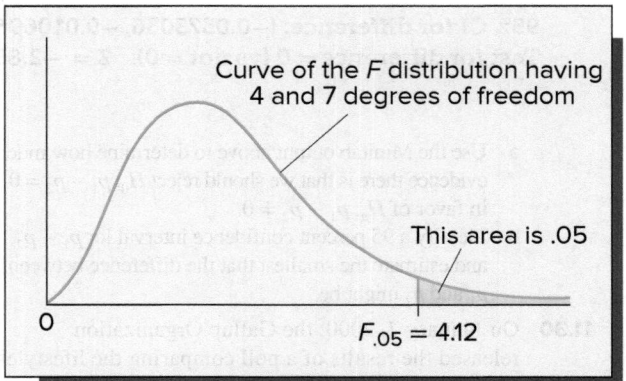

of Table A.7, which gives values of $F_{.05}$, is reproduced in this chapter as Table 11.6. For instance, suppose we wish to find the F point that gives a right-hand tail area of .05 under the curve of the F distribution having 4 numerator and 7 denominator degrees of freedom. To do this, we scan across the top of Table 11.6 until we find the column corresponding to 4 numerator degrees of freedom, and we scan down the left side of the table until we find the row corresponding to 7 denominator degrees of freedom. The table entry in this column and row is the desired F point. We find that the $F_{.05}$ point is 4.12 (see Figure 11.12(b) and Table 11.6).

TABLE 11.6   **A Portion of an F Table: Values of $F_{.05}$**

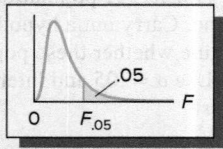

df_2	Numerator Degrees of Freedom (df_1)													
	1	2	3	**4**	5	6	7	8	9	10	12	15	20	24
1	161.4	199.5	215.7	224.6	230.2	234.0	236.8	238.9	240.5	241.9	243.9	245.9	248.0	249.1
2	18.51	19.00	19.16	19.25	19.30	19.33	19.35	19.37	19.38	19.40	19.41	19.43	19.45	19.45
3	10.13	9.55	9.28	9.12	9.01	8.94	8.89	8.85	8.81	8.79	8.74	8.70	8.66	8.64
4	7.71	6.94	6.59	6.39	6.26	6.16	6.09	6.04	6.00	5.96	5.91	5.86	5.80	5.77
5	6.61	5.79	5.41	5.19	5.05	4.95	4.88	4.82	4.77	4.74	4.68	4.62	4.56	4.53
6	5.99	5.14	4.76	4.53	4.39	4.28	4.21	4.15	4.10	4.06	4.00	3.94	3.87	3.84
7	5.59	4.71	4.25	4.12	3.97	3.87	3.79	3.73	3.68	3.64	3.57	3.51	3.44	3.41
8	5.32	4.46	4.07	3.84	3.69	3.58	3.50	3.44	3.39	3.35	3.28	3.22	3.15	3.12
9	5.12	4.26	3.86	3.63	3.48	3.37	3.29	3.23	3.18	3.14	3.07	3.01	2.94	2.90
10	4.96	4.10	3.71	3.48	3.33	3.22	3.14	3.07	3.02	2.98	2.91	2.85	2.77	2.74
11	4.84	3.98	3.59	3.36	3.20	3.09	3.01	2.95	2.90	2.85	2.79	2.72	2.65	2.61
12	4.75	3.89	3.49	3.26	3.11	3.00	2.91	2.85	2.80	2.75	2.69	2.62	2.54	2.51
13	4.67	3.81	3.41	3.18	3.03	2.92	2.83	2.77	2.71	2.67	2.60	2.53	2.46	2.42
14	4.60	3.74	3.34	3.11	2.96	2.85	2.76	2.70	2.65	2.60	2.53	2.46	2.39	2.35
15	4.54	3.68	3.29	3.06	2.90	2.79	2.71	2.64	2.59	2.54	2.48	2.40	2.33	2.29

Source: Data from M. Merrington and C. M. Thompson, "Tables of Percentage Points of the Inverted Beta (F) Distribution," *Biometrika*, Vol. 33 (1943), pp. 73–88.

11.5 Comparing Two Population Variances by Using Independent Samples

LO11-6
Compare two population variances when the samples are independent.

We have seen that we often wish to compare two population means. In addition, it is often useful to compare two population variances. For example, we might compare the variances of the fills that would be produced by two processes that are supposed to fill jars with 16 ounces of strawberry preserves. Or, as another example, we might wish to compare the variance of the chemical yields obtained when using Catalyst XA-100 with that obtained when using Catalyst ZB-200. Here the catalyst that produces yields with the smaller variance is giving more consistent (or predictable) results.

If σ_1^2 and σ_2^2 are the population variances that we wish to compare, one approach is to test the null hypothesis H_0: $\sigma_1^2 = \sigma_2^2$. We might test H_0 versus an alternative hypothesis of, for instance, H_a: $\sigma_1^2 > \sigma_2^2$. To perform this test, we use the test statistic $F = s_1^2/s_2^2$, where s_1^2 and s_2^2 are the variances of independent samples of sizes n_1 and n_2 randomly selected from the two populations. The **sampling distribution of $F = s_1^2/s_2^2$** is the probability distribution that describes the population of all possible values of s_1^2/s_2^2. It can be shown that, if each sampled population is normally distributed, and if the null hypothesis H_0: $\sigma_1^2 = \sigma_2^2$ is true, then the sampling distribution of $F = s_1^2/s_2^2$ is an F distribution having $df_1 = n_1 - 1$ numerator degrees of freedom and $df_2 = n_2 - 1$ denominator degrees of freedom. A value of the test statistic $F = s_1^2/s_2^2$ that is significantly larger than 1 would result from an s_1^2 that is significantly larger than s_2^2. Intuitively, this would imply that we should reject H_0: $\sigma_1^2 = \sigma_2^2$ in favor of H_a: $\sigma_1^2 > \sigma_2^2$. The exact procedures for carrying out this test and for testing H_0: $\sigma_1^2 = \sigma_2^2$ versus H_a: $\sigma_1^2 < \sigma_2^2$ are given in the following summary box.

Testing the Equality of Two Population Variances versus a One-Sided Alternative Hypothesis

Suppose we randomly select independent samples from two normally distributed populations—populations 1 and 2. Let s_1^2 be the variance of the random sample of n_1 observations from population 1, and let s_2^2 be the variance of the random sample of n_2 observations from population 2.

1   In order to test H_0: $\sigma_1^2 = \sigma_2^2$ versus H_a: $\sigma_1^2 > \sigma_2^2$, define the test statistic

$$F = \frac{s_1^2}{s_2^2}$$

and define the corresponding p-value to be the area to the right of F under the curve of the F distribution having $df_1 = n_1 - 1$ numerator degrees of freedom and $df_2 = n_2 - 1$ denominator degrees of freedom. We can reject H_0 at level of significance α if and only if

a $F > F_\alpha$ or, equivalently,

b p-value $< \alpha$.

Here F_α is based on $df_1 = n_1 - 1$ and $df_2 = n_2 - 1$ degrees of freedom.

2   In order to test H_0: $\sigma_1^2 = \sigma_2^2$ versus H_a: $\sigma_1^2 < \sigma_2^2$, define the test statistic

$$F = \frac{s_2^2}{s_1^2}$$

and define the corresponding p-value to be the area to the right of F under the curve of the F distribution having $df_1 = n_2 - 1$ numerator degrees of freedom and $df_2 = n_1 - 1$ denominator degrees of freedom. We can reject H_0 at level of significance α if and only if

a $F > F_\alpha$ or, equivalently,

b p-value $< \alpha$.

Here F_α is based on $df_1 = n_2 - 1$ and $df_2 = n_1 - 1$ degrees of freedom.

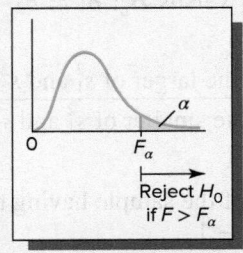

EXAMPLE 11.9 The Jar Fill Case: Comparing Process Consistencies

A jelly and jam producer has decided to purchase one of two jar-filling processes—process 1 or process 2—to fill 16-ounce jars with strawberry preserves. Process 1 is slightly more expensive than process 2, but the maker of process 1 claims that σ_1^2, the variance of all fills

Process 1	Process 2
15.9841	15.9622
16.0150	15.9736
15.9964	15.9753
15.9916	15.9802
15.9949	15.9820
16.0003	15.9860
15.9884	15.9885
16.0016	15.9897
16.0260	15.9903
16.0216	15.9920
16.0065	15.9928
15.9997	15.9934
15.9909	15.9973
16.0043	16.0014
15.9881	16.0016
16.0078	16.0053
15.9934	16.0053
16.0150	16.0098
16.0057	16.0102
15.9928	16.0252
15.9987	16.0316
16.0131	16.0331
15.9981	16.0384
16.0025	16.0386
15.9898	16.0401

 Preserves

that would be produced by process 1, is less than σ_2^2, the variance of all fills that would be produced by process 2. To test $H_0: \sigma_1^2 = \sigma_2^2$ versus $H_a: \sigma_1^2 < \sigma_2^2$ the jelly and jam producer measures the fills of 25 randomly selected jars produced by each process. The jar fill measurements for each process are given in the page margin. A histogram (not shown here) for each sample is bell-shaped and symmetrical, the sample sizes are $n_1 = 25$ and $n_2 = 25$, and the sample variances are $s_1^2 = .0001177$ and $s_2^2 = .0004847$. Therefore, we compute the test statistic

$$F = \frac{s_2^2}{s_1^2} = \frac{.0004847}{.0001177} = 4.1168$$

and we compare this value with F_α based on $df_1 = n_2 - 1 = 25 - 1 = 24$ numerator degrees of freedom and $df_2 = n_1 - 1 = 25 - 1 = 24$ denominator degrees of freedom. If we test H_0 versus H_a at the .05 level of significance, then Table A.7 (a portion of which is shown below) tells us that when $df_1 = 24$ and $df_2 = 24$, we have $F_{.05} = 1.98$.

df_2 \ df_1	1	2	3	4	5	6	7	8	9	10	12	15	20	24
23	4.28	3.42	3.03	2.80	2.64	2.53	2.44	2.37	2.32	2.27	2.20	2.13	2.05	2.01
24	4.26	3.40	3.01	2.78	2.62	2.51	2.42	2.36	2.30	2.25	2.18	2.11	2.03	1.98

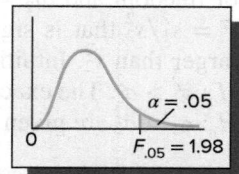

Because $F = 4.1168$ is greater than $F_{.05} = 1.98$, we can reject $H_0: \sigma_1^2 = \sigma_2^2$ in favor of $H_a: \sigma_1^2 < \sigma_2^2$. That is, we conclude (at an α of .05) that the variance of all fills that would be produced by process 1 is less than the variance of all fills that would be produced by process 2. That is, process 1 produces more consistent fills.

The p-value for testing H_0 versus H_a is the area to the right of $F = 4.1168$ under the curve of the F distribution having 24 numerator degrees of freedom and 24 denominator degrees of freedom (see the page margin). The Excel output in Figure 11.13 tells us that this p-value equals 0.0004787. Because this p-value is less than .001, we have extremely strong evidence to support rejecting H_0 in favor of H_a. That is, there is extremely strong evidence that process 1 produces jar fills that are more consistent (less variable) than the jar fills produced by process 2.

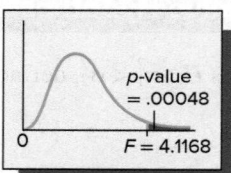

We now present a procedure for testing the null hypothesis $H_0: \sigma_1^2 = \sigma_2^2$ versus the two-sided alternative hypothesis $H_a: \sigma_1^2 \neq \sigma_2^2$.

Testing the Equality of Two Population Variances (Two-Sided Alternative)

Suppose we randomly select independent samples from two normally distributed populations and define all notation as in the previous box. Then, in order to test $H_0: \sigma_1^2 = \sigma_2^2$ versus $H_a: \sigma_1^2 \neq \sigma_2^2$, define the test statistic

$$F = \frac{\text{the larger of } s_1^2 \text{ and } s_2^2}{\text{the smaller of } s_1^2 \text{ and } s_2^2}$$

and let

$df_1 = \{$the size of the sample having the larger variance$\} - 1$

$df_2 = \{$the size of the sample having the smaller variance$\} - 1$

Also, define the corresponding p-value to be twice the area to the right of F under the curve of the F distribution having df_1 numerator degrees of freedom and

df_2 denominator degrees of freedom. We can reject H_0 at level of significance α if and only if

1 $F > F_{\alpha/2}$ or, equivalently,

2 p-value $< \alpha$.

Here $F_{\alpha/2}$ is based on df_1 and df_2 degrees of freedom.

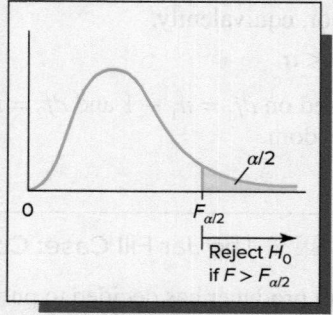

FIGURE 11.13	Excel Output for Testing $H_0: \sigma_1^2 = \sigma_2^2$ versus $H_a: \sigma_1^2 < \sigma_2^2$

F-Test Two-Sample for Variances

	Process 2	Process 1
Mean	16.001756	16.00105
Variance	0.0004847	0.0001177
Observations	25	25
df	24	24
F	4.1167856	
P(F<=f) one-tail	0.0004787	
F Critical one-tail	1.9837596	

FIGURE 11.14 **Minitab and JMP Outputs for Testing $H_0: \sigma_1^2 = \sigma_2^2$ versus $H_a: \sigma_1^2 \neq \sigma_2^2$ in the Catalyst Comparison Case**

(a) The Minitab output

Test for Equal Variances: ZB-200, XA-100

F-Test (Normal Distribution)

Test Statistic = 1.25,

p-value = 0.831

(b) The JMP output

Test	F Ratio	DFNum	DFDen	p-Value
F Test 2-sided	1.2544	4	4	0.834

EXAMPLE 11.10 The Catalyst Comparison Case: Process Improvement

Consider the catalyst comparison situation and suppose the production supervisor wishes to use the sample data in Table 11.1 to determine whether σ_1^2, the variance of the chemical yields obtained by using Catalyst XA-100, differs from σ_2^2, the variance of the chemical yields obtained by using Catalyst ZB-200. Table 11.1 tells us that $n_1 = 5$, $n_2 = 5$, $s_1^2 = 386$, and $s_2^2 = 484.2$. Therefore, we can reject $H_0: \sigma_1^2 = \sigma_2^2$ in favor of $H_a: \sigma_1^2 \neq \sigma_2^2$ at the .05 level of significance if

$$F = \frac{\text{the larger of } s_1^2 \text{ and } s_2^2}{\text{the smaller of } s_1^2 \text{ and } s_2^2} = \frac{484.2}{386} = 1.2544$$

is greater than $F_{\alpha/2} = F_{.05/2} = F_{.025}$. Here, because sample 2 has the larger sample variance ($s_2^2 = 484.2$) and sample 1 has the smaller sample variance ($s_1^2 = 386$), the numerator degrees of freedom for $F_{.025}$ is $df_1 = n_2 - 1 = 5 - 1 = 4$ and the denominator degrees of freedom for $F_{.025}$ is $df_2 = n_1 - 1 = 5 - 1 = 4$. Table A.8 tells us that the appropriate $F_{.025}$ point equals 9.60 (see the page margin). Because $F = 1.2544$ is not greater than 9.60, we cannot reject H_0 at the .05 level of significance. Furthermore, the p-value for the hypothesis test is twice the area to the right of $F = 1.2544$ under the curve of the F distribution having 4 numerator degrees of freedom and 4 denominator degrees of freedom (this p-value is illustrated in the page margin). The Minitab and JMP output in Figure 11.14 tells us that the p-value equals 0.831. Because the p-value is large, we have little evidence that the variabilities of the yields produced by Catalysts XA-100 and ZB-200 differ.

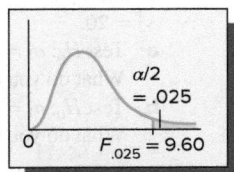

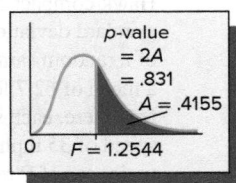

It has been suggested that the F test of $H_0: \sigma_1^2 = \sigma_2^2$ be used to choose between the equal variances and unequal variances t-based procedures when comparing two means (as described in Section 11.1). Certainly the F test is one approach to making this choice. However, studies have shown that the validity of the F test is very sensitive to violations of the normality assumption—much more sensitive, in fact, than the equal variances procedure is to violations of the equal variances assumption. While opinions vary, some statisticians believe that this is a serious problem and that the F test should never be used to choose between the equal variances and unequal variances procedures. Others feel that performing the test for this purpose is reasonable if the test's limitations are kept in mind.

As an example for those who believe that using the F test is reasonable, we found in Example 11.10 that we do not reject $H_0: \sigma_1^2 = \sigma_2^2$ at the .05 level of significance in the context of the catalyst comparison situation. Further, the p-value related to the F test, which equals 0.831, tells us that there is little evidence to suggest that the population variances differ. It follows that it might be reasonable to compare the mean yields of the catalysts by using the equal variances procedures (as we have done in Examples 11.1 and 11.2).

Exercises for Sections 11.4 and 11.5

CONCEPTS

11.31 When is the population of all possible values of s_1^2/s_2^2 described by an F distribution?

11.32 Intuitively explain why a value of s_1^2/s_2^2 that is substantially greater than 1 provides evidence that σ_1^2 is not equal to σ_2^2.

METHODS AND APPLICATIONS

11.33 Use Table 11.6 to find the $F_{.05}$ point for each of the following:

a $df_1 = 3$ and $df_2 = 14$.
b $df_1 = 6$ and $df_2 = 10$.
c $df_1 = 2$ and $df_2 = 11$.
d $df_1 = 7$ and $df_2 = 5$.

Use Tables A.6, A.7, A.8, and A.9 to find the following F_α points:

e $F_{.10}$ with $df_1 = 4$ and $df_2 = 7$.
f $F_{.01}$ with $df_1 = 3$ and $df_2 = 25$.
g $F_{.025}$ with $df_1 = 7$ and $df_2 = 17$.
h $F_{.05}$ with $df_1 = 9$ and $df_2 = 3$.

11.34 Suppose two independent random samples of sizes $n_1 = 9$ and $n_2 = 7$ that have been taken from two normally distributed populations having variances σ_1^2 and σ_2^2 give sample variances of $s_1^2 = 100$ and $s_2^2 = 20$.

a Test $H_0: \sigma_1^2 = \sigma_2^2$ versus $H_a: \sigma_1^2 \neq \sigma_2^2$ with $\alpha = .05$. What do you conclude?
b Test $H_0: \sigma_1^2 = \sigma_2^2$ versus $H_a: \sigma_1^2 > \sigma_2^2$ with $\alpha = .05$. What do you conclude?

11.35 The stopping distances of a random sample of 16 Fire-Hawk compact cars have a mean of 57.2 feet and a standard deviation of 4.81 feet. The stopping distances of a random sample of 16 Lance compact cars have a mean of 62.7 feet and a standard deviation of 7.56 feet. Here, each stopping distance is measured from a speed of 35 mph. If σ_1^2 and σ_2^2 denote the population variances of Fire-Hawk and Lance stopping distances,

respectively, test $H_0: \sigma_1^2 = \sigma_2^2$ versus $H_a: \sigma_1^2 < \sigma_2^2$ by setting α equal to .05 and using a critical value (assume normality).

11.36 The National Golf Association's consumer advocacy group wishes to compare the variabilities in durability of two brands of golf balls: Champ golf balls and Master golf balls. The advocacy group randomly selects 10 balls of each brand and places each ball into a machine that exerts the force produced by a 250-yard drive. The number of simulated drives needed to crack or chip each ball is recorded, and the results are given to the right. Assuming normality, test to see if the variabilities in durability differ for the two brands at the .10, .05, and .01 levels of significance. What do you conclude? **DS** GolfBrands

Champ Golf Balls	
270	334
290	315
301	307
305	325
298	331

Master Golf Balls	
364	302
325	342
350	348
359	327
396	355

DS GolfBrands

11.37 A marketing research firm wishes to compare the prices charged by two supermarket chains—Miller's and Albert's. The research firm, using a standardized one-week shopping plan (grocery list), makes identical purchases at 10 of each chain's stores. The stores for each chain are randomly selected, and all purchases are made during a single week. It is found that the mean and the standard deviation of the shopping expenses at the 10 Miller's stores are $121.92 and $1.40, respectively. It is also found that the mean and the standard deviation of the shopping expenses at the 10 Albert's stores are $114.81 and $1.84, respectively. Assuming normality, test to see if the corresponding population variances differ by setting α equal to .05. Is it reasonable to use the equal variances procedure to compare population means? Explain.

Chapter Summary

This chapter has explained **how to compare two populations** by using confidence intervals and hypothesis tests. First we discussed how to compare **two population means** by using **independent samples.** Here the measurements in one sample are not related to the measurements in the other sample. When the population variances are unknown, **t-based** inferences are appropriate if the populations are normally distributed or the sample sizes are large. Both **equal variances and unequal variances t-based procedures** exist. We learned that, because it can be difficult to compare the population variances, many statisticians believe that it is almost always best to use the unequal variances procedure.

Sometimes samples are not independent. We learned that one such case is what is called a **paired difference experiment.** Here we obtain two different measurements on the same sample units, and we can compare two population means by using a confidence interval or by conducting a hypothesis test that employs the differences between the pairs of measurements. We next explained how to compare **two population proportions** by using **large, independent samples.** Finally, we concluded this chapter by discussing the **F distribution** and how it is used to compare **two population variances** by using independent samples.

Glossary of Terms

F distribution: A continuous probability curve having a shape that depends on two parameters—the numerator degrees of freedom, df_1, and the denominator degrees of freedom, df_2.

independent samples experiment: An experiment in which there is no relationship between the measurements in the different samples.

paired difference experiment: An experiment in which two different measurements are taken on the same units and inferences are made using the differences between the pairs of measurements.

sampling distribution of $\hat{p}_1 - \hat{p}_2$: The probability distribution that describes the population of all possible values of $\hat{p}_1 - \hat{p}_2$, where $\hat{p}_1$ is the sample proportion for a random sample taken from one population and $\hat{p}_2$ is the sample proportion for a random sample taken from a second population.

sampling distribution of s_1^2/s_2^2: The probability distribution that describes the population of all possible values of s_1^2/s_2^2, where s_1^2 is the sample variance of a random sample taken from one population and s_2^2 is the sample variance of a random sample taken from a second population.

sampling distribution of $\bar{x}_1 - \bar{x}_2$: The probability distribution that describes the population of all possible values of $\bar{x}_1 - \bar{x}_2$, where $\bar{x}_1$ is the sample mean of a random sample taken from one population and $\bar{x}_2$ is the sample mean of a random sample taken from a second population.

Important Formulas and Tests

Sampling distribution of $\bar{x}_1 - \bar{x}_2$ (independent random samples): page 478

t-based confidence interval for $\mu_1 - \mu_2$ when $\sigma_1^2 = \sigma_2^2$: page 479

t-based confidence interval for $\mu_1 - \mu_2$ when $\sigma_1^2 \neq \sigma_2^2$: page 482

t test about $\mu_1 - \mu_2$ when $\sigma_1^2 = \sigma_2^2$: page 481

t test about $\mu_1 - \mu_2$ when $\sigma_1^2 \neq \sigma_2^2$: page 482

Confidence interval for μ_d: page 488

A hypothesis test about μ_d: page 489

Sampling distribution of $\hat{p}_1 - \hat{p}_2$ (independent random samples): page 494

Large sample confidence interval for $p_1 - p_2$: page 495

Large sample hypothesis test about $p_1 - p_2$: page 496

Sampling distribution of s_1^2/s_2^2 (independent random samples): page 501

A hypothesis test about the equality of σ_1^2 and σ_2^2 (One-Sided Alternative): page 501

A hypothesis test about the equality of σ_1^2 and σ_2^2 (Two-Sided Alternative): page 502

Supplementary Exercises

connect **11.38** In the book *Essentials of Marketing Research,* William R. Dillon, Thomas J. Madden, and Neil H. Firtle discuss evaluating the effectiveness of a test coupon. Samples of 500 test coupons and 500 control coupons were randomly delivered to shoppers. The results indicated that 35 of the 500 control coupons were redeemed, while 50 of the 500 test coupons were redeemed.

a In order to consider the test coupon for use, the marketing research organization required that the proportion of all shoppers who would redeem the test coupon be statistically shown to be greater than the proportion of all shoppers who would redeem the control coupon. Assuming that the two samples of shoppers are independent, carry out a hypothesis test at the .01 level of significance that will show whether this requirement is met by the test coupon. Explain your conclusion.

b Use the sample data to find a point estimate and a 95 percent interval estimate of the difference between the proportions of all shoppers who would redeem the test coupon and the control coupon. What does this interval say about whether the test coupon should be considered for use? Explain.

c Carry out the test of part *a* at the .10 level of significance. What do you conclude? Is your result statistically significant? Compute a 90 percent interval estimate instead of the 95 percent interval estimate of part *b*. Based on the interval estimate, do you feel that this result is practically important? Explain.

11.39 A marketing manager wishes to compare the mean prices charged for two brands of CD players. The manager conducts a random survey of retail outlets and obtains independent random samples of prices. The six retail outlets at which prices for the Onkyo CD player are obtained have a mean price of \$189 with a standard deviation of \$12. The twelve retail outlets at which prices for the JVC CD player are obtained have a mean price of \$145 with a standard deviation of \$10. Assuming normality and equal variances:

a Use an appropriate hypothesis test to determine whether the mean prices for the two brands differ. How much evidence is there that the mean prices differ?

b Use an appropriate 95 percent confidence interval to estimate the difference between the mean prices of the two brands of CD players. Do you think that the difference has practical importance?

c Use an appropriate hypothesis test to provide evidence supporting the claim that the mean price of the Onkyo CD player is more than $30 higher than the mean price for the JVC CD player. Set α equal to .05.

11.40 Standard deviation of returns is often used as a measure of a mutual fund's volatility (risk). A larger standard deviation of returns is an indication of higher risk. According to Morningstar.com (June 17, 2010), the American Century Equity Income Institutional Fund, a large cap fund, has a standard deviation of returns equal to 14.11 percent. Morningstar.com also reported that the Fidelity Small Cap Discovery Fund has a standard deviation of returns equal to 28.44 percent. Each standard deviation was computed using a sample of size 36. Perform a hypothesis test to determine if the variance of returns for the Fidelity Small Cap Discovery Fund is larger than the variance of returns for the American Century Equity Income Institutional Fund. Perform the test at the .05 level of significance, and assume normality.

Exercises 11.41 and 11.42 deal with the following situation:

In an article in the *Journal of Retailing,* Kumar, Kerwin, and Pereira study factors affecting merger and acquisition activity in retailing by comparing "target firms" and "bidder firms" with respect to several financial and marketing-related variables. If we consider two of the financial variables included in the study, suppose a random sample of 36 "target firms" gives a mean earnings per share of $1.52 with a standard deviation of $0.92, and that this sample gives a mean debt-to-equity ratio of 1.66 with a standard deviation of 0.82. Furthermore, an independent random sample of 36 "bidder firms" gives a mean earnings per share of $1.20 with a standard deviation of $0.84, and this sample gives a mean debt-to-equity ratio of 1.58 with a standard deviation of 0.81.

11.41

a Set up the null and alternative hypotheses needed to test whether the mean earnings per share for all "target firms" differs from the mean earnings per share for all "bidder firms." Test these hypotheses at the .10, .05, .01, and .001 levels of significance. How much evidence is there that these means differ? Explain.

b Calculate a 95 percent confidence interval for the difference between the mean earnings per share for "target firms" and "bidder firms." Interpret the interval.

11.42

a Set up the null and alternative hypotheses needed to test whether the mean debt-to-equity ratio for all "target firms"

differs from the mean debt-to-equity ratio for all "bidder firms." Test these hypotheses at the .10, .05, .01, and .001 levels of significance. How much evidence is there that these means differ? Explain.

b Calculate a 95 percent confidence interval for the difference between the mean debt-to-equity ratios for "target firms" and "bidder firms." Interpret the interval.

c Based on the results of this exercise and Exercise 11.41, does a firm's earnings per share or the firm's debt-to-equity ratio seem to have the most influence on whether a firm will be a "target" or a "bidder"? Explain.

11.43 What impact did the September 11 terrorist attack have on U.S. airline demand? An analysis was conducted by Ito and Lee, "Assessing the Impact of the September 11 Terrorist Attacks on U.S. Airline Demand," in the *Journal of Economics and Business* (January–February 2005). They found a negative short-term effect of over 30 percent and an ongoing negative impact of over 7 percent. Suppose that we wish to test the impact by taking a random sample of 12 airline routes before and after 9/11. Passenger miles (millions of passenger miles) for the same routes were tracked for the 12 months prior to and the 12 months immediately following 9/11. Assume that the population of all possible paired differences is normally distributed.

a Set up the null and alternative hypotheses needed to determine whether there was a reduction in mean airline passenger demand.

b Below we present the Minitab output for the paired differences test. Use the output and critical values to test the hypotheses at the .10, .05, and .01 levels of significance. Has the population mean airline demand been reduced?

Paired T-Test and CI: Before911, After911

Paired T for Before911 - After911

	N	Mean	StDev	SE Mean
Before911	12	117.333	26.976	7.787
After911	12	87.583	25.518	7.366
Difference	12	29.7500	10.3056	2.9750

T-Test of mean difference = 0 (vs > 0):
T-Value = 10.00 P-Value = 0.000

c Use the *p*-value to test the hypotheses at the .10, .05, and .01 levels of significance. How much evidence is there against the null hypothesis?

Appendix 11.1 ■ Two-Sample Hypothesis Testing Using Excel

Test for the difference between means, equal variances, in Figure 11.2(a) (data file: Catalyst.xlsx):

- Enter the data from Table 11.1 into two columns: yields for catalyst XA-100 in column A and yields for catalyst ZB-200 in column B, with labels XA-100 and ZB-200.

- Select **Data : Data Analysis : t-Test: Two-Sample Assuming Equal Variances** and click OK in the Data Analysis dialog box.

- In the t-Test dialog box, enter A1:A6 in the "Variable 1 Range" window.

- Enter B1:B6 in the "Variable 2 Range" window.

- Enter 0 (zero) in the "Hypothesized Mean Difference" box.

- Place a checkmark in the Labels checkbox.

- Enter 0.05 into the Alpha box.

- Under output options, select "New Worksheet Ply" to have the output placed in a new worksheet and enter the name Output for the new worksheet.

- Click OK in the t-Test dialog box.

- The output will be displayed in a new worksheet.

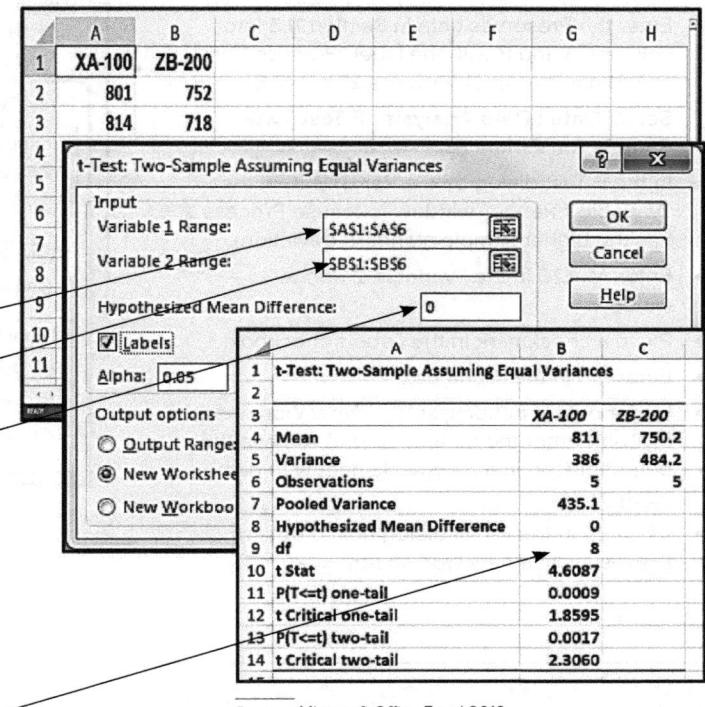

Source: Microsoft Office Excel 2016

Note: The *t*-test assuming unequal variances can be done by selecting **Data : Data Analysis : t-Test : Two-Sample Assuming Unequal Variances.**

Test for paired differences in Figure 11.9 (data file: Repair.xlsx):

- Enter the data from Table 11.2) into two columns: costs for Garage 1 in column A and costs for Garage 2 in column B, with labels Garage1 and Garage2.

- Select **Data : Data Analysis : t-Test: Paired Two Sample for Means** and click OK in the Data Analysis dialog box.

- In the t-Test dialog box, enter A1:A8 into the "Variable 1 Range" window.

- Enter B1:B8 into the "Variable 2 Range" window.

- Enter 0 (zero) in the "Hypothesized Mean Difference" box.

- Place a checkmark in the Labels checkbox.

- Enter 0.05 into the Alpha box.

- Under output options, select "New Worksheet Ply" to have the output placed in a new worksheet and enter the name Output for the new worksheet.

- Click OK in the t-Test dialog box.

- The output will be displayed in a new worksheet.

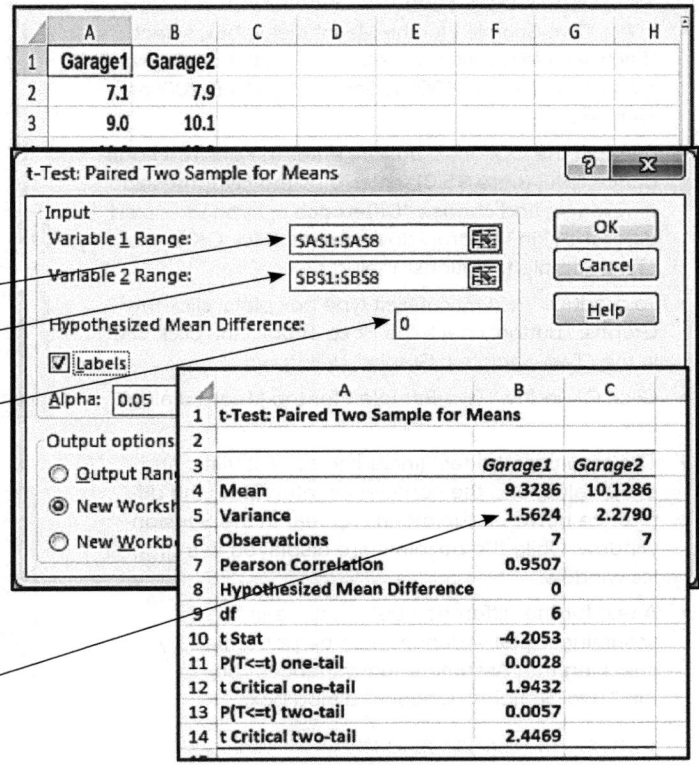

Source: Microsoft Office Excel 2016

Hypothesis test for the equality of two variances in Figure 11.13 (data file: Preserves.xlsx):

- Enter the Preserves data in Section 11.5 into columns A and B with the label "Process 1" in cell A1 and the label "Process 2" in cell B1.

- Select **Data : Data Analysis : F-Test Two-Sample for Variances.**

- In the F-Test dialog box, enter B1:B26 in the "Variable 1 Range" window (because Process 2 has the higher sample standard deviation).

- Enter A1:A26 in the "Variable 2 Range" window.

- Place a checkmark in the Labels checkbox.

- Enter .05 in the Alpha box.

- Under output options, select "New Worksheet Ply" and enter the name "Output" to have the output placed in a new worksheet with name "Output."

- Click OK in the F-test dialog box to obtain the results in a new worksheet.

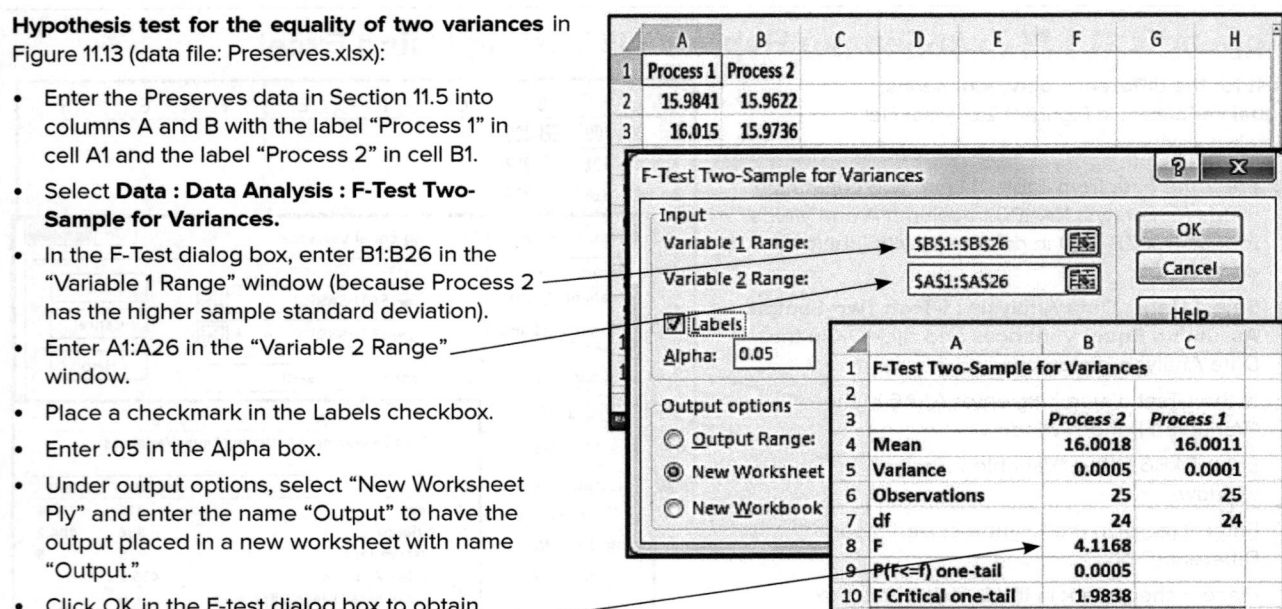

Source: Microsoft Office Excel 2016

Appendix 11.2 ■ Two-Sample Hypothesis Testing Using Minitab

Test for the difference between means, unequal variances, in Figure 11.4 (data file: Catalyst.MTW)

- In the Woksheet window, enter the data from Table 11.1 into two columns with variable names XA-100 and ZB-200.

- Select **Stat : Basic Statistics : 2-Sample t**

- In the "Two-Sample t for the Mean" dialog box, select "Each sample is in its own column" from the drop-down menu and enter 'XA-100' as Sample 1 and 'ZB-200' as Sample 2.

- Click on the Options... button; enter the desired confidence level (here 95.0); enter 0 as the Hypothesized difference; and choose "Difference ≠ hyphothesized difference" in the drop-down menu. Click OK in the "Two-Sample t: Options" dialog box.

- To produce yield by catalyst type box plots, click the Graphs... button; check the Boxplot box; and click OK in the "Two-Sample t: Graphs" dialog box.

- Click OK in the "Two-Sample t for the Mean" dialog box.

- The results of the test (including the t statistic and the p-value) and the confidence interval for the difference between the means appear in the Session window, while the boxplots are displayed in a graphics window.

- A test for the difference between means, assuming equal variances, can be performed by checking the "Assume equal variances" box in the "Two-Sample t: Options" dialog box.

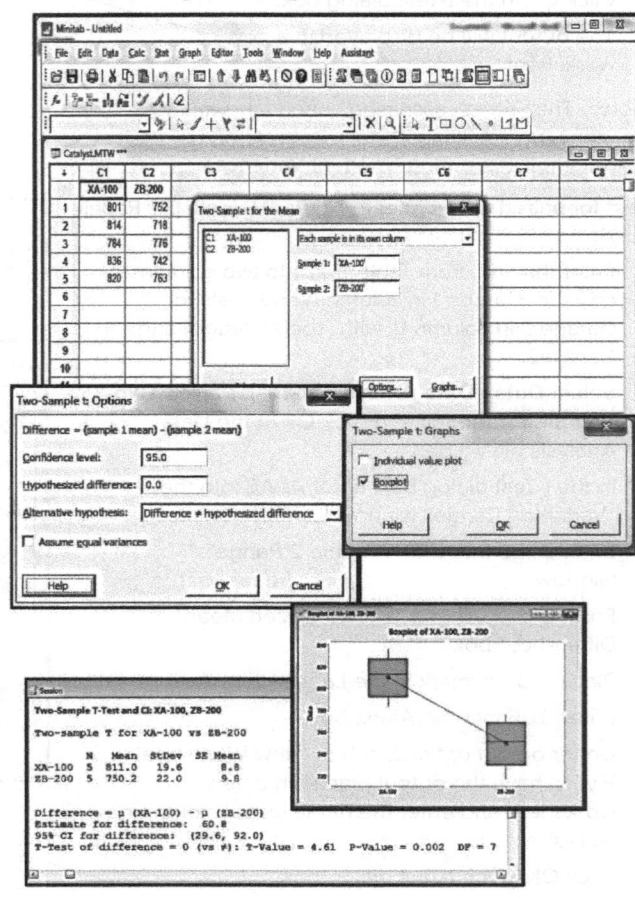

Source: Minitab 18

Test for equality of variances in Figure 11.14(a) (data file: Catalyst.MTW):

- In the Worksheet window, enter the data from Table 11.1 into two columns with variable names XA-100 and ZB-200.

- Select **Stat : Basic Statistics : 2 Variances**

- In the "Two-Sample Variance" dialog box, select "Each sample is in its own column" from the drop-down menu and enter 'ZB-200' as Sample 1 and 'XA-100' as Sample 2 (because the ZB-200 yields have the higher sample variance).

- Click on the Options... button; select (Sample 1 variance)/(Sample 2 variance) in the Ratio menu; select the desired confidence level; choose 1 as the Hypothesized ratio; check the "Use test and confidence intervals based on normal distribution" box; and click OK in the "Two-Sample Variance: Options" dialog box.

- Click OK in the "Two-Sample Variance" dialog box.

- The F statistic and the p-value will be displayed in the Session window.

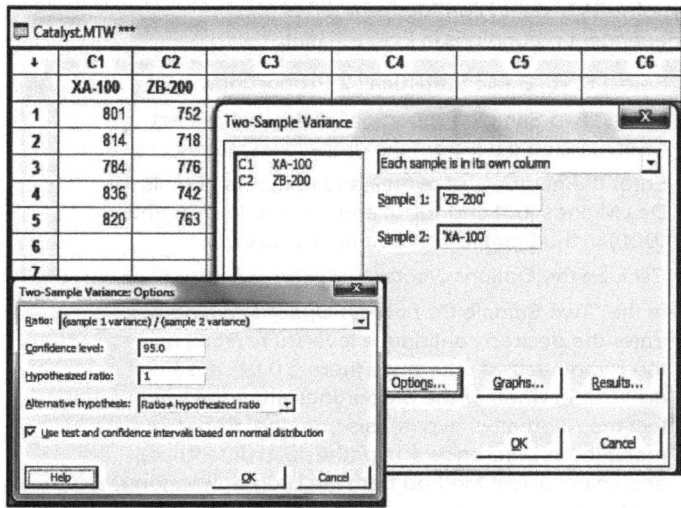

Source: Minitab 18

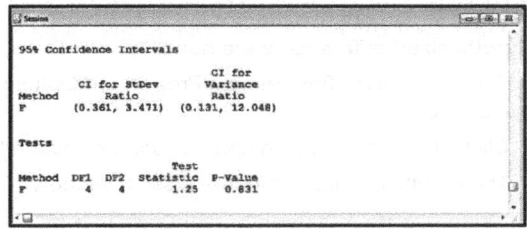

Source: Minitab 18

Test for paired differences in Figure 11.8 (data file: Repair.MTW):

- In the Worksheet window, enter the data from Table 11.2 into columns C1 and C2 with variable names Garage1 and Garage2.

- Select **Stat : Basic Statistics : Paired t**

- In the "Paired t for the Mean" dialog box, select "Each sample is in a column" in the drop-down menu and enter Garage1 and Garage 2 in the Sample 1 and Sample 2 boxes.

- Click the Options... button.

- In the "Paired t: Options" dialog box, select the desired confidence level, choose 0 as the Hypothesized difference, select "Difference < hypothesized difference" as the Alternative hypothesis, and click OK.

- To produce a box plot of differences with a graphical summary of the test, click the Graphs... button, check the "Boxplot of differences" box, and click OK in the "Paired t: Graphs" dialog box.

- Click OK in the "Paired t for the Mean" dialog box. The results of the paired t test are given in the Session window, and the boxplot is displayed in a graphics window.

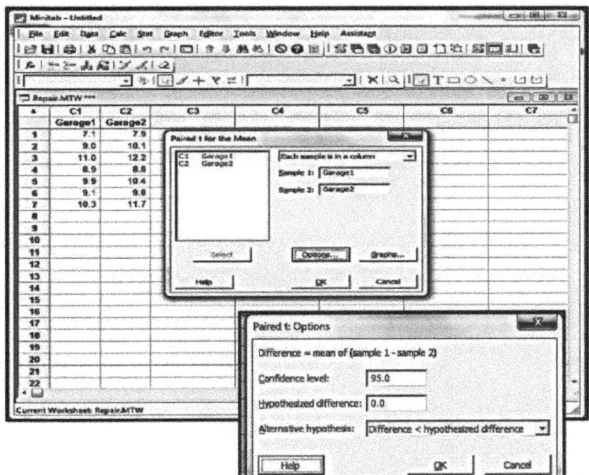

Source: Minitab 18

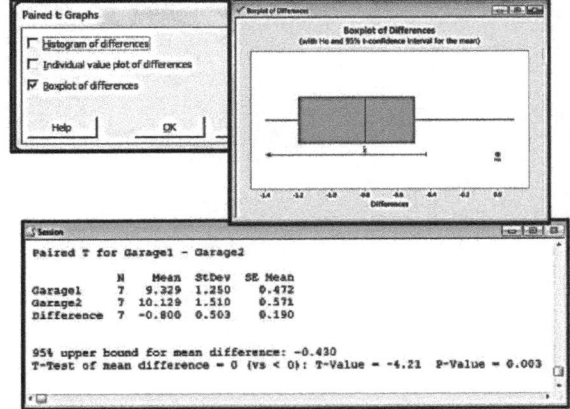

Source: Minitab 18

Hypothesis test and confidence interval for two Independent proportions in Figure 11.11(a):

- Select **Stat : Basic Statistics : 2 Proportions**

- In the "Two-Sample Proportion" dialog box, select "Summarized data" from the drop-down menu.

- Enter the numbers of events and numbers of trials for Des Moines (631 and 1000) and for Toledo (798 and 1000) in the Sample 1 and Sample 2 boxes.

- Click on the Options... button.

- In the "Two-Sample Proportion: Options" dialog box, enter the desired confidence level (here 95.0) and the Hypothesized difference (here 0.0 because we are testing whether the proportions are equal), select the Alternative hypothesis from the drop-down menu (here Difference ≠ hypothesized difference), and select a Test Method from the drop-down menu. We have chosen "Use the pooled estimate of the proportion" because our hypothesized difference is 0. We would not use the pooled estimate if the hypothesized difference were nonzero.

- Click OK in the "Two-Sample Proportion: Options" dialog box.

- Click OK in the "Two-Sample Proportion" dialog box.

- The results will appear in the Session window.

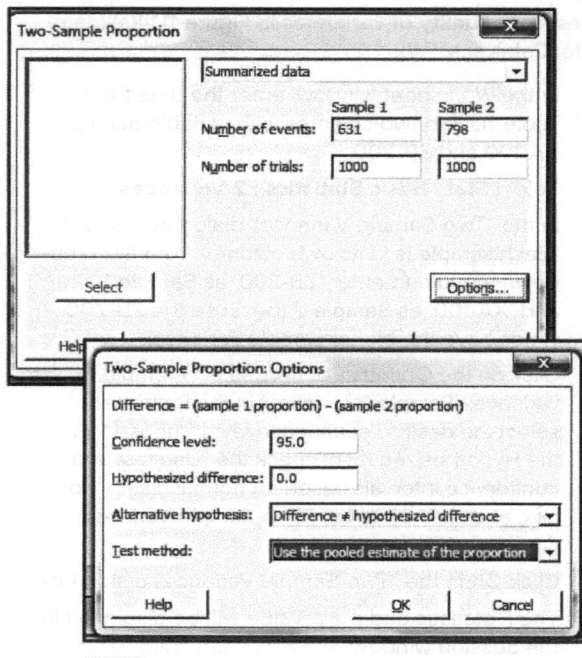

Source: Minitab 18

```
Session

Test and CI for Two Proportions

Sample    X      N    Sample p
1        631   1000   0.631000
2        798   1000   0.798000


Difference = p (1) - p (2)
Estimate for difference:  -0.167
95% CI for difference:  (-0.205906, -0.128094)
Test for difference = 0 (vs ≠ 0):  Z = -8.27  P-Value = 0.000
```

Source: Minitab 18

Appendix 11.3 ■ Two-Sample Hypothesis Testing Using JMP

Test for the difference between means assuming equal variances in Figure 11.7 (data file: AutoLoan. jmp):

- After opening the data file, select **Analyze : Fit Y by X.**

- In the dialog box, select the column "Label" and click the "X, Factor" button and select the column "Data" and click the "Y, Response" button. Click "OK."

- From the red triangle menu beside "Oneway Analysis ..." select **Means and Std Dev** and **Means/Anova/Pooled t.**

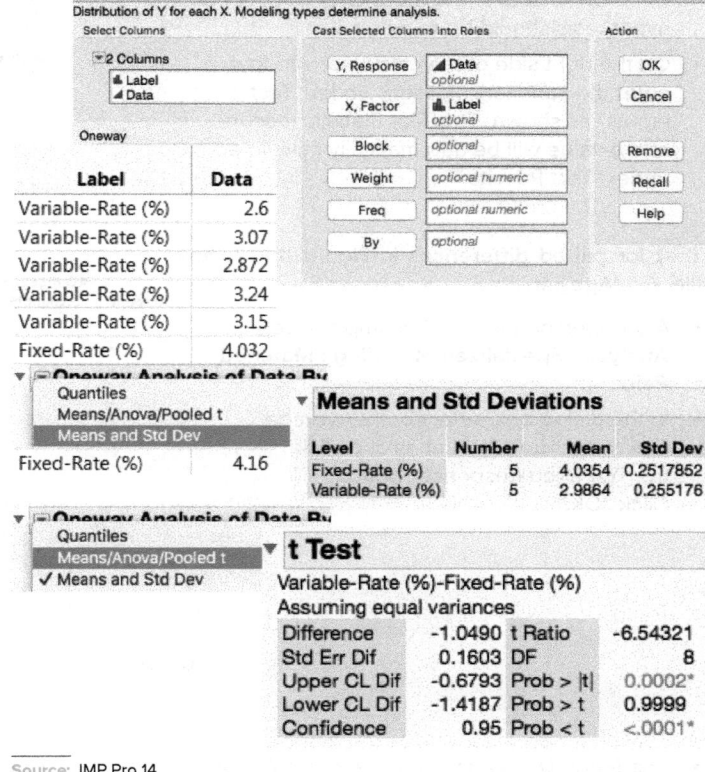

Source: JMP Pro 14

- **NOTE:** The **t Test** option in the same red triangle menu provides the *p*-values and confidence interval for the difference in the means <u>assuming unequal variances</u>.

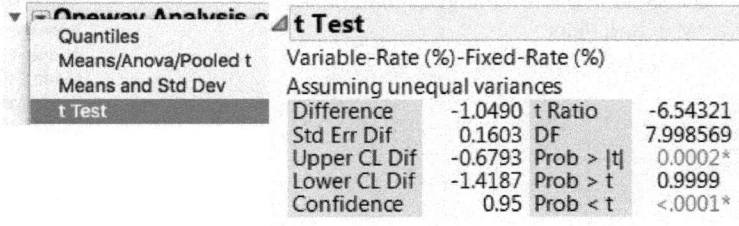

Source: JMP Pro 14

Test for the difference between means assuming unequal variances (see above for example output):

- <u>If you have the raw data</u>, see the preceding **NOTE** about the **t Test** command.

- If you only have summary statistics, select **Help : Sample Data** then select "Hypothesis Test for Two Means" from the **Calculators** under Teaching resources.

- Select Summary Statistics and click "OK."

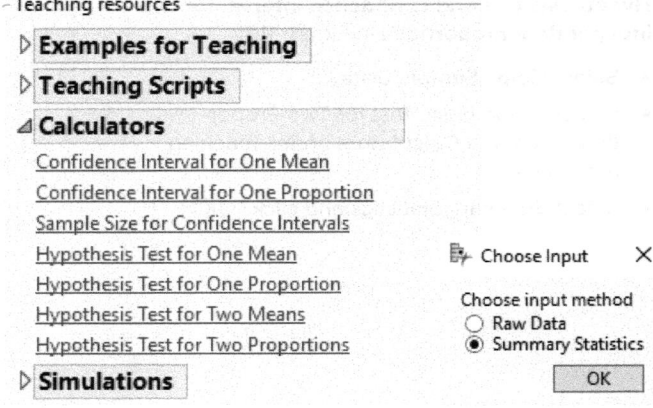

Source: JMP Pro 14

- On the left of the calculator window, select "*t*-test," "Unequal Variances," and "(Mean 2 – Mean 1) is greater than the hypothesized difference." You can also check the "Reveal Decision" box to show whether or not the null hypothesis should be rejected.

- On the right side of the calculator window, enter the summary statistics under "Test Inputs" as shown. The Test Statistic (*t*-score) and *p*-value will be automatically calculated under "Test Results."

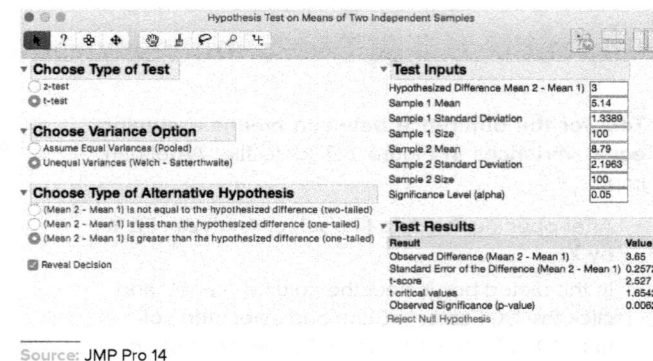

Source: JMP Pro 14

Test for paired differences in Figure 11.10 (data file: AcctAge.jmp):

- After opening the AcctAge.jmp file, select **Analyze : Specialized Modeling : Matched Pairs.**

- In the dialog box, select both "Average age of account…" columns and click the "Y, Paired Response" button. Click "OK."

Account	Average age of account in …	Average age of account in …
1	27	35
2	19	24
3	40	47
4	30	28
5	33	41
6	25	33

Source: JMP Pro 14

- The **Mean Difference**, test statistic (**t-Ratio**), *p*-value (**Prob > t**) appear in the Matched Pairs window. Additional output options are available in the red triangle menu beside "Matched Pairs."

Matched Pairs

Difference: Average age of account in 2016 (days)-Average age of account in 2017 (days)

Average age of account in 2016 (days)	34	t-Ratio	3.61211
Average age of account in 2017 (days)	27	DF	9
Mean Difference	7	Prob > |t|	0.0056*
Std Error	1.93793	Prob > t	0.0028*
Upper 95%	11.3839	Prob < t	0.9972
Lower 95%	2.61611		
N	10		
Correlation	0.80459		

Source: JMP Pro 14

Hypothesis test and confidence interval for two independent proportions in Figure 11.11:

- Select **Help : Sample Data.**

- Select "Hypothesis Test for Two Proportions" from the **Calculators** under Teaching resources.

- Select Summary Statistics and click "OK."

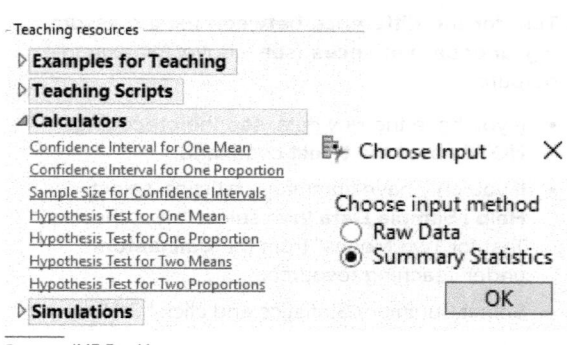

Source: JMP Pro 14

- On the left of the calculator window, select "(p2-p1) is not equal to the hypothesized difference" and "Use Pooled Estimate of Variances." You can also check the "Reveal Decision" box to show whether or not the null hypothesis should be rejected.

- On the right side of the calculator window, enter the summary statistics under "Test Inputs" as shown. The Test Statistic **(z-score)** and *p*-value will be automatically calculated under "Test Results."

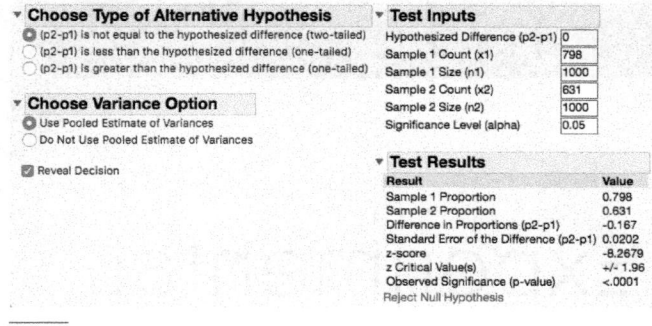

Source: JMP Pro 14

Test for equality of two variances in Figure 11.14 (data file: Catalyst.jmp):

- After opening the Catalyst.jmp file, select **Analyze : Fit Y by X.**

- In the dialog box, select the column "Label" and click the "X, Factor" button and select the column "Data" and click the "Y, Response" button. Click "OK."

- From the red triangle menu beside "Oneway Analysis ..." select **Unequal Variances.** The "F Test 2-sided" output is the last row under "Test."

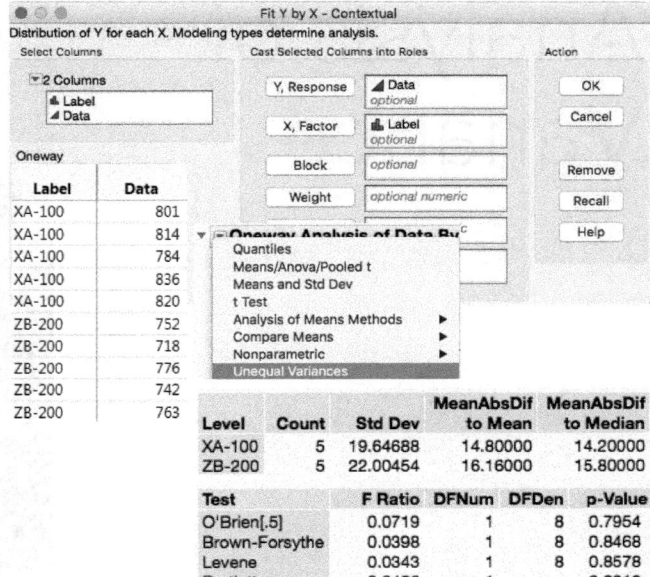

Source: JMP Pro 14

Design Elements: (CD): ©Comstock Images/Alamy; (All Others): ©McGraw-Hill Education

## CHAPTER 12

Experimental Design and Analysis of Variance

©Hal Bergman/Media Bakery

Learning Objectives

After mastering the material in this chapter, you will be able to:

LO12-1 Explain the basic terminology and concepts of experimental design.

LO12-2 Compare several different population means by using a one-way analysis of variance.

LO12-3 Compare treatment effects and block effects by using a randomized block design.

LO12-4 Assess the effects of two factors on a response variable by using a two-way analysis of variance.

LO12-5 Describe what happens when two factors interact.

Chapter Outline

n Chapter 11 we learned that business improvement often involves making **comparisons.** In that chapter we presented several confidence intervals and several hypothesis testing procedures for comparing two population means. However, business improvement often requires that we compare more than two population means. For instance, we might compare the mean sales obtained by using three different advertising campaigns in order to improve a company's marketing process. Or, we might compare the mean production output obtained by using four different manufacturing process designs to improve productivity.

In this chapter we extend the methods presented in Chapter 11 by considering statistical procedures for **comparing two or more population means.** Each of the methods we discuss is called an **analysis of variance (ANOVA)** procedure. We also present some basic concepts of **experimental design,** which involves deciding how to collect data in a way that allows us to most effectively compare population means.

We explain the methods of this chapter in the context of three cases:

The Oil Company Case: An oil company wishes to develop a reasonably priced gasoline that will deliver improved mileages. The company uses **one-way analysis of variance** to compare the effects of three types of gasoline on mileage in order to find the gasoline type that delivers the highest mean mileage.

The Cardboard Box Case: A paper company performs an experiment to investigate the effects of four production methods on the number of defective cardboard boxes produced in an hour. The company uses a **randomized block ANOVA** to determine which production method yields the smallest mean number of defective boxes.

The Supermarket Case: A commercial bakery supplies many supermarkets. In order to improve the effectiveness of its supermarket shelf displays, the company wishes to compare the effects of shelf display height (bottom, middle, or top) and width (regular or wide) on monthly demand. The bakery employs **two-way analysis of variance** to find the display height and width combination that produces the highest monthly demand.

12.1 Basic Concepts of Experimental Design

LO12-1
Explain the basic terminology and concepts of experimental design.

In many statistical studies a variable of interest, called the **response variable** (or **dependent variable**), is identified. Then data are collected that tell us about how one or more **factors** (or **independent variables**) influence the variable of interest. If we cannot control the factor(s) being studied, we say that the data obtained are **observational.** For example, suppose that in order to study how the size of a home relates to the sales price of the home, a real estate agent randomly selects 50 recently sold homes and records the square footages and sales prices of these homes. Because the real estate agent cannot control the sizes of the randomly selected homes, we say that the data are observational.

If we can control the factors being studied, we say that the data are **experimental.** Furthermore, in this case the values, or **levels,** of the factor (or combination of factors) are called **treatments.** The purpose of most experiments is **to compare and estimate the effects of the different treatments on the response variable.** For example, suppose that an oil company wishes to study how three different gasoline types (A, B, and C) affect the mileage obtained by a popular compact automobile model. Here the response variable is gasoline mileage, and the company will study a single factor—gasoline type. Because the oil company can control which gasoline type is used in the compact automobile, the data that the oil company will collect are experimental. Furthermore, the treatments—the levels of the factor gasoline type—are gasoline types A, B, and C.

In order to collect data in an experiment, the different treatments are assigned to objects (people, cars, animals, or the like) that are called **experimental units.** For example, in the gasoline mileage situation, gasoline types A, B, and C will be compared by conducting mileage tests using a compact automobile. The automobiles used in the tests are the experimental units.

In general, when a treatment is applied to more than one experimental unit, it is said to be **replicated.** Furthermore, when the analyst controls the treatments employed and how they are applied to the experimental units, a **designed experiment** is being carried out. A commonly used, simple experimental design is called the **completely randomized experimental design.**

> In a **completely randomized experimental design,** independent random samples of experimental units are assigned to the treatments.

As illustrated in the following examples, we can sometimes assign *independent* random samples of experimental units to the treatments by assigning *different* random samples of experimental units to different treatments.

Ⓒ **EXAMPLE 12.1** The Oil Company Case: Comparing Gasoline Types

©Rubberball/Getty Images

North American Oil Company is attempting to develop a reasonably priced gasoline that will deliver improved gasoline mileages. As part of its development process, the company would like to compare the effects of three types of gasoline (A, B, and C) on gasoline mileage. For testing purposes, North American Oil will compare the effects of gasoline types A, B, and C on the gasoline mileage obtained by a popular compact model called the Lance. Suppose the company has access to 1,000 Lances that are representative of the population of all Lances, and suppose the company will utilize a completely randomized experimental design that employs samples of size five. In order to accomplish this, five Lances will be randomly selected from the 1,000 available Lances. These autos will be assigned to gasoline type A. Next, five *different* Lances will be randomly selected from the remaining 995 available Lances. These autos will be assigned to gasoline type B. Finally, five *different* Lances will be randomly selected from the remaining 990 available Lances. These autos will be assigned to gasoline type C.

Each randomly selected Lance is test driven using the appropriate gasoline type (treatment) under normal conditions for a specified distance, and the gasoline mileage for each test drive is measured. We let x_{ij} denote the jth mileage obtained when using gasoline type i. The mileage data obtained are given in Table 12.1. Here we assume that the set of gasoline mileage observations obtained by using a particular gasoline type is a sample randomly selected from the infinite population of all Lance mileages that could be obtained using that gasoline type. Examining the box plots shown next to the mileage data, we see some evidence that gasoline type B yields the highest gasoline mileages.

TABLE 12.1   **The Gasoline Mileage Data**   ⒟ⓢ GasMile2

Gasoline Type A	Gasoline Type B	Gasoline Type C
$x_{A1} = 34.0$	$x_{B1} = 35.3$	$x_{C1} = 33.3$
$x_{A2} = 35.0$	$x_{B2} = 36.5$	$x_{C2} = 34.0$
$x_{A3} = 34.3$	$x_{B3} = 36.4$	$x_{C3} = 34.7$
$x_{A4} = 35.5$	$x_{B4} = 37.0$	$x_{C4} = 33.0$
$x_{A5} = 35.8$	$x_{B5} = 37.6$	$x_{C5} = 34.9$

Ⓒ **EXAMPLE 12.2** The Supermarket Case: Studying the Effect
of Display Height

The Tastee Bakery Company supplies a bakery product to many supermarkets in a metropolitan area. The company wishes to study the effect of the shelf display height employed by the supermarkets on monthly sales (measured in cases of 10 units each) for

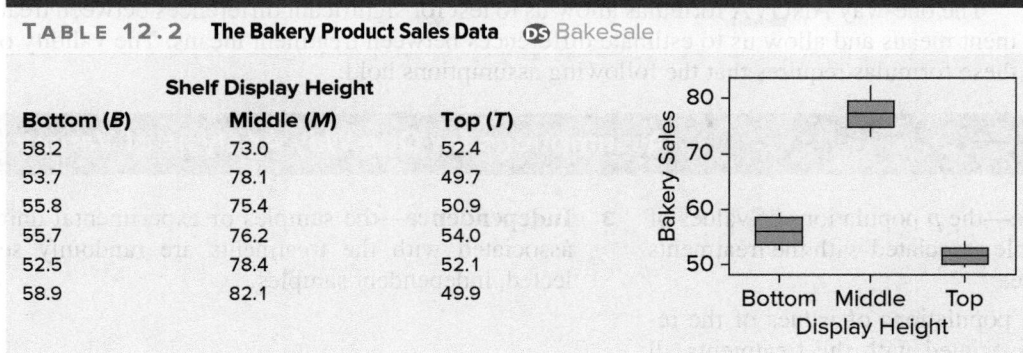

Shelf Display Height		
Bottom (B)	**Middle (M)**	**Top (T)**
58.2	73.0	52.4
53.7	78.1	49.7
55.8	75.4	50.9
55.7	76.2	54.0
52.5	78.4	52.1
58.9	82.1	49.9

TABLE 12.2   The Bakery Product Sales Data   🖸 BakeSale

this product. Shelf display height, the factor to be studied, has three levels—bottom (B), middle (M), and top (T)—which are the treatments. To compare these treatments, the bakery uses a completely randomized experimental design. For each shelf height, six supermarkets (the experimental units) of equal sales potential are randomly selected, and each supermarket displays the product using its assigned shelf height for a month. At the end of the month, sales of the bakery product (the response variable) at the 18 participating stores are recorded, giving the data in Table 12.2. Here we assume that the set of sales amounts for each display height is a sample randomly selected from the population of all sales amounts that could be obtained (at supermarkets of the given sales potential) at that display height. Examining the box plots that are shown next to the sales data, we seem to have evidence that a middle display height gives the highest bakery product sales.

12.2 One-Way Analysis of Variance

LO12-2
Compare several different population means by using a one-way analysis of variance.

Suppose we wish to study the effects of p **treatments** (treatments 1, 2, . . . , p) on a **response variable.** For any particular treatment, say treatment i, we define μ_i and σ_i to be the mean and standard deviation of the population of all possible values of the response variable that could potentially be observed when using treatment i. Here we refer to μ_i as **treatment mean i.** The goal of **one-way analysis of variance** (often called **one-way ANOVA**) is to estimate and compare the effects of the different treatments on the response variable. We do this by **estimating and comparing the treatment means** $\mu_1, \mu_2, . . . , \mu_p$. Here we assume that a sample has been randomly selected for each of the p treatments by employing a completely randomized experimental design. We let n_i denote the size of the sample that has been randomly selected for treatment i, and we let x_{ij} denote the jth value of the response variable that is observed when using treatment i. It then follows that the point estimate of μ_i is $\bar{x}_i$, the average of the sample of n_i values of the response variable observed when using treatment i. It further follows that the point estimate of σ_i is s_i, the standard deviation of the sample of n_i values of the response variable observed when using treatment i.

For example, consider the gasoline mileage situation. We let μ_A, μ_B, and μ_C denote the means and σ_A, σ_B, and σ_C denote the standard deviations of the populations of all possible gasoline mileages using gasoline types A, B, and C. To estimate these means and standard deviations, North American Oil has employed a completely randomized experimental design and has obtained the samples of mileages in Table 12.1. The means of these samples—$\bar{x}_A = 34.92$, $\bar{x}_B = 36.56$, and $\bar{x}_C = 33.98$—are the point estimates of μ_A, μ_B, and μ_C. The standard deviations of these samples—$s_A = .7662$, $s_B = .8503$, and $s_C = .8349$—are the point estimates of σ_A, σ_B, and σ_C. Using these point estimates, we will (later in this section) test to see whether there are any statistically significant differences between the treatment means μ_A, μ_B, and μ_C. If such differences exist, we will estimate the magnitudes of these differences. This will allow North American Oil to judge whether these differences have practical importance.

The one-way ANOVA formulas allow us to test for significant differences between treatment means and allow us to estimate differences between treatment means. The validity of these formulas requires that the following assumptions hold:

Assumptions for One-Way Analysis of Variance

1 **Constant variance**—the p populations of values of the response variable associated with the treatments have equal variances.

2 **Normality**—the p populations of values of the response variable associated with the treatments all have normal distributions.

3 **Independence**—the samples of experimental units associated with the treatments are randomly selected, independent samples.

The one-way ANOVA results are not very sensitive to violations of the equal variances assumption. Studies have shown that this is particularly true when the sample sizes employed are equal (or nearly equal). Therefore, a good way to make sure that unequal variances will not be a problem is to take samples that are the same size. In addition, it is useful to compare the sample standard deviations $s_1, s_2, \ldots, s_p$ to see if they are reasonably equal. As a general rule, *the one-way ANOVA results will be approximately correct if the largest sample standard deviation is no more than twice the smallest sample standard deviation.* The variations of the samples can also be compared by constructing a box plot for each sample (as we have done for the gasoline mileage data in Table 12.1). Several statistical tests also employ the sample variances to test the equality of the population variances [see Bowerman and O'Connell (1990) for two of these tests]. However, these tests have some drawbacks—in particular, their results are very sensitive to violations of the normality assumption. Because of this, there is controversy as to whether these tests should be performed.

The normality assumption says that each of the p populations is normally distributed. This assumption is not crucial. It has been shown that the one-way ANOVA results are approximately valid for mound-shaped distributions. It is useful to construct a box plot and/ or a stem-and-leaf display for each sample. If the distributions are reasonably symmetric, and if there are no outliers, the ANOVA results can be trusted for sample sizes as small as 4 or 5. As an example, consider the gasoline mileage study of Example 12.1. The box plots of Table 12.1 suggest that the variability of the mileages in each of the three samples is roughly the same. Furthermore, the sample standard deviations $s_A = .7662$, $s_B = .8503$, and $s_C = .8349$ are reasonably equal (the largest is not even close to twice the smallest). Therefore, it is reasonable to believe that the constant variance assumption is satisfied. Moreover, because the sample sizes are the same, unequal variances would probably not be a serious problem anyway. Many small, independent factors influence gasoline mileage, so the distributions of mileages for gasoline types A, B, and C are probably mound-shaped. In addition, the box plots of Table 12.1 indicate that each distribution is roughly symmetric with no outliers. Thus, the normality assumption probably approximately holds. Finally, because North American Oil has employed a completely randomized design, the independence assumption probably holds. This is because the gasoline mileages in the different samples were obtained for *different* Lances.

Testing for significant differences between treatment means

As a preliminary step in one-way ANOVA, we wish to determine whether there are any statistically significant differences between the treatment means $\mu_1, \mu_2, \ldots, \mu_p$. To do this, we test the null hypothesis

$$H_0: \mu_1 = \mu_2 = \cdots = \mu_p$$

This hypothesis says that all the treatments have the same effect on the mean response. We test H_0 versus the alternative hypothesis

$$H_a: \text{At least two of } \mu_1, \mu_2, \ldots, \mu_p \text{ differ}$$

FIGURE 12.1   Comparing Between-Treatment Variability and Within-Treatment Variability

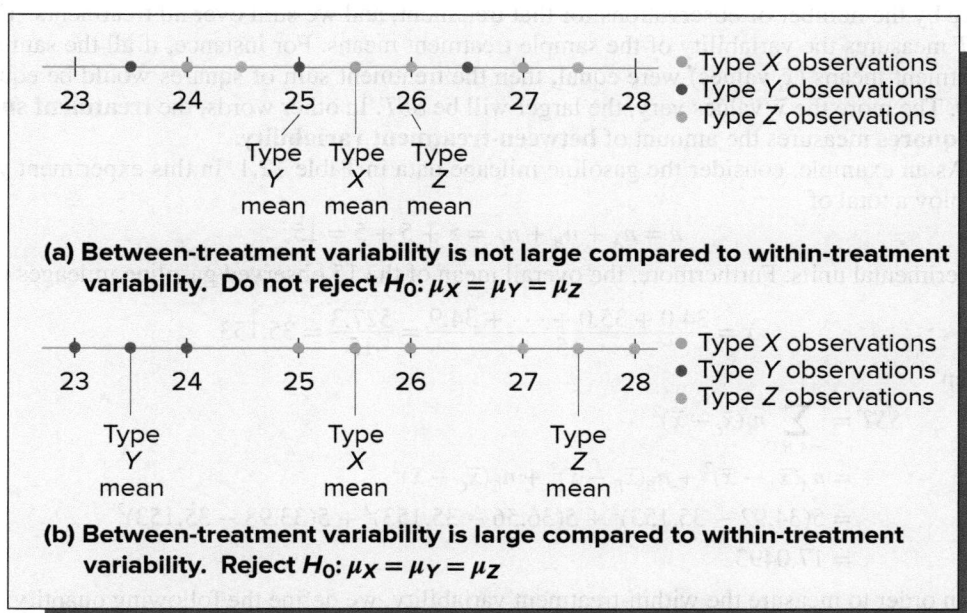

(a) Between-treatment variability is not large compared to within-treatment variability. Do not reject $H_0: \mu_X = \mu_Y = \mu_Z$

(b) Between-treatment variability is large compared to within-treatment variability. Reject $H_0: \mu_X = \mu_Y = \mu_Z$

This alternative says that at least two treatments have different effects on the mean response.

To carry out such a test, we compare what we call the **between-treatment variability** to the **within-treatment variability.** For instance, suppose we wish to study the effects of three gasoline types (X, Y, and Z) on mean gasoline mileage, and consider Figure 12.1(a). This figure depicts three independent random samples of gasoline mileages obtained using gasoline types X, Y, and Z. Observations obtained using gasoline type X are plotted as blue dots (•), observations obtained using gasoline type Y are plotted as red dots (•), and observations obtained using gasoline type Z are plotted as green dots (•). Furthermore, the sample treatment means are labeled as "type X mean," "type Y mean," and "type Z mean." We see that the variability of the sample treatment means—that is, the **between-treatment variability**—is not large compared to the variability within each sample (the **within-treatment variability**). In this case, the differences between the sample treatment means could quite easily be the result of sampling variation. Thus we would not have sufficient evidence to reject

$$H_0: \mu_X = \mu_Y = \mu_Z$$

Next look at Figure 12.1(b), which depicts a different set of three independent random samples of gasoline mileages. Here the variability of the sample treatment means (the between-treatment variability) is large compared to the variability within each sample. This would probably provide enough evidence to tell us to reject $H_0: \mu_X = \mu_Y = \mu_Z$ in favor of H_a: At least two of μ_X, μ_Y, and μ_Z differ. We would conclude that at least two of gasoline types X, Y, and Z have different effects on mean mileage.

In order to numerically compare the between-treatment and within-treatment variability, we can define several **sums of squares** and **mean squares.** To begin, we define n to be the total number of experimental units employed in the one-way ANOVA, and we define $\bar{x}$ to be the overall mean of all observed values of the response variable. Then we define the following:

The **treatment sum of squares** is

$$SST = \sum_{i=1}^{p} n_i (\bar{x}_i - \bar{x})^2$$

In order to compute *SST*, we calculate the difference between each sample treatment mean $\bar{x}_i$ and the overall mean $\bar{x}$, we square each of these differences, we multiply each squared difference by the number of observations for that treatment, and we sum over all treatments. The *SST* measures the variability of the sample treatment means. For instance, if all the sample treatment means ($\bar{x}_i$ values) were equal, then the treatment sum of squares would be equal to 0. The more the $\bar{x}_i$ values vary, the larger will be *SST*. In other words, the **treatment sum of squares** measures the amount of **between-treatment variability.**

As an example, consider the gasoline mileage data in Table 12.1. In this experiment we employ a total of

$$n = n_A + n_B + n_C = 5 + 5 + 5 = 15$$

experimental units. Furthermore, the overall mean of the 15 observed gasoline mileages is

$$\bar{x} = \frac{34.0 + 35.0 + \cdots + 34.9}{15} = \frac{527.3}{15} = 35.153$$

Then

$$SST = \sum_{i=A,B,C} n_i(\bar{x}_i - \bar{x})^2$$
$$= n_A(\bar{x}_A - \bar{x})^2 + n_B(\bar{x}_B - \bar{x})^2 + n_C(\bar{x}_C - \bar{x})^2$$
$$= 5(34.92 - 35.153)^2 + 5(36.56 - 35.153)^2 + 5(33.98 - 35.153)^2$$
$$= 17.0493$$

In order to measure the within-treatment variability, we define the following quantity:

The **error sum of squares** is

$$SSE = \sum_{j=1}^{n_1}(x_{1j} - \bar{x}_1)^2 + \sum_{j=1}^{n_2}(x_{2j} - \bar{x}_2)^2 + \cdots + \sum_{j=1}^{n_p}(x_{pj} - \bar{x}_p)^2$$

Here x_{1j} is the *j*th observed value of the response in the first sample, x_{2j} is the *j*th observed value of the response in the second sample, and so forth. The formula above says that we compute *SSE* by calculating the squared difference between each observed value of the response and its corresponding sample treatment mean and by summing these squared differences over all the observations in the experiment.

The *SSE* measures the variability of the observed values of the response variable around their respective sample treatment means. For example, if there were no variability within each sample, the error sum of squares would be equal to 0. The more the values within the samples vary, the larger will be *SSE*.

As an example, in the gasoline mileage study, the sample treatment means are $\bar{x}_A = 34.92$, $\bar{x}_B = 36.56$, and $\bar{x}_C = 33.98$. It follows that

$$SSE = \sum_{j=1}^{n_A}(x_{Aj} - \bar{x}_A)^2 + \sum_{j=1}^{n_B}(x_{Bj} - \bar{x}_B)^2 + \sum_{j=1}^{n_C}(x_{Cj} - \bar{x}_C)^2$$
$$= [(34.0 - 34.92)^2 + (35.0 - 34.92)^2 + (34.3 - 34.92)^2 + (35.5 - 34.92)^2$$
$$+ (35.8 - 34.92)^2] + [(35.3 - 36.56)^2 + (36.5 - 36.56)^2 + (36.4 - 36.56)^2$$
$$+ (37.0 - 36.56)^2 + (37.6 - 36.56)^2] + [(33.3 - 33.98)^2 + (34.0 - 33.98)^2$$
$$+ (34.7 - 33.98)^2 + (33.0 - 33.98)^2 + (34.9 - 33.98)^2]$$
$$= 8.028$$

Finally, we define a sum of squares that measures the total amount of variability in the observed values of the response:

The **total sum of squares** is

$$SSTO = SST + SSE$$

The variability in the observed values of the response must come from one of two sources—the between-treatment variability or the within-treatment variability. It follows that the total sum of squares equals the sum of the treatment sum of squares and the error sum of squares. Therefore, the **SST and SSE are said to partition the total sum of squares.** For the gasoline mileage study

$$SSTO = SST + SSE = 17.0493 + 8.028 = 25.0773$$

In order to decide whether there are any statistically significant differences between the treatment means, it makes sense to compare the amount of between-treatment variability to the amount of within-treatment variability. This comparison suggests the following F test:

An *F* Test for Differences between Treatment Means

Suppose that we wish to compare p treatment means $\mu_1, \mu_2, \ldots, \mu_p$ and consider testing

$$H_0: \mu_1 = \mu_2 = \cdots = \mu_p \qquad \text{versus} \qquad H_a: \text{At least two of } \mu_1, \mu_2, \ldots, \mu_p \text{ differ}$$
$$\text{(all treatment means are equal)} \qquad\qquad \text{(at least two treatment means differ)}$$

To perform the hypothesis test, define the **treatment mean square** to be $MST = SST/(p-1)$ and define the **error mean square** to be $MSE = SSE/(n-p)$. Also, define the F statistic

$$F = \frac{MST}{MSE} = \frac{SST/(p-1)}{SSE/(n-p)}$$

and its p-value to be the area under the F curve with $p-1$ and $n-p$ degrees of freedom to the right of F. We can reject H_0 in favor of H_a at level of significance α if either of the following equivalent conditions holds:

1 $F > F_\alpha$ **2** p-value $< \alpha$

Here the F_α point is based on $p-1$ numerator and $n-p$ denominator degrees of freedom.

A large value of F results when SST, which measures the between-treatment variability, is large compared to SSE, which measures the within-treatment variability. If F is large enough, this implies that H_0 should be rejected. The critical value F_α tells us when F is large enough to allow us to reject H_0 at level of significance α. When F is large, the associated p-value is small. If this p-value is less than α, we can reject H_0 at level of significance α.

Ⓒ **EXAMPLE 12.3** The Oil Company Case: Comparing Gasoline Types

Consider the North American Oil Company data in Table 12.1. The company wishes to determine whether any of gasoline types A, B, and C have different effects on mean Lance gasoline mileage. That is, we wish to see whether there are any statistically significant differences between μ_A, μ_B, and μ_C. To do this, we test the null hypothesis $H_0: \mu_A = \mu_B = \mu_C$, which says that gasoline types A, B, and C have the same effects on mean gasoline mileage. We test H_0 versus the alternative H_a: At least two of μ_A, μ_B, and μ_C differ, which says that at least two of gasoline types A, B, and C have different effects on mean gasoline mileage.

Because we have previously computed SST to be 17.0493 and SSE to be 8.028, and because we are comparing $p = 3$ treatment means, we have

$$MST = \frac{SST}{p-1} = \frac{17.0493}{3-1} = 8.525$$

and

$$MSE = \frac{SSE}{n-p} = \frac{8.028}{15-3} = 0.669$$

TABLE 12.3   Analysis of Variance (ANOVA) Table for Testing H_0: $\mu_A = \mu_B = \mu_C$ in the Oil Company Case ($p = 3$ Gasoline Types, $n = 15$ Observations)

Source	Degrees of Freedom	Sums of Squares	Mean Squares	F Statistic	p-Value
Treatments	$p - 1 = 3 - 1$ $= 2$	$SST = 17.0493$	$MST = \frac{SST}{p-1}$ $= \frac{17.0493}{3-1}$ $= 8.525$	$F = \frac{MST}{MSE}$ $= \frac{8.525}{0.669}$ $= 12.74$	0.001
Error	$n - p = 15 - 3$ $= 12$	$SSE = 8.028$	$MSE = \frac{SSE}{n-p}$ $= \frac{8.028}{15-3}$ $= 0.669$		
Total	$n - 1 = 15 - 1$ $= 14$	$SSTO = 25.0773$			

It follows that

$$F = \frac{MST}{MSE} = \frac{8.525}{0.669} = 12.74$$

In order to test H_0 at the .05 level of significance, we use $F_{.05}$ with $p - 1 = 3 - 1 = 2$ numerator and $n - p = 15 - 3 = 12$ denominator degrees of freedom. Table A.7 tells us that this F point equals 3.89, so we have

$$F = 12.74 > F_{.05} = 3.89$$

Therefore, we reject H_0 at the .05 level of significance. This says we have strong evidence that at least two of the treatment means μ_A, μ_B, and μ_C differ. In other words, we conclude that at least two of gasoline types A, B, and C have different effects on mean gasoline mileage.

The results of an analysis of variance are often summarized in what is called an **analysis of variance table.** This table gives the sums of squares (SST, SSE, $SSTO$), the mean squares (MST and MSE), and the F statistic and its related p-value for the ANOVA. The table also gives the degrees of freedom associated with each source of variation—treatments, error, and total. Table 12.3 gives the ANOVA table for the gasoline mileage problem. Notice that in the column labeled "Sums of Squares," the values of SST and SSE sum to $SSTO$.

Figure 12.2 gives the Minitab and Excel output of an analysis of variance of the gasoline mileage data. Note that the upper portion of the Minitab output and the lower portion of the Excel output give the ANOVA table of Table 12.3. Also, note that each output gives the value $F = 12.74$ and the related p-value, which equals .001(rounded). Because this p-value is less than .05, we reject H_0 at the .05 level of significance.

Pairwise comparisons

If the one-way ANOVA F test says that at least two treatment means differ, then we investigate which treatment means differ and we estimate how large the differences are. We do this by making what we call **pairwise comparisons** (that is, we compare treatment means *two at a time*). One way to make these comparisons is to compute point estimates of and confidence intervals for **pairwise differences.** For example, in the oil company case we might estimate the pairwise differences $\mu_A - \mu_B$, $\mu_A - \mu_C$, and $\mu_B - \mu_C$. Here, for instance, the pairwise difference $\mu_A - \mu_B$ can be interpreted as the change in mean mileage achieved by changing from using gasoline type B to using gasoline type A.

FIGURE 12.2 Minitab and Excel Output of an Analysis of Variance of the Oil Company Gasoline Mileage Data in Table 12.1

(a) The Minitab output

Analysis of Variance

Source	DF	Adj SS	Adj MS	F-Value	P-Value
Factor	2 [1]	17.049 [4]	8.5247 [7]	12.74 [9]	0.001 [10]
Error	12 [2]	8.028 [5]	0.6690 [8]		
Total	14 [3]	25.077 [6]			

Means

GasType	N	Mean	StDev	95% CI
Type A	5	34.920 [11]	0.766	(34.123, 35.717)
Type B	5	36.560 [12]	0.850	(35.763, 37.357)
Type C	5	33.980 [13]	0.835	(33.183, 34.777)

Pooled StDev = 0.817924

Grouping Information Using the Tukey Method and 95% Confidence

GasType	N	Mean	Grouping
Type B	5	36.560	A
Type A	5	34.920	B
Type C	5	33.980	B

Means that do not share a letter are significantly different.

Tukey Simultaneous Tests for Differences of Means

Difference of Levels	Difference of Means	SE of Difference	95% CI	T-Value	Adjusted P-Value
Type B – Type A	1.640	0.517	(0.261, 3.019)	3.17	0.020
Type C – Type A	−0.940	0.517	(−2.319, 0.439)	−1.82	0.206
Type C – Type B	−2.580	0.517	(−3.959, −1.201)	−4.99	0.001

Individual confidence level = 97.94%

(b) The Excel output

SUMMARY

Groups	Count	Sum	Average	Variance
Type A	5	174.6	34.92 [11]	0.587
Type B	5	182.8	36.56 [12]	0.723
Type C	5	169.9	33.98 [13]	0.697

ANOVA

Source of Variation	SS	df	MS	F	P-Value	F crit
Between Groups	17.0493 [4]	2 [1]	8.5247 [7]	12.7424 [9]	0.0011 [10]	3.8853 [14]
Within Groups	8.0280 [5]	12 [2]	0.6690 [8]			
Total	25.0773 [6]	14 [3]				

[1] $p-1$ [2] $n-p$ [3] $n-1$ [4] SST [5] SSE [6] $SSTO$ [7] MST [8] MSE [9] F statistic [10] p-value related to F [11] $\bar{x}_A$ [12] $\bar{x}_B$ [13] $\bar{x}_C$ [14] $F_{.05}$

There are two approaches to calculating confidence intervals for pairwise differences. The first involves computing the usual, or **individual, confidence interval** for each pairwise difference. Here, if we are computing $100(1 - \alpha)$ percent confidence intervals, we are $100(1 - \alpha)$ percent confident that each individual pairwise difference is contained in its respective interval. That is, the confidence level associated with each (individual) comparison is $100(1 - \alpha)$ percent, and we refer to α as the **comparisonwise error rate.** However, we are less than $100(1 - \alpha)$ percent confident that all of the pairwise differences are simultaneously contained in their respective intervals. A more conservative approach is to compute **simultaneous confidence intervals.** Such intervals make us $100(1 - \alpha)$ percent confident that all of the pairwise differences are simultaneously contained in their respective intervals. That is, when we compute simultaneous intervals, the overall confidence level associated with all the comparisons being made in the experiment is $100(1 - \alpha)$ percent, and we refer to α as the **experimentwise error rate.**

Several kinds of simultaneous confidence intervals can be computed. In this book we present what is called the **Tukey formula** for simultaneous intervals. We do this because, *if we are interested in studying all pairwise differences between treatment means, the Tukey formula yields the most precise (shortest) simultaneous confidence intervals.*

1 Consider the **pairwise difference** $\mu_i - \mu_h$, which can be interpreted to be the change in the mean value of the response variable associated with changing from using treatment h to using treatment i. Then, a **point estimate of the difference** $\mu_i - \mu_h$ **is** $\bar{x}_i - \bar{x}_h$, where $\bar{x}_i$ and $\bar{x}_h$ are the sample treatment means associated with treatments i and h.

2 A **Tukey simultaneous 100(1−α) percent confidence interval for** $\mu_i - \mu_h$ is

$$\left[(\bar{x}_i - \bar{x}_h) \pm q_\alpha \sqrt{\frac{MSE}{m}}\right]$$

Here, the value q_α is obtained from Table A.10, which is a **table of percentage points of the**

studentized range. In this table q_α is listed corresponding to values of p and $n - p$. Furthermore, we assume that the sample sizes n_i and n_h are equal to the same value, which we denote as m. If n_i and n_h are not equal, we replace $q_\alpha\sqrt{MSE/m}$ by $(q_\alpha/\sqrt{2})\sqrt{MSE\,[(1/n_i) + (1/n_h)]}$.

3 A **point estimate of the treatment mean** μ_i **is** $\bar{x}_i$ and an **individual 100(1 − α) percent confidence interval for** μ_i is

$$\left[\bar{x}_i \pm t_{\alpha/2}\sqrt{\frac{MSE}{n_i}}\right]$$

Here, the $t_{\alpha/2}$ point is based on $n - p$ degrees of freedom.

C **EXAMPLE 12.4** The Oil Company Case: Comparing Gasoline Types

Part 1: Using confidence intervals   In the gasoline mileage study, we are comparing $p = 3$ treatment means (μ_A, μ_B, and μ_C). Furthermore, each sample is of size $m = 5$, there are a total of $n = 15$ observed gas mileages, and the MSE found in Table 12.3 is .669. Because $q_{.05} = 3.77$ is the entry found in Table A.10 corresponding to $p = 3$ and $n - p = 12$, a Tukey simultaneous 95 percent confidence interval for $\mu_B - \mu_A$ is

$$\left[(\bar{x}_B - \bar{x}_A) \pm q_{.05}\sqrt{\frac{MSE}{m}}\right] = \left[(36.56 - 34.92) \pm 3.77\sqrt{\frac{.669}{5}}\right]$$
$$= [1.64 \pm 1.379]$$
$$= [.261, 3.019]$$

Similarly, Tukey simultaneous 95 percent confidence intervals for $\mu_A - \mu_C$ and $\mu_B - \mu_C$ are, respectively,

$$[(\bar{x}_A - \bar{x}_C) \pm 1.379] \qquad \text{and} \qquad [(\bar{x}_B - \bar{x}_C) \pm 1.379]$$
$$= [(34.92 - 33.98) \pm 1.379] \qquad\qquad = [(36.56 - 33.98) \pm 1.379]$$
$$= [-0.439, 2.319] \qquad\qquad\qquad = [1.201, 3.959]$$

These intervals make us simultaneously 95 percent confident that (1) changing from gasoline type A to gasoline type B increases mean mileage by between .261 and 3.019 mpg, (2) changing from gasoline type C to gasoline type A might decrease mean mileage by as much as .439 mpg or might increase mean mileage by as much as 2.319 mpg, *and* (3) changing from gasoline type C to gasoline type B increases mean mileage by between 1.201 and 3.959 mpg. The first and third of these intervals make us 95 percent confident that μ_B is at least .261 mpg greater than μ_A and at least 1.201 mpg greater than μ_C. Therefore, we have strong evidence that gasoline type B yields the highest mean mileage of the gasoline types tested. Furthermore, noting that $t_{.025}$ based on $n - p = 12$ degrees of freedom is 2.179, it follows that an individual 95 percent confidence interval for μ_B is

$$\left[\bar{x}_B \pm t_{.025}\sqrt{\frac{MSE}{n_B}}\right] = \left[36.56 \pm 2.179\sqrt{\frac{.669}{5}}\right]$$
$$= [35.763, 37.357]$$

This interval says we can be 95 percent confident that the mean mileage obtained by using gasoline type B is between 35.763 and 37.357 mpg. Notice that this confidence interval is shown on the Minitab output of Figure 12.2. This output also shows the 95 percent confidence intervals for μ_A and μ_C and gives Tukey simultaneous 95 percent intervals for $\mu_B - \mu_A$, $\mu_C - \mu_A$, and $\mu_C - \mu_B$. Note that the last two Tukey intervals on the output are the "negatives" of the Tukey intervals that we hand calculated for $\mu_A - \mu_C$ and $\mu_B - \mu_C$.

Part 2: Using hypothesis testing (optional) We next consider testing $H_0: \mu_i - \mu_h = 0$ versus $H_a: \mu_i - \mu_h \neq 0$. The test statistic t for performing this test is calculated by dividing $\bar{x}_i - \bar{x}_h$ by $\sqrt{MSE[(1/n_i) + (1/n_h)]}$. For example, consider testing $H_0: \mu_B - \mu_A = 0$ versus $H_a: \mu_B - \mu_A \neq 0$. Since $\bar{x}_B - \bar{x}_A = 36.56 - 34.92 = 1.64$ and $\sqrt{MSE[(1/n_B) + (1/n_A)]} = \sqrt{.669[(1/5) + (1/5)]} = .5173$, the test statistic t equals $1.64/.5173 = 3.17$. This test statistic value is given in the leftmost portion of the following Excel add-in (MegaStat) output, as is the test statistic value for testing $H_0: \mu_B - \mu_C = 0$ ($t = 4.99$) and the test statistic value for testing $H_0: \mu_A - \mu_C = 0$ ($t = 1.82$):

Tukey simultaneous comparison t-values (d.f. = 12)

		Type C 33.98	Type A 34.92	Type B 36.56	critical values for experimentwise error rate:	
Type C	33.98				0.05	2.67
Type A	34.92	1.82			0.01	3.56
Type B	36.56	4.99	3.17			

If we wish to use the **Tukey simultaneous comparison procedure** having an experimentwise error rate of α, we reject $H_0: \mu_i - \mu_h = 0$ in favor of $H_a: \mu_i - \mu_h \neq 0$ if the absolute value of t is greater than the critical value $q_\alpha/\sqrt{2}$. Table A.10 tells us that $q_{.05}$ is 3.77 and $q_{.01}$ is 5.04. Therefore, the critical values for experimentwise error rates of .05 and .01 are, respectively, $3.77/\sqrt{2} = 2.67$ and $5.04/\sqrt{2} = 3.56$ (see the right portion of the MegaStat output). Suppose we set α equal to .05. Then, because the test statistic value for testing $H_0: \mu_B - \mu_A = 0$ ($t = 3.17$) and the test statistic value for testing $H_0: \mu_B - \mu_C = 0$ ($t = 4.99$) are greater than the critical value 2.67, we reject both null hypotheses. The Minitab output in Figure 12.2 gives the t statistics and **Tukey adjusted p-values** for testing $H_0: \mu_B - \mu_A = 0$, $H_0: \mu_C - \mu_A = 0$, and $H_0: \mu_C - \mu_B = 0$. The last two t statistics are the negatives of the t statistics in the MegaStat output for testing $H_0: \mu_A - \mu_C = 0$ and $H_0: \mu_B - \mu_C = 0$. We will not discuss how the Tukey adjusted p-values are calculated. However, because the Tukey adjusted p-value of .02 for testing $H_0: \mu_B - \mu_A = 0$ and the Tukey adjusted p-value of .001 for testing $H_0: \mu_C - \mu_B = 0$ are less than .05, we reject both hypotheses using an experimentwise error rate of .05. This, along with the fact that $\bar{x}_B = 36.56$ is greater than $\bar{x}_A = 34.92$ and $\bar{x}_C = 33.98$, leads us to conclude that gasoline type B yields the highest mean mileage.

In general, when we use a completely randomized experimental design, it is important to compare the treatments by using experimental units that are essentially the same with respect to the characteristic under study. For example, in the oil company case we have used cars of the same type (Lances) to compare the different gasoline types, and in the supermarket case we have used grocery stores of the same sales potential for the bakery product to compare the shelf display heights (the reader will analyze the data for this case in the exercises). Sometimes, however, it is not possible to use experimental units that are essentially the same with respect to the characteristic under study. One approach to dealing with this situation is to employ **a randomized block design.** This experimental design is discussed in Section 12.3.

To conclude this section, we note that if we fear that the normality and/or equal variances assumptions for one-way analysis of variance do not hold, we can use the nonparametric Kruskal–Wallis H test to compare several populations. See Chapter 18.

Exercises for Section 12.2

CONCEPTS ◼ connect

12.1 Define the meaning of the terms *response variable, factor, treatments,* and *experimental units.*

12.2 Explain the assumptions that must be satisfied in order to validly use the one-way ANOVA formulas.

12.3 Explain the difference between the between-treatment variability and the within-treatment variability when performing a one-way ANOVA.

12.4 Explain why we conduct pairwise comparisons of treatment means.

METHODS AND APPLICATIONS

12.5 **THE SUPERMARKET CASE** ◙ BakeSale

Consider Example 12.2, and let μ_B, μ_M, and μ_T represent the mean monthly sales when using the bottom, middle, and top shelf display heights, respectively. Figure 12.3 gives the Minitab output of a one-way ANOVA of the bakery sales study data in Table 12.2. Using the computer output in Figure 12.3:

a Test the null hypothesis that μ_B, μ_M, and μ_T are equal by setting $\alpha = .05$. On the basis of this test, can we conclude that the bottom, middle, and top shelf display heights have different effects on mean monthly sales?

b Consider the pairwise differences $\mu_M - \mu_B$, $\mu_T - \mu_B$, and $\mu_T - \mu_M$. Find a point estimate of

and a Tukey simultaneous 95 percent confidence interval for each pairwise difference. Interpret the meaning of each interval in practical terms. Which display height maximizes mean sales?

c Find 95 percent confidence intervals for μ_B, μ_M, and μ_T. Interpret each interval.

12.6 A study compared three different display panels for use by air traffic controllers. Each display panel was tested in a simulated emergency condition; 12 highly trained air traffic controllers took part in the study. Four controllers were randomly assigned to each display panel. The time (in seconds) needed to stabilize the emergency condition was recorded. The results of the study are given in Table 12.4. Let μ_A, μ_B, and μ_C represent the mean times to stabilize the emergency condition when using display panels A, B, and C, respectively. Figure 12.4 gives the JMP output of a one-way ANOVA of the display panel data. Using the computer output ◙ Display

a Test the null hypothesis that μ_A, μ_B, and μ_C are equal by setting $\alpha = .05$. On the basis of this test, can we conclude that display panels A, B, and C have different effects on the mean time to stabilize the emergency condition?

b Consider the pairwise differences $\mu_A - \mu_B$, $\mu_A - \mu_C$, and $\mu_B - \mu_C$. Find a point estimate of and a Tukey simultaneous 95 percent confidence interval for each pairwise difference. Interpret the results by

FIGURE 12.3 **Minitab Output of a One-Way ANOVA of the Bakery Sales Data in Table 12.2 (for Exercise 12.5)**

Analysis of Variance

Source	DF	Adj SS	Adj MS	F-Value	P-Value
Display Height	2	2273.88	1136.94	184.57	0.000
Error	15	92.40	6.16		
Total	17	2366.28			

Means

Display Height	N	Mean	StDev	95% CI
Bottom	6	55.80	2.48	(53.64, 57.96)
Middle	6	77.20	3.10	(75.04, 79.36)
Top	6	51.500	1.648	(49.340, 53.660)

Pooled StDev = 2.48193

Grouping Information Using the Tukey Method and 95% Confidence

Display

Height	N	Mean	Grouping
Middle	6	77.20	A
Bottom	6	55.80	B
Top	6	51.500	C

Means that do not share a letter are significantly different.

Tukey Simultaneous Tests for Differences of Means

Difference of Levels	Difference of Means	SE of Difference	95% CI	T-Value	Adjusted P-Value
Middle - Bottom	21.40	1.43	(17.68, 25.12)	14.93	0.000
Top - Bottom	−4.30	1.43	(−8.02, −0.58)	−3.00	0.023
Top - Middle	−25.70	1.43	(−29.42, −21.98)	−17.94	0.000

Individual confidence level = 97.97%

describing the effects of changing from using each display panel to using each of the other panels. Which display panel minimizes the time required to stabilize the emergency condition?

12.7 A consumer preference study compares the effects of three different bottle designs (A, B, and C) on sales of a popular fabric softener. A completely randomized design is employed. Specifically, 15 supermarkets of equal sales potential are selected, and 5 of these supermarkets are randomly assigned to each bottle design. The number of bottles sold in 24 hours at each supermarket is recorded. The data obtained are displayed in Table 12.5. Let μ_A, μ_B, and μ_C represent mean daily sales using bottle designs A, B, and C, respectively. Figure 12.5 gives the Excel output of a one-way ANOVA of the bottle design study data. Using the computer output **DS** BottleDes

a Test the null hypothesis that μ_A, μ_B, and μ_C are equal by setting $\alpha = .05$. That is, test for statistically significant differences between these treatment means at the .05 level of significance. Based on this test, can we conclude that bottle designs A, B, and C have different effects on mean daily sales?

TABLE 12.4
Display Panel Study Data **DS** Display

Display Panel		
A	**B**	**C**
21	24	40
27	21	36
24	18	35
26	19	32

FIGURE 12.4   JMP Output of a One-Way ANOVA of the Display Panel Study Data in Table 12.4 (for Exercise 12.6)

Analysis of Variance

Source	DF	Sum of Squares	Mean Square	F Ratio
Model	2	500.16667	250.083	30.1104
Error	9	74.75000	8.306	Prob > F
C. Total	11	574.91667		0.0001*

Effect Tests

Source	Nparm	DF	Sum of Squares	F Ratio	Prob > F
Display	2	2	500.16667	30.1104	0.0001*

Least Squares Means Estimates

Display	Estimate	Std Error	DF	Lower 95%	Upper 95%
A	24.500000	1.4409680	9	21.240304	27.759696
B	20.500000	1.4409680	9	17.240304	23.759696
C	35.750000	1.4409680	9	32.490304	39.009696

Tukey HSD All Pairwise Comparisons
Quantile = 2.79201, Adjusted DF = 9.0, Adjustment = Tukey

Display	-Display	Difference	Std Error	t Ratio	Prob > |t|	Lower 95%	Upper 95%
A	B	4.0000	2.037837	1.96	0.1772	−1.6897	9.68966
A	C	−11.2500	2.037837	−5.52	0.0010*	−16.9397	−5.56034
B	C	−15.2500	2.037837	−7.48	<.0001*	−20.9397	−9.56034

FIGURE 12.5   Excel Output of a One-Way ANOVA of the Bottle Design Study Data (for Exercise 12.7)

TABLE 12.5
Bottle Design Study Data
 DS BottleDes

SUMMARY

Groups	Count	Sum	Average	Variance
DESIGN A	5	83	16.6	5.3
DESIGN B	5	164	32.8	9.2
DESIGN C	5	124	24.8	8.2

ANOVA

Source of Variation	SS	df	MS	F	P-Value	F crit
Between Groups	656.1333	2	328.0667	43.35683	3.23E-06	3.88529
Within Groups	90.8	12	7.566667			
Total	746.9333	14				

Bottle Design		
A	**B**	**C**
16	33	23
18	31	27
19	37	21
17	29	28
13	34	25

b Consider the pairwise differences $\mu_B - \mu_A$, $\mu_C - \mu_A$, and $\mu_C - \mu_B$. Find a point estimate of and a Tukey simultaneous 95 percent confidence interval for each pairwise difference. Interpret the results in practical terms. Which bottle design maximizes mean daily sales?

c Find and interpret a 95 percent confidence interval for each of the treatment means μ_A, μ_B, and μ_C.

12.8 In order to compare the durability of four different brands of golf balls (Alpha, Best, Century, and Divot), the National Golf Association randomly selects five balls of each brand and places each ball into a machine that exerts the force produced by a 250-yard drive. The number of simulated drives needed to crack or chip each ball is recorded. The results are given in Table 12.6. The Excel output of a one-way ANOVA of these data is shown in Figure 12.6. Using the computer output, test for statistically significant differences between the treatment means μ_{ALPHA}, μ_{BEST}, μ_{CENTURY}, and μ_{DIVOT}. Set $\alpha = .05$. **DS** GolfBall

TABLE 12.6

Golf Ball Durability Test Results
DS GolfBall

Brand	
Alpha	**Best**
281	270
220	334
274	307
242	290
251	331
Century	**Divot**
218	364
244	302
225	325
273	337
249	355

12.9 Using the computer output, perform pairwise comparisons of the treatment means in Exercise 12.8 by (1) Using Tukey simultaneous 95 percent confidence intervals; (2) Optionally using t statistics and critical values (see the right side of Figure 12.6). (3) Which brands are most durable? (4) Find and interpret a 95 percent confidence interval for each of the treatment means.

12.10 Table 12.7 presents the yields (in bushels per one-third acre plot) for corn hybrid types X, W, and Y. **DS** CornYield

a Test for statistically significant differences between the treatment means μ_X, μ_W, and μ_Y. Set $\alpha = .05$.

b Perform pairwise comparisons of the treatment means by using Tukey simultaneous 95 percent confidence intervals. Which corn hybrid type maximizes mean yield?

c Find and interpret a 95 percent confidence interval for each of the treatment means.

TABLE 12.7

Corn Yield Data
DS CornYield

Corn Hybrid Type		
X	**W**	**Y**
27.6	22.6	25.4
26.5	21.5	24.5
27.0	22.1	26.3
27.2	21.8	24.8
26.8	22.4	25.1

12.11 An oil company wishes to study the effects of four different gasoline additives on mean gasoline mileage. The company randomly selects four groups of six automobiles each and assigns a group of six automobiles to each additive type (W, X, Y, and Z). Here all 24 automobiles employed in the experiment are the same make and model. Each of the six automobiles assigned to a gasoline additive is test driven using the appropriate additive, and the gasoline mileage for the test drive is recorded. The results of the experiment are given in Table 12.8. **DS** GasAdditive

a Test for statistically significant differences between the treatment means μ_W, μ_X, μ_Y, and μ_Z. Set $\alpha = .05$.

b Perform pairwise comparisons of the treatment means by using Tukey simultaneous 95 percent confidence intervals. Which additives give the highest mean gasoline mileage?

c Find and interpret a 95 percent confidence interval for each of the treatment means.

TABLE 12.8

Gas Additive Data
DS GasAdditive

Gas Additive Type			
W	**X**	**Y**	**Z**
31.2	27.6	35.7	34.5
32.6	28.1	34.0	36.2
30.8	27.4	35.1	35.2
31.5	28.5	33.9	35.8
32.0	27.5	36.1	34.9
30.1	28.7	34.8	35.3

FIGURE 12.6 **Excel Output of a One-Way ANOVA of the Golf Ball Durability Data (for Exercises 12.8 and 12.9)**

SUMMARY

Groups	Count	Sum	Average	Variance
Alpha	5	1268	253.6	609.3
Best	5	1532	306.4	740.3
Century	5	1209	241.8	469.7
Divot	5	1683	336.6	605.3

Tukey simultaneous comparison t-values (d.f. = 16)

		Century 241.8	Alpha 253.6	Best 306.4	Divot 336.6
Century	241.8				
Alpha	253.6	0.76			
Best	306.4	4.15	3.39		
Divot	336.6	6.09	5.33	1.94	

ANOVA

Source of Variation	SS	df	MS	F	P-Value	F crit
Between Groups	29860.4	3	9953.4667	16.420798	3.853E-05	3.2388715
Within Groups	9698.4	16	606.15			
Total	39558.8	19				

Critical values for experimentwise error rate:

0.05	2.86
0.01	3.67

12.3 The Randomized Block Design

LO12-3
Compare treatment
effects and block effects
by using a randomized
block design.

Not all experiments employ a completely randomized design. For instance, suppose that when we employ a completely randomized design, we fail to reject the null hypothesis of equality of treatment means because the within-treatment variability (which is measured by the *SSE*) is large. This could happen because differences between the experimental units are concealing true differences between the treatments. We can often remedy this by using what is called a **randomized block design.**

Ⓒ **EXAMPLE 12.5** The Cardboard Box Case: Comparing Production Methods

The Universal Paper Company manufactures cardboard boxes. The company wishes to investigate the effects of four production methods (methods 1, 2, 3, and 4) on the number of defective boxes produced in an hour. To compare the methods, the company could utilize a completely randomized design. For each of the four production methods, the company would select several (say, as an example, three) machine operators, train each operator to use the production method to which he or she has been assigned, have each operator produce boxes for one hour, and record the number of defective boxes produced. The three operators using any one production method would be *different* from those using any other production method. That is, the completely randomized design would utilize a total of 12 machine operators. However, the abilities of the machine operators could differ substantially. These differences might tend to conceal any real differences between the production methods. To overcome this disadvantage, the company will employ a **randomized block experimental design.** This involves randomly selecting three machine operators and training each operator thoroughly to use all four production methods. Then each operator will produce boxes for one hour using each of the four production methods. The order in which each operator uses the four methods should be random. We record the number of defective boxes produced by each operator using each method. The advantage of the randomized block design is that the defective rates obtained by using the four methods result from employing the *same* three operators. Thus any true differences in the effectiveness of the methods would not be concealed by differences in the operators' abilities.

When Universal Paper employs the randomized block design, it obtains the 12 defective box counts in Table 12.9. We let x_{ij} denote the number of defective boxes produced by machine operator j using production method i. For example, $x_{32} = 5$ says that 5 defective boxes were produced by machine operator 2 using production method 3 (see Table 12.9). In addition to the 12 defective box counts, Table 12.9 gives the sample mean of these 12 observations, which is $\bar{x} = 7.5833$, and also gives **sample treatment means** and **sample block means.** The sample treatment means are the average defective box counts obtained when using production methods 1, 2, 3, and 4. Denoting these sample treatment means as $\bar{x}_{1.}, \bar{x}_{2.}, \bar{x}_{3.},$ and $\bar{x}_{4.}$, we see from Table 12.9 that $\bar{x}_{1.} = 10.3333, \bar{x}_{2.} = 10.3333, \bar{x}_{3.} = 5.0,$ and $\bar{x}_{4.} = 4.6667$. Because $\bar{x}_{3.}$ and $\bar{x}_{4.}$ are less than $\bar{x}_{1.}$ and $\bar{x}_{2.}$, we estimate that the mean number of defective boxes produced per hour by production method 3 or 4 is less than the mean number of defective boxes produced per hour by production method 1 or 2. The sample block means are the average defective box counts obtained by machine operators 1, 2, and 3. Denoting these

---

TABLE 12.9 Numbers of Defective Cardboard Boxes Obtained by Production Methods 1, 2, 3, and 4 and Machine Operators 1, 2, and 3 Ⓓ CardBox

Treatment (Production Method)	Block (Machine Operator) 1	2	3	Sample Treatment Mean
1	9	10	12	10.3333
2	8	11	12	10.3333
3	3	5	7	5.0
4	4	5	5	4.6667
Sample Block Mean	6.0	7.75	9.0	$\bar{x} = 7.5833$

sample block means as $\bar{x}_{\cdot1}$, $\bar{x}_{\cdot2}$, and $\bar{x}_{\cdot3}$, we see from Table 12.9 that $\bar{x}_{\cdot1} = 6.0$, $\bar{x}_{\cdot2} = 7.75$, and $\bar{x}_{\cdot3} = 9.0$. Because $\bar{x}_{\cdot1}$, $\bar{x}_{\cdot2}$, and $\bar{x}_{\cdot3}$ differ, we have evidence that the abilities of the machine operators differ and thus that using the machine operators as blocks is reasonable.

In general, a **randomized block design** compares p treatments (for example, production methods) by using b blocks (for example, machine operators). Each block is used exactly once to measure the effect of each and every treatment. The advantage of the randomized block design over the completely randomized design is that we are comparing the treatments by using the *same* experimental units. Thus any true differences in the treatments will not be concealed by differences in the experimental units.

In order to analyze the data obtained in a randomized block design, we define

x_{ij} = the value of the response variable observed when block j uses treatment i

$\bar{x}_{i\cdot}$ = the mean of the b values of the response variable observed when using treatment i

$\bar{x}_{\cdot j}$ = the mean of the p values of the response variable observed when using block j

$\bar{x}$ = the mean of the total of the bp values of the response variable that we have observed in the experiment

The ANOVA procedure for a randomized block design partitions the **total sum of squares** **(SSTO)** into three components: the **treatment sum of squares (SST),** the **block sum of squares (SSB),** and the **error sum of squares (SSE).** The formula for this partitioning is

$$SSTO = SST + SSB + SSE$$

We define each of these sums of squares and show how they are calculated for the defective cardboard box data as follows (note that $p = 4$ and $b = 3$):

Step 1: Calculate *SST*, which measures the amount of between-treatment variability:

$$
\begin{aligned}
SST &= b\sum_{i=1}^{p} (\bar{x}_{i\cdot} - \bar{x})^2 \\
&= 3[(\bar{x}_{1\cdot} - \bar{x})^2 + (\bar{x}_{2\cdot} - \bar{x})^2 + (\bar{x}_{3\cdot} - \bar{x})^2 + (\bar{x}_{4\cdot} - \bar{x})^2] \\
&= 3[(10.3333 - 7.5833)^2 + (10.3333 - 7.5833)^2 \\
&\quad + (5.0 - 7.5833)^2 + (4.6667 - 7.5833)^2] \\
&= 90.9167
\end{aligned}
$$

Step 2: Calculate *SSB*, which measures the amount of variability due to the blocks:

$$
\begin{aligned}
SSB &= p\sum_{j=1}^{b} (\bar{x}_{\cdot j} - \bar{x})^2 \\
&= 4[(\bar{x}_{\cdot1} - \bar{x})^2 + (\bar{x}_{\cdot2} - \bar{x})^2 + (\bar{x}_{\cdot3} - \bar{x})^2] \\
&= 4[(6.0 - 7.5833)^2 + (7.75 - 7.5833)^2 + (9.0 - 7.5833)^2] \\
&= 18.1667
\end{aligned}
$$

Step 3: Calculate *SSTO*, which measures the total amount of variability:

$$
\begin{aligned}
SSTO &= \sum_{i=1}^{p}\sum_{j=1}^{b} (x_{ij} - \bar{x})^2 \\
&= (9 - 7.5833)^2 + (10 - 7.5833)^2 + (12 - 7.5833)^2 \\
&\quad + (8 - 7.5833)^2 + (11 - 7.5833)^2 + (12 - 7.5833)^2 \\
&\quad + (3 - 7.5833)^2 + (5 - 7.5833)^2 + (7 - 7.5833)^2 \\
&\quad + (4 - 7.5833)^2 + (5 - 7.5833)^2 + (5 - 7.5833)^2 \\
&= 112.9167
\end{aligned}
$$

Step 4: Calculate *SSE*, which measures the amount of variability due to the error:

$$
\begin{aligned}
SSE &= SSTO - SST - SSB \\
&= 112.9167 - 90.9167 - 18.1667 \\
&= 3.8333
\end{aligned}
$$

TABLE 12.10   Randomized Block ANOVA Table for the Defective Box Data

Source of Variation	Degrees of Freedom	Sum of Squares	Mean Square	F
Treatments	$p - 1 = 3$	$SST = 90.9167$	$MST = \dfrac{SST}{p-1} = 30.3056$	$F(\text{treatments}) = \dfrac{MST}{MSE} = 47.4348$
Blocks	$b - 1 = 2$	$SSB = 18.1667$	$MSB = \dfrac{SSB}{b-1} = 9.0833$	$F(\text{blocks}) = \dfrac{MSB}{MSE} = 14.2174$
Error	$(p-1)(b-1) = 6$	$SSE = 3.8333$	$MSE = \dfrac{SSE}{(p-1)(b-1)} = .6389$	
Total	$pb - 1 = 11$	$SSTO = 112.9167$		

These sums of squares are shown in Table 12.10, which is the ANOVA table for a randomized block design. This table also gives the degrees of freedom, mean squares, and F statistics used to test the hypotheses of interest in a randomized block experiment, as well as the values of these quantities for the defective cardboard box data.

Of main interest is the test of the null hypothesis H_0 that **no differences exist between the treatment effects** on the mean value of the response variable versus the alternative hypothesis H_a that **at least two treatment effects differ.** We can reject H_0 in favor of H_a at level of significance α if $F(\text{treatments})$ is greater than the F_α point based on $p - 1$ numerator and $(p - 1)(b - 1)$ denominator degrees of freedom. In the defective cardboard box case, $F_{.05}$ based on $p - 1 = 3$ numerator and $(p - 1)(b - 1) = 6$ denominator degrees of freedom is 4.76 (see Table A.7). Because $F(\text{treatments}) = 47.4348$ (see Table 12.10) is greater than $F_{.05} = 4.76$, we reject H_0 at the .05 level of significance. Therefore, we have strong evidence that at least two production methods have different effects on the mean number of defective boxes produced per hour.

It is also of interest to test the null hypothesis H_0 that **no differences exist between the block effects** on the mean value of the response variable versus the alternative hypothesis H_a that **at least two block effects differ.** We can reject H_0 in favor of H_a at level of significance α if $F(\text{blocks})$ is greater than the F_α point based on $b - 1$ numerator and $(p - 1)(b - 1)$ denominator degrees of freedom. In the defective cardboard box case, $F_{.05}$ based on $b - 1 = 2$ numerator and $(p - 1)(b - 1) = 6$ denominator degrees of freedom is 5.14 (see Table A.7). Because $F(\text{blocks}) = 14.2174$ (see Table 12.10) is greater than $F_{.05} = 5.14$, we reject H_0 at the .05 level of significance. Therefore, we have strong evidence that at least two machine operators have different effects on the mean number of defective boxes produced per hour.

Figure 12.7 on the next page gives the Minitab and Excel outputs of a randomized block ANOVA of the defective cardboard box data. The p-value of .000 (<.001) related to $F(\text{treatments})$ provides extremely strong evidence of differences in production method effects. The p-value of .0053 related to $F(\text{blocks})$ provides very strong evidence of differences in machine operator effects.

If, in a randomized block design, we conclude that at least two treatment effects differ, we can perform pairwise comparisons to determine how they differ.

Point Estimates and Confidence Intervals in a Randomized Block ANOVA

Consider the **difference between the effects of treatments i and h on the mean value of the response variable.** Then

1   A **point estimate** of this difference is $\bar{x}_{i\bullet} - \bar{x}_{h\bullet}$.

2   A **Tukey simultaneous $100(1 - \alpha)$ percent confidence interval** for this difference is

$$\left[(\bar{x}_{i\bullet} - \bar{x}_{h\bullet}) \pm q_\alpha \frac{s}{\sqrt{b}} \right]$$

Here the value q_α is obtained from Table A.10, which is a table of percentage points of the studentized range. In this table q_α is listed corresponding to values of p and $(p - 1)(b - 1)$.

FIGURE 12.7   Minitab and Excel Outputs of a Randomized Block ANOVA of the Defective Box Data

(a) The Minitab Output

Rows: Method Columns: Operator

	1	2	3	All	Method	Mean	Operator	Mean
1	9.000	10.000	12.000	10.333	1	10.3333 **12**	1	6.00 **16**
2	8.000	11.000	12.000	10.333	2	10.3333 **13**	2	7.75 **17**
3	3.000	5.000	7.000	5.000	3	5.0000 **14**	3	9.00 **18**
4	4.000	5.000	5.000	4.667	4	4.6667 **15**		
All	6.000	7.750	9.000	7.583				

Two-way ANOVA: Rejects versus Method, Operator

Source	DF	SS	MS	F	P
Method	3	90.917 **1**	30.3056 **5**	47.43 **8**	0.000 **9**
Operator	2	18.167 **2**	9.0833 **6**	14.22 **10**	0.005 **11**
Error	6	3.833 **3**	0.6389 **7**		
Total	11	112.917 **4**			

(b) The Excel Output

ANOVA: Two-Factor Without Replication

Summary	Count	Sum	Average	Variance
Method1	3	31	10.3333 **12**	2.3333
Method2	3	31	10.3333 **13**	4.3333
Method3	3	5	5 **14**	4
Method4	3	14	4.6667 **15**	0.3333
Operator1	4	24	6 **16**	8.6667
Operator2	4	31	7.75 **17**	10.25
Operator3	4	36	9 **18**	12.6667

ANOVA

Source of Variation	SS	df	MS	F	P-value	F crit
Method	90.9167 **1**	3	30.3056 **5**	47.4348 **8**	0.0001 **9**	4.7571
Operator	18.1667 **2**	2	9.0833 **6**	14.2174 **10**	0.0053 **11**	5.1433
Error	3.8333 **3**	6	0.6389 **7**			
Total	112.9167 **4**	11				

1 SST	**2** SSB	**3** SSE	**4** SSTO	**5** MST	**6** MSB	**7** MSE	**8** F(treatments)	**9** p-value for F(treatments)
10 F(blocks)	**11** p-value for F(blocks)	**12** $\bar{x}_{1\bullet}$	**13** $\bar{x}_{2\bullet}$	**14** $\bar{x}_{3\bullet}$	**15** $\bar{x}_{4\bullet}$	**16** $\bar{x}_{\bullet 1}$	**17** $\bar{x}_{\bullet 2}$	**18** $\bar{x}_{\bullet 3}$

C EXAMPLE 12.6 The Cardboard Box Case: Comparing Production Methods

We have previously concluded that we have extremely strong evidence that at least two production methods have different effects on the mean number of defective boxes produced per hour. We have also seen that the sample treatment means are $\bar{x}_{1\bullet} = 10.3333$, $\bar{x}_{2\bullet} = 10.3333$, $\bar{x}_{3\bullet} = 5.0$, and $\bar{x}_{4\bullet} = 4.6667$. Because $\bar{x}_{4\bullet}$ is the smallest sample treatment mean, we will use Tukey simultaneous 95 percent confidence intervals to compare the effect of production method 4 with the effects of production methods 1, 2, and 3. To compute these intervals, we first note that $q_{.05} = 4.90$ is the entry in Table A.10 corresponding to $p = 4$ and $(p - 1)(b - 1) = 6$. Also, note that the *MSE* found in the randomized block ANOVA table is .6389 (see Figure 12.7), which implies that $s = \sqrt{.6389} = .7993$. It follows that a Tukey simultaneous 95 percent confidence interval for the difference between the effects of production methods 4 and 1 on the mean number of defective boxes produced per hour is

$$\left[(\bar{x}_{4\bullet} - \bar{x}_{1\bullet}) \pm q_{.05}\frac{s}{\sqrt{b}}\right] = \left[(4.6667 - 10.3333) \pm 4.90\left(\frac{.7993}{\sqrt{3}}\right)\right]$$

$$= [-5.6666 \pm 2.2615]$$

$$= [-7.9281, -3.4051]$$

Furthermore, it can be verified that a Tukey simultaneous 95 percent confidence interval for the difference between the effects of production methods 4 and 2 on the mean number of defective boxes produced per hour is also $[-7.9281, -3.4051]$. Therefore, we can be 95 percent confident that changing from production method 1 or 2 to production method 4 decreases the mean number of defective boxes produced per hour by a machine operator by between 3.4051 and 7.9281 boxes. A Tukey simultaneous 95 percent confidence interval for the difference between the effects of production methods 4 and 3 on the mean number of defective boxes produced per hour is

$$[(\bar{x}_{4\bullet} - \bar{x}_{3\bullet}) \pm 2.2615] = [(4.6667 - 5) \pm 2.2615]$$
$$= [-2.5948, 1.9282]$$

This interval tells us (with 95 percent confidence) that changing from production method 3 to production method 4 might decrease the mean number of defective boxes produced per hour by as many as 2.5948 boxes or might increase this mean by as many as 1.9282 boxes. In other words, because this interval contains 0, we cannot conclude that the effects of production methods 4 and 3 differ.

Exercises for Section 12.3

CONCEPTS

12.12 In your own words, explain why we sometimes employ the randomized block design.

12.13 Describe what *SSTO*, *SST*, *SSB*, and *SSE* measure.

12.14 How can we test to determine if the blocks we have chosen are reasonable?

METHODS AND APPLICATIONS

12.15 A marketing organization wishes to study the effects of four sales methods on weekly sales of a product. The organization employs a randomized block design in which three salesman use each sales method. The results obtained are given in Figure 12.8, along with the JMP output of a randomized block ANOVA of these data. Using the computer output   SaleMeth

    a Test the null hypothesis H_0 that no differences exist between the effects of the sales methods (treatments) on mean weekly sales. Set $\alpha = .05$. Can we

conclude that the different sales methods have different effects on mean weekly sales?

    b Test the null hypothesis H_0 that no differences exist between the effects of the salesmen (blocks) on mean weekly sales. Set $\alpha = .05$. Can we conclude that the different salesmen have different effects on mean weekly sales?

    c Use Tukey simultaneous 95 percent confidence intervals to make pairwise comparisons of the sales method effects on mean weekly sales. Which sales method(s) maximize mean weekly sales?

12.16 A consumer preference study involving three different bottle designs (A, B, and C) for the jumbo size of a new liquid laundry detergent was carried out using a randomized block experimental design, with supermarkets as blocks. Specifically, four supermarkets were supplied with all three bottle designs, which were priced the same. Table 12.11 gives the number of bottles of each design sold in a 24-hour period at

---

FIGURE 12.8   **The Sales Method Data and the JMP Output of a Randomized Block ANOVA (for Exercise 12.15)**   SaleMeth

Sales Method, *i*	Salesman, *j* A	B	C
1	32	29	30
2	32	30	28
3	28	25	23
4	25	24	23

Least Squares Means Estimates

Method	Estimate	Salesman	Estimate
1	30.333333	A	29.250000
2	30.000000	B	27.000000
3	25.333333	C	26.000000
4	24.000000		

Analysis of Variance

Source	DF	Sum of Squares	Mean Square	F Ratio
Model	5	115.75000	23.1500	26.8839
Error	6	5.16667	0.8611	Prob > F
C. Total	11	120.91667		0.0005*

Effect Tests

Source	Nparm	DF	Sum of Squares	F Ratio	Prob > F
Method	3	3	93.583333	36.2258	0.0003*
Salesman	2	2	22.166667	12.8710	0.0068*

Tukey HSD All Pairwise Comparisons

Quantile = 3.46171, Adjusted DF = 60, Adjustment = Tukey

Method	-Method	Difference	Std Error	t Ratio	Prob>ltl	Lower 95%	Upper 95%
1	2	0.333333	0.7576768	0.44	0.9692	-2.28952	2.956187
1	3	5.000000	0.7576768	6.60	0.0024*	2.37715	7.622854
1	4	6.333333	0.7576768	8.36	0.0007*	3.71048	8.956187
2	3	4.666667	0.7576768	6.16	0.0034*	2.04381	7.289521
2	4	6.000000	0.7576768	7.92	0.0009*	3.37715	8.622854
3	4	1.333333	0.7576768	1.76	0.3741	-1.28952	3.956187

TABLE 12.11 **Results of a Bottle Design Experiment**
DS BottleDes2

Bottle Design, *i*	Supermarket, *j*			
	1	2	3	4
A	16	14	1	6
B	33	30	19	23
C	23	21	8	12

TABLE 12.12 **Results of a Keyboard Experiment**
DS Keyboard

Data Entry Specialist	Keyboard Brand		
	A	B	C
1	77	67	63
2	71	62	59
3	74	63	59
4	67	57	54

each supermarket. If we use these data, *SST*, *SSB*, and *SSE* can be calculated to be 586.1667, 421.6667, and 1.8333, respectively. DS BottleDes2

a Test the null hypothesis H_0 that no differences exist between the effects of the bottle designs on mean daily sales. Set $\alpha = .05$. Can we conclude that the different bottle designs have different effects on mean sales?

b Test the null hypothesis H_0 that no differences exist between the effects of the supermarkets on mean daily sales. Set $\alpha = .05$. Can we conclude that the different supermarkets have different effects on mean sales?

c Use Tukey simultaneous 95 percent confidence intervals to make pairwise comparisons of the bottle design effects on mean daily sales. Which bottle design(s) maximize mean sales?

12.17 To compare three brands of computer keyboards, four data entry specialists were randomly selected. Each specialist used all three keyboards to enter the same kind of text material for 10 minutes, and the number of words entered per minute was recorded. The data obtained are given in Table 12.12. If we use these data, *SST*, *SSB*, and *SSE* can be calculated to be 392.6667, 143.5833, and 2.6667, respectively. DS Keyboard

a Test the null hypothesis H_0 that no differences exist between the effects of the keyboard brands on the mean number of words entered per minute. Set $\alpha = .05$.

b Test the null hypothesis H_0 that no differences exist between the effects of the data entry specialists on the mean number of words entered per minute. Set $\alpha = .05$.

c Use Tukey simultaneous 95 percent confidence intervals to make pairwise comparisons of the keyboard brand effects on the mean number of words entered per minute. Which keyboard brand maximizes the mean number of words entered per minute?

12.18 The Coca-Cola Company introduced New Coke in 1985. Within three months of this introduction, negative consumer reaction forced Coca-Cola to reintroduce the original formula of Coke as Coca-Cola Classic. Suppose that two years later, in 1987, a marketing research firm in Chicago compared the sales of Coca-Cola Classic, New Coke, and Pepsi in public building vending machines. To do this, the marketing research firm randomly selected 10 public buildings in Chicago having both a Coke machine (selling Coke Classic and New Coke) and a Pepsi machine. The data—in number of cans sold over a given period of time—and a Minitab randomized block ANOVA of the data are given in Figure 12.9. Using the computer output DS Coke

a Test the null hypothesis H_0 that no differences exist between the mean sales of Coca-Cola Classic, New Coke, and Pepsi in Chicago public building vending machines. Set $\alpha = .05$.

b Make pairwise comparisons of the mean sales of Coca-Cola Classic, New Coke, and Pepsi in Chicago public building vending machines by using Tukey simultaneous 95 percent confidence intervals.

c By the mid-1990s the Coca-Cola Company had discontinued making New Coke and had returned to making only its original product. Is there evidence in the 1987 study that this might happen? Explain your answer.

FIGURE 12.9 **The Coca-Cola Data and a Minitab Output of a Randomized Block ANOVA of the Data (for Exercise 12.18)**

	Building									
	1	2	3	4	5	6	7	8	9	10
Coke Classic	45	136	134	41	146	33	71	224	111	87
New Coke	6	114	56	14	39	20	42	156	61	140
Pepsi	24	90	100	43	51	42	68	131	74	107

DS Coke

Two-way ANOVA: Cans versus Drink, Building

Source	DF	SS	MS	F	P
Drink	2	7997.6	3998.80	5.78	0.011
Building	9	55573.5	6174.83	8.93	0.000
Error	18	12443.7	691.32		
Total	29	76014.8			

Descriptive Statistics: Cans

Variable	Drink	Mean
Cans	Coke Classic	102.8
	New Coke	64.8
	Pepsi	73.0

12.4 Two-Way Analysis of Variance

Many response variables are affected by more than one factor. Because of this we must often conduct experiments in which we study the effects of several factors on the response. In this section we consider studying the effects of **two factors** on a response variable. To begin, recall that in Example 12.2 we discussed an experiment in which the Tastee Bakery Company investigated the effect of shelf display height on monthly demand for one of its bakery products. This one-factor experiment is actually a simplification of a two-factor experiment carried out by the Tastee Bakery Company. We discuss this two-factor experiment in the following example.

LO12-4
Assess the effects of two factors on a response variable by using a two-way analysis of variance.

 EXAMPLE 12.7 The Supermarket Case: Comparing Display Heights and Widths

The Tastee Bakery Company supplies a bakery product to many metropolitan supermarkets. The company wishes to study the effects of two factors—**shelf display height** and **shelf display width**—on **monthly demand** (measured in cases of 10 units each) for this product. The factor "display height" is defined to have three levels: B (bottom), M (middle), and T (top). The factor "display width" is defined to have two levels: R (regular) and W (wide). The **treatments** in this experiment are **display height and display width combinations.** These treatments are

$$BR \quad BW \quad MR \quad MW \quad TR \quad TW$$

Here, for example, the notation BR denotes the treatment "bottom display height and regular display width." For each display height and width combination the company randomly selects a sample of $m = 3$ metropolitan area supermarkets (all supermarkets used in the study will be of equal sales potential). Each supermarket sells the product for one month using its assigned display height and width combination, and the month's demand for the product is recorded. The six samples obtained in this experiment are given in Table 12.13. We let $x_{ij,k}$ denote the monthly demand obtained at the kth supermarket that used display height i and display width j. For example, $x_{MW,2} = 78.4$ is the monthly demand obtained at the second supermarket that used a middle display height and a wide display.

In addition to giving the six samples, Table 12.13 gives the **sample treatment mean** for each display height and display width combination. For example, $\bar{x}_{BR} = 55.9$ is the mean of the sample of three demands observed at supermarkets using a bottom display height and a regular display width. The table also gives the sample mean demand for each level of display height (B, M, and T) and for each level of display width (R and W). Specifically,

$\bar{x}_{B\bullet} = 55.8$ = the mean of the six demands observed when using a bottom display height
$\bar{x}_{M\bullet} = 77.2$ = the mean of the six demands observed when using a middle display height
$\bar{x}_{T\bullet} = 51.5$ = the mean of the six demands observed when using a top display height
$\bar{x}_{\bullet R} = 60.8$ = the mean of the nine demands observed when using a regular display width
$\bar{x}_{\bullet W} = 62.2$ = the mean of the nine demands observed when using a wide display

Finally, Table 12.13 gives $\bar{x} = 61.5$, which is the overall mean of the total of 18 demands observed in the experiment. Because $\bar{x}_{M\bullet} = 77.2$ is considerably larger than $\bar{x}_{B\bullet} = 55.8$ and $\bar{x}_{T\bullet} = 51.5$, we estimate that mean monthly demand is highest when using a middle display height. Because $\bar{x}_{\bullet R} = 60.8$ and $\bar{x}_{\bullet W} = 62.2$ do not differ by very much, we estimate there is little difference between the effects of a regular display width and a wide display on mean monthly demand.

Figure 12.10 presents a graphical analysis of the bakery demand data. In this figure we plot, for each display width (R and W), the change in the sample treatment mean demand associated with changing the display height from bottom (B) to middle (M) to top (T). Note that, for either a regular display width (R) or a wide display (W), the middle display height (M) gives the highest mean monthly demand. Also, note that, for either a bottom, middle, or top display height, there is little difference between the effects of a regular display

©101dalmatians/Getty Images

TABLE 12.13	Six Samples of Monthly Demands for a Bakery Product DS BakeSale2		
	Display Width		
Display Height	**R**	**W**	
B	58.2	55.7	
	53.7	52.5	
	55.8	58.9	
	$\bar{X}_{BR} = 55.9$	$\bar{X}_{BW} = 55.7$	$\bar{x}_{B\bullet} = 55.8$
M	73.0	76.2	
	78.1	78.4	
	75.4	82.1	
	$\bar{X}_{MR} = 75.5$	$\bar{X}_{MW} = 78.9$	$\bar{x}_{M\bullet} = 77.2$
T	52.4	54.0	
	49.7	52.1	
	50.9	49.9	
	$\bar{X}_{TR} = 51.0$	$\bar{X}_{TW} = 52.0$	$\bar{x}_{T\bullet} = 51.5$
	$\bar{x}_{\bullet R} = 60.8$	$\bar{x}_{\bullet W} = 62.2$	$\bar{x} = 61.5$

FIGURE 12.10   Graphical Analysis of the Bakery Demand Data

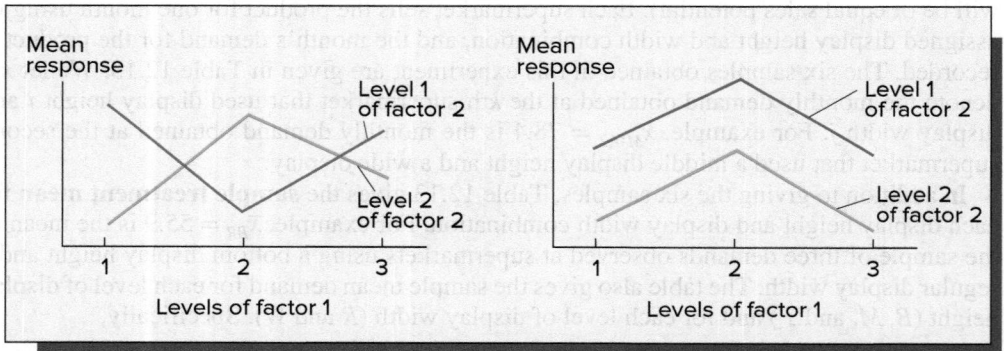

LO12-5
Describe what happens when two factors interact.

width and a wide display on mean monthly demand. This sort of graphical analysis is useful for determining whether a condition called **interaction** exists. In general, for two factors that might affect a response variable, we say that **interaction exists if the relationship between the mean response and one factor depends on the other factor.** This is clearly true in the leftmost figure below:

Specifically, this figure shows that at levels 1 and 3 of factor 1, level 1 of factor 2 gives the highest mean response, while at level 2 of factor 1, level 2 of factor 2 gives the highest mean response. On the other hand, the **parallel** line plots in the rightmost figure indicate a lack of interaction between factors 1 and 2. Because the sample mean plots in Figure 12.10 look nearly parallel, we might intuitively conclude that there is little or no interaction between display height and display width.

Suppose we wish to study the effects of two factors on a response variable. We assume that the first factor, which we refer to as **factor 1**, has a **levels** (levels 1, 2, . . . , a). Further, we assume that the second factor, which we will refer to as **factor 2**, has b **levels** (levels 1, 2, . . . , b). Here a **treatment** is considered to be a **combination of a level of factor 1 and a level of factor 2.** It follows that there are a total of ab treatments, and we assume that we will employ a *completely randomized experimental design* in which we will assign m randomly selected experimental units to each treatment. This procedure results in our observing m values of the response variable for each of the ab treatments, and in this case we say that we are performing a **two-factor factorial experiment.**

In addition to graphical analysis, **two-way analysis of variance (two-way ANOVA)** is a useful tool for analyzing the data from a two-factor factorial experiment. To explain the ANOVA approach for analyzing such an experiment, we define

$x_{ij,k}$ = the kth value of the response variable observed when using level i of factor 1 and level j of factor 2

$\bar{x}_{ij}$ = the mean of the m values observed when using the ith level of factor 1 and the jth level of factor 2

$\bar{x}_{i\bullet}$ = the mean of the bm values observed when using the ith level of factor 1

$\bar{x}_{\bullet j}$ = the mean of the am values observed when using the jth level of factor 2

$\bar{x}$ = the mean of the abm values that we have observed in the experiment

The ANOVA procedure for a two-factor factorial experiment partitions the **total sum of squares (SSTO)** into four components: the **factor 1 sum of squares, SS(1)**; the **factor 2 sum of squares, SS(2)**; the **interaction sum of squares, SS(int)**; and the **error sum of squares, SSE.** The formula for this partitioning is as follows:

$$SSTO = SS(1) + SS(2) + SS(\text{int}) + SSE$$

We define each of these sums of squares and show how they are calculated for the bakery demand data as follows (note that $a = 3$, $b = 2$, and $m = 3$):

Step 1: Calculate $SSTO$, which measures the total amount of variability:

$$SSTO = \sum_{i=1}^{a}\sum_{j=1}^{b}\sum_{k=1}^{m}(x_{ij,k} - \bar{x})^2$$

$$= (58.2 - 61.5)^2 + (53.7 - 61.5)^2 + \cdots + (49.9 - 61.5)^2 = 2{,}366.28$$

Step 2: Calculate $SS(1)$, which measures the amount of variability due to the different levels of factor 1:

$$SS(1) = bm\sum_{i=1}^{a}(\bar{x}_{i\bullet} - \bar{x})^2$$

$$= 2 \cdot 3[(\bar{x}_{B\bullet} - \bar{x})^2 + (\bar{x}_{M\bullet} - \bar{x})^2 + (\bar{x}_{T\bullet} - \bar{x})^2]$$

$$= 6[(55.8 - 61.5)^2 + (77.2 - 61.5)^2 + (51.5 - 61.5)^2] = 2{,}273.88$$

Step 3: Calculate $SS(2)$, which measures the amount of variability due to the different levels of factor 2:

$$SS(2) = am\sum_{j=1}^{b}(\bar{x}_{\bullet j} - \bar{x})^2$$

$$= 3 \cdot 3[(\bar{x}_{\bullet R} - \bar{x})^2 + (\bar{x}_{\bullet W} - \bar{x})^2]$$

$$= 9[(60.8 - 61.5)^2 + (62.2 - 61.5)^2] = 8.82$$

Step 4: Calculate $SS(\text{int})$, which measures the amount of variability due to the interaction between factors 1 and 2:

$$SS(\text{int}) = m\sum_{i=1}^{a}\sum_{j=1}^{b}(\bar{x}_{ij} - \bar{x}_{i\bullet} - \bar{x}_{\bullet j} + \bar{x})^2$$

$$= 3[(\bar{x}_{BR} - \bar{x}_{B\bullet} - \bar{x}_{\bullet R} + \bar{x})^2 + (\bar{x}_{BW} - \bar{x}_{B\bullet} - \bar{x}_{\bullet W} + \bar{x})^2$$

$$+ (\bar{x}_{MR} - \bar{x}_{M\bullet} - \bar{x}_{\bullet R} + \bar{x})^2 + (\bar{x}_{MW} - \bar{x}_{M\bullet} - \bar{x}_{\bullet W} + \bar{x})^2$$

$$+ (\bar{x}_{TR} - \bar{x}_{T\bullet} - \bar{x}_{\bullet R} + \bar{x})^2 + (\bar{x}_{TW} - \bar{x}_{T\bullet} - \bar{x}_{\bullet W} + \bar{x})^2]$$

$$= 3[(55.9 - 55.8 - 60.8 + 61.5)^2 + (55.7 - 55.8 - 62.2 + 61.5)^2$$

$$+ (75.5 - 77.2 - 60.8 + 61.5)^2 + (78.9 - 77.2 - 62.2 + 61.5)^2$$

$$+ (51.0 - 51.5 - 60.8 + 61.5)^2 + (52.0 - 51.5 - 62.2 + 61.5)^2] = 10.08$$

TABLE 12.14 Two-Way ANOVA Table for the Bakery Demand Data

Source of Variation	Degrees of Freedom	Sum of Squares	Mean Square	F
Factor 1	$a - 1 = 2$	$SS(1) = 2{,}273.88$	$MS(1) = \dfrac{SS(1)}{a-1} = 1136.94$	$F(1) = \dfrac{MS(1)}{MSE} = 185.6229$
Factor 2	$b - 1 = 1$	$SS(2) = 8.82$	$MS(2) = \dfrac{SS(2)}{b-1} = 8.82$	$F(2) = \dfrac{MS(2)}{MSE} = 1.44$
Interaction	$(a-1)(b-1) = 2$	$SS(\text{int}) = 10.08$	$MS(\text{int}) = \dfrac{SS(\text{int})}{(a-1)(b-1)} = 5.04$	$F(\text{int}) = \dfrac{MS(\text{int})}{MSE} = .8229$
Error	$ab(m-1) = 12$	$SSE = 73.50$	$MSE = \dfrac{SSE}{ab(m-1)} = 6.125$	
Total	$abm - 1 = 17$	$SSTO = 2{,}366.28$		

Step 5: Calculate SSE, which measures the amount of variability due to the error:

$$SSE = SSTO - SS(1) - SS(2) - SS(\text{int})$$
$$= 2{,}366.28 - 2{,}273.88 - 8.82 - 10.08 = 73.50$$

These sums of squares are shown in Table 12.14, which is called a **two-way analysis of variance (ANOVA) table.** This table also gives the degrees of freedom, mean squares, and F statistics used to test the hypotheses of interest in a two-factor factorial experiment, as well as the values of these quantities for the shelf display data.

We first test the null hypothesis H_0 that **no interaction exists between factors 1 and 2** versus the alternative hypothesis H_a that **interaction does exist.** We can reject H_0 in favor of H_a at level of significance α if $F(\text{int})$ is greater than the F_α point based on $(a - 1)(b - 1)$ numerator and $ab(m - 1)$ denominator degrees of freedom. In the supermarket case, $F_{.05}$ based on $(a - 1)(b - 1) = 2$ numerator and $ab(m - 1) = 12$ denominator degrees of freedom is 3.89 (see Table A.7). Because $F(\text{int}) = .8229$ (see Table 12.14) is less than $F_{.05} = 3.89$, we cannot reject H_0 at the .05 level of significance. We conclude that little or no interaction exists between shelf display height and shelf display width. That is, we conclude that the relationship between mean demand for the bakery product and shelf display height depends little (or not at all) on the shelf display width. Further, we conclude that the relationship between mean demand and shelf display width depends little (or not at all) on the shelf display height. Therefore, we can test the significance of each factor separately.

To test the significance of **factor 1,** we test the null hypothesis H_0 **that no differences exist between the effects of the different levels of factor 1** on the mean response versus the alternative hypothesis H_a **that at least two levels of factor 1 have different effects.** We can reject H_0 in favor of H_a at level of significance α if $F(1)$ is greater than the F_α point based on $a - 1$ numerator and $ab(m - 1)$ denominator degrees of freedom. In the supermarket case, $F_{.05}$ based on $a - 1 = 2$ numerator and $ab(m - 1) = 12$ denominator degrees of freedom is 3.89. Because $F(1) = 185.6229$ (see Table 12.14) is greater than $F_{.05} = 3.89$, we can reject H_0 at the .05 level of significance. Therefore, we have strong evidence that at least two of the bottom, middle, and top display heights have different effects on mean monthly demand.

To test the significance of **factor 2,** we test the null hypothesis H_0 **that no differences exist between the effects of the different levels of factor 2** on the mean response versus the alternative hypothesis H_a **that at least two levels of factor 2 have different effects.** We can reject H_0 in favor of H_a at level of significance α if $F(2)$ is greater than the F_α point based on $b - 1$ numerator and $ab(m - 1)$ denominator degrees of freedom. In the supermarket case, $F_{.05}$ based on $b - 1 = 1$ numerator and $ab(m - 1) = 12$ denominator degrees of freedom is 4.75. Because $F(2) = 1.44$ (see Table 12.14) is less than $F_{.05} = 4.75$, we cannot reject H_0 at the .05 level of significance. Therefore, we do not

FIGURE 12.11   **Minitab and Excel Outputs of a Two-Way ANOVA of the Bakery Demand Data**

(a) The Minitab Output

Rows : Height   Columns : Width

Cell Contents : Demand : Mean

	Regular	Wide	All		Height	Mean		Width	Mean
Bottom	55.90	55.70	55.80		Bottom	55.8 [16]		Regular	60.8 [19]
Middle	75.50	78.90	77.20		Middle	77.2 [17]		Wide	62.2 [20]
Top	51.00	52.00	51.50		Top	51.5 [18]			
All	60.80	62.20	61.50						

Two-way ANOVA: Demand versus Height, Width

Source	DF	SS		MS		F		P	
Height	2	2273.88	[1]	1136.94	[6]	185.62	[10]	0.000	[11]
Width	1	8.82	[2]	8.82	[7]	1.44	[12]	0.253	[13]
Interaction	2	10.08	[3]	5.04	[8]	0.82	[14]	0.462	[15]
Error	12	73.50	[4]	6.12	[9]				
Total	17	2366.28	[5]						

(b) The Excel Output

ANOVA: Two-Factor With Replication

SUMMARY		Regular	Wide	Total	
Bottom					
Count		3	3	6	
Sum		167.7	167.1	334.8	
Average		55.9	55.7	55.8	[16]
Variance		5.07	10.24	6.136	
Middle					
Count		3	3	6	
Sum		226.5	236.7	463.2	
Average		75.5	78.9	77.2	[17]
Variance		6.51	8.89	9.628	
Top					
Count		3	3	6	
Sum		153.0	156.0	309.0	
Average		51.0	52.0	51.5	[18]
Variance		1.8	4.2	2.7	
Total					
Count		9	9		
Sum		547.2	559.8		
Average		60.8 [19]	62.2 [20]		
Variance		129.405	165.277		

ANOVA

Source of Variation	SS		df	MS		F		P-value		F crit
Height	2273.88	[1]	2	1136.94	[6]	185.6229	[10]	0.0000	[11]	3.8853
Width	8.82	[2]	1	8.82	[7]	1.4400	[12]	0.2533	[13]	4.7472
Interaction	10.08	[3]	2	5.04	[8]	0.8229	[14]	0.4625	[15]	3.8853
Within	73.5	[4]	12	6.125	[9]					
Total	2366.28	[5]	17							

[1] SS(1)   [2] SS(2)   [3] SS(int)   [4] SSE   [5] SSTO   [6] MS(1)   [7] MS(2)   [8] MS(int)   [9] MSE   [10] $F(1)$ [11] p-value for $F(1)$ [12] $F(2)$

[13] p-value for $F(2)$ [14] $F(\text{int})$ [15] p-value for $F(\text{int})$ [16] $\bar{x}_{B\bullet}$ [17] $\bar{x}_{M\bullet}$ [18] $\bar{x}_{T\bullet}$ [19] $\bar{x}_{\bullet R}$ [20] $\bar{x}_{\bullet W}$

have strong evidence that the regular display width and the wide display have different effects on mean monthly demand.

Noting that Figure 12.11 gives Minitab and Excel outputs of a two-way ANOVA for the bakery demand data, we next discuss how to make pairwise comparisons.

Point Estimates and Confidence Intervals in Two-Way ANOVA

1 Consider the **difference between the effects of levels i and i' of factor 1 on the mean value of the response variable.**

   a   A **point estimate** of this difference is $\bar{x}_{i\cdot} - \bar{x}_{i'\cdot}$.

   b   A **Tukey simultaneous $100(1 - \alpha)$ percent confidence interval** for this difference (in the set of all possible paired differences between the effects of the different levels of factor 1) is

   $$\left[(\bar{x}_{i\cdot} - \bar{x}_{i'\cdot}) \pm q_\alpha \sqrt{MSE\left(\frac{1}{bm}\right)}\right]$$

   where q_α is obtained from Table A.10, which is a table of percentage points of the studentized range. Here q_α is listed corresponding to values of a and $ab(m - 1)$.

2 Consider the **difference between the effects of levels j and j' of factor 2 on the mean value of the response variable.**

   a   A **point estimate** of this difference is $\bar{x}_{\cdot j} - \bar{x}_{\cdot j'}$.

   b   A **Tukey simultaneous $100(1 - \alpha)$ percent confidence interval** for this difference (in the set of all possible paired differences between the effects of the different levels of factor 2) is

   $$\left[(\bar{x}_{\cdot j} - \bar{x}_{\cdot j'}) \pm q_\alpha \sqrt{MSE\left(\frac{1}{am}\right)}\right]$$

   where q_α is obtained from Table A.10 and is listed corresponding to values of b and $ab(m - 1)$.

3 Let μ_{ij} denote the **mean value of the response variable obtained when using level i of factor 1 and level j of factor 2.** A point estimate of μ_{ij} is $\bar{x}_{ij}$, and an **individual $100(1 - \alpha)$ percent confidence interval** for μ_{ij} is

   $$\left[x_{ij} \pm t_{\alpha/2}\sqrt{\frac{MSE}{m}}\right]$$

   where the $t_{\alpha/2}$ point is based on $ab(m - 1)$ degrees of freedom.

C **EXAMPLE 12.8** The Supermarket Case: Comparing Display Heights and Widths

We have previously concluded that at least two of the bottom, middle, and top display heights have different effects on mean monthly demand. Because $\bar{x}_{M\cdot} = 77.2$ is greater than $\bar{x}_{B\cdot} = 55.8$ and $\bar{x}_{T\cdot} = 51.5$, we will use Tukey simultaneous 95 percent confidence intervals to compare the effect of a middle display height with the effects of the bottom and top display heights. To compute these intervals, we first note that $q_{.05} = 3.77$ is the entry in Table A.10 corresponding to $a = 3$ and $ab(m - 1) = 12$. Also note that the MSE found in the two-way ANOVA table is 6.125 (see Table 12.14). It follows that a Tukey simultaneous 95 percent confidence interval for the difference between the effects of a middle and bottom display height on mean monthly demand is

$$\left[(\bar{x}_{M\cdot} - \bar{x}_{B\cdot}) \pm q_{.05}\sqrt{MSE\left(\frac{1}{bm}\right)}\right] = \left[(77.2 - 55.8) \pm 3.77\sqrt{6.125\left(\frac{1}{2(3)}\right)}\right]$$

$$= [21.4 \pm 3.8091]$$

$$= [17.5909, 25.2091]$$

This interval says we are 95 percent confident that changing from a bottom display height to a middle display height will increase the mean demand for the bakery product by between 17.5909 and 25.2091 cases per month. Similarly, a Tukey simultaneous 95 percent confidence interval for the difference between the effects of a middle and top display height on mean monthly demand is

$$[(\bar{x}_{M\cdot} - \bar{x}_{T\cdot}) \pm 3.8091] = [(77.2 - 51.5) \pm 3.8091]$$

$$= [21.8909, 29.5091]$$

This interval says we are 95 percent confident that changing from a top display height to a middle display height will increase mean demand for the bakery product by between 21.8909 and 29.5091 cases per month. Together, these intervals make us 95 percent confident that a middle shelf display height is, on average, at least 17.5909 cases sold per month better than a bottom shelf display height and at least 21.8909 cases sold per month better than a top shelf display height.

Next, recall that previously conducted F tests suggest that there is little or no interaction between display height and display width and that there is little difference between using a regular display width and a wide display. However, noting that $\bar{x}_{MW} = 78.9$ is slightly larger than $\bar{x}_{MR} = 75.5$, we now find an individual 95 percent confidence interval for μ_{MW}, the mean demand obtained when using a middle display height and a wide display:

$$\left[\bar{x}_{MW} \pm t_{.025}\sqrt{\frac{MSE}{m}}\right] = \left[78.9 \pm 2.179\sqrt{\frac{6.125}{3}}\right]$$
$$= [75.7865, 82.0135]$$

Here $t_{.025} = 2.179$ is based on $ab(m - 1) = 12$ degrees of freedom. This interval says that, when we use a middle display height and a wide display, we can be 95 percent confident that mean demand for the bakery product will be between 75.7865 and 82.0135 cases per month.

If we conclude that (substantial) interaction exists between factors 1 and 2, the effects of changing the level of one factor will depend on the level of the other factor. In this case, we cannot analyze the levels of the two factors separately. One simple alternative procedure is to use one-way ANOVA (see Section 12.2) to compare all of the treatment means (the μ_{ij}'s) with the possible purpose of finding the best combination of levels of factors 1 and 2.

Exercises for Section 12.4

CONCEPTS ■ connect

12.19 What is a treatment in the context of a two-factor factorial experiment?

12.20 Explain what we mean when we say that interaction exists between two factors.

METHODS AND APPLICATIONS

12.21 A study compared three display panels used by air traffic controllers. Each display panel was tested for four different simulated emergency conditions. Twenty-four highly trained air traffic controllers were used in the study. Two controllers were randomly assigned to each display panel–emergency condition combination. The time (in seconds) required to stabilize the emergency condition was recorded. Figure 12.12 gives the resulting data and the JMP output of a two-way ANOVA of the data. Using the computer output ⓭ Display2

a Interpret the interaction plot in Figure 12.12. Then test for interaction with $\alpha = .05$.

b Test the significance of display panel effects with $\alpha = .05$.

c Test the significance of emergency condition effects with $\alpha = .05$.

d Make pairwise comparisons of display panels A, B, and C by using Tukey simultaneous 95 percent confidence intervals.

e Make pairwise comparisons of emergency conditions 1, 2, 3, and 4 by using Tukey simultaneous 95 percent confidence intervals.

f Which display panel minimizes the time required to stabilize an emergency condition? Does your answer depend on the emergency condition? Why?

g Calculate a 95 percent (individual) confidence interval for the mean time required to stabilize emergency condition 4 using display panel B.

12.22 A telemarketing firm has studied the effects of two factors on the response to its television advertisements. The first factor is the time of day at which the ad is run, while the second is the position of the ad within the hour. The data in Figure 12.13, which were obtained by using a completely randomized experimental design, give the number of calls placed to an 800 number following a sample broadcast of the advertisement. If we use Excel to analyze these data, we obtain the output in Figure 12.13. Using the computer output ⓭ TelMktResp

a Perform graphical analysis to check for interaction between time of day and position of advertisement. Explain your conclusion. Then test for interaction with $\alpha = .05$.

b Test the significance of time of day effects with $\alpha = .05$.

c Test the significance of position of advertisement effects with $\alpha = .05$.

d Make pairwise comparisons of the morning, afternoon, and evening times by using Tukey simultaneous 95 percent confidence intervals.

e Make pairwise comparisons of the four ad positions by using Tukey simultaneous 95 percent confidence intervals.

f Which time of day and advertisement position maximizes consumer response? Compute a 95 percent (individual) confidence interval for the mean number of calls placed for this time of day/ad position combination.

FIGURE 12.12   The Display Panel Data and the JMP Output of a Two-Way ANOVA (for Exercise 12.21) DS Display2

Emergency Condition

Display Panel	1	2	3	4
A	17	25	31	14
	14	24	34	13
B	15	22	28	9
	12	19	31	10
C	21	29	32	15
	24	28	37	19

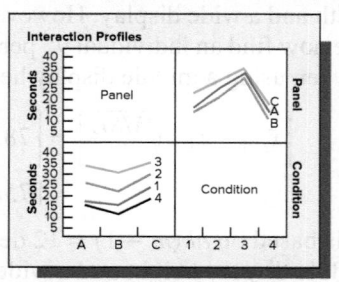

Least Squares Means Estimates

Panel	Estimate	Condition	Estimate
A	21.500000	1	17.166667
B	18.250000	2	24.500000
C	25.625000	3	32.166667
		4	13.333333

Analysis of Variance

Source	DF	Sum of Squares	Mean Square	F Ratio
Model	11	1482.4583	134.769	32.6713
Error	12	49.5000	4.125	Prob > F
C. Total	13	1531.9583		<.0001*

Effect Tests

Source	Nparm	DF	Sum of Squares	F Ratio	Prob > F
Panel	2	2	218.5833	26.4949	<.0001*
Condition	3	3	1247.4583	100.8047	<.0001*
Panel*Condition	6	6	16.4167	0.6633	0.6809

Tukey HSD All Pairwise Comparisons

Quantile = 2.66776, Adjusted DF = 12.0, Adjustment = Tukey

Panel	-Panel	Difference	Std Error	t Ratio	Prob > |t|	Lower 95%	Upper 95%
A	B	3.25000	1.015505	3.20	0.0194*	0.5409	5.95912
A	C	−4.12500	1.015505	−4.06	0.0042*	−6.8341	−1.41588
B	C	−7.37500	1.015505	−7.26	<.0001*	−10.0841	−4.66588

Tukey HSD All Pairwise Comparisons

Quantile = 2.9688, Adjusted DF = 12.0, Adjustment = Tukey

Condition	-Condition	Difference	Std Error	t Ratio	Prob > |t|	Lower 95%	Upper 95%
1	2	−7.3333	1.172604	−6.25	0.0002*	−10.8146	−3.8521
1	3	−15.0000	1.172604	−12.79	<.0001*	−18.4812	−11.5188
1	4	3.8333	1.172604	3.27	0.0297*	0.3521	7.3146
2	3	−7.6667	1.172604	−6.54	0.0001*	−11.1479	−4.1854
2	4	11.1667	1.172604	9.52	<.0001*	7.6854	14.6479
3	4	18.8333	1.172604	16.06	<.0001*	15.3521	22.3146

12.23 A small builder of speculative homes builds three basic house designs and employs two foremen. The builder has used each foreman to build two houses of each design and has obtained the profits given in Table 12.15 (the profits are given in thousands of dollars). Figure 12.14 presents the Minitab output of a two-way ANOVA of the house profitability data. DS HouseProf

a (1) Interpret the Minitab interaction plot in Figure 12.14. (2) Test for interaction with $\alpha = .05$. (3) Can we (separately) test for the significance of house design and foreman effects? Explain why or why not.

b Which house design/foreman combination gets the highest profit? Compute a 95 percent (individual) confidence interval for mean profit when the best house design/foreman combination is employed.

12.24 A research team at a school of agriculture carried out an experiment to study the effects of two fertilizer types (A and B) and four wheat types (M, N, O, and P) on crop yields (in bushels per one-third acre plot). The data in Table 12.16 were obtained by using a completely randomized experimental design. DS Wheat

a Graphically check for interaction.

b Test for significant interaction, for significant fertilizer type effects, and for significant wheat type effects. Set $\alpha = .05$ for each test.

c Use Tukey simultaneous 95 percent confidence intervals (1) to make a pairwise comparison of the two fertilizer types and (2) to make pairwise comparisons of the wheat types.

d Find a 95 percent (individual) confidence interval for the mean crop yield resulting from using the best fertilizer type/wheat type combination.

TABLE 12.15   Results of the House Profitability Study DS HouseProf

Foreman	House Design		
	A	B	C
1	10.2	12.2	19.4
	11.1	11.7	18.2
2	9.7	11.6	13.6
	10.8	12.0	12.7

TABLE 12.16   Results of a Two-Factor Wheat Yield Experiment DS Wheat

Fertilizer Type	Wheat Type			
	M	N	O	P
A	19.4	25.0	24.8	23.1
	20.6	24.0	26.0	24.3
	20.0	24.5	25.4	23.7
B	22.6	25.6	27.6	25.4
	21.6	26.8	26.4	24.5
	22.1	26.2	27.0	26.3

FIGURE 12.13   The Telemarketing Data and the Excel Output of a Two-Way ANOVA (for Exercise 12.22)   ⒹⓈ TelMktResp

Position of Advertisement

Time of Day	On the Hour	On the Half-Hour	Early in Program	Late in Program
10:00 morning	42	36	62	51
	37	41	68	47
	41	38	64	48
4:00 afternoon	62	57	88	67
	60	60	85	60
	58	55	81	66
9:00 evening	100	97	127	105
	96	96	120	101
	103	101	126	107

ANOVA: Two-Factor With Replication

Summary	Hour	Half-Hour	Early	Late	Total
Morning					
Count	3	3	3	3	12
Sum	120	115	194	146	575
Average	40	38.3	64.7	48.7	47.9
Variance	7	6.3	9.3	4.3	123.7
Afternoon					
Count	3	3	3	3	12
Sum	180	172	254	193	799
Average	60	57.3	84.7	64.3	66.6
Variance	4	6.3	12.3	14.3	132.4
Evening					
Count	3	3	3	3	12
Sum	299	294	373	313	1279
Average	99.67	98	124.3	104.3	106.6
Variance	12.33	7	14.3	9.3	128.3
Total					
Count	9	9	9	9	
Sum	599	581	821	652	
Average	66.56	64.56	91.22	72.44	
Variance	697.53	701.78	700.69	625.03	

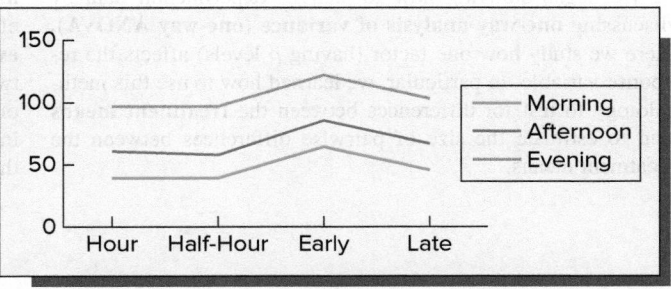

ANOVA

Source of Variation	SS	df	MS	F	P-value	F crit
Sample	21560.89	2	10780.444	1209.02	8.12E-25	3.403
Columns	3989.42	3	1329.806	149.14	1.19E-15	3.009
Interaction	25.33	6	4.222	0.47	0.8212	2.508
Within	214	24	8.917			
Total	25789.64	35				

FIGURE 12.14   Minitab Output of a Two-Way ANOVA of the House Profitability Data (for Exercise 12.23)

Rows: Foreman   Columns: Design

	A	B	C	All
1	10.65	11.95	18.80	13.80
2	10.25	11.80	13.15	11.73
All	10.45	11.88	15.98	12.77

Cell Contents: Profit : Mean

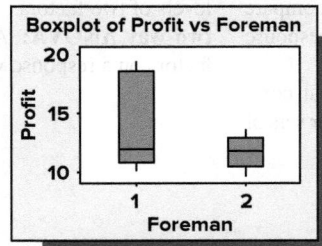

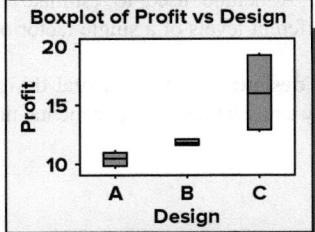

Two-way ANOVA: Profit versus Foreman, Design

Source	DF	SS	MS	F	P
Foreman	1	12.813	12.8133	32.85	0.001
Design	2	65.822	32.9108	84.39	0.000
Interaction	2	19.292	9.6458	24.73	0.001
Error	6	2.340	0.3900		
Total	11	100.267			

Foreman	Mean	Design	Mean
1	13.8000	A	10.450
2	11.7333	B	11.875
		C	15.975

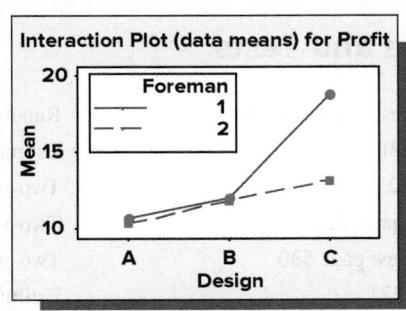

Chapter Summary

We began this chapter by introducing some basic concepts of **experimental design.** We saw that we carry out an experiment by setting the values of one or more **factors** before the values of the **response variable** are observed. The different values (or levels) of a factor are called **treatments,** and the purpose of most experiments is to compare and estimate the effects of the various treatments on the response variable. We saw that the different treatments are assigned to **experimental units,** and we discussed the **completely randomized experimental design.** This design assigns independent, random samples of experimental units to the treatments.

We began studying how to analyze experimental data by discussing **one-way analysis of variance (one-way ANOVA).** Here we study how one factor (having p levels) affects the response variable. In particular, we learned how to use this methodology to test for differences between the **treatment means** and to estimate the size of pairwise differences between treatment means.

Sometimes, even if we randomly select the experimental units, differences between the experimental units conceal differences between the treatments. In such a case, we learned that we can employ a **randomized block design.** Each **block** (experimental unit or set of experimental units) is used exactly once to measure the effect of each and every treatment. Because we are comparing the treatments by using the same experimental units, any true differences between the treatments will not be concealed by differences between the experimental units.

The last technique we studied in this chapter was **two-way analysis of variance (two-way ANOVA).** Here we study the effects of two factors by carrying out a **two-factor factorial experiment.** If there is little or **no interaction** between the two factors, then we are able to study the significance of each of the two factors separately. On the other hand, if substantial interaction exists between the two factors, we study the nature of the differences between the treatment means.

Glossary of Terms

analysis of variance table: A table that summarizes the sums of squares, mean squares, F statistic(s), and p-value(s) for an analysis of variance.

completely randomized experimental design: An experimental design in which independent, random samples of experimental units are assigned to the treatments.

experimental units: The entities (objects, people, and so on) to which the treatments are assigned.

factor: A variable that might influence the response variable; an independent variable.

interaction: When the relationship between the mean response and one factor depends on the level of the other factor.

one-way ANOVA: A method used to estimate and compare the effects of the different levels of a single factor on a response variable.

randomized block design: An experimental design that compares p treatments by using b blocks (experimental units or sets of

experimental units). Each block is used exactly once to measure the effect of each and every treatment.

replication: When a treatment is applied to more than one experimental unit.

response variable: The variable of interest in an experiment; the dependent variable.

treatment: A value (or level) of a factor (or combination of factors).

treatment mean: The mean value of the response variable obtained by using a particular treatment.

two-factor factorial experiment: An experiment in which we randomly assign m experimental units to each combination of levels of two factors.

two-way ANOVA: A method used to study the effects of two factors on a response variable.

Important Formulas and Tests

One-way ANOVA sums of squares: pages 519–520

One-way ANOVA F-test: page 521

One-way ANOVA table: page 522

Estimation in one-way ANOVA: page 524

Randomized block sums of squares: page 530

Randomized block F tests: page 531

Randomized block ANOVA table: page 531

Estimation in a randomized block ANOVA: page 531

Two-way ANOVA sums of squares: pages 537–538

Two-way ANOVA F tests: page 538

Two-way ANOVA table: page 538

Estimation in two-way ANOVA: page 540

Supplementary Exercises

connect | **12.25** An experiment is conducted to study the effects of two sales approaches—high-pressure (H) and low-pressure (L)—and to study the effects of two sales pitches (1 and 2) on the weekly sales of a product. The data in Table 12.17 are obtained by using a completely randomized design, and Figure 12.15 gives the Excel output of a two-way ANOVA of the sales experiment data. Using Table 12.17 and the computer output of Figure 12.15 **DS** SaleMeth2

a Perform graphical analysis to check for interaction between sales pressure and sales pitch.

b Test for interaction by setting $\alpha = .05$.

c Test for differences in the effects of the levels of sales pressure by setting $\alpha = .05$.

d Test for differences in the effects of the levels of sales pitch by setting $\alpha = .05$.

12.26 The loan officers at a large bank can use three different methods for evaluating loan applications. Loan decisions can be based on (1) the applicant's balance sheet (B), (2) examination of key financial ratios (F), or (3) use of a new decision support system (D). In order to compare these three methods, four of the bank's loan officers are randomly selected. Each officer employs each of the evaluation methods for one month (the methods are employed in randomly selected orders). After a year has passed, the percentage of bad loans for each loan officer and evaluation method is determined. The data obtained by using this randomized block design are given in Table 12.18. Completely analyze the data using randomized block ANOVA. **DS** LoanEval

12.27 A drug company wishes to compare the effects of three different drugs (X, Y, and Z) that are being developed to reduce cholesterol levels. Each drug is administered to six patients at the recommended dosage for six months. At the end of this period the reduction in cholesterol level is recorded for each patient. The results are given in Table 12.19. Using these data we obtain $SSTO = 2547.8$, $SSE = 395.7$, $\bar{x}_X = 23.67$, $\bar{x}_Y = 39.17$, and $\bar{x}_Z = 12.50$. Completely analyze these data using one-way ANOVA. **DS** CholRed

TABLE 12.19
Reduction of Cholesterol Levels
DS CholRed

	Drug	
X	**Y**	**Z**
22	40	15
31	35	9
19	47	14
27	41	11
25	39	21
18	33	5

12.28 In an article in *Accounting and Finance* (the journal of the Accounting Association of Australia and New Zealand), Church and Schneider report on a study concerning auditor objectivity. A sample of 45 auditors was randomly divided into three groups: (1) the 15 auditors in group 1 designed an audit program for accounts receivable and evaluated an audit program for accounts payable designed by somebody else; (2) the 15 auditors in group 2 did the reverse; (3) the 15 auditors in group 3 (the control group) evaluated the audit programs for both accounts. All 45 auditors were then instructed to spend an additional 15 hours investigating suspected irregularities in either or both of the audit programs. The mean additional number of hours allocated to the accounts receivable audit program by the

TABLE 12.17 Results of the Sales Approach Experiment **DS** SaleMeth2

	Sales Pitch	
Sales Pressure	**1**	**2**
H	32	32
	29	30
	30	28
L	28	25
	25	24
	23	23

TABLE 12.18 Results of a Loan Evaluation Experiment **DS** Loan Eval

	Loan Evaluation Method		
Loan Officer	**B**	**F**	**D**
1	8	5	4
2	6	4	3
3	5	2	1
4	4	1	0

FIGURE 12.15 Excel Output of a Two-Way ANOVA of the Sales Approach Data (for Exercise 12.25)

ANOVA: Two-Factor With Replication

SUMMARY	Pitch 1	Pitch 2	Total
High Pressure			
Count	3	3	6
Sum	91	90	181
Average	30.3333	30	30.1667
Variance	2.3333	4	2.5667
Low Pressure			
Count	3	3	6
Sum	76	72	148
Average	25.3333	24	24.6667
Variance	6.3333	1	3.4667
Total			
Count	6	6	
Sum	167	162	
Average	27.8333	27	
Variance	10.9667	12.8	

ANOVA

Source of Variation	SS	df	MS	F	P-value	F crit
Pressure	90.75	1	90.75	26.5610	0.0009	5.3177
Pitch	2.0833	1	2.0833	0.6098	0.4574	5.3177
Interaction	0.75	1	0.75	0.2195	0.6519	5.3177
Within	27.3333	8	3.4167			
Total	120.917	11				

TABLE 12.20 **Results of a Two-Factor Gasoline Mileage Experiment** ⒹⓈ PremGas

Premium Gasoline Type	Gasoline Additive Type			
	Q	R	S	T
D	27.4	33.0	33.5	30.8
	28.6	32.0	32.3	29.7
E	33.3	35.6	33.4	29.6
	34.5	34.4	33.1	30.6
F	33.0	34.7	33.0	28.6
	33.5	33.3	32.0	29.8

auditors in groups 1, 2, and 3 were $\bar{x}_1 = 6.7$, $\bar{x}_2 = 9.7$, and $\bar{x}_3 = 7.6$. Furthermore, a one-way ANOVA of the data shows that $SST = 71.51$ and $SSE = 321.3$.

a Define appropriate treatment means μ_1, μ_2, and μ_3. Then test for statistically significant differences between these treatment means. Set $\alpha = .05$. Can we conclude that the different auditor groups have different effects on the mean additional time allocated to investigating the accounts receivable audit program?

b Perform pairwise comparisons of the treatment means by computing a Tukey simultaneous 95 percent confidence interval for each of the pairwise differences $\mu_1 - \mu_2$, $\mu_1 - \mu_3$, and $\mu_2 - \mu_3$. Interpret the results. What do your results imply about the objectivity of auditors? What are the practical implications of this result?

12.29   An experiment is carried out to study the effects of three premium gasoline types (*D*, *E*, and *F*) and of four different gasoline additive types (*Q*, *R*, *S*, and *T*) on the gasoline mileage (in miles per gallon) obtained by a test vehicle. The data in Table 12.20 were obtained by using a completely randomized experimental design. ⒹⓈ PremGas

FIGURE 12.16   **Line Plot for Exercise 12.30**

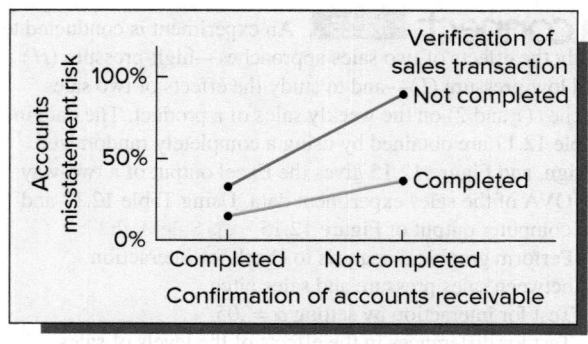

Source: Data from C. E. Brown and I. Solomon, "Configural Information Processing in Auditing: The Role of Domain-Specific Knowledge," *The Accounting Review* 66, no.1 (January 1991), p. 105 (Figure 1).

a Graphically check for interaction.
b Test for the significance of interaction by setting $\alpha = .05$.
c Find a 95 percent (individual) confidence interval for the mean mileage obtained by the best combination of a premium gasoline type and an additive type.

12.30   In an article in the *Accounting Review*, Brown and Solomon study the effects of two factors—confirmation of accounts receivable and verification of sales transactions—on account misstatement risk by auditors. Both factors had two levels—completed or not completed—and a line plot of the treatment mean misstatement risks is shown in Figure 12.16. This line plot makes it appear that interaction exists between the two factors. In your own words, explain what the nature of the interaction means in practical terms.

Appendix 12.1 ■ Experimental Design and Analysis of Variance Using Excel

One-way ANOVA in Figure 12.2(b) (data file: GasMile2 .xlsx):

- Enter the gasoline mileage data from Table 12.1 as follows: type the label "Type A" in cell A1 with its five mileage values in cells A2 to A6; type the label "Type B" in cell B1 with its five mileage values in cells B2 to B6; type the label "Type C" in cell C1 with its five mileage values in cells C2 to C6.

- Select **Data : Data Analysis : Anova : Single Factor**, and click OK in the Data Analysis dialog box.

- In the "Anova: Single Factor" dialog box, enter A1:C6 into the "Input Range" window.

- Select the "Grouped by: Columns" option.

- Place a checkmark in the "Labels in first row" checkbox.

- Enter 0.05 into the Alpha box.

- Under output options, select "New Worksheet Ply" to have the output placed in a new worksheet and enter the name "Output" for the new worksheet.

- Click OK in the "Anova: Single Factor" dialog box to display the ANOVA in a new worksheet.

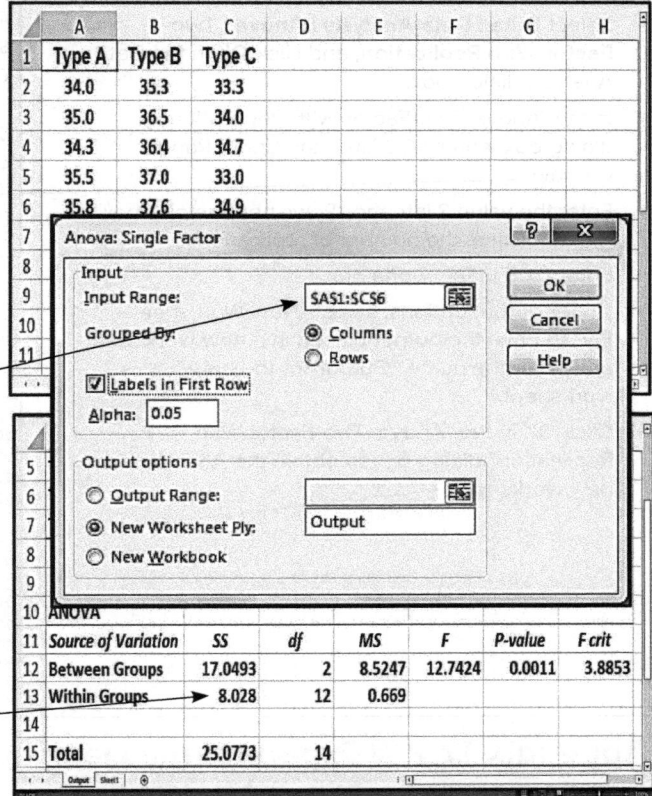

Source: Microsoft Office Excel 2016

Randomized block ANOVA similar to Figure 12.8 (data file: SaleMeth.xlsx):

- Enter the sales methods data from Figure 12.8 as shown in the screen.

- Select **Data : Data Analysis : Anova: Two-Factor Without Replication** and click OK in the Data Analysis dialog box.

- In the "Anova: Two-Factor Without Replication" dialog box, enter A1:D5 into the "Input Range" window.

- Place a checkmark in the "Labels" checkbox.

- Enter 0.05 in the Alpha box.

- Under output options, select "New Worksheet Ply" to have the output placed in a new worksheet and enter the name "Output" for the new worksheet.

- Click OK in the "Anova: Two-Factor Without Replication" dialog box to obtain the ANOVA in a new worksheet.

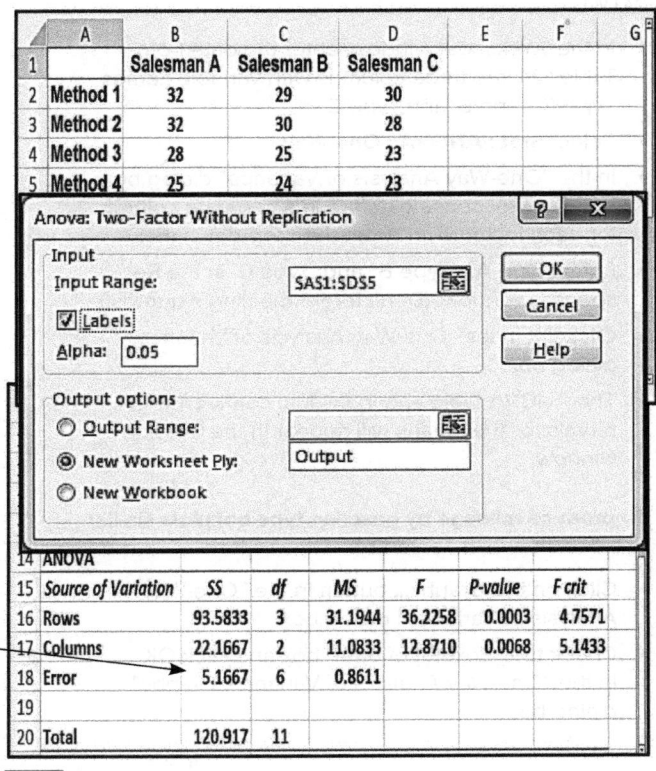

Source: Microsoft Office Excel 2016

Two-way ANOVA in Figure 12.15 (data file: SaleMeth2 .xlsx):

- Enter the sales approach experiment data from Table 12.17 as shown in the screen.

- Select **Data : Data Analysis : Anova : Two-Factor With Replication**, and click OK in the Data Analysis dialog box.

- In the "Anova: Two-Factor With Replication" dialog box, enter A1:C7 into the "Input Range" window.

- Enter the value 3 into the "Rows per Sample" box. (This indicates the number of replications.)

- Enter 0.05 in the Alpha box.

- Under output options, select "New Worksheet Ply" to have the output placed in a new worksheet and enter the name "Output"for the new worksheet.

- Click OK in the "Anova: Two-Factor With Replication" dialog box to obtain the ANOVA in a new worksheet.

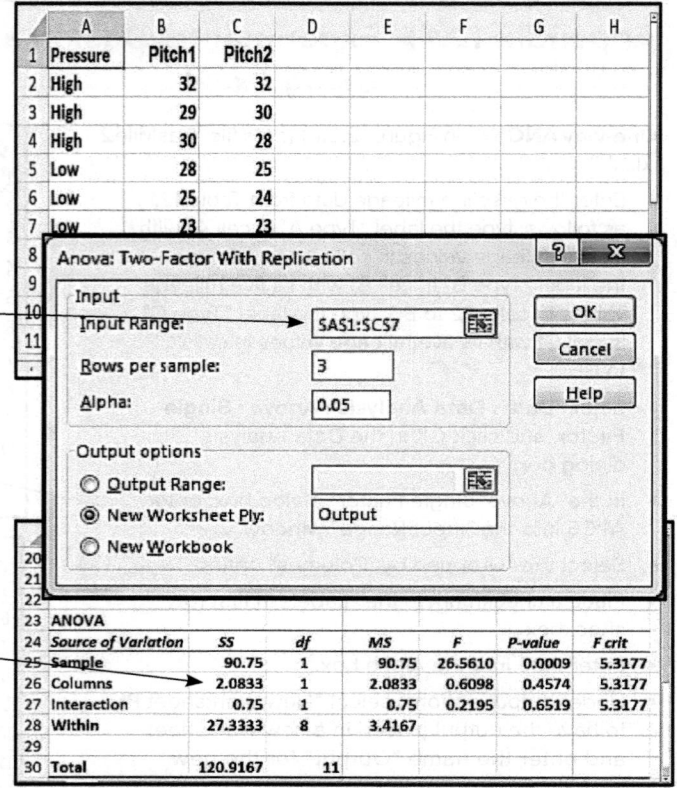

Source: Microsoft Office Excel 2016

Appendix 12.2 ■ Experimental Design and Analysis of Variance Using Minitab

One-way ANOVA in Figure 12.2(a) (data file: GasMile2 .MTW):

- In the Worksheet window, enter the data from Table 12.1 into three columns with variable names Type A, Type B, and Type C.

- Select **Stat : ANOVA : One-Way.**

- In the "One-Way Analysis of Variance" dialog box, choose "Response data are in a separate column for each factor level" from the drop-down menu.

- Enter 'Type A,', 'Type B,' and 'Type C' in the Responses window. (Don't forget the single quotes.)

- Click OK in the "One-Way Analysis of Variance" dialog box.

- The ANOVA table and individual confidence intervals for the means will appear in the Session window.

To produce mileage by gasoline type boxplots similar to those shown in Table 12.1:

- Click on the Graphs... button in the "One-Way Analysis of Variance" dialog box.

- Check the "Boxplot of data" box and click OK in the "One-Way Analysis of Variance: Graphs" dialog box.

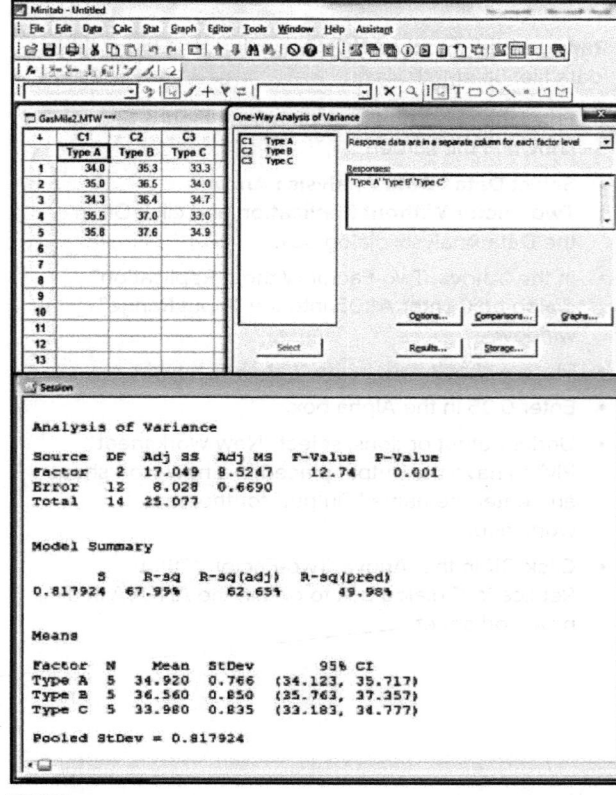

Source: Minitab 18

To produce Tukey pairwise comparisons:

- Click on the Comparisons... button in the "One-Way Analysis of Variance" dialog box.
- Enter 5, denoting 5%, in the "Error rate for comparisons" box and check Tukey as the desired comparison procedure.
- In the Results section of the dialog box, check "Tests" to have the Tukey simultaneous confidence intervals and tests of differences in means computed.
- Click OK in the "One-Way Analysis of Variance: Comparisons" dialog box.
- The Tukey multiple comparisons will appear in the Session window.

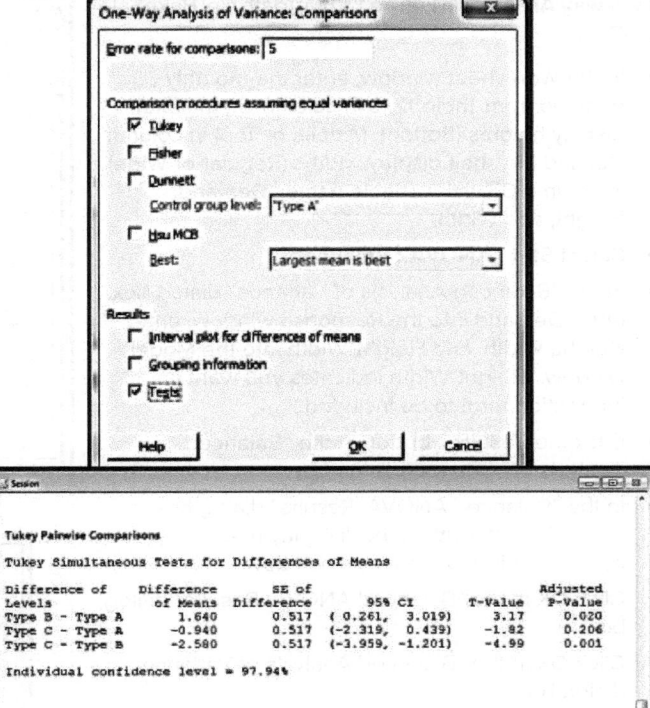

Source: Minitab 18

Randomized Block ANOVA in Figure 12.7(a) (data file: CardBox.MTW):

- In the Worksheet window, enter the number of defective boxes from Table 12.9 in column C1; the production method (1, 2, 3, or 4) in column C2; and the machine operator in column C3 with variable names Rejects, Method, and Operator.
- Select **Stat : ANOVA : Balanced ANOVA.**
- In the "Balanced Analysis of Variance" dialog box, enter Rejects into the Response window, enter Method and Operator into the Model window, and enter Operator into the Random Factors window (because the three operators were a random sample of the company's operators).
- Click OK in the "Balanced Analysis of Variance" dialog box.
- The ANOVA table will appear in the Session window.

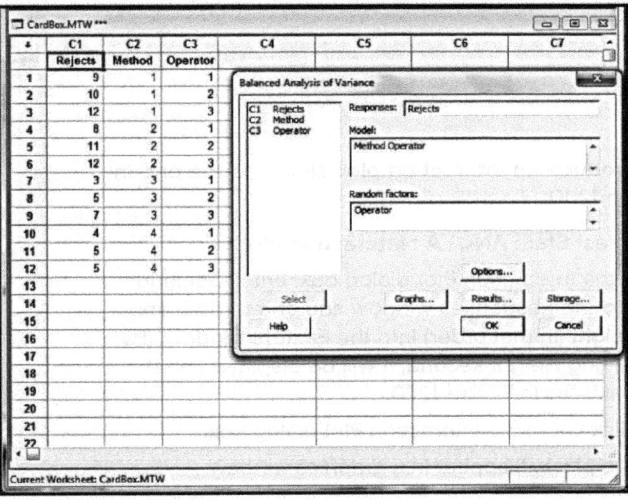

Source: Minitab 18

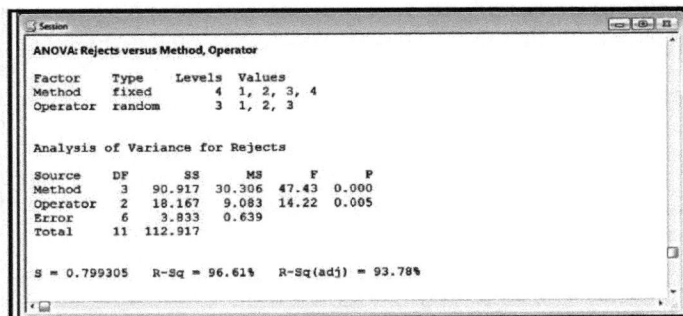

Source: Minitab 18

Two-way ANOVA in Figure 12.11(a) (data file: BakeSale 2.MTW):

- In the Worksheet window, enter the monthly demand from Table 12.13 in column C1; the shelf display heights (Bottom, Middle, or Top) in column C2; and the shelf display widths (Regular or Wide) in column C3 with variable names Demand, Height, and Width.
- Select **Stat : ANOVA : Balanced.**
- In the "Balanced Analysis of Variance" dialog box, enter Demand into the Response window; enter Height, Width, and Height*Width into the Model window. (Height*Width indicates you want an interaction term to be included.)
- Click the Results... button in the "Balanced Analysis of Variance" dialog box.
- In the "Balanced ANOVA: Results" dialog box, enter Height*Width in the "Display means corresponding to the terms" window.
- Click OK in the "Balanced ANOVA: Results" dialog box.
- Click OK in the "Balanced Analysis of Variance" dialog box.
- The ANOVA table and the mean demands will appear in the Session window.

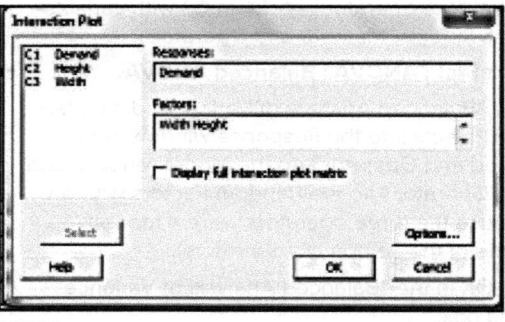

Source: Minitab 18

To produce an interaction plot similar to the one in Figure 12.10:

- Select **Stat : ANOVA : Interaction Plot.**
- In the Interaction Plot dialog box, enter Demand into the Responses window and enter Width and Height (in that order) into the Factors window. (By putting Height second, it will be displayed on the x-axis, as in Figure 12.10.)
- Click OK in the Interaction Plot dialog box.
- The plot will appear in a graphics window.

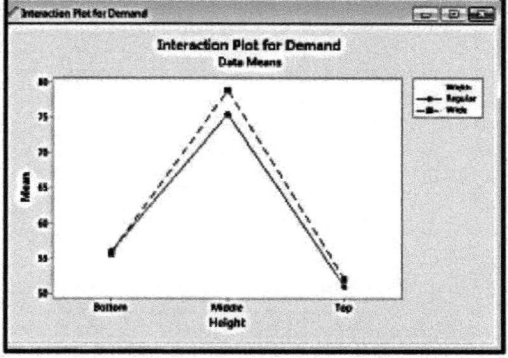

Source: Minitab 18

Appendix 12.3 ■ Experimental Design and Analysis of Variance Using JMP

One-Way Analysis of Variance in Figure 12.4(a) (data file: Display.jmp):

- After opening the Display.jmp file, sele **Analyze: Fit Model**.

- Select the "Time" column and click "Y." Select the "Display" column and click "Add" under the "Construct Model Effects" box. Select "Minimal Report" as the "Emphasis." Click "Run."

- The Analysis of Variance (ANOVA) Table with F test is given in the output. To calculate the pairwise comparisons, click the red triangle menu beside "Response Time" and select **Estimates : Multiple Comparisons**. Choose "Least Squares Means Estimates," "Display," and "All Pairwise Comparisons—Tukey HSD." Click "OK." The three pairwise comparisons calculated using the Tukey HSD method will be added to the output window.

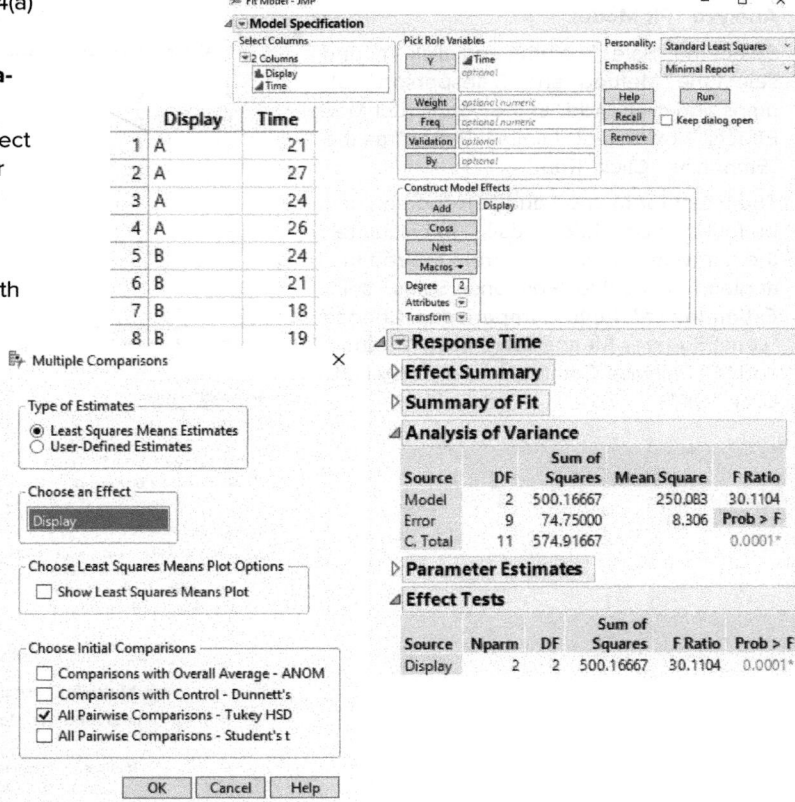

Display	Time
1 A	21
2 A	27
3 A	24
4 A	26
5 B	24
6 B	21
7 B	18
8 B	19

Analysis of Variance

Source	DF	Sum of Squares	Mean Square	F Ratio
Model	2	500.16667	250.083	30.1104
Error	9	74.75000	8.306	Prob > F
C. Total	11	574.91667		0.0001*

Effect Tests

Source	Nparm	DF	Sum of Squares	F Ratio	Prob > F
Display	2	2	500.16667	30.1104	0.0001*

Multiple Comparisons for Display

Least Squares Means Estimates

Display	Estimate	Std Error	DF	Lower 95%	Upper 95%	Arithmetic Mean Estimate
A	24.500000	1.4409680	9	21.240304	27.759696	24.500000
B	20.500000	1.4409680	9	17.240304	23.759696	20.500000
C	35.750000	1.4409680	9	32.490304	39.009696	35.750000

Tukey HSD All Pairwise Comparisons

Quantile = 2.79201, Adjusted DF = 9.0, Adjustment = Tukey

All Pairwise Differences

Display	-Display	Difference	Std Error	t Ratio	Prob>|t|	Lower 95%	Upper 95%
A	B	4.0000	2.037837	1.96	0.1772	-1.6897	9.68966
A	C	-11.2500	2.037837	-5.52	0.0010*	-16.9397	-5.56034
B	C	-15.2500	2.037837	-7.48	<.0001*	-20.9397	-9.56034

Source: JMP Pro 14

- To obtain a display of ordered pairwise differences, click on the gray triangle beside "Effects Details." From the red triangle menu beside "Display," select "LSMeans Tukey HSD." From the red triangle menu beside "LSMeans Difference Tukey HSD," select "Ordered Differences Report." All calculated differences are positive and ordered from largest to smallest.

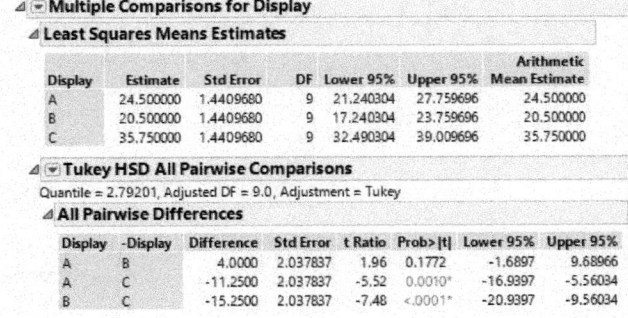

Effect Details

Display

Least Squares Means Table

Level	Least Sq Mean	Std Error	Mean
A	24.500000	1.4409680	24.5000
B	20.500000	1.4409680	20.5000
C	35.750000	1.4409680	35.7500

LSMeans Differences Tukey HSD

- ✓ Crosstab Report
- ✓ Connecting Letters Report
- Save Connecting Letters Table
- Ordered Differences Report

Effect Details

Display

- ✓ LSMeans Table
- LSMeans Plot
- LSMeans Contrast...
- LSMeans Student's t
- LSMeans Tukey HSD
- LSMeans Dunnett
- Test Slices
- Power Analysis

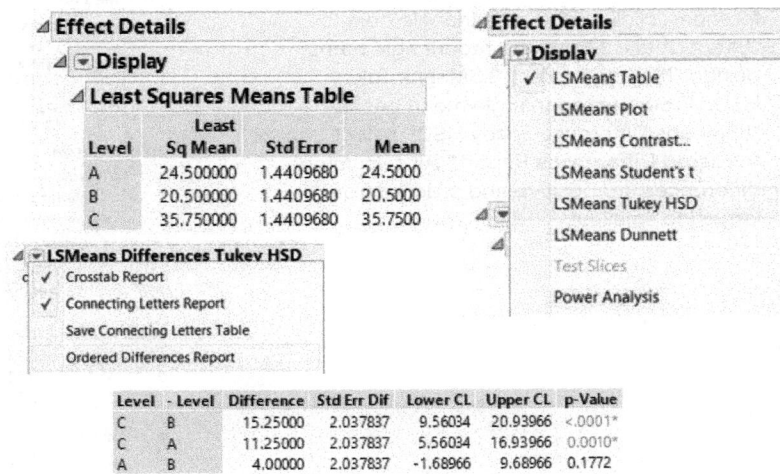

Level	- Level	Difference	Std Err Dif	Lower CL	Upper CL	p-Value
C	B	15.25000	2.037837	9.56034	20.93966	<.0001*
C	A	11.25000	2.037837	5.56034	16.93966	0.0010*
A	B	4.00000	2.037837	-1.68966	9.68966	0.1772

Source: JMP Pro 14

Randomized Block Design in Figure 12.8 (data file: SaleMeth.jmp):

- After opening the SaleMeth.jmp file, select **Analyze : Fit Model**.

- Select the "Sales" column and click "Y" and select the "Method" and "Salesperson" columns and click "Add" in the "Construct Model Effects" box. Select "Minimal Report" as the "Emphasis." Click "Run."

- The F test for Method and Salesperson can be found under "Effect Tests." To calculate the pairwise comparisons, from the red triangle menu next to "Response Sales," select **Estimates : Multiple Comparisons**. Choose "Least Squares Means Estimates," "Method," and "All Pairwise Comparisons—Tukey HSD." Click "OK."

	Method	Salesperson	Sales
1	1	A	32
2	1	B	29
3	1	C	30
4	2	A	32
5	2	B	30
6	2	C	28
7	3	A	28
8	3	B	25

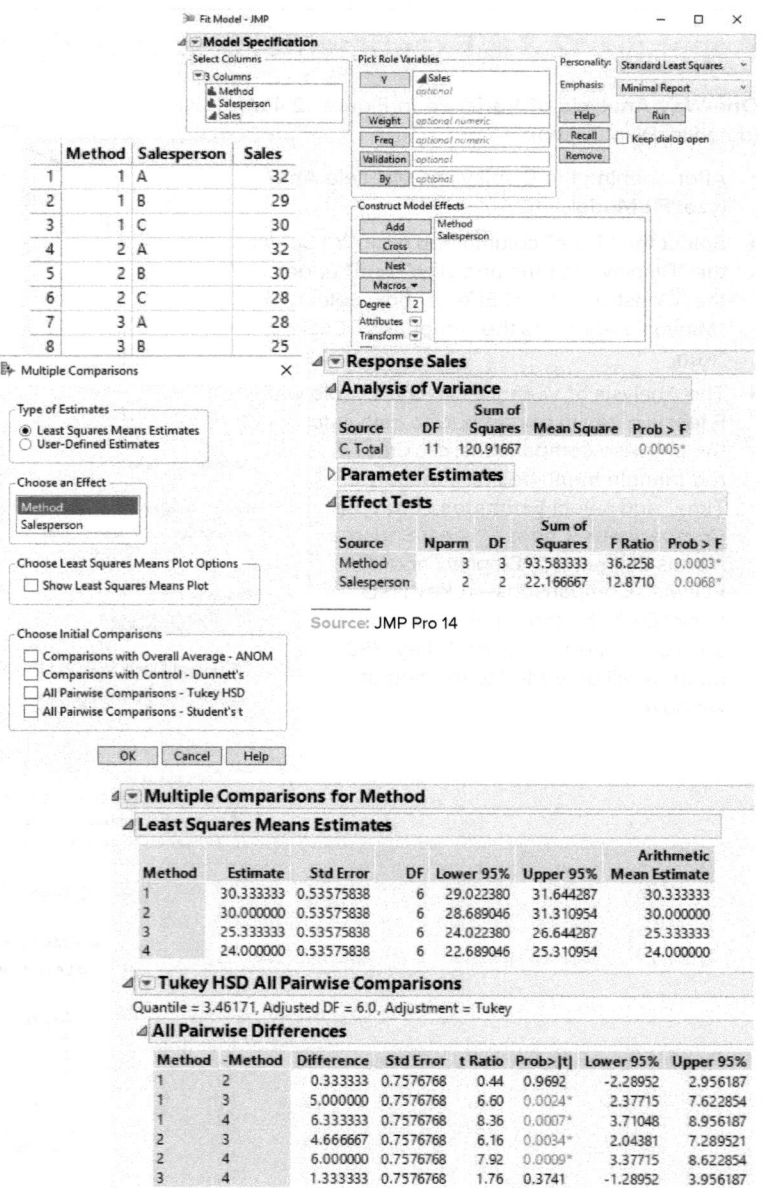

Response Sales

Analysis of Variance

Source	DF	Sum of Squares	Mean Square	Prob > F
C. Total	11	120.91667		0.0005*

Parameter Estimates

Effect Tests

Source	Nparm	DF	Sum of Squares	F Ratio	Prob > F
Method	3	3	93.583333	36.2258	0.0003*
Salesperson	2	2	22.166667	12.8710	0.0068*

Source: JMP Pro 14

Multiple Comparisons for Method

Least Squares Means Estimates

Method	Estimate	Std Error	DF	Lower 95%	Upper 95%	Arithmetic Mean Estimate
1	30.333333	0.53575838	6	29.022380	31.644287	30.333333
2	30.000000	0.53575838	6	28.689046	31.310954	30.000000
3	25.333333	0.53575838	6	24.022380	26.644287	25.333333
4	24.000000	0.53575838	6	22.689046	25.310954	24.000000

Tukey HSD All Pairwise Comparisons

Quantile = 3.46171, Adjusted DF = 6.0, Adjustment = Tukey

All Pairwise Differences

Method	-Method	Difference	Std Error	t Ratio	Prob>|t|	Lower 95%	Upper 95%
1	2	0.333333	0.7576768	0.44	0.9692	-2.28952	2.956187
1	3	5.000000	0.7576768	6.60	0.0024*	2.37715	7.622854
1	4	6.333333	0.7576768	8.36	0.0007*	3.71048	8.956187
2	3	4.666667	0.7576768	6.16	0.0034*	2.04381	7.289521
2	4	6.000000	0.7576768	7.92	0.0009*	3.37715	8.622854
3	4	1.333333	0.7576768	1.76	0.3741	-1.28952	3.956187

Source: JMP Pro 14

- To obtain a display of ordered pairwise differences, click on the gray triangle next to "Effect Details." From the red triangle menu beside "Method," select "LSMeans Tukey HSD." From the red triangle menu beside "LSMeans Difference Tukey HSD," select "Ordered Differences Report." All calculated differences are positive and ordered from largest to smallest.

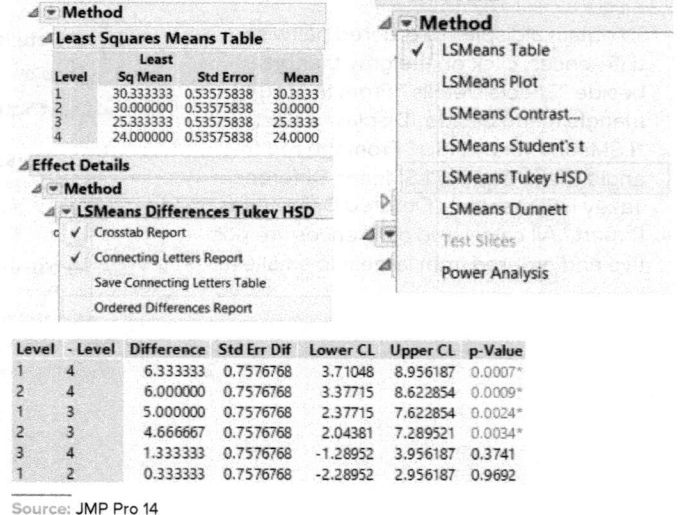

Effect Details

Method

Least Squares Means Table

Level	Least Sq Mean	Std Error	Mean
1	30.333333	0.53575838	30.3333
2	30.000000	0.53575838	30.0000
3	25.333333	0.53575838	25.3333
4	24.000000	0.53575838	24.0000

Effect Details

Method

LSMeans Differences Tukey HSD

- ✓ Crosstab Report
- ✓ Connecting Letters Report
- Save Connecting Letters Table
- Ordered Differences Report

Effect Details

Method

- ✓ LSMeans Table
- LSMeans Plot
- LSMeans Contrast...
- LSMeans Student's t
- LSMeans Tukey HSD
- LSMeans Dunnett
- Test Slices
- Power Analysis

Level	- Level	Difference	Std Err Dif	Lower CL	Upper CL	p-Value
1	4	6.333333	0.7576768	3.71048	8.956187	0.0007*
2	4	6.000000	0.7576768	3.37715	8.622854	0.0009*
1	3	5.000000	0.7576768	2.37715	7.622854	0.0024*
2	3	4.666667	0.7576768	2.04381	7.289521	0.0034*
3	4	1.333333	0.7576768	-1.28952	3.956187	0.3741
1	2	0.333333	0.7576768	-2.28952	2.956187	0.9692

Source: JMP Pro 14

Two-Way ANOVA in Figure 12.12 (data file: Display2.jmp):

- After opening the Display2.jmp file, select **Analyze : Fit Model**.

- Select the "Time" column and click "Y." Select the "Panel" and "Condition" columns and click "Add" in the "Construct Model Effects" box. Select both the "Panel" and "Condition" columns and click "Cross" in the "Construct Model Effects" box to create the interaction term. Select "Minimal Report" as the "Emphasis." Click "Run."

- The F test for "Panel," "Condition," and their interaction "Panel*Condition" can be found under "Effect Tests." To calculate the pairwise comparisons for "Panel," from the red triangle menu beside "Response Time," select **Estimates : Multiple Comparisons**. Choose "Least Squares Means Estimates," "Panel," and "All Pairwise Comparisons—Tukey HSD." Then click "OK."

- Repeat the previous steps to calculate the pairwise comparisons for "Condition."

- To obtain an Interaction Plot, from the red triangle menu beside "Response Time," select **Factor Profiling : Interaction Plots**.

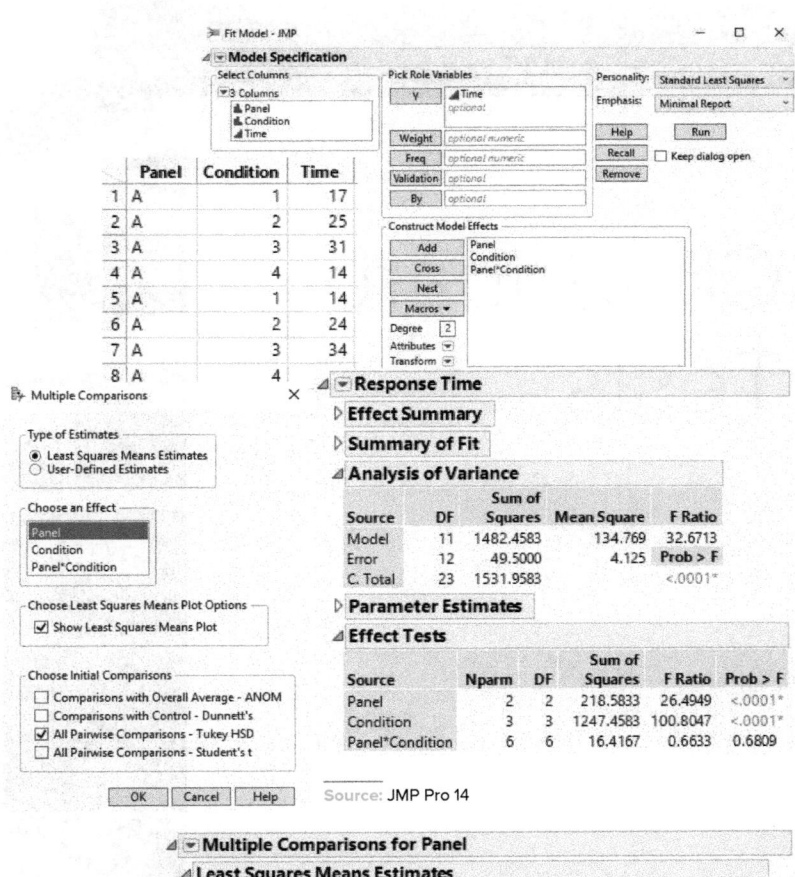

Source: JMP Pro 14

Multiple Comparisons for Panel

Least Squares Means Estimates

Panel	Estimate	Std Error	DF	Lower 95%	Upper 95%	Arithmetic Mean Estimate
A	21.500000	0.71807033	12	19.935459	23.064541	21.500000
B	18.250000	0.71807033	12	16.685459	19.814541	18.250000
C	25.625000	0.71807033	12	24.060459	27.189541	25.625000

Tukey HSD All Pairwise Comparisons

Quantile = 2.66776, Adjusted DF = 12.0, Adjustment = Tukey

All Pairwise Differences

Panel	-Panel	Difference	Std Error	t Ratio	Prob>|t|	Lower 95%	Upper 95%
A	B	3.25000	1.015505	3.20	0.0194*	0.5409	5.95912
A	C	-4.12500	1.015505	-4.06	0.0042*	-6.8341	-1.41588
B	C	-7.37500	1.015505	-7.26	<.0001*	-10.0841	-4.66588

Multiple Comparisons for Condition

Least Squares Means Estimates

Condition	Estimate	Std Error	DF	Lower 95%	Upper 95%	Arithmetic Mean Estimate
1	17.166667	0.82915620	12	15.360091	18.973243	17.166667
2	24.500000	0.82915620	12	22.693424	26.306576	24.500000
3	32.166667	0.82915620	12	30.360091	33.973243	32.166667
4	13.333333	0.82915620	12	11.526757	15.139909	13.333333

Tukey HSD All Pairwise Comparisons

Quantile = 2.9688, Adjusted DF = 12.0, Adjustment = Tukey

All Pairwise Differences

Condition	-Condition	Difference	Std Error	t Ratio	Prob>|t|	Lower 95%	Upper 95%
1	2	-7.3333	1.172604	-6.25	0.0002*	-10.8146	-3.8521
1	3	-15.0000	1.172604	-12.79	<.0001*	-18.4812	-11.5188
1	4	3.8333	1.172604	3.27	0.0297*	0.3521	7.3146
2	3	-7.6667	1.172604	-6.54	0.0001*	-11.1479	-4.1854
2	4	11.1667	1.172604	9.52	<.0001*	7.6854	14.6479
3	4	18.8333	1.172604	16.06	<.0001*	15.3521	22.3146

Source: JMP Pro 14

Interaction Profiles

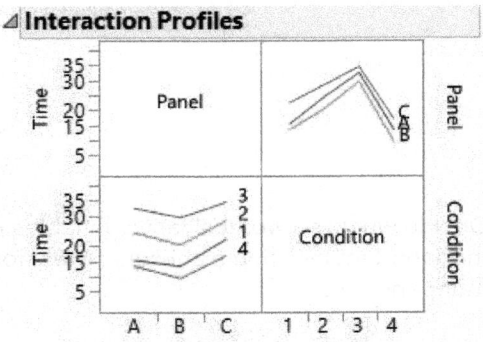

Source: JMP Pro 14

Design Elements: (CD): ©Comstock Images/Alamy; (All Others): ©McGraw-Hill Education

## CHAPTER 13

Chi-Square Tests

©Andrey Rudakov/Bloomberg/Getty Images

Learning Objectives

After mastering the material in this chapter, you will be able to:

LO13-1 Test hypotheses about multinomial probabilities by using a chi-square goodness-of-fit test.

LO13-2 Perform a goodness-of-fit test for normality.

LO13-3 Decide whether two qualitative variables are independent by using a chi-square test for independence.

Chapter Outline

n this chapter we present two useful hypothesis tests based on the **chi-square distribution.** (We have discussed the chi-square distribution in Section 10.6). First, we consider the **chi-square test of goodness-of-fit.** This test evaluates whether data falling into several categories do so with a hypothesized set of probabilities. Second, we discuss the **chi-square test for independence.** Here data are classified on two dimensions and are summarized in a **contingency table.** The test for independence then evaluates whether the cross-classified variables are independent of each other. If we conclude that the variables are not independent, then we have established that the variables in question are related, and we must then investigate the nature of the relationship.

13.1 Chi-Square Goodness-of-Fit Tests

Multinomial probabilities

Sometimes we collect count data in order to study how the counts are distributed among several **categories** or **cells.** As an example, we might study consumer preferences for four different brands of a product. To do this, we select a random sample of consumers, and we ask each survey participant to indicate a brand preference. We then count the number of consumers who prefer each of the four brands. Here we have four categories (brands), and we study the distribution of the counts in each category in order to see which brands are preferred.

LO13-1
Test hypotheses about multinomial probabilities by using a chi-square goodness-of-fit test.

We often use categorical data to carry out a statistical inference. For instance, suppose that a major wholesaler in Cleveland, Ohio, carries four different brands of microwave ovens. Historically, consumer behavior in Cleveland has resulted in the market shares shown in Table 13.1. The wholesaler plans to begin doing business in a new territory—Milwaukee, Wisconsin. To study whether its policies for stocking the four brands of ovens in Cleveland can also be used in Milwaukee, the wholesaler compares consumer preferences for the four ovens in Milwaukee with the historical market shares observed in Cleveland. A random sample of 400 consumers in Milwaukee gives the preferences shown in Table 13.2.

To compare consumer preferences in Cleveland and Milwaukee, we must consider a **multinomial experiment.** This is similar to the binomial experiment. However, a binomial experiment concerns count data that can be classified into two categories, while a multinomial experiment is more general and concerns count data that are classified into two or more categories. Specifically, the assumptions for the multinomial experiment are as follows:

The Multinomial Experiment

1 We perform an experiment in which we carry out n identical trials and in which there are k possible outcomes on each trial.

2 The probabilities of the k outcomes are denoted $p_1, p_2, \ldots, p_k$ where $p_1 + p_2 + \cdots + p_k = 1$. These probabilities stay the same from trial to trial.

3 The trials in the experiment are independent.

4 The results of the experiment are observed frequencies (counts) of the number of trials that result in each of the k possible outcomes. The frequencies are denoted $f_1, f_2, \ldots, f_k$. That is, f_1 is the number of trials resulting in the first possible outcome, f_2 is the number of trials resulting in the second possible outcome, and so forth.

TABLE 13.1 Market Shares for Four Microwave Oven Brands in Cleveland, Ohio DS MicroWav

Brand	Market Share
1	20%
2	35%
3	30%
4	15%

TABLE 13.2 Brand Preferences for Four Microwave Ovens in Milwaukee, Wisconsin DS MicroWav

Brand	Observed Frequency (Number of Consumers Sampled Who Prefer the Brand)
1	102
2	121
3	120
4	57

Notice that the scenario that defines a multinomial experiment is similar to the one that defines a binomial experiment. In fact, a binomial experiment is simply a multinomial experiment where k equals 2 (there are two possible outcomes on each trial).

In general, the probabilities $p_1, p_2, \ldots, p_k$ are unknown, and we estimate their values. Or, we compare estimates of these probabilities with a set of specified values. We now look at such an example.

Ⓒ EXAMPLE 13.1 The Microwave Oven Case: Studying Consumer Preferences

Suppose the microwave oven wholesaler wishes to compare consumer preferences in Milwaukee with the historical market shares in Cleveland. If the consumer preferences in Milwaukee are substantially different, the wholesaler will consider changing its policies for stocking the ovens. Here we will define

p_1 = the proportion of Milwaukee consumers who prefer brand 1

p_2 = the proportion of Milwaukee consumers who prefer brand 2

p_3 = the proportion of Milwaukee consumers who prefer brand 3

p_4 = the proportion of Milwaukee consumers who prefer brand 4

Remembering that the historical market shares for brands 1, 2, 3, and 4 in Cleveland are 20 percent, 35 percent, 30 percent, and 15 percent, we test the null hypothesis

$$H_0: p_1 = .20, \quad p_2 = .35, \quad p_3 = .30, \quad \text{and} \quad p_4 = .15$$

which says that consumer preferences in Milwaukee are consistent with the historical market shares in Cleveland. We test H_0 versus

$$H_a: \text{the previously stated null hypothesis is not true}$$

To test H_0 we must compare the "observed frequencies" given in Table 13.2 with the "expected frequencies" for the brands calculated on the assumption that H_0 is true. For instance, if H_0 is true, we would expect $400(.20) = 80$ of the 400 Milwaukee consumers surveyed to prefer brand 1. Denoting this expected frequency for brand 1 as E_1, the expected frequencies for brands 2, 3, and 4 when H_0 is true are $E_2 = 400(.35) = 140$, $E_3 = 400(.30) = 120$, and $E_4 = 400(.15) = 60$. Recalling that Table 13.2 gives the observed frequency for each brand, we have $f_1 = 102$, $f_2 = 121$, $f_3 = 120$, and $f_4 = 57$. We now compare the observed and expected frequencies by computing a **chi-square statistic** as follows:

$$\chi^2 = \sum_{i=1}^{k=4} \frac{(f_i - E_i)^2}{E_i}$$

$$= \frac{(102 - 80)^2}{80} + \frac{(121 - 140)^2}{140} + \frac{(120 - 120)^2}{120} + \frac{(57 - 60)^2}{60}$$

$$= \frac{484}{80} + \frac{361}{140} + \frac{0}{120} + \frac{9}{60} = 8.7786$$

Clearly, the more the observed frequencies differ from the expected frequencies, the larger χ^2 will be and the more doubt will be cast on the null hypothesis. If the chi-square statistic is large enough (beyond a critical value), then we reject H_0.

To find an appropriate critical value, it can be shown that, when the null hypothesis is true, the sampling distribution of χ^2 is approximately a χ^2 distribution with $k - 1 = 4 - 1 = 3$ degrees of freedom. If we wish to test H_0 at the .05 level of significance, we reject H_0 if and only if

$$\chi^2 > \chi^2_{.05}$$

Chi-square curve with $k - 1 = 3$ degrees of freedom

.05

0     $\chi^2_{.05} = 7.81473$

Reject H_0

FIGURE 13.1   Minitab and JMP outputs of a Goodness-of-Fit Test for the Microwave Oven Case

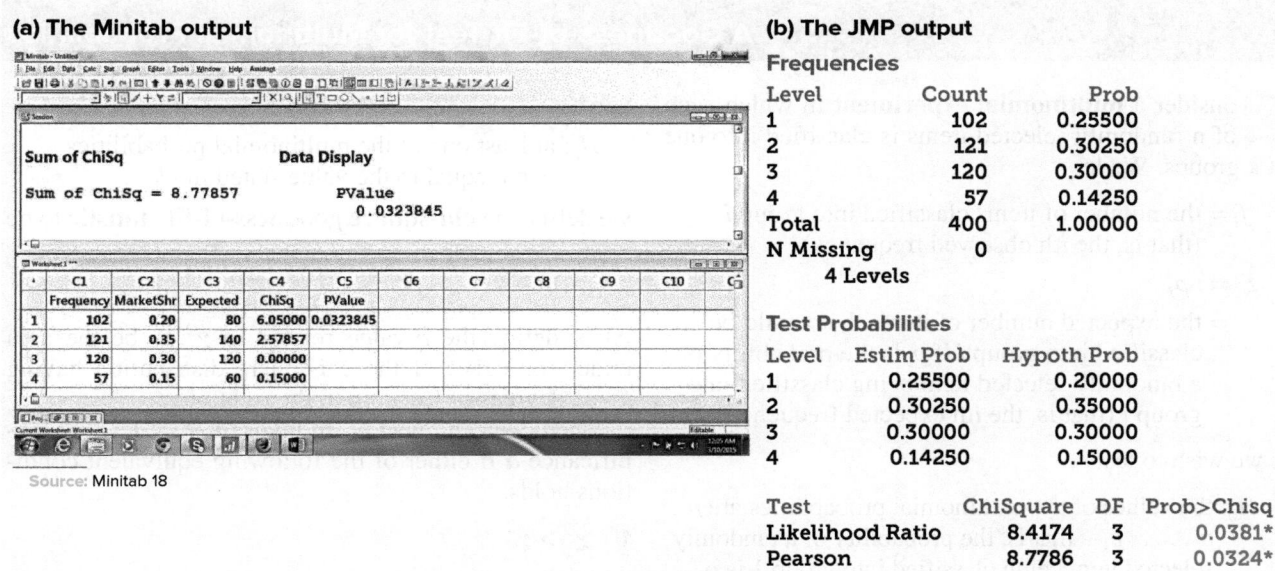

(a) The Minitab output

Sum of ChiSq   Data Display

Sum of ChiSq = 8.77857   PValue
                                    0.0323845

	C1	C2	C3	C4	C5	C6	C7	C8	C9	C10	
	Frequency	MarketShr	Expected	ChiSq	PValue						
1	102	0.20	80	6.05000	0.0323845						
2	121	0.35	140	2.57857							
3	120	0.30	120	0.00000							
4	57	0.15	60	0.15000							

Source: Minitab 18

(b) The JMP output

Frequencies

Level	Count	Prob
1	102	0.25500
2	121	0.30250
3	120	0.30000
4	57	0.14250
Total	400	1.00000
N Missing	0	
4 Levels		

Test Probabilities

Level	Estim Prob	Hypoth Prob
1	0.25500	0.20000
2	0.30250	0.35000
3	0.30000	0.30000
4	0.14250	0.15000

Test	ChiSquare	DF	Prob>Chisq
Likelihood Ratio	8.4174	3	0.0381*
Pearson	8.7786	3	0.0324*

Because Table A.5 tells us that the $\chi^2_{.05}$ point corresponding to $k - 1 = 3$ degrees of freedom equals 7.81473, we find that

$$\chi^2 = 8.7786 > \chi^2_{.05} = 7.81473$$

and we reject H_0 at the .05 level of significance. Alternatively, the p-value for this hypothesis test is the area under the curve of the chi-square distribution having 3 degrees of freedom to the right of $\chi^2 = 8.7786$. This p-value can be calculated to be .0323845. Because this p-value is less than .05, we can reject H_0 at the .05 level of significance. Although there is no single Minitab dialog box that produces a chi-square goodness-of-fit test, Figure 13.1(a) shows the output of a Minitab session that computes the chi-square statistic and its related p-value for the microwave oven case.

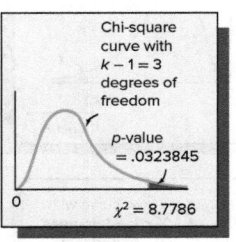

Chi-square curve with $k - 1 = 3$ degrees of freedom

p-value = .0323845

$\chi^2 = 8.7786$

We conclude that the preferences of all consumers in Milwaukee for the four brands of ovens are not consistent with the historical market shares in Cleveland. Based on this conclusion, the wholesaler should consider changing its stocking policies for microwave ovens when it enters the Milwaukee market. To study how to change its policies, the wholesaler might compute a 95 percent confidence interval for, say, the proportion of consumers in Milwaukee who prefer brand 2. Because $\hat{p}_2 = 121/400 = .3025$, this interval is (see Section 9.4)

$$\left[\hat{p}_2 \pm z_{.025} \sqrt{\frac{\hat{p}_2(1 - \hat{p}_2)}{n_2}} \right] = \left[.3025 \pm 1.96 \sqrt{\frac{.3025(1 - .3025)}{400}} \right]$$

$$= [.2575, .3475]$$

Because this entire interval is below .35, it suggests that (1) the market share for brand 2 ovens in Milwaukee will be smaller than the 35 percent market share that this brand commands in Cleveland, and (2) fewer brand 2 ovens (on a percentage basis) should be stocked in Milwaukee. Notice here that by restricting our attention to one particular brand (brand 2), we are essentially combining the other brands into a single group. It follows that we now have two possible outcomes—"brand 2" and "all other brands." Therefore, we have a binomial experiment, and we can employ the methods of Section 9.4, which are based on the binomial distribution.

In the following box we give a general chi-square goodness-of-fit test for multinomial probabilities:

A Goodness-of-Fit Test for Multinomial Probabilities

Consider a **multinomial experiment** in which each of n randomly selected items is classified into one of k groups. We let

f_i = the number of items classified into group i (that is, the ith observed frequency)

$E_i = np_i$

= the expected number of items that would be classified into group i if p_i is the probability of a randomly selected item being classified into group i (that is, the ith expected frequency)

If we wish to test

H_0: the values of the multinomial probabilities are p_1, $p_2, \ldots , p_k$—that is, the probability of a randomly selected item being classified into group 1 is p_1, the probability of a randomly selected item being classified into group 2 is p_2, and so forth

versus

H_a: at least one of the multinomial probabilities is not equal to the value stated in H_0

we define the **chi-square goodness-of-fit statistic** to be

$$\chi^2 = \sum_{i=1}^{k} \frac{(f_i - E_i)^2}{E_i}$$

Also, define the p-value related to χ^2 to be the area under the curve of the chi-square distribution having $k - 1$ degrees of freedom to the right of χ^2.

Then, we can reject H_0 in favor of H_a at level of significance α if either of the following equivalent conditions holds:

1 $\chi^2 > \chi_\alpha^2$

2 p-value < α

Here the χ_α^2 point is based on $k - 1$ degrees of freedom.

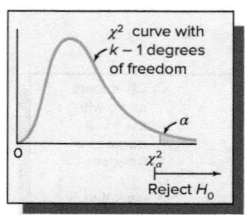

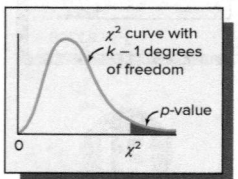

This test is based on the fact that it can be shown that, when H_0 is true, the sampling distribution of χ^2 is approximately a chi-square distribution with $k - 1$ degrees of freedom, if the sample size n is large. **It is generally agreed that n should be considered large if all of the "expected cell frequencies" (E_i values) are at least 5.** Furthermore, recent research implies that this condition on the E_i values can be somewhat relaxed. For example, Moore and McCabe (1993) indicate that **it is reasonable to use the chi-square approximation if the number of groups (k) exceeds 4, the average of the E_i values is at least 5, and the smallest E_i value is at least 1.** Notice that in Example 13.1 all of the E_i values are much larger than 5. Therefore, the chi-square test is valid.

A special version of the chi-square goodness-of-fit test for multinomial probabilities is called a **test for homogeneity.** This involves testing the null hypothesis that all of the multinomial probabilities are equal. For instance, in the microwave oven case we would test

$$H_0: p_1 = p_2 = p_3 = p_4 = .25$$

which would say that no single brand of microwave oven is preferred to any of the other brands (equal preferences). If this null hypothesis is rejected in favor of

$$H_a: \text{At least one of } p_1, p_2, p_3, \text{ and } p_4 \text{ exceeds .25}$$

we would conclude that there is a preference for one or more of the brands. Here each of the expected cell frequencies equals .25(400) = 100. Remembering that the observed cell frequencies are $f_1 = 102, f_2 = 121, f_3 = 120,$ and $f_4 = 57$, the chi-square statistic is

$$
\begin{aligned}
\chi^2 &= \sum_{i=1}^{4} \frac{(f_i - E_i)^2}{E_i} \\
&= \frac{(102 - 100)^2}{100} + \frac{(121 - 100)^2}{100} + \frac{(120 - 100)^2}{100} + \frac{(57 - 100)^2}{100} \\
&= .04 + 4.41 + 4 + 18.49 = 26.94
\end{aligned}
$$

Because $\chi^2 = 26.94$ is greater than $\chi_{.05}^2 = 7.81473$ (see Table A.5 with $k - 1 = 4 - 1 = 3$ degrees of freedom), we reject H_0 at level of significance .05. We conclude that preferences for the four brands are not equal and that at least one brand is preferred to the others.

Normal distributions

We have seen that many statistical methods are based on the assumption that a random sample has been selected from a normally distributed population. We can check the validity of the normality assumption by using frequency distributions, stem-and-leaf displays, histograms, and normal plots. Another approach is to use a chi-square goodness-of-fit test to check the normality assumption. We show how this can be done in the following example.

LO13-2
Perform a goodness-of-fit test for normality.

Ⓒ EXAMPLE 13.2 The Car Mileage Case: Testing Normality

Consider the sample of 50 gas mileages given in Table 1.7. A histogram of these mileages (see Figure 2.9) is symmetrical and bell-shaped. This suggests that the sample of mileages has been randomly selected from a normally distributed population. In this example we use a chi-square goodness-of-fit test to check the normality of the mileages.

To perform this test, we first divide the number line into intervals (or categories). One way to do this is to use the class boundaries of the histogram in Figure 2.9. Table 13.3 gives these intervals and also gives observed frequencies (counts of the number of mileages in each interval), which have been obtained from the histogram of Figure 2.9. The chi-square test is done by comparing these observed frequencies with the expected frequencies in the rightmost column of Table 13.3. To explain how the expected frequencies are calculated, we first use the sample mean $\bar{x} = 31.56$ and the sample standard deviation $s = .798$ of the 50 mileages as point estimates of the population mean μ and population standard deviation σ. Then, for example, consider p_1, the probability that a randomly selected mileage will be in the first interval (less than 30.0) in Table 13.3, if the population of all mileages is normally distributed. We estimate p_1 to be

$$p_1 = P(\text{mileage} < 30.0) = P\left(z < \frac{30.0 - 31.56}{.798}\right)$$

$$= P(z < -1.95) = .0256$$

It follows that $E_1 = 50p_1 = 50(.0256) = 1.28$ is the expected frequency for the first interval under the normality assumption. Next, if we consider p_2, the probability that a randomly selected mileage will be in the second interval in Table 13.3 if the population of all mileages is normally distributed, we estimate p_2 to be

$$p_2 = P(30.0 \leq \text{mileage} < 30.5) = P\left(\frac{30.0 - 31.56}{.798} \leq z < \frac{30.5 - 31.56}{.798}\right)$$

$$= P(-1.95 \leq z < -1.33) = .0918 - .0256 = .0662$$

It follows that $E_2 = 50p_2 = 50(.0662) = 3.31$ is the expected frequency for the second interval under the normality assumption. The other expected frequencies are computed similarly. In general, p_i is the probability that a randomly selected mileage will be in interval i if the

---

TABLE 13.3	Observed and Expected Cell Frequencies for a Chi-Square Goodness-of-Fit Test for Testing the Normality of the Population of Gasoline Mileages　Ⓓ GasMiles		
Interval	Observed Frequency (f_i)	p_i If the Population of Mileages Is Normally Distributed	Expected Frequency, $E_i = np_i = 50p_i$
Less than 30.0	1	$p_1 = P(\text{mileage} < 30.0) = .0256$	$E_1 = 50(.0256) = 1.28$
30.0 < 30.5	3	$p_2 = P(30.0 \leq \text{mileage} < 30.5) = .0662$	$E_2 = 50(.0662) = 3.31$
30.5 < 31.0	8	$p_3 = P(30.5 \leq \text{mileage} < 31.0) = .1502$	$E_3 = 50(.1502) = 7.51$
31.0 < 31.5	11	$p_4 = P(31.0 \leq \text{mileage} < 31.5) = .2261$	$E_4 = 50(.2261) = 11.305$
31.5 < 32.0	11	$p_5 = P(31.5 \leq \text{mileage} < 32.0) = .2407$	$E_5 = 50(.2407) = 12.035$
32.0 < 32.5	9	$p_6 = P(32.0 \leq \text{mileage} < 32.5) = .1722$	$E_6 = 50(.1722) = 8.61$
32.5 < 33.0	5	$p_7 = P(32.5 \leq \text{mileage} < 33.0) = .0831$	$E_7 = 50(.0831) = 4.155$
Greater than 33.0	2	$p_8 = P(\text{mileage} > 33.0) = .0359$	$E_8 = 50(.0359) = 1.795$

population of all possible mileages is normally distributed with mean 31.56 and standard deviation .798, and E_i is the expected number of the 50 mileages that would be in interval i if the population of all possible mileages has this normal distribution.

It seems reasonable to reject the null hypothesis

$$H_0: \text{the population of all mileages is normally distributed}$$

in favor of the alternative hypothesis

$$H_a: \text{the population of all mileages is not normally distributed}$$

if the observed frequencies in Table 13.3 differ substantially from the corresponding expected frequencies in Table 13.3. We compare the observed frequencies with the expected frequencies under the normality assumption by computing the chi-square statistic

$$\chi^2 = \sum_{i=1}^{8} \frac{(f_i - E_i)^2}{E_i}$$

$$= \frac{(1 - 1.28)^2}{1.28} + \frac{(3 - 3.31)^2}{3.31} + \frac{(8 - 7.51)^2}{7.51} + \frac{(11 - 11.305)^2}{11.305}$$

$$+ \frac{(11 - 12.035)^2}{12.035} + \frac{(9 - 8.61)^2}{8.61} + \frac{(5 - 4.155)^2}{4.155} + \frac{(2 - 1.795)^2}{1.795}$$

$$= .43242$$

Because we have estimated $m = 2$ parameters (μ and σ) in computing the expected frequencies (E_i values), it can be shown that the sampling distribution of χ^2 is approximately a chi-square distribution with $k - 1 - m = 8 - 1 - 2 = 5$ degrees of freedom. Therefore, we can reject H_0 at level of significance α if

$$\chi^2 > \chi_\alpha^2$$

where the χ_α^2 point is based on $k - 1 - m = 8 - 1 - 2 = 5$ degrees of freedom. If we wish to test H_0 at the .05 level of significance, Table A.5 tells us that $\chi_{.05}^2 = 11.0705$. Therefore, because

$$\chi^2 = .43242 < \chi_{.05}^2 = 11.0705$$

we cannot reject H_0 at the .05 level of significance, and we cannot reject the hypothesis that the population of all mileages is normally distributed. Therefore, for practical purposes it is probably reasonable to assume that the population of all mileages is approximately normally distributed and that inferences based on this assumption are valid. Finally, the p-value for this test, which is the area under the chi-square curve having 5 degrees of freedom to the right of $\chi^2 = .43242$, can be shown to equal .994. Because this p-value is large (much greater than .05), we have little evidence to support rejecting the null hypothesis (normality).

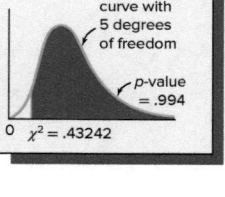

Chi-square curve with 5 degrees of freedom

p-value $= .994$

$0 \quad \chi^2 = .43242$

Note that although some of the expected cell frequencies in Table 13.3 are not at least 5, the number of classes (groups) is 8 (which exceeds 4), the average of the expected cell frequencies is at least 5, and the smallest expected cell frequency is at least 1. Therefore, it is probably reasonable to consider the result of this chi-square test valid. If we choose to base the chi-square test on the more restrictive assumption that all of the expected cell frequencies are at least 5, then we can combine adjacent cell frequencies as follows:

Original f_i Values	Original p_i Values	Original E_i Values	Combined E_i Values	Combined p_i Values	Combined f_i Values
1	.0256	1.28			
3	.0662	3.31	12.1	.2420	12
8	.1502	7.51			
11	.2261	11.305	11.305	.2261	11
11	.2407	12.035	12.035	.2407	11
9	.1722	8.61	8.61	.1722	9
5	.0831	4.155			
2	.0359	1.795	5.95	.1190	7

When we use these combined cell frequencies, the chi-square approximation is based on $k - 1 - m = 5 - 1 - 2 = 2$ degrees of freedom. We find that $\chi^2 = .30102$ and that the p-value $= .860$. Because this p-value is much greater than .05, we cannot reject the hypothesis of normality at the .05 level of significance.

In Example 13.2 we based the intervals employed in the chi-square goodness-of-fit test on the class boundaries of a histogram for the observed mileages. Another way to establish intervals for such a test is to compute the sample mean $\bar{x}$ and the sample standard deviation s and to use intervals based on the Empirical Rule as follows:

Interval 1: less than $\bar{x} - 2s$

Interval 2: $\bar{x} - 2s < \bar{x} - s$

Interval 3: $\bar{x} - s < \bar{x}$

Interval 4: $\bar{x} < \bar{x} + s$

Interval 5: $\bar{x} + s < \bar{x} + 2s$

Interval 6: greater than $\bar{x} + 2s$

However, care must be taken to ensure that each of the expected frequencies is large enough (using the previously discussed criteria).

No matter how the intervals are established, we use $\bar{x}$ as an estimate of the population mean μ and we use s as an estimate of the population standard deviation σ when we calculate the expected frequencies (E_i values). Because we are estimating $m = 2$ population parameters, the critical value χ_α^2 is based on $k - 1 - m = k - 1 - 2 = k - 3$ degrees of freedom, where k is the number of intervals employed.

In the following box we summarize how to carry out this chi-square test:

A Goodness-of-Fit Test for a Normal Distribution

1 We will test the following null and alternative hypotheses:

   H_0: the population has a normal distribution

   H_a: the population does not have a normal distribution

2 Select a random sample of size n and compute the sample mean $\bar{x}$ and sample standard deviation s.

3 Define k intervals for the test. Two reasonable ways to do this are to use the classes of a histogram of the data or to use intervals based on the Empirical Rule.

4 Record the observed frequency (f_i) for each interval.

5 Calculate the expected frequency (E_i) for each interval under the normality assumption. Do this by computing the probability that a normal variable having mean $\bar{x}$ and standard deviation s is within the interval

and by multiplying this probability by n. Make sure that each expected frequency is large enough. If necessary, combine intervals to make the expected frequencies large enough.

6 Calculate the chi-square statistic

$$\chi^2 = \sum_{i=1}^{k} \frac{(f_i - E_i)^2}{E_i}$$

and define the p-value for the test to be the area under the curve of the chi-square distribution having $k - 3$ degrees of freedom to the right of χ^2.

7 Reject H_0 in favor of H_a at level of significance α if either of the following equivalent conditions holds:

   a $\chi^2 > \chi_\alpha^2$ **b** p-value $< \alpha$

Here, the χ_α^2 point is based on $k - 3$ degrees of freedom.

While chi-square goodness-of-fit tests are often used to verify that it is reasonable to assume that a random sample has been selected from a normally distributed population, such tests can also check other distribution forms. For instance, we might verify that it is reasonable to assume that a random sample has been selected from a Poisson distribution. In general, **the number of degrees of freedom for the chi-square goodness-of-fit test will** equal $k - 1 - m$, where k is the number of intervals or categories employed in the test and m is the number of population parameters that must be estimated to calculate the needed expected frequencies.

Exercises for Section 13.1

CONCEPTS ▤ connect

13.1 Describe the characteristics that define a multinomial experiment.

13.2 Give the conditions that the expected cell frequencies must meet in order to validly carry out a chi-square goodness-of-fit test.

13.3 Explain the purpose of a goodness-of-fit test.

13.4 When performing a chi-square goodness-of-fit test, explain why a large value of the chi-square statistic provides evidence that H_0 should be rejected.

13.5 Explain two ways to obtain intervals for a goodness-of-fit test of normality.

METHODS AND APPLICATIONS

13.6 The shares of the U.S. automobile market held in 1990 by General Motors, Japanese manufacturers, Ford, Chrysler, and other manufacturers were, respectively, 36%, 26%, 21%, 9%, and 8%. Suppose that a new survey of 1,000 new-car buyers shows the following purchase frequencies:

GM	Japanese	Ford	Chrysler	Other
193	384	170	90	163

a Show that it is appropriate to carry out a chi-square test using these data. ⑤ AutoMkt

b Test to determine whether the current market shares differ from those of 1990. Use $\alpha = .05$.

13.7 Last rating period, the percentages of viewers watching several channels between 11 P.M. and 11:30 P.M. in a major TV market were as follows: ⑤ TVRate

WDUX (News)	WWTY (News)	WACO (Cheers Reruns)	WTJW (News)	Others
15%	19%	22%	16%	28%

Suppose that in the current rating period, a survey of 2,000 viewers gives the following frequencies:

WDUX (News)	WWTY (News)	WACO (Cheers Reruns)	WTJW (News)	Others
182	536	354	151	777

a Show that it is appropriate to carry out a chi-square test using these data.

b Test to determine whether the viewing shares in the current rating period differ from those in the last rating period at the .10 level of significance. What do you conclude?

13.8 In the *Journal of Marketing Research* (November 1996), Gupta studied the extent to which the purchase behavior of **scanner panels** is representative of overall brand preferences. A scanner panel is a sample of households whose purchase data are recorded when a magnetic identification card is presented at a store checkout. The table below gives peanut butter purchase data collected by the A. C. Nielson Company using a panel of 2,500 households in Sioux Falls, South Dakota. The data were collected over 102 weeks. The table also gives the market shares obtained by recording all peanut butter purchases at the same stores during the same period. ⑤ ScanPan

a Show that it is appropriate to carry out a chi-square test.

b Test to determine whether the purchase behavior of the panel of 2,500 households is consistent with the purchase behavior of the population of all peanut butter purchasers. Use the Excel output of an appropriate chi-square test in the table below. Assume here that purchase decisions by panel members are reasonably independent, and set $\alpha = .05$.

Table and Excel Output for Exercise 13.8

Brand	Size	Number of Purchases by Household Panel	Market Shares	Goodness-of-Fit Test O (obs)	E (expected)	O − E	(O − E)²/E	% of chisq
Jif	18 oz.	3,165	20.10%					
Jif	28	1,892	10.10	3165	3842.115	−677.115	119.331	13.56
Jif	40	726	5.42	1892	1930.615	−38.615	0.772	0.09
Peter Pan	10	4,079	16.01	726	1036.033	−310.033	92.777	10.54
Skippy	18	6,206	28.56	4079	3060.312	1018.689	339.092	38.52
Skippy	28	1,627	12.33	6206	5459.244	746.756	102.147	11.60
Skippy	40	1,420	7.48	1627	2356.880	−729.880	226.029	25.68
Total		19,115		1420	1429.802	−9.802	0.067	0.01
				19115	19115.000	0.000	880.216	100.00

880.22 chisquare   6 df   0.0000   p-value

Source: S. Gupta et al., *The Journal of Marketing Research*, published by the American Marketing Association, Vol. 33. "Do Household Scanner Data Provide Representative Inferences from Brand Choices? A Comparison with Store Data," p. 393 (Table 6).

13.9 The purchase frequencies for six different brands of a digital camera are observed at a discount store over one month: ⊙Ⓢ DigCam

Brand	1	2	3	4	5	6
Purchase Frequency	131	273	119	301	176	200

a Carry out a test of homogeneity for these data with $\alpha = .025$.

b Interpret the result of your test.

13.10 A wholesaler has recently developed a computerized sales invoicing system. Prior to implementing this system, a manual system was used. Historically, the manual system produced 87% of invoices with 0 errors, 8% of invoices with 1 error, 3% of invoices with 2 errors, 1% of invoices with 3 errors, and 1% of invoices with more than 3 errors. After implementation of the computerized system, a random sample of 500 invoices showed 479 invoices with 0 errors, 10 invoices with 1 error, 8 invoices with 2 errors, 2 invoices with 3 errors, and 1 invoice with more than 3 errors. ⊙Ⓢ Invoice2

Ei	fi	$(f - E)^2/E$
435	479	4.4506
40	10	22.5000
15	8	3.2667
5	2	1.8000
5	1	3.2000

Chi-Square
35.21724

p-value
0.0000001096

a Show that it is appropriate to carry out a chi-square test using these data.

b Show how the expected frequencies (the E_is) on the partial Excel output as given above have been calculated.

c Use the partial Excel output to determine whether the error percentages for the computerized system differ from those for the manual system at the .05 level of significance. What do you conclude?

13.11 THE e-Billing CASE

Consider the sample of 65 payment times given in Table 2.4. Use these data to carry out a chi-square goodness-of-fit test to test whether the population of all payment times is normally distributed by doing the following: ⊙Ⓢ PayTime

a It can be shown that $\bar{x} = 18.1077$ and that $s = 3.9612$ for the payment time data. Use these values to compute the intervals (*continues in right column*)

(1) Less than $\bar{x} - 2s$	(4) $\bar{x} < \bar{x} + s$
(2) $\bar{x} - 2s < \bar{x} - s$	(5) $\bar{x} + s < \bar{x} + 2s$
(3) $\bar{x} - s < \bar{x}$	(6) Greater than $\bar{x} + 2s$

b Assuming that the population of all payment times is normally distributed, find the probability that a randomly selected payment time will be contained in each of the intervals found in part *a*. Use these probabilities to compute the expected frequency under the normality assumption for each interval.

c Verify that the average of the expected frequencies is at least 5 and that the smallest expected frequency is at least 1. What does this tell us?

d Formulate the null and alternative hypotheses for the chi-square test of normality.

e For each interval given in part *a*, find the observed frequency. Then calculate the chi-square statistic needed for the chi-square test of normality.

f Use the chi-square statistic to test normality at the .05 level of significance. What do you conclude?

13.12 THE MARKETING RESEARCH CASE

Consider the sample of 60 bottle design ratings given in Table 1.6. Use these data to carry out a chi-square goodness-of-fit test to determine whether the population of all bottle design ratings is normally distributed. Use $\alpha = .05$, and note that $\bar{x} = 30.35$ and $s = 3.1073$ for the 60 bottle design ratings. ⊙Ⓢ Design

13.13 THE BANK CUSTOMER WAITING TIME CASE

Consider the sample of 100 waiting times given in Table 1.9. Use these data to carry out a chi-square goodness-of-fit test to determine whether the population of all waiting times is normally distributed. Use $\alpha = .10$, and note that $\bar{x} = 5.46$ and $s = 2.475$ for the 100 waiting times. ⊙Ⓢ WaitTime

13.14 The table below gives a frequency distribution describing the number of errors found in thirty 1,000-line samples of computer code. Suppose that we wish to determine whether the number of errors can be described by a Poisson distribution with mean $\mu = 4.5$. Using the Poisson probability tables, fill in the table. Then perform an appropriate chi-square goodness-of-fit test at the .05 level of significance. What do you conclude about whether the number of errors can be described by a Poisson distribution with $\mu = 4.5$? Explain. ⊙Ⓢ CodeErr

Number of Errors	Observed Frequency	Probability Assuming Errors Are Poisson Distributed with $\mu = 4.5$	Expected Frequency
0–1	6	_____	_____
2–3	5	_____	_____
4–5	7	_____	_____
6–7	8	_____	_____
8 or more	4	_____	_____

LO13-3

Decide whether two
qualitative variables are
independent by using
a chi-square test for
independence.

13.2 A Chi-Square Test for Independence

We have spent considerable time in previous chapters studying relationships between variables. One way to study the relationship between two variables is to classify multinomial count data on two scales (or dimensions) by setting up a *contingency table*.

 EXAMPLE 13.3 The Brokerage Firm Case: Studying Client Satisfaction

A brokerage firm sells several kinds of investment products—a stock fund, a bond fund, and a tax-deferred annuity. The company is examining whether customer satisfaction depends on the type of investment product purchased. To do this, 100 clients are randomly selected from the population of clients who have purchased shares in exactly one of the funds. The company records the fund type purchased by these clients and asks each sampled client to rate his or her level of satisfaction with the fund as high, medium, or low. Figure 13.2(a) gives the survey results for the first three randomly selected clients, and all of the survey results are shown in Table 2.16. To begin to analyze these data, it is helpful to construct a **contingency table.** Such a table classifies the data on two dimensions—type of fund and degree of client satisfaction. Figures 13.2(b), (c) and (d) give Excel, Minitab, and JMP outputs of a contingency table of fund type versus level of satisfaction. This table consists of a row for each fund type and a column for each level of satisfaction. Together, the rows and columns form a "cell" for each fund type–satisfaction level combination. That is, there is a cell for each "contingency" with respect to fund type and satisfaction level. The Excel, Minitab, and JMP outputs give a **cell frequency** for each cell. On the Minitab and JMP outputs, this is the top number given in the cell. The cell frequency is a count (observed frequency) of the number of surveyed clients with the cell's fund type–satisfaction level combination. For instance, 15 of the surveyed clients invest in the bond fund and report high satisfaction, while 24 of the surveyed clients invest in the tax-deferred annuity and report medium satisfaction. In addition to the cell frequencies, each output also gives

Row totals (at the far right of each table): These are counts of the numbers of clients who invest in each fund type. These row totals tell us that 30 clients invest in the bond fund, 30 clients invest in the stock fund, and 40 clients invest in the tax-deferred annuity.

Column totals (at the bottom of each table): These are counts of the numbers of clients who report high, medium, and low satisfaction. These column totals tell us that 40 clients report high satisfaction, 40 clients report medium satisfaction, and 20 clients report low satisfaction.

Overall total (the bottom-right entry in each table): This tells us that a total of 100 clients were surveyed.

Besides the row and column totals, the Minitab and JMP outputs give **row and column percentages** (below the row and column totals). For example, 30.00 percent of the surveyed clients invest in the bond fund, and 20.00 percent of the surveyed clients report low satisfaction. Furthermore, in addition to a cell frequency, the Minitab output gives a **row percentage** and a **column percentage** for each cell (these are below the cell frequency in each cell). For instance, looking at the "bond fund–high satisfaction cell," we see that the 15 clients in this cell make up 50.0 percent of the 30 clients who invest in the bond fund, and they make up 37.5 percent of the 40 clients who report high satisfaction. The JMP output gives the same information but first gives the column percentage and then gives the row percentage. We will explain the last number that appears in each cell of the Minitab and JMP outputs later in this section.

Looking at the contingency tables, it appears that the level of client satisfaction may be related to the fund type. We see that higher satisfaction ratings seem to be reported by stock and bond fund investors, while holders of tax-deferred annuities report lower satisfaction ratings. To carry out a formal statistical test we can test the null hypothesis H_0: fund type and level of client satisfaction are independent versus the alternative hypothesis H_a: fund type and level of client satisfaction are dependent. To perform this test, we compare the counts (or **observed cell frequencies**) in the contingency table with the counts we would expect if we assume that fund type and level of satisfaction are independent. Because these latter counts

FIGURE 13.2   Partial Survey Results, and the Excel, Minitab, and JMP Outputs of a Contingency Table of Fund Type versus Level of Client Satisfaction   ⓓⓢ Invest

(a) Survey results for the first three randomly selected clients

Client	Fund Type	Level of Satisfaction
1	BOND	HIGH
2	STOCK	HIGH
3	TAXDEF	MED

(b) The Excel output

a = Chi-square statistic.
b = p-value for chi-square.

Source: Microsoft Office Excel 2016

(c) The Minitab output

Rows: FundType   Columns: SatRating

	High	Med	Low	All
Bond	15	12	3	30
	50.00	40.00	10.00	100.00
	37.50	30.00	15.00	30.00
	12	12	6	30
Stock	24	4	2	30
	80.00	13.33	6.67	100.00
	60.00	10.00	10.00	30.00
	12	12	6	30
TaxDef	1	24	15	40
	2.50	60.00	37.50	100.00
	2.50	60.00	75.00	40.00
	16	16	8	40
All	40	40	20	100
	40.00	40.00	20.00	100.00
	100.00	100.00	100.00	100.00
	40	40	20	100

Pearson Chi-Square = 46.438, DF = 4
P-Value = 0.000

Cell Contents: Count
% of Row
% of Column
Expected count

(d) The JMP output

Contingency Table

Count Total % Col % Row % Expected	High	SatRating Med	Low	Total
Bond	15	12	3	30
	15.00	12.00	3.00	30.00
	37.50	30.00	15.00	
	50.00	40.00	10.00	
	12	12	6	
Stock	24	4	2	30
	24.00	4.00	2.00	30.00
	60.00	10.00	10.00	
	80.00	13.33	6.67	
	12	12	6	
TaxDef	1	24	15	40
	1.00	24.00	15.00	40.00
	2.50	60.00	75.00	
	2.50	60.00	37.50	
	16	16	8	
Total	40	40	20	100
	40.00	40.00	20.00	

Tests

N	DF	-LogLike	RSquare (U)
100	4	27.699274	0.2626

Test	ChiSquare	Prob>Chisq
Likelihood Ratio	55.399	<.0001*
Pearson	46.438	<.0001*

are computed by assuming independence, we call them the **expected cell frequencies under the independence assumption.** We illustrate how to calculate these expected cell frequencies by considering the cell corresponding to the bond fund and high client satisfaction. We first use the data in the contingency table to compute an estimate of the probability that a randomly selected client invests in the bond fund. Denoting this probability as p_B, we estimate p_B by dividing the row total for the bond fund by the total number of clients surveyed. That is, denoting the row total for the bond fund as r_B and letting n denote the total number of clients surveyed, the estimate of p_B is $r_B/n = 30/100 = .3$. Next we compute an estimate of the probability that a randomly selected client will report high satisfaction. Denoting this probability as p_H, we estimate p_H by dividing the column total for high satisfaction by the total number of clients surveyed. That is, denoting the column total for high satisfaction as c_H, the estimate of p_H is $c_H/n = 40/100 = .4$. Next, assuming that investing in the bond fund and reporting high satisfaction are **independent,** we compute an estimate of the probability that a randomly selected client invests in the bond fund and reports high satisfaction. Denoting this probability as p_{BH}, we can compute its estimate by recalling from Section 4.4 that if two events A and B are statistically independent, then $P(A \cap B)$ equals $P(A)P(B)$. It follows that, if we assume that investing in the bond fund and reporting high satisfaction are independent, we can compute an estimate of p_{BH} by multiplying the estimate of p_B by the estimate of p_H. That is, the estimate of p_{BH} is $(r_B/n)(c_H/n) = (.3)(.4) = .12$. Finally, we compute an estimate of the expected cell frequency under the independence assumption. Denoting the expected cell frequency as E_{BH}, the estimate of E_{BH} is

$$\hat{E}_{BH} = n\left(\frac{r_B}{n}\right)\left(\frac{c_H}{n}\right) = 100(.3)(.4) = 12$$

This estimated expected cell frequency is given in the Minitab and JMP outputs of Figures 13.2(c) and (d) as the last number under the observed cell frequency for the bond fund–high satisfaction cell.

Noting that the expression for $\hat{E}_{BH}$ can be written as

$$\hat{E}_{BH} = n\left(\frac{r_B}{n}\right)\left(\frac{c_H}{n}\right) = \frac{r_B c_H}{n}$$

we can generalize to obtain a formula for the estimated expected cell frequency for any cell in the contingency table. Letting $\hat{E}_{ij}$ denote the estimated expected cell frequency corresponding to row i and column j in the contingency table, we see that

$$\hat{E}_{ij} = \frac{r_i c_j}{n}$$

where r_i is the row total for row i and c_j is the column total for column j. For example, for the fund type–satisfaction level contingency table, we obtain

$$\hat{E}_{SL} = \frac{r_S c_L}{n} = \frac{30(20)}{100} = \frac{600}{100} = 6$$

and

$$\hat{E}_{TM} = \frac{r_T c_M}{n} = \frac{40(40)}{100} = \frac{1,600}{100} = 16$$

These (and the other estimated expected cell frequencies under the independence assumption) are the last numbers below the observed cell frequencies in the Minitab and JMP outputs of Figure 13.2(c) and (d). Intuitively, these estimated expected cell frequencies tell us what the contingency table would look like if fund type and level of client satisfaction were independent. A table of estimated expected cell frequencies is also given below the contingency table on the Excel output of Figure 13.2(b).

To test the null hypothesis of independence, we will compute a chi-square statistic that compares the observed cell frequencies with the estimated expected cell frequencies calculated assuming independence. Letting f_{ij} denote the observed cell frequency for cell ij, we compute

$$\chi^2 = \sum_{\text{all cells}} \frac{(f_{ij} - \hat{E}_{ij})^2}{\hat{E}_{ij}}$$

$$= \frac{(f_{BH} - \hat{E}_{BH})^2}{\hat{E}_{BH}} + \frac{(f_{BM} - \hat{E}_{BM})^2}{\hat{E}_{BM}} + \cdots + \frac{(f_{TL} - \hat{E}_{TL})^2}{\hat{E}_{TL}}$$

$$= \frac{(15 - 12)^2}{12} + \frac{(12 - 12)^2}{12} + \frac{(3 - 6)^2}{6} + \frac{(24 - 12)^2}{12} + \frac{(4 - 12)^2}{12}$$

$$+ \frac{(2 - 6)^2}{6} + \frac{(1 - 16)^2}{16} + \frac{(24 - 16)^2}{16} + \frac{(15 - 8)^2}{8}$$

$$= 46.4375$$

If the value of the chi-square statistic is large, this indicates that the observed cell frequencies differ substantially from the expected cell frequencies calculated by assuming independence. Therefore, the larger the value of chi-square, the more doubt is cast on the null hypothesis of independence.

To find an appropriate critical value, we let r denote the number of rows in the contingency table and we let c denote the number of columns. Then, it can be shown that, when the null hypothesis of independence is true, the sampling distribution of χ^2 is approximately a χ^2 distribution with $(r - 1)(c - 1) = (3 - 1)(3 - 1) = 4$ degrees of freedom. If we test H_0 at the .05 level of significance, we reject H_0 if and only if

$$\chi^2 > \chi^2_{.05}$$

Because Table A.5 tells us that the $\chi^2_{.05}$ point corresponding to $(r - 1)(c - 1) = 4$ degrees of freedom equals 9.48773, we have

$$\chi^2 = 46.4375 > \chi^2_{.05} = 9.48773$$

and we reject H_0 at the .05 level of significance. We conclude that fund type and level of client satisfaction are not independent.

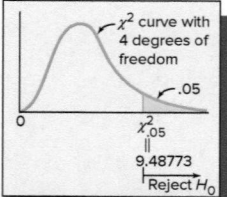

χ^2 curve with 4 degrees of freedom

.05

$\chi^2_{.05}$

9.48773

Reject H_0

BI

In the following box we summarize how to carry out a chi-square test for independence:

A Chi-Square Test for Independence

Suppose that each of n randomly selected elements is classified on two dimensions, and suppose that the result of the two-way classification is a **contingency table having r rows and c columns.** Let

f_{ij} = the cell frequency corresponding to row i and column j of the contingency table (that is, the number of elements classified in row i and column j)

r_i = the row total for row i in the contingency table

c_j = the column total for column j in the contingency table

$$\hat{E}_{ij} = \frac{r_i c_j}{n}$$

= the estimated expected number of elements that would be classified in row i and column j of the contingency table if the two classifications are statistically independent

If we wish to test

H_0: the two classifications are statistically independent

versus

H_a: the two classifications are statistically dependent

we define the test statistic

$$\chi^2 = \sum_{\text{all cells}} \frac{(f_{ij} - \hat{E}_{ij})_2}{\hat{E}_{ij}}$$

Also, define the p-value related to χ^2 to be the area under the curve of the chi-square distribution having $(r-1)(c-1)$ degrees of freedom to the right of χ^2. Then, we can reject H_0 in favor of H_a at level of significance α if either of the following equivalent conditions holds:

1 $\chi^2 > \chi^2_\alpha$

2 p-value $< \alpha$

Here the χ^2_α point is based on $(r-1)(c-1)$ degrees of freedom.

This test is based on the fact that it can be shown that, when the null hypothesis of independence is true, the sampling distribution of χ^2 is approximately a chi-square distribution with $(r-1)(c-1)$ degrees of freedom, if the sample size n is large. **It is generally agreed that n should be considered large if all of the estimated expected cell frequencies ($\hat{E}_{ij}$ values) are at least 5.** Moore and McCabe (1993) indicate that **it is reasonable to use the chi-square approximation if the number of cells (rc) exceeds 4, the average of the $\hat{E}_{ij}$ values is at least 5, and the smallest $\hat{E}_{ij}$ value is at least 1.** Notice that in Figure 13.2 all of the estimated expected cell frequencies are greater than 5.

C EXAMPLE 13.4 The Brokerage Firm Case: Studying Client Satisfaction

Again consider the Excel and Minitab outputs of Figure 13.2, which give the contingency table of fund type versus level of client satisfaction. Both outputs give the chi-square statistic ($= 46.438$) for testing the null hypothesis of independence, as well as the related p-value. We see that this p-value is less than .001. It follows, therefore, that we can reject

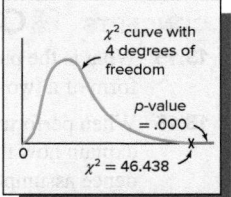

H_0: fund type and level of client satisfaction are independent

at the .05 level of significance, because the p-value is less than .05.

In order to study the nature of the dependency between the classifications in a contingency table, it is often useful to plot the row and/or column percentages. As an example, Figure 13.3 on the next page gives plots of the row percentages in the contingency table of Figure 13.2(c). For instance, looking at the column in this contingency table corresponding to a high level of satisfaction, the contingency table tells us that 40.00 percent of the surveyed clients report a high level of satisfaction. If fund type and level of satisfaction really are independent, then we would expect roughly 40 percent of the clients in each of the three categories—bond fund participants, stock fund participants, and tax-deferred annuity holders—to report a high level of satisfaction. That is, we would expect the row percentages in the "high satisfaction" column to be roughly 40 percent in each row. However, Figure 13.3(a) gives a plot of the percentages of clients reporting a high level of satisfaction for each investment type (that is, the figure plots the three row percentages in the column corresponding to "high satisfaction"). We see that these percentages vary considerably. Noting that the dashed line in the figure is the 40 percent reporting a high level of satisfaction for the overall group, we see that the percentage

FIGURE 13.3   Plots of Row Percentages versus Investment Type for the Contingency Tables in Figure 13.2

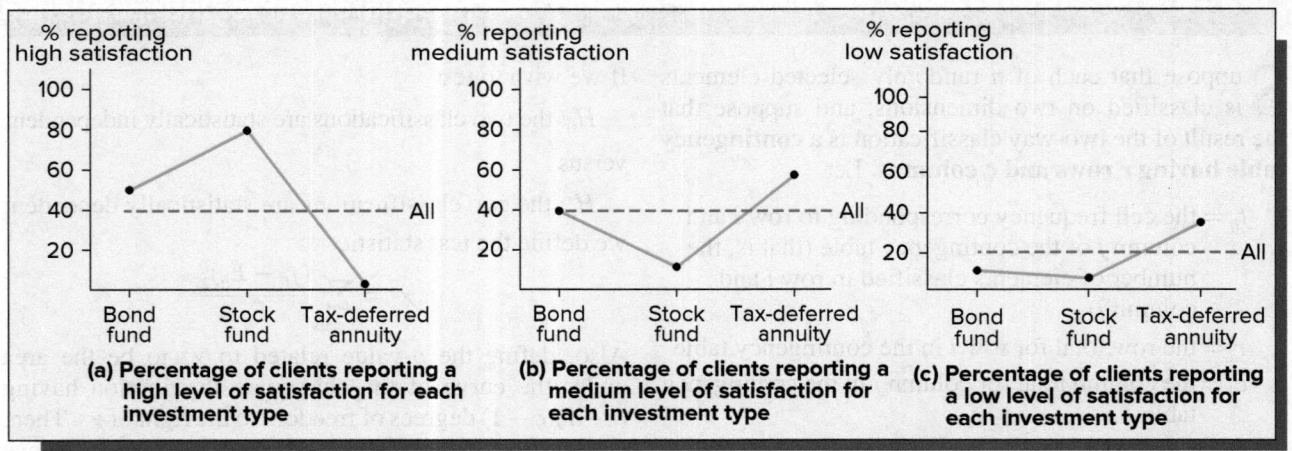

(a) Percentage of clients reporting a high level of satisfaction for each investment type

(b) Percentage of clients reporting a medium level of satisfaction for each investment type

(c) Percentage of clients reporting a low level of satisfaction for each investment type

of stock fund participants reporting high satisfaction is 80 percent. This is far above the 40 percent we would expect if independence exists. On the other hand, the percentage of tax-deferred annuity holders reporting high satisfaction is only 2.5 percent—way below the expected 40 percent if independence exists. In a similar fashion, Figures 13.3(b) and (c) plot the row percentages for the medium and low satisfaction columns in the contingency table. These plots indicate that stock fund participants report medium and low levels of satisfaction less frequently than the overall group of clients, and that tax-deferred annuity participants report medium and low levels of satisfaction more frequently than the overall group of clients.

To conclude this section, we note that the chi-square test for independence can be used to test the equality of several population proportions. We will show how this is done in Exercise 13.21.

Exercises for Section 13.2

CONCEPTS   ■ connect

13.15 What is the purpose behind summarizing data in the form of a two-way contingency table?

13.16 When performing a chi-square test for independence, explain how the "cell frequencies under the independence assumption" are calculated. For what purpose are these frequencies calculated?

METHODS AND APPLICATIONS

13.17 A marketing research firm wishes to study the relationship between wine consumption and whether a person likes to watch professional tennis on television. One hundred randomly selected people are asked whether they drink wine and whether they watch tennis. The following results are obtained:

ⒹⓈ WineCons

	Watch Tennis	Do Not Watch Tennis	Totals
Drink Wine	16	24	40
Do Not Drink Wine	4	56	60
Totals	20	80	100

a For each row and column total, calculate the corresponding row or column percentage.

b For each cell, calculate the corresponding cell, row, and column percentages.

c Test the hypothesis that whether people drink wine is independent of whether people watch tennis. Set $\alpha = .05$.

d Given the results of the chi-square test, does it make sense to advertise wine during a televised tennis match (assuming that the ratings for the tennis match are high enough)? Explain.

13.18 In recent years major efforts have been made to standardize accounting practices in different countries; this is called *harmonization*. In an article in *Accounting and Business Research,* Emmanuel N. Emenyonu and Sidney J. Gray studied the extent to which accounting practices in France, Germany, and the U.K. are harmonized.   ⒹⓈ DeprMeth

a Depreciation method is one of the accounting practices studied by Emenyonu and Gray. Three methods were considered—the straight-line method (S), the declining balance method

TABLE 13.4   Depreciation Methods Used by a Sample of 78 Firms (for Exercise 13.18)   ⓓⓢ DeprMeth

Depreciation Methods	France	Germany	UK	Total
A. Straight line (S)	15	0	25	40
B. Declining Bal (D)	1	1	1	3
C. (D & S)	10	25	0	35
Total companies	26	26	26	78

Source: Data from E. N. Emenyonu and S. J. Gray, "EC Accounting Harmonisation: An Empirical Study of Measurement Practices in France, Germany, and the UK," *Accounting and Business Research* 23, no. 89 (1992), pp. 49–58.

Chi-Square Test for Independence

	France	Germany	UK	Total
A. Straight line (S)	15	0	25	40
B. Declining Bal (D)	1	1	1	3
C. (D & S)	10	25	0	35
Total	26	26	26	78

50.89 chisquare   4 df   0.0000   p-value

TABLE 13.5   A Contingency Table of the Results of the Accidents Study (for Exercise 13.19, parts *a* and *b*)   ⓓⓢ Accident

	On-the-Job Accident		
Smoker	Yes	No	Row Total
Heavy	12	4	16
Moderate	9	6	15
Nonsmoker	13	22	35
Column total	34	32	66

Source: Data from D. R. Cooper and C. W. Emory, *Business Research Methods* (5th ed.) (Burr Ridge, IL: Richard D. Irwin, 1995), p. 451.

FIGURE 13.4   Minitab Output of a Chi-Square Test for Independence in the Accident Study

Expected counts are below observed counts

	Accident	No Accident	Total
Heavy	12 8.24	4 7.76	16
Moderate	9 7.73	6 7.27	15
Nonsmoker	13 18.03	22 16.97	35
Total	34	32	66

Chi-Sq = 6.860, DF = 2, P-Value = 0.032

(*D*), and a combination of *D & S* (sometimes European firms start with the declining balance method and then switch over to the straight-line method when the figure derived from straight line exceeds that from declining balance). The data in Table 13.4 summarize the depreciation methods used by a sample of 78 French, German, and U.K. firms. Use these data and the results of the chi-square analysis in Table 13.4 to test the hypothesis that depreciation method is independent of a firm's location (country) at the .05 level of significance.

b Perform a graphical analysis to study the relationship between depreciation method and country. What conclusions can be made about the nature of the relationship?

13.19 In the book *Business Research Methods* (5th ed.), Donald R. Cooper and C. William Emory discuss studying the relationship between on-the-job accidents and smoking. Cooper and Emory describe the study as follows: ⓓⓢ Accident

Suppose a manager implementing a smoke-free workplace policy is interested in whether smoking affects worker accidents. Since the company has complete reports of on-the-job accidents, she draws a sample of names of workers who were involved in accidents during the last year. A similar sample from among workers who had no reported accidents in the last year is drawn. She interviews members of both groups to determine if they are smokers or not.

The sample results are given in Table 13.5.
a For each row and column total in Table 13.5, find the corresponding row/column percentage.
b For each cell in Table 13.5, find the corresponding cell, row, and column percentages.
c Use the Minitab output in Figure 13.4 to test the hypothesis that the incidence of on-the-job accidents is independent of smoking habits. Set $\alpha = .01$.
d Is there a difference in on-the-job accident occurrences between smokers and nonsmokers? Explain.

13.20 In the book *Essentials of Marketing Research,* William R. Dillon, Thomas J. Madden, and Neil A. Firtle discuss the relationship between delivery time and computer-assisted ordering. A sample of 40 firms shows that 16 use computer-assisted ordering, while 24 do not. Furthermore, past data are used to categorize each firm's delivery times as below the industry average, equal to the industry average, or above the industry average. The results obtained are given in Table 13.6.   ⓓⓢ DelTime

TABLE 13.6   **A Contingency Table Relating Delivery Time and Computer-Assisted Ordering (for Exercise 13.20)**   DS DelTime

Computer-Assisted Ordering	Below Industry Average	Equal to Industry Average	Above Industry Average	Row Total
	Delivery Time			
No	4	12	8	24
Yes	10	4	2	16
Column total	14	16	10	40

TABLE 13.7   **A Summary of the Results of a TV Viewership Study (for Exercise 13.21)**   DS TVView

Watch 11 P.M. News?	18 or Less	19 to 35	36 to 54	55 or Older	Total
		Age Group			
Yes	37	48	56	73	214
No	213	202	194	177	786
Total	250	250	250	250	1,000

a   Test the hypothesis that delivery time performance is independent of whether computer-assisted ordering is used. What do you conclude by setting $\alpha = .05$?

b   Verify that a chi-square test is appropriate.

c   Is there a difference between delivery-time performance between firms using computer-assisted ordering and those not using computer-assisted ordering?

d   Carry out graphical analysis to investigate the relationship between delivery-time performance and computer-assisted ordering. Describe the relationship.

13.21   A television station wishes to study the relationship between viewership of its 11 P.M. news program and viewer age (18 years or less, 19 to 35, 36 to 54, 55 or older). A sample of 250 television viewers in each age group is randomly selected, and the number who watch the station's 11 P.M. news is found for each sample. The results are given in Table 13.7.
DS TVView

a   Let p_1, p_2, p_3, and p_4 be the proportions of all viewers in each age group who watch the station's 11 P.M. news. If these proportions are equal, then whether a viewer watches the station's 11 P.M. news is independent of the viewer's age group. Therefore, we can test the null hypothesis H_0 that p_1, p_2, p_3, and p_4 are equal by carrying out a chi-square test for independence. Perform this test by setting $\alpha = .05$.

b   Compute a 95 percent confidence interval for the difference between p_1 and p_4.

Chapter Summary

In this chapter we presented two hypothesis tests that employ the **chi-square distribution.** In Section 13.1 we discussed a **chi-square test of goodness-of-fit.** Here we considered a situation in which we study how count data are distributed among various categories. In particular, we considered a **multinomial experiment** in which randomly selected items are classified into several groups, and we saw how to perform a goodness-of-fit test for the multinomial probabilities associated with these groups. We also explained how to perform a goodness-of-fit test for normality. In

Section 13.2 we presented a **chi-square test for independence.** Here we classify count data on two dimensions, and we summarize the cross-classification in the form of a **contingency table.** We use the cross-classified data to test whether the two classifications are **statistically independent,** which is really a way to see whether the classifications are related. We also learned that we can use graphical analysis to investigate the nature of the relationship between the classifications.

Glossary of Terms

chi-square test for independence: A test to determine whether two classifications are independent.
contingency table: A table that summarizes data that have been classified on two dimensions or scales.
goodness-of-fit test for multinomial probabilities: A test to determine whether multinomial probabilities are equal to a specific set of values.

goodness-of-fit test for normality: A test to determine if a sample has been randomly selected from a normally distributed population.
homogeneity (test for)**:** A test of the null hypothesis that all multinomial probabilities are equal.
multinomial experiment: An experiment that concerns count data that are classified into more than two categories.

Important Formulas and Tests

A goodness-of-fit test for multinomial probabilities: page 558
A test for homogeneity: page 558

A goodness-of-fit test for a normal distribution: page 561
A chi-square test for independence: page 567

Supplementary Exercises

connect | **13.22** A large supermarket conducted a consumer preference study by recording the brand of wheat bread purchased by customers in its stores. The supermarket carries four brands of wheat bread. In a random sample of 200 purchasers, the numbers of purchasers preferring Brands *A, B, C,* and *D* of the wheat bread were, respectively, 51, 82, 27, and 40. **(1)** Test the null hypothesis that the four brands are equally preferred by setting α equal to .05. **(2)** Find a 95 percent confidence interval for the proportion of all purchasers who prefer Brand *B*.   **DS** BreadPref

13.23 An occupant traffic study was carried out to aid in the remodeling of a large building on a university campus. The building has five entrances, and the choice of entrance was recorded for a random sample of 300 persons entering the building. The numbers of persons using Entrances 1, 2, 3, 4, and 5 were, respectively, 30, 91, 97, 40, and 42. **(1)** Test the null hypothesis that the five entrances are equally used by setting α equal to .05. **(2)** Find a 95 percent confidence interval for the proportion of all people who use Entrance 3.
DS EntrPref

13.24 In an article in *Accounting and Business Research,* Meier, Alam, and Pearson studied auditor lobbying on several proposed U.S. accounting standards that affect banks and savings and loan associations. As part of this study, the authors investigated auditors' positions regarding proposed changes in accounting standards that would increase client firms' reported earnings. It was hypothesized that auditors would favor such proposed changes because their clients' managers would receive higher compensation (salary, bonuses, and so on) when client earnings were reported to be higher. Table 13.8 summarizes auditor and client positions (in favor or opposed) regarding proposed changes in accounting standards that would increase client firms' reported earnings. Here the auditor and client positions are cross-classified versus the size of the client firm.
DS AuditPos

a Test to determine whether auditor positions regarding earnings-increasing changes in accounting standards depend on the size of the client firm. Use $\alpha = .05$.
b Test to determine whether client positions regarding earnings-increasing changes in accounting standards depend on the size of the client firm. Use $\alpha = .05$.
c Carry out a graphical analysis to investigate a possible relationship between (1) auditor positions and the size of the client firm and (2) client positions and the size of the client firm.
d Does the relationship between position and the size of the client firm seem to be similar for both auditors and clients? Explain.

13.25 In the book *Business Research Methods* (5th ed.), Donald R. Cooper and C. William Emory discuss a market researcher for an automaker who is studying consumer preferences for styling features of larger sedans. Buyers, who were classified as "first-time" buyers or "repeat" buyers, were asked to express their preference for one of two types of styling—European styling or Japanese styling. Of 40 first-time buyers, 8 preferred European styling and 32 preferred Japanese styling. Of 60 repeat buyers, 40 preferred European styling and 20 preferred Japanese styling.
a Set up a contingency table for these data.
b Test the hypothesis that buyer status (repeat versus first-time) and styling preference are independent at the .05 level of significance. What do you conclude?
c Carry out a graphical analysis to investigate the nature of any relationship between buyer status and styling preference. Describe the relationship.

13.26 Again consider the situation of Exercise 13.24. Table 13.9 summarizes auditor positions regarding proposed changes in accounting standards that would decrease client firms' reported earnings. Determine whether the relationship between auditor position and the size of the client firm is the same for earnings-decreasing changes in accounting standards as it is for earnings-increasing changes in accounting standards. Justify your answer using both a statistical test and a graphical analysis.   **DS** AuditPos2

TABLE 13.8   Auditor and Client Positions Regarding Earnings-Increasing Changes in Accounting Standards (for Exercise 13.24)   DS AuditPos

(a) Auditor positions

	Large Firms	Small Firms	Total
In Favor	13	130	143
Opposed	10	24	34
Total	23	154	177

(b) Client positions

	Large Firms	Small Firms	Total
In Favor	12	120	132
Opposed	11	34	45
Total	23	154	177

Source: Heidi Hylton Meier, Pervaiz Alam, and Michael A. Pearson, "Auditor Lobbying for Accounting Standards: The Case of Banks and Savings and Loan Associations," *Accounting and Business Research* 23, no. 92 (1993), pp. 477–487.

TABLE 13.9   Auditor Positions Regarding Earnings-Decreasing Changes in Accounting Standards (for Exercise 13.26)   DS AuditPos2

	Large Firms	Small Firms	Total
In Favor	27	152	179
Opposed	29	154	183
Total	56	306	362

Source: Heidi Hylton Meier, Pervaiz Alam, and Michael A. Pearson, "Auditor Lobbying for Accounting Standards: The Case of Banks and Savings and Loan Associations," *Accounting and Business Research* 23, no. 92 (1993), pp. 477–487.

TABLE 13.10   Results of the Coupon Redemption Study (for Exercise 13.27)   DS Coupon

Coupon Redemption Level	Midtown	North Side	South Side	Total
High	69	97	52	218
Medium	101	93	76	270
Low	30	10	72	112
Total	200	200	200	600

FIGURE 13.5   Minitab Output of a Chi-Square Test for Independence in the Coupon Redemption Study

Expected counts are below observed counts

	Midtown	North	South	Total
High	69	97	52	218
	72.67	72.67	72.67	
Medium	101	93	76	270
	90.00	90.00	90.00	
Low	30	10	72	112
	37.33	37.33	37.33	
Total	200	200	200	600

Chi-Sq = 71.476, DF = 4, P-Value = 0.000

TABLE 13.11   A Sample of 65 Customer Satisfaction Ratings (for Exercise 13.28)   DS VideoGame

39	46	42	40	45	44	44	44	45
45	44	46	46	46	41	46	46	
38	40	40	41	43	38	48	39	
42	39	47	43	47	43	44	41	
42	40	44	39	43	36	41	44	
41	42	43	43	41	44	45	42	
38	45	45	46	40	44	44	47	
42	44	45	45	43	45	44	43	

13.27 The manager of a chain of three discount drug stores wishes to investigate the level of discount coupon redemption at its stores. All three stores have the same sales volume. Therefore, the manager will randomly sample 200 customers at each store with regard to coupon usage. The survey results are given in Table 13.10. Test the hypothesis that redemption level and location are independent with $\alpha = .01$. Use the Minitab output in Figure 13.5.   DS Coupon

13.28 THE VIDEO GAME SATISFACTION RATING CASE

Consider the sample of 65 customer satisfaction ratings given in Table 13.11. Carry out a chi-square goodness-of-fit test of normality for the population of all customer satisfaction ratings. Recall that we previously calculated $\bar{x} = 42.95$ and $s = 2.6424$ for the 65 ratings.   DS VideoGame

Appendix 13.1 ■ Chi-Square Tests Using Excel

Chi-square goodness-of-fit test in Exercise 13.10 (data file: Invoice2.xlsx):

- In the first row of the spreadsheet, enter the following column headings in order: Percent, Expected, Number, and ChiSqContribution. Also enter the heading "p-value" in cell C9.

- Beginning in cell A2, enter the "percentage of invoice figures" from Exercise 13.10 as decimal fractions into column A.

- Compute expected values. Enter the formula =500*A2 into cell B2 and press Enter. Copy this formula through cell B6 by double-clicking the drag handle (in the lower right corner) of cell B2.

- Enter the "number of invoice figures" from Exercise 13.10 into cells C2 through C6.

- Compute cell chi-square contributions. In cell D2, enter the formula =(C2 − B2)^2/B2 and press Enter. Copy this formula through cell D6 by double-clicking the drag handle (in the lower right corner) of cell D2.

- Compute the chi-square statistic in cell D7. Use the mouse to select the range of cells D2:D7 and click the "Σ AutoSum" button on the Excel ribbon.

- Click on an empty cell, say cell A9, and select the Insert Function button f_x on the Excel ribbon.

- In the Insert Function dialog box, select Statistical from the "Or select a category:" menu, select CHISQ.DIST from the "Select a function:" menu, and click OK.

- In the "CHISQ.DIST Function Arguments" dialog box, enter D7 into the "X" window, enter 3 into the "Deg_freedom" window, and enter 1 into the "Cumulative" window.

- Click OK in the "CHISQ.DIST Function Arguments" dialog box. The left-tail area related to the value of the chi-square statistic is entered into cell A9.

- To calculate the p-value related to the chi-square statistic, enter the cell formula =1−A9 into cell D9. This produces the p-value in cell D9.

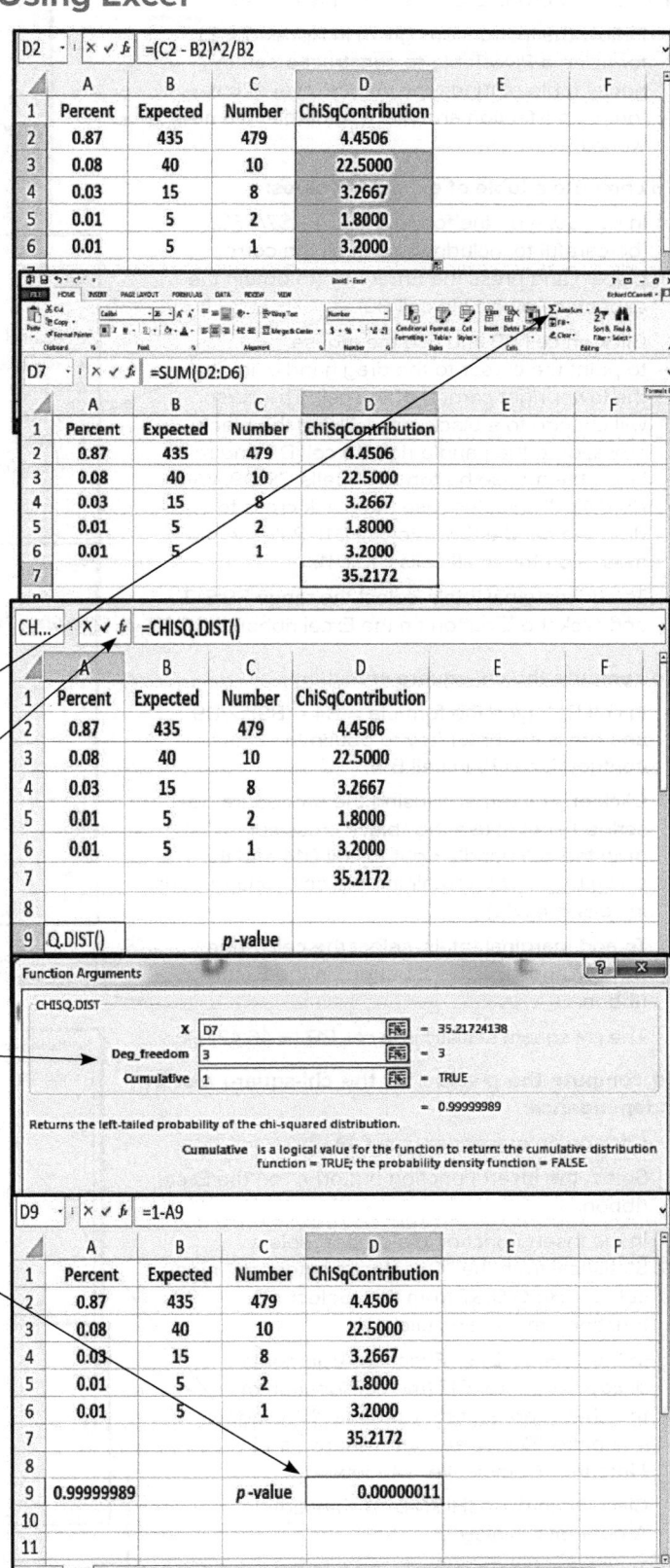

Source: Microsoft Office Excel 2016

Contingency table and chi-square test of independence in Figure 13.2(b) (data file: Invest.xlsx):

- Follow the instructions given in Appendix 2.1 for using a PivotTable to construct a contingency table of fund type versus level of customer satisfaction and place the table in a new worksheet.

To compute a table of expected values:

- In cell B9, type the formula =$E4*B$7/E7 (be careful to include the $ in all the correct places) and press the Enter key (to obtain the expected value 12 in cell B9).

- Click on cell B9 and use the mouse to point the cursor to the drag handle (in the lower right corner) of the cell. The cursor will change to a black cross. Using the black cross, drag the handle right to cell D9 and release the mouse button to fill cells C9:D9. With B9:D9 still selected, use the black cross to drag the handle down to cell D11. Release the mouse button to fill cells B10:D11.

- To add marginal totals, select the range B9:E12 and click the Σ button on the Excel ribbon.

To compute the chi-square statistic:

- In cell B14, type the formula =(B4 − B9)^2/B9 and press the Enter key to obtain the cell contribution 0.75 in cell B14.

- Click on cell B14 and (using the procedure described above) use the "black cross cursor" to drag the cell handle right to cell D14 and then down to cell D16 (obtaining the cell contributions in cells B14:D16).

- To add marginal totals, select the cell range B14:E17 and click the Σ button on the Excel ribbon.

- The chi-square statistic is in cell E17 (= 46.4375).

To compute the *p*-value for the chi-square test of independence:

- Click on an empty cell, say C19.

- Select the Insert Function button f_x on the Excel ribbon.

- In the Insert Function dialog box, select Statistical from the "Or select a category:" menu, select CHISQ.DIST from the "Select a function:" menu, and click OK.

- In the "CHISQ.DIST Function Arguments" dialog box, enter E17 (the cell location of the chi-square statistic) into the "X" window, 4 into the "Deg_freedom" window, and 1 into the "Cumulative" window.

- Click OK in the "CHISQ.DIST Function Arguments" dialog box.

- To produce the *p*-value related to the chi-square statistic, type the cell formula =1.0 − C19 into cell C20.

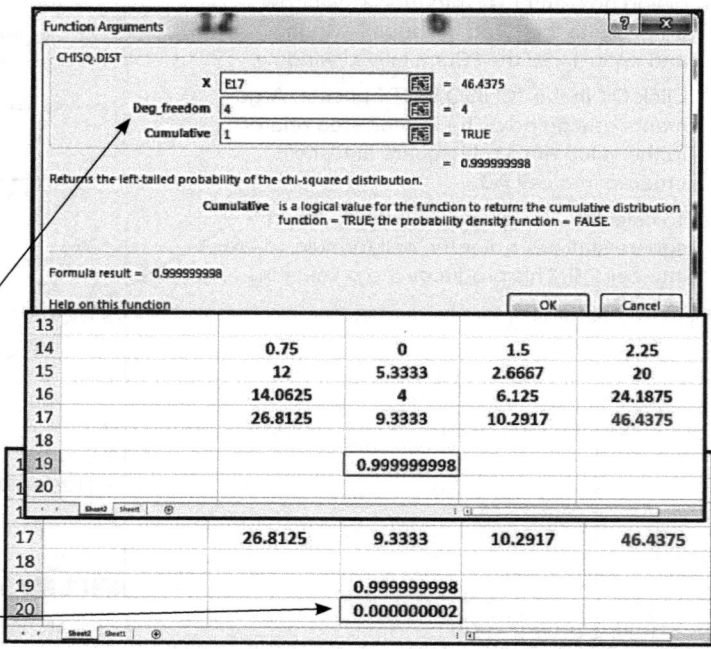

Source: Microsoft Office Excel 2016

Source: Microsoft Office Excel 2016

The *p*-value related to the chi-square statistic can also be obtained by using the CHISQ.TEST function. To do this, we enter the range of the values of the observed cell frequencies (here, B4:D6) into the "Actual_range" window, and we enter the range of the expected cell frequencies (here, B9:D11) into the "Expected_range" window. When we click OK in the CHISQ.TEST Function Arguments dialog box, this function returns the *p*-value in a single spreadsheet cell.

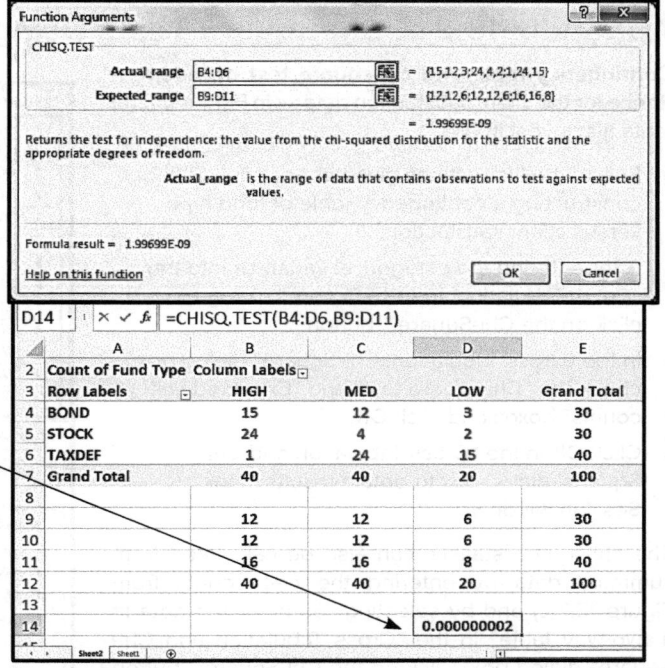

Source: Microsoft Office Excel 2016

The dialog box and spreadsheet show:

Function Arguments — CHISQ.TEST

Actual_range: B4:D6 = {15,12,3;24,4,2;1,24,15}
Expected_range: B9:D11 = {12,12,6;12,12,6;16,16,8}

= 1.99699E-09

Returns the test for independence: the value from the chi-squared distribution for the statistic and the appropriate degrees of freedom.

Actual_range is the range of data that contains observations to test against expected values.

Formula result = 1.99699E-09

D14 =CHISQ.TEST(B4:D6,B9:D11)

	A	B	C	D	E
2	Count of Fund Type	Column Labels			
3	Row Labels	HIGH	MED	LOW	Grand Total
4	BOND	15	12	3	30
5	STOCK	24	4	2	30
6	TAXDEF	1	24	15	40
7	Grand Total	40	40	20	100
8					
9		12	12	6	30
10		12	12	6	30
11		16	16	8	40
12		40	40	20	100
13					
14				0.000000002	

Sheet2 Sheet1

Appendix 13.2 ■ Chi-Square Tests Using Minitab

Contingency table and chi-square test of independence for the client satisfaction data as in Figure 13.2(c) (data file: Invest.MTW):

- Follow the instructions given in Appendix 2.2 for constructing a contingency table of fund type versus client satisfaction.

- After entering the categorical variables into the "Cross Tabulation and Chi-Square" dialog box, click on the Chi-Square... button.

- In the "Cross Tabulation: Chi-Square" dialog box, check the "Chi-square test" and "Expected cell counts" boxes and click OK.

- Click OK in the "Cross Tabulation and Chi-Square" dialog box to obtain results in the Session window.

The chi-square statistic can also be calculated from summary data by entering the cell counts from Figure 13.2(c) and by selecting "Summarized data in a two-way table" in the "Cross Tabulation and Chi-Square" dialog box. Click on the Chi-Square... button and continue with the instructions above.

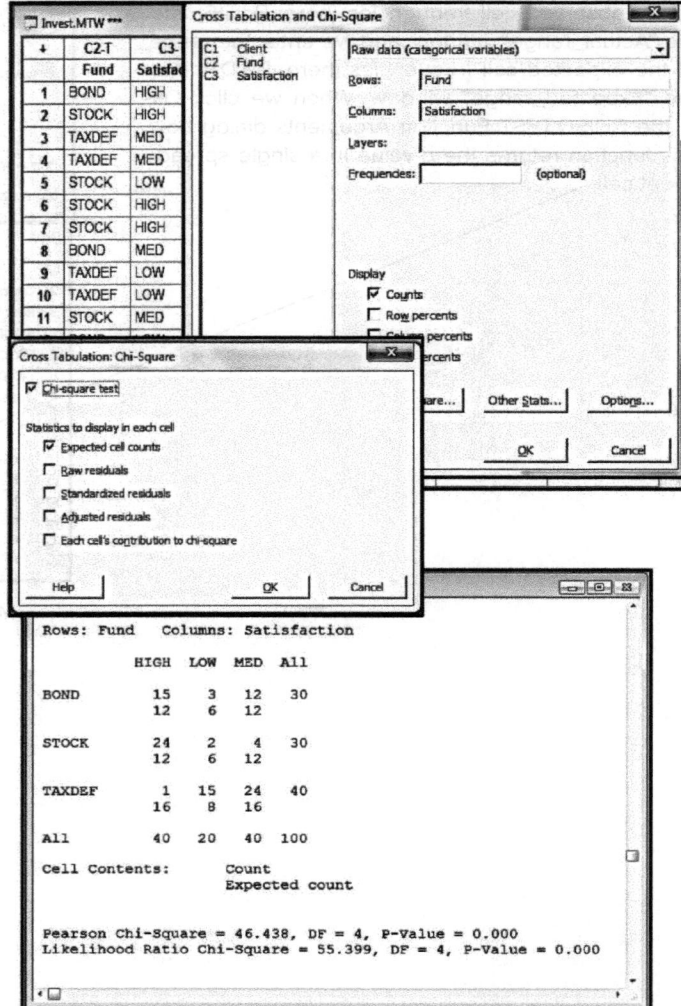

Source: Minitab 18

Chi-square test for goodness-of-fit in Figure 13.1(a) (data file: MicroWav.MTW):

- Enter the microwave oven data from Tables 13.1 and 13.2 with Frequency in column C1 and MrktShr (entered in decimal form) in column C2.

To compute the chi-square statistic:

- Select **Calc : Calculator.**

- In the Calculator dialog box, enter Expected into the "Store results in variable" box.

- In the Expression window, enter 400*MrktShr and click OK to compute the expected values.

- Select **Calc : Calculator.**

- Enter ChiSq into the "Store results in variable" box.

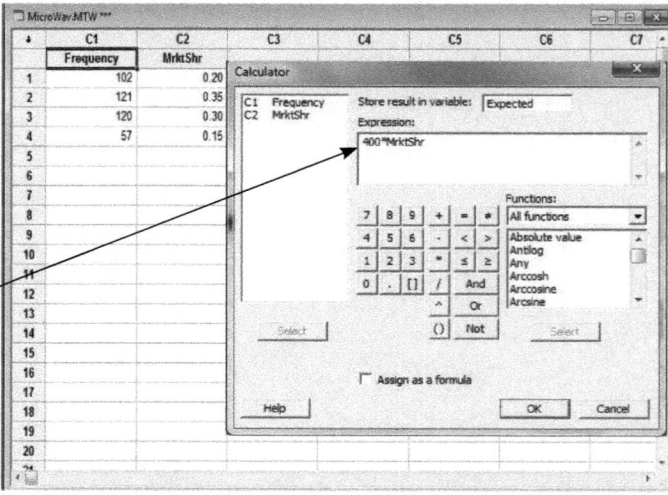

Source: Minitab 18

- In the Expression window, enter the formula (Frequency - Expected)**2/Expected and click OK to compute the cell chi-square contributions.
- Select **Calc : Column Statistics.**
- In the Column Statistics dialog box, click on Sum.
- Enter ChiSq in the "Input Variable" box.
- Enter K1 in the "Store results in" box and click OK to compute the chi-square statistic and store it as the constant K1.
- The chi-square statistic will be displayed in the Session window.

To compute the *p*-value for the test:

Begin by computing the probability of obtaining a value of the chi-square statistic that is less than or equal to the computed value (=8.77857):

- Select **Calc : Probability Distributions : Chi-Square.**
- In the Chi-Square Distribution dialog box, click on "Cumulative probability."
- Enter 3 in the "Degrees of freedom" box.
- Click the "Input constant" option and enter K1.
- Enter K2 in the "Optional storage" box.
- Click OK in the Chi-Square Distribution dialog box. The desired probability is computed and stored as the constant K2.

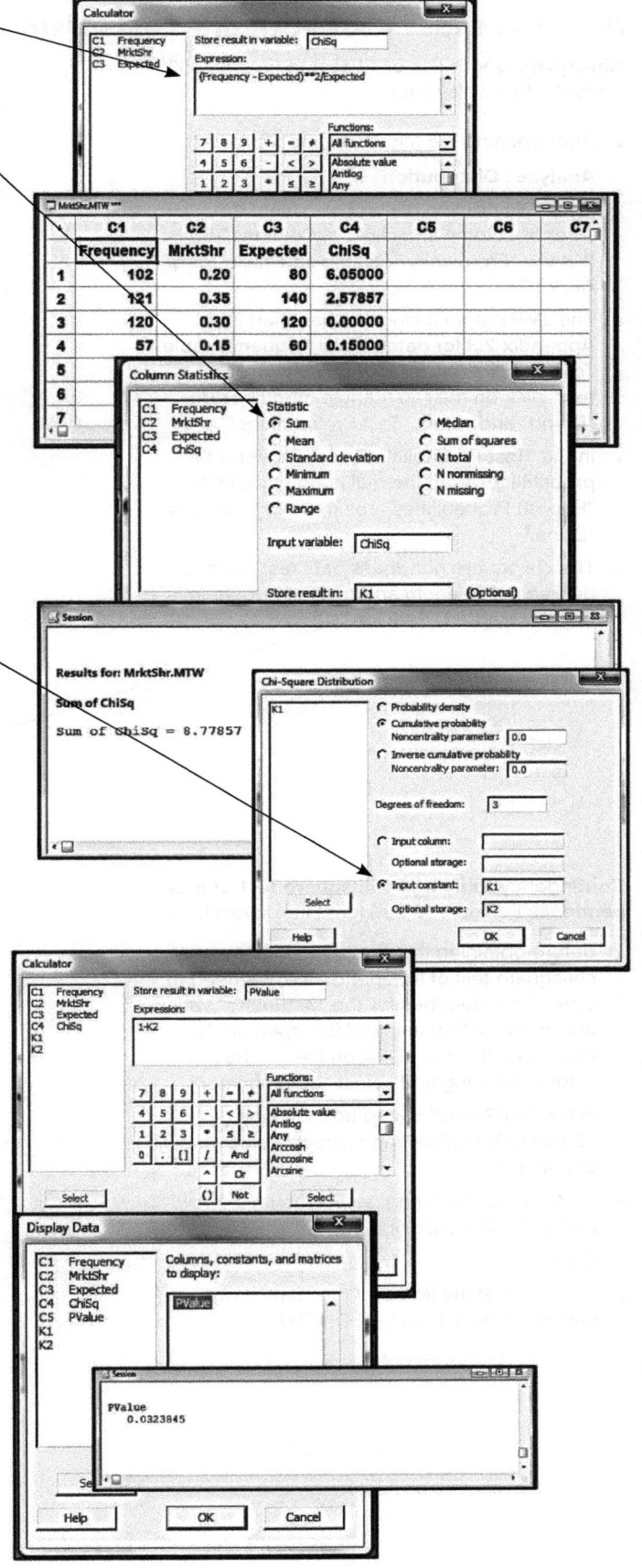

- Select **Calc : Calculator.**
- In the Calculator dialog box, enter PValue into the "Store result in variable" box.
- In the Expression window, enter the formula 1 - K2 and click OK to compute the *p*-value for the chi-square statistic.

To display the *p*-value:

- Select **Data : Display Data.**
- Enter PValue in the "Columns, constants, and matrices to display" window and click OK.
- The *p*-value is displayed in the Session window.

Source: Minitab 18

Appendix 13.3 ■ Chi-Square Tests Using JMP

Chi-square goodness of fit test in Figure 13.1(b) (data file: MicroWav.jmp):

- After opening the MicroWav.jmp file, select **Analyze : Distribution**.

- In the Distribution dialog box, select the "Brand" column and click "Y, Columns." Select the "Frequency" column and click "Freq." Click "OK."

- The JMP output includes a bar chart (see Appendix 2.3 for details) and frequency table. To conduct a **chi- square goodness of fit test**, click on the red triangle menu beside "Brand" and select "Test Probabilities."

- In the "Test Probabilities" section, enter the probabilities from the null hypothesis in the "Hypoth Probabilities" column and then click "Done."

- The chi-square goodness of fit test statistic, degrees of freedom and *p*-value appear in the row marked "Pearson."

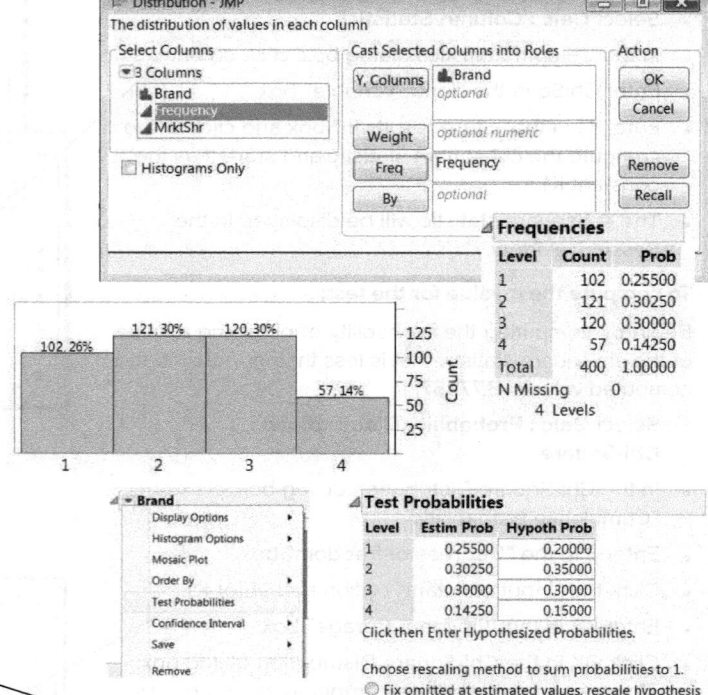

Source: JMP Pro 14

Contingency table and Chi-square test of independence in Figure 13.2(d) (data file: Invest.jmp):

- Before obtaining the contingency table and chi-square test of independence, we need to order the categories for the "SatRating" variable in the JMP data file. After opening the Invest.jmp file, right click on the "SatRating" column heading and select "Column Info."

- In the "SatRating" dialog box, click on "Column Properties" and select "Value Ordering."

- Under Value Ordering, select the level "Low" and click "Move Down."

- Click "OK".

- The order of the levels of the "SatRating" variable is now: HIGH, MED, LOW.

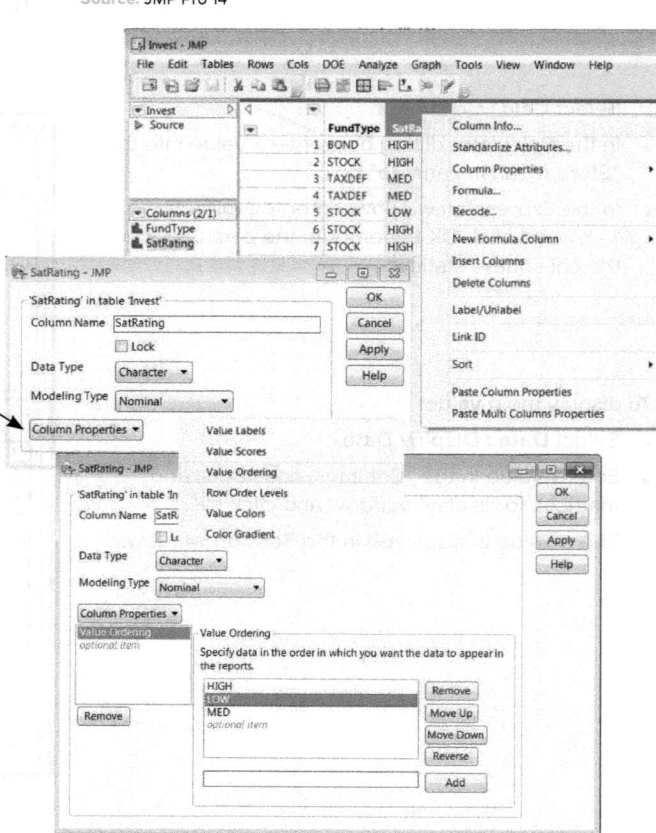

Source: JMP Pro 14

- To obtain the contingency table and chi- square test of independence, select **Analyze : Fit Y by X**.

- In the Fit Y by X dialog box, select the "SatRating" column and click "Y, Response" and select the "FundType" column and click "X, Factor." Click "OK."

- The JMP output will include a Mosaic Plot, **Contingency Table** and **Tests**. Click the red triangle menu beside "Contingency Table" to change the summaries in the table (for example, remove "Total %" or add "Expected").

- The chi-square test of independence test statistic, degrees of freedom and *p*-value appear in the row marked "Pearson."

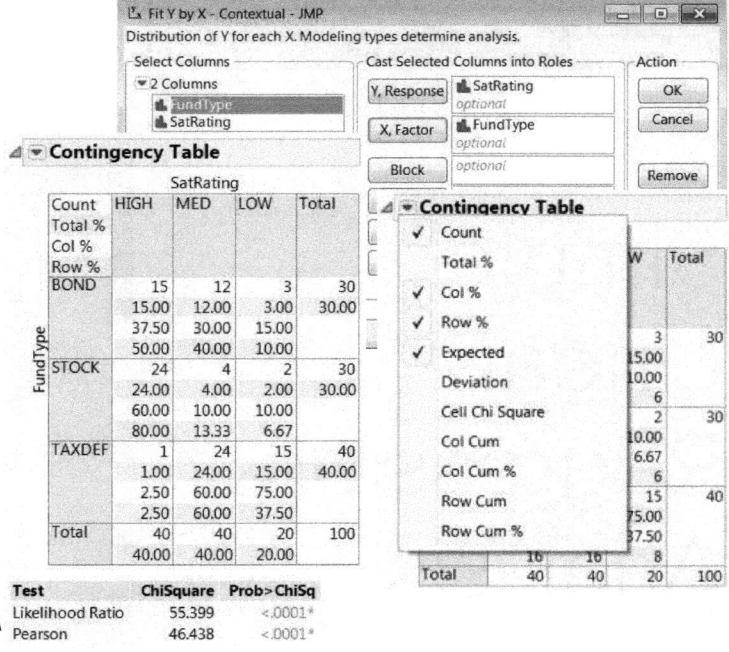

Source: JMP Pro 14

Contingency table and chi-square test of independence with summary data in Figure 13.4 and Exercise 13.9 (data file: Accident.jmp):

- After opening the Accident.jmp file, select **Analyze : Fit Y by X**.

- In the Fit Y by X dialog box, select "Smoker" and click "X, Factor" and select "On-the-Job Accident" and click "Y, Response." To enter the cell counts, select "Count" and click "Freq." Click "OK."

- The output (mosaic plot, contingency table and tests) is the same as in the example above.

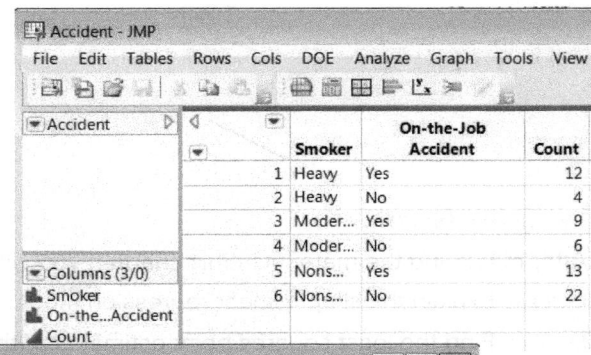

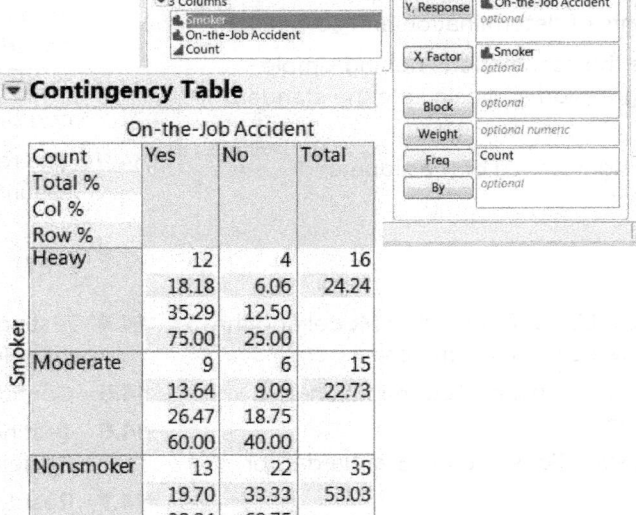

Source: JMP Pro 14

Design Elements: (CD): ©Comstock Images/Alamy; (All Others): ©McGraw-Hill Education

## CHAPTER 14

Simple Linear Regression Analysis

©Paul Bradbury/OJO Images/Getty Images RF

Learning Objectives

After mastering the material in this chapter, you will be able to:

LO14-1 Explain the simple linear regression model.

LO14-2 Find the least squares point estimates of the slope and y-intercept.

LO14-3 Calculate and interpret the simple coefficients of determination and correlation.

LO14-4 Describe the assumptions behind simple linear regression and calculate the standard error.

LO14-5 Test the significance of the slope and y-intercept.

LO14-6 Test the significance of a simple linear regression model by using an F test (Optional).

LO14-7 Calculate and interpret a confidence interval for a mean value and a prediction interval for an individual value.

LO14-8 Test hypotheses about the population correlation coefficient (Optional).

*LO14-9 Use residual analysis to check the assumptions of simple linear regression.

Chapter Outline

14.1 The Simple Linear Regression Model and the Least Squares Point Estimates

14.2 Simple Coefficients of Determination and Correlation

14.3 Model Assumptions and the Standard Error

14.4 Testing the Significance of the Slope and y-Intercept

14.5 Confidence and Prediction Intervals

14.6 Testing the Significance of the Population Correlation Coefficient (Optional)

*14.7 Residual Analysis

*Section 14.7, "Residual Analysis," can be delayed until needed for the study of Section 15.11.

anagers often make decisions by studying the relationships between variables, and process improvements can often be made by understanding how changes in one or more variables affect the process output. **Regression analysis** is a statistical technique in which we use observed data to relate a variable of interest, which is called the **dependent** (or **response**) **variable,** to one or more **independent** (or **predictor**) **variables.** The objective is to build a **regression model,** or **prediction equation,** that can be used to **describe, predict,** and **control** the dependent variable on the basis of the independent variables. For example, a company might wish to improve its marketing process. After collecting data concerning the demand for a product, the product's price, and the advertising expenditures made to promote the product, the company might use regression analysis to develop an equation to predict demand on the basis of price and advertising expenditure. Predictions of demand for various price–advertising expenditure combinations can then be used to evaluate potential changes in the company's marketing strategies.

In the next two chapters we give a thorough presentation of regression analysis. We begin in this chapter by presenting **simple linear regression** analysis. Using this technique is appropriate when we are relating a dependent variable to a single independent variable and when *a straight-line model* describes the relationship between these two variables. We explain many of the methods of this chapter in the context of two new cases:

The Tasty Sub Shop Case: A business entrepreneur uses simple linear regression analysis to predict the yearly revenue for a potential restaurant site on the basis of the number of residents living near the site. The entrepreneur then uses the prediction to assess the profitability of the potential restaurant site.

The QHIC Case: The marketing department at Quality Home Improvement Center (QHIC) uses simple linear regression analysis to predict home upkeep expenditure on the basis of home value. Predictions of home upkeep expenditures are used to help determine which homes should be sent advertising brochures promoting QHIC's products and services.

14.1 The Simple Linear Regression Model and the Least Squares Point Estimates

LO14-1
Explain the simple linear regression model.

The simple linear regression model

The **simple linear regression model** assumes that the relationship between the **dependent variable, which is denoted** y**,** and the **independent variable, denoted** x**,** can be approximated by a straight line. We can tentatively decide whether there is an approximate straight-line relationship between y and x by making a **scatter diagram,** or **scatter plot,** of y versus x. First, data concerning the two variables are observed in pairs. To construct the scatter plot, each value of y is plotted against its corresponding value of x. If the y values tend to increase or decrease in a straight-line fashion as the x values increase, and if there is a scattering of the (x, y) points around the straight line, then it is reasonable to describe the relationship between y and x by using the simple linear regression model. We illustrate this in the following case study.

 EXAMPLE 14.1 The Tasty Sub Shop Case: Predicting Yearly Revenue for a Potential Restaurant Site

Part 1: Purchasing a Tasty Sub Shop Franchise The Tasty Sub Shop is a restaurant chain that sells franchises to business entrepreneurs. Like Quiznos and Subway, the Tasty Sub Shop does not construct a standard, recognizable building to house each of its restaurants. Instead, the entrepreneur wishing to purchase a Tasty Sub franchise finds a suitable site, which consists of a suitable geographical location and suitable store space to rent. Then, when Tasty Sub approves the site, an architect and a contractor are hired to remodel the store rental space and thus "build" the Tasty Sub Shop restaurant. Franchise regulations allow Tasty Sub (and other chains) to help entrepreneurs understand the factors that affect restaurant profitability and to provide basic guidance in evaluating potential restaurant sites. However, in order to prevent restaurant chains from overpredicting profits and thus

TABLE 14.1   **The Tasty Sub Shop Revenue Data**
DS TastySub1

Restaurant	Population Size, x (Thousands of Residents)	Yearly Revenue, y (Thousands of Dollars)
1	20.8	527.1
2	27.5	548.7
3	32.3	767.2
4	37.2	722.9
5	39.6	826.3
6	45.1	810.5
7	49.9	1040.7
8	55.4	1033.6
9	61.7	1090.3
10	64.6	1235.8

FIGURE 14.1   **Excel Output of a Scatter Plot of y versus x**

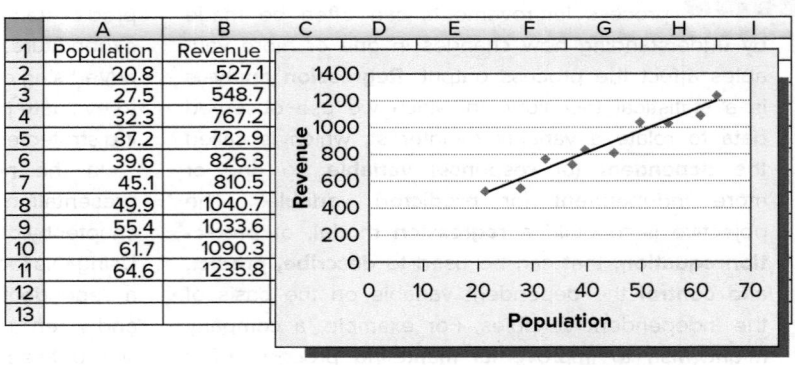

misleading potential franchise owners, these regulations make each individual entrepreneur responsible for predicting the profits of his or her potential restaurant sites.

In this case study we consider a business entrepreneur who has found several potential sites for a Tasty Sub Shop restaurant. Similar to most existing Tasty Sub restaurant sites, each of the entrepreneur's sites is a store rental space located in an outdoor shopping area that is close to one or more residential areas. For a Tasty Sub restaurant built on such a site, yearly revenue is known to partially depend on (1) the number of residents living near the site and (2) the amount of business and shopping near the site. Referring to the number of residents living near a site as *population size* and to the yearly revenue for a Tasty Sub restaurant built on the site as *yearly revenue*, the entrepreneur will—in this chapter—try to predict the **dependent (response) variable** yearly revenue (y) on the basis of the **independent (predictor) variable** population size (x). (In the next chapter the entrepreneur will also use the amount of business and shopping near a site to help predict yearly revenue.) To predict yearly revenue on the basis of population size, the entrepreneur randomly selects 10 existing Tasty Sub restaurants that are built on sites similar to the sites that the entrepreneur is considering. The entrepreneur then asks the owner of each existing restaurant what the restaurant's revenue y was last year and estimates—with the help of the owner and published demographic information—the number of residents, or population size x, living near the site. The values of y (measured in thousands of dollars) and x (measured in thousands of residents) that are obtained are given in Table 14.1. In Figure 14.1 we give an Excel output of a scatter plot of y versus x. This plot shows (1) a tendency for the yearly revenues to increase in a straight-line fashion as the population sizes increase and (2) a scattering of points around the straight line. A **regression model** describing the relationship between y and x must represent these two characteristics. We now develop such a model.

Part 2: The Simple Linear Regression Model   The **simple linear regression model** relating y to x can be expressed as follows:

$$y = \beta_0 + \beta_1 x + \varepsilon$$

This model says that the values of y can be represented by a *mean level* ($\mu_y = \beta_0 + \beta_1 x$) that changes in a straight-line fashion as x changes, combined with random fluctuations (described by the error term ε) that cause the values of y to deviate from the mean level. Here:

1   The **mean level** $\mu_y = \beta_0 + \beta_1 x$ is the mean yearly revenue corresponding to a particular population size x. That is, noting that different Tasty Sub restaurants could potentially be built near different populations of the same size x, the mean level $\mu_y = \beta_0 + \beta_1 x$ is the mean of the yearly revenues that would be obtained by all such restaurants. In addition, because $\mu_y = \beta_0 + \beta_1 x$ is the equation of a straight line, the mean yearly revenues that correspond to increasing values of the population size x lie on a straight line. For example, Table 14.1 tells us that 32,300 residents live near restaurant 3 and 45,100 residents

FIGURE 14.2   **The Simple Linear Regression Model Relating Yearly Revenue (y) to Population (x)**

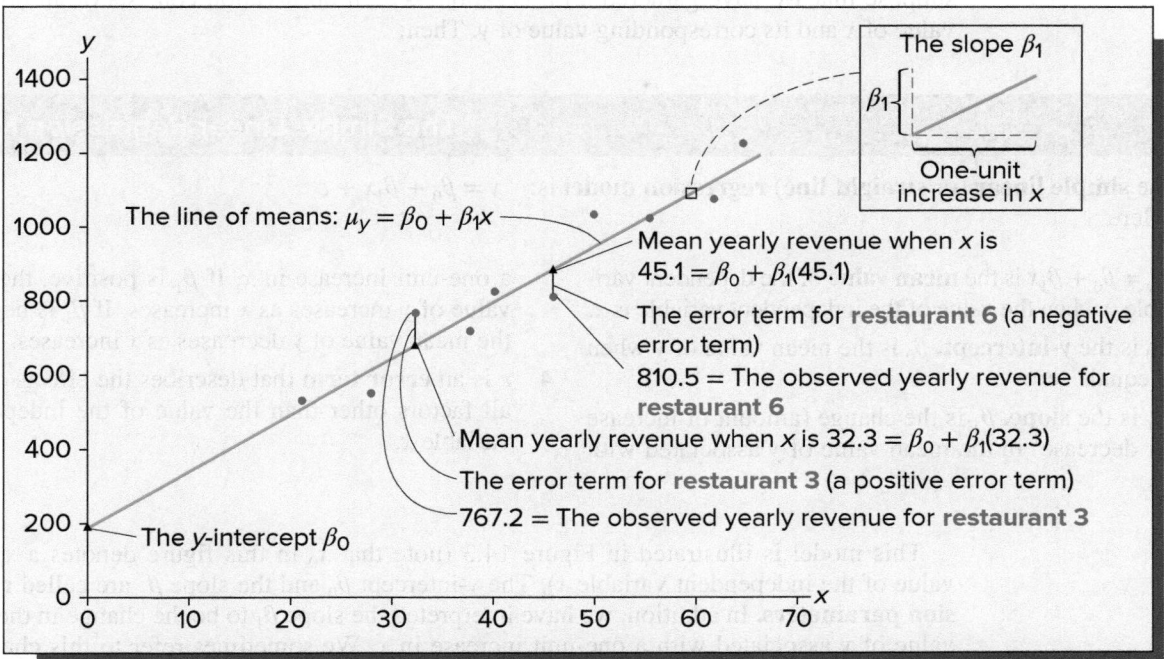

live near restaurant 6. It follows that the mean yearly revenue for all Tasty Sub restaurants that could potentially be built near populations of 32,300 residents is $\beta_0 + \beta_1(32.3)$. Similarly, the mean yearly revenue for all Tasty Sub restaurants that could potentially be built near populations of 45,100 residents is $\beta_0 + \beta_1(45.1)$. Figure 14.2 depicts these two mean yearly revenues as triangles that lie on the straight line $\mu_y = \beta_0 + \beta_1 x$, which we call the **line of means.** The unknown parameters β_0 and β_1 are the **y-intercept** and the **slope** of the line of means. When we estimate β_0 and β_1 in the next subsection, we will be able to estimate mean yearly revenue μ_y on the basis of the population size x.

2   The **y-intercept** β_0 of the line of means can be understood by considering Figure 14.2. As illustrated in this figure, the y-intercept β_0 is the mean yearly revenue for all Tasty Sub restaurants that could potentially be built near populations of zero residents. However, because it is unlikely that a Tasty Sub restaurant would be built near a population of zero residents, this interpretation of β_0 is of dubious practical value. There are many regression situations where the y-intercept β_0 lacks a practical interpretation. In spite of this, statisticians have found that β_0 is almost always an important component of the line of means and thus of the simple linear regression model.

3   The **slope** β_1 of the line of means can also be understood by considering Figure 14.2. As illustrated in this figure, the slope β_1 is the change in mean yearly revenue that is associated with a one-unit increase (that is, a 1,000 resident increase) in the population size x.

4   The **error term** ε of the simple linear regression model accounts for any factors affecting yearly revenue other than the population size x. Such factors would include the amount of business and shopping near a restaurant and the skill of the owner as an operator of the restaurant. For example, Figure 14.2 shows that the error term for restaurant 3 is positive. Therefore, the observed yearly revenue $y = 767.2$ for restaurant 3 is above the corresponding mean yearly revenue for all restaurants that have $x = 32.3$. As another example, Figure 14.2 also shows that the error term for restaurant 6 is negative. Therefore, the observed yearly revenue $y = 810.5$ for restaurant 6 is below the corresponding mean yearly revenue for all restaurants that have $x = 45.1$. Of course, because we do not know the true values of β_0 and β_1, the relative positions of the quantities pictured in Figure 14.2 are only hypothetical.

With the Tasty Sub Shop example as background, we are ready to define the **simple linear regression model relating the dependent variable y to the independent variable x.** We suppose that we have gathered n observations—each observation consists of an observed value of x and its corresponding value of y. Then:

The Simple Linear Regression Model

The **simple linear** (or **straight line**) **regression model** is: $y = \beta_0 + \beta_1 x + \varepsilon$

Here

1 $\mu_y = \beta_0 + \beta_1 x$ is the **mean value** of the dependent variable y when the value of the independent variable is x.

2 β_0 is the **y-intercept.** β_0 is the mean value of y when x equals zero.

3 β_1 is the **slope.** β_1 is the change (amount of increase or decrease) in the mean value of y associated with

a one-unit increase in x. If β_1 is positive, the mean value of y increases as x increases. If β_1 is negative, the mean value of y decreases as x increases.

4 ε is an **error term** that describes the effects on y of all factors other than the value of the independent variable x.

This model is illustrated in Figure 14.3 (note that x_0 in this figure denotes a specific value of the independent variable x). The y-intercept β_0 and the slope β_1 are called **regression parameters.** In addition, we have interpreted the slope β_1 to be the change in the mean value of y associated with a one-unit increase in x. We sometimes refer to this change as *the effect of the independent variable x on the dependent variable y.* However, we cannot prove that a *change in an independent variable causes a change in the dependent variable.* Rather, regression can be used only to establish that the two variables move together and that the independent variable contributes information for predicting the dependent variable. For instance, regression analysis might be used to establish that as liquor sales have increased over the years, college professors' salaries have also increased. However, this does not prove

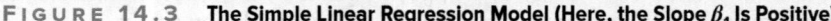

FIGURE 14.3   The Simple Linear Regression Model (Here, the Slope β_1 Is Positive)

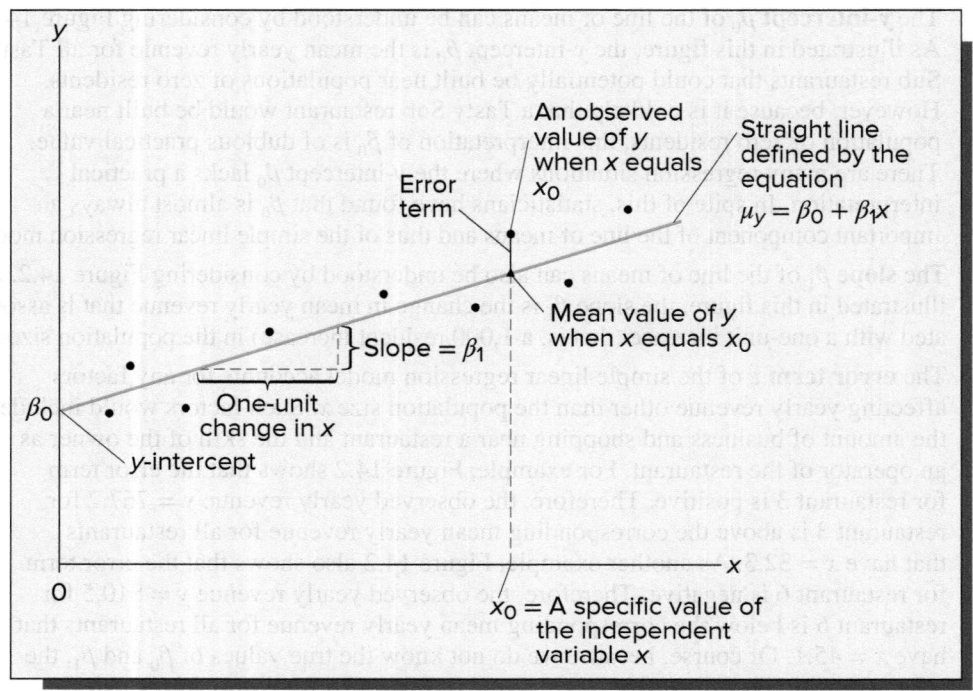

that increases in liquor sales cause increases in college professors' salaries. Rather, both variables are influenced by a third variable—long-run growth in the national economy.

The least squares point estimates

Suppose that we have gathered n observations (x_1, y_1), (x_2, y_2), . . . , (x_n, y_n), where each observation consists of a value of an independent variable x and a corresponding value of a dependent variable y. Also, suppose that a scatter plot of the n observations indicates that the simple linear regression model relates y to x. In order to estimate the y-intercept β_0 and the slope β_1 of the line of means of this model, we could visually draw a line—called an **estimated regression line**—through the scatter plot. Then, we could read the y-intercept and slope off the estimated regression line and use these values as the point estimates of β_0 and β_1. Unfortunately, if different people visually drew lines through the scatter plot, their lines would probably differ from each other. What we need is the "best line" that can be drawn through the scatter plot. Although there are various definitions of what this best line is, one of the most useful best lines is the *least squares line*.

To understand the least squares line, we let

$$\hat{y} = b_0 + b_1 x$$

denote the general equation of an estimated regression line drawn through a scatter plot. Here, because we will use this line to predict y on the basis of x, we call $\hat{y}$ *the predicted value of y* when the value of the independent variable is x. In addition, b_0 is the y-intercept and b_1 is the slope of the estimated regression line. When we determine numerical values for b_0 and b_1, these values will be the point estimates of the y-intercept β_0 and the slope β_1 of the line of means. To explain which estimated regression line is the least squares line, we begin with the Tasty Sub Shop situation. Figure 14.4 shows an estimated regression line drawn through a scatter plot of the Tasty Sub Shop revenue data. In this figure the red dots represent the 10 observed yearly revenues and the black squares represent the 10 predicted yearly revenues given by the estimated regression line. Furthermore, the line segments drawn between the red dots and black squares represent *residuals*, which are the differences between the observed and predicted yearly revenues. Intuitively, if a

LO14-2

Find the least squares point estimates of the slope and y-intercept.

FIGURE 14.4   An Estimated Regression Line Drawn through the Tasty Sub Shop Revenue Data

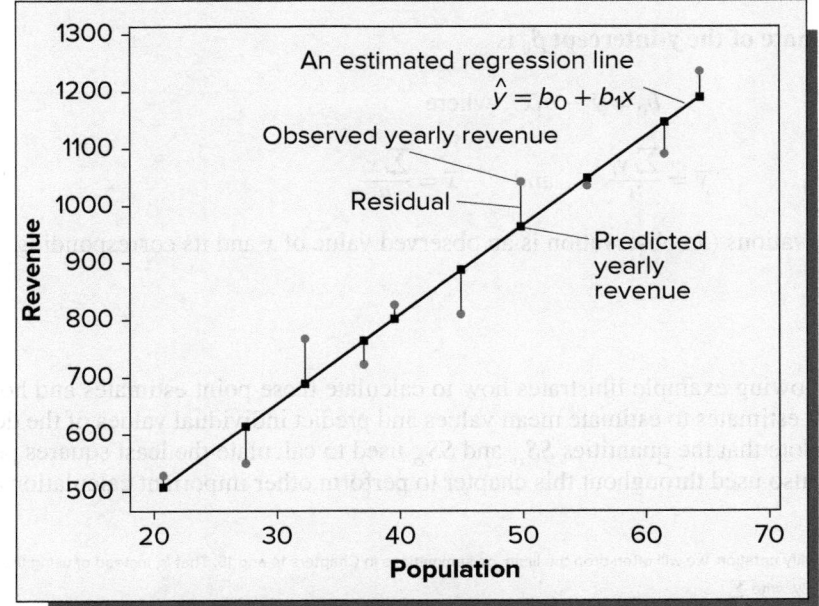

particular estimated regression line provides a good "fit" to the Tasty Sub Shop revenue data, it will make the predicted yearly revenues "close" to the observed yearly revenues, and thus the residuals given by the line will be small. The *least squares line* is the line that minimizes the sum of squared residuals. That is, the least squares line is the line positioned on the scatter plot so as to minimize the sum of the squared vertical distances between the observed and predicted yearly revenues.

To define the least squares line in a general situation, consider an arbitrary observation (x_i, y_i) in a sample of n observations. For this observation, the **predicted value of the dependent variable y** given by an estimated regression line is

$$\hat{y}_i = b_0 + b_1 x_i$$

Furthermore, the difference between the observed and predicted values of y, $y_i - \hat{y}_i$, is the **residual** for the observation, and the **sum of squared residuals** for all n observations is

$$SSE = \sum_{i=1}^{n}(y_i - \hat{y}_i)^2$$

The **least squares line** is the line that minimizes SSE. To find this line, we find the values of the y-intercept b_0 and slope b_1 that give values of $\hat{y}_i = b_0 + b_1 x_i$ that minimize SSE. These values of b_0 and b_1 are called the **least squares point estimates** of β_0 and β_1. Using calculus, it can be shown that these estimates are calculated as follows:[1]

The Least Squares Point Estimates

For the simple linear regression model:

1 The **least squares point estimate of the slope β_1** is

$$b_1 = \frac{SS_{xy}}{SS_{xx}} \quad \text{where}$$

$$SS_{xy} = \sum(x_i - \bar{x})(y_i - \bar{y}) = \sum x_i y_i - \frac{(\sum x_i)(\sum y_i)}{n} \quad \text{and} \quad SS_{xx} = \sum(x_i - \bar{x})^2 = \sum x_i^2 - \frac{(\sum x_i)^2}{n}$$

2 The **least squares point estimate of the y-intercept β_0** is

$$b_0 = \bar{y} - b_1\bar{x} \quad \text{where}$$

$$\bar{y} = \frac{\sum y_i}{n} \quad \text{and} \quad \bar{x} = \frac{\sum x_i}{n}$$

Here n is the number of observations (an observation is an observed value of x and its corresponding value of y).

The following example illustrates how to calculate these point estimates and how to use these point estimates to estimate mean values and predict individual values of the dependent variable. Note that the quantities SS_{xy} and SS_{xx} used to calculate the least squares point estimates are also used throughout this chapter to perform other important calculations.

---

[1]In order to simplify notation, we will often drop the limits on summations in Chapters 14 and 15. That is, instead of using the summation $\sum_{i=1}^{n}$ we will simply write $\sum$.

Ⓒ **EXAMPLE 14.2** The Tasty Sub Shop Case: The Least Squares Estimates

Part 1: Calculating the Least Squares Point Estimates   Again consider the Tasty Sub Shop problem. To compute the least squares point estimates of the regression parameters β_0 and β_1, we first calculate the following preliminary summations:

y_i	x_i	x_i^2	$x_i y_i$
527.1	20.8	$(20.8)^2 = 432.64$	$(20.8)(527.1) = 10963.68$
548.7	27.5	$(27.5)^2 = 756.25$	$(27.5)(548.7) = 15089.25$
767.2	32.3	$(32.3)^2 = 1,043.29$	$(32.3)(767.2) = 24780.56$
722.9	37.2	$(37.2)^2 = 1,383.84$	$(37.2)(722.9) = 26891.88$
826.3	39.6	$(39.6)^2 = 1,568.16$	$(39.6)(826.3) = 32721.48$
810.5	45.1	$(45.1)^2 = 2,034.01$	$(45.1)(810.5) = 36553.55$
1040.7	49.9	$(49.9)^2 = 2,490.01$	$(49.9)(1040.7) = 51930.93$
1033.6	55.4	$(55.4)^2 = 3,069.16$	$(55.4)(1033.6) = 57261.44$
1090.3	61.7	$(61.7)^2 = 3,806.89$	$(61.7)(1090.3) = 67271.51$
1235.8	64.6	$(64.6)^2 = 4,173.16$	$(64.6)(1235.8) = 79832.68$
$\sum y_i = 8603.1$	$\sum x_i = 434.1$	$\sum x_i^2 = 20,757.41$	$\sum x_i y_i = 403,296.96$

Using these summations, we calculate SS_{xy} and SS_{xx} as follows:

$$SS_{xy} = \sum x_i y_i - \frac{\left(\sum x_i\right)\left(\sum y_i\right)}{n}$$

$$= 403,296.96 - \frac{(434.1)(8603.1)}{10}$$

$$= 29,836.389$$

$$SS_{xx} = \sum x_i^2 - \frac{\left(\sum x_i\right)^2}{n}$$

$$= 20,757.41 - \frac{(434.1)^2}{10}$$

$$= 1913.129$$

It follows that the least squares point estimate of the slope β_1 is

$$b_1 = \frac{SS_{xy}}{SS_{xx}} = \frac{29,836.389}{1913.129} = 15.596$$

Furthermore, because

$$\bar{y} = \frac{\sum y_i}{n} = \frac{8603.1}{10} = 860.31 \quad \text{and} \quad \bar{x} = \frac{\sum x_i}{n} = \frac{434.1}{10} = 43.41$$

the least squares point estimate of the y-intercept β_0 is

$$b_0 = \bar{y} - b_1 \bar{x} = 860.31 - (15.596)(43.41) = 183.31$$

(where we have used more decimal place accuracy than shown to obtain the result 183.31).

Because $b_1 = 15.596$, we estimate that mean yearly revenue at Tasty Sub restaurants increases by 15.596 (that is, by \$15,596) for each one-unit (1,000 resident) increase in the population size x. Because $b_0 = 183.31$, we estimate that mean yearly revenue for all Tasty Sub restaurants that could potentially be built near populations of zero residents is \$183,310. However, because it is unlikely that a Tasty Sub restaurant would be built near a population of zero residents, this interpretation is of dubious practical value.

The least squares line

$$\hat{y} = b_0 + b_1 x = 183.31 + 15.596x$$

is sometimes called the *least squares prediction equation*. In Table 14.2 we summarize using this prediction equation to calculate the predicted yearly revenues and the residuals

TABLE 14.2   Calculation of *SSE* Obtained by Using the Least Squares Point Estimates

y_i	x_i	$\hat{y}_i = 183.31 + 15.596x_i$	$y_i - \hat{y}_i$
527.1	20.8	$183.31 + 15.596(20.8) = 507.69$	19.41
548.7	27.5	$183.31 + 15.596(27.5) = 612.18$	−63.48
767.2	32.3	687.04	80.16
722.9	37.2	763.46	−40.56
826.3	39.6	800.89	25.41
810.5	45.1	886.67	−76.17
1040.7	49.9	961.53	79.17
1033.6	55.4	1047.30	−13.70
1090.3	61.7	1145.55	−55.25
1235.8	64.6	1190.78	45.02

$$SSE = \sum(y_i - \hat{y}_i)^2 = (19.41)^2 + (-63.48)^2 + \cdots + (45.02)^2 = 30{,}460.21$$

Note: The predictions and residuals in this table are taken from Minitab, which uses values of b_0 and b_1 that are more precise than the rounded values we have calculated by hand. If you use the formula $\hat{y}_i = 183.31 + 15.596x_i$, your figures may differ slightly from those given here.

for the 10 observed Tasty Sub restaurants. For example, because the population size for restaurant 1 was 20.8, the predicted yearly revenue for restaurant 1 is

$$\hat{y}_1 = 183.31 + 15.596(20.8) = 507.69$$

It follows, because the observed yearly revenue for restaurant 1 was $y_1 = 527.1$, that the residual for restaurant 1 is

$$y_1 - \hat{y}_1 = 527.1 - 507.69 = 19.41$$

If we consider all of the residuals in Table 14.2 and add their squared values, we find that *SSE*, the sum of squared residuals, is 30,460.21. This *SSE* value will be used throughout this chapter. Figure 14.5 gives the Minitab output of the least squares line. Note that this output gives (within rounding) the least squares estimates we have calculated ($b_0 = 183.3$ and $b_1 = 15.60$). In general, we will rely on Excel, Minitab, and JMP to compute the least squares estimates (and to perform many other regression calculations).

Part 2: Estimating a Mean Yearly Revenue and Predicting an Individual Yearly Revenue We define the **experimental region** to be the range of the previously observed population sizes. Referring to Table 14.2, we see that the experimental region consists of

FIGURE 14.5   The Minitab Output of the Least Squares Line

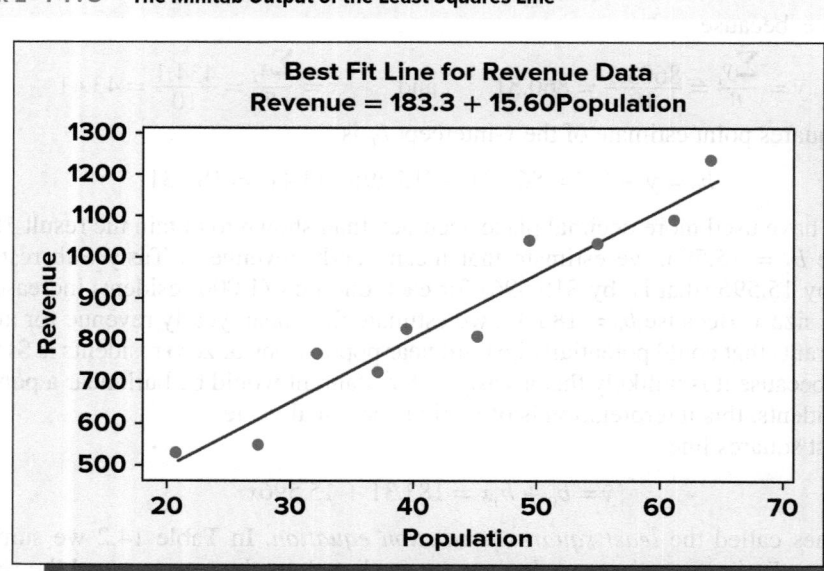

the range of population sizes from 20.8 to 64.6. The simple linear regression model relates yearly revenue y to population size x for values of x that are in the experimental region. For such values of x, the least squares line is the estimate of the line of means. It follows that the point on the least squares line corresponding to a population size of x

$$\hat{y} = b_0 + b_1x$$

is the point estimate of $\beta_0 + \beta_1x$, the mean yearly revenue for all Tasty Sub restaurants that could potentially be built near populations of size x. In addition, we predict the error term ε to be 0. Therefore, $\hat{y}$ is also the *point prediction* of an *individual value* $y = \beta_0 + \beta_1x + \varepsilon$, which is the yearly revenue for a single (individual) Tasty Sub restaurant that is built near a population of size x. Note that the reason we predict the error term ε to be zero is that, because of several *regression assumptions* to be discussed in the next section, ε has a 50 percent chance of being positive and a 50 percent chance of being negative.

For example, suppose that one of the business entrepreneur's potential restaurant sites is near a population of 47,300 residents. Because $x = 47.3$ is in the experimental region,

$$\hat{y} = 183.31 + 15.596(47.3)$$

$$= 921.0 \text{ (that is, \$921,000)}$$

is

1   The **point estimate** of the mean yearly revenue for all Tasty Sub restaurants that could potentially be built near populations of 47,300 residents.

2   The **point prediction** of the yearly revenue for a single Tasty Sub restaurant that is built near a population of 47,300 residents.

Figure 14.6 illustrates $\hat{y} = 921.0$ as a square on the least squares line. Moreover, suppose that the yearly rent and other fixed costs for the entrepreneur's potential restaurant will be $257,550 and that (according to Tasty Sub corporate headquarters) the yearly food and other

FIGURE 14.6   **Point Estimation and Point Prediction, and the Danger of Extrapolation**

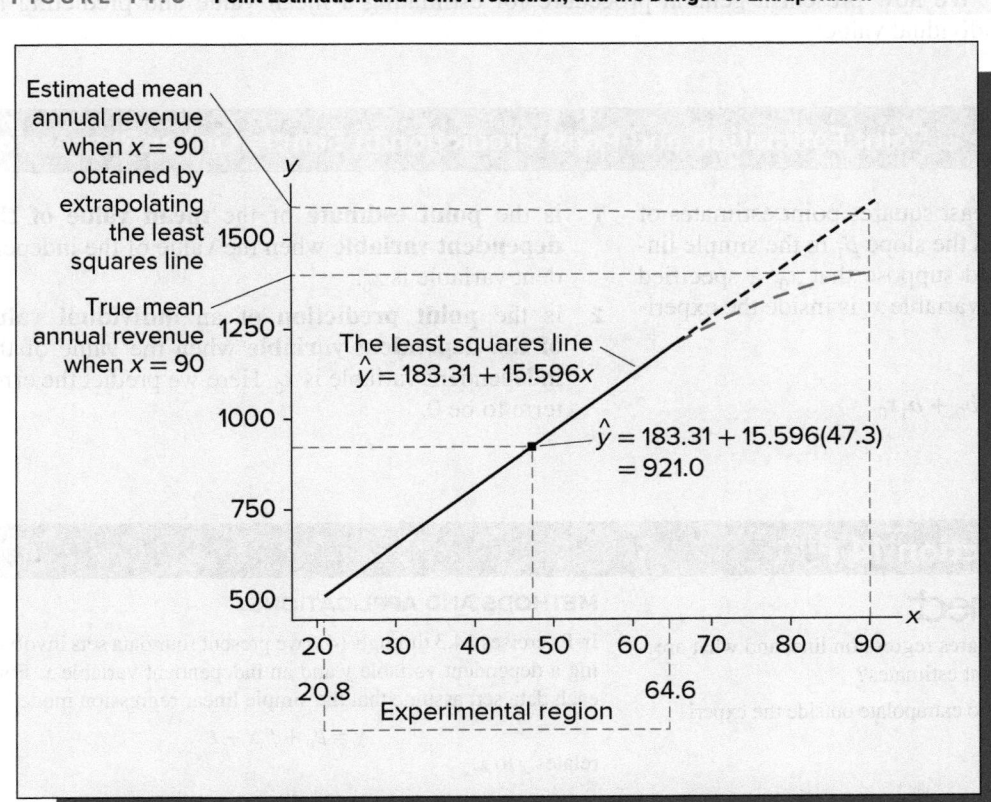

variable costs for the restaurant will be 60 percent of the yearly revenue. Because we predict that the yearly revenue for the restaurant will be $921,000, it follows that we predict that the yearly total operating cost for the restaurant will be $257,550 + .6($921,000) = $810,150. In addition, if we subtract this predicted yearly operating cost from the predicted yearly revenue of $921,000, we predict that the yearly profit for the restaurant will be $110,850. Of course, these predictions are point predictions. In Section 14.5 we will predict the restaurant's yearly revenue and profit *with confidence*.

To conclude this example, note that Figure 14.6 illustrates the potential danger of using the least squares line to predict outside the experimental region. In the figure, we extrapolate the least squares line beyond the experimental region to obtain a prediction for a population size of $x = 90$. As shown in Figure 14.6, for values of x in the experimental region (that is, between 20.8 and 64.6) the observed values of y tend to increase in a straight-line fashion as the values of x increase. However, for population sizes greater than $x = 64.6$, we have no data to tell us whether the relationship between y and x continues as a straight-line relationship or, possibly, becomes a curved relationship. If, for example, this relationship becomes the sort of curved relationship shown in Figure 14.6, then extrapolating the straight-line prediction equation to obtain a prediction for $x = 90$ would overestimate mean yearly revenue (see Figure 14.6).

The previous example illustrates that when we are using a least squares regression line, we should not estimate a mean value or predict an individual value unless the corresponding value of x is in the **experimental region**—the range of the previously observed values of x. Often the value $x = 0$ is not in the experimental region. In such a situation, it would not be appropriate to interpret the y-intercept b_0 as the estimate of the mean value of y when x equals 0. For example, consider the Tasty Sub Shop problem. Figure 14.6 illustrates that the population size $x = 0$ is not in the experimental region. Therefore, it would not be appropriate to use $b_0 = 183.31$ as the point estimate of the mean yearly revenue for all Tasty Sub restaurants that could potentially be built near populations of zero residents. Because it is not meaningful to interpret the y-intercept in many regression situations, we often omit such interpretations.

We now present a general procedure for estimating a mean value and predicting an individual value:

Point Estimation and Point Prediction in Simple Linear Regression

Let b_0 and b_1 be the least squares point estimates of the y-intercept β_0 and the slope β_1 in the simple linear regression model, and suppose that x_0, a specified value of the independent variable x, is inside the experimental region. Then

$$\hat{y} = b_0 + b_1 x_0$$

1   is the **point estimate** of the **mean value of the dependent variable** when the value of the independent variable is x_0.

2   is the **point prediction** of an **individual value of the dependent variable** when the value of the independent variable is x_0. Here we predict the error term to be 0.

Exercises for Section 14.1

CONCEPTS ⬛ connect

14.1   What is the least squares regression line, and what are the least squares point estimates?

14.2   Why is it dangerous to extrapolate outside the experimental region?

METHODS AND APPLICATIONS

In Exercises 14.3 through 14.6 we present four data sets involving a dependent variable y and an independent variable x. For each data set, assume that the simple linear regression model

$$y = \beta_0 + \beta_1 x + \varepsilon$$

relates y to x.

Week	Average Hourly Temperature, x (°F)	Natural Gas Consumption, y (MMcf)
1	28.0	12.4
2	28.0	11.7
3	32.5	12.4
4	39.0	10.8
5	45.9	9.4
6	57.8	9.5
7	58.1	8.0
8	62.5	7.5

DS GasCon1

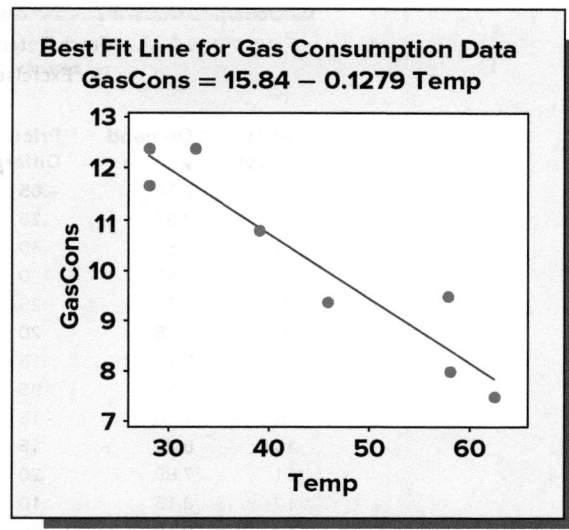

Best Fit Line for Gas Consumption Data
GasCons = 15.84 − 0.1279 Temp

14.3   THE NATURAL GAS CONSUMPTION CASE
DS GasCon1

Above we give the average hourly outdoor temperature (x) in a city during a week and the city's natural gas consumption (y) during the week for each of eight weeks (the temperature readings are expressed in degrees Fahrenheit and the natural gas consumptions are expressed in millions of cubic feet of natural gas—denoted MMcf). The output to the right of the data is obtained when Minitab is used to fit a least squares line to the natural gas consumption data.

a (1) Find the least squares point estimates b_0 and b_1 on the computer output and report their values. (2) Interpret b_0 and b_1. (3) Is an average hourly temperature of 0°F in the experimental region? (4) What does this say about the interpretation of b_0?

b Use the facts that $SS_{xy} = -179.6475$; $SS_{xx} = 1,404.355$; $\bar{y} = 10.2125$; and $\bar{x} = 43.98$ to hand calculate (within rounding) b_0 and b_1.

c Use the least squares line to compute a point estimate of the mean natural gas consumption for all weeks having an average hourly temperature of 40°F and compute a point prediction of the natural gas consumption for an individual week having an average hourly temperature of 40°F.

14.4   THE FRESH DETERGENT CASE   DS Fresh

Enterprise Industries produces Fresh, a brand of liquid laundry detergent. In order to study the relationship between price and demand for the large bottle of Fresh, the company has gathered data concerning demand for Fresh over the last 30 sales periods (each sales period is four weeks). Here, for each sales period,

   y = demand for the large bottle of Fresh (in hundreds of thousands of bottles) in the sales period, and

   x = the difference between the average industry price (in dollars) of competitors' similar detergents and the price (in dollars) of Fresh as offered by Enterprise Industries in the sales period.

The data and Minitab output from fitting a least squares regression line to the data are given in Table 14.3.

a Find the least squares point estimates b_0 and b_1 on the computer output and report their values.

b Interpret b_0 and b_1. Does the interpretation of b_0 make practical sense?

c Write the equation of the least squares line.

d Use the least squares line to compute (1) a point estimate of the mean demand in all sales periods when the price difference is .10 and (2) a point prediction of the demand in an individual sales period when the price difference is .10.

14.5   THE SERVICE TIME CASE   DS SrvcTime

Accu-Copiers, Inc., sells and services the Accu-500 copying machine. As part of its standard service contract, the company agrees to perform routine service on this copier. To obtain information about the time it takes to perform routine service, Accu-Copiers has collected data for 11 service calls. The data and Excel output from fitting a least squares regression line to the data are given in Table 14.4.

a Find the least squares point estimates b_0 and b_1 on the computer output and report their values. Interpret b_0 and b_1. Does the interpretation of b_0 make practical sense?

b Use the least squares line to compute a point estimate of the mean time to service four copiers and a point prediction of the time to service four copiers on a single call.

14.6   THE DIRECT LABOR COST CASE   DS DirLab

An accountant wishes to predict direct labor cost (y) on the basis of the batch size (x) of a product produced in a job shop. Data for 12 production

TABLE 14.3 **Fresh Detergent Demand Data and the Least Squares Line**
(for Exercise 14.4) Ⓓ Fresh

Sales Period	Demand, y	Price Difference, x	Sales Period	Demand, y	Price Difference, x
1	7.38	−.05	24	8.50	.10
2	8.51	.25	25	8.75	.50
3	9.52	.60	26	9.21	.60
4	7.50	0	27	8.27	−.05
5	9.33	.25	28	7.67	0
6	8.28	.20	29	7.93	.05
7	8.75	.15	30	9.26	.55
8	7.87	.05			
9	7.10	−.15			
10	8.00	.15			
11	7.89	.20			
12	8.15	.10			
13	9.10	.40			
14	8.86	.45			
15	8.90	.35			
16	8.87	.30			
17	9.26	.50			
18	9.00	.50			
19	8.75	.40			
20	7.95	−.05			
21	7.65	−.05			
22	7.27	−.10			
23	8.00	.20			

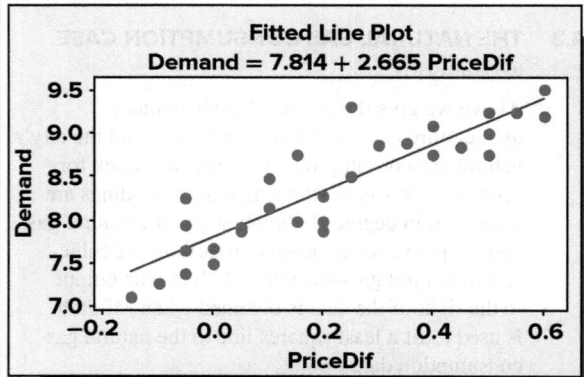

Fitted Line Plot
Demand = 7.814 + 2.665 PriceDif

TABLE 14.4 **The Service Time Data and the Least Squares Line (for Exercise 14.5)** Ⓓ SrvcTime

Service Call	Number of Copiers Serviced, x	Number of Minutes Required, y
1	4	109
2	2	58
3	5	138
4	7	189
5	1	37
6	3	82
7	4	103
8	5	134
9	2	68
10	4	112
11	6	154

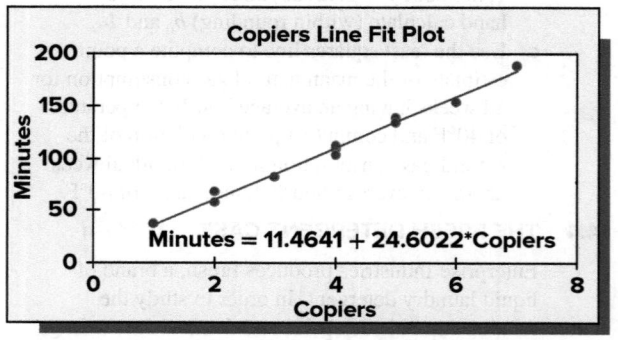

Copiers Line Fit Plot
Minutes = 11.4641 + 24.6022*Copiers

runs are given in Table 14.5, along with the Excel output from fitting a least squares regression line to the data.

a By using the formulas illustrated in Example 14.2 and the data provided, verify that (within rounding) $b_0 = 18.488$ and $b_1 = 10.146$, as shown on the Excel output.

b Interpret the meanings of b_0 and b_1. Does the interpretation of b_0 make practical sense?

c Write the least squares prediction equation.

d Use the least squares line to obtain (**1**) a point estimate of the mean direct labor cost for all batches of size 60 and (**2**) a point prediction of the direct labor cost for an individual batch of size 60.

DirLab

TABLE 14.5   **The Direct Labor Cost Data and the Least Squares Line (for Exercise 14.6)**

Direct Labor Cost, y ($100s)	Batch Size, x
71	5
663	62
381	35
138	12
861	83
145	14
493	46
548	52
251	23
1024	100
435	41
772	75

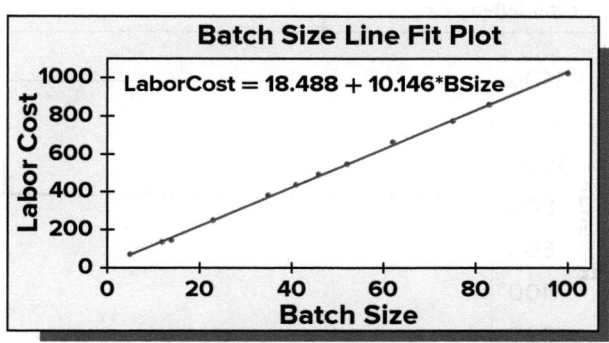

14.2 Simple Coefficients of Determination and Correlation

LO14-3
Calculate and interpret the simple coefficients of determination and correlation.

The simple coefficient of determination

The **simple coefficient of determination** is a measure of the potential usefulness of a simple linear regression model. To introduce this quantity, which is denoted r^2 (pronounced **r squared**), suppose we have observed n values of the dependent variable y. Also, suppose that we are asked to predict y without using a predictor (independent) variable x. In such a case the only reasonable prediction of a specific value of y, say y_i, would be $\bar{y}$, which is simply the average of the n observed values $y_1, y_2, \ldots, y_n$. Here the error of prediction in predicting y_i would be $y_i - \bar{y}$. For example, because the mean of the $n = 10$ Tasty Sub Shop yearly revenues in Table 14.1 is $\bar{y} = (\Sigma y_i/n) = (8603.1/10) = 860.31$, the error of prediction when we predict the ith yearly revenue, y_i, by $\bar{y}$ is $y_i - \bar{y} = y_i - 860.31$. The second column in Table 14.6

TABLE 14.6   **Calculation of the Values of $y_i - \bar{y}$ and $y_i - \hat{y}_i$ and of the Total, Unexplained, and Explained Variations for the Tasty Sub Shop Revenue Data**

y_i	$y_i - \bar{y} = y_i - 860.31$	x_i	$\hat{y}_i = 183.31 + 15.596x_i$	$y_i - \hat{y}_i$
527.1	−333.21	20.8	183.31 + 15.596(20.8) = 507.69	19.41
548.7	−311.61	27.5	183.31 + 15.596(27.5) = 612.18	−63.48
767.2	−93.11	32.3	687.04	80.16
722.9	−137.41	37.2	763.46	−40.56
826.3	−34.01	39.6	800.89	25.41
810.5	−49.81	45.1	886.67	−76.17
1040.7	180.39	49.9	961.53	79.17
1033.6	173.29	55.4	1047.30	−13.70
1090.3	229.99	61.7	1145.55	−55.25
1235.8	375.49	64.6	1190.78	45.02

Total variation = $\sum(y_i - \bar{y})^2 = (-333.21)^2 + (-311.61)^2 + \cdots + (375.49)^2 = 495{,}776.51$

Unexplained variation = $SSE = \sum(y_i - \hat{y}_i)^2 = (19.41)^2 + (-63.48)^2 + \cdots + (45.02)^2 = 30{,}460.21$

Explained variation = Total variation − Unexplained variation

$\qquad\qquad = 495{,}776.51 - 30{,}460.21 = 465{,}316.30$

FIGURE 14.7   **The Reduction in the Prediction Errors Accomplished by Employing the Predictor Variable x**

(a) Prediction errors for the Tasty Sub Shop case when we do not use the information contributed by x

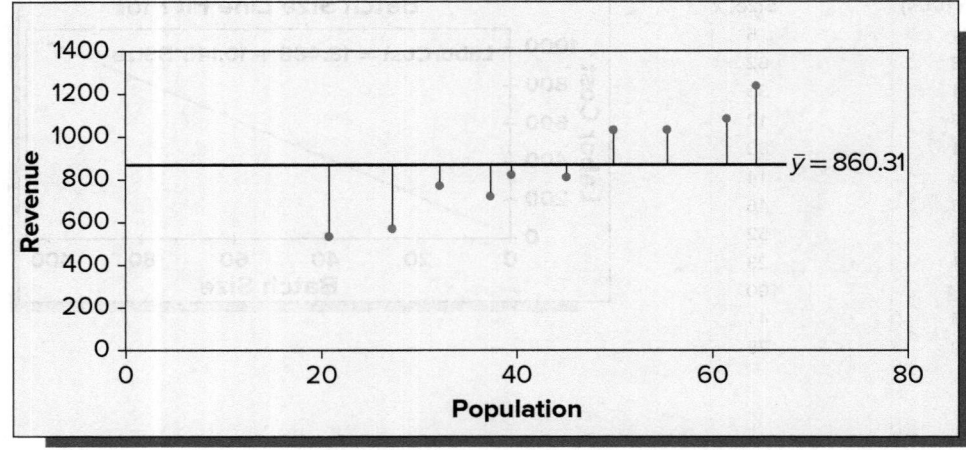

(b) Prediction errors for the Tasty Sub Shop case when we use the information contributed by x by using the least squares line

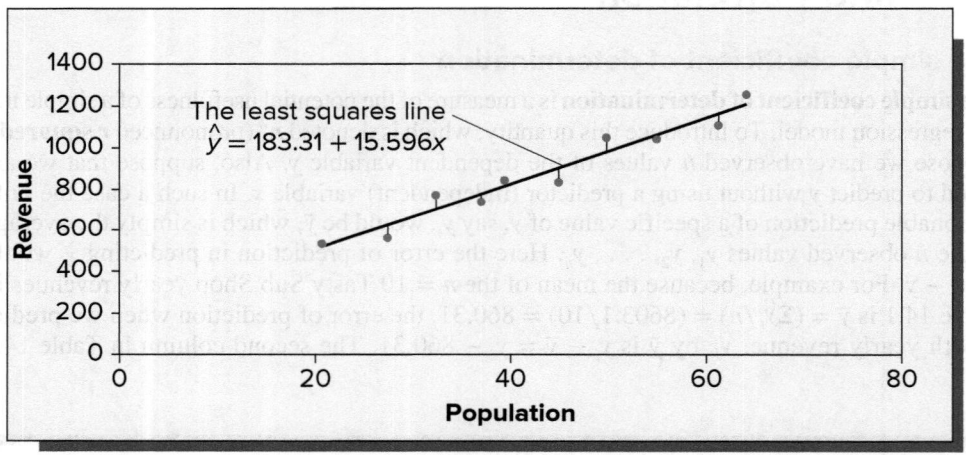

gives the values of $y_i - 860.31$ for the 10 yearly revenues, and Figure 14.7(a) graphically illustrates these prediction errors.

In reality, of course, we would probably not be asked to predict y without using a predictor variable x. Suppose, therefore, that we have observed the n values $x_1, x_2, \ldots, x_n$ of a predictor variable x that correspond to the n observed values $y_1, y_2, \ldots, y_n$ of the dependent variable y. In this case the prediction of y_i is $\hat{y}_i = b_0 + b_1 x_i$ and the error of prediction is $y_i - \hat{y}_i$. For example, recalling that the least squares prediction equation in the Tasty Sub Shop case is $\hat{y}_i = 183.31 + 15.596 x_i$, the last column in Table 14.6 gives the values of $y_i - \hat{y}_i$ for the 10 yearly revenues, and Figure 14.7(b) graphically illustrates these (smaller) prediction errors. Together, Figures 14.7(a) and (b) show the reduction in the prediction errors accomplished by employing the predictor variable x (and the least squares line).

In general, the predictor variable x decreases the prediction error in predicting y_i from $(y_i - \bar{y})$ to $(y_i - \hat{y}_i)$, or by an amount equal to

$$(y_i - \bar{y}) - (y_i - \hat{y}_i) = (\hat{y}_i - \bar{y})$$

It can be shown that

$$\sum (y_i - \bar{y})^2 - \sum (y_i - \hat{y}_i)^2 = \sum (\hat{y}_i - \bar{y})^2$$

Here:

- The sum of squared prediction errors obtained when we do not employ the predictor variable x, $\Sigma(y_i - \bar{y})^2$, is called the **total variation.** Intuitively, this quantity measures the total amount of variation exhibited by the observed values of y. To calculate the total variation in the Tasty Sub Shop case, we square the 10 values of $y_i - \bar{y}$ in Table 14.6 and add together the 10 squared values, obtaining 495,776.51. This is shown in Table 14.6.

- The sum of squared prediction errors obtained when we use the predictor variable x, $\Sigma(y_i - \hat{y}_i)^2$, is called the **unexplained variation (this is another name for *SSE*).** Intuitively, this quantity measures the amount of variation in the values of y that is not explained by the predictor variable. To calculate the unexplained variation (SSE) in the Tasty Sub Shop case, we square the 10 values of $y_i - \hat{y}_i$ in Table 14.6 and add together the 10 squared values, obtaining 30,460.21. This is also shown in Table 14.6 (and has been previously demonstrated in Table 14.2).

- The quantity $\Sigma(\hat{y}_i - \bar{y})^2$ is called the **explained variation.** Using the verbal names for $\Sigma(y_i - \bar{y})^2$, $\Sigma(y_i - \hat{y}_i)^2$, and $\Sigma(\hat{y}_i - \bar{y})^2$ and the fact that it can be shown that $\Sigma(y_i - \bar{y})^2 - \Sigma(y_i - \hat{y}_i)^2 = \Sigma(\hat{y}_i - \bar{y})^2$, we can say that

$$\text{Total variation} - \text{Unexplained variation} = \text{Explained variation}$$

It follows that the explained variation is the reduction in the sum of squared prediction errors that has been accomplished by using the predictor variable x to predict y. It also follows that

$$\text{Total variation} = \text{Explained variation} + \text{Unexplained variation}$$

Intuitively, this equation implies that the explained variation represents the amount of the total variation in the observed values of y that is explained by the predictor variable x (and the simple linear regression model). For example, the explained variation for the Tasty Sub Shop simple linear regression model is

$$\text{Explained variation} = \text{Total variation} - \text{Unexplained variation}$$
$$= 495{,}776.51 - 30{,}460.21 = 465{,}316.30$$

We now define the **simple coefficient of determination** to be

$$r^2 = \frac{\text{Explained variation}}{\text{Total variation}}$$

That is, r^2 is the proportion of the total variation in the n observed values of y that is explained by the simple linear regression model. For example, r^2 in the Tasty Sub Shop situation is

$$r^2 = \frac{\text{Explained variation}}{\text{Total variation}} = \frac{465{,}316.30}{495{,}776.51} = .939$$

This value of r^2 says that the Tasty Sub Shop simple linear regression model explains 93.9 percent of the total variation in the 10 observed yearly revenues. In general, neither the explained variation nor the total variation can be negative (both quantities are sums of squares). Therefore, r^2 is always greater than or equal to 0. Because the explained variation must be less than or equal to the total variation, r^2 cannot be greater than 1. The nearer r^2 is to 1, the larger is the proportion of the total variation that is explained by the simple linear regression model, and the greater is the potential utility of the model in predicting y. Note, however, that a value of r^2 close to 1 does not guarantee that the model will predict accurately enough for the practical needs of the situation being analyzed. We will see in Section 14.5 that perhaps the best way to assess prediction accuracy is to calculate a *prediction interval*. If the value of r^2 is not reasonably close to 1, the independent variable in the model is extremely unlikely to provide accurate predictions of y. In such a case, a different predictor variable must be found in order to accurately predict y. It is also possible that no regression model employing a single predictor variable will accurately predict y. In this case the model must be improved by including more than one independent variable. We show how to do this in Chapter 15. We summarize as follows:

The Simple Coefficient of Determination, r^2

For the simple linear regression model

1   **Total variation** $= \sum (y_i - \bar{y})^2$

2   **Explained variation** $= \sum (\hat{y}_i - \bar{y})^2$

3   **Unexplained variation** $= \sum (y_i - \hat{y}_i)^2$

4   **Total variation = Explained variation
    + Unexplained variation**

5   **The simple coefficient of determination is**

$$r^2 = \frac{\text{Explained variation}}{\text{Total variation}}$$

6   r^2 is the proportion of the total variation in the n observed values of the dependent variable that is explained by the simple linear regression model.

The simple correlation coefficient, r

People often claim that two variables are correlated. For example, a college admissions officer might feel that the academic performance of college students (measured by grade point average) is correlated with the students' scores on a standardized college entrance examination. This means that college students' grade point averages are related to their college entrance exam scores. One measure of the relationship between two variables y and x is the **simple correlation coefficient.** We define this quantity as follows:

The Simple Correlation Coefficient

The **simple correlation coefficient between y and x,** denoted by r, is

$$r = +\sqrt{r^2} \quad \text{if } b_1 \text{ is positive} \qquad \text{and} \qquad r = -\sqrt{r^2} \quad \text{if } b_1 \text{ is negative}$$

where b_1 is the slope of the least squares line relating y to x. This correlation coefficient **measures the strength of the linear relationship between y and x.**

Because r^2 is always between 0 and 1, the correlation coefficient r is between -1 and 1. A value of r near 0 implies little linear relationship between y and x. A value of r close to 1 says that y and x have a strong tendency to move together in a straight-line fashion with a positive slope and, therefore, that y and x are highly related and **positively correlated.** A value of r close to -1 says that y and x have a strong tendency to move together in a straight-line fashion with a negative slope and, therefore, that y and x are highly related and **negatively correlated.** Figure 14.8 illustrates these relationships. Notice that when $r = 1$, y and x have a perfect linear relationship with a positive slope, whereas when $r = -1$, y and x have a perfect linear relationship with a negative slope.

For example, in the Tasty Sub Shop case, we have found that $b_1 = 15.596$ and $r^2 = .939$. It follows that the simple correlation coefficient between y (yearly revenue) and x (population size) is

$$r = +\sqrt{r^2} = +\sqrt{.939} = .969$$

This simple correlation coefficient says that x and y have a strong tendency to move together in a linear fashion with a positive slope. We have seen this tendency in Figure 14.1, which indicates that y and x are positively correlated.

If we have computed the least squares slope b_1 and r^2, the method given in the previous box provides the easiest way to calculate r. The simple correlation coefficient can also be calculated using the formula

$$r = \frac{SS_{xy}}{\sqrt{SS_{xx}\,SS_{yy}}}$$

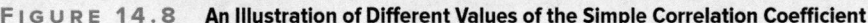

FIGURE 14.8 An Illustration of Different Values of the Simple Correlation Coefficient

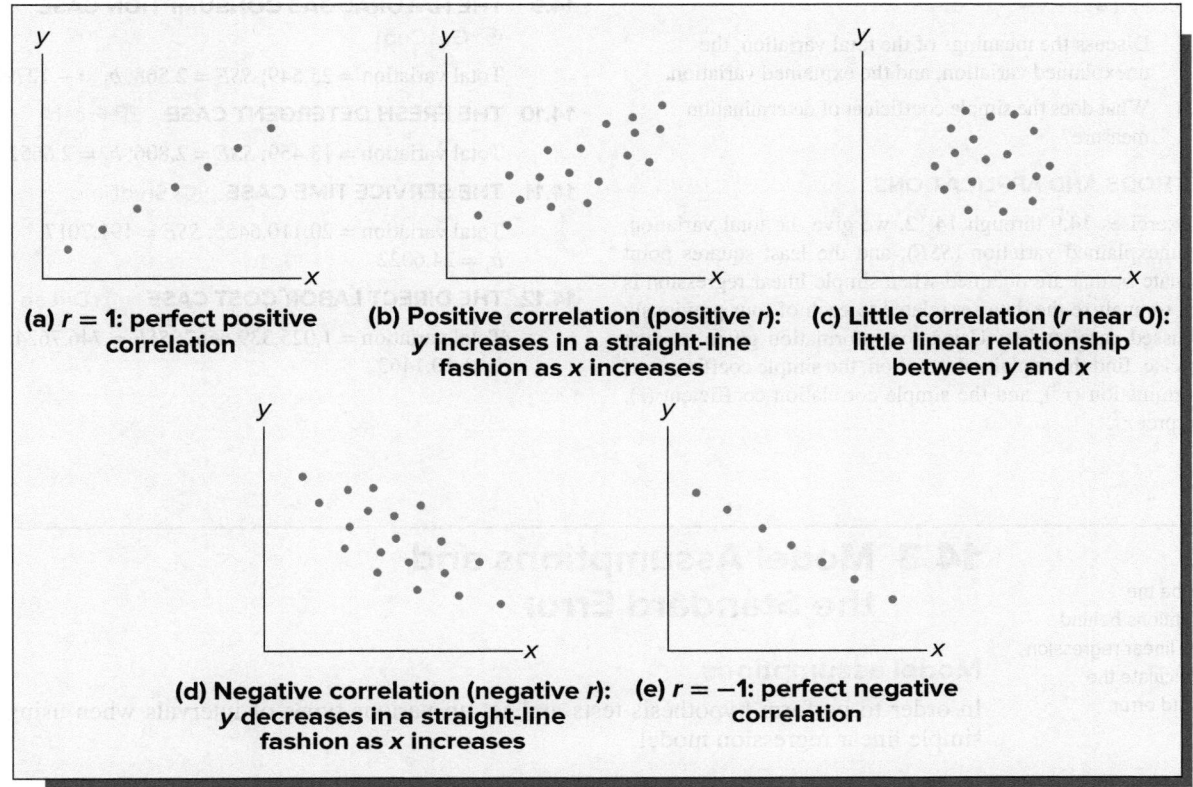

Here SS_{xy} and SS_{xx} have been defined in Section 14.1, and SS_{yy} denotes the total variation, which has been defined in this section. Furthermore, this formula for r automatically gives r the correct ($+$ or $-$) sign. For instance, in the Tasty Sub Shop case, $SS_{xy} = 29{,}836.389$, $SS_{xx} = 1913.129$, and $SS_{yy} = 495{,}776.51$ (see Example 14.2). Therefore:

$$r = \frac{SS_{xy}}{\sqrt{SS_{xx}\,SS_{yy}}}$$

$$= \frac{29{,}836.389}{\sqrt{(1{,}913.129)(495{,}776.51)}} = .969$$

It is important to make two points. First, **the value of the simple correlation coefficient is not the slope of the least squares line.** If we wish to find this slope, we should use the previously given formula for b_1. (It can be shown that b_1 and r are related by the equation $b_1 = (SS_{yy}/SS_{xx})^{1/2}r$.) Second, **high correlation does not imply that a cause-and-effect relationship exists.** When r indicates that y and x are highly correlated, this says that y and x have a strong tendency to move together in a straight-line fashion. The correlation does not mean that changes in x cause changes in y. Instead, some other variable (or variables) could be causing the apparent relationship between y and x. For example, suppose that college students' grade point averages and college entrance exam scores are highly positively correlated. This does not mean that earning a high score on a college entrance exam causes students to receive a high grade point average. Rather, other factors such as intellectual ability, study habits, and attitude probably determine both a student's score on a college entrance exam and a student's college grade point average. In general, while the simple correlation coefficient can show that variables tend to move together in a straight-line fashion, scientific theory must be used to establish cause-and-effect relationships.

Exercises for Section 14.2

CONCEPTS

14.7 Discuss the meanings of the total variation, the unexplained variation, and the explained variation.

14.8 What does the simple coefficient of determination measure?

METHODS AND APPLICATIONS

In Exercises 14.9 through 14.12, we give the total variation, the unexplained variation (*SSE*), and the least squares point estimate b_1 that are obtained when simple linear regression is used to analyze the data set related to each of four previously discussed case studies. Using the information given in each exercise, find the explained variation, the simple coefficient of determination (r^2), and the simple correlation coefficient (r). Interpret r^2.

14.9 THE NATURAL GAS CONSUMPTION CASE ⓄⓈ GasCon1

Total variation = 25.549; *SSE* = 2.568; $b_1 = -.12792$

14.10 THE FRESH DETERGENT CASE ⓄⓈ Fresh

Total variation = 13.459; *SSE* = 2.806; $b_1 = 2.6652$

14.11 THE SERVICE TIME CASE ⓄⓈ SrvcTime

Total variation = 20,110.5455; *SSE* = 191.7017; $b_1 = 24.6022$

14.12 THE DIRECT LABOR COST CASE ⓄⓈ DirLab

Total variation = 1,025,339.6667; *SSE* = 746.7624; $b_1 = 10.1463$

LO14-4
Describe the assumptions behind simple linear regression and calculate the standard error.

14.3 Model Assumptions and the Standard Error

Model assumptions

In order to perform hypothesis tests and set up various types of intervals when using the simple linear regression model

$$y = \beta_0 + \beta_1 x + \varepsilon$$

we need to make certain assumptions about the error term ε. At any given value of x, there is a population of error term values that could potentially occur. These error term values describe the different potential effects on y of all factors other than the value of x. Therefore, these error term values explain the variation in the y values that could be observed when the independent variable is x. Our statement of the simple linear regression model assumes that μ_y, the mean of the population of all y values that could be observed when the independent variable is x, is $\beta_0 + \beta_1 x$. This model also implies that $\varepsilon = y - (\beta_0 + \beta_1 x)$, so this is equivalent to assuming that the mean of the corresponding population of potential error term values is 0. In total, we make four assumptions (called the **regression assumptions**) about the simple linear regression model. These assumptions can be stated in terms of potential y values or, equivalently, in terms of potential error term values. Following tradition, we begin by stating these assumptions in terms of potential error term values:

The Regression Assumptions

1 At any given value of x, the population of potential error term values has a **mean equal to 0.**

2 **Constant Variance Assumption**
At any given value of x, the population of potential error term values has a variance that does not depend on the value of x. That is, the different populations of potential error term values corresponding to different values of x have **equal variances.** We denote the **constant variance as σ^2.**

3 **Normality Assumption**
At any given value of x, the population of potential error term values has a **normal distribution.**

4 **Independence Assumption**
Any one value of the error term ε is **statistically independent** of any other value of ε. That is, the value of the error term ε corresponding to an observed value of y is statistically independent of the value of the error term corresponding to any other observed value of y.

FIGURE 14.9   **An Illustration of the Model Assumptions**

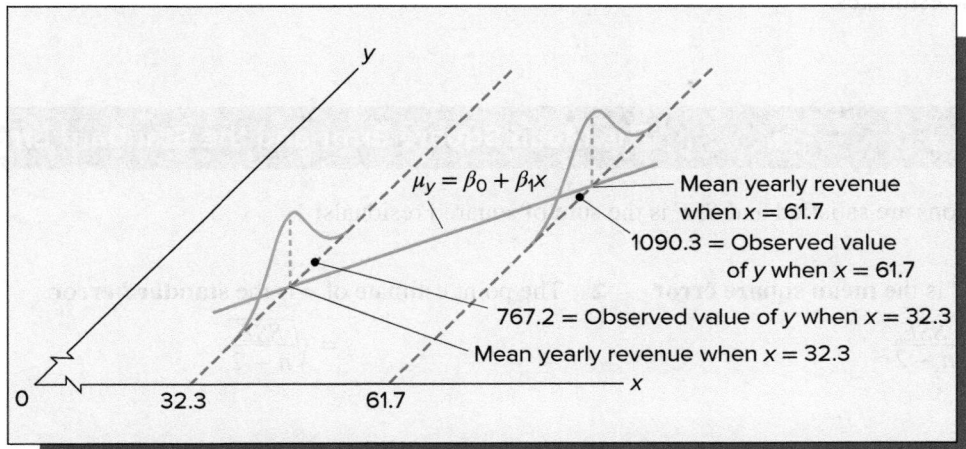

Taken together, the first three assumptions say that, at any given value of x, the population of potential error term values is **normally distributed** with **mean zero** and a **variance** σ^2 **that does not depend on the value of** x. Because the potential error term values cause the variation in the potential y values, these assumptions imply that the population of all y values that could be observed when the independent variable is x is **normally distributed** with **mean** $\beta_0 + \beta_1 x$ and **a variance** σ^2 **that does not depend on** x. These three assumptions are illustrated in Figure 14.9 in the context of the Tasty Sub Shop problem. Specifically, this figure depicts the populations of yearly revenues corresponding to two values of the population size x—32.3 and 61.7. Note that these populations are shown to be normally distributed with different means (each of which is on the line of means) and with the same variance (or spread).

The independence assumption is most likely to be violated when time series data are being utilized in a regression study. For example, the natural gas consumption data in Exercise 14.3 are time series data. Intuitively, the independence assumption says that there is no pattern of positive error terms being followed (in time) by other positive error terms, and there is no pattern of positive error terms being followed by negative error terms. That is, there is no pattern of higher-than-average y values being followed by other higher-than-average y values, and there is no pattern of higher-than-average y values being followed by lower-than-average y values.

It is important to point out that the regression assumptions very seldom, if ever, hold exactly in any practical regression problem. However, it has been found that regression results are not extremely sensitive to mild departures from these assumptions. In practice, only pronounced departures from these assumptions require attention. In Section 14.7 we show how to check the regression assumptions. Prior to doing this, we will suppose that the assumptions are valid in our examples.

In Section 14.1 we stated that, when we predict an individual value of the dependent variable, we predict the error term to be 0. To see why we do this, note that the regression assumptions state that, at any given value of the independent variable, the population of all error term values that can potentially occur is normally distributed with a mean equal to 0. Because we also assume that successive error terms (observed over time) are statistically independent, each error term has a 50 percent chance of being positive and a 50 percent chance of being negative. Therefore, it is reasonable to predict any particular error term value to be 0.

The mean square error and the standard error

To present statistical inference formulas in later sections, we need to be able to compute point estimates of σ^2 and σ, the constant variance and standard deviation of the error term populations. The point estimate of σ^2 is called the **mean square error** and the point estimate of σ is called the **standard error.** In the following box, we show how to compute these estimates:

The Mean Square Error and the Standard Error

I f the regression assumptions are satisfied and *SSE* is the sum of squared residuals:

1   The point estimate of σ^2 is the **mean square error**

$$s^2 = \frac{SSE}{n - 2}$$

2   The point estimate of σ is the **standard error**

$$s = \sqrt{\frac{SSE}{n - 2}}$$

In order to understand these point estimates, recall that σ^2 is the variance of the population of y values (for a given value of x) around the mean value μ_y. Because $\hat{y}$ is the point estimate of this mean, it seems natural to use

$$SSE = \sum(y_i - \hat{y}_i)^2$$

to help construct a point estimate of σ^2. We divide *SSE* by $n - 2$ because it can be proven that doing so makes the resulting s^2 an unbiased point estimate of σ^2. Here we call $n - 2$ the **number of degrees of freedom** associated with *SSE*.

Ⓒ **EXAMPLE 14.3** The Tasty Sub Shop Case: The Standard Error

Consider the Tasty Sub Shop situation, and recall that in Table 14.2 we have calculated the sum of squared residuals to be $SSE = 30,460.21$. It follows, because we have observed $n = 10$ yearly revenues, that the point estimate of σ^2 is the mean square error

$$s^2 = \frac{SSE}{n - 2} = \frac{30,460.21}{10 - 2} = 3807.526$$

This implies that the point estimate of σ is the standard error

$$s = \sqrt{s^2} = \sqrt{3807.526} = 61.7052$$

Exercises for Section 14.3

CONCEPTS **connect**

14.13 What four assumptions do we make about the simple linear regression model?

14.14 What is estimated by the mean square error, and what is estimated by the standard error?

METHODS AND APPLICATIONS

14.15 THE NATURAL GAS CONSUMPTION CASE
Ⓓ GasCon1

When a least squares line is fit to the 8 observations in the natural gas consumption data, we obtain $SSE = 2.568$. Calculate s^2 and s.

14.16 THE FRESH DETERGENT CASE 🅳🆂 Fresh

When a least squares line is fit to the 30 observations in the Fresh detergent data, we obtain $SSE = 2.806$. Calculate s^2 and s.

14.17 THE SERVICE TIME CASE 🅳🆂 SrvcTime

When a least squares line is fit to the 11 observations in the service time data, we obtain $SSE = 191.7017$. Calculate s^2 and s.

14.18 THE DIRECT LABOR COST CASE 🅳🆂 DirLab

When a least squares line is fit to the 12 observations in the labor cost data, we obtain $SSE = 746.7624$. Calculate s^2 and s.

14.4 Testing the Significance of the Slope and *y*-Intercept

LO14-5

Test the significance of the slope and *y*-intercept.

A *t* test for the significance of the slope

A simple linear regression model is not likely to be useful unless there is a **significant relationship between *y* and *x*.** In order to judge the significance of the relationship between *y* and *x*, we test the null hypothesis

$$H_0: \beta_1 = 0$$

which says that there is no change in the mean value of *y* associated with an increase in *x*, versus the alternative hypothesis

$$H_a: \beta_1 \neq 0$$

which says that there is a (positive or negative) change in the mean value of *y* associated with an increase in *x*. It would be reasonable to conclude that *x* is significantly related to *y* if we can be quite certain that we should reject H_0 in favor of H_a.

In order to test these hypotheses, recall that we compute the least squares point estimate b_1 of the true slope β_1 by using a sample of *n* observed values of the dependent variable *y*. Different samples of *n* observed *y* values would yield different values of the least squares point estimate b_1. It can be shown that, if the regression assumptions hold, then the population of all possible values of b_1 is normally distributed with a mean of β_1 and with a standard deviation of

$$\sigma_{b_1} = \frac{\sigma}{\sqrt{SS_{xx}}}$$

The standard error *s* is the point estimate of σ, so it follows that a point estimate of σ_{b_1} is

$$s_{b_1} = \frac{s}{\sqrt{SS_{xx}}}$$

which is called the **standard error of the estimate b_1.** Furthermore, if the regression assumptions hold, then the population of all values of

$$\frac{b_1 - \beta_1}{s_{b_1}}$$

has a *t* distribution with $n - 2$ degrees of freedom. It follows that, if the null hypothesis $H_0: \beta_1 = 0$ is true, then the population of all possible values of the test statistic

$$t = \frac{b_1}{s_{b_1}}$$

has a *t* distribution with $n - 2$ degrees of freedom. Therefore, we can test the significance of the regression relationship as follows:

Testing the Significance of the Regression Relationship: Testing the Significance of the Slope

Null Hypothesis H_0: $\beta_1 = 0$ **Test Statistic** $t = \dfrac{b_1}{s_{b_1}}$ where $s_{b_1} = \dfrac{s}{\sqrt{SS_{xx}}}$ **Assumptions** The regression assumptions

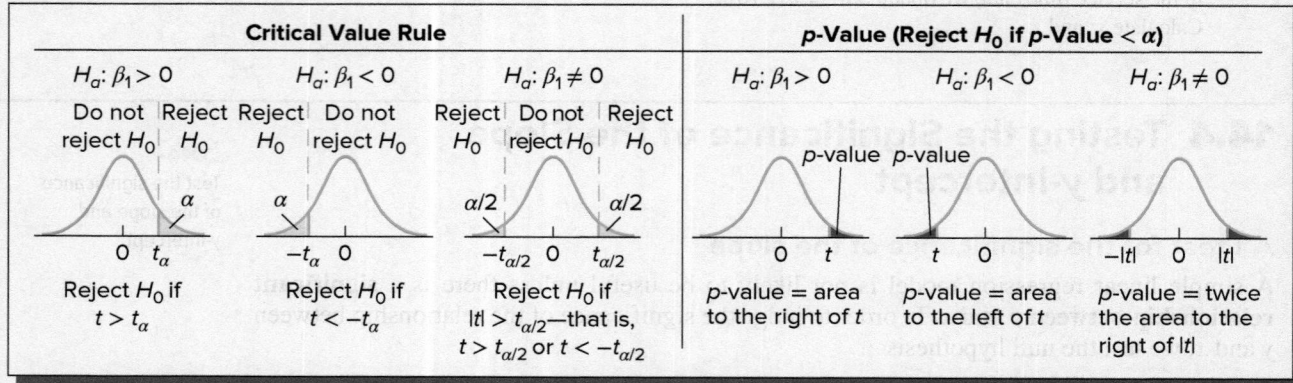

Critical Value Rule			p-Value (Reject H_0 if p-Value $< \alpha$)		
H_a: $\beta_1 > 0$	H_a: $\beta_1 < 0$	H_a: $\beta_1 \neq 0$	H_a: $\beta_1 > 0$	H_a: $\beta_1 < 0$	H_a: $\beta_1 \neq 0$
Reject H_0 if $t > t_\alpha$	Reject H_0 if $t < -t_\alpha$	Reject H_0 if $\lvert t \rvert > t_{\alpha/2}$—that is, $t > t_{\alpha/2}$ or $t < -t_{\alpha/2}$	p-value = area to the right of t	p-value = area to the left of t	p-value = twice the area to the right of $\lvert t \rvert$

Here $t_{\alpha/2}$, t_α, and all p-values are based on $n - 2$ degrees of freedom. **If we can reject H_0: $\beta_1 = 0$ at a given value of α, then we conclude that the slope (or, equivalently, the regression relationship) is significant at the α level.**

We usually use the two-sided alternative H_a: $\beta_1 \neq 0$ for this test of significance. However, sometimes a one-sided alternative is appropriate. For example, in the Tasty Sub Shop problem we can say that if the slope β_1 is not 0, then it must be positive. A positive β_1 would say that mean yearly revenue increases as the population size x increases. Because of this, it would be appropriate to decide that x is significantly related to y if we can reject H_0: $\beta_1 = 0$ in favor of the one-sided alternative H_a: $\beta_1 > 0$. Although this test would be slightly more effective than the usual two-tailed test, there is little practical difference between using the one-tailed or two-tailed test. Furthermore, computer packages (such as Excel, Minitab, and JMP) present results for the two-tailed test. For these reasons we will emphasize the two-tailed test in future discussions.

It should also be noted that

1 **If we can decide that the slope is significant at the .05 significance level,** then we have concluded that x is significantly related to y by using a test that allows only a .05 probability of concluding that x is significantly related to y when it is not. **This is usually regarded as strong evidence that the regression relationship is significant.**

2 **If we can decide that the slope is significant at the .01 significance level, this is usually regarded as very strong evidence that the regression relationship is significant.**

3 **The smaller the significance level α at which H_0 can be rejected, the stronger is the evidence that the regression relationship is significant.**

C EXAMPLE 14.4 The Tasty Sub Shop Case: Testing the Significance of the Slope

Again consider the Tasty Sub Shop revenue model. For this model $SS_{xx} = 1913.129$, $b_1 = 15.596$, and $s = 61.7052$ (see Examples 14.2 and 14.3). Therefore,

$$s_{b_1} = \frac{s}{\sqrt{SS_{xx}}} = \frac{61.7052}{\sqrt{1913.129}} = 1.411$$

and

$$t = \frac{b_1}{s_{b_1}} = \frac{15.596}{1.411} = 11.05$$

FIGURE 14.10   Excel and Minitab Outputs of a Simple Linear Regression Analysis of the Tasty Sub Shop Revenue Data

(a) The Excel Output

Regression Statistics

Multiple R	0.9688
R Square	0.9386 [9]
Adjusted R Square	0.9309
Standard Error	61.7052 [8]
Observations	10

ANOVA	df	SS	MS	F	Significance F
Regression	1	465316.3004 [10]	465316.3004	122.2096 [13]	0.0000 [14]
Residual	8	30460.2086 [11]	3807.5261		
Total	9	495776.5090 [12]			

	Coefficients	Standard Error	t Stat	P-value [7]	Lower 95%	Upper 95%
Intercept	183.3051 [1]	64.2741 [3]	2.8519 [5]	0.0214	35.0888	331.5214
Population	15.5956 [2]	1.4107 [4]	11.0548 [6]	0.0000	12.3424 [19]	18.8488 [19]

(b) The Minitab Output

Analysis of Variance

Source	DF	Adj SS	Adj MS	F-Value	P-Value
Regression	1	465316 [10]	465316	122.21 [13]	0.000 [14]
Population	1	465316	465316	122.21	0.000
Error	8	30460 [11]	3808		
Total	9	495777 [12]			

Model Summary

S	R-sq	R-sq(adj)	R-sq(pred)
61.7052 [8]	93.86% [9]	93.09%	90.74%

Coefficients

Term	Coef	SE Coef	T-Value	P-Value [7]	VIF
Constant	183.3 [1]	64.3 [3]	2.85 [5]	0.021	
Population	15.60 [2]	1.41 [4]	11.05 [6]	0.000	1.00

Regression Equation

Revenue = 183.3 + 15.60 Population

Variable	Setting	Fit [15]	SE Fit [16]	95% CI [17]	95% PI [18]
Population	47.3	920.977	20.2699	(874.234, 967.719)	(771.204, 1070.75)

[1] b_0 = point estimate of the y-intercept [2] b_1 = point estimate of the slope [3] s_{b_0} = standard error of the estimate b_0 [4] s_{b_1} = standard error of the estimate b_1 [5] t for testing significance of the y-intercept [6] t for testing significance of the slope [7] p-values for t statistics [8] s = standard error [9] r^2 [10] Explained variation [11] SSE = Unexplained variation [12] Total variation [13] F(model) statistic [14] p-value for F(model) [15] $\hat{y}$ = point prediction when x = 47.3 [16] $s_{\hat{y}}$ = standard error of the estimate $\hat{y}$ [17] 95% confidence interval when x = 47.3 [18] 95% prediction interval when x = 47.3 [19] 95% confidence interval for the slope β_1

Figure 14.10 presents the Excel and Minitab outputs of a simple linear regression analysis of the Tasty Sub Shop revenue data. Note that b_0 (labeled as [1] on the outputs), b_1 (labeled [2]), s (labeled [8]), r^2 (labeled [9]), the explained variation (labeled [10]), the unexplained variation (labeled [11]), the total variation (labeled [12]), s_{b_1} (labeled [4]), and t (labeled [6]) are given on each of these outputs. (The other quantities on the outputs will be discussed later.) In order to test H_0: $\beta_1 = 0$ versus H_a: $\beta_1 \neq 0$ at the $\alpha = .05$ level of significance, we compare $|t| = 11.05$ with $t_{\alpha/2} = t_{.025} = 2.306$, which is based on $n - 2 = 10 - 2 = 8$ degrees of freedom. Because $|t| = 11.05$ is greater than $t_{.025} = 2.306$, we reject H_0: $\beta_1 = 0$ and conclude that there is strong evidence that the slope (regression relationship) is significant. The p-value for testing H_0 versus H_a is twice the area to the right of $|t| = 11.05$ under the curve of the t distribution having $n - 2 = 8$ degrees of freedom. Both the Excel and Minitab outputs in Figure 14.10 tell us that this p-value is less than .001 (see [7] on the outputs). It follows that we can reject H_0 in favor of H_a at level of significance .05, .01, or .001, which implies that we have extremely strong evidence that the regression relationship between x and y is significant.

If the regression assumptions hold, a **100(1 − α) percent confidence interval for the true slope β_1** is $[b_1 \pm t_{\alpha/2}\, s_{b_1}]$. Here $t_{\alpha/2}$ is based on $n - 2$ degrees of freedom.

Ⓒ **EXAMPLE 14.5** The Tasty Sub Shop Case: A Confidence Interval for the Slope

The Excel output in Figure 14.10(a) tells us that $b_1 = 15.5956$ and $s_{b_1} = 1.4107$. Thus, for instance, because $t_{.025}$ based on $n - 2 = 10 - 2 = 8$ degrees of freedom equals 2.306, a 95 percent confidence interval for β_1 is

$$[b_1 \pm t_{.025}s_{b_1}] = [15.5956 \pm 2.306(1.4107)]$$

$$= [12.342, 18.849]$$

(where we have used more decimal place accuracy than shown to obtain the final result). This interval says we are 95 percent confident that, if the population size increases by one thousand residents, then mean yearly revenue will increase by at least $12,342 and by at most $18,849. Also, because the 95 percent confidence interval for β_1 does not contain 0, we can reject H_0: $\beta_1 = 0$ in favor of H_a: $\beta_1 \neq 0$ at level of significance .05. Note that the 95 percent confidence interval for β_1 is given on the Excel output but not on the Minitab output (see Figure 14.10).

Testing the significance of the y-intercept

We can also test the significance of the y-intercept β_0. We do this by testing the null hypothesis H_0: $\beta_0 = 0$ versus the alternative hypothesis H_a: $\beta_0 \neq 0$. **If we can reject H_0 in favor of H_a by setting the probability of a Type I error equal to α, we conclude that the intercept β_0 is significant at the α level.** To carry out the hypothesis test, we use the test statistic

$$t = \frac{b_0}{s_{b_0}} \quad \text{where} \quad s_{b_0} = s\sqrt{\frac{1}{n} + \frac{\bar{x}^2}{SS_{xx}}}$$

Here the critical value and p-value conditions for rejecting H_0 are the same as those given previously for testing the significance of the slope, except that t is calculated as b_0/s_{b_0}. For example, if we consider the Tasty Sub Shop problem and the Excel output in Figure 14.10(a), we see that $b_0 = 183.3051$, $s_{b_0} = 64.2741$, $t = 2.8519$, and p-value = .0214. Because $t = 2.8519 > t_{.025} = 2.306$ and p-value < .05, we can reject H_0: $\beta_0 = 0$ in favor of H_a: $\beta_0 \neq 0$ at the .05 level of significance. This provides strong evidence that the y-intercept β_0 of the line of means does not equal 0 and thus is significant. Therefore, we should include β_0 in the Tasty Sub Shop revenue model.

In general, if we fail to conclude that the intercept is significant at a level of significance of .05, it might be reasonable to drop the y-intercept from the model. However, it is common practice to include the y-intercept whether or not H_0: $\beta_0 = 0$ is rejected. In fact, experience suggests that it is definitely safest, when in doubt, to include the intercept β_0.

LO14-6

Test the significance of a simple linear regression model by using an F test (Optional).

An F test for the significance of the slope (Optional)

A second way to test the null hypothesis H_0: $\beta_1 = 0$ (the regression relationship between x and y is not significant) versus H_a: $\beta_1 \neq 0$ (the regression relationship between x and y is significant) is to use a hypothesis test based on the F distribution. The following summary box describes this F test. If we can reject H_0 at level of significance α, we often say that **the simple linear regression model is significant at level of significance α.**

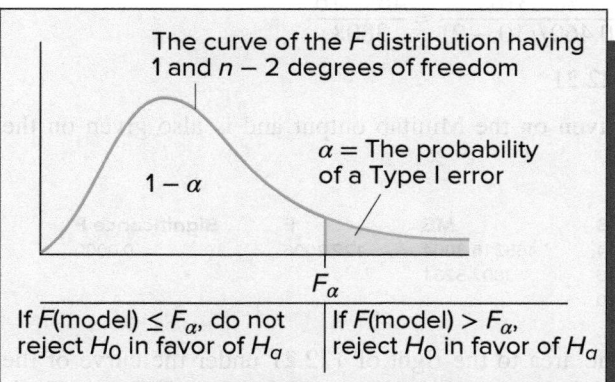

FIGURE 14.11(a) The F Test Critical Value

The curve of the F distribution having 1 and n − 2 degrees of freedom

$1 − \alpha$

α = The probability of a Type I error

F_{α}

If $F(\text{model}) \leq F_{\alpha}$, do not reject H_0 in favor of H_a | If $F(\text{model}) > F_{\alpha}$, reject H_0 in favor of H_a

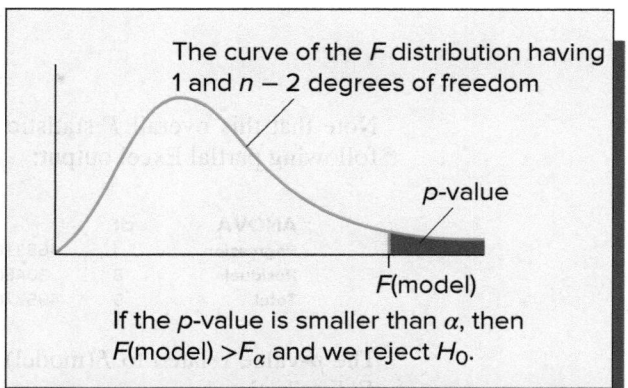

FIGURE 14.11(b) The F Test p-Value

The curve of the F distribution having 1 and n − 2 degrees of freedom

p-value

F(model)

If the p-value is smaller than α, then $F(\text{model}) > F_{\alpha}$ and we reject H_0.

An F Test for the Simple Linear Regression Model

Suppose that the regression assumptions hold, and define the **overall F statistic** to be

$$F(\text{model}) = \frac{\text{Explained variation}}{(\text{Unexplained variation})/(n-2)}$$

Also define the p-value related to $F(\text{model})$ to be the area under the curve of the F distribution (having 1 numerator and $n-2$ denominator degrees of freedom) to the right of $F(\text{model})$.

We can reject $H_0: \beta_1 = 0$ in favor of $H_a: \beta_1 \neq 0$ at level of significance α if either of the following equivalent conditions holds:

1 $F(\text{model}) > F_{\alpha}$
2 p-value $< \alpha$

Here the point F_{α} is based on 1 numerator and $n-2$ denominator degrees of freedom.

The first condition in the box says we should reject $H_0: \beta_1 = 0$ (and conclude that the relationship between x and y is significant) when $F(\text{model})$ is large. This is intuitive because a large overall F statistic would be obtained when the explained variation is large compared to the unexplained variation. This would occur if x is significantly related to y, which would imply that the slope β_1 is not equal to 0. Figure 14.11(a) illustrates that we reject H_0 when $F(\text{model})$ is greater than F_{α}. As can be seen in Figure 14.11(b), when $F(\text{model})$ is large, the related p-value is small. When the p-value is small enough [resulting from an $F(\text{model})$ statistic that is large enough], we reject H_0. Figure 14.11(b) illustrates that the second condition in the box (p-value $< \alpha$) is an equivalent way to carry out this test.

Ⓒ EXAMPLE 14.6 The Tasty Sub Shop Case: An F Test for the Model

Consider the Tasty Sub Shop problem and the following partial Minitab output of the simple linear regression analysis relating yearly revenue y to population size x.

Analysis of Variance

Source	DF	Adj SS	Adj MS	F-Value	P-Value
Regression	1	465316	465316	122.21	0.000
Population	1	465316	465316	122.21	0.000
Error	8	30460	3808		
Total	9	495777			

Looking at this output, we see that the explained variation is 465,316 and the unexplained variation is 30,460. It follows that

$$F(\text{model}) = \frac{\text{Explained variation}}{(\text{Unexplained variation})/(n-2)}$$

$$= \frac{465{,}316}{30{,}460/(10-2)} = \frac{465{,}316}{3808}$$

$$= 122.21$$

Note that this overall F statistic is given on the Minitab output and is also given on the following partial Excel output:

ANOVA	df	SS	MS	F	Significance F
Regression	1	465316.3004	465316.3004	122.2096	0.0000
Residual	8	30460.2086	3807.5261		
Total	9	495776.5090			

The p-value related to $F(\text{model})$ is the area to the right of 122.21 under the curve of the F distribution having 1 numerator and 8 denominator degrees of freedom. This p-value is given on both the Minitab output and the Excel output (labeled "Significance F") and is less than .001. If we wish to test the significance of the regression relationship with level of significance $\alpha = .05$, we use the critical value $F_{.05}$ based on 1 numerator and 8 denominator degrees of freedom. Using Table A.7, we find that $F_{.05} = 5.32$. Because $F(\text{model}) = 122.21 > F_{.05} = 5.32$, we can reject $H_0: \beta_1 = 0$ in favor of $H_a: \beta_1 \neq 0$ at level of significance .05. Alternatively, because the p-value is smaller than .05, .01, and .001, we can reject H_0 at level of significance .05, .01, or .001. Therefore, we have extremely strong evidence that $H_0: \beta_1 = 0$ should be rejected and that the regression relationship between x and y is significant. That is, we might say that we have extremely strong evidence that the simple linear model relating y to x is significant.

Testing the significance of the regression relationship between y and x by using the overall F statistic and its related p-value is equivalent to doing this test by using the t statistic and its related p-value. Specifically, it can be shown that $(t)^2 = F(\text{model})$ and that $(t_{\alpha/2})^2$ based on $n - 2$ degrees of freedom equals F_α based on 1 numerator and $n - 2$ denominator degrees of freedom. It follows that the critical value conditions

$$|t| > t_{\alpha/2} \quad \text{and} \quad F(\text{model}) > F_\alpha$$

are equivalent. Furthermore, the p-values related to t and $F(\text{model})$ can be shown to be equal. Because these tests are equivalent, it would be logical to ask why we have presented the F test. There are two reasons. First, most standard regression computer packages include the results of the F test as a part of the regression output. Second, the F test has a useful generalization in multiple regression analysis (where we employ more than one predictor variable). The F test in multiple regression is not equivalent to a t test. This is further explained in Chapter 15.

Exercises for Section 14.4

CONCEPTS

14.19 What do we conclude if we can reject $H_0: \beta_1 = 0$ in favor of $H_a: \beta_1 \neq 0$ by setting
    a α equal to .05?   **b** α equal to .01?

14.20 Give an example of a practical application of the confidence interval for β_1.

METHODS AND APPLICATIONS

In Exercises 14.21 through 14.26, we refer to Excel, Minitab, and JMP outputs of simple linear regression analyses of data sets related to six case studies (four case studies introduced in the exercises for Section 14.1 and two new case studies). Using the appropriate output for each case study, do the

following parts a through i, and, if you have studied the optional F test, also do parts j through m.

a Find the least squares point estimates b_0 and b_1 of β_0 and β_1 on the output and report their values.

b Find the explained variation, the unexplained variation (SSE), the total variation, r^2, and s on the computer output and report their values.

c Find s_{b_1} and the t statistic for testing the significance of the slope on the output and report their values. Show (within rounding) how t has been calculated by using b_1 and s_{b_1} from the computer output.

d Using the t statistic and an appropriate critical value, test $H_0: \beta_1 = 0$ versus $H_a: \beta_1 \neq 0$ by setting α equal to .05. Is the slope (regression relationship) significant at the .05 level?

e Find the *p*-value for testing $H_0: \beta_1 = 0$ versus $H_a: \beta_1 \neq 0$ on the output and report its value. Using the *p*-value, determine whether we can reject H_0 by setting α equal to .10, .05, .01, and .001. How much evidence is there that the slope (regression relationship) is significant?

f Calculate the 95 percent confidence interval for β_1 using numbers on the output. Interpret the interval.

g Find s_{b_0} and the *t* statistic for testing the significance of the *y*-intercept on the output and report their values. Show (within rounding) how *t* has been calculated by using b_0 and s_{b_0} from the computer output.

h Find the *p*-value for testing $H_0: \beta_0 = 0$ versus $H_a: \beta_0 \neq 0$ on the computer output and report its value. Using the *p*-value, determine whether we can reject H_0 by setting α equal to .10, .05, .01, and .001. What do you conclude about the significance of the *y*-intercept?

i Using the data set and *s* from the computer output, hand calculate (within rounding) SS_{xx}, s_{b_0}, and s_{b_1}.

j Use the explained variation and the unexplained variation as given on the computer output to calculate (within rounding) the *F*(model) statistic.

k Utilize the *F*(model) statistic and the appropriate critical value to test $H_0: \beta_1 = 0$ versus $H_a: \beta_1 \neq 0$ by setting α equal to .05. What do you conclude about the regression relationship between *y* and *x*?

l Find the *p*-value related to *F*(model) on the computer output and report its value. Using the *p*-value, test the significance of the regression model at the .10, .05, .01, and .001 levels of significance. What do you conclude?

m Show that the *F*(model) statistic is (within rounding) the square of the *t* statistic for testing $H_0: \beta_1 = 0$ versus $H_a: \beta_1 \neq 0$. Also, show that the $F_{.05}$ critical value is the square of the $t_{.025}$ critical value.

14.21 THE NATURAL GAS CONSUMPTION CASE
   🗂 GasCon1

The Excel and Minitab outputs of a simple linear regression analysis of the data set for this case (see Exercise 14.3) are given in Figures 14.12 and 14.13. Recall that labeled Excel and Minitab outputs are in Figure 14.10.

14.22 THE FRESH DETERGENT CASE   🗂 Fresh

The JMP output of a simple linear regression analysis of the data set for this case (see Table 14.3) is given in Figure 14.14.

14.23 THE SERVICE TIME CASE   🗂 SrvcTime

The Excel output of a simple linear regression analysis of the data set for this case (see Table 14.4) is given in Figure 14.15.

14.24 THE DIRECT LABOR COST CASE   🗂 DirLab

The Excel output of a simple linear regression analysis of the data set for this case (see Table 14.5) is given in Figure 14.16.

14.25 THE SALES VOLUME CASE   🗂 SalesPlot

Obtain a computer output of a simple linear regression analysis of the sales volume data in Figure 3.22(a) using Excel, Minitab or JMP. In Figure 3.22(a), *y* is the sales volume (in thousands of units) of a product and *x* is the amount spent (in units of $10,000) on advertising the product. When the output is obtained, analyze it as previously described.

14.26 THE REAL ESTATE SALES PRICE CASE   🗂 RealEst

Obtain a computer output of a simple linear regression analysis of the real estate sales price data in Table 14.7 using Excel, Minitab, or JMP. In Table 14.7, *y* is the sales price of a house (in thousands of dollars) and *x* is the size of the house (in hundreds of square feet). When the output is obtained, analyze it as previously described.

TABLE 14.7
The Real Estate Sales Price Data
🗂 RealEst

y	x
180	23
98.1	11
173.1	20
136.5	17
141	15
165.9	21
193.5	24
127.8	13
163.5	19
172.5	25

Source: Data from R. L. Andrews and J. T. Ferguson, "Integrating Judgement with a Regression Appraisal," *The Real Estate Appraiser and Analyst* 52, no. 2 (1986).

FIGURE 14.12   **Excel Output of a Simple Linear Regression Analysis of the Natural Gas Consumption Data (for Exercise 14.21)**

Regression Statistics

Multiple R	0.9484
R Square	0.8995
Adjusted R Square	0.8827
Standard Error	0.6542
Observations	8

ANOVA	df	SS	MS	F	Significance F
Regression	1	22.9808	22.9808	53.6949	0.0003
Residual	6	2.5679	0.4280		
Total	7	25.5488			

	Coefficients	Standard Error	t Stat	P-value	Lower 95%	Upper 95%
Intercept	15.8379	0.8018	19.7535	1.09E-06	13.8760	17.7997
TEMP	−0.1279	0.0175	−7.3277	0.0003	−0.1706	−0.0852

FIGURE 14.13   Minitab Output of a Simple Linear Regression Analysis of the Natural Gas Consumption Data
                (for Exercise 14.21)

Analysis of Variance

Source	DF	Adj SS	Adj MS	F-Value	P-Value
Regression	1	22.9808	22.9808	53.69	0.000
Temp	1	22.9808	22.9808	53.69	0.000
Error	6	2.5679	0.4280		
Lack-of-Fit	5	2.3229	0.4646	1.90	0.500
Pure Error	1	0.2450	0.2450		
Total	7	25.5488			

Model Summary

S	R-sq	R-sq(adj)	R-sq(pred)
0.654209	89.95%	88.27%	81.98%

Coefficients

Term	Coef	SE Coef	T-Value	P-Value	VIF
Constant	15.838	0.802	19.75	0.000	
Temp	−0.1279	0.0175	−7.33	0.000	1.00

Regression Equation

GasCons = 15.838 − 0.1279 Temp

Variable	Setting	Fit	SE Fit	95% CI	95% PI
Temp	40	10.7210	0.241483	(10.1301, 11.3119)	(9.01462, 12.4274)

Note: Minitab carries out a "lack-of-fit" F test if the data set has at least one x value for which there are at least two observed y values. For example, two weeks in the natural gas consumption data set on page 541 have an average hourly temperature of 28°F, and so there are two observed y values for the x value 28. If the p-value associated with the lack-of-fit F statistic is greater than .05 (it is .5 in this figure), then the lack-of-fit F test says that the fitted model shows little or no lack-of-fit to the observed data. In the authors' opinion, Minitab sometimes carries out the lack-of-fit F test when there are not enough data to get meaningful results. We will not further discuss this test.

FIGURE 14.14   JMP Output of a Simple Linear Regression Analysis of the Fresh Detergent Demand Data
                (for Exercise 14.22)

Summary of Fit

RSquare	0.791516
RSquare Adj	0.78407
Root Mean Square Error	0.316561
Mean of Response	8.382667
Observations (or Sum Wgts	30

Analysis of Variance

Source	DF	Sum of Squares	Mean Square	F Ratio
Model	1	10.652685	10.6527	106.3028
Error	28	2.805902	0.1002	Prob > F
C. Total	29	13.45857		<.0001*

Parameter Estimates

Term	Estimate	Std Error	t Ratio	Prob>|t|	Lower 95%	Upper 95%
Intercept	7.8140876	0.079884	97.82	<.0001*	7.650452	7.9777232
PriceDif	2.6652145	0.2585	10.31	<.0001*	2.1357021	3.1947269

	Predicted Demand	Lower 95% Mean Demand	Upper 95% Mean Demand	Lower 95% Indiv Demand	Upper 95% Indiv Demand	StdErr Pred Demand	StdErr Indiv Demand
31	8.0806090242	7.9478783815	8.213339667	7.4187185048	8.7424995436	0.0647970025	0.3231244931

14.27 In an article in *Public Roads* (1983), Bissell, Pilkington, Mason, and Woods study bridge safety y (measured in accident rates per 100 million vehicles) and x, the **difference** between the width of the bridge and the width of the roadway approach (road plus shoulder). The values of x analyzed by the authors were −6, −4, −2, 0, 2, 4, 6, 8, 10, and 12. The corresponding values of y were 120, 103, 87, 72, 58, 44, 31, 20, 12, and 7. Find and interpret a point estimate of and a 95 percent confidence interval for the slope, β_1, of a simple linear regression model describing these data.
Ⓓ AutoAcc

FIGURE 14.15 **Excel Output of a Simple Linear Regression Analysis of the Service Time Data (for Exercise 14.23)**

Regression Statistics

Multiple R	0.9952
R Square	0.9905
Adjusted R Square	0.9894
Standard Error	4.6152
Observations	11

ANOVA	df	SS	MS	F	Significance F
Regression	1	19918.8438	19918.844	935.149	2.094E-10
Residual	9	191.7017	21.300184		
Total	10	20110.5455			

	Coefficients	Standard Error	t Stat	P-value	Lower 95%	Upper 95%
Intercept	11.4641	3.4390	3.3335	0.0087	3.6845	19.2437
Copiers	24.6022	0.8045	30.5802	2.09E-10	22.7823	26.4221

FIGURE 14.16 **Excel Output of a Simple Linear Regression Analysis of the Direct Labor Cost Data (for Exercise 14.24)**

Regression Statistics

Multiple R	0.9996
R Square	0.9993
Adjusted R Square	0.9992
Standard Error	8.6415
Observations	12

ANOVA	df	SS	MS	F	Significance F
Regression	1	1024592.9043	1024592.9043	13720.4677	5.04E-17
Residual	10	746.7624	74.6762		
Total	11	1025339.6667			

	Coefficients	Standard Error	t Stat	P-value	Lower 95%	Upper 95%
Intercept	18.4875	4.6766	3.9532	0.0027	8.0674	28.9076
BatchSize (x)	10.1463	0.0866	117.1344	5.04E-17	9.9533	10.3393

14.5 Confidence and Prediction Intervals

If the regression relationship between y and x is significant, then

$$\hat{y} = b_0 + b_1 x_0$$

is the **point estimate of the mean value of y** when the value of the independent variable x is x_0. We have also seen that $\hat{y}$ is the **point prediction of an individual value of y** when the value of the independent variable x is x_0. In this section we will assess the accuracy of $\hat{y}$ as both a point estimate and a point prediction. To do this, we will find a **confidence interval for the mean value of y** and a **prediction interval for an individual value of y.**

Because each possible sample of n values of the dependent variable gives values of b_0 and b_1 that differ from the values given by other samples, different samples give different values of $\hat{y} = b_0 + b_1 x_0$. If the regression assumptions hold, a confidence interval for the mean value of y is based on the estimated standard deviation of the normally distributed population of all possible values of $\hat{y}$. This estimated standard deviation is called the **standard error of $\hat{y}$,** is denoted $s_{\hat{y}}$, and is given by the formula

$$s_{\hat{y}} = s \sqrt{\frac{1}{n} + \frac{(x_0 - \bar{x})^2}{SS_{xx}}}$$

LO14-7

Calculate and interpret a confidence interval for a mean value and a prediction interval for an individual value.

Here, s is the standard error (see Section 14.3), $\bar{x}$ is the average of the n previously observed values of x, and $SS_{xx} = \Sigma x_i^2 - (\Sigma x_i)^2/n$.

As explained above, a confidence interval for the mean value of y is based on the standard error $s_{\hat{y}}$. A prediction interval for an individual value of y is based on a more complex standard error: the estimated standard deviation of the normally distributed population of all possible values of $y - \hat{y}$. Here $y - \hat{y}$ is the prediction error obtained when predicting y by $\hat{y}$. We refer to this estimated standard deviation as the **standard error of $y - \hat{y}$** and denote it as $s_{(y - \hat{y})}$. If the regression assumptions hold, the formula for $s_{(y - \hat{y})}$ is

$$s_{(y - \hat{y})} = s \sqrt{1 + \frac{1}{n} + \frac{(x_0 - \bar{x})^2}{SS_{xx}}}$$

Intuitively, the "extra 1" under the radical in the formula for $s_{(y - \hat{y})}$ accounts for the fact that there is *more uncertainty* in predicting an individual value $y = \beta_0 + \beta_1 x_0 + \varepsilon$ than in estimating the mean value $\beta_0 + \beta_1 x_0$ (because we must predict the error term ε when predicting an individual value). Therefore, as shown in the following summary box, the prediction interval for an individual value of y is longer than the confidence interval for the mean value of y.

A Confidence Interval and a Prediction Interval

If the regression assumptions hold,

1 A **$100(1 - \alpha)$ percent confidence interval for the mean value of y** when x equals x_0 is

$$\left[\hat{y} \pm t_{\alpha/2} \, s \sqrt{\frac{1}{n} + \frac{(x_0 - \bar{x})^2}{SS_{xx}}} \right]$$

2 A **$100(1 - \alpha)$ percent prediction interval for an individual value of y** when x equals x_0 is

$$\left[\hat{y} \pm t_{\alpha/2} \, s \sqrt{1 + \frac{1}{n} + \frac{(x_0 - \bar{x})^2}{SS_{xx}}} \right]$$

Here, $t_{\alpha/2}$ is based on $(n - 2)$ degrees of freedom.

The summary box tells us that both the formula for the confidence interval and the formula for the prediction interval use the quantity $1/n + (x_0 - \bar{x})^2/SS_{xx}$. We will call this quantity the **distance value,** because it is a measure of the distance between x_0, the value of x for which we will make a point estimate or a point prediction, and $\bar{x}$, the average of the previously observed values of x. The farther that x_0 is from $\bar{x}$, which represents the center of the experimental region, the larger is the distance value, and thus the longer are both the confidence interval $[\hat{y} \pm t_{\alpha/2} s \sqrt{\text{distance value}}]$ and the prediction interval $[\hat{y} \pm t_{\alpha/2} s \sqrt{1 + \text{distance value}}]$. Said another way, when x_0 is farther from the center of the data, $\hat{y} = b_0 + b_1 x_0$ is likely to be less accurate as both a point estimate and a point prediction.

Ⓒ EXAMPLE 14.7 The Tasty Sub Shop Case: Predicting Revenue and Profit

In the Tasty Sub Shop problem, recall that one of the business entrepreneur's potential sites is near a population of 47,300 residents. Also, recall that

$$\hat{y} = b_0 + b_1 x_0$$
$$= 183.31 + 15.596(47.3)$$
$$= 921.0 \text{ (that is, \$921,000)}$$

is the point estimate of the mean yearly revenue for all Tasty Sub restaurants that could potentially be built near populations of 47,300 residents and is the point prediction of the yearly revenue for a single Tasty Sub restaurant that is built near a population of 47,300 residents. Using the information in Example 14.2, we compute

$$\text{distance value} = \frac{1}{n} + \frac{(x_0 - \bar{x})^2}{SS_{xx}}$$

$$= \frac{1}{10} + \frac{(47.3 - 43.41)^2}{1913.129}$$

$$= .1079$$

Because $s = 61.7052$ (see Example 14.3) and because $t_{\alpha/2} = t_{.025}$ based on $n - 2 = 10 - 2 = 8$ degrees of freedom equals 2.306, it follows that a 95 percent confidence interval for the mean yearly revenue when $x = 47.3$ is

$$[\hat{y} \pm t_{\alpha/2}\, s\, \sqrt{\text{distance value}}]$$

$$= [921.0 \pm 2.306(61.7052)\sqrt{.1079}]$$

$$= [921.0 \pm 46.74]$$

$$= [874.3, 967.7]$$

This interval says we are 95 percent confident that the mean yearly revenue for all Tasty Sub restaurants that could potentially be built near populations of 47,300 residents is between $874,300 and $967,700.

Because the entrepreneur would be operating a single Tasty Sub restaurant that is built near a population of 47,300 residents, the entrepreneur is interested in obtaining a prediction interval for the yearly revenue of such a restaurant. A 95 percent prediction interval for this revenue is

$$[\hat{y} \pm t_{\alpha/2}\, s\, \sqrt{1 + \text{distance value}}]$$

$$= [921.0 \pm 2.306(61.7052)\sqrt{1.1079}]$$

$$= [921.0 \pm 149.77]$$

$$= [771.2, 1070.8]$$

This interval says that we are 95 percent confident that the yearly revenue for a single Tasty Sub restaurant that is built near a population of 47,300 residents will be between $771,200 and $1,070,800. Moreover, recall that the yearly rent and other fixed costs for the entrepreneur's potential restaurant will be $257,550 and that (according to Tasty Sub corporate headquarters) the yearly food and other variable costs for the restaurant will be 60 percent of the yearly revenue. Using the lower end of the 95 percent prediction interval [771.2, 1070.8], we predict that (1) the restaurant's yearly operating cost will be $257,550 + .6($771,200) = $720,270 and (2) the restaurant's yearly profit will be $771,200 − $720,270 = $50,930. Using the upper end of the 95 percent prediction interval [771.2, 1070.8], we predict that (1) the restaurant's yearly operating cost will be $257,550 + .6($1,070,800) = $900,030 and (2) the restaurant's yearly profit will be $1,070,800 − $900,030 = $170,770. Combining the two predicted profits, it follows that we are 95 percent confident that the potential restaurant's yearly profit will be between $50,930 and $170,770. If the entrepreneur decides that this is an acceptable range of potential yearly profits, then the entrepreneur might decide to purchase a Tasty Sub franchise for the potential restaurant site. In Chapter 15 we will use a *multiple regression model* to reduce the range of the predicted yearly profits for the potential Tasty Sub restaurant.

Below we repeat the bottom of the Minitab output in Figure 14.10(b). This output gives (within rounding) the point estimate and prediction $\hat{y} = 921.0$, the 95 percent confidence interval for the mean value of y when x equals 47.3, and the 95 percent prediction interval for an individual value of y when x equals 47.3.

Variable	Setting	Fit	SE Fit	95% CI	95% PI
Population	47.3	920.977	20.2699	(874.234, 967.719)	(771.204, 1070.75)

Although the Minitab output does not directly give the distance value, it does give $s_{\hat{y}} = s\sqrt{\text{distance value}}$ under the heading "SE Fit." A little algebra shows that this implies that

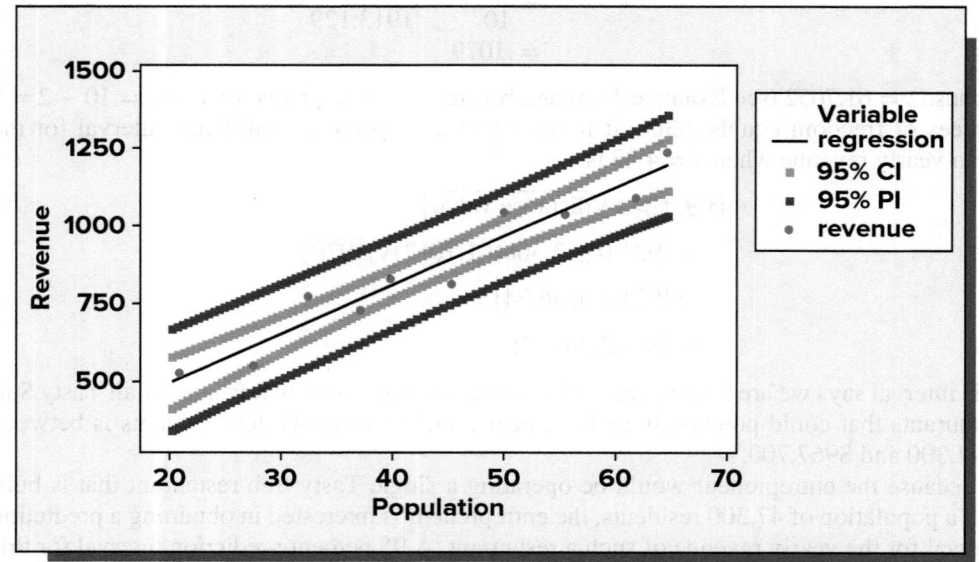

FIGURE 14.17   Minitab Output of 95% Confidence and Prediction Intervals for the Tasty
                Sub Shop Case

the distance value equals $(s_{\hat{y}}/s)^2$. Specifically, because $s_{\hat{y}} = 20.2699$ and $s = 61.7052$, the distance value equals $(20.2699/61.7052)^2 = .1079$.

To conclude this example, note that Figure 14.17 illustrates the Minitab output of the 95 percent confidence and prediction intervals corresponding to all values of x in the experimental region. Here $\bar{x} = 43.41$ can be regarded as the center of the experimental region. Notice that the farther x_0 is from $\bar{x} = 43.41$, the larger is the distance value and, therefore, the longer are the 95 percent confidence and prediction intervals. These longer intervals are undesirable because they give us less information about mean and individual values of y.

Exercises for Section 14.5

CONCEPTS connect

14.28 What is the difference between a confidence interval and a prediction interval?

14.29 What does the distance value measure? How does the distance value affect a confidence or prediction interval?

METHODS AND APPLICATIONS

14.30 THE NATURAL GAS CONSUMPTION CASE DS GasCon1

The following partial Minitab regression output for the natural gas consumption data relates to predicting the city's natural gas consumption (in MMcf) in a week that has an average hourly temperature of 40°F.

Variable	Setting	Fit	SE Fit
Temp	40	10.7210	0.241483
	95% CI		95% PI
	(10.1301, 11.3119)		(9.01462, 12.4274)

a Report (as shown on the computer output) a point estimate of and a 95 percent confidence interval for the mean natural gas consumption for all weeks having an average hourly temperature of 40°F.

b Report (as shown on the computer output) a point prediction of and a 95 percent prediction interval for the natural gas consumption in a single week that has an average hourly temperature of 40°F.

c Remembering that $s = .6542$; $SS_{xx} = 1.404.355$; $\bar{x} = 43.98$; and $n = 8$, hand calculate the distance value when $x_0 = 40$. Remembering that the distance value equals $(s_{\hat{y}}/s)^2$, use s and $s_{\hat{y}}$ from the computer output to calculate (within rounding) the distance value using this formula.

d Remembering that for the natural gas consumption data $b_0 = 15.838$ and $b_1 = -.1279$, calculate (within rounding) the confidence interval of part *a* and the prediction interval of part *b*.

e Suppose that next week the city's average hourly temperature will be 40°F. Also, suppose that the city's natural gas company will use the point prediction $\hat{y} = 10.721$ and order 10.721 MMcf of natural gas to be shipped to the city by a pipeline transmission system. The company will have to pay a fine to the transmission system if the city's actual gas usage y differs from the order of 10.721 MMcf by more than 10.5 percent—that is, is outside of the range [10.721 ± .105(10.721)] = [9.595, 11.847]. Discuss why the

95 percent prediction interval for *y*, [9.015, 12.427], says that *y* might be outside of the allowable range and thus does not make the company 95 percent confident that it will avoid paying a fine.

Note: In the exercises of Chapter 15, we will use multiple regression analysis to predict *y* accurately enough so that the company is likely to avoid paying a fine.

14.31 THE FRESH DETERGENT CASE DS Fresh

The following partial JMP regression output for the Fresh detergent data relates to predicting demand for future sales periods in which the price difference will be .10.

	Predicted Demand	Lower 95% Mean Demand	Upper 95% Mean Demand
31	8.0806090242	7.9478783815	8.213339667

	StdErr Indiv Demand	Lower 95% Indiv Demand	Upper 95% Mean Demand
	0.3231244931	7.4187185048	8.7424995436

a Report (as shown on the computer output) a point estimate of and a 95 percent confidence interval for the mean demand for Fresh in all sales periods when the price difference is .10.

b Report (as shown on the computer output) a point prediction of and a 95 percent prediction interval for the demand for Fresh in an individual sales period when the price difference is .10.

c. **StdErr Indiv Demand** on the JMP output equals $s\sqrt{1 + \text{distance value}}$. Using this information, find 99 percent confidence and prediction intervals for mean and individual demands when $x = .10$.

14.32 THE SERVICE TIME CASE DS SrvcTime

The following partial MegaStat regression output for the service time data relates to predicting service times for four copiers.

Predicted values for: Minutes (y)

		95% Confidence Interval	
Copiers (x)	Predicted	lower	upper
4	109.873	106.721	113.025

95% Prediction Interval		
lower	upper	Leverage
98.967	120.779	0.091

a Report (as shown on the computer output) a point estimate of and a 95 percent confidence interval for the mean time to service four copiers.

b Report (as shown on the computer output) a point prediction of and a 95 percent prediction interval for the time to service four copiers on a single call.

c For this case: $n = 11$, $b_0 = 11.4641$, $b_1 = 24.6022$, and $s = 4.615$. Using this information and a distance value (called **Leverage** on the MegaStat output), hand calculate (within rounding) the confidence interval of part *a* and the prediction interval of part *b*.

d If we examine the service time data, we see that there was at least one call on which Accu-Copiers serviced each of 1, 2, 3, 4, 5, 6, and 7 copiers. The 95 percent confidence intervals for the mean service times on these calls might be used to schedule future service calls.

To understand this, note that a person making service calls will (in, say, a year or more) make a very large number of service calls. Some of the person's individual service times will be below, and some will be above, the corresponding mean service times. However, because the very large number of individual service times will average out to the mean service times, it seems fair to both the efficiency of the company and to the person making service calls to schedule service calls by using estimates of the mean service times. Therefore, suppose we wish to schedule a call to service four copiers. The upper limit of the 95 percent confidence interval for the mean time to service four copiers tells us the largest that this mean service time might be. If we use this upper limit as the number of minutes to allow when scheduling a call to service four copiers, how many minutes will we allow?

14.33 THE DIRECT LABOR COST CASE DS DirLab

The following partial MegaStat regression output for the direct labor cost data relates to predicting direct labor cost when the batch size is 60.

Predicted values for: LaborCost (y)

		95% Confidence Interval	
BatchSize (x)	Predicted	lower	upper
60	627.263	621.054	633.472

95% Prediction Interval		
lower	upper	Leverage
607.032	647.494	0.104

a Report (as shown on the computer output) a point estimate of and a 95 percent confidence interval for the mean direct labor cost of all batches of size 60.

b Report (as shown on the computer output) a point prediction of and a 95 percent prediction interval for the actual direct labor cost of an individual batch of size 60.

c For this case: $n = 12$, $b_0 = 18.4875$, $b_1 = 10.1463$, and $s = 8.6415$. Use this information and the distance value (called **Leverage**) on the computer output to compute 99 percent confidence and prediction intervals for the mean and individual labor costs when $x = 60$.

14.34 THE SALES VOLUME CASE DS SalesPlot

Use the sales volume data in Figure 3.22(a) to find (**1**) a point estimate of and a 95 percent confidence interval for the mean sales volume in all sales regions when advertising expenditure is 10 (that is, $100,000) and (**2**) a point prediction of and a 95 percent prediction interval for the sales volume in an individual sales region when advertising expenditure is 10.

14.35 THE REAL ESTATE SALES PRICE CASE
DS RealEst

Use the real estate sales price data in Table 14.7 to find (**1**) a point estimate of and a 95 percent confidence interval for the mean sales price of all houses having a size of 20 (that is, 2,000 square feet) and (**2**) a point prediction of and a 95 percent prediction interval for the sales price of a single house having a size of 20.

LO14-8
Test hypotheses
about the population
correlation coefficient
(Optional).

14.6 Testing the Significance of the Population Correlation Coefficient (Optional)

We have seen that the simple correlation coefficient measures the linear relationship between the observed values of x and the observed values of y that make up the sample. A similar coefficient of linear correlation can be defined for the population of *all possible combinations of observed values of x and y.* We call this coefficient the **population correlation coefficient** and denote it by the symbol ρ (pronounced ***rho***). We use r as the point estimate of ρ. In addition, we can carry out a hypothesis test. Here we test the null hypothesis $H_0: \rho = 0$, **which says there is no linear relationship between x and y,** against the alternative $H_a: \rho \neq 0$, **which says there is a positive or negative linear relationship between x and y.** This test employs the test statistic

$$t = \frac{r\sqrt{n-2}}{\sqrt{1-r^2}}$$

and is based on the assumption that the population of all possible observed combinations of values of x and y has a **bivariate normal probability distribution.** See Wonnacott and Wonnacott (1981) for a discussion of this distribution. It can be shown that the preceding test statistic t and the p-value used to test $H_0: \rho = 0$ versus $H_a: \rho \neq 0$ are equal to, respectively, the test statistic $t = b_1/s_{b_1}$ and the p-value used to test $H_0: \beta_1 = 0$ versus $H_a: \beta_1 \neq 0$, where β_1 is the slope in the simple linear regression model. Keep in mind, however, that although the mechanics involved in these hypothesis tests are the same, these tests are based on different assumptions (remember that the test for significance of the slope is based on the regression assumptions). If the bivariate normal distribution assumption for the test concerning ρ is badly violated, we can use a nonparametric approach to correlation. One such approach is **Spearman's rank correlation coefficient.**

Ⓒ EXAMPLE 14.8 The Tasty Sub Shop Case: The Correlation Between x and y

Again consider testing the significance of the slope in the Tasty Sub Shop problem. Recall that in Example 14.4 we found that $t = 11.05$ and that the p-value related to this t statistic is less than .001. We therefore (if the regression assumptions hold) can reject $H_0: \beta_1 = 0$ at level of significance .05, .01, or .001, and we have extremely strong evidence that x is significantly related to y. This also implies (if the population of all possible observed combinations of x and y has a bivariate normal probability distribution) that we can reject $H_0: \rho = 0$ in favor of $H_a: \rho \neq 0$ at level of significance .05, .01, or .001. It follows that we have extremely strong evidence of a linear relationship, or correlation, between x and y. Furthermore, because we have previously calculated r to be .969, we estimate that x and y are positively correlated.

Exercises for Section 14.6

CONCEPTS connect

14.36 Explain what is meant by the population correlation coefficient ρ.

14.37 Explain how we test $H_0: \rho = 0$ versus $H_a: \rho \neq 0$. What do we conclude if we reject $H_0: \rho = 0$?

METHODS AND APPLICATIONS

14.38 THE NATURAL GAS CONSUMPTION CASE ⒹⓈ GasCon1

Assuming that the bivariate normal probability distribution assumption holds, use information in

Figure 14.12 to test $H_0: \rho = 0$ versus $H_a: \rho \neq 0$ by setting α equal to .05, .01, and .001. What do you conclude about how x and y are related?

14.39 THE REAL ESTATE SALES PRICE CASE ⒹⓈ RealEst

Assuming that the bivariate normal probability distribution assumption holds, use the data in Table 14.7 to test $H_0: \rho = 0$ versus $H_a: \rho \neq 0$ by setting α equal to .05, .01, and .001. What do you conclude about how x and y are related?

14.7 Residual Analysis

LO14-9
Use residual analysis to
check the assumptions
of simple linear
regression.

As discussed in Section 14.3, four regression assumptions must approximately hold if statistical inferences made using the simple linear regression model

$$y = \beta_0 + \beta_1 x + \varepsilon$$

are to be valid. The first three regression assumptions say that, at any given value of the independent variable x, the population of error terms that could potentially occur

1 Has mean zero.
2 Has a constant variance σ^2 (a variance that does not depend upon x).
3 Is normally distributed.

The fourth regression assumption says that any one value of the error term is statistically independent of any other value of the error term. To assess whether the regression assumptions hold in a particular situation, note that the simple linear regression model $y = \beta_0 + \beta_1 x + \varepsilon$ implies that the error term ε is given by the equation $\varepsilon = y - (\beta_0 + \beta_1 x)$. The point estimate of this error term is the **residual**

$$e = y - \hat{y} = y - (b_0 + b_1 x)$$

where $\hat{y} = b_0 + b_1 x$ is the predicted value of the dependent variable y. Therefore, because the n residuals are the point estimates of the n error terms in the regression analysis, we can use the residuals to check the validity of the regression assumptions.

Residual plots

One useful way to analyze residuals is to plot them versus various criteria. The resulting plots are called **residual plots.** To construct a residual plot, we compute the residual for each observed y value. The calculated residuals are then plotted versus some criterion. To validate the regression assumptions, **we make residual plots against (1) values of the independent variable x; (2) values of $\hat{y}$, the predicted value of the dependent variable; and (3) the time order in which the data have been observed (if the regression data are time series data).**

C EXAMPLE 14.9 The QHIC Case: Constructing Residual Plots

Quality Home Improvement Center (QHIC) operates five stores in a large metropolitan area. The marketing department at QHIC wishes to study the relationship between x, home value (in thousands of dollars), and y, yearly expenditure on home upkeep (in dollars). A random sample of 40 homeowners is taken and survey participants are asked to estimate their expenditures during the previous year on the types of home upkeep products and services offered by QHIC. Public records of the county auditor are used to obtain the previous year's assessed values of the homeowner's homes. Figure 14.18(a) gives the resulting values of x (see Value) and y (see Upkeep), and Figure 14.18(b) gives a scatter plot of these values. The least squares point estimates of the y-intercept β_0 and the slope β_1 of the simple linear regression model describing the QHIC data are $b_0 = -348.3921$ and $b_1 = 7.2583$. Moreover, Figure 14.18(a) presents the predicted home upkeep expenditures and residuals that are given by the regression model. Here each residual is computed as

$$e = y - \hat{y} = y - (b_0 + b_1 x) = y - (-348.3921 + 7.2583x)$$

For instance, for the first home, when $y = 1{,}412.08$ and $x = 237.00$, the residual is

$$e = 1{,}412.08 - (-348.3921 + 7.2583(237))$$

$$= 1{,}412.08 - 1{,}371.816 = 40.264$$

FIGURE 14.18   The QHIC Data, Residuals, and Residual Plots

(a) The QHIC Data and Residuals   QHIC

Home	Value	Upkeep	Predicted	Residual
1	237.00	1,412.080	1,371.816	40.264
2	153.08	797.200	762.703	34.497
3	184.86	872.480	993.371	−120.891
4	222.06	1,003.420	1,263.378	−259.958
5	160.68	852.900	817.866	35.034
6	99.68	288.480	375.112	−86.632
7	229.04	1,288.460	1,314.041	−25.581
8	101.78	423.080	390.354	32.726
9	257.86	1,351.740	1,523.224	−171.484
10	96.28	378.040	350.434	27.606
11	171.00	918.080	892.771	25.309
12	231.02	1,627.240	1,328.412	298.828
13	228.32	1,204.760	1,308.815	−104.055
14	205.90	857.040	1,146.084	−289.044
15	185.72	775.000	999.613	−224.613
16	168.78	869.260	876.658	−7.398
17	247.06	1,396.000	1,444.835	−48.835
18	155.54	711.500	780.558	−69.058
19	224.20	1,475.180	1,278.911	196.269
20	202.04	1,413.320	1,118.068	295.252
21	153.04	849.140	762.413	86.727
22	232.18	1,313.840	1,336.832	−22.992
23	125.44	602.060	562.085	39.975
24	169.82	642.140	884.206	−242.066
25	177.28	1,038.800	938.353	100.447
26	162.82	697.000	833.398	−136.398
27	120.44	324.340	525.793	−201.453
28	191.10	965.100	1,038.662	−73.562
29	158.78	920.140	804.075	116.065
30	178.50	950.900	947.208	3.692
31	272.20	1,670.320	1,627.307	43.013
32	48.90	125.400	6.537	118.863
33	104.56	479.780	410.532	69.248
34	286.18	2,010.640	1,728.778	281.862
35	83.72	368.360	259.270	109.090
36	86.20	425.600	277.270	148.330
37	133.58	626.900	621.167	5.733
38	212.86	1,316.940	1,196.602	120.338
39	122.02	390.160	537.261	−147.101
40	198.02	1,090.840	1,088.889	1.951

(b) Scatter Plot of Upkeep versus Value

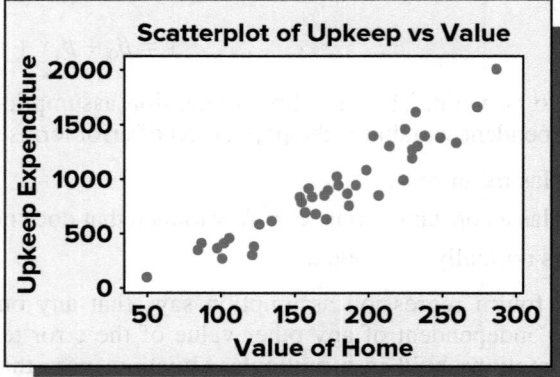

(c) Minitab Residual Plot versus Value

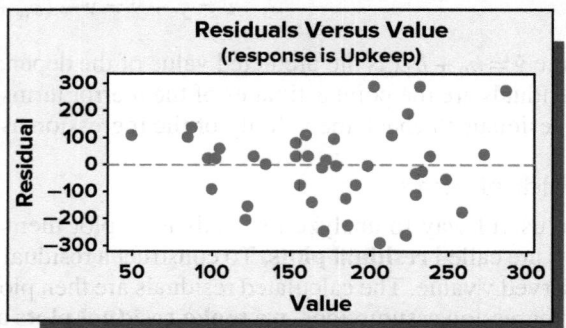

(d) Minitab Residual Plot versus Predicted Upkeep

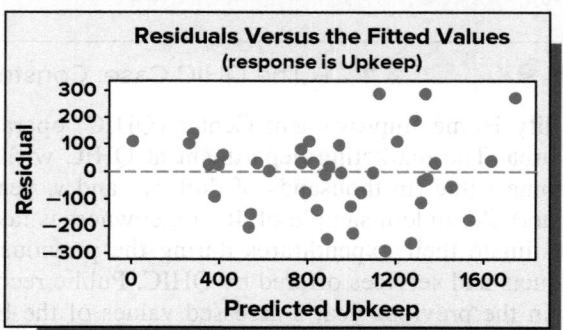

The Minitab output in Figure 14.18(c) and (d) gives plots of the residuals for the QHIC simple linear regression model against values of x (Value) and $\hat{y}$ (Predicted Upkeep). To understand how these plots are constructed, recall that for the first home $y = 1,412.08$, $x = 237.00$, $\hat{y} = 1,371.816$, and the residual is 40.264. It follows that the point plotted in Figure 14.18(c) corresponding to the first home has a horizontal axis coordinate equal to the x value 237.00 and a vertical axis coordinate equal to the residual 40.264. It also follows that the point plotted in Figure 14.18(d) corresponding to the first home has a horizontal axis coordinate equal to the $\hat{y}$ value 1,371.816, and a vertical axis coordinate equal to the residual 40.264. Finally, note that the QHIC data are cross-sectional data, not time series data. Therefore, we cannot make a residual plot versus time.

FIGURE 14.19   **Residual Plots and the Constant Variance Assumption**

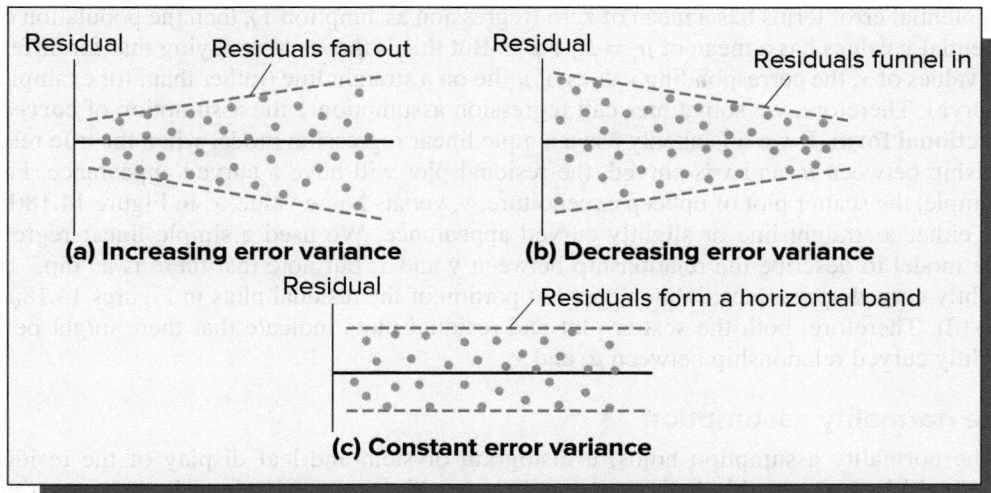

(a) Increasing error variance   **(b) Decreasing error variance**

(c) Constant error variance

The constant variance assumption

To check the validity of the constant variance assumption, we examine plots of the residuals against values of x, $\hat{y}$, and time (if the regression data are time series data). When we look at these plots, the pattern of the residuals' fluctuation around 0 tells us about the validity of the constant variance assumption. A residual plot that "fans out" [as in Figure 14.19(a)] suggests that the error terms are becoming more spread out as the horizontal plot value increases and that the constant variance assumption is violated. Here we would say that an **increasing error variance** exists. A residual plot that "funnels in" [as in Figure 14.19(b)] suggests that the spread of the error terms is decreasing as the horizontal plot value increases and that again the constant variance assumption is violated. In this case we would say that a **decreasing error variance** exists. A residual plot with a "horizontal band appearance" [as in Figure 14.19(c)] suggests that the spread of the error terms around 0 is not changing much as the horizontal plot value increases. Such a plot tells us that the constant variance assumption (approximately) holds.

As an example, consider the QHIC case and the residual plot in Figure 14.18(c). This plot appears to fan out as x increases, indicating that the spread of the error terms is increasing as x increases. That is, an increasing error variance exists. This is equivalent to saying that the variance of the population of potential yearly upkeep expenditures for houses worth x (thousand dollars) appears to increase as x increases. The reason is that the model $y = \beta_0 + \beta_1 x + \varepsilon$ says that the variation of y is the same as the variation of ε. For example, the variance of the population of potential yearly upkeep expenditures for houses worth \$200,000 would be larger than the variance of the population of potential yearly upkeep expenditures for houses worth \$100,000. Increasing variance makes some intuitive sense because people with more expensive homes generally have more discretionary income. These people can choose to spend either a substantial amount or a much smaller amount on home upkeep, thus causing a relatively large variation in upkeep expenditures.

Another residual plot showing the increasing error variance in the QHIC case is Figure 14.18(d). This plot tells us that the residuals appear to fan out as $\hat{y}$ (predicted y) increases, which is logical because $\hat{y}$ is an increasing function of x. Also, note that the scatter plot of y versus x in Figure 14.18(b) shows the increasing error variance—the y values appear to fan out as x increases. In fact, one might ask why we need to consider residual plots when we can simply look at scatter plots of y versus x. One answer is that, in general, because of possible differences in scaling between residual plots and scatter plots of y versus x, one of these types of plots might be more informative in a particular situation. Therefore, we should consider both types of plots.

When the constant variance assumption is violated, statistical inferences made using the simple linear regression model may not be valid. Later in this section we discuss how we can make statistical inferences when a nonconstant error variance or other violations of the regression assumptions exist.

The assumption of correct functional form

If for each value of x in the simple linear regression model, $y = \beta_0 + \beta_1 x + \varepsilon$, the population of potential error terms has a mean of zero (regression assumption 1), then the population of potential y values has a mean of $\mu_y = \beta_0 + \beta_1 x$. But this is the same as saying that for different values of x, the corresponding values of μ_y lie on a straight line (rather than, for example, a curve). Therefore, we sometimes call regression assumption 1 the assumption of **correct functional form.** If we mistakenly use a simple linear regression model when the true relationship between μ_y and x is curved, the residual plot will have a curved appearance. For example, the scatter plot of upkeep expenditure, y, versus home value, x, in Figure 14.18(b) has either a straight-line or slightly curved appearance. We used a simple linear regression model to describe the relationship between y and x, but note that there is a "dip," or slightly curved appearance, in the upper left portion of the residual plots in Figures 14.18(c) and (d). Therefore, both the scatter plot and residual plots indicate that there might be a slightly curved relationship between μ_y and x.

The normality assumption

If the normality assumption holds, a histogram or stem-and-leaf display of the residuals should look reasonably bell-shaped and reasonably symmetric about 0, and a **normal plot** of the residuals should have a straight-line appearance. To construct a normal plot, we first arrange the residuals in order from smallest to largest. Letting the ordered residuals be denoted as $e_{(1)}, e_{(2)}, \ldots, e_{(n)}$, we denote the ith residual in the ordered listing as $e_{(i)}$. We plot $e_{(i)}$ on the vertical axis against the normal point $z_{(i)}$ on the horizontal axis. Here $z_{(i)}$ is defined to be the point on the horizontal axis under the standard normal curve so that the area under this curve to the left of $z_{(i)}$ is $(3i - 1)/(3n + 1)$. Note that the area $(3i - 1)/(3n + 1)$ is employed in regression by statistical software packages. Because this area equals $[i - (1/3)]/[n + (1/3)]$, it is only a slight modification of the area $i/(n + 1)$, which we used in our general discussion of normal plots in Section 7.6. For example, recall in the QHIC case that there are $n = 40$ residuals in Figure 14.18(a). It follows that, when $i = 1, (3i - 1)/(3n + 1) = [3(1) - 1]/[3(40) + 1] = .0165$. Using Table A.3 to look up the normal point $z_{(1)}$, which has a standard normal curve area to its left of .0165, we find that $z_{(1)} = -2.13$. Because the smallest residual in Figure 14.18(a) is -289.044, the first point plotted is $e_{(1)} = -289.044$ on the vertical axis versus $z_{(1)} = -2.13$ on the horizontal axis. Plotting the other ordered residuals $e_{(2)}, e_{(3)}, \ldots, e_{(40)}$ against their corresponding normal points in the same way, we obtain the Excel add-in (MegaStat) normal plot in Figure 14.20(a). An equivalent Minitab normal plot of the residuals is given in Figure 14.20(b). To obtain this normal plot, Minitab essentially reverses the roles of the vertical and horizontal axes and plots $e_{(1)}$ on the horizontal axis

FIGURE 14.20 Normal Plots for the QHIC Simple Linear Regression Model

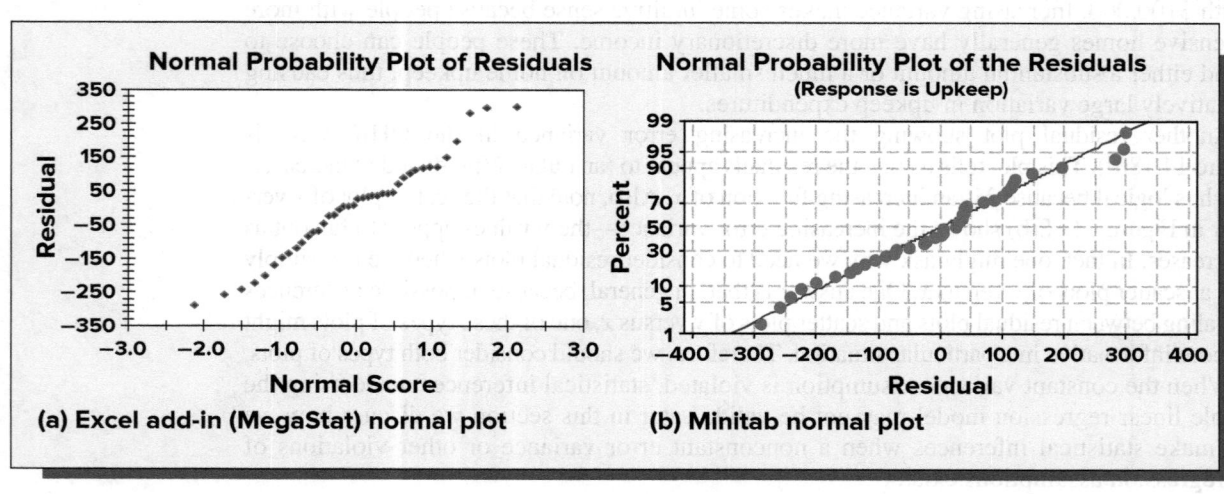

versus the percentage $P_{(i)} = [(3i - 1)/(3n + 1)](100)$ on the vertical axis. Moreover, Minitab scales the $P_{(i)}$ values on the vertical axis so that the resulting normal plot has the same shape as the normal plot we would obtain if we simply plotted $e_{(i)}$ on the horizontal axis versus $z_{(i)}$ on the vertical axis. Examining Figure 14.20, we see that both normal plots have some curvature (particularly in the upper right-hand portion). Therefore, there is a possible violation of the normality assumption.

It is important to realize that violations of the constant variance and correct functional form assumptions can often cause a histogram and/or stem-and-leaf display of the residuals to look nonnormal and can cause the normal plot to have a curved appearance. Because of this, it is usually a good idea to use residual plots to check for nonconstant variance and incorrect functional form before making any final conclusions about the normality assumption.

Transforming the dependent variable: A possible remedy for violations of the constant variance, correct functional form, and normality assumptions

In general, if a data or residual plot indicates that the error variance of a regression model increases as an independent variable or the predicted value of the dependent variable increases, then we can sometimes remedy the situation by transforming the dependent variable. One transformation that works well is to take each y value to a fractional power. As an example, we might use a transformation in which we take the square root (or one-half power) of each y value. Letting y^* denote the value obtained when the transformation is applied to y, we would write the **square root transformation** as $y^* = y^{.5}$. Another commonly used transformation is the **quartic root transformation.** Here we take each y value to the one-fourth power. That is, $y^* = y^{.25}$. If we consider a transformation that takes each y value to a fractional power (such as .5, .25, or the like), as the power approaches 0, the transformed value y^* approaches the natural logarithm of y (commonly written ln y). In fact, we sometimes use the **logarithmic transformation** $y^* = \ln y$, which takes the natural logarithm of each y value.

For example, consider the QHIC upkeep expenditures in Figure 14.18(a). In Figures 14.21, 14.22, and 14.23 we show the plots that result when we take the square root, quartic root, and natural logarithmic transformations of the upkeep expenditures and plot the transformed values versus the home values. To interpret these plots, note that when we take a fractional power (including the natural logarithm) of the dependent variable, the transformation not only tends to equalize the error variance but also tends to "straighten out" certain types of nonlinear data plots. Specifically, if a data plot indicates that the dependent variable is increasing at an increasing rate as the independent variable increases [this is true of the QHIC data plot in Figure 14.18(b)], then a fractional power transformation tends to

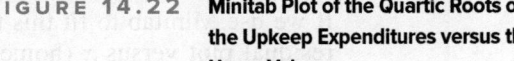

FIGURE 14.21   **Minitab Plot of the Square Roots of the Upkeep Expenditures versus the Home Values**

FIGURE 14.22   **Minitab Plot of the Quartic Roots of the Upkeep Expenditures versus the Home Values**

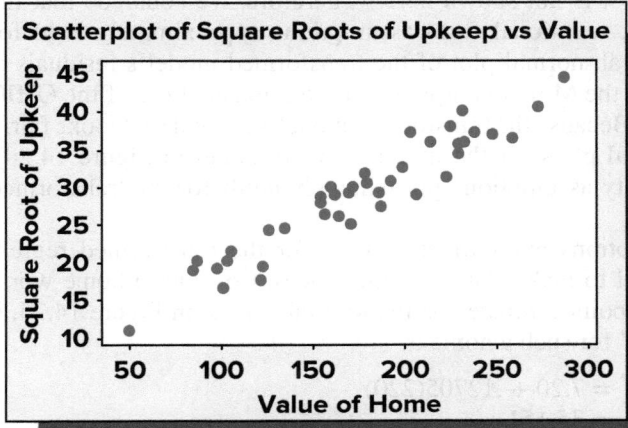

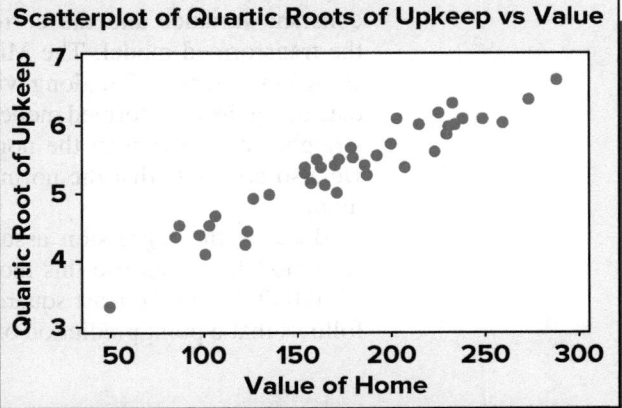

FIGURE 14.23 **Minitab Plot of the Natural Logarithms of the Upkeep Expenditures versus the Home Values**

FIGURE 14.24 **Minitab Output of a Residual Plot versus x (Home Value) for the Upkeep Expenditure Model**
$y^* = \beta_0 + \beta_1 x + \varepsilon$ where $y^* = y^{.5}$

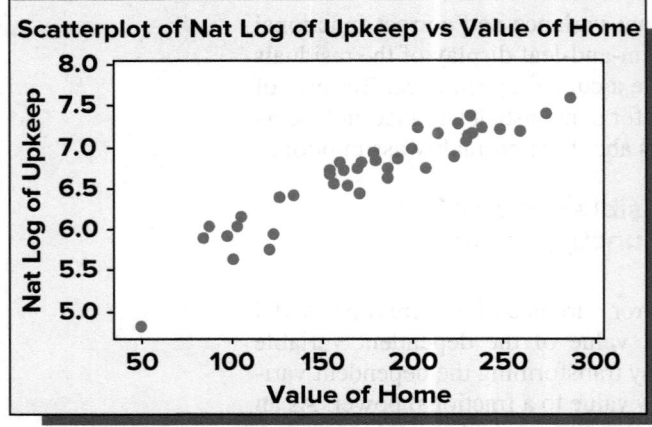

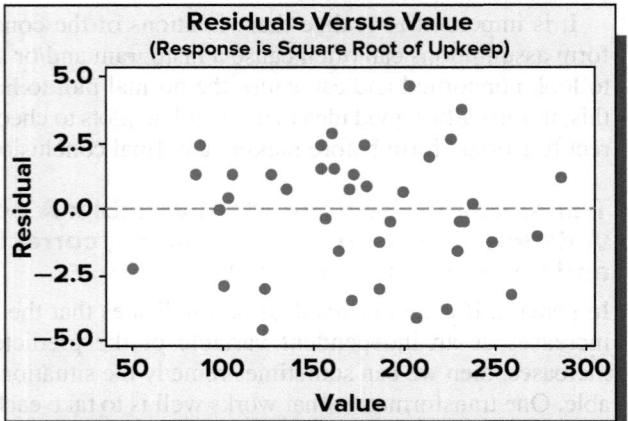

straighten out the data plot. A fractional power transformation can also help to remedy a violation of the normality assumption. Because we cannot know which fractional power to use before we actually take the transformation, we recommend taking all of the square root, quartic root, and natural logarithm transformations and seeing which one best equalizes the error variance and (possibly) straightens out a nonlinear data plot. This is what we have done in Figures 14.21, 14.22, and 14.23, and examining these figures, it seems that the square root transformation best equalizes the error variance and straightens out the curved data plot in Figure 14.18(b). Note that the natural logarithm transformation seems to "overtransform" the data—the error variance tends to decrease as the home value increases and the data plot seems to "bend down." The plot of the quartic roots indicates that the quartic root transformation also seems to overtransform the data (but not by as much as the logarithmic transformation). In general, as the fractional power gets smaller, the transformation gets stronger. Different fractional powers are best in different situations.

Because the plot in Figure 14.21 of the square roots of the upkeep expenditures versus the home values has a straight-line appearance, we consider the model

$$y^* = \beta_0 + \beta_1 x + \varepsilon \qquad \text{where} \qquad y^* = y^{.5}$$

If we use Minitab to fit this transformed model to the QHIC data, we obtain the Minitab residual plot versus x (home value) in Figure 14.24. This residual plot has a horizontal band appearance, as does the transformed model's residual plot versus the predicted value of $y^* = y^{.5}$. (The latter residual plot is not shown here.) Therefore, we conclude that the constant variance and the correct functional form assumptions approximately hold for the transformed model. The Minitab normal plot of the transformed model's residuals is shown in Figure 14.25, along with the Minitab output of a regression analysis of the QHIC data using the transformed model. Because the transformed model's normal plot looks fairly straight (straighter than the normal plots for the untransformed model in Figure 14.20), we also conclude that the normality assumption approximately holds for the transformed model.

Because the regression assumptions approximately hold for the transformed regression model, we can use this model to make statistical inferences. Consider a home worth $220,000. Using the least squares point estimates on the Minitab output in Figure 14.25, it follows that a point prediction of y^* for such a home is

$$\hat{y}^* = 7.20 + .12705(220)$$
$$= 35.151$$

FIGURE 14.25   **Minitab Output of a Regression Analysis of the Upkeep Expenditure Data by Using the Model $y^* = \beta_0 + \beta_1 x + \varepsilon$ where $y^* = y^{.5}$, and a Normal Plot versus x**

Analysis of Variance

Source	DF	Adj SS	Adj MS	F-Value	P-Value
Regression	1	2016.8	2016.84	373.17	0.000
Value	1	2016.8	2016.84	373.17	0.000
Error	38	205.4	5.40		
Total	39	2222.2			

Model Summary

S	R-sq	R-sq (adj)	R-sq (pred)
2.32479	90.76%	90.51%	89.82%

Coefficients

Term	Coef	SE Coef	T-Value	P-Value	VIF
Constant	7.20	1.21	5.98	0.000	
Value	0.12705	0.00658	19.32	0.000	1.00

Regression Equation
SqRtUpkeep = 7.20 + 0.12705 Value

Variable	Setting	Fit	SE Fit	95% CI	95% PI
Value	220	35.1511	0.474033	(34.1914, 36.1107)	(30.3479, 39.9542)

Normal Probability Plot of the Residuals
(Response is Square Root of Upkeep)

This point prediction is given at the bottom of the Minitab output, as is the 95 percent prediction interval for y^*, which is [30.348, 39.954]. It follows that a point prediction of the upkeep expenditure for a home worth $220,000 is $(35.151)^2 = \$1,235.59$ and that a 95 percent prediction interval for this upkeep expenditure is $[(30.348)^2, (39.954)^2] = [\$921.00, \$1596.32]$. Suppose that QHIC wishes to send an advertising brochure to any home that has a predicted upkeep expenditure of at least $500. It follows that a home worth $220,000 would be sent an advertising brochure. This is because the predicted yearly upkeep expenditure for such a home is (as just calculated) $1,235.59. Other homes can be evaluated in a similar fashion.

The independence assumption

The independence assumption is most likely to be violated when the regression data are **time series data**—that is, data that have been collected in a time sequence. For such data the time-ordered error terms can be **autocorrelated.** Intuitively, we say that error terms occurring over time have **positive autocorrelation** when positive error terms tend to be followed over time by positive error terms and when negative error terms tend to be followed over time by negative error terms. Positive autocorrelation in the error terms is depicted in Figure 14.26(a), which illustrates that **positive autocorrelation can produce a cyclical error term pattern over time.** Because the residuals are point estimates of the error terms, if a plot of the residuals versus the data's time sequence has a cyclical appearance, we have evidence that the error terms are positively autocorrelated and thus that the independence assumption is violated.

Another type of autocorrelation that sometimes exists is **negative autocorrelation,** where positive error terms tend to be followed over time by negative error terms and negative error terms tend to be followed over time by positive error terms. **Negative autocorrelation can produce an alternating error term pattern over time** [see Figure 14.26(b)] and is suggested by an alternating pattern in a plot of the time-ordered residuals. However, if a plot of the time-ordered residuals has a random pattern, the plot does not provide evidence of autocorrelation, and thus it is reasonable to conclude that the independence assumption holds.

For example, Figure 14.27(a) presents data concerning weekly sales at Pages Bookstore (Sales), Pages weekly advertising expenditure (Adver), and the weekly advertising expenditure of Pages main competitor (Compadv). Here the sales values are expressed in thousands of dollars, and the advertising expenditure values are expressed in hundreds of dollars. Figure 14.27(a) also gives the Minitab output of the residuals that are obtained when a simple linear regression analysis is performed relating Pages sales to

FIGURE 14.26   Positive and Negative Autocorrelation

(a) Positive Autocorrelation in the Error
    Terms: Cyclical Pattern

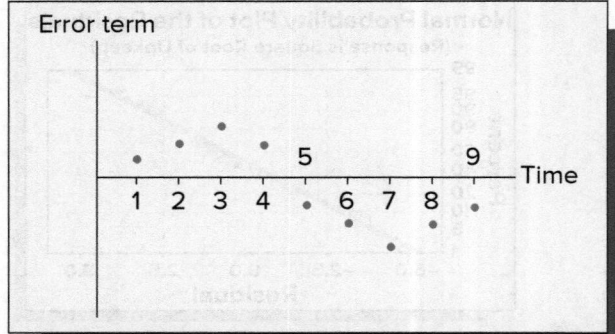

(b) Negative Autocorrelation in the Error
    Terms: Alternating Pattern

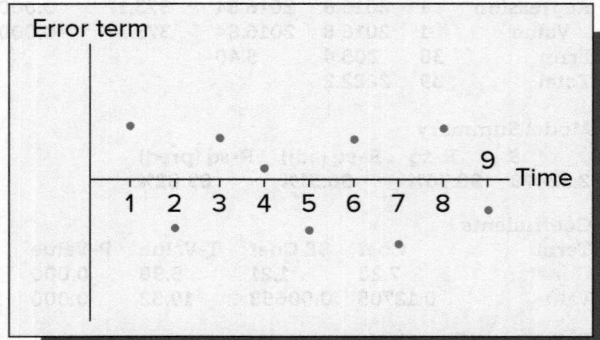

FIGURE 14.27   Pages Bookstore Sales and Advertising Data, Residuals, and a Residual Plot

(a) The Data and Minitab Residuals    BookSales

Obs	Adver	Compadv	Sales	Pred	Resid
1	18	10	22	18.7	3.3
2	20	10	27	23.0	4.0
3	20	15	23	23.0	− 0.0
4	25	15	31	33.9	− 2.9
5	28	15	45	40.4	4.6
6	29	20	47	42.6	4.4
7	29	20	45	42.6	2.4
8	28	25	42	40.4	1.6
9	30	35	37	44.7	− 7.7
10	31	35	39	46.9	− 7.9
11	34	35	45	53.4	− 8.4
12	35	30	52	55.6	− 3.6
13	36	30	57	57.8	− 0.8
14	38	25	62	62.1	− 0.1
15	41	20	73	68.6	4.4
16	45	20	84	77.3	6.7

Durbin-Watson = 0.65

(b) A Residual Plot versus Time

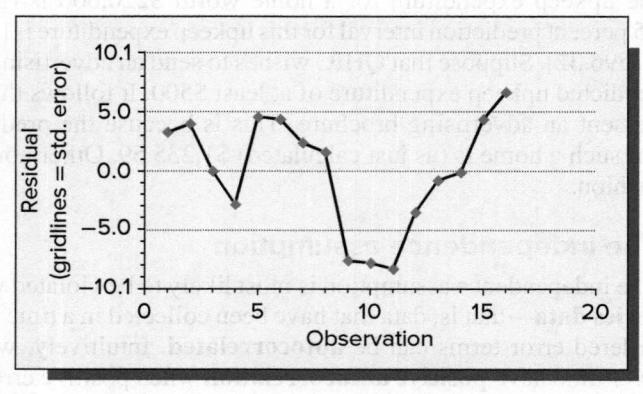

Pages advertising expenditure. These residuals are plotted versus time in Figure 14.27(b). Examining the residual plot, we see that there are (1) positive residuals (actual sales higher than predicted sales) in 6 of the first 8 weeks, when the competitor's advertising expenditure is lower; (2) negative residuals (actual sales lower than predicted sales) in the next 5 weeks, when the competitor's advertising expenditure is higher; and (3) positive residuals again in 2 of the last 3 weeks, when the competitor's advertising expenditure is lower. Overall, the residual plot seems to have a cyclical pattern, which suggests that the error terms in the simple linear regression model are positively autocorrelated and that the independence assumption is violated. Moreover, the competitor's advertising expenditure seems to be causing the positive autocorrelation. Finally, note that the simple linear regression model describing Pages sales has a standard error, s, of 5.038. The residual plot in Figure 14.27(b) includes grid lines that are placed one and two standard errors above and below the residual mean of 0. Such grid lines may help us to better diagnose potential violations of the regression assumptions.

When the independence assumption is violated, various remedies can be employed. One approach is to identify which independent variable left in the error term (for example, competitors' advertising expenditure) is causing the error terms to be autocorrelated. We can then remove this independent variable from the error term and insert it directly into the regression model, forming a **multiple regression model.** (Multiple regression models are discussed in Chapter 15.)

The Durbin–Watson test

One type of positive or negative autocorrelation is called **first-order autocorrelation.** It says that ε_t, the error term in time period t, is related to ε_{t-1}, the error term in time period $t-1$. To check for first-order autocorrelation, we can use the **Durbin–Watson statistic**

$$d = \frac{\sum_{t=2}^{n}(e_t - e_{t-1})^2}{\sum_{t=1}^{n}e_t^2}$$

where $e_1, e_2, \ldots, e_n$ are the time-ordered residuals.

Intuitively, small values of d lead us to conclude that there is positive autocorrelation. This is because, if d is small, the differences $(e_t - e_{t-1})$ are small. This indicates that the adjacent residuals e_t and e_{t-1} are of the same magnitude, which in turn says that the adjacent error terms ε_t and ε_{t-1} are positively correlated. Consider testing the null hypothesis H_0 **that the error terms are not autocorrelated** versus the alternative hypothesis H_a **that the error terms are positively autocorrelated.** Durbin and Watson have shown that there are points (denoted $d_{L,\alpha}$ and $d_{U,\alpha}$) such that if α is the probability of a Type I error, then

1 If $d < d_{L,\alpha}$, we reject H_0.

2 If $d > d_{U,\alpha}$, we do not reject H_0.

3 If $d_{L,\alpha} \leq d \leq d_{U,\alpha}$, the test is inconclusive.

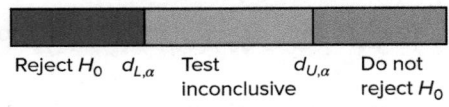

So that the Durbin–Watson test may be easily done, tables containing the points $d_{L,\alpha}$ and $d_{U,\alpha}$ have been constructed. These tables give the appropriate $d_{L,\alpha}$ and $d_{U,\alpha}$ points for various values of α; k, the number of independent variables used by the regression model; and n, the number of observations. Tables A.11, A.12, and A.13 give these points for $\alpha = .05$, $\alpha = .025$, and $\alpha = .01$. A portion of Table A.11 is given in Table 14.8. Note that when we are considering a simple linear regression model, which uses *one* independent variable, we look up the points $d_{L,\alpha}$ and $d_{U,\alpha}$ under the heading "$k = 1$." Other values of k are used when we study multiple regression models in Chapter 15. Using the residuals in Figure 14.27(a), the Durbin–Watson statistic for the simple linear regression model relating Pages sales to Pages advertising expenditure is calculated to be

$$d = \frac{\sum_{t=2}^{16}(e_t - e_{t-1})^2}{\sum_{t=1}^{16}e_t^2}$$

$$= \frac{(4.0 - 3.3)^2 + (0.0 - 4.0)^2 + \cdots + (6.7 - 4.4)^2}{(3.3)^2 + (4.0)^2 + \cdots + (6.7)^2}$$

$$= .65$$

A Minitab output of the Durbin–Watson statistic is given at the bottom of Figure 14.27(a). To test for positive autocorrelation, we note that there are $n = 16$ observations and the regression model uses $k = 1$ independent variable. Therefore, if we set $\alpha = .05$, Table 14.8 tells us

TABLE 14.8   Critical Values for the Durbin–Watson d Statistic (α = .05)

	$k=1$		$k=2$		$k=3$		$k=4$	
n	$d_{L,.05}$	$d_{U,.05}$	$d_{L,.05}$	$d_{U,.05}$	$d_{L,.05}$	$d_{U,.05}$	$d_{L,.05}$	$d_{U,.05}$
15	1.08	1.36	0.95	1.54	0.82	1.75	0.69	1.97
16	1.10	1.37	0.98	1.54	0.86	1.73	0.74	1.93
17	1.13	1.38	1.02	1.54	0.90	1.71	0.78	1.90
18	1.16	1.39	1.05	1.53	0.93	1.69	0.82	1.87
19	1.18	1.40	1.08	1.53	0.97	1.68	0.86	1.85
20	1.20	1.41	1.10	1.54	1.00	1.68	0.90	1.83

that $d_{L,.05} = 1.10$ and $d_{U,.05} = 1.37$. Because $d = .65$ is less than $d_{L,.05} = 1.10$, we reject the null hypothesis of no autocorrelation. That is, we conclude (at an α of .05) that there is positive (first-order) autocorrelation.

It can be shown that the Durbin–Watson statistic d is always between 0 and 4. Large values of d (and hence small values of $4 - d$) lead us to conclude that there is negative autocorrelation because if d is large, this indicates that the differences $(e_t - e_{t-1})$ are large. This says that the adjacent error terms ε_t and ε_{t-1} are negatively autocorrelated. Consider testing the null hypothesis H_0 **that the error terms are not autocorrelated** versus the alternative hypothesis H_a **that the error terms are negatively autocorrelated.** Durbin and Watson have shown that based on setting the probability of a Type I error equal to α, the points $d_{L,\alpha}$ and $d_{U,\alpha}$ are such that

1 If $(4 - d) < d_{L,\alpha}$, we reject H_0.

2 If $(4 - d) > d_{U,\alpha}$, we do not reject H_0.

3 If $d_{L,\alpha} \le (4 - d) \le d_{U,\alpha}$, the test is inconclusive.

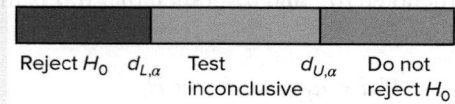

As an example, for the Pages sales simple linear regression model, we see that

$$(4 - d) = (4 - .65) = 3.35 > d_{U,.05} = 1.37$$

Therefore, on the basis of setting α equal to .05, we do not reject the null hypothesis of no autocorrelation. That is, there is no evidence of negative (first-order) autocorrelation.

We can also use the Durbin–Watson statistic to test for positive or negative autocorrelation. Specifically, consider testing the null hypothesis H_0 **that the error terms are not autocorrelated** versus the alternative hypothesis H_a **that the error terms are positively or negatively autocorrelated.** Durbin and Watson have shown that, based on setting the probability of a Type I error equal to α, we perform both the above described test for positive autocorrelation and the above described test for negative autocorrelation by using the critical values $d_{L,\alpha/2}$ and $d_{U,\alpha/2}$ for each test. If **either test** says to reject H_0, then we reject H_0. If **both tests** say to not reject H_0, then we do not reject H_0. Finally, if **either test** is inconclusive, then the overall test is inconclusive. For example, consider testing for positive or negative autocorrelation in the Pages sales model. If we set α equal to .05, then $\alpha/2 = .025$, and we need to find the points $d_{L,.025}$ and $d_{U,.025}$ when $n = 16$ and $k = 1$. Looking up these points in Table A.12, we find that $d_{L,.025} = .98$ and $d_{U,.025} = 1.24$. Because $d = .65$ is less than $d_{L,.025} = .98$, we reject the null hypothesis of no autocorrelation. That is, we conclude (at an α of .05) that there is first-order autocorrelation.

Although we have used the Pages sales model in these examples to demonstrate the Durbin–Watson tests for (1) positive autocorrelation, (2) negative autocorrelation, and (3) positive or negative autocorrelation, we must in practice choose one of these Durbin–Watson tests in a particular situation. Because positive autocorrelation is more common in real time series data than negative autocorrelation, the Durbin–Watson test for positive autocorrelation is used more often than the other two tests. Also, note that each Durbin–Watson test assumes that the population of all possible residuals at any time t has a normal distribution.

Exercises for Section 14.7

CONCEPTS connect

14.40 In regression analysis, what should the residuals be plotted against?

14.41 What patterns in residual plots indicate violations of the regression assumptions?

14.42 How do we check the normality assumption?

METHODS AND APPLICATIONS

14.43 **THE SERVICE TIME CASE** SrvcTime

The residuals given by the service time prediction equation $\hat{y} = 11.4641 + 24.6022x$ are listed in Table 14.9(a), and residual plots versus x and $\hat{y}$ are given in Figures 14.28(a) and (b). Do the plots indicate any violations of the regression assumptions? Explain.

14.44 **THE SERVICE TIME CASE** SrvcTime

The residuals given by the service time prediction equation $\hat{y} = 11.4641 + 24.6022x$ are listed in Table 14.9(a).

a In this exercise we construct a normal plot of these residuals. To construct this plot, we must first arrange the residuals in order from smallest to largest. These ordered residuals are given in Table 14.9(b). Denoting the ith ordered residual as $e_{(i)}$ ($i = 1, 2, \ldots , 11$), we next compute for each value of i the point $z_{(i)}$. These computations are summarized in Table 14.9(b). Show how $z_{(4)} = -.46$ and $z_{(10)} = 1.05$ have been obtained.

b The ordered residuals (the $e_{(i)}$'s are plotted against the $z_{(i)}$'s in Figure 14.28(c). Does this figure indicate a violation of the normality assumption? Explain.

TABLE 14.9	Service Time Model Residuals and Normal Plot Calculations

(a) Predicted values and residuals using
$\hat{y} = 11.4641 + 24.6022x$

Observation	Copiers	Minutes	Predicted	Residual
1	4	109.0	109.9	−0.9
2	2	58.0	60.7	−2.7
3	5	138.0	134.5	3.5
4	7	189.0	183.7	5.3
5	1	37.0	36.1	0.9
6	3	82.0	85.3	−3.3
7	4	103.0	109.9	−6.9
8	5	134.0	134.5	−0.5
9	2	68.0	60.7	7.3
10	6	112.0	109.9	2.1
11	4	154.0	159.1	−5.1

(b) Ordered residuals and normal plot calculations

i	Ordered Residual, $e_{(i)}$	$\dfrac{3i-1}{3n+1}$	$z_{(i)}$
1	−6.9	.0588	−1.565
2	−5.1	.1470	−1.05
3	−3.3	.2353	−.72
4	−2.7	.3235	−.46
5	−0.9	.4118	−.22
6	−0.5	.5000	0
7	0.9	.5882	.22
8	2.1	.6765	.46
9	3.5	.7647	.72
10	5.3	.8529	1.05
11	7.3	.9412	1.565

FIGURE 14.28	Service Time Model Residual and Normal Plots

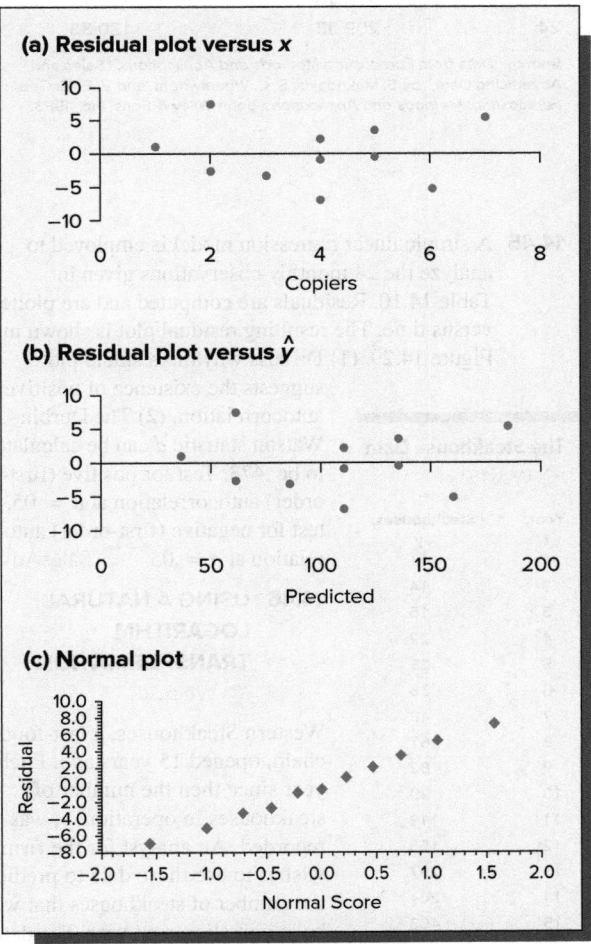

(a) Residual plot versus x

(b) Residual plot versus $\hat{y}$

(c) Normal plot

TABLE 14.10   Sales and Advertising Data for Exercise 14.45 **DS** SalesAdv

Month	Monthly Total Sales, y	Advertising Expenditures, x
1	202.66	116.44
2	232.91	119.58
3	272.07	125.74
4	290.97	124.55
5	299.09	122.35
6	296.95	120.44
7	279.49	123.24
8	255.75	127.55
9	242.78	121.19
10	255.34	118.00
11	271.58	121.81
12	268.27	126.54
13	260.51	129.85
14	266.34	122.65
15	281.24	121.64
16	286.19	127.24
17	271.97	132.35
18	265.01	130.86
19	274.44	122.90
20	291.81	117.15
21	290.91	109.47
22	264.95	114.34
23	228.40	123.72
24	209.33	130.33

Source: Data from *Forecasting Methods and Applications*, "Sales and Advertising Data," by S. Makridakis, S. C. Wheelwright, and V. E. McGee, *Forecasting: Methods and Applications*. John Wiley & Sons, Inc., 1983.

FIGURE 14.29   Residual Plot for Exercise 14.45

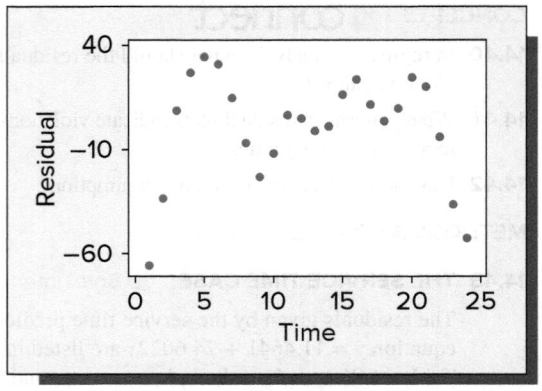

FIGURE 14.30   Plot of Western Steakhouse Openings versus Year (for Exercise 14.46)

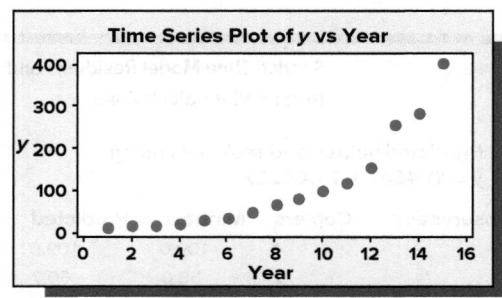

14.45 A simple linear regression model is employed to analyze the 24 monthly observations given in Table 14.10. Residuals are computed and are plotted versus time. The resulting residual plot is shown in Figure 14.29. (**1**) Discuss why the residual plot suggests the existence of positive autocorrelation. (**2**) The Durbin–Watson statistic d can be calculated to be .473. Test for positive (first-order) autocorrelation at $\alpha = .05$, and test for negative (first-order) autocorrelation at $\alpha = .05$ **DS** SalesAdv

The Steakhouse Data
DS WestStk

Year, t	Steakhouses, y
1	11
2	14
3	16
4	22
5	28
6	36
7	46
8	67
9	82
10	99
11	119
12	156
13	257
14	284
15	403

14.46 USING A NATURAL LOGARITHM TRANSFORMATION
DS WestStk

Western Steakhouses, a fast-food chain, opened 15 years ago. Each year since then the number of steakhouses in operation, y, was recorded. An analyst for the firm wishes to use these data to predict the number of steakhouses that will be in operation next year. The data

are given in the left page margin, and a plot of the data is given in Figure 14.30. Examining the data plot, we see that the number of steakhouse openings has increased over time at an increasing rate and with increasing variation. A plot of the natural logarithms of the steakhouse values versus time (see the right side of Figure 14.31) has a straight-line appearance with constant variation. Therefore, we consider the model $\ln y_t = \beta_0 + \beta_1 t + \varepsilon_t$. If we use JMP, we find that the least squares point estimates of β_0 and β_1 are $b_0 = 2.0701$ and $b_1 = .25688$ and that a plot of the residuals from the model versus the predicted values of $\ln y_t$ has a horizontal band appearance. We also find that a point prediction of and a 95 percent prediction interval for the natural logarithm of the number of steakhouses in operation next year (year 16) are 6.1802 and [5.9945, 6.3659]. See the JMP output in Figure 14.31.

a Use the least squares point estimates to calculate the point prediction.

b By exponentiating the point prediction and prediction interval—that is, by calculating $e^{6.1802}$ and $[e^{5.9945}, e^{6.3659}]$—find a point prediction of and a 95 percent prediction interval for the number of steakhouses in operation next year.

FIGURE 14.31   **A Plot of the Logged Steakhouse Values versus Time, and the JMP Output of a Regression Analysis of the Steakhouse Data Using the Model ln $y_t = \beta_0 + \beta_1 t + \varepsilon_t$ (for Exercise 14.46)**

Summary of Fit

RSquare	0.996004
RSquare Adj	0.995696
Root Mean Square Error	0.075516
Mean of Response	4.125163
Observations (or Sum Wgts)	15

Analysis of Variance

Source	DF	Sum of Squares	Mean Square	F Ratio
Model	1	18.476514	18.4765	3239.969
Error	13	0.074135	0.0057	Prob > F
C. Total	14	18.550649		<.0001*

Parameter Estimates

Term	Estimate	Std Error	t Ratio	Prob>|t|	Lower 95%	Upper 95%
Intercept	2.0701197	0.041032	50.45	<.0001*	1.9814749	2.1587646
Year	0.2568804	0.004513	56.92	<.0001*	0.2471308	0.2666301

	Predicated Iny	Lower 95% Mean Iny	Upper 95% Mean Iny	Lower 95% Indiv Iny	Upper 95% Indiv Iny
16	6.1802064901	6.0915616085	6.2688513716	5.9945362618	6.3658767184

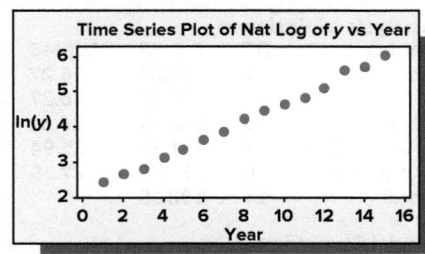

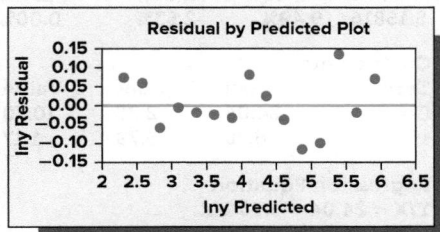

FIGURE 14.32   **The Laptop Service Time Scatter Plot (for Exercise 14.47)** DS SrvcTime2

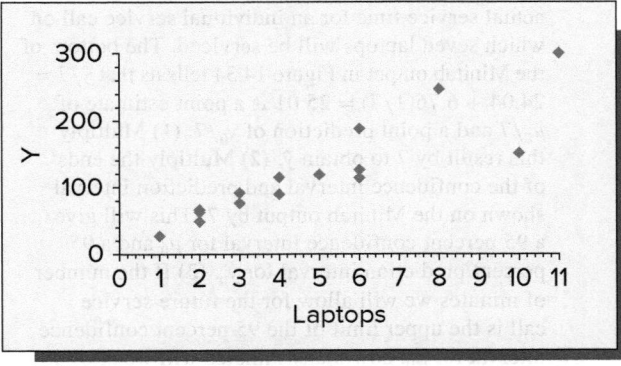

FIGURE 14.33   **Excel Output of a Residual Plot versus x for the Laptop Service Time Model $y = \beta_0 + \beta_1 x + \varepsilon$ (for Exercise 14.47)**

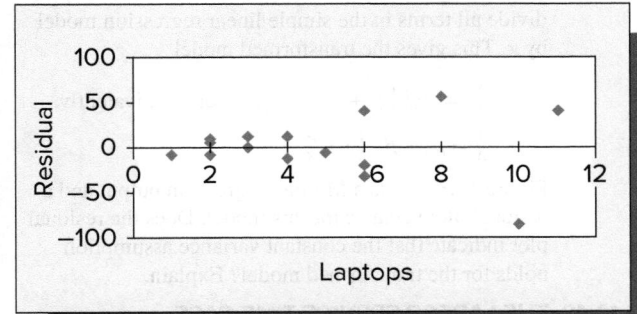

c   The model ln $y_t = \beta_0 + \beta_1 t + \varepsilon_t$ is called a **growth curve model** because it implies that

$$y_t = e^{(\beta_0 + \beta_1 t + \varepsilon_t)} = (e^{\beta_0})(e^{\beta_1 t})(e^{\varepsilon_t})$$
$$= \alpha_0 \alpha_1^t \eta_t$$

where $\alpha_0 = e^{\beta_0}$, $\alpha_1 = e^{\beta_1}$, and $\eta_t = e^{\varepsilon_t}$. Here $\alpha_1 = e^{\beta_1}$ is called the **growth rate** of the y values. Noting that the least squares point estimate of β_1 is $b_1 = .25688$, estimate the growth rate α_1.

d   We see that $y_t = \alpha_0 \alpha_1^t \eta_t = (\alpha_0 \alpha_1^{t-1})\alpha_1 \eta_t \approx (y_{t-1})\alpha_1 \eta_t$. This says that y_t is expected to be approximately α_1 times y_{t-1}. Noting this, interpret the growth rate of part c.

14.47   THE LAPTOP SERVICE TIME CASE
DS SrvcTime2

The page margin shows data concerning the time, y, required to perform service (in minutes) and the number of laptop computers serviced, x, for 15 service calls. Figure 14.32 gives a plot of y versus x, and Figure 14.33 gives the Excel output of a plot of the residuals versus x for a simple linear regression model. What regression assumption appears to be violated? Explain.

14.48   THE LAPTOP SERVICE TIME CASE
DS SrvcTime2

Figure 14.33 shows the residual plot versus x for the simple linear regression of the laptop service time data. This

The Laptop Service Time Data
DS SrvcTime2

Service Time, y	Laptops Serviced, x
92	3
63	2
126	6
247	8
49	2
90	4
119	5
114	6
67	2
115	4
188	6
298	11
77	3
151	10
27	1

FIGURE 14.34   **Minitab Output of a Regression Analysis of the Laptop Service Time Data Using the Model**
$y/x = \beta_0 + \beta_1(1/x) + \varepsilon/x$ **and a Residual Plot versus x (for Exercises 14.48 and 14.49)**

Analysis of Variance

Source	DF	Adj SS	Adj MS	F-Value	P-Value
Regression	1	36.27	36.27	1.36	0.264
1/X	1	36.27	36.27	1.36	0.264
Error	13	345.89	26.61		
Lack-of-Fit	7	181.56	25.94	0.95	0.535
Pure Error	6	164.33	27.39		
Total	14	382.15			

Model Summary

S	R-sq	R-sq(adj)	R-sq(pred)
5.15816	9.49%	2.53%	0.00%

Coefficients

Term	Coef	SE Coef	T-Value	P-Value	VIF
Constant	24.04	2.25	10.70	0.000	
1/X	6.76	5.79	1.17	0.264	1.00

Regression Equation
Y/X = 24.04 + 6.76 1/X

Variable	Setting	Fit	SE Fit	95% CI	95% PI
1/X	0.143	25.0079	1.65360	(21.4355, 28.5802)	(13.3057, 36.7100)

plot fans out, indicating that the error term ε tends to become larger in magnitude as x increases. To remedy this violation of the constant variance assumption, we divide all terms in the simple linear regression model by x. This gives the transformed model

$$\frac{y}{x} = \beta_0\left(\frac{1}{x}\right) + \beta_1 + \frac{\varepsilon}{x} \quad \text{or, equivalently,}$$

$$\frac{y}{x} = \beta_0 + \beta_1\left(\frac{1}{x}\right) + \frac{\varepsilon}{x}$$

Figure 14.34 gives a Minitab regression output and a residual plot versus x for this model. Does the residual plot indicate that the constant variance assumption holds for the transformed model? Explain.

14.49 THE LAPTOP SERVICE TIME CASE
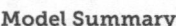 SrvcTime2

Consider a future service call on which seven laptops will be serviced. Let μ_0 represent the mean

service time for all service calls on which seven laptops will be serviced, and let y_0 represent the actual service time for an individual service call on which seven laptops will be serviced. The bottom of the Minitab output in Figure 14.34 tells us that $\hat{y}/7 = 24.04 + 6.76(1/7) = 25.01$ is a point estimate of $\mu_0/7$ and a point prediction of $y_0/7$. **(1)** Multiply this result by 7 to obtain $\hat{y}$. **(2)** Multiply the ends of the confidence interval and prediction interval shown on the Minitab output by 7. This will give a 95 percent confidence interval for μ_0 and a 95 percent prediction interval for y_0. **(3)** If the number of minutes we will allow for the future service call is the upper limit of the 95 percent confidence interval for μ_0, how many minutes will we allow?

Chapter Summary

This chapter has discussed **simple linear regression analysis,** which relates a **dependent variable** to a single **independent** (predictor) **variable.** We began by considering the **simple linear regression model,** which employs two parameters: the **slope** and **y-intercept.** We next discussed how to compute the **least squares point estimates** of these parameters and how to use these estimates to calculate a **point estimate of the mean value of the dependent variable** and a **point prediction of an individual value** of the dependent variable. We continued by considering the **simple coefficient of determination,** which is a measure of the potential utility of the simple linear

regression model. Then, after considering the assumptions behind the simple linear regression model, we discussed **testing the significance of the regression relationship (slope),** calculating a **confidence interval** for the mean value of the dependent variable, and calculating a **prediction interval** for an individual value of the dependent variable. We concluded this chapter by giving a discussion of using **residual analysis** to detect violations of the regression assumptions. We learned that we can sometimes remedy violations of these assumptions by **transforming** the dependent variable.

Glossary of Terms

dependent variable: The variable that is being described, predicted, or controlled.

distance value: A measure of the distance between a particular value x_0 of the independent variable x and $\bar{x}$, the average of the previously observed values of x (the center of the experimental region).

error term: The difference between an individual value of the dependent variable and the corresponding mean value of the dependent variable.

experimental region: The range of the previously observed values of the independent variable.

explained variation: A quantity that measures the amount of the total variation in the observed values of y that is explained by the predictor variable x.

independent (predictor) variable: A variable used to describe, predict, and control the dependent variable.

least squares point estimates: The point estimates of the slope and y-intercept of the simple linear regression model that minimize the sum of squared residuals.

negative autocorrelation: The situation in which positive error terms tend to be followed over time by negative error terms and negative error terms tend to be followed over time by positive error terms.

normal plot: A residual plot that is used to check the normality assumption.

positive autocorrelation: The situation in which positive error terms tend to be followed over time by positive error terms and negative error terms tend to be followed over time by negative error terms.

residual: The difference between the observed value of the dependent variable and the corresponding predicted value of the dependent variable.

residual plot: A plot of the residuals against some criterion. The plot is used to check the validity of one or more regression assumptions.

simple coefficient of determination r^2: The proportion of the total variation in the observed values of the dependent variable that is explained by the simple linear regression model.

simple correlation coefficient: A measure of the linear association between two variables.

simple linear regression model: An equation that describes the straight-line relationship between a dependent variable and an independent variable.

slope (of the simple linear regression model): The change in the mean value of the dependent variable that is associated with a one-unit increase in the value of the independent variable.

total variation: A quantity that measures the total amount of variation exhibited by the observed values of the dependent variable y.

unexplained variation: A quantity that measures the amount of the total variation in the observed values of y that is not explained by the predictor variable x.

y-intercept (of the simple linear regression model): The mean value of the dependent variable when the value of the independent variable is 0.

Important Formulas and Tests

Simple linear regression model: page 584

Least squares point estimates of β_0 and β_1: pages 585–587

Least squares line (prediction equation): page 586

The predicted value of y: page 586

The residual: pages 586 and 615

Sum of squared residuals: pages 586 and 595

Point estimate of a mean value of y: pages 589, 590

Point prediction of an individual value of y: pages 589 and 590

Explained variation: page 596

Unexplained variation: page 596

Total variation: page 596

Simple coefficient of determination r^2: page 596

Simple correlation coefficient: page 596

Mean square error: page 600

Standard error: page 600

Standard error of the estimate b_1: page 601

A t test for the significance of the slope: page 602

Confidence interval for the slope: page 604

Testing the significance of the y-intercept: page 604

An F test for the significance of the slope: page 605

Standard error of $\hat{y}$: page 609

Standard error of $y - \hat{y}$: page 610

Confidence interval for a mean value of y: page 610

Prediction interval for an individual value of y: page 610

Testing the significance of the population correlation coefficient: page 614

Normal plot calculations: page 618

Durbin–Watson test: pages 623–624

Supplementary Exercises

14.50 The data in Table 14.11 concerning the relationship between smoking and lung cancer death are presented in a course of The Open University, *Statistics in Society*, Unit C4, The Open University Press, Milton Keynes, England, 1983. The original source of the data is *Occupational Mortality: The Registrar General's Decennial Supplement for England and Wales, 1970–1972*, Her Majesty's Stationery Office, London, 1978. In the table, a smoking index greater (less) than

TABLE 14.11   The Smoking and Lung Cancer Death Data (for Exercise 14.50)   DS Smoking

Occupational Group	Smoking Index, x	Lung Cancer Death Index, y
Farmers, foresters, and fishermen	77	84
Miners and quarrymen	137	116
Gas, coke, and chemical makers	117	123
Glass and ceramics makers	94	128
Furnace, forge, foundry, and rolling mill workers	116	155
Electrical and electronics workers	102	101
Engineering and allied trades	111	118
Woodworkers	93	113
Leather workers	88	104
Textile workers	102	88
Clothing workers	91	104
Food, drink, and tobacco workers	104	129
Paper and printing workers	107	86
Makers of other products	112	96
Construction workers	113	144
Painters and decorators	110	139
Drivers of stationary engines, cranes, etc.	125	113
Laborers not included elsewhere	133	146
Transport and communications workers	115	128
Warehousemen, storekeepers, packers, and bottlers	105	115
Clerical workers	87	79
Sales workers	91	85
Service, sport, and recreation workers	100	120
Administrators and managers	76	60
Professionals, technical workers, and artists	66	51

New Jersey Bank Data   DS NJBank

Percent Minority, x	Residents per Branch, y
23.3	3,073
13.0	2,095
17.8	2,905
23.4	3,330
7.3	1,321
26.5	2,557
48.8	3,474
10.7	3,068
33.2	3,683
3.7	1,998
24.9	2,607
18.1	3,154
12.6	2,609
8.2	2,253
4.7	2,317
28.1	3,307
16.7	2,511
12.0	2,333
2.4	2,568
25.6	3,048
2.8	2,349

Source: Data from P. D'Ambrosio and S. Chambers, "No Checks and Balances," Asbury Park Press, September 10, 1995.

100 indicates that men in the occupational group smoke more (less) than average when compared to all men of the same age. Similarly, a lung cancer death index greater (less) than 100 indicates that men in the occupational group have a greater (less) than average lung cancer death rate when compared to all men of the same age. In Figure 14.35 we present a plot of the lung cancer death index versus the smoking index.   DS Smoking

a Use simple linear regression analysis to determine if there is a significant relationship between x and y.

b Does the slope of the hypothetical line relating the two indexes when the smoking index is less than 100 seem to equal the slope of the hypothetical line relating the two indexes when the smoking index is greater than 100? Use simple linear regression to make a more precise determination. What practical conclusion might you make?

14.51   In New Jersey, banks have been charged with withdrawing from counties having a high percentage of minorities. To substantiate this charge, P. D' Ambrosio and S. Chambers (1995) present the data concerning the percentage of minority population, x, and the number of county residents per bank branch, y, in each of New Jersey's 21 counties. Using simple linear regression analysis: DS NJBank

a Determine if there is a significant relationship between x and y.

b Describe the exact nature of any relationship that exists between x and y. (Hint: Estimate β_1 by a point estimate and a confidence interval.)

14.52   On January 28, 1986, the space shuttle *Challenger* exploded soon after takeoff, killing all seven astronauts aboard.

The temperature at the Kennedy Space Center at liftoff was 31°F. Before the launch, several scientists argued that the launch should be delayed because the shuttle's O-rings might harden in the cold and leak. Other scientists used a data plot to argue that there was no relationship between temperature and O-ring failure. On the basis of this data plot and other considerations, *Challenger* was launched to its disastrous, last flight.

Figure 14.36 shows a plot of the number of O-ring failures versus temperature for all 24 previous launches, 17 of which had no O-ring failures. Unfortunately, the scientists arguing that there was no relationship between temperature and O-ring failure mistakenly believed that data based on *zero* O-ring failures were *meaningless* and constructed a plot based only on the 7 previous launches that had at least one O-ring failure. That data plot is the plot enclosed in a rectangle in Figure 14.36.

a Discuss the difference between what the complete and incomplete data plots tell us.

b Even though the figure using only seven launches is incomplete, what about it should have cautioned the scientists not to make the launch?

14.53   **THE OXFORD HOME BUILDER CASE**   DS OxHome

To determine what types of homes would attract residents of the college community of Oxford, Ohio, a builder of speculative homes contacted a statistician at a local college. The statistician went to a local real estate agency and obtained the data in Table 14.12. This table presents the sales price y, square footage x_1, number of rooms x_2, number of bedrooms x_3, and age x_4, for each of 60 single-family residences recently sold in

FIGURE 14.35 A Plot of the Lung Cancer Death Index versus the Smoking Index (for Exercise 14.50)

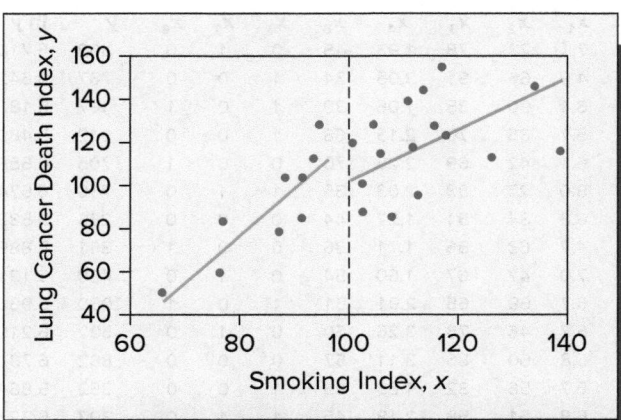

FIGURE 14.36 A Plot of the Number of O-ring Failures versus Temperature (for Exercise 14.52)

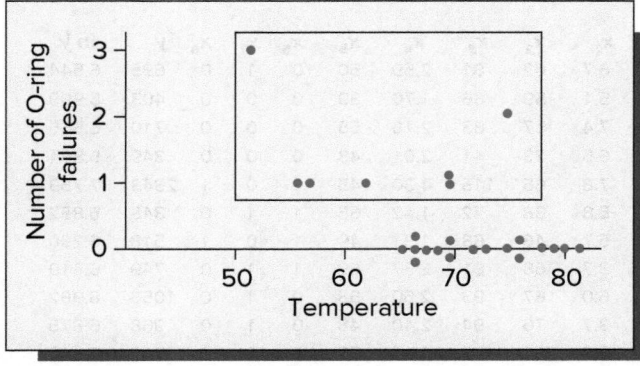

TABLE 14.12 Measurements Taken on 63 Single-Family Residences (for Exercise 14.53) DS OxHome

Residence	Sales Price, y (×$1,000)	Square Feet, x_1	Rooms, x_2	Bedrooms, x_3	Age, x_4	Residence	Sales Price, y (×$1,000)	Square Feet, x_1	Rooms, x_2	Bedrooms, x_3	Age, x_4
1	110	1,008	5	2	35	31	115	1,020	6	3	16
2	101	1,290	6	3	36	32	129	1,053	5	2	24
3	104	860	8	2	36	33	123	1,728	6	3	26
4	102.8	912	5	3	41	34	71	416	3	1	42
5	107	1,204	6	3	40	35	107	1,040	5	2	9
6	113	1,204	5	3	10	36	153	1,496	6	3	30
7	164	1,764	8	4	64	37	189	1,936	8	4	39
8	175	1,600	7	3	19	38	123	1,904	7	4	32
9	141	1,255	5	3	16	39	149	1.080	5	2	24
10	95	864	5	3	37	40	145	1,768	8	4	74
11	79	720	4	2	41	41	169	1,503	6	3	14
12	102	1,008	6	3	35	42	183	1,736	7	3	16
13	213	1,950	8	3	52	43	169	1,695	6	3	12
14	308	2,086	7	3	12	44	233	2,186	8	4	12
15	173	2,011	9	4	76	45	103	888	5	2	34
16	123	1,465	6	3	102	46	113.4	1,120	6	3	29
17	120	1,232	5	2	69	47	125	1,400	5	3	33
18	205	1,736	7	3	67	48	297	2,165	7	3	2
19	161.8	1,296	6	3	11	49	123	1,536	6	3	36
20	253	1,996	7	3	9	50	203	1,972	8	3	37
21	178.8	1,874	5	2	14	51	92	1,120	5	3	27
22	163	1,580	5	3	11	52	113	1,664	7	3	79
23	191	1,920	5	3	14	53	109.8	925	5	3	20
24	151	1,430	9	3	16	54	133	1,288	5	3	2
25	141	1,486	6	3	27	55	149	1,400	5	3	2
26	129	1,008	5	2	35	56	83	1,376	6	3	103
27	138	1,282	5	3	20	57	285	2,038	12	4	62
28	73	1,134	5	2	74	58	139	1,572	6	3	29
29	288	2,400	9	4	15	59	283	1,993	6	3	4
30	187.4	1,701	5	3	15	60	113	1,130	5	2	21

TABLE 14.13 The Liver Operation Survival Time Data: The Training Data Set
DS SurvTimeT

x_1	x_2	x_3	x_4	x_5	x_6	x_7	x_8	y	$\ln y$
6.7	62	81	2.59	50	0	1	0	695	6.544
5.1	59	66	1.70	39	0	0	0	403	5.999
7.4	57	83	2.16	55	0	0	0	710	6.565
6.5	73	41	2.01	48	0	0	0	349	5.854
7.8	65	115	4.30	45	0	0	1	2343	7.759
5.8	38	72	1.42	65	1	1	0	348	5.852
5.7	46	63	1.91	49	1	0	1	518	6.250
3.7	68	81	2.57	69	1	1	0	749	6.619
6.0	67	93	2.50	58	0	1	0	1056	6.962
3.7	76	94	2.40	48	0	1	0	968	6.875
6.3	84	83	4.13	37	0	1	0	745	6.613
6.7	51	43	1.86	57	0	1	0	257	5.549
5.8	96	114	3.95	63	1	0	0	1573	7.361
5.8	83	88	3.95	52	1	0	0	858	6.754
7.7	62	67	3.40	58	0	0	1	702	6.554
7.4	74	68	2.40	64	1	1	0	809	6.695
6.0	85	28	2.98	36	1	1	0	682	6.526
3.7	51	41	1.55	39	0	0	0	205	5.321
7.3	68	74	3.56	59	1	0	0	550	6.309
5.6	57	87	3.02	63	0	0	1	838	6.731
5.2	52	76	2.85	39	0	0	0	359	5.883
3.4	83	53	1.12	67	1	1	0	353	5.866
6.7	26	68	2.10	30	0	0	1	599	6.395
5.8	67	86	3.40	49	1	1	0	562	6.332
6.3	59	100	2.95	36	1	1	0	651	6.478
5.8	61	73	3.50	62	1	1	0	751	6.621
5.2	52	86	2.45	70	0	1	0	545	6.302
11.2	76	90	5.59	58	1	0	1	1965	7.583
5.2	54	56	2.71	44	1	0	0	477	6.167
5.8	76	59	2.58	61	1	1	0	600	6.396
3.2	64	65	0.74	53	0	1	0	443	6.094
8.7	45	23	2.52	68	0	1	0	181	5.198
5.0	59	73	3.50	57	0	1	0	411	6.019
5.8	72	93	3.30	39	1	0	1	1037	6.944
5.4	58	70	2.64	31	1	1	0	482	6.179
5.3	51	99	2.60	48	0	1	0	634	6.453
2.6	74	86	2.05	45	0	0	0	678	6.519
4.3	8	119	2.85	65	1	0	0	362	5.893
4.8	61	76	2.45	51	1	1	0	637	6.457
5.4	52	88	1.81	40	1	0	0	705	6.558
5.2	49	72	1.84	46	0	0	0	536	6.283
3.6	28	99	1.30	55	0	0	1	582	6.366
8.8	86	88	6.40	30	1	1	0	1270	7.147
6.5	56	77	2.85	41	0	1	0	538	6.288
3.4	77	93	1.48	69	0	1	0	482	6.178
6.5	40	84	3.00	54	1	1	0	611	6.416
4.5	73	106	3.05	47	1	1	0	960	6.867
4.8	86	101	4.10	35	1	0	1	1300	7.170
1.1	67	77	2.86	66	1	0	0	581	6.365
1.9	82	103	4.55	50	0	1	0	1078	6.983
6.6	77	46	1.95	50	0	1	0	405	6.005
6.4	85	40	1.21	58	0	0	1	579	6.361
6.4	59	85	2.33	63	0	1	0	550	6.310
8.8	78	72	3.20	56	0	0	0	651	6.478

TABLE 14.14 The Liver Operation Survival Time Data: The Validation Data Set
DS SurvTimeV

x_1	x_2	x_3	x_4	x_5	x_6	x_7	x_8	y	$\ln y$
7.1	23	78	1.93	45	0	1	0	302	5.710
4.9	66	91	3.05	34	1	0	0	767	6.642
6.4	90	35	1.06	39	1	0	1	487	6.188
5.7	35	70	2.13	68	1	0	0	242	5.489
6.1	42	69	2.25	70	0	0	1	705	6.558
8.0	27	83	2.03	35	1	1	0	716	6.574
6.8	34	51	1.27	44	0	0	0	266	5.583
4.7	63	36	1.71	36	0	0	1	361	5.889
7.0	47	67	1.60	54	0	1	0	460	6.131
6.7	69	65	2.91	61	1	0	1	1060	6.966
6.7	46	78	3.26	50	0	1	0	502	6.219
5.8	60	86	3.11	57	0	0	0	882	6.782
6.7	56	32	1.53	69	1	0	0	352	5.864
6.8	51	58	2.18	45	1	1	0	307	5.727
7.2	95	82	4.68	62	1	0	0	1227	7.112
7.4	52	67	3.28	53	0	0	0	508	6.230
5.3	53	62	2.42	39	1	0	0	419	6.038
3.5	58	84	1.74	42	1	1	0	536	6.294
6.8	74	79	2.25	36	0	0	1	902	6.805
4.4	47	49	2.42	35	0	0	0	189	5.242
7.0	66	118	4.69	59	0	0	0	1433	7.269
6.7	61	57	3.87	66	1	0	0	815	6.703
5.6	75	103	3.11	57	1	1	0	1144	7.042
6.9	58	88	3.46	37	1	1	0	571	6.347
6.2	62	57	1.25	53	1	0	1	591	6.382
4.7	97	27	1.77	36	0	1	0	533	6.279
6.8	69	60	2.90	56	1	1	0	534	6.280
6.0	73	58	1.22	36	0	0	1	374	5.924
5.9	50	62	3.19	69	0	1	0	222	5.403
5.5	88	74	3.21	36	1	1	0	881	6.781
3.8	55	52	1.41	69	1	0	0	470	6.153
4.3	99	83	3.93	35	1	0	0	913	6.817
6.6	48	54	2.94	59	1	0	1	527	6.267
6.2	42	63	1.85	32	0	0	1	676	6.516
5.0	60	105	3.17	34	1	0	0	850	6.745
5.8	62	82	3.18	45	1	1	0	569	6.344
5.2	56	49	1.48	54	0	1	0	182	5.204
5.7	70	59	2.28	65	0	0	1	421	6.043
4.7	64	48	1.30	56	1	1	0	245	5.501
7.8	74	49	2.58	33	0	0	1	611	6.415
4.8	52	45	2.71	63	0	1	0	338	5.823
4.9	72	90	3.51	45	0	1	0	875	6.774
4.6	73	57	2.82	38	1	0	0	750	6.620
5.9	78	70	4.28	55	0	0	0	935	6.841
4.6	69	70	3.17	43	1	1	0	583	6.368
6.1	53	52	1.84	35	0	1	0	319	5.765
5.9	88	98	3.33	36	1	1	0	1158	7.054
4.7	66	68	1.80	35	0	1	1	553	6.315
10.4	62	85	4.65	50	0	1	0	1041	6.948
5.8	70	64	2.52	49	0	1	0	589	6.378
5.4	64	81	1.36	62	0	1	0	599	6.395
6.9	90	33	2.78	48	1	0	0	655	6.485
7.9	45	55	2.46	43	0	1	0	377	5.932
4.5	68	60	2.07	59	0	0	0	642	6.465

Source of both Tables 14.13 and 14.14: M. H. Kutner, C. S. Nachtsheim, J. Neter, and W. Li, CD for the book *Applied Linear Statistical Models*, 5th ed (Burr Ridge, IL: McGraw-Hill/Irwin, 2005).

the community. Use simple linear regression analysis to relate y to each of x_1, x_2, x_3, and x_4 and determine which simple linear relationships are significant. In the Supplementary Exercises of Chapter 15, the reader will see how *multiple regression analysis* (regression analysis that employs more than one independent variable) helped the builder to build houses that sold better.

14.54   THE LIVER OPERATION SURVIVAL TIME CASE Ⓓⓢ SurvTimeT

When developing a regression model to predict a dependent variable y, it is best, if there are enough data, to "build" the model using one data set (called the **training data set**) and then "validate" the model by using it to analyze a different data set (called the **validation data set**). To illustrate this, Kutner, Nachtsheim, Neter, and Li (2005) consider 108 observations described by the dependent variable y = survival time (in days) after undergoing a particular liver operation and the independent variables x_1 = blood

clotting score, x_2 = prognostic index, x_3 = enzyme function test score, x_4 = liver function test score, x_5 = age (in years), $x_6 = 1$ for a female patient and 0 for a male patient, $x_7 = 1$ for a patient who is a moderate drinker and 0 otherwise, and $x_8 = 1$ for a patient who is a heavy drinker and 0 otherwise. Because 108 observations are a fairly large number of observations, Kutner and his fellow authors randomly divided these observations into two equal halves, obtaining the training data set in Table 14.13 and the validation data set in Table 14.14. In the exercises of Section 15.10 of Chapter 15 the reader will use both data sets to develop a multiple regression model that can be used to predict survival time, y, for future patients. As a preliminary step in doing this, **(1)** use the training data set to plot y versus each of x_1, x_2, x_3, and x_4, and **(2)** use the training data set to plot ln y versus each of x_1, x_2, x_3, and x_4. **(3)** By comparing the two sets of data plots, determine whether the y values or the ln y values seem to exhibit more of a straight-line appearance and more constant variation as x_1, x_2, x_3, and x_4 increase.

Appendix 14.1 ■ Simple Linear Regression Analysis Using Excel

Simple linear regression in Figure 14.12 (data file: GasCon1.xlsx):

- Enter the natural gas consumption data—the temperatures in column A with label Temp and the gas consumptions in column B with label GasCons.

- Select **Data : Data Analysis : Regression** and click OK in the Data Analysis dialog box.

- In the Regression dialog box:

- Enter B1 : B9 into the "Input Y Range" window.

- Enter A1 : A9 into the "Input X Range" window.

- Place a checkmark in the Labels checkbox.

- Be sure that the "Constant is Zero" checkbox is NOT checked.

- Select the "New Worksheet Ply" option and enter the name "Regression Output" into the New Worksheet window.

- Click OK in the Regression dialog box to obtain the regression results in a new worksheet.

To produce residual analysis similar to that in Section 14.7:

- In the Regression dialog box, place a checkmark in the Residuals checkbox to request predicted values and residuals.

- Place a checkmark in the Residual Plots checkbox.

- Place a checkmark in the Normal Probability Plots checkbox.

- Click OK in the Regression dialog box.

Move the plots to chart sheets to format them for effective viewing. Additional residual plots—residuals versus predicted values and residuals versus time—can be produced using the Excel charting features.

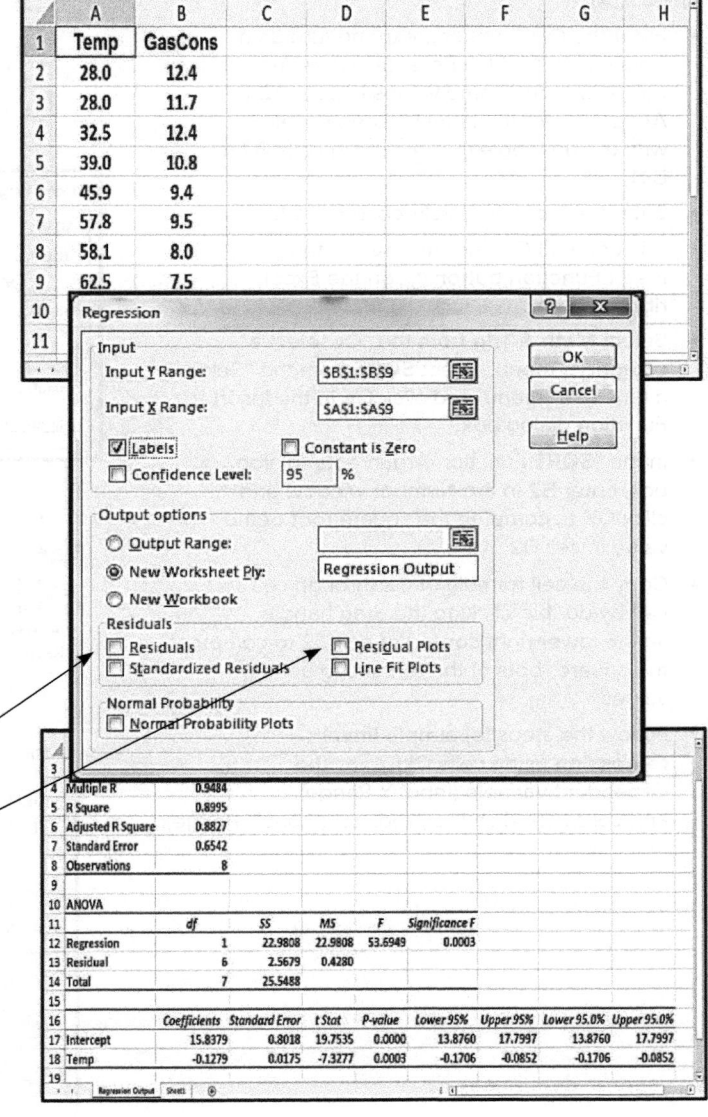

	A	B
1	Temp	GasCons
2	28.0	12.4
3	28.0	11.7
4	32.5	12.4
5	39.0	10.8
6	45.9	9.4
7	57.8	9.5
8	58.1	8.0
9	62.5	7.5

4	Multiple R	0.9484				
5	R Square	0.8995				
6	Adjusted R Square	0.8827				
7	Standard Error	0.6542				
8	Observations	8				
9						
10	ANOVA					
11		df	SS	MS	F	Significance F
12	Regression	1	22.9808	22.9808	53.6949	0.0003
13	Residual	6	2.5679	0.4280		
14	Total	7	25.5488			
15						

		Coefficients	Standard Error	t Stat	P-value	Lower 95%	Upper 95%	Lower 95.0%	Upper 95.0%
16									
17	Intercept	15.8379	0.8018	19.7535	0.0000	13.8760	17.7997	13.8760	17.7997
18	Temp	-0.1279	0.0175	-7.3277	0.0003	-0.1706	-0.0852	-0.1706	-0.0852
19									

Source: Microsoft Office Excel 2016

To compute a point prediction of gas consumption when temperature is 40°F (data file: GasCon1.xlsx):

- The Excel Analysis ToolPak does not provide an option for computing point or interval predictions. A point prediction can be computed from the regression results using Excel cell formulas.

- In the regression output, the estimated intercept and slope parameters from cells A17:B18 have been copied to cells D2:E3 and the predictor value 40 has been placed in cell E5.

- In cell E6, enter the Excel formula =E2+E3*E5 (=10.721) to compute the prediction.

E6		× ✓ fx	=E2 + E3*E5		

	A	B	C	D	E	F
1	SUMMARY OUTPUT					
2				Intercept	15.8379	
3		Regression Statistics		Temp	-0.1279	
4	Multiple R		0.9484			
5	R Square		0.8995	New Temp	40	
6	Adjusted R Square		0.8827	Prediction	10.721	
7	Standard Error		0.6542			
8	Observations		8			
9						
10	ANOVA					
11		df	SS	MS	F	Significance F
12	Regression	1	22.9808	22.9808	53.6949	0.0003
13	Residual	6	2.5679	0.4280		
14	Total	7	25.5488			
15						
16		Coefficients	Standard Error	t Stat	P-value	Lower 95%	Up
17	Intercept	15.8379	0.8018	19.7535	0.0000	13.8760
18	Temp	-0.1279	0.0175	-7.3277	0.0003	-0.1706

Source: Microsoft Office Excel 2016

Simple linear regression with a transformed dependent variable similar to Figure 14.25 (data file: QHIC.xlsx):

- Enter the QHIC upkeep expenditure data from Figure 14.18(a). Enter the label Value in cell A1 with the home values in cells A2 to A41 and enter the label Upkeep in cell B1 with the upkeep expenditures in cells B2 to B41.

- Enter the label SqRtUpkeep in cell C1.

- Click on cell C2 and then select the Insert Function button f_x on the Excel ribbon.

- Select **Math & Trig** from the "Or select a category:" menu, select **SQRT** from the "Select a function:" menu, and click OK in the Insert Function dialog box.

- In the "SQRT Function Arguments" dialog box, enter B2 in the Number window and click OK to compute the square root of the value in cell B2.

- Copy the cell formula of C2 through cell C41 by double-clicking the drag handle (in the lower right corner) of cell C2 to compute the square roots of the remaining upkeep values.

- Follow the steps for **simple linear regression** using cells C1:C41 as the dependent variable (Input Y Range) and cells A1:A41 as the predictor (Input X Range).

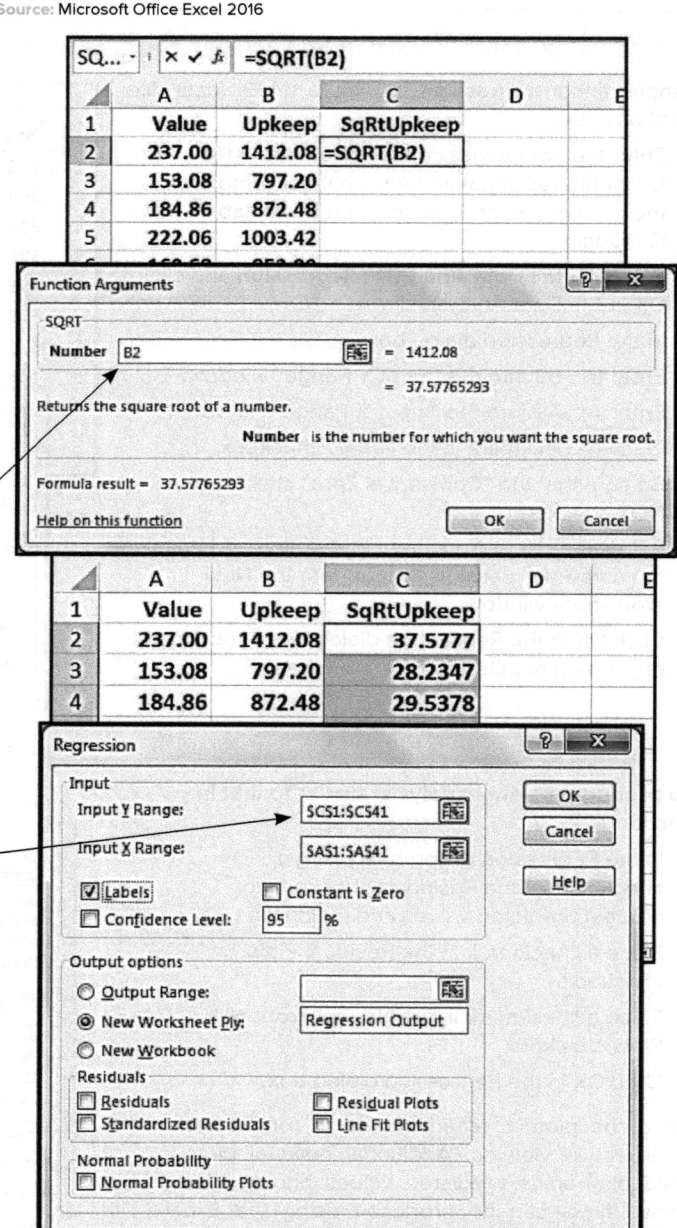

Source: Microsoft Office Excel 2016

Appendix 14.2 ■ Simple Linear Regression Analysis Using Minitab

Simple linear regression in Figure 14.13 (data file: GasCon1.MTW):

- Enter the gas consumption data in the data window with average hourly temperatures in C1 and weekly gas consumptions in C2; use variable names Temp and GasCons.

- Select **Stat : Regression : Regression : Fit Regression Model**

- In the Regression dialog box, select GasCons as the Response and Temp as the Continuous predictor.

- Click OK in the Regression dialog box; the regression results will appear in the Session window.

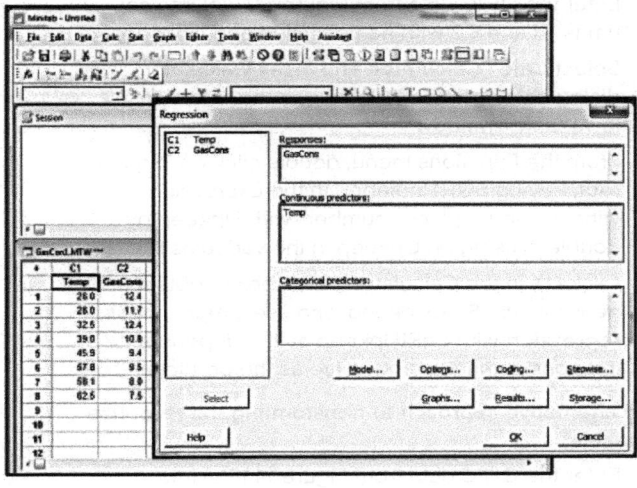

Source: Minitab 18

To compute a prediction of natural gas consumption when temperature is 40° F:

- After fitting the regression model, you have a new option in the **Stat : Regression** sequence.

- Select **Stat : Regression : Regression : Predict**

- In the "Predict" dialog box, GasCons will appear as the Response; choose "Enter individual values" from the drop-box selections, and enter 40 as the temperature.

- The default will be to compute a 95% prediction interval. (Click on the Options button should you desire a different choice.)

- Click OK in the "Predict" dialog box. The prediction interval will be added to the results in the Session window.

To produce **residual analysis** similar to that in Section 14.7:

- In the Regression dialog box, click on the Graphs... button.

- In the "Regression: Graphs" dialog box, select "Regular" from the "Residuals for plots" drop-down menu.

- Click on the "Four in one option" and enter Temp in the "Residuals versus the variables" window.

- Click OK in the "Regression: Graphs" dialog box and in the "Regression" dialog box. The graphs will appear in two high-resolution graphics windows.

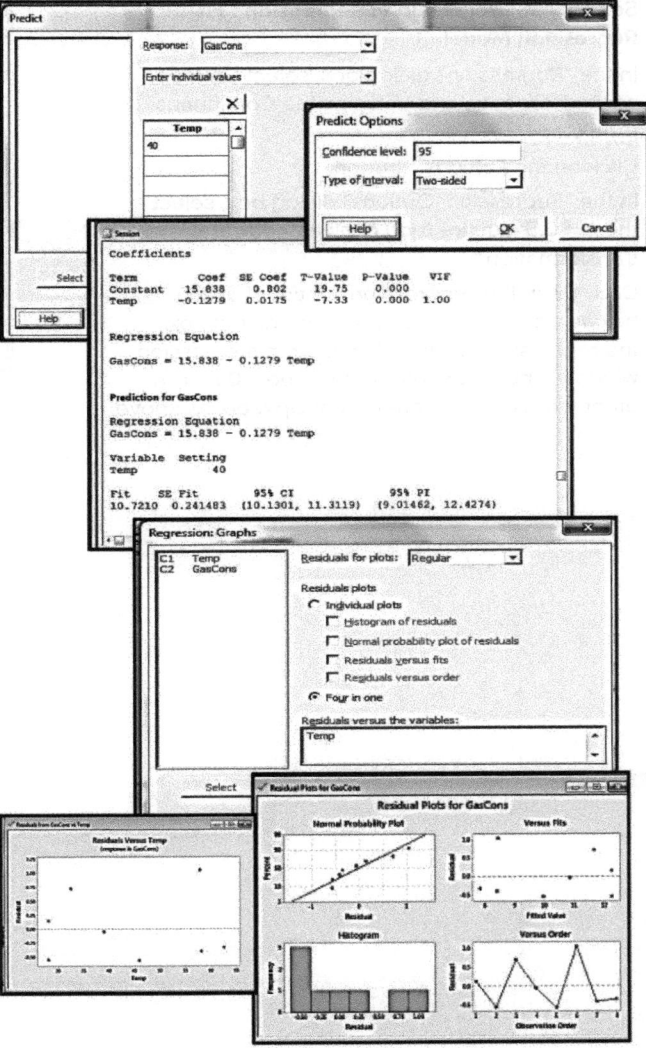

Source: Minitab 18

Simple linear regression with a transformed response as in Figure 14.25 (data file: QHIC.MTW):

- Enter the QHIC data from Figure 14.18(a) into columns C1 and C2 with names Value and Upkeep.

- Select **Calc : Calculator,** and in the Calculator dialog box, enter SqRtUpkeep in the "Store result in variable" window.

- From the Functions menu, double click on "Square root," giving SQRT(number) in the Expression window, and replace "number" with Upkeep by double-clicking on Upkeep in the variables list.

- Click OK in the Calculator dialog box to obtain a new column, SqRtUpkeep, and use **simple linear regression** with SqRtUpkeep as the dependent (response) variable and Value as the predictor.

An alternative approach to transforming the response variable:

- Enter the QHIC data from Figure 14.18(a) into columns C1 and C2 with names Value and Upkeep.

- Select **Stat : Regression : Regression : Fit Regression Model.**

- In the "Regression" dialog box, select Upkeep as the Response and Value as the Continuous predictor.

- Click on the Options... button.

- In the "Regression: Options" dialog box, select the $\lambda = 0.5$ (square root) choice under "Box-Cox transformation."

- Click OK in the "Regression: Options" dialog box as well as in the "Regression" dialog box; the regression results will appear in the Session window. The model predicts Upkeep^0.5 or, in other words, the square root of Upkeep, as above.

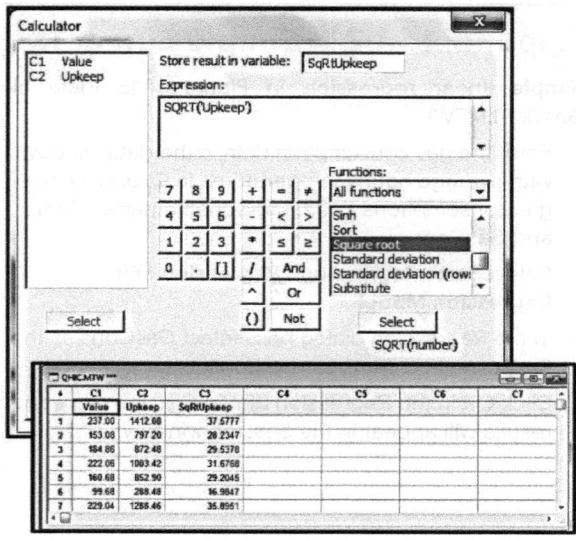

Source: Minitab 18

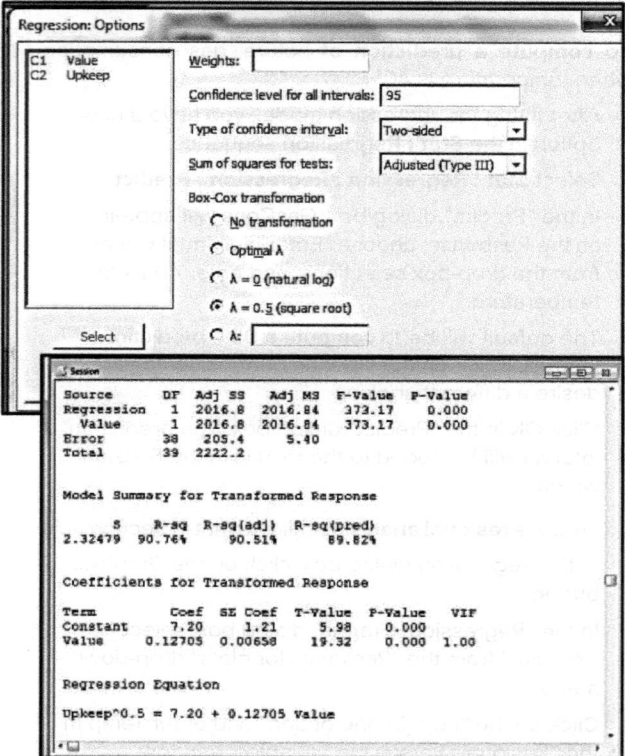

Source: Minitab 18

Appendix 14.3 ■ Simple Linear Regression Analysis Using JMP

Simple linear regression in Figure 14.14 (data file: Fresh.jmp):

- After opening the Fresh.jmp data file, select **Analyze : Fit Model**.

- In the Fit Model dialog box, select the "Demand" column and click the "Y" button and select the "PriceDif" column and click the "Add" button.

- Set Personality to "Standard Least Squares" and set Emphasis to "Minimal Report." Click "Run."

- Option-Control-Click (Mac)/Alt-Right-Click (Windows) on the word "Intercept" in the "Parameter Estimates" section to access options for that section. Click the checkboxes for "Lower 95%" and "Upper 95%" to obtain confidence intervals for the intercept and slope.

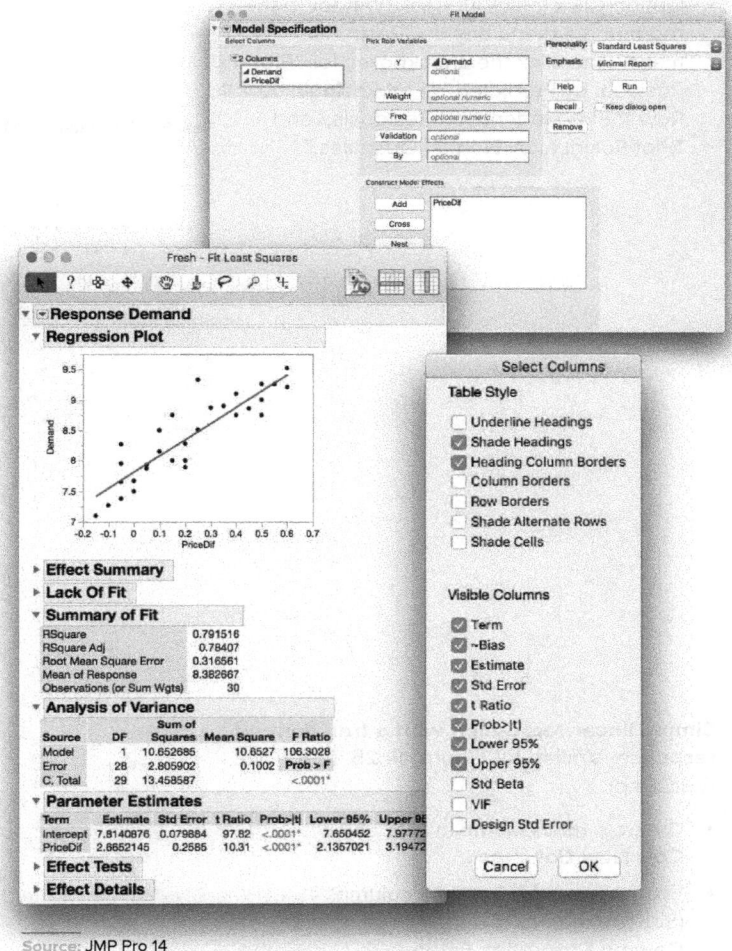

Source: JMP Pro 14

To compute a prediction of Demand when PriceDif is 0.1:

- Add a new row to the data table with PriceDif entered as 0.1 and no value entered for Demand. Redo the steps above for **simple linear regression**. (Note: It is best to enter "prediction rows" like this one *before* running your analysis to avoid repeating steps in JMP. You can add as many prediction rows as desired.)

- From the red triangle menu beside "Response Demand" select **Save Columns : Prediction Formula** to produce the Predicted Demand value at the bottom of Figure 14.14. You can Option-Click (Mac)/Alt-Click (Windows) the same red triangle menu to access a dialog of options allowing you to select multiple **Save Columns** at once. Click the checkboxes for "Mean Confidence Interval," "Indiv Confidence Interval," "Std Error of Predicted," and "Std Error of Individual" to produce the other values at the bottom of Figure 14.14. These values will all appear in the JMP Data Table.

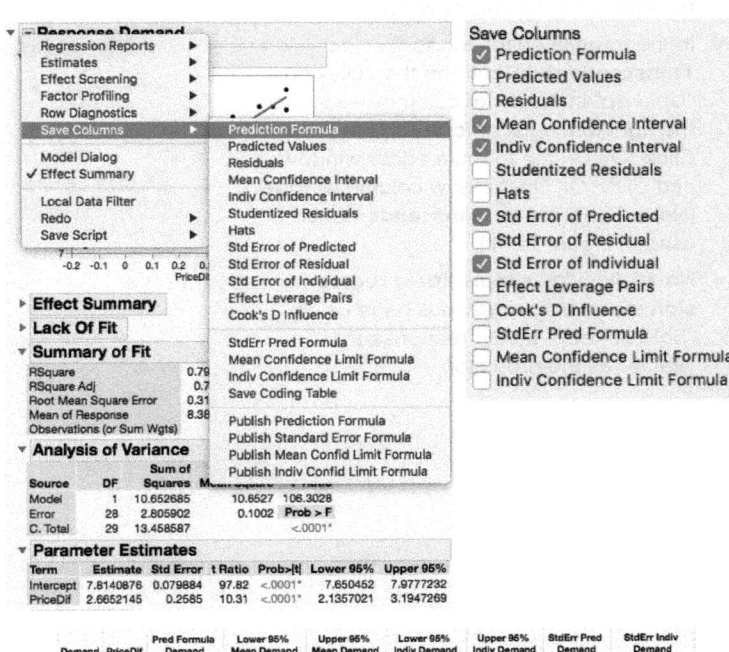

Source: JMP Pro 14

To produce residual plots similar to those in Section 14.7:

- Option-Click (Mac)/Alt-Click (Windows) the red triangle menu beside "Response Demand" and click the checkboxes for "Plot Residual by Predicted," "Plot Residual by Row," "Plot Studentized Residuals," and "Plot Residual by Normal Quantiles."

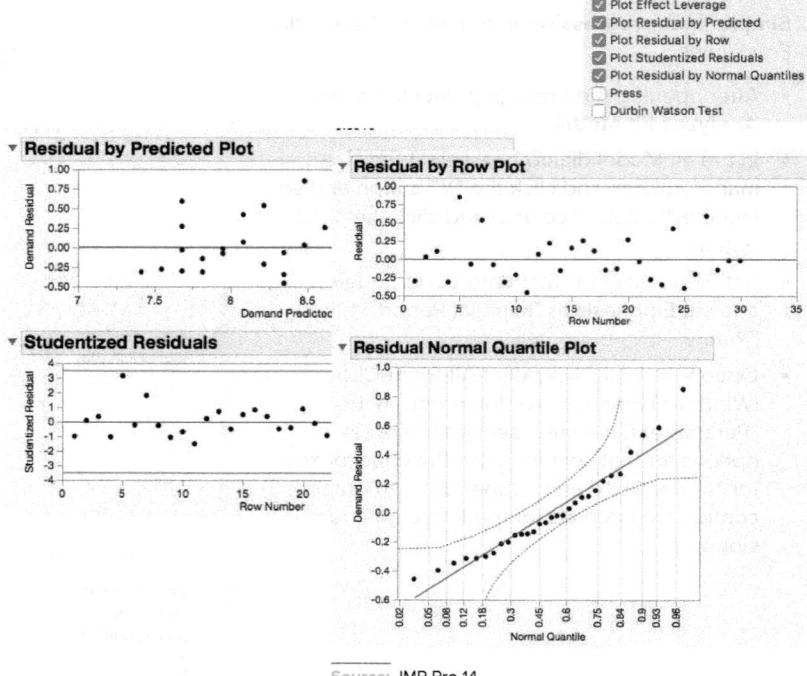

Source: JMP Pro 14

Simple linear regression with a transformed response similar to Figure 14.25 (data file: QHIC.jmp):

- Create a new column in JMP using **Cols:New Columns...**

- Enter a name for the new column (SqRtUpkeep for example) and select **Column Properties:Formula** to access the formula editor for this new column.

- In the formula editor window, select **Transcendental:Root** and then click on "Upkeep" in the list of columns (found to the right of the list of formula categories). Click "OK" in the formula editor window and click "OK" in the new column window. (Note: The formula **Transcendental:Ln** is used in Figure 14.31.)

- Now follow the **simple linear regression** steps on the previous page using "SqRtUpkeep" as the response and "Value" as the predictor.

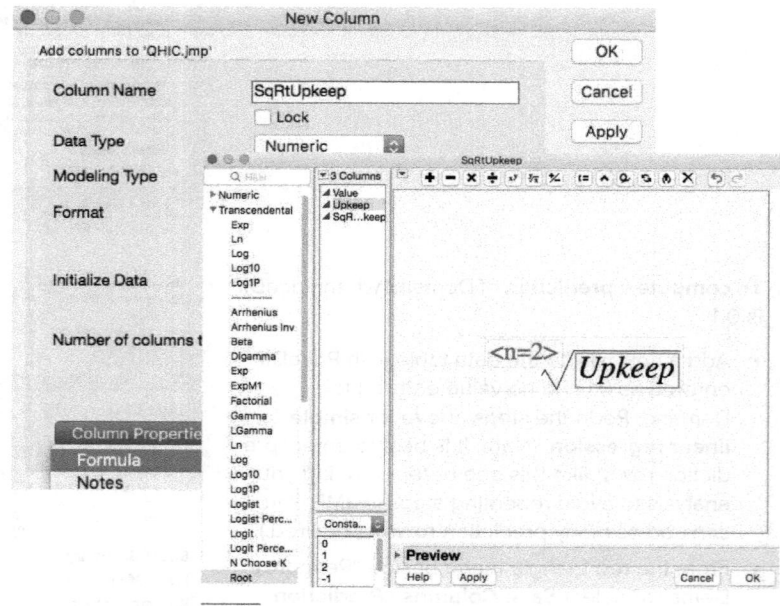

Source: JMP Pro 14

(Note: JMP's **Analyze:Fit Y by X** platform provides another method of applying common transformations to data. Use the **Fit Special...** subcommand to select from the built-in transformations.)

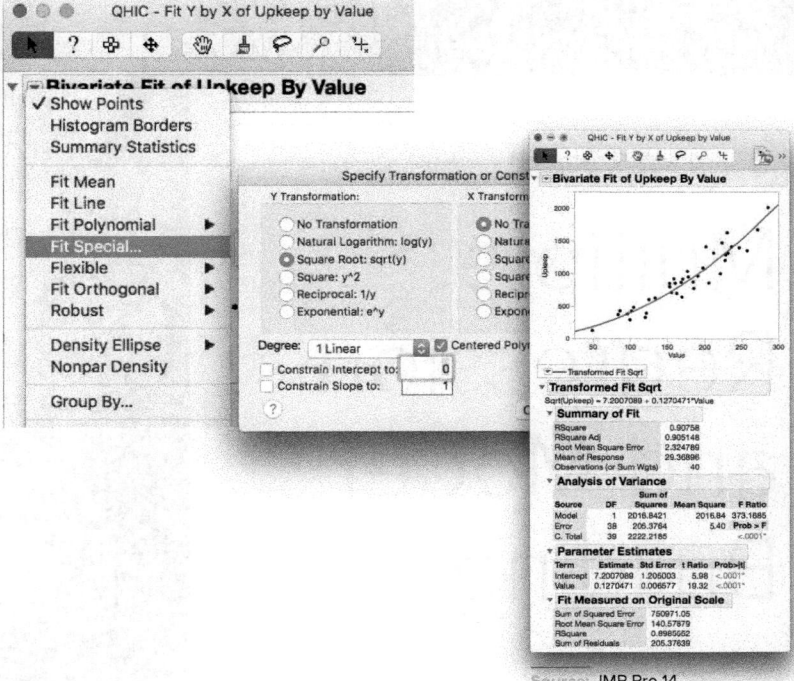

Source: JMP Pro 14

Design Elements: (CD): ©Comstock Images/Alamy; (All Others): ©McGraw-Hill Education

©Tetra Images/Alamy RF

CHAPTER 15

Multiple Regression and Model Building

Learning Objectives

After mastering the material in this chapter, you will be able to:

LO15-1 Explain the multiple regression model and the related least squares point estimates.

LO15-2 Calculate and interpret the multiple and adjusted multiple coefficients of determination.

LO15-3 Explain the assumptions behind multiple regression and calculate the standard error.

LO15-4 Test the significance of a multiple regression model by using an *F* test.

LO15-5 Test the significance of a single independent variable.

LO15-6 Find and interpret a confidence interval for a mean value and a prediction interval for an individual value.

LO15-7 Use dummy variables to model qualitative independent variables. (Optional)

LO15-8 Use squared and interaction variables. (Optional)

LO15-9 Describe multicollinearity and build and validate a multiple regression model. (Optional)

LO15-10 Use residual analysis and outlier detection to check the assumptions of multiple regression. (Optional)

Note: After completing Section 15.7, the reader may study Sections 15.8, 15.9, 15.10, and 15.11 in any order without loss of continuity.

ften we can more accurately describe, predict, and control a dependent variable by using a regression model that employs more than one independent variable. Such a model is called a **multiple regression model,** which is the subject of this chapter.

In order to explain the ideas of this chapter, we consider the following cases:

The Tasty Sub Shop Case: The business entrepreneur more accurately predicts the yearly revenue for a potential restaurant site by using a multiple regression model that employs as independent variables (1) the number of residents living near the site and (2) a rating of the amount of business and shopping near the site. The entrepreneur uses the more accurate predictions given by the multiple regression model to more accurately assess the profitability of the potential restaurant site.

The Sales Representative Case: A sales manager evaluates the performance of sales representatives by using a multiple regression model that predicts sales performance on the basis of five independent variables. Salespeople whose actual performance is far worse than predicted performance will get extra training to help improve their sales techniques.

15.1 The Multiple Regression Model and the Least Squares Point Estimates

LO15-1
Explain the multiple regression model and the related least squares point estimates.

Regression models that employ more than one independent variable are called **multiple regression models.** We begin our study of these models by considering the following example.

 EXAMPLE 15.1 The Tasty Sub Shop Case: A Multiple Regression Model

Part 1: The Data and a Regression Model Consider the Tasty Sub Shop problem in which the business entrepreneur wishes to predict yearly revenue for potential Tasty Sub restaurant sites. In Chapter 14 we used the number of residents, or population size x, living near a site to predict y, the yearly revenue for a Tasty Sub Shop built on the site. We now consider predicting y on the basis of the population size and a second predictor variable—the business rating. The business rating for a restaurant site reflects the amount of business and shopping near the site. This rating is expressed as a whole number between 1 and 10. Sites having only limited business and shopping nearby do not provide many potential customers—shoppers or local employees likely to eat in a Tasty Sub Shop—so they receive ratings near 1. However, sites located near substantial business and shopping activity do provide many potential customers for a Tasty Sub Shop, so they receive much higher ratings. The best possible rating for business activity is 10.

The business entrepreneur has collected data concerning yearly revenue (y), population size (x_1), and business rating (x_2) for 10 existing Tasty Sub restaurants that are built on sites similar to the site the entrepreneur is considering. These data are given in Table 15.1.

Figure 15.1 presents a scatter plot of y versus x_1. This plot shows that y tends to increase in a straight-line fashion as x_1 increases. Figure 15.2 shows a scatter plot of y versus x_2. This plot shows that y tends to increase in a straight-line fashion as x_2 increases. Together, the scatter plots in Figures 15.1 and 15.2 imply that a reasonable multiple regression model relating y (yearly revenue) to x_1 (population size) and x_2 (business rating) is

$$y = \beta_0 + \beta_1 x_1 + \beta_2 x_2 + \varepsilon$$

This model says that the values of y can be represented by a **mean level** ($\mu_y = \beta_0 + \beta_1 x_1 + \beta_2 x_2$) that changes as x_1 and x_2 change, combined with random fluctuations (described by the

TABLE 15.1   The Tasty Sub Shop Revenue Data   DS TastySub2

Restaurant	Population Size, x_1 (Thousands of Residents)	Business Rating, x_2	Yearly Revenue, y (Thousands of Dollars)
1	20.8	3	527.1
2	27.5	2	548.7
3	32.3	6	767.2
4	37.2	5	722.9
5	39.6	8	826.3
6	45.1	3	810.5
7	49.9	9	1040.7
8	55.4	5	1033.6
9	61.7	4	1090.3
10	64.6	7	1235.8

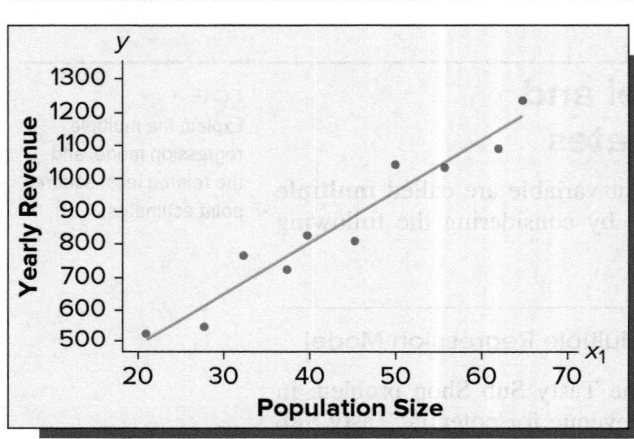

FIGURE 15.1   Plot of y (Yearly Revenue) versus x_1 (Population Size)

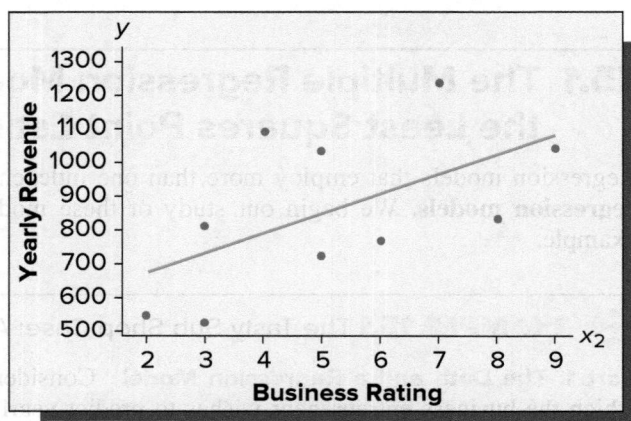

FIGURE 15.2   Plot of y (Yearly Revenue) versus x_2 (Business Rating)

error term ε) that cause the values of y to deviate from the mean level. Here

1   The mean level $\mu_y = \beta_0 + \beta_1 x_1 + \beta_2 x_2$ is the mean yearly revenue for all Tasty Sub restaurants that could potentially be built near populations of size x_1 and business/shopping areas having a rating of x_2. Furthermore, the equation

$$\mu_y = \beta_0 + \beta_1 x_1 + \beta_2 x_2$$

is the equation of a plane—called the **plane of means**—in three-dimensional space. The plane of means is the shaded plane illustrated in Figure 15.3. Different mean yearly revenues corresponding to different population size–business rating combinations lie on the plane of means. For example, Table 15.1 tells us that restaurant 3 is built near a population of 32,300 residents and a business/shopping area having a rating of 6. It follows that

$$\beta_0 + \beta_1(32.3) + \beta_2(6)$$

is the mean yearly revenue for all Tasty Sub restaurants that could potentially be built near populations of 32,300 residents and business/shopping areas having a rating of 6.

2   β_0, β_1, and β_2 are (unknown) regression parameters that relate mean yearly revenue to x_1 and x_2. Specifically:

- β_0 (the *intercept of the model*) is the mean yearly revenue for all Tasty Sub restaurants that could potentially be built near populations of zero residents and business/shopping

FIGURE 15.3   A Geometrical Interpretation of the Regression Model Relating y to x_1 and x_2

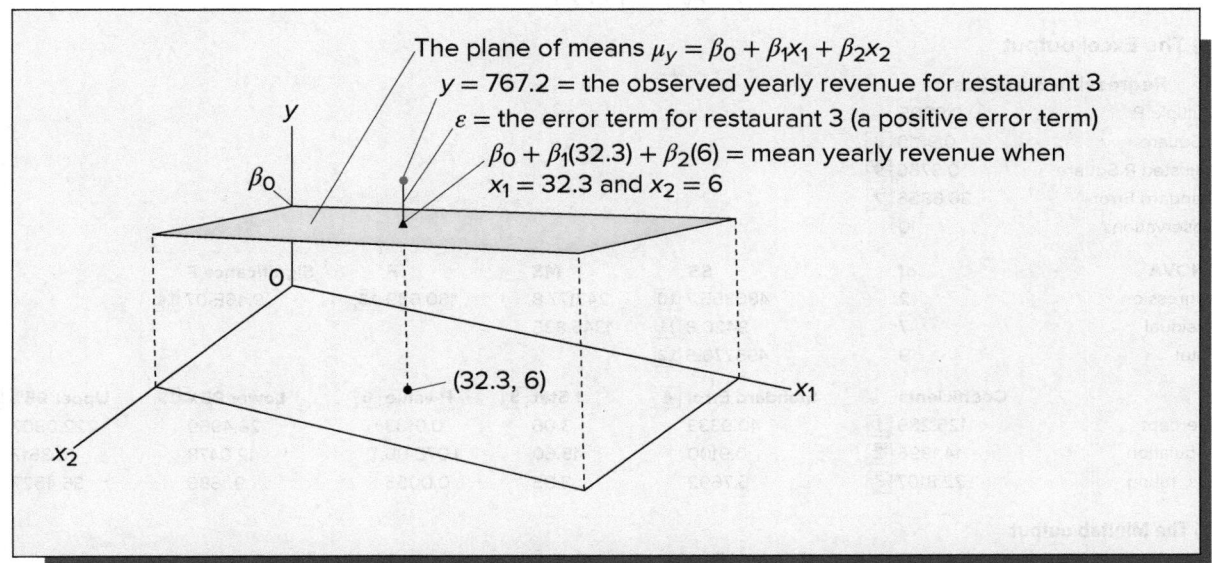

The plane of means $\mu_y = \beta_0 + \beta_1 x_1 + \beta_2 x_2$

$y = 767.2 =$ the observed yearly revenue for restaurant 3

$\varepsilon =$ the error term for restaurant 3 (a positive error term)

$\beta_0 + \beta_1(32.3) + \beta_2(6) =$ mean yearly revenue when $x_1 = 32.3$ and $x_2 = 6$

areas having a rating of 0. This interpretation, however, is of dubious practical value, because we have not observed any Tasty Sub restaurants that are built near populations of zero residents and business/shopping areas having a rating of zero. (The lowest business rating is 1.)

- β_1 (the **regression parameter for the variable x_1**) is the change in mean yearly revenue that is associated with a one-unit (1,000 resident) increase in the population size (x_1) when the business rating (x_2) does not change. Intuitively, β_1 is the slope of the plane of means in the x_1 direction.

- β_2 (the **regression parameter for the variable x_2**) is the change in mean yearly revenue that is associated with a one-unit increase in the business rating (x_2) when the population size (x_1) does not change. Intuitively, β_2 is the slope of the plane of means in the x_2 direction.

3   ε is an error term that describes the effect on y of all factors other than x_1 and x_2. One such factor is the skill of the owner as an operator of the restaurant under consideration. For example, Figure 15.3 shows that the error term for restaurant 3 is positive. This implies that the observed yearly revenue for restaurant 3, $y = 767.2$, is greater than the mean yearly revenue for all Tasty Sub restaurants that could potentially be built near populations of 32,300 residents and business/shopping areas having a rating of 6. In general, positive error terms cause their respective observed yearly revenues to be greater than the corresponding mean yearly revenues. On the other hand, negative error terms cause their respective observed yearly revenues to be less than the corresponding mean yearly revenues.

Part 2: The Least Squares Point Estimates   If b_0, b_1, and b_2 denote point estimates of β_0, β_1, and β_2, then the point prediction of an observed yearly revenue $y = \beta_0 + \beta_1 x_1 + \beta_2 x_2 + \varepsilon$ is

$$\hat{y} = b_0 + b_1 x_1 + b_2 x_2$$

which we call a *predicted yearly revenue*. Here, because the regression assumptions (to be discussed in Section 15.3) imply that the error term ε has a 50 percent chance of being positive and a 50 percent chance of being negative, we predict ε to be zero. Now, consider the 10 Tasty Sub restaurants in Table 15.1. If any particular values of b_0, b_1, and b_2 are good point estimates, they will make the predicted yearly revenue for each restaurant fairly close to the observed yearly revenue for the restaurant. This will make the restaurant's

FIGURE 15.4   Excel and Minitab Outputs of a Regression Analysis of the Tasty Sub Shop Revenue Data in Table 15.1 Using the Model $y = \beta_0 + \beta_1 x_1 + \beta_2 x_2 + \varepsilon$

(a) The Excel output

Regression Statistics

Multiple R	0.9905
R Square	0.9810 8
Adjusted R Square	0.9756 9
Standard Error	36.6856 7
Observations	10

ANOVA	df	SS	MS	F	Significance F
Regression	2	486355.7 10	243177.8	180.689 13	9.46E-07 14
Residual	7	9420.8 11	1345.835		
Total	9	495776.5 12			

	Coefficients	Standard Error 4	t Stat 5	P-value 6	Lower 95% 19	Upper 95% 19
Intercept	125.289 1	40.9333	3.06	0.0183	28.4969	222.0807
population	14.1996 2	0.9100	15.60	1.07E-06	12.0478	16.3517
bus_rating	22.8107 3	5.7692	3.95	0.0055	9.1686	36.4527

(b) The Minitab output

Analysis of Variance

Source	DF	Adj SS	AdJ MS	F-Value	P-Value
Regression	2	486356 10	243178	180.69 13	0.000 14
Population	1	327678	327678	243.48	0.000
Bus_Rating	1	21039	21039	15.63	0.006
Error	7	9421 11	1346		
Total	9	495777 12			

Model Summary

S	R-sq	R-sq(adj)	R-sq(pred)
36.6856 7	98.10% 8	97.56% 9	96.31%

Coefficients

Term	Coef	SE Coef 4	T-Value 5	P-Value 6	VIF
Constant	125.3 1	40.9	3.06	0.018	
Population	14.200 2	0.910	15.60	0.000	1.18
Bus_Rating	22.81 3	5.77	3.95	0.006	1.18

Regression Equation

Revenue = 125.3 + 14.200 Population + 22.81 Bus_Rating

Variable	Setting	Fit 15	SE Fit 16	95% CI 17	95% PI 18
Population	47.3	956.606	15.0476	(921.024, 992.188)	(862.844, 1050.37)
Bus_Rating	7				

---

1 b_0 2 b_1 3 b_2 4 S_{b_j} = standard error of the estimate b_j 5 t statistics   6 p-values for t statistics   7 s = standard error
8 R^2 9 Adjusted R^2 10 Explained variation   11 SSE = Unexplained variation   12 Total variation   13 F(model) statistic
14 p-value for F(model)   15 $\hat{y}$ = point prediction when $x_1 = 47.3$ and $x_2 = 7$ 16 $s_{\hat{y}}$ = standard error of the estimate $\hat{y}$
17 95% confidence interval when $x_1 = 47.3$ and $x_2 = 7$ 18 95% prediction interval when $x_1 = 47.3$ and $x_2 = 7$ 19 95% confidence interval for β_j

---

residual—the difference between the restaurant's observed and predicted yearly revenues—fairly small (in magnitude). We define the **least squares point estimates** to be the values of b_0, b_1, and b_2 that minimize SSE, the sum of squared residuals for the 10 restaurants.

The formula for the least squares point estimates of the parameters in a multiple regression model is expressed using a branch of mathematics called **matrix algebra.** This formula is presented in Bowerman, O'Connell, and Koehler (2005). In the main body of this book, we will rely on Excel, JMP, and Minitab to compute the needed estimates. For example, consider the Excel and Minitab outputs in Figure 15.4. The Excel output tells us that the least squares point estimates of β_0, β_1, and β_2 in the Tasty Sub Shop revenue model are $b_0 = 125.289$, $b_1 = 14.1996$, and $b_2 = 22.8107$ (see 1, 2, and 3). The point estimate $b_1 = 14.1996$ of β_1 says we estimate that mean yearly revenue increases by \$14,199.60 when the population size increases by 1,000 residents and the business rating does not change. The point estimate

TABLE 15.2   The Point Predictions and Residuals Using the Least Squares Point Estimates, $b_0 = 125.289$, $b_1 = 14.1996$, and $b_2 = 22.8107$

Restaurant	Population Size, x_1 (Thousands of Residents)	Business Rating, x_2	Yearly Revenue, y (Thousands of Dollars)	Predicted Yearly Revenue $\hat{y} = 125.289 + 14.1996x_1 + 22.8107x_2$	Residual, $y - \hat{y}$
1	20.8	3	527.1	489.07	38.03
2	27.5	2	548.7	561.40	−12.70
3	32.3	6	767.2	720.80	46.40
4	37.2	5	722.9	767.57	−44.67
5	39.6	8	826.3	870.08	−43.78
6	45.1	3	810.5	834.12	−23.62
7	49.9	9	1040.7	1039.15	1.55
8	55.4	5	1033.6	1026.00	7.60
9	61.7	4	1090.3	1092.65	−2.35
10	64.6	7	1235.8	1202.26	33.54

$$SSE = (38.03)^2 + (-12.70)^2 + \cdots + (33.54)^2 = 9420.8$$

$b_2 = 22.8107$ of β_2 says we estimate that mean yearly revenue increases by \$22,810.70 when there is a one-unit increase in the business rating and the population size does not change.

The equation

$$\hat{y} = b_0 + b_1x_1 + b_2x_2$$
$$= 125.289 + 14.1996x_1 + 22.8107x_2$$

is called the **least squares prediction equation.** In Table 15.2 we summarize using this prediction equation to calculate the predicted yearly revenues and the residuals for the 10 observed Tasty Sub restaurants. For example, because the population size and business rating for restaurant 1 were 20.8 and 3, the predicted yearly revenue for restaurant 1 is

$$\hat{y} = 125.289 + 14.1996(20.8) + 22.8107(3)$$
$$= 489.07$$

It follows, because the observed yearly revenue for restaurant 1 was $y = 527.1$, that the residual for restaurant 1 is

$$y - \hat{y} = 527.1 - 489.07 = 38.03$$

If we consider all of the residuals in Table 15.2 and add their squared values, we find that **SSE, the sum of squared residuals,** is 9420.8. This *SSE* value is given on the Excel and Minitab outputs in Figure 15.4 (see ⑪) and will be used throughout this chapter.

Part 3: Estimating Means and Predicting Individual Values   The least squares prediction equation is the equation of a plane—called the **least squares plane**—in three-dimensional space. The least squares plane is the estimate of the plane of means. It follows that the point on the least squares plane corresponding to the population size x_1 and the business rating x_2

$$\hat{y} = b_0 + b_1x_1 + b_2x_2$$
$$= 125.289 + 14.1996x_1 + 22.8107x_2$$

is the point estimate of $\beta_0 + \beta_1x_1 + \beta_2x_2$, the mean yearly revenue for all Tasty Sub restaurants that could potentially be built near populations of size x_1 and business/shopping areas having a rating of x_2. In addition, because we predict the error term to be 0, $\hat{y}$ is also the point prediction of $y = \beta_0 + \beta_1x_1 + \beta_2x_2 + \varepsilon$, the yearly revenue for a single Tasty Sub restaurant that is built near a population of size x_1 and a business/shopping area having a rating of x_2.

For example, suppose that one of the business entrepreneur's potential restaurant sites is near a population of 47,300 residents and a business/shopping area having a rating of 7. It follows that

$$\hat{y} = 125.289 + 14.1996(47.3) + 22.8107(7)$$
$$= 956.606 \text{ (that is, \$956,606)}$$

is

1 The **point estimate** of the mean yearly revenue for all Tasty Sub restaurants that could potentially be built near populations of 47,300 residents and business/shopping areas having a rating of 7, and

2 The **point prediction** of the yearly revenue for a single Tasty Sub restaurant that is built near a population of 47,300 residents and a business/shopping area having a rating of 7.

Notice that $\hat{y} = 956.606$ is given at the bottom of the Minitab output in Figure 15.4 (see 15). Moreover, recall that the yearly rent and other fixed costs for the entrepreneur's potential restaurant will be \$257,550 and that (according to Tasty Sub corporate headquarters) the yearly food and other variable costs for the restaurant will be 60 percent of the yearly revenue. Because we predict that the yearly revenue for the restaurant will be \$956,606, it follows that we predict that the yearly total operating cost for the restaurant will be \$257,550 + .6(\$956,606) = \$831,514. In addition, if we subtract this predicted yearly operating cost from the predicted yearly revenue of \$956,606, we predict that the yearly profit for the restaurant will be \$125,092. Of course, these predictions are point predictions. In Section 15.6 we will predict the restaurant's yearly revenue and profit *with confidence*.

The Tasty Sub Shop revenue model expresses the dependent variable as a function of two independent variables. In general, we can use a multiple regression model to express a dependent variable as a function of any number of independent variables. For example, in the past, natural gas utilities serving the Cincinnati, Ohio, area have predicted daily natural gas consumption by using four independent (predictor) variables—average temperature, average wind velocity, average sunlight, and change in average temperature from the previous day. The general form of a multiple regression model expresses the dependent variable y as a function of k independent variables $x_1, x_2, \ldots, x_k$. We express this general form in the following box.

The Multiple Regression Model

The **multiple regression model relating** y **to** $x_1, x_2, \ldots, x_k$ is

$$y = \beta_0 + \beta_1 x_1 + \beta_2 x_2 + \cdots + \beta_k x_k + \varepsilon$$

Here

1 $\mu_y = \beta_0 + \beta_1 x_1 + \beta_2 x_2 + \cdots + \beta_k x_k$ is the mean value of the dependent variable y when the values of the independent variables are $x_1, x_2, \ldots, x_k$.

2 $\beta_0, \beta_1, \beta_2, \ldots, \beta_k$ are (unknown) **regression parameters** relating the mean value of y to $x_1, x_2, \ldots, x_k$.

3 ε is an **error term** that describes the effects on y of all factors other than the values of the independent variables $x_1, x_2, \ldots, x_k$.

If $b_0, b_1, b_2, \ldots, b_k$ denote point estimates of $\beta_0, \beta_1, \beta_2, \ldots, \beta_k$, then

$$\hat{y} = b_0 + b_1 x_1 + b_2 x_2 + \cdots + b_k x_k$$

is the **point estimate of the mean value of the dependent variable** when the values of the independent variables are $x_1, x_2, \ldots, x_k$. In addition, because we predict the error term ε to be 0, $\hat{y}$ is also the **point prediction of an individual value of the dependent variable** when the values of the independent variables are $x_1, x_2, \ldots, x_k$. Now, assume that we have obtained n observations, where each observation consists of an observed value of the dependent variable y and corresponding observed values of the independent variables $x_1, x_2, \ldots, x_k$. For the ith observation, let y_i and $\hat{y}_i$ denote the observed and predicted values of the dependent variable, and define the residual to be $e_i = y_i - \hat{y}_i$. It then follows that the **least squares point estimates** are the values of $b_0, b_1, b_2, \ldots, b_k$ that minimize the sum of squared residuals:

$$SSE = \sum_{i=1}^{n} (y_i - \hat{y}_i)^2$$

As illustrated in Example 15.1, we use Excel and Minitab to find the least squares point estimates.

To conclude this section, consider an arbitrary independent variable, which we will denote as x_j, in a multiple regression model. We can then interpret the parameter β_j to be the change in the mean value of the dependent variable that is associated with a one-unit increase in x_j when the other independent variables in the model do not change. This interpretation is based, however, on the assumption that x_j can increase by one unit without the other independent variables in the model changing. In some situations (as we will see) this assumption is not reasonable.

Exercises for Section 15.1

CONCEPTS   ■ connect

15.1   In the multiple regression model, what sum of squared deviations do the least squares point estimates minimize?

15.2   When using the multiple regression model, how do we obtain a point estimate of the mean value of y and a point prediction of an individual value of y for a given set of x values?

METHODS AND APPLICATIONS

15.3   **THE NATURAL GAS CONSUMPTION CASE**   DS GasCon2

Consider the situation in which a gas company wishes to predict weekly natural gas consumption for its city. In the exercises of Chapter 14, we used the single predictor variable x, average hourly temperature, to predict y, weekly natural gas consumption. We now consider predicting y on the basis of average hourly temperature and a second predictor variable—the chill index. The chill index for a given average hourly temperature expresses the combined effects of all other major weather-related factors that influence natural gas consumption, such as wind velocity, sunlight, cloud cover, and the passage of weather fronts. The chill index is expressed as a whole number between 0 and 30. A weekly chill index near 0 indicates that, given the average hourly temperature during the week, all other major weather-related factors will only slightly increase weekly natural gas consumption. A weekly chill index near 30 indicates that, given the average hourly temperature during the week, other weather-related factors will greatly increase weekly natural gas consumption. The natural gas company has collected data concerning weekly natural gas consumption (y, in MMcF), average hourly temperature (x_1, in degrees Fahrenheit), and the chill index (x_2) for the last eight weeks. The data are given in Table 15.3, and scatter plots of y versus x_1 and y versus x_2 are given below the data. Moreover, Figure 15.5 gives Excel and Minitab outputs of a regression analysis of these data using the model
$$y = \beta_0 + \beta_1 x_1 + \beta_2 x_2 + \varepsilon$$

a   Using the Excel or Minitab output (depending on the package used in your class), find (on the computer output) b_1 and b_2, the least squares point estimates of β_1 and β_2, and report their values. Then interpret b_1 and b_2.

b   Calculate **(1)** a point estimate of the mean natural gas consumption for all weeks that have an average hourly temperature of 40 and a chill index of 10, and **(2)** a point prediction of the amount of natural

gas consumed in a single week that has an average hourly temperature of 40 and a chill index of 10. Find this point estimate (prediction), which is given at the bottom of the Minitab output, and verify that it equals (within rounding) your calculated value.

15.4   **THE REAL ESTATE SALES PRICE CASE**   DS RealEst2

A real estate agency collects the data in Table 15.4 concerning

y = sales price of a house (in thousands of dollars)

x_1 = home size (in hundreds of square feet)

x_2 = rating (an overall "niceness rating" for the house expressed on a scale from 1 [worst] to 10 [best], and provided by the real estate agency)

The agency wishes to develop a regression model that can be used to predict the sales prices of future houses it will list. Use Excel, JMP, or Minitab to perform a regression analysis of the real estate sales price data in Table 15.4 using the model
$$y = \beta_0 + \beta_1 x_1 + \beta_2 x_2 + \varepsilon$$

a   Find (on the computer output) and report the values of b_1 and b_2, the least squares point estimates of β_1 and β_2. Interpret b_1 and b_2.

TABLE 15.3
The Natural Gas Consumption Data
DS GasCon2

y	x_1	x_2
12.4	28.0	18
11.7	28.0	14
12.4	32.5	24
10.8	39.0	22
9.4	45.9	8
9.5	57.8	16
8.0	58.1	1
7.5	62.5	0

Temp, x_1

Chill, x_2

TABLE 15.4
The Real Estate Sales Price Data
DS RealEst2

y	x_1	x_2
180	23	5
98.1	11	2
173.1	20	9
136.5	17	3
141	15	8
165.9	21	4
193.5	24	7
127.8	13	6
163.5	19	7
172.5	25	2

Source: R. L. Andrews and J. T. Ferguson, "Integrating Judgement with a Regression Appraisal," *The Real Estate Appraiser and Analyst* 52, no. 2 (1986).

FIGURE 15.5   Excel and Minitab Outputs of a Regression Analysis of the Natural Gas Consumption Data
Using the Model $y = \beta_0 + \beta_1 x_1 + \beta_2 x_2 + \varepsilon$

(a) The Excel output

Regression Statistics

Multiple R	0.9867
R Square	0.9736
Adjusted R Square	0.9631
Standard Error	0.3671
Observations	8

ANOVA

	df	SS	MS	F	Significance F
Regression	2	24.8750	12.4375	92.3031	0.0001
Residual	5	0.6737	0.1347		
Total	7	25.5488			

	Coefficients	Standard Error	t Stat	P-value	Lower 95%	Upper 95%
Intercept	13.1087	0.8557	15.3193	2.15E-05	10.9091	15.3084
TEMP	−0.0900	0.0141	−6.3942	0.0014	−0.1262	−0.0538
CHILL	0.0825	0.0220	3.7493	0.0133	0.0259	0.1391

(b) The Minitab output

Analysis of Variance

Source	DF	Adj SS	AdJ MS	F-Value	P-Value
Regression	2	24.8750	12.4375	92.30	0.000
Temp	1	5.5093	5.5093	40.89	0.001
Chill	1	1.8942	1.8942	14.06	0.013
Error	5	0.6737	0.1347		
Total	7	25.5488			

Model Summary

S	R-sq	R-sq(adj)	R-sq(pred)
0.367078	97.36%	96.31%	93.41%

Coefficients

Term	Coef	SE Coef	T-Value	P-Value	VIF
Constant	13.109	0.856	15.32	0.000	
Temp	−0.0900	0.0141	−6.39	0.001	2.07
Chill	0.0825	0.0220	3.75	0.013	2.07

Regression Equation

GasCons = 13.109 − 0.0900 Temp + 0.0825 Chill

Variable	Setting	Fit	SE Fit	95% CI	95% PI
Temp	40	10.3331	0.170472	(9.89492, 10.7713)	(9.29274, 11.3735)
Chill	10				

b Calculate (1) a point estimate of the mean sales price of all houses having 2,000 square feet and a niceness rating of 8, and (2) a point prediction of the sales price of a single house having 2,000 square feet and a niceness rating of 8. Find this point estimate (prediction), which is given at the bottom of the Minitab output, and verify that it equals (within rounding) your calculated value.

15.5   THE BONNER FROZEN FOODS ADVERTISING CASE DS Bonner

Bonner Frozen Foods, Inc., has designed an experiment to study the effects of radio and television advertising and print advertising on sales of one of its frozen foods lines. Bonner has used five levels of radio and television advertisements (x_1) in combination with five levels of print advertisements (x_2) in 25 sales regions of equal sales potential. Table 15.5 shows the advertising mix

used in each region last August along with the resulting sales, y. Advertising amounts are recorded in $1,000 units, while sales are recorded in units of $10,000. Use Excel, JMP, or Minitab to perform a regression analysis of the Bonner sales volume data in Table 15.5 using the model

$$y = \beta_0 + \beta_1 x_1 + \beta_2 x_2 + \varepsilon$$

a (1) Find (on the computer output) and report the values of b_1 and b_2, the least squares point estimates of β_1 and β_2. (2) Interpret b_1 and b_2. (3) Which of the two types of advertising—radio and television advertising or print advertising—seems to give the greatest increase in mean sales volume when the expenditure on this type of advertising is increased by $1,000 and the expenditure on the other type of advertising remains constant?

b (1) Calculate a point estimate of the mean sales volume in all sales regions when $2,000 is spent on radio and television advertising and $5,000 is spent on print advertising. **(2)** Calculate a point prediction of the sales volume in an individual sales region when $2,000 is spent on radio and television advertising and $5,000 is spent on print advertising.

15.6   THE FRESH DETERGENT CASE   DS Fresh2

Enterprise Industries produces Fresh, a brand of liquid laundry detergent. In order to manage its inventory more effectively and make revenue projections, the company would like to better predict demand for Fresh. To develop a prediction model, the company has gathered data concerning demand for Fresh over the last 30 sales periods (each sales period is defined to be a four-week period). The demand data are presented in Table 15.6. Here, for each sales period,

> y = the demand for the large size bottle of Fresh (in hundreds of thousands of bottles) in the sales period
>
> x_1 = the price (in dollars) of Fresh as offered by Enterprise Industries in the sales period
>
> x_2 = the average industry price (in dollars) of competitors' similar detergents in the sales period
>
> x_3 = Enterprise Industries' advertising expenditure (in hundreds of thousands of dollars) to promote Fresh in the sales period

Figure 15.6 gives the JMP output of a regression analysis of the Fresh Detergent demand data in Table 15.6 using the model

$$y = \beta_0 + \beta_1 x_1 + \beta_2 x_2 + \beta_3 x_3 + \varepsilon$$

a Find (on the computer output) and report the values of b_1, b_2, and b_3, the least squares point estimates of β_1, β_2, and β_3. Interpret b_1, b_2, and b_3.

b Consider the demand for Fresh Detergent in a future sales period when Enterprise Industries' price for Fresh will be $x_1 = 3.70$, the average price of competitors' similar detergents will be $x_2 = 3.90$, and Enterprise Industries' advertising expenditure for Fresh will be $x_3 = 6.50$. The point prediction of this demand is given at the bottom of Figure 15.6 in the JMP output. Report this point prediction and show (within rounding) how it has been calculated.

15.7   THE HOSPITAL LABOR NEEDS CASE   DS HospLab

Table 15.7 presents data concerning the need for labor in 16 U.S. Navy hospitals. Here, y = monthly labor hours required; x_1 = monthly X-ray exposures; x_2 = monthly occupied bed days (a hospital has one occupied bed day if one bed is occupied for an

TABLE 15.5
The Bonner Frozen Foods Sales Volume Data   DS Bonner

x_1	x_2	y
1	1	3.27
1	2	8.38
1	3	11.28
1	4	14.50
1	5	19.63
2	1	5.84
2	2	10.01
2	3	12.46
2	4	16.67
2	5	19.83
3	1	8.51
3	2	10.14
3	3	14.75
3	4	17.99
3	5	19.85
4	1	9.46
4	2	12.61
4	3	15.50
4	4	17.68
4	5	21.02
5	1	12.23
5	2	13.58
5	3	16.77
5	4	20.56
5	5	21.05

TABLE 15.6   Historical Data Concerning Demand for Fresh Detergent   DS Fresh2

Sales Period	Price for Fresh, x_1	Average Industry Price, x_2	Advertising Expenditure for Fresh, x_3	Demand for Fresh, y	Sales Period	Price for Fresh, x_1	Average Industry Price, x_2	Advertising Expenditure for Fresh, x_3	Demand for Fresh, y
1	3.85	3.80	5.50	7.38	16	3.80	4.10	6.80	8.87
2	3.75	4.00	6.75	8.51	17	3.70	4.20	7.10	9.26
3	3.70	4.30	7.25	9.52	18	3.80	4.30	7.00	9.00
4	3.70	3.70	5.50	7.50	19	3.70	4.10	6.80	8.75
5	3.60	3.85	7.00	9.33	20	3.80	3.75	6.50	7.95
6	3.60	3.80	6.50	8.28	21	3.80	3.75	6.25	7.65
7	3.60	3.75	6.75	8.75	22	3.75	3.65	6.00	7.27
8	3.80	3.85	5.25	7.87	23	3.70	3.90	6.50	8.00
9	3.80	3.65	5.25	7.10	24	3.55	3.65	7.00	8.50
10	3.85	4.00	6.00	8.00	25	3.60	4.10	6.80	8.75
11	3.90	4.10	6.50	7.89	26	3.65	4.25	6.80	9.21
12	3.90	4.00	6.25	8.15	27	3.70	3.65	6.50	8.27
13	3.70	4.10	7.00	9.10	28	3.75	3.75	5.75	7.67
14	3.75	4.20	6.90	8.86	29	3.80	3.85	5.80	7.93
15	3.75	4.10	6.80	8.90	30	3.70	4.25	6.80	9.26

FIGURE 15.6   JMP Output of a Regression Analysis of the Fresh Detergent Demand Data Using
the Model $y = \beta_0 + \beta_1 x_1 + \beta_2 x_2 + \beta_3 x_3 + \varepsilon$

Summary of Fit

RSquare	0.893613
RSquare Adj	0.881338
Root Mean Square Error	0.234669
Mean of Response	8.382667
Observations (or Sum Wgts)	30

Analysis of Variance

Source	DF	Sum of Squares	Mean Square	F Ratio
Model	3	12.026774	4.00892	72.7973
Error	26	1.431813	0.05507	Prob > F
C. Total	29	13.458587		<.0001*

Parameter Estimates

Term	Estimate	Std Error	t Ratio	Prob>|t|	Lower 95%	Upper 95%
Intercept	7.5891025	2.445021	3.10	0.0046*	2.5632892	12.614916
Price(X1)	−2.357723	0.637947	−3.70	0.0010*	−3.669042	−1.046404
IndPrice(X2)	1.6122144	0.295353	5.46	<.0001*	1.0051083	2.2193205
AdvExp(X3)	0.5011517	0.125874	3.98	0.0005*	0.2424145	0.7598889

	Predicted Demand	Lower 95% Mean Demand	Upper 95% Mean Demand	Lower 95% Indiv Demand	Upper 95% Indiv Demand	StdErr Pred Demand	StdErr Indiv Demand
31	8.4106503477	8.3143172822	8.5069834132	7.9187553487	8.9025453468	0.04686533	0.2393033103

entire day); and x_3 = average length of patient's stay (in days). Figure 15.7 gives the Excel output of a regression analysis of the data using the model

$$y = \beta_0 + \beta_1 x_1 + \beta_2 x_2 + \beta_3 x_3 + \varepsilon$$

Note that the variables x_1, x_2, and x_3 are denoted as XRay, BedDays, and LengthStay on the output.

a Find (on the computer output) and report the values of b_1, b_2, and b_3, the least squares point estimates of β_1, β_2, and β_3. Interpret b_1, b_2, and b_3. Note that the negative value of b_3 (= −413.7578) might say that, when XRay and BedDays stay constant, an increase in LengthStay implies less

TABLE 15.7   Hospital Labor Needs Data DS HospLab

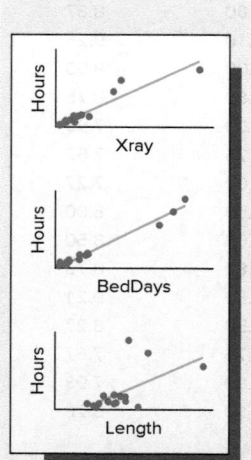

Hospital	Monthly X-Ray Exposures, x_1	Monthly Occupied Bed Days, x_2	Average Length of Stay, x_3	Monthly Labor Hours Required, y
1	2,463	472.92	4.45	566.52
2	2,048	1,339.75	6.92	696.82
3	3,940	620.25	4.28	1,033.15
4	6,505	568.33	3.90	1,603.62
5	5,723	1,497.60	5.50	1,611.37
6	11,520	1,365.83	4.60	1,613.27
7	5,779	1,687.00	5.62	1,854.17
8	5,969	1,639.92	5.15	2,160.55
9	8,461	2,872.33	6.18	2,305.58
10	20,106	3,655.08	6.15	3,503.93
11	13,313	2,912.00	5.88	3,571.89
12	10,771	3,921.00	4.88	3,741.40
13	15,543	3,865.67	5.50	4,026.52
14	34,703	12,446.33	10.78	11,732.17
15	39,204	14,098.40	7.05	15,414.94
16	86,533	15,524.00	6.35	18,854.45

Source: *Procedures and Analysis for Staffing Standards Development Regression Analysis Handbook* (San Diego, CA: Navy Manpower and Material Analysis Center, 1979).

FIGURE 15.7   Excel Output of a Regression Analysis of the Hospital Labor Needs Data
               Using the Model $y = \beta_0 + \beta_1 x_1 + \beta_2 x_2 + \beta_3 x_3 + \varepsilon$

(a) The Excel output

Regression Statistics

Multiple R	0.9981
R Square	0.9961
Adjusted R Square	0.9952
Standard Error	387.1598
Observations	16

ANOVA	df	SS	MS	F	Significance F
Regression	3	462327889.4	154109296.5	1028.1309	9.92E-15
Residual	12	1798712.2	149892.7		
Total	15	464126601.6			

	Coefficients	Standard Error	t Stat	P-value	Lower 95%	Upper 95%
Intercept	1946.8020	504.1819	3.8613	0.0023	848.2840	3045.3201
XRay (x1)	0.0386	0.0130	2.9579	0.0120	0.0102	0.0670
BedDays (x2)	1.0394	0.0676	15.3857	2.91E-09	0.8922	1.1866
LengthStay (x3)	−413.7578	98.5983	−4.1964	0.0012	−628.5850	−198.9306

(b) Prediction using MegaStat

Predicted values for: LaborHours

				95% Confidence Interval		95% Prediction Interval		
XRay (x1)	BedDays (x2)	LengthStay (x3)	Predicted	lower	upper	lower	upper	Leverage
56194	14077.88	6.89	15,896.2473	15,378.0313	16,414.4632	14,906.2361	16,886.2584	0.3774

patient turnover and thus fewer start-up hours needed for the initial care of new patients.

  b Consider a hospital for which XRay = 56,194, BedDays = 14,077.88, and LengthStay = 6.89. A point prediction of the labor hours corresponding to this combination of values of the independent

variables is given on the MegaStat output. Report this point prediction and show (within rounding) how it has been calculated.

  c If the actual number of labor hours used by the hospital was $y = 17,207.31$, how does this y value compare with the point prediction?

15.2 R^2 and Adjusted R^2

The multiple coefficient of determination, R^2

In this section we discuss several ways to assess the utility of a multiple regression model. We first discuss a quantity called the **multiple coefficient of determination,** which is denoted R^2. The formulas for R^2 and several other related quantities are given in the following box.

LO15-2
Calculate and interpret the multiple and adjusted multiple coefficients of determination.

The Multiple Coefficient of Determination, R^2

For the multiple regression model:

1 **Total variation** $= \sum (y_i - \bar{y})^2$

2 **Explained variation** $= \sum (\hat{y}_i - \bar{y})^2$

3 **Unexplained variation** $= \sum (y_i - \hat{y}_i)^2$

4 **Total variation = Explained variation + Unexplained variation**

5 The **multiple coefficient of determination** is
$$R^2 = \frac{\text{Explained variation}}{\text{Total variation}}$$

6 R^2 is the proportion of the total variation in the n observed values of the dependent variable that is explained by the overall regression model.

7 **Multiple correlation coefficient** $= R = \sqrt{R^2}$

As an example, consider the Tasty Sub Shop revenue model

$$y = \beta_0 + \beta_1 x_1 + \beta_2 x_2 + \varepsilon$$

and the following Minitab output:

Analysis of Variance					
Source	DF	Adj SS	Adj MS	F-Value	P-Value
Regression	2	486356	243178	180.69	0.000
Population	1	327678	327678	243.48	0.000
Bus_Rating	1	21039	21039	15.63	0.006
Error	7	9421	1346		
Total	9	495777			

S	R-sq	R-sq(adj)	R-sq(pred)
36.6856	98.10%	97.56%	96.31%

This output tells us that the total variation, explained variation, and unexplained variation for the model are, respectively, 495,777, 486,356, and 9,421. The output also tells us that the multiple coefficient of determination is

$$R^2 = \frac{\text{Explained variation}}{\text{Total variation}} = \frac{486,356}{495,777} = .981 \quad (98.1\% \text{ on the output})$$

which implies that the multiple correlation coefficient is $R = \sqrt{.981} = .9905$. The value of $R^2 = .981$ says that the two independent variable Tasty Sub Shop revenue model explains 98.1 percent of the total variation in the 10 observed yearly revenues. Note this R^2 value is larger than the r^2 of .939 for the simple linear regression model that uses only the population size to predict yearly revenue. Also note that the quantities given on the Minitab output are given on the following Excel output.

Regression Statistics	
Multiple R	0.9905
R Square	0.9810
Adjusted R Square	0.9756
Standard Error	36.6856
Observations	10

ANOVA	df	SS	MS	F	Significance F
Regression	2	486355.7	243177.8	180.689	9.46E-07
Residual	7	9420.8	1345.835		
Total	9	495776.5			

Adjusted R^2

Even if the independent variables in a regression model are unrelated to the dependent variable, they will make R^2 somewhat greater than 0. To avoid overestimating the importance of the independent variables, many analysts recommend calculating an *adjusted* multiple coefficient of determination.

Adjusted R^2

The **adjusted multiple coefficient of determination** (**adjusted R^2**) is

$$\bar{R}^2 = \left(R^2 - \frac{k}{n-1} \right) \left(\frac{n-1}{n-(k+1)} \right)$$

where R^2 is the multiple coefficient of determination, n is the number of observations, and k is the number of independent variables in the model under consideration.

To briefly explain this formula, note that it can be shown that subtracting $k/(n-1)$ from R^2 helps avoid overestimating the importance of the k independent variables. Furthermore, multiplying $[R^2 - (k/(n-1))]$ by $(n-1)/(n-(k+1))$ makes $\overline{R}^2$ equal to 1 when R^2 equals 1.

As an example, consider the Tasty Sub Shop revenue model

$$y = \beta_0 + \beta_1 x_1 + \beta_2 x_2 + \varepsilon$$

Because we have seen that $R^2 = .981$, it follows that

$$\overline{R}^2 = \left(R^2 - \frac{k}{n-1}\right)\left(\frac{n-1}{n-(k+1)}\right)$$

$$= \left(.981 - \frac{2}{10-1}\right)\left(\frac{10-1}{10-(2+1)}\right)$$

$$= .9756$$

which is given on the Minitab and Excel outputs.

If R^2 is less than $k/(n-1)$, which can happen, then $\overline{R}^2$ will be negative. In this case, statistical software systems set $\overline{R}^2$ equal to 0. Historically, R^2 and $\overline{R}^2$ have been popular measures of model utility—possibly because they are unitless and between 0 and 1. In general, we desire R^2 and $\overline{R}^2$ to be near 1. However, sometimes even if a regression model has an R^2 and an $\overline{R}^2$ that are near 1, the model is still not able to predict accurately. We will discuss assessing a model's ability to predict accurately, as well as using R^2 and $\overline{R}^2$ to help choose a regression model, as we proceed through the rest of this chapter. Note that we will explain the quantity on the Minitab output called "R-sq (pred)" in Section 15.10.

15.3 Model Assumptions and the Standard Error

LO15-3
Explain the assumptions behind multiple regression and calculate the standard error.

Model assumptions

In order to perform hypothesis tests and set up various types of intervals when using the multiple regression model

$$y = \beta_0 + \beta_1 x_1 + \beta_2 x_2 + \cdots + \beta_k x_k + \varepsilon$$

we need to make certain assumptions about the error term ε. At any given combination of values of $x_1, x_2, \ldots, x_k$, there is a population of error term values that could potentially occur. These error term values describe the different potential effects on y of all factors other than the combination of values of $x_1, x_2, \ldots, x_k$. Therefore, these error term values explain the variation in the y values that could be observed at the combination of values of $x_1, x_2, \ldots, x_k$. We make the following four assumptions about the potential error term values.

Assumptions for the Multiple Regression Model

1 At any given combination of values of $x_1, x_2, \ldots, x_k$, the population of potential error term values has a mean equal to 0.

2 **Constant variance assumption:** At any given combination of values of $x_1, x_2, \ldots, x_k$, the population of potential error term values has a variance that does not depend on the combination of values of $x_1, x_2, \ldots, x_k$. That is, the different populations of potential error term values corresponding to different combinations of values of $x_1, x_2, \ldots, x_k$ have equal variances. We denote the constant variance as σ^2.

3 **Normality assumption:** At any given combination of values of $x_1, x_2, \ldots, x_k$, the population of potential error term values has a **normal distribution.**

4 **Independence assumption:** Any one value of the error term ε is **statistically independent** of any other value of ε. That is, the value of the error term ε corresponding to an observed value of y is statistically independent of the error term corresponding to any other observed value of y.

Taken together, the first three assumptions say that, at any given combination of values of $x_1, x_2, \ldots, x_k$, the population of potential error term values is normally distributed with mean 0 and a variance σ^2 that does not depend on the combination of values of $x_1, x_2, \ldots, x_k$. Because the potential error term values cause the variation in the potential y values, the first three assumptions imply that, at any given combination of values of $x_1, x_2, \ldots, x_k$, the population of y values that could be observed is normally distributed with mean $\beta_0 + \beta_1 x_1 + \beta_2 x_2 + \cdots + \beta_k x_k$ and a variance σ^2 that does not depend on the combination of values of $x_1, x_2, \ldots, x_k$. Furthermore, the independence assumption says that, when time series data are utilized in a regression study, there are no patterns in the error term values. In Section 15.11 we show how to check the validity of the regression assumptions. That section can be read at any time after Section 15.7. As in simple linear regression, only pronounced departures from the assumptions must be remedied.

The mean square error and the standard error

To present statistical inference formulas in later sections, we need to be able to compute point estimates of σ^2 and σ (the constant variance and standard deviation of the different error term populations). We show how to do this in the following box:

The Mean Square Error and the Standard Error

Suppose that the multiple regression model

$$y = \beta_0 + \beta_1 x_1 + \beta_2 x_2 + \cdots + \beta_k x_k + \varepsilon$$

utilizes k independent variables and thus has $(k + 1)$ parameters $\beta_0, \beta_1, \beta_2, \ldots, \beta_k$. Then, if the regression assumptions are satisfied, and if SSE denotes the sum of squared residuals for the model:

1   A point estimate of σ^2 is the **mean square error**

$$s^2 = \frac{SSE}{n - (k + 1)}$$

2   A point estimate of σ is the **standard error**

$$s = \sqrt{\frac{SSE}{n - (k + 1)}}$$

In order to explain these point estimates, recall that σ^2 is the variance of the population of y values (for given values of $x_1, x_2, \ldots, x_k$) around the mean value μ_y. Because $\hat{y}$ is the point estimate of this mean, it seems natural to use $SSE = \Sigma(y_i - \hat{y}_i)^2$ to help construct a point estimate of σ^2. We divide SSE by $n - (k + 1)$ because it can be proven that doing so makes the resulting s^2 an unbiased point estimate of σ^2. We call $n - (k + 1)$ the **number of degrees of freedom** associated with SSE.

We will see in Section 15.6 that if a particular regression model gives a small standard error, then the model will give short prediction intervals and thus accurate predictions of individual y values. For example, Table 15.2 shows that SSE for the Tasty Sub Shop revenue model

$$y = \beta_0 + \beta_1 x_1 + \beta_2 x_2 + \varepsilon$$

is 9420.8. Because this model utilizes $k = 2$ independent variables and thus has $k + 1 = 3$ parameters (β_0, β_1, and β_2), a point estimate of σ^2 is the mean square error

$$s^2 = \frac{SSE}{n - (k + 1)} = \frac{9420.8}{10 - 3} = \frac{9420.8}{7} = 1345.835$$

and a point estimate of σ is the standard error $s = \sqrt{1345.835} = 36.6856$. Note that $SSE = 9420.8$, $s^2 = 1345.835$, and $s = 36.6856$ are given on the Excel and Minitab outputs in Figure 15.4. Also note that the s of 36.6856 for the two independent variable model is less than the s of 61.7052 for the simple linear regression model that uses only the population size to predict yearly revenue (see Example 14.3).

15.4 The Overall *F* Test

Another way to assess the utility of a regression model is to test the significance of the regression relationship between y and $x_1, x_2, \ldots, x_k$. For the multiple regression model, we test the null hypothesis $H_0: \beta_1 = \beta_2 = \ldots = \beta_k = 0$, which says that **none of the independent variables $x_1, x_2, \ldots, x_k$ is significantly related to y (the regression relationship is not significant),** versus the alternative hypothesis H_a: At least one of $\beta_1, \beta_2, \ldots, \beta_k$ does not equal 0, which says that **at least one of the independent variables is significantly related to y (the regression relationship is significant).** If we can reject H_0 at level of significance α, we say that **the multiple regression model is significant at level of significance α.** We carry out the test as follows:

An *F* Test for the Multiple Regression Model

Suppose that the regression assumptions hold and that the multiple regression model has $(k + 1)$ parameters, and consider testing

$$H_0: \beta_1 = \beta_2 = \cdots = \beta_k = 0$$

versus

H_a: At least one of $\beta_1, \beta_2, \ldots, \beta_k$ does not equal 0.

We define the **overall *F* statistic** to be

$$F(\text{model}) = \frac{(\text{Explained variation})/k}{(\text{Unexplained variation})/[n - (k + 1)]}$$

Also define the *p*-value related to $F(\text{model})$ to be the area under the curve of the F distribution (having k and $[n - (k + 1)]$ degrees of freedom) to the right of $F(\text{model})$. Then, we can reject H_0 in favor of H_a at level of significance α if either of the following equivalent conditions holds:

1 $F(\text{model}) > F_\alpha$

2   *p*-value $< \alpha$

Here, the point F_α is based on k numerator and $n - (k + 1)$ denominator degrees of freedom.

Condition 1 is intuitively reasonable because a large value of $F(\text{model})$ would be caused by an explained variation that is large relative to the unexplained variation. This would occur if at least one independent variable in the regression model significantly affects y, which would imply that H_0 is false and H_a is true.

Ⓒ **EXAMPLE 15.2** The Tasty Sub Shop Case: The Overall *F* Test

Consider the Tasty Sub Shop revenue model

$$y = \beta_0 + \beta_1 x_1 + \beta_2 x_2 + \varepsilon$$

and the following Minitab output.

Analysis of Variance					
Source	DF	Adj SS	Adj MS	F-Value	P-Value
Regression	2	486356	243178	180.69	0.000
Population	1	327678	327678	243.48	0.000
Bus_Rating	1	21039	21039	15.63	0.006
Error	7	9421	1346		
Total	9	495777			

This output tells us that the explained and unexplained variations for this model are, respectively, 486,356 and 9,421. It follows, because there are $k = 2$ independent variables, that

$$F(\text{model}) = \frac{(\text{Explained variation})/k}{(\text{Unexplained variation})/[n - (k + 1)]}$$

$$= \frac{486{,}356/2}{9421/[10 - (2 + 1)]} = \frac{243{,}178}{1345.8}$$

$$= 180.69$$

Note that this overall F statistic is given on the Minitab output and is also given on the following Excel output:

ANOVA	df	SS	MS	F	Significance F
Regression	2	486355.7	243177.8	180.689	9.46E-07
Residual	7	9420.8	1345.835		
Total	9	495776.5			

The p-value related to F(model) is the area to the right of 180.69 under the curve of the F distribution having $k = 2$ numerator and $n - (k + 1) = 10 - 3 = 7$ denominator degrees of freedom. Both the Minitab and Excel outputs say this p-value is less than .001.

If we wish to test the significance of the regression model at level of significance $\alpha = .05$, we use the critical value $F_{.05}$ based on 2 numerator and 7 denominator degrees of freedom. Using Table A.7, we find that $F_{.05} = 4.74$. Because F(model) $= 180.69 > F_{.05} = 4.74$, we can reject H_0 in favor of H_a at level of significance .05. Alternatively, because the p-value is smaller than .05, .01, and .001, we can reject H_0 at level of significance .05, .01, and .001. Therefore, we have extremely strong evidence that the Tasty Sub Shop revenue model is significant. That is, we have extremely strong evidence that at least one of the independent variables x_1 and x_2 in the model is significantly related to y.

If the overall F test tells us that at least one independent variable in a regression model is significant, we next attempt to decide which independent variables are significant. In the next section we discuss one way to do this.

Exercises for Sections 15.2, 15.3, and 15.4

CONCEPTS ▦ connect

15.8 a What do R^2 and $\overline{R}^2$ measure?   **b** How do R^2 and $\overline{R}^2$ differ?

15.9 What is estimated by the mean square error, and what is estimated by the standard error?

15.10 What is the purpose of the overall F test?

METHODS AND APPLICATIONS

In Exercises 15.11 to 15.15 we refer to Excel, JMP, and Minitab outputs of regression analyses of the data sets related to five case studies introduced in Section 15.1. Above each referenced output we give the regression model and the number of observations, n, used to perform the regression analysis under consideration. Using the appropriate model, sample size n, and output:

1 Report the total variation, unexplained variation, and explained variation as shown on the output.

2 Report R^2 and $\overline{R}^2$ as shown on the output. Interpret R^2 and $\overline{R}^2$. Show how $\overline{R}^2$ has been calculated from R^2 and other numbers.

3 Report SSE, s^2, and s as shown on the output. Calculate s^2 from SSE and other numbers.

4 Calculate the F(model) statistic by using the explained and unexplained variations (as shown on the output) and other relevant quantities. Find F(model) on the output to check your answer (within rounding).

5 Use the F(model) statistic and the appropriate critical value to test the significance of the linear regression model under consideration by setting α equal to .05.

6 Find the p-value related to F(model) on the output. Using the p-value, test the significance of the linear regression model by setting $\alpha = .10, .05, .01,$ and $.001$. What do you conclude?

15.11 THE NATURAL GAS CONSUMPTION CASE ⒹⓈ GasCon2

Model: $y = \beta_0 + \beta_1 x_1 + \beta_2 x_2 + \varepsilon$ Sample size: $n = 8$

The output follows.

```
Analysis of Variance
Source          DF    Adj SS    Adj MS   F-Value  P-Value
Regression       2   24.8750   12.4375    92.30    0.000
    Temp         1    5.5093    5.5093    40.89    0.001
    Chill        1    1.8942    1.8942    14.06    0.013
Error            5    0.6737    0.1347
Total            7   25.5488

      S      R-sq    R-sq(adj)   R-sq(pred)
0.367078   97.36%     96.31%       93.41%
```

15.12 THE REAL ESTATE SALES PRICE CASE DS RealEst2

Model: $y = \beta_0 + \beta_1 x_1 + \beta_2 x_2 + \varepsilon$ Sample size: $n = 10$

Obtain a computer output by using Excel, JMP, or Minitab to perform a regression analysis of the real estate sales price data in Table 15.4. Then analyze the output as previously described.

15.13 THE BONNER FROZEN FOODS ADVERTISING CASE DS Bonner

Model: $y = \beta_0 + \beta_1 x_1 + \beta_2 x_2 + \varepsilon$ Sample size: $n = 25$

Obtain a computer output by using Excel, JMP, or Minitab to perform a regression analysis of the Bonner Frozen Foods sales volume data in Table 15.5. Then analyze the output as previously described.

15.14 THE FRESH DETERGENT CASE DS Fresh2

Model: $y = \beta_0 + \beta_1 x_1 + \beta_2 x_2 + \beta_3 x_3 + \varepsilon$ Sample size: $n = 30$

Summary of Fit

RSquare	0.893613
RSquare Adj	0.881338
Root Mean Square Error	0.234669
Mean of Response	8.382667
Observations (or Sum Wgts)	30

Analysis of Variance

Source	df	Sum of Squares	Mean Square	F Ratio
Model	3	12.026774	4.00892	72.7973
Error	26	1.431813	0.05507	Prob > F
C. Total	29	13.458587		<.0001*

15.15 THE HOSPITAL LABOR NEEDS CASE DS HospLab

Model: $y = \beta_0 + \beta_1 x_1 + \beta_2 x_2 + \beta_3 x_3 + \varepsilon$ Sample size: $n = 16$

Regression Statistics

Multiple R	0.9981
R Square	0.9961
Adjusted R Square	0.9952
Standard Error	387.1598
Observations	16

ANOVA	df	SS	MS	F	Significance F
Regression	3	462327889.4	154109296.5	1028.1309	9.92E-15
Residual	12	1798712.2	149892.7		
Total	15	464126601.6			

15.5 Testing the Significance of an Independent Variable

LO15-5
Test the significance of a single independent variable.

Consider the multiple regression model

$$y = \beta_0 + \beta_1 x_1 + \beta_2 x_2 + \cdots + \beta_k x_k + \varepsilon$$

In order to gain information about which independent variables significantly affect y, we can test the significance of a single independent variable. We arbitrarily refer to this variable as x_j

and assume that it is multiplied by the parameter β_j. For example, if $j = 1$, we are testing the significance of x_1, which is multiplied by β_1; if $j = 2$, we are testing the significance of x_2, which is multiplied by β_2. To test the significance of x_j, we test the null hypothesis H_0: $\beta_j = 0$. We usually test H_0 versus the alternative hypothesis H_a: $\beta_j \neq 0$. **It is reasonable to conclude that x_j is significantly related to y in the regression model under consideration if H_0 can be rejected in favor of H_a at a small level of significance.** Here the phrase *in the regression model under consideration* is very important. This is because it can be shown that whether x_j is significantly related to y in a particular regression model can depend on what other independent variables are included in the model. This issue will be discussed in detail in Section 15.10.

Testing the significance of x_j in a multiple regression model is similar to testing the significance of the slope in the simple linear regression model (recall we test H_0: $\beta_1 = 0$ in simple regression). It can be proven that, if the regression assumptions hold, the population of all possible values of the least squares point estimate b_j is normally distributed with mean β_j and standard deviation σ_{b_j}. The point estimate of σ_{b_j} is called the **standard error of the estimate b_j** and is denoted s_{b_j}. The formula for s_{b_j} involves matrix algebra and is discussed in Bowerman, O'Connell, and Koehler (2005). In our discussion here, we will rely on Excel and Minitab to compute s_{b_j}. It can be shown that, if the regression assumptions hold, then the population of all possible values of

$$\frac{b_j - \beta_j}{s_{b_j}}$$

has a t distribution with $n - (k + 1)$ degrees of freedom. It follows that, if the null hypothesis H_0: $\beta_j = 0$ is true, then the population of all possible values of the test statistic

$$t = \frac{b_j}{s_{b_j}}$$

has a t distribution with $n - (k + 1)$ degrees of freedom. Therefore, we can test the significance of x_j as follows:

Testing the Significance of the Independent Variable x_j

Null Hypothesis H_0: $\beta_j = 0$ **Test Statistic** $t = \dfrac{b_j}{s_{b_j}}$ $df = n - (k + 1)$ **Assumptions** The regression assumptions

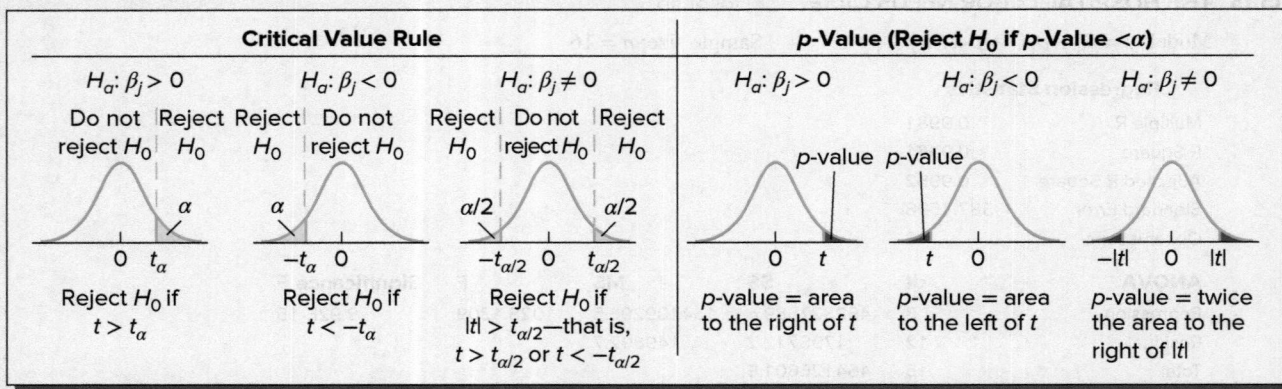

As in testing H_0: $\beta_1 = 0$ in simple linear regression, we usually use the two-sided alternative hypothesis H_a: $\beta_j \neq 0$. Excel, JMP, and Minitab present the results for the two-sided test.

It is customary to test the significance of each and every independent variable in a regression model. Generally speaking,

1   If we can reject H_0: $\beta_j = 0$ at the .05 level of significance, we have strong evidence that the independent variable x_j is significantly related to y in the regression model.

FIGURE 15.8 *t* Statistics and *p*-Values for Testing the Significance of the Intercept, x_1, and x_2 in the Tasty Sub Shop Revenue Model $y = \beta_0 + \beta_1 x_1 + \beta_2 x_2 + \varepsilon$

(a) Calculation of the *t* statistics

Independent Variable	Null Hypothesis	b_j	s_{b_j}	$t = \dfrac{b_j}{s_{b_j}}$	*p*-Value
Intercept	$H_0: \beta_0 = 0$	$b_0 = 125.289$	$s_{b_0} = 40.9333$	$t = \dfrac{b_0}{s_{b_0}} = \dfrac{125.289}{40.9333} = 3.06$	.0183
x_1	$H_0: \beta_1 = 0$	$b_1 = 14.1996$	$s_{b_1} = 0.9100$	$t = \dfrac{b_1}{s_{b_1}} = \dfrac{14.1996}{.9100} = 15.6$	<.001
x_2	$H_0: \beta_2 = 0$	$b_2 = 22.8107$	$s_{b_2} = 5.7692$	$t = \dfrac{b_2}{s_{b_2}} = \dfrac{22.8107}{5.7692} = 3.95$	.0055

(b) The Excel output

	Coefficients	Standard Error	t Stat	P-value	Lower 95%	Upper 95%
Intercept	125.289	40.9333	3.06	0.0183	28.4969	222.0807
population	14.1996	0.9100	15.60	1.07E-06	12.0478	16.3515
bus_rating	22.8107	5.7692	3.95	0.0055	9.1686	36.4527

(c) The Minitab Output

Coefficients

Terms	Coef	SE Coef	T-Value	P-Value	VIF
Constant	125.3	40.9	3.06	0.018	
Population	14.200	0.910	15.60	0.000	1.18
Bus_Rating	22.81	5.77	3.95	0.006	1.18

Analysis of Variance

Source	DF	Adj SS	Adj MS	F-Value	P-Value
Regression	2	486356	243178	180.69	0.000
Population	1	327678	327678	243.48	0.000
Bus_Rating	1	21039	21039	15.63	0.006
Error	7	9421	1346		
Total	9	495777			

2 If we can reject $H_0: \beta_j = 0$ at the .01 level of significance, we have very strong evidence that x_j is significantly related to y in the regression model.

3 The smaller the significance level α at which H_0 can be rejected, the stronger is the evidence that x_j is significantly related to y in the regression model.

© **EXAMPLE 15.3** The Tasty Sub Shop Case: *t* Statistics and Related *p*-Values

Again consider the Tasty Sub Shop revenue model

$$y = \beta_0 + \beta_1 x_1 + \beta_2 x_2 + \varepsilon$$

Figure 15.8(a) summarizes the calculation of the *t* statistics and related *p*-values for testing the significance of the intercept and each of the independent variables x_1 and x_2. Here the values of b_j, s_{b_j}, t, and the *p*-value have been obtained from the Excel output of Figure 15.8(b). If we wish to carry out tests at the .05 level of significance, we use the critical value $t_{.05/2} = t_{.025} = 2.365$, which is based on $n - (k + 1) = 10 - 3 = 7$ degrees of freedom. Looking at Figure 15.8(a), we see that

1 For the intercept, $|t| = 3.06 > 2.365$.

2 For x_1, $|t| = 15.6 > 2.365$.

3 For x_2, $|t| = 3.95 > 2.365$.

Because in each case $|t| > t_{.025}$, we reject each of the null hypotheses in Figure 15.8(a) at the .05 level of significance. Furthermore, because the *p*-value related to x_1 is less than .001, we can reject $H_0: \beta_1 = 0$ at the .001 level of significance. Also, because the *p*-value related to x_2 is less than .01, we can reject $H_0: \beta_2 = 0$ at the .01 level of significance. On the basis of these results, we have extremely strong evidence that in the above model x_1 (population size) is significantly related to y. We also have very strong evidence that in this model x_2 (business rating) is significantly related to y. Note that the left side of the Minitab output in Figure 15.8(c) gives the same (within rounding) *t* statistic and associated *p*-value results that the Excel output gives.

We next note that to the right of the Minitab output of the t statistics and associated p-values in Figure 15.8(c), we show again the Minitab output of the previously discussed F(model) statistic, which equals 180.69. Under this F(model) statistic, we see (1) an F statistic associated with "Population" (x_1), which equals 243.48 and is the square of the t statistic of 15.60 associated with "Population" and (2) an F statistic associated with "Bus_Rating" (x_2), which equals 15.63 and is the square of the t statistic of 3.95 associated with "Bus_ Rating." In general, the F statistic associated with a particular independent variable x_j in a multiple regression model using k independent variables is the square of the t statistic associated with x_j in this model. The F statistic associated with x_j can be calculated directly by the formula $(SSE_j - SSE)/s^2$. Here, SSE and s^2 are the unexplained variation and mean square error for the regression model using all k independent variables, and SSE_j is the unexplained variation for the regression model using all k independent variables *except for x_j*. The p-value for the F statistic associated with x_j is the area under the curve of the F distribution having 1 numerator and $n - (k + 1)$ denominator degrees of freedom to the right of this F statistic. This p-value equals the p-value for the t statistic associated with x_j. Because both p-values give the same information, we will not further discuss the F statistic associated with x_j.

We next consider how to calculate a confidence interval for a regression parameter.

A Confidence Interval for the Regression Parameter β_j

If the regression assumptions hold, a **$100(1 - \alpha)$ percent confidence interval for β_j** is

$$[b_j \pm t_{\alpha/2}s_{b_j}]$$

Here $t_{\alpha/2}$ is based on $n - (k + 1)$ degrees of freedom.

Ⓒ EXAMPLE 15.4 The Tasty Sub Shop Case: A Confidence Interval for β_1

Consider the Tasty Sub Shop revenue model

$$y = \beta_0 + \beta_1 x_1 + \beta_2 x_2 + \varepsilon$$

The Excel output in Figure 15.8(b) tells us that $b_1 = 14.1996$ and $s_{b_1} = .91$. It follows, because $t_{.025}$ based on $n - (k + 1) = 10 - 3 = 7$ degrees of freedom equals 2.365, that a 95 percent confidence interval for β_1 is (see the Excel output)

$$[b_1 \pm t_{.025}s_{b_1}] = [14.1996 \pm 2.365(.91)]$$
$$= [12.048, 16.352]$$

This interval says we are 95 percent confident that, if the population size increases by 1,000 residents and the business rating does not change, then mean yearly revenue will increase by between $12,048 and $16,352. Furthermore, because this 95 percent confidence interval does not contain 0, we can reject $H_0: \beta_1 = 0$ in favor of $H_a: \beta_1 \neq 0$ at the .05 level of significance.

Exercises for Section 15.5

CONCEPTS **connect**

15.16 What do we conclude about x_j if we can reject $H_0: \beta_j = 0$ in favor of $H_a: \beta_j \neq 0$ by setting
a α equal to .05?
b α equal to .01?

15.17 Give an example of a practical application of the confidence interval for β_j.

METHODS AND APPLICATIONS

In Exercises 15.18 through 15.22 we refer to Excel, JMP, and Minitab outputs of regression analyses of the data sets related to five case studies introduced in Section 15.1. The outputs are given in Figure 15.9, or you are asked to obtain on output. Using the appropriate output, do the following for **each parameter β_j** in the model under consideration:

FIGURE 15.9   *t* **Statistics and** *p*-**Values for Four Case Studies**

(a) Minitab output for the natural gas consumption case (sample size: $n = 8$**)**

Coefficients

Term	Coef	SE Coef	T-Value	P-Value	VIF
Constant	13.109	0.856	15.32	0.000	
Temp	−0.0900	0.0141	−6.39	0.001	2.07
Chill	0.0825	0.0220	3.75	0.013	2.07

(b) JMP output for the Fresh detergent case (sample size: $n = 30$**)**

Parameter Estimates

Term	Estimate	Std Error	t Ratio	Prob>|t|	Lower 95%	Upper 95%
Intercept	7.5891025	2.445021	3.10	0.0046*	2.5632892	12.614916
Price (X1)	−2.357723	0.637947	−3.70	0.0010*	−3.669042	−1.046404
IndPrice (X2)	1.6122144	0.295353	5.46	<.0001*	1.0051083	2.2193205
AdvExp (X3)	0.5011517	0.125874	3.98	0.0005*	0.2424145	0.7598889

(c) Excel output for the hospital labor needs case (sample size: $n = 16$**)**

	Coefficients	Standard Error	t Stat	P-value	Lower 95%	Upper 95%
Intercept	1946.8020	504.1819	3.8613	0.0023	848.2840	3045.3201
XRay (x1)	0.0386	0.0130	2.9579	0.0120	0.0102	0.0670
BedDays (x2)	1.0394	0.0676	15.3857	2.91E-09	0.8922	1.1866
LengthStay (x3)	−413.7578	98.5983	−4.1964	0.0012	−628.5850	−198.9306

1. Find b_j, s_{b_j}, and the t statistic for testing $H_0\colon \beta_j = 0$ on the output and report their values. Show how t has been calculated by using b_j and s_{b_j}.

2. Using the t statistic and appropriate critical values, test $H_0\colon \beta_j = 0$ versus $H_a\colon \beta_j \neq 0$ by setting α equal to .05. Which independent variables are significantly related to y in the model with $\alpha = .05$?

3. Find the p-value for testing $H_0\colon \beta_j = 0$ versus $H_a\colon \beta_j \neq 0$ on the output. Using the p-value, determine whether we can reject H_0 by setting α equal to .10, .05, .01, and .001. What do you conclude about the significance of the independent variables in the model?

4. Calculate the 95 percent confidence interval for β_j. Discuss one practical application of this interval.

15.18   **THE NATURAL GAS CONSUMPTION CASE**
      DS GasCon2

Use the Minitab output in Figure 15.9(a) to do (1) through (4) for each of β_0, β_1, and β_2.

15.19   **THE REAL ESTATE SALES PRICE CASE**
      DS RealEst2

Obtain a computer output by using the model $y = \beta_0 + \beta_1 x_1 + \beta_2 x_2 + \varepsilon$ and Excel, JMP, or

Minitab to perform a regression analysis of the real estate sales price data in Table 15.4. Use the output to do (1) through (4) for each of β_0, β_1, and β_2.

15.20   **THE BONNER FROZEN FOODS ADVERTISING CASE**   **DS** Bonner

Obtain a computer output by using the model $y = \beta_0 + \beta_1 x_1 + \beta_2 x_2 + \varepsilon$ and Excel, JMP, or Minitab to perform a regression analysis of the Bonner Frozen Foods sales volume data in Table 15.5. Use the output to do (1) through (4) for each of β_0, β_1, and β_2.

15.21   **THE FRESH DETERGENT CASE**   **DS** Fresh2

Use the Excel output in Figure 15.9(b) to do (1) through (4) for each of β_0, β_1, β_2, and β_3.

15.22   **THE HOSPITAL LABOR NEEDS CASE**
      DS HospLab

Use the Excel output in Figure 15.9(c) to do (1) through (4) for each of β_0, β_1, β_2, and β_3.

15.6 Confidence and Prediction Intervals

In this section we show how to use the multiple regression model to find a **confidence interval for a mean value of *y*** and a **prediction interval for an individual value of *y*.** We first present an example of these intervals, and we then discuss (in an optional technical note) the formulas used to compute the intervals.

LO15-6

Find and interpret a confidence interval for a mean value and a prediction interval for an individual value.

EXAMPLE 15.5 The Tasty Sub Shop Case: Estimating Mean Revenue
and Predicting Revenue and Profit

In the Tasty Sub Shop problem, recall that one of the business entrepreneur's potential sites
is near a population of 47,300 residents and a business/shopping area having a rating of 7.
Also, recall that

$$\hat{y} = b_0 + b_1x_1 + b_2x_2$$
$$= 125.289 + 14.1996(47.3) + 22.8107(7)$$
$$= 956.606 \quad \text{(that is, \$956,606)}$$

is

1   The **point estimate** of the mean yearly revenue for all Tasty Sub restaurants that could
    potentially be built near populations of 47,300 residents and business/shopping areas
    having a rating of 7, and

2   The **point prediction** of the yearly revenue for a single Tasty Sub restaurant that is built
    near a population of 47,300 residents and a business/shopping area having a rating of 7.

This point estimate and prediction are given at the bottom of the Minitab output in Figure 15.4,
which we repeat here as follows:

Fit	SE Fit	95% CI	95% PI
956.606	15.0476	(921.024, 992.188)	(862.844, 1050.37)

In addition to giving $\hat{y} = 956.606$, the Minitab output also gives a 95 percent confidence
interval and a 95 percent prediction interval. The 95 percent confidence interval, [921.024,
992.188], says that we are 95 percent confident that the mean yearly revenue for all Tasty
Sub restaurants that could potentially be built near populations of 47,300 residents and
business/shopping areas having a rating of 7 is between \$921,024 and \$992,188. The
95 percent prediction interval, [862.844, 1050.37], says that we are 95 percent confident
that the yearly revenue for a single Tasty Sub restaurant that is built near a population
of 47,300 residents and a business/shopping area having a rating of 7 will be between
\$862,844 and \$1,050,370.

   Now, recall that the yearly rent and other fixed costs for the entrepreneur's potential restaurant will be \$257,550 and that (according to Tasty Sub corporate headquarters) the yearly
food and other variable costs for the restaurant will be 60 percent of the yearly revenue.
Using the lower end of the 95 percent prediction interval [862.844, 1050.37], we predict that
(1) the restaurant's yearly operating cost will be \$257,550 + .6(862,844) = \$775,256 and
(2) the restaurant's yearly profit will be \$862,844 − \$775,256 = \$87,588. Using the upper
end of the 95 percent prediction interval [862.844, 1050.37], we predict that (1) the restaurant's yearly operating cost will be \$257,550 + .6(1,050,370) = \$887,772 and (2) the
restaurant's yearly profit will be \$1,050,370 − \$887,772 = \$162,598. Combining the two
predicted profits, it follows that we are 95 percent confident that the potential restaurant's
yearly profit will be between \$87,588 and \$162,598. If the entrepreneur decides that this
is an acceptable range of potential yearly profits, then the entrepreneur might decide to
purchase a Tasty Sub franchise for the potential restaurant site.

A technical note (optional)

In general

$$\hat{y} = b_0 + b_1x_1 + b_2x_2 + \cdots + b_kx_k$$

is the **point estimate of the mean value of the dependent variable** y when the values of the
independent variables are $x_1, x_2, \ldots, x_k$ and is the **point prediction of an individual value
of the dependent variable** y when the values of the independent variables are $x_1, x_2, \ldots, x_k$.
Furthermore:

A Confidence Interval and a Prediction Interval

If the regression assumptions hold,

1 A **100(1 − α) percent confidence interval for the mean value of** y when the values of the independent variables are $x_1, x_2, \ldots, x_k$ is

$$[\hat{y} \pm t_{\alpha/2} \, s \sqrt{\text{distance value}}]$$

2 A **100(1 − α) percent prediction interval for an individual value of** y when the values of the independent variables are $x_1, x_2, \ldots, x_k$ is

$$[\hat{y} \pm t_{\alpha/2} \, s \sqrt{1 + \text{distance value}}]$$

Here $t_{\alpha/2}$ is based on $n - (k + 1)$ degrees of freedom and s is the standard error. Furthermore, the formula for the **distance value** (also sometimes called the **leverage value**) involves matrix algebra and is given in Bowerman, O'Connell, and Koehler (2005). In practice, we can obtain the distance value from the outputs of statistical software packages (such as Minitab, JMP, or MegaStat).

Intuitively, the **distance value** is a measure of the distance of the combination of values x_1, $x_2, \ldots, x_k$ from the center of the observed data. The farther that this combination is from the center of the observed data, the larger is the distance value, and thus the longer are both the confidence interval and the prediction interval.

Minitab gives $s_{\hat{y}} = s \sqrt{\text{distance value}}$ under the heading "SE Fit." Because the Minitab output also gives s, the distance value can be found by calculating $(s_{\hat{y}}/s)^2$. For example, the Minitab output on the previous page and in Figure 15.4 tells us that $\hat{y} = 956.606$ (see "Fit") and $s_{\hat{y}} = 15.0476$ (see "SE Fit"). Therefore, because s for the two-variable Tasty Sub Shop revenue model equals 36.6856 (see Figure 15.4), the distance value equals $(15.0476/36.6856)^2 = .1682454$. It follows that the 95 percent confidence and prediction intervals given on the Minitab output have been calculated (within rounding) as follows:

$[\hat{y} \pm t_{.025} \, s \sqrt{\text{distance value}}]$ $[\hat{y} \pm t_{.025} \, s \sqrt{1 + \text{distance value}}]$

$= [956.606 \pm 2.365(36.6856)\sqrt{.1682454}]$ $= [956.606 \pm 2.365(36.6856)\sqrt{1.1682454}]$

$= [956.606 \pm 35.59]$ $= [956.606 \pm 93.78]$

$= [921.0, 992.2]$ $= [862.8, 1050.4]$

Here $t_{\alpha/2} = t_{.025} = 2.365$ is based on $n - (k + 1) = 10 - 3 = 7$ degrees of freedom.

Exercises for Section 15.6

CONCEPTS

15.23 What is the difference between a confidence interval and a prediction interval?

15.24 What does the distance value measure? How does the distance value affect a confidence or prediction interval? (Note: You must read the optional technical note to answer this question.)

METHODS AND APPLICATIONS

15.25 THE NATURAL GAS CONSUMPTION CASE

    GasCon2

The partial Minitab regression output (in the right column) for the natural gas consumption data relates to predicting the city's natural gas consumption (in MMcF) in a week that has an average hourly temperature of 40°F and a chill index of 10.

Fit	SE Fit	95% CI
10.3331	0.170472	(9.89492, 10.7713)
		95% PI
		(9.29274, 11.3735)

a Report (as shown on the computer output) a point estimate of and a 95 percent confidence interval for the mean natural gas consumption for all weeks having an average hourly temperature of 40°F and a chill index of 10.

b Report (as shown on the computer output) a point prediction of and a 95 percent prediction interval for the natural gas consumption in a single week that has an average hourly temperature of 40°F and a chill index of 10.

c Suppose that next week the city's average hourly temperature will be 40°F and the city's chill index will be 10. Also, suppose the city's natural gas company will use the point prediction $\hat{y} = 10.3331$ and order 10.3331 MMcF of natural gas to be shipped to the city by a pipeline transmission system. The gas company will have to pay a fine to the transmission system if the city's actual gas usage y differs from the order of 10.3331 MMcF by more than 10.5 percent—that is, is outside of the range $[10.3331 \pm .105(10.3331)] = [9.248, 11.418]$. Discuss why the 95 percent prediction interval for y, $[9.29274, 11.3735]$, says that y is likely to be inside the allowable range and thus makes the gas company 95 percent confident that it will avoid paying a fine.

d Find 99 percent confidence and prediction intervals for the mean and actual natural gas consumption referred to in parts a and b. Hint: $n = 8$ and $s = .367078$. Optional technical note needed.

15.26 THE REAL ESTATE SALES PRICE CASE

     (DS) RealEst2

Use the real estate sales price data in Table 15.4 and the model $y = \beta_0 + \beta_1 x_1 + \beta_2 x_2 + \varepsilon$ to find **(1)** a point estimate of and a 95 percent confidence interval for the mean sales price of all houses having 2,000 square feet and a niceness rating of 8 and **(2)** a point prediction of and a 95 percent prediction interval for the sales price of a single house having 2,000 square feet and a niceness rating of 8. Hint: Use JMP, MegaStat, or Minitab.

15.27 THE BONNER FROZEN FOODS ADVERTISING CASE   (DS) Bonner

Use the Bonner Frozen Foods sales volume data in Table 15.5 and the model $y = \beta_0 + \beta_1 x_1 + \beta_2 x_2 + \varepsilon$ to find **(1)** a point estimate of and a 95 percent confidence interval for the mean sales volume in all sales regions when \$2,000 is spent on radio and television advertising and \$5,000 is spent on print advertising and **(2)** a point prediction of and a 95 percent prediction interval for the sales volume in an individual sales region when \$2,000 is spent on radio and television advertising and \$5,000 is spent on print advertising. Hint: Use JMP, MegaStat, or Minitab.

15.28 THE FRESH DETERGENT CASE   (DS) Fresh2

Consider the demand for Fresh Detergent in a future sales period when Enterprise Industries' price for Fresh will be $x_1 = 3.70$, the average price of competitors' similar detergents will be $x_2 = 3.90$, and Enterprise Industries' advertising expenditure

for Fresh will be $x_3 = 6.50$. A 95 percent prediction interval for this demand is given on the following JMP output:

	Predicted Demand	Lower 95% Mean Demand	Upper 95% Mean Demand
31	8.4106503477	8.3143172822	8.5069834132
	StdErr Indiv Demand	**Lower 95% Indiv Demand**	**Upper 95% Indiv Demand**
	0.2393033103	7.9187553487	8.9025453468

a **(1)** Find and report the 95 percent prediction interval on the output. **(2)** If Enterprise Industries plans to have in inventory the number of bottles implied by the upper limit of this interval, it can be very confident that it will have enough bottles to meet demand for Fresh in the future sales period. How many bottles is this? **(3)** If we multiply the number of bottles implied by the lower limit of the prediction interval by the price of Fresh (\$3.70), we can be very confident that the resulting dollar amount will be the minimum revenue from Fresh in the future sales period. What is this dollar amount?

b Calculate a 99 percent prediction interval for the demand for Fresh in the future sales period. Use the fact that **StdErr Indiv Demand** equals $s(1 + \text{distance value})^{.5}$.

15.29 THE HOSPITAL LABOR NEEDS CASE

     (DS) HospLab

Consider a hospital for which XRay $= 56,194$, BedDays $= 14,077.88$, and LengthStay $= 6.89$. A 95 percent prediction interval for the labor hours corresponding to this combination of values of the independent variables is given in the following MegaStat output:

	95% Confidence Interval	
Predicted	**lower**	**upper**
15,896.2473	15,378.0313	16,414.4632

95% Prediction Interval		
lower	**upper**	**Leverage**
14,906.2361	16,886.2584	0.3774

Find and report the prediction interval on the output. Then, use this interval to determine if the actual number of labor hours used by the hospital ($y = 17,207.31$) is unusually low or high.

15.7 The Sales Representative Case: Evaluating Employee Performance

Suppose the sales manager of a company wishes to evaluate the performance of the company's sales representatives. Each sales representative is solely responsible for one sales territory, and the manager decides that it is reasonable to measure the performance, y, of a sales representative by using the yearly sales of the company's product in the representative's sales territory. The manager feels that sales performance y substantially depends on five independent variables:

x_1 = number of months the representative has been employed by the company (Time)

x_2 = sales of the company's product and competing products in the sales territory, a measure of sales potential (MktPoten)

x_3 = dollar advertising expenditure in the territory (Adver)

x_4 = weighted average of the company's market share in the territory for the previous four years (MktShare)

x_5 = change in the company's market share in the territory over the previous four years (Change)

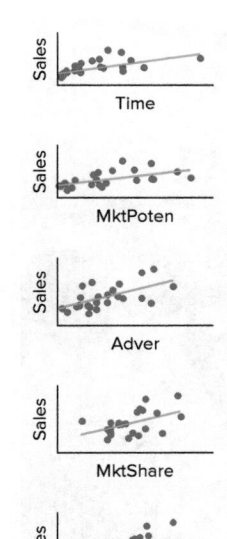

In Table 15.8 we present values of y and x_1 through x_5 for 25 randomly selected sales representatives. To understand the values of y (sales) and x_2 (MktPoten) in the table, note that sales of the company's product or any competing product are measured in hundreds of units of the product sold. Therefore, for example, the first sales figure of 3,669.88 in Table 15.8 means that the first randomly selected sales representative sold 366,988 units of the company's product during the year.

In the page margin are plots of y versus x_1 through x_5. Because each plot has an approximate straight-line appearance, it is reasonable to relate y to x_1 through x_5 by using the regression model

$$y = \beta_0 + \beta_1 x_1 + \beta_2 x_2 + \beta_3 x_3 + \beta_4 x_4 + \beta_5 x_5 + \varepsilon$$

The main objective of the regression analysis is to help the sales manager evaluate sales performance by comparing actual performance to predicted performance. The manager has randomly selected the 25 representatives from all the representatives the company considers to be effective and wishes to use a regression model based on effective representatives to evaluate questionable representatives.

Figure 15.10 gives a partial Excel output of a regression analysis of the sales representative performance data using the five independent variable model. This output tells us that the least

TABLE 15.8   **Sales Representative Performance Data**

DS SalePerf

Sales	Time	MktPoten	Adver	MktShare	Change
3,669.88	43.10	74,065.11	4,582.88	2.51	0.34
3,473.95	108.13	58,117.30	5,539.78	5.51	0.15
2,295.10	13.82	21,118.49	2,950.38	10.91	−0.72
4,675.56	186.18	68,521.27	2,243.07	8.27	0.17
6,125.96	161.79	57,805.11	7,747.08	9.15	0.50
2,134.94	8.94	37,806.94	402.44	5.51	0.15
5,031.66	365.04	50,935.26	3,140.62	8.54	0.55
3,367.45	220.32	35,602.08	2,086.16	7.07	−0.49
6,519.45	127.64	46,176.77	8,846.25	12.54	1.24
4,876.37	105.69	42,053.24	5,673.11	8.85	0.31
2,468.27	57.72	36,829.71	2,761.76	5.38	0.37
2,533.31	23.58	33,612.67	1,991.85	5.43	−0.65
2,408.11	13.82	21,412.79	1,971.52	8.48	0.64
2,337.38	13.82	20,416.87	1,737.38	7.80	1.01
4,586.95	86.99	36,272.00	10,694.20	10.34	0.11
2,729.24	165.85	23,093.26	8,618.61	5.15	0.04
3,289.40	116.26	26,878.59	7,747.89	6.64	0.68
2,800.78	42.28	39,571.96	4,565.81	5.45	0.66
3,264.20	52.84	51,866.15	6,022.70	6.31	−0.10
3,453.62	165.04	58,749.82	3,721.10	6.35	−0.03
1,741.45	10.57	23,990.82	860.97	7.37	−1.63
2,035.75	13.82	25,694.86	3,571.51	8.39	−0.43
1,578.00	8.13	23,736.35	2,845.50	5.15	0.04
4,167.44	58.54	34,314.29	5,060.11	12.88	0.22
2,799.97	21.14	22,809.53	3,552.00	9.14	−0.74

Source: This data set is from a research study published in "An Analytical Approach for Evaluation of Sales Territory Performance," *Journal of Marketing*, January 1972, 31–37 (authors are David W. Cravens, Robert B. Woodruff, and Joseph C. Stamper). We have updated the situation in our case study to be more modern.

FIGURE 15.10   **Partial Excel Output of a Sales Representative Performance Regression Analysis**

(a) The Excel output

Regression Statistics

Multiple R	0.9566
R Square	0.9150
Adjusted R Square	0.8926
Standard Error	430.2319
Observations	25

F	Significance F
40.9106	0.0000

	Coefficients	t Stat	P-value
Intercept	−1113.7879	−2.6526	0.0157
Time	3.6121	3.0567	0.0065
MktPoten	0.0421	6.2527	0.0000
Adver	0.1289	3.4792	0.0025
MktShare	256.9555	6.5657	0.0000
Change	324.5334	2.0634	0.0530

(b) Prediction using MegaStat

Predicted values for: Sales

	95% Prediction Interval	
Predicted	**lower**	**upper**
4,181.74333	3,233.59431	5,129.89235

Leverage

0.109

©Chris Ryan/AGE Fotostock RF

squares point estimates of the model parameters are $b_0 = -1,113.7879$, $b_1 = 3.6121$, $b_2 = .0421$, $b_3 = .1289$, $b_4 = 256.9555$, and $b_5 = 324.5334$. In addition, because the output tells us that the p-values associated with Time, MktPoten, Adver, and MktShare are all less than .01, we have very strong evidence that these variables are significantly related to y and, thus, are important in this model. Because the p-value associated with Change is .0530, we have close to strong evidence that this variable is also important.

Consider a questionable sales representative for whom Time = 85.42, MktPoten = 35,182.73, Adver = 7,281.65, MktShare = 9.64, and Change = .28. The point prediction of the sales, y, corresponding to this combination of values of the independent variables is

$$\hat{y} = -1,113.7879 + 3.6121(85.42) + .0421(35,182.73)$$
$$+ .1289(7,281.65) + 256.9555(9.64) + 324.5334(.28)$$
$$= 4,181.74 \text{ (that is, 418,174 units)}$$

In addition to giving this point prediction, the Excel output tells us that a 95 percent prediction interval for y is [3233.59, 5129.89]. Furthermore, suppose that the actual sales y for the questionable representative were 3,087.52. This actual sales figure is less than the point prediction $\hat{y} = 4,181.74$ and is less than the lower bound of the 95 percent prediction interval for y, [3233.59, 5129.89]. Therefore, we conclude that there is strong evidence that the actual performance of the questionable representative is less than predicted performance. We should investigate the reason for this. Perhaps the questionable representative needs special training.

15.8 Using Dummy Variables to Model Qualitative Independent Variables (Optional)

While the levels (or values) of a quantitative independent variable are numerical, the levels of a **qualitative** independent variable are defined by describing them. For instance, the type of sales technique used by a door-to-door salesperson is a qualitative independent variable. Here we might define three different levels—high pressure, medium pressure, and low pressure.

We can model the effects of the different levels of a qualitative independent variable by using what we call **dummy variables** (also called **indicator variables**). Such variables are usually defined so that they take on two values—either 0 or 1. To see how we use dummy variables, we begin with an example.

EXAMPLE 15.6 The Electronics World Case: Comparing Three Kinds of Store Locations

Part 1: The Data and Data Plots   Suppose that Electronics World, a chain of stores that sells audio and video equipment, has gathered the data in Table 15.9. These data concern store sales volume in July of last year (y, measured in thousands of dollars), the number of households in the store's area (x, measured in thousands), and the location of the store (on a suburban street or in a suburban shopping mall—a qualitative independent variable). Figure 15.11 gives a data plot of y versus x. Stores having a street location are plotted as solid dots, while stores having a mall location are plotted as asterisks. Notice that the line relating y to x for mall locations has a higher y-intercept than does the line relating y to x for street locations.

Part 2: A Dummy Variable Model   In order to model the effects of the street and shopping mall locations, we define a dummy variable denoted D_M as follows:

$$D_M = \begin{cases} 1 & \text{if a store is in a mall location} \\ 0 & \text{otherwise} \end{cases}$$

Using this dummy variable, we consider the regression model

$$y = \beta_0 + \beta_1 x + \beta_2 D_M + \varepsilon$$

TABLE 15.9	The Electronics World Sales Volume Data		

DS Electronics1

Store	Number of Households, x	Location	Sales Volume, y
1	161	Street	157.27
2	99	Street	93.28
3	135	Street	136.81
4	120	Street	123.79
5	164	Street	153.51
6	221	Mall	241.74
7	179	Mall	201.54
8	204	Mall	206.71
9	214	Mall	229.78
10	101	Mall	135.22

FIGURE 15.11   Plot of the Sales Volume Data and a Geometrical Interpretation of the Model $y = \beta_0 + \beta_1 x + \beta_2 D_M + \varepsilon$

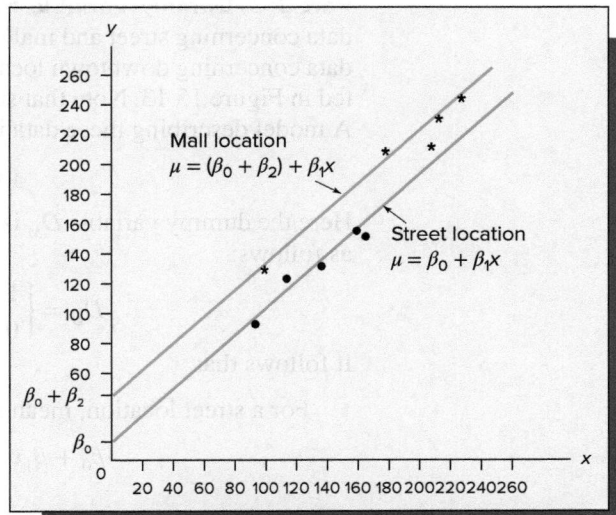

This model and the definition of D_M imply that

1   For a street location, mean sales volume equals

$$\beta_0 + \beta_1 x + \beta_2 D_M = \beta_0 + \beta_1 x + \beta_2(0)$$
$$= \beta_0 + \beta_1 x$$

2   For a mall location, mean sales volume equals

$$\beta_0 + \beta_1 x + \beta_2 D_M = \beta_0 + \beta_1 x + \beta_2(1)$$
$$= (\beta_0 + \beta_2) + \beta_1 x$$

Thus, the dummy variable allows us to model the situation illustrated in Figure 15.11. Here, the lines relating mean sales volume to x for street and mall locations have different y intercepts, β_0 and $(\beta_0 + \beta_2)$, and the same slope β_1. Note that β_2 is the difference between the mean monthly sales volume for stores in mall locations and the mean monthly sales volume for stores in street locations, when all these stores have the same number of households in their areas. That is, we can say that β_2 represents the effect on mean sales of a mall location compared to a street location. The Excel output in Figure 15.12 tells us that the least

FIGURE 15.12   Excel Output of a Regression Analysis of the Sales Volume Data Using the Model $y = \beta_0 + \beta_1 x + \beta_2 D_M + \varepsilon$

Regression Statistics	
Multiple R	0.9913
R Square	0.9827
Adjusted R Square	0.9778
Standard Error	7.3288
Observations	10

ANOVA	df	SS	MS	F	Significance F
Regression	2	21411.7977	10705.8989	199.3216	6.75E-07
Residual	7	375.9817	53.7117		
Total	9	21787.7795			

	Coefficients	Standard Error	t Stat	P-value	Lower 95%	Upper 95%
Intercept	17.3598	9.4470	1.8376	0.1087	−4.9788	39.6985
Households (x)	0.8510	0.0652	13.0439	3.63E-06	0.6968	1.0053
DummyMall	29.2157	5.5940	5.2227	0.0012	15.9881	42.4434

squares point estimate of β_2 is $b_2 = 29.2157$. This says that for any given number of house-holds in a store's area, we estimate that the mean monthly sales volume in a mall location is $29,215.70 greater than the mean monthly sales volume in a street location.

Part 3: A Dummy Variable Model for Comparing Three Locations In addition to the data concerning street and mall locations in Table 15.9, Electronics World has also collected data concerning downtown locations. The complete data set is given in Table 15.10 and plotted in Figure 15.13. Note that stores having a downtown location are plotted as open circles. A model describing these data is

$$y = \beta_0 + \beta_1 x + \beta_2 D_M + \beta_3 D_D + \varepsilon$$

Here the dummy variable D_M is as previously defined and the dummy variable D_D is defined as follows:

$$D_D = \begin{cases} 1 & \text{if a store is in a downtown location} \\ 0 & \text{otherwise} \end{cases}$$

It follows that

1   For a street location, mean sales volume equals

$$\beta_0 + \beta_1 x + \beta_2 D_M + \beta_3 D_D = \beta_0 + \beta_1 x + \beta_2(0) + \beta_3(0)$$
$$= \beta_0 + \beta_1 x$$

2   For a mall location, mean sales volume equals

$$\beta_0 + \beta_1 x + \beta_2 D_M + \beta_3 D_D = \beta_0 + \beta_1 x + \beta_2(1) + \beta_3(0)$$
$$= (\beta_0 + \beta_2) + \beta_1 x$$

3   For a downtown location, mean sales volume equals

$$\beta_0 + \beta_1 x + \beta_2 D_M + \beta_3 D_D = \beta_0 + \beta_1 x + \beta_2(0) + \beta_3(1)$$
$$= (\beta_0 + \beta_3) + \beta_1 x$$

Thus, the dummy variables allow us to model the situation illustrated in Figure 15.13. Here the lines relating mean sales volume to x for street, mall, and downtown locations have

TABLE 15.10	The Complete Electronics World Sales Volume Data DS Electronics2		
Store	**Number of Households,** x	**Location**	**Sales Volume,** y
1	161	Street	157.27
2	99	Street	93.28
3	135	Street	136.81
4	120	Street	123.79
5	164	Street	153.51
6	221	Mall	241.74
7	179	Mall	201.54
8	204	Mall	206.71
9	214	Mall	229.78
10	101	Mall	135.22
11	231	Downtown	224.71
12	206	Downtown	195.29
13	248	Downtown	242.16
14	107	Downtown	115.21
15	205	Downtown	197.82

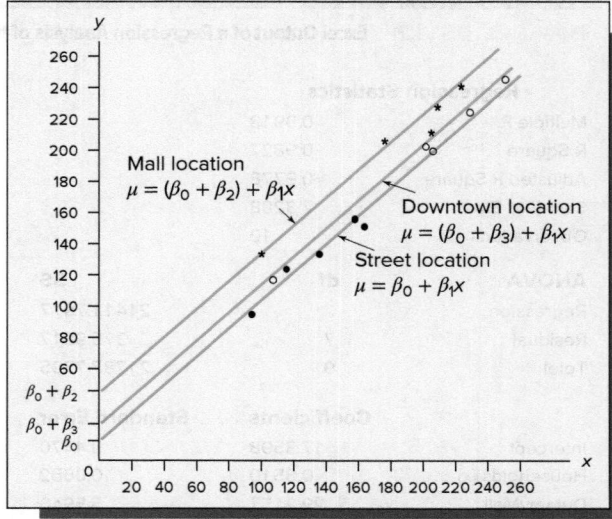

FIGURE 15.13   Plot of the Complete Electronics World Sales Volume Data and a Geometrical Interpretation of the Model $y = \beta_0 + \beta_1 x + \beta_2 D_M + \beta_3 D_D + \varepsilon$

FIGURE 15.14   Excel Output of a Regression Analysis of the Complete Sales Volume Data Using the Model
$$y = \beta_0 + \beta_1 x + \beta_2 D_M + \beta_3 D_D + \varepsilon$$

Regression Statistics

Multiple R	0.9934
R Square	0.9868
Adjusted R Square	0.9833
Standard Error	6.3494
Observations	15

ANOVA	df	SS	MS	F	Significance F
Regression	3	33268.6953	11089.5651	275.0729	1.27E-10
Residual	11	443.4650	40.3150		
Total	14	33712.1603			

	Coefficients	Standard Error	t Stat	P-value	Lower 95%	Upper 95%
Intercept	14.9777	6.1884	2.4203	0.0340	1.3570	28.5984
Households (x)	0.8686	0.0405	21.4520	2.52E-10	0.7795	0.9577
DummyMall	28.3738	4.4613	6.3600	5.37E-05	18.5545	38.1930
DummyDtown	6.8638	4.7705	1.4388	0.1780	-3.6360	17.3635

different y-intercepts, β_0, $(\beta_0 + \beta_2)$, and $(\beta_0 + \beta_3)$, and the same slope β_1. Note that β_2 represents the effect on mean sales of a mall location compared to a street location, and β_3 represents the effect on mean sales of a downtown location compared to a street location. Furthermore, the difference between β_2 and β_3, $\beta_2 - \beta_3$, represents the effect on mean sales of a mall location compared to a downtown location.

Part 4: Comparing the Three Locations   Figure 15.14 gives the Excel output of a regression analysis of the sales volume data using the dummy variable model. This output tells us that the least squares point estimate of β_2 is $b_2 = 28.3738$. It follows that for any given number of households in a store's area, we estimate that the mean monthly sales volume in a mall location is $28,373.80 greater than the mean monthly sales volume in a street location. Furthermore, because the Excel output tells us that a 95 percent confidence interval for β_2 is [18.5545, 38.193], we are 95 percent confident that for any given number of households in a store's area, the mean monthly sales volume in a mall location is between $18,554.50 and $38,193 greater than the mean monthly sales volume in a street location. The Excel output also shows that the t statistic for testing $H_0: \beta_2 = 0$ versus $H_a: \beta_2 \neq 0$ equals 6.36 and that the related p-value is less than .001. Therefore, we have very strong evidence that there is a difference between the mean monthly sales volumes in mall and street locations.

We next note that the Excel output shows that the least squares point estimate of β_3 is $b_3 = 6.8638$. Therefore, we estimate that for any given number of households in a store's area, the mean monthly sales volume in a downtown location is $6,863.80 greater than the mean monthly sales volume in a street location. Furthermore, the Excel output shows that a 95 percent confidence interval for β_3 is [-3.636, 17.3635]. This says we are 95 percent confident that for any given number of households in a store's area, the mean monthly sales volume in a downtown location is between $3,636 less than and $17,363.50 greater than the mean monthly sales volume in a street location. The Excel output also shows that the t statistic and p-value for testing $H_0: \beta_3 = 0$ versus $H_a: \beta_3 \neq 0$ are $t = 1.4388$ and p-value = .178. Therefore, we do not have strong evidence that there is a difference between the mean monthly sales volumes in downtown and street locations.

Finally, note that, because $b_2 = 28.3738$ and $b_3 = 6.8638$, the point estimate of $\beta_2 - \beta_3$ is $b_2 - b_3 = 28.3738 - 6.8638 = 21.51$. Therefore, we estimate that mean monthly sales volume in a mall location is $21,510 higher than mean monthly sales volume in a downtown location. Near the end of this section we show how to compare the mall and downtown locations by using a confidence interval and a hypothesis test. We will find that there is very strong evidence that the mean monthly sales volume in a mall location is higher than the mean monthly sales volume in a downtown location. In summary, the mall location seems to give a higher mean monthly sales volume than either the street or downtown location.

BI

FIGURE 15.15   **Minitab Output of a Regression Analysis of the Complete Sales Volume Data Using the Model** $y = \beta_0 + \beta_1 x + \beta_2 D_M + \beta_3 D_D + \varepsilon$

```
Analysis of Variance
Source            DF    Adj SS    Adj MS    F-Value   P-Value
Regression         3   33268.7   11089.6    275.07    0.000
   Households      1   18552.4   18552.4    460.19    0.000
   DMall           1    1630.7    1630.7     40.45    0.000
   DDowntown       1      83.5      83.5      2.07    0.178
Error             11     443.5      40.3
Total             14   33712.2

Model Summary
     S      R-sq    R-sq(adj)   R-sq(pred)
6.34941   98.68%     98.33%      97.68%

Coefficients
Term          Coef   SE Coef   T-Value   P-Value    VIF
Constant     14.98     6.19      2.42     0.034
Households   0.8686   0.0405    21.45     0.000    1.45
DMall        28.37     4.46      6.36     0.000    1.65
DDowntown     6.86     4.77      1.44     0.178    1.88

Variable     Setting      Fit     SE Fit        95% CI                  95% PI
Households       200   217.069   2.91432   (210.655, 223.484)   (201.692, 232.446)
DMall              1
DDowntown          0
```

Part 5: Predicting a Future Sales Volume   Suppose that Electronics World wishes to predict the sales volume in a future month for an individual store that has 200,000 households in its area and is located in a shopping mall. The point prediction of this sales volume is (note $D_M = 1$ and $D_D = 0$ when a store is in a shopping mall)

$$\hat{y} = b_0 + b_1(200) + b_2(1) + b_3(0)$$
$$= 14.98 + .8686(200) + 28.37(1)$$
$$= 217.069$$

This point prediction is given at the bottom of the Minitab output in Figure 15.15. The corresponding 95 percent prediction interval, which is [201.692, 232.446], says we are 95 percent confident that the sales volume in a future sales period for an individual mall store that has 200,000 households in its area will be between \$201,692 and \$232,446.

Part 6: Interaction Models   Consider the Electronics World data for street and mall locations given in Table 15.9 and the model

$$y = \beta_0 + \beta_1 x + \beta_2 D_M + \beta_3 x D_M + \varepsilon$$

This model uses the *cross-product*, or *interaction*, *term* $x D_M$ and implies that

1   For a street location, mean sales volume equals (because $D_M = 0$)

$$\beta_0 + \beta_1 x + \beta_2(0) + \beta_3 x(0) = \beta_0 + \beta_1 x$$

2   For a mall location, mean sales volume equals (because $D_M = 1$)

$$\beta_0 + \beta_1 x + \beta_2(1) + \beta_3 x(1) = (\beta_0 + \beta_2) + (\beta_1 + \beta_3)x$$

As illustrated in Figure 15.16(a), if we use this model, then the straight lines relating mean sales volume to x for street and mall locations have *different y-intercepts* and *different slopes*. The *different slopes* imply that this model assumes *interaction* between x and store location. In general, we say that **interaction** exists between two independent variables if the relationship between (for example, the slope of the line relating) the mean value of the dependent variable and one of the independent variables depends upon the value (or level) of the other independent variable. We have drawn the lines in Figure 15.16(a) so that they represent one possible type of interaction in the sales volume situation. Specifically, note that the

FIGURE 15.16   Geometrical Interpretations and Partial Excel Output for the Model
$$y = \beta_0 + \beta_1 x + \beta_2 D_M + \beta_3 x D_M + \varepsilon$$

(a) One type of interaction

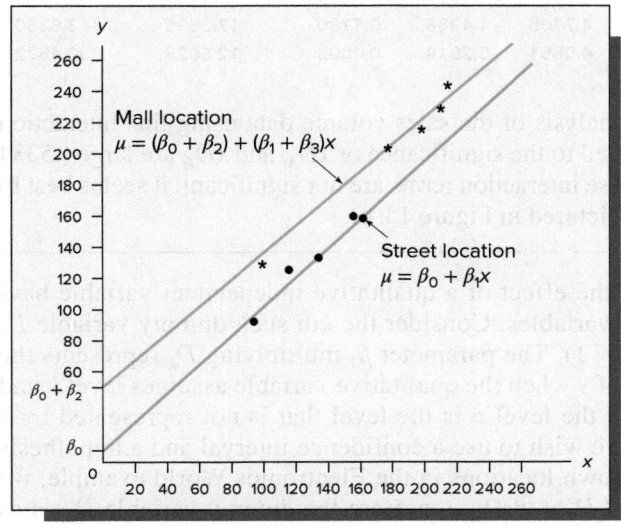

(b) A second type of interaction

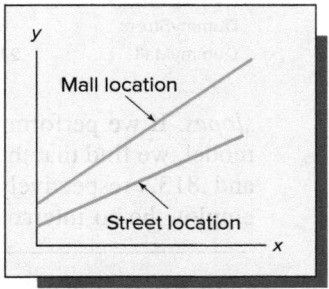

(c) Partial Excel output

R Square	0.9836
Adjusted R Square	0.9755
Standard Error	7.7092

	Coefficients	t Stat	P-value
Intercept	7.9004	0.4090	0.6967
Households	0.9207	6.5792	0.0006
DM	42.7297	1.7526	0.1302
XDM	−0.0917	−0.5712	0.5886

differently sloped lines representing the mean sales volumes in street and mall locations get closer together as x increases. This says that the difference between the mean sales volumes in a mall and street location gets smaller as the number of households in a store's area increases. Such interaction might be logical because, as the number of households in a given area increases, the *concentration* and thus proportion of households that are very near and thus familiar with a street location might increase. Thus, the difference between the always high proportion of customers who would shop at a mall location and the increasing proportion of customers who would shop at a street location would get smaller. There would then be a decreasing difference between the mean sales volumes in the mall and street locations. Of course, the "opposite" type of interaction, in which the differently sloped lines representing the mean sales volumes in street and mall locations get farther apart as x increases, is also possible. As illustrated in Figure 15.16(b), this type of interaction would say that the difference between the mean sales volumes in a mall and street location gets larger as the number of households in a store's area increases. Figure 15.16(c) gives a partial Excel output of a regression analysis of the sales volume data using the interaction model. Here D_M and $x D_M$ are labeled as DM and XDM, respectively, on the output. The Excel output tells us that the p-value related to the significance of $x D_M$ is large (.5886), indicating that this interaction term is not significant. It follows that the no-interaction model seems best. As illustrated in Figure 15.11, this model implies that the lines representing the mean sales volumes in street and mall locations are **parallel.** Therefore, this model assumes that the difference between the mean sales volumes in a mall and street location does not depend on the number of households in a store's area. Note, however, that although interaction does not exist in this situation, it does in others. For example, studies using dummy variable regression models have shown that the difference between the salaries of males and females in an organization tend to increase as an employee's time with the organization increases.

Lastly, consider the Electronics World data for street, mall, and downtown locations given in Table 15.10. In modeling these data, if we believe that interaction exists between the number of households in a store's area and store location, we might consider using the model

$$y = \beta_0 + \beta_1 x + \beta_2 D_M + \beta_3 D_D + \beta_4 x D_M + \beta_5 x D_D + \varepsilon$$

Similar to Figure 15.16, this model implies that the straight lines relating mean sales volume to x for the street, mall, and downtown locations have *different y-intercepts* and *different*

FIGURE 15.17   Partial Excel Output for the Model $y = \beta_0 + \beta_1 x + \beta_2 D_S + \beta_3 D_M + \varepsilon$

	Coefficients	Standard Error	t Stat	P-value	Lower 95%	Upper 95%
Intercept	21.8415	8.5585	2.5520	0.0269	3.0044	40.6785
Households (x)	0.8686	0.0405	21.4520	2.52E-10	0.7795	0.9577
DummyStreet	−6.8638	4.7705	−1.4388	0.1780	−17.3635	3.6360
DummyMall	21.5100	4.0651	5.2914	0.0003	12.5628	30.4572

slopes. If we perform a regression analysis of the sales volume data using this interaction model, we find that the *p*-values related to the significance of xD_M and xD_D are large (.5334 and .8132, respectively). Because these interaction terms are not significant, it seems best to employ the no-interaction model as pictured in Figure 15.13.

In general, if we wish to model the effect of a qualitative independent variable having *a* levels, we use $a - 1$ dummy variables. Consider the *k*th such dummy variable D_k ($k =$ one of the values 1, 2, . . . , $a - 1$). The parameter β_k multiplying D_k represents the mean difference between the level of *y* when the qualitative variable assumes level *k* and when it assumes the level *a* (where the level *a* is the level that is not represented by a dummy variable). For example, if we wish to use a confidence interval and a hypothesis test to compare the mall and downtown locations in the Electronics World example, we can use the model $y = \beta_0 + \beta_1 x + \beta_2 D_S + \beta_3 D_M + \varepsilon$. Here the dummy variable D_M is as previously defined, and D_S is a dummy variable that equals 1 if a store is in a street location and 0 otherwise. Because this model does not use a dummy variable to represent the downtown location, the parameter β_2 expresses the effect on mean sales of a street location compared to a downtown location, and the parameter β_3 expresses the effect on mean sales of a mall location compared to a downtown location. Figure 15.17 gives a partial Excel output of a regression analysis using this model. Because the least squares point estimate of β_3 is $b_3 = 21.51$, we estimate that for any given number of households in a store's area, the mean monthly sales volume in a mall location is $21,510 higher than the mean monthly sales volume in a downtown location. The Excel output tells us that a 95 percent confidence interval for β_3 is [12.5628, 30.4572]. Therefore, we are 95 percent confident that for any given number of households in a store's area, the mean monthly sales volume in a mall location is between $12,562.80 and $30,457.20 greater than the mean monthly sales volume in a downtown location. The Excel output also shows that the *t* statistic and *p*-value for testing $H_0: \beta_3 = 0$ versus $H_a: \beta_3 \neq 0$ in this model are, respectively, 5.2914 and .0003. Therefore, we have very strong evidence that there is a difference between the mean monthly sales volumes in mall and downtown locations.

Exercises for Section 15.8

CONCEPTS connect

15.30 What is a qualitative independent variable?

15.31 How do we use dummy variables to model the effects of a qualitative independent variable?

15.32 What does the parameter multiplied by a dummy variable express?

METHODS AND APPLICATIONS

15.33 Neter, Kutner, Nachtsheim, and Wasserman (1996) relate the speed, *y*, with which a particular insurance innovation is adopted to the size of the insurance firm, *x*, and the type of firm. The dependent variable *y* is measured by the number of months elapsed between the time the first firm adopted the innovation and the time the firm being considered adopted the innovation. The size of the firm, *x*, is measured by the total assets of the firm, and the type of firm—a qualitative independent variable—is either a mutual company or a stock company. The data in Table 15.11 are observed. InsInnov

a   Discuss why the data plot in Figure 15.18 indicates that the model

$$y = \beta_0 + \beta_1 x + \beta_2 D_S + \varepsilon$$

might appropriately describe the observed data.

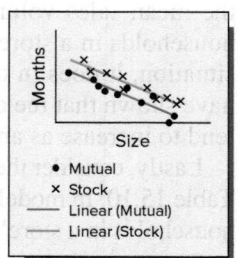

FIGURE 15.18

Plot of the Insurance Innovation Data

TABLE 15.11   The Insurance Innovation Data   DS InsInnov

Firm	Number of Months Elapsed, y	Size of Firm (Millions of Dollars), x	Type of Firm	Firm	Number of Months Elapsed, y	Size of Firm (Millions of Dollars), x	Type of Firm
1	17	151	Mutual	11	28	164	Stock
2	26	92	Mutual	12	15	272	Stock
3	21	175	Mutual	13	11	295	Stock
4	30	31	Mutual	14	38	68	Stock
5	22	104	Mutual	15	31	85	Stock
6	0	277	Mutual	16	21	224	Stock
7	12	210	Mutual	17	20	166	Stock
8	19	120	Mutual	18	13	305	Stock
9	4	290	Mutual	19	30	124	Stock
10	16	238	Mutual	20	14	246	Stock

Here D_S equals 1 if the firm is a stock company and 0 if the firm is a mutual company.

b The model of part *a* implies that the mean adoption time of an insurance innovation by mutual companies having an asset size x equals

$$\beta_0 + \beta_1 x + \beta_2(0) = \beta_0 + \beta_1 x$$

and that the mean adoption time by stock companies having an asset size x equals

$$\beta_0 + \beta_1 x + \beta_2(1) = \beta_0 + \beta_1 x + \beta_2$$

The difference between these two means equals the model parameter β_2. In your own words, interpret the practical meaning of β_2.

c Figure 15.19 presents the Excel output of a regression analysis of the insurance innovation data using the model of part *a*. **(1)** Using the output, test $H_0: \beta_2 = 0$ versus $H_a: \beta_2 \neq 0$ by setting $\alpha = .05$ and .01. **(2)** Interpret the practical meaning of the result of this test. **(3)** Also, use the computer output to find, report, and interpret a 95 percent confidence interval for β_2.

15.34 If we add the interaction term xD_S to the model of part *a* of Exercise 15.33, we find that the *p*-value related to this term is .9821. What does this imply?   DS InsInnov

15.35   THE FLORIDA POOL HOME CASE   DS PoolHome

Table 3.21 gives the selling price (Price, expressed in thousands of dollars), the square footage (SqrFt), the number of bathrooms (Bathrms), and the niceness rating (Niceness, expressed as an integer from 1 to 7) of 80 homes randomly selected from all homes sold in a Florida city during the last six months. (The random selections were made from homes having between 2,000 and 3,500 square feet.) Table 3.21 also gives values of the dummy variable Pool?, which equals 1 if a home has a pool and 0 otherwise. Use Excel, JMP, or Minitab to perform a regression analysis of these data using the model

$$\text{Price} = \beta_0 + \beta_1 \times \text{SqrFt} + \beta_2 \times \text{Bathrms} + \beta_3 \times \text{Niceness} + \beta_4 \times \text{Pool?} + \varepsilon$$

FIGURE 15.19   Excel Output of a Regression Analysis of the Insurance Innovation Data Using the Model $y = \beta_0 + \beta_1 x + \beta_2 D_S + \varepsilon$

Regression Statistics

Multiple R	0.9461
R Square	0.8951
Adjusted R Square	0.8827
Standard Error	3.2211
Observations	20

ANOVA

	df	SS	MS	F	Significance F
Regression	2	1,504.4133	752.2067	72.4971	4.77E-09
Residual	17	176.3867	10.3757		
Total	19	1,680.8			

	Coefficients	Standard Error	t Stat	P-value	Lower 95%	Upper 95%
Intercept	33.8741	1.8139	18.6751	9.15E-13	30.0472	37.7010
Size of Firm (x)	−0.1017	0.0089	−11.4430	2.07E-09	−0.1205	−0.0830
DummyStock	8.0555	1.4591	5.5208	3.74E-05	4.9770	11.1339

a Noting that β_4 is the effect on mean sales price of a home having a pool, find (on the computer output) a point estimate of this effect. If the average current purchase price of the pools in the sample is \$32,500, find a point estimate of the percentage of a pool's cost that a customer buying a pool can expect to recoup when selling his (or her) home.

b If we add various combinations of the interaction terms SqrFt × Pool?, Bathrooms × Pool?, and Niceness × Pool? to the above model, we find that the p-values related to these terms are greater than .05. What does this imply?

15.36 THE FRESH DETERGENT CASE   DS Fresh3

Recall from Exercise 15.6 that Enterprise Industries has observed the historical data in Table 15.6 concerning y (demand for Fresh liquid laundry detergent), x_1 (the price of Fresh), x_2 (the average industry price of competitors' similar detergents), and x_3 (Enterprise Industries' advertising expenditure for Fresh). To ultimately increase the demand for Fresh, Enterprise Industries' marketing department is comparing the effectiveness of three different advertising campaigns. These campaigns are denoted as campaigns A, B, and C. Campaign A consists entirely of television commercials, campaign B consists of a balanced mixture of television and radio commercials, and campaign C consists of a balanced mixture of television, radio, newspaper, and magazine ads. To conduct the study, Enterprise Industries has randomly selected one advertising campaign to be used in each of the 30 sales periods in Table 15.6. Although logic would indicate that each of campaigns A, B, and C should be used in 10 of the 30 sales periods, Enterprise Industries has made previous commitments to the advertising media involved in the study. As a result, campaigns A, B, and C were randomly assigned to, respectively, 9, 11, and 10 sales periods. Furthermore, advertising was done in only the first three weeks of each sales period, so that the carryover effect of the campaign used in a sales period to the next sales period would be minimized. Table 15.12 lists the campaigns used in the sales periods.

TABLE 15.12
Advertising Campaigns Used by Enterprise Industries
DS Fresh3

Sales Period	Advertising Campaign
1	B
2	B
3	B
4	A
5	C
6	A
7	C
8	C
9	B
10	C
11	A
12	C
13	C
14	A
15	B
16	B
17	B
18	A
19	B
20	B
21	C
22	A
23	A
24	A
25	A
26	B
27	C
28	B
29	C
30	C

To compare the effectiveness of advertising campaigns A, B, and C, we define two dummy variables. Specifically, we define the dummy variable D_B to equal 1 if campaign B is used in a sales period and 0 otherwise. Furthermore, we define the dummy variable D_C to equal 1 if campaign C is used in a sales period and 0 otherwise. Figures 15.20 and 15.21 present the JMP output of a regression analysis of the Fresh demand data by using the model

$$y = \beta_0 + \beta_1 x_1 + \beta_2 x_2 + \beta_3 x_3 + \beta_4 D_B + \beta_5 D_C + \varepsilon$$

a Because this model does not use a dummy variable to represent advertising campaign A, the parameter β_4 represents the effect on mean demand of advertising campaign B compared to advertising campaign A, and the parameter β_5 represents the effect on mean demand of advertising campaign C compared to advertising campaign A. (1) Use the regression output to find and report a point estimate of each of the above effects and to test the significance of each of the above effects. (2) Find and report a 95 percent confidence interval for each of the above effects. (3) Interpret your results.

b The prediction results in Figure 15.21 correspond to a future period when Fresh's price will be $x_1 = 3.70$, the average price of similar detergents will be $x_2 = 3.90$, Fresh's advertising expenditure will be $x_3 = 6.50$, and advertising campaign C will be used. (1) Show (within rounding) how $\hat{y} = 8.61621$ is calculated. (2) Find, report, and interpret a 95 percent confidence interval for mean demand and a 95 percent prediction interval for an individual demand when $x_1 = 3.70$, $x_2 = 3.90$, $x_3 = 6.50$, and campaign C is used.

c Consider the alternative model

$$y = \beta_0 + \beta_1 x_1 + \beta_2 x_2 + \beta_3 x_3 + \beta_4 D_A + \beta_5 D_C + \varepsilon$$

Here D_A equals 1 if advertising campaign A is used and equals 0 otherwise. Because this model does not use a dummy variable to represent advertising campaign B, the parameter β_5 in this model represents the effect on mean demand of advertising campaign C compared to advertising campaign B. Use Excel, JMP, or Minitab to perform a regression analysis of the data in Tables 15.6 and 15.12 by using the alternative model. Use the computer output to (1) test the significance of the effect represented by β_5 and (2) find a 95 percent confidence interval for β_5. (3) Interpret your results.

15.37 THE FRESH DETERGENT CASE   DS Fresh3

Add the independent variables $x_3 D_B$ and $x_3 D_C$ to the first model of Exercise 15.36 and use Excel, JMP, or Minitab to determine whether these independent variables are significant. What do your results imply?

FIGURE 15.20   **JMP Output of a Dummy Variable Regression Model Analysis of the Fresh Demand Data**

Summary of Fit

RSquare	0.959727
RSquare Adj	0.951337
Root Mean Square Error	0.15028
Mean of Response	8.382667
Observations (or Sum Wgts)	30

Analysis of Variance

Source	DF	Sum of Squares	Mean Square	F Ratio
Model	5	12.916568	2.58331	114.3862
Error	24	0.542019	0.02258	Prob > F
C. Total	29	13.458587		<.0001*

Parameter Estimates

Term	Estimate	Std Error	t Ratio	Prob>|t|	Lower 95%	Upper 95%
Intercept	8.715414	1.584933	5.50	<.0001*	5.4442726	11.986555
Price(X1)	−2.768024	0.414437	−6.68	<.0001*	−3.62338	−1.912669
IndPrice(X2)	1.6666921	0.191332	8.71	<.0001*	1.2718024	2.0615818
AdvExp(X3)	0.4927425	0.080646	6.11	<.0001*	0.326298	0.659187
DB	0.269496	0.06945	3.88	0.0007*	0.1261574	0.4128347
DC	0.4395621	0.070335	6.25	<.0001*	0.2943979	0.5847264

FIGURE 15.21   **JMP Output of a Prediction of Fresh Demand Using the Dummy Variable Regression Model of Figure 15.20**

	Predicted Demand	Lower 95% Mean Demand	Upper 95% Mean Demand	Lower 95% Indiv Demand	Upper 95% Indiv Demand
31	8.616211576	8.5137990349	8.718624117	8.289578079	8.9428450729

15.9 Using Squared and Interaction Variables (Optional)

LO15-8

Use squared and interaction variables. (Optional)

Using squared variables

One useful form of the multiple regression model is what we call the **quadratic regression model.** Assuming that we have obtained n observations—each consisting of an observed value of y and a corresponding value of x—the model is as follows:

The Quadratic Regression Model

The **quadratic regression model** relating y to x is

$$y = \beta_0 + \beta_1 x + \beta_2 x^2 + \varepsilon$$

where

1   $\beta_0 + \beta_1 x + \beta_2 x^2$ is μ_y, the mean value of the dependent variable y when the value of the independent variable is x.

2   β_0, β_1, and β_2 are (unknown) **regression parameters** relating the mean value of y to x.

3   ε is an error term that describes the effects on y of all factors other than x and x^2.

The quadratic equation $\mu_y = \beta_0 + \beta_1 x + \beta_2 x^2$ that relates μ_y to x is the equation of a **parabola.** Two parabolas are shown in Figure 15.22(a) and (b) and help to explain the meanings of the parameters β_0, β_1, and β_2. Here β_0 is the **y-intercept** of the parabola (the value of μ_y when $x = 0$). Furthermore, β_1 is the **shift parameter** of the parabola: the value of β_1 shifts the parabola to the left or right. Specifically, increasing the value of β_1 shifts the parabola to the left. Lastly, β_2 is the **rate of curvature** of the parabola. If β_2 is greater than 0, the parabola opens upward [see Figure 15.22(a)]. If β_2 is less than 0, the parabola opens

FIGURE 15.22 **The Mean Value of the Dependent Variable Changing in a Quadratic Fashion as x Increases ($\mu_y = \beta_0 + \beta_1 x + \beta_2 x^2$)**

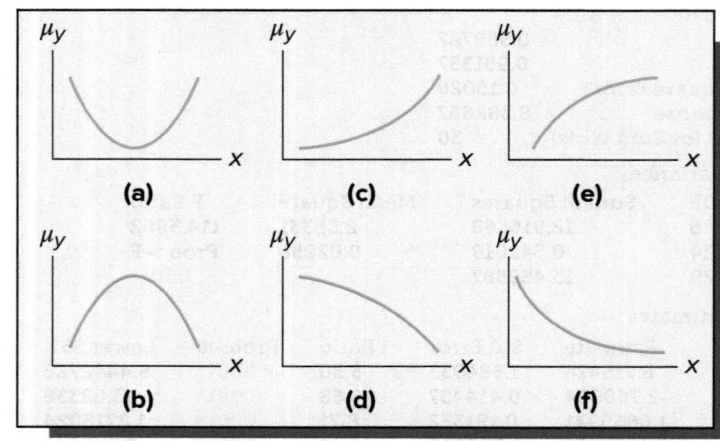

downward [see Figure 15.22(b)]. If a scatter plot of y versus x shows points scattered around a parabola, or a part of a parabola [some typical parts are shown in Figure 15.22(c), (d), (e), and (f)], then the quadratic regression model might appropriately relate y to x.

ⓒ **EXAMPLE 15.7** The Gasoline Additive Case: Maximizing Mileage

TABLE 15.13

The Gasoline Additive Data

ⅅⓢ GasAdd

Additive Units (x)	Gasoline Mileage (y)
0	25.8
0	26.1
0	25.4
1	29.6
1	29.2
1	29.8
2	32.0
2	31.4
2	31.7
3	31.7
3	31.5
3	31.2
4	29.4
4	29.0
4	29.5

An oil company wishes to improve the gasoline mileage obtained by cars that use its regular unleaded gasoline. Company chemists suggest that an additive, ST-3000, be blended with the gasoline. In order to study the effects of this additive, mileage tests are carried out in a laboratory using test equipment that simulates driving under prescribed conditions. The amount of additive ST-3000 blended with the gasoline is varied, and the gasoline mileage for each test run is recorded. Table 15.13 gives the results of the test runs. Here the dependent variable y is gasoline mileage (in miles per gallon) and the independent variable x is the amount of additive ST-3000 used (measured as the number of units of additive added to each gallon of gasoline). One of the study's goals is to determine the number of units of additive that should be blended with the gasoline to maximize gasoline mileage. The company would also like to predict the maximum mileage that can be achieved using additive ST-3000.

Figure 15.23(a) gives a scatter plot of y versus x. Because the scatter plot has the appearance of a quadratic curve (that is, part of a parabola), it seems reasonable to relate y to x by using the quadratic model

$$y = \beta_0 + \beta_1 x + \beta_2 x^2 + \varepsilon$$

Figure 15.23(b) and (c) gives the Minitab output of a regression analysis of the data using this quadratic model. Here the squared term x^2 is denoted as UnitsSq on the output. The Minitab output tells us that the least squares point estimates of the model parameters are $b_0 = 25.715$, $b_1 = 4.976$, and $b_2 = -1.0190$. These estimates give us the least squares prediction equation

$$\hat{y} = 25.715 + 4.976x - 1.0190x^2$$

Intuitively, this is the equation of the best quadratic curve that can be fitted to the data plotted in Figure 15.23(a). The Minitab output also tells us that the p-values related to x and x^2 are less than .001. This implies that we have very strong evidence that each of these model components is significant. The fact that x^2 seems significant confirms the graphical evidence that there is a quadratic relationship between y and x. Once we have such confirmation, we usually retain the linear term x in the model no matter what the size of its p-value. The reason is that geometrical considerations indicate that it is best to use both x and x^2 to model a quadratic relationship.

The oil company wishes to find the value of x that results in the highest predicted mileage. Using calculus, it can be shown that the value $x = 2.44$ maximizes predicted gas mileage.

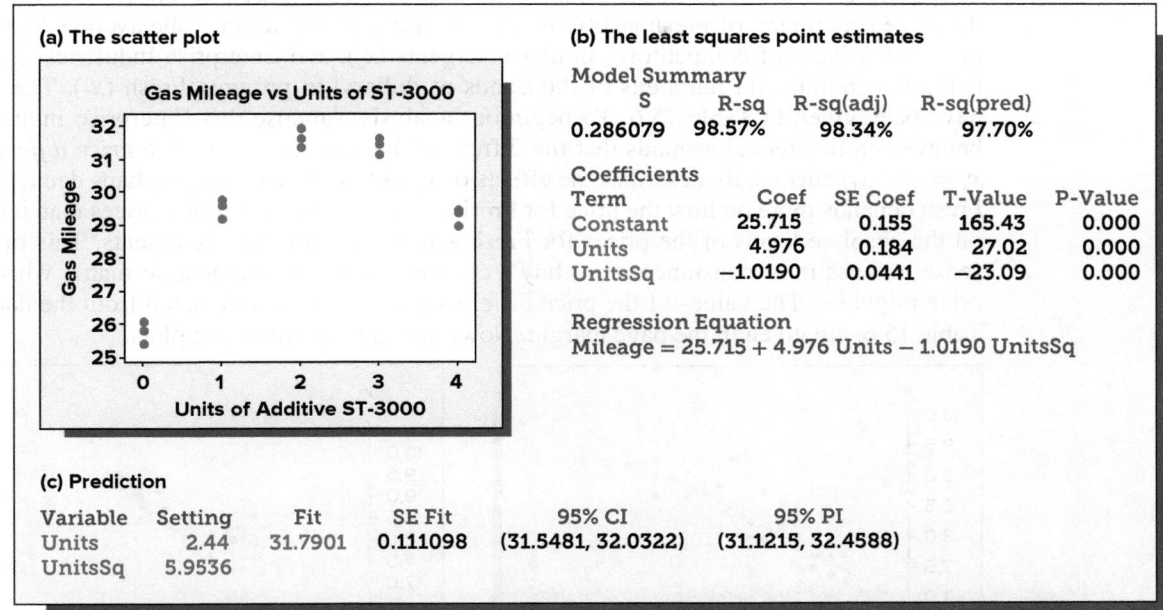

FIGURE 15.23   Minitab Scatter Plot and Quadratic Model Regression Analysis of the Gasoline Additive Data

(a) The scatter plot

Gas Mileage vs Units of ST-3000

(b) The least squares point estimates

Model Summary

S	R-sq	R-sq(adj)	R-sq(pred)
0.286079	98.57%	98.34%	97.70%

Coefficients

Term	Coef	SE Coef	T-Value	P-Value
Constant	25.715	0.155	165.43	0.000
Units	4.976	0.184	27.02	0.000
UnitsSq	−1.0190	0.0441	−23.09	0.000

Regression Equation
Mileage = 25.715 + 4.976 Units − 1.0190 UnitsSq

(c) Prediction

Variable	Setting	Fit	SE Fit	95% CI	95% PI
Units	2.44	31.7901	0.111098	(31.5481, 32.0322)	(31.1215, 32.4588)
UnitsSq	5.9536				

Therefore, the oil company can maximize predicted mileage by blending 2.44 units of additive ST-3000 with each gallon of gasoline. This will result in a predicted gas mileage equal to

$$\hat{y} = 25.715 + 4.976(2.44) - 1.0190(2.44)^2$$
$$= 31.7901 \text{ miles per gallon}$$

This predicted mileage is the point estimate of the mean mileage that would be obtained by all gallons of the gasoline (when blended as just described) and is the point prediction of the mileage that would be obtained by an individual gallon of the gasoline. Note that $\hat{y} = 31.7901$ is given at the bottom of the Minitab output in Figure 15.23(c). In addition, the Minitab output tells us that a 95 percent confidence interval for the mean mileage that would be obtained by all gallons of the gasoline is [31.5481, 32.0322]. If the test equipment simulates driving conditions in a particular automobile, this confidence interval implies that an owner of the automobile can be 95 percent confident that he or she will average between 31.5481 mpg and 32.0322 mpg when using a very large number of gallons of the gasoline. The Minitab output also tells us that a 95 percent prediction interval for the mileage that would be obtained by an individual gallon of the gasoline is [31.1215, 32.4588].

Using interaction variables

Multiple regression models often contain **interaction variables.** We form an interaction variable by multiplying two independent variables together. For instance, if a regression model includes the independent variables x_1 and x_2, then we can form the interaction variable x_1x_2. It is appropriate to employ an interaction variable if the relationship between the dependent variable y and one of the independent variables depends upon the value of the other independent variable. In the following example we consider a multiple regression model that uses a linear variable, a squared variable, and an interaction variable.

 EXAMPLE 15.8 The Fresh Detergent Case: Predicting Demand

Enterprise Industries produces Fresh, a brand of liquid laundry detergent. In order to manage its inventory more effectively and make revenue projections, the company would like to

Ⓓ Fresh2

Price Difference,
$x_4 = x_2 - x_1$
(Dollars)

−.05
.25
.60
0
.25
.20
.15
.05
−.15
.15
.20
.10
.40
.45
.35
.30
.50
.50
.40
−.05
−.05
−.10
.20
.10
.50
.60
−.05
0
.05
.55

better predict demand for Fresh. To develop a prediction model, the company has gathered data concerning demand for Fresh over the last 30 sales periods (each sales period is defined to be a four-week period). For each sales period, these data consist of recorded values of the demands (in hundreds of thousands of bottles) for the large size bottle of Fresh (y), the price (in dollars) of Fresh as offered by Enterprise Industries (x_1), the average industry price (in dollars) of competitors' similar detergents (x_2), and Enterprise Industries' advertising expenditure (in hundreds of thousands of dollars) to promote Fresh (x_3). The data have been given in Table 15.6. To begin our analysis, suppose that Enterprise Industries believes on theoretical grounds that the difference between x_1 and x_2 (the *price difference* $x_4 = x_2 - x_1$) adequately describes the effects of x_1 and x_2 on y. That is, perhaps demand for Fresh depends more on how the price for Fresh compares to competitors' prices than it does on the absolute levels of the prices for Fresh and other competing detergents. This makes sense because most consumers must buy a certain amount of detergent no matter what the price might be. The values of the price difference $x_4 = x_2 - x_1$, calculated from the data in Table 15.6, are given in the page margin. Now, consider the following plots:

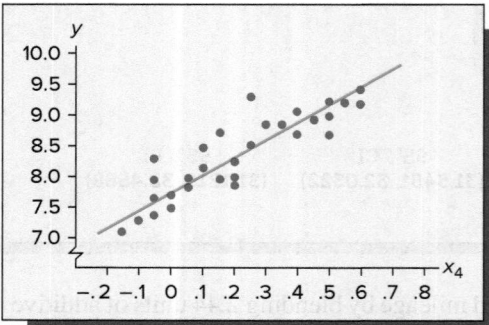 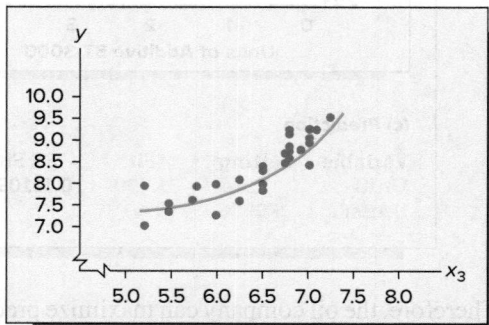

Because the plot on the left shows a linear relationship between y and x_4, we should use x_4 to predict y. Because the plot on the right shows a quadratic relationship between y and x_3, we should use x_3 and x_3^2 to predict y. Moreover, if x_4 and x_3 interact, then we should use the interaction variable $x_4 x_3$ to predict y. This gives the model

$$y = \beta_0 + \beta_1 x_4 + \beta_2 x_3 + \beta_3 x_3^2 + \beta_4 x_4 x_3 + \varepsilon.$$

Figure 15.24 presents the JMP output obtained when this model is fit to the Fresh demand data. The p-values for testing the significance of the intercept and the independent variables are all below .05. Therefore, we have strong evidence that each of these terms should be included in the model. In particular, because the p-value related to $x_4 x_3$ is .0361, we have strong evidence that x_4 and x_3 interact. We will examine the nature of this interaction in the discussion to come.

Suppose that Enterprise Industries wishes to predict demand for Fresh in a future sales period when the price difference will be $.20 (that is, 20 cents) and when advertising expenditure will be $650,000. Using the least squares point estimates in Figure 15.24(a), the point prediction is

$$\hat{y} = 29.1133 + 11.1342(.20) - 7.6080(6.50) + 0.6712(6.50)^2 - 1.4777(.20)(6.50)$$
$$= 8.32725 \ (832,725 \text{ bottles}).$$

Figure 15.24(b) gives this point prediction along with the 95 percent confidence interval for mean demand and the 95 percent prediction interval for an individual demand when x_4 equals 0.20 and x_3 equals 6.50.

To investigate the nature of the interaction between x_3 and x_4, consider the prediction equation

$$\hat{y} = 29.1133 + 11.1342 x_4 - 7.6080 x_3 + 0.6712 x_3^2 - 1.4777 x_4 x_3$$

obtained from the least squares point estimates in Figure 15.24(a). Also, consider the six combinations of price difference x_4 and advertising expenditure x_3 obtained by combining the x_4 values .10 and .30 with the x_3 values 6.0, 6.4, and 6.8. When we use the prediction equation to predict the demands for Fresh corresponding to these six combinations, we obtain the predicted demands ($\hat{y}$ values) shown in Figure 15.25(a). (Note that we consider

FIGURE 15.24 **JMP Output of a Regression Analysis of the Fresh Demand Data by Using the Interaction Model** $y = \beta_0 + \beta_1 x_4 + \beta_2 x_3 + \beta_3 x_3^2 + \beta_4 x_4 x_3 + \varepsilon$

Summary of Fit

RSquare	0.920913
RSquare Adj	0.908259
Root Mean Square Error	0.206339
Mean of Response	8.382667
Observations (or Sum Wgts)	30

Analysis of Variance

Source	DF	Sum of Squares	Mean Square	F Ratio
Model	4	12.394190	3.09855	72.7771
Error	25	1.064397	0.04258	Prob > F
C. Total	29	13.458587		<.0001*

Parameter Estimates

Term	Estimate	Std Error	t Ratio	Prob>ltl	Lower 95%	Upper 95%
Intercept	29.113287	7.483206	3.89	0.0007*	13.701335	44.525238
PriceDif(X4)	11.134226	4.445854	2.50	0.0192*	1.9778181	20.290634
AdvExp(X3)	−7.608007	2.469109	−3.08	0.0050*	−12.69323	−2.522782
AdvExp(X3)*AdvExp(X3)	0.6712472	0.202701	3.31	0.0028*	0.253777	1.0887173
PriceDif(X4)*AdvExp(X3)	−1.477717	0.667165	−2.21	0.0361*	−2.851768	−0.103666

	Predicted Demand	Lower 95% Mean Demand	Upper 95% Mean Demand	Lower 95% Indiv Demand	Upper 95% Indiv Demand
31	8.3272497715	8.2112139058	8.4432856372	7.8867292725	8.7677702705

two x_4 values because there is a *linear* relationship between y and x_4, and we consider *three* x_3 values because there is a *quadratic* relationship between y and x_3.) Now

1   If we fix x_3 at 6.0 in Figure 15.25(a) and plot the corresponding $\hat{y}$ values 7.86 and 8.31 versus the x_4 values .10 and .30, we obtain the two squares connected by the lowest line in Figure 15.25(b). Similarly, if we fix x_3 at 6.4 and plot the corresponding $\hat{y}$ values 8.08 and 8.42 versus the x_4 values .10 and .30, we obtain the two squares connected by the middle line in Figure 15.25(b). Also, if we fix x_3 at 6.8 and plot the corresponding $\hat{y}$ values 8.52 and 8.74 versus the x_4 values .10 and .30, we obtain the two squares connected by the highest line in Figure 15.25(b). Examining the three lines relating $\hat{y}$ to x_4, we see that the slopes of these lines decrease as x_3 increases from 6.0 to 6.4 to 6.8. This says that as the price difference x_4 increases from .10 to .30 (that is, as Fresh becomes less expensive compared to its competitors), the *rate of increase* of predicted demand $\hat{y}$ is slower when advertising expenditure x_3 is higher than when advertising expenditure x_3 is lower. Moreover, this might be logical because it says that when a higher advertising expenditure makes more customers aware of Fresh's cleaning abilities and thus causes customer demand for Fresh to be higher, there is less opportunity for an increased price difference to increase demand for Fresh.

FIGURE 15.25 **Interaction between x_4 and x_3**

(a) Predicted demands ($\hat{y}$ values)

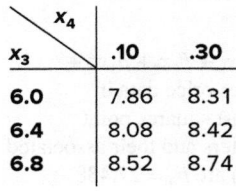

(b) Plots of $\hat{y}$ versus x_4 for different x_3 values

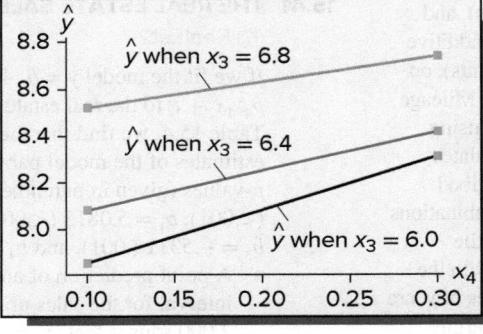

(c) Plots of $\hat{y}$ versus x_3 for different x_4 values

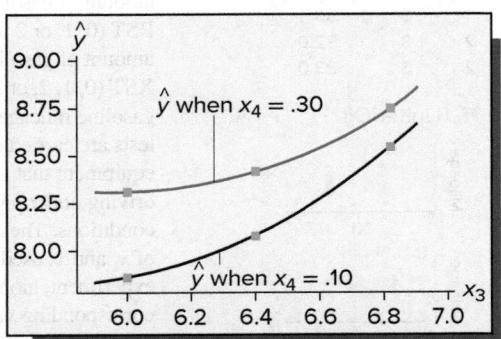

2   If we fix x_4 at .10 in Figure 15.25(a) and plot the corresponding $\hat{y}$ values 7.86, 8.08, and 8.52 versus the x_3 values 6.0, 6.4, and 6.8, we obtain the three squares connected by the lower quadratic curve in Figure 15.25(c). Similarly, if we fix x_4 at .30 and plot the corresponding $\hat{y}$ values 8.31, 8.42, and 8.74 versus the x_3 values 6.0, 6.4, and 6.8, we obtain the three squares connected by the higher quadratic curve in Figure 15.25(c). The nonparallel quadratic curves in Figure 15.25(c) say that as advertising expenditure x_3 increases from 6.0 to 6.8, the rate of increase of predicted demand $\hat{y}$ is slower when the price difference x_4 is larger (that is, $x_4 = .30$) than when the price difference x_4 is smaller (that is, $x_4 = .10$). Moreover, this might be logical because it says that when a larger price difference causes customer demand for Fresh to be higher, there is less opportunity for an increased advertising expenditure to increase demand for Fresh.

To summarize the nature of the interaction between x_4 and x_3, we might say that a higher value of each of these independent variables somewhat weakens the impact of the other independent variable on predicted demand. In Exercise 15.41 we will consider a situation where a higher value of each of two independent variables somewhat strengthens the impact of the other independent variable on the predicted value of the dependent variable. Moreover, if the p-value related to x_4x_3 in the Fresh detergent situation had been large and thus we had removed x_4x_3 from the model (that is, *no interaction*), then the plotted lines in Figure 15.25(b) would have been *parallel* and the plotted quadratic curves in Figure 15.25(c) would have been *parallel*. This would say that as each independent variable increases, predicted demand increases at the same rate whether the other independent variable is larger or smaller.

Exercises for Section 15.9

CONCEPTS ▦ connect

RST Units (x_1)	XST Units (x_2)	Gas Mileage (y, mpg)
0	0	27.4
0	0	28.0
0	0	28.6
1	0	29.6
1	0	30.6
2	0	28.6
2	0	29.8
0	1	32.0
0	1	33.0
1	1	33.3
1	1	34.5
0	2	32.3
0	2	33.5
1	2	34.4
1	2	35.0
1	2	35.6
2	2	33.3
2	2	34.0
2	2	34.7
1	3	33.4
2	3	32.0
2	3	33.0

Ⓓ UnitedOil

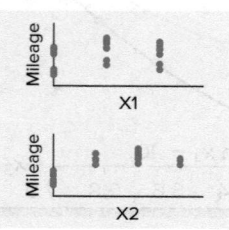

15.38 When does a scatter plot suggest the use of the quadratic regression model?

15.39 How do we model the interaction between two independent variables?

METHODS AND APPLICATIONS

15.40 United Oil Company is attempting to develop a regular grade gasoline that will deliver higher gasoline mileages than can be achieved by its current regular grade gasoline. As part of its development process, United Oil wishes to study the effect of two independent variables—x_1, amount of gasoline additive RST (0, 1, or 2 units), and x_2, amount of gasoline additive XST (0, 1, 2, or 3 units), on gasoline mileage, y. Mileage tests are carried out using equipment that simulates driving under prescribed conditions. The combinations of x_1 and x_2 used in the experiment, along with the corresponding values of y, are given in the page margin.

a   Discuss why the data plots given in the page margin indicate that the model $y = \beta_0 + \beta_1x_1 + \beta_2x_1^2 + \beta_3x_2 + \beta_4x_2^2 + \varepsilon$ might appropriately relate y to x_1 and x_2. If we fit this model to the data in the page margin, we find that the least squares point estimates of the model parameters and their associated p-values (given in parentheses) are $b_0 = 28.1589$ (<.001), $b_1 = 3.3133$ (<.001), $b_2 = -1.4111$ (<.001), $b_3 = 5.2752$ (<.001), and $b_4 = -1.3964$ (<.001). Moreover, consider the mean mileage obtained by all gallons of the gasoline when it is made with one unit of RST and two units of XST (a combination that the data in the page margin indicates would maximize mean mileage). A point estimate of and a 95 percent confidence interval for this mean mileage are 35.0261 and [34.4997, 35.5525]. Using the above model, show how the point estimate is calculated.

b   If we add the independent variable x_1x_2 to the model in part *a*, we find that the p-value related to x_1x_2 is .9777. What does this imply?   Ⓓ UnitedOil

15.41 THE REAL ESTATE SALES PRICE CASE
   Ⓓ RealEst2

If we fit the model $y = \beta_0 + \beta_1x_1 + \beta_2x_2 + \beta_3x_2^2 + \beta_4x_1x_2 + \varepsilon$ to the real estate sales price data in Table 15.4, we find that the least squares point estimates of the model parameters and their associated p-values (given in parentheses) are $b_0 = 27.438$ (<.001), $b_1 = 5.0813$ (<.001), $b_2 = 7.2899$ (<.001), $b_3 = -.5311$ (.001), and $b_4 = .11473$ (.014).

a   A point prediction of and a 95 percent prediction interval for the sales price of a house having 2,000 square feet ($x_1 = 20$) and a niceness rating

Predicted Sales Prices

x_2 \ x_1	13	22
2	108.933	156.730
5	124.124	175.019
8	129.756	183.748

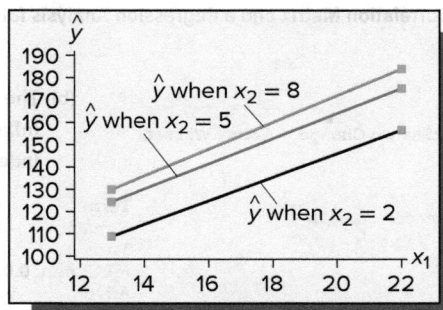

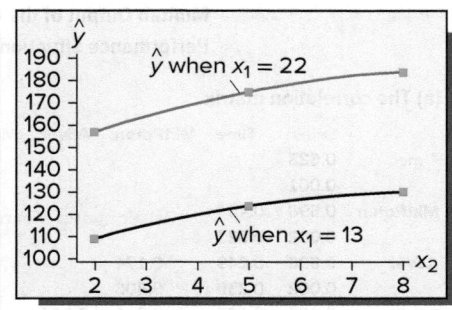

of 8 ($x_2 = 8$) are 171.751 ($171,751) and [168.836, 174.665]. Using the above model, show how the point prediction is calculated.

b Above we give model predictions of sales prices of houses for six combinations of x_1 and x_2, along with plots of the predictions needed to interpret the interaction between x_1 and x_2. Carefully interpret this interaction.

15.42 THE BONNER FROZEN FOODS ADVERTISING CASE DS Bonner

Use Excel, JMP, or Minitab to perform a regression analysis of the Bonner Frozen Foods sales volume data in Table 15.5 by using the model $y = \beta_0 + \beta_1 x_1 + \beta_2 x_2 + \beta_3 x_1 x_2 + \varepsilon$.

a Determine if the interaction term $x_1 x_2$ is important in the model.

b Find **(1)** a point estimate of and a 95 percent confidence interval for the mean sales volume in all sales regions when $2,000 is spent on radio and television advertising and $5,000 is spent on print advertising and **(2)** a point prediction of and a 95 percent prediction interval for the sales volume in an individual sales region when $2,000 is spent on radio and television advertising and $5,000 is spent on print advertising.

15.43 United Oil Company is also attempting to develop a premium grade gasoline that will deliver higher gasoline mileages than can be achieved by its current premium grade gasoline. As part of its development process, United Oil wishes to study the effect of two independent variables—x_1, amount of gasoline additive

WST (0, 1, or 2 units), and x_2, amount of gasoline additive YST (0, 1, 2, or 3 units), on gasoline mileage, y. Mileage tests are carried out using equipment that simulates driving under prescribed conditions. The combinations of x_1 and x_2 used in the experiment, along with the corresponding values of y, are given in the page margin.

a Use Excel, JMP, or Minitab to perform a regression analysis of the data in the page margin by using the model $y = \beta_0 + \beta_1 x_1 + \beta_2 x_2 + \beta_3 x_1^2 + \beta_4 x_2^2 + \beta_5 x_1 x_2 + \beta_6 x_1^2 x_2^2 + \varepsilon$. Which terms appear to be important in the model?

b Find a point estimate of and a 95 percent confidence interval for the mean mileage obtained by all gallons of the gasoline when it is made with one unit of WST and one unit of YST (a combination that the data in the page margin indicates would maximize mean mileage).
DS UnitedOil2

WST Units (x_1)	YST Units (x_2)	Gas Mileage (y, mpg)
0	0	28.0
0	0	28.6
0	0	27.4
1	0	33.3
1	0	34.5
0	1	33.0
0	1	32.0
1	1	35.6
1	1	34.4
1	1	35.0
2	1	34.0
2	1	33.3
2	1	34.7
0	2	33.5
0	2	32.3
1	2	33.4
2	2	33.0
2	2	32.0
1	3	29.6
1	3	30.6
2	3	28.6
2	3	29.8

DS UnitedOil2

15.10 Multicollinearity, Model Building, and Model Validation (Optional)

LO15-9
Describe multicollinearity and build and validate a multiple regression model. (Optional)

Multicollinearity

Recall the sales representative performance data in Table 15.8. These data consist of values of the dependent variable y (SALES) and of the independent variables x_1 (TIME), x_2 (MKTPOTEN), x_3 (ADVER), x_4 (MKTSHARE), and x_5 (CHANGE). The complete sales representative performance data analyzed by Cravens, Woodruff, and Stomper (1972) consist of the data presented in Table 15.8 and data concerning three additional independent variables. These three additional variables are x_6 = number of accounts handled by the representative (ACCTS); x_7 = average workload per account, measured by using a weighting based

FIGURE 15.26   **Minitab Output of the Correlation Matrix and a Regression Analysis for the Sales Representative Performance Situation**

(a) The correlation matrix

	Sales	Time	MktPoten	Adver	MktShare	Change	Accts	WkLoad
Time	0.623							
	0.001							
MktPoten	0.598	0.454						
	0.002	0.023						
Adver	0.596	0.249	0.174					
	0.002	0.230	0.405					
MktShare	0.484	0.106	−0.211	0.264				
	0.014	0.613	0.312	0.201				
Change	0.489	0.251	0.268	0.377	0.085			
	0.013	0.225	0.195	0.064	0.685			
Accts	0.754	0.758	0.479	0.200	0.403	0.327		
	0.000	0.000	0.016	0.338	0.046	0.110		
WkLoad	−0.117	−0.179	−0.259	−0.272	0.349	−0.288	−0.199	
	0.577	0.391	0.212	0.188	0.087	0.163	0.341	
Rating	0.402	0.101	0.359	0.411	−0.024	0.549	0.229	−0.277
	0.046	0.631	0.078	0.041	0.911	0.004	0.272	0.180

Cell contents: Pearson correlation
                P-Value

(b) The t statistics, p-values, and variance inflation factors for the eight-independent-variables model

Term	Coef	SE Coef	T-Value	P-Value	VIF
Constant	−1508	779	−1.94	0.071	
Time	2.01	1.93	1.04	0.313	3.34
MktPoten	0.03720	0.00820	4.54	0.000	1.98
Adver	0.1510	0.0471	3.21	0.006	1.91
MktShare	199.0	67.0	2.97	0.009	3.24
Change	291	187	1.56	0.139	1.60
Accts	5.55	4.78	1.16	0.262	5.64
WkLoad	19.8	33.7	0.59	0.565	1.82
Rating	8	129	0.06	0.950	1.81

TABLE 15.14

Values of ACCTS, WKLOAD, and RATING

DS SalePerf2

Accounts x_6	Work-load x_7	Rating x_8
74.86	15.05	4.9
107.32	19.97	5.1
96.75	17.34	2.9
195.12	13.40	3.4
180.44	17.64	4.6
104.88	16.22	4.5
256.10	18.80	4.6
126.83	19.86	2.3
203.25	17.42	4.9
119.51	21.41	2.8
116.26	16.32	3.1
142.28	14.51	4.2
89.43	19.35	4.3
84.55	20.02	4.2
119.51	15.26	5.5
80.49	15.87	3.6
136.58	7.81	3.4
78.86	16.00	4.2
136.58	17.44	3.6
138.21	17.98	3.1
75.61	20.99	1.6
102.44	21.66	3.4
76.42	21.46	2.7
136.58	24.78	2.8
88.62	24.96	3.9

on the sizes of the orders by the accounts and other workload-related criteria (WKLOAD); and $x_8 =$ an aggregate rating on eight dimensions of the representative's performance, made by a sales manager and expressed on a 1–7 scale (RATING).

Table 15.14 gives the observed values of x_6, x_7, and x_8, and Figure 15.26(a) presents the Minitab output of a **correlation matrix** for the sales representative performance data. Examining the first column of this matrix, we see that the simple correlation coefficient between SALES and WKLOAD is −.117 and that the p-value for testing the significance of the relationship between SALES and WKLOAD is .577. This indicates that there is little or no relationship between SALES and WKLOAD. However, the simple correlation coefficients between SALES and the other seven independent variables range from .402 to .754, with associated p-values ranging from .046 to .000. This indicates the existence of potentially useful relationships between SALES and these seven independent variables.

While simple correlation coefficients (and scatter plots) give us a preliminary understanding of the data, they cannot be relied upon alone to tell us which independent variables are significantly related to the dependent variable. One reason for this is a condition called *multicollinearity*. **Multicollinearity** is said to exist among the independent variables in a regression situation if these independent variables are related to or dependent upon each other. One way to investigate multicollinearity is to examine the correlation matrix. To understand this, note that all of the simple correlation coefficients not located in the first column of this matrix measure the **simple correlations between the independent variables.** For example, the simple correlation coefficient between ACCTS and TIME is .758, which says that the ACCTS values increase as the TIME values increase. Such a relationship makes sense because it is logical that the longer a sales representative has been with the company, the more accounts he or she handles. Statisticians often regard multicollinearity in a data set to be severe if at least one simple correlation coefficient between the independent variables is at least .9. Because the largest such simple correlation coefficient in Figure 15.26(a) is .758, this is not true for the sales representative performance data. Note, however, that even moderate multicollinearity can be a potential problem. This will be demonstrated later using the sales representative performance data.

Another way to measure multicollinearity is to use **variance inflation factors.** Consider a regression model relating a dependent variable y to a set of independent variables $x_1, \ldots,$ $x_{j-1}, x_j, x_{j+1}, \ldots, x_k$. The **variance inflation factor for the independent variable** x_j in this set is denoted VIF_j and is defined by the equation

$$VIF_j = \frac{1}{1 - R_j^2}$$

where R_j^2 is the multiple coefficient of determination for the regression model that relates x_j to all the other independent variables $x_1, \ldots, x_{j-1}, x_{j+1}, \ldots, x_k$ in the set. For example,

Figure 15.26(b) gives the Minitab output of the t statistics, p-values, and variance inflation factors for the sales representative performance model that relates y to all eight independent variables. The largest variance inflation factor is $VIF_6 = 5.64$. To calculate VIF_6, Minitab first calculates the multiple coefficient of determination for the regression model that relates x_6 to x_1, x_2, x_3, x_4, x_5, x_7, and x_8 to be $R_6^2 = .822673$. It then follows that

$$VIF_6 = \frac{1}{1 - R_6^2} = \frac{1}{1 - .822673} = 5.64$$

In general, if $R_j^2 = 0$, which says that x_j is not related to the other independent variables, then the variance inflation factor VIF_j equals 1. On the other hand, if $R_j^2 > 0$, which says that x_j is related to the other independent variables, then $(1 - R_j^2)$ is less than 1, making VIF_j greater than 1. Generally, the multicollinearity between independent variables is considered **(1)** severe if the largest variance inflation factor is greater than 10 and **(2)** moderately strong if the largest variance inflation factor is greater than 5. Moreover, if the mean of the variance inflation factors is substantially greater than 1 (sometimes a difficult criterion to assess), multicollinearity might be problematic. In the sales representative performance model, the largest variance inflation factor, $VIF_6 = 5.64$, is greater than 5. Therefore, we might classify the multicollinearity as being moderately strong.

The Picket Fence Display

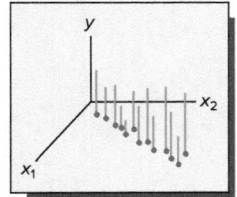

The reason that VIF_j is called the variance inflation factor is that it can be shown that, when VIF_j is greater than 1, then the standard deviation σ_{b_j} of the population of all possible values of the least squares point estimate b_j is likely to be inflated beyond its value when $R_j^2 = 0$. If σ_{b_j} is greatly inflated, two slightly different samples of values of the dependent variable can yield two substantially different values of b_j. To intuitively understand why strong multicollinearity can significantly affect the least squares point estimates, consider the so-called "picket fence" display in the page margin. This figure depicts two independent variables (x_1 and x_2) exhibiting strong multicollinearity (note that as x_1 increases, x_2 increases). The heights of the pickets on the fence represent the y observations. If we assume that the model $y = \beta_0 + \beta_1 x_1 + \beta_2 x_2 + \varepsilon$ adequately describes this data, then calculating the least squares point estimates amounts to fitting a plane to the points on the top of the picket fence. Clearly, this plane would be quite unstable. That is, a slightly different height of one of the pickets (a slightly different y value) could cause the slant of the fitted plane (and the least squares point estimates that determine this slant) to change radically. It follows that, when strong multicollinearity exists, sampling variation can result in least squares point estimates that differ substantially from the true values of the regression parameters. In fact, some of the least squares point estimates may have a sign (positive or negative) that differs from the sign of the true value of the parameter (we will see an example of this in the exercises). Therefore, when strong multicollinearity exists, it is dangerous to interpret the individual least squares point estimates.

The most important problem caused by multicollinearity is that, even when multicollinearity is not severe, it can hinder our ability to use the t statistics and related p-values to assess the importance of the independent variables. Recall that we can reject $H_0: \beta_j = 0$ in favor of $H_a: \beta_j \neq 0$ at level of significance α if and only if the absolute value of the corresponding t statistic is greater than $t_{\alpha/2}$ based on $n - (k + 1)$ degrees of freedom, or, equivalently, if and only if the related p-value is less than α. Thus the larger (in absolute value) the t statistic is and the smaller the p-value is, the stronger is the evidence that we should reject $H_0: \beta_j = 0$ and the stronger is the evidence that the independent variable x_j is significant. When multicollinearity exists, the sizes of the t statistic and of the related p-value **measure the additional importance of the independent variable x_j over the combined importance of the other independent variables in the regression model.** Because two or more correlated independent variables contribute redundant information, multicollinearity often causes the t statistics obtained by relating a dependent variable to a set of correlated independent variables to be smaller (in absolute value) than the t statistics that would be obtained if separate regression analyses were run, where each separate regression analysis relates the dependent variable to a smaller set (for example, only one) of the correlated independent variables. Thus multicollinearity can cause some of the correlated independent variables to appear less important—in terms of having small absolute t statistics and large p-values—than they really are. Another way to understand this is to note that because multicollinearity inflates σ_{b_j}, it inflates the point estimate s_{b_j} of σ_{b_j}. Because $t = b_j/s_{b_j}$, an inflated value of s_{b_j} can (depending on the size of b_j) cause t to be small (and the related p-value to be large). This would suggest that x_j is not significant even though x_j may really be important.

For example, Figure 15.26(b) tells us that when we perform a regression analysis of the sales representative performance data using a model that relates y to all eight independent variables, the p-values related to TIME, MKTPOTEN, ADVER, MKTSHARE, CHANGE, ACCTS, WKLOAD, and RATING are, respectively, .313, .000, .006, .009, .139, .262, .565, and .950. By contrast, recall from Figure 15.10 that when we perform a regression analysis of the sales representative performance data using a model that relates y to the first five independent variables, the p-values related to TIME, MKTPOTEN, ADVER, MKTSHARE, and CHANGE are, respectively, .0065, .0000, .0025, .0000, and .0530. Note that TIME (p-value = .0065) is significant at the .01 level and CHANGE (p-value = .0530) is significant at the .06 level in the five independent variable model. However, when we consider the model that uses all eight independent variables, neither TIME (p-value = .313) nor CHANGE (p-value = .139) is significant at the .10 level. The reason that TIME and CHANGE seem more significant in the five independent variable model is that, because this model uses fewer variables, TIME and CHANGE contribute less overlapping information and thus have more additional importance in this model.

Comparing regression models on the basis of R^2, s, adjusted R^2, prediction interval length, and the C statistic

We have seen that when multicollinearity exists in a model, the p-value associated with an independent variable in the model measures the additional importance of the variable over the combined importance of the other variables in the model. Therefore, it can be difficult to use the p-values to determine which variables to retain and which variables to remove from a model. The implication of this is that we need to evaluate more than the *additional importance* of each independent variable in a regression model. We also need to evaluate how well the independent variables *work together* to accurately describe, predict, and control the dependent variable. One way to do this is to determine if the *overall* model gives a high R^2 and $\overline{R}^2$, a small s, and short prediction intervals.

It can be proven that **adding any independent variable to a regression model, even an unimportant independent variable, will decrease the unexplained variation and will increase the explained variation.** Therefore, because the total variation $\Sigma(y_i - \overline{y})^2$ depends only on the observed y values and thus remains unchanged when we add an independent variable to a regression model, it follows that **adding any independent variable to a regression model will increase R^2,** which equals the explained variation divided by the total variation. This implies that R^2 cannot tell us (by decreasing) that adding an independent variable is undesirable. That is, although we wish to obtain a model with a large R^2, there are better criteria than R^2 that can be used to *compare* regression models.

One better criterion is the standard error $s = \sqrt{SSE/[n - (k + 1)]}$. When we add an independent variable to a regression model, the number of model parameters $(k + 1)$ increases by one, and thus the number of degrees of freedom $n - (k + 1)$ decreases by one. If the decrease in $n - (k + 1)$, which is used in the denominator to calculate s, is proportionally more than the decrease in SSE (the unexplained variation) that is caused by adding the independent variable to the model, then s will increase. **If s increases, this tells us that we should not add the independent variable to the model,** because the new model would give longer prediction intervals and thus less accurate predictions. If s decreases, the new model is likely to give shorter prediction intervals but (as we will see) is not guaranteed to do so. Thus, it can be useful to compare the lengths of prediction intervals for different models. Also, it can be shown that the standard error s decreases if and only if $\overline{R}^2$ (adjusted R^2) increases. It follows that, if we are comparing regression models, the model that gives the smallest s gives the largest $\overline{R}^2$.

Ⓒ EXAMPLE 15.9 The Sales Representative Case: Model Comparisons

Figure 15.27 gives the Minitab output resulting from calculating R^2, $\overline{R}^2$, and s for **all possible regression models** based on all possible combinations of the eight independent variables in the sales representative performance situation (the values of Cp on the output will be

FIGURE 15.27   **Minitab Output of the Best Two Sales Representative Performance Regression Models of Each Size**

Vars	R-Sq	R-Sq (adj)	R-Sq (pred)	Mallows Cp	S	Time	MktPoten	Adver	MktShare	Change	Accts	WLoad	Rating
1	56.8	55.0	43.3	67.6	881.09						X		
1	38.8	36.1	25.0	104.6	1049.3	X							
2	77.5	75.5	70.0	27.2	650.39			X			X		
2	74.6	72.3	65.7	33.1	691.11	X	X						
3	84.9	82.7	79.2	14.0	545.51	X	X	X					
3	82.8	80.3	76.4	18.4	582.64	X	X				X		
4	90.0	88.1	86.0	5.4	453.84	X	X	X			X		
4	89.6	87.5	84.6	6.4	463.95	X	X	X	X				
5	91.5	89.3	86.9	4.4	430.23	X	X	X	X	X			
5	91.2	88.9	87.1	5.0	436.75	X	X	X	X		X		
6	92.0	89.4	85.4	5.4	428.00	X	X	X	X	X	X		
6	91.6	88.9	86.8	6.1	438.20		X	X	X	X	X	X	
7	92.2	89.0	84.1	7.0	435.67	X	X	X	X	X	X	X	
7	92.0	88.8	83.5	7.3	440.30	X	X	X	X	X	X		X
8	92.2	88.3	81.8	9.0	449.03	X	X	X	X	X	X	X	X

explained after we complete this example). The Minitab output gives the two best models of each size in terms of s and $\overline{R}^2$—the two best one-variable models, the two best two-variable models, the two best three-variable models, and so on. Examining the output, we see that the three models having the smallest values of s and largest values of $\overline{R}^2$ are

1   The six-variable model that contains

TIME, MKTPOTEN, ADVER, MKTSHARE, CHANGE, ACCTS

and has $s = 428.00$ and $\overline{R}^2 = 89.4$; we refer to this model as Model 1.

2   The five-variable model that contains

TIME, MKTPOTEN, ADVER, MKTSHARE, CHANGE

and has $s = 430.23$ and $\overline{R}^2 = 89.3$; we refer to this model as Model 2.

3   The seven-variable model that contains

TIME, MKTPOTEN, ADVER, MKTSHARE, CHANGE, ACCTS, WKLOAD

and has $s = 435.67$ and $\overline{R}^2 = 89.0$; we refer to this model as Model 3.

To see that s can increase when we add an independent variable to a regression model, note that s increases from 428.00 to 435.67 when we add WKLOAD to Model 1 to form Model 3. In this case, although it can be verified that adding WKLOAD decreases the unexplained variation from 3,297,279.3342 to 3,226,756.2751, this decrease has not been enough to offset the change in the denominator of $s^2 = SSE/[n - (k + 1)]$, which decreases from $25 - 7 = 18$ to $25 - 8 = 17$. To see that prediction interval lengths might increase even though s decreases, consider adding ACCTS to Model 2 to form Model 1. This decreases s from 430.23 to 428.00. However, consider a questionable sales representative for whom TIME = 85.42, MKTPOTEN = 35,182.73, ADVER = 7,281.65, MKTSHARE = 9.64, CHANGE = .28, and ACCTS = 120.61. The 95 percent prediction interval given by Model 2 for sales corresponding to this combination of values of the independent variables is [3,233.59, 5,129.89] and has length $5,129.89 - 3,233.59 = 1896.3$. The 95 percent prediction interval given by Model 1 for such sales is [3,193.86, 5,093.14] and has length $5,093.14 - 3,193.86 = 1,899.28$. In other words, even though Model 2 has a slightly larger s, it gives a slightly shorter prediction interval. (For those who have studied the formula for a prediction interval, $[\hat{y} \pm t_{\alpha/2} s \sqrt{1 + \text{distance value}}]$, Model 2 gives a slightly shorter 95 percent prediction interval because it uses one less variable and thus can be verified to give slightly smaller values of $t_{.025}$ and the distance value.) In addition, the extra independent variable ACCTS in Model 1 can be verified to have a p-value of .2881. Therefore, we conclude that Model 2 is better than Model 1 and is, in fact, the "best" sales representative performance model (using only linear independent variables).

Another quantity that can be used for comparing regression models is called the **C statistic** (also often called the **C_p statistic**). To show how to calculate the C statistic, suppose that we wish to choose an appropriate set of independent variables from p potential independent variables. We first calculate the mean square error, which we denote as s_p^2, for the model using all p potential independent variables. Then, if SSE denotes the unexplained variation for another particular model that has k independent variables, it follows that the C statistic for this model is

$$C = \frac{SSE}{s_p^2} - [n - 2(k + 1)]$$

For example, consider the sales representative case. It can be verified that the mean square error for the model using all $p = 8$ independent variables is 201,621.21 and that the SSE for the model using the first $k = 5$ independent variables (Model 2 in the previous example) is 3,516,812.7933. It follows that the C statistic for this latter model is

$$C = \frac{3,516,812.7933}{201,621.21} - [25 - 2(5 + 1)] = 4.4$$

Because the C statistic for a given model is a function of the model's SSE, and because we want SSE to be small, **we want C to be small.** Although adding an unimportant independent variable to a regression model will decrease SSE, adding such a variable can increase C. This can happen when the decrease in SSE caused by the addition of the extra independent variable is not enough to offset the decrease in $n - 2(k + 1)$ caused by the addition of the extra independent variable (which increases k by 1). It should be noted that although adding an unimportant independent variable to a regression model can increase both s^2 and C, there is no exact relationship between s^2 and C.

While we want C to be small, it can be shown from the theory behind the C statistic that **we also wish to find a model for which the C statistic roughly equals $k + 1$,** the number of parameters in the model. **If a model has a C statistic substantially greater than $k + 1$, it can be shown that this model has substantial *bias* and is undesirable.** Thus, although we want to find a model for which C is as small as possible, if C for such a model is substantially greater than $k + 1$, we may prefer to choose a different model for which C is slightly larger and more nearly equal to the number of parameters in that (different) model. **If a particular model has a small value of C and C for this model is less than $k + 1$, then the model should be considered desirable.** Finally, it should be noted that for the model that includes all p potential independent variables (and thus utilizes $p + 1$ parameters), it can be shown that $C = p + 1$.

If we examine Figure 15.27, we see that Model 2 of the previous example has the smallest C statistic. The C statistic for this model (denoted Cp on the Minitab output) equals 4.4. Because $C = 4.4$ is less than $k + 1 = 6$, the model is not biased. Therefore, this model should be considered best with respect to the C statistic. (Note that we will discuss the meaning of "R-sq(pred)" in Figure 15.27 later in this section.)

Stepwise regression and backward elimination

In some situations it is useful to employ an **iterative model selection procedure,** where at each step a single independent variable is added to or deleted from a regression model, and a new regression model is evaluated. We begin by discussing one such procedure—**stepwise regression.**

Stepwise regression begins by considering all of the one-independent-variable models and choosing the model for which the p-value related to the independent variable in the model is the smallest. If this p-value is less than α_{entry}, an α value for "entering" a variable, the independent variable is the first variable entered into the stepwise regression model and stepwise regression continues. Stepwise regression then considers the remaining independent variables not in the stepwise model and chooses the independent variable which, when paired with the first independent variable entered, has the smallest p-value. If this p-value is less than α_{entry}, the new variable is entered into the stepwise model. Moreover, the stepwise procedure checks to see if the p-value related to the first variable entered into the stepwise model is less than α_{stay}, an α value for allowing a variable to stay in the stepwise model. This is done because multicollinearity could have changed the p-value of the first variable entered into the stepwise model. The stepwise procedure continues this process and concludes when no new independent variable can be entered into the stepwise model. It is common practice to set both α_{entry} and α_{stay} equal to .05 or .10.

For example, again consider the sales representative performance data. We let x_1, x_2, x_3, x_4, x_5, x_6, x_7, and x_8 be the eight potential independent variables employed in the stepwise procedure. Figure 15.28(a) gives the Minitab output of the stepwise regression employing these independent variables where both α_{entry} and α_{stay} have been set equal to .10. The stepwise procedure (1) adds ACCTS (x_6) on the first step; (2) adds ADVER (x_3) and retains ACCTS

FIGURE 15.28   Minitab Iterative Procedures for the Sales Representative Case

(a) Stepwise regression ($\alpha_{entry} = \alpha_{stay} = .10$)

Stepwise Selection of Terms

Candidate terms: Time, MktPoten, Adver, MktShare, Change, Accts, WkLoad, Rating

	-----Step 1---		------Step 2----		------Step 3-----		------Step 4-----	
	Coef	P	Coef	P	Coef	P	Coef	P
Constant	709		50		−327		−1442	
Accts	21.72	0.000	19.05	0.000	15.55	0.000	9.21	0.004
Adver			0.2265	0.000	0.2161	0.000	0.1750	0.000
MktPoten					0.02192	0.019	0.03822	0.000
MktShare							190.1	0.001
S	881.093		650.392		582.636		453.836	
R-sq	56.85%		77.51%		82.77%		90.04%	
R-sq(adj)	54.97%		75.47%		80.31%		88.05%	
R-sq(pred)	43.32%		70.04%		76.41%		85.97%	
Mallows' Cp	67.56		27.16		18.36		5.43	

α to enter = 0.1, α to remove = 0.1

(b) Backward elimination ($\alpha_{stay} = .05$)

Backward Elimination of Terms

Candidate terms: Time, MktPoten, Adver, MktShare, Change, Accts, WkLoad, Rating

	------Step 1-----		------Step 2-----		------Step 3-----		------Step 4-----		------Step 5-----	
	Coef	P	Coef	P	Coef	P	Coef	P	Coef	P
Constant	−1508		−1486		−1165		−1114		−1312	
Time	2.01	0.313	1.97	0.287	2.27	0.198	3.61	0.006	3.82	0.007
MktPoten	0.03720	0.000	0.03729	0.000	0.03828	0.000	0.04209	0.000	0.04440	0.000
Adver	0.1510	0.006	0.1520	0.003	0.1407	0.002	0.1289	0.003	0.1525	0.001
MktShare	199.0	0.009	198.3	0.007	221.6	0.000	257.0	0.000	259.5	0.000
Change	291	0.139	296	0.090	285	0.093	325	0.053		
Accts	5.55	0.262	5.61	0.234	4.38	0.288				
WkLoad	19.8	0.565	19.9	0.550						
Rating	8	0.950								
S	449.026		435.674		428.004		430.232		463.948	
R-sq	92.20%		92.20%		92.03%		91.50%		89.60%	
R-sq(adj)	88.31%		88.99%		89.38%		89.26%		87.52%	
R-sq(pred)	81.75%		84.09%		85.35%		86.90%		84.57%	
Mallows' Cp	9.00		7.00		5.35		4.44		6.35	

α to remove = 0.05

on the second step; (3) adds MKTPOTEN (x_2) and retains ACCTS and ADVER on the third step; and (4) adds MKTSHARE (x_4) and retains ACCTS, ADVER, and MKTPOTEN on the fourth step. The procedure terminates after step 4 when no more independent variables can be added. Therefore, the stepwise procedure arrives at the model that utilizes x_2, x_3, x_4, and x_6. Note that this model is not the model using x_1, x_2, x_3, x_4, and x_5 that was obtained by evaluating all possible regression models and that has the smallest C statistic of 4.4. In general, stepwise regression can miss finding the best regression model but is useful in data mining, where a massive number of independent variables exist and all possible regression models cannot be evaluated.

In contrast to stepwise regression, **backward elimination** is an iterative model selection procedure that begins by considering the model that contains all of the potential independent variables and then attempts to remove independent variables one at a time from this model. On each step, an independent variable is removed from the model if it has the largest p-value of any independent variable remaining in the model and if its p-value is greater than α_{stay}, an α value for allowing a variable to stay in the model. Backward elimination terminates when all the p-values for the independent variables remaining in the model are less than α_{stay}. For example, Figure 15.28(b) gives the Minitab output of a backward elimination of the sales representative performance data. Here the backward elimination uses $\alpha_{stay} = .05$, begins with the model using all eight independent variables, and removes (in order) RATING (x_8), then WKLOAD (x_7), then ACCTS (x_6), and finally CHANGE (x_5). The procedure terminates when no independent variable remaining can be removed—that is, when no independent variable has a related p-value greater than $\alpha_{stay} = .05$—and arrives at a model that uses TIME (x_1), MKTPOTEN (x_2), ADVER (x_3), and MKTSHARE (x_4). Similar to stepwise regression, backward elimination has not arrived at the model using x_1, x_2, x_3, x_4, and x_5 that was obtained by evaluating all possible regression models and that has the smallest C statistic of 4.4. However, note that the model found in step 4 by backward elimination is the model using x_1, x_2, x_3, x_4, and x_5 and is the final model that would have been obtained by backward elimination if α_{stay} had been set at .10.

The sales representative performance example brings home two important points. First, the models obtained by backward elimination and stepwise regression depend on the choices of α_{entry} and α_{stay} (whichever is appropriate). Second, it is best not to think of these methods as "automatic model-building procedures." Rather, they should be regarded as processes that allow us to find and evaluate a variety of model choices.

Some advanced model-building methods: Using squared and interaction variables and the partial F test

We have concluded that perhaps the best sales representative performance model using only linear independent variables is the model using TIME, MKTPOTEN, ADVER, MKT-SHARE, and CHANGE. We have also seen in Section 15.9 that using squared variables (which model quadratic curvature) and interaction variables can improve a regression model.

In the page margin we present the 5 squared variables and the 10 (pairwise) interaction variables that can be formed using TIME, MKTPOTEN, ADVER, MKT-SHARE, and CHANGE. Consider having Minitab evaluate all possible models involving these squared and interaction variables, where the five linear variables are included in each possible model. If you have Minitab, do this and find the best model of each size in terms of s, we obtain the output in Figure 15.29. Examining the output, we see that the model that uses 12 squared and interaction variables (or a total of 17 variables, including the 5 linear variables) has the smallest s (=174.6) of any model. Unfortunately, although we would like to test the significance of the independent variables in this model, extreme multicollinearity (relationships between the independent variables) exists when using squared and interaction variables. Thus, the usual t tests for assessing the significance of individual independent variables might not be reliable. As an alternative, we will use a *partial F test*. Specifically, considering the model with the smallest s of 174.6 and a total of 17 variables to be a *complete model*, we will use this test to assess whether at least one variable in the *subset* of 12 squared and interaction variables in this model is significant.

Squared Variables and Interaction Variables

SQT	=	TIME*TIME
SQMP	=	MKTPOTEN*MKTPOTEN
SQA	=	ADVER*ADVER
SQMS	=	MKTSHARE*MKTSHARE
SQC	=	CHANGE*CHANGE
TMP	=	TIME*MKTPOTEN
TA	=	TIME*ADVER
TMS	=	TIME*MKTSHARE
TC	=	TIME*CHANGE
MPA	=	MKTPOTEN*ADVER
MPMS	=	MKTPOTEN*MKTSHARE
MPC	=	MKTPOTEN*CHANGE
AMS	=	ADVER*MKTSHARE
AC	=	ADVER*CHANGE
MSC	=	MKTSHARE*CHANGE

FIGURE 15.29   **Minitab Comparisons Using Additional Squared and Interaction Variables**

Response is Sales
The following variables are included in all models: Time MktPoten Adver MktShare Change

Total vars	Squared and Interaction Vars	R-Sq	R-Sq (adj)	R-Sq (pred)	Mallows Cp	S	S Q M T	S Q M P	S Q M A	Q T C	S T P A	T M A S	T M P C	M P M A S	M A S C	A S C	M S C C
6	1	94.2	92.2	87.0	43.2	365.87						X					
7	2	95.8	94.1	91.8	29.7	318.19	X					X					
8	3	96.5	94.7	90.9	25.8	301.61	X					X	X				
9	4	97.0	95.3	91.5	22.5	285.54	X				X X		X				
10	5	97.5	95.7	88.7	20.3	272.05	X				X X		X	X			
11	6	98.1	96.5	89.2	16.4	244.00	X	X			X X		X		X		
12	7	98.7	97.4	91.3	13.0	210.70	X X				X X		X		X X		
13	8	99.0	97.8	89.3	12.3	193.95	X X		X		X X		X		X X		
14	9	99.2	98.0	94.2	12.7	185.45	X X	X			X X		X		X X X		
15	10	99.3	98.2	93.4	13.3	175.70	X X	X			X X		X X X X X				
16	11	99.4	98.2	93.3	14.6	177.09	X X	X	X X X		X X X X X						
17	12	99.5	98.2	88.4	15.8	174.60	X X		X X X X X		X X X X X						
18	13	99.5	98.1	8.7	17.5	183.22	X X X		X X X X X		X X X X X						
19	14	99.6	97.9	4.6	19.1	189.77	X X		X X X X X X X X X X X X								
20	15	99.6	97.4	0.0	21.0	210.78	X X X X X X X X X X X X X X X										

The Partial F Test: An F Test for a Portion of a Regression Model

Suppose that the regression assumptions hold, and consider a **complete model** that uses k independent variables. To assess whether at least one of the independent variables in a subset of k^* independent variables in this model is significant, we test the null hypothesis

H_0: All of the β_j coefficients corresponding to the independent variables in the subset are zero

which says that none of the independent variables in the subset are significant. We test H_0 versus

H_a: At least one of the β_j coefficients corresponding to the independent variables in the subset is not equal to zero

which says that at least one of the independent variables in the subset is significant. Let SSE_C denote the unexplained variation for the complete model, and let SSE_R denote the unexplained variation for the **reduced model** that uses all k independent variables **except** for the k^* independent variables in the subset. Also, define

$$F(\text{partial}) = \frac{(SSE_R - SSE_C)/k^*}{SSE_C/[n - (k + 1)]}$$

and define the p-value related to $F(\text{partial})$ to be the area under the curve of the F distribution (having k^* and $[n - (k + 1)]$ degrees of freedom) to the right of $F(\text{partial})$. Then, we can reject H_0 in favor of H_a at level of significance α if either of the following equivalent conditions holds:

1   $F(\text{partial}) > F_\alpha$

2   p-value $< \alpha$

Here the point F_α is based on k^* numerator and $n - (k + 1)$ denominator degrees of freedom.

Using Excel, JMP, or Minitab, we find that the unexplained variation for the *complete model* that uses all $k = 17$ variables is $SSE_C = 213{,}396.12$ and the unexplained variation for the *reduced model* that does not use the $k^* = 12$ squared and interaction variables (and thus uses only the 5 linear variables) is $SSE_R = 3{,}516{,}859.2$. It follows that

$$F(\text{partial}) = \frac{(SSE_R - SSE_C)/k^*}{SSE_C/[n - (k + 1)]}$$

$$= \frac{(3{,}516{,}859.2 - 213{,}396.12)/12}{213{,}396.12/[25 - (17 + 1)]}$$

$$= \frac{3{,}303{,}463.1/12}{213{,}396.12/7} = 9.03$$

Because $F_{.05}$ based on $k^* = 12$ numerator and $n - (k + 1) = 7$ denominator degrees of freedom is 3.57 (see Table A.7), and because $F(\text{partial}) = 9.03$ is greater than $F_{.05} = 3.57$, we reject H_0 and conclude (at an α of .05) that at least one of the 12 squared and interaction variables in the 17-variable model is significant. In the exercises, the reader will do further analysis and use another partial F test and the C statistic to find a model that is perhaps better than the 17-variable model.

Model validation, the PRESS statistic, and R^2(predict)

When developing a regression model to predict a dependent variable y, it is best, if there are enough data, to "build" the model using one data set (called the **training data set**) and then "validate" the model by using it to analyze a different data set (called the **validation data set**). To illustrate this, Kutner, Nachtsheim, Neter, and Li (2005) consider 108 observations described by the dependent variable y = survival time (in days) after undergoing a particular liver operation and the independent variables x_1 = blood clotting score, x_2 = prognostic index, x_3 = enzyme function test score, x_4 = liver function test score, x_5 = age (in years), $x_6 = 1$ for a female patient and 0 for a male patient, $x_7 = 1$ for a patient who is a moderate drinker and 0 otherwise, and $x_8 = 1$ for a patient who is a heavy drinker and 0 otherwise. Because 108 observations are a fairly large number of observations, Kutner and his fellow authors randomly divided these observations into two equal halves, obtaining the training data set in Table 14.13 and the validation data set in Table 14.14. A regression analysis relating y to x_1, x_2, x_3, and x_4 using the training data set had a residual plot that was curved and fanned out, suggesting the need for a natural logarithm transformation (see Sections 14.7 and 15.11). Using all possible regressions on the training data set, the models with the smallest C statistic, smallest s^2, and largest $\overline{R}^2$ were the following models 1, 2, and 3 (see Table 15.15):

Model 1: $\ln y = \beta_0 + \beta_1 x_1 + \beta_2 x_2 + \beta_3 x_3 + \beta_8 x_8 + \varepsilon$

Model 2: $\ln y = \beta_0 + \beta_1 x_1 + \beta_2 x_2 + \beta_3 x_3 + \beta_6 x_6 + \beta_8 x_8 + \varepsilon$

Model 3: $\ln y = \beta_0 + \beta_1 x_1 + \beta_2 x_2 + \beta_3 x_3 + \beta_5 x_5 + \beta_8 x_8 + \varepsilon$

Each model was then fitted to the training data set and used to predict each observation in the validation data set. The accuracy of these predictions was evaluated by calculating the **mean square of the prediction error**

$$\text{MSPR} = \frac{\sum_{i=1}^{n^*}(y_i' - \hat{y}_i)^2}{n^*}$$

Here, n^* (=54) is the number of observations in the validation data set, y_i' is the value of the dependent variable for the ith observation (in this case, the natural logarithm of the survival time for the ith patient) in the validation data set, and $\hat{y}_i$ is the prediction of y_i' made using the model that was fitted to the training data set. A model that predicts the validation data set dependent variable values accurately has a small value of MSPR, and the values of MSPR for Models 1, 2, and 3 are shown in Table 15.15. Each of Models 1, 2, and 3 was also fitted to the validation data set, but we have not filled the values of C, s^2, and $\overline{R}^2$ for these models into Table 15.15. The reason is that in Exercise 15.48 the reader will use all possible regressions on the validation data set, fill in the blank spaces, and choose a best model.

To conclude this section, note that if there are not enough observations to split into a training data set and a validation data set, and thus if we have only a training data set, we can still assess to some extent how well a model would predict observations not used to build the model

TABLE 15.15 **Comparisons of Models 1, 2, and 3**

	Model 1 Training	Model 1 Validation	Model 2 Training	Model 2 Validation	Model 3 Training	Model 3 Validation
C	5.7508		5.5406		5.7874	
s^2	0.0445		0.0434		0.0427	
$\overline{R}^2$	0.8160		0.8205		0.8234	
MSPR	0.0773		0.0764		0.0794	

by calculating the *PRESS statistic* and R^2*(predict)*. To calculate these statistics, we first calculate the *deleted residual* for each observation. For a particular observation, observation i, the **deleted residual** is found by subtracting from y_i the point prediction $\hat{y}_{(i)}$ computed using least squares point estimates based on all observations other than observation i. The **PRESS statistic** is then defined to be the sum of the squared deleted residuals. **R^2(predict),** which is called *R-Sq (pred)* on Minitab regression outputs, is defined to be $1-[\text{PRESS}/(\text{Total variation})]$. Like C, s^2, and $\overline{R}^2$, PRESS and R^2(predict) are useful model comparison criteria. We desire PRESS to be small and R^2(predict) to be large. These two statistics are regarded as being good indicators of how well a regression model will predict future observations (observations not in the data set). If we examine the output for all possible regression models for the sales representative performance data in Figure 15.27, we see that the model using only linear terms that has the smallest C statistic ($C = 4.4$) has the second largest R^2(predict) [R^2(predict) = 86.9].

Forward selection with simultaneous validation

Suppose that a Toyota dealer wishes to predict the sales prices of used Toyota Corollas and has gathered 1,436 observations concerning sales prices of used Corollas and 36 independent variables. Sixty percent (or 862) of the observations are randomly selected as the training data set and the remaining 40 percent (or 574) of the observations are used as the validation data set. Some of the observations are shown in the JMP output of Figure 15.30, along with a *validation column* that specifies which observations are part of the training data set and which observations are part of the validation data set. Also shown on this output are the independent variable values for used Toyota Corolla 1437, which has just been obtained by the dealer and has not yet been sold. To find a model describing these data, we will use a JMP procedure called *forward selection with simultaneous validation*. What JMP does in implementing this procedure is to use the training set to add one independent variable at a time to the regression model, where the next variable added to the model is the variable not currently in the model that has the smallest p-value when added to the model. As each new independent variable is added, the model obtained is used to predict the response variable values for each observation in the training data set and for each observation in the validation data set. *RSquare* is then calculated using the observations in the training data set, and *RSquare Validation* is calculated using the observations in the validation data set. The stepwise procedure stops if *RSquare Validation* reaches a value that is not exceeded in 10 additional steps of adding independent variables.

When using this procedure, JMP does not use the standard 0,1 dummy variables to model qualitative independent variables. Rather, JMP uses what are called $-1, 0, 1$ dummy variables. For example, consider Figure 15.31(a), which shows all of the quantitative independent variables, including the dummy variables used to model the qualitative independent variables. For example, then, Automatic {0-1} is a dummy variable that equals -1 if a Corolla has an automatic transmission and 1 if the Corolla does not have an automatic transmission. When a qualitative independent variable has more than two levels, JMP defines dummy variables with the help of a decision tree analysis (see Chapter 5), so as to obtain the potentially most significant dummy variables. For example, the independent variable fuel type has three levels, and the two dummy variables JMP uses to model these levels are (1) Fuel_Type {CNG & Petrol-Diesel}, which equals 1 if the fuel type is CNG (natural gas) or Petrol (gasoline) and equals -1 if the fuel type is Diesel, and (2) Fuel_Type {CNG-Petrol}, which equals 1 if the fuel type is CNG, -1 if the fuel type is Petrol, and 0 otherwise (if the fuel type is Diesel). The nine dummy variables modeling the ten colors are defined similarly. For example, Color {White & Beige–Violet & Green & Blue} equals 1 if the car color is white or beige, -1 if the car color is violet or green or red, and 0 otherwise.

Figure 15.31(b) shows the steps of adding independent variables to the model and shows that the last independent variable added before *RSquare Validation* no longer increases is Gears at step 33. (Note that the *RSqure* and *RSquare Validation* values are given to only four decimal places. Also, note that a single dummy variable that is one

FIGURE 15.30   **The JMP Output of a Portion of the Used Toyota Corolla Sales Price Data**   DS Toyota Corolla

Id	Model	Price	Age_08_04	Mfg_Month	Mfg_Year	KM	Fuel_Type	HP	Met_Color
1	TOYOTA Corolla 2.0 D4D HATCHB TERRA 2/3-Doors	13500	23	10	2002	46986	Diesel	90	1
2	TOYOTA Corolla 2.0 D4D HATCHB TERRA 2/3-Doors	13750	23	10	2002	72937	Diesel	90	1
3	TOYOTA Corolla 2.0 D4D HATCHB TERRA 2/3-Doors	13950	24	9	2002	41711	Diesel	90	1
8	TOYOTA Corolla 2.0 D4D 90 3DR TERRA 2/3-Doors	18600	30	3	2002	75889	Diesel	90	1
9	TOYOTA Corolla 1800 T SPORT VVTI 2/3-Doors	21500	27	6	2002	19700	Petrol	192	0
10	TOYOTA Corolla 1.9 D HATCHB TERRA 2/3-Doors	12950	23	10	2002	71138	Diesel	69	0
1437	TOYOTA Corolla 2.0 D4D 90 3DR TERRA 2/3-Doors		27	6	2002	94612	Diesel	90	

	Color	Automatic	CC	Doors	Cylinders	Gears	Quarterly_Tax	Weight	Mfg_Guarantee	BOVAG_Guarantee	Guarantee_Period
1	Blue	0	2000	3	4	5	210	1165	0	1	3
2	Silver	0	2000	3	4	5	210	1165	0	1	3
3	Blue	0	2000	3	4	5	210	1165	1	1	3
8	Grey	0	2000	3	4	5	210	1245	1	1	3
9	Red	0	1800	3	4	5	100	1185	0	1	3
10	Blue	0	1900	3	4	5	185	1105	0	1	3
1437	Grey	0	2000	3	4	5	210	1245	0	1	3

	ABS	Airbag_1	Airbag_2	Airco	Automatic_airco	Boardcomputer	CD_Player	Central_Lock	Powered_Windows
1	1	1	1	0	0	1	0	1	1
2	1	1	1	1	0	1	1	1	0
3	1	1	1	0	0	1	0	0	0
8	1	1	1	1	0	1	1	1	1
9	1	1	0	1	0	0	0	1	1
10	1	1	1	1	0	1	0	0	0
1437	1	1	1	1	0	1	0	1	1

	Power_Steering	Radio	Mistlamps	Sport_Model	Backseat_Divider	Metallic_Rim	Radio_cassette	Tow_Bar	Validation
1	1	0	0	0	1	0	0	0	Validation
2	1	0	0	0	1	0	0	0	Validation
3	1	0	0	0	1	0	0	0	Training
8	1	0	0	0	1	0	0	0	Training
9	1	1	0	0	0	1	1	0	Training
10	1	0	0	0	1	0	0	0	Validation
1437	1	0	0	1	1	0	0	0	Validation

of the nine dummy variables modeling color was added at step 10 and that both of the dummy variables Fuel_Type {CNG–Petrol} and Fuel_Type {CNG & Petrol-Diesel} modeling fuel type were added at step 11 - see the jump in 8 from 12 to 14. In general, a *group* of dummy variables modeling a qualitative independent variable is added if the partial F test *p*-value for the significance of the group is the smallest.) At step 34—and through step 39—*RSquare Validation* does not increase beyond its .9036 value at step 33, and therefore JMP chooses the best model to be the model at step 33. Note that in doing the forward selection with simultaneous validation, we used JMP's *Combine* options which allows a single dummy variable or a group of dummy variables to be added at a single step. JMP's *Whole Effects* option allows only the entire group of dummy variables model-ing a qualitative independent variable to be added at a single step. We will not show the JMP stepwise output obtained using the *Whole Effects* option. However, Figures 15.32

FIGURE 15.31   The JMP Output of the Potential Quantitative Independent Variables, Including the Dummy Variables, and Forward Selection with Simultaneous Validation in the Used Toyota Corolla Sales Price Example

(a) The Variables

Age_08_04
Mfg_Month
Mfg_Year
KM
Fuel_Type{CNG&Petrol-Diesel}
Fuel_Type{CNG-Petrol}
HP
Met_Color{0-1}
Color{White&Beige&Violet&Green&Red-Blue&Black&Silver&Grey&Yellow}
Color{White&Beige-Violet&Green&Red}
Color{White-Beige}
Color{Violet&Green-Red}
Color{Violet-Green}
Color{Blue&Black&Silver-Grey&Yellow}
Color{Blue-Black&Silver}
Color{Black-Silver}
Color{Grey-Yellow}
Automatic{0-1}
CC
Doors
Cylinders
Gears
Quarterly_Tax
Weight
Mfg_Guarantee
BOVAG_Guarantee
Guarantee_Period
ABS
Airbag_1
Airbag_2
Airco
Automatic_airco
Boardcomputer
CD_Player
Central_Lock
Powered_Windows
Power_Steering
Radio
Mistlamps
Sport_Model
Backseat_Divider
Metallic_Rim
Radio_cassette
Tow_Bar

(b) Forward selection with simultaneous validation

Step History

Step	Parameter	Action	"Sig Prob"	RSquare	RSquare Validation	P
1	Mfg_Year	Entered	0.0000	0.7834	0.7831	2
2	Automatic_airco	Entered	0.0000	0.8326	0.8339	3
3	HP	Entered	0.0000	0.8569	0.8392	4
4	KM	Entered	0.0000	0.8675	0.8472	5
5	Weight	Entered	0.0000	0.8865	0.8895	6
6	Powered_Windows	Entered	0.0000	0.8896	0.8933	7
7	Quarterly_Tax	Entered	0.0000	0.8930	0.8942	8
8	Guarantee_Period	Entered	0.0000	0.8961	0.8940	9
9	BOVAG_Guarantee	Entered	0.0000	0.8993	0.8956	10
10	Color{White&Beige&Violet&Green&Red-Blue&Black&Silver&Grey&Yellow}	Entered	0.0001	0.9011	0.8965	11
11	Sport_Model	Entered	0.0003	0.9026	0.8975	12
12	Fuel_Type{CNG-Petrol}	Entered	0.0015	0.9041	0.9008	14
13	Boardcomputer	Entered	0.0014	0.9053	0.9003	15
14	ABS	Entered	0.0110	0.9060	0.9006	16
15	Age_08_04	Entered	0.0209	0.9066	0.9010	17
16	Automatic{0-1}	Entered	0.0185	0.9072	0.9005	18
17	Metallic_Rim	Entered	0.0171	0.9078	0.9007	19
18	Airco	Entered	0.0585	0.9082	0.9010	20
19	Mfg_Guarantee	Entered	0.0712	0.9086	0.9023	21
20	Backseat_Divider	Entered	0.0614	0.9089	0.9029	22
21	Color{Blue&Black&Silver-Grey&Yellow}	Entered	0.0912	0.9092	0.9027	23
22	Central_Lock	Entered	0.0833	0.9096	0.9017	24
23	Doors	Entered	0.1099	0.9098	0.9013	25
24	Color{White&Beige-Violet&Green&Red}	Entered	0.1189	0.9101	0.9020	26
25	Tow_Bar	Entered	0.1151	0.9104	0.9029	27
26	Airbag_1	Entered	0.3079	0.9105	0.9025	28
27	Color{Grey-Yellow}	Entered	0.3464	0.9106	0.9024	29
28	CD_Player	Entered	0.4018	0.9107	0.9029	30
29	Airbag_2	Entered	0.3797	0.9107	0.9030	31
30	Color{Violet&Green-Red}	Entered	0.4227	0.9108	0.9032	32
31	CC	Entered	0.5240	0.9108	0.9034	33
32	Met_Color{0-1}	Entered	0.5621	0.9109	0.9036	34
33	Gears	Entered	0.6239	0.9109	0.9036	35
34	Color{Violet-Green}	Entered	0.6316	0.9109	0.9036	36
35	Mistlamps	Entered	0.6821	0.9110	0.9034	37
36	Color{Black-Silver}	Entered	0.7004	0.9110	0.9032	39
37	Power_Steering	Entered	0.8402	0.9110	0.9032	40
38	Radio	Entered	0.9038	0.9110	0.9032	41
39	Mfg_Month	Entered	0.9997	0.9110	0.9032	42
40	Best	Specific	.	0.9109	0.9036	35

and 15.33 show JMP outputs of regression analysis and prediction using the model from the *Combine* option and the model from the *Whole Effects* option. We see that, although the *Combine* model uses more independent variables and has a smaller value of the C statistic, both models give similar values of *RSquare* and *RSquare Validation* and similar predictions of the sales price of used Toyota Corolla 1437.

FIGURE 15.32   The JMP Output of Regression Analyses of the Used Toyota Corolla Sales Price Data Using the *Combine* and *Whole Effects* Models

(a) The *Combine* model

Summary of Fit

RSquare	0.910911
RSquare Adj	0.907248
Root Mean Square Error	1104.616
Mean of Response	10796.6
Observations (or Sum Wgts)	862

Analysis of Variance

Source	DF	Sum of Squares	Mean Square	F Ratio	Cp
Model	34	1.0318e+10	303460227	248.7019	29.160669
Error	827	1009085920	1220176.4	Prob > F	**RSquare**
C. Total	861	1.1327e+10		<.0001*	**Validation**
					0.9036

Parameter Estimates

Term	Estimate	t Ratio	Prob>|t|	Term	Estimate	t Ratio	Prob>|t|
Intercept	−2371739	−8.29	<.0001*	Quarterly_Tax	12.33489	5.69	<.0001*
Age_08_04	−26.14582	−2.20	0.0283*	Weight	6.6815394	4.87	<.0001*
Mfg_Year	1186.0134	8.30	<.0001*	Mfg_Guarantee	183.60428	2.16	0.0310*
KM	−0.017259	−11.61	<.0001*	BOVAG_Guarantee	606.95314	4.14	<.0001*
Fuel_Type{CNG&Petrol-Diesel}	−253.059	−1.66	0.0978	Guarantee_Period	78.978396	5.02	<.0001*
Fuel_Type{CNG-Petrol}	−876.5524	−4.07	<.0001*	ABS	−359.9936	−2.45	0.0143*
HP	29.120103	7.92	<.0001*	Airbag_1	297.38844	1.08	0.2794
Met_Color[0]	27.264735	0.60	0.5455	Airbag_2	−124.4065	−0.87	0.3844
Color{White&Beige&Violet&Green& Red-Blue&Black&Silver&Grey&Yellow}	−208.2795	−1.92	0.0551	Airco	190.39279	1.90	0.0579
				Automatic_airco	2387.7747	12.37	<.0001*
Color{White&Beige-Violet&Green&Red}	−237.463	−1.74	0.0816	Boardcomputer	−403.7215	−3.07	0.0022*
Color{Violet&Green-Red}	−42.72422	−0.63	0.5295	CD_Player	112.46846	1.00	0.3172
Color{Blue&Black&Silver-Grey&Yellow}	54.246956	0.33	0.7432	Central_Lock	−312.6974	−1.73	0.0832
Color{Grey-Yellow}	291.86664	0.90	0.3698	Powered_Windows	535.79599	2.93	0.0035*
Automatic[0]	−244.7515	−2.76	0.0058*	Sport_Model	419.86302	4.30	<.0001*
CC	−0.049143	−0.63	0.5312	Backseat_Divider	−239.8786	−1.67	0.0962
Doors	83.116693	1.83	0.0671	Metallic_Rim	259.59634	2.52	0.0119*
Gears	100.91757	0.49	0.6239	Tow_Bar	−139.8901	−1.57	0.1175

(b) The *Whole Effects* model

Summary of Fit

RSquare	0.908549
RSquare Adj	0.905475
Root Mean Square Error	1115.126
Mean of Response	10796.6
Observations (or Sum Wgts)	862

Analysis of Variance

Source	DF	Sum of Squares	Mean Square	F Ratio	Cp
Model	28	1.0291e+10	367531907	295.5611	38.932443
Error	833	1035840234	1243505.7	Prob > F	**RSquare**
C. Total	861	1.1327e+10		<.0001*	**Validation**
					0.9023

Parameter Estimates

Term	Estimate	t Ratio	Prob>|t|	Term	Estimate	t Ratio	Prob>|t|
Intercept	−2369638	−8.21	<.0001*	Guarantee_Period	80.021674	5.08	<.0001*
Age_08_04	−26.50997	−2.21	0.0272*	ABS	−385.9352	−2.61	0.0091*
Mfg_Year	1184.9488	8.22	<.0001*	Airbag_1	295.48363	1.07	0.2858
KM	−0.017346	−11.67	<.0001*	Airbag_2	−93.62222	−0.65	0.5154
Fuel_Type[CNG]	−1067.774	−4.11	<.0001*	Airco	184.79899	1.83	0.0674
Fuel_Type[Diesel]	306.01122	1.49	0.1353	Automatic_airco	2439.5204	12.57	<.0001*
HP	28.57111	7.72	<.0001*	Boardcomputer	−372.9491	−2.82	0.0049*
Automatic[0]	−244.3866	−2.74	0.0063*	CD_Player	125.87424	1.12	0.2624
CC	−0.054992	−0.70	0.4870	Central_Lock	−322.3784	−1.79	0.0744
Doors	66.73783	1.47	0.1427	Powered_Windows	570.542	3.12	0.0019*
Gears	100.3322	0.48	0.6284	Sport_Model	423.35845	4.30	<.0001*
Quarterly_Tax	12.812515	5.90	<.0001*	Backseat_Divider	−242.6011	−1.67	0.0949
Weight	6.8640934	4.97	<.0001*	Metallic_Rim	243.22818	2.37	0.0179*
Mfg_Guarantee	165.31113	1.96	0.0507	Tow_Bar	−146.1009	−1.66	0.0981
BOVAG_Guarantee	611.67575	4.13	<.0001*				

F I G U R E   1 5 . 3 3   **Prediction of the Sales Price of Used Toyota Corolla 1437 Using the *Combine* Model and the *Whole Effects* Model**

(a) Prediction using the *Combine* model

Pred Formula Price	Lower 95% Mean Price	Upper 95% Mean Price	Lower 95% Indiv Price	Upper 95% Indiv Price
15779.429458	15353.665825	16205.193091	13569.840754	17989.018162

(b) Prediction using the *Whole effects* model

Pred Formula Price	Lower 95% Mean Price	Upper 95% Mean Price	Lower 95% Indiv Price	Upper 95% Indiv Price
15642.417558	15234.592309	16050.242808	13415.960944	17868.874173

Exercises for Section 15.10

CONCEPTS connect

15.44 What is *multicollinearity*? What problems can be caused by multicollinearity?

15.45 List the criteria and model selection procedures we use to compare regression models.

METHODS AND APPLICATIONS

15.46 THE HOSPITAL LABOR NEEDS CASE

 HospLab2

Recall that Table 15.7 presents data concerning the need for labor in 16 U.S. Navy hospitals. This table gives values of the dependent variable Hours (monthly labor hours) and of the independent variables Xray (monthly X-ray exposures), BedDays (monthly occupied bed days—a hospital has one occupied bed day if one bed is occupied for an entire day), and

Length (average length of patient's stay, in days). The data in Table 15.7 are part of a larger data set analyzed by the Navy. The complete data set includes two additional independent variables—Load (average daily patient load) and Pop (eligible population in the area, in thousands)—values of which are given in the page margin. Figure 15.34 gives JMP outputs of multicollinearity analysis and model building for the complete hospital labor needs data set.

a **(1)** Find the three largest simple correlation coefficients among the independent variables and the three largest variance inflation factors in Figure 15.34(a) and (b). **(2)** Discuss why these statistics imply that the independent variables BedDays, Load, and Pop are most strongly involved in multicollinearity and thus contribute possibly redundant information for predicting Hours. Note that, although we have reasoned in Exercise 15.7(a)

F I G U R E   1 5 . 3 4   **JMP Output of Multicollinearity and a Model Building Analysis in the Hospital Labor Needs Case**

(a) The correlation matrix

	Hours	Xray	BedDays	Length	Load	Pop
Hours	1.0000	0.9425	0.9889	0.5603	0.9886	0.9465
Xray	0.9425	1.0000	0.9048	0.4243	0.9051	0.9124
BedDays	0.9889	0.9048	1.0000	0.6609	0.9999	0.9328
Length	0.5603	0.4243	0.6609	1.0000	0.6610	0.4515
Load	0.9886	0.9051	0.9999	0.6610	1.0000	0.9353
Pop	0.9465	0.9124	0.9328	0.4515	0.9353	1.0000

(b) Regression analysis

Parameter Estimates

Term	Estimate	Std Error	t Ratio	Prob>|t|	Lower 95%	Upper 95%	VIF
Intercept	2270.4153	670.7861	3.38	0.0069*	775.81071	3765.0199	
Xray	0.0411168	0.013677	3.01	0.0132*	0.0106421	0.0715915	8.0773378
BedDays	1.4127901	1.92532	0.73	0.4799	−2.877091	5.702671	8684.2076
Length	−467.8605	131.6268	−3.55	0.0052*	−761.1434	−174.5777	4.212708
Load	−9.297186	60.80999	−0.15	0.8815	−144.7903	126.19591	9334.456
Pop	−3.222942	4.47357	−0.72	0.4878	−13.19068	6.7447921	23.00542

(c) The two best models of each size

All Possible Models

Model	Number	RSquare	RMSE	Cp
BedDays	1	0.9779	856.707	52.3131
Load	1	0.9773	867.672	53.9698
BedDays, Length	2	0.9933	489.126	9.4667
Length, Load	2	0.9927	509.821	11.1488
Xray, BedDays, Length	3	0.9961	387.160	3.2582
Xray, Length, Load	3	0.9955	415.472	4.9649
Xray, BedDays, Length, Pop	4	0.9965	381.555	4.0234
Xray, BedDays, Length, Load	4	0.9964	390.876	4.5190
Xray, BedDays, Length, Load, Pop	5	0.9966	399.712	6.0000

(d) Stepwise regression ($\alpha_{entry} = \alpha_{stay} = .10$)

Step History

Step	Parameter	Action	"Sig Prob"	Seq SS	RSquare	Cp	p
1	All	Removed	.	.	0.0000	2891	1
2	BedDays	Entered	0.0000	4.539e+8	0.9779	52.313	2
3	Length	Entered	0.0001	7165085	0.9933	9.4667	3
4	Xray	Entered	0.0120	1311468	0.9961	3.2582	4

Load	Pop
15.57	18.0
44.02	9.5
20.42	12.8
18.74	36.7
49.20	35.7
44.92	24.0
55.48	43.3
59.28	46.7
94.39	78.7
128.02	180.5
96.00	60.9
131.42	103.7
127.21	126.8
409.20	169.4
463.70	331.4
510.22	371.6

that a negative coefficient (that is, least squares point estimate) for Length might be intuitively reasonable, the negative coefficients for Load and Pop [see Figure 15.34(b)] are not intuitively reasonable and are a further indication of strong multicollinearity. We conclude that a final regression model for predicting Hours may not need all three of the potentially redundant independent variables BedDays, Load, and Pop.

b   Figure 15.34(c) indicates that the two best hospital labor needs models are the model using Xray, BedDays, Pop, and Length, which we will call Model 1, and the model using Xray, BedDays, and Length, which we will call Model 2. (1) Which model gives the smallest value of s and the largest value of $\overline{R}^2$? (2) Which model gives the smallest value of C? (3) Consider a questionable hospital for which Xray = 56,194, BedDays = 14,077.88, Pop = 329.7, and Length = 6.89. The 95 percent prediction intervals given by Models 1 and 2 for labor hours corresponding to this combination of values of the independent variables are, respectively, [14,888.43, 16,861.30] and [14,906.24, 16,886.26]. Which model gives the shortest prediction interval?

c   (1) Which model is chosen by stepwise regression in Figure 15.34(d)? (2) If we start with all five potential independent variables and use backward elimination with an α_{stay} of .05, the procedure removes (in order) Load and Pop and then stops. Which model is chosen by backward elimination? (3) Overall, which model seems best? (4) Which of BedDays, Load, and Pop does this best model use?

15.47   THE SALES REPRESENTATIVE CASE   DS SalePerf

Consider Figure 15.29. The model using 12 squared and interaction variables has the smallest s. However, if we desire a somewhat simpler model, note that s

does not increase substantially until we move from a model having seven squared and interaction variables to a model having six such variables. Moreover, we might subjectively conclude that the s of 210.70 for the model using 7 squared and interaction variables is not that much larger than the s of 174.6 for the model using 12 squared and interaction variables. (1) Using the fact that the unexplained variations for these respective models are 532,733.88 and 213,396.12, perform a partial F test to assess whether at least one of the extra five squared and interaction variables is significant. If none of the five extra variables are significant, we might consider the simpler model to be best. (2) Note from Figure 15.29 that the seven squared and interaction variables model has the third smallest C statistic. What other models seem good based on the C statistic?

15.48   THE LIVER OPERATION SURVIVAL TIME CASE
DS SurvTimeT   and   DS SurvTimeV

Use JMP, MegaStat, or Minitab to perform all possible regressions on the validation data set in Table 14.14, where the dependent variable is ln y and the independent variables are $x_1, x_2, \ldots , x_8$. (1) Are Models 1, 2, and 3 in Table 15.15 the three best models for the validation data set in terms of C, s^2, and $\overline{R}^2$? (2) Fill in the blank spaces for C, s^2, and $\overline{R}^2$ in Table 15.15. (3) Perform a regression analysis of the training data set and a regression analysis of the validation data set using each of Models 1, 2, and 3. The sign (positive or negative) of one of the least squares point estimates in one of the models changes as we move from the training data set to the validation data set. This inconsistency should eliminate one of the models. Which model is this? (4) Choose a "best" model, justify your choice, and specify a prediction equation for predicting the natural logarithm of survival time. Note that if you use Minitab, your choice of a best model will be aided by considering the values of R^2(predict) when Models 1, 2, and 3 are fitted to the training data set and to the validation data set in (3).

LO15-10
Use residual analysis and outlier detection to check the assumptions of multiple regression. (Optional)

15.11 Residual Analysis and Outlier Detection in Multiple Regression (Optional)

Basic residual analysis

For a multiple regression model, we plot the residuals given by the model against (1) values of each independent variable, (2) values of the predicted value of the dependent variable, and (3) the time order in which the data have been observed (if the regression data are time series data). A fanning-out pattern on a residual plot indicates an increasing error variance; a funneling-in pattern indicates a decreasing error variance. Both violate the constant variance assumption. A curved pattern on a residual plot indicates that the functional form of the regression model is incorrect. If the regression data are time series data, a cyclical pattern on the residual plot versus time suggests positive autocorrelation, while an alternating pattern suggests negative autocorrelation. Both violate the independence assumption. On the other hand, if all residual plots have (at least approximately) a horizontal band appearance, then it is reasonable to believe that the constant variance, correct functional form, and independence assumptions approximately hold. To check the normality assumption, we can

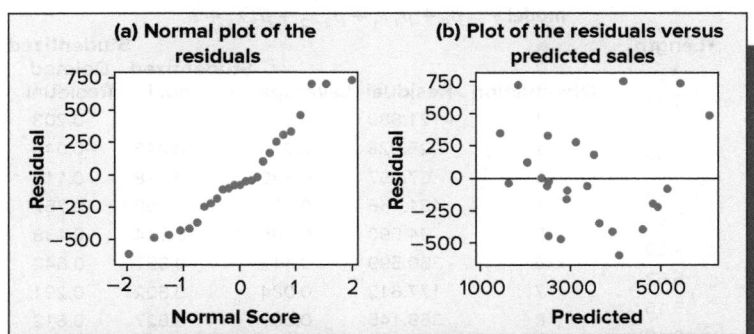

FIGURE 15.35   **Residual Plots for the Sales Representative Performance Model**

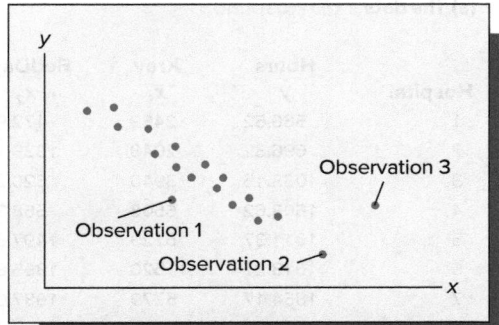

FIGURE 15.36   **Outlying Observations**

construct a histogram, stem-and-leaf display, and normal plot of the residuals. The histogram and stem-and-leaf display should look bell-shaped and symmetric about 0; the normal plot should have a straight-line appearance.

To illustrate these ideas, consider the sales representative performance data in Table 15.8. Figure 15.10 gives a partial Excel output of a regression analysis of these data using the model that relates y to $x_1, x_2, x_3, x_4,$ and x_5. The least squares point estimates on the output give the prediction equation

$$\hat{y} = -1{,}113.7879 + 3.6121x_1 + .0421x_2 + .1289x_3 + 256.9555x_4 + 324.5334x_5$$

Using this prediction equation, we can calculate the predicted sales values and residuals given in the page margin. For example, observation 10 corresponds to a sales representative for whom $x_1 = 105.69$, $x_2 = 42{,}053.24$, $x_3 = 5{,}673.11$, $x_4 = 8.85$, and $x_5 = .31$. If we insert these values into the prediction equation, we obtain a predicted sales value of $\hat{y}_{10} = 4{,}143.597$. Because the actual sales for the sales representative are $y_{10} = 4{,}876.370$, the residual e_{10} equals the difference between $y_{10} = 4{,}876.370$ and $\hat{y}_{10} = 4{,}143.597$, which is 732.773. The normal plot of the residuals in Figure 15.35(a) has an approximate straight-line appearance. The plot of the residuals versus predicted sales in Figure 15.35(b) has a horizontal band appearance, as do the plots of the residuals versus the independent variables (these plots are not given here). We conclude that the regression assumptions approximately hold for the sales representative performance model.

Outliers

An observation that is well separated from the rest of the data is called an **outlier,** and an observation may be an outlier with respect to its y value and/or its x values. We illustrate these ideas by considering Figure 15.36, which is a hypothetical plot of the values of a dependent variable y against an independent variable x. Observation 1 in this figure is outlying with respect to its y value, but not with respect to its x value. Observation 2 is outlying with respect to its x value, but because its y value is consistent with the regression relationship displayed by the nonoutlying observations, it is not outlying with respect to its y value. Observation 3 is an outlier with respect to its x value and its y value.

It is important to identify outliers because (as we will see) outliers can have adverse effects on a regression analysis and thus are candidates for removal from a data set. Moreover, in addition to using data plots, we can use more sophisticated procedures to detect outliers. For example, suppose that the U.S. Navy wishes to develop a regression model based on efficiently run Navy hospitals to evaluate the labor needs of questionably run Navy hospitals. Figure 15.37(a) gives labor needs data for 17 Navy hospitals. Specifically, this table gives values of the dependent variable Hours (y, monthly labor hours required) and of the independent variables Xray (x_1, monthly X-ray exposures), BedDays (x_2, monthly occupied bed days—a hospital has one occupied bed day if one bed is occupied for an entire day), and Length (x_3, average length of patient's stay, in days). When we perform a regression analysis of these data using the model relating y to $x_1, x_2,$ and x_3, we obtain the MegaStat output of

Obs	Predicted	Residual
1	3,504.990	164.890
2	3,901.180	−427.230
3	2,774.866	−479.766
4	4,911.872	−236.312
5	5,415.196	710.764
6	2,026.090	108.850
7	5,126.127	−94.467
8	3,106.925	260.525
9	6,055.297	464.153
10	4,143.597	732.773
11	2,503.165	−34.895
12	1,827.065	706.245
13	2,478.083	−69.973
14	2,351.344	−13.964
15	4,797.688	−210.738
16	2,904.099	−174.859
17	3,362.660	−73.260
18	2,907.376	−106.596
19	3,625.026	−360.826
20	4,056.443	−602.823
21	1,409.835	331.615
22	2,494.101	−458.351
23	1,617.561	−39.561
24	4,574.903	−407.463
25	2,488.700	311.270

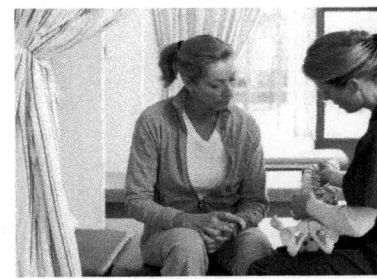

©UpperCut Images/SuperStock RF

FIGURE 15.37   Hospital Labor Needs Data, Outlier Diagnostics, and Residual Plots

(a) The data DS HospLab3

Hospital	Hours y	Xray x_1	BedDays x_2	Length x_3
1	566.52	2463	472.92	4.45
2	696.82	2048	1339.75	6.92
3	1033.15	3940	620.25	4.28
4	1603.62	6505	568.33	3.90
5	1611.37	5723	1497.60	5.50
6	1613.27	11520	1365.83	4.60
7	1854.17	5779	1687.00	5.62
8	2160.55	5969	1639.92	5.15
9	2305.58	8461	2872.33	6.18
10	3503.93	20106	3655.08	6.15
11	3571.89	13313	2912.00	5.88
12	3741.40	10771	3921.00	4.88
13	4026.52	15543	3865.67	5.50
14	10343.81	36194	7684.10	7.00
15	11732.17	34703	12446.33	10.78
16	15414.94	39204	14098.40	7.05
17	18854.45	86533	15524.00	6.35

(b) MegaStat outlier diagnostics for the model $y = \beta_0 + \beta_1 x_1 + \beta_2 x_2 + \beta_3 x_3 + \varepsilon$

Observation	Residual	Leverage	Studentized Residual	Studentized Deleted Residual
1	−121.889	0.121	−0.211	−0.203
2	−25.028	0.226	−0.046	−0.044
3	67.757	0.130	0.118	0.114
4	431.156	0.159	0.765	0.752
5	84.590	0.085	0.144	0.138
6	−380.599	0.112	−0.657	−0.642
7	177.612	0.084	0.302	0.291
8	369.145	0.083	0.627	0.612
9	−493.181	0.085	−0.838	−0.828
10	−687.403	0.120	−1.192	−1.214
11	380.933	0.077	0.645	0.630
12	−623.102	0.177	−1.117	−1.129
13	−337.709	0.064	−0.568	−0.553
14	1,630.503	0.146	2.871	4.558
15	−348.694	0.682	−1.005	−1.006
16	281.914	0.785	0.990	0.989
17	−406.003	0.863	−1.786	−1.975

Source: "Hospital Labor Needs Data" from *Procedures and Analysis for Staffing Standards Development: Regression Analysis Handbook,* © 1979.

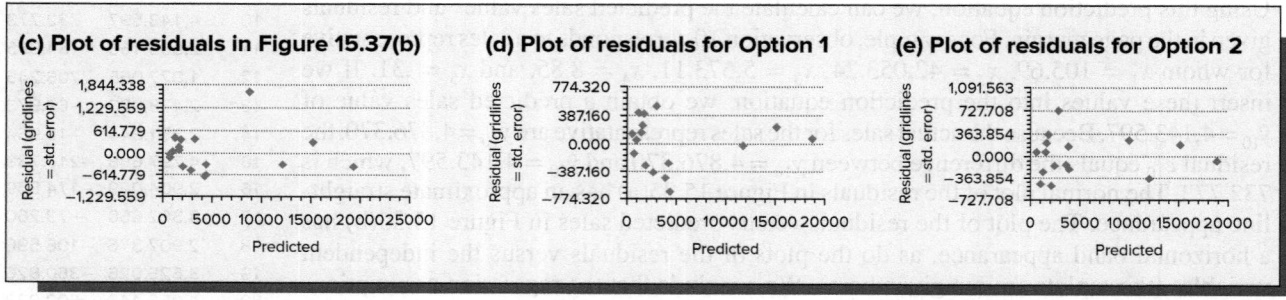

(c) Plot of residuals in Figure 15.37(b)

(d) Plot of residuals for Option 1

(e) Plot of residuals for Option 2

residuals and outlier diagnostics shown in Figure 15.37(b), as well as the residual plot shown in Figure 15.37(c). (Minitab gives the same diagnostics, and at the end of this section we will give formulas for most of these diagnostics.) We now explain the meanings of the diagnostics.

Leverage values

The **leverage value** for an observation is the **distance value** that has been discussed in the optional technical note at the end of Section 15.6. This value is a measure of the distance between the observation's *x* values and the center of all of the observed *x* values. **If the leverage value for an observation is large, the observation is outlying with respect to its *x* values and thus would have substantial leverage in determining the least squares prediction equation.** For example, each of observations 2 and 3 in Figure 15.36 is an outlier with respect to its *x* value and thus would have substantial leverage in determining the position of the least squares line. Moreover, because observations 2 and 3 have inconsistent *y* values, they would pull the least squares line in opposite directions. A **leverage value is considered to be large if it is greater than twice the average of all of the leverage values,** which can be shown to be equal to $2(k + 1)/n$. (MegaStat shades such a leverage value in dark blue.) For example, because there are $n = 17$ observations in Figure 15.37(a) and because the model relating *y* to x_1, x_2, and x_3 utilizes $k = 3$ independent variables, twice the average leverage value is $2(k + 1)/n = 2(3 + 1)/17 = .4706$. Looking at Figure 15.37(b), we see that the leverage values for hospitals 15, 16, and 17 are, respectively, .682, .785, and .863. Because these leverage values are greater than .4706, we conclude that **hospitals 15, 16, and 17 are outliers with respect to their *x* values.** Intuitively, this is because Figure 15.37(a) indicates that x_2 (monthly occupied bed days) is substantially larger for hospitals 15, 16, and 17 than for hospitals 1 through 14. Also note that both x_1 (monthly X-ray exposures) and x_2 (monthly occupied bed days) are substantially larger for hospital 14 than for hospitals 1 through 13. To summarize, we might classify hospitals 1 through 13 as small to medium sized hospitals and hospitals 14, 15, 16, and 17 as larger hospitals.

Residuals and studentized residuals

To identify outliers with respect to their y values, we can use residuals. Any residual that is substantially different from the others is suspect. For example, note from Figure 15.37(b) that the residual for hospital 14, $e_{14} = 1630.503$, seems much larger than the other residuals. Assuming that the labor hours of 10,343.81 for hospital 14 has not been misrecorded, the residual of 1630.503 says that the labor hours are 1630.503 hours more than predicted by the regression model. If we divide an observation's residual by the residual's standard error, we obtain a **studentized residual.** For example, Figure 15.37(b) tells us that the studentized residual for hospital 14 is 2.871. **If the studentized residual for an observation is greater than 2 in absolute value, we have some evidence that the observation is an outlier with respect to its y value.**

Deleted residuals and studentized deleted residuals

Consider again Figure 15.36, and suppose that we use observation 3 to help determine the least squares line. Doing this might draw the least squares line toward observation 3, causing the point prediction $\hat{y}_3$ given by the line to be near y_3 and thus the usual residual $y_3 - \hat{y}_3$ to be small. This would falsely imply that observation 3 is not an outlier with respect to its y value. Moreover, this sort of situation shows the need for computing a **deleted residual.** For a particular observation, observation i, the deleted residual is found by subtracting from y_i the point prediction $\hat{y}_{(i)}$ computed using least squares point estimates based on all n observations except for observation i. Standard statistical software packages calculate the deleted residual for each observation and divide this residual by its standard error to form the **studentized deleted residual.** The experience of the authors leads us to suggest that one should conclude that **an observation is an outlier with respect to its y value if (and only if) the studentized deleted residual is greater in absolute value than $t_{.005}$, which is based on $n - k - 2$ degrees of freedom.** (MegaStat shades such a studentized deleted residual in dark blue.) For the hospital labor needs model, $n - k - 2 = 17 - 3 - 2 = 12$, and therefore $t_{.005} = 3.055$. The studentized deleted residual for hospital 14, which equals 4.558 [see Figure 15.37(b)], is greater in absolute value than $t_{.005} = 3.055$. Therefore, we conclude that **hospital 14 is an outlier with respect to its y value.**

An example of dealing with outliers

One option for dealing with the fact that hospital 14 is an outlier with respect to its y value is to assume that hospital 14 has been run inefficiently. Because we need to develop a regression model using efficiently run hospitals, based on this assumption we would remove hospital 14 from the data set. If we perform a regression analysis using a model relating y to x_1, x_2, and x_3 with hospital 14 removed from the data set (we call this **Option 1**), we obtain a standard error of $s = 387.16$. This s is considerably smaller than the large standard error of 614.779 caused by hospital 14's large residual when we use all 17 hospitals to relate y to x_1, x_2, and x_3.

   A second option is motivated by the fact that *large organizations sometimes exhibit inherent inefficiencies.* To assess whether there might be a *general large hospital inefficiency,* we define a dummy variable D_L that equals 1 for the larger hospitals 14–17 and 0 for the smaller hospitals 1–13. If we fit the resulting regression model $y = \beta_0 + \beta_1 x_1 + \beta_2 x_2 + \beta_3 x_3 + \beta_4 D_L + \varepsilon$ to all 17 hospitals (we call this **Option 2**), we obtain a b_4 of 2871.78 and a p-value for testing H_0: $\beta_4 = 0$ of .0003. This indicates the existence of a large hospital inefficiency that is estimated to be an extra 2871.78 hours per month. In addition, the dummy variable model's s is 363.854, which is slightly smaller than the s of 387.16 obtained using Option 1. The studentized deleted residual for hospital 14 using the dummy variable model tells us what would happen if we removed hospital 14 from the data set and predicted y_{14} by using a newly fitted dummy variable model. In the exercises the reader will show that **the prediction obtained, which uses a large hospital inefficiency estimate based on the remaining large hospitals 15,**

Obs	Cook's D
1	0.002
2	0.000
3	0.001
4	0.028
5	0.000
6	0.014
7	0.002
8	0.009
9	0.016
10	0.049
11	0.009
12	0.067
13	0.006
14	0.353
15	0.541
16	0.897
17	5.033

Hosp	Dffits
1	−0.07541
2	−0.02404
3	0.04383
4	0.32657
5	0.04213
6	−0.22799
7	0.08818
8	0.18406
9	−0.25179
10	−0.44871
11	0.18237
12	−0.52368
13	−0.14509
14	1.88820
15	−1.47227
16	1.89295
17	−4.96226

	Cook's D Influence (Option 2)
1	0.0699061232
2	0.0034984938
3	0.0223768125
4	0.0022035895
5	0.0009130147
6	0.0525707595
7	0.0074281942
8	0.0184548837
9	0.0049568588
10	0.009288363
11	0.1152212446
12	0.0245065598
13	0.0074872707
14	0.8768968657
15	1.4216976841
16	0.8978682961
17	0.7377364159

16, and 17, indicates that hospital 14's labor hours are not unusually large. This justifies leaving hospital 14 in the data set when using the dummy variable model. In summary, both Options 1 and 2 seem reasonable. The reader will further compare these options in the exercises.

Cook's D, Dfbetas, and Dffits

If a particular observation, observation i, is an outlier with respect to its y and/or x values, it might significantly *influence* the least squares point estimates of the model parameters. To detect such influence, we compute **Cook's distance measure** (or **Cook's D**) for observation i, which we denote as D_i. To understand D_i, let $F_{.50}$ denote the 50th percentile of the F distribution based on $(k + 1)$ numerator and $n - (k + 1)$ denominator degrees of freedom. It can be shown that **if D_i is greater than $F_{.50}$, then removing observation i from the data set would significantly change (as a group) the least squares point estimates of the model parameters.** In this case we say that **observation i is influential.** For example, suppose that we relate y to x_1, x_2, and x_3 using all $n = 17$ observations in Figure 15.37(a). Noting that $k + 1 = 4$ and $n - (k + 1) = 13$, we find (using Excel) that $F_{.50} = .8845$. The MegaStat output in the page margin tells us that both $D_{16} = .897$ and $D_{17} = 5.033$ are greater than $F_{.50} = .8845$ (see the dark blue shading). It follows that removing either hospital 16 or 17 from the data set would significantly change (as a group) the least squares estimates of the model parameters.

To assess whether a particular least squares point estimate would significantly change, we can use an advanced statistical software package such as SAS, which gives the following **difference in estimate of β_j statistics (Dfbetas)** for hospitals 16 and 17:

	INTERCEP	X1	X2	X3
Obs	**Dfbetas**	**Dfbetas**	**Dfbetas**	**Dfbetas**
16	0.9880	−1.4289	1.7339	−1.1029
17	0.0294	−3.0114	1.2688	0.3155

Examining the Dfbetas statistics, we see that hospital 17's Dfbetas for the independent variable x_1 (monthly X-ray exposures) equals −3.0114, which is *negative and greater in absolute value than 2*, a sometimes used critical value for Dfbetas statistics. This implies that removing hospital 17 from the data set would *significantly decrease* the least squares point estimate of the effect, β_1, of monthly X-ray exposures on monthly labor hours. One possible consequence might then be that our model would *significantly underpredict* the monthly labor hours for a hospital which [like hospital 17—see Figure 15.37(a)] has a particularly large number of monthly X-ray exposures. In fact, consider the Minitab output in the page margin of the **difference in fits statistic (Dffits).** Dffits for hospital 17 equals −4.96226, which is *negative and greater in absolute value than the critical value 2.* This implies that removing hospital 17 from the data set would *significantly decrease* the point prediction of the monthly labor hours for a hospital that has the same values of x_1, x_2, and x_3 as does hospital 17. Moreover, although it can be verified that using the previously discussed Option 1 or Option 2 to deal with hospital 14's large residual substantially reduces Cook's D, Dfbetas for x_1, and Dffits for hospital 17, these or similar statistics remain or become somewhat significant for the large hospitals 15, 16, and 17 (see the JMP ouput on the page margin of the values of Cook's D when using option 2). The practical implication is that if we wish to predict monthly labor hours for questionably run *large* hospitals, it is very important to keep all of the efficiently run large hospitals 15, 16, and 17 in the data set. (Furthermore, it would be desirable to add information for additional efficiently run large hospitals to the data set).

A technical note (Optional)

Let h_i and e_i denote the leverage value and residual for observation i based on a regression model using k independent variables. Then, for observation i, the studentized residual is e_i divided by $s\sqrt{1 - h_i}$, the deleted residual is e_i divided by $(1 - h_i)$, the studentized deleted residual (denoted t_i) is $e_i\sqrt{n - k - 2}$ divided by $\sqrt{SSE(1 - h_i) - e_i^2}$, Cook's distance measure is $e_i^2 h_i$ divided by $(k + 1)s^2(1 - h_i)^2$, and the difference in fits statistic is $t_i[h_i/(1 - h_i)]^{1/2}$. The formula for the difference in estimate of β_j statistics is very complicated and will not be given here.

Exercises for Section 15.11

CONCEPTS ᴹᶜ connect

15.49 Discuss how we use residual plots to check the regression assumptions for multiple regression.

15.50 List the tools that we use to identify an outlier with respect to its y value and/or x values.

METHODS AND APPLICATIONS

15.51 **The Tasty Sub Shop Case**
Use the following residual plots from the model relating *Revenue* to *Population* and *Business Rating* to check for violations of the regression assumptions.
DS TastySub2

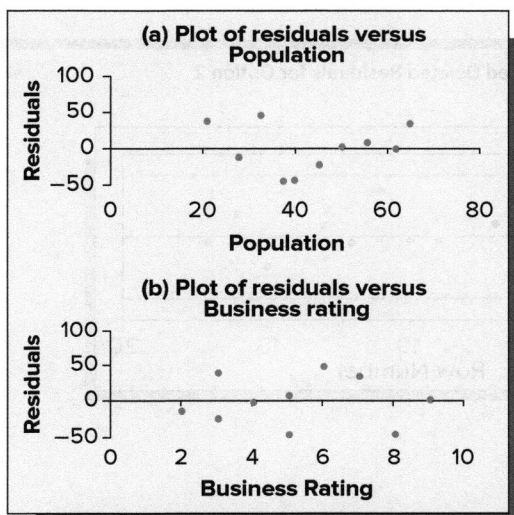

(a) Plot of residuals versus Population

(b) Plot of residuals versus Business rating

15.52 THE HOSPITAL LABOR NEEDS CASE
DS HospLab   **DS** HospLab4

(1) The studentized deleted residuals for Options 1 and 2 are given on the right page margin (see **SDR1** and **SDR2**). Figures 15.38, 15.39, and 15.40 give JMP plots of the studentized deleted residuals in Figure 15.37 (b) and the studentized residuals for Options 1 and 2. The red band critical values are based on $n-k-2$ degrees of freedom and what is called *95 percent simultaneous Bonferroni inference*

(which we will not discuss). It suffices to know that the red band critical values are plus and minus $t_{.025/n}$. For example, Option 2 uses 17 observations and a model with $k = 4$ parameters. Therefore, $n - k - 2 = 11$ and the red band critical values are plus and minus $t_{.025/17} = t_{.00147} = 3.800413$. Analyze the plots and state what they say.
(2) Is hospital 14 an outlier with respect to its y value when using Option 2? **(3)** Consider a questionable large hospital ($D_L = 1$) for which Xray = 56,194, BedDays = 14,077.88, and Length = 6.89. Also, consider the labor needs in an efficiently run large hospital described by this combination of values of the independent variables. The 95 percent prediction intervals for these labor needs given by the models of Options 1 and 2 are, respectively, [14,906.24, 16,886.26] and [15,175.04, 17,030.01]. By comparing these prediction intervals, by analyzing the residual plots for Options 1 and 2 given in Figure 15.37(d) and (e), and by using your conclusions regarding the studentized deleted residuals, recommend which option should be used. **(4)** What would you conclude if the questionable large hospital used 17,207.31 monthly labor hours?

Obs	SDR1	SDR2
1	−0.333	−1.439
2	0.404	0.233
3	0.161	−0.750
4	1.234	0.202
5	0.425	0.213
6	−0.795	−1.490
7	0.677	0.617
8	1.117	1.010
9	−1.078	−0.409
10	−1.359	−0.400
11	1.461	2.571
12	−2.224	−0.624
13	−0.685	0.464
14		1.406
15	−0.137	−2.049
16	1.254	1.108
17	0.597	−0.639

15.53 Recall that Figure 14.27(a) gives $n = 16$ weekly values of Pages Bookstore sales (y), Pages' advertising expenditure (x_1), and competitor's advertising expenditure (x_2). When we fit the model $y = \beta_0 + \beta_1x_1 + \beta_2x_2 + \varepsilon$ to the data, we find that the Durbin–Watson statistic is $d = 1.6\overline{3}$. Use the partial Durbin–Watson table in the page margin to test for positive autocorrelation by setting α equal to .05.

	$k = 2$	
n	$d_{L,.05}$	$d_{U,.05}$
15	0.95	1.54
16	0.98	1.54
17	1.02	1.54
18	1.05	1.53

FIGURE 15.38   **JMP Plot of Studentized Deleted Residuals in Figure 15.37 (b)**

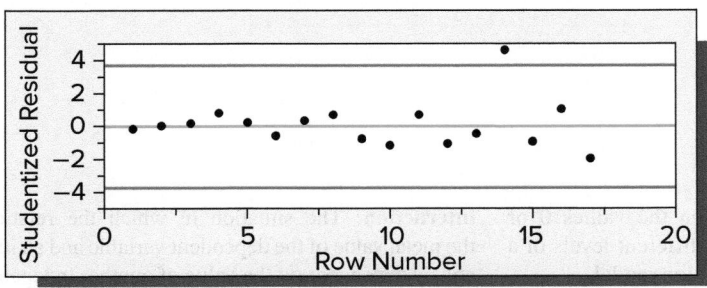

FIGURE 15.39   JMP Plot of Studentized Deleted Residuals for Option 1

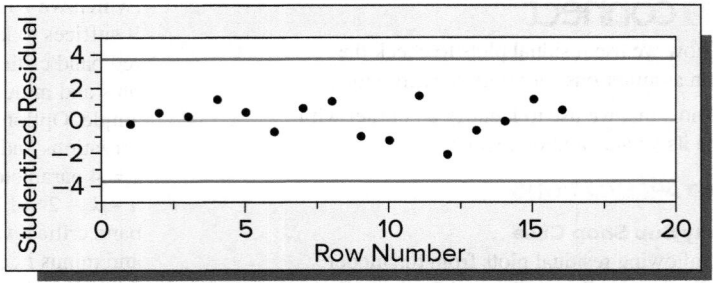

FIGURE 15.40   JMP Plot of Studentized Deleted Residuals for Option 2

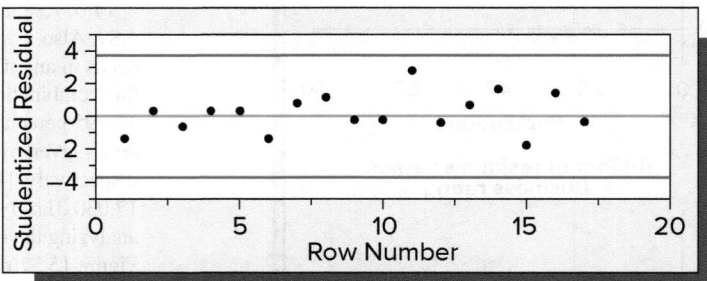

Chapter Summary

This chapter has discussed **multiple regression analysis.** We began by considering the **multiple regression model,** the least **squares point estimates** of the model parameters, and some ways to judge overall model utility—the **multiple coefficient of determination,** and the **adjusted multiple coefficient of determination.** We next presented the assumptions behind the multiple regression model, the **standard error,** and the **overall *F* test.** Then we considered testing the significance of a single independent variable in a multiple regression model, calculating a **confidence interval** for the mean value of the dependent variable, and calculating a **prediction interval** for an individual value of the dependent variable. We continued this chapter by explaining the use of **dummy variables** to model **qualitative** independent variables and the use of **squared and interaction variables.** We then considered **multicollinearity,** which can adversely affect the ability of the *t* statistics and associated *p*-values to assess the importance of the independent variables in a regression model. For this reason, we need to determine if the overall model gives a **high R^2,** a **small *s*,** a **high adjusted R^2,** **short prediction intervals,** and a **small C.** We explained how to compare regression models on the basis of these criteria, and we considered **stepwise regression, backward elimination,** and the **partial *F* test.** We concluded this chapter by discussing the use of **residual analysis** (including the detection of **outliers**) to check the assumptions for multiple regression models.

Glossary of Terms

dummy variable: A variable that takes on the values 0 or 1 and is used to describe the effects of the different levels of a qualitative independent variable in a regression model.

interaction: The situation in which the relationship between the mean value of the dependent variable and an independent variable is dependent on the value of another independent variable.

multicollinearity: The situation in which the independent variables used in a regression analysis are related to each other.

multiple regression model: An equation that describes the relationship between a dependent variable and more than one independent variable.

stepwise regression (and **backward elimination**): Iterative model building techniques for selecting important predictor variables.

Important Formulas and Tests

The least squares point estimates: page 644

The multiple regression model: page 646

Point estimate of a mean value of y: page 646

Point prediction of an individual value of y: page 646

Total variation: page 651

Explained variation: page 651

Unexplained variation: page 651

Multiple coefficient of determination: page 651

Multiple correlation coefficient: page 651

Adjusted multiple coefficient of determination: page 652

Mean square error: page 654

Standard error: page 654

An F test for the multiple regression model: page 655

Testing the significance of an independent variable: page 658

Confidence interval for β_j: page 660

Confidence interval for a mean value of y: pages 661 and 663

Prediction interval for an individual value of y: pages 661 and 663

Distance value (in multiple regression): page 663

The quadratic regression model: page 675

Variance inflation factor: page 682

C statistic: page 686

Partial F test: page 689

PRESS statistic: page 691

R^2(predict): page 691

Studentized deleted residual: pages 699 and 700

Cook's D, Dfbetas, and Dffits: page 700

Supplementary Exercises

connect 15.54 **THE OXFORD HOME BUILDER CASE** ⬥ OxHome

Part I The trend in home building in recent years has been to emphasize open spaces and great rooms, rather than smaller living rooms and family rooms. A builder of speculative homes in the college community of Oxford, Ohio, had been building such homes, but his homes had been taking many months to sell and had been selling for substantially less than the asking price. In order to determine what types of homes would attract residents of the community, the builder contacted a statistician at a local college. The statistician went to a local real estate agency and obtained the data in Table 14.12. This table presents the sales price y, square footage x_1, number of rooms x_2, number of bedrooms x_3, and age x_4 for each of 60 single-family residences recently sold in the community. Use Excel, JMP, or Minitab to perform a regression analysis of these data using the model

$$y = \beta_0 + \beta_1 x_1 + \beta_2 x_2 + \beta_3 x_3 + \beta_4 x_4 + \varepsilon$$

Discuss why the least squares point estimates of β_2 and β_3 suggest that it might be more profitable when building a house of a specified square footage (1) to include both a (smaller) living room and family room rather than a (larger) great room and (2) to not increase the number of bedrooms (at the cost of another type of room) that would normally be included in a house of the specified square footage. Note: Based on the statistical results, the builder realized that there are many families with children in a college town and that the parents in such families would rather have one living area for the children (the family room) and a separate living area for themselves (the living room). The builder started modifying his open-space homes accordingly and greatly increased his profits.

Part II In order to improve the model of Part I we will try to add squared and interaction terms to this model. We will call $x_1, x_2, x_3,$ and x_4 by the computer names *square feet, rooms, bedrooms,* and *age.* The four possible squared variables will be called *sqft_sq, rooms_sq, bedrooms_sq,* and *age_sq.* The six possible (pairwise) interaction variables will be called *sqft ∗ rooms, sqft ∗ bedrooms, sqft ∗ age, rooms ∗ bedrooms, rooms ∗ age,* and *bedrooms ∗ age.* Consider having Minitab evaluate all possible models involving these squared and interaction variables, where the four linear variables are included in each possible model. If we have Minitab do this and find the best two models of each size in terms of s, we obtain the output in Figure 15.41. Examining the Minitab output, we might choose as the best model the model that uses *rooms_sq, bedrooms_sq, sqft ∗ rooms, sqft ∗ age,* and *rooms ∗ age,* in addition to the four linear variables. A partial Minitab regression output obtained by performing a regression analysis of the data in Table 14.12 by using this best model is shown in Figure 15.42.

a Using Figure 15.41, discuss why it is reasonable to choose the model of Figure 15.42 as "best".

b Using the "best" model, calculate **(1)** a point estimate of the mean sales price for all new homes having 2,300 square feet, four bedrooms, and eight total rooms and **(2)** a point estimate of the mean sales price for all new homes having 2,300 square feet, four bedrooms, and seven total rooms. Subtract these point estimates. Based on your results, would you recommend that the builder of a new home having 2,300 square feet, four bedrooms, a kitchen, and a dining room include a large great room in the home or instead include both a (smaller) living room and family room? Justify your answer.

FIGURE 15.41   Model Comparisons Using Additional Squared and Interaction Variables in the Oxford Home Builder Case (for Exercise 15.64)

```
Response is sales price
The following variables are included in all models: square feet
rooms bedrooms age

                                                r
                                          s     o
                                          q     o         b
                            b             f     m         e
                            e    s        t     s         d
                            d    q        *     *    r    r
                       r    r    f    b   s  b   o    o    o
                  s    o    o    t    e   q  o   o    o    o
                  q    o    a    *    d   f  d   m    m    m
                  f    m    m    g    r   r  t   r    s    s
                  t    s    s    e    o   o  *   t    s    *
                  _    _    _    o    o   a  o   o    a    a
         R-Sq  R-Sq Mallows     s s s m m g m g g
 Vars R-Sq (adj) (pred)  Cp    S   q q q s s e s e e
```

Vars	R-Sq	R-Sq (adj)	R-Sq (pred)	Mallows Cp	S
1	83.7	82.2	79.4	18.9	24.331
1	83.7	82.2	78.2	19.2	24.384
2	85.6	83.9	80.9	13.4	23.142
2	85.3	83.7	79.4	14.4	23.318
3	87.3	85.6	81.6	8.3	21.907
3	86.9	85.1	75.9	9.9	22.255
4	88.0	86.1	82.9	7.5	21.519
4	87.8	85.9	82.8	8.3	21.695
5	88.6	86.5	84.3	6.9	21.172
5	88.4	86.4	83.5	7.6	21.319
6	88.9	86.6	83.5	7.8	21.118
6	88.7	86.4	83.0	8.4	21.269
7	89.0	86.5	81.2	9.3	21.223
7	88.9	86.4	82.6	9.7	21.315
8	89.1	86.3	81.0	11.0	21.384
8	89.0	86.2	80.1	11.2	21.428
9	89.1	86.0	79.8	13.0	21.613
9	89.1	86.0	74.4	13.0	21.614
10	89.1	85.7	73.2	15.0	21.852

FIGURE 15.42   Partial Minitab Regression Output Using the Best Oxford Home Builder Case Model (for Exercise 15.64)

Model Summary

S	R-sq	R-sq(adj)	R-sq(pred)
21.1715	88.60%	86.54%	84.25%

Coefficients

Term	Coef	SE Coef	T-Value	P-Value	VIF
Constant	17.1	52.5	0.33	0.745	
square feet	−0.0039	0.0456	−0.08	0.933	50.04
rooms	−14.5	14.2	−1.02	0.313	60.39
bedrooms	88.3	36.0	2.46	0.018	67.50
age	−0.428	0.714	−0.60	0.552	38.15
rooms_sq	−2.53	1.54	−1.64	0.107	146.18
bedrooms_sq	−20.19	6.47	−3.12	0.003	70.37
sqft*rooms	0.02847	0.00714	3.99	0.000	143.98
sqft*age	−0.002813	0.000777	−3.62	0.001	125.06
rooms*age	0.630	0.197	3.19	0.002	164.83

FIGURE 15.43   Minitab Regression Analysis of the Bakery Sales Data Using Model 1:
$$y = \beta_B + \beta_M D_M + \beta_T D_T + \varepsilon$$
(for Exercise 15.55)

Analysis of Variance

Source	DF	Adj SS	Adj MS	F-Value	P-Value
Regression	2	2273.88	1136.94	184.57	0.000
DMiddle	1	1373.88	1373.88	223.03	0.000
DTop	1	55.47	55.47	9.00	0.009
Error	15	92.40	6.16		
Total	17	2366.28			

Model Summary

S	R-sq	R-sq(adj)	R-sq(pred)
2.48193	96.10%	95.57%	94.38%

Coefficients

Term	Coef	SE Coef	T-Value	P-Value	VIF
Constant	55.80	1.01	55.07	0.000	
DMiddle	21.40	1.43	14.93	0.000	1.33
DTop	−4.30	1.43	−3.00	0.009	1.33

Fit	95% CI	95% PI
77.2	(75.0403, 79.3597)	(71.4860, 82.9140)

15.55 THE SUPERMARKET CASE DS BakeSale

The Tastee Bakery Company supplies a bakery product to many supermarkets in a metropolitan area. The company wishes to study the effect of the height of the shelf display employed by the supermarkets on monthly sales, y (measured in cases of 10 units each), for this product. Shelf display height has three levels—bottom (B), middle (M), and top (T). For each shelf display height, six supermarkets of equal sales potential are randomly selected, and each supermarket displays the product using its assigned shelf height for a month. At the end of the month, sales of the bakery product at the 18 participating stores are recorded, and the data in Table 15.16 are obtained. To compare the population mean sales amounts μ_B, μ_M, and μ_T that would be obtained by using the bottom, middle, and top display heights, we use the following dummy variable regression model: $y = \beta_B + \beta_M D_M + \beta_T D_T + \varepsilon$, which we call Model 1. Here, D_M equals 1 if a middle display height is used and 0 otherwise; D_T equals 1 if a top display height is used and 0 otherwise.[1]

a Because the expression $\beta_B + \beta_M D_M + \beta_T D_T$ represents mean monthly sales for the bakery product, the definitions of the dummy variables imply, for example, that $\mu_T = \beta_B + \beta_M (0) + \beta_T (1) = \beta_B + \beta_T$. (1) In a similar fashion, show that $\mu_B = \beta_B$ and $\mu_M = \beta_B + \beta_M$. (2) By appropriately subtracting the expressions for μ_B, μ_M, and μ_T, show that $\mu_M - \mu_B = \beta_M$, $\mu_T - \mu_B = \beta_T$, and $\mu_M - \mu_T = \beta_M - \beta_T$.

b Use the overall F statistic in Figure 15.43 to test H_0: $\beta_M = \beta_T = 0$, or, equivalently, H_0: $\mu_B = \mu_M = \mu_T$. Interpret the practical meaning of the result of this test.

TABLE 15.16
Bakery Sales Study Data (Sales in Cases)
DS BakeSale

Shelf Display Height

Bottom (B)	Middle (M)	Top (T)
58.2	73.0	52.4
53.7	78.1	49.7
55.8	75.4	50.9
55.7	76.2	54.0
52.5	78.4	52.1
58.9	82.1	49.9

c Consider the following two differences in means: $\mu_M - \mu_B = \beta_M$ and $\mu_T - \mu_B = \beta_T$. Use information in Figure 15.43 to (1) find a point estimate of, (2) test the significance of, and (3) find a 95 percent confidence interval for each difference. (Hint: Use the confidence interval formula.) Interpret your results.

d Consider the following alternative model: $y = \beta_T + \beta_B D_B + \beta_M D_M + \varepsilon$, which we call Model 2. Here, D_B equals 1 if a bottom display height is used and 0 otherwise. This

model implies that $\mu_M - \mu_T = \beta_M$. Use Excel, JMP, or Minitab to perform a regression analysis of the data in Table 15.16 using this model. Then, (1) find a point estimate of, (2) test the significance of, and (3) find a 95 percent confidence interval for $\mu_M - \mu_T = \beta_M$. Interpret your results.

e Show by hand calculation that both Models 1 and 2 give the same point estimate $\hat{y} = 77.2$ of mean monthly sales when using a middle display height.

f Figure 15.43 gives the fit for a middle display height. Find (1) a 95 percent confidence interval for mean sales when using a middle display height, and (2) a 95 percent prediction interval for individual sales during a month at a supermarket that employs a middle display height.

15.56 THE FRESH DETERGENT CASE DS Fresh3

Recall from Exercise 15.36 that Enterprise Industries has advertised Fresh liquid laundry detergent by using three different advertising campaigns—advertising campaign A (television commercials), advertising campaign B (a balanced mixture of television and radio commercials), and advertising campaign C (a balanced mixture of television, radio, newspaper, and magazine ads). To compare the effectiveness of these advertising campaigns, consider using two models, Model 1 and Model 2, that are shown with corresponding partial Excel outputs in Figure 15.44. In these models y is demand for Fresh; x_4 is the price difference; x_3 is Enterprise Industries' advertising expenditure for Fresh; D_A equals 1 if advertising campaign A is used in a sales period and 0 otherwise; D_B equals 1 if advertising campaign B is used in a sales period and 0 otherwise; and D_C equals 1 if advertising campaign C is used in a sales period and 0 otherwise. Moreover, in Model 1 the parameter β_5 represents the effect on mean demand of advertising campaign B compared to advertising campaign A, and the parameter β_6 represents the effect on mean demand of advertising campaign C compared to advertising campaign A. In Model 2 the parameter β_6 represents the effect on mean demand of advertising campaign C compared to advertising campaign B.

a Compare advertising compaigns A, B, and C by finding 95 percent confidence intervals for (1) β_5 and β_6 in Model 1 and (2) β_6 in Model 2. Interpret the intervals.

b Using Model 1 or Model 2, a point prediction of Fresh demand when $x_4 = .20$, $x_3 = 6.50$, and campaign C will be used is 8.50068 (that is, 850,068 bottles). Show (by hand calculation) that Model 1 and Model 2 give the same point prediction.

FIGURE 15.44 Excel Output for the Fresh Detergent Case (for Exercise 15.56)

(a) Partial Excel output for Model 1:
$$y = \beta_0 + \beta_1 x_4 + \beta_2 x_3 + \beta_3 x_3^2 + \beta_4 x_4 x_3 + \beta_5 D_B + \beta_6 D_C + \varepsilon$$

	Coefficients	Lower 95%	Upper 95%
Intercept	25.612696	15.6960	35.5294
X4	9.0587	2.7871	15.3302
X3	−6.5377	−9.8090	−3.2664
X3SQ	0.5844	0.3158	0.8531
X4X3	−1.1565	−2.0992	−0.2137
DB	0.2137	0.0851	0.3423
DC	0.3818	0.2551	0.5085

(b) Partial Excel output for Model 2:
$$y = \beta_0 + \beta_1 x_4 + \beta_2 x_3 + \beta_3 x_3^2 + \beta_4 x_4 x_3 + \beta_5 D_A + \beta_6 D_C + \varepsilon$$

	Coefficients	Lower 95%	Upper 95%
Intercept	25.8264	15.9081	35.7447
X4	9.05868	2.7871	15.3302
X3	−6.5377	−9.8090	−3.2664
X3SQ	0.58444	0.3158	0.8531
X4X3	−1.1565	−2.0992	−0.2137
DA	−0.2137	−0.3423	−0.0851
DC	0.16809	0.0363	0.2999

[1] In general, the regression approach of this exercise produces the same comparisons of several population means that are produced by **one-way analysis of variance** (see Section 12.2).

TABLE 15.17

The Least Squares Point Estimates for Model 3

$b_0 = 28.6873 \ (<.0001)$
$b_1 = 10.8253 \ (.0036)$
$b_2 = -7.4115 \ (.0002)$
$b_3 = .6458 \ (<.0001)$
$b_4 = -1.4156 \ (.0091)$
$b_5 = -.4807 \ (.5179)$
$b_6 = -.9351 \ (.2758)$
$b_7 = .10722 \ (.3480)$
$b_8 = .20349 \ (.1291)$

c Consider the alternative model

$$y = \beta_0 + \beta_1 x_4 + \beta_2 x_3 + \beta_3 x_3^2 + \beta_4 x_4 x_3 + \beta_5 D_B + \beta_6 D_C + \beta_7 x_3 D_B + \beta_8 x_3 D_C + \varepsilon$$

which we call Model 3. The least squares point estimates of the model parameters and their associated p-values (given in parentheses) are as shown in Table 15.17. Let $\mu_{[d,a,A]}, \mu_{[d,a,B]},$ and $\mu_{[d,a,C]}$ denote the mean demands for Fresh when the price difference is d, the advertising expenditure is a, and we use advertising campaigns A, B, and C, respectively. The model of this part implies that

$$\mu_{[d,a,A]} = \beta_0 + \beta_1 d + \beta_2 a + \beta_3 a^2 + \beta_4 da + \beta_5(0) + \beta_6(0) + \beta_7 a(0) + \beta_8 a(0)$$

$$\mu_{[d,a,B]} = \beta_0 + \beta_1 d + \beta_2 a + \beta_3 a^2 + \beta_4 da + \beta_5(1) + \beta_6(0) + \beta_7 a(1) + \beta_8 a(0)$$

$$\mu_{[d,a,C]} = \beta_0 + \beta_1 d + \beta_2 a + \beta_3 a^2 + \beta_4 da + \beta_5(0) + \beta_6(1) + \beta_7 a(0) + \beta_8 a(1)$$

(1) Using these equations, verify that $\mu_{[d,a,C]} - \mu_{[d,a,A]}$ equals $\beta_6 + \beta_8 a$. (2) Using the least squares point estimates, show that a point estimate of $\mu_{[d,a,C]} - \mu_{[d,a,A]}$ equals .3266 when $a = 6.2$ and equals .4080 when $a = 6.6$. (3) Verify that $\mu_{[d,a,C]} - \mu_{[d,a,B]}$ equals $\beta_6 - \beta_5 + \beta_8 a - \beta_7 a$. (4) Using the least squares point estimates, show that a point estimate of $\mu_{[d,a,C]} - \mu_{[d,a,B]}$ equals .14266 when $a = 6.2$ and equals .18118 when $a = 6.6$. (5) Discuss why these results imply that the larger that advertising expenditure a is, then the larger is the improvement in mean sales that is obtained by using advertising campaign C rather than advertising campaign A or B.

d If we use MegaStat, we can use Models 1, 2, and 3 to predict demand for Fresh in a future sales period when the price difference will be $x_4 = .20$, the advertising expenditure will be $x_3 = 6.50$, and campaign C will be used. The prediction results using Model 1 or Model 2 are given below and the prediction results using Model 3 are also given below.

Model 1 or 2	95% Prediction Interval	
Predicted	**lower**	**upper**
8.50068	8.21322	8.78813

Model 3	95% Prediction Interval	
Predicted	**lower**	**upper**
8.51183	8.22486	8.79879

Which model gives the shortest 95 percent prediction interval for Fresh demand?

e Using all of the results in this exercise, discuss why there might be a small amount of interaction between advertising expenditure and advertising campaign.

15.57 THE FRESH DETERGENT CASE DS Fresh3

The unexplained variation for Model 1 of the previous exercise,

$y = \beta_0 + \beta_1 x_4 + \beta_2 x_3 + \beta_3 x_3^2 + \beta_4 x_4 x_3 + \beta_5 D_B + \beta_6 D_C + \varepsilon$,

is .3936. If we set both β_5 and β_6 in this model equal to 0 (that is, if we eliminate the dummy variable portion of this model), the resulting reduced model has an unexplained variation of 1.0644. Using an α of .05, perform a partial F test of H_0: $\beta_5 = \beta_6 = 0$. (Hint: $n = 30$, $k = 6$, and $k^* = 2$.) If we reject H_0, we conclude that at least two of advertising campaigns A, B,

and C have different effects on mean demand. Many statisticians believe that rejection of H_0 by using the partial F test makes it more legitimate to make pairwise comparisons of advertising campaigns A, B, and C, as we did in part a of the previous exercise. Here, the partial F test is regarded as a *preliminary test of significance*.

15.58 United Oil is attempting to develop a middle-grade gasoline that will deliver high gasoline mileages. As part of its development process, United Oil will study the effect of three middle-grade gasoline types (A, B, and C) and the amounts 0, 1, 2, and 3 units of additive VST. Mileage tests are run, with the results shown on the page margin. (1) Develop an appropriate model describing these data. (2) Use the model to find a point estimate of and a 95 percent confidence interval for mean mileage when the middle-grade gasoline is made with gasoline type B and one unit of additive VST. Hint: Try squared and interaction variables.
DS UnitedOil3

Gas Type	VST Units	Gas Mileage (mpg)
A	0	27.9
A	0	28.4
A	0	27.5
B	0	33.1
B	0	34.6
A	1	32.9
A	1	32.1
B	1	35.7
B	1	34.4
B	1	35.1
C	1	34.0
C	1	33.4
C	1	34.6
A	2	33.4
A	2	32.1
B	2	33.3
C	2	33.1
C	2	32.1
B	3	29.5
B	3	30.6
C	3	28.5
C	3	29.9

DS UnitedOil3

15.59 THE QHIC CASE
DS QHIC

Consider the QHIC data in Figure 14.18. When we performed a regression analysis of these data by using the simple linear regression model, plots of the model's residuals versus x (home value) and $\hat{y}$ (predicted upkeep expenditure) both fanned out and had a "dip," or slightly curved appearance (see Figure 14.18). In order to remedy the indicated violations of the constant variance and correct functional form assumptions, we transformed the dependent variable by taking the square roots of the upkeep expenditures. An alternative approach consists of two steps. First, the slightly curved appearance of the residual plots implies that it is reasonable to add the squared term x^2 to the simple linear regression model. This gives the quadratic regression model $y = \beta_0 + \beta_1 x + \beta_2 x^2 + \varepsilon$. The upper residual plot in the Minitab output of Figure 15.45 shows that a plot of the model's residuals versus x fans out, indicating a violation of the constant variance assumption. To remedy this violation, we (in the second step) divide all terms in the quadratic model by x, which gives the transformed model and associated Minitab regression output shown to the right of the residual plots. Here, the lower residual plot is the residual plot versus x, for the transformed model.

a Does the lower residual plot indicate the constant variance assumption holds for the transformed model? Explain.

b Consider a home worth $220,000. We let μ_0 represent the mean yearly upkeep expenditure for all homes worth $220,000, and we let y_0 represent the yearly upkeep expenditure for an individual home worth $220,000. (1) The bottom of the Minitab regression output tells us that $\hat{y}/220 = 5.635$ is a point estimate of $\mu_0/220$ and a point prediction of $y_0/220$. Multiply this result by 220 to obtain $\hat{y}$. (2) Multiply the ends of the confidence interval and prediction interval shown on the Minitab output by 220. This will give a 95 percent confidence interval for μ_0 and a 95 percent prediction interval for y_0. (3) Suppose that QHIC

FIGURE 15.45   Residual Plots and Transformed Model Regression for the QHIC Case (for Exercise 15.59)

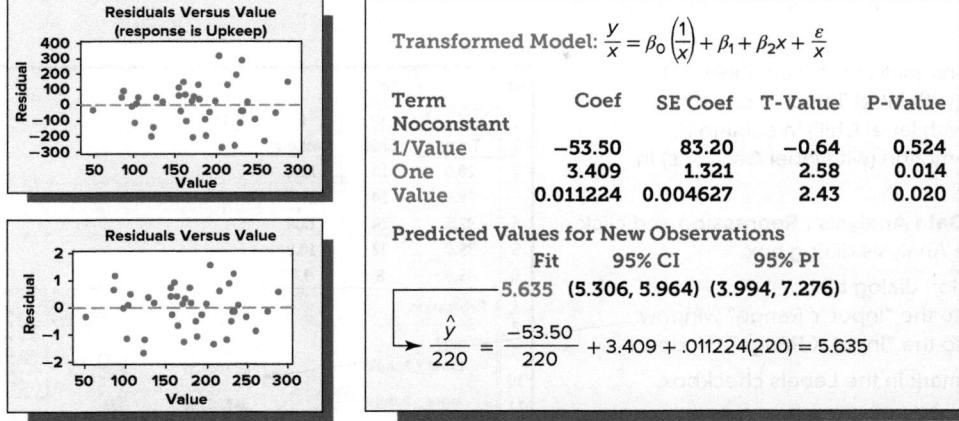

Transformed Model: $\dfrac{y}{x} = \beta_0\left(\dfrac{1}{x}\right) + \beta_1 + \beta_2 x + \dfrac{\varepsilon}{x}$

Term	Coef	SE Coef	T-Value	P-Value
Noconstant				
1/Value	−53.50	83.20	−0.64	0.524
One	3.409	1.321	2.58	0.014
Value	0.011224	0.004627	2.43	0.020

Predicted Values for New Observations

Fit	95% CI	95% PI
5.635	(5.306, 5.964)	(3.994, 7.276)

$$\dfrac{\hat{y}}{220} = \dfrac{-53.50}{220} + 3.409 + .011224(220) = 5.635$$

has decided to send a special, more expensive advertising brochure to any home whose value makes QHIC 95 percent confident that the mean upkeep expenditure for all homes having this value is at least $1,000. Should QHIC send a special brochure to a home worth $220,000?

15.60   Although natural gas is currently inexpensive and nuclear power currently (and perhaps deservedly) does not have a good reputation, it is possible that more nuclear power plants will be constructed in the future. Table 15.18 presents data concerning the construction costs of light water reactor (LWR) nuclear power plants. The dependent variable C, construction cost, is expressed in millions of dollars, adjusted to a 1976 base. This dependent variable and 10 potential predictor variables are summarized in Table 15.18. Preliminary analysis of the data and economic theory indicate that variation in cost increases as cost increases. This suggests transforming cost by taking its natural logarithm. **(1)** Build a multiple regression model to predict ln C by using all possible regressions and stepwise regression. **(2)** Use residual analysis to check your model. **(3)** State which variables are important in predicting the cost of constructing an LWR plant, and **(4)** State a prediction equation that can be used to predict ln C.

TABLE 15.18   Nuclear Power Plant Construction Costs and Variable Definitions (for Exercise 15.60)   DS PowerCosts

Plant	C	D	T1	T2	S	PR	NE	CT	BW	N	PT
1	460.05	68.58	14	46	687	0	1	0	0	14	0
2	452.99	67.33	10	73	1,065	0	0	1	0	1	0
3	443.22	67.33	10	85	1,065	1	0	1	0	1	0
4	652.32	68	11	67	1,065	0	1	1	0	12	0
5	642.23	68	11	78	1,065	1	1	1	0	12	0
6	345.39	67.92	13	51	514	0	1	1	0	3	0
7	272.37	68.17	12	50	822	0	0	0	0	5	0
8	317.21	68.42	14	59	457	0	0	0	0	1	0
9	457.12	68.42	15	55	822	1	0	0	0	5	0
10	690.19	68.33	12	71	792	0	1	1	1	2	0
11	350.63	68.58	12	64	560	0	0	0	0	3	0
12	402.59	68.75	13	47	790	0	1	0	0	6	0
13	412.18	68.42	15	62	530	0	0	1	0	2	0
14	495.58	68.92	17	52	1,050	0	0	0	0	7	0
15	394.36	68.92	13	65	850	0	0	0	1	16	0
16	423.32	68.42	11	67	778	0	0	0	0	3	0
17	712.27	69.5	18	60	845	0	1	0	0	17	0
18	289.66	68.42	15	76	530	1	0	1	0	2	0
19	881.24	69.17	15	67	1,090	0	0	0	0	1	0
20	490.88	68.92	16	59	1,050	1	0	0	0	8	0
21	567.79	68.75	11	70	913	0	0	1	1	15	0
22	665.99	70.92	22	57	828	1	1	0	0	20	0
23	621.45	69.67	16	59	786	0	0	1	0	18	0
24	608.8	70.08	19	58	821	1	0	0	0	3	0
25	473.64	70.42	19	44	538	0	0	1	0	19	0
26	697.14	71.08	20	57	1,130	0	0	1	0	21	0
27	207.51	67.25	13	63	745	0	0	0	0	8	1
28	288.48	67.17	9	48	821	0	0	1	0	7	1
29	284.88	67.83	12	63	886	0	0	0	1	11	1
30	280.36	67.83	12	71	886	1	0	0	1	11	1
31	217.38	67.25	13	72	745	1	0	0	0	8	1
32	270.71	67.83	7	80	886	1	0	0	1	11	1

Variable Definitions:

C   Cost (in millions of dollars), adjusted to a 1976 base

D   Date construction permit issues

T1   Time between application for and issue of permit

T2   Time between issue of operating license and construction permit

S   Power plant net capacity (MWe)

PR   Prior existence of an LWR on same site (=1)

NE   Plant constructed in northeast region of USA (=1)

CT   Use of cooling tower (=1)

BW   Nuclear steam supply system manufactured by Babcock-Wilcox (=1)

N   Cumulative number of power plants constructed by each architect-engineer

PT   Partial turnkey plant (=1)

Source: D.R. Cox and E.J. Snell, *Applied Statistics: Principles and Examples* (Chapman and Hall 1981).

Appendix 15.1 ■ Multiple Regression Analysis Using Excel

Multiple regression in Figure 15.5(a) (data file: GasCon2 .xlsx):

- Enter the gas consumption data from Table 15.3: temperatures (with label Temp) in column A, chill indexes (with label Chill) in column B, and gas consumption (with label GasCons) in column C.

- Select **Data : Data Analysis : Regression** and click OK in the Data Analysis dialog box.

- In the Regression dialog box:
  Enter C1:C9 into the "Input Y Range" window.
  Enter A1:B9 into the "Input X Range" window.

- Place a checkmark in the Labels checkbox.

- Be sure that the "Constant is Zero" checkbox is NOT checked.

- Select the "New Worksheet Ply" Output option.

- Click OK in the Regression dialog box to obtain the regression output in a new worksheet.

Note: The independent variables must be in adjacent columns because the "Input X Range" must span the range of the values for all of the independent variables.

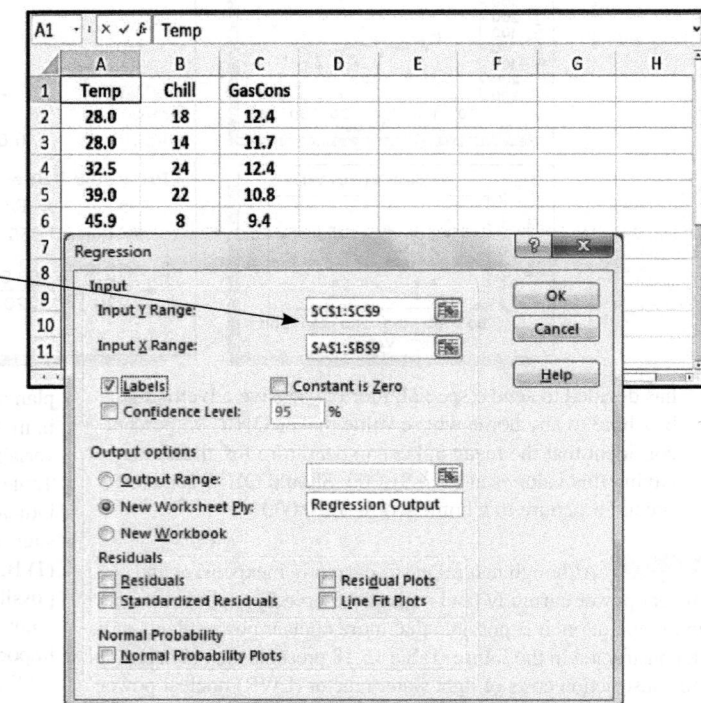

Source: Microsoft Office Excel 2016

To compute a point prediction of natural gas consumption when temperature is 40°F and the chill index is 10:

- The Excel Analysis ToolPak does not provide an option for computing point or interval predictions. A point prediction can be computed from the regression results using Excel cell formulas as follows.

- The estimated regression coefficients and their labels are in cells A17:B19 of the output worksheet and the predictor values 40 and 10 have been placed in cells F2 and F3.

- In cell F4, enter the Excel formula
  =B17+B18*F2+B19*F3
  to compute the point prediction (=10.3331).

Source: Microsoft Office Excel 2016

Sales volume multiple regression with indicator (dummy) variables in Figure 15.14 (data file: Electronics2 .xlsx):

- Enter the sales volume data from Table 15.10: sales volumes (with label Sales) in column A, store locations (with label Location) in column B, and number of households (with label Households) in column C. (The order of the columns is chosen to arrange for an adjacent block of predictor variables.)

- Enter the labels DM and DD in cells D1 and E1.

- Following the definition of the dummy variables DM and DD in Example 15.6, enter the appropriate values of 0 and 1 for these two variables into columns D and E.

- Select **Data : Data Analysis : Regression** and click OK in the Data Analysis dialog box.

- In the Regression dialog box: Enter A1:A16 into the "Input Y Range" window. Enter C1:E16 into the "Input X Range" window.

- Place a checkmark in the Labels checkbox.

- Select the "New Worksheet Ply" Output option.

- Click OK in the Regression dialog box to obtain the regression results in a new worksheet.

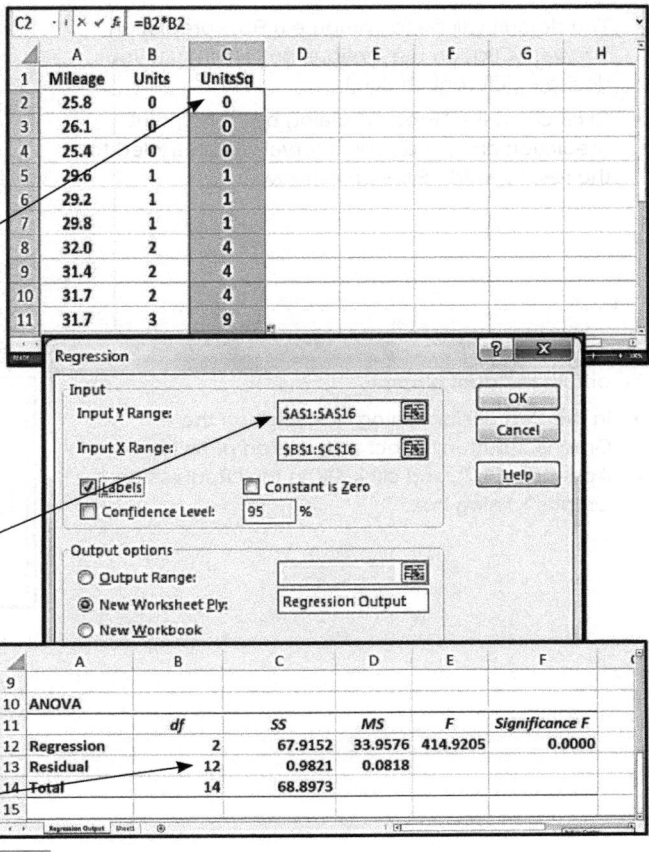

Source: Microsoft Office Excel 2016

Gasoline additive multiple linear regression with a quadratic term similar to Figure 15.23 (data file: GasAdd .xlsx):

- Enter the gas mileage data from Table 15.13: mileages (with label Mileage) in column A and units of additive (with label Units) in column B. (Units are listed second in order to be adjacent to the squared units predictor.)

- Enter UnitsSq into cell C1.

- Click on cell C2, and enter the formula =B2*B2. Press "Enter" to compute the squared value of Units for the first observation.

- Copy the cell formula of C2 through cell C16 (by double-clicking the drag handle in the lower right corner of cell C2) to compute the squared units for the remaining observations.

- Select **Data : Data Analysis : Regression** and click OK in the Data Analysis dialog box.

- In the Regression dialog box: Enter A1:A16 into the "Input Y Range" window. Enter B1:C16 into the "Input X Range" window.

- Place a checkmark in the Labels checkbox.

- Select the "New Worksheet Ply" Output option.

- Click OK in the Regression dialog box to obtain the regression output in a new worksheet.

Source: Microsoft Office Excel 2016

Appendix 15.2 ■ Multiple Regression Analysis Using Minitab

Multiple linear regression in Figure 15.5(b) (data file: GasCon2.MTW):

- Enter the gas consumption data from Table 15.3 in the data window with average hourly temperatures in C1, chill indexes in C2, and weekly gas consumption in C3; use variable names Temp, Chill, and GasCons.

- Select **Stat : Regression : Regression : Fit Regression Model.**

- In the Regression dialog box, select GasCons as the Response; select Temp and Chill as the Continuous predictors.

- Click OK in the Regression dialog box; the regression results will appear in the Session window.

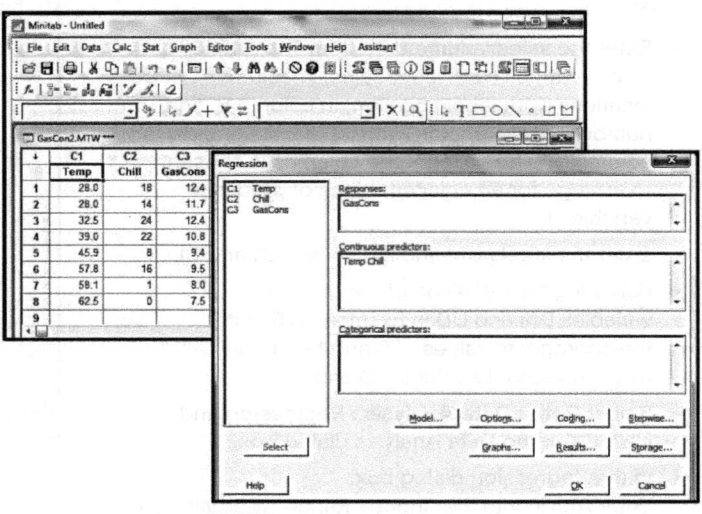

Source: Minitab 18

To compute a prediction of natural gas consumption when temperature is 40° F and the chill index is 10:

- After fitting the regression model, you have a new option in the **Stat : Regression** sequence.

- Select **Stat : Regression : Regression : Predict.**

- In the "Predict" dialog box, GasCons will appear as the Response; choose "Enter individual values" from the drop-box selections; enter 40 as the temperature and 10 as the chill index.

- The default will be to compute a 95% prediction interval. (Click on the Options button should you desire a different choice.)

- Click OK in the "Predict" dialog box. The point prediction and prediction interval will be added to the results in the Session window.

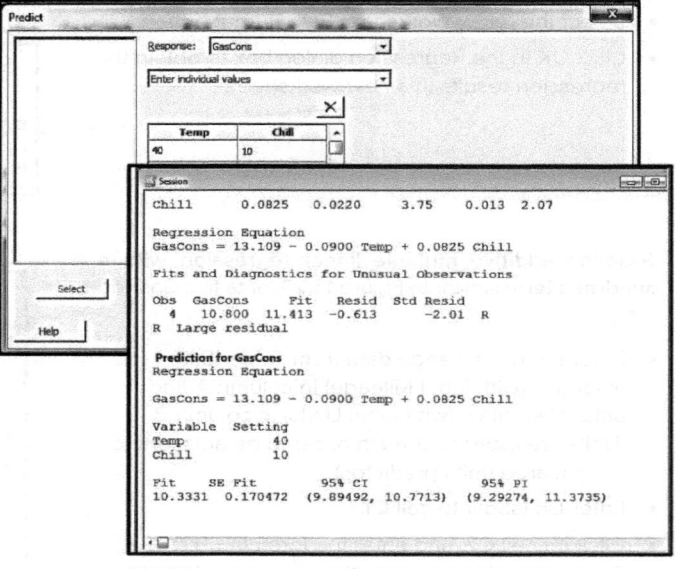

Source: Minitab 18

To obtain **residual plots:**

- In the Regression dialog box, click on the Graphs... button; select the desired plots (see Appendix 14.2), and click OK in the "Regression: Graphs" dialog box.

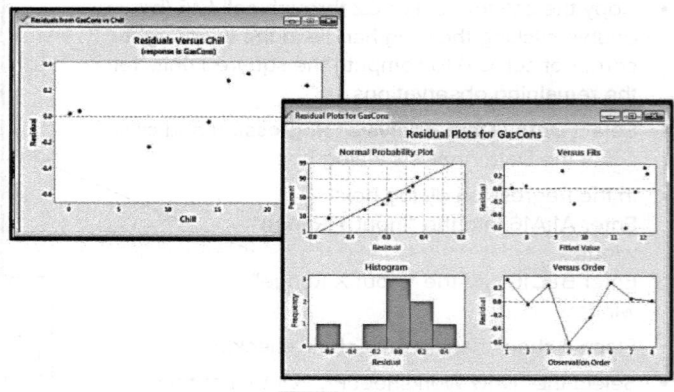

Source: Minitab 18

To obtain outlying and influential observation diagnostics:

- Click the Storage button, and in the "Regression: Storage" dialog box, place checks in the following boxes: Fits (for predicted values), Residuals, Standardized residuals (for studentized residuals), Deleted t residuals (for studentized deleted residuals), Leverages, and Cook's distance.

- Click OK in both the "Regression: Storage" dialog box and the Regression dialog box.

Multiple linear regression with indicator (dummy) variables in Figure 15.15 (data file: Electronics2.MTW):

- Enter the sales volume data from Table 15.10 in the data window with sales volume in C1, location in C2, and number of households in C3. Use variable names Sales, Location, and Households.

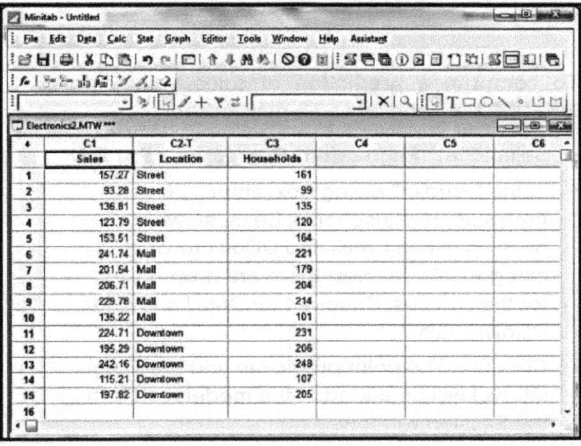

Source: Minitab 18

To create indicator/dummy variables:

- Select **Calc : Make Indicator Variables.**

- In the "Make Indicator Variables" dialog box, enter Location into the "Indicator variables for" window.

- The "Store indicator variables in columns" window lists the distinct values of Location in alphabetical order and provides them default names like "Location_Downtown" in the corresponding column. You can use these assigned names or type in your own, as we have. DDowntown, our first indicator variable, will have 1s in all rows corresponding to Downtown locations and 0s elsewhere. Similarly, DMall will have 1s in all rows corresponding to Mall locations and 0s elsewhere. DStreet is defined in the same fashion.

- Click OK in the "Make Indicator Variables" dialog box to create the new indicator variables.

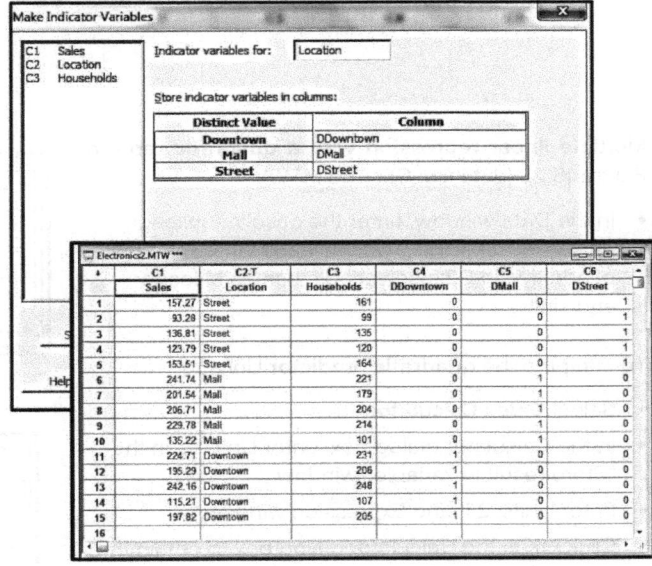

Source: Minitab 18

To fit the multiple regression model:

- Select **Stat : Regression : Regression : Fit Regression Model.**
- In the Regression dialog box, enter Sales in the Response window, Households in the Continuous predictors window, and DMall and DDowntown in the Categorical predictors window.

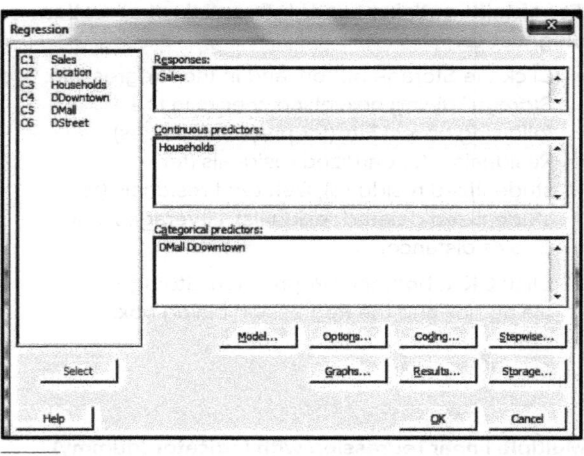

Source: Minitab 18

To compute a prediction of sales volume for 200,000 households and a mall location:

- Select **Stat : Regression : Regression : Predict.**
- In the "Predict" dialog box, enter 200 as the value of Households. Because we have specified DMall and DDowntown as categorical variables, there are drop-boxes for their values. Choose 1 for DMall and 0 for DDowntown.
- Click OK. The point prediction and prediction interval will be added to the model results in the Session window.

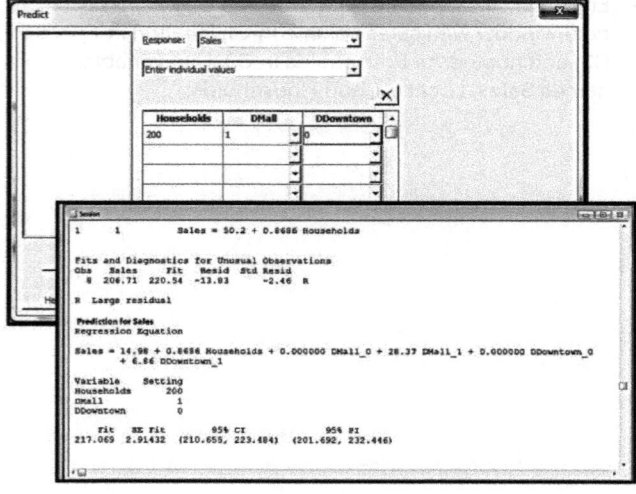

Source: Minitab 18

Multiple linear regression with a quadratic term in Figure 15.23 (data file: GasAdd.MTW):

- In the Data window, enter the gasoline mileage data from Table 15.13 with mileages in C1 and units of additive in C2; use variable names Mileage and Units.

To compute the **quadratic predictor** UnitsSq:

- Select **Calc : Calculator.**
- In the Calculator dialog box, enter UnitsSq in the "Store result in variable" window.
- Enter Units^2 in the Expression window.
- Click OK in the Calculator dialog box to obtain the squared units in C3.

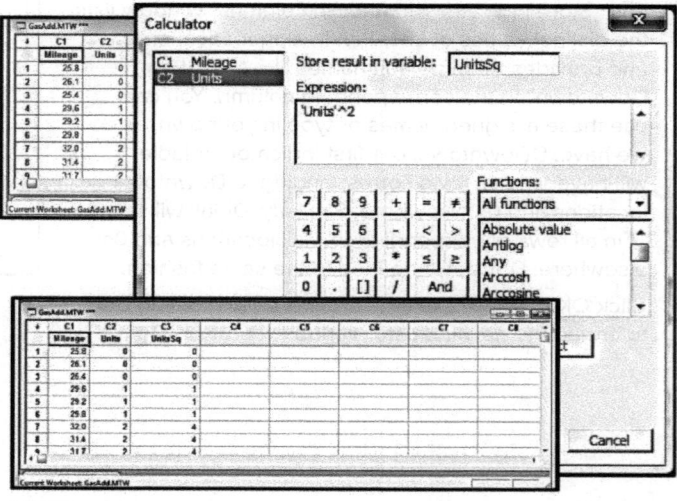

Source: Minitab 18

To fit the quadratic regression model:

- Select **Stat : Regression : Regression : Fit Regression Model.**
- In the Regression dialog box, enter Mileage as the Response variable and Units and UnitsSq as the Continuous predictor variables.
- Click OK in the Regression dialog box.

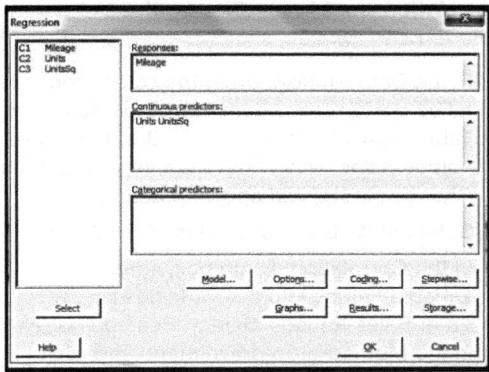

Source: Minitab 18

To compute a **prediction** for mileage when 2.44 units of additive are used:

- Select **Stat : Regression : Regression : Predict.**
- In the "Predict"dialog box, Mileage will appear as the Response; choose "Enter individual values" from the drop-box selections; enter 2.44 as the units and 5.9536 (=2.44^2) as the units squared.
- The default will be to compute a 95% prediction interval. (Click on the Options button should you desire a different choice.)
- Click OK in the "Predict" dialog box. The point prediction and prediction interval will be added to the results in the Session window.

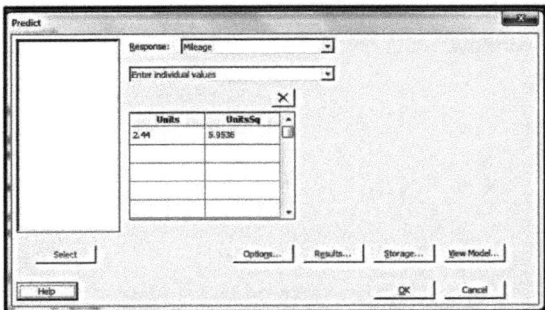

Source: Minitab 18

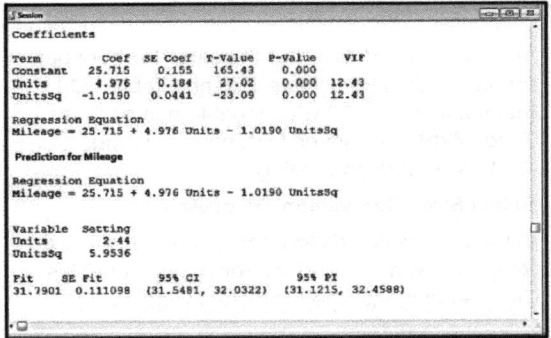

Source: Minitab 18

Alternate approach to specifying a quadratic model:

- Select **Stat : Regression : Regression : Fit Regression Model.**
- In the Regression dialog box, enter Mileage as the Response variable and Units as the Continuous predictor variables.
- Click on Model . . . in the Regression dialog box.
- In the "Regression: Model" dialog box, choose 2 in the "Terms through order" drop-down box. This specifies our choice to use units squared as a predictor in the model. Click OK and continue as above.

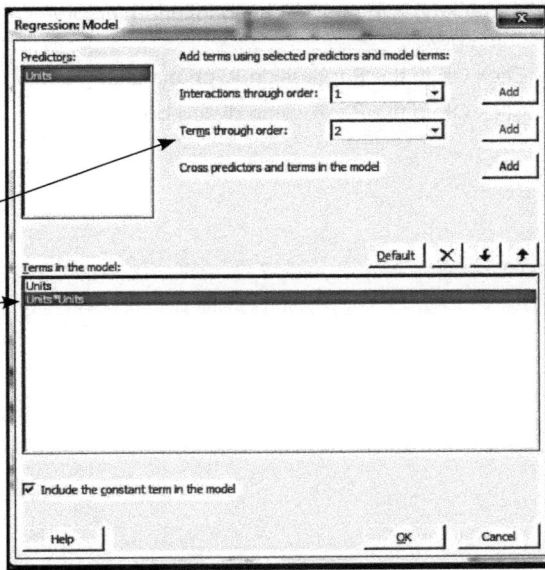

Source: Minitab 18

Correlation matrix in Figure 15.26(a) (data file: SalePerf2.MTW):

- In the Data window, enter the sales representative performance data from Table 15.8 and Table 15.14 into columns C1–C9 with variable names Sales, Time, MktPoten, Adver, MktShare, Change, Accts, WkLoad, and Rating.
- Select **Stat : Basic Statistics : Correlation.**
- In the Correlation dialog box, enter all the variable names into the Variables box or simply enter Sales-Ratings. Select "Pearson correlation" in the Method drop-down menu and click on the Display p-values box.
- Click OK in the Correlation dialog box; the correlation matrix will be displayed in the Session window.

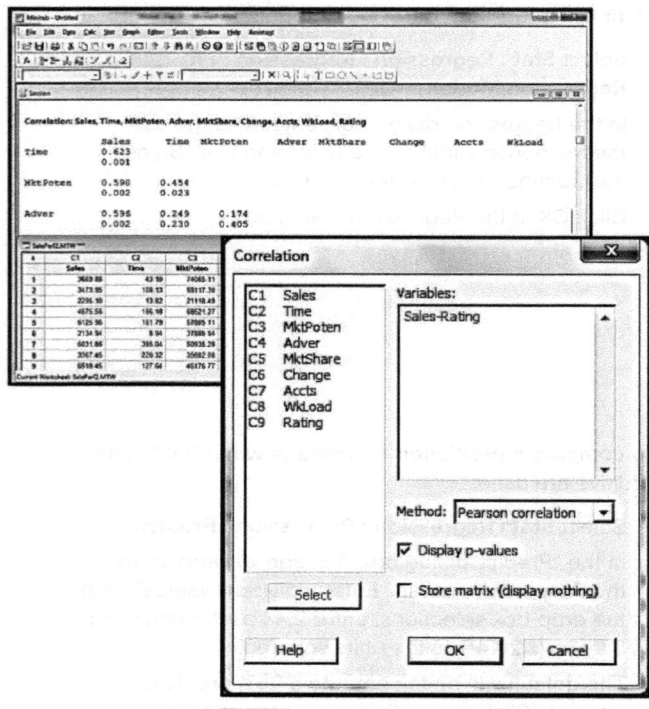

Source: Minitab 18

Variance inflation factors (VIF) in Figure 15.26(b) (data file: SalePerf2.MTW):

- In the Data window, enter the sales territory performance data from Table 15.8 and Table 15.14 into columns C1–C9 with variable names Sales, Time, MktPoten, Adver, MktShare, Change, Accts, WkLoad, and Rating.
- Select **Stat : Regression : Regression.**
- In the Regression dialog box, enter Sales into the Response window and the remaining variables Time—Rating into the Predictors window.
- Click the Options . . . button.
- In the Regression—Options dialog box, place a checkmark in the "Variance inflation factors" checkbox.
- Click OK in the Regression—Options dialog box.
- Click OK in the Regression dialog box.

Best subsets regression in Figure 15.27 (data file: SalePerf2.MTW):

- In the Data window, enter the sales representative performance data from Table 15.8 and Table 15.14 into columns C1–C9; use variable names Sales, Time, MktPoten, Adver, MktShare, Change, Accts, WkLoad, and Rating.

- Select **Stat : Regression : Best Subsets.**

- In the Best Subsets Regression dialog box, enter Sales in the Response window.

- Enter Time-Rating into the "Free predictors" window.

- Click on the Options . . . button.

- In the "Best Subsets Regression: Options" dialog box, enter 2 in the "Models of each size to print" window.

- Click OK in the "Best Subsets Regression: Options" and in the Best Subsets Regression dialog boxes. The results appear in the Session window.

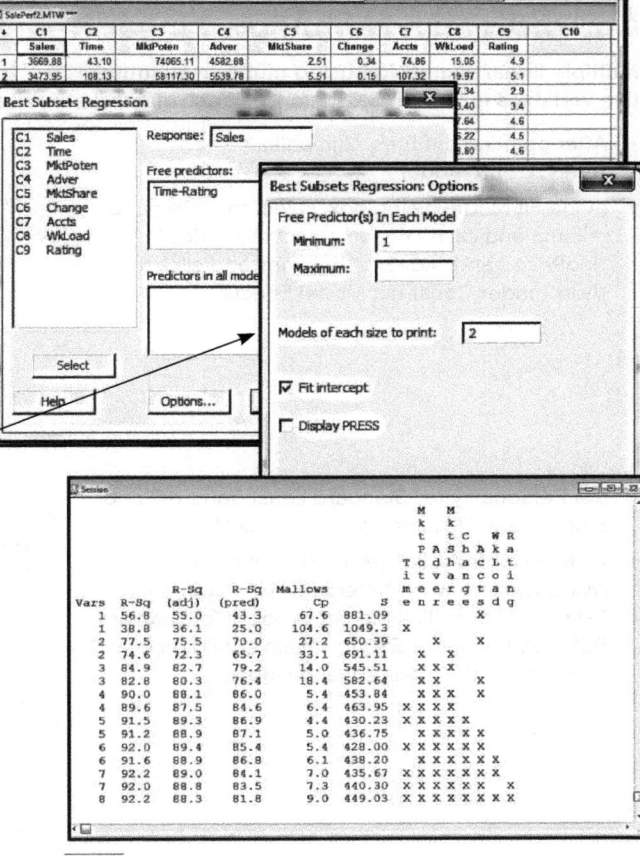

Source: Minitab 18

Stepwise regression in Figure 15.28(a) (data file: SalePerf2 .MTW):

- In the Data window, enter the sales representative performance data from Table 15.8 and Table 15.14 into columns C1–C9; use variable names Sales, Time, MktPoten, Adver, MktShare, Change, Accts, WkLoad, and Rating.

- Select **Stat : Regression : Regression : Fit Regression Model.**

- In the Regression dialog box, enter Sales in the Response window.

- Enter Time-Rating into the "Free predictors" window.

- Click on the Stepwise . . . button in the Regression dialog box.

- In the "Regression: Stepwise" dialog box, enter 0.10 in the "Alpha to enter" and the "Alpha to remove" windows.

- Check the "Display the table of model selection details" box and choose "Include details for each step" in the corresponding drop-down box.

- Click OK in the "Regression: Stepwise" and in the Regression dialog boxes.

- The results are given in the Session window.

- Note that **backward elimination** may be chosen in the "Method" drop-down menu.

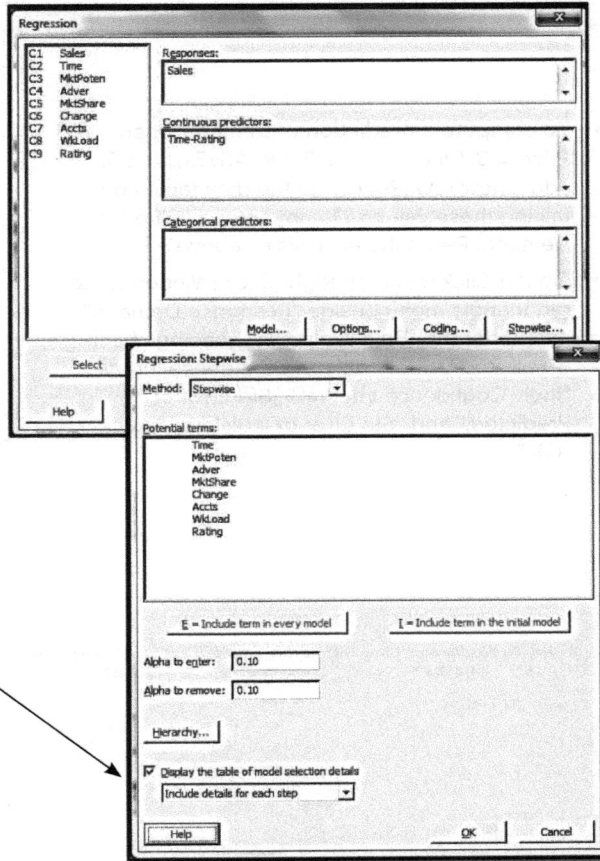

Source: Minitab 18

Appendix 15.3 ■ Multiple Linear Regression Analysis Using JMP

Multiple linear regression using multiple quantitative variables in Figure 15.6 (data file: Fresh2.jmp):

- After opening the Fresh2.jmp file, select **Analyze : Fit Model**.

- In the Fit Model dialog box, select the "Demand" column and click "Y" and select the "Price," "IndPrice," and "AdvExp" columns and click the "Add" under Construct Model Effects.

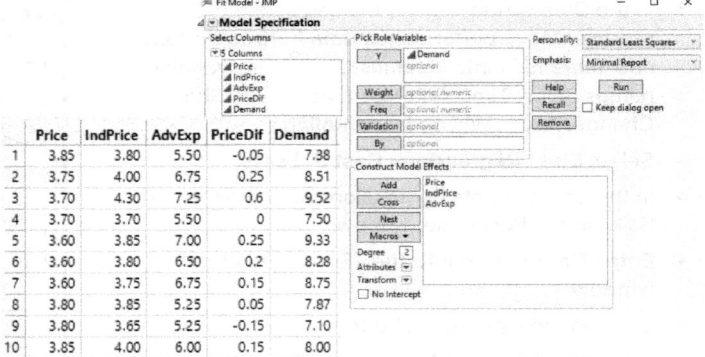

	Price	IndPrice	AdvExp	PriceDif	Demand
1	3.85	3.80	5.50	-0.05	7.38
2	3.75	4.00	6.75	0.25	8.51
3	3.70	4.30	7.25	0.6	9.52
4	3.70	3.70	5.50	0	7.50
5	3.60	3.85	7.00	0.25	9.33
6	3.60	3.80	6.50	0.2	8.28
7	3.60	3.75	6.75	0.15	8.75
8	3.80	3.85	5.25	0.05	7.87
9	3.80	3.65	5.25	-0.15	7.10
10	3.85	4.00	6.00	0.15	8.00

Source: JMP Pro 14

- Set Personality to "Standard Least Squares" and Emphasis to "Minimal Report." Click "Run."

- Option-Control-Click (Mac)/Alt-Right-Click (Windows) the word "Intercept" in "Parameter Estimates" and click the checkboxes for "Lower 95%" and "Upper 95%" to obtain confidence intervals for the intercept and slopes.

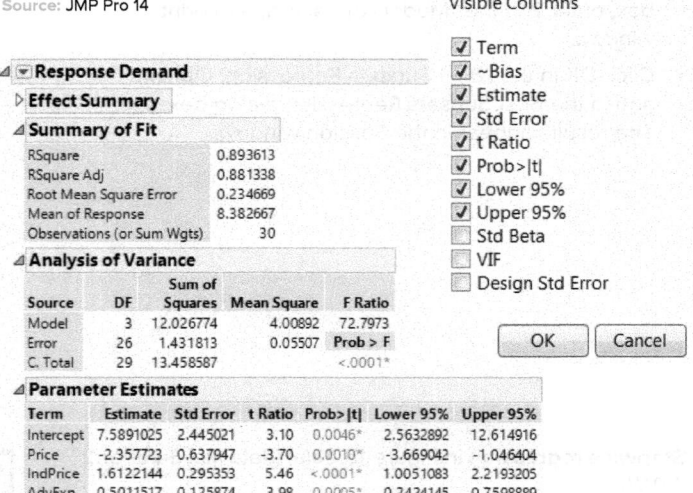

⊿ ▾ Response Demand

▷ Effect Summary

⊿ Summary of Fit

RSquare	0.893613
RSquare Adj	0.881338
Root Mean Square Error	0.234669
Mean of Response	8.382667
Observations (or Sum Wgts)	30

⊿ Analysis of Variance

Source	DF	Sum of Squares	Mean Square	F Ratio
Model	3	12.026774	4.00892	72.7973
Error	26	1.431813	0.05507	Prob > F
C. Total	29	13.458587		<.0001*

⊿ Parameter Estimates

Term	Estimate	Std Error	t Ratio	Prob>|t|	Lower 95%	Upper 95%
Intercept	7.5891025	2.445021	3.10	0.0046*	2.5632892	12.614916
Price	-2.357723	0.637947	-3.70	0.0010*	-3.669042	-1.046404
IndPrice	1.6122144	0.295353	5.46	<.0001*	1.0051083	2.2193205
AdvExp	0.5011517	0.125874	3.98	0.0005*	0.2424145	0.7598889

Visible Columns:
- ☑ Term
- ☑ ~Bias
- ☑ Estimate
- ☑ Std Error
- ☑ t Ratio
- ☑ Prob>|t|
- ☑ Lower 95%
- ☑ Upper 95%
- ☐ Std Beta
- ☐ VIF
- ☐ Design Std Error

- **To compute a prediction** of Demand when Price = 3.7, IndPrice = 3.9 and AdvExp = 6.5, add a new row (Row 31) to the data table containing these values. Do not enter a value for Demand. Rerun the analysis as above.

	Price	IndPrice	AdvExp	Price Dif	Demand
31	3.70	3.90	6.50	0.2	•

Source: JMP Pro 14

- Option-Click (Mac)/Alt-Right Click (Windows) the red triangle menu beside "Response Demand" and under the "Save Columns" heading check "Predicted Values," "Mean Confidence Interval," "Indiv Confidence Interval," "Std Error of Predicted," and "Std Error of Individual." Click "OK."

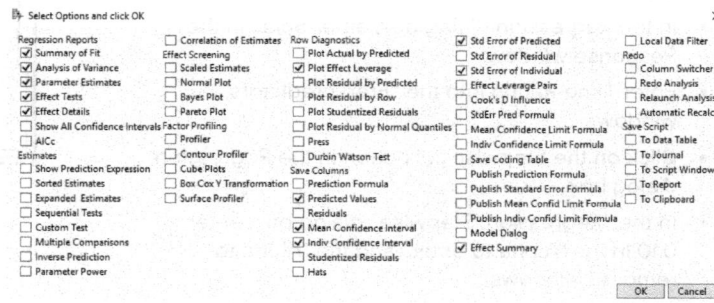

Source: JMP Pro 14

Demand	Predicted Demand	Lower 95% Mean Demand	Upper 95% Mean Demand	Lower 95% Indiv Demand	Upper 95% Indiv Demand	StdErr Pred Demand	StdErr Indiv Demand
•	8.4106503477	8.3143172822	8.5069834132	7.9187553487	8.9025453468	0.04686533	0.2393033103

Source: JMP Pro 14

- To obtain residual plots used to check the regression assumptions, option-Click (Mac)/Alt-Right Click (Windows) the red triangle menu beside "Response Demand" and under the Row Diagnostics heading check "Plot Actual by Predicted," "Plot Residual by Predicted," "Plot Residual by Row," and "Plot Residual by Normal Quantiles." Click "OK."

- To plot residuals versus an independent variable, from the red triangle menu by "Response Demand," select "Save Columns: Residuals." A residual column is added to the data table. Use **Graph Builder** from the main menu to create additional graph(s).

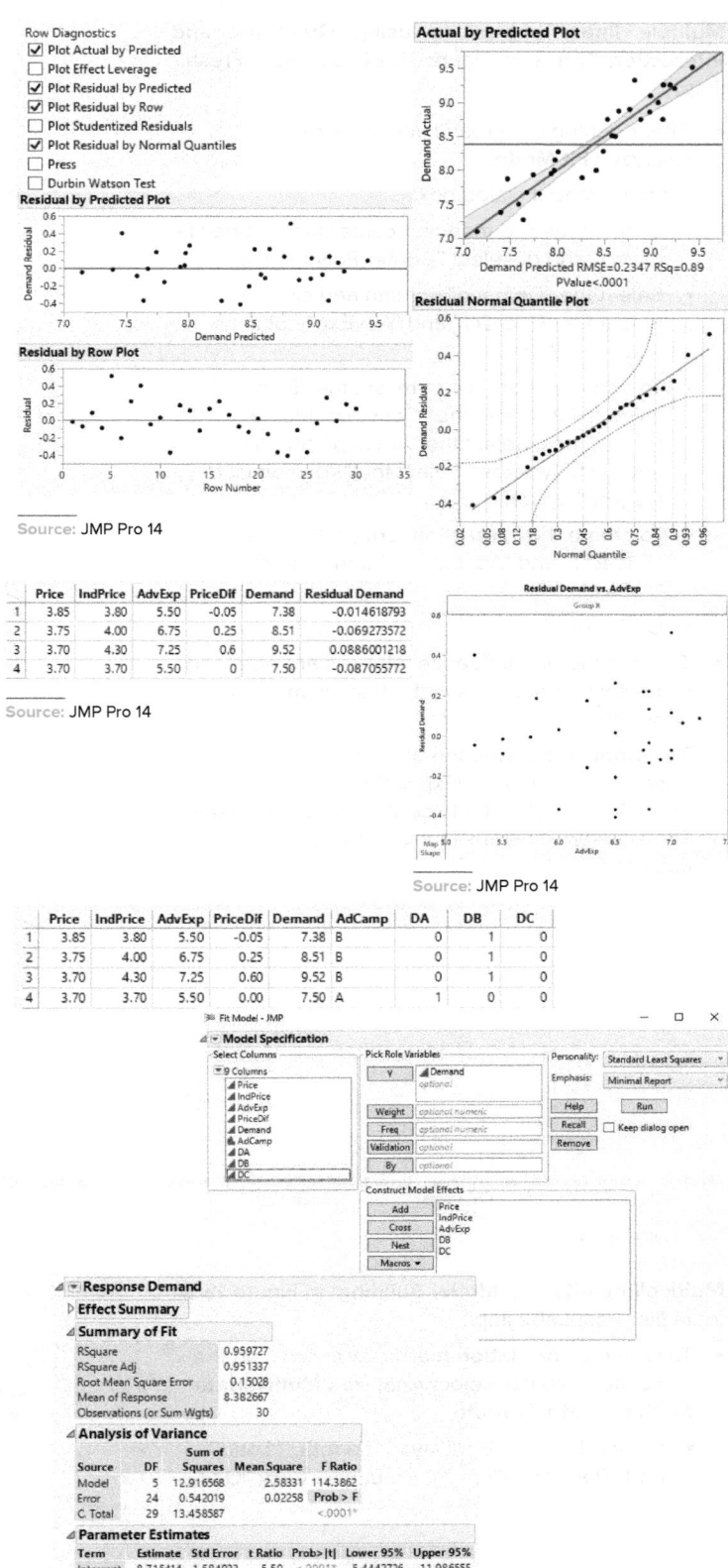

Source: JMP Pro 14

	Price	IndPrice	AdvExp	PriceDif	Demand	Residual Demand
1	3.85	3.80	5.50	-0.05	7.38	-0.014618793
2	3.75	4.00	6.75	0.25	8.51	-0.069273572
3	3.70	4.30	7.25	0.6	9.52	0.0886001218
4	3.70	3.70	5.50	0	7.50	-0.087055772

Source: JMP Pro 14

Multiple linear regression using Dummy Variables in Figures 15.20 and 15.21 (data file: Fresh3.jmp)

- After opening the Fresh3.jmp file, select **Analyze : Fit Model**.

- In the Fit Model dialog box, select the "Demand" column and click the "Y" button and select the "Price," "IndPrice," "AdvExp," "DB," and "DC" columns and click the "Add" button. Click "Run."

- The parameter confidence intervals are obtained in the same manner as in the first example in this appendix.

- **To compute a prediction** of Demand when Price = 3.7, IndPrice = 3.9, AdvExp = 6.5, DB = 0 and DC = 1 (because advertising campaign C is being used) add a new row (row 31) to the data table containing these values. Do not enter a value for Demand. Rerun the analysis.

	Price	IndPrice	AdvExp	PriceDif	Demand	AdCamp	DA	DB	DC
1	3.85	3.80	5.50	-0.05	7.38	B	0	1	0
2	3.75	4.00	6.75	0.25	8.51	B	0	1	0
3	3.70	4.30	7.25	0.60	9.52	B	0	1	0
4	3.70	3.70	5.50	0.00	7.50	A	1	0	0

Summary of Fit

RSquare	0.959727
RSquare Adj	0.951337
Root Mean Square Error	0.15028
Mean of Response	8.382667
Observations (or Sum Wgts)	30

Analysis of Variance

Source	DF	Sum of Squares	Mean Square	F Ratio
Model	5	12.916568	2.58331	114.3862
Error	24	0.542019	0.02258	Prob > F
C. Total	29	13.458587		<.0001*

Parameter Estimates

Term	Estimate	Std Error	t Ratio	Prob>|t|	Lower 95%	Upper 95%
Intercept	8.715414	1.584933	5.50	<.0001*	5.4442726	11.986555
Price	-2.768024	0.414437	-6.68	<.0001*	-3.62338	-1.912669
IndPrice	1.6666921	0.191332	8.71	<.0001*	1.2718024	2.0615818
AdvExp	0.4927425	0.080646	6.11	<.0001*	0.326298	0.659187
DB	0.269496	0.06945	3.88	0.0007*	0.1261574	0.4128347
DC	0.4395621	0.070335	6.25	<.0001*	0.2943979	0.5847264

Source: JMP Pro 14

Predicted Demand	Lower 95% Mean Demand	Upper 95% Mean Demand	Lower 95% Indiv Demand	Upper 95% Indiv Demand	StdErr Pred Demand	StdErr Indiv Demand
8.616211576	8.5137990349	8.718624117	8.289578079	8.9428450729	0.0496209179	0.1582604412

Source: JMP Pro 14

Multiple linear regression using Quadratic and Interaction Terms in Figure 15.24 (data file: Fresh2.jmp)

- After opening the Fresh2.jmp file, select **Analyze : Fit Model**.
- In the Fit Model dialog box:
  - Click on the red triangle beside "Model Specification" and deselect "Center Polynomials."
  - Select the "Demand" column and click "Y." Select the "PriceDif" and "AdvExp" columns and click "Add."
  - To add the quadratic term, set the "Degree" value to "2" under the "Construct Model Effects" box. Select the "AdvExp" column and click "Macros." Select the "Polynomial to Degree" option.
  - To create the interaction term, select the "PriceDif" and "AdvExp" columns and click "Cross."
  - Click "Run."
- The parameter confidence intervals are obtained in the same manner as in the first example in this appendix.
- **To compute a prediction** of Demand when PriceDif = 0.2 and AdvExp = 6.5, add a new row (row 31) to the data table containing these values. Then, follow the steps in the first example in this appendix.

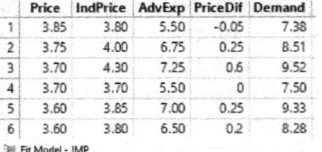

	Price	IndPrice	AdvExp	PriceDif	Demand
1	3.85	3.80	5.50	-0.05	7.38
2	3.75	4.00	6.75	0.25	8.51
3	3.70	4.30	7.25	0.6	9.52
4	3.70	3.70	5.50	0	7.50
5	3.60	3.85	7.00	0.25	9.33
6	3.60	3.80	6.50	0.2	8.28

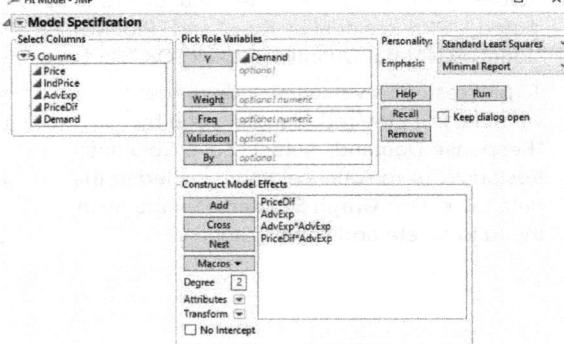

Source: JMP Pro 14

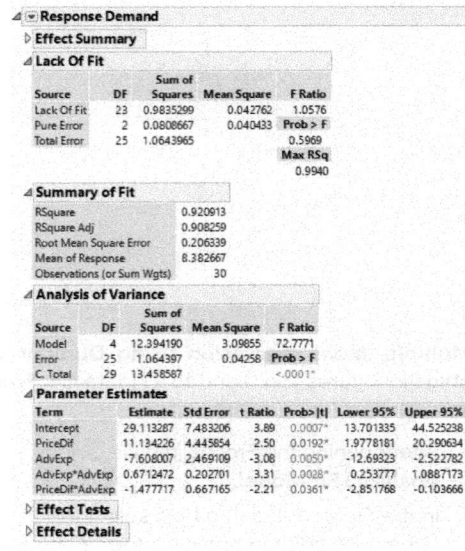

Source: JMP Pro 14

Predicted Demand	Lower 95% Mean Demand	Upper 95% Mean Demand	Lower 95% Indiv Demand	Upper 95% Indiv Demand	StdErr Pred Demand	StdErr Indiv Demand
8.3272497715	8.2112139058	8.4432856372	7.8867292725	8.7677702705	0.0563407107	0.2138928152

Source: JMP Pro 14

Multicollinearity and Model Building: In Figure 15.34 (data file: HospLab2.jmp):

- To create a **correlation matrix**, after opening the HospLab2.jmp file, select **Analyze : Multivariate Methods : Multivariate**.
  - Select "Hours," "BedDays," "Length," "Load," and "Pop" and Click "Y, Columns." Click "OK."

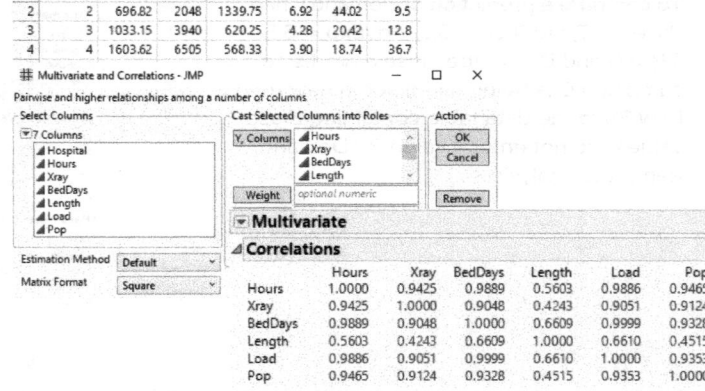

Hospital	Hours	Xray	BedDays	Length	Load	Pop
1	566.52	2463	472.92	4.45	15.57	18.0
2	696.82	2048	1339.75	6.92	44.02	9.5
3	1033.15	3940	620.25	4.28	20.42	12.8
4	1603.62	6505	568.33	3.90	18.74	36.7

Source: JMP Pro 14

- To obtain **VIF's**, after opening the HospLab2. jmp file, select **Analyze : Fit Model**.
  - Select the "Hours" columns and click "Y."
  - Select "Xray," "BedDays," "Length," "Load," and "Pop" and click "Add" in the Construct Model Effects. Click "Run."
  - Option-Control-Click (Mac)/Alt-Right-Click (Windows) the word "Intercept" in "Parameter Estimates." Click the "Lower 95%," "Upper 95%" and "VIF" checkboxes. Click "OK."

- To run **All Possible Models**, after opening the HospLab2.jmp file, select **Analyze : Fit Model**.
  - Select the "Hours" column and click "Y."
  - Select the "Xray," "BedDays," "Length," "Load," and "Pop" columns and click "Add" in the Construct Model Effects box.
  - Set **Personality** to "Stepwise." Click "Run."
  - Option-Click (Mac)/Alt-Right Click (Windows) the red triangle menu beside "Stepwise Fit for Hours" and click the checkbox "All Possible Models." Click "OK."
  - In the "All Possible Models" box, enter "5" as the maximum number of terms and "2" as the number of best models to see. Click "OK."
  - To modify the columns displayed in this report, Option-Right Click (Mac)/Alt-Right Click (Windows) on a column's title and select the desired columns.

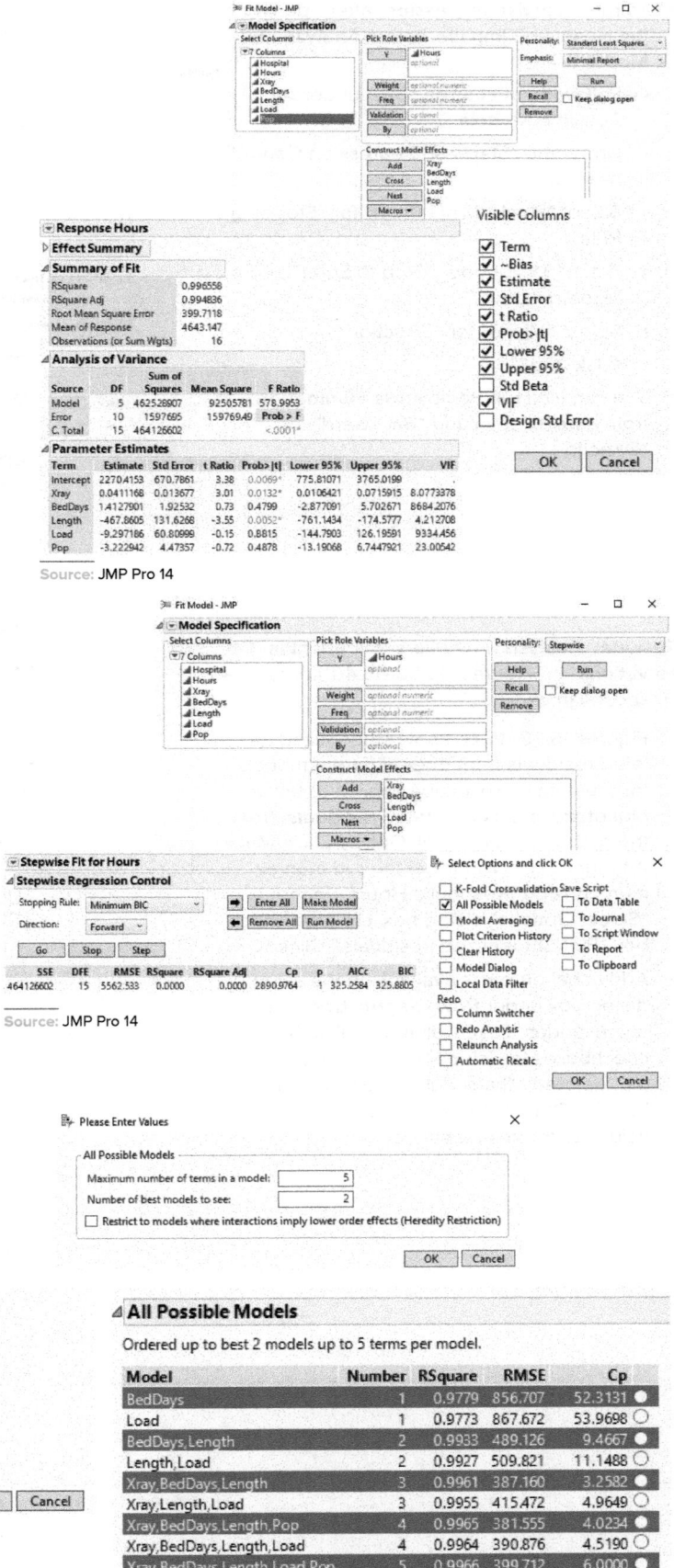

Source: JMP Pro 14

Source: JMP Pro 14

Source: JMP Pro 14

- To run **stepwise regression**, after opening the HospLab2.jmp file, select **Analyze : Fit Model.**

  - Follow the first three steps under **All Possible Models**.

  - Under the "Stepwise Regression Control" heading:

  - Select "P-value Threshold" for "Stopping Rule."

  - Enter "0.1" for both "Prob to Enter" and "Prob to Leave."

  - Select "Mixed" for "Direction."

  - Click "Go."

- You can also run **Backwards Elimination** from here by selecting "Backward" for "Direction."

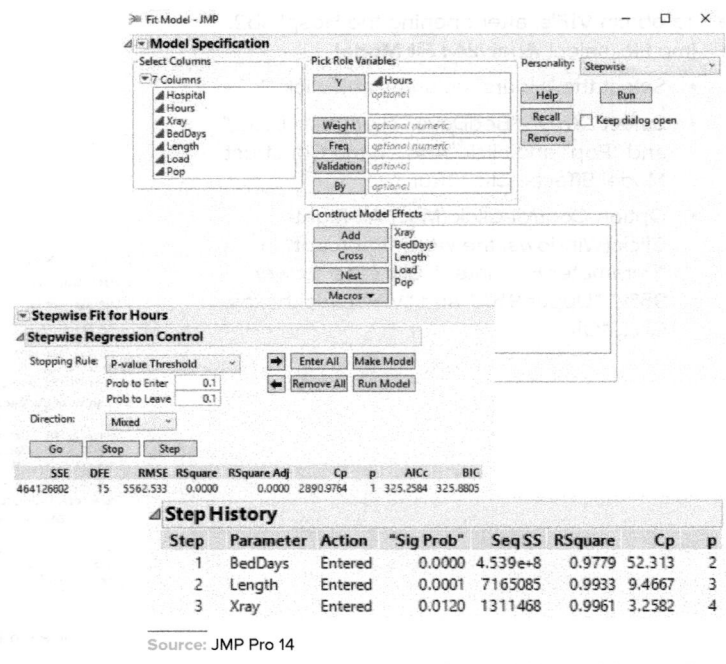

Source: JMP Pro 14

Residual Analysis / Outliers / Influential Observations in Figures 15.38–15.40 (data file: HospLab4.jmp):

- Figures 15.38–15.40 show studentized deleted residuals from three different models for the data in HospLab4.jmp. To obtain a plot of **studentized deleted residuals**, from the output of a fit model, Option-Click (Mac)/Alt-Right Click (Windows) the red triangle menu beside "Response Hours." In the "Select Options" dialog box, click the checkbox "Plot Studentized Residuals." Click "OK."

- Additional output is available in the same dialog box under **Row Diagnostics**. In addition, under "Save Columns," click the checkboxes for "studentized residuals" (not deleted), "hats" (leverage value), and "Cook's D" (influence value) to save these values to the data table.

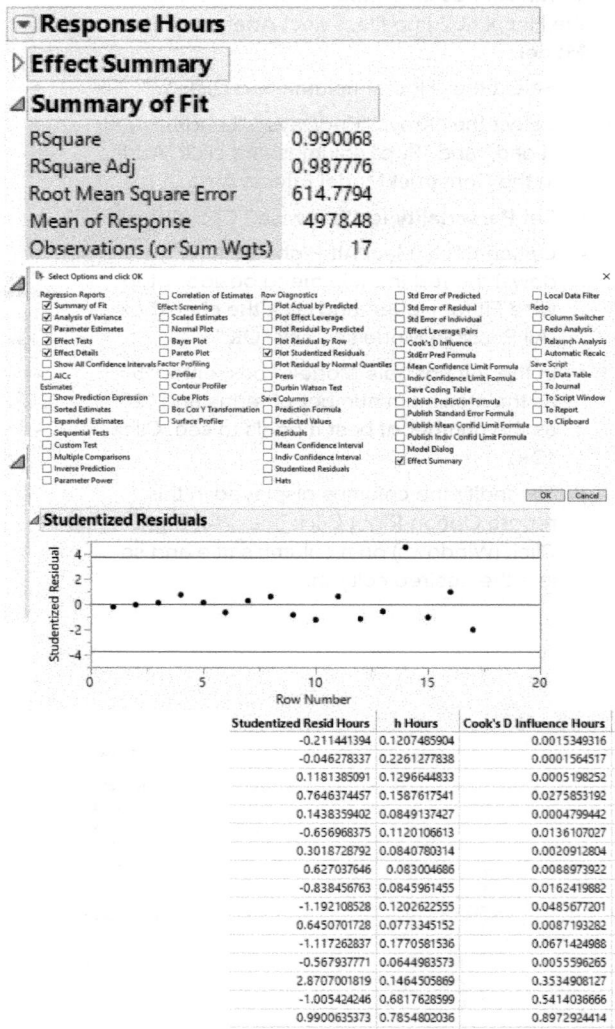

Source: JMP Pro 14

Forward Selection With Simultaneous Validation in Figures 15.31, 15.32, and 15.33 (data file: ToyotaCorolla.jmp):

- After opening the ToyotaCorolla.jmp file, select **Analyze : Fit Model**.

- In the Fit Model Dialog box:
  - Select the "Price" column and click "Y."
  - Select the "Validation" column and click "Validation."
  - Select all columns except "ID," "Model," "Price," and "Validation" and click "Add."
  - Select "Stepwise" for Personality.
  - Click "Run."

- For **Stepwise Regression Control**, choose "Max Validation RSquare" for "Stopping Rule," "Forward" for "Direction," and "Combine" for "Rules." Click "Go."

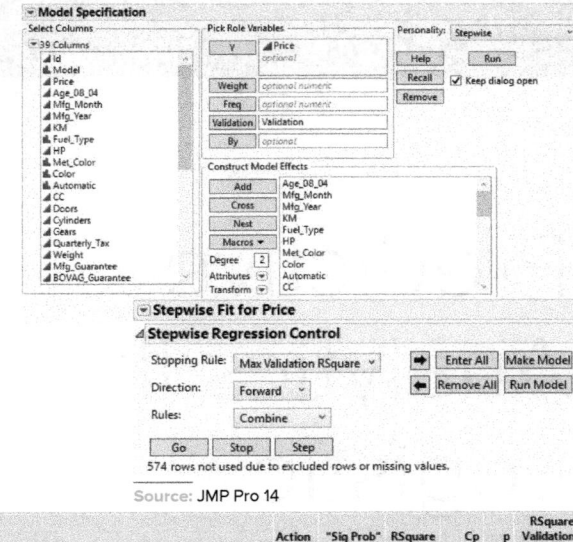

Source: JMP Pro 14

Step	Parameter	Action	"Sig Prob"	RSquare	Cp	p	RSquare Validation
1	Mfg_Year	Entered	0.0000	0.7834	1138	2	0.7831
2	Automatic_airco	Entered	0.0000	0.8326	686.64	3	0.8339
3	HP	Entered	0.0000	0.8569	465.18	4	0.8392
4	KM	Entered	0.0000	0.8675	369.66	5	0.8472
5	Weight	Entered	0.0000	0.8865	196.53	6	0.8895
6	Powered_Windows	Entered	0.0000	0.8896	169.65	7	0.8933
7	Quarterly_Tax	Entered	0.0000	0.8930	139.84	8	0.8942
8	Guarantee_Period	Entered	0.0000	0.8961	113.66	9	0.8940
9	BOVAG_Guarantee	Entered	0.0000	0.8993	86.145	10	0.8956
10	Color{White&Beige&Violet&Green&Red-Blue&Black&Silver&Grey&Yellow}	Entered	0.0001	0.9011	71.494	11	0.8965
38	Radio	Entered	0.9038	0.9110	40	41	0.9032
39	Mfg_Month	Entered	0.9997	0.9110	42	42	0.9032
40	Best	Specific	.	0.9109	29.161	35	0.9036

- To fit the final model after forward selection is complete, in the **Stepwise Regression Control**, click "Run Model." The first set of output is for the "Combine" Rules.

Summary of Fit

RSquare	0.910911
RSquare Adj	0.907248
Root Mean Square Error	1104.616
Mean of Response	10796.6
Observations (or Sum Wgts)	862

Analysis of Variance

Source	DF	Sum of Squares	Mean Square	F Ratio
Model	34	1.0318e+10	303460227	248.7019
Error	827	1009085920	12201764	Prob > F
C. Total	861	1.1327e+10		<.0001*

Crossvalidation

Source	RSquare	RASE	Freq
Training Set	0.9109	1082.0	862
Validation Set	0.9036	1125.1	574

Parameter Estimates

Term	Estimate	Std Error	t Ratio	Prob>|t|
Intercept	-2371739	286063.8	-8.29	<.0001*
Age_08_04	-26.14582	11.89861	-2.20	0.0283*
Mfg_Year	1186.0134	142.8682	8.30	<.0001*
KM	-0.017259	0.001487	-11.61	0.0978
Fuel_Type[CNG&Petrol-Diesel]	-253.059	152.685	-1.66	0.0978
Fuel_Type[CNG-Petrol]	-876.5534	215.564	-4.07	<.0001*
HP	29.120103	3.678955	7.92	<.0001*
Met_Color[0]	27.264735	45.08671	0.60	0.5455
Color{White&Beige&Violet&Green&Red-Blue&Black&Silver&Grey&Yellow}	-208.2795	108.4149	-1.92	0.0551
Color{White&Beige-Violet&Green&Red}	-237.463	136.2054	-1.74	0.0816
Color{Violet&Green-Red}	-42.73422	67.91333	-0.63	0.5295

- To do this analysis using the "Whole Effects" Rules, run the forward selection with the "Whole Effects" Rules and when the forward selection has finished, click on "Run Model" in the **Stepwise Regression Control** dialog box.

Summary of Fit

RSquare	0.908549
RSquare Adj	0.905475
Root Mean Square Error	1115.126
Mean of Response	10796.6
Observations (or Sum Wgts)	862

Analysis of Variance

Source	DF	Sum of Squares	Mean Square	F Ratio
Model	28	1.0291e+10	367531907	295.5611
Error	833	1035840234	1243505.7	Prob > F
C. Total	861	1.1327e+10		<.0001*

Crossvalidation

Source	RSquare	RASE	Freq
Training Set	0.9085	1096.2	862
Validation Set	0.9023	1132.9	574

Parameter Estimates

Term	Estimate	Std Error	t Ratio	Prob>|t|
Intercept	-2369638	288644.2	-8.21	<.0001*
Age_08_04	-26.50997	11.98311	-2.21	0.0272*
Mfg_Year	1184.9488	144.1581	8.22	<.0001*
KM	-0.017346	0.001486	-11.67	<.0001*
Fuel_Type[CNG]	-1067.774	259.5604	-4.11	<.0001*
Fuel_Type[Diesel]	306.01122	204.7073	1.49	0.1353
HP	28.57111	3.699054	7.72	<.0001*
Automatic[0]	-244.3866	89.22187	-2.74	0.0063*
CC	-0.054992	0.079074	-0.70	0.4870
Doors	66.73783	45.48316	1.47	0.1427
Gears	100.3322	207.2298	0.48	0.6284
Quarterly_Tax	12.812515	2.17029	5.90	<.0001*

Source: JMP Pro 14

- **To compute a prediction** for vehicle 1437, create a new row (Row 1437) in the data table using the values associated with this vehicle and then follow the steps outlined in the first example in this appendix.

Predicted Price	Lower 95% Mean Price	Upper 95% Mean Price	Lower 95% Indiv Price	Upper 95% Indiv Price
15779.429458	15353.665825	16205.193091	13569.840754	17989.018162

Predicted Price	Lower 95% Mean Price	Upper 95% Mean Price	Lower 95% Indiv Price	Upper 95% Indiv Price
15642.417558	15234.592309	16050.242808	13415.960944	17868.874173

Source: JMP Pro 14

Design Elements: (CD): ©Comstock Images/Alamy; (All Others): ©McGraw-Hill Education

Predictive Analytics II: Logistic Regression, Discriminate Analysis, and Neural Networks

©Iakov Filimonov/123RF

Learning Objectives

After mastering the material in this chapter, you will be able to:

LO16-1 Use a logistic model to estimate probabilities and odds ratios.

LO16-2 Use linear discriminate analysis to classify observations and estimate probabilities.

LO16-3 Use neural network modeling to estimate probabilities and predict values of quantitative response variables.

Chapter Outline

n this chapter we will discuss three parametric predictive analytics that can be used to classify a qualitative response variable into two or more groups. These predictive analytics are logistic regression, linear discriminate analysis, and neural networks. We will also see that neural networks are particularly useful when analyzing big data and can be used to predict the values of a quantitative response variable.

16.1 Logistic Regression

LO16-1
Use a logistic model to estimate probabilities and odds ratios.

Basic methods

Suppose that in a study of the effectiveness of offering a price reduction on a given product, 300 households having similar incomes were selected. A coupon offering a price reduction, x, on the product, as well as advertising material for the product, was sent to each household. The coupons offered different price reductions (10, 20, 30, 40, 50, and 60 dollars), and 50 homes were assigned at random to each price reduction. The table in the page margin summarizes the number, y, and proportion, $\hat{p}$, of households redeeming coupons for each price reduction, x (expressed in units of $10). On the left side of Figure 16.1, we plot the $\hat{p}$ values versus the x values and draw a curve through the plotted points. A theoretical curve having the shape of the curve in Figure 16.1 is the **logistic curve**

x	y	Proportion
1	4	.08
2	7	.14
3	20	.40
4	35	.70
5	44	.88
6	46	.92

Ⓓ PrcRed

$$p(x) = \frac{e^{(\beta_0 + \beta_1 x)}}{1 + e^{(\beta_0 + \beta_1 x)}}$$

where $p(x)$ denotes the probability that a household receiving a coupon having a price reduction of x will redeem the coupon. The Minitab and JMP outputs in Figures 16.1 and 16.2 tell us that the point estimates of β_0 and β_1 are $b_0 = -3.746$ and $b_1 = 1.111$. (The point estimates in logistic regression are usually obtained by an advanced statistical procedure called *maximum likelihood estimation*.) Using these estimates, it follows that, for example,

$$\hat{p}(5) = \frac{e^{(-3.746+1.111(5))}}{1 + e^{(-3.746+1.111(5))}} = \frac{6.10434}{1 + 6.10434} = .8593$$

That is, $\hat{p}(5) = .8593$ is the point estimate of the probability that a household receiving a coupon having a price reduction of $50 will redeem the coupon. The Minitab and JMP outputs in Figures 16.1 and 16.2 give the values of $\hat{p}(x)$ for $x = 1, 2, 3, 4, 5$, and 6. The curve in Figure 16.1 is a plot of the estimated probabilities, and the dashed bands around the curve are graphical confidence intervals for the true probabilities.

FIGURE 16.1 **Minitab Output of a Logistic Regression of the Price Reduction Data**

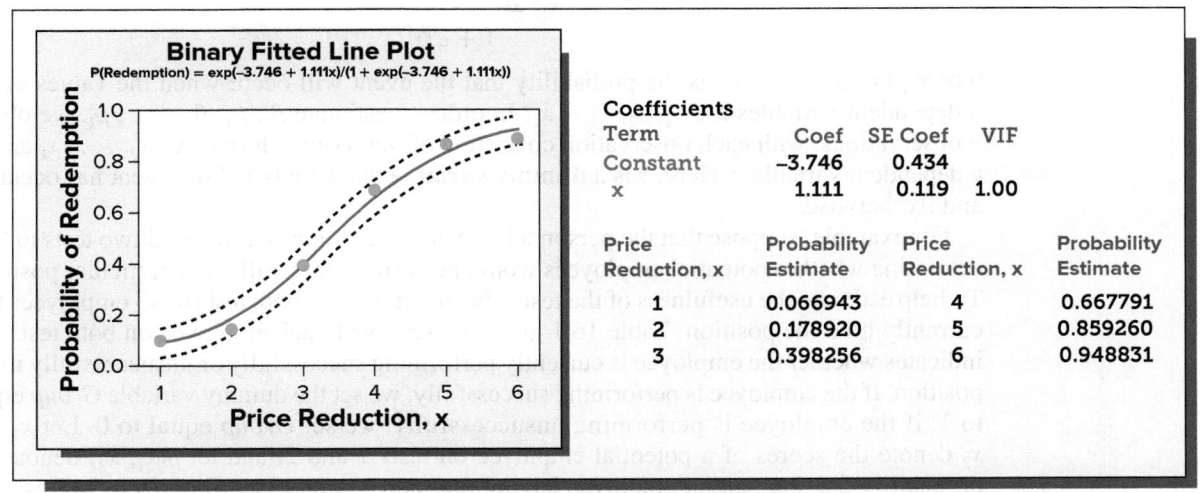

Binary Fitted Line Plot
P(Redemption) = exp(−3.746 + 1.111x)/(1 + exp(−3.746 + 1.111x))

Coefficients

Term	Coef	SE Coef	VIF
Constant	−3.746	0.434	
x	1.111	0.119	1.00

Price Reduction, x	Probability Estimate	Price Reduction, x	Probability Estimate
1	0.066943	4	0.667791
2	0.178920	5	0.859260
3	0.398256	6	0.948831

FIGURE 16.2   JMP Output of a Logistic Regression of the Price Reduction Data

Maximum Likelihood without Validation

Model Summary

Response	Proportion
Distribution	Binomial
Estimation Method	Logistic Regression
Validation Method	None
Probability Model Link	Logit

Measure

Number of rows	6
Sum of Frequencies	6
-LogLikelihood	1.8274301
BIC	7.2383792
AICc	11.65486
Generalized RSquare	0.5916941

Parameter Estimates for Original Predictors

Term	Estimate	Std Error	Wald ChiSquare	Prob> ChiSquare	Lower 95%	Upper 95%
Intercept	−3.745541	3.0713514	1.4872021	0.2227	−9.765279	2.2741976
x	1.1109353	0.8440294	1.732457	0.1881	−0.543332	2.7652025

	x	Proportion	Probability (Proportion=1)	Probability (Proportion=0)
1	1	0.08	0.0669442201	0.9330557799
2	2	0.14	0.1789217439	0.8210782561
3	3	0.4	0.3982566094	0.6017433906
4	4	0.7	0.6677887461	0.3322112539
5	5	0.88	0.8592574379	0.1407425621
6	6	0.92	0.9488297707	0.0511702293

The **general logistic regression model** relates the probability that an event (such as redeeming a coupon) will occur to k independent variables $x_1, x_2, \ldots, x_k$. This general model is

$$p(x_1, x_2, \ldots, x_k) = \frac{e^{(\beta_0+\beta_1 x_1+\beta_2 x_2+ \cdots +\beta_k x_k)}}{1 + e^{(\beta_0+\beta_1 x_1+\beta_2 x_2+ \cdots +\beta_k x_k)}}$$

where $p(x_1, x_2, \ldots, x_k)$ is the probability that the event will occur when the values of the independent variables are $x_1, x_2, \ldots, x_k$. In order to estimate $\beta_0, \beta_1, \beta_2, \ldots, \beta_k$, we obtain n observations, with each observation consisting of observed values of $x_1, x_2, \ldots, x_k$ and of a dependent variable y. Here, y is a **dummy variable** that equals 1 if the event has occurred and 0 otherwise.

For example, suppose that the personnel director of a firm has developed two tests to help determine whether potential employees would perform successfully in a particular position. To help estimate the usefulness of the tests, the director gives both tests to 43 employees that currently hold the position. Table 16.1 gives the scores of each employee on both tests and indicates whether the employee is currently performing successfully or unsuccessfully in the position. If the employee is performing successfully, we set the dummy variable *Group* equal to 1; if the employee is performing unsuccessfully, we set *Group* equal to 0. Let x_1 and x_2 denote the scores of a potential employee on tests 1 and 2, and let $p(x_1, x_2)$ denote the probability that a potential employee having the scores x_1 and x_2 will perform successfully

FIGURE 16.3   Minitab Output of a Logistic Regression of the Performance Data

```
Deviance Table
Source          DF    Adj Dev   Adj Mean   Chi-Square   P-Value
Regression       2    31.483    15.7416      31.48       0.000
   Test 1        1    20.781    20.7806      20.78       0.000
   Test 2        1     3.251     3.2511       3.25       0.071
Error           40    27.918     0.6980
Total           42    59.401


Model Summary
Deviance          Deviance
  R-Sq            R-Sq(adj)     AIC
 53.00%            49.63%      33.92


Coefficients
Term        Coef    SE Coef       95% CI        Z-Value   P-Value   VIF
Constant   -56.2     17.5    ( -90.4, -22.0)     -3.22     0.001
Test 1      0.483    0.158   (  0.174, 0.793)     3.06     0.002    1.03
Test 2      0.165    0.102   (-0.035, 0.365)      1.62     0.106    1.03


Odds Ratios for Continuous Predictors

           Odds Ratio          95% CI

Test 1       1.6214      (1.1901, 2.2091)
Test 2       1.1797      (0.9658, 1.4409)

Goodness-of-Fit Tests

Test              DF    Chi-Square   P-Value
Deviance          40      27.92       0.925
Pearson           40      24.69       0.973
Hosmer-Lemeshow    8       5.03       0.754


                       Fitted
Variable  Setting   Probability     SE Fit          95% CI
Test 1      93       0.934390     0.0568790   (0.697984, 0.988734)
Test 2      84
```

TABLE 16.1
The Performance
Data DS PerfTest

Group	Test 1	Test 2
1	96	85
1	96	88
1	91	81
1	95	78
1	92	85
1	93	87
1	98	84
1	92	82
1	97	89
1	95	96
1	99	93
1	89	90
1	94	90
1	92	94
1	94	84
1	90	92
1	91	70
1	90	81
1	86	81
1	90	76
1	91	79
1	88	83
1	87	82
0	93	74
0	90	84
0	91	81
0	91	78
0	88	78
0	86	86
0	79	81
0	83	84
0	79	77
0	88	75
0	81	85
0	85	83
0	82	72
0	82	81
0	81	77
0	86	76
0	81	84
0	85	78
0	83	77
0	81	71

Source: Performance data
from T. E. Dielman, *Applied
Regression Analysis for
Business and Economics*,
2nd ed., Brooks/Cole, 1996.

in the position. We can estimate the relationship between $p(x_1, x_2)$ and x_1 and x_2 by using the logistic regression model

$$p(x_1, x_2) = \frac{e^{(\beta_0 + \beta_1 x_1 + \beta_2 x_2)}}{1 + e^{(\beta_0 + \beta_1 x_1 + \beta_2 x_2)}}$$

The Minitab output in Figure 16.3 tells us that the point estimates of β_0, β_1, and β_2 are $b_0 = -56.2$, $b_1 = .483$, and $b_2 = .165$. Consider, therefore, a potential employee who scores a 93 on test 1 and an 84 on test 2. It follows that a point estimate of the probability that the potential employee will perform successfully in the position is

$$\hat{p}(93, 84) = \frac{e^{(-56.2 + .483(93) + .165(84))}}{1 + e^{(-56.2 + .483(93) + .165(84))}} = .9344$$

This point estimate is given at the bottom of the Minitab output in Figure 16.3, as is a 95 percent confidence interval for the true probability $p(93, 84)$, which is [.6980, .9887].

We next note that at the top of the Minitab output there are three chi-square statistics and associated p-values. We will not discuss how these statistics are calculated (the calculations are complicated). However, the first chi-square statistic of 31.48 is related to the overall model and is used to test H_0: $\beta_1 = \beta_2 = 0$. The p-value associated with 31.48 is the area under the chi-square curve having $k = 2$ degrees of freedom to the right of 31.48. Because this p-value is less than .001 (see the Minitab output), we have extremely strong evidence that we should reject H_0: $\beta_1 = \beta_2 = 0$ and conclude that at least one of β_1 or β_2 does not equal zero. The p-value for testing H_0: $\beta_1 = 0$ versus H_a: $\beta_1 \neq 0$ is the area under the chi-square curve having one degree of freedom to the right of the second chi-square statistic of 20.78 on the Minitab output. Because this p-value is less than .001, we have extremely strong evidence that we should reject H_0: $\beta_1 = 0$ and conclude that the score on test 1 is related to the probability of a potential employee's success. The p-value for testing H_0: $\beta_2 = 0$ versus H_a: $\beta_2 \neq 0$ is the area under the chi-square curve having one degree of freedom to the right of the third chi-square statistic of 3.25 on the Minitab output. Because this p-value is .071, we do not have strong evidence that the score on test 2 is related to the probability of a potential employee's success. In Exercise 16.3 we will consider a logistic regression model that uses only the score on test 1 to estimate the probability of a potential employee's success.

There is a second way shown in the middle of the Minitab output (under "Coefficients") to test H_0: $\beta_1 = 0$ and to test H_0: $\beta_2 = 0$. This second method gives somewhat different results from the first. Specifically, this second way says that (i) the p-value for testing H_0: $\beta_1 = 0$ is the area under the chi-square curve having one degree of freedom to the right of the square of $z = (b_1/s_{b_1}) = (.483/.158) = 3.06$ and (ii) the p-value for testing H_0: $\beta_2 = 0$ is the area under the chi-square curve having one degree of freedom to the right of the square of $z = (b_2/s_{b_2}) = (.165/.102) = 1.62$. Because these two p-values are .002 and .106, we have very strong evidence that the score on test 1 is related to the probability of a potential employee's success but we do not have strong evidence that the score on test 2 is related to the probability of a potential employee's success. Although slightly different from the results of the previous paragraph, the results are fairly similar to those found previously.

The **odds** of success for a potential employee is defined to be the probability of success divided by the probability of failure for the employee. That is,

$$\text{odds} = \frac{p(x_1, x_2)}{1 - p(x_1, x_2)}$$

For the potential employee who scores a 93 on test 1 and an 84 on test 2, we estimate that the odds of success are $.9344/(1 - .9344) = 14.2$. That is, we estimate that the odds of success for the potential employee are about 14 to 1. It can be shown that $e^{b_1} = e^{.483} = 1.62$ is a point estimate of the **odds ratio for x_1,** which is the proportional change in the odds (for any potential employee) that is associated with an increase of one in x_1 when x_2 stays constant. This point estimate of the odds ratio for x_1 is shown on the Minitab output and says that, for every one point increase in the score on test 1 when the score on test 2 stays constant, we estimate that a potential employee's odds of success increase by 62 percent. Furthermore, the 95 percent confidence interval for the odds ratio for x_1, [1.19, 2.21], does not contain 1. Therefore, as with the chi-square tests of H_0: $\beta_1 = 0$, we conclude that there is strong evidence that the score on test 1 is related to the probability of success for a potential employee. Similarly, it can be shown that $e^{b_2} = e^{.165} = 1.18$ is a point estimate of the **odds ratio for x_2,** which is the proportional change in the odds (for any potential employee) that is associated with an increase of one in x_2 when x_1 stays constant. This point estimate of the odds ratio for x_2 is shown on the Minitab output and says that, for every one point increase in the score on test 2 when the score on test 1 stays constant, we estimate that a potential employee's odds of success increases by 18 percent. However, the 95 percent confidence interval for the odds ratio for x_2, [.97, 1.44], contains 1. Therefore, as with the chi-square tests of H_0: $\beta_2 = 0$, we cannot conclude that there is strong evidence that the score on test 2 is related to the probability of success for a potential employee.

To conclude our interpretation of the Minitab output, consider the Hosmer-Lemeshow goodness-of-fit chi-square statistic value of 5.03 near the bottom of the output. Without going into how this statistic is calculated, we see that the p-value associated with this statistic is .754, which is much greater than .05. This implies that there is no evidence of a lack of fit of the logistic regression model to the performance data. [In general, the Hosmer-Lemeshow chi-square statistic is the only one of the goodness-of-fit chi-square statistics on a Minitab logistic regression output that is valid if, as in the performance data, there are not various combinations of values of the independent variables that are repeated a reasonably large number of (say, 10 or more) times.]

To generalize some of the previous discussions, consider again the **general logistic regression model**

$$p(x_1, x_2, \ldots, x_k) = \frac{e^{(\beta_0 + \beta_1 x_1 + \beta_2 x_2 + \cdots + \beta_k x_k)}}{1 + e^{(\beta_0 + \beta_1 x_1 + \beta_2 x_2 + \cdots + \beta_k x_k)}}$$

where $p(x_1, x_2, \ldots, x_k)$ is the probability that the event under consideration will occur when the values of the independent variables are $x_1, x_2, \ldots, x_k$. The **odds** of the event occurring is defined to be $p(x_1, x_2, \ldots, x_k)/(1 - p(x_1, x_2, \ldots, x_k))$, which is the probability that the event will occur divided by the probability that the event will not occur. It can be shown that the odds equals $e^{(\beta_0 + \beta_1 x_1 + \beta_2 x_2 + \cdots + \beta_k x_k)}$. The natural logarithm of the odds is $(\beta_0 + \beta_1 x_1 + \beta_2 x_2 + \cdots + \beta_k x_k)$, which is called the **logit**. If $b_0, b_1, b_2, \ldots, b_k$ are the point estimates of $\beta_0, \beta_1, \beta_2, \ldots, \beta_k$, the point estimate of the logit, denoted $\widehat{\ell g}$, is $(b_0 + b_1 x_1 + b_2 x_2 + \cdots + b_k x_k)$. It follows that the point estimate of the probability that the event will occur is

$$\hat{p}(x_1, x_2, \ldots, x_k) = \frac{e^{\widehat{\ell g}}}{1 + e^{\widehat{\ell g}}} = \frac{e^{(b_0 + b_1 x_1 + b_2 x_2 + \cdots + b_k x_k)}}{1 + e^{(b_0 + b_1 x_1 + b_2 x_2 + \cdots + b_k x_k)}}$$

Finally, consider an arbitrary independent variable x_j. It can be shown that e^{b_j} is the point estimate of the **odds ratio for x_j**, which is the proportional change in the odds that is associated with a one unit increase in x_j when the other independent variables stay constant.

As a third example of logistic regression, we consider the credit card upgrade example we presented in our discussion of classification trees in Section 5.1. In this example a bank wishes to predict whether or not an existing holder of its Silver credit card will upgrade, for a specified annual fee, to its Platinum credit card. To do this, the bank carries out a pilot study that randomly selects 40 of its existing Silver card holders and offers each Silver card holder an upgrade, for the annual fee, to its Platinum card. The results of the study are shown in the JMP output in Figure 16.4. Here, the response variable *Upgrade* equals 1 if the Silver card holder decided to upgrade and 0 otherwise. Moreover, the predictor variable *Purchases* is last year's purchases (in thousands of dollars) by the Silver card holder, and the predictor variable *PlatProfile* equals 1 if the Silver card holder conforms to the bank's *Platinum profile* and 0 otherwise, Here, the term *Platinum profile* is the name we are using for a real concept used by a major credit card company with which we have spoken. To explain this concept, we assume that when an individual applies for one of the bank's credit cards, the bank has the person fill out a questionnaire related to how he or she will use the credit card. This helps the bank assess which type of credit card might best meet the needs of the individual. For example, a Platinum card might be best for someone who travels frequently and for whom the card's extra reward points and luxury travel perks justify paying the card's annual fee. Based on the questionnaire filled out by the individual, the bank assesses whether the Silver or Platinum card might best meet his or her needs and explains (with a recommendation) the advantages and disadvantages of both cards. Because of the Platinum card's annual fee, many customers whose needs might be best met by the Platinum card choose the Silver card instead. However, the bank keeps track of each Silver card holder's annual purchases and the consistency with which he or she uses the Silver card. Periodically, the bank offers some Silver card holders an upgrade, for the annual fee, to the Platinum card. A Silver card holder who has used the card consistently over the past two years and whose original questionnaire indicates that his or her needs might be best met by a Platinum card is said by the bank to *conform* to its Platinum profile. In this case, the predictor variable *PlatProfile* for the holder is set equal to 1. If a Silver card holder does not conform to the bank's Platinum profile, *PlatProfile* is set equal to 0. Of course, a Silver card holder's needs might change over time, and thus, in addition to using the predictor variable *PlatProfile*, the bank will also use the previously defined predictor variable *Purchases* (last year's purchases) to predict whether a Silver card holder will upgrade. Note that the bank will only consider Silver card holders who have an excellent payment history. Also, note that in sampling the 40 existing Silver card holders, the bank purposely randomly selected 20 Silver card holders who conformed to the Platinum profile and randomly selected 20 Silver card holders who did not conform to the Platinum profile.

Figure 16.5 gives the JMP output of a logistic regression analysis of the credit card upgrade data in Figure 16.4. Examining the output, we see that both of the predictor variables *Purchases* and *PlatProfile* are related to the probability of a Silver card holder upgrading. The odds ratio estimate of 1.2335 for *Purchases* says that for each increase of $1,000 in last year's purchases by a Silver card holder, we estimate that the

FIGURE 16.4   **The JMP Output of the Card Upgrade Data, Upgrade Probability Estimates, and Classification**
CardUpgrade

	UpGrade	Purchases	PlatProfile	Lin[1]	Prob[1]	Prob[0]	Most Likely UpGrade
1	0	7.471	0	−8.384769762	0.0002282662	0.9997717338	0
2	0	21.142	0	−5.516174805	0.0040050951	0.9959949049	0
3	1	39.925	1	2.5729429497	0.9290998029	0.0709001971	1
4	1	32.45	1	1.0044587638	0.7319343211	0.2680656789	1
5	1	48.95	1	4.4666646591	0.9886448599	0.0113551401	1
6	0	28.52	1	0.1798242687	0.5448353131	0.4551646869	1 *
7	0	7.822	0	−8.3111192	0.0002457084	0.9997542916	0
8	0	26.548	0	−4.381830255	0.0123480737	0.9876519263	0
9	1	48.831	1	4.4416948105	0.9883610957	0.0116389043	1
10	0	17.584	0	−6.262752294	0.0019023668	0.9980976332	0
11	1	49.82	1	4.6492173336	0.9905216111	0.0094783889	1
12	1	50.45	0	0.6335421879	0.6532922081	0.3467077919	1
13	0	28.175	0	−4.040435771	0.0172857526	0.9827142474	0
14	0	16.2	0	−6.553157928	0.0014235771	0.9985764229	0
15	1	52.978	1	5.3118625589	0.9950914817	0.0049085183	1
16	1	58.945	1	6.563922109	0.9985916429	0.0014083571	1
17	1	40.075	1	2.6044175487	0.931145345	0.068854655	1
18	1	42.38	0	−1.059791241	0.2573493508	0.7426506492	0 *
19	1	38.11	1	2.1921003012	0.8995378689	0.1004621311	1
20	1	26.185	0	−0.310130323	0.4230829288	0.5769170712	0 *
21	0	52.81	0	1.1287425462	0.7556067658	0.2443932342	1 *
22	1	34.521	1	1.4390180613	0.8083025465	0.1916974535	1
23	0	34.75	0	−2.660799179	0.0653265194	0.9346734806	0
24	1	46.254	1	3.9009611989	0.9801783778	0.0198216222	1
25	0	24.811	0	−4.746306112	0.008608955	0.991391045	0
26	0	4.792	0	−8.946906101	0.0001301222	0.9998698778	0
27	1	55.92	1	5.9291843616	0.997346409	0.002653591	1
28	0	38.62	0	−1.848754524	0.1360191965	0.8639808035	0
29	0	12.742	0	−7.278752352	0.0006895701	0.9993104299	0
30	0	31.95	0	−3.248325028	0.0373871218	0.9626128782	0
31	1	51.211	1	4.9410917821	0.9929039228	0.0070960772	1
32	1	30.92	1	0.6834178535	0.6645010995	0.3354989005	1
33	0	23.527	0	−5.01572868	0.006589093	0.993410907	0
34	0	30.225	0	−3.610282917	0.026332065	0.973667935	0
35	0	28.387	0	0.1519167909	0.5379063233	0.4620936767	1 *
36	0	27.48	0	−4.18626808	0.0149752478	0.9850247522	0
37	1	41.95	1	2.9978500368	0.9524769041	0.0475230959	1
38	1	34.995	1	1.5384777943	0.8232433323	0.1767566677	1
39	0	34.964	1	1.5319730438	0.8222948107	0.1777051893	1 *
40	0	7.998	0	−8.274189004	0.0002549498	0.9997450502	0
41		42.571	1	3.1281548769	0.9580392816	0.0419607184	1
42		51.835	0	0.9241576524	0.7158884981	0.2841115019	1

FIGURE 16.5   **JMP Output of a Logistic Regression of the Credit Card Upgrade Data**

Whole Model Test

Model	-LogLikelihood	DF	ChiSquare	Prob>ChiSq
Difference	18.492960	2	36.98592	<.0001*
Full	9.182906			
Reduced	27.675866			

RSquare (U)	0.6682

Lack Of Fit

Source	DF	-LogLikelihood	ChiSquare
Lack Of Fit	37	9.1829064	18.36581
Saturated	39	0.0000000	Prob>ChiSq
Fitted	2	9.1829064	0.9956

Parameter Estimates

Term	Estimate	Std Error	ChiSquare	Prob>ChiSq	Lower 95%	Upper 95%
Intercept	-9.9524146	4.0818938	5.94	0.0148*	-20.989597	-4.1146535
Purchases	0.20983066	0.0907079	5.35	0.0207*	0.07153188	0.44477918
PlatProfile	4.14786846	1.6101814	6.64	0.0100*	1.64251771	8.57651011

Effect Likelihood Ratio Tests

Source	Nparm	DF	L-R ChiSquare	Prob>ChiSq
Purchases	1	1	11.5458696	0.0007*
PlatProfile	1	1	12.8442451	0.0003*

Odds Ratios

For UpGrade odds of 1 versus 0
Tests and confidence intervals on odds ratios are likelihood ratio based.

Unit Odds Ratios

Per unit change in regressor

Term	Odds Ratio	Lower 95%	Upper 95%	Reciprocal
Purchases	1.233469	1.074152	1.560146	0.8107215
PlatProfile	63.29893	5.168165	5305.557	0.0157981

Silver card holder's odds of upgrading increase by about 23 percent. The odds ratio estimate of 63.2989 for *PlatProfile* says that we estimate that the odds of upgrading for a Silver card holder who conforms to the bank's Platinum profile are about 63 times larger than the odds of upgrading for a Silver card holder who does not conform to the bank's Platinum profile, if both Silver card holders had the same amount of purchases last year. The bottom of the JMP output in Figure 16.5 shows that

- The upgrade probability estimate for a Silver card holder who had purchases of \$42,571 last year and conforms to the bank's Platinum profile is

$$\frac{e^{(-9.9524+.2098(42.571)+4.1479\,(1))}}{1+e^{(-9.9524+.2098(42.571)+4.1479\,(1))}} = \frac{e^{3.1282}}{1+e^{3.1282}} = .9580$$

- The upgrade probability estimate for a Silver card holder who had purchases of \$51,835 last year and does not conform to the bank's Platinum profile is

$$\frac{e^{(-9.9524+.2098(51.835)+4.1479\,(0))}}{1+e^{(-9.9524+.2098(51.835)+4.1479\,(0))}} = \frac{e^{.9242}}{1+e^{.9242}} = .7159$$

FIGURE 16.6	The Confusion Matrix for the Credit Card Upgrade Data

Confusion Matrix

Training

Actual UpGrade	Predicted Count 1	0
1	17	2
0	4	17

FIGURE 16.7	The Confusion Matrix for the Performance Data

Confusion Matrix

Training

Actual Group	Predicted Count 1	0
1	19	4
0	4	16

JMP classifies any Silver card holder as a 1 (that is, as an upgrader) if the upgrade probability estimate for the holder is at least .5 and as a 0 otherwise. The two Silver card holders at the bottom of Figure 16.4 (holders 41 and 42) are two Silver card holders to whom the bank is considering sending an upgrade offer. Because the upgrade probability estimates for these Silver card holders are .9580 and .7159, JMP classifies both card holders as 1s (upgraders). These two classifications, as well as the classifications for the 40 sampled Silver card holders, are given in Figure 16.4 under the column labeled *Most Likely Upgrade*. If we compare the *Upgrade* column with the *Most Likely Upgrade* column, we note that 6 classifications (see the asterisks) out of 40 classifications are inaccurate. This gives a misclassification rate of 6/40 = .15. More specifically, the **Confusion Matrix** in Figure 16.6 says that (1) out of 19 Silver card holders who are 1s (upgraders), 17 are accurately classified as 1s and 2 are inaccurately classified as 0s, for an upgrader misclassification rate of 2/19 = .1053; and (2) out of 21 Silver card holders who are 0s (nonupgraders), 17 are accurately classified as 0s and 4 are inaccurately classified as 1s, for a nonupgrader misclassification rate of 4/21 = .1905. Because the logistic regression model has classified the 40 observed Silver card holders fairly accurately, we can have reasonable confidence in the ability of this model to find accurate upgrade probability estimates and classifications for Silver card holders (for example, holders 41 and 42) to whom the bank is considering sending an upgrade offer. Finally, although we did not give a confusion matrix for the logistic regression of the performance data in Table 16.1, Minitab (like JMP) gives a confusion matrix. Because the confusion matrix in Figure 16.7 indicates that the logistic regression of the performance data has classified the 43 observed employees fairly accurately, we can be reasonably confident that this logistic model will classify future potential employees (like the future potential employee described at the bottom of Figure 16.3) fairly accurately. In general, however, logistic regression does not always provide accurate classifications. We demonstrate this in the example of the following subsection.

Forward selection with simultaneous validation for logistic regression

Consider a cell phone company that has randomly selected a sample of 3,332 observations concerning customer values of the response variable Churn and 17 predictor variables. Here, Churn equals True if a customer churned—left the cell phone company for another cell phone company—and equals False otherwise. The 17 predictor variables describe aspects of the customer's account and recent cell phone bill and are summarized across the top of Figure 16.8, which is a JMP output of twelve observations in the data, Because the sample of 3,332 observations is very large, we begin by having JMP form a *Validation* column (shown in Figure 16.8) by randomly selecting 67 percent (or 2,221) of the observations as the training data set and 33 percent (or 1,111) of the observations as the validation data set. To build a logistic regression model describing the churn data, we will use **forward selection with simultaneous validation.** What JMP does in implementing this technique is use the training data set to add one predictor variable at a time to the logistic model, where the next variable added to the model is the variable not currently in the model that has the smallest *p*-value when added to the model.

FIGURE 16.8 **The JMP Output of Part of the Churn Data and a Validation Column** ⓄⓈ Churn

	Churn	AcctLength	IntlPlan	VMPlan	NVMailMsgs	DayMinutes	DayCalls	DayCharge	EveMinutes	EveCalls	EveCharges
1	False	107	no	yes	26	161.6	123	27.47	195.5	103	16.62
2	False	137	no	no	0	243.4	114	41.38	121.2	110	10.3
3	False	84	yes	no	0	299.4	71	50.9	61.9	88	5.26
4	False	75	yes	no	0	166.7	113	28.34	148.3	122	12.61
10	True	65	no	no	0	129.1	137	21.95	228.5	83	19.42
11	False	74	no	no	0	187.7	127	31.91	163.4	148	13.89
12	False	168	no	no	0	128.8	96	21.9	104.9	71	8.92
13	False	95	no	no	0	156.6	88	26.62	247.6	75	21.05
14	False	62	no	no	0	120.7	70	20.52	307.2	76	26.11
15	True	161	no	no	0	332.9	67	56.59	317.8	97	27.01
16	False	85	no	yes	27	196.4	139	33.39	280.9	90	23.88
99	True	77	no	no	0	251.8	72	42.81	205.7	126	17.48
3333	—	87	no	yes	10	153	110	28.35	190.7	104	18.75

	NightMin	NightCalls	NightCharge	IntlMin	IntlCalls	IntlCharge	NCustServiceCalls	Validation
1	254.4	103	11.45	13.7	3	3.7	1	Training
2	162.6	104	7.32	12.2	5	3.29	0	Training
3	196.9	89	8.86	6.6	7	1.78	2	Training
4	186.9	121	8.41	10.1	3	2.73	3	Validation
10	208.8	111	9.4	12.7	6	3.43	4	Training
11	196	94	8.82	9.1	5	2.46	0	Training
12	141.1	128	6.35	11.2	2	3.02	1	Validation
13	192.3	115	8.65	12.3	5	3.32	3	Validation
14	203	99	9.14	13.1	6	3.54	4	Training
15	160.6	128	7.23	5.4	9	1.46	4	Validation
16	89.3	75	4.02	13.8	4	3.73	1	Training
99	275.2	109	12.38	9.8	7	2.65	2	Validation
3333	205	81	8.81	11.7	3	3.87	5	.

As each new predictor variable is added, the model obtained is used to find a Churn probability estimate that predicts the response variable value (True or False—that is, 1 or 0) for each observation in the training data set and for each observation in the validation data set. *RSquare* is then calculated using the observations in the training data set, and *RSquare Validation* is calculated using the observations in the validation data set. The stepwise procedure stops if *RSquare Validation* reaches a value that is not exceeded in 10 additional steps of adding predictor variables. Figure 16.9 shows that after 7 steps of adding predictor variables, *RSquare Validation* decreases in step 8 from .2061 to .1965. Because (it can be verified) *RSquare Validation* never increases beyond .2061 up through step 17 of adding predictor variables, JMP chooses the 7 predictor variable model. Figure 16.9 gives a Churn probability estimate of .2099 for customer 3333, who has just been sent a new cell phone bill, and thus customer 3333 is predicted not to churn. If customer 3333 had been predicted to churn, then the cell phone company would take early action to try to convince customer 3333 not to churn. No such action seems necessary, but note that the confusion matrices in Figure 16.9 for the training and validation data sets show that the logistic model classifies churners very inaccurately. In the next two sections, we will see that linear discriminate analysis and neural networks classify churners much more accurately.

FIGURE 16.9   The JMP Output of Forward Selection with Simultaneous Validation in the Churn Example, Prediction Using the Logistic Model, and the Confusion Matrices

Step History

Step	Parameter	Action	L-R ChiSquare	"Sig Prob"	RSquare	p	AICc	BIC	RSquare Validation
1	IntlPlan{no-yes}	Entered	112.662	0.0000	0.0613	2	1729.89	1741.3	0.0627
2	NCustServiceCalls	Entered	99.85023	0.0000	0.1156	3	1632.05	1649.15	0.1168
3	DayCharge	Entered	110.714	0.0000	0.1758	4	1523.34	1546.15	0.1600
4	VMPlan{yes-no}	Entered	27.42537	0.0000	0.1907	5	1497.92	1526.43	0.1770
5	EveMinutes	Entered	18.47418	0.0000	0.2008	6	1481.46	1515.66	0.1955
6	IntlMin	Entered	11.09874	0.0009	0.2068	7	1472.38	1512.26	0.2025
7	IntlCalls	Entered	10.79877	0.0010	0.2127	8	1463.59	1509.17	0.2061
8	NVMailMsgs	Entered	9.064382	0.0026	0.2176	9	1456.54	1507.81	0.1965

-LogLikelihood	p	RSquare	AICc	BIC	RSquare Validation	Avg Log Error Validation
723.7629	8	0.2127	1463.59	1509.17	0.2061	0.328492

	Lin[False]	Prob[False]	Prob[True]	Most Likely Churn
3333	1.3256555604	0.7901211055	0.2098788945	False

Training

Actual Churn	Predicted Count False	True
False	1848	51
True	254	68

Validation

Actual Churn	Predicted Count False	True
False	922	28
True	126	35

Exercises for Section 16.1

CONCEPTS connect

16.1 What two values does the dependent variable equal in logistic regression? What do these values represent?

16.2 Define the odds of an event, and the odds ratio for x_j.

METHODS AND APPLICATIONS

16.3 If we use the logistic regression model

$$p(x_1) = \frac{e^{(\beta_0+\beta_1 x_1)}}{1 + e^{(\beta_0+\beta_1 x_1)}}$$

to analyze the performance data in Table 16.1, we find that the point estimates of the model parameters and their associated p-values (given in parentheses) are $b_0 = -43.37(.001)$ and $b_1 = .4897(.001)$. **(1)** Find a point estimate of the probability of success for a potential employee who scores a 93 on test 1. **(2)** Using $b_1 = .4897$, find a point estimate of the odds ratio for x_1. **(3)** Interpret this point estimate.

16.4 Mendenhall and Sincich (1993) present data that can be used to investigate allegations of gender discrimination in the hiring practices of a particular firm. These data are given in the page margin. In this table, y is a dummy variable that equals 1 if a potential employee was hired and 0 otherwise; x_1 is the number of years of education of the potential employee; x_2 is the number of years of experience of the potential employee; and x_3 is a dummy variable that equals 1 if the potential employee was a male and 0 if the potential employee was a female.

a Fit the logistic regression model

$$p(x_1, x_2, x_3) = \frac{e^{(\beta_0+\beta_1 x_1+\beta_2 x_2+\beta_3 x_3)}}{1 + e^{(\beta_0+\beta_1 x_1+\beta_2 x_2+\beta_3 x_3)}}$$

to the gender discrimination data.

b Evaluate the significance of each independent variable in the model.

c Interpret the confusion matrix.

d Consider a potential employee having 4 years of education and 5 years of experience. Find **(1)** a point estimate of the probability that the potential employee will be hired if the potential employee is a male, and **(2)** a point estimate of the probability that the potential employee will be hired if the potential employee is a female.

e Identify and interpret the odds ratio estimates for x_1 (Education), x_2 (Experience), and x_3 (Gender).

f Discuss whether there seems to have been gender discrimination in the firm's hiring practices.

16.5 Read Exercise 5.3 and recall that Figure 5.11(c) gives data concerning **(1)** the dependent variable *Coupon*, which equals 1 if a Williams Apparel customer redeems a coupon and 0 otherwise; **(2)** the independent variable

y	x_1	x_2	x_3
0	6	2	0
0	4	0	1
1	6	6	1
1	6	3	1
0	4	1	0
1	8	3	0
0	4	2	1
0	4	4	0
0	6	1	0
1	8	10	0
0	4	2	1
0	8	5	0
0	4	2	0
0	6	7	0
1	4	5	1
0	6	4	0
0	8	0	1
1	6	1	1
0	4	7	0
0	4	1	1
0	4	5	0
0	6	0	1
1	8	5	1
0	4	9	0
0	8	1	0
0	6	1	1
1	4	10	1
1	6	12	0

DS Gender

Source: William Mendenhall and Terry Sincich, *A Second Course in Business Statistics: Regression Analysis*, 4th edition, Prentice Hall.

FIGURE 16.10   Partial JMP Output of a Logistic Regression of the Coupon Redemption Data

Parameter Estimates

Term	Estimate	Std Error	ChiSquare	Prob>ChiSq	Lower 95%	Upper 95%
Intercept	−15.420089	7.3001666	4.46	0.0347*	−38.23488	−6.1553121
Purchases	0.30451643	0.1467644	4.31	0.0380*	0.11288633	0.75560222
Card	6.94342334	3.0816921	5.08	0.0243*	2.86246555	16.6677408

Effect Likelihood Ratio Tests

Source	Nparm	DF	L-R ChiSquare	Prob>ChiSq
Purchases	1	1	18.0832669	<.0001*
Card	1	1	20.8066786	<.0001*

For Coupon odds of 1 versus 0

Unit Odds Ratios

Per unit change in regressor

Term	Odds Ratio	Lower 95%	Upper 95%	Reciprocal
Purchases	1.355969	1.119505	2.128893	0.7374799
Card	1036.312	17.50463	17326381	0.000965

Confusion Matrix

Training

Actual Coupon	Predicted Count 1	0
1	17	1
0	3	19

	Coupon	Purchases	Card	Lin[1]	Prob[1]	Prob[0]	Most Likely Coupon
41		43.97	1	4.9129214572	0.9927026612	0.0072973388	1
42		52.48	0	0.5609329147	0.6366683712	0.3633316288	1

Purchases, which is last year's purchases (in hundreds of dollars) by the customer; and **(3)** the independent variable *Card*, which equals 1 if the customer has a Williams Apparel credit card and 0 otherwise. Figure 16.10 gives part of a JMP output of a logistic regression of the coupon redemption data.

DS CoupRedemp

a   Evaluate the significance of each independent variable in the model.

b   Identify and interpret the odds ratio estimates for *Purchases* and *Card*.

c   Interpret the confusion matrix.

d   Estimate the probability that a customer who spent $4,397 last year and has a Williams credit card will redeem the coupon.

e   Estimate the probability that a customer who spent $5,248 last year and does not have a Williams credit card will redeem the coupon.

16.2 Linear Discriminate Analysis

LO16-2

Use linear discriminate analysis to classify observations and estimate probabilities.

Figure 16.11 shows the employee performance data and the results of a **linear discriminate analysis,** which predicts whether each of the 43 employees in the historical data, as well as future potential employee 44, would perform successfully (a 1) or unsuccessfully (a 0) in a particular position. To show how linear discriminate analysis works, note that Figure 16.12 shows that (1) the means of the scores on Test 1 and Test 2 for the 20 employees who were actually unsuccessful in the position are 84.75 and 79.1 and (2) the means of the scores of the 23 employees who were actually successful in the position are 92.43 and 84.78. Now, consider future potential employee 44 in Figure 16.11. Figure 16.11 says that the combination (93, 84) of test scores for potential employee 44 is (1) a *squared distance* (see the column headed *SqDist[0] 2*) of 4.8511 from the combination (84.75, 79.1) of mean test scores for the 20 unsuccessful employees and (2) a *squared distance* (see the column headed *SqDist[1]*) of .0518 from the combination (92.43, 84.78) of mean test scores for the 23 successful employees. Here,

FIGURE 16.11   The Performance Data and the Results of a Linear Discriminate Analysis   ⓓⓢ PerfTest

	Group	Test 1	Test 2	SqDist[0]	SqDist[0] 2	SqDist[1]	Prob[0]	Prob[1]	Pred Group
1	1	96	85	745.61314773	8.8403255752	0.876295627	0.01831	0.98169	1
2	1	96	88	758.36061206	9.7632508112	1.0378249679	0.01258	0.98742	1
3	1	91	81	671.65205619	2.6388577133	0.5416130496	0.25949	0.74051	1
4	1	95	78	706.38037694	7.5968603132	2.3797389383	0.06859	0.93141	1
5	1	92	85	699.27732397	4.0993590618	0.0166017198	0.11493	0.88507	1
6	1	93	87	719.19061877	5.730913481	0.1702407241	0.05840	0.94160	1
7	1	98	84	765.54697243	11.918234685	2.2673670654	0.00796	0.99204	1
8	1	92	82	686.93496628	3.5815404607	0.2601790138	0.15967	0.84033	1
9	1	97	89	774.60888785	11.671299266	1.7217566397	0.00686	0.99314	1
10	1	95	96	783.82210904	14.091357859	4.3058611136	0.00744	0.99256	1
11	1	99	93	816.77614698	17.275077643	4.369704187	0.00157	0.99843	1
12	1	89	90	688.23133688	4.5419497422	2.1011536964	0.22787	0.77213	1
13	1	94	90	743.94291241	8.2599537163	0.9675669128	0.02543	0.97457	1
14	1	92	94	740.05945165	9.4078694618	3.0409244347	0.03979	0.96021	1
15	1	94	84	718.0186338	5.9847533022	0.2151582888	0.05291	0.94709	1
16	1	90	92	708.4388057	6.4676781908	2.5489667302	0.12354	0.87646	1
17	1	91	70	632.31375173	6.657196593	7.3517368777	0.58596	0.41404	0 *
18	1	90	81	660.48405627	1.8695721038	0.7426455917	0.36275	0.63725	1
19	1	86	81	617.21072693	0.1911000036	2.9454460976	0.79854	0.20146	0 *
20	1	90	76	641.46796482	2.561045806	2.7031124522	0.51775	0.48225	0 *
21	1	91	79	663.87379477	2.7436223496	1.1539749492	0.31113	0.68887	1
22	1	88	83	646.73443905	1.0343574412	1.3404699688	0.53819	0.46181	0 *
23	1	87	82	631.8862751	0.5264208313	2.0566501421	0.68246	0.31754	0 *
24	0	93	74	668.65818308	6.4381471995	4.1768566544	0.24404	0.75596	1 *
25	0	90	84	672.72816772	2.2891444596	0.4008220524	0.28006	0.71994	1 *
26	0	91	81	671.65205619	2.6388577133	0.5416130496	0.25949	0.74051	1 *
27	0	91	78	660.08897112	2.9003117398	1.5644629711	0.33896	0.66104	1 *
28	0	88	78	626.83902062	0.8465041643	2.4216098502	0.68731	0.31269	0
29	0	86	86	638.33316258	1.6059705001	3.0913234357	0.67758	0.32242	0
30	0	79	81	546.86728138	2.638654628	12.185227783	0.99162	0.00838	0
31	0	83	84	598.85473287	1.206709785	6.1106144386	0.92070	0.07930	0
32	0	79	77	530.70596753	2.2433929013	12.805160583	0.99494	0.00506	0
33	0	88	75	615.73622614	1.5682487751	3.9047503561	0.76283	0.23717	0
34	0	81	85	583.39283893	2.6007314355	9.1914737605	0.96427	0.03573	0
35	0	85	83	615.0192116	0.5152729169	3.7323398991	0.83321	0.16679	0
36	0	82	72	541.87357559	1.9224231856	10.842229571	0.98857	0.01143	0
37	0	82	81	576.17527013	0.7505004435	7.3861191437	0.96503	0.03497	0
38	0	81	77	550.25191686	0.9919136098	9.6130449882	0.98675	0.01325	0
39	0	86	76	597.82674248	0.5146807093	4.5380199616	0.88202	0.11798	0
40	0	81	84	579.00684051	2.156246039	9.0007869956	0.96839	0.03161	0
41	0	85	78	594.84787342	0.0514998926	4.5375600332	0.90405	0.09595	0
42	0	81	71	528.31682332	2.7058982602	12.849821429	0.99377	0.00623	0
43	0	83	77	570.35733432	0.2999024534	6.9803975287	0.96578	0.03422	0
44		93	84	706.48621673	4.8510505409	0.051773679	0.08320	0.91680	1

FIGURE 16.12   The Group Means and p-Values for Test 1 and Test 2

Group Means

Count	Group	Test 1	Test 2	Column	F Ratio	Prob>F
20	0	84.750000	79.100000	Test 1	27.384	0.0000056
23	1	92.434783	84.782609	Test 2	2.369	0.1316083
43	All	88.860465	82.139535			

the formulas for the squared distances involve matrix algebra and are beyond the scope of this course. Intuitively, however, these formulas combine the sort of distance measure discussed in Section 3.9 on cluster analysis with the standard deviations and correlations that we will briefly consider when we discuss the assumptions behind linear discriminate analysis. Moreover, what is important to understand is that, because $SqDist[1] = .0518$ is smaller than $SqDist[0]\ 2 = 4.8511$, potential employee 44's test scores are closer (or nearer) to the mean test scores for the successful employees than to the mean test scores for the unsuccessful employees. Therefore, linear discriminate analysis predicts that potential employee 44 will be successful (that is, classifies potential employee 44 as a 1)—see the column in Figure 16.11 headed *PredGroup*. In addition, using $SqDist[0]\ 2 = 4.8511$ and $SqDist[1] = .0518$, linear discriminate analysis estimates (1) the probability that potential employee 44 will be unsuccessful to be $1/(1 + e^{[-.5(.0518-4.8511)]}) = .0832$ (see the column headed *Prob[0]*) and (2) the probability that potential employee 44 will be successful to be $1/(1 + e^{[-.5(4.8511-.0518)]}) = .9168$ (see the column headed *Prob[1]*). (Note that the *SqDist[0]* values in Figure 16.11 are theoretical quantities that we will not discuss and are used to calculate the values of *SqDist[0]* 2 and *SqDist[1]*.)

In order to evaluate how confident we can be that linear discriminate will accurately classify future potential employees as being successful or unsuccessful, first note from Figure 16.12 that the *p*-value measuring the significance of Test 1 in the linear discriminate analysis is .0000056, indicating that we have extremely strong evidence that Test 1 is significant. However, the *p*-value measuring the significance of Test 2 is .1316083, which indicates that Test 2 may not be significant. (In Exercise 16.7 the reader will carry out a linear discriminate analysis using only Test 1.) Figure 16.13 shows the confusion matrix obtained when linear discriminate analysis is used to classify each of the 43 observed employees in Figure 16.11 as being successful (a 1) or unsuccessful (a 0). Here, an employee is classified as being successful (a 1) if and only if the employee's SqDist[1] is less than or equal to the employee's SqDist[0] 2, or, equivalently, if and only if the employee's Prob[1] is at least .5. Because the confusion matrix indicates that 9 out of the 43 employees (a fairly small 20.93 percent of the employees) have been misclassified, we can be reasonably confident that linear discriminate analysis will classify future potential employees (such as potential employee 44) fairly accurately.

To carry out linear discriminate analysis when the observed data has been divided into a training data set and a validation data set, we calculate the means of the predictor variable values for the observations belonging to a particular category *by using only the observations belonging to that category that are in the training data set*. For example, consider the churn data and customer 3333 in Figure 16.8. To classify customer 3333 as a churner or non-churner, we first note that (as explained in the next paragraph) linear discriminate analysis assumes that all of the predictor variables are quantitative. Because 2 of the 17 predictor variables (IntlPlan and VMPlan) are qualitative, we will omit those variables from the analysis. Therefore JMP first calculates the squared distance between the 15 predictor variable values describing customer 3333 and the means of the 15 predictor variable values for the customers who churned in the training data set. (It can be verified that out of 2,221 customers in the training data set, 322 churned.) Next, linear discriminate analysis calculates the squared distance between the 15 predictor variable values describing customer 3333 and the means of the 15 predictor variable values for the customers who did not churn in the training data set. Using the

---

FIGURE 16.13   **Misclassifications and the Confusion Matrix**

Score Summaries

Source	Count	Number Misclassified	Percent Misclassified	Entropy RSquare	-2LogLikelihood
Training	43	9	20.9302	0.52613	28.1484

Training

	Predicted	
Actual	Count	
Group	0	1
0	16	4
1	5	18

FIGURE 16.14   **A Linear Discriminate Analysis of the Churn Data**   Churn

Probabilities to Each Group

Row	Actual	False	True
3330	False	0.31668	0.68332
3331	False	0.68761	0.31239
3332	False	0.45347	0.54653
3333		0.17764	0.82236

Score Summaries

		Number	Percent		
Source	Count	Misclassified	Misclassified	Entropy RSquare	-2LogLikelihood
Training	2221	640	28.8158	−0.3951	2565.05
Validation	1111	317	28.5329	−0.4310	

Training

Actual Churn	Predicted Count	
	False	True
False	1348	551
True	89	233

Validation

Actual Churn	Predicted Count	
	False	True
False	681	269
True	48	113

two calculated squared distances, linear discriminate analysis then calculates Prob[True], an estimate of the probability that customer 3333 will churn. As shown in Figure 16.14, this estimated probability is .82236, which means that customer 3333 is predicted to churn. Figure 16.14 shows that when all of the customers in the training and validation data sets are classified, the percent misclassification rates are about 29 percent for each data set. Looking at the confusion matrices, and comparing these confusion matrices with the confusion matrices in Figure 16.9 that are obtained using logistic regression, we can see that linear discriminate analysis does a worse job classifying the non-churners but a much better job classifying the churners. In the next section we will see that neural networks classify both the churners and non-churners more accurately than logistic regression or linear discriminate analysis.

Theoretically, linear discriminate analysis assumes that for each particular category being considered, the joint probability distribution of the predictor variables is a *multivariate normal distribution* described by standard deviations and correlations that are the same for any one category as they are for any other category. A thorough explanation of what the multivariate normal distribution is and how to check the assumptions is beyond the scope of this book. However, it has been found that linear discriminate analysis often classifies fairly accurately even when the assumptions are not valid. In fact, although a predictor variable that is qualitative cannot have a normal distribution, linear discriminate analysis is sometimes used successfully with qualitative predictor variables (see Exercises 16.8 and 16.9). Finally, note that linear discriminate analysis is perhaps the easiest classification method to use when the response variable is to be classified into more than two categories. See Exercise 16.10.

Exercises for Section 16.2

CONCEPTS   connect

16.6   Discuss two ways that the squared distances can be used in linear discriminate analysis.

METHODS AND APPLICATIONS

16.7   Using Test 1, perform a linear discriminate analysis of the performance data in Figure 16.11 and classify potential employee 44.

16.8   Using Purchases and PlatProfile, perform a linear discriminate analysis of the card upgrade data in Figure 16.4 and classify Silver card holders 41 and 42.

16.9   Using Purchases and Card, perform a linear discriminate analysis of the coupon redemption data in Figure 5.11(c) and classify customers 41 and 42.

16.10   Figure 16.15 is the JMP output of a linear discriminate analysis of the performance data when the personnel director reclassifies the 43 observed employees as performing either successfully (1), somewhat successfully (.5), or unsuccessfully (0). Use the squared distances and probabilities to explain the classification of future potential employee 44. Also, interpret the group means, *p*-values, and confusion matrix in Figure 16.16.

FIGURE 16.15   A Linear Discriminate Analysis in the Three Group Performance Exercise   DS PerfTest

	Group	Test 1	Test 2	Validation	SqDist[0]	SqDist[.5]	SqDist[0] 2	SqDist[1]	Prob[.5]	Prob[0]	Prob[1]	Pred Group
1	1	96	85	Validation	1502.7309843	7.5918287944	23.454667453	0.507826289	0.02814	0.00001	0.97185	1
2	1	96	88	Training	1528.0977213	9.4014399395	25.988752476	0.4430771348	0.01122	0.00000	0.98878	1
3	1	91	81	Validation	1353.6536035	0.642420003	8.4298158556	2.9979387559	0.75282	0.01533	0.23184	.5
4	1	95	78	Validation	1422.708977	4.6801424789	17.430646563	3.3577221769	0.34027	0.00058	0.65915	1
5	1	92	85	Validation	1408.9085315	2.3441527732	12.519409323	0.8124496219	0.31676	0.00196	0.68129	1
6	1	93	87	Training	1448.7540688	4.3412453122	16.421379974	0.1718937895	0.11057	0.00026	0.88917	1
7	1	98	84	Training	1542.7450438	11.170875062	29.636013483	1.9355096461	0.00978	0.00000	0.99022	1
8	1	92	82	Training	1384.4495664	1.4423134768	10.893096149	1.7849706247	0.54013	0.00479	0.45508	.5
9	1	97	89	Validation	1560.8215521	12.12920129	30.379900646	1.1579768805	0.00413	0.00000	0.99587	1
10	1	95	96	Validation	1575.3995748	16.027985652	33.125333001	3.4594035548	0.00186	0.00000	0.99814	1
11	1	99	93	Validation	1645.2104843	20.821244055	42.881699635	4.5747229023	0.00030	0.00000	0.99970	1
12	1	89	90	Training	1383.231788	3.8366155656	10.953641996	3.3452030978	0.43346	0.01235	0.55419	1
13	1	94	90	Training	1497.8960915	7.7824481266	22.008952193	0.3506614661	0.02376	0.00002	0.97623	1
14	1	92	94	Training	1486.7298285	9.4940724445	21.842750626	2.3392883954	0.02719	0.00006	0.97276	1
15	1	94	84	Training	1447.0827447	4.0833528446	16.860909157	0.4002867827	0.13684	0.00023	0.86293	1
16	.5	90	92	Training	1423.1756206	5.9320035193	14.953908062	2.8029426801	0.17267	0.00190	0.82543	1
17	.5	91	70	Training	1271.3665788	4.7315231134	9.8625147488	13.959696297	0.92014	0.07074	0.00912	.5
18	.5	90	81	Training	1331.0829621	0.2154728549	6.5809731801	3.9590664463	0.83660	0.03469	0.12871	.5
19	.5	86	81	Training	1243.1698037	0.8770912655	1.5550094814	10.172984211	0.58069	0.41375	0.00556	.5
20	.5	90	76	Training	1292.5148123	0.9091992939	6.0672431568	7.7767267173	0.90243	0.06845	0.02912	.5
21	.5	91	79	Training	1337.9515926	0.6451596655	7.9495729331	4.2502519511	0.83976	0.02178	0.13846	.5
22	.5	88	83	Training	1302.6002082	0.3153568348	4.3200486903	5.5855265704	0.82869	0.11189	0.05943	.5
23	.5	87	82	Training	1272.7114639	0.4226819801	2.7639870159	7.7057133206	0.74829	0.23209	0.01961	.5
24	.5	93	74	Training	1345.7941078	3.4621626896	12.40291055	7.4150391303	0.86956	0.00995	0.12049	.5
25	.5	90	84	Training	1355.2114969	0.786881832	7.8768560346	2.6561151241	0.70343	0.02031	0.27626	.5
26	.5	91	81	Validation	1353.6536035	0.642420003	8.4298158556	2.9979387559	0.75282	0.01533	0.23184	.5
27	.5	91	78	Training	1330.2240428	0.7699851017	7.8329070769	4.9998641538	0.86964	0.02545	0.10491	.5
28	.5	88	78	Validation	1263.3482195	0.325244422	3.1224798152	8.7193479897	0.79241	0.19568	0.01192	.5
29	.5	86	86	Training	1283.51715	1.9625612135	3.8479358918	8.1345203269	0.69674	0.27143	0.03183	.5
30	.5	79	81	Validation	1098.4439935	11.157140446	1.8817899704	30.169557261	0.00959	0.99041	0.00000	0
31	0	83	84	Training	1203.5590302	4.1402745215	1.2769800329	15.726031683	0.19273	0.80668	0.00059	0
32	0	79	77	Validation	1068.0373396	12.159987591	1.9186719456	33.671551471	0.00594	0.99406	0.00000	0
33	0	88	75	Training	1240.7846712	1.3188218149	3.3915833307	11.587285682	0.73495	0.26072	0.00433	.5
34	0	81	85	Validation	1170.444394	7.4606514899	1.9950572392	21.197771562	0.06106	0.93887	0.00006	0
35	0	85	83	Validation	1237.6480535	1.7942846849	1.5332899585	11.228678936	0.46548	0.53036	0.00416	0
36	0	82	72	Validation	1092.1087234	9.062165042	1.8790795165	29.533438239	0.02682	0.97318	0.00000	0
37	0	82	81	Validation	1159.0476965	5.3297608809	0.3200969876	20.17795318	0.07552	0.92444	0.00005	0
38	0	81	77	Validation	1108.0915088	7.9267683683	0.5292437779	26.662182572	0.02416	0.97584	0.00000	0
39	0	86	76	Training	1204.8800509	1.8492147351	1.3196764887	14.269041513	0.43381	0.56532	0.00087	0
40	0	81	84	Validation	1162.3622203	7.2308530212	1.523767478	21.59275986	0.05450	0.94546	0.00004	0
41	0	85	78	Training	1198.6048626	2.0129700451	0.5445188563	14.571298128	0.32407	0.67532	0.00061	0
42	0	81	71	Training	1064.7836019	11.733112969	2.8866406236	34.217247771	0.01185	0.98815	0.00000	0
43	0	83	77	Training	1149.0934408	4.6413119472	0.0875784113	20.600576474	0.09305	0.90691	0.00003	0
44		93	84	.	1423.7595217	2.903824041	14.259484826	0.6088328176	0.24075	0.00082	0.75843	1

FIGURE 16.16   The Group Means, p-Values for the Significances of Test 1 and Test 2, and Confusion Matrix for the Three Group Performance Exercise

Group Means

Count	Group	Test 1	Test 2
15	.5	88.733333	80.400000
13	0	82.846154	78.461538
15	1	94.200000	87.066667
43	All	88.860465	82.139535

Column	F Ratio	Prob>F
Test 1	36.190	0.0000000
Test 2	9.628	0.0003995

Training

Actual Group	Predicted Count		
	.5	0	1
.5	13	1	1
0	1	12	0
1	2	0	13

16.3 Neural Networks

The parametric regression techniques discussed thus far in this book were developed for use with data sets having fewer than 1,000 observations and fewer than 50 predictor variables. Yet it is not uncommon in a large data mining project to analyze a data set involving millions of observations and hundreds or even thousands of predictor variables. Examples include sales transaction data used for marketing research, credit card analysis data, Internet e-mail filtering data, and computerized medical record data. In addition to the nonparametric predictive analytics techniques discussed in Chapter 5 (classification trees, regression trees, k-nearest neighbors, and naive Bayes' classification), the tremendous growth in available data has led to the development of parametric regression techniques that can be used to analyze extremely large data sets. In this section we will discuss one such technique: neural network modeling.

The idea behind neural network modeling is to represent the response variable as a nonlinear function of linear combinations of the predictor variables. The simplest but most widely used neural network model is called the **single-hidden-layer, feedforward neural network.** This model, which is also sometimes called the **single-layer perceptron,** is motivated (like all neural network models) by the connections of the neurons in the human brain. As illustrated in Figure 16.17, this model involves

1   An **input layer** consisting of the predictor variables $x_1, x_2, \ldots, x_k$ under consideration.

FIGURE 16.17   **The Single Layer Perceptron**

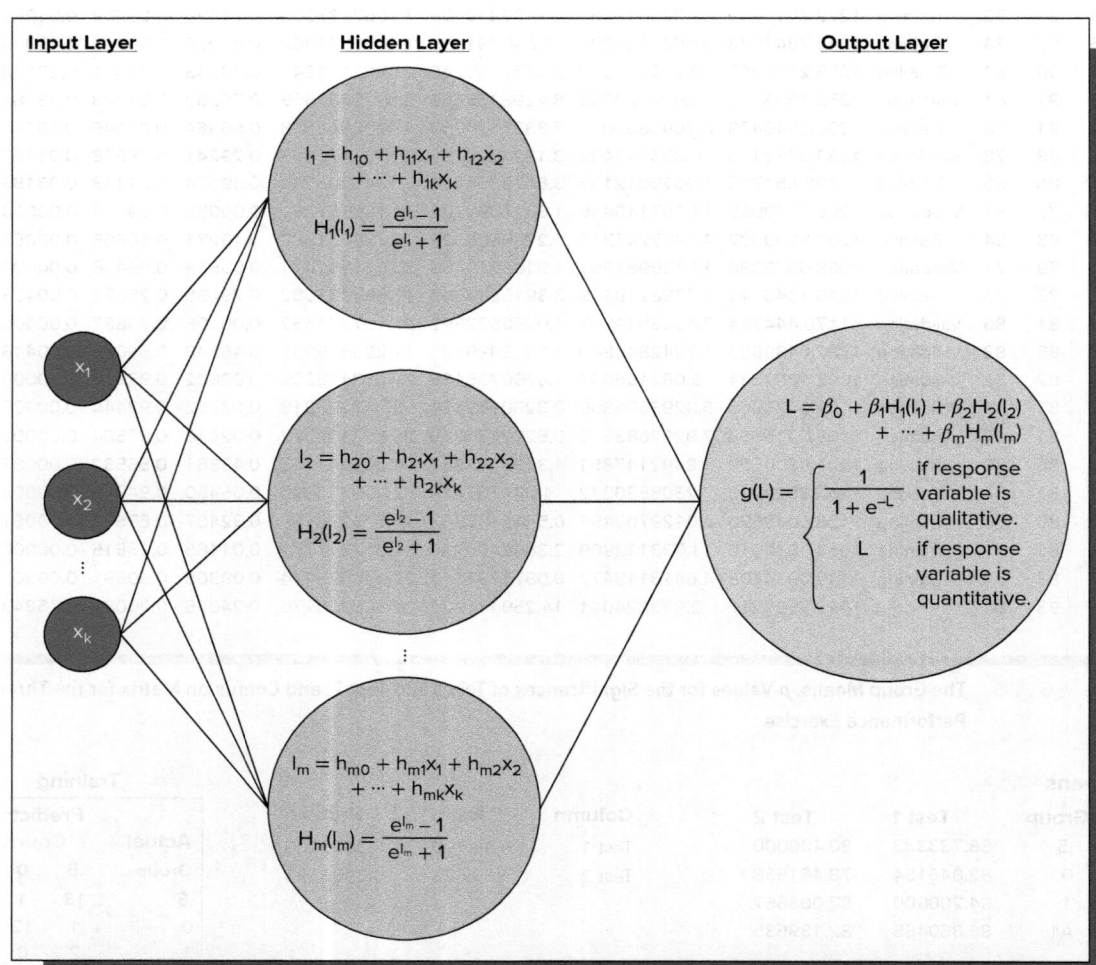

Input Layer   Hidden Layer   Output Layer

$$I_1 = h_{10} + h_{11}x_1 + h_{12}x_2 + \cdots + h_{1k}x_k$$

$$H_1(I_1) = \frac{e^{I_1} - 1}{e^{I_1} + 1}$$

$$I_2 = h_{20} + h_{21}x_1 + h_{22}x_2 + \cdots + h_{2k}x_k$$

$$H_2(I_2) = \frac{e^{I_2} - 1}{e^{I_2} + 1}$$

$$I_m = h_{m0} + h_{m1}x_1 + h_{m2}x_2 + \cdots + h_{mk}x_k$$

$$H_m(I_m) = \frac{e^{I_m} - 1}{e^{I_m} + 1}$$

$$L = \beta_0 + \beta_1 H_1(I_1) + \beta_2 H_2(I_2) + \cdots + \beta_m H_m(I_m)$$

$$g(L) = \begin{cases} \dfrac{1}{1 + e^{-L}} & \text{if response variable is qualitative.} \\ L & \text{if response variable is quantitative.} \end{cases}$$

x_1 x_2 x_k

2   A single **hidden layer** consisting of m **hidden nodes.** At the vth hidden node, for $v = 1, 2, \ldots, m$, we form a linear combination ℓ_v of the k predictor variables:

$$\ell_v = h_{v0} + h_{v1}x_1 + h_{v2}x_2 + \cdots + h_{vk}x_k$$

Here, $h_{v0}, h_{v1}, \ldots, h_{vk}$ are unknown parameters that must be estimated from the sample data. Having formed ℓ_v, we then specify a **hidden node function** $H_v(\ell_v)$ of ℓ_v. This hidden node function, which is also called an **activation function,** is usually nonlinear. The activation function used by JMP is

$$H_v(\ell_v) = \frac{e^{\ell_v} - 1}{e^{\ell_v} + 1}$$

[Noting that $(e^{2x} - 1)/(e^{2x} + 1)$ is the hyperbolic tangent function of the variable x, it follows that $H_v(\ell_v)$ is the hyperbolic tangent function of $x = .5\,\ell_v$.] For example, at nodes $1, 2, \ldots, m$, we specify

Node 1: $\ell_1 = h_{10} + h_{11}x_1 + h_{12}x_2 + \cdots + h_{1k}x_k \qquad H_1(\ell_1) = \dfrac{e^{\ell_1} - 1}{e^{\ell_1} + 1}$

Node 2: $\ell_2 = h_{20} + h_{21}x_1 + h_{22}x_2 + \cdots + h_{2k}x_k \qquad H_2(\ell_2) = \dfrac{e^{\ell_2} - 1}{e^{\ell_2} + 1}$

$\qquad \vdots \qquad\qquad\qquad\qquad \vdots \qquad\qquad\qquad\qquad\qquad \vdots$

Node m: $\ell_m = h_{m0} + h_{m1}x_1 + h_{m2}x_2 + \cdots + h_{mk}x_k \qquad H_m(\ell_m) = \dfrac{e^{\ell_m} - 1}{e^{\ell_m} + 1}$

3   An **output layer** where we form a linear combination L of the m hidden node functions:

$$L = \beta_0 + \beta_1 H_1(\ell_1) + \beta_2 H_2(\ell_2) + \cdots + \beta_m H_m(\ell_m)$$

and then specify an output layer function $g(L)$ of L. If the response variable is qualitative (for example, whether or not a Silver card holder upgrades), JMP specifies $g(L)$ to be the logistic function:

$$g(L) = \frac{e^L}{1 + e^L} = \frac{1}{1 + e^{-L}}$$

If the response variable is quantitative, JMP specifies $g(L)$ to be simply L. When we have obtained $g(L)$, we write the neural network model describing the response variable y as

$$y = g(L) + \varepsilon$$

Moreover, we assume that the mean of all possible values of the error term ε corresponding to any particular combination of values of the predictor variables $x_1, x_2, \ldots, x_k$ is zero.

Because a neural network model employs many parameters (the h_{vj}s and $\beta_0, \beta_1, \ldots, \beta_m$), we say that a neural network model is **overparametrized.** A neural network model's parameters are usually uninterpretable, which is a major drawback of neural network modeling. An advantage of the neural network approach is that a neural network model will often give more accurate predictions than will a standard regression model. Because of the many parameters employed by a neural network model, there is the danger that we will **overfit** the model. This means that we will fit the model too closely to the observed data points and thus fail to capture real patterns in the data that would help us to accurately predict future response variable values. Neural network modeling does not use t statistics and associated p-values to help decide which parameters to retain in a model. Rather, to avoid overfitting, neural network modeling finds parameter estimates that minimize a **penalized least squares criterion.** This criterion equals the sum of the squared residuals plus a *penalty*. The **penalty** equals a positive constant λ times the sum of the squared values of the parameter estimates. (Note that the absolute values of the parameter estimates are sometimes used instead of the squared values.) Thus, the penalty is not imposed on the number of parameters in the model but on the total magnitude of the parameter estimates. The **penalty weight** λ, which is often chosen to be between .001 and .1, controls the tradeoff between overfitting and underfitting. If λ is large, at least some of the parameter estimates will be relatively small in magnitude. We will soon see that the value of λ is chosen by using a validation data set. First, however, note that the neural network approach to balance overfitting and underfitting can be extremely useful in a large data mining project with hundreds or thousands of predictor variables. This is because so many predictor variables would make it virtually impossible to use most of the model-building techniques previously discussed in this book to decide which predictor variables to retain in a model.

Because a neural network model is nonlinear, there is not an equation that can be used (as there is for a linear regression model) to calculate parameter estimates. Instead, for whatever penalty weight λ that is used, a randomly chosen set of "starter values" is used to begin a *nonlinear search* for parameter estimates that minimize the penalized least squares criterion. A poor choice of starter values may lead to convergence to a **local minimum**—that is, to parameter estimates that do not really minimize the penalized least squares criterion. This is a particular problem with neural network models because these models have many parameters to estimate. For this reason, it is common practice to fit the model many times (typically between 10 and 50 times) using different sets of randomly chosen starter values for each fit. The set of parameter estimates that yields the smallest value of the penalized least squares criterion (in the 10 to 50 *tours* through the data) is the final set of parameter estimates for the model corresponding to the λ value being used.

JMP automates the entire estimation process. JMP begins by using **random holdback**— its default procedure—that randomly selects 67 percent of the available data as the **training data set** and uses the remaining 33 percent of the data as the **validation data set.** JMP then fits neural network models having three hidden nodes to the training data set for a range of λ values and uses the fitted model corresponding to each λ value to predict the response variable values in the validation data set. JMP chooses the optimal λ value to be the λ value producing the model that minimizes the total prediction error in the validation data sets. JMP also summarizes information about how well the model corresponding to the optimal λ value fits the training data set and predicts the response variable values in the validation data set. If we conclude from this information that the tentative final model is adequate, we use it to predict future response variable values. Note that the user has the option to specify a different number of hidden nodes, different percentages in random holdback for the training and validation data sets, or a separately defined validation column that explicity says which observations are in the training data set and which observations are in the validation data set. We have found that JMP's default settings usually work fairly well. Also, note that because of both randomly selected starter values for the parameter estimates and the random choice of the training and validation data sets, one will not get the same parameter estimates if one repeatedly uses JMP to analyze the same overall data set. If the data set being analyzed is very large (as it should be for neural network modeling to be the most useful), the effects of the random selections should not be significant. However, when doing neural network modeling with relatively small data sets, these effects can be quite significant. Therefore, when analyzing relatively small data sets, we simply run the JMP procedure over again if initial JMP training and validation set results do not indicate that we have an adequate model.

To illustrate neural network modeling, we consider the credit card upgrade data in Figure 16.18. Here, instead of using JMP's random holdback procedure, we have separately defined a validation column based on randomly selecting 60 percent of the 40 observations as the training data set and 40 percent of the 40 observations as the validation data set. The response variable *Upgrade* equals 1 if the Silver card holder decided to upgrade and 0 otherwise. Moreover, the predictor variable *Purchases* is last year's purchases (in thousands of dollars) by the Silver card holder, and the predictor variable *PlatProfile* equals 1 if the Silver card holder conforms to the bank's Platinum profile and 0 otherwise. Each of the response variable *Upgrade* (with values 1 and 0) and the predictor variable *PlatProfile* (with values 1 and 0) is qualitative, and we must specify this when using JMP. However, in order to interpret and use the JMP neural network modeling output, we must define a **quantitative dummy variable** $JD_{PlatProfile}$ that (1) equals 1 when the qualitative predictor variable *PlatProfile* equals 0 (that is, when the Silver card holder does not conform to the bank's Platinum profile) and (2) equals 0 when the qualitative predictor variable *PlatProfile* equals 1 (that is, when the Silver card holder does not conform to the bank's Platinum profile). We then

FIGURE 16.18 **The JMP Output of the Card Upgrade Data, the Validation Column, Upgrade Probability Estimates, and Classification**   DS CardUpgrade

	UpGrade	Purchases	PlatProfile	Validation	H1_1	H1_2	H1_3	Probability (UpGrade=0)	Probability (UpGrade=1)	Most Likely UpGrade
1	0	7.471	0	Validation	0.3283922339	0.6371926641	0.4904056025	0.9972963541	0.0027036459	0
2	0	21.142	0	Training	0.2977108893	0.3494852724	0.3912570326	0.9842429722	0.0157570278	0
3	1	39.925	1	Training	−0.132207681	−0.369750351	−0.30322582	0.0485455876	0.9514544124	1
4	1	32.45	1	Training	−0.113888984	−0.173884203	−0.240820201	0.1426304937	0.8573695063	1
5	1	48.95	1	Training	−0.154204058	−0.568060608	−0.375209678	0.0145335153	0.9854664847	1
6	0	28.52	1	Validation	−0.104225915	−0.063878268	−0.207154563	0.2428678987	0.7571321013	1*
7	0	7.822	0	Training	0.32761277	0.6312289936	0.487997303	0.9971902248	0.0028097752	0
8	0	26.548	0	Training	0.285402285	0.2081153472	0.3491908402	0.9642923211	0.0357076789	0
9	1	48.831	1	Validation	−0.153914965	−0.565765296	−0.374287051	0.0147472664	0.9852527336	1
10	0	17.584	0	Validation	0.3057585128	0.4349517069	0.4180908016	0.9905174355	0.0094825645	0
11	1	49.82	1	Validation	−0.156316789	−0.584573951	−0.381932283	0.0130776371	0.9869223629	1
12	1	50.45	0	Training	0.2298921492	−0.436723406	0.1478373105	0.3619536103	0.6380463897	1
13	0	28.175	0	Training	0.2816791365	0.1634756758	0.3362394792	0.9540427804	0.0459572196	0
14	0	16.2	0	Training	0.3088772093	0.4662951245	0.4283381854	0.9921504496	0.0078495504	0
15	1	52.978	1	Validation	−0.163973613	−0.640572551	−0.405993654	0.009075762	0.990924238	1
16	1	58.945	1	Training	−0.178386586	−0.730008871	−0.449941095	0.0049132257	0.9950867743	1
17	1	40.075	1	Validation	−0.132574401	−0.373425155	−0.304453896	0.0475184034	0.9524815966	1
18	1	42.38	0	Training	0.2488223861	−0.23436029	0.2181560058	0.6634049837	0.3365950163	0*
19	1	38.11	1	Validation	−0.127767534	−0.324389853	−0.288287164	0.0630146232	0.9369853768	1
20	1	26.185	1	Training	−0.098475173	0.0024031962	−0.18691162	0.3226230077	0.6773769923	1
21	0	52.81	0	Training	0.2243224122	−0.489386982	0.1269530632	0.2883864788	0.7116135212	1*
22	1	34.521	1	Training	−0.118972613	−0.230326282	−0.258336199	0.1060445957	0.8939554043	1
23	0	34.75	0	Training	0.2665475963	−0.021924396	0.282632539	0.8745812573	0.1254187427	0
24	1	46.254	1	Validation	−0.147648141	−0.513907635	−0.354127358	0.0204214906	0.9795785094	1
25	0	24.811	0	Validation	0.2893676646	0.2548304198	0.3628719987	0.9726678493	0.0273321507	0
26	0	4.792	0	Training	0.3343266909	0.6802500817	0.5085399917	0.9979605203	0.0020394797	0
27	1	55.92	1	Training	−0.171089014	−0.687271164	−0.4279148	0.0066238301	0.9933761699	1
28	0	38.62	0	Training	0.257578895	−0.131166264	0.2502072665	0.7851081341	0.2148918659	0
29	0	12.742	0	Validation	0.3166403603	0.5396159234	0.4534624478	0.9949898329	0.0050101671	0
30	0	31.95	0	Validation	0.2730080204	0.0575937152	0.305701188	0.9176040513	0.0823959487	0
31	1	51.211	1	Validation	−0.159691752	−0.60999916	−0.392597065	0.0110954034	0.9889045966	1
32	1	30.92	1	Training	−0.110129484	−0.13141697	−0.227777609	0.1765528873	0.8234471127	1
33	0	23.527	0	Training	0.2922925329	0.2886267381	0.372885985	0.9775164167	0.0224835833	0
34	0	30.225	0	Training	0.2769759223	0.1062848837	0.319738518	0.9368225804	0.0631774196	0
35	0	28.387	1	Training	−0.10389854	−0.060112494	−0.206006114	0.2470202787	0.7529797213	1*
36	0	27.48	0	Validation	0.2832705866	0.1826376259	0.3417877964	0.9587446482	0.0412553518	0
37	1	41.95	1	Training	−0.137155302	−0.418350743	−0.319718488	0.0364786693	0.9635213307	1
38	1	34.995	1	Training	−0.120135262	−0.243043996	−0.262321865	0.0990166808	0.9009833192	1
39	0	34.964	1	Validation	−0.120059233	−0.242214743	−0.262061471	0.0994622409	0.9005377591	1*
40	0	7.998	0	Validation	0.3272217596	0.6282102249	0.4867869163	0.9971350558	0.0028649442	0
41		42.571	1	.	−0.138671214	−0.432804209	−0.324738223	0.0334660985	0.9665339015	1
42		51.835	0	.	0.2266252687	−0.46802718	0.1355958367	0.3173449846	0.6826550154	1

FIGURE 16.19   JMP Output of Neural Network Estimation for the Credit Card Upgrade Data   ⓓⓢ CardUpgrade

Neural   Validation Column: Validation   Model NTanH(3)

Estimates

Parameter	Estimate
H1_1:Purchases	−0.00498
H1_1: PlatProfile:0	0.393252
H1_1:Intercept	0.32598

$$\hat{\ell}_1 = \hat{h}_{10} + \hat{h}_{11}(Purchases) + \hat{h}_{12}(JD_{PlatProfile})$$
$$= .32598 - .00498(51.835) + .393252(1)$$
$$= .4611$$

$$H_1(\hat{\ell}_1) = \frac{e^{.4611} - 1}{e^{.4611} + 1}$$
$$= .226625$$

Parameter	Estimate
H1_2:Purchases	−0.05685
H1_2: PlatProfile:0	0.219115
H1_2:Intercept	1.712453

$$\hat{\ell}_2 = \hat{h}_{20} + \hat{h}_{21}(Purchases) + \hat{h}_{22}(JD_{PlatProfile})$$
$$= 1.712453 - .05685(51.835) + .219115(1)$$
$$= -1.0153$$

$$H_2(\hat{\ell}_2) = \frac{e^{-1.0153} - 1}{e^{-1.0153} + 1}$$
$$= -.468027$$

Parameter	Estimate
H1_3:Purchases	−0.01804
H1_3: PlatProfile:0	0.556931
H1_3:Intercept	0.651032

$$\hat{\ell}_3 = \hat{h}_{30} + \hat{h}_{31}(Purchases) + \hat{h}_{32}(JD_{PlatProfile})$$
$$= .651032 - .01804(51.835) + .556931(1)$$
$$= .2729$$

$$H_3(\hat{\ell}_3) = \frac{e^{.2729} - 1}{e^{.2729} + 1}$$
$$= .135596$$

Parameter	Estimate
Upgrade(0):H1_1	2.299942
Upgrade(0):H1_2	4.493357
Upgrade(0):H1_3	4.160657
Upgrade(0):Intercept	0.25162

$$\hat{L} = b_0 + b_1 H_1(\hat{\ell}_1) + b_2 H_2(\hat{\ell}_2) + b_3 H_3(\hat{\ell}_3)$$
$$= .25162 + 2.299942(.226625)$$
$$+ 4.493357(-.468027) + 4.160657(.135596)$$
$$= -.7660$$

$$g(\hat{L}) = \frac{1}{1 + e^{-(-.7660)}}$$
$$= .3173$$

	Upgrade	Purchases	PlatProfile	Validation	H1_1	H1_2	H1_3	Probability (Upgrade=0)	Probability (Upgrade=1)	Most Likely Upgrade
41		42.571	1	.	−0.138671214	−0.432804209	−0.324738223	0.0334660985	0.9665339015	1
42		51.835	0	.	0.2266252687	−0.46802718	0.1355958367	0.3173449846	0.6826550154	1

define the following linear combinations of the quantitative predictor variables *Purchases* and $JD_{PlatProfile}$ and the following hidden node functions at the three hidden nodes used by JMP:

$$\text{Node 1: } \ell_1 = h_{10} + h_{11}(Purchases) + h_{12}(JD_{PlatProfile}) \qquad H_1(\ell_1) = \frac{e^{\ell_1} - 1}{e^{\ell_1} + 1}$$

$$\text{Node 2: } \ell_2 = h_{20} + h_{21}(Purchases) + h_{22}(JD_{PlatProfile}) \qquad H_2(\ell_2) = \frac{e^{\ell_2} - 1}{e^{\ell_2} + 1}$$

$$\text{Node 3: } \ell_3 = h_{30} + h_{31}(Purchases) + h_{33}(JD_{PlatProfile}) \qquad H_3(\ell_3) = \frac{e^{\ell_3} - 1}{e^{\ell_3} + 1}$$

At the output layer, we form the following linear combination *L* of the three hidden node functions and the following output layer function:

$$L = \beta_0 + \beta_1 H_1(\ell_1) + \beta_2 H_2(\ell_2) + \beta_3 H_3(\ell_3) \qquad g(L) = \frac{1}{1 + e^{-L}}$$

Figure 16.19 shows an annotated JMP output of neural network estimation for the credit card upgrade data. Silver card holders 41 and 42 at the bottom of the output are two Silver card holders who have not yet been sent an upgrade offer and for whom we wish to estimate the probability of upgrading. Silver card holder 42 had purchases last year of \$51,835 (*Purchases* = 51.835) and did not conform to the bank's Platinum profile (*PlatProfile* = 0). Because *PlatProfile* = 0, we have $JD_{PlatProfile}$ = 1. Figure 16.19 shows the parameter estimates for the neural network model based on the training data set and how they are used to estimate the probability that Silver card holder 42 would upgrade. Note that because the response variable *Upgrade* is qualitative, the output layer function is $g(L) = 1/(1 + e^{-L})$. The final result obtained in the calculations, $g(\hat{L}) = .3173$, is an estimate of the probability that Silver card holder 42 would not upgrade (*Upgrade* = 0). This implies that the estimate of the probability that Silver card holder 42 would upgrade is $1 - .3173 = .6827$. If we predict a Silver card holder would upgrade if and only if his or her upgrade probability is at least .5, then Silver card holder 42 is predicted to upgrade (as is Silver card holder 41). JMP uses the model fit to the training data set to calculate an upgrade probability estimate for each of the percent of the Silver card holders in the training data set and for each of the 40 percent of the Silver card holders in the validation data set. If a particular Silver card holder's upgrade probability estimate is at least .5, JMP predicts an upgrade for the card holder and assigns a "most likely" qualitative value of 1 to the card holder. Otherwise, JMP assigns a "most likely" qualitative value of 0 to the card holder. In Figure 16.18 we show the results of the classification. The confusion matrices in Figure 16.20 summarize the neural network model's levels of success. As explained there, in the training data set the model predicts 10 of the 12 nonupgrades and 11 of the 12 upgrades. In the validation data set, the model predicts 7 of the 9 nonupgrades and all 7 of the 7 upgrades. The corresponding success rates are high (see the "confusion rates"), and thus we have evidence that the neural network model will have good success in predicting whether future Silver card holders (such as card holders 41 and 42) will upgrade.

Figures 16.21 and 16.22 show the JMP neural network analysis of the churn data in Figure 16.8. The misclassification rates and confusion matrices in Figure 16.21 indicate that neural network analysis classifies both the non-churners and churners more accurately than either logistic regression or linear discriminate analysis. Figure 16.22 shows that neural network analysis estimates that the probability that customer 3333 will churn is .76703. Therefore, customer 3333 is classified as a churner.

Figure 16.23 shows the JMP output of neural network modeling of the Fresh detergent demand data in Table 15.6. Here, *Demand* is the demand for the large bottle of Fresh (in hundreds of thousands of bottles) in a sales period, *PriceDif* is the difference between the average industry price (in dollars) of competitors' similar detergents and the price (in dollars) of Fresh as offered by Enterprise Industries in the sales period, and *AdvExp* is Enterprise Industries' advertising expenditure (in hundreds of thousands of dollars) to promote Fresh in the sales period. Here, we have used JMP's random holdback procedure that randomly selects 67 percent of the available data as the training data set and the remaining 33 percent of the data as the validation data set. Note that the R^2 value is reasonably high for both the training data set ($R^2 = .9506$) and the validation data set ($R^2 = .8662$). On the output, we show

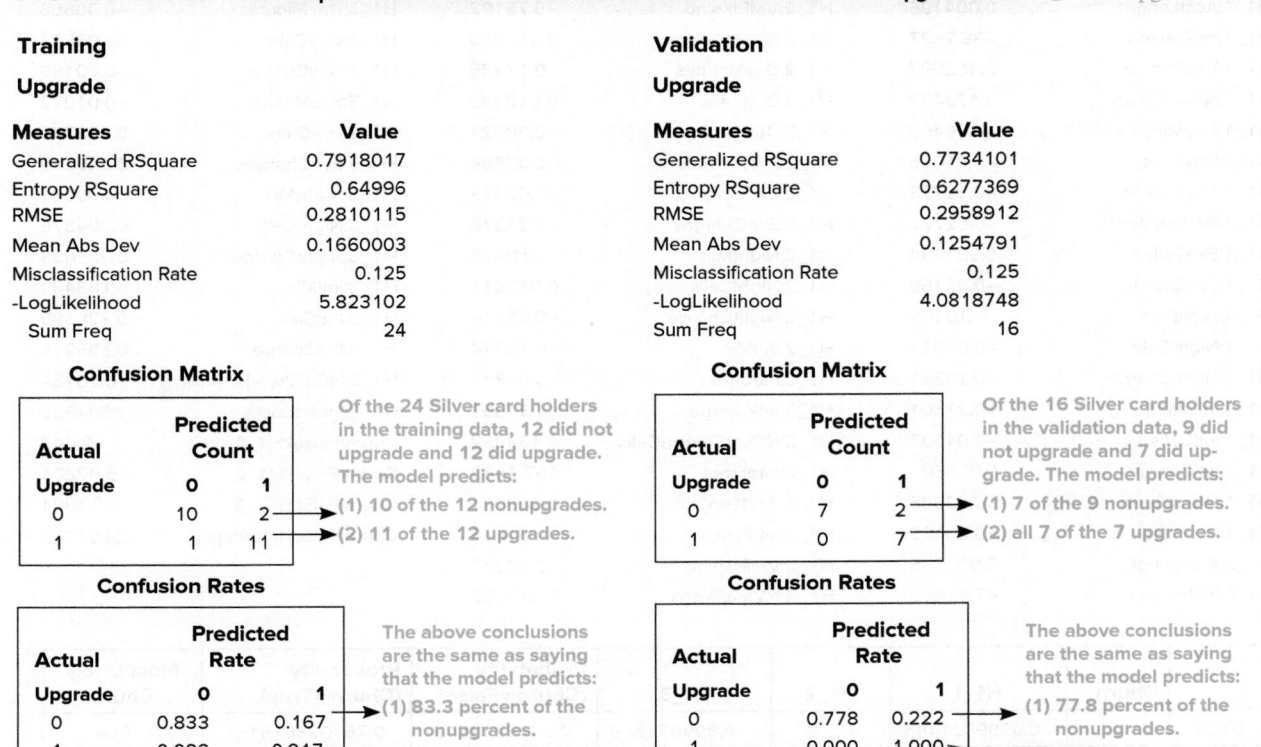

FIGURE 16.20 JMP Output of Neural Network Confusion Matrices for the Credit Card Upgrade Data CardUpgrade

FIGURE 16.21 The JMP Neural Network Misclassification Rates and Confusion Matrices for the Churn Data DS Churn

Neural Validation Column: Validation Model NTanH(3)

Training

Churn

Measures	Value
Generalized RSquare	0.6918409
Entropy RSquare	0.59613
RMSE	0.2094637
Mean Abs Dev	0.0884714
Misclassification Rate	0.0535795
-LogLikelihood	371.26744
Sum Freq	2221

Confusion Matrix

Actual Churn	Predicted Count False	Predicted Count True
False	1865	34
True	85	237

Confusion Rates

Actual Churn	Predicted Rate False	Predicted Rate True
False	0.982	0.018
True	0.264	0.736

Validation

Churn

Measures	Value
Generalized RSquare	0.6466946
Entropy RSquare	0.5468839
RMSE	0.2282956
Mean Abs Dev	0.0960989
Misclassification Rate	0.070207
-LogLikelihood	208.30454
Sum Freq	1111

Confusion Matrix

Actual Churn	Predicted Count False	Predicted Count True
False	917	33
True	45	116

Confusion Rates

Actual Churn	Predicted Rate False	Predicted Rate True
False	0.965	0.035
True	0.280	0.720

FIGURE 16.22 The Neural Network Model Parameter Estimates and the Churn Probability Estimate for Customer 3333

Estimates

Parameter	Estimate
H1_1:AcctLength	0.004196
H1_1:IntlPlan:no	−38.7637
H1_1:VMPlan:no	2.452997
H1_1:NVMailMsgs	0.178437
H1_1:DayMinutes	−0.06469
H1_1:DayCalls	−0.01086
H1_1:DayCharge	−0.39728
H1_1:EveMinutes	−0.02772
H1_1:EveCalls	0.031434
H1_1:EveCharges	−0.37168
H1_1:NightMin	−0.0189
H1_1:NightCalls	−0.00915
H1_1:NightCharge	−0.23751
H1_1:IntlMin	−0.27501
H1_1:IntlCalls	−0.04307
H1_1:IntlCharge	0.075501
H1_1:NCustServiceCalls	14.62885
H1_1:Intercept	18.65425
H1_2:AcctLength	0.003366
H1_2:IntlPlan:no	47.11669

Estimates

Parameter	Estimate
H1_2:VMPlan:no	−9.72192
H1_2:NVMailMsgs	0.316786
H1_2:DayMinutes	−0.17136
H1_2:DayCalls	0.012932
H1_2:DayCharge	−0.98027
H1_2:EveMinutes	−0.07884
H1_2:EveCalls	−0.00373
H1_2:EveCharges	−0.97276
H1_2:NightMin	−0.0478
H1_2:NightCalls	0.012511
H1_2:NightCharge	−0.93418
H1_2:IntlMin	−0.10974
H1_2:IntlCalls	−0.05847
H1_2:IntlCharge	−1.31552
H1_2:NCustServiceCalls	0.169969
H1_2:Intercept	107.1662
H1_3:AcctLength	0.003061
H1_3:IntlPlan:no	−6.01656
H1_3:VMPlan:no	−0.10347
H1_3:NVMailMsgs	0.017032

Estimates

Parameter	Estimate
H1_3:DayMinutes	−0.00668
H1_3:DayCalls	0.000814
H1_3:DayCharge	−0.00183
H1_3:EveMinutes	−0.01022
H1_3:EveCalls	0.006781
H1_3:EveCharges	0.040649
H1_3:NightMin	−0.00931
H1_3:NightCalls	0.004876
H1_3:NightCharge	0.232053
H1_3:IntlMin	−0.63432
H1_3:IntlCalls	0.328599
H1_3:IntlCharge	0.596915
H1_3:NCustServiceCalls	−0.25764
H1_3:Intercept	2.504963
Churn(False):H1_1	−2.442
Churn(False):H1_2	16.07801
Churn(False):H1_3	17.4881
Churn(False):Intercept	2.657303

	Churn	H1_1	H1_2	H1_3	Probability (Churn=False)	Probability (Churn=True)	Most Likely Churn
3333		0.9999821446	1	−0.999820549	0.2329713849	0.7670286151	True

FIGURE 16.23   **JMP Output of Neural Network Modeling of the Fresh Detergent Demand Data**   DS Fresh2

Neural	Validation: Random Holdback	Model NTanH(3)	
Training		**Validation**	
Demand		**Demand**	
Measures	**Value**	**Measures**	**Value**
RSquare	0.9506379	RSquare	0.866165
RMSE	0.1390637	RMSE	0.1593834
Mean Abs Dev	0.1162525	Mean Abs Dev	0.12958
−LogLikelihood	−11.07769	−LogLikelihood	−4.175038
SSE	0.3867744	SSE	0.2540308
Sum Freq	20	Sum Freq	10

Estimates

Parameter	Estimate
H1_1:PriceDif	−5.94909
H1_1:AdvExp	4.315497
H1_1:Intercept	−27.4393
H1_2:PriceDif	2.566626
H1_2:AdvExp	−2.72164
H1_2:Intercept	17.35052
H1_3:PriceDif	0.90711
H1_3:AdvExp	0.396032
H1_3:Intercept	−2.69945
Demand_1:H1_1	3.822497
Demand_2:H1_2	5.84664
Demand_3:H1_3	8.211096
Demand_4:Intercept	8.552722

$$\hat{P}_1 = \hat{h}_{10} + \hat{h}_{11}(PriceDif) + \hat{h}_{12}(AdvExp) \qquad H_1(\hat{P}_1) = \frac{e^{\hat{P}_1} - 1}{e^{\hat{P}_1} + 1}$$

$$\hat{P}_2 = \hat{h}_{20} + \hat{h}_{21}(PriceDif) + \hat{h}_{22}(AdvExp) \qquad H_2(\hat{P}_2) = \frac{e^{\hat{P}_2} - 1}{e^{\hat{P}_2} + 1}$$

$$\hat{P}_3 = \hat{h}_{30} + \hat{h}_{31}(PriceDif) + \hat{h}_{32}(AdvExp) \qquad H_3(\hat{P}_3) = \frac{e^{\hat{P}_3} - 1}{e^{\hat{P}_3} + 1}$$

$$\hat{L} = b_0 + b_1 H_1(\hat{P}_1) + b_2 H_2(\hat{P}_2) + b_3 H_3(\hat{P}_3) \qquad g(\hat{L}) = \hat{L}$$

	Demand	PriceDif	AdvExp	Predicted Demand	H1_1	H1_2	H1_3
1	7.38	−0.05	5.5	7.4443811953	−0.93582353	0.8098452223	−0.275971951
2	8.51	0.25	6.75	8.6653410513	0.101149291	−0.187218623	0.099935036
30	9.26	0.55	6.8	9.0076615453	−0.593452936	0.1268132998	0.2413782177
31	.	0.2	6.5	8.2125811712	−0.281410743	0.0863706613	0.0280806154

the estimates $H_1(\hat{\ell}_1)$, $H_2(\hat{\ell}_2)$, and $H_3(\hat{\ell}_3)$ of the hidden node functions. Because the response variable *Demand* is quantitative, the output layer function $g(L)$ is simply L. In the exercises the reader will verify the calculation of $\hat{L} = 8.21258$ (that is, 821,258 bottles). This is the prediction of Fresh demand when the price difference will be 20 cents and the advertising expenditure for Fresh will be $650,000.

Exercises for Section 16.3

CONCEPTS   connect

16.11   When is neural network modeling most useful?

16.12   How do we define the output layer function $g(L)$?

METHODS AND APPLICATIONS

16.13   Consider the JMP output in Figure 16.23.

a   Identify the values of $H_1(\hat{\ell}_1)$, $H_2(\hat{\ell}_2)$, and $H_3(\hat{\ell}_3)$ for future sales period 31 on the output.

b   Hand calculate $H_1(\hat{\ell}_1)$ for future sales period 31.

c   Hand calculate $\hat{L} = 8.21258$ using the values of $H_1(\hat{\ell}_1)$, $H_2(\hat{\ell}_2)$, and $H_3(\hat{\ell}_3)$ for future sales period 31.

d   Use JMP to rerun the analysis and compare your new results with Figure 16.23.

FIGURE 16.24   **The JMP Output of Neural Network Modeling of the Coupon Redemption Data**   CoupRedemp

Neural	Validation: Random Holdback	Model NTanH(3)	
Training		**Validation**	

Coupon		Coupon	
Measures	**Value**	**Measures**	**Value**
Generalized RSquare	0.7620955	Generalized RSquare	0.9759065
Entropy RSquare	0.6121493	Entropy RSquare	0.950239
RMSE	0.2953379	RMSE	0.0776634
Mean Abs Dev	0.1309155	Mean Abs Dev	0.030409
Misclassification Rate	0.1153846	Misclassification Rate	0
-LogLikelihood	6.9599138	-LogLikelihood	0.4757503
Sum Freq	26	Sum Freq	14

Estimates

Parameter	Estimate
H1_1:Purchases	−0.06443
H1_1:Card:0	1.470573
H1_1:Intercept	6.178974
H1_2:Purchases	−0.045
H1_2:Card:0	2.178333
H1_2:Intercept	0.016893
H1_3:Purchases	0.73072
H1_3:Card:0	−8.34011
H1_3:Intercept	−24.6195
Coupon(0):H1_1	18.08562
Coupon(0):H1_2	−0.33348
Coupon(0):H1_3	−18.009
Coupon(0):Intercept	−0.42799

Confusion Matrix

	Predicted	
Actual	Count	
Coupon	0	1
0	11	3
1	0	12

Confusion Matrix

	Predicted	
Actual	Count	
Coupon	0	1
0	8	0
1	0	6

Confusion Rates

	Predicted	
Actual	Rate	
Coupon	0	1
0	0.786	0.214
1	0.000	1.000

Confusion Rates

	Predicted	
Actual	Rate	
Coupon	0	1
0	1.000	0.000
1	0.000	1.000

	Coupon	Purchases	Card	H1_1	H1_2	H1_3	Probability (Coupon=0)	Probability (Coupon=1)	Most Likely Coupon
41		43.97	1	0.7341853154	−0.968650671	0.9999997386	0.0078781038	0.9921218962	1
42		52.48	0	0.9723772551	−0.082889751	0.9909043164	0.340852896	0.659147104	1

16.14 Read Exercise 5.3 and recall that Figure 5.11(c) gives data concerning (1) the dependent variable *Coupon*, which equals 1 if a Williams Apparel customer redeems a coupon and 0 otherwise; (2) the independent variable *Purchases*, which is last year's purchases (in hundreds of dollars) by the customer; and (3) the independent variable *Card*, which equals 1 if the customer has a Williams Apparel credit card and 0 otherwise. The JMP output of neural network modeling of the coupon redemption data is given in Figure 16.24. Here, we had JMP use random holdback—its default of randomly selecting 67 percent of the data as the training data set and using the remaining 33 percent of the data as the validation data set. CoupRedemp

a Interpret the confusion matrices for both the training data set and the validation data set.

b Identify the values of $H_1(\hat{\ell}_1)$, $H_2(\hat{\ell}_2)$, and $H_3(\hat{\ell}_3)$ for Williams Apparel customer 42 on the output.

c Hand calculate $H_1(\hat{\ell}_1)$ for customer 42 by using the equations that follow:

$$\hat{\ell}_1 = \hat{h}_{10} + \hat{h}_{11}(Purchases) + \hat{h}_{12}(JD_{Card})$$

$$H_1(\hat{\ell}_1) = \frac{e^{\hat{\ell}_1} - 1}{e^{\hat{\ell}_1} + 1}$$

Here, JD_{Card} equals 1 when *Card* equals 0 and equals 0 when *Card* equals 1.

d Hand calculate $\hat{L}$ and $g(\hat{L})$ using the values of $H_1(\hat{\ell}_1)$, $H_2(\hat{\ell}_2)$, and $H_3(\hat{\ell}_3)$ for customer 42 on the output and the equations:

$$\hat{L} = b_0 + b_1 H_1(\hat{\ell}_1) + b_2 H_2(\hat{\ell}_2) + b_3 H_3(\hat{\ell}_3)$$

$$g(\hat{L}) = \frac{1}{1 + e^{-\hat{L}}}$$

e What is the estimate of the probability that customer 42 would redeem the coupon?

f Use JMP to rerun the analysis and compare your new results with the results in Figure 16.24.

16.15 Use JMP to rerun the neural network analysis of the churn data in Figure 16.8 and compare your new results with the results in Figures 16.21 and 16.22.

16.16 Suppose that a Toyota dealer wishes to predict the sales prices of used Toyota Corollas and has gathered 1,436 observations concerning sales prices of used Corollas and 36 predictor variables. Sixty percent (or 862) observations are randomly selected as the training data set and the remaining 40 percent (or 574) observations are used as the validation data set. Some of the observations are shown in Figure 16.25, along with the neural network model prediction of the sales price of used Toyota Corolla 1437, which has just been obtained by the dealer and has not yet been sold.

a. Show how the predicted sales price of $16,984.86 has been calculated by using the values of H1_1, H1_2 and H1_3 shown in Figure 16.25 and the appropriate neural network model parameter estimates, which are given in Figure 16.26.

b. Use JMP to rerun the analysis and compare your new results with the results in Figure 16.25 and 16.26.

FIGURE 16.25   **The JMP Output of a Portion of the Used Toyota Corolla Sales Price Data and the Predicted Sales Price of Used Toyota Corolla 1437**   DS ToyotaCorolla

	Model	Price	Age_08_04	Mfg_Month	Mfg_Year	KM	Fuel_Type	HP	Met_Color
1	TOYOTA Corolla 2.0 D4D HATCHB TERRA 2/3-Doors	13500	23	10	2002	46986	Diesel	90	1
2	TOYOTA Corolla 2.0 D4D HATCHB TERRA 2/3-Doors	13750	23	10	2002	72937	Diesel	90	1
3	TOYOTA Corolla 2.0 D4D HATCHB TERRA 2/3-Doors	13950	24	9	2002	41711	Diesel	90	1
8	TOYOTA Corolla 2.0 D4D 90 3DR TERRA 2/3-Doors	18600	30	3	2002	75889	Diesel	90	1
9	TOYOTA Corolla 1800 T SPORT VVTI 2/3-Doors	21500	27	6	2002	19700	Petrol	192	0
10	TOYOTA Corolla 1.9 D HATCHB TERRA 2/3-Doors	12950	23	10	2002	71138	Diesel	69	0
1437	TOYOTA Corolla 2.0 D4D 90 3DR TERRA 2/3-Doors		27	6	2002	94612	Diesel	90	1

	Color	Automatic	CC	Doors	Cylinders	Gears	Quartly_Tax	Weight	Mfg_Guarantee	BOVAG_Guarantee	Guarantee_Period
1	Blue	0	2000	3	4	5	210	1165	0	1	3
2	Silver	0	2000	3	4	5	210	1165	0	1	3
3	Blue	0	2000	3	4	5	210	1165	1	1	3
8	Grey	0	2000	3	4	5	210	1245	1	1	3
9	Red	0	1800	3	4	5	100	1185	0	1	3
10	Blue	0	1900	3	4	5	185	1105	0	1	3
1437	Grey	0	2000	3	4	5	210	1245	0	1	3

	ABS	Airbag_1	Airbag_2	Airco	Automatic_airco	Boardcomputer	CD_Player	Central_Lock	Powered_Windows
1	1	1	1	0	0	1	0	1	1
2	1	1	1	1	0	1	1	1	0
3	1	1	1	0	0	1	0	0	0
8	1	1	1	1	0	1	1	1	1
9	1	1	0	1	0	0	0	1	1
10	1	1	1	1	0	1	0	0	0
1437	1	1	1	1	0	1	0	1	1

	Power_Steering	Radio	Mistlamps	Sport_Model	Backseat_Divider	Metallic_Rim	Radio_cassette	Tow_Bar	Validation
1	1	0	0	0	1	0	0	0	Validation
2	1	0	0	0	1	0	0	0	Validation
3	1	0	0	0	1	0	0	0	Training
8	1	0	0	0	1	0	0	0	Training
9	1	1	0	0	0	1	1	0	Training
10	1	0	0	0	1	0	0	0	Validation
1437	1	0	0	1	1	0	0	0	.

	Predicted Price	H1_1	H1_2	H1_3
1437	16984.85944	−0.028008838	0.0189808018	−0.0950465

FIGURE 16.26   The JMP Output of the Fit Statistics and the Parameter Estimates for the Neural Network Model Describing the Toyota Corolla Sales Price Data

Training

Price

Measures	Value
RSquare	0.937451
RMSE	906.58549
Mean Abs Dev	691.15597
-LogLikelihood	7093.0738
SSE	708475424
Sum Freq	862

Estimates

Parameter	Estimate
H1_1:Age_08_04	5.138e-5
H1_1:Mfg_Month	−0.00677
H1_1:Mfg_Year	−0.00349
H1_1:KM	−5.94e-7
H1_1:Fuel_Type:CNG	−0.09375
H1_1:Fuel_Type:Diesel	−0.08646
H1_1:HP	0.002549
H1_1:Met_Color:0	−0.03856
H1_1:Color:Beige	0.020215
H1_1:Color:Black	0.128634
H1_1:Color:Blue	−0.00369
H1_1:Color:Green	0.011431
H1_1:Color:Grey	−0.03742
H1_1:Color:Red	−0.07679
H1_1:Color:Silver	0.08814
H1_1:Color:Violet	−0.02694
H1_1:Color:White	−0.22597
H1_1:Automatic:0	0.111076
H1_1:CC	−3.78e-5
H1_1:Doors	−0.00508
H1_1:Cylinders	−2.5e-11
H1_1:Gears	−0.17951
H1_1:Quarterly_Tax	−0.00012
H1_1:Weight	0.001927
H1_1:Mfg_Guarantee	0.068056
H1_1:BOVAG_Guarantee	−0.05106
H1_1:Guararrtee_Period	−0.00174
H1_1:ABS	−0.25978
H1_1:Airbag_1	−0.14027
H1_1:Airbag_2	−0.20608
H1_1:Airco	0.039239
H1_1:Automatic_airco	−0.00281
H1_1:Boardcomputer	0.108221
H1_1:CD_Player	−0.00789
H1_1:Central_Lock	0.014145
H1_1:Powered_Windows	0.071723
H1_1:Power_Steering	−0.39587
H1_1:Radio	0.059401
H1_1:Mistlamps	−0.03013
H1_1:Sport_Model	0.059734
H1_1:Backseat_Divider	0.107749
H1_1:Metallic_Rim	0.153663
H1_1:Radio_cassette	0.05923

Validation

Price

Measures	Value
RSquare	0.909691
RMSE	1089.254
Mean Abs Dev	819.31995
-LogLikelihood	−4828.5953
SSE	681036260
Sum Freq	574

Estimates

Parameter	Estimate
H1_1:Tow_Bar	0.05408
H1_1:Intercept	6.044444
H1_2:Age_08_04	0.00179
H1_2:Mfg_Month	−0.00599
H1_2:Mfg_Year	−0.01938
H1_2:KM	8.158e-7
H1_2:Fuel_Type:CNG	0.03933
H1_2:Fuel_Type:Diesel	0.036832
H1_2:HP	−0.00282
H1_2:Met_Color:0	−0.01854
H1_2:Color:Beige	−0.03661
H1_2:Color:Black	0.0584
H1_2:Color:Blue	−0.01276
H1_2:Color:Green	0.021156
H1_2:Color:Grey	−0.06846
H1_2:Color:Red	−0.02358
H1_2:Color:Silver	0.101673
H1_2:Color:Violet	−0.09835
H1_2:Color:White	−0.06953
H1_2:Automatic:0	0.057123
H1_2:CC	−4.94e-6
H1_2:Doors	−0.00856
H1_2:Cylinders	2.41e-11
H1_2:Gears	0.075998
H1_2:Quarterly_Tax	−0.00007
H1_2:Weight	0.000351
H1_2:Mfg_Guarantee	−0.00393
H1_2:BOVAG_Guarantee	−0.1995
H1_2:Guarantee_Period	−0.01332
H1_2:ABS	−0.12055
H1_2:Airbag_1	−0.19005
H1_2:Airbag_2	−0.12743
H1_2:Airco	0.05019
H1_2:Automatic_airco	−0.07937
H1_2:Boardcomputer	0.061677
H1_2:CD_Player	−0.01558
H1_2:Central_Lock	0.003244
H1_2:Powered_Windows	0.030411
H1_2:Power_Steering	−0.15861
H1_2:Radio	0.029771
H1_2:Mistlamps	0.014592
H1_2:Sport_Model	−0.0421
H1_2:Backseat_Divider	0.061236

Estimates

Parameter	Estimate
H1_2:Metallic_Rim	0.015115
H1_2:Radio_cassette	0.029827
H1_2:Tow_Bar	−0.00685
H1_2:Intercept	38.85837
H1_3:Age_08_04	−0.0014
H1_3:Mfg_Month	0.003486
H1_3:Mfg_Year	0.015594
H1_3:KM	−1.68e-6
H1_3:Fuel_Type:CNG	−0.1401
H1_3:Fuel_Type:Diesel	−0.15341
H1_3:HP	0.007106
H1_3:Met_Color:0	−0.01133
H1_3:Color:Beige	0.062601
H1_3:Colon:Black	0.039921
H1_3:Color:Blue	0.017611
H1_3:Color:Green	−0.01752
H1_3:Color:Grey	0.067837
H1_3:Color:Red	−0.0412
H1_3:Color:Silver	−0.07054
H1_3:Color:Violet	0.134743
H1_3:Color:White	−0.1222
H1_3:Automatic:0	0.027593
H1_3:CC	−0.00003
H1_3:Doors	0.008627
H1_3:Cylinders	−8.8e-12
H1_3:Gears	−0.31193
H1_3:Quarterly_Tax	−0.00016
H1_3:Weight	0.001112
H1_3:Mfg_Guarantee	0.073101
H1_3:BOVAG_Guarantee	0.257445
H1_3:Guarantee_Period	0.017982
H1_3:ABS	−0.07955
H1_3:Airbag_1	0.149054
H1_3:Airbag_2	−0.00818
H1_3:Airco	−0.04164
H1_3:Automatic_airco	0.088074
H1_3:Boardcomputer	0.020769
H1_3:CD_Player	0.014714
H1_3:Central_Lock	0.012489
H1_3:Powered_Windows	0.018621
H1_3:Power_Steering	−0.16082
H1_3:Radio	0.015189
H1_3:Mistlamps	−0.05791
H1_3:Sport_Model	0.125821
H1_3:Backseat_Divider	0.016685
H1_3:Metallic_Rim	0.13599
H1_3:Radio_cassette	0.015227
H1_3:Tow_Bar	0.070434
H1_3:Intercept	−31.9492
Price_1:H1_1	105099.6
Price_2:H1_2	−158660
Price_3:H1_3	−99420.8
Price_4:Intercept	13490.47

Chapter Summary

In this chapter we have studied three parametric predictive analytics—**logistic regression, discriminate analysis** and **neural networks.** We have seen that **logistic regression** uses a fairly simple equation that involves the exponential function. This equation estimates the probability that an observation described by a specified set of predictor variable values will fall into a particular class. The second analytic—**linear discriminate analysis**—classifies an observation and estimates the probability that the observation will fall into a particular class by calculating the *squared distance* between each class's predictor variable value means and an observation's predictor variable values. The observation is then put into the class for which the squared distance is the smallest. Linear discriminate analysis is perhaps the easiest classification analytic to use

when there are more than two classes. The third analytic—**neural network modeling**—represents the response variable as a nonlinear function of linear combinations of predictor variables. This analytic is particularly useful for very large data sets having a very large number of predictor variables. One reason is that, whereas classical multiple regression, logistic regression, and linear discriminate analysis use *p*-values and other model building techniques to help decide which of the potential predictor variables to use in a model, neural network modeling does not require this decision. This is because neural networks use all of the potential predictor variables and a **penalized least squares** criterion that controls the **total magnitude** of the parameter estimates and thus the tradeoff between overfitting and underfitting.

Glossary of Terms

odds: The odds of the event that an observation will fall into a particular class is the probability that this event will occur divided by the probability that this event with not occur.

odds ratio for the predictor variable x_j: the proportional change in the odds that is associated with a one unit increase in x_j when the other predictor variables stay constant.

Important Formulas and Graphics

logistic regression model: page 724

neural networks: page 738

Supplementary Exercises

16.17 Figure 16.27 shows the performance data with a specified validation column and gives the neural network success probability estimates. Figure 16.28 gives the neural network model fit statistics and parameter estimates.

a. Find and report the neural network success probability estimate for potential employee 44.

b. Use JMP to rerun the analysis and compare your new results with the results in Figures 16.27 and 16.28.

16.18 Figure 16.29 gives the neural network analysis of the GPA data, which were discussed in Section 5.2. Some of these data are shown in Figure 5.16.

a. Find and report the predicted GPA of new applicant 706.

b. Use JMP to rerun the analysis and compare your new results with the results given in Figure 16.29.

FIGURE 16.27   The Performance Data, Validation Column, and Success Probability Estimates ⊙ PerfTest

	Group	Test 1	Test 2	Validation	H1_1	H1_2	H1_3	Probability(Group=0)	Probability(Group=1)	Most Likely Group
1	1	96	85	Validation	−0.999997111	0.2604411069	−0.389527231	0.0072221324	0.9927778676	1
2	1	96	88	Training	−0.99999387	0.1381004005	−0.794616078	0.0018794508	0.9981205492	1
3	1	91	81	Validation	−0.999367877	0.8180094249	0.0433414921	0.1794958492	0.8205041508	1
4	1	95	78	Validation	−0.999998207	0.6088624977	0.7893478759	0.264056993	0.735943007	1
5	1	92	85	Validation	−0.999519983	0.6846601553	−0.644002699	0.0300700193	0.9699299807	1
6	1	93	87	Training	−0.999779225	0.5440981987	−0.809280825	0.011636359	0.988363641	1
7	1	98	84	Training	−0.999999826	0.0234819703	−0.010175299	0.005124779	0.994875221	1
8	1	92	82	Training	−0.999773726	0.7466878877	−0.092130919	0.1078112957	0.8921887043	1
9	1	97	89	Validation	−0.999997806	−0.046322449	−0.839534454	0.0007402945	0.9992597055	1
10	1	95	96	Validation	−0.999836433	−0.058379959	−0.994706026	0.0005179694	0.9994820306	1
11	1	99	93	Validation	−0.999999536	−0.463771755	−0.959495201	0.0000865343	0.9999134657	1
12	1	89	90	Training	−0.925065093	0.7832091657	−0.973293276	0.0368677866	0.9631322134	1
13	1	94	90	Training	−0.999869534	0.3270780752	−0.936536575	0.0033838595	0.9966161405	1
14	1	92	94	Training	−0.995424924	0.4260453543	−0.992373909	0.0048842026	0.9951157974	1
15	1	94	84	Training	−0.999971016	0.5332952675	−0.348658771	0.0267765353	0.9732234647	1
16	1	90	92	Training	−0.964827791	0.6781837392	−0.986909649	0.0182441151	0.9817558849	1
17	1	91	70	Training	−0.999959903	0.9244226595	0.9868654162	0.6914490317	0.3085509683	0
18	1	90	81	Training	−0.997732251	0.8600624166	−0.045035967	0.183803971	0.816196029	1
19	1	86	81	Training	−0.68258399	0.9531237404	−0.378927643	0.4965276669	0.5034723331	1
20	1	90	76	Training	−0.999352135	0.9062694201	0.791680088	0.5856227304	0.4143772696	0
21	1	91	79	Training	−0.999617113	0.8442630383	0.4556138425	0.3552952309	0.6447047691	1
22	1	88	83	Training	−0.952815316	0.904085262	−0.58519697	0.1089910622	0.8910089378	1
23	1	87	82	Training	−0.873511146	0.9327957058	−0.488865521	0.2053721239	0.7946278761	1
24	0	93	74	Training	−0.99999152	0.8219704233	0.945720004	0.5638779897	0.4361220103	0
25	0	90	84	Training	−0.9951947	0.8229828544	−0.615455623	0.0594829567	0.9405170433	1
26	0	91	81	Validation	−0.999367877	0.8180094249	0.0433414921	0.1794958492	0.8205041508	1
27	0	91	78	Training	−0.999702013	0.8560485014	0.6143998287	0.4421933148	0.5578066852	1
28	0	88	78	Validation	−0.986299948	0.9362542112	0.4224359755	0.4583109118	0.5416890882	1
29	0	86	86	Training	−0.204189801	0.9291564257	−0.908661128	0.8063045558	0.1936954442	0
30	0	79	81	Validation	0.9986226682	0.9935207826	−0.768990482	0.9997947768	0.0002052232	0
31	0	83	84	Training	0.8975710557	0.9740422062	−0.870877913	0.9995275704	0.0004724296	0
32	0	79	77	Validation	0.9962494708	0.9953851175	−0.120477854	0.9999418924	0.0000581076	0
33	0	88	75	Training	−0.993519323	0.9502544201	0.8086941717	0.6484389693	0.3515610307	0
34	0	81	85	Validation	0.9935039097	0.9839567013	−0.939967392	0.9996922442	0.0003077558	0
35	0	85	83	Validation	0.0558780465	0.9581300218	−0.733210514	0.9646461051	0.0353538949	0
36	0	82	72	Validation	0.5329907033	0.9929034365	0.8524976417	0.9998926259	0.0001073741	0
37	0	82	81	Validation	0.9381774275	0.9848030321	−0.636666213	0.9997713543	0.0002286457	0
38	0	81	77	Validation	0.9527053893	0.9918439278	0.0557464798	0.9999470331	0.0000529669	0
39	0	86	76	Training	−0.897802424	0.9691124206	0.6182557319	0.6992676985	0.3007323015	0
40	0	81	84	Validation	0.9916604473	0.9852553021	−0.907570996	0.999709943	0.000290057	0
41	0	85	78	Training	−0.516019409	0.9724352452	0.1832543757	0.8881538502	0.1118461498	0
42	0	81	71	Training	0.8033843016	0.9950961431	0.8855642754	0.999976869	0.000023131	0
43	0	83	77	Training	0.5241251522	0.9856050404	0.2285647348	0.9996062954	0.0003937046	0
44		93	84	.	−0.999895937	0.6276638439	−0.423829791	0.0353312193	0.9646687807	1

FIGURE 16.28   **Neural Network Analysis of the Performance Data**

Neural	Validation Column: Validation		Model NTanH(3)		

Training

Group

Measures	Value
Generalized RSquare	0.626044
Entropy RSquare	0.4636501
RMSE	0.3366058
Mean Abs Dev	0.2206477
Misclassification Rate	0.1538462
-LogLikelihood	9.2913182
Sum Freq	26

Validation

Group

Measures	Value
Generalized RSquare	0.8482741
Entropy RSquare	0.7326986
RMSE	0.2509753
Mean Abs Dev	0.1106463
Misclassification Rate	0.1176471
-LogLikelihood	3.0786188
Sum Freq	17

Estimates

Parameter	Estimate
H1_1:Test 1	−1.27828
H1_1:Test 2	0.250734
H1_1:Intercept	87.95447
H1_2:Test 1	−0.28563
H1_2:Test 2	−0.08506
H1_2:Intercept	35.18356
H1_3:Test 1	0.17687
H1_3:Test 2	−0.44839
H1_3:Intercept	20.31141
Group(0):H1_1	5.401823
Group(0):H1_2	4.582419
Group(0):H1_3	1.952507
Group(0):Intercept	0.045555

Confusion Matrix

Actual	Predicted Count	
Group	0	1
0	8	2
1	2	14

Confusion Matrix

Actual	Predicted Count	
Group	0	1
0	8	2
1	0	7

Confusion Rates

Actual	Predicted Rate	
Group	0	1
0	0.800	0.200
1	0.125	0.875

Confusion Rates

Actual	Predicted Rate	
Group	0	1
0	0.800	0.200
1	0.000	1.000

FIGURE 16.29   **Neural Network Analysis of the GPA Data**   DS GPA

Neural	Validation Column: Validation		Model NTanH(3)		

Training

GPA

Measures	Value
RSquare	0.2385852
RMSE	0.570192
Mean Abs Dev	0.4478795
-LogLikelihood	301.71907
SSE	114.44187
Sum Freq	352

Validation

GPA

Measures	Value
RSquare	0.25548
RMSE	0.5299264
Mean Abs Dev	0.4073364
-LogLikelihood	276.72422
SSE	99.130147
Sum Freq	353

Estimates

Parameter	Estimate
H1_1:ACT	−0.60614
H1_1:H. S. Rank	−0.07549
H1_1:Intercept	20.81756
H1_2:ACT	−0.09771
H1_2:H. S. Rank	−0.07838
H1_2:Intercept	9.047618
H1_3:ACT	−0.00882
H1_3: H. S. Rank	−0.11898
H1_3:Intercept	9.526342
GPA(1):H1_1	0.210576
GPA(2):H1_2	−1.86742
GPA(3):H1_3	0.869946
GPA(4):Intercept	3.332661

	GPA	ACT	H. S. Rank	Validation	H1_1	H1_2	H1_3	Predicted GPA
706	.	26	91	.	−0.71914728	−0.302973105	−0.644163059	3.1866162845

Appendix 16.1 ■ Logistic Regression Analysis Using Minitab

Logistic regression in Figure 16.3 (data file: PerfTest .MTW):

- In the Data window, enter the performance data in Table 16.1 with Group in C1, the score on test 1 in C2, and the score on test 2 in C3. Use variable names Group, Test1, and Test2.

- Select **Stat : Regression : Binary Logistic Regression.**

- Choose "Response in binary form/frequency format" in the first drop-down menu, enter Group as the Response, choose 1 as the Response event (since Group = 1 indicates the event of success), and enter Test1 and Test2 as the Continuous predictors. Leave the other windows blank.

- Click on the Results... button in the Binary Logistic Regression dialog box.

- In the "Binary Logistic Regression: Results dialog box," choose "Expanded tables" in the drop-down menu for "Display of results."

- Click OK in the "Binary Logistic Regression: Results" and the Binary Logistic Regression dialog boxes.

- The logistic model (including confidence intervals and *z* statistics for the model coefficients) appears in the Session window.

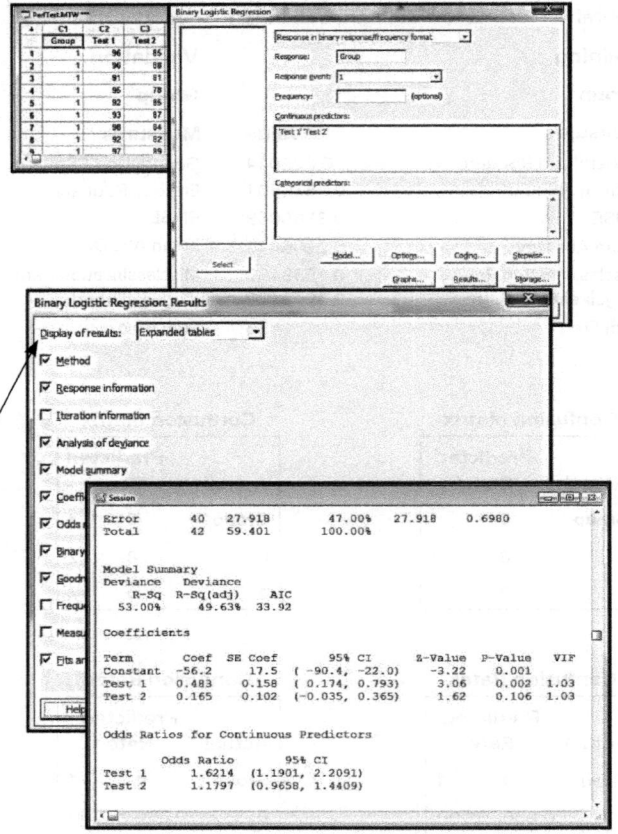

Source: Minitab 18

Appendix 16.2 ■ Logistic Regression, Discriminate Analysis, and Neural Networks using JMP

Logistic regression in Figure 16.2 (data file: PrcRed.jmp):

- After opening the PrcRed.jmp file, select **Analyze : Fit Model.**

- In the Fit Model dialog box, select the "Proportion" column and click the "Y" button and select the "x" column and click the "Add" button.

- Set Personality to "Generalized Regression" and set Distribution to "Binomial." Click "Run."

- From the red triangle menu beside "Logistic Regression" select "Save Columns : Save Prediction Formula" to produce the Probability columns at the bottom of Figure 16.2.

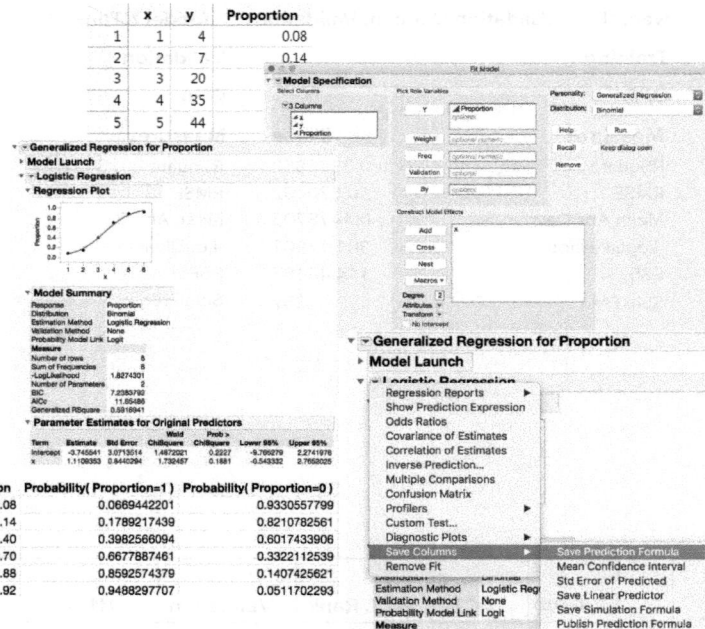

Source: JMP Pro 14

Logistic regression in Figures 16.4–16.6 (data file: CardUpgrade.jmp):

- After opening the CardUpgrade.jmp file, select **Analyze : Fit Model.**

- In the Fit Model dialog box, select the "UpGrade" column and click the "Y" button. Select the "Purchases" and "PlatProfile" columns and click the "Add" button.

- Set Personality to "Nominal Logistic" and set Target Level to "1." Click "Run."

- From the red triangle menu beside "Nominal Logistic Fit for UpGrade," select "Save Probability Formula" to obtain the four output columns in Figure 16.4. Select "Confidence Intervals" and "Odds Ratios" to reveal the output in Figure 16.5. Select "Confusion Matrix" for the output in Figure 16.6.

Logistic regression using forward selection with simultaneous validation in Figures 16.9 (data file: Churn.jmp):

- After opening the Churn.jmp file, select **Analyze : Fit Model.**

- In the Fit Model dialog box, select the "Churn" column and click the "Y" button. Select "AcctLength" through "NCustService-Calls" and click the "Add" button. Select the "Validation" column and click the "Validation" button.

- Set Personality to "Stepwise" and click "Run."

- In the Fit Stepwise dialog, set Stopping Rule to "Max Validation RSquare," set Direction to "Forward," and set Rules to "Combine." Click "Go" to obtain the Step History in Figure 16.9.

- Once the stepwise procedure ends, click "Run Model" to obtain full details on the final stepwise model.

- In the Fit Model report, from the red triangle menu beside "Ordinal Logistic Fit for Churn," select "Confusion Matrix" and "Save Probability Formula" to obtain the remaining output in Figure 16.9.

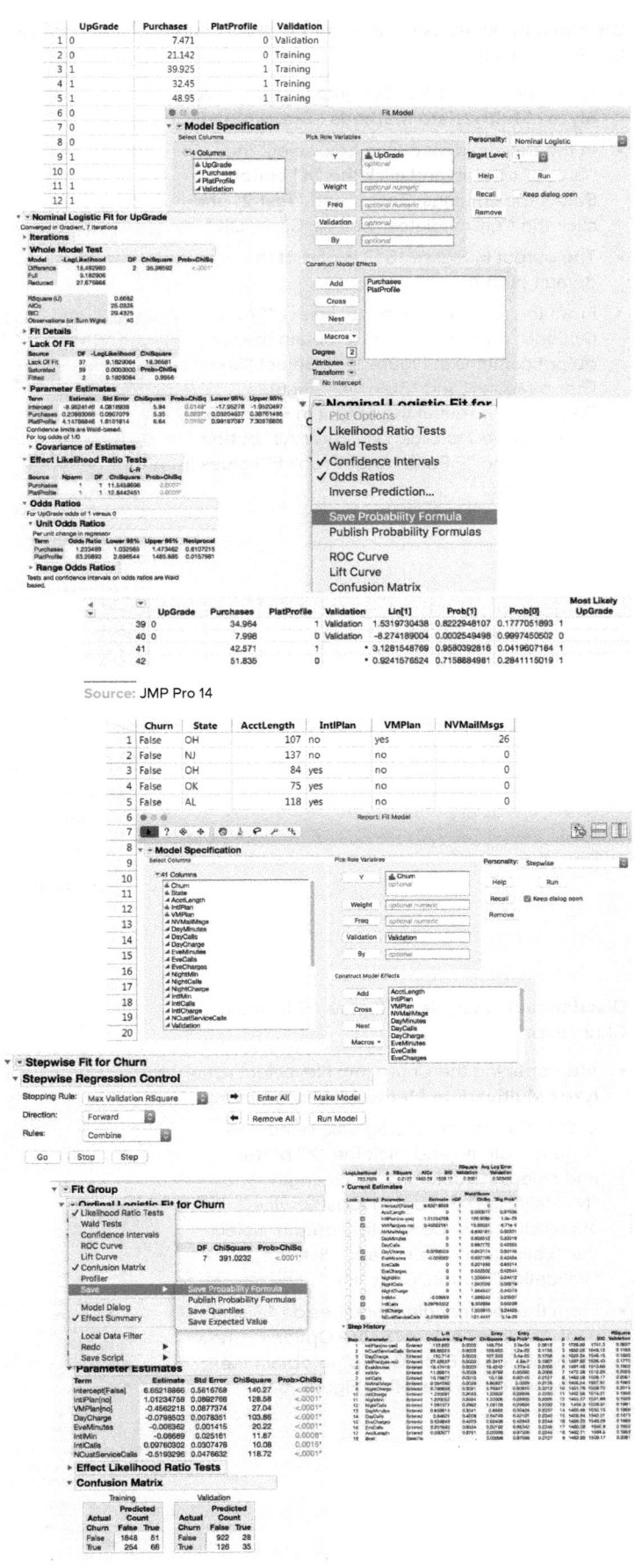

Source: JMP Pro 14

Source: JMP Pro 14

Discriminant analysis in Figures 16.11–16.13 (data file: PerfTest.jmp):

- After opening the PerfTest.jmp file, select **Analyze : Multivariate Methods : Discriminant.**

- In the Discriminant dialog box, select the "Group" column and click the "X" button. Select the columns "Test 1" and "Test 2" and click the "Y, Covariates" button. Click "OK."

- The output in Figure 16.13 is part of the default output.

- From the red triangle menu, select "Score Options : Save Formulas" to obtain the six output columns in Figure 16.11. Select "Show Group Means" and "Stepwise Variable Selection" to reveal the output in Figure 16.12. (You will need to click the "Enter All" button to obtain the "F Ratio" and "Prob>F" values in Figure 16.12.)

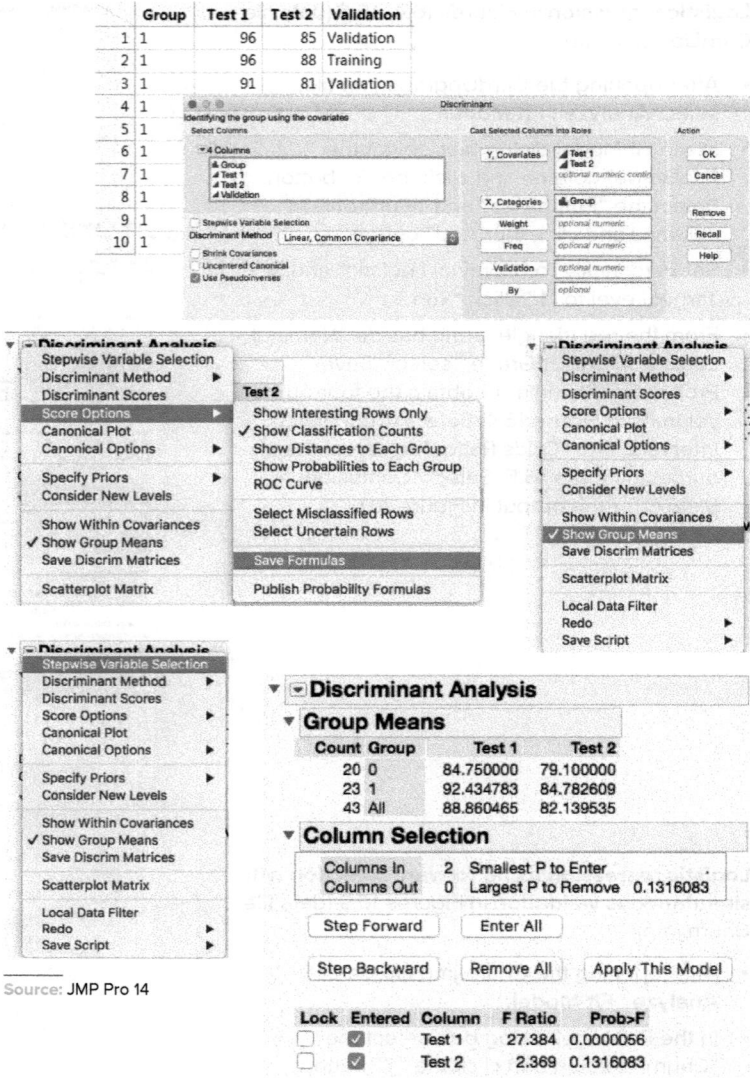

Source: JMP Pro 14

Discriminant analysis in Figure 16.14 (data file: Churn.jmp):

- After opening the Churn.jmp file, select **Analyze : Multivariate Methods : Discriminant.**

- In the Discriminant dialog box, select the "Churn" column and click the "X" button and select the columns "AcctLength" and "NVMailMsgs" through "NCustServiceCalls" and click the "Y, Covariates" button. Select the "Validation" column and click the "Validation" button. Click "OK."

- From the red triangle menu beside "Discriminant Analysis," select "Score Options : Show Probabilities to Each Group" to obtain the probabilities on the left in Figure 16.14.

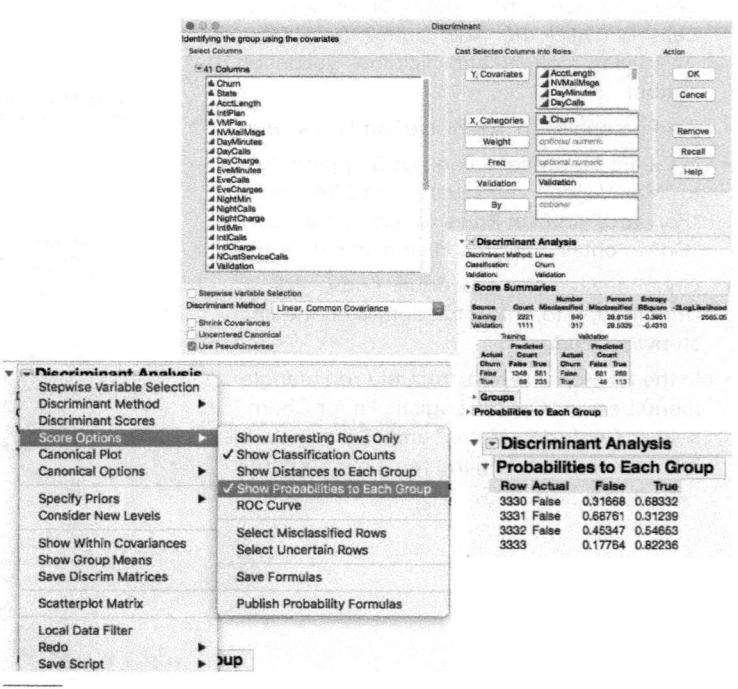

Source: JMP Pro 14

Neural network analysis in Figures 16.18–16.20 (data file: CardUpgrade.jmp):

- NOTE: To reproduce the output in Figures 16.18–16.20 you will need to *manually* create a validation column that exactly matches the column named Validation in Figure 16.18. The following instructions demonstrate one method of *automatically* creating such a validation column.

- After opening the CardUpgrade.jmp data file, select **Analyze : Predictive Modeling : Make Validation Column.**

- Enter 0.6 in the "Training Set" box and 0.4 in the "Validation Set" box then click the "Fixed Random" button. A new column "Validation" is now added to the data file.

- Select **Analyze : Predictive Modeling : Neural.** Select the "UpGrade" column and click the "Y, Response" button. Select the "Purchases" and "PlatProfile" columns and click the "X, Factor" button. Select the "Validation" column and click the "Validation" button. Click "OK."

- In the Model Launch dialog, click the "Go" button to reveal output like that of Figure 16.20.

- From the red triangle menu beside "Model NTanH(3)," select "Show Estimates" to reveal parameter estimates like those in Figure 16.19 and select "Save Formulas" to create the output columns like those in Figure 16.18.

- By adding values for "Purchases" and "Plat-Profile" in rows 41 and 42 (see Figure 16.18), JMP will calculate probabilities and "Most Likely Upgrade" values for these new customers.

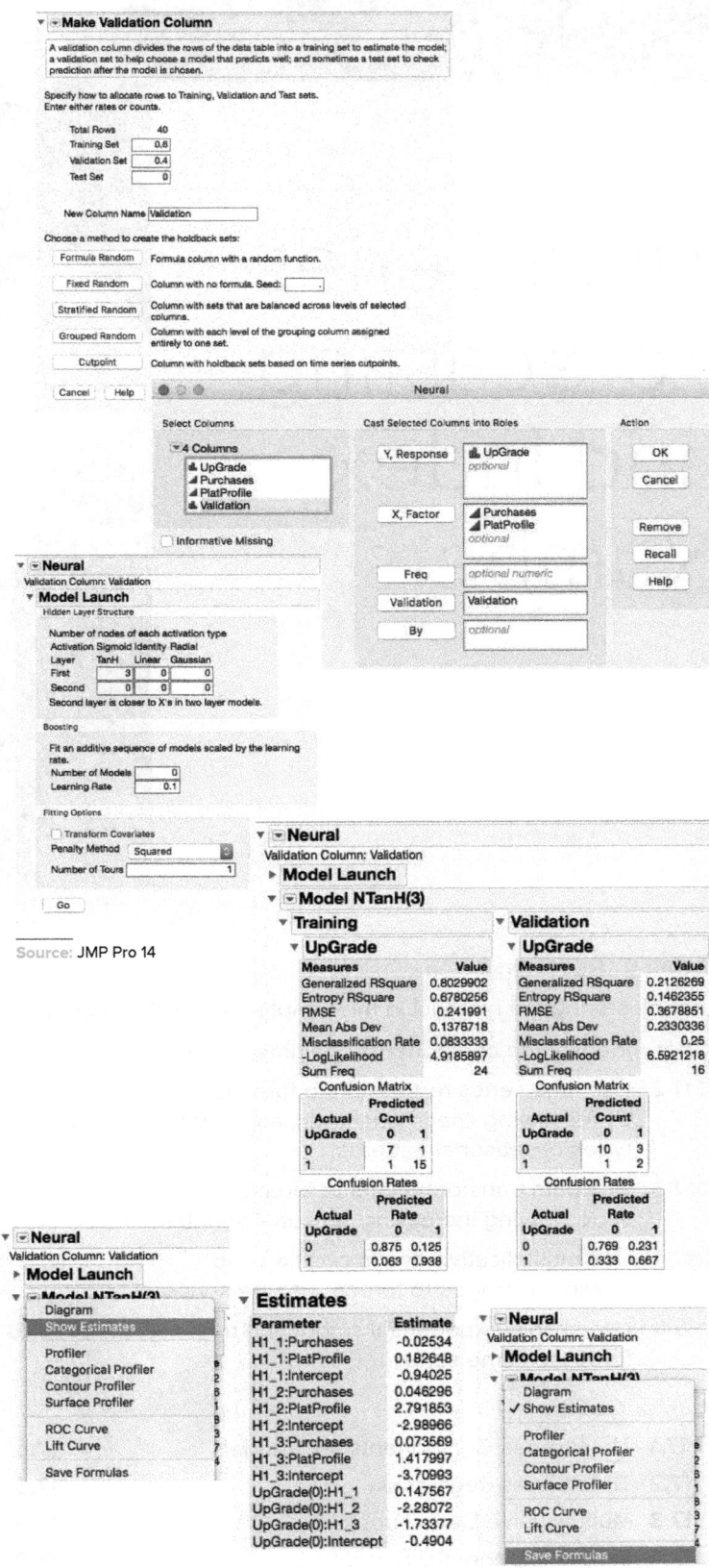

Source: JMP Pro 14

Design Elements:
(CD): ©Comstock Images/Alamy; (All Others): ©McGraw-Hill Education

Source: JMP Pro 14

## CHAPTER 17

Time Series Forecasting and Index Numbers

©Tetra Images/Getty Images RF

Learning Objectives

After mastering the material in this chapter, you will be able to:

LO17-1 Identify the components of a time series.

LO17-2 Use time series regression to forecast time series having linear, quadratic, and certain types of seasonal patterns.

LO17-3 Use data transformations to forecast time series having increasing seasonal variation.

LO17-4 Use multiplicative decomposition and moving averages to forecast a time series.

LO17-5 Use simple exponential smoothing to forecast a time series.

LO17-6 Use double exponential smoothing to forecast a time series.

LO17-7 Use the multiplicative Winters' method to forecast a time series.

LO17-8 Use the Box–Jenkins methodology to forecast a time series.

LO17-9 Compare time series models by using forecast errors.

LO17-10 Use index numbers to compare economic data over time.

Chapter Outline

Note: After completing Section 17.2, the reader may study Sections 17.3, 17.4, 17.6, 17.7, and 17.8 in any order without loss of continuity. Section 17.5 requires background from Sections 17.1, 17.2, and 17.4. Section 17.8 may be covered at any time.

time series is a set of observations on a variable of interest that has been collected in **time order.** In this chapter we discuss developing and using **univariate time series models,** which forecast future values of a time series **solely on the basis of past values of the time series.** Often univariate time series models forecast future time series values by extrapolating the **trend** and/or **seasonal patterns** exhibited by the past values of the time series. To illustrate these ideas, we consider several cases in this chapter, including:

The Calculator Sales Case: By extrapolating an upward trend in past sales of the Bismark X-12 electronic calculator, Smith's Department Stores, Inc., forecasts future sales of this calculator. The forecasts help the department store chain to better implement its inventory and financial policies.

The Traveler's Rest Case: By extrapolating an upward trend and the seasonal behavior of its past hotel room occupancies, Traveler's Rest, Inc., forecasts future hotel room occupancies. The forecasts help the hotel chain to more effectively hire help and acquire supplies.

17.1 Time Series Components and Models

LO17-1
Identify the components of a time series.

In order to identify patterns in time series data, it is often convenient to think of such data as consisting of several components: **trend, cycle, seasonal variations,** and **irregular fluctuations. Trend** refers to the upward or downward movement that characterizes a time series over time. Thus trend reflects the long-run growth or decline in the time series. Trend movements can represent a variety of factors. For example, long-run movements in the sales of a particular industry might be determined by changes in consumer tastes, increases in total population, and increases in per capita income. **Cycle** refers to recurring up-and-down movements around trend levels. These fluctuations can last from 2 to 10 years or even longer measured from peak to peak or trough to trough. One of the common cyclical fluctuations found in time series data is the *business cycle*, which is represented by fluctuations in the time series caused by recurrent periods of prosperity and recession. **Seasonal variations** are periodic patterns in a time series that complete themselves within a calendar year or less and then are repeated on a regular basis. Often seasonal variations occur yearly. For example, soft drink sales and hotel room occupancies are annually higher in the summer months, while department store sales are annually higher during the winter holiday season. Seasonal variations can also last less than one year. For example, daily restaurant patronage might exhibit within-week seasonal variation, with daily patronage higher on Fridays and Saturdays. **Irregular fluctuations** are erratic time series movements that follow no recognizable or regular pattern. Such movements represent what is "left over" in a time series after trend, cycle, and seasonal variations have been accounted for.

Time series that exhibit trend, seasonal, and cyclical components are illustrated in Figure 17.1. In Figure 17.1(a) a time series of sales observations that has an essentially

FIGURE 17.1 Time Series Exhibiting Trend, Seasonal, and Cyclical Components

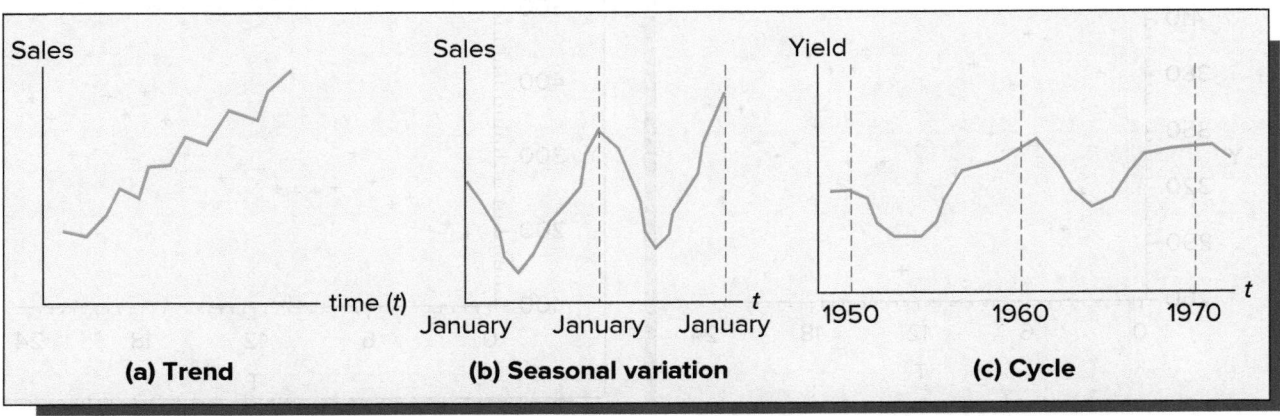

(a) Trend (b) Seasonal variation (c) Cycle

straight-line or linear trend is plotted. Figure 17.1(b) portrays a time series of sales observations that contains a seasonal pattern that repeats annually. Figure 17.1(c) exhibits a time series of agricultural yields that is cyclical, repeating a cycle about once every 10 years.

Time series models attempt to identify significant patterns in the components of a time series. Then, assuming that these patterns will continue into the future, time series models extrapolate these patterns to forecast future time series values. In Section 17.2 we discuss forecasting by **time series regression models,** and in Section 17.3 we discuss forecasting by using an intuitive method called **multiplicative decomposition.** Both of these approaches assume that the time series components remain essentially constant over time. If the time series components might be changing over time, it is appropriate to forecast by using **exponential smoothing.** This approach is discussed in Sections 17.4 and 17.5. A more sophisticated approach to forecasting a time series with components that might be changing over time is to use a model obtained by the **Box–Jenkins methodology.** This more advanced technique is introduced in Section 17.6.

17.2 Time Series Regression

LO17-2
Use time series regression to forecast time series having linear, quadratic, and certain types of seasonal patterns.

Modeling trend components

We begin this section with two examples.

C EXAMPLE 17.1 The Cod Catch Case: No Trend Regression

TABLE 17.1

Cod Catch (in Tons)
🖐 CodCatch

Month	Year 1	Year 2
Jan.	362	276
Feb.	381	334
Mar.	317	394
Apr.	297	334
May	399	384
June	402	314
July	375	344
Aug.	349	337
Sept.	386	345
Oct.	328	362
Nov.	389	314
Dec.	343	365

The Bay City Seafood Company owns a fleet of fishing trawlers and operates a fish processing plant. In order to forecast its minimum and maximum possible revenues from cod sales and plan the operations of its fish processing plant, the company desires to make both point forecasts and prediction interval forecasts of its monthly cod catch (measured in tons). The company has recorded monthly cod catch for the previous two years (years 1 and 2). The cod history is given in Table 17.1. A time series plot shows that the cod catches appear to randomly fluctuate around a constant average level. (See the plot in Figure 17.2.) Because the company subjectively believes that this data pattern will continue in the future, it seems reasonable to use the **"no trend"** regression model

$$y_t = \beta_0 + \varepsilon_t$$

to forecast cod catch in future months. It can be shown that for the no trend regression model the least squares point estimate b_0 of β_0 is $\bar{y}$, the average of the n observed time series values. Because the average $\bar{y}$ of the $n = 24$ observed cod catches is 351.29, it follows that $\hat{y}_t = b_0 = 351.29$ is the point prediction of the cod catch (y_t) in any future month. Furthermore,

---

FIGURE 17.2   **Plot of Cod Catch versus Time**

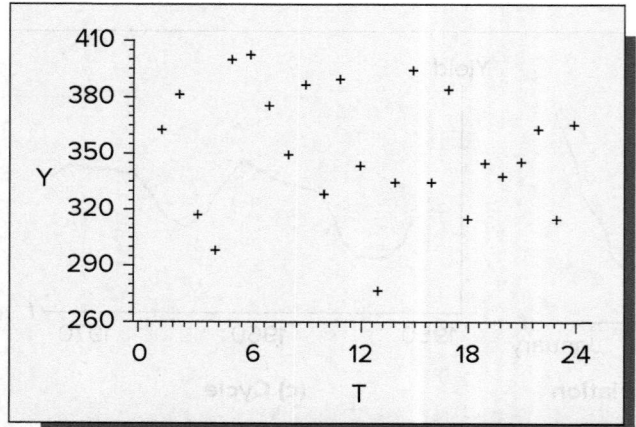

FIGURE 17.3   **Plot of Calculator Sales versus Time**

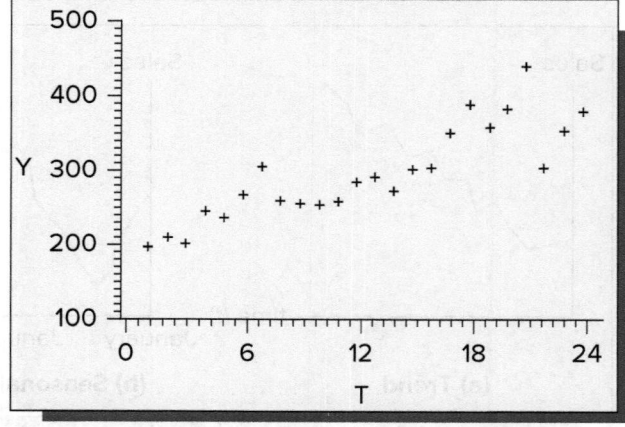

it can be shown that a $100(1 - \alpha)$ percent prediction interval for any future y_t value described by the no trend model is $[\hat{y}_t \pm t_{\alpha/2} s\sqrt{1 + (1/n)}]$. Here s is the sample standard deviation of the n observed time series values, and $t_{\alpha/2}$ is based on $n - 1$ degrees of freedom. For example, because s can be calculated to be 33.82 for the $n = 24$ cod catches, and because $t_{.025}$ based on $n - 1 = 23$ degrees of freedom is 2.069, it follows that a 95 percent prediction interval for the cod catch in any future month is $[351.29 \pm 2.069(33.82)\sqrt{1 + (1/24)}]$, or $[279.92, 422.66]$.

Ⓒ **EXAMPLE 17.2** The Calculator Sales Case: Inventory Policy

For the last two years Smith's Department Stores, Inc., has carried a new type of electronic calculator called the Bismark X-12. Sales of this calculator have generally increased over these two years. Smith's inventory policy attempts to ensure that stores will have enough Bismark X-12 calculators to meet practically all demand for the Bismark X-12, while at the same time ensuring that Smith's does not needlessly tie up its money by ordering many more calculators than can be sold. In order to implement this inventory policy in future months, Smith's requires both point predictions and prediction intervals for total monthly Bismark X-12 demand.

The monthly calculator demand data for the last two years are given in Table 17.2. A time series plot of the demand data is shown in Figure 17.3. The demands appear to randomly fluctuate around an average level that increases over time in a linear fashion. Furthermore, Smith's believes that this trend will continue for at least the next year. Thus it is reasonable to use the **"linear trend"** regression model

$$y_t = \beta_0 + \beta_1 t + \varepsilon_t$$

to forecast calculator sales in future months. Notice that this model is just a simple linear regression model where the time period t plays the role of the independent variable. The least squares point estimates of β_0 and β_1 can be calculated to be $b_0 = 198.0290$ and $b_1 = 8.0743$. [See Figure 17.4(a).] Therefore, for example, point forecasts of Bismark X-12 demand in January and February of year 3 (time periods 25 and 26) are, respectively,

$$\hat{y}_{25} = 198.0290 + 8.0743(25) = 399.9 \quad \text{and}$$

$$\hat{y}_{26} = 198.0290 + 8.0743(26) = 408.0$$

Note that Figure 17.4(b) gives these point forecasts. In addition, it can be shown using either the formulas for simple linear regression or a computer software package [see Figure 17.4(c)] that a 95 percent prediction interval for demand in time period 25 is $[328.6, 471.2]$ and that a 95 percent prediction interval for demand in time period 26 is

TABLE 17.2
Calculator Sales Data
ⒹⓈ CalcSale

Month	Year 1	Year 2
Jan.	197	296
Feb.	211	276
Mar.	203	305
Apr.	247	308
May	239	356
June	269	393
July	308	363
Aug.	262	386
Sept.	258	443
Oct.	256	308
Nov.	261	358
Dec.	288	384

FIGURE 17.4　**Excel Analysis of the Calculator Sales Data Using the Linear Trend Regression Model**

(a) The Excel output

ANOVA

	df	SS	MS	F	Significance F
Regression	1	74974.3567	74974.3567	74.7481	1.5893E-08
Residual	22	22066.6016	1003.0273		
Total	33	97040.9583			

	Coefficients	Standard Error	t Stat	P-value	Lower 95%	Upper 95%
Intercept	198.0290	13.3444	14.8398	6.0955E-13	170.3543	225.7036
T	8.0743	8.9339	8.6457	1.5893E-08	6.1375	10.0112

(b) Prediction using Excel

	A	B	C	D
24	358	23		
25	384	24		
26	399.8877	25	USING TREND	
27	407.962	26		

(c) Prediction using an Excel add-in (MegaStat)

Predicted values for: Sales

t	Predicted	95% Confidence Intervals		95% Prediction Intervals	
		lower	upper	lower	upper
25	399.9	372.2	427.6	328.6	471.2
26	408.0	378.6	437.3	336.0	479.9

[336.0, 479.9]. These prediction intervals can help Smith's implement its inventory policy. For instance, if Smith's stocks 471 Bismark X-12 calculators in January of year 3, we can be reasonably sure that monthly demand will be met.

Example 17.1 illustrates that the intercept β_0 can be used to model a lack of trend over time, and Example 17.2 illustrates that the expression $(\beta_0 + \beta_1 t)$ can model a linear trend over time. In addition, as will be illustrated in Exercise 17.39, the expression $(\beta_0 + \beta_1 t + \beta_2 t^2)$ can model a quadratic trend over time.

Modeling seasonal components

We next consider how to forecast time series described by trend and seasonal components.

Ⓒ EXAMPLE 17.3 The Bike Sales Case: Inventory Policy

©Brand X Pictures/Superstock RF

Table 17.3 presents quarterly sales of the TRK-50 mountain bike for the previous four years at a bicycle shop in Switzerland. The time series plot in Figure 17.5 shows that the bike sales exhibit a linear trend and a strong seasonal pattern, with bike sales being higher in the spring and summer quarters than in the winter and fall quarters. If we let y_t denote the number of TRK-50 mountain bikes sold in time period t at the Swiss bike shop, then a regression model describing y_t is

$$y_t = \beta_0 + \beta_1 t + \beta_{Q2}Q_2 + \beta_{Q3}Q_3 + \beta_{Q4}Q_4 + \varepsilon_t$$

Here the expression $(\beta_0 + \beta_1 t)$ models the linear trend evident in Figure 17.5. Q_2, Q_3, and Q_4 are dummy variables defined for quarters 2, 3, and 4. Specifically, Q_2 equals 1 if quarterly bike sales were observed in quarter 2 (spring) and 0 otherwise; Q_3 equals 1 if quarterly bike sales were observed in quarter 3 (summer) and 0 otherwise; Q_4 equals 1 if quarterly bike sales were observed in quarter 4 (fall) and 0 otherwise. Note that we have not defined a dummy variable for quarter 1 (winter). It follows that the regression parameters β_{Q2}, β_{Q3}, and β_{Q4} compare quarters 2, 3, and 4 with quarter 1. Intuitively, for example, β_{Q4} is the difference, excluding trend, between the level of the time series (y_t) in quarter 4 (fall) and the level of the time series in quarter 1 (winter). A positive β_{Q4} would imply that, excluding trend, bike sales in the fall can be expected to be higher than bike sales in the winter. A negative β_{Q4} would imply that, excluding trend, bike sales in the fall can be expected to be lower than bike sales in the winter.

| TABLE 17.3 | Quarterly Sales of the TRK-50 Mountain Bike   Ⓓⓢ BikeSales |

Year	Quarter	t	Sales, y_t
1	1 (Winter)	1	10
	2 (Spring)	2	31
	3 (Summer)	3	43
	4 (Fall)	4	16
2	1	5	11
	2	6	33
	3	7	45
	4	8	17
3	1	9	13
	2	10	34
	3	11	48
	4	12	19
4	1	13	15
	2	14	37
	3	15	51
	4	16	21

FIGURE 17.5   Time Series Plot of TRK-50 Bike Sales

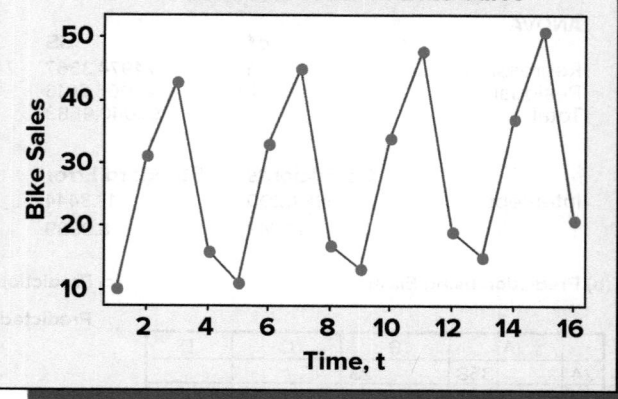

FIGURE 17.6   Minitab Output of an Analysis of the Quarterly Bike Sales Using
               Dummy Variable Regression

Term	Coef	SE Coef	T-Value	P-Value
Constant	8.7500	0.4281	20.44	0.000
Time	0.50000	0.03769	13.27	0.000
Q2	21.0000	0.4782	43.91	0.000
Q3	33.5000	0.4827	69.41	0.000
Q4	4.5000	0.4900	9.18	0.000

$S = 0.674200$ R-Sq = 99.8%   R-Sq(adj) = 99.8%

Regression Equation
BikeSales = 8.75 + 0.500 Time + 21.0 Q2 + 33.5 Q3 + 4.50 Q4

Values of Predictors for New Obs					Predicted Values for New Observations				
New Obs	Time	Q2	Q3	Q4	New Obs	Fit	SE Fit	95% CI	95% PI
1	17.0	0	0	0	1	17.250	0.506	(16.137, 18.363)	(15.395, 19.105)
2	18.0	1	0	0	2	38.750	0.506	(37.637, 39.863)	(36.895, 40.605)
3	19.0	0	1	0	3	51.750	0.506	(50.637, 52.863)	(49.895, 53.605)
4	20.0	0	0	1	4	23.250	0.506	(22.137, 24.363)	(21.395, 25.105)

Figure 17.6 gives the Minitab output of a regression analysis of the quarterly bike sales using the dummy variable model. The Minitab output tells us that the linear trend and the seasonal dummy variables are significant (every t statistic has a related p-value less than .01). Also, notice that the least squares point estimates of β_{Q2}, β_{Q3}, and β_{Q4} are, respectively, $b_{Q2} = 21$, $b_{Q3} = 33.5$, and $b_{Q4} = 4.5$. It follows that, excluding trend, expected bike sales in quarter 2 (spring), quarter 3 (summer), and quarter 4 (fall) are estimated to be, respectively, 21, 33.5, and 4.5 bikes greater than expected bike sales in quarter 1 (winter). Furthermore, using all of the least squares point estimates in Figure 17.6, we can compute point forecasts of bike sales in quarters 1 through 4 of next year (periods 17 through 20) as follows:

$$\hat{y}_{17} = b_0 + b_1(17) + b_{Q2}(0) + b_{Q3}(0) + b_{Q4}(0) = 8.75 + .5(17) = 17.250$$

$$\hat{y}_{18} = b_0 + b_1(18) + b_{Q2}(1) + b_{Q3}(0) + b_{Q4}(0) = 8.75 + .5(18) + 21 = 38.750$$

$$\hat{y}_{19} = b_0 + b_1(19) + b_{Q2}(0) + b_{Q3}(1) + b_{Q4}(0) = 8.75 + .5(19) + 33.5 = 51.750$$

$$\hat{y}_{20} = b_0 + b_1(20) + b_{Q2}(0) + b_{Q3}(0) + b_{Q4}(1) = 8.75 + .5(20) + 4.5 = 23.250$$

These point forecasts are given at the bottom of the Minitab output, as are 95 percent prediction intervals for y_{17}, y_{18}, y_{19}, and y_{20}. The upper limits of these prediction intervals suggest that the bicycle shop can be reasonably sure that it will meet demand for the TRK-50 mountain bike if the numbers of bikes it stocks in quarters 1 through 4 are, respectively, 19, 41, 54, and 25 bikes.

C **EXAMPLE 17.4** The Traveler's Rest Case: Predicting Hotel Room
                 Occupancy

Table 17.4 on the next page presents a time series of hotel room occupancies observed by Traveler's Rest, Inc., a corporation that operates four hotels in a midwestern city. The analysts in the operating division of the corporation were asked to develop a model that could be used to obtain short-term forecasts (up to one year) of the number of occupied rooms in the hotels. These forecasts were needed by various personnel to assist in hiring additional help during the summer months, ordering materials that have long delivery lead times, budgeting of local advertising expenditures, and so on. The available historical data consisted of the number of occupied rooms during each day for the previous 14 years. Because it was desired to obtain monthly forecasts, these data were reduced to monthly averages by dividing each monthly total by the number of days in the month. The monthly

TABLE 17.4   **Monthly Hotel Room Averages**   TravRest

Year	Jan.	Feb.	Mar.	Apr.	May	June	July	Aug.	Sept.	Oct.	Nov.	Dec.
1	501	488	504	578	545	632	728	725	585	542	480	530
2	518	489	528	599	572	659	739	758	602	587	497	558
3	555	523	532	623	598	683	774	780	609	604	531	592
4	578	543	565	648	615	697	785	830	645	643	551	606
5	585	553	576	665	656	720	826	838	652	661	584	644
6	623	553	599	657	680	759	878	881	705	684	577	656
7	645	593	617	686	679	773	906	934	713	710	600	676
8	645	602	601	709	706	817	930	983	745	735	620	698
9	665	626	649	740	729	824	937	994	781	759	643	728
10	691	649	656	735	748	837	995	1040	809	793	692	763
11	723	655	658	761	768	885	1067	1038	812	790	692	782
12	758	709	715	788	794	893	1046	1075	812	822	714	802
13	748	731	748	827	788	937	1076	1125	840	864	717	813
14	811	732	745	844	833	935	1110	1124	868	860	762	877

LO17-3

Use data transformations to forecast time series having increasing seasonal variation.

room averages for the previous 14 years are the time series values given in Table 17.4. A time series plot of these values in Figure 17.7(a) shows that the monthly room averages follow a strong trend and have a seasonal pattern with one major and several minor peaks during the year. Note that the major peak each year occurs during the high summer travel months of June, July, and August.

Although the quarterly bike sales and monthly hotel room averages both exhibit seasonal variation, they exhibit different kinds of seasonal variation. The quarterly bike sales plotted in Figure 17.5 exhibit *constant seasonal variation*. In general, **constant seasonal variation** is seasonal variation where the magnitude of the seasonal swing does not depend on the level of the time series. On the other hand, **increasing seasonal variation** is seasonal variation where the magnitude of the seasonal swing increases as the level of the time series increases. Figure 17.7(a) shows that the monthly hotel room averages exhibit increasing seasonal variation. If a time series exhibits increasing seasonal variation, one approach is to first use a **fractional power transformation** (see Section 14.7) that produces a transformed time series that exhibits constant seasonal variation. Therefore, consider taking the square roots, quartic roots, and natural logarithms of the monthly hotel room averages in Table 17.4. If we do this and plot the resulting three sets of transformed values versus time, we find that the quartic root transformation best equalizes the seasonal variation. Figure 17.7(b) presents a plot of the quartic roots of the monthly hotel room averages versus time. Letting y_t denote the hotel room average observed in time period t, it follows that a regression model describing the quartic root of y_t is

$$y_t^{.25} = \beta_0 + \beta_1 t + \beta_{M1} M_1 + \beta_{M2} M_2 + \cdots + \beta_{M11} M_{11} + \varepsilon_t$$

FIGURE 17.7   **Plots for the Monthly Hotel Room Averages in Table 17.4**

(a) Plot of the monthly hotel room averages versus time

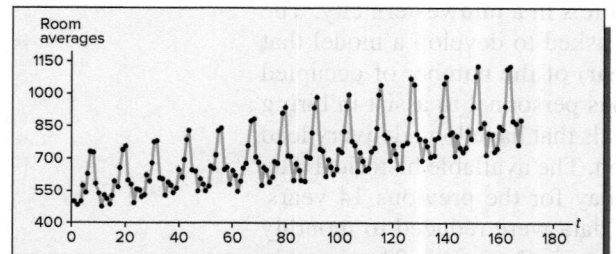

(b) Plot of the quartic roots of the monthly hotel averages versus time

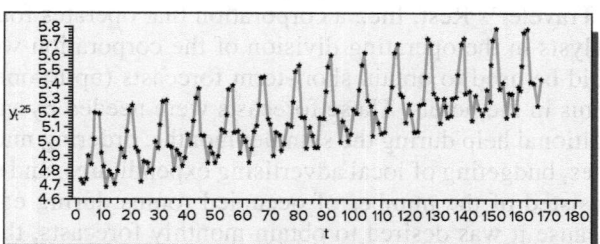

FIGURE 17.8   Excel Output of an Analysis of the Quartic Roots of the Room Averages Using Dummy Variable Regression

(a) The Excel output

	Coefficients	Standard Error	t Stat	P-value
Intercept	4.8073	0.0085	568.0695	4.06E-259
t	0.0035	0.0000	79.0087	3.95E-127
M1	−0.0525	0.0106	−4.9709	1.75E-06
M2	−0.1408	0.0106	−13.3415	1.59E-27
M3	−0.1071	0.0106	−10.1509	7.016E-19
M4	0.0499	0.0105	4.7284	5.05E-06
M5	0.0254	0.0105	2.4096	0.0171
M6	0.1902	0.0105	18.0311	6.85E-40
M7	0.3825	0.0105	36.2663	1.28E-77
M8	0.4134	0.0105	39.2009	2.41E-82
M9	0.0714	0.0105	6.7731	2.47E-10
M10	0.0506	0.0105	4.8029	3.66E-06
M11	−0.1419	0.0105	−13.4626	7.47E-28

(b) Prediction of $y_t^{.25}$ using an Excel add-in (MegaStat)

		95% Prediction Intervals	
t	Predicted	lower	upper
169	5.3489	5.2913	5.4065
170	5.2641	5.2065	5.3217
171	5.3013	5.2437	5.3589
172	5.4618	5.4042	5.5194
173	5.4409	5.3833	5.4984
174	5.6091	5.5515	5.6667
175	5.8049	5.7473	5.8625
176	5.8394	5.7818	5.8969
177	5.5009	5.4433	5.5585
178	5.4837	5.4261	5.5412
179	5.2946	5.2370	5.3522
180	5.4400	5.3825	5.4976

The expression $(\beta_0 + \beta_1 t)$ models the linear trend evident in Figure 17.7(b). Furthermore, M_1, $M_2, \ldots, M_{11}$ are dummy variables defined for months January (month 1) through November (month 11). For example, M_1 equals 1 if a monthly room average was observed in January, and 0 otherwise; M_2 equals 1 if a monthly room average was observed in February, and 0 otherwise. Note that we have not defined a dummy variable for December (month 12). It follows that the regression parameters $\beta_{M1}, \beta_{M2}, \ldots, \beta_{M11}$ compare January through November with December. Intuitively, for example, β_{M1} is the difference, excluding trend, between the level of the time series ($y_t^{.25}$) in January and the level of the time series in December. A positive β_{M1} would imply that, excluding trend, the value of the time series in January can be expected to be greater than the value in December. A negative β_{M1} would imply that, excluding trend, the value of the time series in January can be expected to be smaller than the value in December.

Figure 17.8 gives relevant portions of the Excel output of a regression analysis of the hotel room data using the quartic root dummy variable model. The Excel output tells us that the linear trend and the seasonal dummy variables are significant (every t statistic has a related p-value less than .05). In addition, although not shown on the output, $R^2 = .988$. Now consider time period 169, which is January of next year and which therefore implies that $M_1 = 1$ and that all the other dummy variables equal 0. Using the least squares point estimates in Figure 17.8(a), we compute a point forecast of $y_{169}^{.25}$ to be

$$b_0 + b_1(169) + b_{M1}(1) = 4.8073 + 0.0035(169) + (-.0525)(1)$$
$$= 5.3489$$

Note that this point forecast is given in Figure 17.8(b) [see time period 169]. It follows that a point forecast of y_{169} is

$$(5.3489)^4 = 818.57$$

Furthermore, Figure 17.8(b) shows that a 95 percent prediction interval for $y_{169}^{.25}$ is [5.2913, 5.4065]. It follows that a 95 percent prediction interval for y_{169} is

$$[(5.2913)^4, (5.4065)^4] = [783.88, 854.41]$$

This interval says that Traveler's Rest, Inc., can be 95 percent confident that the monthly hotel room average in period 169 will be no less than 783.88 rooms per day and no more than 854.41 rooms per day. Lastly, note that Figure 17.8(b) also gives point forecasts of and 95 percent prediction intervals for the quartic roots of the hotel room averages in February through December of next year (time periods 170 through 180).

The validity of the regression methods just illustrated requires that the independence assumption be satisfied. However, when time series data are analyzed, this assumption is often violated. It is quite common for the time-ordered error terms to exhibit **positive or negative autocorrelation.** In Section 14.7 we discussed positive and negative autocorrelation, and we saw that we can use residual plots to check for these kinds of autocorrelation.

One type of positive or negative autocorrelation is called **first-order autocorrelation.** It says that ε_t, the error term in time period t, is related to ε_{t-1}, the error term in time period $t-1$, by the equation

$$\varepsilon_t = \phi\varepsilon_{t-1} + a_t$$

Here, ϕ (pronounced *"phi"*) is the correlation coefficient that measures the relationship between error terms separated by one time period (ϕ must be estimated), and a_t, which is called a **random shock,** is a value that is assumed to have been randomly selected from a normal distribution with mean zero and a variance that is the same for each and every time period. Moreover, the random shocks in different time periods are assumed to be statistically independent of each other. To check for positive or negative first-order autocorrelation, we can use the **Durbin–Watson statistic d,** which was discussed in Chapters 14 and 15. For example, it can be verified that this statistic shows no evidence of positive or negative first-order autocorrelation in the error terms of the calculator sales model or in the error terms of the bike sales model. However, the Durbin–Watson statistic for the dummy variable regression model describing the quartic roots of the hotel room averages can be calculated to be $d = 1.26$. Because the dummy variable regression model uses $k = 12$ independent variables, and since Tables A.11, A.12, and A.13 do not give the **Durbin–Watson critical points** corresponding to $k = 12$, we cannot test for autocorrelation using these tables. However, it can be shown that $d = 1.26$ is quite small and indicates **positive autocorrelation** in the error terms. One approach to dealing with first-order autocorrelation in the error terms is to predict future values of the error terms by using the model $\varepsilon_t = \phi\varepsilon_{t-1} + a_t$. Of course, the error term ε_t could be related to more than just the previous error term ε_{t-1}. It could be related to any number of previous error terms. The **autoregressive error term model of order q**

$$\varepsilon_t = \phi_1\varepsilon_{t-1} + \phi_2\varepsilon_{t-2} + \cdots + \phi_q\varepsilon_{t-q} + a_t$$

relates ε_t, the error term in time period t, to the previous error terms $\varepsilon_{t-1}, \varepsilon_{t-2}, \ldots, \varepsilon_{t-q}$. Here $\phi_1, \phi_2, \ldots, \phi_q$ are unknown parameters that must be estimated, and a_t is a random shock that is assumed to satisfy the previously discussed assumptions. The **Box–Jenkins methodology** can be used to systematically identify an autoregressive error term model that relates ε_t to an appropriate number of past error terms. We then combine the time series regression model with the identified error term model to forecast future time series values. How this is done is discussed in Bowerman, O'Connell, and Koehler (1995).

The Box–Jenkins methodology can also be used to identify what is called an **autoregressive–moving average model,** which expresses the time series observation y_t by the equation

$$y_t = \beta_0 + \beta_1 y_{t-1} + \beta_2 y_{t-2} + \cdots + \beta_p y_{t-p}$$
$$+ a_t - \theta_1 a_{t-1} - \theta_2 a_{t-2} - \cdots - \theta_q a_{t-q}$$

Here, $y_{t-1}, y_{t-2}, \ldots, y_{t-p}$ are past time series values, $a_{t-1}, a_{t-2}, \ldots, a_{t-q}$ are past random shocks, and $\beta_0, \beta_1, \beta_2, \ldots, \beta_p, \theta_1, \theta_2, \ldots, \theta_q$ are unknown parameters that must be estimated. While such models are not intuitive, they tend to provide accurate forecasts and are an excellent alternative when the error terms of a time series regression model are found to be autocorrelated. For an introduction to the Box–Jenkins methodology see Section 17.6. Also note that we will discuss exponential smoothing models and methods in Sections 17.4 and 17.5. Exponential smoothing is somewhat similar to the Box–Jenkins methodology in that both techniques are designed to forecast time series described by trend and/or seasonal components that may be changing over time. In fact, some exponential smoothing models are theoretically equivalent to Box–Jenkins models. The two types of models are developed quite differently. Whereas there is a very precise and systematic procedure for developing Box–Jenkins models, we will see in Sections 17.4 and 17.5 that exponential smoothing models (like time series regression models) are developed by visually interpreting a time series plot and intuitively identifying the trend and/or seasonal components that describe the time series. Nevertheless, exponential smoothing models, similar to Box–Jenkins models, can be an effective forecasting approach when the error terms of a time series regression model are found to be autocorrelated. In Exercise 17.28 we will use an exponential smoothing technique called **Winters' method** to forecast the monthly hotel room averages.

To conclude this section, it is important to note that whenever we observe time series data we should determine whether trend and/or seasonal effects exist. For example, recall that the Fresh demand data in Table 15.6 are time series data observed over 30 consecutive four-week sales periods. Although we predicted demand for Fresh detergent on the basis of price difference and advertising expenditure, it is also possible that this demand is affected by a linear or quadratic trend over time and/or by seasonal effects (for example, more laundry detergent might be sold in summer sales periods when children are home from school). If we try using trend equations and dummy variables to search for trend and seasonal effects, we find that these effects do not exist in the Fresh demand data. However, in the Supplementary Exercises (see Exercise 17.45), we present a situation where we use trend equations and seasonal dummy variables, as well as **causal variables** such as price difference and advertising expenditure, to predict demand for a fishing lure.

Exercises for Section 17.2

CONCEPTS   connect

17.1 Discuss how we model no trend and a linear trend.

17.2 Discuss the difference between constant seasonal variation and increasing seasonal variation.

METHODS AND APPLICATIONS

17.3 THE LUMBER PRODUCTION CASE   DS LumberProd

In this exercise we consider annual U.S. lumber production over 30 years. The data were obtained from the U.S. Department of Commerce *Survey of Current Business* and are presented in Table 17.5. (The lumber production values are given in millions of board feet.) Plot the lumber production values versus time and discuss why the plot indicates that the model $y_t = \beta_0 + \varepsilon_t$ might appropriately describe these values.

17.4 THE LUMBER PRODUCTION CASE   DS LumberProd

Referring to the situation of Exercise 17.3, the mean and the standard deviation of the lumber production values can be calculated to be $\bar{y} = 35{,}651.9$ and $s = 2{,}037.3599$. Find a point forecast of and a 95 percent prediction interval for any future lumber production value.

17.5 THE WATCH SALES CASE   DS WatchSale

The past 20 monthly sales figures for a new type of watch sold at Lambert's Discount Stores are given in Table 17.6.

a Plot the watch sales values versus time and discuss why the plot indicates that the model

$$y_t = \beta_0 + \beta_1 t + \varepsilon_t$$

might appropriately describe these values.

b The least squares point estimates of β_0 and β_1 can be calculated to be $b_0 = 290.0895$ and $b_1 = 8.6677$.

(1) Use b_0 and b_1 to show (calculate) that a point forecast of watch sales in period 21 is $\hat{y}_{21} = 472.1$.
(2) Use the formulas of simple linear regression analysis or a computer software package to show that a 95 percent prediction interval for watch sales in period 21 is [421.5, 522.7].

17.6 THE AIR CONDITIONER SALES CASE   DS ACSales

Bargain Department Stores, Inc., is a chain of department stores in the Midwest. Quarterly sales of the "Bargain 8000-Btu Air Conditioner" over the past three years are as given in the left-hand portion of Table 17.7 on the next page.

TABLE 17.5		**TABLE 17.6**	
Annual Total U.S. Lumber Production		**Watch Sales Values**	
DS LumberProd		DS WatchSale	
		298	356
35,404	35,733	302	371
37,462	35,791	301	399
32,901	34,592	351	392
33,178	38,902	336	425
34,449	37,858	361	411
38,044	32,926	407	455
36,762	35,697	351	457
36,742	34,548	357	465
33,385	32,087	346	481
34,171	37,515		
36,124	38,629		
38,658	32,019		
32,901	35,710		
36,356	36,693		
37,166	37,153		

In Tables 17.5 and 17.6 time order is given by reading down the columns from left to right.

TABLE 17.7   **Air Conditioner Sales and a Dummy Variable Regression Analysis of the Sales Data**   ⑤ ACSales

Year	Quarter	Sales	Term	Coef	SE Coef	T-Value	P-Value	
1	1	2,915	Constant	2624.5	100.4	26.15	0.000	S = 92.4244
	2	8,032	T	382.82	34.03	11.25	0.000	R-Sq = 100.0%
	3	10,411	TSq	−11.354	2.541	−4.47	0.004	R-Sq(adj) = 99.9%
	4	2,427	Q2	4629.74	76.08	60.86	0.000	
2	1	4,381	Q3	6738.85	77.38	87.09	0.000	
	2	9,138	Q4	−1565.32	79.34	−19.73	0.000	

Regression Equation
Sales = 2625 + 383 T − 11.4 TSq + 4630 Q2 + 6739 Q3 − 1565 Q4

Year	Quarter	Sales	Time	Fit	SE Fit	95% CI	95% PI
2	3	11,386	13	5682.4	112.6	(5406.9, 5957.9)	(5325.9, 6038.8)
	4	3,382	14	10388.4	142.8	(10039.0, 10737.8)	(9972.2, 10804.6)
3	1	5,105	15	12551.0	177.2	(12117.4, 12984.7)	(12061.9, 13040.2)
	2	9,894	16	4277.7	213.9	(3754.4, 4801.1)	(3707.6, 4847.8)
	3	12,300					
	4	4,013					

a   Plot sales versus time and discuss why the plot indicates that the model

$$y_t = \beta_0 + \beta_1 t + \beta_2 t^2 + \beta_{Q2} Q_2 + \beta_{Q3} Q_3 + \beta_{Q4} Q_4 + \varepsilon_t$$

might appropriately describe the sales values. In this model Q_2, Q_3, and Q_4 are appropriately defined dummy variables for quarters 2, 3, and 4.

b   The right-hand portion of Table 17.7 is the Minitab output of a regression analysis of the air conditioner sales data using the model in part *a*. **(1)** Define the dummy variables Q_2, Q_3, and Q_4. **(2)** Use the Minitab output to find, report, and interpret the least squares point estimates of β_{Q2}, β_{Q3}, and β_{Q4}.

c   At the bottom of the Minitab output are point and prediction interval forecasts of air conditioner sales in the four quarters of year 4. Find and report these forecasts and hand calculate the point forecasts.

17.7   Table 17.8 gives the monthly international passenger totals over the last 11 years for an airline company. A plot of these passenger totals reveals an upward trend with increasing seasonal variation, and the natural

logarithmic transformation is found to best equalize the seasonal variation [see Figure 17.9(a) and (b)]. Figure 17.9(c) gives the JMP output of a regression analysis of the monthly international passenger totals by using the model

$$\ln y_t = \beta_0 + \beta_1 t + \beta_{M1} M_1 + \beta_{M2} M_2 + \cdots + \beta_{M11} M_{11} + \varepsilon_t$$

Here $M_1, M_2, \ldots, M_{11}$ are appropriately defined dummy variables for January (month 1) through November (month 11). Let y_{133} denote the international passenger totals in month 133 (January of next year). The JMP output tells us that a point forecast of and a 95 percent prediction interval for $\ln y_{133}$ are, respectively, 6.08610 and [5.96593, 6.20627]. **(1)** Using the least squares point estimates on the JMP output, calculate the point forecast. **(2)** By calculating $e^{6.08610}$ and $[e^{5.96593}, e^{6.20627}]$, find a point forecast of and a 95 percent prediction interval for y_{133}.   ⑤ AirPass

17.8   Use the Durbin–Watson statistic given at the bottom of the JMP output in Figure 17.9(c) to assess whether there is positive autocorrelation.

TABLE 17.8   **Monthly International Passenger Totals (Thousands of Passengers)**   ⑤ AirPass

Year	Jan.	Feb.	Mar.	Apr.	May	June	July	Aug.	Sept.	Oct.	Nov.	Dec.
1	112	118	132	129	121	135	148	148	136	119	104	118
2	115	126	141	135	125	149	170	170	158	133	114	140
3	145	150	178	163	172	178	199	199	184	162	146	166
4	171	180	193	181	183	218	230	242	209	191	172	194
5	196	196	236	235	229	243	264	272	237	211	180	201
6	204	188	235	227	234	264	302	293	259	229	203	229
7	242	233	267	269	270	315	364	347	312	274	237	278
8	284	277	317	313	318	374	413	405	355	306	271	306
9	315	301	356	348	355	422	465	467	404	347	305	336
10	340	318	362	348	363	435	491	505	404	359	310	337
11	360	342	406	396	420	472	548	559	463	407	362	405

Source: *FAA Statistical Handbook of Civil Aviation* (several annual issues). These data were originally presented by Box and Jenkins (1976). We have updated the situation in this exercise to be more modern.

FIGURE 17.9   Analysis of the Monthly International Passenger Totals

(a) Plot of the passenger totals

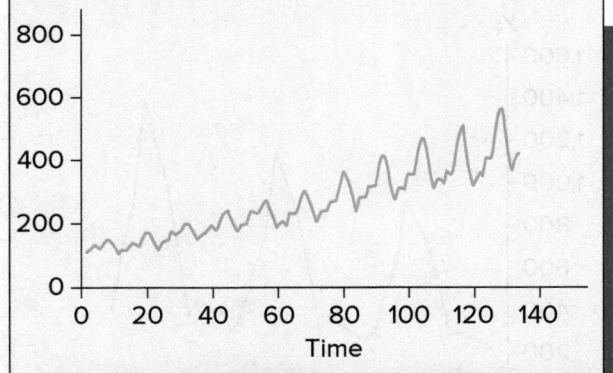

(b) Plot of the natural logarithms of the passenger totals

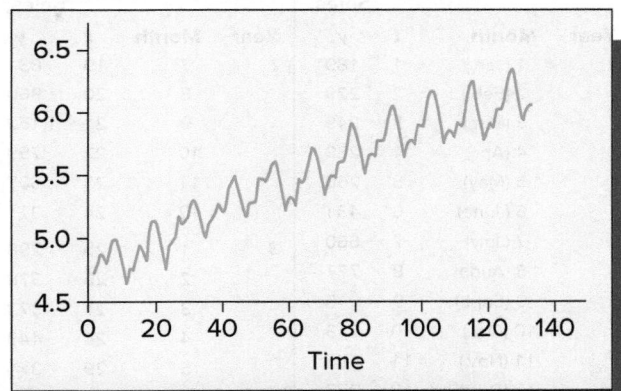

(c) JMP Output of a regression analysis of the monthly international passenger totals using the dummy variable model

Term	Estimate	Std Error	t Ratio	Prob>|t|	Lower 95%	Upper 95%
Intercept	4.6961806	0.01973	238.02	<.0001*	4.6571135	4.7352477
Time	0.0103075	0.000132	78.30	<.0001*	0.0100468	0.0105681
Jan	0.0190283	0.024515	0.78	0.4392	−0.029513	0.0675699
Feb	0.0014978	0.024507	0.06	0.9514	−0.047029	0.0500247
Mar	0.1379529	0.024501	5.63	<.0001*	0.0894393	0.1864665
Apr	0.0958338	0.024495	3.91	0.0002*	0.0473321	0.1443355
May	0.0917785	0.024489	3.75	0.0003*	0.0432873	0.1402696
Jun	0.2143158	0.024485	8.75	<.0001*	0.1658337	0.2627978
Jul	0.3146865	0.024481	12.85	<.0001*	0.2662122	0.3631609
Aug	0.3075925	0.024478	12.57	<.0001*	0.2591245	0.3560605
Sep	0.1665222	0.024475	6.80	<.0001*	0.1180591	0.2149854
Oct	0.0253066	0.024473	1.03	0.3032	−0.023153	0.0737662
Nov	−0.115595	0.024472	−4.72	<.0001*	−0.164053	−0.067137

	Predicted LnPass	Lower 95% Indiv LnPass	Upper 95% Indiv LnPass
133	6.0860998094	5.9659307831	6.2062688358
134	6.0788767584	5.9587077321	6.1990457848
135	6.2256392823	6.105470256	6.3458083087
136	6.1938276716	6.0736586452	6.3139966979
137	6.2000797644	6.079910738	6.3202487907
138	6.3329245192	6.2127554928	6.4530935455
139	6.4436027223	6.323433696	6.5637717487
140	6.4468161495	6.3266471231	6.5669851758
141	6.3160533289	6.1958843026	6.4362223553
142	6.1851451622	6.0649761358	6.3053141885
143	6.0545510203	5.934381994	6.1747200467
144	6.1804534526	6.0602844263	6.300622479

S = 0.0573917   R-Sq = 98.3%   R-Sq(adj) = 98.1%   Durbin-Watson statistic = 0.420944

17.3 Multiplicative Decomposition

When a time series exhibits increasing (or decreasing) seasonal variation, we can use the **multiplicative decomposition method** to decompose the time series into its **trend, seasonal, cyclical,** and **irregular** components. This is illustrated in the following example.

LO17-4
Use multiplicative decomposition and moving averages to forecast a time series.

 EXAMPLE 17.5 The Tasty Cola Case: Predicting Soft Drink Sales

The Discount Soda Shop, Inc., owns and operates 10 drive-in soft drink stores. Discount Soda has been selling Tasty Cola, a soft drink introduced just three years ago and gaining in popularity. Periodically, Discount Soda orders Tasty Cola from the regional distributor. To better implement its inventory policy, Discount Soda needs to forecast monthly Tasty Cola sales (in hundreds of cases).

Discount Soda has recorded monthly Tasty Cola sales for the previous three years. This time series is given in Table 17.9 and is plotted in Figure 17.10, both on the next page. Notice that, in addition to having a linear trend, the Tasty Cola sales time series possesses seasonal variation, with sales of the soft drink greatest in the summer and early fall months and lowest in the winter months. Because the seasonal variation seems to be increasing, we will see as we progress through this example that it might be reasonable to conclude that y_t, the sales of Tasty Cola in period t, is described by the **multiplicative model**

$$y_t = TR_t \times SN_t \times CL_t \times IR_t$$

TABLE 17.9   **Monthly Sales of Tasty Cola (in Hundreds of Cases)** ⊙ TastyCola

Year	Month	t	y_t	Year	Month	t	y_t
1	1 (Jan.)	1	189	2	7	19	831
	2 (Feb.)	2	229		8	20	960
	3 (Mar.)	3	249		9	21	1,152
	4 (Apr.)	4	289		10	22	759
	5 (May)	5	260		11	23	607
	6 (June)	6	431		12	24	371
	7 (July)	7	660	3	1	25	298
	8 (Aug.)	8	777		2	26	378
	9 (Sept.)	9	915		3	27	373
	10 (Oct.)	10	613		4	28	443
	11 (Nov.)	11	485		5	29	374
	12 (Dec.)	12	277		6	30	660
2	1	13	244		7	31	1,004
	2	14	296		8	32	1,153
	3	15	319		9	33	1,388
	4	16	370		10	34	904
	5	17	313		11	35	715
	6	18	556		12	36	441

FIGURE 17.10   **Time Series Plot of the Tasty Cola Sales Data**

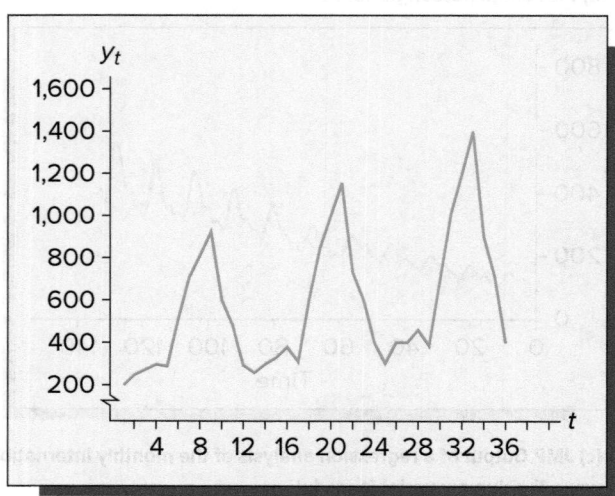

Here TR_t, SN_t, CL_t, and IR_t represent the trend, seasonal, cyclical, and irregular components of the time series in time period t.

Table 17.10 summarizes the calculations needed to find estimates—denoted tr_t, sn_t, cl_t, and ir_t—of TR_t, SN_t, CL_t, and IR_t. As shown in the table, we begin by calculating **moving averages** and **centered moving averages.** The purpose behind computing these averages is to eliminate seasonal variations and irregular fluctuations from the data. The first moving average of the first 12 Tasty Cola sales values is

$$\frac{189 + 229 + 249 + 289 + 260 + 431 + 660 + 777 + 915 + 613 + 485 + 277}{12}$$

$$= 447.833$$

Here we use a "12-period moving average" because the Tasty Cola time series data are monthly (12 time periods or "seasons" per year). If the data were quarterly, we would compute a "4-period moving average." The second moving average is obtained by dropping the first sales value (y_1) from the average and by including the next sales value (y_{13}) in the average. Thus we obtain

$$\frac{229 + 249 + 289 + 260 + 431 + 660 + 777 + 915 + 613 + 485 + 277 + 244}{12}$$

$$= 452.417$$

The third moving average is obtained by dropping y_2 from the average and by including y_{14} in the average. We obtain

$$\frac{249 + 289 + 260 + 431 + 660 + 777 + 915 + 613 + 485 + 277 + 244 + 296}{12}$$

$$= 458$$

Successive moving averages are computed similarly until we include y_{36} in the last moving average. Note that we use the term *moving average* here because, as we calculate these averages, we move along by dropping the most remote observation in the previous average and by including the "next" observation in the new average.

The first moving average corresponds to a time that is midway between periods 6 and 7, the second moving average corresponds to a time that is midway between periods 7 and 8, and so forth. In order to obtain averages corresponding to time periods in the original Tasty

TABLE 17.10   Tasty Cola Sales and the Multiplicative Decomposition Method

t Time Period	y_t Tasty Cola Sales	First Step: 12-Period Moving Average	$tr_t \times cl_t$: Centered Moving Average	$sn_t \times ir_t$: $\frac{y_t}{tr_t \times cl_t}$	sn_t: Table 17.11	d_t: $\frac{y_t}{sn_t}$	tr_t: 380.163 +9.489t	$tr_t \times sn_t$: Multiply tr_t by sn_t	$cl_t \times ir_t$: $\frac{y_t}{tr_t \times sn_t}$	cl_t: 3-Period Moving Average	ir_t: $\frac{cl_t \times ir_t}{cl_t}$
1 (Jan)	189				.493	383.37	389.652	192.10	.9839		
2	229				.596	384.23	399.141	237.89	.9626	.9902	.9721
3	249				.595	418.49	408.630	243.13	1.0241	1.0010	1.0231
4	289				.680	425	418.119	284.32	1.0165	1.0396	.9778
5	260				.564	460.99	427.608	241.17	1.0781	1.0315	1.0452
6	431	447.833			.986	437.12	437.097	430.98	1.0000	1.0285	.9723
7	660	452.417	450.125	1.466	1.467	449.9	446.586	655.14	1.0074	1.0046	1.0028
8	777	458	455.2085	1.707	1.693	458.95	456.075	772.13	1.0063	1.0004	1.0059
9	915	563.833	460.9165	1.985	1.990	459.79	465.564	926.47	.9876	.9937	.9939
10	613	470.583	467.208	1.312	1.307	469.01	475.053	620.89	.9873	.9825	1.0049
11	485	475	472.7915	1.026	1.029	471.33	489.542	498.59	.9727	.9648	1.0082
12	277	485.417	480.2085	.577	.600	461.67	494.031	296.42	.9345	.9634	.9700
13 (Jan)	244	499.667	492.542	.495	.493	494.97	503.520	248.24	.9829	.9618	1.0219
14	296	514.917	507.292	.583	.596	496.64	513.009	305.75	.9681	.9924	.9755
15	319	534.667	524.792	.608	.595	536.13	522.498	310.89	1.0261	1.0057	1.0203
16	370	546.833	540.75	.684	.680	544.12	531.987	361.75	1.0228	1.0246	.9982
17	313	557	551.9165	.567	.564	554.97	541.476	305.39	1.0249	1.0237	1.0012
18	556	564.833	560.9165	.991	.986	563.89	550.965	543.25	1.0235	1.0197	1.0037
19	831	569.333	567.083	1.465	1.467	566.46	560.454	822.19	1.0107	1.0097	1.0010
20	960	576.167	572.75	1.676	1.693	567.04	569.943	964.91	.9949	1.0016	.9933
21	1,152	580.667	578.417	1.992	1.990	578.89	579.432	1,153.07	.9991	.9934	1.0057
22	759	586.75	583.7085	1.300	1.307	580.72	588.921	769.72	.9861	.9903	.9958
23	607	591.833	589.2915	1.030	1.029	589.89	598.410	615.76	.9858	.9964	.9894
24	371	600.5	596.1665	.622	.600	618.33	607.899	364.74	1.0172	.9940	1.0233
25 (Jan)	298	614.917	607.7085	.490	.493	604.46	617.388	304.37	.9791	1.0027	.9765
26	378	631	622.9585	.607	.596	634.23	626.877	373.62	1.0117	.9920	1.0199
27	373	650.667	640.8335	.582	.595	626.89	636.366	378.64	.9851	1.0018	.9833
28	443	662.75	656.7085	.675	.680	651.47	645.855	439.18	1.0087	1.0030	1.0057
29	374	671.75	667.25	.561	.564	663.12	655.344	369.61	1.0119	1.0091	1.0028
30	660	677.583	674.6665	.978	.986	669.37	664.833	655.53	1.0068	1.0112	.9956
31	1,004				1.467	684.39	674.322	989.23	1.0149	1.0059	1.0089
32	1,153				1.693	681.04	683.811	1,157.69	.9959	1.0053	.9906
33	1,388				1.990	697.49	693.300	1,379.67	1.0060	.9954	1.0106
34	904				1.307	691.66	702.789	918.55	.9842	.9886	.9955
35	715				1.029	694.85	712.278	732.93	.9755	.9927	.9827
36	441				.600	735	721.707	433.06	1.0183		

Cola time series, we calculate **centered moving averages.** The centered moving averages are two-period moving averages of the previously computed 12-period moving averages. Thus the first centered moving average is

$$\frac{447.833 + 452.417}{2} = 450.125$$

The second centered moving average is

$$\frac{452.417 + 458}{2} = 455.2085$$

Successive centered moving averages are calculated similarly. The 12-period moving averages and centered moving averages for the Tasty Cola sales time series are given in Table 17.10.

If the original moving averages had been computed using an odd number of time series values, the centering procedure would not have been necessary. For example, if we had three seasons per year, we would compute three-period moving averages. Then, the first moving average would correspond to period 2, the second moving average would correspond to period 3, and so on. However, most seasonal time series are quarterly, monthly, or weekly, so the centering procedure is necessary.

TABLE 17.11   Estimation of the Seasonal Factors

		$sn_t \times ir_t = y_t/(tr_t \times cl_t)$			$sn_t =$
		Year 1	Year 2	$\overline{sn}_t$	$1.0008758(\overline{sn}_t)$
1	Jan.	.495	.490	.4925	.493
2	Feb.	.583	.607	.595	.596
3	Mar.	.608	.582	.595	.595
4	Apr.	.684	.675	.6795	.680
5	May	.567	.561	.564	.564
6	June	.991	.978	.9845	.986
7	July	1.466	1.465	1.4655	1.467
8	Aug.	1.707	1.676	1.6915	1.693
9	Sep.	1.985	1.992	1.9885	1.990
10	Oct.	1.312	1.300	1.306	1.307
11	Nov.	1.026	1.030	1.028	1.029
12	Dec.	.577	.622	.5995	.600

FIGURE 17.11   Plot of Tasty Cola Sales and Deseasonalized Sales

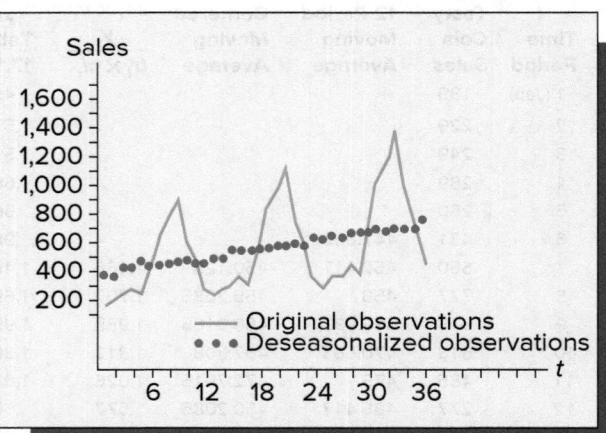

The centered moving average in time period t is considered to equal $tr_t \times cl_t$, the estimate of $TR_t \times CL_t$, because the averaging procedure is assumed to have removed seasonal variations (note that each moving average is computed using exactly one observation from each season) and (short-term) irregular fluctuations. The (longer-term) trend effects and cyclical effects—that is, $tr_t \times cl_t$—remain.

Because the model

$$y_t = TR_t \times SN_t \times CL_t \times IR_t$$

implies that

$$SN_t \times IR_t = \frac{y_t}{TR_t \times CL_t}$$

it follows that the estimate $sn_t \times ir_t$ of $SN_t \times IR_t$ is

$$sn_t \times ir_t = \frac{y_t}{tr_t \times cl_t}$$

Noting that the values of $sn_t \times ir_t$ are calculated in Table 17.10, we can find sn_t by grouping the values of $sn_t \times ir_t$ by months and calculating an average, $\overline{sn}_t$, for each month. These monthly averages are given for the Tasty Cola data in Table 17.11. The monthly averages are then normalized so that they sum to the number of time periods in a year. Denoting the number of time periods in a year by L (for instance, $L = 4$ for quarterly data, $L = 12$ for monthly data), we accomplish the normalization by multiplying each value of $\overline{sn}_t$ by the quantity

$$\frac{L}{\sum \overline{sn}_t} = \frac{12}{.4925 + .595 + \cdots + .5995}$$

$$= \frac{12}{11.9895} = 1.0008758$$

This normalization process results in the estimate $sn_t = 1.0008758(\overline{sn}_t)$, which is the estimate of SN_t. These calculations are summarized in Table 17.11.

Having calculated the values of sn_t and placed them in Table 17.10, we next define the **deseasonalized observation** in time period t to be

$$d_t = \frac{y_t}{sn_t}$$

Deseasonalized observations are computed to better estimate the trend component TR_t. Dividing y_t by the estimated seasonal factor removes the seasonality from the data and allows us to better understand the nature of the trend. The deseasonalized observations are calculated in Table 17.10 and are plotted in Figure 17.11. Because the deseasonalized observations have a straight-line appearance, it seems reasonable to assume a linear trend

$$TR_t = \beta_0 + \beta_1 t$$

We estimate TR_t by fitting a straight line to the deseasonalized observations. That is, we compute the least squares point estimates of the parameters in the simple linear regression model relating the dependent variable d_t to the independent variable t:

$$d_t = \beta_0 + \beta_1 t + \varepsilon_t$$

We obtain $b_0 = 380.163$ and $b_1 = 9.489$. It follows that the estimate of TR_t is

$$tr_t = b_0 + b_1 t = 380.163 + 9.489t$$

The values of tr_t are calculated in Table 17.10. Note that, for example, although $y_{22} = 759$, Tasty Cola sales in period 22 (October of year 2), are larger than $tr_{22} = 588.921$ (the estimated trend in period 22), $d_{22} = 580.72$ is smaller than $tr_{22} = 588.921$. This implies that, on a deseasonalized basis, Tasty Cola sales were slightly down in October of year 2. This might have been caused by a slightly colder October than usual.

Thus far, we have found estimates sn_t and tr_t of SN_t and TR_t. Because the model

$$y_t = TR_t \times SN_t \times CL_t \times IR_t$$

implies that

$$CL_t \times IR_t = \frac{y_t}{TR_t \times SN_t}$$

it follows that the estimate of $CL_t \times IR_t$ is

$$cl_t \times ir_t = \frac{y_t}{tr_t \times sn_t}$$

Moreover, experience has shown that, when considering either monthly or quarterly data, we can average out ir_t and thus calculate the estimate cl_t of CL_t by computing a three-period moving average of the $cl_t \times ir_t$ values.

Finally, we calculate the estimate ir_t of IR_t by using the equation

$$ir_t = \frac{cl_t \times ir_t}{cl_t}$$

The calculations of the values cl_t and ir_t for the Tasty Cola data are summarized in Table 17.10. Because there are only three years of data, and because most of the values of cl_t are near 1, we cannot discern a well-defined cycle. Furthermore, examining the values of ir_t, we cannot detect a pattern in the estimates of the irregular factors.

Traditionally, the estimates tr_t, sn_t, cl_t, and ir_t obtained by using the multiplicative decomposition method are used to describe the time series. However, we can also use these estimates to forecast future values of the time series. If there is no pattern in the irregular component, we predict IR_t to equal 1. Therefore, the point forecast of y_t is

$$\hat{y}_t = tr_t \times sn_t \times cl_t$$

if a well-defined cycle exists and can be predicted. The point forecast is

$$\hat{y}_t = tr_t \times sn_t$$

if a well-defined cycle does not exist or if CL_t cannot be predicted, as in the Tasty Cola example. Because values of $tr_t \times sn_t$ have been calculated in column 9 of Table 17.10, these values are the point forecasts of the $n = 36$ historical Tasty Cola sales values. Furthermore, we present in Table 17.12 on the next page point forecasts of future Tasty Cola sales in the 12 months of year 4. Recalling that the estimated trend equation is $tr_t = 380.163 + 9.489t$ and that the estimated seasonal factor for August is 1.693 (see Table 17.11), it follows, for example, that the point forecast of Tasty Cola sales in period 44 (August of year 4) is

$$\hat{y}_{44} = tr_{44} \times sn_{44}$$
$$= (380.163 + 9.489(44))(1.693)$$
$$= 797.699(1.693)$$
$$= 1,350.50$$

Although there is no theoretically correct prediction interval for y_t, a fairly accurate **approximate $100(1 - \alpha)$ percent prediction interval for y_t** is obtained by computing an

TABLE 17.12   Forecasts of Future Values of Tasty Cola Sales Calculated Using the Multiplicative Decomposition Method

t	sn_t	$tr_t = 380.163 + 9.489t$	Point Prediction, $\hat{y}_t = tr_t \times sn_t$	Approximate 95% Prediction Interval	y_t
37	.493	731.273	360.52	[333.72, 387.32]	352
38	.596	740.762	441.48	[414.56, 468.40]	445
39	.595	750.252	446.40	[419.36, 473.44]	453
40	.680	759.741	516.62	[489.45, 543.79]	541
41	.564	769.231	433.85	[406.55, 461.15]	457
42	.986	778.720	767.82	[740.38, 795.26]	762
43	1.467	788.209	1,156.30	[1,128.71, 1,183.89]	1,194
44	1.693	797.699	1,350.50	[1,322.76, 1,378.24]	1,361
45	1.990	807.188	1,606.30	[1,578.41, 1,634.19]	1,615
46	1.307	816.678	1,067.40	[1,039.35, 1,095.45]	1,059
47	1.029	826.167	850.12	[821.90, 878.34]	824
48	.600	835.657	501.39	[473, 529.78]	495

interval that is centered at $\hat{y}_t$ and that has a length equal to the length of the $100(1 - \alpha)$ percent prediction interval for the **deseasonalized observation** d_t. Here the interval for d_t is obtained by using the model

$$d_t = TR_t + \varepsilon_t$$
$$= \beta_0 + \beta_1 t + \varepsilon_t$$

For instance, using Minitab to predict d_t on the basis of this model, we find that a 95 percent prediction interval for d_{44} is [769.959, 825.439]. Because this interval has a **half-length** equal to $(825.439 - 769.959)/2 = 55.48/2 = 27.74$, it follows that an approximate 95 percent prediction interval for y_{44} is

$$[\hat{y}_{44} \pm 27.74] = [1,350.50 \pm 27.74]$$
$$= [1,322.76, 1,378.24]$$

In Table 17.12 we give the approximate 95 percent prediction intervals (calculated by the above method) for Tasty Cola sales in the 12 months of year 4.

Next, suppose we actually observe Tasty Cola sales in year 4, and these sales are as given in the right-most column of Table 17.12. In Figure 17.12 we plot the observed and forecasted

FIGURE 17.12   A Plot of the Observed and Forecasted Tasty Cola Sales Values

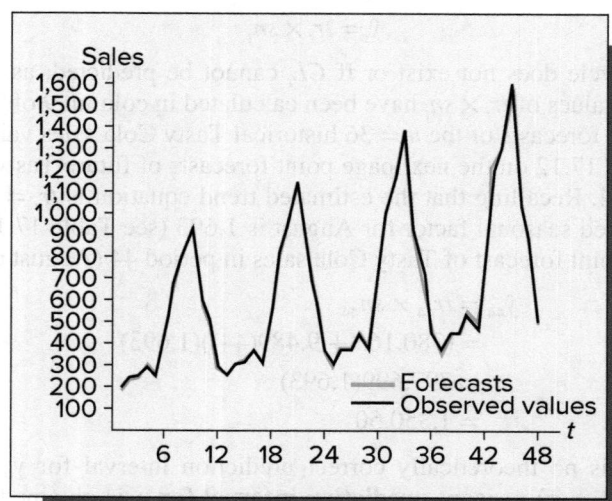

sales for all 48 sales periods. In practice, the comparison of the observed and forecasted sales in years 1 through 3 would be used by the analyst to determine whether the forecasting equation adequately fits the historical data. An adequate fit (as indicated by Figure 17.12, for example) might prompt an analyst to use this equation to calculate forecasts for future time periods. One reason that the Tasty Cola forecasting equation

$$\hat{y}_t = tr_t \times sn_t$$
$$= (380.163 + 9.489t)sn_t$$

provides reasonable forecasts is that this equation *multiplies* tr_t by sn_t. Therefore, as the average level of the time series (determined by the trend) increases, the seasonal swing of the time series increases, which is consistent with the data plots in Figures 17.10 and 17.12. For example, note from Table 17.11 that the estimated seasonal factor for August is 1.693. The forecasting equation yields a prediction of Tasty Cola sales in August of year 1 equal to

$$\hat{y}_8 = [380.163 + 9.489(8)]1.693$$
$$= (456.075)(1.693)$$
$$= 772.13$$

This implies a seasonal swing of $772.13 - 456.075 = 316.055$ (hundreds of cases) above 456.075, the estimated trend level. The forecasting equation yields a prediction of Tasty Cola sales in August of year 2 equal to

$$\hat{y}_{20} = [380.163 + 9.489(20)]1.693$$
$$= (569.943)(1.693)$$
$$= 964.91$$

which implies an increased seasonal swing of $964.91 - 569.943 = 394.967$ (hundreds of cases) above 569.943, the estimated trend level. In general, then, the forecasting equation is appropriate for forecasting a time series with a seasonal swing that is proportional to the average level of the time series as determined by the trend—that is, a time series exhibiting increasing seasonal variation.

We next note that the U.S. Bureau of the Census has developed the **Census II method,** which is a sophisticated version of the multiplicative decomposition method discussed in this section. The initial version of Census II was primarily developed by Julius Shiskin in the late 1950s when a computer program was written to perform the rather complex calculations. Several modifications have been made to the first version of the method over the years. Census II continues to be widely used by a variety of businesses and government agencies.

Census II first adjusts the original data for "trading day variations." That is, the data are adjusted to account for the fact that, for example, different months or quarters will consist of different numbers of business days or "trading days." The method then uses an iterative procedure to obtain estimates of the seasonal component (SN_t), the trading day component, the so-called trend-cycle component ($TR_t \times CL_t$), and the irregular component (IR_t). The iterative procedure makes extensive use of moving averages and a method for identifying and replacing extreme values in order to eliminate randomness. For a good explanation of the details involved here and in the Census II method as a whole, see Makridakis, Wheelwright, and McGee (1983). After carrying out a number of tests to check the correctness of the estimates, the method estimates the trend-cycle, seasonal, and irregular components.

Minitab carries out a modified version of the multiplicative decomposition method discussed in this section. We believe that Minitab's modified version (at the time of the writing of this book) makes some conceptual errors that can result in biased estimates of the time series components. Therefore, we will not present Minitab output of multiplicative decomposition. The Excel add-in MegaStat estimates the seasonal factors and the trend line exactly as described in this section. MegaStat does not estimate the cyclical and irregular components.

However, because it is often reasonable to make forecasts by using estimates of the seasonal factors and trend line, MegaStat can be used to do this.

Exercises for Section 17.3

CONCEPTS ■■ connect

17.9 Explain how the multiplicative decomposition model estimates seasonal factors.

17.10 Explain how the multiplicative decomposition method estimates the trend effect.

17.11 Discuss how the multiplicative decomposition method makes point forecasts of future time series values.

METHODS AND APPLICATIONS

Exercises 17.12 through 17.16 are based on the following situation: International Machinery, Inc., produces a tractor and wishes to use **quarterly** tractor sales data observed in the last four years to predict quarterly tractor sales next year. The following MegaStat output gives the tractor sales data and the estimates of the seasonal factors and trend line for the data: ⒹⓈ IntMach

t	Year	Quarter	Sales, y	Centered Moving Average	Ratio to CMA	Seasonal Indexes	Sales, y Deseasonalized
1	1	1	293			1.191	245.9
2	1	2	392			1.521	257.7
3	1	3	221	275.125	0.803	0.804	275.0
4	1	4	147	302.000	0.487	0.484	303.9
5	2	1	388	325.250	1.193	1.191	325.7
6	2	2	512	338.125	1.514	1.521	336.6
7	2	3	287	354.125	0.810	0.804	357.1
8	2	4	184	381.500	0.482	0.484	380.4
9	3	1	479	405.000	1.183	1.191	402.0
10	3	2	640	417.375	1.533	1.521	420.7
11	3	3	347	435.000	0.798	0.804	431.8
12	3	4	223	462.125	0.483	0.484	461.0
13	4	1	581	484.375	1.199	1.191	487.7
14	4	2	755	497.625	1.517	1.521	496.3
15	4	3	410			0.804	510.2
16	4	4	266			0.484	549.9

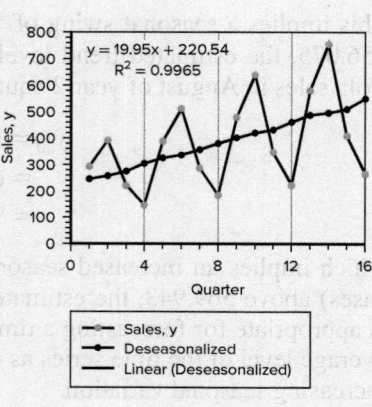

$y = 19.95x + 220.54$
$R^2 = 0.9965$

Sales, y — Deseasonalized — Linear (Deseasonalized)

Calculation of Seasonal Indexes

	1	2	3	4	
1			0.803	0.487	
2	1.193	1.514	0.810	0.482	
3	1.183	1.533	0.798	0.483	
4	1.199	1.517			
mean:	1.192	1.522	0.804	0.484	4.001
adjusted:	1.191	1.521	0.804	0.484	4.000

17.12 Find and identify the four seasonal factors for quarters 1, 2, 3, and 4.

17.13 What type of trend is indicated by the plot of the deseasonalized data?

17.14 What is the equation of the estimated trend that has been calculated using the deseasonalized data?

17.15 Compute a point forecast of tractor sales (based on trend and seasonal factors) for each of the quarters next year.

17.16 Compute an approximate 95 percent prediction interval forecast of tractor sales for each of the quarters next year. Use the fact that the half-lengths of 95 percent prediction intervals for the deseasonalized sales values in the four quarters of next year are, respectively, 14, 14.4, 14.6, and 15.

17.17 If we use the multiplicative decomposition method to analyze the quarterly bicycle sales data given in Table 17.3, we find that the quarterly seasonal factors are .46, 1.22, 1.68, and .64. Furthermore, if we use a statistical software package to fit a straight line to the deseasonalized sales values, we find that the estimate of the trend is $tr_t = 22.61 + .59t$.

In addition, we find that the half-lengths of 95 percent prediction intervals for the deseasonalized sales values in the four quarters of the next year are, respectively, 2.80, 2.85, 2.92, and 2.98. ⒹⓈ BikeSales

a Calculate point predictions of bicycle sales in the four quarters of the next year.

b Calculate approximate 95 percent prediction intervals for bicycle sales in the four quarters of the next year.

17.4 Simple Exponential Smoothing

LO17-5

Use simple exponential smoothing to forecast a time series.

In ongoing forecasting systems, forecasts of future time series values are made each period for succeeding periods. At the end of each period the estimates of the time series parameters and the forecasting equation need to be updated to account for the most recent observation. This updating accounts for possible changes in the parameters that may occur over time. In addition, such changes may imply that unequal weights should be applied to the time series observations when the estimates of the parameters are updated.

In this section we assume that a time series is appropriately described by the no trend equation

$$y_t = \beta_0 + \varepsilon_t$$

When the parameter β_0 remains constant over time, we have seen that it is reasonable to forecast future values of y_t by using regression analysis (see Example 17.1). In such a case, the least squares point estimate of β_0 is

$$b_0 = \bar{y} = \text{the average of the observed time series values}$$

When we compute the point estimate b_0 we are **equally weighting** each of the previously observed time series values $y_1, y_2, \ldots, y_n$.

When the value of the parameter β_0 may be changing over time, the equal weighting scheme might not be appropriate. Instead, it may be desirable to weight recent observations more heavily than remote observations. **Simple exponential smoothing** is a forecasting method that applies unequal weights to the time series observations. This unequal weighting is accomplished by using a **smoothing constant** that determines how much weight is attached to each observation. The most recent observation is given the most weight. More distantly past observations are given successively smaller weights. The procedure allows the forecaster to update the estimate of β_0 so that changes in the value of this parameter can be detected and incorporated into the forecasting equation. We illustrate simple exponential smoothing in the following example.

Ⓒ EXAMPLE 17.6 The Cod Catch Case: Simple Exponential Smoothing

Consider the cod catch data of Example 17.1, which are given in Table 17.1. The plot of these data (in Figure 17.2) suggests that the no trend model

$$y_t = \beta_0 + \varepsilon_t$$

may appropriately describe the cod catch series. It is also possible that the parameter β_0 could be slowly changing over time.

We begin the simple exponential smoothing procedure by calculating an initial estimate of the average level β_0 of the series. This estimate is denoted S_0 and is computed by averaging the first six time series values. We obtain

$$S_0 = \frac{\sum\limits_{t=1}^{6} y_t}{6} = \frac{362 + 381 + \cdots + 402}{6} = 359.667$$

Note that, because simple exponential smoothing attempts to track changes over time in the average level β_0 by using newly observed values to update the estimates of β_0, we use only six of the $n = 24$ time series observations to calculate the initial estimate of β_0. If we do this, then 18 observations remain to tell us how β_0 may be changing over time. Experience has shown that, in general, it is reasonable to calculate initial estimates in exponential smoothing procedures by using the first half of the historical data. However, it can be shown that, in simple exponential smoothing, using six observations is reasonable (it would not, however, be reasonable to use a very small number of observations because doing so might make the initial estimate so different from the true value of β_0 that the exponential smoothing procedure would be adversely affected).

Next, assume that at the end of time period $T-1$ we have an estimate S_{T-1} of β_0. Then, assuming that in time period T we obtain a new observation y_T, we can update S_{T-1} to S_T, which is an estimate made in period T of β_0. We compute the updated estimate by using the so-called **smoothing equation**

$$S_T = \alpha y_T + (1 - \alpha)S_{T-1}$$

Here α is a **smoothing constant** between 0 and 1. (α will be discussed in more detail later.) The updating equation says that S_T, the estimate made in time period T of β_0, equals a fraction α (for example, .1) of the newly observed time series observation y_T plus a fraction $(1 - \alpha)$ (for example, .9) of S_{T-1}, the estimate made in time period $T-1$ of β_0. The more the average level of the process is changing, the more a newly observed time series value should influence our estimate, and thus the larger the smoothing constant α should be set. We will soon see how to use historical data to determine an appropriate value of α.

We will now begin with the initial estimate $S_0 = 359.667$ and update this initial estimate by applying the smoothing equation to the 24 observed cod catches. To do this, we arbitrarily set α equal to .02, and to judge the appropriateness of this choice of α we calculate "one-period-ahead" forecasts of the historical cod catches as we carry out the smoothing procedure. Because the initial estimate of β_0 is $S_0 = 359.667$, it follows that 359.667 is the forecast made at time 0 for y_1, the value of the time series in period 1. Because we see from Table 17.13 that $y_1 = 362$, we have a forecast error of $362 - 359.667 = 2.333$. Using $y_1 = 362$, we can update S_0 to S_1, an estimate made in period 1 of the average level of the time series, by using the equation

$$S_1 = \alpha y_1 + (1 - \alpha)S_0$$
$$= .02(362) + .98(359.667) = 359.713$$

TABLE 17.13 One-Period-Ahead Forecasting of the Historical Cod Catch Time Series Using Simple Exponential Smoothing with $\alpha = .02$

Year	Month	Actual Cod Catch, y_T	Smoothed Estimate, S_T	Forecast Made Last Period	Forecast Error
			($S_0 = 359.667$)		
1	Jan.	362	359.713	359.667	2.333
	Feb.	381	360.139	359.713	21.287
	Mar.	317	359.276	360.139	−43.139
	Apr.	297	358.031	359.276	−62.276
	May	399	358.850	358.031	40.969
	June	402	359.713	358.850	43.150
	July	375	360.019	359.713	15.287
	Aug.	349	359.799	360.019	−11.019
	Sept.	386	360.323	359.799	26.201
	Oct.	328	359.676	360.323	−32.323
	Nov.	389	360.263	359.676	29.324
	Dec.	343	359.917	360.263	−17.263
2	Jan.	276	358.239	359.917	−83.917
	Feb.	334	357.754	358.239	−24.239
	Mar.	394	358.479	357.754	36.246
	Apr.	334	357.990	358.479	−24.479
	May	384	358.510	357.990	26.010
	June	314	357.620	358.510	−44.510
	July	344	357.347	357.620	−13.620
	Aug.	337	356.940	357.347	−20.347
	Sept.	345	356.701	356.940	−11.940
	Oct.	362	356.807	356.701	5.299
	Nov.	314	355.951	356.807	−42.807
	Dec.	365	356.132	355.951	9.049

Because this implies that 359.713 is the forecast made in period 1 for y_2, and because we see from Table 17.13 that $y_2 = 381$, we have a forecast error of $381 - 359.713 = 21.287$. Using $y_2 = 381$, we can update S_1 to S_2, an estimate made in period 2 of β_0, by using the equation

$$S_2 = \alpha y_2 + (1 - \alpha)S_1$$
$$= .02(381) + .98(359.713) = 360.139$$

Because this implies that 360.139 is the forecast made in period 2 for y_3, and because we see from Table 17.13 that $y_3 = 317$, we have a forecast error of $317 - 360.139 = -43.139$. This procedure is continued through the entire 24 periods of historical data. The results are summarized in Table 17.13. Using the results in this table, we find that, for $\alpha = .02$, the mean of the squared forecast errors is 1161.14. To find a "good" value of α, we evaluate the mean of the squared forecast errors for values of α ranging from .02 to .30 in increments of .02. (In exponential smoothing applications, where the level of the time series is changing fairly slowly, the value of the smoothing constant is usually found to be between .01 and .30.) When we do this, we find that $\alpha = .02$ minimizes the mean of the squared forecast errors. Because this minimizing value of α is small, it appears to be best to apply small weights to new observations, which tells us that the level of the time series is not changing very much.

In general, simple exponential smoothing is carried out as follows:

Simple Exponential Smoothing

1 Suppose that the time series $y_1, \ldots, y_n$ is described by the equation $y_t = \beta_0 + \varepsilon_t$, where the average level β_0 of the process may be slowly changing over time. Then the estimate S_T of β_0 made in time period T is given by the **smoothing equation**

$$S_T = \alpha y_T + (1 - \alpha)S_{T-1}$$

where α is a smoothing constant between 0 and 1 and S_{T-1} is the estimate of β_0 made in time period $T - 1$.

2 A point forecast made in time period T for any future value of the time series is S_T.

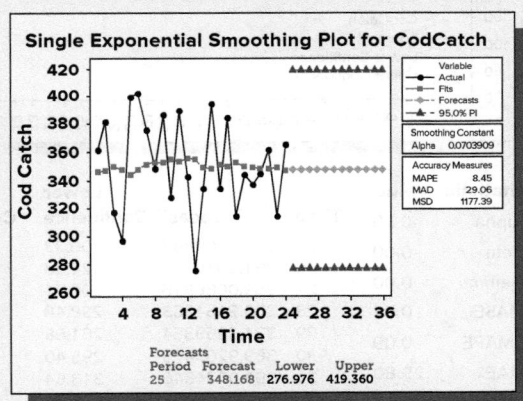

C EXAMPLE 17.7 The Cod Catch Case: Forecasting

In Example 17.6 we saw that $\alpha = .02$ is a "good" value of the smoothing constant when forecasting the 24 observed cod catches in Table 17.13. Therefore, we will use simple exponential smoothing with $\alpha = .02$ to forecast future monthly cod catches. From Table 17.13 we see that $S_{24} = 356.132$ is the estimate made in month 24 of the average level β_0 of the monthly cod catches. It follows that the point forecast made in month 24 of any future monthly cod catch is 356.132 tons of cod.

Computer software packages can be used to implement exponential smoothing. These packages choose the smoothing constant (or constants) in different ways and also compute approximate prediction intervals in different ways. For example, the right side of the summary box gives the Minitab output of using simple exponential smoothing to forecast in month 24 the cod catches in future months. Note that Minitab has selected the smoothing constant $\alpha = .0703909$ and tells us that the point forecast and the 95 percent prediction interval forecast of the cod catch in any future month are, respectively, 348.168 and [276.976, 419.360].

Like Minitab, Excel and JMP will carry out simple exponential smoothing, and all three packages will carry out two advanced exponential smoothing methods to be discussed in the next section. These methods are **Holt–Winters' double exponential smoothing,** which forecasts a time series described by possibly changing level and linear trend components, and **additive Winters' method,** which forecasts a time series described by possibly changing level, linear trend, and constant seasonal variation components. A third advanced method to be discussed in the next section—**multiplicative Winters' method**—is carried out by Minitab but not (at the time of the writing of this book) by Excel or JMP. This method forecasts a time series described by possibly changing level, linear trend, and increasing seasonal variation components. It follows that if we wish to use Excel or JMP to forecast such a time series, we would first use a natural logarithm or fractional power transformation to transform the increasing seasonal variation into constant seasonal variation, and then we would use additive Winters' method.

Whereas Minitab and JMP require the user to specify the exponential smoothing method that he or she wishes to use, Excel automatically selects a method from the three methods it carries out.

FIGURE 17.13   **Excel and Minitab Forecasting Outputs**

(a) Excel additive Winters' method for the cod catches

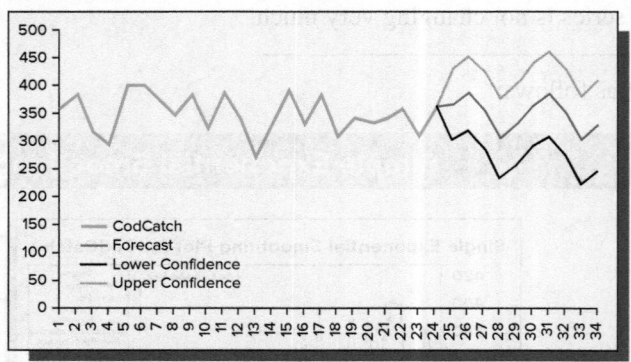

Statistic	Value
Alpha	0.25
Beta	0.00
Gamma	0.00
MASE	0.64
SMAPE	0.09
MAE	29.80
RMSE	33.15

Time	Forecast	Lower Confidence	Upper Confidence
25	371.1209887	306.33	435.91
26	391.2018254	324.38	458.02
27	359.0086915	290.21	427.81
28	309.2154383	238.48	379.96
29	334.3259354	261.68	406.97
30	369.9263478	295.40	444.45
31	390.0071846	313.64	466.37
32	357.8140506	279.64	435.99
33	308.0207974	228.07	387.98
34	333.1312946	251.42	414.84

(b) Excel Holt–Winters' double exponential smoothing for the cod catches

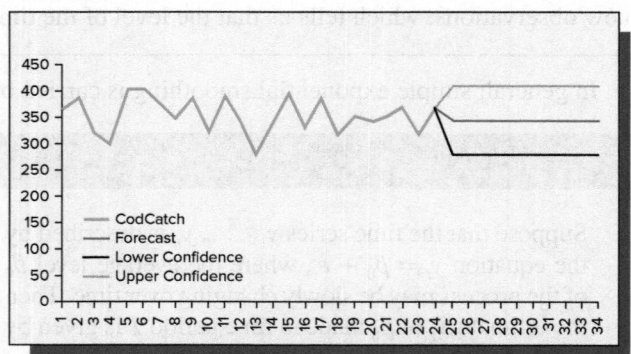

Statistic	Value
Alpha	0.00
Beta	0.00
Gamma	0.00
MASE	0.40
SMAPE	0.05
MAE	18.43
RMSE	22.12

Time	Forecast	Lower Confidence	Upper Confidence
25	341.1421138	277.15	405.13
26	340.2662033	276.28	404.25
27	339.3902929	275.40	403.38
28	338.5143825	274.53	402.50
29	337.6384721	273.65	401.63
30	336.7625616	272.77	400.75
31	335.8866512	271.89	399.88
32	335.0107408	271.01	399.01
33	334.1348304	270.13	398.13
34	333.25892	269.26	397.26

(c) Excel additive Winters' method for bike sales

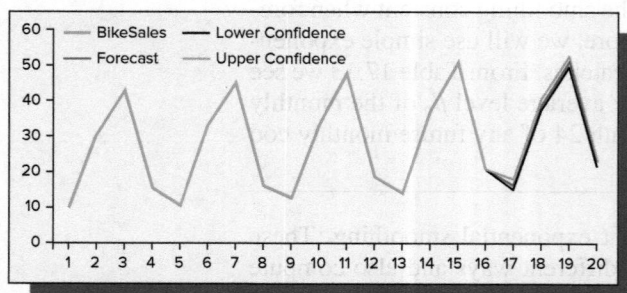

Statistic	Value
Alpha	0.25
Beta	0.00
Gamma	0.00
MASE	0.05
SMAPE	0.03
MAE	0.87
RMSE	1.09

Time	Forecast	Lower Confidence	Upper Confidence
17	17.07724383	15.45	18.71
18	39.12788968	37.45	40.81
19	51.18507665	49.45	52.92
20	23.73316931	21.95	25.51

(d) Minitab multiplicative Winters' method for the hotel room averages

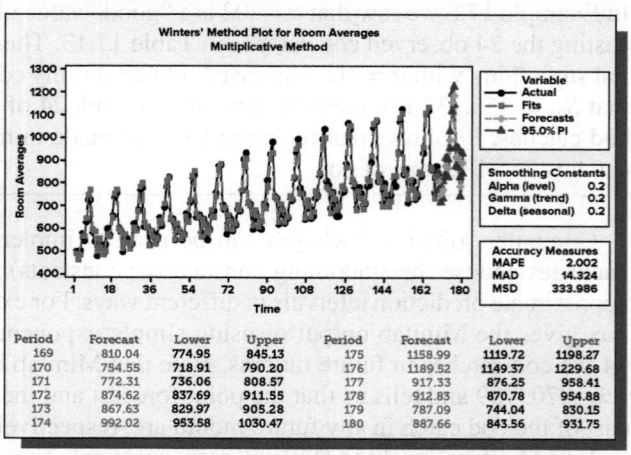

Period	Forecast	Lower	Upper	Period	Forecast	Lower	Upper
169	810.04	774.95	845.13	175	1158.99	1119.72	1198.27
170	754.55	718.91	790.20	176	1189.52	1149.37	1229.68
171	772.31	736.06	808.57	177	917.33	876.25	958.41
172	874.62	837.69	911.55	178	912.83	870.78	954.88
173	867.63	829.97	905.28	179	787.09	744.04	830.15
174	992.02	953.58	1030.47	180	887.66	843.56	931.75

For example, as a preview of advanced exponential smoothing methods, Figure 17.13(a) shows the forecasts obtained when we input the cod catch data into Excel. Because the forecasts (shown in red) have a pattern that repeats itself every five months (this pattern can be shown to continue with more than the 10 forecasts shown), Excel has identified a five-month seasonal pattern in the cod catches and has used additive Winters' method. Although Excel might be right, we do not see a seasonal pattern in the historical data, and so we used Excel's option that allows us to tell Excel not to identify any seasonality. When we do this, Excel produces the forecasts shown in Figure 17.13(b). These forecasts trend slightly downward by about .88 tons of cod per month. Excel has used Holt–Winters' double exponential smoothing to obtain these forecasts. As a preview of exponential smoothing forecasting of an obviously seasonal time series, Figure 17.13(c) shows the Excel output of using additive Winters' method to forecast the bike sales time series in Table 17.3, and Figure 17.13(d) shows the Minitab output of using multiplicative Winters' method to forecast the hotel room average time series in Table 17.4.

Exercises for Section 17.4

CONCEPTS ■ connect

17.18 In general, when is it appropriate to use exponential smoothing?

17.19 What is the purpose of a smoothing constant in exponential smoothing?

METHODS AND APPLICATIONS

17.20 THE COD CATCH CASE ⓓⓢ CodCatch

Consider Table 17.13 (page 776). Verify (calculate) that S_3, an estimate made in period 3 of β_0, is 359.276. Also verify (calculate) that the one-period-ahead forecast error for period 4 is −62.276, as shown in Table 17.13.

17.21 THE COD CATCH CASE ⓓⓢ CodCatch

Consider Example 17.7 (page 777). Suppose that we observe a cod catch in January of year 3 of $y_2 = 328$. Update $S_{24} = 356.132$ to S_{25}, a point forecast made in month 25 of any future monthly cod catch. Use $\alpha = .02$ as in Example 17.7.

17.22 Figure 17.14(a) gives the weekly revenues for the past 20 weeks at Prince's Eyeglasses Repair, which is a small business operated by Brenda Prince. Use the JMP output in Figure 17.14(b) to identify a point forecast and a 95 percent prediction interval forecast of weekly revenue in weeks 21 and 22. ⓓⓢ Prince's Repair

---

FIGURE 17.14　**Weekly Revenue at Prince's Eyeglasses Repair, and the JMP Output of Forecasting by Simple Exponential Smoothing** ⓓⓢ Prince's Repair

(a) Weekly revenue (in dollars)

Week	Revenue
1	1800
2	2200
3	1800
4	2300
5	2000
6	1900
7	2300
8	2500
9	1900
10	1600
11	1700
12	1500
13	1600
14	1400
15	1700
16	2300
17	1900
18	2400
19	2000
20	2100

(b) The JMP output of forecasting by simple exponential smoothing

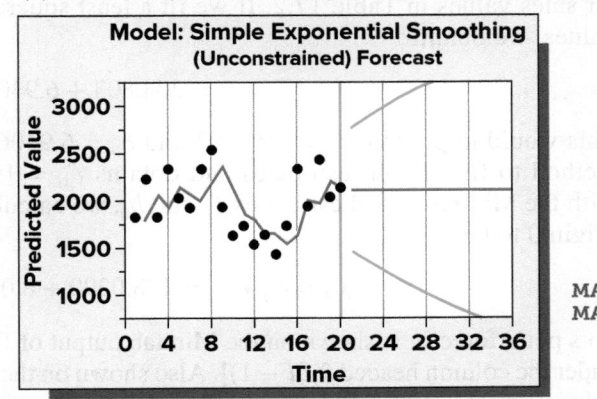

	MAPE	13.8357121
	MAE	273.3462

Parameter Estimates

Term	Estimate	Std Error	t Ratio	Prob>|t|
Level Smoothing Weight	0.52756702	0.2298670	2.30	0.0340*

	Predicted Sales	Upper CL (0.95) Sales	Lower CL (0.95) Sales
21	2092.8954216	2740.0395535	1445.7512896
22	2092.8954216	2824.5767815	1361.2140617
23	2092.8954216	2900.3108644	1285.4799787
24	2092.8954216	2969.5263487	1216.2644945

LO17-6
Use double exponential
smoothing to forecast
a time series.

17.5 Holt–Winters' Models

Holt–Winters' double exponential smoothing

Various extensions of simple exponential smoothing can be used to forecast time series that are described by models that are different from the model $y_t = \beta_0 + \varepsilon_t$. For example, **Holt–Winters' double exponential smoothing** can be used to forecast time series that are described by the linear trend model

$$y_t = \beta_0 + \beta_1 t + \varepsilon_t$$

Here we assume that β_0 and β_1 (and thus the linear trend) may be changing over time. To implement Holt–Winters' double exponential smoothing, we let ℓ_{T-1} denote the estimate of the level $\beta_0 + \beta_1(T - 1)$ of the time series in time period $T - 1$, and we let b_{T-1} denote the estimate of the slope β_1 of the time series in time period $T - 1$. Then, if we observe a new time series value y_T in time period T, the estimate of the level $\beta_0 + \beta_1 T$ of the time series in time period T uses the **smoothing constant α** and is

$$\ell_T = \alpha y_T + (1 - \alpha)[\ell_{T-1} + b_{T-1}]$$

This equation says that ℓ_T equals a fraction α of the newly observed time series value y_T plus a fraction $(1 - \alpha)$ of $[\ell_{T-1} + b_{T-1}]$, which is the estimate of the level of the time series in time period T, as calculated using the estimates ℓ_{T-1} and b_{T-1} computed in time period $T - 1$. Furthermore, the estimate of the slope β_1 of the time series in time period T uses the **smoothing constant γ** and is

$$b_T = \gamma[\ell_T - \ell_{T-1}] + (1 - \gamma)b_{T-1}$$

This equation says that b_T equals a fraction γ of $[\ell_T - \ell_{T-1}]$, which is an estimate of the difference between the levels of the time series in periods T and $T - 1$, plus a fraction $(1 - \gamma)$ of b_{T-1}, the estimate of the slope made in time period $T - 1$.

To use the updating equations, we first obtain initial estimates ℓ_0 and b_0 of the level and the slope of the time series in time period 0. One way to do this is to fit a least squares trend line to the first part (say, one-half) of the historical data and let the y-intercept and slope of the trend line be ℓ_0 and b_0. For example, consider the 24 observed calculator sales values in Table 17.2. If we fit a least squares trend line to the first 12 of those values, we obtain

$$\hat{y}_t = 204.803 + 6.9406t$$

This would imply that $\ell_0 = 204.803$ and $b_0 = 6.9406$. Minitab uses a more complicated method to find initial estimates and obtains $\ell_0 = 198.0290$ and $b_0 = 8.0743$. Starting with the Minitab initial estimates ℓ_0 and b_0, we calculate a point forecast of y_1 from time origin 0 to be

$$\hat{y}_1(0) = \ell_0 + b_0 = 198.0290 + 8.0743 = 206.103$$

This point forecast is shown on the Minitab output of Figure 17.15(a) [it is the first number under the column headed $\hat{y}_T(T - 1)$]. Also shown on the output are the actual calculator sales value $y_1 = 197$ and the forecast error, which is

$$y_1 - \hat{y}_1(0) = 197 - 206.103 = -9.103$$

We next choose values of the smoothing constants α and γ. A reasonable choice (and the default option of Minitab) is to let each of α and γ be .2. Then, using $y_1 = 197$ and the equation for ℓ_T, it follows that the estimate of the level of the time series in time period 1 is

$$\ell_1 = \alpha y_1 + (1 - \alpha)[\ell_0 + b_0]$$
$$= .2(197) + .8[198.0290 + 8.0743]$$
$$= 204.283$$

FIGURE 17.15   The Minitab Output of Double Exponential Smoothing for the Calculator Sales Data

(a) The updated level and slope estimates when $\alpha = .2$ and $\gamma = .2$

$\ell_0 = 198.0290 \quad b_0 = 8.0743$

Time	Sales	Level	Slope	Forecast	Error
T	y_T	ℓ_T	b_T	$\hat{y}_T(T-1)$	$y_T - \hat{y}_T(T-1)$
1	197	204.283	7.7102	206.103	−9.1033
2	211	211.794	7.6705	211.993	−0.9929
3	203	216.172	7.0119	219.465	−16.4648
4	247	227.947	7.9646	223.184	23.8162
5	239	236.529	8.0881	235.912	3.0884
6	269	249.494	9.0634	244.617	24.3827
7	308	268.446	11.0411	258.557	49.4427
8	262	275.990	10.3416	279.487	−17.4869
9	258	280.665	9.2084	286.331	−28.3312
10	256	283.099	7.8535	289.873	−33.8733
11	261	284.962	6.6554	290.952	−29.9521
12	288	290.894	6.5107	291.617	−3.6171
13	296	297.123	6.4545	297.404	−1.4043
14	276	298.062	5.3514	303.578	−27.5780
15	305	303.731	5.4148	303.414	1.5862
16	308	308.917	5.3690	309.146	−1.1459
17	356	322.629	7.0376	314.286	41.7143
18	393	342.333	9.5709	329.666	63.3339
19	363	354.123	10.0148	351.904	11.0962
20	386	368.510	10.8893	364.138	21.8621
21	443	392.120	13.4333	379.400	63.6004
22	308	386.042	9.5312	405.553	−97.5529
23	358	388.059	8.0282	395.574	−37.5735
24	384	393.670	7.5447	396.087	−12.0870

(b) Point and 95 percent prediction interval forecasts when $\alpha = .2$ and $\gamma = .2$

Period	Forecast	Lower	Upper
25	401.214	337.813	464.616
26	408.759	344.037	473.482
27	416.304	350.159	482.449
28	423.849	356.186	491.512
29	431.393	362.124	500.663
30	438.938	367.979	509.898

(c) Point and 95 percent prediction interval forecasts when $\alpha = .496$ and $\gamma = .142$

Period	Forecast	Lower	Upper
25	383.677	319.135	448.220
26	389.121	316.066	462.177
27	394.565	312.109	477.022
28	400.010	307.534	492.486
29	405.454	302.521	508.386
30	410.898	297.191	524.604

Furthermore, using the equation for b_T, the estimate of the slope of the time series in time period 1 is

$$b_1 = \gamma[\ell_1 - \ell_0] + (1 - \gamma)b_0$$
$$= .2[204.283 - 198.0290] + .8(8.0743)$$
$$= 7.7102$$

It follows that a point forecast made in time period 1 of y_2 is

$$\hat{y}_2(1) = \ell_1 + b_1 = 204.283 + 7.7102 = 211.993$$

Because the actual calculator sales value in period 2 is $y_2 = 211$, the forecast error is

$$y_2 - \hat{y}_2(1) = 211 - 211.933 = -.993$$

The Minitab output in Figure 17.15(a) shows the entire process of using the double exponential smoothing updating equations to find new period-by-period estimates of the level and slope of the time series. The output also shows the one-period-ahead forecasts and forecast errors, which are utilized to evaluate the effectiveness of the double exponential smoothing procedure. At the end of the updating process, Minitab uses $\ell_{24} = 393.670$ and $b_{24} = 7.5447$ to calculate point forecasts of future calculator sales values. For example, point forecasts of y_{25} and y_{26} made from time origin 24 are

$$\hat{y}_{25}(24) = \ell_{24} + b_{24} = 393.670 + 7.5447 = 401.214$$

FIGURE 17.16   The Minitab Graphical Forecasts of Calculator Sales When $\alpha = .2$ and $\gamma = .2$

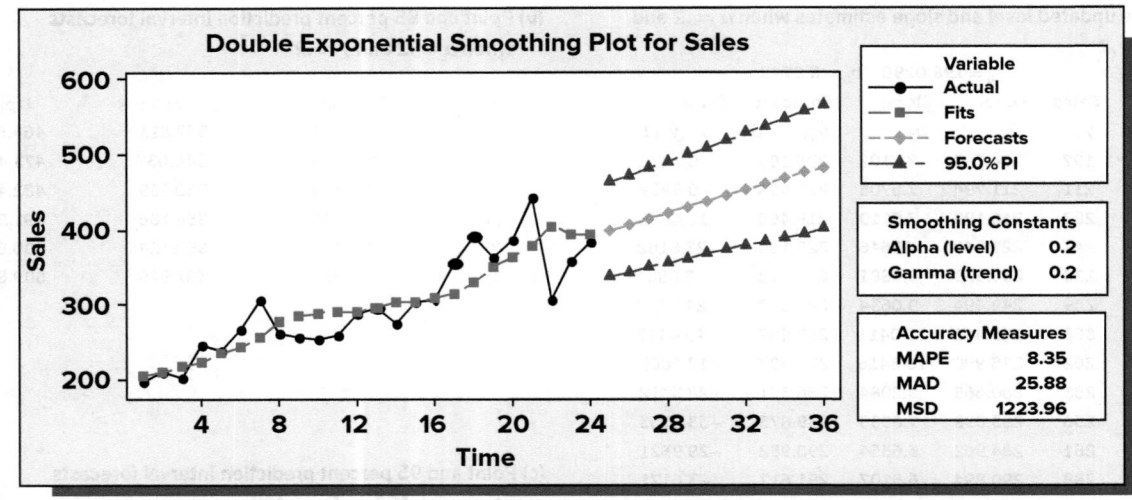

and

$$\hat{y}_{26}(24) = \ell_{24} + 2b_{24} = 393.670 + 2(7.5447) = 408.759$$

These point forecasts, as well as point forecasts of y_{27} through y_{30}, are shown on the Minitab output in Figure 17.15(b). Also shown are 95 percent prediction interval forecasts of y_{25} through y_{30}.

Figure 17.16 shows a Minitab output that graphically illustrates the forecasts when $\alpha = .2$ and $\gamma = .2$. Generally speaking, choosing $\alpha = .2$ and $\gamma = .2$ gives reasonable results, but Minitab will choose its own values of α and γ. If we have Minitab do this, it chooses $\alpha = .496$ and $\gamma = .142$. The forecasts given by this choice of α and γ are given in Figure 17.15(c) and are graphically illustrated in Figure 17.17. To evaluate the choice of a particular set of values for α and γ, Minitab gives the **mean of the absolute forecast**

FIGURE 17.17   The Minitab Graphical Forecasts When $\alpha = .496$ and $\gamma = .142$

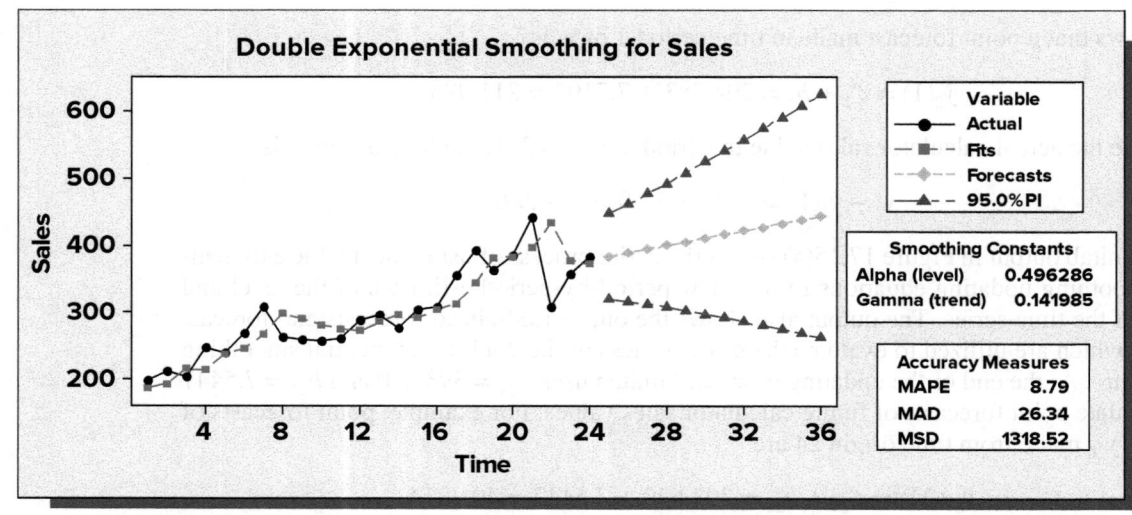

errors (the MAD) and the **mean of the squared forecast errors (the MSD)** for the 24 historical calculator sales values. Comparing Figures 17.16 and 17.17, we see that $\alpha = .2$ and $\gamma = .2$ give a smaller MAD and MSD than do $\alpha = .496$ and $\gamma = .142$. Therefore, we might conclude that we should use the forecasts of y_{25} through y_{30} based on $\alpha = .2$ and $\gamma = .2$. On the other hand, we might believe that the lower sales values at the end of the observed data signal that the sales values will not continue to increase as fast as they have been increasing. In this case, we might use the lower forecasts given by $\alpha = .496$ and $\gamma = .142$ (see Figure 17.17).

Multiplicative Winters' method (requires material from Section 17.3)

LO17-7
Use the multiplicative Winters' method to forecast a time series.

The **Multiplicative Winters' method** can be used to forecast time series that are described by the model

$$y_t = (\beta_0 + \beta_1 t) \times SN_t + \varepsilon_t$$

Here we assume that β_0 and β_1 (and thus the linear trend) and SN_t (which represents the seasonal pattern) may be changing over time. To implement the multiplicative Winters' method, we let ℓ_{T-1} denote the estimate of the deseasonalized level $\beta_0 + \beta_1(T-1)$ of the time series in time period $T-1$, and we let b_{T-1} denote the estimate of the slope β_1 of the time series in time period $T-1$. Then, suppose that we observe a new time series value y_T in time period T, and let sn_{T-L} denote the "most recent" estimate of the seasonal factor for the season corresponding to time period T. Here L denotes the number of seasons in a year ($L = 12$ for monthly data, and $L = 4$ for quarterly data), and thus $T - L$ denotes the time period occurring one year prior to time period T. Furthermore, the subscript $T - L$ of sn_{T-L} denotes the fact that the time series value observed in time period $T - L$ was the most recent time series value observed in the season being analyzed and thus was the most recent time series value used to help find sn_{T-L}. Then, the estimate of the deseasonalized level $\beta_0 + \beta_1 T$ of the time series in time period T uses the smoothing constant α and is

$$\ell_T = \alpha \frac{y_T}{sn_{T-L}} + (1 - \alpha)[\ell_{T-1} + b_{T-1}]$$

where y_T/sn_{T-L} is the deseasonalized observation in time period T. The estimate of the slope β_1 of the time series in time period T uses the smoothing constant γ and is

$$b_T = \gamma[\ell_T - \ell_{T-1}] + (1 - \gamma)b_{T-1}$$

The new estimate of the seasonal factor SN_T in time period T uses the smoothing constant δ and is

$$sn_T = \delta \frac{y_T}{\ell_T} + (1 - \delta)sn_{T-L}$$

where y_T/ℓ_T is an estimate of the newly observed seasonal variation.

To use the updating equations, we first obtain initial estimates ℓ_0, b_0, and sn_0 of the deseasonalized level, slope, and seasonal factors of the time series in time period 0. One way to do this is to use the multiplicative decomposition method (see Section 17.3) to analyze the first part (say, one-half) of the historical data. Here, if there are less than five years of historical data, it is probably best to base the initial estimates on all of the historical data. Then, we regard the y-intercept and slope of the trend line fit to the deseasonalized data as the initial estimates ℓ_0 and b_0. Furthermore, we regard the multiplicative decomposition method's seasonal factors as the initial estimates of the seasonal factors in time period 0. For example, consider the 36 Tasty Cola sales values in Table 17.9. Using the multiplicative decomposition method results summarized in Tables 17.10 and 17.11, we obtain the initial estimates $\ell_0 = 380.163$

Initial Seasonal Factor Estimates

Month	sn_0
Jan.	.493
Feb.	.596
Mar.	.595
Apr.	.680
May	.564
June	.986
July	1.467
Aug.	1.693
Sept.	1.990
Oct.	1.307
Nov.	1.029
Dec.	.600

and $b_0 = 9.489$ and the initial seasonal factor estimates given in the page margin. Starting with these initial estimates, we calculate a point forecast of y_1 from time origin 0 to be

$$\hat{y}_1(0) = (\ell_0 + b_0)sn_0$$
$$= (380.163 + 9.489)(.493)$$
$$= 192.098$$

Here we have used the initial January seasonal factor estimate $sn_0 = .493$ because y_1 is Tasty Cola sales in January of year 1. The actual value of y_1 is 189, so the forecast error is

$$y_1 - \hat{y}_1(0) = 189 - 192.098 = -3.098$$

We next choose values of the smoothing constants α, γ, and δ. A reasonable choice (and the default option of Minitab) is to let each of α, γ, and δ be .2. Then, using $y_1 = 189$ and the equation for ℓ_T, it follows that the estimate of the deseasonalized level of the time series in time period 1 is

$$\ell_1 = \alpha \frac{y_1}{sn_0} + (1 - \alpha)[\ell_0 + b_0]$$
$$= .2\left[\frac{189}{.493}\right] + .8[380.163 + 9.489]$$
$$= 388.395$$

Here we have used the initial January seasonal factor estimate $sn_0 = .493$ as the most recent Winters' method estimate of the January seasonal factor. Using the equation for b_T, the estimate of the slope of the time series in time period 1 is

$$b_1 = \gamma[\ell_1 - \ell_0] + (1 - \gamma)b_0$$
$$= .2[388.395 - 380.163] + .8(9.489)$$
$$= 9.238$$

Using the equation for sn_T, the new estimate of the January seasonal factor in time period 1 is

$$sn_1 = \delta \frac{y_1}{\ell_1} + (1 - \delta)sn_0$$
$$= .2\left[\frac{189}{388.395}\right] + .8(.493)$$
$$= .492$$

It follows that a point forecast made in period 1 of y_2 is

$$\hat{y}_2(1) = (\ell_1 + b_1)sn_0$$
$$= (388.395 + 9.238)(.596)$$
$$= 236.989$$

Here we have used the initial February seasonal factor estimate $sn_0 = .596$ because y_2 is the Tasty Cola sales in February of year 1. The actual value of y_2 is 229, so the forecast error is

$$y_2 - \hat{y}_2(1) = 229 - 236.989 = -7.989$$

The Minitab output in Figure 17.18(a) shows the entire process of using the Winters' method updating equations to find new period-by-period estimates of the level, slope, and seasonal factors of the time series. The output also shows the one-period-ahead forecasts and forecast errors, which are utilized to evaluate the effectiveness of the Winters' method procedure. Minitab does not find initial estimates by using the multiplicative decomposition method. We will not discuss how Minitab obtains initial estimates, but note from Figure 17.18(a) that the values of ℓ_1 and b_1 obtained by Minitab ($\ell_1 = 278.768$

FIGURE 17.18   The Minitab Output of Winters' Method for the Tasty Cola Sales Data, When $\alpha = .2, \gamma = .2,$ and $\delta = .2$

(a) The updated level, slope, and seasonal factor estimates

Time T	Sales y_T	Level ℓ_T	Slope b_T	Seasonal sn_T	Forecast $\hat{y}_T(T-1)$	Error $y_T - \hat{y}_T(T-1)$
1	189	278.768	44.9736	0.48896	106.67	82.334
2	229	343.270	48.8794	0.56818	175.93	53.065
3	249	401.836	50.8167	0.57606	221.63	27.371
4	289	449.774	50.2409	0.65605	298.49	-9.492
5	260	492.009	48.6398	0.55787	282.62	-22.624
6	431	520.567	44.6235	0.94880	529.30	-98.301
7	660	541.448	39.8750	1.42638	835.48	-175.485
8	777	556.089	34.8280	1.64516	992.39	-215.395
9	915	562.722	29.1891	1.95207	1201.68	-286.680
10	613	565.315	23.8699	1.28544	790.62	-177.623
11	485	563.116	18.6561	1.01787	622.78	-137.777
12	277	552.752	12.8521	0.60770	369.04	-92.044
13	244	552.287	10.1887	0.47953	276.56	-32.557
14	296	554.174	8.5282	0.56137	319.59	-23.586
15	319	560.914	8.1706	0.57459	324.15	-5.151
16	370	568.063	7.9664	0.65511	373.35	-3.349
17	313	573.035	7.3675	0.55554	321.35	-8.352
18	556	581.523	7.5916	0.95026	550.68	5.315
19	831	587.811	7.3308	1.42385	840.30	-9.301
20	960	592.820	6.8664	1.64000	979.10	-19.101
21	1152	597.777	6.4846	1.94709	1170.63	-18.631
22	759	601.501	5.9325	1.28072	776.74	-17.742
23	607	605.216	5.4890	1.01488	618.29	-11.287
24	371	610.664	5.4807	0.60767	371.13	-0.126
25	298	617.205	5.6927	0.48019	295.46	2.542
26	378	632.989	7.7111	0.56853	349.67	28.326
27	373	642.391	8.0493	0.57580	368.14	4.859
28	443	655.597	9.0806	0.65923	426.11	16.891
29	374	666.385	9.4222	0.55668	369.26	4.743
30	660	679.556	10.1717	0.95445	642.19	17.807
31	1004	692.808	10.7879	1.42891	982.07	21.934
32	1153	703.487	10.7660	1.63980	1153.90	-0.898
33	1388	713.974	10.7103	1.94648	1390.71	-2.712
34	904	720.918	9.9571	1.27537	928.12	-24.118
35	715	725.603	8.9026	1.00898	741.75	-26.753
36	441	732.750	8.5514	0.60650	446.34	-5.336

(b) Point and 95 percent prediction interval forecasts

Period	Forecast	Lower	Upper
37	355.96	240.98	470.95
38	426.31	309.52	543.10
39	436.69	317.90	555.49
40	505.60	384.60	626.60
41	431.71	308.32	555.10
42	748.35	622.39	874.31
43	1132.57	1003.88	1261.26
44	1313.74	1182.16	1445.32
45	1576.09	1441.48	1710.70
46	1043.59	905.81	1181.37
47	834.24	693.17	975.32
48	506.65	362.17	651.14

and $b_1 = 44.9736$) are very different from the values that we obtained by hand calculation ($\ell_1 = 388.395$ and $b_1 = 9.238$). In addition, the one-period-ahead forecast errors obtained by Minitab are generally quite large in periods 1 through 12 but then become reasonably small for periods 13 through 36. To further illustrate the Winters' method updating equations, note from Figure 17.18(a) that $\ell_{35} = 725.603$ and $b_{35} = 8.9026$. Because the most recent estimate of the December seasonal factor is $sn_{24} = .60767$, the point forecast made in period 35 of y_{36} (sales in December of year 3) is

$$\hat{y}_{36}(35) = (\ell_{35} + b_{35})sn_{24}$$
$$= (725.603 + 8.9026)(.60767)$$
$$= 446.34$$

The actual sales value in period 36 is $y_{36} = 441$, so the forecast error is

$$y_{36} - \hat{y}_{36}(35) = 441 - 446.34 = -5.34$$

The updated estimates ℓ_{36}, b_{36}, and sn_{36} are calculated as follows:

$$\ell_{36} = \alpha \frac{y_{36}}{sn_{24}} + (1 - \alpha)[\ell_{35} + b_{35}]$$

$$= .2\left[\frac{441}{.60767}\right] + .8[725.603 + 8.9026]$$

$$= 732.75$$

$$b_{36} = \gamma[\ell_{36} - \ell_{35}] + (1 - \gamma)b_{35}$$
$$= .2[732.75 - 725.603] + .8(8.9026)$$
$$= 8.5514$$

and

$$sn_{36} = \delta \frac{y_{36}}{\ell_{36}} + (1 - \delta)sn_{24}$$

$$= .2\left[\frac{441}{732.75}\right] + .8(.60767)$$

$$= .6065$$

We are now at the end of the historical data, so we can forecast future Tasty Cola sales values. Figure 17.18(b) gives the point and 95 percent prediction interval forecasts of future sales values in periods 37 through 48, and Figure 17.19 graphically portrays the forecasts. To see how the point forecasts are calculated, note that, for example, the most recent estimates of the January and July seasonal factors are $sn_{25} = .48019$ and $sn_{31} = 1.42891$. Therefore, point forecasts made in period 36 of Tasty Cola sales in periods 37 and 43 (January and July of year 4) are

$$\hat{y}_{37}(36) = (\ell_{36} + b_{36})sn_{25} = (732.75 + 8.5514)(.48019) = 355.96$$

and

$$\hat{y}_{43}(36) = (\ell_{36} + 7b_{36})sn_{31} = [732.75 + 7(8.5514)](1.42891) = 1{,}132.57$$

The reason that the 95 percent prediction intervals are so wide is that they can be shown to be functions of the historical forecast errors, which are very large in periods 1 through 12. The mean absolute forecast error in periods 13 through 36 can be calculated to be 12.98 and is more representative of the Winters' method's accuracy than is the mean absolute forecast error in all 36 periods, which is 46.93 (see Figure 17.19). Therefore, to obtain more reasonable prediction intervals, we might multiply the lengths of the prediction intervals by $12.98/46.93 \approx .28$. For example, Figure 17.18(b) tells us that the 95 percent prediction

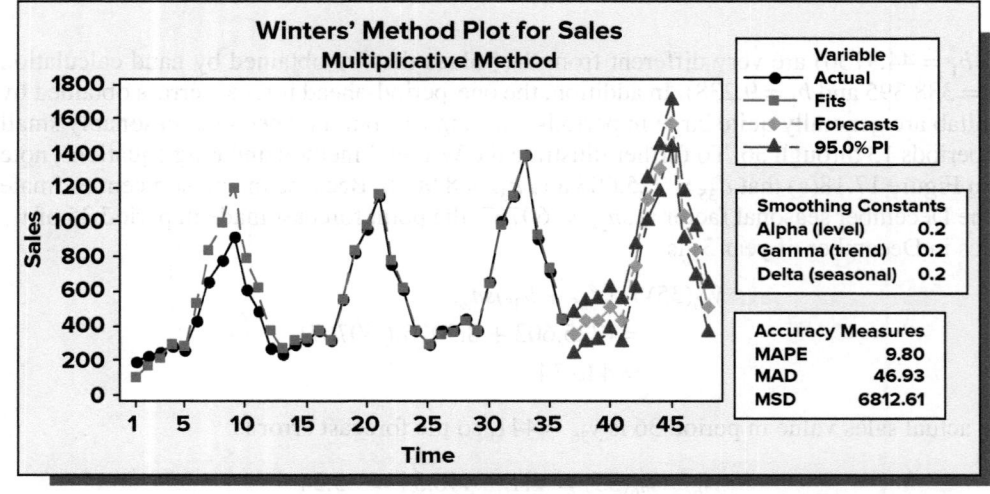

FIGURE 17.19   Minitab Output of Using Winters' Method to Forecast Tasty Cola Sales

FIGURE 17.20   Forecasting the Hotel Room Averages and Bike Sales

(a) Minitab multiplicative Winters' method for the hotel room averages

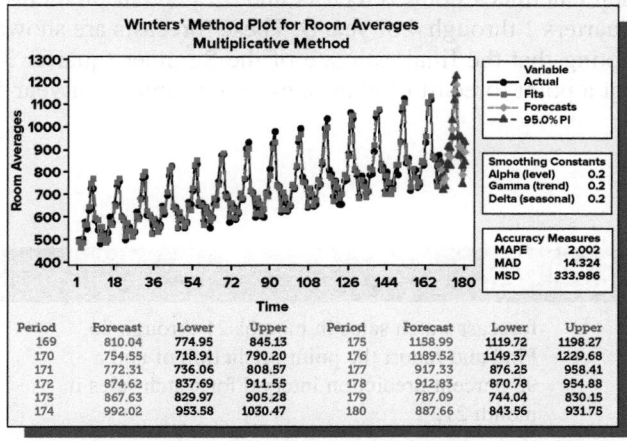

(b) Excel additive Winters' method for bike sales

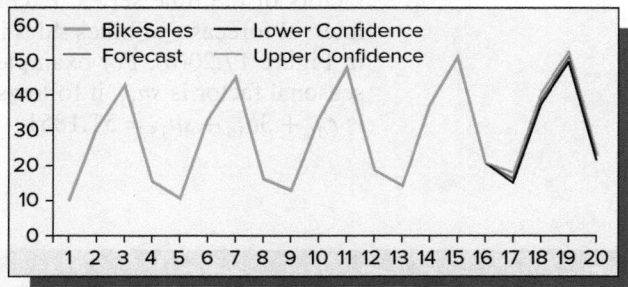

Statistic	Value	Time	Forecast	Lower Confidence	Upper Confidence
Alpha	0.25				
Beta	0.00	17	17.07724383	15.45	18.71
Gamma	0.00	18	39.12788968	37.45	40.81
MASE	0.05	19	51.18507665	49.45	52.92
SMAPE	0.03	20	23.73316931	21.95	25.51
MAE	0.87				
RMSE	1.09				

interval for y_{37} is [240.98, 470.95], which has length $470.95 - 240.98 = 229.97$. Multiplying this length by .28, we obtain $(229.97)(.28) = 64.39$. Surrounding the point forecast 355.96 by a new half-length of $64.39/2 = 32.2$, we obtain a new 95 percent prediction interval of $[355.96 \pm 32.2] = [323.76, 388.16]$. The other 95 percent prediction intervals can be modified similarly.

The wide prediction intervals in Figure 17.18(b) result from a combination of a short historical series (36 sales values) and Minitab obtaining inaccurate initial estimates of the level, slope, and seasonal factors. When the historical series is long, Minitab usually obtains reasonable prediction intervals. For example, see Figure 17.20(a), which shows the Minitab output of using multiplicative Winters' method to forecast the monthly hotel room averages. Finally, note that Minitab will not choose its own values of α, γ, and δ. However, the user can simply experiment with different combinations of values of these smoothing constants until a combination is found that produces the "best" results.

Additive Winters' method

Additive Winters' method can be used to forecast time series that can be described by the model $y_t = \beta_0 + \beta_1 t + SN_t + \varepsilon_t$. Here we assume that β_0 and β_1 (and thus the linear trend) and SN_t (which represents the seasonal pattern) may be changing over time. The first step in implementing additive Winters' method is to find initial estimates of the model parameters. One way to do this is to calculate the least squares point estimates of the parameters of a dummy variable regression model describing the first part of the observed time series data. (If there are less than five years of data, we would use all of the data.) Then, starting with the initial estimates, we use updating equations to find new period-by-period estimates of the level, slope, and seasonal factors in the time series. The appropriate updating equations are modifications of the multiplicative Winters' method updating equations, where we deseasonalize and find seasonal factors by using *subtraction operations* rather than *division operations*. Specifically, the new estimate of the deseasonalized level $\beta_0 + \beta_1 T$ of the time series in time period T is $\ell_T = \alpha [y_T - sn_{T-L}] + (1 - \alpha) [\ell_{T-1} + b_{T-1}]$, where $y_T - sn_{T-L}$ is the deseasonalized observation in time period T. The new estimate of the slope β_T of the time series in time period T is $b_T = \gamma [\ell_T - \ell_{T-1}] + (1 - \gamma)b_{T-1}$. The new estimates of the seasonal factor SN_T in time period T is $sn_T = \delta[y_T - \ell_T] + (1 - \delta)sn_{T-L}$, where $y_T - \ell_T$ is an estimate of the newly observed seasonal variation.

For example, if we input the 16 bike sales values in Table 17.3 into Excel, Excel finds initial parameter estimates and uses the additive Winters' method updating equations to arrive at final estimates ℓ_{16}, b_{16}, sn_{13}, sn_{14}, sn_{15}, and sn_{16} of the level, slope, and quarterly seasonal factors of the time series. Excel then calculates point forecasts and 95 percent prediction interval forecasts of bikes sales in quarters 1 through 4 of year 5. These forecasts are shown in Figure 17.20(b). For example, noting that the final estimate of the Summer (quarter 3) seasonal factor is sn_{15}, it follows that a point forecast of bike sales in the summer of year 5 is $\ell_{16} + 3b_{16} + sn_{15} = 51.1851$.

Exercises for Section 17.5

CONCEPTS

17.23 When do we use double exponential smoothing?

17.24 When do we use the multiplicative Winters' method?

METHODS AND APPLICATIONS

17.25 Consider Figure 17.15 on page 781. Calculate ℓ_2 and b_2 from ℓ_1, b_1, and y_2. Also, calculate $\hat{y}_{27}(24)$ from ℓ_{24} and b_{24}.

17.26 Consider Figure 17.18 on page 785. Calculate ℓ_{35}, b_{35}, and sn_{35} from ℓ_{34}, b_{34}, y_{35}, and sn_{23}. Also, calculate $\hat{y}_{38}(36)$ from ℓ_{36}, b_{36}, and sn_{26}.

17.27 THE WATCH SALES CASE DS WatchSale

Figure 17.21 gives the JMP output of using double exponential smoothing in month 20 to

forecast watch sales in months 21 through 26. Find and report the point prediction of and a 95 percent prediction interval for watch sales in month 21.

17.28 THE TRAVELER'S REST CASE DS TravRest

Use the Minitab output in Figure 17.20(a) to find and report the point prediction of and a 95 percent prediction interval for the monthly hotel room average in month 176.

17.29 Figure 17.22 gives the Minitab output of using multiplicative Winters' method to forecast the monthly international passenger totals in Table 17.8. Using the Minitab output, find and report a point prediction of a 95 percent prediction interval for the monthly international passenger totals in month 133.

FIGURE 17.21   **JMP Holt–Winters' Double Exponential Smoothing for Watch Sales**

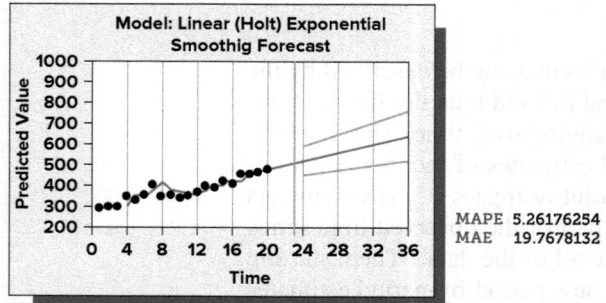

FIGURE 17.22   **Minitab Multiplicative Winters' Method for Monthly International Passenger Totals**

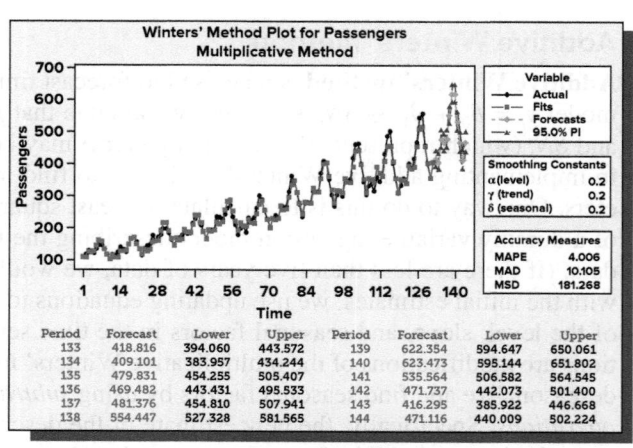

17.6 The Box–Jenkins Methodology

17-8
Use the Box–Jenkins
methodology to forecast
a time series.

In this section we discuss the **Box–Jenkins methodology.** This methodology, developed by G. E. P. Box and G. M. Jenkins (1976), uses an approach to describe the trend and seasonal effects in time series data that is quite different from the approach taken by regression (or exponential smoothing). The Box–Jenkins methodology begins by determining if the time series under consideration is *stationary*. Intuitively, a time series is **stationary** if the statistical properties (for example, the level and the variance) of the time series are essentially constant through time. If we have observed n values $y_1, y_2, \ldots, y_n$ of a time series, we can use a plot of these values (against time) to help us determine whether the time series is stationary. If the n values seem to fluctuate with constant variation around a constant level, then it is reasonable to believe that the time series is stationary. If the n values do not fluctuate around a constant level or do not fluctuate with constant variation, then it is reasonable to believe that the time series is nonstationary. In this case, the Box–Jenkins methodology tells us to *transform* the nonstationary time series values into stationary time series values.

For example, recall that Table 17.8 gives monthly international passenger totals for the past 11 years for an airline. The partial Minitab output in Figure 17.23, in the page margin, gives the first 15 and the last 2 of these monthly passenger totals (see Pass). The plot of these passenger totals is shown in Figure 17.23 and indicates that the passenger totals exhibit an upward linear trend with seasonal variation that increases with the level of the totals. In terms of Box–Jenkins modeling, the **increasing seasonal variation** implies that the time series is **nonstationary with respect to its variance** and we should use a variance-stabilizing transformation. Figure 17.23 also gives the natural logarithms (see LnPass) of the monthly passenger totals for the time periods shown in the figure and shows that the natural logarithms exhibit an upward linear trend with constant (variance-stabilized) seasonal variation. That is, letting y_t^* denote the variance-stabilized passenger total in month t, it follows that $y_t^* = \ln y_t$. The linear trend and seasonal variation in the y_t^* values indicate that, although these values are stationary with respect to their variance, they are **nonstationary with respect to their level.** To transform these values into stationary values with respect to their level, the Box–Jenkins methodology uses one of three types of *differencing: regular differencing, seasonal differencing,* or *combined regular and seasonal differencing.*

The **regular difference** of the passenger totals' natural logarithms in time period t, denoted z_t, is the difference between y_t^*, the natural logarithm in time period t, and y_{t-1}^*, the natural logarithm in time period $t-1$. That is, $z_t = y_t^* - y_{t-1}^*$. For example, Figure 17.23 tells us that the natural logarithms in months 1, 2, 3, 131, and 132 are $y_1^* = 4.71850$, $y_2^* = 4.77068$, $y_3^* = 4.88280$, $y_{131}^* = 5.89164$, and $y_{132}^* = 6.00389$. Therefore, the regular differences of the natural logarithms are:

$$z_2 = y_2^* - y_1^* = 4.77068 - 4.71850 = .05218$$
$$z_3 = y_3^* - y_2^* = 4.88280 - 4.77068 = .11212$$
$$\vdots$$
$$z_{132} = y_{132}^* - y_{131}^* = 6.00389 - 5.89164 = .11225$$

Figure 17.24 on the next page gives the regular differences (see Reg) of the natural logarithms for the time periods shown in the figure. Note that, because Minitab does calculations using more precise decimal place accuracy than it sometimes shows, the results in Figure 17.24 are slightly different from the results that we hand calculate. To calculate the **seasonal differences** of the natural logarithms in time period t, we use the equation $z_t = y_t^* - y_{t-L}^*$, where L is the number of seasons in a year. Because the passenger totals data are monthly, L equals 12, and thus we use the equation $z_t = y_t^* - y_{t-12}^*$. Figure 17.24 tells us that

FIGURE 17.23   The Passenger Totals   AirPass

t	Pass	LnPass
1	112	4.71850
2	118	4.77068
3	132	4.88280
4	129	4.85981
5	121	4.79579
6	135	4.90527
7	148	4.99721
8	148	4.99721
9	136	4.91265
10	119	4.77912
11	104	4.64439
12	118	4.77068
13	115	4.74493
14	126	4.83628
15	141	4.94876
131	362	5.89164
132	405	6.00389

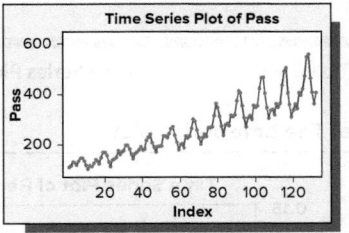

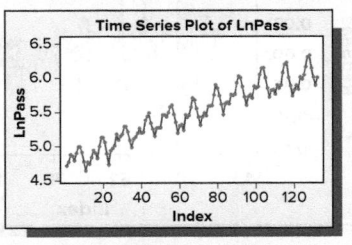

FIGURE 17.24   **The Passenger Totals' Natural Logarithms, Differences, and Two Difference Plots**

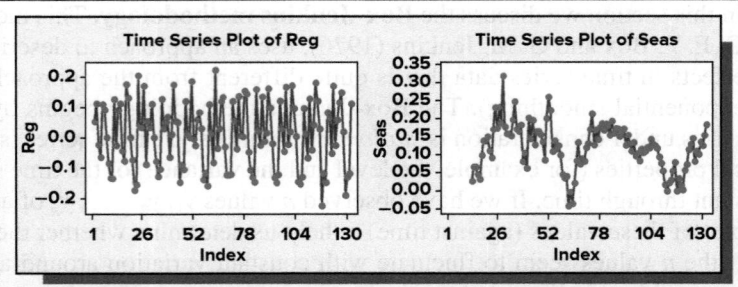

t	LnPass	Reg	Seas	RegSeas
1	4.71850	*	*	*
2	4.77068	0.052186	*	*
3	4.88280	0.112117	*	*
12	4.77068	0.126294	*	*
13	4.74493	−0.025752	0.026433	*
14	4.83628	0.091350	0.065597	0.039164
15	4.94876	0.112478	0.065958	0.000361
131	5.89164	−0.117169	0.155072	0.029581
132	6.00389	0.112243	0.183804	0.028732

$y_1^* = 4.71850$, $y_2^* = 4.77068$, $y_{13}^* = 4.74493$, and $y_{14}^* = 4.83628$. It follows that the first two seasonal differences are:

$$z_{13} = y_{13}^* - y_1^* = 4.74493 - 4.71850 \quad \text{and} \quad z_{14} = y_{14}^* - y_2^* = 4.83628 - 4.77068$$
$$= .02643 \hspace{6cm} = .0656$$

To carry out **combined regular and seasonal differencing,** we take the regular differences of the seasonal differences. For example, the regular difference of the just calculated two seasonal differences is $.0656 - .02643 = .03917$. Figure 17.24 gives the seasonal differences (see Seas) and the combined regular and seasonal differences (see RegSeas) of the natural logarithms for the time periods shown in the figure. Moreover, Figures 17.24 and 17.25(a) show time series plots of the regular differences, the seasonal differences, and the combined regular and seasonal differences of the natural logarithms. Examining the regular difference plot in Figure 17.24, it might seem at first glance that the regular differences are fluctuating with constant variation around a constant level. However, the periodic "dips" in the plot indicate that the regular differences exhibit a seasonal pattern and thus are nonstationary. The seasonal difference plot in Figure 17.24 shows that the seasonal differences do not exhibit any long-term upward or downward trend movements or any seasonal pattern. However, because these seasonal differences exhibit repeated, short-term upward and downward trend movements, we say that they exhibit **changing (or stochastic) trend effects.** These stochastic trend effects imply that the seasonal differences are nonstationary. Finally, Figure 17.25(a) shows that the combined regular and seasonal differences do not seem to exhibit either trend effects or a seasonal pattern. Because these combined differences seem to fluctuate with (approximately) constant variation around a constant level, we will regard them as being stationary.

Tentatively identifying a Box–Jenkins model

The combined regular and seasonal difference in time period t, z_t, is the regular difference of the seasonal difference $y_t^* - y_{t-12}^*$ and thus can be expressed as $z_t = y_t^* - y_{t-12}^* - (y_{t-1}^* - y_{t-13}^*)$.

FIGURE 17.25   **Time Series Plot and the SAC and SPAC of the Combined Regular and Seasonal Differences**

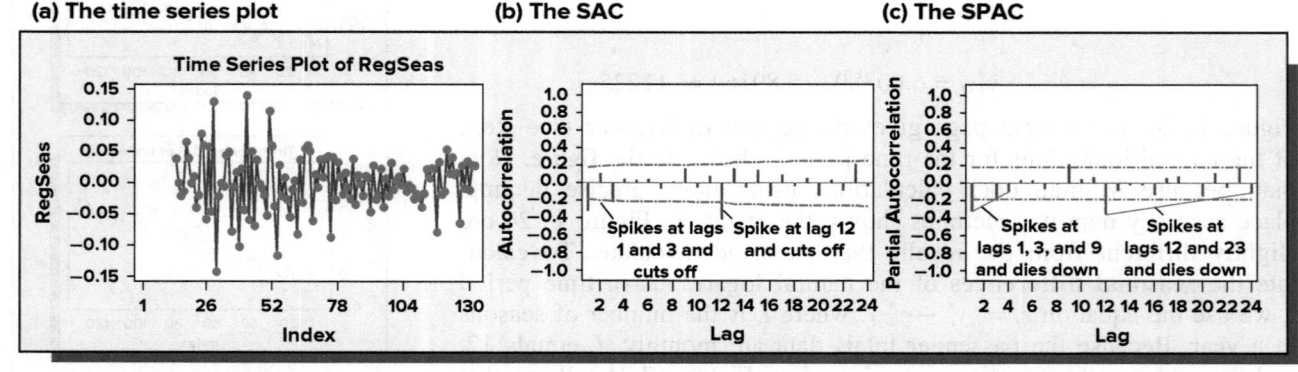

(a) The time series plot   (b) The SAC   (c) The SPAC

In order to tentatively identify a Box–Jenkins model describing z_t, we will consider various types of relationships between the observed z_t values. In general, for any time series of z_t values, we define the **sample autocorrelation at lag k,** denoted r_k, to be the simple correlation coefficient between z_t values separated by k time units. For example, r_1, r_2, and r_3 are the simple correlation coefficients between z_t values separated by, respectively, one, two, and three time units. To illustrate calculating (for instance) r_3, consider the combined regular and seasonal differences of the passenger totals' natural logarithms. In Figure 17.24 we have computed these combined differences to be $z_{14} = .039164$, $z_{15} = .000361$, . . . , $z_{131} = .029581$, and $z_{132} = .028732$. The mean of these combined differences can be shown to equal $\bar{z} = .001322$. Inserting these z_t values and $\bar{z}$ into the following equation

$$r_3 = \frac{(z_{14} - \bar{z})(z_{17} - \bar{z}) + (z_{15} - \bar{z})(z_{18} - \bar{z}) + \cdots + (z_{129} - \bar{z})(z_{132} - \bar{z})}{(z_{14} - \bar{z})^2 + (z_{15} - \bar{z})^2 + \cdots + (z_{132} - \bar{z})^2}$$

it follows that r_3, the simple correlation coefficient between z_t values separated by three time units, can be calculated to be $-.216$. In general, we define the **sample autocorrelation function (SAC)** to be a listing, or graph, of the sample autocorrelations at lags $k = 1, 2, \ldots$. Figure 17.25(b) presents the Minitab output of the SAC of the combined differences of the passenger totals' natural logarithms. The vertical lines on the output represent the sample autocorrelations. For example, the vertical lines at lags 1, 2, and 3 represent r_1, r_2, and r_3, which can be computed to be $-.317$, $.109$, and (as just demonstrated) $-.216$. In order to begin to interpret the meaning of these and the other sample autocorrelations in the SAC, we identify the existence of **spikes** in the SAC. We say that **a spike exists at lag k in the SAC** if the sample autocorrelation r_k is large enough in magnitude to conclude that ρ_k, the population autocorrelation of all possible z_t values separated by k time units, does not equal 0. One frequently used convention is to conclude that a spike exists at lag k in the SAC if r_k is at least as large in magnitude as twice its estimated standard deviation. If we examine the Minitab output of the SAC in Figure 17.25(b), we see that the "center line" on the plot of the r_k values is positioned at 0. Furthermore, for any lag k the dashed line above the center line is two estimated standard deviations greater than 0, and the dashed line below the center line is two estimated standard deviations less than 0. If the vertical line representing r_k extends at least as far up or down as the two standard deviation dashed lines corresponding to lag k, then we conclude that a spike exists at lag k in the SAC. Examining Figure 17.25(b), we conclude that spikes exist at lags 1, 3, and 12 in the SAC.

The **sample partial autocorrelation function (SPAC)** of a time series of z_t values is a listing, or graph, of the **sample partial autocorrelations** at lags $k = 1, 2, \ldots$. It is beyond the scope of this text to give a precise definition of the **sample partial autocorrelation at lag k.** However, this quantity, which is denoted r_{kk}, may intuitively be thought of as the sample autocorrelation of z_t values separated by k time units **with the effects of the intervening z_t values eliminated.** Figure 17.25(c) gives the Minitab output of the SPAC of the combined differences of the passenger totals' natural logarithms. Examining this output, we see that, for example, the vertical lines at lags 1 and 3 represent r_{11} and r_{33}, the sample partial autocorrelations of the combined differences separated by, respectively, one and three time units. In general, if r_{kk} is at least as large in magnitude as twice its estimated standard deviation, we say that **a spike exists at lag k in the SPAC.** This implies that ρ_{kk}, the population partial autocorrelation of all possible z_t values separated by k time units, does not equal 0. The r_{kk} values are plotted in the Minitab output of Figure 17.25(c). We conclude that a spike exists at lag k in the SPAC if the vertical line representing r_{kk} extends at least as far up or down as the two standard deviation dashed lines corresponding to lag k. Because this occurs at lags 1, 3, 9, 12, and 23, we conclude that spikes occur at these lags in the SPAC. We will now see that the spikes in the SAC and SPAC, as well as the overall behavior of the SAC and SPAC, help us to **tentatively identify** a Box–Jenkins model.

Box–Jenkins models describe a **future time series value** z_t by using **past time series values** $z_{t-1}, z_{t-2}, \ldots$ (which are called **autoregressive terms**) and/or **past random shocks** $a_{t-1}, a_{t-2}, \ldots$ (which are called **moving average terms**). In addition, every Box–Jenkins model describes z_t by using the **future random shock** a_t. The future and past random shocks

TABLE 17.14   **Guidelines for Tentatively Identifying a Box–Jenkins Model**

Guideline 1: If at the nonseasonal level:	**Guideline 3:** If at the nonseasonal level:
SAC: Has spikes at one or more lags and then abruptly cuts off (has no spikes) after a certain lag, and	SAC: Dies down (r_k's steadily decrease) fairly quickly, and
SPAC: Dies down (r_{kk}'s steadily decrease) fairly quickly, then	SPAC: Has spikes at one or more lags and then abruptly cuts off (has no spikes) after a certain lag, then
Use: **Nonseasonal moving average terms** $a_{t-1}, a_{t-2}, \ldots$ corresponding to the lags at which spikes exist in the SAC.	Use: **Nonseasonal autoregressive terms** $z_{t-1}, z_{t-2}, \ldots$ corresponding to the lags at which spikes exist in the SPAC.
Guideline 2: If at the seasonal level:	**Guideline 4:** If at the seasonal level:
SAC: Has a spike at lag 12 and cuts off after lag 12 (has a small autocorrelation at lag 24), and	SAC: Dies down fairly quickly (at lags 12 and 24), and
SPAC: Dies down fairly quickly (at lags 12 and 24), then	SPAC: Has a spike at lag 12 and cuts off after lag 12 (has a small partial autocorrelation at lag 24), then
Use: The **seasonal moving average term** a_{t-12}.	Use: The **seasonal autoregressive term** z_{t-12}.

$a_t, a_{t-1}, a_{t-2}, \ldots$ are similar to the error terms in a regression model and are assumed to be randomly and independently selected from a normal distribution having mean zero and a variance that is the same for each and every time period. It is not at all intuitive how past random shocks would be used to predict future time series values, but we will soon see that such random shocks can be quite useful. Guidelines 1 through 4 in Table 17.14 are based on the theory behind Box–Jenkins models and show how to use the SAC and SPAC to tentatively identify which autoregressive terms and/or moving average terms should be included in a Box–Jenkins model. To use the guidelines, we define (for monthly seasonal data) the **nonseasonal level** of the SAC and SPAC to be lags 1 through 9 and the **seasonal level** of the SAC and SPAC to be lags 12 and 24.

We first illustrate the use of Guideline 1 in Table 17.14. To do this, consider Figure 17.25, which gives the SAC and SPAC of the combined differences of the passenger totals' natural logarithms. At the nonseasonal level, the SAC has spikes at lags 1 and 3 and then cuts off (that is, has no spikes) after lag 3, and the SPAC dies down fairly quickly (that is, has partial autocorrelations that steadily decrease fairly quickly), with the exception of a possibly unimportant spike at the high nonseasonal lag 9. This is (approximately) the SAC and SPAC nonseasonal behavior described in Guideline 1. Therefore, because the SAC has spikes at lags 1 and 3, Guideline 1 says that we should describe the combined difference z_t by using the nonseasonal moving average terms a_{t-1} and a_{t-3}. Moreover, because the spike at lag 3 is barely a spike, and because the Box–Jenkins methodology seeks to find simple (*parsimonious*) models, we will begin by attempting to describe z_t by using a_{t-1} (and not a_{t-3}). Then, we will *diagnostically check* the model to see if we made the right decision.

Further considering the SAC and SPAC in Figure 17.25, we see that, at the seasonal level, the SAC has a spike at lag 12 and cuts off after lag 12 (that is, has a small autocorrelation at lag 24) and the SPAC dies down fairly quickly (at lags 12 and 24). This is the SAC and SPAC seasonal behavior described in Guideline 2, and thus Guideline 2 says that we should describe z_t by using the seasonal moving average term a_{t-12}. Because we have also concluded from Guideline 1 that we should describe z_t by using a_{t-1}, it follows that a tentative Box–Jenkins model describing z_t is

$$z_t = a_t - \theta_1 a_{t-1} - \theta_{12} a_{t-12} + \theta_1 \theta_{12} a_{t-13}$$

Here, θ_1 and θ_2 are unknown parameters that must be estimated from sample data, and the theory behind Box–Jenkins models tells us to use the minus signs and the **multiplicative component** $\theta_1 \theta_{12} a_{t-13}$. Moreover, if we insert the expression for the combined difference, $z_t = y_t^* - y_{t-12}^* - (y_{t-1}^* - y_{t-13}^*)$, into this model and solve for y_t^*, we find that a tentative Box–Jenkins model describing y_t^* is

$$y_t^* = y_{t-12}^* + (y_{t-1}^* - y_{t-13}^*) + a_t - \theta_1 a_{t-1} - \theta_{12} a_{t-12} + \theta_1 \theta_{12} a_{t-13}$$

FIGURE 17.26   **Analysis of the Passenger Totals' Box–Jenkins Model**

(a) Estimation and diagnostic checking

Type	Coef	SE Coef	T	P
MA 1	0.3407	0.0868	3.93	0.000
SMA 12	0.6299	0.0766	8.23	0.000

Modified Box-Pierce (Ljung-Box)
Chi-Square statistic

Lag	12	24	36	48
Chi-Square	7.5	19.6	30.5	38.7
DF	10	22	34	46
P-Value	0.679	0.607	0.638	0.770

(b) Residuals

t	LnPass	FITS	RESI
120	5.82008	5.86184	−0.041754
121	5.88610	5.85600	0.030104
122	5.83481	5.83182	0.002987
123	6.00635	5.98014	0.026218
124	5.98141	5.97049	0.010927
125	6.04025	6.00109	0.039167
126	6.15698	6.18944	−0.032462
127	6.30628	6.27991	0.026361
128	6.32615	6.30195	0.024195
129	6.13773	6.15349	−0.015763
130	6.00881	6.01104	−0.002232
131	5.89164	5.87364	0.018005
132	6.00389	5.99133	0.012558

(c) Forecasting

Period	Forecast	95% Limits Lower	Upper
133	6.03771	5.96718	6.10823
134	5.99099	5.90652	6.07546
135	6.14666	6.05023	6.24308
136	6.12046	6.01341	6.22751
137	6.15698	6.04026	6.27369
138	6.30256	6.17692	6.42819
139	6.42828	6.29432	6.56224
140	6.43857	6.29677	6.58037
141	6.26527	6.11604	6.41450
142	6.13438	5.97807	6.29069
143	6.00539	5.84231	6.16846
144	6.11358	5.94401	6.28316

Estimation, diagnostic checking, and forecasting

The upper portion of the Minitab output in Figure 17.26(a) tells us that the least squares point estimates of θ_1 and θ_{12} are $\hat{\theta}_1 = .3407$ and $\hat{\theta}_{12} = .6299$. Moreover, because the p-values for testing $H_0: \theta_1 = 0$ and $H_0: \theta_{12} = 0$ are each less than .001, we conclude that each of θ_1 and θ_{12} is important in the model. The lower portion of Figure 17.26(a) gives a **chi-square analysis** that is used to **diagnostically check** the adequacy of the model. As will be discussed in more detail near the end of this section, it can be shown that because all of the p-values associated with the **Ljung–Box chi-square statistic** values in Figure 17.26(a) are greater than .05, it is reasonable to conclude that the model's residuals are not significantly autocorrelated, and thus that the model is adequate. To calculate a point forecast of y^*_{133}, next month's logged passenger total, we insert $t = 133$, as well as the least squares point estimates $\hat{\theta}_1 = .3407$ and $\hat{\theta}_{12} = .6299$, into the model. This gives the point forecast

$$\hat{y}^*_{133}(132) = y^*_{121} + (y^*_{132} - y^*_{120}) + \hat{a}_{133} - .3407\hat{a}_{132} - .6299\hat{a}_{121} + (.3407)(.6299)\hat{a}_{120}$$
$$= 5.88610 + (6.00389 - 5.82008) + 0 - .3407(.012558)$$
$$\quad - .6299(.030104) + (.3407)(.6299)(-.041754)$$
$$= 6.03771$$

To compute this point forecast, we use the natural logarithms [see LnPass in Figure 17.26(b)] $y^*_{120} = 5.82008$, $y^*_{121} = 5.88610$, and $y^*_{132} = 6.00389$ of the passenger totals in months 120, 121, and 132. In addition, we predict the future random shock a_{133} to be $\hat{a}_{133} = 0$, and we predict the past random shocks a_{132}, a_{121}, and a_{120} by $\hat{a}_{132} = .012558$, $\hat{a}_{121} = .030104$, and $\hat{a}_{120} = -.041754$. These are Minitab-computed residuals for time periods 132, 121, and 120 and are shown in Figure 17.26(b). For month 120, for example, the logged passenger total is $y^*_{120} = 5.82008$ (see LnPass), Minitab uses an advanced procedure to calculate the predicted logged passenger total to be $\hat{y}^*_{120} = 5.86184$ (see FITS), and the residual $\hat{a}_{120}$ is $y^*_{120} - \hat{y}^*_{120} = 5.82008 - 5.86184 = -.041754$ (see RESI). The point forecast of $\hat{y}^*_{133}$ (132) = 6.03771 and a 95 percent prediction interval for y^*_{133} are given in Figure 17.26(c). Exponentiating the point forecast and the ends of the prediction interval, we obtain $e^{6.03771} = 418,933$ and $[e^{5.96718},$ $e^{6.10823}] = [390,403, 449,542]$. Therefore, the airline forecasts that it will fly 418,933 customers in month 133 (January of year 12). The airline is 95 percent confident it will fly between 390,403 and 449,542 customers in month 133.

As another example, consider the seasonal differences $(z_t = y^*_t - y^*_{t-12})$ of the quartic roots $(y^*_t = y_t^{.25})$ of the hotel room averages in Table 17.4. It can be verified that these seasonal differences fluctuate with constant variation around a constant level and are, therefore, stationary. Figures 17.27 and 17.28 give the SAC and SPAC of these seasonal differences. At the nonseasonal level, the SAC dies down fairly quickly, and the SPAC has spikes at lags 1 and 3 and (with the exception of a smaller spike at lag 5) cuts off after lag 3. This is (approximately)

FIGURE 17.27   The Minitab Output of the SAC and SPAC of the Hotel Room Seasonal Differences

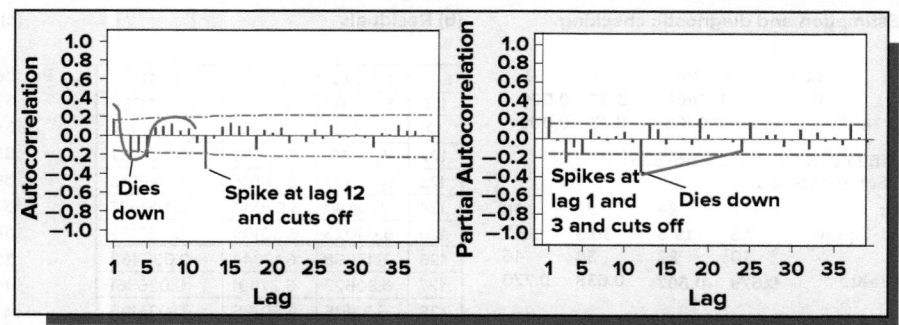

FIGURE 17.28   The JMP Output of the SAC and SPAC of the Hotel Room Seasonal Differences

Lag	AutoCorr	Ljung-Box Q	p-Value	Lag	Partial
0	1.0000	.	.	0	1.0000
1	0.1743	4.8322	0.0279*	1	0.1743
2	0.0143	4.8650	0.0878	2	-0.0166
3	-0.2598	15.7378	0.0013*	3	-0.2676
4	-0.1686	20.3473	0.0004*	4	-0.0861
5	-0.2306	29.0264	<.0001*	5	-0.1993
6	0.0805	30.0918	<.0001*	6	0.0931
7	0.0975	31.6635	<.0001*	7	0.0207
8	0.1212	34.1081	<.0001*	8	-0.0144
9	0.0432	34.4215	<.0001*	9	0.0253
10	0.0710	35.2729	0.0001*	10	0.0705
11	-0.0795	36.3484	0.0001*	11	-0.0341
12	-0.3476	57.0252	<.0001*	12	-0.3491
13	-0.0295	57.1752	<.0001*	13	0.1485
14	0.0906	58.5996	<.0001*	14	0.1022
15	0.1318	61.6348	<.0001*	15	-0.0496
16	0.0950	63.2222	<.0001*	16	0.0013
17	0.0947	64.8134	<.0001*	17	0.0035
18	-0.1493	68.7947	<.0001*	18	-0.0603
19	0.0546	69.3301	<.0001*	19	0.2145
20	0.0264	69.4561	<.0001*	20	0.0446
21	0.0823	70.6927	<.0001*	21	0.0020
22	-0.0749	71.7252	<.0001*	22	0.0058
23	-0.0023	71.7262	<.0001*	23	-0.0258
24	-0.0634	72.4767	<.0001*	24	-0.1273
25	0.0665	73.3089	<.0001*	25	0.1689

the SAC and SPAC nonseasonal behavior described in Guideline 3 in Table 17.14. Therefore, because the SPAC has spikes at lags 1 and 3, Guideline 3 says that we should describe the seasonal difference z_t by using the nonseasonal autoregressive terms z_{t-1} and z_{t-3}. Moreover, Minitab requires us to use the **intervening term** z_{t-2} because we use z_{t-1} and z_{t-3}. At the seasonal level in Figures 17.27 and 17.28, the SAC has a spike at lag 12 and cuts off after lag 12, and the SPAC dies down fairly quickly. This is the SAC and SPAC seasonal behavior described in Guideline 2, and thus Guideline 2 says that we should describe z_t by using the seasonal moving average term a_{t-12}. Because Guideline 3 says that we should also describe z_t by using z_{t-1}, z_{t-2}, and z_{t-3}, a tentative Box–Jenkins model describing z_t is

$$z_t = \delta + \phi_1 z_{t-1} + \phi_2 z_{t-2} + \phi_3 z_{t-3} - \theta_{12} a_{t-12} + a_t$$

FIGURE 17.29 **The Minitab Output of Estimation, Diagnostic Checking, and Forecasting for the Hotel Room Occupancy Box–Jenkins Model**

(a) Estimation and diagnostic checking

Type		Coef	SE Coef	T	P
AR	1	0.2392	0.0794	3.01	0.003
AR	2	0.1322	0.0809	1.63	0.104
AR	3	−0.2642	0.0801	−3.30	0.001
SMA	12	0.5271	0.0735	7.17	0.000
Constant		0.038056	0.001046	36.39	0.000

Modified Box-Pierce (Ljung-Box)
Chi-Square statistic

Lag	12	24	36	48
Chi-Square	9.2	24.3	35.2	43.2
DF	7	19	31	43
P-Value	0.242	0.186	0.276	0.461

(b) Forecasting

Period	Forecast	95% Limits Lower	95% Limits Upper
169	5.38091	5.33070	5.43111
170	5.26784	5.21622	5.31946
171	5.27552	5.22303	5.32801
172	5.42917	5.37584	5.48249
173	5.40940	5.35592	5.46289
174	5.59285	5.53915	5.64654
175	5.82033	5.76663	5.87404
176	5.85258	5.79888	5.90629
177	5.47779	5.42406	5.53152
178	5.47843	5.42470	5.53215
179	5.28889	5.23516	5.34262
180	5.46090	5.40717	5.51463

Here, we have included a **constant term** δ (to be explained momentarily) in the model, and δ, as well as ϕ_1, ϕ_2, ϕ_3, and θ_{12}, are unknown parameters that must be estimated from sample date. Inserting the seasonal difference $z_t = y_t^* - y_{t-12}^*$ into this model and solving for y_t^*, we find that a tentative Box–Jenkins model describing y_t^* is

$$y_t^* = \delta + y_{t-12}^* + \phi_1(y_{t-1}^* - y_{t-13}^*) + \phi_2(y_{t-2}^* - y_{t-14}^*) + \phi_3(y_{t-3}^* - y_{t-15}^*) - \theta_{12}a_{t-12} + a_t$$

The constant term δ in the tentative model says that there is a **deterministic trend** of δ as we move from y_{t-12}^*, the quartic root of the hotel room average one year ago, to y_t^*, the quartic root of the current hotel room average. We will soon see that δ is important in the hotel room average model, but first note that the reason we did not use any multiplicative terms in this model is that the nonseasonal components $\phi_1 z_{t-1}$, $\phi_2 z_{t-2}$, and $\phi_3 z_{t-3}$ are autoregressive components and the seasonal component $-\theta_{12}a_{t-12}$ is a moving average component. We tentatively identified the seasonal moving average term a_{t-12} because, at the seasonal level in Figures 17.27 and 17.28, the SAC has a spike at lag 12 and cuts off and the SPAC dies down fairly quickly (Guideline 2). Suppose, however, that at the seasonal level in Figures 17.27 and 17.28, the SAC had died down fairly quickly and the SPAC had possessed a spike at lag 12 and then cut off. In this case, Guideline 4 would tell us to use the seasonal autoregressive term z_{t-12}, and the tentatively identified model for z_t would be

$$z_t = \delta + \phi_1 z_{t-1} + \phi_2 z_{t-2} + \phi_3 z_{t-3} + \phi_{12} z_{t-12}$$
$$- \phi_1 \phi_{12} z_{t-13} - \phi_2 \phi_{12} z_{t-14} - \phi_3 \phi_{12} z_{t-15} + a_t$$

This model uses three **multiplicative terms** (with negative signs), which are appropriate when *both* the nonseasonal and seasonal components of the model are autoregressive.

Figures 17.29 and 17.30 give the Minitab and JMP outputs of estimation, forecasting, and diagnostic checking for the tentatively identified Box–Jenkins hotel room average model. Examining these outputs, we see that:

- The Minitab parameter estimates are slightly different from, but very close to, the JMP parameter estimates. Note that the estimate of δ on the JMP output is .0372126 and is found under the heading **Constant Estimates.** Also note that the *p*-value for testing H_0: $\delta = 0$ is less than .001, indicating that the constant term δ belongs in the model. (It can be verified that a constant term δ should not be included in the passenger totals model.)

- The Minitab forecasts of the quartic roots of the hotel room averages in months 169 through 180 are slightly different from, but very close to, the JMP forecasts. In the exercises we will help the reader to use the Box–Jenkins hotel room average model and the Minitab parameter estimates to hand calculate the point forecast 5.38091 for month 169.

FIGURE 17.30   The JMP Output of Estimation, Diagnostic Checking, and Forecasting for the Hotel Room Average
                Box–Jenkins Model

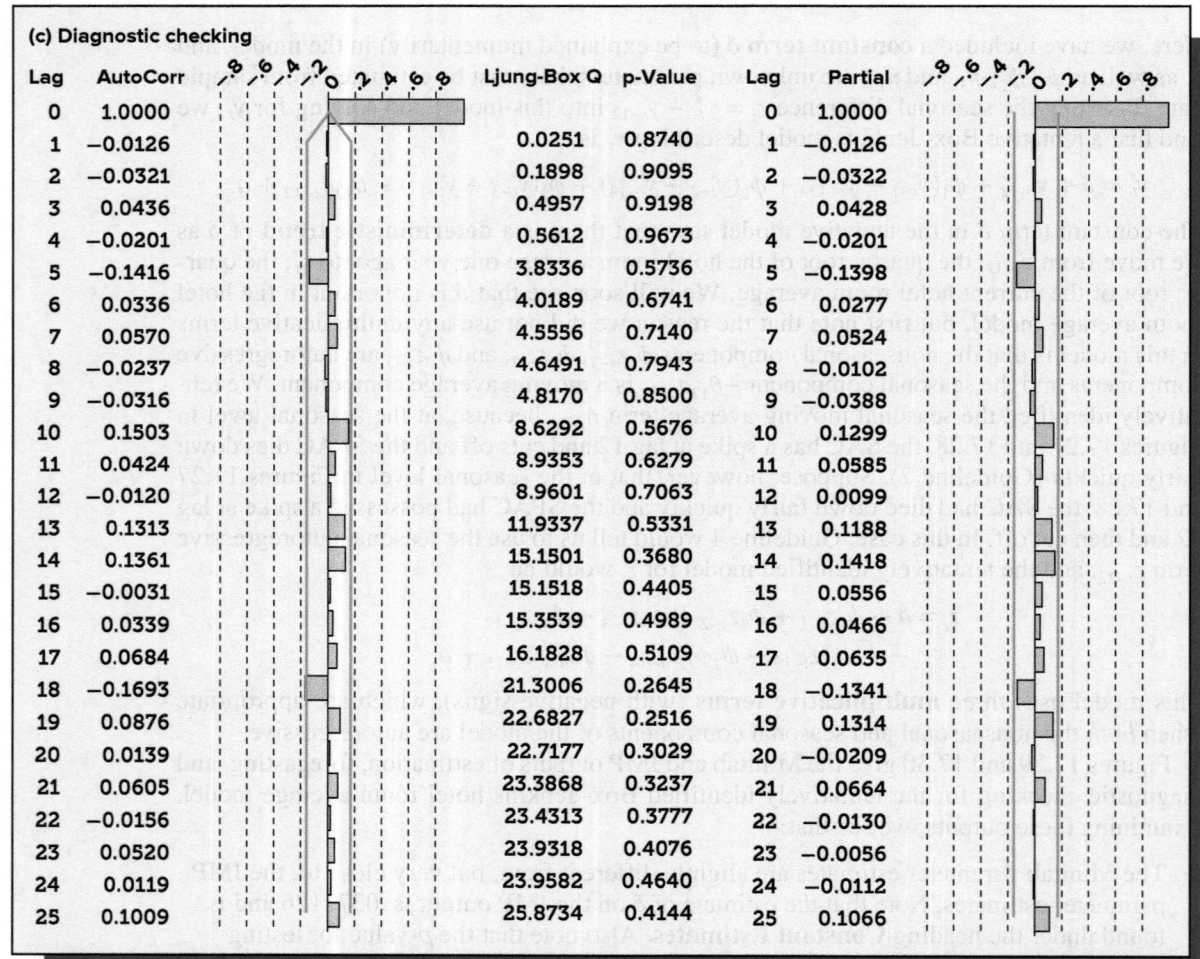

(a) Estimation

Term	Factor	Lag	Estimate	Std Error	t Ratio	Prob>|t|
AR1,1	1	1	0.2523500	0.0805691	3.13	0.0021
AR1,2	1	2	0.1382223	0.0805845	1.72	0.0884
AR1,3	1	3	−0.2680146	0.0781208	−3.43	0.0008*
MA2,12	2	12	0.5686512	0.0806188	7.05	<.0001*
Intercept	1	0	0.0424103	0.0010957	38.71	<.0001*

Constant   Mu
Estimate 0.04241031
0.0372126

(b) Forecasting

Row	Predicted QrOccup	Upper CL (0.95) QrOccup	Lower CL (0.95) QrOccup
169	5.379776292	5.4291250239	5.3304275602
170	5.2687197675	5.319615527	5.217824008
171	5.2759662765	5.3278281371	5.224104416
172	5.429086857	5.4817222306	5.3764514833
173	5.4101171381	5.4629221439	5.3573121323
174	5.5937079262	5.6467472734	5.5406685789
175	5.820368721	5.8734111121	5.7673263299
176	5.8533443392	5.9063899948	5.8002986836
177	5.4793907065	5.5324593502	5.4263220627
178	5.4794889124	5.5325585402	5.4264192846
179	5.2889318717	5.3420016835	5.23586206
180	5.4591805935	5.5122514859	5.4061097012

(c) Diagnostic checking

Lag	AutoCorr	Ljung-Box Q	p-Value	Lag	Partial
0	1.0000			0	1.0000
1	−0.0126	0.0251	0.8740	1	−0.0126
2	−0.0321	0.1898	0.9095	2	−0.0322
3	0.0436	0.4957	0.9198	3	0.0428
4	−0.0201	0.5612	0.9673	4	−0.0201
5	−0.1416	3.8336	0.5736	5	−0.1398
6	0.0336	4.0189	0.6741	6	0.0277
7	0.0570	4.5556	0.7140	7	0.0524
8	−0.0237	4.6491	0.7943	8	−0.0102
9	−0.0316	4.8170	0.8500	9	−0.0388
10	0.1503	8.6292	0.5676	10	0.1299
11	0.0424	8.9353	0.6279	11	0.0585
12	−0.0120	8.9601	0.7063	12	0.0099
13	0.1313	11.9337	0.5331	13	0.1188
14	0.1361	15.1501	0.3680	14	0.1418
15	−0.0031	15.1518	0.4405	15	0.0556
16	0.0339	15.3539	0.4989	16	0.0466
17	0.0684	16.1828	0.5109	17	0.0635
18	−0.1693	21.3006	0.2645	18	−0.1341
19	0.0876	22.6827	0.2516	19	0.1314
20	0.0139	22.7177	0.3029	20	−0.0209
21	0.0605	23.3868	0.3237	21	0.0664
22	−0.0156	23.4313	0.3777	22	−0.0130
23	0.0520	23.9318	0.4076	23	−0.0056
24	0.0119	23.9582	0.4640	24	−0.0112
25	0.1009	25.8734	0.4144	25	0.1066

To discuss diagnostic checking, we see that the JMP output in Figure 17.30(c) gives the sample autocorrelations and sample partial autocorrelations of the residuals obtained when the tentative Box–Jenkins model is fit to the hotel room average data. (Minitab also gives this information, but we have not shown it.) Letting $r_\ell(\hat{a})$ and $r_{\ell\ell}(\hat{a})$ denote the sample autocorrelation and sample partial autocorrelation of residuals separated by ℓ time units, we see, for example, that $r_1(\hat{a}) = -.0126$, $r_2(\hat{a}) = -.0321$, $r_{11}(\hat{a}) = -.0126$, and $r_{22}(\hat{a}) = -.0322$. The previously mentioned **Ljung–Box chi-square statistic** used in diagnostic checking measures, for a specified value of K, the overall size of the first K sample autocorrelations of the residuals. The first step in calculating this chi-square statistic, which is denoted as Q, is to determine n', the total number of appropriately differenced time series values used to fit the tentatively identified model. Because we used seasonal differencing to obtain stationary hotel room average values $z_{13}, z_{14}, \ldots, z_{168}$, it follows that $n' = 156$. Having determined n', Q is then calculated for a specified K by (1) squaring each of $r_1(\hat{a}), r_2(\hat{a}), \ldots, r_K(\hat{a})$; (2) dividing each $r_\ell^2(\hat{a})$ by $(n' - \ell)$; (3) adding up the $r_\ell^2(\hat{a})/(n' - \ell)$ values from step (2); and (4) multiplying the sum in step (3) by $n'(n' + 1)$. For example, if we specify a K of 12, Minitab calculates Q to be 9.2 [see Figure 17.29(a)], and JMP calculates Q to be 8.9601 [see Figure 17.30(c)]. To assess the size of Q, we consider its p-value, which is the area under the curve of the chi-square distribution having $K - n_p$ degrees of freedom to the right of Q. Here, n_p is the number of parameters in the model under consideration. If the p-value associated with Q for a particular K is greater than .05, then Q is small enough to conclude that the first K sample autocorrelations of the residuals do not provide evidence of significant residual autocorrelation. For the hotel room average data, Minitab computes Q for $K = 12, 24, 36$ and 48, and JMP computes Q for $K = 1, 2, \ldots, 30$. Also, note that JMP uses K, rather than $K - n_p$, for the degrees of freedom when calculating the p-value associated with Q. In spite of these differences, Figures 17.29(a) and 17.30(c) show that all of the p-values associated with the different values of Q are greater than .05. This indicates that we do not have evidence that the residuals of the tentatively identified model are significantly autocorrelated, and thus we will conclude that this model is adequate.

To conclude this subsection, recall that we have seen that if a time series fluctuates with constant variation around a constant level, then the time series should be considered to be stationary. It can also be shown that if the SAC cuts off or dies down fairly quickly at the nonseasonal level and cuts off or dies down fairly quickly at the seasonal level, then the time series should be considered to be stationary. However, if the SAC has large sample autocorrelations that die down very slowly at either level, or at both levels, then the time series should be considered to be nonstationary. Also, note that the SAC and SPAC behaviors described in Guidelines 1 through 4 are the most commonly occurring SAC and SPAC behaviors in practice. Theoretically, the SAC and SPAC cannot both cut off at the nonseasonal level and cannot both cut off at the seasonal level, so we should avoid making such interpretations. The SAC and SPAC can theoretically both die down fairly quickly at the seasonal level, but the authors have never seen this occur. What does sometimes occur is that the SAC and SPAC both die down fairly quickly at the nonseasonal level. In this case, we should use both one or more nonseasonal autoregressive terms and one or more nonseasonal moving average terms—usually, simple terms such as both z_{t-1} and a_{t-1}—in a Box–Jenkins model. See Bowerman, O'Connell, and Kochler (1995) for a more complete discussion of the Box–Jenkins methodology.

Exercises for Section 17.6

CONCEPTS   **connect**

17.30 Discuss the purpose of differencing.

17.31 Explain how we use the SAC and SPAC.

METHODS AND APPLICATIONS

17.32 The Minitab least squares point estimates of the parameters of the tentative hotel room average model are $\hat{\delta} = .038056$, $\hat{\phi}_1 = .2392$, $\hat{\phi}_2 = .1322$, $\hat{\phi}_3 = -.2642$, and $\hat{\theta}_{12} = .5271$. Hand calculate the point forecast of y_{169}^* to be $\hat{y}_{169}^*(168) = 5.38091$ [see Figure 17.29(b)]. Hint: Recalling that $y_t^* = y_t^{.25}$, use the facts that $y_{157}^* = 5.33648$, $y_{168}^* - y_{156}^* = .10212$, $y_{167}^* - y_{155}^* = .07934$, $y_{166}^* - y_{154}^* = -.00628$, and the residual for period 157 is .057297. Find a point

forecast of and a 95 percent prediction interval for y_{169}.
   ⊙⊙ TravRest

17.33 Table 17.15 shows the miles flown by an airline company, and Figure 17.31 shows a plot of these data and the SAC and SPAC of the combined regular and seasonal differences of the miles flown. **(1)** Discuss which of the guidelines in Table 17.14 say that a tentative Box–Jenkins model describing $z_t = y_t - y_{t-12} - (y_{t-1} - y_{t-13})$ is $z_t = \phi_1 z_{t-1} + \phi_2 z_{t-2} + \phi_3 z_{t-3} + a_t$. **(2)** Use Minitab or JMP to obtain the SAC and SPAC in Figures 17.31(b) and (c). **(3)** Use Minitab or JMP to fit and diagnostically check the tentatively identified model and to obtain the forecasts shown in Figure 17.32. Also, try a constant term δ and show that it is not appropriate. (Note: Minitab and JMP will give slightly different forecasts.)   ⊙⊙ MilesFlown

TABLE 17.15   **Miles Flown by an Airline Company (in Units of 1,000 Miles Flown)** 💿 MilesFlown

Year / Month	1	2	3	4	5
Jan.	6,827	7,269	8,350	8,186	8,334
Feb.	6,178	6,775	7,829	7,444	7,899
Mar.	7,084	7,819	8,829	8,484	9,994
Apr.	8,162	8,371	9,948	9,864	10,078
May	8,462	9,069	10,638	10,252	10,801
June	9,644	10,248	11,253	12,282	12,950
July	10,466	11,030	11,424	11,637	12,222
Aug.	10,748	10,882	11,391	11,577	12,246
Sept.	9,963	10,333	10,665	12,417	13,281
Oct.	8,194	9,109	9,396	9,637	10,366
Nov.	6,848	7,685	7,775	8,094	8,730
Dec.	7,027	7,602	7,933	9,280	9,614

Source: Data from Kendall and Ord (1990).

FIGURE 17.32

Forecasts for Exercise 17.33

Period	Forecast
61	8824.4
62	8385.8
63	10329.2
64	10547.0
65	11251.7
66	13313.2
67	12682.3
68	12679.7
69	13665.6
70	10818.1
71	9154.9
72	10013.2

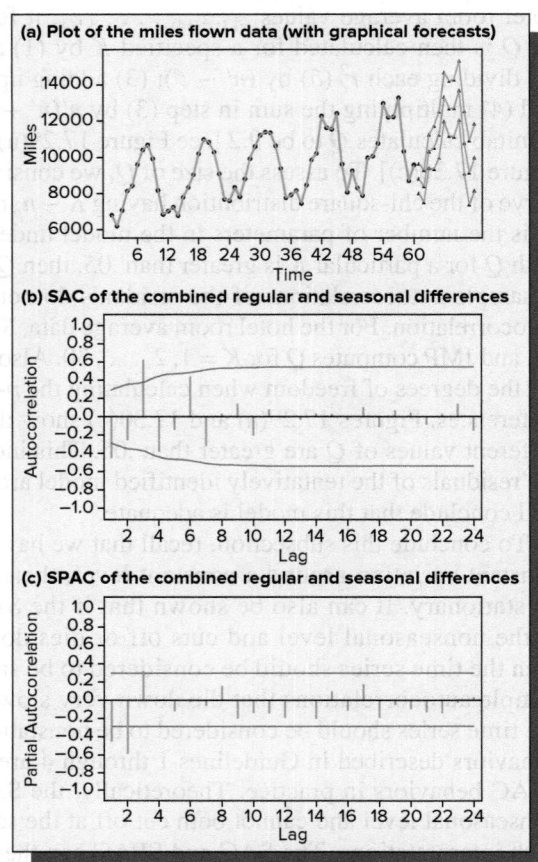

FIGURE 17.31   **A Plot of the Miles Flown Data (with Graphical Forecasts) and the SAC and SPAC of the Combined Regular and Seasonal Differences**

(a) Plot of the miles flown data (with graphical forecasts)

(b) SAC of the combined regular and seasonal differences

(c) SPAC of the combined regular and seasonal differences

LO17-9
Compare time series models by using forecast errors.

TABLE 17.16

Method 1
Forecast Errors

y_t	$\hat{y}_t$	$y_t - \hat{y}_t$
352	360.52	−8.52
445	441.48	3.52
453	446.40	6.6
541	516.62	24.38
457	433.85	23.15
762	767.82	−5.82
1,194	1,156.30	37.7
1,361	1,350.50	10.5
1,615	1,606.30	8.7
1,059	1,067.40	−8.4
824	850.12	−26.12
495	501.39	−6.39

17.7 Forecast Error Comparisons

Consider comparing the forecasting accuracies of two particular forecasting methods—Method 1 and Method 2. To do this, we use each method to compute point forecasts ($\hat{y}_t$) of the future values (y_t) of a time series. We then wait for the future values to occur, record them, and compute the forecast errors ($y_t - \hat{y}_t$). Specifically, suppose that the results we obtain when we use Method 1 and Method 2 to forecast 12 future values of a time series are as shown in Tables 17.16 and 17.17. Three criteria by which to compare Method 1 and Method 2 are the **mean absolute deviation (MAD),** the **mean squared deviation (MSD),** and the **mean absolute percentage error (MAPE).**

    To calculate the MAD, we find the absolute value of each forecast error and then average the resulting absolute values. For example, if we find the absolute value of each of the 12 forecast errors given by Method 1 in Table 17.16, sum the 12 absolute values, and divide the sum by 12, we find that the MAD is 14.15. By contrast, if we calculate the MAD of the Method 2 forecast errors in Table 17.17, we find that the MAD is 25.6.

    To calculate the MSD, we find the squared value of each forecast error and then average the resulting squared values. For example, if we find the squared value of each of the 12 forecast errors given by Method 1 in Table 17.16, sum the 12 squared values, and divide the sum by 12, we find that the MSD is 307.80. By contrast, if we calculate the MSD of the Method 2 forecast errors in Table 17.17, we find that the MSD is 892.44.

    To calculate the MAPE, we find the percentage error for each forecast, $[(y_t - \hat{y}_t)/y_t] \times 100\%$, and we then average the absolute values of the percentage errors. For example, the percentage error for the first forecast given by Method 1 is $[(y_1 - \hat{y}_1)/y_1] \times 100\% = [(352 - 360.52/352] \times 100\% = -2.42\%$. If we average the absolute values of the 12 percentage errors given by Method 1, we find that the MAPE is 2.06%. By contrast, if we calculate the MAPE of the Method 2 forecast errors, we find that the MAPE is 3.24%.

In general, we want a forecasting method that gives small values of the MAD, MSD, and MAPE. Note, however, that the MSD is the average of the *squared forecast errors*. It follows that the MSD, unlike the MAD and MAPE, penalizes a forecasting method much more for large forecast errors than for small forecast errors. Therefore, the forecasting method that gives the smallest MSD may not be the forecasting method that gives the smallest MAD, or MAPE. Furthermore, the forecaster who uses the MSD to choose a forecasting method would prefer several smaller forecast errors to one large error. In our example, Method 1 gives smaller values of the MAD, MSD, and MAPE than does Method 2. Moreover, for those who have studied Sections 17.3 and 17.5, the point forecasts in Tables 17.14 and 17.15 are point forecasts given by the multiplicative decomposition method (Method 1) and multiplicative Winters' method (Method 2). These point forecasts are point forecasts of future Tasty Cola sales in months 37 through 48. Although the multiplicative decomposition method does better in this situation, the multiplicative Winters' method would do better in other situations.

TABLE 17.17
Method 2 Forecast Errors

y_t	$\hat{y}_t$	$y_t - \hat{y}_t$
352	355.96	−3.96
445	426.31	18.69
453	436.69	16.31
541	505.60	35.4
457	431.71	25.29
762	748.35	13.65
1,194	1,132.57	61.43
1,361	1,313.74	47.26
1,615	1,576.09	38.91
1,059	1,043.59	15.41
824	834.24	−10.24
495	506.65	−11.65

Exercises for Section 17.7

CONCEPTS

17.34 What is the MAD? What is the MSD? What is the MAPE? How do we use these quantities?

17.35 Why does the MSD penalize a forecasting method much more for large forecast errors than for small forecast errors?

METHODS AND APPLICATIONS

Exercises 17.36 and 17.37 compare two forecasting methods—method A and method B. Suppose that method A gives the point forecasts 57, 61, and 70 of three future time series values. Method B gives the point forecasts 59, 65, and 73 of these three future values. The three future values turn out to be 60, 64, and 67.

17.36 Calculate the MAD, MSD, and MAPE for method A. Calculate the MAD, MSD, and MAPE for method B.

17.37 Which method—method A or method B—gives the smallest MAD? The smallest MSD? The smallest MAPE?

17.8 Index Numbers

We often wish to compare a value of a time series relative to another value of the time series. For instance, according to the U.S. Bureau of Labor Statistics, energy prices increased by 4.7 percent from 1995 to 1996, while apparel prices decreased by .2 percent from 1995 to 1996. In order to make such comparisons, we must describe the time series. We have seen (in Section 17.3) that time series decomposition can be employed to describe a time series. Another way to describe time-related data is to use *index numbers*.

When we compare time series values to the same previous value, we say that the previous value is in the **base time period,** and successive comparisons of time series values to the value in the base period form a sequence of **index numbers.** More formally, a **simple index number** (or **simple index**) is defined as follows:

LO17-10
Use index numbers to compare economic data over time.

A **simple index** is obtained by dividing the current value of a time series by the value of the time series in the base time period and by multiplying this ratio by 100. That is, if y_t denotes the current value and if y_0 denotes the value in the base time period, then the **simple index number** is

$$\frac{y_t}{y_0} \times 100$$

TABLE 17.18   Price of a Gallon of Regular Gasoline (in Dollars): 2003 to 2011   DS PriceGas1

Year	2003	2004	2005	2006	2007	2008	2009	2010	2011
Price per Gallon	1.59	1.88	2.30	2.59	2.80	3.27	2.35	2.78	3.63
Index (Base Year = 2003)	100.0	118.24	144.65	162.89	176.10	205.66	147.80	174.84	228.30

Source: U.S. Energy Information Administration.

The time series values used to construct an index are often *quantities* or *prices*. For instance, in Table 17.18 we give the price of a gallon of regular gasoline in the United States (in dollars) for the years 2003 through 2011. If we consider 2003 to be the base year, we compute an index for each succeeding year by dividing the price per gallon for each year by 1.59 (the price per gallon for the base year 2003) and multiplying by 100. For example, for 2007 the simple index is $(2.80/1.59) \times 100 = 176.10$, while the simple index for 2008 is $(3.27/1.59) \times 100 = 205.66$. Table 17.18 gives the remaining index values for 2003 through 2011. Notice that (by definition) the index for the base year will always equal 100.0 (as it does here).

Although the simple index is not written with a percentage sign, comparisons of the index with the base year are percentage comparisons. For instance, the index of 205.66 for 2008 tells us that the price per gallon in 2008 was up 105.66 percent compared to the 2003 base year. In general, *if we are comparing the index to the base year*, the difference between the index and 100 gives the percentage change from the base year. It is important to point out that other period-to-period percentage comparisons cannot be made by subtracting indexes. For instance, the percentage difference between the prices per gallon in 2007 and 2008 is *not* $205.66 - 176.10 = 29.56$ percent. Rather, the percentage difference is

$$\frac{205.66 - 176.10}{176.10} \times 100 = 16.79$$

This says that the price per gallon in 2008 was up 16.79 percent relative to 2007.

A simple index is computed by using the values of one time series. Often, however, we compute an index based on the accumulated values of more than one time series. Such an index is called an **aggregate index.** As an example, food prices are often compared with an aggregate index based on a "market basket" of commonly bought grocery items. For instance, consider a market basket consisting of six items—a gallon of whole milk, a one pound jar of peanut butter, a pound of red delicious apples, a dozen eggs, a pound loaf of white bread, and a pound of ground beef. Table 17.19 lists average city prices for these items in June 2006 and in June 2011 according to the Consumer Price Index Detailed Report.

One way to compare prices would be to compute a simple index for each individual item in the market basket. However, we can create an aggregate price index by totaling

TABLE 17.19   2006 and 2011 Prices for a Market Basket of Grocery Items   DS MkBskt

Grocery Item	2006 Price	2011 Price
1 gal. of whole milk	$3.00	$3.62
1 lb. jar of peanut butter	$1.74	$1.96
1 lb. of red delicious apples	$1.05	$1.32
1 dozen eggs	$1.24	$1.68
1 lb. loaf of white bread	$1.07	$1.49
1 lb. of ground beef	$2.24	$2.77
Totals	$10.34	$12.84

Source: http://www.bls.gov/cpi/cpid1106.pdf.

the prices for each year and then computing a simple index of the yearly price totals. Using the data in Table 17.19, we obtain $(12.84/10.34) \times 100 = 124.18$. This index tells us that prices of the market basket grocery items in 2011 have increased by 24.18 percent over the prices of these items in the base year 2006. Notice that this percentage increase does not necessarily apply to each individual grocery item, nor does this index necessarily apply to any of the individual grocery items. It applies only to the aggregate of grocery items in the market basket.

In general, we compute an aggregate price index as follows:

An **aggregate price index** is

$$\left(\frac{\Sigma p_t}{\Sigma p_0}\right) \times 100$$

where Σp_t is the sum of the prices in the current time period and Σp_0 is the sum of the prices in the base year.

A disadvantage of this aggregate price index is that it does not take into account the fact that some items in the market basket are purchased more frequently than others. To remedy this deficiency, we can weight each price by the quantity of that item purchased in a given period (say yearly). Then we can total the weighted prices for each year and compute a simple index of the weighted price totals. To illustrate, Table 17.20 gives the 2006 and 2011 prices of the market basket items and also gives estimates of the quantity of each item purchased in a year by a typical family. The table also gives the price multiplied by the quantity for each item, which is simply the total yearly cost of purchasing the item. These costs are totaled for each year. Looking at Table 17.20, we see that a typical family in 2006 spent $824.09 purchasing the market basket items during the year, while the family spent $1049.68 purchasing the market basket items during 2011. We now compute a simple index of the total costs, which is $(1049.68/824.09) \times 100 = 127.37$.

This type of index is called a **weighted aggregate price index.** Two versions of this kind of index are commonly used. The first version is called a **Laspeyres index.** Here the quantities that are specified for the base year are also employed for all succeeding time periods. This is the assumption we have made in Table 17.20. Notice that the quantities for 2011 are the same as those specified for 2006. In general,

A **Laspeyres index** is

$$\frac{\Sigma p_t q_0}{\Sigma p_0 q_0} \times 100$$

where p_0 represents a base period price, q_0 represents a base period quantity, and p_t represents a current period price.

TABLE 17.20 2006 and 2011 Prices and Quantities for a Market Basket of Grocery Items   DS MkBskt

Grocery Item	2006 (Base Year)			2011		
	Price, p_0	Quantity, q	$p_0 \times q$ = cost	Price, p_t	Quantity, q	$p_t \times q$ = cost
1 gal. of whole milk	$3.00	52	$156.00	$3.62	52	$188.24
1 lb. jar of peanut butter	$1.74	13	$22.62	$1.96	13	$25.48
1 lb. of red delicious apples	$1.05	55	$57.75	$1.32	55	$72.60
1 dozen eggs	$1.24	72	$89.28	$1.68	72	$120.96
1 lb. loaf of white bread	$1.07	156	$166.92	$1.49	156	$232.44
1 lb. of ground beef	$2.24	148	$331.52	$2.77	148	$409.96
Totals	$10.34		$824.09	$12.84		$1049.68

TABLE 17.21   2006 and 2011 Prices and 2011 Quantities for a Market Basket of Grocery Items 📀 MkBsktR

Grocery Item	2006 Price, p_0	2011 Quantity, q_t	$p_0 \times q_t$ = cost	2011 Price, p_t	2011 Quantity, q_t	$p_t \times q_t$ = cost
1 gal. of whole milk	$3.00	40	$120.00	$3.62	40	$144.80
1 lb. jar of peanut butter	$1.74	21	$36.54	$1.96	21	$41.16
1 lb. of red delicious apples	$1.05	82	$86.10	$1.32	82	$108.24
1 dozen eggs	$1.24	60	$74.40	$1.68	60	$100.80
1 lb. loaf of white bread	$1.07	175	$187.25	$1.49	175	$260.75
1 lb. of ground beef	$2.24	148	$331.52	$2.77	148	$409.96
Totals	$10.34	Total	$835.81	$12.84	Total	$1065.71

Because the Laspeyres index employs the base period quantities in all succeeding time periods, this index allows for ready comparison of prices for identical quantities of goods purchased. Such an index is useful as long as the base quantities provide a reasonable representation of consumption patterns in succeeding time periods. However, sometimes purchasing patterns can change drastically as consumer preferences change or as dramatic price changes occur. If consumption patterns in the current period are very different from the quantities specified in the base period, then a Laspeyres index can be misleading because it relates to quantities of goods that few people would purchase.

A second version of the weighted aggregate price index is called a **Paasche index.** Here we update the quantities so that they reflect consumption patterns in the current time period.

> A **Paasche index** is
> $$\frac{\sum p_t q_t}{\sum p_0 q_t} \times 100$$
> where p_0 represents a base period price, p_t represents a current period price, and q_t represents a current period quantity.

As an example, Table 17.21 presents revised quantities for the grocery items in our previously discussed market basket. These quantities reflect increased consumption of apples, peanut butter, and bread, and decreased consumption of milk and eggs. We calculate a 2006 cost of $835.81 for the items in the market basket and a 2011 cost of $1065.71 for the items in the market basket. Therefore, the Paasche index is $(1065.71/835.81) \times 100 = 127.51$.

Because the Paasche index uses quantities from the current period, it reflects current buying habits. However, quantity data for the current period can be difficult to obtain. Furthermore, although each period is compared to the base period, it is difficult to compare the index at other points in time. This is because different quantities are used in different periods, and thus changes in the index are affected by changes in both prices and quantities.

Economic indexes

Several commonly quoted economic indexes are compiled monthly by the U.S. Bureau of Labor Statistics. Two important indexes are the **Consumer Price Index** (the **CPI**) and the **Producer Price Index** (the **PPI**). These are both *Laspeyres indexes.*

The CPI monitors the price of a market basket of goods and services that would be purchased by typical nonfarm consumers. Actually, there are two Consumer Price Indexes. The CPI-U, the Consumer Price Index for all Urban Workers, is often reported by the press as an indicator of price changes. Figure 17.33, which gives a plot of the monthly CPI-U from February 1947 until April 2011, shows the general increasing trend in prices over this period. The U.S. Bureau of the Census periodically changes the base period for the CPI. The plot in Figure 17.33 uses a base period of 1982–1984. Here, assume that

FIGURE 17.33   Plot of the Monthly CPI-U from February 1947 until April 2011

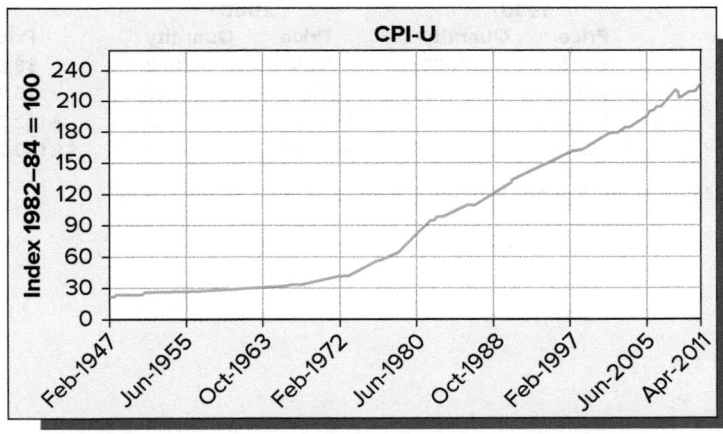

Source: Data from: www.data360.org/pdf_print_group.aspx Print_Group_Id=169.

the base period cost used to compute the CPI-U is the average of the 36 monthly market basket costs over the 1982–1984 base period. The April 2011 CPI-U, computed with a 1982–1984 base index of 100, was 224.91. This says that the cost of purchasing the market basket of goods and services was 124.91 percent higher in April 2011 than it was in the 1982–1984 base period. A second CPI, the CPI-W (Consumer Price Index for Urban Wage Earners and Clerical Workers) is often used to determine wage increases that are written into labor contracts.

The PPI tracks the prices of goods sold by wholesalers. An increase in the PPI is often regarded as an indication that retail prices will soon rise.

Exercises for Section 17.8

CONCEPTS connect

17.38 Explain the difference between a simple index and an aggregate index.

17.39 Explain the difference between a Laspeyres index and a Paasche index.

METHODS AND APPLICATIONS

17.40 Referring to the discussion of Figure 17.33, the CPI-U for March 2011 was 223.47, and the CPI-U for October 1963 was 30. Interpret these CPI-Us.

17.41 Recall that Table 17.18 gives the price of a gallon of regular gasoline in the United States (in dollars) for the years 2003 through 2011. Table 17.22 gives the price of a gallon of regular gasoline in the United States (in dollars) for the years 1990 through 2002.
  DS PriceGas2
  a By using 1990 as the base year, construct a simple index for the prices of a gallon of regular gasoline in Tables 17.22 and 17.18.
  b Plot the indices you have constructed versus time (1990 through 2011). Describe the pattern that you see.

17.42 Suppose that Table 17.23 on the next page gives the yearly automobile operating expenses in a particular

region of the United States for the years 1990, 2000, and 2011.   DS AutoExp
  a Using 1990 as the base year, construct the Laspeyres index for these operating expenses. Describe how the operating expenses have changed over time.
  b Using 1990 as the base year, construct the Paasche index for the operating expenses.
  c Another index, the **value index,** is given by the formula $(\Sigma p_t q_t / \Sigma p_0 q_0) \times$ 100. This index measures the changes in both the prices and quantities involved. Using 1990 as the base year, construct the value index for the operating expenses. Compare the three indices you have constructed.

TABLE 17.22

Gas Prices
DS PriceGas2

Year	Price per Gallon ($)
1990	1.16
1991	1.14
1992	1.13
1993	1.11
1994	1.11
1995	1.15
1996	1.23
1997	1.23
1998	1.06
1999	1.17
2000	1.51
2001	1.46
2002	1.36

TABLE 17.23   Yearly Automobile Operating Expenses in 1990, 2000, and 2011   Autoexp

Item	1990 Price	1990 Quantity	2000 Price	2000 Quantity	2011 Price	2011 Quantity
Gallon of gasoline	$1.16	1000	$1.51	882	$3.63	682
Quart of oil	$2.00	15	$3.50	15	$5.75	15
Tire	$125.00	2	$132.00	2	$142.00	2
Insurance	$750.00	1	$900.00	1	$1,100.00	1

Note: The quantities are based on 15,000 miles driven per year and a 30,000 mile tire lifetime.

Chapter Summary

In this chapter we have discussed using **univariate time series** models to forecast future time series values. We began by seeing that it can be useful to think of a time series as consisting of **trend, seasonal, cyclical, and irregular components.** If these components remain *constant* over time, then it is appropriate to describe and forecast the time series by using a **time series regression model.** We discussed using such models to describe **no trend,** a **linear trend,** and **constant seasonal variation** (by utilizing dummy variables). We also considered various transformations that transform **increasing seasonal variation** into constant seasonal variation. As an alternative to using a transformation and dummy variables to model increasing seasonal variation, we discussed using the **multiplicative decomposition** method. We then turned to a consideration of **exponential smoothing,** which is appropriate to use if the components of a time series may be *changing* over time. Specifically, we discussed **simple exponential smoothing, Holt–Winters' double exponential smoothing,** and **multiplicative Winters' method.** We next discussed the Box–Jenkins methodology, and then we explained how to compare forecasting methods by using the **mean absolute deviation (MAD),** the **mean squared deviation (MSD),** and the **mean absolute percentage error (MAPE).** We concluded this chapter by showing how to use **index numbers** to describe time-related data. Our discussion included the construction of a **simple index,** an **aggregate index,** a **Laspeyres index,** and a **Paasche index.**

Glossary of Terms

cyclical variation: Recurring up-and-down movements of a time series around trend levels that last more than one calendar year (often 2 to 10 years) from peak to peak or trough to trough.

deseasonalized time series: A time series that has had the effect of seasonal variation removed.

double exponential smoothing: An exponential smoothing procedure that can be used to forecast a time series described by a linear trend model with parameters that may be changing over time.

exponential smoothing: A forecasting method that weights recent observations more heavily than distantly past observations.

index number: A number that compares a value of a time series relative to another value of the time series.

irregular component: What is "left over" in a time series after trend, cycle, and seasonal variations have been accounted for.

MAD: The mean of the absolute values of a set of forecast errors.

MAPE: The mean of the absolute values of the percentage errors of a set of forecasts.

moving averages: Averages of successive groups of time series observations.

MSD: The mean of the squares of a set of forecast errors.

multiplicative Winters' method: An exponential smoothing procedure that can be used to forecast a time series described by a linear trend and increasing (or decreasing) seasonal variation with parameters that may be changing over time.

seasonal variation: Periodic patterns in a time series that complete themselves within a calendar year and are then repeated yearly.

simple exponential smoothing: An exponential smoothing procedure that can be used to forecast a time series described by a no trend model with an average level that may be changing over time.

smoothing constant: A number that determines how much weight is attached to each observation when using exponential smoothing.

time series: A set of observations that has been collected in time order.

trend: The long-run upward or downward movement that characterizes a time series over a period of time.

univariate time series model: A model that predicts future values of a time series solely on the basis of past values of the time series.

Important Formulas and Tests

No trend: page 758

Linear trend: page 759

Modeling constant seasonal variation by using dummy variables: pages 760–763

The multiplicative decomposition model: pages 767–773

Simple exponential smoothing: page 777

Double exponential smoothing: pages 780–783

Box–Jenkins methodology: page 789

Multiplicative Winters' method: pages 783–787

Mean absolute deviation (MAD): page 796

Mean absolute percentage error (MAPE): page 796

Mean squared deviation (MSD): page 796

Simple index: page 799

Aggregate price index: pages 800–801

Laspeyres index: page 801

Paasche index: page 802

Supplementary Exercises

connect **17.43** The State University Credit Union, a savings institution open to the faculty and staff of State University, handles savings accounts and makes loans to members. In order to plan its investment strategies, the credit union requires both point and prediction interval forecasts of monthly loan requests (in thousands of dollars) to be made by the faculty and staff in future months. The credit union has recorded monthly loan requests for its past two years of operation. These loan requests are as follows: **DS** Loans

Year 1	297	249	340	406	464	481
	549	553	556	642	670	712
Year 2	808	809	867	855	965	921
	956	990	1019	1021	1033	1127

Note: Time order is given by reading across the rows (left to right) from top to bottom.

If we use Minitab to fit the **quadratic trend model**

$$y_t = \beta_0 + \beta_1 t + \beta_2 t^2 + \varepsilon_t$$

to these data, we obtain the following partial Minitab output.

Term	Coef	SE Coef	T-Value	P-Value
Constant	199.62	20.85	9.58	0.000
Time	50.937	3.842	13.26	0.000
TimeSQ	−0.5677	0.1492	−3.80	0.001

S = 31.2469   R-Sq = 98.7%   R-Sq(adj) = 98.6%

Regression Equation
Y = 200 + 50.9 Time − 0.568 TimeSQ

Predicted Values for New Observations

New Obs	Time	TimeSQ	Fit	SE Fit
1	25.0	625	1118.21	20.85
2	26.0	676	1140.19	24.44

New Obs	95% CI	95% PI
1	(1074.85, 1161.56)	(1040.09, 1196.32)
2	(1089.37, 1191.01)	(1057.70, 1222.68)

a Does the quadratic term t^2 seem important in the model? Justify your answer.

b At the bottom of the Minitab output are point and prediction interval forecasts of loan requests in months 25 and 26. **(1)** Find and report these forecasts. **(2)** Then (using the least squares point estimates of β_0, β_1, and β_2 on the computer output) calculate the point forecasts.

17.44 The Olympia Paper Company, Inc., makes Absorbent Paper Towels. The company would like to develop a prediction model that can be used to give point forecasts and prediction interval forecasts of weekly sales over 100,000 rolls, in units of 10,000 rolls, of Absorbent Paper Towels. With a reliable model, Olympia Paper can more effectively plan its production schedule, plan its budget, and estimate requirements for producing and storing this product. For the past 120 weeks the company has recorded weekly sales of Absorbent Paper Towels. The 120 sales figures, $y_1, y_2, \ldots, y_{120}$, are given in Table 17.24, and the Minitab output resulting from using simple exponential smoothing to forecast future values of the sales figures is given in Figure 17.34 on the next page. Note that Minitab chooses a smoothing constant of 1.35442, which is

TABLE 17.24
Absorbent Paper Towels Sales   **DS** Towels

y_t	y_t	y_t
15.0000	9.2835	11.4986
14.4064	7.7219	13.2778
14.9383	6.8300	13.5910
16.0374	8.2046	13.4297
15.6320	8.5289	13.3125
14.3975	8.8733	12.7445
13.8959	8.7948	11.7979
14.0765	8.1577	11.7319
16.3750	7.9128	11.6523
16.5342	8.7978	11.3718
16.3839	9.0775	10.5502
17.1006	9.3234	11.4741
17.7876	10.4739	11.5568
17.7354	10.6943	11.7986
17.0010	9.8367	11.8867
17.7485	8.1803	11.2951
18.1888	7.2509	12.7847
18.5997	5.0814	13.9435
17.5859	1.8313	13.6859
15.7389	−0.9127	14.1136
13.6971	−1.3173	13.8949
15.0059	−0.6021	14.2853
16.2574	0.1400	16.3867
14.3506	1.4030	17.0884
11.9515	1.9280	15.8861
12.0328	3.5626	14.8227
11.2142	1.9615	15.9479
11.7023	4.8463	15.0982
12.5905	6.5454	13.8770
12.1991	8.0141	14.2746
10.7752	7.9746	15.1682
10.1129	8.4959	15.3818
9.9330	8.4539	14.1863
11.7435	8.7114	13.9996
12.2590	7.3780	15.2463
12.5009	8.1905	17.0179
11.5378	9.9720	17.2929
9.6649	9.6930	16.6366
10.1043	9.4506	15.3410
10.3452	11.2088	15.6453

Note: Time order is given by reading down the columns from left to right.

FIGURE 17.34   Minitab Output of Simple Exponential Smoothing for the Absorbent Paper Towels Sales
                Time Series ⒹⓈ Towels

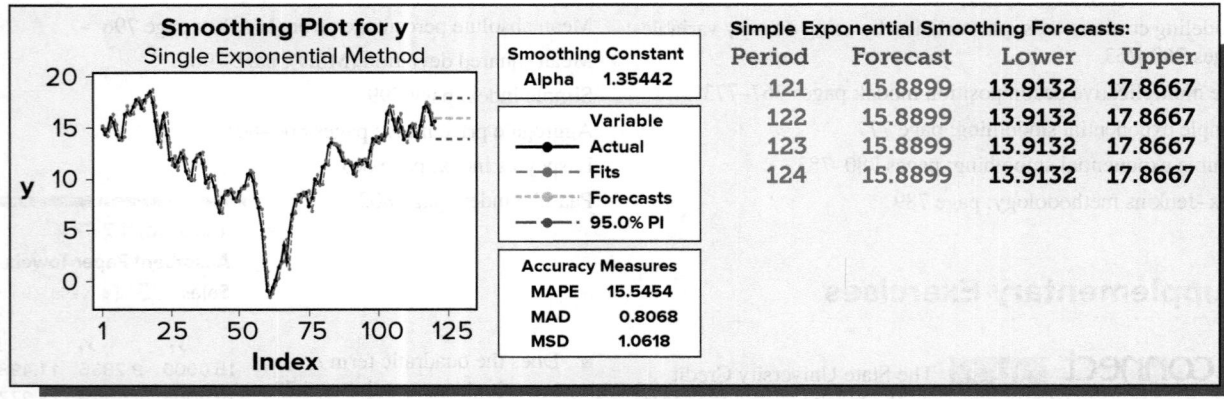

Simple Exponential Smoothing Forecasts:			
Period	Forecast	Lower	Upper
121	15.8899	13.9132	17.8667
122	15.8899	13.9132	17.8667
123	15.8899	13.9132	17.8667
124	15.8899	13.9132	17.8667

not (as is the usual case) between 0 and 1. In general, if the level of a time series is changing very quickly (as the plot of the sales values in Figure 17.34 indicates is true) Minitab will often choose a smoothing constant greater than 1. Use the Minitab output to find a point forecast of and a 95 percent prediction interval for the number of rolls of Absorbent Paper Towels that will be sold in a future week.

17.45   Alluring Tackle, Inc., a manufacturer of fishing equipment, makes the Bass Grabber, a type of fishing lure. The company would like to develop a prediction model that can be used to obtain point forecasts and prediction interval forecasts of the sales of the Bass Grabber. The sales (in tens of thousands of lures) of the Bass Grabber in sales period t, where each sales period is defined to last four weeks, are denoted by the symbol y_t and are believed to be partially determined by one or more of the independent variables x_1 = the price in period t of the Bass Grabber as offered by Alluring Tackle (in dollars); x_2 = the average industry price in period t of competitors' similar lures (in dollars); and x_3 = the advertising expenditure in period t of Alluring Tackle to promote the Bass Grabber (in tens of thousands of dollars). The data in Table 17.25 have been observed over the past 30 sales periods, and a plot of these data indicates that sales of the Bass Grabber have been increasing in a linear fashion over time and have been seasonal, with sales of the lure being largest in the spring and summer, when most recreational fishing takes place. Alluring Tackle believes that this pattern will continue in the future. Hence, remembering that each year consists of 13, four-week seasons, a possible regression model for predicting y_t would relate y_t to x_1, x_2, x_3, t, and the seasonal dummy variables S_2, S_3, . . . , S_{13}.

Here, for example, S_2 equals 1 if sales period t is the second four-week season, and 0 otherwise. As another example, S_{13} equals 1 if sales period t is the 13th four-week season, and 0 otherwise. If we calculate the least squares point estimates of the parameters of the model, we obtain the following prediction equation (the t statistic for the importance of each independent

variable is given in parentheses under the independent variable):
ⒹⓈ BassGrab

$$\hat{y}_t = .1776 + .4071x_1 - .7837x_2 + .9934x_3 + .0435t$$
$$\phantom{\hat{y}_t =} (0.05) \quad (0.42) \quad (-1.51) \quad (4.89) \quad (6.49)$$

$$+ .7800S_2 + 2.373S_3 + 3.488S_4 + 3.805S_5$$
$$ (3.16) \quad (9.28) \quad (12.88) \quad (13.01)$$

$$+ 5.673S_6 + 6.738S_7 + 6.097S_8 + 4.301S_9$$
$$ (19.41) \quad (23.23) \quad (21.47) \quad (14.80)$$

$$+ 3.856S_{10} + 2.621S_{11} + .9969S_{12} - 1.467S_{13}$$
$$ (13.89) \quad (9.24) \quad (3.50) \quad (-4.70)$$

a For sales period 31, which is the fifth season of the year, x_1 will be 3.80, x_2 will be 3.90, and x_3 will be 6.80. Using these values, it can be shown that a point prediction and a 95 percent prediction interval for sales of the Bass Grabber are, respectively, 10.578 and [9.683, 11.473]. Using the given prediction equation, calculate (within rounding) the point prediction 10.578.

TABLE 17.25   **Bass Grabber Sales** ⒹⓈ BassGrab

y_t	x_1	x_2	x_3	y_t	x_1	x_2	x_3
4.797	3.85	3.80	5.50	8.199	3.80	4.10	6.80
6.297	3.75	4.00	6.75	9.630	3.70	4.20	7.10
8.010	3.70	4.30	7.25	9.810	3.80	4.30	7.00
7.800	3.70	3.70	5.50	11.913	3.70	4.10	6.80
9.690	3.60	3.85	7.00	12.879	3.80	3.75	6.50
10.871	3.60	3.80	6.50	12.065	3.80	3.75	6.25
12.425	3.60	3.75	6.75	10.530	3.75	3.65	6.00
10.310	3.80	3.85	5.25	9.845	3.70	3.90	6.50
8.307	3.80	3.65	5.25	9.524	3.55	3.65	7.00
8.960	3.85	4.00	6.00	7.354	3.60	4.10	6.80
7.969	3.90	4.10	6.50	4.697	3.65	4.25	6.80
6.276	3.90	4.00	6.25	6.052	3.70	3.65	6.50
4.580	3.70	4.10	7.00	6.416	3.75	3.75	5.75
5.759	3.75	4.20	6.90	8.253	3.80	3.85	5.80
6.586	3.75	4.10	6.80	10.057	3.70	4.25	6.80

Note: Time order is given by reading down the y_t columns from left to right.

FIGURE 17.35   A Plot of Weekly Thermostat Sales
ⒹⓈ ThermoSales

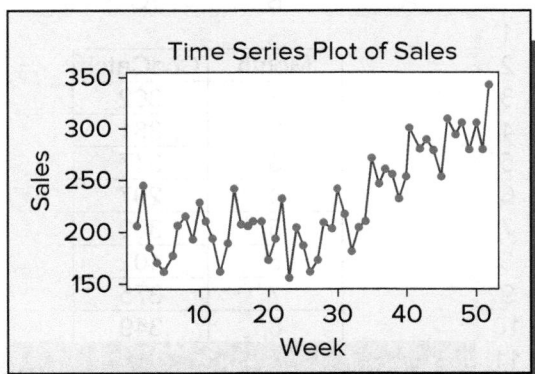

b Some *t* statistics indicate that some of the independent variables might not be important. Using the regression techniques of Chapters 14 and 15, try to find a better model for predicting sales of the Bass Grabber.

17.46 Weekly sales over the previous year of a new thermostat designed for residential use are shown below.

206 245 185 169 162 177 207 216 193 230 212 192 162

189 244 209 207 211 210 173 194 234 156 206 188 162

172 210 205 244 218 182 206 211 273 248 262 258 233

255 303 282 291 280 255 312 296 307 281 308 280 345

Note: Time order is given by reading across the rows (left to right) from top to bottom.

Figure 17.35 is a plot of these data; it suggests they might be described by a trend line with a changing *y*-intercept and slope. Use an appropriate forecasting technique to find point predictions of and 95 percent prediction intervals for thermostat sales in each of the next five weeks. [The source of the data in this exercise is Brown (1962).] ⒹⓈ ThermoSales

TABLE 17.26   Quarterly Sales of Tiger Sports Drink (1000s of Cases) ⒹⓈ TigerDrink

Quarter	Year							
	1	2	3	4	5	6	7	8
1	72	77	81	87	94	102	106	115
2	116	123	131	140	147	162	170	177
3	136	146	158	167	177	191	200	218
4	96	101	109	120	128	134	142	149

17.47 Table 17.26 gives quarterly sales (in thousands of cases) of Tiger Sports Drink over the previous eight years. Plot the data and use an appropriate forecasting technique to find point predictions of and 95 percent prediction intervals for the sales of Tiger Sports Drink in each quarter of next year.
ⒹⓈ TigerDrink

17.48 Consider the Absorbent Paper Towels sales values in Table 17.24. Figure 17.34 shows that these sales values are nonseasonal and nonstationary. At the nonseasonal level (lags 1 through 9) the SAC of the regular differences of the sales values has a spike at lag 1 and cuts off after lag 1, and the SPAC of these regular differences dies down fairly quickly. **(1)** Discuss which of the guidelines in Table 17.14 say that a tentative Box–Jenkins model describing the regular difference $z_t = y_t - y_{t-1}$ is $z_t = a_t - \theta_1 a_{t-1}$ and thus that a tentative model describing y_t is $y_t = y_{t-1} + a_t - \theta_1 a_{t-1}$. **(2)** Use Minitab or JMP to fit and diagnostically check this model (which is equivalent to a simple exponential smoothing model) and to obtain the following forecasts:

Box–Jenkins Model Forecasts:

Period	Forecast	Lower	Upper
121	15.8899	13.8532	17.9267
122	15.8899	12.4609	19.3189
123	15.8899	11.4891	20.2908
124	15.8899	10.6960	21.0839

Appendix 17.1 ■ Time Series Analysis Using Excel

How to create a forecast sheet as in Figure 17.13(a) for the cod catch data in Table 17.1 (data file: CodCatch.xlsx)

- In the Worksheet window, enter the data into two columns with variable names Month and CodCatch. Enter months as numbers 1 to 24 (two years).

- Select both data series.

- On the **Data** tab, in the Forecast group, click **Forecast Sheet.**

- In the **Create Forecast Worksheet** box, pick either a line chart or a column chart for the visual representation of the forecast.

- In the **Forecast End box,** pick an end date, and then click Create. This allows to choose how many time periods to forecast.

- Excel creates a new worksheet that contains both a table of the historical and predicted values and a chart that plots these points. You'll find the new worksheet just to the left ("in front of") the sheet where you entered the data series.

- If you want to change any advanced settings for your forecast, click **Options.** See the descriptions below for the various options:

Seasonality: You can override the automatic detection by choosing **Set Manually** and then picking a number that represents the number of repeating seasons. *Note: A value of 0 will cause Excel not to identify any seasonality.*

Include Forecast Statistics: Check this box if you want additional statistical information on the forecasts included in a new worksheet. Doing this adds a table of statistics generated such as the smoothing coefficients (Alpha, Beta, Gamma), and error metrics (MASE, SMAPE, MAE, RMSE).

	A	B	C
1			
2		Month	CodCatch
3		1	362
4		2	381
5		3	317
6		4	297
7		5	399
8		6	402
9		7	375
10		8	349

Source: Microsoft Office Excel 2016

Source: Microsoft Office Excel 2016

Source: Microsoft Office Excel 2016

Appendix 17.2 ■ Time Series Analysis Using Minitab

Simple exponential smoothing in the simple exponential smoothing summary box in Section 17.4 (data file: CodCatch.MTW):

- In the Data window, enter the cod catch data from Table 17.1.
- Select **Stat : Time Series : Single Exp Smoothing.**
- In the "Single Exponential Smoothing" dialog box, enter CodCatch in the Variable window.
- To request that Minitab select the smoothing constant, select the "Optimal ARIMA" option under "Weight to Use in Smoothing." To choose your own smoothing constant, select the "Use" option and enter a smoothing weight.
- Check the "Generate forecasts" box and enter 12 in the "Number of forecasts:" and 24 in the "Starting from origin:" boxes.
- Click OK in the "Single Exponential Smoothing" dialog box to see the forecast results in the Session window and a graphical summary in a high-resolution graphics window.

Double exponential smoothing can be performed by choosing "Double Exp Smoothing" from the Time Series menu and by following the remainder of the preceding steps.

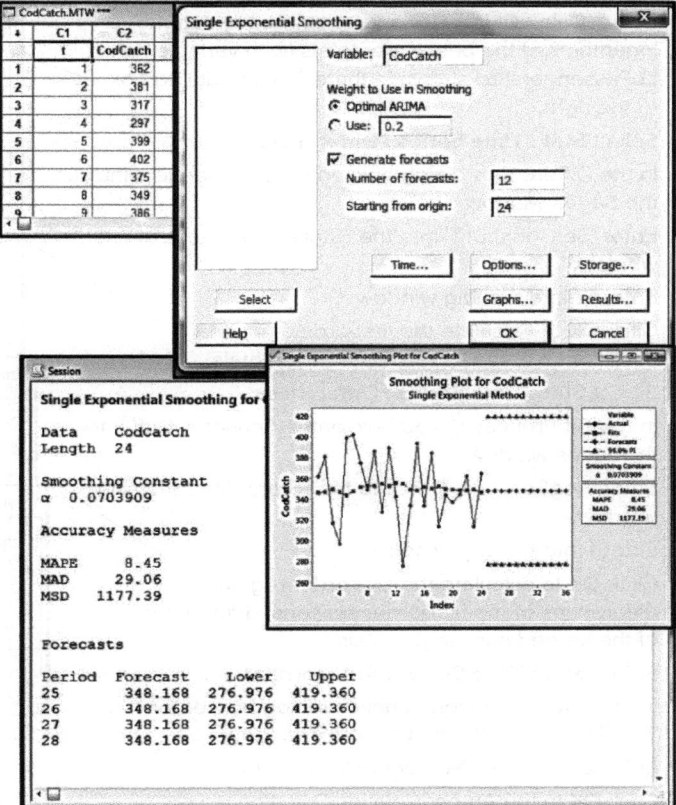

Source: Minitab 18

Multiplicative Winters' method in Figure 17.13(d) and Figure 17.20(a) (data file: TastyCola.MTW):

- In the Data window, enter the TastyCola data from Table 17.9.
- Select **Stat : Time Series : Winters' Method.**
- In the "Winters' Method" dialog box, enter Sales in the Variable window.
- Enter 12 in the "Seasonal length" box.
- Click the Multiplicative option under "Method Type."
- Use the default values for "Weights to Use in Smoothing" (0.2 for Level, Trend, and Seasonal).
- Check the "Generate forecasts" box and enter 12 in the "Number of forecasts:" and 36 in the "Starting from origin:" boxes.
- Click OK in the "Winters' Method" dialog box to see the forecast results in the Session window and a graphical summary in a high-resolution graphics window.

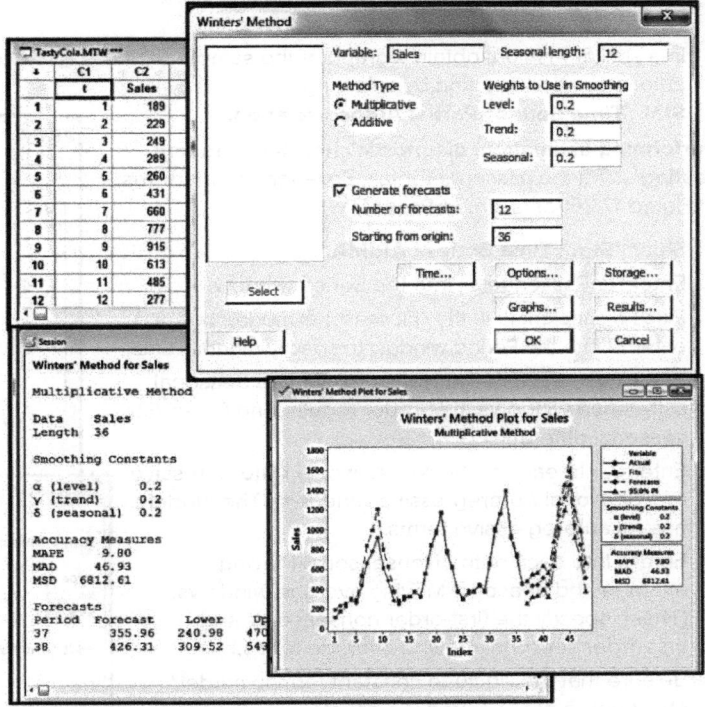

Source: Minitab 18

Computing the SAC and SPAC of the combined regular and seasonal differences of the logged passenger totals in Figure 17.25 (data file: AirPass.MTW):

- Select **Calc : Calculator** and calculate the natural logarithms of the passenger totals (with variable name LnPassengers) as shown in the calculator dialog box to the right.

- Select **Stat : Time Series : Differences.**

- In the Differences dialog box, enter LnPassengers into the Series window.

- Enter 'Seasonal Diff' into the "Store differences in" window.

- Enter 12 into the Lag window.

- Click OK to calculate the first-order seasonal differences of the logged passenger totals.

- Select **Stat : Time Series : Differences.**

- In the Differences dialog box, enter 'Seasonal Diff' into the series window.

- Enter 'RegSeasonal Diff' into the "Store differences in" window.

- Enter 1 into the Lag window.

- Click OK to calculate the first-order regular differences of the first-order seasonal differences of the logged passenger totals.

- Select **Stat : Time Series : Autocorrelation.**

- In the Autocorrelation Function dialog box, enter 'RegSeasonal Diff' into the series window.

- Enter 24 into the "Number of lags" window.

- Click OK in the Autocorrelation Function dialog box to obtain a graph of the sample autocorrelation function of the combined regular and seasonal differences of the logged passenger totals.

- In a similar fashion, obtain a graph of the sample partial autocorrelation function by selecting **Stat : Time Series : Partial Autocorrelation.**

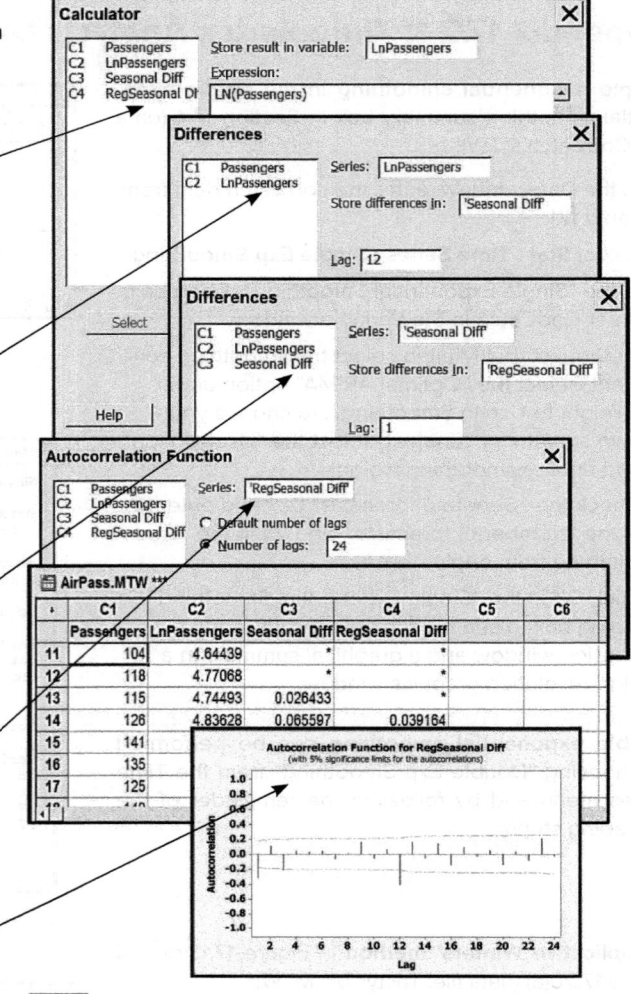

Source: Minitab 18

Performing estimation, diagnostic checking, and forecasting using the passenger totals Box–Jenkins model as in Figure 17.26 (data file: AirPass.MTW):

- Select **Stat : Time Series : ARIMA.**

- Enter "LnPassengers" into the Series window.

- Place a checkmark in the "Fit seasonal model" box and enter 12 into the Period window to specify monthly data.

- Enter 1 into each of the Nonseasonal and Seasonal Difference windows (first-order regular and first-order seasonal differencing).

- Enter 0 into each of the Nonseasonal Autoregressive and Seasonal Autoregressive windows. (The model has no autoregressive terms.)

- Enter 1 into each of the Nonseasonal Moving average and Seasonal Moving average windows. (These specify the first-order nonseasonal and first-order seasonal moving-average terms.)

- Be sure that the "Include constant term in model" checkbox is not checked (no constant in model).

- Click on the Forecasts... button and enter 12 into the "Lead" window to forecast 12 months into the future.

- Click OK in the ARIMA dialog box.

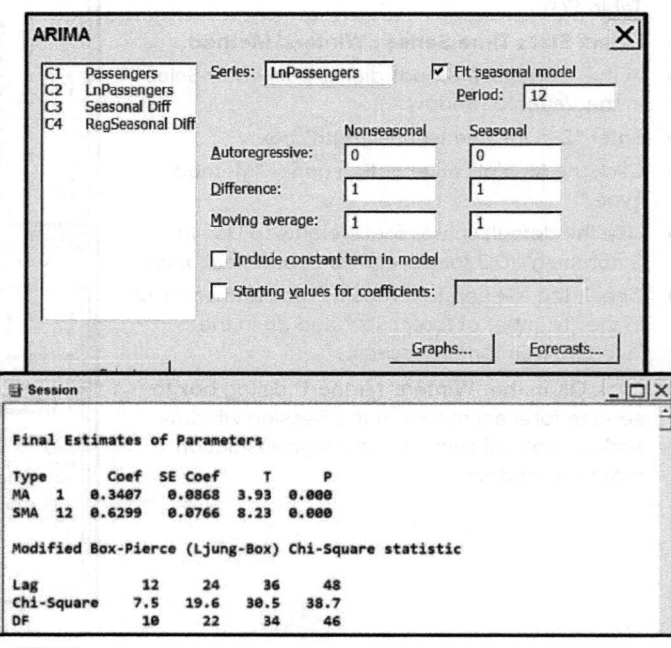

Source: Minitab 18

Appendix 17.3 ■ Time Series Analysis Using JMP

Simple Exponential Smoothing in Figure 17.14a (data file: Prince'sRepair.jmp):

- After opening the Prince'sRepair.jmp file, select **Analyze : Specialized Modeling : Time Series**.

- Select the "Revenue" column and click the "Y, Time Series" button. Select the "Week" column and click the "X, Time ID" button. Click "OK."

- From the red triangle menu beside "Time Series Revenue," select **Smoothing Model : Simple Exponential Smoothing.** Click "Estimate" to fit the model.

- To get a table of forecasts and associated 95% confidence limits for the next 4 weeks, select "Save Columns" from the red triangle menu beside "Model: Simple Exponential Smoothing." A new data table will appear containing the forecasts and associated confidence intervals.

	Week	Revenue
1	1	1800
2	2	2200
3	3	1800
4	4	2300

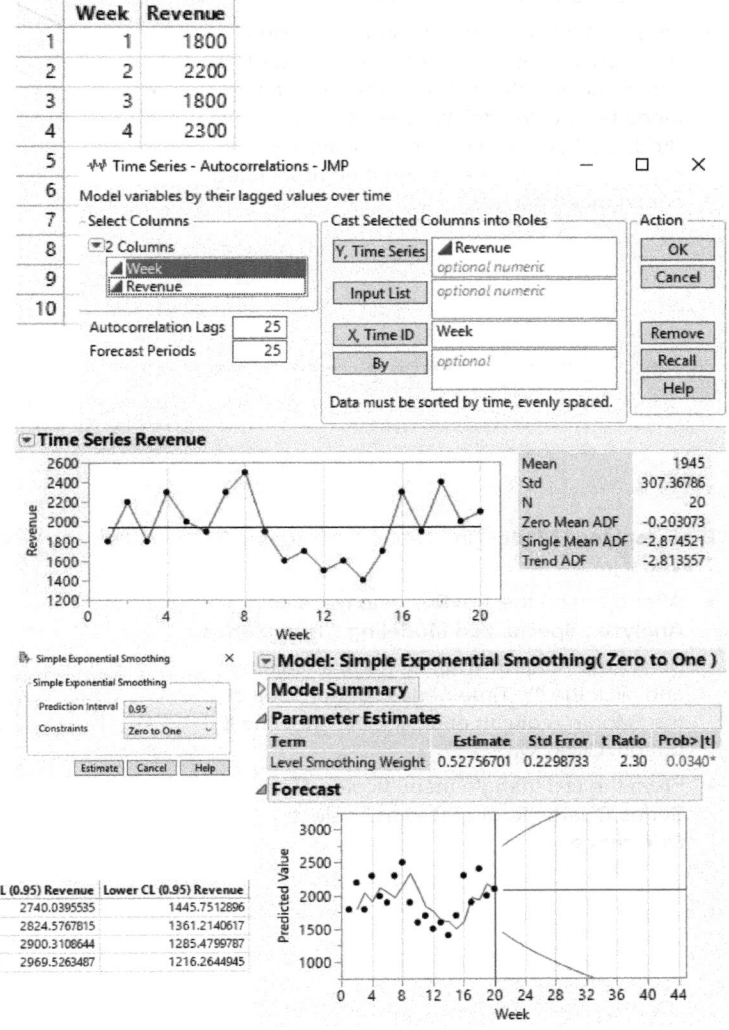

Actual Revenue	Week	Predicted Revenue	Std Err Pred Revenue	Residual Revenue	Upper CL (0.95) Revenue	Lower CL (0.95) Revenue
•	21	2092.8954216	330.18164468	•	2740.0395535	1445.7512896
•	22	2092.8954216	373.31367601	•	2824.5767815	1361.2140617
•	23	2092.8954216	411.95422428	•	2900.3108644	1285.4799787
•	24	2092.8954216	447.26889576	•	2969.5263487	1216.2644945

Source: JMP Pro 14

Source: JMP Pro 14

Additive Winters Method in Table 17.3 (data file: BikeSales.jmp):

- After opening the BikeSales.jmp file, select **Analyze : Specialized Modeling: Time Series**.

- Select the "Sales" column and click the "Y, Time Series" button. Select the "t" column and click the "X, Time ID" button. Click "OK."

- From the red triangle menu beside "Time Series Sales," select **Smoothing Model : Winters Methods**.

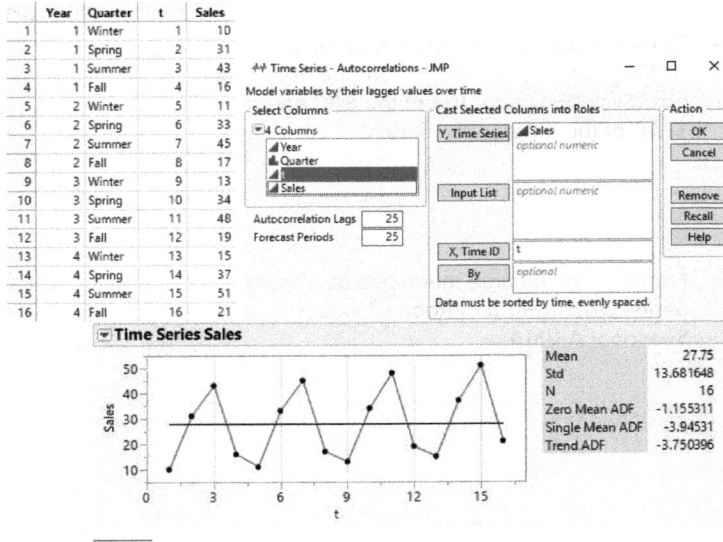

Source: JMP Pro 14

- In the Winters Method dialog box, enter "4" for the "Observations per Period" since we have quarterly data. Click "Estimate."

- To get a table of forecasts and associated 95% confidence limits for the next 4 quarters, select "Save Columns" from the red triangle menu beside "Model: Winters Method (Additive)." A new data table will appear containing the forecasts and their associated confidence intervals.

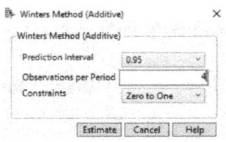

Model: Winters Method (Additive)

▷ Model Summary

△ Parameter Estimates

Term	Estimate	Std Error	t Ratio	Prob>|t|
Level Smoothing Weight	0.1240123	0.173840	0.71	0.4959
Trend Smoothing Weight	1.0000000	1.672999	0.60	0.5666
Seasonal Smoothing Weight	1.0000000	0.318370	3.14	0.0138*

△ Forecast

Actual Sales	t	Predicted Sales	Std Err Pred Sales	Residual Sales	Upper CL (0.95) Sales	Lower CL (0.95) Sales
•	17	17.672446808	0.8416187438	•	19.321989234	16.022904381
•	18	39.662861907	0.8671190212	•	41.362383959	37.963339855
•	19	53.580577669	0.9219193526	•	55.387506397	51.773648942
•	20	23.580577669	1.0120417188	•	25.564142989	21.59701235

Source: JMP Pro 14

Source: JMP Pro 14

Box–Jenkins Model in Table 17.4 (data file: TravRest.jmp):

- After opening the TravRest.jmp file, select **Analyze : Specialized Modeling : Time Series**.

- Select the "Quartic Root Of Rooms" column and click the "Y, Time Series" button. Select the "Month" column and click the "X, Time ID" button. Click "OK."

- From the red triangle menu beside "Time Series Quartic Root of Rooms," select **Difference**.

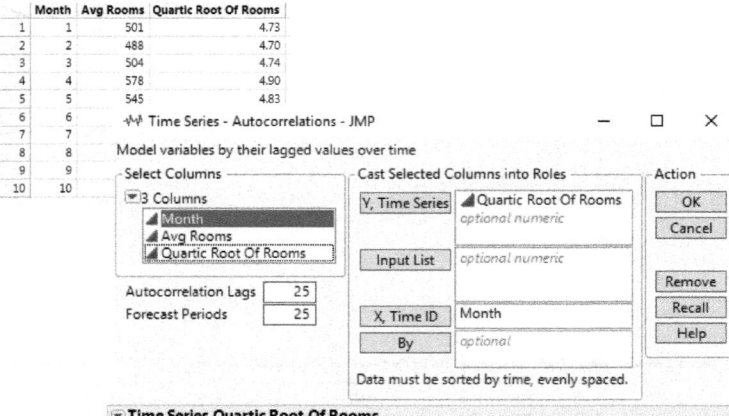

	Month	Avg Rooms	Quartic Root Of Rooms
1	1	501	4.73
2	2	488	4.70
3	3	504	4.74
4	4	578	4.90
5	5	545	4.83
6	6		
7	7		
8	8		
9	9		
10	10		

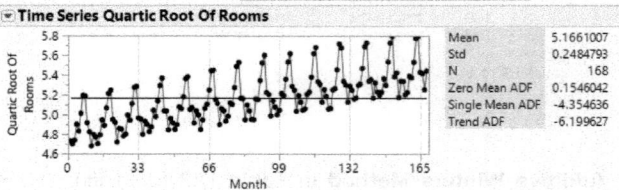

- To take a seasonal difference, select a "Differencing Order" of "1" from the Seasonal box. Click "Estimate" to get the SAC and SPAC of the differenced series.

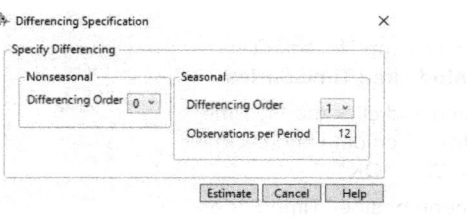

- From the red triangle menu beside "Time Series Quartic Root of Rooms," select **Seasonal ARIMA**.

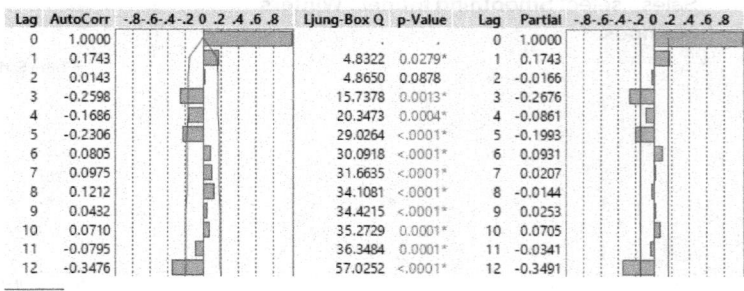

Lag	AutoCorr	-.8-.6-.4-.2 0 .2 .4 .6 .8	Ljung-Box Q	p-Value	Lag	Partial	-.8-.6-.4-.2 0 .2 .4 .6 .8
0	1.0000		.	.	0	1.0000	
1	0.1743		4.8322	0.0279*	1	0.1743	
2	0.0143		4.8650	0.0878	2	-0.0166	
3	-0.2598		15.7378	0.0013*	3	-0.2676	
4	-0.1686		20.3473	0.0004*	4	-0.0861	
5	-0.2306		29.0264	<.0001*	5	-0.1993	
6	0.0805		30.0918	<.0001*	6	0.0931	
7	0.0975		31.6635	<.0001*	7	0.0207	
8	0.1212		34.1081	<.0001*	8	-0.0144	
9	0.0432		34.4215	<.0001*	9	0.0253	
10	0.0710		35.2729	0.0001*	10	0.0705	
11	-0.0795		36.3484	0.0001*	11	-0.0341	
12	-0.3476		57.0252	<.0001*	12	-0.3491	

Source: JMP Pro 14

- The analysis indicated that a seasonal difference should be used. To take this seasonal difference, enter "1" for the "Differencing Order" in the "Seasonal ARIMA" box. The analysis of the SAC and SPAC also indicated the need for three nonseasonal autoregressive terms and one seasonal moving average term. To fit this model, enter "3" for the "Autoregressive Order" in the "ARIMA" box and enter "1" for the "Moving Average Order" in the "Seasonal ARIMA" box. Click "Estimate."

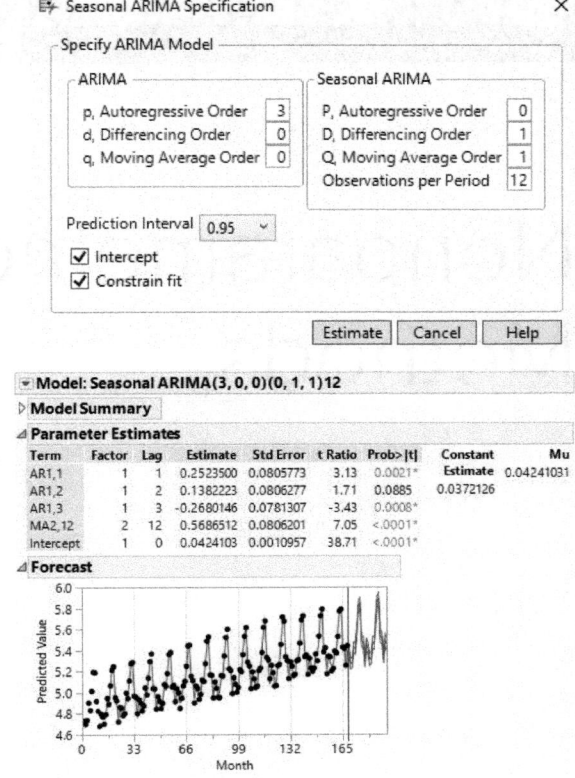

Seasonal ARIMA Specification

Specify ARIMA Model

ARIMA
p, Autoregressive Order	3
d, Differencing Order	0
q, Moving Average Order	0

Seasonal ARIMA
P, Autoregressive Order	0
D, Differencing Order	1
Q, Moving Average Order	1
Observations per Period	12

Prediction Interval 0.95

☑ Intercept
☑ Constrain fit

Estimate   Cancel   Help

Model: Seasonal ARIMA(3, 0, 0)(0, 1, 1)12

▷ Model Summary

Parameter Estimates

Term	Factor	Lag	Estimate	Std Error	t Ratio	Prob>|t|		Constant	Mu
AR1,1	1	1	0.2523500	0.0805773	3.13	0.0021*		Estimate	0.04241031
AR1,2	1	2	0.1382223	0.0806277	1.71	0.0885		0.0372126	
AR1,3	1	3	-0.2680146	0.0781307	-3.43	0.0008*			
MA2,12	2	12	0.5686512	0.0806201	7.05	<.0001*			
Intercept	1	0	0.0424103	0.0010957	38.71	<.0001*			

Forecast

[Forecast plot: Predicted Value vs Month, x-axis labeled 0, 33, 66, 99, 132, 165; y-axis from 4.6 to 6.0]

Source: JMP Pro 14

- From the red triangle menu beside "Model: Seasonal ARIMA" and choose "Save Columns" to create a table of forecasts and their associated confidence intervals.

Month	Predicted Quartic Root Of Rooms	Upper CL (0.95) Quartic Root Of Rooms	Lower CL (0.95) Quartic Root Of Rooms
169	5.38	5.43	5.33
170	5.27	5.32	5.22
171	5.28	5.33	5.22
172	5.43	5.48	5.38
173	5.41	5.46	5.36
174	5.59	5.65	5.54
175	5.82	5.87	5.77
176	5.85	5.91	5.80
177	5.48	5.53	5.43
178	5.48	5.53	5.43
179	5.29	5.34	5.24
180	5.46	5.51	5.41

Source: JMP Pro 14

- To view the SAC and SPAC of the residuals from the model, expand the "Residuals" area by clicking the gray triangle next to "Residuals."

Residuals

Lag	AutoCorr	-.8-.6-.4-.2 0 .2 .4 .6 .8	Ljung-Box Q	p-Value	Lag	Partial	-.8-.6-.4-.2 0 .2 .4 .6 .8
0	1.0000				0	1.0000	
1	-0.0126		0.0251	0.8740	1	-0.0126	
2	-0.0321		0.1898	0.9095	2	-0.0322	
3	0.0436		0.4957	0.9198	3	0.0428	
4	-0.0201		0.5612	0.9673	4	-0.0201	
5	-0.1416		3.8336	0.5736	5	-0.1398	
6	0.0336		4.0189	0.6741	6	0.0277	
7	0.0570		4.5556	0.7140	7	0.0524	
8	-0.0237		4.6491	0.7943	8	-0.0102	
9	-0.0316		4.8170	0.8500	9	-0.0388	
10	0.1503		8.6292	0.5676	10	0.1299	
11	0.0424		8.9353	0.6279	11	0.0585	
12	-0.0120		8.9601	0.7063	12	0.0099	
13	0.1313		11.9337	0.5331	13	0.1188	
14	0.1361		15.1501	0.3680	14	0.1418	
15	-0.0031		15.1518	0.4405	15	0.0556	
16	0.0339		15.3539	0.4989	16	0.0466	
17	0.0684		16.1828	0.5109	17	0.0635	
18	-0.1693		21.3006	0.2645	18	-0.1341	
19	0.0876		22.6827	0.2516	19	0.1314	
20	0.0139		22.7177	0.3029	20	-0.0209	
21	0.0605		23.3868	0.3237	21	0.0664	
22	-0.0156		23.4313	0.3777	22	-0.0130	
23	0.0520		23.9318	0.4076	23	-0.0056	
24	0.0119		23.9582	0.4640	24	-0.0112	
25	0.1009		25.8734	0.4144	25	0.1066	

Source: JMP Pro 14

Design Elements: (CD): ©Comstock Images/Alamy; (All Others): ©McGraw-Hill Education

## CHAPTER 18

Nonparametric Methods

©DaniloAndjus/Getty Images

Learning Objectives

After mastering the material in this chapter, you will be able to:

LO18-1 Use the sign test to test a hypothesis about a population median.

LO18-2 Compare the locations of two distributions using a rank sum test for independent samples.

LO18-3 Compare the locations of two distributions using a signed ranks test for paired samples.

LO18-4 Compare the locations of three or more distributions using a Kruskal–Wallis test for independent samples.

LO18-5 Measure and test the association between two variables by using Spearman's rank correlation coefficient.

Chapter Outline

R ecall from Chapter 3 that the manufacturer of a DVD recorder has randomly selected a sample of 20 purchasers who have owned the recorder for one year. Each purchaser in the sample is asked to rank his or her satisfaction with the recorder along the following 10-point scale:

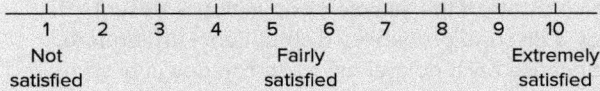

| 1 | 2 | 3 | 4 | 5 | 6 | 7 | 8 | 9 | 10 |
| Not
satisfied | | | | Fairly
satisfied | | | | | Extremely
satisfied |

The stem-and-leaf display below gives the 20 ratings obtained.

```
 1   0
 2
 3   0
 4
 5   00
 6
 7   0
 8   000000
 9   00000
10   0000
```

Let μ denote the mean rating that would be given by all purchasers who have owned the DVD recorder for one year, and suppose we wish to show that μ exceeds 7. To do this, we will test $H_0: \mu \leq 7$ versus $H_a: \mu > 7$. The mean and the standard deviation of the sample of 20 ratings are $\bar{x} = 7.7$ and $s = 2.4301$, and the test statistic t is

$$t = \frac{\bar{x} - 7}{s/\sqrt{n}} = \frac{7.7 - 7}{2.4301/\sqrt{20}} = 1.2882$$

Because $t = 1.2882$ is less than $t_{.10} = 1.328$ (based on 19 degrees of freedom), we cannot reject $H_0: \mu \leq 7$ by setting α equal to .10. That is, the t test does not provide even mildly strong evidence that μ exceeds 7. But how appropriate is the t test in this situation? The t test is, in fact, not appropriate for two reasons:

1 The t test assumes that, when the sample size n is small (less than 30), the sampled population is normally distributed (or, at least, mound-shaped and not highly skewed to the right or left). The stem-and-leaf display of the ratings indicates the population of all DVD recorder ratings might be highly skewed to the left.

2 The ratings of 1 and 3 in the stem-and-leaf display are **outliers** (see Figure 3.17). These outliers affect both the sample mean and the sample standard deviation. First, the sample mean of 7.7 is "pulled down" by the low ratings and thus is smaller than the sample median, which is 8. Although there is not much difference here between the mean and the median, the outlier and overall skewness indicate that the median might be a better measure of central tendency. More important, however, is the fact that the low ratings inflate the sample standard

deviation s. As a result, although the sample mean of 7.7 is greater than 7, the inflated s of 2.4301 makes the denominator of the t statistic large enough to cause us to not reject $H_0: \mu \leq 7$. Intuitively, therefore, even if the population mean DVD recorder rating really does exceed 7, the t test is not **powerful enough** to tell us that this is true.

In addition, some statisticians would consider the t test to be inappropriate for a third reason. The variable DVD recorder rating is an *ordinal variable*. Recall from Section 1.6 that an **ordinal variable** is a **qualitative variable** with a meaningful **ordering,** or **ranking,** of the categories. In general, when the measurements of an ordinal variable are numerical, statisticians debate whether the ordinal variable is "somewhat quantitative." Statisticians who argue that DVD recorder rating is not somewhat quantitative would reason, for instance, that the difference between 10 ("extremely satisfied") and 6 ("fairly satisfied") may not be the same as the difference between 5 ("fairly satisfied") and 1 ("not satisfied"). In other words, although each difference is four rating points, the two differences may not be the same qualitatively. Other statisticians would argue that as soon as respondents see equally spaced numbers (even though the numbers are described by words), their responses are influenced enough to make the ordinal variable somewhat quantitative. In general, the choice of words associated with the numbers probably substantially affects whether an ordinal variable may be considered somewhat quantitative. However, in practice, numerical ordinal ratings are often analyzed as though they are quantitative. For example, although a teacher's effectiveness rating given by a student and a student's course grade are both ordinal variables with the possible measurements 4 ("excellent"), 3 ("good"), 2 ("average"), 1 ("poor"), and 0 ("unsatisfactory"), a teacher's effectiveness **average** and a student's grade point **average** are calculated. Furthermore, some statisticians would argue that when there are "fairly many" numerical ordinal ratings (for example, the 10 ratings in the DVD recorder example), it is even more reasonable to consider the ratings somewhat quantitative and thus to analyze means and variances. However, for statisticians who feel that numerical ordinal ratings should never be considered quantitative, analyzing the means and standard deviations of these ratings—and thus performing t tests—would always be considered inappropriate.

In general, consider the one-sample t test (see Section 10.3), the two independent sample t tests (see Section 11.1), the paired difference t test (see Section 11.2), and the one-way analysis of variance F test (see Section 12.2). All of these procedures assume that the sampled populations are normally distributed (or mound-shaped and not highly skewed to the right or left). When this assumption is not satisfied, we can use techniques that do not require assumptions about the shapes of the probability distributions of the sampled populations. These techniques are often called **nonparametric methods,** and we discuss several of these methods in this chapter. Specifically, we consider four nonparametric tests that can be used in place of the previously mentioned t and F tests. These four nonparametric tests are the **sign test,** the **Wilcoxon ranks sum test,** the **Wilcoxon signed ranks test,** and the **Kruskal–Wallis H test.** These tests require no assumptions about the probability distributions of the sampled populations. In addition, these nonparametric tests are usually better than the t and F tests at correctly finding statistically significant differences in the presence of outliers and extreme skewness. Therefore, we say that the nonparametric tests can be **more powerful** than the t and F tests. For example, we will find in Section 18.1 that, although the t test does not allow us to conclude that the population **mean** DVD recorder rating exceeds 7, the nonparametric sign test does allow us to conclude that the population **median** DVD recorder rating exceeds 7.

Each nonparametric test discussed in this chapter assumes that each sampled population under consideration is described by a continuous probability distribution. However, in most situations, each nonparametric technique is slightly **statistically**

conservative if the sampled population is described by a discrete probability distribution. This means, for example, that a nonparametric hypothesis test has a slightly smaller chance of falsely rejecting the null hypothesis than the specified α value would seem to indicate, if the sampled population is described by a discrete probability distribution. Furthermore, because each nonparametric technique is based essentially on **ranking** the observed sample values, and not on the exact sizes of the sample values, each nonparametric technique can be used to analyze any type of data that can be ranked. This includes ordinal data (for example, teaching effectiveness ratings and DVD recorder ratings) in addition to quantitative data.

To conclude this introduction, we note that t and F tests are more powerful (better at correctly finding statistically significant differences) than nonparametric tests when the sampled populations are normally distributed (or mound-shaped and not highly skewed to the right or left). In addition, nonparametric tests are largely limited to simple settings. For example, there is a nonparametric measure of correlation between two variables— **Spearman's rank correlation coefficient**—which is discussed at the end of this chapter. However, nonparametric tests do not extend easily to multiple regression and complex experimental designs. This is one reason why we have stressed t and F procedures in this book. These procedures can be extended to more advanced statistical methods.

LO18-1

Use the sign test to test a hypothesis about a population median.

18.1 The Sign Test: A Hypothesis Test about the Median

If a population is highly skewed to the right or left, then the population median might be a better measure of central tendency than the population mean. Furthermore, if the sample size is small and the population is highly skewed or clearly non–mound-shaped, then the t test for the population mean that we have presented in Section 10.3 might not be valid. For these reasons, when we have taken a small sample and when we believe that the sampled population might be far from being normally distributed, it is sometimes useful to use a hypothesis test about the population median. This test, called the **sign test,** is valid for any sample size and population shape. To illustrate the sign test, we consider the following example.

EXAMPLE 18.1 The CD Player Case: Testing a Product Lifetime

The leading compact disc player is advertised to have a median lifetime (or time to failure) of 6,000 hours of continuous play. The developer of a new compact disc player wishes to show that the median lifetime of the new player exceeds 6,000 hours of continuous play. To this end, the developer randomly selects 20 new players and tests them in continuous play until each fails. Figure 18.1(a) presents the 20 lifetimes obtained (expressed in hours and

FIGURE 18.1   The Compact Disc Player Lifetime Data and Associated Statistical Analyses

(a) The compact disc player lifetime data 🔵 CompDisc

| 5 | 947 | 2,142 | 4,867 | 5,840 | 6,085 | 6,238 | 6,411 | 6,507 | 6,687 |
| 6,827 | 6,985 | 7,082 | 7,176 | 7,285 | 7,410 | 7,563 | 7,668 | 7,724 | 7,846 |

(c) Minitab output of the sign test of $H_0: M_d = 6{,}000$ **versus** $H_a: M_d > 6{,}000$

Sign test of median = 6000 versus > 6000

	N	Below	Equal	Above	P	Median
LifeTime	20	5	0	15	0.0207	6757

(d) Excel add-in (MegaStat) output of the sign test of $H_0: M_d = 6{,}000$ **versus** $H_a: M_d > 6{,}000$

Sign Test

6000 hypothesized value	5 below		binomial
6757 median Life Time	0 equal	.0207	p-value (one-tailed, upper)
20 n	15 above		

(b) A stem-and-leaf display

0	005
0	947
1	
1	
2	142
2	
3	
3	
4	
4	867
5	
5	840
6	085 238 411
6	507 687 827 985
7	082 176 285 410
7	563 668 724 846

arranged in increasing order), and Figure 18.1(b) shows a stem-and-leaf display of these lifetimes. The stem-and-leaf display and the three low lifetimes of 5, 947, and 2,142 suggest that the population of all lifetimes might be highly skewed to the left. In addition, the sample size is small. Therefore, it might be reasonable to use the sign test.

In order to show that the population median lifetime, M_d, of the new compact disc player exceeds 6,000 (hours), recall that this median divides the population of ordered lifetimes into two equal parts. It follows that, if more than half of the individual population lifetimes exceed 6,000, then the population median, M_d, exceeds 6,000. Let p denote the proportion of the individual population lifetimes that exceed 6,000. Then, we can reject $H_0: M_d = 6{,}000$ in favor of $H_a: M_d > 6{,}000$ if we can reject $H_0: p = .5$ in favor of $H_a: p > .5$. Let x denote the total number of lifetimes that exceed 6,000 in a random sample of 20 lifetimes. If $H_0: p = .5$ is true, then x is a binomial random variable where $n = 20$ and $p = .5$. This says that if $H_0: p = .5$ is true, then we would expect $\mu_x = np = 20(.5) = 10$ of the 20 lifetimes to exceed 6,000. Considering the 20 lifetimes we have actually observed, we note that 15 of these 20 lifetimes exceed 6,000. The p-value for testing $H_0: p = .5$ versus $H_a: p > .5$ is the probability, computed assuming that $H_0: p = .5$ is true, of observing a sample result that is at least as contradictory to H_0 as the sample result we have actually observed. Because any number of lifetimes out of 20 lifetimes that is greater than or equal to 15 is at least this contradictory, we have

$$p\text{-value} = P(x \geq 15) = \sum_{x=15}^{20} \frac{20!}{x!(20-x)!}(.5)^x(.5)^{20-x}$$

Using the binomial distribution table in Table A.1, or the JMP output on the page margin, we find that

$$p\text{-value} = P(x \geq 15)$$
$$= P(x = 15) + P(x = 16) + P(x = 17) + P(x = 18)$$
$$+ P(x = 19) + P(x = 20)$$
$$= .0148 + .0046 + .0011 + .0002 + .0000 + .0000$$
$$= .0207$$

x	P(X=x)
0	9.5367432e-7
1	0.0000190735
2	0.0001811981
3	0.0010871887
4	0.0046205521
5	0.0147857666
6	0.0369644165
7	0.073928833
8	0.1201343536
9	0.1601791382
10	0.176197052
11	0.1601791382
12	0.1201343536
13	0.073928833
14	0.0369644165
15	0.0147857666
16	0.0046205521
17	0.0010871887
18	0.0001811981
19	0.0000190735
20	9.5367432e-7

This says that if $H_0: p = .5$ is true, then the probability that at least 15 out of 20 lifetimes would exceed 6,000 is only .0207. Because it is difficult to believe that such a small chance would occur, we have strong evidence against $H_0: p = .5$ and in favor of $H_a: p > .5$. That is, we have strong evidence that $H_0: M_d = 6{,}000$ is false and $H_a: M_d > 6{,}000$ is true. This implies that it is reasonable to conclude that the median lifetime of the new compact disc player exceeds the advertised median lifetime of the market's leading compact disc player.

Figure 18.1(c) and (d) present the Minitab and Excel add-in (MegaStat) outputs of the sign test of $H_0: M_d = 6,000$ versus $H_a: M_d > 6,000$. In addition, the outputs tell us that a point estimate of the population median lifetime is the sample median of 6,757 hours.

We summarize how to carry out the sign test in the following box:

The Sign Test for a Population Median

Suppose we have randomly selected a sample of size n from a population, and suppose we wish to test the null hypothesis $H_0: M_d = M_0$ versus one of $H_a: M_d < M_0$, $H_a: M_d > M_0$, or $H_a: M_d \neq M_0$ where M_d denotes the population median. Define the test statistic S as follows:

If the alternative is $H_a: M_d < M_0$, then S = the number of sample measurements less than M_0.

If the alternative is $H_a: M_d > M_0$, then S = the number of sample measurements greater than M_0.

If the alternative is $H_a: M_d \neq M_0$, then S = the larger of S_1 and S_2

where S_1 = the number of sample measurements less than M_0, and

S_2 = the number of sample measurements greater than M_0.

Furthermore, define x to be a binomial variable with parameters n and $p = .5$. Then, we can test $H_0: M_d = M_0$ versus a particular alternative hypothesis at level of significance α by using the appropriate p-value.

Alternative Hypothesis	p-Value (reject H_0 if p-value $< \alpha$)
$H_a: M_d > M_0$	The probability that x is greater than or equal to S
$H_a: M_d < M_0$	The probability that x is greater than or equal to S
$H_a: M_d \neq M_0$	Twice the probability that x is greater than or equal to S

Here we can use Table A.1 to find the p-value.

When we take a large sample, we can use the normal approximation to the binomial distribution to implement the sign test. Here, when the null hypothesis $H_0: M_d = M_0$ (or $H_0: p = .5$) is true, the binomial variable x is approximately normally distributed with mean $np = n(.5) = .5n$ and standard deviation $\sqrt{np(1-p)} = \sqrt{n(.5)(1-.5)} = .5\sqrt{n}$. The test is based on the test statistic

$$z = \frac{(S - .5) - .5n}{.5\sqrt{n}}$$

where S is as defined in the previous box and where we subtract .5 from S as a correction for continuity. This motivates the following test:

The Large Sample Sign Test for a Population Median

Suppose we have taken a large sample (for this test, $n \geq 10$ will suffice). Define S as in the previous box, and define the test statistic

$$z = \frac{(S - .5) - .5n}{.5\sqrt{n}}$$

We can test $H_0: M_d = M_0$ versus a particular alternative hypothesis at level of significance α by using the appropriate critical value rule, or, equivalently, the corresponding p-value.

Alternative Hypothesis	Critical Value Rule: Reject H_0 if	p-Value (reject H_0 if p-value $< \alpha$)
$H_a: M_d > M_0$	$z > z_\alpha$	The area under the standard normal curve to the right of z
$H_a: M_d < M_0$	$z > z_\alpha$	The area under the standard normal curve to the right of z
$H_a: M_d \neq M_0$	$z > z_{\alpha/2}$	Twice the area under the standard normal curve to the right of z

EXAMPLE 18.2 The CD Player Case: Testing a Product Lifetime

Consider Example 18.1. Because the sample size $n = 20$ is greater than 10, we can use the large sample sign test to test $H_0: M_d = 6{,}000$ versus $H_a: M_d > 6{,}000$. Because $S = 15$ is the number of compact disc player lifetimes that exceed $M_0 = 6{,}000$, the test statistic z is

$$z = \frac{(S - .5) - .5n}{.5\sqrt{n}} = \frac{(15 - .5) - .5(20)}{.5\sqrt{20}} = 2.01$$

The p-value for the test is the area under the standard normal curve to the right of $z = 2.01$, which is $1 - .9778 = .0222$. Because this p-value is less than .05, we have strong evidence that $H_a: M_d > 6{,}000$ is true. Also, note that the large sample, approximate p-value of .0222 given by the normal distribution is fairly close to the exact p-value of .0207 given by the binomial distribution [see Figure 18.1(c)].

To conclude this section, we consider the DVD recorder rating example discussed in the chapter introduction, and we let M_d denote the median rating that would be given by all purchasers who have owned the DVD recorder for one year. Below we present the Minitab output of the sign test of $H_0: M_d = 7.5$ versus $H_a: M_d > 7.5$:

```
Sign test of median = 7.500 versus > 7.500
             N  Below  Equal  Above       P  Median
DVD Rating  20      5      0     15  0.0207   8.000
```

Because the p-value of .0207 is less than .05, we have strong evidence that the population median rating exceeds 7.5. Furthermore, note that the sign test has reached this conclusion by showing that **more than 50 percent** of all DVD recorder ratings exceed 7.5. It follows, because a rating exceeding 7.5 is the same as a rating being at least 8 (because of the discrete nature of the ratings), that we have strong evidence that the population median rating is at least 8.

Exercises for Section 18.1

CONCEPTS connect

18.1 What is a nonparametric test? Why would such a test be particularly useful when we must take a small sample?

18.2 When we perform the sign test, we use the sample data to compute a p-value. What probability distribution is used to compute the p-value? Explain why.

METHODS AND APPLICATIONS

18.3 Consider the following sample of five chemical yields:
DS ChemYield

   801　　814　　784　　836　　820

   a Use this sample to test $H_0: M_d = 800$ versus $H_a: M_d \neq 800$ by setting $\alpha = .01$.
   b Use this sample to test $H_0: M_d = 750$ versus $H_a: M_d > 750$ by setting $\alpha = .05$.

18.4 Consider the following sample of seven bad debt ratios: DS BadDebt

   7%　4%　6%　7%　5%　4%　9%

Use this sample and the following Minitab output to test the null hypothesis that the median bad debt ratio equals 3.5 percent versus the alternative hypothesis that the median bad debt ratio exceeds 3.5 percent by setting α equal to .05.

```
Sign test of median = 3.500 versus > 3.500
          N  Below  Equal  Above
Ratio     7      0      0      7

          P  Median
     0.0078   6.000
```

18.5 A local newspaper randomly selects 20 patrons of the Springwood Restaurant on a given Saturday night and has each patron rate the quality of his or her meal as 5 (excellent), 4 (good), 3 (average), 2 (poor), or 1 (unsatisfactory). When the results are summarized, it is found that there are 16 ratings of 5, 3 ratings of 4, and 1 rating of 3. Let M_d denote the population median rating that would be given by all possible patrons of the restaurant on the Saturday night.
   a Test $H_0: M_d = 4.5$ versus $H_a: M_d > 4.5$ by setting $\alpha = .05$.
   b Reason that your conclusion in part a implies that we have very strong evidence that the median rating that would be given by all possible patrons is 5.

18.6 Suppose that a particular type of plant has a median growing height of 20 inches in a specified time period when the best plant food currently on the market is used as directed. A developer of a new plant food wishes to show that the new plant food increases the median growing height. If a stem-and-leaf display indicates that the population of all growing heights using the new plant food is markedly nonnormal, it would be appropriate to use the sign test to test $H_0: M_d = 20$ versus $H_a: M_d > 20$. Here M_d denotes the population median growing height when the new plant food is used. Suppose that 13 out of 15 sample plants grown using the new plant food reach a height of more than 20 inches. Test $H_0: M_d = 20$ versus $H_a: M_d > 20$ by using the large sample sign test.

TABLE 18.1   Results of a Taste Test of Coke versus Pepsi   ⓈCokePep

Customer	Preference (Coke or Pepsi)	Value (Sign)
1	Coke	+1
2	Pepsi	−1
3	Pepsi	−1
4	Coke	+1
5	Coke	+1
6	Pepsi	−1
7	Coke	+1
8	Coke	+1
9	Pepsi	−1

Sign Test	0 hypothesized value	4 below	binomial
9 n	1 median Value (sign)	0 equal	1.0000 p-value (two-tailed)
		5 above	

18.7 A common application of the sign test deals with analyzing consumer preferences. For instance, suppose that a blind taste test is administered to nine randomly selected convenience store customers. Each participant is asked to express a preference for either Coke or Pepsi after tasting unidentified samples of each soft drink. The sample results are expressed by recording a +1 for each consumer who prefers Coke and a −1 for each consumer who prefers Pepsi. Note that sometimes, rather than recording either a +1 or a −1, we simply record the sign + or −, hence the name "sign test." A zero is recorded if a consumer is unable to rank the two brands, and these observations are eliminated from the analysis.

The null hypothesis in this application says that there is no difference in preferences for Coke and Pepsi. If this null hypothesis is true, then the number of +1 values in the population of all preferences should equal the number of −1 values, which implies that the median preference $M_d = 0$ (and that the proportion p

of +1 values equals .5). The alternative hypothesis says that there is a significant difference in preferences (or that there is a significant difference in the number of +1 values and −1 values in the population of all preferences). This implies that the median preference does not equal 0 (and that the proportion p of +1 values does not equal .5).   ⓈCokePep

a Table 18.1 gives the results of the taste test administered to the nine randomly selected consumers. If we consider testing $H_0: M_d = 0$ versus $H_a: M_d \neq 0$ where M_d is the median of the (+1 and −1) preference rankings, determine the values of S_1, S_2, and S for the sign test needed to test H_0 versus H_a. Identify the value of S on the Excel add-in (MegaStat) output.

b Use the value of S to find the p-value for testing $H_0: M_d = 0$ versus $H_a: M_d \neq 0$. Then use the p-value to test H_0 versus H_a by setting α equal to .10, .05, .01, and .001. How much evidence is there of a difference in the preferences for Coke and Pepsi? What do you conclude?

LO18-2

Compare the locations of two distributions using a rank sum test for independent samples.

18.2 The Wilcoxon Rank Sum Test

Recall that in Section 11.1 we presented t tests for comparing two population means in an independent samples experiment. If the sampled populations are far from normally distributed and the sample sizes are small, these tests are not valid. In such a case, a nonparametric method should be used to compare the populations.

We have seen that the mean of a population measures the **central tendency,** or **location,** of the probability distribution describing the population. Thus, for instance, if a t test provides strong evidence that μ_1 is greater than μ_2, we might conclude that the probability distribution of population 1 is *shifted to the right* of the probability distribution of population 2. The nonparametric test for comparing the locations of two populations is not (necessarily) a test about the difference between population means. Rather, it is a more general test to detect whether the probability distribution of population 1 is shifted to the right (or left) of the probability distribution of population 2.[1] Furthermore, **the nonparametric test is valid for any shapes that might describe the sampled populations.**

In this section we present the **Wilcoxon rank sum test** (also called the **Mann–Whitney test**), which is used to compare the locations of two populations when **independent samples**

---

[1]To be precise, we say that the probability distribution of population 1 is shifted to the right (left) of the probability distribution of population 2 if there is more than a 50 percent chance that a randomly selected observation from population 1 will be greater than (less than) a randomly selected observation from population 2.

are selected. To perform this test, we first combine all of the observations in both samples into a single set, and we rank these observations from smallest to largest, with the smallest observation receiving rank 1, the next smallest observation receiving rank 2, and so forth. The sum of the ranks of the observations in each sample is then calculated. If the probability distributions of the two populations are identical, we would expect the sum of the ranks for sample 1 to roughly equal the sum of the ranks for sample 2. However, if, for example, the sum of the ranks for sample 1 is substantially larger than the sum of the ranks for sample 2, this would suggest that the probability distribution of population 1 is shifted to the right of the probability distribution of population 2. We explain how to carry out the Wilcoxon rank sum test in the following box:

The Wilcoxon Rank Sum Test

Let D_1 and D_2 denote the probability distributions of populations 1 and 2, and assume that we randomly select independent samples of sizes n_1 and n_2 from populations 1 and 2. Rank the $n_1 + n_2$ observations in the two samples from the smallest (rank 1) to the largest (rank $n_1 + n_2$). Here, if two or more observations are equal, we assign to each "tied" observation a rank equal to the average of the consecutive ranks that would otherwise be assigned to the tied observations. Let T_1 denote the sum of the ranks of the observations in sample 1, and let T_2 denote the sum of the ranks of the observations in sample 2. Furthermore, define the **test statistic T** to be T_1 if $n_1 \leq n_2$ and to be T_2 if $n_1 > n_2$. Then, we can test

$$H_0: D_1 \text{ and } D_2 \text{ are identical probability distributions}$$

versus a particular alternative hypothesis at level of significance α by using the appropriate critical value rule.

Alternative Hypothesis	Critical Value Rule: Reject H_0 if
H_a: D_1 is shifted to the right of D_2	$T \geq T_U$ if $n_1 \leq n_2$
	$T \leq T_L$ if $n_1 > n_2$
H_a: D_1 is shifted to the left of D_2	$T \leq T_L$ if $n_1 \leq n_2$
	$T \geq T_U$ if $n_1 > n_2$
H_a: D_1 is shifted to the right or left of D_2	$T \leq T_L$ or $T \geq T_U$

The first two alternative hypotheses above are **one-sided,** while the third alternative hypothesis is **two-sided.** The critical values T_U and T_L are given in Table A.14 for values of n_1 and n_2 from 3 to 10.

Table 18.2 repeats a portion of Table A.14. This table gives the critical value (T_U or T_L) for testing a one-sided alternative hypothesis at level of significance $\alpha = .05$ and also gives the critical values (T_U or T_L) for testing a two-sided alternative hypothesis at level of significance $\alpha = .10$. The critical values are tabulated according to n_1 and n_2, the sizes of the samples taken from populations 1 and 2, respectively. For instance, as shown in Table 18.2, if we have taken a sample of size $n_1 = 10$ from population 1, and if we have taken a sample of size $n_2 = 7$ from

TABLE 18.2 A Portion of the Wilcoxon Rank Sum Table: Critical Values for $\alpha = .05$ (One-Sided); $\alpha = .10$ (Two-Sided)

n_1	3		4		5		6		7		8		9		10	
n_2	T_L	T_U	T_L	T_U	T_L	T_U	T_L	T_U	T_L	T_U	T_L	T_U	T_L	T_U	T_L	T_U
3	6	15	7	17	7	20	8	22	9	24	9	27	10	29	11	31
4	7	17	12	24	13	27	14	30	15	33	16	36	17	39	18	42
5	7	20	13	27	19	36	20	40	22	43	24	46	25	50	26	54
6	8	22	14	30	20	40	28	50	30	54	32	58	33	63	35	67
7	9	24	15	33	22	43	30	54	39	66	41	71	43	76	46	80
8	9	27	16	36	24	46	32	58	41	71	52	84	54	90	57	95
9	10	29	17	39	25	50	33	63	43	76	54	90	66	105	69	111
10	11	31	18	42	26	54	35	67	46	80	57	95	69	111	83	127

$\alpha = .05$ one-sided; $\alpha = .10$ two-sided

population 2, then for a one-sided test with $\alpha = .05$, we use $T_U = 80$ or $T_L = 46$. Similarly, if $n_1 = 10$ and $n_2 = 7$, we use $T_U = 80$ and $T_L = 46$ for a two-sided test with $\alpha = .10$.

EXAMPLE 18.3 Studying Circuit Court Litigation Times

The State Court Administrator for the State of Oregon commissioned a study of two circuit court jurisdictions within the state to examine the effect of administrative rule differences on litigation processing time. The two jurisdictions of interest are Coos County and Lane County. Samples of 10 cases were selected at random from each jurisdiction. However, records for three of the cases selected from Lane County were incomplete, and the cases had to be discarded from the analysis, leaving $n_1 = 10$ cases for Coos County and $n_2 = 7$ cases for Lane County. Each selected case was examined to determine the total elapsed time (in days) required for processing the case, from filing to completion. The processing times are given in Figure 18.2(a). Because the corresponding box plots indicate that the population of all possible processing times for each county might be skewed to the right, we will perform the Wilcoxon rank sum test. It was theorized before the samples were taken that the administrative rules in Lane County were somewhat inefficient. Therefore, we will test

H_0: the probability distributions of all possible processing times for Coos County and Lane County are identical

versus

H_a: the probability distribution of all possible processing times for Coos County is shifted to the left of the probability distribution of all possible processing times for Lane County (note that this alternative hypothesis intuitively implies that the Coos County processing times are "systematically less than" the Lane County processing times)

To perform the test, we rank the $n_1 + n_2 = 10 + 7 = 17$ processing times in the two samples as shown in Figure 18.2(a). Note that, because there are two processing times of 145 that are tied as the sixth and seventh smallest processing times, we assign each of these an average rank of 6.5. The sum of the ranks of the processing times in sample 1 (Coos County) is $T_1 = 72.5$, and

---

FIGURE 18.2   Analysis of the Coos County and Lane County Litigation Processing Times

(a) The Coos County and Lane County litigation processing times   (DS) Court

Coos County		Lane County		Box Plots of Coos and Lane
Time	**Rank**	**Time**	**Rank**	
48	1	109	4	
97	2	145	6.5	
103	3	196	10	
117	5	273	13	
145	6.5	289	14	
151	8	417	16	
179	9	505	17	
220	11		$T_2 = 80.5$	
257	12			
294	15			
	$T_1 = 72.5$			

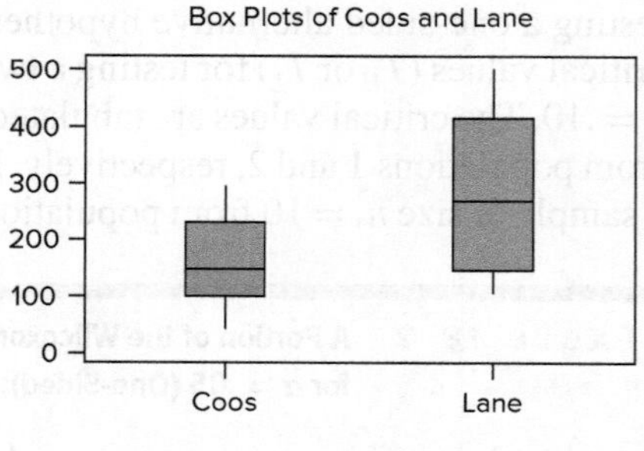

(b) Minitab output of the Wilcoxon rank sum test for the litigation processing times

Coos   N = 10   Median = 148.0
Lane   N = 7   Median = 273.0
Point estimate for ETA1-ETA2 is **−98.0**
95.5 Percent CI for ETA1-ETA2 is (**−248.0,7.9**)
W = **72.5**
Test of ETA1 = ETA2 vs ETA1 < ETA2 is **significant at 0.0486**
The test is significant at 0.0485 (adjusted for ties)

the sum of the ranks of the processing times in sample 2 (Lane County) is $T_2 = 80.5$. Because $n_1 = 10$ is greater than $n_2 = 7$, the summary box tells us that the test statistic T is $T_2 = 80.5$. Because we are testing a "shifted left" alternative hypothesis, and because n_1 is greater than n_2, the summary box also tells us that we can reject H_0 in favor of H_a at the .05 level of significance if T is greater than or equal to T_U. Because $T = 80.5$ is greater than $T_U = 80$ (see Table 18.2), we conclude at the .05 level of significance that the Coos County processing times are shifted to the left of, and thus are "systematically less than," the Lane County processing times. This supports the theory that the Lane County administrative rules are somewhat inefficient.

Figure 18.2(b) presents the Minitab output of the Wilcoxon rank sum test for the litigation processing times. In general, Minitab gives T_1, the sum of the ranks of the observations in sample 1, as the test statistic, which Minitab denotes as W. If, as in the present example, n_1 is greater than n_2 and thus the correct test statistic is T_2, **we can obtain T_2 by subtracting T_1 from $(n_1 + n_2)(n_1 + n_2 + 1)/2$.** This last quantity can be proven to equal the sum of the ranks of the $(n_1 + n_2)$ observations in both samples. In the present example, this quantity equals $(10 + 7)(10 + 7 + 1)/2 = (17)(18)/2$, or 153. Therefore, because the Minitab output tells us that $T_1 = 72.5$, the correct test statistic T_2 is $(153 - 72.5) = 80.5$. In addition to giving T_1, Minitab gives two p-values related to the hypothesis test. The first p-value—.0486—is calculated assuming that there are no ties. Because there is a tie, the second p-value—.0485—is adjusted accordingly and is more correct (although there is little difference in this situation).

In general, the Wilcoxon rank sum test tests the equality of the population medians if the distributions of the sampled populations have the same shapes and equal variances. Minitab tells us that under these assumptions a point estimate of the difference in the population medians is −98.0 (days), and a 95.5 percent confidence interval for the difference in the population medians is [−248.0, 7.9]. Note that the point estimate of the difference in the population medians, which is −98.0, is not equal to the difference in the sample medians, which is $148.0 − 273.0 = −125.0$. In the present example, the box plots in Figure 18.2 indicate that the variances of the two populations are not equal. In fact, in most situations it is a bit too much to ask that the sampled populations have exactly the same shapes and equal variances (although we will see in Exercise 18.12 that this might be approximately true in some situations).

As another example, suppose that on a given Saturday night a local newspaper randomly selects 20 patrons from each of two restaurants and has each patron rate the quality of his or her meal as 5 (excellent), 4 (good), 3 (average), 2 (poor), or 1 (unsatisfactory). The following results are obtained:

Rating	Restaurant 1 Patrons	Restaurant 2 Patrons	Total Patrons	Ranks Involved	Average Rank	Restaurant 1 Rank Sum	Restaurant 2 Rank Sum
5	15	5	20	21–40	30.5	(15)(30.5) = 457.5	(5)(30.5) = 152.5
4	4	11	15	6–20	13	(4)(13) = 52	(11)(13) = 143
3	1	2	3	3, 4, 5	4	(1)(4) = 4	(2)(4) = 8
2	0	1	1	2	2	(0)(2) = 0	(1)(2) = 2
1	0	1	1	1	1	(0)(1) = 0	(1)(1) = 1
						$T_1 = 513.5$	$T_2 = 306.5$

Suppose that we wish to test

H_0: The probability distributions of all possible Saturday night meal
       ratings for restaurants 1 and 2 are identical

versus

H_a: The probability distribution of all possible Saturday night meal ratings
       for restaurant 1 is shifted to the right or left of the probability distribution
       of all possible Saturday night meal ratings for restaurant 2.

Because there are only five numerical ordinal ratings, there are many ties. The above table shows how we determine the sum of the ranks for each sample. Because $n_1 = 20$ and $n_2 = 20$, we cannot obtain critical values by using Table A.14 (which gives critical values for sample sizes up to $n_1 = 10$ and $n_2 = 10$). However, we can use a large sample, normal approximation, which is valid if both n_1 and n_2 are at least 10. The normal approximation involves making two modifications. First, we replace the test statistic T in the previously given summary box by a standardized value of the test statistic. This standardized value, denoted z, is calculated by subtracting the mean $\mu_T = n_i(n_1 + n_2 + 1)/2$ from the test statistic T and by then dividing the resulting difference by the standard deviation $\sigma_T = \sqrt{n_1 n_2(n_1 + n_2 + 1)/12}$. Here n_i in the expression for μ_T equals n_1 if the test statistic T is T_1 and equals n_2 if T is T_2. Second, when testing a one-sided alternative hypothesis, we replace the critical values T_U and T_L by the normal points z_α and $-z_\alpha$. When testing a two-sided alternative hypothesis, we replace T_U and T_L by $z_{\alpha/2}$ and $-z_{\alpha/2}$. For the current example, $n_1 = n_2$, and thus the test statistic T is $T_1 = 513.5$. Furthermore,

$$\mu_T = \frac{n_1(n_1 + n_2 + 1)}{2} = \frac{20(20 + 20 + 1)}{2} = 410$$

$$\sigma_T = \sqrt{\frac{n_1 n_2(n_1 + n_2 + 1)}{12}} = \sqrt{\frac{20(20)(41)}{12}} = 36.968455$$

and

$$z = \frac{T - \mu_T}{\sigma_T} = \frac{513.5 - 410}{36.968455} = 2.7997$$

Because we are testing a "shifted right or left" (that is, a two-sided) alternative hypothesis, the summary box tells us that we reject the null hypothesis if $T \leq T_L$ or $T \geq T_U$. Stated in terms of standardized values, we reject the null hypothesis if $z < -z_{\alpha/2}$ or $z > z_{\alpha/2}$ (here we use strict inequalities to be consistent with other normal distribution critical value conditions). If we set $\alpha = .01$, we use the critical values $-z_{.005} = -2.575$ and $z_{.005} = 2.575$. Because $z = 2.7997$ is greater than $z_{.005} = 2.575$, we reject the null hypothesis at the .01 level of significance. Therefore, we have very strong evidence that there is a systematic difference between the Saturday night meal ratings at restaurants 1 and 2. Looking at the original data, we would estimate that Saturday night meal ratings are higher at restaurant 1.

To conclude this section, we make two comments. First, when there are ties, there is an adjusted formula for σ_T that takes the ties into account.[2] If (as in the restaurant example) we ignore the formula, the results we obtain are statistically conservative. Therefore, if we reject the null hypothesis by using the unadjusted formula, we would reject the null hypothesis by using the adjusted formula. Second, the Excel add-in (MegaStat) calculates p-values by using the large sample, normal approximation (and a *continuity correction*), even if the sample sizes n_1 and n_2 are small (less than 10). This will be illustrated in Exercise 18.12.

Exercises for Section 18.2

CONCEPTS connect

18.8 Explain the circumstances in which we use the Wilcoxon rank sum test.

18.9 Identify the parametric test corresponding to the Wilcoxon rank sum test. What assumption is needed for the validity of this parametric test (and not needed for the Wilcoxon rank sum test)?

METHODS AND APPLICATIONS

18.10 A loan officer at a bank wishes to compare the new car loan rates charged at banks in Ohio with the new car loan rates of Ohio credit unions. Two independent random samples of bank rates and credit union rates in Ohio are obtained with the following results (all rates are fixed rates):

Bank Rates:	4.25	3.50	3.25	4.00	4.10	3.75	3.50	4.25
Credit Union Rates:	3.25	2.25	2.75	3.00	2.50	3.00	2.40	2.50

Because both samples are small, the bank officer is uncertain about the shape of the distributions of bank and credit union new car loan rates. Therefore, the Wilcoxon rank sum test will be used to compare the two types of loan rates. CarLoan

a Let D_1 be the distribution of bank rates and let D_2 be the distribution of credit union rates. Carry out the Wilcoxon rank sum test to determine whether D_1 and D_2 are identical versus the alternative that D_1 is shifted to the right or left of D_2. Use $\alpha = .05$.

b Carry out the Wilcoxon rank sum test to determine whether D_1 is shifted to the right of D_2. Use $\alpha = .025$. What do you conclude?

18.11 A company collected employee absenteeism data (in hours per year) at two of its manufacturing plants. The data were obtained by randomly selecting a sample from all of the employees at the first plant, and by randomly selecting another independent sample from all of the employees at the second plant. For each randomly selected employee, absenteeism records

---
[2]The adjusted formula is quite complicated and will not be given here.

were used to determine the exact number of hours the employee has been absent during the past year. The following results were obtained: ⓭ Absent

Plant 1:	10	131	53	37	59	29	45	26	39	36
Plant 2:	21	46	33	31	49	33	39	19	12	35

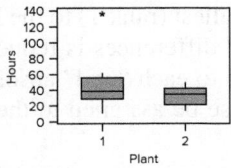

Use a Wilcoxon rank sum test and the following Minitab output to determine whether absenteeism is different at the two plants. Use $\alpha = .05$.

```
            N  Median
Plant 1    10   38.00
Plant 2    10   33.00
Point estimate for ETA1-ETA2 is 7.00
95.5 Percent CI for ETA1-ETA2 is (−6.99, 24.01)
W = 120.5
Test of ETA1 = ETA2 vs ETA1 not = ETA2  is
significant at 0.2568
The test is significant at 0.2565 (adjusted for ties)
```

18.12 THE CATALYST COMPARISON CASE

The following table presents samples of hourly yields for catalysts XA-100 and ZB-200. We analyzed these data using a two independent sample t test in Example 11.2.   ⓭ Catalyst

Catalyst XA-100	Catalyst ZB-200
801	752
814	718
784	776
836	742
820	763

Box Plots of Yield by Catalyst

Wilcoxon/Kruskal–Wallis Tests (Rank Sums)

Level	Count	Score Sum	Expected Score	Score Mean	(Mean-Mean0)/Std0
XA-100	5	40.000	27.500	8.00000	2.507
ZB-200	5	15.000	27.500	3.00000	−2.507

2-Sample Test, Normal Approximation

S	Z	Prob>|Z|
15	−2.50672	0.0122*

a Use a Wilcoxon rank sum test and the JMP output to test for systematic differences in the yields of the two catalysts. Use $\alpha = .05$.

b The p-value on the JMP output has been calculated by finding twice the area under the standard normal curve to the right of

$$z = \frac{39.5 - \mu_T}{\sigma_T} = \frac{39.5 - 27.5}{4.79} = 2.51$$

Here we have used a continuity correction and changed $T_1 = 40$ to 39.5. Verify the calculations of μ_T, σ_T, and the p-value.

c Assume that the second yield for catalyst ZB-200 is invalid. Use the remaining data to determine if we can conclude that the XA-100 yields are systematically higher than the ZB-200 yields. Set $\alpha = .05$.

18.13 Moore (2000) reports on a study by Boo (1997), who asked 303 randomly selected people at fairs:

How often do you think people become sick because of food they consume prepared at outdoor fairs and festivals?

The possible responses were 5 (always), 4 (often), 3 (more often than not), 2 (once in a while), and 1 (very rarely). The following data were obtained:

Response	Females	Males	Total
5	2	1	3
4	23	5	28
3	50	22	72
2	108	57	165
1	13	22	35

Source: H. C. Boo, "Consumers' Perceptions and Concerns about Safety and Healthfulness of Food Served at Fairs and Festivals," M.S. thesis, Purdue University, 1997.

The computer output below presents the results of a Wilcoxon rank sum test that attempts to determine if men and women systematically differ in their responses. Here the normal approximation has been used to calculate the p-value of .0009. What do you conclude?

Gender	N	Sum of Scores
Female	196	31996.5
Male	107	14059.5

T = 14059.5   z = −3.33353
Test significant at 0.0009

18.3 The Wilcoxon Signed Ranks Test

In Section 11.2 we presented a t test for comparing two population means in a paired difference experiment. If the sample size is small and the population of paired differences is far from normally distributed, this test is not valid and we should use a nonparametric test. In this section we present the **Wilcoxon signed ranks test,** which is a nonparametric test for comparing two populations when a **paired difference experiment** has been carried out.

LO18-3

Compare the locations of two distributions using a signed ranks test for paired samples.

The Wilcoxon Signed Ranks Test

Let D_1 and D_2 denote the probability distributions of populations 1 and 2, and assume that we have randomly selected n matched pairs of observations from populations 1 and 2. Calculate the paired differences of the n matched pairs by subtracting each paired population 2 observation from the corresponding population 1 observation, and rank the absolute values of the n paired differences from the smallest (rank 1) to the largest (rank n). Here paired differences equal to 0 are eliminated, and the number n of paired differences is reduced accordingly. Furthermore, if two or more absolute paired differences are equal, we assign to each "tied" absolute paired difference a rank equal to the average of the consecutive ranks that would otherwise be assigned to the tied absolute paired differences. Let

$$T^- = \text{the sum of the ranks associated with the negative paired differences}$$

and

$$T^+ = \text{the sum of the ranks associated with the positive paired differences}$$

We can test

$$H_0: D_1 \text{ and } D_2 \text{ are identical probability distributions}$$

versus a particular alternative hypothesis at level of significance α by using the appropriate test statistic and the corresponding critical value rule.

Alternative Hypothesis	Test Statistic	Critical Value Rule: Reject H_0 if
H_a: D_1 is shifted to the right of D_2	T^-	$T^- \leq T_0$
H_a: D_1 is shifted to the left of D_2	T^+	$T^+ \leq T_0$
H_a: D_1 is shifted to the right or left of D_2	$T = \text{the smaller of } T^- \text{ and } T^+$	$T \leq T_0$

The first two alternative hypotheses above are **one-sided,** while the third alternative hypothesis is **two-sided.** Values of T_0 are given in Table A.15 for values of n from 5 to 50.

Table 18.3 repeats a portion of Table A.15. This table gives the critical value T_0 for testing one-sided and two-sided alternative hypotheses at several different values of α. The critical values are tabulated according to n, the number of paired differences. For instance, Table 18.3 shows that, if we are analyzing 10 paired differences, then the critical value for testing a one-sided alternative hypothesis at the .01 level of significance is equal to $T_0 = 5$. This table also shows that we would use the critical value $T_0 = 5$ for testing a two-sided alternative hypothesis at level of significance $\alpha = .02$.

TABLE 18.3   **A Portion of the Wilcoxon Signed Ranks Table**

One-Sided	Two-Sided	$n = 5$	$n = 6$	$n = 7$	$n = 8$	$n = 9$	$n = 10$
$\alpha = .05$	$\alpha = .10$	1	2	4	6	8	11
$\alpha = .025$	$\alpha = .05$		1	2	4	6	8
$\alpha = .01$	$\alpha = .02$			0	2	3	5
$\alpha = .005$	$\alpha = .01$				0	2	3
		$n = 11$	$n = 12$	$n = 13$	$n = 14$	$n = 15$	$n = 16$
$\alpha = .05$	$\alpha = .10$	14	17	21	26	30	36
$\alpha = .025$	$\alpha = .05$	11	14	17	21	25	30
$\alpha = .01$	$\alpha = .02$	7	10	13	16	20	24
$\alpha = .005$	$\alpha = .01$	5	7	10	13	16	19

(C) EXAMPLE 18.4 The Auto Insurance Case: Comparing Mean Repair Costs

Again consider the automobile repair cost data, which are given in Figure 18.3(a). We analyzed these data using a paired sample t test in Example 11.3. If we fear that the population of all possible paired differences of repair cost estimates at garages 1 and 2 may be far from normally distributed, we can perform the Wilcoxon signed ranks test. Here we test

H_0: the probability distributions of the populations of all possible repair cost estimates at garages 1 and 2 are identical

versus

H_a: the probability distribution of repair cost estimates at garage 1 is shifted to the left of the probability distribution of repair cost estimates at garage 2

To perform this test, we find the absolute value of each paired difference, and we assign ranks to the absolute differences [see Figure 18.3(a)]. Because of the form of the alternative hypothesis (see the preceding summary box), we use the test statistic

T^+ = the sum of the ranks associated with the positive paired differences

Because .1 is the only positive paired difference, and because the rank associated with this difference equals 1, we find that $T^+ = 1$. The alternative hypothesis is one-sided, and we are analyzing $n = 7$ paired differences. Therefore, Table 18.3 tells us that we can test H_0 versus H_a at the .05, .025, and .01 levels of significance by setting the critical value T_0 equal to 4, 2, and 0, respectively. The critical value condition is $T^+ \leq T_0$. It follows that, because $T^+ = 1$ is less than or equal to 4 and 2, but is not less than or equal to 0, we can reject H_0 in favor of H_a at the .05 and .025 levels of significance, but not at the .01 level of significance.

FIGURE 18.3 Analysis of the Repair Cost Estimates at Two Garages

(a) Sample of $n = 7$ paired differences of the repair cost estimates at garages 1 and 2 (DS) Repair
(cost estimates in hundreds of dollars)

Sample of $n = 7$ Damaged Cars	Repair Cost Estimates at Garage 1	Repair Cost Estimates at Garage 2	Sample of $n = 7$ Paired Differences	Absolute Paired Differences	Ranks
Car 1	$ 7.1	$ 7.9	$d_1 = -.8$	.8	4
Car 2	9.0	10.1	$d_2 = -1.1$	1.1	5
Car 3	11.0	12.2	$d_3 = -1.2$	1.2	6
Car 4	8.9	8.8	$d_4 = .1$	.1	1
Car 5	9.9	10.4	$d_5 = -.5$	.5	2
Car 6	9.1	9.8	$d_6 = -.7$	.7	3
Car 7	10.3	11.7	$d_7 = -1.4$	1.4	7

(b) Minitab output of the Wilcoxon signed ranks test

Test of median = 0.0 versus median < 0.0

	N	N for Test	Wilcoxon Statistic	P	Estimated Median
G1 - G2	7	7	1.0	0.017	-0.8250

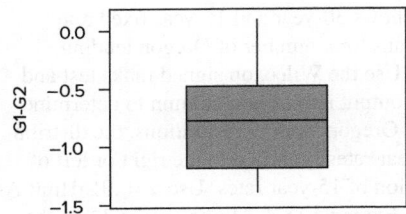

Therefore, we have strong evidence that the probability distribution of repair cost estimates at garage 1 is shifted to the left of the probability distribution of repair cost estimates at garage 2. That is, the repair cost estimates at garage 1 seem to be systematically lower than the repair cost estimates at garage 2. Figure 18.3(b) presents the Minitab output of the Wilcoxon signed ranks test for this repair cost comparison. In general, Minitab gives T^+ as the "Wilcoxon statistic," even if T^- is the appropriate test statistic. It can be shown that **T^- can be obtained by subtracting T^+ from $n(n + 1)/2$,** where n is the total number of paired differences being analyzed.

Notice that in Example 18.4 the nonparametric Wilcoxon signed ranks test would not allow us to reject H_0 in favor of H_a at the .01 level of significance. On the other hand, the *parametric* paired difference t test performed in Example 11.5 did allow us to reject H_0: $\mu_1 - \mu_2 = 0$ in favor of H_a: $\mu_1 - \mu_2 < 0$ at the .01 level of significance. In general, **a parametric test is often *more powerful*** than the analogous nonparametric test. That is, the parametric test often allows us to reject H_0 at smaller values of α. Therefore, if the assumptions for the parametric test are satisfied—for example, if, when we are using small samples, the sampled populations are approximately normally distributed—it is preferable to use the parametric test. **The advantage of nonparametric tests is that they can be used without assuming that the sampled populations have the shapes of any particular probability distributions.** As an example, this can be important when reporting statistical conclusions to U.S. federal agencies. Federal guidelines specify that, when reporting statistical conclusions, the validity of the assumptions behind the statistical methods used must be fully justified. If, for instance, there are insufficient data to justify the assumption that the sampled populations are approximately normally distributed, then we must use a nonparametric method to make conclusions.

Finally, if the sample size n is at least 25, we can use a large sample approximation of the Wilcoxon signed ranks test. This is done by making two modifications. First, we replace the test statistic $(T^-$ or $T^+)$ by a standardized value of the test statistic. This standardized value is calculated by subtracting the mean $n(n + 1)/4$ from the test statistic $(T^-$ or $T^+)$ and then dividing the resulting difference by the standard deviation $\sqrt{n(n + 1)(2n + 1)/24}$. Second, when testing a one-sided alternative hypothesis, we replace the critical value T_0 by the normal point $-z_\alpha$. When testing a two-sided alternative hypothesis, we replace T_0 by $-z_{\alpha/2}$.

Exercises for Section 18.3

CONCEPTS connect

18.14 Explain the circumstances in which we use the Wilcoxon signed ranks test.

18.15 Identify the parametric test corresponding to the Wilcoxon signed ranks test. What assumption is needed for the validity of the parametric test (and not needed for the Wilcoxon signed ranks test)?

METHODS AND APPLICATIONS

18.16 Table 18.4 shows 30-year and 15-year fixed rate mortgage loans for a number of Oregon lending institutions. Use the Wilcoxon signed ranks test and the Minitab output in the right column to determine whether, for Oregon lending institutions, the distribution of 30-year rates is shifted to the right or left of the distribution of 15-year rates. Use $\alpha = .01$. Hint: As discussed in Example 18.4, Minitab gives T^+ as the

Wilcoxon statistic. Find T^- by subtracting T^+ from $n(n + 1)/2$, where $n = 9$. **MortDiff**

Test of median = 0.0 versus median not = 0.0

	N	N for Test	Wilcoxon Statistic	P	Estimated Median
30Yr – 15Yr	9	9	45.0	0.009	0.2355

18.17 A consumer advocacy group is concerned about the ability of tax preparation firms to correctly prepare complex returns. To test the performance of tax preparers in two different tax preparation firms—Quick Tax and Discount Tax—the group designed 10 tax cases for families with gross annual incomes between $100,000 and $200,000. In a "tax-off" competition, the advocacy group randomly assigned pairs of preparers from the two firms to the 10 cases and asked each preparer to compute the tax liability for his or her assigned case. The preparers' returns were collected, and

Lending Institution	30-Year	15-Year	Difference
1	4.715	4.599	0.116
2	4.648	4.367	0.281
3	4.740	4.550	0.190
4	4.597	4.362	0.235
5	4.425	4.162	0.263
6	4.880	4.583	0.297
7	4.900	4.800	0.100
8	4.675	4.394	0.281
9	4.790	4.540	0.250

TABLE 18.4   Mortgage Loan Interest Rates for Nine Randomly Selected Oregon Lending Institutions Ⓓ MortDiff

Annual Percentage Rate

TABLE 18.5   Pretest and Posttest Leadership Scores Ⓓ Leader

Manager	Pretest Score	Posttest Score	Difference
1	35	54	−19
2	27	43	−16
3	51	53	−2
4	38	50	−12
5	32	42	−10
6	44	58	−14
7	33	35	−2
8	26	39	−13
9	40	47	−7
10	50	48	2
11	36	41	−5
12	31	37	−6

the group computed the difference between each preparer's computed tax and the actual tax that should have been computed. The data at the botom of the page consist of the resulting two sets of tax computation errors, one for preparers from Quick Tax and the other for preparers from Discount Tax. Fully interpret the Minitab output at the botom of the page of a Wilcoxon signed ranks test analysis of these data.   Ⓓ TaxErr

18.18   A human resources director wishes to assess the benefits of sending a company's managers to an innovative management course. Twelve of the company's managers are randomly selected to attend the course, and a psychologist interviews each participating manager before and after taking the course. Based on these interviews, the psychologist rates the manager's leadership ability on a 1-to-100 scale. The pretest and posttest leadership scores for each of the 12 managers are given in Table 18.5.
Ⓓ Leader
    a   Let D_1 be the distribution of leadership scores before taking the course, and let D_2 be the distribution of leadership scores after taking the

course. Carry out the Wilcoxon signed ranks test to test whether D_1 and D_2 are identical (that is, the course has no effect on leadership scores) versus the alternative that D_2 is shifted to the right or left of D_1 (that is, the course affects leadership scores). Use $\alpha = .05$.
    b   Carry out the Wilcoxon signed ranks test to determine whether D_2 is shifted to the right of D_1. Use $\alpha = .05$. What do you conclude?

18.19   Recall that in Exercise 11.18 we compared preexposure and postexposure attitude scores for an advertising study by using a paired difference t test. The data obtained and related JMP output are shown in Table 18.6. Use the Wilcoxon signed ranks test and the JMP output to determine whether the distributions of preexposure and postexposure attitude scores are different. Use $\alpha = .05$. Note that the test statistic S used by JMP is a modification of the statistics discussed in this section and the p-value for the two sided alternative being tested is Prob>ISI.   Ⓓ AdStudy

Tax Case	Quick Tax Errors	Discount Tax Errors	Difference
1	857	156	701
2	920	200	720
3	1,090	202	888
4	1,594	390	1,204
5	1,820	526	1,294
6	1,943	749	1,194
7	1,987	911	1,076
8	2,008	920	1,088
9	2,083	2,145	−62
10	2,439	2,602	−163

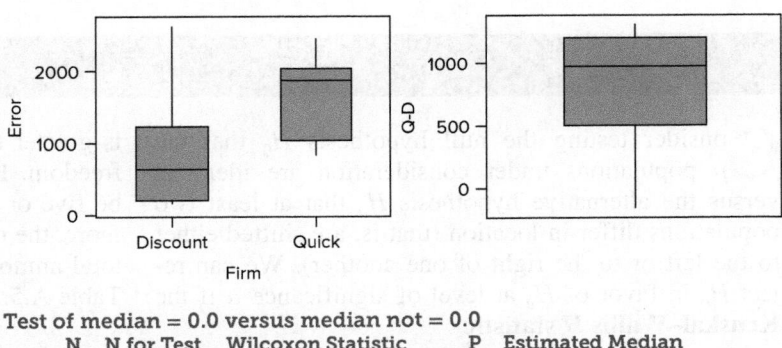

Test of median = 0.0 versus median not = 0.0

	N	N for Test	Wilcoxon Statistic	P	Estimated Median
Q-D	10	10	52.0	0.014	898.0

TABLE 18.6 Preexposure and Postexposure Attitude Scores for an Advertising Study ⓓⓢ AdStudy

Subject	Preexposure Attitudes (A_1)	Postexposure Attitudes (A_2)	Attitude Change (d_i)
1	50	53	−3
2	25	27	−2
3	30	38	−8
4	50	55	−5
5	60	61	−1
6	80	85	−5
7	45	45	0
8	30	31	−1
9	65	72	−7
10	70	78	−8

Wilcoxon Signed Rank

	Postexposure Attitude (A2)- Preexposure Attitudes (A1)
Test Statistic S	27.000
Prob>|S|	0.0039*
Prob>S	0.0020*
Prob>S	0.9980

Source: Attitude scores from W. R. Dillon, et al., *Essentials of Marketing Research*, p. 435. McGraw-Hill, 1993.

18.4 Comparing Several Populations Using the Kruskal–Wallis *H* Test

In this section we present the Kruskal–Wallis *H* test, a nonparametric technique for comparing the locations of three or more populations. This test requires no assumptions about the population probability distributions and assumes we use independent samples chosen randomly.

In general, suppose we wish to use the Kruskal–Wallis *H* test to compare the locations of p populations by using p independent samples of observations randomly selected from these populations. We first rank all of the observations in the p samples from smallest to largest. If n_i denotes the number of observations in the ith sample, we are ranking a total of $n = (n_1 + n_2 + \cdots + n_p)$ observations. Furthermore, we assign tied observations the average of the consecutive ranks that would otherwise be assigned to the tied observations. Next, we calculate the sum of the ranks of the observations in each sample. Letting T_i denote the rank sum for the ith sample, we obtain the rank sums $T_1, T_2, \ldots, T_p$. For example, consider the gasoline mileage case in Chapter 12, and suppose that North American Oil wishes to use the $p = 3$ independent samples of gasoline mileages to compare the locations of the populations of all gasoline mileages that could be obtained by using gasoline types A, B, and C. The gasoline mileage data are repeated in Table 18.7, along with the ranking (given in parentheses) of each observation in each sample. If we sum the ranks in each sample, we find that $T_1 = 37.5$, $T_2 = 63$, and $T_3 = 19.5$. Note that, although the box plots in Table 18.7 do not indicate any serious violations of the normality or equal variances assumptions, the samples are quite small, and thus we cannot be sure that these assumptions approximately hold. Therefore, it is reasonable to compare gasoline types A, B, and C by using the Kruskal–Wallis *H* test.

The Kruskal–Wallis *H* Test

Consider testing the null hypothesis H_0 that the p populations under consideration are identical versus the alternative hypothesis H_a that at least two populations differ in location (that is, are shifted either to the left or to the right of one another). We can reject H_0 in favor of H_a at level of significance α if the **Kruskal–Wallis *H* statistic**

$$H = \frac{12}{n(n+1)} \sum_{i=1}^{p} \frac{T_i^2}{n_i} - 3(n+1)$$

is greater than the χ_α^2 point based on $p - 1$ degrees of freedom. Here, for this test to be valid, there should be five or more observations in each sample. Furthermore, the number of ties should be small relative to the total number of observations. Values of χ_α^2 are given in Table A.5.

TABLE 18.7   **The Gasoline Mileage Samples and Rank Sums**   ⊙ GasMile2

Gasoline Type *A*	Gasoline Type *B*	Gasoline Type *C*
34.0 (3.5)	35.3 (9)	33.3 (2)
35.0 (8)	36.5 (13)	34.0 (3.5)
34.3 (5)	36.4 (12)	34.7 (6)
35.5 (10)	37.0 (14)	33.0 (1)
35.8 (11)	37.6 (15)	34.9 (7)
$T_1 = 37.5$	$T_2 = 63$	$T_3 = 19.5$

FIGURE 18.4   **Minitab Output of the Kruskal–Wallis *H* Test in the Gasoline Mileage Case**

```
Kruskal–Wallis Test on Mileage
Type    N    Median    Ave Rank      Z
A       5    35.00        7.5     -0.31
B       5    36.50       12.6      2.82
C       5    34.00        3.9     -2.51
Overall 15                8.0

H = 9.56    DF = 2    P = 0.008
H = 9.57    DF = 2    P = 0.008 (adjusted for ties)
```

In the gasoline mileage case, $\chi^2_{.05}$ based on $p - 1 = 2$ degrees of freedom is 5.99147 (see Table A.5). Furthermore, because $n = n_1 + n_2 + n_3 = 15$, the Kruskal–Wallis *H* statistic is

$$H = \frac{12}{15(15 + 1)} \left[\frac{(37.5)^2}{5} + \frac{(63)^2}{5} + \frac{(19.5)^2}{5} \right] - 3(15 + 1)$$

$$= \frac{12}{240} \left[\frac{1,406.25}{5} + \frac{3,969}{5} + \frac{380.25}{5} \right] - 48 = 9.555$$

Because $H = 9.555 > \chi^2_{.05} = 5.99147$, we can reject H_0 at the .05 level of significance. Therefore, we have strong evidence that at least two of the three populations of gasoline mileages differ in location. Figure 18.4 presents the Minitab output of the Kruskal–Wallis *H* test in this gasoline mileage case.

To conclude this section, we note that, if the Kruskal–Wallis *H* test leads us to conclude that the *p* populations differ in location, there are various procedures for comparing pairs of populations. A simple procedure is to use the Wilcoxon rank sum test to compare pairs of populations. For example, if we use this test to make separate, **two-sided** comparisons of (1) gasoline types *A* and *B*, (2) gasoline types *A* and *C*, and (3) gasoline types *B* and *C*, and if we set α equal to .05 for each comparison, we find that the mileages given by gasoline type *B* differ systematically from the mileages given by gasoline types *A* and *C*. Examining the mileages in Table 18.7, we would estimate that gasoline type *B* gives the highest mileages. One problem, however, with using the Wilcoxon rank sum test to make pairwise comparisons is that it is difficult to know how to set α for each comparison. Therefore, some practitioners prefer to make **simultaneous** pairwise comparisons (such as given by the Tukey simultaneous confidence intervals discussed in Chapter 12). Gibbons (1985) discusses a nonparametric approach for making simultaneous pairwise comparisons.

Exercises for Section 18.4

CONCEPTS   ▣ **connect**

18.20 Explain the circumstances in which we use the Kruskal–Wallis *H* test.

18.21 Identify the parametric test corresponding to the Kruskal–Wallis *H* test.

18.22 What are the assumptions needed for the validity of the parametric test identified in Exercise 18.21 that are not needed for the Kruskal–Wallis *H* test?

TABLE 18.8   Display Panel Study Data (Time, in Seconds, Required to Stabilize Air Traffic Emergency Condition)   DS Display3

Display Panel		
A	B	C
21	24	40
27	21	36
24	18	35
26	19	32
25	20	37

TABLE 18.9   Bottle Design Study Data (Sales during a 24-Hour Period)   DS BottleDes

Bottle Design		
A	B	C
16	33	23
18	31	27
19	37	21
17	29	28
13	34	25

TABLE 18.10   Bakery Sales Study Data (Sales in Cases)   DS BakeSale

Shelf Display Height		
Bottom (B)	Middle (M)	Top (T)
58.2	73.0	52.4
53.7	78.1	49.7
55.8	75.4	50.9
55.7	76.2	54.0
52.5	78.4	52.1
58.9	82.1	49.9

FIGURE 18.5   Minitab Output of the Kruskal–Wallis H Test for the Bakery Sales Data

Kruskal–Wallis Test on Bakery Sales

Display	N	Median	Ave Rank	Z
Bottom	6	55.75	9.2	−0.19
Middle	6	77.15	15.5	3.37
Top	6	51.50	3.8	−3.18
Overall	18		9.5	

H = 14.36   DF = 2   P = 0.001

TABLE 18.11   Golf Ball Durability Test Results   DS GolfBall

Brand			
Alpha	Best	Century	Divot
281	270	218	364
220	334	244	302
274	307	225	325
242	290	273	337
251	331	249	355

Wilcoxon/Kruskal–Wallis Tests (Rank Sums)

Level	Count	Score Sum	Expected Score	Score Mean	(Mean-Mean0)/Std0
Alpha	5	34.000	52.500	6.8000	−1.571
Best	5	67.000	52.500	13.4000	1.222
Century	5	24.000	52.500	4.8000	−2.444
Divot	5	85.000	52.500	17.0000	2.793

1-Way Test, ChiSquare Approximation

ChiSquare	DF	Prob>ChiSq
13.8343	3	0.0031*

Small sample sizes. Refer to statistical tables for tests, rather than large-sample approximations.

METHODS AND APPLICATIONS

In each of Exercises 18.23 through 18.26, use the given independent samples to perform the Kruskal–Wallis H test of the null hypothesis H_0 that the corresponding populations are identical versus the alternative hypothesis H_a that at least two populations differ in location. Note that we analyzed each of these data sets using the one-way ANOVA F test in the exercises of Chapter 13.

18.23 Use the Kruskal–Wallis H test to compare display panels A, B, and C using the data in Table 18.8. Use $\alpha = .05$.   DS Display3

18.24 Use the Kruskal–Wallis H test to compare bottle designs A, B, and C using the data in Table 18.9. Use $\alpha = .01$.   DS BottleDes

18.25 Use the Kruskal–Wallis H test and the Minitab output in Figure 18.5 to compare the bottom (B), middle (M), and top (T) display heights using the data in Table 18.10. Use $\alpha = .05$. Then, repeat the analysis if the first sales value for the middle display height is found to be incorrect and must be removed from the data set.   DS BakeSale

18.26 Use the Kruskal–Wallis H test to compare golf ball brands Alpha, Best, Century, and Divot using the data in Table 18.11. Use $\alpha = .01$ and the JMP output on the right side of Table 18.11.   DS GolfBall

18.5 Spearman's Rank Correlation Coefficient

LO18-5

Measure and test the association between two variables by using Spearman's rank correlation coefficient.

In Section 14.6 we showed how to test the significance of a population correlation coefficient. This test is based on the assumption that the population of all possible combinations of values of x and y has a bivariate normal probability distribution. If we fear that this assumption is badly violated, we can use a nonparametric approach. One such approach is **Spearman's rank correlation coefficient,** which is denoted r_s.

To illustrate, suppose that Electronics World, a chain of stores that sells audio and video equipment, has gathered the data in Table 18.12. The company wishes to study the relationship between store sales volume in July of last year (y, measured in thousands of dollars) and the number of households in the store's area (x, measured in thousands). Spearman's rank correlation coefficient is found by first ranking the values of x and y separately (ties are treated by averaging the tied ranks). To calculate r_s, we use the formula given in Section 14.2 for r and replace the x and y values in that formula by their ranks. If there are no ties in the ranks, this formula can be calculated by the simple equation

$$r_s = 1 - \frac{6 \, \Sigma d_i^2}{n(n^2 - 1)}$$

where d_i is the difference between the x-rank and the y-rank for the ith observation (if there are few ties in the ranks, this formula is approximately valid). For example, Table 18.12 gives the ranks of x and y, the difference between the ranks, and the squared difference for each of the $n = 15$ stores in the Electronics World example. Because the sum of the squared differences is 32, we calculate r_s to be

$$r_s = 1 - \frac{6(32)}{15(225 - 1)} = .9429$$

Equivalently, if we have Minitab (1) find the ranks of the x (household) values (which we call the *HRanks*) and the ranks of the y (sales) values (which we call the *SRanks*) and (2) use the formula given in Section 14.2 for r to calculate the correlation coefficient between the *HRanks* and *SRanks*, we obtain the following output:

Pearson correlation of HRank and SRank = 0.943

TABLE 18.12   **Electronics World Sales Volume Data and Ranks for 15 Stores**
          DS Electronics

Store	Number of Households, x	Sales Volume, y	x-Rank	y-Rank	Difference, d	d^2
1	161	157.27	6	7	−1	1
2	99	93.28	1	1	0	0
3	135	136.81	5	5	0	0
4	120	123.79	4	3	1	1
5	164	153.51	7	6	1	1
6	221	241.74	13	14	−1	1
7	179	201.54	8	10	−2	4
8	204	206.71	9	11	−2	4
9	214	229.78	12	13	−1	1
10	101	135.22	2	4	−2	4
11	231	224.71	14	12	2	4
12	206	195.29	11	8	3	9
13	248	242.16	15	15	0	0
14	107	115.21	3	2	1	1
15	205	197.82	10	9	1	1
						$\Sigma d_i^2 = 32$

This large positive value of r_s says that there is a strong positive rank correlation between the numbers of households and sales volumes in the sample.

In general, let ρ_s denote the **population rank correlation coefficient**—the rank correlation coefficient for the population of all possible (x, y) values. We can test the significance of ρ_s by using **Spearman's rank correlation test.**

Spearman's Rank Correlation Test

L et r_s denote Spearman's rank correlation coefficient. Then, we can test H_0: $\rho_s = 0$ versus a particular alternative hypothesis at level of significance α by using the appropriate critical value rule.

Alternative Hypothesis	**Critical Value Rule:** Reject H_0 if
H_a: $\rho_s > 0$	$r_s > r_\alpha$
H_a: $\rho_s < 0$	$r_s < -r_\alpha$
H_a: $\rho_s \neq 0$	$\lvert r_s \rvert > r_{\alpha/2}$

Table A.16 gives the critical values r_α, $-r_\alpha$, and $r_{\alpha/2}$ for sample sizes from 5 to 30. Note that for this test to be valid the number of ties encountered in ranking the observations should be small relative to the number of observations.

A portion of Table A.16 is reproduced here as Table 18.13. To illustrate using this table, suppose in the Electronics World example that we wish to test H_0: $\rho_s = 0$ versus H_a: $\rho_s > 0$ by setting $\alpha = .05$. Because there are $n = 15$ stores, Table 18.13 tells us that we use the critical value $r_{.05} = .441$. Because $r_s = .9429$ is greater than this critical value, we can reject H_0: $\rho_s = 0$ in favor of H_a: $\rho_s > 0$ by setting $\alpha = .05$. Therefore, we have strong evidence that in July of last year the sales volume of an Electronics World store was positively correlated with the number of households in the store's area.

To illustrate testing a two-sided alternative hypothesis, consider Table 18.14. This table presents the rankings of $n = 12$ midsize cars given by two automobile magazines. Here each magazine has ranked the cars from 1 (best) to 12 (worst) on the basis of overall ride. Because the two magazines sometimes have differing views, we cannot theorize about whether their rankings would be positively or negatively correlated. Therefore, we will test H_0: $\rho_s = 0$ versus H_a: $\rho_s \neq 0$. The summary box tells us that to perform this test at level of significance α, we use the critical value $r_{\alpha/2}$. To look up $r_{\alpha/2}$ in Table A.16 (or Table 18.13), we replace the table column head α by $\alpha/2$. For example, consider setting $\alpha = .05$. Then, because $\alpha/2 = .025$, we look in Table 18.13 for the value .025. Because there are $n = 12$ cars, we

TABLE 18.13	Critical Values for Spearman's Rank Correlation Coefficient		
n	$\alpha = .05$	$\alpha = .025$	$\alpha = .01$	$\alpha = .005$
10	.564	.648	.745	.794
11	.523	.623	.736	.818
12	.497	.591	.703	.780
13	.475	.566	.673	.745
14	.457	.545	.646	.716
15	.441	.525	.623	.689
16	.425	.507	.601	.666
17	.412	.490	.582	.645
18	.399	.476	.564	.625
19	.388	.462	.549	.608
20	.377	.450	.534	.591

TABLE 18.14	Rankings of 12 Midsize Cars by Two Automobile Magazines

DS CarRank

Car	Magazine 1 Ranking	Magazine 2 Ranking
1	5	7
2	1	1
3	4	5
4	7	4
5	6	6
6	8	10
7	9	8
8	12	11
9	2	3
10	3	2
11	10	12
12	11	9

find that $r_{.025} = .591$. Spearman's rank correlation coefficient for the car ranking data can be calculated to be .8951 (see the JMP output on the page margin). Because $r_s = .8951$ is greater than $r_{.025} = .591$, we reject H_0 at the .05 level of significance. Therefore, we conclude that the midsize car ride rankings given by the two magazines are correlated. Furthermore, because $r_s = .8951$, we estimate that these rankings are positively correlated.

To conclude this section, we make two comments. First, the car ranking example illustrates that Spearman's rank correlation coefficient and test can be used when the raw measurements of the x and/or y variables are themselves **ranks.** Ranks are measurements of an ordinal variable, and Spearman's nonparametric approach applies to ordinal variables. Second, it can be shown that if the sample size n is at least 10, then we can carry out an approximation to Spearman's rank correlation test by replacing r_s by the t statistic

$$t = \frac{r_s\sqrt{n-2}}{\sqrt{1-r_s^2}}$$

and by replacing the critical values r_α, $-r_\alpha$, and $r_{\alpha/2}$ by the t points t_α, $-t_\alpha$, and $t_{\alpha/2}$ (with $n-2$ degrees of freedom). Because Table A.16 gives r_α points for sample sizes up to $n = 30$, this approximate procedure is particularly useful if the sample size exceeds 30. In this case, we can use the z points z_α, $-z_\alpha$, and $z_{\alpha/2}$ in place of the corresponding t points.

Exercises for Section 18.5

CONCEPTS connect

18.27 Explain the circumstances in which we use Spearman's rank correlation coefficient.

18.28 Write the formula that we use to compute Spearman's rank correlation coefficient when
  a There are no (or few) ties in the ranks of the x and y values.
  b There are many ties in the ranks of the x and y values.

METHODS AND APPLICATIONS

18.29 A sales manager ranks 10 people at the end of their training on the basis of their sales potential. A year later, the number of units sold by each person is determined. The following data and Excel add-in (MegaStat) output are obtained. Note that the manager's ranking of 1 is "best."
  SalesRank

Person	1	2	3	4	5	6	7	8	9	10
Manager's Ranking, x	7	4	2	6	1	10	3	5	9	8
Units Sold, y	770	630	820	580	720	440	690	810	560	470

	MgrRank, x	UnitSold, y
MgrRank, x	1.000	
UnitSold, y	−.721	1.000

10 sample size

±.632 critical value .05 (two-tail)
±.765 critical value .01 (two-tail)

  a Find r_s on the Excel add-in (MegaStat) output and use Table 18.13 to find the critical value for testing $H_0: \rho_s = 0$ versus $H_a: \rho_s \neq 0$ at the .05 level of significance. Do we reject H_0?
  b The MegaStat output gives approximate critical values for $\alpha = .05$ and $\alpha = .01$. Do these approximate critical values, which are based on the t distribution, differ by much from the exact critical values in Table 18.13 (recall that $n = 10$)?

18.30 Use the following Minitab output to find r_s, and then test $H_0: \rho_s = 0$ versus $H_a: \rho_s > 0$ for the service time data below. CopyServ

Copiers Serviced, x	4	2	5	7	1	3	4	5	2	4	6
Minutes Required, y	109	58	138	189	37	82	103	134	68	112	154

Pearson correlation of CRank and MRank = **0.986**

18.31 Compute r_s and test $H_0: \rho_s = 0$ versus $H_a: \rho_s > 0$ for the direct labor cost data below. DirLab

Batch Size, x	5	62	35	12	83	14	46	52	23	100	41	75
Direct Labor Cost, y	71	663	381	138	861	145	493	548	251	1,024	435	772

Chapter Summary

The validity of many of the inference procedures presented in this book requires that various assumptions be met. Often, for instance, a normality assumption is required. In this chapter we have learned that, when the needed assumptions are not met, we must employ a **nonparametric method.** Such a method does not require any assumptions about the shape(s) of the distribution(s) of the sampled population(s).

We first presented the **sign test,** which is a hypothesis test about a population median. This test is useful when we have taken a sample from a population that may not be normally distributed. We next presented two nonparametric tests for comparing the locations of two populations. The first such test, the **Wilcoxon rank sum test,** is appropriate when an **independent samples experiment** has been carried out. The second,

the **Wilcoxon signed ranks test,** is appropriate when a **paired difference experiment** has been carried out. Both of these tests can be used without assuming that the sampled populations have the shapes of any particular probability distributions. We then discussed the **Kruskal–Wallis H test,** which is a nonparametric test for comparing the locations of several populations by using independent samples. This test, which employs the chi-square distribution, can be used when the assumptions for one-way analysis of variance do not hold. Finally, we presented a nonparametric approach for testing the significance of a population correlation coefficient. Here we saw how to compute **Spearman's rank correlation coefficient,** and we discussed how to use this quantity to test the significance of the population correlation coefficient.

Glossary of Terms

Kruskal–Wallis H test: A nonparametric test for comparing the locations of three or more populations by using independent random samples.

nonparametric test: A hypothesis test that requires no assumptions about the distribution(s) of the sampled population(s).

sign test: A hypothesis test about a population median that requires no assumptions about the sampled population.

Spearman's rank correlation coefficient: A correlation coefficient computed using the ranks of the observed values of two variables x and y.

Wilcoxon rank sum test: A nonparametric test for comparing the locations of two populations when an independent samples experiment has been carried out.

Wilcoxon signed ranks test: A nonparametric test for comparing the locations of two populations when a paired difference experiment has been carried out.

Important Formulas and Tests

Sign test for a population median: page 818

Large sample sign test: page 818

Wilcoxon rank sum test: page 821

Wilcoxon rank sum test (large sample approximation): pages 823–824

Wilcoxon signed ranks test: page 826

Wilcoxon signed ranks test (large sample approximation): page 828

Kruskal–Wallis H test: page 830

Kruskal–Wallis H statistic: page 830

Spearman's rank correlation coefficient: page 833

Spearman's rank correlation test: page 834

Supplementary Exercises

connect **18.32** Again consider the price comparison situation in which weekly expenses were compared at two chains—Miller's and Albert's. Recall that independent random samples at the two chains yielded the following weekly expenses: **DS** ShopExp

Miller's

$119.25	$121.32	$122.34	$120.14	$122.19
$123.71	$121.72	$122.42	$123.63	$122.44

Albert's

$111.99	$114.88	$115.11	$117.02	$116.89
$116.62	$115.38	$114.40	$113.91	$111.87

Because the sample sizes are small, there might be reason to doubt that the populations of expenses at the two chains are

normally distributed. Therefore, use a Wilcoxon rank sum test to determine whether expenses at Miller's and at Albert's differ. Use $\alpha = .05$.

18.33 A drug company wishes to compare the effects of three different drugs (X, Y, and Z) that are being developed to reduce cholesterol levels. Each drug is administered to six patients at the recommended dosage for six months. At the end of this period the reduction in cholesterol level is recorded for each patient. The results are given in Table 18.15. Assuming that the three samples are independent, use a nonparametric test to see whether the effects of the three drugs differ. Use $\alpha = .05$. **DS** CholRed

TABLE 18.15 Reduction of Cholesterol Levels Using Three Drugs DS CholRed

	Drug	
X	Y	Z
22	40	15
31	35	9
19	47	14
27	41	11
25	39	21
18	33	5

TABLE 18.16 2009 and 2010 Median Prices for Used Homes (in Thousands of Dollars) for Six Randomly Selected Housing Markets DS HomePrice

Housing Market	2009 Median Price	2010 Median Price
1	$243	$234
2	115	112
3	200	190
4	154	153
5	380	372
6	190	169

18.34 Table 18.16 gives the median 2009 and 2010 prices for used homes (in thousands of dollars) for six randomly selected U.S. housing markets. Use a nonparametric test to attempt to show that housing prices decreased from 2009 to 2010. Use $\alpha = .05$ and explain your conclusion. DS HomePrice

18.35 During 2017 a company implemented a number of policies aimed at reducing the ages of its customers' accounts. In order to assess the effectiveness of these measures, the company randomly selects 10 customer accounts. The average age of each account is determined for each of the years 2016 and 2017. These data are given in Table 18.17. Use a nonparametric technique to attempt to show that average account ages have decreased from 2016 to 2017. Use $\alpha = .05$. DS AcctAge

18.36 The following data concern the divorce rate (y) per 1,000 women and the percentage of the female population in the labor force (x): DS Divorce

Year	Divorce Rate, y	% of Females in Labor Force, x*
1890	3.0	18.9
1900	4.1	20.6
1910	4.7	25.4
1920	8.0	23.7
1930	7.5	24.8
1940	8.8	27.4
1950	10.3	31.4
1960	9.2	34.8
1970	14.9	42.6

*15 years old and over 1890–1930; 14 and over 1940–1960; 16 and over thereafter.

Source: U.S. Department of Commerce, Bureau of the Census, *Bicentennial Statistics*, Washington, D.C., 1976.

Use a nonparametric technique to attempt to show that x and y are positively correlated. Use $\alpha = .05$.

18.37 A loan officer wishes to compare the interest rates being charged for 48-month fixed-rate auto loans and 48-month variable-rate auto loans. Two independent, random samples of

TABLE 18.17 Average Account Ages in 2016 and 2017 for 10 Randomly Selected Accounts DS AcctAge

Account	Average Age of Account in 2016 (Days)	Average Age of Account in 2017 (Days)
1	35	27
2	24	19
3	47	40
4	28	30
5	41	33
6	33	25
7	35	31
8	51	29
9	18	15
10	28	21

auto loan rates are selected. A sample of five *48-month fixed-rate* auto loans had the following loan rates: DS AutoLoan

4.032% 3.85% 4.385% 3.75% 4.16%

while a sample of five *48-month variable-rate* auto loans had loan rates as follows:

2.6% 3.07% 2.872% 3.24% 3.15%

Perform a nonparametric test to determine whether loan rates for 48-month fixed-rate auto loans differ from loan rates for 48-month variable-rate auto loans. Use $\alpha = .05$. Explain your conclusion.

18.38 A large bank wishes to limit the median debt-to-equity ratio for its portfolio of commercial loans to 1.5. The bank randomly selects 15 of its commercial loan accounts. Audits result in the following debt-to-equity ratios: DS DebtEq

1.31	1.05	1.45	1.21	1.19
1.78	1.37	1.41	1.22	1.11
1.46	1.33	1.29	1.32	1.65

Can it be concluded that the median debt-to-equity ratio is less than 1.5 at the .05 level of significance? Explain.

18.39   **Internet Exercise**

Did labor force participation rates (LFPR) for women increase between 1968 and 1972? The Data and Story Library (DASL) contains LFPR figures for 1968 and 1972, for each of 19 cities. Go to the DASL website (http://lib.stat.cmu.edu/DASL/) and retrieve the Women in the Labor Force data set (http://lib.stat.cmu.edu/DASL/Datafiles/LaborForce.html). Produce appropriate graphical (histogram, stem-and-leaf, box plot) and numerical summaries of the LFPR data and conduct the following nonparametric statistical analyses:

a   Do the data provide sufficient evidence to conclude that the LFPR for women *increased* between 1968 and 1972? Conduct a nonparametric, two-sample, independent samples Wilcoxon rank sum test at the 0.01 level of significance. Clearly state the hypotheses and your conclusion. Report the *p*-value for your test.

b   Consider, as an alternative to the foregoing independent sample analysis, a paired sample procedure, the nonparametric Wilcoxon signed ranks test. Test once more the hypothesis of part *a*, this time using the Wilcoxon signed ranks test applied to the differences in LFPRs [1972–1968]. Again, clearly state your conclusion and *p*-value.

c   Between the two tests of parts *a* and *b*, which is the more appropriate for the current data situation? Why?

Appendix 18.1 ■ Nonparametric Methods Using Minitab

Sign test for the median in Figure 18.1(c) (data file: CompDisc.MTW):

- Enter the compact disc data from Figure 18.1(a) into column C1 with variable name LifeTime.

- **Select Stat : Nonparametrics : 1-Sample Sign.**

- In the "1-Sample Sign" dialog box, enter LifeTime into the Variables window.

- Select "Test median," and enter the number 6000 into the Test median window.

- Click on the "Alternative" arrow button and select "greater than" from the pull-down menu.

- Click OK in the "1-Sample Sign" dialog box to obtain the sign test results in the Session window.

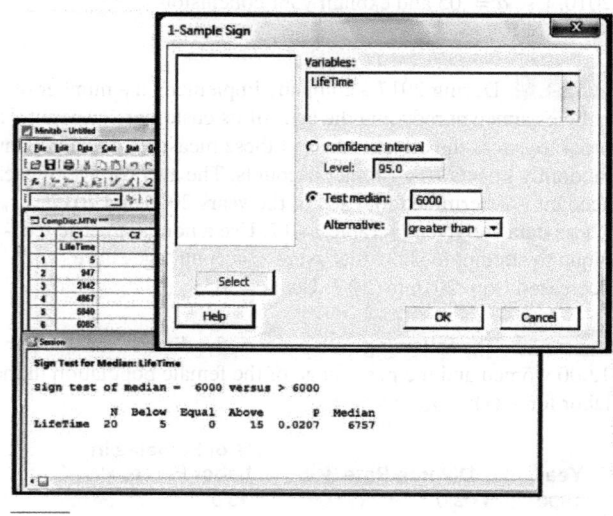

Source: Minitab 18

Wilcoxon (also known as Mann–Whitney) rank sum test for two independent samples in Figure 18.2(b) (data file: Court.MTW):

- Enter the litigation data from Figure 18.2(a) with Coos County in column C1 and Lane County in C2; use variable names Coos and Lane.

- Select **Stat : Nonparametrics : Mann-Whitney.**

- In the "Mann-Whitney" dialog box, enter Coos in the First Sample window and Lane in the Second Sample window.

- Type 95 in the Confidence level window.

- Click on the "Alternative" arrow button and select "less than" from the pull-down menu.

- Click OK in the "Mann-Whitney" dialog box to obtain the test results in the Session window.

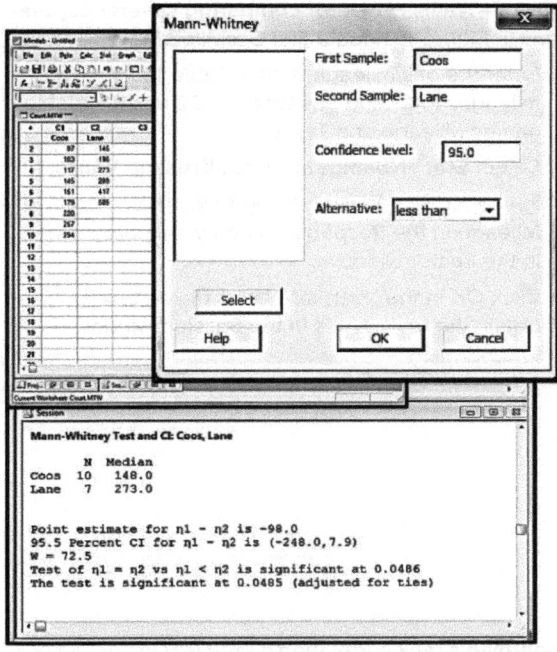

Source: Minitab 18

Wilcoxon signed ranks test for paired differences in Figure 18.3(b) (data file: Repair.MTW):

- Enter the garage repair data from Figure 18.3(a) with garage 1 cost estimates in column C1 and garage 2 cost estimates in column C2, using variable names Garage1 and Garage2.

- Select **Calc : Calculator.**

- In the "Calculator" dialog box, enter G1-G2 in the "Store result in variable" window.

- Enter 'Garage1' - 'Garage2' into the Expression window.

- In the "Calculator" dialog box, click OK to store the repair cost differences in the column named G1-G2.

- Select **Stat : Nonparametrics : 1-Sample Wilcoxon.**

- In the "1-Sample Wilcoxon" dialog box, enter G1-G2 into the Variables window.

- In the "1-Sample Wilcoxon" dialog box, select "Test median," and enter the number 0.0 into the Test median window.

- Click on the "Alternative" arrow button and select "less than" from the pull-down menu.

- Click OK in the "1-Sample Wilcoxon" dialog box to obtain the test results in the Session window.

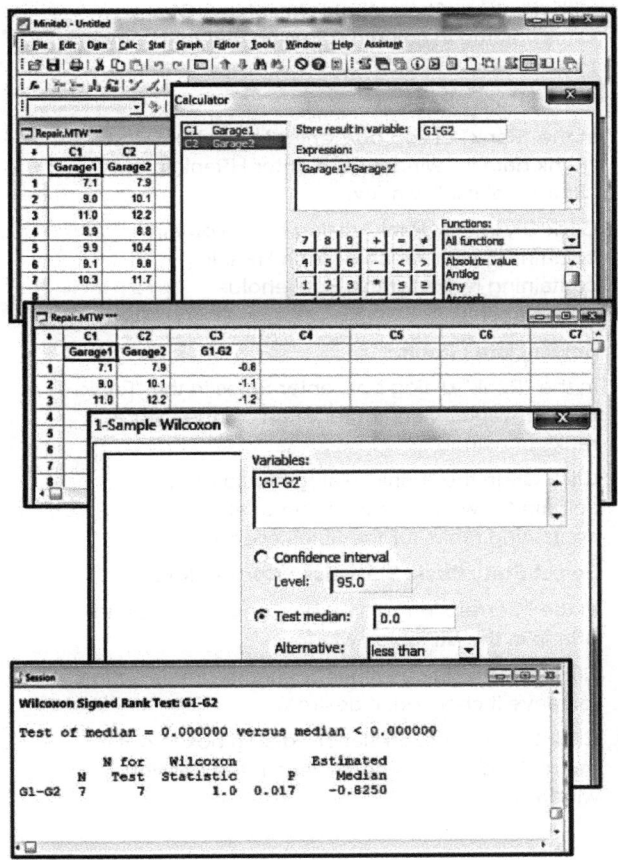

Source: Minitab 18

Kruskal–Wallis *H* test for comparing several populations in Figure 18.4 (data file: GasMile2.MTW):

- Enter the gas mileage data in Table 18.7 with gas mileages in C1 and gas types in C2 using variable names Mileage and Type.

- Select **Stat : Nonparametrics : Kruskal–Wallis.**

- In the "Kruskal-Wallis" dialog box, enter Mileage in the Response window and Type in the Factor window.

- Click OK in the "Kruskal-Wallis" dialog box to obtain the test results in the Session window.

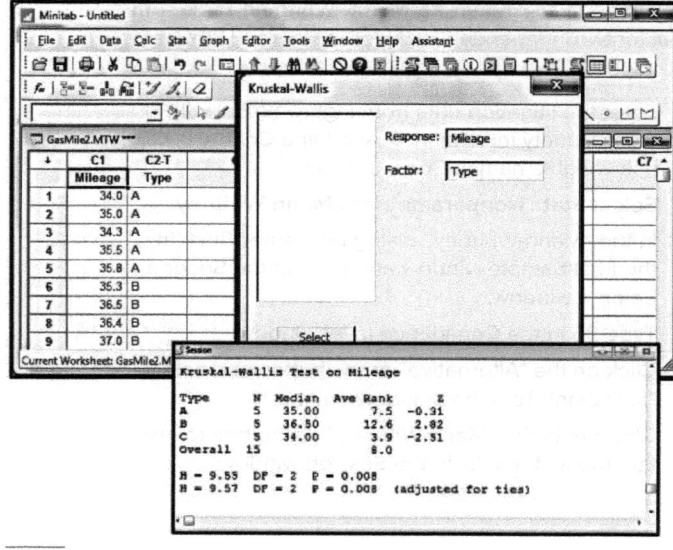

Source: Minitab 18

Spearman's rank correlation coefficient in Section 18.5 (data file: Electronics.MTW):

- Enter the electronics sales data from Table 18.12 with number of households in column C1 and sales volume in column C2 using variable names Households and Sales.

- Select **Data : Rank.**

- In the "Rank" dialog box, enter Households in the "Rank data in" window and enter HRank in the "Store ranks in" window.

- Click OK in the "Rank" dialog box to obtain column C3 with variable name HRank containing ranks for the Households observations.

- Select **Data : Rank.**

- In the "Rank" dialog box, enter Sales in the "Rank data in" window and enter SRank in the "Store ranks in" window.

- Click OK in the "Rank" dialog box to obtain column C4 with variable name SRank containing ranks for the Sales observations.

- Select **Stat : Basic Statistics : Correlation.**

- In the "Correlation" dialog box, enter HRank and SRank in the Variables window.

- Click on "Display p-values" to uncheck this option (or leave it checked, if desired).

- Click OK in the "Correlation" dialog box to obtain the rank correlation coefficient in the Session window.

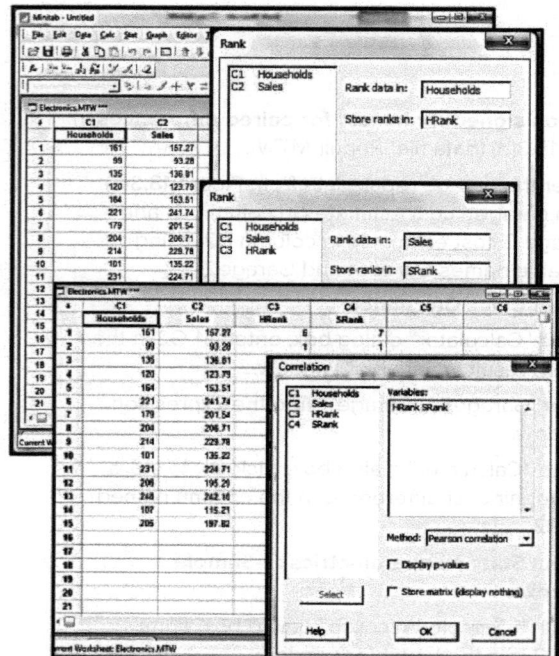

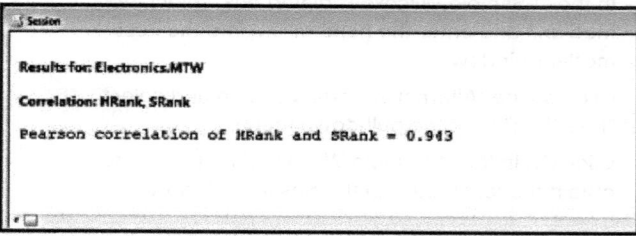

Source: Minitab 18

Appendix 18.2 ▪ Nonparametric Methods Using JMP

Sign test for the median in Example 18.1 (data file: CompDisc.jmp):

- After opening the CompDisc.jmp file, select **Tables : Sort**.

- In the Sort dialog box, select the "LifeTime" column and click the "By" button. Click "OK."

- A new window will appear with the values of "Lifetime" sorted from smallest to largest. Either count the number of observations greater than the median value from the null hypothesis (6000) or follow the steps below to use JMP to determine this number.

  - Select **Rows : Row Selection : Select Where** from the JMP Menu.

  - Select the "LifeTime" column.

  - Choose "is greater than" from the drop-down menu box.

  - Enter "6000" in the input box.

  - Click the "Add Condition" button.

  - Click "OK."

- The number of "Selected" (highlighted) rows is the test statistic *S*. Use the instructions from Appendix 6.3 to calculate the binomial probability $P(S \geq 15)$ with probability of success "0.5."

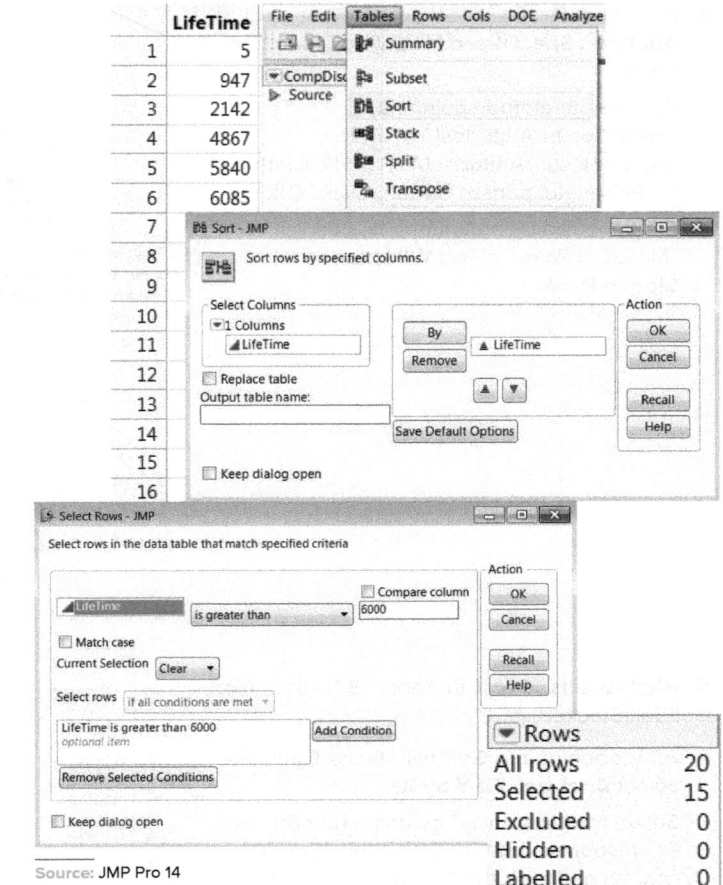

Source: JMP Pro 14

Wilcoxon/Kruskal–Wallis test in Exercise 18.12 (data file: Catalyst(Stacked).JMP):

- After opening the Catalyst (Stacked).jmp file, select **Analyze : Fit Y by X.**

- Select the "Yield" column and click the "Y, Response" button. Select the "Catalyst" column and click the "X, Factor" button. Click "OK."

- From the red triangle menu beside "Oneway Analysis of Yield by Catalyst," select **Nonparametric : Wilcoxon Test.**

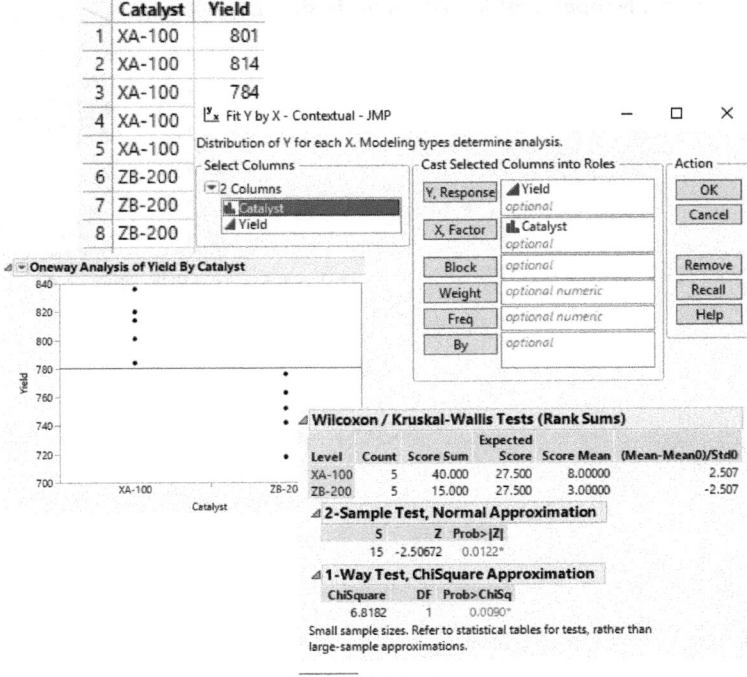

Source: JMP Pro 14

Wilcoxon signed ranks test in Table 18.6 (data file: AdStudy.jmp):

- After opening the AdStudy.jmp file, select **Analyze : Specialized Modeling : Matched Pairs.**

- Select both attitude columns "PreexposureAttitudes(A1)" and "PostexposureAttitudes(A2)" and click the "Y, Paired Response" button. Click "OK."

- From the red triangle menu beside "Matched Pairs," select **Wilcoxon Signed Rank**.

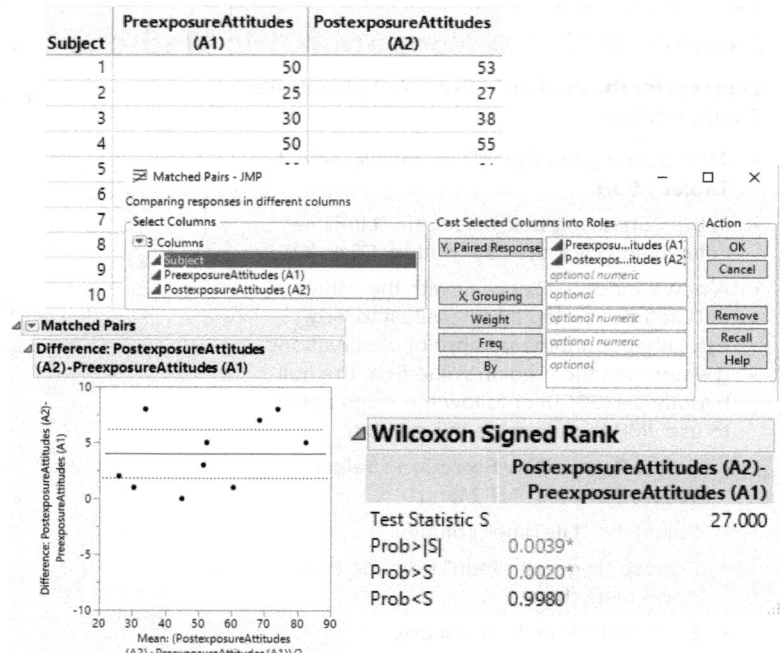

Subject	PreexposureAttitudes (A1)	PostexposureAttitudes (A2)
1	50	53
2	25	27
3	30	38
4	50	55

Wilcoxon Signed Rank

	PostexposureAttitudes (A2)- PreexposureAttitudes (A1)
Test Statistic S	27.000
Prob>|S|	0.0039*
Prob>S	0.0020*
Prob<S	0.9980

Source: JMP Pro 14

Kruskal–Wallis *H* test in Table 18.11 (data file: GolfBall(Stacked).jmp):

- After opening the GolfBall(Stacked).jmp file, select **Analyze : Fit Y by X.**

- Select the "Durability" column and click the "Y, Response" button. Select the "Brand" column and click the "X, Factor" button. Click "OK."

- From the red triangle menu beside "Oneway Analysis of Durability By Brand," select **Nonparametric : Wilcoxon Test.**

	Brand	Durability
1	Alpha	281
2	Alpha	220
3	Alpha	274
4	Alpha	242
5	Alpha	251
6	Best	270
7	Best	334

Oneway Analysis of Durability By Brand

Wilcoxon / Kruskal-Wallis Tests (Rank Sums)

Level	Count	Score Sum	Expected Score	Score Mean	(Mean-Mean0)/Std0
Alpha	5	34.000	52.500	6.8000	-1.571
Best	5	67.000	52.500	13.4000	1.222
Century	5	24.000	52.500	4.8000	-2.444
Divot	5	85.000	52.500	17.0000	2.793

1-Way Test, ChiSquare Approximation

ChiSquare	DF	Prob>ChiSq
13.8343	3	0.0031*

Small sample sizes. Refer to statistical tables for tests, rather than large-sample approximations.

Source: JMP Pro 14

Spearman rank correlation coefficient in Table 18.14 (data file: CarRank.jmp):

- After opening the CarRank.jmp file, select **Analyze : Multivariate Methods : Multivariate.**

- Select both ranking columns ("Mag. 1 Ranking" and "Mag. 2 Ranking") and click "Y, Columns." Click "OK."

- From the red triangle menu beside "Multivariate," select **Nonparametric Correlations : Spearman's ρ.** Note: Since both ranking variables contain rank data, the parametric correlation coefficient under "Correlations" is the same as the Spearman ρ value.

	Mag. 1 Ranking	Mag. 2 Ranking
1	5	7
2	1	1
3	4	5
4	7	4

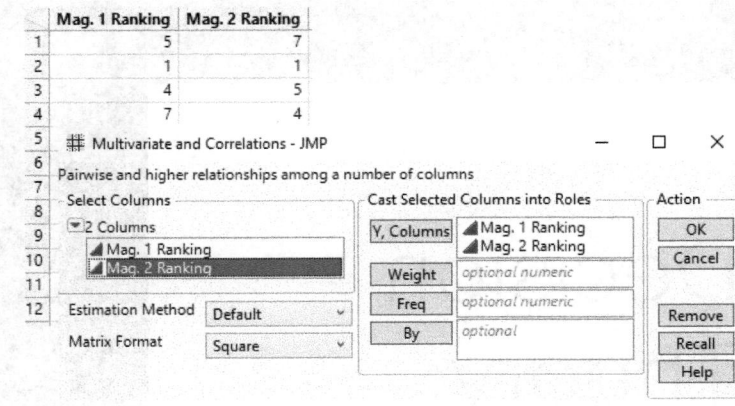

Multivariate
Correlations

	Mag. 1 Ranking	Mag. 2 Ranking
Mag. 1 Ranking	1.0000	0.8951
Mag. 2 Ranking	0.8951	1.0000

The correlations are estimated by Row-wise method.

Scatterplot Matrix

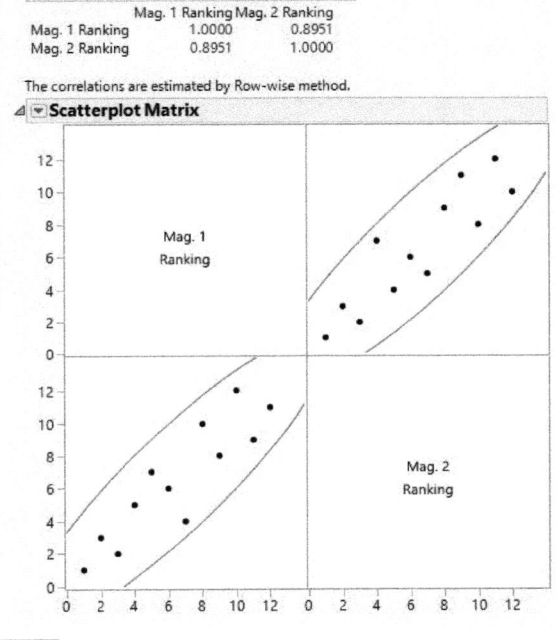

Source: JMP Pro 14

Nonparametric: Spearman's ρ

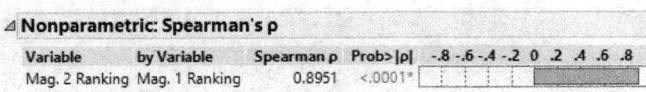

Variable	by Variable	Spearman ρ	Prob>|ρ|	-.8 -.6 -.4 -.2 0 .2 .4 .6 .8
Mag. 2 Ranking	Mag. 1 Ranking	0.8951	<.0001*	

Design Elements: (CD): ©Comstock Images/Alamy; (All Others): ©McGraw-Hill Education

## CHAPTER 19

Decision Theory

©Corbis/Glow Images

Learning Objectives

After mastering the material in this chapter, you will be able to:

LO19-1 Make decisions under uncertainty and under risk and assess the value of perfect information.

LO19-2 Make decisions using posterior analysis and assess the value of sample information.

LO19-3 Make decisions using utility theory.

Chapter Outline

very day businesses and the people who run them face a myriad of decisions. For instance, a manufacturer might need to decide where to locate a new factory and might also need to decide how large the new facility should be. Or, an investor might decide where to invest money from among several possible investment choices. In this chapter we study some probabilistic methods that can help a decision maker to make intelligent decisions. In Section 19.1 we introduce **decision theory.** We discuss the elements of a decision problem, and we present strategies for making decisions when we face various levels of uncertainty. We also show how to construct a **decision tree,** which is a diagram that can help us analyze a decision problem, and we show how the concept of **expected value** can help us make decisions. In Section 19.2 we show how to use **sample information** to help make decisions, and we demonstrate how to assess the worth of sample information in order to decide whether the sample information should be obtained. We conclude this chapter with Section 19.3, which introduces using **utility theory** to help make decisions.

Many of this chapter's concepts are presented in the context of the following case.

The Oil Company Case: An oil company uses decision theory to help decide whether to drill for oil on a particular site. The company can perform a seismic experiment at the site to obtain information about the site's potential, and the company uses decision theory to decide whether to drill based on the various possible survey results. In addition, decision theory is employed to determine whether the seismic experiment should be carried out.

19.1 Introduction to Decision Theory

LO19-1
Make decisions under uncertainty and under risk and assess the value of perfect information.

Suppose that a real estate developer is proposing the development of a condominium complex on an exclusive parcel of lakefront property. The developer wishes to choose between three possible options—building a large complex, building a medium-sized complex, and building a small complex. The profitability of each option depends on the level of demand for condominium units after the complex has been built. For simplicity, the developer considers only two possible levels of demand—high or low; the developer must choose whether to build a large, medium, or small complex based on her beliefs about whether demand for condominium units will be high or low.

The real estate developer's situation requires a decision. **Decision theory** is a general approach that helps decision makers make intelligent choices. A decision theory problem typically involves the following elements:

1 **States of nature:** a set of potential future conditions that affects the results of the decision. For instance, the level of demand (high or low) for condominium units will affect profits after the developer chooses to build a large, medium, or small complex. Thus, we have two states of nature—high demand and low demand.

2 **Alternatives:** several alternative actions for the decision maker to choose from. For example, the real estate developer can choose between building a large, medium, or small condominium complex. Therefore, the developer has three alternatives—large, medium, and small.

3 **Payoffs:** a payoff for each alternative under each potential state of nature. The payoffs are often summarized in a **payoff table.** For instance, Table 19.1 gives a payoff table

TABLE 19.1 A Payoff Table for the Condominium Complex Situation Condo

Alternatives	States of Nature	
	Low Demand	High Demand
Small complex	$8 million	$8 million
Medium complex	$5 million	$15 million
Large complex	–$11 million	$22 million

for the condominium complex situation. This table gives the profit[1] for each alternative under the different states of nature. For example, the payoff table tells us that, if the developer builds a large complex and if demand for units turns out to be high, a profit of $22 million will be realized. However, if the developer builds a large complex and if demand for units turns out to be low, a loss of $11 million will be suffered.

Once the states of nature have been identified, the alternatives have been listed, and the payoffs have been determined, we evaluate the alternatives by using a **decision criterion.** How this is done depends on the **degree of uncertainty** associated with the states of nature. Here we consider three possibilities:

1   **Certainty:** we know for certain which state of nature will actually occur.
2   **Uncertainty:** we have no information about the likelihoods of the various states of nature.
3   **Risk:** the likelihood (probability) of each state of nature can be estimated.

Decision making under certainty

In the unlikely event that we know for certain which state of nature will actually occur, we simply choose the alternative that gives the best payoff for that state of nature. For instance, in the condominium complex situation, if we know that demand for units will be high, then the payoff table (see Table 19.1) tells us that the best alternative is to build a large complex and that this choice will yield a profit of $22 million. On the other hand, if we know that demand for units will be low, then the payoff table tells us that the best alternative is to build a small complex and that this choice will yield a profit of $8 million.

Of course, we rarely (if ever) know for certain which state of nature will actually occur. However, analyzing the payoff table in this way often provides insight into the nature of the problem. For instance, examining the payoff table tells us that, if we know that demand for units will be low, then building either a small complex or a medium complex will be far superior to building a large complex (which would yield an $11 million loss).

Decision making under uncertainty

This is the exact opposite of certainty. Here we have no information about how likely the different states of nature are. That is, we have no idea how to assign probabilities to the different states of nature.

In such a case, several approaches are possible; we will discuss two commonly used methods. The first is called the **maximin criterion.**

> **Maximin:** Find the worst possible payoff for each alternative, and then choose the alternative that yields the maximum worst possible payoff.

For instance, to apply the maximin criterion to the condominium complex situation, we proceed as follows (see Table 19.1):

1   If a small complex is built, the worst possible payoff is $8 million.
2   If a medium complex is built, the worst possible payoff is $5 million.
3   If a large complex is built, the worst possible payoff is −$11 million.

Because the maximum of these worst possible payoffs is $8 million, the developer should choose to build a small complex.

The maximin criterion is a *pessimistic approach* because it considers the worst possible payoff for each alternative. When an alternative is chosen using the maximin criterion, the actual payoff obtained may be higher than the maximum worst possible payoff. However, using the maximin criterion assures a "guaranteed minimum" payoff.

A second approach is called the **maximax criterion.**

---

[1]Here profits are really present values representing current dollar values of expected future income minus costs.

Maximax: Find the best possible payoff for each alternative, and then choose the alternative that yields the maximum best possible payoff.

To apply the maximax criterion to the condominium complex situation, we proceed as follows (see Table 19.1):

1 If a small complex is built, the best possible payoff is $8 million.
2 If a medium complex is built, the best possible payoff is $15 million.
3 If a large complex is built, the best possible payoff is $22 million.

Because the maximum of these best possible payoffs is $22 million, the developer should choose to build a large complex.

The maximax criterion is an *optimistic approach* because we always choose the alternative that yields the highest possible payoff. This is a "go for broke" strategy, and the actual payoff obtained may be far less than the highest possible payoff. For example, in the condominium complex situation, if a large complex is built and demand for units turns out to be low, an $11 million loss will be suffered (instead of a $22 million profit).

Decision making under risk

In this case we can estimate the probability of occurrence for each state of nature. Thus, we have a situation in which we have more information about the states of nature than in the case of uncertainty and less information than in the case of certainty. Here a commonly used approach is to use the **expected monetary value criterion.** This involves computing the expected monetary payoff for each alternative and choosing the alternative with the largest expected payoff.

The expected value criterion can be employed by using *prior probabilities*. As an example, suppose that in the condominium complex situation the developer assigns prior probabilities of .7 and .3 to high and low demands, respectively, as shown in the **decision tree diagram** of Figure 19.1. We find the expected monetary value for each alternative by multiplying

FIGURE 19.1   A Decision Tree for the Condominium Complex Situation

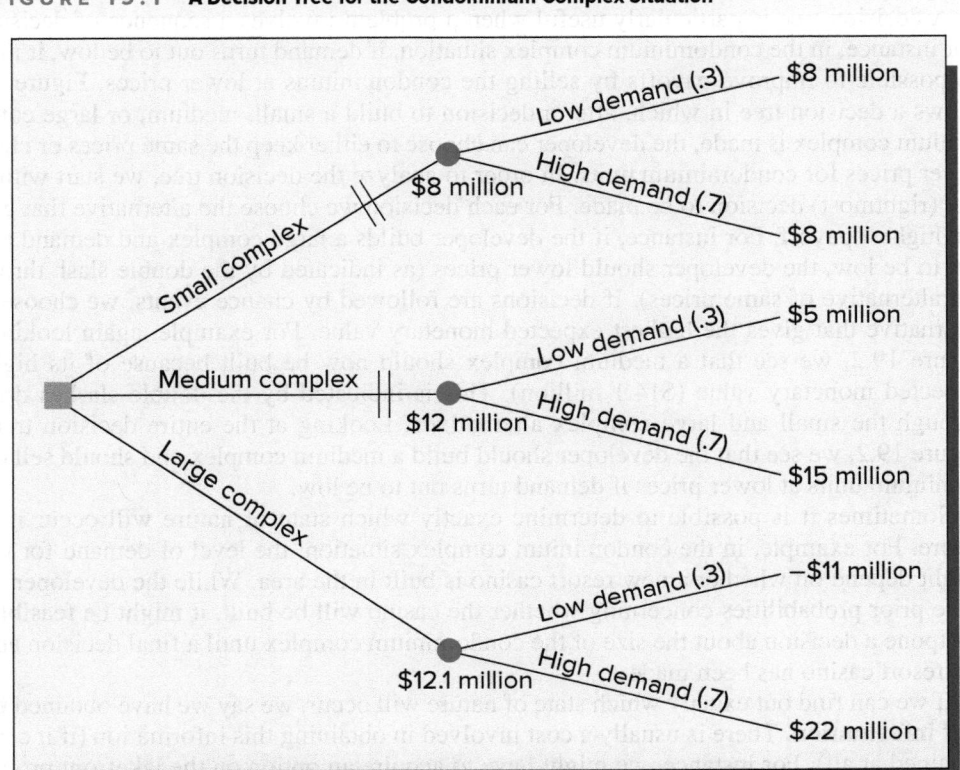

the probability of occurrence for each state of nature by the payoff associated with the state of nature and by summing these products. Referring to the payoff table in Table 19.1, the expected monetary values are as follows:

Small complex: Expected value = .3($8 million) + .7($8 million) = $8 million

Medium complex: Expected value = .3($5 million) + .7($15 million) = $12 million

Large complex: Expected value = .3(−$11 million) + .7($22 million) = $12.1 million

Choosing the alternative with the highest expected monetary value, the developer would choose to build a large complex.

Remember that the expected payoff is not necessarily equal to the actual payoff that will be realized. Rather, the expected payoff is the long-run average payoff that would be realized if many identical decisions were made. For instance, the expected monetary payoff of $12.1 million for a large complex is the average payoff that would be obtained if many large condominium complexes were built. Thus, the expected monetary value criterion is best used when many similar decisions will be made.

Using a decision tree

It is often convenient to depict the alternatives, states of nature, payoffs, and probabilities (in the case of risk) in the form of a **decision tree** or **tree diagram.** The diagram is made up of **nodes** and **branches.** We use square nodes to denote decision points and circular nodes to denote chance events. The branches emanating from a decision point represent alternatives, and the branches emanating from a circular node represent the possible states of nature. As we have seen, Figure 19.1 presents a decision tree for the condominium complex situation (in the case of risk as described previously). Notice that the payoffs are shown at the rightmost end of each branch and that the probabilities associated with the various states of nature are given in parentheses corresponding to each branch emanating from a chance node. The expected monetary values for the alternatives are shown below the chance nodes. The double slashes placed through the small complex and medium complex branches indicate that these alternatives would not be chosen (because of their lower expected payoffs) and that the large complex alternative would be selected.

A decision tree is particularly useful when a problem involves a sequence of decisions. For instance, in the condominium complex situation, if demand turns out to be low, it might be possible to improve payoffs by selling the condominiums at lower prices. Figure 19.2 shows a decision tree in which, after a decision to build a small, medium, or large condominium complex is made, the developer can choose to either keep the same prices or charge lower prices for condominium units. In order to analyze the decision tree, we start with the last (rightmost) decision to be made. For each decision we choose the alternative that gives the highest payoff. For instance, if the developer builds a large complex and demand turns out to be low, the developer should lower prices (as indicated by the double slash through the alternative of same prices). If decisions are followed by chance events, we choose the alternative that gives the highest expected monetary value. For example, again looking at Figure 19.2, we see that a medium complex should now be built because of its highest expected monetary value ($14.1 million). This is indicated by the double slashes drawn through the small and large complex alternatives. Looking at the entire decision tree in Figure 19.2, we see that the developer should build a medium complex and should sell condominium units at lower prices if demand turns out to be low.

Sometimes it is possible to determine exactly which state of nature will occur in the future. For example, in the condominium complex situation, the level of demand for units might depend on whether a new resort casino is built in the area. While the developer may have prior probabilities concerning whether the casino will be built, it might be feasible to postpone a decision about the size of the condominium complex until a final decision about the resort casino has been made.

If we can find out exactly which state of nature will occur, we say we have obtained **perfect information.** There is usually a cost involved in obtaining this information (if it can be obtained at all). For instance, we might have to acquire an option on the lakefront property

FIGURE 19.2 A Decision Tree with Sequential Decisions

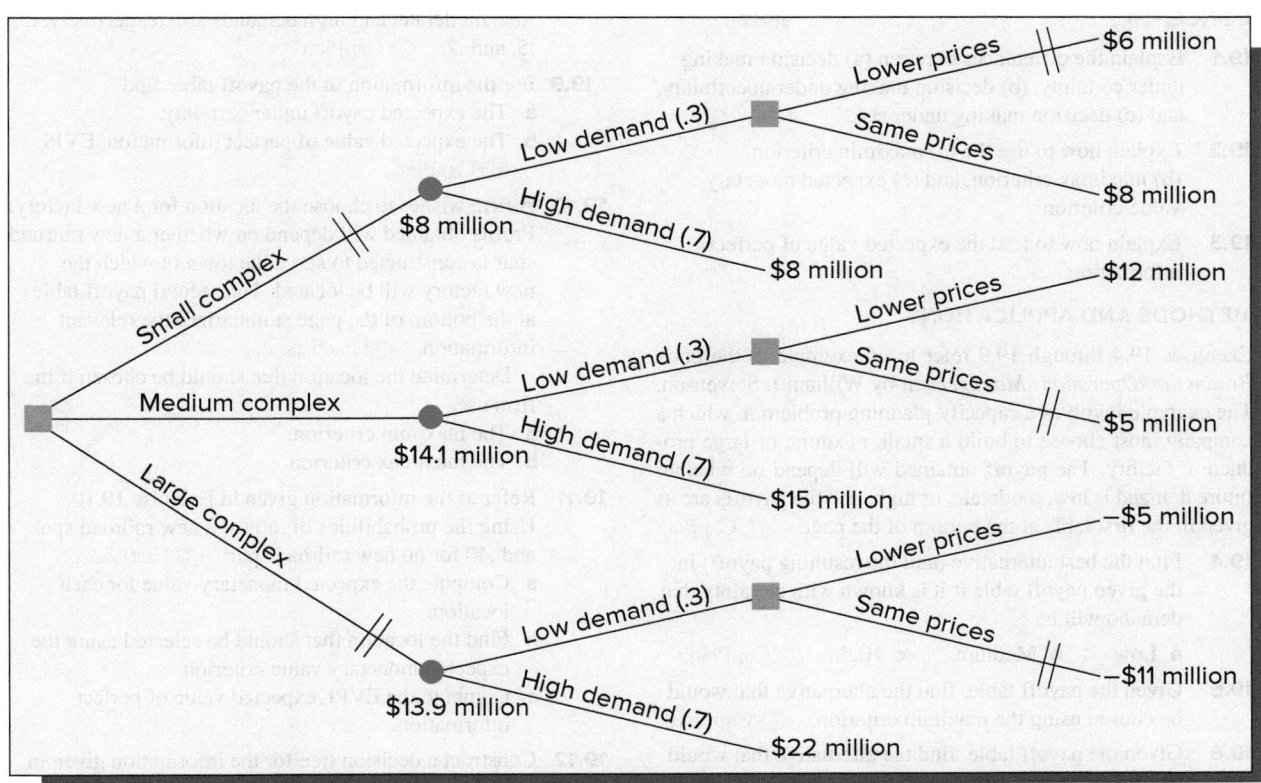

on which the condominium complex is to be built in order to postpone a decision about the size of the complex. Or perfect information might be acquired by conducting some sort of research that must be paid for. A question that arises here is whether it is worth the cost to obtain perfect information. We can answer this question by computing the **expected value of perfect information,** which we denote as the **EVPI.** The EVPI is defined as follows:

> EVPI = expected payoff under certainty − expected payoff under risk

For instance, if we consider the condominium complex situation depicted in the decision tree of Figure 19.1, we found that the expected payoff under risk is $12.1 million (which is the expected payoff associated with building a large complex). To find the expected payoff under certainty, we find the highest payoff under each state of nature. Referring to Table 19.1, we see that if demand is low, the highest payoff is $8 million (when we build a small complex); we see that if demand is high, the highest payoff is $22 million (when we build a large complex). Because the prior probabilities of high and low demand are, respectively, .7 and .3, the expected payoff under certainty is .7($22 million) + .3($8 million) = $17.8 million. Therefore, the expected value of perfect information is $17.8 million − $12.1 million = $5.7 million. This is the maximum amount of money that the developer should be willing to pay to obtain perfect information. That is, the land option should be purchased if it costs $5.7 million or less. Then, if the casino is not built (and demand is low), a small condominium complex should be built; if the casino is built (and demand is high), a large condominium complex should be built. On the other hand, if the land option costs more than $5.7 million, the developer should choose the alternative having the highest expected payoff (which would mean building a large complex—see Figure 19.1).

Finally, another approach to dealing with risk involves assigning what we call **utilities** to monetary values. These utilities reflect the decision maker's attitude toward risk: that is, does the decision maker avoid risk or is he or she a risk taker? Here the decision maker chooses the alternative that **maximizes expected utility.** We will introduce this approach in Section 19.3.

Exercises for Section 19.1

CONCEPTS

19.1 Explain the differences between (a) decision making under certainty, (b) decision making under uncertainty, and (c) decision making under risk.

19.2 Explain how to use the (a) maximin criterion, (b) maximax criterion, and (c) expected monetary value criterion.

19.3 Explain how to find the expected value of perfect information.

METHODS AND APPLICATIONS

Exercises 19.4 through 19.9 refer to an example in the book *Production/Operations Management* by William J. Stevenson. The example involves a capacity-planning problem in which a company must choose to build a small, medium, or large production facility. The payoff obtained will depend on whether future demand is low, moderate, or high, and the payoffs are as given in the first table at the bottom of the page. ⑥ CapPlan

19.4 Find the best alternative (and the resulting payoff) in the given payoff table if it is known with certainty that demand will be

    a Low.   **b** Medium.   **c** High. ⑥ CapPlan

19.5 Given the payoff table, find the alternative that would be chosen using the maximin criterion. ⑥ CapPlan

19.6 Given the payoff table, find the alternative that would be chosen using the maximax criterion. ⑥ CapPlan

19.7 Suppose that the company assigns prior probabilities of .3, .5, and .2 to low, moderate, and high demands, respectively. ⑥ CapPlan
    a Find the expected monetary value for each alternative (small, medium, and large).
    b What is the best alternative if we use the expected monetary value criterion?

19.8 Construct a decision tree for the information in the payoff table assuming that the prior probabilities of

low, moderate, and high demands are, respectively, .3, .5, and .2. ⑥ CapPlan

19.9 For the information in the payoff table find
    a The expected payoff under certainty.
    b The expected value of perfect information, EVPI.
    ⑥ CapPlan

19.10 A firm wishes to choose the location for a new factory. Profits obtained will depend on whether a new railroad spur is constructed to serve the town in which the new factory will be located. The second payoff table at the bottom of the page summarizes the relevant information. ⑥ FactLoc
    Determine the location that should be chosen if the firm uses
    a The maximin criterion.
    b The maximax criterion.

19.11 Refer to the information given in Exercise 19.10. Using the probabilities of .60 for a new railroad spur and .40 for no new railroad spur ⑥ FactLoc
    a Compute the expected monetary value for each location.
    b Find the location that should be selected using the expected monetary value criterion.
    c Compute the EVPI, expected value of perfect information.

19.12 Construct a decision tree for the information given in Exercises 19.10 and 19.11. ⑥ FactLoc

19.13 Figure 19.3 gives a decision tree presented in the book *Production/Operations Management* by William J. Stevenson. Use this tree diagram to do the following:
    a Find the expected monetary value for each of the alternatives (subcontract, expand, and build).
    b Determine the alternative that should be selected in order to maximize the expected monetary value.

Payoff Table for Exercises 19.4–19.9

	Possible Future Demand		
Alternatives	**Low**	**Moderate**	**High**
Small facility	$10*	$10	$10
Medium facility	7	12	12
Large facility	−4	2	16

*Present value in $ millions.

Source: W. J. Stevenson, *Production/Operations Management*, 5th ed. (Burr Ridge, IL: Richard D. Irwin, 1996), p. 73.

Payoff Table for Exercises 19.10–19.12

Alternatives	**New Railroad Spur Built**	**No New Railroad Spur**
Location A	$1*	$14
Location B	2	10
Location C	4	6

*Profits in $ millions.

FIGURE 19.3   **Decision Tree for Exercise 19.13**

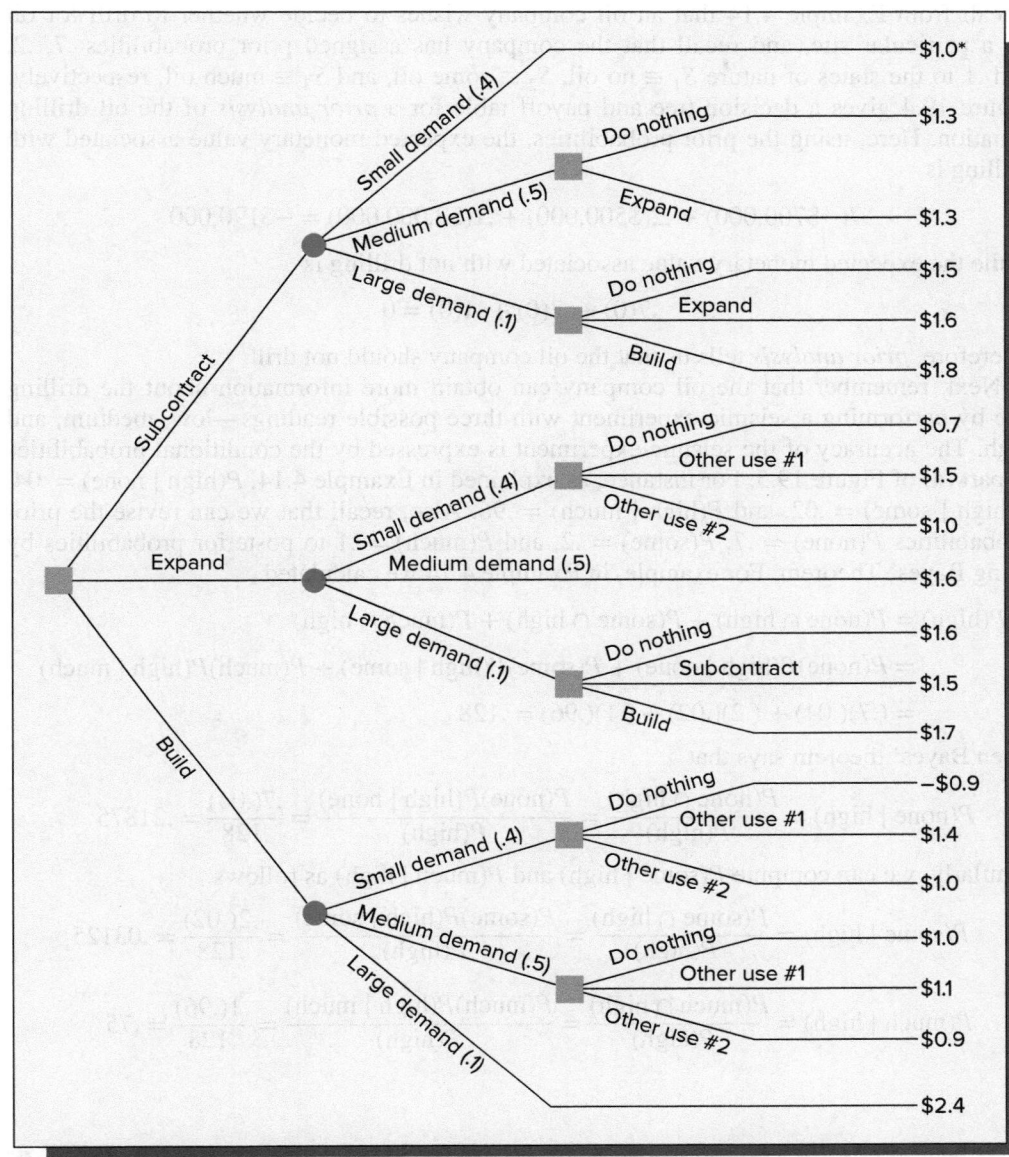

*Net present value in millions.

Source: Decision tree from W. J. Stevenson, *Production/Operations Management*, 6/e, p. 228, and problem from p. 73 © 1999 McGraw-Hill Companies, Inc.

19.2 Decision Making Using Posterior Probabilities

We have seen that the *expected monetary value criterion* tells us to choose the alternative having the highest expected payoff. In Section 19.1 we computed expected payoffs by using *prior probabilities*. When we use the expected monetary value criterion to choose the best alternative based on expected values computed using prior probabilities, we call this **prior decision analysis.** Often, however, sample information can be obtained to help us make decisions. In such a case, we compute expected values by using *posterior probabilities*, and we call the analysis **posterior decision analysis.** In the following example we demonstrate how to carry out posterior analysis.

LO19-2
Make decisions using posterior analysis and assess the value of sample information.

EXAMPLE 19.1 The Oil Company Case: Deciding Whether to Drill for Oil

Recall from Example 4.14 that an oil company wishes to decide whether to drill for oil on a particular site, and recall that the company has assigned prior probabilities .7, .2, and .1 to the states of nature $S_1 \equiv$ no oil, $S_2 \equiv$ some oil, and $S_3 \equiv$ much oil, respectively. Figure 19.4 gives a decision tree and payoff table for a *prior analysis* of the oil drilling situation. Here, using the prior probabilities, the expected monetary value associated with drilling is

$$.7(-\$700,000) + .2(\$500,000) + .1(\$2,000,000) = -\$190,000$$

while the expected monetary value associated with not drilling is

$$.7(0) + .2(0) + .1(0) = 0$$

Therefore, *prior analysis* tells us that the oil company should not drill.

Next, remember that the oil company can obtain more information about the drilling site by performing a seismic experiment with three possible readings—low, medium, and high. The accuracy of the seismic experiment is expressed by the conditional probabilities in part (a) of Figure 19.5. For instance, as explained in Example 4.14, $P(\text{high} \mid \text{none}) = .04$, $P(\text{high} \mid \text{some}) = .02$, and $P(\text{high} \mid \text{much}) = .96$. Also, recall that we can revise the prior probabilities $P(\text{none}) = .7$, $P(\text{some}) = .2$, and $P(\text{much}) = .1$ to posterior probabilities by using Bayes' Theorem. For example, in Example 4.14 we calculated

$$P(\text{high}) = P(\text{none} \cap \text{high}) + P(\text{some} \cap \text{high}) + P(\text{much} \cap \text{high})$$

$$= P(\text{none})P(\text{high} \mid \text{none}) + P(\text{some})P(\text{high} \mid \text{some}) + P(\text{much})P(\text{high} \mid \text{much})$$

$$= (.7)(.04) + (.2)(.02) + (.1)(.96) = .128$$

Then Bayes' theorem says that

$$P(\text{none} \mid \text{high}) = \frac{P(\text{none} \cap \text{high})}{P(\text{high})} = \frac{P(\text{none})P(\text{high} \mid \text{none})}{P(\text{high})} = \frac{.7(.04)}{.128} = .21875$$

Similarly, we can compute $P(\text{some} \mid \text{high})$ and $P(\text{much} \mid \text{high})$ as follows.

$$P(\text{some} \mid \text{high}) = \frac{P(\text{some} \cap \text{high})}{P(\text{high})} = \frac{P(\text{some})P(\text{high} \mid \text{some})}{P(\text{high})} = \frac{.2(.02)}{.128} = .03125$$

$$P(\text{much} \mid \text{high}) = \frac{P(\text{much} \cap \text{high})}{P(\text{high})} = \frac{P(\text{much})P(\text{high} \mid \text{much})}{P(\text{high})} = \frac{.1(.96)}{.128} = .75$$

FIGURE 19.4   **A Decision Tree and Payoff Table for a Prior Analysis of the Oil Company Case**

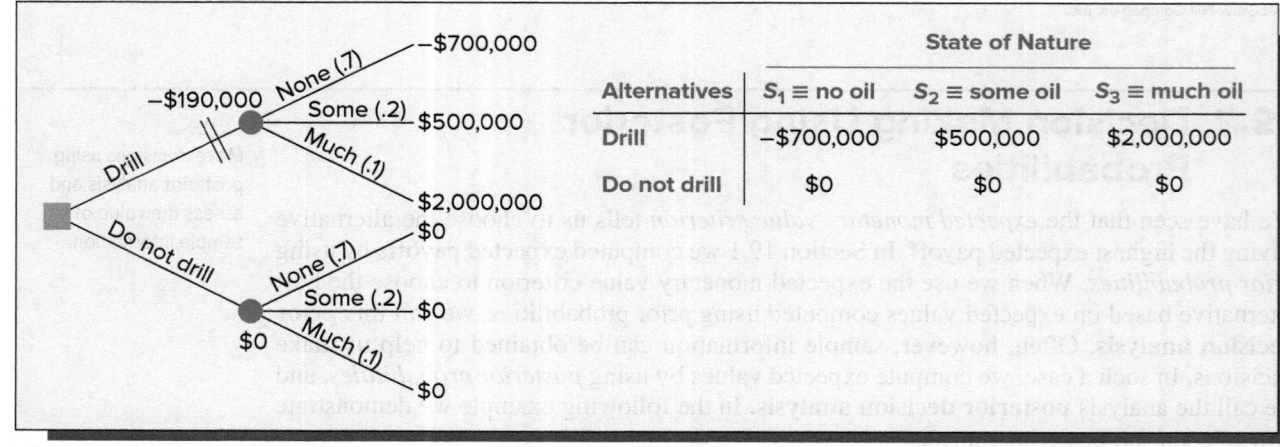

Alternatives	State of Nature		
	$S_1 \equiv$ no oil	$S_2 \equiv$ some oil	$S_3 \equiv$ much oil
Drill	-$700,000	$500,000	$2,000,000
Do not drill	$0	$0	$0

FIGURE 19.5   A Tree Diagram and Probability Revision Tables for Bayes' Theorem in the Oil Company Case

(a) A tree diagram illustrating the prior and conditional probabilities

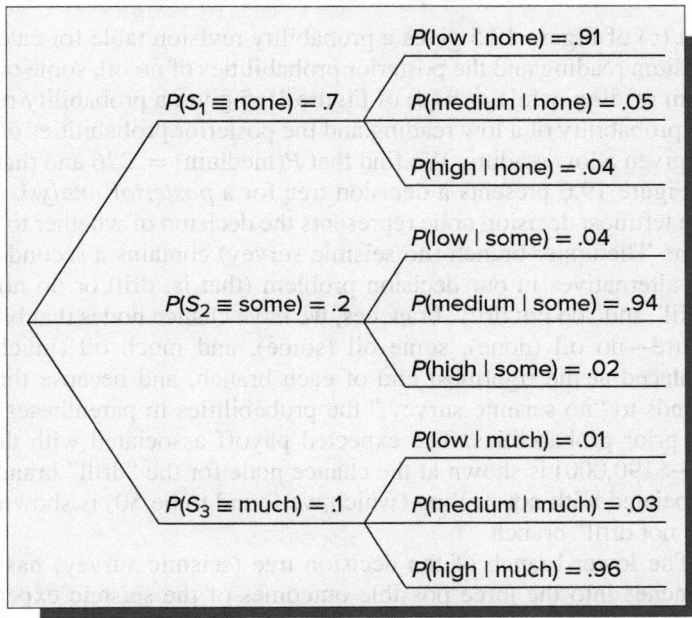

(b) A probability revision table for calculating the probability of a high reading and the posterior probabilities of no oil (S_1), some oil (S_2), and much oil (S_3) given a high reading

S_j	$P(S_j)$	$P(\text{high} \mid S_j)$	$P(S_j \cap \text{high}) = P(S_j)P(\text{high} \mid S_j)$	$P(S_j \mid \text{high}) = P(S_j \cap \text{high})/P(\text{high})$
$S_1 \equiv$ none	$P(\text{none}) = .7$	$P(\text{high} \mid \text{none}) = .04$	$P(\text{none} \cap \text{high}) = .7(.04) = .028$	$P(\text{none} \mid \text{high}) = .028/.128 = .21875$
$S_2 \equiv$ some	$P(\text{some}) = .2$	$P(\text{high} \mid \text{some}) = .02$	$P(\text{some} \cap \text{high}) = .2(.02) = .004$	$P(\text{some} \mid \text{high}) = .004/.128 = .03125$
$S_3 \equiv$ much	$P(\text{much}) = .1$	$P(\text{high} \mid \text{much}) = .96$	$P(\text{much} \cap \text{high}) = .1(.96) = .096$	$P(\text{much} \mid \text{high}) = .096/.128 = .75$
Total	1		$P(\text{high}) = .028 + .004 + .096 = .128$	1

(c) A probability revision table for calculating the probability of a medium reading and the posterior probabilities of no oil (S_1), some oil (S_2), and much oil (S_3) given a medium reading

S_j	$P(S_j)$	$P(\text{medium} \mid S_j)$	$P(S_j \cap \text{medium}) =$ $P(S_j)P(\text{medium} \mid S_j)$	$P(S_j \mid \text{medium}) =$ $P(S_j \cap \text{medium})/P(\text{medium})$
$S_1 \equiv$ none	$P(\text{none}) = .7$	$P(\text{medium} \mid \text{none}) = .05$	$P(\text{none} \cap \text{medium}) = .7(.05) = .035$	$P(\text{none} \mid \text{medium}) = .035/.226 = .15487$
$S_2 \equiv$ some	$P(\text{some}) = .2$	$P(\text{medium} \mid \text{some}) = .94$	$P(\text{some} \cap \text{medium}) = .2(.94) = .188$	$P(\text{some} \mid \text{medium}) = .188/.226 = .83186$
$S_3 \equiv$ much	$P(\text{much}) = .1$	$P(\text{medium} \mid \text{much}) = .03$	$P(\text{much} \cap \text{medium}) = .1(.03) = .003$	$P(\text{much} \mid \text{medium}) = .003/.226 = .01327$
Total	1		$P(\text{medium}) = .035 + .188 + .003 = .226$	1

(d) A probability revision table for calculating the probability of a low reading and the posterior probabilities of no oil (S_1), some oil (S_2), and much oil (S_3) given a low reading

S_j	$P(S_j)$	$P(\text{low} \mid S_j)$	$P(S_j \cap \text{low}) = P(S_j)P(\text{low} \mid S_j)$	$P(S_j \mid \text{low}) = P(S_j \cap \text{low})/P(\text{low})$
$S_1 \equiv$ none	$P(\text{none}) = .7$	$P(\text{low} \mid \text{none}) = .91$	$P(\text{none} \cap \text{low}) = .7(.91) = .637$	$P(\text{none} \mid \text{low}) = .637/.646 = .98607$
$S_2 \equiv$ some	$P(\text{some}) = .2$	$P(\text{low} \mid \text{some}) = .04$	$P(\text{some} \cap \text{low}) = .2(.04) = .008$	$P(\text{some} \mid \text{low}) = .008/.646 = .01238$
$S_3 \equiv$ much	$P(\text{much}) = .1$	$P(\text{low} \mid \text{much}) = .01$	$P(\text{much} \cap \text{low}) = .1(.01) = .001$	$P(\text{much} \mid \text{low}) = .001/.646 = .00155$
Total	1		$P(\text{low}) = .637 + .008 + .001 = .646$	1

These calculations are summarized in the **probability revision table** in Figure 19.5(b). This table also shows that

$$P(\text{high}) = P(\text{none} \cap \text{high}) + P(\text{some} \cap \text{high}) + P(\text{much} \cap \text{high})$$

$$= .028 + .004 + .096 = .128$$

Part (c) of Figure 19.5 gives a probability revision table for calculating the probability of a medium reading and the posterior probabilities of no oil, some oil, and much oil given a medium reading, while part (d) of Figure 19.5 gives a probability revision table for calculating the probability of a low reading and the posterior probabilities of no oil, some oil, and much oil given a low reading. We find that $P(\text{medium}) = .226$ and that $P(\text{low}) = .646$.

Figure 19.6 presents a decision tree for a *posterior analysis* of the oil drilling problem. The leftmost decision node represents the decision of whether to conduct the seismic experiment. The upper branch (no seismic survey) contains a second decision node representing the alternatives in our decision problem (that is, drill or do not drill). At the ends of the "drill" and "do not drill" branches, we have chance nodes that branch into the three states of nature—no oil (none), some oil (some), and much oil (much). The appropriate payoff is placed at the rightmost end of each branch, and because this uppermost branch corresponds to "no seismic survey," the probabilities in parentheses for the states of nature are the prior probabilities. The expected payoff associated with drilling (which we found to be −$190,000) is shown at the chance node for the "drill" branch, and the expected payoff associated with not drilling (which we found to be $0) is shown at the chance node for the "do not drill" branch.

The lower branch of the decision tree (seismic survey) has an extra chance node that branches into the three possible outcomes of the seismic experiment—low, medium, and high. The probabilities of these outcomes are shown on their respective branches. From the low, medium, and high branches, the tree branches into alternatives (drill and do not drill) and from alternatives into states of nature (none, some, and much). However, the probabilities in parentheses written beside the none, some, and much branches are the posterior probabilities that we computed in the probability revision tables in Figure 19.5. This is because advancing to the end of a particular branch in the lower part of the decision tree is conditional; that is, it depends on obtaining a particular experimental result (low, medium, or high).

We can now use the decision tree to determine the alternative (drill or do not drill) that should be selected given that the seismic experiment has been performed and has resulted in a particular outcome. First, suppose that the seismic experiment results in a high reading. Looking at the branch of the decision tree corresponding to a high reading, the expected monetary values associated with the "drill" and "do not drill" alternatives are

Drill: $.21875(-\$700,000) + .03125(\$500,000) + .75(\$2,000,000) = \$1,362,500$

Do not drill: $.21875(0) + .03125(0) + .75(0) = \$0$

These expected monetary values are placed on the decision tree corresponding to the "drill" and "do not drill" alternatives. They tell us that, if the seismic experiment results in a high reading, then the company should drill and the expected payoff will be $1,362,500. The double slash placed through the "do not drill" branch (at the very bottom of the decision tree) blocks off that branch and indicates that the company should drill if a high reading is obtained.

Next, suppose that the seismic experiment results in a medium reading. Looking at the branch corresponding to a medium reading, the expected monetary values are

Drill: $.15487(-\$700,000) + .83186(\$500,000) + .01327(\$2,000,000) = \$334,061$

Do not drill: $.15487(\$0) + .83186(\$0) + .01327(\$0) = \$0$

Therefore, if the seismic experiment results in a medium reading, the oil company should drill, and the expected payoff will be $334,061.

FIGURE 19.6   A Decision Tree for a Posterior Analysis of the Oil Company Case

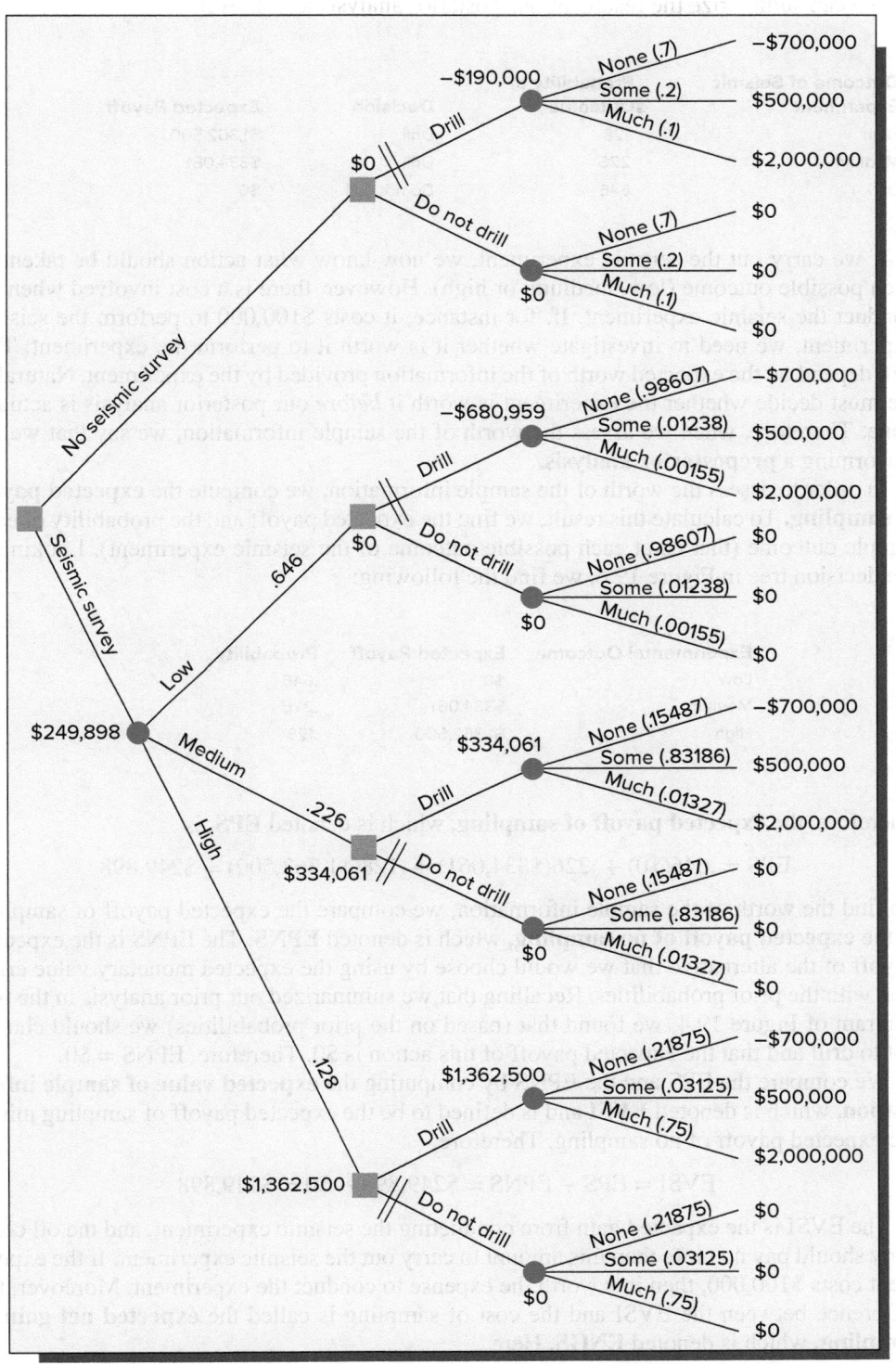

Finally, suppose that the seismic experiment results in a low reading. Looking at the branch corresponding to a low reading, the expected monetary values are

Drill: .98607(−$700,000) + .01238($500,000) + .00155($2,000,000) = −$680,959

Do not drill: .98607($0) + .01238($0) + .00155($0) = $0

Therefore, if the seismic experiment results in a low reading, the oil company should not drill on the site.

We can summarize the results of our posterior analysis as follows:

Outcome of Seismic Experiment	Probability of Outcome	Decision	Expected Payoff
High	.128	Drill	$1,362,500
Medium	.226	Drill	$334,061
Low	.646	Do not drill	$0

If we carry out the seismic experiment, we now know what action should be taken for each possible outcome (low, medium, or high). However, there is a cost involved when we conduct the seismic experiment. If, for instance, it costs $100,000 to perform the seismic experiment, we need to investigate whether it is worth it to perform the experiment. This will depend on the expected worth of the information provided by the experiment. Naturally, we must decide whether the experiment is worth it *before* our posterior analysis is actually done. Therefore, when we assess the worth of the sample information, we say that we are performing a **preposterior analysis.**

In order to assess the worth of the sample information, we compute the **expected payoff of sampling.** To calculate this result, we find the expected payoff and the probability of each sample outcome (that is, at each possible outcome of the seismic experiment). Looking at the decision tree in Figure 19.6, we find the following:

Experimental Outcome	Expected Payoff	Probability
Low	$0	.646
Medium	$334,061	.226
High	$1,362,500	.128

Therefore, the **expected payoff of sampling,** which is denoted **EPS,** is

$$\text{EPS} = .646(\$0) + .226(\$334,061) + .128(\$1,362,500) = \$249,898$$

To find the worth of the sample information, we compare the expected payoff of sampling to the **expected payoff of no sampling,** which is denoted **EPNS.** The EPNS is the expected payoff of the alternative that we would choose by using the expected monetary value criterion with the prior probabilities. Recalling that we summarized our prior analysis in the tree diagram of Figure 19.4, we found that (based on the prior probabilities) we should choose not to drill and that the expected payoff of this action is $0. Therefore, EPNS = $0.

We compare the EPS and the EPNS by computing the **expected value of sample information,** which is denoted **EVSI** and is defined to be the expected payoff of sampling minus the expected payoff of no sampling. Therefore,

$$\text{EVSI} = \text{EPS} - \text{EPNS} = \$249,898 - \$0 = \$249,898$$

The EVSI is the expected gain from conducting the seismic experiment, and the oil company should pay no more than this amount to carry out the seismic experiment. If the experiment costs $100,000, then it is worth the expense to conduct the experiment. Moreover, the difference between the EVSI and the cost of sampling is called the **expected net gain of sampling,** which is denoted **ENGS.** Here

$$\text{ENGS} = \text{EVSI} - \$100,000 = \$249,898 - \$100,000 = \$149,898$$

As long as the ENGS is greater than $0, it is worthwhile to carry out the seismic experiment. That is, the oil company should carry out the seismic experiment before it chooses whether or not to drill. Then, as discussed earlier, our posterior analysis says that if the experiment gives a medium or high reading, the oil company should drill, and if the experiment gives a low reading, the oil company should not drill.

Exercises for Section 19.2

CONCEPTS ■ connect

19.14 Explain what is meant by each of the following and describe the purpose of each:
 a Prior analysis.   b Posterior analysis.
 c Preposterior analysis.

19.15 Define and interpret each of the following:
 a Expected payoff of sampling, EPS.
 b Expected payoff of no sampling, EPNS.
 c Expected value of sample information, EVSI.
 d Expected net gain of sampling, ENGS.

METHODS AND APPLICATIONS

Exercises 19.16 through 19.21 refer to the following situation:

In the book *Making Hard Decisions: An Introduction to Decision Analysis*, 2nd ed., Robert T. Clemen presents an example in which an investor wishes to choose between investing money in (1) a high-risk stock, (2) a low-risk stock, or (3) a savings account. The payoffs received from the two stocks will depend on the behavior of the stock market—that is, whether the market goes up, stays the same, or goes down over the investment period. In addition, in order to obtain more information about the market behavior that might be anticipated during the investment period, the investor can hire an economist as a consultant who will predict the future market behavior. The results of the consultation will be one of the following three possibilities: (1) "economist says up," (2) "economist says flat" (the same), or (3) "economist says down." The conditional probabilities that express the ability of the economist to accurately forecast market behavior are given in the following table: ⓓ InvDec

| | **True Market State** | | |
Economist's Prediction	Up	Flat	Down
"Economist says up"	.80	.15	.20
"Economist says flat"	.10	.70	.20
"Economist says down"	.10	.15	.60

For instance, using this table we see that P("economist says up" | market up) = .80. Figure 19.7 gives an incomplete decision tree for the investor's situation. Notice that this decision tree gives all relevant payoffs and also gives the prior probabilities of up, flat, and down, which are, respectively, 0.5, 0.3, and 0.2. Use the information provided here, and any needed information on the decision tree of Figure 19.7, to complete Exercises 19.16 through 19.21.

19.16 Identify and list each of the following for the investor's decision problem: ⓓ InvDec
 a The investor's alternative actions.
 b The states of nature.
 c The possible results of sampling (that is, of information gathering).

19.17 Write out the payoff table for the investor's decision problem. ⓓ InvDec

19.18 Carry out a prior analysis of the investor's decision problem. That is, determine the investment choice that should be made and find the expected monetary value of that choice assuming that the investor does not

consult the economist about future stock market behavior. ⓓ InvDec

19.19 Set up probability revision tables to ⓓ InvDec
 a Find the probability that the "economist says up" and find the posterior probabilities of market up, market flat, and market down given that the "economist says up."
 b Find the probability that the "economist says flat," and find the posterior probabilities of market up, market flat, and market down given that the "economist says flat."
 c Find the probability that the "economist says down," and find the posterior probabilities of market up, market flat, and market down given that the "economist says down."
 d Reproduce the decision tree of Figure 19.7 and insert the probabilities you found in parts a, b, and c in their appropriate locations.

19.20 Carry out a posterior analysis of the investor's decision problem. That is, determine the investment choice that should be made and find the expected monetary value of that choice assuming ⓓ InvDec
 a The economist says "market up."
 b The economist says "market flat."
 c The economist says "market down."

19.21 Carry out a preposterior analysis of the investor's decision problem by finding ⓓ InvDec
 a The expected monetary value associated with consulting the economist; that is, find the EPS.
 b The expected monetary value associated with not consulting the economist; that is, find the EPNS.
 c The expected value of sample information, EVSI.
 d The maximum amount the investor should be willing to pay for the economist's consulting advice.

Exercises 19.22 through 19.28 refer to the following situation:

A firm designs and manufactures automatic electronic control devices that are installed at customers' plant sites. The control devices are shipped by truck to customers' sites; while in transit, the devices sometimes get out of alignment. More specifically, a device has a prior probability of .10 of getting out of alignment during shipment. When a control device is delivered to the customer's plant site, the customer can install the device. If the customer installs the device, and if the device is in alignment, the manufacturer of the control device will realize a profit of $15,000. If the customer installs the device, and if the device is out of alignment, the manufacturer must dismantle, realign, and reinstall the device for the customer. This procedure costs $3,000, and therefore the manufacturer will realize a profit of $12,000. As an alternative to customer installation, the manufacturer can send two engineers to the customer's plant site to check the alignment of the control device, to realign the device if necessary before installation, and to supervise the installation. Because it is less costly to realign the device before it is installed, sending the engineers costs $500. Therefore, if the engineers are sent to assist with the installation, the manufacturer realizes a profit of $14,500 (this is true whether or not the engineers must realign the device at the site).

FIGURE 19.7   **An Incomplete Decision Tree for the Investor's Decision Problem of Exercises 19.16 through 19.21**

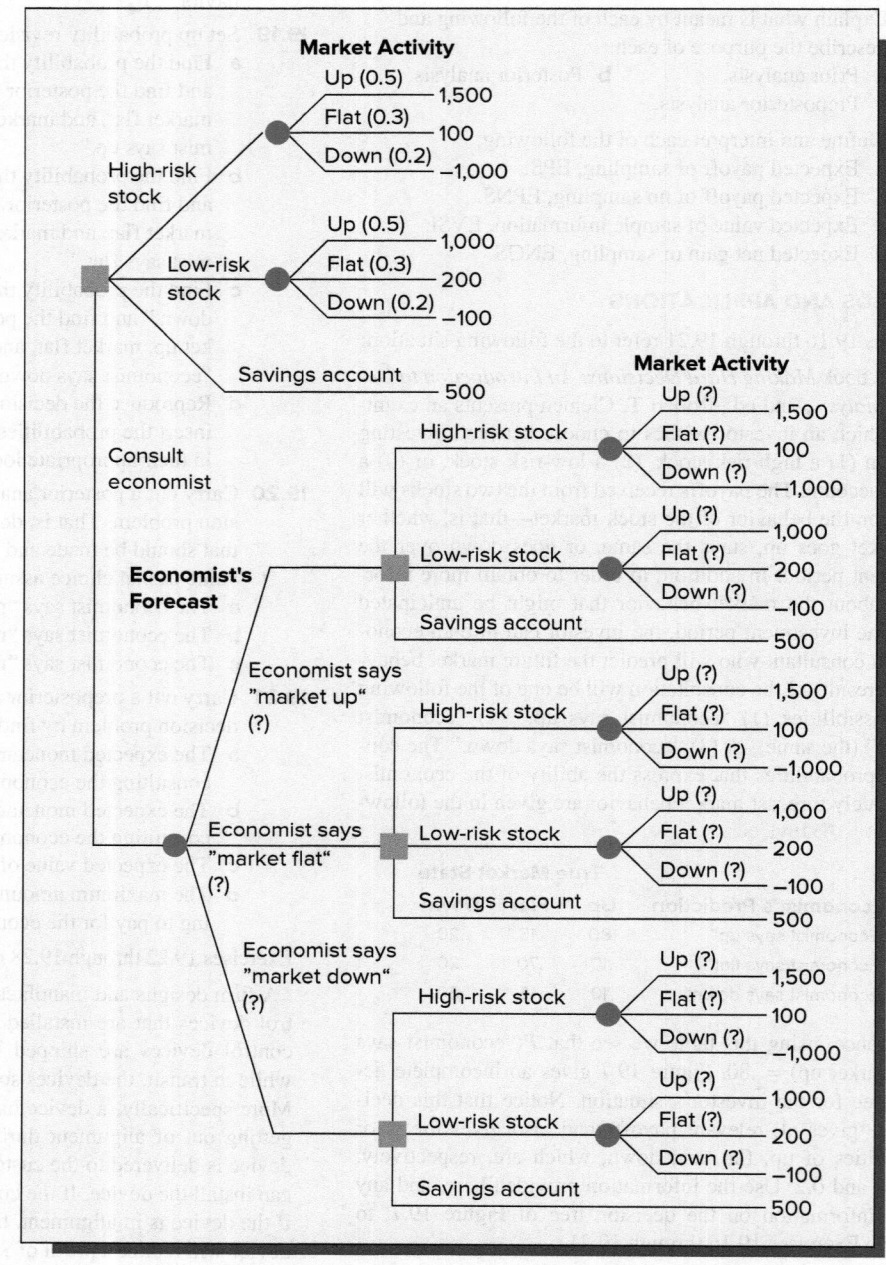

Source: R. T. Clemen, *Making Hard Decisions: An Introduction to Decision Analysis*, 2nd ed., p. 77. Brooks/Cole, 1996.

Before a control device is installed, a piece of test equipment can be used by the customer to check the device's alignment. The test equipment has two readings, "in" or "out" of alignment. If the control device is in alignment, there is a .8 probability that the test equipment will read "in." If the control device is out of alignment, there is a .9 probability that the test equipment will read "out."

19.22 Identify and list each of the following for the control device situation:
   a The firm's alternative actions.
   b The states of nature.
   c The possible results of sampling (that is, of information gathering).

19.23 Write out the payoff table for the control device situation.

19.24 Construct a decision tree for a prior analysis of the control device situation. Then determine whether the engineers should be sent, assuming that the piece of test equipment is not employed to check the device's alignment. Also find the expected monetary value associated with the best alternative action.

19.25 Set up probability revision tables to
  a Find the probability that the test equipment "reads in," and find the posterior probabilities of in alignment and out of alignment given that the test equipment "reads in."
  b Find the probability that the test equipment "reads out," and find the posterior probabilities of in alignment and out of alignment given that the test equipment "reads out."

19.26 Construct a decision tree for a posterior and preposterior analysis of the control device situation.

19.27 Carry out a posterior analysis of the control device problem. That is, decide whether the engineers should be sent, and find the expected monetary value associated with either sending or not sending (depending on which is best) the engineers assuming
  a The test equipment "reads in."
  b The test equipment "reads out."

19.28 Carry out a preposterior analysis of the control device problem by finding
  a The expected monetary value associated with using the test equipment; that is, find the EPS.
  b The expected monetary value associated with not using the test equipment; that is, find the EPNS.
  c The expected value of sample information, EVSI.
  d The maximum amount that should be paid for using the test equipment.

19.3 Introduction to Utility Theory

LO19-3

Make decisions using utility theory.

Suppose that a decision maker is trying to decide whether to invest in one of two opportunities—Investment 1 or Investment 2—or not to invest in either of these opportunities. As shown in Table 19.2(a), (b), and (c), the expected profits associated with Investment 1, Investment 2, and no investment are $32,000, $28,000, and $0. Thus, if

---

TABLE 19.2   **Three Possible Investments and Their Expected Utilities**

(a) Investment 1 profits

Profit	Probability
$50,000	.7
$10,000	.1
−$20,000	.2

Expected profit = 50,000(.7) + 10,000(.1) + (−20,000)(.2) = 32,000

(b) Investment 2 profits

Profit	Probability
$40,000	.6
$30,000	.2
−$10,000	.2

Expected profit = 40,000(.6) + 30,000(.2) + (−10,000)(.2) = 28,000

(c) No investment profit

Profit	Probability
$0	1

Expected profit = 0(1) = 0

(d) Utilities

Profit	Utility
$50,000	1.00
$40,000	.95
$30,000	.90
$10,000	.75
$0	.60
−$10,000	.45
−$20,000	0.00

(e) A utility curve

Profit (in units of $1,000)

(f) Investment 1 utilities

Utility	Probability
1.00	.7
.75	.1
0.00	.2

Expected utility = 1(.7) + .75(.1) + 0(.2) = .775

(g) Investment 2 utilities

Utility	Probability
.95	.6
.90	.2
.45	.2

Expected utility = .95(.6) + .90(.2) + .45(.2) = .84

(h) No investment utility

Utility	Probability
.60	1

Expected utility = .60(1) = .60

the decision maker uses expected profit as a decision criterion, and decides to choose no more than one investment, the decision maker should choose Investment 1. However, as discussed earlier, the expected profit for an investment is the long-run average profit that would be realized if many identical investments could be made. If the decision maker will make only a limited number of investments (perhaps because of limited capital), he or she will not realize the expected profit. For example, a single undertaking of Investment 1 will result in either a profit of $50,000, a profit of $10,000, or a loss of $20,000. Some decision makers might prefer a single undertaking of Investment 2, because the potential loss is only $10,000. Other decision makers might be unwilling to risk $10,000 and would choose no investment.

There is a way to combine the various profits, probabilities, and the decision maker's individual attitude toward risk to make a decision that is best for the decision maker. The method is based on a theory of utility discussed by J. Von Neumann and O. Morgenstern in *Theory of Games and Economic Behavior* (Princeton University Press, Princeton, N.J., 1st ed., 1944; 2nd ed., 1947). This theory says that if a decision maker agrees with certain assumptions about rational behavior (we will not discuss the assumptions here), then the decision maker should replace the profits in the various investments by **utilities** and choose the investment that gives the **highest expected utility.** To find the utility of a particular profit, we first arrange the profits from largest to smallest. The utility of the largest profit is 1 and the utility of the smallest profit is 0. The utility of any particular intermediate profit is the probability, call it u, such that the decision maker is **indifferent** between (1) getting the particular intermediate profit with certainty and (2) playing a lottery (or game) in which the probability is u of getting the highest profit and the probability is $1 - u$ of getting the smallest profit. Table 19.2(d) arranges the profits in Table 19.2(a), (b), and (c) in increasing order and gives a specific decision maker's utility for each profit. The utility of .95 for $40,000 means that the decision maker is indifferent between (1) getting $40,000 with certainty and (2) playing a lottery in which the probability is .95 of getting $50,000 and the probability is .05 of losing $20,000. The utilities for the other profits are interpreted similarly. Table 19.2(f), (g), and (h) show the investments with profits replaced by utilities. Because Investment 2 has the highest expected utility, the decision maker should choose Investment 2.

Table 19.2(e) shows a plot of the specific decision maker's utilities versus the profits. The curve connecting the plot points is the **utility curve** for the decision maker. This curve is an example of a **risk averter's curve.** In general, a risk averter's curve portrays a rapid increase in utility for initial amounts of money followed by a gradual leveling off for larger amounts of money. This curve is appropriate for many individuals or businesses because the marginal value of each additional dollar is not as great once a large amount of money has been earned. A risk averter's curve is shown in the page margin, as are a **risk seeker's curve** and a **risk neutral's curve.** The risk seeker's curve represents an individual who is willing to take large risks to have the opportunity to make large profits. The risk neutral curve represents an individual for whom each additional dollar has the same value. It can be shown that this individual should choose the investment having the highest expected profit.

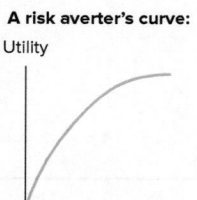

A risk averter's curve:

Utility

Dollar amount

A risk seeker's curve:

Utility

Dollar amount

A risk neutral's curve:

Utility

Dollar amount

Exercises for Section 19.3

CONCEPTS   connect

19.29 What is a utility?

19.30 What is a risk averter? A risk seeker? A risk neutral?

METHODS AND APPLICATIONS

19.31 Suppose that a decision maker has the opportunity to invest in an oil well drilling operation that has a .3 chance of yielding a profit of $1,000,000, a .4 chance of yielding a profit of $400,000, and a .3

chance of yielding a profit of −$100,000. Also, suppose that the decision maker's utilities for $400,000 and $0 are .9 and .7. Explain the meanings of these utilities.

19.32 Consider Exercise 19.31. Find the expected utility of the oil well drilling operation. Find the expected utility of not investing. What should the decision maker do if he/she wishes to maximize expected utility?

Chapter Summary

In Section 19.1 we presented an introduction to decision theory. We saw that a decision problem involves **states of nature, alternatives, payoffs,** and **decision criteria,** and we considered three degrees of uncertainty—**certainty, uncertainty,** and **risk.** In the case of *certainty,* we know which state of nature will actually occur. Here we simply choose the alternative that gives the best payoff. In the case of *uncertainty,* we have no information about the likelihood of the different states of nature. Here we discussed two commonly used decision criteria—the **maximin criterion** and the **maximax criterion.** In the case of *risk,* we are able to estimate the probability of occurrence for each state of nature. In this case we learned how to use the **expected monetary value criterion.** We also learned how

to construct a **decision tree** in Section 19.1, and we saw how to use such a tree to analyze a decision problem. In Section 19.2 we learned how to make decisions by using posterior probabilities. We explained how to perform a **posterior analysis** to determine the best alternative for each of several sampling results. Then we showed how to carry out a **preposterior analysis,** which allows us to assess the worth of sample information. In particular, we saw how to obtain the **expected value of sample information.** This quantity is the expected gain from sampling, which tells us the maximum amount we should be willing to pay for sample information. We concluded this chapter with Section 19.3, which introduced using **utility theory** to help make decisions.

Glossary of Terms

alternatives: Several alternative actions for a decision maker to choose from.

certainty: When we know for certain which state of nature will actually occur.

decision criterion: A rule used to make a decision.

decision theory: An approach that helps decision makers to make intelligent choices.

decision tree: A diagram consisting of nodes and branches that depicts the information for a decision problem.

expected monetary value criterion: A decision criterion in which one computes the expected monetary payoff for each alternative and then chooses the alternative yielding the largest expected payoff.

expected net gain of sampling: The difference between the expected value of sample information and the cost of sampling. If this quantity is positive, it is worth it to perform sampling.

expected value of perfect information: The difference between the expected payoff under certainty and the expected payoff under risk.

expected value of sample information: The difference between the expected payoff of sampling and the expected payoff of no sampling. This measures the expected gain from sampling.

maximax criterion: A decision criterion in which one finds the best possible payoff for each alternative and then chooses the alternative that yields the maximum best possible payoff.

maximin criterion: A decision criterion in which one finds the worst possible payoff for each alternative and then chooses the alternative that yields the maximum worst possible payoff.

payoff table: A tabular summary of the payoffs in a decision problem.

perfect information: Information that tells us exactly which state of nature will occur.

posterior decision analysis: Using a decision criterion based on posterior probabilities to choose the best alternative in a decision problem.

preposterior analysis: When we assess the worth of sample information before performing a posterior decision analysis.

prior decision analysis: Using a decision criterion based on prior probabilities to choose the best alternative in a decision problem.

risk: When the likelihood (probability) of each state of nature can be estimated.

states of nature: A set of potential future conditions that will affect the results of a decision.

uncertainty: When we have no information about the likelihoods of the various states of nature.

utility: A measure of monetary value based on an individual's attitude toward risk.

Important Formulas

Maximin criterion: page 846

Maximax criterion: page 847

Expected monetary value criterion: page 847

Decision tree diagram: pages 847, 848

Expected value of perfect information: page 849

Probability revision table: page 854

Expected payoff of sampling: page 856

Expected payoff of no sampling: page 856

Expected value of sample information: page 856

Expected net gain of sampling: page 856

Expected utility: page 860

Supplementary Exercises

19.33 In the book *Making Hard Decisions: An Introduction to Decision Analysis*, Robert T. Clemen presents a decision tree for a research and development decision (note that payoffs are given in millions of dollars, which is denoted by M). Based on this decision tree (shown in Figure 19.8), answer the following:

a Should development of the research project be continued or stopped? Justify your answer by using relevant calculations, and explain your reasoning.

b If development is continued and if a patent is awarded, should the new technology be licensed, or should the

company develop production and marketing to sell the product directly? Justify your answer by using relevant calculations and explain your reasoning.

19.34 In the book *Production/Operations Management*, William J. Stevenson presents a decision tree concerning a firm's decision about the size of a production facility. This decision tree is given in Figure 19.9 (payoffs are given in millions of dollars). Use the decision tree to determine which alternative (build small or build large) should be chosen in order to

FIGURE 19.8   A Decision Tree for a Research and Development Decision for Exercise 19.33

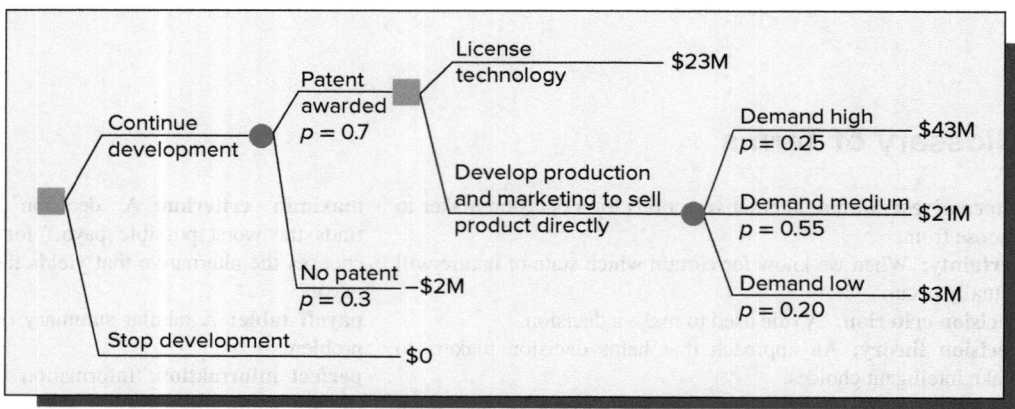

Source: R. T. Clemen, *Making Hard Decisions: An Introduction to Decision Analysis*, 2nd ed., p. 77. Brooks/Cole, 1996.

FIGURE 19.9   A Decision Tree for a Production Facility Decision for Exercises 19.34 and 19.35

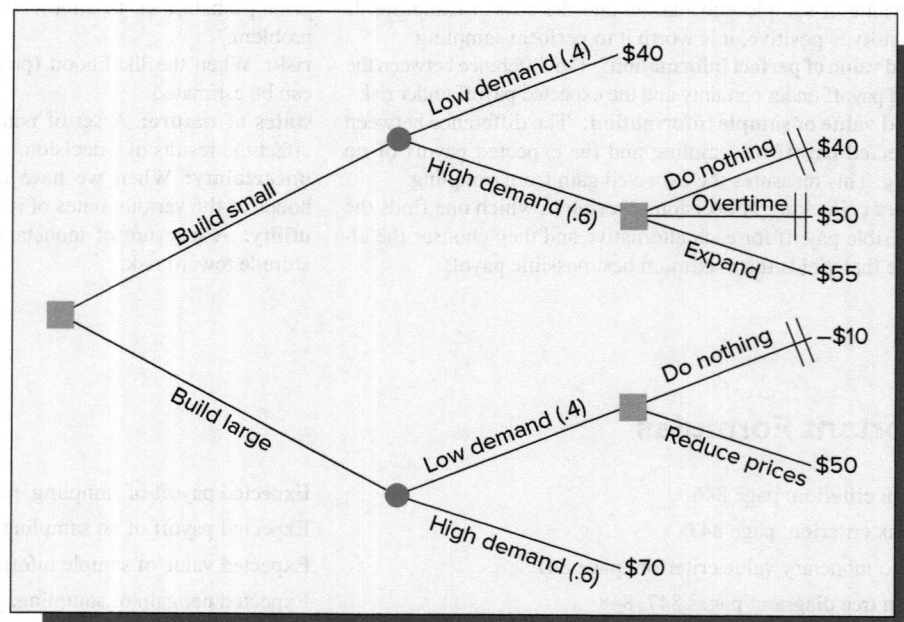

Source: Decision tree from W. J. Stevenson, *Production/Operations Management*, 6th ed., p. 70, © 1999 McGraw-Hill Companies, Inc.

maximize the expected monetary payoff. What is the expected monetary payoff associated with the best alternative?

19.35 Consider the decision tree in Figure 19.9 and the situation described in Exercise 19.34. Suppose that a marketing research study can be done to obtain more information about whether demand will be high or low. The marketing research study will result in one of two outcomes: "favorable" (indicating that demand will be high) or "unfavorable" (indicating that demand will be low). The accuracy of marketing research studies like the one to be carried out can be expressed by the conditional probabilities in the following table:

Study Outcome	True Demand	
	High	Low
Favorable	.9	.2
Unfavorable	.1	.8

For instance, $P(\text{favorable} \mid \text{high}) = .9$ and $P(\text{unfavorable} \mid \text{low}) = .8$. Given the prior probabilities and payoffs in Figure 19.9, do the following:

a Carry out a posterior analysis. Find the best alternative (build small or build large) for each possible study result (favorable or unfavorable), and find the associated expected payoffs.

b Carry out a preposterior analysis. Determine the maximum amount that should be paid for the marketing research study.

19.36 THE OIL COMPANY CASE DS DrillTst
Again consider the oil company case that was described in Example 19.1. Recall that the oil company wishes to decide whether to drill and that the prior probabilities of no oil, some oil, and much oil are $P(\text{none}) = .7$, $P(\text{some}) = .2$, and $P(\text{much}) = .1$. Suppose that, instead of performing the seismic survey to obtain more information about the site, the oil company can perform a cheaper magnetic experiment having two possible results: a high reading and a low reading. The past performance of the magnetic experiment can be summarized as follows:

Magnetic Experiment Result	State of Nature		
	None	Some	Much
Low reading	.8	.4	.1
High reading	.2	.6	.9

Here, for example, $P(\text{low} \mid \text{none}) = .8$ and $P(\text{high} \mid \text{some}) = .6$. Recalling that the payoffs associated with no oil, some oil, and much oil are $-\$700,000$, $\$500,000$, and $\$2,000,000$, respectively, do the following:

a Draw a decision tree for this decision problem.

b Carry out a posterior analysis. Find the best alternative (drill or do not drill) for each possible result of the magnetic experiment (low or high), and find the associated expected payoffs.

c Carry out a preposterior analysis. Determine the maximum amount that should be paid for the magnetic experiment.

19.37 In an exercise in the book *Production/Operations Management*, 5th ed. (1996), William DS ThmPark J. Stevenson considers a theme park whose lease is about to expire. The theme park's management wishes to decide whether to renew its lease for another 10 years or relocate near the site of a new motel complex. The town planning board is debating whether to approve the motel complex. A consultant estimates the payoffs of the theme park's alternatives under each state of nature as shown in the following payoff table:

Theme Park Options	Motel Approved	Motel Rejected
Renew lease	$500,000	$4,000,000
Relocate	$5,000,000	$100,000

a What alternative should the theme park choose if it uses the maximax criterion? What is the resulting payoff of this choice?

b What alternative should the theme park choose if it uses the maximin criterion? What is the resulting payoff of this choice?

19.38 Again consider the situation described in Exercise 19.37, and suppose that management believes there is a .35 probability that the motel complex will be approved.

a Draw a decision tree for the theme park's decision problem.

b Which alternative should be chosen if the theme park uses the maximum expected monetary value criterion? What is the expected monetary payoff for this choice?

c Suppose that management is offered the option of a temporary lease while the planning board decides whether to approve the motel complex. If the lease costs $100,000, should the theme park's management sign the lease? Justify your answer.

Appendix **A**

Statistical Tables

TABLE A.1 **A Binomial Probability Table:**
Binomial Probabilities (n between 2 and 6)

n = 2

p

x↓	.05	.10	.15	.20	.25	.30	.35	.40	.45	.50	
0	.9025	.8100	.7225	.6400	.5625	.4900	.4225	.3600	.3025	.2500	2
1	.0950	.1800	.2550	.3200	.3750	.4200	.4550	.4800	.4950	.5000	1
2	.0025	.0100	.0225	.0400	.0625	.0900	.1225	.1600	.2025	.2500	0
	.95	.90	.85	.80	.75	.70	.65	.60	.55	.50	x↑

n = 3

p

x↓	.05	.10	.15	.20	.25	.30	.35	.40	.45	.50	
0	.8574	.7290	.6141	.5120	.4219	.3430	.2746	.2160	.1664	.1250	3
1	.1354	.2430	.3251	.3840	.4219	.4410	.4436	.4320	.4084	.3750	2
2	.0071	.0270	.0574	.0960	.1406	.1890	.2389	.2880	.3341	.3750	1
3	.0001	.0010	.0034	.0080	.0156	.0270	.0429	.0640	.0911	.1250	0
	.95	.90	.85	.80	.75	.70	.65	.60	.55	.50	x↑

n = 4

p

x↓	.05	.10	.15	.20	.25	.30	.35	.40	.45	.50	
0	.8145	.6561	.5220	.4096	.3164	.2401	.1785	.1296	.0915	.0625	4
1	.1715	.2916	.3685	.4096	.4219	.4116	.3845	.3456	.2995	.2500	3
2	.0135	.0486	.0975	.1536	.2109	.2646	.3105	.3456	.3675	.3750	2
3	.0005	.0036	.0115	.0256	.0469	.0756	.1115	.1536	.2005	.2500	1
4	.0000	.0001	.0005	.0016	.0039	.0081	.0150	.0256	.0410	.0625	0
	.95	.90	.85	.80	.75	.70	.65	.60	.55	.50	x↑

n = 5

p

x↓	.05	.10	.15	.20	.25	.30	.35	.40	.45	.50	
0	.7738	.5905	.4437	.3277	.2373	.1681	.1160	.0778	.0503	.0313	5
1	.2036	.3281	.3915	.4096	.3955	.3602	.3124	.2592	.2059	.1563	4
2	.0214	.0729	.1382	.2048	.2637	.3087	.3364	.3456	.3369	.3125	3
3	.0011	.0081	.0244	.0512	.0879	.1323	.1811	.2304	.2757	.3125	2
4	.0000	.0005	.0022	.0064	.0146	.0284	.0488	.0768	.1128	.1563	1
5	.0000	.0000	.0001	.0003	.0010	.0024	.0053	.0102	.0185	.0313	0
	.95	.90	.85	.80	.75	.70	.65	.60	.55	.50	x↑

n = 6

p

x↓	.05	.10	.15	.20	.25	.30	.35	.40	.45	.50	
0	.7351	.5314	.3771	.2621	.1780	.1176	.0754	.0467	.0277	.0156	6
1	.2321	.3543	.3993	.3932	.3560	.3025	.2437	.1866	.1359	.0938	5
2	.0305	.0984	.1762	.2458	.2966	.3241	.3280	.3110	.2780	.2344	4
3	.0021	.0146	.0415	.0819	.1318	.1852	.2355	.2765	.3032	.3125	3
4	.0001	.0012	.0055	.0154	.0330	.0595	.0951	.1382	.1861	.2344	2
5	.0000	.0001	.0004	.0015	.0044	.0102	.0205	.0369	.0609	.0938	1
6	.0000	.0000	.0000	.0001	.0002	.0007	.0018	.0041	.0083	.0156	0
	.95	.90	.85	.80	.75	.70	.65	.60	.55	.50	x↑

(table continued)

TABLE A.1 *(continued)*
Binomial Probabilities (*n* between 7 and 10)

n = 7   *p*

x↓	.05	.10	.15	.20	.25	.30	.35	.40	.45	.50	
0	.6983	.4783	.3206	.2097	.1335	.0824	.0490	.0280	.0152	.0078	7
1	.2573	.3720	.3960	.3670	.3115	.2471	.1848	.1306	.0872	.0547	6
2	.0406	.1240	.2097	.2753	.3115	.3177	.2985	.2613	.2140	.1641	5
3	.0036	.0230	.0617	.1147	.1730	.2269	.2679	.2903	.2918	.2734	4
4	.0002	.0026	.0109	.0287	.0577	.0972	.1442	.1935	.2388	.2734	3
5	.0000	.0002	.0012	.0043	.0115	.0250	.0466	.0774	.1172	.1641	2
6	.0000	.0000	.0001	.0004	.0013	.0036	.0084	.0172	.0320	.0547	1
7	.0000	.0000	.0000	.0000	.0001	.0002	.0006	.0016	.0037	.0078	0
	.95	.90	.85	.80	.75	.70	.65	.60	.55	.50	x↑

n = 8   *p*

x↓	.05	.10	.15	.20	.25	.30	.35	.40	.45	.50	
0	.6634	.4305	.2725	.1678	.1001	.0576	.0319	.0168	.0084	.0039	8
1	.2793	.3826	.3847	.3355	.2670	.1977	.1373	.0896	.0548	.0313	7
2	.0515	.1488	.2376	.2936	.3115	.2965	.2587	.2090	.1569	.1094	6
3	.0054	.0331	.0839	.1468	.2076	.2541	.2786	.2787	.2568	.2188	5
4	.0004	.0046	.0185	.0459	.0865	.1361	.1875	.2322	.2627	.2734	4
5	.0000	.0004	.0026	.0092	.0231	.0467	.0808	.1239	.1719	.2188	3
6	.0000	.0000	.0002	.0011	.0038	.0100	.0217	.0413	.0703	.1094	2
7	.0000	.0000	.0000	.0001	.0004	.0012	.0033	.0079	.0164	.0313	1
8	.0000	.0000	.0000	.0000	.0000	.0001	.0002	.0007	.0017	.0039	0
	.95	.90	.85	.80	.75	.70	.65	.60	.55	.50	x↑

n = 9   *p*

x↓	.05	.10	.15	.20	.25	.30	.35	.40	.45	.50	
0	.6302	.3874	.2316	.1342	.0751	.0404	.0207	.0101	.0046	.0020	9
1	.2985	.3874	.3679	.3020	.2253	.1556	.1004	.0605	.0339	.0176	8
2	.0629	.1722	.2597	.3020	.3003	.2668	.2162	.1612	.1110	.0703	7
3	.0077	.0446	.1069	.1762	.2336	.2668	.2716	.2508	.2119	.1641	6
4	.0006	.0074	.0283	.0661	.1168	.1715	.2194	.2508	.2600	.2461	5
5	.0000	.0008	.0050	.0165	.0389	.0735	.1181	.1672	.2128	.2461	4
6	.0000	.0001	.0006	.0028	.0087	.0210	.0424	.0743	.1160	.1641	3
7	.0000	.0000	.0000	.0003	.0012	.0039	.0098	.0212	.0407	.0703	2
8	.0000	.0000	.0000	.0000	.0001	.0004	.0013	.0035	.0083	.0176	1
9	.0000	.0000	.0000	.0000	.0000	.0000	.0001	.0003	.0008	.0020	0
	.95	.90	.85	.80	.75	.70	.65	.60	.55	.50	x↑

n = 10   *p*

x↓	.05	.10	.15	.20	.25	.30	.35	.40	.45	.50	
0	.5987	.3487	.1969	.1074	.0563	.0282	.0135	.0060	.0025	.0010	10
1	.3151	.3874	.3474	.2684	.1877	.1211	.0725	.0403	.0207	.0098	9
2	.0746	.1937	.2759	.3020	.2816	.2335	.1757	.1209	.0763	.0439	8
3	.0105	.0574	.1298	.2013	.2503	.2668	.2522	.2150	.1665	.1172	7
4	.0010	.0112	.0401	.0881	.1460	.2001	.2377	.2508	.2384	.2051	6
5	.0001	.0015	.0085	.0264	.0584	.1029	.1536	.2007	.2340	.2461	5
6	.0000	.0001	.0012	.0055	.0162	.0368	.0689	.1115	.1596	.2051	4
7	.0000	.0000	.0001	.0008	.0031	.0090	.0212	.0425	.0746	.1172	3
8	.0000	.0000	.0000	.0001	.0004	.0014	.0043	.0106	.0229	.0439	2
9	.0000	.0000	.0000	.0000	.0000	.0001	.0005	.0016	.0042	.0098	1
10	.0000	.0000	.0000	.0000	.0000	.0000	.0000	.0001	.0003	.0010	0
	.95	.90	.85	.80	.75	.70	.65	.60	.55	.50	x↑

TABLE A.1 *(continued)*
Binomial Probabilities (*n* equal to 12, 14, and 15)

n = 12   p

x↓	.05	.10	.15	.20	.25	.30	.35	.40	.45	.50	
0	.5404	.2824	.1422	.0687	.0317	.0138	.0057	.0022	.0008	.0002	12
1	.3413	.3766	.3012	.2062	.1267	.0712	.0368	.0174	.0075	.0029	11
2	.0988	.2301	.2924	.2835	.2323	.1678	.1088	.0639	.0339	.0161	10
3	.0173	.0852	.1720	.2362	.2581	.2397	.1954	.1419	.0923	.0537	9
4	.0021	.0213	.0683	.1329	.1936	.2311	.2367	.2128	.1700	.1208	8
5	.0002	.0038	.0193	.0532	.1032	.1585	.2039	.2270	.2225	.1934	7
6	.0000	.0005	.0040	.0155	.0401	.0792	.1281	.1766	.2124	.2256	6
7	.0000	.0000	.0006	.0033	.0115	.0291	.0591	.1009	.1489	.1934	5
8	.0000	.0000	.0001	.0005	.0024	.0078	.0199	.0420	.0762	.1208	4
9	.0000	.0000	.0000	.0001	.0004	.0015	.0048	.0125	.0277	.0537	3
10	.0000	.0000	.0000	.0000	.0000	.0002	.0008	.0025	.0068	.0161	2
11	.0000	.0000	.0000	.0000	.0000	.0000	.0001	.0003	.0010	.0029	1
12	.0000	.0000	.0000	.0000	.0000	.0000	.0000	.0000	.0001	.0002	0
	.95	.90	.85	.80	.75	.70	.65	.60	.55	.50	x↑

n = 14   p

x↓	.05	.10	.15	.20	.25	.30	.35	.40	.45	.50	
0	.4877	.2288	.1028	.0440	.0178	.0068	.0024	.0008	.0002	.0001	14
1	.3593	.3559	.2539	.1539	.0832	.0407	.0181	.0073	.0027	.0009	13
2	.1229	.2570	.2912	.2501	.1802	.1134	.0634	.0317	.0141	.0056	12
3	.0259	.1142	.2056	.2501	.2402	.1943	.1366	.0845	.0462	.0222	11
4	.0037	.0349	.0998	.1720	.2202	.2290	.2022	.1549	.1040	.0611	10
5	.0004	.0078	.0352	.0860	.1468	.1963	.2178	.2066	.1701	.1222	9
6	.0000	.0013	.0093	.0322	.0734	.1262	.1759	.2066	.2088	.1833	8
7	.0000	.0002	.0019	.0092	.0280	.0618	.1082	.1574	.1952	.2095	7
8	.0000	.0000	.0003	.0020	.0082	.0232	.0510	.0918	.1398	.1833	6
9	.0000	.0000	.0000	.0003	.0018	.0066	.0183	.0408	.0762	.1222	5
10	.0000	.0000	.0000	.0000	.0003	.0014	.0049	.0136	.0312	.0611	4
11	.0000	.0000	.0000	.0000	.0000	.0002	.0010	.0033	.0093	.0222	3
12	.0000	.0000	.0000	.0000	.0000	.0000	.0001	.0005	.0019	.0056	2
13	.0000	.0000	.0000	.0000	.0000	.0000	.0000	.0001	.0002	.0009	1
14	.0000	.0000	.0000	.0000	.0000	.0000	.0000	.0000	.0000	.0001	0
	.95	.90	.85	.80	.75	.70	.65	.60	.55	.50	x↑

n = 15   p

x↓	.05	.10	.15	.20	.25	.30	.35	.40	.45	.50	
0	.4633	.2059	.0874	.0352	.0134	.0047	.0016	.0005	.0001	.0000	15
1	.3658	.3432	.2312	.1319	.0668	.0305	.0126	.0047	.0016	.0005	14
2	.1348	.2669	.2856	.2309	.1559	.0916	.0476	.0219	.0090	.0032	13
3	.0307	.1285	.2184	.2501	.2252	.1700	.1110	.0634	.0318	.0139	12
4	.0049	.0428	.1156	.1876	.2252	.2186	.1792	.1268	.0780	.0417	11
5	.0006	.0105	.0449	.1032	.1651	.2061	.2123	.1859	.1404	.0916	10
6	.0000	.0019	.0132	.0430	.0917	.1472	.1906	.2066	.1914	.1527	9
7	.0000	.0003	.0030	.0138	.0393	.0811	.1319	.1771	.2013	.1964	8
8	.0000	.0000	.0005	.0035	.0131	.0348	.0710	.1181	.1647	.1964	7
9	.0000	.0000	.0001	.0007	.0034	.0116	.0298	.0612	.1048	.1527	6
10	.0000	.0000	.0000	.0001	.0007	.0030	.0096	.0245	.0515	.0916	5
11	.0000	.0000	.0000	.0000	.0001	.0006	.0024	.0074	.0191	.0417	4
12	.0000	.0000	.0000	.0000	.0000	.0001	.0004	.0016	.0052	.0139	3
13	.0000	.0000	.0000	.0000	.0000	.0000	.0001	.0003	.0010	.0032	2
14	.0000	.0000	.0000	.0000	.0000	.0000	.0000	.0000	.0001	.0005	1
15	.0000	.0000	.0000	.0000	.0000	.0000	.0000	.0000	.0000	.0000	0
	.95	.90	.85	.80	.75	.70	.65	.60	.55	.50	x↑

(table continued)

TABLE A.1 *(continued)*
Binomial Probabilities (*n* equal to 16 and 18)

n = 16   **p**

x↓	.05	.10	.15	.20	.25	.30	.35	.40	.45	.50	
0	.4401	.1853	.0743	.0281	.0100	.0033	.0010	.0003	.0001	.0000	16
1	.3706	.3294	.2097	.1126	.0535	.0228	.0087	.0030	.0009	.0002	15
2	.1463	.2745	.2775	.2111	.1336	.0732	.0353	.0150	.0056	.0018	14
3	.0359	.1423	.2285	.2463	.2079	.1465	.0888	.0468	.0215	.0085	13
4	.0061	.0514	.1311	.2001	.2252	.2040	.1553	.1014	.0572	.0278	12
5	.0008	.0137	.0555	.1201	.1802	.2099	.2008	.1623	.1123	.0667	11
6	.0001	.0028	.0180	.0550	.1101	.1649	.1982	.1983	.1684	.1222	10
7	.0000	.0004	.0045	.0197	.0524	.1010	.1524	.1889	.1969	.1746	9
8	.0000	.0001	.0009	.0055	.0197	.0487	.0923	.1417	.1812	.1964	8
9	.0000	.0000	.0001	.0012	.0058	.0185	.0442	.0840	.1318	.1746	7
10	.0000	.0000	.0000	.0002	.0014	.0056	.0167	.0392	.0755	.1222	6
11	.0000	.0000	.0000	.0000	.0002	.0013	.0049	.0142	.0337	.0667	5
12	.0000	.0000	.0000	.0000	.0000	.0002	.0011	.0040	.0115	.0278	4
13	.0000	.0000	.0000	.0000	.0000	.0000	.0002	.0008	.0029	.0085	3
14	.0000	.0000	.0000	.0000	.0000	.0000	.0000	.0001	.0005	.0018	2
15	.0000	.0000	.0000	.0000	.0000	.0000	.0000	.0000	.0001	.0002	1
	.95	**.90**	**.85**	**.80**	**.75**	**.70**	**.65**	**.60**	**.55**	**.50**	**x↑**

n = 18   **p**

x↓	.05	.10	.15	.20	.25	.30	.35	.40	.45	.50	
0	.3972	.1501	.0536	.0180	.0056	.0016	.0004	.0001	.0000	.0000	18
1	.3763	.3002	.1704	.0811	.0338	.0126	.0042	.0012	.0003	.0001	17
2	.1683	.2835	.2556	.1723	.0958	.0458	.0190	.0069	.0022	.0006	16
3	.0473	.1680	.2406	.2297	.1704	.1046	.0547	.0246	.0095	.0031	15
4	.0093	.0700	.1592	.2153	.2130	.1681	.1104	.0614	.0291	.0117	14
5	.0014	.0218	.0787	.1507	.1988	.2017	.1664	.1146	.0666	.0327	13
6	.0002	.0052	.0301	.0816	.1436	.1873	.1941	.1655	.1181	.0708	12
7	.0000	.0010	.0091	.0350	.0820	.1376	.1792	.1892	.1657	.1214	11
8	.0000	.0002	.0022	.0120	.0376	.0811	.1327	.1734	.1864	.1669	10
9	.0000	.0000	.0004	.0033	.0139	.0386	.0794	.1284	.1694	.1855	9
10	.0000	.0000	.0001	.0008	.0042	.0149	.0385	.0771	.1248	.1669	8
11	.0000	.0000	.0000	.0001	.0010	.0046	.0151	.0374	.0742	.1214	7
12	.0000	.0000	.0000	.0000	.0002	.0012	.0047	.0145	.0354	.0708	6
13	.0000	.0000	.0000	.0000	.0000	.0002	.0012	.0045	.0134	.0327	5
14	.0000	.0000	.0000	.0000	.0000	.0000	.0002	.0011	.0039	.0117	4
15	.0000	.0000	.0000	.0000	.0000	.0000	.0000	.0002	.0009	.0031	3
16	.0000	.0000	.0000	.0000	.0000	.0000	.0000	.0000	.0001	.0006	2
17	.0000	.0000	.0000	.0000	.0000	.0000	.0000	.0000	.0000	.0001	1
	.95	**.90**	**.85**	**.80**	**.75**	**.70**	**.65**	**.60**	**.55**	**.50**	**x↑**

TABLE A.1 *(concluded)*
Binomial Probabilities (*n* equal to 20)

n = 20   **p**

x↓	.05	.10	.15	.20	.25	.30	.35	.40	.45	.50	
0	.3585	.1216	.0388	.0115	.0032	.0008	.0002	.0000	.0000	.0000	20
1	.3774	.2702	.1368	.0576	.0211	.0068	.0020	.0005	.0001	.0000	19
2	.1887	.2852	.2293	.1369	.0669	.0278	.0100	.0031	.0008	.0002	18
3	.0596	.1901	.2428	.2054	.1339	.0716	.0323	.0123	.0040	.0011	17
4	.0133	.0898	.1821	.2182	.1897	.1304	.0738	.0350	.0139	.0046	16
5	.0022	.0319	.1028	.1746	.2023	.1789	.1272	.0746	.0365	.0148	15
6	.0003	.0089	.0454	.1091	.1686	.1916	.1712	.1244	.0746	.0370	14
7	.0000	.0020	.0160	.0545	.1124	.1643	.1844	.1659	.1221	.0739	13
8	.0000	.0004	.0046	.0222	.0609	.1144	.1614	.1797	.1623	.1201	12
9	.0000	.0001	.0011	.0074	.0271	.0654	.1158	.1597	.1771	.1602	11
10	.0000	.0000	.0002	.0020	.0099	.0308	.0686	.1171	.1593	.1762	10
11	.0000	.0000	.0000	.0005	.0030	.0120	.0336	.0710	.1185	.1602	9
12	.0000	.0000	.0000	.0001	.0008	.0039	.0136	.0355	.0727	.1201	8
13	.0000	.0000	.0000	.0000	.0002	.0010	.0045	.0146	.0366	.0739	7
14	.0000	.0000	.0000	.0000	.0000	.0002	.0012	.0049	.0150	.0370	6
15	.0000	.0000	.0000	.0000	.0000	.0000	.0003	.0013	.0049	.0148	5
16	.0000	.0000	.0000	.0000	.0000	.0000	.0000	.0003	.0013	.0046	4
17	.0000	.0000	.0000	.0000	.0000	.0000	.0000	.0000	.0002	.0011	3
18	.0000	.0000	.0000	.0000	.0000	.0000	.0000	.0000	.0000	.0002	2
	.95	.90	.85	.80	.75	.70	.65	.60	.55	.50	x↑

Source: Data from Binomial Probability Table from *Statistical Thinking for Managers*, 3rd Edition, by D. K. Hildebrand & L. Ott. South-Western, 1991.

TABLE A.2 A Poisson Probability Table
Poisson Probabilities (*μ* between .1 and 2.0)

μ

x	.1	.2	.3	.4	.5	.6	.7	.8	.9	1.0
0	.9048	.8187	.7408	.6703	.6065	.5488	.4966	.4493	.4066	.3679
1	.0905	.1637	.2222	.2681	.3033	.3293	.3476	.3595	.3659	.3679
2	.0045	.0164	.0333	.0536	.0758	.0988	.1217	.1438	.1647	.1839
3	.0002	.0011	.0033	.0072	.0126	.0198	.0284	.0383	.0494	.0613
4	.0000	.0001	.0003	.0007	.0016	.0030	.0050	.0077	.0111	.0153
5	.0000	.0000	.0000	.0001	.0002	.0004	.0007	.0012	.0020	.0031
6	.0000	.0000	.0000	.0000	.0000	.0000	.0001	.0002	.0003	.0005

μ

x	1.1	1.2	1.3	1.4	1.5	1.6	1.7	1.8	1.9	2.0
0	.3329	.3012	.2725	.2466	.2231	.2019	.1827	.1653	.1496	.1353
1	.3662	.3614	.3543	.3452	.3347	.3230	.3106	.2975	.2842	.2707
2	.2014	.2169	.2303	.2417	.2510	.2584	.2640	.2678	.2700	.2707
3	.0738	.0867	.0998	.1128	.1255	.1378	.1496	.1607	.1710	.1804
4	.0203	.0260	.0324	.0395	.0471	.0551	.0636	.0723	.0812	.0902
5	.0045	.0062	.0084	.0111	.0141	.0176	.0216	.0260	.0309	.0361
6	.0008	.0012	.0018	.0026	.0035	.0047	.0061	.0078	.0098	.0120
7	.0001	.0002	.0003	.0005	.0008	.0011	.0015	.0020	.0027	.0034
8	.0000	.0000	.0001	.0001	.0001	.0002	.0003	.0005	.0006	.0009

(table continued)

TABLE A.2 *(continued)*

Poisson Probabilities (μ between 2.1 and 5.0)

					μ					
x	2.1	2.2	2.3	2.4	2.5	2.6	2.7	2.8	2.9	3.0
0	.1225	.1108	.1003	.0907	.0821	.0743	.0672	.0608	.0550	.0498
1	.2572	.2438	.2306	.2177	.2052	.1931	.1815	.1703	.1596	.1494
2	.2700	.2681	.2652	.2613	.2565	.2510	.2450	.2384	.2314	.2240
3	.1890	.1966	.2033	.2090	.2138	.2176	.2205	.2225	.2237	.2240
4	.0992	.1082	.1169	.1254	.1336	.1414	.1488	.1557	.1622	.1680
5	.0417	.0476	.0538	.0602	.0668	.0735	.0804	.0872	.0940	.1008
6	.0146	.0174	.0206	.0241	.0278	.0319	.0362	.0407	.0455	.0504
7	.0044	.0055	.0068	.0083	.0099	.0118	.0139	.0163	.0188	.0216
8	.0011	.0015	.0019	.0025	.0031	.0038	.0047	.0057	.0068	.0081
9	.0003	.0004	.0005	.0007	.0009	.0011	.0014	.0018	.0022	.0027
10	.0001	.0001	.0001	.0002	.0002	.0003	.0004	.0005	.0006	.0008
11	.0000	.0000	.0000	.0000	.0000	.0001	.0001	.0001	.0002	.0002

					μ					
x	3.1	3.2	3.3	3.4	3.5	3.6	3.7	3.8	3.9	4.0
0	.0450	.0408	.0369	.0334	.0302	.0273	.0247	.0224	.0202	.0183
1	.1397	.1304	.1217	.1135	.1057	.0984	.0915	.0850	.0789	.0733
2	.2165	.2087	.2008	.1929	.1850	.1771	.1692	.1615	.1539	.1465
3	.2237	.2226	.2209	.2186	.2158	.2125	.2087	.2046	.2001	.1954
4	.1733	.1781	.1823	.1858	.1888	.1912	.1931	.1944	.1951	.1954
5	.1075	.1140	.1203	.1264	.1322	.1377	.1429	.1477	.1522	.1563
6	.0555	.0608	.0662	.0716	.0771	.0826	.0881	.0936	.0989	.1042
7	.0246	.0278	.0312	.0348	.0385	.0425	.0466	.0508	.0551	.0595
8	.0095	.0111	.0129	.0148	.0169	.0191	.0215	.0241	.0269	.0298
9	.0033	.0040	.0047	.0056	.0066	.0076	.0089	.0102	.0116	.0132
10	.0010	.0013	.0016	.0019	.0023	.0028	.0033	.0039	.0045	.0053
11	.0003	.0004	.0005	.0006	.0007	.0009	.0011	.0013	.0016	.0019
12	.0001	.0001	.0001	.0002	.0002	.0003	.0003	.0004	.0005	.0006
13	.0000	.0000	.0000	.0000	.0001	.0001	.0001	.0001	.0002	.0002

					μ					
x	4.1	4.2	4.3	4.4	4.5	4.6	4.7	4.8	4.9	5.0
0	.0166	.0150	.0136	.0123	.0111	.0101	.0091	.0082	.0074	.0067
1	.0679	.0630	.0583	.0540	.0500	.0462	.0427	.0395	.0365	.0337
2	.1393	.1323	.1254	.1188	.1125	.1063	.1005	.0948	.0894	.0842
3	.1904	.1852	.1798	.1743	.1687	.1631	.1574	.1517	.1460	.1404
4	.1951	.1944	.1933	.1917	.1898	.1875	.1849	.1820	.1789	.1755
5	.1600	.1633	.1662	.1687	.1708	.1725	.1738	.1747	.1753	.1755
6	.1093	.1143	.1191	.1237	.1281	.1323	.1362	.1398	.1432	.1462
7	.0640	.0686	.0732	.0778	.0824	.0869	.0914	.0959	.1002	.1044
8	.0328	.0360	.0393	.0428	.0463	.0500	.0537	.0575	.0614	.0653
9	.0150	.0168	.0188	.0209	.0232	.0255	.0281	.0307	.0334	.0363
10	.0061	.0071	.0081	.0092	.0104	.0118	.0132	.0147	.0164	.0181
11	.0023	.0027	.0032	.0037	.0043	.0049	.0056	.0064	.0073	.0082
12	.0008	.0009	.0011	.0013	.0016	.0019	.0022	.0026	.0030	.0034
13	.0002	.0003	.0004	.0005	.0006	.0007	.0008	.0009	.0011	.0013
14	.0001	.0001	.0001	.0001	.0002	.0002	.0003	.0003	.0004	.0005
15	.0000	.0000	.0000	.0000	.0001	.0001	.0001	.0001	.0001	.0002

T A B L E A . 2 *(concluded)*

Poisson Probabilities (μ between 5.5 and 20.0)

x	5.5	6.0	6.5	7.0	7.5	8.0	8.5	9.0	9.5	10.0
0	.0041	.0025	.0015	.0009	.0006	.0003	.0002	.0001	.0001	.0000
1	.0225	.0149	.0098	.0064	.0041	.0027	.0017	.0011	.0007	.0005
2	.0618	.0446	.0318	.0223	.0156	.0107	.0074	.0050	.0034	.0023
3	.1133	.0892	.0688	.0521	.0389	.0286	.0208	.0150	.0107	.0076
4	.1558	.1339	.1118	.0912	.0729	.0573	.0443	.0337	.0254	.0189
5	.1714	.1606	.1454	.1277	.1094	.0916	.0752	.0607	.0483	.0378
6	.1571	.1606	.1575	.1490	.1367	.1221	.1066	.0911	.0764	.0631
7	.1234	.1377	.1462	.1490	.1465	.1396	.1294	.1171	.1037	.0901
8	.0849	.1033	.1188	.1304	.1373	.1396	.1375	.1318	.1232	.1126
9	.0519	.0688	.0858	.1014	.1144	.1241	.1299	.1318	.1300	.1251
10	.0285	.0413	.0558	.0710	.0858	.0993	.1104	.1186	.1235	.1251
11	.0143	.0225	.0330	.0452	.0585	.0722	.0853	.0970	.1067	.1137
12	.0065	.0113	.0179	.0263	.0366	.0481	.0604	.0728	.0844	.0948
13	.0028	.0052	.0089	.0142	.0211	.0296	.0395	.0504	.0617	.0729
14	.0011	.0022	.0041	.0071	.0113	.0169	.0240	.0324	.0419	.0521
15	.0004	.0009	.0018	.0033	.0057	.0090	.0136	.0194	.0265	.0347
16	.0001	.0003	.0007	.0014	.0026	.0045	.0072	.0109	.0157	.0217
17	.0000	.0001	.0003	.0006	.0012	.0021	.0036	.0058	.0088	.0128
18	.0000	.0000	.0001	.0002	.0005	.0009	.0017	.0029	.0046	.0071
19	.0000	.0000	.0000	.0001	.0002	.0004	.0008	.0014	.0023	.0037
20	.0000	.0000	.0000	.0000	.0001	.0002	.0003	.0006	.0011	.0019
21	.0000	.0000	.0000	.0000	.0000	.0001	.0001	.0003	.0005	.0009
22	.0000	.0000	.0000	.0000	.0000	.0000	.0001	.0001	.0002	.0004
23	.0000	.0000	.0000	.0000	.0000	.0000	.0000	.0000	.0001	.0002

μ

x	11.0	12.0	13.0	14.0	15.0	16.0	17.0	18.0	19.0	20.0
0	.0000	.0000	.0000	.0000	.0000	.0000	.0000	.0000	.0000	.0000
1	.0002	.0001	.0000	.0000	.0000	.0000	.0000	.0000	.0000	.0000
2	.0010	.0004	.0002	.0001	.0000	.0000	.0000	.0000	.0000	.0000
3	.0037	.0018	.0008	.0004	.0002	.0001	.0000	.0000	.0000	.0000
4	.0102	.0053	.0027	.0013	.0006	.0003	.0001	.0001	.0000	.0000
5	.0224	.0127	.0070	.0037	.0019	.0010	.0005	.0002	.0001	.0001
6	.0411	.0255	.0152	.0087	.0048	.0026	.0014	.0007	.0004	.0002
7	.0646	.0437	.0281	.0174	.0104	.0060	.0034	.0019	.0010	.0005
8	.0888	.0655	.0457	.0304	.0194	.0120	.0072	.0042	.0024	.0013
9	.1085	.0874	.0661	.0473	.0324	.0213	.0135	.0083	.0050	.0029
10	.1194	.1048	.0859	.0663	.0486	.0341	.0230	.0150	.0095	.0058
11	.1194	.1144	.1015	.0844	.0663	.0496	.0355	.0245	.0164	.0106
12	.1094	.1144	.1099	.0984	.0829	.0661	.0504	.0368	.0259	.0176
13	.0926	.1056	.1099	.1060	.0956	.0814	.0658	.0509	.0378	.0271
14	.0728	.0905	.1021	.1060	.1024	.0930	.0800	.0655	.0514	.0387
15	.0534	.0724	.0885	.0989	.1024	.0992	.0906	.0786	.0650	.0516
16	.0367	.0543	.0719	.0866	.0960	.0992	.0963	.0884	.0772	.0646
17	.0237	.0383	.0550	.0713	.0847	.0934	.0963	.0936	.0863	.0760
18	.0145	.0255	.0397	.0554	.0706	.0830	.0909	.0936	.0911	.0844
19	.0084	.0161	.0272	.0409	.0557	.0699	.0814	.0887	.0911	.0888
20	.0046	.0097	.0177	.0286	.0418	.0559	.0692	.0798	.0866	.0888
21	.0024	.0055	.0109	.0191	.0299	.0426	.0560	.0684	.0783	.0846
22	.0012	.0030	.0065	.0121	.0204	.0310	.0433	.0560	.0676	.0769
23	.0006	.0016	.0037	.0074	.0133	.0216	.0320	.0438	.0559	.0669
24	.0003	.0008	.0020	.0043	.0083	.0144	.0226	.0328	.0442	.0557
25	.0001	.0004	.0010	.0024	.0050	.0092	.0154	.0237	.0336	.0446
26	.0000	.0002	.0005	.0013	.0029	.0057	.0101	.0164	.0246	.0343
27	.0000	.0001	.0002	.0007	.0016	.0034	.0063	.0109	.0173	.0254
28	.0000	.0000	.0001	.0003	.0009	.0019	.0038	.0070	.0117	.0181
29	.0000	.0000	.0001	.0002	.0004	.0011	.0023	.0044	.0077	.0125
30	.0000	.0000	.0000	.0001	.0002	.0006	.0013	.0026	.0049	.0083
31	.0000	.0000	.0000	.0000	.0001	.0003	.0007	.0015	.0030	.0054
32	.0000	.0000	.0000	.0000	.0001	.0001	.0004	.0009	.0018	.0034
33	.0000	.0000	.0000	.0000	.0000	.0001	.0002	.0005	.0010	.0020

Source: Computed by D. K. Hildebrand. Found in D. K. Hildebrand and L. Ott, *Statistical Thinking for Managers*, 3rd ed. (Boston, MA: PWS-KENT Publishing Company, 1991).

TABLE A.3 Cumulative Areas under the Standard Normal Curve

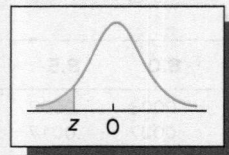

z	0.00	0.01	0.02	0.03	0.04	0.05	0.06	0.07	0.08	0.09
−3.9	0.00005	0.00005	0.00004	0.00004	0.00004	0.00004	0.00004	0.00004	0.00003	0.00003
−3.8	0.00007	0.00007	0.00007	0.00006	0.00006	0.00006	0.00006	0.00005	0.00005	0.00005
−3.7	0.00011	0.00010	0.00010	0.00010	0.00009	0.00009	0.00008	0.00008	0.00008	0.00008
−3.6	0.00016	0.00015	0.00015	0.00014	0.00014	0.00013	0.00013	0.00012	0.00012	0.00011
−3.5	0.00023	0.00022	0.00022	0.00021	0.00020	0.00019	0.00019	0.00018	0.00017	0.00017
−3.4	0.00034	0.00032	0.00031	0.00030	0.00029	0.00028	0.00027	0.00026	0.00025	0.00024
−3.3	0.00048	0.00047	0.00045	0.00043	0.00042	0.00040	0.00039	0.00038	0.00036	0.00035
−3.2	0.00069	0.00066	0.00064	0.00062	0.00060	0.00058	0.00056	0.00054	0.00052	0.00050
−3.1	0.00097	0.00094	0.00090	0.00087	0.00084	0.00082	0.00079	0.00076	0.00074	0.00071
−3.0	0.00135	0.00131	0.00126	0.00122	0.00118	0.00114	0.00111	0.00107	0.00103	0.00100
−2.9	0.0019	0.0018	0.0018	0.0017	0.0016	0.0016	0.0015	0.0015	0.0014	0.0014
−2.8	0.0026	0.0025	0.0024	0.0023	0.0023	0.0022	0.0021	0.0021	0.0020	0.0019
−2.7	0.0035	0.0034	0.0033	0.0032	0.0031	0.0030	0.0029	0.0028	0.0027	0.0026
−2.6	0.0047	0.0045	0.0044	0.0043	0.0041	0.0040	0.0039	0.0038	0.0037	0.0036
−2.5	0.0062	0.0060	0.0059	0.0057	0.0055	0.0054	0.0052	0.0051	0.0049	0.0048
−2.4	0.0082	0.0080	0.0078	0.0075	0.0073	0.0071	0.0069	0.0068	0.0066	0.0064
−2.3	0.0107	0.0104	0.0102	0.0099	0.0096	0.0094	0.0091	0.0089	0.0087	0.0084
−2.2	0.0139	0.0136	0.0132	0.0129	0.0125	0.0122	0.0119	0.0116	0.0113	0.0110
−2.1	0.0179	0.0174	0.0170	0.0166	0.0162	0.0158	0.0154	0.0150	0.0146	0.0143
−2.0	0.0228	0.0222	0.0217	0.0212	0.0207	0.0202	0.0197	0.0192	0.0188	0.0183
−1.9	0.0287	0.0281	0.0274	0.0268	0.0262	0.0256	0.0250	0.0244	0.0239	0.0233
−1.8	0.0359	0.0351	0.0344	0.0336	0.0329	0.0322	0.0314	0.0307	0.0301	0.0294
−1.7	0.0446	0.0436	0.0427	0.0418	0.0409	0.0401	0.0392	0.0384	0.0375	0.0367
−1.6	0.0548	0.0537	0.0526	0.0516	0.0505	0.0495	0.0485	0.0475	0.0465	0.0455
−1.5	0.0668	0.0655	0.0643	0.0630	0.0618	0.0606	0.0594	0.0582	0.0571	0.0559
−1.4	0.0808	0.0793	0.0778	0.0764	0.0749	0.0735	0.0721	0.0708	0.0694	0.0681
−1.3	0.0968	0.0951	0.0934	0.0918	0.0901	0.0885	0.0869	0.0853	0.0838	0.0823
−1.2	0.1151	0.1131	0.1112	0.1093	0.1075	0.1056	0.1038	0.1020	0.1003	0.0985
−1.1	0.1357	0.1335	0.1314	0.1292	0.1271	0.1251	0.1230	0.1210	0.1190	0.1170
−1.0	0.1587	0.1562	0.1539	0.1515	0.1492	0.1469	0.1446	0.1423	0.1401	0.1379
−0.9	0.1841	0.1814	0.1788	0.1762	0.1736	0.1711	0.1685	0.1660	0.1635	0.1611
−0.8	0.2119	0.2090	0.2061	0.2033	0.2005	0.1977	0.1949	0.1922	0.1894	0.1867
−0.7	0.2420	0.2389	0.2358	0.2327	0.2296	0.2266	0.2236	0.2206	0.2177	0.2148
−0.6	0.2743	0.2709	0.2676	0.2643	0.2611	0.2578	0.2546	0.2514	0.2482	0.2451
−0.5	0.3085	0.3050	0.3015	0.2981	0.2946	0.2912	0.2877	0.2843	0.2810	0.2776
−0.4	0.3446	0.3409	0.3372	0.3336	0.3300	0.3264	0.3228	0.3192	0.3156	0.3121
−0.3	0.3821	0.3783	0.3745	0.3707	0.3669	0.3632	0.3594	0.3557	0.3520	0.3483
−0.2	0.4207	0.4168	0.4129	0.4090	0.4052	0.4013	0.3974	0.3936	0.3897	0.3859
−0.1	0.4602	0.4562	0.4522	0.4483	0.4443	0.4404	0.4364	0.4325	0.4286	0.4247
−0.0	0.5000	0.4960	0.4920	0.4880	0.4840	0.4801	0.4761	0.4721	0.4681	0.4641

TABLE A.3 Cumulative Areas under the Standard Normal Curve (*concluded*)

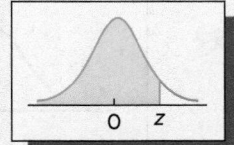

z	0.00	0.01	0.02	0.03	0.04	0.05	0.06	0.07	0.08	0.09
0.0	0.5000	0.5040	0.5080	0.5120	0.5160	0.5199	0.5239	0.5279	0.5319	0.5359
0.1	0.5398	0.5438	0.5478	0.5517	0.5557	0.5596	0.5636	0.5675	0.5714	0.5753
0.2	0.5793	0.5832	0.5871	0.5910	0.5948	0.5987	0.6026	0.6064	0.6103	0.6141
0.3	0.6179	0.6217	0.6255	0.6293	0.6331	0.6368	0.6406	0.6443	0.6480	0.6517
0.4	0.6554	0.6591	0.6628	0.6664	0.6700	0.6736	0.6772	0.6808	0.6844	0.6879
0.5	0.6915	0.6950	0.6985	0.7019	0.7054	0.7088	0.7123	0.7157	0.7190	0.7224
0.6	0.7257	0.7291	0.7324	0.7357	0.7389	0.7422	0.7454	0.7486	0.7518	0.7549
0.7	0.7580	0.7611	0.7642	0.7673	0.7704	0.7734	0.7764	0.7794	0.7823	0.7852
0.8	0.7881	0.7910	0.7939	0.7967	0.7995	0.8023	0.8051	0.8078	0.8106	0.8133
0.9	0.8159	0.8186	0.8212	0.8238	0.8264	0.8289	0.8315	0.8340	0.8365	0.8389
1.0	0.8413	0.8438	0.8461	0.8485	0.8508	0.8531	0.8554	0.8577	0.8599	0.8621
1.1	0.8643	0.8665	0.8686	0.8708	0.8729	0.8749	0.8770	0.8790	0.8810	0.8830
1.2	0.8849	0.8869	0.8888	0.8907	0.8925	0.8944	0.8962	0.8980	0.8997	0.9015
1.3	0.9032	0.9049	0.9066	0.9082	0.9099	0.9115	0.9131	0.9147	0.9162	0.9177
1.4	0.9192	0.9207	0.9222	0.9236	0.9251	0.9265	0.9279	0.9292	0.9306	0.9319
1.5	0.9332	0.9345	0.9357	0.9370	0.9382	0.9394	0.9406	0.9418	0.9429	0.9441
1.6	0.9452	0.9463	0.9474	0.9484	0.9495	0.9505	0.9515	0.9525	0.9535	0.9545
1.7	0.9554	0.9564	0.9573	0.9582	0.9591	0.9599	0.9608	0.9616	0.9625	0.9633
1.8	0.9641	0.9649	0.9656	0.9664	0.9671	0.9678	0.9686	0.9693	0.9699	0.9706
1.9	0.9713	0.9719	0.9726	0.9732	0.9738	0.9744	0.9750	0.9756	0.9761	0.9767
2.0	0.9772	0.9778	0.9783	0.9788	0.9793	0.9798	0.9803	0.9808	0.9812	0.9817
2.1	0.9821	0.9826	0.9830	0.9834	0.9838	0.9842	0.9846	0.9850	0.9854	0.9857
2.2	0.9861	0.9864	0.9868	0.9871	0.9875	0.9878	0.9881	0.9884	0.9887	0.9890
2.3	0.9893	0.9896	0.9898	0.9901	0.9904	0.9906	0.9909	0.9911	0.9913	0.9916
2.4	0.9918	0.9920	0.9922	0.9925	0.9927	0.9929	0.9931	0.9932	0.9934	0.9936
2.5	0.9938	0.9940	0.9941	0.9943	0.9945	0.9946	0.9948	0.9949	0.9951	0.9952
2.6	0.9953	0.9955	0.9956	0.9957	0.9959	0.9960	0.9961	0.9962	0.9963	0.9964
2.7	0.9965	0.9966	0.9967	0.9968	0.9969	0.9970	0.9971	0.9972	0.9973	0.9974
2.8	0.9974	0.9975	0.9976	0.9977	0.9977	0.9978	0.9979	0.9979	0.9980	0.9981
2.9	0.9981	0.9982	0.9982	0.9983	0.9984	0.9984	0.9985	0.9985	0.9986	0.9986
3.0	0.99865	0.99869	0.99874	0.99878	0.99882	0.99886	0.99889	0.99893	0.99897	0.99900
3.1	0.99903	0.99906	0.99910	0.99913	0.99916	0.99918	0.99921	0.99924	0.99926	0.99929
3.2	0.99931	0.99934	0.99936	0.99938	0.99940	0.99942	0.99944	0.99946	0.99948	0.99950
3.3	0.99952	0.99953	0.99955	0.99957	0.99958	0.99960	0.99961	0.99962	0.99964	0.99965
3.4	0.99966	0.99968	0.99969	0.99970	0.99971	0.99972	0.99973	0.99974	0.99975	0.99976
3.5	0.99977	0.99978	0.99978	0.99979	0.99980	0.99981	0.99981	0.99982	0.99983	0.99983
3.6	0.99984	0.99985	0.99985	0.99986	0.99986	0.99987	0.99987	0.99988	0.99988	0.99989
3.7	0.99989	0.99990	0.99990	0.99990	0.99991	0.99991	0.99992	0.99992	0.99992	0.99992
3.8	0.99993	0.99993	0.99993	0.99994	0.99994	0.99994	0.99994	0.99995	0.99995	0.99995
3.9	0.99995	0.99995	0.99996	0.99996	0.99996	0.99996	0.99996	0.99996	0.99997	0.99997

TABLE A.4 A t Table: Values of t_α for df = 1 through 48

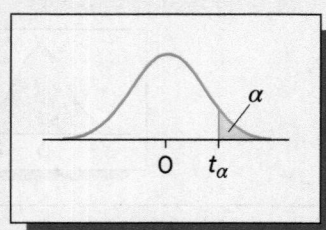

df	$t_{.100}$	$t_{.05}$	$t_{.025}$	$t_{.01}$	$t_{.005}$	$t_{.001}$	$t_{.0005}$
1	3.078	6.314	12.706	31.821	63.657	318.309	636.619
2	1.886	2.920	4.303	6.965	9.925	22.327	31.599
3	1.638	2.353	3.182	4.541	5.841	10.215	12.924
4	1.533	2.132	2.776	3.747	4.604	7.173	8.610
5	1.476	2.015	2.571	3.365	4.032	5.893	6.869
6	1.440	1.943	2.447	3.143	3.707	5.208	5.959
7	1.415	1.895	2.365	2.998	3.499	4.785	5.408
8	1.397	1.860	2.306	2.896	3.355	4.501	5.041
9	1.383	1.833	2.262	2.821	3.250	4.297	4.781
10	1.372	1.812	2.228	2.764	3.169	4.144	4.587
11	1.363	1.796	2.201	2.718	3.106	4.025	4.437
12	1.356	1.782	2.179	2.681	3.055	3.930	4.318
13	1.350	1.771	2.160	2.650	3.012	3.852	4.221
14	1.345	1.761	2.145	2.624	2.977	3.787	4.140
15	1.341	1.753	2.131	2.602	2.947	3.733	4.073
16	1.337	1.746	2.120	2.583	2.921	3.686	4.015
17	1.333	1.740	2.110	2.567	2.898	3.646	3.965
18	1.330	1.734	2.101	2.552	2.878	3.610	3.922
19	1.328	1.729	2.093	2.539	2.861	3.579	3.883
20	1.325	1.725	2.086	2.528	2.845	3.552	3.850
21	1.323	1.721	2.080	2.518	2.831	3.527	3.819
22	1.321	1.717	2.074	2.508	2.819	3.505	3.792
23	1.319	1.714	2.069	2.500	2.807	3.485	3.768
24	1.318	1.711	2.064	2.492	2.797	3.467	3.745
25	1.316	1.708	2.060	2.485	2.787	3.450	3.725
26	1.315	1.706	2.056	2.479	2.779	3.435	3.707
27	1.314	1.703	2.052	2.473	2.771	3.421	3.690
28	1.313	1.701	2.048	2.467	2.763	3.408	3.674
29	1.311	1.699	2.045	2.462	2.756	3.396	3.659
30	1.310	1.697	2.042	2.457	2.750	3.385	3.646
31	1.309	1.696	2.040	2.453	2.744	3.375	3.633
32	1.309	1.694	2.037	2.449	2.738	3.365	3.622
33	1.308	1.692	2.035	2.445	2.733	3.356	3.611
34	1.307	1.691	2.032	2.441	2.728	3.348	3.601
35	1.306	1.690	2.030	2.438	2.724	3.340	3.591
36	1.306	1.688	2.028	2.434	2.719	3.333	3.582
37	1.305	1.687	2.026	2.431	2.715	3.326	3.574
38	1.304	1.686	2.024	2.429	2.712	3.319	3.566
39	1.304	1.685	2.023	2.426	2.708	3.313	3.558
40	1.303	1.684	2.021	2.423	2.704	3.307	3.551
41	1.303	1.683	2.020	2.421	2.701	3.301	3.544
42	1.302	1.682	2.018	2.418	2.698	3.296	3.538
43	1.302	1.681	2.017	2.416	2.695	3.291	3.532
44	1.301	1.680	2.015	2.414	2.692	3.286	3.526
45	1.301	1.679	2.014	2.412	2.690	3.281	3.520
46	1.300	1.679	2.013	2.410	2.687	3.277	3.515
47	1.300	1.678	2.012	2.408	2.685	3.273	3.510
48	1.299	1.677	2.011	2.407	2.682	3.269	3.505

TABLE A.4 (concluded)
A t Table: Values of t_α for df = 49 through 100, 120, and ∞

df	$t_{.100}$	$t_{.05}$	$t_{.025}$	$t_{.01}$	$t_{.005}$	$t_{.001}$	$t_{.0005}$
49	1.299	1.677	2.010	2.405	2.680	3.265	3.500
50	1.299	1.676	2.009	2.403	2.678	3.261	3.496
51	1.298	1.675	2.008	2.402	2.676	3.258	3.492
52	1.298	1.675	2.007	2.400	2.674	3.255	3.488
53	1.298	1.674	2.006	2.399	2.672	3.251	3.484
54	1.297	1.674	2.005	2.397	2.670	3.248	3.480
55	1.297	1.673	2.004	2.396	2.668	3.245	3.476
56	1.297	1.673	2.003	2.395	2.667	3.242	3.473
57	1.297	1.672	2.002	2.394	2.665	3.239	3.470
58	1.296	1.672	2.002	2.392	2.663	3.237	3.466
59	1.296	1.671	2.001	2.391	2.662	3.234	3.463
60	1.296	1.671	2.000	2.390	2.660	3.232	3.460
61	1.296	1.670	2.000	2.389	2.659	3.229	3.457
62	1.295	1.670	1.999	2.388	2.657	3.227	3.454
63	1.295	1.669	1.998	2.387	2.656	3.225	3.452
64	1.295	1.669	1.998	2.386	2.655	3.223	3.449
65	1.295	1.669	1.997	2.385	2.654	3.220	3.447
66	1.295	1.668	1.997	2.384	2.652	3.218	3.444
67	1.294	1.668	1.996	2.383	2.651	3.216	3.442
68	1.294	1.668	1.995	2.382	2.650	3.214	3.439
69	1.294	1.667	1.995	2.382	2.649	3.213	3.437
70	1.294	1.667	1.994	2.381	2.648	3.211	3.435
71	1.294	1.667	1.994	2.380	2.647	3.209	3.433
72	1.293	1.666	1.993	2.379	2.646	3.207	3.431
73	1.293	1.666	1.993	2.379	2.645	3.206	3.429
74	1.293	1.666	1.993	2.378	2.644	3.204	3.427
75	1.293	1.665	1.992	2.377	2.643	3.202	3.425
76	1.293	1.665	1.992	2.376	2.642	3.201	3.423
77	1.293	1.665	1.991	2.376	2.641	3.199	3.421
78	1.292	1.665	1.991	2.375	2.640	3.198	3.420
79	1.292	1.664	1.990	2.374	2.640	3.197	3.418
80	1.292	1.664	1.990	2.374	2.639	3.195	3.416
81	1.292	1.664	1.990	2.373	2.638	3.194	3.415
82	1.292	1.664	1.989	2.373	2.637	3.193	3.413
83	1.292	1.663	1.989	2.372	2.636	3.191	3.412
84	1.292	1.663	1.989	2.372	2.636	3.190	3.410
85	1.292	1.663	1.988	2.371	2.635	3.189	3.409
86	1.291	1.663	1.988	2.370	2.634	3.188	3.407
87	1.291	1.663	1.988	2.370	2.634	3.187	3.406
88	1.291	1.662	1.987	2.369	2.633	3.185	3.405
89	1.291	1.662	1.987	2.369	2.632	3.184	3.403
90	1.291	1.662	1.987	2.368	2.632	3.183	3.402
91	1.291	1.662	1.986	2.368	2.631	3.182	3.401
92	1.291	1.662	1.986	2.368	2.630	3.181	3.399
93	1.291	1.661	1.986	2.367	2.630	3.180	3.398
94	1.291	1.661	1.986	2.367	2.629	3.179	3.397
95	1.291	1.661	1.985	2.366	2.629	3.178	3.396
96	1.290	1.661	1.985	2.366	2.628	3.177	3.395
97	1.290	1.661	1.985	2.365	2.627	3.176	3.394
98	1.290	1.661	1.984	2.365	2.627	3.175	3.393
99	1.290	1.660	1.984	2.365	2.626	3.175	3.392
100	1.290	1.660	1.984	2.364	2.626	3.174	3.390
120	1.289	1.658	1.980	2.358	2.617	3.160	3.373
∞	1.282	1.645	1.960	2.326	2.576	3.090	3.291

Source: Provided by J. B. Orris using Excel.

TABLE A.5 A Chi-Square Table: Values of χ_α^2

df	$\chi_{.995}^2$	$\chi_{.99}^2$	$\chi_{.975}^2$	$\chi_{.95}^2$	$\chi_{.90}^2$	$\chi_{.10}^2$	$\chi_{.05}^2$	$\chi_{.025}^2$	$\chi_{.01}^2$	$\chi_{.005}^2$
1	.0000393	.0001571	.0009821	.0039321	.0157908	2.70554	3.84146	5.02389	6.63490	7.87944
2	.0100251	.0201007	.0506356	.102587	.210720	4.60517	5.99147	7.37776	9.21034	10.5966
3	.0717212	.114832	.215795	.341846	.584375	6.25139	7.81473	9.34840	11.3449	12.8381
4	.206990	.297110	.484419	.710721	1.063623	7.77944	9.48773	11.1433	13.2767	14.8602
5	.411740	.554300	.831211	1.145476	1.61031	9.23635	11.0705	12.8325	15.0863	16.7496
6	.675727	.872085	1.237347	1.63539	2.20413	10.6446	12.5916	14.4494	16.8119	18.5476
7	.989265	1.239043	1.68987	2.16735	2.83311	12.0170	14.0671	16.0128	18.4753	20.2777
8	1.344419	1.646482	2.17973	2.73264	3.48954	13.3616	15.5073	17.5346	20.0902	21.9550
9	1.734926	2.087912	2.70039	3.32511	4.16816	14.6837	16.9190	19.0228	21.6660	23.5893
10	2.15585	2.55821	3.24697	3.94030	4.86518	15.9871	18.3070	20.4831	23.2093	25.1882
11	2.60321	3.05347	3.81575	4.57481	5.57779	17.2750	19.6751	21.9200	24.7250	26.7569
12	3.07382	3.57056	4.40379	5.22603	6.30380	18.5494	21.0261	23.3367	26.2170	28.2995
13	3.56503	4.10691	5.00874	5.89186	7.04150	19.8119	22.3621	24.7356	27.6883	29.8194
14	4.07468	4.66043	5.62872	6.57063	7.78953	21.0642	23.6848	26.1190	29.1413	31.3193
15	4.60094	5.22935	6.26214	7.26094	8.54675	22.3072	24.9958	27.4884	30.5779	32.8013
16	5.14224	5.81221	6.90766	7.96164	9.31223	23.5418	26.2962	28.8454	31.9999	34.2672
17	5.69724	6.40776	7.56418	8.67176	10.0852	24.7690	27.5871	30.1910	33.4087	35.7185
18	6.26481	7.01491	8.23075	9.39046	10.8649	25.9894	28.8693	31.5264	34.8053	37.1564
19	6.84398	7.63273	8.90655	10.1170	11.6509	27.2036	30.1435	32.8523	36.1908	38.5822
20	7.43386	8.26040	9.59083	10.8508	12.4426	28.4120	31.4104	34.1696	37.5662	39.9968
21	8.03366	8.89720	10.28293	11.5913	13.2396	29.6151	32.6705	35.4789	38.9321	41.4010
22	8.64272	9.54249	10.9823	12.3380	14.0415	30.8133	33.9244	36.7807	40.2894	42.7956
23	9.26042	10.19567	11.6885	13.0905	14.8479	32.0069	35.1725	38.0757	41.6384	44.1813
24	9.88623	10.8564	12.4011	13.8484	15.6587	33.1963	36.4151	39.3641	42.9798	45.5585
25	10.5197	11.5240	13.1197	14.6114	16.4734	34.3816	37.6525	40.6465	44.3141	46.9278
26	11.1603	12.1981	13.8439	15.3791	17.2919	35.5631	38.8852	41.9232	45.6417	48.2899
27	11.8076	12.8786	14.5733	16.1513	18.1138	36.7412	40.1133	43.1944	46.9630	49.6449
28	12.4613	13.5648	15.3079	16.9279	18.9392	37.9159	41.3372	44.4607	48.2782	50.9933
29	13.1211	14.2565	16.0471	17.7083	19.7677	39.0875	42.5569	45.7222	49.5879	52.3356
30	13.7867	14.9535	16.7908	18.4926	20.5992	40.2560	43.7729	46.9792	50.8922	53.6720
40	20.7065	22.1643	24.4331	26.5093	29.0505	51.8050	55.7585	59.3417	63.6907	66.7659
50	27.9907	29.7067	32.3574	34.7642	37.6886	63.1671	67.5048	71.4202	76.1539	79.4900
60	35.5346	37.4848	40.4817	43.1879	46.4589	74.3970	79.0819	83.2976	88.3794	91.9517
70	43.2752	45.4418	48.7576	51.7393	55.3290	85.5271	90.5312	95.0231	100.425	104.215
80	51.1720	53.5400	57.1532	60.3915	64.2778	96.5782	101.879	106.629	112.329	116.321
90	59.1963	61.7541	65.6466	69.1260	73.2912	107.565	113.145	118.136	124.116	128.299
100	67.3276	70.0648	74.2219	77.9295	82.3581	118.498	124.342	129.561	135.807	140.169

Source: Data from C. M. Thompson, "Tables of the Percentage Points of the χ^2 Distribution," Biometrika 32 (1941), pp. 188–189.

TABLE A.6 An F Table: Values of $F_{.10}$

Numerator Degrees of Freedom (df_1)

df_2 \ df_1	1	2	3	4	5	6	7	8	9	10	12	15	20	24	30	40	60	120	∞
1	39.86	49.50	53.59	55.83	57.24	58.20	58.91	59.44	59.86	60.19	60.71	61.22	61.74	62.00	62.26	62.53	62.79	63.06	63.33
2	8.53	9.00	9.16	9.24	9.29	9.33	9.35	9.37	9.38	9.39	9.41	9.42	9.44	9.45	9.46	9.47	9.47	9.48	9.49
3	5.54	5.46	5.39	5.34	5.31	5.28	5.27	5.25	5.24	5.23	5.22	5.20	5.18	5.18	5.17	5.16	5.15	5.14	5.13
4	4.54	4.32	4.19	4.11	4.05	4.01	3.98	3.95	3.94	3.92	3.90	3.87	3.84	3.83	3.82	3.80	3.79	3.78	3.76
5	4.06	3.78	3.62	3.52	3.45	3.40	3.37	3.34	3.32	3.30	3.27	3.24	3.21	3.19	3.17	3.16	3.14	3.12	3.10
6	3.78	3.46	3.29	3.18	3.11	3.05	3.01	2.98	2.96	2.94	2.90	2.87	2.84	2.82	2.80	2.78	2.76	2.74	2.72
7	3.59	3.26	3.07	2.96	2.88	2.83	2.78	2.75	2.72	2.70	2.67	2.63	2.59	2.58	2.56	2.54	2.51	2.49	2.47
8	3.46	3.11	2.92	2.81	2.73	2.67	2.62	2.59	2.56	2.54	2.50	2.46	2.42	2.40	2.38	2.36	2.34	2.32	2.29
9	3.36	3.01	2.81	2.69	2.61	2.55	2.51	2.47	2.44	2.42	2.38	2.34	2.30	2.28	2.25	2.23	2.21	2.18	2.16
10	3.29	2.92	2.73	2.61	2.52	2.46	2.41	2.38	2.35	2.32	2.28	2.24	2.20	2.18	2.16	2.13	2.11	2.08	2.06
11	3.23	2.86	2.66	2.54	2.45	2.39	2.34	2.30	2.27	2.25	2.21	2.17	2.12	2.10	2.08	2.05	2.03	2.00	1.97
12	3.18	2.81	2.61	2.48	2.39	2.33	2.28	2.24	2.21	2.19	2.15	2.10	2.06	2.04	2.01	1.99	1.96	1.93	1.90
13	3.14	2.76	2.56	2.43	2.35	2.28	2.23	2.20	2.16	2.14	2.10	2.05	2.01	1.98	1.96	1.93	1.90	1.88	1.85
14	3.10	2.73	2.52	2.39	2.31	2.24	2.19	2.15	2.12	2.10	2.05	2.01	1.96	1.94	1.91	1.89	1.86	1.83	1.80
15	3.07	2.70	2.49	2.36	2.27	2.21	2.16	2.12	2.09	2.06	2.02	1.97	1.92	1.90	1.87	1.85	1.82	1.79	1.76
16	3.05	2.67	2.46	2.33	2.24	2.18	2.13	2.09	2.06	2.03	1.99	1.94	1.89	1.87	1.84	1.81	1.78	1.75	1.72
17	3.03	2.64	2.44	2.31	2.22	2.15	2.10	2.06	2.03	2.00	1.96	1.91	1.86	1.84	1.81	1.78	1.75	1.72	1.69
18	3.01	2.62	2.42	2.29	2.20	2.13	2.08	2.04	2.00	1.98	1.93	1.89	1.84	1.81	1.78	1.75	1.72	1.69	1.66
19	2.99	2.61	2.40	2.27	2.18	2.11	2.06	2.02	1.98	1.96	1.91	1.86	1.81	1.79	1.76	1.73	1.70	1.67	1.63
20	2.97	2.59	2.38	2.25	2.16	2.09	2.04	2.00	1.96	1.94	1.89	1.84	1.79	1.77	1.74	1.71	1.68	1.64	1.61
21	2.96	2.57	2.36	2.23	2.14	2.08	2.02	1.98	1.95	1.92	1.87	1.83	1.78	1.75	1.72	1.69	1.66	1.62	1.59
22	2.95	2.56	2.35	2.22	2.13	2.06	2.01	1.97	1.93	1.90	1.86	1.81	1.76	1.73	1.70	1.67	1.64	1.60	1.57
23	2.94	2.55	2.34	2.21	2.11	2.05	1.99	1.95	1.92	1.89	1.84	1.80	1.74	1.72	1.69	1.66	1.62	1.59	1.55
24	2.93	2.54	2.33	2.19	2.10	2.04	1.98	1.94	1.91	1.88	1.83	1.78	1.73	1.70	1.67	1.64	1.61	1.57	1.53
25	2.92	2.53	2.32	2.18	2.09	2.02	1.97	1.93	1.89	1.87	1.82	1.77	1.72	1.69	1.66	1.63	1.59	1.56	1.52
26	2.91	2.52	2.31	2.17	2.08	2.01	1.96	1.92	1.88	1.86	1.81	1.76	1.71	1.68	1.65	1.61	1.58	1.54	1.50
27	2.90	2.51	2.30	2.17	2.07	2.00	1.95	1.91	1.87	1.85	1.80	1.75	1.70	1.67	1.64	1.60	1.57	1.53	1.49
28	2.89	2.50	2.29	2.16	2.06	2.00	1.94	1.90	1.87	1.84	1.79	1.74	1.69	1.66	1.63	1.59	1.56	1.52	1.48
29	2.89	2.50	2.28	2.15	2.06	1.99	1.93	1.89	1.86	1.83	1.78	1.73	1.68	1.65	1.62	1.58	1.55	1.51	1.47
30	2.88	2.49	2.28	2.14	2.05	1.98	1.93	1.88	1.85	1.82	1.77	1.72	1.67	1.64	1.61	1.57	1.54	1.50	1.46
40	2.84	2.44	2.23	2.09	2.00	1.93	1.87	1.83	1.79	1.76	1.71	1.66	1.61	1.57	1.54	1.51	1.47	1.42	1.38
60	2.79	2.39	2.18	2.04	1.95	1.87	1.82	1.77	1.74	1.71	1.66	1.60	1.54	1.51	1.48	1.44	1.40	1.35	1.29
120	2.75	2.35	2.13	1.99	1.90	1.82	1.77	1.72	1.68	1.65	1.60	1.55	1.48	1.45	1.41	1.37	1.32	1.26	1.19
∞	2.71	2.30	2.08	1.94	1.85	1.77	1.72	1.67	1.63	1.60	1.55	1.49	1.42	1.38	1.34	1.30	1.24	1.17	1.00

Denominator Degrees of Freedom (df_2)

Source: Data from M. Merrington and C. M. Thompson, "Tables of Percentage Points of the Inverted Beta (F-Distribution," Biometrika 33 (1943), pp. 73–88.

TABLE A.7　An F Table: Values of $F_{.05}$

df_2 \ df_1	1	2	3	4	5	6	7	8	9	10	12	15	20	24	30	40	60	120	∞
1	161.4	199.5	215.7	224.6	230.2	234.0	236.8	238.9	240.5	241.9	243.9	245.9	248.0	249.1	250.1	251.1	252.2	253.3	254.3
2	18.51	19.00	19.16	19.25	19.30	19.33	19.35	19.37	19.38	19.40	19.41	19.43	19.45	19.45	19.46	19.47	19.48	19.49	19.50
3	10.13	9.55	9.28	9.12	9.01	8.94	8.89	8.85	8.81	8.79	8.74	8.70	8.66	8.64	8.62	8.59	8.57	8.55	8.53
4	7.71	6.94	6.59	6.39	6.26	6.16	6.09	6.04	6.00	5.96	5.91	5.86	5.80	5.77	5.75	5.72	5.69	5.66	5.63
5	6.61	5.79	5.41	5.19	5.05	4.95	4.88	4.82	4.77	4.74	4.68	4.62	4.56	4.53	4.50	4.46	4.43	4.40	4.36
6	5.99	5.14	4.76	4.53	4.39	4.28	4.21	4.15	4.10	4.06	4.00	3.94	3.87	3.84	3.81	3.77	3.74	3.70	3.67
7	5.59	4.74	4.35	4.12	3.97	3.87	3.79	3.73	3.68	3.64	3.57	3.51	3.44	3.41	3.38	3.34	3.30	3.27	3.23
8	5.32	4.46	4.07	3.84	3.69	3.58	3.50	3.44	3.39	3.35	3.28	3.22	3.15	3.12	3.08	3.04	3.01	2.97	2.93
9	5.12	4.26	3.86	3.63	3.48	3.37	3.29	3.23	3.18	3.14	3.07	3.01	2.94	2.90	2.86	2.83	2.79	2.75	2.71
10	4.96	4.10	3.71	3.48	3.33	3.22	3.14	3.07	3.02	2.98	2.91	2.85	2.77	2.74	2.70	2.66	2.62	2.58	2.54
11	4.84	3.98	3.59	3.36	3.20	3.09	3.01	2.95	2.90	2.85	2.79	2.72	2.65	2.61	2.57	2.53	2.49	2.45	2.40
12	4.75	3.89	3.49	3.26	3.11	3.00	2.91	2.85	2.80	2.75	2.69	2.62	2.54	2.51	2.47	2.43	2.38	2.34	2.30
13	4.67	3.81	3.41	3.18	3.03	2.92	2.83	2.77	2.71	2.67	2.60	2.53	2.46	2.42	2.38	2.34	2.30	2.25	2.21
14	4.60	3.74	3.34	3.11	2.96	2.85	2.76	2.70	2.65	2.60	2.53	2.46	2.39	2.35	2.31	2.27	2.22	2.18	2.13
15	4.54	3.68	3.29	3.06	2.90	2.79	2.71	2.64	2.59	2.54	2.48	2.40	2.33	2.29	2.25	2.20	2.16	2.11	2.07
16	4.49	3.63	3.24	3.01	2.85	2.74	2.66	2.59	2.54	2.49	2.42	2.35	2.28	2.24	2.19	2.15	2.11	2.06	2.01
17	4.45	3.59	3.20	2.96	2.81	2.70	2.61	2.55	2.49	2.45	2.38	2.31	2.23	2.19	2.15	2.10	2.06	2.01	1.96
18	4.41	3.55	3.16	2.93	2.77	2.66	2.58	2.51	2.46	2.41	2.34	2.27	2.19	2.15	2.11	2.06	2.02	1.97	1.92
19	4.38	3.52	3.13	2.90	2.74	2.63	2.54	2.48	2.42	2.38	2.31	2.23	2.16	2.11	2.07	2.03	1.98	1.93	1.88
20	4.35	3.49	3.10	2.87	2.71	2.60	2.51	2.45	2.39	2.35	2.28	2.20	2.12	2.08	2.04	1.99	1.95	1.90	1.84
21	4.32	3.47	3.07	2.84	2.68	2.57	2.49	2.42	2.37	2.32	2.25	2.18	2.10	2.05	2.01	1.96	1.92	1.87	1.81
22	4.30	3.44	3.05	2.82	2.66	2.55	2.46	2.40	2.34	2.30	2.23	2.15	2.07	2.03	1.98	1.94	1.89	1.84	1.78
23	4.28	3.42	3.03	2.80	2.64	2.53	2.44	2.37	2.32	2.27	2.20	2.13	2.05	2.01	1.96	1.91	1.86	1.81	1.76
24	4.26	3.40	3.01	2.78	2.62	2.51	2.42	2.36	2.30	2.25	2.18	2.11	2.03	1.98	1.94	1.89	1.84	1.79	1.73
25	4.24	3.39	2.99	2.76	2.60	2.49	2.40	2.34	2.28	2.24	2.16	2.09	2.01	1.96	1.92	1.87	1.82	1.77	1.71
26	4.23	3.37	2.98	2.74	2.59	2.47	2.39	2.32	2.27	2.22	2.15	2.07	1.99	1.95	1.90	1.85	1.80	1.75	1.69
27	4.21	3.35	2.96	2.73	2.57	2.46	2.37	2.31	2.25	2.20	2.13	2.06	1.97	1.93	1.88	1.84	1.79	1.73	1.67
28	4.20	3.34	2.95	2.71	2.56	2.45	2.36	2.29	2.24	2.19	2.12	2.04	1.96	1.91	1.87	1.82	1.77	1.71	1.65
29	4.18	3.33	2.93	2.70	2.55	2.43	2.35	2.28	2.22	2.18	2.10	2.03	1.94	1.90	1.85	1.81	1.75	1.70	1.64
30	4.17	3.32	2.92	2.69	2.53	2.42	2.33	2.27	2.21	2.16	2.09	2.01	1.93	1.89	1.84	1.79	1.74	1.68	1.62
40	4.08	3.23	2.84	2.61	2.45	2.34	2.25	2.18	2.12	2.08	2.00	1.92	1.84	1.79	1.74	1.69	1.64	1.58	1.51
60	4.00	3.15	2.76	2.53	2.37	2.25	2.17	2.10	2.04	1.99	1.92	1.84	1.75	1.70	1.65	1.59	1.53	1.47	1.39
120	3.92	3.07	2.68	2.45	2.29	2.17	2.09	2.02	1.96	1.91	1.83	1.75	1.66	1.61	1.55	1.50	1.43	1.35	1.25
∞	3.84	3.00	2.60	2.37	2.21	2.10	2.01	1.94	1.88	1.83	1.75	1.67	1.57	1.52	1.46	1.39	1.32	1.22	1.00

Numerator Degrees of Freedom (df_1)

Denominator Degrees of Freedom (df_2)

Source: Data from M. Merrington and C. M. Thompson, "Tables of Percentage Points of the Inverted Beta (F)-Distribution," Biometrika 33 (1943), pp. 73–88.

TABLE A.8 An *F* Table: Values of $F_{.025}$

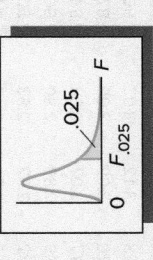

Numerator Degrees of Freedom (df_1)

df_2	1	2	3	4	5	6	7	8	9	10	12	15	20	24	30	40	60	120	∞
1	647.8	799.5	864.2	899.6	921.8	937.1	948.2	956.7	963.3	968.6	976.7	984.9	993.1	997.2	1,001	1,006	1,010	1,014	1,018
2	38.51	39.00	39.17	39.25	39.30	39.33	39.36	39.37	39.39	39.40	39.41	39.43	39.45	39.46	39.46	39.47	39.48	39.49	39.50
3	17.44	16.04	15.44	15.10	14.88	14.73	14.62	14.54	14.47	14.42	14.34	14.25	14.17	14.12	14.08	14.04	13.99	13.95	13.90
4	12.22	10.65	9.98	9.60	9.36	9.20	9.07	8.98	8.90	8.84	8.75	8.66	8.56	8.51	8.46	8.41	8.36	8.31	8.26
5	10.01	8.43	7.76	7.39	7.15	6.98	6.85	6.76	6.68	6.62	6.52	6.43	6.33	6.28	6.23	6.18	6.12	6.07	6.02
6	8.81	7.26	6.60	6.23	5.99	5.82	5.70	5.60	5.52	5.46	5.37	5.27	5.17	5.12	5.07	5.01	4.96	4.90	4.85
7	8.07	6.54	5.89	5.52	5.29	5.12	4.99	4.90	4.82	4.76	4.67	4.57	4.47	4.42	4.36	4.31	4.25	4.20	4.14
8	7.57	6.06	5.42	5.05	4.82	4.65	4.53	4.43	4.36	4.30	4.20	4.10	4.00	3.95	3.89	3.84	3.78	3.73	3.67
9	7.21	5.71	5.08	4.72	4.48	4.32	4.20	4.10	4.03	3.96	3.87	3.77	3.67	3.61	3.56	3.51	3.45	3.39	3.33
10	6.94	5.46	4.83	4.47	4.24	4.07	3.95	3.85	3.78	3.72	3.62	3.52	3.42	3.37	3.31	3.26	3.20	3.14	3.08
11	6.72	5.26	4.63	4.28	4.04	3.88	3.76	3.66	3.59	3.53	3.43	3.33	3.23	3.17	3.12	3.06	3.00	2.94	2.88
12	6.55	5.10	4.47	4.12	3.89	3.73	3.61	3.51	3.44	3.37	3.28	3.18	3.07	3.02	2.96	2.91	2.85	2.79	2.72
13	6.41	4.97	4.35	4.00	3.77	3.60	3.48	3.39	3.31	3.25	3.15	3.05	2.95	2.89	2.84	2.78	2.72	2.66	2.60
14	6.30	4.86	4.24	3.89	3.66	3.50	3.38	3.29	3.21	3.15	3.05	2.95	2.84	2.79	2.73	2.67	2.61	2.55	2.49
15	6.20	4.77	4.15	3.80	3.58	3.41	3.29	3.20	3.12	3.06	2.96	2.86	2.76	2.70	2.64	2.59	2.52	2.46	2.40
16	6.12	4.69	4.08	3.73	3.50	3.34	3.22	3.12	3.05	2.99	2.89	2.79	2.68	2.63	2.57	2.51	2.45	2.38	2.32
17	6.04	4.62	4.01	3.66	3.44	3.28	3.16	3.06	2.98	2.92	2.82	2.72	2.62	2.56	2.50	2.44	2.38	2.32	2.25
18	5.98	4.56	3.95	3.61	3.38	3.22	3.10	3.01	2.93	2.87	2.77	2.67	2.56	2.50	2.44	2.38	2.32	2.26	2.19
19	5.92	4.51	3.90	3.56	3.33	3.17	3.05	2.96	2.88	2.82	2.72	2.62	2.51	2.45	2.39	2.33	2.27	2.20	2.13
20	5.87	4.46	3.86	3.51	3.29	3.13	3.01	2.91	2.84	2.77	2.68	2.57	2.46	2.41	2.35	2.29	2.22	2.16	2.09
21	5.83	4.42	3.82	3.48	3.25	3.09	2.97	2.87	2.80	2.73	2.64	2.53	2.42	2.37	2.31	2.25	2.18	2.11	2.04
22	5.79	4.38	3.78	3.44	3.22	3.05	2.93	2.84	2.76	2.70	2.60	2.50	2.39	2.33	2.27	2.21	2.14	2.08	2.00
23	5.75	4.35	3.75	3.41	3.18	3.02	2.90	2.81	2.73	2.67	2.57	2.47	2.36	2.30	2.24	2.18	2.11	2.04	1.97
24	5.72	4.32	3.72	3.38	3.15	2.99	2.87	2.78	2.70	2.64	2.54	2.44	2.33	2.27	2.21	2.15	2.08	2.01	1.94
25	5.69	4.29	3.69	3.35	3.13	2.97	2.85	2.75	2.68	2.61	2.51	2.41	2.30	2.24	2.18	2.12	2.05	1.98	1.91
26	5.66	4.27	3.67	3.33	3.10	2.94	2.82	2.73	2.65	2.59	2.49	2.39	2.28	2.22	2.16	2.09	2.03	1.95	1.88
27	5.63	4.24	3.65	3.31	3.08	2.92	2.80	2.71	2.63	2.57	2.47	2.36	2.25	2.19	2.13	2.07	2.00	1.93	1.85
28	5.61	4.22	3.63	3.29	3.06	2.90	2.78	2.69	2.61	2.55	2.45	2.34	2.23	2.17	2.11	2.05	1.98	1.91	1.83
29	5.59	4.20	3.61	3.27	3.04	2.88	2.76	2.67	2.59	2.53	2.43	2.32	2.21	2.15	2.09	2.03	1.96	1.89	1.81
30	5.57	4.18	3.59	3.25	3.03	2.87	2.75	2.65	2.57	2.51	2.41	2.31	2.20	2.14	2.07	2.01	1.94	1.87	1.79
40	5.42	4.05	3.46	3.13	2.90	2.74	2.62	2.53	2.45	2.39	2.29	2.18	2.07	2.01	1.94	1.88	1.80	1.72	1.64
60	5.29	3.93	3.34	3.01	2.79	2.63	2.51	2.41	2.33	2.27	2.17	2.06	1.94	1.88	1.82	1.74	1.67	1.58	1.48
120	5.15	3.80	3.23	2.89	2.67	2.52	2.39	2.30	2.22	2.16	2.05	1.94	1.82	1.76	1.69	1.61	1.53	1.43	1.31
∞	5.02	3.69	3.12	2.79	2.57	2.41	2.29	2.19	2.11	2.05	1.94	1.83	1.71	1.64	1.57	1.48	1.39	1.27	1.00

Denominator Degrees of Freedom (df_2)

Source: Data from M. Merrington and C. M. Thompson, "Tables of Percentage Points of the Inverted Beta (*F*)-Distribution," *Biometrika* 33 (1943), pp. 73–88.

TABLE A.9 An F Table: Values of $F_{.01}$

Numerator Degrees of Freedom (df_1)

df_2	1	2	3	4	5	6	7	8	9	10	12	15	20	24	30	40	60	120	∞
1	4,052	4,999.5	5,403	5,625	5,764	5,859	5,928	5,982	6,022	6,056	6,106	6,157	6,209	6,235	6,261	6,287	6,313	6,339	6,366
2	98.50	99.00	99.17	99.25	99.30	99.33	99.36	99.37	99.39	99.40	99.42	99.43	99.45	99.46	99.47	99.47	99.47	99.49	99.50
3	34.12	30.82	29.46	28.71	28.24	27.91	27.67	27.49	27.35	27.23	27.05	26.87	26.69	26.60	26.50	26.41	26.32	26.22	26.13
4	21.20	18.00	16.69	15.98	15.52	15.21	14.98	14.80	14.66	14.55	14.37	14.20	14.02	13.93	13.84	13.75	13.65	13.56	13.46
5	16.26	13.27	12.06	11.39	10.97	10.67	10.46	10.29	10.16	10.05	9.89	9.72	9.55	9.47	9.38	9.29	9.20	9.11	9.02
6	13.75	10.92	9.78	9.15	8.75	8.47	8.26	8.10	7.98	7.87	7.72	7.56	7.40	7.31	7.23	7.14	7.06	6.97	6.88
7	12.25	9.55	8.45	7.85	7.46	7.19	6.99	6.84	6.72	6.62	6.47	6.31	6.16	6.07	5.99	5.91	5.82	5.74	5.65
8	11.26	8.65	7.59	7.01	6.63	6.37	6.18	6.03	5.91	5.81	5.67	5.52	5.36	5.28	5.20	5.12	5.03	4.95	4.86
9	10.56	8.02	6.99	6.42	6.06	5.80	5.61	5.47	5.35	5.26	5.11	4.96	4.81	4.73	4.65	4.57	4.48	4.40	4.31
10	10.04	7.56	6.55	5.99	5.64	5.39	5.20	5.06	4.94	4.85	4.71	4.56	4.41	4.33	4.25	4.17	4.08	4.00	3.91
11	9.65	7.21	6.22	5.67	5.32	5.07	4.89	4.74	4.63	4.54	4.40	4.25	4.10	4.02	3.94	3.86	3.78	3.69	3.60
12	9.33	6.93	5.95	5.41	5.06	4.82	4.64	4.50	4.39	4.30	4.16	4.01	3.86	3.78	3.70	3.62	3.54	3.45	3.36
13	9.07	6.70	5.74	5.21	4.86	4.62	4.44	4.30	4.19	4.10	3.96	3.82	3.66	3.59	3.51	3.43	3.34	3.25	3.17
14	8.86	6.51	5.56	5.04	4.69	4.46	4.28	4.14	4.03	3.94	3.80	3.66	3.51	3.43	3.35	3.27	3.18	3.09	3.00
15	8.68	6.36	5.42	4.89	4.56	4.32	4.14	4.00	3.89	3.80	3.67	3.52	3.37	3.29	3.21	3.13	3.05	2.96	2.87
16	8.53	6.23	5.29	4.77	4.44	4.20	4.03	3.89	3.78	3.69	3.55	3.41	3.26	3.18	3.10	3.02	2.93	2.84	2.75
17	8.40	6.11	5.18	4.67	4.34	4.10	3.93	3.79	3.68	3.59	3.46	3.31	3.16	3.08	3.00	2.92	2.83	2.75	2.65
18	8.29	6.01	5.09	4.58	4.25	4.01	3.84	3.71	3.60	3.51	3.37	3.23	3.08	3.00	2.92	2.84	2.75	2.66	2.57
19	8.18	5.93	5.01	4.50	4.17	3.94	3.77	3.63	3.52	3.43	3.30	3.15	3.00	2.92	2.84	2.76	2.67	2.58	2.49
20	8.10	5.85	4.94	4.43	4.10	3.87	3.70	3.56	3.46	3.37	3.23	3.09	2.94	2.86	2.78	2.69	2.61	2.52	2.42
21	8.02	5.78	4.87	4.37	4.04	3.81	3.64	3.51	3.40	3.31	3.17	3.03	2.88	2.80	2.72	2.64	2.55	2.46	2.36
22	7.95	5.72	4.82	4.31	3.99	3.76	3.59	3.45	3.35	3.26	3.12	2.98	2.83	2.75	2.67	2.58	2.50	2.40	2.31
23	7.88	5.66	4.76	4.26	3.94	3.71	3.54	3.41	3.30	3.21	3.07	2.93	2.78	2.70	2.62	2.54	2.45	2.35	2.26
24	7.82	5.61	4.72	4.22	3.90	3.67	3.50	3.36	3.26	3.17	3.03	2.89	2.74	2.66	2.58	2.49	2.40	2.31	2.21
25	7.77	5.57	4.68	4.18	3.85	3.63	3.46	3.32	3.22	3.13	2.99	2.85	2.70	2.62	2.54	2.45	2.36	2.27	2.17
26	7.72	5.53	4.64	4.14	3.82	3.59	3.42	3.29	3.18	3.09	2.96	2.81	2.66	2.58	2.50	2.42	2.33	2.23	2.13
27	7.68	5.49	4.60	4.11	3.78	3.56	3.39	3.26	3.15	3.06	2.93	2.78	2.63	2.55	2.47	2.38	2.29	2.20	2.10
28	7.64	5.45	4.57	4.07	3.75	3.53	3.36	3.23	3.12	3.03	2.90	2.75	2.60	2.52	2.44	2.35	2.26	2.17	2.06
29	7.60	5.42	4.54	4.04	3.73	3.50	3.33	3.20	3.09	3.00	2.87	2.73	2.57	2.49	2.41	2.33	2.23	2.14	2.03
30	7.56	5.39	4.51	4.02	3.70	3.47	3.30	3.17	3.07	2.98	2.84	2.70	2.55	2.47	2.39	2.30	2.21	2.11	2.01
40	7.31	5.18	4.31	3.83	3.51	3.29	3.12	2.99	2.89	2.80	2.66	2.52	2.37	2.29	2.20	2.11	2.02	1.92	1.80
60	7.08	4.98	4.13	3.65	3.34	3.12	2.95	2.82	2.72	2.63	2.50	2.35	2.20	2.12	2.03	1.94	1.84	1.73	1.60
120	6.85	4.79	3.95	3.48	3.17	2.96	2.79	2.66	2.56	2.47	2.34	2.19	2.03	1.95	1.86	1.76	1.66	1.53	1.38
∞	6.63	4.61	3.78	3.32	3.02	2.80	2.64	2.51	2.41	2.32	2.18	2.04	1.88	1.79	1.70	1.59	1.47	1.32	1.00

Denominator Degrees of Freedom (df_2)

Source: Data from M. Merrington and C. M. Thompson, "Tables of Percentage Points of the Inverted Beta (F)-Distribution," *Biometrika* 33 (1943), pp. 73–88.

TABLE A.10 Percentage Points of the Studentized Range

(Note: r is the "first value" and v is the "second value" referred to in Chapter 13.)

Entry is $q_{.05}$

v	2	3	4	5	6	7	8	9	10	11	12	13	14	15	16	17	18	19	20
1	18.0	27.0	32.8	37.1	40.4	43.1	45.4	47.4	49.1	50.6	52.0	53.2	54.3	55.4	56.3	57.2	58.0	58.8	59.6
2	6.08	8.33	9.80	10.9	11.7	12.4	13.0	13.5	14.0	14.4	14.7	15.1	15.4	15.7	15.9	16.1	16.4	16.6	16.8
3	4.50	5.91	6.82	7.50	8.04	8.48	8.85	9.18	9.46	9.72	9.95	10.2	10.3	10.5	10.7	10.8	11.0	11.1	11.2
4	3.93	5.04	5.76	6.29	6.71	7.05	7.35	7.60	7.83	8.03	8.21	8.37	8.52	8.66	8.79	8.91	9.03	9.13	9.23
5	3.64	4.60	5.22	5.67	6.03	6.33	6.58	6.80	6.99	7.17	7.32	7.47	7.60	7.72	7.83	7.93	8.03	8.12	8.21
6	3.46	4.34	4.90	5.30	5.63	5.90	6.12	6.32	6.49	6.65	6.79	6.92	7.03	7.14	7.24	7.34	7.43	7.51	7.59
7	3.34	4.16	4.68	5.06	5.36	5.61	5.82	6.00	6.16	6.30	6.43	6.55	6.66	6.76	6.85	6.94	7.02	7.10	7.17
8	3.26	4.04	4.53	4.89	5.17	5.40	5.60	5.77	5.92	6.05	6.18	6.29	6.39	6.48	6.57	6.65	6.73	6.80	6.87
9	3.20	3.95	4.41	4.76	5.02	5.24	5.43	5.59	5.74	5.87	5.98	6.09	6.19	6.28	6.36	6.44	6.51	6.58	6.64
10	3.15	3.88	4.33	4.65	4.91	5.12	5.30	5.46	5.60	5.72	5.83	5.93	6.03	6.11	6.19	6.27	6.34	6.40	6.47
11	3.11	3.82	4.26	4.57	4.82	5.03	5.20	5.35	5.49	5.61	5.71	5.81	5.90	5.98	6.06	6.13	6.20	6.27	6.33
12	3.08	3.77	4.20	4.51	4.75	4.95	5.12	5.27	5.39	5.51	5.61	5.71	5.80	5.88	5.95	6.02	6.09	6.15	6.21
13	3.06	3.73	4.15	4.45	4.69	4.88	5.05	5.19	5.32	5.43	5.53	5.63	5.71	5.79	5.86	5.93	5.99	6.05	6.11
14	3.03	3.70	4.11	4.41	4.64	4.83	4.99	5.13	5.25	5.36	5.46	5.55	5.64	5.71	5.79	5.85	5.91	5.97	6.03
15	3.01	3.67	4.08	4.37	4.59	4.78	4.94	5.08	5.20	5.31	5.40	5.49	5.57	5.65	5.72	5.78	5.85	5.90	5.96
16	3.00	3.65	4.05	4.33	4.56	4.74	4.90	5.03	5.15	5.26	5.35	5.44	5.52	5.59	5.66	5.73	5.79	5.84	5.90
17	2.98	3.63	4.02	4.30	4.52	4.70	4.86	4.99	5.11	5.21	5.31	5.39	5.47	5.54	5.61	5.67	5.73	5.79	5.84
18	2.97	3.61	4.00	4.28	4.49	4.67	4.82	4.96	5.07	5.17	5.27	5.35	5.43	5.50	5.57	5.63	5.69	5.74	5.79
19	2.96	3.59	3.98	4.25	4.47	4.65	4.79	4.92	5.04	5.14	5.23	5.31	5.39	5.46	5.53	5.59	5.65	5.70	5.75
20	2.95	3.58	3.96	4.23	4.45	4.62	4.77	4.90	5.01	5.11	5.20	5.28	5.36	5.43	5.49	5.55	5.61	5.66	5.71
24	2.92	3.53	3.90	4.17	4.37	4.54	4.68	4.81	4.92	5.01	5.10	5.18	5.25	5.32	5.38	5.44	5.49	5.55	5.59
30	2.89	3.49	3.85	4.10	4.30	4.46	4.60	4.72	4.82	4.92	5.00	5.08	5.15	5.21	5.27	5.33	5.38	5.43	5.47
40	2.86	3.44	3.79	4.04	4.23	4.39	4.52	4.63	4.73	4.82	4.90	4.98	5.04	5.11	5.16	5.22	5.27	5.31	5.36
60	2.83	3.40	3.74	3.98	4.16	4.31	4.44	4.55	4.65	4.73	4.81	4.88	4.94	5.00	5.06	5.11	5.15	5.20	5.24
120	2.80	3.36	3.68	3.92	4.10	4.24	4.36	4.47	4.56	4.64	4.71	4.78	4.84	4.90	4.95	5.00	5.04	5.09	5.13
∞	2.77	3.31	3.63	3.86	4.03	4.17	4.29	4.39	4.47	4.55	4.62	4.68	4.74	4.80	4.85	4.89	4.93	4.97	5.01

r

(table continued)

TABLE A.10 Percentage Points of the Studentized Range
(Note: r is the "first value" and v is the "second value" referred to in Chapter 13.) *(concluded)*

Entry is $q_{.01}$

r

v	2	3	4	5	6	7	8	9	10	11	12	13	14	15	16	17	18	19	20
1	90.0	135	164	186	202	216	227	237	246	253	260	266	272	277	282	286	290	294	298
2	14.0	19.0	22.3	24.7	26.6	28.2	29.5	30.7	31.7	32.6	33.4	34.1	34.8	35.4	36.0	36.5	37.0	37.5	37.9
3	8.26	10.6	12.2	13.3	14.2	15.0	15.6	16.2	16.7	17.1	17.5	17.9	18.2	18.5	18.8	19.1	19.3	19.5	19.8
4	6.51	8.12	9.17	9.96	10.6	11.1	11.5	11.9	12.3	12.6	12.8	13.1	13.3	13.5	13.7	13.9	14.1	14.2	14.4
5	5.70	6.97	7.80	8.42	8.91	9.32	9.67	9.97	10.2	10.5	10.7	10.9	11.1	11.2	11.4	11.6	11.7	11.8	11.9
6	5.24	6.33	7.03	7.56	7.97	8.32	8.61	8.87	9.10	9.30	9.49	9.65	9.81	9.95	10.1	10.2	10.3	10.4	10.5
7	4.95	5.92	6.54	7.01	7.37	7.68	7.94	8.17	8.37	8.55	8.71	8.86	9.00	9.12	9.24	9.35	9.46	9.55	9.65
8	4.74	5.63	6.20	6.63	6.96	7.24	7.47	7.68	7.87	8.03	8.18	8.31	8.44	8.55	8.66	8.76	8.85	8.94	9.03
9	4.60	5.43	5.96	6.35	6.66	6.91	7.13	7.32	7.49	7.65	7.78	7.91	8.03	8.13	8.23	8.32	8.41	8.49	8.57
10	4.48	5.27	5.77	6.14	6.43	6.67	6.87	7.05	7.21	7.36	7.48	7.60	7.71	7.81	7.91	7.99	8.07	8.15	8.22
11	4.39	5.14	5.62	5.97	6.25	6.48	6.67	6.84	6.99	7.13	7.25	7.36	7.46	7.56	7.65	7.73	7.81	7.88	7.95
12	4.32	5.04	5.50	5.84	6.10	6.32	6.51	6.67	6.81	6.94	7.06	7.17	7.26	7.36	7.44	7.52	7.59	7.66	7.73
13	4.26	4.96	5.40	5.73	5.98	6.19	6.37	6.53	6.67	6.79	6.90	7.01	7.10	7.19	7.27	7.34	7.42	7.48	7.55
14	4.21	4.89	5.32	5.63	5.88	6.08	6.26	6.41	6.54	6.66	6.77	6.87	6.96	7.05	7.12	7.20	7.27	7.33	7.39
15	4.17	4.83	5.25	5.56	5.80	5.99	6.16	6.31	6.44	6.55	6.66	6.76	6.84	6.93	7.00	7.07	7.14	7.20	7.26
16	4.13	4.78	5.19	5.49	5.72	5.92	6.08	6.22	6.35	6.46	6.56	6.66	6.74	6.82	6.90	6.97	7.03	7.09	7.15
17	4.10	4.74	5.14	5.43	5.66	5.85	6.01	6.15	6.27	6.38	6.48	6.57	6.66	6.73	6.80	6.87	6.94	7.00	7.05
18	4.07	4.70	5.09	5.38	5.60	5.79	5.94	6.08	6.20	6.31	6.41	6.50	6.58	6.65	6.72	6.79	6.85	6.91	6.96
19	4.05	4.67	5.05	5.33	5.55	5.73	5.89	6.02	6.14	6.25	6.34	6.43	6.51	6.58	6.65	6.72	6.78	6.84	6.89
20	4.02	4.64	5.02	5.29	5.51	5.69	5.84	5.97	6.09	6.19	6.29	6.37	6.45	6.52	6.59	6.65	6.71	6.76	6.82
24	3.96	4.54	4.91	5.17	5.37	5.54	5.69	5.81	5.92	6.02	6.11	6.19	6.26	6.33	6.39	6.45	6.51	6.56	6.61
30	3.89	4.45	4.80	5.05	5.24	5.40	5.54	5.65	5.76	5.85	5.93	6.01	6.08	6.14	6.20	6.26	6.31	6.36	6.41
40	3.82	4.37	4.70	4.93	5.11	5.27	5.39	5.50	5.60	5.69	5.77	5.84	5.90	5.96	6.02	6.07	6.12	6.17	6.21
60	3.76	4.28	4.60	4.82	4.99	5.13	5.25	5.36	5.45	5.53	5.60	5.67	5.73	5.79	5.84	5.89	5.93	5.98	6.02
120	3.70	4.20	4.50	4.71	4.87	5.01	5.12	5.21	5.30	5.38	5.44	5.51	5.56	5.61	5.66	5.71	5.75	5.79	5.83
∞	3.64	4.12	4.40	4.60	4.76	4.88	4.99	5.08	5.16	5.23	5.29	5.35	5.40	5.45	5.49	5.54	5.57	5.61	5.65

Source: Data from Henry Scheffe, *The Analysis of Variance*, pp. 414–416, John Wiley & Sons, 1959.

TABLE A.11 Critical Values for the Durbin–Watson d Statistic ($\alpha = .05$)

n	k=1 $d_{L,05}$	k=1 $d_{U,05}$	k=2 $d_{L,05}$	k=2 $d_{U,05}$	k=3 $d_{L,05}$	k=3 $d_{U,05}$	k=4 $d_{L,05}$	k=4 $d_{U,05}$	k=5 $d_{L,05}$	k=5 $d_{U,05}$
15	1.08	1.36	0.95	1.54	0.82	1.75	0.69	1.97	0.56	2.21
16	1.10	1.37	0.98	1.54	0.86	1.73	0.74	1.93	0.62	2.15
17	1.13	1.38	1.02	1.54	0.90	1.71	0.78	1.90	0.67	2.10
18	1.16	1.39	1.05	1.53	0.93	1.69	0.82	1.87	0.71	2.06
19	1.18	1.40	1.08	1.53	0.97	1.68	0.86	1.85	0.75	2.02
20	1.20	1.41	1.10	1.54	1.00	1.68	0.90	1.83	0.79	1.99
21	1.22	1.42	1.13	1.54	1.03	1.67	0.93	1.81	0.83	1.96
22	1.24	1.43	1.15	1.54	1.05	1.66	0.96	1.80	0.86	1.94
23	1.26	1.44	1.17	1.54	1.08	1.66	0.99	1.79	0.90	1.92
24	1.27	1.45	1.19	1.55	1.10	1.66	1.01	1.78	0.93	1.90
25	1.29	1.45	1.21	1.55	1.12	1.66	1.04	1.77	0.95	1.89
26	1.30	1.46	1.22	1.55	1.14	1.65	1.06	1.76	0.98	1.88
27	1.32	1.47	1.24	1.56	1.16	1.65	1.08	1.76	1.01	1.86
28	1.33	1.48	1.26	1.56	1.18	1.65	1.10	1.75	1.03	1.85
29	1.34	1.48	1.27	1.56	1.20	1.65	1.12	1.74	1.05	1.84
30	1.35	1.49	1.28	1.57	1.21	1.65	1.14	1.74	1.07	1.83
31	1.36	1.50	1.30	1.57	1.23	1.65	1.16	1.74	1.09	1.83
32	1.37	1.50	1.31	1.57	1.24	1.65	1.18	1.73	1.11	1.82
33	1.38	1.51	1.32	1.58	1.26	1.65	1.19	1.73	1.13	1.81
34	1.39	1.51	1.33	1.58	1.27	1.65	1.21	1.73	1.15	1.81
35	1.40	1.52	1.34	1.58	1.28	1.65	1.22	1.73	1.16	1.80
36	1.41	1.52	1.35	1.59	1.29	1.65	1.24	1.73	1.18	1.80
37	1.42	1.53	1.36	1.59	1.31	1.66	1.25	1.72	1.19	1.80
38	1.43	1.54	1.37	1.59	1.32	1.66	1.26	1.72	1.21	1.79
39	1.43	1.54	1.38	1.60	1.33	1.66	1.27	1.72	1.22	1.79
40	1.44	1.54	1.39	1.60	1.34	1.66	1.29	1.72	1.23	1.79
45	1.48	1.57	1.43	1.62	1.38	1.67	1.34	1.72	1.29	1.78
50	1.50	1.59	1.46	1.63	1.42	1.67	1.38	1.72	1.34	1.77
55	1.53	1.60	1.49	1.64	1.45	1.68	1.41	1.72	1.38	1.77
60	1.55	1.62	1.51	1.65	1.48	1.69	1.44	1.73	1.41	1.77
65	1.57	1.63	1.54	1.66	1.50	1.70	1.47	1.73	1.44	1.77
70	1.58	1.64	1.55	1.67	1.52	1.70	1.49	1.74	1.46	1.77
75	1.60	1.65	1.57	1.68	1.54	1.71	1.51	1.74	1.49	1.77
80	1.61	1.66	1.59	1.69	1.56	1.72	1.53	1.74	1.51	1.77
85	1.62	1.67	1.60	1.70	1.57	1.72	1.55	1.75	1.52	1.77
90	1.63	1.68	1.61	1.70	1.59	1.73	1.57	1.75	1.54	1.78
95	1.64	1.69	1.62	1.71	1.60	1.73	1.58	1.75	1.56	1.78
100	1.65	1.69	1.63	1.72	1.61	1.74	1.59	1.76	1.57	1.78

Source: Data from J. Durbin and G. S. Watson, "Testing for Serial Correlation in Least Squares Regression, II," *Biometrika* 30 (1951), pp. 159–178.

TABLE A.12 Critical Values for the Durbin–Watson d Statistic ($\alpha = .025$)

n	k=1 $d_{L,025}$	k=1 $d_{U,025}$	k=2 $d_{L,025}$	k=2 $d_{U,025}$	k=3 $d_{L,025}$	k=3 $d_{U,025}$	k=4 $d_{L,025}$	k=4 $d_{U,025}$	k=5 $d_{L,025}$	k=5 $d_{U,025}$
15	0.95	1.23	0.83	1.40	0.71	1.61	0.59	1.84	0.48	2.09
16	0.98	1.24	0.86	1.40	0.75	1.59	0.64	1.80	0.53	2.03
17	1.01	1.25	0.90	1.40	0.79	1.58	0.68	1.77	0.57	1.98
18	1.03	1.26	0.93	1.40	0.82	1.56	0.72	1.74	0.62	1.93
19	1.06	1.28	0.96	1.41	0.86	1.55	0.76	1.72	0.66	1.90
20	1.08	1.28	0.99	1.41	0.89	1.55	0.79	1.70	0.70	1.87
21	1.10	1.30	1.01	1.41	0.92	1.54	0.83	1.69	0.73	1.84
22	1.12	1.31	1.04	1.42	0.95	1.54	0.86	1.68	0.77	1.82
23	1.14	1.32	1.06	1.42	0.97	1.54	0.89	1.67	0.80	1.80
24	1.16	1.33	1.08	1.43	1.00	1.54	0.91	1.66	0.83	1.79
25	1.18	1.34	1.10	1.43	1.02	1.54	0.94	1.65	0.86	1.77
26	1.19	1.35	1.12	1.44	1.04	1.54	0.96	1.65	0.88	1.76
27	1.21	1.36	1.13	1.44	1.06	1.54	0.99	1.64	0.91	1.75
28	1.22	1.37	1.15	1.45	1.08	1.54	1.01	1.64	0.93	1.74
29	1.24	1.38	1.17	1.45	1.10	1.54	1.03	1.63	0.96	1.73
30	1.25	1.38	1.18	1.46	1.12	1.54	1.05	1.63	0.98	1.73
31	1.26	1.39	1.20	1.47	1.13	1.55	1.07	1.63	1.00	1.72
32	1.27	1.40	1.21	1.47	1.15	1.55	1.08	1.63	1.02	1.71
33	1.28	1.41	1.22	1.48	1.16	1.55	1.10	1.63	1.04	1.71
34	1.29	1.41	1.24	1.48	1.17	1.55	1.12	1.63	1.06	1.70
35	1.30	1.42	1.25	1.48	1.19	1.55	1.13	1.63	1.07	1.70
36	1.31	1.43	1.26	1.49	1.20	1.56	1.15	1.63	1.09	1.70
37	1.32	1.43	1.27	1.49	1.21	1.56	1.16	1.62	1.10	1.70
38	1.33	1.44	1.28	1.50	1.23	1.56	1.17	1.62	1.12	1.70
39	1.34	1.44	1.29	1.50	1.24	1.56	1.19	1.63	1.13	1.69
40	1.35	1.45	1.30	1.51	1.25	1.57	1.20	1.63	1.15	1.69
45	1.39	1.48	1.34	1.53	1.30	1.58	1.25	1.63	1.21	1.69
50	1.42	1.50	1.38	1.54	1.34	1.59	1.30	1.64	1.26	1.69
55	1.45	1.52	1.41	1.56	1.37	1.60	1.33	1.64	1.30	1.69
60	1.47	1.54	1.44	1.57	1.40	1.61	1.37	1.65	1.33	1.69
65	1.49	1.55	1.46	1.59	1.43	1.62	1.40	1.66	1.36	1.69
70	1.51	1.57	1.48	1.60	1.45	1.63	1.42	1.66	1.39	1.70
75	1.53	1.58	1.50	1.61	1.47	1.64	1.45	1.67	1.42	1.70
80	1.54	1.59	1.52	1.62	1.49	1.65	1.47	1.67	1.44	1.70
85	1.56	1.60	1.53	1.63	1.51	1.65	1.49	1.68	1.46	1.71
90	1.57	1.61	1.55	1.64	1.53	1.66	1.50	1.69	1.48	1.71
95	1.58	1.62	1.56	1.65	1.54	1.67	1.52	1.69	1.50	1.71
100	1.59	1.63	1.57	1.65	1.55	1.67	1.53	1.70	1.51	1.72

Source: Data from J. Durbin and G. S. Watson, "Testing for Serial Correlation in Least Squares Regression, II," *Biometrika* 30 (1951), pp. 159–178.

TABLE A.13 Critical Values for the Durbin–Watson d Statistic ($\alpha = .01$)

n	$d_{L,.01}$	$d_{U,.01}$	$d_{L,.01}$	$d_{U,.01}$	$d_{L,.01}$	$d_{U,.01}$	$d_{L,.01}$	$d_{U,.01}$	$d_{L,.01}$	$d_{U,.01}$
	k = 1		k = 2		k = 3		k = 4		k = 5	
15	0.81	1.07	0.70	1.25	0.59	1.46	0.49	1.70	0.39	1.96
16	0.84	1.09	0.74	1.25	0.63	1.44	0.53	1.66	0.44	1.90
17	0.87	1.10	0.77	1.25	0.67	1.43	0.57	1.63	0.48	1.85
18	0.90	1.12	0.80	1.26	0.71	1.42	0.61	1.60	0.52	1.80
19	0.93	1.13	0.83	1.26	0.74	1.41	0.65	1.58	0.56	1.77
20	0.95	1.15	0.86	1.27	0.77	1.41	0.68	1.57	0.60	1.74
21	0.97	1.16	0.89	1.27	0.80	1.41	0.72	1.55	0.63	1.71
22	1.00	1.17	0.91	1.28	0.83	1.40	0.75	1.54	0.66	1.69
23	1.02	1.19	0.94	1.29	0.86	1.40	0.77	1.53	0.70	1.67
24	1.04	1.20	0.96	1.30	0.88	1.41	0.80	1.53	0.72	1.66
25	1.05	1.21	0.98	1.30	0.90	1.41	0.83	1.52	0.75	1.65
26	1.07	1.22	1.00	1.31	0.93	1.41	0.85	1.52	0.78	1.64
27	1.09	1.23	1.02	1.32	0.95	1.41	0.88	1.51	0.81	1.63
28	1.10	1.24	1.04	1.32	0.97	1.41	0.90	1.51	0.83	1.62
29	1.12	1.25	1.05	1.33	0.99	1.42	0.92	1.51	0.85	1.61
30	1.13	1.26	1.07	1.34	1.01	1.42	0.94	1.51	0.88	1.61
31	1.15	1.27	1.08	1.34	1.02	1.42	0.96	1.51	0.90	1.60
32	1.16	1.28	1.10	1.35	1.04	1.43	0.98	1.51	0.92	1.60
33	1.17	1.29	1.11	1.36	1.05	1.43	1.00	1.51	0.94	1.59
34	1.18	1.30	1.13	1.36	1.07	1.43	1.01	1.51	0.95	1.59
35	1.19	1.31	1.14	1.37	1.08	1.44	1.03	1.51	0.97	1.59
36	1.21	1.32	1.15	1.38	1.10	1.44	1.04	1.51	0.99	1.59
37	1.22	1.32	1.16	1.38	1.11	1.45	1.06	1.51	1.00	1.59
38	1.23	1.33	1.18	1.39	1.12	1.45	1.07	1.52	1.02	1.58
39	1.24	1.34	1.19	1.39	1.14	1.45	1.09	1.52	1.03	1.58
40	1.25	1.34	1.20	1.40	1.15	1.46	1.10	1.52	1.05	1.58
45	1.29	1.38	1.24	1.42	1.20	1.48	1.16	1.53	1.11	1.58
50	1.32	1.40	1.28	1.45	1.24	1.49	1.20	1.54	1.16	1.59
55	1.36	1.43	1.32	1.47	1.28	1.51	1.25	1.55	1.21	1.59
60	1.38	1.45	1.35	1.48	1.32	1.52	1.28	1.56	1.25	1.60
65	1.41	1.47	1.38	1.50	1.35	1.53	1.31	1.57	1.28	1.61
70	1.43	1.49	1.40	1.52	1.37	1.55	1.34	1.58	1.31	1.61
75	1.45	1.50	1.42	1.53	1.39	1.56	1.37	1.59	1.34	1.62
80	1.47	1.52	1.44	1.54	1.42	1.57	1.39	1.60	1.36	1.62
85	1.48	1.53	1.46	1.55	1.43	1.58	1.41	1.60	1.39	1.63
90	1.50	1.54	1.47	1.56	1.45	1.59	1.43	1.61	1.41	1.64
95	1.51	1.55	1.49	1.57	1.47	1.60	1.45	1.62	1.42	1.64
100	1.52	1.56	1.50	1.58	1.48	1.60	1.46	1.63	1.44	1.65

Source: Data from J. Durbin and G. S. Watson, "Testing for Serial Correlation in Least Squares Regression, II," Biometrika 30 (1951), pp. 159–178.

TABLE A.14 A Wilcoxon Rank Sum Table: Values of T_L and T_U

(a) $\alpha = .025$ One-Sided; $\alpha = .05$ Two-Sided

n_2 \ n_1	3 T_L	3 T_U	4 T_L	4 T_U	5 T_L	5 T_U	6 T_L	6 T_U	7 T_L	7 T_U	8 T_L	8 T_U	9 T_L	9 T_U	10 T_L	10 T_U
3	5	16	6	18	6	21	7	23	7	26	8	28	8	31	9	33
4	6	18	11	25	12	28	12	32	13	35	14	38	15	41	16	44
5	6	21	12	28	18	37	19	41	20	45	21	49	22	53	24	56
6	7	23	12	32	19	41	26	52	28	56	29	61	31	65	32	70
7	7	26	13	35	20	45	28	56	37	68	39	73	41	78	43	83
8	8	28	14	38	21	49	29	61	39	73	49	87	51	93	54	98
9	8	31	15	41	22	53	31	65	41	78	51	93	63	108	66	114
10	9	33	16	44	24	56	32	70	43	83	54	98	66	114	79	131

(b) $\alpha = .05$ One-Sided; $\alpha = .10$ Two-Sided

n_2 \ n_1	3 T_L	3 T_U	4 T_L	4 T_U	5 T_L	5 T_U	6 T_L	6 T_U	7 T_L	7 T_U	8 T_L	8 T_U	9 T_L	9 T_U	10 T_L	10 T_U
3	6	15	7	17	7	20	8	22	9	24	9	27	10	29	11	31
4	7	17	12	24	13	27	14	30	15	33	16	36	17	39	18	42
5	7	20	13	27	19	36	20	40	22	43	24	46	25	50	26	54
6	8	22	14	30	20	40	28	50	30	54	32	58	33	63	35	67
7	9	24	15	33	22	43	30	54	39	66	41	71	43	76	46	80
8	9	27	16	36	24	46	32	58	41	71	52	84	54	90	57	95
9	10	29	17	39	25	50	33	63	43	76	54	90	66	105	69	111
10	11	31	18	42	26	54	35	67	46	80	57	95	69	111	83	127

Source: Data from F. Wilcoxon and R. A. Wilcox, "Some Rapid Approximate Statistical Procedures" (New York: American Cyanamid Company, 1964), pp. 20–23.

TABLE A.15 A Wilcoxon Signed Ranks Table: Values of T_0

One-Sided	Two-Sided	$n = 5$	$n = 6$	$n = 7$	$n = 8$	$n = 9$	$n = 10$
$\alpha = .05$	$\alpha = .10$	1	2	4	6	8	11
$\alpha = .025$	$\alpha = .05$		1	2	4	6	8
$\alpha = .01$	$\alpha = .02$			0	2	3	5
$\alpha = .005$	$\alpha = .01$				0	2	3

One-Sided	Two-Sided	$n = 11$	$n = 12$	$n = 13$	$n = 14$	$n = 15$	$n = 16$
$\alpha = .05$	$\alpha = .10$	14	17	21	26	30	36
$\alpha = .025$	$\alpha = .05$	11	14	17	21	25	30
$\alpha = .01$	$\alpha = .02$	7	10	13	16	20	24
$\alpha = .005$	$\alpha = .01$	5	7	10	13	16	19

One-Sided	Two-Sided	$n = 17$	$n = 18$	$n = 19$	$n = 20$	$n = 21$	$n = 22$
$\alpha = .05$	$\alpha = .10$	41	47	54	60	68	75
$\alpha = .025$	$\alpha = .05$	35	40	46	52	59	66
$\alpha = .01$	$\alpha = .02$	28	33	38	43	49	56
$\alpha = .005$	$\alpha = .01$	23	28	32	37	43	49

One-Sided	Two-Sided	$n = 23$	$n = 24$	$n = 25$	$n = 26$	$n = 27$	$n = 28$
$\alpha = .05$	$\alpha = .10$	83	92	101	110	120	130
$\alpha = .025$	$\alpha = .05$	73	81	90	98	107	117
$\alpha = .01$	$\alpha = .02$	62	69	77	85	93	102
$\alpha = .005$	$\alpha = .01$	55	61	68	76	84	92

One-Sided	Two-Sided	$n = 29$	$n = 30$	$n = 31$	$n = 32$	$n = 33$	$n = 34$
$\alpha = .05$	$\alpha = .10$	141	152	163	175	188	201
$\alpha = .025$	$\alpha = .05$	127	137	148	159	171	183
$\alpha = .01$	$\alpha = .02$	111	120	130	141	151	162
$\alpha = .005$	$\alpha = .01$	100	109	118	128	138	149

One-Sided	Two-Sided	$n = 35$	$n = 36$	$n = 37$	$n = 38$	$n = 39$
$\alpha = .05$	$\alpha = .10$	214	228	242	256	271
$\alpha = .025$	$\alpha = .05$	195	208	222	235	250
$\alpha = .01$	$\alpha = .02$	174	186	198	211	224
$\alpha = .005$	$\alpha = .01$	160	171	183	195	208

One-Sided	Two-Sided	$n = 40$	$n = 41$	$n = 42$	$n = 43$	$n = 44$	$n = 45$
$\alpha = .05$	$\alpha = .10$	287	303	319	336	353	371
$\alpha = .025$	$\alpha = .05$	264	279	295	311	327	344
$\alpha = .01$	$\alpha = .02$	238	252	267	281	297	313
$\alpha = .005$	$\alpha = .01$	221	234	248	262	277	292

One-Sided	Two-Sided	$n = 46$	$n = 47$	$n = 48$	$n = 49$	$n = 50$
$\alpha = .05$	$\alpha = .10$	389	408	427	446	466
$\alpha = .025$	$\alpha = .05$	361	379	397	415	434
$\alpha = .01$	$\alpha = .02$	329	345	362	380	398
$\alpha = .005$	$\alpha = .01$	307	323	339	356	373

Source: Data from F. Wilcoxon and R. A. Wilcox, "Some Rapid Approximate Statistical Procedures" (New York: American Cyanamid Company, 1964), p. 28.

TABLE A.16 Critical Values for Spearman's Rank Correlation Coefficient

n	α = .05	α = .025	α = .01	α = .005	n	α = .05	α = .025	α = .01	α = .005
5	.900	—	—	—	18	.399	.476	.564	.625
6	.829	.886	.943	—	19	.388	.462	.549	.608
7	.714	.786	.893	—	20	.377	.450	.534	.591
8	.643	.738	.833	.881	21	.368	.438	.521	.576
9	.600	.683	.783	.833	22	.359	.428	.508	.562
10	.564	.648	.745	.794	23	.351	.418	.496	.549
11	.523	.623	.736	.818	24	.343	.409	.485	.537
12	.497	.591	.703	.780	25	.336	.400	.475	.526
13	.475	.566	.673	.745	26	.329	.392	.465	.515
14	.457	.545	.646	.716	27	.323	.385	.456	.505
15	.441	.525	.623	.689	28	.317	.377	.448	.496
16	.425	.507	.601	.666	29	.311	.370	.440	.487
17	.412	.490	.582	.645	30	.305	.364	.432	.478

Source: Data from E. G. Olds, "Distribution of Sums of Squares of Rank Differences for Small Samples," *Annals of Mathematical Statistics*, 1938, p. 9.

TABLE A.17 A Table of Areas under the Standard Normal Curve

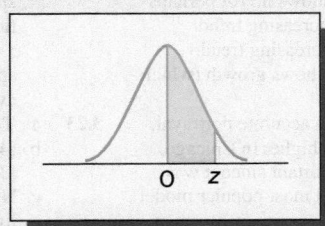

z	.00	.01	.02	.03	.04	.05	.06	.07	.08	.09
0.0	.0000	.0040	.0080	.0120	.0160	.0199	.0239	.0279	.0319	.0359
0.1	.0398	.0438	.0478	.0517	.0557	.0596	.0636	.0675	.0714	.0753
0.2	.0793	.0832	.0871	.0910	.0948	.0987	.1026	.1064	.1103	.1141
0.3	.1179	.1217	.1255	.1293	.1331	.1368	.1406	.1443	.1480	.1517
0.4	.1554	.1591	.1628	.1664	.1700	.1736	.1772	.1808	.1844	.1879
0.5	.1915	.1950	.1985	.2019	.2054	.2088	.2123	.2157	.2190	.2224
0.6	.2257	.2291	.2324	.2357	.2389	.2422	.2454	.2486	.2517	.2549
0.7	.2580	.2611	.2642	.2673	.2704	.2734	.2764	.2794	.2823	.2852
0.8	.2881	.2910	.2939	.2967	.2995	.3023	.3051	.3078	.3106	.3133
0.9	.3159	.3186	.3212	.3238	.3264	.3289	.3315	.3340	.3365	.3389
1.0	.3413	.3438	.3461	.3485	.3508	.3531	.3554	.3577	.3599	.3621
1.1	.3643	.3665	.3686	.3708	.3729	.3749	.3770	.3790	.3810	.3830
1.2	.3849	.3869	.3888	.3907	.3925	.3944	.3962	.3980	.3997	.4015
1.3	.4032	.4049	.4066	.4082	.4099	.4115	.4131	.4147	.4162	.4177
1.4	.4192	.4207	.4222	.4236	.4251	.4265	.4279	.4292	.4306	.4319
1.5	.4332	.4345	.4357	.4370	.4382	.4394	.4406	.4418	.4429	.4441
1.6	.4452	.4463	.4474	.4484	.4495	.4505	.4515	.4525	.4535	.4545
1.7	.4554	.4564	.4573	.4582	.4591	.4599	.4608	.4616	.4625	.4633
1.8	.4641	.4649	.4656	.4664	.4671	.4678	.4686	.4693	.4699	.4706
1.9	.4713	.4719	.4726	.4732	.4738	.4744	.4750	.4756	.4761	.4767
2.0	.4772	.4778	.4783	.4788	.4793	.4798	.4803	.4808	.4812	.4817
2.1	.4821	.4826	.4830	.4834	.4838	.4842	.4846	.4850	.4854	.4857
2.2	.4861	.4864	.4868	.4871	.4875	.4878	.4881	.4884	.4887	.4890
2.3	.4893	.4896	.4898	.4901	.4904	.4906	.4909	.4911	.4913	.4916
2.4	.4918	.4920	.4922	.4925	.4927	.4929	.4931	.4932	.4934	.4936
2.5	.4938	.4940	.4941	.4943	.4945	.4946	.4948	.4949	.4951	.4952
2.6	.4953	.4955	.4956	.4957	.4959	.4960	.4961	.4962	.4963	.4964
2.7	.4965	.4966	.4967	.4968	.4969	.4970	.4971	.4972	.4973	.4974
2.8	.4974	.4975	.4976	.4977	.4977	.4978	.4979	.4979	.4980	.4981
2.9	.4981	.4982	.4982	.4983	.4984	.4984	.4985	.4985	.4986	.4986
3.0	.4987	.4987	.4987	.4988	.4988	.4989	.4989	.4989	.4990	.4990

Source: Data from A. Hald, *Statistical Tables and Formulas* (New York: Wiley, 1952), abridged from Table 1.

ANSWERS TO MOST ODD-NUMBERED EXERCISES

Chapter 1

1.3 (2) cross-sectional
(3) time series
1.7 $494,000; $447,000
1.13 a. 33,276; 3,427; 8,178; 51,259; 60,268; 58,586; 9,998; 14,346; 24,200; and 7,351.
b. Most should fall between 36 and 48. We estimate $46/65 = 70.8\%$ of scores will be at least 42.
1.15 No, since this is a voluntary response sample, it is probably not representative of the population of all viewers.
1.23 Ordinal; nominative; ordinal; nominative; ordinal; nominative

Chapter 2

2.5 a. 144
b. 36
2.7 a.

Rating	Frequency	Relative Frequency
Outstanding	14	$14/30 = .467$
Very Good	10	$10/30 = .333$
Good	5	$5/30 = .167$
Average	1	$1/30 = .033$
Poor	0	$0/30 = .000$
	30	1.000

2.21 a. Between 40 and 46.
b. Slightly skewed with a tail to the left.
2.23 a. Between 45 and 60.
b. Roughly symmetric and mound-shaped.
2.25 a. Sales appear to cluster into two groups, one centered near $175M and the other in the high $700M's.
b. Growth values are skewed with a tail to the right.
2.31 Most growth rates are no more than 71%, but 4 companies had growth rates of 87% or more.
2.39 Payment times are slightly right skewed, while ratings are strongly left skewed.
2.41 In general, Ruth hit more home runs than Maris. 1961 was a highly unusual year for Maris.
2.43 b. Ratings are slightly left skewed.
c. No. 19 of 65 customers (29.2%) gave scores below 42.
2.47 a. 17
b. 14
c. If you have purchased Rola previously, you are more likely to prefer Rola. If you have not purchased Rola previously, you are more likely to prefer Koka.

2.49 a. 22
b. 4
c. People who drink more cola seem more likely to prefer Rola.
2.51 b. 1st row: 79.7%; 20.3%; 100% 2nd row: 65.8%; 34.2%; 100%
c. 1st column: 50.2%; 49.8%; 100% 2nd column: 33.0%; 67.0%; 100%
d. Viewers who think TV violence has increased are more likely to think TV quality has declined.
2.55 There is a positive linear association between copiers and minutes.
2.61 a. No, the graph shows no (or perhaps a very slight) increasing trend.
b. Yes, a strong increasing trend.
c. The line graph shows growth (which did occur).
d. Neither gives an accurate portrayal.
2.65 The ozone level is higher in Chicago.
2.69 Liberty is still important since it was the dealer's second most popular model in 2011.
2.71 The majority of the cars (23 of 32) received a design rating of 3.
2.73 All three regions have a majority of models rated 3 in design. Europe has the only models with ratings of 5, but like the Pacific Rim, it also has models rated 2, unlike the US.
2.75 The US and Europe have similar mechanical quality ratings, although the US ratings are slightly lower than the European ones. The Pacific Rim cars overwhelmingly receive ratings of 3.
2.77 See 2.73.
2.81 e. 72%
f. 50%
2.83 Distribution has one high outlier and, with or without the outlier, is skewed right.

Chapter 3

3.3 a. 9.6, 10, 10
b. 103.33, 100, 90
3.5 a. Yes, $\bar{x} = 42.954 > 42$.
b. $\bar{x} <$ median $= 43$ since there is a slight left skew.
3.7 a. Yes, $\bar{x} = 52.167 > 50$.
b. Median $= 52 \approx \bar{x}$ since the distribution is almost symmetric.
3.9 The graphs show revenue is right skewed.
3.11 The graphs show expenses are slightly right skewed.
3.13 The graphs show incomes are skewed.
3.15 a. They would argue that the mean team income is $6.09 million, so a hard salary cap is not needed.

b. 13 of the teams made money, while 17 of them lost money.
c. The owners would cite that 17 teams lost money and that the median income is $-$2.3 million to argue that the teams need a hard salary cap.
3.19 range $= 10$; $\sigma^2 = 11.6$; $\sigma = 3.4059$
3.21 a. OSU: $z = 1.324$ (OSU's total spending is 1.324 standard deviations above average); . . . ; Michigan: $z = -.775$ (Michigan's total spending is .775 std devs below average)
b. OSU: $z = 2.389$ (OSU's scholarship spending is 2.389 standard deviations above average); . . . ; Michigan: $z = -0.681$ (Michigan's scholarship spending is .681 std devs below average)
3.23 a. The rule is appropriate.
b. [46.773, 57.561]; [41.379, 62.955]; [35.985, 68.349]
c. No, since 45 is greater than the lower limit of 35.985.
d. 66.67%, 96.67%, and 100% fall within the intervals, suggesting our inferences are valid.
3.25 a. The rule is reasonably appropriate.
b. [40.3076, 45.5924]; [37.6652, 48.2348]; [35.0228, 50.8772]
c. Yes, since its lower limit exceeds 35.
d. 63%; 98.46% ; and 100% > Yes.
3.27 a. [$-$72.99, 94.85]; [$-$5.72, 31.72]; [$-$47.87, 116.77]
c. 383.9%; 72.0%; 119.5%. Fund 1 is the riskiest; Fund 2 is least risky.
3.31 a. 192
b. 152
c. 141
d. 171
e. 132
f. 30
3.33 a. 10, 15, 17, 21, 29
b. 20, 29, 31, 33, 35
c. Payment times are right skewed with no outliers; design ratings are left skewed with 2 low outliers.
3.35 The top 75% of the 30-year rates are higher than the bottom 75% of the 15-year rates. The variability is similar.
3.39 a. It is a strong positive linear relationship.
b. $\hat{y} = 134.4751$ minutes.
3.43 2.1
3.45 a. 4.6 lb
b. 3.829
3.47 51.5; 81.6102; 9.0338
3.51 .4142
3.53 a. $R_g = -4.207\%$
b. $84,204.46

3.57 a. A to renters of B; B to renters of A; B to renters of C; C to renters of B; A to renters of B & C; C to renters of A & C; C to renters of A & B; A and C to renters of B; B and C to renters of A; B to renters of C & E; C to renters of B & E. (Require lift > 1.)

b. (i) The support for C and E is 5 since 5 of the customers rented both movies. The support for C, E, and B is 4 since 4 of the customers rented all three movies.
(ii) Of the 5 customers who rented both C and E, 4 also rented B. Therefore the confidence percentage is 4/5 = 80%.
(iii) The lift ratio is the confidence percentage divided by support percentage for the recommended movie = .80/.70 = 1.143

3.59 Text mining is the science of discovering knowledge, insights, and patterns from a collection of textual documents or databases.

3.61 Factor 1: What people like most about owning a cat; Factor 2: Having dogs pull sleds in the snow.

3.63 a. Majority of food types above Axis I are all Asian cuisines, while those below Axis I are all Western cuisines. Food types increase in spicy rating moving left to right.

b. The food types appear to cluster into five groups: Asian, not spicy and moderate in calories; American and Cantonese, bland, heavy, and high in calories; Szechuan, highly spicy but light; European and Mediterranean, spicy and hearty; Mexican, spicy yet lighter.

c. Choose restaurants belonging to as many different clusters as possible.

3.67 Staff quality; merchandise pricing; store's attractiveness; merchandise display; store size.

3.69 A variable's communality is the percentage of the variance of that variable that is explained by the factors in the model.

3.71 The median marketing base salary is higher than the median research base salary, and the variability of base marketing salary is much greater than the variability of base research salary.

3.73 c. The distribution of revenue is skewed to the right with two high outliers. It is still skewed to the right if the outliers are removed. The distribution of number of employees is skewed to the right with three high outliers. Without the outliers, the distribution is almost symmetric.

Chapter 4

4.3 b1. *AA*
b2. *AA, BB, CC*
b3. *AB, AC, BA, BC, CA, CB*
b4. *AA, AB, AC, BA, CA*
b5. *AA, AB, BA, BB*
c. 1/9, 1/3, 2/3, 5/9, 4/9

4.5 b1. *PPPN, PPNP, PNPP, NPPP*
b2. Outcomes with ≤2 *P*'s (11)
b3. Outcomes with ≥1 *P* (15)
b4. *PPPP, NNNN*
c. 1/4, 11/16, 15/16, 1/8

4.7 .15
4.11 a1. .25
a2. .40
a3. .10
b1. .55
b2. .45
b3. .45
4.13 a. 5/8
b. 21/40
c. 19/40
d. 3/8
e. 31/40
4.15 .343
4.19 a. .6
b. .4
c. Dependent
4.21 .55
4.23 .1692
4.25 .31
4.27 b. .40
c. Yes, $P(FRAUD|FIRE) = P(FRAUD)$
4.29 a. .874
b. .996
c. .004
4.31 .999946
4.33 a. .0295
b. .9705
c. Probably not
4.37 .0976; .6098; .2927
4.39 a. .0892
b. No. Too many paying customers would lose credit.
4.41 .1549, .8318, .0133
4.43 .2466, .6164, .1370
4.51 .001
4.53 1/56
4.55 1/9; 1/9; 4/9
4.57 .04; .56; .26; .32
4.59 .9029
4.61 .9436
4.63 .721
4.65 .362
4.67 .502
4.69 Slight dependence
4.71 a. .2075
b. .25
c. .105
d. .42
e. Yes since P(bonus) < P(bonus|training)
4.73 a. .1860
b. It helps the case since .186 > .015
c. .625
d. Much more support than before since $P(S \mid A) = 0.625 > P(S) = 0.10$
e. A fairly strong case since $P(S \mid A) = 0.8333 > P(S) = 0.25$

Chapter 5

5.1 Probability; if the estimated probability is above the preselected threshold, then the item is assigned to that specific class.
5.3 a. The overall misclassification rate for the entire model = 6/40 or 15%

b. Customer 41: Prob(Coupon = 0) = .0397 and Prob(Coupon = 1) = .9603; "Most Likely Coupon = 1" Customer 42: Prob(Coupon = 0) = .5972 and Prob(Coupon = 1) .4028; "Most Likely Coupon = 0" Send Customer 41 a catalog with a coupon.

5.5 b. Male: .8902 or 89%; Female: .2229 or 22%
5.7 Predicted demand in sales period 31: 8.8411 hundred thousand bottles; Predicted demand in sales period 32: 8.0271 hundred thousand bottles.
5.11 a. $k = 3$
b. 2/3 of customer 4's nearest neighbors redeemed their coupons
c. Both are classified as "redeemers"
5.13 Averaging the sales prices of the seven nearest Toyota Corollas in the training data set to Toyota Corolla #1437
5.15 a. Customer 41: Naive Prob 1 = .9346, Naive Prob 0 = .0654; Customer 42: Naive Prob 1 = .5306, Naive Prob 0 = .4694.
b. Redeemers
5.17 3.2581

Chapter 6
6.3 a. Discrete
b. Discrete
c. Continuous
d. Discrete
e. Discrete
f. Continuous
g. Continuous
6.5 $p(x) \geq 0$, each x
$\sum_{all\ x} p(x) = 1$
6.9 $\mu_x = 2.1$, $\sigma_x^2 = 1.49$, $\sigma_x = 1.22$
6.11 b. $500
6.13 a.

x	$p(x)$
$400	.995
−$49,600	.005

b. $150
c. $1,250
6.15 −$4.20
6.17 3.86
6.21 a. $p(x) = \frac{5!}{x!\,(5-x)!}(.3)^x(.7)^{5-x}$
$x = 0, 1, 2, 3, 4, 5$
c. .1323
d. .9692
e. .8369
f. .0308
g. .1631
h. $\mu_x = 1.5$, $\sigma_x^2 = 1.05$, $\sigma_x = 1.024695$
i. [−.54939, 3.54939], .9692
6.23 a. $p(x) = \frac{15!}{x!\,(15-x)!}(.9)^x(.1)^{15-x}$
b1. .4509
b2. .9873
b3. .5491
b4. .1837
b5. .0022
c. No, $P(x \leq 9)$ is very small

6.25 a1. .0625
a2. .3125
b1. .4119
b2. .2517
b3. .0059
c. No, $P(x < 5)$ is very small
6.27 a. .9996, .0004
b. .4845, .5155
c. $p = 1/35$
d. .000040019
6.31 a. $\mu_x = 2, \sigma_x^2 = 2,$
$\sigma_x = 1.414$
b. $[-.828, 4.828], .9473$
$[-2.242, 6.242], .9955$
6.33 a. .7852
b. .2148
6.35 a. Approximately zero
b. Rate of comas unusually high.
6.39 a. 0
b. .0714
c. .4286
d. .4286
e. .0714
f. .9286
g. .5
h. .9286
6.41 a. .1273   b. .8727
6.43 $\dfrac{\binom{160}{6}\binom{40}{1}}{\binom{200}{7}} + \dfrac{\binom{160}{7}\binom{40}{0}}{\binom{200}{7}} \approx .5767$
6.47 a. $\mu_x = 10, \sigma_x = 1.077,$
$\mu_y = 10, \sigma_y = 1.095, \sigma_{xy}^2 = .6$
b. $\sigma_p^2 = .89, \sigma_p = .943$
6.49 b. 87,000
c. 75%
d. [46,454, 127,546]; 95%
6.51 a. .7373
b1. .01733
b2. .42067
b3. .61291
b4. .02361
c. No. The probability is very small.
6.53 a. .2231
b. .9344
c. .9913
d. .0025
6.55 .0025. Claim is probably not true.
6.57 .0037. Business failures are probably increasing.
6.59 .3328. The claim seems reasonable.

Chapter 7
7.7 $h = 1/125$
7.9 a. 3, 3, 1.73205
b. [1.268, 4.732], .57733
7.11 a. $f(x) = 1/20$ for $120 \le x \le 140$
c. .5
d. .25
7.13 $c = 1/6$
7.15 1.0
7.19 a. -1, one σ below μ
b. -3, three σ below μ
c. 0, equals μ
d. 2, two σ above μ
e. 4, four σ above μ
7.21 a. 2.33
b. 1.645
c. 2.05

d. -2.33
e. -1.645
f. -1.28
7.23 a. 696   f. 335.5
b. 664.5   g. 696
c. 304   h. 700
d. 283   i. 300
e. 717
7.25 a1. .9830
a2. .0033
a3. .0456
b. 947
7.27 a. .0013
b. Claim probably not true
7.29 .0424
7.31 a. 10%, 90%, -13.968
b. $-1.402, 26.202$
7.33 a. A: .3085
B: .4013
B is investigated more often
b. A: .8413
B: .6915
A is investigated more often
c. B
d. Investigate if cost variance exceeds $5,000; .5987
7.35 $\mu = 700, \sigma = 100$
7.37 Both np and $n(1 - p) \ge 5$
7.39 a. $np = 80$ and $n(1 - p) = 120$
both ≥ 5
b1. .0558
b2. .9875
b3. .0125
b4. .0025
b5. .0015
7.41 a1. $np = 200$ and $n(1 - p) = 800$
both ≥ 5
a2. 200, 12.6491
a3. Less than .001
b. No
7.43 a. Less than .001
b. No
7.49 a. $3e^{-3x}$ for $x \ge 0$
c. .9502
d. .4226
e. .0025
f. 1/3, 1/9, 1/3
g. .9502
7.51 a. $(2/3)e^{-(2/3)x}$ for $x \ge 0$
c1. .8647
c2. .2498
c3. .0695
c4. .2835
7.53 a1. .1353
a2. .2325
a3. .2212
b. Probably not, probability is .2212
7.55 That the data come from a normal population.
7.59 .0062
7.61 a. .8944   b. 73
7.63 a. .8944
b. .7967
c. .6911
7.65 298
7.67 .9306
7.71 2/3
7.73 a. .0062
b. .6915
c. 3.3275%
7.75 .7745

Chapter 8
8.3 a. 10, .16, .4
b. 500, .0025, .05
c. 3, .0025, .05
d. 100, .000625, .025
8.5 a. Normally distributed
No, sample size is large (≥ 30)
b. $\mu_{\bar{x}} = 20, \sigma_{\bar{x}} = .5$
c. .0228
d. .1093
8.7 30, 40, 50, 50, 60, 70
8.9 2/3
8.11 a. Normal distribution because $n \ge 30$
b. 6, .247
c. .0143
d. 1.43%, conclude $\mu < 6$
8.13 a. .2206
b. .0027
c. Yes
8.19 a. .5, .001, .0316
b. .1, .0009, .03
c. .8, .0004, .02
d. .98, .0000196, .004427
8.21 a. Approximately normal
b. .9, .03
c. .0228
d. .8664
e. .6915
8.23 a. .0122
b. Yes.
8.25 No; yes.
8.27 a. .0294
b. Yes.
8.29 $\sigma_{\bar{x}} = 11.63\%$; $[-26.76, 19.76]$
8.31 a. $\mu = 0.55$; $\sigma = 0.8646$
b. Approximately normal with mean 0.55 and $\sigma_{\bar{x}} = 0.08646$
c. 0.0104

Chapter 9
9.5 It becomes shorter.
9.7 a. [50.2343, 54.099]; [49.6272, 54.7062]
b. Yes. All values in the interval exceed 50.
c. No. Some values in the interval are below 50.
9.9 a. [42.31, 43.59]; [42.107, 43.793]
b. Yes. All values in the interval exceed 42.
c. Yes. All values exceed 42.
9.11 a. [76.132, 89.068]
b. [85.748, 100.252]
c. The intervals overlap, so there is some doubt, that the means differ.
9.15 1.363, 2.201, 4.025
1.440, 2.447, 5.208
9.17 a. [3.442, 8.558]
9.19 a. [6.832, 7.968]
b. Yes, 95% interval is below 8.
9.21 a. [786.609, 835.391]
b. Yes, 95% interval is above 750
9.23 [4.969, 5.951]; Yes
9.29 a. $n = 262$
b. $n = 452$
9.31 a. $n = 47$
b. $n = 328$
9.33 $n = 54$
9.35 a. $p = .5$
b. $p = .3$
c. $p = .8$

9.37 Part a. [.304, .496],
[.286, .514],
[.274, .526]
Part b. [.066, .134],
[.060, .140],
[.055, .145]
Part c. [.841, .959],
[.830, .970],
[.823, .977]
Part d. [.464, .736],
[.439, .761],
[.422, .778]

9.39 a. [.570, .630]
b. Yes, the interval is above 0.5.

9.43 a. [.611, .729]
b. Yes, interval above .6

9.45 [.264, .344]
Yes, 95% interval exceeds .20.

9.47 $n = 1426$

9.51 a. $532; [$514.399, $549.601];
$5,559,932;
[$5,375,983.95, $5,743,880.05]
b. Claim is very doubtful

9.53 a. [63.59, 72.49]
b. [52.29, 61.19]
c. Yes. The interval for financial firms is below the one for industrial firms.

9.55 a. [56.47, 59.13]. Yes, the interval is below 60.
b. $n = 144$

9.57 a. [25.1562, 27.2838]
b. Yes, not much more than 25

Chapter 10

10.3 a. $H_0: \mu \leq 42$ versus $H_a: \mu > 42$
b. Type I: decide $\mu > 42$ when it isn't. Type II: decide $\mu \leq 42$ when it isn't.

10.5 a. $H_0: \mu = 16$ oz.; $H_a: \mu \neq 16$ oz.
b. Type I: Reject H_0 and decide to incur the cost of adjusting the filler when it is not necessary; Type II: Do not reject H_0 and do not adjust the filler even though we should.

10.13 a. $H_0: \mu \leq 42$ versus $H_a: \mu > 42$
b. $z = 2.91$. Since this exceeds the critical values 1.28, 1.645, and 2.33, can reject H_0 at $\alpha = .1, .05,$ and .01. Fail to reject H_0 at $\alpha = .001$.
c. p-value $= .002$. Same conclusion as part (b)
d. Very strong

10.15 a. $H_0: \mu \leq 60$ versus $H_a: \mu > 60$
b. $z = 2.41$; p-value $= .008$. Since $z > 1.645$ and p-value $< .05$, reject H_0 and shut down

10.17 $z = 3.09$ and p-value $= .001$. Since $z > 1.645$ and p-value $< .05$, shut down the plant.

10.19 a. $H_0: \mu = 16$ versus $H_a: \mu \neq 16$
b. $\bar{x} = 16.05$: (1) $z = 3.00$;
(2) p-value $= .003$;
(3) critical values ± 2.575;
(4) [16.007, 16.093]; reject H_0 and decide to readjust.

$\bar{x} = 15.96$: (1) $z = -2.40$;
(2) p-value $= .016$;
(3) critical values ± 2.575;
(4) [15.917, 16.003]; fail to reject H_0 so don't readjust.

$\bar{x} = 16.02$: (1) $z = 1.20$;
(2) p-value $= .230$;
(3) critical values ± 2.575;
(4) [15.977, 16.063]; fail to reject H_0 so don't readjust.

$\bar{x} = 15.94$: (1) $z = -3.60$;
(2) p-value $= .000$;
(3) critical values ± 2.575;
(4) [15.897, 15.983]; reject H_0 and decide to readjust.

10.23 $t = -4.30$; reject H_0 at .10, .05, and .01 but not at .001.

10.25 (1) $t = 2.2 > 1.699 = t_{.05}$ (for $df = 29$)
(2) Reject H_0; *strong* evidence that the mean breaking strength is greater than 50.

10.27 a. (1) $H_0: \mu \leq 3.5$ versus $H_a: \mu > 3.5$
(2) Type I: decide Ohio mean higher when it isn't. Type II: decide Ohio mean isn't higher when it is.
b. (1) $t = 3.62 > 3.143$; reject H_0
(2) The p-value provides very strong evidence for $\mu > 3.5$.

10.29 a. $H_0: \mu \leq 42$ versus $H_a: \mu > 42$
b. (1) $t = 2.899 > 2.386$; reject H_0.
(2) The p-value provides very strong evidence for $\mu > 42$.

10.31 a. $H_0: \mu = 750$ versus $H_a: \mu \neq 750$
b. (1) $t = 6.94 > 4.604$; reject H_0.
(2) The p-value provides very strong evidence for $\mu \neq 750$.

10.33 Since $t = -4.97$ and p-value $= .000$, there is extremely strong evidence that $\mu < 18.8$

10.37 a. $H_0: p \leq .5$ versus $H_a: p > .5$
b. $z = 1.19$. Do not reject H_0 at any α. There is little evidence.

10.39 a. $H_0: p \leq .18$ versus $H_a: p > .18$
b. (1) p-value $= .0329$.
(2) Reject H_0 at $\alpha = .10$ and .05 but not at .01 or .001.
(3) There is strong evidence.
c. Possibly.

10.41 (1) $H_0: p = .73$ versus $H_a: p \neq .73$
(2) $z = -.80$ and p-value $= .4238$ provide insufficient evidence to reject H_0 at any α.

10.45 a. .9279; .8315; .6772; .4840; .2946; .1492; .0618; .0207; .0055; .00118
b. No. Must increase n.
c. The power increases.

10.47 246

10.51 $\chi^2 = 11.62 < 17.7083$. Reject H_0.

10.53 (1) $\chi^2 = 6.72 < 13.8484$. Reject H_0.
(2) [.000085357, .000270944]; [.009239, .016460].
(3) $\mu \pm 3\sigma = 3 \pm 3(.0165) =$ [2.9505, 3.0495]. Yes, this is inside the specification limits.

10.55 $\chi^2 = 16.0286$. Cannot reject H_0 at either .05 or .01.

10.57 a. $H_0: \mu \geq 25$ versus $H_a: \mu < 25$
b. $t = -2.63$; reject H_0 at all α's except .001.
c. Very strong evidence.
d. Yes, if you believe the standard should be met.

10.59 a. $t = 2.50$. Reject H_0 at $\alpha = .10, .05, .01$ but not .001; very strong.
b. $t = 1.11$. Do not reject H_0 at any α.

10.61 a. Reject H_0 at $\alpha = .10$ and .05 but not at .01 or .001.
b. Strong evidence

10.63 (1) $z = 5.81$; p-value $< .001$
(2) Extremely strong evidence
(3) Yes; sales would increase by about 50%

Chapter 11

11.5 $t = 3.39$; reject H_0 at all α's except .001. There is very strong evidence that $\mu_1 - \mu_2 > 20$.

11.7 95% CI: [23.50, 36.50]. Yes, the entire interval is above 20.
Upper tail test: $t = 3.39$ ($df = 11$) Reject H_0 at all α's except .001. Very strong evidence that $\mu_1 - \mu_2 > 20$.
Two tail test: $t = 3.39$. Reject H_0 at all α's except .001. There is very strong evidence that $\mu_1 - \mu_2 \neq 20$.

11.9 a. $H_0: \mu_1 - \mu_2 \leq 0$ versus $H_a: \mu_1 - \mu_2 > 0$
b. $t = 1.97$. Reject H_0 at $\alpha = .10$ and .05 but not .01 or .001. Strong evidence.
c. [−12.01, 412.01]. A's mean could be anywhere from $12.01 lower to $412.01 higher than B's.

11.11 a. $H_0: \mu_1 - \mu_2 = 0$ versus $H_a: \mu_1 - \mu_2 \neq 0$
b. Reject H_0 at $\alpha = .10, .05$ but not .01 or .001; strong evidence
c. [$1.10, $100.90]

11.17 a. [100.141, 106.859]; [98.723, 108.277]
b. $t = 2.32$; reject H_0 at $\alpha = .05$ but not .01; strong
c. $t = -4.31$; reject H_0 at $\alpha = .05, .01$; extremely strong.

11.19 a. $t = 6.18$; decide there is a difference.
b. A 95% confidence interval is [2.01, 4.49], so we can estimate the minimum to be 2.01 and the maximum to be 4.49.

11.21 a. $H_0: \mu_d = 0$ versus $H_a: \mu_d \neq 0$
b. $t = 3.89$; reject H_0 at all α except .001; yes
c. p-value $= .006$; reject H_0 at all α except .001; very strong evidence

11.25 $z = -10.14$; reject H_0 at each value of α; extremely strong evidence

11.27 a. $H_0: p_1 - p_2 = 0$ versus $H_a: p_1 - p_2 \neq 0$
b. $z = 3.63$; reject H_0 at each value of α
c. $H_0: p_1 - p_2 \leq .05$ versus $H_a: p_1 - p_2 > .05$
$z = 1.99$ and p-value $= .0233$; strong evidence
d. [.0509, .1711]; yes

11.29 a. p-value $= .004$; very strong evidence
b. [−.057, −.011]; −.011

11.33 a. 3.34   e. 2.96
b. 3.22   f. 4.68
c. 3.98   g. 3.16
d. 4.88   h. 8.81

11.35 $F = 2.47 > 2.40$. Reject H_0 at $\alpha = .05$.

11.37 $F = 1.73 < 4.03$. Cannot claim $\sigma_1 \neq \sigma_2$. Yes, the assumption of equal variances is reasonable.

11.39 a. $t = 8.251$; reject $H_0: \mu_O - \mu_{JVC} = 0$ at $\alpha = .001$

b. [\$32.69, \$55.31]; probably

c. $t = 2.627$; reject $H_0: \mu_O - \mu_{JVC} \leq 30$ at $\alpha = .05$

11.41 a. $H_0: \mu_T - \mu_B = 0$ versus $H_a: \mu_T - \mu_B \neq 0$

$t = 1.54$; cannot reject H_0 at any value of α; little to no evidence.

b. [$-.09$, $.73$]

11.43 a. $H_0: \mu_d \leq 0$ versus $H_a: \mu_d > 0$

b. $t = 10.00$; reject H_0 at all levels of α

c. p-value $= .000$; reject H_0 at all levels of α; extremely strong evidence

Chapter 12 (Answers to several Even-Numbered Exercises also given)

12.1 Factor = independent variables in a designed experiment.
treatments = values of a factor (or combination of factors).
experimental units = entities to which treatments are assigned.
response variable = the dependent variable (or variable of interest).

12.3 Between-treatment variability measures differences in sample means, while within-treatment variability measures differences of measurements in the same samples.

12.5 a. $F = 184.57$, p-value $= .000$; reject H_0 and decide shelf height affects sales.

b. Point estimate of $\mu_M - \mu_B$ is 21.4; [17.681, 25.119], $\mu_T - \mu_B$: -4.3; [-8.019, $-.581$], $\mu_T - \mu_M$: -25.7; [-29.419, -21.981]. Middle.

c. μ_B: [53.65, 57.96]
μ_M: [75.04, 79.36]
μ_T: [49.34, 53.66]

12.7 a. $F = 43.36$, p-value $= .000$; reject H_0; designs affect sales

b. B − A: [11.56, 20.84]
C − A: [3.56, 12.84]
C − B: [-12.64, -3.36]
Design B.

c. μ_A: [13.92, 19.28]
μ_B: [30.12, 35.48]
μ_C: [22.12, 27.48]

12.8 $F = 16.42$; p-value $< .001$; reject H_0; brands differ

12.9 (1) Divot − Alpha: [38.41, 127.59]
Divot − Century: [50.21, 139.39]
Divot − Best: [-14.39, 74.79]
Century − Alpha: [-56.39, 32.79]
Century − Best: [-109.19, -20.01]
Best − Alpha: [8.29, 97.39]

12.11 a. $F = 127.22$, p-value $= .000$
Reject H_0. There are significant differences between mean mileages.

b. W − X: [2.19, 4.61]
W − Y: [-4.77, -2.35]
W − Z: [-5.16, -2.74]
X − Y: [-8.17, -5.75]
X − Z: [-8.56, -6.14]

Y − Z: [-1.60, 0.82]
Additives Y and Z give the highest mean mileages.

c. μ_W: [30.733, 32.007]
μ_X: [27.333, 28.607]
μ_Y: [34.293, 35.567]
μ_Z: [34.683, 35.957]

12.15 a. $F = 36.2258$; p-value $= .0003$; reject H_0; sales methods differ

b. $F = 12.8710$; p-value $= .0068$; reject H_0; salesman effects differ

c. Method 1 − Method 2: [-2.2895, 2.9562]
Method 1 − Method 3: [2.3372, 7.6229]
Method 1 − Method 4: [3.7105, 8.9562]
Method 2 − Method 3: [2.0438, 7.2895]
Method 2 − Method 4: [3.3772, 8.6229]
Method 3 − Method 4: [-1.2895, 3.9562]
Methods 1 and 2

12.17 a. $F = 441.75$ and p-value $= .000$; reject H_0; keyboard brand effects differ.

b. $F = 107.69$ and p-value $= .000$; reject H_0; specialist effects differ.

c. A − B: [8.55, 11.45]
A − C: [12.05, 14.95]
B − C: [2.05, 4.95]
Keyboard A

12.18 a. $F = 5.78$; p-value $= .0115$; reject H_0; soft drink brands affect sales.

b. Coke Classic − New Coke: [7.98, 68.02]
Coke Classic − Pepsi: [$-.22$, 59.82]
New Coke − Pepsi: [-38.22, 21.82]

c. Yes, mean sales of Coke Classic were significantly higher than those of New Coke.

12.19 A combination of a level of factor 1 and a level of factor 2.

12.20 See leftmost graph following Figure 12.10.

12.21 a. Plot suggests little interaction. $F = .66$ and p-value $= .681$; do not reject H_0. Conclude no interaction.

b. $F = 26.49$ and p-value $= .000$; reject H_0; display panel effects differ.

c. $F = 100.80$ and p-value $= .000$; reject H_0; emergency condition effects differ.

d. A − B: [.49, 5.91]
A − C: [-6.81, -1.39]
B − C: [-10.01, -4.59]

e. $\mu_1 - \mu_2$: $[(\bar{x}_1 - \bar{x}_2) \pm q_{.05} \cdot \sqrt{\frac{MSE}{3?2}}] =$

$[(17.17 - 24.50) \pm 4.20 \cdot \sqrt{\frac{4.125}{6}}] =$

$[-7.33 \pm 3.48] = [-10.81, -3.85]$

$\mu_1 - \mu_3$: [-18.48, -11.52]
$\mu_1 - \mu_4$: [.36, 7.32]
$\mu_2 - \mu_3$: [-11.15, -4.19]
$\mu_2 - \mu_4$: [7.69, 14.65]
$\mu_3 - \mu_4$: [15.36, 22.32]
$\mu_2 - \mu_3$: [-11.15, -4.19]
$\mu_2 - \mu_4$: [7.69, 14.65]

$\mu_3 - \mu_4$: [15.36, 22.32]

f. Panel B. No, there is no interaction.

g. [6.37, 12.63]

12.23 a. The graph indicates there is interaction between foreman and design. F(int) $= 24.73$ and p-value $= .001 < \alpha = .05$. Claim there is interaction. No, we can't. The effect of house design depends on the foreman.

b. Foreman 1 and design C lead to the highest profit. 95% CI: [17.72, 19.88]

Chapter 13

13.7 a. Each $E_i \geq 5$

b. $\chi^2 = 300.605$; reject H_0

13.9 a. $\chi^2 = 137.14$; reject H_0

b. Differences between brand preferences

13.11 a1. [$-\infty$, 10.185]
a2. [10.185, 14.147]
a3. [14.147, 18.108]
a4. [18.108, 22.069]
a5. [22.069, 26.030]
a6. [26.030, ∞]

b. 1.368, 8.154, 20.478, 20.478, 8.154, 1.368

c. Can use χ^2 test

e. 1, 9, 30, 15, 8, 2
$\chi^2 = 5.4997$

f. Fail to reject; normal

13.13 Fail to reject H_0; normal

13.17 a.

	40%
	60%
20%	80%

b.

16%	24%
40%	60%
80%	30%
4%	56%
6.67%	93.33%
20%	70%

c. $\chi^2 = 16.667$; reject H_0

d. Yes

13.19 a.

	24.24%
	22.73%
	53.03%
51.515%	48.485%

b. For Heavy/Yes cell: cell: 18.18%; row: 75%; column: 35.29%

c. $\chi^2 = 6.86$; fail to reject H_0

d. Possibly; can reject H_0 at $\alpha = .05$

13.21 a. $\chi^2 = 16.384$; reject H_0

b. [$-.216$, $-.072$]

13.23 1. $\chi^2 = 65.91$; reject H_0:
2. [.270, .376]

13.25 b. $\chi^2 = 20.941$; reject H_0

c. First time buyers prefer Japanese styling while repeat buyers prefer European.

13.27 $\chi^2 = 71.476$; reject H_0

Chapter 14

14.3 a. $b_0 = 15.84$, $b_1 = -.1279$. b_0 is the estimated mean consumption at $0°$. b_1 is the estimated mean change in consumption for each $1°$ increase in temperature. No; not a valid interpretation.

c. Both are 10.724 MMcF.

14.5 a. $b_0 = 11.4641$, $b_1 = 24.6022$. b_0 is the estimated mean service time when 0 copiers require service. No, it is not. b_1 is the estimated mean change in service time for each additional copier.

b. 109.9 minutes in each case.

14.9 Explained variation = 22.981; $r^2 = .8995$; $r = -.9484$

14.11 Explained variation = 19,918.8438; $r^2 = .9905$. $r = .9952$

14.15 $s^2 = .428$; $s = .6542$

14.17 $s^2 = 21.3002$; $s = 4.6152$

14.21 a. $b_0 = 15.8379$; $b_1 = -.1279$

b. Explained variation = 22.9808; SSE = 2.5679; total variation = 25.5488; $r^2 = .8995$; $s = .6542$

c. $s_{b_1} = .0175$; $t = -7.3277$

d. $t < -t_{.025} = -2.447$ Reject H_0. Yes, significant at .05.

e. p-value = .0003 Reject H_0 at all α's. There is extremely strong evidence.

f. $[-.1706, -.0852]$

g. $s_{b_0} = .8018$ and $t = 19.7535$

h. p-value = .0000 < every α. We have extremely strong evidence.

i. $SS_{xx} = 16874.8 - \dfrac{(351.8)^2}{8} = 1404.395$; $s_{b_0} = (.6542) \times \sqrt{\dfrac{1}{8} + \dfrac{(43.975)^2}{1404.365}} = .8018$; $s_{b_1} = \dfrac{.6542}{\sqrt{1404.395}} = .0175$

j. $F(\text{model}) = \dfrac{22.9808}{2.5679/6} = 53.6955 \approx 53.6949$

k. Since $F(\text{model}) > 5.99$, we reject H_0; the regression relationship is significant at $\alpha = 0.05$.

l. p-value = .0003. The regression relationship is significant at every α.

m. $t^2 = (-7.3277)^2 = 53.6952 \approx 53.6949$; $(t_{.025})^2 = (2.447)^2 = 5.9878$ or 5.99.

14.23 $b_0 = 11.4641$; $b_1 = 24.6022$

b. Explained variation = 19918.8438; SSE = 191.7017; total variation = 20,110.5455; $r^2 = .9905$; $s = 4.6152$

c. $s_{b_1} = 0.8045$; $t = 30.5802$

d. $t > 2.262$ Reject H_0. Yes, significant at 0.05.

e. p-value = .0000 Reject H_0 at all α's. There is extremely strong evidence.

f. [22.7824, 26.4220]

g. $s_{b_0} = 3.4390$ and $t = 3.3335$

h. p-value = .0087 < every α except 0.001. We have very strong evidence.

i. $SS_{xx} = 201 - \dfrac{(43)^2}{11} = 32.9091$; $s_{b_0} = (4.6152)\sqrt{\dfrac{1}{11} + \dfrac{(3.90909)^2}{32.9091}} = 3.4390$; $s_{b_1} = \dfrac{4.6152}{\sqrt{32.9091}} = .8045$

j. $F(\text{model}) = \dfrac{19,918.8438}{191.7017/9} = 935.149$

k. Since $F(\text{model}) > 5.12$, we can reject H_0; the regression relationship is significant at $\alpha = .05$.

l. p-value = .0000. The regression relationship is significant at every α.

m. $t^2 = (30.5802)^2 = 935.149$; $(t_{.025})^2 = (2.262)^2 = 5.12$.

14.25 a. $b_0 = 66.2121$; $b_1 = 4.4303$

b. Explained variation = 1619.2758; SSE = 222.8242; total variation = 1842.1; $r^2 = .8790$; $s = 5.2776$

c. $s_{b_1} = .5810$; $t = 7.6247$

d. $t > 2.306$ Reject H_0. Yes, significant at 0.05.

e. p-value < .0001 Reject H_0 at all α's. There is extremely strong evidence.

f. [3.0903, 5.7703]

g. $s_{b_0} = 5.767$ and $t = 11.4818$

h. p-value = .0000 < every α. We have extremely strong evidence.

i. $SS_{xx} = 985 - \dfrac{(95)^2}{10} = 82.5$; $s_{b_0} = (5.2776)\sqrt{\dfrac{1}{10} + \dfrac{(9.5)^2}{82.5}} = 5.7667$; $s_{b_1} = \dfrac{5.2776}{\sqrt{82.5}} = .5810$

j. $F(\text{model}) = \dfrac{1619.2758}{222.8242/8} = 58.1364$

k. Since $F(\text{model}) > 5.32$, we can reject H_0; the regression relationship is significant at $\alpha = .05$.

l. p-value = .0000. The regression relationship is significant at every α.

m. $t^2 = (7.6247)^2 = 58.136$; $(t_{.025})^2 = (2.306)^2 = 5.32$.

14.27 $b_1 = -6.442$; $[-7.1194, -5.7646]$

14.31 a. 8.0806; [7.9479, 8.2133]

b. 8.0806; [7.4187, 8.7425]

c. 99% CI for mean demand: (7.9016, 8.2597); 99% PI for individual demand: (7.1877, 8.9735)

14.33 a. 627.263; [621.054, 633.472]

b. 627.263; [607.032, 647.494]

c. Distance value = .104; 99% CI: [618.435, 636.097]; PI: [598.492, 656.040]

14.35 (1) $\hat{y} = 162.0304$; distance = .1071; [154.0400, 170.0208]

(2) $\hat{y} = 162.0304$; [136.3403, 187.7206]

14.39 $t = 7.644$; p-value < .0001, so conclude that $\rho \neq 0$ at all α's. Extremely strong evidence of a linear relationship.

14.43 No. There are no systematic patterns in the residual plots.

14.45 (1) The residual plot has a cyclical appearance; this suggests positive autocorrelation.

(2) Since $d = .473 < d_{L,.05} = 1.27$, conclude there is positive autocorrelation. Since $3.527 > d_{U,.05} = 1.45$, conclude that there is no negative autocorrelation.

14.47 The residuals fan out, indicating the variance is not constant.

14.49 (1) $\hat{y} = 175.07$ minutes

(2) CI: [150.05, 200.06]; PI: [93.14, 256.97]

(3) Allow 200.06 minutes.

14.51 a. From Excel, the p-value associated with b_1 is 0.000197, providing extremely strong evidence of a significant relationship between x and y.

b. $b_1 = 35.288$; 95% CI for β_1 is [19.221, 51.355]. For every 1% rise in the minority population, we estimate the number of residents per branch bank increases between about 19 and 51. This is evidence of underservice.

14.53 (1) price vs sq ft: extremely significant; p-value associated with $b_1 \approx 0$.

(2) price vs rooms: extremely significant; p-value associated with $b_1 \approx 0$.

(3) price vs bedrooms: extremely significant; p-value associated with $b_1 = .000265$.

(4) price vs age: strong relationship; p-value associated with $b_1 = .0222$.

Chapter 15

15.3 a. $b_1 = -.0900$; $b_2 = .0825$

b. 10.334 in each case.

15.5 a. $b_1 = 1.314$; $b_2 = 3.136$

b. 19.098 or $19,098 in each case.

15.7 a. $b_1 = .0386$, $b_2 = 1.0394$, $b_3 = -413.7578$

c. 17207.31 is not in the 95% prediction interval.

15.11 (1) Total variation = 25.5488; Unexplained variation = .6737; Explained variation = 24.8750

(2) $R^2 = 97.36\%$; $\overline{R}^2 = 96.31\%$

(3) $SSE = .6737$; $s^2 = .1347$; $s = .367078$

(4) $F(\text{model}) = \dfrac{24.875/2}{.6737/[8-3]} = 92.31 \approx 92.30$

(5) $92.30 > 5.79$; at least one of the variables is related to y.

(6) p-value = .000; extremely strong evidence that $H_0: \beta_1 = \beta_2 = 0$ is false.

15.13 (1) Total variation = 598.63; Unexplained variation = 20.39; Explained variation = 578.23

(2) $R^2 = 96.59\%$; $\overline{R}^2 = 96.28\%$.

(3) $SSE = 20.39$; $s^2 = .9268$; $s = .9628$

(4) $F(\text{model}) = \dfrac{578.23/2}{20.39/[25-3]} = 311.94 \approx 311.87$

(5) $311.87 > 3.44$; at least one of the variables is related to y.

(6) p-value = .000; extremely strong evidence that $H_0: \beta_1 = \beta_2 = 0$ is false.

15.15 (1) Total variation = 464,126,601.6; Unexplained variation = 1,798,712.2; Explained variation = 462,327,889.4

(2) $R^2 = 99.61\%$; $\overline{R}^2 = 99.52\%$

(3) $SSE = 1,798,712.2$; $s^2 = 149,892.7$; $s = 387.16$

(4) $F(\text{model}) = \dfrac{462,327,889.4/3}{1,798,712.2/[16-4]} = 1028.13$

(5) $1028.13 > 3.49$; at least one of the variables is related to y.

(6) p-value = .0000; extremely strong evidence that $H_0: \beta_1 = \beta_2 = \beta_3 = 0$ is false.

15.19 (1) $b_0 = 29.347$; $s_{b_0} = 4.891$; $t = 6.000$

$b_1 = 5.613$; $s_{b_1} = .2285$; $t = 24.56$

$b_2 = 3.834$; $s_{b_2} = .4332$; $t = 8.85$

(2) $t = 6.00 > 2.365$; conclude $\beta_0 \neq 0$ at $\alpha = .05$.

$t = 24.56 > 2.365$; conclude that $\beta_1 \neq 0$ at $\alpha = .05$.

$t = 8.85 > 2.365$; conclude that $\beta_2 \neq 0$ at $\alpha = .05$.

(3) p-value $= .0005$; conclude $\beta_0 \neq 0$ at all α's.

p-value $= .000$; conclude that $\beta_1 \neq 0$ at all α's.

p-value $= .000$; conclude that $\beta_2 \neq 0$ at all α's.

(4) [17.780, 40.914]; [5.073, 6.153]; [2.810, 4.858]

15.21 (1) $b_0 = 7.5891$; $s_{b_0} = 2.4450$; $t = 3.1039$

$b_1 = -2.3577$; $s_{b_1} = .6379$; $t = -3.6958$

$b_2 = 1.6122$; $s_{b_2} = .2954$; $t = 5.4586$

$b_3 = 0.5012$; $s_{b_3} = .1259$; $t = 3.9814$

(2) $t = 3.104 > 2.056$; conclude that $\beta_0 \neq 0$ at $\alpha = .05$.

$t = -3.700 < -2.056$; conclude that $\beta_1 \neq 0$ at $\alpha = .05$.

$t = 5.459 > 2.056$; conclude that $\beta_2 \neq 0$ at $\alpha = .05$.

$t = 3.981 > 2.056$; conclude that $\beta_3 \neq 0$ at $\alpha = .05$.

(3) p-value $= .0046$; conclude that $\beta_0 \neq 0$ at all α's except .001.

p-value $= .0010$; conclude that $\beta_1 \neq 0$ at all α's except .001.

p-value $= .0000$; conclude that $\beta_2 \neq 0$ at all α's.

p-value $= .0005$; conclude that $\beta_2 \neq 0$ at all α's.

(4) [2.562, 12.616]; [−3.669, −1.046]; [1.005, 2.220]; [.242, .760]

15.25 a. 10.3331; [9.895, 10.771]

b. 10.3331; [9.293, 11.374]

c. PI is entirely within the no-fine interval.

d. [9.6456, 11.0206]; [8.7012, 11.9650]

15.27 (1) 19.1012; [18.354, 19.848]

(2) 19.1012; [16.969, 21.233]

15.29 17207.31 is above the upper limit of the PI; this y-value is unusually high.

15.33 a. The plot suggests parallel lines with different y-intercepts.

b. β_2 equals the difference between the mean innovation adoption times of stock and mutual companies of the same size.

c. (1) p-value $< .001$; reject H_0 at both α's.

(2) Innovation adoption times differ between company types.

(3) [4.9770, 11.1339]

15.35 a. The pool coefficient is $25,862$; 79.6%.

b. There is no interaction between pool and any other independent variable.

15.37 The large p-values imply that there is no interaction between expenditure and campaign type.

15.41 b. For houses with higher niceness ratings, price increases more quickly with size.

15.43 a. All the terms are important in the model. The p-values corresponding to each of $b_1, \ldots, b_6$ are all less than .001.

b. $\hat{y} = 35.0694$; [34.3788, 35.7600].

15.47 (1) F(partial) $= 2.095 < 3.97$. Conclude we can drop all five variables in question.

(2) Consider the models having 8 and 9 squared or interaction terms. They have the smallest C values and $C < k + 1$.

15.51 Both residual plots have horizontal band appearance.

15.53 $d = 1.63 > d_u = 1.54$; conclude there is no positive autocorrelation.

15.55 b. $F = 184.57$; p-value $= .000$; conclude there are sales differences associated with shelf height.

c. (1) $b_M = 21.40$; $b_T = -4.30$

(2) $t_{b_M} = 14.93 > 4.037$; extremely strong evidence that $\beta_M \neq 0$; $t_{b_T} = -3.00 < -2.947$; very strong evidence $\beta_T \neq 0$

(3) [18.353, 24.447];[−7.347, −1.253]

d. (1) $b_M = 25.7$

(2) $t_{b_M} = 17.935 > 4.037$; extremely strong evidence that $\beta_M \neq 0$.

(3) [22.6463, 28.7537]

f. (1) [75.0403, 79.3597]

(2) [71.4860, 82.9140]

15.57 F(partial) $= \dfrac{(1.0644 - .3936)/2}{.3936/[30 - (6 + 1)]} =$

$\dfrac{.3354}{.0171} = 19.6140 > 3.42$; conclude that at least one of the additional variables is significant.

15.59 a. Yes, the residuals fall in a horizontal band about 0.

b. (1) $\hat{y} = 1239.70$

(2) [1167.32, 1312.08]; [878.68, 1600.72]

(3) Yes, it should since the entire confidence interval is above 1,000.

Chapter 16

16.1 1 if the event of interest has occurred; 0 otherwise.

16.3 .8977; 1.6318; for a one point increase in the score, we estimate that the odds of success increase by about 63%.

16.5 a. *Purchases* is statistically significant (p-value $= .0380 < .05$) as is *Card* (p-value $= .0243$).

b. *Purchases* odds ratio $= 1.2335$; for each additional $100 spent by a customer (holding PlatProfile fixed), the odds of the person redeeming the coupon increase by about 23%. *Card* odds ratio $= 1036.312$; a card holder is 1036 times more likely to redeem the coupon than a customer without a card (assuming purchases are the same).

c. The misclassification rate for redeemers is 1/18 or .0556. The misclassification rate for non-redeemers is 3/22 or .1364. The overall misclassification rate is 4/40 or .10.

d. .9927

e. .6367

16.7 Prob[1] $= .9077$; we classify this person as a "1": they *would* be successful in the position.

16.9 Customer 41: Prob[1] $= .9883$; redeemer. Customer 42: Prob[1] $= .4529$; non-redeemer.

16.11 Neural network modeling is most useful for extremely large data sets (millions of rows/observations and thousands of columns/predictors).

16.13 a. −.2814; .0864; and .0281

b. −.2814

c. 8.2129

d. Each run of a neural network model will result in different values.

16.15 Each run of a neural network model will result in different values.

16.17 a. .9647

b. Each run of a neural network model will result in different values.

Chapter 17

17.3 The data display no trend.

17.5 a. The plot suggests linear growth.

b. 95% PI for y_{21}: $\hat{y}_t \pm t_{.025} s \times$

$$\sqrt{1 + \frac{1}{n} + \frac{(x_0 - \bar{x}^2)}{SS_{xx}}} =$$

$472.1 \pm 2.101(21.8395) \times$

$$\sqrt{1 + \frac{1}{20} + \frac{110.25}{665}} = [421.5,$$

522.7]. (This can be obtained from software.)

17.7 (1) Forecast of $\ln y_{133} = 4.69618 + .0103075(133) + .01903 = 6.086108$

(2) $\hat{y}_{133} = e^{6.0861} = 439.7$; PI: $[e^{5.96593}, e^{6.20627}] = [389.92, 495.85]$

17.13 Positive linear trend.

17.15 $\hat{y}_{17} = 666.6$; $\hat{y}_{18} = 881.6$; $\hat{y}_{19} = 482.1$; $\hat{y}_{20} = 299.9$

17.17 a. 15.01; 40.54; 56.82; 22.02

b. [12.21, 17.81]; [37.69, 43.39]; [53.90, 59.74]; [19.04, 25.00]

17.21 $S_{26} = .02(328) + (1 - .02)356.69 = 356.12$

17.25 $l_2 = .2(211) + .8(204.283 + 7.7102) = 211.794$

$b_2 = .2(211.794 - 204.283) + 0.8(7.7102) = 7.6704$

$\hat{y}_{27} = l_{24} + 3b_{24} = 393.670 + 3(7.5447) = 416.304$

17.27 488.0297; [440.9014, 535.1580]

17.29 418.816; [394.060, 443.572]

17.37

Year	$/Gal	Index
1990	1.16	1
1991	1.14	.98
⋮	⋮	⋮
2010	2.78	2.4
2011	3.63	3.13

17.39 a. Yes, its p-value $= .001$.

b. $1118.21 = 199.62 + 50.937(25) - .5677(625)$; $1140.19 = 199.62 + 50.937(26) - .5677(676)$

17.41 a. $\hat{y} = .1776 + .4071(3.80) - .7837(3.90) + .9934(6.80) + .0435(31) + 3.805(1) = 10.577 \approx 10.578$

b. The model using all the seasonal terms, x_3, and time provides a good fit. All independent variables are significant.

17.43 The data suggest a linear trend and increasing quarterly seasonal variability. We can equalize the seasonal variation by using the transformed model $\ln(y_t) = \beta_0 + \beta_1 t + \beta_2 S_2 + \beta_3 S_3 + \beta_4 S_4 + \epsilon_t$. Exponentiating yields the following predictions for y_t:

t	$\hat{y}$	95% PI lower	95% PI upper
33	121.53	117.99	125.18
34	193.36	187.72	199.16
35	230.78	224.06	237.71
36	162.22	157.49	167.09

Chapter 18
18.3 a. $S = 4$; p-value $= .375$; do not reject H_0
 b. $S = 5$; p-value $= .031$; reject H_0
18.5 a. p-value $= .0059$; reject H_0
18.7 a. $S_1 = 4$, $S_2 = 5$, $S = 5$
 b. p-value $= 1.0$; do not reject H_0 at any α; conclude no difference.
18.11 $T_1 = 120.5$; do not reject H_0; conclude no difference.
18.13 Gender differences exist.
18.17 Using the Minitab output, we can reject at .02 but not .01.
18.19 p-value $= .0039$; reject H_0; attitude scores differ.
18.23 Reject H_0; conclude panels differ.
18.25 $H = 14.36$; reject H_0; sales differ at different heights. After removing the value, $H = 13.35$; same conclusion.
18.29 a. $r_s = -.721$; .648; yes
 b. No
18.31 $r_s = 1.0$; reject H_0.
18.33 $H = 14.36$; p-value $= .001$; drugs differ.
18.35 $T = 1.0 \leq 11 = T_0$; account ages have decreased.
18.37 $T = T_2 = 15 < 21 = T_L$. Reject H_0 and conclude the loan rates do differ.

Chapter 19
19.5 Small facility
19.7 a. $10 million, $10.5 million, $3 million
 b. Medium facility

19.9 a. $12.2 million
 b. $1.7 million
19.11 a. A: 6.2, B: 5.2, C: 4.8
 b. Location A
 c. $1.8 million
19.13 a. Subcontract: $1.23
 Expand: $1.57
 Build: $1.35
 b. Expand
19.19 a. .485, .8247, .0928, .0825
 b. .300, .1667, .700, .1333
 c. .215, .2326, .2093, .5581
19.21 a. 822
 b. 580
 c. 242
 d. 242
19.25 a. .73, .9863, .0137
 b. .27, .6667, .3333
19.27 a. Do not send, $14,958.90
 b. Send, $14,500
19.33 a. Should be continued since EMV (continue) $= $15.5
 b. Should be licensed since EMV (develop) $= $22.9 < $23.
19.35 a. $P(F) = .62$, $P(H|F) = .871$, $P(L|F) = .129$; $P(U) = .38$, $P(H|U) = .158$, $P(L|U) = .842$
 If favorable, build large.
 If unfavorable, also build large.
 b. EVSI $= 0$. Don't pay for advice.
19.37 a. Relocate; $5,000,000
 b. Renew lease; $500,000

Chapter 20 (Online)
20.9 b. Yes. No subgroup range is above 12.76.
 c. No. $\bar{x}$ is above 146.35 for subgroups 4, 10, 22 and below 136.21 for subgroup 19.
 d. Yes. All values are within limits.
20.11 a. $CL_{\bar{x}} = 841.45$,
 $UCL_{\bar{x}} = 845.21$,
 $LCL_{\bar{x}} = 837.69$,
 $CL_R = 5.16$,
 $UCL_R = 11.78$, no LCL_R
 Yes, both charts out of control
 c. $CL_{\bar{x}} = 841.40$,
 $UCL_{\bar{x}} = 844.96$,
 $LCL_{\bar{x}} = 837.84$,
 $CL_R = 4.88$,
 $UCL_R = 11.14$,
 no LCL_R
 d. R chart in control; yes, can use $\bar{x}$ chart

e. No, $\bar{x}$ chart out of control; process mean is changing
 f. $CL_{\bar{x}} = 840.46$,
 $UCL_{\bar{x}} = 844.29$,
 $LCL_{\bar{x}} = 836.63$,
 $CL_R = 5.25$,
 $UCL_R = 11.98$,
 no LCL_R
 g. Yes, all within control limits
20.13 a. Run down; 9 points in B or beyond; 3 points in A; run up.
 b. Alternating pattern.
20.19 a. [132.31, 149.15]
 b. min $= 132.31$; max $= 149.15$
 c. Yes. The tolerance limits are inside the specification limits.
 d. 3.30
20.21 a. [.6518, 1.0416]
 b. 1.0416 lb.
 c. No; weights might be as low as .6518 lb.
 d. .0681 or 6.81%
20.23 a. [50.9189, 54.2561]
 b. Can be reduced by .4189 lb.; .4189(1,000,000)($2) $= $837,800
20.25 .8736
20.29 UCL $= .19$, LCL $= .01$
20.31 a. UCL $= .6217$, LCL $= .4323$
 b. In control, no assignable causes
20.33 a. UCL $= .085$, LCL $= .019$
 b. UCL $= .057$, LCL $= .005$; yes
20.41 a. $UCL_{\bar{x}} = 5.35$, $LCL_{\bar{x}} = 3.51$
 b. $UCL_R = 3.38$
 c. In control
20.43 a. $UCL_{\bar{x}} = 5.08$, $LCL_{\bar{x}} = 3.92$ $UCL_R = 2.135$
 b. Yes, in control
 c. [3.20, 5.80]
 d. Yes, capable. Sigma level $= 3.45$

REFERENCES

Abraham, B., and J. Ledolter. *Statistical Methods for Forecasting.* New York, NY: John Wiley & Sons, 1983.

Akaah, Ishmael P., and Edward A. Riordan. "Judgments of Marketing Professionals about Ethical Issues in Marketing Research: A Replication and Extension." *Journal of Marketing Research,* February 1989, pp. 112–20.

Anderson, D. R., D. J. Sweeney, T. A. Williams, J. Camm, and J. J. Cochran. *Essentials of Statistics for Business and Economics.* 8th ed. Stamford, CT: Cengage Learning, 2015.

Anderson, David R., Dennis J. Sweeney, Thomas A. Williams, Jeffrey D. Camm, and James J. Cochran. *Statistics for Business and Economics.* 12th ed. Belmont, CA: Cengage Learning, 2014.

Ashton, Robert H., John J. Willingham, and Robert K. Elliott. "An Empirical Analysis of Audit Delay." *Journal of Accounting Research* 25, no. 2 (Autumn 1987), pp. 275–92.

Axcel, Amir. *Complete Business Statistics.* 3rd ed. Burr Ridge, IL: Irwin/McGraw-Hill, 1996.

Bayus, Barry L. "The Consumer and Durable Replacement Buyer." *Journal of Marketing* 55 (January 1991), pp. 42–51.

Beattie, Vivien, and Michael John Jones. "The Use and Abuse of Graphs in Annual Reports: Theoretical Framework and Empirical Study." *Accounting and Business Research* 22, no. 88 (Autumn 1992), pp. 291–303.

Berenson, M., D. Levine, and K. Szabat. *Basic Business Statistics: Concepts and Applications.* Boston, MA: Pearson, 2015.

Berenson, Mark L., David L. Levine, and Kathryn A. Szabat, *Basic Business Statistics: Concepts and Applications.* 13th ed. Boston, MA: Pearson Education, Inc., 2015.

Bissell, H. H., I. I. Pilkington, J. M. Mason, and D. L. Woods. "Roadway Cross Section: An Alinement." *Public Roads* 46, no. 4 (1983).

Blauw, Jan Nico, and Willem E. During. "Total Quality Control in Dutch Industry." *Quality Progress* (February 1990), pp. 50–51.

Blodgett, Jeffrey G., Donald H. Granbois, and Rockney G. Walters. "The Effects of Perceived Justice on Complainants' Negative Word-of-Mouth Behavior and Repatronage Intentions." *Journal of Retailing* 69, no. 4 (Winter 1993), pp. 399–428.

Bowerman, Bruce L., and Richard T. O'Connell. *Linear Statistical Models: An Applied Approach.* 2nd ed.

Boston, MA: PWS-KENT Publishing Company, 1990, pp. 457, 460–64, 729–974.

Bowerman, Bruce L., Richard T. O'Connell, and Anne B. Koehler. *Forecasting, Time Series, and Regression.* 4th ed. Belmont, CA: Brooks Cole, 2005.

Bowerman, Bruce L., Richard T. O'Connell, and Emily S. Murphree. *Experimental Design: Unified Concepts, Practical Applications, and Computer Implementation.* New York, NY: Business Expert Press, 2015.

Bowerman, Bruce L., Richard T. O'Connell, and Emily S. Murphree. *Regression Analysis: Unified Concepts, Practical Applications, and Computer Implementation.* New York, NY: Business Expert Press, 2015.

Box, G. E. P., and G. M. Jenkins. *Time Series Analysis: Forecasting and Control.* 2nd ed. San Francisco, CA: Holden-Day, 1976.

Brown, R. G. *Smoothing, Forecasting and Prediction of Discrete Time Series.* Englewood Cliffs, NJ: Prentice Hall, 1962.

Carey, John, Robert Neff, and Lois Therrien. "The Prize and the Passion." *BusinessWeek* (Special 1991 bonus issue: The Quality Imperative), January 15, 1991, pp. 58–59.

Carslaw, Charles A. P. N., and Steven E. Kaplan. "An Examination of Audit Delay: Further Evidence from New Zealand." *Accounting and Business Research* 22, no. 85 (1991), pp. 21–32.

Cateora, Philip R. *International Marketing.* 9th ed. Homewood, IL: Irwin/McGraw-Hill, 1993, p. 262.

CEEM Information Services. "Is ISO 9000 for You?" 1993.

Clemen, Robert T. *Making Hard Decisions: An Introduction to Decision Analysis.* 2nd ed. Belmont, CA: Duxbury Press, 1996, p. 443.

Conlon, Edward J., and Thomas H. Stone. "Absence Schema and Managerial Judgment." *Journal of Management* 18, no. 3 (1992), pp. 435–54.

Cooper, Donald R., and C. William Emory. *Business Research Methods.* 5th ed. Homewood, IL: Richard D. Irwin, 1995, pp. 434–38, 450–51, 458–68.

Cravens, David W., Robert B. Woodruff, and Joe C. Stamper. "An Analytical Approach for Evaluating Sales Territory Performance." *The Journal of Marketing* (1972), pp. 31–37.

Cuprisin, Tim. "Inside TV & Radio." *The Milwaukee Journal Sentinel,* April 26, 1995.

Dawson, Scott. "Consumer Responses to Electronic Article Surveillance Alarms." *Journal of Retailing* 69, no. 3 (Fall 1993), pp. 353–62.

Deming, W. Edwards. *Out of the Crisis.* Cambridge, MA: Massachusetts Institute of Technology Center for Advanced Engineering Study, 1986, pp. 18–96, 312–14.

Dielman, Terry. *Applied Regression Analysis for Business and Economics.* Belmont, CA: Duxbury Press, 1996.

Dillon, William R., Thomas J. Madden, and Neil H. Firtle. *Essentials of Marketing Research.* Homewood, IL: Richard D. Irwin Inc., 1993, pp. 382–84, 416–17, 419–20, 432–33, 445, 462–64, 524–27.

Dondero, Cort. "SPC Hits the Road." *Quality Progress,* January 1991, pp. 43–44.

Draper, N., and H. Smith. *Applied Regression Analysis.* 2nd ed. New York, NY: John Wiley & Sons, 1981.

Farnum, Nicholas R. *Modern Statistical Quality Control and Improvement.* Belmont, CA: Duxbury Press, 1994, p. 55.

Fitzgerald, Neil. "Relations Overcast by Cloudy Conditions." *CA Magazine,* April 1993, pp. 28–35.

Garvin, David A. *Managing Quality.* New York, NY: Free Press/Macmillan, 1988.

Gibbons, J. D. *Nonparametric Statistical Inference.* 2nd ed. New York, NY: McGraw-Hill, 1985.

Gitlow, Howard, Shelly Gitlow, Alan Oppenheim, and Rosa Oppenheim. *Tools and Methods for the Improvement of Quality.* Homewood, IL: Richard D. Irwin, 1989, pp. 14–25, 533–53.

Gunter, B. "Process Capability Studies Part 3: The Tale of Charts." *Quality Progress* (June 1991), pp. 77–82.

Gupta, Sachin, Pradeep Chintagunta, Anil Kaul, and Dick R. Wittinik. "Do Household Scanner Data Provide Representative Inferences from Brand Choices? A Comparison of Store Data." *Journal of Marketing Research* 33, no. 4 (1996), pp. 383–398.

Guthrie, James P., Curtis M. Grimm, and Ken G. Smith. "Environmental Change and Management Staffing: A Reply." *Journal of Management* 19, no. 4 (1993), pp. 889–96.

Hoexter, R. and M. Julian. "Legal Eagles Become Quality Hawks." *Quality Progress,* January 1994, pp. 31–33.

Ito, Harumi, and Darin Lee. "Assessing the Impact of the September 11 Terrorist Attacks on US Airline Demand." *Journal of Economics and Business* 57, no. 1 (2005), pp. 75–95.

Kuhn, Susan E. "A Closer Look at Mutual Funds: Which Ones Really Deliver?" *Fortune,* October 7, 1991, pp. 29–30.

Kumar, V., Roger A. Kerin, and Arun Pereira. "An Empirical Assessment of Merger and Acquisition Activity in Retailing." *Journal of Retailing* 67, no. 3 (Fall 1991), pp. 321–38.

Kutner, Michael H., Christopher J. Nachtsheim, John Neter, and William Li. *Applied Linear Statistical Models.* 5th ed. Burr Ridge, IL: McGraw-Hill, Irwin, 2005.

Magee, Robert P. *Advanced Managerial Accounting.* New York, NY: Harper & Row, 1986, p. 223.

Mahmood, Mo Adam, and Gary J. Mann. "Measuring the Organizational Impact of Information Technology Investment: An Exploratory Study." *Journal of Management Information Systems* 10, no. 1 (Summer 1993), pp. 97–122.

Makridakis, Spyros, Steven C. Wheelwright, and Victor E. McGee. *Forecasting: Methods and Applications.* New York, NY: J. Wiley, 1983.

Martocchio, Joseph J. "The Financial Cost of Absence Decisions." *Journal of Management* 18, no. 1 (1992), pp. 133–52.

McCabe, W.J. "Examining Processes Improves Operations." *Quality Progress,* July 1989, pp. 26–32.

McCabe, William J. "Examining Processes Improves Operations." *Quality Progress* 22, no. 7 (1989), pp. 26–32.

Meier, Heidi Hylton, Pervaiz Alam, and Michael A. Pearson. "Auditor Lobbying for Accounting Standards: The Case of Banks and Savings and Loan Associations." *Accounting and Business Research* 23, no. 92 (1993), pp. 477–487.

Mendenhall, W., and J. Reinmuth. *Statistics for Management Economics.* 4th ed. Boston, MA: PWS-KENT Publishing Company, 1982.

Mendenhall, W., and T. Sincich. *A Second Course in Statistics: Regression Analysis.* 4th ed. Boston, MA: Prentice Hall, 1993.

The Miami University Report. Miami University, Oxford, OH, vol. 8, no. 26, 1989.

Moore, David S. *The Basic Practice of Statistics.* 2nd ed. New York, NY: W. H. Freeman and Company, 2000.

Moore, David S., and George P. McCabe. *Introduction to the Practice of Statistics.* 2nd ed. New York, NY: W. H. Freeman, 1993.

Morris, Michael H., Ramon A. Avila, and Jeffrey Allen. "Individualism and the Modern Corporation: Implications for Innovation and Entrepreneurship." *Journal of Management* 19, no. 3 (1993), pp. 595–612.

Nunnally, Bennie H., Jr., and D. Anthony Plath. *Cases in Finance.* Burr Ridge, IL: Richard D. Irwin, 1995, pp. 12-1–12-7.

Olmsted, Dan, and Gigi Anders. "Turned Off." *USA Weekend,* June 2–4, 1995.

Ott, Lyman. *An Introduction to Statistical Methods and Data Analysis.* 2nd ed. Boston, MA: PWS-Kent, 1987.

Schaeffer, R. L., William Mendenhall, and Lyman Ott. *Elementary Survey Sampling.* 3rd ed. Boston, MA: Duxbury Press, 1986.

Schargel, F. P. "Teaching TQM in an Inner City High School." *Quality Progress,* September 1994, pp. 87–90.

Scherkenbach, William. *The Deming Route to Quality and Productivity: Road Maps and Roadblocks.* Washington, DC.: CEEPress Books, 1986.

Seigel, James C. "Managing with Statistical Models." SAE Technical Paper 820520. Warrendale, PA: Society for Automotive Engineers, Inc., 1982.

Shmueli, G. P. C. Bruce, M. L. Stephens, and N. R. Patel. *Data Mining for Business Analytics: Concepts, Techniques, and Applications with JMP Pro.* Hoboken, NJ: John Wiley and Sons, 2017.

Sichelman, Lew. "Random Checks Find Loan Application Fibs." *The Journal-News* (Hamilton, Ohio), September 26, 1992 (originally published in *The Washington Post*).

Siegel, Andrew F. *Practical Business Statistics.* 2nd ed. Homewood, IL: Richard D. Irwin, 1990, p. 588.

Silk, Alvin J., and Ernst R. Berndt. "Scale and Scope Effects on Advertising Agency Costs." *Marketing Science* 12, no. 1 (Winter 1993), pp. 53–72.

Stevenson, William J. *Production/Operations Management.* 6th ed. Homewood, IL: Irwin/McGraw-Hill, 1999, p. 228.

Thomas, Anisya S., and Kannan Ramaswamy. "Environmental Change and Management Staffing: A Comment." *Journal of Management* 19, no. 4 (1993), pp. 877–87.

"The Tools of Quality Part V: Check Sheets," from *QI Tools: Data Collection Workbook,* p. 12. Juran Institute Inc., 1989.

U.S. Department of Commerce, Bureau of the Census. *Bicentennial Statistics,* Washington, DC, 1976.

Von Neumann, J., and O. Morgenstern. *Theory of Games and Economic Behavior.* 2nd ed. Princeton, NJ: Princeton University Press, 1947.

Walton, Mary. *The Deming Management Method.* New York, NY: Dodd, Mead & Company, 1986.

Weinberger, Marc G., and Harlan E. Spotts. "Humor in U.S. versus U.K. TV Commercials: A Comparison." *Journal of Advertising* 18, no. 2 (1989), pp. 39–44.

Wright, Thomas A., and Douglas G. Bonett. "Role of Employee Coping and Performance in Voluntary Employee Withdrawal: A Research Refinement and Elaboration." *Journal of Management* 19, no. 1 (1993), pp. 147–61.